American Herbal Products Association's

BOTANICAL SAFETY HANDBOOK

Second Edition

Edited by

Zoë Gardner

Michael McGuffin

Expert Advisory Council

Roy Upton

Soaring Bear

David Winston

Daniel Gagnon

Aviva Jill Romm

Tieraona Low Dog

Mary Hardy

Lyle Craker

Reviewers

David Bechtel

Leslie Beyer

Bill J. Gurley

Proofreaders

Bill Schoenbart

Constance A. Parks

AHPA
AMERICAN HERBAL PRODUCTS ASSOCIATION

CRC Press
Taylor & Francis Group
Boca Raton London New York

CRC Press is an imprint of the
Taylor & Francis Group, an **informa** business

First published 2013 by CRC Press

Published 2019 by CRC Press
Taylor & Francis Group
6000 Broken Sound Parkway NW, Suite 300
Boca Raton, FL 33487-2742

Version Date: 20130125

ISBN-13: 978-1-4665-1694-6 (hbk)

Library of Congress Cataloging-in-Publication Data

American Herbal Products Association botanical safety handbook / editors: Zoe Gardner, Michael McGuffin. -- 2nd ed.
p. cm.
Botanical safety handbook
Includes bibliographical references and index.
ISBN 978-1-4665-1694-6 (hardcover : alk. paper)
1. Herbs--Toxicology--Handbooks, manuals, etc. 2. Materia medica, Vegetable--Toxicology--Handbooks, manuals, etc. I. Gardner, Z. E. (Zoë E.) II. McGuffin, Michael. III. American Herbal Products Association. IV. Title: Botanical safety handbook.

RA1250.A45 2013
615.3'21--dc23 2012037920

Visit the Taylor & Francis Web site at
http://www.taylorandfrancis.com

and the CRC Press Web site at
http://www.crcpress.com

DEDICATION

This book is dedicated to the memory of Mary Frances Picciano, Ph.D., former Senior Nutrition Research Scientist at the Office of Dietary Supplements, National Institutes of Health. Her interest, foresight, and willingness to have the Office of Dietary Supplements support a partnership with the American Herbal Products Association and the University of Massachusetts made this work possible.

CONTENTS

PREFACE TO THE FIRST EDITION

Increased attention on herbal products, both in the marketplace and in the legislative arena, has created a need for wider public access to data regarding the safety of botanicals. The passage of the Dietary Supplement Health and Education Act in October, 1994, furthered the need for such information, as this law authorizes the use of cautionary labeling for dietary supplements, including those that contain herbs.

The American Herbal Products Association (AHPA), through its Standards Committee, convened a special SubCommittee (hereinafter "the Committee") to address this need. The Committee members identified considerable safety data in varied texts and journals and discovered that some attempts to classify herbs had been undertaken in several other countries. No comprehensive compilation or review of this data for botanical ingredients sold in the North American marketplace, however, was available in a useful format.

The goal of the present work is to find a rational platform for the evaluation of herb safety, neither assuming that all natural substances are inherently safe, as some popular references suggest, nor blindly accepting reports of toxicity from uncritical sources. In undertaking this task, the Editors met with information that presented significant challenges. Many authors utilize unreferenced data, perpetuate historical inaccuracies or display inherent biases against the use of botanicals. Also, contemporary reviews of the toxicity of many herbs are not available. Nonetheless, the Editors are confident that the body of information presented here is largely accurate. It is our sincere hope that readers of this work will find it to be a valuable reference and will address all useful criticisms to our attention.

In sponsoring this effort, the American Herbal Products Association (AHPA) addresses the common interest of industry, the public, and regulatory agencies in assuring safe access to a wide range of herbs and herbal products. This document provides accurate data to guide manufacturers and consumers in safe utilization of herbal products. As the most broadly established trade association in the herbal marketplace, AHPA has, by supporting and sponsoring the creation of this work, furthered the herb industry's leadership role in promoting the responsible use of herbs.

PREFACE

This second edition of AHPA's *Botanical Safety Handbook* represents a significant modification from the first edition, published in 1997. At the same time, the second edition reflects the continued commitment of the American Herbal Products Association (AHPA) to provide accurate information about the safe use of herbs in a practical and accessible format.

The original edition classified botanical ingredients in four safety classifications to differentiate those that can be safely consumed when used appropriately from those for which some contraindication or other restriction is known, as well as those that should be used only under the guidance of a qualified expert. A handful of entries in that edition were also placed in a separate class if the editors had insufficient data for classification. The revised edition largely retains this safety classification system, except that if the review process did not provide enough information to make a knowledgeable decision on any specific herb, that species was removed from the text.

The present edition also includes a separate classification system to address what is known about the potential for an herb to interact with any drugs. Each of the herbs listed here is identified in one of three interaction classes to differentiate between those for which no clinically relevant interactions are expected and those for which clinically relevant interactions are biologically plausible or are, in fact, known to occur.

Botanical products continue to be broadly used throughout the world. In the United States, most herbs are sold in loose form or as tablets, capsules, or tinctures, and regulated as dietary supplements (this product class also includes vitamins, minerals, amino acids, and numerous other ingredients). Many herbs are also common flavorings for foods, or are used in teas. In addition, a handful of herbs provide active ingredients in non-prescription drugs. The U.S. marketplace for herbal products in the supplement category has increased significantly in the years since publication of the first edition, and the retail value of this product category grew from $3 billion in 1996 (Muth et al. 1999) to $5 billion in 2010 (Anon. 2011).

An even more significant change in the past 15 years has been the emergence of the Internet and online scientific databases as tools for accessing scientific information. The first edition of this book relied almost entirely on secondary references (i.e., books and other summaries of traditional or scientific information), and the editors of that document used their personal collections of such texts to compile the information needed to make safety determinations for the plants addressed in it. On the other hand, the process for compiling information for this second edition, as described in the introduction, involved a much more thorough review of primary references (i.e., published research papers, case reports, and other original literature). Thus, while the first edition included just 280 references to evaluate the over 500 plants addressed therein, this revision cites 301 references just in its treatment of St. John's wort herb (*Hypericum perforatum*), ginkgo leaf and seed (*Ginkgo biloba*), and garlic bulb (*Allium sativum*).

Any attempt to provide a summary of safety information on botanicals will encounter certain prejudices and inaccuracies in the published record. One such prejudice, often repeated in reviews of herbal medicines and dietary supplements, is the view that consumers have been led to believe the myth that "anything natural is safe" (Barnes 2003; Dasgupta and Bernard 2006). While one survey of consumer attitudes in Canada found that 7 percent of respondents completely agree that there is no risk associated with products made with natural ingredients (Anon. 2005), there are no published analyses of consumer beliefs that indicate that there is broad acceptance of any such assumption.

It is, however, true that many of the plants that enjoy broad culinary and traditional therapeutic usage are generally safe. We can safely season our food with any number of herbs to make a meal more flavorful. We can appreciate a delicious cup of peppermint leaf or rose hips tea, or safely take an herbal supplement containing dandelion root, saw palmetto berries, or any number of other herbs. Although allergies and individual reactions have been recorded for a few herbs that are widely used in foods and supplements, such individual concerns are also seen with many other foods, and do not diminish the safety profile of the many herbs that are widely regarded as safe.

On the other hand, and as everyone knows, there are any number of plants that are highly toxic, even deadly. Every savvy North American hiker knows to stay away from poison ivy (*Toxicodendron* spp.) when walking in the woods. The death sentence imposed on Socrates by an Athenian jury 2,400 years ago was carried out with a fatal dose of poison hemlock (*Conium maculatum*). The poison curare, a blend of several equatorial rain forest plants (e.g., species of *Chondrodendron*, *Curarea* and *Strychnos*) is used by some South American hunter cultures to make their arrows more deadly (Schultes and Raffauf 1990). And in the "concrete jungle" of Los Angeles, two young boys died in 2000 from ingesting a few leaves of the ubiquitous oleander (*Nerium oleander*) (Garrison 2000). Federal law and good common sense, however, prevent the use of any such highly toxic plants in products that are readily available to consumers.

The revised edition of the *American Herbal Products Association's Botanical Safety Handbook* fills the need for a reference that neither promulgates the myth that all herbs are always safe, since they are "natural," nor accepts without review every case report or conceptual theory that draws an unsubstantiated or illogical conclusion of harm from an herb or herbal product. In assembling this revision, significant effort has gone into sorting out references that are factual from those that are inaccurate. Texts that communicate that all natural substances are inherently safe would not have been included here, though in fact no such documents were encountered. More effort was needed to avoid blind acceptance of reports that purport to identify herbal safety concerns with unreferenced statements or incomplete records of specific herbal preparations, which are unfortunately quite common, even in peer-reviewed scientific journals. Such references may nonetheless be included in this text to provide readers with a complete record, though efforts were made to highlight any perceived flaws.

Even as the consumer market for herbal supplement products expands and scientific information becomes more accessible, the goals of the second edition of the AHPA's *Botanical Safety Handbook* are essentially the same as those of the original edition. Companies that market herbal products are bound by federal regulations to disclose known safety concerns that may result from a product's use. Health care providers, especially those lacking in training or experience in the use of herbs, are in need of accurate data if they are to provide guidance to their patients who use herbs. And consumers of herbs and herbal products need readily understandable information to assist them in making safe and appropriate health care choices. AHPA's *Botanical Safety Handbook, 2nd edition* is designed to provide the information needed by each of these audiences.

It should be recognized, however, that this reference is not an herbal user's guide. Numerous excellent references exist that provide information on the uses and benefits of herbs. Readers of the present document are advised to seek out these references, or to consult with experts qualified by training and experience, for advice on when and how to use herbs for their health benefits.

The editors are confident that the body of information presented in this second edition of the AHPA *Botanical Safety Handbook* is largely accurate, and hope that readers of this work will find it to be a valuable reference. Useful criticisms will nonetheless be welcome, and should be addressed to the attention of the editors.

LITERATURE CITED

Anon. (Ipsos-Reid). 2005. Baseline Natural Health Products Survey Among Consumers: Final Report. Health Canada, Natural Health Products Directorate.

Anon. 2011. Organic wins big in 2010. *Nutr. Bus. J.* 16(6):1-8.

Barnes J. 2003. Quality, efficacy and safety of complementary medicines: Fashions, facts and the future. Part II: Efficacy and safety. *Br. J. Clin. Pharmacol.* 55:331-340.

Dasgupta, A. and D.W. Bernard. 2006. Herbal remedies: Effects on clinical laboratory tests. *Arch. Pathol. Lab. Med.* 130:521–528.

Garrison, J. 2000. Two Toddlers Died from Oleander Poisoning, Coroner Says. *Los Angeles Times*: July 26, 2000.

Muth, M.K., D.W. Anderson, J.L. Domanico, J.B. Smith, and B. Wendling. 1999. Economic characterization of the dietary supplement industry. Contract No. 223-96-2290: Task Order 3. Final Report. Research Triangle Park, NC: Research Triangle Institute.

Schultes, R.E., and R.F. Raffauf. 1990. *The healing forest.* Portland, OR: Dioscorides Press.

ACKNOWLEDGMENTS

Appreciation was expressed in the first edition of the *Botanical Safety Handbook* to three individuals who were essential to AHPA's decision to take on the task of creating this document. This appreciation is restated now to Daniel Gagnon, for his vision and persistence; to John Hallagan, for the encouragement borne of his own experience; and to the late William Appler, for so clearly seeing the value of this text when it was just an idea.

The members of this edition's Expert Advisory Council met together on a regular basis for nearly five years, all on their own time, and without any financial compensation. The expertise and experience embodied in these individuals are unsurpassed, and without them the work could not have proceeded beyond a collection of references, as it was through their efforts that these references were evaluated and organized into the present text. Biographies of each of these individuals follow.

Additional specific guidance was occasionally solicited from a number of other experts, and thanks are due to Dennis Awang, Dan Bensky, Paul Bergner, Mary Bove, Eric Brand, Josef Brinckmann, Francis Brinker, Chanchal Cabrera, Todd Caldecott, John Chen, Sigrun Chrubasik, Emily Cohen, Cynthia Copple, Amanda McQuade Crawford, De-Qiang Dou, Lana Dvorkin-Camiel, Andrew Ellis, Thomas Avery Garran, Christopher Hobbs, David Hoffmann, Prashanti de Jager, K.P. Khalsa, Vasant Lad, Reinhard Länger, Wilson Lau, Phyllis Light, Russell Molyneux, Vikram Naharwar, Robert Newman, Xie Peishan, Sebastian Pole, Bill Schoenbart, Atreya Smith, Ed Smith, James Snow, Alan Tillotson, Jonathan Treasure, Nancy Turner, Donnie Yance, Eric Yarnell, and Yifang Zhang.

Thanks are also due to the generations of herbalists and scientists around the world whose research and experience have provided the basis for our understanding of the safety of medicinal plants. Their work and publications have created a significant foundation for our understanding of the safety of the botanicals reviewed in the present text.

Appreciation is also due to Joseph Betz, Ph.D. and the late Mary Frances Picciano, Ph.D. of the Office of Dietary Supplements (ODS) at the National Institutes of Health. Dr. Betz shared his ideas on the makeup of the Expert Advisory Council and on the importance of addressing the potential for an herb to modify the effect of a drug taken concomitantly, commonly referred to as an herb–drug interaction (a topic that was outside of the scope of the first edition). Dr. Picciano facilitated ODS's significant financial support of the revision process and ensured that the planned revision met high academic and scientific standards.

A number of research assistants helped to acquire and manage the thousands of documents reviewed in this project. A work of this scope would not have been possible without the enthusiastic assistance of Jamie Blair, Brittney Laramee, Annie Winkler, Ryan Rogan, Rye Zemelsky, Kathleen Broadhurst, Jennifer Kehoe, Margo Voskanian, Jennifer Hast, and Abigail Haines. Thanks are also due to Constance Parks and Bill Schoenbart for their detailed reading and editing of the manuscript.

FINANCIAL SUPPORT

This work was made possible with financial support from the Office of Dietary Supplements at the National Institutes of Health, and in-kind contributions by the Medicinal Plants Program at the University of Massachusetts, Amherst.

Financial contributions to support this effort were also made by numerous AHPA member companies and individuals identified as follows.

VISIONARIES

Herbs Etc.

BENEFACTORS

Aveda — the art and science of pure flower and plant essences

Herbalife

Body Building.com

Nature's Way

Celestial Seasonings

New Chapter

Herb Pharm

Traditional Medicinals

CHAMPIONS

Amin Talati, LLC
CRC Insurance Services, Inc.
Frontier Natural Products Co-op
GNC (General Nutrition Centers, Inc.)
Michael McGuffin
Mountain Rose Herbs
Pacific Nutritional, Inc.
Rainbow Light
Systemic Formulas
Vitality Works

SPONSORS

Alkemists Laboratories
Animal Essentials Inc.
Arise & Shine Herbal Products, Inc.
Bent Creek Institute, Inc.
Emerson Ecologics
Gaia Herbs
Golden Flower Chinese Herbs
Herbalist & Alchemist
Iovate Health Sciences International Inc.
Motherlove Herbal Co.
NOW Foods
Nuherbs Co.
Paragon Laboratories
Jim Prochnow
Ridgecrest Herbals
Sabinsa Corporation

ADVOCATES AND SUPPORTERS

American Botanical Council
Cindy Angerhofer
Beehive Botanicals, Inc.
Jon Benninger
BI Nutraceuticals
Bighorn Botanicals
Ed Croom
Steven Dentali
Dicentra
Earth Mama Angel Baby
EuroPharma USA
Evergreen Herbs, LLC
James Fischer
Dagmar Goldschmidt
Jackie Greenfield
Howard Miller
Mountain Meadow Herbs, Inc.
Pacific Botanicals
Pharmline Inc.
St. Francis Herb Farm Inc.
Strategic Sourcing Inc.
Mary Beth Watkins

EDITORS

Zoë Gardner, M.S., Ph.D.(c)
Research & Development Manager, Traditional Medicinals

Fascinated by the connection between people and plants, Zoë Gardner has been studying, researching, and teaching on the production, conservation, quality, safety, and appropriate use of medicinal plants since 1998. After completing her undergraduate degree in environmental studies at the Audubon Expedition Institute, Zoë helped to establish the Medicinal Plant Program at the University of Massachusetts, earning her master's degree there in plant & soil sciences. More recently, Zoë joined the Research & Development Department at Traditional Medicinals, a leading producer of botanical dietary supplements. A self-proclaimed "herb nerd," Zoë is currently completing her Ph.D. on medicinal plant quality and safety.

Michael McGuffin
President, American Herbal Products Association

Michael McGuffin has been active in the herbal industry since 1975, having owned and managed both retail and manufacturing businesses in this field. He is the managing editor of *Botanical Safety Handbook, 1st edition* (1997) and of *Herbs of Commerce, 2nd edition* (2000). He serves on the boards of the American Herbal Pharmacopoeia and United Plant Savers; on the Advisory Board of the USC School of Pharmacy Regulatory Science Program; and as chair of the U.S. Technical Advisory Group for ISO/TC 249, the International Organization for Standardization's Technical Committee on Traditional Chinese Medicine (provisional title). Michael maintains active involvement with regulatory agencies, and served on FDA's Food Advisory Committee Working Group on Good Manufacturing Practices for Dietary Supplements (1998–1999), FDA's Food Advisory Committee's Dietary Supplements Subcommittee (2003–2005), and California's Office of Environmental Health Hazard Analysis Food Warning Workgroup (2008–2010).

EXPERT ADVISORY COUNCIL

Roy Upton, RH (AHG), DAyu
Executive Director, American Herbal Pharmacopoeia

Roy Upton has been trained in traditional Ayurvedic, Chinese and Western herbal traditions, has studied Native American and Caribbean ethnobotanical traditions, and is a professional member of the American Herbalists Guild. He is the executive director and editor of the *American Herbal Pharmacopoeia* and a member of the Standards Committee of the American Herbal Products Association and advisory committees for the American Botanical Council, AOAC International, and NSF International. Along with being an author and lecturer, Roy was co-founder and past president of the American Herbalists Guild and is the herbalist and director of the California-based herbal company Planetary Herbals.

Soaring Bear, Ph.D.
Technical Information Specialist (Ret.), National Library of Medicine

Dr. Soaring Bear has been collecting herbal data, with a focus on toxicology, since the early 1970s. He earned a B.S. in biochemistry with honors and a Ph.D. in pharmacology from the University of Arizona. His doctoral research on structure–activity relationships and chemical interactions provides him with a unique perspective on bioactivity of herbs. He created herbmed.org and edited over ten thousand quick summaries and links into Medline. His work at the National Library of Medicine in the Medical Subject Headings (MeSH) section included significant revisions of the herbal, alternative medicine, and chemistry sections of MeSH, which improves the quality of millions of searches done every day on the pubmed.gov database.

David Winston, RH (AHG)
President, Herbalist & Alchemist

David Winston is an herbalist and ethnobotanist with over 40 years of training in Cherokee, Chinese, and Western/Eclectic herbal traditions. David is a founding/professional member of the American Herbalists Guild and has been in clinical practice for over 33 years. He is an herbal consultant to many physicians and other healthcare professionals throughout the U.S. and Canada. David is also the president of Herbalist & Alchemist, Inc. and founder and director of David Winston's Center for Herbal Studies and the Herbal Therapeutics Research Library. He is the author of numerous texts on herbal medicine, is an internationally known lecturer, and teaches frequently at medical schools, symposia, and herb conferences.

Daniel Gagnon, M.S., RH (AHG)
President, Herbs, Etc.

Daniel Gagnon, owner of Herbs, Etc., has been a practicing herbalist since 1976. Daniel has studied medical herbalism, pharmacognosy, and related subjects at the Santa Fe College of Natural Medicine, the College of Santa Fe, and the University of New Mexico. He received his B.S. in herbal medicine from the North American College of Botanical Medicine and his M.S. in herbal medicine from the Scottish School of Herbal Medicine. Daniel was a faculty member of the North American College of Botanical Medicine for over 10 years and currently serves as an herbal consultant to healthcare providers. He is the author of several books on herbal medicine.

Aviva Jill Romm, M.D.
Director, Integrative Medicine for Women and Children

Before becoming a physician, Dr. Aviva Romm was a recognized expert in midwifery, women's health, and women's and pediatric botanical medicine, and she practiced as a homebirth midwife and herbalist for over 20 years. A graduate of the Yale School of Medicine, she completed her internship in internal medicine (Yale) and her residency in family medicine (Tufts). She was the president of the American Herbalists Guild, the founder and director of Herbal Medicine for Women, a distance learning program, and is the medical director for the American Herbal Pharmacopoeia. Dr. Romm has been active in establishing standards for botanical medicine practice and education in the United States, and is the author of numerous texts on botanical medicine, pregnancy, and children's health.

Tieraona Low Dog, M.D.
Director of the Fellowship, Arizona Center for Integrative Medicine, University of Arizona College of Medicine

Dr. Tieraona Low Dog's extensive career in natural medicine began more than 25 years ago. A graduate of the University of New Mexico School of Medicine, Tieraona has served as president of the American Herbalists Guild and is currently the Director of the Fellowship at the Arizona Center for Integrative Medicine at the University of Arizona School of Medicine. She has been involved in national health policy and regulatory issues, serving previously on the White House Commission of Complementary and Alternative Medicine and as a member of the Advisory Council for the National Center for Complementary and Alternative Medicine, and is currently chair of the U.S. Pharmacopeia Dietary Supplements and Botanicals Expert Committee.

Mary Hardy, M.D.
Medical Director, UCLA Center for Integrative Oncology

Dr. Mary Hardy is an integrative medicine physician who received her Doctor of Medicine degree from Louisiana State University School of Medicine and studied medical ethics at Harvard Divinity School and Loma Linda University. She has studied with herbal practitioners in the United States and Europe and visited traditional healers in Peru, Kenya, South Africa, Morocco and China. She is the complementary and alternative medicine expert for a number of research projects conducted by the Southern California Evidence-Based Practice Center at the RAND Corporation, co-chairperson of the Clinical Practice Committee of the Academic Consortium of Integrative Medicine, and associate director of the UCLA Botanical Research Center, funded by the National Institutes of Health.

Lyle Craker, Ph.D.
Professor, Medicinal Plant Program, University of Massachusetts Amherst

Dr. Lyle Craker has been a researcher in the field of medicinal plants for over 30 years. With a Ph.D. in agronomy from the University of Minnesota, he is founding and past editor of the *Journal of Herbs, Spices, and Medicinal Plants*, founding and current executive editor of the *Journal of Medicinally Active Plants*, past chairman of the International Society for Horticultural Science (ISHS) Section on Medicinal and Aromatic Plants, organizer of the Herb, Spice, & Medicinal Plant Working Group within the American Society for Horticultural Science (ASHS), and an organizing member of the International Council on Medicinal and Aromatic Plants and the American Council for Medicinally Active Plants. He is an advisory board member of the American Botanical Council and serves on the board of the AHPA Foundation for Education and Research on Botanicals.

INTRODUCTION

The second edition of AHPA's *Botanical Safety Handbook* provides information on a number of safety factors that may affect an individual's decision to ingest any of the herbal* substances listed in this work. The information was prepared through a process that involved identification of relevant publications on each botanical, as well as a review by experts qualified by training and experience in the traditional and therapeutic use of herbs and herbal products.

Each of the botanical ingredients† included in this text is classified into one or more Safety Class, and also into an Interaction Class, details of which are described below. These classifications, as well as a synopsis of pertinent information from reviewed references, are presented in a Quick Reference Summary, which provides basic data needed to understand safety issues associated with each botanical. This summary is followed by a section titled Review Details in which more in-depth information is presented when available. Thorough descriptions of the templates and contents of each of these sections are provided later in this introduction.

DETERMINATION OF HERB SAFETY

In developing this document, the voices and experience of various organizations and individuals were considered. A primary source of guidance and inspiration for the first edition of this text was the work of the World Health Organization (WHO). In 1991, WHO's Programme on Traditional Medicines presented Guidelines for the Assessment of Herbal Medicines at the Sixth International Conference of Drug Regulatory Authorities. These guidelines, which were subsequently reviewed and adopted by WHO, propose that the safety of herbal medicine be assessed according to the following principle:

> *...that if the product has been traditionally used without demonstrated harm, no specific restrictive regulatory action should be undertaken unless new evidence demands a revised risk-benefit assessment.* (WHO 1991)

The editors of the first edition adopted this principle from the WHO Guidelines and this view has been maintained for the compilation of the present work.

In his classic text, *The Problem of Poisonous Plants*, J.M. Kingsbury provides further direction by calling attention to the fact that there are many instances in which a plant contains a measurable amount of a toxic substance, though the plant may be poisonous only if consumed in excessive quantities. He notes:

> *In order for a plant to be functionally poisonous, it must not only contain a toxic secondary compound, but also possess effective means of presenting that compound to an animal in sufficient concentrations, and the compound must be capable of overcoming whatever physiological or biochemical defenses the animal may possess against it. Thus the presence of a known poison principle, even in toxicologically significant amounts, in a plant does not automatically mean that either man or a given species of animal will ever be effectively poisoned by the plant.* (Kingsbury 1979)

In examining the relevance of Kingsbury's position, it is of interest to revisit the means by which concerns for the safety of herbs arise. Toxicity studies are often conducted by feeding abnormally high quantities of an herb or isolated constituent of an herb to laboratory animals. For example, Bensky and Gamble report in their monograph on mulberry leaves that "long-term use of 250 times the normal human dose in mice produced both liver and kidney damage" (Bensky and Gamble 1986). Data based on excessive consumption have little relevance to the practical use of herbal supplements, and such findings are clearly not pertinent to normal human consumption patterns. In addition, information is sometimes available that identifies an LD_{50} for an herb, herbal preparation, or isolated compound (i.e., the "lethal dose" at which 50 percent of test animals are killed by the studied substance), but often fails to specify the concentration or form of the specific material used. Such incomplete data cannot be accurately applied to safety evaluations of human consumption.

Significant toxicity data exist for isolated constituents of a wide variety of commonly available foods, as well as herbs. Potatoes, as a member of the Solanaceae family, contain trace amounts of the toxic glycoalkaloid solanine, especially in green parts of the potato tuber (Turner and Szczawinski 1991). Although the symptoms of solanine poisoning are serious, potatoes themselves are generally considered to be a safe food. While consumption of

* The terms "herbal" and "botanical" are used interchangeably throughout this work.

† Or occasionally, groups of ingredients. Examples are listings for more than one plant part from a specific taxa, when the safety concerns for these are not different, or groups of two or more species within a genus, when these have common safety profiles.

as little as five grams of nutmeg can cause marked hallucinations (Sangalli and Chiang 2000), no safety concerns prevent us from enjoying a sprinkle of this characteristic flavor on our holiday eggnog. Similarly, no safety concern is associated with a candy flavored with peppermint oil, though as many as 26 toxins are reported to have been observed in the plant (Duke 1989). Safety concerns for herbal products need not be extrapolated from constituent profiles with any more alarm than is appropriate for foods.

In following the principles espoused by WHO and incorporating the ideas delineated by Kingsbury, it is imperative that herb safety be assessed according to the intended use of the substance within the historical context of its use. In establishing safety classifications, this work has intentionally refrained from automatically applying information on the toxicity of isolated constituents or considering excessive or irresponsible consumption patterns. The decision to place an herb in a restrictive safety class was made only if the use of the herb in a normal dosage range is documented as presenting a safety concern, or if the amount of a harmful or potentially harmful constituent obtainable from the crude plant is of sufficient quantity to be problematic.

ADDRESSING POTENTIAL DRUG INTERACTIONS

The issue of herb–drug interactions was specifically excluded from the first edition of this work, since at that time very little accurate information had been developed on this subject. In the years since then, this topic has been much more prominently studied. Some early publications on the subject were largely speculative, but researchers have now begun to develop scientifically based data that have measured actual effects of several herbs on the metabolism of selected drugs or on drug-metabolizing enzymes. At the same time, emerging research on many specific botanicals has confirmed that no drug interactions should be expected with these herbs.

Drug interactions are generally divided into two categories: pharmacodynamic interactions, in which the physiological effects of drugs or botanicals interact (including additive and opposing effects), and pharmacokinetic interactions, in which an interaction affects a drug's absorption, metabolism, or excretion, and changes the amount and duration of a drug's bioavailability (see CYP450 and P-gp interactions profile in Appendix 3). While pharmacodynamic interactions are generally predictable based on the pharmacological effects of drugs and botanicals, pharmacokinetic interactions, until identified through testing or well-documented case reports, generally cannot be predicted.

This work focuses primarily on pharmacokinetic interactions, although a small number of pharmacodynamic interactions are also listed, especially when such interactions may have significant health consequences (e.g., additive effects on heart medications or antiplatelet drugs). In pharmacokinetic herb–drug interactions, the severity of an interaction is generally based on the toxicity of the drug being used or the consequences if the therapeutic dose is not achieved. When herbs are used with drugs that have a narrow therapeutic window (i.e., small difference between the effective dose and the toxic dose, such as with digoxin, warfarin, lithium, cyclosporine, phenytoin, and theophylline), supervision by a qualified healthcare practitioner is strongly advised.

Both pharmacodynamic and pharmacokinetic interactions may have positive effects, such as increasing the efficacy or bioavailability of drugs or botanicals. Such positive therapeutic interactions are not covered in this text, unless the interaction also poses a safety concern.

SELECTION OF THE EXPERT ADVISORY COUNCIL

Methods and considerations for safety evaluations that are outlined in the U.S. Institute of Medicine's (IOM) *Framework for Evaluating the Safety of Dietary Supplements* provided guidance in the literature collection and review processes that went into the creation of this text (IOM 2005). This IOM document also highlights the importance of using experts from a number of fields related to dietary supplements. Consistent with this advice, an advisory panel of qualified experts was assembled at the outset of this project. All members were selected for their extensive knowledge and experience in areas such as medicine, clinical herbalism, pharmacology, biochemistry, or traditional herbal medicine systems (e.g., traditional Chinese medicine or Ayurvedic medicine). When knowledge of a particular topic or botanical was not found in the Expert Advisory Council, the experience and opinions of outside experts were solicited.

LITERATURE REVIEW METHODS

Systematic literature searches were conducted in several electronic databases from January 2007 to May 2010, using search terms developed in cooperation with a technical information specialist from the National Library of Medicine, as follows:

- PubMed ([Latin name] OR [standardized common name]) AND (adverse effects OR adverse reaction OR safety OR tolerability OR drug interactions OR herb–drug interactions OR poisoning OR toxic OR toxicity OR toxicology OR drug toxicity OR teratogen* OR contraindicat* OR cytochrome OR p450 OR pregnancy OR lactation OR breast feeding OR breast milk OR pharmacodynamics OR "[MeSH term]/adverse effects" OR "[MeSH term]/toxicity" OR (Case Reports[ptyp]))
- EMBASE and Biological Abstracts ([Latin name] OR [standardized common name]) AND (adverse drug reaction OR safety OR tolerability OR drug tolerability OR herb drug interaction OR drug interaction OR drug contraindication AND contraindication OR poisoning OR intoxication OR drug toxicity OR toxic OR toxicity OR toxicology OR teratogen OR teratogenic OR teratogenicity OR cytochrome OR pregnancy OR lactation)
- TOXNET ([Latin name] OR [standardized common name]) AND (teratogen* OR adverse effects OR safety OR tolerability OR drug interactions OR poisoning OR toxicity OR cytochrome OR contraindications OR pregnancy OR lactation) NOT PubMed

Literature selected from these searches for review included meta-analyses, systematic reviews, other reviews, clinical trials, human, animal and *in vitro* pharmacological studies related to safety (including drug interaction studies), toxicity studies including reproductive and developmental toxicity studies, epidemiological studies, and ethnobotanical surveys. Articles on combination products and homeopathic products were generally excluded. No language restrictions were imposed, so publications in other languages were included whenever possible, but the review focused on English language publications.

Besides this extensive literature review, numerous other publications were consulted. These consisted primarily of authoritative references on the traditional use of herbal medicines, and also included regulatory documents, ethnobotanical records, pharmacopoeial texts, and writings on toxicology, food ingredients, and other relevant topics.

Literature was identified, obtained, and summarized by the research editor. Full literature summaries were presented to the Expert Advisory Council and other experts as needed, for review and assignment of safety and interaction ratings, contraindications, and precautions. Ratings were assigned by the Expert Advisory Council bearing in mind the history of use of the botanical under review.

No formal assessment of the validity of each reference was undertaken in this process, although the levels of evidence afforded by different types of publications (i.e., case report vs. randomized, placebo-controlled double-blind study) were actively considered during the review process. In addition, it was observed that some identified publications were of limited value, especially those that lack sufficient detail about the specific herbal preparation addressed, and case reports that postulate a causal relationship between a specific herbal ingredient and a reported adverse effect, without consideration for confounding factors such as patient history or concomitant drug use. Some such references were nonetheless retained, though the editors attempted to call attention to their perceived flaws.

Additional articles in scientific journals that were published subsequent to the 2007 to 2010 review were also considered for several entries during the editing stages that followed the process described above.

THE REVIEW PROCESS AND CLASSIFICATION

The herbal ingredients included in this edition are very nearly the same as those included the first edition, published in 1997. Some other herbs were added in order to include ingredients that have become more prominent in the U.S. marketplace in the interim. A few herbs addressed in the first edition are not included here, usually because no relevant contemporary publications were found and evidence from historical sources was lacking or insufficient.

Classifications are included for each part of the plant* identified in an entry, and are for dried plant material, unless otherwise stated. Classifications address only the identified part of the herb in its whole, cut, or powdered form, as a raw material or as an ingredient in a finished product (tablets, capsules, teas, etc.); or as a decoction, tincture, or extract prepared from that plant part by a traditional process. Concentrated extracts, extracts with added compounds, or compounds isolated from botanicals may be expected to have different physiological effects and safety and interaction considerations than the source

* Plant parts identified as "herb" consist of the leaf and stem of the identified plant, and this term is generally used only for non-woody plants. A plant part identified as "above-ground parts" means all of the plant above the ground, so it generally includes not only leaf and stem, but also flowers, fruits, and seeds, depending on the state of maturity of the plant at the time of harvest. All other plant parts (e.g., bark, leaf, root) are each identified with the generally used botanical term.

botanical, and classifications should not be extrapolated to other such ingredients without additional review.

Classifications are generally based on data that are associated with use of the specific herb and in the quantities generally consumed for a health-promoting or therapeutic effect. Any cautions may therefore be somewhat overstated for an herb that appears in the market in a smaller amount as part of a combination product, or for herbs that are used as flavorings in less than therapeutic quantities.

Each herb is placed in two classes based on all of the information included, along with the experience of the Expert Advisory Council. The first is the **Safety Class**, which evaluates the safety of a particular herb. The second is the **Interaction Class**, which provides information on what is currently known about the potential for an herb to alter the effect of prescription or non-prescription drugs when the herb and drug are used concomitantly. Central to the appropriate application of this document is the understanding that classifications are based on an assumption of rational, informed use of herbs and herbal products.

Classes are defined below, and are followed by bullet points which list criteria and considerations for inclusion in each particular class.

SAFETY CLASSES

Class 1. Herbs that can be safely consumed when used appropriately.

- History of safe traditional use
- No case reports of significant adverse events with high probability of causality
- No significant adverse events in clinical trials
- No identified concerns for use during pregnancy or lactation
- No innately toxic constituents
- Toxicity associated with excessive use is not a basis for exclusion from this class
- Minor or self-limiting side effects are not bases for exclusion from this class

Class 2.* Herbs for which the following use restrictions apply, unless otherwise directed by an expert qualified in the use of the described substance:

2a: For external use only

- Toxicity demonstrated with crude preparation taken orally at traditional dose
- Adverse event data in humans with probability of causality of toxicity (e.g., hepatotoxicity, nephrotoxicity, neurotoxicity) associated with oral use

2b: Not to be used during pregnancy

- Traditional use contraindicates
- Traditional use as an abortifacient or uterine stimulant
- Relevant adverse event data in humans exist and have probability of causality
- Data in animals suggesting teratogenicity or other adverse effects on the fetus or mother, with reasonable application to humans
- For plants with common food uses, standard dose is in excess of typical food amounts

2c: Not to be used while nursing

- Traditional use contraindicates
- Relevant adverse event data in humans exists and has probability of causality
- Potential hepatotoxicity or neurotoxicity
- Bioavailability of constituents of concern in breast milk has been demonstrated

2d: Other specific use restrictions as noted

- Information exists that use may be unsafe for specific populations
- Dosage level outside of a standard range known to cause adverse effects

Class 3. Herbs to be used only under the supervision of a qualified expert. The following labeling is recommended for Class 3 herbs: "To be used only under the supervision of an expert qualified in the appropriate use of this substance." Labeling must include proper use information: dosage, contraindications, potential adverse effects and drug interactions, and any other relevant information related to the safe use of the substance.

- Narrow therapeutic range
- Identified safety concerns in many populations

Interaction classes

Class A. Herbs for which no clinically relevant interactions are expected

- No case reports of suspected interactions with probability of causality
- No clinically relevant interactions in human pharmacological studies, if any

Class B. Herbs for which clinically relevant interactions are biologically plausible

- Human or animal pharmacological study data suggest potential for clinically relevant interaction.

* Herbs placed in any of the subparts of Class 2 may also be placed in other of these subparts.

- Multiple case reports have suggested a potential interaction concern.
- Cell culture or biochemical assays establish a basis for biologically plausible mechanism of interaction.

Class C. Herbs for which clinically relevant interactions are known to occur

- Human pharmacological study has demonstrated interaction with a specific drug or supplement.
- Human pharmacological study has demonstrated clinically relevant effects on drug metabolizing enzymes or drug transporter proteins.
- Case reports of suspected interactions have a probability of causality.

Limitations of Scope

This work specifically excludes the following data, conditions, and related products:

- Excessive consumption. Safety and interaction classifications given here are for normally consumed amounts, and cannot be assumed to have relevance for any quantity. Also, any concerns that are significant only in excessive or immoderate use are not relevant to assignation of classifications, though these may be referred to in an Editors' Note.
- Safety or toxicity concerns based on isolated constituents. As is the case with many common foods, some herbs are known to contain constituents that, in isolation, exhibit toxicity. Data based solely on constituents are not considered relevant to safety classification except in those cases where such compounds are known to accumulate, or where consumption patterns are sufficient to provide cause for health concerns. The presence of a constituent has been identified in a Notice if knowledge of the constituent is relevant to the safe use of an herb.
- Toxicity data based solely upon intravenous or intraperitoneal administration. The majority of herbal products consumed by the public are taken orally and with adequate dosage instructions. The physiological effects of injectable preparations are not relevant to oral consumption. Information associated with other forms of administration was reviewed but was not considered as a sole basis for classifications, and classifications should be assumed to address oral administration, unless otherwise stated.
- Traditional Chinese and Ayurvedic contraindications. In Chinese and Ayurvedic therapeutic traditions, most herbs have contraindications based on an individual's constitutional strengths and weaknesses, seasons, climate, and other factors that can only be understood in the context of the specific tradition. These traditional concerns have not been included in the text unless they can also be interpreted in a modern biomedical context, such as contraindication in pregnancy.
- Gastrointestinal disturbances. Reports of nausea or emesis from excessive doses, or occasional and/or minor gastrointestinal disturbances, have been noted but have not been considered in establishing safety classification, unless frequency or severity of such reactions warrants consideration.
- Idiosyncratic reactions. Any plant substance, whether used as food or medicine, has the potential to stimulate a negative response in unpredictably sensitive individuals. Safety classifications do not take into account such idiosyncratic responses, unless there is evidence to suggest that such an idiosyncratic reaction may be predictable.
- Allergic reactions. Certain plants in the Asteraceae, Apiaceae, and other plant families possess a relatively high degree of allergenicity, and specific mention of this is provided in the text for certain plants, such as feverfew herb (*Tanacetum parthenium*) and *Echinacea* spp. flowering tops. A plant's allergic potential, however, is not generally considered a basis for restrictive safety classification. Persons with a known allergy to ragweeds are nonetheless advised to observe caution in the consumption of all plants of the Asteraceae family, especially flowering parts.
- Contact dermatitis. The primary focus of this work is on herbal products for oral ingestion. Except in cases where there is a history of external therapeutic use, coupled with a record of associated dermatitis (e.g., mustard plasters), such concerns are beyond the scope of this document.
- Well-known toxic plants with known safety concerns that are not broadly traded. Many of the plants which are listed in standard toxicological texts as highly poisonous are not included in this document. Although isolates and constituents of some of these might be included in prescription drugs, they are not found in products which are otherwise accessible in a retail setting. Among the plants excluded are *Adonis vernalis*,

Claviceps purpurea, Chondrodendron tomentosum, Colchicum autumnale, Conium maculatum, Croton tiglium, Datura spp., *Gelsemium sempervirens, Hyoscyamus niger, Nicotiana* spp., *Rauwolfia* spp., *Stramonium* spp., *Strophanthus kombe,* and *Strychnos nux-vomica.*

- Homeopathic herbal preparations. Homeopathic products are classified as over-the-counter or prescription drugs and are regulated under the *Homeopathic Pharmacopoeia of the United States.* Safety concerns that arise for an herb in crude form may not apply to homeopathic preparations of the same herb, and this document does not address herbal products in homeopathic forms.
- Essential oils. Essential oils are concentrations of specific volatile compounds. While many essential oils have a well-documented history of safe use by appropriately skilled persons, they often present toxicological concerns that are absent or moderate in the crude plant materials from which the oil is derived. Except for a small number of essential oils that have a history of internal use, the classification of essential oils is beyond the scope of this document.
- Herbal products to which chemically defined active substances, including chemically defined isolated constituents of an herb, have been added. Safety of such products should be determined by manufacturers and marketers prior to market introduction.
- Environmental factors, additives, or contaminants. Classifications do not consider potential adulteration of botanical materials, although known adulterations that present health risks may be listed in an Editors' Note. Safety concerns of this sort must be addressed by the manufacturing practices of suppliers and manufacturers, who are responsible for assuring that herbal products are not contaminated or adulterated.

LITERATURE CITED

Bensky, D., and A. Gamble. 1986. *Chinese herbal medicine: Materia medica.* Seattle: Eastland Press.

Duke, J.A. 1989. *CRC handbook of medicinal herbs.* Boca Raton, FL: CRC Press.

IOM. 2005. *Dietary supplements; A framework for evaluating safety.* Institute of Medicine and National Research Council of the National Academies. Washington, D.C.: National Academies Press.

Kingsbury, J.M. 1979. *The problem of poisonous plants.* New York: Columbia University Press.

Sangalli, B., and W. Chiang. 2000. Toxicology of nutmeg abuse. *J. Toxicol. Clin. Toxicol.* 38 (6):671-678.

Turner, N., and A. Szczawinski. 1991. *Common poisonous plants and mushrooms of North America.* Portland, OR: Timber Press.

WHO. 1991. WHO Guidelines for the assessment of herbal medicines. WHO/TRM/91.4. Geneva: World Health Organization.

ORGANIZATION OF DATA

Listings are alphabetically arranged by Latin name. More than one species of a genus are combined into a single listing in those cases where two or more species are used interchangeably, or where the issues relevant to safe use are the same or nearly the same for related species. Some herbs supply more than one useful part. These parts are listed and classified together only in those cases where the safety issues of all parts are sufficiently similar; otherwise, separate listings are included for each plant part.

Following the Latin name is the botanical family name. In instances where synonymous Latin names may be encountered in relevant references, one or more of these may be listed as a Synonym (**Syn).**

It is not unusual for a plant to have many common names, a fact which can confound the understanding of an herb's uses and potential safety concerns. AHPA published *Herbs of Commerce, 2nd edition* (McGuffin et al. 2000) to address this concern by assigning a single common or usual name to each herb, denoted in each listing in the current text as its standardized common name (**SCN**). Additional familiar common names are listed as other common names (**OCN**), though this field is generally not intended to be exhaustive. Ayurvedic names (**AN**) and pinyin names (**PN**) for botanical ingredients commonly used in Ayurvedic or traditional Chinese medicine are also included; note that Ayurvedic names tend to identify the plant itself, while pinyin names usually identify a specific plant part. With occasional exceptions, nomenclature in this work is derived from *Herbs of Commerce, 2nd edition*.

Following the plant's names is the **Part** of the plant for which the safety and interactions classifications that follow are made. Occasional specific information is included for those herbs that require special processing.

The remainder of each listing is divided into two sections, the **Quick Reference Summary**, which provides a concise, clinically relevant summary of the scientific information and traditional knowledge on the safety of each species or set of species, and the **Review Details** section, which provides details on the information presented in the summary.

Each entry's **Quick Reference Summary** includes the following elements. Each of the fields printed below in bold are always included, and state "None known" in the absence of any information relevant to the entry. All other fields are optional, and are included only for those entries for which information in the described area is relevant.

- Safety Class: Each entry is assigned one or more of the Safety Classes described earlier in this introduction.
- Interaction Class: Each herb is also assigned an Interaction Class as described previously.
- Contraindications: Any situations, conditions, or populations in which the botanical should not be used are listed here.
- Other Precautions: Special considerations for use are identified in this field. These may include, for example, common idiosyncratic effects (e.g., allergic reactions), adverse effects that may be undesirable but are not typically dangerous, or other conditions that require some specific caution, as stated.
- Drug and Supplement Interactions: This section gives details on known or suspected interactions in order to provide further information on any possible or probable interactions noted in an Interaction Class B or C. Note, however, that possible interactions that have low levels of evidence, or drugs for which a lack of interactions has been demonstrated, are generally listed under Pharmacological Considerations.
- Standard Dose: Quantitative dosage information is included here only for those plants listing a recommendation that excessive dosage be avoided. The dose is usually given in the quantity and form for direct consumption or for preparation as a tea or decoction and is based on the herb in its dried (dehydrated) form, unless otherwise stated. Equivalent dosage in the form of tinctures and extracts must be calculated based on the concentration of such extracts on a dry weight basis. Standard Dose should not be taken to be the equivalent of a dosage limitation. Rather, this dosage should be seen as related to the concept of "serving size." Although Standard Dose may be relevant to the determination of appropriate dosage limits, a thorough examination of other specific factors would be required prior to setting such levels.
- Notice: Certain plant constituents, such as caffeine or pyrrolizidine alkaloids, and herbs with known physiological actions, such as emetics and nervous system stimulants, may present safety considerations in numerous species. Rather than address such concerns in detail for each individual species, a Notice identifies these constituents

or actions and directs the reader to a thorough discussion of each such subject in Appendix 1, 2, or 3.

- Editors' Notes. Supplemental information relevant to the safe use of an herb, such as specific labeling recommendations, information regarding preparation, content of a chemical compound of potential concern, exceptions to use restrictions, possible adulteration, and other information are all included in this section, if required. Some discussion of the details, quality, or applicability of cited references may also be included here.
- Adverse Events and Side Effects. Recorded adverse changes in health, including any abnormal signs or symptoms, that have been reported to have occurred in association with the use of a particular herb are listed at this field. Side effects are defined as predictable effects of an herb that are not the principal effect for which the herb was taken (e.g., some people experience heartburn after ingestion of ginger). Adverse events, which include any health-related event associated with the use of a product that is perceived as harmful to the user, may or may not be related to an herb that was being taken at the time of an event. While some adverse events temporally associated with usage may be attributable to the herb consumed (e.g., nausea, vomiting, and central nervous system disturbances with overdose of raw *Ginkgo biloba* seed), many adverse events identified in case reports are not likely to be related to the associated herb. Sufficient detail is often lacking in case reports to determine whether a particular herb was likely the cause of any adverse event, and a case report cannot be considered to be in and of itself evidence that the reported adverse event was caused by the identified herb. All case reports in this text refer to human cases, unless they are listed under animal studies or otherwise specified.
- Pharmacological Considerations. If the physiological effects or other pharmacological activity of an herb may be relevant to the safe use of that herb, this information is reported here. Preference is given to data from human and animal use, although in vitro data that may be relevant to clinical use is also listed here. Low-level evidence for potential drug interactions is also typically included here.
- Pregnancy and Lactation. As available, information on the safety of herbs during pregnancy or while nursing is provided in this field. For a number of the Class 1 herbs, substantial data or traditional use suggests that these may be safely used in pregnancy and lactation. For other botanicals, less data and clinical experience are available regarding their use in pregnancy or lactation. The absence of formal data and clinical experience regarding the use of a botanical in pregnancy or lactation, in and of itself, was not justification to contraindicate the botanical in these conditions. In such cases, the editors and Expert Advisory Council have used their best judgment in conjunction with the available literature to make the most appropriate determination. The following statement is included in those entries for which data and clinical experience for the botanical were lacking or less robust than desired:

No information on the safety of this herb in pregnancy or lactation was identified in the scientific or traditional literature. Although this review did not identify any concerns for use while pregnant or nursing, safety has not been conclusively established.

The **Review Details** section for each entry is divided into five primary fields, each of which has its own organization. The reader will observe considerable redundancy when reading an herb's Quick Reference Summary and its Review Details sections together, as each of these is designed to be complete in itself. Thus, while the Quick Reference Summary provides enough information to understand an herb's safety and interaction profile, the Review Details section provides a more in-depth discussion of the data that was reviewed for the entry.

Some of the specific elements of this section are always present (again shown in bold font below) and when there is no relevant information known for a specific entry, that fact is affirmatively stated (e.g., "No clinical trials of drug or supplement interactions were identified."). All other elements are optional, and are again included only for those entries where information in the described area is relevant to the listing.

I. Drug and Supplement Interactions
   - Clinical trials of drug or supplement interactions
   - Case reports of suspected drug or supplement interactions
   - Animal trials of drug or supplement interactions

II. Adverse Events
   - Adverse events reported in clinical trials
   - Case reports of adverse events

III. Pharmacology and Pharmacokinetics
   - Human pharmacological studies
   - Animal pharmacological studies
   - In vitro pharmacological studies

IV. Pregnancy and Lactation
V. Toxicity Studies
- Acute toxicity
- Short-term toxicity
- Subchronic toxicity
- Chronic toxicity
- Genotoxicity
- Cytotoxicity

Each entry closes with a listing of the **Literature Cited** for that particular entry.

LITERATURE CITED

McGuffin, M., J. Kartesz, A. Leung, and A.O. Tucker. 2000. *Herbs of commerce. 2nd ed*. Silver Spring, MD: American Herbal Products Association.

DISCLAIMER

The editors and the Expert Advisory Council of the *Botanical Safety Handbook* have endeavored to ensure that the information contained in this document accurately represents contemporary knowledge on the safe use of herbal ingredients. In developing this work, particular care was given to identifying references that provide accurate information, and efforts were made to present a balanced view of all available scientific information.

The safe oral consumption of any substance can depend to a great deal on the health of an individual consumer, as well as to the quantity of the substance consumed. In addition, idiosyncratic or allergic reactions are often unpredictable. Any person who consumes an herb listed in this reference based on its classifications does so at his or her own risk, and should consult a healthcare provider in the event of an adverse response.

There is no obligation at this time for AHPA members to adopt the specific information contained here in their product labeling. Rather, this document is presented as a guideline, providing data to assist member and nonmember manufacturers in developing labels that fully inform consumers. Verification of all data and classifications for the purpose of label development is the responsibility of the manufacturer.

Abies balsamea (L.) Mill. — Pinaceae

SCN: balsam fir
OCN: American silver fir

Part: bark, needles, sap, wood

Quick Reference Summary

Safety Class: 1
Interaction Class: A

Contraindications
None known.

Other Precautions
None known.

Drug and Supplement Interactions
None known.

Adverse Events and Side Effects
None known.

Pharmacological Considerations
None known.

Pregnancy and Lactation
No information on the safety of balsam fir in pregnancy or lactation was identified in the scientific or traditional literature. Although this review did not identify any concerns for use while pregnant or nursing, safety has not been conclusively established.

Review Details

I. Drug and Supplement Interactions

Clinical Trials of Drug or Supplement Interactions
No clinical trials of drug or supplement interactions were identified.

Case Reports of Suspected Drug or Supplement Interactions
No case reports of suspected drug or supplement interactions were identified.

Animal Trials of Drug or Supplement Interactions
No animal trials of drug or supplement interactions were identified.

II. Adverse Events

Case Reports of Adverse Events
No case reports of adverse events were identified.

III. Pharmacology and Pharmacokinetics

Human Pharmacological Studies
No relevant human pharmacological studies were identified.

Animal Pharmacological Studies
No relevant animal pharmacological studies were identified.

In Vitro Pharmacological Studies
An ethanol extract of an unidentified part of balsam fir demonstrated in vitro mechanism-based inhibition of the drug metabolizing isoenzyme CYP3A4 (Tam et al. 2011).

IV. Pregnancy and Lactation
No information on the safety of balsam fir during pregnancy or lactation was identified.

V. Toxicity Studies
No toxicity studies were identified.

Literature Cited

Tam, T.W., R. Liu, J.T. Arnason, A. Krantis, W.A. Staines, P.S. Haddad, and B.C. Foster. 2011. Cree antidiabetic plant extracts display mechanism-based inactivation of CYP3A4. *Can. J. Physiol. Pharmacol.* 89(1):13-23.

A

Achillea millefolium L. — Asteraceae

SCN: yarrow
OCN: milfoil

Part: herb

Quick Reference Summary

Safety Class: 1
Interaction Class: A

Contraindications

None known.

Other Precautions

Persons with allergies to other members of the Asteraceae family (such as feverfew, chamomile, or *Echinacea* species) should exercise caution with yarrow, as allergic cross-reactivity is common to Asteraceae plants (Hausen 1996; Paulsen et al. 1993).

Drug and Supplement Interactions

None known.

Notice

Thujone (trace amounts) (Bradley 1992); *see* Appendix 1.

Editors' Notes

Use of yarrow as a food additive in the United States is subject to a limitation that the finished food or beverage is thujone-free (CFR 2011). Dietary ingredients for use in dietary supplements, however, are specifically excluded from the federal food additive definition (U.S.C. 2010).

Thujone is present in yarrow only in trace amounts (Leung and Foster 1996). Some concerns regarding the safety of thujone have been based on the effects of absinthe, an alcoholic beverage that historically contained thujone. Recent research, however, indicates that the alcohol content, rather than the thujone content, of absinthe was responsible for the reported adverse effects (Lachenmeier et al. 2006, 2008).

Adverse Events and Side Effects

Cases of contact allergy to yarrow plants have been reported, and allergic cross-reactivity to plants in the Asteraceae family has been documented (Davies and Kersey 1986; Guin and Skidmore 1987; Hausen 1996; Paulsen et al. 1993).

Pharmacological Considerations

In vitro studies with yarrow have reported inhibition of some CYP enzymes (Scott et al. 2006), increase in bile flow (Benedek et al. 2006), and estrogenic activity (Innocenti et al. 2007). One animal study showed some adverse effects on sperm at high doses (1.2 g/kg daily) but not at lower doses (Dalsenter et al. 2004).

Pregnancy and Lactation

Information on the safety of yarrow during pregnancy and lactation is limited. One animal study showed a decrease in fetal weight in offspring of rats administered high (2.8 g/kg) doses of yarrow, but no adverse effects were seen at lower doses (Boswell-Ruys et al. 2003).

No information on the safety of yarrow during lactation was identified. While this review did not identify any concerns for use while nursing, safety has not been conclusively established.

Review Details

I. Drug and Supplement Interactions

Clinical Trials of Drug or Supplement Interactions

No clinical trials of drug or supplement interactions were identified.

Case Reports of Suspected Drug or Supplement Interactions

No case reports of suspected drug or supplement interactions were identified.

Animal Trials of Drug or Supplement Interactions

No animal trials of drug or supplement interactions were identified.

II. Adverse Events

Case Reports of Adverse Events

Cases of contact allergy to yarrow have been documented (Davies and Kersey 1986; Guin and Skidmore 1987) and are believed to be caused primarily by the sesquiterpene lactone α-peroxyachifolid (Hausen et al. 1991).

In patch testing of Asteraceae-sensitive individuals, approximately 1.5% of 3800 test subjects were sensitive to yarrow (Hausen 1996). Similarly, patch testing of 686 subjects revealed 32 with sensitivity to several species of Asteraceae plants, including yarrow (Paulsen et al. 1993).

III. Pharmacology and Pharmacokinetics

Human Pharmacological Studies

No relevant human pharmacological studies were identified.

Animal Pharmacological Studies

No adverse effects on the male reproductive system were observed in male rats orally administered up to 600 mg/kg daily of an aqueous yarrow extract for 90 days. An increase in the percentage of abnormal sperm was observed in rats treated with 1.2 g/kg daily (Dalsenter et al. 2004).

In Vitro Pharmacological Studies

Inhibition of CYP450 isoenzymes CYP2C19, CYP19, and CYP3A4 by a methanolic extract of yarrow was observed in vitro (Scott et al. 2006).

A dose-dependent increase in bile flow was observed in isolated perfused rat livers treated with a polar fraction of yarrow (Benedek et al. 2006).

A methanol and water extract of yarrow demonstrated estrogenic activity in estrogen receptor-positive human breast cancer cells (MCF-7). Activation of estrogen receptors α and β was seen (Innocenti et al. 2007).

IV. Pregnancy and Lactation

A decrease in fetal weight was observed in offspring of rats administered 2.8 g/kg daily of an ethanolic extract of yarrow on gestational days 8 to 15, but no effects on fetuses were seen when yarrow was administered on GD 1 to 8. No changes in pre-implantation or post-implantation loss were observed (Boswell-Ruys et al. 2003).

No information on the safety of yarrow during lactation was identified.

V. Toxicity Studies

Acute Toxicity

The LD_{50} of both orally and subcutaneously administered yarrow extract (2% in propylene glycol and water) in mice is 1 g/kg (Provital 1998).

Chronic Toxicity

No signs of toxicity were observed in rats administered up to 1.2 g/kg daily of a yarrow aqueous extract for 90 days (Cavalcanti et al. 2006).

Genotoxicity

A weakly genotoxic effect of a yarrow aqueous extract was reported in *Drosophila melanogaster* (Graf et al. 1994).

Literature Cited

Benedek, B., N. Geisz, W. Jager, T. Thalhammer, and B. Kopp. 2006. Choleretic effects of yarrow (*Achillea millefolium* s.l.) in the isolated perfused rat liver. *Phytomedicine* 13(9-10):702-706.

Boswell-Ruys, C.L., H.E. Ritchie, and P.D. Brown-Woodman. 2003. Preliminary screening study of reproductive outcomes after exposure to yarrow in the pregnant rat. *Birth Defects Res. B Dev. Reprod. Toxicol.* 68(5):416-420.

Bradley, P.R. 1992. *British herbal compendium: A handbook of scientific information on widely used plant drugs.* Bournemouth, Dorset: British Herbal Medicine Association.

Cavalcanti, A.M., C.H. Baggio, C.S. Freitas, et al. 2006. Safety and antiulcer efficacy studies of *Achillea millefolium* L. after chronic treatment in Wistar rats. *J. Ethnopharmacol.* 107(2):277-284.

CFR. 2011. *Code of federal regulations*, Title 21 Part 172.510, 2011 ed. Food additives permitted for direct addition to food for human consumption. Flavoring agents and related substances. Natural flavoring substances and natural substances used in conjunction with flavors. Washington, DC: U.S. Government Printing Office.

Dalsenter, P.R., A.M. Cavalcanti, A.J. Andrade, S.L. Araujo, and M.C. Marques. 2004. Reproductive evaluation of aqueous crude extract of *Achillea millefolium* L. (Asteraceae) in Wistar rats. *Reprod. Toxicol.* 18(6):819-823.

Davies, M.G., and P.J. Kersey. 1986. Contact allergy to yarrow and dandelion. *Contact Dermat.* 14(4):256-257.

Graf, U., A. Alonso Moraga, R. Castro, and E. Diaz Carrillo. 1994. Genotoxicity testing of different types of beverages in the *Drosophila* wing somatic mutation and recombination test. *Food Chem. Toxicol.* 32(5):423-430.

Guin, J.D., and G. Skidmore. 1987. Compositae dermatitis in childhood. *Arch. Dermatol.* 123(4):500-502.

Hausen, B.M. 1996. A 6-year experience with compositae mix. *Am. J. Contact Dermat.* 7(2):94-99.

Hausen, B.M., J. Breuer, J. Weglewski, and G. Rücker. 1991. alpha-Peroxyachifolid and other new sensitizing sesquiterpene lactones from yarrow (*Achillea millefolium* L., Compositae). *Contact Dermat.* 24(4):274-280.

Innocenti, G., E. Vegeto, S. Dall'Acqua, et al. 2007. In vitro estrogenic activity of *Achillea millefolium* L. *Phytomedicine* 14(2-3):147-152.

Lachenmeier, D., D. Nathan-Maister, T. Breaux, et al. 2008. Chemical composition of vintage preban absinthe with special reference to thujone, fenchone, pinocamphone, methanol, copper, and antimony concentrations. *J. Agric. Food Chem.* 59(9):3073-3081.

Lachenmeier, D.W., J. Emmert, T. Kuballa, and G. Sartor. 2006. Thujone—Cause of absinthism? *Forensic Sci. Int.* 158(1):1-8.

Leung, A.Y., and S. Foster. 1996. *Encyclopedia of common natural ingredients used in food, drugs, and cosmetics.* 2nd ed. New York: Wiley.

Paulsen, E., K.E. Andersen, and B.M. Hausen. 1993. Compositae dermatitis in a Danish dermatology department in one year. I. Results of routine patch testing with the sesquiterpene lactone mix supplemented with aimed patch testing with extracts and sesquiterpene lactones of Compositae plants. *Contact Dermat.* 29(1):6-10.

Provital, S.A. 1998. Raw materials documentation for yarrow (*Achillea millefolium*) extract. Unpublished data. *Cited in* Fiume, M. 2001. Final report on the safety assessment of yarrow (*Achillea millefolium*) extract. *Int. J. Toxicol.* 20(Suppl. 2):79-84.

Scott, I.M., R.I. Leduc, A.J. Burt, et al. 2006. The inhibition of human cytochrome P450 by ethanol extracts of North American botanicals. *Pharmaceut. Biol.* 44(5):315-327.

U.S.C. 2010. United States Code, Title 21, Part 321 (s)(6). Current as of January 7, 2011. Washington, DC: U.S. Government Printing Office.

A

Achyranthes bidentata Blume — Amaranthaceae

SCN: achyranthes
PN: *niu xi* (root)
OCN: ox knee
Part: root

Quick Reference Summary

Safety Class: 2b, 2d
Interaction Class: A

Contraindications

Not for use in pregnancy except under the supervision of a qualified healthcare practitioner (Bensky et al. 2004; Chen and Chen 2004).

Not for use in excessive menstruation (Bensky et al. 2004; Chen and Chen 2004).

Other Precautions

None known.

Notice

Uterine stimulant (Bensky et al. 2004; Chen and Chen 2004); *see* Appendix 2.

Editors' Note

Multiple species are traded under the name *niu xi,* and all are contraindicated in pregnancy (Bensky et al. 2004).

Adverse Events and Side Effects

None known.

Pharmacological Considerations

None known.

Pregnancy and Lactation

Texts on traditional Chinese medicine indicate that achyranthes should not be used during pregnancy (Bensky et al. 2004; Chen and Chen 2004). Dilation of the cervical os was observed in association with achyranthes use in women who had abortions (Chen and Chen 2004). Animal studies of achyranthes have indicated anti-implantation, antifertility, and uterine stimulating activity (Che 1988; Chen and Chen 2004; Zhu and Che 1987).

No information on the safety of achyranthes during lactation was identified. While this review did not identify any concerns for use while nursing, safety has not been conclusively established.

Review Details

I. Drug and Supplement Interactions

Clinical Trials of Drug or Supplement Interactions

No clinical trials of drug or supplement interactions were identified.

Case Reports of Suspected Drug or Supplement Interactions

No case reports of suspected drug or supplement interactions were identified.

Animal Trials of Drug or Supplement Interactions

No animal trials of drug or supplement interactions were identified.

II. Adverse Events

Case Reports of Adverse Events

No case reports of adverse events were identified.

III. Pharmacology and Pharmacokinetics

Human Pharmacological Studies

No relevant human pharmacological studies were identified.

Animal Pharmacological Studies

In mice intraperitoneally administered polysaccharides from achyranthes at doses of 50, 100, or 200 mg/kg daily for 15 days, the 50 mg/kg dose inhibited the growth of introduced lung cancer tumors while the 200 mg/kg dose increased tumor growth. Tumor growth in mice administered 100 mg/kg was equivalent to that of the untreated control group (Jin et al. 2007).

Also see Pregnancy and Lactation for this entry.

In Vitro Pharmacological Studies

No relevant in vitro pharmacological studies were identified.

IV. Pregnancy and Lactation

Dilation of the cervical os was observed in association with achyranthes use in women who had abortions (Chen and Chen 2004).

A benzene extract of achyranthes saponins orally administered to mice at doses of 50 or 80 mg/kg reduced female fertility and implantation. The chloroform extract administered at doses of 80 or 120 mg/kg reduced fertility but did not affect implantation (Che 1988).

A dose-dependent decrease in fertility was observed in mice orally administered achyranthes saponins at doses of 125 to 1000 mg/kg (ED_{50} was 218 mg/kg). Implantation was prevented in mice orally administered 500 mg/kg of achyranthes saponins 5 days after mating, although no such activity was observed in rats administered 500 mg/kg. No abortifacient activity was observed in rats orally administered 2 g/kg daily achyranthes saponins on days 14 to 19 after mating (Zhu and Che 1987).

Administration of achyranthes to mated female mice at doses of 250 to 500 mg/kg for 20 days resulted in a decrease in fertility and increased risk of miscarriage (Chen and Chen 2004).

Studies in rabbits and rats indicated that achyranthes stimulates uterine contractions (dose and route of administration not specified in English language translation) (Chen and Chen 2004).

No information on the safety of achyranthes during lactation was identified.

V. Toxicity Studies

Acute Toxicity

The LD_{50} of an orally administered combination of the compounds ecdysterone and inoteosterone in mice is 9 g/kg (Chen and Chen 2004).

Short-Term Toxicity

No abnormalities in blood parameters, liver, or kidneys were observed in mice orally administered an achyranthes decoction at a dose of 60 g/kg daily for 30 days (Chen and Chen 2004).

Literature Cited

Bensky, D., S. Clavey, and E. Stöger. 2004. *Chinese herbal medicine: Materia medica.* 3rd ed. Seattle: Eastland Press.

Che, X. 1988. Anti-fertility effects of *Achyranthes bidentata* in mice. *Xi'an Yike Daxue Xuebao* 9(2):119-121.

Chen, J.K., and T.T. Chen. 2004. *Chinese medical herbology and pharmacology.* City of Industry, CA: Art of Medicine Press.

Jin, L.Q., Z.J. Zheng, Y. Peng, et al. 2007. Opposite effects on tumor growth depending on dose of *Achyranthes bidentata* polysaccharides in C57BL/6 mice. *Int. Immunopharmacol.* 7(5):568-577.

Zhu, H., and X. Che. 1987. Study on antifertility effect of *Achyranthes bidentata* saponins on rats and mice. *Xi'an Yike Daxue Xuebao* 8(3):246-249.

Aconitum carmichaelii Debeaux

Ranunculaceae

SCN: Sichuan aconite

PN: *chuan wu* (prepared main root); *fu zi* (prepared lateral root)

OCN: Japanese aconite

Part: prepared main and lateral root

Quick Reference Summary

Safety Class: 3

Interaction Class: A

Contraindications

Not for use except under the supervision of an expert qualified in the appropriate use of this substance (Bensky et al. 2004; Bisset 1981; Chan 2009; Fitzpatrick et al. 1994; Lin et al. 2004).

The unprepared root should never be taken internally (Bensky et al. 2004).

Other Precautions

References on traditional Chinese medicine indicate that alcohol should not be consumed with Sichuan aconite, as the absorption of the toxic constituents of Sichuan aconite will be greatly enhanced (Bensky et al. 2004; Chen and Chen 2004).

Drug and Supplement Interactions

Extreme caution is advised for use of Sichuan aconite in combination with antiarrhythmic medications (Chen and Chen 2004).

Editors' Notes

Both prepared and unprepared Sichuan aconite are available commercially. Sichuan aconite contains aconitine, a toxic alkaloid that affects the heart and the central nervous system (Bensky et al. 2004). Due to aconitine content, the unprepared root is highly toxic and is the primary herb associated with serious adverse events in traditional Chinese medicine hospitals in Hong Kong (Chan 2002; Chan et al. 1994a, 1994c; Poon et al. 2006). Processing of Sichuan aconite root greatly reduces the content of aconitine (Chen and Chen 2004). The prepared root, that has been processed to reduce toxicity, is the subject of this entry.

Sichuan aconite may be prepared in several ways, the most common of which is to cook the herb at boiling temperature for several hours. Such processing reduced the toxicity of Sichuan aconite to between 1/2000 and 1/4000 of the toxicity of the unprocessed herb (Chen and Chen 2004). Heat processing at temperatures above 120°C for 50 minutes decreased the diester alkaloids, such as mesaconitine, aconitine, and hypaconitine, and increased monoester alkaloids, such as benzoylmesaconine, benzoylaconine, and benzoylhypaconine, whereas heating to 105°C preserved the diester alkaloids (Taki et al. 1998).

A text on traditional Chinese medicine notes that while prepared Sichuan aconite is recognized to be toxic, if the appropriate dosage of the prepared root is combined with other appropriate ingredients such as ginger and licorice, and the patient is carefully instructed on the method of proper decoction, little likelihood of toxicity exists (Bensky et al. 2004).

Other species of *Aconitum* are also in trade (Bensky et al. 2004), and all should be considered class 3.

Adverse Events and Side Effects

Cases of aconite poisoning have been reported; some have been fatal. Characteristic symptoms of poisoning include nausea, vomiting, generalized paresthesia (numbness), irregular heartbeat, and cold extremities (Bisset 1981; Fitzpatrick et al. 1994).

A review of case reports of poisoning from various species of aconite indicated that the risk of poisoning is higher with inadequately processed aconite root, improperly prepared extracts (i.e., patients not boiling root as long as directed when making decoctions), large doses, and alcohol-based extracts (Lin et al. 2004).

Pharmacological Considerations

See Adverse Events and Side Effects above.

Pregnancy and Lactation

Traditional Chinese medicine texts contraindicate the use of prepared Sichuan aconite in pregnancy (Bensky et al. 2004; Chen and Chen 2004).

No information on the safety of prepared Sichuan aconite during lactation was identified. While this review did not identify any concerns for use while nursing, safety has not been conclusively established, and this substance is not recommended for use except under the supervision of an expert qualified in its appropriate use.

Review Details

I. Drug and Supplement Interactions

Clinical Trials of Drug or Supplement Interactions

No clinical trials of drug or supplement interactions were identified.

Case Reports of Suspected Drug or Supplement Interactions

No case reports of suspected drug or supplement interactions were identified.

Animal Trials of Drug or Supplement Interactions

No animal trials of drug or supplement interactions were identified.

II. Adverse Events

Case Reports of Adverse Events

Aconite poisoning has been reported to occur after mistaken use of the unprepared herb, inappropriate preparation, or overdose. Poisoning may affect the nervous system (dizziness, blurred vision, mydriasis, loss of vision, and numbness of mouth, limbs, or whole body), digestive system (severe nausea and vomiting), and circulatory system (palpitations, low blood pressure, cold extremities, chest pain, bradycardia, sinus tachycardia, ventricular ectopics, ventricular arrhythmias, and junctional rhythm) (Bisset 1981; Chan 2009; Fitzpatrick et al. 1994).

Toxic effects are caused by the alkaloid aconitine (Fu et al. 2006; Lin et al. 2004). Aconitine and other alkaloids activate the sodium channel and have widespread effects on the excitable membranes of cardiac, neural, and muscle tissues. Muscarinic activation also causes hypotension and bradyarrhythmias (Chang and But 1986).

A number of cases of aconite poisoning have been reported, most with typical clinical symptoms of aconite poisoning; some cases were fatal (But et al. 1994; Chan 2002; Chan et al. 1993, 1994a, 1994b, 1994c; Fatovich 1992; Fujita et al. 2007; Kolev et al. 1996; Lowe et al. 2005; Smith et al. 2005; Tai et al. 1992a, 1992b). Severe poisoning has been reported after consumption of as little as 0.2 mg of the compound aconitine or decoctions prepared from a Chinese herbal prescription containing 6 g of prepared Sichuan aconite (But et al. 1994). The toxic dose range is reported to be between 15 and 60 g of dried root prepared as a decoction, and is dependent on the time of harvest, method of preparation, and length of decocting time (Bensky et al. 2004).

A review of case reports of poisoning from various species of aconite indicated that the risk of poisoning is higher with inadequately processed aconite root, improperly prepared extracts (i.e., patients not boiling root as long as directed when making decoctions), large doses, and alcohol-based extracts (Lin et al. 2004).

III. Pharmacology and Pharmacokinetics

Human Pharmacological Studies

No relevant human pharmacological studies of prepared Sichuan aconite were identified.

Animal Pharmacological Studies

A dose-dependent decrease in plasma glucose levels was seen in diabetic rats orally administered up to 50 mg/kg prepared Sichuan aconite. The plasma glucose-lowering effect was eliminated by blockage of μ-opioid receptors (Liou et al. 2006).

In mice orally administered 1 mg/kg daily of the compound aconitine, various types of arrhythmias were observed including ventricular fibrillation, ventricular tachycardia, and bundle branch block. The arrhythmias occurred within 30 minutes of administration of aconitine,

varies according to origin. The essential oil of plants from India contains up to 75% β-asarone (*see* Alkenylbenzenes in Appendix 1), while the oil of calamus from Japan and eastern Russia contains 10–40%, oil from European plants contains approximately 13%, and that from North America contains almost no β-asarone (Keller and Stahl 1982, 1983; Raina et al. 2003; Stahl and Keller 1981; Subramanian et al. 2008). Since varieties of calamus may not be well differentiated in commerce, the caution stated for the Asian and European varieties should be considered relevant to any sample that is not positively identified as the North American variety.

Animal and in vitro studies have indicated that the compound β-asarone has carcinogenic, mutagenic, and chromosome-damaging properties (Abel 1987; Balachandran et al. 1991; FAO/WHO 1981; Goggelmann and Schimmer 1983; Habermann 1971; Hasheminejad and Caldwell 1994).

All varieties of calamus are prohibited in foods in the United States (CFR 2011).

Adverse Events and Side Effects

None known.

Pharmacological Considerations

A reference text on traditional Chinese medicine indicated that overdose and long-term use of calamus should be avoided (Bensky et al. 2004).

Pregnancy and Lactation

A study with calamus essential oil in chicken eggs showed no adverse effects on embryo development (Yabiku et al. 1979). No other information on the safety of calamus in pregnancy or lactation was identified.

While this review did not identify any concerns for use while pregnant or nursing, safety has not been conclusively established, and this substance is not recommended for use except under the supervision of an expert qualified in its appropriate use.

Review Details

I. Drug and Supplement Interactions

Clinical Trials of Drug or Supplement Interactions

No clinical trials of drug or supplement interactions were identified.

Case Reports of Suspected Drug or Supplement Interactions

No case reports of suspected drug or supplement interactions were identified.

Animal Trials of Drug or Supplement Interactions

Extracts of calamus have been shown to potentiate pentobarbitone-induced sleeping time (Dandiya et al. 1959; Hazra et al. 2007; Panchal et al. 1989).

II. Adverse Events

Case Reports of Adverse Events

A 19-year-old man experienced diaphoresis, persistent vomiting, and mild leukocytosis after ingesting an 8-inch-long calamus rhizome (Vargas et al. 1998).

III. Pharmacology and Pharmacokinetics

Human Pharmacological Studies

No relevant human pharmacological studies were identified.

Animal Pharmacological Studies

An ethanolic extract of calamus demonstrated immunomodulatory potential by inhibiting proliferation of mitogen- and antigen-stimulated human peripheral blood mononuclear cells. The extract inhibited growth of several mouse and human cell lines (Mehrotra et al. 2003).

Animal studies have indicated that calamus has a depressant effect on the central nervous system (Agarwal et al. 1956; Dandiya et al. 1958, 1959; Dandiya and Cullumbine 1959; Dasgupta et al. 1977).

In Vitro Pharmacological Studies

Negative ionotropic and chronotropic effects were observed in frog heart preparations treated with concentrations of 100 μg/ml of an alcohol extract of calamus (Panchal et al. 1989).

IV. Pregnancy and Lactation

No teratogenic effects were observed in chicken eggs injected with calamus essential oil at doses of 0.12, 0.60, 3.00, 15.00, or 75.00 mg/egg (Yabiku et al. 1979). Similarly, no teratogenic effects were observed in chicken eggs injected with α-asarone at doses up to 4 mg/egg. In eggs injected with β-asarone, 43% of embryos survived a dose of 0.04 mg/egg, while none survived a dose of 4 mg/egg (Yabiku et al. 1979).

No information on the safety of calamus in lactation was identified.

V. Toxicity Studies

Acute Toxicity

The LD_{50} of orally administered calamus oil in rats was 0.77 g/kg for oil from Jammu (~75% β-asarone) (Jenner et al. 1964), 4.3 g/kg for oil from Kashmir (~5% β-asarone) (WHO 1981), and 3.5 g/kg for oil from Europe (~5% β-asarone) (WHO 1981).

The LD_{50} of intraperitoneally administered European calamus oil in mice was 1.1 g/kg for oil containing β-asarone and 1.7 g/kg for oil with no β-asarone (Yabiku et al. 1979). In rats, the intraperitoneal LD_{50} of the essential oil was 299 mg/kg (Yabiku et al. 1979). The LD_{50} of intraperitoneally administered β-asarone in mice was 0.184 g/kg (Yabiku et al. 1979).

A

Chronic Toxicity

In rats fed diets containing 0, 500, 1000, 2500, or 5000 ppm (0, 0.05, 0.1, 0.25, and 0.5%) Jammu calamus essential oil (~75% β-asarone) daily for 2 years, all of the 5000 ppm group died within 45 weeks, all of the 2500 ppm group died within 68 weeks, and all of the 1000 ppm group died within 104 weeks. Gross abnormalities were observed, including liver damage, fluid in the pleural and or peritoneal cavity, and tumorous masses in the intestines. Cardiac atrophy was observed in both test and control animals but was more severe in test animals (Taylor 1967, 1981).

In rats fed diets containing 0.1, 0.5, 1.0, or 2.0% European calamus essential oil (~5% β-asarone) daily for 2 years, leiomyosarcomas, hepatocellular adenomas, and hepatocellular adenocarcinomas were observed at the 1 and 2% dose levels. Other dose-dependent adverse effects on the livers were observed, with effects at the 0.1% dose being similar to those in the controls, or slightly increased. Dose-dependent changes observed in the heart included myocardial atrophy, fibrosis, fatty degeneration, and fatty infiltration (Taylor 1981).

In rats fed diets containing 0, 400, 800, or 2000 ppm (0, 0.04, 0.08, or 0.2%) β-asarone for 2 years, none of the animals receiving 2000 ppm β-asarone survived more than 84 weeks, and mortality was increased at the 800 ppm dose. Gross pathological changes were observed and included serous fluid in the abdominal and pleural cavities, liver and kidney changes, and tumorous masses in the intestinal tract. Occurrence of tumors was dose-related. Changes in the heart included myocardial atrophy, fibrosis, thrombosis, fatty degeneration, and fatty infiltration (Taylor 1981).

Genotoxicity

No mutagenic activity of a calamus extract was observed in *Salmonella typhimurium* strains TA97a, TA100, TA102, and TA104. Dose-dependent antimutagenic activity was observed at concentrations of 25 to 100 µg/plate (Aqil et al. 2008).

The compound *cis*-asarone has shown mutagenic activity in vitro (Goggelmann and Schimmer 1983), although this activity has been characterized as weak in comparison with other mutagenic/carcinogenic natural substances (Wichtl 2004).

No mutagenic activity of β-asarone was observed in *Salmonella typhimurium* strains TA98, TA100, TA1535, TA1537, or TA1538 at concentrations of 2 to 200 µg/plate with metabolic activation. Tests without metabolic activation were not completed (Hsia et al. 1979).

No mutagenic activity of α-asarone was observed in the Ames test with *Salmonella typhimurium* at concentrations of up to 5000 ppm with or without activation. In a related study, β-asarone was not mutagenic at 50 ppm, but did show mutagenic activity at a concentration of 5000 ppm with activation (Yabiku et al. 1979).

Cytotoxicity

An ethanol extract of calamus demonstrated cytotoxic activity in a brine shrimp lethality test (Padmaja et al. 2002).

Literature Cited

Abel, G. 1987. Chromosome-damaging effect of beta-asarone on human lymphocytes. *Planta Med.* 53(3):251-253.

Agarwal, S.L., P. Dandiya, K. Singh, and R. Arora. 1956. A note on the preliminary studies of certain pharmacological actions of *Acorus calamus*. *J. Am. Pharm. Assn.* 45:655-6.

Aqil, F., M. Zahin, and I. Ahmad. 2008. Antimutagenic activity of methanolic extracts of four Ayurvedic medicinal plants. *Indian J. Exp. Biol.* 46(9):668-672.

Balachandran, B., S.N. Sivaswamy, and V.M. Sivaramakrishnan. 1991. Genotoxic effects of some foods and food components in Swiss mice. *Indian J. Med. Res.* 94:378-383.

Bensky, D., S. Clavey, and E. Stöger. 2004. *Chinese herbal medicine: Materia medica*. 3rd ed. Seattle: Eastland Press.

CFR. 2011. *Code of federal regulations*, Title 21 Part 189.110, 2011 ed. Substances prohibited from use in human food. Calamus and its derivatives. Washington, DC: U.S. Government Printing Office.

Chadha, Y. 1988. *The wealth of India: A dictionary of Indian raw materials and industrial products*. Delhi: Council of Scientific and Industrial Research.

Dandiya, P.C., R. Baxter, and H. Cullumbine. 1958. Studies on *Acorus calamus*. I. Phytochemical investigation. *Can. Pharm. J.* 91:607.

Dandiya, P.C., and H. Cullumbine. 1959. Studies on *Acorus calamus*. III. Some pharmacological actions of the volatile oil. *J. Pharmacol. Exp. Ther.* 125:353-359.

Dandiya, P.C., H. Cullumbine, and E.A. Sellers. 1959. Studies on *Acorus calamus*. IV. Investigations on mechanism of action in mice. *J. Pharmacol. Exp. Ther.* 126:334-337.

Dasgupta, S.R., B. Patra, and S. Sikdar. 1977. Preliminary studies of the effect of a chloroform extracted factor from *Acorus calamus* on the behavior of conscious Rhesus monkeys. *Sci. Culture* 43:218.

De Smet, P.A. 1985. A multidisciplinary overview of intoxicating snuff rituals in the western hemisphere. *J. Ethnopharmacol.* 13(1):3-49.

FAO/WHO. 1981. β-Asarone. FAO/WHO Expert Committee on Food Additives. Toxicological evaluation of certain food additives. WHO Food Additives Series 16. Geneva.

Goggelmann, W., and O. Schimmer. 1983. Mutagenicity testing of beta-asarone and commercial calamus drugs with *Salmonella typhimurium*. *Mutat. Res.* 121(3-4):191-194.

Habermann, R.T. 1971. Carcinogenicity of beta-asarone in rats in a two-year feeding study. *Cited in* FAO/WHO. 1981. β-Asarone. FAO/WHO Expert Committee on Food Additives. Toxicological evaluation of certain food additives. WHO Food Additives Series 16. Geneva.

Hanson, K., M. Gayton-Ely, L. Holland, P. Zehr, and B. Söderberg. 2005. Rapid assessment of beta-asarone content of *Acorus calamus* by micellar electrokinetic capillary chromatography. *Electrophoresis* 26(4-5):943-946.

Hasheminejad, G., and J. Caldwell. 1994. Genotoxicity of the alkenylbenzenes alpha- and beta-asarone, myristicin and elimicin as determined by the UDS assay in cultured rat hepatocytes. *Food Chem. Toxicol.* 32 (3):223-231.

Hazra, R., K. Ray, and D. Guha. 2007. Inhibitory role of *Acorus calamus* in ferric chloride-induced epileptogenesis in rat. *Hum. Exp. Toxicol.* 26(12):947-953.

Hsia, M.T.S., J.A. Adamovics, and B.L. Kreamer. 1979. Microbial mutagenicity studies of insect growth regulators and other potential insecticidal compounds in *Salmonella typhimurium*. *Chemosphere* 8(8):521-529.

Jenner, P.M., E.C. Hagan, J.M. Taylor, E.L. Cook, and O.G. Fitzhugh. 1964. Food flavourings and compounds of related structure. I. Acute oral toxicity. *Food Cosmet. Toxicol.* 2(3):327-343.

Keller, K., and E. Stahl. 1982. Kalamus: Inhaltsstoffe und β-Asarongehalt bei verschiedenen Herkünften. *Dtsch. Apoth. Ztg.* 122:2463-2466.

Keller, K., and E. Stahl. 1983. Composition of the essential oil from beta-asarone free calamus. *Planta Med.* 47(2):71-74.

Kumar, V.S., R.K. Srivastava, A. Krishna, et al. 2000. Cultivation, chemistry, biology and utilization of bach (*Acorus calamus*): A review. *J. Med. Aromatic Plant Sci.* 22(2-3):338-348.

Leung, A.Y., and S. Foster. 1996. *Encyclopedia of common natural ingredients used in food, drugs, and cosmetics*. 2nd ed. New York: Wiley.

Martindale, W., and J.E.F. Reynolds. 1996. *The extra pharmacopoeia*. 31st ed. London: Pharmaceutical Press.

Mehrotra, S., K.P. Mishra, R. Maurya, et al. 2003. Anticellular and immunosuppressive properties of ethanolic extract of *Acorus calamus* rhizome. *Int. Immunopharmacol.* 3(1):53-61.

Motley, T. 1994. The ethnobotany of sweet flag, *Acorus calamus* (Araceae). *Econ. Bot.* 48(4):397-412.

Oprean, R., M. Tamas, and L. Roman. 1998. Comparison of GC-MS and TLC techniques for asarone isomers determination. *J. Pharm. Biomed. Anal.* 18(1-2):227-234.

Padmaja, R., P.C. Arun, D. Prashanth, et al. 2002. Brine shrimp lethality bioassay of selected Indian medicinal plants. *Fitoterapia* 73:508-510.

Panchal, G.M., H. Venkatakrishna-Bhatt, R.B. Doctor, and S. Vajpayee. 1989. Pharmacology of *Acorus calamus* L. *Indian J. Exp. Biol.* 27(6):561-567.

Raina, V., S. Srivastava, and K. Syamasunder. 2003. Essential oil composition of *Acorus calamus* L. from the lower region of the Himalayas. *Flav. Frag. J.* 18(1):18-20.

Stahl, E., and K. Keller. 1981. The classification of commercial *Acorus calamus* drugs. *Plant Med.* 43(2):128-140.

Subramanian, L., S. Murali, and P.M. Murali. 2004. Analysis of asarones from commercial samples of *Acorus calamus* L. *Proc. Natl. Acad. Sci. India B (Biol. Sci.)* 74(1):75-78.

Subramanian, R., V. Ozaa, P. Parmara, and S. Mehtab. 2008. Rapid determination of β-asarone-free *Acorus calamus* cytotypes by HPTLC *Curr. Trends Biotechnol. Pharm.* 2(4):506-513.

Taylor, J.M. 1967. Toxicity of oil of calamus (Jammu variety). *Toxicol. Exp. Pharmacol.* 10:405.

Taylor, J.M. 1981. Personal communication to the World Health Organization concerning unpublished studies on beta-asarone and calamus oils. *Cited in* WHO. 1981. β-Asarone. Toxicological evaluation of certain food additives. WHO Food Additives Series 16. Joint FAO/WHO Expert Committee on Food Additives.

Vargas, C.P., L.R. Wolf, S.R. Gamm, and K. Koontz. 1998. Getting to the root (*Acorus calamus*) of the problem. *J. Toxicol. Clin. Toxicol.* 36(3):259-260.

WHO. 1981. β-Asarone. *In* Toxicological evaluation of certain food additives. WHO Food Additives Series 16. Joint FAO/WHO Expert Committee on Food Additives.

Wichtl, M. 2004. *Herbal drugs and phytopharmaceuticals: A handbook for practice on a scientific basis*. 3rd ed. Boca Raton, FL: CRC Press.

Widmer, V., A. Schibli, and E. Reich. 2005. Quantitative determination of β-asarone in calamus by high-performance thin-layer chromatography. *J. AOAC Int.* 88(5):1562-1567.

Yabiku, H.Y., S. Oga, and F.M. Lajolo. 1979. Toxic effects of *Acorus calamus* oil. Preliminary study with rats and chicken embryos. *An. Farm. Quim. Sao Paulo* 19(2):252-258.

Acorus calamus L. — Acoraceae

SCN: calamus
AN: *vacha*
OCN: acorus; sweet calamus; sweetflag
Part: rhizome of the diploid asarone-free variety

Quick Reference Summary

Safety Class: 1
Interaction Class: A

Contraindications

None known.

Other Precautions

None known.

Drug and Supplement Interactions

None known.

Editors' Notes

Calamus grows wild in India, China, Europe, and North America, and the phytochemical profile of the plant material varies according to origin. The essential oil of plants from India contains up to 75% β-asarone (*see* Alkenylbenzenes in Appendix 1), while the oil of calamus from Japan and eastern Russia contains 10–40%, oil from European plants contains approximately 13%, and that from North America contains almost no β-asarone (Keller and Stahl 1982, 1983; Raina et al. 2003; Stahl and Keller 1981; Subramanian et al. 2008). Since varieties of calamus may not be well differentiated in commerce, the caution stated for the Asian and European varieties should be considered relevant to any sample that is not positively identified as the North American variety.

All varieties of calamus are prohibited in foods in the United States (CFR 2011).

Adverse Events and Side Effects

None known.

Pharmacological Considerations

A reference text on traditional Chinese medicine indicated that overdose and long-term use of calamus should be avoided (Bensky et al. 2004).

Pregnancy and Lactation

A study in chicken eggs showed no teratogenic activity of calamus essential oil (Yabiku et al. 1979). No other information on the safety of calamus in pregnancy or lactation was identified.

Review Details

I. Drug and Supplement Interactions

Clinical Trials of Drug or Supplement Interactions

No clinical trials of drug or supplement interactions were identified.

Case Reports of Suspected Drug or Supplement Interactions

No case reports of suspected drug or supplement interactions were identified.

Animal Trials of Drug or Supplement Interactions

Extracts of calamus have been shown to potentiate pentobarbitone-induced sleeping time (Dandiya et al. 1959; Hazra et al. 2007; Panchal et al. 1989).

II. Adverse Events

Case Reports of Adverse Events

A 19-year-old man experienced diaphoresis, persistent vomiting, and mild leukocytosis after ingesting an 8-inch-long calamus rhizome (variety unspecified) (Vargas et al. 1998).

III. Pharmacology and Pharmacokinetics

Human Pharmacological Studies

No relevant human pharmacological studies were identified.

Animal Pharmacological Studies

An ethanolic extract of calamus demonstrated immunomodulatory potential by inhibiting proliferation of mitogen- and antigen-stimulated human peripheral blood mononuclear cells. The extract inhibited growth of several mouse and human cell lines (Mehrotra et al. 2003).

Animal studies have indicated that calamus has a depressant effect on the central nervous system (Agarwal et al. 1977; Dandiya et al. 1958, 1959; Dandiya and Cullumbine 1959; Dasgupta et al. 1977).

In Vitro Pharmacological Studies

Negative ionotropic and chronotropic effects were observed in frog heart preparations treated with concentrations of 100 μg/ml of an alcohol extract of calamus (Panchal et al. 1989).

IV. Pregnancy and Lactation

No teratogenic effects were observed in chicken eggs injected with calamus essential oil at doses of 0.12, 0.60, 3.00, 15.00, or 75.00 mg/egg (Yabiku et al. 1979).

No other information on the safety of calamus in pregnancy or lactation was identified.

V. Toxicity Studies

Acute Toxicity

The LD_{50} of intraperitoneally administered calamus oil from Europe in mice was 1.1 g/kg for oil containing β-asarone, and 1.7 g/kg for oil with no β-asarone (Yabiku et al. 1979).

Genotoxicity

No mutagenic activity of a calamus extract was observed in *Salmonella* strains TA97a, TA100, TA102, and TA104. Dose-dependent antimutagenic activity was observed at concentrations of 25–100 μg/plate (Aqil et al. 2008).

Cytotoxicity

An ethanol extract of calamus demonstrated cytotoxic activity in a brine shrimp lethality test (Padmaja et al. 2002).

Literature Cited

Agarwal, B.L., R.K. Agarwal, and D.N. Misra. 1977. Malignant arrhythmias induced by accidental aconite poisoning. *Indian Heart J.* 29(5):246-248.

Aqil, F., M. Zahin, and I. Ahmad. 2008. Antimutagenic activity of methanolic extracts of four Ayurvedic medicinal plants. *Indian J. Exp. Biol.* 46(9):668-672.

Bensky, D., S. Clavey, and E. Stöger. 2004. *Chinese herbal medicine: Materia medica*. 3rd ed. Seattle: Eastland Press.

CFR. 2011. *Code of federal regulations*, Title 21 Part 189.110, 2011 ed. Substances prohibited from use in human food. Calamus and its derivatives. Washington, DC: U.S. Government Printing Office.

Dandiya, P.C., R. Baxter, and H. Cullumbine. 1958. Studies on *Acorus calamus*. I. Phytochemical investigation. *Can. Pharm. J.* 91:607.

Dandiya, P.C., and H. Cullumbine. 1959. Studies on *Acorus calamus*. III. Some pharmacological actions of the volatile oil. *J. Pharmacol. Exp. Ther.* 125:353-359.

Dandiya, P.C., H. Cullumbine, and E.A. Sellers. 1959. Studies on *Acorus calamus*. IV. Investigations on mechanism of action in mice. *J. Pharmacol. Exp. Ther.* 126:334-337.

Dasgupta, S.R., B. Patra, and S. Sikdar. 1977. Preliminary studies of the effect of a chloroform extracted factor from *Acorus calamus* on the behavior of conscious Rhesus monkeys. *Sci. Culture* 43:218.

Hazra, R., K. Ray, and D. Guha. 2007. Inhibitory role of *Acorus calamus* in ferric chloride-induced epileptogenesis in rat. *Hum. Exp. Toxicol.* 26(12):947-953.

Keller, K., and E. Stahl. 1982. Kalamus: Inhaltsstoffe und β-Asarongehalt bei verschiedenen Herkünften. *Dtsch. Apoth. Ztg.* 122:2463-2466.

Keller, K., and E. Stahl. 1983. Composition of the essential oil from beta-asarone free calamus. *Planta Med.* 47(2):71-74.

Mehrotra, S., K.P. Mishra, R. Maurya, et al. 2003. Anticellular and immunosuppressive properties of ethanolic extract of *Acorus calamus* rhizome. *Int. Immunopharmacol.* 3(1):53-61.

Padmaja, R., P.C. Arun, D. Prashanth, et al. 2002. Brine shrimp lethality bioassay of selected Indian medicinal plants. *Fitoterapia* 73:508-510.

Panchal, G.M., H. Venkatakrishna-Bhatt, R.B. Doctor, and S. Vajpayee. 1989. Pharmacology of *Acorus calamus* L. *Indian J. Exp. Biol.* 27(6):561-567.

Raina, V., S. Srivastava, and K. Syamasunder. 2003. Essential oil composition of *Acorus calamus* L. from the lower region of the Himalayas. *Flav. Frag. J.* 18(1):18-20.

Stahl, E., and K. Keller. 1981. The classification of commercial *Acorus calamus* drugs. *Plant Med.* 43(2):128-140.

Subramanian, R., V. Ozaa, P. Parmara, and S. Mehtab. 2008. Rapid determination of β-asarone-free *Acorus calamus* cytotypes by HPTLC. *Curr. Trends Biotechnol. Pharm.* 2(4):506-513.

Vargas, C.P., L.R. Wolf, S.R. Gamm, and K. Koontz. 1998. Getting to the root (*Acorus calamus*) of the problem. *J. Toxicol. Clin. Toxicol.* 36(3):259-260.

Yabiku, H.Y., S. Oga, and F.M. Lajolo. 1979. Toxic effects of *Acorus calamus* oil. Preliminary study with rats and chicken embryos. *An. Farm. Quim. Sao Paulo* 19(2):252-258.

Acorus gramineus Sol. ex Aiton

Acoraceae

SCN: grass-leaf sweetflag
Syn: *Acorus tatarinowii* Schott
PN: *shi chang pu* (rhizome)
OCN: grass-leaf calamus
Part: rhizome

Quick Reference Summary

Safety Class: 3
Interaction Class: A

Contraindications

Not for use except under the supervision of an expert qualified in the appropriate use of this substance (Bensky et al. 2004).

Other Precautions

None known.

Drug and Supplement Interactions

See Pharmacological Considerations.

Notice

Alkenylbenzenes (β-asarone and α-asarone, 0.08–0.8%) (Chang and But 1986; Chen and Chen 2004; Cho et al. 2002; Sugimoto et al. 1997a, 1997b). (See Appendix 1.)

Editors' Notes

Different varieties of grass-leaf sweetflag and material of different geographical origin have varying levels of β-asarone (*see* Alkenylbenzenes in Appendix 1) (Bensky et al. 2004; Sugimoto et al. 1997b). Plant material with low β-asarone is strongly preferred, while material high in β-asarone should be used only when absolutely necessary, only for acute conditions, and only for short periods (Bensky et al. 2004).

Adverse Events and Side Effects

None known.

Pharmacological Considerations

Increases in pentobarbital-induced sleeping time have been observed in mice after oral administration or aroma inhalation of grass-leaf sweetflag (Koo et al. 2003; Liao et al. 1998).

Pregnancy and Lactation

A study in chicken eggs showed no teratogenic activity of compounds from grass-leaf sweetflag (Yabiku et al. 1979). No other information on the safety of grass-leaf sweetflag in pregnancy or lactation was identified.

While this review did not identify any concerns for use while pregnant or nursing, safety has not been conclusively established, and this substance is not recommended for use except under the supervision of an expert qualified in its appropriate use.

A

Review Details

I. Drug and Supplement Interactions

Clinical Trials of Drug or Supplement Interactions

No clinical trials of drug or supplement interactions were identified.

Case Reports of Suspected Drug or Supplement Interactions

No case reports of suspected drug or supplement interactions were identified.

Animal Trials of Drug or Supplement Interactions

A dose-dependent increase in pentobarbital-induced sleeping time was observed in mice orally administered an aqueous extract of 0.5 to 5.0 g/kg of grass-leaf sweetflag (Liao et al. 1998).

Inhalation of the aroma of grass-leaf sweetflag also prolonged the pentobarbital-induced sleeping time in mice (Koo et al. 2003).

II. Adverse Events

Case Reports of Adverse Events

No case reports of adverse events were identified.

III. Pharmacology and Pharmacokinetics

Human Pharmacological Studies

No relevant human pharmacological studies were identified.

Animal Pharmacological Studies

A dose-dependent decrease in locomotor activity was observed in mice orally administered an aqueous extract of 0.5 to 5.0 g/kg of grass-leaf sweetflag (Liao et al. 1998).

In Vitro Pharmacological Studies

No relevant in vitro pharmacological studies were identified.

IV. Pregnancy and Lactation

No teratogenic effects were observed in chicken eggs injected with α-asarone at doses up to 4 mg/egg. In eggs injected with β-asarone, 43% of embryos survived a dose of 0.04 mg/egg, while none survived a dose of 4 mg/egg (Yabiku 1980).

No information on the safety of grass-leaf sweetflag in lactation was identified.

V. Toxicity Studies

Also see entry for *Acorus calamus* rhizome of the asarone-containing triploid or tetraploid varieties for more information on the toxicity of β-asarone.

Acute Toxicity

The LD_{50} of intraperitoneally administered aqueous extract of grass-leaf sweetflag in mice was 53 g/kg (Chen and Chen 2004). The LD_{50} of intraperitoneally administered α-asarone in mice was 339 mg/kg. Toxic effects included seizures, convulsions, and slowed respiration (Chen and Chen 2004).

Literature Cited

Bensky, D., S. Clavey, and E. Stöger. 2004. *Chinese herbal medicine: Materia medica*. 3rd ed. Seattle: Eastland Press.

Chang, H.-M., and P.P.H. But. 1986. *Pharmacology and applications of Chinese materia medica*. English ed. Singapore, Philadelphia: World Scientific.

Chen, J.K., and T.T. Chen. 2004. *Chinese medical herbology and pharmacology*. City of Industry, CA: Art of Medicine Press.

Cho, J., Y.H. Kim, J.Y. Kong, C.H. Yang, and C.G. Park. 2002. Protection of cultured rat cortical neurons from excitotoxicity by asarone, a major essential oil component in the rhizomes of *Acorus gramineus*. *Life Sci.* 71(5):591-599.

Koo, B.S., K.S. Park, J.H. Ha, et al. 2003. Inhibitory effects of the fragrance inhalation of essential oil from *Acorus gramineus* on central nervous system. *Biol. Pharm. Bull.* 26(7):978-982.

Liao, J.F., S.Y. Huang, Y.M. Jan, L.L. Yu, and C.F. Chen. 1998. Central inhibitory effects of water extract of Acori graminei rhizoma in mice. *J. Ethnopharmacol.* 61(3):185-193.

Sugimoto, N., M. Mikage, H. Ohtsubo, F. Kiuchi, and Y. Tsuda. 1997a. Pharmacognostical investigations of acori rhizomes: I. Histological and chemical studies of rhizomes of *A. calamus* and *A. gramineus* distributed in Japan. *Nat. Med.* 51(3):259-264.

Sugimoto, N., H. Ohtsubo, M. Mikiage, et al. 1997b. Pharmacognostical investigation of Acori rhizomes: II. Histological and chemical studies of Acori rhizomes in Asian markets. *Nat. Med.* 51(4):316-324.

Yabiku, H.K. 1980. Calamus oil—Toxicological aspects and their control in alcoholic beverages. *Cited in* WHO. 1981. β-Asarone. Toxicological evaluation of certain food additives. WHO Food Additives Series 16. Joint FAO/WHO Expert Committee on Food Additives.

Yabiku, H.Y., S. Oga, and F.M. Lajolo. 1979. Toxic effects of *Acorus calamus* oil. Preliminary study with rats and chicken embryos. *An. Farm. Quim. Sao Paulo* 19(2):252-258.

Actaea spp.

Ranunculaceae

Actaea cimicifuga L.
SCN: Chinese cimicifuga
Syn: *Cimicifuga foetida* L.
PN: *sheng ma* (rhizome)
OCN: skunk bugbane

Actaea dahurica **(Turcz. ex Fisch. & C.A. Mey.) Franch.**
SCN: Chinese cimicifuga
Syn: *Cimicifuga dahurica* (Turcz. ex Fisch. & C.A. Mey.) Maxim.
PN: *sheng ma* (rhizome)
OCN: Dahurian bugbane

Actaea heracleifolia **(Kom.) J. Compton**
SCN: Chinese cimicifuga
Syn: *Cimicifuga heracleifolia* Kom.
PN: *sheng ma* (rhizome)
OCN: large-leaf bugbane
Part: rhizome

Quick Reference Summary

Safety Class: 2d
Interaction Class: A

Contraindications

A typical therapeutic use of Chinese cimicifuga is in the initial stages of measles or measles with incomplete eruptions. Use after full eruption of measles is not recommended (Bensky et al. 2004; Chen and Chen 2004).

Other Precautions

None known.

Drug and Supplement Interactions

None known.

Adverse Events and Side Effects

A reference text on traditional Chinese medicine notes that overdose (standard dose listed as decoction of 3–9 g) of Chinese cimicifuga may cause headaches, dizziness, vomiting, tremors, gastroenteritis, and pathogenic erections (Bensky et al. 2004).

Pharmacological Considerations

None known.

Pregnancy and Lactation

No information on the safety of Chinese cimicifuga in pregnancy or lactation was identified in the scientific or traditional literature. Although this review did not identify any concerns for use while pregnant or nursing, safety has not been conclusively established.

Review Details

I. Drug and Supplement Interactions

Clinical Trials of Drug or Supplement Interactions

No clinical trials of drug or supplement interactions were identified.

Case Reports of Suspected Drug or Supplement Interactions

No case reports of suspected drug or supplement interactions were identified.

Animal Trials of Drug or Supplement Interactions

No animal trials of drug or supplement interactions were identified.

II. Adverse Events

Case Reports of Adverse Events

Overdoses of Chinese cimicifuga (standard dose listed as decoction of 3–9 g) may cause nausea, vomiting, and gastroenteritis. High doses have been noted to cause headache, tremors, tetanic contraction of limbs, lassitude, vertigo, and abnormal erections. Extreme overdoses may result in hypotension, dyspnea, delirium, and respiratory arrest (Bensky et al. 2004; Chen and Chen 2004).

III. Pharmacology and Pharmacokinetics

Human Pharmacological Studies

No relevant human pharmacological studies were identified.

Animal Pharmacological Studies

The compound isoferulic acid, isolated from Chinese cimicifuga, has been shown to lower plasma glucose in diabetic rats in a dose-dependent manner with effects on plasma glucose observed at doses of 5 mg/kg (intravenous) and higher (Liu et al. 1999).

In Vitro Pharmacological Studies

An extract of Chinese cimicifuga demonstrated antibacterial activity in Gram-positive and Gram-negative bacteria (Moskalenko 1986).

IV. Pregnancy and Lactation

No information on the safety of Chinese cimicifuga in pregnancy or lactation was identified.

A

V. Toxicity Studies

Acute Toxicity

The LD_{50} of a methanolic extract of Chinese cimicifuga orally administered to mice could not be determined at doses up to 10 g/kg. The LD_{50} of the same extract administered intraperitoneally to mice was 8.5 g/kg (Shibata et al. 1975).

Literature Cited

Bensky, D., S. Clavey, and E. Stöger. 2004. *Chinese herbal medicine: Materia medica*. 3rd ed. Seattle: Eastland Press.

Chen, J.K., and T.T. Chen. 2004. *Chinese medical herbology and pharmacology*. City of Industry, CA: Art of Medicine Press.

Liu, I.M., T.C. Chi, F.L. Hsu, C.F. Chen, and J.T. Cheng. 1999. Isoferulic acid as active principle from the rhizoma of *Cimicifuga dahurica* to lower plasma glucose in diabetic rats. *Planta Med.* 65(8):712-714.

Moskalenko, S.A. 1986. Preliminary screening of far-eastern ethnomedicinal plants for antibacterial activity. *J. Ethnopharmacol.* 15(3):231-259.

Shibata, M., Y. Yamatake, Y. Amagaya, and M. Fukushima. 1975. Pharmacological studies on the Chinese crude drug "Shoma." I. Acute toxicity and anti-inflammatory action of *Cimicifuga rhizoma, Cimicifuga dahurica* Maxim. (author's transl.). *Yakugaku Zasshi* 95(5):539-546.

Actaea racemosa L.

Ranunculaceae

SCN: black cohosh
Syn: *Cimicifuga racemosa* (L.) Nutt.
OCN: black bugbane; black snakeroot; rheumatism weed
Part: rhizome

Quick Reference Summary

Safety Class: 2B
Interaction Class: A

Contraindications

Not for use in pregnancy except under the supervision of a qualified healthcare practitioner.

Other Precautions

None known.

Drug and Supplement Interactions

None known.

Editors' Notes

Analyses of black cohosh products associated with liver toxicity in Canada found that several products were not black cohosh but instead a closely related species (Painter et al. 2010). An analysis of products on the American market indicated that 3 of 11 products tested contained Asian species of *Actaea* in place of or in addition to *Actaea racemosa* (Jiang et al. 2006).

Adverse Events and Side Effects

Broad attention has been paid to case reports of hepatotoxicity in persons taking black cohosh. An assessment by the European Medicines Agency (EMEA) of 7 published and 42 unpublished cases of hepatotoxicity reported in persons taking black cohosh products concluded, "Overall, all discussed cases of literature and pharmacovigilance reports are poorly documented" but nevertheless found three "possible" and two "probable" cases (EMEA 2006). Based on some of the same and also additional case reports of hepatotoxicity, Britain's Medicines and Healthcare Products Regulatory Agency and the Australian Therapeutic Goods Administration require cautionary labels on products containing black cohosh (MHRA 2006; TGA 2006). The U.S. Pharmacopoeia has also recommended cautionary labeling (Mahady et al. 2008). Prior to the emergence of these case reports, reviews of clinical trials and other safety data have indicated that black cohosh is generally safe (Huntley and Ernst 2003; Low Dog et al. 2003).

A review of 69 published and unpublished case reports of black cohosh-associated hepatotoxicity indicated that there was an excluded, unlikely, unrelated, or unassessable causality for black cohosh in 68 of 69 cases and "little, if any, supportive evidence for a significant hepatotoxic risk of black cohosh" (Teschke et al. 2009). Although one animal study on black cohosh rhizome has identified a biologically plausible mechanism of hepatotoxicity at high dose levels (Lüde et al. 2007), no changes in liver enzyme levels have been observed in several human studies (Bai et al. 2007; Nasr and Nafeh 2009; Osmers et al. 2005; van Breemen et al. 2009). Healthcare practitioners and consumers should be aware of the possible association between products containing black cohosh and hepatotoxicity.

Occasional gastrointestinal discomfort has been reported to occur with black cohosh use (Bradley 1992).

Pharmacological Considerations

While preclinical human, animal, and in vitro studies gave mixed results on the estrogenic activity of black cohosh,

6- and 12-month human clinical trials indicated a lack of estrogenic effects (Huntley 2004; Liske et al. 2002; Mahady 2003, 2005; Raus et al. 2006; Reed et al. 2008).

Pregnancy and Lactation

Black cohosh has been contraindicated by some references for use during pregnancy due to the reported emmenagogic effect (Brinker 2001; Dugoua et al. 2006). Contemporary herbal practitioners, however, have not observed this effect and have used black cohosh early in pregnancy to prevent miscarriage (Upton 2002).

In this work, the contraindication for use in pregnancy is based on concerns regarding the recent cases of hepatotoxicity reported in association with black cohosh use, as the implications of these case reports and possible mechanisms of hepatotoxicity have yet to be fully understood.

Black cohosh is traditionally used late in pregnancy, in or around labor as a partus preparator (Ellingwood 1919; Felter 1891; Felter and Lloyd 1898; McFarlin et al. 1999; Scudder 1903).

While one reference (Dugoua et al. 2006) noted low-level evidence suggesting potential hormonal activity of black cohosh could be a cause for concern during lactation, further studies indicated a lack of estrogenic activity of black cohosh (*see* Human pharmacological studies).

Review Details

I. Drug and Supplement Interactions

Clinical Trials of Drug or Supplement Interactions

A trial of single doses of digoxin (0.4 mg) before and after 14 days of black cohosh administration (40 mg/day) indicated no interaction between black cohosh and digoxin (Gurley et al. 2006).

Black cohosh (1090 mg, twice daily for 28 days) was shown to inhibit CYP 2D6 in a nonclinically significant manner (Gurley et al. 2005).

Case Reports of Suspected Drug or Supplement Interactions

No cases of suspected drug interactions were identified.

Animal Trials of Drug or Supplement Interactions

No animal trials of drug or supplement interactions were identified.

II. Adverse Events

Adverse Events Reported in Clinical Trials

In a systematic review of clinical trials, published case reports and pharmacovigilance reporting center data, Huntley and Ernst (2003) concluded that black cohosh is generally safe. If products are taken for a limited amount of time, the risk of adverse events is slight and the events are usually mild and transient, with gastrointestinal upset and rashes being the most common events reported. The reviewers noted that some serious adverse events have been reported, including hepatic and circulatory conditions, but due to limited information on cases, causality could not be determined (Huntley and Ernst 2003).

Similarly, a review of human clinical trials (including over 2800 participants), postmarketing surveillance, and uncontrolled case reports on black cohosh indicated a low incidence (5.4%) of adverse events in persons taking black cohosh. Of the adverse events reported, 97% were minor and the severe events were attributed to causes other than black cohosh use (Low Dog et al. 2003).

Case Reports of Adverse Events

As of October 2009, a total of 83 cases of liver toxicity associated with black cohosh use had been reported to drug monitoring agencies and or published in the literature (Mahady et al. 2009). Using rating scales that generally include categories of "unassessable, unlikely, possible, probable, or highly probable," critical analyses of these case reports conclude that most cases are "unlikely" or "possible" in relation to black cohosh, while only two cases have been categorized as "probable" and none as "highly probable." Reviewers noted that lack of detail in most of the cases made causality assessment difficult (EMEA 2006; Mahady et al. 2008; Teschke and Schwarzenboeck 2009). Synopses of the published case reports are as follows.

A case of autoimmune hepatitis was reported in a 57-year-old woman with a history of polymyositis, diabetes, high blood pressure, and obstructive sleep apnea. Medications being taken were labetalol, fosinopril, verapramil, metformin, insulin, aspirin, and aminosalicylic acid. The woman had been taking black cohosh (product and dose unspecified) for approximately 1 week (Cohen et al. 2004).

Fulminant liver failure was reported in a 50-year-old woman who had been taking 500 mg black cohosh daily for 5 months (Levitsky et al. 2005a). The initial case report did not include any listed comedications or significant medical history, although in a published erratum, the physicians indicated that the patient had been consuming alcohol on a regular basis and was taking the drug valaciclovir at the time of the original incident (Levitsky et al. 2005b).

Fulminant liver failure was reported in a 54-year-old woman who had been taking 1000 mg of an unspecified black cohosh product daily for 8 months. The woman had a history of fibromyalgia, osteoarthritis, depression, and hypothyroidism and was also taking fluoxetine, propoxyphene, acetaminophen, and levothyroxine (Lynch et al. 2006).

Acute hepatitis was reported in a 47-year-old woman who had been taking 40 mg of an *isopropanol* extract of black cohosh daily for 6 days (Whiting et al. 2002).

A

Hepatitis was reported in a 50-year-old woman with a history of gallstones, gastroesophageal *reflux* disease, and anxiety who took 40 mg of black cohosh (product unspecified) daily for 2 weeks along with lansoprazole (Nisbet and O'Connor 2007).

Acute liver failure was reported in a 41-year-old woman who had been taking black cohosh for 2 weeks; the dose and product used were not specified. In this case, the woman developed giant cell hepatitis that increased in severity after cessation of black cohosh (Dunbar and Solga 2007).

Elevated liver enzymes were reported in a 50-year-old woman with a history of chronic fatigue and high blood pressure. The woman had been taking black cohosh, although the dose, duration, and product used were not specified (Joy et al. 2008).

Elevated liver enzymes were reported in a 51-year-old woman with a history of gallstones, fatty liver, and asthma who had been taking black cohosh for 2 months. The black cohosh dose and product used were not specified (Joy et al. 2008).

Liver failure was reported in a 51-year-old woman who had been taking 20 mg of black cohosh intermittently for 3 years, with titration according to her menopausal symptoms. The woman had a history of obesity, gastric bypass surgery, and alcoholism (Chow et al. 2008).

Liver injury was reported in a 42-year-old woman with a history of hypothyroidism who had been taking black cohosh (product and dose not specified) for 6 months along with levothyroxine. Liver enzyme levels remained elevated for several months after cessation of black cohosh (Guzman et al. 2009).

Elevated liver enzymes were reported in a 53-year-old woman with a history of irritable bowel syndrome. Medications included dicyclomine and nonsteroidal anti-inflammatory drugs as needed and a product containing black cohosh and soy protein (dose, duration, and product used not specified) (Guzman et al. 2009).

Chronic hepatitis was reported in a 58-year-old woman with a history of high blood pressure, hypothyroidism, diabetes, and high cholesterol. The woman had been taking 80 mg of black cohosh extract daily for 1 year along with irbesartan, levothyroxin, simvastatin, and insulin. The liver injury was initially thought to be induced by simvastatin, although discontinuation of simvastatin did not result in a reduction of liver enzyme levels (Pierard et al. 2009).

Cutaneous vasculitis was reported in two patients taking 80 mg black cohosh daily for 2–4 months (one woman was taking a product that contained black cohosh and "other vitamin, mineral, and herbal supplements"). The vasculitis resolved after treatment and cessation of black cohosh. No other medications or significant medical history were reported in either patient, and both patients declined a rechallenge (Ingraffea et al. 2007).

A renal transplant rejection occurred 16 years post-transplant in a woman taking black cohosh, alfalfa (*Medicago sativa*), azathioprine, and cyclosporine. The woman had undergone treatment for breast cancer several months prior to the incident (Light and Light 2003).

Other adverse events associated with black cohosh use include a case of asthenia (Minciullo et al. 2006) and a case of an erythematous rash (Meyer et al. 2007), each in women who had been taking black cohosh for 1 year.

III. Pharmacology and Pharmacokinetics

Human Pharmacological Studies

Human clinical trials have indicated a lack of estrogenic effects of black cohosh. In a clinical trial of 400 postmenopausal women, administration of 40 mg of black cohosh extract daily for 1 year did not produce any endometrial hyperplasia or other adverse endometrial outcomes, and no change in endometrial thickness was observed (Raus et al. 2006). Likewise, in a clinical trial of peri- and postmenopausal women, 24 weeks of black cohosh (39 or 127 mg/day) administration did not produce any changes in vaginal cytology and no systemic estrogenic effects were observed (Liske et al. 2002).

No changes in vaginal cytology profiles, vaginal dryness, menstrual cyclicity, or hormone profiles were observed in women ages 45 to 55 orally administered 160 mg black cohosh daily for 12 months (Reed et al. 2008).

No estrogenic activity was observed in postmenopausal women orally administered 80 mg black cohosh daily for 12 weeks. Measures of estrogenicity included estrogenic markers in serum, pS2 levels, and cellular morphology in nipple aspirate fluid (Ruhlen et al. 2007).

No changes in mammographic breast density, breast cell proliferation, or endometrial thickness were observed in postmenopausal women who consumed 40 mg daily of isopropanolic extract of black cohosh for 6 months (Hirschberg et al. 2007).

In a retrospective cohort study of breast cancer survivors, a delay in breast cancer recurrence was observed in women using black cohosh, as compared to women not taking black cohosh (Zepelin et al. 2007).

In postmenopausal women taking 40 mg black cohosh extract daily for 4 months, no changes in total hepatic blood flow, bilirubin, γ-glutamyltransferase, alkaline phosphatase, alanine aminotransferase, aspartate aminotransferase, serum albumin, or prothrombin time and concentration were observed (Nasr and Nafeh 2009). No changes in liver enzyme levels were observed in other studies with women taking 40 mg black cohosh daily for 3 or 4 months, or in women given single doses up to 128 mg (Bai et al. 2007; Osmers et al. 2005; van Breemen et al. 2009).

A human study with the compound 23-epi-26-deoxyactein , administered at doses of 1.4, 2.8, or 5.6 mg, indicated that no phase I or phase II metabolites of this compound were found. Based on the lack of metabolites, 23-epi-26-deoxyactein is believed to not be metabolized by cytochrome

P450 enzymes and thus is unlikely to be a source of interactions caused by competition for CYP enzymes (van Breemen et al. 2010).

Animal Pharmacological Studies

Reviews examining the estrogenic effects of black cohosh have indicated that data from animal studies is mixed but currently leans toward lack of estrogenic activity (Huntley 2004; Mahady 2003, 2005). Reviewers have noted that older studies tend to demonstrate estrogenic effects while new studies demonstrate a lack of estrogenicity. Increased uterine weight and decreased luteinizing hormone (LH) levels are indications of estrogenic activity (Upton 2002).

Studies in rats and mice (typically ovariectomized animals) have indicated increased uterine weight and or decreased luteinizing hormone (LH) levels in animals administered black cohosh (Düker et al. 1991; Eagon et al. 1999; Foldes 1959; Gizicky 1944; Jarry and Harnischfeger 1985; Jarry et al. 1985). Other studies have shown no increase in uterine weight (Einer-Jensen et al. 1996; Kretzschmar et al. 2005). In rats, black cohosh reduced bone mineral density loss with no effect on uterine weight or gene expression, suggesting that black cohosh is an organ-specific selective estrogen receptor modulator (Seidlova-Wuttke et al. 2003).

In rats orally administered up to 600 mg of a black cohosh extract or up to 40 mg of a lipophilic fraction of black cohosh extract, no uterotrophic activity was observed (Bolle et al. 2007).

In mice genetically predisposed to breast cancer that were fed diets containing black cohosh at a level equivalent to a human 40 mg daily dose, no differences were detected in the incidence or onset of mammary tumors. In tumor-bearing mice, an increase in the incidence of lung metastases was observed in animals on the same black cohosh-containing diet (Davis et al. 2008).

In rats with estrogen-dependent mammary tumors, treatment with black cohosh did not stimulate cancerous growths (Freudenstein et al. 2002). Likewise, in rats with endometrial cancer treated with black cohosh alone or with tamoxifen, animals in both the black cohosh and black cohosh plus tamoxifen treatment groups had fewer metastases and smaller tumor mass than the untreated control animals (Nisslein and Freudenstein 2004).

In Vitro Pharmacological Studies

Reviews examining the estrogenic effects of black cohosh have indicated that data from in vitro studies is mixed but leans toward lack of estrogenic activity (Huntley 2004; Mahady 2003, 2005). The authors noted that older studies tend to demonstrate estrogenic effects while new studies demonstrate a lack of estrogenicity.

In vitro experiments have examined effects of black cohosh on proliferation of various cancer cell lines, mostly on estrogen receptor-positive breast cancer lines. A number of studies have shown that black cohosh did not cause cell proliferation and in some cases even inhibited proliferation (Bodinet and Freudenstien 2002; Dixon Shanies and Shaikh 1999; Einbond et al. 2007, 2008; Freudenstein and Bodinet 1999; Gaube et al. 2007; Hostanska et al. 2004; Lupu et al. 2003; Nesselhut et al. 1993; Rice et al. 2007; Zava et al. 1998), while others have indicated a proliferative effect (Harnischfeger and Cillien 1996; Liu et al. 2001a; Lohning et al. 2000). Similarly, studies examining the binding of black cohosh to estrogen receptors have shown no binding (Düker et al. 1991; Eagon et al. 1996; Harnischfeger and Cillien 1996; Liu et al. 2001b; Zava et al. 1998; Zierau et al. 2002), while others have shown binding (Jarry et al. 1985, 1999, 2003; Liu et al. 2001a).

In mouse mammary tumor cells, treatment with black cohosh increased the cytotoxicity of doxorubicin and docetaxel and decreased the cytotoxicity of cisplatin, but did not alter the effects of radiation or 4-hydroperoxycyclophosphamide (an analog of cyclophosphamide that is active in cell culture) (Rockwell et al. 2005).

In human liver cancer cells (HepG2), an ethanol extract of black cohosh impaired mitochondrial β-oxidation at a concentration of 10 μg/ml and demonstrated cytotoxic activity at a concentration of 75 μg/ml (Lüde et al. 2007).

In estrogen receptor-positive breast cancer cells (MCF-7), black cohosh alone did not show any stimulatory effect on cell growth, while a dose-dependent inhibition of estrogen proliferative effect with black cohosh was noted. A combination of black cohosh with increasing tamoxifen concentrations further inhibited breast cancer cell growth (Al-Akoum et al. 2007).

Black cohosh was shown to be a partial agonist of the μ-opiate receptor in hamster ovary cells expressing human opiate receptors (Rhyu et al. 2006).

IV. Pregnancy and Lactation

Limited scientific information is available on the use of black cohosh during pregnancy and labor. Traditional use indicates black cohosh as a partus preparator (Upton 2002).

Reviewing the available literature, Dugoua et al. (2006) concluded that low-level evidence showed concerns for use during pregnancy due to labor inducing effects, hormonal effects, emmenagogue properties, and anovulatory effects (Dugoua et al. 2006). Those authors noted that low-level evidence suggested that potential hormonal activity of black cohosh could be a cause for concern during breast-feeding. Human studies conducted since that review was completed have indicated a lack of estrogenic activity of black cohosh (*see* Human pharmacological studies).

One reference (Dugoua et al. 2006) noted low-level evidence suggesting potential hormonal activity of black cohosh could be a cause for concern during lactation, while other studies indicated a lack of estrogenic activity of black cohosh (*see* Human pharmacological studies).

A

Review Details

V. Toxicity Studies

Acute Toxicity

The LD_{50} of the compound acteina was greater than 500 mg/kg in mice after intraperitoneal administration, was 1000 mg/kg in rats after oral administration, and was 70 mg/kg in rabbits after intravenous administration. Toxic doses of acteina could not be determined in rabbits administered the compound orally or subcutaneously (Genazzani and Sorrentino 1962).

Short-Term Toxicity

In a study of an ethanolic extract of black cohosh administered by gavage to rats in daily doses from 1 to 1000 mg/kg for 21 days, some dose-dependent changes in liver cell mitochondria were observed beginning at the 10 mg/kg dose. At 10 mg/kg, a slight amount of mitochondrial swelling and an enlargement of bile canaliculi was observed. At the 100 or 300 mg/kg dose, more distinct mitochondrial swelling and alterations in mitochondrial morphology such as vacuoles in the matrix was observed. At 1000 mg/kg, effects included microvesicular steatosis of the hepatocytes, and glycogen depletion (Lüde et al. 2007).

No changes in liver morphology or hepatic function indices were observed in rats orally administered 300 mg/kg of black cohosh extract daily for 30 days (Mazzanti et al. 2008).

Chronic Toxicity

In rats administered black cohosh extract up to 5000 mg/kg/day, slight and reversible increases in some organ weights were noted in animals administered the highest doses, but no toxic effects were noted at any of the dose levels (Korn 1991).

Genotoxicity

No evidence of mutagenicity was demonstrated in the Ames test (Beuscher 1996; Boblitz et al. 2000; Schaper and Brümmer 1990).

Literature Cited

Al-Akoum, M., S. Dodin, and A. Akoum. 2007. Synergistic cytotoxic effects of tamoxifen and black cohosh on MCF-7 and MDA-MB-231 human breast cancer cells: An in vitro study. *Can. J. Physiol. Pharmacol.* 85(11):1153-1159.

Bai, W., H.-H. Henneicke-von Zepelin, S. Wang, et al. 2007. Efficacy and tolerability of a medicinal product containing an isopropanolic black cohosh extract in Chinese women with menopausal symptoms: A randomized, double blind, parallel-controlled study versus tibolone. *Maturitas* 58(1):31-41.

Beuscher, N. 1996. European phytotherapy: *Cimicifuga racemosa* L.—black cohosh. *Q. Rev. Nat. Med.* (Spring):19-27.

Boblitz, N., E. Liske, and P. Wüstenberg. 2000. Black cohosh: Efficacy, effect and safety of *Cimicifuga racemosa* in gynecology. *Dtsch. Apoth. Ztg.* 140(24):107-114.

Bodinet, C., and J. Freudenstien. 2002. Influence of *Cimicifuga racemosa* on the proliferation of estrogen receptor-positive human breast cancer cells. *Breast Cancer Res. Treat.* 76(1):1-10.

Bolle, P., S. Mastrangelo, F. Perrone, and M.G. Evandri. 2007. Estrogen-like effect of a *Cimicifuga racemosa* extract subfraction as assessed by in vivo, ex vivo and in vitro assays. *J. Steroid Biochem. Mol. Biol.* 107(3-5):262-269.

Bradley, P.R. 1992. *British herbal compendium: A handbook of scientific information on widely used plant drugs*. Bournemouth, Dorset: British Herbal Medicine Association.

Brinker, F. 2001. *Herb contraindications and drug interactions*. 3rd ed. Sandy, OR: Eclectic Medical Publications.

Chow, E.C., M. Teo, J.A. Ring, and J.W. Chen. 2008. Liver failure associated with the use of black cohosh for menopausal symptoms. *Med. J. Aust.* 188 (7):420.

Cohen, S.M., A.M. O'Connor, J. Hart, N.H. Merel, and H.S. Te. 2004. Autoimmune hepatitis associated with the use of black cohosh: a case study. *Menopause* 11(5):575-577.

Davis, V.L., M.J. Jayo, A. Ho, et al. 2008. Black cohosh increases metastatic mammary cancer in transgenic mice expressing c-erbB2. *Cancer Res.* 68(20):8377-8383.

Dixon Shanies, D., and N. Shaikh. 1999. Growth inhibition of human breast cancer cells by herbs and phytoestrogens. *Oncol. Rep.* 6:1383-1387.

Dugoua, J.J., D. Seely, D. Perri, G. Koren, and E. Mills. 2006. Safety and efficacy of black cohosh (*Cimicifuga racemosa*) during pregnancy and lactation. *Can. J. Clin. Pharmacol.* 13(3):e257-e261.

Düker, E.M., L. Kopanski, H. Jarry, and W. Wuttke. 1991. Effects of extracts from *Cimicifuga racemosa* on gonadotropin release in menopausal women and ovariectomized rats. *Planta Med.* 57:424-427.

Dunbar, K., and S.F. Solga. 2007. Black cohosh, safety, and public awareness. *Liver Int.* 27(7):1017.

Eagon, C.L., M.S. Elm, and P.K. Eagon. 1996. Estrogenicity of traditional Chinese and Western herbal remedies. *Proc. Am. Assoc. Cancer Res.* 37:284.

Eagon, P.K., N.B. Tress, H.A. Ayer, et al. 1999. Medicinal botanicals with hormonal activity. *Proc. Am. Assoc. Cancer Res.* 40:161-162.

Einbond, L.S., T. Su, H.A. Wu, et al. 2007. Gene expression analysis of the mechanisms whereby black cohosh inhibits human breast cancer cell growth. *Anticancer Res.* 27(2):697-712.

Einbond, L.S., Y. Wen-Cai, K. He, et al. 2008. Growth inhibitory activity of extracts and compounds from *Cimicifuga* species on human breast cancer cells. *Phytomedicine* 15(6-7):504-511.

Einer-Jensen, N., J. Zhao, K.P. Andersen, and K. Kristoffersen. 1996. *Cimicifuga* and *Melbrosia* lack oestrogenic effects in mice and rats. *Maturitas* 25(2):149-153.

Ellingwood, F. 1919. *American materia medica therapeutics and pharmacognosy*. 11th ed. Chicago: Ellingwood's Therapeutist.

EMEA. 2006. Assessment of case reports connected to herbal medicinal products containing *Cimicifugae racemosae rhizoma* (black cohosh, root). EMEA/HMPC/269258/2006. European Medicines Agency. London.

Felter, H.W. 1891. Eclectic medicines—I. Macrotys. *Med. Gleaner* 3(5):109-111.

Felter, H.W., and J.U. Lloyd. 1898. *King's American dispensatory*. 18th ed., 3rd rev. 2 vols. Cincinnati: Ohio Valley Co.

Foldes, J. 1959. The actions of an extract of *Cimicifuga racemosa*. *Arzneimittelforschung* 13:623-624.

Freudenstein, J., and C. Bodinet. 1999. Influence of an isopropanolic aqueous extract of *Cimicifuga racemosa* rhizoma on the proliferation of MCF-7 cells (abstract). Paper read at 23rd Int. LOF-Symposium on Phyto-Oestrogens, at University of Gent, Belgium.

Freudenstein, J., C. Dasenbrock, and T. Nisslein. 2002. Lack of promotion of estrogen-dependent mammary gland tumors in vivo by an isopropanolic *Cimicifuga racemosa* extract. *Cancer Res.* 62(12):3448-3852.

Gaube, F., S. Wolfl, L. Pusch, T.C. Kroll, and M. Hamburger. 2007. Gene expression profiling reveals effects of *Cimicifuga racemosa* (L.) Nutt. (black cohosh) on the estrogen receptor positive human breast cancer cell line MCF-7. *BMC Pharmacol.* 7(1):11.

Genazzani, E., and L. Sorrentino. 1962. Vascular action of acteina: Active constituent of *Actaea racemosa* L. *Nature* 194:544-555.

Gizicky, H.U. 1944. Arzneipflanzen in ihren Beziehungen zum weiblichen Genitalsystem. Versuche anweissen ratten und mausen mit *Cimicifuga racemosa*. *Z. Ges. Exp. Med.* 113:635-644.

Gurley, B.J., G.W. Barone, D.K. Williams, et al. 2006. Effect of milk thistle (*Silybum marianum*) and black cohosh (*Cimicifuga racemosa*) supplementation on digoxin pharmacokinetics in humans. *Drug Metab. Dispos.* 34(1):69-74.

Gurley, B.J., S.F. Gardner, M.A. Hubbard, et al. 2005. In vivo effects of goldenseal, kava kava, black cohosh, and valerian on human cytochrome P450 1A2, 2D6, 2E1, and 3A4/5 phenotypes. *Clin. Pharmacol. Ther.* 77(5):415-426.

Guzman, G., E.R. Kallwitz, C. Wojewoda, et al. 2009. Liver injury with features mimicking autoimmune hepatitis following the use of black cohosh. *Case Rep. Med.* 2009:1-8.

Harnischfeger, G., and N. Cillien. 1996. Influence of *Cimicifuga racemosa* extract fractions on the proliferation of human carcinoma cells in vitro with regard to their estrogen receptor sensitivity. Paper presented at 44th Annual Congress of Georg August Universitat, at Gottingen, Germany.

Hirschberg, A.L., M. Edlund, G. Svane, et al. 2007. An isopropanolic extract of black cohosh does not increase mammographic breast density or breast cell proliferation in postmenopausal women. *Menopause* 14(1):89-96.

Hostanska, K., T. Nisslein, J. Freudenstein, J. Reichling, and R. Saller. 2004. *Cimicifuga racemosa* extract inhibits proliferation of estrogen receptor-positive and negative human breast carcinoma cell lines by induction of apoptosis. *Breast Cancer Res. Treat.* 84(2):151-160.

Huntley, A. 2004. The safety of black cohosh (*Actaea racemosa, Cimicifuga racemosa*). *Expert Opin. Drug Safety* 3(6):615-623.

Huntley, A., and E. Ernst. 2003. A systematic review of the safety of black cohosh. *Menopause* 10(1):58-64.

Ingraffea, A., K. Donohue, C. Wilkel, and V. Falanga. 2007. Cutaneous vasculitis in two patients taking an herbal supplement containing black cohosh. *J. Am. Acad. Dermatol.* 56(5 Suppl):S124-S126.

Jarry, H., and G. Harnischfeger. 1985. Endocrine effects of constituents of *Cimicifuga racemosa*. 1. The effect on serum levels of pituitary hormones in ovariectomized rats. *Planta Med.* 51(1):46-49.

Jarry, H., G. Harnischfeger, and E. Duker. 1985. The endocrine effects of constituents of *Cimicifuga racemosa*. 2. In vitro binding of constituents to estrogen receptors. *Planta Med.* 51(4):316-319.

Jarry, H., S. Leonhardt, C. Duls, et al. 1999. Organ-specific effects of *Cimicifuga racemosa* in brain and uterus (abstract). Paper read at 23rd International LOF-Symposium on Phyto-Oestrogens, at University of Gent, Belgium.

Jarry, H., M. Metten, B. Spengler, V. Christoffel, and W. Wuttke. 2003. In vitro effects of the *Cimicifuga racemosa* extract BNO 1055. *Maturitas* 44(Suppl 1):S31-S38.

Jiang, B., F. Kronenberg, P. Nuntanakorn, M.H. Qiu, and E.J. Kennelly. 2006. Evaluation of the botanical authenticity and phytochemical profile of black cohosh products by high performance liquid chromatography with selected ion monitoring liquid-chromatography mass spectrometry. *J. Agric. Food. Chem.* 54:3242-3253.

Joy, D., J. Joy, and P. Duane. 2008. Black cohosh: A cause of abnormal postmenopausal liver function tests. *Climacteric* 11(1):84-88.

Korn, W.D. 1991. Six month oral toxicity study with remifemin-granulate in rats followed by an 8-week recovery period. Hannover, Germany. International Bioresearch.

Kretzschmar, G., T. Nisslein, O. Zierau, and G. Vollmer. 2005. No estrogen-like effects of an isopropanolic extract of Rhizoma Cimicifugae racemosae on uterus and vena cava of rats after 17 day treatment. *J. Steroid Biochem. Mol. Biol.* 97(3):271-277.

Levitsky, J., T.A. Alli, J. Wisecarver, and M.F. Sorrell. 2005a. Fulminant liver failure associated with the use of black cohosh. *Dig. Dis. Sci.* 50(3):538-539.

Levitsky, J., T.A. Alli, J. Wisecarver, and M.F. Sorrell. 2005b. Fulminant liver failure associated with the use of black cohosh. *Dig. Dis. Sci.* 53 (3):869.

Light, T.D., and J.A. Light. 2003. Acute renal transplant rejection possibly related to herbal medications. *Am. J. Transplant.* 3(12):1608-1609.

Liske, E., W. Hanggi, H.H. Henneicke-von Zepelin, et al. 2002. Physiological investigation of a unique extract of black cohosh (Cimicifugae racemosae rhizoma): A 6-month clinical study demonstrates no systemic estrogenic effect. *J. Women's Health Gend. Based Med.* 11(2):163-174.

Liu, Z., Z. Yang, M. Zhu, and J. Huo. 2001a. Estrogenicity of black cohosh (*Cimicifuga racemosa*) and its effects on estrogen receptor level in human breast cancer MCF-7 cells. *Wei Sheng Yan Jiu* 30(2):77-80.

Liu, Z., B. Yu, J.S. Huo, C.Q. Lu, and J.S. Chen. 2001b. Estrogenic effects of *Cimicifuga racemosa* (black cohosh) in mice and on estrogen receptors in MCF-7 cells. *J. Med. Food* 4:171-178.

Lohning, A., E.J. Verspohl, and H. Winterhoff. 2000. *Cimicifuga racemosa*: In vitro findings using MCF-7 cells (abstract). Paper read at Phytopharmakaforschung, at Bonn, Germany.

Low Dog, T.L., K.L. Powell, and S.M. Weisman. 2003. Critical evaluation of the safety of *Cimicifuga racemosa* in menopause symptom relief. *Menopause* 10(4):299-313.

A

Lüde, S., M. Török, S. Dieterle, et al. 2007. Hepatic effects of *Cimicifuga racemosa* extract in vivo and in vitro. *Cell Mol. Life Sci.* 64(21):2848-2857.

Lupu, R., I. Mehmi, E. Atlas, et al. 2003. Black cohosh, a menopausal remedy, does not have estrogenic activity and does not promote breast cancer cell growth. *Int. J. Oncol.* 23(5):1407-1412.

Lynch, C.R., M.E. Folkers, and W.R. Hutson. 2006. Fulminant hepatic failure associated with the use of black cohosh: A case report. *Liver Transplant.* 12(6):989-992.

Mahady, G., T. Low Dog, D.N. Sarma, and G.I. Giancaspro. 2009. Suspected black cohosh hepatotoxicity—Causality assessment versus safety signal. *Maturitas* 64(2):139-140.

Mahady, G.B. 2003. Is black cohosh estrogenic? *Nutr. Rev.* 61(5 Pt 1):183-186.

Mahady, G.B. 2005. Black cohosh (*Actaea/Cimicifuga racemosa*): Review of the clinical data for safety and efficacy in menopausal symptoms. *Treat. Endocrinol.* 4(3):177-184.

Mahady, G.B., T. Low Dog, M.L. Barrett, et al. 2008. United States Pharmacopeia review of the black cohosh case reports of hepatotoxicity. *Menopause* 15(4):628.

Mazzanti, G., A. Di Sotto, A. Franchitto, et al. 2008. Effects of *Cimicifuga racemosa* extract on liver morphology and hepatic function indices. *Phytomedicine* 15(11):1021-1024.

McFarlin, B.L., M.H. Gibson, J. O'Rear, and P. Harman. 1999. A national survey of herbal preparation use by nurse-midwives for labor stimulation. Review of the literature and recommendations for practice. *J. Nurse Midwifery* 44(3):205-216.

Meyer, S., T. Vogt, E.C. Obermann, M. Landthaler, and S. Karrer. 2007. Cutaneous pseudolymphoma induced by *Cimicifuga racemosa*. *Dermatology* 214(1):94-96.

MHRA. 2006. Black cohosh. UK Public Assessment Report.

Minciullo, P.L., A. Saija, M. Patafi, et al. 2006. Muscle damage induced by black cohosh (*Cimicifuga racemosa*). *Phytomedicine* 13(1-2):115-118.

Nasr, A., and H. Nafeh. 2009. Influence of black cohosh (*Cimicifuga racemosa*) use by postmenopausal women on total hepatic perfusion and liver functions. *Fertil. Steril.* 92(5):1780-1782.

Nesselhut, T., C. Schellhase, R. Dietrich, and W. Kuhn. 1993. Study on the proliferative potential of phytopharmacological agents with estrogen-like effect in breast cancer cells. *Arch. Gynecol. Obstet.* 254:817-818.

Nisbet, B.C., and R.E. O'Connor. 2007. Black cohosh-induced hepatitis. *Del. Med. J.* 79(11):441-444.

Nisslein, T., and J. Freudenstein. 2004. Concomitant administration of an isopropanolic extract of black cohosh and tamoxifen in the in vivo tumor model of implanted RUCA-I rat endometrial adenocarcinoma cells. *Toxicol. Lett.* 150(3):271-275.

Osmers, R., M. Friede, E. Liske, et al. 2005. Efficacy and safety of isopropanolic black cohosh extract for climacteric symptoms. *Obstet. Gynecol.* 105(5):1074-1083.

Painter, D., S. Perwaiz, and M. Murty. 2010. Black cohosh products and liver toxicity: update. *Can. Adverse React. Newsl.* 20(1):1-2.

Pierard, S., J.C. Coche, P. Lanthier, et al. 2009. Severe hepatitis associated with the use of black cohosh: A report of two cases and an advice for caution. *Eur. J. Gastroenterol. Hepatol.* 21(8):941.

Raus, K., C. Brucker, C. Gorkow, and W. Wuttke. 2006. First-time proof of endometrial safety of the special black cohosh extract (*Actaea* or *Cimicifuga racemosa* extract) CR BNO 1055. *Menopause* 13(4):678-691.

Reed, S.D., K.M. Newton, A.Z. Lacroix, et al. 2008. Vaginal, endometrial, and reproductive hormone findings: Randomized, placebo-controlled trial of black cohosh, multibotanical herbs, and dietary soy for vasomotor symptoms: The Herbal Alternatives for Menopause (HALT) study. *Menopause* 15(1):51-58.

Rhyu, M.R., J. Lu, D.E. Webster, et al. 2006. Black cohosh (*Actaea racemosa, Cimicifuga racemosa*) behaves as a mixed competitive ligand and partial agonist at the human mu opiate receptor. *J. Agric. Food Chem.* 54(26):9852-9857.

Rice, S., A. Amon, and S.A. Whitehead. 2007. Ethanolic extracts of black cohosh (*Actaea racemosa*) inhibit growth and oestradiol synthesis from oestrone sulphate in breast cancer cells. *Maturitas* 56(4):359-367.

Rockwell, S., Y. Liu, and S.A. Higgins. 2005. Alteration of the effects of cancer therapy agents on breast cancer cells by the herbal medicine black cohosh. *Breast Cancer Res. Treat.* 90(3):233-239.

Ruhlen, R.L., J. Haubner, J.K. Tracy, et al. 2007. Black cohosh does not exert an estrogenic effect on the breast. *Nutr. Cancer* 59(2):269-277.

Schaper and Brümmer. 1990. Unpublished internal research, *In* Low Dog, T., et al. Critical evaluation of the safety of *Cimicifuga racemosa* in menopause symptom relief. *Menopause* 10(4):299-313.

Scudder, J. 1903. *Specific medications and specific medicines*. Cincinnati: Scudder Bros.

Seidlova-Wuttke, D., H. Jarry, T. Becker, V. Christoffel, and W. Wuttke. 2003. Pharmacology of *Cimicifuga racemosa* extract BNO 1055 in rats: Bone, fat and uterus. *Maturitas* 44(Suppl 1):S39-S50.

Teschke, R., R. Bahre, A. Genthner, et al. 2009. Suspected black cohosh hepatotoxicity—Challenges and pitfalls of causality assessment. *Maturitas* 63(4):302-314.

Teschke, R., and A. Schwarzenboeck. 2009. Suspected hepatotoxicity by *Cimicifugae racemosae rhizoma* (black cohosh, root): Critical analysis and structured causality assessment. *Phytomedicine* 16(1):72-84.

TGA. 2006. Therapeutic Goods Administration. Black cohosh (*Cimifuga racemosa*). New labeling and consumer information for medicines containing black cohosh (*Cimifuga racemosa*).

Upton, R. 2002. *Black cohosh rhizome*. Santa Cruz, CA: American Herbal Pharmacopoeia.

van Breemen, R.B., W. Liang, S. Banuvar, et al. 2010. Pharmacokinetics of 23-epi-26-deoxyactein in women after oral administration of a standardized extract of black cohosh. *Clin. Pharmacol. Ther.* 87(2):219-225.

Whiting, P.W., A. Clouston, and P. Kerlin. 2002. Black cohosh and other herbal remedies associated with acute hepatitis. *Med. J. Aust.* 177(8):440-443.

Zava, D.T., C.M. Dollbaum, and M. Blen. 1998. Estrogen and progestin bioactivity of foods, herbs, and spices. *Proc. Soc. Exp. Biol.* 217:369-378.

Zepelin, H.H., H. Meden, K. Kostev, et al. 2007. Isopropanolic black cohosh extract and recurrence-free survival after breast cancer. *Int. J. Clin. Pharmacol. Ther.* 45(3):143-154.

Zierau, O., C. Bodinet, S. Kolba, M. Wulf, and G. Vollmer. 2002. Antiestrogenic activities of *Cimicifuga racemosa* extracts. *J. Steroid Biochem. Mol. Biol.* 80(1):125-130.

A

Adenophora spp. — Campanulaceae

Adenophora stricta Miq.
SCN: adenophora
PN: *nan sha shen* (root)
OCN: ladybells
Adenophora triphylla (Thunb. ex Murray) A. DC.
SCN: adenophora
Syn: *Adenophora tetraphylla* Fisch.
PN: *nan sha shen* (root)
OCN: ladybells
Part: root

Quick Reference Summary

Safety Class: 1
Interaction Class: A

Contraindications

None known.

Other Precautions

None known.

Drug and Supplement Interactions

None known.

Adverse Events and Side Effects

Allergic reactions to adenophora have been reported (Bensky et al. 2004).

Pharmacological Considerations

None known.

Pregnancy and Lactation

No information on the safety of adenophora in pregnancy or lactation was identified in the scientific or traditional literature. Although this review did not identify any concerns for use while pregnant or nursing, safety has not been conclusively established.

Review Details

I. Drug and Supplement Interactions

Clinical Trials of Drug or Supplement Interactions

No clinical trials of drug or supplement interactions were identified.

Case Reports of Suspected Drug or Supplement Interactions

No case reports of suspected drug or supplement interactions were identified.

Animal Trials of Drug or Supplement Interactions

No animal trials of drug or supplement interactions were identified.

II. Adverse Events

Case Reports of Adverse Events

Inappropriate use of adenophora may cause headaches, weakness, apathy, aversion to cold, distended abdomen, vomiting, or delayed menstruation (Bensky et al. 2004).

Allergic reactions to adenophora, including drug rash and asthma, have been reported (Bensky et al. 2004).

III. Pharmacology and Pharmacokinetics

Human Pharmacological Studies

No relevant human pharmacological studies were identified.

Animal Pharmacological Studies

No relevant animal pharmacological studies were identified.

In Vitro Pharmacological Studies

Methanol and ethanol extracts of adenophora exhibited weak estrogenic effects in recombinant yeast system assays (Kang et al. 2006; Kim et al. 2008).

IV. Pregnancy and Lactation

No information on the use of adenophora during pregnancy or lactation was identified.

V. Toxicity Studies

No toxicity studies were identified.

Literature Cited

Bensky, D., S. Clavey, and E. Stöger. 2004. *Chinese herbal medicine: Materia medica.* 3rd ed. Seattle: Eastland Press.

Kang, S.C., C.M. Lee, H. Choi, et al. 2006. Evaluation of oriental medicinal herbs for estrogenic and antiproliferative activities. *Phytother. Res.* 20(11):1017.

Kim, I.G., S.C. Kang, K.C. Kim, E.S. Choung, and O.P. Zee. 2008. Screening of estrogenic and antiestrogenic activities from medicinal plants. *Env. Toxicol. Pharmacol.* 25(1):75-82.

A

Adiantum spp. — Adiantiaceae

Adiantum capillus-veneris L.
SCN: maidenhair fern
OCN: southern maidenhair; Venus' hair fern
Adiantum pedatum L.
SCN: maidenhair fern
OCN: northern maidenhair
Part: herb

Quick Reference Summary

Safety Class: 2b
Interaction Class: A

Contraindications

Not for use in pregnancy except under the supervision of a qualified healthcare practitioner (List and Hörhammer 1973; Taylor 2005).

Other Precautions

None known.

Drug and Supplement Interactions

None known.

Adverse Events and Side Effects

"Large" doses may be emetic (Chadha 1988; List and Hörhammer 1973).

Pharmacological Considerations

None known.

Pregnancy and Lactation

Maidenhair fern has traditionally been used to promote menstruation (Taylor 2005). Use in pregnancy is not recommended except under the supervision of a qualified healthcare practitioner.

No information on the safety of maidenhair fern during lactation was identified. While this review did not identify any concerns for use while nursing, safety has not been conclusively established.

Review Details

I. Drug and Supplement Interactions

Clinical Trials of Drug or Supplement Interactions

No clinical trials of drug or supplement interactions were identified.

Case Reports of Suspected Drug or Supplement Interactions

No case reports of suspected drug or supplement interactions were identified.

Animal Trials of Drug or Supplement Interactions

No animal trials of drug or supplement interactions were identified.

II. Adverse Events

Case Reports of Adverse Events

No case reports of adverse events were identified.

III. Pharmacology and Pharmacokinetics

Human Pharmacological Studies

No relevant human pharmacological studies were identified.

Animal Pharmacological Studies

No relevant animal pharmacological studies were identified.

In Vitro Pharmacological Studies

No relevant in vitro pharmacological studies were identified.

IV. Pregnancy and Lactation

Maidenhair fern has traditionally been used to promote menstruation (Taylor 2005).

No information on the safety of maidenhair fern during lactation was identified.

V. Toxicity Studies

No toxicity studies were identified.

Literature Cited

Chadha, Y. 1988. *The wealth of India: A dictionary of Indian raw materials and industrial products*. Delhi: Council of Scientific and Industrial Research.

List, P.H., and H. Hörhammer. 1973. *Hagers handbuch der pharmazeutischen praxis*. Berlin: Springer.

Taylor, L. 2005. *The healing power of rainforest herbs*. Garden City Park, NY: Square One Publishers.

Aesculus hippocastanum L. — Hippocastanaceae

SCN: horse chestnut

Part: seed

Quick Reference Summary

Safety Class: 1
Interaction Class: A

Contraindications
None known.

Other Precautions
None known.

Drug and Supplement Interactions
None known.

Adverse Events and Side Effects
Systematic reviews and meta-analyses of clinical trials with horse chestnut indicate that horse chestnut is generally well tolerated with few associated adverse events (Pittler and Ernst 2006; Siebert et al. 2002). Mild adverse events reported in clinical trials were gastrointestinal complaints, dizziness, nausea, headache, and itching (Pittler and Ernst 2006).

Allergic reactions, including anaphylactic reactions, to horse chestnut have been reported (Jaspersen-Schib et al. 1996; Sirtori 2001).

Cases of kidney and liver damage have been reported in association with injections of purified extracts of horse chestnut (Grasso and Corvaglia 1976; Hellberg et al. 1975; Klose and Pistor 1976; Takegoshi et al. 1986; Voigt and Junger 1978). No such reactions have been reported or are expected with oral use of horse chestnut (Mills and Bone 2005).

Pharmacological Considerations
In vitro studies examining the effects of horse chestnut on the drug-metabolizing isoenzymes CYP3A4, CYP1A2, CYP2D6, CYP2E1, and CYP2C19, and P-glycoprotein (P-gp) drug efflux transporters showed some effects on these isoenzymes and transporters, but none were thought to be clinically relevant (Brandin et al. 2007; Hellum et al. 2007, 2009; Hellum and Nilsen 2007, 2008).

Pregnancy and Lactation
No adverse effects on fetal development were reported in several clinical studies on the use of horse chestnut extract in pregnant women (Alter 1973; Steiner 1990; Steiner and Hillemanns 1986, 1990). In animal studies, decreased fetal weight was observed in rabbits orally administered high doses (300 mg/kg) of horse chestnut. No adverse effects were observed at the same dose in rats or lower doses (100 mg/kg) in rats or rabbits (Liehn et al. 1972).

No information on the safety of horse chestnut during lactation was identified. While this review did not identify any concerns for use while nursing, safety has not been conclusively established.

Review Details

I. Drug and Supplement Interactions

Clinical Trials of Drug or Supplement Interactions
No clinical trials of drug or supplement interactions were identified.

Case Reports of Suspected Drug or Supplement Interactions
No case reports of suspected drug or supplement interactions were identified.

Animal Trials of Drug or Supplement Interactions
No animal trials of drug or supplement interactions were identified.

II. Adverse Events

Adverse Events Reported in Clinical Trials
A systematic review of double-blind, controlled studies of horse chestnut indicated that 14 of the identified studies reported on adverse events. Of the 14, 4 reported that there were no treatment-related adverse events in the horse chestnut groups. In 6 studies, gastrointestinal complaints, dizziness, nausea, headache, and pruritus were reported with a frequency ranging from 1 to 36% of treated patients. The remaining 4 studies reported that horse chestnut treatments were well tolerated. The review characterized adverse events in the horse chestnut groups as mild and infrequent (Pittler and Ernst 2006).

A meta-analysis of randomized controlled trials (totaling 1051 patients) and large-scale observational studies (totaling 10,725 patients) of horse chestnut extract indicated that no serious adverse events were reported in the trials and studies and that treatment with horse chestnut extract did not significantly increase mild adverse events (Siebert et al. 2002).

Case Reports of Adverse Events
Cases of acute renal failure have been reported in children who were intravenously administered injections of 0.5 mg/kg (approximately 2 to 3 times the recommended dose) of aescin (a mixture of saponins from horse chestnut) for an

average of 4 days. The cases were primarily seen in children ages 2 to 10 and were observed after 3 to 4 days of normal renal function. The overall mortality rate was 10.7% (Grasso and Corvaglia 1976; Hellberg et al. 1975; Klose and Pistor 1976; Voigt and Junger 1978). Commenting on these cases, a text on the use of drugs in patients with renal insufficiency notes that the cases involved patients administered either the standard dose or overdose, and that some were also taking other drugs. Also, most of the patients were polytraumatic after accidents or had undergone severe surgery, conditions that alone may lead to acute renal failure (Seyffart 1991). Aescin may cause hemolysis after injection, and the liberated hemoglobin can deposit in the kidneys, leading to renal failure. Such effects are not expected with oral use of horse chestnut (Mills and Bone 2005).

Hepatic injury was reported in a 37-year-old man who had received a single intramuscular injection of 65 mg of a standardized horse chestnut extract prior to surgery. Liver tests performed 17 days after injection revealed moderate elevation of total bilirubin, ALP, GGTP and mild eosinophilia. The lymphocyte stimulation test was positive, and the liver biopsy demonstrated marked cholestasis with zonal necrosis in the centrilobular areas but showed little or no changes in the portal tracts (Takegoshi et al. 1986).

Allergic reactions, including anaphylactic reactions, to horse chestnut have been reported (Jaspersen-Schib et al. 1996; Sirtori 2001).

III. Pharmacology and Pharmacokinetics

Human Pharmacological Studies

In trials assessing the renal effects of intravenously administered aescin (a mixture of saponins from horse chestnut; 10 to 25 mg daily for 3 to 10 days), no impairment of renal function was observed in patients with healthy kidneys, and no worsening of function was observed in patients with renal impairment (Ascher 1977; Bastian and Valilensieck 1976; Sirtori 2001; Wilhelm and Feldmeier 1975).

Animal Pharmacological Studies

No relevant animal pharmacological studies were identified.

In Vitro Pharmacological Studies

Inhibition of the drug-metabolizing isoenzyme CYP3A4 and the P-gp drug efflux transporters was observed in cDNA baculovirus-expressed CYP3A4 and Caco-2 cells treated with horse chestnut extract. The effects were less than that of St. John's wort (Hellum and Nilsen 2008). Some inhibition of CYP2D6 was observed in cDNA baculovirus-expressed CYP2D6, although the activity was not considered to be clinically relevant (Hellum and Nilsen 2007).

General inhibitory potential of the drug-metabolizing isoenzymes CYP1A2, CYP2D6, and CYP3A4 were observed in primary human hepatocytes treated with an extract of horse chestnut. The activity was less than that of other botanicals considered to have clinically relevant CYP interactions (Hellum et al. 2007).

A twofold induction of the drug-metabolizing isoenzyme CYP1A2 was observed in human colon cancer cells (LS180) treated with horse chestnut extract. No effects on CYP3A4 or the transporter protein MDR1 were observed (Brandin et al. 2007).

No significant effects of horse chestnut extract were observed in the drug-metabolizing isoenzymes CYP2C19 and CYP2E1 in cultured human hepatocytes (Hellum et al. 2009).

IV. Pregnancy and Lactation

Several clinical studies on the use of horse chestnut extract in pregnant women have been completed. Dosages were 480 to 600 mg daily (standardized to 100 mg aescin) for 2 to 4 weeks. In these studies, no adverse effects on fetal development were reported (Alter 1973; Steiner 1990; Steiner and Hillemanns 1986; Steiner and Hillemanns 1990).

No teratogenic effects were observed in the offspring of rats orally administered 100 or 300 mg/kg or rabbits orally administered 100 mg/kg horse chestnut extract during pregnancy. In rabbits administered 300 mg/kg, a significant reduction in the weight of fetuses was observed (Liehn et al. 1972). No teratogenic or embryotoxic effects of horse chestnut extract were observed in rats intravenously administered 9 or 30 mg/kg on days 6 to 15 of pregnancy, or rabbits administered the same doses on days 6 to 18 of pregnancy (Liehn et al. 1972).

No information on the safety of horse chestnut during lactation was identified.

V. Toxicity Studies

Acute Toxicity

The LD_{50} of orally administered horse chestnut extract is 990 mg/kg in mice, 2150 mg/kg in rats, 1120 mg/kg in guinea pigs, 1530 mg/kg in rabbits, 10,700 mg/kg in hamsters, and 10,600 mg/kg in chicks (Liehn et al. 1972; Williams and Olsen 1984). In dogs, no oral LD_{50} could be determined, as dogs vomited the test substance at doses over 130 mg/kg (Liehn et al. 1972). The LD_{50} of intravenously administered horse chestnut extract is 138 mg/kg in mice, 165 mg/kg in rats, 465 mg/kg in guinea pigs, and 180 mg/kg in rabbits. The LD_{50} of intraperitoneally administered horse chestnut in mice was 342 mg/kg (Liehn et al. 1972).

Subchronic Toxicity

In rats intravenously administered 9, 30, or 90 mg/kg horse chestnut extract daily for 8 weeks, no adverse effects were reported at the 9 mg/kg dose. At the 90 mg/kg dose, 8 of the 30 test animals died during the first several days, although the rest of the animals in that treatment group developed normal body weights (Liehn et al. 1972).

No toxic effects or organ damage were observed in dogs orally administered 20, 40, or 80 mg/kg (8 times the human dose) or rats orally administered 100, 200, or 400 mg/kg (40 times the human dose) of horse chestnut extract 5 days per week for 34 weeks (Liehn et al. 1972).

Genotoxicity

In the Ames test for mutagenicity, horse chestnut extract was weakly mutagenic with metabolic activation by S9 but showed no mutagenicity without activation. Fluid extracts of horse chestnut showed no mutagenic activity without activation, and weak mutagenic activity with activation (Schimmer et al. 1994). The authors of this study indicated that the mutagenic effects were likely due to the compound quercetin, and quercetin is not considered to have clinically relevant mutagenic activity in humans (Harwood et al. 2007).

Literature Cited

Alter, H. 1973. Zur medikamentösen therapie der varikosis. *Z. Allg. Med.* 49:1301-1304.

Ascher, P. 1977. Renale Funktionsgroessen unter Na-Aescinat bei nierengesunden und nierenkranken Patienten. *Therapiewoche* 52:3-10.

Bastian, H.P., and W. Valilensieck. 1976. Nierenfunktion unter parenteraler Aescin-Behandlung. *Med. Klin.* 71:1295-1299.

Brandin, H., E. Viitanen, O. Myrberg, and A.K. Arvidsson. 2007. Effects of herbal medicinal products and food supplements on induction of CYP1A2, CYP3A4 and MDR1 in the human colon carcinoma cell line LS180. *Phytother. Res.* 21(3):239-244.

Grasso, A., and E. Corvaglia. 1976. Due casi di sospetta tubulonefrosi tossica da escina. *Gazz. Med. Ital.* 135:581-584.

Harwood, M., B. Danielewska-Nikiel, J.F. Borzelleca, et al. 2007. A critical review of the data related to the safety of quercetin and lack of evidence of in vivo toxicity, including lack of genotoxic/carcinogenic properties. *Food Chem. Toxicol.* 45(11):2179-2205.

Hellberg, K., W. Ruschewski, and R. de Vivie. 1975. Medikamentoes bedingtes post-operatives Nierenversagen nach herzchirurgischen Eingriffen. *Thoraxchirurgie* 23:396-399.

Hellum, B.H., Z. Hu, and O.G. Nilsen. 2007. The induction of CYP1A2, CYP2D6 and CYP3A4 by six trade herbal products in cultured primary human hepatocytes. *Basic Clin. Pharmacol. Toxicol.* 100(1):23-30.

Hellum, B.H., Z. Hu, and O.G. Nilsen. 2009. Trade herbal products and induction of CYP2C19 and CYP2E1 in cultured human hepatocytes. *Basic Clin. Pharmacol. Toxicol.* 105(1):58-63.

Hellum, B.H., and O.G. Nilsen. 2007. The in vitro inhibitory potential of trade herbal products on human CYP2D6-mediated metabolism and the influence of ethanol. *Basic Clin. Pharmacol. Toxicol.* 101(5):350-358.

Hellum, B.H., and O.G. Nilsen. 2008. In vitro inhibition of CYP3A4 metabolism and P-glycoprotein-mediated transport by trade herbal products. *Basic Clin. Pharmacol. Toxicol.* 102(5):466-475.

Jaspersen-Schib, R., L. Theus, M. Guirguis-Oeschger, B. Gossweiler, and P.J. Meier-Abt. 1996. Acute poisonings with toxic plants in Switzerland between 1966 and 1994. *Schweiz. Med. Wochenschr.* 126(25):1085-1098.

Klose, P., and K. Pistor. 1976. Posttraumatisches nierenversagen bei 2 Kindern nach beta-aescin-therapie. *Munch. Med. Wschr.* 719-720.

Liehn, H.D., P.A. Franco, H. Hampel, and G. Hofrichter. 1972. A toxicological study of extractum hippocastani semen. *Panminerva Med.* 14(3):84.

Mills, S., and K. Bone. 2005. *The essential guide to herbal safety*. St. Louis: Elsevier.

Pittler, M.H., and E. Ernst. 2006. Horse chestnut seed extract for chronic venous insufficiency. *Cochrane Database Syst. Rev.* (1):CD003230.

Schimmer, O., A. Krüger, H. Paulini, and F. Haefele. 1994. An evaluation of 55 commercial plant extracts in the Ames mutagenicity test. *Pharmazie* 49:448-451.

Seyffart, G. 1991. *Drug dosage in renal insufficiency*. Boston: Kluwer.

Siebert, U., M. Brach, G. Sroczynski, and K. Berla. 2002. Efficacy, routine effectiveness, and safety of horsechestnut seed extract in the treatment of chronic venous insufficiency. A meta-analysis of randomized controlled trials and large observational studies. *Int. Angiol.* 21(4):305-315.

Sirtori, C.R. 2001. Aescin: Pharmacology, pharmacokinetics and therapeutic profile. *Pharmacol. Res.* 44(3):183-193.

Steiner, M. 1990. Untersuchungen zur ödemvermindernden und ödemprotektiven wirkung von roßkastaniensamenextrakt. *Phlebol. Proktol.* 19:239-242.

Steiner, M., and H. Hillemanns. 1986. Untersuchungen zur ödemprotektiven Wirkung eines Venentherapeutikums. *Munch. Med. Wschr.* 128:551-555.

Steiner, M., and H. Hillemanns. 1990. Venostasin retard in the management of venous problems during pregnancy. *Phlebology* 5:41-44.

Takegoshi, K., T. Tohyama, and K. Okuda. 1986. A case of Venoplant-induced hepatic injury. *Gastroenterol. Jap.* 21(1):62-65.

Voigt, E., and H. Junger. 1978. Acute posttraumatic renal failure following therapy with antibiotics and beta-aescin. *Anaesthesist* 27(2):81.

Wilhelm, R., and C. Feldmeier. 1975. Postoperative und posttraumatische Oedemprophylaxe und therapie. *Med. Klin.* 70:2079-2083.

Williams, M.C., and J.D. Olsen. 1984. Toxicity of seeds of three *Aesculus* spp. to chicks and hamsters. *Am. J. Vet. Res.* 45(3):539-542.

A

Aframomum melegueta K. Schum.

Zingiberaceae

SCN: grains-of-paradise (seed)
Syn: *Amomum melegueta* Roscoe
AN: *brihadela*
OCN: Guinea grains (seed); melegueta pepper (seed)
Part: fruit, seed

Quick Reference Summary

Safety Class: 1
Interaction Class: A

Contraindications

None known.

Other Precautions

None known.

Drug and Supplement Interactions

None known.

Notice

Contains piperine (Githens 1948); *see* Appendix 3.

Editors' Note

Concerns for this herb are based on the relatively higher doses used for therapeutic purposes in contrast to lower amounts generally used in cooking, and have not been associated with its use as a spice.

Adverse Events and Side Effects

In one human study, consumption of single doses of 350 mg of grains-of-paradise seed led to temporary ocular changes (Igwe et al. 1999).

Pharmacological Considerations

None known.

Pregnancy and Lactation

No information on the safety of grains-of-paradise in pregnancy or lactation was identified in the scientific or traditional literature. Although this review did not identify any concerns for use while pregnant or nursing, safety has not been conclusively established.

Review Details

I. Drug and Supplement Interactions

Clinical Trials of Drug or Supplement Interactions

No clinical trials of drug or supplement interactions were identified.

Case Reports of Suspected Drug or Supplement Interactions

No case reports of suspected drug or supplement interactions were identified.

Animal Trials of Drug or Supplement Interactions

No animal trials of drug or supplement interactions were identified.

II. Adverse Events

Case Reports of Adverse Events

No case reports of adverse events were identified.

III. Pharmacology and Pharmacokinetics

Human Pharmacological Studies

Ingestion of 350 mg of grains-of-paradise seed by healthy males increased the near point of convergence in the eyes by 17% and reduced the amplitude of ocular accommodation by 9.2% without affecting pupil size or visual acuity. The increased near point of convergence leads to doubling of vision while the reduction or loss of accommodation leads to blurring of vision, with these effects synergizing to transiently impair vision. The authors indicated that traditional practice of long-term, excessive consumption of grains-of-paradise may contribute to premature presbyopia that is common among the Igbo peoples of eastern Nigeria (Igwe et al. 1999).

Animal Pharmacological Studies

No relevant animal pharmacological studies were identified.

In Vitro Pharmacological Studies

No relevant in vitro pharmacological studies were identified.

IV. Pregnancy and Lactation

No information on the use of grains-of-paradise during pregnancy or lactation was identified.

V. Toxicity Studies

The LD_{50} of intraperitoneally administered grains-of-paradise ethanol extract in mice was 2.1 g/kg (Okoli et al. 2007).

Literature Cited

Githens, T. 1948. *Drug plants of Africa.* Philadelphia: University of Pennsylvania Press.

Igwe, S., I. Emeruwa, and J. Modie. 1999. Ocular toxicity of *Aframomum melegueta* (alligator pepper) on healthy Igbos of Nigeria. *J. Ethnopharmacol.* 65(3):203-206.

Okoli, C.O., P.A. Akah, S.V. Nwafor, U.U. Ihemelandu, and C. Amadife. 2007. Anti-inflammatory activity of seed extracts of *Aframomum melegueta. J. Herbs Spices Med. Plants* 13(1):11-21.

Agastache rugosa (Fisch. & C.A. Mey.) Kuntze — Lamiaceae

SCN: Chinese giant hyssop
PN: *huo xiang* (herb)
Part: herb

Quick Reference Summary

Safety Class: 1
Interaction Class: A

Contraindications

None known.

Other Precautions

None known.

Drug and Supplement Interactions

None known.

Adverse Events and Side Effects

None known.

Pharmacological Considerations

None known.

Pregnancy and Lactation

Texts on traditional Chinese medicine do not indicate any cautions for use of Chinese giant hyssop during pregnancy or lactation (Bensky et al. 2004; Chen and Chen 2004). Although this review did not identify any concerns for use while pregnant or nursing, safety has not been conclusively established.

Review Details

I. Drug and Supplement Interactions

Clinical Trials of Drug or Supplement Interactions

No clinical trials of drug or supplement interactions were identified.

Case Reports of Suspected Drug or Supplement Interactions

No case reports of suspected drug or supplement interactions were identified.

Animal Trials of Drug or Supplement Interactions

No animal trials of drug or supplement interactions were identified.

II. Adverse Events

Case Reports of Adverse Events

No case reports of adverse events were identified.

III. Pharmacology and Pharmacokinetics

Human Pharmacological Studies

No relevant human pharmacological studies were identified.

Animal Pharmacological Studies

No relevant animal pharmacological studies were identified.

In Vitro Pharmacological Studies

No relevant in vitro pharmacological studies were identified.

IV. Pregnancy and Lactation

Texts on traditional Chinese medicine do not indicate any cautions for use of Chinese giant hyssop during pregnancy or lactation (Bensky et al. 2004; Chen and Chen 2004).

V. Toxicity Studies

No toxicity studies were identified.

Literature Cited

Bensky, D., S. Clavey, and E. Stöger. 2004. *Chinese herbal medicine: Materia medica.* 3rd ed. Seattle: Eastland Press.

Chen, J.K., and T.T. Chen. 2004. *Chinese medical herbology and pharmacology.* City of Industry, CA: Art of Medicine Press.

A

Agathosma spp. Rutaceae

Agathosma betulina (P.J. Bergius) Pillans
SCN: buchu
Syn: *Barosma betulina* (Bergius) Bartl. & H.L. Wendl.
OCN: round buchu; short buchu
Agathosma crenulata (L.) Pillans
SCN: buchu
Syn: *Barosma crenulata* (L.) Hook.
OCN: ovate buchu
Agathosma serratifolia (Curtis) Spreeth
SCN: buchu
Syn: *Barosma serratifolia* (Curtis) Willd.
OCN: long buchu
Part: leaf

Quick Reference Summary

Safety Class: 2b
Interaction Class: A

Contraindications

Not for use in pregnancy except under the supervision of a qualified healthcare practitioner (Bradley 1992; Collins and Graven 1996; Kaiser et al. 1975).

Other Precautions

Use with caution in persons with kidney inflammation (McGuffin et al. 1997).

Drug and Supplement Interactions

None known.

Notice

Diuretic (Felter and Lloyd 1898; Moolla and Viljoen 2008; Remington and Wood 1918). (See Appendix 2.)

Adverse Events and Side Effects

None known.

Pharmacological Considerations

None known.

Pregnancy and Lactation

Buchu contains the compound pulegone (2.4–4.5% of *A. betulina* essential oil and 31.6–73.2% of *A. crenulata* essential oil) (Collins and Graven 1996; Kaiser et al. 1975). Pulegone is considered to be the primary compound in European pennyroyal (*Mentha pulegium*) responsible for the abortifacient activity of this plant (Anderson et al. 1996). Although no reports of abortifacient activity of buchu were identified, use of buchu during pregnancy is not recommended.

No information on the safety of buchu during lactation was identified. While this review did not identify any concerns for use while nursing, safety has not been conclusively established.

Review Details

I. Drug and Supplement Interactions

Clinical Trials of Drug or Supplement Interactions

No clinical trials of drug or supplement interactions were identified.

Case Reports of Suspected Drug or Supplement Interactions

No case reports of suspected drug or supplement interactions were identified.

Animal Trials of Drug or Supplement Interactions

No animal trials of drug or supplement interactions were identified.

II. Adverse Events

Case Reports of Adverse Events

No case reports of adverse events were identified.

III. Pharmacology and Pharmacokinetics

Human Pharmacological Studies

No relevant human pharmacological studies were identified.

Animal Pharmacological Studies

No relevant animal pharmacological studies were identified.

In Vitro Pharmacological Studies

No relevant in vitro pharmacological studies were identified.

IV. Pregnancy and Lactation

Buchu contains the compound pulegone (2.4–4.5% of *A. betulina* essential oil and 31.6–73.2% of *A. crenulata* essential oil) (Collins and Graven 1996; Kaiser et al. 1975). Pulegone is considered to be the primary compound in European pennyroyal (*Mentha pulegium*) responsible for the abortifacient activity of that plant (Anderson et al. 1996). Although no reports of abortifacient activity of buchu were identified, use of buchu during pregnancy is not recommended.

No information on the safety of buchu during lactation was identified.

V. Toxicity Studies

Cytotoxicity

At concentrations up to 100 µg/ml, no cytotoxic activity of the essential oil of buchu species (*A. betulina* and *A. crenulata*) was observed in the MTT cellular viability assay (Viljoen et al. 2006).

Literature Cited

Anderson, I.B., W.H. Mullen, J.E. Meeker, et al. 1996. Pennyroyal toxicity: Measurement of toxic metabolite levels in two cases and review of the literature. *Ann. Intern. Med.* 124(8):726-734.

Bradley, P.R. 1992. *British herbal compendium: A handbook of scientific information on widely used plant drugs*. Bournemouth, UK: British Herbal Medicine Association.

Collins, N.F., and E.H. Graven. 1996. Chemotaxonomy of commercial buchu species (*Agathosma betulina* and *A. crenulata*). *J. Essen. Oil Res.* 8:229-235.

Felter, H.W., and J.U. Lloyd. 1898. *King's American dispensatory*. 18th ed., 3rd rev. 2 vols. Cincinnati: Ohio Valley Co.

Kaiser, R., D. Lamparsky, and P. Schudel. 1975. Analysis of buchu leaf oil. *J. Agric. Food Chem.* 23:943-950.

McGuffin, M., C. Hobbs, R. Upton, and A. Goldberg. 1997. *Botanical safety handbook*. Boca Raton, FL: CRC Press.

Moolla, A., and A.M. Viljoen. 2008. 'Buchu'—*Agathosma betulina* and *Agathosma crenulata* (Rutaceae): A review. *J. Ethnopharmacol.* 119(3):413-419.

Remington, J.P., and H.C. Wood. 1918. *The dispensatory of the United States of America*. 20th ed. Philadelphia: Lippincott.

Viljoen, A.M., A. Moolla, S.F. Van Vuuren, et al. 2006. The biological activity and essential oil composition of 17 *Agathosma* (Rutaceae) species. *J. Essen. Oil Res.* 18:2-16.

Agrimonia eupatoria L. — Rosaceae

SCN: agrimony
OCN: church steeples

Part: herb

Quick Reference Summary

Safety Class: 1
Interaction Class: A

Contraindications

None known.

Other Precautions

None known.

Drug and Supplement Interactions

Notice

Tannins (4.0–10.0%) (Wichtl 2004); *see* Appendix 1.

Adverse Events and Side Effects

None known.

Pharmacological Considerations

None known.

Pregnancy and Lactation

No information on the safety of agrimony in pregnancy or lactation was identified in the scientific or traditional literature. Although this review did not identify any concerns for use while pregnant or nursing, safety has not been conclusively established.

Review Details

I. Drug and Supplement Interactions

Clinical Trials of Drug or Supplement Interactions

No clinical trials of drug or supplement interactions were identified.

Case Reports of Suspected Drug or Supplement Interactions

No case reports of suspected drug or supplement interactions were identified.

Animal Trials of Drug or Supplement Interactions

No animal trials of drug or supplement interactions were identified.

II. Adverse Events

Case Reports of Adverse Events

No case reports of adverse events were identified.

III. Pharmacology and Pharmacokinetics

Human Pharmacological Studies

No relevant human pharmacological studies were identified.

Animal Pharmacological Studies

No relevant animal pharmacological studies were identified.

In Vitro Pharmacological Studies

No relevant in vitro pharmacological studies were identified.

IV. Pregnancy and Lactation

No information on the safety of agrimony in pregnancy or lactation was identified.

V. Toxicity Studies

Genotoxicity

No mutagenic activity of agrimony methanolic extract was observed in the Ames test with or without metabolic activation (Bilia et al. 1993).

Literature Cited

Bilia, A.R., E. Palme, S. Catalano, L. Pistelli, and I. Morelli. 1993. Constituents and biological assay of *Agrimonia eupatoria*. *Fitoterapia* 64(6):549-550.

Wichtl, M. 2004. *Herbal drugs and phytopharmaceuticals: A handbook for practice on a scientific basis*. 3rd ed. Boca Raton, FL: CRC Press.

Albizia julibrissin Durazz. — Fabaceae

SCN: silk tree
PN: *he huan pi* (bark)
OCN: mimosa tree
Part: bark

Quick Reference Summary

Safety Class: 2b
Interaction Class: A

Contraindications

Not for use in pregnancy except under the supervision of a qualified healthcare practitioner (Bensky et al. 2004; Chen and Chen 2004).

Other Precautions

None known.

Drug and Supplement Interactions

None known.

Adverse Events and Side Effects

None known.

Pharmacological Considerations

None known.

Pregnancy and Lactation

Texts on traditional Chinese medicine indicate that silk tree bark should be used with caution in pregnancy (Bensky et al. 2004; Chen and Chen 2004). One text indicates that silk tree bark stimulates uterine contractions (Chen and Chen 2004).

No information on the safety of silk tree bark during lactation was identified. While this review did not identify any concerns for use while nursing, safety has not been conclusively established.

Review Details

I. Drug and Supplement Interactions

Clinical Trials of Drug or Supplement Interactions

No clinical trials of drug or supplement interactions were identified.

Case Reports of Suspected Drug or Supplement Interactions

No case reports of suspected drug or supplement interactions were identified.

Animal Trials of Drug or Supplement Interactions

No animal trials of drug or supplement interactions were identified.

II. Adverse Events

Case Reports of Adverse Events

No case reports of adverse events were identified.

III. Pharmacology and Pharmacokinetics

Human Pharmacological Studies

No relevant human pharmacological studies were identified.

Animal Pharmacological Studies

No relevant animal pharmacological studies were identified.

In Vitro Pharmacological Studies

The compound julibrin II (4′-*O*-methylpyridoxine) was found to induce arrhythmias in isolated frog hearts. Other related compounds also isolated from silk tree did not induce arrhythmias (Higuchi et al. 1992). The compound 4′-*O*-methylpyridoxine is also found in ginkgo seed and is believed to be the compound associated with seizures reported after excessive consumption of ginkgo seed. The seizures are thought to be due to vitamin B_6 deficiency, which can be caused by 4′-*O*-methylpyridoxine (van Beek and Montoro 2009; Wada et al. 1985). No cases of seizures have been reported in association with traditional use of silk tree.

IV. Pregnancy and Lactation

Texts on traditional Chinese medicine indicate that silk tree bark should be used with caution in pregnancy (Bensky et al. 2004; Chen and Chen 2004) and indicate that silk tree bark stimulates uterine contractions (Chen and Chen 2004).

No information on the safety of silk tree bark during lactation was identified.

V. Toxicity Studies

Silk tree bark contains the compound 4′-*O*-methylpyridoxine, a compound that can cause vitamin B_6 deficiency symptoms and is believed to be responsible for the toxic effects (seizures and gastrointestinal symptoms) of ginkgo seed consumption (see *Ginkgo biloba* seed entry) (Mooney et al. 2009; Wada et al. 1985). No toxic effects of silk tree bark have been reported in association with use in traditional Chinese medicine (Bensky et al. 2004; Chen and Chen 2004).

Cytotoxicity

Saponin fractions of silk tree exhibited cytotoxic activity in vitro (Ikeda et al. 1997).

Literature Cited

Bensky, D., S. Clavey, and E. Stöger. 2004. *Chinese herbal medicine: Materia medica*. 3rd ed. Seattle: Eastland Press.

Chen, J.K., and T.T. Chen. 2004. *Chinese medical herbology and pharmacology*. City of Industry, CA: Art of Medicine Press.

Higuchi, H., J. Kinjo, and T. Nohara. 1992. An arrhythmic-inducing glycoside from *Albizia julibrissin* Durazz, IV. *Chem. Pharm. Bull.* 40(3):829-831.

Ikeda, T., S. Fujiwara, K. Araki, et al. 1997. Cytotoxic glycosides from *Albizia julibrissin*. *J. Nat. Prod.* 60(2):102-107.

Mooney, S., J.E. Leuendorf, C. Hendrickson, and H. Hellmann. 2009. Vitamin B_6: A long known compound of surprising complexity. *Molecules* 14(1):329-351.

van Beek, T.A., and P. Montoro. 2009. Chemical analysis and quality control of *Ginkgo biloba* leaves, extracts, and phytopharmaceuticals. *J. Chromatogr. A* 1216(11):2002-2032.

Wada, K., S. Ishigaki, K. Ueda, M. Sakata, and M. Haga. 1985. An antivitamin B_6, 4′-methoxypyridoxine from the seed of *Ginkgo biloba* L. *Chem. Pharm. Bull.* 33:3555-3557.

Albizia julibrissin Durazz. — Fabaceae

SCN: silk tree
PN: *he huan hua* (flower)
OCN: mimosa tree
Part: flower

Quick Reference Summary

Safety Class: 1
Interaction Class: A

Contraindications

None known.

Other Precautions

None known.

Drug and Supplement Interactions

None known.

Adverse Events and Side Effects

None known.

Pharmacological Considerations

None known.

Pregnancy and Lactation

Texts on traditional Chinese medicine do not indicate any cautions for use of silk tree flower during pregnancy or lactation (Bensky et al. 2004; Chen and Chen 2004). Although this review did not identify any concerns for use while pregnant or nursing, safety has not been conclusively established.

A

Review Details

I. Drug and Supplement Interactions

Clinical Trials of Drug or Supplement Interactions

No clinical trials of drug or supplement interactions were identified.

Case Reports of Suspected Drug or Supplement Interactions

No case reports of suspected drug or supplement interactions were identified.

Animal Trials of Drug or Supplement Interactions

No animal trials of drug or supplement interactions were identified.

II. Adverse Events

Case Reports of Adverse Events

No case reports of adverse events were identified.

III. Pharmacology and Pharmacokinetics

Human Pharmacological Studies

No relevant human pharmacological studies were identified.

Animal Pharmacological Studies

No relevant animal pharmacological studies were identified.

In Vitro Pharmacological Studies

No relevant in vitro pharmacological studies were identified.

IV. Pregnancy and Lactation

Texts on traditional Chinese medicine do not indicate any cautions for use of silk tree flower during pregnancy or lactation (Bensky et al. 2004; Chen and Chen 2004).

V. Toxicity Studies

No toxicity studies were identified.

Literature Cited

Bensky, D., S. Clavey, and E. Stöger. 2004. *Chinese herbal medicine: Materia medica*. 3rd ed. Seattle: Eastland Press.

Chen, J.K., and T.T. Chen. 2004. *Chinese medical herbology and pharmacology*. City of Industry, CA: Art of Medicine Press.

Alcea rosea L. — Malvaceae

SCN: hollyhock
Syn: *Althaea rosea* (L.) Cav.

Part: root

Quick Reference Summary

Safety Class: 1
Interaction Class: A

Contraindications

None known.

Other Precautions

None known.

Drug and Supplement Interactions

Other drugs should be taken one hour prior to consumption of hollyhock or several hours after consumption, as mucilaginous plants such as hollyhock may slow the absorption of orally administered drugs (Brinker 2001; De Smet 1993; Mills and Bone 2005).

Notice

Mucilages (Tomoda et al. 1983; Turowska et al. 1966); *see* Appendix 3.

Adverse Events and Side Effects

None known.

Pharmacological Considerations

None known.

Pregnancy and Lactation

No information on the safety of hollyhock in pregnancy or lactation was identified in the scientific or traditional literature. Although this review did not identify any concerns for use while pregnant or nursing, safety has not been conclusively established.

Review Details

I. Drug and Supplement Interactions

Clinical Trials of Drug or Supplement Interactions

No clinical trials of drug or supplement interactions were identified.

Case Reports of Suspected Drug or Supplement Interactions

No case reports of suspected drug or supplement interactions were identified.

Animal Trials of Drug or Supplement Interactions

No animal trials of drug or supplement interactions were identified.

II. Adverse Events

Case Reports of Adverse Events

No case reports of adverse events were identified.

III. Pharmacology and Pharmacokinetics

Human Pharmacological Studies

No relevant human pharmacological studies were identified.

Animal Pharmacological Studies

No relevant animal pharmacological studies were identified.

In Vitro Pharmacological Studies

No relevant in vitro pharmacological studies were identified.

IV. Pregnancy and Lactation

No information on the safety of hollyhock in pregnancy or lactation was identified.

V. Toxicity Studies

No toxicity studies were identified.

Literature Cited

Brinker, F. 2001. *Herb contraindications and drug interactions*. 3rd ed. Sandy, OR: Eclectic Medical Publications.

De Smet, P.A.G.M. 1993. *Adverse effects of herbal drugs, Vol. 2*. New York: Springer.

Mills, S., and K. Bone. 2005. *The essential guide to herbal safety*. St. Louis: Elsevier.

Tomoda, M., K. Shimada, and N. Shimizu. 1983. Plant mucilages. XXXII. A representative mucilage, "*Althaea*-mucilage R", from the roots of *Althaea rosea*. *Chem. Pharm. Bull.* 31(8):2677-2684.

Turowska, I., S. Kohlmunzer, and Z. Maga. 1966. Studies on the correlation between the concentration of the mucilaginous elements and the value of *Althaea rosea* as a mucilaginous raw material: I. *Acta Biol. Cracov Ser. Bot.* 9:111.

Alchemilla xanthochlora Rothm. — Rosaceae

SCN: lady's mantle
Syn: *Alchemilla vulgaris* auct. non L.

Part: herb

Quick Reference Summary

Safety Class: 1

Interaction Class: A

Contraindications

None known.

Other Precautions

None known.

Drug and Supplement Interactions

None known.

Notice

Tannins (6–16%) (Fraisse et al. 1999; Wichtl 2004); *see* Appendix 1.

Adverse Events and Side Effects

None known.

Pharmacological Considerations

None known.

Pregnancy and Lactation

No information on the safety of lady's mantle in pregnancy or lactation was identified in the scientific or traditional literature. Although this review did not identify any concerns for use while pregnant or nursing, safety has not been conclusively established.

A

Review Details

I. Drug and Supplement Interactions

Clinical Trials of Drug or Supplement Interactions

No clinical trials of drug or supplement interactions were identified.

Case Reports of Suspected Drug or Supplement Interactions

No case reports of suspected drug or supplement interactions were identified.

Animal Trials of Drug or Supplement Interactions

No animal trials of drug or supplement interactions were identified.

II. Adverse Events

Case Reports of Adverse Events

No case reports of adverse events were identified.

III. Pharmacology and Pharmacokinetics

Human Pharmacological Studies

No relevant human pharmacological studies were identified.

Animal Pharmacological Studies

No relevant animal pharmacological studies were identified.

In Vitro Pharmacological Studies

No relevant in vitro pharmacological studies were identified.

IV. Pregnancy and Lactation

No information on the safety of lady's mantle during pregnancy or lactation was identified.

V. Toxicity Studies

Genotoxicity

In the Ames test for mutagenicity with *Salmonella typhimurium* strains TA98 and TA100, some mutagenic activity of an ethanol extract of lady's mantle was observed in TA98 but not in TA100. Mutagenic activity in this and other plant extracts tested in the study was consistent with quercetin levels in the plants and the mutagenic activity was attributed to the quercetin (Schimmer et al. 1988). Reviews of quercetin have indicated that although mutagenic effects have been observed in vitro, such effects are not seen in vivo (Harwood et al. 2007).

Literature Cited

Fraisse, D., A. Carnat, A.P. Carnat, and J.L. Lamaison. 1999. Standardization of the aerial parts of *Alchemilla. Ann. Pharm. Fr.* 57(5):401-405.

Harwood, M., B. Danielewska-Nikiel, J.F. Borzelleca, et al. 2007. A critical review of the data related to the safety of quercetin and lack of evidence of *in vivo* toxicity, including lack of genotoxic/carcinogenic properties. *Food Chem. Toxicol.* 45(11):2179-2205.

Schimmer, O., F. Hafele, and A. Kruger. 1988. The mutagenic potencies of plant extracts containing quercetin in *Salmonella typhimurium* TA98 and TA100. *Mutat. Res.* 206(2):201-208.

Wichtl, M. 2004. *Herbal drugs and phytopharmaceuticals: A handbook for practice on a scientific basis.* 3rd ed. Boca Raton, FL: CRC Press.

Aletris farinosa L.

Liliaceae

SCN: aletris
OCN: blazing star; colic root; star grass; true unicorn

Part: rhizome, root

Quick Reference Summary

Safety Class: 1
Interaction Class: A

Contraindications

None known.

Other Precautions

None known.

Drug and Supplement Interactions

None known.

Adverse Events and Side Effects

None known.

Pharmacological Considerations

None known.

Pregnancy and Lactation

One early animal study indicates a uterine depressant effect of aletris and reports that aletris may antagonize the oxytocin-stimulating effects of the pituitary gland (Butler and Costello 1944). There is no evidence to suggest this would result in a clinically relevant interaction with oxytocin administration during labor.

No information on the safety of aletris during lactation was identified. While this review did not identify any concerns for use while nursing, safety has not been conclusively established.

Review Details

I. Drug and Supplement Interactions

Clinical Trials of Drug or Supplement Interactions

No clinical trials of drug or supplement interactions were identified.

Case Reports of Suspected Drug or Supplement Interactions

No case reports of suspected drug or supplement interactions were identified.

Animal Trials of Drug or Supplement Interactions

No animal trials of drug or supplement interactions were identified.

II. Adverse Events

Case Reports of Adverse Events

No case reports of adverse events were identified.

III. Pharmacology and Pharmacokinetics

Human Pharmacological Studies

No relevant human pharmacological studies were identified.

Animal Pharmacological Studies

No relevant animal pharmacological studies were identified.

In Vitro Pharmacological Studies

A hydroalcoholic extract of aletris showed minimal binding to human breast cancer cells rich in estrogen (MCF-7) or progesterone (T47D) receptors (Zava et al. 1998).

IV. Pregnancy and Lactation

A fluid extract of aletris exhibited depressant activity on the uterus of anesthetized cats, and in isolated rat uteri. An antagonistic effect against the stimulating action of the oxytocic principle of the posterior lobe of the pituitary was also observed (Butler and Costello 1944).

No information on the safety of aletris during lactation was identified.

V. Toxicity Studies

No toxicity studies were identified.

Literature Cited

Butler, C.L., and C.H. Costello. 1944. Pharmacological studies. I. *Aletris farinosa*. *J. Am. Pharm. Assoc.* 33:177-183.

Zava, D.T., C.M. Dollbaum, and M. Blen. 1998. Estrogen and progestin bioactivity of foods, herbs, and spices. *Proc. Soc. Exp. Biol. Med.* 217(3):369-378.

Alisma plantago-aquatica L. ssp. *orientale* (Sam.) Sam. — Alismataceae

SCN: Asian water plantain
Syn: *Alisma orientale* (Sam.) Juz.
PN: *ze xie* (rhizome)
OCN: alisma
Part: rhizome

Quick Reference Summary

Safety Class: 1
Interaction Class: A

Contraindications

None known.

Other Precautions

Prolonged use may cause gastrointestinal irritation or disturb electrolyte balance (Bensky et al. 2004).

Drug and Supplement Interactions

None known.

Notice

Diuretic (Bensky et al. 2004; Chen and Chen 2004; Zhu 1998); *see* Appendix 2.

Adverse Events and Side Effects

Long-term use or overdose of Asian water plantain can disturb electrolyte balance (Bensky et al. 2004).

Nausea, vomiting, abdominal pain, diarrhea, and disturbance of liver function have been reported in association with Asian water plantain use (Bensky et al. 2004).

Allergic skin rash after exposure to Asian water plantain has been reported (Bensky et al. 2004).

Pharmacological Considerations

One in vitro study has indicated that Asian water plantain may inhibit the drug-transporting protein P-glycoprotein (Fong et al. 2007).

Pregnancy and Lactation

No information on the safety of Asian water plantain in pregnancy or lactation was identified in the scientific or traditional literature. Although this review did not identify any concerns for use while pregnant or nursing, safety has not been conclusively established.

Review Details

I. Drug and Supplement Interactions

Clinical Trials of Drug or Supplement Interactions

No clinical trials of drug or supplement interactions were identified.

Case Reports of Suspected Drug or Supplement Interactions

No case reports of suspected drug or supplement interactions were identified.

Animal Trials of Drug or Supplement Interactions

No animal trials of drug or supplement interactions were identified.

II. Adverse Events

Case Reports of Adverse Events

Long-term use or overdose of Asian water plantain can disturb the electrolyte balance and may cause hematuria and, in severe cases, acidosis (Bensky et al. 2004).

Nausea, vomiting, abdominal pain, diarrhea, and disturbance of liver function were reported in association with Asian water plantain use (Bensky et al. 2004).

Allergic skin rash after exposure to Asian water plantain was reported (Bensky et al. 2004).

III. Pharmacology and Pharmacokinetics

Human Pharmacological Studies

No relevant human pharmacological studies were identified.

Animal Pharmacological Studies

No relevant animal pharmacological studies were identified.

In Vitro Pharmacological Studies

In multidrug-resistant cancer cells, extracts of Asian water plantain showed a synergistic growth inhibitory effect with cancer drugs that are P-glycoprotein (P-gp) substrates including actinomycin D, puromycin, paclitaxel, vinblastine, and doxorubicin. The authors of the study concluded that Asian water plantain may contain components that are effective inhibitors of P-gp (Fong et al. 2007).

IV. Pregnancy and Lactation

No information on the safety of alisma in pregnancy or lactation was identified.

V. Toxicity Studies

Acute Toxicity

The LD_{50} of an intravenously injected alcohol extract of Asian water plantain in rats was 0.98 g/kg and 1.27 g/kg after intraperitoneal injection (Chen and Chen 2004).

Subchronic Toxicity

No adverse effects on internal organs were observed in rats orally administered an extract of Asian water plantain at doses of 1 or 2 g/kg daily for 3 months (Chen and Chen 2004).

Literature Cited

Bensky, D., S. Clavey, and E. Stöger. 2004. *Chinese herbal medicine: Materia medica*. 3rd ed. Seattle: Eastland Press.

Chen, J.K., and T.T. Chen. 2004. *Chinese medical herbology and pharmacology*. City of Industry, CA: Art of Medicine Press.

Fong, W.F., C. Wang, G.Y. Zhu, et al. 2007. Reversal of multidrug resistance in cancer cells by *Rhizoma Alismatis* extract. *Phytomedicine* 14(2-3):160-165.

Zhu, Y.P. 1998. *Chinese materia medica chemistry, pharmacology and applications*. Boca Raton, FL: CRC Press.

Alkanna tinctoria (L.) Tausch

Boraginaceae

SCN: alkanet

Part: root

Quick Reference Summary

Safety Class: 2a, 2b, 2c
Interaction Class: A

Contraindications

For external use only (Roeder et al. 1984).

Not for use in pregnancy or lactation except under the supervision of a qualified healthcare practitioner (Roeder et al. 1984).

Other Precautions

None known.

Drug and Supplement Interactions

None known.

Notice

Pyrrolizidine alkaloids (Roeder et al. 1984); *see* Appendix 1.

Editors' Notes

The American Herbal Products Association has established a trade requirement (AHPA 2011) that all products with botanical ingredients that contain toxic pyrrolizidine alkaloids, including alkanet, are not offered for sale for internal use and display the following cautionary label: "For external use only. Do not apply to broken or abraded skin. Do not use when nursing."

Alkanet contains unsaturated pyrrolizidine alkaloids, although the concentrations of alkaloids in various parts of the plant are not known (Roeder et al. 1984).

Adverse Events and Side Effects

None known.

Pharmacological Considerations

None known.

Pregnancy and Lactation

No information on the safety of alkanet in pregnancy or lactation was identified in the scientific or traditional literature. Based on the presence of pyrrolizidine alkaloids, alkanet should not be used in pregnancy or lactation.

Review Details

I. Drug and Supplement Interactions

Clinical Trials of Drug or Supplement Interactions

No clinical trials of drug or supplement interactions were identified.

Case Reports of Suspected Drug or Supplement Interactions

No case reports of suspected drug or supplement interactions were identified.

Animal Trials of Drug or Supplement Interactions

No animal trials of drug or supplement interactions were identified.

II. Adverse Events

Case Reports of Adverse Events

No case reports of adverse events were identified.

III. Pharmacology and Pharmacokinetics

Human Pharmacological Studies

No relevant human pharmacological studies were identified.

Animal Pharmacological Studies

No relevant animal pharmacological studies were identified.

In Vitro Pharmacological Studies

No relevant in vitro pharmacological studies were identified.

IV. Pregnancy and Lactation

No information on the safety of alkanet in pregnancy or lactation was identified. Based on the presence of pyrrolizidine alkaloids, alkanet should not be used in pregnancy or lactation.

V. Toxicity Studies

Subchronic Toxicity

No toxic effects were observed in mice fed a diet containing 1% of the compound alkannin daily for 15 weeks. The compound was not detected in abdominal fat or in urine of the treated animals (Majláthová 1971).

Genotoxicity

No mutagenic activity of the compound alkannin was observed in *Salmonella typhimurium* strains TA98 or TA1535 with or without metabolic activation (Papageorgiou 1978).

Literature Cited

AHPA. July 2011. Code of Ethics & Business Conduct. Silver Spring, MD: American Herbal Products Association.

Majláthová, L. 1971. [Feeding trial with alkannin on mice]. *Nahrung* 15(5):505-508.

Papageorgiou, V.P. 1978. Wound healing properties of naphthaquinone pigments from *Alkanna tinctoria*. *Experientia* 34(11):1499.

Roeder, E., H. Wiedenfeld, and R. Schraut. 1984. Pyrrolizidine alkaloids from *Alkanna tinctoria*. *Phytochemistry* 23(9):2125-2126.

Allium sativum L.

Liliaceae

SCN: garlic
AN: *lashuna; rasona*
PN: *da suan* (bulb)
Part: bulb

Quick Reference Summary

Safety Class: 1
Interaction Class: C

Contraindications

None known.

Other Precautions

Use of two or more grams of garlic daily in persons with coagulation disorders should be under the supervision of a qualified healthcare professional (Bordia et al. 1996; Kiesewetter et al. 1993a; Kiesewetter et al. 1993b; Steiner and Li 2001).

Patients undergoing surgery are advised to stop consuming garlic seven days before surgery (Bordia et al. 1996; Kiesewetter et al. 1993a; Kiesewetter et al. 1993b; Steiner and Li 2001).

Persons with acute or chronic stomach inflammation should use caution when taking raw or minimally processed garlic products (De Smet 1992; Felter and Lloyd 1898; Nelson et al. 2006). Persons with acid-reflux should avoid garlic due to the reported relaxing effects on the lower esophageal sphincter of aromatic compounds (Brinker 2001).

External applications of raw garlic may cause skin irritation, contact dermatitis, or chemical burn (Friedman et al. 2006).

Drug and Supplement Interactions

Doses equivalent to two or more grams daily of fresh garlic may reduce plasma levels of saquinavir (Piscitelli et al. 2002).

Also see Pharmacological Considerations for this entry.

Editors' Note

The classifications and concerns for this herb are based on relatively higher doses used for therapeutic purposes in contrast to lower amounts generally used in cooking, and have not been associated with its use as a spice.

Adverse Events and Side Effects

In a systematic review of garlic clinical trials, adverse effects included "smelly" breath and body odor. Other effects, noted as possible but not proven, included flatulence, esophageal and abdominal pain, small intestinal obstruction, dermatitis, rhinitis, asthma, and bleeding (Mulrow et al. 2000). Garlic may cause gastrointestinal disturbance in sensitive individuals (Bradley 1992; De Smet 1992; Felter and Lloyd 1898; Nelson et al. 2006).

Several cases of abnormal bleeding have been reported in association with garlic use (Burnham 1995; Carden et al. 2002; German et al. 1995; Rose et al. 1990).

Allergic reactions to garlic, including anaphylactic reactions, have been reported (Perez-Pimiento et al. 1999). Topical exposure to raw garlic is recognized to cause various types of allergic and non-allergic skin irritation in sensitive individuals (Jappe et al. 1999).

Pharmacological Considerations

Persons taking heparin, clopidogrel, or aspirin, and doses equivalent to two or more grams daily of fresh garlic should be monitored for abnormal bleeding (Bordia et al. 1996; Kiesewetter et al. 1993a; Kiesewetter et al. 1993b; Steiner and Li 2001). Human studies have shown no additive effects of garlic and warfarin on platelet aggregation (Abdul et al. 2008; Macan et al. 2006).

Human studies have shown varying effects of garlic preparations on platelet aggregation, with some studies showing no effects and others showing inhibition of aggregation (Beckert et al. 2007; Bordia et al. 1996; Kiesewetter et al. 1993a; Kiesewetter et al. 1993b; Scharbert et al. 2007; Steiner and Li 2001).

In human studies, garlic oil has been shown to inhibit the drug metabolizing isoenzyme CYP2E1 (Gurley et al. 2002; Gurley et al. 2005).

Pregnancy and Lactation

No maternal or fetal adverse events were observed in one human study of garlic use during the third trimester of pregnancy (Ziaei et al. 2001).

Ingestion of garlic by nursing mothers has been shown to change the odor of human breast milk (Mennella and Beauchamp 1991), in some cases leading to increased nursing

(Mennella and Beauchamp 1991; Mennella and Beauchamp 1993). As with some commonly consumed vegetables, such as broccoli and cabbage, garlic consumption by nursing mothers has been associated with colic in some breastfeeding infants (Lust et al. 1996).

One text on Ayurvedic medicine contraindicates garlic in pregnancy (Chadha 1988); however, the long history of use as a food has established that consumption of garlic in reasonable quantities is generally safe, though caution may be indicated in children.

Review Details

I. Drug and Supplement Interactions

Clinical Trials of Drug or Supplement Interactions

Coadministration of 3.6 g saquinavir and 2.4 g garlic daily for 20 days resulted in a decrease in plasma levels of saquinavir (Piscitelli et al. 2002).

No changes in INR (a standardized scale used to report the results of blood coagulation tests), platelet aggregation, clotting factor activity, or plasma levels of warfarin, were observed in healthy volunteers orally administered two garlic tablets (each the equivalent of 2 g fresh garlic) daily for two weeks (Abdul et al. 2008).

In patients on warfarin therapy, supplementation with 10 ml daily of aged garlic extract did not require adjustments in warfarin dose. No significant changes in hematological parameters, including INR, platelet count, and incidences of hemorrhages, were observed (Macan et al. 2006).

Single doses of acetaminophen (1 g) administered during three months supplementation of aged garlic extract (equivalent to 6 cloves/day) showed that peak plasma concentrations of acetaminophen and acetaminophen metabolites were increased and renal clearance was decreased by garlic coadministration (Gwilt et al. 1994).

No effects on ritonavir (400 mg, single dose) were observed after ingestion of 10 mg garlic extract (equivalent to 1 g of fresh garlic) daily for four days (Gallicano et al. 2003).

In renal transplant patients on cyclosporine therapy, ingestion of one clove of garlic daily did not affect cyclosporine levels (Jabbari et al. 2005).

In breast cancer patients undergoing treatment with docetaxel (30 mg/m^2/week), coadministration with 1200 mg garlic daily did not significantly affect docetaxel pharmacokinetics (Cox et al. 2006).

Case Reports of Suspected Drug or Supplement Interactions

Two cases of increased international normalized ratio (INR) (a standardized scale used to report the results of blood coagulation tests) and clotting time have been reported in persons taking garlic and the anticoagulant, warfarin (doses unspecified) (Sunter 1991).

A case of decreased INR was reported in an 82-year-old man taking garlic (600 mg daily) and the anticoagulant, fluindione (Pathak et al. 2003).

A case of platelet dysfunction in an 87-year-old patient was reported after chronic consumption of approximately 2 g daily of garlic cloves (De Smet 1993).

Two cases of gastrointestinal toxicity were reported in patients taking garlic and the protease inhibitor, ritonavir (Laroche et al. 1999).

Animal Trials of Drug or Supplement Interactions

In rabbits, single doses of the antibiotic isoniazid (30 mg/kg) before and after 14 days supplementation with an aqueous garlic extract significantly decreased the bioavailability of isoniazid with no change in rate of elimination (Dhamija et al. 2006). An identical study completed with the antibiotic rifampicin (24 mg/kg) indicated no change in rifampicin bioavailability (Dhamija et al. 2006).

II. Adverse Events

Adverse Events Reported in Clinical Trials

In a systematic review of garlic clinical trials, adverse effects included "smelly" breath and body odor. Other effects, noted as possible but not proven, included flatulence, esophageal and abdominal pain, small intestinal obstruction, dermatitis, rhinitis, asthma, and bleeding (Mulrow et al. 2000).

Case Reports of Adverse Events

Several cases of abnormal bleeding have been reported in association with garlic ingestion. These cases include spinal epidural hematoma with platelet dysfunction (Rose et al. 1990), post-operative bleeding (Burnham 1995; German et al. 1995) and a periorbital hematoma (Carden et al. 2002).

Allergic reactions to garlic, including anaphylactic reactions, have been reported (Perez-Pimiento et al. 1999). Topical exposure to raw garlic has caused irritant, allergic and protein contact dermatitis, contact urticaria, induction of pemphigus (blisters) and combinations thereof (Jappe et al. 1999).

III. Pharmacology and Pharmacokinetics

Human Pharmacological Studies

In patients with stage II peripheral arterial occlusive disease, 800 mg garlic daily for 12 weeks was found to reduce spontaneous thrombocyte aggregation and plasma viscosity (Kiesewetter et al. 1993b).

In patients with cerebrovascular risk factors and constantly increased platelet aggregation, administration of 800 mg garlic daily for four weeks resulted in inhibition of circulating platelet aggregates and of spontaneous platelet aggregation. No significant changes were observed in the placebo group (Kiesewetter et al. 1993a).

In blood samples from patients with coronary artery disease who were administered garlic oil, inhibition of chemically induced platelet aggregation was observed. Administration of different single doses indicated a lack of dose-dependent effect on platelet aggregation after acute administration, while lower doses administered long-term had anti-platelet activity (Bordia et al. 1996).

In blood samples taken from healthy volunteers orally administered 2.4, 4.8, or 7.2 g aged garlic extract daily for six weeks, platelet aggregation induced by ADP was slowed only at the highest dose level. When aggregation was induced by collagen, the middle and high doses had a slower rate than the low dose, while for epinepherine-induced aggregation, the low and middle doses were more effective in reducing aggregation than the high dose (Steiner and Li 2001).

In healthy volunteers orally administered garlic capsules at the manufacturer's recommended dose (amount not specified) daily for two weeks, no effects on platelet function or other hematological parameters were observed, including prothrombin time, partial thromboplastin time, thrombin time, bleeding time, the collagen/epinephrine assay, or the collagen/adenosine diphosphate assay. Aspirin was used as a positive control and markedly inhibited platelet function (Beckert et al. 2007).

In healthy volunteers fed 4.2 g raw garlic in food as either a single dose or a daily dose for one week, platelet function was not impaired by the single or repeated doses (Scharbert et al. 2007).

In blood samples taken from healthy volunteers orally administered 5 ml of aged garlic extract daily for 13 weeks, inhibition of the rate of ADP-induced aggregation was observed. Plasma levels of thromboxane B_2 and 6-ketoprostaglandin $F_{1\alpha}$ were unchanged (Rahman 2007).

Garlic oil (1500 mg/day for 28 days) has been shown to cause a significant inhibition of CYP2E1 activity in humans (Gurley et al. 2002; Gurley et al. 2005). No effects of garlic on CYP3A4, CYP1A2, or CYP2D6 have been observed (Gurley et al. 2002; Gurley et al. 2005; Markowitz et al. 2003).

Animal Pharmacological Studies

In rats orally administered 5 or 50 mg/kg diallyl trisulfide-rich garlic oil, prolongation of bleeding time and thrombin time and enhanced anticoagulation factor activity were observed. At 50 mg/kg, garlic oil increased plasma fibrinogen concentration and affected the levels of several hematological parameters such as erythrocyte count, hemoglobin, and platelets (Chan et al. 2007).

In rats administered 1 g/kg odorless garlic extract daily for two weeks, activation of the coagulation system in animals with induced thrombosis was suppressed in the garlic group, as evaluated by prothrombin time and activated partial thromboplastin time (Fukao et al. 2007).

In Vitro Pharmacological Studies

Studies available but omitted due to availability of human data.

IV. Pregnancy and Lactation

In a study of garlic in pregnant women, garlic intake of 800 mg daily for 8 weeks in the third trimester of pregnancy produced no adverse effects on infants or mothers, aside from "odor due to garlic" reported in 34% of mothers taking garlic (Ziaei et al. 2001).

Ingestion of garlic by nursing mothers has been shown to change the odor of human breast milk (Mennella and Beauchamp 1991). In studies of infant breastfeeding from nursing mothers who had taken a single dose of 1.5 grams of garlic, sulfur-containing volatile compounds were transmitted to breast milk. Infant nursing behavior on this milk ranged from unchanged to a significantly greater amount of time attached to the nipple (Mennella and Beauchamp 1991; Mennella and Beauchamp 1993).

As with some commonly consumed vegetables, such as broccoli and cabbage (Lust et al. 1996), garlic consumption by nursing mothers has been associated with colic in some breastfeeding infants, although no colic was reported in the studies detailed above (Mennella and Beauchamp 1991; Mennella and Beauchamp 1993).

V. Toxicity Studies

Acute Toxicity

The LD_{50} of intraperitoneally administered garlic extract in male rats was 30 ml/kg; female rats survived this dose of garlic (Nakagawa et al. 1984). Male and female rats survived doses of 30 ml/kg garlic administered orally or subcutaneously (Nakagawa et al. 1984).

Short-Term Toxicity

In rats intraperitoneally administered 500 mg/kg garlic extract for four weeks, changes in lung and liver tissue were observed. Side effects of garlic treatment included soft feces, dehydration and increase of white and red blood cells (Alnaqeeb et al. 1996).

In rats orally administered 5 ml/kg of fresh garlic juice daily for 21 days, serious stomach injury occurred along with swelling of the liver, hypertrophy of the adrenal glands and spleen, and changes in red blood cell count (Nakagawa et al. 1980).

Raw garlic powder administered by gavage (direct feeding tube) into dog stomachs caused severe damage of gastrointestinal mucosa. Boiled garlic powder administered in the same manner caused reddening of the stomach mucosa, while aged garlic extract did not cause any adverse effects (Hoshino et al. 2001).

Rats administered 1, 2.5 or 5 g/kg of fresh garlic showed a significant deterioration in liver function after one week of garlic ingestion. Lower doses of 0.1 and 0.25 g/kg did not produce any adverse effects (Rana et al. 2006).

Subchronic Toxicity

In rats administered 2000 mg/kg daily of garlic extract five times weekly for 6 months, no adverse effects were observed (Sumiyoshi et al. 1984). In mice administered 100 mg/kg daily of garlic extract for three months, heart, liver, and spleen weights were slightly reduced as compared to control (al-Bekairi et al. 1990).

Genotoxicity

In the Ames test for mutagenicity, no mutagenic effects were shown for garlic tincture. Garlic juice, however, in the absence of metabolic activation, induced significant levels of chromosomal damage (Schimmer et al. 1994).

Literature Cited

Abdul, M.I.M., X. Jiang, K.M. Williams, et al. 2008. Pharmacodynamic interaction of warfarin with cranberry but not with garlic in healthy subjects. *Br. J. Pharmacol.* 154(8):1691-1700.

al-Bekairi, A., A. Shah, and S. Qureshi. 1990. Effect of *Allium sativum* on epididymal spermatozoa, estradiol-treated mice and general toxicity. *J. Ethnopharmacol.* 29(2):117-125.

Alnaqeeb, M.A., M. Thomson, T. Bordia, and M. Ali. 1996. Histopathological effects of garlic on liver and lung of rats. *Toxicol. Lett.* 85(3):157-164.

Beckert, B.W., M.J. Concannon, S.L. Henry, D.S. Smith, and C.L. Puckett. 2007. The effect of herbal medicines on platelet function: An in vivo experiment and review of the literature. *Plast Reconstr. Surg.* 120(7):2044-2050.

Bordia, A., S.K. Verma, and K.C. Srivastava. 1996. Effect of garlic on platelet aggregation in humans: A study in healthy subjects and patients with coronary artery disease. *Prostaglandins Leukot. Essent. Fatty Acids* 55(3):201-205.

Bradley, P.R. 1992. *British herbal compendium: A handbook of scientific information on widely used plant drugs.* Bournemouth, Dorset: British Herbal Medicine Association.

Brinker, F. 2001. *Herb contraindications and drug interactions.* 3rd ed. Sandy, OR: Eclectic Medical Publications.

Burnham, B. 1995. Garlic as a possible risk for postoperative bleeding. *Plast. Reconstr. Surg.* 95(1):213.

Carden, S., W. Good, P. Carden, and R.M. Good 2002. Garlic and the strabismus surgeon. *Clin. Experiment. Ophthalmol.* 30(4):303-304.

Chadha, Y. 1988. *The wealth of India: A dictionary of Indian raw materials and industrial products.* Delhi: Council of Scientific and Industrial Research.

Chan, K.C., M.C. Yin, and W.J. Chao. 2007. Effect of diallyl trisulfide-rich garlic oil on blood coagulation and plasma activity of anticoagulation factors in rats. *Food Chem. Toxicol.* 45(3):502-507.

Cox, M.C., J. Low, J. Lee, et al. 2006. Influence of garlic (*Allium sativum*) on the pharmacokinetics of docetaxel. *Clin. Cancer Res.* 12(15):4636-4640.

De Smet, P.A.G.M. 1992. *Adverse effects of herbal drugs Volume 1.* Berlin; New York: Springer.

De Smet, P.A.G.M. 1993. *Adverse effects of herbal drugs Volume 2.* Berlin; Heidelberg; New York: Springer.

Dhamija, P., S. Malhotra, and P. Pandhi. 2006. Effect of oral administration of crude aqueous extract of garlic on pharmacokinetic parameters of isoniazid and rifampicin in rabbits. *Pharmacology* 77(2):100-104.

Felter, H.W., and J.U. Lloyd. 1898. *King's American dispensatory.* 18th ed., 3rd rev. 2 vols. Cincinnati: Ohio Valley Co.

Friedman, T., A. Shalom, and M. Westreich. 2006. Self-inflicted garlic burns: Our experience and literature review. *Int. J. Dermatol.* 45(10):1161-1163.

Fukao, H., H. Yoshida, Y. Tazawa, and T. Hada. 2007. Antithrombotic effects of odorless garlic powder both in vitro and in vivo. *Biosci. Biotechnol. Biochem.* 71(1):84-90.

Gallicano, K., B. Foster, and S. Choudhri. 2003. Effect of short-term administration of garlic supplements on single-dose ritonavir pharmacokinetics in healthy volunteers. *Br. J. Clin. Pharmacol.* 55(2):199-202.

German, K., U. Kumar, and H. Blackford. 1995. Garlic and the risk of TURP bleeding. *Br. J. Urol.* 76(4):518.

Gurley, B., S. Gardner, M. Hubbard, et al. 2002. Cytochrome P450 phenotypic ratios for predicting herb-drug interactions in humans. *Clin. Pharmacol. Ther.* 72(3):276-287.

Gurley, B.J., S.F. Gardner, M.A. Hubbard, et al. 2005. Clinical assessment of effects of botanical supplementation on cytochrome P450 phenotypes in the elderly: St John's wort, garlic oil, *Panax ginseng* and *Ginkgo biloba*. *Drug. Aging* 22(6):525-539.

Gwilt, P., C. Lear, M. Tempero, et al. 1994. The effect of garlic extract on human metabolism of acetaminophen. *Cancer Epidemiol. Biomarkers. Prev.* 3:155-160.

Hoshino, T., N. Kashimoto, and S. Kasuga. 2001. Effects of garlic preparations on the gastrointestinal mucosa. *J. Nutr.* 131(3s):1109s-1113s.

Jabbari, A., H. Argani, A. Ghorbanihaghjo, and R. Mahdavi. 2005. Comparison between swallowing and chewing of garlic on levels of serum lipids, cyclosporine, creatinine and lipid peroxidation in renal transplant recipients. *Lipids Health Dis.* 4(11)1-4.

Jappe, U., B. Bonnekoh, B. Hausen, and H. Gollnick. 1999. Garlic-related dermatoses: Case report and review of the literature. *Am. J. Contact Derm.* 10(1):37-39.

Kiesewetter, H., F. Jung, E. Jung, et al. 1993a. Effect of garlic on platelet aggregation in patients with increased risk of juvenile ischaemic attack. *Eur. J. Clin. Pharmacol.* 45(4):333-336.

Kiesewetter, H., F. Jung, E.M. Jung, et al. 1993b. Effects of garlic coated tablets in peripheral arterial occlusive disease. *Clin. Investig.* 71(5):383-386.

Laroche, M., S. Choudhri, K. Gallicano, and B. Foster. 1999. Severe gastrointestinal toxicity with concomitant ingestion of ritonavir and garlic [Abstract only]. *Can. J. Infect. Dis.* 9(Suppl A):471P.

Lust, K.D., J. Brown, and W. Thomas. 1996. Maternal intake of cruciferous vegetables and other foods and colic symptoms in exclusively breast-fed infants. *J. Am. Diet. Assoc.* 96(1):46-48.

Macan, H., R. Uykimpang, M. Alconcel, et al. 2006. Aged garlic extract may be safe for patients on warfarin therapy. *J. Nutr.* 136(3 Suppl):793S-795S.

Markowitz, J., C. DeVane, K. Chavin, et al. 2003. Effects of garlic (*Allium sativum* L.) supplementation on cytochrome P450 2D6 and 3A4 activity in healthy volunteers. *Clin. Pharmacol. Ther.* 74(2):170-177.

Mennella, J., and G. Beauchamp. 1991. Maternal diet alters the sensory qualities of human milk and the nursling's behavior. *Pediatr. Dermatol.* (88):737-744.

Mennella, J.A., and G.K. Beauchamp. 1993. The effects of repeated exposure to garlic-flavored milk on the nursling's behavior. *Pediatr. Res.* 34(6):805-808.

Mulrow, C., V. Lawrence, R. Ackermann, et al. 2000. Garlic: Effects on cardiovascular risks and disease, protective effects against cancer, and clinical adverse effects. *Evid. Rep. Technol. Assess.* (20):1-4.

Nakagawa, S., K. Masamoto, H. Sumiyoshi, and H. Harada. 1984. Acute toxicity test of garlic extract. *J. Toxicol. Sci.* 9(1):57-60.

Nakagawa, S., K. Masamoto, H. Sumiyoshi, K. Kunihiro, and T. Fuwa. 1980. Effect of raw and extracted-aged garlic juice on growth of young rats and their organs after peroral administration. *J. Toxicol. Sci.* 5(1):91-112.

Nelson, L., R.D. Shih, M.J. Balick, and K.F. Lampe. 2006. *Handbook of poisonous and injurious plants.* New York: Springer.

Pathak, A., P. Leger, H. Bagheri, et al. 2003. Garlic interaction with fluindione: A case report. *Therapie* 58(4):380-381.

Perez-Pimiento, A., I. Moneo, M. Santaolalla, et al. 1999. Anaphylactic reaction to young garlic. *Allergy* 54(6):626-629.

Piscitelli, S.C., A.H. Burstein, N. Welden, K.D. Gallicano, and J. Falloon. 2002. The effect of garlic supplements on the pharmacokinetics of saquinavir. *Clin. Infect. Dis.* 34(2):234-238.

Rahman, K. 2007. Effects of garlic on platelet biochemistry and physiology. *Molec. Nutr. Food. Res.* 51(11):1335-1344.

Rana, S.V., R. Pal, K. Vaiphei, and K. Singh. 2006. Garlic hepatotoxicity: Safe dose of garlic. *Trop. Gastroenterol.* 27(1):26-30.

Rose, K., P. Croissant, C. Parliament, and M. Levin. 1990. Spontaneous spinal epidural hematoma with associated platelet dysfunction from excessive garlic ingestion: A case report. *Neurosurg.* 26(5):880-882.

Scharbert, G., M.L. Kalb, M. Duris, C. Marschalek, and S.A. Kozek-Langenecker. 2007. Garlic at dietary doses does not impair platelet function. *Anesthes. Analges.* 105(5):1214-1218.

Schimmer, O., A. Kruger, H. Paulini, and F. Haefele. 1994. An evaluation of 55 commercial plant extracts in the Ames mutagenicity test. *Pharmazie* 49:448-451.

Steiner, M., and W. Li. 2001. Aged garlic extract, a modulator of cardiovascular risk factors: A dose-finding study on the effects of AGE on platelet functions. *J. Nutr.* 131(3s):980s-984s.

Sumiyoshi, H., A. Kanezawa, K. Masamoto, et al. 1984. Chronic toxicity test of garlic extract in rats. *J. Toxicol. Sci.* 9(1):61-75.

Sunter, W. 1991. Warfarin and garlic. *Pharm. J.* (246):722.

Ziaei, S., S. Hantoshzadeh, P. Rezasoltani, and M. Lamyian. 2001. The effect of garlic tablet on plasma lipids and platelet aggregation in nulliparous pregnants at high risk of preeclampsia. *Eur. J. Obstet. Gynecol. Reprod. Biol.* 99(2):201-206.

Aloe spp. — Liliaceae

Aloe ferox Mill.
SCN: cape aloe

Aloe perryi Baker
SCN: Perry's aloe
OCN: Socotrine aloe; Zanzibar aloe

Aloe vera (L.) Burm. f.
SCN: aloe vera
Syn: *Aloe barbadensis* Mill.
AN: *ghrita kumari*; *kanyasara* (dried juice of leaf)
PN: *lu hui* (dried concentrated juice of leaf)
OCN: aloe; Barbados aloe; Curaçao aloe
Part: dried juice from the pericyclic region of the leaf (commonly called aloe latex)

Quick Reference Summary

Safety Class: 2b, 2c, 2d
Interaction Class: A

Contraindications

Aloe latex should not be used in pregnancy or lactation except under the supervision of a qualified healthcare practitioner (Bensky et al. 2004; Chen and Chen 2004; Mills and Bone 2005).

Internal use of aloe latex is contraindicated in persons with the following conditions: intestinal obstruction, abdominal pain of unknown origin, or any inflammatory condition of the intestines (i.e., appendicitis, colitis, Crohn's disease, irritable bowel syndrome) (Bensky et al. 2004; Bradley 1992; Chadha 1988; Roth et al. 1984; Weiss and Meuss 2001; Wichtl 2004); melanosis coli; hemorrhoids (Bradley 1992; Felter and Lloyd 1898; List and Hörhammer 1973; Roth et al. 1984); liver disease; kidney dysfunction (Wichtl 2004); and severe dehydration (with water and electrolyte depletion).

Aloe latex should not be used during menstruation (Wichtl 2004) or in children less than 12 years of age (Bradley 1992; De Smet 1993).

Not for use in excess of 8 days (Bensky et al. 2004; Bradley 1992; Chadha 1988; De Smet 1993; Felter and Lloyd 1898; Leung and Foster 1996; Martindale and Reynolds 1996; Roth et al. 1984; Weiss and Meuss 2001; Wichtl 2004).

Other Precautions

None known.

Drug and Supplement Interactions

See Pharmacological Considerations below.

Notice

Stimulant laxative (Bensky et al. 2004; Bradley 1992; Chadha 1988; De Smet 1993; Felter and Lloyd 1898; Leung and Foster 1996; Martindale and Reynolds 1996; Weiss and Meuss 2001; Wichtl 2004; Williamson 2003); *see* Appendix 2.

Standard Dose

50–300 mg in a single dose at bedtime (Bradley 1992; Williamson 2003).

Editors' Notes

Aloe latex is prepared by drying the bitter yellow juice from the pericyclic region of the leaf and is a distinct product from aloe vera gel. The latex contains the laxative compound aloin, whereas the gel (*see* next entry) does not (Leung and Foster 1996).

The American Herbal Products Association has established a trade requirement (AHPA 2011) that products containing this herb in sufficient quantity to warrant such labeling bear the following label statement:

> NOTICE: Do not use this product if you have abdominal pain or diarrhea. Consult a health care provider prior to use if you are pregnant or nursing. Discontinue use in the event of diarrhea or watery stools. Do not exceed recommended dose. Not for long-term use.

Aloe latex is generally considered to be obsolete as a laxative, as it is a less desirable agent for laxation than other available substances (Hoover 1970; Martindale and Reynolds 1996).

Adverse Events and Side Effects

Adverse effects due to long-term use include electrolyte depletion, potassium depletion, damage to the colon, kidney malfunction, and heart palpitations (Wichtl 2004). In some epidemiological studies, long-term use of stimulant laxatives has been associated with colorectal cancer (Siegers 1992; Siegers et al. 1993), while other epidemiological studies have shown no correlation with cancer (Wald 2003).

Two cases of acute hepatitis have been reported in persons taking 500 mg of an extract of *Aloe vera* in tablet or capsule form for three to four weeks. No additional information on the preparation of the aloe ingredient was identified in either case (Kanat et al. 2006; Rabe et al. 2005).

Overdose (stated at as little as 1.0 g per day for several days) can cause colonic perforation and bleeding diarrhea, as well as kidney damage, and, according to one reference, death (Bradley 1992; Leung and Foster 1996; List and Hörhammer 1973; Martindale and Reynolds 1996; Wichtl 2004; Williamson 2003).

Pharmacological Considerations

Concomitant use of aloe latex is cautioned with antiarrhythmic drugs and botanicals containing cardiac glycosides, as long-term use of aloe latex as a laxative can cause potassium loss, leading to increased toxicity of these drugs and botanicals (Brinker 2001; De Smet 1993).

Concomitant internal use of aloe latex is cautioned with thiazide diuretics, corticosteroids, and licorice, and long-term use of aloe latex as a laxative may increase the potassium loss induced by these drugs and botanicals (Brinker 2001).

Use of stimulant laxatives, such as aloe latex, may reduce the gastrointestinal transit time and thus reduce the absorption of orally administered drugs (Brinker 2001; De Smet 1993).

Pregnancy and Lactation

Oral use of aloe latex is contraindicated during pregnancy in traditional Chinese medicine (Bensky et al. 2004; Chen and Chen 2004). While most stimulant laxatives have traditionally been contraindicated in pregnancy due to concerns regarding stimulation of the uterus, a number of stimulant laxatives, including aloe latex, have shown a lack of adverse effects on pregnancy or the fetus when used according to the recommended dosage schedule (De Smet 1993; ESCOP 2003). Thus, these laxatives are now considered appropriate for use during pregnancy (De Smet 1993; ESCOP 2003; Prather 2004). Due to the potential genotoxicity of certain anthraquinones, however, it is recommended that use of certain anthranoid laxatives, including aloe latex, be avoided in the first trimester of pregnancy or used under professional supervision (ESCOP 2003).

Limited animal studies have provided mixed results on use of aloe latex in pregnancy, with one study showing no adverse effects and another showing some fetal abnormalities (ESCOP 2003; Nath et al. 1992).

Use of aloe latex while trying to become pregnant is best avoided (Mills and Bone 2005).

Use during lactation is cautioned (Bensky et al. 2004). Eclectic physicians traditionally administered aloe latex to breast-feeding mothers to produce a laxative effect in the nursing child (Felter and Lloyd 1898), although more recent publications indicate that laxative effects have not been observed in breast-feeding infants after maternal ingestion of other anthraquinone laxatives (Committee on Drugs 2001; Faber and Strenge-Hesse 1988; Tyson et al. 1937).

Review Details

I. Drug and Supplement Interactions

Clinical Trials of Drug or Supplement Interactions

No clinical trials of drug or supplement interactions were identified.

Case Reports of Suspected Drug or Supplement Interactions

Use of stimulant laxatives may reduce the gastrointestinal transit time and thus absorption of orally administered drugs (Brinker 2001; Mills and Bone 2005).

Animal Trials of Drug or Supplement Interactions

No animal trials of drug or supplement interactions were identified.

II. Adverse Events

Case Reports of Adverse Events

Adverse events reported in association with aloe latex ingestion include nausea, vomiting, nose bleeds, abdominal pain, diarrhea, bloody stool, hematuria, albuminuria, and, after long-term administration, colitis (Bensky et al. 2004).

Two cases of acute hepatitis have been reported in persons taking *Aloe vera* internally (500 mg daily) for 3 to 4 weeks. No additional information on the preparation of the aloe ingredient was identified in either case (Kanat et al. 2006; Rabe et al. 2005).

Overdose of products containing aloe latex (stated at as little as 1.0 g per day for several days) can cause colonic perforation and bleeding diarrhea, as well as kidney damage, and, according to one reference, death (Bradley 1992; Leung and Foster 1996; List and Hörhammer 1973; Martindale and Reynolds 1996; Wichtl 2004; Williamson 2003).

One case of kidney failure was reported in a 47-year-old man with a history of hypertension and abdominal pain who had taken three doses of a traditional remedy believed to be *Aloe ferox* for constipation (Luyckx et al. 2002).

III. Pharmacology and Pharmacokinetics

Human Pharmacological Studies

Retrospective and prospective epidemiological studies of anthranoid-containing laxatives have correlated long-term use of these laxatives with colorectal cancer (Siegers 1992; Siegers et al. 1993). A review of chronic use of stimulant laxatives concluded that while stimulant laxatives can cause structural damage to surface cells in the colon, no causal relationship between laxative use and colorectal cancer could be established (Wald 2003).

Animal Pharmacological Studies

Animal pharmacological studies were identified but omitted due to availability of human data.

In Vitro Pharmacological Studies

In vitro pharmacological studies were identified but omitted due to availability of human data.

IV. Pregnancy and Lactation

Aloe latex extract (up to 1000 mg/kg) administered to pregnant rats did not demonstrate any embryotoxic, fetotoxic, or teratogenic activity (ESCOP 2003). Some fetal abnormalities were noted in fetuses of pregnant albino rats fed an extract of ground aloe vera leaf on days 0 through 9 of gestation (Nath et al. 1992).

Oral use of aloe latex is traditionally contraindicated during pregnancy (Bensky et al. 2004; Chen and Chen 2004), as side effects of stimulant laxatives, such as aloe latex, are reported to include induction of uterine contractions (De Smet 1993). Use during lactation is cautioned (Bensky et al. 2004; Brinker 2001).

Eclectic physicians traditionally administered aloe latex to breast-feeding mothers to produce a laxative effect in the nursing child (Felter and Lloyd 1898), although more recent publications indicate that laxative effects have not been observed in breast-feeding infants after maternal ingestion of other anthraquinone laxatives such as senna and cascara sagrada (Committee on Drugs 2001; Faber and Strenge-Hesse 1988; Tyson et al. 1937).

V. Toxicity Studies

Acute Toxicity

The LD_{50} of orally administered aloe vera extract (0.50% total solids; 43.70% alcohol) in mice has been reported at 120 mg/kg (Lagarto Parra et al. 2001). This LD_{50} value is questionable, as oral administration of single doses of 0.5, 1, or 3 g/kg of aloe vera showed no signs of toxicity at either the 0.5 or 1 g/kg dose. At the 3 g/kg dose, a decrease in nervous system activity was noted (Shah et al. 1989).

Subchronic Toxicity

In rats fed whole leaf powder of aloe (species unspecified) at doses of 2, 4, or 8 g/kg daily for 90 days, some incidences of pigmentation of various organs was observed (Zhou et al. 2003).

Genotoxicity

Aqueous extracts of cape aloe (5 to 100 mg/ml) gave mixed results for mutagenicity, with mutagenicity reported for tests in the *Bacillus subtilis* rec-assay but no mutagenicity shown in *Salmonella* strains TA98 or TA100 (Morimoto et al. 1982).

Some genotoxicity but no cytotoxicity was noted for aloe vera pulp in a bacterial transformation assay with the pulp added to *E. coli* and exposed to UV light (Paes-Leme et al. 2005).

Literature Cited

AHPA. July 2011. Code of Ethics & Business Conduct. Silver Spring, MD: American Herbal Products Association.

Bensky, D., S. Clavey, and E. Stöger. 2004. *Chinese herbal medicine: Materia medica*. 3rd ed. Seattle: Eastland Press.

Bradley, P.R. 1992. *British herbal compendium: A handbook of scientific information on widely used plant drugs*. Bournemouth, Dorset: British Herbal Medicine Association.

Brinker, F. 2001. *Herb contraindications and drug interactions*. 3rd ed. Sandy, OR: Eclectic Medical Publications.

Chadha, Y. 1988. *The wealth of India: A dictionary of Indian raw materials and industrial products*. Delhi: Council of Scientific and Industrial Research.

Chen, J.K., and T.T. Chen. 2004. *Chinese medical herbology and pharmacology*. City of Industry, CA: Art of Medicine Press.

Committee on Drugs. 2001. American Academy of Pediatrics. The transfer of drugs and other chemicals into human milk. *Pediatrics* 108(3):776-789.

De Smet, P.A.G.M. 1993. *Adverse effects of herbal drugs, Volume 2*. Berlin, Heidelberg, New York: Springer.

ESCOP. 2003. *ESCOP monographs: The scientific foundation for herbal medicinal products*. 2nd ed. Exeter, U.K.: European Scientific Cooperative on Phytotherapy.

Faber, P., and A. Strenge-Hesse. 1988. Relevance of rhein excretion into breast milk. *Pharmacology* 36(1):212-220.

Felter, H.W., and J.U. Lloyd. 1898. *King's American dispensatory*. 18th ed., 3rd rev. 2 vols. Cincinnati: Ohio Valley Co.

Hoover, J.M. 1970. *Remington's pharmaceutical sciences*. 14th ed. Easton, PA: Mack Publishing Company.

Kanat, O., A. Ozet, and S. Ataergin. 2006. *Aloe vera*-induced acute toxic hepatitis in a healthy young man. *Eur. J. Intern. Med.* 17(8):589.

Lagarto Parra, A., R. Silva Yhebra, I. Guerra Sardinas, and L. Iglesias Buela. 2001. Comparative study of the assay of *Artemia salina* L. and the estimate of the medium lethal dose (LD_{50} value) in mice, to determine oral acute toxicity of plant extracts. *Phytomedicine* 8(5):395-400.

Leung, A.Y., and S. Foster. 1996. *Encyclopedia of common natural ingredients used in food, drugs, and cosmetics*. 2nd ed. New York: Wiley.

List, P.H., and H. Hörhammer. 1973. *Hagers handbuch der pharmazeutischen praxis*. Vollst. 4. Neuausg. ed. Berlin, Heidelberg, New York: Springer.

Luyckx, V.A., R. Ballantine, M. Claeys, et al. 2002. Herbal remedy-associated acute renal failure secondary to Cape aloes. *Am. J. Kidney Dis.* 39(3):E13.

Martindale, W., and J.E.F. Reynolds. 1996. *The extra pharmacopoeia*. 31st ed. edited by James E.F. Reynolds; deputy editor, Kathleen Parfitt; assistant editors, Anne V. Parsons, Sean C. Sweetman. London: Pharmaceutical Press.

Mills, S., and K. Bone. 2005. *The essential guide to herbal safety*. St. Louis: Elsevier.

Morimoto, I., F. Watanabe, T. Osawa, and T. Okitsu. 1982. Mutagenicity screening of crude drugs with *Bacillus subtilis* rec-assay and *Salmonella* microsome reversion assay. *Mutat. Res.* 97:81-102.

Nath, D., N. Sethi, R. Singh, and A. Jain. 1992. Commonly used Indian abortifacient plants with special reference to their teratologic effects in rats. *J. Ethnopharmacol.* 36(2):147-154.

Paes-Leme, A.A., E.S. Motta, J.C. De Mattos, et al. 2005. Assessment of *Aloe vera* (L.) genotoxic potential on *Escherichia coli* and plasmid DNA. *J. Ethnopharmacol.* 102(2):197-201.

Rabe, C., A. Musch, P. Schirmacher, W. Kruis, and R. Hoffman. 2005. Acute hepatitis induced by an *Aloe vera* preparation: A case report. *World J. Gastroenterol.* 11(2):303-304.

Roth, L., M. Daunderer, and K. Kormann. 1984. *Giftpflanzen-pflanzengifte: Vorkommen, wirkung, therapie*. 2. Aufl. ed. Landsberg, Germany: Ecomed.

Shah, A.H., S. Qureshi, M. Tariq, and A.M. Ageel. 1989. Toxicity studies on six plants used in the traditional Arab system of medicine. *Phytother. Res.* 3:25-29.

Siegers, C.P. 1992. Anthranoid laxatives and colorectal cancer. *Trends Pharmacol. Sci.* 13(6):229-231.

Siegers, C.P., E. von Hertzberg-Lottin, M. Otte, and B. Schneider. 1993. Anthranoid laxative abuse—a risk for colorectal cancer? *Gut* 34(8):1099-1101.

Tyson, R.M., E.A. Shrader, and H.H. Perlman. 1937. Drugs transmitted through breast milk: Part I: Laxatives. *J. Pediatr.* 11(6):824-832.

Wald, A. 2003. Is chronic use of stimulant laxatives harmful to the colon? *J. Clin. Gastroenterol.* 36(5):386-389.

Weiss, R.F., and A.R. Meuss. 2001. *Weiss's herbal medicine*. Classic ed. New York: Thieme.

Wichtl, M. 2004. *Herbal drugs and phytopharmaceuticals: A handbook for practice on a scientific basis*. 3rd ed. Boca Raton, FL: CRC Press.

Williamson, E.M. 2003. *Potter's herbal cyclopedia*. Saffron Walden, Essex: C.W. Daniel Co.

Zhou, Y., Y. Feng, H. Wang, and H. Yang. 2003. 90-day subchronic toxicity study of aloe whole-leaf powder. *Wei Sheng Yan Jiu* 32(6):590-593.

Aloe vera (L.) Burm. f. — Liliaceae

SCN: aloe vera
Syn: *Aloe barbadensis* Mill.
OCN: aloe; Barbados aloe; Curaçao aloe

Part: mucilaginous leaf gel from parenchymatous leaf cells (commonly called aloe vera inner leaf; often referred to as aloe vera gel)

Quick Reference Summary

Safety Class: 1
Interaction Class: A

Contraindications

None known.

Other Precautions

None known.

Drug and Supplement Interactions

Other drugs should be taken 1 hour prior to consumption of aloe vera leaf gel or several hours after consumption, as it may slow the absorption of orally administered drugs (Brinker 1997).

Notice

Mucilages (~30%) (Roboz and Haagen-Smit 1948); *see* Appendix 3.

Editors' Note

Aloe vera leaf gel is commonly consumed in quantity as a cleansing juice preparation. It does not act as a strong cathartic like the dried juice from the pericyclic region of the leaf (*see* previous entry), which is a distinct product that contains the laxative compound aloin (Leung and Foster 1996).

Adverse Events and Side Effects

Reviews of topical use of aloe vera leaf gel have indicated that topical use is generally very well tolerated with few adverse effects (Vogler and Ernst 1999). Contact dermatitis has been reported after topical application of aloe vera leaf gel on some individuals (Ferreira et al. 2007; Hogan 1988; Hunter and Frumkin 1991).

Two cases of Henoch-Schönlein purpura (a disease that involves purple spots on the skin, joint pain, and gastrointestinal and kidney problems) have been reported in persons taking aloe vera leaf juice or extract (Cao et al. 2005; Cholongitas et al. 2005).

Pharmacological Considerations

Use on surgical wounds has been cautioned as wounds being healed by second intention (left open to heal from the base outward) treated with aloe vera leaf gel have been observed to have increased healing time in one study (Schmidt and Greenspoon 1991), although other studies have indicated beneficial effects of aloe vera leaf gel or extract on wound healing or psoriasis (Fulton 1990; Syed et al. 1996).

Pregnancy and Lactation

Among five animal studies of aloe vera leaf gel in pregnancy, one study showed some effects in preventing pregnancy, while four studies showed no effects on pregnancy (Kamboj and Dhawan 1982).

No information on the safety of aloe vera leaf gel during lactation was identified. While this review did not identify any concerns for use while nursing, safety has not been conclusively established.

Review Details

I. Drug and Supplement Interactions

Clinical Trials of Drug or Supplement Interactions

No clinical trials of drug or supplement interactions were identified.

Case Reports of Suspected Drug or Supplement Interactions

Mass bleeding during surgery was reported in a patient administered the anesthetic sevoflurane (0.5–1.2%) who had been taking an aloe vera product for 2 weeks prior to surgery (4 tablets daily). The physicians reporting this case indicated that sevoflurane is known to reduce platelet aggregation and that approximately 1% of patients in sevoflurane clinical trials experience hemorrhages, and also stated that the patient did not know if the aloe preparation was a fresh herb product, a dehydrated ingredient, or an extract (Lee et al. 2004). The physicians also referenced a study that indicated significant inhibition of prostaglandin synthesis in rats administered aloe vera leaf gel extracts (Vazquez et al. 1996).

Animal Trials of Drug or Supplement Interactions

No animal trials of drug or supplement interactions were identified.

II. Adverse Events

Adverse Events Reported in Clinical Trials

A systematic review of 10 controlled clinical trials of topical or oral use of aloe vera leaf gel (or in one case of an extract, not further described) indicated no withdrawals due to adverse effects of aloe vera. Adverse effects reported included burning after topical application, contact dermatitis, and mild itching. Adverse effects were noted as reversible, and aloe vera was characterized as very well tolerated (Vogler and Ernst 1999).

Case Reports of Adverse Events

Cases of contact dermatitis have been reported after topical applications of aloe vera leaf gel (Ferreira et al. 2007; Hogan 1988; Hunter and Frumkin 1991). A case of hypersensitivity to aloe was reported in a woman who had used aloe vera orally (3 teaspoons daily) and topically (Morrow et al. 1980).

Two cases of Henoch–Schönlein purpura (a disease that involves purple spots on the skin, joint pain, gastrointestinal problems, and glomerulonephritis) have been reported in individuals that were using aloe. One patient had taken "some juice extracted from four to five leaflets of *Aloe vera*" (he had taken the same remedy several months prior with no reaction) 24 hours prior to developing a rash, abdominal pain followed by rapid deterioration of renal function and abnormal levels of serum creatinine (Cholongitas et al. 2005). The second patient had taken "extracts of *Aloe vera*" (parts consumed and dose consumed unspecified) for approximately 1 week. He developed general edema and palpable purpura that eventually resolved (Cao et al. 2005).

One case of thyroid dysfunction was reported in a woman who had used aloe vera plant juice orally (10 ml daily) and topically (Pigatto and Guzzi 2005).

III. Pharmacology and Pharmacokinetics

Human Pharmacological Studies

Several clinical trials have studied the effects of topical applications of aloe on healing of various types of wounds and skin conditions. A speeded healing time was observed in patients treated with aloe after dermabrasion (a surgical scraping of the skin) therapy for acne (Fulton 1990) and for healing in psoriasis patients (Syed et al. 1996). A slowed healing time was observed in postsurgical complicated wounds treated topically with aloe (Schmidt and Greenspoon 1991). In allergic patch testing of 700 patients, no reactions to a concentrated (10×) aloe vera leaf gel were observed (Reider et al. 2005).

Animal Pharmacological Studies

Animal pharmacological studies were identified but omitted due to availability of human data.

In Vitro Pharmacological Studies

In vitro pharmacological studies were identified but omitted due to availability of human data.

IV. Pregnancy and Lactation

A review of published and unpublished Indian research on plants for fertility regulation indicated that, in one study, aloe vera leaf extracts (100 to 500 mg/kg daily) administered to female rats on postcoital days 1 through 7 had some activity in preventing pregnancy. Contrarily, four similar studies indicated no effect on pregnancy (Kamboj and Dhawan 1982).

No information on the safety of aloe vera leaf gel during lactation was identified.

V. Toxicity Studies

Short-Term Toxicity

No clinical signs of toxicity, mortality, or other adverse events were observed in rats orally administered 2 ml aloe vera leaf in water daily for 14 days (MDS 2000).

Subchronic Toxicity

No significant adverse effects were noted when technical grade acemannan (a polysaccharide fraction of aloe gel) was administered orally to rats for 14 days as 5% of the diet and for 6 months at up to 2000 mg/kg daily and to beagle dogs for 90 days at up to 1500 mg/kg daily (Fogelman et al. 1992).

No changes in organ weights or body weight were reported in male mice fed 100 mg/kg daily aloe vera extract for 90 days. Degeneration of sex organs was reported in 20% of mice (Shah et al. 1989).

Chronic Toxicity

No harmful effects were reported in rats that ingested aloe vera leaf powder as 1% of their diet from 42 days old until natural death (Ikeno et al. 2002).

Genotoxicity

Some genotoxicity but no cytotoxicity was noted for aloe vera leaf pulp in a bacterial transformation assay with the pulp added to *E. coli* and exposed to UV light (Paes-Leme et al. 2005).

Literature Cited

Brinker, F. 1997. Interactions of pharmaceutical and botanical medicines. *J. Naturopathic Med.* 7(2):14-20.

Cao, D., C.H. Yoon, B.S. Shin, et al. 2005. Effects of aloe, aloesin, or propolis on the pharmacokinetics of benzo[*a*]pyrene and 3-OH-benzo[*a*]pyrene in rats. *J. Toxicol. Env. Health A* 68(23-24):2227-2238.

Cholongitas E., Katsoudas, S., and Dourakis, S. 2005. Henoch–Schonlein purpura associated with *Aloe vera* administration. *Eur. J. Intern. Med.* 16(1):59-60.

Ferreira, M., M. Teixeira, E. Silva, and M. Selores. 2007. Allergic contact dermatitis to *Aloe vera*. *Contact Dermat.* 57(4):278-279.

Fogelman, R.W., T.E. Shellenberger, M.F. Balmer, R.H. Carpenter, and B.H. McAnalley. 1992. Subchronic oral administration of acemannan in the rat and dog. *Vet. Hum. Toxicol.* 34(2):144-147.

Fulton, J.E. 1990. The stimulation of postdermabrasion wound healing with stabilized *Aloe vera* gel-polyethylene oxide dressing. *J. Dermatol. Surg. Oncol.* 16(5):460-467.

Hogan, D.J. 1988. Widespread dermatitis after topical treatment of chronic leg ulcers and stasis dermatitis. *Can. Med. Assoc. J.* 138(4):336-338.

Hunter, D., and A. Frumkin. 1991. Adverse reactions to vitamin E and *Aloe vera* preparations after dermabrasion and chemical peel. *Cutis* 47(3):193-196.

Ikeno, Y., G.B. Hubbard, S. Lee, B.P. Yu, and J.T. Herlihy. 2002. The influence of long-term *Aloe vera* ingestion on age-related disease in male Fischer 344 rats. *Phytother. Res.* 16(8):712-718.

Kamboj, V.P., and B.N. Dhawan. 1982. Research on plants for fertility regulation in India. *J. Ethnopharmacol.* 6(2):191-226.

Lee, A., P.T. Chui, C.S. Aun, T. Gin, and A.S. Lau. 2004. Possible interaction between sevoflurane and *Aloe vera*. *Ann. Pharmacother.* 38(10):1651-1654.

Leung, A.Y., and S. Foster. 1996. *Encyclopedia of common natural ingredients used in food, drugs, and cosmetics*. 2nd ed. New York: Wiley.

MDS. 2000. MDS Pharma Services. Single dose toxicity study by the oral route in the rat. Unpublished data. *In* Bergfeld et al. 2004. *Safety assessment of aloe*. Cosmetic Ingredient Review.

Morrow, D.M., M.J. Rapaport, and R.A. Strick. 1980. Hypersensitivity to aloe. *Arch. Dermatol.* 116(9):1064-1065.

Paes-Leme, A.A., E.S. Motta, J.C. De Mattos, et al. 2005. Assessment of *Aloe vera* (L.) genotoxic potential on *Escherichia coli* and plasmid DNA. *J. Ethnopharmacol.* 102(2):197-201.

Pigatto, P.D., and G. Guzzi. 2005. Aloe linked to thyroid dysfunction. *Arch. Med. Res.* 36(5):608.

Reider, N., A. Issa, T. Hawranek, et al. 2005. Absence of contact sensitization to *Aloe vera* (L.) Burm. f. *Contact Dermat.* 53(6):332-334.

Roboz, E., and A.J. Haagen-Smit. 1948. A mucilage from *Aloe vera*. *J. Am. Chem. Soc.* 70(10):3248-3249.

Syed, T.A., S.A. Ahmad, A.H. Holt, et al. 1996. Management of psoriasis with *Aloe vera* extract in a hydrophilic cream: A placebo-controlled, double-blind study. *Trop. Med. Int. Health* 1(4):505-509.

Schmidt, J.M., and J.S. Greenspoon. 1991. Aloe vera dermal wound gel is associated with a delay in wound healing. *Obstet. Gynecol.* 78(1):115-117.

Shah, A.H., S. Qureshi, M. Tariq, and A.M. Ageel. 1989. Toxicity studies on six plants used in the traditional Arab system of medicine. *Phytother. Res.* 3:25-29.

Vazquez, B., G. Avila, D. Segura, and B. Escalante. 1996. Antiinflammatory activity of extracts from *Aloe vera* gel. *J. Ethnopharmacol.* 55(1):69-75.

Vogler, B.K., and E. Ernst. 1999. *Aloe vera*: A systematic review of its clinical effectiveness. *Br. J. Gen. Pract.* 49(447):823-828.

Aloysia citriodora Palau — Verbenaceae

SCN: lemon verbena
Syn: *Aloysia triphylla* (L'Hér.) Britton; *Lippia citriodora* Kunth, nom. illeg.
OCN: verbena
Part: leaf

Quick Reference Summary

Safety Class: 1
Interaction Class: A

Contraindications

None known.

Other Precautions

None known.

Drug and Supplement Interactions

None known.

Adverse Events and Side Effects

None known.

Pharmacological Considerations

None known.

Pregnancy and Lactation

No information on the safety of lemon verbena in pregnancy or lactation was identified in the scientific or traditional literature. Although this review did not identify any concerns for use while pregnant or nursing, safety has not been conclusively established.

Review Details

I. Drug and Supplement Interactions

Clinical Trials of Drug or Supplement Interactions

No clinical trials of drug or supplement interactions were identified.

Case Reports of Suspected Drug or Supplement Interactions

No case reports of suspected drug or supplement interactions were identified.

Animal Trials of Drug or Supplement Interactions

No animal trials of drug or supplement interactions were identified.

II. Adverse Events

Case Reports of Adverse Events

No case reports of adverse events were identified.

III. Pharmacology and Pharmacokinetics

Human Pharmacological Studies

No relevant human pharmacological studies were identified.

Animal Pharmacological Studies

No relevant animal pharmacological studies were identified.

In Vitro Pharmacological Studies

A hexane extract of lemon verbena was observed to have an antagonistic effect on the activity of the β-receptor in rat hearts (Vargas Solis 2000).

IV. Pregnancy and Lactation

No information on the safety of lemon verbena in pregnancy or lactation was identified.

V. Toxicity Studies

Acute Toxicity

The LC_{50} of lemon balm essential oil in the brine shrimp lethality assay was 1279 μg/ml (Oliva et al. 2007).

Genotoxicity

Antigenotoxic activity of aqueous extracts of lemon verbena was observed in the comet assay in mice (Zamorano-Ponce et al. 2004; Zamorano-Ponce et al. 2006).

Literature Cited

Oliva, M.D.L.M., N. Gallucci, J.A. Zygadlo, and M.S. Demo. 2007. Cytotoxic activity of Argentinean essential oils on *Artemia salina*. *Pharm. Biol.* 45(4):259-262.

Vargas Solis, R.C. 2000. Inhibitory effect of *Aloysia triphylla* hexanic extract on Wistar rat heart. *Rev. Mex. Cien. Farm.* 31(3):23-25.

Zamorano-Ponce, E., J. Fernandez, G. Vargas, P. Rivera, and M.A. Carballo. 2004. Protective activity of cedron (*Aloysia triphylla*) infusion over genetic damage induced by cisplatin evaluated by the comet assay technique. *Toxicol. Lett.* 152(1):85-90.

Zamorano-Ponce, E., C. Morales, D. Ramos, et al. 2006. Anti-genotoxic effect of *Aloysia triphylla* infusion against acrylamide-induced DNA damage as shown by the comet assay technique. *Mutat. Res.* 603(2):145-150.

Alpinia galanga (L.) Sw. — Zingiberaceae

SCN: greater galangal
Syn: *Maranta galanga* L.
AN: *kulanjana*
PN: *da gao liang jiang*
OCN: galanga; Java galanga; Siamese galanga
Part: rhizome

Quick Reference Summary

Safety Class: 1
Interaction Class: A

Contraindications

None known.

Other Precautions

None known.

Drug and Supplement Interactions

None known.

Editors' Notes

Although use of this species as a food additive in the U.S. is limited to its function as a flavoring substance in alcoholic beverages (CFR 2011a), the related *A. officinarum* is identified in the same reference as generally recognized as safe as a spice or flavoring (CFR 2011b). Also, dietary ingredients for use in dietary supplements are specifically excluded from the federal food additive definition (U.S.C. 2010).

Adverse Events and Side Effects

None known.

Pharmacological Considerations

None known.

Pregnancy and Lactation

No information on the safety of greater galangal in pregnancy or lactation was identified in the scientific or traditional literature. Although this review did not identify any concerns for use while pregnant or nursing, safety has not been conclusively established.

Review Details

I. Drug and Supplement Interactions

Clinical Trials of Drug or Supplement Interactions

No clinical trials of drug or supplement interactions were identified.

Case Reports of Suspected Drug or Supplement Interactions

No case reports of suspected drug or supplement interactions were identified.

Animal Trials of Drug or Supplement Interactions

No animal trials of drug or supplement interactions were identified.

II. Adverse Events

Case Reports of Adverse Events

Allergic contact dermatitis, confirmed by patch testing (both fresh and dried greater galangal), was reported after topical application of greater galangal (Hong and Chang 2006).

III. Pharmacology and Pharmacokinetics

Human Pharmacological Studies

No relevant human pharmacological studies were identified.

Animal Pharmacological Studies

A reduction in blood glucose levels was observed in healthy rabbits but not in diabetic rabbits administered methanol or aqueous extracts or powdered rhizome of greater galangal (Akhtar et al. 2002).

A

In Vitro Pharmacological Studies

Inhibition of platelet-activating factor binding to rabbit platelets was observed after treatment with greater galangal (Jantan et al. 2005).

Inhibition of the drug metabolizing isoenzymes CYP2D6 and CYP3A4 were observed in human liver microsomes treated with methanol extracts of greater galangal (Subehan et al. 2006; Usia et al. 2006).

IV. Pregnancy and Lactation

No information on the safety of greater galangal in pregnancy or lactation was identified.

V. Toxicity Studies

Acute Toxicity

No significant mortality was observed in mice orally administered single doses of up to 3 g/kg of an ethanol extract of greater galangal (Qureshi et al. 1992).

The LD_{50} of an ethanol extract of greater galangal in the brine shrimp lethality assay was 109 μg/ml (Khattak et al. 2005).

Subchronic Toxicity

An increase in red blood cell count but no changes in organ weights and no signs of toxicity were observed in mice orally administered 100 mg/kg of an ethanol extract of greater galangal daily for 90 days (Qureshi et al. 1992).

Genotoxicity

Single-strand breaks in DNA were observed in human fibroblast, mammary tumor, and lung adenocarcinoma cell lines treated with an aqueous extract of greater galangal. Breaks were observed after exposure to greater galangal at concentrations of 100 μg/ml or more (Muangnoi et al. 2007).

Cytotoxicity

Cytotoxic effects were observed in human fibroblast, mammary tumor, and lung adenocarcinoma cell lines treated with an aqueous or methanol and dichloromethane extract of greater galangal (Muangnoi et al. 2007; Nam et al. 2005).

Literature Cited

Akhtar, M.S., M.A. Khan, and M.T. Malik. 2002. Hypoglycaemic activity of *Alpinia galanga* rhizome and its extracts in rabbits. *Fitoterapia* 73(7-8):623-628.

CFR. 2011a. *Code of federal regulations*, Title 21 Part 172.510, 2011 ed. Food additives permitted for direct addition to food for human consumption. Natural flavoring substances and natural substances used in conjunction with flavors. Washington, DC: U.S. Government Printing Office.

CFR. 2011b. *Code of federal regulations*, Title 21 Part 182.10, 2011 ed. Substances generally recognized as safe. Spices and other natural seasonings and flavorings. Washington, DC: U.S. Government Printing Office.

Hong, S.J., and C.H. Chang. 2006. Erythema multiforme-like generalized allergic contact dermatitis caused by *Alpinia galanga*. *Contact Dermat.* 54(2):118-120.

Jantan, I., I.A. Rafi, and J. Jalil. 2005. Platelet-activating factor (PAF) receptor-binding antagonist activity of Malaysian medicinal plants. *Phytomedicine* 12(1-2):88-92.

Khattak, S., R. Saeed ur, H.U. Shah, W. Ahmad, and M. Ahmad. 2005. Biological effects of indigenous medicinal plants *Curcuma longa* and *Alpinia galanga*. *Fitoterapia* 76(2):254-257.

Muangnoi, P., M. Lu, J. Lee, et al. 2007. Cytotoxicity, apoptosis and DNA damage induced by *Alpinia galanga* rhizome extract. *Planta Med.* 73(8):748-754.

Nam, J.-W., S.-J. Kim, A.-R. Han, S.K. Lee, and E.-K. Seo. 2005. Cytotoxic phenylpropanoids from the rhizomes of *Alpinia galanga*. *J. Appl. Pharmacol.* 13(4):263-266.

Qureshi, S., A.H. Shah, and A.M. Ageel. 1992. Toxicity studies on *Alpinia galanga* and *Curcuma longa*. *Planta Med.* 58(2):124-127.

Subehan, T. Usia, H. Iwata, S. Kadota, and Y. Tezuka. 2006. Mechanism-based inhibition of CYP3A4 and CYP2D6 by Indonesian medicinal plants. *J. Ethnopharmacol.* 105(3):449-455.

U.S.C. 2010. United States Code, Title 21, Part 321 (s)(6). Current as of January 7, 2011. Washington, DC: U.S. Government Printing Office.

Usia, T., H. Iwata, A. Hiratsuka, et al. 2006. CYP3A4 and CYP2D6 inhibitory activities of Indonesian medicinal plants. *Phytomedicine* 13(1-2):67-73.

Alpinia officinarum Hance — Zingiberaceae

SCN: lesser galangal
PN: *gao liang jiang* (rhizome)
OCN: Chinese galangal; Chinese ginger
Part: rhizome

Quick Reference Summary

Safety Class: 1
Interaction Class: A

Contraindications

None known.

Other Precautions

None known.

Drug and Supplement Interactions

None known.

Adverse Events and Side Effects

None known.

Pharmacological Considerations

None known.

Pregnancy and Lactation

No information on the safety of lesser galangal in pregnancy or lactation was identified in the scientific or traditional literature. Although this review did not identify any concerns for use while pregnant or nursing, safety has not been conclusively established.

Review Details

I. Drug and Supplement Interactions

Clinical Trials of Drug or Supplement Interactions

No clinical trials of drug or supplement interactions were identified.

Case Reports of Suspected Drug or Supplement Interactions

No case reports of suspected drug or supplement interactions were identified.

Animal Trials of Drug or Supplement Interactions

No animal trials of drug or supplement interactions were identified.

II. Adverse Events

Case Reports of Adverse Events

No case reports of adverse events were identified.

III. Pharmacology and Pharmacokinetics

Human Pharmacological Studies

No relevant human pharmacological studies were identified.

Animal Pharmacological Studies

No relevant animal pharmacological studies were identified.

In Vitro Pharmacological Studies

Several compounds isolated from lesser galangal were shown to inhibit platelet-activating factor receptor binding (Fan et al. 2007).

IV. Pregnancy and Lactation

No information on the safety of lesser galangal in pregnancy or lactation was identified.

V. Toxicity Studies

Acute Toxicity

The LD_{50} of an orally administered alcohol extract of lesser galangal in mice was 4.2 ml/kg (Chen and Chen 2004).

No fatalities were reported in mice orally administered 120 g/kg daily of an aqueous extract of lesser galangal daily for 7 days (Chen and Chen 2004).

Cytotoxicity

Cytotoxic effects were observed in human breast cancer and lung cancer cell lines treated with a methanol and dichloromethane extract of lesser galangal (Lee and Houghton 2005).

Literature Cited

Chen, J.K., and T.T. Chen. 2004. *Chinese medical herbology and pharmacology*. City of Industry, CA: Art of Medicine Press.

Fan, G.J., Y.H. Kang, Y.N. Han, and B.H. Han. 2007. Platelet-activating factor (PAF) receptor binding antagonists from *Alpinia officinarum*. *Bioorg. Med. Chem. Lett.* 17(24):6720-6722.

Lee, C.C., and P. Houghton. 2005. Cytotoxicity of plants from Malaysia and Thailand used traditionally to treat cancer. *J. Ethnopharmacol.* 100(3):237-243.

Althaea officinalis L.

Malvaceae

SCN: marshmallow
OCN: althaea; althea

Part: flower, leaf, root

Quick Reference Summary

Safety Class: 1

Interaction Class: A

Contraindications

None known.

Other Precautions

Should be taken with at least 250 ml (8 oz.) of liquid.

Drug and Supplement Interactions

Other drugs should be taken 1 hour prior to consumption of marshmallow or several hours after consumption, as marshmallow may slow the absorption of orally administered drugs (Brinker 2001; De Smet 1993; Mills and Bone 2005).

Notice

Mucilages (~10%) (Evans 2002); *see* Appendix 3.

Adverse Events and Side Effects

None known.

Pharmacological Considerations

None known.

Pregnancy and Lactation

No information on the safety of marshmallow in pregnancy or lactation was identified in the scientific or traditional literature. Although this review did not identify any concerns for use while pregnant or nursing, safety has not been conclusively established.

Review Details

I. Drug and Supplement Interactions

Clinical Trials of Drug or Supplement Interactions

No clinical trials of drug or supplement interactions were identified.

Case Reports of Suspected Drug or Supplement Interactions

No case reports of suspected drug or supplement interactions were identified.

Animal Trials of Drug or Supplement Interactions

No animal trials of drug or supplement interactions were identified.

II. Adverse Events

Case Reports of Adverse Events

No case reports of adverse events were identified.

III. Pharmacology and Pharmacokinetics

Human Pharmacological Studies

No relevant human pharmacological studies were identified.

Animal Pharmacological Studies

In rats administered 10, 30, or 100 mg/kg marshmallow, plasma glucose levels dropped to 74%, 81%, and 65%, respectively, of control values 7 hours after administration (Tomoda et al. 1987).

In Vitro Pharmacological Studies

No relevant in vitro pharmacological studies were identified.

Literature Cited

Brinker, F. 2001. *Herb contraindications and drug interactions*. 3rd ed. Sandy, OR: Eclectic Medical Publications.

De Smet, P.A.G.M. 1993. *Adverse effects of herbal drugs Volume 2*. Berlin, Heidelberg, New York: Springer.

Evans, W. 2002. *Trease and Evans pharmacognosy*. 15th ed. New York: Saunders.

Mills, S., and K. Bone. 2005. *The essential guide to herbal safety*. St. Louis: Elsevier.

Tomoda, M., N. Shimizu, Y. Oshima, et al. 1987. Hypoglycemic activity of twenty plant mucilages and three modified products. *Planta Med.* 53(1):8-12.

Amomum tsao-ko Crevost & Lemarié — Zingiberaceae

SCN: tsao-ko amomum

PN: *cao guo* (fruit)

OCN: *tsao ko*

Part: fruit

Quick Reference Summary

Safety Class: 1

Interaction Class: A

Contraindications

None known.

Other Precautions

None known.

Drug and Supplement Interactions

None known.

Adverse Events and Side Effects

None known.

Pharmacological Considerations

None known.

Pregnancy and Lactation

No information on the safety of tsao-ko amomum in pregnancy or lactation was identified in the scientific or traditional literature. Although this review did not identify any concerns for use while pregnant or nursing, safety has not been conclusively established.

Review Details

I. Drug and Supplement Interactions

Clinical Trials of Drug or Supplement Interactions

No clinical trials of drug or supplement interactions were identified.

Case Reports of Suspected Drug or Supplement Interactions

No case reports of suspected drug or supplement interactions were identified.

Animal Trials of Drug or Supplement Interactions

No animal trials of drug or supplement interactions were identified.

II. Adverse Events

Case Reports of Adverse Events

No case reports of adverse events were identified.

III. Pharmacology and Pharmacokinetics

Human Pharmacological Studies

A decease in the expression of platelet-activating factor was observed in patients with *Helicobacter pylori* infection orally administered 0.5 ml tsao-ko amomum essential oil three times daily for 4 weeks (Huang et al. 2008).

Animal Pharmacological Studies

No relevant animal pharmacological studies were identified.

In Vitro Pharmacological Studies

No relevant in vitro pharmacological studies were identified.

IV. Pregnancy and Lactation

No information on the use of tsao-ko amomum during pregnancy or lactation was identified.

V. Toxicity Studies

Acute Toxicity

The LD_{50} of the compound geraniol orally administered in rats is 4.8 g/kg and intravenously administered in rabbits is 50 mg/kg (Chen and Chen 2004).

Literature Cited

Chen, J.K., and T.T. Chen. 2004. *Chinese medical herbology and pharmacology*. City of Industry, CA: Art of Medicine Press.

Huang, G.D., Y.H. Huang, M.Z. Xiao, et al. 2008. Effect of volatile oil of amomum on expressions of platelet activating factor and mastocarcinoma-related peptide in the gastric membrane of chronic gastritis patients with helicobacter-pylori infection. *Chin. J. Integr. Med.* 14(1):23-27.

Andrographis paniculata (Burm. f.) Nees — Acanthaceae

SCN: andrographis
AN: *bhunimba; mahatikta*
PN: *chuan xin lian* (herb)
OCN: creat; green chiretta; Indian chiretta
Part: herb

Quick Reference Summary

Safety Class: 2b
Interaction Class: A

Contraindications

Not for use in pregnancy except under the supervision of a qualified healthcare practitioner (But 1988; Chang and But 1986; Chen and Chen 2004; Zoha et al. 1989).

Other Precautions

None known.

Drug and Supplement Interactions

None known.

Notice

Abortifacient (Chang and But 1986; Chen and Chen 2004); *see* Appendix 2.

Adverse Events and Side Effects

A systematic review of human studies of andrographis concluded that "short-term therapy is associated with an encouraging safety profile, although adverse events have been demonstrated at higher doses" (Coon and Ernst 2004).

Large oral doses are reported to cause gastric discomfort and loss of appetite (Chang and But 1986). Allergic reactions, including an anaphylactic reaction, to andrographis have been reported (Coon and Ernst 2004).

Pharmacological Considerations

Animal studies have demonstrated that andrographis may modify glucose regulation; thus diabetic persons using insulin or oral hypoglycemic drugs should continue to monitor glucose levels during andrographis use (Borhanuddin et al. 1994; Husen et al. 2004; Zhang and Tan 2000).

Results of animal studies on the effects of andrographis on female and male fertility and embryo implantation are conflicting. Some studies indicated that high doses of andrographis prevented implantation and caused abortion (But 1988; Zhang and Tan 1997), while other studies with similar doses showed no adverse effects on female fertility or pregnancy (Shamsuzzoha et al. 1978, 1979). Results of studies of andrographis or the compound andrographolide on male animals vary from no observed effects on fertility in two studies, one with andrographis and the other with andrographolide (Burgos et al. 1997; Sattayasai et al. 2010), to a reduction in sperm counts and lack of sperm motility in a third study using andrographolide (Akbarsha and Murugaian 2000). Thus, until further evidence is available, the use of andrographis by women or men attempting to conceive is cautioned.

Pregnancy and Lactation

Information on the safety of andrographis during pregnancy is limited. Conflicting results on effects on fertility and pregnancy have been reported in animal studies, with some studies indicating no adverse effects on fertility in mice (Dhammaupakorn and Chaichantipyuth 1989; Shamsuzzoha et al. 1978, 1979) and other studies indicating antifertility or abortifacient activity (Chen and Chen 2004; But 1988; Zoha et al. 1989).

No information on the safety of andrographis during lactation was identified. While this review did not identify any concerns for use while nursing, safety has not been conclusively established.

Also see Pharmacological Considerations for this entry.

Review Details

I. Drug and Supplement Interactions

Clinical Trials of Drug or Supplement Interactions

No clinical trials of drug or supplement interactions were identified.

Case Reports of Suspected Drug or Supplement Interactions

No case reports of suspected drug or supplement interactions were identified.

Animal Trials of Drug or Supplement Interactions

A prolongation of pentobarbitone-induced sleeping time was observed in mice intraperitoneally administered 100 to 300 mg/kg of an andrographis extract (Mandal et al. 2001).

II. Adverse Events

Adverse Events Reported in Clinical Trials

A systematic review of human studies of andrographis concluded that "short-term therapy is associated with an encouraging safety profile, although adverse events have been demonstrated at higher doses" (Coon and Ernst 2004). The study utilizing the higher doses was a phase I trial of the compound andrographolide in HIV-positive and normal volunteers. The study was terminated early due to a large number of adverse events. The doses used in this trial, 5 or 10 mg/kg daily of andrographolide, were 6 to 12 times higher than doses used in other studies. Adverse events included allergic reaction (one anaphylactic reaction), fatigue, headache, rash, diarrhea, nausea, decreased taste, dry tongue, reduced libido, reduced short-term memory, eyes sensitive to light, decreased memory, dizziness, heartburn, tender lymph nodes, and lymphadenopathy. There was no placebo group in the study (Calabrese et al. 2000). Urticaria was reported in 2 of 60 subjects in a trial of standardized andrographis extract (Melchoir et al. 1996). Several additional human studies either reported no adverse events or did not provide information on adverse events (Coon and Ernst 2004).

Case Reports of Adverse Events

As of June 2003, no adverse events in persons taking andrographis had been reported to any national drug safety monitoring agencies including those in the United Kingdom, Germany, and Austria (Coon and Ernst 2004). One case of anaphylactic shock and two cases of anaphylactic reactions had been reported to the World Health Organization. All cases were from Sweden, and no details on the case reports were available in the published literature (Coon and Ernst 2004). Data obtained from manufacturers of andrographis products indicated that one company had received five reports of allergic reactions to products that contained andrographis (Coon and Ernst 2004).

III. Pharmacology and Pharmacokinetics

Human Pharmacological Studies

No relevant human pharmacological studies were identified.

Animal Pharmacological Studies

No testicular toxicity was observed in male rats orally administered a standardized extract of andrographis at doses up to 1000 mg/kg daily for 60 days (Burgos et al. 1997). A decrease in sperm counts and lack of motility in spermatozoa were observed in 3-month-old male rats orally administered 25 or 50 mg/kg of the compound andrographolide daily for 48 days (Akbarsha and Murugaian 2000). Conversely, no significant changes in sperm morphology and motility were observed in male mice orally administered 50 mg/kg of the compound andrographolide daily for up to eight weeks. After four weeks of treatment, an increase in serum testosterone levels was observed as compared to control. Male mice receiving andrographis mounted females more quickly and with greater frequency than untreated males (Sattayasai et al. 2010). Reductions in alkaline phosphatase activity and total protein were observed in the reproductive organs of male rats orally administered an andrographis powder suspension at a dose of 10 or 20 mg per animal daily for 24 or 48 days (Akbarsha and Manivannan 1993). Reduced libido was observed in male mice fed a diet supplemented with 0.75% andrographis stem powder for up to 4 weeks (Shamsuzzoha et al. 1979).

Inhibition of fasting serum glucose levels and glucose tolerance test levels was observed in diabetic rats orally administered an ethanolic extract of andrographis at doses of 100, 200, or 400 mg/kg (Zhang and Tan 2000). Prevention of hyperglycemia but no effects on adrenaline-induced hyperglycemia were observed in rabbits orally administered 10 mg/kg of an aqueous extract of andrographis (Borhanuddin et al. 1994). Similarly, a significant reduction in blood glucose levels was observed in hyperglycemic rats orally administered an aqueous extract of andrographis. The hyperglycemic effect was enhanced when the freeze-dried extract was used (Husen et al. 2004).

No significant changes in CYP450 enzymes were observed in rats administered 1 g/kg andrographis extract or 10 mg/kg of the compound andrographolide, each in single doses (Choudhury et al. 1987).

Dose-dependent choleretic action of the compound andrographolide was observed in rats and guinea pigs administered doses of 1.5 to 12 mg/kg (Shukla et al. 1992).

In Vitro Pharmacological Studies

In human platelets, the compound andrographolide inhibited platelet-activating-factor-induced aggregation in a dose-dependent manner (Amroyan et al. 1999). In mouse platelets, a standardized aqueous extract of andrographis, and the compounds andrographolide and 14-deoxy-11,12-didehydroandrographolide inhibited thrombin-induced platelet aggregation in a dose-dependent manner (Thisoda et al. 2006).

Methanolic and ethanolic extracts of andrographis inhibited CYP3A4 and CYP2D6 in vitro (Subehan et al. 2006; Usia et al. 2006). The compound andrographolide induced CYP1A1 and CYP1A2 but had no effect on CYP1B1 in vitro (Jaruchotikamol et al. 2007).

The compound bisandrographolide A has been shown to activate transient receptor potential V4 (TRPV4) with an EC_{50} (half maximal effective concentration) of 790 to 950 nM (Smith et al. 2006).

IV. Pregnancy and Lactation

After administration of 2 g/kg daily as part of the diet for 6 weeks, no pregnancies occurred in fertile female mice when the females were mated with male mice (not administered andrographis) known to be fertile (Zoha et al. 1989).

No antifertility effects were observed in mice administered powdered andrographis leaf or root as 1% of the diet (~2 g/kg daily) for 14 days prior to mating, and 21 days during mating (Shamsuzzoha et al. 1978). Similarly, no antifertility effects were observed in female mice administered andrographis stem powder for 28 days prior to mating (Shamsuzzoha et al. 1979).

Intraperitoneal administration of an aqueous extract of whole andrographis plants (dose not specified in English language abstract) prevented implantation in mice and caused abortion in mice during the first, second, and third trimesters of pregnancy (But 1988; Chen and Chen 2004).

In rats orally administered 200, 600, or 2000 mg/kg (30 to 300 times higher than the standard human dose) of an andrographis extract for days 0–19 of pregnancy, no change in progesterone levels and no progesterone-mediated pregnancy termination were observed (Panossian et al. 1999).

No teratogenicity or toxicity was observed in pregnant mice orally administered 200 or 400 mg/kg andrographis leaf powder suspension on alternate days for 4 weeks (Dhammaupakorn and Chaichantipyuth 1989).

No information on the safety of andrographis during lactation was identified.

V. Toxicity Studies

Acute Toxicity

The LD_{50} of an orally or subcutaneously administered alcoholic extract of andrographis in mice could not be determined at doses up to 15 g/kg (Sithisomwongse et al. 1989). The LD_{50} of an intraperitoneally administered alcoholic extract was 14.98 g/kg (Sithisomwongse et al. 1989).

The LD_{50} of an aqueous-methanolic andrographis whole plant extract intraperitoneally administered in mice could not be determined at doses up to 1 g/kg (Nakannishi et al. 1965).

The LD_{50} of the compound andrographolide orally administered in mice could not be determined at doses up to 40 g/kg (Chang and But 1986).

Literature Cited

Akbarsha, M.A., and B. Manivannan. 1993. Biochemical changes in the testis and male accessory organs of albino rats on treatment with *Andrographis paniculata* (Nees). *Indian J. Comp. An. Physiol.* 11(2):103-108.

Akbarsha, M.A., and P. Murugaian. 2000. Aspects of the male reproductive toxicity/male antifertility property of andrographolide in albino rats: Effect on the testis and the cauda epididymidal spermatozoa. *Phytother. Res.* 14(6):432-435.

Amroyan, E., E. Gabrielian, A. Panossian, G. Wikman, and H. Wagner. 1999. Inhibitory effect of andrographolide from *Andrographis paniculata* on PAF-induced platelet aggregation. *Phytomedicine* 6(1):27-31.

Borhanuddin, M., M. Shamsuzzoha, and A.H. Hussain. 1994. Hypoglycaemic effects of *Andrographis paniculata* Nees on non-diabetic rabbits. *Bangladesh Med. Res. Counc. Bull.* 20(1):24-26.

Burgos, R.A., E.E. Caballero, N.S. Sanchez, et al. 1997. Testicular toxicity assessment of *Andrographis paniculata* dried extract in rats. *J. Ethnopharmacol.* 58(3):219-224.

But, P. 1988. Chinese medicine for birth control. *Abstr. Chin. Med.* 2(2):247-269.

Calabrese, C., S.H. Berman, J.G. Babish, et al. 2000. A phase I trial of andrographolide in HIV positive patients and normal volunteers. *Phytother. Res.* 14(5):333-338.

Chang, H.-M., and P.P.H. But. 1986. *Pharmacology and applications of Chinese materia medica*. English ed. Singapore, Philadelphia: World Scientific.

Chen, J.K., and T.T. Chen. 2004. *Chinese medical herbology and pharmacology*. City of Industry, CA: Art of Medicine Press.

Choudhury, B.R., S.J. Haque, and M.K. Poddar. 1987. In vivo and in vitro effects of kalmegh (*Andrographis paniculata*) extract and andrographolide on hepatic microsomal drug metabolizing enzymes. *Planta Med.* 53(2):135-140.

Coon, J.T., and E. Ernst. 2004. *Andrographis paniculata* in the treatment of upper respiratory tract infections: A systematic review of safety and efficacy. *Planta Med.* 70(4):293-298.

Dhammaupakorn, P., and C. Chaichantipyuth. 1989. *Cited in* Mills, S., and K. Bone. 2005. *The essential guide to herbal safety*. St. Louis: Elsevier.

Husen, R., A.H. Pihie, and M. Nallappan. 2004. Screening for antihyperglycaemic activity in several local herbs of Malaysia. *J. Ethnopharmacol.* 95(2-3):205-208.

Jaruchotikamol, A., K. Jarukamjorn, W. Sirisangtrakul, et al. 2007. Strong synergistic induction of CYP1A1 expression by andrographolide plus typical CYP1A inducers in mouse hepatocytes. *Toxicol. Appl. Pharmacol.* 224(2):156-162.

Mandal, S.C., A.K. Dhara, and B.C. Maiti. 2001. Studies on psychopharmacological activity of *Andrographis paniculata* extract. *Phytother. Res.* 15(3):253-256.

Melchoir, J., S. Palm, and G. Wikman. 1996. Controlled clinical study of standardised *Andrographis paniculata* extract in common cold—a pilot trial. *Phytomedicine* 3:315-318.

Nakannishi, K., S. Sasaki, A. Kiang, et al. 1965. Phytochemical and pharmacological screening. *Chem. Pharm. Bull.* 13:822.

Panossian, A., A. Kochikian, E. Gabrielian, et al. 1999. Effect of *Andrographis paniculata* extract on progesterone in blood plasma of pregnant rats. *Phytomedicine* 6(3):157-161.

Sattayasai, J., S. Srisuwan, T. Arkaravichien, and C. Aromdee. 2010. Effects of andrographolide on sexual functions, vascular reactivity and serum testosterone level in rodents. *Food. Chem. Toxicol.* 48 (7):1934-1938.

Shamsuzzoha, M., M.S. Rahman, and M.M. Ahmed. 1979. Antifertility activity of a medicinal plant of the genus *Andrografis* (sic) Wall (family Acanthaceae). Part II. *Bangladesh Med. Res. Counc. Bull.* 5(1):14-18.

Shamsuzzoha, M., M.S. Rahman, M.M. Ahmed, and A.K. Islam. 1978. Antifertility effect in mice of medicinal plant of family Acanthaceae. *Lancet* 312(8095):900.

Shukla, B., P.K. Visen, G.K. Patnaik, and B.N. Dhawan. 1992. Choleretic effect of andrographolide in rats and guinea pigs. *Planta Med.* 58(2):146-149.

Sithisomwongse, N., J. Phengchata, and S. Cheewapatana. 1989. Acute and chronic toxicity of *Andrographis paniculata* Nees. *Thai J. Pharm. Sci.* 14(2):109-117.

Smith, P.L., K.N. Maloney, R.G. Pothen, J. Clardy, and D.E. Clapham. 2006. Bisandrographolide from *Andrographis paniculata* activates TRPV4 channels. *J. Biol. Chem.* 281(40):29897.

Subehan, T. Usia, H. Iwata, S. Kadota, and Y. Tezuka. 2006. Mechanism-based inhibition of CYP3A4 and CYP2D6 by Indonesian medicinal plants. *J. Ethnopharmacol.* 105(3):449-455.

Thisoda, P., N. Rangkadilok, N. Pholphana, et al. 2006. Inhibitory effect of *Andrographis paniculata* extract and its active diterpenoids on platelet aggregation. *Eur. J. Pharmacol.* 553(1-3):39-45.

Usia, T., H. Iwata, A. Hiratsuka, et al. 2006. CYP3A4 and CYP2D6 inhibitory activities of Indonesian medicinal plants. *Phytomedicine* 13(1-2):67-73.

Zhang, C.Y., and B.K. Tan. 1997. Mechanisms of cardiovascular activity of *Andrographis paniculata* in the anaesthetized rat. *J. Ethnopharmacol.* 56(2):97-101.

Zhang, X.F., and B.K. Tan. 2000. Anti-diabetic property of ethanolic extract of *Andrographis paniculata* in streptozotocin-diabetic rats. *Acta Pharmacol. Sin.* 21(12):1157-1164.

Zoha, M.S., A.H. Hussain, and S.A. Choudhury. 1989. Antifertility effect of *Andrographis paniculata* in mice. *Bangladesh Med. Res. Counc. Bull.* 15(1):34-37.

Anemarrhena asphodeloides Bunge

Liliaceae

A

SCN: anemarrhena
PN: *zhi mu* (rhizome)

Part: rhizome

Quick Reference Summary

Safety Class: 2d
Interaction Class: A

Contraindications

Not for use in persons with diarrhea (Bensky et al. 2004; Chen and Chen 2004).

Other Precautions

None known.

Drug and Supplement Interactions

None known.

Adverse Events and Side Effects

None known.

Pharmacological Considerations

Animal studies have demonstrated that anemarrhena may modify glucose regulation; thus diabetic persons using insulin or oral hypoglycemic drugs should continue to monitor glucose levels during anemarrhena use (Chen and Chen 2004; Jia et al. 2003; Miura et al. 2001; Nakashima et al. 1993; Takahashi et al. 1985).

Pregnancy and Lactation

No information on the safety of anemarrhena in pregnancy or lactation was identified in the scientific or traditional literature. Although this review did not identify any concerns for use while pregnant or nursing, safety has not been conclusively established.

Review Details

I. Drug and Supplement Interactions

Clinical Trials of Drug or Supplement Interactions

No clinical trials of drug or supplement interactions were identified.

Case Reports of Suspected Drug or Supplement Interactions

No case reports of suspected drug or supplement interactions were identified.

Animal Trials of Drug or Supplement Interactions

No animal trials of drug or supplement interactions were identified.

II. Adverse Events

Case Reports of Adverse Events

No case reports of adverse events were identified.

III. Pharmacology and Pharmacokinetics

Human Pharmacological Studies

No relevant human pharmacological studies were identified.

Animal Pharmacological Studies

Aqueous extracts of anemarrhena have been shown to dose-dependently lower glucose levels in diabetic mice (Miura et al. 2001; Nakashima et al. 1993; Takahashi et al. 1985).

In Vitro Pharmacological Studies

Steroidal saponins isolated from anemarrhena inhibited platelet aggregation and activated partial thromboplastin time in human blood (Zhang et al. 1999). Another saponin inhibited PAF-induced platelet aggregation in rabbit blood (Dong and Han 1991).

An ethanol extract of anemarrhena stimulated insulin secretion in the islets of pancreases isolated from normal and diabetic rats at concentrations of 3.3 mM (normal) and 16.7 mM (diabetic) (Hoa et al. 2004).

IV. Pregnancy and Lactation

No information on the safety of anemarrhena in pregnancy or lactation was identified.

V. Toxicity Studies

Acute Toxicity

In rabbits intravenously administered an extract (solvent type not specified in English language translation), a dose of 7 ml led to respiratory failure and death, while a dose of 1 to 3 ml caused difficulty breathing, and a dose of 0.5 ml showed no adverse effects (Chen and Chen 2004).

Literature Cited

Bensky, D., S. Clavey, and E. Stöger. 2004. *Chinese herbal medicine: Materia medica.* 3rd ed. Seattle: Eastland Press.

Chen, J.K., and T.T. Chen. 2004. *Chinese medical herbology and pharmacology*. City of Industry, CA: Art of Medicine Press.

A

Dong, J.X., and G.Y. Han. 1991. A new active steroidal saponin from *Anemarrhena asphodeloides*. *Planta Med.* 57(5):460-462.

Hoa, N.K., D.V. Phan, N.D. Thuan, and C.G. Ostenson. 2004. Insulin secretion is stimulated by ethanol extract of *Anemarrhena asphodeloides* in isolated islet of healthy Wistar and diabetic Goto-Kakizaki rats. *Exp. Clin. Endocrinol. Diabetes* 112(9):520-525.

Jia, W., W. Gaoz, and L. Tang. 2003. Antidiabetic herbal drugs officially approved in China. *Phytother. Res.* 17(10):1127-1134.

Miura, T., H. Ichiki, N. Iwamoto, et al. 2001. Antidiabetic activity of the rhizoma of *Anemarrhena asphodeloides* and active components, mangiferin and its glucoside. *Biol. Pharm. Bull.* 24(9):1009-1011.

Nakashima, N., I. Kimura, M. Kimura, and H. Matsuura. 1993. Isolation of pseudoprototimosaponin AIII from rhizomes of *Anemarrhena asphodeloides* and its hypoglycemic activity in streptozotocin-induced diabetic mice. *J. Nat. Prod.* 56(3):345-350.

Takahashi, M., C. Konno, and H. Hikino. 1985. Isolation and hypoglycemic activity of anemarans A, B, C and D, glycans of *Anemarrhena asphodeloides* rhizomes. *Planta Med.* 51(2):100-102.

Zhang, J., Z. Meng, M. Zhang, et al. 1999. Effect of six steroidal saponins isolated from anemarrhenae rhizoma on platelet aggregation and hemolysis in human blood. *Clin. Chim. Acta* 289(1-2):79-88.

Anemopsis californica (Nutt.) Hook. & Arn. — Saururaceae

SCN: yerba mansa

Part: rhizome, root

Quick Reference Summary

Safety Class: 1
Interaction Class: A

Contraindications

None known.

Other Precautions

None known.

Drug and Supplement Interactions

None known.

Adverse Events and Side Effects

None known.

Pharmacological Considerations

None known.

Pregnancy and Lactation

No information on the safety of yerba mansa in pregnancy or lactation was identified in the scientific or traditional literature. Although this review did not identify any concerns for use while pregnant or nursing, safety has not been conclusively established.

Review Details

I. Drug and Supplement Interactions

Clinical Trials of Drug or Supplement Interactions

No clinical trials of drug or supplement interactions were identified.

Case Reports of Suspected Drug or Supplement Interactions

No case reports of suspected drug or supplement interactions were identified.

Animal Trials of Drug or Supplement Interactions

No animal trials of drug or supplement interactions were identified.

II. Adverse Events

Case Reports of Adverse Events

No case reports of adverse events were identified.

III. Pharmacology and Pharmacokinetics

Human Pharmacological Studies

No relevant human pharmacological studies were identified.

Animal Pharmacological Studies

No relevant animal pharmacological studies were identified.

In Vitro Pharmacological Studies

Aqueous and ethanol extracts of yerba mansa inhibited the growth of human breast cancer cells (MCF-7/AZ, estrogen receptor positive) in a concentration-dependent manner and had no effect on the growth of colon cancer cells (HCT8/E11) (Daniels et al. 2006).

IV. Pregnancy and Lactation

No information on the safety of yerba mansa in pregnancy or lactation was identified.

V. Toxicity Studies

No toxicity studies were identified.

Literature Cited

Daniels, A.L., S. Van Slambrouck, R.K. Lee, et al. 2006. Effects of extracts from two Native American plants on proliferation of human breast and colon cancer cell lines in vitro. *Oncol. Rep.* 15(5):1327-1331.

Anethum graveolens L. — Apiaceae

SCN: dill
Syn: *Peucedanum graveolens* (L.) Benth. & Hook. f.
AN: *shatapushpa*
Part: fruit (commonly known as "seed"), herb

Quick Reference Summary

Safety Class: 1
Interaction Class: A

Contraindications

None known.

Other Precautions

None known.

Drug and Supplement Interactions

None known.

Notice

Diuretic (Mahran et al. 1991); *see* Appendix 2.

Adverse Events and Side Effects

Allergic reactions to dill, confirmed by patch testing, have been reported (Monteseirin et al. 2003; Monteseirin et al. 2002).

Pharmacological Considerations

None known.

Pregnancy and Lactation

No information on the safety of dill in pregnancy was identified in the scientific or traditional literature. While this review did not identify any concerns for pregnant women, safety has not been conclusively established.

Dill seed has been traditionally been used to promote milk production in nursing women (Jain et al. 2004; Mahran et al. 1992).

Review Details

I. Drug and Supplement Interactions

Clinical Trials of Drug or Supplement Interactions

No clinical trials of drug or supplement interactions were identified.

Case Reports of Suspected Drug or Supplement Interactions

No case reports of suspected drug or supplement interactions were identified.

Animal Trials of Drug or Supplement Interactions

No animal trials of drug or supplement interactions were identified.

II. Adverse Events

Case Reports of Adverse Events

Allergic contact dermatitis to dill, confirmed by patch testing, has been reported (Monteseirin et al. 2002, 2003).

III. Pharmacology and Pharmacokinetics

Human Pharmacological Studies

No relevant human pharmacological studies were identified.

Animal Pharmacological Studies

A decrease in serum glucose and insulin levels was observed in rats with corticosteroid-induced diabetes orally administered 100 mg/kg of dill leaf extract daily for 15 days (Panda 2008).

In female rats orally administered 0.045 or 0.45 g/kg of an aqueous extract of dill seed or 0.5 or 5 mg/kg of an ethanol extract of dill seed daily for 10 days, a significant increase was observed in the duration of the estrus cycle, the diestrus phase, and the progesterone concentration in the high-dose groups of both extracts. The stereological study did not reveal any significant changes in the volumes of ovaries, or primary, secondary, or graafian follicles (Monsefi et al. 2006).

In Vitro Pharmacological Studies

No relevant in vitro pharmacological studies were identified.

A

IV. Pregnancy and Lactation

No information on the safety of dill in pregnancy was identified. While this review did not identify any concerns for pregnant women, safety has not been conclusively established.

Dill seed has traditionally been used in India to promote milk production in nursing women (Jain et al. 2004).

V. Toxicity Studies

Acute Toxicity

The LD_{50} of intraperitoneally administered dill seed extracts in mice was 3.04 g/kg for the aqueous extract and 6.89 g/kg for the ethanolic extract (Hosseinzadeh et al. 2002).

The LD_{50} of dill seed oil in rats was 4.6 g/kg (Opdyke and Letizia 1982).

Guinea pigs intravenously administered 35 mg/kg of dill seed essential oil experienced anaphylactic shock and died. The same oil intravenously administered to cats at doses of 5 to 10 mg/kg resulted in lowered blood pressure and increased respiratory volume (Shipochliev 1968).

Genotoxicity

Dose-dependent induction of chromosomal aberrations and sister chromatid exchanges were observed in human lymphocytes treated with dill leaf or seed essential oil. No significant effects of dill seed essential oil were observed in the *Drosophila melanogaster* somatic mutation and recombination test (SMART) in vivo (Lazutka et al. 2001).

No genotoxic effects were observed in bone marrow cells of mice administered 1 g/kg dill seed essential oil daily (duration of dosing not specified in English language abstract). The seed oil protected against the genotoxic activity of benzo[*a*]pyrene (Morkunas 2002).

Cytotoxicity

Cytotoxic effects of dill seed and leaf essential oils were observed in human lymphocytes (Lazutka et al. 2001).

Literature Cited

Hosseinzadeh, H., G.R. Karimi, and M. Ameri. 2002. Effects of *Anethum graveolens* L. seed extracts on experimental gastric irritation models in mice. *BMC Pharmacol.* 2:21.

Jain, A., S.S. Katewa, B.L. Chaudhary, and P. Galav. 2004. Folk herbal medicines used in birth control and sexual diseases by tribals of southern Rajasthan, India. *J. Ethnopharmacol.* 90(1):171-177.

Lazutka, J.R., J. Mierauskiene, G. Slapsyte, and V. Dedonyte. 2001. Genotoxicity of dill (*Anethum graveolens* L.), peppermint (*Mentha piperita* L.) and pine (*Pinus sylvestris* L.) essential oils in human lymphocytes and *Drosophila melanogaster*. *Food Chem. Toxicol.* 39(5):485-492.

Mahran, G.H., H.A. Kadry, Z.G. Isaac, et al. 1991. Investigation of diuretic drug plants. 1. Phytochemical screening and pharmacological evaluation of *Anethum graveolens* L., *Apium graveolens* L., *Daucus carota* L. and *Eruca sativa* Mill. *Phytother. Res.* 5(4):169-172.

Mahran, G.H., H.A. Kadry, C.K. Thabet, et al. 1992. GC/MS analysis of volatile oil of fruits of *Anethum graveolens*. *Int. J. Pharmacog.* 30(2):139-144.

Monsefi, M., M. Ghasemi, and A. Bahaoddini. 2006. The effects of *Anethum graveolens* L. on female reproductive system. *Phytother. Res.* 20(10):865-868.

Monteseirin, J., J.L. Perez-Formoso, M. Hernandez, et al. 2003. Contact urticaria from dill. *Contact Dermat.* 48(5):275.

Monteseirin, J., J.L. Perez-Formoso, M.C. Sanchez-Hernandez, et al. 2002. Occupational contact dermatitis to dill. *Allergy* 57(9):866-867.

Morkunas, V. 2002. [Investigation of the genetic toxicology of dill essential oil and benzo(*a*)pyrene in mouse bone marrow by micronucleus test]. *Biologija* 4:14-16.

Opdyke, C., and C. Letizia. 1982. Flavor and fragrance raw materials. *Food Chem. Toxicol.* 20:633.

Panda, S. 2008. The effect of *Anethum graveolens* L. (dill) on corticosteroid induced diabetes mellitus: Involvement of thyroid hormones. *Phytother. Res.* 22(12):1695-1697.

Shipochliev, T. 1968. [Pharmacological study of several essential oils. I. Effect on the smooth muscle]. *Vet. Med. Nauk.* 5(6):63-69.

Angelica dahurica (Fisch. ex Hoffm.) Benth. & Hook. f. ex Franch. & Sav. Apiaceae

SCN: fragrant angelica
PN: *bai zhi* (root)
OCN: Dahurian angelica
Part: root

Quick Reference Summary

Safety Class: 1
Interaction Class: A

Contraindications

None known.

Other Precautions

None known.

Drug and Supplement Interactions

None known.

Editors' Note

Other species of *Angelica* are commonly traded as *bai zhi* (Bensky et al. 2004; Chang and But 1986).

Adverse Events and Side Effects

Topical allergic reactions and dermatitis caused by contact with the fresh herb have been reported (Bensky et al. 2004). Although furanocoumarin compounds, such as those present in fragrant angelica, have photosensitizing effects (a heightened response of the skin to sunlight or other ultraviolet light) after contact with skin, no cases of photosensitivity due to oral ingestion of fragrant angelica are known (Bensky et al. 2004); such a reaction is theoretically possible.

Pharmacological Considerations

An animal study indicated that fragrant angelica inhibited the drug-metabolizing isoenzymes CYP2C, CYP3A, CYP2D1, and CYP2C9 (Ishihara et al. 2000). The relevance of this information to human use is not known.

Pregnancy and Lactation

No information on the safety of fragrant angelica in pregnancy or lactation was identified in the scientific or traditional literature. Although this review did not identify any concerns for use while pregnant or nursing, safety has not been conclusively established.

Review Details

I. Drug and Supplement Interactions

Clinical Trials of Drug or Supplement Interactions

No clinical trials of drug or supplement interactions were identified.

Case Reports of Suspected Drug or Supplement Interactions

No case reports of suspected drug or supplement interactions were identified.

Animal Trials of Drug or Supplement Interactions

Coumarins isolated from fragrant angelica administered orally to rats in single doses of 25, 50, or 100 mg/kg increased the duration of hypnosis induced by pentobarbital sodium. The 100 mg/kg dose also prolonged the latent period and shortened the duration of hypnosis induced by barbital sodium (Liu et al. 2006).

II. Adverse Events

Case Reports of Adverse Events

Large doses (30–60 g) of fragrant angelica may cause slowed heart rate, increased blood pressure, increased depth of respiration, nausea, vomiting, dizziness, difficulty breathing, sweating, or numbness of the limbs (Bensky et al. 2004; Chen and Chen 2004). Cases of gross overdose have been associated with seizures and convulsions (Chen and Chen 2004).

Although several sources indicate that persons ingesting fragrant angelica should limit exposure to the sun, due to the furanocoumarin content of the herb, no cases of photosensitivity associated with oral ingestion are known (Bensky et al. 2004). Topical allergic reactions and dermatitis caused by contact with the fresh herb have been reported (Bensky et al. 2004). Two cases of phytophotodermatitis associated with accidental skin contact with a multi-herb decoction including fragrant angelica have been reported (Zhang and Zhu 2011).

III. Pharmacology and Pharmacokinetics

Human Pharmacological Studies

No relevant human pharmacological studies were identified.

Animal Pharmacological Studies

Inhibition of the drug-metabolizing isoenzymes CYP2C, CYP3A, and CYP2D1 was observed in rats orally administered a single dose of 1 g/kg of fragrant angelica. At the same dose level, fragrant angelica prolonged the half-life and reduced the clearance of intravenously administered tolbutamide (a CYP2C9 substrate) (Ishihara et al. 2000).

The furanocoumarin compound phellopterin was shown to be a partial agonist of the central benzodiazepine receptors. Two other naturally occurring furanocoumarins, byakangelicol and imperatorin, exhibited significantly less potent binding activity to the benzodiazepine site of the rat brain GABA-A receptor (Dekermendjian et al. 1996).

In Vitro Pharmacological Studies

Infusions of fragrant angelica inhibited the drug-metabolizing isoenzyme CYP3A4. Activity varied according to horticultural variety, and the infusions were significantly more active than the 40% ethanol extract of the different products (Guo et al. 2001).

Of 11 furanocoumarin compounds isolated from fragrant angelica, 2 showed strong abilities to induce alkaline phosphatase with EC_{50} values of 1.1 and 0.8 μg/ml, respectively, whereas the other 9 furanocoumarins were weakly or only slightly active (Piao et al. 2006).

Fragrant angelica inhibited melanogenesis in B16 melanoma cells (Cho et al. 2006).

Of eight furanocoumarin compounds isolated from fragrant angelica, one strongly inhibited (IC_{50} of 0.36 μ*M*) binding of diazepam to central nervous system benzodiazepine receptors, while other structurally similar compounds had minimal effect on receptor binding (Bergendorff et al. 1997).

A

IV. Pregnancy and Lactation

No information on the safety of fragrant angelica in pregnancy or lactation was identified.

V. Toxicity Studies

Acute Toxicity

The LD_{50} of an orally administered decoction of fragrant angelica in mice was 53.82 g/kg (Chen and Chen 2004).

Literature Cited

Bensky, D., S. Clavey, and E. Stöger. 2004. *Chinese herbal medicine: Materia medica*. 3rd ed. Seattle: Eastland Press.

Bergendorff, O., K. Dekermendjian, M. Nielsen, et al. 1997. Furanocoumarins with affinity to brain benzodiazepine receptors in vitro. *Phytochemistry* 44(6):1121-1124.

Chang, H.-M., and P.P.H. But. 1986. *Pharmacology and applications of Chinese materia medica*. English ed. Singapore, Philadelphia: World Scientific.

Chen, J.K., and T.T. Chen. 2004. *Chinese medical herbology and pharmacology*. City of Industry, CA: Art of Medicine Press.

Cho, Y.H., J.H. Kim, S.M. Park, et al. 2006. New cosmetic agents for skin whitening from *Angelica dahurica*. *J. Cosmet. Sci.* 57(1):11-21.

Dekermendjian, K., J. Ai, M. Nielsen, et al. 1996. Characterisation of the furanocoumarin phellopterin as a rat brain benzodiazepine receptor partial agonist in vitro. *Neurosci. Lett.* 219 (3):151-154.

Guo, L.Q., M. Taniguchi, Q.Y. Chen, K. Baba, and Y. Yamazoe. 2001. Inhibitory potential of herbal medicines on human cytochrome P450-mediated oxidation: Properties of umbelliferous or citrus crude drugs and their relative prescriptions. *Jpn. J. Pharmacol.* 85(4):399-408.

Ishihara, K., H. Kushida, M. Yuzurihara, et al. 2000. Interaction of drugs and Chinese herbs: Pharmacokinetic changes of tolbutamide and diazepam caused by extract of *Angelica dahurica*. *J. Pharm. Pharmacol.* 52(8):1023-1029.

Liu, Z., J. Li, and J. Wu. 2006. [Influence of coumarins from Radix Angelicae Dahuricae on the hypnotic effects of pentobarbital sodium and barbital sodium in mice]. *Med. J. Wuhan Univ.* 27(1):63-65.

Piao, X.L., H.H. Yoo, H.Y. Kim, et al. 2006. Estrogenic activity of furanocoumarins isolated from *Angelica dahurica*. *Arch. Pharm. Res.* 29(9):741-745.

Zhang, R., and W. Zhu. 2011. Phytophotodermatitis due to Chinese herbal medicine decoction. *Indian J. Dermatol.* 56(3):329-321.

Angelica pubescens Maxim. — Apiaceae

SCN: pubescent angelica
PN: *du huo* (root)

Part: root

Quick Reference Summary

Safety Class: 2d
Interaction Class: A

Contraindications

Avoid prolonged exposure to sunlight (Chen and Chen 2004).

Other Precautions

None known.

Drug and Supplement Interactions

None known.

Notice

Photosensitizing (Chang and But 1986; Chen and Chen 2004); *see* Appendix 2.

Editors' Note

Other species of *Angelica* and plants of other genera are commonly traded as *du huo*.

Adverse Events and Side Effects

Gastric discomfort, nausea, vomiting, dizziness, headache, and numb tongue have been reported in association with pubescent angelica use, although detailed information on these cases is lacking (Bensky et al. 2004).

Pharmacological Considerations

Overdoses (standard therapeutic dose is 3–9 g as a decoction) of pubescent angelica have been associated with increased sensitivity to sunlight (Chen and Chen 2004).

Pregnancy and Lactation

No information on the safety of pubescent angelica in pregnancy or lactation was identified in the scientific or traditional literature. Although this review did not identify any concerns for use while pregnant or nursing, safety has not been conclusively established.

A

Review Details

I. Drug and Supplement Interactions

Clinical Trials of Drug or Supplement Interactions

No clinical trials of drug or supplement interactions were identified.

Case Reports of Suspected Drug or Supplement Interactions

No case reports of suspected drug or supplement interactions were identified.

Animal Trials of Drug or Supplement Interactions

No animal trials of drug or supplement interactions were identified.

II. Adverse Events

Case Reports of Adverse Events

During the use of pubescent angelica in the treatment of bronchitis, adverse events of gastric discomfort, nausea, vomiting, dizziness, headache, and numb tongue have been reported. Details on these reports are lacking (Bensky et al. 2004).

Overdose of pubescent angelica has been associated with increased photosensitivity (Chen and Chen 2004). Overdose of the dried root or "large" doses (standard therapeutic dose is 3–9 g as a decoction) of the fresh root of pubescent angelica have been reported to cause rapid breathing, agitation, hallucinations, delirium, tetanic spasms, and, in severe cases, death. Details on these cases, including doses used and relevant medical history, are lacking (Bensky et al. 2004).

III. Pharmacology and Pharmacokinetics

Human Pharmacological Studies

No relevant human pharmacological studies were identified.

Animal Pharmacological Studies

No relevant animal pharmacological studies were identified.

In Vitro Pharmacological Studies

The compound osthole, isolated from pubescent angelica, inhibited platelet aggregation and ATP release induced by ADP, arachidonic acid, PAF, collagen, ionophore A23187, and thrombin in washed rabbit platelets. It showed a weak activity in platelet-rich plasma (Ko et al. 1989).

IV. Pregnancy and Lactation

No information on the safety of pubescent angelica in pregnancy or lactation was identified.

V. Toxicity Studies

Acute Toxicity

The LD_{50} of the compound xanthotoxin intramuscularly administered in rats is 160 mg/kg, while that of bergapten is 945 mg/kg (Chen and Chen 2004).

Literature Cited

Bensky, D., S. Clavey, and E. Stöger. 2004. *Chinese herbal medicine: Materia medica*. 3rd ed. Seattle: Eastland Press.

Chang, H.-M., and P.P.H. But. 1986. *Pharmacology and applications of Chinese materia medica*. English ed. Singapore, Philadelphia, PA: World Scientific.

Chen, J.K., and T.T. Chen. 2004. *Chinese medical herbology and pharmacology*. City of Industry, CA: Art of Medicine Press.

Ko, F.N., T.S. Wu, M.J. Liou, T.F. Huang, and C.M. Teng. 1989. Inhibition of platelet thromboxane formation and phosphoinositides breakdown by osthole from *Angelica pubescens*. *Thromb. Haemost.* 62(3):996-999.

Angelica sinensis (Oliv.) Diels

Apiaceae

SCN: dong quai
Syn: *Angelica polymorpha* Maxim. var. *sinensis* Oliv.
PN: *dang gui shen* (root body); *dang gui tou* (root head); *dang gui wei* (root tail)
OCN: Chinese angelica; tang kuei
Part: root

Quick Reference Summary

Safety Class: 1
Interaction Class: C

Contraindications

None known.

Other Precautions

Patients undergoing surgery are advised to stop use of dong quai 7 days before surgery due to reported inhibition of platelet aggregation (Li and Yang 1982; Tu and Huang 1984; Yan et al. 1987).

Drug and Supplement Interactions

Human case reports and an animal study have indicated a possible interaction between warfarin and dong quai (Ellis and Stephens 1999; Lo et al. 1995; Page and Lawrence 1999).

Adverse Events and Side Effects

Adverse events reported in persons taking dong quai include aggravation of endometriosis, severe bleeding of gums (high doses for many months), overstimulation of menstrual cycle, increased or excessive menstrual flow, edema and breast tenderness, headaches, increased irritability, gynecomastia, and occupational asthma (Goh and Loh 2001; Lee et al. 2001; Upton 2003). A causal association between these adverse events and dong quai was not determined in any of the cases. No significant adverse events are reported in the traditional Chinese medicine literature (Bensky et al. 2004; Chen and Chen 2004).

Pharmacological Considerations

Inhibition of platelet aggregation by dong quai extract has been observed in human and animal studies (Dong et al. 2004; Li and Yang 1982; Tu and Huang 1984; Yan et al. 1987).

Although conflicting results on the estrogenicity of dong quai have been reported in animal and in vitro studies (Circosta et al. 2006; Dixon-Shanies and Shaikh 1999; Lau et al. 2005), a human study of dong quai in menopausal women demonstrated no estrogenic activity (Hirata et al. 1997), and a binding and transcription study showed no activity of dong quai in estrogen receptors ERα or ERβ (Amato et al. 2002).

Induction of drug-metabolizing enzymes CYP2D6 and CYP3A was observed in rats administered dong quai extracts (Tang et al. 2006).

Pregnancy and Lactation

Limited information on the safety of dong quai during pregnancy or lactation is available. Both a stimulating and relaxing effect on the uterus have been reported for constituents of dong quai in vitro (Zhu 1998). In traditional Chinese medicine practices, dong quai is used in combination with other herbs at various stages of pregnancy (Upton 2003). In the West, dong quai is commonly used as a single herb; some texts contraindicate dong quai during pregnancy (Brinker 2001; Mills and Bone 2005), but no contraindication during pregnancy is noted in either of the Chinese herbal references cited in this work (Bensky et al. 2004; Chen and Chen 2004).

Cases of rashes and high blood pressure during lactation have been reported (Nambiar et al. 1999; Upton 2003).

Review Details

I. Drug and Supplement Interactions

Clinical Trials of Drug or Supplement Interactions

No clinical trials of drug or supplement interactions were identified.

Case Reports of Suspected Drug or Supplement Interactions

A woman taking warfarin, digoxin, and furosemide had an increased INR (a standardized scale used to report the results of blood coagulation tests; increased INR indicates slowed blood clotting) after having taken one or two tablets (565 mg each) daily of dong quai. The woman's INR returned to the normal range after cessation of dong quai (Page and Lawrence 1999). Increased INR and bleeding were reported in a woman taking warfarin and dong quai (dose, duration, and product unspecified) (Ellis and Stephens 1999).

Animal Trials of Drug or Supplement Interactions

In rabbits orally administered 4 g/kg daily dong quai and subcutaneously administered warfarin, an increase in prothrombin time was observed in a steady-state model of warfarin administration, whereas in a single-dose model, no significant changes in prothrombin time were observed. Dong quai alone did not produce any changes in prothrombin time (Lo et al. 1995).

II. Adverse Events

Adverse Events Reported in Clinical Trials

A review of the available traditional and scientific data indicated that dong quai is a very safe herb with a low probability of side effects (Upton 2003).

Case Reports of Adverse Events

A case of gynecomastia was reported in a man who had taken dong quai capsules (dose unspecified) daily for 1 month. The man was on long-term treatment with phenytoin and folate. The gynecomastia resolved completely several months after cessation of the dong quai (Goh and Loh 2001). Several case reports have indicated an association of phenytoin use with gynecomastia (Ikeda et al. 1998; Monson and Scott 1987; Rossi et al. 1983).

Results of a survey of members of the American Herbalists Guild indicated that adverse events observed during the use of dong quai included aggravation of endometriosis, overstimulation of menstrual cycle, increased or excessive menstrual flow, edema and breast tenderness, headaches (sometimes severe), increased irritability, and severe bleeding gums (high doses for many months) (Upton 2003).

A case of occupational asthma was reported in a pharmacist who had worked in an herb shop processing herbal materials. The patient had a positive response to dong quai ("Angelica radix," species unspecified) in a skin prick test, and bronchoprovocation tests showed an early asthmatic response to dong quai extract (Lee et al. 2001).

III. Pharmacology and Pharmacokinetics

Human Pharmacological Studies

In patients with ulcerative colitis, intravenous administration of 40 ml daily of dong quai extract inhibited platelet activation (Dong et al. 2004). Intravenous administration of 200 ml daily of a 25% solution of dong quai for 20 days to patients with acute ischemic stroke resulted in decreases in a number of hematological parameters including platelet adhesion rate, platelet electrophoresis time, and whole blood adhesion, and an increase in prothrombin time (Tu and Huang 1984). A reduction in whole blood viscosity was observed for approximately 3 hours in people orally administered a dong quai extract (Terasawa et al. 1985).

In a human study of dong quai in menopausal women, involving administration of 4.5 g daily of dong quai for 24 weeks, no changes in serum hormone levels, vaginal cytology, or endometrial thickness were observed, indicating a lack of estrogenic effects (Hirata et al. 1997).

Animal Pharmacological Studies

In rats administered 3 g/kg dong quai aqueous or ethanolic extracts, both extracts significantly induced the drug-metabolizing isoenzymes CYP2D6 and CYP3A (Tang et al. 2006).

Injection of 1 g/ml dong quai prevented platelet aggregation and erythrocyte stasis in alveolar capillaries induced by the injection of measles vaccine (Yan et al. 1987). Administration of two doses of 6 or 8 ml each (1.5 hours between doses) of an aqueous extract of dong quai to fasting rats resulted in a decrease in time to thrombus formation and an increase in prothrombin time (Li and Yang 1982).

No increase in uterine weight was observed in ovariectomized mice orally administered 500 µl daily of an ethanolic extract of dong quai for 4 days (Amato et al. 2002). In ovariectomized rats administered 100 or 300 mg/kg daily of a dried ethanolic extract of dong quai or 1 µg/rat estradiol, a significant increase in uterine weight was observed in rats administered dong quai, but the increase was significantly less than that observed in rats administered estradiol. Mean luteinizing hormone levels were reduced in rats treated with dong quai, and no changes were observed in follicle-stimulating hormone levels. These results suggest an estrogenic effect of dong quai (Circosta et al. 2006).

In Vitro Pharmacological Studies

In a study of a dong quai ethanolic extract on MCF-7 (ER positive) human breast cancer cells, dong quai induced the growth of MCF-7 cells by 16-fold over that of untreated control cells. In a transient gene expression assay system, dong quai failed to show transactivation of either estrogen receptor ERα or ERβ cDNA (Amato et al. 2002). An aqueous extract of dong quai dose-dependently stimulated the proliferation of MCF-7 and BT-20 (ER negative) breast cancer cells (Lau et al. 2005). An extract of dong quai inhibited growth of T-47D (ER positive) and MCF-7 human breast cancer cells (Dixon-Shanies and Shaikh 1999).

IV. Pregnancy and Lactation

A survey of 593 pregnant women in Hong Kong indicated that 12% had used "Chinese angelica" during their pregnancy. This was the second most popular herb for use during pregnancy in the survey. No data on maternal or infant health was collected in the survey (Ong et al. 2005).

Subcutaneous administration of 0.1 to 0.4 ml daily of an aqueous dong quai extract to mice for 5 days did not affect fertility or show any teratogenic effects (Matsui et al. 1967).

Dong quai essential oil applied to isolated uteri produced a relaxant effect while the nonvolatile fractions produced a stimulant effect, potentiating uterine contraction (Zhu 1998).

Three unpublished cases of rashes have been reported in breast-feeding newborn infants whose mothers were taking dong quai (dose, duration, and preparation unreported) (Upton 2003).

A breast-feeding woman and her 3-week-old infant developed high blood pressure after the woman consumed an unspecified amount of dong quai prepared in a soup. Blood pressure of both the woman and her infant returned to normal 48 hours after cessation of dong quai (Nambiar et al. 1999).

V. Toxicity Studies

Acute Toxicity

The LD_{50} of orally administered dong quai concentrated extract (8:1 to 16:1) in rats is 100 g/kg (Zschocke et al. 1998); of intravenously administered aqueous extract in mice is 100 g/kg (Wei 1987); and of orally administered 50% ethanol extract in mice is over 40 g/kg (Yang and Chen 1992). The LD_{50} of 4:1 concentrated extract of dong quai in rats is 100 g/kg (route of administration unspecified) (Zhu 1987). No toxic effects of dong quai extract were observed in rats or mice after single doses of 6 g/kg (Tanaka et al. 1983).

The LD_{50} of intravenously administered ferulic acid in mice is reported to be 856.6 mg/kg (Ozaki and Ma 1990); of intraperitoneally administered ligustilide in mice is approximately 410 mg/kg (Xie and Tao 1985), and of orally administered 3-*n*-butylidene phthalide in rats is 2.45 g/kg (Opdyke 1979).

In a review of the toxicology literature on dong quai, intravenous injection of the volatile fraction of dong quai was reported to cause kidney degeneration (Mei et al. 1991).

Short-Term Toxicity

After 21 days of oral administration of 1.5 or 3 g/kg dong quai extract daily to rats, increases in serum free cholesterol levels and kidney CYP450 enzymes were observed (Tanaka et al. 1983).

A

Genotoxicity

No genotoxic effects of dong quai were observed in a salmonella genotoxicity assay (NTP 2008).

Literature Cited

Amato, P., S. Christophe, and P.L. Mellon. 2002. Estrogenic activity of herbs commonly used as remedies for menopausal symptoms. *Menopause* 9(2):145-150.

Bensky, D., S. Clavey, and E. Stöger. 2004. *Chinese herbal medicine: Materia medica*. 3rd ed. Seattle: Eastland Press.

Brinker, F. 2001. *Herb contraindications and drug interactions*. 3rd ed. Sandy, OR: Eclectic Medical Publications.

Chen, J.K., and T.T. Chen. 2004. *Chinese medical herbology and pharmacology*. City of Industry, CA: Art of Medicine Press.

Circosta, C., R.D. Pasquale, D.R. Palumbo, S. Samperi, and F. Occhiuto. 2006. Estrogenic activity of standardized extract of *Angelica sinensis*. *Phytother. Res.* 20(8):665-669.

Dixon-Shanies, D., and N. Shaikh. 1999. Growth inhibition of human breast cancer cells by herbs and phytoestrogens. *Oncol. Rep.* 6(6):1383-1387.

Dong, W.G., S.P. Liu, H.H. Zhu, H.S. Luo, and J.P. Yu. 2004. Abnormal function of platelets and role of *Angelica sinensis* in patients with ulcerative colitis. *World J. Gastroenterol.* 10(4):606-609.

Ellis, G., and M. Stephens. 1999. Untitled. *Br. Med. J.* 319:650.

Goh, S., and K. Loh. 2001. Gynaecomastia and the herbal tonic "dong quai." *Singapore Med. J.* 42(3):155-156.

Hirata, J.D., L.M. Swiersz, B. Zell, R. Small, and B. Ettinger. 1997. Does dong quai have estrogenic effects in postmenopausal women? A double-blind, placebo-controlled trial. *Fertil. Steril.* 68 (6):981-986.

Ikeda, A., H. Hattori, A. Odani, J. Kimura, and H. Shibasaki. 1998. Gynaecomastia in association with phenytoin and zonisamide in a patient having a CYP2C subfamily mutation. *J. Neurol. Neurosurg. Psychiat.* 65(5):803-804.

Lau, C.B., T.C. Ho, T.W. Chan, and S.C. Kim. 2005. Use of dong quai (*Angelica sinensis*) to treat peri- or postmenopausal symptoms in women with breast cancer: Is it appropriate? *Menopause* 12(6):734-740.

Lee, S.K., H.K. Cho, S.H. Cho, et al. 2001. Occupational asthma and rhinitis caused by multiple herbal agents in a pharmacist. *Ann. Allergy Asthma Immunol.* 86(4):469-474.

Li, C., and S. Yang. 1982. The influence of yimucao, chishao, danggui, sanleng, erzhu and zelan upon the blood coagulation of rats. *Zhongxiyi Jiehe Zazhi* 69(2):111-112.

Lo, A.C.T., K. Chan, J.H.K. Yeung, and K.S. Woo. 1995. Danggui (*Angelica sinensis*) affects the pharmacodynamics but not the pharmacokinetics of warfarin in rabbits. *Eur. J. Drug Metab. Pharmacokinet.* 20(1):55-60.

Matsui, A., J. Rogers, Y. Woo, and W. Cutting. 1967. Effects of some natural products on fertility in mice. *Med. Pharmacol. Exp. Int. J. Exp. Med.* 16(5):414-424.

Mills, S., and K. Bone. 2005. *The essential guide to herbal safety*. St. Louis: Elsevier.

Monson, J.P., and D.F. Scott. 1987. Gynaecomastia induced by phenytoin in men with epilepsy. *Br. Med. J.* 294(6572):612.

Nambiar, S., R. Schwartz, and A. Constantino. 1999. Hypertension in mother and baby linked to ingestion of Chinese herbal medicine. *West. J. Med.* 171:152.

NTP. 2008. Dong quai: Genetic toxicology. National Toxicology Program.

Ong, C.O., L.Y. Chan, P.B. Yung, and T.N. Leung. 2005. Use of traditional Chinese herbal medicine during pregnancy: A prospective survey. *Acta Obstet. Gynecol. Scand.* 84(7):699-700.

Opdyke, D.L.J. 1979. *Monographs on fragrance raw materials*. New York: Pergamon Press.

Page, R.L. 2nd, and J.D. Lawrence. 1999. Potentiation of warfarin by dong quai. *Pharmacotherapy* 19(7):870-876.

Rossi, L., U. Bonuccelli, G. Marcacci, et al. 1983. Gynecomastia in epileptics treated with phenobarbital, phenytoin and fluoresone: Two case reports. *Ital. J. Neurol. Sci.* 4(2):207-210.

Tanaka, S., A. Takahashi, and K. Onoda. 1983. [Toxicological studies on biological effects of the herbal drug extracts in rats and mice—peony root, peach kernel, Japanese angelica root and *Cnidium* rhizome]. *Yakugaku Zasshi* 103(9):937-955.

Tang, J.C., J.N. Zhang, Y.T. Wu, and Z.X. Li. 2006. Effect of the water extract and ethanol extract from traditional Chinese medicines *Angelica sinensis* (Oliv.) Diels, *Ligusticum chuanxiong* Hort. and *Rheum palmatum* L. on rat liver cytochrome P450 activity. *Phytother. Res.* 20(12):1046-1051.

Terasawa, K., A. Imadaya, H. Tosa, et al. 1985. Chemical and clinical evaluation of crude drugs derived from *Angelica acutilobae* and *Angelica sinensis*. *Fitoterapia* 56(4):201-208.

Tu, J., and H. Huang. 1984. Effects of radix Angelicae sinensis on hemorrheology in patients with acute ischemic stroke. *Zhongyi Zazhi* 4(3):225-228.

Upton, R. 2003. *Dang gui root: Angelica sinensis (Oliv.) Diels: Standards of analysis, quality control, and therapeutics, American Herbal Pharmacopoeia and therapeutic compendium*. Scotts Valley, CA: American Herbal Pharmacopoeia.

Yan, T., A. Hou, G. Zhou, et al. 1987. Pharmacological effects of *Angelica* injection and its treatment of infantile viral pneumonia. *Zhongguo Zhongxiyi Jiehe Zazhi* 7(3):161-162.

Zhu, D.P. 1987. Dong quai. *Am. J. Chin. Med.* 15(3-4):117-125.

Zhu, Y.P. 1998. *Chinese materia medica chemistry, pharmacology and applications*. Boca Raton, FL: CRC Press.

Zschocke, S., J.-H. Liu, H. Stuppner, and R. Bauer. 1998. Comparative study of roots of *Angelica sinensis* and related umbelliferous drugs by thin layer chromatography, high-performance liquid chromatography, and liquid chromatography-mass spectrometry. *Phytochem. Anal.* 9(6):283-290.

Angelica spp. — Apiaceae

Angelica archangelica **L.**
SCN: angelica
Syn: *Angelica officinalis* Moench; *Archangelica officinalis* (Moench) Hoffm.
AN: *chanda; chirakabheda*
OCN: archangel; European angelica

Angelica atropurpurea **L.**
SCN: purple angelica
OCN: alexanders; American angelica; great angelica; purple-stem angelica
Part: fruit (commonly known as "seed"), root

Quick Reference Summary

Safety Class: 2b
Interaction Class: A

Contraindications

Not for use in pregnancy except under the supervision of a qualified healthcare practitioner (Felter and Lloyd 1898).

Other Precautions

None known.

Drug and Supplement Interactions

None known.

Notice

Emmenagogue (Felter and Lloyd 1898; Puri 1971); *see* Appendix 2.

Adverse Events and Side Effects

Although furanocoumarin compounds, such as those present in angelica or purple angelica, have photosensitizing effects (a heightened response of the skin to sunlight or other ultraviolet light) after contact with the skin, and some sources indicate that persons taking angelica or purple angelica should avoid prolonged exposure to sunlight (Blumenthal et al. 1998; Williamson 2003), no cases of phototoxicity associated with internal use of angelica or purple angelica were identified in the literature, although such a reaction is theoretically possible.

Pharmacological Considerations

None known.

Pregnancy and Lactation

Angelica and purple angelica have traditionally been used as emmenagogues (Felter and Lloyd 1898; Puri 1971). Based on this, use during pregnancy is not recommended except under the supervision of a qualified healthcare practitioner.

No information on the safety of angelica in lactation was identified. While this review did not identify any concerns for use while nursing, safety has not been conclusively established.

Review Details

I. Drug and Supplement Interactions

Clinical Trials of Drug or Supplement Interactions

No clinical trials of drug or supplement interactions were identified.

Case Reports of Suspected Drug or Supplement Interactions

No case reports of suspected drug or supplement interactions were identified.

Animal Trials of Drug or Supplement Interactions

No animal trials of drug or supplement interactions were identified.

II. Adverse Events

Case Reports of Adverse Events

No case reports of adverse events were identified.

III. Pharmacology and Pharmacokinetics

Human Pharmacological Studies

No sensitization reactions were observed after 1% angelica root or seed oil in petroleum was applied to volunteers in a 48 hour closed patch test (Opdyke 1979).

Animal Pharmacological Studies

Among seven plants generally recognized to cause phytophotodermatitis, angelica had the highest concentration of coumarin compounds on the interior surface of the leaf. The concentration was greater than that of *Psoralea bituminosa,* although *P. bituminosa* had the highest concentration of coumarin on the leaf surface, as compared to other species (Zobel and Brown 1991).

A mild reaction was observed in rabbits topically treated with 500 mg of angelica seed essential oil (Opdyke 1974).

In Vitro Pharmacological Studies

No relevant in vitro pharmacological studies were identified.

A

IV. Pregnancy and Lactation

Among other uses, angelica has traditionally been used as an emmenagogue (Felter and Lloyd 1898).

No information on the safety of angelica during lactation was identified.

V. Toxicity Studies

Acute Toxicity

The LD_{50} of orally administered angelica root essential oil in rats is 11.16 g/kg (Skramlik 1959), and in mice is 2.2 g/kg (Opdyke 1975).

The dermal LD_{50} for angelica root essential oil in rabbits could not be determined at doses up to 5 g/kg (Opdyke 1975).

Genotoxicity

No genotoxic effects of angelica were observed in the micronucleus test in mouse bone marrow and peripheral blood cells (Salikhova et al. 1993). Antimutagenic activity of an alcohol extract of angelica was observed in the micronucleus test in murine bone marrow cells (Salikhova et al. 1993; Salikhova and Poroshenko 1995).

Literature Cited

Blumenthal, M., W. Busse, A. Goldberg, et al. 1998. *The complete German Commission E monographs*. Austin, TX: American Botanical Council.

Felter, H.W., and J.U. Lloyd. 1898. *King's American dispensatory*. 18th ed., 3rd rev. 2 vols. Cincinnati: Ohio Valley Co.

Opdyke, D. 1974. Fragrance raw materials monographs: Angelica seed oil. *Food Cosmet. Toxicol.* 12(7-8):821.

Opdyke, D. 1975. Angelica root oil. *Food Cosmet. Toxicol.* 13(6):713.

Opdyke, D.L.J. 1979. *Monographs on fragrance raw materials*. New York: Pergamon Press.

Puri, H.S. 1971. Comparative study of folk lore vegetable drugs of Europe and India. *Acta Phytother.* 18:21-23.

Salikhova, R.A., N. Dulatova, and G.G. Poroshenko. 1993. [Studies of *Angelica archangelica* L. antimutagen properties by the micronucleus test]. *Byull. Eksper. Biol. I Med.* 115(4):371-372.

Salikhova, R.A., and G.G. Poroshenko. 1995. Antimutagenic properties of *Angelica archangelica* L. *Vestn. Ross. Akad. Med. Nauk.* 1:58-61.

Skramlik, V.E.V. 1959. Über die giftigkeit und verträglichkeit von ätherischen ölen. *Pharmazie* 14:435-445.

Williamson, E.M. 2003. *Potter's herbal cyclopedia*. Saffron Walden, Essex: C.W. Daniel Co.

Zobel, A.M., and S.A. Brown. 1991. Dermatitis-inducing psoralens on the surfaces of seven medicinal plant species. *J. Toxicol. Cutan. Ocular Toxicol.* 10(3):223-231.

Anthriscus cerefolium (L.) Hoffm. — Apiaceae

SCN: chervil
OCN: garden chervil

Part: herb

Quick Reference Summary

Safety Class: 2b
Interaction Class: A

Contraindications

Not for use in pregnancy except under the supervision of a qualified healthcare practitioner (Chadha 1988; List and Hörhammer 1973).

Other Precautions

None known.

Drug and Supplement Interactions

None known.

Notice

Emmenagogue (List and Hörhammer 1973; Remington and Wood 1918); *see* Appendix 2.

Editors' Note

The classifications and concerns for this herb are based on relatively higher doses used for therapeutic purposes in contrast to lower amounts generally used in cooking, and have not been associated with its use as a spice.

Adverse Events and Side Effects

None known.

Pharmacological Considerations

None known.

Pregnancy and Lactation

Based on traditional use as an emmenagogue (Chadha 1988; List and Hörhammer 1973; Remington and Wood 1918), two herbal reference texts indicate that chervil should not be used during pregnancy (Chadha 1988; List and Hörhammer 1973). No other information on the use of chervil during pregnancy was identified.

No information on the use of chervil during lactation was identified. While this review did not identify any concerns for use while nursing, safety has not been conclusively established.

Review Details

I. Drug and Supplement Interactions

Clinical Trials of Drug or Supplement Interactions

No clinical trials of drug or supplement interactions were identified.

Case Reports of Suspected Drug or Supplement Interactions

No case reports of suspected drug or supplement interactions were identified.

Animal Trials of Drug or Supplement Interactions

No animal trials of drug or supplement interactions were identified.

II. Adverse Events

Case Reports of Adverse Events

No case reports of adverse events were identified.

III. Pharmacology and Pharmacokinetics

Human Pharmacological Studies

No relevant human pharmacological studies were identified.

Animal Pharmacological Studies

No relevant animal pharmacological studies were identified.

In Vitro Pharmacological Studies

No relevant in vitro pharmacological studies were identified.

IV. Pregnancy and Lactation

Based on traditional use as an emmenagogue (Chadha 1988; List and Hörhammer 1973; Remington and Wood 1918), two herbal reference texts indicate that chervil should not be used during pregnancy (Chadha 1988; List and Hörhammer 1973). No other information on the use of chervil during pregnancy or lactation was identified.

V. Toxicity Studies

No toxicity studies were identified.

Literature Cited

Chadha, Y. 1988. *The wealth of India: A dictionary of Indian raw materials and industrial products*. Delhi: Council of Scientific and Industrial Research.

List, P.H., and H. Hörhammer. 1973. *Hagers handbuch der pharmazeutischen praxis*. Vollst. 4. Neuausg. ed. Berlin: Springer.

Remington, J.P., and H.C. Wood. 1918. *The dispensatory of the United States of America*. 20th ed. Philadelphia: Lippincott.

Apium graveolens L. — Apiaceae

SCN: celery
OCN: wild celery

Part: fruit (commonly known as "seed")

Quick Reference Summary

Safety Class: 2b
Interaction Class: A

Contraindications

Not for use in pregnancy except under the supervision of a qualified healthcare practitioner (Bradley 1992).

Other Precautions

Celery seed should be used with caution by individuals with inflammation of the kidneys or a history of irritation of the kidneys (Bradley 1992; Wichtl 2004).

Drug and Supplement Interactions

None known.

Notice

Photosensitizing (Ahluwalia et al. 1988; Garg et al. 1978, 1979; Schimmer 1983); *see* Appendix 2.

Editors' Note

The classifications and concerns for this herb are based on relatively higher doses used for therapeutic purposes in contrast to lower amounts generally used in cooking, and have not been associated with its use as a spice.

Adverse Events and Side Effects

Allergic reactions, including anaphylactic reactions, to celery have been reported (Ballmer-Weber et al. 2000; Barg et al. 2008; Darsow et al. 2004; Luttkopf et al. 2000; Pauli et al. 1988; Vilke 2002).

Phototoxicity has been reported, primarily in persons handling fresh plants or ingesting large quantities of celery root followed by exposure to high-intensity UV radiation (UV therapy or tanning salon) (Birmingham et al. 1961; Jeanmougin et al. 2005; Ljunggren 1990; Schimmer 1983). No reports of phototoxicity in association with seed ingestion were identified. Furanocoumarin compounds present in celery seed, stem, and leaf are responsible for the phototoxicity (Beattie et al. 2007; Beier et al. 1983; Birmingham et al. 1961; Schimmer 1983).

Pharmacological Considerations

None known.

Pregnancy and Lactation

No adverse effects of celery seed essential oil on fertility were observed in two animal studies (Garg et al. 1970; Sharma et al. 1983). Apiol, a compound found in celery essential oil, has been reported to have abortifacient activity (Kochmann 1931; van Itallie et al. 1932; Wichtl 2004).

No information on the safety of celery seed during lactation was identified. While this review did not identify any concerns for use while nursing, safety has not been conclusively established.

Review Details

I. Drug and Supplement Interactions

Clinical Trials of Drug or Supplement Interactions

No clinical trials of drug or supplement interactions were identified.

Case Reports of Suspected Drug or Supplement Interactions

No case reports of suspected drug or supplement interactions were identified.

Animal Trials of Drug or Supplement Interactions

No animal trials of drug or supplement interactions were identified.

II. Adverse Events

Case Reports of Adverse Events

Anaphylactic reactions to celery have been reported, including two cases of food-dependent, exercise-induced anaphylaxis after ingestion of celery (Barg et al. 2008; Vilke 2002). One case of anaphylactic reaction to the drug oseltamivir (Tamiflu) occurred in a patient with celery-carrot-mugwort-spice syndrome (Hirschfeld et al. 2008).

Allergic reactions to celery have been confirmed in double-blind studies and by patch testing (Ballmer-Weber et al. 2000; Darsow et al. 2004; Ermertcan et al. 2007; Luttkopf et al. 2000; Pauli et al. 1988).

Phototoxicicity was reported in a woman who consumed large amounts of celery root prior to visiting a suntan parlor (Ljunggren 1990). A similar reaction was reported in a patient undergoing photochemotherapy (UVA) treatment (Jeanmougin et al. 2005).

Cases of mild to severe photodermatitis have been reported in persons handling fresh celery plants (Beier et al. 1983; Birmingham et al. 1961; Finkelstein et al. 1994; Seligman et al. 1987).

III. Pharmacology and Pharmacokinetics

Human Pharmacological Studies

Birch-celery syndrome has been described, with individuals sensitized to birch (*Betula* spp.) pollen or mugwort (*Artemisia vulgaris*) pollen frequently displaying type I allergic symptoms after ingestion of celery (Breiteneder et al. 1995; Luttkopf et al. 2000).

No phototoxic reactions were observed in healthy volunteers administered 200 g raw or cooked celery stem or parsnip followed by exposure to UVA light (Beattie et al. 2007).

Animal Pharmacological Studies

In female rats administered 200 mg/kg of an ethanol extract of celery seed daily for 10 days, a lengthening of the estrus phase was observed, while administration for 20 or 30 days resulted in a decrease in other menstrual phases (Barethia 2008).

In Vitro Pharmacological Studies

No relevant in vitro pharmacological studies were identified.

IV. Pregnancy and Lactation

No effects on implantation were observed in rats orally administered 100 mg/kg of petroleum ether, alcoholic, or aqueous extracts of celery on days 1 to 7 of pregnancy (Garg et al. 1970). No adverse effects on fertility were reported in rats orally administered 250 mg/kg of an ethanol extract of celery daily on days 1 to 7 of pregnancy (Sharma et al. 1983).

Uterine stimulant activity of celery (plant part not specified in English language translation) has been reported (Kreitmair 1936).

No information on the safety of celery seed during lactation was identified.

V. Toxicity Studies

Acute Toxicity

The LD_{50} of orally administered celery seed essential oil in rats could not be determined at doses up to 5 g/kg (Opdyke 1974). The dermal LD_{50} of celery seed essential oil applied to the skin in rabbits could not be determined at doses up to 5 g/kg (Opdyke 1974).

Literature Cited

Ahluwalia, V.K., D.R. Boyd, A.K. Jain, C.H. Khanduri, and N.D. Sharma. 1988. Furanocoumarin glucosides from the seeds of *Apium graveolens*. *Phytochemistry* 27(4):1181-1183.

Ballmer-Weber, B.K., S. Vieths, D. Luttkopf, P. Heuschmann, and B. Wuthrich. 2000. Celery allergy confirmed by double-blind, placebo-controlled food challenge: A clinical study in 32 subjects with a history of adverse reactions to celery root. *J. Allergy Clin. Immunol.* 106(2):373-378.

Barethia, R. 2008. Effect of *Apium graveolens* on the oestrus cycle in rats. *J. Exp. Zool. India* 11(2):311-312.

Barg, W., A. Wolanczyk-Medrala, A. Obojski, et al. 2008. Food-dependent exercise-induced anaphylaxis: Possible impact of increased basophil histamine releasability in hyperosmolar conditions. *J. Investig. Allergol. Clin. Immunol.* 18(4):312-315.

Beattie, P.E., M.J. Wilkie, G. Smith, J. Ferguson, and S.H. Ibbotson. 2007. Can dietary furanocoumarin ingestion enhance the erythemal response during high-dose UVA1 therapy? *J. Am. Acad. Dermatol.* 56(1):84-87.

Beier, R.C., G.W. Ivie, E.H. Oertli, and D.L. Holt. 1983. HPLC analysis of linear furocoumarins (psoralens) in healthy celery (*Apium graveolens*). *Food Chem. Toxicol.* 21(2):163-165.

Birmingham, D.J., M.M. Key, G.E. Tubich, and V.B. Perone. 1961. Phototoxic bullae among celery harvesters. *Arch. Dermatol.* 83(1):73.

Bradley, P.R. 1992. *British herbal compendium: A handbook of scientific information on widely used plant drugs.* Bournemouth, UK: British Herbal Medicine Association.

Breiteneder, H., K. Hoffmann-Sommergruber, G. O'Riordain, et al. 1995. Molecular characterization of Api g 1, the major allergen of celery (*Apium graveolens*), and its immunological and structural relationships to a group of 17-kDa tree pollen allergens. *Eur. J. Biochem.* 233(2):484-489.

Darsow, U., J. Laifaoui, K. Kerschenlohr, et al. 2004. The prevalence of positive reactions in the atopy patch test with aeroallergens and food allergens in subjects with atopic eczema: A European multicenter study. *Allergy* 59(12):1318-1325.

Ermertcan, A.T., S. Ozturkcan, M.T. Sahin, C. Bilac, and D.B. Bilac. 2007. Acute irritant contact dermatitis due to *Apium graveolens*. *Contact Dermat.* 57(2):122-123.

Finkelstein, E., U. Afek, E. Gross, et al. 1994. An outbreak of phytophotodermatitis due to celery. *Int. J. Dermatol.* 33(2):116-118.

Garg, S.K., S.K. Saksena, and R.R. Chaudhury. 1970. Antifertility screening of plants. VI. Effect of five indigenous plants on early pregnancy in albino rats. *Indian J. Med. Res.* 58(9):1285-1289.

Garg, S.K., S.R. Gupta, and N.D. Sharma. 1978. Apiumetin: A new furanocoumarin from the seeds of *Apium graveolens*. *Phytochemistry* 17:2135–2136.

Garg, S.K., S.R. Gupta, and N.D. Sharma. 1979. Apiumoside, a new furanocoumarin glucoside from the seeds of *Apium graveolens*. *Phytochemistry* 18:1764-1765.

Hirschfeld, G., L. Weber, A. Renkl, K. Scharffetter-Kochanek, and J.M. Weiss. 2008. Anaphylaxis after oseltamivir (Tamiflu) therapy in a patient with sensitization to star anise and celery-carrot-mugwort-spice syndrome. *Allergy* 63(2):243-244.

Jeanmougin, M., C. Varroud-Vial, and L. Dubertret. 2005. Phototoxic side-effect following celery ingestion during puva therapy. *Ann. Dermatol. Venereol.* 132(6-7 Pt 1):566-567.

Kochmann, M. 1931. *Anthemis nobilis* und Apiol, sind sie Abortivmittel? *Arch. Toxicol.* 2(1):35-36.

Kreitmair, H. 1936. Pharmacological trials with some domestic plants. *E. Merck's Jahresber*. 50:102-111.

Ljunggren, B. 1990. Severe phototoxic burn following celery ingestion. *Arch. Dermatol.* 126(10):1334-1336.

Luttkopf, D., B.K. Ballmer-Weber, B. Wuthrich, and S. Vieths. 2000. Celery allergens in patients with positive double-blind placebo-controlled food challenge. *J. Allergy Clin. Immunol.* 106(2):390-399.

Opdyke, D.L.J. 1974. Fragrance raw materials monographs. *Food Chem. Toxicol.* 12:849.

Pauli, G., J.C. Bessot, P.A. Braun, et al. 1988. Celery allergy clinical and biological study of 20 cases. *Ann. Allergy* 60(3):243-246.

Schimmer, O. 1983. Determination of the phototoxic and photomutagenic potency of drug and commercial preparations containing furocoumarin using a *Chlamydomonas* test system. *Planta Med.* 47(2):79-82.

Seligman, P.J., C.G. Mathias, M.A. O'Malley, et al. 1987. Phytophotodermatitis from celery among grocery store workers. *Arch. Dermatol.* 123(11):1478-1482.

Sharma, B.B., M.D. Varshney, D.N. Gupta, and A.O. Prakash. 1983. Antifertility screening of plants. Part I. Effect of ten indigenous plants on early pregnancy in albino rats. *Int. J. Crude Drug Res.* 21(4):183-187.

van Itallie, L., A. Harmsma, and L.W. van Esveld. 1932. Abortifacients, particularly apiole. *Arch. Exp. Pathol. Pharmakol.* 165:84-100.

Vilke, G.M. 2002. Food-dependent exercise-induced anaphylaxis. *Prehosp. Emerg. Care* 6(3):348-350.

Wichtl, M. 2004. *Herbal drugs and phytopharmaceuticals: A handbook for practice on a scientific basis.* 3rd ed. Boca Raton, FL: CRC Press.

A

Apocynum spp. — Apocynaceae

Apocynum androsaemifolium **L.**
SCN: spreading dogbane
OCN: common dogbane

Apocynum cannabinum **L.**
SCN: Indian hemp
OCN: Canada hemp; hemp dogbane
Part: root

Quick Reference Summary

Safety Class: 3
Interaction Class: B

Contraindications

Not for use except under the supervision of an expert qualified in the appropriate use of this substance (Felter and Lloyd 1898; Wood and LaWall 1926).

Other Precautions

None known.

Drug and Supplement Interactions

Based on the presence of cardiac glycosides in spreading dogbane and Indian hemp, caution should be used in patients taking cardiac medications (Genkina et al. 1974; Grundmann and Gerlach 1967; Lee et al. 1972).

Notice

Emetic (Felter and Lloyd 1898; Wood and LaWall 1926); *see* Appendix 2.

Diuretic (Felter and Lloyd 1898; Wood and LaWall 1926); *see* Appendix 2.

Adverse Events and Side Effects

May cause irritation of the gastrointestinal tract, nausea, vomiting, and increased perspiration (Felter and Lloyd 1898; Wood and LaWall 1926).

Pharmacological Considerations

Medical texts from the late 1800s and early 1900s indicate that use of spreading dogbane almost invariably produces nausea or vomiting (sometimes copious), copious perspiration, diuretic activity, and "acts powerfully upon the heart, slowing its action and raising arterial tension" (Felter and Lloyd 1898; Graham 1909; Wood and LaWall 1926).

Pregnancy and Lactation

No information on the safety of spreading dogbane and Indian hemp in pregnancy or lactation was identified in the scientific or traditional literature. Based on the pharmacological activity, use during pregnancy is not recommended except under the supervision of a qualified healthcare practitioner.

While this review did not identify any concerns for use while nursing, safety has not been conclusively established, and this substance is not recommended for use except under the supervision of an expert qualified in its appropriate use.

Review Details

I. Drug and Supplement Interactions

Clinical Trials of Drug or Supplement Interactions

No clinical trials of drug or supplement interactions were identified.

Case Reports of Suspected Drug or Supplement Interactions

No case reports of suspected drug or supplement interactions were identified.

Animal Trials of Drug or Supplement Interactions

No animal trials of drug or supplement interactions were identified.

II. Adverse Events

Case Reports of Adverse Events

No case reports of adverse events were identified.

III. Pharmacology and Pharmacokinetics

Human Pharmacological Studies

No relevant human pharmacological studies were identified.

Animal Pharmacological Studies

Slowing of heart rate and increased urination were observed in cats, dogs, and frogs treated with varying doses of spreading dogbane (Graham 1909; Wood and LaWall 1926).

In Vitro Pharmacological Studies

Slowing of heart rate was observed in excised rabbit hearts treated with varying doses of spreading dogbane (Graham 1909).

IV. Pregnancy and Lactation

No information on the safety of spreading dogbane or Indian hemp in pregnancy or lactation was identified.

V. Toxicity Studies

Acute Toxicity

The LD_{50} of a tincture of spreading dogbane administered intravenously to frogs is 5.1 ml/kg (Graham 1909).

Cytotoxicity

Fractions of a water-alcohol extract of spreading dogbane exhibited cytotoxic activity in human nasopharynx carcinoma cells (Kupchan et al. 1964).

Literature Cited

Felter, H.W., and J.U. Lloyd. 1898. *King's American dispensatory*. 18th ed., 3rd rev. 2 vols. Cincinnati: Ohio Valley Co.

Genkina, G.L., K.K. Khodzhaev, T.T. Shakirov, and N.K. Abubakirov. 1974. An investigation of the roots of *Apocynum androsaemifolium* and *A. cannabinum* for their cardenolide content. *Chem. Nat. Compd.* 8(3):316-318.

Graham, J.C. 1909. The pharmacology of *Apocynum cannabinum*. *Biochem. J.* 4(9):385-404.

Grundmann, W., and H. Gerlach. 1967. Cardiac glycosides from the roots of *Apocynum cannabinum*: Quantitative determination of glycosides in photometry cymarin apocannoside. *Pharm. Zentralh.* 106(8):501-508.

Kupchan, S.M., R.J. Hemingway, and R.W. Doskotch. 1964. Tumor inhibitors. IV. Apocannoside and cymarin, the cytotoxic principles of *Apocynum cannabinum* L. *J. Med. Chem.* 7:803-804.

Lee, P.K., D.P. Carew, and J. Rosazza. 1972. *Apocynum cannabinum* tissue culture. Growth and chemical analysis. *Lloydia* 35(2):150-156.

Wood, H., and C. LaWall. 1926. *The dispensatory of the United States of America*. Philadelphia: J.B. Lippincott.

Aralia californica S. Watson

Araliaceae

SCN: California spikenard
OCN: elk clover

Part: rhizome, root

Quick Reference Summary

Safety Class: 1
Interaction Class: A

Contraindications

None known.

Other Precautions

None known.

Editors' Note

A reference text on medicinal plants of western America does not list any cautions or contraindications for use of California spikenard (Moore 2003).

Drug and Supplement Interactions

None known.

Adverse Events and Side Effects

None known.

Pharmacological Considerations

None known.

Pregnancy and Lactation

No information on the safety of California spikenard in pregnancy or lactation was identified in the scientific or traditional literature. Although this review did not identify any concerns for use while pregnant or nursing, safety has not been conclusively established.

Review Details

I. Drug and Supplement Interactions

Clinical Trials of Drug or Supplement Interactions

No clinical trials of drug or supplement interactions were identified.

Case Reports of Suspected Drug or Supplement Interactions

No case reports of suspected drug or supplement interactions were identified.

Animal Trials of Drug or Supplement Interactions

No animal trials of drug or supplement interactions were identified.

II. Adverse Events

Case Reports of Adverse Events

No case reports of adverse events were identified.

III. Pharmacology and Pharmacokinetics

Human Pharmacological Studies

No relevant human pharmacological studies were identified.

Animal Pharmacological Studies

No relevant animal pharmacological studies were identified.

In Vitro Pharmacological Studies

No relevant in vitro pharmacological studies were identified.

IV. Pregnancy and Lactation

No information on the use of California spikenard during pregnancy or lactation was identified.

V. Toxicity Studies

No toxicity studies were identified.

Literature Cited

Moore, M. 1993. *Medicinal plants of the Pacific West*. Santa Fe: Red Crane Books.

Aralia nudicaulis L. — Araliaceae

SCN: small spikenard
OCN: false sarsaparilla; wild sarsaparilla

Part: rhizome

Quick Reference Summary

Safety Class: 1
Interaction Class: A

Contraindications

None known.

Other Precautions

None known.

Drug and Supplement Interactions

None known.

Editors' Note

No cautions for use of small spikenard are reported in historical American medical texts (Felter and Lloyd 1898; Remington and Wood 1918).

Adverse Events and Side Effects

None known.

Pharmacological Considerations

None known.

Pregnancy and Lactation

No information on the safety of small spikenard in pregnancy or lactation was identified in the scientific or traditional literature. Although this review did not identify any concerns for use while pregnant or nursing, safety has not been conclusively established.

Review Details

I. Drug and Supplement Interactions

Clinical Trials of Drug or Supplement Interactions

No clinical trials of drug or supplement interactions were identified.

Case Reports of Suspected Drug or Supplement Interactions

No case reports of suspected drug or supplement interactions were identified.

Animal Trials of Drug or Supplement Interactions

No animal trials of drug or supplement interactions were identified.

II. Adverse Events

Case Reports of Adverse Events

No case reports of adverse events were identified.

III. Pharmacology and Pharmacokinetics

Human Pharmacological Studies

No relevant human pharmacological studies were identified.

Animal Pharmacological Studies

No relevant animal pharmacological studies were identified.

In Vitro Pharmacological Studies

No relevant in vitro pharmacological studies were identified.

IV. Pregnancy and Lactation

No information on the use of small spikenard during pregnancy or lactation was identified.

V. Toxicity Studies

No toxicity studies were identified.

Literature Cited

Felter, H.W., and J.U. Lloyd. 1898. *King's American dispensatory*. 18th ed., 3rd rev. 2 vols. Cincinnati: Ohio Valley Co.

Remington, J.P., and H.C. Wood. 1918. *The dispensatory of the United States of America*. 20th ed. Philadelphia: Lippincott.

Aralia racemosa L.

Araliaceae

SCN: spikenard
OCN: American spikenard

Part: rhizome

Quick Reference Summary

Safety Class: 2b
Interaction Class: A

Contraindications

Not for use in pregnancy except under the supervision of a qualified healthcare practitioner (Densmore 1928; Herrick 1977).

Other Precautions

None known.

Drug and Supplement Interactions

None known.

Adverse Events and Side Effects

None known.

Pharmacological Considerations

None known.

Pregnancy and Lactation

Spikenard is reported to have been used as an abortifacient by the Chippewa and Iroquois (Densmore 1928; Herrick 1977). Based on this information, use during pregnancy is not recommended except under the supervision of a qualified healthcare practitioner.

No information on the safety of spikenard use during lactation was identified. While this review did not identify any concerns for use while nursing, safety has not been conclusively established.

Review Details

I. Drug and Supplement Interactions

Clinical Trials of Drug or Supplement Interactions

No clinical trials of drug or supplement interactions were identified.

Case Reports of Suspected Drug or Supplement Interactions

No case reports of suspected drug or supplement interactions were identified.

Animal Trials of Drug or Supplement Interactions

No animal trials of drug or supplement interactions were identified.

II. Adverse Events

Case Reports of Adverse Events

No case reports of adverse events were identified.

III. Pharmacology and Pharmacokinetics

Human Pharmacological Studies

No relevant human pharmacological studies were identified.

Animal Pharmacological Studies

No relevant animal pharmacological studies were identified.

In Vitro Pharmacological Studies

No relevant in vitro pharmacological studies were identified.

IV. Pregnancy and Lactation

Spikenard is reported to have been used as an abortifacient by the Chippewa and Iroquois (Densmore 1928; Herrick 1977).

No information on the safety of spikenard use during lactation was identified.

V. Toxicity Studies

No toxicity studies were identified.

Literature Cited

Densmore, F. 1928. Uses of plants by the Chippewa Indians. *Smithsonian Inst. Bur. Am. Ethnol. Annu. Rep.* 44:273-379.

Herrick, J.W. 1977. Iroquois medical botany. Ph.D. Thesis. State University of New York, Albany.

Arctium lappa L. — Asteraceae

SCN: burdock
PN: *niu bang zi* (fruit)
OCN: *gobo; goboshi;* great burdock
Part: root, seed

Quick Reference Summary

Safety Class: 1
Interaction Class: A

Contraindications

None known.

Other Precautions

None known.

Drug and Supplement Interactions

None known.

Adverse Events and Side Effects

An anaphylactic reaction to burdock root has been reported (Sasaki et al. 2003). Several cases of contact dermatitis have been reported after topical applications of burdock root plasters (Rodriguez et al. 1995).

Cases of atropine poisoning have been reported in persons that took adulterated burdock root products (Bryson et al. 1978; Fletcher and Cantwell 1978; Gandolfo and Accascina 1953; Rhoads et al. 1984). Atropine is not a known constituent of burdock (Leung and Foster 1996; Mills and Bone 2005).

Pharmacological Considerations

None known.

Pregnancy and Lactation

One animal study indicated no adverse effects of burdock on fetal development (Matsui et al. 1967). No other information on the safety of this herb in pregnancy or lactation was identified in the scientific or traditional literature. Although this review did not identify any concerns for use while pregnant or nursing, safety has not been conclusively established.

Review Details

I. Drug and Supplement Interactions

Clinical Trials of Drug or Supplement Interactions

No clinical trials of drug or supplement interactions were identified.

Case Reports of Suspected Drug or Supplement Interactions

No case reports of suspected drug or supplement interactions were identified.

Animal Trials of Drug or Supplement Interactions

No animal trials of drug or supplement interactions were identified.

II. Adverse Events

Case Reports of Adverse Events

Several cases of atropine poisoning have been reported in persons consuming tea labeled as burdock root (Bryson et al. 1978; Fletcher and Cantwell 1978; Gandolfo and Accascina 1953; Rhoads et al. 1984). Atropine is not a known constituent of burdock; its presence was a result of contamination (Leung and Foster 1996; Mills and Bone 2005).

An anaphylactic reaction to cooked burdock root has been reported. The patient tested positively to skin prick tests with raw and boiled burdock root, and raw and boiled carrot (Sasaki et al. 2003). Allergic contact dermatitis has been reported in three patients who topically applied burdock root plasters. All three patients tested positively in closed but not open patch tests, and none reported allergies to other Asteraceae family plants. A positive patch test for nickel was reported in one patient. Tests with a sesquiterpene lactone mix were not completed (Rodriguez et al. 1995).

III. Pharmacology and Pharmacokinetics

Human Pharmacological Studies

No relevant human pharmacological studies were identified.

Animal Pharmacological Studies

Some earlier reviews indicated a hypoglycemic effect of burdock root (Bever and Zahnd 1979; Farnsworth and Segelman 1971).

No adverse effects on fertility were seen in mice subcutaneously administered 0.05 to 0.2 ml of an aqueous burdock extract (part unspecified) twice daily for 5 days (Matsui et al. 1967).

In Vitro Pharmacological Studies

An extract of burdock root weakly inhibited the CYP isoenzymes CYP3A4, CYP19, and CYP2C19 (Scott et al. 2006).

An aqueous extract of burdock seed inhibited platelet-activating factor binding at a concentration of 200 μg/ml (Iwakami et al. 1992).

IV. Pregnancy and Lactation

No teratogenic effects were seen in offspring of pregnant mice subcutaneously administered 0.05 to 0.2 ml of an aqueous burdock extract (part unspecified) twice daily for 5 days (Matsui et al. 1967).

No information on the safety of burdock root or seed during lactation was identified.

V. Toxicity Studies

Acute Toxicity

The LD_{50} of orally administered alcohol extract of burdock root in rats is 700 mg/kg (Sharma et al. 1978).

Subchronic Toxicity

No toxic effects were observed in rats fed burdock root as 30% of the diet for 4 months (Hirono et al. 1978).

Literature Cited

Bever, B., and G. Zahnd. 1979. Plants with oral hypoglycemic activity. *Q. J. Crude Drug Res.* 17:139-196.

Bryson, P.D., A.S. Watanabe, B.H. Rumack, and R.C. Murphy. 1978. Burdock root tea poisoning. Case report involving a commercial preparation. *J. Am. Med. Assoc.* 239(20):2157.

Farnsworth, N., and A. Segelman. 1971. Hypoglycemic plants. *Tile Till* 57:52-56.

Fletcher, G.F., and J.D. Cantwell. 1978. Burdock root tea poisoning. *J. Am. Med. Assoc.* 240(15):1586.

Gandolfo, N., and G. Accascina. 1953. Atropine poisoning by ingestion of decoction of burdock roots. *Rend. Ist. Sup. Sanit.* 16(10-11-12):844-851.

Hirono, I., H. Mori, K. Kato, et al. 1978. Safety examination of some edible plants, Part 2. *J. Environ. Pathol. Toxicol.* 1(1):71-74.

Iwakami, S., J.B. Wu, Y. Ebizuka, and U. Sankawa. 1992. Platelet activating factor (PAF) antagonists contained in medicinal plants: Lignans and sesquiterpenes. *Chem. Pharm. Bull. (Tokyo)* 40(5):1196-1198.

Leung, A.Y., and S. Foster. 1996. *Encyclopedia of common natural ingredients used in food, drugs, and cosmetics*. 2nd ed. New York: Wiley.

Matsui, A.S., J. Rogers, Y.K. Woo, and W.C. Cutting. 1967. Effects of some natural products on fertility in mice. *Med. Pharmacol. Exp. Int. J. Exp. Med.* 16(5):414-424.

Mills, S., and K. Bone. 2005. *The essential guide to herbal safety*. St. Louis: Elsevier.

Rhoads, P.M., T.G. Tong, W. Banner, Jr., and R. Anderson. 1984. Anticholinergic poisonings associated with commercial burdock root tea. *J. Toxicol. Clin. Toxicol.* 22(6):581-584.

Rodriguez, P., J. Blanco, S. Juste, et al. 1995. Allergic contact dermatitis due to burdock (*Arctium lappa*). *Contact Dermat.* 33(2):134-135.

Sasaki, Y., Y. Kimura, T. Tsunoda, and H. Tagami. 2003. Anaphylaxis due to burdock. *Int. J. Dermatol.* 42(6):472-473.

Scott, I.M., R.I. Leduc, A.J. Burt, et al. 2006. The inhibition of human cytochrome P450 by ethanol extracts of North American botanicals. *Pharm. Biol.* 44(5):315-327.

Sharma, M.L., N. Chandokhe, B.J. Ghatak, et al. 1978. Pharmacological screening of Indian medicinal plants. *Indian J. Exp. Biol.* 16(2):228-240.

Arctostaphylos uva-ursi (L.) Spreng. — Ericaceae

SCN: uva-ursi
OCN: bearberry; kinnickinick

Part: leaf

Quick Reference Summary

Safety Class: 1
Interaction Class: A

Contraindications

None known.

Other Precautions

None known.

Drug and Supplement Interactions

None known.

Notice

Tannins (6.0–27.5%, usually around 10%) (Leung and Foster 1996; Wichtl 2004); *see* Appendix 1.

Editors' Notes

Although some herbal reference texts indicate that uva-ursi is contraindicated in kidney disorders (Bradley 1992; ESCOP 2003), a recent review of the literature found no evidence to support this contraindication (Upton et al. 2008). Some evidence of kidney damage has been observed in rats chronically administered the compound hydroquinone, which may have been the basis for contraindication in kidney disorders (Hoffman-Bohm and Simon 1992; Kari et al. 1992; Shibata et al. 1991). The kidney damage was seen primarily in aged male rats, a population that is predisposed to kidney damage (DeCaprio 1999; McGregor 2007).

One herbal reference text notes that "hydroquinone appears to be nontoxic with the administration of uva-ursi leaf tea, however hydroquinone must remain in suspicion with regard to having mutagenic and possibly carcinogenic effects" (Wichtl 2004). Reviews of the compound hydroquinone indicate that although genotoxic effects were observed in vitro and in animal tests using intraperitoneal or subcutaneous administration, such effects were generally not seen in animals orally administered the compound (DeCaprio 1999; McGregor 2007).

Uva-ursi contains arbutin and other hydroquinone glucosides, present at up to 12% of the dried leaf (Hegnauer 1966). The exposure to hydroquinone, however, depends on multiple factors including concentration in the leaf, concentration in the extract or other dosage form, and the absorption and distribution of hydroquinone in the body, which have yet to be fully investigated (Upton et al. 2008).

Adverse Events and Side Effects

A case of maculopathy was reported after long-term consumption of uva-ursi (Wang and Del Priore 2004).

Ingestion of uva-ursi has been reported to change urine to a brown-green color due to hydroquinone metabolites (Upton et al. 2008).

Pharmacological Considerations

Although several herbal reference texts indicate that concomitant use of uva-ursi and medications that can acidify the urine may result in a decrease in efficacy of uva-ursi with acidic urine or in conjunction with remedies that produce acidic urine (Blumenthal et al. 1998; Bradley 1992; ESCOP 2003; Weiss and Meuss 2001), a review of uva-ursi indicated that this concern appears to be theoretical, with no data available to support such an interaction (Upton et al. 2008).

Pregnancy and Lactation

No studies on the safety of uva-ursi in pregnancy were identified. A number of reproductive toxicity studies have been completed on the compounds arbutin and hydroquinone. The no-observed-effect level (NOEL) for arbutin in rats was 100 mg/kg daily (Itabashi et al. 1988). The NOEL for hydroquinone in rats was 100 mg/kg daily (Murphy et al. 1992). In rabbits, the NOEL for maternal toxicity of hydroquinone was 25 mg/kg, while the developmental toxicity NOEL was 75 mg/kg (Murphy et al. 1992).

In a two-generation reproductive toxicity study, no significant changes in rat pup development were observed (Blacker et al. 1993).

No information on the safety of uva-ursi during lactation was identified. While this review did not identify any concerns for use while nursing, safety has not been conclusively established.

Review Details

I. Drug and Supplement Interactions

Clinical Trials of Drug or Supplement Interactions

No clinical trials of drug or supplement interactions were identified.

Case Reports of Suspected Drug or Supplement Interactions

No case reports of drug or supplement interactions were identified. Several reference texts indicate that concomitant use of uva-ursi with medications that can acidify the urine may result in a decrease of effectiveness of uva-ursi (Bradley 1992; ESCOP 2003; Weiss and Meuss 2001). The acidity of cranberry, another botanical commonly use to treat urinary tract infections, is often cited as a reason to avoid taking cranberry and uva-ursi together. Studies on the ability of cranberry to acidify the urine generally indicate that any such effects are mild or nonexistent (Upton 2002).

Animal Trials of Drug or Supplement Interactions

Studies of uva-ursi in rats, mice, and rabbits have indicated that uva-ursi may potentiate the anti-inflammatory effects of dexamethasone, prednisolone, and indomethacin (Matsuda et al. 1990, 1991, 1992).

II. Adverse Events

Adverse Events Reported in Clinical Trials

No adverse effects or abnormal values in hematology or urinalysis were reported in men or women administered 300 or 500 mg daily of purified hydroquinone for 3 to 5 months (Carlson and Brewer 1953).

Case Reports of Adverse Events

A case of bull's-eye maculopathy was reported in a 56-year-old woman who had been taking uva-ursi tea "regularly" (dose and frequency not reported) for 3 years. The woman reported a decrease in visual acuity in the third year. The reporting authors noted that hydroquinone compounds are known inhibitors of tyrosine kinase and thus melanin synthesis, and that decreased melanin in the eye might account for the maculopathy (Wang and Del Priore 2004).

III. Pharmacology and Pharmacokinetics

Human Pharmacological Studies

No adverse effects were observed in healthy volunteers administered uva-ursi standardized extract or gastric juice-resistant tablet containing the same extract (Paper et al. 1993).

No adverse event data were reported in studies of urinary arbutin levels in persons taking uva-ursi (Quintus et al. 2005; Siegers et al. 1997).

Animal Pharmacological Studies

Some extracts of uva-ursi may have inhibitory effects on melanin synthesis, melasma, and other abnormal pigmentations in animals (Maeda and Fukuda 1996; Matsuda et al. 1992, 1996; Ortiz et al. 1999).

In Vitro Pharmacological Studies

Aqueous and methanolic extracts of uva-ursi inhibited the drug-metabolizing CYP isoenzymes CYP3A5, CYP3A7, CYP3A4, and CYP2C19; the aqueous extract also inhibited CYP19. For all isoenzymes, the aqueous extract had a stronger inhibitory effect than the methanolic extract (Chauhan et al. 2007).

Induction of MDR1 and the drug-metabolizing isoenzymes CYP1A2 and CYP3A4 was observed in vitro in human colon carcinoma cells (Brandin et al. 2007). Conversely, inhibition of CYP3A4 and CYP2C19 by ethanol extracts of uva-ursi was observed in a high-throughput screening assay (Scott et al. 2006).

Uva-ursi potentiated the effects of β-lactam antibiotics against methicillin-resistant *Staphylococcus aureus* (Shimizu et al. 2001).

IV. Pregnancy and Lactation

In pregnant rats subcutaneously administered 100 or 400 mg/kg daily of the compound arbutin, no signs of toxicity in parent animals or fetal toxicity were observed. The maximal no-observed-effect level (NOEL) dose was 100 mg/kg per day (equivalent to a human dose of ~7000 mg) (Itabashi et al. 1988).

In a two-generation reproductive toxicity study in rats, the compound hydroquinone was administered daily at doses up to 150 mg/kg 10 weeks prior to cohabitation, during cohabitation, and until scheduled termination, and no adverse effects were observed on feed consumption, survival, or reproductive parameters for the first or second generation (F0 or F1) parental animals. Mild, transient tremors were observed shortly after dosing at 150 mg/kg/day in several F0 and F1 animals and in a single F0 male at 50 mg/kg/day (Blacker et al. 1993).

In pregnant rats orally administered hydroquinone up to 300 mg/kg daily on days 6 to 15 of pregnancy, no changes in fertility, pregnancy outcomes, or fetal development were observed except that a slightly reduced fetal weight was seen at the highest dose level (Krasavage et al. 1992). In pregnant rats orally administered up to 300 mg/kg of the compound hydroquinone daily on days 6 to 15 of pregnancy, the NOEL for both maternal and developmental toxicity was 100 mg/kg, whereas 300 mg/kg was the no-observable-adverse-effect level (NOAEL) (Murphy et al. 1992).

In rabbits orally administered up to 150 mg/kg of the compound hydroquinone on days 6 to 15 of pregnancy, the NOEL for maternal toxicity was 25 mg/kg daily, and for developmental toxicity was 75 mg/kg daily. Under the conditions of this study, hydroquinone at 150 mg/kg/day produced minimal developmental alterations in the presence of maternal toxicity (Murphy et al. 1992).

No effects of uva-ursi on isolated rat or guinea pig uteruses were observed after application of 1 to 2 mg of crude drug per square centimeter (Shipochliev 1981).

No information on the safety of uva-ursi during lactation was identified.

V. Toxicity Studies

Acute Toxicity

The LD_{50} of orally administered hydroquinone is 200 mg/kg in rabbits, 80 mg/kg in cats, and could not be determined at doses up to 375 mg/kg in rats (Clayton and Clayton 1991; Topping et al. 2007). In rats, doses between 285 and 375 mg/kg caused animals to exhibit minor to moderate tremors and minor convulsions within the first hour after dosing (Topping et al. 2007). The LD_{50} of subcutaneously administered hydroquinone in mice is 160 mg/kg (Clayton and Clayton 1991). Based on these values, the LD_{50} of the compound hydroquinone in humans can be estimated at a mean of 21 g, suggesting that the potential for toxicity associated with typical doses of uva-ursi is low (Upton et al. 2008).

Subchronic Toxicity

In rats orally administered the compound hydroquinone 5 days a week for 13 weeks at doses of 0 to 200 mg/kg daily, doses of 64 or 200 mg/kg hydroquinone resulted in acutely observable behavioral effects including tremors and reduced activity. Oral doses of 64 mg/kg or more of hydroquinone resulted in acute neurobehavioral effects indicative of CNS stimulation; however, subchronic exposure to dose levels that produced repetitive CNS stimulation by hydroquinone did not result in an exacerbation of acute stimulatory effects over time nor morphological changes in the CNS or PNS or nephrotoxicity. Nephrotoxic effects observed in other studies in Fischer 344 rats after hydroquinone exposure were not observed in this study with Sprague-Dawley rats (Topping et al. 2007).

Chronic Toxicity

Several chronic toxicity studies of the compound hydroquinone in rats have shown nephropathy after long-term dietary or other oral exposure, with male rats being significantly more affected than female rats (Hoffman-Bohm and Simon 1992; Kari et al. 1992; Shibata et al. 1991). Reviews of the toxicology of hydroquinone note that chronic progressive nephropathy is characteristic of aged male rats (DeCaprio 1999; McGregor 2007).

A chronic toxicity test of hydroquinone in mice and rats indicated some adverse effects after long-term administration as 0.8% of the diet for 2 years. In the hydroquinone group, renal tubular hyperplasia occurred in 100% of male rats, 30% of male mice, and 7% of female rats; adenomas occurred in 47% of male rats and 10% of male mice (Shibata et al. 1991).

After chronic exposure to hydroquinone, nephropathy occurred in 80% of male rats, 27% of female rats, and not at all in mice. Mice developed squamous cell hyperplasia in the stomach; 67% of male mice developed hypertrophy of the liver and 47% developed hepatocellular adenoma. No differences in survival, however, were noted between treated animals and controls (Hoffman-Bohm and Simon 1992).

In a chronic toxicity test of hydroquinone orally administered to rats at doses up to 50 mg/kg and mice at doses up to 100 mg/kg for 2 years, at the highest dose, male rats had severe nephropathy. Renal tubular adenomas were seen in 7% of low-dose rats and 14% of high-dose rats, but not in female rats or control animals. Mononuclear cell leukemia increased with dosage in female rats (Kari et al. 1992).

In rats orally administered 25 or 50 mg/kg of hydroquinone 5 days a week for 2 years, there was some evidence of carcinogenic activity of hydroquinone for male and female F344/N rats, and B6C3F1 mice, as shown by marked increases in tubular cell adenomas of the kidney (male F344/N rats), increases in mononuclear cell leukemia (female F344/N rats), and increases in hepatocellular neoplasms (female B6C3F1 mice). No evidence was found for carcinogenic activity of hydroquinone in male B6C3F1 mice administered 50 or 100 mg/kg. Administration of hydroquinone was associated with thyroid follicular cell hyperplasia in both male and female mice and anisokaryosis, multinucleated hepatocytes, and basophilic foci of the liver in male mice (NTP 1989).

Genotoxicity

No mutagenic activity of uva-ursi was seen in the Ames test with *Salmonella typhimurium* or in the *Bacillus subtilis* rec-assay (ESCOP 2003; Hoffman-Bohm and Simon 1992). No mutagenicity of the compound arbutin was observed in rats subcutaneously administered 100 to 400 mg/kg daily (Itabashi et al. 1988).

In Ames and micronucleus tests of urine from human volunteers administered the compound arbutin at doses of 420 mg, no mutagenic or genotoxic activity was observed (Siegers et al. 1997). No mutagenicity of arbutin was observed in a gene mutation assay at a concentration of 0.01 *M*, whereas the compound hydroquinone caused an increase in mutation frequency at a concentration of 0.01 *M* (Mueller and Kasper 1996).

Reviews of the compound hydroquinone indicate that although genotoxic effects were observed in vitro and in animal tests using intraperitoneal or subcutaneous administration, such effects were generally not seen in animals orally administered the compound (DeCaprio 1999; McGregor 2007).

Literature Cited

Blacker, A.M., R.E. Schroeder, J.C. English, et al. 1993. A 2-generation reproduction study with hydroquinone in rats. *Fund. Appl. Toxicol.* 21(4):420-424.

Blumenthal, M., W. Busse, A. Goldberg, et al. 1998. *The complete German Commission E monographs*. Austin, TX: American Botanical Council.

Bradley, P.R. 1992. *British herbal compendium: A handbook of scientific information on widely used plant drugs*. Bournemouth, Dorset: British Herbal Medicine Association.

Brandin, H., E. Viitanen, O. Myrberg, and A.K. Arvidsson. 2007. Effects of herbal medicinal products and food supplements on induction of CYP1A2, CYP3A4 and MDR1 in the human colon carcinoma cell line LS180. *Phytother. Res.* 21(3):239-244.

Carlson, A.J., and N.R. Brewer. 1953. Toxicity studies on hydroquinone. *Proc. Soc. Exp. Biol. Med.* 84(3):684-688.

Chauhan, B., C. Yu, A. Krantis, et al. 2007. In vitro activity of uva-ursi against cytochrome P450 isoenzymes and P-glycoprotein. *Can. J. Physiol. Pharmacol.* 85(11):1099-1107.

Clayton, G., and F. Clayton. 1991. *Patty's industrial hygiene and toxicology*. 4th ed. New York: Wiley.

DeCaprio, A.P. 1999. The toxicology of hydroquinone—Relevance to occupational and environmental exposure. *Crit. Rev. Toxicol.* 29(3):283-330.

ESCOP. 2003. *ESCOP monographs: The scientific foundation for herbal medicinal products*. 2nd ed. Exeter, U.K.: European Scientific Cooperative on Phytotherapy.

Hegnauer, R. 1966. *Ericaceae*. Basel: Birkhäuser.

Hoffman-Bohm, K., and P. Simon. 1992. *Arctostaphylos*. Edited by Hansel, R., Keller, K., Rimpler, H. Schneider, G. *Hagers handbuch der pharmazeutichen praxis*. Berlin: Springer.

Itabashi, M., H. Aihara, T. Inoue, et al. 1988. Reproduction study of arbutin in rats by subcutaneous administration. *Iyakuhin Kenkyu* 19:282-297.

Kari, F.W., J. Bucher, S.L. Eustis, J.K. Haseman, and J.E. Huff. 1992. Toxicity and carcinogenicity of hydroquinone in F344/N rats and B6c3f1 mice. *Food Chem. Toxicol.* 30(9):737-747.

Krasavage, W.J., A.M. Blacker, J.C. English, and S.J. Murphy. 1992. Hydroquinone: A developmental toxicity study in rats. *Fund. Appl. Toxicol.* 18(3):370-375.

Leung, A.Y., and S. Foster. 1996. *Encyclopedia of common natural ingredients used in food, drugs, and cosmetics.* 2nd ed. New York: Wiley.

Maeda, K., and M. Fukuda. 1996. Arbutin: Mechanism of its depigmenting action in human melanocyte culture. *J. Pharmacol. Exp. Ther.* 276:765-769.

Matsuda, H., M. Higashino, Y. Nakai, et al. 1996. Studies of cuticle drugs from natural sources. IV. Inhibitory effects of some *Arctostaphylos* plants on melanin biosynthesis. *Biol. Pharm. Bull.* 19(1):153-156.

Matsuda, H., S. Nakamura, H. Shiomoto, T. Tanaka, and M. Kubo. 1992. Pharmacological studies on leaf of *Arctostaphylos uva-ursi* (L.) Spreng. IV. Effect of 50% methanolic extract from *Arctostaphylos uva-ursi* (L.) Spreng. (bearberry leaf) on melanin synthesis. *Yakugaku Zasshi* 112(4):276-282.

Matsuda, H., H. Nakata, T. Tanaka, and M. Kubo. 1990. [Pharmacological study on *Arctostaphylos uva-ursi* (L.) Spreng. II. Combined effects of arbutin and prednisolone or dexamethasone on immuno-inflammation]. *Yakugaku Zasshi* 110(1):68-76.

Matsuda, H., T. Tanaka, and M. Kubo. 1991. Pharmacological studies on leaf of *Arctostaphylos uva-ursi* (L.) Spreng. III. Combined effect of arbutin and indomethacin on immuno-inflammation. *Yakugaku Zasshi* 111(4-5):253-258.

McGregor, D. 2007. Hydroquinone: An evaluation of the human risks from its carcinogenic and mutagenic properties. *Crit. Rev. Toxicol.* 37:887-914.

Mueller, L., and P. Kasper. 1996. The mutagenic potential of arbutine, a naturally occurring hydroquinone glycoside. *Mutat. Res.* 360:291-292.

Murphy, S.J., R.E. Schroeder, A.M. Blacker, W.J. Krasavage, and J.C. English. 1992. A study of developmental toxicity of hydroquinone in the rabbit. *Fund. Appl. Toxicol.* 19(2):214-221.

NTP. 1989. Toxicology and carcinogenesis studies of hydroquinone in F-344/N rats and B6C3F1 mice. NIH Publication No. 90-2821. Research Triangle Park, NC: National Institutes of Health.

Ortiz, Y., B. Elba, E. Del Pino Ma, G. Guzman, and I. Arias. 1999. Malasma: An aleatory, double blind, comparative study to evaluate the efficacy and safety of uva ursi: Uva ursi and lactic acid vs hydroquinone. *Dermatol. Rev. Mex.* 43:245-254.

Paper, D., J. Koehler, and G. Franz. 1993. Bioavailability of drug preparations containing a leaf extract from *Arctostaphylos uva-ursi (Uvae ursi folium). Planta Med.* 59:A589.

Quintus, J., K.A. Kovar, P. Link, and H. Hamacher. 2005. Urinary excretion of arbutin metabolites after oral administration of bearberry leaf extracts. *Planta Med.* 71(2):147-152.

Scott, I.M., R.I. Leduc, A.J. Burt, et al. 2006. The inhibition of human cytochrome P450 by ethanol extracts of North American botanicals. *Pharm. Biol.* 44(5):315-327.

Shibata, M.A., M. Hirose, H. Tanaka, et al. 1991. Induction of renal-cell tumors in rats and mice, and enhancement of hepatocellular tumor-development in mice after long-term hydroquinone treatment. *Jpn. J. Cancer Res.* 82(11):1211-1219.

Shimizu, M., S. Shiota, T. Mizushima, et al. 2001. Marked potentiation of activity of beta-lactams against methicillin-resistant *Staphylococcus aureus* by corilagin. *Antimicrob. Agents Chemother.* 45(11):3198-3201.

Shipochliev, T. 1981. [Uterotonic action of extracts from a group of medicinal plants]. *Vet. Med. Nauk.* 18(4):94-98.

Siegers, C.P., J.P. Siegers, R. Pentz, C. Bodinet, and J. Freudenstein. 1997. Metabolism of arbutin from uvae ursi-extracts in humans. *Pharm. Pharmacol. Lett.* 7(2-3):90-92.

Topping, D.C., L.G. Bernard, J.L. O'Donoghue, and J.C. English. 2007. Hydroquinone: Acute and subchronic toxicity studies with emphasis on neurobehavioral and nephrotoxic effects. *Food Chem. Toxicol.* 45(1):70-78.

Upton, R. 2002. *Cranberry fruit:* Vaccinium macrocarpon *Aiton: Standards of analysis, quality control, and therapeutics, American Herbal Pharmacopoeia and therapeutic compendium.* Santa Cruz, CA: American Herbal Pharmacopoeia.

Upton, R., A. Graff, and D. Swisher. 2008. *Uva ursi leaf,* Arctostaphylos uva-ursi *(L.) Spreng: Standards of analysis, quality control, and therapeutics.* Scotts Valley, CA: American Herbal Pharmacopoeia.

Wang, L., and L.V. Del Priore. 2004. Bull's-eye maculopathy secondary to herbal toxicity from uva ursi. *Am. J. Ophthalmol.* 137(6):1135-1137.

Weiss, R.F., and A.R. Meuss. 2001. *Weiss's herbal medicine.* Classic ed. New York: Thieme.

Wichtl, M. 2004. *Herbal drugs and phytopharmaceuticals: A handbook for practice on a scientific basis.* 3rd ed. Boca Raton, FL: CRC Press.

A

Arisaema spp. — Araceae

Arisaema amurense Maxim.
SCN: Chinese arisaema
PN: *tian nan xing* (rhizome)

Arisaema erubescens **(Wall.) Schott**
SCN: Chinese arisaema
Syn: *Arisaema consanguineum* Schott
PN: *tian nan xing* (rhizome)

Arisaema heterophyllum **Blume**
SCN: Chinese arisaema
PN: *tian nan xing* (rhizome)
Part: prepared rhizome

Quick Reference Summary

Safety Class: 3
Interaction Class: A

Contraindications

Not for use except under the supervision of an expert qualified in the appropriate use of this substance (Bensky et al. 2004; Chen and Chen 2004).

Other Precautions

None known.

Drug and Supplement Interactions

None known.

Standard Dose

The standard dose is 3 to 10 g of the processed rhizome as a decoction (Bensky et al. 2004; Chen and Chen 2004).

Editors' Notes

Before use, Chinese arisaema must be processed to reduce toxicity. Processing typically consists of soaking the herb in multiple changes of water with alum added in later soakings, followed by boiling with fresh ginger and alum. Processing with bile, according to traditional methods, is an alternate method that reduces toxicity of the herb (Bensky et al. 2004). Unprocessed Chinese arisaema is considered toxic and inappropriate for internal use (Bensky et al. 2004; Chen and Chen 2004). The prepared rhizome, that has been processed to reduce toxicity, is the subject of this entry.

Adverse Events and Side Effects

Adverse reactions to Chinese arisaema are generally associated with the incompletely processed herb or with overdose. Irritating calcium oxalate crystals are generally responsible for reactions, which may include numbness of the tongue and mouth, itching and burning sensations, swelling, salivation, and loss of taste (Bensky et al. 2004; Chen and Chen 2004; Wu and Zhong 2008).

Topical reactions may occur if fresh Chinese arisaema comes in contact with the skin. This reaction may include swelling, itching, and pain (Bensky et al. 2004; Chen and Chen 2004).

Pharmacological Considerations

None known.

Pregnancy and Lactation

While one text on traditional Chinese medicine indicates that Chinese arisaema should be used with extreme caution during pregnancy (Chen and Chen 2004), another text indicates that this herb is contraindicated in pregnancy (Bensky et al. 2004).

No information on the safety of Chinese arisaema during lactation was identified.

While this review identified equivocal concerns for pregnant women and none while nursing, safety has not been conclusively established, and this substance is not recommended for use except under the supervision of an expert qualified in its appropriate use.

Review Details

I. Drug and Supplement Interactions

Clinical Trials of Drug or Supplement Interactions

No clinical trials of drug or supplement interactions were identified.

Case Reports of Suspected Drug or Supplement Interactions

No case reports of suspected drug or supplement interactions were identified.

Animal Trials of Drug or Supplement Interactions

No animal trials of drug or supplement interactions were identified.

II. Adverse Events

Case Reports of Adverse Events

Adverse reactions to Chinese arisaema are generally associated with the incompletely processed herb or with overdose (standard dose listed as a decoction of 3 to 10 g). Reactions may include numbness of the tongue and mouth, itching, burning sensations, swelling, edema, salivation, loss of taste, difficulty opening the mouth, slurred speech, erosion and necrosis of the oral and pharyngeal mucosa, headache, nausea, vomiting, dizziness, and palpitations. In severe cases, muscle spasms, convulsions, respiratory depression, or respiratory paralysis may occur (Bensky et al. 2004; Chen and Chen 2004).

III. Pharmacology and Pharmacokinetics

Human Pharmacological Studies

No relevant human pharmacological studies were identified.

Animal Pharmacological Studies

No relevant animal pharmacological studies were identified.

In Vitro Pharmacological Studies

No relevant in vitro pharmacological studies were identified.

IV. Pregnancy and Lactation

While one text on traditional Chinese medicine indicates that Chinese arisaema should be used with extreme caution during pregnancy (Chen and Chen 2004), another text indicates that this herb is contraindicated in pregnancy (Bensky et al. 2004).

No information on the safety of Chinese arisaema during lactation was identified.

V. Toxicity Studies

Acute Toxicity

The LD_{50} of a decoction of processed Chinese arisaema orally administered in mice could not be determined at doses up to 150 g/kg. After intraperitoneal administration, the LD_{50} of the same extract in mice is 13.5 g/kg (Chen and Chen 2004).

Literature Cited

Bensky, D., S. Clavey, and E. Stöger. 2004. *Chinese herbal medicine: Materia medica*. 3rd ed. Seattle: Eastland Press.

Chen, J.K., and T.T. Chen. 2004. *Chinese medical herbology and pharmacology*. City of Industry, CA: Art of Medicine Press.

Wu, H., and L.Y. Zhong. 2008. Study on irritation of calcium oxalate crystal in Araceae plants. *Zhongguo Zhong Yao Za Zhi* 33(4):380-384.

Arisaema triphyllum (L.) Schott — Araceae

SCN: Jack-in-the-pulpit
OCN: Indian turnip

Part: dried tuber

Quick Reference Summary

Safety Class: 2d
Interaction Class: A

Contraindications

Do not exceed recommended dose (Felter 1922; Felter and Lloyd 1898).

Other Precautions

None known.

Drug and Supplement Interactions

None known.

Standard Dose

The standard dose is 1 to 5 drops of tincture every 30 minutes to 1 hour (Felter 1922).

Editors' Note

Jack-in-the-pulpit contains water-insoluble calcium oxalate crystals that are destroyed by processing (heating or drying) (Felter and Lloyd 1898; Keating 2004; Nakata and McConn 2000; Nelson et al. 2006; Weber 1891).

Adverse Events and Side Effects

The calcium oxalate crystals, present in the unprocessed plant, are microscopic needlelike structures that mechanically irritate the skin and mucous membranes, causing a painful burning sensation of the lips and mouth after ingestion. Ingestion may cause an inflammatory reaction, often with edema and blistering, sometimes resulting in hoarseness and difficulty swallowing (Nelson et al. 2006).

Pharmacological Considerations

None known.

Pregnancy and Lactation

No information on the safety of Jack-in-the-pulpit in pregnancy or lactation was identified in the scientific or traditional literature. Although this review did not identify any concerns for use while pregnant or nursing, safety has not been conclusively established.

Review Details

I. Drug and Supplement Interactions

Clinical Trials of Drug or Supplement Interactions

No clinical trials of drug or supplement interactions were identified.

Case Reports of Suspected Drug or Supplement Interactions

No case reports of suspected drug or supplement interactions were identified.

Animal Trials of Drug or Supplement Interactions

No animal trials of drug or supplement interactions were identified.

II. Adverse Events

Case Reports of Adverse Events

No case reports of adverse events were identified.

III. Pharmacology and Pharmacokinetics

Human Pharmacological Studies

No relevant human pharmacological studies were identified.

Animal Pharmacological Studies

No relevant animal pharmacological studies were identified.

In Vitro Pharmacological Studies

No relevant in vitro pharmacological studies were identified.

IV. Pregnancy and Lactation

No information on the safety of Jack-in-the-pulpit during pregnancy or lactation was identified.

V. Toxicity Studies

No toxicity studies were identified.

Literature Cited

Felter, H.W. 1922. *The Eclectic materia medica, pharmacology and therapeutics*. Cincinnati, OH: Scudder.

Felter, H.W., and J.U. Lloyd. 1898. *King's American dispensatory*. 18th ed., 3rd rev. 2 vols. Cincinnati: Ohio Valley Co.

Keating, R.C. 2004. Systematic occurrence of raphide crystals in Araceae. *Ann. Missouri Bot. Gard.* 91(3):495-504.

Nakata, P.A., and M.M. McConn. 2000. Isolation of *Medicago truncatula* mutants defective in calcium oxalate crystal formation. *Plant Physiol.* 124(3):1097-1104.

Nelson, L., R.D. Shih, M.J. Balick, and K.F. Lampe. 2006. *Handbook of poisonous and injurious plants*. 2nd ed. Berlin: Springer.

Weber, R.A. 1891. Raphides, the cause of the acridity of certain plants. *J. Am. Chem. Soc.* 13(7):215-217.

Armoracia rusticana P. Gaertn. et al. — Brassicaceae

SCN: horseradish
Syn: *Armoracia lapathifolia* Gilib.
Part: rhizome, root

Quick Reference Summary

Safety Class: 1
Interaction Class: A

Contraindications

None known.

Other Precautions

Use with caution in persons with stomach or intestinal ulcers, or kidney disorders (Blumenthal et al. 1998).

Drug and Supplement Interactions

None known.

Editors' Note

Although the German Commission E indicates that horseradish should not be used in children under the age of 4 (Blumenthal et al. 1998), no basis for this caution is provided and no other information from the traditional or scientific literature was found to support this contraindication.

Adverse Events and Side Effects

Horseradish may cause gastrointestinal discomfort (Blumenthal et al. 2000).

Pharmacological Considerations

None known.

Pregnancy and Lactation

No information on the safety of horseradish in pregnancy or lactation was identified in the scientific or traditional literature. Although this review did not identify any concerns for use while pregnant or nursing, safety has not been conclusively established.

Review Details

I. Drug and Supplement Interactions

Clinical Trials of Drug or Supplement Interactions

No clinical trials of drug or supplement interactions were identified.

Case Reports of Suspected Drug or Supplement Interactions

No case reports of suspected drug or supplement interactions were identified.

Animal Trials of Drug or Supplement Interactions

No animal trials of drug or supplement interactions were identified.

II. Adverse Events

Case Reports of Adverse Events

No case reports of adverse events were identified.

III. Pharmacology and Pharmacokinetics

Human Pharmacological Studies

No relevant human pharmacological studies were identified.

Animal Pharmacological Studies

No relevant animal pharmacological studies were identified.

In Vitro Pharmacological Studies

No relevant in vitro pharmacological studies were identified.

IV. Pregnancy and Lactation

No information on the safety of horseradish in pregnancy or lactation was identified.

V. Toxicity Studies

Genotoxicity

A slight mutagenic effect of horseradish paste was observed in bacterial mutagenicity assays. The effect was significantly less than that of cabbage or Brussels sprouts (Kassie et al. 1996). Antimutagenic activity of horseradish was observed in rat bone marrow cells (Agabeili and Kasimova 2005).

Literature Cited

Agabeili, R.A., and T.E. Kasimova. 2005. Antimutagenic activity of *Armoracia rusticana*, *Zea mays* and *Ficus carica* plant extracts and their mixture. *Tsitol. I Genet.* 39(3):75-79.

Blumenthal, M., W. Busse, A. Goldberg, et al. 1998. *The complete German Commission E monographs*. Austin, TX: American Botanical Council.

Blumenthal, M., A. Goldberg, and J. Brinckmann. 2000. *Herbal medicine: Expanded Commission E monographs*. Newton, MA: Integrative Medicine.

Kassie, F., W. Parzefall, S. Musk, et al. 1996. Genotoxic effects of crude juices from *Brassica* vegetables and juices and extracts from phytopharmaceutical preparations and spices of cruciferous plants origin in bacterial and mammalian cells. *Chem. Biol. Interact.* 102(1):1-16.

Arnica spp.

Asteraceae

Arnica latifolia Bong.
SCN: arnica
Arnica montana L.
SCN: arnica
OCN: European arnica; leopard's bane; mountain tobacco
Part: flower, rhizome, whole plant

Quick Reference Summary

Safety Class: 3 (internal use), 2d (external use)
Interaction Class: A

Contraindications

Not for internal use except under the supervision of an expert qualified in the appropriate use of this substance (De Smet 1992; Felter and Lloyd 1898; Leung and Foster 1996; List and Hörhammer 1973; Wichtl 2004).

Do not use on open wounds or broken skin (Felter and Lloyd 1898; Mitchell 1983).

Other Precautions

Persons with allergies to other members of the Asteraceae family (such as sunflower, marigold, or feverfew) should exercise caution with arnica, as allergic cross-reactivity to Asteraceae plants is common (Hausen 1996).

Drug and Supplement Interactions

None known.

Editors' Notes

Homeopathic preparations of arnica are commonly used and are distinct from herbal preparations of arnica that are the subject of this entry (Leivers 2005).

Arnica contains trace amounts of the nontoxic pyrrolizidine alkaloids tussilagine and isotussilagine (Passreiter et al. 1992).

Other species of arnica (*A. angustifolia, A. chamissonis, A. chamissonis* Less. ssp. *foliosa, A. cordifolia*, and *A. sororia*) are used interchangeably in trade (McGuffin et al. 1997).

Adverse Events and Side Effects

Allergic contact dermatitis to arnica has been reported and is associated in some cases with sensitivity to other species of plants in the Asteraceae and Lauraceae families (Brinkhaus et al. 2006; Hausen 1980, 1985, 1992, 1996; Hausen and Schulz 1978; Machet et al. 1993; Paulsen et al. 2008; Pirker et al. 1992; Schempp et al. 2002; Schwarzkopf et al. 2006).

Pharmacological Considerations

None known.

Pregnancy and Lactation

Overdoses of arnica have led to miscarriage (Blaschek et al. 2002; Merdinger 1938). Uterine stimulation has been observed in rats and guinea pigs (Brunzell and Wester 1947; Kreitmair 1936).

No information on the safety of arnica during lactation was identified. While this review did not identify any concerns for use while nursing, safety has not been conclusively established, and this substance is not recommended for internal use except under the supervision of an expert qualified in its appropriate use.

Review Details

I. Drug and Supplement Interactions

Clinical Trials of Drug or Supplement Interactions

No clinical trials of drug or supplement interactions were identified.

Case Reports of Suspected Drug or Supplement Interactions

No case reports of suspected drug or supplement interactions were identified.

Animal Trials of Drug or Supplement Interactions

No animal trials of drug or supplement interactions were identified.

II. Adverse Events

Case Reports of Adverse Events

Arnica contains sesquiterpene lactones that produce contact dermatitis in sensitive persons (Hausen 1978, 1980, 1996; Herrmann et al. 1978). Studies on sensitivity to Asteraceae species have indicated that up to 1.4 % of the population may be prone to contact allergy to arnica (de Groot et al. 1988; Hausen 1996; Paulsen et al. 1993; Reider et al. 2001). Cross-reactivity to arnica and other species of Asteraceae, including sunflower (*Helianthus annuus*) and marigolds (*Tagetes* spp.), and to species of Lauraceae have been reported (Hausen 1996; Hausen and Schulz 1978; Machet et al. 1993; Paulsen et al. 2008; Pirker et al. 1992). Other cases of contact allergy to arnica or products containing arnica have been reported (Hausen 1980, 1985; Hörmann and Korting 1994, 1995; Pirker et al. 1992; Rudzki and Grzywa 1977; Schempp et al. 2002; Schwarzkopf et al. 2006; Spettoli et al. 1998).

Topical application of a cream containing 1.5% arnica was reported to trigger a case of leukemia-related Sweet's syndrome (Delmonte et al. 1998).

A young man mistakenly consumed tea made from an unknown amount of arnica flower and leaf. Two hours later he experienced myalgia, headache, chills, and developed hyperthermia, tachycardia, hypotension, and elevated serum levels of creatinine, aspartate aminotransferase, and alanine aminotransferase (Topliff and Grande 2000). Stomach cramping followed by death was reported in a man who had consumed 70 g of arnica tincture (Blaschek et al. 2002). Poisoning with arnica has been reported to cause death due to circulatory paralysis with secondary respiratory arrest (product and dose not specified) (Hänsel et al. 1993).

An older herbal text indicated that internal use of "large doses" of arnica may cause "heat in the throat, nausea, vomiting, purging, spasmodic contractions of the limbs, difficulty of respiration, and sometimes inflammation of the alimentary canal, and coma" (Felter and Lloyd 1898).

III. Pharmacology and Pharmacokinetics

Human Pharmacological Studies

No relevant human pharmacological studies were identified.

Animal Pharmacological Studies

In an irritation study using New Zealand albino rabbits, the primary irritation index of arnica in a base of soybean oil and tocopherol was 0 (Henkel Corp. 1997). A mixture of arnica extract, butylene glycol, and water, applied topically, was not irritating to rabbits (Ichimaru Pharcos Co. 1995). Open application of arnica absolute and arnica resinoid was not irritating to mouse skin (RIFM 1996a, 1996b).

A raw extract, an ether extract, and a tincture of arnica produced sensitization reactions in Pirbright white guinea pigs (Hausen 1978). A mixture containing arnica extract, butylene glycol, and water was not phototoxic to guinea pigs, and arnica absolute and arnica resinoid were not phototoxic to hairless mice (Ichimaru Pharcos Co. 1995; RIFM 1996a, 1996b).

In Vitro Pharmacological Studies

The compounds helenalin and 11a,13-dihydrohelenalin have been shown to inhibit collagen-induced platelet aggregation, thromboxane formation, and 5-hydroxytryptamine secretion in a concentration-dependent manner in human platelets (Schröder et al. 1990).

IV. Pregnancy and Lactation

Miscarriages have been reported after overdoses of arnica infusion (made from 20 g arnica flower) or arnica flower tincture (44 ml) (Blaschek et al. 2002; Merdinger 1938).

No change in uterine tone or contractility was observed in isolated pregnant rat uteruses treated with arnica tincture (Blaschek et al. 2002). Uterine stimulation was observed in rats intragastrically administered a hot aqueous extract of arnica flower (dose not specified) (Kreitmair 1936) and in guinea pigs administered arnica tincture (dose not specified) (Brunzell and Wester 1947).

A survey of medical records of women who had used herbs as abortifacients indicated that consumption of arnica (preparation and dose unspecified) led to two cases of multiple organ system failure (Ciganda and Laborde 2003).

No information on the safety of arnica during lactation was identified.

V. Toxicity Studies

Acute Toxicity

The LD_{50} of orally administered arnica extract in mice is 123 mg/kg (RTECS 1996), and could not be determined in rats at doses up to 5 g/kg (CTFA 1981). The LD_{50} of intraperitoneally administered arnica extract in mice is 31 mg/kg (RTECS 1996).

The dermal LD_{50} of arnica resinoid in rabbits is greater than 5 g/kg; slight irritant effects were observed at this dose level (RIFM 1996b).

Genotoxicity

In the Ames test for mutagenicity, ethanol extracts of arnica produced a two- to fourfold increase in the number of revertants, as compared to controls with *S. typhimurium* TA98 with and without metabolic activation and with *S. typhimurium* TA100 with metabolic activation; an increase was not seen with TA100 without metabolic activation (Goggelmann and Schimmer 1986). The authors indicated that the effect was likely due to the flavonol compounds in arnica, noting that "the origin of the plant is important for the presence of essential components" and that results can differ based on the district of growth and the preparation of the extract (Goggelmann and Schimmer 1986).

Literature Cited

Blaschek, W., S. Ebel, E. Hackenthal, et al. 2002. *Hagers handbuch der drogen und arzneistoffe. HagerROM.* Heidelberg: Springer.

Brinkhaus, B., J.M. Wilkens, R. Ludtke, et al. 2006. Homeopathic arnica therapy in patients receiving knee surgery: Results of three randomised double-blind trials. *Complement Ther. Med.* 14(4):237-246.

Brunzell, A., and S. Wester. 1947. *Arnica chamissonis* and *Arnica montana* compared. *Svensk Farm. Tidskr.* 51:645-651.

Ciganda, C., and A. Laborde. 2003. Herbal infusions used for induced abortion. *J. Toxicol. Clin. Toxicol.* 41(3):235-239.

Cosmetic, Toiletry, and Fragrance Association (CTFA). 1981. Acute oral toxicity, skin irritation, sensitization, and ocular irritation testing on *Arnica montana* extract. Unpublished data.

de Groot, A.C., D.P. Bruynzeel, J.D. Bos, et al. 1988. The allergens in cosmetics. *Arch. Dermatol.* 124(10):1525-1529.

De Smet, P.A.G.M. 1992. *Adverse effects of herbal drugs, volume 1.* Berlin, New York: Springer.

Delmonte, S., C. Brusati, A. Parodi, and A. Rebora. 1998. Leukemia-related Sweet's syndrome elicited by pathergy to *Arnica. Dermatology* 197(2):195-196.

Felter, H.W., and J.U. Lloyd. 1898. *King's American dispensatory.* 18th ed., 3rd rev. 2 vols. Cincinnati: Ohio Valley Co.

Goggelmann, W., and O. Schimmer. 1986. Mutagenic activity of phytotherapeutical drugs. *Prog. Clin. Biol. Res.* 206:63-72.

Hänsel, R., K. Keller, H. Rimpler, and G. Schneider, eds. 1993. *Hagers handbuch der pharmazeutischen praxis.* 5th ed. Berlin: Springer.

Hausen, B.M. 1978. Identification of the allergens of *Arnica montana* L. *Contact Dermat.* 4(5):308.

Hausen, B.M. 1980. [*Arnica* allergy]. *Hautarzt* 31(1):10-17.

Hausen, B.M. 1985. *Gaillardia* allergy. *Derm. Beruf. Umwelt.* 33(2):62-65.

Hausen, B.M. 1992. *In* De Smet PAGM. 1992. *Adverse effects of herbal drugs, volume 1.* New York: Springer.

Hausen, B.M. 1996. A 6-year experience with compositae mix. *Am. J. Contact Dermat.* 7(2):94-99.

Hausen, B.M., and K.H. Schulz. 1978. Polyvalent contact allergy in a florist. *Derm. Beruf. Umwelt.* 26(5):175-176.

Henkel Corp. 1997. *Cited in* Fiume M. 2001. Final report on the safety assessment of *Arnica montana* extract and *Arnica montana. Int. J. Toxicol.* 20(Suppl. 2):1-11.

Herrmann, H.D., G. Willuhn, and B.M. Hausen. 1978. Helenalinmethacrylate, a new pseudoguaianolide from the flowers of *Arnica montana* L. and the sensitizing capacity of their sesquiterpene lactones. *Planta Med.* 34(3):299-304.

Hörmann, H., and H. Korting. 1994. Akute allergische Kontaktdermatitis auf Arnika-Tinktur. *Dermatosen* 42:246-249.

Hörmann, H., and H. Korting. 1995. Allergic acute contact dermatitis due to arnica tincture self-medication. *Phytomedicine* 3:315-317.

Ichimaru Pharcos Co. 1995. Specifications of arnica liquid (*Arnica montana* extract and butylene glycol water). Unpublished data submitted by CTFA.

Kreitmair, H. 1936. Pharmacological trials with some domestic plants. *E Merck's Jahr. Neuer. Geb. Pharmakother. Pharm.* 50:102-110.

Leivers, K. 2005. Unravelling the confusion around arnica's herbal and homoeopathic use. *Pharm. J.* 275:289-291.

Leung, A.Y., and S. Foster. 1996. *Encyclopedia of common natural ingredients used in food, drugs, and cosmetics.* 2nd ed. New York: Wiley.

A

List, P.H., and H. Hörhammer. 1973. *Hagers handbuch der pharmazeutischen praxis*. Vollst. 4. Neuausg. ed. Berlin, Heidelberg, New York: Springer.
Machet, L., L. Vaillant, A. Callens, et al. 1993. Allergic contact dermatitis from sunflower (*Helianthus annuus*) with cross-sensitivity to arnica. *Contact Dermat.* 28(3):184-185.
McGuffin, M., C. Hobbs, R. Upton, and A. Goldberg. 1997. *Botanical safety handbook*. Boca Raton, FL: CRC Press.
Merdinger, O. 1938. *MMW* 1496. *Cited in* Mills S., Bone K. 2005. *The essential guide to herbal safety*. St. Louis: Elsevier.
Mitchell, H. 1983. *British herbal pharmacopoeia*. Bournemouth, U.K.: British Herbal Medicine Association.
Passreiter, C.M., G. Willuhn, and E. Roder. 1992. Tussilagine and isotussilagine: Two pyrrolizidine alkaloids in the genus *Arnica*. *Planta Med.* 58(6):556-557.
Paulsen, E., K.E. Andersen, and B.M. Hausen. 1993. Compositae dermatitis in a Danish dermatology department in one year (I). Results of routine patch testing with the sesquiterpene lactone mix supplemented with aimed patch testing with extracts and sesquiterpene lactones of Compositae plants. *Contact Dermat.* 29(1):6-10.
Paulsen, E., L.P. Christensen, and K.E. Andersen. 2008. Cosmetics and herbal remedies with Compositae plant extracts—Are they tolerated by Compositae-allergic patients? *Contact Dermat.* 58(1):1523.
Pirker, C., T. Moslinger, D.Y. Koller, M. Gotz, and R. Jarisch. 1992. Cross-reactivity with *Tagetes* in *Arnica* contact eczema. *Contact Dermat.* 26(4):217-219.
Reider, N., P. Komericki, B.M. Hausen, P. Fritsch, and W. Aberer. 2001. The seamy side of natural medicines: Contact sensitization to arnica (*Arnica montana* L.) and marigold (*Calendula officinalis* L.). *Contact Dermat.* 45(5):269-272.
RIFM. 1996a. Monograph 1121—Arnica resinoid. Hackensack, NJ: Research Institute for Fragrance Materials. *Cited in* Fiume, M. 2001. Final report on the safety assessment of *Arnica montana* extract and *Arnica montana*. *Int. J. Toxicol.* 20(Suppl. 2):1-11.
RIFM. 1996b. Monograph 1240—Arnica absolute. Hackensack, NJ: Research Institute for Fragrance Materials. *Cited in* Fiume M. 2001. Final report on the safety assessment of *Arnica montana* extract and *Arnica montana*. *Int. J. Toxicol.* 20(Suppl. 2):1-11.
RTECS. 1996. Registry of the toxic effects of chemical substances (RTECS). Bethesda, MD: National Library of Medicine. *Cited in* Fiume M. 2001. Final report on the safety assessment of *Arnica montana* extract and *Arnica montana*. *Int. J. Toxicol.* 20(Suppl. 2):1-11.
Rudzki, E., and Z. Grzywa. 1977. Dermatitis from *Arnica montana*. *Contact Dermat.* 3:281-282.
Schempp, C.M., E. Schopf, and J.C. Simon. 2002. Plant-induced toxic and allergic dermatitis (phytodermatitis). *Hautarzt* 53(2):93-97.
Schröder, H., W. Lösche, H. Strobach, et al. 1990. Helenalin and 11a,13-dihydrohelenalin, two constituents from *Arnica montana* L., inhibit human platelet function via thiol-dependent pathways. *Thromb. Res.* 57:839-845.
Schwarzkopf, S., P.L. Bigliardi, and R.G. Panizzon. 2006. Allergic contact dermatitis from Arnica. *Rev. Med. Suisse* 2(91):2884-2885.
Spettoli, E., S. Silvani, P. Lucente, L. Guerra, and C. Vincenzi. 1998. Contact dermatitis caused by sesquiterpene lactones. *Am. J. Contact Dermat.* 9(1):49-50.
Topliff, A., and G. Grande. 2000. Significant toxicity after the ingestion of arnica. *J. Toxicol. Clin. Toxicol.* 38(5):518.
Wichtl, M. 2004. *Herbal drugs and phytopharmaceuticals: A handbook for practice on a scientific basis*. 3rd ed. Boca Raton, FL: CRC Press.

Artemisia abrotanum L. — Asteraceae

SCN: southernwood
OCN: lad's love
Part: herb

Quick Reference Summary

Safety Class: 2b
Interaction Class: A

Contraindications

Not for use in pregnancy except under the supervision of a qualified healthcare practitioner (McGuffin et al. 1997).

Other Precautions

None known.

Drug and Supplement Interactions

None known.

Notice

Emmenagogue (Madaus 1938; Steinegger and Hänsel 1972); *see* Appendix 2.

Adverse Events and Side Effects

None known.

Pharmacological Considerations

None known.

Pregnancy and Lactation

Use of southernwood during pregnancy is not recommended (McGuffin et al. 1997). Southernwood has been reported to have emmenagogue activity (Madaus 1938; McGuffin et al. 1997; Steinegger and Hänsel 1972).

No information on the safety of southernwood during lactation was identified. While this review did not identify any concerns for use while nursing, safety has not been conclusively established.

Review Details

I. Drug and Supplement Interactions

Clinical Trials of Drug or Supplement Interactions

No clinical trials of drug or supplement interactions were identified.

Case Reports of Suspected Drug or Supplement Interactions

No case reports of suspected drug or supplement interactions were identified.

Animal Trials of Drug or Supplement Interactions

No animal trials of drug or supplement interactions were identified.

II. Adverse Events

Adverse Events Reported in Clinical Trials

In a small study, a nasal spray preparation of southernwood essential oil was generally well tolerated in adults with allergic rhinitis. Immediately after local nasal administration, all patients reported a slight to moderate nasal stinging sensation that lasted from 5 to 20 s. The sensation was not experienced as unpleasant. During repeated and prolonged use, none of the patients reported untoward effects such as mucosal irritation or injury, bleeding, or dry nose (Remberg et al. 2004).

Case Reports of Adverse Events

No case reports of adverse events were identified.

III. Pharmacology and Pharmacokinetics

Human Pharmacological Studies

No relevant human pharmacological studies were identified.

Animal Pharmacological Studies

No relevant animal pharmacological studies were identified.

In Vitro Pharmacological Studies

No relevant in vitro pharmacological studies were identified.

IV. Pregnancy and Lactation

Use of southernwood during pregnancy is not recommended (McGuffin et al. 1997).

No information on the safety of southernwood during lactation was identified.

V. Toxicity Studies

No toxicity studies were identified.

Literature Cited

Madaus, G. 1938. *Lehrbuch der biologischen heilmittel*. Leipzig: Thieme.

McGuffin, M., C. Hobbs, R. Upton, and A. Goldberg. 1997. *Botanical safety handbook*. Boca Raton, FL: CRC Press.

Remberg, P., L. Björk, T. Hedner, and O. Sterner. 2004. Characteristics, clinical effect profile and tolerability of a nasal spray preparation of *Artemisia abrotanum* L. for allergic rhinitis. *Phytomedicine* 11(1):36-42.

Steinegger, E., and R. Hänsel. 1972. *Lehrbuch der pharmakognosie auf phytochemischer grundlage*. New York: Springer.

Artemisia absinthium L. — Asteraceae

SCN: wormwood

Part: herb

Quick Reference Summary

Safety Class: 2b, 2c, 2d
Interaction Class: A

Contraindications

Not for use during pregnancy or lactation (Blagojevic et al. 2006; McGuffin et al. 1997).

Not for long-term use; do not exceed recommended dose (Leung and Foster 1996; Pinto-Scognamiglio 1967; Wichtl 2004).

Other Precautions

None known.

Drug and Supplement Interactions

None known.

Standard Dose

Not to exceed 1.5 g of dried herb in tea, two to three times daily (Wichtl 2004).

Notice

Thujone (α-thujone 0–20% of essential oil, β-thujone 1.3–46% of essential oil; essential oil content of plant is 0.2 to 1.5%) (Blagojevic et al. 2006; Lawrence 1995; Wichtl 2004); *see* Appendix 1

Editors' Notes

Several herbal reference texts do not substantiate toxicological concerns for wormwood, except with long-term use (Leung and Foster 1996), excessive doses, or use of the essential oil (Felter and Lloyd 1898; Weiss and Meuss 2001; Wichtl 2004).

Regulatory restrictions against the use of wormwood requiring finished food products to be thujone-free exist in the United States and other countries (CFR 2011; Leung and Foster 1996; Martindale and Reynolds 1996).

Wormwood contains α- and β-thujone, compounds that may cause convulsions and that bind to GABA receptors, resulting in excitation of the autonomic nervous system (Höld et al. 2000). High amounts of thujone and thujone metabolites in the body may provoke convulsions and unconsciousness (Lee and Balick 2005; Olsen 2000). Nicotinic acid receptor, muscarinic cholinergic receptor, and 5-HT_3 receptor activity has also been reported (Deiml et al. 2004; Lee and Balick 2005).

Adverse Events and Side Effects

No reports of adverse events or side effects of wormwood used at standard doses were identified. Overdose of wormwood essential oil has resulted in seizures, rhabdomyolysis (breakdown of muscle fibers), vomiting, gastrointestinal cramps, retention of urine, and, in severe cases, confusion and renal lesions (Berlin and Smilkstein 1996; Weisbord et al. 1997; Wichtl 2004).

Pharmacological Considerations

Thujone is a porphyrogenic terpenoid and thus may be hazardous to patients with an underlying defect in hepatic heme synthesis (Bonkovsky et al. 1992).

Pregnancy and Lactation

While safety classes 2b and 2c in this text indicate that use in pregnancy or lactation is acceptable under the supervision of an expert qualified in the use of the described substance, the editors of this text know of no reason for which wormwood would be used in pregnancy or lactation. Based on the bioactivity of thujone and other compounds, wormwood should not be used during pregnancy or lactation.

Limited information on the safety of wormwood use during pregnancy and lactation is available. One study in rats indicated anti-implantation activity of wormwood (Rao et al. 1988).

Review Details

I. Drug and Supplement Interactions

Clinical Trials of Drug or Supplement Interactions

No clinical trials of drug or supplement interactions were identified.

Case Reports of Suspected Drug or Supplement Interactions

No case reports of suspected drug or supplement interactions were identified.

Animal Trials of Drug or Supplement Interactions

No animal trials of drug or supplement interactions were identified.

II. Adverse Events

Case Reports of Adverse Events

Acute renal failure caused by rhabdomyolysis resulting from tonic clonic seizures was reported in a man who ingested 10 ml of wormwood essential oil. The patient had no history of neuromuscular disease, kidney disease, or history of alcohol dependence. The manufacturer confirmed that the wormwood essential oil consumed was pure and not adulterated (Weisbord et al. 1997). Ingestion of 60 ml of wormwood essential oil led to altered mental status, seizures, and secondary complications including hyperthermia, rhabdomyolysis, and aspiration (Berlin and Smilkstein 1996).

Bradyarrhythmia was reported in a man with acute absinthe intoxication. The authors of the report noted that although tachyarrhythmias are frequently developed in acute alcohol intoxication, bradyarrhythmias are exceptional in this context (Benezet-Mazuecos and de la Fuente 2006).

A condition known as absinthism was observed in chronic consumers of the alcoholic beverage absinthe, which contains wormwood extract. The condition was described as a form of alcoholism that included delirium, hallucinations, tremors, and seizures (Lee and Balick 2005). While the compound thujone was once thought to be the primary cause of the psychotropic activity and toxicity of absinthe, recent analyses of absinthe indicate that the thujone content of historical and contemporary samples is insignificant and that other ingredients, such as the coloring agents copper sulfate or antimony chloride, may have been responsible for the adverse effects of absinthe (Blaschek et al. 2002; Lachenmeier et al. 2008).

III. Pharmacology and Pharmacokinetics

Human Pharmacological Studies

A decrease in attention performance was associated with thujone content of beverages administered to healthy volunteers at concentrations of 0, 10, or 100 mg/l thujone in alcohol, leading to an average thujone dose of 0, 0.026, or 0.26 mg/kg (doses were adjusted to provide equal amounts of alcohol in each dose). The thujone concentration was associated with a temporary counteracting of the anxiolytic effect of alcohol (Dettling et al. 2004).

Animal Pharmacological Studies

No relevant animal pharmacological studies were identified.

In Vitro Pharmacological Studies

No relevant in vitro pharmacological studies were identified.

IV. Pregnancy and Lactation

Anti-implantation activity was reported in rats administered an ethanol extract of wormwood (Rao et al. 1988).

Wormwood is listed as an ingredient in two herbal abortifacient formulas traditionally used in Iran (Madari and Jacobs 2004). Toxic effects and death after consumption of large amounts of wormwood (extract not specified but believed to be the essential oil) have been reported in the context of using the plant as an abortifacient (Gessner 1974).

V. Toxicity Studies

Acute Toxicity

The LD_{50} value of wormwood liquid extract administered orally in mice is 2.5 g/kg (Lagarto Parra et al. 2001); that of an ethanol extract administered intraperitoneally in rats is 1 g/kg (Sharma et al. 1978); and that of the essential oil administered orally in rats is 96 mg/kg (Opdyke 1975).

No mortality was observed in rabbits orally administered hexane, chloroform, and water extracts of wormwood at doses up to 1.6 g/kg (Khattak et al. 1985).

The doses of intraperitoneally administered α- and β-thujone that produce convulsions have been reported as 590 mg/kg in mice (Wenzel and Ross 1957) and 100 mg/kg in rats (Sampson and Fernandez 1939).

Subchronic Toxicity

A 13-week repeated-dose toxicity study of wormwood extract performed in male and female rats indicated that the no-observed-adverse-effect level was equivalent to 1.27 g/kg daily in male rats and 2.06 g/kg daily in female rats (Muto et al. 2003).

Chronic Toxicity

Diminished life expectancy and growth but no increase in liver damage or tumor formation were observed in rats orally administered a total of 15.3 g of powdered wormwood over an 18-month period (Schmahl 1956).

Chronic administration of thujone is reported to lead to fatty degeneration of the liver (Pinto-Scognamiglio 1967).

Genotoxicity

No mutagenic activity was reported in an Ames test of wormwood extract (Schimmer et al. 1994).

Literature Cited

Benezet-Mazuecos, J., and A. de la Fuente. 2006. Electrocardiographic findings after acute absinthe intoxication. *Int. J. Cardiol.* 113(2):e48-e50.

Berlin, R., and M. Smilkstein. 1996. Wormwood oil@toxic.ing. *J. Toxicol. Clin. Toxicol.* 34(5):543.

Blagojevic, P., N. Radulovic, R. Palic, and G. Stojanovic. 2006. Chemical composition of the essential oils of Serbian wild-growing *Artemisia absinthium* and *Artemisia vulgaris*. *J. Agric. Food Chem.* 54(13):4780-4789.

Blaschek, W., S. Ebel, E. Hackenthal, et al. 2002. *Hagers handbuch der drogen und arzneistoffe. HagerROM.* Heidelberg: Springer.

Bonkovsky, H.L., E.E. Cable, J.W. Cable, et al. 1992. Porphyrogenic properties of the terpenes camphor, pinene, and thujone (with a note on historic implications for absinthe and the illness of Vincent van Gogh). *Biochem. Pharmacol.* 43(11):2359-2368.

CFR. 2011. *Code of federal regulations*, Title 21 Part 172.510, 2011 ed. Food additives permitted for direct addition to food for human consumption. Natural flavoring substances and natural substances used in conjunction with flavors. Washington, DC: U.S. Government Printing Office.

Deiml, T., R. Haseneder, W. Zieglglinsberger, et al. 2004. α-Thujone reduces $5\text{-}HT_3$ receptor activity by an effect on the agonist-reduced desensitization. *Neuropharmacology* 46:192-201.

Dettling, A., H. Grass, A. Schuff, et al. 2004. Absinthe: Attention performance and mood under the influence of thujone. *J. Stud. Alcohol* 65(5):573-581.

Felter, H.W., and J.U. Lloyd. 1898. *King's American dispensatory*. 18th ed., 3rd rev. 2 vols. Cincinnati: Ohio Valley Co.

Gessner, O. 1974. *Cited in* Mills, S., Bone K. 2005. *The essential guide to herbal safety*. St. Louis: Elsevier.

Höld, K.M., N.S. Sirisoma, T. Ikeda, T. Narahashi, and J.E. Casida. 2000. Alpha-thujone (the active component of absinthe): Gamma-aminobutyric acid type A receptor modulation and metabolic detoxification. *Proc. Natl. Acad. Sci. U.S.A.* 97(8):3826-3831.

Khattak, S.G., S.N. Gilani, and M. Ikram. 1985. Antipyretic studies on some indigenous Pakistani medicinal plants. *J. Ethnopharmacol.* 14(1):45-51.

Lachenmeier, D., D. Nathan-Maister, T. Breaux, et al. 2008. Chemical composition of vintage preban absinthe with special reference to thujone, fenchone, pinocamphone, methanol, copper, and antimony concentrations. *J. Agric. Food Chem.* 59(9):3073-3081.

Lagarto Parra, A., R. Silva Yhebra, I. Guerra Sardinas, and L. Iglesias Buela. 2001. Comparative study of the assay of *Artemia salina* L. and the estimate of the medium lethal dose (LD_{50} value) in mice, to determine oral acute toxicity of plant extracts. *Phytomedicine* 8(5):395-400.

Lawrence, B.M. 1995. Essential oils 1992–1994. In *Natural flavor and fragrance materials*. Carol Stream, IL: Allured Publishing Corp.

Lee, R.A., and M.J. Balick. 2005. Absinthe: La fee vert. *Explore J. Sci. Healing* 1(3):217-219.

Leung, A.Y., and S. Foster. 1996. *Encyclopedia of common natural ingredients used in food, drugs, and cosmetics*. 2nd ed. New York: Wiley.

Madari, H., and R.S. Jacobs. 2004. An analysis of cytotoxic botanical formulations used in the traditional medicine of ancient Persia as abortifacients. *J. Nat. Prod.* 67(8):1204-1210.

Martindale, W., and J.E.F. Reynolds. 1996. *The extra pharmacopoeia*. 31st ed. edited by James E.F. Reynolds London: Pharmaceutical Press.

McGuffin, M., C. Hobbs, R. Upton, and A. Goldberg. 1997. *Botanical safety handbook*. Boca Raton, FL: CRC Press.

A

Muto, T., T. Watanabe, M. Okamura, et al. 2003. Thirteen-week repeated dose toxicity study of wormwood (*Artemisia absinthium*) extract in rats. *J. Toxicol. Sci.* 28(5):471-478.

Olsen, R. 2000. Absinthe and the GABA receptors. *Proc. Natl. Acad. Sci. U.S.A.* 97:4417-4418.

Opdyke, D. 1975. Monographs on fragrance raw materials. *Food Cosmet. Toxicol.* 13:721-722.

Pinto-Scognamiglio, W. 1967. Connaissances actuelles sur l'activité pharmacodynamique de la thujone, aromatisant naturel d'un emploi entendu. *Boll. Chim. Farm.* 106:292-300.

Rao, V., A. Menezes, and M. Gadelha. 1988. Antifertility screening of some indigenous plants of Brasil. *Fitoterapia* 59:17-20.

Sampson, W.L., and L. Fernandez. 1939. Experimental convulsions in the rat. *J. Pharm. Exp. Ther.* 65:275.

Schimmer, O., A. Kruger, H. Paulini, and F. Haefele. 1994. An evaluation of 55 commercial plant extracts in the Ames mutagenicity test. *Pharmazie* 49(6):448-451.

Schmahl, D. 1956. Failure of cancerogenic effect of *Artemesia absinthium* fed to rats. *Z. Krebsforsch.* 61(3):227-229.

Sharma, M.L., N. Chandokhe, B.J. Ghatak, et al. 1978. Pharmacological screening of Indian medicinal plants. *Indian J. Exp. Biol.* 16(2):228-240.

Weisbord, S.D., J.B. Soule, and P.L. Kimmel. 1997. Poison on line—acute renal failure caused by oil of wormwood purchased through the Internet. *New Engl. J. Med.* 337:825-827.

Weiss, R.F., and A.R. Meuss. 2001. *Weiss's herbal medicine.* Classic ed. New York: Thieme.

Wenzel, D., and C. Ross. 1957. Central stimulating properties of some terpenones. *J. Am. Pharm. Assoc.* 46:77.

Wichtl, M. 2004. *Herbal drugs and phytopharmaceuticals: A handbook for practice on a scientific basis.* 3rd ed. Boca Raton, FL: CRC Press.

Artemisia annua L.

Asteraceae

SCN: sweet wormwood
PN: *qing hao* (above-ground parts)
OCN: annual wormwood; sweet Annie
Part: above-ground parts

Quick Reference Summary

Safety Class: 2b
Interaction Class: A

Contraindications

Not for use in pregnancy except under the supervision of a qualified healthcare practitioner (Al Kadi 2007; Boareto et al. 2008).

Other Precautions

None known.

Drug and Supplement Interactions

None known.

Editors' Note

Sweet wormwood contains the compound artemisinin. Artemisinin and compounds derived from sweet wormwood are widely used as antimalarial drugs (Meshnick 2002).

Adverse Events and Side Effects

No adverse effects from sweet wormwood are reported in texts on traditional Chinese medicine for normal doses prepared as a decoction (Bensky et al. 2004; Chen and Chen 2004), though use of an extract tablet (dosage not provided) is reported to have caused gastrointestinal side effects in 3.4% of patients (Bensky et al. 2004).

A case of hepatitis was reported in association with a supplement containing the compound artemisinin (Malhotra et al. 2009).

No serious adverse events have been reported from artemisinin or artemisinin derivatives used as antimalarial drugs. Mild adverse events associated with the use of these compounds include nausea, vomiting, and diarrhea, although these have also been noted as symptoms of malaria (Gordi and Lepist 2004; McGready et al. 1998; Meshnick 2002; Meshnick et al. 1996; Ribeiro and Olliaro 1998).

Pharmacological Considerations

None known.

Pregnancy and Lactation

A review of antimalarial drugs indicated that artemisinin-based compounds are not recommended for use during the first trimester of pregnancy, due to lack of safety data, but the compounds may be used during the second and third trimesters (Al Kadi 2007). An animal study published after that review indicated that an increase in postimplantation losses was observed in animals given relatively high amounts (35 or 70 mg/kg) of the compound artemisinin but not in animals given amounts comparable to the standard human dose (7 mg/kg) (Boareto et al. 2008).

No information on the safety of sweet wormwood during lactation was identified. While this review did not identify any concerns for use while nursing, safety has not been conclusively established.

Review Details

I. Drug and Supplement Interactions

Clinical Trials of Drug or Supplement Interactions

No clinical trials of drug or supplement interactions were identified.

Case Reports of Suspected Drug or Supplement Interactions

No case reports of suspected drug or supplement interactions were identified.

Animal Trials of Drug or Supplement Interactions

In studies with the compound artemisinin in rodent malaria (a normally susceptible strain of *Plasmodium berghei*), marked potentiative synergism was found with the drugs mefloquine, tetracycline, and spiramycin. Some synergism was also observed with primaquine. Combinations of artemisinin with dapsone, sulfadiazine, sulfadoxine, pyrimethamine, pyrimethamine/sulfadoxine, and cycloguanil showed antagonism. A high degree of potentiation was shown between artemisinin and primaquine with a primaquine-resistant strain, while the combination with mefloquine showed enhanced potentiation with a mefloquine-resistant strain. Combinations of artemisinin with mefloquine, primaquine, tetracycline or clindamycin showed marked potentiation with an artemisinin-resistant strain (Chawira et al. 1987).

II. Adverse Events

Case Reports of Adverse Events

Reviews of animal toxicity studies of the compound artemisinin and artemisinin derivatives indicate that neurotoxic effects have been observed, notably in the brain stem. These reviews acknowledge, however, that such toxicity has not been observed after widespread use in humans. As of 1996, an estimated 2 million malaria patients had been treated with artemisinin and its derivatives, with no serious adverse events reported. Common nonserious adverse events include nausea, vomiting, and diarrhea, although these have also been noted as symptoms of malaria (Gordi and Lepist 2004; McGready et al. 1998; Meshnick 2002; Meshnick et al. 1996; Ribeiro and Olliaro 1998). Occasional neutropenia, reticulocytopenia, elevated liver enzymes, transient bradycardia, and prolonged QT intervals have also been reported (Ribeiro and Olliaro 1998).

Hepatitis was reported in a 52-year-old man who had been taking two capsules, three times daily for 1 week, of a supplement containing 100 mg per capsule of the compound artemisinin. Laboratory analyses confirmed the presence of artemisinin and the absence of adulterants (Malhotra et al. 2009).

III. Pharmacology and Pharmacokinetics

Human Pharmacological Studies

No relevant human pharmacological studies were identified.

Animal Pharmacological Studies

No relevant animal pharmacological studies were identified.

In Vitro Pharmacological Studies

Flavonol compounds isolated from sweet wormwood were found to potentiate the activity of berberine and norfloxacin against a multidrug-resistant strain of *Staphylococcus aureus* (Stermitz et al. 2002).

IV. Pregnancy and Lactation

In rats orally administered 7, 35, or 70 mg/kg of the compound artemisinin daily from gestational days 7 to 13 or 14 to 20, an increase in postimplantation losses was observed at the 35 and 75 mg/kg dose levels. The losses were correlated with a trend to lower maternal progestogens and a maternal testosterone decrease. Results indicate toxicity for both periods of treatment, with lower sensitivity at later stages of pregnancy (Boareto et al. 2008).

A review of antimalarial drugs indicated that, due to lack of safety data, artemisinin-based compounds are not recommended for use during the first trimester of pregnancy but the compounds may be used during the second and third trimesters (Al Kadi 2007).

No information on the safety of sweet wormwood during lactation was identified.

V. Toxicity Studies

Acute Toxicity

The LD_{50} of the compound artemisinin in mice is 4.23 g/kg after oral administration, 3.84 g/kg after intramuscular administration, 1.59 g/kg after intraperitoneal administration, and 9 g/kg after subcutaneous administration (Zhu 1998). The LD_{50} of the compound artemisinin in rats is 5.58 g/kg after oral administration and 2.57 g/kg after intramuscular administration (Zhu 1998).

Literature Cited

Al Kadi, H.O. 2007. Antimalarial drug toxicity: A review. *Chemotherapy* 53(6):385-391.

Bensky, D., S. Clavey, and E. Stöger. 2004. *Chinese herbal medicine: Materia medica*. 3rd ed. Seattle: Eastland Press.

Boareto, A.C., J.C. Muller, A.C. Bufalo, et al. 2008. Toxicity of artemisinin (*Artemisia annua* L.) in two different periods of pregnancy in Wistar rats. *Reprod. Toxicol.* 25(2):239-246.

A

Chawira, A.N., D.C. Warhurst, B.L. Robinson, and W. Peters. 1987. The effect of combinations of qinghaosu (artemisinin) with standard antimalarial drugs in the suppressive treatment of malaria in mice. *Trans. R. Soc. Trop. Med. Hyg.* 81(4):554-558.

Chen, J.K., and T.T. Chen. 2004. *Chinese medical herbology and pharmacology*. City of Industry, CA: Art of Medicine Press.

Gordi, T., and E.I. Lepist. 2004. Artemisinin derivatives: Toxic for laboratory animals, safe for humans? *Toxicol. Lett.* 147(2):99-107.

Malhotra, U., R. Rakita, F. Fernandez, et al. 2009. Hepatitis temporally associated with an herbal supplement containing artemisinin—Washington, 2008. *J. Am. Med. Assoc.* 302(13):1412-1414.

McGready, R., T. Cho, J.J. Cho, et al. 1998. Artemisinin derivatives in the treatment of falciparum malaria in pregnancy. *Trans. R. Soc. Trop. Med. Hyg.* 92(4):430-433.

Meshnick, S.R. 2002. Artemisinin: Mechanisms of action, resistance and toxicity. *Int. J. Parasitol.* 32(13):1655-1660.

Meshnick, S.R., T.E. Taylor, and S. Kamchonwongpaisan. 1996. Artemisinin and the antimalarial endoperoxides: From herbal remedy to targeted chemotherapy. *Microbiol. Rev.* 60(2):301-15.

Ribeiro, I.R., and P. Olliaro. 1998. Safety of artemisinin and its derivatives. A review of published and unpublished clinical trials. *Med. Trop.* 58(3 Suppl.):50-53.

Stermitz, F.R., L.N. Scriven, G. Tegos, and K. Lewis. 2002. Two flavonols from *Artemisia annua* which potentiate the activity of berberine and norfloxacin against a resistant strain of *Staphylococcus aureus*. *Planta Med.* 68(12):1140-1141.

Zhu, Y.-P. 1998. *Chinese materia medica: Chemistry, pharmacology and applications* Amsterdam: Harwood Academic Publishers.

Artemisia dracunculus L. 'Sativa' — Asteraceae

SCN: French tarragon
OCN: estragon; tarragon

Part: herb

Quick Reference Summary

Safety Class: 1
Interaction Class: A

Contraindications

None known.

Other Precautions

None known.

Drug and Supplement Interactions

None known.

Notice

Alkenylbenzenes (estragole 60–81% of the essential oil, essential oil content is 0.9–5.3%) (Arabhosseini et al. 2006; De Vincenzi et al. 2000; Ribnicky et al. 2004; Wichtl 2004); *see* Appendix 1.

Adverse Events and Side Effects

None known.

Pharmacological Considerations

An animal study and traditional human use indicate that French tarragon may modify glucose regulation (Ribnicky et al. 2006; Swanston-Flatt et al. 1991). People with diabetes are advised to monitor their blood sugar closely and discuss the use of this herb with a qualified healthcare practitioner prior to use.

In vitro studies indicate that extracts of French tarragon may inhibit platelet aggregation (Shahriyary and Yazdanparast 2007; Tognolini et al. 2006).

Pregnancy and Lactation

No information on the safety of French tarragon in pregnancy or lactation was identified in the scientific or traditional literature. Although this review did not identify any concerns for use while pregnant or nursing, safety has not been conclusively established.

Review Details

I. Drug and Supplement Interactions

Clinical Trials of Drug or Supplement Interactions

No clinical trials of drug or supplement interactions were identified.

Case Reports of Suspected Drug or Supplement Interactions

No case reports of suspected drug or supplement interactions were identified.

Animal Trials of Drug or Supplement Interactions

No animal trials of drug or supplement interactions were identified.

II. Adverse Events

Case Reports of Adverse Events

No case reports of adverse events were identified.

III. Pharmacology and Pharmacokinetics

Human Pharmacological Studies

No relevant human pharmacological studies were identified.

Animal Pharmacological Studies

In diabetic mice orally administered 500 mg/kg of an ethanol extract of French tarragon daily for 7 days, a reduction in blood glucose levels was observed (Ribnicky et al. 2006).

In Vitro Pharmacological Studies

In guinea pig and rat plasma, French tarragon essential oil exhibited antiplatelet activity against ADP-, arachidonic acid-, and U46619-induced platelet aggregation. The essential oil also demonstrated an ability to destabilize clot retraction (Tognolini et al. 2006).

Inhibition of platelet adhesion, aggregation, and secretion was observed in human platelets treated with a methanol extract of French tarragon (Shahriyary and Yazdanparast 2007).

IV. Pregnancy and Lactation

No information on the safety of French tarragon during pregnancy or lactation was identified.

V. Toxicity Studies

Acute Toxicity

The LD_{50} of an ethanol extract of French tarragon orally administered to rats could not be determined at doses up to 5 g/kg (Ribnicky et al. 2004).

Subchronic Toxicity

In rats orally administered 10, 100, or 1000 mg/kg of an ethanol extract of French tarragon daily for 90 days, clinical chemistry, organ weights, and microscopic examination of organs indicated no adverse effects (Ribnicky et al. 2004).

Genotoxicity

In the Ames test for mutagenicity with *Salmonella typhimurium* strains TA98, TA100, TA1535, TA1537, and TA102, no biologically relevant mutagenic activity of an ethanol extract of French tarragon was observed with or without metabolic activation by S9 (Ribnicky et al. 2004).

The essential oil of French tarragon exhibited mutagenic activity in the *Bacillus subtilis* rec-assay but not in the *Salmonella*/microsome reversion assay (Zani et al. 1991).

Literature Cited

Arabhosseini, A., S. Padhye, T.A. van Beek, et al. 2006. Loss of essential oil of tarragon *Artemisia dracunculus* L. due to drying. *J. Sci. Food Agric.* 86(15):2543-2550.

De Vincenzi, M., M. Silano, F. Maialetti, and B. Scazzocchio. 2000. Constituents of aromatic plants: II. Estragole. *Fitoterapia* 71(6):725-729.

Ribnicky, D.M., A. Poulev, J. O'Neal, et al. 2004. Toxicological evaluation of the ethanolic extract of *Artemisia dracunculus* L. for use as a dietary supplement and in functional foods. *Food Chem. Toxicol.* 42(4):585-598.

Ribnicky, D.M., A. Poulev, M. Watford, W.T. Cefalu, and I. Raskin. 2006. Antihyperglycemic activity of Tarralin, an ethanolic extract of *Artemisia dracunculus* L. *Phytomedicine* 13(8):550-557.

Shahriyary, L., and R. Yazdanparast. 2007. Inhibition of blood platelet adhesion, aggregation and secretion by *Artemisia dracunculus* leaves extracts. *J. Ethnopharmacol.* 114(2):194-198.

Swanston-Flatt, S.K., P.R. Flatt, C. Day, and C.J. Bailey. 1991. Traditional dietary adjuncts for the treatment of diabetes mellitus. *Proc. Nutr. Soc.* 50:641-651.

Tognolini, M., E. Barocelli, V. Ballabeni, et al. 2006. Comparative screening of plant essential oils: Phenylpropanoid moiety as basic core for antiplatelet activity. *Life Sci.* 78(13):1419-1432.

Wichtl, M. 2004. *Herbal drugs and phytopharmaceuticals: A handbook for practice on a scientific basis*. 3rd ed. Boca Raton, FL: CRC Press.

Zani, F., G. Massimo, S. Benvenuti, et al. 1991. Studies on the genotoxic properties of essential oils with *Bacillus subtilis* rec-assay and *Salmonella*/microsome reversion assay. *Planta Med.* 57(3):237-241.

Artemisia spp.

Asteraceae

Artemisia douglasiana **Bess.**
SCN: western mugwort

Artemisia lactiflora **Wall. ex DC.**
SCN: white mugwort

Artemisia vulgaris **L.**
SCN: mugwort
AN: *nagadamani*
Part: herb

Quick Reference Summary

Safety Class: 2b
Interaction Class: A

Contraindications

Not for use in pregnancy except under the supervision of a qualified healthcare practitioner (Jamwal and Anand 1962; Riddle 1997; Saha et al. 1961; Wichtl 2004).

Other Precautions

None known.

Drug and Supplement Interactions

None known.

Notice

Emmenagogue (Saha et al. 1961; Williamson 2003); *see* Appendix 2.

Thujone (~64% of the essential oil; *A. vulgaris* contains 0.03–0.3% essential oil) (Misra and Singh 1986); *see* Appendix 1.

Adverse Events and Side Effects

A review of contact dermatitis due to plant exposures indicated that mugwort was a relatively common cause of dermatitis (Kurz and Rapaport 1979; Saito et al. 1982).

Pharmacological Considerations

None known.

Pregnancy and Lactation

Some reports indicate that mugwort has been used as an abortifacient or in abortifacient formulas (Jamwal and Anand 1962; Riddle 1997; Saha et al. 1961; Wichtl 2004). Based on this information, use in pregnancy is not recommended except under the supervision of a qualified healthcare practitioner.

No information on the safety of mugwort during lactation was identified. While this review did not identify any concerns for use while nursing, safety has not been conclusively established.

Review Details

I. Drug and Supplement Interactions

Clinical Trials of Drug or Supplement Interactions

No clinical trials of drug or supplement interactions were identified.

Case Reports of Suspected Drug or Supplement Interactions

No case reports of suspected drug or supplement interactions were identified.

Animal Trials of Drug or Supplement Interactions

No animal trials of drug or supplement interactions were identified.

II. Adverse Events

Case Reports of Adverse Events

A review of contact dermatitis due to plant exposures indicated that mugwort was a relatively common cause of dermatitis (Kurz and Rapaport 1979; Saito et al. 1982).

A severe allergic reaction was reported in a florist with a preexisting sunflower allergy after occupational contact with mugwort. Skin testing (prick and scratch) revealed strong sensitization to mugwort and French tarragon (Kurzen et al. 2003).

Skin prick testing in a patient that experienced anaphylaxis after pine nut ingestion indicated cross-reactivity to mugwort (Rodrigues-Alves et al. 2008). In a patient sensitized to lettuce, CAP inhibition testing indicated that the patient was also allergic to mugwort (Vila et al. 1998).

III. Pharmacology and Pharmacokinetics

Human Pharmacological Studies

No relevant human pharmacological studies were identified.

Animal Pharmacological Studies

In nondiabetic mice orally administered 250 mg/kg of a methanol extract of mugwort, an increase in the blood glucose level was observed in the oral glucose tolerance test (Villasenor and Lamadrid 2006).

In Vitro Pharmacological Studies

No relevant in vitro pharmacological studies were identified.

IV. Pregnancy and Lactation

Some reports indicate that mugwort has been used as an abortifacient or in abortifacient formulas (Jamwal and Anand 1962; Riddle 1997; Saha et al. 1961; Wichtl 2004). No stimulant action of mugwort was observed in isolated guinea pig uteruses (Saha et al. 1961).

No information on the safety of mugwort during lactation was identified.

V. Toxicity Studies

No toxicity studies were identified.

Literature Cited

Jamwal, K.S., and K.K. Anand. 1962. Preliminary screening of some reputed abortifacient indigenous plants. *Indian J. Pharm.* 24:218-220.

Kurz, G., and M.J. Rapaport. 1979. External/internal allergy to plants (*Artemesia*). *Contact Dermat.* 5(6):407-409.

Kurzen, M., C. Bayerl, and S. Goerdt. 2003. Occupational allergy to mugwort. *J. Dtsch. Dermatol. Ges.* 1(4):285-290.

Misra, L.N., and S.P. Singh. 1986. α-Thujone, the major component of the essential oil from *Artemisia vulgaris* growing wild in Nilgiri Hills. *J. Nat. Prod.* 49(5):941.

Riddle, J. 1997. *Eve's herbs: A history of contraception and abortion in the West.* Cambridge, MA: Harvard University Press.

Rodrigues-Alves, R., A. Pregal, M.C. Pereira-Santos, et al. 2008. Anaphylaxis to pine nut: Cross-reactivity to *Artemisia vulgaris*? *Allergol. Immunopathol. (Madrid)* 36(2):113-116.

Saha, J.C., S. Kasinathan, J.C. Saha, and S. Kasinathan. 1961. Ecbolic properties of Indian medicinal plants. *Indian J. Med. Res.* 49:130-151.

Saito, F., O. Urushibata, and T. Murao. 1982. Contact dermatitis from plants for the last 6 years. *Hifu* 24(2):238-249.

Vila, L., G. Sanchez, M.L. Sanz, et al. 1998. Study of a case of hypersensitivity to lettuce (*Lactuca sativa*). *Clin. Exp. Allergy* 28(8):1031-1035.

Villasenor, I.M., and M.R. Lamadrid. 2006. Comparative anti-hyperglycemic potentials of medicinal plants. *J. Ethnopharmacol.* 104(1-2):129-131.

Wichtl, M. 2004. *Herbal drugs and phytopharmaceuticals: A handbook for practice on a scientific basis.* 3rd ed. Boca Raton, FL: CRC Press.

Williamson, E.M. 2003. *Potter's herbal cyclopedia.* Saffron Walden, Essex: C.W. Daniel Co.

Artemisia spp. — Asteraceae

Artemisia capillaris Thunb.
SCN: yin-chen wormwood
PN: *yin chen hao* (above-ground parts, herb)
OCN: capillaris; capillary artemisia

***Artemisia scoparia* Waldst. & Kit.**
SCN: yin-chen wormwood
PN: *yin chen hao* (herb)
Part: above-ground parts, herb

Quick Reference Summary

Safety Class: 1
Interaction Class: A

Contraindications

None known.

Other Precautions

None known.

Drug and Supplement Interactions

None known.

Notice

Thujone (α-thujone 0.7% of *A. scoparia* essential oil, 0.1% of *A. capillaris* essential oil; β-thujone 0.3% of *A. scoparia* essential oil, trace amount in *A. capillaris* essential oil) (Cha et al. 2005); *see* Appendix 1.

Adverse Events and Side Effects

None known.

Pharmacological Considerations

None known.

Pregnancy and Lactation

No information on the safety of yin-chen wormwood herb or above-ground parts in pregnancy or lactation was identified in the scientific or traditional literature. Although this review did not identify any concerns for use while pregnant or nursing, safety has not been conclusively established.

Review Details

I. Drug and Supplement Interactions

Clinical Trials of Drug or Supplement Interactions

No clinical trials of drug or supplement interactions were identified.

Case Reports of Suspected Drug or Supplement Interactions

No case reports of suspected drug or supplement interactions were identified.

Animal Trials of Drug or Supplement Interactions

No animal trials of drug or supplement interactions were identified.

II. Adverse Events

Case Reports of Adverse Events

Rare cases of adverse events have been reported in persons taking relatively high (15 to 30 g) doses of yin-chen wormwood. Symptoms included headache, dizziness, numbness and tremor of the upper extremities, heart palpitations, fainting, and an oppressive feeling in the chest. Details on the exact doses, preparation, and relevant medical history were not reported (Bensky et al. 2004).

III. Pharmacology and Pharmacokinetics

Human Pharmacological Studies

No relevant human pharmacological studies were identified.

A

Animal Pharmacological Studies

No relevant animal pharmacological studies were identified.

In Vitro Pharmacological Studies

Of 75 compounds isolated from yin-chen wormwood, 15 showed antiplatelet aggregation activity (Wu et al. 2001).

IV. Pregnancy and Lactation

Inhibition of implantation was observed in female rats orally administered 10 mg/kg of the compound scoparone. At doses of 25 or 50 mg/kg, the number of pups born, ratio of live/dead pups, and their weight gain were adversely affected, with skeletal abnormalities observed in some pups (Chandhoke 1979).

No information on the safety of yin-chen wormwood during pregnancy or lactation was identified.

V. Toxicity Studies

Acute Toxicity

The LD_{50} of the compound capillarin intraperitoneally administered to mice is 262.5 mg/kg (Chen and Chen 2004).

Short-Term Toxicity

Oral administration of the compound scoparone to male rats at doses of 10, 25, or 50 mg/kg daily for 75 days produced a decrease in weight gain and dose-related mortality (Chandhoke 1979).

Genotoxicity

No mutagenic activity of an aqueous extract of yin-chen wormwood was observed in the Ames mutagenicity assay with *Salmonella typhimurium* strains TA98 or TA100 with or without metabolic activation by S9 (Yin et al. 1991).

In the mouse micronucleus and chromosomal aberration assays, some mutagenic activity was observed in mice intraperitoneally administered an aqueous extract of yin-chen wormwood at doses of 4.5, 9, or 18 g/kg. In the micronucleus assay, an increase of polychromatic erythrocytes was observed at the 4.5, 9, and 18 g/kg dose levels. An increase in the incidence of chromosomal aberrations was observed at the 4.5, 9, and 18 g/kg dose levels, but not at the 0.9 g/kg dose level (Yin et al. 1991).

Literature Cited

Bensky, D., S. Clavey, and E. Stöger. 2004. *Chinese herbal medicine: Materia medica*. 3rd ed. Seattle: Eastland Press.

Cha, J.-D., M.-R. Jeong, S.-I. Jeong, et al. 2005. Chemical composition and antimicrobial activity of the essential oils of *Artemisia scoparia* and *A. capillaris*. *Planta Med.* 71(2):186-190.

Chandhoke, N. 1979. Scoparone effect on reproductive processes in rats. *Indian J. Exp. Biol.* 17(8):740-742.

Chen, J.K., and T.T. Chen. 2004. *Chinese medical herbology and pharmacology*. City of Industry, CA: Art of Medicine Press.

Wu, T.S., Z.J. Tsang, P.L. Wu, et al. 2001. New constituents and antiplatelet aggregation and anti-HIV principles of *Artemisia capillaris*. *Bioorg. Med. Chem.* 9(1):77-83.

Yin, X.J., D.X. Liu, H.C. Wang, and Y. Zhou. 1991. A study on the mutagenicity of 102 raw pharmaceuticals used in Chinese traditional medicine. *Mutat. Res.* 260(1):73-82.

Asclepias asperula (Decne.) Woodson — Asclepiadaceae

SCN: inmortal

Part: root

Quick Reference Summary

Safety Class: 2b
Interaction Class: A

Contraindications

Not for use in pregnancy except under the supervision of a qualified healthcare practitioner (Conway and Slocumb 1979; Moore 2003).

Other Precautions

Use with heart medications is cautioned (Moore 2003).

Drug and Supplement Interactions

None known.

Adverse Events and Side Effects

Overdose may cause nausea and vomiting (standard dose is listed as 2 teaspoons of root daily) (Moore 2003).

Pharmacological Considerations

None known.

Pregnancy and Lactation

Inmortal has been used to facilitate childbirth and to induce abortions (Conway and Slocumb 1979). Based on this, use during pregnancy is not recommended except under the supervision of a qualified healthcare practitioner.

No information on the safety of inmortal during lactation was identified. While this review did not identify any concerns for use while nursing, safety has not been conclusively established.

Review Details

I. Drug and Supplement Interactions

Clinical Trials of Drug or Supplement Interactions

No clinical trials of drug or supplement interactions were identified.

Case Reports of Suspected Drug or Supplement Interactions

No cases of suspected drug or supplement interactions were identified.

Animal Trials of Drug or Supplement Interactions

No animal trials of drug or supplement interactions were identified.

II. Adverse Events

Case Reports of Adverse Events

No case reports of adverse events associated with inmortal were identified.

III. Pharmacology and Pharmacokinetics

Human Pharmacological Studies

No relevant human pharmacological studies were identified.

Animal Pharmacological Studies

No relevant animal pharmacological studies were identified.

In Vitro Pharmacological Studies

Some cardenolides isolated from inmortal have exhibited binding affinity for a physiological receptor, porcine kidney Na^+,K^+-ATPase. The order of binding affinities for the cardenolides from highest to lowest was uzarigenin, desglucouzarin, uzarin, and 6′-*O*-(*E*-4-hydroxycinnamoyl)desglucouzarin (Abbott et al. 1998).

IV. Pregnancy and Lactation

Hot and cold water extracts of inmortal have been used in childbirth and large doses of decoctions have been used to induce abortions. Large doses are reported to cause nausea (Conway and Slocumb 1979).

No information on the safety of inmortal during lactation was identified.

V. Toxicity Studies

No toxicity studies were identified.

Literature Cited

Abbott, A.J., C.G. Holoubek, and R.A. Martin. 1998. Inhibition of Na^+,K^+-ATPase by the cardenolide 6′-*O*-(*E*-4-hydroxycinnamoyl) desglucouzarin. *Biochem. Biophys. Res. Commun.* 251(1):256-259.

Conway, G., and J. Slocumb. 1979. Plants used as abortifacients and emmenagogues by Spanish New Mexicans. *J. Ethnopharmacol.* 1:241-261.

Kingsbury, J.M. 1964. *Poisonous plants of the United States and Canada*. Englewood Cliffs, NJ: Prentice-Hall.

List, P.H., and H. Hörhammer. 1973. *Hagers handbuch der pharmazeutischen praxis*. Vollst. 4. Neuausg. ed. Berlin, Heidelberg, New York: Springer.

Martindale, W., and J.E.F. Reynolds. 1996. *The extra pharmacopoeia*. 31st ed. London: Pharmaceutical Press.

Moore, M. 2003. *Medicinal plants of the Mountain West*. Revised and expanded edition. Santa Fe: Museum of New Mexico Press.

Turner, N., and A. Szczawinski. 1991. *Common poisonous plants and mushrooms of North America*. Portland, OR: Timber Press.

Asclepias tuberosa L. — Asclepiadaceae

SCN: pleurisy root

Part: root

Quick Reference Summary

Safety Class: 2b

Interaction Class: A

Contraindications

Not for use in pregnancy except under the supervision of a qualified healthcare practitioner (List and Hörhammer 1973; Moore 2003).

Other Precautions

May cause nausea and vomiting (List and Hörhammer 1973).

Drug and Supplement Interactions

None known.

Notice

Emetic (List and Hörhammer 1973); *see* Appendix 2.

Editors' Note

While the standardized common name of the root of *A. tuberosa* is pleurisy root, that of the plant itself is butterfly weed (McGuffin et al. 2000).

Adverse Events and Side Effects

While one toxicology text indicates that ingestion of pleurisy root may cause irritation of the mouth, throat, and gastrointestinal tract (Lewis 1998), a pharmacy text written in the late 1800s, when pleurisy root was commonly used, did not list such adverse effects (Felter and Lloyd 1898).

Pharmacological Considerations

None known.

Pregnancy and Lactation

Uterine stimulation and uterotonic activity has been reported in animal studies (Costello and Butler 1949). One reference indicates that pleurisy root is probably inappropriate for use in "delicate pregnancies" (Moore 2003). Based on these reports, use in pregnancy is not recommended except under the supervision of a qualified healthcare practitioner.

No information on the safety of pleurisy root during lactation was identified. While this review did not identify any concerns for use while nursing, safety has not been conclusively established.

Review Details

I. Drug and Supplement Interactions

Clinical Trials of Drug or Supplement Interactions

No clinical trials of drug or supplement interactions were identified.

Case Reports of Suspected Drug or Supplement Interactions

No cases of suspected drug or supplement interactions were identified.

Animal Trials of Drug or Supplement Interactions

No animal trials of drug or supplement interactions were identified.

II. Adverse Events

Case Reports of Adverse Events

No case reports of adverse events associated with pleurisy root were identified.

III. Pharmacology and Pharmacokinetics

Human Pharmacological Studies

No relevant human pharmacological studies were identified.

Animal Pharmacological Studies

In an older study, estrogenic activity of a pleurisy root extract was observed in castrated rats (Costello and Butler 1949).

In Vitro Pharmacological Studies

Inhibition of ouabain binding to membrane Na^+,K^+-ATPase and cross-reactivity with digoxin antibody in a radioimmunoassay from a pleurisy root aqueous extract were observed in vitro (Longerich et al. 1993).

IV. Pregnancy and Lactation

In vivo uterine stimulant action was observed in rabbits, dogs, and cats intravenously administered an ethanol- and isopropanol-based extract of pleurisy root (Costello and Butler 1949). Increased uterine tone was observed in cats and rabbits injected with an extract of pleurisy root (Hassan 1952).

One reference indicates that pleurisy root is probably inappropriate for use in "delicate pregnancies" (Moore 2003).

No information on the safety of pleurisy root during lactation was identified.

V. Toxicity Studies

Acute Toxicity

An alcohol extract of pleurisy root administered intravenously to rabbits and intraperitoneally to rats at doses of 0.04 ml/kg induced partial paralysis (Hassan 1952).

Short-Term Toxicity

Rats intraperitoneally administered 10 mg daily of an alcohol extract of pleurisy root developed diarrhea and continuous tremors after 5 days (Hassan 1952).

Literature Cited

Costello, C., and C. Butler. 1949. The estrogenic and uterine stimulating activity of *Asclepias tuberosa*. *J. Am. Pharm. Assoc. Sci. Ed.* 39:233-237.

Felter, H.W., and J.U. Lloyd. 1898. *King's American dispensatory*. 18th ed., 3rd rev. 2 vols. Cincinnati: Ohio Valley Co.

Hassan, W.E., Jr. 1952. Studies on species of *Asclepias*. VI. Toxicology, pathology, and pharmacology. *J. Am. Pharm. Assoc.* 41(6):298-300.

Kingsbury, J.M. 1964. *Poisonous plants of the United States and Canada*. Englewood Cliffs, NJ: Prentice-Hall.

Lewis, R. 1998. *Lewis' dictionary of toxicology*. Boca Raton, FL: CRC Press.

List, P.H., and H. Hörhammer. 1973. *Hagers handbuch der pharmazeutischen praxis*. Vollst. 4. Neuausg. ed. Berlin, Heidelberg; New York: Springer.

Longerich, L., E. Johnson, and M.H. Gault. 1993. Digoxin-like factors in herbal teas. *Clin. Invest. Med.* 16(3):210-218.

Martindale, W., and J.E.F. Reynolds. 1996. *The extra pharmacopoeia*. 31st ed. London: Pharmaceutical Press.

McGuffin, M., J. Kartesz, A. Leung, and A.O. Tucker. 2000. *Herbs of commerce*. 2nd ed. Silver Spring, MD: American Herbal Products Association.

Mills, S., and K. Bone. 2005. *The essential guide to herbal safety*. St. Louis: Elsevier.

Moore, M. 2003. *Medicinal plants of the Mountain West*. Revised and expanded edition. Santa Fe: Museum of New Mexico Press.

Turner, N., and A. Szczawinski. 1991. *Common poisonous plants and mushrooms of North America*. Portland, OR: Timber Press.

Asparagus adscendens Roxb. — Liliaceae

SCN: *Asparagus adscendens*
AN: *shveta mushali*

Part: root

Quick Reference Summary

Safety Class: 1
Interaction Class: A

Contraindications

None known.

Other Precautions

None known.

Drug and Supplement Interactions

None known.

Adverse Events and Side Effects

None known.

Pharmacological Considerations

None known.

Pregnancy and Lactation

No information on the safety of *Asparagus adscendens* in pregnancy or lactation was identified in the scientific or traditional literature. Although this review did not identify any concerns for use while pregnant or nursing, safety has not been conclusively established.

Asparagus adscendens has traditionally been used to promote milk secretion in lactating women (Kapoor 2001).

Review Details

I. Drug and Supplement Interactions

Clinical Trials of Drug or Supplement Interactions

No clinical trials of drug or supplement interactions were identified.

Case Reports of Suspected Drug or Supplement Interactions

No case reports of suspected drug or supplement interactions were identified.

Animal Trials of Drug or Supplement Interactions

No animal trials of drug or supplement interactions were identified.

II. Adverse Events

Case Reports of Adverse Events

No case reports of adverse events were identified.

III. Pharmacology and Pharmacokinetics

Human Pharmacological Studies

No relevant human pharmacological studies were identified.

Animal Pharmacological Studies

No relevant animal pharmacological studies were identified.

In Vitro Pharmacological Studies

An aqueous extract of *Asparagus adscendens* produced an increase in glucose-dependent insulinotropic actions in pancreatic β-cells. The extract also produced an increase in glucose uptake in adipocytes (Mathews et al. 2006).

IV. Pregnancy and Lactation

No information on the safety of *Asparagus adscendens* during pregnancy was identified.

Asparagus adscendens has traditionally been used to promote milk secretion in lactating women (Kapoor 2001).

V. Toxicity Studies

No toxicity studies were identified.

Literature Cited

Kapoor, L.D. 2001. *Handbook of Ayurvedic medicinal plants*. Boca Raton, FL: CRC Press.

Mathews, J.N., P.R. Flatt, and Y.H. Abdel-Wahab. 2006. *Asparagus adscendens* (shweta musali) stimulates insulin secretion, insulin action and inhibits starch digestion. *Br. J. Nutr.* 95(3):576-581.

A

Asparagus cochinchinensis (Lour.) Merr. — Liliaceae

SCN: Chinese asparagus
PN: *tian men dong* (root tuber)

Part: root tuber

Quick Reference Summary

Safety Class: 1
Interaction Class: A

Contraindications
None known.

Other Precautions
None known.

Drug and Supplement Interactions
None known.

Adverse Events and Side Effects
None known.

Pharmacological Considerations
None known.

Pregnancy and Lactation
Texts on traditional Chinese medicine do not indicate any cautions for use of Chinese asparagus during pregnancy or lactation (Bensky et al. 2004; Chen and Chen 2004). Although this review did not identify any concerns for use while pregnant or nursing, safety has not been conclusively established.

Review Details

I. Drug and Supplement Interactions

Clinical Trials of Drug or Supplement Interactions
No clinical trials of drug or supplement interactions were identified.

Case Reports of Suspected Drug or Supplement Interactions
No case reports of suspected drug or supplement interactions were identified.

Animal Trials of Drug or Supplement Interactions
No animal trials of drug or supplement interactions were identified.

II. Adverse Events

Case Reports of Adverse Events
No case reports of adverse events were identified.

III. Pharmacology and Pharmacokinetics

Human Pharmacological Studies
No relevant human pharmacological studies were identified.

Animal Pharmacological Studies
No relevant animal pharmacological studies were identified.

In Vitro Pharmacological Studies
No relevant in vitro pharmacological studies were identified.

IV. Pregnancy and Lactation
Texts on traditional Chinese medicine do not indicate any cautions for use of Chinese asparagus during pregnancy or lactation (Bensky et al. 2004; Chen and Chen 2004).

V. Toxicity Studies
No toxicity studies were identified.

Literature Cited

Bensky, D., S. Clavey, and E. Stöger. 2004. *Chinese herbal medicine: Materia medica*. 3rd ed. Seattle: Eastland Press.

Chen, J.K., and T.T. Chen. 2004. *Chinese medical herbology and pharmacology*. City of Industry, CA: Art of Medicine Press.

Asparagus officinalis L.

Liliaceae

SCN: asparagus

Part: rhizome

Quick Reference Summary

Safety Class: 1
Interaction Class: A

Contraindications

None known.

Other Precautions

None known.

Drug and Supplement Interactions

None known.

Notice

Diuretic (Blumenthal et al. 1998; Fyfe and Scudder 1909; Jeaffreson 1855; Van Wyk and Wink 2004); *see* Appendix 2.

Adverse Events and Side Effects

Occupational allergic contact dermatitis from the above-ground parts of asparagus has been reported (Rademaker and Yung 2000; Tabar et al. 2004). Allergic reactions, including anaphylactic reactions, have been reported after ingestion of the above-ground parts of asparagus (Tabar et al. 2004). No relevance to asparagus root of these records of side effects for the above-ground parts has been established.

Pharmacological Considerations

See Notice.

Pregnancy and Lactation

No information on the safety of asparagus in pregnancy or lactation was identified in the scientific or traditional literature. Although this review did not identify any concerns for use while pregnant or nursing, safety has not been conclusively established.

Review Details

I. Drug and Supplement Interactions

Clinical Trials of Drug or Supplement Interactions

No clinical trials of drug or supplement interactions were identified.

Case Reports of Suspected Drug or Supplement Interactions

No case reports of suspected drug or supplement interactions were identified.

Animal Trials of Drug or Supplement Interactions

No animal trials of drug or supplement interactions were identified.

II. Adverse Events

Case Reports of Adverse Events

Occupational allergic contact dermatitis to asparagus was reported in a farm worker who worked with asparagus plants (Rademaker and Yung 2000).

Allergic reactions, including anaphylactic reactions, to asparagus have been reported and confirmed by patch testing and IgE immunoblotting (Tabar et al. 2004).

III. Pharmacology and Pharmacokinetics

Human Pharmacological Studies

No relevant human pharmacological studies were identified.

Animal Pharmacological Studies

No relevant animal pharmacological studies were identified.

In Vitro Pharmacological Studies

No relevant in vitro pharmacological studies were identified.

IV. Pregnancy and Lactation

No information on the safety of asparagus in pregnancy or lactation was identified.

V. Toxicity Studies

Cytotoxicity

Several steroidal compounds isolated from the rhizome of asparagus had cytotoxic effects in human and mouse tumor cell lines (Huang et al. 2008).

Literature Cited

Blumenthal, M., W. Busse, A. Goldberg, et al. 1998. *The complete German Commission E monographs*. Austin, TX: American Botanical Council.

Fyfe, J.W., and J.M. Scudder. 1909. *Specific diagnosis and specific medication*. Cincinnati, OH: Scudder Brothers.

Huang, X.F., Y.Y. Lin, and L.Y. Kong. 2008. Steroids from the roots of *Asparagus officinalis* and their cytotoxic activity. *J. Integr. Plant Biol.* 50(6):717-722.

Jeaffreson, S.J. 1855. On asparagus as a diuretic. *Br. Med. J.* 3(123):439.

Rademaker, M., and A. Yung. 2000. Contact dermatitis to *Asparagus officinalis*. *Aust. J. Dermatol.* 41(4):262-263.

Tabar, A.I., M.J. Alvarez-Puebla, B. Gomez, et al. 2004. Diversity of asparagus allergy: Clinical and immunological features. *Clin. Exp. Allergy* 34(1):131-136.

Van Wyk, B., and M. Wink. 2004. *Medicinal plants of the world: An illustrated scientific guide to important medicinal plants and their uses*. Portland, OR: Timber Press.

Asparagus racemosus Willd. — Liliaceae

SCN: shatavari
AN: *shatavari*
Part: rhizome

Quick Reference Summary

Safety Class: 1
Interaction Class: A

Contraindications

None known.

Other Precautions

None known.

Drug and Supplement Interactions

None known.

Notice

Diuretic (Kapoor 2001; Williamson 2002); *see* Appendix 2.

Adverse Events and Side Effects

None known.

Pharmacological Considerations

An animal study indicated that shatavari may have estrogenic activity (Pandey et al. 2005).

Pregnancy and Lactation

Shatavari is commonly used as a reproductive system tonic in India and has traditionally been used to prevent threatened miscarriage (Caldecott 2006). A review of the literature on shatavari indicated that, in Ayurveda, this herb is generally considered safe for use during pregnancy (Goyal et al. 2003). An animal study indicated adverse effects in developing fetuses of pregnant mice administered a methanol extract of shatavari during pregnancy (Goel et al. 2006). Shatavari is traditionally prepared in milk, ghee, or water and used throughout pregnancy. A methanol extract would provide larger amounts of steroidal compounds than traditional methods, which may account for the adverse effects reported in the animal study.

Shatavari has traditionally been used to promote milk secretion in lactating women (Joglekar et al. 1967; Kumar et al. 2008; Nadkarni 1954; Patel and Kanitkar 1969).

Review Details

I. Drug and Supplement Interactions

Clinical Trials of Drug or Supplement Interactions

No clinical trials of drug or supplement interactions were identified.

Case Reports of Suspected Drug or Supplement Interactions

No case reports of suspected drug or supplement interactions were identified.

Animal Trials of Drug or Supplement Interactions

No animal trials of drug or supplement interactions were identified.

II. Adverse Events

Case Reports of Adverse Events

No case reports of adverse events were identified.

III. Pharmacology and Pharmacokinetics

Human Pharmacological Studies

No relevant human pharmacological studies were identified.

Animal Pharmacological Studies

In rats orally administered 3 mg/kg of an alcohol extract of shatavari daily on days 1 to 15 of pregnancy, macroscopic findings revealed a prominence of the mammary glands, a dilated vaginal opening, and a transversely situated uterine horn. A thickening of the vaginal epithelium was observed. The authors of the study indicated that those results, along with other reported changes in female reproductive tissues, suggest an estrogenic effect of shatavari on the female mammary gland and genital organs (Pandey et al. 2005).

A reduction in chemically induced tumor incidence was observed in rats administered a chloroform-methanol extract of shatavari (Rao 1981).

In Vitro Pharmacological Studies

The ethanol extract, and hexane, chloroform, and ethyl acetate fractions of shatavari extracts concentration-dependently stimulated insulin secretion in isolated perfused rat pancreas, isolated rat islet cells, and clonal beta-cells (Hannan et al. 2007).

IV. Pregnancy and Lactation

Increases in resorptions of fetuses, gross malformations, intrauterine growth retardation, and small placental size were observed in rats orally administered 100 mg/kg daily of a methanolic extract of shatavari daily for 60 days. A smaller litter size and decrease in pup body weight and length were associated with the treatment (Goel et al. 2006).

The saponin fraction of shatavari was reported to inhibit oxytocin in isolated rat uteri (Gaitonde and Jetmalani 1969).

Shatavari has traditionally been used to prevent threatened miscarriage (Caldecott 2006). A review of the literature on shatavari indicated that, in Ayurveda, this herb is generally considered safe for use during pregnancy (Goyal et al. 2003).

Shatavari has traditionally been used to promote milk secretion in lactating women (Joglekar et al. 1967; Kumar et al. 2008; Nadkarni 1954; Patel and Kanitkar 1969).

V. Toxicity Studies

Acute Toxicity

No adverse effects were observed in mice orally administered 1 g/kg of a shatavari extract (Debelmas and Hache 1976; Rege et al. 1999).

Short-Term Toxicity

No adverse effects were observed in mice and rats orally administered 1 g/kg of an aqueous extract of shatavari daily for 15 to 30 days (Rege et al. 1999).

Literature Cited

Caldecott, T. 2006. *Ayurveda: The divine science of life.* New York: Mosby.

Debelmas, A.M., and J. Hache. 1976. Toxicity of several medicinal plants of Nepal including some behavioral and central nervous system effects. *Plant Med. Phytother.* 10:128-138.

Gaitonde, B.B., and M.H. Jetmalani. 1969. Antioxytocic action of saponin isolated from *Asparagus racemosus* Willd. (shatavari) on uterine muscle. *Arch. Int. Pharmacodyn. Ther.* 179(1):121-129.

Goel, R.K., T. Prabha, M.M. Kumar, et al. 2006. Teratogenicity of *Asparagus racemosus* Willd. root, a herbal medicine. *Indian J. Exp. Biol.* 44(7):570-573.

Goyal, R.K., J. Singh, and H. Lal. 2003. *Asparagus racemosus*—an update. *Indian J. Med. Sci.* 57(9):408-414.

Hannan, J.M., L. Marenah, L. Ali, et al. 2007. Insulin secretory actions of extracts of *Asparagus racemosus* root in perfused pancreas, isolated islets and clonal pancreatic beta-cells. *J. Endocrinol.* 192 (1):159-168.

Joglekar, G.V., R.H. Ahuja, and J.H. Balwani. 1967. Galactogogue effect of *Asparagus racemosus*. Preliminary communication. *Indian Med. J.* 61(7):165.

Kapoor, L.D. 2001. *Handbook of Ayurvedic medicinal plants.* Boca Raton, FL: CRC Press.

Kumar, S., R.K. Mehla, and A.K. Dang. 2008. Use of shatavari (*Asparagus racemosus*) as a galactopoietic and therapeutic herb-A review. *Agric. Rev.* 29(2):132-138.

Nadkarni, A.K. 1954. *Indian materia medica.* Vol. 1. Bombay: Popular Book Dept.

Pandey, S.K., A. Sahay, R.S. Pandey, and Y.B. Tripathi. 2005. Effect of *Asparagus racemosus* rhizome (shatavari) on mammary gland and genital organs of pregnant rat. *Phytother. Res.* 19(8):721-724.

Patel, A.B., and U.K. Kanitkar. 1969. *Asparagus racemosus* Willd., form bordi, as a galactogogue in buffaloes. *Indian Vet. J.* 46(8):718-721.

Rao, A.R. 1981. Inhibitory action of *Asparagus racemosus* on DMBA-induced mammary carcinogenesis in rats. *Int. J. Cancer* 28(5):607-610.

Rege, N.N., U.M. Thatte, and S.A. Dahanukar. 1999. Adaptogenic properties of six rasayana herbs used in Ayurvedic medicine. *Phytother. Res.* 13(4):275-291.

Williamson, E.M. 2002. *Major herbs of Ayurveda.* Edinburgh, New York: Churchill Livingstone.

Astragalus mongholicus Bunge — Fabaceae

SCN: astragalus
Syn: *Astragalus membranaceus* (Fisch. ex Link) Bunge var. *mongholicus* (Bunge) P. K. Hsiao
PN: *huang qi* (root)
OCN: membranous milkvetch, Mongolian milkvetch
Part: root

Quick Reference Summary

Safety Class: 1
Interaction Class: B

Contraindications

None known.

Other Precautions

None known.

Drug and Supplement Interactions

Theoretically, astragalus could be incompatible with immune suppressant drugs such as cyclosporine and corticosteroids (Upton 1999). A limited number of human studies have indicated that astragalus increases the therapeutic effects of corticosteroids and cyclophosphamide in patients with autoimmune disease (Cai et al. 2006; Pan et al. 2008; Su et al. 2007), while animal studies have indicated a reversal of cyclophosphamide-induced immune suppression (Chu et al. 1988b, 1988c).

Astragalus has been shown to potentiate the therapeutic effects of alpha recombinant interferon 1 (rIFN-α-1) in humans (Qian et al. 1990), acyclovir in mice (Zuo et al. 1995), and interleukin 2 (rIL-2) and interferon in vitro (Chu et al. 1988a; Hou et al. 1981).

Editors' Note

The above-ground parts of certain chemically distinct species of astragalus are known to cause a disease called locoism which can cause reproductive abnormalities in livestock. The compound swainsonine, which causes locoism, is not present in *Astragalus mongholicus* (Rios and Waterman 1997).

Adverse Events and Side Effects

None known.

Pharmacological Considerations

None known.

Pregnancy and Lactation

No information on the safety of this herb in pregnancy or lactation was identified in the scientific or traditional literature. Although this review did not identify any concerns for use while pregnant or nursing, safety has not been conclusively established.

Review Details

I. Drug and Supplement Interactions

Clinical Trials of Drug or Supplement Interactions

Astragalus has been shown to potentiate the therapeutic effects of alpha recombinant interferon 1 (rIFN-α-1) in patients with human papillomavirus type 16 or herpes simplex virus type 2 (Qian et al. 1990).

A purified fraction of astragalus was shown to reverse cyclophosphamide-induced immune suppression in rats, enhancing the ability of the rats to reject xenogeneic grafts (Chu et al. 1988b).

In patients with systemic lupus erythematosus, concomitant administration (intraperitoneally) of astragalus and standard corticosteroid/immunosuppressant drug therapy (drug not specified in English language abstract) was shown to enhance the inhibitory function of the corticosteroid on apoptosis and regulate the ratio and function of T lymphocyte subsets as compared to standard corticosteroid treatment alone (Cai et al. 2006). In patients with systemic lupus erythematosus complicated by kidney damage, intravenous administration of 20 ml of astragalus 12 days per month for 3 months with cyclophosphamide administered once a month, a 4.3% infection rate was observed in the astragalus group, while the rate was 25% in the control group, indicating that astragalus used together with cyclophosphamide is more effective than cyclophosphamide alone in decreasing infection rate and urine protein and improving immune function for patients with lupus nephritis (Su et al. 2007).

Case Reports of Suspected Drug or Supplement Interactions

The T-cell-stimulating activity of astragalus could, theoretically, contradict the effects of immune suppressant drugs such as cyclosporine and corticosteroids.

Animal Trials of Drug or Supplement Interactions

A partially purified fraction of astragalus coadministered to mice with the immune suppressant cyclophosphamide after transplantation of mononuclear cells resulted in transplant rejection, indicating a reversal of cyclophosphamide action (Chu et al. 1988b).

Astragalus has been shown to potentiate the therapeutic effects of alpha recombinant interferon 1 (rIFN-α-1) in humans (Qian et al. 1990), acyclovir in mice (Zuo et al. 1995), and interleukin 2 (rIL-2) and interferon in vitro (Chu et al. 1988a; Hou et al. 1981).

II. Adverse Events

Case Reports of Adverse Events

Allergic reactions including skin eruptions and pruritus have been reported in persons taking astragalus (Bensky et al. 2004). Overdoses may cause headache, tightness in the chest, insomnia, dizziness, and hypertension. Anaphylactic reactions have been reported in patients intravenously administered extracts of astragalus (Bensky et al. 2004).

III. Pharmacology and Pharmacokinetics

Human Pharmacological Studies

CD25 expression on T cells was increased in healthy volunteers administered extracts equivalent to 1.23 g of astragalus daily for 7 days (Zwickey et al. 2007).

Animal Pharmacological Studies

No relevant animal pharmacological studies were identified.

In Vitro Pharmacological Studies

An aqueous extract of astragalus was shown to enhance immune function by increasing proliferation of spleen cells,

B cell IgG production, macrophage cytokine production of IL-6 and TNF, and enhancing induction of cytotoxic T cells (Yoshida et al. 1997).

IV. Pregnancy and Lactation

No information on the safety of astragalus during pregnancy or lactation was identified.

V. Toxicity Studies

Acute Toxicity

The LD_{50} of astragalus extract in mice has been reported as 40 g/kg (Chang and But 1987), although no toxicity was observed after administration of a concentrated aqueous extract to rats at doses equivalent to 100 g/kg (routes of administration unspecified in English translations of studies) (Chang and But 1987; Wagner et al. 1997). Intraperitoneal administration of 50 g/kg astragalus extract in mice led to prostration, paralysis, dyspnea, and cyanosis; in some animals, contracture of the extremities was observed before the death of the animals (Chang and But 1987).

Subchronic Toxicity

No signs of clinical toxicity were observed in rats intraperitoneally administered up to 39.9 g/kg astragalus extract for 12 weeks or in dogs intraperitoneally administered astragalus extract up to 19.5 g/kg for 17 weeks (Yu et al. 2007).

Mutagenicity

Astragalus has been reported to have no mutagenic effects (Wagner et al. 1997).

Literature Cited

Bensky, D., S. Clavey, and E. Stöger. 2004. *Chinese herbal medicine: Materia medica*. 3rd ed. Seattle: Eastland Press.

Cai, X.Y., Y.L. Xu, and X.J. Lin. 2006. Effects of radix Astragali injection on apoptosis of lymphocytes and immune function in patients with systemic lupus erythematosus. *Zhongguo Zhong Xi Yi Jie He Za Zhi* 26(5):443-445.

Chang, H., and P. But. 1987. *Pharmacology and applications of Chinese materia medica, volume 2*. Singapore: World Scientific.

Chu, D.T., J. Lepe-Zuniga, W.L. Wong, R. LaPushin, and G.M. Mavligit. 1988a. Fractionated extract of *Astragalus membranaceus*, a Chinese medicinal herb, potentiates LAK cell cytotoxicity generated by low dose of recombinant interleukin-2. *J. Clin. Lab. Immunol.* 26(4):183-187.

Chu, D.T., W.L. Wong, and G.M. Mavligit. 1988b. Immunotherapy with Chinese medicinal herbs. II. Reversal of cyclophosphamide-induced immune suppression by administration of fractionated *Astragalus membranaceus* in vivo. *J. Clin. Lab. Immunol.* 25(3):125-129.

Chu, D.T., W.L. Wong, and G.M. Mavligit. 1988c. Immunotherapy with Chinese medicinal herbs. I: Reversal of cyclophosphamide-induced immune suppression by administration of fractionated *Astragalus membranaceus* in vivo. *J. Clin. Lab. Immunol.* 25(3):119-123.

Hou, Y., Z. Zhang, S. Su, and S. Duan. 1981. Interferon induction and lymphocyte transformation stimulated by *Astragalus membranaceus* in mouse spleen cell cultures. *Zhonghua Weisheng Wuxue Hemian Yixue Zazhi* 1(2):137-139.

Pan, H.F., X.H. Fang, W.X. Li, D.Q. Ye, G.C. Wu, and X.P. Li. 2008. Radix Astragali: A promising new treatment option for systemic lupus erythematosus. *Med. Hypotheses* 71(2):311-312.

Qian, Z., S. Mao, X. Cai, X. Zhang, F. Gao, M. Lu, X. Shao, Y. Li, X. Yang, and Y. Zhuo. 1990. Viral etiology of chronic cervicitis and its therapeutic response to α-recombinant interferon. *Chin. Med. J. (Engl.)* 103(8):647-651.

Rios, J., and P. Waterman. 1997. A review of the pharmacology and toxicology of astragalus. *Phytother. Res.* 11:411-418.

Su, L., J.C. Mao, and J.H. Gu. 2007. Effect of intravenous drip infusion of cyclophosphamide with high-dose *Astragalus* injection in treating lupus nephritis. *Zhong Xi Yi Jie He Xue Bao* 5(3):272-275.

Upton, R. 1999. Astragalus root: *Astragalus membranaceus* and *Astragalus membranaceus* var. *mongholicus*: Analytical, quality control, and therapeutic monograph. Santa Cruz, CA: American Herbal Pharmacopoeia.

Wagner, H., R. Bauer, X. Peigen, C. Jianming, and H. Michler. 1997. Radix Astragali [huang qi]: Chinese drug monographs and analysis. Bayer, Wald: Verlag für Ganzheitliche Medizin Dr Erich Wühr GmbH. Volume 1(8). Available from American Botanical Council, Austin, TX.

Yoshida, Y., M.Q. Wang, J.N. Liu, B.E. Shan, and U. Yamashita. 1997. Immunomodulating activity of Chinese medicinal herbs and *Oldenlandia diffusa* in particular. *Int. J. Immunopharmacol.* 19(7):359-370.

Yu, S.Y., H.T. Ouyang, J.Y. Yang, X.L. Huang, T. Yang, J.P. Duan, J.P. Cheng, Y.X. Chen, Y.J. Yang, and P. Qiong. 2007. Subchronic toxicity studies of Radix Astragali extract in rats and dogs. *J. Ethnopharmacol.* 110(2):352-355.

Zuo, L., X. Dong, and X. Sun. 1995. The curative effects of *Astragalus membranaceus* Bunge (A-6) in combination with acyclovir on mice infected with HSV-1. *Virol. Sin.* 110(2).

Zwickey, H., J. Brush, C.M. Iacullo, E. Connelly, W.L. Gregory, A. Soumyanath, and R. Buresh. 2007. The effect of *Echinacea purpurea*, *Astragalus membranaceus* and *Glycyrrhiza glabra* on CD25 expression in humans: A pilot study. *Phytother. Res.* 21(11):1109-1112.

A

Atractylodes lancea (Thunb.) DC. — Asteraceae

SCN: cang-zhu atractylodes
Syn: *Atractylodes chinensis* (Bunge) Koidz.; *Atractylis ovata* Thunb.

PN: *cang zhu* (rhizome)
Part: rhizome

Quick Reference Summary

Safety Class: 1
Interaction Class: A

Contraindications

None known.

Other Precautions

None known.

Drug and Supplement Interactions

None known.

Adverse Events and Side Effects

None known.

Pharmacological Considerations

None known.

Pregnancy and Lactation

No information on the use of cang-zhu atractylodes during pregnancy or lactation was identified in the scientific or traditional literature. While this review did not identify any concerns for pregnant or lactating women, safety has not been conclusively established.

Review Details

I. Drug and Supplement Interactions

Clinical Trials of Drug or Supplement Interactions

No clinical trials of drug or supplement interactions were identified.

Case Reports of Suspected Drug or Supplement Interactions

No case reports of suspected drug or supplement interactions were identified.

Animal Trials of Drug or Supplement Interactions

No animal trials of drug or supplement interactions were identified.

II. Adverse Events

Adverse Events Reported in Clinical Trials

Extracts of cang-zhu atractylodes have been used in clinical studies to treat infants with rickets. In one study, oral doses of 0.066 ml of the essential oil were administered to 2- or 3-year-olds three times daily for 1 to 2 weeks (Chen and Chen 2004). In a second study, infants were treated with a syrup containing the equivalent of 4.5 g of cang-zhu atractylodes twice daily for 15 days (Chen and Chen 2004). No adverse event information was reported in English language translations.

Case Reports of Adverse Events

No case reports of adverse events were identified.

III. Pharmacology and Pharmacokinetics

Human Pharmacological Studies

No relevant human pharmacological studies were identified.

Animal Pharmacological Studies

No relevant animal pharmacological studies were identified.

In Vitro Pharmacological Studies

No relevant in vitro pharmacological studies were identified.

IV. Pregnancy and Lactation

No information on the safety of cang-zhu atractylodes during pregnancy or lactation was identified.

V. Toxicity Studies

No toxicity studies were identified.

Literature Cited

Chen, J.K., and T.T. Chen. 2004. *Chinese medical herbology and pharmacology*. City of Industry, CA: Art of Medicine Press.

Atractylodes macrocephala Koidz.

Asteraceae

SCN: bai-zhu atractylodes
PN: *bai zhu* (rhizome)

Part: rhizome

Quick Reference Summary

Safety Class: 1
Interaction Class: A

Contraindications

None known.

Other Precautions

None known.

Drug and Supplement Interactions

None known.

Adverse Events and Side Effects

None known.

Pharmacological Considerations

In one human study, a lengthening of prothrombin time (indicating a slowing of blood clotting) was observed in patients treated with bai-zhu atractylodes. Information on the magnitude of effect on prothrombin time was not listed in the English language translation (Chen and Chen 2004).

Pregnancy and Lactation

Bai-zhu atractylodes is used in formulas during pregnancy in traditional Chinese medicine to "calm the fetus" (Bensky et al. 2004).

No information on the safety of bai-zhu atractylodes in lactation was identified. While this review did not identify any concerns for use while nursing, safety has not been conclusively established.

Review Details

I. Drug and Supplement Interactions

Clinical Trials of Drug or Supplement Interactions

No clinical trials of drug or supplement interactions were identified.

Case Reports of Suspected Drug or Supplement Interactions

No case reports of suspected drug or supplement interactions were identified.

Animal Trials of Drug or Supplement Interactions

No animal trials of drug or supplement interactions were identified.

II. Adverse Events

Case Reports of Adverse Events

Overdose (standard dose is a decoction of 6–15 g) or inappropriate use has been associated with vomiting of blood, nosebleeds, bloody stools, elevated temperature, and agitation (Bensky et al. 2004).

III. Pharmacology and Pharmacokinetics

Human Pharmacological Studies

A lengthening of prothrombin time was observed in healthy volunteers administered a 5% decoction of bai-zhu atractylodes three times daily for 4 days. Information on the magnitude of the effect on prothrombin time was not listed in the English language translation (Chen and Chen 2004).

Animal Pharmacological Studies

No relevant animal pharmacological studies were identified.

In Vitro Pharmacological Studies

Induction of the drug-metabolizing isoenzyme CYP3A4 was observed in human liver cells and rat liver microsomes treated with extracts of bai-zhu atractylodes (Dong et al. 2008a, 2008b).

IV. Pregnancy and Lactation

Bai-zhu atractylodes is used during pregnancy in traditional Chinese medicine to "calm the fetus" (Bensky et al. 2004).

Sesquiterpene lactones isolated from bai-zhu atractylodes inhibited contractions induced by acetylcholine and oxytocin in isolated rat uteri (Zhang et al. 2000).

No information on the safety of bai-zhu atractylodes during lactation was identified.

V. Toxicity Studies

Acute Toxicity

The LD_{50} of a decoction of bai-zhu atractylodes intraperitoneally administered in mice is 13.3 g/kg (Chen and Chen 2004).

Cytotoxicity

Pro-oxidant and cytotoxic activities of the compound atractylenolide have been observed (Wang et al. 2006).

Literature Cited

Bensky, D., S. Clavey, and E. Stöger. 2004. *Chinese herbal medicine: Materia medica*. 3rd ed. Seattle: Eastland Press.

Chen, J.K., and T.T. Chen. 2004. *Chinese medical herbology and pharmacology*. City of Industry, CA: Art of Medicine Press.

Dong, H.Y., J.W. Shao, J.F. Chen, et al. 2008a. Transcriptional regulation of cytochrome P450 3A4 by four kinds of traditional Chinese medicines. *Zhongguo Zhong Yao Za Zhi* 33(9):1014-1017, 1089.

Dong, H.Y., J.W. Shao, T. Wang, Y.H. Guo, and L.Y. Yan. 2008b. Effects on the activities and mRNA expression of CYP3A in rat's liver by four kinds of extracts from anti-cancer traditional Chinese medicines. *Zhong Yao Cai* 31(1):68-71.

Wang, C.C., S.Y. Lin, H.C. Cheng, and W.C. Hou. 2006. Pro-oxidant and cytotoxic activities of atractylenolide I in human promyeloleukemic HL-60 cells. *Food Chem. Toxicol.* 44(8):1308-1315.

Zhang, Y.Q., S.B. Xu, Y.C. Lin, et al. 2000. Antagonistic effects of 3 sesquiterpene lactones from *Atractylodes macrocephala* Koidz on rat uterine contraction in vitro. *Acta Pharmacol. Sin.* 21(1):91-96.

Atropa belladonna L. — Solanaceae

SCN: belladonna
OCN: deadly nightshade

Part: leaf

Quick Reference Summary

Safety Class: 3
Interaction Class: C

Contraindications

Not for use except under the supervision of an expert qualified in the appropriate use of this substance (Caksen et al. 2003; Felter and Lloyd 1898; Frohne and Pfänder 2000; Hsu et al. 1995; Kingsbury 1964; Lee 2007; Nelson et al. 2006; Witthaus 1911).

Other Precautions

None known.

Drug and Supplement Interactions

Interaction considerations are the same as those for the drug atropine (Caksen et al. 2003; Lee 2007).

Adverse Events and Side Effects

In overdose, belladonna elicits symptoms typical of anticholinergic poisoning, with symptoms colloquially described as "hot as a hare, dry as a bone, red as a beet, mad as a hatter, blind as a bat" (Lee 2007). The most common poisoning symptoms are redness of the face, dilation of the pupils, increased heart rate, meaningless speech, hallucinations, dry mouth, and agitation (Caksen et al. 2003; Lee 2007).

Pharmacological Considerations

The compound atropine, which has anticholinergic effects in the central and peripheral nervous system, is primarily responsible for the effects of belladonna (Leung and Foster 1996).

Pregnancy and Lactation

A retrospective study of women who had used belladonna in the first trimester of pregnancy indicated that belladonna use was associated with malformations; however, interpretation of data was difficult and no causality could be determined (Heinonen et al. 1977).

The compound atropine has been shown to rapidly cross the placenta (Kanto et al. 1981; Kivalo and Saarikoski 1977; Onnen et al. 1979) and has been used to test placental function in high-risk obstetric patients (Hellman and Fillisti 1965). Atropine has also been used to reduce gastric secretions before cesarean section with no fetal or neonatal effects observed (Diaz et al. 1980; Roper and Salem 1981).

Although the compound atropine is reported to cross into breast milk, and neonates are sensitive to anticholinergic agents, no adverse effects have been reported in nursing infants whose mothers were taking atropine (Briggs et al. 2008; Dart 2004). The American Academy of Pediatrics lists the compound atropine as being compatible with breastfeeding (AAP 2001).

Review Details

I. Drug and Supplement Interactions

Clinical Trials of Drug or Supplement Interactions

No clinical trials of drug or supplement interactions were identified.

Case Reports of Suspected Drug or Supplement Interactions

No case reports of suspected drug or supplement interactions were identified.

Animal Trials of Drug or Supplement Interactions

No animal trials of drug or supplement interactions were identified.

II. Adverse Events

Case Reports of Adverse Events

The overdose symptoms of belladonna are those of anticholinergic poisoning, with symptoms colloquially described as "hot as a hare, dry as a bone, red as a beet, mad as a hatter, blind as a bat" (Lee 2007).

Intoxication with belladonna results in dry mouth with dysphonia, difficulty swallowing, increased heart rate, and urinary retention. Elevation of body temperature may be accompanied by flushed, dry skin. Mydriasis, blurred vision, excitement and delirium, headache, and confusion may be observed (Nelson et al. 2006).

In a review of 49 cases of belladonna poisoning in children, occurring over a 7-year period, cases were categorized into mild/moderate or severe intoxication. Cases with encephalopathy were considered severe, while those without were considered mild/moderate. In the mild/moderate group, the most common symptoms were tachycardia, flushing, mydriasis, meaningless speech, hyperglycemia, and agitation. In the severe group, the most common symptoms were meaningless speech, coma, mydriasis, lethargy, tachycardia, flushing, and convulsions (Caksen et al. 2003). The review did not indicate the part of the plant consumed, although the berries are most commonly responsible for poisoning in children (Lee 2007).

Belladonna poisoning has been reported after eating a rabbit that had been feeding on belladonna (Firth and Bentley 1921).

Intoxication with belladonna was reported in several families that had consumed tea labeled as mate (*Ilex paraguariensis*) and later discovered to be belladonna leaf. Symptoms of poisoning were consistent with belladonna intoxication, including agitation, flushing, mydriasis, hallucinations, disorientation, tachycardia, nausea, and vomiting (Hsu et al. 1995).

An early toxicology text describes a series of 379 cases of poisoning due to belladonna preparations (eye drops, plasters, and liniments) (Witthaus 1911). A review article on belladonna notes that medical use of belladonna has declined significantly, and such poisonings have become rare (Lee 2007).

Numerous other cases of belladonna intoxication have been reported in the literature and include adverse effects from accidental or intentional consumption of the leaf, berry, or root of belladonna (e.g., Ceha et al. 1997; Gabel 1968; Gibbons 1954; Lange and Toft 1990; Minors 1948; Montoya et al. 2009; Schneider et al. 1996; Southgate et al. 2000; Trabattoni et al. 1984).

III. Pharmacology and Pharmacokinetics

Human Pharmacological Studies

In healthy volunteers orally administered single doses of 1, 2, or 5 ml of belladonna tincture (0.1 mg/ml alkaloid concentration, atropine/scopolamine = 20:1), RR interval and high-frequency spectral power of heart rate variability increased markedly after the 1 or 2 ml doses. Decreases in RR interval, high-frequency spectral power of heart rate variability, and noninvasive baroreflex sensitivity were observed after the 5 ml dose. The activity of the 2 ml dose was characterized as vagotonic, while that of the 5 ml dose was considered vagolytic. No significant effects on blood pressure were observed (Bettermann et al. 2001).

Animal Pharmacological Studies

In mice treated with belladonna tincture or atropine, the anticholinergic effects of belladonna were greater than those suggested by the alkaloid content of the tincture (Mazzanti et al. 1988).

In Vitro Pharmacological Studies

No relevant in vitro pharmacological studies were identified.

IV. Pregnancy and Lactation

A retrospective study of 554 women who had used belladonna in the first trimester of pregnancy indicated that belladonna was associated with eye and ear malformations, and also associated with increased risk of respiratory tract anomalies and hypospadias. The authors of the study indicated that interpretation of the data was difficult and that no causality could be determined. Among women who used belladonna any time in pregnancy, 1355 exposures were reported and no association with belladonna and malformations was observed in the group (Heinonen et al. 1977).

The compound atropine has been shown to rapidly cross the placenta (Kanto et al. 1981; Kivalo and Saarikoski 1977; Onnen et al. 1979) and has been used to test placental function in high-risk obstetric patients by producing fetal vagal blockade and subsequent tachycardia (Hellman and Fillisti 1965). Atropine has also been used to reduce gastric secretions before cesarean section with no fetal or neonatal effects observed (Diaz et al. 1980; Roper and Salem 1981).

Although the compound atropine is reported to cross into breast milk, and neonates are sensitive to anticholinergic agents, no adverse effects have been reported in nursing infants whose mothers were taking atropine (Briggs et al. 2008; Dart 2004). The American Academy of Pediatrics lists the compound atropine as being compatible with breastfeeding (AAP 2001).

V. Toxicity Studies

Acute Toxicity

The LD_{50} of the compound atropine orally administered in rats is 500 mg/kg, and in mice is 75 mg/kg; that after intravenous administration in rats is 73 mg/kg and in mice is 30 mg/kg. The LD_{50} of atropine in rats is 920 mg/kg after intramuscular administration and 280 mg/kg after intraperitoneal administration (Sax and Lewis 1992).

A

Literature Cited

AAP. 2001. The transfer of drugs and other chemicals into human milk. American Academy of Pediatrics Committee on Drugs. *Pediatrics* 108(3):776-789.

Bettermann, H., D. Cysarz, A. Portsteffen, and H.C. Kummell. 2001. Bimodal dose-dependent effect on autonomic, cardiac control after oral administration of *Atropa belladonna*. *Auton. Neurosci.* 90(1-2):132-137.

Briggs, G.G., R.K. Freeman, and S.J. Yaffe. 2008. *Drugs in pregnancy and lactation: A reference guide to fetal and neonatal risk.* 8th ed. Philadelphia: Lippincott, Williams & Wilkins.

Caksen, H., D. Odabas, S. Akbayram, et al. 2003. Deadly nightshade (*Atropa belladonna*) intoxication: An analysis of 49 children. *Hum. Exp. Toxicol.* 22(12):665-668.

Ceha, L.J., C. Presperin, E. Young, M. Allswede, and T. Erickson. 1997. Anticholinergic toxicity from nightshade berry poisoning responsive to physostigmine. *J. Emerg. Med.* 15(1):65-69.

Dart, R.C. 2004. *Medical toxicology*. Philadelphia: Lippincott, Williams & Wilkins.

Diaz, D.M., S.F. Diaz, and G.F. Marx. 1980. Cardiovascular effects of glycopyrrolate and belladonna derivatives in obstetric patients. *Bull. N.Y. Acad. Med.* 56(2):245.

Felter, H.W., and J.U. Lloyd. 1898. *King's American dispensatory*. 18th ed., 3rd rev. 2 vols. Cincinnati: Ohio Valley Co.

Firth, D., and J.R. Bentley. 1921. Belladonna poisoning from eating rabbit. *Lancet* 2:901.

Frohne, D., and H.J. Pfänder. 2000. *A colour atlas of poisonous plants: A handbook for pharmacists, doctors, toxicologists, biologists and veterinarians*. 2nd ed. London: Manson.

Gabel, M.C. 1968. Purposeful ingestion of belladonna for hallucinatory effects. *J. Pediatr.* 72(6):864-866.

Gibbons, W.D. 1954. Deadly nightshade poisoning. *Med. J. Aust.* 41:254-255.

Heinonen, O.P., D. Slone, and S. Shapiro. 1977. *Birth defects and drugs in pregnancy*. Littleton, MA: Publishing Sciences Group.

Hellman, L.M., and L.P. Fillisti. 1965. Analysis of the atropine test for placental transfer in gravidas with toxemia and diabetes. *Am. J. Obstet. Gynecol.* 91:797.

Hsu, C.K., P. Leo, D. Shastry, et al. 1995. Anticholinergic poisoning associated with herbal tea. *Arch. Int. Med.* 155(20):2245-2248.

Kanto, J., R. Virtanen, E. Iisalo, K. Maenpaa, and P. Liukko. 1981. Placental transfer and pharmacokinetics of atropine after a single maternal intravenous and intramuscular administration. *Acta Anaesthesiol. Scand.* 25(2):85-88.

Kingsbury, J.M. 1964. *Poisonous plants of the United States and Canada*. Englewood Cliffs, NJ: Prentice-Hall.

Kivalo, I., and S. Saarikoski. 1977. Placental transmission of atropine at full-term pregnancy. *Br. J. Anaes.* 49(10):1017-1021.

Lange, A., and P. Toft. 1990. Poisoning with deadly nightshade (*Atropa belladonna*). *Ugeskr. Laeger* 152:1096.

Lee, M.R. 2007. Solanaceae IV: *Atropa belladonna*, deadly nightshade. *J. R. Coll. Physicians Edinb.* 37(1):77-84.

Leung, A.Y., and S. Foster. 1996. *Encyclopedia of common natural ingredients used in food, drugs, and cosmetics*. 2nd ed. New York: Wiley.

Mazzanti, G., B. Tita, P. Bolle, M. Bonanomi, and D. Piccinelli. 1988. A comparative study of behavioural and autonomic effects of atropine and *Atropa belladonna*. *Pharm. Res. Commun.* 20(Suppl. 5):49-53.

Minors, E.H. 1948. Five cases of belladonna poisoning. *Br. Med. J.* 2:518-519.

Montoya, A.M., N. Mavrakanas, and J.S. Schutz. 2009. Acute anticholinergic syndrome from *Atropa belladonna* mistaken for blueberries. *Eur. J. Ophthalmol.* 19:170-172.

Nelson, L., R.D. Shih, M.J. Balick, and K.F. Lampe. 2006. *Handbook of poisonous and injurious plants*. 2nd ed. Berlin: Springer.

Onnen, I., G. Barrier, P. d'Athis, C. Sureau, and G. Olive. 1979. Placental transfer of atropine at the end of pregnancy. *Eur. J. Clin. Pharmacol.* 15(6):443-446.

Roper, R.E., and M.G. Salem. 1981. Effects of glycopyrrolate and atropine combined with antacid on gastric acidity. *Br. J. Anaesth.* 53(12):1277-1280.

Sax, N.I., and R.J. Lewis. 1992. *Dangerous properties of industrial materials.* 8th ed. New York: Van Nostrand Reinhold.

Schneider, F., P. Lutun, P. Kintz, et al. 1996. Plasma and urine concentrations of atropine after the ingestion of cooked deadly nightshade berries. *J. Toxicol. Clin. Toxicol.* 34(1):113-117.

Southgate, H.J., M. Egerton, and E.A. Dauncey. 2000. Lessons to be learned: A case study approach. Unseasonal severe poisoning of two adults by deadly nightshade (*Atropa belladonna*). *J. R. Soc. Health* 120(2):127-130.

Trabattoni, G., D. Visintini, G.M. Terzano, and A. Lechi. 1984. Accidental poisoning with deadly nightshade berries: A case report. *Hum. Toxicol.* 3(6):513-516.

Witthaus, R.A. 1911. *Manual of toxicology.* 2nd ed. New York: Baillière Tindall and Cox.

Avena spp.

Poaceae

Avena fatua L.
SCN: wild oat
Part: spikelets

Avena sativa **L.**
SCN: oat
Part: spikelets

Quick Reference Summary

Safety Class: 1
Interaction Class: A

Contraindications

None known.

A

Other Precautions

None known.

Drug and Supplement Interactions

None known.

Adverse Events and Side Effects

Allergic contact dermatitis to oat has been reported and confirmed by patch testing (De Paz Arranz et al. 2002; Pazzaglia et al. 2000).

Pharmacological Considerations

Animal studies have indicated that oat extracts may antagonize the effects of morphine and nicotine (Connor et al. 1975).

Pregnancy and Lactation

No information on the safety of oat or wild oat in pregnancy or lactation was identified in the scientific or traditional literature. Although this review did not identify any concerns for use while pregnant or nursing, safety has not been conclusively established.

Review Details

I. Drug and Supplement Interactions

Clinical Trials of Drug or Supplement Interactions

No clinical trials of drug or supplement interactions were identified.

Case Reports of Suspected Drug or Supplement Interactions

No case reports of suspected drug or supplement interactions were identified.

Animal Trials of Drug or Supplement Interactions

Alcohol extracts of oat antagonized the effects of morphine and nicotine injections in mice. The extracts did not affect the seizure threshold to bemegride or nicotine or the sleeping time induced by barbitone sodium (Connor et al. 1975).

II. Adverse Events

Case Reports of Adverse Events

Allergic contact dermatitis to oats has been reported and confirmed by patch testing (De Paz Arranz et al. 2002; Pazzaglia et al. 2000).

III. Pharmacology and Pharmacokinetics

Human Pharmacological Studies

In sensitization studies with oat proteins (1, 3, and 5%) in children with atopic dermatitis, skin prick tests to oat proteins were positive in 19% of children, whereas atopy patch tests were positive in 15% (Boussault et al. 2007).

In patients with biopsy-confirmed dermatitis herpetiformis, no adverse effects were observed after addition of oats to the strict gluten-free diet of the patients (maintained for an average of 15.8 years prior to the study). Serological tests for antigliadin, antireticulin, and antiendomysial antibodies were negative before oats were introduced into the diet and after they were discontinued, and villous architecture remained normal (Hardman et al. 1997).

In patients with celiac disease and dermatitis herpetiformis, no adverse effects were seen after ingestion of 2.5 g of the protein avenin daily (equivalent to 300 g of oats, or about 10 bowls of oatmeal) for 5 days. Biopsies of the small intestines indicated that that avenin did not change the ratio of the villous height to the crypt depth, the height of enterocytes, or intraepithelial lymphocyte counts (Hardman et al. 1999).

In a study introducing oats into the diets of patients with celiac disease, the addition of 50 g of oats daily for 12 weeks was generally well tolerated, although one patient developed partial villous atrophy and a rash after the introduction of oats. Several other patients showed positive levels of interferon gamma mRNA after introduction of oats (Lundin et al. 2003). Other studies have indicated that moderate quantities of oats are safe for persons with gluten intolerance (Janatuinen et al. 1995, 2002; Picarelli et al. 2001; Storsrud et al. 2003).

Animal Pharmacological Studies

In irritation tests using 2 to 200 % colloidal extracts of oats, no ocular or cutaneous toxicity were observed, nor any sensitization or photosensitization (Fabre 2004).

In Vitro Pharmacological Studies

No relevant in vitro pharmacological studies were identified.

IV. Pregnancy and Lactation

No information on the use of oat or wild oat during pregnancy or lactation was identified.

V. Toxicity Studies

No toxicity studies were identified.

Literature Cited

Boussault, P., C. Leaute-Labreze, E. Saubusse, et al. 2007. Oat sensitization in children with atopic dermatitis: Prevalence, risks and associated factors. *Allergy* 62(11):1251-1256.

Connor, J., T. Connor, P.B. Marshall, A. Reid, and M.J. Turnbull. 1975. The pharmacology of *Avena sativa*. *J. Pharm. Pharmacol.* 27(2):92-98.

De Paz Arranz, S., A. Perez Montero, L.Z. Remon, and M.I. Molero. 2002. Allergic contact urticaria to oatmeal. *Allergy* 57(12):1215.

Fabre, B. 2004. *Avena sativa*, demande d'inscription en usage topique. Pierre Fabre Innovation Développement. Concept paper 2004. *Cited in* EMEA. 2008. Assessment report on *Avena sativa* L., herba and *Avena sativa* L., fructus. Doc. Ref. EMEA/HMPC/202967/2007. London: European Medicines Agency.

Hardman, C., L. Fry, A. Tatham, and H.J. Thomas. 1999. Absence of toxicity of avenin in patients with dermatitis herpetiformis. *N. Engl. J. Med.* 340(4):321.

Hardman, C.M., J.J. Garioch, J.N. Leonard, et al. 1997. Absence of toxicity of oats in patients with dermatitis herpetiformis. *N. Engl. J. Med.* 337(26):1884-1887.

Janatuinen, E.K., T.A. Kemppainen, R.J. Julkunen, et al. 2002. No harm from five year ingestion of oats in coeliac disease. *Gut* 50(3):332-335.

Janatuinen, E.K., P.H. Pikkarainen, T.A. Kemppainen, et al. 1995. A comparison of diets with and without oats in adults with celiac disease. *N. Engl. J. Med.* 333(16):1033-1037.

Lundin, K.E., E.M. Nilsen, H.G. Scott, et al. 2003. Oats induced villous atrophy in coeliac disease. *Gut* 52(11):1649-1652.

Pazzaglia, M., M. Jorizzo, G. Parente, and A. Tosti. 2000. Allergic contact dermatitis due to *Avena* extract. *Contact Dermat.* 42(6):364.

Picarelli, A., M. Di Tola, L. Sabbatella, et al. 2001. Immunologic evidence of no harmful effect of oats in celiac disease. *Am. J. Clin. Nutr.* 74(1):137-140.

Storsrud, S., M. Olsson, R. Arvidsson Lenner, et al. 2003. Adult coeliac patients do tolerate large amounts of oats. *Eur. J. Clin. Nutr.* 57(1):163-169.

Azadirachta indica A. Juss. — Meliaceae

SCN: neem
Syn: *Melia azadirachta* L.
AN: *nimba*
OCN: bead tree; margosa
Part: bark

Quick Reference Summary

Safety Class: 1
Interaction Class: A

Contraindications

None known.

Other Precautions

None known.

Drug and Supplement Interactions

None known.

Adverse Events and Side Effects

None known.

Pharmacological Considerations

Changes in hormone levels and sperm production were observed in male rats intraperitoneally administered extracts of neem stem bark (Raji et al. 2003).

Pregnancy and Lactation

Other than one animal study that indicated anti-implantation activity of large doses (500 mg/kg) of neem bark extract (Bhargava and Prakash 2000), no information on the safety of neem in pregnancy or lactation was identified in the scientific or traditional literature. While this review did not identify any concerns thought to be clinically relevant for pregnant or nursing women, safety has not been conclusively established.

Review Details

I. Drug and Supplement Interactions

Clinical Trials of Drug or Supplement Interactions

No clinical trials of drug or supplement interactions were identified.

Case Reports of Suspected Drug or Supplement Interactions

No case reports of suspected drug or supplement interactions were identified.

Animal Trials of Drug or Supplement Interactions

No animal trials of drug or supplement interactions were identified.

II. Adverse Events

Case Reports of Adverse Events

No case reports of adverse events were identified.

III. Pharmacology and Pharmacokinetics

Human Pharmacological Studies

No relevant human pharmacological studies were identified.

Animal Pharmacological Studies

In rats intraperitoneally administered up to 150 mg/kg of an ethanol extract of neem bark daily for 10 weeks, dose-dependent decreases in weights of the testis, epididymis, and seminal vesicles, and an increased weight of the adrenal gland were observed. The treatment caused a dose-dependent reduction in serum testosterone and LH but no change in FSH levels. Male rats were unable to impregnate female rats throughout the duration of the study, and female rats were later bred successfully with untreated male rats (Raji et al. 2003).

In Vitro Pharmacological Studies

No relevant in vitro pharmacological studies were identified.

IV. Pregnancy and Lactation

In female rats orally administered 500 mg/kg neem bark extract daily for 10 days, significant anti-implantation activity was observed. At that dose, fetal resorptions were observed although no changes in the number of corpora lutea were seen (Bhargava and Prakash 2000).

No information on the safety of neem during lactation was identified.

V. Toxicity Studies

No toxicity studies were identified.

Literature Cited

Bhargava, V., and A.O. Prakash. 2000. Effect of a herbal preparation from neem bark on early and late pregnancy in rats. *Indian Drugs* 37(4):178-181.

Raji, Y., U.S. Udoh, O.O. Mewoyeka, F.C. Ononye, and A.F. Bolarinwa. 2003. Implication of reproductive endocrine malfunction in male antifertility efficacy of *Azadirachta indica* extract in rats. *Afr. J. Med. Med. Sci.* 32(2):159-165.

Azadirachta indica A. Juss. — Meliaceae

SCN: neem
Syn: *Melia azadirachta* L.
AN: *nimba*
OCN: bead tree; margosa
Part: leaf

Quick Reference Summary

Safety Class: 2d
Interaction Class: A

Contraindications

Not for internal use by men attempting to conceive (Aladakatti and Ahamed 2005; Aladakatti et al. 2001; Awasthy 2001; Ghosesawar et al. 2003; Kataria et al. 2000; Kaushic and Upadhyay 1995; Mishra and Singh 2005; Parveen et al. 1993).

Other Precautions

None known.

Drug and Supplement Interactions

None known.

Adverse Events and Side Effects

Ventricular fibrillation followed by heart attack was reported in a woman several hours after she consumed 1 liter of neem leaf extract. Causality could not be determined, and details on the concentration and type of extract were not reported (Sivashanmugham et al. 1984).

Pharmacological Considerations

Animal studies investigating the use of neem leaf as a male contraceptive have shown changes in the tissues responsible for sperm formation, changes in sperm shape and motility, and decreases in sperm counts (Aladakatti and Ahamed 2005; Aladakatti et al. 2001; Awasthy 2001; Ghosesawar et al. 2003; Kataria et al. 2000; Kaushic and Upadhyay 1995; Mishra and Singh 2005; Parveen et al. 1993). Doses studied ranged from 20 to 2000 mg/kg. Changes in tissues and sperm formation were shown to return to normal after cessation of treatment (Joshi et al. 1996; Mishra and Singh 2005). No human studies on the safety or efficacy of neem leaf as a male contraceptive have been completed.

Animal studies have demonstrated that neem leaf may modify glucose regulation (Chandra et al. 2008; Chattopadhyay 1999; Kar et al. 2003; Khosla et al. 2000). People with diabetes are advised to monitor their blood sugar closely and discuss the use of this herb with a qualified healthcare practitioner prior to use.

An animal study showed a decrease in thyroid hormone levels in animals treated with high doses (100 mg/kg) but not lower doses (40 mg/kg) of neem leaf (Panda and Kar 2000).

Pregnancy and Lactation

No adverse effects on fetal development were observed in rats fed the compound azadirachtin (0.5 mg/kg) for two generations, or in rats fed large doses (1.5 g/kg) of azadirachtin daily on days 6 to 15 of pregnancy.

No information on the safety of neem leaf during lactation was identified. While this review did not identify any concerns for use while nursing, safety has not been conclusively established.

Review Details

I. Drug and Supplement Interactions

Clinical Trials of Drug or Supplement Interactions

No clinical trials of drug or supplement interactions were identified.

Case Reports of Suspected Drug or Supplement Interactions

No case reports of suspected drug or supplement interactions were identified.

Animal Trials of Drug or Supplement Interactions

No animal trials of drug or supplement interactions were identified.

II. Adverse Events

Adverse Events Reported in Clinical Trials

No adverse effects, including changes in liver or kidney function parameters, were reported in HIV/AIDS patients orally administered 1 g of fractionated acetone-water extracts of neem leaf daily for 12 weeks (Mbah et al. 2007).

Case Reports of Adverse Events

Ventricular fibrillation and cardiac arrest were reported in a 24-year-old woman 2 hours after she consumed 1 liter of neem leaf extract (type and concentration of extract unspecified) (Sivashanmugham et al. 1984).

III. Pharmacology and Pharmacokinetics

Human Pharmacological Studies

No relevant human pharmacological studies were identified.

Animal Pharmacological Studies

Effects on Male Reproductive System

In male rats orally administered 50, 100, or 200 mg/kg aqueous extract of neem leaf daily for 28 days, animals administered the 200 mg/kg dose had marked alterations both in histological appearance and in the level of sialic acid in the epididymal duct, while those administered 50 or 100 mg/kg had no appreciable alterations in histological appearance of the epididymis. In treated animals, testes showed both normal and affected seminiferous tubules in the same sections. After 42 days of withdrawal of the treatment, the alterations in the reproductive organs returned to control levels (Mishra and Singh 2005).

No effects on spermatogenesis were observed in male rats orally administered 100 mg/kg daily of an ethanol extract of neem leaf for 21 days. In untreated females mated with treated males, anti-implantation effects were observed (Choudhary et al. 1990).

In rats orally administered 100 mg neem leaf powder daily (average rat weight 185 g) for 24 days, a decrease in the total sperm count, sperm motility, and forward velocity in vas deferens fluid was observed. The percentage of abnormal sperm increased and the fructose content decreased (Ghosesawar et al. 2003).

Decreases in the weight of the seminal vesicle and ventral prostate, sperm motility, sperm count, and relative percentage of normal sperm were observed in male rats orally administered 20, 40, or 60 mg of neem leaf powder (average rat weight 240 g) daily for 24 days (Parveen et al. 1993).

In male rats orally administered 100 mg/animal neem leaf powder daily for 48 days, decreases in the total sperm count, sperm motility, and forward velocity were observed. The percentage of abnormal sperm increased and the fructose content of caudal semen of the epididymis decreased (Aladakatti et al. 2001).

In male rats orally administered 500 mg/kg neem leaf daily for 48 days, atrophic seminiferous tubules with widening intercellular spaces, degeneration of Leydig cells, and pathological changes in the spermatozoa of the cauda epididymis were observed (Aladakatti and Ahamed 2005).

Reduction in the diameters of seminiferous tubule and nuclei of the germinal elements as well as atrophy of the spermatogenic elements and Leydig cells were observed in rats orally administered 100 mg/animal neem leaf powder daily (average rat weight 225 g) for 24 days. After cessation of neem administration, a gradual recovery was observed within 8 days (Joshi et al. 1996).

Histological and biochemical changes in the caput and cauda epididymis were observed in rats orally administered 20, 40, or 60 mg/animal of neem leaf powder daily for 24 days. In the treated rats, the height of the epithelium and the diameter of the nucleus in both the regions were reduced, and the lumen of the caput was packed with lymphocytes. Biochemically, a dose-dependent decrease in the protein content and acid phosphatase and increase in the alkaline phosphatase and lactate dehydrogenase activities were observed in both regions, along with a decrease in serum testosterone concentration (Kasturi et al. 1995).

In male rats orally administered 100 mg/animal neem leaf powder daily for 24 days, intracellular spaces and vacuolization were observed in Sertoli cells; in Leydig cells, cytoplasmic inclusions appeared diminished, and the configuration of granular endoplasmic reticulum appeared as a single unbranched tubule, leading the authors to conclude that neem leaf might affect spermatogenesis through antispermatogenic and antiandrogenic properties (Kasturi et al. 2002).

A reduction in sperm count and increased frequency of spermatozoa with abnormal head morphology was observed in mice orally administered 0.5, 1, or 2 g/kg neem leaf extract daily for 6 weeks (Awasthy 2001).

Effects on Female Reproductive System

In ovariectomized rats orally administered 0.5 ml aqueous extract of neem leaf daily for 3 days, an absence of open vagina, absence of epithelial cells in a vaginal smear, and incomplete development of the uterus were observed, indicating an absence of estrogenic activity (Mateenuddin et al. 1986).

Antidiabetic Activity

In diabetic and healthy rats, oral administration of 500 mg/kg of an aqueous neem leaf extract resulted in a decrease in blood glucose levels, with the effect being more pronounced in diabetic than in healthy animals. The effects were similar to that of the drug glibenclamide (Khosla et al. 2000).

A reduction in blood glucose levels was observed in diabetic rats orally administered 500 mg/kg neem extract (Chandra et al. 2008). A reduction in blood glucose levels was observed after 2 weeks of treatment in diabetic rats orally administered 500 mg/kg of an ethanolic neem leaf extract daily (Kar et al. 2003).

A reduction in blood glucose levels was observed in mice orally administered 50 to 400 mg/kg or rats orally administered 0.25 to 8 mg/kg of an alcohol extract of neem leaf (Chattopadhyay 1999).

Other Pharmacological Activity

In mice orally administered 40 or 100 mg/kg neem leaf extract daily for 20 days, animals in the higher dose group had a decrease in serum triiodothyronine (T_3) and increased serum thyroxine (T_4) concentrations. No significant alterations of T_3 or T_4 levels were observed in the lower dose group (Panda and Kar 2000).

Intravenous administration of 100, 300, or 1000 mg/kg of an alcohol extract of neem leaf to rats produced an initial bradycardia followed by cardiac arrhythmia. A significant and dose-related fall in blood pressure was observed and was immediate, sharp, and persistent. Pretreatment with either atropine or mepyramine did not alter the hypotensive activity of the neem leaf extract (Koley and Lal 1994).

In Vitro Pharmacological Studies

In a study of the concentration of neem leaf extract needed to immobilize and kill 100% of human spermatozoa within 20 seconds, the minimum effective spermicidal concentration was approximately 2.85 mg/million sperm. The effect of extracts on morphology and viability of sperm was also studied and no change was observed in morphology of head, midpiece, or tail. The extract made from "tender" leaf was more effective than the extract made from "old" leaf (Khillare and Shrivastav 2003).

IV. Pregnancy and Lactation

In pregnant rats orally administered 0.5, 1, or 1.5 g/kg of the compound azadirachtin daily on gestational days 6 to 15, no significant adverse effects were observed on fetal development, number of implantations, post-implantation loss, or other reproductive parameters. Some minor malformations were observed in high doses but were not compound- or dose-related (Srivastava and Raizada 2001).

No adverse effects on reproductive function or fetal or pup development were observed over two generations of rats fed diets containing the equivalent of 5, 25, or 50 mg/kg of the compound azadirachtin daily (Srivastava and Raizada 2007).

No information on the safety of neem leaf during lactation was identified.

V. Toxicity Studies

Acute Toxicity

The LD_{50} of orally administered aqueous neem leaf extract in mice could not be determined at doses up to 2.5 g/kg (Dorababu et al. 2006).

Acute toxicity, including nervous symptoms (head movements, walking in circles) with dyspnea, an increase in body temperature, hepatic failure, and tympanites, was observed in a sheep that ate 100 g of neem leaf. The symptoms lasted for 12 hours and were followed by death of the animal (Ali and Salih 1982).

Short-Term Toxicity

No adverse effects were observed in rats orally administered 1 g/kg neem leaf extract daily for 28 days. Observed parameters included body and tissue weights, food and water intake, hematological profile, and various liver and kidney function tests (Dorababu et al. 2006).

No adverse effects, including changes in body weight, organ-body weight ratios, or liver enzyme levels were observed in mice subcutaneously administered 0.5 or 1 mg neem leaf aqueous extract daily for 28 days (Haque et al. 2006).

In rats administered 100 mg/kg neem leaf extract daily for 49 days, decreases in appetite, body weight, and pupillary reflex were observed. Histopathological studies revealed congestion in the liver, kidneys, lungs, and brain (Hore et al. 1999).

Subchronic Toxicity

In goats and guinea pigs orally administered 50 or 200 mg/kg fresh or dried neem leaf daily for up to 8 weeks, a progressive decrease in body weight, weakness, inappetence, and loss of condition were observed along with a decrease in heart rate, pulse rate, and respiratory rates. Diarrhea was observed in animals given the fresh leaf. In goats, the higher doses of the plant leaf produced tremors and ataxia during the last few days of treatment. No statistically significant hematological changes were observed. On necropsy of treated goats, there were areas of hemorrhagic erosions. Histopathologically, various degrees of hemorrhage, congestion, and degeneration were seen in the liver, kidney, lung, duodenum, and brain, along with degeneration of the seminiferous tubules (Ali 1987).

Chronic Toxicity

In rats fed diets containing the equivalent of 5, 25, or 50 mg/kg of the compound azadirachtin daily for two generations, no toxicological effects, including enzymatic parameters (AST, ALT, ALP), bilirubin, cholesterol, total protein, or histopathology of liver, brain, kidney, testes, or ovary were observed (Srivastava and Raizada 2007).

Genotoxicity

Chromosome strand breakages or spindle disturbances as well as the regulation of genes responsible for sperm shaping were affected in mice orally administered 0.5, 1, or 2 g/kg neem leaf extract daily for 6 weeks (Awasthy 2001).

An increase in the incidence of structural and mitotic disruptive changes in metaphase chromosomes of bone marrow cells was observed on days 8, 15, and 35 in rats orally administered 0.5, 1, or 2 g/kg neem leaf ethanolic extract daily (Awasthy et al. 1999).

In the Ames test for mutagenicity, a weak antimutagenic effect of the ethanol extract of neem leaf was observed in *Salmonella typhimurium* TA100 (Kusamran et al. 1998).

Literature Cited

Aladakatti, R.H., and R.N. Ahamed. 2005. Ultrastructural changes in Leydig cells and cauda epididymal spermatozoa induced by *Azadirachta indica* leaves in albino rats. *Phytother. Res.* 19(9):756-766.

Aladakatti, R.H., R. Nazeer Ahamed, M. Ahmed, and M.G. Ghosesawar. 2001. Sperm parameters changes induced by *Azadirachta indica* in albino rats. *J. Basic Clin. Physiol. Pharmacol.* 12(1):69-76.

Ali, B.H. 1987. The toxicity of *Azadirachta indica* leaves in goats and guinea pigs. *Vet. Hum. Toxicol.* 29(1):16-19.

Ali, B.H., and A.M.M. Salih. 1982. Suspected *Azadirachta indica* toxicity in a sheep. *Vet. Rec.* 111:494.

Awasthy, K.S. 2001. Genotoxicity of a crude leaf extract of neem in male germ cells of mice. *Cytobios* 106(Suppl. 2):151-164.

Awasthy, K.S., O.P. Chaurasia, and S.P. Sinha. 1999. Prolonged murine genotoxic effects of crude extracted from neem. *Phytother. Res.* 13(1):81-83.

Chandra, A., A.A. Mahdi, R.K. Singh, F. Mahdi, and R. Chander. 2008. Effect of Indian herbal hypoglycemic agents on antioxidant capacity and trace elements content in diabetic rats. *J. Med. Food* 11(3):506-512.

Chattopadhyay, R.R. 1999. A comparative evaluation of some blood sugar lowering agents of plant origin. *J. Ethnopharmacol.* 67(3):367-372.

Choudhary, D.N., J.N. Singh, S.K. Verma, and B.P. Singh. 1990. Antifertility effects of leaf extracts of some plants in male rats. *Indian J. Exp. Biol.* 28(8):714-716.

Dorababu, M., M.C. Joshi, G. Bhawani, et al. 2006. Effect of aqueous extract of neem (*Azadirachta indica*) leaves on offensive and diffensive gastric mucosal factors in rats. *Indian J. Physiol. Pharmacol.* 50(3):241-249.

Ghosesawar, M.G., R.N. Ahamed, M. Ahmed, and R.H. Aladakatti. 2003. *Azadirachta indica* adversely affects sperm parameters and fructose levels in vas deferens fluid of albino rats. *J. Basic Clin. Physiol. Pharmacol.* 14(4):387-395.

Haque, E., I. Mandal, S. Pal, and R. Baral. 2006. Prophylactic dose of neem (*Azadirachta indica*) leaf preparation restricting murine tumor growth is nontoxic, hematostimulatory and immunostimulatory. *Immunopharmacol. Immunotoxicol.* 28(1):33-50.

Hore, S.K., S.K. Maiti, and N. Gupta. 1999. Effect of subacute exposure to neem (*Azadirachta indica*) leaf extract in rats. *Indian Vet. J.* 76(11):1011-1012.

Joshi, A.R., R.N. Ahamed, K.M. Pathan, and B. Manivannan. 1996. Effect of *Azadirachta indica* leaves on testis and its recovery in albino rats. *Indian J. Exp. Biol.* 34(11):1091-1094.

Kar, A., B.K. Choudhary, and N.G. Bandyopadhyay. 2003. Comparative evaluation of hypoglycaemic activity of some Indian medicinal plants in alloxan diabetic rats. *J. Ethnopharmacol.* 84(1):105-108.

Kasturi, M., R.N. Ahamed, K.M. Pathan, B. Manivannan, and R.H. Aladakatti. 2002. Ultrastructural changes induced by leaves of *Azadirachta indica* (Neem) in the testis of albino rats. *J. Basic Clin. Physiol. Pharmacol.* 13(4):311-328.

Kasturi, M., B. Manivannan, R.N. Ahamed, P.D. Shaikh, and K.M. Pathan. 1995. Changes in epididymal structure and function of albino rat treated with *Azadirachta indica* leaves. *Indian J. Exp. Biol.* 33(10):725-729.

Kataria, M., P.K. Gupta, S. Gupta, and V.K. Vijan. 2000. Effect of petroleum ether extracts of neem (*Azadirachta indica*) seed kernel and husk on blood parameters in rats. *Indian J. Toxicol.* 7:73-81.

Kaushic, C., and S. Upadhyay. 1995. Mode of long-term antifertility effect of intrauterine neem treatment (IUNT). *Contraception* 51(3):203-207.

Khillare, B., and T.G. Shrivastav. 2003. Spermicidal activity of *Azadirachta indica* (neem) leaf extract. *Contraception* 68(3):225-229.

Khosla, P., S. Bhanwra, J. Singh, S. Seth, and R.K. Srivastava. 2000. A study of hypoglycaemic effects of *Azadirachta indica* (Neem) in normal and alloxan diabetic rabbits. *Indian J. Physiol. Pharmacol.* 44(1):69-74.

Koley, K.M., and J. Lal. 1994. Pharmacological effects of *Azadirachta indica* (neem) leaf extract on the ECG and blood pressure of rat. *Indian J. Physiol. Pharmacol.* 38(3):223-225.

Kusamran, W.R., A. Tepsuwan, and P. Kupradinun. 1998. Antimutagenic and anticarcinogenic potentials of some Thai vegetables. *Mutat. Res.* 402(1-2):247-258.

Mateenuddin, M., K.K. Khairatkar, K.N. Mendhulkar, and N.L. Sadre. 1986. Assessment of oestrogenecity of neem leaf extract in rats. *Indian J. Physiol. Pharmacol.* 30(1):118-119.

Mbah, A.U., I.J. Udeinya, E.N. Shu, et al. 2007. Fractionated neem leaf extract is safe and increases $CD4^+$ cell levels in HIV/AIDS patients. *Am. J. Ther.* 14(4):369-374.

Mishra, R.K., and S.K. Singh. 2005. Effect of aqueous leaf extract of *Azadirachta indica* on the reproductive organs in male mice. *Indian J. Exp. Biol.* 43(11):1093-1103.

Panda, S., and A. Kar. 2000. How safe is neem extract with respect to thyroid function in male mice? *Pharmacol. Res.* 41(4):419-422.

Parveen, D.S., B. Manivannan, K.M. Pathan, M. Kasturi, and R.N. Ahamed. 1993. Antispermatic activity *Azadirachta indica* leaves in albino rats. *Curr. Sci.* 64:688-689.

Sivashanmugham, R., N. Bhaskar, and N. Banumathi. 1984. Ventricular fibrillation and cardiac arrest due to neem leaf poisoning. *J. Assoc. Physicians India* 32(7):610-611.

Srivastava, M.K., and R.B. Raizada. 2001. Assessment of embryo/fetotoxicity and teratogenicity of azadirachtin in rats. *Food Chem. Toxicol.* 39(10):1023-1027.

Srivastava, M.K., and R.B. Raizada. 2007. Lack of toxic effect of technical azadirachtin during postnatal development of rats. *Food Chem. Toxicol.* 45(3):465-471.

Bacopa monnieri (L.) Pennell

Scrophulariaceae

B

SCN: bacopa
Syn: *Herpestis monniera* (L.) Kunth
AN: *brahmi; mandukaparni*
OCN: herb-of-grace; Indian pennywort; water hyssop
Part: herb

Quick Reference Summary

Safety Class: 1
Interaction Class: A

Contraindications

None known.

Other Precautions

See Pharmacological Considerations.

Drug and Supplement Interactions

None known.

Editors' Note

As a general rule, *Bacopa monnieri* is *brahmi* in Kerala and *mandukaparni* in north and west India, whereas *Centella asiatica* is called *brahmi* in north and west India and *mandukaparni* in Kerala (south India) (McGuffin et al. 2000).

Adverse Events and Side Effects

Human dose escalation and safety studies have indicated that bacopa is generally well tolerated, with only minor and transient adverse events, primarily gastrointestinal discomfort, reported (Pravina et al. 2007; Singh and Dhawan 1997).

Pharmacological Considerations

An animal study with large doses (250 mg/kg) of bacopa demonstrated a decrease in sperm count and viability. All parameters were shown to return to normal several weeks after cessation of treatment (Singh and Singh 2009). No human studies on the safety or efficacy of bacopa as a male contraceptive have been completed.

Pregnancy and Lactation

No information on the safety of bacopa in pregnancy or lactation was identified in the scientific or traditional literature. Although this review did not identify any concerns for use while pregnant or nursing, safety has not been conclusively established.

Review Details

I. Drug and Supplement Interactions

Clinical Trials of Drug or Supplement Interactions

No clinical trials of drug or supplement interactions were identified.

Case Reports of Suspected Drug or Supplement Interactions

No case reports of suspected drug or supplement interactions were identified.

Animal Trials of Drug or Supplement Interactions

No animal trials of drug or supplement interactions were identified.

II. Adverse Events

Case Reports of Adverse Events

No case reports of adverse events were identified.

III. Pharmacology and Pharmacokinetics

Human Pharmacological Studies

In a phase I safety evaluation study, healthy volunteers were orally administered 300 mg of a standardized bacopa extract daily for 15 days, followed by 450 mg daily for the next 15 days. No adverse effects on hematological, biochemical, or electrocardiographic parameters were observed. Mild gastrointestinal disturbances, which subsided spontaneously without the need for stopping treatment, were reported in two volunteers at the 300 mg dose and in one volunteer at the 450 mg dose (Pravina et al. 2007).

In a dose tolerance study, healthy volunteers were orally administered a combination of isolated bacosides A and B at doses of 20 to 200 mg daily for 4 weeks, or a single dose of 300 mg. No adverse effects were reported, and monitoring of clinical, hematological, and biochemical parameters did not reveal any drug-related abnormalities (Singh and Dhawan 1997).

B

Animal Pharmacological Studies

In male rats orally administered 250 mg/kg of bacopa daily for 28 or 56 days, a reversible reduction in sperm motility, viability, and number and changes in sperm morphology were observed along with changes of cells and tissues in the testes. No changes were observed in libido, levels of testosterone, alanine aminotransferase, aspartate aminotransferase, and creatinine in blood serum, hematological parameters, or in liver and kidney histology. All treatment-related effects returned to control levels within 56 days after treatment (Singh and Singh 2009).

In Vitro Pharmacological Studies

No relevant in vitro pharmacological studies were identified.

IV. Pregnancy and Lactation

No information on the safety of bacopa during pregnancy or lactation was identified.

V. Toxicity Studies

Acute Toxicity

The LD_{50} of orally administered bacopa in rats is 5 g/kg for an aqueous extract, 17 g/kg for an alcohol extract (Martis et al. 1992), and 2.2 g/kg for a standardized extract (Joshua Allan et al. 2007). The LD_{50} of intraperitoneally administered bacopa rats is 1 g/kg for the aqueous extract and 15 g/kg for the alcohol extract (Martis et al. 1992).

In the brine shrimp lethality assay, the LC_{50} of various extracts of bacopa is 295 µg/ml for the ethanol extract, 5332 µg/ml for the water extract, 95 µg/ml for the saponin-rich fraction, and 38 µg/ml for the compound bacoside A (D'Souza et al. 2002).

Short-Term Toxicity

In rats orally administered 250, 500, or 1000 mg/kg of a standardized extract of bacopa daily for 14 days, a mild reduction in body weight gain was observed in treated males (Joshua Allan et al. 2007).

Subchronic Toxicity

In rats orally administered 85, 210, or 500 mg/kg of a standardized extract of bacopa daily for 90 days, no evidence of toxicity was observed. Observed parameters included clinical signs, food consumption, body weight gain, relative organ weights, histopathology, and blood biochemistry (Joshua Allan et al. 2007).

Genotoxicity

In the Ames test for mutagenicity in *Salmonella typhimurium*, no mutagenic activity of a standardized extract of bacopa was observed at concentrations of up to 5000 µg/plate (Deb et al. 2008).

In a clastogenicity assay, no significant activity of a standardized extract of bacopa was observed at concentrations up to 62.5 µg/ml. At a concentration of 125 µg/ml, mild clastogenic activity was observed without, but not with, metabolic activation. The extract demonstrated dose-dependent protection against tested clastogens, with maximum protection observed with metabolic activation (Deb et al. 2008).

Literature Cited

D'Souza, P., M. Deepak, P. Rani, et al. 2002. Brine shrimp lethality assay of *Bacopa monnieri*. *Phytother. Res.* 16(2):197-198.

Deb, D.D., P. Kapoor, R.P. Dighe, et al. 2008. In vitro safety evaluation and anticlastogenic effect of BacoMind on human lymphocytes. *Biomed. Environ. Sci.* 21(1):7-23.

Joshua Allan, J., A. Damodaran, N.S. Deshmukh, K.S. Goudar, and A. Amit. 2007. Safety evaluation of a standardized phytochemical composition extracted from *Bacopa monnieri* in Sprague-Dawley rats. *Food Chem. Toxicol.* 45(10):1928-1937.

Martis, G., A. Rao, and K.S. Karanth. 1992. Neuropharmacological activity of *Herpestis monniera*. *Fitoterapia* 63:399-404.

McGuffin, M., J. Kartesz, A. Leung, and A.O. Tucker. 2000. *Herbs of commerce.* 2nd ed. Silver Spring, MD: American Herbal Products Association.

Pravina, K., K.R. Ravindra, K.S. Goudar, et al. 2007. Safety evaluation of BacoMind in healthy volunteers: Å phase I study. *Phytomedicine* 14(5):301-308.

Singh, A., and S.K. Singh. 2009. Evaluation of antifertility potential of brahmi in male mouse. *Contraception* 79(1):71-79.

Singh, H.K., and B.N. Dhawan. 1997. Neuropsychopharmacological effects of the Ayurvedic nootropic *Bacopa monniera* Linn. (brahmi). *Indian J. Pharmacol.* 29:S359-S365.

Ballota nigra L.

Lamiaceae

SCN: black horehound

Part: herb

Quick Reference Summary

Safety Class: 1
Interaction Class: A

Contraindications

None known.

Other Precautions

None known.

Drug and Supplement Interactions

None known.

Adverse Events and Side Effects

None known.

Pharmacological Considerations

None known.

Pregnancy and Lactation

No information on the safety of black horehound in pregnancy or lactation was identified in the scientific or traditional literature. Although this review did not identify any concerns for use while pregnant or nursing, safety has not been conclusively established.

Review Details

I. Drug and Supplement Interactions

Clinical Trials of Drug or Supplement Interactions

No clinical trials of drug or supplement interactions were identified.

Case Reports of Suspected Drug or Supplement Interactions

No case reports of suspected drug or supplement interactions were identified.

Animal Trials of Drug or Supplement Interactions

No animal trials of drug or supplement interactions were identified.

II. Adverse Events

Case Reports of Adverse Events

No case reports of adverse events were identified.

III. Pharmacology and Pharmacokinetics

Human Pharmacological Studies

No relevant human pharmacological studies were identified.

Animal Pharmacological Studies

A decrease in blood glucose levels was observed in healthy rats orally administered 400 mg/kg of an alcohol extract of black horehound daily for 7 days (Nusier et al. 2007a). In a glucose tolerance test, healthy rats administered a single 400 mg/kg dose of the same extract showed a significant decrease in blood glucose levels with a significant increase in serum insulin levels (Nusier et al. 2007a). In diabetic rats, this extract significantly reduced elevated glucose levels associated with the diabetic condition (Nusier et al. 2007b).

In Vitro Pharmacological Studies

No relevant in vitro pharmacological studies were identified.

IV. Pregnancy and Lactation

No information on the use of black horehound during pregnancy or lactation was identified.

V. Toxicity Studies

Acute Toxicity

No changes in aspartate aminotransferase (AST) or alanine aminotransferase (ALT) levels were observed in rats orally administered 400 mg/kg of an alcohol extract of black horehound daily for 7 days (Nusier et al. 2007a).

Literature Cited

Nusier, M.K., H.N. Bataineh, Z.M. Bataineh, and H.M. Daradka. 2007a. Effects of *Ballota nigra* on blood biochemical parameters and insulin in albino rats. *Neuroendocrinol. Lett.* 28(4):473-476.

Nusier, M.K., H.N. Bataineh, Z.M. Bataineh, and H.M. Daradka. 2007b. Effects of *Ballota nigra* on glucose and insulin in alloxan-diabetic albino rats. *Neuroendocrinol. Lett.* 28(4):470-472.

Baptisia tinctoria (L.) R. Br. — Fabaceae

SCN: wild indigo
OCN: false indigo

Part: root

Quick Reference Summary

Safety Class: 2b
Interaction Class: A

Contraindications

Not for use in pregnancy except under the supervision of a qualified healthcare practitioner.

Other Precautions

None known.

Drug and Supplement Interactions

None known.

Adverse Events and Side Effects

"Large" doses can cause vomiting and diarrhea (Felter and Lloyd 1898).

Pharmacological Considerations

None known.

Pregnancy and Lactation

No information on the safety of wild indigo during pregnancy or lactation was identified. Based on the activity and uses of this herb, however, the editors of this text believe that use in pregnancy should be under the supervision of a qualified healthcare practitioner.

Review Details

I. Drug and Supplement Interactions

Clinical Trials of Drug or Supplement Interactions

No clinical trials of drug or supplement interactions were identified.

Case Reports of Suspected Drug or Supplement Interactions

No case reports of suspected drug or supplement interactions were identified.

Animal Trials of Drug or Supplement Interactions

No animal trials of drug or supplement interactions were identified.

II. Adverse Events

Case Reports of Adverse Events

No case reports of adverse events were identified.

III. Pharmacology and Pharmacokinetics

Human Pharmacological Studies

No relevant human pharmacological studies were identified.

Animal Pharmacological Studies

No relevant animal pharmacological studies were identified.

In Vitro Pharmacological Studies

Chromatographically purified fractions of aqueous-ethanolic extracts from wild indigo produced a strong lymphocyte DNA synthesis-stimulating activity (Beuscher et al. 1989).

Arabinogalactan-containing proteins isolated from wild indigo increased proliferation of mouse spleen cells, IgM production of mouse lymphocytes, IL6 production in alveolar mouse macrophage cultures, and NO_2 production in alveolar mouse macrophage cultures (Classen et al. 2006).

IV. Pregnancy and Lactation

No information on the safety of wild indigo during pregnancy or lactation was identified. Based on the activity and uses of this herb, however, the editors of this text believe that use in pregnancy should be under the supervision of a qualified healthcare practitioner.

V. Toxicity Studies

Acute Toxicity

In rats intraperitoneally administered a decoction or infusion of wild indigo, the fatal dose was reported to be 1 ml per 100 g (Macht and Black 1927).

In rabbits (1.5 kg) administered 10 ml of wild indigo infusion via stomach tube, some narcotic effects and a slight paralysis of the hind limbs were observed. Rabbits intravenously administered 3 ml of the same preparation experienced diarrhea and weakness of the limbs (Macht and Black 1927).

In cats (2.5 kg) subcutaneously administered 5 ml of a decoction or infusion of wild indigo, salivation and vomiting resulted, followed by dilation of the pupils, muscle spasms, and decreased respiration and heart rate (Macht and Black 1927).

Literature Cited

Beuscher, N., K.H. Scheit, C. Bodinet, and L. Kopanski. 1989. Immunologically active glycoproteins of *Baptisia tinctoria*. *Planta Med.* 55(4):358-363.

Classen, B., S. Thude, W. Blaschek, M. Wack, and C. Bodinet. 2006. Immunomodulatory effects of arabinogalactan-proteins from *Baptisia* and *Echinacea*. *Phytomedicine* 13(9-10):688-694.

Felter, H.W., and J.U. Lloyd. 1898. *King's American dispensatory*. 18th ed., 3rd rev. 2 vols. Cincinnati: Ohio Valley Co.

Macht, D.I., and J.A. Black. 1927. A pharmacological note on *Baptisia tinctoria*. *J. Am. Pharm. Assoc.* 16:1056-1059.

Bauhinia forficate Link

Fabaceae

SCN: cow's foot
OCN: *pata de vaca*

Part: leaf

Quick Reference Summary

Safety Class: 1
Interaction Class: A

Contraindications

None known.

Other Precautions

None known.

Drug and Supplement Interactions

None known.

Adverse Events and Side Effects

None known.

Pharmacological Considerations

Although one human study showed no effect of cow's foot on serum glucose levels (Russo et al. 1990), several animal studies have demonstrated that cow's foot modifies glucose regulation (Lino et al. 2004; Pepato et al. 2002; Silva et al. 2002). People with diabetes are advised to monitor their blood sugar closely and discuss the use of this herb with a qualified healthcare practitioner prior to use.

Pregnancy and Lactation

Several animal studies indicated no adverse effects on maternal weight gain, reproductive performance, or fetal or placental development in rats treated with aqueous extracts of cow's foot at varying doses from 500 to 1000 mg/kg daily on days 0 to 20 of pregnancy (Calderon et al. 2001; Damasceno et al. 2004; Rudge et al. 2001; Volpato et al. 2008). In one study, a reduction in postimplantation losses was observed in pregnant diabetic rats administered 200 mg/kg daily cow's foot during pregnancy (Volpato et al. 1999).

No information on the safety of cow's foot during lactation was identified. While this review did not identify any concerns for use while nursing, safety has not been conclusively established.

Review Details

I. Drug and Supplement Interactions

Clinical Trials of Drug or Supplement Interactions

No clinical trials of drug or supplement interactions were identified.

Case Reports of Suspected Drug or Supplement Interactions

No case reports of suspected drug or supplement interactions were identified.

Animal Trials of Drug or Supplement Interactions

No animal trials of drug or supplement interactions were identified.

II. Adverse Events

Case Reports of Adverse Events

No case reports of adverse events were identified.

III. Pharmacology and Pharmacokinetics

Human Pharmacological Studies

No acute or chronic effects on plasma glucose levels or glycated hemoglobin were observed in healthy volunteers or volunteers with type II diabetes who consumed infusions of 3 g of cow's foot daily for 56 days (Russo et al. 1990).

Animal Pharmacological Studies

Significant reductions in plasma glucose levels were observed in diabetic rats orally administered aqueous, ethanolic, and hexanic extracts of cow's foot daily for 7 days (Lino et al. 2004). Similarly, a reduction in serum and urinary glucose levels was observed in diabetic rats provided a decoction of cow's foot as the sole source of drinking water (150 g leaf/l water; mean daily dose of 3.52 ml/kg) (Pepato et al. 2002). A significant reduction of blood glucose levels was observed in healthy and in diabetic rats orally administered 500 to 600 mg/kg of the *n*-butanol fraction of cow's foot (Silva et al. 2002).

In Vitro Pharmacological Studies

No relevant in vitro pharmacological studies were identified.

IV. Pregnancy and Lactation

Several studies have examined the safety of cow's foot in pregnant diabetic and healthy rats using an orally administered aqueous extract at doses of 500 mg/kg on gestational days (GD) 0 to 4, 600 mg/kg on GD 5 to 14, and 1000 mg/kg on GD 15 to 20. No teratogenic activity and no effects on maternal weight gain, reproductive performance, or fetal or placental development were observed. Treatment with cow's foot reduced the number of skeletal abnormalities and visceral malformations observed in offspring of

B

diabetic animals, as compared to controls. No effects on maternal blood glucose levels were observed in either the healthy or diabetic rats (Calderon et al. 2001; Damasceno et al. 2004; Rudge et al. 2001; Volpato et al. 2008).

A reduction in postimplantation losses was observed in pregnant diabetic rats orally administered 200 mg/kg daily during pregnancy (Volpato et al. 1999).

No information on the safety of cow's foot during lactation was identified.

V. Toxicity Studies

Short-Term Toxicity

No changes in serum levels of lactate dehydrogenase, creatine kinase, amylase, angiotensin-converting enzyme, or bilirubin were observed in normal or diabetic rats provided a decoction of cow's foot (150 g/l) in place of drinking water daily for 33 days (Pepato et al. 2004).

Genotoxicity

Aqueous extracts of cow's foot demonstrated mutagenic activity in the Ames test for mutagenicity in *Salmonella typhimurium* strains TA98 with or without metabolic activation, and in strain TA100 without metabolic activation. No mutagenic activity was observed when the same extract was tested in TA100 with metabolic activation, or in TA97a and TA104 with or without metabolic activation (Rivera et al. 1994).

Literature Cited

Calderon, I.M., G.T. Volpato, D.C. Damasceno, and M.V. Rudge. 2001. Maternal, reproductive and perinatal repercussions after treatment with *Bauhinia forficata* aqueous extract during pregnancy of non-diabetic and diabetic rats. *J. Perinat. Med.* 29(Suppl. 1):504-505.

Damasceno, D.C., G.T. Volpato, M. Calderon Ide, R. Aguilar, and M.V. Rudge. 2004. Effect of *Bauhinia forficata* extract in diabetic pregnant rats: Maternal repercussions. *Phytomedicine* 11(2-3):196-201.

Lino, C., J.P. Diogenes, B.A. Pereira, et al. 2004. Antidiabetic activity of *Bauhinia forficata* extracts in alloxan-diabetic rats. *Biol. Pharm. Bull.* 27(1):125-127.

Pepato, M.T., A.M. Baviera, R.C. Vendramini, and I.L. Brunetti. 2004. Evaluation of toxicity after one-month's treatment with *Bauhinia forficata* decoction in streptozotocin-induced diabetic rats. *BMC Complement Altern. Med.* 4:7.

Pepato, M.T., E.H. Keller, A.M. Baviera, et al. 2002. Anti-diabetic activity of *Bauhinia forficata* decoction in streptozotocin-diabetic rats. *J. Ethnopharmacol.* 81(2):191-197.

Rivera, I.G., M.T. Martins, P.S. Sanchez, et al. 1994. Genotoxicity assessment through the Ames test of medicinal plants commonly used in Brazil. *Env. Toxicol. Water Qual.* 9(2):87-93.

Rudge, M.V., D.C. Damasceno, G.T. Volpato, and I.M. Calderon. 2001. Treatment of pregnant diabetic rats with *Bauhinia forficata* (Paw-of-cow) aqueous extract: Maternal and fetal repercussions. *J. Perinat. Med.* 29(Suppl. 1):493-494.

Russo, E.M., A.A. Reichelt, J.R. De-Sa, et al. 1990. Clinical trial of *Myrcia uniflora* and *Bauhinia forficata* leaf extracts in normal and diabetic patients. *Braz. J. Med. Biol. Res.* 23(1):11-20.

Silva, F.R., B. Szpoganicz, M.G. Pizzolatti, M.A. Willrich, and E. de Sousa. 2002. Acute effect of *Bauhinia forficata* on serum glucose levels in normal and alloxan-induced diabetic rats. *J. Ethnopharmacol.* 83(1-2):33-37.

Volpato, G.T., D.C. Damasceno, I.D.M. Paranhos Calderon, and M.V.C. Rudge. 1999. Study of *Bauhinia forficata* L. extract on diabetes in pregnant rats. *Rev. Bras. Plant. Med.* 2(1):49-55.

Volpato, G.T., D.C. Damasceno, M.V. Rudge, C.R. Padovani, and I.M. Calderon. 2008. Effect of *Bauhinia forficata* aqueous extract on the maternal-fetal outcome and oxidative stress biomarkers of streptozotocin-induced diabetic rats. *J. Ethnopharmacol.* 116(1):131-137.

Benincasa hispida (Thunb.) Cogn.

Cucurbitaceae

SCN: winter melon
Syn: *Benincasa cerifera* Savi
AN: *kushmanda*
PN: *dong gua pi* (fruit rind)
OCN: wax gourd; white pumpkin
Part: rind

Quick Reference Summary

Safety Class: 1
Interaction Class: A

Contraindications

None known.

Other Precautions

None known.

Drug and Supplement Interactions

None known.

Adverse Events and Side Effects
None known.

Pharmacological Considerations
None known.

Pregnancy and Lactation
Texts on traditional Chinese medicine do not indicate any cautions for use of bitter melon rind during pregnancy or lactation (Bensky et al. 2004; Chen and Chen 2004). Although this review did not identify any concerns for use while pregnant or nursing, safety has not been conclusively established.

Review Details

I. Drug and Supplement Interactions

Clinical Trials of Drug or Supplement Interactions
No clinical trials of drug or supplement interactions were identified.

Case Reports of Suspected Drug or Supplement Interactions
No case reports of suspected drug or supplement interactions were identified.

Animal Trials of Drug or Supplement Interactions
No animal trials of drug or supplement interactions were identified.

II. Adverse Events

Case Reports of Adverse Events
No case reports of adverse events were identified.

III. Pharmacology and Pharmacokinetics

Human Pharmacological Studies
No relevant human pharmacological studies were identified.

Animal Pharmacological Studies
No relevant animal pharmacological studies were identified.

In Vitro Pharmacological Studies
No relevant in vitro pharmacological studies were identified.

IV. Pregnancy and Lactation
Texts on traditional Chinese medicine do not indicate any cautions for use of bitter melon rind during pregnancy or lactation (Bensky et al. 2004; Chen and Chen 2004).

V. Toxicity Studies
No toxicity studies were identified.

Literature Cited

Bensky, D., S. Clavey, and E. Stöger. 2004. *Chinese herbal medicine: Materia medica*. 3rd ed. Seattle: Eastland Press.

Chen, J.K., and T.T. Chen. 2004. *Chinese medical herbology and pharmacology*. City of Industry, CA: Art of Medicine Press.

Benincasa hispida (Thunb.) Cogn. — Cucurbitaceae

SCN: winter melon
Syn: *Benincasa cerifera* Savi
AN: *kushmanda*
PN: *dong gua zi* (seed)
OCN: wax gourd; white pumpkin
Part: seed

Quick Reference Summary

Safety Class: 1
Interaction Class: A

Contraindications
None known.

Other Precautions
None known.

Drug and Supplement Interactions
None known.

Adverse Events and Side Effects
None known.

Pharmacological Considerations
A traditional Chinese medicine text notes that winter melon seed has mild diuretic activity (Chen and Chen 2004).

Pregnancy and Lactation
No information on the safety of winter melon seed in pregnancy or lactation was identified in the scientific or traditional literature. Although this review did not identify any concerns for use while pregnant or nursing, safety has not been conclusively established.

B

Review Details

I. Drug and Supplement Interactions

Clinical Trials of Drug or Supplement Interactions

No clinical trials of drug or supplement interactions were identified.

Case Reports of Suspected Drug or Supplement Interactions

No case reports of suspected drug or supplement interactions were identified.

Animal Trials of Drug or Supplement Interactions

No animal trials of drug or supplement interactions were identified.

II. Adverse Events

Case Reports of Adverse Events

No case reports of adverse events were identified.

III. Pharmacology and Pharmacokinetics

Human Pharmacological Studies

No relevant human pharmacological studies were identified.

Animal Pharmacological Studies

No relevant animal pharmacological studies were identified.

In Vitro Pharmacological Studies

No relevant in vitro pharmacological studies were identified.

IV. Pregnancy and Lactation

No information on the use of winter melon seed during pregnancy or lactation was identified.

V. Toxicity Studies

No toxicity studies were identified.

Literature Cited

Chen, J.K., and T.T. Chen. 2004. *Chinese medical herbology and pharmacology*. City of Industry, CA: Art of Medicine Press.

Berberis vulgaris L.

Berberidaceae

SCN: barberry
OCN: European barberry

Part: root, root bark

Quick Reference Summary

Safety Class: 2b
Interaction Class: A

Contraindications

Not for use in pregnancy except under the supervision of a qualified healthcare practitioner (Jahnke et al. 2006).

Other Precautions

Use of barberry during lactation is not recommended (Chan 1993).

Use is cautioned in persons with severe or acute hepatocellular disease, septic cholecystitis, unconjugated hyperbilirubinemia, intestinal spasms or ileus, or liver cancer (Chan 1993; Mills and Bone 2005).

Drug and Supplement Interactions

None known.

Notice

Berberine (root bark: 6.1%; woody portion of root: 0.4%) (De Smet 1992; Leung and Foster 1996); *see* Appendix 1.

Editors' Notes

Most safety concerns reported for barberry are based on studies of the compound berberine and other alkaloids. Data regarding isolated compounds may not apply directly to products or extracts made from the root or root bark.

The compound berberine has been shown to exhibit a number of bioactivities including cytotoxicity in cancer cell lines (Kettmann et al. 2004; Kim et al. 2005; Orfila et al. 2000) and topoisomerase I and II inhibition (Kim et al. 1998; Mantena et al. 2006; Pasqual et al. 1993). Studies have indicated that the antimicrobial action of berberine may be potentiated by non-antimicrobial compounds found in berberine-containing plants, indicating a synergistic action of compounds (Stermitz et al. 2000).

Adverse Events and Side Effects

None known.

Pharmacological Considerations

Human studies have provided mixed results on the effects of the compound berberine on cyclosporine A, with one study indicating an increase in cyclosporine A bioavailability after repeated coadministration (Wu et al. 2005) and a second study indicating increased bioavailability after a single dose of the compound berberine but no change after repeated dosing of berberine (Xin et al. 2006).

Pregnancy and Lactation

Barberry is contraindicated or cautioned against in pregnancy in several contemporary texts on herbal safety (Brinker 2001; Mills and Bone 2005). These contraindications are based primarily on the uterine stimulant activity of the isolated compound berberine in excised mouse uteri (Furuya 1957; Imaseki et al. 1961) and the potential ability of berberine to displace bilirubin and cause neonatal jaundice (Chan 1993). Although definitive data confirming the safety of barberry during pregnancy is lacking, several reproductive toxicity studies of the isolated compound berberine in mice and rats have shown no adverse effects on the fetus at doses equivalent to more than 75 times the standard human dose (Jahnke et al. 2006).

Berberine-containing plants have been reported to induce uterine stimulation (Furuya 1957; Imaseki et al. 1961); however, no uterine stimulation was noted in rats administered high doses of berberine during pregnancy (Jahnke et al. 2006), and a study of berberine-containing herbal extracts on isolated uteri showed no correlation between uterine stimulation and berberine concentration (Haginiwa and Harada 1962).

Berberine has been shown to be present in the breast milk of lactating women who have taken berberine-containing plants (Chan 1993).

Review Details

I. Drug and Supplement Interactions

Clinical Trials of Drug or Supplement Interactions

In a randomized controlled trial of renal transplant patients taking cyclosporine A and 200 mg of the compound berberine three times daily for 3 months, the minimum blood concentrations of cyclosporine A increased by 88.9% and the ratios of concentration/dose of cyclosporine A in the berberine-treated group increased by 98.4%. In the control group, the increase in trough blood concentrations of cyclosporine A was 64.5% and the increase in the ratio of concentration/dose of cyclosporine A was 69.4%. The authors indicated that the mechanism for this interaction is most likely an inhibition of CYP3A4 by berberine (Wu et al. 2005).

An increase in bioavailability of orally administered cyclosporine A was observed after healthy male volunteers were administered a single dose of 3 mg cyclosporine A followed by a single oral dose of 300 mg of the compound berberine. The increase in area under the time-concentration curve of cyclosporine A was 19.2%; no changes in elimination half-life, maximum blood drug concentration, time to maximum concentration, or apparent oral clearance were observed (Xin et al. 2006). Conversely, no changes in bioavailability of orally administered cyclosporine A were observed after healthy male volunteers were administered 300 mg of the compound berberine twice daily for 10 days followed by a single dose of 6 mg cyclosporine A (Xin et al. 2006).

No effects of oryzanol (a derivative of rice bran oil that includes sterols and ferulic acid) on berberine absorption were observed in healthy volunteers (Li et al. 2000).

Case Reports of Suspected Drug or Supplement Interactions

No case reports of suspected drug or supplement interactions were identified.

Animal Trials of Drug or Supplement Interactions

The compound berberine has been shown to significantly prolong pentobarbital-induced sleeping time in rats (Janbaz and Gilani 2000).

II. Adverse Events

Case Reports of Adverse Events

No case reports of adverse events were identified.

III. Pharmacology and Pharmacokinetics

Human Pharmacological Studies

Intravenous administration of the compound berberine (0.2 mg/kg/min) caused a significant decrease in blood pressure in patients with severe congestive heart failure (Marin-Neto et al. 1988).

Animal Pharmacological Studies

Chronic administration of berberine to rats was shown to significantly decrease bilirubin protein binding at doses of 10 to 29 mg/kg (administered intraperitoneally), although at doses of 2 mg/kg the displacement was not significant (Chan 1993). Berberine is speculated to have possible interactions with drugs that displace bilirubin protein binding (Mills and Bone 2005).

Berberine was shown to promote blood coagulation in mice and rats (Ziablitskii et al. 1996).

In Vitro Pharmacological Studies

Berberine has been shown to increase the efflux of paclitaxel and rhodamine in human digestive tract and liver cancer cells by inducing P-glycoprotein (Blaschek et al. 2002). Berberine may interfere with the action of tetracycline in the treatment of cholera (Khin Maung et al. 1985).

B

IV. Pregnancy and Lactation

In pregnant rats fed the compound berberine chloride dihydrate on gestational days (GD) 6 to 20, some reduction in maternal weight gain was observed, with a lowest-observed-adverse-effect level (LOAEL) of 530 mg/kg daily. Only a mild reduction in fetal weights was observed, and the LOAEL based on fetal weight reduction was 1000 mg/kg (Jahnke et al. 2006). Similarly, in mice administered berberine on GD 6 to 17 at doses up to 1155 mg/kg daily, the maternal LOAEL was determined to be 531 mg/kg daily, and the developmental toxicity level was 1000 mg/kg daily. In mice, 33% of the treated females died. Surviving animals had increased relative water intake, and average fetal body weight per litter decreased 5–6% with no change in live litter size (Jahnke et al. 2006).

Berberine has been shown to stimulate uterine contractions in both pregnant and nonpregnant mice (Furuya 1957; Imaseki et al. 1961). A study of various berberine-containing herbal extracts on isolated uteri, however, indicated that relaxation or stimulation of the uterus did not correlate with the concentration of berberine in the extract, suggesting that not all berberine-containing herbs will have the same effect on the uterus (Haginiwa and Harada 1962).

Berberine has been shown to be present in the breast milk of lactating women who have taken berberine-containing plants (Chan 1993).

V. Toxicity Studies

Acute Toxicity

The LD_{50} of orally administered barberry root extract fraction (containing 80% berbamine and three other isoquinoline alkaloids) in mice is 520 mg/kg, and in rats is 1280 mg/kg (Manolov et al. 1985). The LD_{50} of orally administered barberry root in mice was determined to be equivalent to 2600 mg/kg of the powdered root (Peychev 2005).

The LD_{50} of orally administered berberine in mice is 329 mg/kg (Haginiwa and Harada 1962). The LD_{50} of orally administered berberine sulfate in rats is greater than 1000 mg/kg (Kowalewski et al. 1975).

Genotoxicity

No mutagenic activity of berberine was observed in *Salmonella typhimurium* TA100 and TA98 with or without metabolic activation by S9 mix. Berberine hydrochloride was weakly mutagenic to strain TA98 without S9 mix but showed no mutagenic activity in TA100 without S9 mix (Nozaka et al. 1990).

No genotoxic, mutagenic, or recombinogenic activity of berberine with or without metabolic activation was observed in the SOS chromotest. Berberine did not induce significant cytotoxic, mutagenic or recombinogenic effects during treatments performed under nongrowth conditions; however, in dividing cells, this alkaloid induced cytotoxic and cytostatic effects in proficient and repair-deficient strains of *Saccharomyces cerevisiae*. In dividing cells, the induction of frameshift and mitochondrial mutations, as well as crossing over, indicated that berberine is not a potent mutagenic agent (Pasqual et al. 1993).

Literature Cited

Blaschek, W., S. Ebel, E. Hackenthal, et al. 2002. *Hagers handbuch der drogen und arzneistoffe. HagerROM.* Heidelberg: Springer.

Brinker, F. 2001. *Herb contraindications and drug interactions.* 3rd ed. Sandy, OR: Eclectic Medical Publications.

Chan, E. 1993. Displacement of bilirubin from albumin by berberine. *Biol. Neonate* 63(4):201-208.

De Smet, P.A.G.M. 1992. *Adverse effects of herbal drugs. Volume 1.* New York: Springer.

Furuya, T. 1957. Pharmacological action, including toxicity and excretion of berberine hydrochloride and its oxidation product. *Bull. Osaka Med. School* 3:62-67. *Cited in* De Smet, P.A.G.M. 1992. *Adverse effects of herbal drugs. Volume 1.* New York: Springer.

Haginiwa, J., and M. Harada. 1962. [Pharmacological studies on crude drugs. V. Comparison of berberine type alkaloid-containing plants on their components and several pharmacological actions.] *Jpn. J. Pharmacol.* 82:726-731.

Imaseki, I., Y. Kitabatake, and T. Taguchi. 1961. Studies on the effect of berberine alkaloids on intestine and uterus in mice. *Yakugaku Zasshi* 81:1281-1284.

Jahnke, G.D., C.J. Price, M.C. Marr, C.B. Myers, and J.D. George. 2006. Developmental toxicity evaluation of berberine in rats and mice. *Birth Defects Res. B Dev. Reprod. Toxicol.* 77(3):195-206.

Janbaz, K.H., and A.H. Gilani. 2000. Studies on preventive and curative effects of berberine on chemical-induced hepatotoxicity in rodents. *Fitoterapia* 71(1):25-33.

Kettmann, V., D. Kosfalova, S. Jantova, M. Cernakova, and J. Drimal. 2004. In vitro cytotoxicity of berberine against HeLa and L1210 cancer cell lines. *Pharmazie* 59(7):548-551.

Khin Maung, U., K. Myo, W. Nyunt Nyunt, K. Aye, and U. Tin. 1985. Clinical trial of berberine in acute watery diarrhoea. *Br. Med. J. (Clin. Res. Ed.)* 291(6509):1601-1605.

Kim, H.R., H.Y. Min, Y.H. Jeong, et al. 2005. Cytotoxic constituents from the whole plant of *Corydalis pallida*. *Arch. Pharm. Res.* 28(11):1224-1227.

Kim, S.A., Y. Kwon, J.H. Kim, M.T. Muller, and I.K. Chung. 1998. Induction of topoisomerase II-mediated DNA cleavage by a protoberberine alkaloid, berberrubine. *Biochemistry* 37(46):16316-16324.

Kowalewski, Z., A. Mrozikiewicz, T. Bobkiewicz, K. Drost, and B. Hladon. 1975. [Toxicity of berberine sulfate.] *Acta Pol. Pharm.* 32(1):113-120.

Leung, A.Y., and S. Foster. 1996. *Encyclopedia of common natural ingredients used in food, drugs, and cosmetics.* 2nd ed. New York: Wiley.

Li, B.X., B.F. Yang, X.M. Hao, et al. 2000. Study on the pharmacokinetics of berberine in single dosage and coadministration with oryzanol in rabbits and healthy volunteers. *Chin. Pharm. J.* 35(1)33-35.

Manolov, P., N. Nikolov, M. Markov, and M. Toneva. 1985. [Experimental research on *Berberis vulgaris*.] *Eksp. Med. Morfol.* 24(2):41-45.

Mantena, S.K., S.D. Sharma, and S.K. Katiyar. 2006. Berberine inhibits growth, induces G_1 arrest and apoptosis in human epidermoid carcinoma A431 cells by regulating Cdki-Cdk-cyclin cascade, disruption of mitochondrial membrane potential and cleavage of caspase 3 and PARP. *Carcinogenesis* 27(10):2018-2027.

Marin-Neto, J.A., B.C. Maciel, A.L. Secches, and L. Gallo, Jr. 1988. Cardiovascular effects of berberine in patients with severe congestive heart failure. *Clin. Cardiol.* 11(4):253-260.

Mills, S., and K. Bone. 2005. *The essential guide to herbal safety*. St. Louis: Elsevier.

Nozaka, T., F. Watanabe, S. Tadaki, et al. 1990. Mutagenicity of isoquinoline alkaloids, especially of the aporphine type. *Mutat. Res.* 240(4):267-279.

Orfila, L., M. Rodriguez, T. Colman, et al. 2000. Structural modification of berberine alkaloids in relation to cytotoxic activity in vitro. *J. Ethnopharmacol.* 71(3):449-456.

Pasqual, M.S., C.P. Lauer, P. Moyna, and J.A. Henriques. 1993. Genotoxicity of the isoquinoline alkaloid berberine in prokaryotic and eukaryotic organisms. *Mutat. Res.* 286(2):243-252.

Peychev, L. 2005. Pharmacological investigation on the cardiovascular effects of *Berberis vulgaris* on tested animals. *Pharmacia* 52(1-2):118-121.

Stermitz, F.R., P. Lorenz, J.N. Tawara, L.A. Zenewicz, and K. Lewis. 2000. Synergy in a medicinal plant: Antimicrobial action of berberine potentiated by 5′-methoxyhydnocarpin, a multidrug pump inhibitor. *Proc. Natl. Acad. Sci. U.S.A.* 97(4):1433-1437.

Wu, X., Q. Li, H. Xin, A. Yu, and M. Zhong. 2005. Effects of berberine on the blood concentration of cyclosporin A in renal transplanted recipients: Clinical and pharmacokinetic study. *Eur. J. Clin. Pharmacol.* 61(8):567-572.

Xin, H.W., X.C. Wu, Q. Li, et al. 2006. The effects of berberine on the pharmacokinetics of ciclosporin A in healthy volunteers. *Methods Find. Exp. Clin. Pharmacol.* 28(1):25-29.

Ziablitskii, V.M., V.N. Romanovskaia, R.Z. Umurzakova, A.N. Starosel'skaia, and T. Mikhal'skaia. 1996. [Modification to the functional status of the hemostatic system with the use of berberine sulfate.] *Eksp. Klin. Farmakol.* 59(1):37-39.

B

Betula lenta L. — Betulaceae

SCN: sweet birch
OCN: black birch; cherry birch

Part: bark, leaf

Quick Reference Summary

Safety Class: 1
Interaction Class: A

Contraindications

None known.

Other Precautions

None known.

Drug and Supplement Interactions

None known.

Notice

Salicylates (bark and leaf 0.23–0.6%) (Burdock 1996; Felter and Lloyd 1898; List and Hörhammer 1973; Nowak 1966); *see* Appendix 1.

Adverse Events and Side Effects

None known.

Pharmacological Considerations

None known.

Pregnancy and Lactation

No information on the safety of sweet birch in pregnancy or lactation was identified in the scientific or traditional literature. Although this review did not identify any concerns for use while pregnant or nursing, safety has not been conclusively established.

Review Details

I. Drug and Supplement Interactions

Clinical Trials of Drug or Supplement Interactions

No clinical trials of drug or supplement interactions were identified.

Case Reports of Suspected Drug or Supplement Interactions

No case reports of suspected drug or supplement interactions were identified.

Animal Trials of Drug or Supplement Interactions

No animal trials of drug or supplement interactions were identified.

II. Adverse Events

Case Reports of Adverse Events

No case reports of adverse events were identified.

III. Pharmacology and Pharmacokinetics

Human Pharmacological Studies

No relevant human pharmacological studies were identified.

Animal Pharmacological Studies

No relevant animal pharmacological studies were identified.

In Vitro Pharmacological Studies

No relevant in vitro pharmacological studies were identified.

IV. Pregnancy and Lactation

No information on the safety of sweet birch during pregnancy or lactation was identified.

V. Toxicity Studies

Acute Toxicity

The oral LD_{50} of the compound methyl salicylate (which comprises 97–99% of sweet birch essential oil) is 1110 mg/kg in mice, 887 or 1250 mg/kg in rats, 1060 mg/kg in guinea pigs, 1300 or 2800 mg/kg in rabbits, and 2100 mg/kg in dogs (FAO/WHO 1967). The adult human oral LD_{50} is estimated at 500 mg/kg (FAO/WHO 1967).

Subchronic Toxicity

In dogs orally administered 50, 100, 250, 500, 800, or 1200 mg/kg methyl salicylate for up to 10 weeks, no adverse effects were noted in animals receiving 250 mg/kg or less, but an increasing dose-dependent fatty metamorphosis of the liver was observed at higher test levels (Webb and Hansen 1963).

In rats fed diets containing 0.1, 0.5, or 1.0% methyl salicylate for 17 weeks, both sexes showed a significant reduction in growth rate at the 1.0% level, but histological examination of the major organs revealed no abnormality. In a related experiment, in rats fed a diet containing 2% methyl salicylate for up to 10 weeks, bone growth was reduced and excessive density of bone with reduced chondroclastic and osteoclastic activity was observed (Webb and Hansen 1963).

Chronic Toxicity

In rats fed diets containing 0.1, 0.5, 1.0, or 2.0% methyl salicylate for 2 years, animals fed the highest dose did not survive past 49 weeks. At the 1% level, growth rates were considerably reduced and enlargement of male testes and female hearts and kidneys were noted. Excess cancellous bone formation was seen at the 2.0, 1.0, and 0.5% levels (Webb and Hansen 1963). Another 2-year feeding study revealed no adverse effects, including bone changes, at dietary levels up 0.21% of methyl salicylate (Packman et al. 1961).

In dogs orally administered 0, 50, 150, and 250 mg/kg methyl salicylate daily for 2 years, some growth retardation and liver enlargement was noted at the 150 and 250 mg/kg level, and histology revealed enlarged hepatic parenchymal cells (Webb and Hansen 1963).

Literature Cited

Burdock, G.A. 1996. *Encyclopedia of food and color additives*. Boca Raton, FL: CRC Press.

FAO/WHO. 1967. Methyl salicylate. Toxicological evaluation of some flavouring substances and non-nutritive sweetening agents. FAO Nutrition Meetings Resort Series No. 44A and WHO Food Additives 68.33.

Felter, H.W., and J.U. Lloyd. 1898. *King's American dispensatory*. 18th ed., 3rd rev. 2 vols. Cincinnati: Ohio Valley Co.

List, P.H., and H. Hörhammer. 1973. *Hagers handbuch der pharmazeutischen praxis*. Vollst. 4. Neuausg. ed. Berlin: Springer.

Nowak, G.A. 1966. Cosmetic and medicinal properties of the birch. *Am. Perfum. Cosmet.* 81:37-40.

Packman, E.W., D.D. Abbott, B.M. Wagner, and J.W.E. Harrison. 1961. Chronic oral toxicity of oil of sweet birch (methyl salicylate). *Pharmacologist* 3:62.

Webb, W.K., and W.H. Hansen. 1963. Chronic and subacute toxicology and pathology of methyl salicylate in dogs, rats and rabbits. *Toxicol. Appl. Pharmacol.* 5:576.

Betula spp. — Betulaceae

Betula pendula Roth
SCN: birch
Syn: *Betula verrucosa* Ehrh.
OCN: European white birch; silver birch; weeping birch

Betula pubescens **Ehrh.**
SCN: birch
Syn: *Betula alba* L.
OCN: downy birch; white birch
Part: bark, leaf

Quick Reference Summary

Safety Class: 1
Interaction Class: A

Contraindications

None known.

Other Precautions

None known.

Drug and Supplement Interactions

None known.

Notice

Diuretic (Remington and Wood 1918; Wichtl 2004); *see* Appendix 2.

Adverse Events and Side Effects

None known.

Pharmacological Considerations

None known.

Pregnancy and Lactation

No information on the safety of birch in pregnancy or lactation was identified in the scientific or traditional literature. Although this review did not identify any concerns for use while pregnant or nursing, safety has not been conclusively established.

Review Details

I. Drug and Supplement Interactions

Clinical Trials of Drug or Supplement Interactions

No clinical trials of drug or supplement interactions were identified.

Case Reports of Suspected Drug or Supplement Interactions

No case reports of suspected drug or supplement interactions were identified.

Animal Trials of Drug or Supplement Interactions

No animal trials of drug or supplement interactions were identified.

II. Adverse Events

Case Reports of Adverse Events

No case reports of adverse events were identified.

III. Pharmacology and Pharmacokinetics

Human Pharmacological Studies

No relevant human pharmacological studies were identified.

Animal Pharmacological Studies

No relevant animal pharmacological studies were identified.

In Vitro Pharmacological Studies

No relevant in vitro pharmacological studies were identified.

IV. Pregnancy and Lactation

No information on the use of birch during pregnancy or lactation was identified.

V. Toxicity Studies

Genotoxicity

In the Ames test for mutagenicity, weak mutagenic activity was observed, a frequent response from plant materials containing flavonols (Göggleman and Schimmer 1986).

Literature Cited

Göggleman, W., and O. Schimmer. 1986. Mutagenic activity of phytotherapeutical drugs. In *Genetic toxicology of the diet*, edited by Knudsen, I. New York: Alan Liss.

Remington, J.P., and H.C. Wood. 1918. *The dispensatory of the United States of America.* 20th ed. Philadelphia: Lippincott.

Wichtl, M. 2004. *Herbal drugs and phytopharmaceuticals: A handbook for practice on a scientific basis.* 3rd ed. Boca Raton, FL: CRC Press.

Bixa orellana L. — Bixaceae

SCN: annatto
OCN: *achiote*; lipstick tree

Part: seed

Quick Reference Summary

Safety Class: 1
Interaction Class: A

Contraindications

None known.

Other Precautions

None known.

Drug and Supplement Interactions

None known.

Adverse Events and Side Effects

Allergic reactions to annatto, including anaphylactic reactions and urticaria, have been reported and confirmed by patch testing (Nish et al. 1991).

Pharmacological Considerations

Several animal studies have indicated that annatto may modify glucose regulation (Fernandes et al. 2002; Morrison et al. 1991; Morrison and West 1985; Russell et al. 2005, 2008). People with diabetes are advised to monitor their blood sugar closely and discuss the use of this herb with a qualified healthcare practitioner prior to use.

Pregnancy and Lactation

No adverse effects on pregnancy outcomes were observed in two rat studies on the use of annatto in pregnancy (Paumgartten et al. 2002; van Esch et al. 1959). One of the studies concluded that the maternal and fetal no-observed-adverse-effect level for orally consumed annatto was over 500 mg/kg daily (Paumgartten et al. 2002).

No information on the safety of annatto during lactation was identified. While this review did not identify any concerns for use while nursing, safety has not been conclusively established.

Review Details

I. Drug and Supplement Interactions

Clinical Trials of Drug or Supplement Interactions

No clinical trials of drug or supplement interactions were identified.

Case Reports of Suspected Drug or Supplement Interactions

No case reports of suspected drug or supplement interactions were identified.

Animal Trials of Drug or Supplement Interactions

No animal trials of drug or supplement interactions were identified.

II. Adverse Events

Case Reports of Adverse Events

Allergic reactions to annatto, including anaphylactic reactions and urticaria, have been reported and confirmed by patch testing (Nish et al. 1991).

III. Pharmacology and Pharmacokinetics

Human Pharmacological Studies

In studies of patients with recurrent or chronic urticaria, one study yielded reactions in 26% of the patients treated topically with 25 μl of annatto extract (0.065% bixin) (Mikkelsen et al. 1978), while 10% of patients orally administered 10 mg annatto extract reacted to the annatto (Juhlin 1981).

Animal Pharmacological Studies

In anesthetized dogs intragastrically administered the compound bixin at doses of 50 ml of a chloroform extract dissolved in oil or 15 ml of a chloroform extract dissolved in ethanol, a reduction in the blood sugar increase rate was observed after the oral glucose tolerance test (Morrison and West 1985). A reduction in blood glucose levels was observed in dogs administered a partially purified annatto extract at a dose of 80 mg/kg (Russell et al. 2005). A reduction in fasting glucose levels was observed in healthy and diabetic dogs orally administered an extract of annatto (Russell et al. 2008). Hyperglycemia was observed in anesthetized dogs treated with the compound *trans*-bixin (Morrison et al. 1991).

In rats administered drinking water containing 0.8, 8.5, or 74 mg/kg norbixin daily for 21 days, norbixin induced dose-dependent hyperglycemia. Conversely, in mice administered drinking water containing 0.8, 7.6, 66, or 274 mg/kg norbixin daily for 21 days, norbixin induced dose-dependent hypoglycemia. These effects were not observed in animals treated with annatto extracts containing 50% norbixin. Rats and mice treated with annatto pigments showed hyperinsulinemia and hypoinsulinemia, respectively, indicating that pancreatic beta-cells were affected (Fernandes et al. 2002).

In the local lymph node assay, a threefold increase in the proliferation of auricular lymph node cells was observed in mice treated topically with the compound bixin (1-25% w/v) at concentrations of 5 to 25%. In the mouse ear swelling test, a significant increase in percent ear swelling was observed in animals treated with 5 to 10% bixin. Bixin was classified as a nonirritant at concentrations of 1 to 25% (w/v) but was considered a contact sensitizer. In these same tests, the compound norbixin (1–20% w/v) showed no irritation or contact sensitization (Auttachoat et al. 2005).

In Vitro Pharmacological Studies

No relevant in vitro pharmacological studies were identified.

IV. Pregnancy and Lactation

In pregnant rats orally administered annatto (28% bixin) on days 6 to 15 of pregnancy at doses of 0 to 500 mg/kg, no increase in embryo lethality and no reduction of fetal body weight were observed among annatto-exposed rats, and annatto did not induce any increase in the incidence of externally visible, visceral, or skeletal anomalies in the exposed offspring. The maternal and fetal no-observed-adverse-effect level (NOAEL) for orally consumed annatto was over 500 mg/kg daily (Paumgartten et al. 2002).

No teratogenic effects or other adverse effects on growth or reproduction were observed in rats fed diets containing 0.05% fat-soluble annatto and 0.05% water-soluble extracts of annatto for three generations (van Esch et al. 1959).

No information on the safety of annatto during lactation was identified.

V. Toxicity Studies

Acute Toxicity

The LD_{50} of orally administered annatto in rats is over 50 g/kg for the fat-soluble extract, and over 35 g/kg for the water-soluble extract (Hallagan et al. 1995; van Esch et al. 1959).

Short-Term Toxicity

In rats orally administered 2 g/kg of annatto powder (27% bixin) 20 times over a 4-week period, no changes in hematological or plasma biochemical parameters were observed. A decrease in body weight gain was observed in male rats, and kidney apoptosis in restricted areas without proliferation or tubular segments modification was observed in 20% of female rats. The precise nature of apoptosis was not investigated. The authors concluded that annatto was not toxic to the rat (Bautista et al. 2004).

Disarrangement of mitochondria structure, fusion of mitochondria, and formation of residual bodies were observed in the liver and pancreas of dogs (9 to 16 kg each) orally administered 2 g of a dried chloroform extract of annatto dissolved in ethanol daily for 14 days. Stacking of hepatic mitochondria was also observed in these animals. Mitochondrial changes were not observed in animals fed 3 mg riboflavin in addition to the annatto extract (Morrison et al. 1987).

No carcinogenic activity was observed in rats fed a diet containing up to 1000 ppm annatto (4.23 g/kg bixin daily) for 2 weeks before or 8 weeks after treatment with diethylnitrosamine (Agner et al. 2005).

In a 21-day study, mice were given drinking water containing annatto extract (56 or 351 mg/kg) or the compound norbixin (0.8, 7.6, 66, or 274 mg/kg), and rats were given drinking water containing annatto extract (0.8, 7.5 or 68 mg/kg) or norbixin (0.8, 8.5, or 74 mg/kg). In rats, no toxicity was detected by plasma chemistry. In mice, norbixin induced an increase in plasma alanine aminotransferase activity (Fernandes et al. 2002).

Subchronic Toxicity

In beagle dogs fed a diet containing 2.7% fat-soluble extract of annatto seed for 9 weeks, then fed normal diet for 5 weeks, and then fed only 1.35% extract for 38 weeks, no abnormalities were found in growth, food intake, mortality, liver and kidney function, hematology, or histopathology. One dog in the test group died and had hepatocellular degeneration, although it is not clear whether the death was treatment related (Kay and Calandra 1961).

In beagle dogs fed a diet of up to 10% aqueous extract of annatto for 1 year, or fed 20% aqueous extract of annatto for 16 weeks, then the same amount half in diet and half in gelatin capsules daily for 36 weeks, liver or kidney function tests, hematology, and histopathology of all major tissues showed no abnormalities attributable to the test substance (Kay and Calandra 1961).

In rats fed annatto extract at dietary levels of 0, 0.1, 0.3, or 0.9% for 13 weeks, no treatment-related adverse effects were observed on body weight, food and water consumption, or ophthalmological or hematological data. Blood biochemical analysis revealed changes in rats of both sexes in the 0.9% and 0.3% diet groups, including increased alkaline phosphatase, phospholipid, total protein, albumin, and albumin/globulin ratio. Marked elevation in absolute and relative liver weights was also found in both sexes of the 0.9% and 0.3%, but not the 0.1%, groups. In the 0.9% group, hepatocyte hypertrophy was evident and an additional electron microscopic examination demonstrated this to be linked to abundant mitochondria after exposure. The authors of this study concluded that the no-observed-adverse-effect-level (NOAEL) of annatto was 0.1% of the diet, equivalent to 69 mg/kg daily for males and 76 mg/kg daily for females (Hagiwara et al. 2003).

No changes in food intake, growth, hematological examination, organ weights, or histopathology of major organs were observed in pigs administered a diet of 1%

B

fat-soluble and 1% water-soluble extract of annatto daily for 21 weeks (van Esch et al. 1959).

Cyst formation was observed at the site of injection in mice subcutaneously administered 0.1 ml of fat-soluble extract of annatto three times a week for 17 months (van Esch et al. 1959).

No changes in food intake, growth, hematological examination, organ weights, or histopathology of major organs were observed in rats fed a diet of 2% fat-soluble annatto, and 2% water-soluble annatto daily for 13 weeks (van Esch et al. 1959). No adverse effects were seen in rats orally administered 1 g/kg annatto extract daily for 100 days (Zbinden and Studer 1958).

No adverse effects were observed in rats subcutaneously administered 0.05 ml of fat-soluble annatto extract three times a week, at the same injection site, for 36 weeks. The rats were observed for a total of 24 months (van Esch et al. 1959).

No adverse effects and no accumulation of carotenoid compounds (bixin and norbixin) were observed in rats orally administered diets of 0.1% of annatto extracts containing primarily bixin or norbixin daily for 52 weeks (Philip 1981).

No skin papillomas or other tumors were observed on mice treated topically, twice a week, with 0.025 ml of annatto extract for 3 months (Engelbreth-Holm and Inversen 1955).

Chronic Toxicity

No pathological effects were observed in rats orally administered 26 mg annatto daily in soy oil for 26 months (Engelbreth-Holm and Inversen 1955).

Genotoxicity

In comet assays, no mutagenic activity was observed in rats fed a diet containing up to 1000 ppm annatto (4.23 g/kg bixin daily) for 2 weeks before or 8 weeks after treatment with diethylnitrosamine (Agner et al. 2004, 2005), and no enhancement in DNA breakage was detected in liver or kidney from mice treated with annatto pigments at doses up to 351 mg/kg daily for 21 days (Fernandes et al. 2002).

In the micronucleus test, no mutagenic activity was observed in the bone marrow cells of mice fed diets containing 1330, 5330, or 10,670 ppm annatto daily for 7 days. An increased frequency of micronucleated cells was observed with coadministration of cyclophosphamide in animals fed the highest concentration of annatto, as compared to cyclophosphamide alone (Alves de Lima et al. 2003).

No mutagenic activity was observed in *E. coli* treated with annatto at a concentration of 0.5 g/100 ml (Lück and Rickerl 1960). Other earlier studies have shown a lack of genotoxic effects of annatto in various test systems (Hallagan et al. 1995).

No clastogenic activity was observed in human lymphocytes treated with the compound bixin at concentrations up to 10 µg/ml (Antunes et al. 2005).

No detectable genotoxicity was observed in reverse mutation assays with *E. coli* or *Salmonella typhimurium* TA1538 with or without metabolic activation (Haveland-Smith 1981).

No genotoxic effects of the compound norbixin were observed in murine fibroblasts (Kovary et al. 2001).

Literature Cited

Agner, A.R., L.F. Barbisan, C. Scolastici, and D.M. Salvadori. 2004. Absence of carcinogenic and anticarcinogenic effects of annatto in the rat liver medium-term assay. *Food Chem. Toxicol.* 42(10):1687-1693.

Agner, A.R., A.P. Bazo, L.R. Ribeiro, and D.M. Salvadori. 2005. DNA damage and aberrant crypt foci as putative biomarkers to evaluate the chemopreventive effect of annatto (*Bixa orellana* L.) in rat colon carcinogenesis. *Mutat. Res.* 582(1-2):146-154.

Alves de Lima, R.O., L. Azevedo, L.R. Ribeiro, and D.M. Salvadori. 2003. Study on the mutagenicity and antimutagenicity of a natural food colour (annatto) in mouse bone marrow cells. *Food Chem. Toxicol.* 41(2):189-192.

Antunes, L.M.G., L.M. Pascoal, M.D.L.P. Bianchi, and F.L. Dias. 2005. Evaluation of the clastogenicity and anticlastogenicity of the carotenoid bixin in human lymphocyte cultures. *Mutat. Res.* 585(1-2):113-119.

Auttachoat, W., D.R. Germolec, K.L. White, Jr., and T.L. Guo. 2005. Induction of contact sensitization by annatto extract bixin but not by norbixin in female BALB/c mice. *Toxicol. Sci.* 84(1 Supp):248.

Bautista, A.R., E.L. Moreira, M.S. Batista, M.S. Miranda, and I.C. Gomes. 2004. Subacute toxicity assessment of annatto in rat. *Food Chem. Toxicol.* 42(4):625-629.

Engelbreth-Holm, J., and S. Inversen. 1955. Is vegetable annatto butter colour carcinogenic? *Acta Pathol. Microbiol. Scand.* 37:483-491.

Fernandes, A.C., C.A. Almeida, F. Albano, et al. 2002. Norbixin ingestion did not induce any detectable DNA breakage in liver and kidney but caused a considerable impairment in plasma glucose levels of rats and mice. *J. Nutr. Biochem.* 13(7):411-420.

Hagiwara, A., N. Imai, T. Ichihara, et al. 2003. A thirteen-week oral toxicity study of annatto extract (norbixin), a natural food color extracted from the seed coat of annatto (*Bixa orellana* L.), in Sprague-Dawley rats. *Food Chem. Toxicol.* 41(8):1157-1164.

Hallagan, J.B., D.C. Allen, and J.F. Borzelleca. 1995. The safety and regulatory status of food, drug and cosmetics colour additives exempt from certification. *Food Chem. Toxicol.* 33(6):515-528.

Haveland-Smith, R.B. 1981. Evaluation of the genotoxicity of some natural food colours using bacterial assays. *Mutat. Res.* 91:285-290.

Juhlin, L. 1981. Recurrent urticaria: Clinical investigation of 330 patients. *Br. J. Derm.* 104:369-381.

Kay, J.H., and J.C. Calandra. 1961. Unpublished report by industrial bio-test laboratories *Cited in* WHO. 1982. Annatto extracts. Toxicological evaluation of certain food additives. WHO Food Additives Series, No. 17.

Kovary, K., T.S. Louvain, M.C. Costa e Silva, et al. 2001. Biochemical behaviour of norbixin during *in vitro* DNA damage induced by reactive oxygen species. *Br. J. Nutr.* 85(4):431-440.

Lück, H., and E. Rickerl. 1960. Lebensmittelzusatzstoffe und Mutagene Wirkung. VI. Mitteilung. Prüfung der in Westdeutschland zugelarsenen und ursprünglich vorgeschlagenen Lebensmittel-farbstoffe auf Mutagene Wirkung an *E. coli*. *Z. Lebensm. Untersuch.* 112:157-174.

Mikkelsen, H., J.C. Larsen, and F. Tarding. 1978. Hypersensitivity reactions to food colours with special reference to the natural colour annatto extract (butter colour). *Arch. Toxicol.* (Suppl.):141-143.

Morrison, E.Y., S. Smith-Richardson, M. West, et al. 1987. Toxicity of the hyperglycaemic-inducing extract of the annatto (*Bixa orellana*) in the dog. *West Indian Med. J.* 36(2):99-103.

Morrison, E.Y., H. Thompson, K. Pascoe, M. West, and C. Fletcher. 1991. Extraction of an hyperglycemic principle from the annatto (*Bixa orellana*): A medicinal plant in the West Indies. *Trop. Geogr. Med.* 43(1-2):184-188.

Morrison, E.Y., and M.E. West. 1985. The effect of *Bixa orellana* (annatto) on blood sugar levels in the anaesthetized dog. *West Indian Med. J.* 34(1):38-42.

Nish, W.A., B.A. Whisman, D.W. Goetz, and D.A. Ramirez. 1991. Anaphylaxis to annatto dye: A case report. *Ann. Allergy* 66(2):129-131.

Paumgartten, F.J., R.R. De-Carvalho, I.B. Araujo, et al. 2002. Evaluation of the developmental toxicity of annatto in the rat. *Food Chem. Toxicol.* 40(11):1595-1601.

Philip, J. 1981. Unpublished report by Unilever Research Division to WHO. *Cited in* WHO. 1982. Annatto extracts. Toxicological evaluation of certain food additives WHO Food Additives Series, No 17.

Russell, K.R., E.Y. Morrison, and D. Ragoobirsingh. 2005. The effect of annatto on insulin binding properties in the dog. *Phytother. Res.* 19(5):433-436.

Russell, K.R., F.O. Omoruyi, K.O. Pascoe, and E.Y. Morrison. 2008. Hypoglycaemic activity of *Bixa orellana* extract in the dog. *Methods Find. Exp. Clin. Pharmacol.* 30(4):301-305.

van Esch, G.J., H. van Genderen, and H.H. Vink. 1959. Uber die chronische Verträglichkeit von Annattofarbstoff. *Z. Lebensm. Untersuch.* 111:93-108.

Zbinden, G., and A. Studer. 1958. Tierexperimentelle Untersuchungen Uber die chronische Verträglichkeit von ß-Caratin, Lycopin, 7,7-Dihydro-β-carotin und bixin. *Z. Lebensm. Untersuch.* 108:113-134.

Boerhavia diffusa L. — Nyctaginaceae

SCN: boerhavia
AN: *punarnava*
OCN: spreading hogweed
Part: root

Quick Reference Summary

Safety Class: 1
Interaction Class: A

Contraindications

None known.

Other Precautions

Use in persons with diarrhea is not recommended (Pole 2006).

Drug and Supplement Interactions

None known.

Notice

Diuretic (Chowdhury 1955; Kapoor 2001; Singh et al. 1992; Taylor 2005); *see* Appendix 2.

Adverse Events and Side Effects

None known.

Pharmacological Considerations

Animal studies have demonstrated that boerhavia may modify glucose regulation. People with diabetes are advised to monitor their blood sugar closely and discuss the use of this herb with a qualified healthcare practitioner prior to use.

Pregnancy and Lactation

Although one reference notes, without details, that boerhavia has been used for abortion (Taylor 2005), an animal study indicated no adverse effects at relatively high doses (250 mg/kg) of boerhavia in pregnancy (Singh et al. 1991).

No information on the safety of boerhavia during lactation was identified. While this review did not identify any concerns for use while nursing, safety has not been conclusively established.

Review Details

I. Drug and Supplement Interactions

Clinical Trials of Drug or Supplement Interactions

No clinical trials of drug or supplement interactions were identified.

Case Reports of Suspected Drug or Supplement Interactions

No case reports of suspected drug or supplement interactions were identified.

Animal Trials of Drug or Supplement Interactions

No animal trials of drug or supplement interactions were identified.

II. Adverse Events

Case Reports of Adverse Events

No case reports of adverse events were identified.

III. Pharmacology and Pharmacokinetics

Human Pharmacological Studies

No relevant human pharmacological studies were identified.

Animal Pharmacological Studies

In diabetic rats orally administered 200 mg/kg of a boerhavia extract daily for 4 weeks, a reduction in fasting blood glucose levels was observed (Pari and Satheesh 2004). In diabetic rats orally administered single doses of 100, 200, or 400 mg/kg of an aqueous extract of boerhavia, a reduction in blood glucose levels was observed (Chude et al. 2001).

In Vitro Pharmacological Studies

No relevant in vitro pharmacological studies were identified.

IV. Pregnancy and Lactation

No adverse effects on fetal development in rats were observed after oral administration of 250 mg/kg of an ethanol extract of boerhavia daily throughout pregnancy (Singh et al. 1991).

One reference notes that boerhavia has been used for abortion. Details were not reported on part used, dose, route of administration, and whether the plant was used singly or in a formula (Taylor 2005).

No information on the safety of boerhavia during lactation was identified.

V. Toxicity Studies

Acute Toxicity

The LD_{50} of an aqueous extract of boerhavia orally administered to rats and mice could not be determined at doses up to 2 g/kg (Orisakwe et al. 2003). The oral LD_{50} of an ethanol extract of boerhavia in rats was 1 g/kg (Dhar et al. 1968).

Subchronic Toxicity

No signs of toxicity were observed in rats and mice orally administered 500, 1000, or 2000 mg/kg of an aqueous extract of boerhavia daily for 90 days. The parameters measured included absolute and relative weight of various organs, hematological parameters, and tests for liver function. Rats in the treatment group had an increase in food and fluid intake as compared to control that resulted in a greater weight gain in the group administered boerhavia extract (Orisakwe et al. 2003).

Literature Cited

Chowdhury, A. 1955. *Boerhavia diffusa*: Effect on diuresis and some renal enzymes. *Ann. Biochem. Exp. Med.* 15:119-126.

Chude, M.A., O.E. Orisakwe, O.J. Afonne, et al. 2001. Hypoglycaemic effect of the aqueous extract of *Boerhavia diffusa* leaves. *Indian J. Pharmacol.* 33(3):215-216.

Dhar, M.M., B.N. Dhawan, B.V. Mehrotra, and C. Ray. 1968. Screening of Indian plants for biological activity. III. *Indian J. Exp. Biol.* 6:232-247.

Kapoor, L.D. 2001. *Handbook of Ayurvedic medicinal plants.* Boca Raton, FL: CRC Press.

Orisakwe, O.E., O.J. Afonne, M.A. Chude, E. Obi, and C.E. Dioka. 2003. Sub-chronic toxicity studies of the aqueous extract of *Boerhavia diffusa* leaves. *J. Health Sci.* 49(6):444-447.

Pari, L., and M.A. Satheesh. 2004. Antidiabetic effect of *Boerhavia diffusa*: Effect on serum and tissue lipids in experimental diabetes. *J. Med. Food* 7(4):472-476.

Pole, S. 2006. *Ayurvedic medicine: The principles of traditional practice.* New York: Churchill Livingstone.

Singh, A., R.G. Singh, R.H. Singh, N. Mishra, and N. Singh. 1991. An experimental evaluation of possible teratogenic potential in *Boerhavia diffusa* in albino rats. *Planta Med.* 57(4):315-316.

Singh, S.K.P., B.L. Pandey, and R.G. Singh. 1992. Recent approach in clinical and experimental evaluation of diuretic action of Punarnava. *Indian J. Med. Educ. Res.* 11(1):29-36.

Taylor, L. 2005. *The healing power of rainforest herbs.* Garden City Park, NY: Square One Publishers.

Borago officinalis L.

Boraginaceae

SCN: borage

Part: herb

Quick Reference Summary

Safety Class: 2a
Interaction Class: A

Contraindications

For external use only (Dodson and Stermitz 1986; Herrmann et al. 2002; Huizing and Malingré 1981; Wretensjö and Karlberg 2003).

Other Precautions

None known.

Drug and Supplement Interactions

None known.

Notice

Pyrrolizidine alkaloids (0.0002–0.001%) (Dodson and Stermitz 1986; Herrmann et al. 2002; Huizing and Malingré 1981; Wretensjö and Karlberg 2003); *see* Appendix 1.

Editors' Notes

The American Herbal Products Association has established a trade requirement (AHPA 2011) that all products with botanical ingredients that contain toxic pyrrolizidine alkaloids, including borage herb, are not offered for sale for internal use and display the following cautionary label: "For external use only. Do not apply to broken or abraded skin. Do not use when nursing."

Borage herb contains relatively minor amounts of pyrrolizidine alkaloids (PA), including lycopsamine, amabiline, supinine, 7-acetyllycopsamine, intermedine, acetylintermedine, and thesinine (Herrmann et al. 2002; Larson et al. 1984; Roitman 1983). Of these compounds, all but thesinine are unsaturated PAs, a group of compounds that are generally classified as toxic (De Smet 1993). Within the unsaturated PAs, toxicity ranges from mild to severe. Lycopsamine is a monoesterified, unsaturated PA that is less toxic than diesterified, unsaturated PAs that occur in other PA-containing plants.

Adverse Events and Side Effects

None known.

Pharmacological Considerations

None known.

Pregnancy and Lactation

No information on the safety of borage herb during pregnancy or lactation was identified. Based on the presence of pyrrolizidine alkaloids, use during pregnancy or lactation is not recommended.

Review Details

I. Drug and Supplement Interactions

Clinical Trials of Drug or Supplement Interactions

No clinical trials of drug or supplement interactions were identified.

Case Reports of Suspected Drug or Supplement Interactions

No case reports of suspected drug or supplement interactions were identified.

Animal Trials of Drug or Supplement Interactions

No animal trials of drug or supplement interactions were identified.

II. Adverse Events

Case Reports of Adverse Events

Borage and foxglove (*Digitalis purpurea*) have botanically similar leaves. Adverse cardiovascular effects due to foxglove consumption have been reported in persons that mistook foxglove leaf for borage leaf (Brustbauer and Wenisch 1997; Cardano et al. 2002; Maffe et al. 2009).

III. Pharmacology and Pharmacokinetics

Human Pharmacological Studies

No relevant human pharmacological studies were identified.

Animal Pharmacological Studies

No relevant animal pharmacological studies were identified.

In Vitro Pharmacological Studies

No relevant in vitro pharmacological studies were identified.

IV. Pregnancy and Lactation

No information on the safety of borage herb during pregnancy or lactation was identified.

V. Toxicity Studies

No toxicity studies were identified.

Literature Cited

AHPA. July 2011. Code of Ethics & Business Conduct. Silver Spring, MD: American Herbal Products Association.

Brustbauer, R., and C. Wenisch. 1997. Bradycardiac atrial fibrillation after consuming herbal tea. *Dtsch. Med. Wochenschr.* 122(30):930-932.

Cardano, S., F. Beldi, C. Bignoli, A. Monteverde, and E. Uglietti. 2002. A dangerous "risotto." *Rec. Prog. Med.* 93(4):245-246.

De Smet, P.A.G.M. 1993. *Adverse effects of herbal drugs. Volume 2.* New York: Springer.

Dodson, C.D., and F.R. Stermitz. 1986. Pyrrolizidine alkaloids from borage (*Borago officinalis*) seeds and flowers. *J. Nat. Prod.* 49(4):727-728.

Herrmann, M., H. Joppe, and G. Schmaus. 2002. Thesinine-4'-*O*-beta-D-glucoside the first glycosylated plant pyrrolizidine alkaloid from *Borago officinalis*. *Phytochemistry* 60:399-402.

Huizing, H.J., and T.M. Malingré. 1981. A chemotaxonomical study of some Boraginaceae: Pyrrolizidine alkaloids and phenolic compounds. *Plant Syst. Evol.* 137(1):127-134.

Larson, K.M., M.R. Roby, and F.R. Stermitz. 1984. Unsaturated pyrrolizidines from borage (*Borago officinalis*), a common garden herb. *J. Nat. Prod.* 47(4):747-748.

Maffe, S., L. Cucchi, F. Zenone, et al. 2009. *Digitalis* must be banished from the table: A rare case of acute accidental *Digitalis* intoxication of a whole family. *J. Cardiovasc. Med.* 10(9):727-732.

Roitman, J.N. 1983. *Ingestion of pyrrolizidine alkaloids: A health hazard of global proportions.* Edited by Finley, J.W. and D.E. Schwass, *Xenobiotics in foods and fields.* ACS Symposium Series 234. Washington, DC: American Chemical Society.

Wretensjö, I., and B. Karlberg. 2003. Pyrrolizidine alkaloid content in crude and processed borage oil from different processing stages. *J. Am. Oil Chem. Soc.* 80(10):963-970.

Borago officinalis L. — Boraginaceae

SCN: borage

Part: seed oil

Quick Reference Summary

Safety Class: 1
Interaction Class: A

Contraindications

None known.

Other Precautions

None known.

Drug and Supplement Interactions

None known.

Editors' Notes

Borage seed oil is commonly traded as a source of gamma-linolenic acid (GLA). Processing methods eliminate pyrrolizidine alkaloids (PA) from the finished material (Wretensjö and Karlberg 2003). A study of the effects of the borage seed oil refinement process on the content of pyrrolizidine alkaloids in the oil indicated that processing reduces the PA content by a factor of about 30,000. In finished samples, no PAs were present above the detection limit of 20 ppb (Langer and Franz 1997; Wretensjö and Karlberg 2003). The seeds themselves contain only small amounts of the saturated (nontoxic) pyrrolizidine alkaloid thesinine (De Smet 1993; Dodson and Stermitz 1986).

Adverse Events and Side Effects

None known.

Pharmacological Considerations

None known.

Pregnancy and Lactation

No information on the safety of borage seed oil in pregnancy was identified in the scientific or traditional literature. While this review did not identify any concerns for pregnant women, safety has not been conclusively established.

Gamma-linolenic acid (GLA), one of the primary constituents of borage seed oil, is a compound naturally formed in the human body and is a component of human breast milk (Carter 1988).

Review Details

I. Drug and Supplement Interactions

Clinical Trials of Drug or Supplement Interactions

No clinical trials of drug or supplement interactions were identified.

Case Reports of Suspected Drug or Supplement Interactions

No case reports of suspected drug or supplement interactions were identified.

Animal Trials of Drug or Supplement Interactions

No animal trials of drug or supplement interactions were identified.

II. Adverse Events

Case Reports of Adverse Events

No case reports of adverse events were identified.

III. Pharmacology and Pharmacokinetics

Human Pharmacological Studies

No effects on platelet aggregation were observed in healthy volunteers orally administered 3 g of borage seed oil daily for 6 weeks (Bard et al. 1997).

No changes in hematological or biochemical parameters were observed in adults and children with atopic eczema orally administered 2 g (adult dose) or 1 g (child dose) of borage seed oil daily for 12 weeks. Adverse events in the treatment group were similar to or less than those reported in the control group (Takwale et al. 2003).

Animal Pharmacological Studies

No relevant animal pharmacological studies were identified.

In Vitro Pharmacological Studies

No relevant in vitro pharmacological studies were identified.

IV. Pregnancy and Lactation

No information on the safety of borage seed oil during pregnancy was identified. Gamma-linolenic acid (GLA), one of the primary constituents of borage seed oil, is a compound naturally formed in the human body and is a component of human breast milk (Carter 1988).

V. Toxicity Studies

No toxicity studies were identified.

Literature Cited

Bard, J.M., G. Luc, B. Jude, et al. 1997. A therapeutic dosage (3 g/day) of borage oil supplementation has no effect on platelet aggregation in healthy volunteers. *Fund. Clin. Pharmacol.* 11(2):143-144.

Carter, J.P. 1988. Gamma-linolenic acid as a nutrient. *Food Technol.* 42(6):72-82.

De Smet, P.A.G.M. 1993. *Adverse effects of herbal drugs. Volume 2.* New York: Springer.

Dodson, C.D., and F.R. Stermitz. 1986. Pyrrolizidine alkaloids from borage (*Borago officinalis*) seeds and flowers. *J. Nat. Prod.* 49(4):727-728.

Langer, T., and C. Franz. 1997. Determination of pyrrolizidine alkaloids in commercial samples of borage seed oil products by GC-MS. *Sci. Pharm.* 65:321-328.

Takwale, A., E. Tan, S. Agarwal, et al. 2003. Efficacy and tolerability of borage oil in adults and children with atopic eczema: Randomised, double blind, placebo controlled, parallel group trial. *Br. Med. J.* 327(7428):1385.

Wretensjö, I., and B. Karlberg. 2003. Pyrrolizidine alkaloid content in crude and processed borage oil from different processing stages. *J. Am. Oil Chem. Soc.* 80(10):963-970.

Boswellia spp. — Burseraceae

Boswellia sacra Flueck.
SCN: frankincense (oleo gum resin)
Syn: *Boswellia carterii* Birdw.
PN: *ru xiang* (oleo gum resin)
OCN: (oleo gum resin) bible frankincense; incense; olibanum

Boswellia serrata Roxb.
SCN: Indian frankincense (oleo gum resin)
AN: *shallaki*
OCN: (oleo gum resin): frankincense; Indian olibanum
Part: gum resin

Quick Reference Summary

Safety Class: 2b
Interaction Class: A

Contraindications

Not for use in pregnancy except under the supervision of a qualified healthcare practitioner (Bensky et al. 2004; Chen and Chen 2004).

Other Precautions

None known.

Drug and Supplement Interactions

None known.

Adverse Events and Side Effects

Frankincense may be irritating to the stomach and the bitter taste may cause nausea and vomiting (Bensky et al. 2004; Chen and Chen 2004).

Allergic reactions to frankincense and Indian frankincense have been reported (Acebo et al. 2004; Bensky et al. 2004).

B

Pharmacological Considerations

None known.

Pregnancy and Lactation

While texts on traditional Chinese medicine indicate that frankincense should not be used during pregnancy (Bensky et al. 2004; Chen and Chen 2004), in Yemen, women chew frankincense during pregnancy (Ghazanfar 1994).

No information on the safety of frankincense during lactation was identified. While this review did not identify any concerns for use while nursing, safety has not been conclusively established.

Review Details

I. Drug and Supplement Interactions

Clinical Trials of Drug or Supplement Interactions

No clinical trials of drug or supplement interactions were identified.

Case Reports of Suspected Drug or Supplement Interactions

No case reports of suspected drug or supplement interactions were identified.

Animal Trials of Drug or Supplement Interactions

No animal trials of drug or supplement interactions were identified.

II. Adverse Events

Adverse Events Reported in Clinical Trials

A systematic review of clinical trials of Indian frankincense products indicated that adverse effects related to Indian frankincense were minor and not significantly different from those observed in control treatment or placebo groups. Diarrhea, abdominal pain, and nausea were reported in several of the seven studies reviewed (Frank and Unger 2006). In a study not included in the review, epigastric pain, abdominal fullness, gastroesophageal reflux, diarrhea, and nausea were reported in 18% of patients with ulcerative colitis receiving 1.5 g of Indian frankincense daily for 6 weeks (Gupta et al. 1997).

Case Reports of Adverse Events

Allergic contact dermatitis to Indian frankincense has been reported (Acebo et al. 2004).

A 17-year-old woman developed an epigastric bezoar of frankincense after repeated ingestion of "large quantities" of frankincense over an unspecified period of time (El Fortia et al. 2006).

III. Pharmacology and Pharmacokinetics

Human Pharmacological Studies

No relevant human pharmacological studies were identified.

Animal Pharmacological Studies

Primary skin and eye irritation tests in rabbits indicated that an Indian frankincense extract containing 30% 3-*O*-acetyl-11-keto-β-boswellic acid was nonirritating to the skin and mildly irritating to the eyes (Lalithakumari et al. 2006).

Indian frankincense and fractions of Indian frankincense have exhibited immunomodulatory activity in several animal studies (Khajuria et al. 2008; Pungle et al. 2003; Sharma et al. 1996).

In Vitro Pharmacological Studies

Methanolic extracts of frankincense and Indian frankincense nonselectively inhibited the human drug-metabolizing isoenzymes CYP 1A2, 2C8, 2C9, 2C19, 2D6, and 3A4 (Frank and Unger 2006).

Inhibition of P-glycoprotein (P-gp) was observed in human lymphocytic leukemia cells and porcine brain capillary endothelial cells treated with Indian frankincense extract with inhibitory concentrations in the micromolar range. The authors of the study indicated that inhibition of P-gp at the blood-brain barrier is probably not relevant for the brain availability of other P-gp substrates, due to low plasma levels determined for selected boswellic acids, but the data cannot exclude the possibility of drug interactions caused by modulation of gastrointestinal P-gp (Weber et al. 2006).

IV. Pregnancy and Lactation

While texts on traditional Chinese medicine indicate that frankincense should not be used during pregnancy (Bensky et al. 2004; Chen and Chen 2004), in Yemen, women chew frankincense during pregnancy (Ghazanfar 1994).

No information on the safety of frankincense or Indian frankincense during lactation was identified.

V. Toxicity Studies

Acute Toxicity

The LD_{50} of orally or intraperitoneally administered extracts of Indian frankincense in mice and rats could not be determined at doses up to 2 g/kg (Singh and Atal 1986).

The LD_{50} of an Indian frankincense extract containing 30% 3-*O*-acetyl-11-keto-β-boswellic acid orally administered to rats could not be determined at doses up to 5 g/kg. The acute dermal LD_{50} of the same extract could not be determined at doses up to 2 g/kg (Lalithakumari et al. 2006).

Subchronic Toxicity

No changes in hematology, clinical chemistry, or histopathology were observed in rats fed diets containing 0 to 2.5% of an Indian frankincense extract containing 30%

3-*O*-acetyl-11-keto-β-boswellic acid daily for 90 days (Lalithakumari et al. 2006).

Chronic Toxicity

In rats with chemically induced colitis fed diets containing 0.1 or 1% Indian frankincense hexane and water fractions daily for their lifetime, hepatotoxicity with pronounced hepatomegaly and steatosis was observed. Microarray analyses of hepatic gene expression demonstrated dysregulation of a number of genes, including a large group of lipid metabolism-related genes and detoxifying enzymes, a response consistent with that of hepatotoxic xenobiotics (Kiela et al. 2005).

Genotoxicity

No changes in hepatic DNA fragmentation were observed in rats fed diets containing 0 to 2.5% of an Indian frankincense extract containing 30% 3-*O*-acetyl-11-keto-β-boswellic acid daily for 90 days (Lalithakumari et al. 2006).

Cytotoxicity

Mild to low toxicities of Indian frankincense gum resin and 3-*O*-acetyl-11-keto-β-boswellic acid were observed in undifferentiated keratinocytes (HaCaT and NCTC 2544) and fetal dermal fibroblasts (HFFF2) (Burlando et al. 2008).

B

Literature Cited

Acebo, E., J.A. Raton, S. Sautua, et al. 2004. Allergic contact dermatitis from *Boswellia serrata* extract in a naturopathic cream. *Contact Dermat.* 51(2):91-92.

Bensky, D., S. Clavey, and E. Stöger. 2004. *Chinese herbal medicine: Materia medica.* 3rd ed. Seattle: Eastland Press.

Burlando, B., A. Parodi, A. Volante, and A.M. Bassi. 2008. Comparison of the irritation potentials of *Boswellia serrata* gum resin and of acetyl-11-keto-(beta)-boswellic acid by in vitro cytotoxicity tests on human skin-derived cell lines. *Toxicol. Lett.* 177(2):144-149.

Chen, J.K., and T.T. Chen. 2004. *Chinese medical herbology and pharmacology.* City of Industry, CA: Art of Medicine Press.

El Fortia, M., H. Badi, K. Elalem, O. Kadiki, and Y. Topov. 2006. Olibanum bezoar: Complication of a traditional popular medicine. *East. Mediterr. Health J.* 12(6):927-929.

Frank, A., and M. Unger. 2006. Analysis of frankincense from various *Boswellia* species with inhibitory activity on human drug metabolising cytochrome P450 enzymes using liquid chromatography mass spectrometry after automated on-line extraction. *J. Chromatogr. A* 1112(1-2):255-262.

Ghazanfar, S.A. 1994. *Handbook of Arabian medicinal plants.* Boca Raton, FL: CRC Press.

Gupta, I., A. Parihar, P. Malhotra, et al. 1997. Effects of *Boswellia serrata* gum resin in patients with ulcerative colitis. *Eur. J. Med. Res.* 2(1):37-43.

Khajuria, A., A. Gupta, P. Suden, et al. 2008. Immunomodulatory activity of biopolymeric fraction BOS 2000 from *Boswellia serrata*. *Phytother. Res.* 22(3):340-348.

Kiela, P.R., A.J. Midura, N. Kuscuoglu, et al. 2005. Effects of *Boswellia serrata* in mouse models of chemically induced colitis. *Am. J. Physiol. Gastrointest. Liver Physiol.* 288(4):798-808.

Lalithakumari, K., A.V. Krishnaraju, K. Sengupta, G.V. Subbaraju, and A. Chatterjee. 2006. Safety and toxicological evaluation of a novel, standardized 3-*O*-acetyl-11-keto-(beta)-boswellic acid (AKBA)-enriched *Boswellia serrata* extract (5-Loxin™). *Toxicol. Mech. Methods* 16(4):199-226.

Pungle, P., M. Banavalikar, A. Suthar, M. Biyani, and S. Mengi. 2003. Immunomodulatory activity of boswellic acids of *Boswellia serrata* Roxb. *Indian J. Exp. Biol.* 41(12):1460-1462.

Sharma, M.L., A. Kaul, A. Khajuria, S. Singh, and G.B. Singh. 1996. Immunomodulatory activity of boswellic acids (pentacyclic triterpene acids) from *Boswellia serrata*. *Phytother. Res.*10(2):107-112.

Singh, G.B., and C.K. Atal. 1986. Pharmacology of an extract of salai guggal ex-*Boswellia serrata*, a new nonsteroidal anti-inflammatory agent. *Agents Actions* 18:407-412.

Weber, C.C., K. Reising, W.E. Muller, M. Schubert-Zsilavecz, and M. Abdel-Tawab. 2006. Modulation of Pgp function by boswellic acids. *Planta Med.* 72(6):507-513.

Brassica spp. and *Sinapis* spp.

Brassicaceae

Brassica juncea (L.) Czernov var. ***tumida*** Tsen & Lee
SCN: mustard
PN: *bai jie zi* (seed)
OCN: swollen-stem mustard
Brassica nigra (L.) W.D.J. Koch
SCN: black mustard
AN: *sarshapa*
OCN: brown mustard
Sinapis alba L.
SCN: white mustard
Syn: *Brassica alba* Rabenh.; *Brassica hirta* Moench
PN: *bai jie zi* (seed)
OCN: yellow mustard
Part: seed

Quick Reference Summary

Safety Class: 1 (internal), 2d (external)

Interaction Class: A

Contraindications

Internal Use

None known.

External Use

Not for external use with children under 6 years of age (Felter and Lloyd 1898; List and Hörhammer 1973; Watt and Breyer-Brandwijk 1962).

Not for use in persons with varicose veins, severe circulatory damage, or other venous disorders (Wichtl 2004).

Other Precautions

Internal Use

Mustards may irritate the gastrointestinal tract and should be used with caution in patients with peptic ulcer or gastric hemorrhage (Bensky et al. 2004).

External Use

If left on the skin for more than 15 to 30 minutes, plasters made from mustards can cause blistering that may be accompanied by suppurating, poorly healing ulcerations and necrosis (Wichtl 2004).

Drug and Supplement Interactions

None known.

Adverse Events and Side Effects

The local skin irritant effect of topically applied mustards lasts for 24 to 48 hours (Wichtl 2004). Blistering, often accompanied by suppurating, poorly healing ulcerations and necrosis, may occur if topical black mustard plasters are applied for more than 15 to 30 minutes (Wichtl 2004).

Large doses (standard dose listed as an aqueous extract of 3–9 g) of mustards can cause enteritis, abdominal pain, and diarrhea (Bensky et al. 2004).

Allergic reactions to mustards have been reported after both internal and topical use. Reactions included anaphylactic shock and urticaria, and were confirmed by patch testing (Bensky et al. 2004; Morisset et al. 2003).

Pharmacological Considerations

Several animal studies have demonstrated that black mustard seed may modify glucose regulation. Diabetic persons are advised to discuss the use of this herb with a qualified healthcare practitioner prior to use (Anand et al. 2007, 2009; Yadav et al. 2004).

Pregnancy and Lactation

No information on the safety of mustards in pregnancy or lactation was identified in the scientific or traditional literature. Although this review did not identify any concerns for use while pregnant or nursing, safety has not been conclusively established.

Review Details

I. Drug and Supplement Interactions

Clinical Trials of Drug or Supplement Interactions

No clinical trials of drug or supplement interactions were identified.

Case Reports of Suspected Drug or Supplement Interactions

No case reports of suspected drug or supplement interactions were identified.

Animal Trials of Drug or Supplement Interactions

No animal trials of drug or supplement interactions were identified.

II. Adverse Events

Case Reports of Adverse Events

Toxicosis was reported in cattle that ingested significant amounts of mustard seed (Katamoto et al. 2001; Semalulu and Rousseaux 1989). The compound allyl isothiocyanate, which has an LD_{50} of 10 mg/kg in cattle (Kingsbury 1964) and 339 mg/kg in rats (Jenner et al. 1964), is believed to be responsible for the toxicosis.

Topical application of mustard seed as a plaster has been associated with tachypnea, sweating, dizziness, agitation, and low blood pressure, with symptoms appearing 40 minutes after topical application (Bensky et al. 2004).

III. Pharmacology and Pharmacokinetics

Human Pharmacological Studies

Mustard seed allergy accounts for 1.1% of food allergies in children (Morisset et al. 2003).

In a study of 30 subjects, age 3 to 20, presenting positive prick tests to ground black mustard seed (*B. nigra*), mustard flour (*B. juncea*), metabisulfite-free strong mustard seasoning (*B. juncea*), or a commercialized allergenic extract (*B. nigra*), 27 subjects were screened for mustard-specific immunoglobulin E (IgE) and single- or double-blind placebo-controlled food challenges were completed with up to 1340 mg of the metabisulfite-free seasoning. The mean diameter of the wheal induced by prick tests with the allergenic extract was lower than that induced by the native mustard products, ground black mustard seed, mustard flour, or the strong mustard seasoning. The mean of mustard specific-IgE values was 8.7 KU/l (Morisset et al. 2003).

Animal Pharmacological Studies

A decease in serum glucose and increase in serum insulin were observed in diabetic rats orally administered 200 mg/kg black mustard seed daily for 2 months (Anand et al.

2009). In the oral glucose tolerance test, the effective dose of the aqueous extract of black mustard seed was 200 mg/kg, with the aqueous extract being more effective than chloroform, acetone, or ethanol extracts (Anand et al. 2007). In rats fed a diet containing 10% black mustard seed for 30 days, significant decreases in fasting serum glucose and insulin were observed, although the levels were not normalized (Yadav et al. 2004).

In Vitro Pharmacological Studies

No relevant in vitro pharmacological studies were identified.

IV. Pregnancy and Lactation

No information on the safety of mustards in pregnancy or lactation was identified.

V. Toxicity Studies

Acute Toxicity

The LD_{50} of orally administered black mustard seed extract in rats could not be determined at doses up to 3 g/kg (Anand et al. 2009).

Genotoxicity

Antimutagenic activity of mustard seed was observed in rats exposed to a chemical mutagen after being fed a diet containing 1 to 10% mustard seed daily for 30 days (Polasa et al. 1994).

Literature Cited

Anand, P., K.Y. Murali, V. Tandon, R. Chandra, and P.S. Murthy. 2007. Preliminary studies on antihyperglycemic effect of aqueous extract of *Brassica nigra* (L.) Koch in streptozotocin induced diabetic rats. *Indian J. Exp. Biol.* 45(8):696-701.

Anand, P., Y.K. Murali, V. Tandon, P.S. Murthy, and R. Chandra. 2009. Insulinotropic effect of aqueous extract of *Brassica nigra* improves glucose homeostasis in streptozotocin induced diabetic rats. *Exp. Clin. Endocrinol. Diabetes* 117(6):251-256.

Bensky, D., S. Clavey, and E. Stöger. 2004. *Chinese herbal medicine: Materia medica.* 3rd ed. Seattle: Eastland Press.

Felter, H.W., and J.U. Lloyd. 1898. *King's American dispensatory.* 18th ed., 3rd rev. 2 vols. Cincinnati: Ohio Valley Co.

Jenner, P.M., E.C. Hagan, J.M. Taylor, E.L. Cook, and O.G. Fitzhugh. 1964. Food flavourings and compounds of related structure. I. Acute oral toxicity. *Food Cosmet. Toxicol.* 2(3):327-343.

Katamoto, H., S. Nishiguchi, K. Harada, et al. 2001. Suspected Oriental mustard (*Brassica juncea*) intoxication in cattle. *Vet. Rec.* 149(7):215-216.

Kingsbury, J.M. 1964. *Poisonous plants of the United States and Canada.* Prentice-Hall biological science series. Englewood Cliffs, NJ: Prentice-Hall.

List, P.H., and H. Hörhammer. 1973. *Hagers handbuch der pharmazeutischen praxis.* Vollst. 4. Neuausg. ed. Berlin: Springer.

Morisset, M., D.A. Moneret-Vautrin, F. Maadi, et al. 2003. Prospective study of mustard allergy: First study with double-blind placebo-controlled food challenge trials (24 cases). *Allergy* 58(4):295-299.

Polasa, K., P.U. Kumar, and K. Krishnaswamy. 1994. Effect of *Brassica nigra* on benzo[*a*]pyrene mutagenicity. *Food Chem. Toxicol.* 32(8):777-781.

Semalulu, S.S., and C.G. Rousseaux. 1989. Saskatchewan. Suspected oriental mustard seed (*Brassica juncea*) poisoning in cattle. *Can. Vet. J.* 30(7):595-596.

Watt, J.M., and M.G. Breyer-Brandwijk. 1962. *The medicinal and poisonous plants of southern and eastern Africa.* 2nd ed. Edinburgh: E. & S. Livingstone.

Wichtl, M. 2004. *Herbal drugs and phytopharmaceuticals: A handbook for practice on a scientific basis.* 3rd ed. Boca Raton, FL: CRC Press.

Yadav, S.P., V. Vats, A.C. Ammini, and J.K. Grover. 2004. *Brassica juncea* (Rai) significantly prevented the development of insulin resistance in rats fed fructose-enriched diet. *J. Ethnopharmacol.* 93(1):113-116.

Bupleurum spp. — Apiaceae

Bupleurum chinense DC.
SCN: bupleurum
PN: *chai hu* (root)
OCN: Chinese thoroughwax
Bupleurum scorzonerifolium Willd.
SCN: bupleurum
PN: *chai hu* (root)
OCN: Chinese thoroughwax
Part: root

Quick Reference Summary

Safety Class: 1
Interaction Class: A

Contraindications

None known.

Other Precautions

None known.

Drug and Supplement Interactions

None known.

Adverse Events and Side Effects

An allergic reaction to bupleurum has been reported (Bensky et al. 2004).

Pharmacological Considerations

None known.

Pregnancy and Lactation

No information on the safety of bupleurum in pregnancy or lactation was identified in the scientific or traditional literature. Although this review did not identify any concerns for use while pregnant or nursing, safety has not been conclusively established.

Review Details

I. Drug and Supplement Interactions

Clinical Trials of Drug or Supplement Interactions

No clinical trials of drug or supplement interactions were identified.

Case Reports of Suspected Drug or Supplement Interactions

No case reports of suspected drug or supplement interactions were identified.

Animal Trials of Drug or Supplement Interactions

No animal trials of drug or supplement interactions were identified.

II. Adverse Events

Case Reports of Adverse Events

An allergic reaction to bupleurum has been reported (Bensky et al. 2004).

III. Pharmacology and Pharmacokinetics

Human Pharmacological Studies

No relevant human pharmacological studies were identified.

Animal Pharmacological Studies

No relevant animal pharmacological studies were identified.

In Vitro Pharmacological Studies

The compound saikosaponin inhibited human platelet aggregation induced by ADP, with a potency of inhibition comparable to that of aspirin, and dose-dependently inhibited platelet thromboxane formation from exogenous and endogenous arachidonic acid. Effects were similar to that of epigallocatechin isolated from green tea (Chang and Hsu 1991).

The compounds eugenin and saikochromone inhibited CD28-costimulated activation of human peripheral blood T cells (Chang et al. 2003).

In human G6PD-deficient red blood cells, an aqueous extract of bupleurum demonstrated dose-dependent reduction of reduced glutathione (GSH) and methemoglobin (MetHb) at concentrations of 5–10 mg/ml (Ko et al. 2008).

No estrogenic activity of an ethanol extract of bupleurum was observed in a recombinant yeast system (Kim et al. 2008).

IV. Pregnancy and Lactation

No information on the use of bupleurum during pregnancy or lactation was identified.

V. Toxicity Studies

Acute Toxicity

The LD_{50} of intraperitoneally administered bupleurum essential oil in mice is 1.19 g/kg, while that of intraperitoneally administered saikosaponin is 1.9 g/kg (Chen and Chen 2004).

The LD_{50} of the crude saponin fraction of bupleurum in mice is 4700 mg/kg after oral administration, 1800 mg/kg after subcutaneous administration, 112 mg/kg after intraperitoneal administration, and 70 mg/kg after intravenous administration. In guinea pigs, the LD_{50} of the same fraction is 58.3 mg/kg after intraperitoneal administration (Takagi and Shibata 1969).

No adverse effects were observed in rats 14 days after intranasal administration of bupleurum essential oil made from 10 g of bupleurum root (50 times the human clinical dose) (Xie et al. 2006).

Cytotoxicity

The compound saikosaponin d induced apoptosis in human hepatocellular carcinoma cells (HepG2) through the activation of caspases 3 and 7, resulting in poly(ADP-ribose)-polymerase (PARP) cleavage. DNA fragmentation was clearly noted at more than 6 hours after exposure to saikosaponin d (Chiang et al. 2003).

Literature Cited

Bensky, D., S. Clavey, and E. Stöger. 2004. *Chinese herbal medicine: Materia medica.* 3rd ed. Seattle: Eastland Press.

Chang, W.C., and F.L. Hsu. 1991. Inhibition of platelet activation and endothelial cell injury by flavan-3-ol and saikosaponin compounds. *Prostaglandins Leukotrienes Essent. Fatty Acids* 44(1):51-56.

Chang, W.L., L.W. Chiu, J.H. Lai, and H.C. Lin. 2003. Immunosuppressive flavones and lignans from *Bupleurum scorzonerifolium. Phytochemistry* 64(8):1375-1379.

Chen, J.K., and T.T. Chen. 2004. *Chinese medical herbology and pharmacology.* City of Industry, CA: Art of Medicine Press.

Chiang, L.C., L.T. Ng, L.T. Liu, D.E. Shieh, and C.C. Lin. 2003. Cytotoxicity and anti-hepatitis B virus activities of saikosaponins from *Bupleurum* species. *Planta Med.* 69(8):705-709.

Kim, I.G., S.C. Kang, K.C. Kim, E.S. Choung, and O.P. Zee. 2008. Screening of estrogenic and antiestrogenic activities from medicinal plants. *Env. Toxicol. Pharmacol.* 25(1):75-82.

Ko, C.H., K. Li, P.C. Ng, et al. 2008. Pro-oxidative effects of Chinese herbal medicine on G6PD-deficient erythrocytes in vitro. *Toxicol. in Vitro* 22(5):1222-1227.

Takagi, K., and M. Shibata. 1969. Pharmacological studies on *Bupleurum falcatum* L. I. Acute toxicity and central depressant action of crude saikosides. *Yakugaku Zasshi* 89(5):712-720.

Xie, Y., W. Lu, S. Cao, et al. 2006. Preparation of bupleurum nasal spray and evaluation on its safety and efficacy. *Chem. Pharm. Bull. (Tokyo)* 54(1):48-53.

Buxus sempervirens L. — Buxaceae

SCN: boxwood

Part: leaf

Quick Reference Summary

Safety Class: 3
Interaction Class: A

Contraindications

Not for use except under the supervision of an expert qualified in the appropriate use of this substance (Kingsbury 1964; List and Hörhammer 1973).

Other Precautions

None known.

Drug and Supplement Interactions

None known.

Adverse Events and Side Effects

None known.

Pharmacological Considerations

None known.

Pregnancy and Lactation

No information on the safety of boxwood during pregnancy or lactation was identified. While this review did not identify any concerns for use while pregnant or nursing, safety has not been conclusively established, and this substance is not recommended for use except under the supervision of an expert qualified in its appropriate use.

Review Details

I. Drug and Supplement Interactions

Clinical Trials of Drug or Supplement Interactions

No clinical trials of drug or supplement interactions were identified.

Case Reports of Suspected Drug or Supplement Interactions

No case reports of suspected drug or supplement interactions were identified.

Animal Trials of Drug or Supplement Interactions

No animal trials of drug or supplement interactions were identified.

II. Adverse Events

Case Reports of Adverse Events

After ingestion of an unknown amount of boxwood leaf, a 1-year-old was reported to be apathetic then overexcited (Frohne and Pfänder 2000).

Poisoning has been reported in livestock that have eaten boxwood. A horse died after consuming 1.5 pounds of boxwood (0.15% of the animal's body weight) (Frohne and Pfänder 2000; Kingsbury 1964).

III. Pharmacology and Pharmacokinetics

Human Pharmacological Studies

No relevant human pharmacological studies were identified.

Animal Pharmacological Studies

In anesthetized cats administered 1.25 μmol/kg of the compound buxaminol-E, a small short-lasting increase in blood pressure followed by marked hypotension was observed. Atropine inhibited the hypotensive effect almost completely and methylatropine partially inhibited the effect (Kvaltínová et al. 1991).

In Vitro Pharmacological Studies

No relevant in vitro pharmacological studies were identified.

IV. Pregnancy and Lactation

No information on the use of boxwood during pregnancy or lactation was identified.

V. Toxicity Studies

No toxicity studies were identified.

B

Literature Cited

Frohne, D., and H.J. Pfänder. 2000. *A colour atlas of poisonous plants: A handbook for pharmacists, doctors, toxicologists, biologists and veterinarians.* 2nd ed. London: Manson.

Kingsbury, J.M. 1964. *Poisonous plants of the United States and Canada.* Prentice-Hall biological science series. Englewood Cliffs, NJ: Prentice-Hall.

Kvaltínová, Z., L. Lukovic, J. Machová, and M. Fatranská. 1991. Effect of the steroidal alkaloid buxaminol-E on blood pressure, acetylcholinesterase activity and [$^{\$1}$H]quinuclidinyl benzilate binding in cerebral cortex. *Pharmacology* 43:20-25.

List, P.H., and H. Hörhammer. 1973. *Hagers handbuch der pharmazeutischen praxis.* Vollst. 4. Neuausg. ed. Berlin: Springer.

Calendula officinalis L.

Asteraceae

C

SCN: calendula
OCN: marigold; pot marigold

Part: flower

Quick Reference Summary

Safety Class: 1
Interaction Class: A

Contraindications

None known.

Other Precautions

Persons with allergies to other members of the Asteraceae family (such as feverfew, chamomile, or *Echinacea* species) should exercise caution with calendula, as allergic cross-reactivity to Asteraceae plants is common (Paulsen 2002).

Drug and Supplement Interactions

None known.

Editors' Note

The Cosmetic Ingredient Review deemed calendula "safe as used" as a topical cosmetic product (CIR 2009).

Adverse Events and Side Effects

Allergic reaction to calendula, including anaphylactic reaction, is rare but has been reported (Fiume 2001; Goldman 1974). Irritation and sensitization studies in animals have indicated that calendula is nonirritating and nonsensitizing (Fiume 2001).

Pharmacological Considerations

None known.

Pregnancy and Lactation

An animal study on a water-ethanol extract of calendula at doses up to 1.0 g/kg indicated no adverse effects on fetal development (Silva et al 2009).

No information on the safety of calendula during lactation was identified. While this review did not identify any concerns for use while nursing, safety has not been conclusively established.

Review Details

I. Drug and Supplement Interactions

Clinical Trials of Drug or Supplement Interactions

No clinical trials of drug or supplement interactions were identified.

Case Reports of Suspected Drug or Supplement Interactions

No case reports of suspected drug or supplement interactions were identified.

Animal Trials of Drug or Supplement Interactions

Calendula has been shown to increase the hexobarbital-induced sleeping time in rats (Samochowiec 1983).

II. Adverse Events

Case Reports of Adverse Events

Patch testing with calendula preparations has indicated a 0–2% incidence of reaction to such testing with ointments containing 10% calendula (Bruynzeel et al. 1992), an ether extract of calendula (Reider et al. 2001), an ethanol extract of calendula (de Groot et al. 1988), and calendula absolute (Rodriquez and Mitchell 1977). In one set of patch tests, patients sensitive to calendula were also sensitive to nickel, propolis, fragrance mix, and balsam of Peru (*Myroxylon balsamum* var. *pereirae*). Sensitization to calendula can be assessed only by use of an Asteraceae mix or a sesquiterpene mix (Reider et al. 2001).

Anaphylactic reaction to a calendula infusion has been reported (Goldman 1974). Contact dermatitis from contact with a calendula plant has been reported (Wrangsjö et al. 1990).

III. Pharmacology and Pharmacokinetics

Human Pharmacological Studies

No relevant human pharmacological studies were identified.

Animal Pharmacological Studies

Little to no irritation was observed in animal dermal and ocular irritation studies of a variety of calendula extracts (Fiume 2001). Similarly, in animal sensitization studies, calendula did not produce sensitization reactions (Fiume 2001).

In Vitro Pharmacological Studies

No relevant in vitro pharmacological studies were identified.

IV. Pregnancy and Lactation

In a reproductive toxicity assessment of a hydroalcoholic extract of calendula, female rats were orally administered (by gavage) 0, 0.25, 0.5, or 1.0 g/kg of the extract on days 1 to 6, 7 to 14, or 15 to 19 of gestation. On day 20 of gestation, rats were killed for evaluation of maternal and fetal parameters. The treatment did not cause any changes in the number of corpora lutea, implantation sites, or resorption sites and no significant differences were found in the preimplantation and postimplantation loss rates in the treated groups as compared to the control. The number of live fetuses and the fetal weight was similar in the treatment and control groups. A dose-dependent reduction in maternal weight gain was observed in the group administered calendula on gestational days 15 to 19, although fetal weight was not affected (Silva et al. 2009).

In isolated rabbit and guinea pig uteruses, an aqueous extract of calendula exhibited uterotonic activity (Shipochliev 1981). The relevance of that finding to human use is unknown.

Older studies and references report emmenagogic activity of calendula (Palma 1964; Puri 1971; Saha et al. 1961).

No information on the safety of calendula during lactation was identified.

V. Toxicity Studies

Acute Toxicity

The LD_{50} of orally administered hydroalcoholic extract in rats and mice could not be determined at doses up to 5 g/kg (CTFA 1980; Silva et al. 2007). The LD_{50} of intraperitoneally administered calendula extract in mice is 300 mg/kg (Dhar et al. 1968).

Short-Term Toxicity

In rats and mice administered a hydroalcoholic extract of calendula at doses up to 1 g/kg daily for 30 days, no hematological alterations were observed at doses up to 1 g/kg, and no morphological changes in the brain, kidney, and heart were seen. An increase in blood urea nitrogen and in alanine transaminase levels was observed, and inflammatory sites were found in the lung (associated with oral gavage) and liver (Silva et al. 2007).

Genotoxicity

No mutagenic activity of saponins isolated from calendula was seen in the Ames test using *Salmonella typhimurium* TA98 without and with metabolic activation (Elias et al. 1990). No mutagenic activity of an aqueous extract of calendula was seen in a somatic mutation and recombination test using *Drosophila melanogaster*. Two flavonols, quercetin and rutin, had weak genotoxic activity (Graf et al. 1994).

No chromosomal damage was observed in the bone marrow micronucleus test after mice were orally administered up to 1 g/kg calendula aqueous-ethanolic extract daily for 2 days (Ramos et al. 1998). An aqueous-ethanolic extract of calendula exhibited dose-dependent toxicity and genotoxicity (both mitotic crossing-over and chromosome malsegregation) to *Aspergillus* at concentrations of 0.1 to 1.0 mg solids/ml (Ramos et al. 1998). No mutagenic activity of an aqueous-ethanolic calendula extract at concentrations ranging from 50 to 5000 μg/plate was seen in several *Salmonella* strains with or without S9 activation (Ramos et al. 1998).

At low doses, aqueous and aqueous-ethanolic extracts of calendula provided effects against unscheduled DNA synthesis induced by diethylnitrosamine, whereas at high concentrations without diethylnitrosamine, the same extracts produced some genotoxic effects (Perez-Carreon et al. 2002).

Cytotoxicity

In vitro cytotoxic activity of calendula extracts was observed in MRC5, Hep2, and Ehrlich cell lines. When the same extracts were tested in mice, one was inactive and three were poorly active, whereas the most saponin-rich extract did not produce development of ascites (Boucaud-Maitre et al. 1988).

Literature Cited

Boucaud-Maitre, Y., O. Algernon, and J. Raynaud. 1988. Cytotoxic and antitumoral activity of *Calendula officinalis* extracts. *Pharmazie* 43(3):220-221.

Bruynzeel, D.P., W.G. van Ketel, E. Young, T. van Joost, and G. Smeenk. 1992. Contact sensitization by alternative topical medicaments containing plant extracts. The Dutch Contact Dermatoses Group. *Contact Dermat.* 27(4):278-279.

CIR. 2009. Cosmetic ingredients found safe as used. December 2009. Washington D.C.: Cosmetic Ingredient Review.

CTFA. 1980. Acute oral toxicity of *Calendula officinalis* extract.

de Groot, A.C., D.P. Bruynzeel, J.D. Bos, et al. 1988. The allergens in cosmetics. *Arch. Dermatol.* 124(10):1525-1529.

Dhar, M.L., M.M. Dhar, B.N. Dhawan, B.N. Mehrotra, and C. Ray. 1968. Screening of Indian plants for biological activity: I. *Indian J. Exp. Biol.* 6 (4):232-247.

Elias, R., M. Demeo, E. Vidalollivier, et al. 1990. Antimutagenic activity of some saponins isolated from *Calendula officinalis* L., *Calendula arvensis* L. and *Hedera helix* L. *Mutagenesis* 5 (4):327-331.

Fiume, M. 2001. Final report on the safety assessment of *Calendula officinalis* extract and *Calendula officinalis*. *Int. J. Toxicol.* 20(Suppl. 2):13-20.

Goldman, I. 1974. [Anaphylactic shock after gargling with an infusion of *Calendula*.] *Klin. Med. (Mosk.)* 52 (4):142-143.

Graf, U., A.A. Moraga, R. Castro, and E. Diaz Carrillo. 1994. Genotoxicity testing of different types of beverages in the *Drosophila* wing somatic mutation and recombination test. *Food Chem. Toxicol.* 32 (5):423-430.

Palma, L. 1964. *Le piante medicinali d'Italia.* Torino: SEI.

Paulsen, E. 2002. Contact sensitization from Compositae-containing herbal remedies and cosmetics. *Contact Dermat.* 47 (4):189-198.

Perez-Carreon, J.I., G. Cruz-Jimenez, J.A. Licea-Vega, et al. 2002. Genotoxic and anti-genotoxic properties of *Calendula officinalis* extracts in rat liver cell cultures treated with diethylnitrosamine. *Toxicol. In Vitro* 16 (3):253-258.

Puri, H. 1971. *Cited in* Farnsworth, N.R., A.S. Bingel, G.A. Cordell, FA Crane, H.H. Fong. 1975. Potential value of plants as sources of new antifertility agents I. *J. Pharm. Sci.* 64(4):535-598.

Ramos, A., A. Edreira, A. Vizoso, et al. 1998. Genotoxicity of an extract of *Calendula officinalis* L. *J. Ethnopharmacol.* 61 (1):49-55.

Reider, N., P. Komericki, B.M. Hausen, P. Fritsch, and W. Aberer. 2001. The seamy side of natural medicines: Contact sensitization to arnica (*Arnica montana* L.) and marigold (*Calendula officinalis* L.). *Contact Dermat.* 45 (5):269-272.

Rodriquez, E., and J. Mitchell. 1977 Absence of contact hypersensitivity to some perfume materials derived from Compositae species. *Contact Dermat.* 3 (3):168-169.

Saha, J.C., E.C. Savani, and S. Kasinathan. 1961. Ecbolic properties of Indian medicinal plants. Part 1. *Indian J. Med. Res.* 49:130-151.

Samochowiec, L. 1983. Pharmacological study of saponosides from *Aralia mandshurica* Rupr. et Maxim and *Calendula officinalis* L. *Herba Pol.* 29:151-155.

Shipochliev, T. 1981. Uterotonic action of extracts from a group of medicinal plants. *Vet. Med. Nauk.* 18(4):94-98.

Silva, EJR, J.H. Costa-Silva, L.B. Evêncio, M. do Carmo, C.A. Fraga, M. Cristina, O.C. Coelho, and A.G. Wanderley. 2009. Reproductive assessment of hydroalcohol extract of *Calendula officinalis* L. in Wistar rats *Phytother. Res.* 23:1392-1398.

Silva, E.J., E.S. Goncalves, F. Aguiar, et al. 2007. Toxicological studies on hydroalcohol extract of *Calendula officinalis* L. *Phytother. Res.* 21(4):332-336.

Wrangsjö, K., A.M. Ros, and J.E. Wahlberg. 1990. Contact allergy to Compositae plants in patients with summer-exacerbated dermatitis. *Contact Dermat.* 22 (3):148-154.

Camellia sinensis (L.) Kuntze **Theaceae**

SCN: tea
Syn: *Thea sinensis* L.
OCN: black tea
Part: fully fermented leaf, stem

Quick Reference Summary

Safety Class: 1
Interaction Class: C*

Contraindications

None known.

Other Precautions

Many black tea extracts contain caffeine, a nervous system stimulant. If taken in large amounts, black tea products containing caffeine can cause insomnia, nervousness, and the other well-known symptoms of excess caffeine intake (Donovan and DeVane 2001).

Due to the CNS stimulant effects of caffeine, use of caffeine-containing products is cautioned in persons with heart disorders, as excessive caffeine consumption may increase heart rate or exacerbate arrhythmias; or psychological disorders, as caffeine may aggravate depression or induce anxiety (Brinker 2001).

Drug and Supplement Interactions

Use of caffeine with other central nervous system (CNS) stimulants, including bronchodilators or adrenergic drugs, may cause excessive central nervous system stimulation resulting in nervousness, irritability, insomnia, and possibly convulsions or cardiac arrhythmias (PDR 2006).

Caffeine is metabolized by the isoenzyme CYP1A2. Drugs that inhibit this isoenzyme (including fluvoxamine, ciprofloxacin, cimetidine, amiodarone, fluoroquinolones, furafylline, interferon, methoxsalen, and mibefradil) may slow the metabolism of caffeine resulting in high blood levels of caffeine in persons drinking multiple cups of black tea daily (Carrillo and Benitez 2000).

Notice

Caffeine (0.9–5.0% caffeine) (Wichtl 2004; Williamson 2003); *see* Appendix 1.

Diuretic (Brunton et al. 2006); *see* Appendix 2.

Tannins (12.9%) (List and Hörhammer 1973; Martindale and Reynolds 1996; Wichtl 2004); *see* Appendix 1.

Editors' Note

The American Herbal Products Association has established a trade requirement (AHPA 2011) that dietary supplement products that contain caffeine, whether as a direct ingredient or as a constituent of herbal ingredients, be labeled to disclose the presence of caffeine in the product and the quantity of added caffeine if greater than 25 mg; be formulated and labeled in a manner to recommend a maximum of 200 mg of caffeine per serving, not more often than every 3 to 4 hours; and bear the following or similar statement on the label of any dietary supplement that contains caffeine in sufficient quantity to warrant such labeling:

Too much caffeine may cause nervousness, irritability, sleeplessness, and, occasionally, rapid heartbeat. Not recommended for use by children under 18 years of age.

* For caffeine-free preparations, no interactions are expected.

See Appendix 1 for more specific details on this AHPA trade requirement.

Adverse Events and Side Effects

A review of clinical trials on black tea did not indicate any adverse effects associated with black tea in the trials (Gardner et al. 2007).

Pharmacological Considerations

Black tea consumption may reduce absorption of iron, calcium, and magnesium (Disler et al. 1975; Gardner et al. 2007; Merhav et al. 1985).

Pregnancy and Lactation

No studies on the safety of black tea during pregnancy or lactation were identified. Pregnant women are advised to limit use of caffeinated black tea products to 300 mg caffeine (approximately six cups black tea) daily (PDR 2006). Nursing women are advised to limit use of caffeinated black tea products to two to three cups (~150 mg caffeine) daily (AAP 2001).

Review Details

I. Drug and Supplement Interactions

Clinical Trials of Drug or Supplement Interactions

No clinical trials of interactions with black tea were identified. A number of interaction studies on the compound caffeine have been completed; *see* Caffeine in Appendix 1.

Case Reports of Suspected Drug or Supplement Interactions

No case reports of suspected drug or supplement interactions were identified.

Animal Trials of Drug or Supplement Interactions

No animal trials of interactions with black tea or black tea extract were identified.

II. Adverse Events

Adverse Events Reported in Clinical Trials

A review of clinical trials on black tea did not indicate any adverse effects associated with black tea in the trials (Gardner et al. 2007).

Case Reports of Adverse Events

No case reports of adverse events were identified.

III. Pharmacology and Pharmacokinetics

Human Pharmacological Studies

Human studies have shown increased urinary excretion and decreased calcium and magnesium absorption in women administered the minerals and caffeine at the same time (Bergman et al. 1990; Heaney and Recker 1982), although an epidemiological study of approximately 1000 people over age 30 indicated a higher bone mineral density in persons who had habitually consumed tea for 19 years or more, as compared with nonhabitual tea drinkers (Nesher et al. 2003).

Black tea has been shown to reduce iron absorption, especially in individuals who have an increased risk of anemia (Disler et al. 1975; Merhav et al. 1985). The postulated mechanism of action is the formation of iron complexes with black tea tannins in the gut (South et al. 1997). However, in a review of clinical trials and epidemiological studies of black tea, no change in iron status was noted except in populations that were at risk for anemia (Gardner et al. 2007).

Caffeine is a substrate of the drug-metabolizing isoenzyme CYP1A2 (Nordmark et al. 1999).

Animal Pharmacological Studies

Animal studies were available but omitted due to the presence of human data.

In Vitro Pharmacological Studies

In vitro studies were available but omitted due to the presence of human data.

IV. Pregnancy and Lactation

Caffeine is in the FDA pregnancy category C and has been shown to cross the placenta and achieve blood and tissue concentrations in the fetus. Excessive intake of caffeine by pregnant women has been associated with fetal arrhythmias. Pregnant women are advised to limit caffeine intake to less than 300 mg (approximately six cups black tea) daily (PDR 2006).

Caffeine is listed as a "Maternal Medication Usually Compatible with Breastfeeding" by the American Academy of Pediatrics Committee on Drugs. The Committee noted that maternal consumption of caffeine may cause irritability and poor sleeping patterns in nursing infants, and that maternal consumption of caffeinated beverages should be limited to two to three cups daily (AAP 2001).

Epidemiological studies have indicated an association between high caffeine intake during pregnancy and an increased risk of spontaneous abortions. An analysis concluded that methodological flaws in many of the studies led to biased results and that a causal link between caffeine consumption and abortion cannot yet be confirmed (Signorello and McLaughlin 2004).

V. Toxicity Studies

Acute Toxicity

The LD_{50} of orally administered caffeine in rats is 335 mg/kg (Mills and Bone 2005).

Literature Cited

AAP. 2001. The transfer of drugs and other chemicals into human milk. American Academy of Pediatrics Committee on Drugs. *Pediatrics* 108(3):776-789.

AHPA. July 2011. Code of Ethics & Business Conduct. Silver Spring, MD: American Herbal Products Association.

Bergman, E.A., L.K. Massey, K.J. Wise, and D.J. Sherrard. 1990. Effects of dietary caffeine on renal handling of minerals in adult women. *Life Sci.* 47(6):557-564.

Brinker, F. 2001. *Herb contraindications and drug interactions.* 3rd ed. Sandy, OR: Eclectic Medical Publications.

Brunton, L.L., J.S. Lazo, and K.L. Parker. 2006. *Goodman & Gilman's the pharmacological basis of therapeutics.* 11th ed. New York: McGraw-Hill.

Carrillo, J.A., and J. Benitez. 2000. Clinically significant pharmacokinetic interactions between dietary caffeine and medications. *Clin. Pharmacokinet.* 39(2):127-153.

Disler, P.B., S.R. Lynch, R.W. Charlton, et al. 1975. The effect of tea on iron absorption. *Gut* 16(3):193-200.

Donovan, J.L., and C.L. DeVane. 2001. A primer on caffeine pharmacology and its drug interactions in clinical psychopharmacology. *Psychopharmacol. Bull.* 35(3):30-48.

Gardner, E.J., C.H. Ruxton, and A.R. Leeds. 2007. Black tea—Helpful or harmful? A review of the evidence. *Eur. J. Clin. Nutr.* 61(1):3-18.

Heaney, R.P., and R.R. Recker. 1982. Effects of nitrogen, phosphorus, and caffeine on calcium balance in women. *J. Lab. Clin. Med.* 99(1):46-55.

List, P.H., and H. Hörhammer. 1973. *Hagers handbuch der pharmazeutischen praxis.* Berlin: Springer.

Martindale, W., and J.E.F. Reynolds. 1996. *The extra pharmacopoeia.* London: Pharmaceutical Press.

Merhav, H., Y. Amitai, H. Palti, and S. Godfrey. 1985. Tea drinking and microcytic anemia in infants. *Am. J. Clin. Nutr.* 41(6):1210-1213.

Mills, S., and K. Bone. 2005. *The essential guide to herbal safety.* St. Louis: Elsevier.

Nesher, G., M. Mates, and S. Zevin. 2003. Effect of caffeine consumption on efficacy of methotrexate in rheumatoid arthritis. *Arthr. Rheum.* 48(2):571-572.

Nordmark, A., S. Lundgren, S. Cnattingius, and A. Rane. 1999. Dietary caffeine as a probe agent for assessment of cytochrome P4501A2 activity in random urine samples. *Br. J. Clin. Pharmacol.* 47(4):397-402.

PDR. 2006. *Physicians' desk reference for nonprescription drugs and dietary supplements.* 27th ed. Montvale, NJ: Medical Economics Co.

Signorello, L.B., and J.K. McLaughlin. 2004. Maternal caffeine consumption and spontaneous abortion: A review of the epidemiologic evidence. *Epidemiology* 15(2):229-239.

South, P.K., W.A. House, and D.D. Miller. 1997. Tea consumption does not affect iron absorption in rats unless tea and iron are consumed together. *Nutr. Res.* 17(8):1303-1310.

Wichtl, M. 2004. *Herbal drugs and phytopharmaceuticals: A handbook for practice on a scientific basis.* 3rd ed. Boca Raton, FL: CRC Press.

Williamson, E.M. 2003. *Potter's herbal cyclopedia.* Saffron Walden, Essex: C.W. Daniel Co.

Camellia sinensis (L.) Kuntze

Theaceae

SCN: tea
Syn: *Thea sinensis* L.
OCN: green tea
Part: unfermented leaf, stem

Quick Reference Summary

Safety Class: 1
Interaction Class: C*

Contraindications

None known.

Other Precautions

Green tea and many green tea extracts contain caffeine, a nervous system stimulant. If taken in large amounts, green tea products containing caffeine can cause insomnia, nervousness, and the other well-known symptoms of excess caffeine intake (Donovan and DeVane 2001).

Due to the CNS stimulant effects of caffeine, use of caffeine-containing products is cautioned in persons with heart disorders, as excessive caffeine consumption may increase heart rate or exacerbate arrhythmias; or psychological disorders, as caffeine may aggravate depression or induce anxiety (Brinker 2001).

Ethanol extracts of green tea should be taken with a meal (Sarma et al. 2008).

Drug and Supplement Interactions

Use of caffeine with other central nervous system (CNS) stimulants, including bronchodilators or adrenergic drugs, may cause excessive central nervous system stimulation resulting in nervousness, irritability, insomnia, and possibly convulsions or cardiac arrhythmias (PDR 2006).

Notice

Caffeine (0.9–5.0% caffeine) (Leung and Foster 1996; Martindale and Reynolds 1996; Wichtl 2004); *see* Appendix 1.

Diuretic (Brunton et al. 2006); *see* Appendix 2.

* For caffeine-free preparations, no interactions are expected.

Tannins (22.2%) (Martindale and Reynolds 1996; Wichtl 2004); *see* Appendix 1.

Editors' Notes

Green, unfermented tea is the world's second most consumed beverage, after water (Graham 1992). A typical cup of green tea contains 15 to 40 mg of caffeine (Chin et al. 2008).

The American Herbal Products Association has established a trade requirement (AHPA 2011) that dietary supplement products that contain caffeine, whether as a direct ingredient or as a constituent of herbal ingredients, be labeled to disclose the presence of caffeine in the product and the quantity of added caffeine if greater than 25 mg; be formulated and labeled in a manner to recommend a maximum of 200 mg of caffeine per serving, not more often than every 3 to 4 hours; and bear the following or similar statement on the label of any dietary supplement that contains caffeine in sufficient quantity to warrant such labeling:

Too much caffeine may cause nervousness, irritability, sleeplessness, and, occasionally, rapid heartbeat. Not recommended for use by children under 18 years of age.

See Appendix 1 for more specific details on this AHPA trade requirement.

Adverse Events and Side Effects

Cases of liver toxicity have been reported in persons taking products that contain green tea extracts (Abu el Wafa et al. 2005; Bonkovsky 2006; Duenas Sadornil et al. 2004; Federico et al. 2007; Garcia-Moran et al. 2004; Gloro et al. 2005; Javaid and Bonkovsky 2006; Jimenez-Saenz and Martinez-Sanchez 2006; Mazzanti et al. 2009; Molinari et al. 2006; Pedros et al. 2003; Seddik et al. 2001; Thiolet et al. 2002; Vial et al. 2003). These cases were primarily associated with ethanolic extracts of green tea, although several cases were associated with aqueous extracts (Federico et al. 2007; Jimenez-Saenz and Martinez-Sanchez 2006; Mazzanti et al. 2009). A review of the reported cases indicated that most patients developed severe hepatocellular injury with elevated levels of bilirubin and the liver enzyme alanine aminotransferase, suggesting idiosyncratic drug-induced liver injury rather than immunoallergic injury (Javaid and Bonkovsky 2006). Two case reports involved inadvertent re-exposure to the green tea extract, with a return of liver symptoms (Bonkovsky 2006; Jimenez-Saenz and Martinez-Sanchez 2006).

A systematic review of studies on the effects of green tea on liver disease indicated that 8 of 10 studies examined indicated a significant protective role of green tea against various liver diseases including liver cancer, cirrhosis, and fatty liver disease (Jin et al. 2008).

Pharmacological Considerations

A review of the safety of green tea extracts indicated that human and animal data showed that the maximum plasma concentration of the compound epigallocatechin gallate (EGCG) was significantly higher when administered under fasting conditions than with or after a meal, leading to the recommendation that green tea extracts be taken with a meal (Kapetanovic et al. 2009; Sarma et al. 2008).

Caffeine is a substrate of the drug-metabolizing isoenzyme CYP1A2 (Nordmark et al. 1999). No significant effects of decaffeinated green tea extract were observed on the drug-metabolizing isoenzymes CYP3A4 or CYP2D6 (Donovan et al. 2004). No clinically significant changes in CYP1A2, CYP2D6, or CYP2C19 were observed in humans administered green tea catechins (Chow et al. 2006).

Green tea may reduce the absorption of iron (Samman et al. 2001). Caffeine may decrease calcium and magnesium absorption (Bergman et al. 1990; Heaney and Recker 1982).

Pregnancy and Lactation

Pregnant women are advised to limit caffeine intake to less than 300 mg (approximately eight cups green tea) daily (PDR 2006). Lactating women are advised to limit consumption of caffeinated beverages to two to three cups daily (AAP 2001).

Review Details

I. Drug and Supplement Interactions

Clinical Trials of Drug or Supplement Interactions

No clinical trials of green tea or green tea extract interactions were identified. A number of interaction trials on the compound caffeine have been completed. *See* Caffeine in Appendix 1.

Case Reports of Suspected Drug or Supplement Interactions

Decreased INR (international normalized ratio, a system used to report the results of blood coagulation tests) was reported in a patient taking warfarin and drinking one-half to one gallon of green tea daily (Taylor and Wilt 1999). Warfarin works by interfering with vitamin K metabolism, a vitamin essential for blood clotting, and patients taking warfarin must limit dietary intake of vitamin K. The vitamin K content of green tea is relatively high, although brewed green tea contains less than 0.01% of the leaf content of vitamin K (Booth et al. 1995).

Animal Trials of Drug or Supplement Interactions

In rats orally administered 175 mg/kg of a green tea extract 4 days prior to oral administration of clozapine, a reduction in serum levels of clozapine was observed (Jang et al. 2005).

II. Adverse Events

Adverse Events Reported in Clinical Trials

In a study of green tea standardized ethanolic extract (375 mg catechins daily) in obese volunteers, 1 of the 70 trial participants had an increase in liver transaminases after 3 months of green tea extract ingestion (Chantre and Lairon 2002).

Case Reports of Adverse Events

Cases of hepatotoxicity have been reported in persons taking products that contain green tea extracts (Abu el Wafa et al. 2005; Bonkovsky 2006; Duenas Sadornil et al. 2004; Federico et al. 2007; Garcia-Moran et al. 2004; Gloro et al. 2005; Javaid and Bonkovsky 2006; Jimenez-Saenz and Martinez-Sanchez 2006; Mazzanti et al. 2009; Molinari et al. 2006; Pedros et al. 2003; Seddik et al. 2001; Thiolet et al. 2002; Vanstraelen et al. 2008; Verhelst et al. 2009; Vial et al. 2003). Supplement use duration in these cases ranged from 5 to 120 days and cumulative doses ranged from 6 to 240 g (Bonkovsky 2006). Hepatotoxicity cases were primarily associated with ethanolic extracts of green tea, although several cases were associated with aqueous extracts (Federico et al. 2007; Jimenez-Saenz and Martinez-Sanchez 2006; Mazzanti et al. 2009). Two case reports involved inadvertent re-exposure to the green tea extract, with a return of liver symptoms (Bonkovsky 2006; Jimenez-Saenz and Martinez-Sanchez 2006). One review of the reported cases indicated that most patients developed severe hepatocellular injury with elevated levels of bilirubin and the liver enzyme alanine aminotransferase, suggesting idiosyncratic drug-induced liver injury rather than immunoallergic injury (Javaid and Bonkovsky 2006). A second review of published adverse events and those reported to pharmacovigilance centers indicated that of 34 identified case reports, "probable" causality was found in 7 cases and "possible" causality was found in the remainder of the cases (Sarma et al. 2008), based on the Naranjo causality algorithm scale (Naranjo et al. 1981). Notable diversity was seen in the products involved in these case reports, and product composition was verified in only one report (Sarma et al. 2008).

III. Pharmacology and Pharmacokinetics

Human Pharmacological Studies

A systematic review of studies on the effects of green tea on liver disease indicated that 8 of 10 studies showed a significant protective role of green tea against various liver diseases including liver cancer, cirrhosis, and fatty liver disease (Jin et al. 2008).

In men with positive prostate biopsies, oral administration of a standardized green tea extract containing 1.3 g green tea polyphenols (800 mg EGCG) daily for 12 to 214 days (mean of 34 days) was studied; a decrease in liver enzyme levels, including aspartate aminotransferase, alkaline phosphatase, and amylase, was observed (McLarty et al. 2009).

Human studies have shown increased urinary excretion and decreased calcium and magnesium absorption in women administered the minerals and caffeine at the same time (Bergman et al. 1990; Heaney and Recker 1982), although an epidemiological study of approximately 1000 people over age 30 indicated a higher bone mineral density in persons who had habitually consumed tea for 19 years or more, as compared with nonhabitual tea drinkers (Nesher et al. 2003).

Green tea has been shown to reduce iron absorption, especially in individuals who have an increased risk of anemia (Samman et al. 2001). The postulated mechanism of action is the formation of iron complexes with green tea tannins in the gut (South et al. 1997).

No significant effects of decaffeinated green tea extract (800 mg daily for 14 days) were observed on the drug-metabolizing isoenzymes CYP3A4 or CYP2D6 (Donovan et al. 2004). Green tea catechins were observed to have no effect on CYP1A2, CYP2D6, or CYP2C19, and clinically insignificant inhibition of CYP3A4 (Chow et al. 2006). Caffeine is a substrate of the drug-metabolizing isoenzyme CYP1A2 (Nordmark et al. 1999).

No changes in liver enzyme levels were observed in healthy men orally administered 6 capsules containing 384 mg (119 mg polyphenols) of an aqueous green tea extract daily (2 capsules before each meal) for 3 weeks (Frank et al. 2009).

Animal Pharmacological Studies

Animal studies were identified but omitted due to the presence of human data.

In Vitro Pharmacological Studies

In vitro studies were identified but omitted due to the presence of human data.

IV. Pregnancy and Lactation

Caffeine is in the FDA pregnancy category C and has been shown to cross the placenta and achieve blood and tissue concentrations in the fetus. Excessive intake of caffeine by pregnant women has been associated with fetal arrhythmias. Pregnant women are advised to limit caffeine intake to less than 300 mg (approximately eight cups green tea) daily (PDR 2006).

Caffeine is listed as a "Maternal Medication Usually Compatible with Breastfeeding" by the American Academy of Pediatrics Committee on Drugs. The Committee noted that maternal consumption of caffeine may cause irritability and poor sleeping patterns in nursing infants, and that maternal consumption of caffeinated beverages should be limited to two to three cups daily (AAP 2001).

In rats fed green tea extract and catechins during pregnancy, fetuses were found to have one-hundredth of the catechins found in maternal plasma (Chu et al. 2006). Catechins were identified in several different fetal organs, with eyes having the highest concentration of these compounds (Chu et al. 2007).

V. Toxicity Studies

Acute Toxicity

The LD_{50} of orally administered caffeine in rats is 335 mg/kg (Mills and Bone 2005). The LD_{50} of orally administered EGCG in rats is 2000 mg/kg (Isbrucker et al. 2006b).

In mice orally administered a single dose of 1500 mg/kg of the compound EGCG, plasma levels of alanine aminotransferase (ALT) increased by 138-fold and reduced survival by 85%. A greater increase in ALT levels was observed after doses of 750 mg/kg of EGCG daily for 2 days (Lambert et al. 2010).

Short-Term Toxicity

In rats orally administered 2.5 or 5 g/kg of Pu-erh green or Pu-erh black tea extract daily for 28 days, alanine aminotransferase increased in males at the 5 g/kg dose, and creatinine increased in all animals at the 5 g/kg dose and in males at the 2.5 g/kg dose. Slight bile duct hyperplasia in the liver was also observed. No adverse effects were observed in animals treated with Pu-erh black tea extract (Wang et al. 2010).

Subchronic Toxicity

No signs of toxicity were observed in rats after administration of 500 mg/kg daily EGCG for 13 weeks (Isbrucker et al. 2006b). No adverse effects were observed in prefed dogs administered 500 mg/kg EGCG daily in divided doses for 13 weeks, although morbidity was observed in fasted dogs fed 500 mg/kg in a single dose. This dose was noted by the study authors to be unrealistic as compared to standard human use (Isbrucker et al. 2006b). A no-observed-adverse-effect level (NOAEL) for human consumption of EGCG was estimated to be 500 mg/kg daily (Isbrucker et al. 2006b).

In rats fed diets containing 0.3, 1.25, or 5.0% green tea catechins for 90 days (daily intake of catechins was 184, 792, 3533 mg/kg), no mortality or obvious clinical signs were observed. Increased relative liver weight, alanine transaminase, and alkaline phosphatase were observed at the 5% level in both sexes. Aspartate transaminase was increased in females at the 5% level. Hematology and histopathological observation revealed no toxicological changes. The no-observed-adverse-effect level was estimated to be 764 mg/kg for males and 820 mg/kg for females (Takami et al. 2008).

In beagle dogs orally administered 200, 500, or 1000 mg/kg daily of green tea extract (56–72% EGCG), on an empty stomach, for a planned duration of 9 months, extensive morbidity, mortality, and pathology of many major organs led to termination of the study at 6.5 months and prevented identification of the toxicity mechanisms. In a 13-week follow-up study, animals were fed 175 to 200 mg/kg of the same extract daily on an empty stomach or after daily feeding. Dosing in a fed state resulted in considerably lower and less variable exposure than found under fasted conditions. Toxicity was less frequent and of lesser severity with lower exposure (Kapetanovic et al. 2009).

High concentrations of hydroalcoholic green tea extracts have been shown to exert acute toxicity in rat liver cells *in vitro*. The compound (–)-epigallocatechin-3-gallate (EGCG) was noted as the constituent that was likely responsible for this effect. The bioavailability of catechins is relatively low with oral ingestion, however, suggesting that EGCG is not responsible for the hepatotoxicity reported in human studies (Schmidt et al. 2005).

Administration of a single dose of the green tea phenolic acid gallic acid (100–800 mg/kg), propyl gallate (100 - 200 mg/kg), and the green tea catechin EGCG (100 mg/kg) resulted in a dose-dependent increase in plasma ALT levels in mice, indicative of liver injury (Galati et al. 2006).

Genotoxicity

Studies on the compound EGCG indicated a lack of genotoxicity in tests on *Salmonella typhimurium* and in rats (oral administration up to 2000 mg/kg) (Isbrucker et al. 2006a).

Literature Cited

AAP. 2001. The transfer of drugs and other chemicals into human milk. American Academy of Pediatrics Committee on Drugs. *Pediatrics* 108(3):776-789.

Abu el Wafa, Y., A. Benavente Fernandez, A. Talavera Fabuel, M.A. Perez Ramos, and J.I. Ramos-Clemente. 2005. Acute hepatitis induced by *Camellia sinensis* (green tea). *An. Med. Intern.* 22(6):298.

AHPA. July 2011. Code of Ethics & Business Conduct. Silver Spring, MD: American Herbal Products Association.

Bergman, E.A., L.K. Massey, K.J. Wise, and D.J. Sherrard. 1990. Effects of dietary caffeine on renal handling of minerals in adult women. *Life Sci.* 47(6):557-564.

Bonkovsky, H.L. 2006. Hepatotoxicity associated with supplements containing Chinese green tea (*Camellia sinensis*). *Ann. Intern. Med.* 144(1):68-71.

Booth, S.L., H.T. Madabushi, K.W. Davidson, and J.A. Sadowski. 1995. Tea and coffee brews are not dietary sources of vitamin K-1 (phylloquinone). *J. Am. Diet. Assoc.* 95(1):82-83.

Brinker, F. 2001. *Herb contraindications and drug interactions*. 3rd ed. Sandy, OR: Eclectic Medical Publications.

Brunton, L.L., J.S. Lazo, and K.L. Parker. 2006. *Goodman & Gilman's the pharmacological basis of therapeutics.* 11th ed. New York: McGraw-Hill.

Chantre, P., and D. Lairon. 2002. Recent findings of green tea extract AR25 (Exolise) and its activity for the treatment of obesity. *Phytomedicine* 9(1):3-8.

Chin, J.M., M.L. Merves, B.A. Goldberger, A. Sampson-Cone, and E.J. Cone. 2008. Technical note: Caffeine content of brewed teas. *J. Analyt. Technol.* 32(8):702-704.

Chow, H.H., I.A. Hakim, D.R. Vining, et al. 2006. Effects of repeated green tea catechin administration on human cytochrome P450 activity. *Cancer Epidemioi. Biomarkers Prev.* 15(12):2473-2476.

Chu, K.O., C.C. Wang, C.Y. Chu, et al. 2006. Pharmacokinetic studies of green tea catechins in maternal plasma and fetuses in rats. *J. Pharm. Sci.* 95(6):1372-1381.

Chu, K.O., C.C. Wang, C.Y. Chu, et al. 2007. Uptake and distribution of catechins in fetal organs following in utero exposure in rats. *Human Reprod.* 22(1):280-287.

Donovan, J.L., K.D. Chavin, C.L. Devane, et al. 2004. Green tea (*Camellia sinensis*) extract does not alter cytochrome P450 3A4 or 2D6 activity in healthy volunteers. *Drug Metab. Dispos.* 32(9):906-908.

Donovan, J.L., and C.L. DeVane. 2001. A primer on caffeine pharmacology and its drug interactions in clinical psychopharmacology. *Psychopharmacol. Bull.* 35(3):30-48.

Duenas Sadornil, C., S. Fabregas Puigtio, and R. Durandez. 2004. Hepatotoxicity due to *Camellia sinensis. Med. Clin.* 122(17):677-678.

Federico, A., A. Tiso, and C. Loguercio. 2007. A case of hepatotoxicity caused by green tea. *Free Radicals Biol. Med.* 43(3):474.

Frank, J., T.W. George, J.K. Lodge, et al. 2009. Daily consumption of an aqueous green tea extract supplement does not impair liver function or alter cardiovascular disease risk biomarkers in healthy men. *J. Nutr.* 139(1):58-62.

Galati, G., A. Lin, A.M. Sultan, and P.J. O'Brien. 2006. Cellular and in vivo hepatotoxicity caused by green tea phenolic acids and catechins. *Free Radicals Biol. Med.* 40(4):570-580.

Garcia-Moran, S., F. Saez-Royuela, E. Gento, A. Lopez Morante, and L. Arias. 2004. Acute hepatitis associated with *Camellia thea* and *Orthosiphon stamineus* ingestion. *Gastroenterol. Hepatol.* 27(9):559-560.

Gloro, R., I. Hourmand-Ollivier, B. Mosquet, et al. 2005. Fulminant hepatitis during self-medication with hydroalcoholic extract of green tea. *Eur. J. Gastroenterol. Hepatol.* 17(10):1135-1137.

Graham, H.N. 1992. Green tea composition, consumption, and polyphenol chemistry. *Prev. Med.* 21:334-350.

Heaney, R.P., and R.R. Recker. 1982. Effects of nitrogen, phosphorus, and caffeine on calcium balance in women. *J. Lab. Clin. Med.* 99(1):46-55.

Isbrucker, R.A., J. Bausch, J.A. Edwards, and E. Wolz. 2006a. Safety studies on epigallocatechin gallate (EGCG) preparations. Part 1: Genotoxicity. *Food Chem. Toxicol.* 44(5):626-635.

Isbrucker, R.A., J.A. Edwards, E. Wolz, A. Davidovich, and J. Bausch. 2006b. Safety studies on epigallocatechin gallate (EGCG) preparations. Part 2: Dermal, acute and short-term toxicity studies. *Food Chem. Toxicol.* 44(5):636-650.

Jang, E.H., J.Y. Choi, C.S. Park, et al. 2005. Effects of green tea extract administration on the pharmacokinetics of clozapine in rats. *J. Pharm. Pharmacol.* 57(3):311-316.

Javaid, A., and H.L. Bonkovsky. 2006. Hepatotoxicity due to extracts of Chinese green tea (*Camellia sinensis*): A growing concern. *J. Hepatol.* 45(2):334-335; author reply 335-336.

Jimenez-Saenz, M., and M.d.C. Martinez-Sanchez. 2006. Acute hepatitis associated with the use of green tea infusions. *J. Hepatol.* 44(3):616-617.

Jin, X., R.H. Zheng, and Y.M. Li. 2008. Green tea consumption and liver disease: A systematic review. *Liver Int.* 28(7):990-996.

Kapetanovic, I.M., J.A. Crowell, R. Krishnaraj, et al. 2009. Exposure and toxicity of green tea polyphenols in fasted and non-fasted dogs. *Toxicology* 260(1-3):28-36.

Lambert, J.D., M.J. Kennett, S. Sang, et al. 2010. Hepatotoxicity of high oral dose (−)-epigallocatechin-3-gallate in mice. *Food Chem. Toxicol.* 48(1):409-416.

Leung, A.Y., and S. Foster. 1996. *Encyclopedia of common natural ingredients used in food, drugs, and cosmetics.* 2nd ed. New York: Wiley.

Martindale, W., and J.E.F. Reynolds. 1996. *The extra pharmacopoeia.* London: Pharmaceutical Press.

Mazzanti, G., F. Menniti-Ippolito, P.A. Moro, et al. 2009. Hepatotoxicity from green tea: A review of the literature and two unpublished cases. *Eur. J. Clin. Pharmacol.* 65(4):331-341.

McLarty, J., R.L. Bigelow, M. Smith, et al. 2009. Tea polyphenols decrease serum levels of prostate-specific antigen, hepatocyte growth factor, and vascular endothelial growth factor in prostate cancer patients and inhibit production of hepatocyte growth factor and vascular endothelial growth factor in vitro. *Cancer Prev. Res.* 2(7):673-682.

Mills, S., and K. Bone. 2005. *The essential guide to herbal safety.* St. Louis: Elsevier.

Molinari, M., K.D. Watt, T. Kruszyna, et al. 2006. Acute liver failure induced by green tea extracts: Case report and review of the literature. *Liver Transplant.* 12(12):1892-1895.

Naranjo, C.A., U. Busto, E.M. Sellers, et al. 1981. A method for estimating the probability of adverse drug reactions. *Clin. Pharmacol. Ther.* 30(2):239-245.

Nesher, G., M. Mates, and S. Zevin. 2003. Effect of caffeine consumption on efficacy of methotrexate in rheumatoid arthritis. *Arthr. Rheum.* 48(2):571-572.

Nordmark, A., S. Lundgren, S. Cnattingius, and A. Rane. 1999. Dietary caffeine as a probe agent for assessment of cytochrome P4501A2 activity in random urine samples. *Br. J. Clin. Pharmacol.* 47(4):397-402.

PDR. 2006. *Physicians' desk reference for nonprescription drugs and dietary supplements.* 27th ed. Montvale, NJ: Medical Economics Co.

Pedros, C., G. Cereza, N. Garcia, and J.R. Laporte. 2003. Liver toxicity of *Camellia sinensis* dried ethanolic extract. *Med. Clin.* 121(15):598-599.

Samman, S., B. Sandstrom, M.B. Toft, et al. 2001. Green tea or rosemary extract added to foods reduces nonheme-iron absorption. *Am. J. Clin. Nutr.* 73(3):607-612.

Sarma, D.N., M.L. Barrett, M.L. Chavez, et al. 2008. Safety of green tea extracts: A systematic review by the U.S. Pharmacopeia. *Drug Saf.* 31(6):469-484.

Schmidt, M., H.J. Schmitz, A. Baumgart, et al. 2005. Toxicity of green tea extracts and their constituents in rat hepatocytes in primary culture. *Food Chem. Toxicol.* 43(2):307-314.

Seddik, M., D. Lucidarme, C. Creusy, and B. Filoche. 2001. Is Exolise hepatotoxic? *Gastroenterol. Clin. Biol.* 25(8-9):834-835.

South, P.K., W.A. House, and D.D. Miller. 1997. Tea consumption does not affect iron absorption in rats unless tea and iron are consumed together. *Nutr. Res.* 17(8):1303-1310.

Takami, S., T. Imai, M. Hasumura, et al. 2008. Evaluation of toxicity of green tea catechins with 90-day dietary administration to F344 rats. *Food Chem. Toxicol.* 46(6):2224-2229.

Taylor, J.R., and V.M. Wilt. 1999. Probable antagonism of warfarin by green tea. *Ann. Pharmacother.* 33(Apr):426-428.

Thiolet, C., D. Mennecier, C. Bredin, et al. 2002. Acute cytolysis induced by Chinese tea. *Gastroenterol. Clin. Biol.* 26(10):939-940.
Vanstraelen, S., J. Rahier, and A.P. Geubel. 2008. Jaundice as a misadventure of a green tea (*Camellia sinensis*) lover: A case report. *Acta Gastroenterol. Belg.* 71(4):409-412.
Verhelst, X., P. Burvenich, D. Van Sassenbroeck, et al. 2009. Acute hepatitis after treatment for hair loss with oral green tea extracts (*Camellia sinensis*). *Acta Gastroenterol. Belg.* 72(2):262-264.
Vial, T., G. Bernard, B. Lewden, J. Dumortier, and J. Descotes. 2003. Acute hepatitis due to Exolise, a *Camellia sinensis*-derived drug. *Gastroenterol. Clin. Biol.* 27(12):1166-1167.
Wang, D., R. Xiao, X. Hu, et al. 2010. Comparative safety evaluation of Chinese Pu-erh green tea extract and Pu-erh black tea extract in Wistar rats. *J. Agric. Food Chem.* 58(2):1350-1358.
Wichtl, M. 2004. *Herbal drugs and phytopharmaceuticals: A handbook for practice on a scientific basis*. 3rd ed. Boca Raton, FL: CRC Press.

Cananga odorata (Lam.) Hook. f. & Thomson — Annonaceae

SCN: ylang ylang

Part: flower

Quick Reference Summary

Safety Class: 1
Interaction Class: A

Contraindications
None known.

Other Precautions
None known.

Drug and Supplement Interactions
None known.

Adverse Events and Side Effects
Allergic reactions to ylang ylang have been reported and confirmed by patch testing (Burdock and Carabin 2008).

Pharmacological Considerations
Allergic patch tests and sensitization tests in normal or fragrance-sensitive volunteers treated with ylang ylang essential oil have elicited reactions in 0–6.2% of tested populations (Burdock and Carabin 2008).

Pregnancy and Lactation
No information on the safety of ylang ylang in pregnancy or lactation was identified in the scientific or traditional literature. Although this review did not identify any concerns for use while pregnant or nursing, safety has not been conclusively established.

Review Details

I. Drug and Supplement Interactions

Clinical Trials of Drug or Supplement Interactions

No clinical trials of drug or supplement interactions were identified.

Case Reports of Suspected Drug or Supplement Interactions

No case reports of suspected drug or supplement interactions were identified.

Animal Trials of Drug or Supplement Interactions

No effect on pentobarbital-induced sleeping time was observed in rats exposed to air odorized with ylang ylang essential oil (concentration in air not reported) (Komori et al. 1997, 2006).

II. Adverse Events

Case Reports of Adverse Events

Allergic reactions to ylang ylang essential oil, including occupational contact dermatitis, have been reported and confirmed by patch testing (Burdock and Carabin 2008).

III. Pharmacology and Pharmacokinetics

Human Pharmacological Studies

No sensitization or irritation was observed in a closed patch test in test subjects treated with 10% ylang ylang essential oil in a petroleum base (Opdyke 1979).

In human patch tests, ylang ylang essential oil tested at concentrations between 0.2 and 10% elicited positive skin reactions in 1.3 to 6.2% of volunteers in different testing centers (Frosch et al. 2002).

Tested at concentrations of 5%, ylang ylang absolute caused dermatitis and eczema in approximately 5% of test subjects, while the essential oil elicited reactions in 3.4% of subjects (Itoh et al. 1988).

In patch testing with 65 allergens in over 4900 volunteers, ylang ylang essential oil elicited contact dermatitis in 1.1% of volunteers, and ranked 47th out of the allergens tested (Pratt et al. 2004).

In repeated-insult patch tests, 0 or 5% of subjects became sensitized after being treated with 10% ylang ylang essential oil in a petroleum base. In a rechallenge patch test with the same test substance, no sensitization reactions were observed (Opdyke 1979).

Animal Pharmacological Studies

In irritation tests, ylang ylang essential oil was not irritating when applied undiluted to the back of hairless mice or on intact or abraded rabbit skin (Opdyke 1979).

In Vitro Pharmacological Studies

No relevant in vitro pharmacological studies were identified.

IV. Pregnancy and Lactation

No information on the safety of ylang ylang during pregnancy or lactation was identified.

V. Toxicity Studies

Acute Toxicity

The LD_{50} of orally administered ylang ylang essential oil in rats and the acute dermal LD_{50} in rabbits could not be determined at doses up to 5 g/kg (Opdyke 1979).

C

Literature Cited

Burdock, G.A., and I.G. Carabin. 2008. Safety assessment of ylang-ylang (*Cananga* spp.) as a food ingredient. *Food Chem. Toxicol.* 46(2):433-445.

Frosch, P.J., J.D. Johansen, T. Menne, et al. 2002. Further important sensitizers in patients sensitive to fragrances. *Contact Dermat.* 47(2):78-85.

Itoh, M., K. Hosono, H. Kantoh, et al. 1988. Patch tests results with cosmetic ingredients conducted between 1978 and 1986. *Nippon Koshohin Kagakkaishi* 12:27-41.

Komori, T., T. Matsumoto, E. Motomura, and T. Shiroyama. 2006. The sleep-enhancing effect of valerian inhalation and sleep-shortening effect of lemon inhalation. *Chem. Senses* 31:731-737.

Komori, T., M. Tanida, A. Kikuchi, et al. 1997. Effects of odorant inhalation on pentobarbital-induced sleep time in rats. *Hum. Psychopharmacol.* 12:601.

Opdyke, D.L.J. 1979. *Monographs on fragrance raw materials.* New York: Pergamon.

Pratt, M.D., D.V. Belsito, V.A. DeLeo, et al. 2004. North American Contact Dermatitis Group patch-test results, 2001-2002 study period. *Dermatitis* 15(4):176-183.

Canarium album (Lour.) Räusch. — Burseraceae

SCN: Chinese white olive
PN: *qing guo* (fruit)
OCN: white canary tree
Part: fruit

Quick Reference Summary

Safety Class: 1
Interaction Class: A

Contraindications

None known.

Other Precautions

None known.

Drug and Supplement Interactions

None known.

Editors' Note

In addition to being used medicinally, the fruit of Chinese white olive is consumed as an edible fruit in China (Jin et al. 1999).

Adverse Events and Side Effects

None known.

Pharmacological Considerations

None known.

Pregnancy and Lactation

No information on the safety of Chinese white olive in pregnancy or lactation was identified in the scientific or traditional literature. Although this review did not identify any concerns for use while pregnant or nursing, safety has not been conclusively established.

Review Details

I. Drug and Supplement Interactions

Clinical Trials of Drug or Supplement Interactions

No clinical trials of drug or supplement interactions were identified.

Case Reports of Suspected Drug or Supplement Interactions

No case reports of suspected drug or supplement interactions were identified.

Animal Trials of Drug or Supplement Interactions

No animal trials of drug or supplement interactions were identified.

II. Adverse Events

Case Reports of Adverse Events

No case reports of adverse events were identified.

III. Pharmacology and Pharmacokinetics

Human Pharmacological Studies

No relevant human pharmacological studies were identified.

Animal Pharmacological Studies

No relevant animal pharmacological studies were identified.

In Vitro Pharmacological Studies

No relevant in vitro pharmacological studies were identified.

IV. Pregnancy and Lactation

No information on the use of Chinese white olive during pregnancy or lactation was identified.

V. Toxicity Studies

No toxicity studies were identified.

Literature Cited

Jin, C., S. Yin-Chun, C. Gui-Qin, and W. Wen-Dun. 1999. Ethnobotanical studies on wild edible fruits in southern YunnAN: Folk names; nutritional value and uses. *Econ. Bot.* 53(1):2-14.

Capsella bursa-pastoris (L.) Medik. — Brassicaceae

SCN: shepherd's purse

Part: herb

Quick Reference Summary

Safety Class: 2b
Interaction Class: A

Contraindications

Not for use in pregnancy except under the supervision of a qualified healthcare practitioner (Chadha 1988; Kuroda and Takagi 1968, 1969; Shipochliev 1981; Watt and Breyer-Brandwijk 1962).

Other Precautions

None known.

Drug and Supplement Interactions

None known.

Notice

Uterine stimulant (Williamson 2003); *see* Appendix 2.

Editors' Notes

Early references note, and contemporary herbalists maintain, that extracts made from the fresh herb are more active than those from the dried herb (Felter and Lloyd 1898, 1901). While this does not relate to the safety of the extract itself, one of the primary indications of shepherd's purse is to stop postpartum bleeding, a setting in which the use of a nonefficacious product may present a safety concern.

Adverse Events and Side Effects

None known.

Pharmacological Considerations

None known.

Pregnancy and Lactation

Limited information on the safety of shepherd's purse in pregnancy or lactation was identified. Older animal and in vitro studies indicate that shepherd's purse may stimulate uterine contractions (Kuroda and Takagi 1968, 1969; Shipochliev 1981). Older, primarily ethnobotanical surveys report that shepherd's purse was traditionally used as an emmenagogue and abortifacient (Casey 1960; de Laszlo and Henshaw 1954; Palma 1964; Watt and Breyer-Brandwijk 1962). Based on this information, use during pregnancy is not recommended except under the supervision of a qualified healthcare practitioner.

Shepherd's purse contains glucosinolates, compounds present in many plants in the Brassicaceae family (e.g., cauliflower and Brussels sprouts). These compounds have been shown to cross the placenta and to be present in breast milk (Panter and James 1990). At high concentrations, glucosinolates inhibit iodine uptake by the thyroid, although at lower concentrations cancer-preventive effects have been shown (Stoewsand 1995).

C

Review Details

I. Drug and Supplement Interactions

Clinical Trials of Drug or Supplement Interactions

No clinical trials of suspected drug or supplement interactions were identified.

Case Reports of Suspected Drug or Supplement Interactions

No case reports of drug or supplement interactions were identified.

Animal Trials of Drug or Supplement Interactions

No animal trials of drug or supplement interactions were identified.

II. Adverse Events

Case Reports of Adverse Events

No case reports of adverse events were identified.

III. Pharmacology and Pharmacokinetics

Human Pharmacological Studies

No relevant human pharmacological studies were identified.

Animal Pharmacological Studies

Hypotensive activity of shepherd's purse observed in rats is inhibited by a β-adrenoceptor blocker but not by atropine, dismissing earlier reports of cholinergic activity (Jurisson 1971; Kuroda and Takagi 1968).

In Vitro Pharmacological Studies

No relevant in vitro pharmacological studies were identified.

IV. Pregnancy and Lactation

In isolated animal uterine tissue, an infusion of shepherd's purse exhibited uterine tonic activity at a concentration of 1 to 2 mg/ml. The effect was less than that exhibited by chamomile (Shipochliev 1981). A fraction of an ethanol extract of shepherd's purse intravenously administered to rats increased experimentally induced contractions (Kuroda and Takagi 1969). The same extract exerted contractile activity on isolated rat uteruses (Kuroda and Takagi 1968). Shepherd's purse has been used traditionally to stimulate uterine contractions during childbirth (Bastien 1983). Older, primarily ethnobotanical surveys report that shepherd's purse was traditionally used as an emmenagogue and abortifacient (Casey 1960; de Laszlo and Henshaw 1954; Palma 1964; Watt and Breyer-Brandwijk 1962). The ancient Greek physician Pedanius Dioscorides wrote that shepherd's purse "moves the menstrual and destroys the embrya," an early observation of some of the physiological activity reported above (Dioscorides 512).

Shepherd's purse contains glucosinolates, compounds present in many plants in the Brassicaceae family (e.g., cauliflower and Brussels sprouts). Placental transfer of glucosinolate compounds has been observed (Panter and James 1990), and some adverse developmental effects have been reported in offspring of animals administered high levels of dietary glucosinolates (Panter and James 1990). No teratogenic effects were seen in fetuses of rats subcutaneously administered selected glucosinolates on gestational day 8 and or 9 (Nishie and Daxenbichler 1980).

At high doses, glucosinolate compounds inhibit iodine uptake although cancer-preventive properties have been shown at lower doses (Stoewsand 1995).

Glucosinolate compounds have been detected in breast milk of animals (Cheeke and Shull 1985).

V. Toxicity Studies

Acute Toxicity

The LD_{50} of an intraperitoneally administered ethanol extract of shepherd's purse in rats is 1000 mg/kg (Sharma et al. 1978). Other LD_{50} values of shepherd's purse (type of extract not available in English language information) are 1500 mg/kg for intraperitoneal administration and 3100 mg/kg for subcutaneous administration. Symptoms of toxicity included sedation, dilation of pupils, paralysis of hind limbs, respiratory difficulty, and death (Jurisson 1971). Fractions of an ethanol extract of shepherd's purse in mice showed "low" toxicity (Kuroda and Takagi 1969).

Short-Term Toxicity

In guinea pigs administered shepherd's purse as 20 or 40% of the diet, no effects on fertility were observed at the 20% level; at 40%, temporary infertility was observed in males and females, and a reduction in ovulation was observed. No estrogenic activity was demonstrated (East 1955).

Cytotoxicity

The therapeutic ratio of anti-HIV activity of an aqueous extract of shepherd's purse versus toxicity to MT2 lymphoblastoid cells was less than 1 (Abdel-Malek et al. 1996).

Literature Cited

Abdel-Malek, S., J.W. Bastien, W.F. Mahler, et al. 1996. Drug leads from the Kallawaya herbalists of Bolivia. 1. Background, rationale, protocol and anti-HIV activity. *J. Ethnopharmacol.* 50(3):157-166.

Bastien, J.W. 1983. Pharmacopeia of Qollahuaya Andeans. *J. Ethnopharmacol.* 8(1):97-111.

Casey, R.C. 1960. 298 alleged anti-fertility plants of India. *Indian J. Med. Sci.* 14:590-600.

Chadha, Y. 1988. *The wealth of India: A dictionary of Indian raw materials and industrial products.* Delhi: Council of Scientific and Industrial Research.

Cheeke, P.R., and L. Shull. 1985. *Natural toxicants in feeds and poisonous plants.* Westport, CT: AVI Publishing Company.

de Laszlo, H., and P.S. Henshaw. 1954. Plant materials used by primitive peoples to affect fertility. *Science* 119(3097):626-631.

Dioscorides, P. 512. *De materia medica: The Greek herbal of Dioscorides.* Translated by Goodyer, J. (1959 edition). New York: Hafner Pub. Co.

East, J. 1955. The effect of certain plant preparations on the fertility of laboratory mammals. 3. *Capsella bursa-pastoris* L. *J. Endocrinol.* 12(4):267-272.

Felter, H., and J. Lloyd. 1901. *Syllabus of Eclectic materia medica and therapeutics. Compiled from notes taken from the lectures of Frederick J. Locke, M.D.* Cincinnati, OH: Scudder Brothers Company.

Felter, H.W., and J.U. Lloyd. 1898. *King's American dispensatory.* 18th ed., 3rd rev. 2 vols. Cincinnati: Ohio Valley Co.

Jurisson, S. 1971. [Determination of active substances of *Capsella bursa-pastoris.*] *Tartu Riiliku Ulikooli Toim* 270:71-79.

Kuroda, K., and K. Takagi. 1968. Physiologically active substance in *Capsella bursa-pastoris. Nature* 220 (5168):707-708.

Kuroda, K., and K. Takagi. 1969. Studies on *Capsella bursa-pastoris.* I. General pharmacology of ethanol extract of the herb. *Arch. Int. Pharmacodyn. Ther.* 178(2):382-391.

Nishie, K., and M.E. Daxenbichler. 1980. Toxicology of glucosinolates, related compounds (nitriles, *R*-goitrin, isothiocyanates) and vitamin U found in Cruciferae. *Food Cosmet. Toxicol.* 18 (2):159-172.

Palma, L. 1964. *Le piante medicinali d'Italia.* Torino: SEI.

Panter, K.E., and L.F. James. 1990. Natural plant toxicants in milk: A review. *J. Anim. Sci.* 68(3):892-904.

Sharma, M.L., N. Chandokhe, B.J. Ghatak, et al. 1978. Pharmacological screening of Indian medicinal plants. *Indian J. Exp. Biol.* 16(2):228-240.

Shipochliev, T. 1981. [Uterotonic action of extracts from a group of medicinal plants.] *Vet. Med. Nauk.* 18 (4):94-98.

Stoewsand, G.S. 1995. Bioactive organosulfur phytochemicals in *Brassica oleracea* vegetables—A review. *Food Chem. Toxicol.* 33 (6):537-543.

Watt, J.M., and M.G. Breyer-Brandwijk. 1962. *The medicinal and poisonous plants of southern and eastern Africa.* 2nd ed. Edinburgh: E. & S. Livingstone.

Williamson, E.M. 2003. *Potter's herbal cyclopedia.* Saffron Walden, Essex: C.W. Daniel Co.

Capsicum spp. — Solanaceae

Capsicum annuum L. var. ***annuum***
SCN: cayenne
Syn: *Capsicum frutescens* L.
OCN: cayenne pepper; chili pepper; paprika; red pepper; tabasco pepper

***Capsicum annuum* L. var. *glabriusculum* (Dunal) Heiser & Pickersgill**
SCN: bird pepper
OCN: African bird pepper; piquin
Part: fruit

Quick Reference Summary

Safety Class: 1 (internal use), 2d (external use)
Interaction Class: A

Contraindications

External Use

Not for use in or near eyes (Agrawal et al. 1985; Fett 2003; Johnson 2007).

Other Precautions

Internal Use

In sensitive individuals, excessive doses may cause gastrointestinal irritation or heartburn, or exacerbate gastroesophageal reflux (Chadha 1988; Felter and Lloyd 1898; Martindale and Reynolds 1996; Milke et al. 2006; Myers et al. 1987; Viranuvatti et al. 1972; Watt and Breyer-Brandwijk 1962).

External Use

Cayenne preparations irritate the mucous membranes and injured or broken skin even at very low concentrations, and may cause a painful burning sensation (Fett 2003; Johnson 2007).

Drug and Supplement Interactions

Use of topical cayenne products may exacerbate cough caused by angiotensin-converting enzyme (ACE) inhibitors, although cayenne is not believed to affect the efficacy of these drugs (Stargrove et al. 2008; Yeo et al. 1994, 1995).

Editors' Note

The German Commission E suggests that *Capsicum* should not be applied externally for more than 2 days, with a 14-day time lapse between applications (Blumenthal et al. 2000). However, two systematic reviews of human studies of topically applied products containing the compound capsaicin (0.025–0.075%) as the active ingredient indicated that study periods were generally 4 to 8 weeks in duration (Mason et al. 2004; Zhang and Po 1994).

Adverse Events and Side Effects

Topical application of cayenne may produce a burning sensation, often accompanied by erythema (redness of the skin due to capillary congestion). Repeated topical use generally causes desensitization that is reversible on discontinuation of the drug (Hautkappe et al. 1998; Johnson 2007). Cayenne preparations irritate mucous membranes even at very low concentrations, and may cause a painful burning sensation (Fett 2003).

Ingestion of cayenne may cause or exacerbate gastric irritation, heartburn, or gastroesophageal reflux (GER) in sensitive individuals (Milke et al. 2006).

Pharmacological Considerations

One study in rats indicated that cayenne reduced the bioavailability of aspirin and salicylic acid (Cruz et al. 1999). A study in rabbits indicated that cayenne increased the plasma levels of orally but not intravenously administered theophylline (Bouraoui et al. 1995). The relevance of those data to human use is not known.

Pregnancy and Lactation

Information on the safety of cayenne use in pregnancy is limited. One animal study showed no adverse effects of cayenne after administration to pregnant rats for 5 days (Pellicer et al. 1996).

No information on use of cayenne during lactation was identified. While this review did not identify any concerns for use while nursing, safety has not been conclusively established.

Review Details

I. Drug and Supplement Interactions

Clinical Trials of Drug or Supplement Interactions

Exacerbation of ACE inhibitor-induced cough by capsaicin has been reported in two small studies in humans (Yeo et al. 1994, 1995).

Case Reports of Suspected Drug or Supplement Interactions

Topical application of a capsaicin-containing cream (0.075% capsaicin) induced cough in a woman taking an unspecified ACE inhibitor. The cough occurred just after the cream was applied (Hakas 1990).

Animal Trials of Drug or Supplement Interactions

Coadministration of aspirin or salicylic acid and cayenne extract (300 mg/kg, 10% capsaicin) orally to rats significantly reduced the bioavailability of salicylic acid and aspirin. Chronic oral administration of aspirin or salicylic acid and cayenne extract (100 or 300 mg/kg) resulted in undetectable plasma levels of aspirin and significantly reduced blood levels of salicylic acid (Cruz et al. 1999). The compound capsaicin increased the absorption of acetaminophen in rats (Metwally and Kandil 1985).

Oral coadministration of 5 ml/kg of a cayenne suspension (10% cayenne) and 20 mg/kg theophylline to rabbits significantly increased the plasma levels of theophylline (Bouraoui et al. 1988), whereas oral administration of cayenne (5 ml/kg of a 10% cayenne suspension) and intravenous administration of theophylline (12 mg/kg) did not affect the plasma levels of theophylline (Bouraoui et al. 1995). Dosing with cayenne and theophylline for 7 days in rabbits significantly reduced urinary excretion of the theophylline metabolite 1-methyluric acid (Bouraoui et al. 1995).

Administration of 0.1 µg/kg of the compound capsaicin to rats protected the stomach from mucosal damage by ethanol. Doses of 10 to 30 mg/kg capsaicin, however, aggravated the damage caused by ethanol (Salam et al. 1995).

II. Adverse Events

Adverse Events Reported in Clinical Trials

A systematic review of herbal medicines for lower back pain identified two trials of topical cayenne monopreparations and reported that adverse events in cayenne treatment groups were inflammatory contact eczema, urticaria, minute hemorrhagic spots, and vesiculation or dermatitis. In one trial, 15 adverse events were reported in the cayenne group and 9 in the placebo group (Gagnier et al. 2007). In a clinical trial of a topical capsicum plaster or placebo, localized adverse drug reactions were found in 7.5% of the patients on capsicum and 3.1% on placebo (Frerick et al. 2003).

A review of clinical trials of the compound capsaicin indicated that repeated topical applications of capsaicin to the skin produce an initial burning sensation followed by gradual desensitization that is reversible on discontinuation of the drug (Hautkappe et al. 1998).

A meta-analysis of topical capsaicin for the treatment of chronic pain, including 6 studies with a total of 656 patients, reported that 54% of patients using capsaicin had one or more local adverse events compared with 15% in the placebo group. Coughing was reported in 8% of patients using products with 0.075% capsaicin and none using products with 0.025% capsaicin. One active-controlled trial reported coughing in 1 of 32 patients using 0.025% capsaicin and in 7 using 0.25% capsaicin (Mason et al. 2004).

Case Reports of Adverse Events

Topical reactions to cayenne, including contact dermatitis and urticaria, have been reported (Burnett 1989; Feldman and Levy 2003).

C

III. Pharmacology and Pharmacokinetics

Human Pharmacological Studies

A review of the effects of cayenne on gastric ulcers indicated that large doses of chili pepper administered directly into the stomach of normal fasting subjects generally produced a mild response consisting of a small increase in gastric acid secretion and an increased blood flow (Ayad 1995). Intragastric administration of 3 g cayenne in healthy subjects with varying histories of cayenne consumption caused no detectable changes in 13 of the 20 subjects, mild to moderate edema or increased blood flow in 6 subjects, and a hemorrhage in 1 subject (Viranuvatti et al. 1972). An assessment of cayenne on human gastric mucosa in healthy volunteers indicated that administration of 0.1 to 1.5 g cayenne exhibited dose-dependent parietal secretion, pepsin secretion, potassium loss, and gastric cell exfoliation, and that the gastric effects were similar to those of aspirin (Myers et al. 1987).

Consumption of 20 g of cayenne provided a protective effect against gastrointestinal mucosal damage induced by aspirin (600 mg) in a study of healthy volunteers (Yeoh et al. 1995).

A study on the effects of chronic chili ingestion on gastroesophageal reflux (GER) symptoms indicated that daily ingestion of 3 g of chili (two different varieties of chilies, one with 0.048% capsaicin, the other with 0.088% capsaicin) induced GER and that the magnitude of the induced reflux was related to the kind of chili (Milke et al. 2006).

A trial of guajillo chili (a cultivar of *Capsicum annuum* L. var. *annuum*) in patients with irritable bowel syndrome (IBS) indicated that ingestion of 1 g of guajillo chili (0.112% capsaicin) did not change IBS symptoms but did induce upper abdominal discomfort and lowered the rectal pain threshold. The authors indicated that guajillo chili may induce rectal hyperalgesia (high sensitivity to pain) in persons with IBS (Schmulson et al. 2003).

No exacerbation of hemorrhoidal symptoms was observed after administration of 10 mg of ground cayenne pepper in patients with hemorrhoids (Altomare et al. 2006).

Ingestion of iron with a meal supplemented with 4.2 g of freeze-dried chili significantly inhibited iron absorption (Tuntipopipat et al. 2006).

A "sensory hyperreactivity" cough induced by inhalation of the compound capsaicin has been described (Millqvist 2000; Millqvist and Bende 2001; Millqvist et al. 2000). Cough was reported as an adverse event in a study of topical application of a capsaicin cream (0.075% capsaicin) (Ellison et al. 1997). The effects are believed to be due to aerosolization and subsequent inhalation of particles of dried cream (McKenna et al. 2002). Capsaicin activates the type 1 vanilloid receptor, a sensory receptor that stimulates coughing (Kissin 2008; Knotkova et al. 2008; Morice and Geppetti 2004; Nagy et al. 2004; Pingle et al. 2007; Szolcsanyi 2004).

Animal Pharmacological Studies

Topical application of cayenne tinctures at concentrations of 0.1 to 1% to rabbits indicated that such concentrations were nonirritant (Maruzen Pharmaceuticals 2002).

In Vitro Pharmacological Studies

The compound capsaicin inhibits the drug-metabolizing isoenzyme CYP2E1 (Surh and Sup Lee 1995).

In human ileocarcinoma cells, cayenne increased cell permeability (tight junction gap increase) (Jensen-Jarolim et al. 1998).

IV. Pregnancy and Lactation

A study on the administration of cayenne (2.75 mg capsaicin) daily for 5 days to rats during pregnancy indicated no adverse developmental effects on the fetus. Offspring of the rats showed a delayed response to acute heat stimuli. The authors of this study indicate that the compound capsaicin is capable of crossing the placental barrier (Pellicer et al. 1996).

No effects on parturition time, skeletal development, or weight of offspring was observed in rats subcutaneously administered 50 to 200 mg/kg of the compound capsaicin every other day on gestational days (GD) 7 through 15 (Perfumi and Sparapassi 1999). In rats subcutaneously administered 50 mg/kg capsaicin on GD 14, 16, 18, or 20, or on days 15 and 16, or 16 and 17, a reduced crown-rump length of fetuses was observed only in dams injected on day 18, and a decrease in fetal spontaneous activity and loss of fetal responsiveness to morphine were observed in offspring of mothers treated on days 16 and 17, but not on days 15 and 16. No skeletal or soft tissue malformations, differences in fetal weight or average number of fetuses, or incidences of resorption were observed (Kirby et al. 1982). Capsaicin was shown to cross the placenta in rats subcutaneously administered the compound at doses of 20 to 50 mg/kg (Atkinson and Chaggar 1983).

No information on the safety of cayenne during lactation was identified.

V. Toxicity Studies

Acute Toxicity

The LD_{50} of cayenne extract orally administered to rats is 23.58 ml/kg (Winek et al. 1982). The LD_{50} values of the compound capsaicin in mice via various routes of administration are 190 mg/kg (oral), 7.65 mg/kg (intraperitoneal), and 0.56 mg/kg (intravenous) (Glinsukon et al. 1980).

Short-Term Toxicity

In mice fed a diet of up to 10% cayenne for 4 weeks, only minor effects of cayenne were observed, including a slight glycogen depletion and anisocytosis of hepatocytes. General health, body weight, and food intake were apparently not affected, and no treatment-related changes were observed in any of the tissues examined (Jang et al. 1992).

Subchronic Toxicity

In rats intragastrically administered 0.5 g/kg cayenne powder daily for 60 days, some reduction in growth rate was observed as compared to control but no significant differences from control were observed in rectal temperature, water intake, plasma chemistry, urine dilution and concentration, or relative organ weights (Monsereenusorn 1983).

No significant adverse effects were observed in rats fed a diet of 2% cayenne for 8 weeks. In rats fed a diet of 10%, however, feed intake and growth rate were depressed and exfoliation of the intestinal epithelium into the lumen, cytoplasmic fatty vacuolation, and necrosis of the centrilobular hepatocytes were observed (al-Qarawi and Adam 1999).

Chronic Toxicity

Chronic administration of 20 µl cayenne extract (2.5 mg/ml capsaicin) to the cheek pouch of hamsters induced shrunken eyeballs and closing of the eyelids in 23% of the animals. This effect was not observed in hamsters that received a single application of the potent carcinogen methyl(acetoxymethyl) nitrosamine prior to repeated treatment with the cayenne extract (Agrawal et al. 1985).

In rabbits administered 5 mg/kg of cayenne daily for 12 months, spleen and liver weights were increased as compared to control animals. Microscopic observation of these organs revealed effects at the cellular level (Lee 1963).

Genotoxicity

Mutagenic activity of chilies (species unspecified) and the compound capsaicin was observed in the Ames mutagenicity test with *Salmonella typhimurium* with metabolic activation from S9 mix (Nagabhushan and Bhide 1986).

Cayenne extract and the compound capsaicin exhibited mutagenic activity in *Salmonella typhimurium* strains TA98, TA100, TA1535, TA1537, and TA1538 with metabolic activation. Capsaicin had more potent mutagenic effects than cayenne. Mutagenic activity of the same products was not observed in the micronucleus test or in the 8-azaguanine-resistant mutagenesis assay with Chinese hamster cells. Capsaicin inhibited DNA synthesis in the testes of Swiss mice injected at two dose levels (Nagabhushan and Bhide 1985).

Carcinogenicity

A case-control study of men in India with cancers of the oral cavity indicated that use of red chili powder was associated, in a dose-related manner, with an increased risk of cancers. Tea drinking was also observed to be a risk factor for esophageal cancers and, to a lesser extent, for pharyngeal cancers (Notani and Jayant 1987). Similarly, a case-control study of men and women in Mexico indicated that a dose-related trend of increasing risk of gastric cancer was reported with consumption of chili peppers (López-Carnllo et al. 1994). Conversely, a case-control study of men and women in Italy indicated that consumption of chili peppers was associated with a decreased risk of gastric cancer (Buiatti et al. 1989).

In rats administered cayenne as 10% of the diet for 2 years, neoplastic changes in the liver were observed after 6 months of treatment, with males more prone to tumor development. In female rats, cystic cholangiomas, solid cholangiomas, adenocarcinomas, and hepatomas were observed (Hoch-Ligeti 1951).

Rats fed 4 g/kg cayenne as part of the diet for 12 months were found to have a 35% incidence of adenocarcinoma of the abdomen, with no adenocarcinoma observed in the control group (Balachandran and Sivaramkrishnan 1995). A dose of 80 mg/kg daily cayenne for 30 days was associated with an 83% incidence of intestinal tumors or polyps, a 90% incidence of tumors or polyps in the positive control group treated with the carcinogen 1,2-dimethylhydrazine (DMH), and no such abnormalities in the untreated control group (Nalini et al. 1997). In rats administered cayenne alone, DMH alone, or a combination of the two, incidences of intestinal and colon tumors were 90% in the cayenne group, 93% in the DMH group, and 96% in the combination group. The authors of the study concluded that cayenne promoted the carcinogenesis of DMH (Chitra et al. 1997).

A review of studies on the anticarcinogenic and carcinogenic effects of the compound capsaicin indicated that small amounts of capsaicin had few or no deleterious effects, while heavy ingestion of the compound has been associated with necrosis, ulceration, and carcinogenesis. Capsaicin has also been shown to alter the metabolism of certain carcinogens, acting as a chemopreventive compound (Surh and Lee 1996).

Cytotoxicity

The compounds capsaicin and resiniferatoxin (both vanilloids) inhibit the NADH-electron transport system, activate c-Jun kinase (JNK) but not AP1, and induce apoptosis in transformed cells (Macho et al. 1998). Capsaicin, dihydrocapsaicin, and resiniferatoxin inhibit NADH oxidase and induce apoptosis (Vaillant et al. 1996; Wolvetang et al. 1996).

Literature Cited

Agrawal, R.C., A.V. Sarode, V.S. Lalitha, and S.V. Bhide. 1985. Chilli extract treatment and induction of eye lesions in hamsters. *Toxicol. Lett.* 28(1):1-7.

al-Qarawi, A.A., and S.E. Adam. 1999. Effects of red chilli (*Capsicum frutescens* L.) on rats. *Vet. Hum. Toxicol.* 41(5):293-295.

Altomare, D.F., M. Rinaldi, F. La Torre, et al. 2006. Red hot chili pepper and hemorrhoids: The explosion of a myth: Results of a prospective, randomized, placebo-controlled, crossover trial. *Dis. Colon Rectum* 49 (7):1018-1023.

C

Atkinson, M.E., and J.S. Chaggar. 1983. The effects of prenatal capsaicin on the distribution of substance P in developing primary afferent neurons. *Neurosci. Lett.* 35 (1):25-29.

Ayad, M. 1995. Do chili peppers cause ulcers? A burning question. *Nutr. Bytes* 1(1):1-3.

Balachandran, B., and V.M. Sivaramkrishnan. 1995. Induction of tumours by Indian dietary constituents. *Indian J. Cancer* 32 (3):104-109.

Blumenthal, M., A. Goldberg, and J. Brinckmann. 2000. *Herbal medicine: Expanded Commission E monographs*. Newton, MA: Integrative Medicine.

Bouraoui, A., J.L. Brazier, H. Zouaghi, and M. Rousseau. 1995. Theophylline pharmacokinetics and metabolism in rabbits following single and repeated administration of capsicum fruit. *Eur. J. Drug Metab. Pharmacokinet.* 20(3):173-178.

Bouraoui, A., A. Toumi, H. Ben Mustapha, and J.L. Brazier. 1988. Effects of capsicum fruit on theophylline absorption and bioavailability in rabbits. *Drug Nutr. Interact.* 5(4):345-350.

Buiatti, E., D. Palli, A. DeCarli, et al. 1989. A case-control study of gastric cancer and diet in Italy. *Int. J. Cancer* 44 (4):611-616.

Burnett, J.W. 1989. Capsicum pepper dermatitis. *Cutis* 43 (6):534.

Chadha, Y. 1988. *The wealth of India: A dictionary of Indian raw materials and industrial products*. Delhi: Council of Scientific and Industrial Research.

Chitra, S., P. Viswanathan, N. Nalini, K. Sabitha, and V.P. Menon. 1997. Role of red chilli (capsaicin) in the formation of colonic carcinoma. *Indian J. Pathol. Microbiol.* 40 (1):21-25.

Cruz, L., G. Castaneda-Hernandez, and A. Navarrete. 1999. Ingestion of chilli pepper (*Capsicum annuum*) reduces salicylate bioavailability after oral aspirin administration in the rat. *Can. J. Physiol. Pharmacol.* 77 (6):441-446.

Ellison, N., C.L. Loprinzi, J. Kugler, et al. 1997. Phase III placebo-controlled trial of capsaicin cream in the management of surgical neuropathic pain in cancer patients. *J. Clin. Oncol.* 15(8):2974.

Feldman, H., and P.D. Levy. 2003. Hot pepper-induced urticaria while repairing a digital laceration. *Am. J. Emerg. Med.* 21(2):159.

Felter, H.W., and J.U. Lloyd. 1898. *King's American dispensatory*. 18th ed., 3rd rev. 2 vols. Cincinnati: Ohio Valley Co.

Fett, D.D. 2003. Botanical briefs: Capsicum peppers. *Cutis* 72 (1):21-23.

Frerick, H., W. Keitel, U. Kuhn, et al. 2003. Topical treatment of chronic low back pain with a capsicum plaster. *Pain* 106 (1-2):59-64.

Gagnier, J.J., M.W. van Tulder, B. Berman, and C. Bombardier. 2007. Herbal medicine for low back pain: A Cochrane review. *Spine* 32 (1):82-92.

Glinsukon, T., V. Stitmunnaithum, C. Toskulkao, T. Buranawuti, and V. Tangkrisanavinont. 1980. Acute toxicity of capsaicin in several animal species. *Toxicon* 18(2):215-220.

Hakas, J.F., Jr. 1990. Topical capsaicin induces cough in patient receiving ACE inhibitor. *Ann. Allergy* 65(4):322-323.

Hautkappe, M., M.F. Roizen, A. Toledano, et al. 1998. Review of the effectiveness of capsaicin for painful cutaneous disorders and neural dysfunction. *Clin. J. Pain* 14 (2):97.

Hoch-Ligeti, C. 1951. Production of liver tumours by dietary means; effect of feeding chilies [*Capsicum frutescens* and *annuum* (Linn.)] to rats. *Acta Unio Int. Contra Cancrum* 7(3):606-611.

Jang, J.J., D.E. Devor, D.L. Logsdon, and J.M. Ward. 1992. A 4-week feeding study of ground red chilli (*Capsicum annuum*) in male B6C3F1 mice. *Food Chem. Toxicol.* 30 (9):783-787.

Jensen-Jarolim, E., L. Gajdzik, I. Haberl, et al. 1998. Hot spices influence permeability of human intestinal epithelial monolayers. *J. Nutr.* 128(3):577-581.

Johnson, W. 2007. Final report on the safety assessment of *Capsicum annuum* extract, *Capsicum annuum* fruit extract, *Capsicum annuum* resin, *Capsicum annuum* fruit powder, *Capsicum frutescens* fruit, *Capsicum frutescens* fruit extract, *Capsicum frutescens* resin, and capsaicin. *Int. J. Toxicol.* 26(Suppl. 1):3-106.

Kirby, M.L., T.F. Gale, and T.G. Mattio. 1982. Effects of prenatal capsaicin treatment on fetal spontaneous activity, opiate receptor binding, and acid phosphatase in the spinal cord. *Exp. Neurol.* 76(2):298-308.

Kissin, I. 2008. Vanilloid-induced conduction analgesia: Selective, dose-dependent, long-lasting, with a low level of potential neurotoxicity. *Anesth. Analg.* 107 (1):271-281.

Knotkova, H., M. Pappagallo, and A. Szallasi. 2008. Capsaicin (TRPV1 agonist) therapy for pain relief: Farewell or revival? *Clin. J. Pain* 24(2):142-154.

Lee, S. 1963. Studies on the influence of diets and lipotropic substances upon the various organs and metabolic changes in rabbits on long term feeding with red pepper. *Cited in* Johnson, W. 2007. Final report on the safety assessment of *Capsicum annuum* extract, *Capsicum annuum* fruit extract, *Capsicum annuum* resin, *Capsicum annuum* fruit powder, *Capsicum frutescens* fruit, *Capsicum frutescens* fruit extract, *Capsicum frutescens* resin, and capsaicin. *Int J. Toxicol.* 26(Suppl. 1):3-106.

López-Carillo, L., M.H. Avila, and R. Dubrow. 1994. Chili pepper consumption and gastric cancer in Mexico: A case-control study. *Am. J. Epidemiol.* 139(3):263-271.

Macho, A., M.V. Blazquez, P. Navas, and E. Munoz. 1998. Induction of apoptosis by vanilloid compounds does not require de novo gene transcription and activator protein 1 activity. *Cell Growth Differ.* 9(3):277-286.

Martindale, W., and J.E.F. Reynolds. 1996. *The extra pharmacopoeia*. 31st ed. London: Pharmaceutical Press.

Maruzen Pharmaceuticals. 2002. Data on *Capsicum annuum* fruit extract: UV absorption, concentration of use, and skin irritation. *Cited in* Johnson, W. 2007. Final report on the safety assessment of *Capsicum annuum* extract, *Capsicum annuum* fruit extract, *Capsicum annuum* resin, *Capsicum annuum* fruit powder, *Capsicum frutescens* fruit, *Capsicum frutescens* fruit extract, *Capsicum frutescens* resin, and capsaicin. *Int. J. Toxicol.* 26(Suppl. 1):3-106.

Mason, L., R.A. Moore, S. Derry, J.E. Edwards, and H.J. McQuay. 2004. Systematic review of topical capsaicin for the treatment of chronic pain. *Br. Med. J.* 238:991-994.

McKenna, D., K. Jones, and K. Hughes. 2002. *Botanical medicines: The desk reference for major herbal supplements*. New York: Haworth Press.

Metwally, S.A., and A.M. Kandil. 1985. Effect of capsaicin on gastrointestinal absorption of acetaminophen. *J. Drug Res.* 16(1-2):1-6.

Milke, P., A. Diaz, M.A. Valdovinos, and S. Moran. 2006. Gastroesophageal reflux in healthy subjects induced by two different species of chilli (*Capsicum annum*). *Dig. Dis.* 24(1-2):184-188.

Millqvist, E. 2000. Cough provocation with capsaicin is an objective way to test sensory hyperreactivity in patients with asthma-like symptoms. *Allergy* 55(6):546-550.

Millqvist, E., and M. Bende. 2001. Capsaicin cough sensitivity is decreased in smokers. *Resp. Med.* 95(1):19-21.

Millqvist, E., O. Lowhagen, and M. Bende. 2000. Quality of life and capsaicin sensitivity in patients with sensory airway hyperreactivity. *Allergy* 55(6):540-545.

Monsereenusorn, Y. 1983. Subchronic toxicity studies of capsaicin and capsicum in rats. *Res. Commun. Chem. Pathol. Pharmacol.* 41(1):95-110.

Morice, A.H., and P. Geppetti. 2004. Cough 5: The type 1 vanilloid receptor: A sensory receptor for cough. *Thorax* 59(3):257.

Myers, B.M., J.L. Smith, and D.Y. Graham. 1987. Effect of red pepper and black pepper on the stomach. *Am. J. Gastroenterol.* 82(3):211-214.

Nagabhushan, M., and S.V. Bhide. 1985. Mutagenicity of chili extract and capsaicin in short-term tests. *Env. Mutagen.* 7(6):881-888.

Nagabhushan, M., and S.V. Bhide. 1986. Nonmutagenicity of curcumin and its antimutagenic action versus chili and capsaicin. *Nutr. Cancer* 8 (3):201-210.

Nagy, I., P. Santha, G. Jancso, and L. Urban. 2004. The role of the vanilloid (capsaicin) receptor (TRPV1) in physiology and pathology. *Eur. J. Pharmacol.* 500(1-3):351-369.

Nalini, N., K. Sabitha, S. Chitra, P. Viswanathan, and V.P. Menon. 1997. Histopathological and lipid changes in experimental colon cancer: Effect of coconut kernal (*Cocos nucifera* Linn.) and (*Capsicum annum* Linn.) red chilli powder. *Indian J. Exp. Biol.* 35(9):964-971.

Notani, P.N., and K. Jayant. 1987. Role of diet in upper aerodigestive tract cancers. *Nutr. Cancer* 10(1-2):103-113.

Pellicer, F., O. Picazo, B. Gomez-Tagle, and O.I. de la Roldan. 1996. Capsaicin or feeding with red peppers during gestation changes the thermonociceptive response of rat offspring. *Physiol. Behav.* 60(2):435-438.

Perfumi, M., and L. Sparapassi. 1999. Rat offspring treated prenatally with capsaicin do not show some of the irreversible effects induced by neonatal treatment with neurotoxin. *Pharmacol. Toxicol.* 84(2):66.

Pingle, S.C., J.A. Matta, and G.P. Ahern. 2007. Capsaicin receptor: TRPV1 a promiscuous TRP channel. *Handb. Exp. Pharmacol.* 179:155-171.

Salam, O.M.E.A., G. Mozsik, and J. Szolcsanyi. 1995. Studies on the effect of intragastric capsaicin on gastric ulcer and on the prostacyclin-induced cytoprotection in rats. *Pharmacol. Res.* 32(4):209-215.

Schmulson, M.J., M.A. Valdovinos, and P. Milke. 2003. Chili pepper and rectal hyperalgesia in irritable bowel syndrome. *Am. J. Gastroenterol.* 98(5):1214-1215.

Stargrove, M., J. Treasure, and D. McKee. 2008. *Herb, nutrient, and drug interactions: Clinical implications and therapeutic solutions.* St. Loius, MO: Elsevier.

Surh, Y.J., and S.S. Lee. 1996. Capsaicin in hot chili pepper: Carcinogen, co-carcinogen or anticarcinogen? *Food Chem. Toxicol.* 34(3):313-316.

Surh, Y.J., and S. Sup Lee. 1995. Capsaicin, a double-edged sword: Toxicity, metabolism, and chemopreventive potential. *Life Sci.* 56(22):1845-1855.

Szolcsanyi, J. 2004. Forty years in capsaicin research for sensory pharmacology and physiology. *Neuropeptides* 38(6):377-384.

Tuntipopipat, S., K. Judprasong, C. Zeder, et al. 2006. Chili, but not turmeric, inhibits iron absorption in young women from an iron-fortified composite meal. *J. Nutr.* 136 (12):2970-2974.

Vaillant, F., J.A. Larm, G.L. McMullen, E.J. Wolvetang, and A. Lawen. 1996. Effectors of the mammalian plasma membrane NADH-oxidoreductase system. Short-chain ubiquinone analogues as potent stimulators. *J. Bioenerg. Biomembr.* 28(6):531-540.

Viranuvatti, V., C. Kalayasiri, O. Chearani, and U. Plengvanit. 1972. Effects of capsicum solution on human gastric mucosa as observed gastroscopically. *Am. J. Gastroenterol.* 58(3):225-232.

Watt, J.M., and M.G. Breyer-Brandwijk. 1962. *The medicinal and poisonous plants of southern and eastern Africa.* 2nd ed. Edinburgh: E. & S. Livingstone.

Winek, C.L., D.C. Markie, and S.P. Shanor. 1982. Pepper sauce toxicity. *Drug Chem. Toxicol.* 5(2):89-113.

Wolvetang, E.J., J.A. Larm, P. Moutsoulas, and A. Lawen. 1996. Apoptosis induced by inhibitors of the plasma membrane NADH-oxidase involves Bcl-2 and calcineurin. *Cell Growth Differ.* 7(10):1315-1325.

Yeo, W.W., I. Chadwick, M. Kraskiewicz, P. Jackson, and L. Ramsay. 1995. Resolution of ACE inhibitor cough: Changes in subjective cough and responses to inhaled capsaicin, intradermal bradykinin and substance-P. *Br. J. Clin. Pharmacol.* 40(5):423-429.

Yeo, W.W., K.S. Higgins, G. Foster, P.R. Jackson, and L.E. Ramsay. 1994. Effect of dose adjustment on enalapril-induced cough and the response to inhaled capsaicin. *J. Hyperten.* 12(11):1311.

Yeoh, K.G., J.Y. Kang, I. Yap, et al. 1995. Chili protects against aspirin-induced gastroduodenal mucosal injury in humans. *Dig. Dis. Sci.* 40(3):580-583.

Zhang, W., and L. Po. 1994. The effectiveness of topically applied capsaicin *Eur. J. Clin. Pharmacol.* 46(6):517-522.

Carica papaya L. — Caricaceae

SCN: papaya

Part: leaf

Quick Reference Summary

Safety Class: 2b
Interaction Class: A

Contraindications

Not for use in pregnancy except under the supervision of a qualified healthcare practitioner (Adebiyi et al. 2002).

Other Precautions

None known.

Drug and Supplement Interactions

None known.

Adverse Events and Side Effects

None known.

Pharmacological Considerations

None known.

Pregnancy and Lactation

Animal studies on the compound papain provide conflicting information on the safety of this compound in pregnancy, with some studies demonstrating no adverse effects and other studies indicating an increase in fetal losses and other adverse effects on fetuses (Adebiyi et al. 2002; Devi and Singh 1978, 1979; Schmidt 1995; Singh and Devi 1976). Unripe papaya fruit is traditionally avoided by pregnant women in many Asian countries (Adebiyi et al. 2002). Different parts of the papaya plant have been used to produce abortion and to induce labor (Adebiyi et al. 2002; Kamatenesi-Mugisha and Oryem-Origa 2007).

No information on the safety of papaya leaf during lactation was identified. While this review did not identify any concerns for use while nursing, safety has not been conclusively established.

Review Details

I. Drug and Supplement Interactions

Clinical Trials of Drug or Supplement Interactions

No clinical trials of drug or supplement interactions were identified.

Case Reports of Suspected Drug or Supplement Interactions

No case reports of suspected drug or supplement interactions were identified.

Animal Trials of Drug or Supplement Interactions

No animal trials of drug or supplement interactions were identified.

II. Adverse Events

Case Reports of Adverse Events

Testing of serum samples from 136 patients with well-documented, clinically relevant, immediate-type hypersensitivity against latex proteins ("latex fruit allergy") for IgE antibodies against a panel of different fruits indicated that 50% of test subjects had positive tests to papaya (Brehler et al. 1997). Sensitization to latex-containing fruits such as papaya, avocado, banana, fig, and kiwi has been reported in persons allergic to latex (Diaz-Perales et al. 1999; Marin et al. 2002).

Cases of occupational sensitization to the compound papain have been reported in a number of persons working with this compound. Sensitization has been confirmed by skin prick testing, detection of specific IgE antibodies, and specific bronchoprovocation tests (Van Kampen et al. 2005).

III. Pharmacology and Pharmacokinetics

Human Pharmacological Studies

No relevant human pharmacological studies were identified.

Animal Pharmacological Studies

No relevant animal pharmacological studies were identified.

In Vitro Pharmacological Studies

No relevant in vitro pharmacological studies were identified.

IV. Pregnancy and Lactation

In rats orally administered 800 mg/kg of the compound papain daily on gestational days 0 to 6 or 6 to 15, a reduction in preimplantation losses was observed along with a lack of fetal abnormalities (Schmidt 1995).

Gross and histological examination indicated changes in the placenta of rats orally or intraperitoneally administered the compound papain on gestational days 8 to 17 (Devi and Singh 1978).

Intraperitoneal administration of the compound papain to pregnant rabbits on gestational days 10 to 20 was associated with fetal stunting, visceral and subcutaneous hemorrhages, and edema of various organs (Devi and Singh 1979; Singh and Devi 1976).

A slight increase in postimplantation losses was observed in rats fed diets containing papain on gestational days 4 to 8 (Gopalakrishnan and Rajasekharasetty 1978).

An ethnobotanical survey indicated that papaya leaf, fruit, and root are used in Uganda to induce labor (Kamatenesi-Mugisha and Oryem-Origa 2007). In isolated rat uteruses, papaya latex and the compounds papain and chymopapain induced spasmodic contraction of the uterine muscles (Adaikan and Adebiyi 2004; Adebiyi et al. 2002).

No information on the safety of papaya leaf during lactation was identified.

V. Toxicity Studies

No toxicity studies were identified.

Literature Cited

Adaikan, P.G., and A. Adebiyi. 2004. Mechanisms of the oxytocic activity of papaya proteinases. *Pharm. Biol.* 42(8):646-655.

Adebiyi, A., P.G. Adaikan, and R.N. Prasad. 2002. Papaya (*Carica papaya*) consumption is unsafe in pregnancy: Fact or fable? Scientific evaluation of a common belief in some parts of Asia using a rat model. *Br. J. Nutr.* 88(2):199-203.

Brehler, R., U. Theissen, C. Mohr, and T. Luger. 1997. "Latex-fruit syndrome": Frequency of cross-reacting IgE antibodies. *Allergy* 52(4):404-410.

Devi, S., and S. Singh. 1978. Changes in placenta of rat fetuses induced by maternal administration of papain. *Indian J. Exp. Biol.* 16(12):1256-1260.

Devi, S., and S. Singh. 1979. Teratogenic effect of papain in rabbit fetuses. *J. Anat. Soc. India* 28(1):6-10.

Diaz-Perales, A., C. Collada, C. Blanco, et al. 1999. Cross-reactions in the latex-fruit syndrome: A relevant role of chitinases but not of complex asparagine-linked glycans. *J. Allergy Clin. Immunol.* 104(3. Pt. 1):681-687.

Gopalakrishnan, M., and M.R. Rajasekharasetty. 1978. Effect of papaya (*Carica papaya* Linn.) on pregnancy and estrous cycle in albino rats of Wistar strain. *Indian J. Physiol. Pharmacol.* 22(1):66-70.

Kamatenesi-Mugisha, M., and H. Oryem-Origa. 2007. Medicinal plants used to induce labour during childbirth in western Uganda. *J. Ethnopharmacol.* 109(1):1-9.

Marin, F.A., S.P.D.B.A. Peres, and A. Zuliani. 2002. Latex-fruit allergy. *Rev. Nutr.* 15(1):95-103.

Schmidt, H. 1995. Effect of papain on different phases of prenatal ontogenesis in rats. *Reprod. Toxicol.* 9(1):49-55.

Singh, S., and S. Devi. 1976. Lethality and teratogenicity of papain in rabbit fetuses. *J. Anat. Soc. India* 25(1):31-32.

Van Kampen, V., R. Merget, and T. Bruning. 2005. Occupational allergies to papain. *Pneumologie* 59 (6):405-410.

Carthamus tinctorius L. — Asteraceae

SCN: safflower
AN: *kusumbha*
PN: *hong hua* (flower)
Part: flower

Quick Reference Summary

Safety Class: 2b, 2d
Interaction Class: B

Contraindications

Not for use in pregnancy except under the supervision of a qualified healthcare practitioner (Bahmanfour and Javidnia 2003; Bahmanpour et al. 2003; Bensky et al. 2004; Chadha 1988; Chen and Chen 2004; Leung and Foster 1996; List and Hörhammer 1973; Tang and Eisenbrand 1992).

Not for use in patients with bleeding disorders, hemorrhagic diseases, or peptic ulcers (Bensky et al. 2004; Chang and But 1986; Chen and Chen 2004).

Other Precautions

None known.

Drug and Supplement Interactions

Concomitant use with anticoagulant medications should be under the supervision of a qualified healthcare practitioner (Chen and Chen 2004).

Notice

Abortifacient (List and Hörhammer 1973); *see* Appendix 2.

Uterine stimulant (Bensky et al. 2004; Chen and Chen 2004); *see* Appendix 2.

Editors' Note

The oil extracted from safflower and commonly used in cooking is distinct from the flower that is the subject of this entry. Concerns and considerations for the flower do not apply to the seed oil.

Adverse Events and Side Effects

Allergic reactions, including anaphylactic reactions, to safflower have been reported (Bensky et al. 2004).

Pharmacological Considerations

Safflower, and the food colorant safflower yellow, have been shown to have antiplatelet activity in animal and in vitro studies and may prolong blood coagulation time (Chen and Chen 2004; Zang et al. 2002; Zhu 1998).

Pregnancy and Lactation

Reference texts on traditional Chinese medicine indicate that safflower should not be used in pregnancy (Bensky et al. 2004; Chen and Chen 2004). Animal studies have shown that safflower has stimulating effects on the uterus and increases muscular contractions (Chen and Chen 2004; Shi et al. 1995; Wang 1983). Adverse effects on fetuses and embryos have been observed in pregnant rats administered safflower extracts (Bahmanfour and Javidnia 2003; Bahmanpour et al. 2003; Nobakht et al. 2000).

No information on the safety of safflower during lactation was identified. While this review did not identify any concerns for use while nursing, safety has not been conclusively established.

C

Review Details

I. Drug and Supplement Interactions

Clinical Trials of Drug or Supplement Interactions

No clinical trials of drug or supplement interactions were identified.

Case Reports of Suspected Drug or Supplement Interactions

No case reports of suspected drug or supplement interactions were identified.

Animal Trials of Drug or Supplement Interactions

No animal trials of drug or supplement interactions were identified.

II. Adverse Events

Case Reports of Adverse Events

Overdose (standard dose listed as 3–10 g in decoction) of safflower has been associated with bleeding, nausea, headache, dilated pupils, and an increase in intraocular pressure (Bensky et al. 2004; Chen and Chen 2004).

Allergic reactions, including anaphylactic reactions, to safflower have been reported (Bensky et al. 2004). Occupational asthma was reported in a worker that routinely handled dried safflower flowers (Compes et al. 2006).

III. Pharmacology and Pharmacokinetics

Human Pharmacological Studies

No relevant human pharmacological studies were identified.

Animal Pharmacological Studies

In dogs intravenously administered 10 mg/kg aqueous extract of safflower, a 58% increase in coronary blood flow was observed. The alcohol extract, however, had no effect when administered at doses of 10 to 30 mg/kg (Wang 1983).

In dogs, an alcohol extract of safflower prolonged clotting time and serum thrombin time, and reduced serum prothrombin time. The compound carthamin inhibited ADP-induced platelet aggregation in rabbits. Doses used in these studies were not reported in the available English language translations (Zhu 1998).

In rats administered an extract of safflower, the extract was 73% effective in preventing the formation of thrombi. Dose, duration, and route of administration were not listed in the available English language translation (Chen and Chen 2004).

In Vitro Pharmacological Studies

Alcoholic and aqueous extracts of safflower inhibited ADP- and collagen-induced aggregation of rabbit and rat platelets (Zhu 1998). The food colorant safflower yellow inhibited ADP-induced platelet aggregation in rabbit blood platelets and prolonged the rabbit plasma prothrombin time (Zang et al. 2002).

An increase in the proliferation of estrogen receptor-positive human breast cancer cells (MCF-7) was observed in cells treated with serum from mice that had been administered a safflower extract. In cells treated with serum from animals administered diethylstilbestrol or diethylstilbestrol and safflower, the cells treated with the combination showed less proliferation than those administered diethylstilbestrol alone (Zhao et al. 2007).

In mice intraperitoneally administered 50 to 450 mg/kg of the food colorant safflower yellow daily for 8 days, a decrease in serum lysozyme concentration and phagocytosing functions of both peritoneal macrophages and peripheral leukocytes was observed. The treatment also diminished the production of plaque-forming cells and specific rosette-forming cells, and inhibited the delayed-type hypersensitivity reaction and the activation of T suppressor cells elicited by supraoptimal immunization (Lu et al. 1991).

IV. Pregnancy and Lactation

In mice intraperitoneally administered 1, 10, 25, or 50 mg/kg aqueous extract of safflower on days 7 or 8 of pregnancy as single dose and days 9 and 10 as multidose, congenital malformations were observed. Malformations included exencephaly, spina bifida, tail and limb necrosis, and malformation of the eye and eyelid (Bahmanfour and Javidnia 2003; Bahmanpour et al. 2003).

In mice orally administered 0.2, 0.8, 1.2, 1.6, or 2.0 mg/kg safflower aqueous extract daily on days 0 through 8 of gestation, resorption of embryos was observed in the animals receiving 1.6 and 2 mg/kg daily. In animals receiving 1.2 mg/kg daily, changes in external, internal, and longitudinal diameters as well as cellular orientation and degeneration were observed (Nobakht et al. 2000).

An extract of safflower was found to have a stimulating effect on the H1-receptor and α-adrenergic receptor in isolated rat uteri (Shi et al. 1995). A translation of the Chinese scientific literature indicates that several animal studies have shown that safflower has a stimulating effect on the smooth muscle of the uterus and increases muscle contraction, with the effect being more pronounced in pregnant uteri (Chen and Chen 2004; Wang 1983).

No information on the safety of safflower during lactation was identified.

V. Toxicity Studies

Acute Toxicity

The LD_{50} of the compound carthamin orally administered to mice could not be determined at doses up to 8 g/kg (Zhu 1998). The LD_{50} of the food colorant safflower yellow intraperitoneally administered to mice was 7.54 g/kg (Zang et al. 2002).

Subchronic Toxicity

No abnormalities were observed in the liver, kidney, heart, stomach, or intestines of rats administered up to 1.5 g/kg safflower extract daily for 90 days. Route of administration was not specified in the available English language translation (Chen and Chen 2004).

In rats intraperitoneally administered 20, 60, or 180 mg/kg of the food colorant safflower yellow daily for 90 days, a prolonged blood coagulation time without changes in the normal blood coagulation process was observed in animals administered 60 or 180 mg/kg. The prolonged blood coagulation time was recovered to a normal level on the 28th day after withdrawing the test product. At the 180 mg/kg dose, some kidney injury was observed along with increases in the liver index without obvious pathological changes in liver histology. No adverse effects were observed in other organs (Liu et al. 2004). Conversely, intraperitoneal administration of safflower extract has been shown to reduce renal dysfunction and injury in rats with induced renal ischemia and reperfusion (Gao et al. 2006).

Genotoxicity

No significant mutagenic activity of an aqueous safflower extract was observed in the Ames test in *Salmonella typhimurium* strains TA98 or TA100, nor in micronucleus or chromosomal aberration assays in mice (Yin et al. 1991).

Literature Cited

Bahmanfour, S., and K. Javidnia. 2003. Ocular abnormalities due to toxic effects of the *Carthamus* tincture in the mouse embryo (abstract). *Toxicol. Lett.* 144(Suppl. 1):S114.

Bahmanpour, S., K. Javidnia, and H. Arandi. 2003. Weight and crown-rump length reduction, gross malformation and pregnancy outcome in *Carthamus tinctorius* L.-treated mice. *Arch. Iranian Med.* 6(2):117-120.

Bensky, D., S. Clavey, and E. Stöger. 2004. *Chinese herbal medicine: Materia medica.* 3rd ed. Seattle: Eastland Press.

Chadha, Y. 1988. *The wealth of India: A dictionary of Indian raw materials and industrial products.* Delhi: Council of Scientific and Industrial Research.

Chang, H.-M., and P.P.H. But. 1986. *Pharmacology and applications of Chinese materia medica.* English ed. Singapore, Philadelphia: World Scientific.

Chen, J.K., and T.T. Chen. 2004. *Chinese medical herbology and pharmacology.* City of Industry, CA: Art of Medicine Press.

Compes, E., B. Bartolome, M. Fernandez-Nieto, J. Sastre, and J. Cuesta. 2006. Occupational asthma from dried flowers of *Carthamus tinctorius* (safflower) and *Achillea millefolium* (yarrow). *Allergy* 61(10):1239-1240.

Gao, F., X.H. Wu, C.L. Luo, et al. 2006. Effect of saffor (*Carthamus tinctorius*) injection on renal ischemia/reperfusion injury in rats. *Zhongguo Zhong Yao Za Zhi* 31(21):1814-1818.

Leung, A.Y., and S. Foster. 1996. *Encyclopedia of common natural ingredients used in food, drugs, and cosmetics.* 2nd ed. New York: Wiley.

List, P.H., and H. Hörhammer. 1973. *Hagers handbuch der pharmazeutischen praxis.* Berlin: Springer.

Liu, Z., C. Li, M. Li, D. Li, and K. Liu. 2004. The subchronic toxicity of hydroxysafflor yellow A of 90 days repeatedly intraperitoneal injections in rats. *Toxicology* 203(1-3):139-143.

Lu, Z.W., F. Liu, J. Hu, D. Bian, and F.G. Li. 1991. Suppressive effects of safflower yellow on immune functions. *Zhongguo Yao Li Xue Bao* 12(6):537-542.

Nobakht, M., M. Fattahi, M. Hoormand, et al. 2000. A study on the teratogenic and cytotoxic effects of safflower extract. *J. Ethnopharmacol.* 73(3):453-459.

Shi, M., L. Chang, and G. He. 1995. Stimulating action of *Carthamus tinctorius* L., *Angelica sinensis* (Oliv.) Diels and *Leonurus sibiricus* L. on the uterus. *Zhongguo Zhong Yao Za Zhi* 20(3):173-175, 192.

Tang, W., and G. Eisenbrand. 1992. *Chinese drugs of plant origin: Chemistry, pharmacology, and use in traditional and modern medicine.* New York: Springer.

Wang, Y.S. 1983. *Pharmacology and applications of Chinese materia medica.* Beijing: People's Health Publishers.

Yin, X.J., D.X. Liu, H.C. Wang, and Y. Zhou. 1991. A study on the mutagenicity of 102 raw pharmaceuticals used in Chinese traditional medicine. *Mutat. Res.* 260(1):73-82.

Zang, B.X., W. Wu, W.R. Li, et al. 2002. Study on the anticoagulation effect of gross safflor yellow prepared by silica gel adsorption. *Chin. Pharm. J.* 37(2):106-109.

Zhao, P.W., D.W. Wang, J.Z. Niu, J.F. Wang, and L.Q. Wang. 2007. Evaluation on phytoestrogen effects of ten kinds of Chinese medicine including *Flos Carthami*. *Zhongguo Zhong Yao Za Zhi* 32(5):436-439.

Zhu, Y.-P. 1998. *Chinese materia medica: Chemistry, pharmacology and applications.* Amsterdam: Harwood Academic Publishers.

Carum carvi L. — Apiaceae

SCN: caraway
AN: *krishna jiraka*

Part: fruit (commonly known as "seed")

Quick Reference Summary

Safety Class: 1
Interaction Class: A

Contraindications

None known.

Other Precautions

None known.

Drug and Supplement Interactions

None known.

Adverse Events and Side Effects

None known.

Pharmacological Considerations

Animal studies have demonstrated that caraway may modify glucose regulation (Eddouks et al. 2004; Ene et al. 2008). People with diabetes are advised to monitor their blood sugar closely and discuss the use of this herb with a qualified healthcare practitioner prior to use.

Pregnancy and Lactation

No information on the safety of caraway in pregnancy or lactation was identified in the scientific or traditional literature. Although this review did not identify any concerns for use while pregnant or nursing, safety has not been conclusively established.

Review Details

I. Drug and Supplement Interactions

Clinical Trials of Drug or Supplement Interactions

No clinical trials of drug or supplement interactions were identified.

Case Reports of Suspected Drug or Supplement Interactions

No case reports of suspected drug or supplement interactions were identified.

Animal Trials of Drug or Supplement Interactions

No animal trials of drug or supplement interactions were identified.

II. Adverse Events

Case Reports of Adverse Events

No case reports of adverse events were identified.

III. Pharmacology and Pharmacokinetics

Human Pharmacological Studies

In a patient with occupational allergy to anise seed, a skin prick test showed immediate positive responses to caraway and other spices and foods including coriander, cumin, dill, fennel, and asparagus. Skin prick tests with celery, carrot, birch pollen, and mugwort pollen extracts were negative (Garcia-Gonzalez et al. 2002).

Caraway seed oil was nonirritating when applied for 48 hours at a concentration of 4% in human patch tests (Opdyke 1979).

Animal Pharmacological Studies

A significant decrease in blood glucose was observed in diabetic rats orally administered a single dose of 20 mg/kg aqueous extract of caraway seed. In diabetic animals administered 20 mg/kg of the extract daily for 15 days, blood glucose levels were nearly normalized. No highly significant changes in blood glucose were observed in healthy rats under either the single or repeated treatments (Eddouks et al. 2004). In diabetic rats orally administered 5, 10, 20, 40, or 80 mg/kg caraway seed oil daily for 10 weeks, a significant reduction in blood glucose was observed in animals receiving the 10 mg/kg dose (Ene et al. 2008).

No irritation was observed after caraway seed oil was applied to the back of hairless mice, although irritating effects were observed after application to intact or abraded rabbit skin (Opdyke 1979).

In Vitro Pharmacological Studies

No relevant in vitro studies were identified.

IV. Pregnancy and Lactation

No information on the safety of caraway in pregnancy or lactation was identified.

V. Toxicity Studies

Acute Toxicity

The LD_{50} of orally administered caraway seed oil in rats is 3.5 ml/kg. The dermal LD_{50} of the same product applied to rabbits is 1.78 ml/kg (Opdyke 1979).

Subchronic Toxicity

In rats orally administered 5, 10, 20, 40, or 80 mg/kg caraway seed oil daily for 10 weeks, elevated levels of liver marker enzymes (AST, ALT, ALP) were observed in animals receiving doses of 20 mg/kg or more (Ene et al. 2006).

Genotoxicity

In Ames tests for mutagenicity, no mutagenic activity of water, methanol, or hexane extracts of caraway was observed in *Salmonella typhimurium* strains TA98 and TA100 (Higashimoto et al. 1993), while an ethanol extract showed moderate mutagenic activity in strains TA98 and TA102 (Mahmoud et al. 1992).

Literature Cited

Eddouks, M., A. Lemhadri, and J.B. Michel. 2004. Caraway and caper: Potential anti-hyperglycaemic plants in diabetic rats. *J. Ethnopharmacol.* 94(1):143-148.

Ene, A.C., M.A. Milala, and E.A. Nwankwo. 2006. The effect of different doses of black caraway (*Carum carvi* L.) oil on the liver enzymes of alloxan-induced diabetic rats. *J. Med. Sci.* 6(6):994-998.

Ene, A.C., E.A. Nwankwo, and L.M. Samdi. 2008. Alloxan-induced diabetes in rats and the effects of Black caraway (*Carum carvi* L.) oil on their body weights. *J. Pharmacol. Toxicol.* 3(2):141-146.

Garcia-Gonzalez, J.J., B. Bartolome-Zavala, S. Fernandez-Melendez, et al. 2002. Occupational rhinoconjunctivitis and food allergy because of aniseed sensitization. *Ann. Allergy Asthma Immunol.* 88(5):518-522.

Higashimoto, M., J. Purintrapiban, K. Kataoka, et al. 1993. Mutagenicity and antimutagenicity of extracts of three spices and a medicinal plant in Thailand. *Mutat. Res.* 303(3):135-142.

Mahmoud, I., A. Alkofahi, and A. Abdelaziz. 1992. Mutagenic and toxic activities of several spices and some Jordanian medicinal plants. *Int. J. Pharmacogn.* 30(2):81-85.

Opdyke, D.L.J. 1979. *Monographs on fragrance raw materials.* New York: Pergamon.

Castanea dentata (Marsh.) Borkh. — Fagaceae

SCN: American chestnut

Part: leaf

Quick Reference Summary

Safety Class: 1
Interaction Class: A

Contraindications
None known.

Other Precautions
None known.

Drug and Supplement Interactions
None known.

Notice
Tannins (9%) (Hale-White 1901); *see* Appendix 1.

Adverse Events and Side Effects
None known.

Pharmacological Considerations
None known.

Pregnancy and Lactation
No information on the safety of American chestnut in pregnancy or lactation was identified in the scientific or traditional literature. Although this review did not identify any concerns for use while pregnant or nursing, safety has not been conclusively established.

Review Details

I. Drug and Supplement Interactions

Clinical Trials of Drug or Supplement Interactions
No clinical trials of drug or supplement interactions were identified.

Case Reports of Suspected Drug or Supplement Interactions
No case reports of suspected drug or supplement interactions were identified.

Animal Trials of Drug or Supplement Interactions
No animal trials of drug or supplement interactions were identified.

II. Adverse Events

Case Reports of Adverse Events
No case reports of adverse events were identified.

III. Pharmacology and Pharmacokinetics

Human Pharmacological Studies
No relevant human pharmacological studies were identified.

Animal Pharmacological Studies
No relevant animal pharmacological studies were identified.

In Vitro Pharmacological Studies
No relevant in vitro pharmacological studies were identified.

IV. Pregnancy and Lactation

No information on the use of American chestnut during pregnancy or lactation was identified.

V. Toxicity Studies

No toxicity studies were identified.

Literature Cited

Hale-White, W. 1901. *Materia medica pharmacy, pharmacology and therapeutics.* 5th ed. Philadelphia: P. Blakiston's Son & Co.

Castanea sativa Mill. — Fagaceae

SCN: Spanish chestnut

Part: leaf

Quick Reference Summary

Safety Class: 1
Interaction Class: A

Contraindications

None known.

Other Precautions

None known.

Drug and Supplement Interactions

None known.

Adverse Events and Side Effects

None known.

Pharmacological Considerations

None known.

Pregnancy and Lactation

No information on the safety of American chestnut in pregnancy or lactation was identified in the scientific or traditional literature. Although this review did not identify any concerns for use while pregnant or nursing, safety has not been conclusively established.

Review Details

I. Drug and Supplement Interactions

Clinical Trials of Drug or Supplement Interactions

No clinical trials of drug or supplement interactions were identified.

Case Reports of Suspected Drug or Supplement Interactions

No case reports of suspected drug or supplement interactions were identified.

Animal Trials of Drug or Supplement Interactions

No animal trials of drug or supplement interactions were identified.

II. Adverse Events

Case Reports of Adverse Events

In a woman with hypersensitivity to latex, testing revealed hypersensitivity to banana, avocado, and Spanish chestnut (Beier and Disch 1994).

III. Pharmacology and Pharmacokinetics

Human Pharmacological Studies

No irritant effects of an ethanol-water extract of Spanish chestnut leaf were observed in patch testing in healthy volunteers (Almeida et al. 2008).

Animal Pharmacological Studies

No relevant animal pharmacological studies were identified.

In Vitro Pharmacological Studies

No relevant in vitro pharmacological studies were identified.

IV. Pregnancy and Lactation

No information on the use of Spanish chestnut during pregnancy or lactation was identified.

V. Toxicity Studies

No toxicity studies were identified.

Literature Cited

Almeida, I.F., P. Valentão, P.B. Andrade, et al. 2008. In vivo skin irritation potential of a *Castanea sativa* (chestnut) leaf extract, a putative natural antioxidant for topical application. *Basic Clin. Pharmacol. Toxicol.* 103(5):461-467.

Beier, C., and R. Disch. 1994. Contact urticaria to latex with associated immediate reaction to banana, avocado and chestnut. *Allergologie* 17(6):268-270.

C

Catharanthus roseus (L.) G. Don — Apocynaceae

SCN: Madagascar periwinkle
Syn: *Vinca rosea* L.

Part: herb

Quick Reference Summary

Safety Class: 2b
Interaction Class: A

Contraindications

Not for use in pregnancy except under the supervision of a qualified healthcare practitioner (Gupta 2009; Gupta and Mathur 2009; List and Hörhammer 1973; Mathur et al. 1996).

Other Precautions

None known.

Drug and Supplement Interactions

None known.

Notice

Abortifacient (Gupta 2009; List and Hörhammer 1973; Mathur et al. 1996); *see* Appendix 2.

Editors' Note

The compounds vinblastine and vincristine, isolated from Madagascar periwinkle, are used as chemotherapy agents in the treatment of various cancers (Leveque and Jehl 2007; van Der Heijden et al. 2004). The concentration of these compounds in Madagascar periwinkle is very low (~2 ppm for vinblastine) (van Der Heijden et al. 2004).

Adverse Events and Side Effects

Animal studies have demonstrated adverse effects of high doses of Madagascar periwinkle on testicular cells (Mathur et al. 1996; Mathur and Chaudan 1985).

A case of severe bone marrow depression has been reported in association with use of Madagascar periwinkle (Wu et al. 2004). Animal studies have indicated depletion of white blood cells and bone marrow depression after injections, but not after oral use, of Madagascar periwinkle (Noble et al. 1958; Svoboda et al. 1962).

Pharmacological Considerations

Animal studies have demonstrated that Madagascar periwinkle may modify glucose regulation (Ahmed et al. 2007; Chattopadhyay et al. 1991; Nammi et al. 2003; Singh et al. 2001). People with diabetes are advised to monitor their blood sugar closely and discuss the use of this herb with a qualified healthcare practitioner prior to use.

Pregnancy and Lactation

Studies have indicated that Madagascar periwinkle reduced the number of viable fetuses in pregnant animals (Gupta 2009; Gupta and Mathur 2009; Mathur et al. 1996). Based on this information, use in pregnancy is not recommended except under the supervision of a qualified healthcare practitioner.

No information on the safety of Madagascar periwinkle during lactation was identified. While this review did not identify any concerns for use while nursing, safety has not been conclusively established.

Review Details

I. Drug and Supplement Interactions

Clinical Trials of Drug or Supplement Interactions

No clinical trials of drug or supplement interactions were identified.

Case Reports of Suspected Drug or Supplement Interactions

No case reports of suspected drug or supplement interactions were identified.

C

Animal Trials of Drug or Supplement Interactions

No animal trials of drug or supplement interactions were identified.

II. Adverse Events

Case Reports of Adverse Events

Severe bone marrow depression was reported in a 67-year-old woman with a history of diabetes, hepatitis C, and liver cancer. The woman had consumed unspecified amounts of the juice of Madagascar periwinkle for 5 days prior to the onset of symptoms (Wu et al. 2004).

III. Pharmacology and Pharmacokinetics

Human Pharmacological Studies

No relevant human pharmacological studies were identified.

Animal Pharmacological Studies

Injections of different extracts and extract fractions of Madagascar periwinkle led to depletion of white cells and bone marrow depression in rats. Such effects were reportedly not seen after oral administration (Noble et al. 1958; Svoboda et al. 1962).

In male rats orally administered 150 or 300 mg/kg of an ethanol extract of Madagascar periwinkle, testicular necrosis, hyalinization of tubules and Sertoli cell-only-syndrome was observed. Biochemical studies revealed reductions in glycogen and fructose levels in reproductive tissues (Mathur and Chaudan 1985).

Decreases in the weights of the prostate and seminal vesicles were observed in male mice intramuscularly administered 10 mg of a petroleum ether extract or ethanol extract of Madagascar periwinkle every other day for 20 days (Mathur et al. 1996).

In mice intramuscularly administered 10 mg daily of a petroleum ether extract of Madagascar periwinkle along with 17β-estradiol, inhibition of uterine weight gain was observed as compared to estradiol alone (Gupta 2009).

In diabetic rats orally administered 0.5 or 1 ml/kg of Madagascar periwinkle juice, or 300 or 450 mg/kg of an aqueous extract of Madagascar periwinkle, a reduction in serum glucose levels was observed (Ahmed et al. 2007). In diabetic rats orally administered 500 mg/kg of a dichloromethane-methanol extract of Madagascar periwinkle daily for 7 or 15 days, significant hypoglycemic activity was observed (Singh et al. 2001).

Oral administration of 0.5, 0.75, and 1.0 ml/kg of Madagascar periwinkle juice to healthy or diabetic rabbits resulted in dose-dependent reduction in blood glucose of both healthy and diabetic rabbits. Activity was comparable to the positive control, 40 µg/kg glibenclamide (Nammi et al. 2003).

In Vitro Pharmacological Studies

An aqueous extract of Madagascar periwinkle inhibited the drug-metabolizing isoenzyme CYP2D6 (Usia et al. 2006). The compounds serpentine and ajmalicine were identified as the most potent inhibitors among tested compounds from Madagascar periwinkle (Usia et al. 2005).

A dichloromethane–methanol–water extract of Madagascar periwinkle demonstrated toxicity in Chang liver cells and adipose cells at concentrations of 12.5 µg/ml (van de Venter et al. 2008).

IV. Pregnancy and Lactation

Total inhibition of pregnancy was observed in mice administered 10 mg (per mouse) of a petroleum ether extract or ethanol extract of Madagascar periwinkle on gestational days 7 to 9 (Mathur et al. 1996).

A study in rats indicated abortifacient activity of Madagascar periwinkle administered on days 1 to 10 of pregnancy. Details on dose and product used were lacking (Adhikary et al. 1990).

In mice intramuscularly administered 2, 5, or 10 mg daily of petroleum ether extract of the leaf of Madagascar periwinkle on gestational days 7 to 9, no implantation sites were found in animals given the 10 mg dose, while 20% implantation was observed at the 2 mg dose and 48% implantation was observed at the 5 mg dose, as compared to 85% implantation in the control group (Gupta 2009; Gupta and Mathur 2009).

No information on the safety of Madagascar periwinkle during lactation was identified.

V. Toxicity Studies

Acute Toxicity

The LD_{50} of an aqueous extract of Madagascar periwinkle intraperitoneally administered in mice is 3.16 g/kg (Sarma et al. 1997).

No mortalities were observed in mice orally administered 10 g/kg of an aqueous extract of Madagascar periwinkle (Sarma et al. 1997).

No adverse effects were observed in rats intraperitoneally administered up to 4 g/kg of a water-soluble fraction of ethanolic extract of Madagascar periwinkle (Chattopadhyay et al. 1991).

Genotoxicity

In the Ames test for mutagenicity, with *Salmonella typhimurium* strains TA98 and TA100, a methanol extract of Madagascar periwinkle exhibited some mutagenic activity in strain TA98 with but not without metabolic activation. No mutagenic activity was observed in strain TA100 or in the VitoTOX assay (Elgorashi et al. 2003).

Literature Cited

Adhikary, P., J. Banerji, D. Choudhury, et al. 1990 Anti-implantation activity of some indigenous plants in adult female rats [abstract]. *Indian J. Pharmacol.* 22(1):24-25.

Ahmed, A.U., A.H. Ferdous, S.K. Saha, et al. 2007. Hypoglycemic effect of *Catharanthus roseus* in normal and streptozotocin-induced diabetic rats. *Mymensingh Med. J.* 16(2):143-148.

Chattopadhyay, R.R., S.K. Sarkar, S. Ganguly, R.N. Banerjee, and T.K. Basu. 1991. Hypoglycemic and antihyperglycemic effect of leaves of *Vinca rosea* Linn. *Indian J. Physiol. Pharmacol.* 35(3):145-151.

Elgorashi, E.E., J.L. Taylor, A. Maes, et al. 2003. Screening of medicinal plants used in South African traditional medicine for genotoxic effects. *Toxicol. Lett.* 143(2):195-207.

Gupta, P. 2009. Antiestrogenic activity of petroleum ether extract of the leaves of *Catharanthus roseus* (*Vinca rosea*) in female albino mice. *Asian J. Exp. Sci.* 23(1):313-316.

Gupta, P., and P. Mathur. 2009. Changes in the uterine phosphatase levels in female swiss albino mice treated with chromatographic fractions of petroleum ether extracts of *Vinca rosea* leaves. *J. Herb Med. Toxicol.* 3(2):143-145.

Leveque, D., and F. Jehl. 2007. Molecular pharmacokinetics of *Catharanthus* (*Vinca*) alkaloids. *J. Clin. Pharmacol.* 47(5):579-588.

List, P.H., and H. Hörhammer. 1973. *Hagers handbuch der pharmazeutischen praxis*. Berlin: Springer.

Mathur, P., P. Garg, D.K. Vyas, and D. Jacob. 1996. Antifertility efficacy of various extracts of leaves of *Catharanthus roseus* in Swiss albino mouse. *J. Anim. Morphol. Physiol.* 43(1):29-33.

Mathur, R., and S. Chaudan. 1985. Antifertility efficacy of *Catharanthus roseus* Linn.: A biochemical and histological study. *Acta Eur. Fertil.* 16(3):203-205.

Nammi, S., M.K. Boini, S.D. Lodagala, and R.B. Behara. 2003. The juice of fresh leaves of *Catharanthus roseus* Linn. reduces blood glucose in normal and alloxan diabetic rabbits. *BMC Complement. Altern. Med.* 3:4.

Noble, R.L., C.T. Beer, and J.H. Cutts. 1958. Role of chance observations in chemotherapy: *Vinca rosea*. *Ann N.Y. Acad. Sci.* 76:882-894.

Sarma, J., N. Ahmed, and N. Ahmed. 1997. Preliminary studies on acute toxicity of *Vinca rosea* Linn. in albino mice. *Indian J. Indig. Med.* 19(2):140-142.

Singh, S.N., P. Vats, S. Suri, et al. 2001. Effect of an antidiabetic extract of *Catharanthus roseus* on enzymic activities in streptozotocin induced diabetic rats. *J. Ethnopharmacol.* 76(3):269-277.

Svoboda, G., I.S. Johnson, M. Gorman, and N. Neuss. 1962. Current status of research on the alkaloids of *Vinca rosea* Linn. (*Catharanthus roseus* G. Don). *J. Pharm. Sci.* 51(8):707-720.

Usia, T., H. Iwata, A. Hiratsuka, et al. 2006. CYP3A4 and CYP2D6 inhibitory activities of Indonesian medicinal plants. *Phytomedicine* 13 (1-2):67-73.

Usia, T., T. Watabe, S. Kadota, and Y. Tezuka. 2005. Cytochrome P450 2D6 (CYP2D6) inhibitory constituents of *Catharanthus roseus*. *Biol. Pharm. Bull.* 28(6):1021-1024.

van de Venter, M., S. Roux, L.C. Bungu, et al. 2008. Antidiabetic screening and scoring of 11 plants traditionally used in South Africa. *J. Ethnopharmacol.* 119(1):81-86.

van Der Heijden, R., D.I. Jacobs, W. Snoeijer, D. Hallard, and R. Verpoorte. 2004. The Catharanthus alkaloids: Pharmacognosy and biotechnology. *Curr. Med. Chem.* 11(5):607-628.

Wu, M.L., J.F. Deng, J.C. Wu, F.S. Fan, and C.F. Yang. 2004. Severe bone marrow depression induced by an anticancer herb *Cantharanthus roseus*. *J. Toxicol. Clin. Toxicol.* 42(5):667-671.

Caulophyllum thalictroides (L.) Michx. — Berberidaceae

SCN: blue cohosh
OCN: papoose root
Part: root

Quick Reference Summary

Safety Class: 2b
Interaction Class: A

Contraindications

Not for use during pregnancy except under the supervision of a qualified healthcare practitioner (Dugoua et al. 2008; Flynn et al. 1998; Kennelly et al. 1999).

Other Precautions

None known.

Drug and Supplement Interactions

None known.

Notice

Abortifacient (De Smet 1993); Emmenagogue (De Smet 1993); *see* Appendix 2.

Editors' Note

Blue cohosh contains *N*-methylcytisine, an alkaloid that is similar in action to nicotine in its ability to stimulate intestinal activity, increase respiration, and elevate blood pressure. Blue cohosh also contains caulosaponin, a glycoside that can act as a coronary vasoconstrictor and is believed to be responsible for stimulation of uterine contractions and induction of labor (Ferguson and Edwards 1954; McFarlin et al. 1999; Satchithanandam et al. 2008; Scott and Chin 1943).

Adverse Events and Side Effects

Myocardial infarction, profound congestive heart failure, multiorgan failure, and perinatal stroke have been reported in infants born of mothers using blue cohosh, sometimes in combination with black cohosh or other herbs, before birth (Finkel and Zarlengo 2004; Gunn and Wright 1996; Jones and Lawson 1998).

Pharmacological Considerations

Increased uterine tone and uterine spasms have been reported in human and animal studies of isolated compounds of blue cohosh (Ferguson and Edwards 1954; Pilcher et al. 1916; Pilcher and Mauer 1918; Vinks et al. 1982).

Pregnancy and Lactation

Blue cohosh has been used to augment labor and also for up to several weeks prior to the due date as a parturifacient (Felter and Lloyd 1898) and was officially listed as a labor inducer in the U.S. Pharmacopoeia from 1882–1905. There have been a small number of serious adverse events reported in infants born to mothers taking blue cohosh several weeks prior to birth. A 1999 survey of nurse midwives found that 64% used blue cohosh, often in combination with black cohosh, to augment labor during delivery. This survey noted that blue cohosh is the herb midwives have the least comfort with during pregnancy and a significant number had observed an increased rate of meconium, tachycardia, and need for resuscitation in association with its use (McFarlin et al. 1999). It is unclear, however, whether these adverse events were associated with the use of blue cohosh itself or with the conditions that led to its use, such as post-term pregnancy (Romm 2005).

Eclectic physicians of the early 20th century and some midwives today utilize small amounts of blue cohosh during pregnancy for threatened miscarriage. The safety of blue cohosh in early pregnancy is questionable, however, given that compounds isolated from blue cohosh caused fetal malformations in rat embryo cultures (Flynn et al. 1998; Kennelly et al. 1999).

Given the current information, it is advisable to limit use of blue cohosh in pregnancy only under the supervision of a practitioner who is familiar with its use and who can appropriately monitor both the mother and infant for possible adverse effects. *Also see* Adverse Events and Side Effects *and* Pharmacological Considerations for this entry.

No information on the safety of blue cohosh during lactation was identified. While this review did not identify any concerns for use while nursing, safety has not been conclusively established.

Review Details

I. Drug and Supplement Interactions

Clinical Trials of Drug or Supplement Interactions

No clinical trials of suspected drug or supplement interactions were identified.

Case Reports of Suspected Drug or Supplement Interactions

No case reports of drug or supplement interactions were identified.

Animal Trials of Drug or Supplement Interactions

No animal trials of drug or supplement interactions were identified.

II. Adverse Events

Case Reports of Adverse Events

A survey of 172 nurse midwives indicated that 90 regularly used herbal preparations to stimulate labor. Of these 90, 64% reported using blue cohosh. Adverse events data included in the results of this survey did not separate black cohosh (*Actaea racemosa*) and blue cohosh; the adverse events reported for the two botanicals were nausea, increased meconium-stained fluid, and transient fetal tachycardia. The frequency of these events was not indicated (McFarlin et al. 1999).

Acute myocardial infarction, congestive heart failure, and cardiogenic shock were reported in a newborn infant whose mother had taken blue cohosh. The mother had taken 3 tablets of blue cohosh daily for 3 weeks at the end of her pregnancy, although the recommended dose was 1 tablet daily (Jones and Lawson 1998). A review of the traditional herbal literature indicated that the dose taken in the case constituted an overdose (Bergner 2001).

After an apparently healthy pregnancy and normal labor, a newborn infant was not able to breathe spontaneously and had CNS hypoxic-ischemic damage. The midwife attending the birth had given blue cohosh and black cohosh (dose and duration unspecified) to induce labor. At 3 months of life, the child had lower limb spasticity and required nasogastric tube feeding (Gunn and Wright 1996).

Perinatal stroke was reported in a 26-hour-old infant whose mother had taken a blue cohosh-containing tea (dose and duration unspecified). Urine and stool samples of the infant tested positive for the cocaine metabolite benzoylecgonine. Analyses of the blue cohosh product that the mother had taken and a different preparation of blue cohosh were also reported to test positive for benzoylecgonine. The authors of this report suggested that either benzoylecgonine is a metabolite of both cocaine and blue cohosh or that the blue cohosh was contaminated with cocaine (Finkel and Zarlengo 2004). Comments on this report have indicated that no published evidence exists that blue cohosh contains benzoylecgonine or any compound that could be metabolized to benzoylecgonine (Chan and Nelson 2004; Potterton 2004). In response to these comments, the reporting physicians indicated that maternal cocaine use, an insensitive immunoassay, or incorrect interpretation of chemical analysis of the product were possible explanations for the detection of benzoylecgonine (Finkel and Zarlengo 2004). Another analysis of the case report indicated that a false-positive reading from the analytical test used was likely and that blue cohosh contains a phytochemical that looks like the cocaine metabolite in terms of the tests used but is not related in any way to cocaine (Mills and Bone 2005).

A 21-year-old woman who had taken 10 to 20 doses (dose amount unspecified) of blue cohosh tincture daily for 4 days during weeks 5 or 6 of pregnancy in an attempt to induce abortion was diagnosed with nicotinic toxicity. Symptoms included hypertension, tachycardia, vomiting, diaphoresis, and muscle fasciculations of the abdominal wall. The woman was also taking slippery elm orally and using slippery elm and parsley douches (Rao and Hoffman 2002).

III. Pharmacology and Pharmacokinetics

Human Pharmacological Studies

The compound sparteine was found to cause uterine spasms in a small percentage (~5%) of women who were unable to metabolize the compound (Eichelbaum et al. 1979; Vinks et al. 1982).

Animal Pharmacological Studies

In dogs administered an extract of blue cohosh, no effect on the uterus was observed (Pilcher and Mauer 1918). Intravenous administration of caulosaponin to rats and to in situ rat uteruses at a dose of 5 or 10 mg/kg increased uterine tone and rate of contraction (Ferguson and Edwards 1954).

In Vitro Pharmacological Studies

A methanol extract of blue cohosh showed no effects on CYP450 isoenzymes. A combination of alkaloids isolated from the extract showed inhibition of CYP2C19, CYP3A4, CPY2D6, and CYP1A2 (Madgula et al. 2009).

IV. Pregnancy and Lactation

Also see Case reports of adverse events *and* Pharmacology and Pharmacokinetics for this entry.

In a rat embryo culture system, the compound caulophyllumine was embryotoxic, producing 100% lethality at a concentration of 5 μg/ml, and the compound *N*-methylcytisine was described as potentially teratogenic with neural tube defects observed at a concentration of 80 μg/ml. No teratogenic effects were observed for the compound anagyrine at a concentration of 500 μg/ml, although inhibition of embryo growth and development was observed (Flynn et al. 1998). A second rat embryo culture study on alkaloids of blue cohosh found that *N*-methylcytisine exhibited teratogenic activity, whereas taspine showed high embryotoxicity but no teratogenic activity, and other alkaloids did not produce teratogenic effects (Kennelly et al. 1999).

No information on the safety of blue cohosh during lactation was identified.

V. Toxicity Studies

Acute Toxicity

In mice, the LD_{50} of methylcytisine could not be determined at doses up to 500 mg/kg after oral administration, and is 51 mg/kg after intraperitoneal administration and 21 mg/kg after intravenous administration (Barlow and McLeod 1969).

The LD_{50} of intravenously administered caulosaponin in mice is 12 mg/kg and in rats is 20 mg/kg. In these studies, increases in activity, ataxia, and clonic seizures were observed, and death was attributed to asphyxia (Ferguson and Edwards 1954).

Subchronic Toxicity

Subcutaneous administration of caulosaponin to rats at a dose of 5 mg/kg daily for 60 days did not produce any signs of toxicity (Ferguson and Edwards 1954).

Literature Cited

Barlow, R.B., and L.J. McLeod. 1969. Some studies on cytisine and its methylated derivatives. *Br. J. Pharmacol.* 35(1):161-174.

Bergner, P. 2001. *Caulophyllum*: Cardiotoxic effects of blue cohosh on a fetus. *Med. Herbalism* 12(1):12-14.

Chan, G.M., and L.S. Nelson. 2004. More on blue cohosh and perinatal stroke. *N. Engl. J. Med.* 351(21):2239-2241.

De Smet, P.A.G.M. 1993. *Adverse effects of herbal drugs, Volume 2*. Berlin, Heidelberg, New York: Springer.

Dugoua, J.J., D. Perri, D. Seely, E. Mills, and G. Koren. 2008. Safety and efficacy of blue cohosh (*Caulophyllum thalictroides*) during pregnancy and lactation. *Can. J. Pharmacol.* 15:e66-e73.

Eichelbaum, M., N. Spannbrucker, and H.J. Dengler. 1979. Influence of the defective metabolism of sparteine on its pharmacokinetics. *Eur. J. Clin. Pharmacol.* 16(3):189-194.

Felter, H.W., and J.U. Lloyd. 1898. *King's American dispensatory*. 18th ed., 3rd rev. 2 vols. Cincinnati: Ohio Valley Co.

Ferguson, H.C., and L.D. Edwards. 1954. A pharmacological study of a crystalline glycoside of *Caulophyllum thalictroides*. *J. Am. Pharm. Assoc.* 43:16-21.

Finkel, R.S., and K.M. Zarlengo. 2004. Blue cohosh and perinatal stroke. *N. Engl. J. Med.* 351(3):302-303.

Flynn, T.J., E.J. Kennelly, E.P. Mazzola, T.G. McCloud, and J.M. Betz. 1998. Screening of the dietary supplement blue cohosh for potentially teratogenic alkaloids using rat embryo culture. *Teratology* 57(4/5):219.

Gunn, T.R., and I.M. Wright. 1996. The use of black and blue cohosh in labour. *N.Z. Med. J.* 109(1032):410-411.

Jones, T.K., and B.M. Lawson. 1998. Profound neonatal congestive heart failure caused by maternal consumption of blue cohosh herbal medication. *J. Pediatr.* 132(3, Pt. 1):550-552.

Kennelly, E.J., T.J. Flynn, E.P. Mazzola, et al. 1999. Detecting potential teratogenic alkaloids from blue cohosh rhizomes using an in vitro rat embryo culture. *J. Nat. Prod.* 62(10):1385-1389.

Madgula, V., Z. Ali, T. Smillie, et al. 2009. Alkaloids and saponins as cytochrome P450 inhibitors from blue cohosh (*Caulophyllum thalictroides*) in an in vitro assay. *Planta Med.* 75:329-332.

McFarlin, B.L., M.H. Gibson, J. O'Rear, and P. Harman. 1999. A national survey of herbal preparation use by nurse-midwives for labor stimulation. Review of the literature and recommendations for practice. *J. Nurse Midwifery* 44(3):205-216.

Mills, S., and K. Bone. 2005. *The essential guide to herbal safety*. St. Louis: Elsevier.

Pilcher, D., W. Delzell, and G. Burman. 1916. The action of various "female" remedies on the excised uterus of the guinea-pig. *J. Am. Med. Assoc.* 47:490-492.

Pilcher, J., and R. Mauer. 1918. The action of "female remedies" on intact uteri of animals. *Surg. Gynecol. Obstet.* 27:97-99.

Potterton, D. 2004. More on blue cohosh and perinatal stroke. *N. Engl. J. Med.* 351(21):2239-2241.

Rao, R.B., and R.S. Hoffman. 2002. Nicotinic toxicity from tincture of blue cohosh (*Caulophyllum thalictroides*) used as an abortifacient. *Vet. Hum. Toxicol.* 44(4):221-222.

Romm, A. 2005. Blue cohosh (*Caulophyllum thalictroides*): Safety questioned during pregnancy and labor. 4th Oxford International Conference on Science of Botanicals, Oxford, MS.

Satchithanandam, S., E. Grundel, J. Roach, et al. 2008. Alkaloids and saponins in dietary supplements of blue cohosh (*Caulophyllum thalictroides*). *J. AOAC Int.* 91(1):21-32.

Scott, C., and K. Chin. 1943. The pharmacologic action of *N*-methylcytisine. *Therapeutics* 79:334.

Vinks, A., T. Inaba, S.V. Otton, and W. Kalow. 1982. Sparteine metabolism in Canadian Caucasians. *Clin. Pharmacol. Ther.* 31(1):23-29.

Ceanothus americanus L. — Rhamnaceae

SCN: red root
OCN: Jersey tea; New Jersey tea
Part: root, root bark

Quick Reference Summary

Safety Class: 1
Interaction Class: A

Contraindications

None known.

Other Precautions

None known.

Drug and Supplement Interactions

See *Pharmacological Considerations* below.

Adverse Events and Side Effects

None known.

Pharmacological Considerations

Early human, animal, and in vitro studies indicated that red root causes an increase in the speed of blood coagulation (Groot 1927; Lynch 1966; Lynch et al. 1958; Taylor 1927; Tharaldsen 1929; Tharaldsen and Krawetz 1927).

Pregnancy and Lactation

No information on the safety of red root in pregnancy or lactation was identified in the scientific or traditional literature. Although this review did not identify any concerns for use while pregnant or nursing, safety has not been conclusively established.

Review Details

I. Drug and Supplement Interactions

Clinical Trials of Drug or Supplement Interactions

No clinical trials of drug or supplement interactions were identified.

Case Reports of Suspected Drug or Supplement Interactions

No case reports of suspected drug or supplement interactions were identified.

Animal Trials of Drug or Supplement Interactions

No animal trials of drug or supplement interactions were identified.

II. Adverse Events

Case Reports of Adverse Events

No case reports of adverse events were identified.

III. Pharmacology and Pharmacokinetics

Human Pharmacological Studies

A reduction in blood coagulation time was observed in healthy volunteers orally administered 15 or 30 ml of a hydroalcoholic extract of red root (Groot 1927).

Animal Pharmacological Studies

A reduction in blood coagulation time was observed in dogs intravenously administered a hydroalcoholic extract of red root, while no effects on coagulation time were observed after oral administration (Groot 1927).

Intravenous administration of 4 ml of red root hydroalcoholic extract to dogs was reported to lower blood pressure, while no effects on blood pressure were observed after oral administration (Groot 1927).

In Vitro Pharmacological Studies

A reduction in blood coagulation time was observed in blood treated with various fractions of red root extract (Lynch et al. 1958).

IV. Pregnancy and Lactation

No information on the use of red root during pregnancy or lactation was identified.

V. Toxicity Studies

No toxicity studies were identified.

Literature Cited

Groot, J.T. 1927. Pharmacology of *Ceanothus americanus*. I. Preliminary studies; hemodynamics and the effects of coagulation. *J. Pharmacol.* 30:275-291.

Lynch, T.A. 1966. Blood coagulating principles from *Ceanothus americanus*.

Lynch, T.A., T.S. Miya, and C.J. Carr. 1958. An investigation of the blood coagulating principles from *Ceanothus americanus*. *J. Am. Pharm. Assoc.* 47(11):816-819.

Taylor, G.C. 1927. *Ceanothus americanus* L. as a hemostatic. A resume of recent investigations into the chemistry, pharmacology and clinical use of the drug. *Am. J. Pharm.* 99:214-232.

Tharaldsen, C.E. 1929. Ceanothyn as a hemostatic. *J. Am. Inst. Homeopathy* 22:428-435.

Tharaldsen, C.E., and J. Krawetz. 1927. The blood reactions of the alkaloids of *Ceanothus americanus*. *Am. J. Physiol.* 79:445-452.

Centaurea cyanus L.

Asteraceae

SCN: cornflower
OCN: bachelor's button; cyani

Part: herb

Quick Reference Summary

Safety Class: 1
Interaction Class: A

Contraindications

None known.

Other Precautions

None known.

Drug and Supplement Interactions

None known.

Adverse Events and Side Effects

None known.

Pharmacological Considerations

None known.

Pregnancy and Lactation

No information on the safety of cornflower in pregnancy or lactation was identified in the scientific or traditional literature. Although this review did not identify any concerns for use while pregnant or nursing, safety has not been conclusively established.

Review Details

I. Drug and Supplement Interactions

Clinical Trials of Drug or Supplement Interactions

No clinical trials of drug or supplement interactions were identified.

Case Reports of Suspected Drug or Supplement Interactions

No case reports of suspected drug or supplement interactions were identified.

Animal Trials of Drug or Supplement Interactions

No animal trials of drug or supplement interactions were identified.

II. Adverse Events

Case Reports of Adverse Events

No case reports of adverse events were identified.

III. Pharmacology and Pharmacokinetics

Human Pharmacological Studies

No relevant human pharmacological studies were identified.

Animal Pharmacological Studies

No relevant animal pharmacological studies were identified.

In Vitro Pharmacological Studies

No relevant in vitro pharmacological studies were identified.

IV. Pregnancy and Lactation

No information on the use of cornflower during pregnancy or lactation was identified.

V. Toxicity Studies

Acute Toxicity

In the brine shrimp lethality assay, the LC_{50} of a methanol extract of cornflower is 38 µg/ml, while that of an ether extract of cornflower is 103 µg/ml (Janackovic et al. 2008).

Literature Cited

Janackovic, P., V. Tesevic, P.D. Marin, et al. 2008. Brine shrimp lethality bioassay of selected *Centaurea* L. species (Asteraceae). *Arch. Biol. Sci.* 60(4):681-685.

Centaurium erythraea Rafn

Gentianaceae

SCN: centaury
OCN: common centaury; lesser centaury

Part: flowering top

Quick Reference Summary

Safety Class: 1
Interaction Class: A

Contraindications

None known.

Other Precautions

None known.

Drug and Supplement Interactions

None known.

Adverse Events and Side Effects

None known.

Pharmacological Considerations

An animal study indicated diuretic effects of centaury after several days of administration (Haloui et al. 2000).

Pregnancy and Lactation

No information on the safety of centaury in pregnancy or lactation was identified in the scientific or traditional literature. Although this review did not identify any concerns for use while pregnant or nursing, safety has not been conclusively established.

Review Details

I. Drug and Supplement Interactions

Clinical Trials of Drug or Supplement Interactions

No clinical trials of drug or supplement interactions were identified.

Case Reports of Suspected Drug or Supplement Interactions

No case reports of suspected drug or supplement interactions were identified.

Animal Trials of Drug or Supplement Interactions

No animal trials of drug or supplement interactions were identified.

II. Adverse Events

Case Reports of Adverse Events

No case reports of adverse events were identified.

III. Pharmacology and Pharmacokinetics

Human Pharmacological Studies

No relevant human pharmacological studies were identified.

Animal Pharmacological Studies

In rats orally administered 10 ml/kg of an 8 or 16% aqueous extract of centaury daily for 7 days, enhanced diuresis was observed after 5 days. No increase was recorded for 24 hour urinary excretion of sodium, potassium, and calcium during the first 4 days of treatment with either extract, whereas the effects on those parameters were highly significant thereafter. No changes in plasma electrolytes or urea were observed (Haloui et al. 2000).

In Vitro Pharmacological Studies

No relevant in vitro pharmacological studies were identified.

IV. Pregnancy and Lactation

No information on the use of centaury during pregnancy or lactation was identified.

V. Toxicity Studies

Acute Toxicity

In the brine shrimp lethality assay, the LC_{50} values of compounds isolated from centaury are 8 μg/ml for swertiamarin, 34 μg/ml for sweroside, and 24 μg/ml for gentiopicroside (Kumarasamy et al. 2003a, 2003b).

Genotoxicity

No mutagenic activity of a tincture or fluid extract of centaury was observed in *Salmonella typhimurium* strains TA98 or TA100, while an evaporated extract exhibited weak mutagenic activity in TA98 (Schimmer et al. 1994). Strong antimutagenic activity of the compounds eustomin and demethyleustomin were observed in *Salmonella typhimurium* strains TA98, TA100, and TA102 (Schimmer and Mauthener 1996).

C

Literature Cited

Haloui, M., L. Louedec, J.B. Michel, and B. Lyoussi. 2000. Experimental diuretic effects of *Rosmarinus officinalis* and *Centaurium erythraea*. *J. Ethnopharmacol.* 71(3):465-472.

Kumarasamy, Y., L. Nahar, P.J. Cox, M. Jaspars, and S.D. Sarker. 2003a. Bioactivity of secoiridoid glycosides from *Centaurium erythraea*. *Phytomedicine* 10(4):344-347.

Kumarasamy, Y., L. Nahar, and S.D. Sarker. 2003b. Bioactivity of gentiopicroside from the aerial parts of *Centaurium erythraea*. *Fitoterapia* 74(1-2):151-154.

Schimmer, O., A. Krüger, H. Paulini, and F. Haefele. 1994. An evaluation of 55 commercial plant extracts in the Ames mutagenicity test. *Pharmazie* 49:448-448.

Schimmer, O., and H. Mauthener. 1996. Polymethoxylated xanthones from the herb of *Centaurium erythraea* with strong antimutagenic properties in *Salmonella typhimurium*. *Planta Med.* 62(6):561-564.

Centella asiatica (L.) Urb. — Apiaceae

SCN: gotu kola
Syn: *Hydrocotyle asiatica* L.
AN: *brahmi; mandukaparni*
PN: *ji xue cao* (whole plant)
OCN: Asiatic pennywort; Indian pennywort
Part: herb

Quick Reference Summary

Safety Class: 1
Interaction Class: A

Contraindications

None known.

Other Precautions

None known.

Drug and Supplement Interactions

None known.

Editors' Note

As a general rule, *Centella asiatica* is called "brahmi" in north and west India and "mandukaparni" in Kerala (south India), whereas *Bacopa monnieri* is "brahmi" in Kerala and "mandukaparni" in north and west India.

Adverse Events and Side Effects

Allergic contact dermatitis, confirmed by patch testing, has been reported after topical application of products containing gotu kola (Bilbao et al. 1995; Eun and Lee 1985; Izu et al. 1992; Morante et al. 1998; Santucci et al. 1985; Vena and Angelini 1986).

Pharmacological Considerations

None known.

Pregnancy and Lactation

No adverse effects on fetal development were observed in the offspring of mice administered fresh juice of gotu kola (equivalent to 20 to 80 g/kg of fresh plant) for either 14 or 21 days beginning 7 days prior to mating (Dutta and Basu 1968).

No information on the safety of gotu kola during lactation was identified. While this review did not identify any concerns for use while nursing, safety has not been conclusively established.

Review Details

I. Drug and Supplement Interactions

Clinical Trials of Drug or Supplement Interactions

No clinical trials of drug or supplement interactions were identified.

Case Reports of Suspected Drug or Supplement Interactions

No case reports of suspected drug or supplement interactions were identified.

Animal Trials of Drug or Supplement Interactions

No animal trials of drug or supplement interactions were identified.

II. Adverse Events

Case Reports of Adverse Events

Cases of jaundice and hepatotoxicity were reported in three women taking gotu kola tablets for weight loss for 21 to 60 days (products and doses not specified). The pathological diagnoses were granulomatous hepatitis with marked necrosis and apoptosis, chronic hepatitis with cirrhotic transformation and intense necroinflammatory activity, and granulomatous hepatitis. All patients improved after cessation of gotu kola, and one patient experienced a recurrence of jaundice after 2 weeks of re-exposure (Jorge and Jorge 2005).

Night eating syndrome was observed in a 41-year-old woman who had been taking a tincture of gotu kola (dose unspecified) for approximately 2 years. Onset of the syndrome was temporally associated with the use of gotu kola, and symptoms resolved after cessation of the tincture (O'Brien 2005).

Allergic contact dermatitis, confirmed by patch testing, has been reported after topical application of products containing gotu kola (Bilbao et al. 1995; Eun and Lee 1985; Izu et al. 1992; Morante et al. 1998; Santucci et al. 1985; Vena and Angelini 1986).

III. Pharmacology and Pharmacokinetics

Human Pharmacological Studies

No relevant human pharmacological studies were identified.

Animal Pharmacological Studies

Weak sensitization capacity of gotu kola was observed in guinea pigs after repeated topical application (Hausen 1993).

A reduction in conception rates was observed in mice orally administered fresh juice of gotu kola (equivalent to 20–80 g/kg of fresh plant) for either 14 or 21 days beginning 7 days prior to mating. Conception rates were approximately 45% for treated animals and 85% for untreated controls (Dutta and Basu 1968).

Administration of ethanol, methanol, and *n*-butanol extracts but not a water extract of gotu kola to rats at doses of 28 ml/kg daily for 14 days inhibited in vitro platelet reactivity. The ethanol extract also inhibited dynamic coagulation (Satake et al. 2007).

In Vitro Pharmacological Studies

No relevant in vitro pharmacological studies were identified.

IV. Pregnancy and Lactation

No adverse effects on fetal development were observed in the offspring of mice orally administered fresh juice of gotu kola (equivalent to 20–80 g/kg of fresh plant) for either 14 or 21 days beginning 7 days prior to cohabitation with males (Dutta and Basu 1968).

No information on the safety of gotu kola during lactation was identified.

V. Toxicity Studies

Acute Toxicity

The LD_{50} of orally administered hydroalcoholic extract of gotu kola in rats could not be determined at doses up to 675 mg/kg (De Lucia et al. 1997).

Short-Term Toxicity

No adverse effects were observed in rats orally administered 150 mg/kg of a hydroalcoholic extract of gotu kola daily for 30 days. Plasma glucose and protein levels showed no changes, and no alterations in the stomach, liver, spleen, or kidneys were observed (De Lucia et al. 1997).

No adverse effects were observed in mice orally administered up to 1 mg/kg gotu kola aqueous extract daily for 15 days (Rao et al. 2005).

Genotoxicity

Mutagenic activity of aqueous and ethanolic extracts of gotu kola was observed in *Salmonella typhimurium* strain TA98 with but not without metabolic activation (Mills and Bone 2005; Rivera et al. 1994). Conversely, weak to moderate inhibition of mutagenicity was observed in *Salmonella typhimurium* strains TA98 and TA100 treated with a gotu kola extract (Yen et al. 2001), and a dose-dependent decrease in chemically induced genotoxic damage (chromosomal aberrations and sister chromatid exchanges) was observed in human lymphocytes treated with gotu kola extracts (Siddique et al. 2008).

Literature Cited

Bilbao, I., A. Aguirre, R. Zabala, et al. 1995. Allergic contact dermatitis from butoxyethyl nicotinic acid and *Centella asiatica* extract. *Contact Dermat.* 33(6):435-436.

De Lucia, R., J.A.A. Sertie, E.A. Camargo, and S. Panizza. 1997. Pharmacological and toxicological studies on *Centella asiatica* extract. *Fitoterapia* 68(5):413-416.

Dutta, T., and U.P. Basu. 1968. Crude extract of *Centella asiatica* and products derived from its glycosides as oral antifertility agents. *Indian J. Exp. Biol.* 6(3):181-182.

Eun, H.C., and A.Y. Lee. 1985. Contact dermatitis due to madecassol. *Contact Dermat.* 13(5):310-313.

Hausen, B.M. 1993. *Centella asiatica* (Indian pennywort), an effective therapeutic but a weak sensitizer. *Contact Dermat.* 29(4):175-179.

Izu, R., A. Aguirre, N. Gil, and J.L. Diaz-Perez. 1992. Allergic contact dermatitis from a cream containing *Centella asiatica* extract. *Contact Dermat.* 26(3):192-193.

Jorge, O.A., and A.D. Jorge. 2005. Hepatotoxicity associated with the ingestion of *Centella asiatica*. *Rev. Esp. Enferm. Dig.* 97(2):115-124.

Mills, S., and K. Bone. 2005. *The essential guide to herbal safety*. St. Louis: Elsevier.

Morante, J.M.O., J.J.G. Bujan, M.G. Guemes, I.Y. Bayona, and R.S. Arechavala. 1998. Allergic contact dermatitis from *Centella asiatica* extract: Report of a new case. *Acta Dermo-Sifiliograf.* 89(6):341-343.

O'Brien, B. 2005. Night eating syndrome and gotu kola. *Iranian Med. J.* 98(10):250-251.

Rao, S.B., M. Chetana, and P. Uma Devi. 2005. *Centella asiatica* treatment during postnatal period enhances learning and memory in mice. *Physiol. Behav.* 86(4):449-457.

Rivera, I.G., M.T. Martins, P.S. Sanchez, et al. 1994. Genotoxicity assessment through the Ames test of medicinal-plants commonly used in Brazil. *Env. Toxicol. Water Qual.* 9(2):87-93.

Santucci, B., M. Picardo, and A. Cristaudo. 1985. Contact dermatitis due to centelase. *Contact Dermat.* 13(1):39.

Satake, T., K. Kamiya, Y. An, T. Oishi Nee Taka, and J. Yamamoto. 2007. The anti-thrombotic active constituents from *Centella asiatica*. *Biol. Pharm. Bull.* 30(5):935-940.

Siddique, Y.H., G. Ara, T. Beg, et al. 2008. Antigenotoxic role of *Centella asiatica* L. extract against cyproterone acetate induced genotoxic damage in cultured human lymphocytes. *Toxicol. In Vitro* 22(1):10-7.

Vena, G.A., and G. Angelini. 1986. Contact allergy to centelase. *Contact Dermat.* 15(2):108-109.

Yen, G.C., H.Y. Chen, and H.H. Peng. 2001. Evaluation of the cytotoxicity, mutagenicity and antimutagenicity of emerging edible plants. *Food Chem. Toxicol.* 39(11):1045-1053.

C

Cephaelis ipecacuanha (Brot.) Tussac — Rubiaceae

SCN: ipecac
Syn: *Psychotria ipecacuanha* (Brot.) Stokes
OCN: Brazilian ipecac; ipecacuanha; Rio ipecac
Part: rhizome

Quick Reference Summary

Safety Class: 3
Interaction Class: A

Contraindications

Not for use except under the supervision of an expert qualified in the appropriate use of this substance (CFR 2011; Manno and Manno 1977; Manoguerra and Cobaugh 2005; Quang and Woolf 2000).

Other Precautions

Sufficient doses will cause nausea and vomiting in most people (Lloyd 1897; Manno and Manno 1977; Manoguerra and Cobaugh 2005; Silber 2005; Wichtl 2004; Wood and LaWall 1926).

Drug and Supplement Interactions

None known.

Standard Dose

The emetic dose of ipecac syrup (U.S.P.) is 15 ml (CFR 2011). The standard therapeutic dose for ipecac as an expectorant is achieved at an adult dose of 0.4 to 1.4 ml of ipecac syrup (U.S.P.) (Martindale and Reynolds 1996).

For emetic purposes, a single dose is required for 80–85% of patients, while 10–15% require two doses, and 4–5% of patients do not vomit after administration of ipecac (Manoguerra and Cobaugh 2005).

Notice

Emetic (Leung and Foster 1996; Martindale and Reynolds 1996; Wichtl 2004; Wood and LaWall 1926); *see* Appendix 2.

Editors' Notes

Ipecac syrup sold in the United States must be in packages of less than 1.0 ounce and labeled with contraindications for use in unconscious persons, or following the consumption of "strychnine, corrosives such as alkalies (lye) and strong acids, or petroleum distillates such as kerosene, gasoline, coal oil, fuel oil, paint thinner, or cleaning fluid." Packaging must also direct the user to call a physician, poison control center, or hospital prior to use, and to keep the product out of reach of children. Labeling should also indicate that the standard dose of ipecac syrup is 1 tablespoon (15 ml) in persons over 1 year of age (CFR 2011).

Concentrated preparations such as fluid extracts of ipecac can be extremely toxic (Gilman et al. 1985; Martindale and Reynolds 1996) and should be used only under proper supervision. Ipecac has an irritant effect on the skin and mucous membranes and should be handled with care. Inhalation of airborne dust from ipecac has been associated with induction of asthma and sensitization to ipecac (Persson 1991; Wichtl 2004).

Adverse Events and Side Effects

Sufficient doses of ipecac syrup generally induce vomiting within 15 to 30 minutes after ingestion (Manoguerra and Cobaugh 2005).

Adverse events associated with use of ipecac for induction of vomiting include violent and prolonged vomiting, diarrhea, atypical lethargy, rupture of the stomach, tears of the upper gastrointestinal tract, hemorrhagic gastroenteritis, slowing of the heart rate, brain hemorrhages, and air in the space between the lungs (pneumomediastinum) (Czajka and Russell 1985; Quang and Woolf 2000).

Chronic ipecac abuse by anorexic or bulimic individuals has been associated with cardiac toxicity, generalized muscle weakness, metabolic disturbances, shock, and death (Quang and Woolf 2000).

Pharmacological Considerations

Use of ipecac in the management of poisonings is now discouraged by a number of organizations, with activated charcoal or other treatment protocols being recognized as safer and more effective (Manoguerra and Cobaugh 2005; Meadows-Oliver 2004; Silber 2005).

In 2003, the American Academy of Pediatrics issued a policy statement recommending that ipecac syrup no longer be routinely used in the home as a poison treatment strategy (AAP 2003). A 1997 joint statement issued by American and European toxicology organizations recommended that ipecac syrup not be routinely administered in the management of poisoned patients (Krenzelok et al. 1997). A 2005 statement from the American Association of Poison Control Centers provided further clinical guidelines indicating that ipecac may be appropriate for use in a limited number of situations. These guidelines indicate that, for emetic purposes, ipecac syrup should not be used unless directed by a healthcare professional, and should not be used if:

- A patient is comatose or has altered mental status and the risk of breathing in the stomach contents is high.
- The patient is having convulsions.
- The substance ingested is capable of causing altered mental status or convulsions.
- The substance ingested is a caustic or corrosive agent.
- The substance ingested is a low viscosity petroleum distillate with the potential for pulmonary aspiration.
- The patient has a medical condition that may be exacerbated by vomiting (e.g., severe hypertension, bradycardia, hemorrhagic diathesis) (Manoguerra and Cobaugh 2005).

Other guidelines list additional "relative contraindications" beyond the "absolute contraindications" listed above. These guidelines indicate that ipecac should not be used if:

- The patient is already vomiting.
- More than 1 hour has passed since ingestion of the product of concern.
- The patient is susceptible to bleeding or hemorrhaging (bleeding diathesis).
- An oral antidote to the consumed poison is available.
- The patient is less than 6 months of age.
- The patient is elderly or has a history of heart disease.
- The patient ingested cardiotoxic drugs (calcium-channel blockers, beta blockers) (Quang and Woolf 2000).

Ipecac should not be used in persons with heart disease (Gilman et al. 1985; Mitchell 1983).

Pregnancy and Lactation

No studies on the safety of ipecac during pregnancy or lactation were identified. Based on the emetic effects, use of ipecac during pregnancy is not recommended except under the supervision of a qualified healthcare practitioner.

While this review did not identify any concerns for use while nursing, safety has not been conclusively established, and this substance is not recommended for use except under the supervision of an expert qualified in its appropriate use.

Review Details

I. Drug and Supplement Interactions

Clinical Trials of Drug or Supplement Interactions

No clinical trials of drug or supplement interactions were identified.

Case Reports of Suspected Drug or Supplement Interactions

No case reports of suspected drug or supplement interactions were identified.

Animal Trials of Drug or Supplement Interactions

No animal trials of drug or supplement interactions were identified.

II. Adverse Events

Case Reports of Adverse Events

Adverse events associated with use of ipecac for induction of vomiting include violent and persistent vomiting, gastric rupture, Mallory-Weiss tears, hemorrhagic gastroenteritis, bradycardia, cerebral hemorrhages, and air in the space between the lungs (pneumomediastinum) (Quang and Woolf 2000). Cases of protracted vomiting (in some cases, lasting for several days) have been reported in 2- to 4-year-olds administered 15 to 30 ml of ipecac syrup. Some of these cases were fatal (Manno and Manno 1977).

Chronic ipecac abuse by anorexic or bulimic individuals has been associated with cardiac toxicity characterized by PR prolongation, T-wave abnormalities, QRS abnormalities, congestive cardiomyopathy, and atrial and ventricular dysrhythmia. Generalized muscle weakness, myopathy, diarrhea, mild tremors, edema, dehydration, metabolic disturbances (hypokalemia, hypochloremic acidosis, elevation of creatinine phosphokinase), shock, and death have also been reported after chronic use (Manno and Manno 1977; Quang and Woolf 2000).

Ipecac has an irritant effect on the skin and mucous membranes and should be handled with care. Inhalation of airborne dust from ipecac has been associated with induction of asthma and sensitization to ipecac (Persson 1991; Wichtl 2004).

III. Pharmacology and Pharmacokinetics

Human Pharmacological Studies

A follow-up study with 146 patients that had been administered ipecac syrup indicated that within 4 hours after ipecac-induced emesis, 33.6% had no symptoms and 17.1% experienced protracted emesis. Incidences of diarrhea and atypical lethargy were higher after ipecac-induced emesis than in patients not receiving ipecac syrup (Czajka and Russell 1985).

Animal Pharmacological Studies

In an animal model of anorexia nervosa, rats were intraperitoneally administered ipecac daily for 4 to 10 weeks. The treatment induced myopathy that resolved completely after 6 to 12 weeks after cessation of ipecac (Hopf and Goebel 1993).

In Vitro Pharmacological Studies

No relevant in vitro pharmacological studies were identified.

IV. Pregnancy and Lactation

No studies on the safety of ipecac during pregnancy or lactation were identified.

V. Toxicity Studies

Acute Toxicity

The LD_{50} of the compound emetine intraperitoneally administered in rats is 12.1 mg/kg, while that of the compound cephaeline is 9.9 mg/kg (Radomski et al. 1952).

Literature Cited

AAP. 2003. Policy statement: Poison treatment in the home. American Academy of Pediatrics. *Pediatrics* 112(5):1182-1185.

CFR. 2011. *Code of federal regulations*, Title 21 Part 201.308, 2011 ed. Specific labeling requirements for specific drug products. Ipecac syrup; warnings and directions for use for over-the-counter sale. Washington, DC: U.S. Government Printing Office.

Czajka, P.A., and S.L. Russell. 1985. Nonemetic effects of ipecac syrup. *Pediatrics* 75(6):1101-1104.

Gilman, A.G., L.S. Goodman, and A. Gilman. 1985. *Goodman & Gilman's the pharmacological basis of therapeutics*. 7th ed. New York: MacMillan.

Hopf, N.J., and H.H. Goebel. 1993. Experimental emetine myopathy: Enzyme histochemical, electron microscopic, and immunomorphological studies. *Acta Neuropathol.* 85(4):414-418.

Krenzelok, E.P., M. McGuigan, and P. Lheur. 1997. Position statement: Ipecac syrup. American Academy of Clinical Toxicology; European Association of Poisons Centres and Clinical Toxicologists. *J. Toxicol. Clin. Toxicol.* 35(7):699-709.

Leung, A.Y., and S. Foster. 1996. *Encyclopedia of common natural ingredients used in food, drugs, and cosmetics*. 2nd ed. New York: Wiley.

Lloyd, J.U. 1897. *Cephaelis ipecacuanha*. Cincinnati, OH: Self-published.

Manno, B.R., and J.E. Manno. 1977. Toxicology of ipecac: A review. *Clin. Toxicol.* 10(2):221-242.

Manoguerra, A.S., and D.J. Cobaugh. 2005. Guideline on the use of ipecac syrup in the out-of-hospital management of ingested poisons. *Clin. Toxicol. (Phila.)* 43(1):1-10.

Martindale, W., and J.E.F. Reynolds. 1996. *The extra pharmacopoeia*. 31st ed. London: Pharmaceutical Press.

Meadows-Oliver, M. 2004. Syrup of ipecac: New guidelines from the AAP. *J. Pediatr. Health Care* 18(2):109-110.
Mitchell, H. 1983. *British herbal pharmacopoeia*. Bournemouth, U.K.: British Herbal Medicine Association.
Persson, C.G. 1991. Ipecacuanha asthma: More lessons. *Thorax* 46(6):467-468.
Quang, L.S., and A.D. Woolf. 2000. Past, present, and future role of ipecac syrup. *Curr. Opin. Pediatr.* 12(2):153-162.
Radomski, J.L., E.C. Hagan, H.N. Fuyat, and A.A. Nelson. 1952. The pharmacology of ipecac. *J. Pharm. Exp. Ther.* 104(4):421-426.
Silber, T.J. 2005. Ipecac syrup abuse, morbidity, and mortality: Isn't it time to repeal its over-the-counter status? *J. Adolesc. Health* 37(3):256-260.
Wichtl, M. 2004. *Herbal drugs and phytopharmaceuticals: A handbook for practice on a scientific basis*. 3rd ed. Boca Raton, FL: CRC Press.
Wood, H., and C. LaWall. 1926. *The dispensatory of the United States of America*. Philadelphia: J.B. Lippincott.

Ceratonia siliqua L. — Fabaceae

SCN: carob
OCN: locust bean; St. John's bread
Part: fruit

Quick Reference Summary

Safety Class: 1
Interaction Class: A

Contraindications

None known.

Other Precautions

None known.

Drug and Supplement Interactions

None known.

Adverse Events and Side Effects

None known.

Pharmacological Considerations

None known.

Pregnancy and Lactation

No adverse effects of carob bean gum on fetal development were observed in animal studies (FAO/WHO 1981).

No information on the safety of carob bean gum during lactation was identified. While this review did not identify any concerns for use while nursing, safety has not been conclusively established.

Review Details

I. Drug and Supplement Interactions

Clinical Trials of Drug or Supplement Interactions

No clinical trials of drug or supplement interactions were identified.

Case Reports of Suspected Drug or Supplement Interactions

No case reports of suspected drug or supplement interactions were identified.

Animal Trials of Drug or Supplement Interactions

No animal trials of drug or supplement interactions were identified.

II. Adverse Events

Case Reports of Adverse Events

Occupational asthma from repeated exposure to carob flour has been reported (Scoditti et al. 1996; van der Brempt et al. 1992). An allergic reaction to carob bean gum has been reported (Savino et al. 1999).

III. Pharmacology and Pharmacokinetics

Human Pharmacological Studies

In healthy human volunteers, the addition of carob bean gum (9.5 g/1000 kcal) to the diet for 2 weeks resulted in a significant reduction in the absorption of calcium, iron, and zinc as compared to the control diet, while the absorption of copper remained unchanged (Harmuth-Hoene et al. 1982).

Animal Pharmacological Studies

No relevant animal pharmacological studies were identified.

In Vitro Pharmacological Studies

No relevant in vitro pharmacological studies were identified.

IV. Pregnancy and Lactation

No teratogenic effects of carob bean gum were observed in rats or mice orally administered up to 1300 mg/kg, hamsters administered up to 1000 mg/kg, or rabbits administered up to 196 mg/kg. A dose of 910 mg/kg was lethal to most pregnant rabbits, and a dose of 1300 mg/kg was lethal to 25% of pregnant mice (FAO/WHO 1981).

No information on the safety of carob bean gum during lactation was identified.

V. Toxicity Studies

Acute Toxicity

The LD_{50} of orally administered carob bean gum in rats could not be determined at doses up to 5 g/kg (FAO/WHO 1981).

Chronic Toxicity

No histopathological effects were observed in rats and mice fed diets containing 2.5 or 5.0% locust bean gum daily for 103 weeks. In male mice on the 5.0% diet, a depression in body weight gain was observed (Melnick et al. 1983; NTP 1981).

Genotoxicity

Weak mutagenic activity of a carob extract was observed in *Salmonella typhimurium* strains TA98 and TA100 (Alkofahi et al. 1990).

C

Literature Cited

Alkofahi, A.S., A. Abdelaziz, I. Mahmoud, et al. 1990. Cytotoxicity, mutagenicity and antimicrobial activity of forty Jordanian medicinal plants. *Int. J. Crude Drug Res.* 28(2):139-144.

FAO/WHO. 1981. Carob (Locust) bean gum. International Programme on Chemical Safety. Toxicological evaluation of certain food additives and food contaminants. In *WHO Food Additives Series*.

Harmuth-Hoene, A.E., A. Meier-Ploeger, and C. Leitzmann. 1982. Effect of carob bean flour on the resorption of minerals and trace elements in man. *Z. Ernahrungswiss.* 21(3):202-213.

Melnick, R.L., J. Huff, J.K. Haseman, et al. 1983. Chronic effects of agar, guar gum, gum arabic, locust-bean gum, or tara gum in F344 rats and B6C3F1 mice. *Food Chem. Toxicol.* 21(3):305-311.

NTP. 1981. Carcinogenesis bioassay of locust bean gum (CAS No. 9000-40-2) in F344 rats and B6C3F1 mice (feed study). In *National Toxicology Program Technical Report Series Report No 221*.

Savino, F., M.C. Muratore, L. Silvestro, R. Oggero, and M. Mostert. 1999. Allergy to carob gum in an infant. *J Pediatr. Gastroenterol. Nutr.* 29(4):475-476.

Scoditti, A., P. Peluso, R. Pezzuto, T. Giordano, and A. Melica. 1996. Asthma to carob bean flour. *Ann. Allergy Asthma Immunol.* 77(1):81.

van der Brempt, X., C. Ledent, and M. Mairesse. 1992. Rhinitis and asthma caused by occupational exposure to carob bean flour. *J. Allergy Clin. Immunol.* 90(6, Pt. 1):1008-1010.

Cetraria islandica (L.) Ach. — Parmeliaceae

SCN: Iceland moss

Part: thallus

Quick Reference Summary

Safety Class: 1
Interaction Class: A

Contraindications

None known.

Other Precautions

None known.

Drug and Supplement Interactions

None known.

Adverse Events and Side Effects

None known.

Pharmacological Considerations

None known.

Pregnancy and Lactation

No information on the safety of Iceland moss in pregnancy or lactation was identified in the scientific or traditional literature. Although this review did not identify any concerns for use while pregnant or nursing, safety has not been conclusively established.

Review Details

I. Drug and Supplement Interactions

Clinical Trials of Drug or Supplement Interactions

No clinical trials of drug or supplement interactions were identified.

Case Reports of Suspected Drug or Supplement Interactions

No case reports of suspected drug or supplement interactions were identified.

C

Animal Trials of Drug or Supplement Interactions

No animal trials of drug or supplement interactions were identified.

II. Adverse Events

Adverse Events Reported in Clinical Trials

In clinical studies, pastilles containing amounts equivalent to 0.5 to 5 g of Iceland moss were generally well tolerated when taken daily for 4 or 5 days (Bradley 2006; ESCOP 2003).

Case Reports of Adverse Events

No case reports of adverse events were identified.

III. Pharmacology and Pharmacokinetics

Human Pharmacological Studies

No relevant human pharmacological studies were identified.

Animal Pharmacological Studies

As compared to control animals, significantly less arthritis was observed in mice with antigen-induced arthritis subcutaneously administered 2.5 mg/kg of an aqueous Iceland moss extract twice a week for 14 days before arthritis was induced and continuing for 36 days after (Freysdottir et al. 2008).

In Vitro Pharmacological Studies

Upregulation of IL10 and IL12p40 secretion was observed in human monocyte-derived immature dendritic cells cultured with an aqueous extract from Iceland moss (Freysdottir et al. 2008).

IV. Pregnancy and Lactation

No information on the safety of Iceland moss in pregnancy or lactation was identified.

V. Toxicity Studies

Acute Toxicity

Gastrointestinal upset was observed in mice fed diets containing 50% Iceland moss daily, and the mice died within 6 days (Airaksinen et al. 1986a).

Subchronic Toxicity

In rats fed diets containing 25% Iceland moss (pretreated either by boiling for 10 min, or soaking in a 2% wood ash solution for 2 days) for 3 months, kidneys of some animals had focal tubular damage and an increase in glomerular cell counts. Urinary protein levels were higher as compared to control. A reduction in weight gain was observed (Airaksinen et al. 1986b).

Literature Cited

Airaksinen, M.M., P. Peura, L. Ala-Fossi-Salokangas, et al. 1986a. Toxicity of plant material used as emergency food during famines in Finland. *J. Ethnopharmacol.* 18(3):273-296.

Airaksinen, M.M., P. Peura, and S. Antere. 1986b. Toxicity of Iceland lichen and reindeer lichen. *Arch. Toxicol. Suppl.* 9:406-409.

Bradley, P.R. 2006. *British herbal compendium: A handbook of scientific information on widely used plant drugs. Volume* 2. Bournemouth, UK: British Herbal Medicine Association.

ESCOP. 2003. *ESCOP monographs: The scientific foundation for herbal medicinal products.* 2nd ed. Exeter, U.K.: European Scientific Cooperative on Phytotherapy.

Freysdottir, J., S. Omarsdottir, K. Ingolfsdottir, A. Vikingsson, and E.S. Olafsdottir. 2008. In vitro and in vivo immunomodulating effects of traditionally prepared extract and purified compounds from *Cetraria islandica. Int. Immunopharmacol.* 8(3):423-430.

Chaenomeles speciosa (Sweet) Nakai — Rosaceae

SCN: flowering quince
Syn: *Chaenomeles lagenaria* (Loisel.) Koidz.
PN: *mu gua* (fruit)
OCN: Chinese quince; Japanese quince
Part: fruit

Quick Reference Summary

Safety Class: 1
Interaction Class: A

Contraindications

None known

Other Precautions

None known.

Drug and Supplement Interactions

None known.

Editors' Note

Although one reference text on traditional Chinese medicine contraindicates flowering quince in persons with excessive stomach acid (Bensky et al. 2004), another reference does not indicate any such concern (Chen and Chen 2004).

Adverse Events and Side Effects

Allergic reactions have been reported after topical exposure to flowering quince (Bensky et al. 2004).

Pharmacological Considerations

None known.

Pregnancy and Lactation

No information on the safety of flowering quince in pregnancy or lactation was identified in the scientific or traditional literature. Although this review did not identify any concerns for use while pregnant or nursing, safety has not been conclusively established.

C

Review Details

I. Drug and Supplement Interactions

Clinical Trials of Drug or Supplement Interactions

No clinical trials of drug or supplement interactions were identified.

Case Reports of Suspected Drug or Supplement Interactions

No case reports of suspected drug or supplement interactions were identified.

Animal Trials of Drug or Supplement Interactions

No animal trials of drug or supplement interactions were identified.

II. Adverse Events

Case Reports of Adverse Events

Allergic reactions have been reported after topical exposure to flowering quince (Bensky et al. 2004).

III. Pharmacology and Pharmacokinetics

Human Pharmacological Studies

No relevant human pharmacological studies were identified.

Animal Pharmacological Studies

No relevant animal pharmacological studies were identified.

In Vitro Pharmacological Studies

No relevant in vitro pharmacological studies were identified.

IV. Pregnancy and Lactation

No information on the safety of flowering quince during pregnancy or lactation was identified.

V. Toxicity Studies

No toxicity studies were identified.

Literature Cited

Bensky, D., S. Clavey, and E. Stöger. 2004. *Chinese herbal medicine: Materia medica.* 3rd ed. Seattle: Eastland Press.

Chen, J.K., and T.T. Chen. 2004. *Chinese medical herbology and pharmacology*. City of Industry, CA: Art of Medicine Press.

Chamaelirium luteum (L.) A. Gray — Liliaceae

SCN: false unicorn
OCN: blazing star; fairy wand; helonias
Part: rhizome

Quick Reference Summary

Safety Class: 1
Interaction Class: A

Contraindications

None known.

Other Precautions

None known.

Drug and Supplement Interactions

None known.

Adverse Events and Side Effects

Large doses (standard dose listed as 0.5–1 g of powdered root) of false unicorn may be emetic (Felter and Lloyd 1898).

Pharmacological Considerations

None known.

Pregnancy and Lactation

Historically, false unicorn has been used to treat morning sickness and to reduce the risk of miscarriage in women with a history of miscarriage (Felter and Lloyd 1898).

No information on the safety of false unicorn in lactation was identified. While this review did not identify any concerns for use while nursing, safety has not been conclusively established.

C

Review Details

I. Drug and Supplement Interactions

Clinical Trials of Drug or Supplement Interactions

No clinical trials of drug or supplement interactions were identified.

Case Reports of Suspected Drug or Supplement Interactions

No case reports of suspected drug or supplement interactions were identified.

Animal Trials of Drug or Supplement Interactions

No animal trials of drug or supplement interactions were identified.

II. Adverse Events

Case Reports of Adverse Events

No case reports of adverse events were identified.

III. Pharmacology and Pharmacokinetics

Human Pharmacological Studies

No relevant human pharmacological studies were identified.

Animal Pharmacological Studies

No relevant animal pharmacological studies were identified.

In Vitro Pharmacological Studies

An aqueous-ethanolic extract of false unicorn showed no significant binding activity in assays with estradiol and progesterone receptors (Zava et al. 1998).

IV. Pregnancy and Lactation

Historically, false unicorn has been used to treat morning sickness and to reduce the risk of miscarriage in women with a history of miscarriage (Felter and Lloyd 1898).

No information on the safety of false unicorn during lactation was identified.

V. Toxicity Studies

No toxicity studies were identified.

Literature Cited

Felter, H.W., and J.U. Lloyd. 1898. *King's American dispensatory*. 18th ed., 3rd rev. 2 vols. Cincinnati: Ohio Valley Co.

Zava, D.T., C.M. Dollbaum, and M. Blen. 1998. Estrogen and progestin bioactivity of foods, herbs, and spices. *Proc. Soc. Exp. Biol. Med.* 217(3):369-378.

Chamaemelum nobile (L.) All. — Asteraceae

SCN: Roman chamomile
Syn: *Anthemis nobilis* L.
OCN: dog fennel; English chamomile
Part: flower

Quick Reference Summary

Safety Class: 2b
Interaction Class: A

Contraindications

Not for use in pregnancy except under the supervision of a qualified healthcare practitioner (Kochmann 1931; van Itallie et al. 1932).

Other Precautions

Persons with allergies to other members of the Asteraceae family (such as feverfew or chamomile) should exercise caution with Roman chamomile, as allergic cross-reactivity is common to Asteraceae plants (Hausen 1979).

Drug and Supplement Interactions

None known.

Notice

Abortifacient (List and Hörhammer 1973); *see* Appendix 2.

Emmenagogue (List and Hörhammer 1973; Palma 1964; Saha et al. 1961); *see* Appendix 2.

Adverse Events and Side Effects

Allergic reactions to Roman chamomile have been reported and confirmed by patch testing (Bossuyt and Dooms-Goossens 1994; Giordano-Labadie et al. 2000; Hausen 1979; McGeorge and Steele 1991; Paulsen 2002; Pereira et al. 1997).

Pharmacological Considerations

None known.

Pregnancy and Lactation

References from the 1930s indicate that Roman chamomile was an ingredient in abortifacient formulas sold in

Germany at that time (Kochmann 1931; van Itallie et al. 1932). Emmenagogue activity of Roman chamomile has been reported (Palma 1964; Saha et al. 1961). Based on these reports, use during pregnancy is not recommended except under the supervision of a qualified healthcare practitioner.

No information on the safety of Roman chamomile during lactation was identified. While this review did not identify any concerns for use while nursing, safety has not been conclusively established.

Review Details

I. Drug and Supplement Interactions

Clinical Trials of Drug or Supplement Interactions

No clinical trials of drug or supplement interactions were identified.

Case Reports of Suspected Drug or Supplement Interactions

No case reports of suspected drug or supplement interactions were identified.

Animal Trials of Drug or Supplement Interactions

No animal trials of drug or supplement interactions were identified.

II. Adverse Events

Case Reports of Adverse Events

A student in an aromatherapy class became light-headed, with tachycardia, high blood pressure, and nausea, after inhaling Roman chamomile essential oil (Maddocks-Jennings 2004).

Contact dermatitis from Roman chamomile tea and a topical cream containing Roman chamomile have been reported and confirmed by patch testing (Bossuyt and Dooms-Goossens 1994; Giordano-Labadie et al. 2000; McGeorge and Steele 1991; Pereira et al. 1997). Sesquiterpene lactone compounds in Roman chamomile are generally recognized as being responsible for this effect (Paulsen 2002).

III. Pharmacology and Pharmacokinetics

Human Pharmacological Studies

Among 290 patients with eczema, patch testing with 5% Roman chamomile elicited a positive reaction in only one person (Meneghini et al. 1971). Patch testing with Roman chamomile in individuals with prior allergy to different Compositae plants indicated that 2 of 25 patients were allergic to Roman chamomile (Hausen 1979). A maximization test in 25 healthy volunteers indicated no sensitization reactions to 4% Roman chamomile essential oil (Opdyke 1979).

Animal Pharmacological Studies

In rats intravenously administered 50, 100, or 200 mg/kg of an aqueous extract of Roman chamomile, an increase in urine output was observed at the highest dose. Dose-dependent hypotensive activity was observed at all dose levels (Zeggwagh et al. 2007).

Undiluted Roman chamomile oil applied to the backs of hairless mice produced no irritating effects. Mild irritation was observed after the oil was applied full strength to intact or abraded rabbit skin for 24 hours (Opdyke 1979).

In Vitro Pharmacological Studies

No relevant in vitro pharmacological studies were identified.

IV. Pregnancy and Lactation

References from the 1930s indicate that Roman chamomile was an ingredient in abortifacient formulas sold in Germany at that time (Kochmann 1931; van Itallie et al. 1932). Emmenagogue activity of Roman chamomile has been reported (Palma 1964; Saha et al. 1961).

No information on the safety of Roman chamomile during lactation was identified.

V. Toxicity Studies

Acute Toxicity

The LD_{50} of Roman chamomile could not be determined at doses up to 5 g/kg after oral or topical administration (Opdyke 1979).

Genotoxicity

No mutagenic activity of Roman chamomile essential oil was seen in the *Bacillus subtilis rec* assay and the *Salmonella*/microsome reversion assay (Zani et al. 1991).

Literature Cited

Bossuyt, L., and A. Dooms-Goossens. 1994. Contact sensitivity to nettles and camomile in "alternative" remedies. *Contact Dermat.* 31(2):131-132.

Giordano-Labadie, F., H.P. Schwarze, and J. Bazex. 2000. Allergic contact dermatitis from chamomile used in phytotherapy. *Contact Dermat.* 42(2):247.

Hausen, B.M. 1979. The sensitizing capacity of Compositae plants. III. Test results and cross-reactions in Compositae-sensitive patients. *Dermatologica* 159(1):1-11.

Kochmann, M. 1931. *Anthemis nobilis* und apiol, sind sie abortivmittel? *Arch. Toxicol.* 2(1):35-36.

List, P.H., and H. Hörhammer. 1973. *Hagers handbuch der pharmazeutischen praxis.* Berlin: Springer.

Maddocks-Jennings, W. 2004. Critical incident: Idiosyncratic allergic reactions to essential oils. *Complement Ther. Nurse Midwifery* 10(1):58-60.

McGeorge, B.C.L., and M.C. Steele. 1991. Allergic contact dermatitis of the nipple from Roman chamomile ointment. *Contact Dermat.* 24:139-140.

Meneghini, C.L., F. Rantuccio, and M. Lomuto. 1971. Additives, vehicles and active drugs of topical medicaments as causes of delayed type allergic dermatitis. *Dermatologica* 143:137-147.

Opdyke, D.L.J. 1979. *Monographs on fragrance raw materials.* New York: Pergamon.

Palma, L. 1964. *Cited in* Farnsworth, N.R., A.S. Bingel, G.A. Cordell, F.A. Crane, and H.H. Fong. 1975. Potential value of plants as sources of new antifertility agents I. *J. Pharm. Sci.* 64(4):535-598.

Paulsen, E. 2002. Contact sensitization from Compositae-containing herbal remedies and cosmetics. *Contact Dermat.* 47(4):189-198.

Pereira, F., R. Santos, and A. Pereira. 1997. Contact dermatitis from chamomile tea. *Contact Dermat.* 37(6):307.

Saha, J.C., E.C. Savini, and S. Kasinthan. 1961. *Cited in* Farnsworth, N.R., A.S. Bingel, G.A. Cordell, F.A. Crane, and H.H. Fong. 1975. Potential value of plants as sources of new antifertility agents I. *J. Pharm. Sci.* 64(4):535-598.

van Itallie, L., A. Harmsma, and L.W. van Esveld. 1932. Abortifacients, particularly apiole. *Arch. Exp. Pathol. Pharmakol.* 165:84-100.

Zani, F., G. Massimo, S. Benvenuti, et al. 1991. Studies on the genotoxic properties of essential oils with *Bacillus subtilis* rec-assay and *Salmonella*/microsome reversion assay. *Planta Med.* 57(3):237-241.

Zeggwagh, N.A., J.B. Michel, M. Eddouks, and A. Pereira. 2007. Acute hypotensive and diuretic activities of *Chamaemelum nobile* aqueous extract in normal rats. *Am. J. Pharmacol. Toxicol.* 2(3):140-145.

Chamaesyce hirta (L.) Millsp. — Euphorbiaceae

SCN: pill-bearing spurge
Syn: *Euphorbia hirta* L.; *Euphorbia pilulifera* auct. non L.
OCN: asthma herb; garden euphorbia
Part: herb

Quick Reference Summary

Safety Class: 1
Interaction Class: A

Contraindications

None known.

Other Precautions

None known.

Drug and Supplement Interactions

None known.

Standard Dose

As an expectorant, 2.0 g (Cook and Martin 1948; Powers et al. 1942); the emetic dose may be similar.

Notice

Diuretic (Johnson et al. 1999); *see* Appendix 2.

Adverse Events and Side Effects

May cause nausea and vomiting (Felter and Lloyd 1898; List and Hörhammer 1973).

Pharmacological Considerations

Traditional use and an animal study indicate that pill-bearing spurge is a diuretic (Johnson et al. 1999).

Pregnancy and Lactation

No information on the safety of pill-bearing spurge in pregnancy or lactation was identified in the scientific or traditional literature. Although this review did not identify any concerns for use while pregnant or nursing, safety has not been conclusively established.

Review Details

I. Drug and Supplement Interactions

Clinical Trials of Drug or Supplement Interactions

No clinical trials of drug or supplement interactions were identified.

Case Reports of Suspected Drug or Supplement Interactions

No case reports of suspected drug or supplement interactions were identified.

Animal Trials of Drug or Supplement Interactions

No animal trials of drug or supplement interactions were identified.

II. Adverse Events

Case Reports of Adverse Events

No case reports of adverse events were identified.

III. Pharmacology and Pharmacokinetics

Human Pharmacological Studies

No relevant human pharmacological studies were identified.

Animal Pharmacological Studies

In rats orally administered single doses of 50 or 100 mg/kg of water or ethanol extracts of pill-bearing spurge daily, a time-dependent increase in urine output was observed. The water extract increased the urine excretion of sodium, potassium, and bicarbonate. In contrast, the ethanol extract increased the excretion of bicarbonate, decreased the loss of potassium, and had little effect on renal removal of sodium. The activity of pill-bearing spurge was noted as similar to the drug acetazolamide (Johnson et al. 1999).

In Vitro Pharmacological Studies

No relevant in vitro pharmacological studies were identified.

IV. Pregnancy and Lactation

An ethnobotanical study indicated that pill-bearing spurge is used in Trinidad and Tobago to treat infertility (Lans 2007).

No information on the safety of pill-bearing spurge during lactation was identified.

V. Toxicity Studies

Acute Toxicity

No adverse effects were observed in chickens orally administered 2 g/kg of an aqueous extract of pill-bearing spurge. Parameters observed included liver enzyme levels and microscopic evaluation of livers and kidneys (Hashemi et al. 2008).

Short-Term Toxicity

In rats orally administered 10 g/kg of pill-bearing spurge aqueous extract daily for 14 days, leukocytosis, macrocytic hypochromic anemia, reduction in packed cell volume, and an elevation of liver enzyme levels (ALT and AST) were observed. By the end of the study, 20% of the animals in the pill-bearing spurge group had died (Adedapo et al. 2004).

Literature Cited

Adedapo, A.A., M.O. Abatan, and O.O. Olorunsogo. 2004. Toxic effects of some plants in the genus *Euphorbia* on haematological and biochemical parameters of rats. *Vet. Arhiv.* 74(1):53-62.

Cook, E.F., and E.W. Martin. 1948. *Remington's practice of pharmacy.* 9th ed. Easton, PA: Mack Publishing Company.

Felter, H.W., and J.U. Lloyd. 1898. *King's American dispensatory.* 18th ed., 3rd rev. 2 vols. Cincinnati: Ohio Valley Co.

Hashemi, S.R., I. Zulkifli, M.H. Bejo, A. Farida, and M.N. Somchit. 2008. Acute toxicity study and phytochemical screening of selected herbal aqueous extract in broiler chickens. *Int. J. Pharmacol.* 4(5):352-360.

Johnson, P.B., E.M. Abdurahman, E.A. Tiam, I. Abdu-Aguye, and I.M. Hussaini. 1999. *Euphorbia hirta* leaf extracts increase urine output and electrolytes in rats. *J. Ethnopharmacol.* 65(1):63-69.

Lans, C. 2007. Ethnomedicines used in Trinidad and Tobago for reproductive problems. *J. Ethnobiol. Ethnomed.* 3:13.

List, P.H., and H. Hörhammer. 1973. *Hagers handbuch der pharmazeutischen praxis.* Berlin: Springer.

Powers, J.L., E.H. Wirth, and A.B. Nichols. 1942. *National formulary.* 7th ed. Washington, DC: American Pharmaceutical Association.

Changium smyrnioides Wolff — Apiaceae

SCN: changium
PN: *ming dang shen* (root)

Part: root

Quick Reference Summary

Safety Class: 2b
Interaction Class: A

Contraindications

Not for use in pregnancy except under the supervision of a qualified healthcare practitioner (Bensky et al. 2004).

Other Precautions

None known.

Drug and Supplement Interactions

None known.

Adverse Events and Side Effects

None known.

Pharmacological Considerations

None known.

Pregnancy and Lactation

A reference text on traditional Chinese medicine indicates that changium should not be used during pregnancy (Bensky et al. 2004).

No information on the safety of changium during lactation was identified. While this review did not identify any concerns for use while nursing, safety has not been conclusively established.

Review Details

C

I. Drug and Supplement Interactions

Clinical Trials of Drug or Supplement Interactions

No clinical trials of drug or supplement interactions were identified.

Case Reports of Suspected Drug or Supplement Interactions

No case reports of suspected drug or supplement interactions were identified.

Animal Trials of Drug or Supplement Interactions

No animal trials of drug or supplement interactions were identified.

II. Adverse Events

Case Reports of Adverse Events

No case reports of adverse events were identified.

III. Pharmacology and Pharmacokinetics

Human Pharmacological Studies

No relevant human pharmacological studies were identified.

Animal Pharmacological Studies

No relevant animal pharmacological studies were identified.

In Vitro Pharmacological Studies

No relevant in vitro pharmacological studies were identified.

IV. Pregnancy and Lactation

A reference text on traditional Chinese medicine indicates that changium should not be used during pregnancy (Bensky et al. 2004).

No information on the safety of changium during lactation was identified.

V. Toxicity Studies

No toxicity studies were identified.

Literature Cited

Bensky, D., S. Clavey, and E. Stöger. 2004. *Chinese herbal medicine: Materia medica.* 3rd ed. Seattle: Eastland Press.

Chelidonium majus L. — Papaveraceae

SCN: celandine
OCN: greater celandine
Part: herb

Quick Reference Summary

Safety Class: 2b, 2c, 2d
Interaction Class: A

Contraindications

Not for use in pregnancy or lactation except under the supervision of a qualified healthcare practitioner (BfArM 2008).

Not for use in persons with biliary obstruction, having a history of liver disease, or taking substances contraindicated with liver disease (Benninger et al. 1999; BfArM 2008; Crijns et al. 2002; ESCOP 2003; Greving et al. 1998; Hardeman et al. 2008; Rifai et al. 2006; Stickel et al. 2003).

Not for use in excess of 2 to 4 weeks unless under the supervision of a qualified healthcare practitioner (BfArM 2008; ESCOP 2003; Mills and Bone 1999; Wichtl 2004).

Do not exceed recommended dose (ESCOP 2003).

Other Precautions

If signs of liver damage (yellowing of the skin or eyes, dark urine, abnormally colored stool, pain in the upper abdomen, nausea, loss of appetite, fatigue) are observed during celandine use, celandine use should be stopped, and appropriate medical attention should be sought (BfArM 2008).

Drug and Supplement Interactions

None known.

Standard Dose

The recommended dose is 125–700 mg of standardized hydroalcoholic extract (9–24 mg total alkaloids); 2–4 ml of tincture three times daily; 1–2 ml of fluid extract three times daily (ESCOP 2003).

Notice

Berberine (0.011–0.25%) (Frohne and Pfänder 2000; Kursinszki et al. 2006; Osol and Farrar 1955; Sárközi et al. 2006; Tomè and Colombo 1995); *see* Appendix 1.

Editors' Note

Due to reported cases of liver toxicity, German regulatory authorities recommend that liver function values should be monitored if celandine is used for more than 4 weeks at a time at daily doses greater than 2.5 g or containing over 2.5 mg of alkaloids (calculated as chelidonin) (BfArM 2008; ESCOP 2003; Wichtl 2004).

Adverse Events and Side Effects

Cases of acute hepatitis, often cholestatic in nature, have been reported in persons taking celandine (Benninger et al. 1999; Crijns et al. 2002; Greving et al. 1998; Hardeman et al. 2008; Rifai et al. 2006; Stickel et al. 2003). None of the cases of hepatitis were fatal, and all were reversed on discontinuation of celandine. In some cases the patients were also taking pharmaceutical drugs that have been associated with cholestatic hepatitis (Huang and Liaw 1995; Kullak-Ublick and Meier 2000).

An herbal reference text notes that celandine infusions (hot water extracts) have significantly lower concentrations of alkaloids as compared to more recently available commercial extracts, and that the historical lack of adverse effects may be attributed to the lower alkaloid content in traditional preparations (Wichtl 2004).

Mild gastrointestinal disturbances, such as nausea, upset stomach, and diarrhea, have been reported as side effects of celandine (ESCOP 2003). Severe irritation of the digestive system has been reported in association with consumption of the fresh plant (Kingsbury 1964; Nelson et al. 2006).

Pharmacological Considerations

None known.

Pregnancy and Lactation

No information on the safety of celandine in pregnancy or lactation was identified. In this work, the contraindications for use in pregnancy and lactation are based primarily on concerns regarding the cases of liver toxicity reported in association with celandine use, as the implications of these case reports, and possible mechanisms of liver toxicity, have yet to be fully understood.

Review Details

I. Drug and Supplement Interactions

Clinical Trials of Drug or Supplement Interactions

No clinical trials of drug or supplement interactions were identified.

Case Reports of Suspected Drug or Supplement Interactions

No case reports of suspected drug or supplement interactions were identified.

Animal Trials of Drug or Supplement Interactions

No animal trials of drug or supplement interactions were identified.

II. Adverse Events

Case Reports of Adverse Events

A case series describes 10 incidences of acute hepatitis that were observed over a 2-year period. Cholestasis was observed in 5 of the 10 patients, and in 7 of the 10 cases, the hepatitis was characterized as drug-induced. No liver failure occurred in any of the patients, and cessation of celandine resulted in resolution of hepatitis. Use of celandine was from 1 to 9 months. Prescription and nonprescription drugs and herbal supplements were also being used in some cases, including thyroxine (2 patients), iodide (2 patients), estradiol (2 patients), and amitriptyline (1 patient) (Benninger et al. 1999). Cholestatic hepatitis has been associated with antithyroid medications, estradiol, and amitriptyline (Huang and Liaw 1995; Kullak-Ublick and Meier 2000; Larrey et al. 1988; Randeva et al. 2000).

Cholestatic hepatitis was reported in a 35-year-old woman who had been taking celandine for 4 months. The woman had been taking thyroxine for 5 years and took a course of roxithromycine in the month prior to the hepatitis (Greving et al. 1998). Severe cholestatic hepatitis was reported in a 39-year-old woman, with both incidences temporally related to the use of celandine. In both cases, the woman was taking or had recently finished taking either sulfamethoxazole or penicillin (Stickel et al. 2003).

Noncholestatic hepatitis was reported in a 28-year-old woman who had been taking celandine for 4 months (~4–8 mg of "active principle" per dose). The woman had been taking mestranol and chlormadinone acetate for several years, St. John's wort for 2 months, and had completed a course of tetracycline 6 months prior to the hepatitis (Greving et al. 1998).

Cholestatic hepatitis indicative of drug toxicity was reported in a 69-year-old man that had taken celandine capsules (about 2 daily, each containing 4 mg of chelidonin) (Stickel et al. 2003). A 58-year-old man experienced severe acute cholestatic hepatitis in association with use of a celandine and turmeric product (dose and duration of use not specified) (Rifai et al. 2006). A 42-year-old woman developed jaundice due to acute hepatitis after several weeks of using celandine. Liver functions returned to normal within 2 months of withdrawal from celandine (Crijns et al. 2002). Acute hepatitis was reported in a 58-year-old woman after she began taking capsules of celandine, gentian, and turmeric (Hardeman et al. 2008).

Hemolytic anemia with kidney failure, liver cytolysis, and thrombocytopenia was reported in a 72-year-old woman who had occasionally consumed a tea of celandine (Pinto Garcia et al. 1990).

Overdose of celandine may cause abdominal pain, gastrointestinal cramps, urinary urgency, and hematuria (ESCOP 2003).

Contact dermatitis was reported in a 64-year-old woman who was using celandine juice topically (Etxenagusia et al. 2000).

III. Pharmacology and Pharmacokinetics

Human Pharmacological Studies

No relevant human pharmacological studies were identified.

Animal Pharmacological Studies

No hepatoprotective activity was observed against carbon tetrachloride-induced liver toxicity in rats after animals were intraperitoneally administered 100 or 200 mg/kg of an aqueous extract of celandine (Sever Yilmaz et al. 2007).

Hepatoprotective, antitumor, and antigenotoxic activity was reported in mice with chemically induced liver tumors orally administered 0.1 ml daily (per 18 to 22 g animal) of an ethanol extract of celandine for 60 or 120 days (Biswas et al. 2008).

In Vitro Pharmacological Studies

Alkaloids isolated from celandine (chelerythrine, sanguinarine, berberine, and coptisine) had a significant inhibitory effect in mitochondrial respiration in mouse liver cells (Barreto et al. 2003).

IV. Pregnancy and Lactation

No information on the safety of celandine in pregnancy or lactation was identified.

V. Toxicity Studies

See *Sanguinaria canadensis* for additional toxicity studies on the compound sanguinarine.

Acute Toxicity

The LD_{50} of an intraperitoneally administered decoction of celandine in mice is 9.5 g/kg (Chang and But 1986). The LD_{50} of a subcutaneously administered extract of the alkaloids of celandine in mice is 300 mg/kg (Huang 1993).

The LD_{50} of intraperitoneally administered chelidonine in mice is 1.3 g/kg and in rats is 2 g/kg. Sublethal doses caused ptosis, tremor, sedation, and a decrease in body temperature (Jagiello-Wojtowicz et al. 1989).

Ingestion of 500 g of celandine can cause toxic effects in cattle and horses (Frohne and Pfänder 2000).

Subchronic Toxicity

No changes in liver enzymes were observed in pigs administered the compounds sanguinarine and chelerythrine (3:1 ratio) as 0.1 to 0.0002% of the diet for 90 days (Kosina et al. 2004).

Genotoxicity

DNA-damaging effects of the alkaloids chelidonine and sanguinarine were detected in the comet assay, with sanguinarine being active at a lower concentration than chelidonine (Philchenkov et al. 2008).

Cytotoxicity

In the brine shrimp lethality assay, the LC_{50} of chelidonine was 319 ppm, and for prototropine was 49 ppm. LC_{50} values for ethanol and aqueous extracts of celandine could not be determined at concentrations up to 1000 ppm (Saglam and Arar 2003).

Cytotoxic activity of an ethanolic celandine extract was observed in human lymphoblastoid cells treated with 10 or 50 μg/ml of the extract (Spiridonov et al. 2005).

The compounds chelidonine and sanguinarine induced apoptosis in human acute T-lymphoblastic leukemia MT-4 cells (Philchenkov et al. 2008).

Cytotoxicity tests in rat hepatocytes indicated that EC_{50} values for alkaloids isolated from celandine were as follows: 5 μg/ml for sanguinarine, 8 μg/ml for chelerythrine, 13 μg/ml for coptisine, 100 μg/ml for prototropine, and over 100 μg/ml for chelidonine (ESCOP 2003).

Literature Cited

Barreto, M.C., R.E. Pinto, J.D. Arrabaca, and M.L. Pavao. 2003. Inhibition of mouse liver respiration by *Chelidonium majus* isoquinoline alkaloids. *Toxicol. Lett.* 146(1):37-47.

Benninger, J., H.T. Schneider, D. Schuppan, T. Kirchner, and E.G. Hahn. 1999. Acute hepatitis induced by greater celandine (*Chelidonium majus*). *Gastroenterology* 117(5):1234-1237.

BfArM. 2008. Abwehr von gefahren durch arzneimittel, stufe II: Schöllkraut-haltige arzneimittel zur innerlichen anwendung. April 9, 2008. Bonn: Bundesinstitut für Arzneimittel und Medizinprodukte.

Biswas, S.J., N. Bhattacharjee, and A.R. Khuda-Bukhsh. 2008. Efficacy of a plant extract (*Chelidonium majus* L.) in combating induced hepatocarcinogenesis in mice. *Food Chem. Toxicol.* 46(5):1474-1487.

Chang, H.-M., and P.P.H. But. 1986. *Pharmacology and applications of Chinese materia medica*. English ed. Singapore, Philadelphia: World Scientific.

Crijns, A.P., P.G.A.M. De Smet, M. van den Heuvel, B.W. Schot, and E.B. Haagsma. 2002. Acute hepatitis after use of a herbal preparation with greater celandine (*Chelidonium majus*). *Ned. Tijdschr. Geneeskd.* 146(3):124-128.

ESCOP. 2003. *ESCOP monographs: The scientific foundation for herbal medicinal products*. 2nd ed. Exeter, U.K.: European Scientific Cooperative on Phytotherapy.

Etxenagusia, M.A., M. Anda, I. Gonzalez-Mahave, E. Fernandez, and L. Fernandez de Corres. 2000. Contact dermatitis from *Chelidonium majus* (greater celandine). *Contact Dermat.* 43(1):47.

Frohne, D., and H.J. Pfänder. 2000. *A colour atlas of poisonous plants: A handbook for pharmacists, doctors, toxicologists, biologists and veterinarians*. 2nd ed. London: Manson.

Greving, I., V. Meister, C. Monnerjahn, K.M. Muller, and B. May. 1998. *Chelidonium majus*: A rare reason for severe hepatotoxic reaction. *Pharmacoepidemiol. Drug Safety* 7(Suppl.):S66-S69.

Hardeman, E., L. Van Overbeke, S. Ilegems, and M. Ferrante. 2008. Acute hepatitis induced by greater celandine (*Chelidonium majus*). *Acta Gastroenterol. Belg.* 71(2):281-282.

Huang, K.C. 1993. *The pharmacology of Chinese herbs*. Boca Raton, FL: CRC Press.

Huang, M.J., and Y.F. Liaw. 1995. Clinical associations between thyroid and liver diseases. *J. Gastroenterol. Hepatol.* 10(3):344-350.

Jagiello-Wojtowicz, E., L. Jusiak, J. Szponar, and Z. Kleinrok. 1989. Preliminary pharmacological evaluation of chelidonine in rodents. *Pol. J. Pharmacol. Pharm.* 41(2):125-131.

Kingsbury, J.M. 1964. *Poisonous plants of the United States and Canada*. Prentice-Hall biological science series. Englewood Cliffs, NJ: Prentice-Hall.

Kosina, P., D. Walterova, J. Ulrichova, et al. 2004. Sanguinarine and chelerythrine: Assessment of safety on pigs in ninety days feeding experiment. *Food Chem. Toxicol.* 42(1):85-91.

Kullak-Ublick, G.A., and P.J. Meier. 2000. Mechanisms of cholestasis. *Clin. Liver Dis.* 4(2):357-385.

Kursinszki, L., Á. Sárközi, Á. Kéry, and É. Szöke. 2006. Improved RP-HPLC method for analysis of isoquinoline alkaloids in extracts of *Chelidonium majus*. *Chromatographia* 63:131-135.

Larrey, D., G. Amouyal, D. Pessayre, et al. 1988. Amitriptyline-induced prolonged cholestasis. *Gastroenterology* 94(1):200-203.

Mills, S., and K. Bone. 1999. *Principles and practice of phytotherapy: Modern herbal medicine*. London: Churchill Livingstone.

Nelson, L., R.D. Shih, M.J. Balick, and K.F. Lampe. 2006. *Handbook of poisonous and injurious plants*. 2nd ed. New York: New York Botanical Garden.

Osol, A., and G. Farrar. 1955. *The dispensatory of the United States of America*. 25th ed. Philadelphia: Lippincott.

Philchenkov, A., V. Kaminskyy, M. Zavelevich, and R. Stoika. 2008. Apoptogenic activity of two benzophenanthridine alkaloids from *Chelidonium majus* L. does not correlate with their DNA damaging effects. *Toxicol. In Vitro* 22(2):287-295.

Pinto Garcia, V., P.R. Vicente, A. Barez, et al. 1990. Hemolytic anemia induced by *Chelidonium majus*. Clinical case. *Sangre* 35(5):401-403.

Randeva, H.S., V. Bangar, S. Sailesh, and E.W. Hillhouse. 2000. Fatal cholestatic jaundice associated with amitriptyline. *Int. J. Clin. Pract.* 54(6):405-406.

Rifai, K., P. Flemming, M.P. Manns, and C. Trautwein. 2006. Severe drug hepatitis caused by *Chelidonium*. *Internist* 47(7):749-751.

Saglam, H., and G. Arar. 2003. Cytotoxic activity and quality control determinations on *Chelidonium majus*. *Fitoterapia* 74 (1-2):127-129.

Sárközi, Á., G. Janicsák, L. Kursinszki, and Á. Kéry. 2006. Alkaloid composition of *Chelidonium majus* L. studied by different chromatographic techniques. *Chromatographia* 63:81-86.

Sever Yilmaz, B., H. Ozbek, G. Saltan Citoglu, et al. 2007. Analgesic and hepatoprotective effects of *Chelidonium majus* L. *Ankara Univ. Eczacil. Fakult. Derg.* 36(1):9-20.

Spiridonov, N.A., D.A. Konovalov, and V.V. Arkhipov. 2005. Cytotoxicity of some Russian ethnomedicinal plants and plant compounds. *Phytother. Res.* 19(5):428-432.

Stickel, F., G. Poschl, H.K. Seitz, et al. 2003. Acute hepatitis induced by greater celandine (*Chelidonium majus*). *Scand. J. Gastroenterol.* 38(5):565-568.

Tomè, F., and M.L. Colombo. 1995. Distribution of alkaloids in *Chelidonium majus* and factors affecting their accumulation. *Phytochemistry* 40(1):37-39.

Wichtl, M. 2004. *Herbal drugs and phytopharmaceuticals: A handbook for practice on a scientific basis*. 3rd ed. Boca Raton, FL: CRC Press.

Chimaphila umbellata (L.) W.P.C. Barton — Pyrolaceae

SCN: pipsissewa
OCN: prince's pine
Part: herb

Quick Reference Summary

Safety Class: 1
Interaction Class: A

Contraindications
None known.

Other Precautions
None known.

Drug and Supplement Interactions
None known.

Adverse Events and Side Effects
None known.

Pharmacological Considerations
None known.

Pregnancy and Lactation

No information on the safety of pipsissewa in pregnancy or lactation was identified in the scientific or traditional literature. Although this review did not identify any concerns for use while pregnant or nursing, safety has not been conclusively established.

Review Details

C

I. Drug and Supplement Interactions

Clinical Trials of Drug or Supplement Interactions

No clinical trials of drug or supplement interactions were identified.

Case Reports of Suspected Drug or Supplement Interactions

No case reports of suspected drug or supplement interactions were identified.

Animal Trials of Drug or Supplement Interactions

No animal trials of drug or supplement interactions were identified.

II. Adverse Events

Case Reports of Adverse Events

No case reports of adverse events were identified.

III. Pharmacology and Pharmacokinetics

Human Pharmacological Studies

No relevant human pharmacological studies were identified.

Animal Pharmacological Studies

No relevant animal pharmacological studies were identified.

In Vitro Pharmacological Studies

No relevant in vitro pharmacological studies were identified.

IV. Pregnancy and Lactation

No information on the safety of pipsissewa in pregnancy or lactation was identified.

V. Toxicity Studies

The leaf of pipsissewa contains 18 to 22% of the compound arbutin (Shnyakina et al. 1981; Trubachev 1967). See *Arctostaphylos uva-ursi* Toxicity Studies for toxicity data on arbutin.

Literature Cited

Shnyakina, G.P., V.A. Sedel'nikova, and N.B. Tsygankova. 1981. Arbutin content in the leaves of some plants grown at Dal'nyi Vostok. *Rastitel'nye Resursy* 17(4):568-571.

Trubachev, A.A. 1967. Phytochemical study of *Chimaphila umbellata*. *Tr. Leningrad Khim.* 21:176-182.

Chionanthus virginicus L. — Oleaceae

SCN: fringetree

Part: root bark

Quick Reference Summary

Safety Class: 1
Interaction Class: A

Contraindications

None known.

Other Precautions

None known.

Drug and Supplement Interactions

None known.

Editors' Note

No cautions for use of fringe tree are reported in historical American medical texts (Felter and Lloyd 1898; Remington and Wood 1918).

Adverse Events and Side Effects

None known.

Pharmacological Considerations

None known.

Pregnancy and Lactation

No information on the safety of fringetree in pregnancy or lactation was identified in the scientific or traditional literature. Although this review did not identify any concerns for use while pregnant or nursing, safety has not been conclusively established.

Review Details

I. Drug and Supplement Interactions

Clinical Trials of Drug or Supplement Interactions

No clinical trials of drug or supplement interactions were identified.

Case Reports of Suspected Drug or Supplement Interactions

No case reports of suspected drug or supplement interactions were identified.

Animal Trials of Drug or Supplement Interactions

No animal trials of drug or supplement interactions were identified.

II. Adverse Events

Case Reports of Adverse Events

No case reports of adverse events were identified.

III. Pharmacology and Pharmacokinetics

Human Pharmacological Studies

No relevant human pharmacological studies were identified.

Animal Pharmacological Studies

No relevant animal pharmacological studies were identified.

In Vitro Pharmacological Studies

No relevant in vitro pharmacological studies were identified.

IV. Pregnancy and Lactation

No information on the safety of fringetree in pregnancy or lactation was identified.

V. Toxicity Studies

No toxicity studies were identified.

Literature Cited

Felter, H.W., and J.U. Lloyd. 1898. *King's American dispensatory.* 18th ed., 3rd rev. 2 vols. Cincinnati: Ohio Valley Co.

Remington, J.P., and H.C. Wood. 1918. *The dispensatory of the United States of America.* 20th ed. Philadelphia: Lippincott.

Chrysanthemum morifolium Ramat. — Asteraceae

SCN: chrysanthemum
Syn: *Chrysanthemum sinense* Sabine; *Dendranthema grandiflorum* (Ramat.) Kitam.; *Dendranthema morifolium* (Ramat.) Tzvelev
PN: *ju hua* (flower)
OCN: florist's chrysanthemum; mum
Part: flower

Quick Reference Summary

Safety Class: 1
Interaction Class: A

Contraindications

None known.

Other Precautions

Persons highly sensitive to ragweed (*Ambrosia artemisiifolia*) may be sensitive to chrysanthemum.

Drug and Supplement Interactions

None known.

Adverse Events and Side Effects

Mild allergic skin reactions have been reported after ingestion of chrysanthemum (Bensky et al. 2004). Contact dermatitis has been reported after contact with fresh or dried chrysanthemum (Bensky et al. 2004; Frain-Bell et al. 1979; Goncalo et al. 1996; Sertoli et al. 1985; Sharma et al. 1989).

Pharmacological Considerations

None known.

Pregnancy and Lactation

No information on the safety of chrysanthemum in pregnancy or lactation was identified in the scientific or traditional literature. Although this review did not identify any concerns for use while pregnant or nursing, safety has not been conclusively established.

C

Review Details

I. Drug and Supplement Interactions

Clinical Trials of Drug or Supplement Interactions

No clinical trials of drug or supplement interactions were identified.

Case Reports of Suspected Drug or Supplement Interactions

No case reports of suspected drug or supplement interactions were identified.

Animal Trials of Drug or Supplement Interactions

No animal trials of drug or supplement interactions were identified.

II. Adverse Events

Case Reports of Adverse Events

Mild allergic skin reactions have been reported after ingestion of chrysanthemum (Bensky et al. 2004).

Contact dermatitis to chrysanthemum has been reported after contact with the fresh or dried material (Bensky et al. 2004; Frain-Bell et al. 1979; Goncalo et al. 1996; Sertoli et al. 1985; Sharma et al. 1989). Airborne contact dermatitis has also been reported, primarily in workers in facilities that process chrysanthemums and other ornamental flowers (Sharma and Kaur 1989; Sharma et al. 1989).

III. Pharmacology and Pharmacokinetics

Human Pharmacological Studies

No relevant human pharmacological studies were identified.

Animal Pharmacological Studies

No relevant animal pharmacological studies were identified.

In Vitro Pharmacological Studies

No relevant in vitro pharmacological studies were identified.

IV. Pregnancy and Lactation

No information on the safety of chrysanthemum during pregnancy or lactation was identified.

V. Toxicity Studies

Genotoxicity

Antimutagenic activity of flavonoid compounds from chrysanthemum has been observed in the Ames test for mutagenicity in *Salmonella typhimurium* strains treated with chemical mutagens (Miyazawa and Hisama 2003).

Literature Cited

Bensky, D., S. Clavey, and E. Stöger. 2004. *Chinese herbal medicine: Materia medica*. 3rd ed. Seattle: Eastland Press.

Frain-Bell, W., A. Hetherington, and B.E. Johnson. 1979. Contact allergic sensitivity to chrysanthemum and the photosensitivity dermatitis and actinic reticuloid syndrome. *Br. J. Dermatol.* 101(5):491-501.

Goncalo, S., M. Goncalo, and J. Sequeira. 1996. Contact dermatitis to *Dendranthema morifolium* (Ramat). *Contact Dermat.* 35(5):310-311.

Miyazawa, M., and M. Hisama. 2003. Antimutagenic activity of flavonoids from *Chrysanthemum morifolium*. *Biosci. Biotechnol. Biochem.* 67(10):2091-2099.

Sertoli, A., P. Campolmi, P. Fabbri, N. Gelsomini, and E. Panconesi. 1985. Contact eczema caused by *Chrysanthemum morifolium* Ramat. *G. Ital. Dermatol. Venereol.* 120(5):365-370.

Sharma, S.C., and S. Kaur. 1989. Airborne contact dermatitis from Compositae plants in northern India. *Contact Dermat.* 21(1):1-5.

Sharma, S.C., R.C. Tanwar, and S. Kaur. 1989. Contact dermatitis from chrysanthemums in India. *Contact Dermat.* 21(2):69-71.

Chrysopogon zizanioides (L.) Roberty — Poaceae

SCN: vetiver
Syn: *Andropogon muricatus* Retz.; *Vetiveria zizanoides* (L.) Nash ex Small
AN: *ushira*
Part: root

Quick Reference Summary

Safety Class: 2b
Interaction Class: A

Contraindications

Not for use in pregnancy except under the supervision of a qualified healthcare practitioner (Caldecott 2006; Watt and Breyer-Brandwijk 1962).

Other Precautions

None known.

Drug and Supplement Interactions

None known.

Notice

Abortifacient (Watt and Breyer-Brandwijk 1962); *see* Appendix 2.

Adverse Events and Side Effects

None known.

Pharmacological Considerations

None known.

Pregnancy and Lactation

A text on Ayurvedic medicine indicates that vetiver should not be used during pregnancy (Caldecott 2006). Another reference reports that vetiver has been used as an abortifacient (Watt and Breyer-Brandwijk 1962). An ethnobotanical survey indicated that vetiver was one of a group of plants used to shorten labor and expel the placenta (Lans 2007).

No information on the safety of vetiver during lactation was identified. While this review did not identify any concerns for use while nursing, safety has not been conclusively established.

C

Review Details

I. Drug and Supplement Interactions

Clinical Trials of Drug or Supplement Interactions

No clinical trials of drug or supplement interactions were identified.

Case Reports of Suspected Drug or Supplement Interactions

No case reports of suspected drug or supplement interactions were identified.

Animal Trials of Drug or Supplement Interactions

No animal trials of drug or supplement interactions were identified.

II. Adverse Events

Case Reports of Adverse Events

No case reports of adverse events were identified.

III. Pharmacology and Pharmacokinetics

Human Pharmacological Studies

No relevant human pharmacological studies were identified.

Animal Pharmacological Studies

No relevant animal pharmacological studies were identified.

In Vitro Pharmacological Studies

No relevant in vitro pharmacological studies were identified.

IV. Pregnancy and Lactation

An ethnobotanical survey of medicinal plants used for reproductive health in Trinidad and Tobago listed vetiver as one of a group of plants used to shorten labor and expel the placenta (Lans 2007).

A text on Ayurvedic medicine indicates that vetiver should not be used during pregnancy (Caldecott 2006). Another reference reports that vetiver has been used as an abortifacient, although no details are provided on dose, preparation, or whether the plant was used alone or in a formula (Watt and Breyer-Brandwijk 1962).

No information on the safety of vetiver during lactation was identified.

V. Toxicity Studies

No toxicity studies were identified.

Literature Cited

Caldecott, T. 2006. *Ayurveda: The divine science of life*. New York: Mosby.

Lans, C. 2007. Ethnomedicines used in Trinidad and Tobago for reproductive problems. *J. Ethnobiol. Ethnomed.* 3:13.

Watt, J.M., and M.G. Breyer-Brandwijk. 1962. *The medicinal and poisonous plants of southern and eastern Africa.* 2nd ed. Edinburgh: E. & S. Livingstone.

Cibotium barometz (L.) J. Sm. — Dicksoniaceae

SCN: Scythian lamb
PN: *gou ji* (rhizome)
OCN: golden moss, Tartarian lamb
Part: rhizome

Quick Reference Summary

Safety Class: 1
Interaction Class: A

Contraindications

None known.

C

Other Precautions
None known.

Drug and Supplement Interactions
None known.

Adverse Events and Side Effects
None known.

Pharmacological Considerations
None known.

Pregnancy and Lactation
Texts on traditional Chinese medicine do not indicate any cautions for use of Scythian lamb during pregnancy or lactation (Bensky et al. 2004; Chen and Chen 2004). Although this review did not identify any concerns for use while pregnant or nursing, safety has not been conclusively established.

Review Details

I. Drug and Supplement Interactions

Clinical Trials of Drug or Supplement Interactions
No clinical trials of drug or supplement interactions were identified.

Case Reports of Suspected Drug or Supplement Interactions
No case reports of suspected drug or supplement interactions were identified.

Animal Trials of Drug or Supplement Interactions
No animal trials of drug or supplement interactions were identified.

II. Adverse Events

Case Reports of Adverse Events
No case reports of adverse events were identified.

III. Pharmacology and Pharmacokinetics

Human Pharmacological Studies
No relevant human pharmacological studies were identified.

Animal Pharmacological Studies
No relevant animal pharmacological studies were identified.

In Vitro Pharmacological Studies
No relevant in vitro pharmacological studies were identified.

IV. Pregnancy and Lactation
Texts on traditional Chinese medicine do not indicate any cautions for use of Scythian lamb during pregnancy or lactation (Bensky et al. 2004; Chen and Chen 2004).

V. Toxicity Studies
No toxicity studies were identified.

Literature Cited

Bensky, D., S. Clavey, and E. Stöger. 2004. *Chinese herbal medicine: Materia medica*. 3rd ed. Seattle: Eastland Press.

Chen, J.K., and T.T. Chen. 2004. *Chinese medical herbology and pharmacology*. City of Industry, CA: Art of Medicine Press.

Cichorium intybus L. — Asteraceae

SCN: chicory
AN: *kasni*

Part: roasted root

Quick Reference Summary

Safety Class: 1
Interaction Class: A

Contraindications
None known.

Other Precautions
None known.

Drug and Supplement Interactions
None known.

Adverse Events and Side Effects
Allergic skin reactions to chicory have been reported (Blumenthal et al. 1998), most likely due to the presence of sesquiterpene lactone compounds present in the fresh, unroasted plant material (Friis et al. 1975; Malarz et al. 2002; Pyrek 1985; Ross et al. 1993).

Animal studies have indicated that chicory root may modify glucose regulation (Kaur and Gupta 2002; Pushparaj et al. 2007). People with diabetes are advised to monitor their blood sugar closely and discuss the use of this herb with a qualified healthcare practitioner prior to use.

Pharmacological Considerations

None known.

Pregnancy and Lactation

No information on the safety of chicory in pregnancy or lactation was identified in the scientific or traditional literature. Although this review did not identify any concerns for use while pregnant or nursing, safety has not been conclusively established.

Review Details

I. Drug and Supplement Interactions

Clinical Trials of Drug or Supplement Interactions

No clinical trials of drug or supplement interactions were identified.

Case Reports of Suspected Drug or Supplement Interactions

No case reports of suspected drug or supplement interactions were identified.

Animal Trials of Drug or Supplement Interactions

No animal trials of drug or supplement interactions were identified.

II. Adverse Events

Case Reports of Adverse Events

No case reports of adverse events were identified.

III. Pharmacology and Pharmacokinetics

Human Pharmacological Studies

No relevant human pharmacological studies were identified.

Animal Pharmacological Studies

A decrease in serum glucose levels with no change in serum insulin levels was observed in diabetic rats orally administered 125 mg/kg of an extract of whole chicory plant (Pushparaj et al. 2007).

In Vitro Pharmacological Studies

No relevant in vitro pharmacological studies were identified.

IV. Pregnancy and Lactation

No information on the safety of chicory in pregnancy or lactation was identified.

V. Toxicity Studies

Short-Term Toxicity

No adverse effects, including clinical observations, changes in body weights, food consumption, clinical pathology, gross necropsy, and histology, were observed in rats orally administered up to 1 g/kg chicory root extract daily for 28 days. Based on the results of that study, the researchers indicated that the no-observed-adverse-effect level was 1 g/kg daily (Schmidt et al. 2007).

Genotoxicity

No mutagenic activity of a chicory root extract was observed in the Ames test for mutagenicity in *Salmonella* with or without metabolic activation (Schmidt et al. 2007).

Literature Cited

Blumenthal, M., W. Busse, A. Goldberg, et al. 1998. *The complete German Commission E monographs*. Austin, TX: American Botanical Council.

Friis, B., N. Hjorth, J.T. Vail, and J.C. Mitchell. 1975. Occupational contact dermatitis from *Cichorium* (chicory, endive) and *Lactuca* (lettuce). *Contact Dermat.* 1(5):311-313.

Kaur, N., and A.K. Gupta. 2002. Applications of inulin and oligofructose in health and nutrition. *J. Biosci.* 27(7):703-714.

Malarz, J., A. Stojakowska, and W. Kisiel. 2002. Sesquiterpene lactones in a hairy root culture of *Cichorium intybus*. *Z. Naturforsch. C J. Biosci.* 57(11-12):994-997.

Pushparaj, P.N., H.K. Low, J. Manikandan, B.K. Tan, and C.H. Tan. 2007. Anti-diabetic effects of *Cichorium intybus* in streptozotocin-induced diabetic rats. *J. Ethnopharmacol.* 111(2):430-434.

Pyrek, J.S.T. 1985. Sesquiterpene lactones of *Cichorium intybus* and *Leontodon autumnalis*. *Phytochemistry* 24:186-188.

Ross, J.S., H. Du Peloux Menage, J.L.M. Hawk, and I.R. White. 1993. Sesquiterpene lactone contact sensitivity: Clinical patterns of Compositae dermatitis and relationship to chronic actinic dermatitis. *Contact Dermat.* 29(2):84-87.

Schmidt, B.M., N. Ilic, A. Poulev, and I. Raskin. 2007. Toxicological evaluation of a chicory root extract. *Food Chem. Toxicol.* 45(7):1131-1139.

Cinchona spp. Rubiaceae

Cinchona calisaya Wedd.
SCN: yellow cinchona
Syn: *Cinchona ledgeriana* Moens ex Trimen
OCN: calisaya; yellow quinine

Cinchona officinalis **L.**
SCN: red cinchona
OCN: Jesuit's bark; Peruvian bark; red quinine

Cinchona pubescens **Vahl**
SCN: red cinchona
OCN: red quinine
Part: bark

C

Quick Reference Summary

Safety Class: 2b, 2d
Interaction Class: A

Contraindications

Not for use in pregnancy except under the supervision of a qualified healthcare practitioner (Blaschek et al. 2006; Wichtl 2004).

Do not exceed recommended dose (Martindale and Reynolds 1996).

Other Precautions

The compound quinine may cause serious and potentially life-threatening hypersensitivity reactions in some individuals (Belkin 1967; BfR 2005; Brasic and Baltimore 2001; CFR 2011a; Howard et al. 2003).

Cinchona may cause gastrointestinal irritation; use with caution in persons with gastric or intestinal ulcers (Wichtl 2004).

Drug and Supplement Interactions

None known.

Standard Dose

The standard dose is 1 to 4 g daily as a tea (Wichtl 2004).

Editors' Notes

Red and yellow cinchona are sources of the alkaloid quinine, which has been widely used as an antimalarial drug. Red and yellow cinchona typically contain 5–8% alkaloids (including quinine and related compounds), though may contain up to 15% (Bradley 1992; DAB 1987; McHale 1986; Wichtl 2004).

Red and yellow cinchona are regulated in the United States as allowable in carbonated beverages in certain salt forms as long as the content of quinine does not exceed 83 ppm and as long as the ingredient is identified on the label as "quinine" (CFR 2011b).

U.S. regulations for over-the-counter drugs require the following label for quinine and other *Cinchona* derivatives: Caution: Discontinue use if ringing in the ears, deafness, skin rash, or visual disturbances occur (CFR 2011c). In the United States, products containing quinine sulfate may not be marketed for use related to nocturnal leg cramps (CFR 2011a).

Adverse Events and Side Effects

Symptoms of "cinchonism," due to overdose of the compound quinine, include headache, nausea, distributed vision, tinnitus, delirium, abdominal pain, and diarrhea. Extreme reactions including blindness, deafness, convulsions, paralysis, collapse, and even death may occur in hypersensitive individuals (Gilman et al. 1985; Martindale and Reynolds 1996; Remington and Wood 1918).

The compound quinine has been reported to cause thrombotic thrombocytopenic purpura (a disorder that causes blood clots to form in small blood vessels and leads to a low platelet count) in sensitive individuals. This condition has been associated with use of quinine tablets and tonic water containing quinine. Individuals can become sensitized to cinchona alkaloids and develop sudden reactions after long-term use (Belkin 1967; Brasic and Baltimore 2001; Howard et al. 2003).

Allergic reactions to cinchona are relatively common (Wichtl 2004).

Pharmacological Considerations

None known.

Pregnancy and Lactation

The compound quinine is reported to have been used as an abortifacient, although some sources indicate the compound is generally ineffective for this purpose and toxic to the mother (Dannenberg and Dorfman 1983). Based on this and related information, use during pregnancy is not recommended except under the supervision of a qualified healthcare practitioner.

No information on the safety of red cinchona during lactation was identified. While this review did not identify any concerns for use while nursing, safety has not been conclusively established.

C

Review Details

I. Drug and Supplement Interactions

Clinical Trials of Drug or Supplement Interactions

No clinical trials of drug or supplement interactions were identified.

Case Reports of Suspected Drug or Supplement Interactions

No case reports of suspected drug or supplement interactions were identified.

Animal Trials of Drug or Supplement Interactions

No animal trials of drug or supplement interactions were identified.

II. Adverse Events

Case Reports of Adverse Events

A review of elderly patients with lupus anticoagulant (LA) indicated that 73% were taking cinchona alkaloids. The frequencies of drug usage differed significantly from age- and sex-matched controls. Features suggestive of the antiphospholipid syndrome were observed in 5 patients. Repeat testing showed persistent LA activity in all but 2 of 5 patients in whom the relevant drug had been ceased (Bird et al. 1995).

The compound quinine has been reported to cause thrombotic thrombocytopenic purpura (a disorder that causes blood clots to form in small blood vessels and leads to a low platelet count) in sensitive individuals. This condition has been associated with use of quinine tablets and tonic water containing quinine (Belkin 1967; Brasic and Baltimore 2001; Howard et al. 2003).

Contact urticaria due to cinchona has been reported (Dooms-Goossens et al. 1986).

III. Pharmacology and Pharmacokinetics

Human Pharmacological Studies

No relevant human pharmacological studies were identified.

Animal Pharmacological Studies

Photosensitizing activity of the compound quinine and derivatives has been reported (Spikes 1998; Waddell et al. 1997).

In Vitro Pharmacological Studies

No relevant in vitro pharmacological studies were identified.

IV. Pregnancy and Lactation

The compound quinine is reported to have been used as an abortifacient, although some sources indicate the compound is generally ineffective for this purpose and is generally toxic to the mother (Dannenberg and Dorfman 1983).

The compound quinine is listed in FDA pregnancy category C, "Animal reproduction studies have shown an adverse effect on the fetus and there are no adequate and well-controlled studies in humans, but potential benefits may warrant use of the drug in pregnant women despite potential risks."

No information on the safety of red cinchona during lactation was identified.

V. Toxicity Studies

Acute Toxicity

The LD_{50} of the compound quinine intraperitoneally administered in mice is 245 mg/kg (Carlson and Cretcher 1951).

Literature Cited

Belkin, G.A. 1967. Cocktail purpura. *Ann. Intern. Med.* 66(3):583.

BfR. 2005. Quinine-containing beverages may cause health problems. Updated BfR Health Assessment No. 020/2008, 17 February. German Federal Institute for Risk Assessment.

Bird, M.R., A.I. O'Neill, R.R. Buchanan, K.M. Ibrahim, and J. Des Parkin. 1995. Lupus anticoagulant in the elderly may be associated with both quinine and quinidine usage. *Pathology* 27(2):136-139.

Blaschek, W., S. Ebel, E. Hackenthal, et al., eds. 2006. *Hagers handbuch der drogen und arzneistoffe.* Version 5.0, CD-ROM. Heidelberg: Springer.

Bradley, P.R. 1992. *British herbal compendium: A handbook of scientific information on widely used plant drugs.* Bournemouth, UK: British Herbal Medicine Association.

Brasic, J.R., and M. Baltimore. 2001. Quinine-induced thrombocytopenia in a 64-year-old man who consumed tonic water to relieve nocturnal leg cramps. *Mayo Clin. Proc.* 76(8):863.

Carlson, W.W., and L.H. Cretcher. 1951. The blood distribution and toxicity of several cinchona alkaloid derivatives. *J. Am. Pharm. Assoc.* 40(9):471-473.

CFR. 2011a. *Code of federal regulations*, Title 21 Part 310.546, 2011 ed. Requirements for specific new drugs or devices. Drug products containing active ingredients offered over-the-counter (OTC) for the treatment and/or prevention of nocturnal leg muscle cramps. Washington, DC: U.S. Government Printing Office.

CFR. 2011b. *Code of federal regulations*, Title 21 Part 172.575, 2011 ed. Food additives permitted for direct addition to food for human consumption. Flavoring agents and related substances. Quinine. Washington, DC: U.S. Government Printing Office.

CFR. 2011c. *Code of federal regulations*, Title 21 Part 369.20, 2011 ed. Interpretive statements re warnings on drugs and devices for over-the-counter sale. Warnings and caution statements for drugs. Drugs; recommended warning and caution statements. Washington, DC: U.S. Government Printing Office.

DAB. 1987. DAB 9—Kommentar. Deutsches Arzneibuch 9, Band 2:1157-1164.

Dannenberg, A.L., and S.F. Dorfman. 1983. Use of quinine for self-induced abortion. *South. Med. J.* 76(7):846-849.

Dooms-Goossens, A., H. Deveylder, C. Duron, M. Dooms, and H. Degreef. 1986. Airborne contact urticaria due to cinchona. *Contact Dermat.* 15(4):258.

Gilman, A.G., L. Goodman, T.W. Rall, and F. Murad. 1985. *Goodman and Gilman's the pharmacological basis of therapeutics.* 7th ed. New York: Macmillan.

Howard, M.A., A.B. Hibbard, D.R. Terrell, et al. 2003. Quinine allergy causing acute severe systemic illness: Report of 4 patients manifesting multiple hematologic, renal, and hepatic abnormalities. *Proc. Baylor Univ. Med. Cent.* 16(1):21-26.

Martindale, W., and J.E.F. Reynolds. 1996. *The extra pharmacopoeia.* 31st ed. London: Pharmaceutical Press.

McHale, D. 1986. The *Cinchona* tree. *Biologist* 33:45-53.

Remington, J.P., and H.C. Wood. 1918. *The dispensatory of the United States of America.* 20th ed. Philadelphia: Lippincott.

Spikes, J.D. 1998. Photosensitizing properties of quinine and synthetic antimalarials. *J. Photochem. Photobiol. B* 42(1):1-11.

Waddell, T.G., A.D. Carter, J.T. Arnason, R. Marles, and G. Guillet. 1997. Phototoxicity of quinine derivatives. Structure-activity trends. *Fitoterapia* 68(4):381-383.

Wichtl, M. 2004. *Herbal drugs and phytopharmaceuticals: A handbook for practice on a scientific basis.* 3rd ed. Boca Raton, FL: CRC Press.

Cinnamomum aromaticum Nees — Lauraceae

SCN: cassia
Syn: *Cinnamomum cassia* auct.
PN: *rou gui* (bark); *gui zhi* (twig)
OCN: Chinese cinnamon
Part: bark

Quick Reference Summary

Safety Class: 2b
Interaction Class: A

Contraindications

Not for use in pregnancy except under the supervision of a qualified healthcare practitioner (Bensky et al. 2004; Chen and Chen 2004; List and Hörhammer 1973).

Other Precautions

Use with caution in persons with sensitivity to balsam of Peru tree (Calnan 1976; Hoskins 1984; Pfutzner et al. 2003).

Drug and Supplement Interactions

None known.

Editors' Notes

The classifications and concerns for this herb are based on relatively higher doses used for therapeutic purposes in contrast to lower amounts generally used in cooking, and have not been associated with its use as a spice.

Several species of *Cinnamomum* are commonly traded under the name "cinnamon" (Leung and Foster 1996).

European authorities have raised concern over the coumarin content of cassia (BfR 2006; EFSA 2008). In the 1980s, coumarin was suspected to have genotoxic and carcinogenic effects, although more recent evidence suggests that coumarin is not genotoxic (Lake 1999; Sproll et al. 2008). As of 2008, the European Food Safety Authority indicated that the appropriate total daily intake (TDI) level of coumarin in foods was 0.1 mg/kg, and that intake of three times this amount for 1 to 2 weeks posed no concerns (EFSA 2008). The German Federal Institute for Risk Assessment (BfR) reviewed human clinical data from pharmaceutical use of coumarin and also concluded that 0.1 mg/kg was the acceptable TDI (BfR 2006). Others have questioned this TDI level since it is based on studies of an isolated compound and does not account for activity of other compounds in the whole plant or plant extract that may modify the toxicity of coumarin (Schmidt et al. 2007). The coumarin content of cassia is reported to range from 1250 to 1490 mg/kg, while in cassia essential oil, the coumarin content is 4370 mg/kg (Rychlik 2008).

Adverse Events and Side Effects

Allergic or sensitization reactions to cassia or compounds from cassia are relatively common. These reactions usually affect the skin or mucous membranes (Allen and Blozis 1988; Endo and Rees 2006; Lamey et al. 1990; Miller et al. 1992; Wright 2007).

Pharmacological Considerations

Human and animal studies indicated that cassia may modify glucose levels (Dugoua et al. 2007; Kim et al. 2006; Pham et al. 2007; Verspohl et al. 2005). People with diabetes are advised to monitor their blood sugar closely and discuss the use of this herb with a qualified healthcare practitioner prior to use.

Occupational allergy to cassia and cassia essential oil has been reported and confirmed by patch testing (De Benito and Alzaga 1999; Steger et al. 2000).

Pregnancy and Lactation

Texts on traditional Chinese medicine indicate that cassia should not be used during pregnancy (Bensky et al. 2004;

Chen and Chen 2004). In rats administered the compound cinnamaldehyde (5 to 250 mg/kg daily), an increase in fetal malformations was observed (Mantovani et al. 1989). Based on this information, use during pregnancy is not recommended except under the supervision of a qualified healthcare practitioner.

No information on the safety of cassia during lactation was identified. While this review did not identify any concerns for use while nursing, safety has not been conclusively established.

Review Details

I. Drug and Supplement Interactions

Clinical Trials of Drug or Supplement Interactions

No clinical trials of drug or supplement interactions were identified.

Case Reports of Suspected Drug or Supplement Interactions

No case reports of suspected drug or supplement interactions were identified.

Animal Trials of Drug or Supplement Interactions

No animal trials of drug or supplement interactions were identified.

II. Adverse Events

Adverse Events Reported in Clinical Trials

No significant adverse events were reported in clinical trials with cassia at doses up to 6 g daily for 40 or 60 days (Hlebowicz et al. 2007; Khan et al. 2003).

Case Reports of Adverse Events

Overdoses (aqueous extracts of 18 g or more daily) of cassia have been associated with a sensation of abdominal tension, dysuria, red urine, burning urination, proteinuria, constipation, a sensation of heat in the chest, a thirst for cold beverages, dry and swollen eyes, flushed face, dizziness, blurred vision, and numbness of the tongue (Bensky et al. 2004).

The compound cinnamaldehyde, which comprises approximately 70 to 95% of cassia essential oil, can cause skin sensitization and irritation in sensitive persons. Reactions to cinnamaldehyde occur most commonly with prolonged oral exposure to products such as toothpaste, gum, or hard candies. Such cinnamon-flavored products have been found to be responsible for a number of cases of oral inflammation or lesions (Allen and Blozis 1988; Endo and Rees 2006; Lamey et al. 1990; Miller et al. 1992; Wright 2007). These include stomatitis (an inflammation of the mucous lining of any of the structures in the mouth), orofacial granulomatosis, erythematous patches, erythema multiforme, lichen planus, leukoplakia, varying degrees of superimposed keratosis or ulceration, and plasma cell gingivitis (Allen and Blozis 1988; Anil 2007; Cohen and Bhattacharyya 2000; Endo and Rees 2006, 2007; Lamey et al. 1990; Mihail 1992; Miller et al. 1992; Wright 2007). Some cases have been diagnosed as allergic in nature (Drake and Maibach 1976; Tremblay and Avon 2008). *See* Case reports of adverse events in *Cinnamomum verum* for more information on adverse events related to cinnamaldehyde.

Persons that took 36 g of powdered cassia (standard dose of powdered herb is 1–2 g) in a single dose experienced symptoms such as dizziness, blurred vision, rapid pulse, increased intraocular pressure, cough, decreased urine production, and thirst (Bensky et al. 2004; Zhu 1998).

Occupational allergy to cassia essential oil has been reported and confirmed by patch testing (De Benito and Alzaga 1999).

In workers occupationally exposed to cassia, skin prick tests and specific IgE tests indicated that 11% of 62 workers with high or low exposure to cassia were sensitized to this herb (Steger et al. 2000).

III. Pharmacology and Pharmacokinetics

Human Pharmacological Studies

A review of randomized controlled studies on cassia in diabetes reported that two out of three studies showed that doses of 1 to 6 g daily reduced fasting blood glucose. The third study indicated no change on glucose levels after doses of 1.5 g daily (Dugoua et al. 2007). Studies published since that review indicated no effects on glucose at a dose of 3 g cassia daily, and a reduction in glucose at a dose of 6 g daily (Hlebowicz et al. 2007, 2009).

Cassia and balsam of Peru contain many similar allergenic compounds. Patch testing has demonstrated cross-reactivity in persons with allergies to these two species (Calnan 1976; Hoskins 1984; Pfutzner et al. 2003). Patch testing in a group of 118 patients allergic to balsam of Peru indicated that 14.5% of patients were allergic to cassia. Of 220 patients tested who were not allergic to balsam of Peru, only one tested positively to cassia (Niinimaki 1984).

In healthy volunteers orally administered 2.8 g (containing ~55 mg of oxalates) of cassia daily for 4 weeks, no significant changes in fasting plasma glucose were observed. No significant changes in urinary oxalate levels were observed at the end of the test period (Tang et al. 2008).

Animal Pharmacological Studies

In rats orally administered 5.29 mg/kg of a dried aqueous extract of cassia or 85.7 mg/kg of powdered cassia (equivalent to the dose of cassia used in the extract), a decrease in blood glucose levels was observed in the glucose tolerance test but not in animals challenged with a glucose load (Verspohl et al. 2005).

A dose-dependent reduction in glucose levels was observed in diabetic mice orally administered 50, 100, 150, or 200 mg/kg of a cassia extract daily for 6 weeks (Kim et al. 2006). In diabetic rats orally administered 5, 10, or 20 mg/kg of the compound cinnamaldehyde daily for 45 days, a dose-dependent decrease in plasma glucose levels was observed (Subash Babu et al. 2007).

In Vitro Pharmacological Studies

Antiestrogenic effects of an ethanol extract of cassia were observed in a recombinant yeast system featuring a human estrogen receptor expression plasmid and a reporter plasmid (Kim et al. 2008).

An extract of cassia decreased the dissolution of tetracycline hydrochloride (Miyazaki et al. 1977).

IV. Pregnancy and Lactation

Texts on traditional Chinese medicine indicate that cassia should not be used during pregnancy (Bensky et al. 2004; Chen and Chen 2004).

No information on the safety of cassia during lactation was identified.

V. Toxicity Studies

Acute Toxicity

The LD_{50} of an intravenously administered decoction of cassia in mice is 18.48 g/kg (Chen and Chen 2004). The LD_{50} of orally administered cassia essential oil in rats is 2.8 ml/kg, while the dermal LD_{50} of the same product in rabbits is 0.32 ml/kg (Opdyke 1979).

The LD_{50} of the compound cinnamic acid orally administered to rats, mice, and guinea pigs could not be determined at doses up to 5 g/kg (Hoskins 1984).

The LD_{50} of the compound cinnamaldehyde in mice is 132 mg/kg after intravenous administration, 610 mg/kg after intraperitoneal administration, and 2225 mg/kg after oral administration (Zhu 1998).

Genotoxicity

No mutagenic activity of cassia essential oil was observed in the Ames test for mutagenicity in *Salmonella typhimurium* strain TA100 with or without activation by S9 (Park 2002; Sekizawa and Shibamoto 1982). Cassia essential oil tested positively in the *rec* assay with *Bacillus subtilis* (Sekizawa and Shibamoto 1982).

Literature Cited

Allen, C.M., and G.G. Blozis. 1988. Oral mucosal reactions to cinnamon-flavored chewing gum. *J. Am. Dent. Assoc.* 116(6):664-667.

Anil, S. 2007. Plasma cell gingivitis among herbal toothpaste users: A report of three cases. *J. Contemp. Dent. Pract.* 8(4):60-66.

Bensky, D., S. Clavey, and E. Stöger. 2004. *Chinese herbal medicine: Materia medica.* 3rd ed. Seattle: Eastland Press.

BfR. 2006. High daily intakes of cinnamon: Health risk cannot be ruled out. BfR Health Assessment No. 044/2006. Bundesinstitut für Risikobewertung.

Calnan, C.D. 1976. Cinnamon dermatitis from an ointment. *Contact Dermat.* 2(3):167-170.

Chen, J.K., and T.T. Chen. 2004. *Chinese medical herbology and pharmacology.* City of Industry, CA: Art of Medicine Press.

Cohen, D.M., and I. Bhattacharyya. 2000. Cinnamon-induced oral erythema multiformelike sensitivity reaction. *J. Am. Dent. Assoc.* 131:929-934.

De Benito, V., and R. Alzaga. 1999. Occupational allergic contact dermatitis from cassia (Chinese cinnamon) as a flavouring agent in coffee. *Contact Dermat.* 40(3):165.

Drake, T.E., and H.I. Maibach. 1976. Allergic contact dermatitis and stomatitis caused by a cinnamic aldehyde-flavored toothpaste. *Arch. Dermatol.* 112(2):202-203.

Dugoua, J.J., D. Seely, D. Perri, et al. 2007. From type 2 diabetes to antioxidant activity: A systematic review of the safety and efficacy of common and cassia cinnamon bark. *Can. J. Physiol. Pharmacol.* 85(9):837-847.

EFSA. 2008. Coumarin in flavourings and other food ingredients with flavouring properties. Scientific opinion of the Panel on Food Additives, Flavourings, Processing Aids and Materials in Contact with Food. *EFSA J.* No. 793:1-15.

Endo, H., and T.D. Rees. 2006. Clinical features of cinnamon-induced contact stomatitis. *Compend. Contin. Educ. Dent.* 27(7):403-409.

Endo, H., and T.D. Rees. 2007. Cinnamon products as a possible etiologic factor in orofacial granulomatosis. *Med. Oral Patol. Oral Cir. Bucal* 12(6):E440-E444.

Hlebowicz, J., G. Darwiche, O. Bjorgell, and L.O. Almer. 2007. Effect of cinnamon on postprandial blood glucose, gastric emptying, and satiety in healthy subjects. *Am. J. Clin. Nutr.* 85(6):1552-1526.

Hlebowicz, J., A. Hlebowicz, S. Lindstedt, et al. 2009. Effects of 1 and 3 g cinnamon on gastric emptying, satiety, and postprandial blood glucose, insulin, glucose-dependent insulinotropic polypeptide, glucagon-like peptide 1, and ghrelin concentrations in healthy subjects. *Am. J. Clin. Nutr.* 89(3):815-821.

Hoskins, J.A. 1984. The occurrence, metabolism and toxicity of cinnamic acid and related compounds. *J. Appl. Toxicol.* 4(6):283-292.

Khan, A., M. Safdar, M.M. Ali Khan, K.N. Khattak, and R.A. Anderson. 2003. Cinnamon improves glucose and lipids of people with type 2 diabetes. *Diabetes Care* 26(12):3215-3218.

Kim, I.G., S.C. Kang, K.C. Kim, E.S. Choung, and O.P. Zee. 2008. Screening of estrogenic and antiestrogenic activities from medicinal plants. *Env. Tox. Pharmacol.* 25(1):75-82.

Kim, S.H., S.H. Hyun, and S.Y. Choung. 2006. Anti-diabetic effect of cinnamon extract on blood glucose in *db/db* mice. *J. Ethnopharmacol.* 104(1-2):119-123.

Lake, B.G. 1999. Coumarin metabolism, toxicity and carcinogenicity: Relevance for human risk assessment. *Food Chem. Toxicol.* 37(4):423-453.
Lamey, P.J., M.A. Lewis, T.D. Rees, et al. 1990. Sensitivity reaction to the cinnamonaldehyde component of toothpaste. *Br. Dent. J.* 168(3):115-118.
Leung, A.Y., and S. Foster. 1996. *Encyclopedia of common natural ingredients used in food, drugs, and cosmetics.* 2nd ed. New York: Wiley.
List, P.H., and H. Hörhammer. 1973. *Hagers handbuch der pharmazeutischen praxis.* Berlin: Springer.
Mantovani, A., A.V. Stazi, C. Macri, et al. 1989. Pre-natal (segment II) toxicity study of cinnamic aldehyde in the Sprague-Dawley rat. *Food Chem. Toxicol.* 27(12):781-786.
Mihail, R.C. 1992. Oral leukoplakia caused by cinnamon food allergy. *J. Otolaryngol.* 21(5):366-367.
Miller, R.L., A.R. Gould, and M.L. Bernstein. 1992. Cinnamon-induced stomatitis venenata: Clinical and characteristic histopathologic features. *Oral Surg. Oral Med. Oral Pathol.* 73(6):708-716.
Miyazaki, S., H. Inoue, and T. Nadai. 1977. Effect of antacids on the dissolution behavior of tetracycline and methacycline. *Chem. Pharm. Bull.* 27:2523-2527.
Niinimaki, A. 1984. Delayed-type allergy to spices. *Contact Dermat.* 11(1):34-40.
Opdyke, D.L.J. 1979. *Monographs on fragrance raw materials.* New York: Pergamon.
Park, H.J. 2002. Mutagenicity of the essential oils in Ames test. *Korean J. Pharmacog.* 33(4):372-375.
Pfutzner, W., P. Thomas, A. Niedermeier, et al. 2003. Systemic contact dermatitis elicited by oral intake of balsam of Peru. *Acta Derm. Venereol.* 83(4):294-295.
Pham, A.Q., H. Kourlas, and D.Q. Pham. 2007. Cinnamon supplementation in patients with type 2 diabetes mellitus. *Pharmacotherapy* 27(4):595-599.
Rychlik, M. 2008. Quantification of free coumarin and its liberation from glucosylated precursors by stable isotope dilution assays based on liquid chromatography-tandem mass spectrometric detection. *J. Agric. Food Chem.* 56(3):796-801.
Schmidt, M., M. Mueller, and S.D. Mueller. 2007. Cinnamon and coumarins: Bitter "truth"—extrapolation of risks. *Deutsche Lebensmittel-Rundschau* 103(8):378-383.
Sekizawa, J., and T. Shibamoto. 1982. Genotoxicity of safrole-related chemicals in microbial test systems. *Mutat. Res.* 101(2):127-140.
Sproll, C., W. Ruge, C. Andlauer, R. Godelmann, and D.W. Lachenmeier. 2008. HPLC analysis and safety assessment of coumarin in foods. *Food Chem.* 109(2):462-469.
Steger, A., K. Radon, A. Pethran, and D. Nowak. 2000. Sensitization and lung function in workers occupationally exposed to natural thickening products. *Allergy* 55(4):376-381.
Subash Babu, P., S. Prabuseenivasan, and S. Ignacimuthu. 2007. Cinnamaldehyde—A potential antidiabetic agent. *Phytomedicine* 14(1):15-22.
Tang, M., D.E. Larson-Meyer, and M. Liebman. 2008. Effect of cinnamon and turmeric on urinary oxalate excretion, plasma lipids, and plasma glucose in healthy subjects. *Am. J. Clin. Nutr.* 87(5):1262-1267.
Tremblay, S., and S.L. Avon. 2008. Contact allergy to cinnamon: Case report. *J. Can. Dent. Assoc.* 74(5):445-461.
Verspohl, E.J., K. Bauer, and E. Neddermann. 2005. Antidiabetic effect of *Cinnamomum cassia* and *Cinnamomum zeylanicum in vivo* and *in vitro*. *Phytother. Res.* 19(3):203-206.
Wright, J. 2007. Diagnosis and management of oral lichenoid reactions. *J. Calif. Dent. Assoc.* 35(6):412-416.
Zhu, Y.-P. 1998. *Chinese materia medica: Chemistry, pharmacology and applications.* Amsterdam: Harwood Academic Publishers.

Cinnamomum camphora (L.) J. Presl — Lauraceae

SCN: camphor
Syn: *Camphora camphora* (L.) H. Karst, nom. illeg.; *Camphora officinalis* Nees; *Laurus camphora* L.
AN: *karpura*
Part: distillate of the wood

Quick Reference Summary

Safety Class: 2b, 2d
Interaction Class: A

Contraindications

Not for use in pregnancy except under the supervision of a qualified healthcare practitioner (Bensky et al. 2004; Chen and Chen 2004).

Not for long-term use; do not exceed recommended dose (Manoguerra et al. 2006; McGuffin et al. 1997).

Not for use in children under 2 years of age (Bensky et al. 2004; Guilbert et al. 2007; Uc et al. 2000).

Other Precautions

Use with caution in small children (Bensky et al. 2004; Guilbert et al. 2007; Uc et al. 2000).

Camphor must be used with extreme caution if taken internally (Bensky et al. 2004; Chen and Chen 2004; Geller et al. 1984; Love et al. 2004; Manoguerra et al. 2006).

Drug and Supplement Interactions

None known.

Notice

Alkenylbenzenes (safrole 1.1–43%) (Stubbs et al. 2004); *see* Appendix 1.

Standard Dose

External Use

Products should contain 11% or less of the compound camphor (Love et al. 2004; Manoguerra et al. 2006).

Internal Use

The standard dose is 100 to 200 mg daily (Bensky et al. 2004; Chen and Chen 2004).

Editors' Notes

"Camphor" is the common name for *Cinnamomum camphora* and the name of the compound that gives this plant its distinctive smell. The compound camphor is frequently used in topical personal care products. The compound camphor comprises approximately 50% of camphor essential oil (Stubbs et al. 2004).

Use of camphor as a food additive in the United States is subject to a limitation that the finished food or beverage is safrole-free (CFR 2011). Dietary ingredients for use in dietary supplements, however, are specifically excluded from the federal food additive definition (U.S.C. 2010).

Adverse Events and Side Effects

Numerous cases of poisoning have been reported after accidental or intentional ingestion of products containing the compound camphor. Symptoms of poisoning, which generally begin soon after ingestion, include mucous membrane irritation, nausea, vomiting, generalized seizures, and abdominal pain. Other symptoms include central nervous system depression, headache, dizziness, confusion, anxiety, hallucinations, and twitching. Fatal cases of poisoning have been reported. Poisoning after topical use of products is relatively rare and occurred primarily in small children (Love et al. 2004; Manoguerra et al. 2006).

A review of cases of poisoning from the compound camphor indicated that patients ingesting less than 2 mg/kg were asymptomatic, while 5 to 10 mg/kg resulted in minor symptoms. A dose of 30 mg/kg is generally considered toxic, and severe symptoms are generally observed at 59 mg/kg. The average fatal dose has been reported as 199 mg/kg (Geller et al. 1984; Manoguerra et al. 2006).

Pharmacological Considerations

See Adverse Events and Side Effects.

Pregnancy and Lactation

Texts on traditional Chinese medicine indicate that camphor should not be used during pregnancy (Bensky et al. 2004; Chen and Chen 2004).

An animal study showed no adverse effects of the compound camphor at relatively high doses (Leuschner 1997).

No information on the safety of camphor during lactation was identified. While this review did not identify any concerns for use while nursing, safety has not been conclusively established.

Review Details

I. Drug and Supplement Interactions

Clinical Trials of Drug or Supplement Interactions

No clinical trials of drug or supplement interactions were identified.

Case Reports of Suspected Drug or Supplement Interactions

No case reports of suspected drug or supplement interactions were identified.

Animal Trials of Drug or Supplement Interactions

No animal trials of drug or supplement interactions were identified.

II. Adverse Events

Case Reports of Adverse Events

Numerous cases of poisoning have been reported after accidental or intentional ingestion of products containing the compound camphor. Symptoms of poisoning, which generally begin soon after ingestion, include mucous membrane irritation, nausea, vomiting, abdominal pain, and generalized seizures. Other symptoms include central nervous system depression, headache, dizziness, confusion, anxiety, hallucinations, and twitching. Rare cases of hepatotoxicity have been reported. Some cases of poisoning have been fatal. Poisoning after topical use of products is relatively rare and occurred primarily in small children (Love et al. 2004; Manoguerra et al. 2006).

A review of cases of poisoning from the compound camphor indicated that patients ingesting less than 2 mg/kg were asymptomatic, while 5 to 10 mg/kg resulted in minor symptoms. A dose of 30 mg/kg is generally considered toxic, and severe symptoms are generally observed at 59 mg/kg. The average fatal dose has been reported as 199 mg/kg (Geller et al. 1984; Manoguerra et al. 2006).

Myocarditis (inflammation of the heart muscle) was reported in a 60-year-old man with a history of severe anemia. Symptoms developed after the man consumed 120 mg/kg of the compound camphor (Bhaya and Beniwal 2007).

Seizures were reported in a 4-month-old girl who had been treated topically on her abdomen with an unspecified amount of a topical product containing the compound camphor (Guilbert et al. 2007).

Hepatotoxicity was reported in a 2-month-old undernourished girl that had been topically administered "generous amounts" of a cold remedy containing the compound camphor three times daily for 5 days. The hepatotoxicity resolved after cessation of the cold remedy (Uc et al. 2000).

III. Pharmacology and Pharmacokinetics

Human Pharmacological Studies

No relevant human pharmacological studies were identified.

Animal Pharmacological Studies

No relevant animal pharmacological studies were identified.

In Vitro Pharmacological Studies

No relevant in vitro pharmacological studies were identified.

IV. Pregnancy and Lactation

In studies with rats and rabbits, the compound D-camphor elicited no evidence of teratogenicity when administered orally at doses up to 1000 mg/kg daily in rats and 681 mg/kg daily in rabbits. The developmental no-observed-effect level for rats was above 1000 mg/kg and for rabbits was 681 mg/kg (Leuschner 1997).

No information on the safety of camphor during lactation was identified.

V. Toxicity Studies

Acute Toxicity

The oral LD_{50} of the compound camphor is 1800 mg/kg in guinea pigs, 2000 mg/kg in rabbits, and 800 mg/kg in dogs. The intraperitoneal LD_{50} of the compound camphor is 900 mg/kg in rats and 3000 mg/kg in mice. The subcutaneous LD_{50} of the compound camphor is 2200 mg/kg in mice (Wickstrom 1988).

Genotoxicity

No mutagenic activity of the compound camphor was observed in the *Salmonella*/microsome assay with *Salmonella* strains TA97a, TA98, TA100, and TA102 with or without metabolic activation (Gomes-Carneiro et al. 1998).

Literature Cited

Bensky, D., S. Clavey, and E. Stöger. 2004. *Chinese herbal medicine: Materia medica.* 3rd ed. Seattle: Eastland Press.

Bhaya, M., and R. Beniwal. 2007. Camphor induced myocarditis: A case report. *Cardiovasc. Toxicol.* 7(3):212-214.

CFR. 2011. *Code of federal regulations*, Title 21 Part 172.510, 2011 ed. Food additives permitted for direct addition to food for human consumption. Flavoring agents and related substances. Natural flavoring substances and natural substances used in conjunction with flavors. Washington, DC: U.S. Government Printing Office.

Chen, J.K., and T.T. Chen. 2004. *Chinese medical herbology and pharmacology*. City of Industry, CA: Art of Medicine Press.

Geller, R.J., D.A. Spyker, L.K. Garrettson, and A.D. Rogol. 1984. Camphor toxicity: Development of a triage strategy. *Vet. Hum. Toxicol.* 26(2):8-10.

Gomes-Carneiro, M.R., I. Felzenszwalb, and F.J. Paumgartten. 1998. Mutagenicity testing (±)-camphor, 1,8-cineole, citral, citronellal, (−)-menthol and terpineol with the *Salmonella*/ microsome assay. *Mutat. Res.* 416(1-2):129-136.

Guilbert, J., C. Flamant, F. Hallalel, et al. 2007. Anti-flatulence treatment and status epilepticus: A case of camphor intoxication. *Emerg. Med. J.* 24(12):859-860.

Leuschner, J. 1997. Reproductive toxicity studies of D-camphor in rats and rabbits. *Arzneimittelforschung* 47(2):124-128.

Love, J.N., M. Sammon, and J. Smereck. 2004. Are one or two dangerous? Camphor exposure in toddlers. *J. Emerg. Med.* 27(1):49-54.

Manoguerra, A.S., A.R. Erdman, P.M. Wax, et al. 2006. Camphor poisoning: An evidence-based practice guideline for out-of-hospital management. *Clin. Toxicol.* 44(4):357-370.

McGuffin, M., C. Hobbs, R. Upton, and A. Goldberg. 1997. *Botanical safety handbook*. Boca Raton, FL: CRC Press.

Stubbs, B.J., A. Specht, and D. Brushett. 2004. The essential oil of *Cinnamomum camphora* (L.) Nees and Eberm.—Variation in oil composition throughout the tree in two chemotypes from eastern Australia. *J. Ess. Oil Res.* 16(1):9-14.

Uc, A., W.P. Bishop, and K.D. Sanders. 2000. Camphor hepatotoxicity. *South. Med. J.* 93(6):596-598.

U.S.C. 2010. United States Code, Title 21, Part 321 (s)(6). Current as of January 7, 2011. Washington, DC: U.S. Government Printing Office.

Wickstrom, E. 1988. Camphor. PIM 095. Poisons Information Monographs—Pharmaceuticals. INCHEM. International Programme on Chemical Safety: Inter-Organization Programme of UNEP, ILO, FAO, WHO, UNIDO, UNITAR, and OECD.

Cinnamomum verum J. Presl — Lauraceae

SCN: cinnamon
Syn: *Cinnamomum zeylanicum* Nees
AN: *tvak*
OCN: Ceylon cinnamon; true cinnamon
Part: bark

Quick Reference Summary

Safety Class: 2b
Interaction Class: A

Contraindications

Not for use in pregnancy except under the supervision of a qualified healthcare practitioner (Chadha 1988; Mantovani et al. 1989; Wichtl 2004).

C

Other Precautions

Use with caution in persons with sensitivity to balsam of Peru (Calnan 1976; Hoskins 1984; Pfutzner et al. 2003).

Drug and Supplement Interactions

None known.

Editors' Notes

The classifications and concerns for this herb are based on relatively higher doses used for therapeutic purposes in contrast to lower amounts generally used in cooking, and have not been associated with its use as a spice.

Several species of *Cinnamomum* are commonly traded under the name "cinnamon" (Leung and Foster 1996).

Adverse Events and Side Effects

Allergic or sensitization reactions to cinnamon or compounds from cinnamon are relatively common. These reactions usually affect the skin or mucous membranes (Allen and Blozis 1988; Endo and Rees 2006; Lamey et al. 1990; Miller et al. 1992; Wright 2007).

Pharmacological Considerations

Animal studies have indicated that cinnamon may modify glucose levels (Kannappan et al. 2006; Verspohl et al. 2005). People with diabetes are advised to monitor their blood sugar closely and discuss the use of this herb with a qualified healthcare practitioner prior to use.

Pregnancy and Lactation

A reduction in the number of embryos was observed in mice fed diets containing cinnamon essential oil (Domaracky et al. 2007). In rats administered the compound cinnamaldehyde (5 to 250 mg/kg daily) on days 7 to 17 of pregnancy, an increase in fetal malformations was observed (Mantovani et al. 1989).

No information on the safety of cinnamon during lactation was identified. While this review did not identify any concerns for use while nursing, safety has not been conclusively established.

Review Details

I. Drug and Supplement Interactions

Clinical Trials of Drug or Supplement Interactions

No clinical trials of drug or supplement interactions were identified.

Case Reports of Suspected Drug or Supplement Interactions

No case reports of suspected drug or supplement interactions were identified.

Animal Trials of Drug or Supplement Interactions

No animal trials of drug or supplement interactions were identified.

II. Adverse Events

Case Reports of Adverse Events

No side effects from cinnamon are expected at standard therapeutic doses (2–4 g daily). In overdose, cinnamon may excite the central nervous system causing increases in heart rate, peristalsis, respiration, and perspiration. The excitation is sometimes followed by sedation, with sleepiness (Wichtl 2004).

The compound cinnamaldehyde, which comprises approximately 55 to 75% of cinnamon essential oil, can cause skin sensitization and irritation in sensitive persons. Reactions to cinnamaldehyde occur most commonly with prolonged oral exposure to products such as toothpaste, gum, or hard candies. Such cinnamon-flavored products have been found to be responsible for a number of cases of oral inflammation or lesions (Allen and Blozis 1988; Endo and Rees 2006; Lamey et al. 1990; Miller et al. 1992; Wright 2007). These include stomatitis (an inflammation of the mucous lining of any of the structures in the mouth), orofacial granulomatosis, erythematous patches, erythema multiforme, lichen planus, leukoplakia, varying degrees of superimposed keratosis or ulceration, and plasma cell gingivitis (Allen and Blozis 1988; Anil 2007; Cohen and Bhattacharyya 2000; Endo and Rees 2006, 2007; Lamey et al. 1990; Mihail 1992; Miller et al. 1992; Wright 2007). Some cases have been diagnosed as allergic in nature (Drake and Maibach 1976; Tremblay and Avon 2008).

Cases of allergic contact dermatitis have been reported in persons that added powdered cinnamon to their coffee. In at least one of the cases, the patient held the coffee in his mouth to cool before swallowing, resulting in increased oral contact with cinnamon (De Rossi and Greenberg 1998).

Contact allergy to the compound cinnamal was confirmed in a patient with oral lichen planus (Hoskyn and Guin 2005).

Squamous cell carcinoma of the tongue was reported in a 24-year-old woman who chewed up to 5 packs daily of a sugarless gum flavored with cinnamon aldehydes, compounds that can incite irritation of oral mucosa (Westra et al. 1998). An oral leukoplakic lesion that was clinically thought to be a squamous cell carcinoma was reported in a patient that chewed cinnamon gum. The patient was found to be allergic to cinnamon (Mihail 1992).

Numerous cases of topical contact dermatitis from cinnamon, cinnamon essential oil, and compounds isolated from cinnamon have been reported and confirmed by patch testing (Farkas 1981; Garcia-Abujeta et al. 2005; Goh and Ng 1988; Hartmann and Hunzelmann 2004; Kern 1960; Kirton 1978; Ludera-Zimoch 1981; Sanchez-Perez and Garcia-Diez 1999). Some cases have been due to occupational exposure in bakers and restaurant workers (Ackermann et al. 2009; Fisher 1982; Nixon 1995).

Severe exacerbation of rosacea was reported in a 68-year-old woman with type 2 diabetes who had been taking a 500 mg capsule of cinnamon every other day for 1 week, and then one capsule daily for a second week (Campbell et al. 2008).

First and second degree burns were reported on the rear thigh of an 11-year-old male that had been carrying a vial of cinnamon essential oil in his back pants pocket. The vial broke, resulting in contact with a relatively large amount of cinnamon oil (Sparks 1985).

Oral burning, abdominal pain, and a single case of nausea were reported in 32 adolescent males that had sucked on toothpicks, hard candy, or their fingers that had been dipped in cinnamon essential oil (Perry et al. 1990).

Vomiting, diarrhea, dizziness, and loss of consciousness were reported in a child who ingested 60 ml of cinnamon essential oil (Pilapil 1989).

Self-inflicted bloody tears were reported in a woman who had been applying cinnamon bark wrapped in gauze to her eye for an unspecified amount of time (Awan et al. 2006).

Occupational asthma has been reported in workers regularly exposed to cinnamon dust (Uragoda 1984). In workers occupationally exposed to cinnamon dust, skin and eye irritation as well as hair loss have been reported and are likely due to the irritant nature of cinnamaldehyde (Calnan 1976; Uragoda 1984).

III. Pharmacology and Pharmacokinetics

Human Pharmacological Studies

Cinnamon and balsam of Peru contain many similar allergenic compounds. Patch testing has demonstrated cross-reactivity in persons with allergies to these two species (Calnan 1976; Hoskins 1984; Pfutzner et al. 2003).

Among spice factory workers, pruritus and skin irritation were reported as common in persons working with spice powders, with cinnamon being noted as the most common irritant. Positive patch test reactions to cinnamaldehyde were found in 11 of 25 factory workers who had skin reactions, and skin prick testing elicited positive reactions in 6 workers (Meding 1993).

Animal Pharmacological Studies

In rats orally administered 0.2 or 2.0 ml (per 150–170 g animal) of an aqueous extract of cinnamon daily for 60 days and fed a high-fructose diet, blood glucose levels in the 2.0 ml dose group were similar to untreated animals fed a regular diet. Untreated animals fed a high-fructose diet had elevated blood glucose levels (Kannappan et al. 2006).

In rats orally administered 5.96 mg/kg of a dried aqueous extract of cinnamon, a decrease in blood glucose levels was observed in the glucose tolerance test but not in animals challenged with a glucose load (Verspohl et al. 2005).

In diabetic rats orally administered 5, 10, or 20 mg/kg of the compound cinnamaldehyde daily for 45 days, a dose dependent decrease in plasma glucose levels was observed (Subash Babu et al. 2007).

In maximization tests with guinea pigs, animals sensitized to cinnamaldehyde also reacted to cinnamyl alcohol and cinnamic acid. Animals sensitized to cinnamyl alcohol reacted to cinnamyl alcohol and cinnamaldehyde, but not to cinnamic acid. Cinnamic acid did not sensitize guinea pigs. Compared to the challenge concentration for cinnamaldehyde, a 15 times higher concentration of cinnamyl alcohol and a 25 times higher concentration of cinnamic acid were required to give positive reactions in animals sensitized to cinnamaldehyde (Weibel et al. 1989).

In Vitro Pharmacological Studies

Aqueous extracts of cinnamon provoked dose-related contractile responses of isolated guinea pig tracheal smooth muscle. The authors indicated that, based on that result, reactions to cinnamon experienced in spice factory workers exposed to cinnamon dust may be due to direct irritant reactions (Zuskin et al. 1988).

IV. Pregnancy and Lactation

In rats orally administered the compound cinnamaldehyde at doses of 5, 25, or 250 mg/kg on days 7 to 17 of pregnancy, the incidence of poor cranial ossification was increased in all treated groups, while reduced ossification of the tympanic bulla was increased in the 25 and 250 mg/kg groups. Increases in the incidences of dilated pelvis/reduced papilla in the kidney, dilated ureters, and abnormal sternebrae were detected in the 25 mg/kg group, which had the highest overall prevalence of minor abnormalities. Increases in the incidences of reduced cranial ossification, dilated ureters, and renal variants were observed at 5 mg/kg, a dose at which there was no detectable maternal toxicity (Mantovani et al. 1989).

In mice orally administered a diet containing 0.25% cinnamon essential oil for 2 weeks, a decrease in the number of nuclei and an alteration of the distribution of embryos was observed at gestational day 4 (Domaracky et al. 2007).

No information on the safety of cinnamon during lactation was identified.

V. Toxicity Studies

Acute Toxicity

The LD_{50} of orally administered ethanol extract of cinnamon in mice could not be determined at does up to 3 g/kg (Shah et al. 1998).

The oral LD_{50} of the compound cinnamaldehyde has been reported in different studies as 1.85 2.22, 3.35, or 3.4 g/kg in rats, 1.16 or 3.4 g/kg in guinea pigs, and 0.46 or 2.32 g/kg in mice (Hoskins 1984; Subash Babu et al. 2007).

The LD_{50} of the compound cinnamic acid orally administered in rats, mice, and guinea pigs could not be determined at doses up to 5 g/kg (Hoskins 1984).

Subchronic Toxicity

In mice orally administered 100 mg/kg of an ethanol extract of cinnamon daily for 90 days, no signs of mortality were observed. Hematological studies indicated a reduction in hemoglobin levels. Increases in reproductive organ weights, sperm motility, and sperm count were observed with no spermatotoxic effects (Shah et al. 1998).

Chronic Toxicity

In mice orally administered 100 mg per animal of cinnamon powder daily for 12 months, an increase in squamous papillomas and poorly differentiated carcinomas was observed. Similar results were observed for popularly consumed spices and foods such as chili peppers (*Capsicum annuum*) and large hairtail fish (*Trichiurus lepturus*) (Balachandran and Sivaramkrishnan 1995).

Genotoxicity

In the *Bacillus subtilis rec* assay with strains H17 *rec*$^+$ and M45 *rec*$^-$, no mutagenic activity of an ethanol extract of cinnamon was observed. Petroleum ether and chloroform extracts exhibited mutagenic activity that was not seen after metabolic activation by S9 mix (Ungsurungsie et al. 1984).

"Intermediate" mutagenic activity of an ethanol extract of cinnamon was observed in *Salmonella typhimurium* strain TA102 but not in TA98 (Mahmoud et al. 1992).

In the Ames test for mutagenicity and the *Escherichia coli* WP2 *uvrA* reversion test, no mutagenic activity of cinnamon essential oil was observed. The essential oil exhibited mutagenic activity in the *Bacillus subtilis rec* assay without S9 (Sekizawa and Shibamoto 1982).

Literature Cited

Ackermann, L., K. Aalto-Korte, R. Jolanki, and K. Alanko. 2009. Occupational allergic contact dermatitis from cinnamon including one case from airborne exposure. *Contact Dermat.* 60(2):96-99.

Allen, C.M., and G.G. Blozis. 1988. Oral mucosal reactions to cinnamon-flavored chewing gum. *J. Am. Dent. Assoc.* 116(6):664-667.

Anil, S. 2007. Plasma cell gingivitis among herbal toothpaste users: A report of three cases. *J. Contemp. Dent. Pract.* 8(4):60-66.

Awan, S., H.S. Kazmi, and A.A. Awan. 2006. An unusual case of bloody tears. *J. Ayub Med. Coll. Abbottabad* 18(1):68-69.

Balachandran, B., and V.M. Sivaramkrishnan. 1995. Induction of tumours by Indian dietary constituents. *Indian J. Cancer* 32(3):104-109.

Calnan, C.D. 1976. Cinnamon dermatitis from an ointment. *Contact Dermat.* 2(3):167-170.

Campbell, T.M., R. Neems, and J. Moore. 2008. Severe exacerbation of rosacea induced by cinnamon supplements. *J. Drugs Dermatol.* 7(6):586-587.

Chadha, Y. 1988. *The wealth of India: A dictionary of Indian raw materials and industrial products*. Delhi: Council of Scientific and Industrial Research.

Cohen, D.M., and I. Bhattacharyya. 2000. Cinnamon-induced oral erythema multiformelike sensitivity reaction. *J. Am. Dent. Assoc.* 131:929-934.

De Rossi, S.S., and M.S. Greenberg. 1998. Intraoral contact allergy: A literature review and case reports. *J. Am. Dent. Assoc.* 129(10):1435-1441.

Domaracky, M., P. Rehak, S. Juhas, and J. Koppel. 2007. Effects of selected plant essential oils on the growth and development of mouse preimplantation embryos *in vivo*. *Physiol. Res.* 56(1):97-104.

Drake, T.E., and H.I. Maibach. 1976. Allergic contact dermatitis and stomatitis caused by a cinnamic aldehyde-flavored toothpaste. *Arch. Dermatol.* 112(2):202-203.

Endo, H., and T.D. Rees. 2006. Clinical features of cinnamon-induced contact stomatitis. *Compend. Contin. Educ. Dent.* 27(7):403-409.

Endo, H., and T.D. Rees. 2007. Cinnamon products as a possible etiologic factor in orofacial granulomatosis. *Med. Oral Patol. Oral Cir. Bucal* 12(6):E440-E444.

Farkas, J. 1981. Perioral dermatitis from marjoram, bay leaf and cinnamon. *Contact Dermat.* 7(2):121.

Fisher, A.A. 1982. Hand dermatitis—A "baker's dozen." *Cutis* 29(3):214, 217-218, 221.

Garcia-Abujeta, J.L., C.H. de Larramendi, J.P. Berna, and E.M. Palomino. 2005. Mud bath dermatitis due to cinnamon oil. *Contact Dermat.* 52(4):234.

Goh, C.L., and S.K. Ng. 1988. Bullous contact allergy from cinnamon. *Derm. Beruf. Umwelt.* 36(6):186-187.

Hartmann, K., and N. Hunzelmann. 2004. Allergic contact dermatitis from cinnamon as an odour-neutralizing agent in shoe insoles. *Contact Dermat.* 50(4):253-254.

Hoskins, J.A. 1984. The occurrence, metabolism and toxicity of cinnamic acid and related compounds. *J. Appl. Toxicol.* 4(6):283-292.

Hoskyn, J., and J.D. Guin. 2005. Contact allergy to cinnamal in a patient with oral lichen planus. *Contact Dermat.* 52(3):160-161.

Kannappan, S., T. Jayaraman, P. Rajasekar, M.K. Ravichandran, and C.V. Anuradha. 2006. Cinnamon bark extract improves glucose metabolism and lipid profile in the fructose-fed rat. *Singapore Med. J.* 47(10):858-863.

Kern, A.B. 1960. Contact dermatitis from cinnamon. *Arch. Dermatol.* 81:599-600.

Kirton, V. 1978. Contact urticaria and cinnamic aldehyde. *Contact Dermat.* 4(6):374-375.

Lamey, P.J., M.A. Lewis, T.D. Rees, et al. 1990. Sensitivity reaction to the cinnamonaldehyde component of toothpaste. *Br. Dent. J.* 168(3):115-118.

Leung, A.Y., and S. Foster. 1996. *Encyclopedia of common natural ingredients used in food, drugs, and cosmetics*. 2nd ed. New York: Wiley.

Ludera-Zimoch, G. 1981. Case of urticaria with immediate local and generalized reaction to cinnamon oil and benzaldehyde.] *Przegl. Dermatol.* 68(1):67-70.

Mahmoud, I., A. Alkofahi, and A. Abdelaziz. 1992. Mutagenic and toxic activities of several spices and some Jordanian medicinal plants. *Int. J. Pharmacogn.* 30(2):81-85.

Mantovani, A., A.V. Stazi, C. Macri, et al. 1989. Prenatal (segment II) toxicity study of cinnamic aldehyde in the Sprague-Dawley rat. *Food Chem. Toxicol.* 27(12):781-786.

Meding, B. 1993. Skin symptoms among workers in a spice factory. *Contact Dermat.* 29(4):202-205.

Mihail, R.C. 1992. Oral leukoplakia caused by cinnamon food allergy. *J. Otolaryngol.* 21(5):366-367.
Miller, R.L., A.R. Gould, and M.L. Bernstein. 1992. Cinnamon-induced stomatitis venenata: Clinical and characteristic histopathologic features. *Oral Surg. Oral Med. Oral Pathol.* 73(6):708-716.
Nixon, R. 1995. Cinnamon allergy in a baker. *Australas. J. Dermatol.* 36(1):41.
Perry, P.A., B.S. Dean, and E.P. Krenzelok. 1990. Cinnamon oil abuse by adolescents. *Vet. Hum. Toxicol.* 32(2):162-164.
Pfutzner, W., P. Thomas, A. Niedermeier, et al. 2003. Systemic contact dermatitis elicited by oral intake of balsam of Peru. *Acta Derm. Venereol.* 83(4):294-295.
Pilapil, V.R. 1989. Toxic manifestations of cinnamon oil ingestion in a child. *Clin. Pediatr.* 28(6):276.
Sanchez-Perez, J., and A. Garcia-Diez. 1999. Occupational allergic contact dermatitis from eugenol, oil of cinnamon and oil of cloves in a physiotherapist. *Contact Dermat.* 41(6):346-347.
Sekizawa, J., and T. Shibamoto. 1982. Genotoxicity of safrole-related chemicals in microbial test systems. *Mutat. Res.* 101(2):127-140.
Shah, A.H., A.H. Al-Shareef, A.M. Ageel, and S. Qureshi. 1998. Toxicity studies in mice of common spices, *Cinnamomum zeylanicum* bark and *Piper longum* fruits. *Plant Foods Hum. Nutr.* 52(3):231-239.
Sparks, T. 1985. Cinnamon oil burn. *West. J. Med.* 142(6):835.
Subash Babu, P., S. Prabuseenivasan, and S. Ignacimuthu. 2007. Cinnamaldehyde—A potential antidiabetic agent. *Phytomedicine* 14(1):15-22.
Tremblay, S., and S.L. Avon. 2008. Contact allergy to cinnamon: Case report. *J. Can. Dent. Assoc.* 74(5):445-461.
Ungsurungsie, M., C. Paovalo, and A. Noonai. 1984. Mutagenicity of extracts from Ceylon cinnamon in the rec assay. *Food Chem. Toxicol.* 22(2):109-112.
Uragoda, C.G. 1984. Asthma and other symptoms in cinnamon workers. *Br. J. Indian Med.* 41(2):224-227.
Verspohl, E.J., K. Bauer, and E. Neddermann. 2005. Antidiabetic effect of *Cinnamomum cassia* and *Cinnamomum zeylanicum in vivo* and *in vitro*. *Phytother. Res.* 19(3):203-206.
Weibel, H., J. Hansen, and K.E. Andersen. 1989. Cross-sensitization patterns in guinea pigs between cinnamaldehyde, cinnamyl alcohol and cinnamic acid. *Acta Derm. Venereol.* 69(4):302-307.
Westra, W.H., J.S. McMurray, J. Califano, P.W. Flint, and R.L. Corio. 1998. Squamous cell carcinoma of the tongue associated with cinnamon gum use: A case report. *Head Neck* 20(5):430-433.
Wichtl, M. 2004. *Herbal drugs and phytopharmaceuticals: A handbook for practice on a scientific basis.* 3rd ed. Boca Raton, FL: CRC Press.
Wright, J. 2007. Diagnosis and management of oral lichenoid reactions. *J. Calif. Dent. Assoc.* 35(6):412-416.
Zuskin, E., B. Kanceljak, Z. Skuric, et al. 1988. Immunological and respiratory findings in spice-factory workers. *Env. Res.* 47(1):95-108.

C

Cistanche spp. — Orobanchaceae

Cistanche deserticola Ma
SCN: desert broomrape
PN: *rou cong rong* (stem)

Cistanche salsa (C.A. Mey.) Beck
SCN: broomrape
PN: *rou cong rong* (stem)
Part: stem

Quick Reference Summary

Safety Class: 1
Interaction Class: A

Contraindications
None known.

Other Precautions
None known.

Drug and Supplement Interactions
None known.

Adverse Events and Side Effects
None known.

Pharmacological Considerations
None known.

Pregnancy and Lactation
Texts on traditional Chinese medicine do not indicate any cautions for use of broomrape or desert broomrape during pregnancy or lactation (Bensky et al. 2004; Chen and Chen 2004). Although this review did not identify any concerns for use while pregnant or nursing, safety has not been conclusively established.

Review Details

I. Drug and Supplement Interactions

Clinical Trials of Drug or Supplement Interactions
No clinical trials of drug or supplement interactions were identified.

Case Reports of Suspected Drug or Supplement Interactions
No case reports of suspected drug or supplement interactions were identified.

C

Animal Trials of Drug or Supplement Interactions

No animal trials of drug or supplement interactions were identified.

II. Adverse Events

Case Reports of Adverse Events

No case reports of adverse events were identified.

III. Pharmacology and Pharmacokinetics

Human Pharmacological Studies

No relevant human pharmacological studies were identified.

Animal Pharmacological Studies

No relevant animal pharmacological studies were identified.

In Vitro Pharmacological Studies

No relevant in vitro pharmacological studies were identified.

IV. Pregnancy and Lactation

Texts on traditional Chinese medicine do not indicate any cautions for use of broomrape or desert broomrape during pregnancy or lactation (Bensky et al. 2004; Chen and Chen 2004).

V. Toxicity Studies

No toxicity studies were identified.

Literature Cited

Bensky, D., S. Clavey, and E. Stöger. 2004. *Chinese herbal medicine: Materia medica*. 3rd ed. Seattle: Eastland Press

Chen, J.K., and T.T. Chen. 2004. *Chinese medical herbology and pharmacology*. City of Industry, CA: Art of Medicine Press.

Citrus × aurantifolia (Christm.) Swingle — Rutaceae

SCN: lime
OCN: key lime

Part: peel

Quick Reference Summary

Safety Class: 1
Interaction Class: A

Contraindications

None known.

Other Precautions

None known.

Drug and Supplement Interactions

None known.

Notice

Photosensitizing (Opdyke 1979; Pomeranz and Karen 2007); *see* Appendix 2.

Adverse Events and Side Effects

Lime peel contains furanocoumarin compounds that can cause phytophotodermatitis (a nonimmunological phototoxic reaction of the skin) in sensitive individuals after topical exposure to lime peel followed by exposure to the sun (Opdyke 1979; Pomeranz and Karen 2007). Such reactions are not expected after oral consumption.

Pharmacological Considerations

None known.

Pregnancy and Lactation

No information on the safety of lime in pregnancy or lactation was identified in the scientific or traditional literature. Although this review did not identify any concerns for use while pregnant or nursing, safety has not been conclusively established.

Review Details

I. Drug and Supplement Interactions

Clinical Trials of Drug or Supplement Interactions

No clinical trials of drug or supplement interactions were identified.

Case Reports of Suspected Drug or Supplement Interactions

No case reports of suspected drug or supplement interactions were identified.

Animal Trials of Drug or Supplement Interactions

No animal trials of drug or supplement interactions were identified.

II. Adverse Events

Case Reports of Adverse Events

Several cases of phytophotodermatitis have been reported after topical exposure to lime peel followed by exposure to the sun. Erythematous vesicles and bullae were observed on the hands of a 62-year-old man who had been picking limes (Schmidt 2007). Phytophotodermatitis with a painful, erythematous, blistering rash was observed in a 23-year-old woman who had prepared mojitos (cocktails that contain limes) during 2 days she spent at the beach (Pomeranz and Karen 2007). Cases of phytophotodermatitis have been reported in children after preparing lime juice (Mill et al. 2008).

Allergic reactions to lime peel have also been reported after topical exposure. Such was the case for a 52-year-old woman with a history of eczema and eyelid dermatitis who presented with angular cheilitis after having developed a habit of sucking on the limes in her gin and tonic for up to a minute after finishing the drink (Thomson et al. 2007).

III. Pharmacology and Pharmacokinetics

Human Pharmacological Studies

In sensitization tests with fragrance raw materials, several volunteers that tested positive to patch tests with a fragrance mix also tested positive to lime extract (Roesyanto-Mahadi et al. 1990). Lime essential oil demonstrated phototoxic activity in human patch tests (Opdyke 1979).

Animal Pharmacological Studies

In phototoxicity testing, lime essential oil processed by expression showed phototoxic effects in pigs and hairless mice, while the steam-distilled oil showed no such effects (Opdyke 1979).

In Vitro Pharmacological Studies

No relevant in vitro pharmacological studies were identified.

IV. Pregnancy and Lactation

No information on the safety of lime in pregnancy or lactation was identified.

V. Toxicity Studies

Acute Toxicity

The LD_{50} of orally administered lime essential oil in rats and rabbits could not be determined at doses up to 5 g/kg (Opdyke 1979).

Literature Cited

Mill, J., B. Wallis, L. Cuttle, et al. 2008. Phytophotodermatitis: Case reports of children presenting with blistering after preparing lime juice. *Burns* 34(5):731-733.

Opdyke, D.L.J. 1979. *Monographs on fragrance raw materials*. New York: Pergamon.

Pomeranz, M.K., and J.K. Karen. 2007. Images in clinical medicine. Phytophotodermatitis and limes. *N. Engl. J. Med.* 357(1):e1.

Roesyanto-Mahadi, I.D., A.M. Geursen-Reitsma, T. van Joost, and T.W. van den Akker. 1990. Sensitization to fragrance materials in Indonesian cosmetics. *Contact Dermat.* 22(4):212-217.

Schmidt, J. 2007. Phytophotodermatitis. *Dermatol. Nurs.* 19(5):486.

Thomson, M.A., P.W. Preston, L. Prais, and I.S. Foulds. 2007. Lime dermatitis from gin and tonic with a twist of lime. *Contact Dermat.* 56(2):114-115.

Citrus × aurantium L. — Rutaceae

SCN: bitter orange

PN: *zhi shi* (immature fruit); *zhi qiao* (nearly mature fruit)

OCN: bigarade; marmalade orange; Seville orange; sour orange

Part: fruit

Quick Reference Summary

Safety Class: 1

Interaction Class: C

Contraindications

None known.

Other Precautions

None known.

Drug and Supplement Interactions

In human studies, the fruit juice of bitter orange has been shown to inhibit the drug metabolizing isoenzyme CYP3A4, leading to increased blood levels of certain drugs metabolized by CYP3A4. Human studies have confirmed interactions of the juice with cyclosporine, felodipine, dextromethorphan, and saquinavir (Di Marco et al. 2002;

Edwards et al. 1999; Malhotra et al. 2001; Mouly et al. 2005). *See* Cytochrome P450 in Appendix 3.

Editors' Notes

Three types of bitter orange fruit products are commercially available. The crude (minimally processed) dried immature fruit and nearly mature fruit are the products used in traditional Chinese medicine, and extracts of the fruit with an increased concentration of the compound synephrine (usually 4 to 6%) are also available (*see* next entry).

Synephrine, an alkaloid present in bitter orange, has a stimulant action on the nervous system. Synephrine affects α-adrenergic receptors and, to a lesser extent, β-adrenergic receptors, and it can raise blood pressure and have other effects on cardiac function that may be beneficial for selected patients when a proper dosage is administered (ILS 2004). Synephrine content in dried bitter orange fruit ranges from 0.012 to 0.25% (Rossato et al. 2011).

The interaction class established for this entry is based primarily on studies of bitter orange fruit juice, and on the fact that bergamottin, one of the furocoumarins in the juice that are responsible for its interaction with numerous drugs (Baumgart 2004; Edwards et al. 1999; Malhotra et al. 2001; Messer et al. 2012), is also present in the fruit (Peroutka et al. 2007).

In addition, some furocoumarins are recognized to be phototoxic (Kavli and Volden 1984). However, no cases of phototoxicity in persons using bitter orange fruit have been reported, and two of the citrus furocoumarins, bergaptol and bergamottin, were not found to be phototoxic in vitro (Messer et al. 2012).

Adverse Events and Side Effects

Ingestion of "large amounts" of orange peel (bitter or sweet) has been reported to cause intestinal colic, convulsions, and even death in children (Leung and Foster 1996). A review of this reference and its sources, however, discloses that the reported death was limited to a single case recorded in 1833 that was attributed to "eating the rind of an orange," and that no additional details were provided on the other reported effects (Wood and Bache 1833). The current literature does not indicate any concern for such events at standard therapeutic doses.

Pharmacological Considerations

The juice of bitter orange fruit and extracts of whole bitter orange fruits have been shown to affect the metabolism of certain drugs in ways similar to, but not precisely the same as, grapefruit juice (Malhotra et al. 2001; Di Marco et al. 2002). Animal studies have found that decoctions of the fruit significantly decreased blood levels of the drug tacrolimus in rats, whereas an article described as the peel of the ripe fruit had no effect on the drug (Lin et al. 2011) and increased plasma concentrations of cyclosporine in swine (Hou et al. 2000).

Human studies have confirmed interactions of bitter orange juice with cyclosporine, felodipine, dextromethorphan, and saquinavir (Di Marco et al. 2002; Edwards et al. 1999; Malhotra et al. 2001; Mouly et al. 2005).

Pregnancy and Lactation

Information on the safety of bitter orange fruit during pregnancy is limited. One reference text on traditional Chinese medicine cites caution for both the immature and nearly mature fruit during pregnancy (Bensky et al. 2004), a concern not documented in other references.

No information on the safety of bitter orange fruit in lactation was identified in the scientific or traditional literature. While this review did not identify any concerns for use while nursing, safety has not been conclusively established.

Review Details

I. Drug and Supplement Interactions

Clinical Trials of Drug or Supplement Interactions

In healthy volunteers orally administered 240 ml of bitter orange fruit juice along with felodipine, a 73% increase in the area under the plasma time-concentration curve of felodipine was observed. The effects of bitter orange fruit juice were similar to that of grapefruit juice, in that the felodipine maximum concentration was augmented while the terminal elimination half-life was unchanged. Inactivation of intestinal CYP3A4 was the suggested mechanism of action (Malhotra et al. 2001).

In healthy volunteers orally administered 200 ml of bitter orange fruit juice or grapefruit juice, the bioavailability of orally administered dextromethorphan increased significantly with both juices, but only returned to half the baseline value after three days of washout. The results suggested that both juices are clinically significant inhibitors of intestinal CYP3A4 and P-glycoprotein (Di Marco et al. 2002).

Case Reports of Suspected Drug or Supplement Interactions

No case reports of suspected drug or supplement interactions were identified.

Animal Trials of Drug or Supplement Interactions

Coadministration of cyclosporine (10 mg/kg) and an aqueous extract of crude bitter orange fruit in pigs increased the plasma concentration of cyclosporine by 64% (Hou et al. 2000).

In rats orally administered 1.5 mg/kg of tacrolimus (a substrate of CYP3A4) with or without 2 g/kg of a decoction of the unripe fruits or an article described as the ripe peel of bitter orange, a 72% decrease in the maximum plasma concentration of tacrolimus was observed in rats administered the decoction of the unripe fruits of bitter orange. No change in tacrolimus levels was observed in the rats administered a decoction of the ripe peels. A related in vitro study with the same plant materials indicated that the unripe fruit increased the activity of P-gp, a drug transport protein that transports drugs out of cells. For both studies, the decoctions of the unripe fruits and ripe peels was made by adding 25 g of plant material to 500 ml of water and cooking gently until 100 ml of liquid remained (Lin et al. 2011).

II. Adverse Events

Case Reports of Adverse Events

No case reports of adverse events were identified.

III. Pharmacology and Pharmacokinetics

Human Pharmacological Studies

No relevant human pharmacological studies were identified.

Animal Pharmacological Studies

No relevant animal pharmacological studies were identified.

In Vitro Pharmacological Studies

No relevant in vitro pharmacological studies were identified.

IV. Pregnancy and Lactation

Information on the safety of bitter orange fruit during pregnancy is limited. One reference text on traditional Chinese medicine cites caution for both the immature and nearly mature fruit during pregnancy (Bensky et al. 2004), a concern not documented in other references.

No information on the safety of bitter orange fruit during lactation was identified. While this review did not identify any concerns for use while nursing, safety has not been conclusively established.

V. Toxicity Studies

Acute Toxicity

The LD_{50} of an injection solution of immature bitter orange fruit administered intravenously in mice has been recorded as 72 g/kg, and intravenous injection administered to dogs did not produce any serious reactions at a dose of 21 g/kg (Zhu 1998).

C

Literature Cited

Baumgart, A., M. Schmidt, H.J. Schmitz, and D. Schrenk. 2005. Natural furocoumarins as inducers and inhibitors of cytochrome P450 1A1 in rat hepatocytes. *Biochem. Pharmacol.* 69(4):657-667.

Bensky, D., S. Clavey, and E. Stöger. 2004. *Chinese herbal medicine: Materia medica.* 3rd ed. Seattle: Eastland Press.

Di Marco M.P., D.J. Edwards, I.W. Wainer, and M.P. Ducharme. 2002. The effect of grapefruit juice and seville orange juice on the pharmacokinetics of dextromethorphan: The role of gut CYP3A and P-glycoprotein. *Life Sci.* 71(10):1149-60.

Edwards D.J., M.E. Fitzsimmons, E.G. Schuetz, K. Yasuda, M.P. Ducharme, L.H. Warbasse, P.M. Woster, J.D. Schuetz, and P. Watkins. 1999. 6′,7′-Dihydroxybergamottin in grapefruit juice and Seville orange juice: Effects on cyclosporine disposition, enterocyte CYP3A4, and P-glycoprotein. *Clin. Pharmacol. Ther.* 65(3):237-44.

Hou, Y.C., S.L. Hsiu, C.W. Tsao, Y.H. Wang, and P.D. Chao. 2000. Acute intoxication of cyclosporin caused by coadministration of decoctions of the fruits of *Citrus aurantium* and the pericarps of *Citrus grandis*. *Planta Med.* 66(7):653-655.

ILS. 2004. Bitter orange (*Citrus aurantium* var. *amara*) extracts and constituents (±)-*p*-synephrine [CAS No. 94-07-5] and (±)-*p*-octopamine [CAS No. 104-14-3]: Review of toxicological literature. Research Triangle Park, NC: Integrated Laboratory Systems, Inc.

Kavli, G., and G. Volden. 1984. Phytophotodermatitis. *Photodermatol.* 1(2):65-75.

Leung, A.Y., and S. Foster. 1996. *Encyclopedia of common natural ingredients used in food, drugs, and cosmetics.* 2nd ed. New York: Wiley.

Lin, S.P., P.P. Wu, Y.C. Hou, S.Y. Tsai, M.J. Wang, S.H. Fang, and P.D.L. Chao. 2011. Different influences on tacrolimus pharmacokinetics by coadministrations of zhi ke and zhi shi in rats. *Evid. Based Complement. Alternat. Med.* 2011:751671.

Malhotra S., D.G. Bailey, M.F. Paine, and P.B. Watkins. 2001. Seville orange juice-felodipine interaction: Comparison with dilute grapefruit juice and involvement of furocoumarins. *Clin. Pharmacol. Ther.* 69(1):14-23.

Messer A., N. Raquet, C. Lohr, and D. Schrenk. 2012. Major furocoumarins in grapefruit juice II: Phototoxicity, photogenotoxicity, and inhibitory potency *versus* cytochrome P450 3A4 activity. *Food Chem. Toxicol.* 50:756-60.

Mouly, S.J., C. Matheny, M.F. Paine, J. Lamba, V. Lamba, S.N. Pusek, E.G. Shuetz, P.W. Stewart, and P.B. Watkins. 2005. Variation in oral clearance of saquinavir is predicted by CYP3A5*1 genotype but not by enterocyte content of cytochrome P450 3A5. *Clin. Pharmacol. Ther.* 78(6):605-618.

Rossato L.G., V.M. Costa, R.P. Limberger, M.L. Bastos, and F. Remião. 2011. Synephrine: From trace concentrations to massive consumption in weight-loss. *Food Chem Toxicol* 49(1):8-16.

Wood, B., and F. Bache. 1833. *The dispensatory of the United States of America.* 1st ed. Philadelphia: Lippincott.

Zhu, Y.-P. 1998. *Chinese materia medica: Chemistry, pharmacology and applications.* Amsterdam: Harwood Academic Publishers.

Citrus × aurantium L. — Rutaceae

SCN: bitter orange
OCN: bigarade; marmalade orange; Seville orange; sour orange

Part: fruit concentrated extract (3–8% synephrine)

C

Quick Reference Summary

Safety Class: 2d*
Interaction Class: C†

Contraindications

Bitter orange fruit concentrated extract combined with caffeine or caffeine-containing herbs is not recommended for excessive or long-term use or for use by persons with heart irregularities, hypertension, insomnia, or anxiety (Haller 2005).

Other Precautions

None known.

Drug and Supplement Interactions

In human studies, the fruit juice of bitter orange has been shown to inhibit the drug metabolizing isoenzyme CYP3A4, leading to increased blood levels of certain drugs metabolized by CYP3A4 (Di Marco et al. 2002; Edwards et al. 1999; Malhotra et al. 2001; Mouly et al. 2005). *See* Editors' Notes below and Cytochrome P450 in Appendix 3.

Editors' Notes

Three types of bitter orange fruit products are commercially available. The crude (minimally processed) dried immature fruit and nearly mature fruit are the products used in traditional Chinese medicine (*see* previous entry), and extracts of the fruit with an increased concentration of the compound synephrine (usually 4–6%) are also available. This concentrated extract is marketed primarily as an ingredient in weight loss supplements.

Several of the articles referenced in this entry clearly state that the subject extract is derived from the whole fruit of bitter orange (Calapai et al. 1999; Firenzuoli et al. 2005; Min et al. 2005), while numerous others do not specifically identify the part of the plant from which the extract is derived (Bouchard et al. 2005; Bui et al. 2006; Colker et al. 1999; Gurley et al. 2004; Hansen et al. 2006; Nykamp et al. 2004). Some of the latter articles, however, provide *zhi shi*, which is the pinyin name for the dried immature fruit of this *Citrus* species, as a synonym for bitter orange or *Citrus aurantium* (Bui et al. 2006; Gurley et al. 2004; Hansen et al. 2006). Just one case report identifies bitter orange peel as an ingredient in a weight loss formula, apparently based on the product's labeling (Gange et al. 2006), and some studies refer to one particular brand of bitter orange fruit extract (Douds 1997; Haller et al. 2005). For purposes of this entry, it is assumed that ingredients reviewed in all references are actually extracted from whole bitter orange fruit rather than from the peel.

Synephrine, an alkaloid present in bitter orange fruit, has a stimulant action on the nervous system. Synephrine affects α-adrenergic receptors and, to a lesser extent, β-adrenergic receptors, and it can raise blood pressure and have other effects on cardiac function that may be beneficial for selected patients when a proper dosage is administered (ILS 2004). Synephrine content in dried bitter orange ranges from 0.012 to 0.25% (Rossato et al. 2011). Concentrated extracts of bitter orange fruit are typically standardized to 4 or 6% synephrine (Bui et al. 2006; Calapai et al. 1999; Gurley et al. 2004; Hansen et al. 2006; Kubo et al. 2005; Nguyen et al. 2006).

The interaction class established for this entry is based primarily on the fact that furocoumarins present in bitter orange fruit are known to be CYP3A4 inhibitors (Ishihara et al. 2011). In one study of a bitter orange fruit extract, however, no significant effects on the drug metabolizing enzyme CYP3A4 were observed, though it was also reported that the furocoumarin 6,7-dihydroxybergamottin, a known inhibitor of CYP3A4, was not detected in the tested extract (Gurley et al. 2004). No clinically relevant drug or supplement interactions are expected with bitter orange fruit concentrated extract in which the furocoumarins that are naturally-occurring in the fruit are not present.

Furocoumarins are recognized to be phototoxic (Kavli and Volden 1984). However, no cases of phototoxicity in persons using bitter orange fruit extract have been reported, and two of the citrus furocoumarins, bergaptol and bergamottin, were not found to be phototoxic in vitro (Messer et al. 2012).

Adverse Events and Side Effects

Adverse events, including stroke, variant angina, and myocardial infarction, have been reported in temporal association with consumption of bitter orange concentrated extracts in combination with caffeine or a caffeine-containing ingredient (Bouchard et al. 2005; Gange et al. 2006; Nasir et al. 2004; Nykamp et al. 2004). One of 16 reports

* No contraindications known when used without caffeine or caffeine-containing herbs.

† No clinically relevant drug or supplement interactions are expected with bitter orange fruit concentrated extract in which the furocoumarins that are naturally-occurring in the fruit are not present.

of cardiovascular adverse reactions received by Health Canada between 1998 and 2004 and associated with products containing bitter orange or synephrine was related to a product that was reported to contain neither caffeine nor ephedrine, while the other reported products contained one or both of these alkaloids (Jordan et al. 2004). Instances of tachycardia were reported in a 52-year-old woman on two separate occasions, each on the same day that oral use of a dietary supplement containing 500 mg of an extract of *Citrus aurantium* standardized to 30 mg of synephrine was initiated; other ingredients, if any, were not disclosed (Firenzuoli 2005).

Pharmacological Considerations

Studies on the effects of bitter orange fruit concentrated extract on heart rate and blood pressure are conflicting. No increase in heart rate was associated with acute use of bitter orange fruit concentrated extract (containing 50 mg *p*-synephrine) at 75 minutes (Stohs et al. 2011), though a significant increase was noted with a similar dose after 4 hours (Bui et al. 2006) and 6 hours (Haller et al. 2005). Similarly, one study showed a significant increase in blood pressure after use of bitter orange fruit concentrated extract (Bui et al. 2006), with others showing no such effect (Haller et al. 2005; Min et al. 2005; Stohs et al 2011).

Bitter orange fruit concentrated extract has been shown in one human trial to have no effect on the drug-metabolizing enzyme CYP3A4 (Gurley et al. 2004). Bitter orange juice (also known as Seville orange juice), however, has been shown to significantly inhibit the activity of CYP3A4 (Penzak et al. 2002). Research findings on the juice of bitter orange may not be relevant to products made from the fruit, as the two parts are distinct.

Bitter orange contains furocoumarins, compounds recognized to be phototoxic and to act as inhibitors of certain drug-metabolizing enzymes (Baumgart et al. 2005). However, no cases of phototoxicity in persons using bitter orange have been reported.

Pregnancy and Lactation

Information on the safety of bitter orange fruit concentrated extract in pregnancy is limited. A study of bitter orange fruit concentrated extract in pregnant rats showed no adverse effects on mothers or fetuses (Hansen et al. 2006).

No information on the safety of bitter orange fruit extract in lactation was identified in the scientific or traditional literature. Although this review did not identify any concerns for nursing women, safety has not been conclusively established.

Review Details

I. Drug and Supplement Interactions

Clinical Trials of Drug or Supplement Interactions

No significant effects of bitter orange fruit concentrated extract (700 mg daily with 28 mg synephrine) were observed on CYP3A4 (Gurley et al. 2004).

Case Reports of Suspected Drug or Supplement Interactions

No case reports of suspected drug or supplement interactions were identified.

Animal Trials of Drug or Supplement Interactions

No animal trials of drug or supplement interactions were identified.

II. Adverse Events

Adverse Events Reported in Clinical Trials

No adverse events were reported in a clinical trial of adults taking a supplement containing bitter orange fruit concentrated extract (975 mg with 58 mg synephrine), caffeine, and St. John's wort (*Hypericum perforatum*) for 6 weeks (Colker et al. 1999).

Case Reports of Adverse Events

A stroke was reported in a 38-year-old man with a one-week history of taking a supplement containing bitter orange fruit concentrated extract (6–12 mg synephrine daily) and caffeine (200–400 mg daily) (Bouchard et al. 2005).

Two episodes of tachycardia were reported in a woman who took a dietary supplement for weight loss that contained bitter orange concentrated extract (30 mg synephrine daily); no additional information on the product formulation was provided in the case report. The episodes occurred on two separate days, 1 month apart, on which the woman initiated use of the extract (Firenzuoli et al. 2005).

An acute lateral-wall myocardial infarction was reported in a 55-year-old woman with a 1-year history of use of a weight loss product containing 300 mg *Citrus aurantium* (no other description provided) as well as 30 mg caffeine (misidentified as guararanine [sic]) and 30 mg green tea (no other description). The patient also reported a nearly 40-year tobacco-smoking habit (1½ packs per day), high caffeine intake from cola, coffee, and tea, and a preexisting heart murmur (Nykamp et al. 2004).

A 57-year-old man with a 35-day history of use of a weight loss product containing 125 mg of a combination of *Citrus aurantium* extract standardized to 5% synephrine and green tea leaf extract standardized to 50% epigallocatechin gallate, along with several other ingredients, presented with symptoms typical of variant angina involving the right coronary artery (Gange et al. 2006).

A previously healthy 22-year-old woman, who had recently resumed use of a combination product with numerous caffeine-containing ingredients along with standardized bitter orange extract, experienced a syncopal episode while running. An electrocardiogram revealed sinus tachycardia and QT prolongation (Nasir et al. 2004).

In the six-year period from 1998 to 2004, Health Canada received 16 reports of cardiovascular adverse reactions associated with products containing bitter orange or synephrine. One of these products was reported to contain neither caffeine nor ephedrine, while the others contained one or both of these alkaloids (Jordan et al. 2004).

III. Pharmacology and Pharmacokinetics

Human Pharmacological Studies

No changes in the QT_c interval or blood pressure were observed after a single dose of bitter orange fruit concentrated extract (450 mg with 27 mg synephrine) (Min et al. 2005). No increases in systolic or diastolic blood pressure or heart rate occurred in fasting volunteers 45 minutes after oral consumption of bitter orange fruit concentrated extract containing 50 mg *p*-synephrine with or without the flavonoids naringin and hesperidin (Stohs et al 2011).

Conversely, a single dose of bitter orange fruit concentrated extract containing 46.9 mg synephrine was reported to be associated with increased heart rate of 11.4 ± 10.8 beats per minute after 6 hours, while blood pressure remained unchanged (Haller et al. 2005). In another study of a bitter orange fruit concentrated extract (900 mg with 54 mg synephrine), heart rate increased from pretreatment rate by 4.2 ± 4.5 beats per minute 4 hours after administration, and systolic and diastolic blood pressure increased by 7.3 ± 4.6 mm Hg and 2.6 ± 3.8 mm Hg, respectively (Bui et al. 2006).

Animal Pharmacological Studies

In rats administered 2.5 to 20 mg/kg bitter orange fruit concentrated extract daily (4 or 6% synephrine) for 7 or 15 days, mortality was observed in a dose-dependent manner in animals receiving extracts with either 4 or 6% synephrine. No differences in blood pressure were observed as compared to the control group (Calapai et al. 1999).

In rats administered bitter orange concentrated extract as 5% of the daily diet, adrenaline and dopamine were elevated, and heart weight was slightly lower, as compared to control animals (Kubo et al. 2005).

In Vitro Pharmacological Studies

No relevant in vitro pharmacological studies were identified.

IV. Pregnancy and Lactation

No teratogenic or other adverse effects were found in fetuses of rats administered up to 100 mg/kg daily by gavage of bitter orange concentrated extract (6% synephrine). In combination with caffeine (25 mg/kg), however, a decrease in the number of implanted eggs was observed (Hansen et al. 2006).

No information on the safety of bitter orange fruit concentrated extract during lactation was identified.

V. Toxicity Studies

Short-Term Toxicity

Rats fed 10 g/kg bitter orange fruit concentrated extract daily for 14 days showed no mortality, although clinical abnormalities were observed, including rough coat, decreased activity, congested breathing, dark material around the facial area, decreased defecation, salivation, soft stools and urine/fecal stain (Douds 1997).

Literature Cited

Baumgart, A., M. Schmidt, H.J. Schmitz, and D. Schrenk. 2005. Natural furocoumarins as inducers and inhibitors of cytochrome P450 1A1 in rat hepatocytes. *Biochem. Pharmacol.* 69(4):657-667.

Bouchard, N.C., M.A. Howland, H.A. Greller, R.S. Hoffman, and L.S. Nelson. 2005. Ischemic stroke associated with use of an ephedra-free dietary supplement containing synephrine. *Mayo Clin. Proc.* 80(4):541-545.

Bui, L.T., D.T. Nguyen, and P.J. Ambrose. 2006. Blood pressure and heart rate effects following a single dose of bitter orange. *Ann. Pharmacother.* 40(1):53-57.

Calapai, G., F. Firenzuoli, A. Saitta, et al. 1999. Antiobesity and cardiovascular toxic effects of *Citrus aurantium* extracts in the rat: A preliminary report. *Fitoterapia* 70(6):586-592.

Colker, C.M., D.S. Kalman, G.C. Torina, T. Perlis, and C. Street. 1999. Effects of *Citrus aurantium* extract, caffeine, and St. John's wort on body fat loss, lipid levels, and mood states in overweight healthy adults. *Curr. Therap. Res.* 60(3):145-153.

Douds, D. 1997. An acute oral toxicity study in rats with Advantra Z. Spencerville, OH: Springborn Laboratories, Inc. (Study No. 3443.1; submitted to Nutratech Inc., Fairfield, NJ.)

Firenzuoli, F., L. Gori, and C. Galapai. 2005. Adverse reaction to an adrenergic herbal extract (*Citrus aurantium*). *Phytomedicine* 12(3):247-248.

Gange, C.A., C. Madias, E.M. Felix-Getzik, A.R. Weintraub, and N.A. Estes, 3rd. 2006. Variant angina associated with bitter orange in a dietary supplement. *Mayo Clin. Proc.* 81(4):545-548.

Gurley, B.J., S.F. Gardner, M.A. Hubbard, et al. 2004. In vivo assessment of botanical supplementation on human cytochrome P450 phenotypes: *Citrus aurantium, Echinacea purpurea,* milk thistle, and saw palmetto. *Clin. Pharmacol. Ther.* 76(5):428-440.

Haller, C.A., N.L. Benowitz, and P. Jacob, 3rd. 2005. Hemodynamic effects of ephedra-free weight-loss supplements in humans. *Am. J. Med.* 118(9):998-1003.

Hansen, D.K., K.S. Wall, G. White, and L.S. Pellicore. 2006. Teratogenic potential of *Citrus aurantium* [abstract only]. *Birth Defects Res. A Clin. Mol. Teratol.* 76(5):385.

ILS. 2004. Bitter orange (*Citrus aurantium* var. *amara*) extracts and constituents (±)-*p*-synephrine [CAS No. 94-07-5] and (±)-*p*-octopamine [CAS No. 104-14-3]: Review of toxicological literature. Research Triangle Park, NC: Integrated Laboratory Systems, Inc.

Ishihara, M., H. Toda, N. Sunagane, and T. Ohta. 2011. Furanocoumarins contents and cytochrome P450 3A (CYP3A) inhibitory activities of various processed fruit peel products: Outflow of 6′,7′-Dihydroxybergamottin during processing treatment of peel [article in Japanese]. *Yakugaku Zasshi* 131(5):679-684.

Kavli, G., and G. Volden. 1984. Phytophotodermatitis. *Photodermatol.* 1(2):65-75.

Jordan S., M. Murty, and K. Pilon. 2004. Products containing bitter orange or synephrine: Suspected cardiovascular adverse reactions. *Can. Med. Assoc. J.* 171(8):993-994.

Kubo, K., C. Kiyose, S. Ogino, and M. Saito. 2005. Suppressive effect of *Citrus aurantium* against body fat accumulation and its safety. *J. Clin. Biochem. Nutr.* 36(1):11-17.

Messer A., N. Raquet, C. Lohr, and D. Schrenk. 2012. Major furocoumarins in grapefruit juice II: Phototoxicity, photogenotoxicity, and inhibitory potency vs. cytochrome P450 3A4 activity. *Food Chem. Toxicol.* 50:756-760.

Min, B., D. Cios, J. Kluger, and C.M. White. 2005. Absence of QT_c-interval-prolonging or hemodynamic effects of a single dose of bitter-orange extract in healthy subjects. *Pharmacotherapy* 25(12):1719-1724.

Nasir, J. M., S.J. Durning, M. Ferguson, H.J.S. Barold, and M.C. Haigney. 2004. Exercise-induced syncope associated with QT prolongation and ephedra-free Xenadrine. *Mayo Clin. Proc.* 79(8):1059–1062.

Nguyen, D.T., L.T. Bui, and P.J. Ambrose. 2006. Response of CEDIA amphetamines assay after a single dose of bitter orange. *Ther. Drug Monit.* 28(2):252-254.

Nykamp, D., M. Fackih, and A. Compton. 2004. Possible association of acute lateral-wall myocardial infarction and bitter orange supplement. *Ann. Pharmacother.* 38(5):812-816.

Penzak, S.R., E.P. Acosta, M. Turner, et al. 2002. Effect of Seville orange juice and grapefruit juice on indinavir pharmacokinetics. *J. Clin. Pharmacol.* 42(10):1165-1170.

Rossato L.G., V.M. Costa, R.P. Limberger, M.L. Bastos, and F. Remião. 2011. Synephrine: From trace concentrations to massive consumption in weight-loss. *Food Chem. Toxicol.* 49(1):8-16.

Stohs, S.J., H.G. Preuss, S.C. Keith, P.L. Keith, H. Miller, and G.R. Kaats. 2011. Effects of *p*-synephrine alone and in combination with selected bioflavonoids on resting metabolism, blood pressure, heart rate, and self-reported mood changes. *Int. J. Med. Sci.* 8:295-301.

C

Citrus bergamia Risso & Poit. — Rutaceae

SCN: bergamot orange
Syn: *Citrus* × *aurantium* L. ssp. *bergamia* (Risso & Poit.) Wight & Arn. ex Engl.

Part: peel

Quick Reference Summary

Safety Class: 1
Interaction Class: A

Contraindications

None known.

Other Precautions

None known.

Drug and Supplement Interactions

None known.

Notice

Photosensitizing (Dubertret et al. 1990; Freund 1916; Kejlova et al. 2007; Maibach and Marzulli 1986; Zaynoun et al. 1977a, 1977b); *see* Appendix 2.

Editors' Note

Bergamot orange peel contains the compound bergapten, a furanocoumarin that can cause photodermatitis after topical application followed by exposure to the sun (Dubertret et al. 1990; Freund 1916; Kejlova et al. 2007; Maibach and Marzulli 1986; Zaynoun et al. 1977a, 1977b). Such a reaction is not expected after oral ingestion of bergamot orange. Commercial processing can remove bergapten from bergamot orange products, and bergapten-free products are commercially available.

Adverse Events and Side Effects

Cases of dermatological reactions have been reported after topical application of products containing bergamot orange followed by exposure to the sun (Gruson and Chang 2002; Kaddu et al. 2001; Meyer 1970; Wang et al. 2002; Weisenseel and Woitalla 2005; Zacher and Ippen 1984).

Muscle cramping and twitching and blurred vision were reported in a man who drank approximately 4 liters of Earl Grey tea (black tea with bergamot orange essential oil) daily. Symptoms worsened with time and abated with cessation of Earl Grey (Finsterer 2002).

Pharmacological Considerations

None known.

Pregnancy and Lactation

No information on the safety of bergamot orange in pregnancy or lactation was identified in the scientific or traditional literature. Although this review did not identify any concerns for use while pregnant or nursing, safety has not been conclusively established.

Review Details

I. Drug and Supplement Interactions

Clinical Trials of Drug or Supplement Interactions

No clinical trials of drug or supplement interactions were identified.

Case Reports of Suspected Drug or Supplement Interactions

No case reports of suspected drug or supplement interactions were identified.

Animal Trials of Drug or Supplement Interactions

No animal trials of drug or supplement interactions were identified.

II. Adverse Events

Case Reports of Adverse Events

Cases of photodermatitis have been reported in people topically exposed to bergamot orange essential oil, or products containing the oil, followed by exposure to sunlight (Cocks and Wilson 1998; Gruson and Chang 2002; Kaddu et al. 2001; Knott and Hofmann 2007; Wang et al. 2002; Weisenseel and Woitalla 2005).

A 44-year-old man with a history of drinking up to 4 liters of black tea daily developed muscle cramps and twitching, a tingling sensation of the limbs, and a feeling of pressure in the eyes associated with blurred vision with symptoms beginning 1 week after switching from black tea to Earl Grey tea (black tea with bergamot orange). Symptoms worsened with time and resolved after cessation of Earl Grey. The reporting author noted that the adverse effects of bergamot essential oil in this patient were likely due to the fact that the compound bergapten is a largely selective axolemmal potassium channel blocker that reduces potassium permeability at the nodes of Ranvier in a time-dependent manner, sometimes leading to hyperexcitability of the axonal membrane and phasic alterations of potassium currents, causing muscle twitching and cramping (Finsterer 2002).

Allergic contact dermatitis confirmed by patch testing was reported in two patients repeatedly exposed to perfumes or cosmetics containing bergamot orange (Zacher and Ippen 1984).

III. Pharmacology and Pharmacokinetics

Human Pharmacological Studies

In a study on factors that affect the phototoxic reaction of bergamot orange essential oil, the minimal effective concentrations of bergamot oil ranged from 1% to over 20%, showing some correlation with skin tone, with darker skins requiring a greater concentration of the essential oil to elicit a reaction after exposure to UVA light (Zaynoun et al. 1977b).

Animal Pharmacological Studies

In hairless albino mice treated topically with the compound bergapten at doses of 0 to 50 ppm, a dose-dependent relationship to time of tumor onset was observed and bergapten was reported to have phototumorigenic potential at the lowest tested dose of 5 ppm. The addition of UVA and UVB sunscreens to the treatment vehicle significantly reduced the tumorigenic effects (Young et al. 1990).

No papillomas were observed in mice with chemically initiated tumors subsequently treated topically with undiluted bergamot orange essential oil (Roe and Field 1965).

Studies comparing human, animal, and in vitro reactions to the compound bergapten have been used to develop phototoxicity tests (Girard et al. 1979; Gloxhuber 1970; Kejlova et al. 2007).

In Vitro Pharmacological Studies

Bergamot orange essential oil elicited a phototoxic reaction in human erythrocytes exposed to radiation sources rich in ultraviolet A or B (Placzek et al. 2007).

IV. Pregnancy and Lactation

No information on the safety of bergamot orange in pregnancy or lactation was identified.

V. Toxicity Studies

Acute Toxicity

The LD_{50} of orally administered bergamot orange essential oil in rats could not be determined at doses up to 10 g/kg (Opdyke 1979).

Genotoxicity

The compound bergapten was found to induce lethal and mutagenic photosensitization of bacteria, "dark"-induced frameshift mutagenesis in bacteria, and lethal and clastogenic effects on mammalian cells in tissue cultures (Ashwood-Smith et al. 1980).

Studies with bergapten and bergamot orange essential oil (containing equivalent amounts of bergapten) in *Saccharomyces cerevisiae* using solar simulated radiation (SSR) demonstrated that equal concentrations of bergapten alone or bergapten in bergamot orange essential oil have a similar influence on survival and on the induction of cytoplasmic *petite* mutations, reverse and forward mutations, mitotic gene conversion and genetically aberrant colonies including mitotic crossing-over. No reciprocity was found between SSR dose and bergapten concentration for cytotoxic, mutagenic, or recombinogenic effects. In the presence of chemical UVA or UVB filters, considerable protection against induction of genetic effects was observed (Averbeck et al. 1990).

No mutagenic effects of bergamot orange essential oil were observed in the *Bacillus subtilis rec* assay and *Salmonella*/microsome reversion assay with or without metabolic activation (Zani et al. 1991).

C

Literature Cited

Ashwood-Smith, M.J., G.A. Poulton, M. Barker, and M. Mildenberger. 1980. 5-Methoxypsoralen, an ingredient in several suntan preparations, has lethal, mutagenic and clastogenic properties. *Nature* 285(5764):407-409.

Averbeck, D., S. Averbeck, L. Dubertret, A.R. Young, and P. Morliere. 1990. Genotoxicity of bergapten and bergamot oil in *Saccharomyces cerevisiae*. *J. Photochem. Photobiol. B* 7(2-4):209-229.

Cocks, H., and D. Wilson. 1998. Dangers of the intake of psoralens and subsequent UV exposure producing significant burns. *Burns* 24(1):82.

Dubertret, L., P. Morliere, D. Averbeck, and A.R. Young. 1990. The photochemistry and photobiology of bergamot oil as a perfume ingredient: An overview. *J. Photochem. Photobiol. B* 7(2-4):362-365.

Finsterer, J. 2002. Earl Grey tea intoxication. *Lancet* 359(9316):1484.

Freund, E. 1916. Über bisher noch nicht beschriebene künstliche hautverfärbungen. *Wochenschrift* 63:931-933.

Girard, J., J. Unkovic, J. Delahayes, and C. Lafille. 1979. [Phototoxicity of bergamot oil. Comparison between humans and guinea pigs.] *Dermatologica* 158(4):229-243.

Gloxhuber, C. 1970. Phototoxicity testing of cosmetic materials. *J. Soc. Cosmet. Chem.* 21(Nov.):825-833.

Gruson, L.M., and M.W. Chang. 2002. Berloque dermatitis mimicking child abuse. *Arch. Pediatr. Adolesc. Med.* 156(11):1091-1093.

Kaddu, S., H. Kerl, and P. Wolf. 2001. Accidental bullous phototoxic reactions to bergamot aromatherapy oil. *J. Am. Acad. Dermatol.* 45(3):458-461.

Kejlova, K., D. Jirova, H. Bendova, et al. 2007. Phototoxicity of bergamot oil assessed by in vitro techniques in combination with human patch tests. *Toxicol. In Vitro* 21(7):1298-1303.

Knott, E., and H. Hofmann. 2007. [Purely natural: Phototoxic dermatitis.] *MMW Fortschr. Med.* 149(6):36.

Maibach, H.I., and F.N. Marzulli. 1986. Photoirritation (phototoxicity) from topical agents. *Dermatol. Clin.* 4(2):217-222.

Meyer, J. 1970. Accidents due to tanning cosmetics with a base of bergamot oil. *Bull. Soc. Fr. Dermatol. Syphiligr.* 77(6):881-884.

Opdyke, D.L.J. 1979. *Monographs on fragrance raw materials*. New York: Pergamon.

Placzek, M., W. Fromel, B. Eberlein, K.P. Gilbertz, and B. Przybilla. 2007. Evaluation of phototoxic properties of fragrances. *Acta Derm. Venereol.* 87(4):312-316.

Roe, F.J.C., and W.E.H. Field. 1965. Chronic toxicity of essential oils and certain other products of natural origin. *Food Cosmet. Toxicol.* 3(2):311-323.

Wang, L., B. Sterling, and P. Don. 2002. Berloque dermatitis induced by "Florida water." *Cutis* 70(1):29-30.

Weisenseel, P., and S. Woitalla. 2005. Toxic mustard plaster dematitis and phototoxic dematitis after application of bergamot oil. *MMW Fortschr. Med.* 147(51-52):53, 55.

Young, A.R., S.L. Walker, J.S. Kinley, et al. 1990. Phototumorigenesis studies of 5-methoxypsoralen in bergamot oil: Evaluation and modification of risk of human use in an albino mouse skin model. *J. Photochem. Photobiol. B* 7(2-4):231-250.

Zacher, K.D., and H. Ippen. 1984. Contact dermatitis caused by bergamot oil. *Derm. Beruf. Umwelt.* 32(3):95-97.

Zani, F., G. Massimo, S. Benvenuti, et al. 1991. Studies on the genotoxic properties of essential oils with *Bacillus subtilis* rec-assay and *Salmonella*/microsome reversion assay. *Planta Med.* 57(3):237-241.

Zaynoun, S.T., B.E. Johnson, and W. Frain-Bell. 1977a. A study of oil of bergamot and its importance as a phototoxic agent. I. Characterization and quantification of the photoactive component. *Br. J. Dermatol.* 96(5):475-482.

Zaynoun, S.T., B.E. Johnson, and W. Frain-Bell. 1977b. A study of oil of bergamot and its importance as a phototoxic agent. II. Factors which affect the phototoxic reaction induced by bergamot oil and psoralen derivatives. *Contact Dermat.* 3(5):225-239.

Citrus × limon (L.) Osbeck — Rutaceae

SCN: lemon
AN: *nimbuka*

Part: peel

Quick Reference Summary

Safety Class: 1
Interaction Class: A

Contraindications

None known.

Other Precautions

None known.

Drug and Supplement Interactions

None known.

Adverse Events and Side Effects

None known.

Pharmacological Considerations

Although lemon peel contains furanocoumarin compounds known to cause phototoxicity after topical application followed by exposure to the sun (Naganuma et al. 1985; Opdyke 1979; Zobel and Brown 1991), no case reports of phototoxicity were identified in the literature, and no such reaction is expected after ingestion.

Pregnancy and Lactation

No information on the safety of lemon in pregnancy or lactation was identified in the scientific or traditional literature. Although this review did not identify any concerns for use while pregnant or nursing, safety has not been conclusively established.

C

Review Details

I. Drug and Supplement Interactions

Clinical Trials of Drug or Supplement Interactions

No clinical trials of drug or supplement interactions were identified.

Case Reports of Suspected Drug or Supplement Interactions

No case reports of suspected drug or supplement interactions were identified.

Animal Trials of Drug or Supplement Interactions

No animal trials of drug or supplement interactions were identified.

II. Adverse Events

Case Reports of Adverse Events

Allergic contact dermatitis due to lemon peel oil (among other products) was reported in perfume factory workers routinely exposed to essential oils and related compounds (Schubert 2006). A case of allergy to lemon peel essential oil has been reported and confirmed by patch testing (Audicana and Bernaola 1994).

III. Pharmacology and Pharmacokinetics

Human Pharmacological Studies

No relevant human pharmacological studies were identified.

Animal Pharmacological Studies

The compound oxypeucedanin, a furanocoumarin present in lemon peel essential oil, was found to elicit photopigmentation on colored guinea pig skin without preceding visible erythema. The phototoxic potency of oxypeucedanin was approximately one-quarter of that of the compound bergapten (Naganuma et al. 1985).

In Vitro Pharmacological Studies

Extracts of immature whole lemon fruit inhibited the drug-metabolizing isoenzymes CYP3A4 and CYP2C9 but did not cause significant inhibition of CYP2D6. The inhibition of CYP3A4 was similar to that of grapefruit (Fujita et al. 2008).

IV. Pregnancy and Lactation

No information on the safety of lemon during pregnancy or lactation was identified.

V. Toxicity Studies

Acute Toxicity

The LD_{50} of orally administered lemon essential oil in rats could not be determined at doses up to 5 g/kg (Opdyke 1979).

Literature Cited

Audicana, M., and G. Bernaola. 1994. Occupational contact dermatitis from citrus fruits: Lemon essential oils. *Contact Dermat.* 31(3):183-185.

Fujita, T., A. Kawase, T. Niwa, et al. 2008. Comparative evaluation of 12 immature citrus fruit extracts for the inhibition of cytochrome P450 isoform activities. *Biol. Pharm. Bull.* 31(5):925-930.

Naganuma, M., S. Hirose, and Y. Nakayama. 1985. A study of the phototoxicity of lemon oil. *Arch. Dermatol. Res.* 278(1):31-36.

Opdyke, D.L.J. 1979. *Monographs on fragrance raw materials.* New York: Pergamon.

Schubert, H.J. 2006. Skin diseases in workers at a perfume factory. *Contact Dermat.* 55(2):81-83.

Zobel, A.M., and S.A. Brown. 1991. Dermatitis-inducing psoralens on the surfaces of seven medicinal plant species. *J. Toxicol. Cutan. Ocul. Toxicol.* 10(3):223-231.

Citrus reticulata Blanco

Rutaceae

SCN: tangerine

PN: *chen pi* (dried peel of mature fruit); *qing pi* (dried peel of green fruit)

OCN: mandarin; Mandarin orange; red tangerine

Part: peel

Quick Reference Summary

Safety Class: 1
Interaction Class: A

Contraindications
None known.

Other Precautions
Long-term use of tangerine peel is not recommended (Chen and Chen 2004).

Drug and Supplement Interactions
None known.

Adverse Events and Side Effects
An anaphylactic reaction to tangerine has been reported and confirmed by patch testing (Ebo et al. 2007). Contact dermatitis due to tangerine essential oil in a perfume has also been reported and confirmed by patch testing (Vilaplana 2002).

Pharmacological Considerations
None known.

Pregnancy and Lactation
No information on the safety of tangerine in pregnancy or lactation was identified in the scientific or traditional literature. Although this review did not identify any concerns for use while pregnant or nursing, safety has not been conclusively established.

C

Review Details

I. Drug and Supplement Interactions

Clinical Trials of Drug or Supplement Interactions
No clinical trials of drug or supplement interactions were identified.

Case Reports of Suspected Drug or Supplement Interactions
No case reports of suspected drug or supplement interactions were identified.

Animal Trials of Drug or Supplement Interactions
No animal trials of drug or supplement interactions were identified.

II. Adverse Events

Case Reports of Adverse Events
An anaphylactic reaction to tangerine has been reported and confirmed by patch testing (Ebo et al. 2007). Contact dermatitis due to tangerine essential oil in a perfume has also been reported and confirmed by patch testing (Vilaplana 2002).

III. Pharmacology and Pharmacokinetics

Human Pharmacological Studies
No relevant human pharmacological studies were identified.

Animal Pharmacological Studies
No relevant animal pharmacological studies were identified.

In Vitro Pharmacological Studies
No relevant in vitro pharmacological studies were identified.

IV. Pregnancy and Lactation
No information on the safety of tangerine in pregnancy or lactation was identified.

V. Toxicity Studies

Acute Toxicity
No adverse effects of aged tangerine peel were reported after administration of a 50% decoction to dogs at doses of 3 ml/kg (Chen and Chen 2004).

In mice, the intravenous LD_{50} of the compound tangeretin is 780 mg/kg, while that of the compound hesperidin is 850 mg/kg (Chen and Chen 2004).

Literature Cited

Chen, J.K., and T.T. Chen. 2004. *Chinese medical herbology and pharmacology*. City of Industry, CA: Art of Medicine Press.

Ebo, D.G., O. Ahrazem, G. Lopez-Torrejon, et al. 2007. Anaphylaxis from mandarin (*Citrus reticulata*): Identification of potential responsible allergens. *Int. Arch. Allergy Immunol.* 144(1):39-43.

Vilaplana, J., and C. Romaguera. 2002. Contact dermatitis from the essential oil of tangerine in fragrance. *Contact Derm.* 46 (2):108.

C

Clematis chinensis Osbeck — Ranunculaceae

SCN: Chinese clematis
PN: *wei ling xian* (root and rhizome)

Part: root

Quick Reference Summary

Safety Class: 1
Interaction Class: A

Contraindications

None known.

Other Precautions

Use with caution in pregnancy (Chen and Chen 2004).

Drug and Supplement Interactions

None known.

Editors' Note

The sap of Chinese clematis contains anemonin and protoanemonin, compounds that irritate the skin and mucous membranes (Bensky et al. 2004; Chang and But 1986; Kingsbury 1964; List and Hörhammer 1973; Nelson et al. 2006). These irritating principles are mostly lost on drying and lengthy storage in a dry environment (Bensky et al. 2004; List and Hörhammer 1973).

Adverse Events and Side Effects

No adverse effects have been observed with typical therapeutic use of the properly dried root (Bensky et al. 2004). Long-term or excessive use has been associated with adverse events including gastrointestinal irritation, slowed heart rate, and dilated pupils (Bensky et al. 2004).

Topical application of the fresh herb may cause dermatitis (Bensky et al. 2004; Tan et al. 2008; Wang et al. 2001).

Pharmacological Considerations

None known.

Pregnancy and Lactation

Although one reference text on traditional Chinese medicine indicates that Chinese clematis should be used with caution in pregnancy (Chen and Chen 2004), another reference did not list any concern for use in pregnancy (Bensky et al. 2004). An animal study indicated that the alcohol extract of Chinese clematis had abortifacient activity in mice (Chen and Chen 2004). Information on dose and route of administration was not reported in the available English language translation. In traditional Chinese medicine, herbal products are typically extracted in water and used in formulas rather than single herbs, and thus the relevance of the above animal study to standard therapeutic use is not known.

No information on the safety of Chinese clematis in lactation was identified. While this review did not identify any concerns for use while nursing, safety has not been conclusively established.

Review Details

I. Drug and Supplement Interactions

Clinical Trials of Drug or Supplement Interactions

No clinical trials of drug or supplement interactions were identified.

Case Reports of Suspected Drug or Supplement Interactions

No case reports of suspected drug or supplement interactions were identified.

Animal Trials of Drug or Supplement Interactions

No animal trials of drug or supplement interactions were identified.

II. Adverse Events

Case Reports of Adverse Events

No adverse effects have been observed with typical therapeutic use of the properly processed root. Long-term or excessive use (standard dose is a decoction of 6-9 g of the herb) has been associated with burning sensation, swelling, and ulcerations of the oral cavity, vomiting, abdominal pain, severe diarrhea, difficulty breathing, slowed heart rate, agitation, pale face, cold sweat, and dilated pupils. In severe cases, death was reported about 10 hours after ingestion (Bensky et al. 2004).

Topical application of large doses of Chinese clematis was associated with rash, blistering of the skin, and allergic dermatitis (Bensky et al. 2004). Pruritic erythema with depigmentation and hyperpigmentation was reported after topical application of fresh Chinese clematis (Tan et al. 2008). Severe contact dermatitis with a systemic reaction was reported after contact with fresh Chinese clematis (Wang et al. 2001).

III. Pharmacology and Pharmacokinetics

Human Pharmacological Studies

No relevant human pharmacological studies were identified.

Animal Pharmacological Studies

No relevant animal pharmacological studies were identified.

In Vitro Pharmacological Studies

No relevant in vitro pharmacological studies were identified.

IV. Pregnancy and Lactation

Although one reference text on traditional Chinese medicine indicates that Chinese clematis should be used with caution in pregnancy (Chen and Chen 2004), another reference did not list any concern for use in pregnancy (Bensky et al. 2004). An animal study indicated that the alcohol extract of Chinese clematis had abortifacient activity in mice (Chen and Chen 2004). Information on dose and route of administration was not reported in the available English language translation. In traditional Chinese medicine, herbal products are typically extracted in water and used in formulas rather than single herbs, and thus the relevance of the above animal study to standard therapeutic use is not known.

No information on the safety of Chinese clematis in lactation was identified. While this review did not identify any concerns for use while nursing, safety has not been conclusively established.

V. Toxicity Studies

Acute Toxicity

The LD_{50} of an orally or intraperitoneally administered 50% ethanol extract of Chinese clematis in mice could not be determined at doses up to 50 g/kg (Yang and Chen 1997).

Short-Term Toxicity

In rats orally administered 5 or 10 g/kg of a 50% ethanol extract of Chinese clematis daily for 14 days, an increase in heart and liver wet weights were observed in animals administered the 10 g/kg dose. At the low dose, increased urinary output was observed, and at both doses, an increase in white blood cells and urinary protein and decrease in serum glutamic-pyruvic transaminase were observed (Yang and Chen 1997).

Literature Cited

Bensky, D., S. Clavey, and E. Stöger. 2004. *Chinese herbal medicine: Materia medica.* 3rd ed. Seattle: Eastland Press.

Chang, H.-M., and P.P.H. But. 1986. *Pharmacology and applications of Chinese materia medica.* English ed. Singapore, Philadelphia: World Scientific.

Chen, J.K., and T.T. Chen. 2004. *Chinese medical herbology and pharmacology.* City of Industry, CA: Art of Medicine Press.

Kingsbury, J.M. 1964. *Poisonous plants of the United States and Canada.* Prentice-Hall biological science series. Englewood Cliffs, NJ: Prentice-Hall.

List, P.H., and H. Hörhammer. 1973. *Hagers handbuch der pharmazeutischen praxis.* Berlin: Springer.

Nelson, L., R.D. Shih, M.J. Balick, and K.F. Lampe. 2006. *Handbook of poisonous and injurious plants.* 2nd ed. New York: New York Botanical Garden.

Tan, C., W.Y. Zhu, and Z.S. Min. 2008. Co-existence of contact leukoderma and pigmented contact dermatitis attributed to *Clematis chinensis* Osbeck. *Contact Dermat.* 58(3):177-178.

Wang, N.Z., Z.Y. Xue, and Z.G. Bi. 2001. Severe contact dermatitis and systemic reaction caused by fresh *Clematis chinensis* Osbeck: A case report. *Clin. J. Dermatol.* 30:256-257.

Yang, H.Y., and C.F. Chen. 1997. Subacute toxicity of 15 commonly used Chinese drugs (II). *J. Food Drug Anal.* 5(4):355-380.

Cnicus benedictus L. — Asteraceae

SCN: blessed thistle
Syn: *Centaurea benedicta* (L.) L.

OCN: holy thistle
Part: herb

Quick Reference Summary

Safety Class: 1
Interaction Class: A

Contraindications

None known.

Other Precautions

Persons with allergies to other members of the Asteraceae family (such as feverfew or *Echinacea* species) should exercise caution with blessed thistle, as allergic cross-reactivity is common to Asteraceae plants (Upton 2007; Zeller et al. 1985).

Drug and Supplement Interactions

None known.

Adverse Events and Side Effects

Allergic reactions to blessed thistle have been reported (De Smet 1993). In animal studies, blessed thistle was a sensitizer and showed cross-reactivity to other members of the Asteraceae family (Zeller et al. 1985).

High doses (more than 5.0 g per cup of tea) may irritate the stomach and cause vomiting (Roth et al. 1984).

Pharmacological Considerations

None known.

Pregnancy and Lactation

Limited information on the safety of blessed thistle in pregnancy or lactation was identified in the scientific or traditional literature. Although this review did not identify any concerns for use while pregnant or nursing, safety has not been conclusively established.

Review Details

C

I. Drug and Supplement Interactions

Clinical Trials of Drug or Supplement Interactions

No clinical trials of drug or supplement interactions were identified.

Case Reports of Suspected Drug or Supplement Interactions

No case reports of suspected drug or supplement interactions were identified.

Animal Trials of Drug or Supplement Interactions

No animal trials of drug or supplement interactions were identified.

II. Adverse Events

Case Reports of Adverse Events

No case reports of adverse events were identified.

III. Pharmacology and Pharmacokinetics

Human Pharmacological Studies

No relevant human pharmacological studies were identified.

Animal Pharmacological Studies

Sensitization testing in guinea pigs indicated that blessed thistle is a relatively strong sensitizer. In this testing, cross-reactivity to "a considerable number" of other plants in the Asteraceae family was observed (Zeller et al. 1985). Blessed thistle contains the compound cnicin, a sesquiterpene lactone. Such compounds are responsible for allergic contact dermatitis associated with a number of species of the Asteraceae family (Gordon 1999).

In Vitro Pharmacological Studies

Blessed thistle is reported to contain lithospermic acid, a compound with antigonadotropic activity (Graham and Noble 1955).

IV. Pregnancy and Lactation

Blessed thistle extract did not stimulate contractions in isolated guinea pig uterine tissue (Pilcher et al. 1916).

No information on the safety of blessed thistle during lactation was identified.

V. Toxicity Studies

Acute Toxicity

The LD_{50} of the compound cnicin in mice is 1.6–3.2 µmol/kg (Müller and Schneider 1999).

Genotoxicity

An ethanolic extract of blessed thistle was not mutagenic in the Ames assay with *Salmonella typhimurium* with or without S9 mix (Schimmer et al. 1994).

Literature Cited

De Smet, P.A.G.M. 1993. *Adverse effects of herbal drugs, Volume 2.* New York: Springer.

Gordon, L.A. 1999. Compositae dermatitis. *Australas. J. Dermatol.* 40(3):123-130.

Graham, R., and R. Noble. 1955. Comparison of in vitro activity of various species of *Lithospermum* and other plants to inactivate gonadotrophin. *Endocrinology* 56:239.

Müller, B., and B. Schneider. 1999. Anwendungsbereiche eines Trockenextrakts aus Birkenblätten bei Harnwegserkrankungen: Ergebnisse einer Anwendungsbeobachtung. *In Abstracts of Phytotherapie an der Schwelle zum neuen Jahrtausend;* 10. Jahrestagung der Gesellschaft für Phytotherapie. Abstract P16., at Münster, 11-13 November.

Pilcher, J.D., G.E. Burman, and W.R. Delzell. 1916. The action of the so-called female remedies on the excized uterus of the guinea pig. *Arch. Intern. Med.* 18:557-583.

Roth, L., M. Daunderer, and K. Kormann. 1984. *Giftpflanzen-pflanzengifte: Vorkommen, wirkung, therapie.* Landsberg, Germany: Ecomed.

Schimmer, O., A. Krueger, H. Paulini, and F. Haefele. 1994. An evaluation of 55 commercial plant extracts in the Ames mutagenicity test. *Pharmazie* 49 (6):448-451.

Upton, R. 2007. *Feverfew aerial parts: Tanacetum parthenium (L.) Schultz Bip: Standards of analysis, quality control, and therapeutics.* Santa Cruz, CA: American Herbal Pharmacopoeia.

Zeller, W., M. de Gols, and B.M. Hausen. 1985. The sensitizing capacity of Compositae plants. VI. Guinea pig sensitization experiments with ornamental plants and weeds using different methods. *Arch. Dermatol. Res.* 277(1):28-35.

Cnidium monnieri (L.) Cusson ex Juss. — Apiaceae

SCN: cnidium

PN: *she chuang zi* (seed)

Part: seed

Quick Reference Summary

Safety Class: 1
Interaction Class: A

Contraindications
None known.

Other Precautions
Cnidium should be used with caution by individuals with inflammation of the kidneys or a history of irritation of the kidneys (Yan et al. 2001).

Drug and Supplement Interactions
None known.

Adverse Events and Side Effects
None known.

Pharmacological Considerations
None known.

Pregnancy and Lactation
No information on the use of cnidium during pregnancy or lactation was identified in the scientific or traditional literature. While this review did not identify any concerns for pregnant or lactating women, safety has not been conclusively established.

Review Details

I. Drug and Supplement Interactions

Clinical Trials of Drug or Supplement Interactions
No clinical trials of drug or supplement interactions were identified.

Case Reports of Suspected Drug or Supplement Interactions
No case reports of suspected drug or supplement interactions were identified.

Animal Trials of Drug or Supplement Interactions
No animal trials of drug or supplement interactions were identified.

II. Adverse Events

Case Reports of Adverse Events
No adverse effects have been reported from ingestion of cnidium extracts. Ingestion of the coumarin fraction of cnidium has led to dry mouth, drowsiness, and mild gastric discomfort (Bensky et al. 2004).

III. Pharmacology and Pharmacokinetics

Human Pharmacological Studies
No relevant human pharmacological studies were identified.

Animal Pharmacological Studies
No relevant animal pharmacological studies were identified.

In Vitro Pharmacological Studies
No relevant in vitro pharmacological studies were identified.

IV. Pregnancy and Lactation

No information on the safety of cnidium in pregnancy or lactation was identified.

V. Toxicity Studies

Cytotoxicity
Coumarin compounds (osthol, imperatorin, bergapten, isopimpinellin, and xanthotoxin) isolated from a cnidium extract exhibited cytotoxic activity against human leukemia, cervical carcinoma, and colorectal carcinoma cells in vitro. The compound osthol showed the strongest cytotoxic activity on tumor cell lines. The compound imperatorin showed the highest sensitivity to human leukemia cells and the least cytotoxicity to normal peripheral blood mononuclear cells (Yang et al. 2003).

Literature Cited

Bensky, D., S. Clavey, and E. Stöger. 2004. *Chinese herbal medicine: Materia medica.* 3rd ed. Seattle: Eastland Press.

Yan, F., Z. Liang, C. Jianna, et al. 2001. Analysis of *Cnidium monnieri* fruits in different regions of China. *Talanta* 53(6):1155-1162.

Yang, L.L., M.C. Wang, L.G. Chen, and C.C. Wang. 2003. Cytotoxic activity of coumarins from the fruits of *Cnidium monnieri* on leukemia cell lines. *Planta Med.* 69(12):1091-1095.

Codonopsis spp. — Campanulaceae

Codonopsis pilosula (Franch.) Nannf.
SCN: codonopsis

Syn: *Codonopsis silvestris* Kom.
PN: *dang shen* (root)

OCN: bellflower
Codonopsis tangshen Oliv.
SCN: codonopsis
PN: *dang shen* (root)
OCN: bellflower; Sichuan dang shen
Part: root

Quick Reference Summary

C

Safety Class: 1
Interaction Class: A

Contraindications

None known.

Other Precautions

None known.

Drug and Supplement Interactions

None known.

Adverse Events and Side Effects

No side effects are expected at standard therapeutic doses (Bensky et al. 2004).

Pharmacological Considerations

None known.

Pregnancy and Lactation

No information on the safety of codonopsis in pregnancy or lactation was identified in the scientific or traditional literature. Although this review did not identify any concerns for use while pregnant or nursing, safety has not been conclusively established.

Review Details

I. Drug and Supplement Interactions

Clinical Trials of Drug or Supplement Interactions

No clinical trials of drug or supplement interactions were identified.

Case Reports of Suspected Drug or Supplement Interactions

No case reports of suspected drug or supplement interactions were identified.

Animal Trials of Drug or Supplement Interactions

No animal trials of drug or supplement interactions were identified.

II. Adverse Events

Case Reports of Adverse Events

Doses of codonopsis over 63 g may cause discomfort in the left pectoral area or arrhythmia. These effects subside when the herb is discontinued (Bensky et al. 2004; Liang 1976). Other symptoms of overdose (aqueous extract of 30–60 g codonopsis) include visual impairment, throat pain, vertigo, disorientation, loss of balance, leg spasms, and loss of voice (Bensky et al. 2004). Codonopsis is reported to have a low toxicity, and no side effects are expected within the standard dose range (aqueous extract of 6–9 g daily) (Bensky et al. 2004).

III. Pharmacology and Pharmacokinetics

Human Pharmacological Studies

No relevant human pharmacological studies were identified.

Animal Pharmacological Studies

Administration of "large doses" (dose not specified in English language abstract) of codonopsis in rats was reported to cause a reduction in weight gain and serum triiodothyronine (T_3) levels, and an increase of reverse T_3 and thyrotropin-releasing hormone (TRH) levels (Chen et al. 1989).

In Vitro Pharmacological Studies

No relevant in vitro pharmacological studies were identified.

IV. Pregnancy and Lactation

No information on the safety of codonopsis in pregnancy or lactation was identified.

V. Toxicity Studies

Acute Toxicity

The LD_{50} of intraperitoneally administered aqueous extract of codonopsis in mice is 79 g/kg (Chen and Chen 2004). No toxicity or fatalities were seen in mice orally administered up to 10 g/kg of the same extract (Chen and Chen 2004).

Short-Term Toxicity

No toxic effects were seen in rats subcutaneously administered an extract of codonopsis daily for 13 days (Wang 1983). No toxic effects or changes in serum glutamic-pyruvic transaminase levels were observed in mice intraperitoneally administered the same extract daily for 15 days (Wang 1983).

Literature Cited

Bensky, D., S. Clavey, and E. Stöger. 2004. *Chinese herbal medicine: Materia medica.* 3rd ed. Seattle: Eastland Press.

Chen, J.K., and T.T. Chen. 2004. *Chinese medical herbology and pharmacology.* City of Industry, CA: Art of Medicine Press.

Chen, M.D., A.K. Kuang, and J.L. Chen. 1989. Influence of yang-restoring herb medicines upon metabolism of thyroid hormone in normal rats and a drug administration schedule. *Zhong Xi Yi Jie He Za Zhi* 9(2):93-95.

Liang, J. 1976. *Res. Disc. Chin. Med. Herbol.* 4:33 *Cited in* Chen, J.K., T.T. Chen. 2004. *Chinese medical herbology and pharmacology.* City of Industry, CA: Art of Medicine Press.

Wang, Y. 1983. *Pharmacology and applications of Chinese materia medica.* Beijing: People's Health Publishers.

Coffea arabica L. — Rubiaceae

SCN: coffee
OCN: Arabian coffee

Part: roasted seed kernel

Quick Reference Summary

Safety Class: 1
Interaction Class: C*

Contraindications

None known.

Other Precautions

Coffee contains caffeine, a nervous system stimulant. If taken in large amounts, coffee products containing caffeine can cause insomnia, nervousness, and the other well-known symptoms of excess caffeine intake (Donovan and DeVane 2001).

Due to the CNS stimulant effects of caffeine, use of caffeine-containing products is cautioned in persons with heart disorders, as excessive caffeine consumption may increase heart rate or exacerbate arrhythmias. Caution should also be observed in persons with psychological disorders, as caffeine may aggravate depression or induce anxiety (Brinker 2001).

Drug and Supplement Interactions

Use of caffeine with other central nervous system (CNS) stimulants, including bronchodilators or adrenergic drugs, may cause excessive central nervous system stimulation resulting in nervousness, irritability, insomnia, and possibly convulsions or cardiac arrhythmias (PDR 2006).

Caffeine is metabolized by the isoenzyme CYP1A2. Drugs that inhibit this isoenzyme (including fluvoxamine, ciprofloxacin, cimetidine, amiodarone, fluoroquinolones, furafylline, interferon, methoxsalen, and mibefradil) may slow the metabolism of caffeine. In persons drinking multiple cups of coffee daily, high levels of caffeine could accumulate (Carrillo and Benitez 2000).

Notice

Caffeine (1.5–2.5%) (Leung and Foster 1996; McCusker et al. 2003); *see* Appendix 1.

Diuretic (Carrillo and Benitez 2000; Maughan and Griffin 2003); *see* Appendix 2.

Editors' Note

An average 8 oz cup of coffee contains 75 to 130 mg caffeine. A shot of espresso contains 55 to 76 mg caffeine (McCusker et al. 2003). The caffeine content of a single type of coffee purchased from one store of an international coffee company on six different days was found to vary from 130 to 282 mg per 8 oz cup (McCusker et al. 2003).

The American Herbal Products Association has established a trade requirement (AHPA 2011) that dietary supplement products that contain caffeine, whether as a direct ingredient or as a constituent of herbal ingredients, be labeled to disclose the presence of caffeine in the product and the quantity of added caffeine if greater than 25 mg; be formulated and labeled in a manner to recommend a maximum of 200 mg of caffeine per serving, not more often than every 3 to 4 hours; and bear the following or similar statement on the label of any dietary supplement that contains caffeine in sufficient quantity to warrant such labeling:

> *Too much caffeine may cause nervousness, irritability, sleeplessness, and, occasionally, rapid heartbeat. Not recommended for use by children under 18 years of age.*

See Appendix 1 for more specific details on this AHPA trade requirement.

Adverse Events and Side Effects

Overdose of coffee may result in restlessness, nervousness, excitement, insomnia, flushed face, frequent urination,

* For caffeine-free preparations, no interactions are expected.

stomach ache, nausea, tremors, talkativeness, periods of inexhaustibility, and irritability (Hughes et al. 1991).

Cessation of coffee may result in withdrawal symptoms including headache, drowsiness, fatigue, and irritability. Significant withdrawal symptoms can occur after 3 days of consuming as little as 100 mg caffeine (about one cup coffee) daily (Evans and Griffiths 1999; Juliano and Griffiths 2004).

Epidemiological studies have provided conflicting results on the association between coffee consumption and coronary heart disease risk. While some studies indicate no association, others indicate that drinking five cups of coffee or more daily increases the risk of coronary heart disease (Higdon and Frei 2006; Sofi et al. 2007).

The risk of heart attack has been shown to be higher in persons consuming five to ten cups of coffee daily as compared with those drinking two to three cups daily (Hammar et al. 2003; Palmer et al. 1995; Sesso et al. 1999; Tavani et al. 2001, 2004).

Consumption of boiled coffee has been associated with an increase in cholesterol levels, while no significant effects of filtered coffee have been observed (Jee et al. 2001).

Studies have generally indicated no association between coffee and risk of stroke (Adolfsson et al. 1977; Grobbee et al. 1990; Heyden et al. 1978).

Contrary to earlier studies, more recent reviews and meta-analyses indicate no association between coffee and the risk of certain cancers, including pancreatic, bladder, ovarian, breast, prostate, and gastric cancer (Higdon and Frei 2006; Tavani and La Vecchia 2000).

Pharmacological Considerations

Coffee has been shown to cause an acute rise in blood pressure, although with chronic consumption the rise is minimal (1.2 mm Hg systolic, 0.49 mm Hg diastolic) (Geleijnse 2008; Noordzij et al. 2005; Nurminen et al. 1999).

Clinical trials and prospective studies have indicated no association between coffee and risk of irregular heart rhythms (Chelsky et al. 1990; Frost and Vestergaard 2005; Myers 1991; Wilhelmsen et al. 2002).

Coffee consumption has been associated with a slight decrease in calcium absorption. Studies have indicated no effect of coffee on bone mineral density and mixed results on coffee consumption and the risk of hip fracture in elderly persons (Barger-Lux and Heaney 1995; Hasling et al. 1992; Higdon and Frei 2006).

Coffee has been shown to strongly inhibit dietary iron absorption, though at only about half the inhibitory effect of black tea (Hurrell et al. 1999).

Pregnancy and Lactation

Pregnant women are advised to limit caffeine intake to less than 300 mg (about three cups of coffee) daily (PDR 2006; Sato et al. 1993). Lactating women are advised to limit consumption of caffeinated beverages to two to three cups daily (AAP 2001).

Studies of caffeine have indicated no adverse effects on pregnancy or fetal development at levels equivalent to three to four cups daily. At much higher doses (equivalent to 21 cups of coffee or more daily), no adverse effects on the fetus have been observed (Christian and Brent 2001).

Review Details

I. Drug and Supplement Interactions

Clinical Trials of Drug or Supplement Interactions

In healthy volunteers orally administered theophylline with or without two to seven cups of regular instant coffee during a 24-hour study period, blood levels of theophylline and caffeine were significantly higher in the coffee group. The half-life of theophylline was increased from 6.3 to 8.3 h, and the clearance of theophylline was reduced (Sato et al. 1993).

Case Reports of Suspected Drug or Supplement Interactions

In patients taking lithium, high coffee consumption and caffeine withdrawal have been associated with an increase in lithium tremors (Grandjean and Aubry 2009; Mester et al. 1995). A series of case reports and a related small trial indicated that coffee may inhibit intestinal absorption of thyroxine (T_4) (Benvenga et al. 2008).

Animal Trials of Drug or Supplement Interactions

No animal trials of drug or supplement interactions were identified.

II. Adverse Events

Case Reports of Adverse Events

Overdose of coffee may result in restlessness, nervousness, excitement, insomnia, flushed face, frequent urination, stomach ache, nausea, tremulousness, talkativeness, periods of inexhaustibility, irritability, and psychomotor agitation (Hughes et al. 1991).

Cessation of coffee may result in withdrawal symptoms including headache, drowsiness, fatigue, and irritability. Significant withdrawal symptoms can occur after 3 days of consuming as little as 100 mg caffeine (about one cup coffee) daily (Evans and Griffiths 1999; Juliano and Griffiths 2004).

Occupational asthma and allergy to roasted and green coffee have been reported in persons exposed to coffee dust (Diba and English 2002; Lemiere et al. 1996; Treudler et al. 1997). An anaphylactic reaction to coffee, with skin prick testing confirming caffeine as the responsible compound, has been reported (Infante et al. 2003).

C

III. Pharmacology and Pharmacokinetics

Human Pharmacological Studies

Epidemiological studies have provided conflicting results on the association between coffee consumption and coronary heart disease risk (Higdon and Frei 2006; Sofi et al. 2007). Case-control studies have found a 40 to 60% increased risk of coronary heart disease in people who consumed five or more cups of coffee daily, as compared to non-coffee drinkers (Greenland 1993; Kawachi et al. 1994). Prospective cohort studies, conversely, have generally not found a significant association between coffee consumption and coronary heart disease risk (Kawachi et al. 1994; Myers and Basinski 1992). The difference in results may be explained by the study methods. In case-control studies, the health history and related factors from individuals who have developed a disease are compared with a control population. A prospective cohort study follows a group over time (Coggon et al. 1997).

The risk of heart attack has been shown to be higher in persons consuming five to ten cups of coffee daily as compared with those drinking two to three cups daily (Hammar et al. 2003; Palmer et al. 1995; Sesso et al. 1999; Tavani et al. 2001, 2004). A case-control study identified an increased risk of cardiac arrest in patients with established coronary heart disease who drank ten cups of coffee or more daily (de Vreede-Swagemakers et al. 1999), whereas a similar prospective study found no association between heavy (ten or more cups daily) coffee consumption and survival in persons who had survived a heart attack (Mukamal et al. 2004).

Clinical trials and prospective studies have indicated no association between coffee and risk of cardiac arrhythmias (Chelsky et al. 1990; Frost and Vestergaard 2005; Myers 1991; Wilhelmsen et al. 2002) and generally indicated no association between coffee and risk of stroke (Adolfsson et al. 1977; Grobbee et al. 1990; Heyden et al. 1978). One study, however, did show an increased risk of stroke in nonsmoking men with high blood pressure who consumed three cups of coffee or more daily (Hakim et al. 1998).

An acute rise in blood pressure, usually occurring 30 minutes to 4 hours after coffee consumption, has been observed in a number of studies with healthy volunteers (Nurminen et al. 1999). The blood-pressure raising effects may be more pronounced in persons with high blood pressure (Nurminen et al. 1999). With routine consumption of coffee, most individuals develop tolerance to the blood-pressure raising effect of coffee, while the tolerance is incomplete in some (James 1994; Lovallo et al. 2004). A meta-analysis of studies on the effects of chronic coffee consumption on blood pressure indicated that the average increase in blood pressure was small (1.2 mm Hg systolic, 0.49 mm Hg diastolic), with boiled coffee causing a greater increase than filtered coffee. In the studies reviewed, the average coffee intake was 450 to 1235 ml (225–798 mg caffeine) daily (Noordzij et al. 2005). Some epidemiological studies indicate an inverse, U-shaped relationship between blood pressure and coffee consumption, with low pressure shown in those who did not drink coffee and those who drank more than six or nine cups daily, as compared with those who drank one to two cups daily (Geleijnse 2008).

Unfiltered coffee (Turkish, French press, or boiled coffee) has been associated with an increase in serum cholesterol levels (total and LDL), while reviews indicate that filtered coffee results in very little increase in serum cholesterol levels (Jee et al. 2001). This effect has been attributed to the compounds cafestrol and kahweol that are present in unfiltered coffee (Urgert and Katan 1997).

In clinical and epidemiological studies, coffee has been associated with a dose-dependent increase in plasma total homocysteine levels (Bree et al. 2001; Husemoen et al. 2004; Mennen et al. 2002; Nygard et al. 1997; Stolzenberg-Solomon et al. 1999). Filtered coffee showed greater effects than unfiltered, with an 18% increase in homocysteine levels in persons drinking 34 oz of filtered coffee daily for 2 weeks, as compared to a 10% increase in persons drinking the same volume of unfiltered coffee (Urgert et al. 2000). A clinical trial in healthy volunteers showed that supplementation with 200 μg of folic acid daily prevented elevation of homocysteine levels caused by 2.5 cups coffee daily. Elevated total homocysteine levels have been associated with increased risk of cardiovascular disease such as coronary artery disease, stroke, and congestive heart failure (Mart-Carvajal et al. 2009).

Coffee has been shown to cause a slight decrease in the efficiency of calcium absorption, with an estimated 6 mg decrease in calcium balance per cup of coffee (Barger-Lux and Heaney 1995; Hasling et al. 1992). The majority of epidemiological studies indicate that coffee consumption is not associated with changes in bone mineral density (Higdon and Frei 2006). Epidemiological studies provide conflicting results on coffee consumption and the risk of hip fracture (Higdon and Frei 2006). Among five case-control studies, none found a relationship between coffee consumption and the risk of hip fracture (Cumming and Klineberg 1994; Johnell et al. 1995; Kanis et al. 1999; Nieves et al. 1992; Tavani et al. 1995). In cohort studies, three found no association between coffee consumption and the risk of hip fracture, while three other studies indicated an association in women drinking more than two to four cups of coffee daily (Cummings et al. 1995; Hernandez-Avila et al. 1991; Kiel et al. 1990).

Coffee has been shown to strongly inhibit dietary iron absorption, though at only about half the inhibitory effect of black tea (Hurrell et al. 1999).

While earlier case-control studies indicated an association between coffee consumption and increased incidences of pancreatic, bladder, and ovarian cancers, more recent, better-designed studies have not supported those findings (Higdon and Frei 2006). Prospective cohort studies generally indicate a lack of association between coffee consumption and pancreatic, bladder, ovarian, breast, prostate, and gastric cancer (Tavani and La Vecchia 2000). Reviews and meta-analyses indicate that coffee may decrease the risk of colorectal cancer (Giovannucci 1998; Tavani and La Vecchia 2004) and liver cancer (Higdon and Frei 2006).

Caffeine is a substrate of the drug-metabolizing isoenzyme CYP1A2 (Nordmark et al. 1999).

Animal Pharmacological Studies

Animal studies were identified but omitted due to the availability of human data.

In Vitro Pharmacological Studies

In vitro studies were identified but omitted due to the availability of human data.

IV. Pregnancy and Lactation

Caffeine is in the FDA pregnancy category C and has been shown to cross the placenta and achieve blood and tissue concentrations in the fetus. Excessive intake of caffeine by pregnant women has been associated with fetal arrhythmias. Pregnant women are advised to limit caffeine intake to less than 300 mg (approximately 3 cups of coffee) daily (PDR 2006; Sato et al. 1993).

Caffeine is listed as a "maternal medication usually compatible with breast-feeding" by the American Academy of Pediatrics Committee on Drugs. The committee noted that maternal consumption of caffeine may cause irritability and poor sleeping patterns in nursing infants, and that maternal consumption of caffeinated beverages should be limited to two to three cups daily (AAP 2001).

A review of reproductive and developmental toxicity studies of caffeine indicated that, at levels of 5 to 6 mg/kg of caffeine daily (about three or four cups of coffee), no adverse effects on pregnancy or fetal development are expected. The developmental toxicity no-observed-effect level (NOEL) of caffeine is 30 mg/kg daily (about 21 cups of coffee) and the reproductive toxicity NOEL is between 80 and 120 mg/kg daily (Christian and Brent 2001).

Epidemiological studies have indicated an association between high caffeine intake during pregnancy and an increased risk of spontaneous abortions. An analysis of these studies concluded that methodological flaws in many of these studies led to biased results and that a causal link between caffeine consumption and abortion cannot be confirmed (Signorello and McLaughlin 2004).

V. Toxicity Studies

Acute Toxicity

The LD_{50} of coffee essential oil orally administered in mice could not be determined at doses up to 5 g/kg (Viani 1988).

The LD_{50} of orally administered caffeine in rats is 335 mg/kg (Mills and Bone 2005). Other sources have more broadly estimated the oral LD_{50} of caffeine in rats to be between 261 and 383 mg/kg, or 200 and 400 mg/kg (NTP 1982; OECD 2004).

The lethal oral dose of caffeine in humans has been estimated as 5 to 10 g, equivalent to 50 to 100 cups of coffee (Stavric 1988).

Chronic Toxicity

In rats fed diets containing 6% regular or decaffeinated instant coffee for 2 years, the average coffee intake was 2.9 g/kg daily in males and 3.5 g/kg daily in females, a human equivalent of 70 to 80 cups of coffee daily. In all groups, the total number of neoplasms was either similar to or lower than the total in the control group. Body weights of coffee-treated groups were generally lower than body weights of controls and were inversely proportional to the caffeine content of the coffee samples. Food intake in treated rats was similar to or higher than that of control groups. Differences in blood chemistry and hematology were noted, but they were not considered to indicate toxic or ill effects, an interpretation substantiated by the histological findings. Plasma cholesterol levels in both sexes consistently showed a positive correlation with caffeine intake (Würzner et al. 1977a, 1977b).

In rats provided with coffee at dilutions of 25, 50, or 100% (intake at the 100% level was equivalent to human intake of 37 cups coffee by male rats and 67 cups coffee by female rats) as the sole source of fluid for up to 2 years, elevated levels of cholesterol were observed. A decrease in bone calcium and increase in serum calcium was observed after 1 year but returned to normal without change in treatment. Mean serum alkaline phosphatase, bilirubin, and blood urea nitrogen values were occasionally elevated. Treatment-related increases in relative weights of lungs, kidneys, liver, and epididymides were recorded. An increase in mortality was observed in females receiving 50 or 100% coffee. Levels of tumors were comparable in treatment and control groups (Palm et al. 1984).

In rats fed diets containing 0.5 to 5% instant coffee daily for 2 years, weight gain was impaired at the highest dose level. Liver and kidney hypertrophy were also observed at that dose level (Daubert 1967).

Genotoxicity

A review of genotoxicity studies on coffee indicated that mutagenic activity is seen in vitro with bacteria and fungi and in studies at high concentrations in mammalian cells in vitro. Mutagenic activity, however, is generally lacking when bacteria or mammalian cells are cultured with liver enzymes that provide metabolic activation. Animal studies indicate a lack of mutagenic activity of coffee and caffeine, with a number of studies showing antimutagenic activity of coffee (Nehlig and Debry 1994).

Literature Cited

AAP. 2001. The transfer of drugs and other chemicals into human milk. American Academy of Pediatrics Committee on Drugs. *Pediatrics* 108(3):776-789.

AHPA. July 2011. Code of Ethics & Business Conduct. Silver Spring, MD: American Herbal Products Association.

Adolfsson, R., K. Svardsudd, and G. Tibblin. 1977. 1913 men study—A longitudinal study of the development of stroke in a population. *Scand. J. Soc. Med. Suppl.* 14:122-127.

Barger-Lux, M.J., and R.P. Heaney. 1995. Caffeine and the calcium economy revisited. *Osteoporosis Int.* 5(2):97-102.

Benvenga, S., L. Bartolone, M.A. Pappalardo, et al. 2008. Altered intestinal absorption of L-thyroxine caused by coffee. *Thyroid* 18(3):293-301.

Bree, A., W.M. Verschuren, H.J. Blom, and D. Kromhout. 2001. Lifestyle factors and plasma homocysteine concentrations in a general population sample. *Am. J. Epidemiol.* 154(2):150-154.

Brinker, F. 2001. *Herb contraindications and drug interactions.* 3rd ed. Sandy, OR: Eclectic Medical Publications.

Carrillo, J.A., and J. Benitez. 2000. Clinically significant pharmacokinetic interactions between dietary caffeine and medications. *Clin. Pharmacokinet.* 39(2):127-153.

Chelsky, L.B., J.E. Cutler, K. Griffith, et al. 1990. Caffeine and ventricular arrhythmias: An electrophysiological approach. *J. Am. Med. Assoc.* 264(17):2236.

Christian, M.S., and R.L. Brent. 2001. Teratogen update: Evaluation of the reproductive and developmental risks of caffeine. *Teratology* 64(1):51-78.

Coggon, D., G. Rose, and D.J.P. Barker. 1997. *Epidemiology for the uninitiated.* 4th ed. London: BMJ.

Cumming, R.G., and R.J. Klineberg. 1994. Case-control study of risk factors for hip fractures in the elderly. *Am. J. Epidemiol.* 139(5):493-503.

Cummings, S.R., M.C. Nevitt, W.S. Browner, et al. 1995. Risk factors for hip fracture in white women. *New Engl. J. Med.* 332(12):767-773.

Daubert, B.F. 1967. Effects of long-term administration of coffee and caffeine in rats. Colloque Scientifique International sur le Café 3, Trieste (Italia), June 2-9.

de Vreede-Swagemakers, J.J.M., A.P.M. Gorgels, M.P. Weijenberg, et al. 1999. Risk indicators for out-of-hospital cardiac arrest in patients with coronary artery disease. *J. Clin. Epidemiol.* 52 (7):601-607.

Diba, V.C., and J.S. English. 2002. Contact allergy to green coffee bean dust in a coffee processing plant worker. *Contact Dermat.* 47(1):56.

Donovan, J.L., and C.L. DeVane. 2001. A primer on caffeine pharmacology and its drug interactions in clinical psychopharmacology. *Psychopharmacol. Bull.* 35 (3):30-48.

Evans, S.M., and R.R. Griffiths. 1999. Caffeine withdrawal: A parametric analysis of caffeine dosing conditions. *J. Pharmacol. Exp. Ther.* 289(1):285-294.

Frost, L., and P. Vestergaard. 2005. Caffeine and risk of atrial fibrillation or flutter: The Danish Diet, Cancer, and Health Study. *Am. J. Clin. Nutr.* 81(3):578.

Geleijnse, J.M. 2008. Habitual coffee consumption and blood pressure: An epidemiological perspective. *Vasc. Health Risk Manag.* 4(5):963-970.

Giovannucci, E. 1998. Meta-analysis of coffee consumption and risk of colorectal cancer. *Am. J. Epidemiol.* 147(11):1043-1052.

Grandjean, E.M., and J.M. Aubry. 2009. Lithium: Updated human knowledge using an evidence-based approach: Part III: Clinical safety. *CNS Drugs* 23(5):397-418.

Greenland, S. 1993. A meta-analysis of coffee, myocardial infarction, and coronary death. *Epidemiology* 4(4):366-374.

Grobbee, D.E., E.B. Rimm, E. Giovannucci, et al. 1990. Coffee, caffeine, and cardiovascular disease in men. *New Engl. J. Med.* 323(15):1026.

Hakim, A.A., G.W. Ross, J.D. Curb, et al. 1998. Coffee consumption in hypertensive men in older middle-age and the risk of stroke: The Honolulu Heart Program. *J. Clin. Epidemiol.* 51(6):487-494.

Hammar, N., T. Andersson, L. Alfredsson, et al. 2003. Association of boiled and filtered coffee with incidence of first nonfatal myocardial infarction: The SHEEP and the VHEEP study. *J. Intern. Med.* 253(6):653-659.

Hasling, C., K. Sondergaard, P. Charles, and L. Mosekilde. 1992. Calcium metabolism in postmenopausal osteoporotic women is determined by dietary calcium and coffee intake. *J. Nutr.* 122(5):1119.

Hernandez-Avila, M., G.A. Colditz, M.J. Stampfer, et al. 1991. Caffeine, moderate alcohol intake, and risk of fractures of the hip and forearm in middle-aged women. *Am. J. Clin. Nutr.* 54(1):157-163.

Heyden, S., H.A. Tyroler, G. Heiss, C.G. Hames, and A. Bartel. 1978. Coffee consumption and mortality: Total mortality, stroke mortality, and coronary heart disease mortality. *Arch. Intern. Med.* 138(10):1472-1475.

Higdon, J.V., and B. Frei. 2006. Coffee and health: A review of recent human research. *Crit. Rev. Food Sci. Nutr.* 46(2):101-123.

Hughes, J.R., S.T. Higgins, W.K. Bickel, et al. 1991. Caffeine self-administration, withdrawal, and adverse effects among coffee drinkers. *Arch. Gen. Psychiatry* 48(7):611-617.

Hurrell, R.F., M. Reddy, and J.D. Cook. 1999. Inhibition of non-haem iron absorption in man by polyphenolic-containing beverages. *Br. J. Nutr.* 81(4):289-295.

Husemoen, L.L.N., T.F. Thomsen, M. Fenger, and T. Jørgensen. 2004. Effect of lifestyle factors on plasma total homocysteine concentrations in relation to MTHFR (C677T) genotype. Inter99 (7). *Eur. J. Clin. Nutr.* 58(8):1142-1150.

Infante, S., M.L. Baeza, M. Calvo, et al. 2003. Anaphylaxis due to caffeine. *Allergy* 58(7):681-682.

James, J.E. 1994. Chronic effects of habitual caffeine consumption on laboratory and ambulatory blood pressure levels. *J. Cardiovasc. Risk* 1:159-164.

Jee, S.H., J. He, L.J. Appel, et al. 2001. Coffee consumption and serum lipids: A meta-analysis of randomized controlled clinical trials. *Am. J. Epidemiol.* 153(4):353-362.

Johnell, O., B. Gullberg, and J.A. Kanis. 1995. Risk factors for hip fracture in European women: The MEDOS Study. *J. Bone Miner. Res.* 10:1802-1815.

Juliano, L.M., and R.R. Griffiths. 2004. A critical review of caffeine withdrawal: Empirical validation of symptoms and signs, incidence, severity, and associated features. *Psychopharmacology* 176(1):1-29.

C

Kanis, J., O. Johnell, B. Gullberg, et al. 1999. Risk factors for hip fracture in men from Southern Europe: The MEDOS study. *Osteoporosis Int.* 9(1):45-54.

Kawachi, I., G.A. Colditz, and C.B. Stone. 1994. Does coffee drinking increase the risk of coronary heart disease? Results from a meta-analysis. *Br. Med. J.* 72(3):269-275.

Kiel, D.P., D.T. Felson, M.T. Hannan, J.J. Anderson, and P.W.F. Wilson. 1990. Caffeine and the risk of hip fracture: The Framingham Study. *Am. J. Epidemiol.* 132(4):675.

Lemiere, C., J.L. Malo, M. McCants, and S. Lehrer. 1996. Occupational asthma caused by roasted coffee: Immunologic evidence that roasted coffee contains the same antigens as green coffee, but at a lower concentration. *J. Allergy Clin. Immunol.* 98(2):464-466.

Leung, A.Y., and S. Foster. 1996. *Encyclopedia of common natural ingredients used in food, drugs, and cosmetics.* 2nd ed. New York: Wiley.

Lovallo, W.R., M.F. Wilson, and A.S. Vincent. 2004. Blood pressure response to caffeine shows incomplete tolerance after short-term regular consumption. *Hypertension* 43:760-765.

Mart-Carvajal, A.J., I. Sol, D. Lathyris, and G. Salanti. 2009. Homocysteine lowering interventions for preventing cardiovascular events. *Cochrane Database Syst. Rev.* No. 4:CD006612.

Maughan, R.J., and J. Griffin. 2003. Caffeine ingestion and fluid balance: A review. *J. Hum. Nutr. Dietet.* 16(6):411-420.

McCusker, R.R., B.A. Goldberger, and E.J. Cone. 2003. Technical note: Caffeine content of specialty coffees. *J. Analyt. Toxicol.* 27(7):520-522.

Mennen, L.I., G.P. de Courcy, J.C. Guilland, et al. 2002. Homocysteine, cardiovascular disease risk factors, and habitual diet in the French Supplementation with Antioxidant Vitamins and Minerals Study. *Am. J. Clin. Nutr.* 76(6):1279-1289.

Mester, R., P. Toren, and I. Mizrachi. 1995. Caffeine withdrawal increases lithium blood levels. *Biol. Psychiatry* 37(5):348-350.

Mills, S., and K. Bone. 2005. *The essential guide to herbal safety.* St. Louis: Elsevier.

Mukamal, K.J., M. Maclure, J.E. Muller, J.B. Sherwood, and M.A. Mittleman. 2004. Caffeinated coffee consumption and mortality after acute myocardial infarction. *Am. Heart J.* 147(6):999-1004.

Myers, M.G. 1991. Caffeine and cardiac arrhythmias. *Ann. Intern. Med.* 114(2):147.

Myers, M.G., and A. Basinski. 1992. Coffee and coronary heart disease. *Arch. Intern. Med.* 152(9):1767-1772.

Nehlig, A., and G. Debry. 1994. Potential genotoxic, mutagenic and antimutagenic effects of coffee: A review. *Mutat. Res.* 317(2):145-162.

Nieves, J.W., J.A. Grisso, and J.L. Kelsey. 1992. A case-control study of hip fracture: Evaluation of selected dietary variables and teenage physical activity. *Osteoporosis Int.* 2(3):122-127.

Noordzij, M., C. Uiterwaal, L.R. Arends, et al. 2005. Blood pressure response to chronic intake of coffee and caffeine: A meta-analysis of randomized controlled trials. *J. Hyperten.* 23(5):921-928.

Nordmark, A., S. Lundgren, S. Cnattingius, and A. Rane. 1999. Dietary caffeine as a probe agent for assessment of cytochrome P4501A2 activity in random urine samples. *Br. J. Clin. Pharmacol.* 47(4):397-402.

NTP. 1982. Acute toxicity report on caffeine (3,7-dihydro-1,3,7-trimethyl-1*H*-purine-2,6-dione) in Fischer 344 rats. Final Report, NCTR Tech Rpt. Jefferson, AR: National Center for Toxicological Research.

Nurminen, M.L., L. Niittynen, R. Korpela, and H. Vapaatalo. 1999. Coffee, caffeine and blood pressure: A critical review. *Eur. J. Clin. Nutr.* 53:831–839.

Nygard, O., H. Refsum, P.M. Ueland, et al. 1997. Coffee consumption and plasma total homocysteine: The Hordaland Homocysteine Study. *Am. J. Clin. Nutr.* 65(1):136-143.

OECD. 2004. Caffeine CAS: 58-08-2. In *Screening information dataset (SIDS) high production volume chemicals.* Paris: Organisation for Economic Co-operation and Development.

Palm, P.E., E.P. Arnold, M.S. Nick, J.R. Valentine, and T.E. Doerfler. 1984. Two-year toxicity/carcinogenicity study of fresh-brewed coffee in rats initially exposed in utero. *Toxicol. Appl. Pharmacol.* 74(3):364-382.

Palmer, J.R., L. Rosenberg, R.S. Rao, and S. Shapiro. 1995. Coffee consumption and myocardial infarction in women. *Am. J. Epidemiol.* 141(8):724-731.

PDR. 2006. *Physicians' desk reference for nonprescription drugs and dietary supplements.* 27th ed. Montvale, NJ: Medical Economics Co.

Sato, J., H. Nakata, E. Owada, et al. 1993. Influence of usual intake of dietary caffeine on single-dose kinetics of theophylline in healthy human subjects. *Eur. J. Clin. Pharmacol.* 44(3):295-298.

Sesso, H.D., J.M. Gaziano, J.E. Buring, and C.H. Hennekens. 1999. Coffee and tea intake and the risk of myocardial infarction. *Am. J. Epidemiol.* 149(2):162.

Signorello, L.B., and J.K. McLaughlin. 2004. Maternal caffeine consumption and spontaneous abortion: A review of the epidemiologic evidence. *Epidemiology* 15(2):229-239.

Sofi, F., A.A. Conti, A.M. Gori, et al. 2007. Coffee consumption and risk of coronary heart disease: A meta-analysis. *Nutr. Metab. Cardiovasc. Dis.* 17(3):209-223.

Stavric, B. 1988. Methylxanthines: Toxicity to humans. 2. Caffeine. *Food Chem. Toxicol.* 26(7):645-662.

Stolzenberg-Solomon, R.Z., E.R. Miller 3rd, M.G. Maguire, J. Selhub, and L.J. Appel. 1999. Association of dietary protein intake and coffee consumption with serum homocysteine concentrations in an older population. *Am. J. Clin. Nutr.* 69(3):467-475.

Tavani, A., M. Bertuzzi, S. Gallus, E. Negri, and C. La Vecchia. 2004. Risk factors for non-fatal acute myocardial infarction in Italian women. *Prev. Med.* 39(1):128-134.

Tavani, A., M. Bertuzzi, E. Negri, L. Sorbara, and C. La Vecchia. 2001. Alcohol, smoking, coffee and risk of non-fatal acute myocardial infarction in Italy. *Eur. J. Epidemiol.* 17(12):1131-1137.

Tavani, A., and C. La Vecchia. 2000. Coffee and cancer: A review of epidemiological studies, 1990–1999. *Eur. J. Cancer Prev.* 9(4):241-256.

Tavani, A., and C. La Vecchia. 2004. Coffee, decaffeinated coffee, tea and cancer of the colon and rectum: A review of epidemiological studies, 1990–2003. *Cancer Causes Control* 15(8):743-757.

Tavani, A., E. Negri, and C. LaVecchia. 1995. Coffee intake and risk of hip fracture in women in northern Italy. *Prev. Med.* 24(4):396-400.

Treudler, R., B. Tebbe, and C.E. Orfanos. 1997. Coexistence of type I and type IV sensitization in occupational coffee allergy. *Contact Dermat.* 36(2):109.

Urgert, R., and M.B. Katan. 1997. The cholesterol-raising factor from coffee beans. *Annu. Rev. Nutr.* 17:305-324.

Urgert, R., T. van Vliet, P.L. Zock, and M.B. Katan. 2000. Heavy coffee consumption and plasma homocysteine: A randomized controlled trial in healthy volunteers. *Am. J. Clin. Nutr.* 72(5):1107-1110.

Viani, R. 1988. Physiologically active substances in coffee. In *Coffee, Volume 3: Physiology*, edited by Clarke, R.J., and R. Macrae. New York: Elsevier Applied Science.

Wilhelmsen, L., A. Rosengren, and G. Lappas. 2002. Hospitalizations for atrial fibrillation in the general male population: Morbidity and risk factors. *J. Intern. Med.* 250(5):382-389.

Würzner, H.P., E. Lindström, L. Vuataz, and H. Luginbühl. 1977a. A 2-year feeding study of instant coffees in rats. I. Body weight, food consumption, haematological parameters and plasma chemistry. *Food Cosmet. Toxicol.* 15(1):7-16.

Würzner, H.P., E. Lindström, L. Vuataz, and H. Luginbühl. 1977b. A 2-year feeding study of instant coffees in rats. II. Incidence and types of neoplasms. *Food Cosmet. Toxicol.* 15(4):289-296.

Coix lacryma-jobi L. — Poaceae

SCN: Job's tears
OCN: coix
PN: *yi yi ren* (seed)
Part: seed

Quick Reference Summary

Safety Class: 2b
Interaction Class: A

Contraindications

Not for use in pregnancy except under the supervision of a qualified healthcare practitioner (Chen and Chen 2004; Tzeng et al. 2005).

Other Precautions

Job's tears should be used with caution in persons with spermatorrhea (frequent involuntary emission of the semen without copulation) and polyuria (passage of large volumes of urine) (Chen and Chen 2004).

Drug and Supplement Interactions

None known.

Editors' Note

Job's tears has traditionally been eaten as a food in China, India, and southeast Asia (Arber 2010; Simoons 1991).

Adverse Events and Side Effects

None known.

Pharmacological Considerations

Animal studies have demonstrated that Job's tears may modify glucose regulation (Takahashi et al. 1986; Yeh et al. 2006). People with diabetes are advised to monitor their blood sugar closely and discuss the use of this herb with a qualified healthcare practitioner prior to use.

Pregnancy and Lactation

Although two animal studies of Job's tears indicated a lack of adverse effects on fetal development (Shinoda et al. 2007; Tzeng et al. 2005), one of the studies using very large doses of Job's tears showed some loss of embryos but no adverse effects on fetal development (Tzeng et al. 2005).

Animal and in vitro studies have yielded conflicting data on the effects of Job's tears on uterine contractions. Inhibitory effects were observed on induced contractions in rats (Hsia et al. 2008), whereas studies in isolated rat uteri showed that different extracts induced, inhibited, or had no effects on uterine contractions (Hsia et al. 2008; Tzeng et al. 2005).

A reference text on traditional Chinese medicine indicates that Job's tears should not be used during pregnancy (Chen and Chen 2004).

No information on the safety of Job's tears during lactation was identified. While this review did not identify any concerns for use while nursing, safety has not been conclusively established.

Review Details

I. Drug and Supplement Interactions

Clinical Trials of Drug or Supplement Interactions

No clinical trials of drug or supplement interactions were identified.

Case Reports of Suspected Drug or Supplement Interactions

No case reports of suspected drug or supplement interactions were identified.

C

Animal Trials of Drug or Supplement Interactions

No animal trials of drug or supplement interactions were identified.

II. Adverse Events

Case Reports of Adverse Events

No case reports of adverse events were identified.

III. Pharmacology and Pharmacokinetics

Human Pharmacological Studies

No relevant human pharmacological studies were identified.

Animal Pharmacological Studies

An inhibitory effect of Job's tears hulls on prostaglandin $F_{2\alpha}$-induced uterine contractions was observed in rats (Hsia et al. 2008).

A reduction in plasma glucose levels was observed in diabetic rats fed a diet containing 58% dehulled Job's tears daily for 4 weeks (Yeh et al. 2006). Hypoglycemic activity of Job's tears aqueous extract was observed in normal and diabetic mice after oral administration (Takahashi et al. 1986).

In Vitro Pharmacological Studies

In isolated rat uteri, an aqueous extract of Job's tears enhanced spontaneous uterine contractions. No effects of a methanol extract were observed (Tzeng et al. 2005). An ethanol extract of Job's tears hull inhibited smooth muscle contraction in isolated rat uteri (Hsia et al. 2008).

Downregulation of progesterone biosynthesis was observed in rat granulosa cells treated with an extract of the bran of Job's tears (Hsia et al. 2006).

IV. Pregnancy and Lactation

In pregnant rats orally administered 1 g/kg of either aqueous or methanolic extracts of Job's tears on day 6 of pregnancy, an increase in fetal resorptions and postimplantation losses was observed in the group administered the water extract. No fetal malformations were observed in either group (Tzeng et al. 2005).

No adverse effects on fetal development were observed in rats fed diets containing up to 5% Job's tears during the full course of pregnancy. The authors concluded that the nontoxic dose level was over 3.5 g/kg daily for both mothers and pups (Shinoda et al. 2007).

No information on the safety of Job's tears during lactation was identified.

V. Toxicity Studies

Short-Term Toxicity

No significant adverse changes in organ weights or blood chemistry were observed in rats orally administered aqueous extracts of whole Job's tears at doses of 0.5 to 2 g/kg daily for 14 days (Hayashi et al. 2009).

Literature Cited

Arber, A. 2010. *The gramineae: A study of cereal, bamboo and grass.* Cambridge: Cambridge University Press.

Chen, J.K., and T.T. Chen. 2004. *Chinese medical herbology and pharmacology.* City of Industry, CA: Art of Medicine Press.

Hayashi, H., Y. Ohta, T. Arai, et al. 2009. Preliminary reproduction toxicity screening test of *Coix lacryma-jobi* L. var. *ma-yuen* Stapf by oral administration in rats. *Jpn. J. CAM* 6(2):105-110.

Hsia, S.M., W. Chiang, Y.H. Kuo, and P.S. Wang. 2006. Downregulation of progesterone biosynthesis in rat granulosa cells by adlay (*Coix lachryma-jobi* L. var. *ma-yuen* Stapf.) bran extracts. *Int. J. Impotence Res.* 18(3):264-274.

Hsia, S.M., Y.H. Kuo, W. Chiang, and P.S. Wang. 2008. Effects of adlay hull extracts on uterine contraction and Ca^{2+} mobilization in the rat. *Am. J. Physiol. Endocrinol. Metab.* 295(3):E719-E726.

Shinoda, Y., M. Hirata, K. Sato, et al. 2007. Preliminary reproduction toxicity screening test of *Coix lacryma-jobi* L. var. *ma-yuen* Stapf by oral administration in rats. *Jpn. Pharmacol. Therapeut.* 35(1):67-70.

Simoons, F.J. 1991. *Food in China: A cultural and historical inquiry.* Boca Raton, FL: CRC Press.

Takahashi, M., C. Konno, and H. Hikino. 1986. Isolation and hypoglycemic activity of coixans A, B and C, glycans of *Coix lachryma-jobi* var. *ma-yuen* seeds. *Planta Med.* 52(1):64-65.

Tzeng, H.P., W. Chiang, T.H. Ueng, and S.H. Liu. 2005. The abortifacient effects from the seeds of *Coix lachryma-jobi* L. var. *ma-yuen* Stapf. *J. Toxicol. Env. Health A* 68(17-18):1557-1565.

Yeh, P.H., W. Chiang, and M.T. Chiang. 2006. Effects of dehulled adlay on plasma glucose and lipid concentrations in streptozotocin-induced diabetic rats fed a diet enriched in cholesterol. *Int. J. Vitam. Nutr. Res.* 76(5):299-305.

Cola spp. — Sterculiaceae

Cola acuminata (Pall.) Schott & Endl.
SCN: cola
OCN: abata cola; bissy nut; kola

Cola nitida **(Vent.) A. Chev.**
SCN: cola
Syn: *Sterculia nitida* Vent.
OCN: ghanja cola; kola
Part: seed

Quick Reference Summary

Safety Class: 1
Interaction Class: C*

Contraindications

None known.

Other Precautions

Cola contains caffeine, a nervous system stimulant. If taken in large amounts, cola products containing caffeine can cause insomnia, nervousness, and the other well-known symptoms of excess caffeine intake (Donovan and DeVane 2001).

Due to the central nervous system (CNS) stimulant effects of caffeine, use of caffeine-containing products is cautioned in persons with heart disorders, as excessive caffeine consumption may increase heart rate or exacerbate arrhythmias. Caution is also advised in psychological disorders, as caffeine may aggravate depression or induce anxiety (Brinker 2001).

Drug and Supplement Interactions

Use of caffeine with other CNS stimulants, including bronchodilators or adrenergic drugs, may cause excessive central nervous system stimulation resulting in nervousness, irritability, insomnia, and possibly convulsions or cardiac arrhythmias (PDR 2006).

Notice

Caffeine (0.6–3.8%) (Atawodi et al. 2007; Leung and Foster 1996; Wichtl 2004); *see* Appendix 1.

Diuretic (Brunton et al. 2006); *see* Appendix 2.

Editors' Note

The American Herbal Products Association has established a trade requirement (AHPA 2011) that dietary supplement products that contain caffeine, whether as a direct ingredient or as a constituent of herbal ingredients, be labeled to disclose the presence of caffeine in the product and the quantity of added caffeine if greater than 25 mg; be formulated and labeled in a manner to recommend a maximum of 200 mg of caffeine per serving, not more often than every 3 to 4 hours; and bear the following or similar statement on the label of any dietary supplement that contains caffeine in sufficient quantity to warrant such labeling:

> *Too much caffeine may cause nervousness, irritability, sleeplessness, and, occasionally, rapid heartbeat. Not recommended for use by children under 18 years of age.*

See Appendix 1 for more specific details on this AHPA trade requirement.

Adverse Events and Side Effects

Habitual chewing of cola nuts may cause yellow pigmentation of the gums (Ashri and Gazi 1990), and excessive use may cause gastrointestinal complaints (Wichtl 2004).

Allergic reactions to cola have been reported (Speer 1976).

Pharmacological Considerations

See Other Precautions and Drug and Supplement Interactions above.

Pregnancy and Lactation

In a study of pregnant women who had consumed cola nuts during pregnancy, some association between cola use and infant head and chest circumference were observed (Abidoye et al. 1990), whereas an animal study showed no adverse effects on fetal development (Ajarem and Ahmad 1994). Pregnant women are advised to limit intake of caffeine to 300 mg daily (PDR 2006).

Caffeine is listed as a "Maternal Medication Usually Compatible with Breastfeeding" by the American Academy of Pediatrics Committee on Drugs with an indication that maternal consumption of caffeinated beverages should be limited to 150 mg daily (AAP 2001).

* For caffeine-free preparations, no interactions are expected.

Review Details

I. Drug and Supplement Interactions

Clinical Trials of Drug or Supplement Interactions

No clinical trials of drug or supplement interactions were identified.

Case Reports of Suspected Drug or Supplement Interactions

No case reports of suspected drug or supplement interactions were identified.

Animal Trials of Drug or Supplement Interactions

No animal trials of drug or supplement interactions were identified.

II. Adverse Events

Case Reports of Adverse Events

Bright yellow pigmentation of the gums has been reported in persons who habitually chew on cola nuts (Ashri and Gazi 1990).

Cola nut has been listed as a common food allergen (Speer 1976).

C

III. Pharmacology and Pharmacokinetics

Human Pharmacological Studies

In persons without malaria, administration of 35 g of cola nut was reported to cause malaria-like symptoms in a number of patients, including sleeplessness, lack of concentration, dizziness, and weakness (Alaribe et al. 2003).

Animal Pharmacological Studies

In mice intraperitoneally administered 2.5, 5, or 10 mg/kg of fresh cola nut extract, the 10 mg/kg dose depressed locomotor activities while the 5 mg/kg dose increased locomotor activity and the 2.5 mg/kg dose had no significant effects (Ajarem 1990).

An increase in arterial pressure was observed in rats fed diets containing 33 or 100% kola nuts daily for 7 days (Osim and Udia 1993).

In Vitro Pharmacological Studies

A methanol extract of cola showed a time-, dose-, and estrogen receptor-dependent stimulation of the *pS2* gene in estrogen receptor-positive human breast cancer cells (MCF-7), while an acetone extract showed no transcription stimulation of *pS2* (Fontenot et al. 2007).

In isolated rat hearts, application of an aqueous extract of *Cola nitida* subsp. *rubra* increased the heart metabolic rate at concentrations of 4 to 8 mg/l, while the rate decreased to control levels at 10 mg/l. In hearts treated with *Cola acuminata*, the heart metabolic rate increased after treatment with 4 to 10 mg/l with peak activity at 6 mg/l. Extracts of both species did not cause any change in heart metabolic rate after treatment with a concentration of 2 mg/l. Both extracts caused a dose-dependent increase followed by a decrease in heart rate (Chukwu et al. 2006).

IV. Pregnancy and Lactation

In a study of the effects of cola nut consumption during pregnancy on the anthropometric measurements of newborns, cola consumption was associated with a head circumference of 30 to 35 cm and was also associated with a difference in chest circumference. No correlation between cola nut consumption and birth weight was observed. Most mothers (76%) consumed between 13 and 91 g of cola nuts per week (Abidoye et al. 1990).

No adverse effects on development were reported in the offspring of pregnant mice provided up to 32 mg/l (equivalent to 10 mg/kg daily) of cola extract as the sole source of drinking water during the full course of pregnancy. In the 4-day-old pups, a decrease in body weight was observed as compared to the control group, but no differences in body weight were observed in pups of mothers that had been provided lower concentrations of cola extract (equivalent to 2.5 or 5 mg/kg daily) (Ajarem and Ahmad 1994).

Caffeine is in the FDA pregnancy category C and has been shown to cross the placenta and achieve blood and tissue concentrations in the fetus. Excessive intake of caffeine by pregnant women has been associated with fetal arrhythmias. Pregnant women are advised to limit caffeine intake to less than 300 mg daily (PDR 2006).

Caffeine is listed as a "Maternal Medication Usually Compatible with Breastfeeding" by the American Academy of Pediatrics Committee on Drugs. The Committee noted that maternal consumption of caffeine may cause irritability and poor sleeping patterns in nursing infants, and that maternal consumption of caffeinated beverages should not exceed 150 mg daily (AAP 2001).

Epidemiological studies have indicated an association between high caffeine intake during pregnancy and an increased risk of spontaneous abortions. An analysis concluded that methodological flaws in many of the studies led to biased results and that a causal link between caffeine consumption and abortion cannot yet be confirmed (Signorello and McLaughlin 2004).

V. Toxicity Studies

Acute Toxicity

The LD_{50} of orally administered caffeine in rats is 335 mg/kg (Mills and Bone 2005).

No adverse effects were reported in rats fed a diet of cola nuts (daily intake ~9.2 g; rats weighed ~210 g) (Osim and Udia 1993).

Subchronic Toxicity

In rats orally administered 57 mg/kg (equivalent to a human dose of 1 g/kg) of an aqueous extract of cola nut every other day for 18 weeks, a decrease in total body weight and an increase in absolute weights of liver, kidney, brain, and testis were observed. Total protein, RNA, and DNA of these organs were significantly depressed, and serum levels of total and conjugated bilirubin were significantly decreased (Ikegwuonu et al. 1981).

Literature Cited

AAP. 2001. The transfer of drugs and other chemicals into human milk. American Academy of Pediatrics Committee on Drugs. *Pediatrics* 108(3):776-789.

AHPA. July 2011. Code of Ethics & Business Conduct. Silver Spring, MD: American Herbal Products Association.

Abidoye, R.O. and A.P. Chijioke. 1990. Effect of kolanut (*Cola nitida* Vent.) on the anthropometric measurement of newborn babies in Nigeria. *Nutr. Res.* 10(10):1091-1098.

Ajarem, J.S. 1990. Effects of fresh kola-nut extract (*Cola nitida*) on the locomotor activities of male mice. *Acta Physiol. Pharmacol. Bulg.* 16(4):10-15.

Ajarem, J.S., and M. Ahmad. 1994. Effects of consumption of fresh kola-nut extract by female mice on the post-natal development and behavior of their offspring. *J. King Saud Univ.* 6:41-50.
Alaribe, A.A.A., G.C. Ejezie, and E.N.U. Ezedinachi. 2003. The role of kola nut (*Cola nitida*) in the etiology of malaria morbidity. *Pharm. Biol.* 41(6):458-462.
Ashri, N., and M. Gazi. 1990. More unusual pigmentations of the gingiva. *Oral Surg. Oral Med. Oral Pathol.* 70(4):445-449.
Atawodi, S.E., B. Pfundstein, R. Haubner, et al. 2007. Content of polyphenolic compounds in the Nigerian stimulants *Cola nitida* ssp. *alba*, *Cola nitida* ssp. *rubra* A. Chev, and *Cola acuminata* Schott. & Endl. and their antioxidant capacity. *J. Agric. Food Chem.* 55(24):9824-9828.
Brinker, F. 2001. *Herb contraindications and drug interactions.* 3rd ed. Sandy, OR: Eclectic Medical Publications.
Brunton, L.L., J.S. Lazo, and K.L. Parker. 2006. *Goodman & Gilman's the pharmacological basis of therapeutics.* 11th ed. New York: McGraw-Hill.
Chukwu, L.O., W.O. Odiete, and L.S. Briggs. 2006. Basal metabolic regulatory responses and rhythmic activity of mammalian heart to aqueous kola nut extracts. *Afr. J. Biotechnol.* 5(5):484-486.
Donovan, J.L., and C.L. DeVane. 2001. A primer on caffeine pharmacology and its drug interactions in clinical psychopharmacology. *Psychopharmacol. Bull.* 35(3):30-48.
Fontenot, K., S. Naragoni, M. Claville, and W. Gray. 2007. Characterization of bizzy nut extracts in estrogen-responsive MCF-7 breast cancer cells. *Toxicol. Appl. Pharmacol.* 220(1):25-32.
Ikegwuonu, F.I., T.A. Aire, and S.O. Ogwuegbu. 1981. Effects of kola-nut extract administration on the liver, kidney, brain, testis and some serum constituents of the rat. *J. Appl. Toxicol.* 1(6):292-294.
Leung, A.Y., and S. Foster. 1996. *Encyclopedia of common natural ingredients used in food, drugs, and cosmetics.* 2nd ed. New York: Wiley.
Mills, S., and K. Bone. 2005. *The essential guide to herbal safety.* St. Louis: Elsevier.
Osim, E.E., and P.M. Udia. 1993. Effects of consuming a kola nut (*Cola nitida*) diet on mean arterial pressure in rats. *Int. J. Pharmacogn.* 31(3):193-197.
PDR. 2006. *Physicians' desk reference for nonprescription drugs and dietary supplements.* 27th ed. Montvale, NJ: Medical Economics Co.
Signorello, L.B., and J.K. McLaughlin. 2004. Maternal caffeine consumption and spontaneous abortion: A review of the epidemiologic evidence. *Epidemiology* 15(2):229-239.
Speer, F. 1976. Food allergy: The 10 common offenders. *Am. Fam. Physician* 13(2):106-112.
Wichtl, M. 2004. *Herbal drugs and phytopharmaceuticals: A handbook for practice on a scientific basis.* 3rd ed. Boca Raton, FL: CRC Press.

C

Collinsonia canadensis L. — Lamiaceae

SCN: stoneroot
OCN: citronella; horse balm; richweed
Part: root

Quick Reference Summary

Safety Class: 1
Interaction Class: A

Contraindications

None known.

Other Precautions

None known.

Drug and Supplement Interactions

None known.

Adverse Events and Side Effects

None known.

Pharmacological Considerations

None known.

Pregnancy and Lactation

Eclectic medical texts indicate that stoneroot was used to stop threatened abortions and to treat hemorrhoids in pregnant women (Ellingwood 1919; Felter and Lloyd 1898).

No information on the safety of stoneroot during lactation was identified. While this review did not identify any concerns for use while nursing, safety has not been conclusively established.

Review Details

I. Drug and Supplement Interactions

Clinical Trials of Drug or Supplement Interactions

No clinical trials of drug or supplement interactions were identified.

Case Reports of Suspected Drug or Supplement Interactions

No case reports of suspected drug or supplement interactions were identified.

Animal Trials of Drug or Supplement Interactions

No animal trials of drug or supplement interactions were identified.

II. Adverse Events

Case Reports of Adverse Events

No case reports of adverse events were identified.

III. Pharmacology and Pharmacokinetics

Human Pharmacological Studies

No relevant human pharmacological studies were identified.

Animal Pharmacological Studies

No relevant animal pharmacological studies were identified.

In Vitro Pharmacological Studies

No relevant in vitro pharmacological studies were identified.

IV. Pregnancy and Lactation

Eclectic medical texts indicate that stoneroot was used to stop threatened abortions and to treat hemorrhoids in pregnant women (Ellingwood 1919; Felter and Lloyd 1898).

No information on the safety of stoneroot during lactation was identified.

V. Toxicity Studies

No toxicity studies were identified.

Literature Cited

Ellingwood, F. 1919. *The American materia medica, therapeutics and pharmacognosy*. Evanston, IL: Ellingwood's Therapeutist.

Felter, H.W., and J.U. Lloyd. 1898. *King's American dispensatory*. 18th ed., 3rd rev. 2 vols. Cincinnati: Ohio Valley Co.

Commiphora spp. — Burseraceae

Commiphora madagascariensis Jacq.
SCN: myrrh (oleo gum resin)
Syn: *Commiphora abyssinica* (O. Berg.) Engl., orth. var.; *Commiphora habessinica* (O. Berg.) Engl.
OCN: Abyssinian myrrh; Arabian myrrh; Yemen myrrh

***Commiphora molmol* (Engl.) Engl.**
SCN: myrrh (oleo gum resin)
OCN: molmol; Somalian myrrh

***Commiphora myrrha* (Nees) Engl.**
SCN: myrrh (oleo gum resin)
AN: *bola*
PN: *mo yao* (oleo gum resin)
OCN: common myrrh; Hirabol myrrh
Part: gum resin

Quick Reference Summary

Safety Class: 2b
Interaction Class: A

Contraindications

Not for use during pregnancy except under the supervision of a qualified healthcare practitioner (Bensky et al. 2004; Chen and Chen 2004).

Other Precautions

None known.

Drug and Supplement Interactions

None known.

Editors' Note

Doses over 2 to 4 grams may cause irritation of the kidneys and diarrhea (List and Hörhammer 1973).

Adverse Events and Side Effects

Allergic reactions to myrrh, including contact dermatitis and systemic allergic reactions, have been reported (Al-Suwaidan et al. 1998; Bensky et al. 2004; Bian and Pan 1987; Gallo et al. 1999; Lee and Lam 1993). Adverse events reported in one human study were giddiness, somnolence, mild fatigue, and abdominal discomfort (Sheir 2001).

Pharmacological Considerations

The bitter taste of myrrh may cause nausea and vomiting in persons with sensitive stomachs (Chen and Chen 2004).

Pregnancy and Lactation

In traditional Chinese medicine texts, myrrh is contraindicated in pregnancy (Bensky et al. 2004; Chen and Chen 2004). Based on this information, use during pregnancy is not recommended except under the supervision of a qualified healthcare practitioner.

No adverse effects of myrrh on fetal development were reported in an animal study (Massoud et al. 2000).

No information on the safety of myrrh during lactation was identified. While this review did not identify any concerns for use while nursing, safety has not been conclusively established.

Review Details

I. Drug and Supplement Interactions

Clinical Trials of Drug or Supplement Interactions

No clinical trials of drug or supplement interactions were identified.

Case Reports of Suspected Drug or Supplement Interactions

No case reports of suspected drug or supplement interactions were identified.

Animal Trials of Drug or Supplement Interactions

No animal studies of drug or supplement interactions were identified.

II. Adverse Events

Adverse Events Reported in Clinical Trials

In a human study of persons with schistosomiasis taking myrrh at doses of 10 mg/kg daily for 6 days, adverse events were transient and mild. The most frequently reported events were giddiness, somnolence, mild fatigue, and abdominal pain or discomfort. Myrrh had no significant effects on liver functions, serum creatinine, or electrocardiographic findings, nor on liver and kidney functions (Sheir 2001).

No adverse events were reported in persons with *Fasciola* orally administered a resin and oil combination of myrrh at doses of 12 mg/kg daily for 6 days (Massoud 2001).

Case Reports of Adverse Events

Reports of allergic contact dermatitis to topical products containing myrrh have been confirmed by patch testing (Al-Suwaidan et al. 1998; Gallo et al. 1999; Lee and Lam 1993). Two cases of allergic reactions in persons orally administered Chinese herbal formulas containing myrrh have been reported. Myrrh was identified as the allergen in both cases (Bian and Pan 1987).

III. Pharmacology and Pharmacokinetics

Human Pharmacological Studies

No relevant human pharmacological studies were identified.

Animal Pharmacological Studies

No relevant animal pharmacological studies were identified.

In Vitro Pharmacological Studies

No relevant in vitro pharmacological studies were identified.

IV. Pregnancy and Lactation

In traditional Chinese medicine texts, myrrh is contraindicated in pregnancy (Bensky et al. 2004; Chen and Chen 2004).

No fetal abnormalities were observed in offspring of pregnant rats administered 50 to 200 mg/kg daily of a myrrh volatile oil and resin mixture on gestational days 6 to 15 (Massoud et al. 2000).

No information on the safety of myrrh during lactation was identified.

V. Toxicity Studies

Acute Toxicity

The LD_{50} of orally administered myrrh in mice could not be determined at doses up to 3 g/kg. No signs of toxicity were observed at that dose (Rao et al. 2001).

In goat kids orally administered 1 to 5 g/kg myrrh daily, grinding of teeth, salivation, soft feces, loss of appetite, jaundice, shortness of breath, ataxia, and recumbency were seen, and the animals died between days 5 and 16. Intestinal, hepatic, and renal toxicity were accompanied by anemia, leukopenia, increases in serum ALP activity and in concentrations of bilirubin, cholesterol, triglycerides and creatinine, and decreases in total protein and albumin. A dose of 0.25 g/kg was not toxic (Omer and Adam 1999).

Short-Term Toxicity

In rats administered myrrh at doses of 1000 mg/kg orally, 500 mg/kg intramuscularly, or 250 mg/kg intraperitoneally for 2 weeks, animals exhibited the following symptoms: depression, soft feces, jaundice, ruffled hair, hepatonephropathy, hemorrhagic myositis and patchy peritonitis (at the injection site), and death. These were accompanied by increases in serum ALP and ALT activities, bilirubin, and creatinine concentrations, and decreases in total protein and albumin levels, and macrocytic anemia and leukopenia. Doses of 500 mg/kg administered orally or 250 mg/kg administered intramuscularly were not lethal, and when given daily for 1 week the effects were less marked (Omer et al. 1999).

Subchronic Toxicity

No significant toxicity was observed in mice orally administered 100 mg/kg daily of myrrh for 90 days. At the end of the treatment period, the myrrh group had increased body weight gain and had increased weights of testes, epididymides, and seminal vesicles, although no spermatotoxic effects were found. An increase in red blood cell count and hemoglobin levels was seen (Rao et al. 2001).

Literature Cited

Al-Suwaidan, S.N., M.O.G. Rab, S. Al-Fakhiry, et al. 1998. Allergic contact dermatitis from myrrh, a topical herbal medicine used to promote healing. *Contact Dermat.* 39(3):137-137.

Bensky, D., S. Clavey, and E. Stöger. 2004. *Chinese herbal medicine: Materia medica.* 3rd ed. Seattle: Eastland Press.

Bian, H.Z., and M.S. Pan. 1987. Systemic allergic reaction to myrrh, a report of 2 cases. *Bull. Chin. Mater. Med.* 12:53.

Chen, J.K., and T.T. Chen. 2004. *Chinese medical herbology and pharmacology*. City of Industry, CA: Art of Medicine Press.

Gallo, R., G. Rivara, G. Cattarini, E. Cozzani, and M. Guarrera. 1999. Allergic contact dermatitis from myrrh. *Contact Dermat.* 41(4):230-231.

Lee, T.Y., and T.H. Lam. 1993. Myrrh is the putative allergen in bonesetter's herbs dermatitis. *Contact Dermat.* 29(5):279.

List, P.H., and H. Hörhammer. 1973. *Hagers handbuch der pharmazeutischen praxis*. Berlin: Springer.

Massoud, A. 2001. Preliminary study of therapeutic efficacy of a new fasciolicidal drug derived from *Commiphora molmol* (myrrh). *Am. J. Trop. Med. Hyg.* 65(2):96-99.

Massoud, A.M., I.M. El-Ashmawy, S.A. Hemeda, and O.M. Salama. 2000. Hematological, chromosomal and teratogenic studies of a new schistosomicidal agent derived from myrrh. *Alex. J. Pharm. Sci.* 14(1):61-68.

Omer, S.A., and S.E. Adam. 1999. Toxicity of *Commiphora myrrha* to goats. *Vet. Hum. Toxicol.* 41(5):299-301.

Omer, S.A., S.E. Adam, and H.E. Khalid. 1999. Effects on rats of *Commiphora myrrha* extract given by different routes of administration. *Vet. Hum. Toxicol.* 41(4):193-196.

Rao, R.M., Z.A. Khan, and A.H. Shah. 2001. Toxicity studies in mice of *Commiphora molmol* oleo-gum-resin. *J. Ethnopharmacol.* 76(2):151-154.

Sheir, Z. 2001. A safe, effective, herbal antischistosomal therapy derived from myrrh. *Am. J. Trop. Med. Hyg.* 65(6):700-704.

Commiphora wightii (Arn.) Bhandari — Burseraceae

SCN: guggul (oleo gum resin)
Syn: *Balsamodendron mukul* Hook.; *Balsamodendron wightii* Arn.; *Commiphora mukul* (Hook. ex Stocks) Engl.
AN: *guggulu*
OCN: bdellium tree; false myrrh; Indian bdellium tree
Part: gum resin

Quick Reference Summary

Safety Class: 2b
Interaction Class: B

Contraindications

Not for use during pregnancy except under the supervision of a qualified healthcare practitioner (Chadha 1988).

Other Precautions

None known.

Drug and Supplement Interactions

Reductions in plasma concentrations of propanolol and diltiazem were observed after a single dose of guggulipid (Dalvi et al. 1994).

Notice

Uterine stimulant (Chadha 1988); *see* Appendix 2.

Editors' Notes

Many of the products on the U.S. market are standardized to contain specific levels of guggulsterones (Ulbricht et al. 2005; Urizar and Moore 2003). Preparations characterized on select guggulsterones may be expected to have different physiological effects than traditional preparations of the herb.

Guggulipid is a standardized extract of guggul (Sahni et al. 2005).

Adverse Events and Side Effects

Gastrointestinal upset, most commonly involving loose stools or diarrhea, has been reported as a side effect in human studies of guggul and guggul extracts (Kuppurajan et al. 1978; Malhotra and Ahuja 1971; Malhotra et al. 1977; Nityanand et al. 1989; Singh et al. 2007; Szapary et al. 2003). Such activity is generally seen at doses over 2 to 4 g (List and Hörhammer 1973).

Hypersensitivity drug rashes have been observed in human studies of orally administered guggul and guggul extracts (Gelfand et al. 2005; Satyavati 1991; Shanavaskhan and Binu 1997; Szapary et al. 2003). Allergic contact dermatitis to topical creams containing guggul was confirmed by patch testing (Kölönte et al. 2006; Salavert et al. 2007).

Pharmacological Considerations

Several human studies of guggul and guggulipid have indicated that guggul reduces platelet aggregation and increases serum fibrinolytic activity (Baldav et al. 1980; Bordia and Chuttani 1979; Gaur et al. 1997; Mester et al. 1979), although no reports of bleeding in humans is reported in the available literature, including multiple controlled trials (Ulbricht et al. 2005).

Pregnancy and Lactation

No adverse effects on fetal development have been observed in animal studies (CDIR 1986). An older study in rats indicated a reduction in fertility after repeated dosing of guggul (Amma et al. 1978). Uterine stimulant activity of guggul has been reported (Chadha 1988).

No information on the safety of guggul during lactation was identified. While this review did not identify any concerns for use while nursing, safety has not been conclusively established.

Review Details

I. Drug and Supplement Interactions

Clinical Trials of Drug or Supplement Interactions

Reductions in plasma concentrations of orally administered propanolol and diltiazem were observed in healthy male volunteers administered a single oral 1 g dose of guggulipid (Dalvi et al. 1994).

Case Reports of Suspected Drug or Supplement Interactions

No case reports of suspected drug or supplement interactions were identified.

Animal Trials of Drug or Supplement Interactions

No animal studies of drug or supplement interactions were identified.

II. Adverse Events

Adverse Events Reported in Clinical Trials

A systematic review of clinical trials of gugulipid, guggulu A, and gum guggulu (Bordia and Chuttani 1979; Kuppurajan et al. 1978; Malhotra and Ahuja 1971; Singh et al. 1994; Szapary et al. 2002; Verma and Bordia 1988) indicated that several mild adverse events were reported during these trials, including rash, nausea, vomiting, eructation, hiccup, headache, loose stools, restlessness, and apprehension, although information regarding adverse events experienced during placebo administration was not always provided (Thompson Coon and Ernst 2003).

In a placebo-controlled double-blind human trial of guggulipid, 3% of the 3 g daily dose ("standard dose") group, 15% of the 6 g daily dose ("high dose") group, and none of the placebo group experienced a hypersensitivity drug rash judged by the investigators to be possibly related to the guggulipid. The skin reactions were associated with pruritus and all occurred within 48 hours of the initiation of therapy. Guggulipid was characterized by the investigators as generally well tolerated without statistically significant differences in rates of adverse events among the treatment groups. No significant changes in renal function, levels of liver-associated enzymes, electrolytes, or TSH levels were observed among any of the treatment groups. In the high dose group, a higher incidence of loose stools and diarrhea were reported than in the standard dose and placebo groups (Gelfand et al. 2005; Szapary et al. 2003).

Side effects reported in a human study of 100 mg daily of guggulipid for 24 weeks were headache, mild nausea, belching, and hiccups (Singh et al. 1994). Diarrhea was reported as a side effect in 12 to 15% of subjects in trials of guggul extract fractions. Anxiety and hiccups were also reported as side effects of these preparations (Malhotra and Ahuja 1971; Malhotra et al. 1977).

Unpurified guggul gum may cause itching or inflammation when applied topically. In early human studies of guggul, skin rashes, diarrhea, and irregular menstruation were reported as common side effects. Rashes were controlled by reducing the dose of guggul (Satyavati 1991; Shanavaskhan and Binu 1997).

No side effects were reported in some human studies of ethyl acetate fractions of guggul gum (Agarwal et al. 1986; Beg et al. 1996; Gopal et al. 1986).

Case Reports of Adverse Events

Reports of allergic contact dermatitis to topical creams containing guggul have been confirmed by patch testing. In one case, the patient was also sensitive to nickel and cobalt (Kölönte et al. 2006; Salavert et al. 2007). Cross-reactivity to abietic acid and balsam of Peru (*Myroxylon balsamum* var. *pereirae*) has been reported (Willis 1973).

Loose stools was reported in a man taking 3 g daily of guggulipid for 1 week. Prior to that week, the patient had been taking guggulipid at a dose of 550 mg daily for 5 months with no apparent side effects. Other drugs and supplements used daily were a nicotine patch, aspirin, and saw palmetto with lycopene (Sahni et al. 2005).

III. Pharmacology and Pharmacokinetics

Human Pharmacological Studies

In healthy subjects and subjects with coronary artery disease, administration of a petrol-soluble extract of guggul reduced the platelet adhesion index and increased serum fibrinolytic activity (Bordia and Chuttani 1979). Several other studies have indicated that guggulipid administration inhibits platelet aggregation and increases fibrinolysis (Baldav et al. 1980; Gaur et al. 1997; Mester et al. 1979). In theory, the risk of bleeding may increase, although there are no reports of bleeding in humans in the available literature, including multiple controlled trials (Ulbricht et al. 2005).

Animal Pharmacological Studies

An increase in serum triiodothyronine (T_3) and enhancement of the T_3/T_4 ratio were observed in mice orally administered a guggul extract at a dose of 0.2 g/kg daily for 15 days (Panda and Kar 1999). A ketosteroid isolated from guggul exhibited a strong thyroid-stimulatory action when administered to rats. Administration of 0.1 mg/kg caused an increase in iodine uptake by the thyroid and enhanced activities of thyroid peroxidase and protease (Tripathi et al. 1984).

In Vitro Pharmacological Studies

Guggulsterones activate the estrogen receptor-α isoform, progesterone receptor, and the pregnane X receptor with EC_{50} values in the low micromolar range. Guggulsterone-mediated activation of the pregnane X receptor induced the expression of CYP3A genes in both rodent and human hepatocytes (Brobst et al. 2004).

IV. Pregnancy and Lactation

A reduction in weights of the uterus, ovaries, and cervix as well as increases in sialic acid and glycogen levels in the uterus and ovaries were observed in rats orally administered guggul and the acid fraction of guggul at doses of 20 or 200 mg/kg daily for 7 days. The authors noted that such activity indicated an antifertility effect of this extract (Amma et al. 1978).

No teratogenic effects were observed in rats, monkeys, and dogs administered an ethyl acetate fraction of guggul (CDIR 1986).

Uterine stimulant activity of guggul has been reported (Chadha 1988).

No information on the safety of guggul during lactation was identified.

V. Toxicity Studies

Acute Toxicity

The LD_{50} of orally or intraperitoneally administered ethyl acetate fraction of guggul in mice could not be determined at doses up to 2 g/kg (Shanker and Singh 2001). The LD_{50} of orally administered guggul essential oil in mice is 1.7 g/kg (Bagi et al. 1985).

Short-Term Toxicity

In rats administered 250 mg/kg daily of a guggul petroleum ether fraction for 3 months, a 50% mortality rate was observed; a 20% morality rate was observed in the control group (Malhotra and Ahuja 1971).

Subchronic Toxicity

No mortality was observed in dogs administered 1 g daily of a guggul petroleum ether fraction for 3 months (Malhotra and Ahuja 1971).

Literature Cited

Agarwal, R.C., S.P. Singh, R.K. Saran, et al. 1986. Clinical-trial of gugulipid a new hypolipidemic agent of plant-origin in primary hyperlipidemia. *Indian J. Med. Res.* 84:626-634.

Amma, M.K., N. Malhotra, R.K. Suri, et al. 1978. Effect of oleoresin of gum guggul (*Commiphora mukul*) on the reproductive organs of female rat. *Indian J. Exp. Biol.* 16(9):1021-1023.

Bagi, M.K., H. Kakrani, G. Kalyani, D. Satyanarayana, and F. Manvi. 1985. Preliminary pharmacological studies of essential oil from *Commiphora mukul*. *Fitoterapia* 56(4):245-248.

Baldav, V.S., R.C. Sharma, P.C. Ranka, and M.D. Chittera. 1980. Effect of *Commiphora mukul* (guggul) on fibrinolytic activity and platelet aggregation in coronary artery disease. *Rajasthan Med. J.* 19:84.

Beg, M., K.C. Singhal, and S. Afzaal. 1996. A study of effect of guggulsterone on hyperlipidemia of secondary glomerulopathy. *Indian J. Physiol. Pharmacol.* 40(3):237-240.

Bordia, A., and S.K. Chuttani. 1979. Effect of gum guggulu on fibrinolysis and platelet adhesiveness in coronary heart disease. *Indian J. Med. Res.* 70:992-996.

Brobst, D.E., X. Ding, K.L. Creech, et al. 2004. Guggulsterone activates multiple nuclear receptors and induces CYP3A gene expression through the pregnane X receptor. *J. Pharmacol. Exp. Ther.* 310(2):528-535.

CDIR. 1986. Gugulipid studies. Lucknow, India: Central Drug Research Institute.

Chadha, Y. 1988. *The wealth of India: A dictionary of Indian raw materials and industrial products*. Delhi: Council of Scientific and Industrial Research.

Dalvi, S.S., V.K. Nayak, S.M. Pohujani, et al. 1994. Effect of gugulipid on bioavailability of diltiazem and propranolol. *J. Assoc. Physicians India* 42(6):454-455.

Gaur, S.P.S., R.K. Garg, A.M. Kar, Y.K. Purohit, and A. Gupta. 1997. Gugulipid, a new hypolipidaemic agent, in patients of acute ischaemic stroke: Effect on clinical outcome, platelet function and serum lipids. *Asia Pacific J. Pharmacol.* 12(3-4):65-69.

Gelfand, J.M., G.H. Crawford, B.A. Brod, and P.O. Szazpary. 2005. Adverse cutaneous reactions to guggulipid. *J. Am. Acad. Dermatol.* 52(3, Part 1):533-534.

Gopal, K., R. Saran, S. Nityanand, P. Gupta, and M. Hasan. 1986. Clinical trial of ethyl acetate extract of gum gugulu (gugulipid) in primary hyperlipidemia. *J. Assoc. Physicians India* 34:249-251.

Kölönte, A., B. Guillot, and N. Raison-Peyron. 2006. Allergic contact dermatitis to guggul extract contained in an anticellulite gel-cream. *Contact Dermat.* 54(4):226-227.

Kuppurajan, K., S.S. Rajagopalan, T.K. Rao, and R. Sitaraman. 1978. Effect of guggulu (*Commiphora mukul* Engl.) on serum lipids in obese, hypercholesterolemic and hyperlipemic cases. *J. Assoc. Physicians India* 26(5):367-373.

List, P.H., and H. Hörhammer. 1973. *Hagers handbuch der pharmazeutischen praxis*. Berlin: Springer.

Malhotra, S.C., and M.M. Ahuja. 1971. Comparative hypolipidaemic effectiveness of gum guggulu (*Commiphora mukul*) fraction 'A,' ethyl-*p*-chlorophenoxyisobutyrate and Ciba-13437-Su. *Indian J. Med. Res.* 59(10):1621-1632.

Malhotra, S.C., M.M. Ahuja, and K.R. Sundaram. 1977. Long term clinical studies on the hypolipidaemic effect of *Commiphora mukul* (guggulu) and clofibrate. *Indian J. Med. Res.* 65(3):390-395.

Mester, L., M. Mester, and S. Nityanand. 1979. Inhibition of platelet aggregation by "guggulu" steroids. *Planta Med.* 37(4):367-369.

Nityanand, S., J.S. Srivastava, and O.P. Asthana. 1989. Clinical trials with gugulipid. A new hypolipidaemic agent. *J. Assoc. Physicians India* 37(5):323-328.

Panda, S., and A. Kar. 1999. Gugulu (*Commiphora mukul*) induces triiodothyronine production: Possible involvement of lipid peroxidation. *Life Sci.* 65(12):PL137-141.

Sahni, S., C.A. Hepfinger, and K.A. Sauer. 2005. Guggulipid use in hyperlipidemia: Case report and review of the literature. *Am. J. Health Syst. Pharm.* 62(16):1690-1692.

Salavert, M., S. Amarger, M.C. Le Bouedec, et al. 2007. Allergic contact dermatitis to guggul in a slimming cream. *Contact Dermat.* 56(5):286-287.

Satyavati, G. 1991. Guggulipid: A promising hypolipidemic agent from gum guggul (*Commiphora wightii*). In *Economic and medicinal plant research, Volume 5*. London: Academic Press.

Shanavaskhan, A.E., and S. Binu. 1997. Detoxification techniques of traditional physicians of Kerala, India on some toxic herbal drugs. *Fitoterapia* 68(1):69-74.

Shanker, G., and H. Singh. 2001. Pharmacology profile of standardised extract of gugul. *Indian J. Pharmacol.* 33(3).

Singh, B.B., S.P. Vinjamury, C. Der-Martirosian, et al. 2007. Ayurvedic and collateral herbal treatments for hyperlipidemia: A systematic review of randomized controlled trials and quasi-experimental designs. *Altern. Ther. Health Med.* 13(4):22-28.

Singh, R.B., M.A. Niaz, and S. Ghosh. 1994. Hypolipidemic and antioxidant effects of *Commiphora mukul* as an adjunct to dietary therapy in patients with hypercholesterolemia. *Cardiovasc. Drugs Ther.* 8(4):659-664.

Szapary, P.O., M.L. Wolfe, L.T. Bloedon, et al. 2003. Guggulipid for the treatment of hypercholesterolemia: A randomized controlled trial. *J. Am. Med. Assoc.* 290(6):765-772.

Szapary, P.O., M.L. Wolfe, L.T. Bloedon, et al. 2002. A double blind, randomised, placebo controlled clinical trial of standardized guggul extract in patients with hypercholesterolemia. *Complement. Ther. Med.* 10:112.

Thompson Coon, J.S., and E. Ernst. 2003. Herbs for serum cholesterol reduction: A systematic view. *J. Fam. Pract.* 52(6):468-478.

Tripathi, Y., O. Malhotra, and S. Tripathi. 1984. Thyroid stimulating action of Z-guggulsterone obtained from *Commiphora mukul*. *Planta Med.* 50(1):78-80.

Ulbricht, C., E. Basch, P. Szapary, et al. 2005. Guggul for hyperlipidemia: A review by the Natural Standard Research Collaboration. *Complement. Ther. Med.* 13(4):279-290.

Urizar, N., and D. Moore. 2003. Gugulipid: A natural cholesterol-lowering agent. *Annu. Rev. Nutr.* 23(1):303-313.

Verma, S., and A. Bordia. 1988. Effect of *Commiphora mukul* (gum guggulu) in patients with hyperlipidemia with special reference to HDL cholesterol. *Indian J. Med. Res.* 87:356-360.

Willis, J. 1973. *A dictionary of the flowering plants and ferns.* 8th ed. Cambridge: Cambridge University Press.

Convallaria majalis L. — Liliaceae

SCN: lily-of-the-valley

Part: entire plant

Quick Reference Summary

Safety Class: 3
Interaction Class: B

Contraindications

Not for use except under the supervision of an expert qualified in the appropriate use of this substance (List and Hörhammer 1973; Martindale and Reynolds 1996; Nelson et al. 2006; Roth et al. 1984; Wood and LaWall 1926).

Other Precautions

None known.

Drug and Supplement Interactions

Interaction considerations are the same as those for digoxin and other digitalis glycosides (Weiss and Fintelmann 2001).

Adverse Events and Side Effects

May cause nausea and vomiting (Felter and Lloyd 1898; Wood and LaWall 1926).

Ingestion of "large" amounts of lily-of-the-valley has been associated with changes in heart rhythms and rates. Pain in the oral cavity, nausea, emesis, abdominal pain, cramping, and diarrhea may also occur after ingestion (Nelson et al. 2006).

Pharmacological Considerations

Lily-of-the-valley has been reported to slow heart rate when given in "moderate doses" and to stop the heart at "toxic" doses (Wood and LaWall 1926).

Pregnancy and Lactation

No information on the safety of lily-of-the-valley during pregnancy or lactation was identified. While this review did not identify any concerns for use while pregnant or nursing, safety has not been conclusively established, and this substance is not recommended for use except under the supervision of an expert qualified in its appropriate use.

Review Details

I. Drug and Supplement Interactions

Clinical Trials of Drug or Supplement Interactions

No clinical trials of drug or supplement interactions were identified.

Case Reports of Suspected Drug or Supplement Interactions

No case reports of suspected drug or supplement interactions were identified.

C

Animal Trials of Drug or Supplement Interactions

No animal trials of drug or supplement interactions were identified.

II. Adverse Events

Case Reports of Adverse Events

Dysrhythmia, heart block, hypotension, gastrointestinal distress, nausea, and vomiting were reported in four members of a family that ingested an unspecified amount of lily-of-the-valley leaves, having mistaken them for wild leeks (*Allium tricoccum*) (Edgerton 1989).

Ingestion of "large" amounts of lily-of-the-valley has been associated with dysrhythmia, sometimes accompanied by symptoms including sinus bradycardia, premature ventricular contractions, atrioventricular conduction defects, or ventricular tachydysrhythmias. Pain in the oral cavity, nausea, emesis, abdominal pain, cramping, and diarrhea may also occur after ingestion (Nelson et al. 2006).

Increased heart rate, restlessness, convulsions, and stupor were observed in a 2-year-old that had been mistakenly administered a teaspoon full of lily-of-the-valley fluid extract (Andrew 1898).

A review of lily-of-the-valley exposure cases reported to U.S. poison control centers over a 10-year period identified 2639 exposures to lily-of-the-valley and *Convallaria montana*, 93% of which were in children under 6 years of age. No fatalities were reported, and only 3 patients suffered major outcomes, with symptoms reported in 6% of all cases (Krenzelok et al. 1996).

Suspected lily-of-the-valley poisoning, with third degree atrioventricular block, bradycardia, vomiting, and lethargy, was reported in a dog that had been seen chewing on lily-of-the-valley leaves (Atkinson et al. 2008).

III. Pharmacology and Pharmacokinetics

Human Pharmacological Studies

No relevant human pharmacological studies were identified.

Animal Pharmacological Studies

No relevant animal pharmacological studies were identified.

In Vitro Pharmacological Studies

No relevant in vitro pharmacological studies were identified.

IV. Pregnancy and Lactation

No information on the use of lily-of-the-valley during pregnancy or lactation was identified.

V. Toxicity Studies

No toxicity studies were identified.

Literature Cited

Andrew, J.H. 1898. A case of poisoning by *Convallaria majalis*. *Therapeutic Gazette* 14(1):144.

Atkinson, K.J., D.M. Fine, T.J. Evans, and S. Khan. 2008. Suspected lily-of-the-valley (*Convallaria majalis*) toxicosis in a dog. *J. Vet. Emer. Crit. Care* 18(4):399-403.

Edgerton, P.H. 1989. Symptoms of digitalis-like toxicity in a family after accidental ingestion of lily of the valley plant. *J. Emerg. Nurs.* 15(3):220-223.

Felter, H.W., and J.U. Lloyd. 1898. *King's American dispensatory*. 18th ed., 3rd rev. 2 vols. Cincinnati: Ohio Valley Co.

Krenzelok, E.P., T.D. Jacobsen, and J.M. Aronis. 1996. Lily-of-the-valley *Convallaria majalis* exposures: Are the outcomes consistent with the reputation? 1996 Annual Meeting of the North American Congress of Clinical Toxicology, Portland, Oregon, USA, October 10-15, 1996. *J. Toxicol. Clin. Toxicol.* 34(5):601.

List, P.H., and H. Hörhammer. 1973. *Hagers handbuch der pharmazeutischen praxis*. Berlin: Springer.

Martindale, W., and J.E.F. Reynolds. 1996. *The extra pharmacopoeia.* London: Pharmaceutical Press.

Nelson, L., R.D. Shih, M.J. Balick, and K.F. Lampe. 2006. *Handbook of poisonous and injurious plants.* New York: Springer.

Roth, L., M. Daunderer, and K. Kormann. 1984. *Giftpflanzen-pflanzengifte: Vorkommen, wirkung, therapie.* Landsberg, Germany: Ecomed.

Weiss, R.F., and V. Fintelmann. 2001. *Weiss's herbal medicine.* New York: Thieme.

Wood, H., and C. LaWall. 1926. *The dispensatory of the United States of America.* Philadelphia: Lippincott.

Conyza canadensis (L.) Cronquist — Asteraceae

SCN: Canada fleabane
Syn: *Erigeron canadensis* L.
OCN: Canadian horseweed
Part: herb

Quick Reference Summary

Safety Class: 1
Interaction Class: A

Contraindications

None known.

Other Precautions

None known.

Drug and Supplement Interactions

None known.

Adverse Events and Side Effects

None known.

Pharmacological Considerations

In vitro studies have demonstrated some antiaggregatory effects of compounds from Canada fleabane on blood platelets (Olas et al. 2006; Saluk-Juszczak et al. 2007). The relevance of this in vitro data to human use is not known.

Pregnancy and Lactation

No information on the safety of Canada fleabane in pregnancy or lactation was identified in the scientific or traditional literature. Although this review did not identify any concerns for use while pregnant or nursing, safety has not been conclusively established.

C

Review Details

I. Drug and Supplement Interactions

Clinical Trials of Drug or Supplement Interactions

No clinical trials of drug or supplement interactions were identified.

Case Reports of Suspected Drug or Supplement Interactions

No case reports of suspected drug or supplement interactions were identified.

Animal Trials of Drug or Supplement Interactions

No animal trials of drug or supplement interactions were identified.

II. Adverse Events

Case Reports of Adverse Events

No case reports of adverse events were identified.

III. Pharmacology and Pharmacokinetics

Human Pharmacological Studies

No relevant human pharmacological studies were identified.

Animal Pharmacological Studies

No relevant animal pharmacological studies were identified.

In Vitro Pharmacological Studies

Polysaccharides from Canada fleabane have been observed to have dose-dependent antiaggregatory effects in ADP- and collagen-induced aggregation in human blood platelets (Olas et al. 2006; Saluk-Juszczak et al. 2007). The relevance of this in vitro data to human use is not known.

IV. Pregnancy and Lactation

No information on the use of Canada fleabane during pregnancy or lactation was identified.

V. Toxicity Studies

No toxicity studies were identified.

Literature Cited

Olas, B., J. Saluk-Juszczak, I. Pawlaczyk, et al. 2006. Antioxidant and antiaggregatory effects of an extract from *Conyza canadensis* on blood platelets in vitro. *Platelets* 17(6):354-360.

Saluk-Juszczak, J., B. Olas, I. Pawlaczyk, R. Gancarz, and B. Wachowicz. 2007. Effects of the extract from *Conyza canadensis* on human blood platelet aggregation. *Gen. Physiol. Biophys.* 26(2):150-152.

Coptis chinensis Franch. — Ranunculaceae

SCN: coptis

PN: *huang lian* (rhizome)

OCN: Chinese goldthread

Part: rhizome

C

Quick Reference Summary

Safety Class: 2b
Interaction Class: A

Contraindications

Not for use in pregnancy except under the supervision of a qualified healthcare practitioner (Chan 1993; Jahnke et al. 2006; Yeung et al. 1990).

Other Precautions

Use of coptis during lactation is not recommended (Chan 1993).

Drug and Supplement Interactions

None known.

Notice

Berberine (4–7%) (Chang and But 1986); *see* Appendix 1.

Editors' Note

Most safety concerns reported for coptis are based on studies of the compound berberine and other alkaloids. Data regarding isolated compounds may not apply directly to products or extracts made from coptis.

Adverse Events and Side Effects

Hyperbilirubinemia was reported in an infant with G6PD deficiency (a hereditary enzyme deficiency, commonly called favism) who was administered coptis (Yeo and Tan 1996), although this may have been due to the G6PD condition (Kaplan and Hammerman 2002).

Texts on traditional Chinese medicine indicate that while coptis is regarded as relatively safe, adverse events reported in association with coptis include diarrhea, abdominal distention, reduction of red blood cells, heart palpitations, shortness of breath, dizziness, gastrointestinal discomfort, nausea, and vomiting (Bensky et al. 2004; Chen and Chen 2004).

Allergic reactions to coptis have been reported (Bensky et al. 2004; Chen and Chen 2004).

Pharmacological Considerations

None known.

Pregnancy and Lactation

The use of berberine-containing herbs is contraindicated or cautioned against in pregnancy in several contemporary texts on herbal safety (Brinker 2001; Mills and Bone 2005). These contraindications are based primarily on the uterine stimulant activity of the isolated compound berberine in excised mouse uteruses (Furuya 1957; Imaseki et al. 1961) and the potential ability of berberine to displace bilirubin and cause neonatal jaundice (Chan 1993). While definitive data confirming the safety of coptis during pregnancy is lacking, reproductive toxicity studies on the isolated compound berberine in mice and rats have shown no adverse effects on the fetus at doses equivalent to over 75 times the standard human dose (Jahnke et al. 2006; Price and George 2003).

Surveys and retrospective studies of coptis use during pregnancy in Taiwan indicate that coptis is commonly used by pregnant women (Chuang et al. 2006, 2007). One study indicated that no significant adverse effects on fetal growth were associated with coptis use (Chuang et al. 2006).

Some concerns exist for use of berberine-containing plants during pregnancy, including uterine stimulation (Furuya 1957; Imaseki et al. 1961), although no uterine stimulation was noted in rats administered high doses of berberine during pregnancy (Jahnke et al. 2006), and a study of berberine-containing herbal extracts on isolated uteri showed no correlation between uterine stimulation and berberine concentration (Haginiwa and Harada 1962).

Berberine has been shown to be present in the breast milk of lactating women who have taken berberine-containing plants (Chan 1993).

Review Details

I. Drug and Supplement Interactions

Clinical Trials of Drug or Supplement Interactions

No clinical trials of coptis monopreparations were identified.

Case Reports of Suspected Drug or Supplement Interactions

No case reports of suspected drug or supplement interactions were identified.

Animal Trials of Drug or Supplement Interactions

No animal studies of drug or supplement interactions were identified.

II. Adverse Events

Case Reports of Adverse Events

An infant with glucose-6-phosphate dehydrogenase (G6PD) deficiency developed severe hyperbilirubinemia and transient bilirubin encephalopathy that the reporting authors indicated as being likely related to consumption of Chinese herbs, including coptis. The peak serum bilirubin was 562 μmol/l (Yeo and Tan 1996). A review of G6PD deficiency indicated that this condition is a potential source of severe neonatal hyperbilirubinemia and kernicterus (Kaplan and Hammerman 2002).

Texts on traditional Chinese medicine indicate that while coptis is regarded as relatively safe, adverse effects reported are transient diarrhea, abdominal distention, reduction of red blood cells, heart palpitations, shortness of breath, dizziness, tinnitus, vomiting, nausea, digestive tract discomfort, and allergic reaction. Details on the products used, doses, and durations of use were not reported (Bensky et al. 2004; Chen and Chen 2004).

III. Pharmacology and Pharmacokinetics

Human Pharmacological Studies

No relevant human pharmacological studies were identified.

Animal Pharmacological Studies

Intraperitoneal administration of 0.02 mg/kg of the compound berberine daily for 1 week to adult rats resulted in a significant decrease in mean bilirubin serum protein binding due a displacement effect (Chan 1993).

In Vitro Pharmacological Studies

In vitro, an aqueous extract of coptis was found to displace bilirubin from serum protein binding as assessed by the peroxidase oxidation method. The authors indicated that coptis may increase the risk of brain damage by free bilirubin in jaundiced infants (Yeung et al. 1990).

The compound berberine was found to be 10 times more potent in vitro than phenylbutazone, a known displacer of bilirubin, and approximately 100 times more potent than papaverine (Chan 1993).

IV. Pregnancy and Lactation

In a retrospective survey of pregnant women in Taiwan taking two to three capsules of 300 to 500 mg of coptis daily during pregnancy, no significant adverse effects on fetal growth were observed. A statistically nonsignificant slight decrease in birth weight and increase in risk of low birth weight was reported in women who had taken coptis, especially in women who had taken more than 56 doses of coptis, as compared to a matched group of pregnant women who had not taken coptis (Chuang et al. 2006).

A survey regarding herb use during pregnancy in Taiwan indicated that 24% of women used at least one herbal product. Coptis was the third most commonly used herb, with 4% of women reporting use (Chuang et al. 2007).

A number of cases of kernicterus (brain damage caused by neonatal jaundice) were reported in south Asian countries in the 1970s and 1980s (Upton 2001). Coptis was a common ingredient in traditional formulas used in pregnant women and newborns. It has been shown in animal models to prevent bilirubin from binding to serum protein, an effect which was reported to increase the risk of brain damage by free bilirubin in jaundiced infants (Yeung et al. 1990). A review of traditional Chinese medicine use in cases of neonatal jaundice indicated that no association could be made between maternal use of Chinese goldthread during pregnancy and neonatal jaundice (Fok 2001).

In pregnant rats fed the compound berberine chloride dihydrate on gestational days (GD) 6 to 20, some reduction in maternal weight gain was observed, with a lowest-observed-adverse-effect level (LOAEL) of 530 mg/kg daily. Only a mild reduction in fetal weights was observed, and the LOAEL based on fetal weight reduction was 1000 mg/kg (Jahnke et al. 2006). Similarly, in mice administered berberine on GD 6 to 17 at doses up to 1155 mg/kg daily, the maternal LOAEL was determined to be 531 mg/kg daily, and the developmental toxicity level was 1000 mg/kg daily. In mice, 33% of the treated females died. Surviving animals had increased relative water intake, and average fetal body weight per litter decreased 5–6% with no change in live litter size (Jahnke et al. 2006).

Berberine has been shown to stimulate uterine contractions in both pregnant and nonpregnant mice (Furuya 1957; Imaseki et al. 1961). A study of various berberine-containing herbal extracts on isolated uteri, however, indicated that relaxation or stimulation of the uterus did not correlate with the concentration of berberine in the extract, suggesting that not all berberine-containing herbs will have the same effect on the uterus (Haginiwa and Harada 1962).

Berberine has been shown to be present in the breast milk of lactating women who have taken berberine-containing plants (Chan 1993).

V. Toxicity Studies

Acute Toxicity

The LD_{50} of orally administered berberine in mice is 329 mg/kg (Haginiwa and Harada 1962). The LD_{50} of orally administered berberine sulfate in rats is greater than 1000 mg/kg (Kowalewski et al. 1975).

Genotoxicity

No mutagenic activity of berberine was observed in *Salmonella typhimurium* TA100 and TA98 with or without metabolic activation by S9 mix. Berberine hydrochloride was weakly mutagenic to strain TA98 without S9 mix, but showed no mutagenic activity in TA100 without S9 mix (Nozaka et al. 1990).

No genotoxic, mutagenic, or recombinogenic activity of berberine with or without metabolic activation was observed in the SOS chromotest. Berberine did not induce significant cytotoxic, mutagenic, or recombinogenic effects during treatments performed under nongrowth conditions; however, in dividing cells, the alkaloid induced cytotoxic and cytostatic effects in proficient and repair-deficient strains of *Saccharomyces cerevisiae*. In dividing cells, the induction of frameshift and mitochondrial mutations, as well as crossing-over, indicated that berberine is not a potent mutagenic agent (Pasqual et al. 1993).

Literature Cited

Bensky, D., S. Clavey, and E. Stöger. 2004. *Chinese herbal medicine: Materia medica.* 3rd ed. Seattle: Eastland Press.

Brinker, F. 2001. *Herb contraindications and drug interactions.* 3rd ed. Sandy, OR: Eclectic Medical Publications.

Chan, E. 1993. Displacement of bilirubin from albumin by berberine. *Neonatology* 63(4):201-208.

Chang, H.-M., and P.P.H. But. 1986. *Pharmacology and applications of Chinese materia medica.* English ed. Singapore, Philadelphia: World Scientific.

Chen, J.K., and T.T. Chen. 2004. *Chinese medical herbology and pharmacology.* City of Industry, CA: Art of Medicine Press.

Chuang, C.H., W.S. Hsieh, Y.L. Guo, et al. 2007. Chinese herbal medicines used in pregnancy: A population-based survey in Taiwan. *Pharmacoepidemiol. Drug Saf.* 16(4):464-468.

Chuang, C.H., J.N. Lai, J.D. Wang, P.J. Chang, and P.C. Chen. 2006. Use of coptidis rhizoma and foetal growth: A follow-up study of 9895 pregnancies. *Pharmacoepidemiol. Drug Saf.* 15(3):185-192.

Fok, T.F. 2001. Neonatal jaundice—Traditional Chinese medicine approach. *J. Perinatol.* 21(Suppl. 1):S98-S100; discussion S104-S107.

Furuya, T. 1957. Pharmacological action, including toxicity and excretion of berberine hydrochloride and its oxidation product. *Bull. Osaka Med. School* 3:62-67.

Haginiwa, J., and M. Harada. 1962. Pharmacological studies on crude drugs. V. Comparison of berberine type alkaloid-containing plants on their components and several pharmacological actions. *Yakugaku Zasshi* 82:726.

Imaseki, I., Y. Kitabatakea, and T. Taguchi. 1961. Studies on the effect of berberine alkaloids on intestine and uterus in mice. *Yakugaku Zasshi* 81:1281-1284.

Jahnke, G.D., C.J. Price, M.C. Marr, C.B. Myers, and J.D. George. 2006. Developmental toxicity evaluation of berberine in rats and mice. *Birth Defects Res. B Dev. Reprod. Toxicol.* 77(3):195-206.

Kaplan, M., and C. Hammerman. 2002. Glucose-6-phosphate dehydrogenase deficiency: A potential source of severe neonatal hyperbilirubinaemia and kernicterus. *Semin. Neonatol.* 7:121-128.

Kowalewski, Z., A. Mrozikiewicz, T. Bobkiewicz, K. Drost, and B. Hladon. 1975. Studies of toxicity of berberine sulfate. *Acta Pol. Pharmaceut.* 32(1):113-120.

Mills, S., and K. Bone. 2005. *The essential guide to herbal safety.* St. Louis: Elsevier.

Nozaka, T., F. Watanabe, S.I. Tadaki, et al. 1990. Mutagenicity of isoquinoline alkaloids, especially of the aporphine type. *Mutat. Res.* 240(4):267-279.

Pasqual, M.S., C.P. Lauer, P. Moyna, and J.A.P. Henriques. 1993. Genotoxicity of the isoquinoline alkaloid berberine in prokaryotic and eukaryotic organisms. *Mutat. Res.* 286(2):243-252.

Price, C.J., and J.D. George. 2003. Final study report on the developmental toxicity evaluation for berberine chloride dihydrate (CAS No. 5956-60-5) administered in the feed to Swiss (CD-1™) mice on gestational days 6 through 17. *Gov. Rep. Announc. Index* No. 20:112.

Upton, R. 2001. *Goldenseal root:* Hydrastis canadensis; *Standards of analysis, quality control, and therapeutics*, American Herbal Pharmacopoeia and therapeutic compendium. Santa Cruz, CA: American Herbal Pharmacopoeia.

Yeo, K.L., and V.C. Tan. 1996. Severe hyperbilirubinaemia associated with Chinese herbs—a case report. *Singapore Paediatr. J.* 38(4):180-182.

Yeung, C.Y., F.T. Lee, and H.N. Wong. 1990. Effect of a popular Chinese herb on neonatal bilirubin protein binding. *Biol. Neonate* 58(2):98-103.

Coptis trifolia (L.) Salisb.

Ranunculaceae

SCN: American goldthread
Syn: *Coptis groenlandica* (Oeder) Fernald
OCN: canker root
Part: rhizome

Quick Reference Summary

Safety Class: 2b
Interaction Class: A

Contraindications

Not for use in pregnancy except under the supervision of a qualified healthcare practitioner (Chan 1993; Jahnke et al. 2006).

Other Precautions

Use of American goldthread during lactation is not recommended (Chan 1993).

Drug and Supplement Interactions

None known.

Notice

Berberine (0.8%) (Felter and Lloyd 1898; Schultz 1884) *see* Appendix 1.

Editors' Note

Most safety concerns reported for American goldthread are based on studies of the compound berberine and other alkaloids. Data regarding isolated compounds may not apply directly to products or extracts made from American goldthread.

Adverse Events and Side Effects

None known.

Pharmacological Considerations

None known.

Pregnancy and Lactation

The use of berberine-containing herbs is contraindicated or cautioned against in pregnancy in several contemporary texts on herbal safety (Brinker 2001; Mills and Bone 2005). These contraindications are based primarily on uterine stimulant activity of the isolated compound berberine in excised mouse uteruses (Furuya 1957; Imaseki et al. 1961) and the potential ability of berberine to displace bilirubin and cause neonatal jaundice (Chan 1993).

While definitive data confirming the safety of American goldthread during pregnancy are lacking, several reproductive toxicity studies on the isolated compound berberine in mice and rats have shown no adverse effects on the fetus at doses equivalent to over 75 times the standard human dose (Jahnke et al. 2006; Price and George 2003).

Some concerns exist for use of berberine-containing plants during pregnancy, including uterine stimulation (Furuya 1957; Imaseki et al. 1961), although no uterine stimulation was noted in rats administered high doses of berberine during pregnancy (Jahnke et al. 2006). Moreover, a study of berberine-containing herbal extracts on isolated uteri showed no correlation between uterine stimulation and berberine concentration (Haginiwa and Harada 1962).

Berberine has been shown to be present in the breast milk of lactating women who have taken berberine-containing plants (Chan 1993).

Review Details

I. Drug and Supplement Interactions

Clinical Trials of Drug or Supplement Interactions

No clinical trials of drug or supplement interactions were identified.

Case Reports of Suspected Drug or Supplement Interactions

No case reports of suspected drug or supplement interactions were identified.

Animal Trials of Drug or Supplement Interactions

No animal studies of drug or supplement interactions were identified.

II. Adverse Events

Case Reports of Adverse Events

No case reports of adverse events were identified.

III. Pharmacology and Pharmacokinetics

Human Pharmacological Studies

No relevant human pharmacological studies were identified.

Animal Pharmacological Studies

Intraperitoneal administration of 0.02 mg/kg of the compound berberine daily for 1 week to adult rats resulted in a significant decrease in mean bilirubin serum protein binding due to a displacement effect (Chan 1993).

In Vitro Pharmacological Studies

The compound berberine was found to be 10 times more potent in vitro than phenylbutazone, a known displacer of bilirubin, and approximately 100 times more potent than papaverine (Chan 1993).

IV. Pregnancy and Lactation

In pregnant rats fed the compound berberine on gestational days (GD) 6 to 20, some reduction in maternal weight gain was observed, with a lowest-observed-adverse-effect level (LOAEL) of 530 mg/kg daily. Only a mild reduction in fetal weights was observed, and the LOAEL based on fetal weight reduction was 1000 mg/kg (Jahnke et al. 2006). Similarly, in mice administered berberine on GD 6 to 17 at doses up to 1155 mg/kg daily, the maternal LOAEL was determined to be 531 mg/kg daily, and the developmental toxicity level was 1000 mg/kg daily. In mice, 33% of the treated females died. Surviving animals had increased relative water intake, and average fetal body weight per litter decreased 5–6% with no change in live litter size (Jahnke et al. 2006).

Berberine has been shown to stimulate uterine contractions in both pregnant and nonpregnant mice (Furuya 1957; Imaseki et al. 1961). A study of various berberine-containing herbal extracts on isolated uteri, however, indicated that relaxation or stimulation of the uterus did not correlate with the concentration of berberine in the extract, suggesting that not all berberine-containing herbs will have the same effect on the uterus (Haginiwa and Harada 1962).

Berberine has been shown to be present in the breast milk of lactating women who have taken berberine-containing plants (Chan 1993).

V. Toxicity Studies

Acute Toxicity

The LD_{50} of orally administered berberine in mice is 329 mg/kg (Haginiwa and Harada 1962). The LD_{50} of orally administered berberine sulfate in rats is greater than 1000 mg/kg (Kowalewski et al. 1975).

C

Genotoxicity

No mutagenic activity of berberine was observed in *Salmonella typhimurium* TA100 and TA98 with or without metabolic activation by S9 mix. Berberine hydrochloride was weakly mutagenic to strain TA98 without S9 mix but showed no mutagenic activity in TA 100 without S9 mix (Nozaka et al. 1990).

No genotoxic, mutagenic, or recombinogenic activity of berberine with or without metabolic activation was observed in the SOS chromotest. Berberine did not induce significant cytotoxic, mutagenic, or recombinogenic effects during treatments performed under nongrowth conditions; however, in dividing cells, the alkaloid induced cytotoxic and cytostatic effects in proficient and repair-deficient strains of *Saccharomyces cerevisiae*. In dividing cells, the induction of frameshift and mitochondrial mutations, as well as crossing-over, indicated that berberine is not a potent mutagenic agent (Pasqual et al. 1993).

Literature Cited

Brinker, F. 2001. *Herb contraindications and drug interactions*. 3rd ed. Sandy, OR: Eclectic Medical Publications.

Chan, E. 1993. Displacement of bilirubin from albumin by berberine. *Neonatology* 63(4):201-208.

Felter, H.W., and J.U. Lloyd. 1898. *King's American dispensatory*. 18th ed., 3rd rev. 2 vols. Cincinnati: Ohio Valley Co.

Furuya, T. 1957. Pharmacological action, including toxicity and excretion of berberine hydrochloride and its oxidation product. *Bull. Osaka Med. School* 3:62-67.

Haginiwa, J., and M. Harada. 1962. Pharmacological studies on crude drugs. V. Comparison of berberine type alkaloid-containing plants on their components and several pharmacological actions. *Yakugaku Zasshi* 82:726.

Imaseki, I., Y. Kitabatakea, and T. Taguchi. 1961. Studies on the effect of berberine alkaloids on intestine and uterus in mice. *Yakugaku Zasshi* 81:1281-1284.

Jahnke, G.D., C.J. Price, M.C. Marr, C.B. Myers, and J.D. George. 2006. Developmental toxicity evaluation of berberine in rats and mice. *Birth Defects Res. B Dev. Reprod. Toxicol.* 77(3):195-206.

Kowalewski, Z., A. Mrozikiewicz, T. Bobkiewicz, K. Drost, and B. Hladon. 1975. Studies of toxicity of berberine sulfate. *Acta Pol. Pharmaceut.* 32(1):113-120.

Mills, S., and K. Bone. 2005. *The essential guide to herbal safety*. St. Louis: Elsevier.

Nozaka, T., F. Watanabe, S.I. Tadaki, et al. 1990. Mutagenicity of isoquinoline alkaloids, especially of the aporphine type. *Mutat. Res.* 240(4):267-279.

Pasqual, M.S., C.P. Lauer, P. Moyna, and J.A.P. Henriques. 1993. Genotoxicity of the isoquinoline alkaloid berberine in prokaryotic and eukaryotic organisms. *Mutat. Res.* 286(2):243-252.

Price, C.J., and J.D. George. 2003. Final study report on the developmental toxicity evaluation for berberine chloride dihydrate (CAS No. 5956-60-5) administered in the feed to Swiss (CD-1™) mice on gestational days 6 through 17. *Gov. Rep. Announce. Index* No. 20:112.

Schultz, J. 1884. The alkaloids of *Coptis trifolia*. *Am. J. Pharm.* 56.

Cordia ecalyculata Vell. — Boraginaceae

SCN: chá-de-bugre
Syn: *Cordia salicifolia* Cham.

Part: leaf

Quick Reference Summary

Safety Class: 1
Interaction Class: A

Contraindications

None known.

Other Precautions

None known.

Drug and Supplement Interactions

None known.

Adverse Events and Side Effects

None known.

Pharmacological Considerations

None known.

Pregnancy and Lactation

No information on the safety of chá-de-bugre in pregnancy or lactation was identified in the scientific or traditional literature. Although this review did not identify any concerns for use while pregnant or nursing, safety has not been conclusively established.

Review Details

I. Drug and Supplement Interactions

Clinical Trials of Drug or Supplement Interactions

No clinical trials of drug or supplement interactions were identified.

Case Reports of Suspected Drug or Supplement Interactions

No case reports of suspected drug or supplement interactions were identified.

Animal Trials of Drug or Supplement Interactions

No animal trials of drug or supplement interactions were identified.

II. Adverse Events

Case Reports of Adverse Events

No case reports of adverse events were identified.

III. Pharmacology and Pharmacokinetics

Human Pharmacological Studies

No relevant human pharmacological studies were identified.

Animal Pharmacological Studies

No effects on blood glucose were observed in diabetic rats orally administered an aqueous extract of chá-de-bugre daily for 13 days (Siqueira et al. 2006).

In Vitro Pharmacological Studies

No relevant in vitro pharmacological studies were identified.

IV. Pregnancy and Lactation

No information on the use of chá-de-bugre during pregnancy or lactation was identified.

V. Toxicity Studies

Acute Toxicity

The LD_{50} of intraperitoneally administered aqueous extract of chá-de-bugre in mice is 920 mg/kg whereas the oral LD_{50} could not be determined at doses up to 2 g/kg (Caparroz-Assef et al. 2005).

Subchronic Toxicity

No adverse effects, including changes in the body weight gain, organ weights, or biochemical and hematological parameters, were observed in mice orally administered up to 400 mg/kg aqueous extract of chá-de-bugre daily for 90 days (Caparroz-Assef et al. 2005).

Literature Cited

Caparroz-Assef, S.M., R. Grespan, R.C. Freire Batista, et al. 2005. Toxicity studies of *Cordia salicifolia* extract. *Acta Sci. Health Sci.* 27(1):41-44.

Siqueira, V.L.D., D.A.G. Cortez, C.E. Oliveira, C.V. Nakamura, and R.B. Bazoette. 2006. Pharmacological studies of *Cordia salicifolia* Cham in normal and diabetic rats. *Braz. Arch. Biol. Tech.* 49(2):215-218.

Cordyceps sinensis (Berk.) Sacc.

Clavicipitaceae

SCN: cordyceps
PN: *dong chong xia cao* (fungal fruiting body and body of host larvae); *jinshuibao jiaonang, jinshuibao pian* (fermented Cs-4 mycelium)
OCN: Chinese caterpillar fungus
Part: fungal fruiting body and body of host larvae; mycelium

Quick Reference Summary

Safety Class: 1
Interaction Class: A

Contraindications

None known.

C

Other Precautions

None known.

Drug and Supplement Interactions

None known. *See* Clinical trials of drug or supplement interactions below for human studies demonstrating a lack of interaction with certain drugs.

Editors' Notes

Cordyceps is a fungus that parasitizes caterpillars (larvae) of the species *Hepialus armoricanus*. Commercially marketed cordyceps products include the fruiting body of the fungus with the body of the parasitized caterpillar, and products made from cultured strains of cordyceps mycelium (the nonreproductive portion of the fungus). The preparations Cs-4 and Cs-B_{414} that are the subject of several pharmacological studies are fermentation products from mycelial strains isolated from wild strains of cordyceps (Hockaday 2002).

Wild-collected cordyceps fruiting bodies (with body of host larva attached) have been reported to be adulterated with lead to increase the ingredient's weight, a practice that has led to contamination of cordyceps products and cases of lead poisoning (Wu et al. 1996). Such contamination is not a concern in products made from the mycelium or from cultivated cordyceps.

Adverse Events and Side Effects

A review of clinical trials of cordyceps and Cs-4 indicated that mild upper gastrointestinal tract discomfort has been reported (Zhu et al. 1998).

Pharmacological Considerations

Cordyceps and polysaccharide fractions of cordyceps have been shown to modify blood sugar levels in diabetic mice and rats, with the polysaccharide fraction having more marked effects than the whole extract (Balon et al. 2002; Li et al. 2006; Zhang et al. 2006; Zhao et al. 2002). People with diabetes are advised to monitor their blood sugar closely and discuss the use of this herb with a qualified healthcare practitioner prior to use.

Pregnancy and Lactation

Limited information on the safety of cordyceps in pregnancy is available. One animal study of a cordyceps product in pregnancy indicated no adverse effects on maternal health or fetal development (Jing et al. 1987).

No information on the safety of cordyceps in lactation was identified in the scientific or traditional literature. While this review did not identify any concerns for use while nursing, safety has not been conclusively established.

Review Details

I. Drug and Supplement Interactions

Clinical Trials of Drug or Supplement Interactions

No adverse effects on cyclosporine A efficacy were observed when liver transplant patients taking cyclosporine A (5 mg/kg daily) were administered 3 g daily of cordyceps for 15 days. A significant reduction of cyclosporine A nephrotoxicity was reported (Xu et al. 1995).

No significant adverse events or interactions were reported in a study of Cs-4 in patients with chronic heart failure. Patients were administered 3 to 4 g daily of Cs-4 in addition to their standard prescription medication which included digoxin, hydrochlorothiazide, isosorbide dinitrate, furosemide, lanatoside, dopamine, and dobutamine (Zhang et al. 1995).

In elderly patients with acute infections, administration of 6 g daily of cordyceps reduced the toxicity of amikacin (400 mg daily for 6 days) with no apparent adverse effects on amikacin efficacy. Patients treated with cordyceps and amikacin had significantly lower levels of urinary N-acetyl-β-D-glucosaminidase as compared with those taking amikacin alone (Bao et al. 1994).

Case Reports of Suspected Drug or Supplement Interactions

No case reports of suspected drug or supplement interactions were identified.

Animal Trials of Drug or Supplement Interactions

No adverse effects on methotrexate efficacy were observed in mice inoculated with melanoma cells orally administered 200 mg/kg daily of an aqueous extract of cordyceps with 15 mg/kg daily methotrexate (Nakamura et al. 2003).

Cordyceps reduced the nephrotoxic effects of cyclosporine A in rats, ameliorating glomerular and interstitial injuries. In the acute study, rats were administered 50 mg/kg cyclosporine A and 1000 mg/kg cordyceps daily for 15 days, while in the chronic study the animals received 30 mg/kg cyclosporine A and 500 mg/kg cordyceps daily for 3 months (Zhao and Li 1993).

II. Adverse Events

Adverse Events Reported in Clinical Trials

A review of clinical trials of cordyceps and Cs-4 indicated that adverse events reported in the trials were mild upper gastrointestinal tract discomfort, including nausea, dry mouth, and stomach discomfort (Zhu et al. 1998).

Case Reports of Adverse Events

Side effects associated with the use of cordyceps have been reported as headache, irritability, restlessness, edema, swelling of the face and extremities, nosebleeds, decreased volume of urine, and rapid pulse. Case details including exact product, dose, and duration of use were not available. (Chen and Chen 2004).

One case of constipation, abdominal distention, and decrease in peristalsis in a person taking cordyceps has been reported. Case details were not available (Bensky et al. 2004).

A systemic allergic reaction to Cs-4 has been reported (Xu 1992).

III. Pharmacology and Pharmacokinetics

Human Pharmacological Studies

No relevant human pharmacological studies were identified.

Animal Pharmacological Studies

A polysaccharide-enriched preparation (10 mg/kg daily) of cordyceps and an aqueous extract of cordyceps (100 mg/kg daily) produced hypoglycemic effects on streptozotocin-induced diabetic rats, with the polysaccharide-enriched preparation producing greater effects than the aqueous extract (Zhang et al. 2006). A polysaccharide fraction of cordyceps (200 mg/kg daily) produced a hypoglycemic effect in alloxan-induced diabetic mice and streptozotocin-induced diabetic rats (Li et al. 2006).

Cs-4 lowered the fasting plasma levels of glucose and insulin, improving oral glucose tolerance and increasing the glucose-insulin index in rats (Zhao et al. 2002), and was observed to increase whole body insulin sensitivity in rats (Balon et al. 2002).

In Vitro Pharmacological Studies

The compound cordycepin has been shown to completely inhibit aggregation of human platelets induced by U46619 (a thromboxane A_2 analog) (Cho et al. 2006). Similarly, dose-dependent inhibition of collagen-induced human platelet aggregation by cordycepin in the presence of various concentrations of exogenous $CaCl_2$ has been observed (Cho et al. 2007).

The compound cordycepin produced apoptotic effects in OEC-M1, a human oral squamous cancer cell line (Wu et al. 2007).

IV. Pregnancy and Lactation

No adverse effects on pregnant rats or their fetuses were reported after pregnant rats were administered up to 5 g/kg cultured cordyceps mycelia on gestational days 6 to 16 (Jing et al. 1987).

No information on the safety of cordyceps during lactation was identified.

V. Toxicity Studies

Acute Toxicity

The LD_{50} of intraperitoneally administered cordyceps in mice is 21 g/kg. Symptoms of overdose included generalized inhibition, followed by excitation, convulsions, spasms, and respiratory depression (Chen and Chen 2004). Intraperitoneal administration of a macerate extract of cordyceps was lethal to some mice at a dose of 5 g/kg and was lethal to all treated mice at a dose of 30–50 g/kg (Zhu 1998). A macerate of cordyceps administered intravenously or subcutaneously had an inhibitory effect on mice and rabbits, with large doses inducing rapid respiration and pulse, then spasms resulting in death. The boiled extract was reported as nontoxic (Zhu 1998).

Cordyceps has been well tolerated in mice administered doses of 45 g/kg (route of administration not noted, likely oral) (Chen and Chen 2004).

Subchronic Toxicity

No significant changes in organ weights, blood parameters, or body weight were reported in rats administered 5 or 20 g/kg Cs-4 as part of the diet for 3 months (IMM 1996). No changes in blood parameters or liver or renal function were observed in dogs administered 3 g/kg Cs-4 as part of the diet for 3 months (IMM 1996).

No adverse effects were observed in rabbits orally administered Cs-B_{414} at a dose of 10 g/kg daily for 3 months. No changes in blood parameters, liver or renal function, and organ weights were observed, except an increase in the weight of the testes in treated males, which was associated with a significant increase in sperm count (Huang et al. 1987).

Genotoxicity

No genotoxic or mutagenic effects of Cs-B_{414} were observed in rabbits or in the Ames test with *Salmonella typhimurium* (Zhu et al. 1998).

Literature Cited

Balon, T.W., A.P. Jasman, and J.S. Zhu. 2002. A fermentation product of *Cordyceps sinensis* increases whole-body insulin sensitivity in rats. *J. Altern. Complement. Med.* 8(3):315-323.

Bao, Z., Z. Wu, and F. Zheng. 1994. Amelioration of aminoglycoside nephrotoxicity by *Cordyceps sinensis* in old patients. *Chin J. Integ. Trad. West. Med. (Chung-Kuo Chung Hsi i Chieh Ho Tsa Chih)* 14(5):271-273.

Bensky, D., S. Clavey, and E. Stöger. 2004. *Chinese herbal medicine: Materia medica.* 3rd ed. Seattle: Eastland Press.

Chen, J.K., and T.T. Chen. 2004. *Chinese medical herbology and pharmacology.* City of Industry, CA: Art of Medicine Press.

Cho, H.J., J.Y. Cho, M.H. Rhee, C.R. Lim, and H.J. Park. 2006. Cordycepin (3′-deoxyadenosine) inhibits human platelet aggregation induced by U46619, a TXA_2 analogue. *J. Pharm. Pharmacol.* 58(12):1677-1682.

Cho, H.J., J.Y. Cho, M.H. Rhee, and H.J. Park. 2007. Cordycepin (3′-deoxyadenosine) inhibits human platelet aggregation in a cyclic AMP- and cyclic GMP-dependent manner. *Eur. J. Pharmacol.* 558(1-3):43-51.

Hockaday, T.D.R. 2002. Two herbal preparations, Cordyceps Cs4 and Cogent db: Do they act on blood glucose, insulin sensitivity, and diabetes as "viscous dietary fibers?" *J. Altern. Complement. Med.* 8(4):403-405.

Huang, Y., J. Lu, B. Zhu, et al. 1987. Toxicity study of fermentation *Cordyceps* mycelia B414. *Zhongchengyao Yanjiu* 10:24-25. *Cited in* Zhu, J.S., G.M. Halpern, and K. Jones. 1998. The scientific rediscovery of a precious ancient Chinese herbal regimen: *Cordyceps sinensis*: Part II. *J. Altern. Complement. Med.* 4(4):429-457.

IMM. 1996. Institute of Materia Medica. Clinical application of fermented *Cordyceps sinensis* Cs-4 (Part III): Toxicology. Unpublished report. *Cited in* Zhu, J.S., G.M. Halpern, and K. Jones. 1998. The scientific rediscovery of a precious ancient Chinese herbal regimen: *Cordyceps sinensis*: Part II. *J. Altern. Complement. Med.* 4(4):429-457.

Jing, A., Q. Tao, and Y. Zhang. 1987. Studies on teratogenicity of mycelial powder of *Cephalosporium sinensis* of *Cordyceps*. *Trad. Chin. Mater. Med.* 18(7):45.

Li, S.P., G.H. Zhang, Q. Zeng, et al. 2006. Hypoglycemic activity of polysaccharide, with antioxidation, isolated from cultured *Cordyceps* mycelia. *Phytomedicine* 13(6):428-433.

Nakamura, K., K. Konoha, Y. Yamaguchi, et al. 2003. Combined effects of *Cordyceps sinensis* and methotrexate on hematogenic lung metastasis in mice. *Receptors Channels* 9(5):329-334.

Wu, T.N., K.C. Yang, C.M. Wang, et al. 1996. Lead poisoning caused by contaminated *Cordyceps*, a Chinese herbal medicine: Two case reports. *Sci. Total Env.* 182(1-3):193-195.

Wu, W.C., J.R. Hsiao, Y.Y. Lian, C.Y. Lin, and B.M. Huang. 2007. The apoptotic effect of cordycepin on human OEC-M1 oral cancer cell line. *Cancer Chemother. Pharmacol.* 60(1):103-111.

Xu, F. 1992. Pharmaceutical studies of submerged culture of *Cordyceps* mycelia in China. *Chin. Pharmaceut. J.* 27(4):195-197.

Xu, F., J.B. Huang, L. Jiang, J. Xu, and J. Mi. 1995. Amelioration of cyclosporin nephrotoxicity by *Cordyceps sinensis* in kidney-transplanted recipients. *Nephrol. Dial. Transplant.* 10(1):142-143.

Zhang, G., Y. Huang, Y. Bian, et al. 2006. Hypoglycemic activity of the fungi *Cordyceps militaris, Cordyceps sinensis, Tricholoma mongolicum*, and *Omphalia lapidescens* in streptozotocin-induced diabetic rats. *Appl. Microbiol. Biotechnol.* 72 (6):1152-1156.

Zhang, Z., W. Huang, S. Liao, et al. 1995. Clinical and laboratory studies of Jin Shui Bao in scavenging oxygen free radicals in elderly senescent Xu-Zheng patients. *J. Admin. Trad. Chin. Med.* 5:14-18.

Zhao, C.S., W.T. Yin, J.Y. Wang, et al. 2002. CordyMax Cs-4 improves glucose metabolism and increases insulin sensitivity in normal rats. *J. Altern. Complement. Med.* 8(3):309-314.

Zhao, X., and L. Li. 1993. *Cordyceps sinensis* in protection of the kidney from cyclosporine A nephrotoxicity. *Zhonghua yi xue za zhi* 73(7):410-412, 447.

Zhu, J.S., G.M. Halpern, and K. Jones. 1998. The scientific rediscovery of a precious ancient Chinese herbal regimen: *Cordyceps sinensis*: Part II. *J. Altern. Complement. Med.* 4(4):429-457.

Zhu, Y.-P. 1998. *Chinese materia medica: Chemistry, pharmacology and applications.* Amsterdam: Harwood Academic Publishers.

Coriandrum sativum L. — Apiaceae

SCN: coriander (fruit)
AN: *dhanyaka*
PN: *yuan sui zi* (fruit)
OCN: Chinese parsley; culantro
Part: fruit (commonly known as "seed")

Quick Reference Summary

Safety Class: 1
Interaction Class: A

Contraindications

None known.

Other Precautions

None known.

Drug and Supplement Interactions

None known.

Adverse Events and Side Effects

Allergic reactions to coriander fruit, including anaphylactic reactions, have been reported and confirmed by patch testing (Manzanedo et al. 2004).

Pharmacological Considerations

Coriander fruit has traditionally been used in the treatment of diabetes, and animal studies have demonstrated that it may modify glucose regulation (Aissaoui et al. 2008; Eddouks et al. 2002; Jabeen et al. 2009; Otoom et al. 2006; Srinivasan 2005). People with diabetes are advised to monitor their blood sugar closely and discuss the use of coriander fruit with a qualified healthcare practitioner prior to use.

Diuretic activity of coriander fruit has been observed in animal studies (Aissaoui et al. 2008; Jabeen et al. 2009).

Pregnancy and Lactation

In animal studies, no adverse effects of coriander fruit or oil on fetal development have been observed (Al-Said et al. 1987; Burdock and Carabin 2009; Vollmuth et al. 1990), although one study showed anti-implantation activity of a water extract of coriander fruit (Al-Said et al. 1987). The no-observed-adverse-effect level (NOAEL) of the oil for pregnant animals was estimated as 250 mg/kg daily and the NOAEL for fetuses was 500 mg/kg daily (Burdock and Carabin 2009; Vollmuth et al. 1990).

No information on the safety of coriander during lactation was identified. While this review did not identify any concerns for use while nursing, safety has not been conclusively established.

Review Details

I. Drug and Supplement Interactions

Clinical Trials of Drug or Supplement Interactions

No clinical trials of drug or supplement interactions were identified.

Case Reports of Suspected Drug or Supplement Interactions

No case reports of suspected drug or supplement interactions were identified.

Animal Trials of Drug or Supplement Interactions

No animal trials of drug or supplement interactions were identified.

II. Adverse Events

Case Reports of Adverse Events

Allergic reactions to coriander fruit, including anaphylactic reactions, have been reported and confirmed by patch testing (Manzanedo et al. 2004). Occupational protein contact dermatitis and occupational asthma from coriander exposure have been reported (Kanerva and Soini 2001; Sastre et al. 1996).

III. Pharmacology and Pharmacokinetics

Human Pharmacological Studies

Coriander has traditionally been used in the treatment of diabetes (Eddouks et al. 2002; Otoom et al. 2006; Srinivasan 2005).

Animal Pharmacological Studies

Significant hypoglycemic activity of coriander fruit was observed in rats fed a diet containing 10% coriander seed daily for 90 days (Chithra and Leelamma 1999). An increase in insulin secretion was observed in diabetic mice fed a diet containing coriander fruit (62.5 g/kg approximate daily dose) and given drinking water containing coriander fruit (2.5 g/l) (Gray and Flatt 2007).

Dose-dependent diuretic activity of a coriander fruit extract was observed in rats intraperitoneally administered doses of 30 or 100 mg/kg, with the 30 mg/kg dose characterized as a mild diuretic and the 100 mg/kg dose reported as a "significant" diuretic, although the effects were less than that of furosemide, the control drug used in the study (Jabeen et al. 2009). Diuretic activity was observed in anesthetized rats administered a continuous intravenous infusion of an aqueous extract of coriander fruit at doses of 40 or 100 mg/kg for 2 h. Coriander fruit dose-dependently increased diuresis, excretion of electrolytes, and glomerular filtration rate but was less potent as a diuretic and saluretic than furosemide (Aissaoui et al. 2008).

A reduction in arterial blood pressure, partially blocked by atropine, was observed in anesthetized rats intravenously administered a water–methanol extract of coriander fruit (Jabeen et al. 2009).

In Vitro Pharmacological Studies

No adverse effects on immune function were observed in the plaque-forming cell and host-resistance assays in mice orally administered coriander oil at doses up to 1250 mg/kg daily for 5 days (Gaworski et al. 1994).

IV. Pregnancy and Lactation

In rats orally administered 250, 500, or 1000 mg/kg coriander essential oil daily 7 days prior to mating through 4 days after birth, at the highest dose a decrease in body weight and food consumption were observed along with a decrease in gestation index, length of gestation, and litter size. The only effect on offspring was a decrease in viability of pups at 1000 mg/kg/day. The authors of the study indicated that the maternal no-observed-adverse-effect level (NOAEL) of the oil was 250 mg/kg daily and the developmental NOAEL was 500 mg/kg daily (Burdock and Carabin 2009; Vollmuth et al. 1990).

In pregnant rats orally administered 250 or 500 mg/kg of an aqueous extract of fresh coriander fruit daily for at least 20 days, a dose-dependent anti-implantation effect was observed, but treatment failed to produce complete infertility. On the fifth day of treatment, a significant decrease in the serum progesterone level was noted. The investigators indicated this decrease might be responsible for the anti-implantation effects of the extract. No adverse effects on fetal development were observed, including changes in the weight or length of the fetuses, nor any changes in organs (Al-Said et al. 1987).

No information on the safety of coriander during lactation was identified.

V. Toxicity Studies

Acute Toxicity

The LD_{50} of an orally administered water-methanol extract of coriander fruit in mice could not be determined at doses up to 10 g/kg (Jabeen et al. 2009). The LD_{50} of orally administered coriander essential oil in rats is 4.13 g/kg, whereas the dermal LD_{50} could not be determined at doses up to 5 g/kg in rabbits (Hart 1971).

Short-Term Toxicity

In rats orally administered 160, 400, or 1000 mg/kg coriander essential oil daily for 28 days, increases in absolute and relative kidney weights were observed in the high-dose group and in the mid-dose males. Absolute and relative liver weights were increased in the mid- and high-dose groups. Some histological changes were observed in the kidneys of high-dose males and in the livers of high-dose females. No treatment-related effects on survival, clinical observations, body weights, or food consumption were noted. The study suggested that the no-observed-effect level (NOEL) for coriander essential oil was 160 mg/kg daily for male rats and less than that for female rats (Letizia et al. 2003).

Genotoxicity

Some mutagenic activity of an ethanol extract of coriander fruit was observed in the Ames test in *Salmonella typhimurium* strains TA98 and TA100 (Mahmoud et al. 1992). In streptomycin-dependent strains of *S. typhimurium* TA98, some mutagenic activity of an alcohol extract of coriander was observed with but not without metabolic activation by S9 (Shashikanth and Hosono 1987). No mutagenic activity of hot water, methanol, or hexane extracts of coriander fruit was observed in the Ames test in *S. typhimurium* strains TA98 and TA100 with or without metabolic activation (Bersani et al. 1981; Higashimoto et al. 1993).

No genotoxic activity was observed in cultured rat embryo fibroblast cells treated with an alcohol extract of coriander (plant part not stated) in the comet assay with treatments up to 1020 mg (Heibatullah et al. 2008).

No clastogenic activity of coriander essential oil was observed in the chromosomal aberration test in vitro using a Chinese hamster fibroblast cell line treated with concentrations of the oil up to 0.125 mg/ml (Ishidate et al. 1984).

No genotoxic activity was observed in the yeast *Saccharomyces cerevisiae* treated with coriander essential oil (Bakkali et al. 2005).

Literature Cited

Aissaoui, A., J. El-Hilaly, Z.H. Israili, and B. Lyoussi. 2008. Acute diuretic effect of continuous intravenous infusion of an aqueous extract of *Coriandrum sativum* L. in anesthetized rats. *J. Ethnopharmacol.* 115(1):89-95.

Al-Said, M.S., K.I. Al-Khamis, M.W. Islam, et al. 1987. Post-coital antifertility activity of the seeds of *Coriandrum sativum* in rats. *J. Ethnopharmacol.* 21(2):165-173.

Bakkali, F., S. Averbeck, D. Averbeck, A. Zhiri, and M. Idaomar. 2005. Cytotoxicity and gene induction by some essential oils in the yeast *Saccharomyces cerevisiae*. *Mutat. Res.* 585(1-2):1-13.

Bersani, C., C. Cantoni, and G. Soncini. 1981. Ames test evaluation of mutagenic activity in essences and spices. *Arch. Vet. Ital.* 32:10-11.

Burdock, G.A., and I.G. Carabin. 2009. Safety assessment of coriander (*Coriandrum sativum* L.) essential oil as a food ingredient. *Food Chem. Toxicol.* 47(1):22-34.

Chithra, V., and S. Leelamma. 1999. *Coriandrum sativum*—Mechanism of hypoglycemic action. *Food Chem.* 67(3):229-231.

Eddouks, M., M. Maghrani, A. Lemhadri, M.L. Ouahidi, and H. Jouad. 2002. Ethnopharmacological survey of medicinal plants used for the treatment of diabetes mellitus, hypertension and cardiac diseases in the south-east region of Morocco (Tafilalet). *J. Ethnopharmacol.* 82(2-3):97-103.

Gaworski, C.L., T.A. Vollmuth, M.M. Dozier, et al. 1994. An immunotoxicity assessment of food flavouring ingredients. *Food Chem. Toxicol.* 32(5):409.

Gray, A.M., and P.R. Flatt. 2007. Insulin-releasing and insulin-like activity of the traditional anti-diabetic plant *Coriandrum sativum* (coriander). *Br. J. Nutrit.* 81(03):203-209.

Hart, E.P. 1971. Report to the research institute for fragrance materials. *In* Burdock, G.A., and I.G. Carabin. 2009. Safety assessment of coriander (*Coriandrum sativum* L.) essential oil as a food ingredient. *Food Chem. Toxicol.* 47(1):22-34.

Heibatullah, K., P. Marzieh, I. Arefeh, and M. Ebrahim. 2008. Genotoxicity determinations of coriander drop and extract of *Coriander sativum* in cultured fibroblast of rat embryo by comet assay. *Saudi Pharmaceut. J.* 16:85-88.

Higashimoto, M., J. Purintrapiban, K. Kataoka, et al. 1993. Mutagenicity and antimutagenicity of extracts of three spices and a medicinal plant in Thailand. *Mutat. Res.* 303(3):135-142.

Ishidate, M., T. Sofuni, K. Yoshikawa, et al. 1984. Primary mutagenicity screening of food additives currently used in Japan. *Food Chem. Toxicol.* 22(8):623-636.

Jabeen, Q., S. Bashir, B. Lyoussi, and A.H. Gilani. 2009. Coriander fruit exhibits gut modulatory, blood pressure lowering and diuretic activities. *J. Ethnopharmacol.* 122(1):123-130.

Kanerva, L., and M. Soini. 2001. Occupational protein contact dermatitis from coriander. *Contact Dermat.* 45(6):354-355.

Letizia, C.S., J. Cocchiara, J. Lalko, and A.M. Api. 2003. Fragrance material review on linalool. *Food Chem. Toxicol.* 41:943-964.

Mahmoud, I., A. Alkofahi, and A. Abdelaziz. 1992. Mutagenic and toxic activities of several spices and some Jordanian medicinal plants. *Int. J. Pharmacogn.* 30(2):81-85.

Manzanedo, L., J. Blanco, M. Fuentes, M.L. Caballero, and I. Moneo. 2004. Anaphylactic reaction in a patient sensitized to coriander seed. *Allergy* 59(3):362-363.

Otoom, S.A., S.A. Al-Safi, Z.K. Kerem, and A. Alkofahi. 2006. The use of medicinal herbs by diabetic Jordanian patients. *J. Herb. Pharmacother.* 6(2):31-41.

Sastre, J., M. Olmo, A. Novalvos, D. Ibanez, and C. Lahoz. 1996. Occupational asthma due to different spices. *Allergy* 51(2):117-120.

Shashikanth, K.N., and A. Hosono. 1987. Screening of streptomycin-dependent strains of *Salmonella typhimurium* and *Escherichia coli* for in vitro detection of spice-induced mutagenicity. *Lebensmittel-Wissenschaft Technol.* 20(2):91-94.

Srinivasan, K. 2005. Plant foods in the management of diabetes mellitus: Spices as beneficial antidiabetic food adjuncts. *Int. J. Food. Sci. Nutr.* 56(6):399-414.

Vollmuth, T.A., M.B. Bennett, A.M. Hoberman, and M.S. Christian. 1990. An evaluation of food flavoring ingredients using an in vivo reproductive and developmental toxicity screening test. *Teratology* 41:597-598.

Cornus officinalis Siebold & Zucc. — Cornaceae

C

SCN: Asiatic dogwood
Syn: *Macrocarpium officinale* (Siebold & Zucc.) Nakai
PN: *shan zhu yu* (fruit without seed)
OCN: Asiatic cornel; Asiatic cornelian cherry; Japanese cornel
Part: fruit

Quick Reference Summary

Safety Class: 1
Interaction Class: A

Contraindications

None known.

Other Precautions

Use with caution in persons with painful or difficult urination (Bensky et al. 2004).

Drug and Supplement Interactions

None known.

Adverse Events and Side Effects

None known.

Pharmacological Considerations

None known.

Pregnancy and Lactation

No information on the safety of Asiatic dogwood in pregnancy or lactation was identified in the scientific or traditional literature. Although this review did not identify any concerns for use while pregnant or nursing, safety has not been conclusively established.

Review Details

I. Drug and Supplement Interactions

Clinical Trials of Drug or Supplement Interactions

No clinical trials of drug or supplement interactions were identified.

Case Reports of Suspected Drug or Supplement Interactions

No case reports of suspected drug or supplement interactions were identified.

Animal Trials of Drug or Supplement Interactions

No animal trials of drug or supplement interactions were identified.

II. Adverse Events

Case Reports of Adverse Events

No case reports of adverse events were identified.

III. Pharmacology and Pharmacokinetics

Human Pharmacological Studies

No relevant human pharmacological studies were identified.

Animal Pharmacological Studies

Glucose-modulating activity was observed in diabetic rats administered an ether extract of Asiatic dogwood (Yamahara et al. 1981).

In Vitro Pharmacological Studies

No estrogenic or antiestrogenic activity of an ethanol extract of Asiatic dogwood was observed in a recombinant yeast system featuring a human estrogen receptor expression plasmid and a reporter plasmid (Kim et al. 2008).

IV. Pregnancy and Lactation

No information on the safety of Asiatic dogwood in pregnancy or lactation was identified.

V. Toxicity Studies

Acute Toxicity

In mice, a study of anti-inflammatory activity utilized a dose of 5 g/kg of Asiatic dogwood with no adverse effects indicated (Chen and Chen 2004).

Genotoxicity

No mutagenic activity of water or methanol extracts of Asiatic dogwood were observed in the *Bacillus subtilis rec* assay or the Ames test with *Salmonella typhimurium* strains TA98 and TA100 with or without metabolic activation (Morimoto et al. 1982).

Literature Cited

Bensky, D., S. Clavey, and E. Stöger. 2004. *Chinese herbal medicine: Materia medica*. 3rd ed. Seattle: Eastland Press.

Chen, J.K., and T.T. Chen. 2004. *Chinese medical herbology and pharmacology*. City of Industry, CA: Art of Medicine Press.

Kim, I.G., S.C. Kang, K.C. Kim, E.S. Choung, and O.P. Zee. 2008. Screening of estrogenic and antiestrogenic activities from medicinal plants. *Env. Toxicol. Pharmacol.* 25(1):75-82.

Morimoto, I., F. Watanabe, T. Osawa, T. Okitsu, and T. Kada. 1982. Mutagenicity screening of crude drugs with *Bacillus subtilis* rec-assay and *salmonella*/microsome reversion assay. *Mutat. Res.* 97:81-102.

Yamahara, J., H. Mibu, T. Sawada, et al. 1981. Biologically active principles of crude drugs. Antidiabetic principles of Corni fructus in experimental diabetes induced by streptozotocin. *Yakugaku Zasshi* 101(1):86-90.

C

Corydalis yanhusuo W.T. Wang

Papaveraceae

SCN: *Corydalis yanhusuo*
Syn: *Corydalis turtschaninovii* Bess. f. *yanhusuo* Y.H. Chou & C.C. Hsu
PN: *yan hu suo* (tuber)
OCN: Chinese fumewort
Part: tuber

Quick Reference Summary

Safety Class: 2b
Interaction Class: A

Contraindications

Not for use in pregnancy except under the supervision of a qualified healthcare practitioner (Bensky et al. 2004; Chen and Chen 2004).

Other Precautions

None known.

Drug and Supplement Interactions

None known.

Notice

Uterine stimulant (List and Hörhammer 1973); *see* Appendix 2.

Berberine (0.005–0.057%) (Ding et al. 2007; Zhang et al. 2009); *see* Appendix 1.

Adverse Events and Side Effects

None known.

Pharmacological Considerations

Allergic reactions to *Corydalis yanhusuo* have been reported (Bensky et al. 2004).

Pregnancy and Lactation

Reference texts on traditional Chinese medicine indicate that *Corydalis yanhusuo* should not be used in pregnancy (Bensky et al. 2004; Chen and Chen 2004), with one text making an exception for "exceptional circumstances" (Bensky et al. 2004). Based on this information, use during pregnancy is not recommended except under the supervision of a qualified healthcare practitioner.

No information on the safety of *Corydalis yanhusuo* during lactation was identified. While this review did not identify any concerns for use while nursing, safety has not been conclusively established.

Review Details

I. Drug and Supplement Interactions

Clinical Trials of Drug or Supplement Interactions

No clinical trials of drug or supplement interactions were identified.

Case Reports of Suspected Drug or Supplement Interactions

No case reports of suspected drug or supplement interactions were identified.

Animal Trials of Drug or Supplement Interactions

No animal trials of drug or supplement interactions were identified.

II. Adverse Events

Case Reports of Adverse Events

Although no adverse effects are expected at typical therapeutic doses, persons taking large doses (10–15 g) of *Corydalis yanhusuo* may develop drowsiness, dizziness, and abdominal distention (Bensky et al. 2004).

Allergic reactions to *Corydalis yanhusuo* have been reported, including drug fever, erythema, and pruritus, with nausea, dizziness, shortness of breath, and numbness of lips and extremities (Bensky et al. 2004).

III. Pharmacology and Pharmacokinetics

Human Pharmacological Studies

No relevant human pharmacological studies were identified.

Animal Pharmacological Studies

No relevant animal pharmacological studies were identified.

In Vitro Pharmacological Studies

No relevant in vitro pharmacological studies were identified.

IV. Pregnancy and Lactation

Reference texts on traditional Chinese medicine indicate that *Corydalis yanhusuo* should not be used in pregnancy (Bensky et al. 2004; Chen and Chen 2004), with one text making an exception for "exceptional circumstances" (Bensky et al. 2004).

No information on the safety of *Corydalis yanhusuo* during lactation was identified.

V. Toxicity Studies

Acute Toxicity

The LD_{50} of orally administered *Corydalis yanhusuo* extract in mice was approximately 100 g/kg (Chen and Chen 2004). At this dose, side effects included decreased blood pressure, heart rate, and respiration along with sedation and tremor (Chen and Chen 2004).

The toxic dose of *Corydalis yanhusuo* in humans has been reported as 60–120 g. Symptoms of poisoning are reported to appear 1 to 4 hours after ingestion and include dizziness, facial pallor, drowsiness, weakness, dyspnea, spasms, shock, low blood pressure, and weak pulse. In severe cases of shock, tetanic convulsion and respiratory inhibition have been reported (Bensky et al. 2004).

Literature Cited

Bensky, D., S. Clavey, and E. Stöger. 2004. *Chinese herbal medicine: Materia medica.* 3rd ed. Seattle: Eastland Press.

Chen, J.K., and T.T. Chen. 2004. *Chinese medical herbology and pharmacology*. City of Industry, CA: Art of Medicine Press.

Ding, B., T. Zhou, G. Fan, Z. Hong, and Y. Wu. 2007. Qualitative and quantitative determination of ten alkaloids in traditional Chinese medicine *Corydalis yanhusuo* WT Wang by LC-MS/MS and LC-DAD. *J. Pharmaceut. Biomed. Anal.* 45(2):219-226.

List, P.H., and H. Hörhammer. 1973. *Hagers handbuch der pharmazeutischen praxis*. Berlin: Springer.

Zhang, J., Y. Jin, J. Dong, et al. 2009. Systematic screening and characterization of tertiary and quaternary alkaloids from *Corydalis yanhusuo* W. T. Wang using ultra-performance liquid chromatography-quadrupole-time-of-flight mass spectrometry. *Talanta* 78(2):513-522.

Corylus spp. — Betulaceae

Corylus avellana L.
SCN: European hazel
OCN: European filbert
Corylus cornuta Marsh.
SCN: beaked hazel
OCN: beaked filbert
Part: bark, leaf

Quick Reference Summary

Safety Class: 1
Interaction Class: A

Contraindications

None known.

Other Precautions

None known.

Drug and Supplement Interactions

None known.

Notice

Tannins (Amarowicz et al. 2008; Fraisse et al. 1999); *see* Appendix 1.

Adverse Events and Side Effects

Allergic reactions to the nuts and pollen of European hazel and beaked hazel have been reported (Bozkurt et al. 2005; Peroni et al. 2007; Soyer and Sekerel 2008), although no reports of allergic reactions to the leaf or bark were identified.

Pharmacological Considerations

None known.

Pregnancy and Lactation

No information on the safety of European hazel or beaked hazel in pregnancy or lactation was identified in the scientific or traditional literature. Although this review did not identify any concerns for use while pregnant or nursing, safety has not been conclusively established.

C

Review Details

I. Drug and Supplement Interactions

Clinical Trials of Drug or Supplement Interactions

No clinical trials of drug or supplement interactions were identified.

Case Reports of Suspected Drug or Supplement Interactions

No case reports of suspected drug or supplement interactions were identified.

Animal Trials of Drug or Supplement Interactions

No animal trials of drug or supplement interactions were identified.

II. Adverse Events

Case Reports of Adverse Events

Allergic reactions to the nuts and pollen of European hazel and beaked hazel have been reported (Bozkurt et al. 2005; Peroni et al. 2007; Soyer and Sekerel 2008), although no reports of allergic reactions to the leaf or bark were identified.

III. Pharmacology and Pharmacokinetics

Human Pharmacological Studies

No relevant human pharmacological studies were identified.

Animal Pharmacological Studies

No relevant animal pharmacological studies were identified.

In Vitro Pharmacological Studies

No relevant in vitro pharmacological studies were identified.

IV. Pregnancy and Lactation

No information on the safety of European hazel or beaked hazel during pregnancy or lactation was identified.

V. Toxicity Studies

No toxicity studies were identified.

Literature Cited

Amarowicz, R., G.A. Dykes, and R.B. Pegg. 2008. Antibacterial activity of tannin constituents from *Phaseolus vulgaris, Fagoypyrum esculentum, Corylus avellana* and *Juglans nigra*. *Fitoterapia* 79(3):217-219.

Bozkurt, B., G. Karakaya, and A.F. Kalyoncu. 2005. Food hypersensitivity in patients with seasonal rhinitis in Ankara. *Allergol. Immunopathol. (Madrid)* 33(2):86-92.

Fraisse, D., A. Carnat, A.P. Carnat, and J.L. Lamaison. 1999. Standardization of hazel leaf. *Ann. Pharm. Fr.* 57(5):406-409.

Peroni, D.G., A. Dall'Agnola, G.L. Piacentini, and A.L. Boner. 2007. Worsening of atopic dermatitis by hazelnut essence contained in hydroxyzine syrup. *Acta Paediatr.* 96(11):1710.

Soyer, O.U., and B.E. Sekerel. 2008. Food dependent exercise induced anaphylaxis or exercise induced anaphylaxis? *Allergol. Immunopathol. (Madrid)* 36(4):242-243.

Crataegus spp. — Rosaceae

Crataegus laevigata (Poir.) DC.
SCN: hawthorn
Syn: *Crataegus oxyacantha* auct.
OCN: English hawthorn; May tree; white thorn

Crataegus monogyna Jacq.
SCN: hawthorn
OCN: English hawthorn; one-seed hawthorn
Part: flower, leaf

Quick Reference Summary

Safety Class: 1
Interaction Class: A

Contraindications

None known.

Other Precautions

None known.

Drug and Supplement Interactions

Some experience suggests that the dose of cardiac drugs may be reduced when hawthorn flower and leaf are taken concomitantly (Ammon and Haendel 1981; Bauer and Hölscher 1992; Blesken 1992; van Hellemont and Delfosse 1988). No interaction with digoxin was observed in a human study (Tankanow et al. 2003).

Adverse Events and Side Effects

In meta-analyses and systematic reviews of clinical trials, hawthorn flower and leaf extracts have been characterized as well tolerated in clinical trials, with few adverse events reported (Daniele et al. 2006; Pittler et al. 2003).

Pharmacological Considerations

A preliminary human study indicated a lack of interaction of an extract of dried flowering tops of hawthorn with low-dose insulin, metformin, gliclazide, ACE inhibitors, calcium channel blockers, β-blockers, and diuretics (Walker et al. 2006).

Pregnancy and Lactation

An animal study indicated no adverse effects of an ethanol extract of hawthorn leaf administered at a dose of 2.8 g/kg, noted as 56 times the human dose, daily on days 1–8 or 8–15 of gestation (Yao et al 2008). No adverse effects of hawthorn leaf and flower on pregnancy or fetal development have been observed in a limited number of additional animal studies (ESCOP 2003; Manolov and Daleva 1969).

No information on the safety of hawthorn leaf and flower during lactation was identified. While this review did not identify any concerns for use while nursing, safety has not been conclusively established.

Review Details

I. Drug and Supplement Interactions

Clinical Trials of Suspected Drug or Supplement Interactions

A clinical trial with eight healthy volunteers indicated that administration of 900 mg of hawthorn leaf and flower extract daily for 3 weeks had no effect on the pharmacokinetics of digoxin (Tankanow et al. 2003).

In a randomized trial of type 2 diabetes patients taking prescribed drugs, administration of 1200 mg daily of a 3:1 hawthorn leaf and flower extract for 16 weeks resulted in a moderate decrease in blood pressure with no adverse interactions reported. Prescription drugs being taken by the trial participants were hypoglycemic medications (low-dose insulin, metformin, gliclazide, and "others") and hypotensive medication (ACE inhibitors, calcium channel blockers, β-blockers, diuretics, and "others") (Walker et al. 2006).

Older studies indicated that hawthorn leaf and flower extracts could potentiate the effects of digitalis glycosides (Trunzler and Schuler 1962), or the coronary vasodilating effects of drugs such as caffeine, papaverine, adenosine, sodium nitrate, and epinephrine (Hahn et al. 1960). In the 1960s and 1970s, however, hawthorn leaf and flower extracts were often combined with conventional cardiac drugs including digitalis. Such combinations were reported to reduce the side effects of the drugs and sometimes allowed for a reduction in dose of the drug (Ammon and Haendel 1981; Bauer and Hölscher 1992; Blesken 1992; Upton 1999).

Case Reports of Drug or Supplement Interactions

No case reports of suspected drug or supplement interactions were identified.

Animal Trials of Drug or Supplement Interactions

Coadministration of hawthorn leaf and flower extracts with barbiturates in mice resulted in significantly increased sleeping time as compared to barbiturates alone (Della Loggia et al. 1982).

II. Adverse Events

Adverse Events Reported in Clinical Trials

A meta-analysis of hawthorn leaf and flower extract monopreparations, including 13 randomized, double-blind, and placebo-controlled clinical trials, indicated that adverse events reported were infrequent, mild, and transient. Adverse events reported in persons taking hawthorn included nausea, dizziness, and cardiac and gastrointestinal complaints. The number and nature of adverse events in placebo groups were not identified (Pittler et al. 2003).

A systematic review of adverse events in clinical trials, uncontrolled studies, observational studies, and case reports associated with hawthorn leaf and flower monopreparations, including 22 studies with a total of 7080 participants, concluded that hawthorn is generally well tolerated. Adverse events noted in the trials were similar in the hawthorn and placebo groups. The daily doses of hawthorn preparation in the clinical trials ranged from 160 to 1800 mg and the duration of treatment ranged from 3 to 24 weeks (Daniele et al. 2006).

Case Reports of Adverse Events

No case reports of adverse events were identified.

III. Pharmacology and Pharmacokinetics

Human Pharmacological Studies

No relevant human pharmacological studies were identified.

Animal Pharmacological Studies

No relevant animal pharmacological studies were identified.

In Vitro Pharmacological Studies

No relevant in vitro pharmacological studies were identified.

IV. Pregnancy and Lactation

In a study with rats orally administered 2.8 g/kg daily of an ethanol extract of hawthorn leaf on gestational days 1–8 or 8–15, examination of fetuses showed no adverse effects on fetal development. The dose used was calculated as 56 times the recommended human dose (Yao et al. 2008).

No adverse effects were observed in offspring of mice fed a flavonoid mixture obtained from the leaf of hawthorn at a dose of 100 mg/kg daily for 1 month (Manolov and Daleva 1969). No teratogenic effects were observed in pregnant rats and rabbits orally administered 1.6 g/kg of a hydroethanolic extract of hawthorn leaf and flower. The same study showed no postnatal toxicity in rats or their offspring (ESCOP 2003).

No information on the safety of hawthorn leaf and flower during lactation was identified.

V. Toxicity Studies

Acute Toxicity

No signs of clinical toxicity or fatalities were observed in rats and mice orally administered up to 3000 mg/kg of hawthorn leaf and flower extract in a single dose (Schlegelmilch and Heywood 1994). No adverse effects were observed in rats orally administered 2.8 g/kg of an ethanol extract of hawthorn leaf daily for 8 days (Yao et al. 2008).

The LD_{50} of intraperitoneally administered hawthorn leaf and flower extract in rats is 750 mg/kg and in mice is 1170 mg/kg. Toxicity signs observed after intraperitoneal administration included sedation, piloerection, dyspnea, and tremor (Schlegelmilch and Heywood 1994). No adverse effects were observed in rats orally administered 2.8 g/kg of an ethanol extract of hawthorn leaf daily for 8 days (Yao et al. 2008).

Subchronic Toxicity

Administration of 300 mg/kg of hawthorn leaf and flower extract daily for 26 weeks to rats and dogs did not produce any signs of clinical toxicity (Schlegelmilch and Heywood 1994).

Genotoxicity and Mutagenicity

No evidence of mutagenicity or clastogenicity of extracts of hawthorn leaf and flower were observed using standard tests, including the Ames test, the mouse micronucleus assay, the mouse lymphoma test, and the human lymphoma test (Schlegelmilch and Heywood 1994).

Literature Cited

Ammon, H.P.T., and M. Haendel. 1981. *Crataegus*, toxicology and pharmacology. Part II: Pharmacokinetics. *Planta Med.* 43:209-239.

Bauer, I., and U. Hölscher. 1992. *In* Hänsel, R., K. Keller, H. Rimpler, and G. Schneider, eds. *Hagers handbuch der pharmazeutischen praxis, Volume 4.* Berlin: Springer.

Blesken, R. 1992. *Crataegus* in cardiology. *Fortschr. Med.* 110(15):290-292.

Daniele, C., G. Mazzanti, M.H. Pittler, and E. Ernst. 2006. Adverse-event profile of *Crataegus* spp.: A systematic review. *Drug Saf.* 29(6):523-535.

Della Loggia, R., A. Tubaro, C. Zilli, and C. Redaelli. 1982. The sedative action of hawthorn. *Planta Med.* 45(3):K7.

ESCOP. 2003. *ESCOP monographs: The scientific foundation for herbal medicinal products.* 2nd ed. Exeter, U.K.: European Scientific Cooperative on Phytotherapy.

Hahn, F., F. Klinkhammer, and A. Oberdorf. 1960. Darstellung und pharmakologische untersuchungen eines neuen therapeutischen wirkstoffes aus *Crataegus oxyacantha*. *Arzneimittelforschung* 10:825-826. *Cited in* Upton R., 1999. *Hawthorn leaf with flower, Crataegus* spp. Santa Cruz, CA: American Herbal Pharmacopoeia.

Manolov, P., and L. Daleva. 1969. Pharmacological study of a preparation based on a flavonoid mixture from *Crataegus monogyna*. *Farmatsiya* 19:38-44.

Pittler, M.H., K. Schmidt, and E. Ernst. 2003. Hawthorn extract for treating chronic heart failure: Meta-analysis of randomized trials. *Am. J. Med.* 114(8):665-674.

Schlegelmilch, R., and R. Heywood. 1994. Toxicity of *Crataegus* (hawthorn) extract (WS 1442). *J. Am. Coll. Toxicol.* 13(2):103-111.

Tankanow, R., H.R. Tamer, D.S. Streetman, et al. 2003. Interaction study between digoxin and a preparation of hawthorn (*Crataegus oxyacantha*). *J. Clin. Pharmacol.* 43(6):637-642.

Trunzler, G., and E. Schuler. 1962. [Comparative studies on the effect of a *Crataegus* extract, digitoxin, digoxin and γ-strophanthin on the isolated mammalian heart.] *Arzneimittelforschung* 12:198-202. *Cited in* Upton R., 1999. *Hawthorn leaf with flower, Crataegus* spp. Santa Cruz, CA: American Herbal Pharmacopoeia.

Upton, R. 1999. *Hawthorn leaf with flower, Crataegus spp.: Analytical, quality control, and therapeutic monograph.* Santa Cruz, CA: American Herbal Pharmacopoeia.

van Hellemont, J., and M. Delfosse. 1988. *Compendium de phytothérapie.* Association Pharmaceutique Belge.

Walker, A.F., G. Marakis, E. Simpson, et al. 2006. Hypotensive effects of hawthorn for patients with diabetes taking prescription drugs: A randomised controlled trial. *Br. J. Gen. Pract.* 56(527):437-443.

Yao, M., H.E. Ritchie, P.D. Brown-Woodman, 2008. A reproductive screening test of hawthorn. *J. Ethnopharmacol.* 118(1):127-132.

Crataegus spp. — Rosaceae

Crataegus laevigata (Poir.) DC.
SCN: hawthorn
Syn: *Crataegus oxyacantha* auct.
OCN: English hawthorn; May tree; white thorn

Crataegus monogyna **Jacq.**
SCN: hawthorn
OCN: English hawthorn; one-seed hawthorn

Crataegus pinnatifida **Bunge**
SCN: Chinese hawthorn
PN: *shan zha* (fruit)
OCN: northern shanzha
Part: fruit

Quick Reference Summary

Safety Class: 1
Interaction Class: A

Contraindications

None known.

Other Precautions

None known.

Drug and Supplement Interactions

In the 1960s and 1970s, hawthorn fruit extracts were sometimes combined with conventional cardiac drugs, including digitalis, reportedly allowing for a reduction in dose of the drug (Ammon and Haendel 1981; Bauer and Hölscher 1992; Blesken 1992).

Editors' Note

Hawthorn fruit has been widely consumed as food, prepared as syrups and jams, and eaten fresh, and is considered to have a safety profile similar to commonly consumed fruits (Upton 1999).

Adverse Events and Side Effects

In meta-analyses and systematic reviews of clinical trials, hawthorn fruit extracts have been characterized as well tolerated in clinical trials with few adverse events reported (Daniele et al. 2006).

Pharmacological Considerations

None known.

Pregnancy and Lactation

No information on the safety of Chinese hawthorn fruit in pregnancy or lactation was identified in the scientific or traditional literature. Although this review did not identify any concerns for use while pregnant or nursing, safety has not been conclusively established.

C

Review Details

I. Drug and Supplement Interactions

Clinical Trials of Suspected Drug or Supplement Interactions

In the 1960s and 1970s, hawthorn fruit extracts were sometimes combined with conventional cardiac drugs including digitalis. Such combinations were reported to reduce the side effects of the drugs and sometimes allowed for a reduction in dose of the drug (Ammon and Haendel 1981; Bauer and Hölscher 1992; Blesken 1992; Upton 1999).

Case Reports of Drug or Supplement Interactions

No case reports of suspected drug or supplement interactions were identified.

Animal Trials of Drug or Supplement Interactions

No animal studies of drug or supplement interactions were identified.

II. Adverse Events

Adverse Events Reported in Clinical Trials

A systematic review of adverse events in clinical trials and case reports associated with hawthorn monopreparations, including two trials of fruit extracts, concluded that hawthorn fruit extracts are generally well tolerated. Adverse events noted in the clinical trials were similar in the hawthorn fruit extracts and placebo groups. The duration of treatment was 8 or 10 weeks (Daniele et al. 2006).

Case Reports of Adverse Events

No case reports of adverse events were identified.

III. Pharmacology and Pharmacokinetics

Human Pharmacological Studies

No relevant human pharmacological studies were identified.

Animal Pharmacological Studies

No relevant animal pharmacological studies were identified.

In Vitro Pharmacological Studies

No relevant in vitro pharmacological studies were identified.

IV. Pregnancy and Lactation

No information on the use of hawthorn fruit during pregnancy or lactation was identified.

V. Toxicity Studies

Hawthorn fruit has been widely consumed as food, eaten fresh, and prepared as syrups and jams, and is considered to have a safety profile similar to commonly consumed fruits (Upton 1999).

Literature Cited

Ammon, H.P.T., and M. Haendel. 1981. *Crataegus*, toxicology and pharmacology. Part II: Pharmacokinetics. *Planta Med.* 43:209-239.

Bauer, I., and U. Hölscher. 1992. *In* Hänsel, R., K. Keller, H. Rimpler, and G. Schneider, eds. *Hagers handbuch der pharmazeutischen praxis, Volume 4 (AD).* Berlin: Springer-Verlag.

Blesken, R. 1992. *Crataegus* in cardiology. *Fortschr. Med.* 110(15):290-292.
Daniele, C., G. Mazzanti, M.H. Pittler, and E. Ernst. 2006. Adverse-event profile of Crataegus spp.: A systematic review. *Drug Saf.* 29(6):523-535.
Upton, R. 1999. *Hawthorn berry, Crataegus spp.: Analytical, quality control, and therapeutic monograph.* Santa Cruz, CA: American Herbal Pharmacopoeia.

C

Crocus sativus L. — Iridaceae

SCN: saffron
AN: *kunkuma*
PN: *fan hong hua* (stigma)
OCN: Spanish saffron, true saffron
Part: stigma

Quick Reference Summary

Safety Class: 2b
Interaction Class: A

Contraindications

Not for use in pregnancy except under the supervision of a qualified healthcare practitioner (Bensky et al. 2004; Chadha 1988; Frohne and Pfänder 2000; List and Hörhammer 1973; Tang and Eisenbrand 1992).

Other Precautions

None known.

Drug and Supplement Interactions

None known.

Notice

Abortifacient (Chadha 1988; Frohne and Pfänder 2000; List and Hörhammer 1973; Tang and Eisenbrand 1992; Wichtl 2004); *see* Appendix 2.

Editors' Note

The classifications and concerns for this herb are based on relatively higher doses used for therapeutic purposes in contrast to lower amounts generally used in cooking, and have not been associated with its use as a spice.

Adverse Events and Side Effects

Overdose of saffron has been associated with vomiting, bleeding, vertigo, confusion, and yellowing of the skin and mucous membranes (mimicking jaundice) (Frank 1961; Wichtl 2004).

Pharmacological Considerations

Adverse effects have been associated with ingestion of 5 g of saffron stigmas (Frank 1961), although the standard dose in traditional Chinese medicine is an aqueous extract of 1.5 to 6 g (Bensky et al. 2004). The German Commission E lists the maximum daily dose as 1.5 g (Wichtl 2004). A human study indicated no adverse effects at doses up to 400 mg daily (Modaghegh et al. 2008). No risk is associated with consumption in standard food use quantities or in therapeutic dosage of less than 1.5 g per day (Wichtl 2004).

Pregnancy and Lactation

Saffron is reported to have a stimulating effect on the smooth muscles of the uterus (Wichtl 2004) and has been used as an abortifacient (Chadha 1988; Wichtl 2004). A reference text on traditional Chinese medicine indicates that saffron is contraindicated in pregnancy (Bensky et al. 2004). Based on this information, use during pregnancy is not recommended except under the supervision of a qualified healthcare practitioner.

No information on the safety of saffron during lactation was identified. While this review did not identify any concerns for use while nursing, safety has not been conclusively established.

Review Details

I. Drug and Supplement Interactions

Clinical Trials of Drug or Supplement Interactions

No clinical trials of drug or supplement interactions were identified.

Case Reports of Suspected Drug or Supplement Interactions

No case reports of suspected drug or supplement interactions were identified.

Animal Trials of Drug or Supplement Interactions

No animal trials of drug or supplement interactions were identified.

II. Adverse Events

Case Reports of Adverse Events

Overdose of saffron (maximum daily dose listed as 1.5 g) has been associated with vomiting, uterine bleeding, bloody diarrhea and urine, bleeding from the nose, lips, and eyelids, convulsions, intestinal colic, vertigo, confusion, and yellowing of the skin and mucous membranes (mimicking jaundice) (Bensky et al. 2004; Wichtl 2004).

Ingestion of 5 g of saffron in an attempted abortion resulted in severe purpura with black necrosis of the nose, thrombocytopenia, hypothrombinemia, uremia, and collapse (Frank 1961).

III. Pharmacology and Pharmacokinetics

Human Pharmacological Studies

In a safety evaluation of saffron in healthy volunteers, administration of 200 or 400 mg of saffron daily for 7 days did not cause any clinically important changes in hematological, biochemical, or electrocardiographic parameters. Observed changes regarded as not clinically significant included a decrease in standing systolic blood pressure and mean arterial pressure at the 400 mg dose, along with a slight decrease in red blood cells, hemoglobin, hematocrit, and platelets and an increase in sodium, blood urea nitrogen, and creatinine (Modaghegh et al. 2008).

Animal Pharmacological Studies

No relevant animal pharmacological studies were identified.

In Vitro Pharmacological Studies

An inhibitory effect on the calcium channel was observed in isolated guinea pig hearts treated with an aqueous-ethanol extract of saffron at concentrations of 1 or 5%, but not at 0.1 or 0.5% (Boskabady et al. 2008).

The compound crocetin demonstrated dose-dependent inhibition of platelet aggregation induced by ADP or collagen but not by arachidonic acid (Yang et al. 2008).

IV. Pregnancy and Lactation

In mice orally administered 100 mg/kg daily of a saffron extract on days 6 to 15 of pregnancy, a delay in bone ossification in the hip, metatarsus, metacarpus, fingers, and sternum was observed in fetuses excised on the 18th day of pregnancy. Decreased body weight of the fetuses, with no major congenital malformations, was observed (Golalipoor et al. 2006, 2008).

No information on the safety of saffron during lactation was identified.

V. Toxicity Studies

Acute Toxicity

The LD_{50} of saffron in rodents is reported as 20.7 g/kg (Nair et al. 1995). The LD_{50} of an ethanol extract of saffron in mice is reported as greater than 600 mg/kg (Nair et al. 1991).

The human lethal dose of saffron has been reported as 20 g (Wichtl 2004).

Short-Term Toxicity

In rats orally administered 0.35, 0.70, or 1.05 g/kg of an ethanol extract of saffron daily for 14 days, dose-dependent reductions in hemoglobin level, hematocrit, and total red blood cell count and an increase in total white blood cell count were observed. Dose-dependent increases in AST, ALT, urea, uric acid, and creatinine values were also seen. Microscopic observation revealed mild to severe hepatic and renal tissue injuries (Mohajeri et al. 2007).

Subchronic Toxicity

In rats subcutaneously administered 400 mg/kg of the compound crocin weekly for 13 weeks, acute tubular necrosis was found in all kidney samples from crocin-treated animals, and slight signs of nephrotoxicity were identified by biochemical analysis of the serum (Garcia-Olmo et al. 1999).

Genotoxicity

No mutagenic effects of ethyl acetate, methanol, or aqueous extracts of saffron were observed in the *Salmonella*/microsome assay using *S. typhimurium* strains TA98 and TA100 with or without metabolic activation (Yamamoto et al. 1982). No mutagenic effects of a 2:1 chloroform-methanol extract of saffron were observed in pig kidney cells or in trophoblastic placenta cells treated at concentrations of 100 mg/plate (Rockwell and Raw 1979).

Antimutagenic activity was observed in mice orally administered 20, 40, or 80 mg/kg of an aqueous extract of saffron three times in 5 days before administration of known mutagens (Premkumar et al. 2001, 2006). Antimutagenic effects of saffron were also observed in the mouse bone marrow micronucleus test with oral doses of 25, 50, and 100 mg/kg (Premkumar et al. 2003), and in the Ames test in *Salmonella typhimurium* strain TA98 (Abdullaev et al. 2003).

Literature Cited

Abdullaev, F.I., L. Riveron-Negrete, H. Caballero-Ortega, et al. 2003. Use of in vitro assays to assess the potential antigenotoxic and cytotoxic effects of saffron (*Crocus sativus* L.). *Toxicol. In Vitro* 17(5-6):731-736.

Bensky, D., S. Clavey, and E. Stöger. 2004. *Chinese herbal medicine: Materia medica.* 3rd ed. Seattle: Eastland Press.

Boskabady, M.H., M.N. Shafei, A. Shakiba, and H.S. Sefidi. 2008. Effect of aqueous-ethanol extract from *Crocus sativus* (saffron) on guinea-pig isolated heart. *Phytother. Res.* 22(3):330-334.
Chadha, Y. 1988. *The wealth of India: A dictionary of Indian raw materials and industrial products*. Delhi: Council of Scientific and Industrial Research.
Frank, A. 1961. Purpura resulting from artificial abortion. *Dtsch. Med. Wochenschr.* 86:1618.
Frohne, D., and H.J. Pfänder. 2000. *A colour atlas of poisonous plants: A handbook for pharmacists, doctors, toxicologists, biologists and veterinarians*. 2nd ed. London: Manson.
Garcia-Olmo, D.C., H.H. Riese, J. Escribano, et al. 1999. Effects of long-term treatment of colon adenocarcinoma with crocin, a carotenoid from saffron (*Crocus sativus* L.): An experimental study in the rat. *Nutr. Cancer* 35(2):120-126.
Golalipoor, M.J., V. Khori, and M. Afshar. 2006. Teratogenic effect of saffron on mice. *Reprod. Toxicol.* 22(2):272.
Golalipour, M.J., A.M. Gharravi, S. Ghafari, M. Afshar, and V. Khori. 2008. Effects of *Crocus sativus* on the fetal development of NMRI mice. *Saudi Med. J.* 29(2):309-310.
List, P.H., and H. Hörhammer. 1973. *Hagers handbuch der pharmazeutischen praxis*. Berlin: Springer.
Modaghegh, M.H., M. Shahabian, H.A. Esmaeili, O. Rajbai, and H. Hosseinzadeh. 2008. Safety evaluation of saffron (*Crocus sativus*) tablets in healthy volunteers. *Phytomedicine* 15(12):1032-1037.
Mohajeri, D., G. Mousavi, M. Mesgari, Y. Doustar, and M.H.K. Nouri. 2007. Subacute toxicity of *Crocus sativus* L. (saffron) stigma ethanolic extract in rats. *Am. J. Pharmacol. Toxicol.* 2(4):189-193.
Nair, S.C., S.K. Kurumboor, and J.H. Hasegawa. 1995. Saffron chemoprevention in biology and medicine: A review. *Cancer Biother.* 10(4):257-264.
Nair, S.C., B. Panikkar, and K.R. Panikkar. 1991. Antitumour activity of saffron. *Cancer Lett.* 57(2):109-114.
Premkumar, K., S.K. Abraham, S.T. Santhiya, P.M. Gopinath, and A. Ramesh. 2001. Inhibition of genotoxicity by saffron (*Crocus sativus* L.) in mice. *Drug Chem. Toxicol.* 24(4):421-428.
Premkumar, K., S.K. Abraham, S.T. Santhiya, and A. Ramesh. 2003. Inhibitory effects of aqueous crude extract of saffron (*Crocus sativus* L.) on chemical-induced genotoxicity in mice. *Asia Pac. J. Clin. Nutr.* 12(4):474-476.
Premkumar, K., C. Thirunavukkarasu, S.K. Abraham, S.T. Santhiya, and A. Ramesh. 2006. Protective effect of saffron (*Crocus sativus* L.) aqueous extract against genetic damage induced by antitumor agents in mice. *Hum. Exp. Toxicol.* 25(2):79-84.
Rockwell, P., and I. Raw. 1979. A mutagenic screening of various herbs, spices, and food additives. *Nutr. Cancer* 1:10-15.
Tang, W., and G. Eisenbrand. 1992. *Chinese drugs of plant origin: Chemistry, pharmacology, and use in traditional and modern medicine*. New York: Springer.
Wichtl, M. 2004. *Herbal drugs and phytopharmaceuticals: A handbook for practice on a scientific basis*. 3rd ed. Boca Raton, FL: CRC Press.
Yamamoto, H., T. Mizutani, and H. Nomura. 1982. Studies on the mutagenicity of crude drug extracts. I. *Yakugaku Zasshi* 102:596-601.
Yang, L., Z. Qian, Y. Yang, et al. 2008. Involvement of Ca^{2+} in the inhibition by crocetin of platelet activity and thrombosis formation. *J. Agric. Food Chem.* 56(20):9429-9433.

Cullen corylifolium (L.) Medik. — Fabaceae

SCN: psoralea
Syn: *Psoralea corylifolia* L.
AN: *bakuchi*
PN: *bu gu zhi* (fruit)
OCN: Malaytea scurfpea; scurfy pea
Part: seed

Quick Reference Summary

Safety Class: 2b
Interaction Class: A

Contraindications

Not for use in pregnancy except under the supervision of a qualified healthcare practitioner (Chen et al. 2007).

Other Precautions

None known.

Drug and Supplement Interactions

None known.

Notice

Photosensitizing (Chang and But 1986; Innocenti et al. 1977; Maurice and Cream 1989); *see* Appendix 2.

Adverse Events and Side Effects

Allergic reactions of the skin have been reported after oral use of psoralea (Bensky et al. 2004).

Although psoralea contains compounds known to cause phototoxicity when applied externally (Epstein 1999; Innocenti et al. 1977), no phototoxic reactions from ingestion of psoralea have been reported in the Chinese literature (Bensky et al. 2004). A single case of phototoxicity after ingestion was identified in other literature, and the case involved long-term use of an excessive amount (30 g daily) of psoralea (Maurice and Cream 1989).

Pharmacological Considerations

None known.

Pregnancy and Lactation

Although one text on traditional Chinese medicine indicated that psoralea is sometimes used to prevent miscarriage

(Bensky et al. 2004), another text indicates that psoralea should be used with caution during pregnancy (Chen and Chen 2004). Based on this information, use during pregnancy is not recommended except under the supervision of a qualified healthcare practitioner.

No information on the safety of psoralea during lactation was identified. While this review did not identify any concerns for use while nursing, safety has not been conclusively established.

Review Details

I. Drug and Supplement Interactions

Clinical Trials of Drug or Supplement Interactions

No clinical trials of drug or supplement interactions were identified.

Case Reports of Suspected Drug or Supplement Interactions

No case reports of suspected drug or supplement interactions were identified.

Animal Trials of Drug or Supplement Interactions

No animal trials of drug or supplement interactions were identified.

II. Adverse Events

Case Reports of Adverse Events

Photosensitivity was reported in a 30-year-old man with vitiligo. The man had been drinking an infusion prepared from 30 g of powdered seed daily for approximately 6 months (Maurice and Cream 1989). Note that the 30 g in this case is a gross overdose; the standard dose in traditional Chinese medicine is 4.5 to 9 g, while in Ayurvedic medicine the standard dose is 1 to 3 g (Bensky et al. 2004; Kapoor 2001).

Acute cholestatic hepatitis was reported in a 44-year-old woman who drank an unspecified amount of psoralea with a cup of black tea every hour for 7 weeks (Nam et al. 2005).

A text on traditional Chinese medicine reports that no adverse effects of psoralea are expected within the normal dose range (decoction of 4.5–9 g) (Bensky et al. 2004). In overdose, psoralea has been associated with dizziness, general weakness, blurred vision, rapid breathing, and vomiting. Severe cases of overdose have been associated with vomiting of blood, loss of consciousness, and coma (Bensky et al. 2004).

Allergic reactions have been reported after use of oral and injected preparations of psoralea (Bensky et al. 2004; Chen and Chen 2004).

III. Pharmacology and Pharmacokinetics

Human Pharmacological Studies

No relevant human pharmacological studies were identified.

Animal Pharmacological Studies

The compounds psoralen and angelicin extracted from psoralea demonstrated photosensitizing activity after topical application to guinea pigs (Innocenti et al. 1977).

No changes in liver enzymes or hepatic CYP450 enzymes were observed in mice orally administered the compound isopsoralen daily for 6 days (Bickers et al. 1982).

A reduction in stress-induced increases in MAO-A and MAO-B was observed in mice orally administered 30 or 50 mg/kg of total furanocoumarins from psoralea daily for 21 days, and psoralea reduced biochemical changes induced by chronic mild stress (Chen et al. 2007).

No changes in uterine weight were observed in ovariectomized rats orally administered 25 or 50 mg/kg of a hydroethanolic extract of psoralea daily for 3 months. Rats in a control group administered 5 μg/kg estrogen daily had significantly higher uterine weights, suggesting a lack of estrogenic activity of psoralea (Tsai et al. 2007).

In Vitro Pharmacological Studies

The compounds psoralen and isopsoralen inhibited monoamine oxidase (MAO) activity in rat brain mitochondria, preferentially inhibiting MAO-A activity over MAO-B activity. This inhibition of enzyme activities was found to be dose-dependent and reversible. The IC_{50} values for MAO-A were 15.2 μ*M* for psoralen and 9.0 μ*M* for isopsoralen. The IC_{50} for MAO-B was 61.8 μ*M* for isopsoralen (Kong et al. 2001).

The compounds psoralen and isopsoralen were selective ER-α agonists in a HeLa cell assay. These compounds promoted proliferation of estrogen receptor-positive breast cancer cells (MCF-7). Other compounds in psoralea were ER-β agonists (Xin et al. 2010).

An ethanol extract of psoralea demonstrated estrogenic activity in a recombinant yeast system with both a human estrogen receptor expression plasmid and a reporter plasmid (Zhang et al. 2005).

Inhibition of platelet aggregation induced by arachidonic acid, collagen, and platelet-activating factor was observed in rabbit platelets treated with a methanol extract of psoralea (Tsai et al. 1996).

IV. Pregnancy and Lactation

Although one text on traditional Chinese medicine indicated that psoralea is sometimes used to prevent miscarriage (Bensky et al. 2004), another text indicates that psoralea should be used with caution during pregnancy (Chen and Chen 2004).

In rats subcutaneously administered 60 or 90 mg daily of nonsaponified fractions of psoralea, a reduction in fetal implantations and increase in resorptions was observed (Khan and Samad 1975).

No information on the safety of psoralea during lactation was identified.

C

V. Toxicity Studies

Acute Toxicity

The LD_{50} of psoralea essential oil in mice was 38 g/kg (Zhu 1998). The LD_{50} of compounds from psoralea orally administered to mice was 180 mg/kg for isopsoralen and 2.3 ml/kg for bakuchiol (Zhu 1998).

The LD_{50} of an ether-acetone fraction of the *n*-hexane extract of psoralea in mice could not be determined after oral administration of doses up to 400 mg/kg (Latha and Panikkar 1999).

In male rats orally administered a single dose of 10 g/kg of an ethanol extract of psoralea, a reduction of serum testosterone and FSH levels preceded degeneration of certain developing germ cells at 3 and 7 days after administration (Takizawa et al. 2004).

No adverse effects were observed in mice orally administered 50, 100, or 200 mg/kg of the compound isopsoralen daily for 3 days (Zhu 1998).

Subchronic Toxicity

In rats fed diets containing 0.375, 0.75, 1.5, or 3.0% of an ethanol extract of psoralea daily for 90 days, weights of the testes in the 1.5 and 3.0% groups and weights of the ovaries in the 3.0% group were significantly lower than controls. Histopathological examination revealed seminiferous tubular atrophy and Leydig cell atrophy in the testes, and epithelial cell atrophy in the seminal vesicles and prostate at the 1.5 and 3.0% level in males. In females, a decrease in the number of corpora lutea associated with necrotic follicles in the ovaries was observed at the 1.5 and 3.0% level (Takizawa et al. 2002). To clarify the pathogenetic targets for the testicular effects of psoralea, rats were fed a diet containing 3% psoralea extract for up to 12 weeks. Tubular degeneration associated with Leydig cell atrophy was observed along with a reduction of serum testosterone and FSH levels (Takizawa et al. 2004).

No changes in liver or kidney function, ECG, or histology of internal organs were observed in dogs orally administered 10 to 100 mg/kg of the compound isopsoralen daily for 14 days (Zhu 1998).

In mice orally administered up to 1 ml/kg of the compound bakuchiol daily for 1 to 4 weeks, progressive renal damage was observed but no pathological changes were reported in other organs (Zhu 1998).

Literature Cited

Bensky, D., S. Clavey, and E. Stöger. 2004. *Chinese herbal medicine: Materia medica.* 3rd ed. Seattle: Eastland Press.

Bickers, D.R., H. Mukhtar, S.J. Mulica, and M.A. Pathak. 1982. The effect of psoralens on hepatic and cutaneous drug metabolizing enzymes and cytochrome P-450. *J. Invest. Dermatol.* 79(3):201-205.

Chang, H.-M., and P.P.H. But. 1986. *Pharmacology and applications of Chinese materia medica.* English ed. Singapore, Philadelphia: World Scientific.

Chen, J.K., and T.T. Chen. 2004. *Chinese medical herbology and pharmacology.* City of Industry, CA: Art of Medicine Press.

Chen, Y., H.D. Wang, X. Xia, et al. 2007. Behavioral and biochemical studies of total furocoumarins from seeds of *Psoralea corylifolia* in the chronic mild stress model of depression in mice. *Phytomedicine* 14(7-8):523-529.

Epstein, J.H. 1999. Phototoxicity and photoallergy. *Semin. Cutan. Med. Surg.* 18(4):274-284.

Innocenti, G., F. Dall'Acqua, A. Guiotto, and G. Caporale. 1977. Investigation on skin photo sensitizing activity of various kinds of psoralea. *Planta Med.* 31(2):151-155.

Kapoor, L.D. 2001. *Handbook of Ayurvedic medicinal plants.* Boca Raton, FL: CRC Press.

Khan, Z., and F. Samad. 1975. Anti-fertility properties of the non-saponified fraction of seeds of *Psoralea corylifolia* in the adult female rats. *Pak. J. Sci. Indust. Res.* 18(1-2):54-56.

Kong, L.D., R.X. Tan, A.Y.H. Woo, and C.H.K. Cheng. 2001. Inhibition of rat brain monoamine oxidase activities by psoralen and isopsoralen: Implications for the treatment of affective disorders. *Pharmacol. Toxicol.* 88(2):75-80.

Latha, P.G., and K.R. Panikkar. 1999. Inhibition of chemical carcinogenesis by *Psoralea corylifolia* seeds. *J. Ethnopharmacol.* 68(1-3):295-298.

Maurice, P.D., and J.J. Cream. 1989. The dangers of herbalism. *Br. Med. J.* 299:1204.

Nam, S.W., J.T. Baek, D.S. Lee, et al. 2005. A case of acute cholestatic hepatitis associated with the seeds of *Psoralea corylifolia* (boh-gol-zhee). *Clin. Toxicol.* 43(6):589-591.

Takizawa, T., T. Imai, K. Mitsumori, et al. 2002. Gonadal toxicity of an ethanol extract of *Psoralea corylifolia* in a rat 90-day repeated dose study. *J. Toxicol. Sci.* 27(2):97-105.

Takizawa, T., K. Mitsumori, H. Takagi, et al. 2004. Sequential analysis of testicular lesions and serum hormone levels in rats treated with a *Psoralea corylifolia* extract. *Food Chem. Toxicol.* 42(1):1-7.

Tsai, M.H., G.S. Huang, Y.C. Hung, et al. 2007. *Psoralea corylifolia* extract ameliorates experimental osteoporosis in ovariectomized rats. *Am. J. Chin. Med.* 35(4):669-680.

Tsai, W.J., W.C. Hsin, and C.C. Chen. 1996. Antiplatelet flavonoids from seeds of *Psoralea corylifolia*. *J. Nat. Prod.* 59(7):671-672.

Xin, D., H. Wang, J. Yang, et al. 2010. Phytoestrogens from *Psoralea corylifolia* reveal estrogen receptor-subtype selectivity. *Phytomedicine* 17(2):126-131.

Zhang, C.Z., S.X. Wang, Y. Zhang, J.P. Chen, and X.M. Liang. 2005. *In vitro* estrogenic activities of Chinese medicinal plants traditionally used for the management of menopausal symptoms. *J. Ethnopharmacol.* 98(3):295-300.

Zhu, Y.-P. 1998. *Chinese materia medica: Chemistry, pharmacology and applications* Amsterdam: Harwood Academic Publishers.

Cuminum cyminum L. — Apiaceae

SCN: cumin
AN: *jiraka*

Part: fruit (commonly known as "seed")

Quick Reference Summary

Safety Class: 1
Interaction Class: A

Contraindications

None known.

Other Precautions

Cumin should be used with caution by individuals with inflammation of the kidneys or a history of irritation of the kidneys (Gachkar et al. 2007; Rong and Zi-Tao 2004).

Drug and Supplement Interactions

None known.

Editors' Note

Concerns for this herb are based on relatively higher doses used for therapeutic purposes in contrast to lower amounts generally used in cooking, and have not been associated with its use as a spice.

Adverse Events and Side Effects

Allergic reactions, including anaphylactic reactions, to cumin have been reported (Anliker et al. 2002; Boxer et al. 1997; Stager et al. 1991).

Pharmacological Considerations

An animal study indicated that cumin increased plasma levels of the drug rifampicin (Sachin et al. 2007). Animal studies suggested estrogenic activity of cumin seed (Al-Khamis et al. 1988; Malini and Vanithakumari 1987).

Pregnancy and Lactation

Animal studies have indicated that large doses of cumin have antifertility activity (Al-Khamis et al. 1988; Garg 1976).

No information on the safety of cumin in lactation was identified. While this review did not identify any concerns for use while nursing, safety has not been conclusively established.

Review Details

I. Drug and Supplement Interactions

Clinical Trials of Drug or Supplement Interactions

No clinical trials of drug or supplement interactions were identified.

Case Reports of Suspected Drug or Supplement Interactions

No case reports of suspected drug or supplement interactions were identified.

Animal Trials of Drug or Supplement Interactions

An increase in plasma levels of rifampicin was observed in mice orally administered 16 mg/kg of a water-ethanol extract of cumin with 40 mg/kg of rifampicin. The activity was attributed to a flavonoid glycoside and was hypothesized to be due to a permeation-enhancing effect of the glycoside (Sachin et al. 2007).

II. Adverse Events

Case Reports of Adverse Events

Anaphylactic reactions to cumin have been reported (Boxer et al. 1997). Persons with mugwort-spice syndrome have also tested positive to cumin (Anliker et al. 2002; Stager et al. 1991).

III. Pharmacology and Pharmacokinetics

Human Pharmacological Studies

No relevant human pharmacological studies were identified.

Animal Pharmacological Studies

A reduction in bone loss was observed in ovariectomized rats with estrogen deficiency-induced osteoporosis orally administered 1 g/kg of a methanol extract of cumin daily for 10 weeks (Shirke et al. 2008).

In ovariectomized rats orally administered 10, 20, or 25 μg/kg of an acetone extract of cumin daily for 10 days, estrus was induced. In rats administered the 20 or 25 μg/kg doses, an increase in uterine weight and protein concentration in the endometrium were observed (Malini and Vanithakumari 1987).

In Vitro Pharmacological Studies

No relevant in vitro pharmacological studies were identified.

IV. Pregnancy and Lactation

In rats orally administered 250 or 500 mg/kg of an aqueous extract of cumin, mild abortifacient activity was observed when the extract was administered on gestational days 8 to 12. Cumin also increased the plasma levels of

C

estradiol-17β at the proestrus and estrus stages of the estrus cycle (Al-Khamis et al. 1988).

In rats administered an alcohol extract of cumin, 100% antifertility activity was observed at a dose of 150 mg/kg (Garg 1976).

A survey of mothers in the United Arab Emirates indicated that cumin is one of the spices commonly used to soothe crying infants. Of 998 mothers surveyed, 33 reported using cumin to soothe their infants (Abdulrazzaq et al. 2009).

No information on the safety of cumin during lactation was identified.

V. Toxicity Studies

Acute Toxicity

The LD_{50} of orally administered ethanol extract of cumin could not be determined at doses up to 3 g/kg (Shah et al. 1989).

Short-Term Toxicity

In rats fed diets containing 2 or 10% cumin for 6 weeks, hematological changes were observed in the 10% group after 3 weeks of treatment. These changes included decreases in hemoglobin concentration, red blood cells, packed cell volume, and mean corpuscular hemoglobin concentration, and increases in white blood cells and lymphocytes. After 6 weeks of treatment, changes in erythrocytic series persisted, but the values for white blood cells and lymphocytes decreased. Impairment of growth and enterohepatonephropathy were also observed in the 10% group. No hematological changes were observed in the 2% group (Haroun et al. 2002).

Subchronic Toxicity

In rats orally administered 100 mg/kg of an ethanol extract of cumin daily for 90 days, 30% developed wounds and ulcers on the tail, 15% developed alopecia, and 25% of the animals had died by the end of the 90 days, as compared with 10% that died in the untreated control group. An increase in hemoglobin, and no change in organ weights, was observed (Shah et al. 1989).

Genotoxicity

No mutagenic activity of an aqueous extract of cumin was observed in *Salmonella typhimurium* strains (Sivaswamy et al. 1991). In the mouse micronucleus test, cumin exhibited dose-dependent inhibition of chemically induced genotoxicity (Abraham et al. 1998).

Literature Cited

Abdulrazzaq, Y.M., A. Al Kendi, and N. Nagelkerke. 2009. Soothing methods used to calm a baby in an Arab country. *Acta Paediatr.* 98(2):392-396.

Abraham, S.K., S.P. Singh, and P.C. Kesavan. 1998. In vivo antigenotoxic effects of dietary agents and beverages co-administered with urethane: Assessment of the role of glutathione S-transferase activity. *Mutat. Res.* 413(2):103-110.

Al-Khamis, K.I., M.A. Al-Said, M.W. Islam, et al. 1988. Antifertility anti-implantation and abortifacient activity of the aqueous extract of *Cuminum cyminum*. *Fitoterapia* 59(1):5-10.

Anliker, M.D., S. Borelli, and B. Wuthrich. 2002. Occupational protein contact dermatitis from spices in a butcher: A new presentation of the mugwort-spice syndrome. *Contact Dermat.* 46(2):72-74.

Boxer, M., M. Roberts, and L. Grammer. 1997. Cumin anaphylaxis: A case report. *J. Allergy Clin. Immunol.* 99(5):722-723.

Gachkar, L., D. Yadegari, M.B. Rezaei, et al. 2007. Chemical and biological characteristics of *Cuminum cyminum* and *Rosmarinus officinalis* essential oils. *Food Chem.* 102(3):898-904.

Garg, S.K. 1976. Antifertility screening of plants: Effect of four indigenous plants on early pregnancy in female albino rats. *Indian J. Med. Res.* 64(8):1133-1135.

Haroun, E.M., O.M. Mahmoud, and S.E. Adam. 2002. Effect of feeding *Cuminum cyminum* fruits, *Thymus vulgaris* leaves or their mixture to rats. *Vet. Hum. Toxicol.* 44(2):67-69.

Malini, T., and G. Vanithakumari. 1987. Estrogenic activity of *Cuminum cyminum* in rats. *Indian J. Exp. Biol.* 25(7):442-444.

Rong, L., and J. Zi-Tao. 2004. Chemical composition of the essential oil of *Cuminum cyminum* L. from China. *Flav. Frag. J.* 19(4):311-313.

Sachin, B.S., S.C. Sharma, S. Sethi, et al. 2007. Herbal modulation of drug bioavailability: Enhancement of rifampicin levels in plasma by herbal products and a flavonoid glycoside derived from *Cuminum cyminum*. *Phytother. Res.* 21(2):157-163.

Shah, A.H., S. Qureshi, M. Tariq, and A.M. Ageel. 1989. Toxicity studies on six plants used in the traditional Arab system of medicine. *Phytother. Res.* 3(1):25-29.

Shirke, S.S., S.R. Jadhav, and A.G. Jagtap. 2008. Methanolic extract of *Cuminum cyminum* inhibits ovariectomy-induced bone loss in rats. *Exp. Biol. Med.* 233(11):1403-1410.

Sivaswamy, S.N., B. Balachandran, S. Balanehru, and V.M. Sivaramakrishnan. 1991. Mutagenic activity of south Indian food items. *Indian J. Exp. Biol.* 29(8):730-737.

Stager, J., B. Wuthrich, and S.G. Johansson. 1991. Spice allergy in celery-sensitive patients. *Allergy* 46(6):475-478.

Curculigo orchioides Gaertn. — Hypoxidaceae

SCN: curculigo
PN: *xian mao* (rhizome)
OCN: golden eye grass
Part: rhizome

Quick Reference Summary

Safety Class: 2b, 2d
Interaction Class: A

Contraindications

Not for use in pregnancy except under the supervision of a qualified healthcare practitioner (Tarafder 1983).

Not for long-term use; do not exceed recommended dose (Bensky et al. 2004; Chen and Chen 2004).

Other Precautions

None known.

Drug and Supplement Interactions

See Pharmacological Considerations.

Standard Dose

A decoction of 3 to 10 g (Bensky et al. 2004; Chen and Chen 2004); maximum dose is 12 g daily (Bensky et al. 2004).

Adverse Events and Side Effects

Large doses have been associated with cold sweating, numbness of the limbs, swollen tongue, agitation, and loss of consciousness (Bensky et al. 2004).

Pharmacological Considerations

An animal study indicated that curculigo may modify glucose regulation. People with diabetes are advised to monitor their blood sugar closely and discuss the use of this herb with a qualified healthcare practitioner prior to use (Madhavan et al. 2007).

Animal studies have indicated that curculigo has immunomodulatory activity and may counteract the effects of immunosuppressant drugs (Bafna and Mishra 2006; Zhu 1998).

Conflicting results on estrogenic activity of curculigo have been reported in animal studies, with one study showing estrogenic activity (Vijayanarayana et al. 2007) and another study showing a lack of estrogenic activity (Cao et al. 2008).

Pregnancy and Lactation

An ethnobotanical survey in northern India indicated that curculigo was traditionally used as an abortifacient (Tarafder 1983). Based on this information, use during pregnancy is not recommended except under the supervision of a qualified healthcare practitioner.

No information on the safety of curculigo in lactation was identified. While this review did not identify any concerns for use while nursing, safety has not been conclusively established.

Review Details

I. Drug and Supplement Interactions

Clinical Trials of Drug or Supplement Interactions

No clinical trials of drug or supplement interactions were identified.

Case Reports of Suspected Drug or Supplement Interactions

No case reports of suspected drug or supplement interactions were identified.

Animal Trials of Drug or Supplement Interactions

No animal trials of drug or supplement interactions were identified.

II. Adverse Events

Case Reports of Adverse Events

Large doses have been reported to cause side effects such as cold sweating, numbness of the limbs, swollen tongue, agitation, and loss of consciousness (Bensky et al. 2004).

III. Pharmacology and Pharmacokinetics

Human Pharmacological Studies

No relevant human pharmacological studies were identified.

Animal Pharmacological Studies

In mice with cyclophosphamide-induced immunosuppression, oral administration of 50 to 800 mg/kg methanol extract of curculigo daily for 13 days produced dose-dependent increases in humoral antibody titer, delayed-type hypersensitivity, and levels of white blood cells (Bafna and Mishra 2006).

In cyclophosphamide-induced immunosuppressed mice, oral administration of 10 or 20 g/kg of curculigo increased peritoneal macrophages and counteracted the immunosuppressive effect of cyclophosphamide (Zhu 1998).

In ovariectomized rats orally administered 300, 600, or 1200 mg/kg of an ethanol extract of curculigo daily for 7 days, increases in percentage vaginal cornification, uterine wet weight, uterine glycogen content, and proliferative changes in the uterine endometrium were observed, as compared to control (Vijayanarayana et al. 2007).

A reduction in bone loss with no effects on body weight or uterine weight was observed in ovariectomized rats intragastrically administered 0.5, 1.0, or 2.0 g/kg of curculigo extract daily for 12 weeks (Cao et al. 2008).

A reduction in blood glucose levels was observed in diabetic rats orally administered 0.5 or 1.0 g/kg of alcohol or aqueous extracts of curculigo daily for 21 days (Madhavan et al. 2007).

In Vitro Pharmacological Studies

No relevant in vitro pharmacological studies were identified.

IV. Pregnancy and Lactation

An ethnobotanical survey of tribal peoples in Bihar, India, indicated that a paste of curculigo was traditionally used as an abortifacient (Tarafder 1983).

No information on the safety of curculigo during lactation was identified.

V. Toxicity Studies

Acute Toxicity

The LD_{50} of orally administered curculigo in mice could not be determined at doses up to 150 g/kg (Chen and Chen 2004).

C

Literature Cited

Bafna, A.R., and S.H. Mishra. 2006. Immunostimulatory effect of methanol extract of *Curculigo orchioides* on immunosuppressed mice. *J. Ethnopharmacol.* 104(1-2):1-4.

Bensky, D., S. Clavey, and E. Stöger. 2004. *Chinese herbal medicine: Materia medica.* 3rd ed. Seattle: Eastland Press.

Cao, D.P., Y.N. Zheng, L.P. Qin, et al. 2008. *Curculigo orchioides*, a traditional Chinese medicinal plant, prevents bone loss in ovariectomized rats. *Maturitas* 59(4):373-380.

Chen, J.K., and T.T. Chen. 2004. *Chinese medical herbology and pharmacology.* City of Industry, CA: Art of Medicine Press.

Madhavan, V., R. Joshi, A. Murali, and S.N. Yoganarasimhan. 2007. Antidiabetic activity of *Curculigo orchioides* root tuber. *Pharmaceut. Biol.* 45(1):18-21.

Tarafder, C.R. 1983. Ethnogynecology in relation to plants 2. Plants used for abortion. *J. Econ. Taxon. Bot.* 4(2):507-516.

Vijayanarayana, K., R.S. Rodrigues, K.S. Chandrashekhar, and E.V. Subrahmanyam. 2007. Evaluation of estrogenic activity of alcoholic extract of rhizomes of *Curculigo orchioides*. *J. Ethnopharmacol.* 114(2):241-245.

Zhu, Y.-P. 1998. *Chinese materia medica: Chemistry, pharmacology and applications.* Amsterdam: Harwood Academic Publishers.

Curcuma zedoaria (Christm.) Roscoe — Zingiberaceae

SCN: zedoary
AN: *karchura*
PN: *e zhu* (rhizome)
Part: rhizome

Quick Reference Summary

Safety Class: 2b, 2d
Interaction Class: A

Contraindications

Not for use in pregnancy except under the supervision of a qualified healthcare practitioner (Bensky et al. 2004; Chen and Chen 2004). Not for use in excessive menstruation (Bensky et al. 2004; Chen and Chen 2004).

Other Precautions

None known.

Drug and Supplement Interactions

None known.

Editors' Note

The classifications and concerns for this herb are based on relatively higher doses used for therapeutic purposes in contrast to lower amounts generally used in cooking, and have not been associated with its use as a spice.

Other species of *Curcuma* are used interchangeably as sources of *e zhu* (Bensky et al. 2004; Chen and Chen 2004; PPRC 2005).

Adverse Events and Side Effects

None known.

Pharmacological Considerations

Extracts of zedoary have inhibited platelet binding in vitro (Han et al. 1995). A text on traditional Chinese medicine indicates that zedoary should be used with caution in persons taking anticoagulant or antiplatelet drugs, although this concern is theoretical and no interactions have been documented (Chen and Chen 2004).

Pregnancy and Lactation

A text on traditional Chinese medicine indicates that zedoary should not be used during pregnancy (Bensky et al. 2004; Chen and Chen 2004). Based on this information, use during pregnancy is not recommended except under the supervision of a qualified healthcare practitioner.

No information of the safety of zedoary in lactation was identified. While this review did not identify any concerns for use while nursing, safety has not been conclusively established.

Review Details

I. Drug and Supplement Interactions

Clinical Trials of Drug or Supplement Interactions

No clinical trials of drug or supplement interactions were identified.

Case Reports of Suspected Drug or Supplement Interactions

No case reports of suspected drug or supplement interactions were identified.

Animal Trials of Drug or Supplement Interactions

No animal trials of drug or supplement interactions were identified.

II. Adverse Events

Case Reports of Adverse Events

No case reports of adverse events were identified.

III. Pharmacology and Pharmacokinetics

Human Pharmacological Studies

No relevant human pharmacological studies were identified.

Animal Pharmacological Studies

No relevant animal pharmacological studies were identified.

In Vitro Pharmacological Studies

Inhibition of the drug-metabolizing isoenzyme CYP3A4 was observed in human intestinal epithelial (Caco-2) cells treated with a methanol extract of zedoary. The 50% inhibitory concentration (IC_{50}) of the extract was 0.014 mg/ml (Hou et al. 2007). Extracts of zedoary inhibited the drug-metabolizing isoenzymes CYP1A1, CYP1A2, CYP2B1, CYP2B2, and CYP2E1 (Jeong et al. 2002).

An increase in activity of the drug transporter P-gp was observed in human intestinal epithelial (Caco-2) cells treated with a methanol extract of zedoary at a concentration of 0.1 mg/ml (Hou et al. 2008).

Inhibition of platelet-activating factor binding to rabbit platelets was observed after treatment of platelets with 200 μg/ml of the freeze-dried form of an aqueous extract of zedoary (Han et al. 1995).

IV. Pregnancy and Lactation

Texts on traditional Chinese medicine indicates that zedoary should not be used during pregnancy (Bensky et al. 2004; Chen and Chen 2004).

No information of the safety of zedoary in lactation was identified. While this review did not identify any concerns for use while nursing, safety has not been conclusively established.

V. Toxicity Studies

Acute Toxicity

The LD_{50} of zedoary in mice is 147 g/kg after oral administration and 55 g/kg after intramuscular injection (Chen and Chen 2004).

Short-Term Toxicity

In rats fed a high-protein flour prepared from zedoary, ingestion of 320 g/kg resulted in death of all animals by day 6. The author of that study noted that the traditional method of preparing zedoary in India involves prolonged washing in multiple changes of water, a process that removes most of the protein and other water-soluble nutrients (Latif et al. 1979).

In weanling rats fed 400 g/kg daily of dried zedoary, 40% of the animals died within 4 days. Day-old chicks fed 100 or 200 g/kg of the same dried zedoary had a decrease in body weight but survived the 20-day test period (Latif et al. 1979).

Genotoxicity

No genotoxic activity of a polysaccharide fraction of zedoary was observed in the Ames test for mutagenicity with or without metabolic activation with S9. No clastogenic activity of the same polysaccharide fraction was observed in micronucleus or chromosomal aberration assays performed using Chinese hamster lung fibroblast cells with or without metabolic activation (Kim et al. 2000).

Literature Cited

Bensky, D., S. Clavey, and E. Stöger. 2004. *Chinese herbal medicine: Materia medica.* 3rd ed. Seattle: Eastland Press.

Chen, J.K., and T.T. Chen. 2004. *Chinese medical herbology and pharmacology.* City of Industry, CA: Art of Medicine Press.

Han, B.H., H.O. Yang, Y.H. Kang, and Y.N. Han. 1995. Screening of the inhibitory effects of herbal medicines on the platelet activating factor (PAF) binding: 35 selected herbal medicines based on folk medicinal informations. *J. Pharm. Soc. Korea* 39:10-13.

Hou, X.L., K. Takahashi, N. Kinoshita, et al. 2007. Possible inhibitory mechanism of *Curcuma* drugs on CYP3A4 in 1(alpha),25-dihydroxyvitamin D3 treated Caco-2 cells. *Int. J. Pharmaceut.* 337(1-2):169-177.

Hou, X.L., K. Takahashi, K. Tanaka, et al. 2008. *Curcuma* drugs and curcumin regulate the expression and function of P-gp in Caco-2 cells in completely opposite ways. *Int. J. Pharm.* 358(1-2):224-229.

Jeong, H.G., H.J. You, Y.S. Chang, et al. 2002. Inhibitory effects of medicinal herbs on cytochrome P450 drug metabolizing enzymes. *Korean J. Pharmacog.* 33(1):35-41.

Kim, K.I., J.W. Kim, B.S. Hong, et al. 2000. Antitumor, genotoxicity and anticlastogenic activities of polysaccharide from *Curcuma zedoaria*. *Mol. Cells* 10(4):392-398.

Latif, M.A., T.R. Morris, A.H. Miah, D. Hewitt, and J.E. Ford. 1979. Toxicity of shoti (Indian arrowroot: *Curcuma zedoaria*) for rats and chicks. *Br. J. Nutr.* 41 (1):57-63.

PPRC. 2005. *Pharmacopoeia of the People's Republic of China*. Beijing: People's Medical Publishing House.

C

Curcuma longa L. — Zingiberaceae

SCN: turmeric
Syn: *Curcuma domestica* Valeton
AN: *haridra*
PN: *jiang huang* (rhizome)
OCN: common turmeric; Indian saffron; yellow ginger
Part: rhizome

Quick Reference Summary

Safety Class: 1
Interaction Class: A

Contraindications

None known.

Other Precautions

None known.

Drug and Supplement Interactions

None known.

Editors' Note

Turmeric contains 0.3–5.4% of the compound curcumin (Leung and Foster 1996). Commercially available turmeric products include products with this normal percentage of curcumin and those modified to contain up to 95% curcuminoids (including curcumin and other similar compounds). Products with modified concentrations of selected compounds may be expected to have different physiological effects than traditional preparations of the herb.

Adverse Events and Side Effects

Contact dermatitis has been reported in sensitive persons after topical application of turmeric products (Hata et al. 1997; Nath and Thappa 2007; Sakurane et al. 1999).

Turmeric and the compound curcumin have been generally well tolerated in human studies at doses up to 8 g daily. Adverse events reported included nausea and diarrhea (Hsu and Cheng 2007).

Pharmacological Considerations

Several references on herbal safety contraindicate the use of turmeric in persons with biliary tract obstruction (Brinker 2001; De Smet 1993; Mills and Bone 2005) due to reported bile-stimulating activity (De Smet 1993) observed in rats administered the compound curcumin (Bhat et al. 1984), and gallbladder contraction (50% contraction after 40 mg dose) observed in humans administered the compound curcumin (Rasyid et al. 2002). Such concerns are theoretical, and no clinical evidence for turmeric has been reported to support or refute these concerns.

In one study, a large dose (100 mg/kg) of the compound curcumin caused ulcers in rats but showed no ulcerogenic activity at a smaller dose (50 mg/kg) (Gupta et al. 1980); in other studies, curcumin exhibited potent antiulcer activity at a medium dose (80 mg/kg) (Sivalingam et al. 2007; Swarnakar et al. 2005). Extracts of turmeric have shown antiulcer activity in humans (3 g daily dose) and rats (500 mg/kg dose) (Prucksunand et al. 2001; Rafatullah et al. 1990).

Antiplatelet activity of turmeric and the compound curcumin has been reported in vitro but has not been confirmed in human or animal studies (Jantan et al. 2007; Shah et al. 1999; Srivastava 1989; Srivastava et al. 1995).

Pregnancy and Lactation

Although texts on traditional Chinese medicine contraindicate the use of turmeric rhizome in pregnancy (Bensky et al. 2004; Chen and Chen 2004), no adverse effects on reproduction in mice were observed at high doses (1 g/kg daily) of the compound curcumin in a two-generation study (Ganiger et al. 2007). Earlier studies on reproduction provided mixed results, with some indicating a decrease in implantation (An 1998; Chen 1988; Garg 1974) and other studies showing no adverse effects (Francis 2002; Govindarajan 1980; Vijayalaxmi 1980).

The compound curcumin and its metabolites have been shown to cross into breast milk (Singh et al. 1995, 1996).

Review Details

I. Drug and Supplement Interactions

Clinical Trials of Drug or Supplement Interactions

No clinical trials of drug or supplement interactions were identified.

Case Reports of Suspected Drug or Supplement Interactions

No case reports of suspected drug or supplement interactions were identified.

Animal Trials of Drug or Supplement Interactions

No animal trials of drug or supplement interactions were identified.

II. Adverse Events

Adverse Events Reported in Clinical Trials

A review of clinical trials of turmeric and the compound curcumin indicated that turmeric and curcumin were generally safe and well tolerated at doses up to 8 g daily for 4 months. Adverse events reported in human pharmacological studies (see below) were primarily diarrhea or nausea, with other events reported as headache, rash and yellowish stool (Hsu and Cheng 2007).

Case Reports of Adverse Events

Kumkum, a turmeric-based paste applied topically, has been associated with cases of dermatitis (Nath and Thappa 2007). Cases of contact urticaria from the compounds curcumin and tetrahydrocurcumin have been reported (Fischer and Agner 2004; Lamb and Wilkinson 2003; Liddle et al. 2006; Thompson and Tan 2006). Contact dermatitis with positive patch test for turmeric was reported in a spice shop worker routinely exposed to spice powders (Goh and Ng 1987). Similarly, contact dermatitis was reported in a pasta factory worker exposed to curcumin food coloring (Kiec-Swierczynska and Krecisz 1998). Two cases of contact dermatitis were reported in individuals using a topical ointment containing turmeric. Patch testing of both individuals indicated a sensitivity to both turmeric and curcumin (Hata et al. 1997; Sakurane et al. 1999).

III. Pharmacology and Pharmacokinetics

Human Pharmacological Studies

In a dose escalation study of single doses of a standardized turmeric extract (95% curcuminoids), healthy volunteers were administered doses from 0.5 to 8 g. Tolerance was reported as excellent at all dose levels. Adverse events, including yellow stool, diarrhea, headache, and rash, were reported at doses of 1 g and above (Lao et al. 2006). In a dose escalation study in patients with colorectal cancer, doses between 0.4 and 2.2 g of turmeric extract daily (36–180 mg curcumin) for up to 4 months were well tolerated, and dose-limiting toxicity was not observed. Ingestion of 440 mg of turmeric extract for 29 days was accompanied by a 59% decrease in lymphocytic glutathione *S*-transferase activity, but at higher dose levels that effect was not observed (Sharma et al. 2001). In a human safety study of turmeric essential oil, no effects on hematological, renal, or hepatic parameters were observed in healthy volunteers administered 1.8 ml of turmeric essential oil daily for 1 month and 3 ml daily for 2 months. One volunteer discontinued the study due to an allergic skin rash (Joshi et al. 2003).

Curcuminoids were reported as well tolerated at doses of 2 g daily for 7 weeks (Chainani-Wu et al. 2007), and curcumin was well tolerated at a dose of 2 g daily for 6 months (Hanai et al. 2006). A dose escalation study of curcumin at doses of 1 to 12 g daily in patients at high risk of selected cancers indicated that no toxicity was observed at doses up to 8 g daily for 3 months. The 12 g dose was reported as unacceptable to patients due to the bulky volume of the drug (Cheng et al. 2001).

Dose-dependent contraction of the gallbladder was observed in healthy volunteers administered 20, 40, or 80 mg of the compound curcumin, with higher doses causing greater contraction (Rasyid et al. 2002).

In patients with duodenal or gastric ulcers, administration of 3 g turmeric daily for 9 weeks resolved the ulcers in 75% of the patients (Prucksunand et al. 2001).

Turmeric paste has traditionally been applied topically to prevent unwanted hair growth in women (Rao and Kotagi 1984).

Animal Pharmacological Studies

Administration of the compound curcumin to rats at doses of 100 mg/kg daily for 6 days increased the formation of stomach ulcers, whereas doses of 50 mg/kg had no effect (Gupta et al. 1980). Conversely, an ethanolic extract of turmeric at a dose of 500 mg/kg was shown to have significant antiulcerogenic activity in rats subjected to stress or administered ulcer-causing compounds (Rafatullah et al. 1990).

Significant decreases in testis weight and serum testosterone were observed in rats subcutaneously administered 1 ml of a concentrated ethanol extract of turmeric (equivalent to 50 mg powder) daily for 10 days. No changes in liver, kidney, or spleen morphology were observed (Rao and Kotagi 1984).

In Vitro Pharmacological Studies

The compound curcumin exhibited significant antiplatelet activity in vitro, causing inhibition of arachidonic acid-, collagen- and ADP-induced platelet aggregation with IC_{50} values of 13.8, 22.4, and 16.8 µ*M*, respectively (Jantan et al. 2007). Inhibition of arachidonate-induced platelet aggregation was observed in human blood platelets treated with turmeric extract (Srivastava 1989). In a similar study, curcumin inhibited PAF- and AA-induced platelet aggregation, but not epinephrine-induced aggregation. Curcumin also inhibited the formation of thromboxane A_2 (TXA_2) by platelets at an IC_{50} value of 70 µ*M*, suggesting that curcumin-mediated preferential inhibition of PAF- and AA-induced platelet aggregation involves inhibitory effects on TXA_2 synthesis (Shah et al. 1999). In a third study, curcumin inhibited collagen- and AA-induced platelet aggregation and inhibited thromboxane B_2 production (Srivastava et al. 1995).

The compound ar-tumerone was shown to inhibit in vitro platelet aggregation induced by collagen or arachidonic acid but had no effect on platelet-activating factor or thrombin-induced platelet aggregation (Lee 2006).

In rat hepatocytes, the compound curcumin was cytoprotective at concentrations of 0.05 m*M*, whereas at 5 m*M* the observed protective effect on lipid peroxidation was accompanied with a tendency to increase cellular glutathione depletion and lactate dehydrogenase leakage (Donatus et al. 1990).

IV. Pregnancy and Lactation

In a two-generation reproductive study of mice fed a diet of 1500, 3000, or 10,000 ppm curcumin, no adverse effects were observed in parental animals or on reproduction. A small reduction in preweaning body weight gain of the F_2 pups was observed at the highest dose level. The no-observed-adverse-effect level (NOAEL) for reproductive toxicity of curcumin was determined to be 1043 mg/kg daily (Ganiger et al. 2007).

No significant effects on pregnancy rate, implantations, or embryo survival were observed in mice fed a diet of 0.5% turmeric or 0.015% curcumin for 12 weeks (Govindarajan 1980; Vijayalaxmi 1980). A reduction in implantations and litter size was observed in rabbits administered 100 or 200 mg/kg of turmeric daily during the first week of pregnancy. No teratogenic effects were observed (Garg 1974). No teratogenic or reproductive toxicity was observed in pups of mothers administered 50 mg/kg turmeric daily for 1 year (Francis 2002) or in rabbits administered 600 mg/kg curcumin daily from day 6 to 15 of pregnancy (Govindarajan 1980).

Abortions and prevention of implantation were reported in mice orally administered ethanol extracts of turmeric at very high doses of 15 g/kg (Chen 1988). The essential oil administered intravaginally, subcutaneously, or intraperitoneally produced similar antifertility effects (An 1998).

Texts on traditional Chinese medicine contraindicate the use of turmeric in pregnancy (Bensky et al. 2004; Chen and Chen 2004).

Administration of 4 g/kg turmeric or 0.4 g/kg curcumin for 2 or 3 weeks caused an increase in cytochrome P450 and cytochrome b5 levels in lactating mice and their pups (Singh et al. 1995). One other pharmacological study indicated that the active constituents and/or metabolites of turmeric cross into breast milk (Singh et al. 1996).

V. Toxicity Studies

Acute Toxicity

The LD_{50} of orally administered turmeric oleoresin in mice and rats could not be determined at doses up to 10 g/kg (Francis 2002). No toxic effects were observed after rats, guinea pigs, and monkeys were administered 2.5 g/kg turmeric powder or 0.3 g/kg of an alcoholic turmeric extract (Shankar et al. 1980). No apparent toxicity of the compound curcumin was observed in rats administered doses up to 5 g/kg (Wahlstrom and Blennow 1978).

In the brine shrimp assay, no toxic effects of a turmeric ethanol extract were observed (Mahmoud et al. 1992).

Short-Term Toxicity

In mice and rats fed turmeric as 0, 1, or 5% of the diet or fed a turmeric ethanolic extract as 0, 0.05, or 0.25% for 14 days, mice were found to be more susceptible than rats to symptoms of hepatotoxicity observed in the study, with mice showing adverse effects at the 1% dose level after 14 days (Deshpande et al. 1998).

Histopathological changes were observed in the livers of mice administered whole turmeric at doses of 0.2, 1.0, or 5.0% of the diet or turmeric ethanolic extracts at doses of 0.05 or 0.25% of the diet for 14 days (Kandarkar et al. 1998).

A decrease in the levels of liver lipid peroxides were observed in mice administered 4 mg/kg (10% curcumin) daily of turmeric hydroalcoholic extract for 4 weeks. No toxic effects of the turmeric were observed (Miquel et al. 1995).

Subchronic Toxicity

In mice and rats fed turmeric as 0, 1, or 5% of the diet or fed a turmeric ethanolic extract as 0, 0.05, or 0.25% for 90 days, animals ingesting the highest dose of turmeric had symptoms of hepatotoxicity including focal necrosis or focal necrosis with regeneration (Deshpande et al. 1998).

In mice and rats administered turmeric oleoresin (79–85% curcumin) daily in food at doses of 0, 1000, 5000, 10,000, 25,000, or 50,000 ppm (equivalent to average daily doses of 50, 250, 480, 1300, or 2600 mg/kg in males and 60, 300, 550, 1450, or 2800 mg/kg in females) for 13 weeks, an increase in liver weights was observed at the 5000 ppm dose and above; no histopathological lesions were observed, nor were there any significant differences in hematology, clinical chemistry, or urinalysis parameters, although some animals were noted to have stained fur and discolored feces and urine (NTP 1993).

Dose-related increases in liver and thyroid weight were reported in pigs administered 60, 296, or 1551 mg/kg of turmeric oleoresin daily for 102 days. Inflammation of the bile duct, thyroid hyperplasia, and epithelial changes in the kidney and urinary bladder were observed in the two higher dose groups (Bille et al. 1985).

In mice administered 100 mg/kg daily of a concentrated turmeric extract for 90 days, a significant increase in heart, lung, and caudae epididymis weights were observed (Qureshi et al. 1992).

Chronic Toxicity

In mice and rats administered turmeric oleoresin (79–85% curcumin) daily in food at doses of 0, 1000, 5000, 10,000, 25,000, or 50,000 ppm (equivalent to average daily doses of 50, 250, 480, 1300, or 2600 mg/kg in males and 60, 300, 550, 1450, or 2800 mg/kg in females) for 2 years, an increase in liver weights was observed at the 10,000 ppm dose and above, but no toxic effects were observed. At the highest dose, decreased hematocrit values, hemoglobin concentration, and erythrocyte counts were significantly lower in rats.

Also at that dose, increased incidences of ulcers and chronic inflammation were observed (NTP 1993).

Genotoxicity

No mutagenic effects of turmeric alcoholic extracts, turmeric oleoresin, or curcumin were observed in *Salmonella typhimurium* strains with or without metabolic activation or in the Ames test (Abraham and Kesavan 1984; Jensen 1982; Nagabhushan and Bhide 1986; NTP 1993; Shah and Netrawali 1988).

A single acute dose of 6 g/kg intraperitoneally administered turmeric methanolic extract in mice was reported to cause chromosomal damage (Jain et al. 1987). Changes in chromosome morphology were observed in hamster, mouse, and deer cell lines as well as in human lymphocyte cultures treated with an alcoholic turmeric extract (Goodpasture and Arrighi 1976). The compound curcumin was reported to induce DNA damage in human lymphocytes and gastric mucosa cells in vitro (Blasiak et al. 1999). No chromosomal aberrations were reported in rats fed a diet of up to 0.5% turmeric or mice fed a diet of 0.5% turmeric or 0.15% curcumin for 12 weeks (Vijayalaxmi 1980). Small but significant increases in sister chromatid exchanges and chromosomal aberrations were observed in cultured Chinese hamster ovary cells treated with turmeric oleoresin (NTP 1993).

Literature Cited

Abraham, S.K., and P.C. Kesavan. 1984. Genotoxicity of garlic, turmeric and asafoetida in mice. *Mutat. Res.* 136(1):85-88.

An, Y. 1998. *Cited in* But, P.P.H. 1988. Chinese medicine for birth control. *Abstr. Chin. Med.* 2(2): 247-269.

Bensky, D., S. Clavey, and E. Stöger. 2004. *Chinese herbal medicine: Materia medica.* 3rd ed. Seattle: Eastland Press.

Bhat, B., M.R. Srinivasan, and N. Chandrasekhara. 1984. Influence of curcumin and capsaicin on the composition and secretion of bile in rats. *J. Food Sc. Technol. (Mysore)* 21(4):225-227.

Bille, N., J.C. Larsen, E.V. Hansen, and G. Wurtzen. 1985. Subchronic oral toxicity of turmeric oleoresin in pigs. *Food Chem. Toxicol.* 23(11):967-973.

Blasiak, J., A. Trzeciak, E. Malecka-Panas, et al. 1999. DNA damage and repair in human lymphocytes and gastric mucosa cells exposed to chromium and curcumin. *Teratogen. Carcinogen. Mutagen.* 19(1):19-31.

Brinker, F. 2001. *Herb contraindications and drug interactions.* 3rd ed. Sandy, OR: Eclectic Medical Publications.

Chainani-Wu, N., S. Silverman, Jr., A. Reingold, et al. 2007. A randomized, placebo-controlled, double-blind clinical trial of curcuminoids in oral lichen planus. *Phytomedicine* 14(7-8):437-446.

Chen, J.K., and T.T. Chen. 2004. *Chinese medical herbology and pharmacology.* City of Industry, CA: Art of Medicine Press.

Chen, Z. 1988. *Cited in* But, P.P.H. 1998. Chinese medicine for birth control. *Abstr. Chin. Med.* 2(2): 247-269.

Cheng, A.L., C.H. Hsu, J.K. Lin, et al. 2001. Phase I clinical trial of curcumin, a chemopreventive agent, in patients with high-risk or pre-malignant lesions. *Anticancer Res.* 21(4B):2895-2900.

De Smet, P.A.G.M. 1993. *Adverse effects of herbal drugs, Volume 2.* Berlin: Springer.

Deshpande, S.S., V.S. Lalitha, A.D. Ingle, et al. 1998. Subchronic oral toxicity of turmeric and ethanolic turmeric extract in female mice and rats. *Toxicol. Lett.* 95(3):183-193.

Donatus, I.A., Sardjoko, and N.P.E. Vermeulen. 1990. Cytotoxic and cytoprotective activities of curcumin—Effects on paracetamol-induced cytotoxicity, lipid-peroxidation and glutathione depletion in rat hepatocytes. *Biochem. Pharmacol.* 39(12):1869-1875.

Fischer, L.A., and T. Agner. 2004. Curcumin allergy in relation to yellow chlorhexidine solution used for skin disinfection prior to surgery. *Contact Dermat.* 51(1):39-40.

Francis, F.J. 2002. Colorants. *In* Watson, D., ed. *Colour in food: Improving quality (Food chemical safety, Volume 2).* Cambridge, U.K.: Woodhead Publishing Co.

Ganiger, S., H.N. Malleshappa, H. Krishnappa, et al. 2007. A two generation reproductive toxicity study with curcumin, turmeric yellow, in Wistar rats. *Food Chem. Toxicol.* 45(1):64-69.

Garg, S.K. 1974. Effect of *Curcuma longa* (rhizomes) on fertility in experimental animals. *Planta Med.* 26(3):225-227.

Goh, C.L., and S.K. Ng. 1987. Allergic contact dermatitis to *Curcuma longa* (turmeric). *Contact Dermat.* 17(3):186-186.

Goodpasture, C.E., and F.E. Arrighi. 1976. Effects of food seasonings on cell cycle and chromosome morphology of mammalian cells in vitro with special reference to turmeric. *Food Cos. Toxicol.* 14(1):9.

Govindarajan, V.S. 1980. Turmeric—Chemistry, technology, and quality. *CRC Crit. Rev. Food Sci. Nutr.* 12(3):199-301.

Gupta, B., V.K. Kulshrestha, R.K. Srivastava, and D.N. Prasad. 1980. Mechanisms of curcumin induced gastric ulcer in rats. *Indian J. Med. Res.* 71:806-14.

Hanai, H., T. Iida, K. Takeuchi, et al. 2006. Curcumin maintenance therapy for ulcerative colitis: Randomized, multicenter, double-blind, placebo-controlled trial. *Clin. Gastroenterol. Hepatol.* 4(12):1502-1506.

Hata, M., E. Sasaki, M. Ota, et al. 1997. Allergic contact dermatitis from curcumin (turmeric). *Contact Dermat.* 36(2):107-108.

Hsu, C.H., and A.L. Cheng. 2007. Clinical studies with curcumin. *Adv. Exp. Med. Biol.* 595:471-480.

Jain, A.K., H. Tezuka, T. Kada, and I. Tomita. 1987. Evaluation of genotoxic effects of turmeric in mice. *Curr. Sci.* 56(19):1005-1006.

Jantan, I., S.M. Raweh, H.M. Sirat, et al. 2007. Inhibitory effect of compounds from Zingiberaceae species on human platelet aggregation. *Phytomedicine* 15:306-309.

Jensen, N.J. 1982. Lack of mutagenic effect of turmeric oleoresin and curcumin in the Salmonella mammalian microsome test. *Mutat. Res.* 105(6):393-396.

Joshi, J., S. Ghaisas, A. Vaidya, et al. 2003. Early human safety study of turmeric oil (*Curcuma longa* oil) administered orally in healthy volunteers. *J. Assoc. Physicians India* 51:1055-1060.

Kandarkar, S.V., S.S. Sawant, A.D. Ingle, S.S. Deshpande, and G.B. Maru. 1998. Subchronic oral hepatotoxicity of turmeric in mice—Histopathological and ultrastructural studies. *Indian J. Exp. Biol.* 36(7):675-679.

Kiec-Swierczynska, M., and B. Krecisz. 1998. Occupational allergic contact dermatitis due to curcumin food colour in a pasta factory worker. *Contact Dermat.* 39(1):30-31.

Lamb, S.R., and S.M. Wilkinson. 2003. Contact allergy to tetrahydrocurcumin. *Contact Dermat.* 48(4):227.

Lao, C.D., M.T. Ruffin, D. Normolle, et al. 2006. Dose escalation of a curcuminoid formulation. *BMC Complement. Altern. Med.* 6:10.

Lee, H.S. 2006. Antiplatelet property of *Curcuma longa* L. rhizome-derived ar-turmerone. *Bioresour. Technol.* 97(12):1372-1376.

Leung, A.Y., and S. Foster. 1996. *Encyclopedia of common natural ingredients used in food, drugs, and cosmetics.* 2nd ed. New York: Wiley.

Liddle, M., C. Hull, C. Liu, and D. Powell. 2006. Contact urticaria from curcumin. *Dermatitis* 17(4):196-197.

Mahmoud, I., A. Alkofahi, and A. Abdelaziz. 1992. Mutagenic and toxic activities of several spices and some Jordanian medicinal plants. *Int. J. Pharmacog.* 30:81-85.

Mills, S., and K. Bone. 2005. *The essential guide to herbal safety.* St. Louis: Elsevier.

Miquel, J., M. Martinez, A. Diez, et al. 1995. Effects of turmeric on blood and liver lipoperoxide levels of mice: Lack of toxicity. *Age* 18(4):171-174.

Nagabhushan, M., and S.V. Bhide. 1986. Nonmutagenicity of curcumin and its antimutagenic action versus chili and capsaicin. *Nutr. Cancer* 8(3):201-210.

Nath, A.K., and D.M. Thappa. 2007. Kumkum-induced dermatitis: An analysis of 46 cases. *Clin. Exp. Dermatol.* 32(4):385-387.

NTP. 1993. NTP toxicology and carcinogenesis studies of turmeric oleoresin (CAS No. 8024-37-1) (major component 79%–85% curcumin, CAS No. 458-37-7) in F344/N rats and B6C3F1 mice (feed studies). *Natl. Toxicol. Program Tech. Rep. Ser.* No. 427:1-275.

Prucksunand, C., B. Indrasukhsri, M. Leethochawalit, and K. Hungspreugs. 2001. Phase II clinical trial on effect of the long turmeric (*Curcuma longa* Linn) on healing of peptic ulcer. *Southeast Asian J. Trop. Med. Public Health* 32(1):208-215.

Qureshi, S., A.H. Shah, and A.M. Ageel. 1992. Toxicity studies on *Alpinia galanga* and *Curcuma longa*. *Planta Med.* 58(2):124-127.

Rafatullah, S., M. Tariq, M.A. Al-Yahya, J.S. Mossa, and A.M. Ageel. 1990. Evaluation of turmeric (*Curcuma longa*) for gastric and duodenal antiulcer activity in rats. *J. Ethnopharmacol.* 29(1):25-34.

Rao, A.J., and S.G. Kotagi. 1984. Antiandrogenic action of the plant *Curcuma longa* root extract in male rats. *Ircs. Med. Sci. Biochem.* 12(6):500-501.

Rasyid, A., A.R.A. Rahman, K. Jaalam, and A. Lelo. 2002. Effect of different curcumin dosages on human gall bladder. *Asia Pac. J. Clin. Nutr.* 11(4):314-318.

Sakurane, J., H. Komamura, M. Isonokami, et al. 1999. A case of contact dermatitis due to *Curcuma longa* (turmeric). *Env. Dermatol.* 6(4):237-242.

Shah, B.H., Z. Nawaz, S.A. Pertani, et al. 1999. Inhibitory effect of curcumin, a food spice from turmeric, on platelet-activating factor- and arachidonic acid-mediated platelet aggregation through inhibition of thromboxane formation and Ca^{2+} signaling. *Biochem. Pharmacol.* 58(7):1167-1172.

Shah, R.G., and M.S. Netrawali. 1988. Evaluation of mutagenic activity of turmeric extract containing curcumin, before and after activation with mammalian cecal microbial extract of liver microsomal fraction, in the Ames *Salmonella* test. *Bull. Env. Contam. Toxicol.* 40(3):350-357.

Shankar, T.N., N.V. Shantha, H.P. Ramesh, I.A. Murthy, and V.S. Murthy. 1980. Toxicity studies on turmeric (*Curcuma longa*): Acute toxicity studies in rats, guineapigs and monkeys. *Indian J. Exp. Biol.* 18(1):73-75.

Sharma, R.A., H.R. McLelland, K.A. Hill, et al. 2001. Pharmacodynamic and pharmacokinetic study of oral *Curcuma* extract in patients with colorectal cancer. *Clin. Cancer Res.* 7(7):1894-1900.

Singh, A., S.P. Singh, and R. Bamezai. 1995. Postnatal modulation of hepatic biotransformation system enzymes via translactational exposure of F_1 mouse pups to turmeric and curcumin. *Cancer Lett.* 96(1):87-93.

Singh, A., S.P. Singh, and R. Bamezai. 1996. Effect of arecoline on the curcumin-modulated hepatic biotransformation system enzymes in lactating mice and translactationally exposed F_1 pups. *Nutr. Cancer* 25(1):101-110.

Sivalingam, N., R. Hanumantharaya, M. Faith, et al. 2007. Curcumin reduces indomethacin-induced damage in the rat small intestine. *J. Appl. Toxicol.* 27(6):551-560.

Srivastava, K.C. 1989. Extracts from two frequently consumed spices—cumin (*Cuminum cyminum*) and turmeric (*Curcuma longa*)—inhibit platelet aggregation and alter eicosanoid biosynthesis in human blood platelets. *Prostaglandins Leukotrienes Essent. Fatty Acids* 37(1):57-64.

Srivastava, K.C., A. Bordia, and S.K. Verma. 1995. Curcumin, a major component of food spice turmeric (*Curcuma longa*), inhibits aggregation and alters eicosanoid metabolism in human blood platelets. *Prostaglandins Leukotrienes Essent. Fatty Acids* 52(4):223-227.

Swarnakar, S., K. Ganguly, P. Kundu, et al. 2005. Curcumin regulates expression and activity of matrix metalloproteinases 9 and 2 during prevention and healing of indomethacin-induced gastric ulcer. *J. Biol. Chem.* 280(10):9409.

Thompson, D.A., and B.B. Tan. 2006. Tetrahydrocurcumin-related allergic contact dermatitis. *Contact Dermat.* 55(4):254-255.

Vijayalaxmi. 1980. Genetic effects of turmeric and curcumin in mice and rats. *Mutat. Res.* 79(2):125-132.

Wahlstrom, B., and G. Blennow. 1978. Study on fate of curcumin in rats. *Acta Pharmacol. Toxicol.* 43(2):86-92.

Cuscuta spp. — Cuscutaceae

Cuscuta chinensis Lam.
SCN: Chinese dodder
PN: *tu si zi* (seed)
OCN: Chinese cuscuta

Cuscuta japonica **Choisy**
SCN: Japanese dodder
PN: *tu si zi* (seed)
OCN: Japanese cuscuta
Part: seed

Quick Reference Summary

Safety Class: 1
Interaction Class: A

Contraindications
None known.

Other Precautions
None known.

Drug and Supplement Interactions
None known.

Adverse Events and Side Effects
None known.

Pharmacological Considerations
None known.

Pregnancy and Lactation
Chinese dodder is used in traditional Chinese medicine to "calm the fetus" and is noted as an important herb for habitual or threatened miscarriage (Bensky et al. 2004; Chen and Chen 2004). An animal study indicated that compounds from Chinese dodder prevented miscarriage (Ma et al. 2008).

No information on the safety of Chinese dodder during lactation was identified. While this review did not identify any concerns for use while nursing, safety has not been conclusively established.

C

Review Details

I. Drug and Supplement Interactions

Clinical Trials of Drug or Supplement Interactions
No clinical trials of drug or supplement interactions were identified.

Case Reports of Suspected Drug or Supplement Interactions
No case reports of suspected drug or supplement interactions were identified.

Animal Trials of Drug or Supplement Interactions
No animal trials of drug or supplement interactions were identified.

II. Adverse Events

Case Reports of Adverse Events
Nausea, vomiting, drowsiness, gastric bleeding, and clonic spasms have been reported in association with Chinese dodder use. Information on dose, duration, and relevant medical history were not available (Bensky et al. 2004).

III. Pharmacology and Pharmacokinetics

Human Pharmacological Studies
No relevant human pharmacological studies were identified.

Animal Pharmacological Studies
No relevant animal pharmacological studies were identified.

In Vitro Pharmacological Studies
No relevant in vitro pharmacological studies were identified.

IV. Pregnancy and Lactation
Chinese dodder is used in traditional Chinese medicine to "calm the fetus" and is noted as an important herb for habitual or threatened miscarriage (Bensky et al. 2004; Chen and Chen 2004).

Prevention of miscarriage was observed in pregnant rats orally administered the total flavones of Chinese dodder. Doses and days of administration were not specified in the English language abstract (Ma et al. 2008).

No information on the safety of Chinese dodder during lactation was identified.

V. Toxicity Studies

Acute Toxicity
The LD_{50} of a subcutaneously administered ethanol extract of Chinese dodder in mice is 2.465 g/kg (Zhu 1998). The LD_{50} of orally administered Chinese dodder in mice could not be determined at doses up to 5 g/kg (Akbar et al. 1985). No toxic effects were observed in mice orally administered up to 40 g/kg of an ethanol extract of Chinese dodder (Zhu 1998).

Literature Cited

Akbar, S., M. Nisa, and M. Tariq. 1985. Central nervous system depressant activity of *Cuscuta chinensis*. *Int. J. Crude Drug Res.* 23(2):91-94.

Bensky, D., S. Clavey, and E. Stöger. 2004. *Chinese herbal medicine: Materia medica.* 3rd ed. Seattle: Eastland Press.

Chen, J.K., and T.T. Chen. 2004. *Chinese medical herbology and pharmacology.* City of Industry, CA: Art of Medicine Press.

Ma, H.X., Z.L. You, and R.G. Wang. 2008. Effect of total flavones from *Cuscuta chinensis* on expression of Th type-1/Th type-2 cytokines, serum P and PR in abortion rats model. *Zhong yao cai* 31(8):1201.

Zhu, Y.-P. 1998. *Chinese materia medica: Chemistry, pharmacology and applications* Amsterdam: Harwood Academic Publishers.

Cyathula officinalis K.C. Kuan — Amaranthaceae

SCN: cyathula
PN: *chuan niu xi* (root)

Part: root

C

Quick Reference Summary

Safety Class: 2b, 2d
Interaction Class: A

Contraindications

Not for use in pregnancy except under the supervision of a qualified healthcare practitioner (Bensky et al. 2004; Chen and Chen 2004).

Not for use in excessive menstruation (Bensky et al. 2004; Chen and Chen 2004).

Other Precautions

None known.

Drug and Supplement Interactions

None known.

Adverse Events and Side Effects

None known.

Pharmacological Considerations

Although contraindicated in excessive menstruation, a human clinical trial indicated a therapeutic effect of cyathula in patients with uterine bleeding (Chen and Chen 2004).

Pregnancy and Lactation

Reference texts on traditional Chinese medicine indicate that cyathula should not be used during pregnancy (Bensky et al. 2004; Chen and Chen 2004). An animal study indicated decreased fertility and increased risk of miscarriage after treatment with cyathula (Chen and Chen 2004). Based on this information, use during pregnancy is not recommended except under the supervision of a qualified healthcare practitioner.

No information on the safety of cyathula during lactation was identified. While this review did not identify any concerns for use while nursing, safety has not been conclusively established.

Review Details

I. Drug and Supplement Interactions

Clinical Trials of Drug or Supplement Interactions

No clinical trials of drug or supplement interactions were identified.

Case Reports of Suspected Drug or Supplement Interactions

No case reports of suspected drug or supplement interactions were identified.

Animal Trials of Drug or Supplement Interactions

No animal trials of drug or supplement interactions were identified.

II. Adverse Events

Case Reports of Adverse Events

No case reports of adverse events were identified.

III. Pharmacology and Pharmacokinetics

Human Pharmacological Studies

No relevant human pharmacological studies were identified.

Animal Pharmacological Studies

No relevant animal pharmacological studies were identified.

In Vitro Pharmacological Studies

No relevant in vitro pharmacological studies were identified.

IV. Pregnancy and Lactation

Reference texts on traditional Chinese medicine indicate that cyathula should not be used during pregnancy (Bensky et al. 2004; Chen and Chen 2004).

In mice administered 250 to 500 mg/kg cyathula daily for 20 days, a decrease in egg fertilization and increase in the number of miscarriages were observed (Chen and Chen 2004).

No information on the safety of cyathula during lactation was identified.

V. Toxicity Studies

Genotoxicity

An aqueous extract of cyanthula inhibited chemically induced mutagenicity in the Ames assay (Niikawa et al. 1995).

Literature Cited

Bensky, D., S. Clavey, and E. Stöger. 2004. *Chinese herbal medicine: Materia medica.* 3rd ed. Seattle: Eastland Press.

Chen, J.K., and T.T. Chen. 2004. *Chinese medical herbology and pharmacology.* City of Industry, CA: Art of Medicine Press.

Niikawa, M., A.F. Wu, T. Sato, H. Nagase, and H. Kito. 1995. Effects of Chinese medicinal plant extracts on mutagenicity of Trp-P-1. *Nat. Med.* 49(3):329-331.

Cymbopogon citratus (DC. ex Nees) Stapf — Poaceae

SCN: West Indian lemongrass
Syn: *Andropogon citratus* DC. ex Nees
AN: *bhutrina*
OCN: lemongrass; fever grass
Part: herb

Quick Reference Summary

Safety Class: 1
Interaction Class: A

Contraindications

None known.

Other Precautions

None known.

Drug and Supplement Interactions

None known.

Adverse Events and Side Effects

Allergic contact dermatitis to West Indian lemongrass was reported after ingestion of an infusion in a person who had been repeatedly exposed to the essential oil (Bleasel et al. 2002).

Pharmacological Considerations

An animal study demonstrated that West Indian lemongrass may modify glucose regulation, although a human study indicated no effects on serum glucose levels (Adeneye and Agbaje 2007; Leite et al. 1986). People with diabetes are advised to monitor their blood sugar closely and discuss the use of this herb with a qualified healthcare practitioner prior to use.

Pregnancy and Lactation

Among herbal references reporting on West Indian lemongrass use in different regions, a text on medicinal plants of the Caribbean indicated that West Indian lemongrass has been used as an abortifacient (Morton 1981). However, no other identified references indicated this use of West Indian lemongrass (Ayiku 1992; Blaschek et al. 2006; Teuscher et al. 2006), and no adverse effects on development were observed in mice administered West Indian lemongrass throughout pregnancy at doses equivalent to 10 to 20 times the standard human dose (Souza Formigoni et al. 1986).

Studies of the compound citral indicated that the developmental toxicity no-observed-adverse-effect level (NOAEL) in rats was approximately 60 mg/kg (Delgado et al. 1993). In rats exposed to citral vapor, no adverse effects on fetuses were observed up to or above levels that were toxic to the mothers (Gaworski et al. 1992). Studies of the compound β-myrcene indicated that the NOAEL for fertility and general reproductive performance was 300 mg/kg daily (Paumgartten et al. 1998), whereas that for developmental toxicity was 250 mg/kg (Delgado et al. 1993) and that for the compound citral was below 60 mg/kg (Delgado et al. 1993).

No information on the safety of West Indian lemongrass during lactation was identified. While this review did not identify any concerns for use while nursing, safety has not been conclusively established.

Review Details

I. Drug and Supplement Interactions

Clinical Trials of Drug or Supplement Interactions

No clinical trials of drug or supplement interactions were identified.

Case Reports of Suspected Drug or Supplement Interactions

No case reports of suspected drug or supplement interactions were identified.

Animal Trials of Drug or Supplement Interactions

No animal trials of drug or supplement interactions were identified.

II. Adverse Events

Case Reports of Adverse Events

Allergic contact dermatitis was reported after ingestion of West Indian lemongrass tea by an aromatherapist. The reporting authors noted that persons working with essential oils should be aware of the sensitization potential of these products (Bleasel et al. 2002).

III. Pharmacology and Pharmacokinetics

Human Pharmacological Studies

In healthy volunteers orally administered an infusion made from 2, 4, or 10 g of West Indian lemongrass daily for 2 weeks, no changes in serum glucose, urea, creatinine, cholesterol, triglycerides, lipids, total bilirubin, indirect bilirubin, GOT, GPT, alkaline phosphatase, total protein, albumin, LDH, or CPK were observed. Slight elevations of direct bilirubin and of amylase were observed in some of the volunteers, but without any clinical manifestation. At the 10 g dose, upset stomach was reported, while no adverse events were reported at the lower doses (Leite et al. 1986).

In an evaluation of fragrance sensitization, West Indian lemongrass essential oil elicited reactions in 1.6% of the 1600 fragrance-sensitive volunteers (Frosch et al. 2002).

In sensitization studies, no sensitization of West Indian lemongrass essential oil was observed at test concentrations of 4 or 5% in a petroleum base (Opdyke 1979). In a 48-hour closed patch test, no irritation was observed at a test concentration of 4%.

Animal Pharmacological Studies

A dose-dependent reduction in fasting plasma glucose levels was observed in healthy rats orally administered 125 to 500 mg/kg of an aqueous extract of West Indian lemongrass daily for 42 days (Adeneye and Agbaje 2007).

In the local lymph node assay in mice, weak dermal sensitization potential of West Indian lemongrass was observed (Lalko and Api 2006).

No phototoxic effects of undiluted West Indian lemongrass essential oil were observed in pigs or hairless mice (Opdyke 1979). Applied full strength to abraded rabbit skin, West Indian lemongrass essential oil was moderately irritating (Opdyke 1979).

In Vitro Pharmacological Studies

A reduction in cardiac rate, with no change in contractile force, was observed in isolated rat hearts treated with aqueous extracts of West Indian lemongrass (Gazola et al. 2004).

In human colon carcinoma cells (Caco-2) treated with an aqueous extract of West Indian lemongrass, mild induction of sulfoconjugation was detected in cells exposed to a 5% solution (Okamura and Tamura 2004).

IV. Pregnancy and Lactation

No adverse effects were observed in the offspring of mice orally administered 4 or 8 ml/kg of an aqueous extract of West Indian lemongrass (approximately 10 and 20 times the standard human dose) daily throughout gestation (Souza Formigoni et al. 1986).

A reference on medicinal plants of the Caribbean indicated that West Indian lemongrass has been used as an abortifacient. No information on dose or preparation used was listed (Morton 1981).

In rats orally administered the compound citral at doses of 60, 125, 250, 500, or 1000 mg/kg daily on days 6 to 15 of pregnancy, a transient decrease in weight gain and reduction in maternal body weight (minus uterine weight) was observed at the highest dose. A slight but statistically significant increase in the ratio of resorptions per implantations was observed at the 60 and 125 mg/kg doses. Signs of fetal growth retardation and a higher incidence of minor skeletal abnormalities were found in doses higher than 60 mg/kg. No increase in the frequency of visceral anomalies was found at any dose level, but an increase in fetal spleen weight was observed in doses higher than 125 mg/kg. The reporting authors indicated that the fetal no-observed-adverse-effect level was below 60 mg/kg (Delgado et al. 1993).

In rats exposed to the compound citral at concentrations of 0, 10, or 34 ppm as vapor or 68 ppm as an aerosol-vapor mixture for 6 hours daily on days 6 to 15 of pregnancy, exposure to 68 ppm was maternally toxic. At this dose level, reduced body weight gains, ocular opacity, breathing difficulty, nasal discharge, and salivation were observed. No maternal toxicity was seen at the lower vapor exposure levels. The number of corpora lutea, implantations, resorptions, fetal viability, litter size, and sex ratio were not adversely affected by citral at any exposure level tested, and no exposure-related malformations were observed. At a maternally toxic exposure level, a slight reduction in mean fetal body weight and a slight increase in the incidence of hypoplastic bones were noted (Gaworski et al. 1992).

In rats orally administered 0, 100, 300, or 500 mg/kg of the compound β-myrcene in pregnancy daily for 21 days prior to mating, during mating and pregnancy, and throughout the period of lactation up to postnatal day 21, no adverse effects on development were observed except at the highest dose. At the highest dose, increases in the resorption rate and frequency of fetal skeleton anomalies were observed. No sign of maternal toxicity and no increase in externally visible malformations were observed at any dose level. The authors concluded that the no-observed-adverse-effect level (NOAEL) for fertility and general reproductive performance is 300 mg/kg daily (Paumgartten et al. 1998).

In rats orally administered 0.25, 0.5, 1.0, or 1.5 g/kg of the compound β-myrcene beginning on day 15 of pregnancy through postnatal day 21, no adverse effects on offspring were seen at the lowest dose tested. At doses of 0.5 g/kg and higher, decreased birth weight, increased perinatal mortality, and delays in the day of appearance of landmarks of postnatal development were observed, and fertility was impaired in female offspring exposed to the two highest doses of β-myrcene. The authors concluded that the NOAEL for peri- and postnatal developmental toxicity is 0.250 g/kg (Delgado et al. 1993).

No information on the safety of West Indian lemongrass during lactation was identified.

V. Toxicity Studies

Acute Toxicity

The LD_{50} of orally administered West Indian lemongrass essential oil in rats is 3.25 g/kg (Fandohan et al. 2008). The LD_{50} of orally or topically administered West Indian lemongrass essential oil in rabbits could not be determined at doses up to 5 g/kg (Opdyke 1979).

The LD_{50} of an orally administered aqueous extract of West Indian lemongrass in rats could not be determined at doses up to 5 g/kg (Adeneye and Agbaje 2007). The LD_{50} of an orally administered alcohol extract of West Indian lemongrass in rats is 460 mg/kg (Lagarto Parra et al. 2001).

Short-Term Toxicity

In rats orally administered West Indian lemongrass essential oil at doses of 50, 500, 1000, 1500, 2000, or 3000 mg/kg daily for 14 days, no adverse effects were observed in animals treated with doses of 5 to 1500 mg/kg. At doses of 2000 or 3000 mg/kg, all animals died on the second day (Fandohan et al. 2008).

No adverse effects were observed in rats intragastrically administered 20 ml/kg of an aqueous extract of West Indian lemongrass daily for 6 weeks. A diuretic effect, similar to that of green tea, was noted (Mirza et al. 2001).

Genotoxicity

In the Ames test for mutagenicity in *Salmonella typhimurium* strains TA97a, TA98, TA100, and TA104 with or without metabolic activation by S9, some mutagenic activity of an aqueous extract of West Indian lemongrass was observed in TA104 without metabolic activation (Rivera et al. 1994). Conversely, an ethanol extract of West Indian lemongrass was antimutagenic in *Salmonella typhimurium* strains TA98 and TA100 treated with mutagenic compounds (Vinitketkumnuen et al. 1994).

Literature Cited

Adeneye, A.A., and E.O. Agbaje. 2007. Hypoglycemic and hypolipidemic effects of fresh leaf aqueous extract of *Cymbopogon citratus* Stapf. in rats. *J. Ethnopharmacol.* 112(3):440-444.

Ayiku, M.N.B. 1992. *Ghana herbal pharmacopoeia.* Osu, Accra, Ghana: The Advent Press.

Blaschek, W., S. Ebel, E. Hackenthal, et al. 2006. *Hagers handbuch der drogen und arzneistoffe. HagerROM.* Heidelberg: Springer.

Bleasel, N., B. Tate, and M. Rademaker. 2002. Allergic contact dermatitis following exposure to essential oils. *Australas. J. Dermatol.* 43(3):211-213.

Delgado, I.F., A.C. Nogueira, C.A. Souza, et al. 1993. Peri- and postnatal developmental toxicity of beta-myrcene in the rat. *Food Chem. Toxicol.* 31(9):623-638.

Fandohan, P., B. Gnonlonfin, A. Laleye, et al. 2008. Toxicity and gastric tolerance of essential oils from *Cymbopogon citratus*, *Ocimum gratissimum* and *Ocimum basilicum* in Wistar rats. *Food Chem. Toxicol.* 46(7):2493-2497.

Frosch, P.J., J.D. Johansen, T. Menne, et al. 2002. Further important sensitizers in patients sensitive to fragrances. *Contact Dermat.* 47(5):279-287.

Gaworski, C.L., T.A. Vollmuth, R.G. York, J.D. Heck, and C. Aranyi. 1992. Developmental toxicity evaluation of inhaled citral in Sprague-Dawley rats. *Food Chem. Toxicol.* 30(4):269-275.

Gazola, R., D. Machado, C. Ruggiero, G. Singi, and M. Macedo Alexandre. 2004. *Lippia alba, Melissa officinalis* and *Cymbopogon citratus*: Effects of the aqueous extracts on the isolated hearts of rats. *Pharmacol. Res.* 50(5):477-480.

Lagarto Parra, A., R. Silva Yhebra, I. Guerra Sardinas, and L. Iglesias Buela. 2001. Comparative study of the assay of *Artemia salina* L. and the estimate of the medium lethal dose (LD_{50} value) in mice, to determine oral acute toxicity of plant extracts. *Phytomedicine* 8(5):395-400.

Lalko, J., and A.M. Api. 2006. Investigation of the dermal sensitization potential of various essential oils in the local lymph node assay. *Food Chem. Toxicol.* 44(5):739-746.

Leite, J.R., L. Seabra Mde, E. Maluf, et al. 1986. Pharmacology of lemongrass (*Cymbopogon citratus* Stapf). III. Assessment of eventual toxic, hypnotic and anxiolytic effects on humans. *J. Ethnopharmacol.* 17(1):75-83.

Mirza, M., A. Askari, Z. Yaqueen, Z. Ahmad, and R.B. Qadri. 2001. Diuretic studies on lemon grass tea from *Cymbopogon citratus* (DC) Stapf. in rat. *Pak. J. Sci. Indust. Res.* 44(2):96-100.

Morton, J.F. 1981. *Atlas of medicinal plants of Middle America.* Springfield, IL: C.C Thomas.

Okamura, S., and H. Tamura. 2004. Effect of herbal teas on conjugation reactions in a human colon carcinoma cell line, Caco-2. *J. Health Sci.* 50(2):189-192.

Opdyke, D.L.J. 1979. *Monographs on fragrance raw materials.* New York: Pergamon.

Paumgartten, F.J., R.R. De-Carvalho, C.A. Souza, K. Madi, and I. Chahoud. 1998. Study of the effects of beta-myrcene on rat fertility and general reproductive performance. *Braz. J. Med. Biol. Res.* 31(7):955-965.

Rivera, I.G., M.T. Martins, P.S. Sanchez, et al. 1994. Genotoxicity assessment through the Ames test of medicinal plants commonly used in Brazil. *Env. Toxicol. Water Qual.* 9(2):87-93.

Souza Formigoni, M.L., H.M. Lodder, O. Gianotti Filho, T.M. Ferreira, and E.A. Carlini. 1986. Pharmacology of lemongrass (*Cymbopogon citratus* Stapf). II. Effects of daily two month administration in male and female rats and in offspring exposed "in utero." *J. Ethnopharmacol.* 17(1):65-74.

Teuscher, E., U. Bauermann, and M. Werner. 2006. *Medicinal spices: A handbook of culinary herbs, spices, spice mixtures and their essential oils.* Translated by Brinckmann, J., and M.P. Lindenmaier. Stuttgart: Medpharm Scientific Publishers.

Vinitketkumnuen, U., R. Puatanachokchai, P. Kongtawelert, N. Lertprasertsuke, and T. Matsushima. 1994. Antimutagenicity of lemon grass (*Cymbopogon citratus* Stapf) to various known mutagens in *Salmonella* mutation assay. *Mutat. Res.* 341(1):71-75.

Cynanchum atratum Bunge

Asclepiadaceae

SCN: *Cynanchum atratum*
PN: *bai wei* (root and rhizome)
OCN: black-end swallowwort; swallowwort
Part: root

Quick Reference Summary

Safety Class: 1
Interaction Class: A

Contraindications
None known.

Other Precautions
None known.

Drug and Supplement Interactions
None known.

Adverse Events and Side Effects
None known.

Pharmacological Considerations
None known.

Pregnancy and Lactation
No information on the safety of *Cynanchum atratum* in pregnancy or lactation was identified in the scientific or traditional literature. Although this review did not identify any concerns for use while pregnant or nursing, safety has not been conclusively established.

Review Details

I. Drug and Supplement Interactions

Clinical Trials of Drug or Supplement Interactions
No clinical trials of drug or supplement interactions were identified.

Case Reports of Suspected Drug or Supplement Interactions
No case reports of suspected drug or supplement interactions were identified.

Animal Trials of Drug or Supplement Interactions
No animal trials of drug or supplement interactions were identified.

II. Adverse Events

Case Reports of Adverse Events
In overdose (aqueous extract of 30–45 g), *Cynanchum atratum* has been associated with heart palpitations, nausea, vomiting, dizziness, headache, diarrhea, and salivation (Bensky et al. 2004).

III. Pharmacology and Pharmacokinetics

Human Pharmacological Studies
No relevant human pharmacological studies were identified.

Animal Pharmacological Studies
The compound cynatroside B demonstrated antiacetylcholinesterase activity in mice (Lee et al. 2005). No information was identified on the concentration of cynatroside B in *Cynanchum atratum.*

In Vitro Pharmacological Studies
No relevant in vitro pharmacological studies were identified.

IV. Pregnancy and Lactation

No information on the use of *Cynanchum atratum* during pregnancy or lactation was identified.

V. Toxicity Studies

No toxicity studies were identified.

Literature Cited

Bensky, D., S. Clavey, and E. Stöger. 2004. *Chinese herbal medicine: Materia medica.* 3rd ed. Seattle: Eastland Press.

Lee, K.Y., J.S. Yoon, E.S. Kim, S.Y. Kang, and Y.C. Kim. 2005. Anti-acetylcholinesterase and anti-amnesic activities of a pregnane glycoside, cynatroside B, from *Cynanchum atratum. Planta Med.* 71(1):7-11.

Cynomorium songaricum Rupr.

Cynomoriaceae

SCN: cynomorium
PN: *suo yang* (fleshy stem)

Part: fleshy stem

Quick Reference Summary

Safety Class: 1
Interaction Class: A

Contraindications
None known.

Other Precautions

None known.

Drug and Supplement Interactions

None known.

Adverse Events and Side Effects

None known.

Pharmacological Considerations

None known.

Pregnancy and Lactation

No information on the safety of cynomorium in pregnancy or lactation was identified in the scientific or traditional literature. Although this review did not identify any concerns for use while pregnant or nursing, safety has not been conclusively established.

C

Review Details

I. Drug and Supplement Interactions

Clinical Trials of Drug or Supplement Interactions

No clinical trials of drug or supplement interactions were identified.

Case Reports of Suspected Drug or Supplement Interactions

No case reports of suspected drug or supplement interactions were identified.

Animal Trials of Drug or Supplement Interactions

No animal trials of drug or supplement interactions were identified.

II. Adverse Events

Case Reports of Adverse Events

No case reports of adverse events were identified.

III. Pharmacology and Pharmacokinetics

Human Pharmacological Studies

No relevant human pharmacological studies were identified.

Animal Pharmacological Studies

No relevant animal pharmacological studies were identified.

In Vitro Pharmacological Studies

Estrogenic activity of an ethanol extract of cynomorium was observed in a recombinant yeast system with a human estrogen receptor expression plasmid and a reporter plasmid (Zhang et al. 2005).

IV. Pregnancy and Lactation

No information on the safety of cynomorium during pregnancy or lactation was identified.

V. Toxicity Studies

No toxicity studies were identified.

Literature Cited

Zhang, C.Z., S.X. Wang, Y. Zhang, J.P. Chen, and X.M. Liang. 2005. In vitro estrogenic activities of Chinese medicinal plants traditionally used for the management of menopausal symptoms. *J. Ethnopharmacol.* 98(3):295-300.

Cyperus rotundus L. — Cyperaceae

SCN: cyperus
AN: *musta*
PN: *xiang fu* (rhizome)
OCN: galingale; nut grass; nut sedge
Part: rhizome

Quick Reference Summary

Safety Class: 1
Interaction Class: A

Contraindications

None known.

Other Precautions

None known.

Drug and Supplement Interactions

None known.

Adverse Events and Side Effects

None known.

Pharmacological Considerations

None known.

Pregnancy and Lactation

No information on the safety of cyperus in pregnancy or lactation was identified in the scientific or traditional literature. Although this review did not identify any concerns for use while pregnant or nursing, safety has not been conclusively established.

Review Details

C

I. Drug and Supplement Interactions

Clinical Trials of Drug or Supplement Interactions

No clinical trials of drug or supplement interactions were identified.

Case Reports of Suspected Drug or Supplement Interactions

No case reports of suspected drug or supplement interactions were identified.

Animal Trials of Drug or Supplement Interactions

No animal trials of drug or supplement interactions were identified.

II. Adverse Events

Case Reports of Adverse Events

No case reports of adverse events were identified.

III. Pharmacology and Pharmacokinetics

Human Pharmacological Studies

No relevant human pharmacological studies were identified.

Animal Pharmacological Studies

A decrease in serum glucose levels was observed in diabetic rats orally administered 500 mg/kg of a cyperus extract daily for 7 days (Raut and Gaikwad 2006).

In ovariectomized rats, two subcutaneous doses of 0.2 ml of cyperus essential oil administered at an interval of 6 hours resulted in complete keratinization of the vaginal epithelium within 48 h. After three doses of 0.3 ml each, white cells appeared among keratinized cells (Kilani et al. 2005).

In Vitro Pharmacological Studies

No relevant in vitro pharmacological studies were identified.

IV. Pregnancy and Lactation

No information on the use of cyperus during pregnancy or lactation was identified.

V. Toxicity Studies

Acute Toxicity

The LD_{50} of an intraperitoneally administered cyperus alcohol extract in mice is 1.5 g/kg (Zhu 1998), whereas that of cyperus essential oil in rats is 0.29 ml/kg and that of the compound cyperone in rats is 1.5 g/kg (Chen and Chen 2004).

Short-Term Toxicity

No adverse effects were reported in rats fed diets containing 25% cyperus (duration of trial not specified in English language translation) (Chen and Chen 2004).

Literature Cited

Chen, J.K., and T.T. Chen. 2004. *Chinese medical herbology and pharmacology*. City of Industry, CA: Art of Medicine Press.

Kilani, S., R.B. Ammar, I. Bouhlel, et al. 2005. Investigation of extracts from (Tunisian) *Cyperus rotundus* as antimutagens and radical scavengers. *Env. Toxicol. Pharmacol.* 20(3):478-484.

Raut, N.A., and N.J. Gaikwad. 2006. Antidiabetic activity of hydro-ethanolic extract of *Cyperus rotundus* in alloxan induced diabetes in rats. *Fitoterapia* 77(7-8):585-588.

Zhu, Y.-P. 1998. *Chinese materia medica: Chemistry, pharmacology and applications.* Amsterdam: Harwood Academic Publishers.

Cytisus scoparius (L.) Link — Fabaceae

SCN: Scotch broom
Syn: *Sarothamnus scoparius* (L.) Wimm. ex W.D.J. Koch; *Spartium scoparium* L.
OCN: broom; scoparium
Part: flowering top

Quick Reference Summary

Safety Class: 3
Interaction Class: B

Contraindications

Not for use except under the supervision of an expert qualified in the appropriate use of this substance (Felter and Lloyd 1898; Leung and Foster 1996; Martindale and Reynolds 1996; Williamson 2003).

Other Precautions

None known.

Drug and Supplement Interactions

Based on the presence of small amounts of tyramine, Scotch broom should not be taken with MAO inhibitors (Wichtl 2004). *Also see* Pharmacological Considerations for this entry.

Notice

Abortifacient (List and Hörhammer 1973); *see* Appendix 2.

Diuretic (Felter and Lloyd 1898; Wood and LaWall 1926) *see* Appendix 2.

Uterine stimulant (Seel 1949; Wolfes et al. 1936) ; *see* Appendix 2.

Adverse Events and Side Effects

Large doses of Scotch broom are reported to be emetic and cathartic, while smaller doses are reported to be diuretic (Felter and Lloyd 1898; Wood and LaWall 1926). A staggering gait, impaired vision, and profuse vomiting and sweating have been reported in association with Scotch broom use (Felter and Lloyd 1898).

Pharmacological Considerations

Scotch broom contains relatively small amounts (0.08–0.8%) of the compound tyramine (also found in many fermented foods), a monoamine compound that may accumulate and cause high blood pressure if MAO inhibitors are being taken. No cases of such an interaction from Scotch broom have been documented (Gresser et al. 1996; Schmalfuss and Heider 1931; Wichtl 2004).

The compound sparteine is antiarrhythmic, inhibiting the transport of sodium ions across cell membranes and reducing overstimulation of the conduction system of the heart (Wichtl 2004).

The compound sparteine is a substrate of the drug-metabolizing isoenzyme CYP2D6 (Casarett et al. 2001; Zanger et al. 2004). Serum levels of sparteine may be altered by drugs or supplements that induce or inhibit CYP2D6.

Pregnancy and Lactation

In vitro studies have indicated that Scotch broom has a stimulating effect on the uterus (Seel 1949; Wolfes et al. 1936). The compound sparteine was the subject of human clinical trials as a labor inducer and was found to be as potent as oxytocin (Casarett et al. 2001).

No information on the safety of Scotch broom during lactation was identified. While this review did not identify any concerns for use while nursing, safety has not been conclusively established, and this substance is not recommended for use except under the supervision of an expert qualified in its appropriate use.

Review Details

I. Drug and Supplement Interactions

Clinical Trials of Drug or Supplement Interactions

No clinical trials of drug or supplement interactions were identified.

Case Reports of Suspected Drug or Supplement Interactions

No case reports of suspected drug or supplement interactions were identified.

Animal Trials of Drug or Supplement Interactions

No animal trials of drug or supplement interactions were identified.

II. Adverse Events

Case Reports of Adverse Events

A staggering gait, impaired vision, and profuse vomiting and sweating have been reported in association with Scotch broom use. No details on dose or product were provided (Felter and Lloyd 1898).

III. Pharmacology and Pharmacokinetics

Human Pharmacological Studies

No relevant human pharmacological studies were identified.

Animal Pharmacological Studies

No relevant animal pharmacological studies were identified.

In Vitro Pharmacological Studies

No relevant in vitro pharmacological studies were identified.

IV. Pregnancy and Lactation

In vitro studies have indicated that Scotch broom has a stimulating effect on the uterus (Seel 1949; Wolfes et al. 1936). The compound sparteine was the subject of human clinical trials as a labor inducer and was found to be as potent as oxytocin; however, it caused exaggerated responses in some women, including prolonged uterine contraction and abnormally rapid labor (Casarett et al. 2001).

No information on the safety of Scotch broom during lactation was identified.

V. Toxicity Studies

Acute Toxicity

The LD_{50} of the compound sparteine orally administered to rats is approximately 430 mg/kg (Wink 1994).

Literature Cited

Casarett, L.J., C.D. Klaassen, and J. Doull. 2001. *Casarett and Doull's toxicology: The basic science of poisons*. New York: McGraw-Hill Professional.

Felter, H.W., and J.U. Lloyd. 1898. *King's American dispensatory*. 18th ed., 3rd rev. 2 vols. Cincinnati: Ohio Valley Co.

Gresser, G., L. Witte, V.P. Dedkov, and F.C. Czygan. 1996. A survey of quinolizidine alkaloids and phenylethylamine tyramine in *Cytisus scoparius* (Leguminosae) from different origins. *Z. Naturforsch.* 51c:791-801.

Leung, A.Y., and S. Foster. 1996. *Encyclopedia of common natural ingredients used in food, drugs, and cosmetics*. 2nd ed. New York: Wiley.

List, P.H., and H. Hörhammer. 1973. *Hagers handbuch der pharmazeutischen praxis*. Berlin: Springer.

Martindale, W., and J.E.F. Reynolds. 1996. *The extra pharmacopoeia*. 31st ed. London: Pharmaceutical Press.

Schmalfuss, H., and A. Heider. 1931. Tyramine and hydroxytyramine, the blood-pressure-raising substances of the pod of the common broom *Sarothamnus scoparius* Wimm. *Biochem. Z.* 236:226-230.

Seel, H. 1949. Pharmacological and clinical investigations on German medicinal plants. X. *Sarothamnus scoparius* (*Spartium scoparium*) in the therapy of disturbances of the heart rhythm. *Hippokrates* 20:193-196.

Wichtl, M. 2004. *Herbal drugs and phytopharmaceuticals: A handbook for practice on a scientific basis*. 3rd ed. Boca Raton, FL: CRC Press.

Williamson, E.M. 2003. *Potter's herbal cyclopedia*. Saffron Walden, Essex: C.W. Daniel Co.

Wink, M. 1994. Biological activities and potential application of lupin alkaloids. In Neves-Martins, J.M., and M.L. Beirão da Costa, eds. *Advances in lupin research*. Lisboa: ISA Press.

Wolfes, O., H. Kreitmair, and W. Sieckmann. 1936. Broom and its active principles. *Merck's Jahresber.* 50:111-129.

Wood, H., and C. LaWall. 1926. *The dispensatory of the United States of America*. Philadelphia: Lippincott.

Zanger, U.M., S. Raimundo, and M. Eichelbaum. 2004. Cytochrome P450 2D6: Overview and update on pharmacology, genetics, biochemistry. *Naunyn Schmiedebergs Arch. Pharmacol.* 369(1):23-37.

Daemonorops draco (Willd.) Blume. Arecaceae

SCN: dragon's blood palm
PN: *xue jie* (resin)

Part: resin

D

Quick Reference Summary

Safety Class: 2b, 2d
Interaction Class: A

Contraindications

Not for use in pregnancy except under the supervision of a qualified healthcare practitioner (Bensky et al. 2004; Chen and Chen 2004).

Not for internal use during menstruation (Bensky et al. 2004).

Other Precautions

None known.

Drug and Supplement Interactions

None known.

Adverse Events and Side Effects

Allergic reactions to dragon's blood palm have been reported after ingestion and topical exposure (Bensky et al. 2004).

Pharmacological Considerations

None known.

Pregnancy and Lactation

Reference texts on traditional Chinese medicine indicate that dragon's blood palm should not be used during pregnancy (Chen and Chen 2004) or should not be used internally during pregnancy (Bensky et al. 2004). Based on this information, use during pregnancy is not recommended except under the supervision of a qualified healthcare practitioner.

No information on the safety of dragon's blood palm during lactation was identified. While this review did not identify any concerns for use while nursing, safety has not been conclusively established.

Review Details

I. Drug and Supplement Interactions

Clinical Trials of Drug or Supplement Interactions

No clinical trials of drug or supplement interactions were identified.

Case Reports of Suspected Drug or Supplement Interactions

No case reports of suspected drug or supplement interactions were identified.

Animal Trials of Drug or Supplement Interactions

No animal trials of drug or supplement interactions were identified.

II. Adverse Events

Case Reports of Adverse Events

Allergic reactions to dragon's blood palm have been reported after ingestion and topical exposure. Symptoms have included pruritis, skin wheals and eruptions, and angioneurotic edema of the hands and feet (Bensky et al. 2004).

III. Pharmacology and Pharmacokinetics

Human Pharmacological Studies

No relevant human pharmacological studies were identified.

Animal Pharmacological Studies

No additive effects on Δ^9-THC-induced inhibition of mouse motor performance was observed in mice intraperitoneally coadministered 10 or 100 mg/kg of dragon's blood palm and 10 mg/kg Δ^9-THC (Ford et al. 2001). This research was completed after reports of recreational use of *Cannabis sativa* in combination with dragon's blood palm (Ford et al. 2001).

In Vitro Pharmacological Studies

No relevant in vitro pharmacological studies were identified.

IV. Pregnancy and Lactation

Reference texts on traditional Chinese medicine indicate that dragon's blood palm should not be used during pregnancy (Chen and Chen 2004) or should not be used internally during pregnancy (Bensky et al. 2004).

No information on the safety of dragon's blood palm during lactation was identified.

V. Toxicity Studies

No toxicity studies were identified.

Literature Cited

Bensky, D., S. Clavey, and E. Stöger. 2004. *Chinese herbal medicine: Materia medica*. 3rd ed. Seattle: Eastland Press.

Chen, J.K., and T.T. Chen. 2004. *Chinese medical herbology and pharmacology*. City of Industry, CA: Art of Medicine Press.

Ford, S.L., R.R. Steiner, R. Thiericke, R. Young, and W.H. Soine. 2001. Dragon's blood incense: Misbranded as a drug of abuse? *Forensic Sci. Int.* 115(1-2):1-8.

Daucus carota L. ssp. *carota* — Apiaceae

SCN: wild carrot
OCN: Queen Ann's lace

Part: fruit (commonly known as "seed")

Quick Reference Summary

Safety Class: 2b
Interaction Class: A

Contraindications

Not for use in pregnancy except under the supervision of a qualified healthcare practitioner (Garg and Mathur 1972; Kaliwal et al. 1986; Kaliwal and Appaswamy Rao 1979; Kant et al. 1989; Sharma et al. 1976).

Other Precautions

None known.

Drug and Supplement Interactions

None known.

Notice

Diuretic (Felter and Lloyd 1898; Wood and LaWall 1918); *see* Appendix 2.

Adverse Events and Side Effects

None known.

Pharmacological Considerations

Animal studies have indicated effects of wild carrot on estrus cycles and have shown weak estrogenic activity (Majumder et al. 1997; Sharma et al. 1976).

Pregnancy and Lactation

A number of studies on wild carrot have examined the effects of different extracts on implantation in pregnant or recently mated animals. Dose- and timing-related effects on implantation were observed in some studies, while other studies showed no effect on implantation (Garg and Mathur 1972; Kaliwal and Ahamed 1987; Kaliwal et al. 1986; Kaliwal and Appaswamy Rao 1979; Kant et al. 1989; Lal et al. 1986; Sharma et al. 1976). Based on this information, use during pregnancy is not recommended except under the supervision of a qualified healthcare practitioner.

No information on the safety of wild carrot in lactation was identified. While this review did not identify any concerns for use while nursing, safety has not been conclusively established.

Review Details

I. Drug and Supplement Interactions

Clinical Trials of Drug or Supplement Interactions

No clinical trials of drug or supplement interactions were identified.

Case Reports of Suspected Drug or Supplement Interactions

No case reports of suspected drug or supplement interactions were identified.

Animal Trials of Drug or Supplement Interactions

No animal trials of drug or supplement interactions were identified.

II. Adverse Events

Case Reports of Adverse Events

No case reports of adverse events were identified.

III. Pharmacology and Pharmacokinetics

Human Pharmacological Studies

No relevant human pharmacological studies were identified.

Animal Pharmacological Studies

The fatty acid fraction of wild carrot arrested the normal estrus cycle and reduced the weight of ovaries in mice intraperitoneally administered 3 mg/kg of the fraction daily for 15 days. A similar arrest of the cycle was observed in mice intraperitoneally administered 3 or 10 mg/kg of a petroleum ether extract of wild carrot, but no effects of a 1 mg/kg dose were observed (Majumder et al. 1997).

In the 3-day uterotrophic bioassay in mice, a significant increase in wet uterine weights of mice was observed in animals treated with 60 or 120 mg per animal (average animal weight not stated) of an alcohol extract of wild carrot. In comparison to estradiol-17β, the activity was characterized as "extremely weak" (Sharma et al. 1976). In the 3-day antiestrogenic assay, the same extract and doses significantly inhibited the uterotrophic effect of estradiol-17β (Sharma et al. 1976).

In Vitro Pharmacological Studies

No relevant in vitro pharmacological studies were identified.

IV. Pregnancy and Lactation

In rats orally administered 100 mg/kg of fractions of a water extract, 50 mg/kg of fractions of an alcohol extract, or 20 mg/kg of fractions of a petroleum ether extract of wild carrot seed on days 1 to 7 of pregnancy, a reduction in pregnancies was observed with all fractions, although some fractions were more active than others. No adverse effects on development were observed in offspring that were born (Garg and Mathur 1972). A later study re-examining the most effective fractions and doses indicated no effects of these fractions on implantation (Kant et al. 1989).

In rats subcutaneously administered 0.006 ml/kg of a petroleum ether extract of wild carrot on days 1 to 7 of pregnancy, no implantation sites were observed on day 8 of pregnancy (Kaliwal et al. 1986).

In rats subcutaneously administered 0.06 ml/kg of a petroleum ether extract of wild carrot on days 1 to 7 of pregnancy, anti-implantation activity was observed in nearly all of the treated animals. Administration of progesterone blocked the activity of wild carrot seed (Kaliwal and Ahamed 1987). In rats treated subcutaneously with the same extract at the same dose on days 1 to 3, 1 to 5, 3 and 4, or day 3 of pregnancy, no anti-implantation activity was observed. In animals treated with 0.01 ml/kg, no anti-implantation activity was observed, while a dose of 0.02 to 0.04 ml/kg inhibited some implantation (Kaliwal and Appaswamy Rao 1979).

In rats administered an ethanol extract of wild carrot at doses of 50, 150, or 500 mg/kg on selected days after mating, the maximum inhibition of implantation (80%) was observed in animals treated with 500 mg/kg on days 4 to 6 or 1 to 7 after mating (Kant et al. 1989).

In rats orally administered 2 or 4.5 g/kg of wild carrot powder on days 1 to 10 of pregnancy, no anti-implantation effect was observed (Lal et al. 1986).

In mice orally administered 80 or 120 mg per animal (average animal weight not stated) of an alcohol extract of wild carrot, administration on days 4 to 6 after mating was effective in inhibiting implantation, while administration on days 8 to 10 after mating had no effect (Sharma et al. 1976).

In isolated rat uteri, a chloroform fraction of wild carrot seed inhibited spontaneous activity and oxytocin evoked activity. The activity was dose dependent and a concentration of 0.5 mg/ml of the extract produced a marked inhibition of oxytocin-induced activity (Dhar et al. 1975).

No information on the safety of wild carrot in lactation was identified. While this review did not identify any concerns for use while nursing, safety has not been conclusively established.

V. Toxicity Studies

Acute Toxicity

In the brine shrimp lethality assay, the LD_{50} values of the major compounds from the methanol extract of wild carrot were 0.053 mg/ml for luteolin, 1 mg/ml for luteolin 3′-*O*-β-D-glucopyranoside, and 1 mg/ml for luteolin 4′-*O*-β-D-glucopyranoside (Kumarasamy et al. 2005).

Literature Cited

Dhar, V.J., V.S. Mathur, and S.K. Garg. 1975. Pharmacological studies on *Daucus carota* part I. *Planta Med.* 28(1):12-15.

Felter, H.W., and J.U. Lloyd. 1898. *King's American dispensatory.* 18th ed., 3rd rev. 2 vols. Cincinnati: Ohio Valley Co.

Garg, S.K., and V.S. Mathur. 1972. Effect of chromatographic fractions of *Daucus carota* Linn. (seeds) on fertility in female albino rats. *J. Reprod. Fertil.* 31(1):143-145.

Kaliwal, B.B., and R.N. Ahamed. 1987. Maintenance of implantation by progesterone in carrot seed *Daucus carota* extract treated albino rats. *Indian J. Phys. Nat. Sci. A* 7:10-14.

Kaliwal, B.B., R.N. Ahamed, and M.A. Rao. 1986. Implantation delay and nidation by progesterone in carrot seed *Daucus carota* extract treated albino rats. *Proc. Indian Acad. Sci. Anim. Sci.* 95(2):263-268.

Kaliwal, B.B., and M. Appaswamy Rao. 1979. Dose and durational effect of carrot seed extract *Daucus-carota* on implantation in albino rats. *Comp. Physiol. Ecol.* 4(2):92-97.

Kant, A.K., N.K. Lohiya, and D. Jacob. 1989. A revaluation of the pregnancy interceptory efficacy of carrot *Daucus carota* L. seeds in the rat *J. Adv. Zool.* 10(2):110-113.

Kumarasamy, Y., L. Nahar, M. Byres, A. Delazar, and S.D. Sarker. 2005. The assessment of biological activities associated with the major constituents of the methanol extract of 'wild carrot' (*Daucus carota* L.) seeds. *J. Herb. Pharmacother.* 5(1):61-72.

Lal, R., M. Gandhi, A. Sankaranarayanan, V.S. Mathur, and P.L. Pharma. 1986. Antifertility effect of *Daucus carota* seeds in female albino rats. *Fitoterapia* 57(4):243-246.

Majumder, P.K., S. Dasgupta, R.K. Mukhopadhaya, U.K. Mazumdar, and M. Gupta. 1997. Anti-steroidogenic activity of the petroleum ether extract and fraction 5 (fatty acids) of carrot (*Daucus carota* L.) seeds in mouse ovary. *J. Ethnopharmacol.* 57(3):209-212.

Sharma, M.M., G. Lal, and D. Jacob. 1976. Estrogenic and pregnancy interceptory effects of carrot *Daucus carota* seeds. *Indian J. Exp. Biol.* 14(4):506-508.

Wood, H., and C. LaWall. 1918. *The dispensatory of the United States of America*. 21st ed. Philadelphia: Lippincott.

Dendrobium nobile Lindl. — Orchidaceae

SCN: dendrobium
PN: *shi hu* (stem)
OCN: Chinese orchid
Part: herb

Quick Reference Summary

Safety Class: 1
Interaction Class: A

Contraindications

None known.

Other Precautions

None known.

Drug and Supplement Interactions

None known.

Editors' Notes

Other species of *Dendrobium* are considered acceptable substitutes for *D. nobile*. They include *D. loddigesii*, *D. fimbriatum* var. *occulatum*, *D. chrysanthum*, and *D. officinale*, (Bensky et al. 2004; Chen and Chen 2004). Adulteration of dendrobium with species of *Pholidota* has been reported (Lau et al. 2001; Zhang et al. 2007).

Adverse Events and Side Effects

Gross overdose of dendrobium (standard dose is listed as decoction of 6–15 g) has been associated with seizures (Chen and Chen 2004).

Pharmacological Considerations

None known.

Pregnancy and Lactation

No information on the safety of dendrobium in pregnancy or lactation was identified in the scientific or traditional literature. Although this review did not identify any concerns for use while pregnant or nursing, safety has not been conclusively established.

Review Details

I. Drug and Supplement Interactions

Clinical Trials of Drug or Supplement Interactions

No clinical trials of drug or supplement interactions were identified.

Case Reports of Suspected Drug or Supplement Interactions

No case reports of suspected drug or supplement interactions were identified.

Animal Trials of Drug or Supplement Interactions

No animal trials of drug or supplement interactions were identified.

II. Adverse Events

Case Reports of Adverse Events

Seizures have been reported in association with gross overdose of dendrobium (standard dose is a decoction of 6–15 g) (Chen and Chen 2004).

III. Pharmacology and Pharmacokinetics

Human Pharmacological Studies

No relevant human pharmacological studies were identified.

Animal Pharmacological Studies

No relevant animal pharmacological studies were identified.

In Vitro Pharmacological Studies

No relevant in vitro pharmacological studies were identified.

IV. Pregnancy and Lactation

No information on the safety of dendrobium during pregnancy or lactation was identified.

V. Toxicity Studies

Acute Toxicity

The LD_{50} of dendrobium juice in mice could not be determined at doses up to 10 g/kg (route of administration not specified in English language abstract) (Jiang and Cheng 1999).

D

Genotoxicity

No mutagenic activity of dendrobium was observed in the Ames test for mutagenicity, a mouse myeloid cell micronucleus test, or a mouse teratospermia test (doses administered to mice not specified in English language abstract) (Jiang and Cheng 1999).

Extracts of dendrobium have shown antimutagenic activity in *Salmonella typhimurium* strains treated with mutagenic compounds (Miyazawa et al. 1997; Miyazawa et al. 1999).

Literature Cited

Bensky, D., S. Clavey, and E. Stöger. 2004. *Chinese herbal medicine: Materia medica*. 3rd ed. Seattle: Eastland Press.

Chen, J.K., and T.T. Chen. 2004. *Chinese medical herbology and pharmacology*. City of Industry, CA: Art of Medicine Press.

Jiang, M., and S. Cheng. 1999. *Dendrobium nobile* juice. *Shipin Kexue* 20(11):39-41.

Lau, D.T.W., P.C. Shaw, J. Wang, and P.P.H. But. 2001. Authentication of medicinal *Dendrobium* species by the internal transcribed spacer of ribosomal DNA. *Planta Med.* 67:456-460.

Miyazawa, M., H. Shimamura, S.-I. Nakamura, and H. Kameoka. 1997. Antimutagenic activity of gigantol from *Dendrobium nobile*. *J. Agric. Food Chem.* 45(8):2849-2853.

Miyazawa, M., H. Shimamura, S.-I. Nakamura, et al. 1999. Moscatilin from *Dendrobium nobile*, a naturally occurring bibenzyl compound with potential antimutagenic activity. *J. Agric. Food Chem.* 47(5):2163-2167.

Zhang, Y.B., P.P.H. But, Z.T. Wang, and P.C. Shaw. 2007. Current approaches for the authentication of medicinal *Dendrobium* species and its products. *Plant Genet. Resources* 3(2):144-148.

Digitalis spp.

Scrophulariaceae

Digitalis purpurea L.
SCN: digitalis
OCN: foxglove; purple foxglove
Digitalis lanata Ehrh.
SCN: Grecian foxglove
OCN: digitalis
Part: leaf

Quick Reference Summary

Safety Class: 3
Interaction Class: C

Contraindications

Not for use except under the supervision of an expert qualified in the appropriate use of this substance (Hanzlik 1929; Hauptman and Kelly 1999; Maffe et al. 2009; Ramlakhan and Fletcher 2007; Withering 1785).

Other Precautions

None known.

Drug and Supplement Interactions

Interaction considerations are the same as those for digitoxin, digoxin, and other digitalis glycosides (Hauptman and Kelly 1999; Maffe et al. 2009; Vivo et al. 2008).

Editors' Notes

Grecian foxglove (*D. lanata*) contains the compound digoxin, which is not present in digitalis (*D. purpurea*). Digitalis contains digitoxin and gitoxin in the leaf and digitalin in the seed (Ramlakhan and Fletcher 2007). Grecian foxglove also contains lanatosides (Weiler and Zenk 1976).

U.S. regulations require that digitalis drugs be labeled in a manner which informs consumers of the inappropriateness of use as an antiobesity agent (CFR 2011).

Adverse Events and Side Effects

Cases of digitalis and Grecian foxglove poisoning have been reported after intentional self-poisoning or mistaking the leaves for other edible or medicinal species (Bain 1985; Cardano et al. 2002; Colls 1999; Dickstein and Kunkel 1980; Maffe et al. 2009; Slifman et al. 1998; Thierry et al. 2000). Symptoms of poisoning are similar to those from overdose of therapeutic cardiac glycosides such as digoxin and digitoxin. Symptoms include nausea, vomiting, changes in heartbeat and heart function, abdominal pain, diarrhea, dizziness, headache, confusion, delirium, hallucinations, changes in vision, shortness of breath, and chest pain (Ramlakhan and Fletcher 2007).

Pharmacological Considerations

Digitalis and Grecian foxglove have marked effects on heart function. As with digoxin, digitoxin, and other related compounds, the difference between the effective dose and the toxic dose is quite small (Hauptman and Kelly 1999; Withering 1785).

Pregnancy and Lactation

No information on the use of digitalis or Grecian foxglove during pregnancy or lactation was identified. Studies on the drug digoxin indicate that this compound readily crosses the placenta, and digoxin has been used to treat fetal arrhythmias (Chow et al. 1998; Joglar and Page 1999; Oudijk et al. 2002). Digoxin has also been shown to cross into breast milk, although the concentration is relatively low and the drug is considered compatible with breast-feeding (AAP 2001; Chow et al. 1998; Reinhardt et al. 1982). Safety has not been conclusively established for the use of this herb during pregnancy or lactation, and this substance is not recommended for use except under the supervision of an expert qualified in its appropriate use.

Review Details

I. Drug and Supplement Interactions

Clinical Trials of Drug or Supplement Interactions

No clinical trials of drug or supplement interactions were identified.

Case Reports of Suspected Drug or Supplement Interactions

No case reports of suspected drug or supplement interactions were identified.

Animal Trials of Drug or Supplement Interactions

No animal trials of drug or supplement interactions were identified.

II. Adverse Events

Case Reports of Adverse Events

Digitalis or Grecian foxglove poisoning typically occurs after intentional self-poisoning or mistaking the leaves for other edible or medicinal species including borage (*Borago officinalis*), comfrey species (*Symphytum* spp.), and plantain species (*Plantago* spp.) (Bain 1985; Cardano et al. 2002; Colls 1999; Dickstein and Kunkel 1980; Maffe et al. 2009; Slifman et al. 1998; Thierry et al. 2000). Cases of poisoning result in nonspecific symptoms similar to those from overdose of therapeutic cardiac glycosides such as digoxin and digitoxin. Cardiovascular symptoms may include ECG changes, including ventricular premature contractions, sinus bradycardia or tachycardia, nonparoxysmal junctional tachycardia, atrioventricular (AV) dissociation, first-, second-, or third-degree AV block, and sinoatrial block or arrest. Other symptoms may include nausea or vomiting, abdominal pain, diarrhea, dizziness, headache, confusion, delirium, numbness, psychosis, visual and auditory hallucinations, somnolence, aphasia and seizures, blurred vision, photophobia, a predominance of yellow or green in the vision (xanthopsia), diplopia, transient blindness, palpitations, chest pain, and shortness of breath (Ramlakhan and Fletcher 2007).

Nausea and vomiting, abdominal pain, and cardiovascular shock with sinus bradycardia were reported in a 36-year-old female who ingested an unspecified amount of digitalis in an attempted suicide (Lacassie et al. 2000).

A family composed of father, mother, and daughter, respectively 62, 58, and 35 years of age, experienced persistent nausea, vomiting, lethargy, and intense weakness after having eaten potato dumplings containing digitalis leaves that had been mistaken for borage leaves in their garden. All three had marked sinus bradycardia and other electrocardiographic and biochemical aspects that were indicative of acute cardiac glycoside poisoning. No estimations on amount ingested were reported, but all three patients made a complete recovery. The reporting authors noted that the patients have since decided to purchase their produce from the market (Maffe et al. 2009).

Bradycardia and runs of second degree AV block, which rapidly progressed to a 3.5 s sinus pause, followed by sinus arrest, were reported in a 64-year-old man that ate "a whole *Digitalis purpurea* plant." The ingestion was believed to be with suicidal intention and was fatal despite treatment attempts (Ramlakhan and Fletcher 2007).

Poisonings by digitalis have been reported in a number of animals under human care, including cows, deer, ponies, minks, and turkeys (Corrigall et al. 1978; Parker 1951; Thomas et al. 1987; Wijnberg et al. 1999; Zimowski 1973).

III. Pharmacology and Pharmacokinetics

Human Pharmacological Studies

No relevant human pharmacological studies were identified.

Animal Pharmacological Studies

No relevant animal pharmacological studies were identified.

In Vitro Pharmacological Studies

No relevant in vitro pharmacological studies were identified.

IV. Pregnancy and Lactation

No information on the use of digitalis or Grecian foxglove during pregnancy or lactation was identified. Studies on the drug digoxin indicate that this compound readily crosses the placenta, and digoxin has been used to treat fetal arrhythmias (Chow et al. 1998; Joglar and Page 1999; Oudijk et al. 2002).

Digoxin has also been shown to cross into breast milk, although the concentration is relatively low and the drug is considered compatible with breast-feeding (AAP 2001; Chow et al. 1998; Reinhardt et al. 1982).

V. Toxicity Studies

Acute Toxicity

The LD_{50} of the compound digitoxin is 8.9 mg/kg in female rats and 15.4 mg/kg in male rats (Scott et al. 1971).

Literature Cited

AAP. 2001. Transfer of drugs and other chemicals into human milk. American Academy of Pediatrics Committee on Drugs. *Pediatrics* 108(3):776-789.

Bain, R.J. 1985. Accidental digitalis poisoning due to drinking herbal tea. *Br. Med. J.* 290(6482):1624.

Cardano, S., F. Beldi, C. Bignoli, A. Monteverde, and E. Uglietti. 2002. A dangerous "risotto." *Rec. Prog. Med.* 93(4):245-246.

CFR. 2011. *Code of federal regulations*, Title 21 Part 201.317, 2011 ed. Specific labeling requirements for specific drug products. Digitalis and related cardiotonic drugs for human use in oral dosage forms; required warning. Washington, DC: U.S. Government Printing Office.

Chow, T., J. Galvin, and B. McGovern. 1998. Antiarrhythmic drug therapy in pregnancy and lactation. *Am. J. Cardiol.* 82(4):58I-62I.

Colls, B.M. 1999. A salutary lesson: Three very unwise men. *Br. Med. J.* 318(7200):1729.

Corrigall, W., R.R. Moody, and J.C. Forbes. 1978. Foxglove (*Digitalis purpurea*) poisoning in farmed red deer (*Cervus elaphus*). *Vet. Rec.* 102(6):119-122.

Dickstein, E.S., and F.W. Kunkel. 1980. Foxglove tea poisoning. *Am. J. Med.* 69(1):167.

Hanzlik, P.J. 1929. A new method of estimating the potency of digitalis: Pigeon emesis. *J. Pharm. Exp. Ther.* 35(4):363-391.

Hauptman, P.J., and R.A. Kelly. 1999. Digitalis. *Circulation* 99(9):1265-1270.

Joglar, J.A., and R.L. Page. 1999. Treatment of cardiac arrhythmias during pregnancy: Safety considerations. *Drug Saf.* 20(1):85-94.

Lacassie, E., P. Marquet, S. Martin-Dupont, J.-M. Gaulier, and G. Lachatre. 2000. A non-fatal case of intoxication with foxglove, documented by means of liquid chromatography-electrospray-mass spectrometry. *J. Forensic Sci.* 45(5):1154-1158.

Maffe, S., L. Cucchi, F. Zenone, et al. 2009. Digitalis must be banished from the table: A rare case of acute accidental digitalis intoxication of a whole family. *J. Cardiovasc. Med.* 10(9):727-732.

Oudijk, M.A., J.M. Ruskamp, B.E. Ambachtsheer, et al. 2002. Drug treatment of fetal tachycardias. *Pediat. Drugs* 4(1):49-63.

Parker, W.H. 1951. Foxglove (*Digitalis purpurea*) poisoning in turkeys. *Vet. Rec.* 63(24):416.

Ramlakhan, S.L., and A.K. Fletcher. 2007. It could have happened to Van Gogh: A case of fatal purple foxglove poisoning and review of the literature. *Eur. J. Emerg. Med.* 14(6):356-359.

Reinhardt, D., O. Richter, T. Genz, and S. Potthoff. 1982. Kinetics of the translactal passage of digoxin from breast feeding mothers to their infants. *Eur. J. Pediat.* 138(1):49-52.

Scott, W.J., R.P. Beliles, and H.I. Silverman. 1971. The comparative acute toxicity of two cardiac glycosides in adult and newborn rats. *Toxicol. Appl. Pharmacol.* 20:599-601.

Slifman, N.R., W.R. Obermeyer, B.K. Aloi, et al. 1998. Contamination of botanical dietary supplements by *Digitalis lanata*. *N. Engl. J. Med.* 339(12):806-811.

Thierry, S., F. Blot, J.C. Lachérade, et al. 2000. Poisoning with foxglove extract: Favorable evolution without Fab fragments. *Intens. Care Med.* 26(10):1586.

Thomas, D.L., M.P. Quick, and R.P. Morgan. 1987. Suspected foxglove (*Digitalis purpurea*) poisoning in a dairy cow. *Vet. Rec.* 120(13):300-301.

Vivo, R.P., S.R. Krim, J. Perez, et al. 2008. Digoxin: Current use and approach to toxicity. *Am. J. Med. Sci.* 336(5):423-428.

Weiler, E.W., and M.H. Zenk. 1976. Radioimmunoassay for the determination of digoxin and related compounds in *Digitalis lanata*. *Phytochemistry* 15(10):1537-1545.

Wijnberg, I.D., J.H. van der Kolk, and E.G. Hiddink. 1999. Use of phenytoin to treat digitalis-induced cardiac arrhythmias in a miniature Shetland pony. *Vet. Rec.* 144(10):259-261.

Withering, W. 1785. *An account of the foxglove and some of its medical uses: With practical remarks on dropsy and other diseases*. London: J & J Robinson.

Zimowski, A. 1973. Intoxication of mink with *Digitalis purpurea*.] *Med. Wet.* 29(4):226.

Dimocarpus longan Lour. — Sapindaceae

SCN: longan

Syn: *Euphoria longan* (Lour.) Steud.; *Euphoria longana* Lam.; *Nephelium longan* (Lour.) Hook.; *Nephelium longana* (Lam.) Cambess.

PN: *long yan rou* (fleshy seed cover)

Part: fruit

Quick Reference Summary

Safety Class: 1

Interaction Class: A

Contraindications

None known.

Other Precautions

None known.

Drug and Supplement Interactions

None known.

Editors' Note

Longan is commonly consumed as a fruit in Asian countries (Menzel and Waite 2005).

Adverse Events and Side Effects

Allergic reactions to longan have been reported (Bensky et al. 2004; Rank and Li 2007).

Pharmacological Considerations

None known.

Pregnancy and Lactation

No information on the safety of longan in pregnancy or lactation was identified in the scientific or traditional literature. Although this review did not identify any concerns for use while pregnant or nursing, safety has not been conclusively established.

Review Details

I. Drug and Supplement Interactions

Clinical Trials of Drug or Supplement Interactions

No clinical trials of drug or supplement interactions were identified.

Case Reports of Suspected Drug or Supplement Interactions

No case reports of suspected drug or supplement interactions were identified.

Animal Trials of Drug or Supplement Interactions

No animal trials of drug or supplement interactions were identified.

II. Adverse Events

Case Reports of Adverse Events

Allergic reactions to longan have been reported, with one case producing an allergic drug rash with vertigo and elevated body temperature (Bensky et al. 2004; Rank and Li 2007).

III. Pharmacology and Pharmacokinetics

Human Pharmacological Studies

No relevant human pharmacological studies were identified.

Animal Pharmacological Studies

No relevant animal pharmacological studies were identified.

In Vitro Pharmacological Studies

No relevant in vitro pharmacological studies were identified.

IV. Pregnancy and Lactation

No information on the safety of longan during pregnancy or lactation was identified.

V. Toxicity Studies

No toxicity studies were identified.

Literature Cited

Bensky, D., S. Clavey, and E. Stöger. 2004. *Chinese herbal medicine: Materia medica.* 3rd ed. Seattle: Eastland Press.

Menzel, C.M., and G.K. Waite. 2005. *Litchi and longan: Botany, production, and uses.* Wallingford, U.K.: CABI.

Rank, M.A., and J.T. Li. 2007. A case of food allergy due to longan fruit. *Ann. Allerg. Asthma Immunol.* 98(4):402.

Dioscorea oppositifolia L.

Dioscoreaceae

SCN: Chinese yam
Syn: *Dioscorea opposita* Thunb.
PN: *shan yao* (rhizome)
OCN: cinnamon vine; common yam
Part: rhizome

Quick Reference Summary

Safety Class: 1
Interaction Class: A

Contraindications

None known.

D

Other Precautions

None known.

Drug and Supplement Interactions

None known.

Adverse Events and Side Effects

None known.

Pharmacological Considerations

Animal studies have indicated that Chinese yam may modify glucose regulation (Chen and Chen 2004; Gao et al. 2007). People with diabetes are advised to monitor their blood sugar closely and discuss the use of this herb with a qualified healthcare practitioner prior to use.

Allergic reactions to Chinese yam have been reported after ingestion and topical application (Bensky et al. 2004).

Pregnancy and Lactation

No information on the safety of Chinese yam in pregnancy or lactation was identified in the scientific or traditional literature. Although this review did not identify any concerns for use while pregnant or nursing, safety has not been conclusively established.

Review Details

I. Drug and Supplement Interactions

Clinical Trials of Drug or Supplement Interactions

No clinical trials of drug or supplement interactions were identified.

Case Reports of Suspected Drug or Supplement Interactions

No case reports of suspected drug or supplement interactions were identified.

Animal Trials of Drug or Supplement Interactions

No animal trials of drug or supplement interactions were identified.

II. Adverse Events

Case Reports of Adverse Events

Allergic reactions to Chinese yam have been reported after ingestion and topical application. Symptoms following ingestion included maculopapular rashes over the whole body, itchy throat, pruritus, and a sensation of pressure in the chest (Bensky et al. 2004).

III. Pharmacology and Pharmacokinetics

Human Pharmacological Studies

No relevant human pharmacological studies were identified.

Animal Pharmacological Studies

In mice orally administered 30 g/kg of a decoction of Chinese yam daily for 10 days, blood glucose levels decreased by 10 to 30 mg/dl. The same extract and dose controlled the normally sharp increase of glucose following intraperitoneal administration of glucose (Chen and Chen 2004). An ethanol extract of Chinese yam orally administered to diabetic rats significantly reduced blood insulin and glucose levels (Gao et al. 2007).

In Vitro Pharmacological Studies

No relevant in vitro pharmacological studies were identified.

IV. Pregnancy and Lactation

No information on the safety of Chinese yam during pregnancy or lactation was identified.

V. Toxicity Studies

No toxicity studies were identified.

Literature Cited

Bensky, D., S. Clavey, and E. Stöger. 2004. *Chinese herbal medicine: Materia medica*. 3rd ed. Seattle: Eastland Press.

Chen, J.K., and T.T. Chen. 2004. *Chinese medical herbology and pharmacology*. City of Industry, CA: Art of Medicine Press.

Gao, X., B. Li, H. Jiang, et al. 2007. *Dioscorea opposita* reverses dexamethasone induced insulin resistance. *Fitoterapia* 78(1):12-15.

Dioscorea villosa L. — Dioscoreaceae

SCN: wild yam
OCN: China root; colic root; North American wild yam; rheumatism root

Part: rhizome

D

Quick Reference Summary

Safety Class: 1
Interaction Class: A

Contraindications

None known.

Other Precautions

None known.

Drug and Supplement Interactions

None known.

Editors' Note

Wild yam contains the compound diosgenin, commonly and mistakenly promoted as a natural source of progesterone. Diosgenin has been used as a precursor for the synthesis of progesterone, but the transformation required to form progesterone does not occur in the human body (Dentali 1996).

Adverse Events and Side Effects

"Large" doses of tincture (standard dose listed as 20–60 drops of tincture) are reported to cause vomiting, although smaller doses have been used to alleviate nausea and vomiting (Felter and Lloyd 1898).

An animal study indicated that inflammation of the liver and kidneys was observed in rats administered high doses (790 mg/kg daily) of a wild yam extract for 4 weeks. No inflammation was observed in rats administered the same dose for 2 weeks (Wojcikowski et al. 2008).

Pharmacological Considerations

None known.

Pregnancy and Lactation

No information on the safety of wild yam in pregnancy or lactation was identified in the scientific or traditional literature. Although this review did not identify any concerns for use while pregnant or nursing, safety has not been conclusively established.

Review Details

I. Drug and Supplement Interactions

Clinical Trials of Drug or Supplement Interactions

No clinical trials of drug or supplement interactions were identified.

Case Reports of Suspected Drug or Supplement Interactions

No case reports of suspected drug or supplement interactions were identified.

Animal Trials of Drug or Supplement Interactions

No animal trials of drug or supplement interactions were identified.

II. Adverse Events

Case Reports of Adverse Events

No case reports of adverse events were identified.

III. Pharmacology and Pharmacokinetics

Human Pharmacological Studies

No relevant human pharmacological studies were identified.

Animal Pharmacological Studies

In rats orally administered 50, 150, or 500 mg/kg of wild yam oleyl alcohol eluant extract daily for 4 days, observation of uterine and vaginal parameters indicated no estrogenic effects of the extract (Zava et al. 1998).

In a dermal irritation study in rabbits, no irritation was observed after topical application of a single dose of a 10% solution of wild yam oleyl alcohol eluant extract (Hooker 2004).

In Vitro Pharmacological Studies

No estrogenic activity of a hydroethanolic extract of wild yam was observed in estrogen receptor-positive (MCF-7) human breast cancer cells (Zava et al. 1998).

In normal kidney cells (NRK49F) and tubular epithelial cells (NRK52E), ethyl acetate, methanol, and water-methanol extracts of wild yam exhibited some toxicity. At concentrations of 5 to 50 μg/ml, induction of epithelial to mesenchymal transdifferentiation was observed (Wojcikowski et al. 2009).

IV. Pregnancy and Lactation

No information on the safety of wild yam in pregnancy or lactation was identified.

V. Toxicity Studies

Acute Toxicity

The LD_{50} of orally or dermally administered wild yam oleyl alcohol eluant extract in rats could not be determined at doses up to 2 g/kg (Hooker 2004).

Short-Term Toxicity

In rats orally administered 790 mg/kg wild yam hydroalcoholic extract daily for 28 days, an increase in fibrosis in the kidneys and inflammation in the liver was observed. No adverse effects on the kidneys or liver were observed in animals treated for 14 days (Wojcikowski et al. 2008).

In ovariectomized rats treated with a total of 500 mg of the compound diosgenin over a 45-day period (via an intraperitoneally implanted tricalcium phosphate drug delivery system), increases in body weight and spleen weight were observed along with a slight increase in wet adrenal weight. Histopathological evaluation of the adrenal gland revealed a decrease in the cortical and medullary adrenal areas (Benghuzzi et al. 2003).

In dermal toxicity testing, no adverse effects were observed in rats treated with 1 ml/kg of a 1, 3, or 10% concentration of wild yam oleyl alcohol eluant extract daily for 29 days (Hooker 2004).

Genotoxicity

In the Ames mutagenicity test with *Salmonella typhimurium* strains TA98, TA100, TA1535, TA1537, and TA1538, no mutagenic activity of an oleyl alcohol eluant extract of wild yam was observed with or without metabolic activation (Hooker 2004).

In the bone marrow micronucleus test in rats, no mutagenic activity of an oleyl alcohol eluant extract of wild yam was observed after the animals were administered two doses up to 2 g/kg (Hooker 2004).

Literature Cited

Benghuzzi, H., M. Tucci, R. Eckie, and J. Hughes. 2003. The effects of sustained delivery of diosgenin on the adrenal gland of female rats. *Biomed. Sci. Instrum.* 39:335-340.

Dentali, S. 1996. Clearing up confusion over yams and progesterone. *Altern. Ther. Health Med.* 2(4):19.

Felter, H.W., and J.U. Lloyd. 1898. *King's American dispensatory.* 18th ed., 3rd rev. 2 vols. Cincinnati: Ohio Valley Co.

Hooker, E. 2004. Final report of the amended safety assessment of *Dioscorea villosa* (wild yam) root extract. *Int. J. Toxicol.* 23(Suppl. 2):49-54.

Wojcikowski, K., H. Wohlmuth, D.W. Johnson, and G. Gobe. 2008. *Dioscorea villosa* (wild yam) induces chronic kidney injury via pro-fibrotic pathways. *Food Chem. Toxicol.* 46(9):3122-3131.

Wojcikowski, K., H. Wohlmuth, D.W. Johnson, M. Rolfe, and G. Gobe. 2009. An in vitro investigation of herbs traditionally used for kidney and urinary system disorders: Potential therapeutic and toxic effects. *Nephrology* 14(1):70-79.

Zava, D.T., C.M. Dollbaum, and M. Blen. 1998. Estrogen and progestin bioactivity of foods, herbs, and spices. *Proc. Soc. Exp. Biol. Med.* 217(4):369.

Dipsacus spp. — Dipsacaceae

Dipsacus asper Wall.
SCN: Sichuan teasel
Syn: *Dipsacus asper* auct.
PN: *xu duan* (root)
OCN: Sichuan dipsacus

Dipsacus japonicus Miq.
SCN: Japanese teasel
PN: *xu duan* (root)
OCN: Japanese dipsacus
Part: root

Quick Reference Summary

Safety Class: 1
Interaction Class: A

Contraindications

None known.

Other Precautions

None known.

Drug and Supplement Interactions

None known.

Adverse Events and Side Effects

Allergic reactions to Sichuan teasel have been reported (Bensky et al. 2004; Chen and Chen 2004).

Pharmacological Considerations

None known.

Pregnancy and Lactation

In traditional Chinese medicine, Sichuan teasel is used to "calm the fetus," stop uterine bleeding during pregnancy, and prevent threatened miscarriage (Bensky et al. 2004; Chen and Chen 2004).

No information on the safety of Sichuan teasel in lactation was identified. While this review did not identify any concerns for use while nursing, safety has not been conclusively established.

Review Details

I. Drug and Supplement Interactions

Clinical Trials of Drug or Supplement Interactions

No clinical trials of drug or supplement interactions were identified.

Case Reports of Suspected Drug or Supplement Interactions

No case reports of suspected drug or supplement interactions were identified.

Animal Trials of Drug or Supplement Interactions

No animal trials of drug or supplement interactions were identified.

II. Adverse Events

Case Reports of Adverse Events

Allergic reactions to Sichuan teasel have been reported. Reactions included localized erythema, itching, and a sensation of heat or burning (Bensky et al. 2004; Chen and Chen 2004).

III. Pharmacology and Pharmacokinetics

Human Pharmacological Studies

No relevant human pharmacological studies were identified.

Animal Pharmacological Studies

No relevant animal pharmacological studies were identified.

In Vitro Pharmacological Studies

No relevant in vitro pharmacological studies were identified.

IV. Pregnancy and Lactation

In traditional Chinese medicine, Sichuan teasel is used to "calm the fetus," to stop uterine bleeding during pregnancy, and to prevent threatened miscarriage (Bensky et al. 2004; Chen and Chen 2004).

No information on the safety of Sichuan teasel in lactation was identified. While this review did not identify any concerns for use while nursing, safety has not been conclusively established.

V. Toxicity Studies

No toxicity studies were identified.

Literature Cited

Bensky, D., S. Clavey, and E. Stöger. 2004. *Chinese herbal medicine: Materia medica*. 3rd ed. Seattle: Eastland Press.

Chen, J.K., and T.T. Chen. 2004. *Chinese medical herbology and pharmacology*. City of Industry, CA: Art of Medicine Press.

Drynaria fortunei (Kunze ex Mett.) J. Sm. — Polypodiaceae

SCN: drynaria
PN: *gu sui bu* (rhizome)
Part: rhizome

Quick Reference Summary

Safety Class: 1
Interaction Class: A

Contraindications

None known.

Other Precautions

None known.

Drug and Supplement Interactions

None known.

Adverse Events and Side Effects

None known.

Pharmacological Considerations

In vitro studies have suggested estrogenic activity of drynaria extracts (Chang et al. 2003; Jeong et al. 2005).

Pregnancy and Lactation

No information on the safety of drynaria in pregnancy or lactation was identified in the scientific or traditional literature. Although this review did not identify any concerns for use while pregnant or nursing, safety has not been conclusively established.

Review Details

I. Drug and Supplement Interactions

Clinical Trials of Drug or Supplement Interactions

No clinical trials of drug or supplement interactions were identified.

Case Reports of Suspected Drug or Supplement Interactions

No case reports of suspected drug or supplement interactions were identified.

Animal Trials of Drug or Supplement Interactions

No animal trials of drug or supplement interactions were identified.

II. Adverse Events

Case Reports of Adverse Events

No adverse effects have been reported with use of drynaria at standard therapeutic doses (decoction of 9–21 g) (Bensky et al. 2004). Overdose of drynaria (a decoction of over 100 g) has been associated with dry mouth, excessive speech, fear, palpitations, and a manic-depressive type psychosis (Bensky et al. 2004; Chen and Chen 2004).

III. Pharmacology and Pharmacokinetics

Human Pharmacological Studies

No relevant human pharmacological studies were identified.

Animal Pharmacological Studies

No relevant animal pharmacological studies were identified.

In Vitro Pharmacological Studies

Proliferation of nontransformed osteoblastic cells (MC3T3-E1) induced by an extract of drynaria at concentrations of 30–100 μg/ml was eliminated by the estrogen antagonist tamoxifen (Jeong et al. 2005).

A methanol extract of drynaria exhibited proliferative activity in estrogen receptor-positive human breast cancer cells (MCF-7) and osteoblast-like cells (ROS 17/2.8) after treatment with concentrations of 100 μg/ml (Chang et al. 2003).

IV. Pregnancy and Lactation

No information on the use of drynaria during pregnancy or lactation was identified.

V. Toxicity Studies

No toxicity studies were identified.

Literature Cited

Bensky, D., S. Clavey, and E. Stöger. 2004. *Chinese herbal medicine: Materia medica*. 3rd ed. Seattle: Eastland Press.

Chang, E.J., W.J. Lee, S.H. Cho, and S.W. Choi. 2003. Proliferative effects of flavan-3-ols and propelargonidins from rhizomes of *Drynaria fortunei* on MCF-7 and osteoblastic cells. *Arch. Pharm. Res.* 26(8):620-630.

Chen, J.K., and T.T. Chen. 2004. *Chinese medical herbology and pharmacology*. City of Industry, CA: Art of Medicine Press.

Jeong, J.C., J.W. Lee, C.H. Yoon, et al. 2005. Stimulative effects of Drynariae Rhizoma extracts on the proliferation and differentiation of osteoblastic MC3T3-E1 cells. *J. Ethnopharmacol.* 96(3):489-495.

Dryopteris filix-mas (L.) Schott — Aspidiaceae

SCN: male fern
Syn: *Aspidium filix-mas* (L.) Sw.

Part: rhizome

Quick Reference Summary

Safety Class: 3
Interaction Class: A

Contraindications

Not for use except under the supervision of an expert qualified in the appropriate use of this substance (Felter and Lloyd 1898; Maisch 1883; Weiss and Meuss 2001; Williamson 2003; Wood and LaWall 1918).

Other Precautions

None known.

Drug and Supplement Interactions

None known.

Adverse Events and Side Effects

Symptoms of poisoning from male fern use include vomiting, diarrhea, vertigo, headache, tremors, cold sweats, difficulty breathing, cyanosis, convulsions, stupor, and mental disturbances. Poisonings have included disturbance of vision and sometimes blindness, which was permanent in several cases. Fatal cases of poisonings have been reported

(Felter and Lloyd 1898; Maisch 1883; Wood and LaWall 1918).

Pharmacological Considerations

Male fern was historically used in conjunction with castor oil. Castor oil is believed to increase absorption of the toxic compounds in male fern, and the concomitant use of these products is considered dangerous and is no longer recommended (Leung and Foster 1996; Wood and LaWall 1918).

Pregnancy and Lactation

Based on the toxicity of this species (Felter and Lloyd 1898; Wood and LaWall 1918), use of male fern in pregnancy or lactation is not recommended.

D

Review Details

I. Drug and Supplement Interactions

Clinical Trials of Drug or Supplement Interactions

No clinical trials of drug or supplement interactions were identified.

Case Reports of Suspected Drug or Supplement Interactions

No case reports of suspected drug or supplement interactions were identified.

Animal Trials of Drug or Supplement Interactions

No animal trials of drug or supplement interactions were identified.

II. Adverse Events

Case Reports of Adverse Events

Symptoms of poisoning from male fern use include vomiting, diarrhea, vertigo, headache, tremors, cold sweats, difficulty breathing, cyanosis, convulsions, and mental disturbances. Some cases of poisonings included disturbance of vision and sometimes blindness, which was permanent in several cases. Cases of blindness were believed to be due to spasm of the retinal vessels and subsequent atrophy (Wood and LaWall 1918).

Cattle that grazed on male fern and a related species, *Dryopteris borreri*, down to the roots of the plant were found comatose and blind (Macleod and Greig 1978). Characteristics of *Dryopteris* poisonings in cattle are blindness, a desire to stand in water, and profound drowsiness, but with low rates of mortality (Macleod and Greig 1978).

III. Pharmacology and Pharmacokinetics

Human Pharmacological Studies

No relevant human pharmacological studies were identified.

Animal Pharmacological Studies

No relevant animal pharmacological studies were identified.

In Vitro Pharmacological Studies

No relevant in vitro pharmacological studies were identified.

IV. Pregnancy and Lactation

Based on the toxicity of this species (Felter and Lloyd 1898; Maisch 1883; Wood and LaWall 1918), use of male fern in pregnancy or lactation is not recommended.

V. Toxicity Studies

No toxicity studies were identified.

Literature Cited

Felter, H.W., and J.U. Lloyd. 1898. *King's American dispensatory*. 18th ed., 3rd rev. 2 vols. Cincinnati: Ohio Valley Co.

Leung, A.Y., and S. Foster. 1996. *Encyclopedia of common natural ingredients used in food, drugs, and cosmetics*. 2nd ed. New York: Wiley.

Macleod, N.S., and A. Greig. 1978. Poisoning in cattle associated with *Dryopteris filix-mas* and *D. borreri*. *Vet. Rec.* 102(11):239-240.

Maisch, J.M. 1883. Gleanings in materia medica. *Am. J. Pharm.* 55(Feb.).

Weiss, R.F., and A.R. Meuss. 2001. *Weiss's herbal medicine*. Classic ed. New York: Stuttgart.

Williamson, E.M. 2003. *Potter's herbal cyclopedia*. Saffron Walden, Essex: C.W. Daniel Co.

Wood, H., and C. LaWall. 1918. *The dispensatory of the United States of America*. 21st ed. Philadelphia: Lippincott.

Echinacea spp.

Asteraceae

Echinacea angustifolia **DC.**
SCN: *Echinacea angustifolia*
OCN: narrow-leaf echinacea; Kansas snakeroot; narrow-leaf purple coneflower
Part: root, seed

Echinacea pallida **(Nutt.) Nutt.**
SCN: *Echinacea pallida*
Syn: *Rudbeckia pallida* Nutt.
OCN: pale-flower echinacea; pale purple coneflower
Part: root, seed

Echinacea purpurea **(L.) Moench**
SCN: *Echinacea purpurea*
Syn: *Rudbeckia purpurea* L.
OCN: purple coneflower
Part: above-ground parts, root, seed

E

Quick Reference Summary

Safety Class: 1
Interaction Class: A

Contraindications

None known.

Other Precautions

Allergic reactions may occur in susceptible persons (Huntley et al. 2005). Reactions are most likely to products made from the flowers or flowering tops of *Echinacea* (Mills and Bone 2005; Upton and Graff 2004).

Drug and Supplement Interactions

None known.

Editors' Notes

The references cited in this entry are not always sufficiently specific to clearly identify a particular species of *Echinacea* or a specific plant part or preparation (Ang-Lee et al. 2001; Kemp and Franco 2002; Lee and Werth 2004; Logan and Ahmed 2003; Mullins and Heddle 2002; Soon and Crawford 2001). The relevance of such citations to all *Echinacea* species and preparations should therefore not be assumed.

In addition, some references are focused on just one of the three species addressed in this entry, while others review all three taxa. The unmodified generic name (i.e., *Echinacea*) is used here to indicate information that is relevant to all of the species.

Commission E reports for all *Echinacea* species that they are not to be used in systemic diseases such as tuberculosis, leukosis, collagenosis, multiple sclerosis, AIDS, HIV infections, and other autoimmune diseases (Blumenthal et al. 1998). Concerns regarding the use of *Echinacea* species in these conditions are theoretical, and definitive data supporting or refuting an association of *Echinacea* with autoimmune symptom exacerbation are lacking (Mills and Bone 2005; Upton and Graff 2004).

Allergic reactions to *Echinacea* products have been reported (*see* Adverse Events and Side Effects, below). Some authorities have suggested that the reactions may be due to the presence of *Echinacea* pollen in products, as allergies to pollens in the Asteraceae family are common. Ingestion of products made from the flowering top of *Echinacea* are thus more likely to result in allergic reactions than products made from the root and leaf (Mills and Bone 2005; Upton and Graff 2004).

Adverse Events and Side Effects

In clinical trials and reviews of clinical trials, *Echinacea* products have been noted as being well tolerated, with adverse events reported similar to those of placebo (Huntley et al. 2005). Allergic reactions to *Echinacea* species have been reported (Coeugniet and Elek 1987; George et al. 2006; Huntley et al. 2005; Kemp and Franco 2002; Lee and Werth 2004; Logan and Ahmed 2003; Moell 1951; Mullins 1998; Mullins and Heddle 2002; Röseler 1952; Soon and Crawford 2001).

Pharmacological Considerations

No drug or supplement interactions in humans have been reported in either case reports or clinical trials. Some sources (Ang-Lee et al. 2001; Izzo and Ernst 2001) indicate a theoretical concern for interactions with immunosuppressant drugs, although no clinical studies are available to support or refute this concern.

While several human studies of the effects of various parts and preparations of *Echinacea purpurea* on the CYP450 drug-metabolizing isoenzymes indicate a lack of effects or clinically insignificant effects (Gorski et al. 2004; Gurley et al. 2004; Heinrich et al. 2008), one study suggested that *Echinacea purpurea* induced the drug metabolizing isoenzyme CYP3A (Penzak et al. 2010).

Pregnancy and Lactation

Limited human data has not indicated any safety concerns for use of *Echinacea* species during pregnancy (Gallo et al. 1998; Gallo et al. 1999; Perri et al. 2006; Tsui et al. 2001). Animal studies have given mixed results on the safety of *Echinacea purpurea* in pregnancy (Chow et al. 2006; Maass et al. 2005; Mengs et al. 2000).

No information on the safety of *Echinacea* spp. during lactation was identified. While this review did not identify any concerns for use while nursing, safety has not been conclusively established.

Review Details

I. Drug and Supplement Interactions

Clinical Trials of Drug or Supplement Interactions

No changes in drug serum levels were observed in healthy volunteers orally administered 400 mg of lopinavir with 100 mg ritonavir twice daily with or without 1500 mg per day of a proprietary *Echinacea purpurea* extract (identified by the authors as an 8:1 extract of fresh *E. purpurea*, and by the manufacturer as derived from fresh root and flower) for 14 days. In the same study, the volunteers were administered 8 mg midazolam (a substrate of CYP3A) before and after *Echinacea purpurea* administration. A decrease in plasma levels ($AUC_{0-\infty}$) of midazolam was observed, suggesting induction of the drug metabolizing enzyme CYP3A. Lopinavir is a CYP3A substrate, thus a decrease in plasma levels of this drug might be expected. The lack of changes in lopinavir levels observed in this study was attributed to coadministration with ritonavir, a potent inducer of CYP3A (Penzak et al. 2010).

Case Reports of Suspected Drug or Supplement Interactions

No case reports of suspected drug or supplement interactions were identified.

Animal Trials of Drug or Supplement Interactions

Coadministration of *Echinacea purpurea* root extract and melatonin to mice significantly reduced levels of mature, functional granulocyte progeny in the spleen and bone marrow as compared to either *Echinacea* or melatonin administered alone (Currier et al. 2001).

II. Adverse Events

Adverse Events Reported in Clinical Trials

Echinacea has been generally well tolerated in clinical trials, with most trials reporting similar incidences of adverse reactions for *Echinacea* and placebo (Huntley et al. 2005). One trial of fresh juice of *Echinacea purpurea* aerial parts in children showed a significantly higher incidence of rashes in children taking the *Echinacea* preparation than in children taking placebo (Taylor et al. 2003).

Case Reports of Adverse Events

Allergic reactions have been reported in persons taking *Echinacea* species, with events including anaphylaxis, asthma attacks, thrombotic thrombocytopenic purpura, leukopenia, abdominal pain, nausea, dysuria, arthralgia, myalgia, and dizziness (Coeugniet and Elek 1987; George et al. 2006; Huntley et al. 2005; Kemp and Franco 2002; Lee and Werth 2004; Logan and Ahmed 2003; Moell 1951; Mullins 1998; Mullins and Heddle 2002; Röseler 1952; Soon and Crawford 2001). Most case reports do not distinguish between the different species of *Echinacea*.

Clinical herbalists have reported differing information on the use of *Echinacea* species in autoimmune conditions. Exacerbation of symptoms has been reported in systemic lupus, ulcerative colitis (autoimmune etiology uncertain), glomerular nephritis, and multiple sclerosis. In "some" cases, effects reoccurred on rechallenge. In rheumatoid arthritis, treatment with *Echinacea* species for 10 days did not exacerbate the condition (Upton and Graff 2007). A survey of 25 medical herbalists indicated that 12 had used *Echinacea* species in persons with autoimmune conditions. Of these 12, 11 indicated a beneficial effect and 1 indicated a worsening of symptoms (Upton and Graff 2007).

III. Pharmacology and Pharmacokinetics

Human Pharmacological Studies

In healthy volunteers orally administered 1600 mg *Echinacea purpurea* root daily for 8 days, no clinically significant changes in activity of tolbutamide (CYP2C9) or dextromethorphan (CYP2D6) were observed. *Echinacea purpurea* root reduced plasma levels of caffeine (CYP1A2 induction), while effects on midazolam (CYP3A4) were conflicting, with *Echinacea purpurea* root increasing midazolam systemic clearance (~34% increase) through an induction of hepatic CYP3A4 activity, while inhibiting intestinal CYP3A4 activity (Gorski et al. 2004). A review of drug interaction studies noted the effects of *Echinacea purpurea* root observed in this study pale in comparison to those reported by the same investigators for known CYP3A4 inhibitors (Gurley et al. 2012).

No significant effects on the drug-metabolizing isoenzymes CYP1A2 (caffeine), CYP2D6 (debrisoquine), CYP2E1 (chlorzoxazone), or CYP3A4 (midazolam) activity were observed in healthy volunteers orally administered 1600 mg of a whole plant extract of *Echinacea purpurea* (no additional description provided) daily for 28 days (Gurley et al. 2004).

Research was conducted on a proprietary *Echinacea purpurea* extract, identified as an 8:1 extract of fresh *E. purpurea* root and flower. In healthy volunteers orally administered 8 mg midazolam (a substrate of CYP3A) and 120 mg fexofenadine (a substrate of P-gp) before and after administration of 1500 mg per day of this extract for 14 days, a decrease in plasma levels ($AUC_{0-\infty}$) of midazolam was observed, suggesting induction of the drug metabolizing enzyme CYP3A.

No changes in blood levels of fexofenadine were observed, suggesting a lack of effect on the drug transporter protein P-glycoprotein (P-gp) (Penzak et al. 2010).

Echinacea purpurea root at a repeated daily dose of 1600 mg did not affect CYP2C9, using tolbutamide as a probe drug (Gorski et al. 2004); neither this dose of the root nor 800 mg of an *E. purpurea* extract (plant part not stated; containing 6.6 mg isobutylamides) affected CYP2D6 activity, using dextromethorphan and debrisoquine as probe drugs (Gorski et al. 2004; Gurley et al. 2008).

Animal Pharmacological Studies

No relevant animal pharmacological studies were identified.

In Vitro Pharmacological Studies

In vitro studies on the effect of *Echinacea purpurea* on CYP enzymes have given mixed results. Results from one study on the effects of a preparation of the aerial parts of *Echinacea purpurea* on CYP3A4 differed according to the model substrate used; one substrate (7-benzyloxy-4-trifluoromethylcoumarin) indicated induction, while the other (resorufin benzyl ether) indicated inhibition (Yale and Gulrich 2005). Another study showed significant inhibition by a "commercial grade extract" (no other description provided) of *Echinacea angustifolia* root (Budzinski et al. 2000).

A preparation of the aerial parts of *Echinacea purpurea* moderately inhibited CYP2C9 but did not influence CYP2D6 activity (Yale and Gulrich 2005).

In human liver cells, clinically relevant concentrations of an extract of fresh *Echinacea purpurea* herb (95%) and root (5%) or alkylamides isolated from *Echinacea purpurea* had no effect on expression of the transcription of the drug-metabolizing enzyme CYP3A4 (Modarai et al. 2009).

The effects of *Echinacea purpurea* on CYP enzymes have been shown to vary widely based on the alkylamide content of specific *Echinacea purpurea* preparations. Such variation in effects may explain the differing results obtained in the clinical studies (Modarai et al. 2007).

In male rats fed diets containing 50 mg/kg of *Echinacea purpurea* root extracted in 50% ethanol for 4 or 8 weeks, a reduction in the relative testicle size and changes in histology of the testicles were observed after 8 weeks of echinacea consumption (Skaudickas et al. 2004).

Exposure to extremely high concentrations (8 mg/ml) but not lower concentrations (0.8 mg/ml) of *Echinacea purpurea* root extract adversely affected the function of hamster gametes (Ondrizek et al. 1999).

IV. Pregnancy and Lactation

In a prospective study of women using *Echinacea* species products during pregnancy, no significant differences in minor malformations, neonatal complications, birth weight, or gestational age were observed in the *Echinacea* species group as compared to the control group (Gallo et al. 2000).

A survey of dietary supplement use by pregnant women in California indicated that *Echinacea* (species not specified) was the most commonly used supplement, with 8.9% of respondents reporting use during pregnancy (Tsui et al. 2001).

In pigs fed diets containing 1.2 or 3.6% dried aerial parts of *Echinacea purpurea* on days 85 to 110 of pregnancy and 0.5 or 1.5% on up to day 28 of lactation, no significant differences were found for growth performance, weight loss, blood picture, plasma enzymes, or colostrum composition. Performance of the sucking piglets was not impaired either during lactation or during a 4-week observation period after weaning (Maass et al. 2005). No embryotoxicity or adverse effects on postnatal development were observed in rats or rabbits fed up to 2700 mg/kg per day of an ethanol stabilized fresh juice preparation from *Echinacea purpurea* herb (Mengs et al. 2000).

In mice fed diets containing 45 mg/animal daily of an extract of *Echinacea purpurea* (plant part not stated; average animal weight not specified) from the beginning of pregnancy through days 10–14 of pregnancy, a reduction in number of fetuses per pregnancy was observed (Chow et al. 2006).

No information on the safety of *Echinacea* spp. during lactation was identified.

V. Toxicity Studies

Acute Toxicity

The LD_{50} of intravenously administered polysaccharide fractions of *Echinacea purpurea* is reported to be between 1000 and 2500 mg/kg in mice (Lenk 1989). LD_{50} of intravenously administered *Echinacea purpurea* is 50 ml/kg in mice (Lang and Mengs 1976a, 1976b). Other studies with *Echinacea purpurea* indicated that no lethal dose of the expressed juice of the plant could be found in the highest physically possible oral and intravenous doses of 30,000 mg/kg (oral) and 10,000 mg/kg (intravenous) in mice (Mengs et al. 1991).

Short-Term Toxicity

Daily doses of 800 to 8000 mg/kg of the expressed juice of *Echinacea purpurea* for up to 4 weeks showed no toxic effects in rats (Mengs et al. 1991).

Genotoxicity

No carcinogenicity or mutagenicity was identified in in vitro assays (Kennelly 1985; Mengs et al. 1991; Schimmer et al. 1989).

Literature Cited

Ang-Lee, M.K., J. Moss, and C.S. Yuan. 2001. Herbal medicines and perioperative care. *J. Am. Med. Assoc.* 286(2):208-216.

Blumenthal, M., W.R. Busse, and A.B. Council. 1998. *The complete German Commission E monographs: Therapeutic guide to herbal medicines.* Philadelphia: Lippincott.

Budzinski, J.W., B.C. Foster, S. Vandenhoek, and J.T. Arnason. 2000. An in vitro evaluation of human cytochrome P450 3A4 inhibition by selected commercial herbal extracts and tinctures. *Phytomedicine* 7(4):273-282.

Chow, G., T. Johns, and S.C. Miller. 2006. Dietary *Echinacea purpurea* during murine pregnancy: Effect on maternal hemopoiesis and fetal growth. *Biol. Neonate* 89(2):133-138.

Coeugniet, E., and E. Elek. 1987. Immunomodulation with *Viscum album* and *Echinacea purpurea* extracts. *Onkologie* 10(3 Suppl.):27-33.

Currier, N.L., M. Sicotte, and S.C. Miller. 2001. Deleterious effects of *Echinacea purpurea* and melatonin on myeloid cells in mouse spleen and bone marrow. *J. Leukoc. Biol.* 70(2):274-276.

Gallo, M., W. Au, and G. Koren. 1998. The safety of echinacea use during pregnancy: A prospective controlled cohort study. *Teratology* 57(4/5):283.

Gallo, M., M. Sarkar, W. Au, et al. 1999. The safety of echinacea use during pregnancy: A prospective controlled study. *Clin. Invest. Med.* 22(4 Suppl.).

Gallo, M., M. Sarkar, W. Au, et al. 2000. Pregnancy outcome following gestational exposure to echinacea: A prospective controlled study. *Arch. Intern. Med.* 160(20):3141-3143.

George, L., E. Ioannis, T. Radostina, and M. Antonios. 2006. Severe thrombotic thrombocytopenic purpura (TTP) induced or exacerbated by the immunostimulatory herb echinacea. *Am. J. Hematol.* 81(3):224.

Gorski, J.C., S.M. Huang, A. Pinto, et al. 2004. The effect of echinacea (*Echinacea purpurea* root) on cytochrome P450 activity in vivo. *Clin. Pharmacol. Ther.* 75(1):89-100.

Gurley, B.J., S.F. Gardner, M.A. Hubbard, et al. 2004. In vivo assessment of botanical supplementation on human cytochrome P450 phenotypes: *Citrus aurantium, Echinacea purpurea,* milk thistle, and saw palmetto. *Clin. Pharmacol. Ther.* 76(5):428-440.

Gurley, B.J., A. Swain, M.A. Hubbard, et al. 2008. Clinical assessment of CYP2D6-mediated herb-drug interactions in humans: Effects of milk thistle, black cohosh, goldenseal, kava kava, St. John's wort, and *Echinacea*. *Mol. Nutr. Food Res.* 52(7):755-763.

Gurley, B.J., E.K. Fifer, and Z. Gardner. 2012. Pharmacokinetic herb-drug interactions (Part 2): Drug interactions involving popular botanical dietary supplements and their clinical relevance. *Planta Med.* [In press].

Heinrich, M., M. Modarai, and A. Kortenkamp. 2008. Herbal extracts used for upper respiratory tract infections: Are there clinically relevant interactions with the cytochrome P450 enzyme system? *Planta Med.* 74(6):657-660.

Huntley, A.L., J. Thompson Coon, and E. Ernst. 2005. The safety of herbal medicinal products derived from *Echinacea* species: A systematic review. *Drug Saf.* 28(5):387-400.

Izzo, A.A., and E. Ernst. 2001. Interactions between herbal medicines and prescribed drugs: A systematic review. *Drugs* 61(15):2163-2175.

Kemp, D.E., and K.N. Franco. 2002. Possible leukopenia associated with long-term use of echinacea. *J. Am. Board Fam. Pract.* 15(5):417-419.

Kennelly, J. 1985. Echinacea. York, England: Microtest Research Ltd. *In* Hobbs, C. 1997. *Echinacea: The immune herb.* Capitola, CA: Botanica Press.

Lang, W., and U. Mengs. 1976a. Report on echinacea toxicity in mice. *In* Hobbs, C. 1997. *Echinacea: The immune herb.* Capitola, CA: Botanica Press.

Lang, W., and U. Mengs. 1976b. Report on echinacea toxicity in rats. *In* Hobbs, C. 1997. *Echinacea: The immune herb.* Capitola, CA: Botanica Press.

Lee, A.N., and V.P. Werth. 2004. Activation of autoimmunity following use of immunostimulatory herbal supplements. *Arch. Dermatol.* 140(6):723-727.

Lenk, W. 1989. Acute toxicity of various polysaccharides from *Echinacea purpurea* in the mouse. *Ztschr. Phytother.* 10:49-51.

Logan, J.L., and J. Ahmed. 2003. Critical hypokalemic renal tubular acidosis due to Sjogren's syndrome: Association with the purported immune stimulant echinacea. *Clin. Rheumatol.* 22(2):158-159.

Maass, N., J. Bauer, B.R. Paulicks, B.M. Bohmer, and D.A. Roth-Maier. 2005. Efficiency of *Echinacea purpurea* on performance and immune status in pigs. *J. Anim. Physiol. Anim. Nutr.* 89(7-8):244-252.

Mengs, U., C.B. Clare, and J.A. Poiley. 1991. Toxicity of *Echinacea purpurea*. Acute, subacute and genotoxicity studies. *Arzneimittelforschung* 41(10):1076-1081.

Mengs, U., J. Leuschner, and R. Marshall. 2000. Toxicity studies with echinacin [abstract]. Third International Conference on Phytomedicine, Munich, Germany, 2000 October 11-13. *Phytomedicine* 7(Suppl. 2):32.

Mills, S., and K. Bone. 2005. *The essential guide to herbal safety*. St. Louis: Elsevier.

Modarai, M., J. Gertsch, A. Suter, M. Heinrich, and A. Kortenkamp. 2007. Cytochrome P450 inhibitory action of *Echinacea* preparations differs widely and co-varies with alkylamide content. *J. Pharm. Pharmacol.* 59(4):567-573.

Modarai, M., E. Silva, A. Suter, M. Heinrich, and A. Kortenkamp. 2009. Safety of herbal medicinal products: *Echinacea* and selected alkylamides do not induce CYP3A4 mRNA expression. *Evid. Based Complement. Alternat. Med.*

Moell, O. 1951. Primäre Ergebnisse der Echinacinbehandlung bei entzündlichen Unterleibserkrankungen. *Krankenhausarzt* 24:299-302.

Mullins, R.J. 1998. Echinacea-associated anaphylaxis. *Med. J. Aust.* 168(4):170-171.

Mullins, R.J., and R. Heddle. 2002. Adverse reactions associated with echinacea: The Australian experience. *Ann. Allergy Asthma Immunol.* 88(1):42-51.

Ondrizek, R.R., P.J. Chan, W.C. Patton, and A. King. 1999. An alternative medicine study of herbal effects on the penetration of zona-free hamster oocytes and the integrity of sperm deoxyribonucleic acid. *Fertil. Steril.* 71(3):517-522.

Penzak, S.R., S.M. Robertson, J.D. Hunt, et al. 2010. *Echinacea purpurea* significantly induces cytochrome P450 3A activity but does not alter lopinavir-ritonavir exposure in healthy subjects. *Pharmacother.* 30 (8):797-805.

Perri, D., J.J. Dugoua, E. Mills, and G. Koren. 2006. Safety and efficacy of echinacea (*Echinacea angustifolia*, *E. purpurea* and *E. pallida*) during pregnancy and lactation. *Can. J. Clin. Pharmacol.* 13(3):e262-e267.

Röseler, W. 1952. Erfahrungen mit der Echinacin-Therapie fieberhafter gynäkologischer Erkrankungen. *Medizinische* 3:93-95.

Schimmer, O., A. Erlangen, and B. Nurnberg. 1989. Investigation of the genotoxic potency of a neutral polysaccharide from echinacea tissue cultures in human lymphocyte cultures. *Ztschr. Phytother.* 10:39-42. *In* Hobbs, C. 1997. *Echinacea: The immune herb.* Capitola, CA: Botanica Press..

Skaudickas, D., A. Kondrotas, and K. Baltrusaitis. 2004. The effect of *Echinacea purpurea* extract on sexual glands of male rats. *Medicina* 40(12):1211-1218.

Soon, S.L., and R.I. Crawford. 2001. Recurrent erythema nodosum associated with *Echinacea* herbal therapy. *J. Am. Acad. Dermatol.* 44(2):298-299.

Taylor, J.A., W. Weber, L. Standish, et al. 2003. Efficacy and safety of echinacea in treating upper respiratory tract infections in children: A randomized controlled trial. *J. Am. Med. Assoc.* 290(21):2824-2830.

Tsui, B., C.E. Dennehy, and C. Tsourounis. 2001. A survey of dietary supplement use during pregnancy at an academic medical center. *Am. J. Obstet. Gynecol.* 185(2):433-437.

Upton, R., and A. Graff. 2004. *Echinacea purpurea root: Standards of analysis, quality control, and therapeutics.* American Herbal Pharmacopoeia. Scotts Valley, CA.

Upton, R., and A. Graff. 2007. *Echinacea purpurea aerial parts: Standards of analysis, quality control, and therapeutics.* American Herbal Pharmacopoeia. Scotts Valley, CA.

Yale, S.H., and I. Gulrich. 2005. Analysis of the inhibitory potential of *Ginkgo biloba*, *Echinacea purpurea*, and *Serenoa repens* on the metabolic activity of cytochrome P450 3A4, 2D6, and 2C9. *J. Altern. Complement. Med.* 11(3):433-439.

Yale, S.H., and K. Liu. 2004. *Echinacea purpurea* therapy for the treatment of the common cold: A randomized, double-blind, placebo-controlled clinical trial. *Arch. Intern. Med.* 164(11):1237-1241.

Echinodorus macrophyllus (Knuth) Micheli — Alismataceae

SCN: chapéau de couro
OCN: water plantain
Part: leaf

Quick Reference Summary

Safety Class: 1
Interaction Class: A

Contraindications

None known.

Other Precautions

None known.

Drug and Supplement Interactions

None known.

Adverse Events and Side Effects

None known.

Pharmacological Considerations

An animal study indicated immunosuppressive activity of chapéau de couro (Pinto et al. 2007).

Pregnancy and Lactation

No information on the safety of chapéau de couro in pregnancy or lactation was identified in the scientific or traditional literature. Although this review did not identify any concerns for use while pregnant or nursing, safety has not been conclusively established.

Review Details

I. Drug and Supplement Interactions

Clinical Trials of Drug or Supplement Interactions

No clinical trials of drug or supplement interactions were identified.

Case Reports of Suspected Drug or Supplement Interactions

No case reports of suspected drug or supplement interactions were identified.

Animal Trials of Drug or Supplement Interactions

No animal trials of drug or supplement interactions were identified.

II. Adverse Events

Case Reports of Adverse Events

No case reports of adverse events were identified.

III. Pharmacology and Pharmacokinetics

Human Pharmacological Studies

No relevant human pharmacological studies were identified.

Animal Pharmacological Studies

In mice orally administered 0.5 or 5 mg/kg of an aqueous extract of chapéau de couro daily for 7 days, immunosuppressive activity was observed, including inhibition of B cell antibody production and delayed-type hypersensitivity mediated by T cells, reducing subcutaneous tissue leukocyte infiltration (Pinto et al. 2007).

In Vitro Pharmacological Studies

No relevant in vitro pharmacological studies were identified.

IV. Pregnancy and Lactation

No information on the use of chapéau de couro during pregnancy or lactation was identified.

V. Toxicity Studies

Short-Term Toxicity

In mice administered an aqueous extract of chapéau de couro in drinking water for 6 weeks with approximate intake of 3, 23, or 297 mg/kg of a lyophilized extract, or 2200 mg/kg of a crude extract daily, a decrease in body weight was observed in animals receiving the highest dose. Biochemical analysis of the plasma revealed some minor alterations indicating subclinical hepatic toxicity (da Costa Lopes et al. 2000).

Genotoxicity

No mutagenicity of an aqueous extract of chapéau de couro was observed in the *Salmonella*/microsome assay in strains TA97a, TA98, TA100, or TA102 with or without metabolic activation (da Costa Lopes et al. 2000).

In mice administered an aqueous extract of chapéau de couro in drinking water for 6 weeks with intake of 3, 23, or 297 mg/kg of a lyophilized extract, or 2200 mg/kg of a crude extract daily, no genotoxic effects on liver or blood cells were observed in the comet assay. DNA analyses of the kidney cells detected some genotoxic activity for the highest dose tested but not at lower doses (da Costa Lopes et al. 2000).

Literature Cited

da Costa Lopes, L., F. Albano, G. Augusto Travassos Laranja, et al. 2000. Toxicological evaluation by in vitro and in vivo assays of an aqueous extract prepared from *Echinodorus macrophyllus* leaves. *Toxicol. Lett.* 116(3):189-198.

Pinto, A.C., G.C. Rego, A.M. Siqueira, et al. 2007. Immunosuppressive effects of *Echinodorus macrophyllus* aqueous extract. *J. Ethnopharmacol.* 111(2):435-439.

Eclipta prostrata (L.) L. — Asteraceae

SCN: eclipta
Syn: *Eclipta alba* (L.) Hassk.
AN: *bhringaraja*
PN: *han lian cao* (above-ground parts); *mo han lian* (above-ground parts)
OCN: false daisy
Part: above-ground parts

Quick Reference Summary

Safety Class: 1
Interaction Class: A

Contraindications

None known.

Other Precautions

None known.

Drug and Supplement Interactions

None known.

Adverse Events and Side Effects

None known.

Pharmacological Considerations

None known.

Pregnancy and Lactation

No information on the safety of eclipta in pregnancy or lactation was identified in the scientific or traditional literature. Although this review did not identify any concerns for use while pregnant or nursing, safety has not been conclusively established.

Review Details

I. Drug and Supplement Interactions

Clinical Trials of Drug or Supplement Interactions

No clinical trials of drug or supplement interactions were identified.

Case Reports of Suspected Drug or Supplement Interactions

No case reports of suspected drug or supplement interactions were identified.

Animal Trials of Drug or Supplement Interactions

No animal trials of drug or supplement interactions were identified.

II. Adverse Events

Case Reports of Adverse Events

No case reports of adverse events were identified.

III. Pharmacology and Pharmacokinetics

Human Pharmacological Studies

No relevant human pharmacological studies were identified.

Animal Pharmacological Studies

A significant reduction in blood glucose levels was observed in diabetic rats orally administered 2 or 4 g/kg of a leaf suspension of eclipta daily for 60 days (Ananthi et al. 2003).

Examining the immunomodulatory effects of eclipta in mice, administration of doses from 100 to 500 mg/kg of a methanol extract of eclipta resulted in dose-dependent increases in the phagocytic index and antibody titer (Jayathirtha and Mishra 2004).

In Vitro Pharmacological Studies

No relevant in vitro pharmacological studies were identified.

IV. Pregnancy and Lactation

No information on the use of eclipta during pregnancy or lactation was identified.

V. Toxicity Studies

Acute Toxicity

The LD_{50} values for an aqueous extract of eclipta are 7841 mg/kg after oral administration, 303 mg/kg after intravenous administration, and 328 mg/kg after intraperitoneal administration (Qadri et al. 2001).

E

Literature Cited

Ananthi, J., A. Prakasam, and K.V. Pugalendi. 2003. Antihyperglycemic activity of *Eclipta alba* leaf on alloxan-induced diabetic rats. *Yale J. Biol. Med.* 76(3):97-102.

Jayathirtha, M.G., and S.H. Mishra. 2004. Preliminary immunomodulatory activities of methanol extracts of *Eclipta alba* and *Centella asiatica*. *Phytomedicine* 11(4):361-365.

Qadri, N.M., S. Ahmad, S. Qureshi, and Y. Badar. 2001. Acute toxicological evaluation of the aqueous extract of *Eclipta alba* Hassk. *Pakistan J. Sci. Ind. Res.* 44(1):38-41.

Elettaria cardamomum (L.) Maton var. *cardamomum* — Zingiberaceae

SCN: cardamom

Syn: *Amomum cardamomum* L.; *Elettaria cardamomum* L. var. *miniscula* Burkill; *Elettaria cardamomum* L. var. *minus* Watt

AN: *ela*

OCN: Mysore cardamom

Part: fruit

Quick Reference Summary

Safety Class: 1

Interaction Class: A

Contraindications

None known.

Other Precautions

None known.

Drug and Supplement Interactions

None known.

Adverse Events and Side Effects

Contact and systemic contact-type dermatitis reactions to cardamom have been reported and confirmed by patch testing (Dooms-Goossens et al. 1990; Mobacken and Fregert 1975).

Pharmacological Considerations

None known.

Pregnancy and Lactation

No information on the safety of cardamom in pregnancy or lactation was identified in the scientific or traditional literature. Although this review did not identify any concerns for use while pregnant or nursing, safety has not been conclusively established.

Review Details

E

I. Drug and Supplement Interactions

Clinical Trials of Drug or Supplement Interactions

No clinical trials of drug or supplement interactions were identified.

Case Reports of Suspected Drug or Supplement Interactions

No case reports of suspected drug or supplement interactions were identified.

Animal Trials of Drug or Supplement Interactions

No animal trials of drug or supplement interactions were identified.

II. Adverse Events

Case Reports of Adverse Events

Contact and systemic contact-type dermatitis reactions to cardamom have been reported and confirmed by patch testing (Dooms-Goossens et al. 1990; Mobacken and Fregert 1975). In spice factory workers with a history of skin symptoms, irritant patch test reactions were observed for cardamom in some workers (Meding 1993).

III. Pharmacology and Pharmacokinetics

Human Pharmacological Studies

No irritant activity of 4% cardamom essential oil in petrolatum base was observed in 48-hour closed human patch tests (Opdyke 1979). In sensitization testing, no sensitization reactions were observed after testing with 4% cardamom essential oil in petrolatum base (Opdyke 1979).

Animal Pharmacological Studies

In an irritation test, no irritant effects were observed after cardamom essential oil was applied full strength to abraded or intact rabbit skin (Opdyke 1979).

In Vitro Pharmacological Studies

An extract of cardamom inhibited aggregation in human blood platelets treated with the agonists ADP, epinephrine, collagen, and calcium ionophore A 23187. No inhibition of aggregation was observed in platelets treated with ristocetin (Suneetha and Krishnakantha 2005).

IV. Pregnancy and Lactation

No information on the safety of cardamom in pregnancy or lactation was identified.

V. Toxicity Studies

Acute Toxicity

The LD_{50} of orally administered or topically applied cardamom essential oil in rats could not be determined at doses up to 5 g/kg (Opdyke 1979).

In mice orally administered 3 or 300 mg/kg daily for 7 days, microscopic evaluation indicated morphological perturbation of the heart along with inhibition of glyceraldehyde 3-phosphate dehydrogenase, and increases in the levels of thiobarbituric acid-reactive substances, succinate dehydrogenase, and catalase activity (El Malti et al. 2007).

Genotoxicity

No mutagenic activity of a cardamom aqueous extract was observed in the diploid yeast *Saccharomyces cerevisiae* strain D7 (Chughtai et al. 1998).

Literature Cited

Chughtai, S.R., M.A. Ahmad, N. Khalid, and A.S. Mohmand. 1998. Genotoxicity testing of some spices in diploid yeast. *Pakistan J. Bot.* 30(1):33-38.

Dooms-Goossens, A., R. Dubelloy, and H. Degreef. 1990. Contact and systemic contact-type dermatitis to spices. *Dermatol. Clin.* 8(1):89-93.

El Malti, J., D. Mountassif, and H. Amarouch. 2007. Antimicrobial activity of *Elettaria cardamomum*: Toxicity, biochemical and histological studies. *Food Chem.* 104(4):1560-1568.

Meding, B. 1993. Skin symptoms among workers in a spice factory. *Contact Dermat.* 29(4):202-205.

Mobacken, H., and S. Fregert. 1975. Allergic contact dermatitis from cardamom. *Contact Dermat.* 1(3):175-176.

Opdyke, D.L.J. 1979. *Monographs on fragrance raw materials.* New York: Pergamon.

Suneetha, W.J., and T.P. Krishnakantha. 2005. Cardamom extract as inhibitor of human platelet aggregation. *Phytother. Res.* 19(5):437-440.

Eleutherococcus senticosus (Rupr. & Maxim.) Maxim. — Araliaceae

SCN: eleuthero
Syn: *Acanthopanax senticosus* (Rupr. & Maxim.) Harms
PN: *ci wu jia* (root and stem)
OCN: Siberian ginseng; Ussurian thorny pepperbush
Part: root, root bark

Quick Reference Summary

Safety Class: 1
Interaction Class: A

Contraindications

None known.

Other Precautions

None known.

Drug and Supplement Interactions

None known.

Editors' Note

Eleuthero has been reported to be adulterated with *Periploca sepium* (Martindale and Reynolds 1996), a botanical that contains cardioactive glycosides (Zhu 1998).

Adverse Events and Side Effects

A case of "hairy baby syndrome" was reported in association with a product labeled as "Siberian ginseng" (Koren et al. 1990). This product, however, was determined to be adulterated with *Periploca sepium* (Awang 1991).

Pharmacological Considerations

Insomnia has been rarely observed in association with clinical studies (De Smet 1993).

One source indicates that eleuthero should not be used in patients with high blood pressure (WHO 2002), based on the recommendation of two Russian studies published in the 1960s (Dalinger 1966; Lapchik 1967).

Pregnancy and Lactation

Basic animal studies have shown no adverse effects of eleuthero use during pregnancy or lactation (Curtze 1980; Davydov and Krikorian 2000; Farnsworth et al. 1985).

One reference contraindicates eleuthero during pregnancy, but the reason for this contraindication is not provided, with the reference citing animal studies that have shown no harm during pregnancy (WHO 2002).

E

Review Details

I. Drug and Supplement Interactions

Clinical Trials of Drug or Supplement Interactions

No clinical trials of drug or supplement interactions were identified.

Case Reports of Suspected Drug or Supplement Interactions

Elevated serum digoxin levels were reported in a man taking eleuthero; the digoxin levels decreased and again increased when the man ceased and resumed consumption of eleuthero (McRae 1996). The elevated digoxin levels may have been due to an interference with the digoxin assay, especially since no toxic effects of elevated digoxin were observed (Brinker 2001). Others have questioned the botanical identity of the product the patient was taking, indicating that the product may have been adulterated with *Periploca sepium*, an herb that contains cardioactive glycosides (Zhu 1998).

Animal Trials of Drug or Supplement Interactions

Administration of eleuthero (40–320 mg/kg daily) and the barbiturate hexobarbital was shown to increase the sedative effects of hexobarbital (Medon et al. 1984).

II. Adverse Events

Adverse Events Reported in Clinical Trials

A review of clinical trials (over 4000 patients total) indicated that eleuthero was well tolerated with no significant adverse effects at doses of 2 to 16 ml daily of alcohol extract for up to 60 consecutive days (Farnsworth et al. 1985). Authors of two trials indicated that eleuthero should not be administered to persons with high blood pressure (over 180/90 mm Hg) (Dalinger 1966; Lapchik 1967), although no side effects of eleuthero were reported in these studies, which included more than 2100 participants. In studies on patients with atherosclerosis or with rheumatic heart disease, some adverse events were reported including insomnia, tachycardia, extra systole, hypertonia, headaches, pericardial pain, palpitations, and elevated blood pressure (Koshkareva and Kovinsky 1966; Mikunis et al. 1966).

Case Reports of Adverse Events

Use of eleuthero was reported in association with "ginseng abuse syndrome" (see *Panax ginseng* entry) (Siegel 1979). The study that included this information has been criticized for methodological flaws, notably, that no distinction was made between the effects of Asian ginseng and eleuthero (Mills and Bone 2005).

III. Pharmacology and Pharmacokinetics

Human Pharmacological Studies

Administration of standardized eleuthero extract (970 mg daily) for 2 weeks had no significant effects on the drug-metabolizing isoenzymes CYP2D6 or CYP3A4 (Donovan et al. 2003).

Animal Pharmacological Studies

An aqueous extract of eleuthero administered intraperitoneally to mice lowered plasma sugar levels (Hikino et al. 1986).

No androgenic effects were observed in castrated rats administered eleuthero and testosterone for 7 days (Waller et al. 1992).

In Vitro Pharmacological Studies

Application of eleuthero extract (10%, v/v) to neonatal rat heart muscle cells (cardiomyocytes) resulted in a rapid cessation of heart beating due to calcium overload, although diluted extracts resulted in normal heart beating. Cardiotonic effects were observed when the same extract was applied to adult rat cardiomyocytes (Poindexter et al. 2006).

Animal and in vitro studies have indicated that consumption of eleuthero may cause some interference with certain digoxin immunoassays, falsely increasing serum digoxin levels in the FPIA assay or decreasing levels in the MEIA assay. No effects were observed in the EMIT, Randox, CLIA, or Tina-quant assays (Dasgupta and Reyes 2005; Dasgupta et al. 2003). Filtration of serum with activated charcoal prior to analysis increased the accuracy of test results (Dasgupta and Veras 2006).

IV. Pregnancy and Lactation

A case of neonatal androgenization was reported in the infant of a mother who had taken a product labeled as "Siberian ginseng" (an alternate common name for eleuthero) during her pregnancy (Koren et al. 1990). A subsequent analysis of this product revealed that the product was actually *Periploca sepium*, an herb that contains cardioactive glycosides (Awang 1991).

In two studies, eleuthero extracts (13.5 ml/kg or 10 mg/kg) administered for 9 or 16 days to pregnant rats showed no teratogenic effects (Curtze 1980; Davydov and Krikorian 2000). Likewise, no teratogenic effects were observed in sheep or mink after an ethanol extract of eleuthero was fed to pregnant animals (Farnsworth et al. 1985). No adverse effects were reported in minks fed eleuthero (10 ml/kg) on days 1 to 45 of lactation (Farnsworth et al. 1985).

V. Toxicity Studies

Acute Toxicity

The LD_{50} of orally administered powdered eleuthero root in mice is 31 g/kg (Brekhman 1960, 1963). The LD_{50} of orally administered ethanolic extract of eleuthero in mice is 14.5 g/kg (Brekhman 1970).

Administration of a single oral dose of 3 g/kg of a freeze-dried aqueous extract of eleuthero to mice produced no deaths (Medon et al. 1981).

Subchronic Toxicity

No toxic effects were observed in rats fed 10 mg/kg daily of an ethanol extract of eleuthero for 2 months (Davydov and Krikorian 2000).

Chronic Toxicity

Rats administered 5 ml/kg daily of an ethanolic extract of eleuthero for 320 days showed no adverse reactions (Golotin et al. 1972).

Mice administered aqueous extracts of eleuthero for up to 96 days showed no significant difference in weight gain, liquid consumed, pathogenesis, or mortality, although mice that drank an extract with sugar exhibited aggressive behavior (Lewis et al. 1983).

Genotoxicity

No mutagenic or carcinogenic effects of eleuthero extracts were observed in bacterial mutagenicity assays or in rats (Hirosue et al. 1986).

Literature Cited

Awang, D.V. 1991. Maternal use of ginseng and neonatal androgenization. *J. Am. Med. Assoc.* 266(3):363.

Brekhman, I.I. 1960. Toxicity and general action of *Eleutherococcus*. In *Eleutherococcus root—New stimulating and tonic remedy* [in Russian]. Leningrad: VI Lenin Military Institute of Physical Culture and Sports.

Brekhman, I.I. 1963. Comparative data on pharmacological effect of ginseng, *Eleutherococcus*, *Echinopanax* and *Aralia* roots [in Russian]. *Materials to the studies of ginseng and other therapeutical medicines of the Far East* 5:219-227. *Cited in* Farnsworth, N., et al. 1985. Siberian ginseng (*Eleutherococcus senticosus*). *In* Wagner, H., et al., eds. *Economic and medicinal plant research, Volume 1*, pp. 155-215. London: Academic Press.

Brekhman, I.I. 1970. *Eleutherococcus*. Clinical data. Moscow: USSR Foreign Trade Publications 28524/2. *Cited in* Farnsworth, N., et al. 1985. Siberian ginseng (*Eleutherococcus senticosus*). *In* Wagner, H., et al., eds. *Economic and medicinal plant research, Volume 1*, pp. 155-215. London: Academic Press.

Brinker, F. 2001. *Herb contraindications and drug interactions*. 3rd ed. Sandy, OR: Eclectic Medical Publications.

Curtze, A. 1980. Die Arzneipflanze *Eleutherococcus senticosus* Maxim. in der Bundesrepublik Deutschland. *Kassenarzt* 20:497-503.

Dalinger, O.I. 1966. Effect of *Eleutherococcus* extract on the cardiovascular system and some measures of working capacity in older persons [in Russian]. *Central Nervous System Stimulants Tomsk* 1:112-4. *Cited in* Farnsworth, N., et al. 1985. Siberian ginseng (*Eleutherococcus senticosus*). *In* Wagner, H., et al., eds. *Economic and medicinal plant research, Volume 1,* pp. 155-215. London: Academic Press.

Dasgupta, A., and M.A. Reyes. 2005. Effect of Brazilian, Indian, Siberian, Asian, and North American ginseng on serum digoxin measurement by immunoassays and binding of digoxin-like immunoreactive components of ginseng with Fab fragment of antidigoxin antibody (Digibind). *Am. J. Clin. Pathol.* 124(2):229-236.

Dasgupta, A., and E. Veras. 2006. Effectiveness of activated charcoal and equilibrium dialysis in removing Asian, American, Siberian and Indian ginseng from human serum. *Clin. Chim. Acta* 367(1-2):144-149.

Dasgupta, A., S. Wu, J. Actor, et al. 2003. Effect of Asian and Siberian ginseng on serum digoxin measurement by five digoxin immunoassays. Significant variation in digoxin-like immunoreactivity among commercial ginsengs. *Am. J. Clin. Pathol.* 119(2):298-303.

Davydov, M., and A.D. Krikorian. 2000. *Eleutherococcus senticosus* (Rupr. & Maxim.) Maxim. (Araliaceae) as an adaptogen: A closer look. *J. Ethnopharmacol.* 72(3):345-393.

De Smet, P.A.G.M. 1993. *Adverse effects of herbal drugs, Volume 2.* Berlin: Springer.

Donovan, J.L., C. Lindsay DeVane, K.D. Chavin, R.M. Taylor, and J.S. Markowitz. 2003. Siberian ginseng (*Eleutherococcus senticosus*) effects on CYP2D6 and CYP3A4 activity in normal volunteers. *Drug Metab. Disposition* 31(5):519-522.

Farnsworth, N.R., D.A. Kinghorn, D.D. Soejarto, and D.P. Waller. 1985. Siberian ginseng (*Eleutherococcus senticosus*): Current status as an adaptogen. *Econ. Med. Plant Res.* 1:155-215.

Golotin, V.G., I.I. Brekhman, and A.I. Dobryakova. 1972. Influence of ginseng and *Eleutherococcus* extracts on life expectancy of white rats [in Russian]. *Medicines of the Soviet Far East Vladivostok* 11:37-41. *Cited in* Farnsworth, N., et al. 1985. Siberian ginseng (*Eleutherococcus senticosus*). *In* Wagner, H., et al., eds. *Economic and medicinal plant research, Volume 1,* pp. 155-215. London: Academic Press.

Hikino, H., M. Takahashi, K. Otake, and C. Konno. 1986. Isolation and hypoglycemic activity of eleutherans A, B, C, D, E, F, and G: Glycans of *Eleutherococcus senticosus* roots. *J. Nat. Prod.* 49(2):293-297.

Hirosue, T., M. Matsuzawa, H. Kawai, and et al. 1986. Mutagenicity and subacute toxicity of *Acanthopanax senticosus* extracts in rats. *J. Food Hyg. Soc. Jpn.* 27:380-386.

Koren, G., S. Randor, S. Martin, and D. Danneman. 1990. Maternal ginseng use associated with neonatal androgenization. *J. Am. Med. Assoc.* 264(22):2866.

Koshkareva, K.I., and K.P. Kovinsky. 1966. Application of *Eleutherococcus* extract on the hypochondriacal conditions [in Russian]. *Central Nervous System Stimulants Tomsk* 1:128-130. *Cited in* Farnsworth, N., et al. 1985. Siberian ginseng (*Eleutherococcus senticosus*). *In* Wagner, H., et al., eds. *Economic and medicinal plant research, Volume 1,* pp. 155-215. London: Academic Press.

Lapchik, V.F. 1967. [Article in Russian.] *Visn. Kiiv Univ. Ser. Biol.* 9:131. *Cited in* Farnsworth, N., et al. 1985. Siberian ginseng (*Eleutherococcus senticosus*). *In* Wagner, H., et al., eds. *Economic and medicinal plant research, Volume 1,* pp. 155-215. London: Academic Press.

Lewis, W.H., V.E. Zenger, and R.G. Lynch. 1983. No adaptogen response of mice to ginseng and *Eleutherococcus* infusions. *J. Ethnopharmacol.* 8(2):209-214.

Martindale, W., and J.E.F. Reynolds. 1996. *The extra pharmacopoeia.* 31st ed. London: Pharmaceutical Press.

McRae, S. 1996. Elevated serum digoxin levels in a patient taking digoxin and Siberian ginseng. *Can. Med. Assoc. J.* 155(3):293-295.

Medon, P.J., P.W. Ferguson, and C.F. Watson. 1984. Effects of *Eleutherococcus senticosus* extracts on hexobarbital metabolism in vivo and in vitro. *J. Ethnopharmacol.* 10(2):235-241.

Medon, P.J., E.B. Thompson, and N.R. Farnsworth. 1981. Hypoglycemic effect and toxicity of *Eleutherococcus senticosus* following acute and chronic administration in mice. *Acta Pharmacol. Sin.* 2(4):281-285.

Mikunis, R., V. Serkova, and T. Shirkova. 1966. Influence of *Eleutherococcus* on the immunobiologic reactivity of patients with rheumatic defects of heart. *In* Brekhman, I.I., ed. *Eleutherococcus* and other adaptogens among the Far Eastern plants [in Russian]. *Vladivostok* 7:221-226. *Cited in* Farnsworth, N., et al. 1985. Siberian ginseng (*Eleutherococcus senticosus*). *In* Wagner, H., et al., eds. *Economic and medicinal plant research, Volume 1,* pp. 155-215. London: Academic Press.

Mills, S., and K. Bone. 2005. *The essential guide to herbal safety.* St. Louis: Elsevier.

Poindexter, B.J., A.W. Allison, R.J. Bick, and A. Dasgupta. 2006. Ginseng: Cardiotonic in adult rat cardiomyocytes, cardiotoxic in neonatal rat cardiomyocytes. *Life Sci.* 79(25):2337-2344.

Siegel, R.K. 1979. Ginseng abuse syndrome. Problems with the panacea. *J. Am. Med. Assoc.* 241(15):1614-1615.

Waller, D.P., A.M. Martin, N.R. Farnsworth, and D.V. Awang. 1992. Lack of androgenicity of Siberian ginseng. *J. Am. Med. Assoc.* 267(17):2329.

WHO. 2002. *WHO monographs on selected medicinal plants, Volume 2.* Geneva: WHO.

Zhu, Y.-P. 1998. *Chinese materia medica: Chemistry, pharmacology and applications.* Amsterdam: Harwood Academic Publishers.

Elymus repens (L.) Gould — Poaceae

SCN: couch grass
Syn: *Agropyron repens* (L.) P. Beauv.; *Elytrigia repens* (L.) Desv. ex B.D. Jackson; *Triticum repens* L.
OCN: dog grass; graminis; quack grass; triticum; twitch grass
Part: rhizome

Quick Reference Summary

Safety Class: 1
Interaction Class: A

Contraindications
None known.

Other Precautions
None known.

Drug and Supplement Interactions
None known.

Notice
Diuretic (Mills and Bone 2005); *see* Appendix 2.

Adverse Events and Side Effects
None known.

Pharmacological Considerations
None known.

Pregnancy and Lactation
No information on the safety of couch grass in pregnancy or lactation was identified in the scientific or traditional literature. Although this review did not identify any concerns for use while pregnant or nursing, safety has not been conclusively established.

Review Details

I. Drug and Supplement Interactions

Clinical Trials of Drug or Supplement Interactions
No clinical trials of drug or supplement interactions were identified.

Case Reports of Suspected Drug or Supplement Interactions
No case reports of suspected drug or supplement interactions were identified.

Animal Trials of Drug or Supplement Interactions
No animal trials of drug or supplement interactions were identified.

II. Adverse Events

Case Reports of Adverse Events
No case reports of adverse events were identified.

III. Pharmacology and Pharmacokinetics

Human Pharmacological Studies
No relevant human pharmacological studies were identified.

Animal Pharmacological Studies
No relevant animal pharmacological studies were identified.

In Vitro Pharmacological Studies
No relevant in vitro pharmacological studies were identified.

IV. Pregnancy and Lactation
No information on the safety of couch grass during pregnancy or lactation was identified.

V. Toxicity Studies

Genotoxicity
No mutagenic activity of a fluid extract (1:1, 20% ethanol) of couch grass was observed in the Ames test using *Salmonella typhimurium* strains TA98 and TA100 (Schimmer et al. 1994).

Literature Cited

Mills, S., and K. Bone. 2005. *The essential guide to herbal safety*. St. Louis: Elsevier.

Schimmer, O, A. Krüger, H. Paulini, and F. Haefele. 1994. An evaluation of 55 commercial plant extracts in the Ames mutagenicity test. *Planta Med.* 49(6): 448-451.

Epigaea repens L. — Ericaceae

SCN: trailing arbutus
OCN: gravel root

Part: leaf

Quick Reference Summary

Safety Class: 1
Interaction Class: A

Contraindications

None known.

Other Precautions

None known.

Drug and Supplement Interactions

None known.

Editors' Note

Trailing arbutus has a chemical profile and traditional use similar to uva-ursi (*Arctostaphylos uva-ursi*), and these herbs have been considered as acceptable substitutes for each other (Felter and Lloyd 1898; Remington and Wood 1918).

Adverse Events and Side Effects

None known.

Pharmacological Considerations

None known.

Pregnancy and Lactation

No information on the safety of trailing arbutus in pregnancy or lactation was identified in the scientific or traditional literature. Although this review did not identify any concerns for use while pregnant or nursing, safety has not been conclusively established.

E

Review Details

I. Drug and Supplement Interactions

Clinical Trials of Drug or Supplement Interactions

No clinical trials of drug or supplement interactions were identified.

Case Reports of Suspected Drug or Supplement Interactions

No case reports of suspected drug or supplement interactions were identified.

Animal Trials of Drug or Supplement Interactions

No animal trials of drug or supplement interactions were identified.

II. Adverse Events

Case Reports of Adverse Events

No case reports of adverse events were identified.

III. Pharmacology and Pharmacokinetics

Human Pharmacological Studies

No relevant human pharmacological studies were identified.

Animal Pharmacological Studies

No relevant animal pharmacological studies were identified.

In Vitro Pharmacological Studies

No relevant in vitro pharmacological studies were identified.

IV. Pregnancy and Lactation

No information on the safety of trailing arbutus during pregnancy or lactation was identified.

V. Toxicity Studies

See *Arctostaphylos uva-ursi* for a review of toxicity studies on the compound hydroquinone, which is hydrolyzed from the compound arbutin that is present in trailing arbutus (Felter and Lloyd 1898; Tanaka and Iwamoto 2002).

Literature Cited

Felter, H.W., and J.U. Lloyd. 1898. *King's American dispensatory*. 18th ed., 3rd rev. 2 vols. Cincinnati: Ohio Valley Co.

Remington, J.P., and H.C. Wood. 1918. *The dispensatory of the United States of America*. 20th ed. Philadelphia: Lippincott.

Tanaka, Y., and A. Iwamoto. 2002. Skin care preparation/Skin external preparation comprising extract of plant belonging to genus *Epigaea* of family Ericaceae for whitening skin. Japan: Matsumoto Trading Co.

E

Epilobium angustifolium L. — Onagraceae

SCN: fireweed
Syn: *Chamaenerion angustifolium* (L.) Scop.; *Chamerion angustifolium* (L.) Holub

OCN: great willow herb
Part: herb

Quick Reference Summary

Safety Class: 1
Interaction Class: A

Contraindications

None known.

Other Precautions

None known.

Drug and Supplement Interactions

None known.

Adverse Events and Side Effects

None known.

Pharmacological Considerations

None known.

Pregnancy and Lactation

No information on the safety of fireweed in pregnancy or lactation was identified in the scientific or traditional literature. Although this review did not identify any concerns for use while pregnant or nursing, safety has not been conclusively established.

Review Details

I. Drug and Supplement Interactions

Clinical Trials of Drug or Supplement Interactions

No clinical trials of drug or supplement interactions were identified.

Case Reports of Suspected Drug or Supplement Interactions

No case reports of suspected drug or supplement interactions were identified.

Animal Trials of Drug or Supplement Interactions

No animal trials of drug or supplement interactions were identified.

II. Adverse Events

Case Reports of Adverse Events

No case reports of adverse events were identified.

III. Pharmacology and Pharmacokinetics

Human Pharmacological Studies

No relevant human pharmacological studies were identified.

Animal Pharmacological Studies

In intact or testosterone-stimulated castrated male rats orally administered 40 mg/kg of hexane or water extracts or an ultrafiltration fraction daily for 20 days, the water extract caused a decrease in the weight of seminal vesicles in intact rats, and an increase in the weight of all accessory sexual organs in castrated rats. In intact rats, the ultrafiltration fraction exhibited the same but smaller effects than the water extract, with no effects observed in castrated rats. No effects on sexual organs were observed on rats administered the hexane extract (Hiermann and Bucar 1997).

In Vitro Pharmacological Studies

No relevant in vitro pharmacological studies were identified.

IV. Pregnancy and Lactation

No information on the use of fireweed during pregnancy or lactation was identified.

V. Toxicity Studies

No toxicity studies were identified.

Literature Cited

Hiermann, A., and F. Bucar. 1997. Studies of *Epilobium angustifolium* extracts on growth of accessory sexual organs in rats. *J. Ethnopharmacol.* 55(3):179-183.

Epilobium parviflorum Schreb. — Onagraceae

SCN: small-flower willow herb

Part: herb

Quick Reference Summary

Safety Class: 1
Interaction Class: A

Contraindications

None known.

Other Precautions

None known.

Drug and Supplement Interactions

None known.

Notice

Tannins (4–14%) (Ducrey et al. 1997; Lesuisse et al. 1996; Spilkova et al. 1995); *see* Appendix 1.

Adverse Events and Side Effects

None known.

Pharmacological Considerations

None known.

Pregnancy and Lactation

No information on the safety of small-flower willow herb in pregnancy or lactation was identified in the scientific or traditional literature. Although this review did not identify any concerns for use while pregnant or nursing, safety has not been conclusively established.

E

Review Details

I. Drug and Supplement Interactions

Clinical Trials of Drug or Supplement Interactions

No clinical trials of drug or supplement interactions were identified.

Case Reports of Suspected Drug or Supplement Interactions

No case reports of suspected drug or supplement interactions were identified.

Animal Trials of Drug or Supplement Interactions

No animal trials of drug or supplement interactions were identified.

II. Adverse Events

Case Reports of Adverse Events

No case reports of adverse events were identified.

III. Pharmacology and Pharmacokinetics

Human Pharmacological Studies

No relevant human pharmacological studies were identified.

Animal Pharmacological Studies

No relevant animal pharmacological studies were identified.

In Vitro Pharmacological Studies

Compounds isolated from small-flower willow herb have been shown to inhibit 5α-reductase (Ducrey et al. 1997; Lesuisse et al. 1996).

IV. Pregnancy and Lactation

No information on the use of small-flower willow herb during pregnancy or lactation was identified.

V. Toxicity Studies

Acute Toxicity

The LD_{50} of an intraperitoneally administered ethanol extract of small-flower willow herb in rats is 200 mg/kg (Sharma et al. 1978).

No toxic effects were observed in rats orally administered up to 2 g/kg of a decoction of small-flower willow herb (Montalvo et al. 2003).

Literature Cited

Ducrey, B., A. Marston, S. Göhring, R.W. Hartmann, and K. Hostettmann. 1997. Inhibition of 5α-reductase and aromatase by the ellagitannins oenothein A and oenothein B from *Epilobium* species. *Planta Med.* 63(2):111-114.

Lesuisse, D., J. Berjonneau, C. Ciot, et al. 1996. Determination of oenothein B as the active 5-α-reductase-inhibiting principle of the folk medicine *Epilobium parviflorum*. *J. Nat. Prod.* 59(5):490-492.

Montalvo, R.V., R. Menendéz, V.B. Pavón, T.I.G. Sardiñas, and T.Y.V. Hurtado. 2003. Toxicidad aguda oral de la droga seca de *Epilobium parviflorum* L. *Rev. Cubana Plant Med.* 2003(2).

Sharma, M.L., N. Chandokhe, B.J. Ghatak, et al. 1978. Pharmacological screening of Indian medicinal plants. *Indian J. Exp. Biol.* 16(2):228.

Spilkova, J., M. Johankova, and J. Dusek. 1995. Pharmacognostic evaluation of some species of the genus *Epilobium*. *Ceska Slov. Farm.* 44(4):196-200.

Epimedium spp. — Berberidaceae

Epimedium brevicornum Maxim.
SCN: epimedium
PN: *yin yang huo* (herb, leaf)

Epimedium grandiflorum **C. Morren**
SCN: barrenwort
Syn: *Epimedium macranthum* Morren & Decne.

Epimedium koreanum **Nakai**
SCN: epimedium
PN: *yin yang huo* (herb, leaf)
OCN: Korean epimedium

Epimedium pubescens **Maxim.**
SCN: epimedium
PN: *yin yang huo* (herb, leaf)
OCN: pubescent epimedium

Epimedium sagittatum **(Siebold & Zucc.) Maxim.**
SCN: epimedium
PN: *yin yang huo* (herb, leaf)
OCN: sagittate epimedium

Epimedium wushanense **T.S. Ying**
SCN: epimedium
PN: *yin yang huo* (herb, leaf)
OCN: Wushan epimedium
Part: herb, leaf

Quick Reference Summary

Safety Class: 1
Interaction Class: A

Contraindications

None known.

Other Precautions

None known.

Drug and Supplement Interactions

None known.

Adverse Events and Side Effects

Some persons taking epimedium have experienced dry mouth, stomach discomfort, nausea, and vertigo, which were reported to abate without cessation of the herb (Bensky et al. 2004; Zhu 1998).

Pharmacological Considerations

Animal and in vitro studies have demonstrated estrogenic activity of epimedium (De Naeyer et al. 2005; Wang and Lou 2004; Yap et al. 2005, 2007; Zhang et al. 2005), although one study in menopausal women indicated a reduction in bone loss but no effects on estradiol levels (Zhang et al. 2007). In vitro, epimedium extracts have shown a greater affinity for estrogen receptor α than β (Shen et al. 2007).

Pregnancy and Lactation

No information on the safety of epimedium in pregnancy or lactation was identified in the scientific or traditional literature. Although this review did not identify any concerns for use while pregnant or nursing, safety has not been conclusively established.

Review Details

I. Drug and Supplement Interactions

Clinical Trials of Drug or Supplement Interactions

No clinical trials of drug or supplement interactions were identified.

Case Reports of Suspected Drug or Supplement Interactions

No case reports of suspected drug or supplement interactions were identified.

Animal Trials of Drug or Supplement Interactions

No animal trials of drug or supplement interactions were identified.

II. Adverse Events

Case Reports of Adverse Events

Some persons taking epimedium have experienced dry mouth, stomach discomfort, nausea, and vertigo, which were reported to abate without cessation of the herb (Bensky et al. 2004; Zhu 1998). Other adverse events that have been associated with epimedium use include vomiting, nosebleeds, and abdominal distention (Bensky et al. 2004). Details on products and doses associated with these adverse events were not reported in English language translations.

III. Pharmacology and Pharmacokinetics

Human Pharmacological Studies

In late postmenopausal women, administration of epimedium-derived (*E. brevicornum*) phytoestrogen flavonoid capsules (containing 60 mg icariin, 15 mg daidzein, and 3 mg genistein daily) for 2 years, a reduction in bone loss was observed without a detectable hyperplasia effect on the endometrium. Serum estradiol remained unchanged (Zhang et al. 2007).

Animal Pharmacological Studies

In male rats orally administered 500 mg of fractions of an ethanol extract of epimedium (*E. brevicornum*), an increase in estrogen receptor (ER)-α activity was observed (Yap et al. 2007).

In Vitro Pharmacological Studies

In estrogen receptor (ER)-α-positive breast cancer cells (MCF-7), 100 μg/ml of epimedium (*E. brevicornum*) inhibited the effects of tamoxifen. ER-α expression in the nucleus was observed with a low dose of epimedium (1 μg/ml), but higher doses (10 or 100 μg/ml) markedly reduced nuclear ER-α protein content (Yap et al. 2007).

In the modified MCF-7 cell proliferation assay, the compounds icaritin and desmethylicaritin, but not icariin, dose-dependently stimulated the proliferation of MCF-7/BUS cells. The estrogen receptor-regulated progesterone receptor and PS2 mRNA levels were increased by treatment with icaritin or desmethylicaritin (Wang and Lou 2004).

Epimedium (*E. brevicornum*) was reported to increase estrogen-responsive human breast cancer cell proliferation at low concentrations, and paradoxically cause profound inhibition of growth at higher concentrations (Yap et al. 2005).

In a recombinant yeast system with a human estrogen receptor expression plasmid and a reporter plasmid, estrogenic activity of epimedium (*E. brevicornum*) was observed, with an EC_{50} of 100 μg/ml (Zhang et al. 2005).

Epimedium extracts demonstrated a greater affinity for ER-α than ER-β in human cells transfected with ER-α and ER-β (Shen et al. 2007).

An extract of the polyphenolic compounds from epimedium (*E. brevicornum*) exhibited estrogenic activity in estrogen-responsive bioassays, a yeast cell assay, and the Ishikawa Var-I assay (De Naeyer et al. 2005).

IV. Pregnancy and Lactation

No information on the safety of epimedium in pregnancy or lactation was identified in the scientific or traditional literature. Although this review did not identify any concerns for use while pregnant or nursing, safety has not been conclusively established.

V. Toxicity Studies

Acute Toxicity

The LD_{50} of epimedium extract in mice is 36 g/kg after intraperitoneal administration and 56.8 g/kg after intravenous administration (Chen and Chen 2004). The LD_{50} of flavonoids extracted from epimedium is 3 g/kg after intraperitoneal administration to mice (Leung and Foster 1996).

No adverse effects were observed in mice orally administered 450 g/kg of epimedium daily for 3 days (Chen and Chen 2004).

Literature Cited

Bensky, D., S. Clavey, and E. Stöger. 2004. *Chinese herbal medicine: Materia medica*. 3rd ed. Seattle: Eastland Press.

Chen, J.K., and T.T. Chen. 2004. *Chinese medical herbology and pharmacology*. City of Industry, CA: Art of Medicine Press.

De Naeyer, A., V. Pocock, S. Milligan, and D. De Keukeleire. 2005. Estrogenic activity of a polyphenolic extract of the leaves of *Epimedium brevicornum*. *Fitoterapia* 76(1):35-40.

Leung, A.Y., and S. Foster. 1996. *Encyclopedia of common natural ingredients used in food, drugs, and cosmetics*. 2nd ed. New York: Wiley.

Shen, P., B.L. Guo, Y. Gong, et al. 2007. Taxonomic, genetic, chemical and estrogenic characteristics of *Epimedium* species. *Phytochemistry* 68(10):1448-1458.

Wang, Z.Q., and Y.J. Lou. 2004. Proliferation-stimulating effects of icaritin and desmethylicaritin in MCF-7 cells. *Eur. J. Pharmacol.* 504(3):147-153.

Yap, S.P., P. Shen, M.S. Butler, et al. 2005. New estrogenic prenylflavone from *Epimedium brevicornum* inhibits the growth of breast cancer cells. *Planta Med.* 71(2):114-119.

Yap, S.P., P. Shen, J. Li, L.S. Lee, and E.L. Yong. 2007. Molecular and pharmacodynamic properties of estrogenic extracts from the traditional Chinese medicinal herb, *Epimedium*. *J. Ethnopharmacol.* 113(2):218-224.

Zhang, C.Z., S.X. Wang, Y. Zhang, J.P. Chen, and X.M. Liang. 2005. In vitro estrogenic activities of Chinese medicinal plants traditionally used for the management of menopausal symptoms. *J. Ethnopharmacol.* 98(3):295-300.

Zhang, G., L. Qin, and Y. Shi. 2007. Epimedium-derived phytoestrogen flavonoids exert beneficial effect on preventing bone loss in late postmenopausal women: A 24-month randomized, double-blind and placebo-controlled trial. *J. Bone Miner. Res.* 22(7):1072-1079.

Zhu, Y.-P. 1998. *Chinese materia medica: Chemistry, pharmacology and applications.* Amsterdam: Harwood Academic Publishers.

Equisetum hyemale L. — Equisetaceae

SCN: scouring rush
OCN: rough horsetail
PN: *mu zei* (above-ground parts)
Part: above-ground parts

Quick Reference Summary

Safety Class: 2b
Interaction Class: A

E

Contraindications

Not for use in pregnancy except under the supervision of a qualified healthcare practitioner (Bensky et al. 2004; Chen and Chen 2004).

Other Precautions

Long-term use is not recommended (Chen and Chen 2004).

Drug and Supplement Interactions

None known.

Notice

Diuretic (Chen and Chen 2004; Perez Gutierrez et al. 1985); *see* Appendix 2.

Adverse Events and Side Effects

None known.

Pharmacological Considerations

None known.

Pregnancy and Lactation

Reference texts on traditional Chinese medicine indicate that scouring rush should be used with caution in pregnant women (Bensky et al. 2004; Chen and Chen 2004). Based on this information, use during pregnancy is not recommended except under the supervision of a qualified healthcare practitioner.

No information on the safety of scouring rush in lactation was identified. While this review did not identify any concerns for use while nursing, safety has not been conclusively established.

Review Details

I. Drug and Supplement Interactions

Clinical Trials of Drug or Supplement Interactions

No clinical trials of drug or supplement interactions were identified.

Case Reports of Suspected Drug or Supplement Interactions

No case reports of suspected drug or supplement interactions were identified.

Animal Trials of Drug or Supplement Interactions

No animal trials of drug or supplement interactions were identified.

II. Adverse Events

Case Reports of Adverse Events

No case reports of adverse events were identified.

III. Pharmacology and Pharmacokinetics

Human Pharmacological Studies

No relevant human pharmacological studies were identified.

Animal Pharmacological Studies

Diuretic activity of a chloroform extract of horsetail was observed in mice. The effect of 50 mg/kg of horsetail was greater than that of the drugs hydrochlorothiazide, spironolactone, or furosemide (all 25 mg/kg) in relation to the urine output and excretion of sodium, potassium, and chloride (Perez Gutierrez et al. 1985).

In Vitro Pharmacological Studies

No relevant in vitro pharmacological studies were identified.

IV. Pregnancy and Lactation

Reference texts on traditional Chinese medicine indicate that scouring rush should be used with caution in pregnant women (Bensky et al. 2004; Chen and Chen 2004).

No information on the safety of scouring rush during lactation was identified.

V. Toxicity Studies

No toxicity studies were identified.

Literature Cited

Bensky, D., S. Clavey, and E. Stöger. 2004. *Chinese herbal medicine: Materia medica.* 3rd ed. Seattle: Eastland Press.

Chen, J.K., and T.T. Chen. 2004. *Chinese medical herbology and pharmacology.* City of Industry, CA: Art of Medicine Press.

Perez Gutierrez, R.M., G. Yescas Laguna, and A. Walkowski. 1985. Diuretic activity of Mexican *Equisetum. J. Ethnopharmacol.* 14(2-3):269-272.

Equisetum spp. — Equisetaceae

Equisetum arvense L.
SCN: horsetail
OCN: field horsetail; shave grass; shavetail grass
Equisetum telmateia Ehrh.
SCN: giant horsetail
OCN: shave grass
Part: above-ground parts

Quick Reference Summary

Safety Class: 2d
Interaction Class: A

Contraindications

Not for use in persons with kidney disease (Bradley 1992).

Other Precautions

None known.

Drug and Supplement Interactions

None known.

Notice

Diuretic (Felter and Lloyd 1898; Wichtl 2004); *see* Appendix 2.

Editors' Notes

The herb in powdered form is not recommended for children or for prolonged use due to the inorganic silica content (McGuffin et al. 1997), though decoctions contain mainly organic silica in colloidal form, so they are not problematic in this regard (Weiss and Meuss 1998). A review of horsetail states that toxicity has been reported after "eating large amounts of horsetail, which has occurred in children who have used the stems as blowguns or whistles," but does not cite the source of this information (Hamon and Awang 1992). Chewing fresh or dried horsetail may damage tooth enamel.

Adulteration of horsetail with *E. palustre*, which contains the potentially toxic alkaloid palustrin, is widely reported (Langhammer et al. 1972; Wichtl 2004), and hybrids of *E. arvense* and other *Equisetum* species have been reported, further challenging correct identification. The toxicity of palustrin-containing species in humans is not known, as poisonings have only been reported in grazing animals that have consumed large amounts of *Equisetum* species (Wichtl 2004).

Adverse Events and Side Effects

Low levels of sodium and potassium were reported in a woman with a history of hypertension who had been taking infusions of an unspecified amount of giant horsetail daily for approximately 6 months (Miro et al. 1996).

A reaction resembling seborrheic dermatitis, similar to that sometimes observed in association with nicotine, was reported in a man who had been in contact with fresh horsetail plants and then exposed to secondary tobacco smoke (Sudan 1985).

Allergic reaction to horsetail has been reported and confirmed by patch testing (Agustin-Ubide et al. 2004).

Pharmacological Considerations

An animal study demonstrated that horsetail may modify glucose regulation (Soleimani et al. 2007). People with diabetes are advised to monitor their blood sugar closely and discuss the use of this herb with a qualified healthcare practitioner prior to use.

Animal studies and traditional use have indicated that horsetail has diuretic activity (Perez Gutierrez et al. 1985; Wichtl 2004).

Pregnancy and Lactation

No information on the safety of horsetail or giant horsetail in pregnancy or lactation was identified in the scientific or traditional literature. Although this review did not identify any concerns for use while pregnant or nursing, safety has not been conclusively established.

Review Details

I. Drug and Supplement Interactions

Clinical Trials of Drug or Supplement Interactions

No clinical trials of drug or supplement interactions were identified.

Case Reports of Suspected Drug or Supplement Interactions

No case reports of suspected drug or supplement interactions were identified.

Animal Trials of Drug or Supplement Interactions

No animal trials of drug or supplement interactions were identified.

II. Adverse Events

Case Reports of Adverse Events

Hyponatremia and hypokalemia were reported in an 84-year-old woman with a history of hypertension who had been taking infusions of giant horsetail (dose unspecified) for approximately 6 months (Miro et al. 1996).

A reaction resembling seborrheic dermatitis was reported in a man who had been in contact with fresh horsetail plants and then exposed to secondary tobacco smoke. A re-exposure to the plant elicited a more dramatic and rapid reaction. The reporting author indicated that nicotine present in horsetail was believed to be responsible for the reaction (Sudan 1985).

Allergic reaction to horsetail has been reported and confirmed by patch testing (Agustin-Ubide et al. 2004).

Multiple cases of cattle and horse poisoning by *Equisetum* species have been reported in the literature, with some reports indicating *E. arvense* as the causative species (Groh 1930; Hansen 1928; Hudson 1924; Jones 1901; Pammel 1921; Rapp 1954; Rich 1902). A review of these reports indicated that uncertainty in the identification of the species makes the reports difficult to interpret (Mills and Bone 2005). A 1904 study on the toxicity of various species of *Equisetum* in livestock indicated that *E. arvense* was nontoxic but that the botanically similar *E. palustre* demonstrated toxicity (Long 1924).

Symptoms of poisoning were observed in horses, but not in cattle, that ingested hay containing horsetail. The poisoning was treated successfully with injections of thiamine hydrochloride. Related in vitro testing indicated that horsetail destroyed thiamine by enzymatic activity (Henderson et al. 1952).

III. Pharmacology and Pharmacokinetics

Human Pharmacological Studies

No relevant human pharmacological studies were identified.

Animal Pharmacological Studies

A reduction in blood glucose levels was observed in diabetic rats orally administered 50 mg/kg of a methanol extract of equisetum daily for 5 weeks (Soleimani et al. 2007).

In Vitro Pharmacological Studies

Inhibition of thrombin- but not ADP-induced aggregation was observed in rat platelets treated with aqueous extracts of horsetail, with an IC_{50} value of 6.75 mg/ml (Mekhfi et al. 2004).

IV. Pregnancy and Lactation

No information on the use of horsetail or giant horsetail during pregnancy or lactation was identified.

V. Toxicity Studies

Short-Term Toxicity

In rats fed a 20% casein diet with or without cholesterol, the addition of 4% horsetail powder resulted in dermatitis of the neck, head, and back in 20 to 65% of the rats. The dermatitis resolved after the diet was changed to commercial pellets (Maeda et al. 1997).

No adverse effects on liver or kidney function were observed in voles fed diets containing increasing amounts of horsetail from 0.1 to 0.8 mg/kg daily (Jean and Bergeron 1986).

Subchronic Toxicity

No adverse effects were observed in rats intraperitoneally administered 50 mg/kg of a hydroalcoholic extract of horsetail daily for 8 weeks (Guilherme dos Santos et al. 2005).

Literature Cited

Agustin-Ubide, M.P., C. Martinez-Cocera, A. Alonso-Llamazares, et al. 2004. Diagnostic approach to anaphylaxis by carrot, related vegetables and horsetail (*Equisetum arvense*) in a homemaker. *Allergy* 59(7):786-787.

Bradley, P.R. 1992. *British herbal compendium: A handbook of scientific information on widely used plant drugs*. Bournemouth, UK: British Herbal Medicine Association.

Felter, H.W., and J.U. Lloyd. 1898. *King's American dispensatory*. 18th ed., 3rd rev. 2 vols. Cincinnati: Ohio Valley Co.

Groh, H. 1930. Horsetail a horse-poisoning weed. *Can. Dept. Agric. Dom. Exp. Farms Circ.* 74:1-3.

Guilherme dos Santos, J., Jr., F. Hoffmann Martins do Monte, M. Marcela Blanco, et al. 2005. Cognitive enhancement in aged rats after chronic administration of *Equisetum arvense* L. with demonstrated antioxidant properties in vitro. *Pharmacol. Biochem. Behav.* 81(3):593-600.

Hamon, N.W., and D.V. Awang. 1992. Horsetail. *Can. Pharm. J.* 125:399-401.

Hansen, A.A. 1928. Stock poisoning plants. *North Am. Vet.* 9(5):24-27.

Henderson, J.A., E.V. Evans, and R.A. McIntosh. 1952. The antithiamine action of *Equisetum*. *J. Am. Vet. Med. Assoc.* 120:375-378.

Hudson, R. 1924. Poisoning by horsetail (*Equisetum arvense*). *Vet. J.* 80:40.

Jean, Y., and J.M. Bergeron. 1986. Can voles (*Microtus pennsylvanicus*) be poisoned by secondary metabolites of commonly eaten foods? *Can. J. Zool.* 64(1):158-162.

Jones, L.R. 1901. Are our native horsetails or ferns poisonous? *Proc. Soc. Promotion Agric. Sci.* 22:70-74.

Langhammer, L., K. Blaszkiewitz, and I. Kotzorek. 1972. Detection of toxic adulteration of *Equisetum* herbs. *Dtsch. Apoth. Ztg.* 112(44):1749-1751.

Long, H.C. 1924. *Plants poisonous to live stock*. 2nd ed. Cambridge: Cambridge University Press.

Maeda, H., K. Miyamoto, and T. Sano. 1997. Occurrence of dermatitis in rats fed a cholesterol diet containing field horsetail (*Equisetum arvense* L.). *J. Nutr. Sci. Vitaminol.* 43(5):553-563.

McGuffin, M., C. Hobbs, R. Upton, and A. Goldberg. 1997. *Botanical safety handbook*. Boca Raton, FL: CRC Press.

Mekhfi, H., M. El Haouari, A. Legssyer, et al. 2004. Platelet anti-aggregant property of some Moroccan medicinal plants. *J. Ethnopharmacol.* 94(2-3):317-322.

Mills, S., and K. Bone. 2005. *The essential guide to herbal safety*. St. Louis: Elsevier.

Miro, O., E. Pedrol, S. Nogue, and F. Cardellach. 1996. [Severe hyponatremia and hypopotassemia induced by the consumption of *Equisetum telmateia*]. *Med. Clin.* 106(16):639.

Pammel, L.H. 1921. Equisetosis or horsetail poisoning. *Vet. Med.* 16(4):43.
Perez Gutierrez, R.M., G.Y. Laguna, and A. Walkowski. 1985. Diuretic activity of Mexican *Equisetum*. *J. Ethnopharmacol.* 14(2-3):269-272.
Rapp, W.F. 1954. The toxicity of *Equisetum*. *Am. Fern J.* 44:148-154.
Rich, F.A. 1902. *Equisetum* poisoning. *Am. Vet. Rev.* 26:944-945.
Soleimani, S., F.F. Azarbaizani, and V. Nejati. 2007. The effect of *Equisetum arvense* L. (Equisetaceae) in histological changes of pancreatic beta-cells in streptozotocin-induced diabetic in rats. *Pakistan J. Biol. Sci.* 10(23):4236-4240.
Sudan, B.J. 1985. Seborrhoeic dermatitis induced by nicotine of horsetails (*Equisetum arvense* L.). *Contact Dermat.* 13(3):201-202.
Weiss, R.F., and A.R. Meuss. 1998. *Weiss's herbal medicine*. Classic ed. New York: Thieme.
Wichtl, M. 2004. *Herbal drugs and phytopharmaceuticals: A handbook for practice on a scientific basis*. 3rd ed. Boca Raton, FL: CRC Press.

Eriobotrya japonica (Thunb.) Lindley — Rosaceae

SCN: loquat
PN: *pi pa ye* (leaf)
Part: leaf

Quick Reference Summary

Safety Class: 1
Interaction Class: A

Contraindications

None known.

Other Precautions

None known.

Drug and Supplement Interactions

None known.

Standard Dose

The standard dose is 4.5 to 15 g of dried leaf or 15 to 30 g of fresh leaf daily as a tea (Bensky et al. 2004; Chen and Chen 2004).

Notice

Cyanogenic glycosides (0.06% amygdalin) (Bensky et al. 2004; List and Hörhammer 1973); *see* Appendix 1.

Editors' Notes

The hairs naturally present on loquat leaf should be removed to avoid irritation of the mucous membranes that may aggravate a cough, causing edema and spasms of the larynx (Bensky et al. 2004; Tu 1988). While material imported from China has, as a rule, been properly processed with regard to this concern, added caution should be exercised when utilizing domestic sources.

Adverse Events and Side Effects

A case of severe muscular pain was reported after use of 2 liters daily of loquat tea for two weeks (Saliba et al. 2004).

Pharmacological Considerations

Animal studies have indicated that loquat may modify glucose regulation (Chen et al. 2008; De Tommasi et al. 1991; Li et al. 2007; Noreen et al. 1988; Roman-Ramos et al. 1991). People with diabetes are advised to monitor their blood sugar closely and discuss the use of this herb with a qualified healthcare practitioner prior to use.

Pregnancy and Lactation

No information on the safety of loquat in pregnancy or lactation was identified in the scientific or traditional literature. Although this review did not identify any concerns for use while pregnant or nursing, safety has not been conclusively established.

Review Details

I. Drug and Supplement Interactions

Clinical Trials of Drug or Supplement Interactions

No clinical trials of drug or supplement interactions were identified.

Case Reports of Suspected Drug or Supplement Interactions

No case reports of suspected drug or supplement interactions were identified.

Animal Trials of Drug or Supplement Interactions

No animal trials of drug or supplement interactions were identified.

II. Adverse Events

Case Reports of Adverse Events

Severe myalgia, particularly of the proximal muscles of the arms and legs, was observed in a 39-year-old man who had consumed approximately 2 liters of loquat tea daily for 2 weeks (amount of loquat used to make tea was not specified). The myalgia resolved after a reduction of the loquat dose (Saliba et al. 2004).

III. Pharmacology and Pharmacokinetics

Human Pharmacological Studies

No relevant human pharmacological studies were identified.

Animal Pharmacological Studies

In normal and diabetic rabbits orally administered up to 200 mg/kg of an alcohol extract of loquat, a short-lived reduction in blood glucose was observed in healthy animals, but no significant effect was observed in diabetic animals (Noreen et al. 1988).

Compounds isolated from loquat exhibited hypoglycemic activity in diabetic mice and healthy rats (Chen et al. 2008; De Tommasi et al. 1991), with one compound showing activity after oral doses of 25 or 75 mg/kg (Chen et al. 2008).

A reduction in blood glucose levels was observed in healthy rabbits orally administered extracts of loquat (Roman-Ramos et al. 1991).

Hypoglycemic activity of an ethanol extract of loquat was observed in healthy and diabetic mice orally administered 15 to 60 g/kg. A 30 g/kg dose was reported to be more effective than 100 mg/kg of the drug phenformin. An extract of sesquiterpene compounds from loquat also lowered blood glucose at a dose of 30 g/kg (Li et al. 2007).

In Vitro Pharmacological Studies

An ethanol extract of loquat exhibited estrogenic activity in a recombinant yeast system with a human estrogen receptor expression plasmid and a reporter plasmid (Kang et al. 2006; Kim et al. 2008).

IV. Pregnancy and Lactation

No information on the safety of loquat during pregnancy or lactation was identified.

V. Toxicity Studies

Acute Toxicity

The LD_{50} of an orally administered ethanol extract of loquat in mice is 400.1 g/kg (Li et al. 2007).

No adverse effects were observed in rabbits orally administered up to 200 mg/kg of an alcohol extract of loquat (Noreen et al. 1988).

Literature Cited

Bensky, D., S. Clavey, and E. Stöger. 2004. *Chinese herbal medicine: Materia medica*. 3rd ed. Seattle: Eastland Press.

Chen, J., W.L. Li, J.L. Wu, B.R. Ren, and H.Q. Zhang. 2008. Hypoglycemic effects of a sesquiterpene glycoside isolated from leaves of loquat (*Eriobotrya japonica* (Thunb.) Lindl.). *Phytomedicine* 15(1-2):98-102.

Chen, J.K., and T.T. Chen. 2004. *Chinese medical herbology and pharmacology*. City of Industry, CA: Art of Medicine Press.

De Tommasi, N., F. De Simone, G. Cirino, C. Cicala, and C. Pizza. 1991. Hypoglycemic effects of sesquiterpene glycosides and polyhydroxylated triterpenoids of *Eriobotrya japonica*. *Planta Med.* 57(5):414-416.

Kang, S.C., C.M. Lee, H. Choi, et al. 2006. Evaluation of oriental medicinal herbs for estrogenic and antiproliferative activities. *Phytother. Res.* 20(11):1017-1019.

Kim, I.G., S.C. Kang, K.C. Kim, E.S. Choung, and O.P. Zee. 2008. Screening of estrogenic and antiestrogenic activities from medicinal plants. *Env. Toxicol. Pharmacol.* 25(1):75-82.

Li, W.L., J.L. Wu, B.R. Ren, J. Chen, and C.G. Lu. 2007. Pharmacological studies on anti-hyperglycemic effect of folium eriobotryae. *Am. J. Chin. Med.* 35(4):705-711.

List, P.H., and H. Hörhammer. 1973. *Hagers handbuch der pharmazeutischen praxis*. Berlin: Springer.

Noreen, W., A. Wadood, H.K. Hidayat, and S.A. Wahid. 1988. Effect of *Eriobotrya japonica* on blood glucose levels of normal and alloxan-diabetic rabbits. *Planta Med.* 54(3):196-199.

Roman-Ramos, R., J.L. Flores-Saenz, G. Partida-Hernandez, A. Lara-Lemus, and F. Alarcon-Aguilar. 1991. Experimental study of the hypoglycemic effect of some antidiabetic plants. *Arch. Invest. Med.* 22(1):87-93.

Saliba, W.R., L.H. Goldstein, G.S. Habib, and M.S. Elias. 2004. Toxic myopathy induced by the ingestion of loquat leaf extract. *Ann. Rheum. Dis.* 63(10):1355-1356.

Tu, G. 1988. *Pharmacopoeia of the People's Republic of China*. English ed. Beijing, China: People's Medical Pub. House.

Eriodictyon spp. — Hydrophyllaceae

Eriodictyon californicum (Hook. & Arn.) Torr.
SCN: yerba santa
Syn: *Eriodictyon glutinosum* Benth.

Eriodictyon tomentosum Benth.
SCN: woolly yerba santa
Part: herb

Quick Reference Summary

Safety Class: 1
Interaction Class: A

Contraindications

None known.

Other Precautions

None known.

Drug and Supplement Interactions

None known.

Adverse Events and Side Effects

None known.

Pharmacological Considerations

None known.

Pregnancy and Lactation

No information on the safety of yerba santa or woolly yerba santa in pregnancy or lactation was identified in the scientific or traditional literature. Although this review did not identify any concerns for use while pregnant or nursing, safety has not been conclusively established.

Review Details

I. Drug and Supplement Interactions

Clinical Trials of Drug or Supplement Interactions

No clinical trials of drug or supplement interactions were identified.

Case Reports of Suspected Drug or Supplement Interactions

No case reports of suspected drug or supplement interactions were identified.

Animal Trials of Drug or Supplement Interactions

No animal trials of drug or supplement interactions were identified.

II. Adverse Events

Case Reports of Adverse Events

No case reports of adverse events were identified.

III. Pharmacology and Pharmacokinetics

Human Pharmacological Studies

No relevant human pharmacological studies were identified.

Animal Pharmacological Studies

No relevant animal pharmacological studies were identified.

In Vitro Pharmacological Studies

No relevant in vitro pharmacological studies were identified.

IV. Pregnancy and Lactation

No information on the safety of yerba santa or woolly yerba santa during pregnancy or lactation was identified.

V. Toxicity Studies

Subchronic Toxicity

In rats fed diets containing yerba santa fluid extract at an approximate dose of 537 mg/kg daily for 90 days, no gross pathological changes and no significant changes in hematological parameter, weight gain, or liver or kidney weights were observed (Oser et al. 1965).

Genotoxicity

No mutagenic activity of the compounds eriodictyol, diosmetin, and luteolin was observed in the *Salmonella*/mammalian microsome test (Brown and Dietrich 1979; Brown et al. 1977).

Literature Cited

Brown, J.P., and P.S. Dietrich. 1979. Mutagenicity of plant flavonols in the *Salmonella*/mammalian microsome test: Activation of flavonol glycosides by mixed glycosidases from rat cecal bacteria and other sources. *Mutat. Res.* 66(3):223-240.

Brown, J.P., P.S. Dietrich, and R.J. Brown. 1977. Frameshift mutagenicity of certain naturally occurring phenolic compounds in the '*Salmonella*/microsome' test: Activation of anthraquinone and flavonol glycosides by gut bacterial enzymes. *Biochem. Soc. Trans.* 5(5):1489-1492.

Oser, B.L., S. Carson, and M. Oser. 1965. Toxicological tests on flavoring matters. *Food Cosmet. Toxicol.* 3(4):563-569.

Eryngium spp. — Apiaceae

Eryngium maritimum L.
SCN: eryngo
OCN: sea holly; seaside eryngo
Eryngium planum L.
SCN: plains eryngo
Eryngium yuccifolium Michx.
SCN: button eryngo
Part: root

Quick Reference Summary

Safety Class: 1
Interaction Class: A

Contraindications

None known.

Other Precautions

None known.

Drug and Supplement Interactions

None known.

Adverse Events and Side Effects

"Large" doses of button eryngo may cause nausea or vomiting (Felter and Lloyd 1898; Wood and LaWall 1918).

Pharmacological Considerations

None known.

Pregnancy and Lactation

No information on the safety of eryngo, plains eryngo, or button eryngo in pregnancy or lactation was identified in the scientific or traditional literature. Although this review did not identify any concerns for use while pregnant or nursing, safety has not been conclusively established.

Review Details

I. Drug and Supplement Interactions

Clinical Trials of Drug or Supplement Interactions

No clinical trials of drug or supplement interactions were identified.

Case Reports of Suspected Drug or Supplement Interactions

No case reports of suspected drug or supplement interactions were identified.

Animal Trials of Drug or Supplement Interactions

No animal trials of drug or supplement interactions were identified.

II. Adverse Events

Case Reports of Adverse Events

No case reports of adverse events were identified.

III. Pharmacology and Pharmacokinetics

Human Pharmacological Studies

No relevant human pharmacological studies were identified.

Animal Pharmacological Studies

No relevant animal pharmacological studies were identified.

In Vitro Pharmacological Studies

No relevant in vitro pharmacological studies were identified.

IV. Pregnancy and Lactation

No information on the safety of eryngo, plains eryngo, or button eryngo during pregnancy or lactation was identified.

V. Toxicity Studies

No toxicity studies were identified.

Literature Cited

Felter, H.W., and J.U. Lloyd. 1898. *King's American dispensatory*. 18th ed., 3rd rev. 2 vols. Cincinnati: Ohio Valley Co.

Wood, H., and C. LaWall. 1918. *The dispensatory of the United States of America*. 21st ed. Philadelphia: Lippincott.

Erythrina variegata L. — Fabaceae

SCN: Indian coral tree
Syn: *Erythrina indica* Lam.
AN: *paribhadra*
OCN: tiger's claw
PN: *hai tong pi* (bark)
Part: bark

Quick Reference Summary

Safety Class: 1
Interaction Class: A

Contraindications

None known.

Other Precautions

None known.

Drug and Supplement Interactions

None known.

Adverse Events and Side Effects

None known.

Pharmacological Considerations

None known.

Pregnancy and Lactation

No information on the safety of Indian coral tree in pregnancy or lactation was identified in the scientific or traditional literature. Although this review did not identify any concerns for use while pregnant or nursing, safety has not been conclusively established.

Review Details

I. Drug and Supplement Interactions

Clinical Trials of Drug or Supplement Interactions

No clinical trials of drug or supplement interactions were identified.

Case Reports of Suspected Drug or Supplement Interactions

No case reports of suspected drug or supplement interactions were identified.

Animal Trials of Drug or Supplement Interactions

No animal trials of drug or supplement interactions were identified.

II. Adverse Events

Case Reports of Adverse Events

No case reports of adverse events were identified.

III. Pharmacology and Pharmacokinetics

Human Pharmacological Studies

No relevant human pharmacological studies were identified.

Animal Pharmacological Studies

No relevant animal pharmacological studies were identified.

In Vitro Pharmacological Studies

No relevant in vitro pharmacological studies were identified.

IV. Pregnancy and Lactation

No information on the safety of Indian coral tree during pregnancy or lactation was identified.

V. Toxicity Studies

Acute Toxicity

The LD_{50} of the total alkaloids of Indian coral tree intraperitoneally administered to rats is 127.8 mg/kg (Ghosal et al. 1972).

Literature Cited

Ghosal, S., S.K. Dutta, and S.K. Bhattacharya. 1972. *Erythrina*—Chemical and pharmacological evaluation II: Alkaloids of *Erythrina variegata* L. *J. Pharm. Sci.* 61(8):1274-1277.

Erythroxylum catuaba A. J. da Silva ex Hamet

Erythroxylaceae

SCN: catuaba
OCN: golden trumpet

Part: bark

Quick Reference Summary

Safety Class: 1
Interaction Class: A

E

Contraindications

None known.

Other Precautions

None known.

Drug and Supplement Interactions

None known.

Editors' Notes

A number of different species of plants are traded under the common name "catuaba." These include species of *Anemopaegma, Erythroxylum, Ilex, Micropholis, Phyllanthus, Secondatia, Tetragastris,* and *Trichilia* (Kletter et al. 2004). According to *Herbs of Commerce,* "catuaba" is the standardized common name for *Erythroxylum catuaba* (McGuffin et al. 2000).

Adverse Events and Side Effects

None known.

Pharmacological Considerations

None known.

Pregnancy and Lactation

No information on the safety of *Erythroxylum catuaba* in pregnancy or lactation was identified in the scientific or traditional literature. Although this review did not identify any concerns for use while pregnant or nursing, safety has not been conclusively established.

Review Details

I. Drug and Supplement Interactions

Clinical Trials of Drug or Supplement Interactions

No clinical trials of drug or supplement interactions were identified.

Case Reports of Suspected Drug or Supplement Interactions

No case reports of suspected drug or supplement interactions were identified.

Animal Trials of Drug or Supplement Interactions

No animal trials of drug or supplement interactions were identified.

II. Adverse Events

Case Reports of Adverse Events

No case reports of adverse events were identified.

III. Pharmacology and Pharmacokinetics

Human Pharmacological Studies

No relevant human pharmacological studies were identified.

Animal Pharmacological Studies

No relevant animal pharmacological studies were identified.

In Vitro Pharmacological Studies

No relevant in vitro pharmacological studies were identified.

IV. Pregnancy and Lactation

No information on the safety of catuaba during pregnancy or lactation was identified.

V. Toxicity Studies

No toxicity studies were identified.

Literature Cited

Kletter, C., S. Glasl, A. Presser, et al. 2004. Morphological, chemical and functional analysis of catuaba preparations. *Planta Med.* 70(10):993-1000.

McGuffin, M., J. Kartesz, A. Leung, and A.O. Tucker. 2000. *Herbs of commerce.* 2nd ed. Silver Spring, MD: American Herbal Products Association.

Eschscholzia californica Cham. — Papaveraceae

SCN: California poppy

Part: whole plant in flower

Quick Reference Summary

Safety Class: 2b
Interaction Class: A

Contraindications

Not for use in pregnancy except under the supervision of a qualified healthcare practitioner (Moore 2003).

Other Precautions

None known.

Drug and Supplement Interactions

None known.

Adverse Events and Side Effects

One text indicates that "large" doses may cause nausea. The standard dose listed in this text is 15 to 25 drops of tincture three times daily, or an infusion of 2 to 4 ounces of dried herb (Moore 2003).

Pharmacological Considerations

None known.

Pregnancy and Lactation

Information on the safety of California poppy in pregnancy and lactation is limited. One contemporary herbal text indicates that California poppy should not be used during pregnancy (Moore 2003). Based on this information, use during pregnancy is not recommended except under the supervision of a qualified healthcare practitioner.

E

Review Details

I. Drug and Supplement Interactions

Clinical Trials of Drug or Supplement Interactions

No clinical trials of drug or supplement interactions were identified.

Case Reports of Suspected Drug or Supplement Interactions

No case reports of suspected drug or supplement interactions were identified.

Animal Trials of Drug or Supplement Interactions

No animal trials of drug or supplement interactions were identified.

II. Adverse Events

Case Reports of Adverse Events

No case reports of adverse events were identified.

III. Pharmacology and Pharmacokinetics

Human Pharmacological Studies

No relevant human pharmacological studies were identified.

Animal Pharmacological Studies

No relevant animal pharmacological studies were identified.

In Vitro Pharmacological Studies

An aqueous extract of California poppy inhibited dopamine β-hydroxylase and monoamine oxidase (MAO-B) (Kleber et al. 1995).

An ethanol extract of California poppy was shown to bind to the serotonin 5-HT_{1A} and 5-HT_7 receptors at concentrations of 100 μg/ml (Gafner et al. 2006).

IV. Pregnancy and Lactation

One contemporary herbal text indicates that California poppy should not be used during pregnancy (Moore 2003). An ethnobotanical survey indicated that Native Americans of central California avoided use of California poppy during pregnancy and lactation (Bocek 1984). No other information on the safety of California poppy in pregnancy or lactation was identified.

V. Toxicity Studies

Acute Toxicity

The LD_{50} of orally and intraperitoneally administered aqueous-ethanolic extract of California poppy in mice could not be determined at doses up to 5 g/kg (Rolland et al. 2001).

Literature Cited

Bocek, B.R. 1984. Ethnobotany of Costanoan Indians, California, based on collections by John P. Harrington. *Econ. Bot.* 38(2):240-255.

Gafner, S., B.M. Dietz, K.L. McPhail, et al. 2006. Alkaloids from *Eschscholzia californica* and their capacity to inhibit binding of [^{3}H]8-hydroxy-2-(di-*N*-propylamino)tetralin to 5-HT_{1A} receptors in vitro. *J. Nat. Prod.* 69(3):432.

Kleber, E., W. Schneider, H. Schafer, and E. Elstner. 1995. Modulation of key reactions of the catecholamine metabolism by extracts from *Eschscholtzia californica* and *Corydalis cava*. *Arzneimittel-Forschung* 45(2):127-131.

Moore, M. 2003. *Medicinal plants of the Mountain West.* Revised and expanded edition. Santa Fe: Museum of New Mexico Press.

Rolland, A., J. Fleurentin, M.C. Lanhers, R. Misslin, and F. Mortier. 2001. Neurophysiological effects of an extract of *Eschscholzia californica* Cham. (Papaveraceae). *Phytother. Res.* 15(5):377-381.

Eucalyptus globulus Labill. — Myrtaceae

SCN: eucalyptus
OCN: blue gum; southern blue gum; Tasmanian blue gum

Part: leaf

Quick Reference Summary

Safety Class: 1
Interaction Class: A

Contraindications

None known.

Other Precautions

None known.

Drug and Supplement Interactions

None known.

Notice

Tannins (up to 11.0%) (De Smet 1992; Wichtl 2004); *see* Appendix 1.

Editors' Note

Eucalyptus leaf contains eucalyptus essential oil, which is the subject of a separate entry in this text (*see* next entry).

Adverse Events and Side Effects

Vocal cord dysfunction after exposure to eucalyptus leaf and allergic reaction to eucalyptus leaf have been reported (Galdi et al. 2003; Huggins et al. 2004).

Pharmacological Considerations

None known.

Pregnancy and Lactation

No information on the safety of eucalyptus leaf in pregnancy or lactation was identified in the scientific or traditional literature. Although this review did not identify any concerns for use while pregnant or nursing, safety has not been conclusively established.

See eucalyptus leaf essential oil (next entry) for information on the safety of the essential oil in pregnancy and lactation.

Review Details

I. Drug and Supplement Interactions

Clinical Trials of Drug or Supplement Interactions

No clinical trials of drug or supplement interactions were identified.

Case Reports of Suspected Drug or Supplement Interactions

No case reports of suspected drug or supplement interactions were identified.

Animal Trials of Drug or Supplement Interactions

No animal trials of drug or supplement interactions were identified.

II. Adverse Events

Case Reports of Adverse Events

Exacerbation of asthma was observed in a woman with asthma and rhinoconjunctivitis after ingestion of an infusion of eucalyptus leaf. Specific IgE were positive for eucalyptus pollens and negative for common aeroallergens (Galdi et al. 2003).

Vocal cord dysfunction after exposure to eucalyptus leaf and ammonia but not ammonia alone was observed in a woman undergoing inhalation challenges to identify agents that had been causing her respiratory distress. Negative skin prick test results, total IgE level, and negative IgE eucalyptus-specific antibodies support a nonimmunological mechanism (Huggins et al. 2004).

III. Pharmacology and Pharmacokinetics

Human Pharmacological Studies

No relevant human pharmacological studies of eucalyptus leaf were identified.

Animal Pharmacological Studies

No relevant animal pharmacological studies of eucalyptus leaf were identified.

In Vitro Pharmacological Studies

No relevant in vitro pharmacological studies of eucalyptus leaf were identified.

IV. Pregnancy and Lactation

No information on the safety of eucalyptus leaf during pregnancy or lactation was identified. *See Eucalyptus globulus* leaf essential oil (next entry).

V. Toxicity Studies

See Eucalyptus globulus leaf essential oil.

E

Literature Cited

De Smet, P.A.G.M. 1992. *Adverse effects of herbal drugs, Volume 1*. New York: Springer.

Galdi, E., L. Perfetti, G. Calcagno, M.C. Marcotulli, and G. Moscato. 2003. Exacerbation of asthma related to *Eucalyptus* pollens and to herb infusion containing *Eucalyptus*. *Monaldi Arch. Chest Dis.* 59(3):220-221.

Huggins, J.T., A. Kaplan, B. Martin-Harris, and S.A. Sahn. 2004. *Eucalyptus* as a specific irritant causing vocal cord dysfunction. *Ann. Allergy Asthma Immunol.* 93(3):299-303.

Wichtl, M. 2004. *Herbal drugs and phytopharmaceuticals: A handbook for practice on a scientific basis*. 3rd ed. Boca Raton, FL: CRC Press.

Eucalyptus globulus Labill. — Myrtaceae

SCN: eucalyptus
OCN: blue gum; southern blue gum; Tasmanian blue gum

Part: leaf essential oil

Quick Reference Summary

Safety Class: 2b, 2d
Interaction Class: A

Contraindications

Not for internal use in pregnancy except under the supervision of a qualified healthcare practitioner.

Do not exceed the recommended dose (Ruse 1998; Tibballs 1995).

Other Precautions

Eucalyptus essential oil is a highly concentrated product and should not be used on areas of the face, and especially the nose, in infants and young children.

Drug and Supplement Interactions

None known.

Standard Dose

The German standard dose for the compound eucalyptol is 200 mg three times daily (Juergens et al. 2003). This is equivalent to approximately 0.35 ml of eucalyptus essential oil.

Editors' Notes

Eucalyptus essential oil contains 70 to 80% of the compound eucalyptol (also called 1,8-cineole or 1,8-epoxy-*p*-menthane) (Chennoufi et al. 1980; De Vincenzi et al. 2002; HCPDG 2002; Leung and Foster 1996; Tyler et al. 1988). The oil must be free of other eucalyptus oils containing large amounts of phellandrene (Tyler et al. 1988).

Adverse Events and Side Effects

Cases of poisoning after ingestion of varying amounts of eucalyptus essential oil have been reported (Allan 1910; Anpalahan and Le Couteur 1998; Atkinson 1909; Benjamin 1906; Chun 1951; Craig 1953; Foggie 1911; Gurr and Scroggie 1965; Hindle 1994; Kirkness 1910; Myott 1906; Orr and Edin 1906; Owen 1885; Patel and Wiggins 1980; Sewell 1925; Spoerke et al. 1989; Taylor 1905; Webb and Pitt 1993; Wood 1900). Poisoning may affect the central nervous system (loss of consciousness, hypoventilation, depression of reflexes, and convulsions), the gastrointestinal system (abdominal pain, vomiting, and diarrhea), the respiratory system (respiratory depression, dyspnea, pneumonitis, and bronchospasm), and the cardiovascular system (low blood pressure, arrhythmia) (Craig 1953; Patel and Wiggins 1980; Tibballs 1995). Similar symptoms have been observed after topical application of eucalyptus essential oil (Darben et al. 1998).

Pharmacological Considerations

Older studies in rats indicate that eucalyptus essential oil administered subcutaneously or as an aerosol may increase the metabolism of pentobarbital (Jori and Briatico 1973).

Pregnancy and Lactation

One study in rats indicated no embryotoxicity or fetotoxicity in pregnant mice administered eucalyptus essential oil (Pages et al. 1990). In rats administered the compound eucalyptol during or after pregnancy, eucalyptol was shown to cross the placenta but was reported not to cross into breast milk (Jori and Briatico 1973).

Review Details

I. Drug and Supplement Interactions

Clinical Trials of Drug or Supplement Interactions

No clinical trials of drug or supplement interactions were identified.

Case Reports of Suspected Drug or Supplement Interactions

No case reports of suspected drug or supplement interactions were identified.

Animal Trials of Drug or Supplement Interactions

Subcutaneous or aerosol administration of the compound eucalyptol to rats was shown to increase the metabolism of pentobarbital (Jori and Briatico 1973).

Reductions in duration of efficacy of pentobarbital, zoxazolamine, and amphetamine were seen in rats exposed to an aerosol of eucalyptus essential oil for 2 to 10 minutes daily for 4 days (Jori and Briatico 1973).

II. Adverse Events

Adverse Events Reported in Clinical Trials

In a study of the compound eucalyptol at a dose of 200 mg three times daily in patients with bronchial asthma, side effects considered by the study physician to be possibly attributable to eucalyptol were heartburn and gastritis. These effects were experienced by 2 of 16 in the treatment group and none in the placebo group. No serious adverse events were reported, and no clinically relevant abnormalities in routine blood test parameters were observed (Juergens et al. 2003).

In Germany, eucalyptol is a licensed medicinal product sold in gut-soluble capsules and is reported to be well tolerated at a dose of 600 mg daily. Instructions for use indicate that this product should be taken with cold water about 20 minutes prior to eating to prevent epigastric pain (Juergens et al. 2003).

Case Reports of Adverse Events

Cases of poisoning after ingestion of varying amounts of eucalyptus essential oil have been reported (Allan 1910; Anpalahan and Le Couteur 1998; Atkinson 1909; Benjamin 1906; Chun 1951; Craig 1953; Foggie 1911; Gurr and Scroggie 1965; Hindle 1994; Kirkness 1910; Myott 1906; Orr and Edin 1906; Owen 1885; Patel and Wiggins 1980; Sewell 1925; Spoerke et al. 1989; Taylor 1905; Webb and Pitt 1993; Wood 1900). Poisoning affects the central nervous system (loss of consciousness, hypoventilation, depression of reflexes, and convulsions), the gastrointestinal system (abdominal pain, vomiting, and diarrhea), the respiratory system (respiratory depression, dyspnea, pneumonitis, and bronchospasm), and the cardiovascular system (low blood pressure, arrhythmia) (Craig 1953; Patel and Wiggins 1980; Tibballs 1995). Similar symptoms have been observed after topical application of eucalyptus essential oil (Darben et al. 1998).

Doses of eucalyptus oil causing poisoning vary widely, as a 15 ml dose administered to a 6-year-old resulted in slight drowsiness (Atkinson 1909), while approximately 4 ml given to a 10-year-old boy resulted in death (Foggie 1911), and 4 ml given to a 1-year-old girl resulted in severe poisoning with eventual recovery (Allan 1910). In adults, 10 ml led to poisoning with drowsiness and ataxia (Kirkness 1910), and 4 to 5 ml (Macpherson 1925) and 25 ml led to death (Myott 1906); however, severe poisonings were successfully treated in an adult who had ingested 120 to 220 ml (Gurr and Scroggie 1965), and in a 73-year-old woman who had ingested approximately 180 ml of eucalyptus essential oil (Anpalahan and Le Couteur 1998).

A review of cases of eucalyptus poisoning in children ages 5 months to 9 years categorized poisoning cases into five levels of severity, with information on doses of eucalyptus oil ingested by children (note that the doses are actual and not adjusted to ml/kg based on each child's weight): no effects (0.2 to 5.0 ml, mean 1.7 ml) with symptoms of sputtering or coughing after ingestion; minor poisoning (0.2 to 7.0 ml, mean 2.0 ml) with symptoms of ataxia, vomiting, abdominal pain, miosis; moderate poisoning (0.6 to 5.0 ml, mean 2.5 ml) with symptoms of depression of consciousness, Glasgow coma scale score of 8 to 14; major poisoning (7.5 ml) with symptoms of loss of consciousness, Glasgow coma scale score of 3 to 7 without hypoventiliation; life-threatening poisoning (no dose specified) with symptoms of unconsciousness with hypoventilation (Tibballs 1995). Older case reports indicate that children have recovered after ingestion of 14 and 24 ml of eucalyptus essential oil (Benjamin 1906; Sewell 1925). Tiballs (1995) provides information on the clinical management of the different levels of poisoning.

III. Pharmacology and Pharmacokinetics

Human Pharmacological Studies

No relevant human pharmacological studies were identified.

Animal Pharmacological Studies

Ingestion or inhalation of the compound eucalyptol induces microsome mixed function oxidases in rats (Jori and Briatico 1973). Such activity may speed the clearance of certain drugs such as pentobarbital, aminopyrine, and amphetamine (Katzung 2007) and increase the toxicity of pyrrolizidine compounds (White et al. 1983).

In Vitro Pharmacological Studies

Eucalyptus essential oil weakly inhibited the drug-metabolizing isoenzymes CYP2C8, CYP2C19, and CYP3A4 in vitro with IC_{50} values over 100 μg/ml (Unger and Frank 2004).

Eucalyptus essential oil enhanced the penetration of 5-fluorouracil in excised human skin by a factor of 34 (Williams and Barry 1989).

IV. Pregnancy and Lactation

No embryotoxicity or fetotoxicity was observed in the offspring of pregnant mice subcutaneously administered 135 mg/kg daily eucalyptus essential oil on gestational days 6 to 15 (Pages et al. 1990).

In rats subcutaneously administered the compound eucalyptol on gestational days 10 to 14, the last 4 days of pregnancy, or on postnatal days 2 through 6, treatment enhanced microsomal enzyme activity of the mothers and induced enzyme activity in fetal livers, but did not induce microsomal activity of suckling newborn rats, suggesting that eucalyptol is able to cross the placenta but does not cross into breast milk (Jori and Briatico 1973).

V. Toxicity Studies

Acute Toxicity

The LD_{50} of orally administered eucalyptol in rats has been reported as 1560 mg/kg (Brownlee 1940) and 2480 mg/kg (Jenner et al. 1964). Toxic effects at the 1560 mg/kg dose were rapid cyanosis, stupor, irregular breathing, extreme sensitivity to noise, convulsions, and death from respiratory failure (Brownlee 1940).

The estimated lethal dose in humans ranges from 0.05 to 0.5 ml/kg (Hindle 1994).

Short-Term Toxicity

In rats orally administered the compound eucalyptol via stomach tube at doses up to 1200 mg/kg 5 days per week or encapsulated and added to feed at doses up to 3342 mg/kg daily for 28 days, a dose-related decrease in weight gain and absence of a normal degree of hepatic centrilobular cytoplasmic vacuolization was observed in male rats. Dose-related lesions in the liver, kidneys, and parotid salivary glands were found at all dose levels in male rats fed encapsulated eucalyptol (Wolff 1987b).

In male rats administered 0, 500, or 1000 mg/kg of the compound eucalyptol daily for 28 days, statistically significant decreases in terminal body weight and increases in relative liver and kidney weights were found in both dose groups, whereas relative brain weight was increased only in the highest dose group. Minor focal infiltration of mononuclear cells in the liver was observed in all groups. In kidneys, a dose-related accumulation of eosinophilic protein droplets containing α2μ-globulin in the cytoplasm of proximal tubular epithelial cells was induced (Kristiansen and Madsen 1995).

In mice orally administered the compound eucalyptol via stomach tube at doses up to 1200 mg/kg 5 days per week or encapsulated and added to feed at doses up to 5607 mg/kg daily for 28 days, the liver weight/body weight ratio in males was increased at all but the lowest dose given in encapsulated form; likewise, the brain weight/body weight ratio was increased in females at the highest dose. Microscopic examination revealed a minimal hypertrophy of centrilobular hepatocytes in animals of both sexes fed the encapsulated compound, especially at the two highest dose levels (Wolff 1987a).

Commercial rutin derived from eucalyptus, but not purified rutin, caused the development of cataracts in rats orally administered the compound (Nakagawa et al. 1965).

Chronic Toxicity

Weak tumor-promoting activity was seen in mice treated topically once a week for 33 weeks with eucalyptus oil after application of the carcinogen 9,10-dimethyl-1,2-benzanthracene (DMBA). The tumor-promoting effect was less than that of orange or lemon essential oils (Roe and Field 1965).

Genotoxicity

No mutagenic effects of the compound eucalyptol were observed in *Salmonella typhimurium* strains TS97, TA98, TA100, TA102, TA1535, and TA1537 with or without metabolic activation (Gomes-Carneiro et al. 1998; Haworth et al. 1983). Eucalyptol was not active in *Bacillus subtilis* using the *rec* assay or in CHO K-1 cells using the SCE assay (Sasaki et al. 1989).

Literature Cited

Allan, J. 1910. Poisoning by oil of *Eucalyptus*. *Br. Med. J.* 1:569.

Anpalahan, M., and D.G. Le Couteur. 1998. Deliberate self-poisoning with eucalyptus oil in an elderly woman. *Aust. N. Z. J. Med.* 28(1):58.

Atkinson, T. 1909. Eucalyptus oil. *Br. Med. J.* 2:1656.

Benjamin, J. 1906. Eucalyptus poisoning. *Br. Med. J.* 1:1020.

Brownlee, G. 1940. A pharmacological examination of cineole and phellandrone. *Quart. J. Pharm. Pharmacol.* 13:130-137.

Chennoufi, R., J. Morizue, H. Richard, and F. Sandret. 1980. Etude des huilles essentielles d'*Eucalyptus globulus* au Maroc. *Riv. Ital.* 62:353-357.

E

Chun, L.T. 1951. Accidental poisoning in children; with special reference to kerosene poisoning. *Hawaii Med. J.* 11(2):83-87.

Craig, J.O. 1953. Poisoning by the volatile oils in childhood. *Arch. Dis. Child* 28(142):475-483.

Darben, T., B. Cominos, and C.T. Lee. 1998. Topical eucalyptus oil poisoning. *Australas. J. Dermatol.* 39(4):265-267.

De Vincenzi, M., M. Silano, A. De Vincenzi, F. Maialetti, and B. Scazzocchio. 2002. Constituents of aromatic plants: Eucalyptol. *Fitoterapia* 73(3):269-275.

Foggie, W.E. 1911. Eucalyptus oil poisoning. *Br. Med. J.* 1911:359-360.

Gomes-Carneiro, M.R., I. Felzenszwalb, and F.J.R. Paumgartten. 1998. Mutagenicity testing of (±)-camphor, 1,8-cineole, citral, citronellal, (−)-menthol and terpineol with the *Salmonella*/microsome assay. *Mutat. Res.* 416(1-2):129-136.

Gurr, F.W., and J.G. Scroggie. 1965. Eucalyptus oil poisoning treated by dialysis and mannitol infusion, with an appendix on the analysis of biological fluids for alcohol and eucalyptol. *Australas. Ann. Med.* 14:238-249.

Haworth, S., T. Lawlor, K. Mortelmans, W. Speck, and E. Zeiger. 1983. *Salmonella* mutagenicity test results for 250 chemicals. *Env. Mutagen.* 5(1):1-142.

HCPDG. 2002. Opinion of the Scientific Committee on Food on eucalyptol. EU Health & Consumer Protection Directorate-General. Brussels: European Commission, Scientific Committee on Food.

Hindle, R.C. 1994. Eucalyptus oil ingestion. *N. Z. Med. J.* 107(977):185-186.

Jenner, P.M., E.C. Hagan, J.M. Taylor, E.L. Cook, and O.G. Fitzhugh. 1964. Food flavourings and compounds of related structure. I. Acute oral toxicity. *Food Cosmet. Toxicol.* 2(3):327-343.

Jori, A., and G. Briatico. 1973. Effect of eucalyptol on microsomal enzyme activity of foetal and newborn rats. *Biochem. Pharmacol.* 22(4):543-544.

Juergens, U.R., U. Dethlefsen, G. Steinkamp, et al. 2003. Anti-inflammatory activity of 1,8-cineol (eucalyptol) in bronchial asthma: A double-blind placebo-controlled trial. *Resp. Med.* 97(3):250-256.

Katzung, B. 2007. *Basic & clinical pharmacology.* 10th ed. New York: McGraw-Hill.

Kirkness, E. 1910. Poisoning by oil of Eucalyptus *Br. Med. J.* 1:261.

Kristiansen, E., and C. Madsen. 1995. Induction of protein droplet (a2μ-globulin) nephropathy in male rats after short-term dosage with 1,8-cineole and *l*-limonene. *Toxicol. Lett.* 80(1-3):147-152.

Leung, A.Y., and S. Foster. 1996. *Encyclopedia of common natural ingredients used in food, drugs, and cosmetics*. 2nd ed. New York: Wiley.

Macpherson, J. 1925. The toxicology of eucalyptus oil. *Med. J. Aust.* 2:108-110.

Myott, M. 1906. Case of eucalyptus poisoning. *Br. Med. J.* 1:558.

Nakagawa, Y., M.R. Shetlar, and S.H. Wender. 1965. Physiological effects of commercial rutin samples on rats. *Life Sci.* 4:753-758.

Orr, J., and S. Edin. 1906. Eucalyptus poisoning. *Br. Med. J.* 1:1085.

Owen, F. 1885. Notes on a case of poisoning by eucalyptus. *Med. J. Aust.*15:394-397.

Pages, N., G. Fournier, F. Le Luyer, and M.C. Marques. 1990. Essential oils and their potential teratogenic properties: Preliminary study using the essential oil of *Eucalyptus globulus* in mice. *Plant Med. Phytother.* 24(1):21-26.

Patel, S., and J. Wiggins. 1980. Eucalyptus oil poisoning. *Arch. Dis. Child* 55(5):405-406.

Roe, F.J., and W.E. Field. 1965. Chronic toxicity of essential oils and certain other products of natural origin. *Food Cosmet. Toxicol.* 3(2):311-323.

Ruse, M. 1998. Eucalyptus oil: Inchem. International Programme on Chemical Safety, Poisons Information Monograph 031.

Sasaki, Y.F., H. Imanishi, T. Ohta, and Y. Shirasu. 1989. Modifying effects of components of plant essence on the induction of sister-chromatid exchange in cultured Chinese hamster ovary cells. *Mutat. Res.* 226:103-110.

Sewell, J.S. 1925. Poisoning by eucalyptus oil. *Br. Med. J.* 1925:922.

Spoerke, D.G., S.A. Vandenberg, S.C. Smolinske, K. Kulig, and B.H. Rumack. 1989. Eucalyptus oil: 14 cases of exposure. *Vet. Hum. Toxicol.* 31(2):166-168.

Taylor, H.S. 1905. A case of acute poisoning by eucalyptus oil. *Lancet* 2:963-964.

Tibballs, J. 1995. Clinical effects and management of eucalyptus oil ingestion in infants and young children. *Med. J. Aust.* 163(4):177-180.

Tyler, V., L. Brady, and J. Roberts. 1988. *Pharmacognosy.* 9th ed. Philadelphia: Lea & Febiger.

Unger, M., and A. Frank. 2004. Simultaneous determination of the inhibitory potency of herbal extracts on the activity of six major cytochrome P450 enzymes using liquid chromatography/mass spectrometry and automated online extraction. *Rapid Commun. Mass Spectrom.* 18(19):2273-2281.

Webb, N.J., and W.R. Pitt. 1993. Eucalyptus oil poisoning in childhood: 41 cases in south-east Queensland. *J. Paediatr. Child Health* 29(5):368-371.

White, R.D., R.A. Swick, and P.R. Cheeke. 1983. Effects of microsomal enzyme induction on the toxicity of pyrrolizidine (*Senecio*) alkaloids. *J. Toxicol. Env. Health* 12(4):633-640.

Williams, A.C., and B.W. Barry. 1989. Essential oils as novel human skin penetration enhancers. *Int. J. Pharmaceut.* 57(2):7-9.

Wolff, G.L. 1987a. Twenty-eight day gavage and encapsulated feed study on 1,8-cineole in B6C3F1 hybrid mice: National Toxicology Program. NTP chemical no. 15 – NTP experiment nos: 5014-02 (encapsulated) and 5014-06 (gavage). Final report.

Wolff, G.L. 1987b. Twenty-eight day gavage and encapsulated feed study on 1,8-cineole in Fischer 344 rats: National Toxicology Program. NTP chemical no. 15 – NTP experiment nos: 5014-03 (encapsulated) and 5014-07 (gavage). Final report.

Wood, F. 1900. Poisoning by oleum eucalypti. *Br. Med. J.* 1:194.

Eucommia ulmoides Oliv.

Eucommiaceae

SCN: eucommia
PN: *du zhong* (bark)
OCN: Chinese rubber tree; hardy rubber tree
Part: bark

Quick Reference Summary

Safety Class: 1
Interaction Class: A

Contraindications

None known.

Other Precautions

None known.

Drug and Supplement Interactions

None known.

Adverse Events and Side Effects

Allergic reactions after extensive skin contact with eucommia have been reported (Bensky et al. 2004).

Pharmacological Considerations

None known.

Pregnancy and Lactation

Eucommia is one of the herbs most commonly used in traditional Chinese medicine to stabilize pregnancy and prevent miscarriage (Bensky et al. 2004; Chen and Chen 2004).

No information on the safety of eucommia during lactation was identified. While this review did not identify any concerns for use while nursing, safety has not been conclusively established.

E

Review Details

I. Drug and Supplement Interactions

Clinical Trials of Drug or Supplement Interactions

No clinical trials of drug or supplement interactions were identified.

Case Reports of Suspected Drug or Supplement Interactions

No case reports of suspected drug or supplement interactions were identified.

Animal Trials of Drug or Supplement Interactions

No animal trials of drug or supplement interactions were identified.

II. Adverse Events

Case Reports of Adverse Events

Allergic reactions after extensive skin contact with eucommia have been reported (Bensky et al. 2004).

III. Pharmacology and Pharmacokinetics

Human Pharmacological Studies

No relevant human pharmacological studies were identified.

Animal Pharmacological Studies

No relevant animal pharmacological studies were identified.

In Vitro Pharmacological Studies

No relevant in vitro pharmacological studies were identified.

IV. Pregnancy and Lactation

Eucommia is one of the herbs most commonly used in traditional Chinese medicine to stabilize pregnancy and prevent miscarriage (Bensky et al. 2004; Chen and Chen 2004).

In excised uteruses, eucommia dose-dependently relaxed chemically induced uterine stimulation and contraction. Processed eucommia was more active than unprocessed eucommia (Wang et al. 1989).

No information on the safety of eucommia during lactation was identified.

V. Toxicity Studies

Acute Toxicity

The LD_{50} of eucommia intravenously administered to mice is 574.1 g/kg (Chen and Chen 2004). No deaths were reported in mice intraperitoneally administered 500 g/kg of eucommia daily for 6 days (Chen and Chen 2004).

Short-Term Toxicity

No evidence of toxicity, as determined by clinical appearance, histopathology, and serum chemistry evaluation, was observed in rats orally administered 200, 600, or 1200 mg/kg of eucommia extract daily for 28 days (Lang et al. 2005).

Genotoxicity

In the mouse micronucleus and chromosomal aberration assays, mutagenic activity was observed in mice intraperitoneally administered an aqueous extract of eucommia at doses of 1 to 2 g/kg, with no mutagenic activity observed at a dose of 0.2 g/kg. A dose-dependent increase in the incidence of chromosomal aberrations was observed at doses of 1 to 2 g/kg. In the micronucleus assay, an increase of polychromatic erythrocytes was observed at the 1 to 2 g/kg dose level (Yin et al. 1991).

In the Ames mutagenicity assay with *Salmonella typhimurium* strains TA98 or TA100, an extract of eucommia exhibited some mutagenic activity in TA100 with and without metabolic activation (Yin et al. 1991).

Literature Cited

Bensky, D., S. Clavey, and E. Stöger. 2004. *Chinese herbal medicine: Materia medica.* 3rd ed. Seattle: Eastland Press.

Chen, J.K., and T.T. Chen. 2004. *Chinese medical herbology and pharmacology.* City of Industry, CA: Art of Medicine Press.

Lang, C., Z. Liu, H.W. Taylor, and D.G. Baker. 2005. Effect of *Eucommia ulmoides* on systolic blood pressure in the spontaneous hypertensive rat. *Am. J. Chin. Med.* 33(2):215-230.

Wang, Q., X. Li, S. Liu, and X. Guan. 1989. Experimental studies on the significance of processing *Eucommia ulmoides* Oliv. *Zhongguo Zhongyao Zazhi* 14(11):21.

Yin, X.J., D.X. Liu, H.C. Wang, and Y. Zhou. 1991. A study on the mutagenicity of 102 raw pharmaceuticals used in Chinese traditional medicine. *Mutat. Res.* 260(1):73-82.

E

Euonymus atropurpureus Jacq. — Celastraceae

SCN: wahoo
OCN: eastern burningbush; euonymus

Part: root bark

Quick Reference Summary

Safety Class: 2b, 2c, 2d
Interaction Class: A

Contraindications

Not for use in pregnancy or lactation except under the supervision of a qualified healthcare practitioner (Bliss and Ramstad 1957a, 1957b; Bradley 1992).

Do not exceed recommended dose (Bradley 1992; Lyle 1932).

Other Precautions

None known.

Drug and Supplement Interactions

None known.

Standard Dose

A decoction of 0.3 to 1 g or 0.3 to 1 ml of tincture (Bradley 1992).

Adverse Events and Side Effects

Wahoo may cause nausea (Lyle 1932). A British reference text notes that "in excessive doses" wahoo is "relatively toxic" (Bradley 1992).

Pharmacological Considerations

None known.

Pregnancy and Lactation

A British herbal reference text indicates that wahoo should not be used during pregnancy or lactation (Bradley 1992). Based on this information, use during pregnancy is not recommended except under the supervision of a qualified healthcare practitioner.

Review Details

I. Drug and Supplement Interactions

Clinical Trials of Drug or Supplement Interactions

No clinical trials of drug or supplement interactions were identified.

Case Reports of Suspected Drug or Supplement Interactions

No case reports of suspected drug or supplement interactions were identified.

Animal Trials of Drug or Supplement Interactions

No animal trials of drug or supplement interactions were identified.

II. Adverse Events

Case Reports of Adverse Events

A British reference text notes that "in excessive doses" wahoo is "relatively toxic." The standard dose is reported as a decoction of 0.3 to 1 g or 0.3 to 1 ml of tincture (Bradley 1992).

III. Pharmacology and Pharmacokinetics

Human Pharmacological Studies

No relevant human pharmacological studies were identified.

Animal Pharmacological Studies

No relevant animal pharmacological studies were identified.

In Vitro Pharmacological Studies

No relevant in vitro pharmacological studies were identified.

IV. Pregnancy and Lactation

A British herbal reference text indicates that wahoo should not be used during pregnancy or lactation (Bradley 1992).

V. Toxicity Studies

No toxicity studies were identified.

Literature Cited

Bliss, C.A., and E. Ramstad. 1957a. Cardiac glycosides of *Euonymus atropurpurea* Jacq. I. Detection, separation, and isolation. *J. Am. Pharm. Assoc.* 46(1):15-18.

Bliss, C.A., and E. Ramstad. 1957b. Cardiac glycosides of *Euonymus atropurpurea* Jacq. II. A study of the structure of euatroside and euatromonoside. *J. Am. Pharm. Assoc.* 46(7):423-426.

Bradley, P.R. 1992. *British herbal compendium: A handbook of scientific information on widely used plant drugs, Volume 1*. Bournemouth, UK: British Herbal Medicine Association.

Lyle, T.J. 1932. *Physio-medical therapeutics, materia medica and pharmacy*. London: National Association of Medical Herbalists.

Eupatorium perfoliatum L.

Asteraceae

SCN: boneset

OCN: thoroughwort

Part: herb

Quick Reference Summary

Safety Class: 1

Interaction Class: A

Contraindications

None known.

Other Precautions

None known.

Drug and Supplement Interactions

None known.

Editors' Note

The presence or absence of pyrrolizidine alkaloids (PAs), some of which may cause liver toxicity, in boneset has not been fully investigated. Although many species of the genus *Eupatorium* contain PAs, these compounds have not been confirmed in boneset, and several references indicate a lack of these compounds in boneset (Arzneimittelkommission 1990; De Smet 1993; Hensel et al. 2011; Leung and Foster 1996; Zhang et al. 2008).

Adverse Events and Side Effects

Boneset contains sesquiterpene lactones, compounds that may cause allergic contact dermatitis in sensitive individuals (Herz et al. 1977; Warshaw and Zug 1996).

"Large" doses (decoction of 5.5–7 g) of boneset may be emetic or cathartic (Felter and Lloyd 1898; Wood and LaWall 1926).

Pharmacological Considerations

None known.

Pregnancy and Lactation

No information on the safety of boneset in pregnancy or lactation was identified in the scientific or traditional literature. Although this review did not identify any concerns for use while pregnant or nursing, safety has not been conclusively established.

Review Details

I. Drug and Supplement Interactions

Clinical Trials of Drug or Supplement Interactions

No clinical trials of drug or supplement interactions were identified.

Case Reports of Suspected Drug or Supplement Interactions

No case reports of suspected drug or supplement interactions were identified.

Animal Trials of Drug or Supplement Interactions

No animal trials of drug or supplement interactions were identified.

II. Adverse Events

Case Reports of Adverse Events

No case reports of adverse events were identified.

III. Pharmacology and Pharmacokinetics

Human Pharmacological Studies

No relevant human pharmacological studies were identified.

Animal Pharmacological Studies

No relevant animal pharmacological studies were identified.

In Vitro Pharmacological Studies

No relevant in vitro pharmacological studies were identified.

IV. Pregnancy and Lactation

No information on the safety of boneset in pregnancy or lactation was identified.

V. Toxicity Studies

Cytotoxicity

An ethanol extract of boneset showed cytotoxicity with EC_{50} values of 12 to 14 μg/ml, comparable to a standard cytotoxic agent, chlorambucil (Habtemariam 1998).

Literature Cited

Arzneimittelkommission. 1990. Vorinformation pyrrolizidinalkaloid-haltige humanarzneimittel. *Pharm. Ztg.* 135:2532-2533, 2623-2624.

De Smet, P.A.G.M. 1993. *Adverse effects of herbal drugs, Volume 2*. Berlin: Springer.

Felter, H.W., and J.U. Lloyd. 1898. *King's American dispensatory*. 18th ed., 3rd rev. 2 vols. Cincinnati: Ohio Valley Co.

Habtemariam, S. 1998. Cistifolin, an integrin-dependent cell adhesion blocker from the anti-rheumatic herbal drug, gravel root (rhizome of *Eupatorium purpureum*). *Planta Med.* 64(8):683-685.

Hensel, A., M. Maas, J. Sendker, M. Lechtenberg, F. Petereit, A. Deters, T. Schmidt, T. Stark. 2011. *Eupatorium perfoliatum* L. Phytochemistry, traditional use and current applications. *J. Ethnopharmacol.* 138(3):641-651.

Herz, W., P.S. Kalyanaraman, and G. Ramakrishnan. 1977. Sesquiterpene lactones of *Eupatorium perfoliatum*. *J. Org. Chem.* 42(13):2264-2271.

Leung, A.Y., and S. Foster. 1996. *Encyclopedia of common natural ingredients used in food, drugs, and cosmetics*. 2nd ed. New York: Wiley.

Warshaw, E.M., and K.A. Zug. 1996. Sesquiterpene lactone allergy. *Am. J. Contact Dermat.* 7(1):1-23.

Wood, H., and C. LaWall. 1926. *The dispensatory of the United States of America*. Philadelphia: Lippincott.

Zhang, M.L., M. Wu, J.J. Zhang, et al. 2008. Chemical constituents of plants from the genus *Eupatorium*. *Chem. Biodivers.* 5(1):40-55.

Euphrasia spp. — Scrophulariaceae

Euphrasia rostkoviana F. Hayne
SCN: eyebright
Euphrasia stricta J.P. Wolff ex J.F. Lehm.
SCN: eyebright
Syn: *Euphrasia officinalis* L.
Part: herb

Quick Reference Summary

Safety Class: 1
Interaction Class: A

Contraindications

None known.

Other Precautions

None known.

Drug and Supplement Interactions

None known.

Notice

Tannins (~12%) (Harkiss and Timmins 1973); *see* Appendix 1.

Editors' Note

Species of eyebright are difficult to distinguish botanically (Harkiss and Timmins 1973). *Euphrasia stricta*, *E. rostikoviana*, and hybrids or mixtures of these species are commonly accepted as eyebright (Wichtl 2004).

Adverse Events and Side Effects

None known.

Pharmacological Considerations

An animal study indicated that eyebright may modify glucose regulation (Porchezhian et al. 2000). People with diabetes are advised to monitor their blood sugar closely and discuss the use of this herb with a qualified healthcare practitioner prior to use.

Pregnancy and Lactation

No information on the safety of eyebright in pregnancy or lactation was identified in the scientific or traditional literature. Although this review did not identify any concerns for use while pregnant or nursing, safety has not been conclusively established.

Review Details

I. Drug and Supplement Interactions

Clinical Trials of Drug or Supplement Interactions

No clinical trials of drug or supplement interactions were identified.

Case Reports of Suspected Drug or Supplement Interactions

No case reports of suspected drug or supplement interactions were identified.

Animal Trials of Drug or Supplement Interactions

No animal trials of drug or supplement interactions were identified.

II. Adverse Events

Case Reports of Adverse Events

No case reports of adverse events were identified.

III. Pharmacology and Pharmacokinetics

Human Pharmacological Studies

No relevant human pharmacological studies were identified.

Animal Pharmacological Studies

A hypoglycemic effect was observed in diabetic rats orally administered 600 mg/kg of an aqueous extract of eyebright. No significant hypoglycemic activity was observed in healthy rats (Porchezhian et al. 2000).

In Vitro Pharmacological Studies

No relevant in vitro pharmacological studies were identified.

IV. Pregnancy and Lactation

No information on the use of eyebright during pregnancy or lactation was identified.

V. Toxicity Studies

Acute Toxicity

No adverse effects were observed in rats orally administered up to 6 g/kg of an aqueous extract of eyebright (Porchezhian et al. 2000).

Genotoxicity

No mutagenic activity of an ethanol extract of eyebright was observed in the Ames mutagenicity assay with *Salmonella typhimurium* strains TA98 and TA100 with or without metabolic activation by S9 (Schimmer et al. 1994).

Literature Cited

Harkiss, K.J., and P. Timmins. 1973. Studies in the Scrophulariaceae—Part VIII. Phytochemical investigation of *Euphrasia officinalis*. *Planta Med.* 23:182.

Porchezhian, E., S.H. Ansari, and N.K.K. Shreedharan. 2000. Antihyperglycemic activity of *Euphrasia officinale* leaves. *Fitoterapia* 71(5):522-526.

Schimmer, O., A. Krüger, H. Paulini, and F. Haefele. 1994. An evaluation of 55 commercial plant extracts in the Ames mutagenicity test. *Pharmazie* 49(6):448-451.

Wichtl, M. 2004. *Herbal drugs and phytopharmaceuticals: A handbook for practice on a scientific basis*. 3rd ed. Boca Raton, FL: CRC Press.

Euryale ferox Salib. — Nymphaeaceae

SCN: euryale
AN: *makhanna*
PN: *qian shi* (seed)
OCN: foxnut; gorgon waterlily
Part: seed

Quick Reference Summary

Safety Class: 1
Interaction Class: A

Contraindications

None known.

Other Precautions

Use with caution in persons with difficulty urinating or defecating (Bensky et al. 2004).

Drug and Supplement Interactions

None known.

Adverse Events and Side Effects

None known.

Pharmacological Considerations

None known.

Pregnancy and Lactation

No information on the safety of euryale in pregnancy or lactation was identified in the scientific or traditional literature. Although this review did not identify any concerns for use while pregnant or nursing, safety has not been conclusively established.

E

Review Details

I. Drug and Supplement Interactions

Clinical Trials of Drug or Supplement Interactions

No clinical trials of drug or supplement interactions were identified.

Case Reports of Suspected Drug or Supplement Interactions

No case reports of suspected drug or supplement interactions were identified.

Animal Trials of Drug or Supplement Interactions

No animal trials of drug or supplement interactions were identified.

II. Adverse Events

Case Reports of Adverse Events

A case of allergic reaction to euryale, with tingling pruritus and localized urticaria-like papular rashes, was reported (Bensky et al. 2004).

III. Pharmacology and Pharmacokinetics

Human Pharmacological Studies

No relevant human pharmacological studies were identified.

Animal Pharmacological Studies

No relevant animal pharmacological studies were identified.

In Vitro Pharmacological Studies

No relevant in vitro pharmacological studies were identified.

IV. Pregnancy and Lactation

No information on the use of euryale during pregnancy or lactation was identified.

V. Toxicity Studies

No toxicity studies were identified.

Literature Cited

Bensky, D., S. Clavey, and E. Stöger. 2004. *Chinese herbal medicine: Materia medica*. 3rd ed. Seattle: Eastland Press.

Euterpe oleracea Mart. — Arecaceae

SCN: açaí
OCN: assai palm; cabbage palm

Part: fruit

Quick Reference Summary

Safety Class: 1
Interaction Class: A

Contraindications

None known.

Other Precautions

None known.

Drug and Supplement Interactions

None known.

Editors' Notes

The fruit of açaí is widely consumed as a food by people of the Amazon basin in South America (Brondízio et al. 2002; Janick and Paull 2008; Muñiz-Miret et al. 1996).

Cases of Chagas disease (a parasitic disease that can lead to heart and intestinal complications) have been associated with açaí juice consumption in the Amazon (Nóbrega et al. 2009; Valente et al. 1999). Chagas disease is transmitted by several types of insects. Infections in the Amazon regions are believed to be due to triatomine bugs that, attracted by electrical lights, fell into the machine used to process the juice consumed by the patients (Valente et al. 1999). Use of properly cleaned machinery and fruit should prevent transmission of the disease.

Adverse Events and Side Effects

None known.

Pharmacological Considerations

None known.

Pregnancy and Lactation

No information on the safety of açaí in pregnancy or lactation was identified in the scientific or traditional literature. Although this review did not identify any concerns for use while pregnant or nursing, safety has not been conclusively established.

Review Details

I. Drug and Supplement Interactions

Clinical Trials of Drug or Supplement Interactions

No clinical trials of drug or supplement interactions were identified.

Case Reports of Suspected Drug or Supplement Interactions

No case reports of suspected drug or supplement interactions were identified.

Animal Trials of Drug or Supplement Interactions

No animal trials of drug or supplement interactions were identified.

II. Adverse Events

Case Reports of Adverse Events

No case reports of adverse events were identified.

III. Pharmacology and Pharmacokinetics

Human Pharmacological Studies

No relevant human pharmacological studies were identified.

Animal Pharmacological Studies

No relevant animal pharmacological studies were identified.

In Vitro Pharmacological Studies

No relevant in vitro pharmacological studies were identified.

IV. Pregnancy and Lactation

No information on the use of açaí during pregnancy or lactation was identified.

V. Toxicity Studies

Genotoxicity

Some mutagenic activity of açaí was observed in *Saccharomyces cerevisiae* yeast treated with high concentrations (5, 10, or 15%, w/v) of previously frozen açaí pulp. The mutagenic activity was similar to that of cashew apple (*Anacardium occidentale*), kiwi fruit (*Actinidia chinensis*), and strawberry (*Fragaria vesca*) (Spada et al. 2008).

Literature Cited

Brondízio, E.S., C.A.M. Safar, and A.D. Siqueira. 2002. The urban market of acai fruit (*Euterpe oleracea* Mart.) and rural land use change: Ethnographic insights into the role of price and land tenure constraining agricultural choices in the Amazon estuary. *Urban Ecosys.* 6(1):67-97.

Janick, J., and R.E. Paull. 2008. *The encyclopedia of fruit & nuts.* Wallingford, UK: CABI Publishing.

Muñiz-Miret, N., R. Vamos, M. Hiraoka, F. Montagnini, and R.O. Mendelsohn. 1996. The economic value of managing the acai palm (*Euterpe oleracea* Mart.) in the floodplains of the Amazon estuary, Para, Brazil. *Forest Ecol. Management* 87(1-3):163-173.

Nóbrega, A.A., M.H. Garcia, E. Tatto, et al. 2009. Oral transmission of Chagas disease by consumption of acai palm fruit, Brazil. *Emerg. Infect. Dis.* 15(4):653-655.

Spada, P.D., G.G. de Souza, G.V. Bortolini, J.A. Henriques, and M. Salvador. 2008. Antioxidant, mutagenic, and antimutagenic activity of frozen fruits. *J. Med. Food* 11(1):144-151.

Valente, S.A.S., V.C. Valente, and H. Fraiha Neto. 1999. Considerations on the epidemiology and transmission of Chagas disease in the Brazilian Amazon. *Mem. Inst. Oswaldo Cruz* 94:395-398.

Eutrochium spp. Asteraceae

Eutrochium fistulosum (Barratt) E.E. Lamont
SCN: hollow Joe Pye
Syn: *Eupatorium fistulosum* Barratt
Eutrochium maculatum (L.) E.E. Lamont
SCN: spotted Joe Pye
Syn: *Eupatorium maculatum* L.

Eutrochium purpureum (L.) E.E. Lamont
SCN: Joe Pye; gravel root (root)
Syn: *Eupatorium purpureum* L.
OCN: Joe Pye weed; queen-of-the-meadow
Part: herb, rhizome, root

E

Quick Reference Summary

Safety Class: 2a, 2b, 2c
Interaction Class: A

Contraindications

For external use only (McGuffin et al. 1997).

Not for use in pregnancy or lactation except under the supervision of a qualified healthcare practitioner (De Smet 1993).

Other Precautions

None known.

Drug and Supplement Interactions

None known.

Notice

Pyrrolizidine alkaloids (Mills 1967; Smith and Culvenor 1981); *see* Appendix 1.

Editors' Notes

The American Herbal Products Association has established a trade requirement (AHPA 2011) that all products with botanical ingredients that contain toxic pyrrolizidine alkaloids, including *Eupatorium* species, are not offered for sale for internal use and display the following cautionary label: "For external use only. Do not apply to broken or abraded skin. Do not use when nursing."

The presence or absence of pyrrolizidine alkaloids (PAs), some of which may cause liver toxicity, in Joe Pye has not been fully investigated. Although many species of the genus *Eupatorium* contain PAs, these compounds have not been confirmed in Joe Pye. The one reference that indicates the presence of these compounds is a dissertation from the 1960s that lists Joe Pye as "probably" containing the compound echinatine in the aerial parts, and echinatine and trachelanthamine in the root (Mills 1967; Smith and Culvenor 1981). In 1983, Joe Pye was listed as a PA-containing plant "the toxicity of which has not been, or has been insufficiently, investigated" (Danninger et al. 1983; INCHEM 1988). Echinatine is an unsaturated PA, which is the type that has been associated with liver toxicity, as compared with trachelanthamine, a saturated PA which is the type considered nontoxic.

Adverse Events and Side Effects

None known.

Pharmacological Considerations

None known.

Pregnancy and Lactation

No information on the safety of Joe Pye in pregnancy or lactation was identified. Based on the probable presence of pyrrolizidine alkaloid compounds, use of Joe Pye during pregnancy or lactation is not recommended.

Review Details

I. Drug and Supplement Interactions

Clinical Trials of Drug or Supplement Interactions

No clinical trials of drug or supplement interactions were identified.

Case Reports of Suspected Drug or Supplement Interactions

No case reports of suspected drug or supplement interactions were identified.

Animal Trials of Drug or Supplement Interactions

No animal trials of drug or supplement interactions were identified.

II. Adverse Events

Case Reports of Adverse Events

No case reports of adverse events were identified.

III. Pharmacology and Pharmacokinetics

Human Pharmacological Studies

No relevant human pharmacological studies were identified.

Animal Pharmacological Studies

No relevant animal pharmacological studies were identified.

In Vitro Pharmacological Studies

No relevant in vitro pharmacological studies were identified.

IV. Pregnancy and Lactation

No information on the use of Joe Pye during pregnancy or lactation was identified.

V. Toxicity Studies

No toxicity studies were identified.

Literature Cited

AHPA. July 2011. Code of Ethics & Business Conduct. Silver Spring, MD: American Herbal Products Association.

Danninger, T., U. Hagemann, V. Schmidt, and P.S. Schoenhoefer. 1983. Toxicity of pyrrolizidine alkaloid-containing medicinal plants. *Pharm. Ztg.* 128:289-303.

De Smet, P.A.G.M. 1993. *Adverse effects of herbal drugs, Volume 2.* Berlin: Springer.

INCHEM. 1988. Pyrrolizidine alkaloids. In *International Programme on Chemical Safety, Environmental Health Criteria 80.* Geneva: World Health Organization.

McGuffin, M., C. Hobbs, R. Upton, and A. Goldberg. 1997. *Botanical safety handbook.* Boca Raton, FL: CRC Press.

Mills, F. 1967. Phytochemical constituents of *Eupatorium purpureum* (Joe-Pye weed). Western Reserve Univ., Cleveland, OH.

Smith, L.W., and C.C.J. Culvenor. 1981. Plant sources of hepatotoxic pyrrolizidine alkaloids. *J. Nat. Prod.* 44:129-152.

E

Evernia spp. — Usneaceae

Evernia furfuracea (L.) W. Mann
SCN: tree moss
Evernia prunastri (L.) Ach.
SCN: oak moss
Part: thallus

Quick Reference Summary

Safety Class: 1
Interaction Class: A

Contraindications

None known.

Other Precautions

None known.

Drug and Supplement Interactions

None known.

Notice

May contain thujone (CFR 2011; Furia and Bellanca 1971); *see* Appendix 1.

Editors' Notes

Although some research indicated the presence of the compound thujone in oak moss (Gavin and Tabacchi 1975; Stoll and Scherrer 1937), a review of the chemistry of oak moss indicated that the presence of thujone in oak moss is questionable, and that more recent investigation has not confirmed the presence of the compound (Joulain and Tabacchi 2009).

The hot alcohol extract of *Evernia* produces a toxic ethyl ester and is therefore not for internal use (Furia and Bellanca 1971).

Use of oak moss as a food additive in the United States is subject to a limitation that the finished food or beverage is thujone-free (CFR 2011). Dietary ingredients for use in dietary supplements, however, are specifically excluded from the federal food additive definition. (U.S.C. 2010).

Adverse Events and Side Effects

Allergic reactions to oak moss as a fragrance ingredient have been reported (Goncalo 1987; Johansen et al. 2002; Schnuch et al. 2004).

Pharmacological Considerations

None known.

Pregnancy and Lactation

No information on the safety of oak moss or tree moss in pregnancy or lactation was identified in the scientific or traditional literature. Although this review did not identify any concerns for use while pregnant or nursing, safety has not been conclusively established.

Review Details

I. Drug and Supplement Interactions

Clinical Trials of Drug or Supplement Interactions

No clinical trials of drug or supplement interactions were identified.

Case Reports of Suspected Drug or Supplement Interactions

No case reports of suspected drug or supplement interactions were identified.

Animal Trials of Drug or Supplement Interactions

No animal trials of drug or supplement interactions were identified.

II. Adverse Events

Case Reports of Adverse Events

Oak moss is a common ingredient in perfumes and is included in the standard fragrance mix used to screen for fragrance allergies (Johansen et al. 2002). Of 59,298 persons that tested positively to fragrance mix between 1996 and 2002, 29.9% had reactions to oak moss (Schnuch et al. 2004). In persons with allergic reactions to *Frullania*, many also tested positive to oak moss (Goncalo 1987).

III. Pharmacology and Pharmacokinetics

Human Pharmacological Studies

No irritation was observed in healthy volunteers treated with 10% oak moss concrete in petrolatum in a 48-hour closed patch test (Opdyke 1979). No sensitization was observed in healthy volunteers treated topically with a 10% oak moss concrete in petrolatum (Opdyke 1979).

Animal Pharmacological Studies

In irritation tests, no irritation was observed after oak moss concrete was applied undiluted to the backs of hairless mice or pigs or to intact or abraded rabbit skin for 24 hours under occlusion (Opdyke 1979).

In Vitro Pharmacological Studies

No relevant in vitro pharmacological studies were identified.

IV. Pregnancy and Lactation

No information on the use of oak moss or tree moss during pregnancy or lactation was identified.

V. Toxicity Studies

Acute Toxicity

The LD_{50} of orally administered oak moss concrete in rats is 2.9 g/kg; the dermal LD_{50} of the same product tested in rabbits could not be determined at doses up to 5 g/kg (Opdyke 1979).

Literature Cited

CFR. 2011. *Code of federal regulations*, Title 21 Part 172.510, 2011 ed. Food additives permitted for direct addition to food for human consumption. Flavoring agents and related substances. Natural flavoring substances and natural substances used in conjunction with flavors. Washington, DC: U.S. Government Printing Office.

Furia, T.E., and N. Bellanca. 1971. *Feranoli's handbook of flavor ingredients*. Cleveland, OH: The Chemical Rubber Co.

Gavin, J., and R. Tabacchi. 1975. Isolation and identification of phenolic and monoterpenic compounds from oak moss (*Evernia prunastri*). *Helv. Chim. Acta* 58(1):190-194.

Goncalo, S. 1987. Contact sensitivity to lichens and Compositae in *Frullania* dermatitis. *Contact Dermat.* 16(2):84-86.

Johansen, J.D., S. Heydorn, and T. Menne. 2002. Oak moss extracts in the diagnosis of fragrance contact allergy. *Contact Dermat.* 46(3):157-161.

Joulain, D., and R. Tabacchi. 2009. Lichen extracts as raw materials in perfumery. Part 1: Oakmoss. *Flav. Fragrance J.* 24(2):49-61.

Opdyke, D.L.J. 1979. *Monographs on fragrance raw materials*. New York: Pergamon.

Schnuch, A., H. Lessmann, J. Geier, P.J. Frosch, and W. Uter. 2004. Contact allergy to fragrances: Frequencies of sensitization from 1996 to 2002. Results of the IVDK. *Contact Dermat.* 50(2):65-76.

Stoll, M., and W. Scherrer. 1937. Concrete of oak moss. *Chim. Ind.* 29:205-212.

U.S.C. 2010. United States Code, Title 21, Part 321 (s)(6). Current as of January 7, 2011. Washington, DC: U.S. Government Printing Office.

Evolvulus alsinoides (L.) L. — Convolvulaceae

SCN: dwarf morning glory
AN: *shankhapushpi*

Part: whole plant

Quick Reference Summary

Safety Class: 1
Interaction Class: A

Contraindications
None known.

Other Precautions
None known.

Drug and Supplement Interactions
None known.

Adverse Events and Side Effects
None known.

Pharmacological Considerations
None known.

Pregnancy and Lactation
No information on the safety of dwarf morning glory in pregnancy or lactation was identified in the scientific or traditional literature. Although this review did not identify any concerns for use while pregnant or nursing, safety has not been conclusively established.

Review Details

I. Drug and Supplement Interactions

Clinical Trials of Drug or Supplement Interactions
No clinical trials of drug or supplement interactions were identified.

Case Reports of Suspected Drug or Supplement Interactions
No case reports of suspected drug or supplement interactions were identified.

Animal Trials of Drug or Supplement Interactions
No animal trials of drug or supplement interactions were identified.

II. Adverse Events

Case Reports of Adverse Events
No case reports of adverse events were identified.

III. Pharmacology and Pharmacokinetics

Human Pharmacological Studies
No relevant human pharmacological studies were identified.

Animal Pharmacological Studies
No relevant animal pharmacological studies were identified.

In Vitro Pharmacological Studies
No relevant in vitro pharmacological studies were identified.

IV. Pregnancy and Lactation

No information on the use of dwarf morning glory during pregnancy or lactation was identified.

V. Toxicity Studies

Acute Toxicity
The LD_{50} of an ethanol extract of dwarf morning glory orally administered in mice is 7 g/kg (Agarwala and Dey 1977).

Literature Cited

Agarwala, N., and C.D. Dey. 1977. Behavioral and lethal effects of alcoholic extract of *Evolvulus alsinoides* in albino mice. *Indian J. Physiol. Allied Sci.* 31(2):81-86.

Ferula spp.

Apiaceae

Ferula assa-foetida L.
SCN: asafetida (oleo gum resin)
AN: *hingu*
OCN: asafoetida (oleo gum resin); devil's dung (oleo gum resin); giant fennel

***Ferula foetida* (Bunge) Regel**
SCN: asafetida (oleo gum resin)
AN: *hingu*
OCN: asafoetida (oleo gum resin); devil's dung (oleo gum resin); giant fennel
Part: oleo gum resin from rhizomes and roots

Quick Reference Summary

Safety Class: 2b, 2d
Interaction Class: A

Contraindications

Not for use in pregnancy except under the supervision of a qualified healthcare practitioner (Keshri et al. 1999, 2004; Madari and Jacobs 2004).

Not for use in infants (Kelly et al. 1984).

Other Precautions

None known.

Drug and Supplement Interactions

None known.

Notice

Emmenagogue (Felter and Lloyd 1898); *see* Appendix 2.

Editors' Note

The classifications and concerns for this herb are based on relatively higher doses used for therapeutic purposes in contrast to lower amounts generally used in cooking, and have not been associated with its use as a spice.

Adverse Events and Side Effects

Methemoglobinemia (a disorder characterized by a higher than normal level of methemoglobin in the blood) was reported in a 5-week-old after ingestion of glycerited asafetida. A related in vitro study showed activity of asafetida in infant but not adult hemoglobin (Kelly et al. 1984).

Side effects reported from excessive consumption include swollen lips, gastric burning, belching, flatulence and diarrhea, burning urination, headaches, and dizziness. Details on products, dose, and duration associated with these effects were not available (De Smet 1992; Roth et al. 1984).

Doses of 50 to 100 mg of asafetida have been reported to cause convulsions in persons suffering from nervousness (Roth et al. 1984).

Pharmacological Considerations

None known.

Pregnancy and Lactation

Animal studies have indicated anti-implantation activity of asafetida (Keshri et al. 1999, 2004). An Eclectic medical text lists asafetida as an emmenagogue (Felter and Lloyd 1898). Based on this information, use during pregnancy is not recommended except under the supervision of a qualified healthcare practitioner.

No information on the safety of asafetida during lactation was identified. While this review did not identify any concerns for use while nursing, safety has not been conclusively established.

Review Details

I. Drug and Supplement Interactions

Clinical Trials of Drug or Supplement Interactions

No clinical trials of drug or supplement interactions were identified.

Case Reports of Suspected Drug or Supplement Interactions

No case reports of suspected drug or supplement interactions were identified.

Animal Trials of Drug or Supplement Interactions

No animal trials of drug or supplement interactions were identified.

II. Adverse Events

Adverse Events Reported in Clinical Trials

No adverse effects were observed in healthy volunteers administered 3 g of asafetida with 100 g of butter on bread (Bordia and Arora 1975).

Case Reports of Adverse Events

Methemoglobinemia (a disorder characterized by a higher than normal level of methemoglobin in the blood) was diagnosed in a 5-week-old boy who had been administered glycerited asafetida (dose unspecified). An in vitro study conducted by the reporting physicians indicated that asafetida had an oxidizing effect on infant hemoglobin but not on adult hemoglobin (Kelly et al. 1984).

Side effects reported from excessive consumption include swollen lips, gastric burning, belching, flatulence and diarrhea, burning urination, headaches, and dizziness. Details on products, dose, and duration were not available (De Smet 1992; Roth et al. 1984).

Doses of 50 to 100 mg of asafetida have been reported to cause convulsions in persons suffering from nervousness (Roth et al. 1984).

III. Pharmacology and Pharmacokinetics

Human Pharmacological Studies

No relevant human pharmacological studies were identified.

Animal Pharmacological Studies

No relevant animal pharmacological studies were identified.

In Vitro Pharmacological Studies

No relevant in vitro pharmacological studies were identified.

IV. Pregnancy and Lactation

Dose-dependent anti-implantation activity was observed in rats orally administered 200 to 400 mg/kg of extracts of asafetida on days 1 to 10 of pregnancy. Pregnancy was prevented in 80% of rats administered 400 mg/kg of a methanol extract and in 100% of rats administered 400 mg/kg of the extract in a polyvinyl pyrrolidone 1:2 complex (Keshri et al. 1999).

Anti-implantation activity of asafetida was observed in rats orally administered 400 mg/kg of a methanol extract of asafetida daily on days 1 through 7 of pregnancy (Keshri et al. 2004).

Asafetida is listed as an ingredient in ancient Persian abortifacient formulas, along with other botanicals (Madari and Jacobs 2004).

No information on the safety of asafetida during lactation was identified.

V. Toxicity Studies

Genotoxicity

Weak induction of sister-chromatid exchanges in the spermatogonia was observed after oral administration of 0.5 or 1 g/kg asafetida to mice (Abraham and Kesavan 1984).

No genotoxic activity of an aqueous extract of asafetida was observed in *Drosophila* treated with concentrations of 0.25 or 0.5% (Abraham and Kesavan 1985).

Antimutagenic activity of asafetida has been reported (Soni et al. 1997; Soudamini et al. 1995).

Literature Cited

Abraham, S.K., and P.C. Kesavan. 1984. Genotoxicity of garlic, turmeric and asafoetida in mice. *Mutat. Res.* 136(1):85-88.

Abraham, S.K., and P.C. Kesavan. 1985. A preliminary analysis of the genotoxicity of a few spices in *Drosophila*. *Mutat. Res.* 143(4):219-224.

Bordia, A., and S.K. Arora. 1975. The effect of essential oil (active principle) of asafoetida on alimentary lipemia. *Indian J. Med. Res.* 63(5):707-711.

De Smet, P.A.G.M. 1992. *Adverse effects of herbal drugs, Volume 1*. Berlin: Springer.

Felter, H.W., and J.U. Lloyd. 1898. *King's American dispensatory*. 18th ed., 3rd rev. 2 vols. Cincinnati: Ohio Valley Co.

Kelly, K.J., J. Neu, B.M. Camitta, and G.R. Honig. 1984. Methemoglobinemia in an infant treated with the folk remedy glycerited asafoetida. *Pediatrics* 73(5):717-719.

Keshri, G., M. Bajpai, V. Lakshmi, B.S. Setty, and G. Gupta. 2004. Role of energy metabolism in the pregnancy interceptive action of *Ferula assafoetida* and *Melia azedarach* extracts in rat. *Contraception* 70(5):429-432.

Keshri, G., V. Lakshmi, M.M. Singh, and V.P. Kamboj. 1999. Postcoital antifertility activity of *Ferula assafoetida* extract in female rats. *Pharm. Biol.* 37(4):273-276.

Madari, H., and R.S. Jacobs. 2004. An analysis of cytotoxic botanical formulations used in the traditional medicine of ancient Persia as abortifacients. *J. Nat. Prod.* 67(8):1204-1210.

Roth, L., M. Daunderer, and K. Kormann. 1984. *Giftpflanzen-pflanzengifte: Vorkommen, wirkung, therapie*. Landsberg: Ecomed.

Soni, K.B., M. Lahiri, P. Chackradeo, S.V. Bhide, and R. Kuttan. 1997. Protective effect of food additives on aflatoxin-induced mutagenicity and hepatocarcinogenicity. *Cancer Lett.* 115(2):129-133.

Soudamini, K.K., M.C. Unnikrishnan, K. Sukumaran, and R. Kuttan. 1995. Mutagenicity and anti-mutagenicity of selected spices. *Indian J. Physiol. Pharmacol.* 39(4):347-353.

Filipendula ulmaria (L.) Maxim. — Rosaceae

SCN: meadowsweet
Syn: *Spiraea ulmaria* L.

OCN: queen-of-the-meadow
Part: herb

Quick Reference Summary

Safety Class: 1
Interaction Class: A

Contraindications

None known.

Other Precautions

Use in persons with sensitivity to aspirin or other salicylate-containing drugs is cautioned (ESCOP 2003; Mills and Bone 2005; Wichtl 2004).

Use in persons with G6PD deficiency is cautioned (Mills and Bone 2005).

Drug and Supplement Interactions

None known.

Notice

Salicylates (up to 0.5%) (Hansel et al. 1993; Meier 1987; Meier et al. 1987; Thieme 1965); *see* Appendix 1.

Tannins (10–15%) (Hansel et al. 1993; Haslam et al. 1989; Okuda et al. 1993); *see* Appendix 1.

Adverse Events and Side Effects

None known.

Pharmacological Considerations

Meadowsweet flower has been reported to contain "heparin-like compounds" bound to plant proteins. In animal studies, injections of these compounds showed anticoagulant and fibrinolytic activity (Kudriashov et al. 1990, 1991). The relevance of those data to oral use of meadowsweet in humans is not known.

Pregnancy and Lactation

A study on isolated animal uteri indicated that meadowsweet increased the tone and force of uterine contractions (Barnaulov et al. 1978). The relevance of such in vitro studies to human use is not known.

No information on the safety of meadowsweet during lactation was identified. While this review did not identify any concerns for use while nursing, safety has not been conclusively established.

Review Details

I. Drug and Supplement Interactions

Clinical Trials of Drug or Supplement Interactions

No clinical trials of drug or supplement interactions were identified.

Case Reports of Suspected Drug or Supplement Interactions

No case reports of suspected drug or supplement interactions were identified.

Animal Trials of Drug or Supplement Interactions

No animal trials of drug or supplement interactions were identified.

II. Adverse Events

Case Reports of Adverse Events

No case reports of adverse events were identified.

III. Pharmacology and Pharmacokinetics

Human Pharmacological Studies

No relevant human pharmacological studies were identified.

Animal Pharmacological Studies

No relevant animal pharmacological studies were identified.

In Vitro Pharmacological Studies

Heparin-like compounds bound to plant proteins were reportedly identified in meadowsweet flower. Intramuscular and intravenous administration of the compounds in animals enhanced anticoagulant and fibrinolytic activity. The activity was neutralized by protamine sulfate (Kudriashov et al. 1990, 1991).

IV. Pregnancy and Lactation

After treatment with an aqueous extract of meadowsweet, an increase in the tone and force of contractions was observed in smooth muscle sections of the uterine horns of rats, cats, and guinea pigs (Barnaulov et al. 1978). The relevance of such in vitro studies to human use is not known.

No information on the safety of meadowsweet during lactation was identified.

V. Toxicity Studies

Acute Toxicity

The LD_{50} of an ethanol extract of meadowsweet in rabbits is 1770 mg/kg after intraperitoneal administration and 75.7 mg/kg after intravenous administration (Barnaulov et al. 1977).

The LD_{50} of a 1:20 meadowsweet decoction intraperitoneally administered to mice is 1050 mg/kg in females and 535 mg/kg in males. The LD_{50} of the decoction in rabbits is 141.5 mg/kg after intravenous administration (Barnaulov et al. 1977).

No adverse effects on liver function were observed in rabbits treated with different extracts of meadowsweet (extracts, doses, and duration of treatment not specified in available English translations) (Barnaulov et al. 1979).

Literature Cited

Barnaulov, O.D., I.G. Boldina, V.V. Galushko, et al. 1979. Pharmacological properties of galenic preparations from the flowers of *Filipendula ulmaria. Rastitel'nye Resursy* 15:399-407.

Barnaulov, O.D., T.V. Bukreeva, A.A. Kokarev, and A.I. Shevchenko. 1978. Primary evaluation of the spasmolytic properties of some natural compounds and galenicals. *Rastitel'nye Resursy* 14:573-579.

Barnaulov, O.D., A.V. Kumkov, and N.A. Khalikova. 1977. Chemical composition and primary evaluation of the properties of preparations from *Filipendula ulmaria* (L.) Maxim flowers. *Rastitel'nye Resursy* 13(4):661-669.

ESCOP. 2003. *ESCOP monographs: The scientific foundation for herbal medicinal products*. 2nd ed. Exeter, U.K.: European Scientific Cooperative on Phytotherapy.

Hansel, R., K. Keller, and H. Rimpler. 1993. *Hagers handbuch der pharmazeutischen praxis, Volume 5*. 5th ed. Berlin: Springer.

Haslam, E., T.H. Lilley, Y. Cai, R. Martin, and D. Magnolato. 1989. Traditional herbal medicines—The role of polyphenols. *Planta Med.* 55(1):1-8.

Kudriashov, B.A., I.M. Ammosova, L.A. Liapina, et al. 1991. Heparin from the meadowsweet (*Filipendula ulmaria*) and its properties. *Izv. Akad. Nauk. SSSR Biol.* 6:939-943.

Kudriashov, B.A., L.A. Liapina, and L.D. Azieva. 1990. The content of a heparin-like anticoagulant in the flowers of the meadowsweet (*Filipendula ulmaria*). *Farmakol. Toksikol.* 53(4):39-41.

Meier, B. 1987. Analytik, chromatographisches Verhalten und potentielle Wirksamkeit der Inhaltsstoffe salicylathaltiger Arzneipflanzen Mitteleuropas (Habilitationsschrift). Zürich: Eidgenossische Technische Hochschule.

Meier, D., L. Lehmann, O. Sticher, and A. Bettschart. 1987. Salicylate in arzneipflanzen. *Dtsche. Apoth. Ztg.* 127:2401-2407.

Mills, S., and K. Bone. 2005. *The essential guide to herbal safety*. St. Louis: Elsevier.

Okuda, T., T. Yoshida, and T. Hatano. 1993. Classification of oligomeric hydrolysable tannins and specificity of their occurrence in plants. *Phytochemistry* 32:507-521.

Thieme, H. 1965. Isolation of a new phenolic glycoside from the blossoms of *Filipendula ulmaria* (L.) Maxim. *Pharmazie* 20 (7):436-439.

Wichtl, M. 2004. *Herbal drugs and phytopharmaceuticals: A handbook for practice on a scientific basis*. 3rd ed. Boca Raton, FL: CRC Press.

F

Foeniculum spp.

Apiaceae

Foeniculum vulgare **Mill.**
SCN: fennel
AN: *mishreya; shatapushpa*
PN: *xiao hui xiang* (fruit)

Foeniculum vulgare **Mill. ssp.** ***vulgare*** **var.** ***dulce*** **(Mill.) Batt. & Trab.**
SCN: sweet fennel

Foeniculum vulgare **Mill. ssp.** ***vulgare*** **var.** ***vulgare***
SCN: bitter fennel
Part: fruit (commonly known as "seed")

Quick Reference Summary

Safety Class: 1
Interaction Class: A

Contraindications

None known.

Other Precautions

None known.

Drug and Supplement Interactions

See Pharmacological Considerations.

Notice

Alkenylbenzenes (estragole as 5–10% of essential oil) (ESCOP 2003); *see* Appendix 1.

Adverse Events and Side Effects

A case series of premature thelarche (isolated breast development with no other clinical signs of sexual maturation), with elevated estradiol levels, was reported in four girls ages 5 months to 5 years who had been administered two to three cups of fennel tea daily for 4 months to 2 years (Turkyilmaz et al. 2008).

A case series of methemoglobinemia with respiratory distress, cyanosis, and tachycardia, was reported in four infants, 7 to 9 months old, that had been fed purées of fennel (part and dose used not specified). High nitrate levels in the products administered were believed to be responsible for the methemoglobinemia (Murone et al. 2005).

Allergic reactions to fennel are rare but have been reported (Bensky et al. 2004; De Smet 1992; Ottolenghi et al. 1995). Human studies have indicated an association between birch–mugwort–celery syndrome and fennel allergy (Jensen-Jarolim et al. 1997; Stager et al. 1991).

Pharmacological Considerations

Animal studies and human case reports have indicated estrogenic activity of fennel (Malini et al. 1985; Turkyilmaz et al. 2008).

An animal study indicated that the absorption of ciprofloxin was reduced with coadministration of fennel (Zhu et al. 1999).

Pregnancy and Lactation

Fennel seed is one of the most commonly used botanicals by nursing mothers, and it is also given to colicky infants. Although case reports and in vitro studies indicate some estrogenic activity of fennel, considering the widespread use and relative lack of adverse events reported, the editors of this text believe that fennel tea is safe for use during pregnancy and lactation. Fennel essential oil and alcohol extracts of fennel, however, should not be used during pregnancy or lactation (Wichtl 2004).

A review of the safety of fennel indicated that, while fennel essential oil, like many other essential oils, has been reported to cause excitation of the uterus, this effect is not likely to occur with therapeutic doses of fennel. The review also indicated that no cases of successfully self-induced abortions from fennel or fennel essential oil have been reported (De Smet 1992).

A risk-benefit analysis of fennel in infants, regarding concern for carcinogenicity of the compound estragole, concluded that, based on available clinical and epidemiological data, any cancer risk is negligible and fennel seed is safe for use in infants (Iten and Saller 2004).

Also see Adverse Events and Side Effects for this entry.

Review Details

I. Drug and Supplement Interactions

Clinical Trials of Drug or Supplement Interactions

No clinical trials of drug or supplement interactions were identified.

Case Reports of Suspected Drug or Supplement Interactions

No case reports of drug or supplement interactions were identified.

Animal Trials of Drug or Supplement Interactions

In rats orally administered ciprofloxin with or without 2 g/kg of an aqueous extract of fennel, decreases in the maximum plasma concentration, area under the curve, and urinary recovery of ciprofloxin were observed after coadministration with fennel. The volume of distribution and terminal elimination half-life of ciprofloxin were increased (Zhu et al. 1999).

An increase in pentobarbital-induced sleeping time was observed in mice intraperitoneally administered 50 mg/kg of fennel essential oil (Marcus and Lichtenstein 1982).

II. Adverse Events

Case Reports of Adverse Events

A case series of premature thelarche (isolated breast development with no other clinical signs of sexual maturation) was reported in four girls ages 5 months to 5 years who had been administered two to three cups of fennel tea daily for 4 months to 2 years. Age-specific serum estradiol levels were 15 to 20 times higher than normal. All of the girls had been breast-fed for their first 9 months, and none had a history of prolonged drug intake that would account for the elevation in hormone levels. After cessation of administration, premature thelarche resolved within 3 to 6 months and hormone levels decreased to the normal range (Turkyilmaz et al. 2008).

A case series of methemoglobinemia with respiratory distress, cyanosis, and tachycardia was reported in four infants, 7 to 9 months old, that had been fed purées of fennel (part and dose used not specified). Purées administered to two of the infants were analyzed for nitrate content and found to contain high (2550 mg/kg) levels of nitrates. In two cases, the purées were combined with carrot, a vegetable that is high in nitrates. Nitrates are the most common cause of acquired methemoglobinemia (Murone et al. 2005).

Rare cases of allergic reaction to fennel have been reported (Bensky et al. 2004; De Smet 1992; Ottolenghi et al. 1995). Allergic reactions are reported to have been associated with symptoms such as shortness of breath, facial pallor, low blood pressure, excessive sweating, accelerated pulse, and impaired consciousness (Bensky et al. 2004).

In patch testing of persons with birch-mugwort-celery syndrome, cross-sensitivity to fennel was relatively high as compared to other allergens tested (Jensen-Jarolim et al. 1997; Stager et al. 1991).

III. Pharmacology and Pharmacokinetics

Human Pharmacological Studies

In a study of women with idiopathic hirsutism (excessive male pattern hair growth in women with normal levels of serum androgens), involving topical application of a cream containing 1 or 2% fennel ethanol extract daily for 8 or 12 weeks, a dose-dependent reduction in hair diameter was observed. The authors of the article indicated that estrogenic activity of fennel was responsible for this reduction (Javidnia et al. 2003).

Animal Pharmacological Studies

In rats orally administered an acetone extract of fennel at doses of 0.5 to 2.5 mg/kg daily for 14 days, dose-dependent induction of the estrus cycle and increased weights of mammary glands, endometrium, cervix, and vagina were observed in females. In males, a decrease in total protein concentration in the testes and vas deferens and an increase in the seminal vesicles and prostate gland were observed after doses of 1.5 or 2.5 mg/kg (Malini et al. 1985).

A decrease in the weight of testes, epididymis, seminal vesicle, ventral prostate, and vas deferens was observed in male rats orally administered 250 or 500 mg/kg of fennel aqueous extract daily for 60 days. A decrease in sperm motility was also observed along with a decrease in fertility (Agarwal et al. 2006).

Antithrombotic activity was observed in mice orally administered 30 mg/kg fennel essential oil daily for 5 days. At that dose, prevention of paralysis induced by collagen-epinephrine injection was observed and did not cause any prohemorrhagic side effects (Tognolini et al. 2007).

No effects on blood glucose levels were observed in diabetic mice intraperitoneally administered fixed oil of fennel daily for 7 days (Ozbek et al. 2003).

In Vitro Pharmacological Studies

Fennel essential oil inhibited platelet aggregation induced by arachidonic acid, ADP, thromboxane A_2, and U46619 in guinea pig and rat plasma (Tognolini et al. 2006). The compound anethole, the main component of fennel essential oil, inhibited platelet aggregation induced by arachidonic acid, collagen, ADP, and U46619, and prevented thrombin-induced clot retraction in guinea pig plasma (Tognolini et al. 2007).

Studies on the effects of fennel methanol extracts on the activity of the human drug-metabolizing isoenzyme CYP3A4 have shown differing results, with one study indicating significant inhibition (Subehan et al. 2006, 2007) and another study showing no activity (Usia et al. 2006). No significant effects of a methanol extract of fennel seed were observed on the human drug-metabolizing isoenzyme CYP2D6 (Subehan et al. 2006; Usia et al. 2006).

IV. Pregnancy and Lactation

German authorities have indicated that while fennel essential oil and alcohol extracts should not be taken during pregnancy, water-based extracts of fennel are regarded as safe (Wichtl 2004).

A review of the safety of fennel indicated that, while fennel essential oil, like many other essential oils, has been reported to cause excitation of the gravid uterus, this effect has not been verified and is not likely to occur with therapeutic doses of fennel. The review also indicated that no cases of successfully self-induced abortions from fennel or fennel essential oil have been reported (De Smet 1992).

An in vitro test using limb bud cells from day 13 rat embryos indicated some evidence of toxicity of fennel essential oil to fetal cells but no evidence of teratogenicity. After 5 days of incubation in concentrations of 0.93 mg/ml and above, a significant reduction in the number of stained differentiated foci were observed. This reduction was thought to be due to cell loss rather than to a decrease in cell differentiation (Ostad et al. 2004).

The compound *trans*-anethole, isolated from fennel essential oil, exhibited dose-dependent anti-implantation activity in rats administered doses of 50 to 80 mg/kg on days 1 to 10 of pregnancy. At the 80 mg/kg dose level, administration on days 1 and 2 of pregnancy did not cause any changes in fertility, while with the same dose administered on days 3 to 5 of pregnancy, no implantation occurred. The same dose administered on days 6 to 10 of pregnancy caused a reduction in the number of pregnancies. No malformations were observed in any of the animals born from treated mothers (Dhar 1995).

In isolated rat uteri, treatment with fennel essential oil reduced the intensity of contractions induced by oxytocin and prostaglandin E_2. The oil also reduced the frequency of contractions induced by prostaglandin E_2 but not by oxytocin (Ostad et al. 2001).

A risk-benefit analysis on the use of fennel tea as a colic remedy for infants indicated that studies regarding carcinogenicity of the compound estragole in animal studies were difficult to interpret, as direct translation of animal experimental data to humans is problematic for numerous reasons. The analysis concluded that, based on available clinical and epidemiological data, any cancer risk is negligible and fennel seed is safe for use in infants (Iten and Saller 2004).

V. Toxicity Studies

Acute Toxicity

The LD_{50} of orally administered fennel oil in rats is 3.8 ml/kg, whereas that of bitter fennel oil is 5.42 ml/kg (Opdyke 1979). The dermal LD_{50} of fennel oil and bitter fennel oil in rabbits could not be determined at doses up to 5 ml/kg (Opdyke 1979). The LD_{50} of an ethanol extract of fennel orally administered in mice and rats could not be determined at doses up 3 g/kg (Shah et al. 1991; Tanira et al. 1996).

The LD_{50} of orally administered fennel essential oil in rats has also been reported as 1.3 mg/kg (Ostad et al. 2001). In mice, the intraperitoneal LD_{50} of fennel essential oil is 1.04 ml/kg whereas the oral LD_{50} is 5.52 ml/kg (Ozbek et al. 2003, 2006).

Subchronic Toxicity

No adverse morphological, hematological, or spermatogenic changes were observed in mice orally administered 100 mg/kg of an aqueous extract of fennel daily for 90 days (Shah et al. 1991).

Genotoxicity

Some mutagenic activity of a fennel extract was observed in *Salmonella typhimurium* strains TA98 and TA102 (Mahmoud et al. 1992). Fennel essential oil exhibited mutagenic activity in the Ames test with *Salmonella typhimurium* strains TA98 and TA100, with activity potentiated by metabolic activation with S13 (Marcus and Lichtenstein 1982).

No mutagenic activity of aqueous and methanolic extracts of fennel were observed in the Ames test for mutagenicity in *Salmonella typhimurium* strains TA98 or TA100 with or without metabolic activation (Morimoto et al. 1982; Yamamoto et al. 1982). No mutagenic activity of the same extract was found in the *Bacillus subtilis rec* assay (Morimoto et al. 1982).

In the *Bacillus subtilis* DNA-repair test, sweet fennel essential oil showed some mutagenic activity, while in the chromosomal aberration test in Chinese hamster fibroblast cells (Sekizawa and Shibamoto 1982), fennel essential oil showed no mutagenic activity (Ishidate et al. 1984).

F

Literature Cited

Agarwal, M., T. Gehani, B. Sharma, and A. Chauhan. 2006. Antifertility effect of *Foeniculum vulgare* seeds (aqueous extract) on the reproductive organs of male rats, *Rattus norvegicus*. *J. Exp. Zool. India* 9(2):269-274.

Bensky, D., S. Clavey, and E. Stöger. 2004. *Chinese herbal medicine: Materia medica*. 3rd ed. Seattle: Eastland Press.

De Smet, P.A.G.M. 1992. *Adverse effects of herbal drugs, Volume 1*. Berlin: Springer.

Dhar, S.K. 1995. Anti-fertility activity and hormonal profile of *trans*-anethole in rats. *Indian J. Physiol. Pharmacol.* 39(1):63-67.

ESCOP. 2003. *ESCOP monographs: The scientific foundation for herbal medicinal products*. 2nd ed. Exeter, U.K.: European Scientific Cooperative on Phytotherapy.

Ishidate, M., T. Sofuni, K. Yoshikawa, et al. 1984. Primary mutagenicity screening of food additives currently used in Japan. *Food Chem. Toxicol.* 22(8):623-636.

Iten, F., and R. Saller. 2004. Fennel tea: Risk assessment of the phytogenic monosubstance estragole in comparison to the natural multicomponent mixture. *Forsch. Komplementarmed. Klass. Nat.* 11(2):104-108.

Javidnia, K., L. Dastgheib, S. Mohammadi Samani, and A. Nasiri. 2003. Antihirsutism activity of fennel (fruits of *Foeniculum vulgare*) extract. A double-blind placebo controlled study. *Phytomedicine* 10(6-7):455-458.

Jensen-Jarolim, E., A. Leitner, R. Hirschwehr, et al. 1997. Characterization of allergens in Apiaceae spices: Anise, fennel, coriander and cumin. *Clin. Exp. Allergy* 27(11):1299-1306.

Mahmoud, I., A. Alkofahi, and A. Abdelaziz. 1992. Mutagenic and toxic activities of several spices and some Jordanian medicinal plants. *Int. J. Pharmacogn.* 30(2):81-85.

Malini, T., G. Vanithakumari, N. Megala, et al. 1985. Effect of *Foeniculum vulgare* Mill. seed extract on the genital organs of male and female rats. *Indian J. Physiol. Pharmacol.* 29(1):21-26.

Marcus, C., and E.P. Lichtenstein. 1982. Interactions of naturally occurring food plant components with insecticides and pentobarbital in rats and mice. *J Agric. Food. Chem.* 30(3):563-568.

Morimoto, I., F. Watanabe, T. Osawa, T. Okitsu, and T. Kada. 1982. Mutagenicity screening of crude drugs with *Bacillus subtilis* recassay and *Salmonella*/microsome reversion assay. *Mutat. Res.* 97(2):81.

Murone, A.J., P. Stucki, M.G. Roback, and M. Gehri. 2005. Severe methemoglobinemia due to food intoxication in infants. *Pediatr. Emerg. Care* 21(8):536-538.

Opdyke, D.L.J. 1979. *Monographs on fragrance raw materials*. New York: Pergamon.

Ostad, S.N., B. Khakinegad, and O. Sabzevari. 2004. Evaluation of the teratogenicity of fennel essential oil (FEO) on the rat embryo limb buds culture. *Toxicol. In Vitro* 18 (5):623-627.

Ostad, S.N., M. Soodi, M. Shariffzadeh, N. Khorshidi, and H. Marzban. 2001. The effect of fennel essential oil on uterine contraction as a model for dysmenorrhea, pharmacology and toxicology study. *J. Ethnopharmacol.* 76(3):299-304.

Ottolenghi, A., A. De Chiara, S. Arrigoni, L. Terracciano, and M. De Amici. 1995. Diagnosis of food allergy caused by fruit and vegetables in children with atopic dermatitis. *Pediatr. Med. Chirurg.* 17(6):525.

Ozbek, H., M. Ozturk, I. Bayram, S. Ugras, and G.S. Citoglu. 2003. Hypoglycemic and hepatoprotective effects of *Foeniculum vulgare* Miller seed fixed oil extract in mice and rats. *Eastern J. Med.* 8(2):35-40.

Ozbek, H., A. Tas, F. Ozgokce, et al. 2006. Evaluation of median lethal dose and analgesic activity of *Foeniculum vulgare* Miller essential oil. *Int. J. Pharmacol.* 2(2):181-183.

Sekizawa, J., and T. Shibamoto. 1982. Genotoxicity of safrole-related chemicals in microbial test systems. *Mutat. Res.* 101(2):127.

Shah, A.H., S. Qureshi, and A.M. Ageel. 1991. Toxicity studies in mice of ethanol extracts of *Foeniculum vulgare* fruit and *Ruta chalepensis* aerial parts. *J. Ethnopharmacol.* 34(2-3):167-172.

Stager, J., B. Wuthrich, and S.G.O. Johansson. 1991. Spice allergy in celery-sensitive patients. *Allergy* 46(6):475-478.

Subehan, T. Usia, H. Iwata, S. Kadota, and Y. Tezuka. 2006. Mechanism-based inhibition of CYP3A4 and CYP2D6 by Indonesian medicinal plants. *J. Ethnopharmacol.* 105(3):449-455.

Subehan, S.F. Zaidi, S. Kadota, and Y. Tezuka. 2007. Inhibition on human liver cytochrome P450 3A4 by constituents of fennel (*Foeniculum vulgare*): Identification and characterization of a mechanism-based inactivator. *J. Agric. Food. Chem.* 55(25):10162-10167.

Tanira, M.O.M., A.H. Shah, A. Mohsin, A.M. Ageel, and S. Qureshi. 1996. Pharmacological and toxicological investigations on *Foeniculum vulgare* dried fruit extract in experimental animals. *Phytother. Res.* 10(1):33-36.

Tognolini, M., V. Ballabeni, S. Bertoni, et al. 2007. Protective effect of *Foeniculum vulgare* essential oil and anethole in an experimental model of thrombosis. *Pharmacol. Res.* 56(3):254-260.

Tognolini, M., E. Barocelli, V. Ballabeni, et al. 2006. Comparative screening of plant essential oils: Phenylpropanoid moiety as basic core for antiplatelet activity. *Life Sci.* 78(13):1419-1432.

Turkyilmaz, Z., R. Karabulut, K. Sonmez, and A. Can Basaklar. 2008. A striking and frequent cause of premature thelarche in children: *Foeniculum vulgare*. *J. Pediatr. Surg.* 43(11):2109-2111.

Usia, T., H. Iwata, A. Hiratsuka, et al. 2006. CYP3A4 and CYP2D6 inhibitory activities of Indonesian medicinal plants. *Phytomedicine* 13(1-2):67-73.

Wichtl, M. 2004. *Herbal drugs and phytopharmaceuticals: A handbook for practice on a scientific basis*. 3rd ed. Boca Raton, FL: CRC Press.

Yamamoto, H., T. Mizutani, and H. Nomura. 1982. Studies on the mutagenicity of crude drug extracts. I. *Yakugaku Zasshi* 102(6):596-601.

Zhu, M., P.Y.K. Wong, and R.C. Li. 1999. Effect of oral administration of fennel (*Foeniculum vulgare*) on ciprofloxacin absorption and disposition in the rat. *J. Pharm. Pharmacol.* 51(12):1391-1396.

Forsythia suspensa (Thunb.) Vahl — Oleaceae

SCN: forsythia
PN: *lian qiao* (fruit)
OCN: goldenbells
Part: fruit

Quick Reference Summary

Safety Class: 1
Interaction Class: A

Contraindications

Not for use in pregnancy except under the supervision of a qualified healthcare practitioner (List and Hörhammer 1973; Roth et al. 1984).

Other Precautions

Use with caution in persons with diarrhea (Chen and Chen 2004).

Drug and Supplement Interactions

None known.

Notice

Emmenagogue (List and Hörhammer 1973; Roth et al. 1984); *see* Appendix 2.

Adverse Events and Side Effects

A case of photodermatitis after topical application of forsythia has been reported (Bensky et al. 2004).

Pharmacological Considerations

An in vitro study indicated antiplatelet activity of forsythia. The applicability of the data to human use is not known (Iwakami et al. 1992).

Pregnancy and Lactation

Although forsythia has been reported as an emmenagogue (List and Hörhammer 1973; Roth et al. 1984), no cautions for use during pregnancy are listed in the traditional Chinese medicine literature.

No information on the safety of forsythia during lactation was identified. While this review did not identify any concerns for use while nursing, safety has not been conclusively established.

Review Details

I. Drug and Supplement Interactions

Clinical Trials of Drug or Supplement Interactions

No clinical trials of drug or supplement interactions were identified.

Case Reports of Suspected Drug or Supplement Interactions

No case reports of suspected drug or supplement interactions were identified.

Animal Trials of Drug or Supplement Interactions

No animal trials of drug or supplement interactions were identified.

II. Adverse Events

Case Reports of Adverse Events

A case of photodermatitis was reported after topical application of a formula containing forsythia. Forsythia was determined to be the causative agent (Bensky et al. 2004).

III. Pharmacology and Pharmacokinetics

Human Pharmacological Studies

No relevant human pharmacological studies were identified.

Animal Pharmacological Studies

No relevant animal pharmacological studies were identified.

In Vitro Pharmacological Studies

Antiplatelet activity of an aqueous extract of forsythia was observed in rabbit platelets (Iwakami et al. 1992).

Inhibition of the drug-metabolizing isoenzyme CYP3A4 and very weak induction of CYP2E1 were observed in rats orally administered an extract of forsythia daily for 6 days. Dose and type of extract were not specified in the English language abstract (Yan et al. 2003).

IV. Pregnancy and Lactation

Although forsythia has been reported as an emmenagogue (List and Hörhammer 1973; Roth et al. 1984), no cautions for use during pregnancy are listed in the traditional Chinese medicine literature.

No information on the safety of forsythia during lactation was identified.

V. Toxicity Studies

Acute Toxicity

The LD_{50} of forsythia in mice is 172.2 g/kg after oral administration and 20.96 g/kg after intraperitoneal administration (type of extract not specified in available English language translation) (Chen and Chen 2004).

Genotoxicity

In chromosomal aberration and micronucleus assays in mice, an aqueous extract of forsythia tested positively for mutagenic activity (Yin et al. 1991).

F

Literature Cited

Bensky, D., S. Clavey, and E. Stöger. 2004. *Chinese herbal medicine: Materia medica*. 3rd ed. Seattle: Eastland Press.

Chen, J.K., and T.T. Chen. 2004. *Chinese medical herbology and pharmacology*. City of Industry, CA: Art of Medicine Press.

Iwakami, S., J.B. Wu, Y. Ebizuka, and U. Sankawa. 1992. Platelet activating factor (PAF) antagonists contained in medicinal plants: Lignans and sesquiterpenes. *Chem. Pharm. Bull. (Tokyo)* 40(5):1196-1198.

List, P.H., and H. Hörhammer. 1973. *Hagers handbuch der pharmazeutischen praxis*. Berlin: Springer.

Roth, L., M. Daunderer, and K. Kormann. 1984. *Giftpflanzen-pflanzengifte: Vorkommen, wirkung, therapie*. Landsberg: Ecomed.

Yan, S.L., J.P. Hu, Y.X. Xu, and J.N. Zhang. 2003. Simultaneous evaluation of activity of hepatic CYP450 following administration of Fisch Fructus Forsythiae by cocktail probe drugs. *Chin. Pharm. J.* 38(10):761-763.

Yin, X.J., D.X. Liu, H.C. Wang, and Y. Zhou. 1991. A study on the mutagenicity of 102 raw pharmaceuticals used in Chinese traditional medicine. *Mutat. Res.* 260(1):73-82.

Fouquieria splendens Engelm. — Fouquieriaceae

SCN: ocotillo

Part: stem

Quick Reference Summary

Safety Class: 2b

Interaction Class: A

Contraindications

Not for use in pregnancy except under the supervision of a qualified healthcare practitioner (Moore 1990).

Other Precautions

None known.

Drug and Supplement Interactions

None known.

Adverse Events and Side Effects

None known.

Pharmacological Considerations

None known.

Pregnancy and Lactation

A reference text on medicinal plants of the southwestern United States indicates that ocotillo is not recommended for use in pregnancy (Moore 1990). Based on this information, use during pregnancy is not recommended except under the supervision of a qualified healthcare practitioner.

No information on the safety of ocotillo during lactation was identified. While this review did not identify any concerns for use while nursing, safety has not been conclusively established.

Review Details

I. Drug and Supplement Interactions

Clinical Trials of Drug or Supplement Interactions

No clinical trials of drug or supplement interactions were identified.

Case Reports of Suspected Drug or Supplement Interactions

No case reports of suspected drug or supplement interactions were identified.

Animal Trials of Drug or Supplement Interactions

No animal trials of drug or supplement interactions were identified.

II. Adverse Events

Case Reports of Adverse Events

No case reports of adverse events were identified.

III. Pharmacology and Pharmacokinetics

Human Pharmacological Studies

No relevant human pharmacological studies were identified.

Animal Pharmacological Studies

No relevant animal pharmacological studies were identified.

In Vitro Pharmacological Studies

No relevant in vitro pharmacological studies were identified.

IV. Pregnancy and Lactation

A reference text on medicinal plants of the southwestern United States indicates that ocotillo is not recommended for use in pregnancy (Moore 1990). No details on this concern are provided.

No information on the safety of ocotillo during lactation was identified.

V. Toxicity Studies

No toxicity studies were identified.

Literature Cited

Moore, M. 1990. *Los remedios: Traditional herbal remedies of the Southwest*. Santa Fe, NM: Red Crane Books.

Fragaria spp. — Rosaceae

Fragaria vesca **L.**
SCN: strawberry
OCN: alpine strawberry

Fragaria virginiana **Duchesne**
SCN: strawberry
OCN: Virginian strawberry
Part: leaf

Quick Reference Summary

Safety Class: 1
Interaction Class: A

Contraindications

None known.

Other Precautions

Use with caution in persons with allergy to strawberry fruit (Wichtl 2004).

Drug and Supplement Interactions

None known.

Adverse Events and Side Effects

None known.

Pharmacological Considerations

None known.

Pregnancy and Lactation

No information on the safety of strawberry leaf in pregnancy or lactation was identified in the scientific or traditional literature. Although this review did not identify any concerns for use while pregnant or nursing, safety has not been conclusively established.

Review Details

I. Drug and Supplement Interactions

Clinical Trials of Drug or Supplement Interactions

No clinical trials of drug or supplement interactions were identified.

Case Reports of Suspected Drug or Supplement Interactions

No case reports of suspected drug or supplement interactions were identified.

Animal Trials of Drug or Supplement Interactions

No animal trials of drug or supplement interactions were identified.

II. Adverse Events

Case Reports of Adverse Events

No case reports of adverse events were identified.

III. Pharmacology and Pharmacokinetics

Human Pharmacological Studies

No relevant human pharmacological studies were identified.

Animal Pharmacological Studies

No relevant animal pharmacological studies were identified.

In Vitro Pharmacological Studies

No relevant in vitro pharmacological studies were identified.

IV. Pregnancy and Lactation

No information on the safety of strawberry leaf during pregnancy or lactation was identified.

V. Toxicity Studies

No toxicity studies were identified.

Literature Cited

Wichtl, M. 2004. *Herbal drugs and phytopharmaceuticals: A handbook for practice on a scientific basis*. 3rd ed. Boca Raton, FL: CRC Press.

F

Frangula alnus Mill. — Rhamnaceae

SCN: frangula
Syn: *Rhamnus frangula* L.
OCN: alder buckthorn; buckthorn
Part: bark

Quick Reference Summary

Safety Class: 2b, 2c, 2d
Interaction Class: A

Contraindications

Not for use in pregnancy or lactation except under the supervision of a qualified healthcare practitioner (Bradley 1992; Chadha 1988; Wichtl 2004).

Contraindicated with intestinal obstruction, abdominal pain of unknown origin, or any inflammatory condition of the intestines (i.e., appendicitis, colitis, Crohn's disease, irritable bowel syndrome, and melanosis coli) (Bradley 1992; De Smet 1993; Wichtl 2004).

Not for long-term use in excess of 8 consecutive days (Bradley 1992; De Smet 1993; Leung and Foster 1996; Weiss and Meuss 2001; Wichtl 2004).

Not for use in children younger than 12 years of age (Bradley 1992; De Smet 1993).

Other Precautions

None known.

Drug and Supplement Interactions

See Pharmacological Considerations below.

Notice

Stimulant laxative (Bradley 1992; ESCOP 2003; Felter and Lloyd 1898; Leung and Foster 1996; Weiss and Meuss 2001; Wichtl 2004); *see* Appendix 2.

Standard Dose

The individually correct dose is the smallest dose necessary to produce a comfortable, soft stool (ESCOP 2003). Typically, this is 2 g (a scant teaspoon) infused as tea (Wichtl 2004).

Editors' Notes

The bark of frangula must be aged for 1 to 2 years prior to use to destroy emetic compounds (De Smet 1993; Weiss and Meuss 2001).

The American Herbal Products Association has established a trade requirement (AHPA 2011) that products containing this herb in sufficient quantity to warrant such labeling bear the following label statement:

NOTICE: Do not use this product if you have abdominal pain or diarrhea. Consult a health care provider prior to use if you are pregnant or nursing. Discontinue use in the event of diarrhea or watery stools. Do not exceed recommended dose. Not for long-term use.

Adverse Events and Side Effects

Discoloration of the urine by frangula metabolites may occur but is not clinically significant. Abdominal spasms and pain have been reported (ESCOP 2003).

Pharmacological Considerations

Concomitant use of frangula is cautioned with antiarrhythmic drugs and botanicals containing cardiac glycosides, as long-term use of frangula as a laxative can cause potassium loss, leading to increased toxicity of these drugs and botanicals (Brinker 2001; De Smet 1993; ESCOP 2003).

Concomitant internal use of frangula is cautioned with thiazide diuretics, corticosteroids, and licorice, as long-term use of frangula as a laxative may increase the potassium loss induced by these drugs and botanicals (Brinker 2001; De Smet 1993; ESCOP 2003).

Use of stimulant laxatives, such as frangula, may reduce the gastrointestinal transit time and thus reduce the absorption of orally administered drugs (Brinker 2001; De Smet 1993).

Pregnancy and Lactation

While most stimulant laxatives have traditionally been contraindicated in pregnancy due to concerns regarding stimulation of the uterus, a number of stimulant laxatives, including frangula, have shown a lack of adverse effects on pregnancy or the fetus when used according to the recommended dosage schedule (De Smet 1993; ESCOP 2003). Thus, these laxatives are now considered appropriate for use during pregnancy (De Smet 1993; ESCOP 2003; Prather 2004). Due to the potential genotoxicity of certain anthraquinones, however, it is recommended that use of certain anthranoid laxatives, including frangula, should be avoided in the first trimester of pregnancy or used under professional supervision (ESCOP 2003).

No information on the safety of frangula during lactation was identified. While this review did not identify any concerns for use while nursing, safety has not been conclusively established.

Review Details

I. Drug and Supplement Interactions

Clinical Trials of Drug or Supplement Interactions

No clinical trials of drug or supplement interactions were identified.

Case Reports of Suspected Drug or Supplement Interactions

No case reports of suspected drug or supplement interactions were identified.

Animal Trials of Drug or Supplement Interactions

No animal trials of drug or supplement interactions were identified.

II. Adverse Events

Case Reports of Adverse Events

No case reports of adverse events were identified.

III. Pharmacology and Pharmacokinetics

Human Pharmacological Studies

No relevant human pharmacological studies were identified.

Animal Pharmacological Studies

No relevant animal pharmacological studies were identified.

In Vitro Pharmacological Studies

No relevant in vitro pharmacological studies were identified.

IV. Pregnancy and Lactation

While most stimulant laxatives have traditionally been contraindicated in pregnancy due to concerns regarding stimulation of the uterus, a number of stimulant laxatives, including frangula, have shown a lack of adverse effects on pregnancy or the fetus when used according to the recommended dosage schedule (De Smet 1993; ESCOP 2003). Thus, these laxatives are now considered appropriate for use during pregnancy (De Smet 1993; ESCOP 2003; Prather 2004). Due to the potential genotoxicity of certain anthraquinones, however, it is recommended that use of certain anthranoid laxatives, including frangula, should be avoided in the first trimester of pregnancy or used under professional supervision (ESCOP 2003).

Several herbal texts caution against the use of anthraquinone-containing botanicals in pregnancy due to mutagenic activity of some anthraquinone compounds, but these texts also note that animal tests of various anthraquinones show no teratogenic or embryotoxic activity (Brinker 2001; De Smet 1993; ESCOP 2003).

No information on the safety of frangula during lactation was identified.

V. Toxicity Studies

Genotoxicity

Dose-dependent genotoxic activity of an alcohol extract of frangula was observed in the *Salmonella* microsome mutagen test and the DNA repair test of primary rat hepatocytes (Helmholz et al. 1993).

Literature Cited

AHPA. July 2011. Code of Ethics & Business Conduct. Silver Spring, MD: American Herbal Products Association.

Bradley, P.R. 1992. *British herbal compendium: A handbook of scientific information on widely used plant drugs*. Bournemouth, UK: British Herbal Medicine Association.

Brinker, F. 2001. *Herb contraindications and drug interactions*. 3rd ed. Sandy, OR: Eclectic Medical Publications.

Chadha, Y. 1988. *The wealth of India: A dictionary of Indian raw materials and industrial products*. Delhi: Council of Scientific and Industrial Research.

De Smet, P.A.G.M. 1993. *Adverse effects of herbal drugs, Volume 2*. Berlin: Springer.

ESCOP. 2003. *ESCOP monographs: The scientific foundation for herbal medicinal products*. 2nd ed. Exeter, U.K.: European Scientific Cooperative on Phytotherapy.

Felter, H.W., and J.U. Lloyd. 1898. *King's American dispensatory*. 18th ed., 3rd rev. 2 vols. Cincinnati: Ohio Valley Co.

Helmholz, H., A. Ruge, A. Piasecki, S. Schroder, and J. Westendorf. 1993. Genotoxicity of buckthorn bark. *Pharm. Ztg.* 138(Oct):48-50.

Leung, A.Y., and S. Foster. 1996. *Encyclopedia of common natural ingredients used in food, drugs, and cosmetics*. 2nd ed. New York: Wiley.

Weiss, R.F., and A.R. Meuss. 1998. *Weiss's herbal medicine*. Classic ed. New York: Thieme.

Wichtl, M. 2004. *Herbal drugs and phytopharmaceuticals: A handbook for practice on a scientific basis*. 3rd ed. Boca Raton, FL: CRC Press.

Williamson, E.M. 2003. *Potter's herbal cyclopedia*. Saffron Walden, Essex: C.W. Daniel Co.

Frangula purshiana (DC.) J.G. Cooper — Rhamnaceae

SCN: cascara sagrada
Syn: *Rhamnus purshiana* DC.
OCN: bearberry; cascara; chittem bark; sacred bark
Part: bark

Quick Reference Summary

Safety Class: 2b, 2c, 2d
Interaction Class: A

Contraindications

Not for use in pregnancy or lactation except under the supervision of a qualified healthcare practitioner (Bradley 1992; Chadha 1988; Wichtl 2004).

Contraindicated with intestinal obstruction, abdominal pain of unknown origin, or any inflammatory condition of the intestines (i.e., appendicitis, colitis, Crohn's disease, irritable bowel syndrome, and melanosis coli) (Bradley 1992; De Smet 1993; Martindale and Reynolds 1996; Wichtl 2004).

Not for use in children younger than 12 years of age (Bradley 1992; De Smet 1993).

Not for use in excess of 8 days unless under the supervision of a qualified healthcare practitioner (Bradley 1992; Chadha 1988; De Smet 1993; Leung and Foster 1996; Wichtl 2004).

Other Precautions

None known.

Drug and Supplement Interactions

See Pharmacological Considerations below.

Standard Dose

The individually correct dose is the smallest dose necessary to produce a comfortable, soft stool (ESCOP 2003). Typical adult dosages of cascara sagrada bark or preparations provide the equivalent of 0.6 to 2 g of the bark (Osol and Farrar 1955). An infusion made from 1.5 to 2 g of bark in 150 ml of water is also used (Hansel et al. 1994).

Notice

Stimulant laxative (Bradley 1992; Chadha 1988; De Smet 1993; ESCOP 2003; Felter and Lloyd 1898; Leung and Foster 1996; List and Hörhammer 1973; Martindale and Reynolds 1996; Wichtl 2004; Williamson 2003); *see* Appendix 2.

Editors' Notes

The American Herbal Products Association has established a trade requirement (AHPA 2011) that products containing this herb in sufficient quantity to warrant such labeling bear the following label statement:

> NOTICE: Do not use this product if you have abdominal pain or diarrhea. Consult a health care provider prior to use if you are pregnant or nursing. Discontinue use in the event of diarrhea or watery stools. Do not exceed recommended dose. Not for long-term use.

The bark of cascara sagrada must be aged for 1 year or heat-treated prior to use (De Smet 1993; Wichtl 2004).

Adverse Events and Side Effects

Discoloration of the urine by cascara sagrada metabolites may occur but is not clinically significant. Abdominal spasms and pain have been reported (ESCOP 2003).

F

One case of cholestatic hepatitis was reported in association with cascara sagrada use, although the patient was taking other medications that could have caused the hepatitis (Nadir et al. 2000). Occupational asthma due to cascara sagrada exposure has been confirmed (Giavina-Bianchi et al. 1997).

Pharmacological Considerations

Concomitant use of cascara sagrada is cautioned with antiarrhythmic drugs and botanicals containing cardiac glycosides, as long-term use of cascara sagrada as a laxative can cause potassium loss, leading to increased toxicity of these drugs and botanicals (Brinker 2001; De Smet 1993; ESCOP 2003; Mills and Bone 2005).

Concomitant internal use of cascara sagrada is cautioned with thiazide diuretics, corticosteroids, and licorice, and long-term use of cascara sagrada as a laxative may increase the potassium loss induced by these drugs and botanicals (Brinker 2001; ESCOP 2003; Mills and Bone 2005).

Use of stimulant laxatives, such as cascara sagrada, may reduce the gastrointestinal transit time and thus reduce the absorption of orally administered drugs (Brinker 2001; Mills and Bone 2005).

Pregnancy and Lactation

While most stimulant laxatives have traditionally been contraindicated in pregnancy due to concerns regarding stimulation of the uterus, a number of stimulant laxatives, including cascara sagrada, have shown a lack of adverse effects on pregnancy or the fetus when used according to the recommended dosage schedule (De Smet 1993; ESCOP 2003). Thus, these laxatives are now considered appropriate for use during pregnancy (De Smet 1993; ESCOP 2003; Prather 2004). Due to the potential genotoxicity of certain anthraquinones, however, it is recommended that use of certain anthranoid laxatives, including cascara sagrada, should be avoided in the first trimester of pregnancy or used under professional supervision (ESCOP 2003).

In one large prospective study, a statistical association between cascara sagrada use in pregnancy and an increased number of benign tumors in infants was reported, although the significance of that association is not known (Heinonen et al. 1977).

The American Academy of Pediatrics lists cascara sagrada as being compatible with breast-feeding (AAP 2001).

Review Details

I. Drug and Supplement Interactions

Clinical Trials of Drug or Supplement Interactions

No clinical trials of drug or supplement interactions were identified.

Case Reports of Suspected Drug or Supplement Interactions

No case reports of suspected drug or supplement interactions were identified.

Animal Trials of Drug or Supplement Interactions

No animal trials of drug or supplement interactions were identified.

II. Adverse Events

Case Reports of Adverse Events

A 48-year-old man developed cholestatic hepatitis, complicated by portal hypertension, after having taken single 425 mg capsules of aged cascara sagrada three times daily for 3 days. Other medications included amitriptyline, cimetidine, and baclofen (Nadir et al. 2000). A*mitriptyline* has been associated with rare but severe incidences of *hepatotoxicity, notably, cholestatic injury* (Saeian and Rajender-Reddy 2003; Wen et al. 2008).

A case of IgE-mediated occupational asthma and rhinitis, confirmed by skin testing, was reported in a pharmacy worker (Giavina-Bianchi et al. 1997).

III. Pharmacology and Pharmacokinetics

Human Pharmacological Studies

No relevant human pharmacological studies were identified.

Animal Pharmacological Studies

No induction of aberrant crypt foci (considered putative preneoplastic lesions) or increase in incidence of chemically induced aberrant crypt foci were observed in rats fed diets containing up to 0.1% anthraquinone glycosides from cascara sagrada for 56 days (Mereto et al. 1996).

In Vitro Pharmacological Studies

No relevant in vitro pharmacological studies were identified.

IV. Pregnancy and Lactation

In a prospective study on maternal drug use in pregnancy, 53 of 50,282 mothers took cascara sagrada during the first trimester of pregnancy, and 188 took cascara sagrada anytime in pregnancy. An association between cascara sagrada use and benign tumors in infants was observed in the statistical analyses, although the significance of this association is not known (Heinonen et al. 1977).

The American Academy of Pediatrics lists cascara sagrada as being compatible with breast-feeding (AAP 2001).

V. Toxicity Studies

No toxicity studies on cascara sagrada were identified.

Literature Cited

AAP. 2001. Transfer of drugs and other chemicals into human milk. American Academy of Pediatrics. *Pediatrics* 108(3):776-789.

AHPA. July 2011. Code of Ethics & Business Conduct. Silver Spring, MD: American Herbal Products Association.

Bradley, P.R. 1992. *British herbal compendium: A handbook of scientific information on widely used plant drugs*. Bournemouth, UK: British Herbal Medicine Association.

Brinker, F. 2001. *Herb contraindications and drug interactions*. 3rd ed. Sandy, OR: Eclectic Medical Publications.

Chadha, Y. 1988. *The wealth of India: A dictionary of Indian raw materials and industrial products*. Delhi: Council of Scientific and Industrial Research.

De Smet, P.A.G.M. 1993. *Adverse effects of herbal drugs, Volume 2*. Berlin: Springer.

ESCOP. 2003. *ESCOP monographs: The scientific foundation for herbal medicinal products*. 2nd ed. Exeter, U.K.: European Scientific Cooperative on Phytotherapy.

Felter, H.W., and J.U. Lloyd. 1898. *King's American dispensatory*. 18th ed., 3rd rev. 2 vols. Cincinnati: Ohio Valley Co.

Giavina-Bianchi, P.F., F.F.M. Castro, M.L.S. Machado, and A.J.S. Duarte. 1997. Occupational respiratory allergic disease induced by *Passiflora alata* and *Rhamnus purshiana*. *Ann. Allergy Asthma Immunol.* 79(5):449-454.

Hansel, R., K. Keller, H. Rimpler, and G. Schneider. 1994. *Rhamnus*. In *Hagers handbuch der pharmazeutischen praxis*. Berlin: Springer.

Heinonen, O.P., D. Slone, and S. Shapiro. 1977. *Birth defects and drugs in pregnancy*. Boston: PSG Inc.

Leung, A.Y., and S. Foster. 1996. *Encyclopedia of common natural ingredients used in food, drugs, and cosmetics*. 2nd ed. New York: Wiley.

List, P.H., and H. Hörhammer. 1973. *Hagers handbuch der pharmazeutischen praxis*. Berlin: Springer.

Martindale, W., and J.E.F. Reynolds. 1996. *The extra pharmacopoeia*. 31st ed. London: Pharmaceutical Press.

Mereto, E., M. Ghia, and G. Brambilla. 1996. Evaluation of the potential carcinogenic activity of senna and cascara glycosides for the rat colon. *Cancer Lett.* 101(1):79-83.

Mills, S., and K. Bone. 2005. *The essential guide to herbal safety*. St. Louis: Elsevier.

Nadir, A., D. Reddy, and D.H. Van Thiel. 2000. Cascara sagrada-induced intrahepatic cholestasis causing portal hypertension: Case report and review of herbal hepatotoxicity. *Am. J. Gastroenterol.* 95(12):3634-3637.

Osol, A., and G. Farrar. 1955. *The dispensatory of the United States of America*. 25th ed. Philadelphia: Lippincott.

Saeian, K., and K. Rajender-Reddy. 2003. Hepatotoxicity of psychotropic drugs and drugs of abuse. In *Drug-induced liver disease*, edited by Kaplowitz, N. and L. DeLeve. New York: Marcel Dekker.

Wen, B., L. Ma, and M. Zhu. 2008. Bioactivation of the tricyclic antidepressant amitriptyline and its metabolite nortriptyline to arene oxide intermediates in human liver microsomes and recombinant P450s. *Chem. Biol. Interact.* 173(1):59-67.

Wichtl, M. 2004. *Herbal drugs and phytopharmaceuticals: A handbook for practice on a scientific basis*. 3rd ed. Boca Raton, FL: CRC Press.

Williamson, E.M. 2003. *Potter's herbal cyclopedia*. Saffron Walden, Essex: C.W. Daniel Co.

Fraxinus americana L. — Oleaceae

SCN: white ash

Part: bark

Quick Reference Summary

Safety Class: 2b
Interaction Class: A

Contraindications

Not for use in pregnancy except under the supervision of a qualified healthcare practitioner (Moerman 1998; Rousseau 1947).

Other Precautions

None known.

Drug and Supplement Interactions

None known.

Adverse Events and Side Effects

None known.

Pharmacological Considerations

None known.

Pregnancy and Lactation

Ethnobotanical references indicate that white ash was traditionally used as an abortifacient (Moerman 1998; Rousseau 1947). No modern data were identified to support or refute this activity. Based on this information, use during pregnancy is not recommended except under the supervision of a qualified healthcare practitioner.

Review Details

I. Drug and Supplement Interactions

Clinical Trials of Drug or Supplement Interactions

No clinical trials of drug or supplement interactions were identified.

Case Reports of Suspected Drug or Supplement Interactions

No case reports of suspected drug or supplement interactions were identified.

Animal Trials of Drug or Supplement Interactions

No animal trials of drug or supplement interactions were identified.

II. Adverse Events

Case Reports of Adverse Events

No case reports of adverse events were identified.

III. Pharmacology and Pharmacokinetics

Human Pharmacological Studies

No relevant human pharmacological studies were identified.

Animal Pharmacological Studies

No relevant animal pharmacological studies were identified.

In Vitro Pharmacological Studies

No relevant in vitro pharmacological studies were identified.

IV. Pregnancy and Lactation

No information on the safety of white ash during pregnancy or lactation was identified.

V. Toxicity Studies

No toxicity studies were identified.

Literature Cited

Moerman, D.E. 1998. *Native American ethnobotany*. Portland, OR: Timber Press.

Rousseau, J. 1947. Ethnobotanique Abénakise. *Arch. Folklore* 11:145-182.

Fraxinus excelsior L. — Oleaceae

SCN: European ash

Part: bark

Quick Reference Summary

Safety Class: 1
Interaction Class: A

Contraindications

None known.

Other Precautions

None known.

Drug and Supplement Interactions

None known.

Editors' Note

No cautions for use of European ash are reported in historical medical texts and articles (Bentley and Trimen 1880; Hayes 1853).

Adverse Events and Side Effects

None known.

Pharmacological Considerations

None known.

Pregnancy and Lactation

No information on the safety of European ash in pregnancy or lactation was identified in the scientific or traditional literature. Although this review did not identify any concerns for use while pregnant or nursing, safety has not been conclusively established.

Review Details

I. Drug and Supplement Interactions

Clinical Trials of Drug or Supplement Interactions

No clinical trials of drug or supplement interactions were identified.

Case Reports of Suspected Drug or Supplement Interactions

No case reports of suspected drug or supplement interactions were identified.

Animal Trials of Drug or Supplement Interactions

No animal trials of drug or supplement interactions were identified.

II. Adverse Events

Case Reports of Adverse Events

No case reports of adverse events were identified.

III. Pharmacology and Pharmacokinetics

Human Pharmacological Studies

No relevant human pharmacological studies were identified.

Animal Pharmacological Studies

No relevant animal pharmacological studies were identified.

In Vitro Pharmacological Studies

No relevant in vitro pharmacological studies were identified.

IV. Pregnancy and Lactation

No information on the safety of European ash during pregnancy or lactation was identified.

V. Toxicity Studies

No toxicity studies were identified.

F

Literature Cited

Bentley, R., and H. Trimen. 1880. *Medicinal plants. Vol III, Compositae to Thymelaceae.* London: J.A. Churchill.

Hayes, I. 1853. Therapeutic use of the bark, leaves, seeds, and root of the common Ash (Fraxinus excelsior), Quarterly Summaries of the Improvements and Discoveries in the Medical Sciences. *Am. J. Med. Sci.* 25:492-494.

Fritillaria spp.

Liliaceae

Fritillaria cirrhosa D. Don
SCN: Sichuan fritillary
PN: *chuan bei mu* (bulb)
OCN: tendril-leaf fritillary

Fritillaria thunbergii Miq.
SCN: Zhejiang fritillary
PN: *zhe bei mu* (bulb)
Part: bulb

Quick Reference Summary

Safety Class: 1
Interaction Class: A

Contraindications

None known.

Other Precautions

None known.

Drug and Supplement Interactions

None known.

Editors' Notes

Other species of *Fritillaria* are considered acceptable substitutes for the species listed above. These include *F. unibracteata, F. przewalskii,* and *F. delavayi.* Many other species of *Fritillaria* are used locally, including *F. hupehensis, F. anhuiensis, F. monantha, F. karelinii, F. maximowiczii,* and *F. davidii* (Bensky et al. 2004).

Adulteration with several species has been reported, including *Bolbostemma paniculatum, Tulipa edulis, Cremastra appendiculata,* and the highly toxic *Iphigenia indica* (Bensky et al. 2004).

Adverse Events and Side Effects

None known.

Pharmacological Considerations

None known.

Pregnancy and Lactation

No information on the safety of Sichuan fritillary or Zhejiang fritillary in pregnancy or lactation was identified in the scientific or traditional literature. Although this review did not identify any concerns for use while pregnant or nursing, safety has not been conclusively established.

Review Details

I. Drug and Supplement Interactions

Clinical Trials of Drug or Supplement Interactions

No clinical trials of drug or supplement interactions were identified.

Case Reports of Suspected Drug or Supplement Interactions

No case reports of suspected drug or supplement interactions were identified.

Animal Trials of Drug or Supplement Interactions

No animal trials of drug or supplement interactions were identified.

II. Adverse Events

Case Reports of Adverse Events

No case reports of adverse events were identified.

III. Pharmacology and Pharmacokinetics

Human Pharmacological Studies

No relevant human pharmacological studies were identified.

Animal Pharmacological Studies

No relevant animal pharmacological studies were identified.

In Vitro Pharmacological Studies

Alkaloids isolated from Sichuan fritillary or Zhejiang fritillary have shown antihypertensive and anticholinergic activity (Li et al. 2006).

IV. Pregnancy and Lactation

No information on the safety of Sichuan fritillary or Zhejiang fritillary during pregnancy or lactation was identified.

V. Toxicity Studies

Acute Toxicity

The LD_{50} of an extract of Sichuan fritillary orally administered in mice could not be determined at doses up to 8 g/kg (Chen and Chen 2004).

The LD_{50} of the compound peiminine, isolated from Zhejiang fritillary and intravenously administered, is 11 mg/kg in rabbits and 9 mg/kg in cats (Chen and Chen 2004).

Literature Cited

Bensky, D., S. Clavey, and E. Stöger. 2004. *Chinese herbal medicine: Materia medica*. 3rd ed. Seattle: Eastland Press.

Chen, J.K., and T.T. Chen. 2004. *Chinese medical herbology and pharmacology*. City of Industry, CA: Art of Medicine Press.

Li, H.J., Y. Jiang, and P. Li. 2006. Chemistry, bioactivity and geographical diversity of steroidal alkaloids from the Liliaceae family. *Nat. Prod. Rep.* 23(5):735-752.

Fucus vesiculosus L. — Fucaceae

SCN: bladderwrack
OCN: dyer's fucus; red fucus; rockwrack

Part: thallus

Quick Reference Summary

Safety Class: 2d
Interaction Class: A

Contraindications

Not for use in persons with hyperthyroidism (Bradley 2006; Weiss and Meuss 2001).

Other Precautions

None known.

Drug and Supplement Interactions

None known, although use with thyroid medications should be under the supervision of a qualified healthcare practitioner (Bradley 2006).

Other drugs should be taken 1 hour prior to consumption of bladderwrack or several hours after consumption, as mucilaginous plants such as bladderwrack may slow the absorption of orally administered drugs (Brinker 2001; De Smet 1993; Mills and Bone 2005).

Editors' Notes

A number of seaweeds have been found to contain heavy metal residues (Almela et al. 2006; Rose et al. 2007). Total arsenic has been measured in bladderwrack at levels ranging from 20 to 40 mg/kg (Almela et al. 2006), though the inorganic form of arsenic is reported at much lower levels of 0.2–1.8 mg/kg (Almela et al. 2006). Evidence about the negative impact of inorganic arsenic on fetal health and infant development is emerging (EFSA 2009). Caution is therefore recommended in pregnancy and for young children.

Adverse Events and Side Effects

None known.

Pharmacological Considerations

Bladderwrack contains the element iodine (0.03–0.2% dry weight) (Bradley 2006; Chapman 1970; Wichtl 2004). A number of cautions related to the use of bladderwrack in persons with thyroid conditions are listed in the European literature. These cautions are based primarily on the presence of iodine. Reference texts note that iodine idiosyncrasy, hyperthyroidism, and thyrotoxicosis have been indicated as possible side effects after long-term, uncontrolled use of bladderwrack (Wichtl 2004), and excessive consumption of iodine can lead to goiter (Baker 2004; Pennington 1990). No data exist to support or refute the clinical relevance of these concerns to bladderwrack use.

Pregnancy and Lactation

Although iodine is an essential nutrient that is required for proper prenatal development, infants are reported to be particularly sensitive to the effects of iodine (Baker 2004), and cases of infant goiter after maternal use of iodine during pregnancy or lactation have been reported (Pennington 1990). *Also see* Editors' Notes for this entry.

F

Review Details

I. Drug and Supplement Interactions

Clinical Trials of Drug or Supplement Interactions

No clinical trials of drug or supplement interactions were identified.

Case Reports of Suspected Drug or Supplement Interactions

No case reports of suspected drug or supplement interactions were identified.

Animal Trials of Drug or Supplement Interactions

No animal trials of drug or supplement interactions were identified.

II. Adverse Events

Case Reports of Adverse Events

Kidney toxicity was diagnosed in a woman who had been taking a bladderwrack supplement found to contain high levels (21.3 mg/kg) of arsenic. The woman had been taking 1.2 g of the supplement daily for 3 months and presented with polydipsia, polyuria, proteinuria, and acute kidney injury. A renal biopsy revealed tubular degeneration and lymphomonocytic infiltration, which were attributed to the effects of arsenic (Conz et al. 1998).

III. Pharmacology and Pharmacokinetics

Human Pharmacological Studies

In a pilot study in three premenopausal women with abnormal menstrual cycling patterns and/or menstrual-related disease histories, intake of bladderwrack was associated with increases in menstrual cycle lengths, ranging from an increase of 5.5 to 14 days. Bladderwrack was taken in doses of 0.7 g daily for two menstrual cycles and 1.4 g daily for two additional cycles. Hormone testing in one woman indicated a reduction in 17β-estradiol and an increase in progesterone (Skibola 2004).

Animal Pharmacological Studies

No relevant animal pharmacological studies were identified.

In Vitro Pharmacological Studies

Dose-dependent anticoagulant activity of fucoidan fractions from bladderwrack was observed in human platelet-rich plasma treated with concentrations of 5, 10, and 50 μg/ml (Durig et al. 1997).

IV. Pregnancy and Lactation

Infants are reported to be particularly sensitive to the effects of iodine (Baker 2004), and cases of infant goiter after maternal use of iodine during pregnancy or lactation have been reported (Pennington 1990).

Also see Editors' Notes for this entry.

V. Toxicity Studies

Acute Toxicity

The LD_{50} of orally administered bladderwrack extracts in rats is between 1 and 2 g/kg for a 33% ethanol extract and could not be determined at doses up to 2 g/kg for a 60% ethanol extract. In mice, the oral LD_{50} values are between 1 and 2 g/kg for a 33% ethanol extract and approximately 0.75 g/kg for a 60% ethanol extract. After intraperitoneal administration, the LD_{50} values for a 33% ethanol extract are 0.25 g/kg for female rats, between 1 and 2 g/kg for male rats, and between 0.15 and 0.2 g/kg in mice. For a 60% ethanol extract, the LD_{50} values are 0.5 g/kg in female rats, over 0.5 g/kg in male rats, and 0.25 to 0.5 g/kg in mice (Zaragoza et al. 2008).

Short-Term Toxicity

No signs of toxicity, including relevant changes in organ weights or serum parameters, were observed in rats orally administered 200 or 750 mg/kg of a hydroethanolic extract of bladderwrack daily for 28 days (Zaragoza et al. 2008).

Genotoxicity

No genotoxic activity of an aqueous extract of bladderwrack was observed in chromosome aberration and comet assays in human lymphocytes treated with concentrations of 0.25, 0.5, or 1.0 mg/ml (Leite-Silva et al. 2007).

Literature Cited

Almela, C., M.J. Clemente, D. Velez, and R. Montoro. 2006. Total arsenic, inorganic arsenic, lead and cadmium contents in edible seaweed sold in Spain. *Food Chem. Toxicol.* 44(11):1901-1908.

Baker, D.H. 2004. Iodine toxicity and its amelioration. *Exp. Biol. Med.* 229(6):473-478.

Bradley, P.R. 2006. *British herbal compendium: A handbook of scientific information on widely used plant drugs, Volume 2*. Bournemouth, UK: British Herbal Medicine Association.

Brinker, F. 2001. *Herb contraindications and drug interactions*. 3rd ed. Sandy, OR: Eclectic Medical Publications.

Chapman, V.J. 1970. *Seaweeds and their uses*. London: Methuen Co. Ltd.

Conz, P.A., G. La Greca, P. Benedetti, P.A. Bevilacqua, and L. Cima. 1998. *Fucus vesiculosus*: A nephrotoxic alga? *Nephrol. Dial. Transplant.* 13(2):526-527.

De Smet, P.A.G.M. 1993. *Adverse effects of herbal drugs, Volume 2*. New York: Springer.

Durig, J., T. Bruhn, K.H. Zurborn, et al. 1997. Anticoagulant fucoidan fractions from *Fucus vesiculosus* induce platelet activation in vitro. *Thromb. Res.* 85(6):479-491.

EFSA. 2009. Scientific opinion on arsenic in food. *EFSA J.* 7(10):1351-1550.

Leite-Silva, C., C.L.S. Gusmao, and C.S. Takahashi. 2007. Genotoxic and antigenotoxic effects of *Fucus vesiculosus* extract on cultured human lymphocytes using the chromosome aberration and comet assays. *Genet. Mol. Biol.* 30(1):105-111.

Mills, S., and K. Bone. 2005. *The essential guide to herbal safety*. St. Louis: Elsevier.

Pennington, J.A. 1990. A review of iodine toxicity reports. *J. Am. Diet. Assoc.* 90(11):1571.

Rose, M., J. Lewis, N. Langford, et al. 2007. Arsenic in seaweed—Forms, concentration and dietary exposure. *Food Chem. Toxicol.* 45(7):1263-1267.

Skibola, C.F. 2004. The effect of *Fucus vesiculosus*, an edible brown seaweed, upon menstrual cycle length and hormonal status in three pre-menopausal women: A case report. *BMC Complement. Altern. Med.* 4:10.

Weiss, R.F., and A.R. Meuss. 2001. *Weiss's herbal medicine*. Classic ed. Stuttgart: New York.

Wichtl, M. 2004. *Herbal drugs and phytopharmaceuticals: A handbook for practice on a scientific basis*. 3rd ed. Boca Raton, FL: CRC Press.

Zaragoza, M.C., D. Lopez, M.P. Sáiz, et al. 2008. Toxicity and antioxidant activity in vitro and in vivo of two *Fucus vesiculosus* extracts. *J. Agric. Food Chem.* 56(17):7773-7780.

Galium odoratum (L.) Scop. — Rubiaceae

SCN: sweet woodruff
Syn: *Asperula odorata* L.
OCN: sweet-scented bedstraw; woodruff
Part: herb

Quick Reference Summary

Safety Class: 1
Interaction Class: A

Contraindications

None known.

Other Precautions

None known.

Drug and Supplement Interactions

None known.

Adverse Events and Side Effects

Ingestion of sweet woodruff extracts has been associated with headaches in some individuals (Roth et al. 1984).

Pharmacological Considerations

None known.

Pregnancy and Lactation

No information on the safety of sweet woodruff in pregnancy or lactation was identified in the scientific or traditional literature. Although this review did not identify any concerns for use while pregnant or nursing, safety has not been conclusively established.

Review Details

I. Drug and Supplement Interactions

Clinical Trials of Drug or Supplement Interactions

No clinical trials of drug or supplement interactions were identified.

Case Reports of Suspected Drug or Supplement Interactions

No case reports of suspected drug or supplement interactions were identified.

Animal Trials of Drug or Supplement Interactions

No animal trials of drug or supplement interactions were identified.

II. Adverse Events

Case Reports of Adverse Events

Ingestion of sweet woodruff extracts has been associated with headaches in some individuals. Details on product, dose, and duration were not provided (Roth et al. 1984).

III. Pharmacology and Pharmacokinetics

Human Pharmacological Studies

No relevant human pharmacological studies were identified.

Animal Pharmacological Studies

No relevant animal pharmacological studies were identified.

In Vitro Pharmacological Studies

No relevant in vitro pharmacological studies were identified.

IV. Pregnancy and Lactation

No information on the safety of sweet woodruff during pregnancy or lactation was identified.

V. Toxicity Studies

No toxicity studies were identified.

Literature Cited

Roth, L., M. Daunderer, and K. Kormann. 1984. *Giftpflanzen-pflanzengifte: Vorkommen, wirkung, therapie.* Landsberg: Ecomed.

Galium spp.

Rubiaceae

Galium aparine L.
SCN: cleavers
OCN: bedstraw; clivers; goosegrass

Galium verum **L.**
SCN: lady's bedstraw
OCN: our Lady's bedstraw; yellow bedstraw
Part: herb

Quick Reference Summary

Safety Class: 1
Interaction Class: A

Contraindications

None known.

Other Precautions

None known.

Drug and Supplement Interactions

None known.

Adverse Events and Side Effects

None known.

Pharmacological Considerations

None known.

Pregnancy and Lactation

No information on the safety of cleavers or lady's bedstraw in pregnancy or lactation was identified in the scientific or traditional literature. Although this review did not identify any concerns for use while pregnant or nursing, safety has not been conclusively established.

Review Details

I. Drug and Supplement Interactions

Clinical Trials of Drug or Supplement Interactions

No clinical trials of drug or supplement interactions were identified.

Case Reports of Suspected Drug or Supplement Interactions

No case reports of suspected drug or supplement interactions were identified.

Animal Trials of Drug or Supplement Interactions

No animal trials of drug or supplement interactions were identified.

II. Adverse Events

Case Reports of Adverse Events

No case reports of adverse events were identified.

III. Pharmacology and Pharmacokinetics

Human Pharmacological Studies

No relevant human pharmacological studies were identified.

Animal Pharmacological Studies

No relevant animal pharmacological studies were identified.

In Vitro Pharmacological Studies

No relevant in vitro pharmacological studies were identified.

IV. Pregnancy and Lactation

No information on the safety of cleavers or lady's bedstraw during pregnancy or lactation was identified.

V. Toxicity Studies

Acute Toxicity

The LD_{50} of an intraperitoneally administered ethanol extract of cleavers in rats is 1 g/kg (Sharma et al. 1978).

In toxicity assays in brine shrimp and fruit flies, the water fraction of a methanol extract of cleavers showed no toxicity, whereas the ethyl acetate fraction produced over 80% lethality in both types of organisms (McChesney and Adams 1985).

Literature Cited

McChesney, J.D., and R.P. Adams. 1985. Co-evaluation of plant extracts as petrochemical substitutes and for biologically active compounds. *Econ. Bot.* 39(1):74-86.

Sharma, M.L., N. Chandokhe, B.J. Ghatak, et al. 1978. Pharmacological screening of Indian medicinal plants. *Indian J. Exp. Biol.* 16(2):228.

Ganoderma lucidum (Curtis: Fr.) P. Karst. — Ganodermataceae

SCN: reishi
Syn: *Polyporus lucidus* (Curtis: Fr.) Fr.
PN: *ling zhi* (fruiting body)
OCN: ganoderma
Part: fruiting body, mycelium

Quick Reference Summary

Safety Class: 1
Interaction Class: A

Contraindications

None known.

Other Precautions

None known.

Drug and Supplement Interactions

See Pharmacological Considerations.

Adverse Events and Side Effects

Although one reference text notes that rare side effects, including dryness of the mouth, throat, and nasal area, itchiness, upset stomach, nosebleed, and bloody stools, have been recorded after 3 to 6 months of continuous use of reishi (Leung and Foster 1996), clinical trials and dose-ranging studies have indicated that reishi extracts are generally well tolerated at doses of 5.4 g daily for 3 months (Chen et al. 2006; Gao et al. 2002, 2003, 2004a; Noguchi et al. 2008).

Allergic reactions to reishi are rare but have been reported (Bensky et al. 2004; Leung and Foster 1996).

Fulminant hepatitis was reported in two patients who had been taking water extracts of reishi for several months and recently switched to a powdered reishi preparation (Wanmuang et al. 2007; Yuen et al. 2004).

Pharmacological Considerations

Human and animal studies have provided differing results on the effects of reishi on platelet aggregation. Inhibition of platelet aggregation was observed in healthy persons and persons with atherosclerosis after 3 g daily of reishi for 2 weeks (Tao and Feng 1990), whereas no antiplatelet activity was observed in HIV-positive patients given 900 mg reishi daily (Gau et al. 1990), suggesting that the activity could be dose related. Compounds isolated from reishi have demonstrated antiplatelet activity in vitro (Kawagishi et al. 1993; Mongold et al. 1993). Although no interactions have been reported, reishi should be used with caution in persons taking anticoagulant or antiplatelet medications (Upton 2000).

Human and animal studies have indicated immunomodulating effects of reishi, suggesting that reishi should be used with caution in persons taking immune-suppressant medications (Chen et al. 2006; Gao et al. 2003, 2005; Lin 2005; Upton 2000; Zhu et al. 2007).

Human and animal studies have shown that reishi may modify glucose levels (Gao et al. 2004b; Kimura et al. 1988; Zhang and Lin 2004). People with diabetes are advised to monitor their blood sugar closely and discuss the use of this herb with a qualified healthcare practitioner prior to use.

Pregnancy and Lactation

No information on the safety of reishi in pregnancy or lactation was identified in the scientific or traditional literature. Although this review did not identify any concerns for use while pregnant or nursing, safety has not been conclusively established.

Review Details

I. Drug and Supplement Interactions

Clinical Trials of Drug or Supplement Interactions

No clinical trials of drug or supplement interactions were identified.

Case Reports of Suspected Drug or Supplement Interactions

No case reports of suspected drug or supplement interactions were identified.

Animal Trials of Drug or Supplement Interactions

No animal trials of drug or supplement interactions were identified.

II. Adverse Events

Adverse Events Reported in Clinical Trials

A reishi extract comprised of polysaccharide fractions was reported as well tolerated in patients with type 2 diabetes, advanced-stage cancer, or coronary heart disease, at doses of 5.4 g daily for 12 weeks (Gao et al. 2002, 2004a, 2004b,

2005), with minor adverse events, primarily nausea and diarrhea, reported in one study (Gao et al. 2002).

Case Reports of Adverse Events

A 47-year-old woman with a history of schizophrenia developed fatal fulminant hepatitis after having taken 400 mg of powdered reishi daily for 2 months. She was also taking lithium, perphenazine, and trihexyphenidyl long-term. The woman had previously consumed an extract made from boiled reishi slices (dose not specified) for several years with no adverse effects. The reporting physicians indicated that the reaction could have been idiosyncratic (Wanmuang et al. 2007).

Fulminant hepatitis was diagnosed in a 78-year-old woman who had been taking felodipine, calcium supplements, multivitamins, and reishi. The woman had been taking an extract made from boiled reishi slices (dose not specified) for at least 1 year, and began taking a powdered reishi product (dose not specified) 4 weeks before the onset of symptoms (Yuen et al. 2004).

In rare cases, side effects from reishi have been reported, including dry mouth and nose, itchiness, upset stomach, nausea, vertigo, nosebleeds, constipation or diarrhea, and bloody stool. Products and doses taken were not specified, and one reference noted that such effects were associated with 3 to 6 months of reishi use (Bensky et al. 2004; Leung and Foster 1996).

Allergic reactions to reishi have been reported, primarily after intramuscular administration of reishi extracts (Bensky et al. 2004; Leung and Foster 1996). An instance of skin rash was observed following the consumption of reishi wine (Leung and Foster 1996).

Chronic watery diarrhea and pseudoparasitosis were observed in a 49-year-old man with non-Hodgkin's lymphoma who took reishi (product, dose, and duration not specified). Spores from reishi were mistakenly identified as parasites. The report indicates that the man had a 6-month history of diarrhea and that loose stools were observed several days after beginning reishi, although it is unclear if the reishi was suspected as the cause of the 6 months of diarrhea or if the reishi was thought to worsen existing diarrhea. The man was also taking drugs (not specified) for tuberculosis (Wanachiwanawin et al. 2006).

III. Pharmacology and Pharmacokinetics

Human Pharmacological Studies

Inhibition of first and second phase platelet aggregation was observed in blood taken from healthy volunteers or patients with atherosclerotic disease, who had ingested 3 g of a water soluble extract of reishi daily for 2 weeks (Tao and Feng 1990). No antiplatelet activity of reishi extract was observed in HIV-positive hemophiliac patients treated with 0.9 g (containing 1.35 mg of adenosine) daily (Gau et al. 1990).

No adverse effects, including evidence of liver, renal, or DNA toxicity, were observed in healthy volunteers orally administered 1.44 g daily of a reishi preparation (equivalent to 13.2 g fresh reishi) for 4 weeks (Wachtel-Galor et al. 2004).

Good treatment tolerance and no major adverse events were reported in a dose-ranging study of reishi in men with lower urinary tract symptoms orally administered 0.6, 6, or 60 mg reishi daily for 8 weeks (Noguchi et al. 2008).

A reduction in mean fasting and stimulated glycosylated hemoglobin was observed in patients with type 2 diabetes orally administered 5.4 g of polysaccharide fractions of reishi daily for 12 weeks (Gao et al. 2004b).

No adverse effects were observed in a safety study of reishi in healthy volunteers orally administered 2 g of reishi extract daily for 10 days. No significant changes in hematological, cardiac, or urinary parameters were observed. Blood tests for immunological activity indicated no obvious changes in CD4, CD8, and CD19, with some elevation of CD56 cell counts (Wicks et al. 2007).

In an open-label study in patients with colorectal cancer, ingestion of 5.4 g of reishi daily for 12 weeks produced immune modulating effects. Treatment with reishi tended to increase mitogenic reactivity to phytohemagglutinin, counts of CD3, CD4, CD8, and CD56 lymphocytes, plasma concentrations of interleukin (IL)-2, IL-6, and interferon (IFN)-gamma, and NK activity, whereas plasma concentrations of IL-1 and tumor necrosis factor (TNF)-alpha were decreased. For all of these parameters, no statistical significance was observed when a comparison was conducted between baseline and treatment values. The changes of IL-1 were correlated with those for IL-6, IFN-gamma, CD3, CD4, CD8, and NK activity, and the IL-2 changes were correlated with those for IL-6, CD8, and NK activity (Chen et al. 2006).

In patients with advanced-stage cancer treated with 5.4 g of polysaccharide fractions of reishi daily for 12 weeks, increases in the mean plasma concentrations of IL-2, IL-6, and IFN-gamma were observed. Levels of IL-1 and TNF were decreased. A marked variability was observed in the numbers of each lymphocyte subset at baseline. The mean absolute number of CD56 cells was increased after treatment, and a marginal increase in the numbers of CD3, CD4, and CD8 was observed as compared to baseline, with the CD4:CD8 T cell ratios unchanged. PHA responses were enhanced in most patients, and the mean NK activity increased (Gao et al. 2003).

Animal Pharmacological Studies

In castrated rats, administration of a methanol extract of reishi inhibited testosterone-induced growth of the ventral prostate (Fujita et al. 2005).

A dose-dependent hypoglycemic effect was observed in healthy mice intraperitoneally administered reishi polysaccharides at doses of 25, 50, or 100 mg/kg (Zhang and Lin 2004).

A reduction in blood glucose levels with no elevation of blood insulin was observed in rats orally administered 50 mg per animal (200–250 g average weight) of an aqueous extract of reishi (Kimura et al. 1988).

In mice with immune suppression induced by cyclophosphamide, intraperitoneal administration of 2.5, 25, or 250 mg/kg of reishi polysaccharides daily for 7 days enhanced the function of immunological effector cells. The 2.5 mg/kg dose accelerated recovery of bone marrow cells, red blood cells, white blood cells, splenic natural killer cells, and natural killer T cells, and enhanced T and B cell proliferation responses, cytotoxic T lymphocyte activity, and NK cell and lymphokine activated killer cell activity (Zhu et al. 2007).

Dose-dependent inhibition of the drug-metabolizing isoenzymes CYP2E1, CYP1A2, and CYP3A was observed in rats with immune hepatic injury administered 50 or 200 mg/kg of reishi polysaccharides (Wang et al. 2007).

The compound ganoderol B suppressed testosterone-induced regrowth of the ventral prostate in rats (Liu et al. 2007).

In Vitro Pharmacological Studies

A review of the immunomodulating effects of reishi polysaccharides in vivo and in vitro demonstrated that the polysaccharides promote the function of antigen-presenting cells, mononuclear phagocytes, humoral immunity, and cellular immunity (Tahvonen et al. 2005).

Estrogen receptor (ER)-modulating activity of reishi was observed in human breast cancer cells (MCF-7). Reishi down-regulated the expression of ER-α in ER-positive (MCF-7) cells but did not effect the expression of ER-β in ER-positive and ER-negative (MDA-MB-231) cells. Additionally, reishi inhibited estrogen-dependent and constitutive transactivation activity of the ER through an estrogen response element in a reporter gene assay (Jiang et al. 2006).

Differing effects on viability of peripheral blood mononuclear cells from healthy adults and children and from children undergoing chemotherapy were observed after treatment with either of two crude extracts (solvents not specified) of reishi or polysaccharides isolated from reishi (Gill and Rieder 2008).

The compound ganoderol B inhibited androgen-induced growth of human prostate adenocarcinoma cells (LNCaP) (Liu et al. 2007).

The compound ganodermic acid S inhibited collagen-induced platelet aggregation in human platelets (Mongold et al. 1993). Inhibition of platelet aggregation was observed after application of the adenosine derivative 5′-deoxy-5′-methylsulfinyladenosine from reishi (50 μg/ml) (Kawagishi et al. 1993).

Tested with a variety of antimicrobial drugs, reishi showed antagonistic activity with ampicillin and cefazolin against *Proteus vulgaris*, but no antagonistic effects were observed with the drugs oxytetracycline or chloramphenicol, or in tested species of *Bacillus, Staphylococcus, Escherichia, Klebsiella,* or *Salmonella* (Yoon et al. 1994).

IV. Pregnancy and Lactation

No information on the safety of reishi in pregnancy or lactation was identified.

G

V. Toxicity Studies

Acute Toxicity

The oral LD_{50} of an aqueous extract or polysaccharide fraction of reishi could not be determined at doses up to 5 g/kg in mice (Kim et al. 1986).

The LD_{50} of intraperitoneally administered extracts of reishi have been reported to range from 3.42 to 38.3 g/kg, with effects dependent on the type of extract administered (Leung and Foster 1996).

Short-Term Toxicity

No changes in hematological parameters or organ weights were observed in mice orally administered 5 g/kg of an aqueous extract of reishi daily for 30 days (Kim et al. 1986).

No adverse effects were observed in mice orally administered a dose equivalent to 925.9 mg/kg of reishi freeze-dried extract daily for 14 days (Chiu et al. 2000).

No signs of toxicity were observed in mice orally administered an extract of reishi equivalent to 1.8 g/kg daily for 20 days (Leung and Foster 1996).

Genotoxicity

No genotoxic activity of reishi was observed in the comet assay in lymphocytes of mice orally administered single doses of an extract of reishi equivalent to 220 g of reishi (Chiu et al. 2000).

In the comet assay with human lymphocytes, reishi extract exhibited genoprotective activity at a concentration of 0.0001% but induced DNA damage at concentrations of 0.001% (w/v) and above (Wachtel-Galor et al. 2005).

Literature Cited

Bensky, D., S. Clavey, and E. Stöger. 2004. *Chinese herbal medicine: Materia medica*. 3rd ed. Seattle: Eastland Press.

Chen, X., Z.P. Hu, X.X. Yang, et al. 2006. Monitoring of immune responses to a herbal immuno-modulator in patients with advanced colorectal cancer. *Int. Immunopharmacol.* 6(3):499-508.

Chiu, S.W., Z.M. Wang, T.M. Leung, and D. Moore. 2000. Nutritional value of ganoderma extract and assessment of its genotoxicity and antigenotoxicity using comet assays of mouse lymphocytes. *Food Chem. Toxicol.* 38(2-3):173-178.

G

Fujita, R., J. Liu, K. Shimizu, et al. 2005. Anti-androgenic activities of *Ganoderma lucidum*. *J. Ethnopharmacol.* 102(1):107-112.

Gao, Y., G. Chen, X. Dai, J. Ye, and S. Zhou. 2004a. A phase I/II study of ling zhi mushroom *Ganoderma lucidum* (W. Curt.: Fr.) Lloyd (Aphyllophoromycetideae) extract in patients with coronary heart disease. *Int. J. Med. Mushrooms* 6(4):327-334.

Gao, Y., J. Lan, X. Dai, J. Ye, and S. Zhou. 2004b. A phase I/II study of ling zhi mushroom *Ganoderma lucidum* (W. Curt.: Fr.) Lloyd (Aphyllophoromycetideae) extract in patients with type II diabetes mellitus. *Int. J. Med. Mushrooms* 6 (1):33-39.

Gao, Y., W. Tang, X. Dai, et al. 2005. Effects of water-soluble *Ganoderma lucidum* polysaccharides on the immune functions of patients with advanced lung cancer. *J. Med. Food* 8(2):159-168.

Gao, Y., S. Zhou, G. Chen, X. Dai, and J. Ye. 2002. A phase I/II study of a *Ganoderma lucidum* (Curt.: Fr.) P. Karst. extract (ganopoly) in patients with advanced cancer. *Int. J. Med. Mushrooms* 4 (3):207-214.

Gao, Y., S. Zhou, W. Jiang, M. Huang, and X. Dai. 2003. Effects of ganopoly (a *Ganoderma lucidum* polysaccharide extract) on the immune functions in advanced-stage cancer patients. *Immunol. Invest.* 32(3):201-215.

Gau, J.P., C.K. Lin, S.S. Lee, and S.R. Wang. 1990. The lack of antiplatelet effect of crude extracts from *Ganoderma lucidum* on HIV-positive hemophiliacs. *Am. J. Chin. Med.* 18(3-4):175-179.

Gill, S.K., and M.J. Rieder. 2008. Toxicity of a traditional Chinese medicine, *Ganoderma lucidum,* in children with cancer. *Can. J. Clin. Pharmacol.* 15(2):e275-e285.

Jiang, J., V. Slivova, and D. Sliva. 2006. *Ganoderma lucidum* inhibits proliferation of human breast cancer cells by down-regulation of estrogen receptor and NF-kappaB signaling. *Int. J. Oncol.* 29(3):695-703.

Kawagishi, H., F. Fukuhara, M. Sazuka, et al. 1993. 5′-Deoxy-5′-methylsulphinyladenosine, a platelet aggregation inhibitor from *Ganoderma lucidum*. *Phytochemistry* 32(2):239-241.

Kim, M.J., H.W. Kim, Y.S. Lee, et al. 1986. Studies on safety of *Ganoderma lucidum*. *Korean J. Mycol.* 14(1):49-60.

Kimura, Y., H. Okuda, and S. Arichi. 1988. Effects of the extracts of *Ganoderma lucidum* on blood glucose level in rats. *Planta Med.* 54(4):290-294.

Leung, A.Y., and S. Foster. 1996. *Encyclopedia of common natural ingredients used in food, drugs, and cosmetics.* 2nd ed. New York: Wiley.

Lin, Z.B. 2005. Cellular and molecular mechanisms of immunomodulation by *Ganoderma lucidum*. *J. Pharmacol. Sci.* 99(2):144-153.

Liu, J., K. Shimizu, F. Konishi, S. Kumamoto, and R. Kondo. 2007. The anti-androgen effect of ganoderol B isolated from the fruiting body of *Ganoderma lucidum*. *Bioorg. Med. Chem.* 15(14):4966-4972.

Mongold, J.J., P. Susplugas, C. Taillade, and J.J. Serrano. 1993. Anti-inflammatory activity of *Ribes nigrum* leaf extract in rats. *Plant. Med. Phytother.* 26 (2):109-116.

Noguchi, M., T. Kakuma, K. Tomiyasu, et al. 2008. Effect of an extract of *Ganoderma lucidum* in men with lower urinary tract symptoms: A double-blind, placebo-controlled randomized and dose-ranging study. *Asian J. Androl.* 10(4):651-658.

Tahvonen, R.L., U.S. Schwab, K.M. Linderborg, H.M. Mykkanen, and H.P. Kallio. 2005. Black currant seed oil and fish oil supplements differ in their effects on fatty acid profiles of plasma lipids, and concentrations of serum total and lipoprotein lipids, plasma glucose and insulin. *J. Nutr. Biochem.* 16(6):353-359.

Tao, J., and K.Y. Feng. 1990. Experimental and clinical studies on inhibitory effect of *Ganoderma lucidum* on platelet aggregation. *J. Tongji Med. Univ.* 10(4):240-243.

Upton, R. 2000. *Reishi mushroom* Ganoderma lucidum: *Standards of analysis, quality control, and therapeutics. American Herbal Pharmacopoeia and therapeutic compendium.* Santa Cruz, CA: American Herbal Pharmacopoeia.

Wachtel-Galor, S., S.W. Choi, and I.F. Benzie. 2005. Effect of *Ganoderma lucidum* on human DNA is dose dependent and mediated by hydrogen peroxide. *Redox Rep.* 10(3):145-149.

Wachtel-Galor, S., B. Tomlinson, and I.F.F. Benzie. 2004. *Ganoderma lucidum* ('lingzhi'), a Chinese medicinal mushroom: Biomarker responses in a controlled human supplementation study. *Br. J. Nutr.* 91(2):263-269.

Wanachiwanawin, D., A. Piankijagum, A. Chaiprasert, et al. 2006. *Ganoderma lucidum*: A cause of pseudoparasitosis. *Southeast Asian J. Trop. Med. Public Health* 37(6):1099-1102.

Wang, X., X. Zhao, D. Li, et al. 2007. Effects of *Ganoderma lucidum* polysaccharide on CYP2E1, CYP1A2 and CYP3A activities in BCG-immune hepatic injury in rats. *Biol. Pharm. Bull.* 30(9):1702-1706.

Wanmuang, H., J. Leopairut, C. Kositchaiwat, W. Wananukul, and S. Bunyaratvej. 2007. Fatal fulminant hepatitis associated with *Ganoderma lucidum* (lingzhi) mushroom powder. *J. Med. Assoc. Thailand* 90(1):179-181.

Wicks, S.M., R. Tong, C.Z. Wang, et al. 2007. Safety and tolerability of *Ganoderma lucidum* in healthy subjects: A double-blind randomized placebo-controlled trial. *Am. J. Chin. Med.* 35(3):407-414.

Yoon, S.Y., S.K. Eo, Y.S. Kim, C.K. Lee, and S.S. Han. 1994. Antimicrobial activity of *Ganoderma lucidum* extract alone and in combination with some antibiotics. *Arch. Pharm. Res.* 17(6):438-442.

Yuen, M.F., P. Ip, W.K. Ng, and C.L. Lai. 2004. Hepatotoxicity due to a formulation of *Ganoderma lucidum* (lingzhi). *J. Hepatol.* 41(4):686-687.

Zhang, H.N., and Z.B. Lin. 2004. Hypoglycemic effect of *Ganoderma lucidum* polysaccharides. *Acta Pharmacol. Sin.* 25(2):191-195.

Zhu, X.L., A.F. Chen, and Z.B. Lin. 2007. *Ganoderma lucidum* polysaccharides enhance the function of immunological effector cells in immunosuppressed mice. *J. Ethnopharmacol.* 111(2):219-226.

Garcinia cambogia (Gaertn.) Desr. — Clusiaceae

SCN: garcinia
OCN: brindall berry; Malabar tamarind

Part: fruit

Quick Reference Summary

Safety Class: 1
Interaction Class: A

Contraindications

None known.

Other Precautions

None known.

Drug and Supplement Interactions

None known.

Editors' Notes

Much of the research on garcinia is focused on a standardized extract referred to as HCA-SX, a garcinia extract that contains 60% of the compound hydroxycitric acid (HCA) (Soni et al. 2004). Garcinia fruit rind naturally contains approximately 16 to 26% HCA (Antony et al. 1998; Jayaprakasha and Sakariah 1998).

Adverse Events and Side Effects

Cases of liver toxicity have been reported in patients taking weight-loss products containing garcinia and other botanical and mineral ingredients (Actis et al. 2007; Dara et al. 2008; Jones and Andrews 2007; Shim and Saab 2009; Stevens et al. 2005). The relationship between these cases and garcinia is not clear (Stohs et al. 2009).

Pharmacological Considerations

None known.

Pregnancy and Lactation

Multigeneration animal studies have indicated no adverse effects of HCA-SX on pregnancy or fetal development (Deshmukh 2008a, 2008b).

No information on the safety of garcinia in lactation was identified in the scientific or traditional literature. While this review did not identify any concerns for use while nursing, safety has not been conclusively established.

Review Details

I. Drug and Supplement Interactions

Clinical Trials of Drug or Supplement Interactions

No clinical trials of drug or supplement interactions were identified.

Case Reports of Suspected Drug or Supplement Interactions

No case reports of suspected drug or supplement interactions were identified.

Animal Trials of Drug or Supplement Interactions

No animal trials of drug or supplement interactions were identified.

II. Adverse Events

Case Reports of Adverse Events

Cases of liver toxicity have been reported in patients taking weight-loss products containing garcinia and other botanical and mineral ingredients (Actis et al. 2007; Dara et al. 2008; Jones and Andrews 2007; Shim and Saab 2009; Stevens et al. 2005). The relationship between these cases and garcinia is not clear (Stohs et al. 2009).

III. Pharmacology and Pharmacokinetics

Human Pharmacological Studies

In overweight subjects, no changes in serum testosterone, estrone, and estradiol levels were observed after ingestion of 1667 mg daily of garcinia extract (1000 mg HCA daily) (Hayamizu et al. 2008).

Animal Pharmacological Studies

No relevant animal pharmacological studies were identified.

In Vitro Pharmacological Studies

No relevant in vitro pharmacological studies were identified.

IV. Pregnancy and Lactation

In a two-generation reproductive toxicity study, rats were fed diets containing 1000, 3000, or 10,000 ppm HCA-SX for 10 weeks prior to mating, during mating, and, for females, through gestation and lactation, across two generations. No changes were observed in reproductive performance as evaluated by sexual maturity, fertility and mating, gestation, parturition, litter properties, lactation, and development of the offspring. HCA-SX exposure did not affect feed consumption or body weight at any of the exposure levels (Deshmukh 2008b). Animals from the second generation (F2b) of the study were fed diets containing 1000, 3000, or 10,000 ppm HCA-SX until sexual maturity, at which time they were mated. Pregnant rats were continued on the same diet. A slight lowering of maternal body weight gain during the gestation period was observed in the group receiving 10,000 ppm HCA-SX. No evidence of maternal toxicity or adverse effects on fetal development were observed (Deshmukh 2008a).

V. Toxicity Studies

Acute Toxicity

The oral LD_{50} of garcinia extract (60% HCA) in rats could not be determined at doses up to 5 g/kg (Ohia et al. 2002).

The dermal LD_{50} of garcinia extract (60% HCA) in rabbits could not be determined at doses up to 2 g/kg (Ohia et al. 2002).

G

Short-Term Toxicity

In female rats fed diets containing 7.3% garcinia extract HCA-SX (41% HCA) for 2 or 4 weeks, no changes in reproductive hormones were observed (Kiyose et al. 2006a).

Subchronic Toxicity

In rats fed diets containing 0.2, 2, or 5% of a garcinia extract HCA-SX (60% HCA) for 90 days, an advancing age-induced marginal increase in hepatic lipid peroxidation was observed as compared to control animals. No differences were observed in hematology, clinical chemistry, histopathological evaluation, hepatic DNA fragmentation, or testicular lipid peroxidation. Relative organ weights, including liver, testis, and brain, were comparable to the control group (Shara et al. 2003, 2004).

In obese rats fed diets containing 0.5, 2.4, 4.9, or 7.3% (77, 388, 778, or 1244 mg/kg body weight HCA) garcinia extract for 92 days, animals on the 4.9% or higher diets had testicular atrophy and toxicity (Saito et al. 2005). The design of this study has been questioned for accuracy as a test of testicular toxicity (Burdock et al. 2005).

In male rats fed diets containing 4.9% garcinia extract HCA-SX (41% HCA) for 2 or 4 weeks, a decrease in the serum level of inhibin B and increase in follicle-stimulating hormone was observed. The level of testis meiosis-activating sterol was statistically lower in the testes of rats on the garcinia diet as compared to control (Kiyose et al. 2006b).

In male rats orally administered 100 or 200 mg/kg of an ethanol extract of garcinia seed 6 days per week for 6 weeks, changes in the testes were observed along with an increase in sperm counts (Oluyemi et al. 2007).

Genotoxicity

In the micronucleus test with HCA intraperitoneally administered to mice at doses of 20, 100, 500, 2,500, or 12,500 µmol/kg, HCA was found to increase the number of micronucleated polychromatic erythrocytes (Lee and Lee 2007). A commentary on this study noted that the route of administration (intraperitoneal instead of oral) and the use of DMSO as part of the treatment but not the control group were likely to produce results different than those seen after oral use of HCA (Lau et al. 2008).

No mutagenic activity of the compound (–)-hydroxycitric acid (HCA) was observed in the Ames test for mutagenicity with or without metabolic activation, or in the chromosomal aberration test (Aujoulat 2003; Lee and Lee 2007).

Literature Cited

Actis, G.C., E. Bugianesi, A. Ottobrelli, and M. Rizzetto. 2007. Fatal liver failure following food supplements during chronic treatment with montelukast. *Dig. Liver Dis.* 39(10):953-955.

Antony, J.I.X., P.D. Josan, and M.L. Shankaranarayana. 1998. Quantitative analysis of (–)-hydroxycitric acid and (–)-hydroxycitric acid lactone in *Garcinia* fruits and *Garcinia* products. *J. Food Sci. Technol.* 35:399-402.

Aujoulat, M. 2003. *Cited in* Soni, M.G., G.A. Burdock, H.G. Preuss, S.J. Stohs, S.E. Ohia, and D. Bagchi. 2004. Safety assessment of (–)-hydroxycitric acid and Super CitriMax, a novel calcium/potassium salt. *Food Chem. Toxicol.* 42(9):1513-1529.

Burdock, G., M. Soni, M. Bagchi, and D. Bagchi. 2005. *Garcinia cambogia* toxicity is misleading. *Food Chem. Toxicol.* 43(11):1683-1684; author reply 1685-1686.

Dara, L., J. Hewett, and J.K. Lim. 2008. Hydroxycut hepatotoxicity: A case series and review of liver toxicity from herbal weight loss supplements. *World J. Gastroenterol.* 14(45):6999-7004.

Deshmukh, N.S. 2008a. Safety of a novel calcium/potassium salt of (–)-hydroxycitric acid (HCA-SX): II. Developmental toxicity study in rats. *Toxicol. Mech. Meth.* 18(5):443-451.

Deshmukh, N.S. 2008b. Safety of a novel calcium/potassium salt of hydroxycitric acid (HCA-SX): I. Two-generation reproduction toxicity study. *Toxicol. Mech. Meth.* 18(5):433-442.

Hayamizu, K., H. Tomi, I. Kaneko, et al. 2008. Effects of *Garcinia cambogia* extract on serum sex hormones in overweight subjects. *Fitoterapia* 79(4):255-261.

Jayaprakasha, G.K., and K.K. Sakariah. 1998. Determination of organic acids in *Garcinia cambogia* (Desr.) by high-performance liquid chromatography. *J. Chromatogr. A* 806(2):337-339.

Jones, F.J., and A.H. Andrews. 2007. Acute liver injury associated with the herbal supplement Hydroxycut in a soldier deployed to Iraq. *Am. J. Gastroenterol.* 102(10):2357-2358.

Kiyose, C., K. Kubo, and M. Saito. 2006a. Effect of *Garcinia cambogia* administration on female reproductive organs in rats. *J. Clin. Biochem. Nutr.* 38(3):188-194.

Kiyose, C., S. Ogino, K. Kubo, M. Takeuchi, and M. Saito. 2006b. Relationship between *Garcinia cambogia*-induced impairment of spermatogenesis and meiosis-activating sterol production in rat testis. *J. Clin. Biochem. Nutr.* 38(3):180-187.

Lau, F.C., M. Bagchi, and D. Bagchi. 2008. Refuting "Evaluation of the genotoxicity of (–)-hydroxycitric acid (HCA-SX) isolated from *Garcinia cambogia*." *J. Toxicol. Environ. Health A* 71(5):348-349.

Lee, K.H., and B.M. Lee. 2007. Evaluation of the genotoxicity of (–)-hydroxycitric acid (HCA-SX) isolated from *Garcinia cambogia*. *J. Toxicol. Environ. Health A* 70(5):388-392.

Ohia, S.E., C.A. Opere, A.M. LeDay, et al. 2002. Safety and mechanism of appetite suppression by a novel hydroxycitric acid extract (HCA-SX). *Mol. Cell. Biochem.* 238(1-2):89-103.

Oluyemi, K.A., O.R. Jimoh, O.A. Adesanya, et al. 2007. Effects of crude ethanolic extract of *Garcinia cambogia* on the reproductive system of male Wistar rats (*Rattus novergicus*). *Afr. J. Biotechnol.* 6(10):1236-1238.

Saito, M., M. Ueno, S. Ogino, et al. 2005. High dose of *Garcinia cambogia* is effective in suppressing fat accumulation in developing male Zucker obese rats, but highly toxic to the testis. *Food Chem. Toxicol.* 43(3):411-419.

Shara, M., S.E. Ohia, R.E. Schmidt, et al. 2004. Physico-chemical properties of a novel (−)-hydroxycitric acid extract and its effect on body weight, selected organ weights, hepatic lipid peroxidation and DNA fragmentation, hematology and clinical chemistry, and histopathological changes over a period of 90 days. *Mol. Cell. Biochem.* 260(1-2):171-186.

Shara, M., S.E. Ohia, T. Yasmin, et al. 2003. Dose- and time-dependent effects of a novel (−)-hydroxycitric acid extract on body weight, hepatic and testicular lipid peroxidation, DNA fragmentation and histopathological data over a period of 90 days. *Mol. Cell. Biochem.* 254(1-2):339-346.

Shim, M., and S. Saab. 2009. Severe hepatotoxicity due to Hydroxycut: A case report. *Dig. Dis. Sci.* 54(2):406-408.

Soni, M.G., G.A. Burdock, H.G. Preuss, et al. 2004. Safety assessment of (−)-hydroxycitric acid and Super CitriMax, a novel calcium/potassium salt. *Food Chem. Toxicol.* 42(9):1513-1529.

Stevens, T., A. Qadri, and N.N. Zein. 2005. Two patients with acute liver injury associated with use of the herbal weight-loss supplement Hydroxycut. *Ann. Intern. Med.* 142(6):477-478.

Stohs, S.J., H.G. Preuss, S.E. Ohia, et al. 2009. No evidence demonstrating hepatotoxicity associated with hydroxycitric acid. *World J. Gastroenterol.* 15(32):4087-4089.

Gardenia jasminoides J. Ellis — Rubiaceae

SCN: gardenia
Syn: *Gardenia augusta* Merr.
PN: *zhi zi* (fruit)
OCN: Cape jasmine
Part: fruit

Quick Reference Summary

Safety Class: 1
Interaction Class: A

Contraindications

None known.

Other Precautions

None known.

Drug and Supplement Interactions

None known.

Adverse Events and Side Effects

Allergic skin reactions to gardenia have been reported (Bensky et al. 2004).

Pharmacological Considerations

None known.

Pregnancy and Lactation

No information on the safety of gardenia in pregnancy or lactation was identified in the scientific or traditional literature. Although this review did not identify any concerns for use while pregnant or nursing, safety has not been conclusively established.

Review Details

I. Drug and Supplement Interactions

Clinical Trials of Drug or Supplement Interactions

No clinical trials of drug or supplement interactions were identified.

Case Reports of Suspected Drug or Supplement Interactions

No case reports of suspected drug or supplement interactions were identified.

Animal Trials of Drug or Supplement Interactions

No animal trials of drug or supplement interactions were identified.

II. Adverse Events

Case Reports of Adverse Events

Allergic skin reactions to gardenia have been reported (Bensky et al. 2004).

III. Pharmacology and Pharmacokinetics

Human Pharmacological Studies

No relevant human pharmacological studies were identified.

Animal Pharmacological Studies

In diabetic mice, administration of 200 or 400 mg/kg of the compound geniposide caused a dose-dependent decrease in blood glucose and insulin levels in diabetic mice (Wu et al. 2009).

The compounds geniposide and genipin isolated from gardenia and intravenously administered to mice prolonged the time required for thrombotic occlusion induced by photochemical reaction in the mouse femoral artery (Suzuki et al. 2001).

In Vitro Pharmacological Studies

The compounds geniposide and genipin isolated from gardenia inhibited collagen-induced but did not inhibit arachidonate-induced platelet aggregation in mouse platelets (Suzuki et al. 2001).

G

IV. Pregnancy and Lactation

No information on the safety of gardenia in pregnancy or lactation was identified.

V. Toxicity Studies

Acute Toxicity

The LD_{50} of orally administered alcohol extract of gardenia in mice is 107.4 g/kg (Chen and Chen 2004). The LD_{50} of subcutaneously administered water extract of gardenia in mice is 31.79 g/kg (Zhu 1998).

The LD_{50} of the compound genipin in mice is 237 mg/kg after oral administration, 190 mg/kg after intraperitoneal administration, and 153 mg/kg after intravenous administration (Chen and Chen 2004; Zhu 1998).

The LD_{50} of the compound geniposide in mice is 3 g/kg after oral, intraperitoneal, or intravenous administration (Chen and Chen 2004).

Chronic Toxicity

No carcinogenic activity of the food colorant gardenia blue was observed in rats fed diets containing 2.5 or 5% gardenia blue for 2 years (Imazawa et al. 2000).

Genotoxicity

Mutagenicity tests, including the Ames assay, *rec* assay, and sister-chromatid exchange test, were completed on gardenia yellow, a food colorant derived from ethanol extracts of gardenia, and its components, crocetin, gentiobiose, geniposide, and genipin. Gardenia yellow and its components were found not to be mutagenic in the Ames assay. Gardenia yellow and genipin caused damage of DNA in the *rec* assay. Gardenia yellow and genipin induced sister-chromatid exchanges. Genipin induced a significant increase of tetraploids at all doses tested. Subsequent analysis of a gardenia yellow preparation indicated that genipin was not observed (Ozaki et al. 2002).

No mutagenic or clastogenic activity of the compound crocin was observed in the Ames assay, chromosomal aberration assay, *tk*(+/−) gene forward mutation assay, or comet assay. Post-treatment with crocin dose-dependently suppressed H_2O_2-induced DNA damage (Choi et al. 2008).

The compound crocetin inhibited chemically induced genotoxicity and neoplastic transformation in mouse C3H10T1/2 cells in vitro (Chang et al. 1996).

Literature Cited

Bensky, D., S. Clavey, and E. Stöger. 2004. *Chinese herbal medicine: Materia medica*. 3rd ed. Seattle: Eastland Press.

Chang, W.C., Y.L. Lin, M.J. Lee, S.J. Shiow, and C.J. Wang. 1996. Inhibitory effect of crocetin on benzo[*a*]pyrene genotoxicity and neoplastic transformation in C3H10T1/2 cells. *Anticancer Res.* 16(6B):3063-3068.

Chen, J.K., and T.T. Chen. 2004. *Chinese medical herbology and pharmacology*. City of Industry, CA: Art of Medicine Press.

Choi, H.-Y., Y.-J. Kim, H.-K. Jeon, and J.-C. Ryu. 2008. Study on genotoxicity of crocin, a component of gardenia fruit, in bacterial and mammalian cell systems. *Mol. Cell. Toxicol.* 4(4).

Imazawa, T., A. Nishikawa, F. Furukawa, et al. 2000. Lack of carcinogenicity of gardenia blue colour given chronically in the diet to F344 rats. *Food Chem. Toxicol.* 38(4):313-318.

Ozaki, A., M. Kitano, N. Furusawa, et al. 2002. Genotoxicity of gardenia yellow and its components. *Food Chem. Toxicol.* 40(11):1603-1610.

Suzuki, Y., K. Kondo, Y. Ikeda, and K. Umemura. 2001. Antithrombotic effect of geniposide and genipin in the mouse thrombosis model. *Planta Med.* 67(9):807-810.

Wu, S.Y., G.F. Wang, Z.Q. Liu, et al. 2009. Effect of geniposide, a hypoglycemic glucoside, on hepatic regulating enzymes in diabetic mice induced by a high-fat diet and streptozotocin. *Acta Pharmacol. Sin.* 30(2):202-208.

Zhu, Y.-P. 1998. *Chinese materia medica: Chemistry, pharmacology and applications*. Amsterdam: Harwood Academic Publishers.

Gastrodia elata Blume — Orchidaceae

SCN: gastrodia
PN: *tian ma* (rhizome)

Part: rhizome

Quick Reference Summary

Safety Class: 1
Interaction Class: A

Contraindications

None known.

Other Precautions

None known.

Drug and Supplement Interactions

None known.

Adverse Events and Side Effects

Rare cases of allergic reactions to gastrodia have been reported, including urticaria, purpura, vertigo, acute renal failure, and anaphylactic shock (Bensky et al. 2004). Rash, eczema and hair loss have been reported, even in association with aqueous preparations of as little as 10 g of gastrodia (Chen and Chen 2004).

Pharmacological Considerations

None known.

Pregnancy and Lactation

No information on the safety of gastrodia in pregnancy or lactation was identified in the scientific or traditional literature. Although this review did not identify any concerns for use while pregnant or nursing, safety has not been conclusively established.

G

Review Details

I. Drug and Supplement Interactions

Clinical Trials of Drug or Supplement Interactions

No clinical trials of drug or supplement interactions were identified.

Case Reports of Suspected Drug or Supplement Interactions

No case reports of suspected drug or supplement interactions were identified.

Animal Trials of Drug or Supplement Interactions

No animal trials of drug or supplement interactions were identified.

II. Adverse Events

Case Reports of Adverse Events

Rare cases of allergic reactions to gastrodia have been reported, including urticaria, purpura, vertigo, acute renal failure, and anaphylactic shock (Bensky et al. 2004). Rash, eczema and hair loss have been reported in association with gastrodia use, and side effects are reported to occur even at doses of as little as 10 g as an aqueous preparation (Chen and Chen 2004).

Very large doses (aqueous extracts of 80 g within 3 h) have been reported to cause poisoning, with symptoms including flushing and hot sensation of the face, general weakness, headache, dizziness, visual disturbance, loss of muscle coordination, and loss of consciousness (Bensky et al. 2004). Overdose of gastrodia (standard dose listed as an aqueous extract of 3–10 g) is characterized by headache, nausea, vomiting, facial flushing, drowsiness, and delayed response to light (Chen and Chen 2004).

III. Pharmacology and Pharmacokinetics

Human Pharmacological Studies

No relevant human pharmacological studies were identified.

Animal Pharmacological Studies

No relevant animal pharmacological studies were identified.

In Vitro Pharmacological Studies

Antiplatelet activity of several compounds isolated from gastrodia was observed in platelet-rich plasma from rats (Pyo et al. 2004).

IV. Pregnancy and Lactation

No information on the safety of gastrodia during pregnancy or lactation was identified.

V. Toxicity Studies

Acute Toxicity

The LD_{50} of intraperitoneally administered aqueous extract of gastrodia in mice is 51.4 to 61.4 g/kg (Zhu 1998).

The LD_{50} of the compound gastrodin in mice is 1000 mg/kg after oral administration and 337 mg/kg after intravenous administration (Chen and Chen 2004).

In rabbits intraperitoneally administered 12 g/kg of the compound gastrodin, lethargy, reduced deep-tendon reflexes, loss of appetite, and tachycardia were observed (Chen and Chen 2004).

Subchronic Toxicity

No changes in serum chemistry, liver or kidney function tests, or blood lipids were observed in dogs orally administered 75 mg/kg of the compound gastrodin daily for 60 days, or in mice orally administered 375 mg/kg of the compound *p*-hydroxybenzyl alcohol or 250 mg/kg of the compound gastrodin daily for 60 days (Zhu 1998).

Literature Cited

Bensky, D., S. Clavey, and E. Stöger. 2004. *Chinese herbal medicine: Materia medica*. 3rd ed. Seattle: Eastland Press.

Chen, J.K., and T.T. Chen. 2004. *Chinese medical herbology and pharmacology*. City of Industry, CA: Art of Medicine Press.

Pyo, M.K., J.L. Jin, Y.K. Koo, and H.S. Yun-Choi. 2004. Phenolic and furan type compounds isolated from *Gastrodia elata* and their anti-platelet effects. *Arch. Pharm. Res.* 27(4):381-385.

Zhu, Y.-P. 1998. *Chinese materia medica: Chemistry, pharmacology and applications*. Amsterdam: Harwood Academic Publishers.

G

Gaultheria procumbens L. — Ericaceae

SCN: wintergreen
OCN: checkerberry; teaberry

Part: leaf

Quick Reference Summary

Safety Class: 1
Interaction Class: A

Contraindications

None known.

Other Precautions

None known.

Drug and Supplement Interactions

None known.

Notice

Salicylates (0.5–1.0% in the leaf, 98% in the essential oil) (Chyka et al. 2007; Ribnicky et al. 2003; Williamson 2003); *see* Appendix 1.

Adverse Events and Side Effects

Cases of salicylate poisoning have been reported in persons that accidentally (typically children) or intentionally (typically persons attempting to self-poison) ingested wintergreen essential oil (Chan 1996a; Chyka et al. 2007; Stevenson 1937). Common features of salicylate poisoning include nausea, vomiting, tinnitus, deafness, sweating, hyperventilation, respiratory alkalosis, and metabolic acidosis (Chyka et al. 2007).

Pharmacological Considerations

None known.

Pregnancy and Lactation

A review of reproductive and developmental toxicity studies of the compound methyl salicylate reported that the no-observed-adverse-effect level for reproductive toxicity was 75 to 100 mg/kg daily, similar to that of salicylic acid (Belsito et al. 2007).

No information on the safety of wintergreen in lactation was identified. While this review did not identify any concerns for use while nursing, safety has not been conclusively established.

Review Details

I. Drug and Supplement Interactions

Clinical Trials of Drug or Supplement Interactions

No clinical trials of drug or supplement interactions were identified.

Case Reports of Suspected Drug or Supplement Interactions

No case reports of suspected drug or supplement interactions were identified.

Animal Trials of Drug or Supplement Interactions

No animal trials of drug or supplement interactions were identified.

II. Adverse Events

Case Reports of Adverse Events

A 1937 case report and review of the literature identified 43 cases of wintergreen oil poisoning, almost half of which were in young children that drank bottles of wintergreen oil (Stevenson 1937). A more recent review of cases of methyl salicylate poisoning from wintergreen reported to the American poison control centers between 1985 and 2003 identified 11 cases. The review reports wintergreen oil intake in terms of estimated salicylate intake associated with the poisonings and provides a conversion factor of 1.4 for an aspirin equivalent dose of wintergreen oil. The smallest amount of wintergreen oil reported to cause toxicity was the equivalent of 346 mg/kg of salicylates; the lowest lethal dose of salicylates reported was 432 mg/kg in a 2.5-year-old, although a 1-year-old survived a 486 mg/kg dose with symptoms of severe toxicity. This review also includes evidence-based guidelines for out-of-hospital management of salicylate poisoning (Chyka et al. 2007). Another review indicated that 30 ml of wintergreen oil can be lethal in adults, although doses of 6 ml have been lethal in some adults, and doses of 4 ml are lethal to children (Botma et al. 2001).

Common features of salicylate poisoning include nausea, vomiting, tinnitus, deafness, sweating, hyperventilation, respiratory alkalosis, and metabolic acidosis. Uncommon features of poisoning include coma, fever, hypo- or hyperglycemia, gastrointestinal bleeding, seizures, fluid retention, pulmonary edema, respiratory distress syndrome, cerebral edema, and renal failure (Botma et al. 2001; Chan 1996b; Chyka et al. 2007; Proudfoot 1983).

Laryngeal edema, vomiting, irritability, lethargy, and tachypnea were reported in an 18-month-old who had accidentally ingested an unspecified amount of wintergreen essential oil. Serum salicylate levels at the time of hospital admission (~12 hours after initial ingestion) were 4.8 mmol/l (Botma et al. 2001).

Topical application of products containing the compound methyl salicylate has led to poisonings in some individuals (Chyka et al. 2007).

III. Pharmacology and Pharmacokinetics

Human Pharmacological Studies

No relevant human pharmacological studies were identified.

Animal Pharmacological Studies

No relevant animal pharmacological studies were identified.

In Vitro Pharmacological Studies

Inhibition of the drug-metabolizing isoenzyme CYP3A4 was observed after treatment with an ethanol extract of wintergreen. No effects on CYP19 were observed (Scott et al. 2006).

IV. Pregnancy and Lactation

A review of reproductive and developmental toxicity studies of methyl salicylate indicated that, under conditions of sufficient exposure, there is a pattern of embryotoxicity and teratogenesis that is similar to that caused by comparable doses of salicylic acid. The abnormalities included neural tube defects and malformations of the skeleton and viscera. Studies of orally administered methyl salicylate indicated that the no-observed-adverse-effect level (NOAEL) for reproductive toxicity is 75–100 mg/kg daily, which is a level consistent with data from subchronic and chronic toxicity studies and is also consistent with the reproductive NOAEL for salicylic acid, which has been reported as 80 mg/kg daily (Belsito et al. 2007).

No information on the safety of wintergreen during lactation was identified.

V. Toxicity Studies

Acute Toxicity

The oral LD_{50} of the compound methyl salicylate (which comprises 97 to 99% of wintergreen essential oil) is 1110 mg/kg in mice, 887 or 1250 mg/kg in rats, 1060 mg/kg in guinea pigs, 1300 or 2800 mg/kg in rabbits, and 2100 mg/kg in dogs (FAO/WHO 1967). The adult human oral LD_{50} is estimated at 500 mg/kg (FAO/WHO 1967).

Subchronic Toxicity

In dogs orally administered 50, 100, 250, 500, 800, or 1200 mg/kg methyl salicylate for up to 10 weeks, no adverse effects were noted in animals receiving 250 mg/kg or less, but an increasing dose-dependent fatty metamorphosis of the liver was observed at higher test levels (Webb and Hansen 1963).

In rats fed diets containing 0.1, 0.5, or 1.0% methyl salicylate for 17 weeks, both sexes showed a significant reduction in growth rate at the 1.0% level, but histological examination of the major organs revealed no abnormality. In a related experiment, in rats fed a diet containing 2% methyl salicylate for up to 10 weeks, bone growth was reduced and excessive density of bone with reduced chondroclastic and osteoclastic activity was observed (Webb and Hansen 1963).

Chronic Toxicity

In rats fed diets containing 0.1, 0.5, 1.0, or 2.0% methyl salicylate for 2 years, animals fed the highest dose did not survive past 49 weeks. At the 1% level, growth rates were considerably reduced and enlargement of male testes and female hearts and kidneys were noted. Excess cancellous bone formation was seen at the 2.0, 1.0, and 0.5% levels (Webb and Hansen 1963). Another 2-year feeding study revealed no adverse effects, including bone changes, at dietary levels up to 0.21% of methyl salicylate (Packman et al. 1961).

In dogs orally administered 0, 50, 150, and 250 mg/kg methyl salicylate daily for 2 years, some growth retardation and liver enlargement were noted at the 150 and 250 mg/kg level, and histology revealed enlarged hepatic parenchymal cells (Webb and Hansen 1963).

Genotoxicity

In the *Bacillus subtilis rec* assay, no mutagenic activity of methyl salicylate was observed. In the Ames test for mutagenicity, one study indicated no mutagenic activity; another study indicated no mutagenic activity with S9 from rats, mice, or guinea pigs but some mutagenic activity with S9 from hamsters (Lapczynski et al. 2007). No mutagenic activity was observed in the chromosomal aberration test in a Chinese hamster fibroblast cell line (Lapczynski et al. 2007).

Literature Cited

Belsito, D., D. Bickers, M. Bruze, et al. 2007. A toxicologic and dermatologic assessment of salicylates when used as fragrance ingredients. *Food Chem. Toxicol.* 45(1S1):318-361.

Botma, M., W. Colquhoun-Flannery, and S. Leighton. 2001. Laryngeal oedema caused by accidental ingestion of oil of wintergreen. *Int. J. Pediatr. Otorhinolaryngol.* 58(3):229-232.

Chan, T.Y. 1996a. Potential dangers from topical preparations containing methyl salicylate. *Hum. Exp. Toxicol.* 15(9):747-750.

Chan, T.Y. 1996b. The risk of severe salicylate poisoning following the ingestion of topical medicaments or aspirin. *Br. Med. J.* 72(844):109-112.

Chyka, P.A., A.R. Erdman, G. Christianson, et al. 2007. Salicylate poisoning: An evidence-based consensus guideline for out-of-hospital management. *Clin. Toxicol. (Phila.)* 45(2):95-131.

FAO/WHO. 1967. Methyl salicylate. Toxicological evaluation of some flavouring substances and non-nutritive sweetening agents. FAO Nutrition Meetings Resort Series No. 44A and WHO Food Additives 68.33.

Lapczynski, A., L. Jones, D. McGinty, S.P. Bhatia, C.S. Letizia, and A.M. Api. 2007. Fragrance material review on methyl salicylate. *Food Chem. Toxicol.* 45(1, Suppl. 1):S428-S452.

Packman, E.W., D.D. Abbott, B.M. Wagner, and J.W.E. Harrison. 1961. Chronic oral toxicity of oil of sweet birch (methyl salicylate). *Pharmacologist* 3:62.

Proudfoot, A.T. 1983. Toxicity of salicylates. *Am. J. Med.* 75(5A):99.

Ribnicky, D.M., A. Poulev, and I. Raskin. 2003. The determination of salicylates in *Gaultheria procumbens* for use as a natural aspirin alternative. *J. Nutraceut. Funct. Med. Foods* 4(1):39-52.

Scott, I.M., R.I. Leduc, A.J. Burt, R.J. Marles, J.T. Arnason, and B.C. Foster. 2006. The inhibition of human cytochrome P450 by ethanol extracts of North American botanicals. *Pharmaceut. Biol.* 44(5):315-327.

Stevenson, C.S. 1937. Oil of wintergreen (methyl salicylate) poisoning: Report of three cases, one with autopsy, and a review of the literature. *Am. J. Med. Sci.* 193(6):772-788.

Webb, W.K., and W.H. Hansen.1963. Chronic and subacute toxicology and pathology of methyl salicylate in dogs, rats and rabbits. *Toxicol. Appl. Pharmacol.* 5:576.

Williamson, E.M. 2003. *Potter's herbal cyclopedia.* Saffron Walden, Essex: C.W. Daniel Co.

G

Gelidiella spp. and *Gelidium* spp. — Gelidiaceae

Gelidiella acerosa (Forssk.) Feldman & Hamel
SCN: agar (dried mucilage)
OCN: agar-agar

Gelidium amansii **(Lamouroux) Lamouroux**
SCN: agar (dried mucilage)
OCN: American agar; agar-agar

Gelidium capense **(S.G. Gmelin) P.C. Silva**
SCN: agar (dried mucilage)
Syn: *Gelidium cartilagineum* (L.) Gaillon
OCN: agar-agar; Japanese isinglass; Pacific coast agar

Gelidium crinale **(Turner) Gaillon**
SCN: agar (dried mucilage)
OCN: agar-agar

Gelidium divaricatum **G. Martens**
SCN: agar (dried mucilage)
OCN: agar-agar

Gelidium pacificum **Okamura**
SCN: agar (dried mucilage)
OCN: agar-agar

Gelidium vagum **Okamura**
SCN: agar (dried mucilage)
OCN: agar-agar
Part: thallus

Quick Reference Summary

Safety Class: 1
Interaction Class: A

Contraindications

None known

Other Precautions

When taken as a bulk-forming laxative, agar should not be used in persons with bowel obstruction or with abnormal esophageal or intestinal narrowing.

Take with at least 250 ml (8 oz.) of liquid (CFR 2011a).

Drug and Supplement Interactions

Other drugs or supplements should be taken 1 hour prior to consumption of agar or several hours after consumption, as agar may reduce the absorption of certain drugs and supplements due to the mucilage content and the increased speed of passage through the intestines (Bradley 1992; Brinker 2001); *see* Mucilages in Appendix 3.

Standard Dose

The standard dose is 4 to 16 g taken once or twice daily (Martindale and Reynolds 1996).

Notice

Bulk-forming laxative (CFR 2011a); *see* Appendix 2.

Mucilages; *see* Appendix 3.

Editors' Notes

Specific labeling is required in the United States for all over-the-counter drug products containing agar (CFR 2011a); *see* Bulk-forming laxatives in Appendix 2.

Concerns for this herb are based on relatively higher doses used for therapeutic purposes in contrast to lower amounts generally used in cooking, and have not been associated with its use as a spice.

Agar is commonly used as a thickening agent in foods and is identified as generally recognized as safe by the U.S. FDA in limited quantities for certain specified uses, including in baked goods, confections, and soft candy (CFR 2011b). Although an associated regulation states that other uses in foods would require a food additive regulation (CFR 2011c), dietary ingredients for use in dietary supplements are specifically excluded from the federal food additive definition (U.S.C. 2010).

Adverse Events and Side Effects

Blockages in the small intestine have occurred after ingestion of agar with insufficient quantities of liquids (Osada et al. 2008).

Pharmacological Considerations

None known.

Pregnancy and Lactation

Although animal studies have indicated that very large doses (0.5 or 1 g/kg) of agar have led to postimplantation losses in pregnant rats (Premakumara et al. 1995, 1996; Ratnasooriya et al. 1994), agar is commonly used as a thickening agent in foods and is believed to be safe at standard food and therapeutic levels.

No information on the safety of agar during lactation was identified. While this review did not identify any concerns for use while nursing, safety has not been conclusively established.

Review Details

I. Drug and Supplement Interactions

Clinical Trials of Drug or Supplement Interactions

No clinical trials of drug or supplement interactions were identified.

Case Reports of Suspected Drug or Supplement Interactions

No case reports of suspected drug or supplement interactions were identified.

Animal Trials of Drug or Supplement Interactions

No animal trials of drug or supplement interactions were identified.

II. Adverse Events

Case Reports of Adverse Events

Small bowel obstruction due to primary small bowel bezoars was seen in a 70-year-old woman who had ingested "a large amount" of highly concentrated agar dissolved in boiling water 2 days earlier (Osada et al. 2008).

An allergic reaction to agar, including nasal and asthmatic symptoms, was reported in a bakery worker who routinely used a product containing agar. Skin and sniff testing confirmed agar as the allergen (Criep and Riley 1951).

III. Pharmacology and Pharmacokinetics

Human Pharmacological Studies

No relevant human pharmacological studies were identified.

Animal Pharmacological Studies

No relevant animal pharmacological studies were identified.

In Vitro Pharmacological Studies

No relevant in vitro pharmacological studies were identified.

IV. Pregnancy and Lactation

In pregnant rats orally administered 1 g/kg of an agar extract on days 7 and 8 of pregnancy, an 89% postimplantation loss rate was observed, resulting in fetal death between days 9 and 14 of pregnancy (Premakumara et al. 1995).

The hexane fraction of a methanol-dichloromethane extract of agar exhibited antifertility activity in rats orally administered the compound on days 1 through 7 of pregnancy (Premakumara et al. 1996).

Dose-dependent increases in resorption sites and increases in postimplantation losses were observed in rats orally administered 0.5 or 1 g/kg daily of a methanol-dichloromethane extract of agar on days 1 through 7 of pregnancy. Postimplantation losses of 44% were observed after the 0.5 g/kg dose and 98% after the 1 g/kg dose (Ratnasooriya et al. 1994).

No information on the safety of agar during lactation was identified.

V. Toxicity Studies

No toxicity studies were identified.

Literature Cited

Bradley, P.R. 1992. *British herbal compendium: A handbook of scientific information on widely used plant drugs.* Bournemouth, UK: British Herbal Medicine Association.

Brinker, F. 2001. *Herb contraindications and drug interactions.* 3rd ed. Sandy, OR: Eclectic Medical Publications.

CFR. 2011a. *Code of federal regulations*, Title 21 Part 201.319, 2011 ed. Specific labeling requirements for specific drug products. Water-soluble gums, hydrophilic gums, and hydrophilic mucilloids (including, but not limited to agar, alginic acid, calcium polycarbophil, carboxymethylcellulose sodium, carrageenan, chondrus, glucomannan ((B-1,4 linked) polymannose acetate), guar gum, karaya gum, kelp, methylcellulose, plantago seed (psyllium), polycarbophil tragacanth, and xanthan gum) as active ingredients; required warnings and directions. Washington, DC: U.S. Government Printing Office.

CFR. 2011b. *Code of federal regulations*, Title 21 Part 184.1115, 2011 ed. Direct food substances affirmed as generally recognized as safe. Listing of specific substances affirmed as GRAS. Agar-agar. Washington, DC: U.S. Government Printing Office.

CFR. 2011c. *Code of federal regulations*, Title 21 Part 184.1(b)(2), 2011 ed. Direct food substances affirmed as generally recognized as safe. Substances added directly to human food affirmed as generally recognized as safe (GRAS). Washington, DC: U.S. Government Printing Office.

Criep, L.H., and W.K. Riley. 1951. Allergic manifestations to agar. *J. Am. Med. Assoc.* 145(7):485-486.

Martindale, W., and J.E.F. Reynolds. 1996. *The extra pharmacopoeia.* London: Pharmaceutical Press.

Osada, T., T. Shibuya, T. Kodani, K. Beppu, N. Sakamoto, A. Nagahara, T. Ohkusa, T. Ogihara, and S. Watanabe. 2008. Obstructing small bowel bezoars due to an agar diet: Diagnosis using double balloon enteroscopy. *Intern. Med.* 47(7):617-620.

Premakumara, G.A., W.D. Ratnasooriya, and L.M. Tillekeratne. 1995. Studies on the post-coital contraceptive mechanisms of crude extract of Sri Lankan marine red algae, *Gelidiella acerosa. Contraception* 52(3):203-207.

Premakumara, G.A., W.D. Ratnasooriya, and L.M. Tillekeratne. 1996. Isolation of a non-steroidal contragestative agent from the Sri Lankan marine red alga, *Gelidiella acerosa. Contraception* 54(6):379-383.

Ratnasooriya, W.D., G.A. Premakumara, and L.M. Tillekeratne. 1994. Post-coital contraceptive activity of crude extracts of Sri Lankan marine red algae. *Contraception* 50(3):291-299.

U.S.C. 2010. United States Code, Title 21, Part 321 (s)(6). Current as of January 7, 2011. Washington, DC: U.S. Government Printing Office

Genista tinctoria L. — Fabaceae

SCN: dyer's broom
OCN: dyer's greenwood
Part: flower, herb

Quick Reference Summary

Safety Class: 2b, 2d
Interaction Class: A

Contraindications

Not for use in pregnancy except under the supervision of a qualified healthcare practitioner (Casarett et al. 2001; McGuffin et al. 1997; Wichtl 2004).

Not for use in persons with high blood pressure (Wichtl 2004).

Other Precautions

Exceeding the standard dose may cause diarrhea (Wichtl 2004).

Drug and Supplement Interactions

None known.

Standard Dose

The standard dose is 1 to 2 g as a tea (Wichtl 2004).

Notice

Emetic (Williamson 2003); *see* Appendix 2.

Editors' Note

Although one reference text lists dyer's broom as "toxic" (Roth et al. 1984), this concern is not substantiated by other texts (List and Hörhammer 1973; Madaus 1976; Wichtl 2004; Williamson 2003).

Adverse Events and Side Effects

Dyer's broom has been reported to cause nausea and vomiting (Williamson 2003). Diarrhea has been associated with excessive use of dyer's broom (Wichtl 2004).

Pharmacological Considerations

None known.

Pregnancy and Lactation

No information on the safety of dyer's broom in pregnancy or lactation was identified, although constituents found in dyer's broom are similar to the compound sparteine, a uterine stimulant, found in *Cytisus scoparius* (Casarett et al. 2001; McGuffin et al. 1997; Wichtl 2004). Based on this information, use during pregnancy is not recommended except under the supervision of a qualified healthcare practitioner.

Review Details

I. Drug and Supplement Interactions

Clinical Trials of Drug or Supplement Interactions

No clinical trials of drug or supplement interactions were identified.

Case Reports of Suspected Drug or Supplement Interactions

No case reports of suspected drug or supplement interactions were identified.

Animal Trials of Drug or Supplement Interactions

No animal trials of drug or supplement interactions were identified.

II. Adverse Events

Case Reports of Adverse Events

No case reports of adverse events were identified.

III. Pharmacology and Pharmacokinetics

Human Pharmacological Studies

No relevant human pharmacological studies were identified.

Animal Pharmacological Studies

No relevant animal pharmacological studies were identified.

In Vitro Pharmacological Studies

No relevant in vitro pharmacological studies were identified.

IV. Pregnancy and Lactation

No information on the safety of dyer's broom in pregnancy or lactation was identified, although constituents found in dyer's broom are similar to the compound sparteine, a uterine stimulant, found in *Cytisus scoparius* (Casarett et al. 2001; McGuffin et al. 1997; Wichtl 2004).

V. Toxicity Studies

No toxicity studies were identified.

Literature Cited

Casarett, L.J., C.D. Klaassen, and J. Doull. 2001. *Casarett and Doull's toxicology: The basic science of poisons*. New York: McGraw-Hill Professional.

List, P.H., and H. Hörhammer. 1973. *Hagers handbuch der pharmazeutischen praxis*. Berlin: Springer.

Madaus, G. 1976. *Lehrbuch der biologischen heilmittel*. New York: Hildesheim.

McGuffin, M., C. Hobbs, R. Upton, and A. Goldberg. 1997. *Botanical safety handbook*. Boca Raton, FL: CRC Press.

Roth, L., M. Daunderer, and K. Kormann. 1984. *Giftpflanzen-pflanzengifte: Vorkommen, wirkung, therapie*. Landsberg: Ecomed.

Wichtl, M. 2004. *Herbal drugs and phytopharmaceuticals: A handbook for practice on a scientific basis*. 3rd ed. Boca Raton, FL: CRC Press.

Williamson, E.M. 2003. *Potter's herbal cyclopedia*. Saffron Walden, Essex: C.W. Daniel Co.

Gentiana lutea L. — Gentianaceae

SCN: gentian
OCN: yellow gentian

Part: root

Quick Reference Summary

Safety Class: 1
Interaction Class: A

Contraindications

None known.

Other Precautions

Use with caution in persons with gastrointestinal irritation (Felter and Lloyd 1898), and duodenal ulcers (Bradley 1992).

Drug and Supplement Interactions

None known.

Notice

Emmenagogue (India 1987); *see* Appendix 2.

Editors' Notes

Multiple cases of misidentification of *Veratrum* species as gentian by individuals collecting for home use have been reported in the literature. *Veratrum* species produce adverse cardiovascular and gastrointestinal effects after consumption (Festa et al. 1996; Garnier et al. 1982, 1985; Grobosch et al. 2008; Zagler et al. 2005).

Adverse Events and Side Effects

The German Standard License label states that headaches may occasionally occur in persons sensitive to bitter substances (Wichtl 2004).

Gentian may cause gastrointestinal irritation. Irritating qualities are maximized in tincture form and minimized as a tea (McGuffin et al. 1997).

Pharmacological Considerations

None known.

Pregnancy and Lactation

Gentian has been used as an emmenagogue in the Unani tradition of medicine in India (India 1987), although such activity has not been reported in any literature from Europe where gentian is commonly used.

No information on the use of gentian during lactation was identified. While this review did not identify any concerns for use while nursing, safety has not been conclusively established.

Review Details

I. Drug and Supplement Interactions

Clinical Trials of Drug or Supplement Interactions

No clinical trials of drug or supplement interactions were identified.

Case Reports of Suspected Drug or Supplement Interactions

No case reports of suspected drug or supplement interactions were identified.

Animal Trials of Drug or Supplement Interactions

No animal trials of drug or supplement interactions were identified.

II. Adverse Events

Adverse Events Reported in Clinical Trials

Among 205 patients that took an average of 576 mg of a dry hydroethanolic extract of gentian (equivalent to 2.9 g of dried root) daily for 15 days, 2.4% reported minor adverse events including flatulence, stomach cramps, nausea, and headache (Wegener 1998).

Case Reports of Adverse Events

No case reports of adverse events associated with gentian use were identified.

Multiple cases of misidentification of *Veratrum* species as gentian by individuals collecting for home use have been reported in the literature. *Veratrum* species are botanically similar to and grow in similar habitats to gentian and produce adverse cardiovascular and gastrointestinal effects after consumption (Festa et al. 1996; Garnier et al. 1982, 1985; Grobosch et al. 2008; Zagler et al. 2005).

III. Pharmacology and Pharmacokinetics

Human Pharmacological Studies

No relevant human pharmacological studies were identified.

Animal Pharmacological Studies

No relevant animal pharmacological studies were identified.

In Vitro Pharmacological Studies

Some induction of the drug-metabolizing isoenzyme CYP1A2 and the drug transporter protein MDR1 were observed in LS180 (human colon carcinoma) cells treated with an ethanolic extract of gentian (Brandin et al. 2007).

IV. Pregnancy and Lactation

Gentian has been used as an emmenagogue in the Unani tradition of medicine in India (India 1987), although such activity has not been reported in any literature from Europe where gentian is commonly used.

No information on the use of gentian during lactation was identified.

V. Toxicity Studies

Acute Toxicity

The LD_{50} of an orally administered ethanol extract of gentian (bitterness value 200 Ph. Helv. Units/g) in mice is 25 ml/kg (Leslie 1978).

No signs of toxicity were observed in rabbits treated with 12.6 mg/animal of gentian extract daily for 3 days. Aside from slightly lower erythrocyte levels as compared to the control group, no abnormal serum parameters were observed (Chibanguza et al. 1984).

No signs of toxicity were observed in mice intraperitoneally administered 250 or 500 mg/kg of a methanol extract of gentian (Ozturk et al. 2002).

Genotoxicity

Weak mutagenic activity of gentian and compounds isolated from gentian were observed in the Ames test for mutagenicity in *Salmonella typhimurium* strains TA97, TA98, TA100, and TA2637 with metabolic activation by S9 (Göggelmann and Schimmer 1986; Hänsel et al. 1993; Matsushima et al. 1985; Morimoto et al. 1983). Compounds believed to be responsible for the mutagenic activity were noted as being structurally similar to quercetin, a compound that has mutagenic activity in vitro but is regarded as safe in humans (ESCOP 2003; Harwood et al. 2007).

Literature Cited

Bradley, P.R. 1992. *British herbal compendium: A handbook of scientific information on widely used plant drugs*. Bournemouth, UK: British Herbal Medicine Association.

Brandin, H., E. Viitanen, O. Myrberg, and A.K. Arvidsson. 2007. Effects of herbal medicinal products and food supplements on induction of CYP1A2, CYP3A4 and MDR1 in the human colon carcinoma cell line LS180. *Phytother. Res.* 21(3):239-244.

Chibanguza, G., R. März, and W. Sterner. 1984. Zur Wirksamkeit und Toxizität eines pflanzlichen Sekretolytikums und seiner Einzeldrogen. *Arzneimittel-Forschung* 34(1):32-36.

ESCOP. 2003. *ESCOP monographs: The scientific foundation for herbal medicinal products*. 2nd ed. Exeter, U.K.: European Scientific Cooperative on Phytotherapy.

Felter, H.W., and J.U. Lloyd. 1898. *King's American dispensatory*. 18th ed., 3rd rev. 2 vols. Cincinnati: Ohio Valley Co.

Festa, M., B. Andreetto, M.A. Ballaris, A. Panio, and R. Piervittori. 1996. A case of *Veratrum* poisoning. *Minerva Anestesiol.* 62(5):195-196.

Garnier, R., P. Carlier, J. Hoffelt, and A. Savidan. 1985. Acute dietary poisoning by white hellebore (*Veratrum album* L.). Clinical and analytical data. A propos of 5 cases. *Ann. Med. Intern. (Paris)* 136(2):125-128.

Garnier, R., J. Hoffelt, and P. Carlier. 1982. *Veratrum* poisoning with home made gentian wine; clinical and analytical findings. *Vet. Hum. Toxicol.* 24(Suppl.):138-141, 193.

Göggelmann, W., and O. Schimmer. 1986. Mutagenic activity of phytotherapeutic drugs. In *Genetic toxicology of the diet*, edited by Knudsen, I. New York: Alan R. Liss.

Grobosch, T., T. Binscheck, F. Martens, and D. Lampe. 2008. Accidental intoxication with *Veratrum album*. *J. Anal. Toxicol.* 39(9):768-773.

Hänsel, R., K. Keller, H. Rimpler, and G. Schneider, eds. 1993. *Hagers handbuch der pharmazeutischen praxis, Volume 5*. 5th ed. Berlin: Springer.

Harwood, M., B. Danielewska-Nikiel, J.F. Borzelleca, et al. 2007. A critical review of the data related to the safety of quercetin and lack of evidence of in vivo toxicity, including lack of genotoxic/carcinogenic properties. *Food Chem. Toxicol.* 45(11):2179-2205.

India, G.o.I. 1987. *Standardisation of single drugs of Unani medicine, Part II*. New Delhi: Central Council for Research in Unani Medicine.

Leslie, G.B. 1978. A pharmacometric evaluation of nine Bio-Strath herbal remedies. *Medita* 8:31-37.

Matsushima, T., A. Araki, O. Yagame, et al. 1985. Mutagenicities of xanthone derivatives in *Salmonella typhimurium* TA100, TA98, TA97, and TA2637. *Mutat. Res.* 150(1-2):141-146.

McGuffin, M., C. Hobbs, R. Upton, and A. Goldberg. 1997. *Botanical safety handbook*. Boca Raton, FL: CRC Press.

Morimoto, I., T. Nozaka, F. Watanabe, et al. 1983. Mutagenic activities of gentisin and isogentisin from *Gentianae radix* (Gentianaceae). *Mutat. Res.* 116(2):103-117.

Ozturk, N., K.H. Can Baser, S. Aydin, Y. Ozturk, and I. Calis. 2002. Effects of *Gentiana lutea* ssp. *symphyandra* on the central nervous system in mice. *Phytother. Res.* 16(7).

Wegener, T. 1998. Anwendung eines Trockenextraktes Symptomcomplex. *Z. Phytother.* 19:163-164.

Wichtl, M. 2004. *Herbal drugs and phytopharmaceuticals: A handbook for practice on a scientific basis*. 3rd ed. Boca Raton, FL: CRC Press.

Zagler, B., A. Zelger, C. Salvatore, et al. 2005. Dietary poisoning with *Veratrum album*—A report of two cases. *Wien Klin. Wochenschr.* 117(3):106-108.

G

Gentiana macrophylla Pall. — Gentianaceae

SCN: large-leaf gentian
PN: *qin jiao* (root)

Part: root

Quick Reference Summary

Safety Class: 1
Interaction Class: A

Contraindications

None known.

Other Precautions

See Editors' Note.

Drug and Supplement Interactions

None known.

Editors' Note

The concerns stated for *Gentiana lutea* (*see* previous entry), though not specified in available references, may also be relevant to this species (McGuffin et al. 1997).

Adverse Events and Side Effects

Large-leaf gentian may cause drowsiness or sedation (Chen and Chen 2004).

Pharmacological Considerations

None known.

Pregnancy and Lactation

No information on the safety of large-leaf gentian in pregnancy or lactation was identified in the scientific or traditional literature. Although this review did not identify any concerns for use while pregnant or nursing, safety has not been conclusively established.

Review Details

I. Drug and Supplement Interactions

Clinical Trials of Drug or Supplement Interactions

No clinical trials of drug or supplement interactions were identified.

Case Reports of Suspected Drug or Supplement Interactions

No case reports of suspected drug or supplement interactions were identified.

Animal Trials of Drug or Supplement Interactions

No animal trials of drug or supplement interactions were identified.

II. Adverse Events

Case Reports of Adverse Events

No case reports of adverse events were identified.

III. Pharmacology and Pharmacokinetics

Human Pharmacological Studies

No relevant human pharmacological studies were identified.

Animal Pharmacological Studies

Oral administration of 100 mg of the compound gentianine daily to mice for 4 to 13 days resulted in severe nausea and vomiting (Zhu 1998).

In Vitro Pharmacological Studies

No relevant in vitro pharmacological studies were identified.

IV. Pregnancy and Lactation

No information on the safety of large-leaf gentian during pregnancy or lactation was identified.

V. Toxicity Studies

Acute Toxicity

The LD_{50} of the compound gentianine in mice is 480 mg/kg after oral administration, 350 mg/kg after intraperitoneal administration, and 250 to 300 mg/kg after intravenous administration (Chen and Chen 2004). The LD_{50} of the compound gentianadine is 1250 mg/kg after oral administration (Chang and But 1986).

Literature Cited

Chang, H.-M., and P.P.H. But. 1986. *Pharmacology and applications of Chinese materia medica*. English ed. Philadelphia: World Scientific.

Chen, JK, TT Chen. 2004. *Chinese medical herbology and pharmacology*. City of Industry, CA: Art of Medicine Press.

Felter, H.W., and J.U. Lloyd. 1898. *King's American dispensatory*. 18th ed., 3rd rev. 2 vols. Cincinnati: Ohio Valley Co.

McGuffin, M., C. Hobbs, R. Upton, and A. Goldberg. 1997. *Botanical safety handbook*. Boca Raton, FL: CRC Press.

Zhu, Y.-P. 1998. *Chinese materia medica: Chemistry, pharmacology and applications*. Amsterdam: Harwood Academic Publishers.

Gentiana scabra Bunge — Gentianaceae

SCN: scabrous gentian
PN: *long dan cao* (root and rhizome)

Part: root and rhizome

Quick Reference Summary

Safety Class: 1
Interaction Class: A

Contraindications

None known.

Other Precautions

See Editors' Note.

Drug and Supplement Interactions

None known.

Editors' Note

The concerns stated for *Gentiana lutea* (*see* previous entries), though not specified in available references, may also be relevant to this species (McGuffin et al. 1997).

Adverse Events and Side Effects

When administered after meals or in excessive dosage, scabrous gentian may cause impairment of digestive function and, occasionally, headache, flushing of the face, and vertigo (Bensky et al. 2004; Chang and But 1986; Chen and Chen 2004).

Pharmacological Considerations

Scabrous gentian may cause drowsiness or sedation (Chen and Chen 2004).

Pregnancy and Lactation

No information on the safety of scabrous gentian in pregnancy or lactation was identified in the scientific or traditional literature. Although this review did not identify any concerns for use while pregnant or nursing, safety has not been conclusively established.

G

Review Details

I. Drug and Supplement Interactions

Clinical Trials of Drug or Supplement Interactions

No clinical trials of drug or supplement interactions were identified.

Case Reports of Suspected Drug or Supplement Interactions

No case reports of suspected drug or supplement interactions were identified.

Animal Trials of Drug or Supplement Interactions

No animal trials of drug or supplement interactions were identified.

II. Adverse Events

Case Reports of Adverse Events

In high doses (aqueous extract of 30 g as a single dose), abdominal pain, nausea, vomiting, vertigo, impaired consciousness, and stiff neck have been reported (Bensky et al. 2004).

In a case of overdose, a 150 g dose resulted in the previously listed symptoms within 30 minutes of ingestion. Emergency treatment was necessary, and the patient fully recovered in 2 days (Bensky et al. 2004).

III. Pharmacology and Pharmacokinetics

Human Pharmacological Studies

No relevant human pharmacological studies were identified.

Animal Pharmacological Studies

No relevant animal pharmacological studies were identified.

In Vitro Pharmacological Studies

The compound 2-hydroxy-3-methoxybenzoic acid glucose ester isolated from large-leaved gentian was found to be a potent antagonist of platelet-activating factor (Huh et al. 1998).

IV. Pregnancy and Lactation

No information on the safety of scabrous gentian during pregnancy or lactation was identified.

V. Toxicity Studies

Acute Toxicity

The LD_{50} of the compound gentianine in mice is 460 mg/kg after oral administration, 500 mg/kg after subcutaneous administration, and 250 to 300 mg/kg after intravenous administration (Chen and Chen 2004).

Literature Cited

Bensky, D., S. Clavey, and E. Stöger. 2004. *Chinese herbal medicine: Materia medica*. 3rd ed. Seattle: Eastland Press.

Chang, H.-M., and P.P.H. But. 1986. *Pharmacology and applications of Chinese materia medica*. English ed. Philadelphia: World Scientific.

Chen, J.K., and T.T. Chen. 2004. *Chinese medical herbology and pharmacology*. City of Industry, CA: Art of Medicine Press.

Huh, H., H.K. Kim, and H.K. Lee. 1998. PAF antagonistic activity of 2-hydroxy-3-methoxybenzoic acid glucose ester from *Gentiana scabra*. *Arch. Pharmacol. Res.* 21(4):436-439.

McGuffin, M., C. Hobbs, R. Upton, and A. Goldberg. 1997. *Botanical safety handbook*. Boca Raton, FL: CRC Press.

Geranium maculatum L. — Geraniaceae

G

SCN: cranesbill
OCN: alumroot; storksbill; wild geranium

Part: root

Quick Reference Summary

Safety Class: 1
Interaction Class: A

Contraindications

None known.

Other Precautions

None known.

Drug and Supplement Interactions

None known.

Notice

Tannins (9–28%) (Peacock and Deg 1928; Trimble and Peacock 1891); *see* Appendix 1.

Adverse Events and Side Effects

None known.

Pharmacological Considerations

None known.

Pregnancy and Lactation

No information on the safety of cranesbill in pregnancy or lactation was identified in the scientific or traditional literature. Although this review did not identify any concerns for use while pregnant or nursing, safety has not been conclusively established.

Review Details

I. Drug and Supplement Interactions

Clinical Trials of Drug or Supplement Interactions

No clinical trials of drug or supplement interactions were identified.

Case Reports of Suspected Drug or Supplement Interactions

No case reports of suspected drug or supplement interactions were identified.

Animal Trials of Drug or Supplement Interactions

No animal trials of drug or supplement interactions were identified.

II. Adverse Events

Case Reports of Adverse Events

No case reports of adverse events were identified.

III. Pharmacology and Pharmacokinetics

Human Pharmacological Studies

No relevant human pharmacological studies were identified.

Animal Pharmacological Studies

No relevant animal pharmacological studies were identified.

In Vitro Pharmacological Studies

No relevant in vitro pharmacological studies were identified.

IV. Pregnancy and Lactation

No information on the safety of cranesbill during pregnancy or lactation was identified.

V. Toxicity Studies

No toxicity studies were identified.

Literature Cited

Peacock, J.C., and B.L. Deg. 1928. Further study of the tannin of *Geranium maculatum*. *Am. J. Pharm.* 100:548-557.

Trimble, H., and J.C. Peacock. 1891. *Geranium maculatum*. *Am. J. Pharm.* 265.

Ginkgo biloba L. — Ginkgoaceae

SCN: ginkgo
PN: *yin xing ye* (leaf); *yin guo ye* (leaf); *bai guo ye* (leaf)
OCN: maidenhair tree
Part: leaf

Quick Reference Summary

Safety Class: 1
Interaction Class: B

Contraindications

None known.

Other Precautions

Use in persons with coagulation disorders should be under the supervision of a qualified healthcare practitioner (Bent et al. 2005; Kudolo et al. 2002).

Patients undergoing surgery are advised to stop ginkgo leaf 7 days prior to surgery (Bebbington et al. 2005; Destro et al. 2005; Fessenden et al. 2001; Yagmur et al. 2005).

Drug and Supplement Interactions

Concomitant use with anticoagulants should be under the supervision of a qualified healthcare practitioner (Bent et al. 2005; Kudolo et al. 2002).

Ginkgo leaf may lower plasma concentrations of nifedipine (Smith et al. 2001).

Concomitant use with MAO inhibitors should be under the supervision of a qualified healthcare practitioner (Rojas et al. 2004; Woerdenbag and Van Beek 1997).

Editors' Notes

Ginkgo leaf preparations are generally sold either as a tincture, for which no side effects have been recorded, or as a highly concentrated extract. Ginkgo leaf extracts (concentrated to 24% flavonol glycosides and 6% terpene lactones; referred to herinafter as "24/6") are among the most actively studied botanicals. With over 5 million prescriptions per year reported in Germany alone, only occasional side effects, such as headaches and gastrointestinal upset, have been reported (Hobbs 1991).

Ginkgo leaf contains ginkgolic acids (urushiol type alkylphenols), compounds that are related to those found in poison ivy (*Toxicodendron radicans*) and associated with contact allergies, especially dermatitis. Ginkgolic acids are found in relatively high concentrations in ginkgo seed coats and in low concentrations in the leaf (Blumenthal 1997).

Adverse Events and Side Effects

Reviews and meta-analyses of clinical trials have shown that ginkgo leaf is generally well tolerated, with adverse events similar to those of placebo (Birks and Grimley Evans 2006; DeFeudis 1991; Ernst 2002; Hilton and Stuart 2004; Kleijnen and Knipschild 1992a; Pittler and Ernst 2000; Woerdenbag and Van Beek 1997; Zeng et al. 2005).

Case reports of adverse events associated with ginkgo leaf use include abnormal bleeding (several cases post-surgically) (Bebbington et al. 2005; Benjamin et al. 2001; Castellote Varona and Atienza Morales 2005; Destro et al. 2005; Evans 2000; Farnsworth et al. 1975; Fessenden et al. 2001; Gilbert 1997; Hauser et al. 2002; MacVie and Harney 2005; Matthews 1998; Meisel et al. 2003; Miller and Freeman 2002; Purroy Garcia et al. 2002; Rosenblatt and Mindel 1997; Rowin and Lewis 1996; Schneider et al. 2002; Smolinske 1999; Yagmur et al. 2005) and cases of skin eruptions (Chiu et al. 2002; Pennisi 2006).

Pharmacological Considerations

Although a series of case reports have correlated ginkgo leaf extract use with incidences of abnormal bleeding (Bent et al. 2005), controlled human studies have yielded mixed results on the effects of ginkgo leaf on bleeding. In patients with type 2 diabetes, ginkgo leaf extract has been shown to decrease platelet function by inhibiting platelet aggregation (Kudolo et al. 2002). However, no significant modifications of bleeding time, platelet function, or coagulation factors were observed in healthy male volunteers administered up to 480 mg daily of ginkgo leaf extract (Bal Dit Sollier et al. 2003; Beckert et al. 2007; Kohler et al. 2004).

Human studies have shown a lack of interactions with a number of drugs, including warfarin, clopidogrel, aspirin, ticlopidine, digoxin, talinolol, lopinavir, ritonavir, voriconazole, bupropion, diazepam, donepezil, and fexofenadine (Aruna and Naidu 2007; Engelsen et al. 2002; Fan et al. 2009; Gardner et al. 2007; Kim et al. 2010; Lei et al. 2009a, 2009b; Lu et al. 2006; Mauro et al. 2003; Robertson et al. 2008; Wolf 2006; Yasui-Furukori et al. 2004; Zuo et al. 2010).

A number of human studies have examined the effects of ginkgo leaf preparations on the CYP450 drug-metabolizing isoenzymes. Ginkgo leaf extract has been shown to induce CYP2C19 (Yin et al. 2004). Effects on CYP3A4 have been mixed, with most studies showing no effect and others suggesting induction or inhibition (Gurley et al. 2002, 2005; Markowitz et al. 2003a; Robertson et al. 2008; Uchida et al. 2006). Studies have shown no activity to slight induction of CYP2E1 and CYP2C9 (Gurley et al. 2002, 2005; Mohutsky et al. 2006; Uchida et al. 2006; Yin et al. 2004) and no activity on CYP2D6 or CYP1A2 (Gurley et al. 2002, 2005; Markowitz et al. 2003b).

Pregnancy and Lactation

Information on the safety of ginkgo leaf during pregnancy and lactation is limited. Animal studies on the use of ginkgo leaf during pregnancy have provided conflicting results. One study showed decreased fetal weights from mother rats fed doses of 7 or 14 mg/kg daily of ginkgo leaf extract. In the same study, no adverse effects on fetal development were reported at a lower dose (Pinto et al. 2007). Other studies indicated no adverse effects on mothers or offspring at doses up to 1600 mg/kg daily in rabbits and 900 mg/kg daily in rats (Blaschek et al. 2002; DeFeudis 1991; Li et al. 2003).

No information on the safety of gingko leaf during lactation was identified. While this review did not identify any concerns for use while nursing, safety has not been conclusively established.

Review Details

I. Drug and Supplement Interactions

Clinical Trials of Drug or Supplement Interactions

Trials with nifedipine yielded conflicting results, with significantly lower nifedipine plasma levels reported after ingestion of 120 mg ginkgo (no additional information given) for 18 days (Smith et al. 2001), but an increase in plasma nifedipine levels was reported after a single dose of 240 mg of a ginkgo leaf extract standardized to 24% flavonol glycosides and 6% terpene lactones (Yoshioka et al. 2004).

In healthy volunteers orally administered six ginkgo tablets (each containing the equivalent of 2 g of ginkgo leaf) daily for 7 days, then administered ginkgo with 25 mg warfarin daily, no significant changes in plasma levels of warfarin were observed. No changes were observed in platelet aggregation or international normalized ratio (INR) of prothrombin time (Jiang et al. 2005, 2006). Another study in patients stabilized on warfarin indicated no effects of 100 mg of standardized (24/6) ginkgo extract daily for 4 weeks on INR values (Engelsen et al. 2002).

In healthy volunteers orally administered single doses of 120 mg ginkgo (no specific product description stated) with 100 mg cilostazol or with 75 mg clopidogrel, platelet inhibition with the combination of ginkgo and clopidogrel or cilostazol was not significantly different as compared with clopidogrel or cilostazol alone. With the combination of ginkgo and cilostazol, prolongation of bleeding time was observed as compared to either product alone, although no changes were found in clotting time or platelet count (Aruna and Naidu 2007).

In healthy volunteers orally administered 500 mg aspirin with or without 240 mg standardized ginkgo leaf extract daily for 1 week, no significant differences were observed in bleeding time or agonist-induced platelet aggregation during treatment with ginkgo and aspirin as compared to aspirin alone (Wolf 2006). In adults with peripheral artery disease or risk factors for cardiovascular disease, who were administered 325 mg aspirin alone or with 300 mg of the same ginkgo leaf extract daily for 4 weeks, no significant changes in platelet function analysis, platelet aggregation, or bleeding or bruising episodes were observed (Gardner et al. 2007).

In healthy volunteers orally administered single doses of 250 mg ticlopidine with or without 80 mg standardized (24/6) ginkgo leaf extract, no changes in plasma levels of ticlopidine were observed. Bleeding times were not prolonged with gingko, and coadministration was not associated with additional antiplatelet effects compared with ticlopidine alone (Kim et al. 2010). Similarly, no changes in ticlopidine plasma levels were observed after ticlopidine was administered to healthy volunteers before or after 120 mg ginkgo leaf extract daily for 3 days (Lu et al. 2006).

In healthy volunteers orally administered 0.5 mg digoxin before or after 240 mg of standardized (24/6) ginkgo leaf extract daily for 2 weeks, no changes in serum levels of digoxin were observed (Mauro et al. 2003).

In healthy volunteers orally administered 100 mg talinolol with a single dose of 120 mg ginkgo leaf extract, no changes in plasma levels of talinolol were observed. After repeated dosing for 14 days with 360 mg of the same ginkgo leaf extract daily, plasma levels of talinolol were increased. Talinolol is a substrate of the drug transporter protein P-gp (Fan et al. 2009).

In healthy volunteers administered a lopinavir (400 mg) and ritonavir (100 mg) combination to steady state, the addition of 240 mg ginkgo leaf extract (analyzed as containing 29.2% flavonol glycosides and 5.1% terpene lactones) daily for 2 weeks had no effects on plasma levels of either of the drugs (Robertson et al. 2008).

No significant changes in plasma levels of bupropion were observed in healthy volunteers orally administered 150 mg bupropion before or during administration of 240 mg standardized (24/6) ginkgo leaf extract daily for 14 days (Lei et al. 2009a).

No significant changes in voriconazole plasma levels were observed in healthy volunteers (including both extensive and poor metabolizers of CYP2C19) orally administered 200 mg of voriconazole before or after 240 mg standardized (24/6) ginkgo leaf extract daily for 12 days (Lei et al. 2009b).

No significant changes in plasma levels of diazepam (a CYP2C19 substrate) were observed in healthy volunteers orally administered 10 mg diazepam before or during treatment with 240 mg ginkgo leaf extract daily for 28 days (Zuo et al. 2010).

A study of Alzheimer's patients taking donepezil with 90 mg/day of ginkgo leaf extract for 30 days showed no significant impact on plasma concentration of donepezil (Yasui-Furukori et al. 2004).

Case Reports of Suspected Drug or Supplement Interactions

Several cases of suspected drug interactions have been reported in the literature. Coma was reported in a woman taking trazodone (40 mg daily) concomitantly with standardized (24/6) ginkgo leaf extract (160 mg daily) for two days (Galluzzi et al. 2000). Fatal intracerebral mass bleeding was reported in a patient four weeks after introduction of 600 mg ibuprofen daily. This was in addition to ongoing daily use of 180 mg of a concentrated extract of ginkgo leaf for at least 30 months (Meisel et al. 2003). A breakthrough seizure was reported in a patient taking phenytoin and valproic acid (unspecified doses) and multiple herbal supplements with ginkgo (unspecified dose) (Kupiec and Raj 2005).

Animal Trials of Drug or Supplement Interactions

A ginkgo leaf decoction has been shown to reduce plasma concentrations of orally, but not intravenously, administered cyclosporine in rats (Yang et al. 2006). Similarly, when coadministered in rats with phenobarbital, a ginkgo leaf extract (24.2% flavonol glycosides and 9.2% terpene lactones) was shown to significantly reduce plasma concentrations of phenobarbital and sleeping time induced by phenobarbital (Kubota et al. 2004). Studies in rats with the same ginkgo leaf extract indicated significant reduction in the hypotensive effect of nicardipine (Kubota et al. 2003; Shinozuka et al. 2002). Coadministration of standardized (24/6) ginkgo leaf extract and amikacin to rats resulted in an earlier onset and more significant amikacin-induced ototoxicity (Miman et al. 2002). The applicability of these animal data to humans is unknown.

In rats with induced pulmonary embolism given 20 mg/kg ginkgo leaf extract (24.74% flavonol glycosides and 7.55% terpene lactones) with 25 mg/kg cilostazol, no significant effects on the bleeding time, prothrombin time, and activated partial thromboplastin time were observed as compared to cilostazol alone. Survival was improved in the group that received both ginkgo and cilostazol, suggesting that ginkgo may potentiate the antiplatelet effect of cilostazol without the prolongation of bleeding time or coagulation time (Ryu et al. 2009).

II. Adverse Events

Adverse Events Reported in Clinical Trials

Reviews and meta-analyses of clinical trials with ginkgo leaf extract have shown that no serious side effects were reported in any trials and, if present, side effects were not significantly different from those in patients treated with placebo (Birks and Grimley Evans 2006; DeFeudis 1991; Ernst 2002; Hilton and Stuart 2004; Kleijnen and Knipschild 1992a; Pittler and Ernst 2000; Woerdenbag and Van Beek 1997; Zeng et al. 2005).

Case Reports of Adverse Events

Ginkgo leaf preparations have been associated with a number of cases of abnormal bleeding, including cerebral hemorrhages (Benjamin et al. 2001; Castellote Varona and Atienza Morales 2005; Matthews 1998; Meisel et al. 2003; Purroy Garcia et al. 2002; Smolinske 1999), hematomas (Evans 2000; Gilbert 1997; Hauser et al. 2002; Miller and Freeman 2002; Rowin and Lewis 1996), intraocular bleeding (Farnsworth et al. 1975; Hauser et al. 2002; MacVie and Harney 2005; Rosenblatt and Mindel 1997; Schneider et al. 2002), and postsurgical bleeding (Bebbington et al. 2005; Destro et al. 2005; Fessenden et al. 2001; Yagmur et al. 2005). Spontaneous bleeding was reported in a patient with a history of bleeding who was taking ginkgo leaf extract reported to be standardized to 27% flavonol glycosides and 10% terpene lactones, 75 mg per day for the previous 6 months, along with aspirin, vitamins A, C, D, E, and folic acid (Bent et al. 2005).

A review of all spontaneous reports from 1982 to 1988 of adverse events associated with the most established concentrated ginkgo extract concluded that side effects are rare and that "tolerance was generally excellent" (DeFeudis 1991).

Recurrent seizures in two patients with well-controlled epilepsy were associated with use of "*Ginkgo biloba*," presumably consisting of leaf extracts, for 12 days to two weeks (Granger 2001). A review of gingko hypothesized that the compound ginkgotoxin (4′-*O*-methylpyridoxine) may be responsible for lowering seizure threshold by inhibiting synthesis of vitamin B_6 (Leistner and Drewke 2010).

Cases of skin eruptions have been reported in persons taking ginkgo leaf products (Chiu et al. 2002; Pennisi 2006).

Two sisters with no significant prior psychiatric history, but with a significant paternal history of paranoid schizophrenia, were reported to experience manic episodes after taking a ginkgo product of at least twice the recommended dosage for about two years. After discontinuation of ginkgo and stabilization on medication, the sisters were reported to have relapses while not taking ginkgo (Lamonaca et al. 2001).

III. Pharmacology and Pharmacokinetics

Human Pharmacological Studies

Study results on the influence of ginkgo on platelet aggregation are mixed. No effects on hemostasis, coagulation, bleeding time, platelet function, or fibrinolysis were observed after administration of 120, 240, or 480 mg daily of ginkgo leaf standardized extract to healthy volunteers for 14 days (Bal Dit Sollier et al. 2003). No changes in platelet aggregation or blood coagulation were observed in healthy volunteers administered 240 mg standardized ginkgo leaf extract daily for 7 days (Kohler et al. 2004). A significant reduction in platelet aggregation was observed in type 2 diabetes patients after 3 months of daily administration of 120 mg standardized ginkgo leaf extract, whereas with the same treatment in healthy volunteers a significant reduction in collagen-induced but no change in PAF-mediated platelet aggregation was observed (Kudolo et al. 2002). Other studies indicate that ginkgo is believed to have antiplatelet activity via platelet-activating factor (PAF) antagonism (Braquet 1986; Braquet et al. 1987). PAF is primarily involved with pathological thrombosis formation and therefore may not affect normal mechanisms of platelet aggregation and thus have no effect on bleeding time (Mills and Bone 2005). The extent of the effect of ginkgo on the PAF pathway at the standard therapeutic dose has been debated (Braquet 1993; DeFeudis 1991; Skogh 1998).

In healthy volunteers orally administered ginkgo capsules at the manufacturer's recommended dose (amount not specified) daily for two weeks, no effects on platelet function or other hematological parameters were observed, including prothrombin time, partial thromboplastin time, thrombin time, bleeding time, the collagen/epinephrine assay, or the collagen/adenosine diphosphate assay. Aspirin (325 mg daily) was used as a positive control and markedly inhibited platelet function (Beckert et al. 2007).

A reduction in blood viscosity was observed in healthy volunteers orally administered 80 mg standardized ginkgo leaf extract daily for 6 months (Galduróz et al. 2007).

Human study results on the effects of ginkgo on CYP2C9 are mixed, with one trial of standardized (24/6) ginkgo leaf extract in extensive CYP2C9 metabolizers showing slight but significant induction (Uchida et al. 2006) and another trial with the same extract showing no effect (Mohutsky et al. 2006). Ginkgo has been shown to significantly induce CYP2C19 (Yin et al. 2004). Ginkgo leaf extracts showed no significant effect on CYP2E1 in two studies (Gurley et al. 2002, 2005b) and significant induction of CYP2E1 in another study (Yin et al. 2004). No effect of ginkgo has been observed in CYP2D6 (Gurley et al. 2002, 2005b; Markowitz et al. 2003a) or CYP1A2 (Gurley et al. 2002, 2005b).

Several human studies on the effects of ginkgo leaf extracts on CYP3A4 activity have shown no effect on CYP3A4 metabolism (Gurley et al. 2002, 2005; Markowitz et al. 2003a), although one trial of extensive CYP2C9 metabolizers showed a significant inhibition of CYP3A4 (Uchida et al. 2006).

In healthy volunteers orally administered single doses of midazolam and fexofenadine before and after 240 mg per day for 28 days of a gingko leaf extract analyzed as containing 29.2% flavonol glycosides and 5.1% terpene lactones, a significant decrease (~30%) in plasma levels of midazolam were observed, suggesting induction of the drug-metabolizing isoenzyme CYP3A4. Plasma levels of fexofenadine were unchanged (Robertson et al. 2008).

In healthy volunteers orally administered flurbiprofen after three doses of 120 mg standardized (24/6) ginkgo leaf extract in a 24-hour period, no significant changes in plasma levels of flurbiprofen were observed. These results suggest a lack of activity of ginkgo on the drug-metabolizing isoenzyme CYP2C9 (Greenblatt et al. 2006).

A study on the effects of ginkgo on monoamine oxidase (MAO) activity in humans indicated that 60 mg of a standardized ginkgo leaf extract did not affect MAO activity and that MAO inhibitory compounds in ginkgo did not cross the blood-brain barrier (Fowler et al. 2000).

Animal Pharmacological Studies

A decrease in monoamine oxidase (MAO) activity was observed in mice orally administered 50 mg/kg of a standardized ginkgo leaf extract daily for 7 months (Pardon et al. 2000). Similar inhibition was observed in mice intraperitoneally administered a standardized ginkgo leaf extract at a dose of 20 to 100 mg/kg for 7 days (Wu and Zhu 1999), and in mice orally administered 10 mg/kg of standardized ginkgo leaf extract for 17 days (Rojas et al. 2004). A fourth study, however, showed no effect on MAO activity in mice orally administered 25 to 100 mg/kg of a standardized ginkgo leaf extract daily for 5 days (Porsolt et al. 2000).

Other animal studies were identified but omitted due to availability of human data.

In Vitro Pharmacological Studies

In vitro studies were identified but omitted due to availability of human data.

IV. Pregnancy and Lactation

Oral administration of standardized ginkgo leaf extract to rats (up to 1600 mg/kg daily) and rabbits (up to 900 mg/kg daily) did not produce any signs of teratogenicity or embryotoxicity (Blaschek et al. 2002; DeFeudis 1991). Fetuses of pregnant rats fed 7 or 14 mg/kg daily ginkgo leaf extract during pregnancy were shown to have a lower birth weight as compared to control, while treatment with 3.5 mg/kg ginkgo had no adverse effects on fetuses. No signs of toxicity were observed in mothers fed ginkgo at any of the dose levels (Pinto et al. 2007). Standardized (24/6) ginkgo leaf extracts administered to pregnant rats at doses of 100 or 300 mg/kg daily for 5 days increased the number of hippocampal neurons of fetuses. No adverse effects in the fetuses or the mothers were noted (Li et al. 2003).

Ginkgo extract and ginkgo extract fractions with high levels of ginkgolic acids demonstrated embryotoxic effects when injected into recently fertilized chicken eggs (Floissac and Chopin 1999). Toxic effects were noted to be dose dependent. A significant effect of ginkgolic acids was observed in the LD_{50} values for ginkgo extract fractions, with LD_{50} ranging from 250 mg/egg (1% ginkgolic acid, 16% biflavones), to 1.8 mg/egg (16% ginkgolic acid, 6.7% biflavones) and 3.5 mg/egg (58% ginkgolic acid, 0.02% biflavones) (Baron-Ruppert and Luepke 2001).

One study of pregnant women taking herbal remedies reported to have identified the highly toxic alkaloid colchicine in a commercial ginkgo preparation (Petty et al. 2001). Colchicine, however, has never been previously detected in ginkgo extracts, and studies following the report of colchicine failed to identify the compound in other commercial preparations (Bone 2002; Li et al. 2002a, 2002b).

Direct treatment of hamster oocytes with high concentrations of gingko (0.1 and 1 mg/ml) extract significantly reduced penetration by sperm and the integrity of sperm DNA. The relevance of this study to humans taking ginkgo supplements is doubtful (Ondrizek et al. 1999).

No relevant information on the safety of ginkgo leaf during lactation was identified.

V. Toxicity Studies

Acute Toxicity

The LD_{50} of orally administered standardized ginkgo leaf extract in mice is 7.7 g/kg. The intraperitoneal LD_{50} is 1.9 g/kg in mice and 2.1 g/kg in rats (Blaschek et al. 2002).

An extract of ginkgo leaf with relatively high levels (~30%) of alkylphenols produced immunotoxic effects after subplantar injection in mice (Koch et al. 2000).

Subacute Toxicity

In rats and mice, ginkgo administered orally did not produce any signs of organ damage or impairment of hepatic or renal function when administered for over 27 weeks in doses ranging from 100 to 1600 mg/kg (Salvador 1995).

G

Literature Cited

Aruna, D., and M.U. Naidu. 2007. Pharmacodynamic interaction studies of *Ginkgo biloba* with cilostazol and clopidogrel in healthy human subjects. *Br. J. Clin. Pharmacol.* 63(3):333-338.

Bal Dit Sollier, C., H. Caplain, and L. Drouet. 2003. No alteration in platelet function or coagulation induced by EGb761 in a controlled study. *Clin. Lab. Haematol.* 25(4):251-253.

Baron-Ruppert, G., and N.P. Luepke. 2001. Evidence for toxic effects of alkylphenols from *Ginkgo biloba* in the hen's egg test (HET). *Phytomedicine* 8(2):133-138.

Bebbington, A., R. Kulkarni, and P. Roberts. 2005. *Ginkgo biloba*: Persistent bleeding after total hip arthroplasty caused by herbal self-medication. *J. Arthroplasty* 20(1):125-126.

Beckert, B.W., M.J. Concannon, S.L. Henry, et al. 2007. The effect of herbal medicines on platelet function: An in vivo experiment and review of the literature. *Plast. Reconstr. Surg.* 120 (7):2044-2050.

Benjamin, J., T. Muir, K. Briggs, and B. Pentland. 2001. A case of cerebral haemorrhage—Can *Ginkgo biloba* be implicated? *Postgrad. Med. J.* 77(904):112-113.

Bent, S., H. Goldberg, A. Padula, and A.L. Avins. 2005. Spontaneous bleeding associated with *Ginkgo biloba*: A case report and systematic review of the literature. *J. Gen. Intern. Med.* 20(7):657-661.

Birks, J., and J. Grimley Evans. 2006. *Ginkgo biloba* for cognitive impairment and dementia. *Cochrane Database of Systematic Reviews, Volume* 3. Chichester, UK: Wiley.

Blaschek, W., S. Ebel, E. Hackenthal, et al. 2002. *HagerROM 2002: Hagers handbuch der drogen und arzneistoffe*. Heidelberg: Springer.

Blumenthal, M. 1997. German government limits ginkgolic acid levels. *HerbalGram* 41:29.

Bone, K. 2002. Ginkgo and colchicine: "Curiouser and curiouser," says Alice. Review of comment on Petty. *HerbalGram* 55:24.

Braquet, P. 1986. Proofs of involvement of PAF-acether in various immune disorders using BN 52021 (ginkgolide B): A powerful PAF-acether antagonist isolated from *Ginkgo biloba* L. *Adv. Prost. Thrombox. Leukotr. Res.* 16:179-198.

Braquet, P. 1993. Cedemin, a *Ginkgo biloba* extract, should not be considered as a PAF antagonist. *Am. J. Gastroenterol.* 88(12):2138.

Braquet, P., L. Touqui, T.Y. Shen, et al. 1987. Perspectives in platelet-activating factor research. *Pharmacol. Rev.* 39:97-145.

Castellote Varona, F.J., and M.P. Atienza Morales. 2005. *Ginkgo biloba* and cerebral hemorrhage. *Ann. Med. Intern.* 22(4):199.

Chiu, A., A. Lane, and A. Kimball. 2002. Diffuse morbilliform eruption after consumption of *Ginkgo biloba* supplement. *J. Am. Acad. Dermatol.* 46(1):145-146.

DeFeudis, F.V. 1991. *Ginkgo biloba extract (EGb 761): Pharmacological activities and clinical applications*. Paris: Elsevier.

Destro, M.W., M.B. Speranzini, C. Cavalheiro Filho, T. Destro, and C. Destro. 2005. Bilateral haematoma after rhytidoplasty and blepharoplasty following chronic use of *Ginkgo biloba*. *Br. J. Plast. Surg.* 58(1):100-101.

Engelsen, J., J.D. Nielsen, and K. Winther. 2002. Effect of coenzyme Q_{10} and *Ginkgo biloba* on warfarin dosage in stable, long-term warfarin treated outpatients. A randomised, double blind, placebo-crossover trial. *Thromb. Haemost.* 87(6):1075-1076.

Ernst, E. 2002. The risk-benefit profile of commonly used herbal therapies: Ginkgo, St. John's wort, ginseng, echinacea, saw palmetto, and kava. *Ann. Intern. Med.* 136(1):42-53.

Evans, V. 2000. Herbs and the brain: Friend or foe? The effects of ginkgo and garlic on warfarin use. *J. Neurosci. Nurs.* 32(4):229-232.

Fan, L., G.Y. Tao, G. Wang, et al. 2009. Effects of *Ginkgo biloba* extract ingestion on the pharmacokinetics of talinolol in healthy Chinese volunteers. *Ann. Pharmacother.* 43(5):944-949.

Farnsworth, N.R., A.S. Bingel, G.A. Cordell, F.A. Crane, and H.H. Fong. 1975. Potential value of plants as sources of new antifertility agents I. *J. Pharm. Sci.* 64(4):535-598.

Fessenden, J.M., W. Wittenborn, and L. Clarke. 2001. *Gingko biloba*: A case report of herbal medicine and bleeding postoperatively from a laparoscopic cholecystectomy. *Am. Surg.* 67(1):33-35.

Floissac, M., and S. Chopin. 1999. *Ginkgo biloba* extract is embryotoxic to chick embryos. *FASEB J.* 13:5.

Fowler, J.S., G.J. Wang, N.D. Volkow, et al. 2000. Evidence that *Ginkgo biloba* extract does not inhibit MOA A and B in living human brain. *Life Sci.* 66(9):141-146.

Galduróz, J.C.F., H.K. Antunes, and R.F. Santos. 2007. Gender- and age-related variations in blood viscosity in normal volunteers: A study of the effects of extract of *Allium sativum* and *Ginkgo biloba*. *Phytomedicine* 14(7-8):447-451.

Galluzzi, S., O. Zanetti, G. Binetti, M. Trabucchi, and G.B. Frisoni. 2000. Coma in a patient with Alzheimer's disease taking low dose trazodone and *Gingko biloba*. *J. Neurol. Neurosurg. Psychiat.* 68(5):679-680.

Gardner, C.D., J.L. Zehnder, A.J. Rigby, J.R. Nicholus, and J.W. Farquhar. 2007. Effect of *Ginkgo biloba* (EGb 761) and aspirin on platelet aggregation and platelet function analysis among older adults at risk of cardiovascular disease: A randomized clinical trial. *Blood Coagul. Fibrinolysis* 18(8):787-793.

Gilbert, G.J. 1997. *Ginkgo biloba*. *Neurology* 48(4):1137.

Granger, A.S. 2001. *Ginkgo biloba* precipitating epileptic seizures. *Age Ageing* 30(6):523-525.

Greenblatt, D.J., L.L. von Moltke, Y. Luo, et al. 2006. *Ginkgo biloba* does not alter clearance of flurbiprofen, a cytochrome P450-2C9 substrate. *J. Clin. Pharmacol.* 46(2):214-221.

Gurley, B.J., S.F. Gardner, M.A. Hubbard, et al. 2002. Cytochrome P450 phenotypic ratios for predicting herb-drug interactions in humans. *Clin. Pharmacol. Ther.* 72 (3):276-287.

Gurley, B.J., S.F. Gardner, M.A. Hubbard, et al. 2005a. Clinical assessment of effects of botanical supplementation on cytochrome P450 phenotypes in the elderly: St John's wort, garlic oil, *Panax ginseng* and *Ginkgo biloba*. *Drugs Aging* 22(6):525-539.

Hauser, D., T. Gayowski, and N. Singh. 2002. Bleeding complications precipitated by unrecognized *Gingko biloba* use after liver transplantation. *Transpl. Int.* 15(7):377-379.

Hilton, M., and E. Stuart. 2004. *Ginkgo biloba* for tinnitus. *Cochrane database of systematic reviews*. Chichester: Wiley.

Hobbs, C. 1991. *Ginkgo, elixir of youth: Modern medicine from an ancient tree*. Capitola, CA: Botanica Press.

Jiang, X., E.Y. Blair, and A.J. McLachlan. 2006. Investigation of the effects of herbal medicines on warfarin response in healthy subjects: A population pharmacokinetic-pharmacodynamic modeling approach. *J. Clin. Pharmacol.* 46(11):1370-1378.

Jiang, X., K.M. Williams, W.S. Liauw, et al. 2005. Effect of ginkgo and ginger on the pharmacokinetics and pharmacodynamics of warfarin in healthy subjects. *Br. J. Clin. Pharmacol.* 59(4):425-432.

Kim, B.H., K.P. Kim, K.S. Lim, et al. 2010. Influence of *Ginkgo biloba* extract on the pharmacodynamic effects and pharmacokinetic properties of ticlopidine: An open-label, randomized, two-period, two-treatment, two-sequence, single-dose crossover study in healthy Korean male volunteers. *Clin. Ther.* 32(2):380-390.

Kleijnen, J., and P. Knipschild. 1992a. *Ginkgo biloba*. *Lancet* 340:1136-1139.

Koch, E., H. Jaggy, and S.S. Chatterjee. 2000. Evidence for immunotoxic effects of crude *Ginkgo biloba* L. leaf extracts using the popliteal lymph node assay in the mouse. *Int. J. Immunopharmacol.* 22(3):229-236.

Kohler, S., P. Funk, and M. Kieser. 2004. Influence of a 7-day treatment with *Ginkgo biloba* special extract EGb 761 on bleeding time and coagulation: A randomized, placebo-controlled, double-blind study in healthy volunteers. *Blood Coagul. Fibrinolysis* 15(4):303-309.

Kubota, Y., K. Kobayashi, N. Tanaka, et al. 2003. Interaction of *Ginkgo biloba* extract (GBE) with hypotensive agent, nicardipine, in rats. *In Vivo* 17(5):409-412.

Kubota, Y., K. Kobayashi, N. Tanaka, et al. 2004. Pretreatment with *Ginkgo biloba* extract weakens the hypnosis action of phenobarbital and its plasma concentration in rats. *J. Pharm. Pharmacol.* 56(3):401-405.

Kudolo, G.B., S. Dorsey, and J. Blodgett. 2002. Effect of the ingestion of *Ginkgo biloba* extract on platelet aggregation and urinary prostanoid excretion in healthy and type 2 diabetic subjects. *Thromb. Res.* 108(2-3):151-160.

Kupiec, T., and V. Raj. 2005. Fatal seizures due to potential herb-drug interactions with *Ginkgo biloba*. *J. Anal. Toxicol.* 29(7):755-758.

Lamonaca, G., J. Klesmer, and J.L. Katz. 2001. Manic psychosis associated with high-dose *Gingko biloba*. *Prim. Psychiat.* 8(6):63-64.

Lei, H.P., W. Ji, J. Lin, et al. 2009a. Effects of *Ginkgo biloba* extract on the pharmacokinetics of bupropion in healthy volunteers. *Br. J. Clin. Pharmacol.* 68(2):201-206.

Lei, H.P., G. Wang, L.S. Wang, et al. 2009b. Lack of effect of *Ginkgo biloba* on voriconazole pharmacokinetics in Chinese volunteers identified as CYP2C19 poor and extensive metabolizers. *Ann. Pharmacother.* 43(4):726-731.

Leistner, E., and C. Drewke. 2010. *Ginkgo biloba* and ginkgotoxin. *J. Nat. Prod.* 73(1):86-92.

Li, W., J.F. Fitzloff, N.R. Farnsworth, and H.H. Fong. 2002a. Evaluation of commercial *Ginkgo biloba* dietary supplements for the presence of colchicine by high-performance liquid chromatography. *Phytomedicine* 9(5):442-446.

Li, W., Y. Sun, J.F. Fitzloff, and R.B. van Breemen. 2002b. Evaluation of commercial ginkgo and echinacea dietary supplements for colchicine using liquid chromatography-tandem mass spectrometry. *Chem. Res. Toxicol.* 15(9):1174-1178.

Li, W., F. Trovero, J. Cordier, et al. 2003. Prenatal exposure of rats to *Ginkgo biloba* extract (EGb 761) increases neuronal survival/growth and alters gene expression in the developing fetal hippocampus. *Brain Res. Dev. Brain Res.* 144(2):169-180.

Lu, W.J., J.D. Huang, and M.L. Lai. 2006. The effects of ergoloid mesylates and *Ginkgo biloba* on the pharmacokinetics of ticlopidine. *J. Clin. Pharmacol.* 46(6):628-634.

MacVie, O.P., and B.A. Harney. 2005. Vitreous haemorrhage associated with *Gingko biloba* use in a patient with age related macular disease. *Br. J. Ophthalmol.* 89(10):1378-1379.

Markowitz, J., C. DeVane, K. Chavin, et al. 2003a. Effects of garlic (*Allium sativum* L.) supplementation on cytochrome P450 2D6 and 3A4 activity in healthy volunteers. *Clin. Pharmacol. Ther.* 74(2):170-177.

Markowitz, J.S., J.L. Donovan, C. Lindsay DeVane, L. Sipkes, and K.D. Chavin. 2003b. Multiple-dose administration of *Ginkgo biloba* did not affect cytochrome P-450 2D6 or 3A4 activity in normal volunteers. *J. Clin. Psychopharmacol.* 23(6):576-581.

Matthews, M.K., Jr. 1998. Association of *Ginkgo biloba* with intracerebral hemorrhage. *Neurology* 50(6):1933-1934.

Mauro, V.F., L.S. Mauro, J.F. Kleshinski, et al. 2003. Impact of *Ginkgo biloba* on the pharmacokinetics of digoxin. *Am. J. Ther.* 10(4):247.

Meisel, C., A. Johne, and I. Roots. 2003. Fatal intracerebral mass bleeding associated with *Ginkgo biloba* and ibuprofen. *Atherosclerosis* 167(2):367.

Miller, L.G., and B. Freeman. 2002. Possible subdural hematoma associated with *Ginkgo biloba*. *J. Herb. Pharmcother.* 2(2):57-63.

Mills, S., and K. Bone. 2005. *The essential guide to herbal safety*. St. Louis: Elsevier.
Miman, M.C., O. Ozturan, M. Iraz, T. Erdem, and E. Olmez. 2002. Amikacin ototoxicity enhanced by *Ginkgo biloba* extract (EGb 761). *Hearing Res.* 169:121-129.
Mohutsky, M.A., G.D. Anderson, J.W. Miller, and G.W. Elmer. 2006. *Ginkgo biloba*: Evaluation of CYP2C9 drug interactions in vitro and in vivo. *Am. J. Ther.* 13(1):24-31.
Ondrizek, R.R., P.J. Chan, W.C. Patton, and A. King. 1999. An alternative medicine study of herbal effects on the penetration of zona-free hamster oocytes and the integrity of sperm deoxyribonucleic acid. *Fertil. Steril.* 71(3):517-522.
Pardon, M.C., C. Joubert, F. Perez-Diaz, et al. 2000. In vivo regulation of cerebral monoamine oxidase activity in senescent controls and chronically stressed mice by long-term treatment with *Ginkgo biloba* extract (EGb 761). *Mech. Ageing Dev.* 113(3):157-168.
Pennisi, R.S. 2006. Acute generalised exanthematous pustulosis induced by the herbal remedy *Ginkgo biloba*. *Med. J. Aust.* 184(11):583-584.
Petty, H.R., M. Fernando, A.L. Kindzelskii, et al. 2001. Identification of colchicine in placental blood from patients using herbal medicines. *Chem. Res. Toxicol.* 14(9):1254-1258.
Pinto, R.M., E.S. Fernandes, J.E. Reis, V.M. Peters, and M.D. Guerra. 2007. Intra-uterine growth retardation after prenatal administration of *Ginkgo biloba* to rats. *Reprod. Toxicol.* 23:480-485.
Pittler, M., and E. Ernst. 2000. *Ginkgo biloba* extract for the treatment of intermittent claudication: A meta-analysis of randomized trials. *Am. J. Med.* 108(4):276-281.
Porsolt, R.D., S. Roux, and K. Drieu. 2000. Evaluation of a *Ginkgo biloba* extract (EGb 761) in functional tests for monoamine oxidase inhibition. *Arzneimittelforschung* 50(3):232-235.
Purroy Garcia, F., C. Molina, and J. Alvarez Sabin. 2002. Spontaneous cerebellar haemorrhage associated with *Ginkgo biloba* ingestion. *Med. Clin.* 119(15):596-597.
Robertson, S.M., R.T. Davey, J. Voell, et al. 2008. Effect of *Ginkgo biloba* extract on lopinavir, midazolam and fexofenadine pharmacokinetics in healthy subjects. *Curr. Med. Res. Opin.* 24(2):591-599.
Rojas, P., C. Rojas, M. Ebadi, et al. 2004. EGb761 pretreatment reduces monoamine oxidase activity in mouse corpus striatum during 1-methyl-4-phenylpyridinium neurotoxicity. *Neurochem. Res.* 29(7):1417-1423.
Rosenblatt, M., and J. Mindel. 1997. Spontaneous hyphema associated with ingestion of *Ginkgo biloba* extract. *N. Engl. J. Med.* 336(15):1108.
Rowin, J., and S.L. Lewis. 1996. Spontaneous bilateral subdural hematomas associated with chronic *Ginkgo biloba* ingestion. *Neurology* 46(6):1775-1776.
Ryu, K.H., H.Y. Han, S.Y. Lee, et al. 2009. *Ginkgo biloba* extract enhances antiplatelet and antithrombotic effects of cilostazol without prolongation of bleeding time. *Thromb. Res.* 124(3):328-334.
Salvador, R.L. 1995. Herbal medicine—*Ginkgo*. *Can. Pharm. J.* 52:39-41.
Schneider, C., C. Bord, P. Misse, B. Arnaud, and C.F. Schmitt-Bernard. 2002. Spontaneous hyphema caused by *Ginkgo biloba* extract. *J. Fr. Ophthalmol.* 25(7):731-732.
Shinozuka, K., K. Umegaki, Y. Kubota, et al. 2002. Feeding of *Ginkgo biloba* extract (GBE) enhances gene expression of hepatic cytochrome P-450 and attenuates the hypotensive effect of nicardipine in rats. *Life Sci.* 70(23):2783-2792.
Skogh, M. 1998. Extracts of *Ginkgo biloba* and bleeding or haemorrhage. *Lancet* 352(9134):1145-1146.
Smith, M., K. Lin, and Y. Zheng. 2001. An open trial of nifedipine-herb interactions: Nifedipine with St. John's wort, ginseng or *Ginkgo biloba*. *Clin. Pharmacol. Ther.* 69(2):abstr. PIII-89.
Smolinske, S.C. 1999. Dietary supplement-drug interactions. *J. Am. Med. Womens Assoc.* 54(4):191-195.
Uchida, S., H. Yamada, X.D. Li, et al. 2006. Effects of *Ginkgo biloba* extract on pharmacokinetics and pharmacodynamics of tolbutamide and midazolam in healthy volunteers. *J. Clin. Pharmacol.* 46(11):1290-1298.
Woerdenbag, H., and T. Van Beek. 1997. *Ginkgo biloba*. In *Adverse effects of herbal drugs, Volume 3*, edited by De Smet, P.A.G.M., K. Keller, R. Hansel, and R.F. Chandler. New York: Springer.
Wolf, H.R. 2006. Does *Ginkgo biloba* special extract EGb 761 provide additional effects on coagulation and bleeding when added to acetylsalicylic acid 500 mg daily? *Drugs Res. Dev.* 7(3):163-172.
Wu, W.R., and X.Z. Zhu. 1999. Involvement of monoamine oxidase inhibition in neuroprotective and neurorestorative effects of *Ginkgo biloba* extract against MPTP-induced nigrostriatal dopaminergic toxicity in C57 mice. *Life Sci.* 65 (2):157-164.
Yagmur, E., A. Piatkowski, A. Groger, et al. 2005. Bleeding complication under *Gingko biloba* medication. *Am. J. Hematol.* 79(4):343-344.
Yang, C., P.D.L. Chao, Y.C. Hou, et al. 2006. Marked decrease of cyclosporin bioavailability caused by coadministration of ginkgo and onion in rats. *Food Chem. Toxicol.* 44(9):1572-1578.
Yasui-Furukori, N., H. Furukori, A. Kaneda, S. Kaneko, and T. Tateishi. 2004. The effects of *Ginkgo biloba* extracts on the pharmacokinetics and pharmacodynamics of donepezil. *J. Clin. Pharmacol.* 44(5):538-542.
Yin, O.Q.P., B. Tomlinson, M.M.Y. Waye, A.H.L. Chow, and M.S.S. Chow. 2004. Pharmacogenetics and herb-drug interactions: Experience with *Ginkgo biloba* and omeprazole. *Pharmacogenetics* 14(12):841-850.
Yoshioka, M., N. Ohnishi, T. Koishi, et al. 2004. Studies on interactions between functional foods or dietary supplements and medicines. IV. Effects of *Ginkgo biloba* leaf extract on the pharmacokinetics and pharmacodynamics of nifedipine in healthy volunteers. *Biol. Pharm. Bull.* 27(12):2006-2009.
Zeng, X., M. Liu, Y. Yang, Y. Li, and K. Asplund. 2005. *Ginkgo biloba* for acute ischaemic stroke. In *Cochrane database of systematic reviews*. Chichester, UK: Wiley.
Zuo, X.C., B.K. Zhang, S.J. Jia, et al. 2010. Effects of *Ginkgo biloba* extracts on diazepam metabolism: A pharmacokinetic study in healthy Chinese male subjects. *Eur. J. Clin. Pharmacol.* 66(5):503-509.

Ginkgo biloba L. — Ginkgoaceae

SCN: ginkgo
PN: *yin xing* (seed); *bai guo* (seed); *yin guo* (seed)
OCN: maidenhair tree
Part: seed

Quick Reference Summary

Safety Class: 2d
Interaction Class: A

Contraindications

Do not exceed recommended dose (Bensky et al. 2004; Chang and But 1986; Leung and Foster 1996).

Other Precautions

Eating raw ginkgo seed may cause toxic reactions such as nausea, vomiting, seizures, and other symptoms of central nervous system disturbances (Chen and Chen 2004; Leung and Foster 1996), and even cooked seed should be eaten only in small amounts (Leung and Foster 1996). These concerns are not associated with the seed administered in decoction (Bensky et al. 2004).

Not recommended for long-term use (Bensky et al. 2004; Chang and But 1986; Leung and Foster 1996).

Ginkgo seed should be used with caution in pregnancy (Chen and Chen 2004).

Drug and Supplement Interactions

None known.

Standard Dose

Prepared: 1.5 to 9 g daily (Bensky et al. 2004; Chen and Chen 2004). Unprocessed raw ginkgo seed is generally not used due to its toxicity (Chen and Chen 2004).

Editors' Notes

Safety concerns for ginkgo seed are largely dependent on whether the seed is consumed in raw or cooked form. Consumption of cooked ginkgo seed is well documented (Chadha 1988; Leung and Foster 1996), although case reports of adverse effects from consumption of large amounts of the cooked seed have been reported (Hasegawa et al. 2006; Kajiyama 2002; Miwa 2001). One source advises a limit of no more than 10 boiled or roasted seeds daily (Leung and Foster 1996). The seed is considered toxic if eaten raw and has been reported to cause death in children (Chadha 1988; Leung and Foster 1996). Toxic reactions are not known to occur, however, when ginkgo seed is administered in decoction (Bensky et al. 2004).

Ginkgolic acids are urushiol type alkylphenols, related to compounds in poison ivy (*Rhus toxicodendron*), and are associated with contact allergic responses, especially dermatitis. Ginkgolic acids are found in relatively high concentrations in ginkgo seed coats and in lower concentrations in ginkgo leaf (Blumenthal 1997).

Adverse Events and Side Effects

Adverse reactions, including seizures, have been reported in individuals who have consumed large numbers of raw or cooked ginkgo seeds (Hasegawa et al. 2006; Kajiyama et al. 2002; Miwa 2001), though are not associated with decoctions of the seed (Bensky et al. 2004). Symptoms of overdosage, described as including hyperthermia, nausea, vomiting, shortness of breath, abdominal pain, diarrhea, restlessness, convulsions, and loss of consciousness, among other symptoms, can appear 1 to 12 hours after consuming the seeds, and are associated with doses of 7 to 150 seeds in children and 40 to 300 seeds in adults (Bensky et al. 2004; Chen and Chen 2004). It is also reported that death may occur due to respiratory and circulatory failure 1 to 2 days following gross overdose in the absence of emergency treatment (Bensky et al. 2004; Chen and Chen 2004). The number of ginkgo seeds eaten in reported fatalities ranges from 15 to 574 (Kajiyama et al. 2002).

Contact with ginkgo fruit pulp has been associated with contact dermatitis in a number of individuals (Nakamura 1985; Tomb et al. 1988).

Pharmacological Considerations

None known.

Pregnancy and Lactation

A text on traditional Chinese medicine indicates that ginkgo seed should be used with caution during pregnancy (Chen and Chen 2004).

No information on the safety of gingko seed during lactation was identified. While this review did not identify any concerns for use while nursing, safety has not been conclusively established.

Review Details

I. Drug and Supplement Interactions

Clinical Trials of Drug or Supplement Interactions

No clinical trials of drug or supplement interactions were identified.

Case Reports of Suspected Drug or Supplement Interactions

No case reports of suspected drug or supplement interactions were identified.

Animal Trials of Drug or Supplement Interactions

No animal trials of suspected drug or supplement interactions were identified.

II. Adverse Events

Case Reports of Adverse Events

Several cases of convulsions and vomiting have been reported in association with consumption of cooked ginkgo seeds, especially in Japan where cooked ginkgo seed is used as food. A 34-year-old woman experienced two episodes of convulsions and vomited several times after eating 70–80 microwaved ginkgo seeds (Miwa 2001). A 2-year-old presented with vomiting, diarrhea, irritability, and a symmetrical generalized clonic seizure from consumption of 50–60 roasted ginkgo seeds (Kajiyama et al. 2002). A 2-year-old boy experienced vomiting and afebrile convulsions after eating a large number of roasted ginkgo seeds (Hasegawa et al. 2006). In each of these cases, onset of symptoms occurred several hours after consumption of the ginkgo.

Other symptoms of overdose of ginkgo seed may include abdominal pain, dyspnea, and other symptoms related to central nervous system disturbances (Bensky et al. 2004; Chen and Chen 2004). The compound 4′-*O*-methylpyridoxine, a neurotoxin that can cause vitamin B_6 deficiency symptoms, is believed to be responsible for the toxic effects of ginkgo seed (Wada et al. 1985). Although the seed is recognized to be toxic in large amounts, cooked ginkgo seed is a food item commonly eaten throughout Japan, Korea, and China. The number of seeds that can be eaten safely in a single meal has not been defined, and fatalities have been reported from consumption of from 15 to 574 seeds (Kajiyama et al. 2002), with overdose in children reported from as few as seven seeds (Chen and Chen 2004).

Contact with ginkgo fruit pulp has been associated with contact dermatitis in a number of individuals (Nakamura 1985; Tomb et al. 1988).

III. Pharmacology and Pharmacokinetics

Human Pharmacological Studies

No relevant human pharmacological studies were identified.

Animal Pharmacological Studies

No relevant animal pharmacological studies were identified.

In Vitro Pharmacological Studies

No relevant in vitro pharmacological studies were identified.

IV. Pregnancy and Lactation

A text on traditional Chinese medicine indicates that ginkgo seed should be used with caution during pregnancy (Chen and Chen 2004).

No information on the safety of gingko seed during lactation was identified.

V. Toxicity Studies

Mice fed a "gross overdose" of ginkgo seed for 60 days were observed to have weight loss, poor appetite, liver damage, and glomerulonephritis, and some animals died. Intravenous administration of ginkgo seed extract caused an increase followed by a decrease in blood pressure, dyspnea, seizures, and death (Anonymous 1989, 1995). Decoctions of the seed are recognized as nontoxic (Bensky et al. 2004; Chen and Chen 2004).

Literature Cited

Anonymous. 1989. *Jiang Su Zhong Yi [Jiangsu Chinese Medicine]* 10(8):32. *In* Chen, J.K., and T.T. Chen. 2004. *Chinese medical herbology and pharmacology*. City of Industry, CA: Art of Medicine Press.

Anonymous. 1995. *Chang Yong Zhong Yao Xian Dai Yan Jiu Yu Lin Chuan [Recent study and clinical applications of common traditional Chinese medicine]. In* Chen, J.K., and T.T. Chen. 2004. *Chinese medical herbology and pharmacology*. City of Industry, CA: Art of Medicine Press.

Bensky, D., S. Clavey, and E. Stöger. 2004. *Chinese herbal medicine: Materia medica*. 3rd ed. Seattle: Eastland Press.

Blumenthal, M. 1997. German government limits ginkgolic acid levels. *HerbalGram* 41:29.

Chadha, Y. 1988. *The wealth of India: A dictionary of Indian raw materials and industrial products*. Delhi: Council of Scientific and Industrial Research.

Chang, H.-M., and P.P.H. But. 1986. *Pharmacology and applications of Chinese materia medica*. English ed. Philadelphia: World Scientific.

Chen, J.K., and T.T. Chen. 2004. *Chinese medical herbology and pharmacology*. City of Industry, CA: Art of Medicine Press.

Hasegawa, S., Y. Oda, T. Ichiyama, Y. Hori, and S. Furukawa. 2006. Ginkgo nut intoxication in a 2-year-old male. *Pediatric Neurology* 35(4):275-276.

Kajiyama, Y., K. Fujii, H. Takeuchi, and Y. Manabe. 2002. Ginkgo seed poisoning. *Pediatrics* 109(2):325-327.

Leung, A.Y., and S. Foster. 1996. *Encyclopedia of common natural ingredients used in food, drugs, and cosmetics*. 2nd ed. New York: Wiley.

Miwa, H., M. Iijima, S. Tanaka, and Y. Mizuno. 2001. Generalized convulsions after consuming a large amount of gingko nuts. *Epilepsia* 42 (2):280-281.

Nakamura, T. 1985. Ginkgo tree dermatitis. *Contact Dermat.* 12(5):281-282.

Tomb, R.R., J. Foussereau, and Y. Sell. 1988. Mini-epidemic of contact dermatitis from ginkgo tree fruit (*Ginkgo biloba* L.). *Contact Dermat.* 19(4):281-283.

Wada, K., S. Ishigaki, K. Ueda, M. Sakata, and M. Haga. 1985. An antivitamin B_6, 4′-methoxypyridoxine from the seed of *Ginkgo biloba* L. *Chem. Pharm. Bull.* 33:3555-3557.

Glehnia littoralis F. Schmidt ex Miq. — Apiaceae

SCN: glehnia
PN: *bei sha shen* (root)
OCN: beach silvertop
Part: root

G

Quick Reference Summary

Safety Class: 1
Interaction Class: A

Contraindications

None known.

Other Precautions

None known.

Drug and Supplement Interactions

None known.

Adverse Events and Side Effects

None known.

Pharmacological Considerations

Although furanocoumarins, compounds that can cause photosensitivity after contact with skin, have been identified in glehnia (Mizukami et al. 1993), no cases of photosensitivity have been reported.

Pregnancy and Lactation

No information on the safety of glehnia in pregnancy or lactation was identified in the scientific or traditional literature. Although this review did not identify any concerns for use while pregnant or nursing, safety has not been conclusively established.

Review Details

I. Drug and Supplement Interactions

Clinical Trials of Drug or Supplement Interactions

No clinical trials of drug or supplement interactions were identified.

Case Reports of Suspected Drug or Supplement Interactions

No case reports of suspected drug or supplement interactions were identified.

Animal Trials of Drug or Supplement Interactions

No animal trials of drug or supplement interactions were identified.

II. Adverse Events

Case Reports of Adverse Events

No case reports of adverse events were identified.

III. Pharmacology and Pharmacokinetics

Human Pharmacological Studies

No relevant human pharmacological studies were identified.

Animal Pharmacological Studies

No relevant animal pharmacological studies were identified.

In Vitro Pharmacological Studies

No relevant in vitro pharmacological studies were identified.

IV. Pregnancy and Lactation

No information on the safety of glehnia during pregnancy or lactation was identified.

V. Toxicity Studies

No toxicity studies were identified.

Literature Cited

Mizukami, H., K. Ohbayashi, K. Umetsu, and N. Hiraoka. 1993. Restriction fragment length polymorphisms of medicinal plants and crude drugs. II. Analysis of *Glehnia littoralis* of different geographical origin. *Biol. Pharm. Bull.* 16(6):611.

Glycyrrhiza spp. — Fabaceae

Glycyrrhiza echinata L.
SCN: licorice
OCN: east European licorice

Glycyrrhiza glabra **L.**
SCN: licorice
Syn: *Glycyrrhiza glandulifera* Walst. & Kit.
AN: *yashtimadhu*
PN: *gan cao* (root and rhizome)
OCN: Russian licorice; Spanish licorice; Turkish licorice

Glycyrrhiza uralensis **Fisch. ex DC.**
SCN: Chinese licorice
PN: *gan cao* (root and rhizome)
OCN: licorice; Ural licorice
Part: root and rhizome

Quick Reference Summary

Safety Class: 2b, 2d
Interaction Class: B

Contraindications

Not for use in pregnancy except under the supervision of a qualified healthcare practitioner (Bradley 1992; Strandberg et al. 2001, 2002).

Not for use in persons with hypertension, liver disorders, edema, severe kidney insufficiency, low blood potassium, heart disease with edema, or congestive heart failure (Bensky et al. 2004; Bradley 1992; De Smet 1993; Mills and Bone 2005).

Not for prolonged use or in high doses except under the supervision of a qualified healthcare practitioner (Bensky et al. 2004; Bradley 1992; Chadha 1988; Martindale and Reynolds 1996).

Other Precautions

None known.

Drug and Supplement Interactions

No interactions are expected at standard therapeutic doses. At higher doses or with long-term use, licorice may potentiate potassium depletion of high-ceiling loop diuretics and thiazide diuretics, stimulant laxatives (Mills and Bone 2005), and corticosteroids such as prednisolone (Cheng et al. 2004; De Smet 1993), and may potentiate the action of cardiac glycosides such as digoxin (Kelly 1990).

One reference indicated a theoretical concern that licorice may counteract the effects of oral contraceptives (Mills and Bone 2005).

Editors' Notes

The contraindications and precautions listed above apply to products that contain the compound glycyrrhizin and not to deglycyrrhizinated licorice (DGL) products.

Licorice contains approximately 1 to 7% of the compound glycyrrhizin (Kondo et al. 2007). Many commercial products are standardized to 12% glycyrrhizin, and the European Pharmacopoeia specifies that licorice root should contain a minimum of 4% glycyrrhizin (Council of Europe 2001).

Standard Dose

The standard dose is 1 to 5 g, three times daily for up to 6 weeks, though French regulation limits daily consumption to 5 g for direct consumption or 8 g as a tea; reduction in sodium and increase in potassium intakes are recommended (De Smet 1993; Mitchell et al. 1991).

In traditional Chinese medicine texts, the standard dose is listed as 1.5 to 10 g daily of sliced licorice root prepared as a decoction or infusion (Bensky et al. 2004; Chen and Chen 2004).

Adverse Events and Side Effects

No case reports of adverse effects have been reported in persons using licorice within the recommended dose (less than 50 g daily) and treatment period (less than 6 weeks) (WHO 1999).

In highly sensitive individuals, a daily intake of 100 mg glycyrrhizin (approximately 50 g licorice sweets) may produce adverse effects, and most individuals that consume 400 mg experience adverse effects (Størmer et al. 1993). A 12-week trial of the compound glycyrrhizin in women indicated that a dose of 2 mg/kg glycyrrhizin daily, equivalent to approximately 6 g of licorice daily, was the no-observed-adverse-effect level (NOAEL) of the compound (van Gelderen et al. 2000).

The mineralocorticoid effects of licorice are well documented (*see* Human pharmacological studies) (Stewart et al. 1987). Licorice may cause reversible potassium depletion and sodium retention when consumed in therapeutic dosages over a prolonged period (Bensky et al. 2004; Bradley 1992; Martindale and Reynolds 1996). Most cases of adverse events associated with licorice have been reported in persons consuming excessive amounts of licorice candies. Overdoses of licorice have been associated with temporary paralysis, loss of vision, sodium and fluid retention, hypertension, and decreased serum levels of aldosterone. Licorice-related potassium depletion has resulted in symptoms such as tachycardia, rhabdomyolysis, myopathy, and hypokalemic paralysis. These effects are attributed primarily to the action of the compound glycyrrhizin (Isbrucker and Burdock 2006). Preparations of licorice without this compound, deglycyrrhizinated licorice (DGL), are available.

Pharmacological Considerations

Deglycyrrhizinated licorice (DGL) is usually free of adverse effects (Martindale and Reynolds 1996).

Pregnancy and Lactation

Two retrospective epidemiological studies on licorice consumption by pregnant women suggested that heavy licorice consumption was associated with earlier births (Strandberg et al. 2001, 2002). Another epidemiological study indicated that heavy consumption (over 500 mg daily) of licorice candies by pregnant women was associated with adverse effects on cognitive and psychiatric development in children (Räikkönen et al. 2009).

In animal studies, doses of the compounds glycyrrhizin or glycyrrhetic acid over 100 mg/kg were associated with some adverse developmental effects in the fetus, although no adverse effects were observed at lower doses (FDRL 1971; Hundertmark et al. 2002; Mantovani et al. 1988).

No information on the safety of licorice during lactation was identified. While this review did not identify any concerns for use while nursing, safety has not been conclusively established.

Review Details

I. Drug and Supplement Interactions

Clinical Trials of Drug or Supplement Interactions

Coadministration of glycyrrhizin (50 mg orally) and prednisolone (0.096 mg/kg intravenously) significantly decreased the clearance of and increased the plasma concentrations of prednisolone (Cheng et al. 2004).

Estrogens may react with the mineralocorticoid receptor or inhibit 11β-hydroxysteroid dehydrogenase activity, exacerbating the effects of licorice (Clyburn and DiPette 1995).

Case Reports of Suspected Drug or Supplement Interactions

A case of neuromuscular paralysis due to low potassium was reported in a woman taking excessive amounts of both the diuretic furosemide (60–150 mg per day, three times per week) and licorice (40–60 g daily) (Famularo et al. 1999).

Animal Trials of Drug or Supplement Interactions

No animal trials of drug or supplement interactions were identified.

II. Adverse Events

Adverse Events Reported in Clinical Trials

Adverse events reported in clinical trials are consistent with the known pharmacological effects of licorice, including sodium and fluid retention, increased blood pressure, and decreased serum levels of potassium and aldosterone. Such events have been associated with doses of licorice containing high levels (greater than 2 mg/kg) of the compound glycyrrhizin, and are not common with products containing 2 mg/kg or less of glycyrrhizin (Isbrucker and Burdock 2006).

Case Reports of Adverse Events

No case reports of adverse effects have been reported in persons using licorice within the recommended dose (less than 50 g daily) and treatment period (less than 6 weeks) (WHO 1999).

Adverse events associated with licorice consumption are primarily related to the mineralocorticoid effects (*see* Human pharmacological studies), with symptoms such as sodium and fluid retention, increased blood pressure, and decreased serum levels of potassium (hypokalemia) and aldosterone. Multiple forms of licorice products have been associated with adverse events, including licorice candies, teas, and capsules. Most reports of adverse events are associated with ingestion of at least 35 g per day of licorice candies.

One case of cardiac arrest was reported in a patient with licorice-induced hypokalemia (Bannister et al. 1977). Multiple cases of hypertension (Beretta-Piccoli et al. 1985; Brouwers and van der Meulen 2001; Conn et al. 1968; Cugini et al. 1983; Cuspidi et al. 1981; Dellow et al. 1999; Janse et al. 2005; Pozzoli et al. 1980; Takeda et al. 1979; Wash and Bernard 1975) and one case of hypertension encephalopathy (van den Bosch et al. 2005) have been reported.

Various effects of licorice-induced hypokalemia have been reported, including hypokalemia-induced ventricular tachycardia or ventricular fibrillation (Eriksson et al. 1999; Gerritsen et al. 2009; Harris 2000; van den Bosch et al. 2005), hypokalemic rhabdomyolysis (breakdown of muscle fibers) (Barrella et al. 1997; Piette et al. 1984), hypokalemic myopathy (nonfunctioning muscle fibers) (Famularo et al. 1999; Hussain 2003; Ishikawa et al. 1999; Shintani et al. 1992; Yoshida and Takayama 2003), and hypokalemic paralysis (Cheng et al. 2004; Cumming et al. 1980; de Rohan Chabot et al. 1984; Elinav and Chajek-Shaul 2003; Famularo et al. 1999; Lin et al. 2003; Yasue et al. 2007). In one reported case, a woman had generalized edema and sodium retention without elevated blood pressure, suggesting that hypokalemia is not always associated with blood pressure changes (Negro et al. 2000). One case of myopathy was associated with renal tubular damage in an anorexic individual (Ishikawa et al. 1999).

Other related conditions reported in case reports include mineralocorticoid excess (Doeker and Andler 1999; Hamidon and Jeyabalan 2006) and pseudoaldosteronism (Colloredo et al. 1987; Gomez Fernandez et al. 1981; Kageyama et al. 1997).

One case of vestibular neuronitis (swelling of the inner ear, causing vertigo) (Belhadj-Tahar et al. 2003) was reported. Ischemia with a blood vessel blockage was reported in a hypokalemic patient who had consumed licorice regularly for over 10 years (Lozano et al. 2000).

A case series of five patients with transient visual loss observed over a 10-year period was reported. All of the patients had consumed significant amounts of licorice prior to the visual loss. Potential other causes of the loss were noted, including significant coffee consumption in three of the patients and histories of ischemia or migraine headaches in two of the patients (Dobbins and Saul 2000).

One case of occupational asthma was reported in an herbalist exposed to licorice powder. The allergy was IgE-mediated and confirmed by inhalation challenges (Cartier et al. 2002).

III. Pharmacology and Pharmacokinetics

Human Pharmacological Studies

The compound glycyrrhetinic acid has been shown to inhibit the enzyme 11β-hydroxysteroid dehydrogenase, leading to a decreased conversion of cortisol to cortisone (Stewart et al. 1987). Cortisol has the same affinity for the mineralocorticoid receptors as aldosterone, increasing the mineralocorticoid effects in the body. Such an increase in effect may lead to a syndrome known as pseudoaldosteronism, with symptoms such as sodium and fluid retention, increased blood pressure, and decreased serum levels of potassium and aldosterone (Sigurjonsdottir et al. 2006).

A review of case reports and human studies of licorice indicated significant variability in individual susceptibility to the compound glycyrrhizin. In highly sensitive individuals, a daily intake of 100 mg glycyrrhizin (~50 g licorice sweets) may produce adverse effects, and most individuals that consume 400 mg experience adverse effects (Størmer et al. 1993).

Administration of 100 or 200 g daily of licorice (0.7 or 1.4 g glycyrrhizin) for 1 to 4 weeks has been shown to significantly decrease plasma renin secretion and inhibit the renin-angiotensin-aldosterone system (RAAS) (Epstein et al. 1977).

A human study of 50 to 200 g of licorice daily for 2 or 4 weeks showed a dose-response relationship for blood pressure, with licorice significantly raising blood pressure, even at the 50 g dose (including 75 mg glycyrrhetinic acid). Variation in individual sensitivity was observed, with responses following a normal distribution curve (Sigurjonsdottir et al. 2001). A trial of 100 g daily of licorice (including 150 mg glycyrrhetinic acid) for 4 weeks indicated that patients with hypertension are more sensitive to inhibition of 11β-hydroxysteroid dehydrogenase than persons with regular blood pressure (Sigurjonsdottir et al. 2003). In a similar trial, men were shown to be more sensitive to inhibition of 11β-hydroxysteroid dehydrogenase than women (Sigurjonsdottir et al. 2006).

A reduction in serum potassium and renin and an increase in body weight were observed in persons taking 5 or 11 g of licorice daily (380 or 814 mg of glycyrrhizin). Several of the study participants withdrew due to hypertension, edema, or headaches, although some participants had possibly contributing factors, including a family history of hypertension or concomitant use of oral contraceptives. Doses of 1.4 or 2.8 g of licorice were not associated with any significant physiological changes or study withdrawals (Bernardi et al. 1994).

Testing of varied doses (0–4 mg/kg) of the compound glycyrrhizin for 12 weeks in women to determine a no-observable-adverse-effect level (NOAEL) for blood pressure and potassium effects suggested that 2 mg/kg glycyrrhizin was the NOAEL. This level is approximately equivalent to 6 g of licorice daily (van Gelderen et al. 2000).

The results of one small clinical trial (7 men studied) indicated a significant reduction of serum testosterone levels after 4 days of administration of 7 g daily of a licorice preparation (0.5 g glycyrrhizin, equivalent to ~10 g daily of licorice root) (Armanini et al. 1999). Replications of that trial using the compound glycyrrhizin resulted in a non-significant decrease of testosterone (Josephs et al. 2001). The discrepancy in results between the studies is believed to be caused by the method used for testosterone analysis, with the first study measuring testosterone in serum (total hormone level) and the second measuring testosterone in saliva (free hormone level) (Isbrucker and Burdock 2006). A repeat of the original methodology provided similar results to the earlier study (Armanini et al. 2003).

A trial in women showed that serum testosterone was significantly decreased during the luteal phase of the menstrual cycle after administration of 3.5 g licorice (containing 0.26 g glycyrrhizic acid) daily for two menstrual cycles (Armanini et al. 2004).

Administration of 3.5 g daily of licorice (containing 0.26 g glycyrrhizic acid) to healthy women for 2 months significantly increased urinary levels of parathyroid hormone (PTH), vitamin D, and calcium (Armanini et al. 2002).

Animal Pharmacological Studies

Animal pharmacological studies were identified but omitted due to presence of human data.

In Vitro Pharmacological Studies

In vitro pharmacological studies were identified but omitted due to presence of human data.

IV. Pregnancy and Lactation

Two retrospective epidemiological studies on licorice consumption by pregnant women (total 1144 women) suggested that heavy licorice consumption (≥500 mg/week glycyrrhizin) was correlated with earlier births. Heavy consumption reduced the mean gestational period by 2.5 days (Strandberg et al. 2001, 2002).

A study on cognitive and psychiatric outcomes in children based on maternal consumption of licorice during pregnancy suggested some adverse effects associated with high levels of licorice candy consumption. The study populations were sorted based on whether the mothers had low (0–249 mg) or high (>500 mg) rates of licorice candy intake during pregnancy. Children of mothers with a high licorice candy intake had decrements in verbal and visuospatial abilities and narrative memory, and increases in rule-breaking, aggression, and attention problems. The authors indicated that results of this study are consistent with recognized effects of glucocorticoid drugs (Räikkönen et al. 2009).

Doses of licorice in excess of those normally recommended (3–15 g daily) may affect blood pressure (*see* Human pharmacological studies) and should be avoided by women with preeclampsia (a condition during pregnancy that includes hypertension) (Mills and Bone 2005).

No teratogenic effects were observed in mice, rats, hamsters, and rabbits administered the compound ammoniated glycyrrhizin at doses up to 1000 mg/kg daily starting on the sixth day of gestation (continuing for 10, 10, 5, or 13 days, respectively) (FDRL 1971).

Renal tissue abnormalities were observed in fetuses of some rats fed 100 or 250 mg/kg of the compound disodium glycyrrhizin on days 0 to 20 of gestation. Skeletal abnormalities were observed in fetuses of the two highest dose groups, although the authors noted that similar abnormalities were observed in control groups. No adverse effects were observed in rats fed 10 mg/kg disodium glycyrrhizin under the same protocol (Mantovani et al. 1988).

A study on the effects of glycyrrhetic acid on rat fetal lung development found moderately but significantly reduced levels of 11β-hydroxysteroid dehydrogenase in the lungs of fetuses of rats fed 1000 mg/kg daily glycyrrhetic acid on days 13 through 17, 19, or 21, of pregnancy or postpartum day 1. Fetuses of rats fed lower doses (up to 100 mg/kg) showed no adverse effects (Hundertmark et al. 2002).

No information on the safety of licorice during lactation was identified.

V. Toxicity Studies

Acute Toxicity

The LD_{50} of an orally administered licorice extract (53% glycyrrhizin) is 18 g/kg in male rats and 14.2 g/kg in female rats. In mice, the LD_{50} is greater than 7.5 g/kg (Komiyama et al. 1977).

Subchronic Toxicity

Mice administered the compound glycyrrhizin at amounts of 0.6 or 1.25% of drinking water (~880 or 1830 mg/kg glycyrrhizin daily) died within 10 weeks of treatment (Kobuke et al. 1985).

Chronic Toxicity

No significant changes in body weight, clinical chemistry, or histology were observed in rats fed 100, 300, or 1000 mg/kg glycyrrhetinic acid daily for 1 year. Serum potassium and chloride levels were slightly reduced at the highest dose. In dogs fed the same doses, animals receiving 1000 mg/kg displayed a decrease in weight gain and increases in serum levels of alkaline phosphatase and alanine aminotransferase (Kelloff et al. 1994).

Mice administered the compound glycyrrhizin at amounts up to 0.15% of drinking water (~220 mg/kg glycyrrhizin daily) for 96 weeks did not develop tumors (Kobuke et al. 1985).

Literature Cited

Armanini, D., G. Bonanni, M.J. Mattarello, et al. 2003. Licorice consumption and serum testosterone in healthy man. *Exp. Clin. Endocrinol. Diabetes* 111(6):341-343.

Armanini, D., G. Bonanni, and M. Palermo. 1999. Reduction of serum testosterone in men by licorice. *N. Engl. J. Med.* 341(15):1158.

Armanini, D., C. Fiore, M.J. Mattarello, J. Bielenberg, and M. Palermo. 2002. History of the endocrine effects of licorice. *Exp. Clin. Endocrinol. Diabetes* 110(6):257-261.

Armanini, D., M.J. Mattarello, C. Fiore, et al. 2004. Licorice reduces serum testosterone in healthy women. *Steroids* 69(11-12):763-766.

Bannister, B., R. Ginsburg, and J. Shneerson. 1977. Cardiac arrest due to liquorice-induced hypokalaemia. *Br. Med. J.* 2(6089):738-739.

Barrella, M., G. Lauria, R. Quatrale, and E. Paolino. 1997. Hypokalemic rhabdomyolysis associated with liquorice ingestion: Report of an atypical case. *Ital. J. Neurol. Sci.* 18(4):217-220.

Belhadj-Tahar, H., B. Nassar, Y. Coulais, J.L. Montastruc, and N. Sadeg. 2003. Acute pseudo-aldosteronism syndrome induced by liquorice. *Therapie* 58(4):375-378.

Bensky, D., S. Clavey, and E. Stöger. 2004. *Chinese herbal medicine: Materia medica.* 3rd ed. Seattle: Eastland Press.

Beretta-Piccoli, C., G. Salvade, P.L. Crivelli, and P. Weidmann. 1985. Body-sodium and blood volume in a patient with licorice-induced hypertension. *J. Hypertens.* 3(1):19-23.

Bernardi, M., E. Paola, F. D'Intino, et al. 1994. Effects of prolonged ingestion of graded doses of licorice by healthy volunteers. *Life Sci.* 55(11):863-872.

Bradley, P.R. 1992. *British herbal compendium: A handbook of scientific information on widely used plant drugs.* Bournemouth, Dorset: British Herbal Medicine Association.

Brouwers, A.J., and J. van der Meulen. 2001. 'Licorice hypertension' also caused by licorice tea. *Ned. Tijdschr. Geneeskd.* 145(15):744-747.

Cartier, A., J.L. Malo, and M. Labrecque. 2002. Occupational asthma due to liquorice roots. *Allergy* 57(9):863.

Chadha, Y. 1988. *The wealth of India: A dictionary of Indian raw materials and industrial products.* Delhi: Council of Scientific and Industrial Research.

Chen, J.K., and T.T. Chen. 2004. *Chinese medical herbology and pharmacology.* City of Industry, CA: Art of Medicine Press.

Cheng, C.J., Y.H. Chen, T. Chau, and S.H. Lin. 2004. A hidden cause of hypokalemic paralysis in a patient with prostate cancer. *Support Care Cancer* 12(11):810-812.

Clyburn, E.B., and D.J. DiPette. 1995. Hypertension induced by drugs and other substances. *Semin. Nephrol.* 15(2):72-86.

Colloredo, G., V. Bertone, P. Peci, et al. 1987. Pseudoaldosteronism caused by licorice. Review of the literature and description of 4 clinical cases. *Minerva Med.* 78(2):93-101.

Conn, J.W., D.R. Rovner, and E.L. Cohen. 1968. Licorice-induced pseudoaldosteronism. Hypertension, hypokalemia, aldosteronopenia, and suppressed plasma renin activity. *J. Am. Med. Assoc.* 205(7):492-496.

Council of Europe. 2001. *European pharmacopoeia*. 4th ed. Strasbourg: Council of Europe.

Cugini, P., R. Gentile, A. Zard, and G. Rocchi. 1983. Hypertension in licorice abuse. A case report. *G. Ital. Cardiol.* 13(2):126-128.

Cumming, A.M., K. Boddy, J.J. Brown, et al. 1980. Severe hypokalaemia with paralysis induced by small doses of liquorice. *Postgrad. Med. J.* 56(657):526-529.

Cuspidi, C., M. Gelosa, E. Moroni, and L. Sampieri. 1981. Pseudo-Conn's syndrome after habitual ingestion of liquorice. Report on various clinical cases. *Minerva Med.* 72(13):825-830.

de Rohan Chabot, P., B. Gueguen, D. Patte, et al. 1984. Hypokalemic paralysis after prolonged absorption of a non-alcoholic pastis in a diabetic. *Rev. Neurol. (Paris)* 140(3):207-211.

De Smet, P.A.G.M. 1993. *Adverse effects of herbal drugs, Volume 2*. New York: Springer.

Dellow, E.L., R.J. Unwin, and J.W. Honour. 1999. Pontefract cakes can be bad for you: Refractory hypertension and liquorice excess. *Nephrol. Dial. Transplant.* 14(1):218-220.

Dobbins, K.R., and R.F. Saul. 2000. Transient visual loss after licorice ingestion. *J. Neuroophthalmol.* 20(1):38-41.

Doeker, B.M., and W. Andler. 1999. Liquorice, growth retardation and Addison's disease. *Horm. Res.* 52(5):253-255.

Elinav, E., and T. Chajek-Shaul. 2003. Licorice consumption causing severe hypokalemic paralysis. *Mayo Clin. Proc.* 78(6):767-768.

Epstein, M.T., E.A. Espiner, R.A. Donald, and H. Hughes. 1977. Effect of eating liquorice on the renin-angiotensin-aldosterone axis in normal subjects. *Br. Med. J.* 1(6059):488-890.

Eriksson, J.W., B. Carlberg, and V. Hillorn. 1999. Life-threatening ventricular tachycardia due to liquorice-induced hypokalaemia. *J. Intern. Med.* 245(3):307-310.

Famularo, G., F.M. Corsi, and M. Giacanelli. 1999. Iatrogenic worsening of hypokalemia and neuromuscular paralysis associated with the use of glucose solutions for potassium replacement in a young woman with licorice intoxication and furosemide abuse. *Acad. Emerg. Med.* 6(9):960-964.

FDRL. 1971. Teratogenic evaluation of FDA 71-1 (ammonium glycyrrhizinate). Food and Drug Research Labs. US NTIS Report PB-221793. *Cited in* Isbrucker, R.A., and G.A. Burdock. 2006. Risk and safety assessment on the consumption of licorice root (*Glycyrrhiza* sp.), its extract and powder as a food ingredient, with emphasis on the pharmacology and toxicology of glycyrrhizin. *Regul. Toxicol. Pharmacol.* 46(3):167-192.

Gerritsen, K.G.F., J. Meulenbelt, W. Spiering, et al. 2009. An unusual cause of ventricular fibrillation. *Lancet* 373(9669):1144.

Gomez Fernandez, P., M. Casares, J. Martinez Ara, et al. 1981. Primary pseudohyperaldosteronism produced by chronic licorice consumption. *Rev. Clin. Esp.* 163(4):277-278.

Hamidon, B.B., and V. Jeyabalan. 2006. Exogenously-induced apparent hypermineralocorticoidism associated with ingestion of "asam boi." *Singapore Med. J.* 47(2):156-158.

Harris, J. 2000. Lethal licorice. *Aust. Nurs. J.* 7(Suppl.):1-3.

Hundertmark, S., A. Dill, H. Buhler, et al. 2002. 11beta-Hydroxysteroid dehydrogenase type 1: A new regulator of fetal lung maturation. *Horm. Metab. Res.* 34(10):537-544.

Hussain, R.M. 2003. The sweet cake that reaches parts other cakes can't! *Postgrad. Med. J.* 79(928):115-116.

Isbrucker, R.A., and G.A. Burdock. 2006. Safety and risk assessment on the consumption of licorice root. *Regul. Toxicol. Pharmacol.* 46:168-192.

Ishikawa, S., M. Kato, T. Tokuda, et al. 1999. Licorice-induced hypokalemic myopathy and hypokalemic renal tubular damage in anorexia nervosa. *Int. J. Eat. Disord.* 26(1):111-114.

Janse, A., M. van Iersel, W.H. Hoefnagels, and M.G. Olde Rikker. 2005. The old lady who liked liquorice: Hypertension due to chronic intoxication in a memory-impaired patient. *Neth. J. Med.* 63(4):149-150.

Josephs, R.A., J.S. Guinn, M.L. Harper, and F. Askari. 2001. Liquorice consumption and salivary testosterone concentrations. *Lancet* 358(9293):1613-1614.

Kageyama, K., H. Watanobe, M. Nishie, K. Imamura, and T. Suda. 1997. A case of pseudoaldosteronism induced by a mouth refresher containing licorice. *Endocr. J.* 44(4):631-632.

Kelloff, G.J., J.A. Crowell, C.W. Boone, et al. 1994. Clinical development plan: 18beta-glycyrrhetinic acid. *J. Cell. Biochem. Suppl.* 20:166-175.

Kelly, R.A. 1990. Cardiac glycosides and congestive heart failure. *Am. J. Cardiol.* 65(10):10-16E.

Kobuke, T., K. Inai, S. Nambu, et al. 1985. Tumorigenicity study of disodium glycyrrhizinate administered orally to mice. *Food Chem. Toxicol.* 23(11):979-983.

Komiyama, K., Y. Kawakubo, T. Fukushima, et al. 1977. Acute and subacute toxicity test on the extract from *Glycyrrhiza*. *Oyo Yakuri* 14(4):535-548. *Cited in* Isbrucker, R.A., and G.A. Burdock. 2006. Risk and safety assessment on the consumption of Licorice root (*Glycyrrhiza* sp.), its extract and powder as a food ingredient, with emphasis on the pharmacology and toxicology of glycyrrhizin. *Regul. Toxicol. Pharmacol.* 46(3):167-192.

Kondo, K., M. Shiba, R. Nakamura, T. Morota, and Y. Shoyama. 2007. Constituent properties of licorices derived from *Glycyrrhiza uralensis*, *G. glabra*, or *G. inflata* identified by genetic information. *Biol. Pharm. Bull.* 30(7):1271-1277.

Lin, S.H., S.S. Yang, T. Chau, and M.L. Halperin. 2003. An unusual cause of hypokalemic paralysis: Chronic licorice ingestion. *Am. J. Med. Sci.* 325 (3):153-156.

Lozano, P., D. Flores, S. Martinez, et al. 2000. Upper limb ischemia induced by chronic licorice ingestion. *J. Cardiovasc. Surg. (Torino)* 41(4):631-632.

Mantovani, A., C. Ricciardi, A.V. Stazi, et al. 1988. Teratogenicity study of ammonium glycyrrhizinate in the Sprague-Dawley rat. *Food Chem. Toxicol.* 26(5):435-440.

Martindale, W., and J.E.F. Reynolds. 1996. *The extra pharmacopoeia*. 31st ed. London: Pharmaceutical Press.

Mills, S., and K. Bone. 2005. *The essential guide to herbal safety*. St. Louis: Elsevier.

Mitchell, H., et al. 1991. *British herbal pharmacopoeia, 4th impression*. Bournemouth, U.K.: British Herbal Medicine Association.

Negro, A., E. Rossi, G. Regolisti, and F. Perazzoli. 2000. Liquorice-induced sodium retention. Merely an acquired condition of apparent mineralocorticoid excess? A case report. *Ann. Ital. Med. Int.* 15(4):296-300.

Piette, A.M., D. Bauer, and A. Chapman. 1984. Major hypokalemia with rhabdomyolysis secondary to the intake of a nonalcoholic aniseed aperitif. *Ann. Med. Intern. (Paris)* 135(4):296-298.

Pozzoli, G., G.C. Mariotti, and F. Colombo. 1980. Arterial hypertension caused by ingestion of licorice. *G. Ital. Cardiol.* 10(10):1415-1418.

Räikkönen, K., A.K. Pesonen, K. Heinonen, et al. 2009. Maternal licorice consumption and detrimental cognitive and psychiatric outcomes in children. *Am. J. Epidemiol.* 170(9):1137.

Shintani, S., H. Murase, H. Tsukagoshi, and T. Shiigai. 1992. Glycyrrhizin (licorice)-induced hypokalemic myopathy. *Eur. Neurol.* 32(1):44-51.

Sigurjonsdottir, H.A., M. Axelson, G. Johannsson, et al. 2006. The liquorice effect on the RAAS differs between the genders. *Blood Press.* 15(3):169-172.

Sigurjonsdottir, H.A., L. Franzson, K. Manhem, et al. 2001. Liquorice-induced rise in blood pressure: A linear dose-response relationship. *J. Hum. Hypertens.* 15(8):549-552.

Sigurjonsdottir, H.A., K. Manhem, M. Axelson, and S. Wallerstedt. 2003. Subjects with essential hypertension are more sensitive to the inhibition of 11 beta-HSD by liquorice. *J. Hum. Hypertens.* 17(2):125-131.

Stewart, P.M., A.M. Wallace, R. Valentino, et al. 1987. Mineralocorticoid activity of liquorice: 11-beta-hydroxysteroid dehydrogenase deficiency comes of age. *Lancet* 2(8563):821-824.

Størmer, F., R. Reistad, and J. Alexander. 1993. Glycyrrhizic acid in liquorice—Evaluation of health hazard. *Food Chem. Toxicol.* 31(4):303-312.

Strandberg, T.E., S. Andersson, A.L. Jarvenpaa, and P.M. McKeigue. 2002. Preterm birth and licorice consumption during pregnancy. *Am. J. Epidemiol.* 156(9):803-805.

Strandberg, T.E., A.L. Jarvenpaa, H. Vanhanen, and P.M. McKeigue. 2001. Birth outcome in relation to licorice consumption during pregnancy. *Am. J. Epidemiol.* 153(11):1085-1088.

Takeda, R., S. Morimoto, K. Uchida, et al. 1979. Prolonged pseudoaldosteronism induced by glycyrrhizin. *Endocrinol. Jpn.* 26(5):541-547.

van den Bosch, A.E., J.M. van der Klooster, D.M. Zuidgeest, R.J. Ouwendijk, and A. Dees. 2005. Severe hypokalaemic paralysis and rhabdomyolysis due to ingestion of liquorice. *Neth. J. Med.* 63(4):146-148.

van Gelderen, C.E., J.A. Bijlsma, W. van Dokkum, and T.J. Savelkoul. 2000. Glycyrrhizic acid: The assessment of a no effect level. *Hum. Exp. Toxicol.* 19(8):434-439.

Wash, L.K., and J.D. Bernard. 1975. Licorice-induced pseudoaldosteronism. *Am. J. Hosp. Pharm.* 32(1):73-74.

WHO. 1999. *Monographs on selected medicinal plants, Volume 1.* Geneva: World Health Organization.

Yasue, H., T. Itoh, Y. Mizuno, and E. Harada. 2007. Severe hypokalemia, rhabdomyolysis, muscle paralysis, and respiratory impairment in a hypertensive patient taking herbal medicines containing licorice. *Intern. Med.* 46(9):575-578.

Yoshida, S., and Y. Takayama. 2003. Licorice-induced hypokalemia as a treatable cause of dropped head syndrome. *Clin. Neurol. Neurosurg.* 105(4):286-287.

Gossypium spp. — Malvaceae

Gossypium herbaceum L.
SCN: Levant cotton
AN: *karpasa*
OCN: *algodao*

Gossypium hirsutum L.
SCN: cotton
Part: root bark

Quick Reference Summary

Safety Class: 2b
Interaction Class: A

Contraindications

Not for use in pregnancy except under the supervision of a qualified healthcare practitioner (Conway and Slocumb 1979; Felter and Lloyd 1898; Randel et al. 1992).

Other Precautions

None known.

Drug and Supplement Interactions

None known.

Notice

Abortifacient (Chadha 1988; Conway and Slocumb 1979; De Smet 1993; Felter and Lloyd 1898; Moore 1978); *see* Appendix 2.

Uterine stimulant (Chadha 1988; Felter and Lloyd 1898; List and Hörhammer 1973; Williamson 2003); *see* Appendix 2.

Diuretic (Felter and Lloyd 1898); *see* Appendix 2.

Editors' Notes

Cotton and Levant cotton contain the compound gossypol, found primarily in the seed but also present in lesser amounts in the root bark. Gossypol content in the root and root bark reportedly ranges from 0.16 to 5.1% (Royce et al. 1941; Stipanovic et al. 2006; Wang 1987). Gossypol was widely researched as a male contraceptive. *See* Pharmacological Considerations for more information.

Adverse Events and Side Effects

See Pharmacological Considerations.

Pharmacological Considerations

The compound gossypol has been widely investigated as a potential male contraceptive, with a number of human and animal studies showing a significant reduction or complete elimination of sperm in the semen. Research on gossypol as a contraceptive was stopped for two reasons. First, the compound was found to produce chronic or permanent infertility in 25 to 50% of trial participants. Secondly, some clinical trials in China indicated that gossypol produced or exacerbated hypokalemia (low levels of potassium), although this effect was generally not observed in studies outside of China. More recent animal studies have indicated a possible mechanism of action for the hypokalemia (Coutinho 2002; Qian and Wang 1984; Waites et al. 1998).

Pregnancy and Lactation

Cotton and Levant cotton root barks have traditionally been used as abortifacients (Conway and Slocumb 1979; Felter and Lloyd 1898; Moore 1978). A number of animal studies have indicated that the compound gossypol has embryotoxic activity, while other studies have indicated no such activity and a lack of teratogenic effects (Li et al. 1989; Qian and Wang 1984; Randel et al. 1992; Sein 1986; Lin et al. 1985). Based on this information, use during pregnancy is not recommended except under the supervision of a qualified healthcare practitioner.

No information on the safety of cotton root bark during lactation was identified. While this review did not identify any concerns for use while nursing, safety has not been conclusively established.

Review Details

I. Drug and Supplement Interactions

Clinical Trials of Drug or Supplement Interactions

No clinical trials of drug or supplement interactions were identified.

Case Reports of Suspected Drug or Supplement Interactions

No case reports of suspected drug or supplement interactions were identified.

Animal Trials of Drug or Supplement Interactions

No animal trials of drug or supplement interactions were identified.

II. Adverse Events

Case Reports of Adverse Events

No case reports of adverse events were identified.

III. Pharmacology and Pharmacokinetics

Human Pharmacological Studies

The compound gossypol has been shown to have antifertility activity in men at doses of 0.3 mg/kg. Doses in clinical trials generally range between 7.5 and 15 mg daily. After clinical trials, approximately 60% of men regained fertility by 16 weeks after the end of the trial, whereas azoospermia (lack of sperm in the semen) persisted in about 19% of men (Coutinho 2002). Human studies have indicated that gossypol has no significant effects on hormone levels (De Smet 1993; Qian and Wang 1984).

Animal Pharmacological Studies

No relevant animal pharmacological studies were identified.

In Vitro Pharmacological Studies

No relevant in vitro pharmacological studies were identified.

IV. Pregnancy and Lactation

Cotton and Levant cotton root barks have traditionally been used as abortifacients (Conway and Slocumb 1979; Felter and Lloyd 1898; Moore 1978). An Eclectic medical text reported that cotton acts as an emmenagogue and parturient (Felter and Lloyd 1898).

A reproductive toxicity review indicated that animal studies of the compound gossypol demonstrated no teratogenic activity or abortifacient activity of the compound (Randel et al. 1992). No teratogenicity or embryotoxicity of gossypol was reported in studies with rats and rabbits at 5 to 30 times the clinical dose (Qian and Wang 1984). No teratogenic effects were observed in rats orally administered 40 mg/kg of gossypol (Beaudoin 1985).

Administration of the compound gossypol to rats on days 0 to 8 of pregnancy inhibited ovum implantation and maintenance of pregnancy (Lin et al. 1985).

In mice orally administered 60 or 120 mg/kg of gossypol acetic acid daily on days 6 to 13 of pregnancy, an increase in fetal resorptions and late fetal deaths was observed (Li et al. 1989).

In mice orally administered 50 or 75 mg/kg of gossypol acetic acid daily on days 1 to 15 of pregnancy, a dose-dependent embryotoxic effect was observed. No abnormalities were observed in surviving fetuses (Sein 1986).

No information on the safety of cotton root bark during lactation was identified.

V. Toxicity Studies

Acute Toxicity

The oral LD_{50} of the compound gossypol has been reported as 2400 to 3340 mg/kg in rats, 500 to 950 mg/kg in mice, 360 to 600 mg/kg in rabbits, 280 to 300 mg/kg in guinea pigs, and 500 mg/kg in pigs (Qian and Wang 1984).

Short-Term Toxicity

In rats orally administered 0.4, 2, or 4 ml/kg of a fluid extract of Levant cotton root daily for 30 days, no toxicity was observed at the lowest dose (equivalent to the human therapeutic dose). Dose-dependent changes in weight gain, hematological values, and biochemical blood analyses at the two higher doses were indicative of systemic toxicity (Mello et al. 2008).

Genotoxicity

No mutagenic activity of the compound gossypol was observed in the Ames test for mutagenicity (Li et al. 1989).

Literature Cited

Beaudoin, A.R. 1985. A reproduction and teratology study of gossypol. In *Gossypol*, edited by Segal, S.J. New York: Plenum Press.

Chadha, Y. 1988. *The wealth of India: A dictionary of Indian raw materials and industrial products*. Delhi: Council of Scientific and Industrial Research.

Conway, G.A., and J.C. Slocumb. 1979. Plants used as abortifacients and emmenagogues by Spanish New Mexicans. *J. Ethnopharmacol.* 1(3):241-261.

Coutinho, E.M. 2002. Gossypol: A contraceptive for men. *Contraception* 65(4):259-263.

De Smet, P.A.G.M. 1993. *Adverse effects of herbal drugs, Volume 2*. Berlin: Springer.

Felter, H.W., and J.U. Lloyd. 1898. *King's American dispensatory*. 18th ed., 3rd rev. 2 vols. Cincinnati: Ohio Valley Co.

Li, Y.F., G.M. Booth, and R.E. Seegmiller. 1989. Evidence for embryotoxicity of gossypol in mice and chicks with no evidence of mutagenic activity in the Ames test. *Reprod. Toxicol.* 3(1):59-62.

Lin, Y.C., T. Fukaya, Y. Rikihisa, and A. Walton. 1985. Gossypol in female fertility control: Ovum implantation and early pregnancy inhibited in rats. Life Sci 37(1):39.

List, P.H., and H. Hörhammer. 1973. *Hagers handbuch der pharmazeutischen praxis*. Berlin: Springer.

Mello, J.R.B., F.B. Mello, R.N. Etges, et al. 2008. Pre-clinical toxicity of a phytotherapy containing *Gossypium herbaceum* (cotton plant) in Wistar rats. *Lat. Am. J. Pharm.* 27(1):46-55.

Moore, M. 1978. *Los remedios de la gente*. Santa Fe, NM: M. Moore.

Qian, S., and Z. Wang. 1984. Gossypol: A potential antifertility agent for males. *Ann. Rev. Pharmacol. Toxicol.* 24(1):329-360.

Randel, R.D., C.C. Chase, Jr., and S.J. Wyse. 1992. Effects of gossypol and cottonseed products on reproduction of mammals. *J. An. Sci.* 70(5):1628-1638.

Royce, H.D., J.R. Harrison, and E.R. Hahn. 1941. Cotton-root bark as a source of gossypol. *J. Am. Oil Chem. Soc.* 18(2):27-29.

Sein, G.M. 1986. The embryotoxic and immunodepressive effects of gossypol. *Am. J. Chin. Med.* 14(3-4):110-115.

Stipanovic, R.D., L.S. Puckhaber, and A.A. Bell. 2006. Ratios of (+)- and (−)-gossypol in leaves, stems, and roots of selected accessions of *Gossypium hirsutum* var. Marie Galante (Watt) Hutchinson. *J. Agric. Food Chem.* 54(5):1633-1637.

Waites, G.M., C. Wang, and P.D. Griffin. 1998. Gossypol: Reasons for its failure to be accepted as a safe, reversible male antifertility drug. *Int. J. Androl.* 21(1):8.

Wang, M.Z. 1987. Analysis of gossypol by high performance liquid chromatography. *J. Ethnopharmacol.* 20(1):1-11.

Williamson, E.M. 2003. *Potter's herbal cyclopedia*. Saffron Walden, Essex: C.W. Daniel Co.

Grifola frondosa (Dicks.: Fr.) Gray — Polyporaceae

SCN: maitake
Syn: *Polyporus frondosus* (Dicks.: Fr.) Fr.
OCN: dancing mushroom; hen-of-the-woods
Part: fruiting body, mycelium

Quick Reference Summary

Safety Class: 1
Interaction Class: A

Contraindications

None known.

Other Precautions

None known.

Drug and Supplement Interactions

None known.

Editors' Note

Maitake is commonly consumed as a food (Zhuang and Wasser 2004).

Adverse Events and Side Effects

None known.

Pharmacological Considerations

Animal studies have demonstrated that maitake may modify glucose regulation (Cui et al. 2009; Han and Liu 2009; Hong et al. 2007; Horio and Ohtsuru 2001; Kubo et al. 1994; Lo et al. 2008; Preuss et al. 2007). People with diabetes are advised to monitor their blood sugar closely and discuss the use of this herb with a qualified healthcare practitioner prior to use.

Pregnancy and Lactation

No information on the safety of maitake in pregnancy or lactation was identified in the scientific or traditional literature. Although this review did not identify any concerns for use while pregnant or nursing, safety has not been conclusively established.

Review Details

I. Drug and Supplement Interactions

Clinical Trials of Drug or Supplement Interactions

No clinical trials of drug or supplement interactions were identified.

Case Reports of Suspected Drug or Supplement Interactions

No case reports of suspected drug or supplement interactions were identified.

Animal Trials of Drug or Supplement Interactions

No animal trials of drug or supplement interactions were identified.

II. Adverse Events

Case Reports of Adverse Events

No case reports of adverse events were identified.

III. Pharmacology and Pharmacokinetics

Human Pharmacological Studies

In a phase I/II dose escalation trial in postmenopausal women with a history of breast cancer, an extract of maitake was administered at doses of 0.5, 1.5, 3, or 5 mg/kg twice daily for 3 weeks. No dose-limiting toxicity was encountered. Nausea and joint swelling were reported in one patient and rash and pruritus in another. Increasing doses of maitake increased certain immunologic parameters and depressed others, and the dose-response curves for many endpoints were non-monotonic, with intermediate doses having either immune enhancing or immune suppressant effects compared with both high and low doses (Deng et al. 2009).

Animal Pharmacological Studies

In diabetic mice orally administered 150 or 450 mg/kg of alpha-glucan from maitake, a decrease in fasting plasma glucose levels was observed (Hong et al. 2007).

A reduction in fasting blood glucose levels was observed in diabetic rats fed diets containing 20 percent maitake for 100 days (Horio and Ohtsuru 2001).

A reduction in blood glucose levels was observed in diabetic mice orally administered 1 g of maitake powder. When extracts of maitake were tested, an ethanol extract had hypoglycemic activity, but no activity was observed from an aqueous extract (Kubo et al. 1994).

Oral administration of 1 g daily of maitake fruiting body extract or mycelium extract for 15 days attenuated diabetes-induced changes in blood glucose levels (Lo et al. 2008).

Improved glucose sensitivity was observed in spontaneously hypertensive rats fed diets containing a maitake extract fraction (Preuss et al. 2007).

A reduction in blood glucose levels was observed in diabetic mice orally administered a vanadium-rich extract of maitake daily for 20 days (Cui et al. 2009; Han and Liu 2009).

In Vitro Pharmacological Studies

No relevant in vitro pharmacological studies were identified.

IV. Pregnancy and Lactation

No information on the safety of maitake during pregnancy or lactation was identified.

V. Toxicity Studies

Acute Toxicity

No adverse effects were observed in rats orally administered single doses of 0.5 or 2 g/kg of powdered maitake (Koike et al. 2003).

Literature Cited

Cui, B., L. Han, J. Qu, and Y. Lv. 2009. Hypoglycemic activity of *Grifola frondosa* rich in vanadium. *Biol. Trace Elem. Res.*131 (2):186-191.

Deng, G., H. Lin, A. Seidman, et al. 2009. A phase I/II trial of a polysaccharide extract from *Grifola frondosa* (Maitake mushroom) in breast cancer patients: Immunological effects. *J. Cancer Res. Clin. Oncol.* 135 (9):1215–1221.

Han, C., and T. Liu. 2009. A comparison of hypoglycemic activity of three species of basidiomycetes rich in vanadium. *Biol. Trace Elem. Res.* 127 (2):177-182.

Hong, L., M. Xun, and W. Wutong. 2007. Anti-diabetic effect of an alpha-glucan from fruit body of maitake (*Grifola frondosa*) on KK-Ay mice. *J. Pharm. Pharmacol.* 59 (4):575-582.

Animal Pharmacological Studies

No relevant animal pharmacological studies were identified.

In Vitro Pharmacological Studies

No relevant in vitro pharmacological studies were identified.

IV. Pregnancy and Lactation

No information on the safety of grindelia in pregnancy or lactation was identified.

V. Toxicity Studies

No toxicity studies of grindelia were identified.

Literature Cited

BHP. 1976. *British herbal pharmacopoeia*. Scientific Committee of the British Herbal Medicine Association. Cowling, UK: British Herbal Medicine Association.

Culbreth, D. 1927. *A manual of materia medica and pharmacology*. 7th ed. Philadelphia: Lea & Febiger.

Ellingwood, F. 1919. *The American materia medica, therapeutics and pharmacognosy*. Evanston, IL: Ellingwood's Therapeutist.

Felter, H.W., and J.U. Lloyd. 1898. *King's American dispensatory*. 18th ed., 3rd rev. 2 vols. Cincinnati: Ohio Valley Co.

Frankenberger, W., and S. Benson. 1994. *Selenium in the environment*. Boca Raton, FL: CRC Press.

Remington, J.P., and H.C. Wood. 1918. *The dispensatory of the United States of America*. 20th ed. Philadelphia: Lippincott.

Gymnema sylvestre (Retz.) R. Br. ex Schult. — Asclepiadaceae

SCN: gymnema
AN: *meshashiringi*

Part: leaf

Quick Reference Summary

Safety Class 1
Interaction Class A

Contraindications

None known.

Other Precautions

None known.

Drug and Supplement Interactions

None known.

Adverse Events and Side Effects

Compounds from gymnema temporarily suppress the sensation of sweetness from certain sugars (i.e., sucrose) and other substances with sweet tastes (i.e., saccharin) (Singh et al. 2008). This effect has traditionally been experienced by persons chewing the leaf (Yeh et al. 2003).

Pharmacological Considerations

Human and animal studies have demonstrated that gymnema may modify glucose regulation (Baskaran et al. 1990; Gholap and Kar 2005; Shanmugasundaram et al. 1990; Spasov et al. 2008). People with diabetes are advised to monitor their blood sugar closely and discuss the use of this herb with a qualified healthcare practitioner prior to use.

Pregnancy and Lactation

No information on the safety of gymnema in pregnancy or lactation was identified in the scientific or traditional literature. Although this review did not identify any concerns for use while pregnant or nursing, safety has not been conclusively established.

Review Details

I. Drug and Supplement Interactions

Clinical Trials of Drug or Supplement Interactions

No clinical trials of drug or supplement interactions were identified.

Case Reports of Suspected Drug or Supplement Interactions

No case reports of suspected drug or supplement interactions were identified.

Animal Trials of Drug or Supplement Interactions

No animal trials of drug or supplement interactions were identified.

II. Adverse Events

Case Reports of Adverse Events

No case reports of adverse events were identified.

III. Pharmacology and Pharmacokinetics

Human Pharmacological Studies

Reductions in blood glucose levels were observed in patients with type 1 diabetes orally administered 400 mg of a gymnema extract daily for 10 to 12 months and in patients with type 2 diabetes orally administered 400 mg of a gymnema extract daily for 18 to 20 months (Baskaran et al. 1990; Shanmugasundaram et al. 1990).

Animal Pharmacological Studies

A reduction in blood glucose levels was observed in mice with high blood glucose levels intraperitoneally administered 6.7, 13.4, or 26.8 mg/kg of gymnemic acids (Gholap and Kar 2005).

No significant changes in blood sugar levels were observed in healthy or diabetic rats orally administered gymnema aqueous extract as a single dose of 1000 mg/kg or as a repeated dose of 30 mg/kg daily for four weeks (Galletto et al. 2004).

A reduction in fasting blood glucose levels was observed in diabetic rats orally administered the compound conduritol A isolated from the stem of gymnema (Wei et al. 2008).

In Vitro Pharmacological Studies

A concentration-dependent stimulation of insulin secretion was observed in isolated human islets and the MIN6 β-cell line after treatment with varying amounts of gymnema extract (Liu et al. 2009).

IV. Pregnancy and Lactation

No information on the use of gymnema during pregnancy or lactation was identified.

V. Toxicity Studies

Acute Toxicity

The LD_{50} of water and ethanol extracts of gymnema intraperitoneally administered to mice was 375 mg/kg (Bhakuni and Dhar 1971).

Chronic Toxicity

In rats fed diets containing 0.01, 0.10, or 1.0% gymnema for 52 weeks, no exposure-related changes in body-weight, in the food consumption, in the hematological parameters, or in the serum biochemical examinations were recognized. No histopathological alterations were seen (Ogawa et al. 2004).

G

Literature Cited

Baskaran, K., A.B. Kizar, S.K. Radha, and E.R. Shanmugasundaram. 1990. Antidiabetic effect of a leaf extract from *Gymnema sylvestre* in non-insulin-dependent diabetes mellitus patients. *J. Ethnopharmacol.* 30 (3):295.

Bhakuni, D.S., and M.L. Dhar. 1971. Screening of Indian plants for biological activity: Part III. *Indian J. Exp. Biol.* 9:91-102.

Galletto, R., V.L.D. Siqueira, E.B. Ferreira, A.J.B. Oliveira, and R.B. Bazotte. 2004. Absence of antidiabetic and hypolipidemic effect of *Gymnema sylvestre* in non-diabetic and alloxan-diabetic rats. *Braz. Arch. Biol. Technol.* 47:545-551.

Gholap, S., and A. Kar. 2005. Gymnemic acids from *Gymnema sylvestre* potentially regulates dexamethasone-induced hyperglycemia in mice. *Pharmaceut. Biol.* 43 (2):192-195.

Liu, B., H. Asare-Anane, A. Al-Romaiyan, et al. 2009. Characterisation of the insulinotropic activity of an aqueous extract of *Gymnema sylvestre* in mouse beta-cells and human islets of Langerhans. *Cell. Physiol. Biochem.* 23 (1-3):125-132.

Ogawa, Y., K. Sekita, T. Umemura, et al. 2004. *Gymnema sylvestre* Leaf Extract: A 52-Week Dietary Toxicity Study in Wistar Rats. *J. Food Hyg. Soc. Jap.* 45 (1):8-18.

Shanmugasundaram, E.R., G. Rajeswari, K. Baskaran, et al. 1990. Use of *Gymnema sylvestre* leaf extract in the control of blood glucose in insulin-dependent diabetes mellitus. *J. Ethnopharmacol.* 30 (3):281.

Singh, V.K., S. Umar, S.A. Ansari, and M. Iqbal. 2008. *Gymnema sylvestre* for diabetics. *J. Herbs, Spices, Med. Plants* 14 (1):88-106.

Spasov, A.A., M.P. Samokhina, and A.E. Bulanov. 2008. Antidiabetic properties of *Gymnema sylvestre*. *Pharm. Chem. J* 42 (11):626-629.

Wei, J.H., H.S. Zhen, Q. Qiu, J. Chen, and F. Zhou. 2008. Experimental study of hypoglycemic activity of conduritol A of stems of *Gymnema sylvestre*. *Zhongguo Zhong Yao Za Zhi* 33 (24):2961-2965.

Yeh, G.Y., D.M. Eisenberg, T.J. Kaptchuk, and R.S. Phillips. 2003. Systematic review of herbs and dietary supplements for glycemic control in diabetes. *Diabetes Care* 26 (4):1277.

Hamamelis virginiana L. **Hamamelidaceae**

SCN: witch hazel

Part: bark, leaf

Quick Reference Summary

Safety Class: 1
Interaction Class: A

Contraindications

None known.

Other Precautions

None known.

Drug and Supplement Interactions

None known.

Notice

Tannins (10% bark, 3–10% leaf) (Leung and Foster 1996; Touriño et al. 2008; Wichtl 2004); *see* Appendix 1.

Editors' Note

Witch hazel water is a steam distillate and does not contain tannins (Leung and Foster 1996).

Adverse Events and Side Effects

In sensitive individuals, witch hazel may cause irritation of the stomach (Wichtl 2004).

Allergic contact dermatitis from witch hazel has been reported (Granlund 1994).

Pharmacological Considerations

None known.

Pregnancy and Lactation

No information on the safety of witch hazel in pregnancy or lactation was identified in the scientific or traditional literature. Although this review did not identify any concerns for use while pregnant or nursing, safety has not been conclusively established.

Review Details

I. Drug and Supplement Interactions

Clinical Trials of Drug or Supplement Interactions

No clinical trials of drug or supplement interactions were identified.

Case Reports of Suspected Drug or Supplement Interactions

No case reports of suspected drug or supplement interactions were identified.

Animal Trials of Drug or Supplement Interactions

No animal trials of drug or supplement interactions were identified.

II. Adverse Events

Adverse Events Reported in Clinical Trials

Witch hazel ointment was reported as generally well tolerated in an observational study in children (age 27 days to 11 years) with minor skin injuries (Wolff and Kieser 2007).

Case Reports of Adverse Events

No case reports of adverse events were identified.

III. Pharmacology and Pharmacokinetics

Human Pharmacological Studies

In patients with allergy to chamomile, patch testing with cosmetic and herbal product ingredients indicated that witch hazel tincture was a sensitizing agent (Paulsen et al. 2008).

Among 1032 patients selected from patch testing clinics, none had positive reactions to patch testing with a 25% witch hazel extract (Bruynzeel et al. 1992).

Allergic contact dermatitis from witch hazel has been reported (Granlund 1994).

Animal Pharmacological Studies

No relevant animal pharmacological studies were identified.

In Vitro Pharmacological Studies

No relevant in vitro pharmacological studies were identified.

IV. Pregnancy and Lactation

No information on the safety of witch hazel during pregnancy or lactation was identified.

V. Toxicity Studies

Chronic Toxicity

In rats subcutaneously administered 10 mg daily of a freeze-dried extract of witch hazel leaf weekly for 78 weeks, 3 of 15 male rats and 0 of 15 female rats developed malignant mesenchymoma. This rate of tumors was not considered significant (Kapadia et al. 1978).

No carcinogenic activity of witch hazel water was observed in 2-year carcinogenicity tests. Rats were administered a dose of 0.6 ml and mice were administered 0.2 ml (Haseman and Clark 1990).

Genotoxicity

No mutagenic activity of witch hazel water was reported in mutagenicity assays with Chinese hamster ovary cells with or without metabolic activation, nor in the mouse lymphoma cell forward mutation assay (Galloway et al. 1987; McGregor et al. 1988; Tennant et al. 1987).

Literature Cited

Bruynzeel, D.P., W.G. Van Ketel, E. Young, T. Van Joost, and G. Smeenk. 1992. Contact sensitization by alternative topical medicaments containing plant extracts. *Contact Dermat.* 27(4):278-279.

Galloway, S.M., M.J. Armstrong, C. Reuben, et al. 1987. Chromosome aberrations and sister chromatid exchanges in Chinese hamster ovary cells: Evaluations of 108 chemicals. *Environ. Mol. Mutagen.* 10(Suppl. 10):1-175.

Granlund, H. 1994. Contact allergy to witch hazel. *Contact Dermat.* 31(3):195.

Haseman, J.K., and A.M. Clark. 1990. Carcinogenicity results for 114 laboratory animal studies used to assess the predictivity of four *in vitro* genetic toxicity assays for rodent carcinogenicity. *Environ. Mol. Mutagen.* 16(Suppl. 18):15-31.

Kapadia, G.J., E.B. Chung, B. Ghosh, et al. 1978. Carcinogenicity of some folk medicinal herbs in rats. *J. Nat. Cancer Inst.* 60(3):683-686.

Leung, A.Y., and S. Foster. 1996. *Encyclopedia of common natural ingredients used in food, drugs, and cosmetics*. 2nd ed. New York: Wiley.

McGregor, D.B., A. Brown, P. Cattanach, et al. 1988. Responses of the L5178Y *tk+/tk* mouse lymphoma cell forward mutation assay II: 18 coded chemicals. *Environ. Mol. Mutagen.* 11(1):91-118.

Paulsen, E., L.P. Chistensen, and K.E. Andersen. 2008. Cosmetics and herbal remedies with Compositae plant extracts—Are they tolerated by Compositae-allergic patients? *Contact Dermat.* 58(1):15-23.

Tennant, R.W. B.H. Margolin, M.D. Shelby, et al. 1987. Prediction of chemical carcinogenicity in rodents from *in vitro* genetic toxicity assays. *Science* 236(4804):933-941.

Touriño, S., D. Lizárraga, A. Carreras, et al. 2008. Highly galloylated tannin fractions from witch hazel (*Hamamelis virginiana*) bark: Electron transfer capacity, *in vitro* antioxidant activity, and effects on skin-related cells. *Chem. Res. Toxicol.* 21(3):696-704.

Wichtl, M. 2004. *Herbal drugs and phytopharmaceuticals: A handbook for practice on a scientific basis*. 3rd ed. Boca Raton, FL: CRC Press.

Wolff, H.H., and M. Kieser. 2007. *Hamamelis* in children with skin disorders and skin injuries: Results of an observational study. *Eur. J. Pediatr.* 166(9):943-948.

H

Harpagophytum procumbens (Burch.) DC. ex Meisn. — Pedaliaceae

SCN: devil's claw

Part: secondary tuber

Quick Reference Summary

Safety Class: 1

Interaction Class: A

Contraindications

None known.

Other Precautions

Use with caution in persons with gastric or duodenal ulcers (Bradley 1992; ESCOP 2003; Hänsel et al. 1993; Weiss 1991).

Drug and Supplement Interactions

None known.

Adverse Events and Side Effects

Systematic reviews of clinical trials of devil's claw have reported that devil's claw is generally well tolerated and that adverse effects associated with devil's claw are mild, with gastrointestinal complaints and allergic reactions reported in a small number of trial participants (Brien et al. 2006; Denner 2007; Gagnier et al. 2004; Vlachojannis et al. 2008).

Pharmacological Considerations

Although devil's claw has been studied for anti-inflammatory activity, notably in patients with arthritis, studies have indicated that devil's claw does not affect the biosynthesis of prostanoids, and thus is not expected to produce the adverse effects associated with nonsteroidal anti-inflammatory and glucocorticoid drugs (ESCOP 2003; Loew et al. 1996; Moussard et al. 1992; Whitehouse et al. 1983).

While some references contraindicate devil's claw in persons with gastric or duodenal ulcers, one herbal reference text noted that these contraindications were theoretical in nature, and based on the bitter tonic activity of the herb. The same text notes that bitters should be used with caution in persons with esophageal reflux or hyperacidity (Mills and Bone 2005).

Pregnancy and Lactation

In South Africa, devil's claw has traditionally been administered to pregnant women in low doses (250 mg, three times daily) for pain relief during pregnancy. A reduced dose is used during the postpartum period (Watt and Breyer-Brandwijk 1962).

In vitro studies have indicated that devil's claw caused contractions of the uterus. The authors of the study noted that this activity was consistent with traditional use for induction or acceleration of labor, or for expulsion of retained placentas (Mahomed and Ojewole 2006). Another in vitro study indicated that some mouse embryos incubated in a solution containing devil's claw had misshapen tails but were otherwise normal (Yokoyama et al. 2005).

No information on the safety of devil's claw during lactation was identified. While this review did not identify any concerns for use while nursing, safety has not been conclusively established.

Review Details

I. Drug and Supplement Interactions

Clinical Trials of Drug or Supplement Interactions

No clinical trials of drug or supplement interactions were identified.

Case Reports of Suspected Drug or Supplement Interactions

A case of purpura was reported in a patient taking warfarin and devil's claw. The type of devil's claw preparation and dose and duration of the drug and herb were not specified (Shaw et al. 1997).

Animal Trials of Drug or Supplement Interactions

No animal trials of drug or supplement interactions were identified.

II. Adverse Events

Adverse Events Reported in Clinical Trials

A systematic review of the safety of devil's claw preparations for osteoarthritic and low back pain included 28 clinical trials with a total of 6892 patients. The review included both double-blind and observational studies with doses of 2 to 27 g (most studies used 4.5 g) daily of devil's claw aqueous or ethanol extracts for durations of 3 to 54 weeks. In the double-blind studies, incidences of adverse events with devil's claw were less than or equal to those reported in placebo groups. In the observational studies, adverse events were reported in 3% of participants and were primarily gastrointestinal complaints and allergies. Allergic reactions to devil's claw were characterized as rare (Vlachojannis et al. 2008). Other systematic reviews have reported similar findings (Brien et al. 2006; Denner 2007; Gagnier et al. 2004).

Case Reports of Adverse Events

An allergic reaction to devil's claw, confirmed by provocation testing, was reported in a worker routinely exposed to devil's claw (Altmeyer et al. 1991).

III. Pharmacology and Pharmacokinetics

Human Pharmacological Studies

No relevant human pharmacological studies were identified.

Animal Pharmacological Studies

No relevant animal pharmacological studies were identified.

In Vitro Pharmacological Studies

Moderate inhibition of the drug-metabolizing isoenzymes CYP2C8, CYP2C9, CYP2C19, and CYP3A4 were observed in baculovirus-infected insect cells treated with a hydroalcoholic extract of devil's claw. The same extract caused mild inhibition of CYP1A2 and CYP2D6 (Unger and Frank 2004). Conversely, in another study, no effects on human CYP3A4 were observed (Budzinski et al. 2000).

An extract of devil's claw dose-dependently inhibited lipopolysaccharide-induced synthesis of TNF-α in stimulated primary human monocytes. The compounds harpagide and harpagoside had no effect on LPS-induced TNF-α release (Fiebich et al. 2001).

IV. Pregnancy and Lactation

Increases in the baseline tone and spontaneous, rhythmic, myogenic contractions of estrogen-dominated rat uterine muscle strips were observed after treatment with an aqueous extract of devil's claw. Effective concentrations were 10 to 800 μg/ml. The authors noted that the activity was consistent with traditional use for induction or acceleration of labor, or for expulsion of retained placentas (Mahomed and Ojewole 2006).

Devil's claw has traditionally been administered to pregnant women at doses of approximately 250 mg three times daily for pain relief during pregnancy. A reduced dose is used during the postpartum period (Watt and Breyer-Brandwijk 1962).

In mice embryos cultured from days 11 to 13 of gestation in a solution containing devil's claw, no effects on heartbeat, crown-rump length, embryo weight, or the total number of somites were observed, although short or curly tail was associated with devil's claw treatment (Yokoyama et al. 2005).

No information on the safety of devil's claw during lactation was identified.

V. Toxicity Studies

Acute Toxicity

The LD_{50} of orally administered devil's claw could not be determined at doses up to 13.5 g/kg (Whitehouse et al. 1983). The LD_{50} of a 10% aqueous solution of devil's claw orally administered in mice is 220 ml/kg (Capresse 1980).

The LD_{50} of an intravenously administered purified extract of devil's claw (85% harpagoside) in mice is 511 mg/kg (Erdös et al. 1978), whereas that of pure harpagoside intraperitoneally administered to mice is 1 g/kg (Van Haelen et al. 1983).

Short-Term Toxicity

No adverse effects, including changes in hematology or gross pathology, were observed in mice orally administered 7.5 g/kg of devil's claw daily for 21 days (Whitehouse et al. 1983).

Literature Cited

H

Altmeyer, N., R. Garnier, N. Rosenberg, A.M. Geerolf, and A. Ghaem. 1991. Conjunctivite, rhinite et asthme rythmes par l'exposition professionnelle a l'*Harpagophytum*. *Soc. Med. Hyg. Travail* 289-291.

Bradley, P.R. 1992. *British herbal compendium: A handbook of scientific information on widely used plant drugs*. Bournemouth, UK: British Herbal Medicine Association.

Brien, S., G.T. Lewith, and G. McGregor. 2006. Devil's claw (*Harpagophytum procumbens*) as a treatment for osteoarthritis: A review of efficacy and safety. *J. Altern. Complement. Med.* 12(10):981-993.

Budzinski, J.W., B.C. Foster, S. Vandenhoek, and J.T. Arnason. 2000. An in vitro evaluation of human cytochrome CYP3A4 inhibition by selected commercial herbal extracts and tinctures. *Phytomedicine* 7:273-282.

Capresse, M. 1980. Description, identification et usages therapeutiques de la "griffe du diable": *Harpagophytum procumbens* DC. *J. Pharm. Belg.* 35:143-149.

Denner, S.S. 2007. A review of the efficacy and safety of devil's claw for pain associated with degenerative musculoskeletal diseases, rheumatoid, and osteoarthritis. *Holist. Nurs. Pract.* 21(4):203-207.

Erdös, A., R. Fontaine, H. Friehe, R. Durand, and T. Pöppinghaus. 1978. Contribution to the pharmacology and toxicology of different extracts as well as the harpagoside from *Harpagophytum procumbens* DC. *Planta Med.* 34(1):97.

ESCOP. 2003. *ESCOP monographs: The scientific foundation for herbal medicinal products*. 2nd ed. Exeter, U.K.: European Scientific Cooperative on Phytotherapy.

Fiebich, B.L., M. Heinrich, K.O. Hiller, and N. Kammerer. 2001. Inhibition of TNF-α synthesis in LPS-stimulated primary human monocytes by *Harpagophytum* extract SteiHap 69. *Phytomedicine* 8(1):28-30.

Gagnier, J.J., S. Chrubasik, and E. Manheimer. 2004. *Harpgophytum procumbens* for osteoarthritis and low back pain: A systematic review. *BMC Complement. Altern. Med.* 4:13.

Hänsel, R., K. Keller, H. Rimpler, and G. Schneider, eds. 1993. *Hagers handbuch der pharmazeutischen praxis*. 5th ed. Berlin: Springer.

Loew, D., O. Schuster, and J. Möllerfeld. 1996. Stabilität und biopharmazeutische Qualität. Voraussetzung fur Bioverfügbarkeit und Wirksamkeit von *Harpagophytum procumbens*. In *Phytopharmaka II*, edited by Loew, D. and N. Rietbrock. Darmstadt: Steinkopf.

Mahomed, I.M., and J.A.O. Ojewole. 2006. Oxytocin-like effect of *Harpagophytum procumbens* DC. [Pedaliaceae] secondary root aqueous extract on rat isolated uterus. *Afr. J. Trad. Comp. Alt. Med.* 3(1):82-89.

Mills, S., and K. Bone. 2005. *The essential guide to herbal safety*. St. Louis: Elsevier.

Moussard, C., D. Alber, M.M. Toubin, N. Thevenon, and J.C. Henry. 1992. A drug used in traditional medicine, *Harpagophytum procumbens*: No evidence for NSAID-like effect on whole blood eicosanoid production in human. *Prostaglandins Leukot. Essent. Fatty Acids* 46(4):283.

Shaw, D., C. Leon, S. Kolev, and V. Murray. 1997. Traditional remedies and food supplements: A 5-year toxicological study (1991–1995). *Drug Saf.* 17(5):342-356.

Unger, M., and A. Frank. 2004. Simultaneous determination of the inhibitory potency of herbal extracts on the activity of six major cytochrome P450 enzymes using liquid chromatography/mass spectrometry and automated online extraction. *Rapid Commun. Mass Spectrom.* 18(19):2273-2281.

Van Haelen, M., R. Van Haelen-Fastre, J. Samaey-Fontaine, et al. 1983. Aspects botaniques, constitution chimique et activite pharacologique d'*Harpagophytum procumbens* DC. *Phytotherapy* 5:7-13.

Vlachojannis, J., B.D. Roufogalis, and S. Chrubasik. 2008. Systematic review on the safety of *Harpagophytum* preparations for osteoarthritic and low back pain. *Phytother. Res.* 22(2):149-152.

Watt, J.M., and M.G. Breyer-Brandwijk. 1962. *The medicinal and poisonous plants of southern and eastern Africa*. 2nd ed. Edinburgh: E. & S. Livingstone.

Weiss, R.F. 1991. *Harpagophytum procumbens*, Teufelskralle. In *Lehrbuch der phytotherapie*. Stuttgart: Hippokrates.

Whitehouse, L.W., M. Znamirowska, and C.J. Paul. 1983. Devil's claw (*Harpagophytum procumbens*): No evidence for anti-inflammatory activity in the treatment of arthritic disease. *Can. Med. Assoc. J.* 129(3):249-251.

Yokoyama, A. H. Yokoyama, E.A. Johnston, and M. Akita. 2005. Effects of devil crow (DC) on cultured rat embryos. *Congenit. Anom.* 45(4):A62.

Hedeoma pulegioides (L.) Pers. — Lamiaceae

SCN: American pennyroyal **Part:** herb

Quick Reference Summary

Safety Class: 2b, 2c
Interaction Class: A

Contraindications

Not for use during pregnancy or lactation (De Smet 1992; Sleckman et al. 1983).

Other Precautions

Not recommended for use in persons with liver or kidney disease (Gordon et al. 1982; Mizutani et al. 1987; Speijers 2001; Sztajnkrycer et al. 2003).

Not recommended for use in children or infants (Bakerink et al. 1996).

Drug and Supplement Interactions

None known.

Editors' Notes

American pennyroyal and European pennyroyal (*Mentha pulegium*) are historically interchangeable as sources for pennyroyal oil (De Smet 1992). American pennyroyal is reported to contain less of the toxic compound pulegone (0.0017–0.0057% of dried leaf) than European pennyroyal (0.8–1.9%) (Furia and Bellanca 1971; List and Hörhammer 1973; Lorenzo et al. 2002; Sleckman et al. 1983).

Adverse Events and Side Effects

None known.

Pharmacological Considerations

None known.

Pregnancy and Lactation

Although animal or other studies on the use of American pennyroyal during pregnancy are lacking, the presence of the potentially toxic compound pulegone (*see* Toxicity Studies in *Mentha pulegium* herb and essential oil entries) suggests that American pennyroyal should not be taken during pregnancy (Sleckman et al. 1983).

No information on the safety of American pennyroyal during lactation was identified. While this review did not identify any concerns for use while nursing, safety has not been conclusively established.

H

Review Details

I. Drug and Supplement Interactions

Clinical Trials of Drug or Supplement Interactions

No clinical trials of drug or supplement interactions were identified.

Case Reports of Suspected Drug or Supplement Interactions

No case reports of suspected drug or supplement interactions were identified.

Animal Trials of Drug or Supplement Interactions

No animal trials of drug or supplement interactions were identified.

II. Adverse Events

Case Reports of Adverse Events

No case reports of adverse events were identified.

III. Pharmacology and Pharmacokinetics

Human Pharmacological Studies

No relevant human pharmacological studies were identified.

Animal Pharmacological Studies

No relevant animal pharmacological studies were identified.

In Vitro Pharmacological Studies

No relevant in vitro pharmacological studies were identified.

IV. Pregnancy and Lactation

Although animal or other studies on the use of American pennyroyal during pregnancy are lacking, the presence of the potentially toxic compound pulegone (*see* Toxicity Studies in entries for *Mentha pulegium* herb and essential oil) suggests that American pennyroyal should not be taken during pregnancy (De Smet 1992).

No information on the safety of American pennyroyal during lactation was identified.

V. Toxicity Studies

See Mentha pulegium herb and essential oil Toxicity Studies for information on the toxicity of the compound pulegone.

Literature Cited

Bakerink, J.A., S.M. Gospe, Jr., R.J. Dimand, and M.W. Eldridge. 1996. Multiple organ failure after ingestion of pennyroyal oil from herbal tea in two infants. *Pediatrics* 98(5):944-947.

De Smet, P.A.G.M. 1992. *Adverse effects of herbal drugs, Volume 1.* Berlin: Springer.

Furia, T.E., and N. Bellanca. 1971. *Fenaroli's handbook of flavor ingredients.* Cleveland, OH: The Chemical Rubber Co.

Gordon, W.P., A.J. Forte, R.J. McMurtry, J. Gal, and S.D. Nelson. 1982. Hepatotoxicity and pulmonary toxicity of pennyroyal oil and its constituent terpenes in the mouse. *Toxicol. Appl. Pharmacol.* 65(3):413-424.

List, P.H., and H. Hörhammer. 1973. *Hagers handbuch der pharmazeutischen praxis.* Berlin: Springer.

Lorenzo, D., D. Paz, E. Dellacassa, et al. 2002. Essential oils of *Mentha pulegium* and *Mentha rotundifolia* from Uruguay. *Braz. Arch. Biol. Technol.* 45:519-524.

Mizutani, T., H. Nomura, K. Nakanishi, and S. Fujita. 1987. Effects of drug metabolism modifiers on pulegone-induced hepatotoxicity in mice. *Res. Commun. Chem. Pathol. Pharmacol.* 58(1):75-83.

Sleckman, B.P., J. Sherma, and L.C. Mineo. 1983. Determination of pulegone in *H. pulegioides* and peppermint oil by thin layer chromatography with densitometry. *J. Liq. Chromatog. Relat. Technol.* 6(7):1175-1182.

Speijers, G. 2001. WHO Food Additives Series 46: Pulegone and related substances. Bilthoven, Netherlands: National Institute of Public Health and the Environment.

Sztajnkrycer, M.D., E.J. Otten, G.R. Bond, C.J. Lindsell, and R.J. Goetz. 2003. Mitigation of pennyroyal oil hepatotoxicity in the mouse. *Acad. Emerg. Med.* 10(10):1024-1028.

H

Helianthus annuus L. — Asteraceae

SCN: sunflower

Part: seed

Quick Reference Summary

Safety Class: 1
Interaction Class: A

Contraindications

None known.

Other Precautions

None known.

Drug and Supplement Interactions

None known.

Adverse Events and Side Effects

Allergic reactions to sunflower seed, including anaphylactic reactions, have been reported (Asero et al. 2004; Duran et al. 1997; Fremont et al. 2002; Iwaya et al. 1994; Palma-Carlos et al. 2005).

Cases of bezoars (masses trapped in the gastrointestinal tract) have been reported in persons, primarily children, who have eaten "large" amounts of unshelled sunflower seeds (Bakr et al. 2006; Dent and Levine 1989; Lowry and Shah 2001; Melchreit et al. 1984; Sawnani and McFarlane-Ferreira 2003; Tsou et al. 1997).

A case of black hairy tongue was reported in a man who orally dehulled and consumed sunflower seeds nightly for 2 years (Pipili et al. 2008).

Pharmacological Considerations

None known.

Pregnancy and Lactation

No information on the use of sunflower during pregnancy or lactation was identified. Based on the widespread use of sunflower as a food and the seed oil as a cooking oil, no adverse effects are expected.

Review Details

I. Drug and Supplement Interactions

Clinical Trials of Drug or Supplement Interactions

No clinical trials of drug or supplement interactions were identified.

Case Reports of Suspected Drug or Supplement Interactions

No case reports of suspected drug or supplement interactions were identified.

Animal Trials of Drug or Supplement Interactions

No animal trials of suspected drug or supplement interactions were identified.

II. Adverse Events

Case Reports of Adverse Events

Allergic reactions to sunflower, including anaphylactic reactions, have been reported and confirmed by skin prick tests and immunoblotting (Asero et al. 2004; Duran et al. 1997; Fremont et al. 2002; Iwaya et al. 1994; Palma-Carlos et al. 2005).

Numerous cases of rectal sunflower seed bezoars (a mass trapped in the gastrointestinal tract) have been reported in persons, primarily children, who have eaten "large" amounts of unshelled sunflower seeds (Bakr et al. 2006; Dent and Levine 1989; Lowry and Shah 2001; Melchreit et al. 1984; Sawnani and McFarlane-Ferreira 2003; Tsou et al. 1997).

Occupational asthma due to inhalation of sunflower dust was reported in a worker exposed to sunflower (Vandenplas et al. 1998).

Black hairy tongue, a benign hyperplasia of the surface of the tongue, was observed in a man who ate sunflower seeds for approximately 2 hours every night while watching television. The man removed the hulls by cracking the seeds with his teeth, and spitting out the hulls while keeping the kernels in his mouth. The tongue returned to normal after cessation of sunflower seed ingestion, nightly brushing of the tongue, and treatment with 40% urea solution (Pipili et al. 2008).

III. Pharmacology and Pharmacokinetics

Human Pharmacological Studies

No relevant human pharmacological studies were identified.

Animal Pharmacological Studies

No relevant animal pharmacological studies were identified.

In Vitro Pharmacological Studies

No relevant in vitro pharmacological studies were identified.

IV. Pregnancy and Lactation

No information on the use of sunflower during pregnancy or lactation was identified.

V. Toxicity Studies

Subchronic Toxicity

No adverse effects, including changes in kidney function, serum chemistry, or organ weights, were observed in rats fed diets with sunflower protein isolates at amounts up to 13 g/kg daily for 90 days (Hoernicke et al. 1988).

Genotoxicity

Inconclusive results on the genotoxicity of sunflower seed oil were obtained in the *Drosophila* somatic mutation and recombination test (Rojas-Molina et al. 2005).

Clastogenic activity of a water extract of thermally stressed sunflower oil (heated to 180–220°C) was observed in human lymphocytes (Indart et al. 2007).

Literature Cited

Asero, R., G. Mistrello, D. Roncarolo, and S. Amato. 2004. Airborne allergy to sunflower seed. *J. Invest. Allergol. Clin. Immunol.* 14(3):244-246.

Bakr, A.F., N. Sharma, and J.C. Mathias. 2006. Rectal sunflower seed bezoar. *Acta Paediatr.* 95(7):886-887.

Dent, J.M., and S.I. Levine. 1989. Sunflower seed bezoar presenting as diarrhea. *Arch. Pediatr. Adolescent Med.* 143(6):643-644.

Duran, S., J. Delgado, R. Gamez, et al. 1997. Contact urticaria from sunflower seeds. *Contact Dermat.* 37(4):184.

Fremont, S., Y. Errahali, M. Bignol, M. Metche, and J.P. Nicolas. 2002. Allergenicity of oils. *Allerg. Immunol.* 34(3):91-94.

Hoernicke, E., H.J. Lewerenz, and G. Mieth. 1988. Toxicological animal studies of protein isolates from sunflower seeds. *Nahrung* 32(7):691-700.

Indart, A., M. Viana, S. Clapes, et al. 2007. Clastogenic and cytotoxic effects of lipid peroxidation products generated in culinary oils submitted to thermal stress. *Food Chem. Toxicol.* 45(10):1963-1967.

Iwaya, M., G. Murakami, K. Kurose, et al. 1994. A case of anaphylaxis caused by sunflower seed. *Arerugi* 43(6):696-700.

Lowry, M.H., and A.N. Shah. 2001. Sunflower seed rectal bezoar in an adult. *Gastrointest. Endosc.* 53(3):388-389.

Melchreit, R., G. McGowan, and J.S. Hyams. 1984. "Colonic crunch" sign in sunflower-seed bezoar. *N. Engl. J. Med.* 310(26):1748.

Palma-Carlos, A.G., M.L. Palma-Carlos, and F. Tengarrinha. 2005. Allergy to sunflower seeds. *Eur. Ann. Allergy Clin. Immunol.* 37(5):183-186.

Pipili, C., E. Cholongitas, and D. Ioannidou. 2008. Is sunflower seed implicated in the development of black hairy tongue? *Eur. J. Dermatol.* 18(6):732.

Rojas-Molina, M., J. Campos-Sanchez, M. Analla, A. Munoz-Serrano, and A. Alonso-Moraga. 2005. Genotoxicity of vegetable cooking oils in the *Drosophila* wing spot test. *Environ. Mol. Mutagen.* 45(1):90-95.

Sawnani, H., and Y. McFarlane-Ferreira. 2003. Proctological crunch: Sunflower-seed bezoar. *J. La. State Med. Soc.* 155(3):163-164.

Tsou, V.M., P.R. Bishop, and M.J. Nowicki. 1997. Colonic sunflower seed bezoar. *Pediatrics* 99(6):896-897.

Vandenplas, O., T. Vander Borght, and J.P. Delwiche. 1998. Occupational asthma caused by sunflower-seed dust. *Allergy* 53(9):907-908.

Hemidesmus indicus (L.) W.T. Aiton — Asclepiadaceae

SCN: *Hemidesmus indicus*
Syn: *Periploca indica* L.
AN: *sariva*
OCN: East Indian sarsaparilla
Part: root

Quick Reference Summary

Safety Class: 1
Interaction Class: A

Contraindications

None known.

Other Precautions

None known.

Drug and Supplement Interactions

None known.

Adverse Events and Side Effects

None known.

Pharmacological Considerations

None known.

Pregnancy and Lactation

No information on the safety of *Hemidesmus indicus* in pregnancy or lactation was identified in the scientific or traditional literature. Although this review did not identify any concerns for use while pregnant or nursing, safety has not been conclusively established.

Review Details

I. Drug and Supplement Interactions

Clinical Trials of Drug or Supplement Interactions

No clinical trials of drug or supplement interactions were identified.

Case Reports of Suspected Drug or Supplement Interactions

No case reports of suspected drug or supplement interactions were identified.

Animal Trials of Drug or Supplement Interactions

No animal trials of drug or supplement interactions were identified.

II. Adverse Events

Case Reports of Adverse Events

No case reports of adverse events were identified.

III. Pharmacology and Pharmacokinetics

Human Pharmacological Studies

No relevant human pharmacological studies were identified.

Animal Pharmacological Studies

No relevant animal pharmacological studies were identified.

In Vitro Pharmacological Studies

An ethanol extract of *Hemidesmus indicus* was found to reduce the toxicity of gentamicin to hair cells in organotypic cultures without affecting the uptake of gentamicin. The effective concentrations ranged from 25 to 100 μg/ml (Previati et al. 2007).

In sheep red blood cells, an ethanol extract of *Hemidesmus indicus* was found to suppress cell-mediated and humoral components of the immune system (Atal et al. 1986).

An aqueous extract of *Hemidesmus indicus* inhibited TNF-α and interleukin-8 in human peripheral blood mononuclear cells (Jain and Basal 2003).

IV. Pregnancy and Lactation

No information on the safety of *Hemidesmus indicus* during pregnancy or lactation was identified.

V. Toxicity Studies

Short-Term Toxicity

In rats fed diets containing 25% *Hemidesmus indicus* for 14 days, diffuse hydropic degeneration and focal hepatocellular necrosis were observed in the livers (Arseculeratne et al. 1985). Conversely, a number of studies have demonstrated a hepatoprotective effect of alcohol extracts of *Hemidesmus indicus* administered at doses of 100 to 500 mg/kg for up to 30 days (Baheti et al. 2006; Prabakan et al. 2000; Saravanan and Nalini 2007a, 2007b, 2007c, 2008).

Literature Cited

Arseculeratne, S.N., A.A. Gunatilaka, and R.G. Panabokke. 1985. Studies of medicinal plants of Sri Lanka. Part 14: Toxicity of some traditional medicinal herbs. *J. Ethnopharmacol.* 13(3):323-335.

Atal, C.K., M.L. Sharma, A. Kaul, and A. Khajuria. 1986. Immunomodulating agents of plant origin. I: Preliminary screening. *J. Ethnopharmacol.* 18(2):133-141.

Baheti, J.R., R.K. Goyal, and G.B. Shah. 2006. Hepatoprotective activity of *Hemidesmus indicus* R. Br. in rats. *Indian J. Exp. Biol.* 44(5):399-402.

Jain, A., and E. Basal. 2003. Inhibition of *Propionibacterium acnes*-induced mediators of inflammation by Indian herbs. *Phytomedicine* 10(1):34-38.

Prabakan, M., R. Anandan, and T. Devaki. 2000. Protective effect of *Hemidesmus indicus* against rifampicin and isoniazid-induced hepatotoxicity in rats. *Fitoterapia* 71(1):55-59.

Previati, M., E. Corbacella, L. Astolfi, et al. 2007. Ethanolic extract from *Hemidesmus indicus* (Linn) displays otoprotectant activities on organotypic cultures without interfering on gentamicin uptake. *J. Chem. Neuroanat.* 34(3-4):128-133.

Saravanan, N., and N. Nalini. 2007a. Antioxidant effect of *Hemidesmus indicus* on ethanol-induced hepatotoxicity in rats. *J. Med. Food.* 10(4):675-682.

Saravanan, N., and N. Nalini. 2007b. Impact of *Hemidesmus indicus* R.Br. extract on ethanol-mediated oxidative damage in rat kidney. *Redox Rep.* 12(5):229-235.

Saravanan, N., and N. Nalini. 2007c. Inhibitory effect of *Hemidesmus indicus* and its active principle 2-hydroxy-4-methoxybenzoic acid on ethanol-induced liver injury. *Fundam. Clin. Pharmacol.* 21(5):507-514.

Saravanan, N., and N. Nalini. 2008. *Hemidesmus indicus* protects against ethanol-induced liver toxicity. *Cell Mol. Biol. Lett.* 13(1):20-37.

Hepatica spp.

Ranunculaceae

Hepatica nobilis Schreb. var. ***acuta*** (Pursh) Steyerm.
SCN: liverwort
Syn: *Hepatica acutiloba* DC.
OCN: American liverleaf; sharp-lobe hepatica

Hepatica nobilis Schreb. var. ***obtusa*** (Pursh) Steyerm.
SCN: liverwort
Syn: *Anemone hepatica* L.; *Hepatica americana* (DC.) Ker Gawl.
OCN: American liverleaf; round-lobe hepatica

Part: herb

Quick Reference Summary

Safety Class: 2b
Interaction Class: A

Contraindications

Not for use in pregnancy except under the supervision of a qualified healthcare practitioner (Blumenthal et al. 1998; Moerman 1998).

Other Precautions

In large doses, irritation of the kidneys and urinary tract may be caused by the compound protoanemonin (Blumenthal et al. 1998).

Drug and Supplement Interactions

None known.

Editors' Notes

Liverwort, like several other species in the Ranunculaceae family, contains protoanemonin, a strongly irritating vesicant oil (Muhe 1947; Turner 1984). Protoanemonin, however, has a solubility of about 1% in water (Windholz 1983). Although the fresh herb can cause subepidermal blistering of the skin, the compound and the related concern are destroyed on drying (Epstein 1990; List and Hörhammer 1973; Mitchell 1979).

Adverse Events and Side Effects

None known.

Pharmacological Considerations

None known.

Pregnancy and Lactation

Ethnobotanical records indicate that liverwort was traditionally used as a contraceptive, to treat side or abdominal pain during pregnancy, to treat amenorrhea, and to induce labor in middle-aged women. Information on typical doses and durations used was not reported (Moerman 1998). Based on this information, use during pregnancy is not recommended except under the supervision of a qualified healthcare practitioner.

No information on the safety of liverwort during lactation was identified. While this review did not identify any concerns for use while nursing, safety has not been conclusively established.

Review Details

I. Drug and Supplement Interactions

Clinical Trials of Drug or Supplement Interactions

No clinical trials of drug or supplement interactions were identified.

Case Reports of Suspected Drug or Supplement Interactions

No case reports of suspected drug or supplement interactions were identified.

Animal Trials of Drug or Supplement Interactions

No animal trials of drug or supplement interactions were identified.

II. Adverse Events

Case Reports of Adverse Events

No case reports of adverse events were identified.

III. Pharmacology and Pharmacokinetics

Human Pharmacological Studies

No relevant human pharmacological studies were identified.

Animal Pharmacological Studies

No relevant animal pharmacological studies were identified.

In Vitro Pharmacological Studies

No relevant in vitro pharmacological studies were identified.

IV. Pregnancy and Lactation

Ethnobotanical records indicate that liverwort was traditionally used as a contraceptive, to treat side or abdominal pain during pregnancy, to treat amenorrhea, and to induce labor in middle-aged women. Information on typical doses and durations used were not reported (Moerman 1998).

No information on the safety of liverwort during lactation was identified.

V. Toxicity Studies

No toxicity studies were identified.

Literature Cited

Blumenthal, M., W. Busse, A. Goldberg, et al. 1998. *The complete German Commission E monographs*. Austin, TX: American Botanical Council.

Epstein, W. 1990. House and garden plants. In *Irritant contact dermatitis*, edited by Jackson, E.M. and R. Goldner. New York: M. Dekker.

List, P.H., and H. Hörhammer. 1973. *Hagers handbuch der pharmazeutischen praxis*. Berlin: Springer.

Mitchell, J.R.A. 1979. *Botanical dermatology: Plants and plant products injurious to the skin*. Vancouver: Greengrass.

Moerman, D.E. 1998. *Native American ethnobotany*. Portland, OR: Timber Press.

Muhe, R. 1947. Anemone oil. *Pharmazie* 2:333-334.

Turner, N.J. 1984. Counter-irritant and other medicinal uses of plants in Ranunculaceae by native peoples in British Columbia and neighbouring areas. *J. Ethnopharmacol.* 11(2):181-201.

Windholz, M. 1983. *The Merck index*. 10th ed. Rahway, N.J.: Merck.

Heuchera micrantha Douglas ex Lindl. — Saxifragaceae

SCN: alumroot

Part: root

Quick Reference Summary

Safety Class: 1

Interaction Class: A

Contraindications

None known.

Other Precautions

None known.

Drug and Supplement Interactions

None known.

Notice

Tannins (dried root, 9.3–19.6%) (Felter and Lloyd 1898; Osol and Farrar 1955); *see* Appendix 1.

Adverse Events and Side Effects

None known.

Pharmacological Considerations

None known.

Pregnancy and Lactation

No information on the safety of alumroot in pregnancy or lactation was identified in the scientific or traditional literature. Although this review did not identify any concerns for use while pregnant or nursing, safety has not been conclusively established.

Review Details

I. Drug and Supplement Interactions

Clinical Trials of Drug or Supplement Interactions

No clinical trials of drug or supplement interactions were identified.

Case Reports of Suspected Drug or Supplement Interactions

No case reports of suspected drug or supplement interactions were identified.

Animal Trials of Drug or Supplement Interactions

No animal trials of drug or supplement interactions were identified.

II. Adverse Events

Case Reports of Adverse Events

No case reports of adverse events were identified.

III. Pharmacology and Pharmacokinetics

Human Pharmacological Studies

No relevant human pharmacological studies were identified.

Animal Pharmacological Studies

No relevant animal pharmacological studies were identified.

In Vitro Pharmacological Studies

No relevant in vitro pharmacological studies were identified.

IV. Pregnancy and Lactation

No information on the safety of alumroot during pregnancy or lactation was identified.

V. Toxicity Studies

No toxicity studies were identified.

Literature Cited

Felter, H.W., and J.U. Lloyd. 1898. *King's American dispensatory*. 18th ed., 3rd rev. 2 vols. Cincinnati: Ohio Valley Co.

Osol, A., and G. Farrar. 1955. *The dispensatory of the United States of America*. 25th ed. Philadelphia: Lippincott.

Hibiscus sabdariffa L. — Malvaceae

SCN: hibiscus
AN: *ambashthaki*
OCN: roselle
Part: flower

Quick Reference Summary

Safety Class: 1
Interaction Class: A

Contraindications

None known.

Other Precautions

None known.

Drug and Supplement Interactions

See Pharmacological Considerations.

Adverse Events and Side Effects

None known.

Pharmacological Considerations

Hibiscus may increase the elimination rate of acetaminophen without affecting the plasma levels or other parameters of acetaminophen. Investigators suggested that acetaminophen should be taken at least 3 hours before or after the ingestion of hibiscus (Kolawole and Maduenyi 2004).

A human study indicated that hibiscus has mixed effects on the urinary excretion of diclofenac, with excretion increased in some patients and reduced in others (Fakeye et al. 2007). Another human study indicated that hibiscus juice may reduce the bioavailability of chloroquine, with effects of hibiscus similar to those of lemonade and a tamarind beverage. Investigators recommended not taking

H

chloroquine tablets with acidic beverages (Mahmoud et al. 1994).

Pregnancy and Lactation

No adverse effects of large doses of hibiscus on pregnancy or fetal development were observed in several animal studies. In animals administered very large doses (1.5 or 3 g/kg) of hibiscus during pregnancy or lactation, a decrease in maternal food and water intake and associated delay in puberty in offspring were observed (Iyare and Adegoke 2008a, 2008b; Iyare and Iyare 2008).

Review Details

I. Drug and Supplement Interactions

Clinical Trials of Drug or Supplement Interactions

An increase in the elimination rate of acetaminophen was observed in healthy volunteers orally administered 1 liter of sweetened aqueous extract of hibiscus (made from 30 g of hibiscus/liter) 1.5 hours prior to oral administration of acetaminophen (1000 mg). No changes in the maximum plasma concentration, time to maximum concentration, or area under the time-concentration curve were observed. The authors of the study indicated that acetaminophen should be taken at least 3 hours before the ingestion of hibiscus so that the therapeutic activity of acetaminophen will not be shortened (Kolawole and Maduenyi 2004).

Changes in urinary excretion of diclofenac were observed in healthy volunteers pretreated with 300 ml of an aqueous hibiscus extract (containing 8.18 mg anthocyanins) daily for 3 days. Urinary excretion was increased in some volunteers and decreased in others (Fakeye et al. 2007).

A reduction in serum levels of chloroquine was observed in healthy volunteers orally administered 600 mg chloroquine with 300 ml of a juice made from hibiscus (amount of plant material used was not specified). The effects of hibiscus were similar to those of lemonade and a tamarind beverage, with the acidity of all beverages believed to be the cause of reduced absorption. The authors of the study recommended not taking chloroquine tablets with acidic beverages (Mahmoud et al. 1994).

Case Reports of Suspected Drug or Supplement Interactions

No case reports of suspected drug or supplement interactions were identified.

Animal Trials of Drug or Supplement Interactions

No animal trials of drug or supplement interactions were identified.

II. Adverse Events

Case Reports of Adverse Events

No case reports of adverse events were identified.

III. Pharmacology and Pharmacokinetics

Human Pharmacological Studies

In a tolerability study in hypertensive patients, an infusion prepared from 10 g hibiscus taken daily for 4 weeks was well tolerated (Herrera-Arellano et al. 2004).

Animal Pharmacological Studies

Intraperitoneal administration of 100, 200, or 400 mg/kg of an aqueous extract of hibiscus reduced the spontaneous motor activity and increased the duration of pentobarbital-induced sleeping time in mice (Amos et al. 2003).

A reduction in blood glucose after rice starch and or sucrose challenges was observed in rats orally administered 0.5 g/animal (344–442 g average animal weight) of hibiscus extract. No significant changes in blood glucose levels were observed when hibiscus was administered outside of the challenge tests (Preuss et al. 2007).

In Vitro Pharmacological Studies

No relevant in vitro pharmacological studies were identified.

IV. Pregnancy and Lactation

An aqueous extract of hibiscus dose-dependently inhibited the rate and amplitude of uterine contractions in rats after intra-arterial injection of 1, 5, 10, 50, or 100 mg/kg. The uterine response to hibiscus was not affected by administration of either atropine or propranolol. At the 500 mg/kg dose, hibiscus produced a slight reduction of contraction amplitude in oxytocin-precontracted uteruses (Fouda et al. 2007).

In rats provided with drinking water containing an aqueous extract of hibiscus at a concentration that provided doses equivalent to 1.5 or 3.0 g/kg extract daily throughout pregnancy, a decrease in maternal food and fluid intake was observed; in offspring, an increase in postnatal weight gain and delay in onset of puberty were observed (Iyare and Adegoke 2008c). A follow-up study indicated that the delay in puberty onset in the offspring was associated with elevated maternal plasma levels of sodium and corticosterone during pregnancy (Iyare and Adegoke 2008b).

In rats provided with drinking water containing 0.6 or 1.8 g/100 ml of aqueous extract of hibiscus throughout pregnancy and through postnatal day 20, no adverse effects on early postnatal growth of offspring were observed (Iyare and Iyare 2008).

In lactating rats provided with drinking water containing 0.6 or 1.8 g/100 ml of aqueous extract of hibiscus throughout lactation (21 days), a decrease in maternal food and fluid intake was observed along with increased postnatal weight gain and delayed onset of puberty in the offspring (Iyare and Adegoke 2008a).

V. Toxicity Studies

Acute Toxicity

The LD_{50} in rats was 5 g/kg of an orally administered hibiscus aqueous extract prepared from dried hibiscus calyx and described as yielding 22% of the dry plant material (Onyenekwe et al. 1999; Orisakwe et al. 2004). The LD_{50} of an intraperitoneally administered aqueous extract of hibiscus could not be determined at doses up to 5 g/kg (Amos et al. 2003).

Subchronic Toxicity

No changes in testicular weights were observed in rats orally administered 1.15, 2.30, or 4.60 g/kg of an aqueous extract of hibiscus daily for 12 weeks. A decrease in the epididymal sperm counts and disintegration of sperm cells were observed in the 4.6 g/kg group. Distortion of tubules and a disruption of normal epithelial organization were observed in the 1.15 g/kg group, and hyperplasia of the testes with thickening of the basement membrane was observed in the 2.3 g/kg group (Orisakwe et al. 2004).

In rats orally administered 300 or 2000 mg/kg (~10,000 and 66,000 times the average human dose) of an aqueous, aqueous-ethanolic, or ethanolic extract of hibiscus daily for 90 days, all animals at the 2000 mg/kg dose had severe diarrhea by day 3. By day 8, all in the ethanol extract group had died, and by day 28 all from the aqueous-ethanolic extract group had died. In the 300 mg/kg group, animals developed mild diarrhea, all of the animals in the aqueous-ethanolic extract group died by day 40, and 80% of the animals in the aqueous extract group died by day 60. No mortality was observed in the animals in the ethanol extract group (Fakeye et al. 2009).

In rats orally administered 1, 3, 5, 10, or 15 doses of 250 mg/kg each of a hydromethanolic extract of hibiscus daily, serum aspartate aminotransferase (AST) and alanine aminotransferase (ALT) levels were increased in all the treatments, as compared to the control group. Serum levels of alkaline phosphatase and lactate dehydrogenase were not significantly affected. In the group that received 15 doses, serum levels of albumin were elevated. Histopathological studies indicated no pathological changes in livers or hearts in any of the treatment groups (Akindahunsi and Olaleye 2003).

Genotoxicity

No chromosomal aberrations were observed in lymphocytes of rats that had been intraperitoneally administered up to 200 mg/kg, twice at a 24-hour interval, of a hydroethanolic extract of hibiscus (Sowemimo et al. 2007).

Mutagenic activity of a food colorant derived from hibiscus was observed in the Ames assay in *Salmonella typhimurium* strains TA98 and TA100 with or without activation by S9. The compounds responsible for the mutagenic activity were identified as kaempferol and/or quercetin (Takeda and Yasui 1985). Although quercetin has been shown to have mutagenic activity in vitro, in vivo tests indicate a lack of mutagenicity in animals (Harwood et al. 2007).

Literature Cited

Akindahunsi, A.A., and M.T. Olaleye. 2003. Toxicological investigation of aqueous-methanolic extract of the calyces of *Hibiscus sabdariffa* L. *J. Ethnopharmacol.* 89(1):161-164.

Amos, S., L. Binda, B.A. Chindo, et al. 2003. Neuropharmacological effects of *Hibiscus sabdariffa* aqueous extract. *Pharmaceut. Biol.* 41(5):325-329.

Fakeye, T.O., A.O. Adegoke, O.C. Omoyeni, and A.A. Famakinde. 2007. Effects of water extract of *Hibiscus sabdariffa*, Linn. (Malvaceae) 'Roselle' on excretion of a diclofenac formulation. *Phytother. Res.* 21(1):96-98.

Fakeye, T.O., A. Pal, D.U. Bawankule, N.P. Yadav, and S.P. Khanuja. 2009. Toxic effects of oral administration of extracts of dried calyx of *Hibiscus sabdariffa* Linn. (Malvaceae). *Phytother. Res.* 23(3):412-416.

Fouda, A.M., M.H. Daba, and G.M. Dahab. 2007. Inhibitory effects of aqueous extract of *Hibiscus sabdariffa* on contractility of the rat bladder and uterus. *Can. J. Physiol. Pharmacol.* 85(10):1020-1031.

Harwood, M., B. Danielewska-Nikiel, J.F. Borzelleca, et al. 2007. A critical review of the data related to the safety of quercetin and lack of evidence of in vivo toxicity, including lack of genotoxic/carcinogenic properties. *Food Chem. Toxicol.* 45(11):2179-2205.

Herrera-Arellano, A., S. Flores-Romero, M.A. Chavez-Soto, and J. Tortoriello. 2004. Effectiveness and tolerability of a standardized extract from *Hibiscus sabdariffa* in patients with mild to moderate hypertension: A controlled and randomized clinical trial. *Phytomedicine* 11(5):375-382.

Iyare, E.E., and O.A. Adegoke. 2008a. Maternal consumption of an aqueous extract of *Hibiscus sabdariffa* during lactation accelerates postnatal weight and delays onset of puberty in female offspring. *Niger. J. Physiol. Sci.* 23(1-2):89-94.

Iyare, E.E., and O.A. Adegoke. 2008b. Mechanism of the delayed puberty onset in offspring of rats that consumed aqueous extract of *Hibiscus sabdariffa* during pregnancy. *Niger. J. Physiol. Sci.* 23(1-2):71-77.

Iyare, E.E., and O.A. Adegoke. 2008c. Postnatal weight gain and onset of puberty in rats exposed to aqueous extract of *Hibiscus sabdariffa* in utero. *Pak. J. Nutr.* 7(1):98-101.

Iyare, E.E., and F.E. Iyare. 2008. Effect of prenatal and postnatal exposure to an aqueous extract of *Hibiscus sabdariffa* (HS) on postnatal growth in Sprague-Dawley rats. *Pak. J. Nutr.* 7(2):255-257.

Kolawole, J.A., and A. Maduenyi. 2004. Effect of zobo drink (*Hibiscus sabdariffa* water extract) on the pharmacokinetics of acetaminophen in human volunteers. *Eur. J. Drug Metab. Pharmacokinet.* 29(1):25-29.

Mahmoud, B.M., H.M. Ali, M.M. Homeida, and J.L. Bennett. 1994. Significant reduction in chloroquine bioavailability following coadministration with the Sudanese beverages aradaib, karkadi and lemon. *J. Antimicrob. Chemother.* 33:1005-1009.

Onyenekwe, P.C., E.O. Ajani, D.A. Ameh, and K.S. Gamaniel. 1999. Antihypertensive effect of roselle (*Hibiscus sabdariffa*) calyx infusion in spontaneously hypertensive rats and a comparison of its toxicity with that in Wistar rats. *Cell Biochem. Funct.* 17(3):199–206.

Orisakwe, O.E., D.C. Husaini, and O.J. Afonne. 2004. Testicular effects of sub-chronic administration of *Hibiscus sabdariffa* calyx aqueous extract in rats. *Reprod. Toxicol.* 18(2):295-298.

Preuss, H.G., B. Echard, D. Bagchi, and S. Stohs. 2007. Inhibition by natural dietary substances of gastrointestinal absorption of starch and sucrose in rats and pigs: 1. Acute studies. *Int. J. Med. Sci.* 4(4):196-202.

Sowemimo, A.A., F.A. Fakoya, I. Awopetu, O.R. Omobuwajo, and S.A. Adesanya. 2007. Toxicity and mutagenic activity of some selected Nigerian plants. *J. Ethnopharmacol.* 113(3):427-432.

Takeda, N., and Y. Yasui. 1985. Identification of mutagenic substances in roselle color, elderberry color and safflower yellow. *Agric. Biol. Chem.* 49:1851-1852.

Hoodia gordonii (Masson) Sweet ex Decne — Apocynaceae

SCN: *Hoodia gordonii*
Syn: *Stapelia gordonii* Masson

Part: fleshy stem

Quick Reference Summary

Safety Class: 1
Interaction Class: A

Contraindications

None known.

Other Precautions

None known.

Drug and Supplement Interactions

None known.

Editors' Note

Numerous cases of adulteration of *Hoodia gordonii* with *Opuntia* species and *Caralluma fimbriata have been reported (Avula et al. 2007; Rumalla et al. 2008).*

Adverse Events and Side Effects

None known.

Pharmacological Considerations

An in vitro study indicated that *Hoodia gordonii* inhibited the drug-metabolizing isoenzyme CYP3A4 but had no effects on other CYP450 isoenzymes (Madgula et al. 2008). The relevance of those data to human use is not known.

Pregnancy and Lactation

No information on the safety of *Hoodia gordonii* in pregnancy or lactation was identified in the scientific or traditional literature. Although this review did not identify any concerns for use while pregnant or nursing, safety has not been conclusively established.

Review Details

I. Drug and Supplement Interactions

Clinical Trials of Drug or Supplement Interactions

No clinical trials of drug or supplement interactions were identified.

Case Reports of Suspected Drug or Supplement Interactions

No case reports of suspected drug or supplement interactions were identified.

Animal Trials of Drug or Supplement Interactions

No animal trials of drug or supplement interactions were identified.

II. Adverse Events

Case Reports of Adverse Events

A report on adverse events associated with natural health products in Italy from the Italian Pharmacovigilance System indicated that one case of acute hepatitis was reported in a person taking a *Hoodia gordonii* product along with other products including deanol, acetoaminobenzoate, iodocasein, metformin, triiodoacetic acid, and furosemide. Details on the *Hoodia gordonii* product, dose, and duration taken and the patient medical history were not provided (Menniti-Ippolito et al. 2008).

III. Pharmacology and Pharmacokinetics

Human Pharmacological Studies

No relevant human pharmacological studies were identified.

Animal Pharmacological Studies

No relevant animal pharmacological studies were identified.

In Vitro Pharmacological Studies

Inhibition of the drug-metabolizing isoenzyme CYP3A4 was observed in human liver microsomes treated with a steroidal glycoside compound (P57) from *Hoodia gordonii*, with a 50% inhibitory concentration (IC_{50}) of 45 μ*M*. No effects on the activity of CYP1A2, CYP2C9, or CYP2D6 were observed (Madgula et al. 2008).

In human colon cancer epithelial cells (Caco-2), the compound P57 exhibited a higher transport in the secretory direction than in the absorptive direction, and efflux was inhibited by selective inhibitors of multidrug resistance-associated proteins MRP1/MRP2 (MK-571) and P-gp (Madgula et al. 2008).

IV. Pregnancy and Lactation

No information on the safety of *Hoodia gordonii* in pregnancy or lactation was identified.

V. Toxicity Studies

Short-Term Toxicity

No adverse effects were observed in rats fed a diet containing 2% of an aqueous homogenate of dried *Hoodia gordonii* daily for 3 weeks (Tulp et al. 2001).

H

Literature Cited

Avula, B., Y.H. Wang, R.S. Pawar, Y.J. Shukla, and I.A. Khan. 2007. Chemical fingerprinting of *Hoodia* species and related genera: Chemical analysis of oxypregnane glycosides using high-performance liquid chromatography with UV detection in *Hoodia gordonii*. *J. AOAC Int.* 90(6):1526-1531.

Madgula, V.L., B. Avula, R.S. Pawar, et al. 2008. In vitro metabolic stability and intestinal transport of P57AS3 (P57) from *Hoodia gordonii* and its interaction with drug metabolizing enzymes. *Planta Med.* 74(10):1269-1275.

Menniti-Ippolito, F., G. Mazzanti, C. Santuccio, et al. 2008. Surveillance of suspected adverse reactions to natural health products in Italy. *Pharmacoepidemiol. Drug Saf.* 17(6):626-635.

Rumalla, C.S., B. Avula, Y.J. Shukla, et al. 2008. Chemical fingerprint of *Hoodia* species, dietary supplements, and related genera by using HPTLC. *J. High Res. Chromatog.* 31(22):3959-3964.

Tulp, O.L., N.A. Harbi, J. Mihalov, and A. DerMaderosian. 2001. Effect of *Hoodia* plant on weight loss in congenic obese LA/Ntul//-cp rats. *FASEB J.* 15(4):A404.

Hordeum vulgare L.

Poaceae

SCN: barley
AN: *yava*
PN: *mai ya* (dried germinated ripe seed)
Part: sprouted seed

Quick Reference Summary

Safety Class: 1
Interaction Class: A

Contraindications

None known.

Other Precautions

None known.

Drug and Supplement Interactions

None known.

Adverse Events and Side Effects

Allergic reactions to barley, including anaphylactic reactions, have been reported (Bonadonna et al. 1999; Curioni et al. 1999; Gutgesell and Fuchs 1995).

Pharmacological Considerations

Human and animal studies have demonstrated that barley may modify glucose regulation (Chang and But 1986; Chen and Chen 2004). People with diabetes are advised to monitor their blood sugar closely and discuss the use of this herb with a qualified healthcare practitioner prior to use.

Pregnancy and Lactation

No information on the safety of sprouted barley in pregnancy was identified in the scientific or traditional literature. While this review did not identify any concerns for pregnant women, safety has not been conclusively established.

Sprouted barley is used in large doses (typically 30–60 g, sometimes up to 120 g daily) to reduce or stop lactation. However, small doses (up to 9 g) are considered to promote lactation (Bensky et al. 2004; Chen and Chen 2004). Large doses of sprouted barley should not be used by women who are lactating and wish to continue to do so.

III. Pharmacology and Pharmacokinetics

Human Pharmacological Studies

Single oral doses of 50, 250, or 750 mg of the compound 8-PN were well tolerated in postmenopausal women (Rad et al. 2006).

Animal Pharmacological Studies

In rats, the compound 8-PN was shown to have estrogenic activity in reproductive tissue about 20,000-fold weaker compared to 17β-estradiol (Schaefer et al. 2003). Doses of 0.67 to 18 mg/kg of 8-PN daily administered subcutaneously to mice for 28 days completely inhibited ovariectomy-induced bone loss while exhibiting minimal, dose-independent, trophic effects on uterus and endometrium (Humpel et al. 2005).

Five dogs, four of which were greyhounds, developed hyperthermia, restlessness, panting, vomiting, signs of abdominal pain, and seizures after ingestion of spent hops, and four of the five dogs died despite aggressive therapeutic measures (Duncan et al. 1997).

Lupulones, notably the compound colupulone, induced the drug-metabolizing isoenzyme CYP3A in mice (Mannering et al. 1992).

In Vitro Pharmacological Studies

A methanol extract of hops showed significant competitive binding to estrogen receptors alpha (ERα) and beta (ERβ), exhibited estrogenic activity in cultured endometrial cells, induced alkaline phosphatase activity, upregulated presenelin-2 (an estrogen-inducible gene) in S30 breast cancer cells, and upregulated progesterone receptor mRNA. For estrogen receptor binding, IC_{50} (the concentration that provided 50% inhibition) was 30 μg/ml for ERα and 27 μg/ml for ERβ, while for alkaline phosphatase the IC_{50} was 13 μg/ml (Liu et al. 2001).

A fraction of a CO_2 extract of hops showed an estrogenic potential equivalent to that of red clover. The extract demonstrated competitive ER binding, activation of transiently transfected ERE-luciferase, and alkaline phosphatase enzyme induction (Overk et al. 2005).

A hydroethanolic extract of hops bound to estrogen receptors in the estrogen receptor (ER)-positive human breast cancer cell line MCF-7. The same extract did not bind to progesterone receptors in the human breast cancer cell line T-47D (Zava et al. 1998). The extract stimulated the proliferation of T-47D (ER positive) cells but had no effect on MDA486 (ER negative) cells (Zava et al. 1998). An extract of hops was shown to have growth-inhibitory activity on estrogen receptor-positive breast cancer cell lines MCF-7 and T-47D (Dixon-Shanies and Shaikh 1999).

Binding and transcription studies with 8-PN demonstrated that the compound bound to both receptors, with a more than twofold preference for ERα over ERβ. 8-PN was shown to be 10-fold more potent than coumestrol and 100-fold stronger than genistein but 70 times weaker than 17β-estradiol (Schaefer et al. 2003). A similar study indicated that 8-PN exhibited a slight preference for ERβ over ERα (Milligan et al. 2000). The compound 8-PN was shown to have proliferative activity in MCF-7 cells (Effenberger et al. 2005).

IV. Pregnancy and Lactation

No information on the safety of hops during pregnancy or lactation was identified.

V. Toxicity Studies

Acute Toxicity

The LD_{50} of an orally administered ethanolic extract of hops in mice is 3.5 g/kg and in rats is 2.7 g/kg. The LD_{50} of an orally administered ethanol extract of hops in mice is 2.7 g/kg and in rats is 0.415 g/kg (Hansel and Wagener 1967). The LD_{50} of orally administered lupulone in mice is 525 mg/kg (Hansel and Wagener 1967).

No adverse effects were observed in rats orally administered a single dose of 2 g/kg of hops polyphenols (Nagasako-Akazome et al. 2007).

Short-Term Toxicity

No signs of toxicity were observed in rats subcutaneously administered 30 mg/kg of the compound 8-PN daily for 2 weeks (Miyamoto et al. 1998).

No adverse effects on major organ functions or protein, lipid, or carbohydrate metabolism were observed in rats orally administered the compound xanthohumol ad libitum in drinking water at a concentration of 0.0005 *M* for 28 days (Vanhoecke et al. 2005).

In mice administered a diet containing 2 or 4% lupulone for 40 days, leukocytic infiltration into the lungs and some cases of bronchopneumonia were observed (Chin and Anderson 1950).

Subchronic Toxicity

No adverse effects on blood chemistry, organ weights, or histology were observed in rats orally administered hops polyphenols at doses up to 2 g/kg daily for 90 days (Nagasako-Akazome et al. 2007).

Oral administration of the compounds hexahydroisohumulone at doses of 10, 25, 35, or 50 mg/kg or tetrahydroisohumulone at doses of 25, 50, or 100 mg/kg daily for 14 weeks to beagle dogs showed that the compounds were generally well tolerated. At the high dose levels, both compounds induced vomiting, and much of the material administered was excreted in the feces. The no-observed-adverse-effect level (NOAEL) was 50 mg/kg for hexahydroisohumulone and 100 mg/kg for tetrahydroisohumulone (Chappel et al. 1998).

Genotoxicity

A slight mutagenic effect of hops polyphenols was observed at concentrations of 5000 µg/plate in the Ames test in the absence of S9 mix for TA98 and in the presence of S9 mix for TA1537. No mutagenic activity of the same extract was observed in the micronucleus test (Nagasako-Akazome et al. 2007).

Cytotoxicity

Flavonoids isolated from hops were shown to have cytotoxic effects at high concentrations (at or above 10–4 *M*) (Effenberger et al. 2005).

Literature Cited

Chadwick, L.R., G.F. Pauli, and N.R. Farnsworth. 2006. The pharmacognosy of *Humulus lupulus* L. (hops) with an emphasis on estrogenic properties. *Phytomedicine* 13(1-2):119-131.

Chappel, C.I., S.Y. Smith, and M. Chagnon. 1998. Subchronic toxicity study of tetrahydroisohumulone and hexahydroisohumulone in the beagle dog. *Food Chem. Toxicol.* 36(11):915-922.

Chin, Y., and H. Anderson. 1950. Toxicology and pharmacology of lupulon. *Arch. Int. Pharmcodyn. Ther.* 82:1-15.

Dixon-Shanies, D., and N. Shaikh. 1999. Growth inhibition of human breast cancer cells by herbs and phytoestrogens. *Oncol. Rep.* 6(6):1383-1387.

Duncan, K.L., W.R. Hare, and W.B. Buck. 1997. Malignant hyperthermia-like reaction secondary to ingestion of hops in five dogs. *J. Am. Vet. Med. Assoc.* 210(1):51-54.

Effenberger, K.E., S.A. Johnsen, D.G. Monroe, T.C. Spelsberg, and J.J. Westendorf. 2005. Regulation of osteoblastic phenotype and gene expression by hop-derived phytoestrogens. *J. Steroid Biochem. Mol. Biol.* 96(5):387-399.

Estrada, J.L., F. Gozalo, C. Cecchini, and E. Casquete. 2002. Contact urticaria from hops (*Humulus lupulus*) in a patient with previous urticaria-angioedema from peanut, chestnut and banana. *Contact Dermat.* 46(2):127.

Godnic-Cvar, J., E. Zuskin, J. Mustajbegovic, et al. 1999. Respiratory and immunological findings in brewery workers. *Am. J. Ind. Med.* 35(1):68-75.

Gora, A., C. Skorska, Z. Prazmo, et al. 2004a. Exposure to bioaerosols: Allergic reactions and respiratory function in Polish hop growers. *Am. J. Ind. Med.* 46(4):371-374.

Gora, A., C. Skorska, J. Sitkowska, et al. 2004b. Exposure of hop growers to bioaerosols. *Ann. Agric. Environ. Med.* 11(1):129-138.

Hansel, R., and H.H. Wagener. 1967. Attempts to identify sedative-hypnotic active substances in hops. *Arzneimittelforschung* 17(1):79-81.

Horvat, O., A. Raskovic, V. Jakovljevic, J. Sabo, and J. Berenji. 2007. Interaction of alcoholic extracts of hops with cocaine and paracetamol in mice. *Eur. J. Drug Metab. Pharmacokinet.* 32(1):39-44.

Humpel, M., P. Isaksson, O. Schaefer, et al. 2005. Tissue specificity of 8-prenylnaringenin: Protection from ovariectomy induced bone loss with minimal trophic effects on the uterus. *J. Steroid Biochem. Mol. Biol.* 97(3):299-305.

Liu, J., J.E. Burdette, H. Xu, et al. 2001. Evaluation of estrogenic activity of plant extracts for the potential treatment of menopausal symptoms. *J. Agric. Food Chem.* 49(5):2472-2479.

Mannering, G.J., J.A. Shoeman, and L.B. Deloria. 1992. Identification of the antibiotic hops component, colupulone, as an inducer of hepatic cytochrome P-4503A in the mouse. *Drug Metab. Dispos.* 20(2):142-147.

Milligan, S., J. Kalita, V. Pocock, et al. 2002. Oestrogenic activity of the hop phyto-oestrogen, 8-prenylnaringenin. *Reproduction* 123(2):235-242.

Milligan, S.R., J.C. Kalita, A. Heyerick, et al. 1999. Identification of a potent phytoestrogen in hops (*Humulus lupulus* L.) and beer. *J. Clin. Endocrinol. Metab.* 84(6):2249-2252.

Milligan, S.R., J.C. Kalita, V. Pocock, et al. 2000. The endocrine activities of 8-prenylnaringenin and related hop (*Humulus lupulus* L.) flavonoids. *J. Clin. Endocrinol. Metab.* 85(12):4912-4915.

Miyamoto, M., Y. Matsushita, A. Kikokawa, et al. 1998. Prenylflavonoids: A new class of non-steroidal phytoestrogen (part 2). Estrogenic effects of 8-isopentenylnaringenin on bone metabolism. *Planta Med.* 64(8):769.

Nagasako-Akazome, Y., D. Honma, M. Tagashira, et al. 2007. Safety evaluation of polyphenols extracted from hop bracts. *Food Chem. Toxicol.* 45(8):1383-1392.

Newmark, F.M. 1978. Hops allergy and terpene sensitivity: An occupational disease. *Ann. Allergy* 41(5):311-312.

Overk, C.R., P. Yao, L.R. Chadwick, et al. 2005. Comparison of the in vitro estrogenic activities of compounds from hops (*Humulus lupulus*) and red clover (*Trifolium pratense*). *J. Agric. Food Chem.* 53(16):6246-6253.

Pradalier, A., C. Campinos, and C. Trinh. 2002. Systemic urticaria induced by hop. *Allerg. Immunol.* 34(9):330-332.

Rad, M., M. Humpel, O. Schaefer, et al. 2006. Pharmacokinetics and systemic endocrine effects of the phyto-oestrogen 8-prenylnaringenin after single oral doses to postmenopausal women. *Br. J. Clin. Pharmacol.* 62(3):288-296.

Raskovic, A., O. Horvat, V. Jakovljevic, J. Sabo, and R. Vasic. 2007. Interaction of alcoholic extracts of hops with pentobarbital and diazepam in mice. *Eur. J. Drug Metab. Pharmacokinet.* 32(1):45-49.

Schaefer, O., M. Humpel, K.H. Fritzemeier, R. Bohlmann, and W.D. Schleuning. 2003. 8-Prenyl naringenin is a potent ERalpha selective phytoestrogen present in hops and beer. *J. Steroid Biochem. Mol. Biol.* 84(2-3):359-360.

Schiller, H., A. Forster, C. Vonhoff, et al. 2006. Sedating effects of *Humulus lupulus* L. extracts. *Phytomedicine* 13(8):535-541.

Spiewak, R., and J. Dutkiewicz. 2002. Occupational airborne and hand dermatitis to hop (*Humulus lupulus*) with non-occupational relapses. *Ann. Agric. Environ. Med.* 9(2):249-252.

Spiewak, R., A. Gora, and J. Dutkiewicz. 2001. Work-related skin symptoms and type I allergy among eastern-Polish farmers growing hops and other crops. *Ann. Agric. Environ. Med.* 8(1):51-56.

Stricker, W.E., E. Anorve-Lopez, and C.E. Reed. 1986. Food skin testing in patients with idiopathic anaphylaxis. *J. Allergy Clin. Immunol.* 77(3):516-519.

Vanhoecke, B.W., F. Delporte, E. Van Braeckel, et al. 2005. A safety study of oral tangeretin and xanthohumol administration to laboratory mice. *In Vivo* 19:103-107.

Zanoli, P., M. Rivasi, M. Zavatti, F. Brusiani, and M. Baraldi. 2005. New insight in the neuropharmacological activity of *Humulus lupulus* L. *J. Ethnopharmacol.* 102(1):102-106.

Zanoli, P., and M. Zavatti. 2008. Pharmacognostic and pharmacological profile of *Humulus lupulus* L. *J. Ethnopharmacol.* 116(3):383-396.

Zava, D.T., C.M. Dollbaum, and M. Blen. 1998. Estrogen and progestin bioactivity of foods, herbs, and spices. *Proc. Soc. Exp. Biol. Med.* 217(3):369-378.

Hydrangea arborescens L. — Hydrangeaceae

SCN: hydrangea
OCN: seven barks; wild hydrangea

Part: root

Quick Reference Summary

H

Safety Class: 1
Interaction Class: A

Contraindications

None known.

Other Precautions

None known

Drug and Supplement Interactions

None known.

Notice

Cyanogenic glycosides (hydrangin, ~1–3%) (Bondurant 1887; Cook and Martin 1948; Palmer 1963); *see* Appendix 1.

Editors' Note

The flower and leaf of hydrangea are reported to have caused toxic symptoms in humans. This is assumed to be due to the cyanogenic glycoside hydrangin, which is also reported to be found in the root (Bondurant 1887; Leung and Foster 1996; List and Hörhammer 1973).

Adverse Events and Side Effects

Allergic contact dermatitis to *Hydrangea* species has been reported, primarily in agricultural workers (Avenel-Audran et al. 2000; Bruynzeel 1986, 1991; De Rooij et al. 2006; Kuligowski et al. 1992; Meijer et al. 1990; Rademaker 2003).

Overdose of hydrangea (standard dose listed as 2 g) has been reported to cause dizziness and a sense of oppression of the chest (Felter and Lloyd 1898; Powers et al. 1942).

Pharmacological Considerations

None known.

Pregnancy and Lactation

No information on the safety of hydrangea in pregnancy or lactation was identified in the scientific or traditional literature. Although this review did not identify any concerns for use while pregnant or nursing, safety has not been conclusively established.

Review Details

I. Drug and Supplement Interactions

Clinical Trials of Drug or Supplement Interactions

No clinical trials of drug or supplement interactions were identified.

Case Reports of Suspected Drug or Supplement Interactions

No case reports of suspected drug or supplement interactions were identified.

Animal Trials of Drug or Supplement Interactions

No animal trials of drug or supplement interactions were identified.

II. Adverse Events

Case Reports of Adverse Events

Allergic contact dermatitis to *Hydrangea* species has been reported, primarily in agricultural workers (Avenel-Audran et al. 2000; Bruynzeel 1986, 1991; De Rooij et al. 2006; Kuligowski et al. 1992; Meijer et al. 1990; Rademaker 2003).

Overdose of hydrangea has been reported to cause dizziness and a sense of oppression of the chest (Felter and Lloyd 1898).

III. Pharmacology and Pharmacokinetics

Human Pharmacological Studies

No relevant human pharmacological studies were identified.

Animal Pharmacological Studies

No relevant animal pharmacological studies were identified.

In Vitro Pharmacological Studies

No relevant in vitro pharmacological studies were identified.

IV. Pregnancy and Lactation

No information on the safety of hydrangea during pregnancy or lactation was identified.

V. Toxicity Studies

No toxicity studies were identified.

Literature Cited

Avenel-Audran, M., B.M. Hausen, J. le Sellin, G. Ledieu, and J.L. Verret. 2000. Allergic contact dermatitis from hydrangea—Is it so rare? *Contact Dermat.* 43(4):189-191.

Bondurant, C.S. 1887. Analysis of *Hydrangea arborescens*. *Am. J. Pharm.* 59(3):122-124.

Bruynzeel, D.P. 1986. Allergic contact dermatitis to hydrangea. *Contact Dermat.* 14(2):128.

Bruynzeel, D.P. 1991. Contact dermatitis from hydrangea. *Contact Dermat.* 24(1):78.

Cook, E.F., and E.W. Martin. 1948. *Remington's practice of pharmacy*. 9th ed. Easton, PA: Mack Publishing Company.

De Rooij, J., D.P. Bruynzeel, and T. Rustemeyer. 2006. Occupational allergic contact dermatitis from hydrangea. *Contact Dermat.* 54(1):65-66.

Felter, H.W., and J.U. Lloyd. 1898. *King's American dispensatory*. 18th ed., 3rd rev. 2 vols. Cincinnati: Ohio Valley Co.

Kuligowski, M.E., A. Chang, and J.H. Leemreize. 1992. Allergic contact hand dermatitis from hydrangea: Report of a 10th case. *Contact Dermat.* 26(4):269-270.

Leung, A.Y., and S. Foster. 1996. *Encyclopedia of common natural ingredients used in food, drugs, and cosmetics*. 2nd ed. New York: Wiley.

List, P.H., and H. Hörhammer. 1973. *Hagers handbuch der pharmazeutischen praxis*. Berlin: Springer.

Meijer, P., P.J. Coenraads, and B.M. Hausen. 1990. Allergic contact dermatitis from hydrangea. *Contact Dermat.* 23(1):59-60.

Palmer, K.H. 1963. The structure of hydrangin. *Can. J. Chem.* 41(9):2387-2389.

Powers, J.L., E.H. Wirth, and A.B. Nichols. 1942. *National formulary*. 7th ed. Washington, D.C.: American Pharmaceutical Association.

Rademaker, M. 2003. Occupational contact dermatitis to hydrangea. *Australas. J. Dermatol.* 44(3):220-221.

Hydrastis canadensis L. — Ranunculaceae

SCN: goldenseal
OCN: yellow puccoon; yellow root
Part: rhizome, root

Quick Reference Summary

Safety Class: 2b
Interaction Class: C

Contraindications

Not for use in pregnancy except under the supervision of a qualified healthcare practitioner (Bradley 1992; List and Hörhammer 1973; Mills and Bone 2005).

Other Precautions

Use of goldenseal during lactation is not recommended (Mills and Bone 2005).

Drug and Supplement Interactions

Goldenseal may slow the metabolism of drugs metabolized by the drug-metabolizing isoenzymes CYP3A4 and CYP2D6 (Gurley et al. 2005, 2008a, 2008b). *See* Appendix 3.

Notice

Berberine (0.5–6%) (Upton 2001); *see* Appendix 1.

Editors' Notes

Most safety concerns reported for goldenseal are based on studies of the compound berberine and other alkaloids. Data regarding isolated compounds may not apply directly to products or extracts made from the whole rhizome and root. Berberine has been shown to exhibit a number of bioactivities including cytotoxicity in cancer cell lines (Kettmann et al. 2004; Kim et al. 2005; Orfila et al. 2000) and topoisomerase I and II inhibition (Kim et al. 1998; Mantena et al. 2006).

Studies have indicated that the antimicrobial action of berberine may be potentiated by nonantimicrobial compounds found in berberine-containing plants, indicating a synergistic action of compounds (Stermitz et al. 2000).

Adverse Events and Side Effects

No cases of adverse events associated with the use of goldenseal were identified.

Pharmacological Considerations

Human studies have shown that goldenseal may inhibit drug-metabolizing isoenzymes CYP3A4 and CYP2D6, slowing the clearance of drugs metabolized by these enzymes and increasing plasma levels of these drugs (Gurley et al. 2005, 2008a, 2008b). *See* Cytochrome CYP450 in Appendix 3.

Pregnancy and Lactation

Goldenseal is contraindicated in pregnancy in several contemporary texts on herbal safety (Brinker 2001; Mills et al. 2006; Mills and Bone 2005). These contraindications are based primarily on uterine stimulant activity of the isolated compound berberine in excised mouse uteruses (Furuya 1957; Imaseki et al. 1961) and the potential ability of berberine to displace bilirubin and cause neonatal jaundice (Chan 1993). Although definitive data confirming the safety of goldenseal during pregnancy is lacking, several reproductive toxicity studies of goldenseal and the isolated compound berberine in mice and rats have shown no adverse effects on the fetus at doses equivalent to over 45 times the standard human dose (Jahnke et al. 2006; NTP 2002; Price 2003; Yao et al. 2005).

Some concerns exist for use of berberine-containing plants during pregnancy, including uterine stimulation (Furuya 1957; Imaseki et al. 1961), although no adverse effects were observed in rats administered high doses of goldenseal during pregnancy (Yao et al. 2005).

Berberine has been shown to be present in the breast milk of lactating women who have taken berberine-containing plants (Chan 1993).

Review Details

I. Drug and Supplement Interactions

Clinical Trials of Drug or Supplement Interactions

No effects on digoxin were observed after oral administration of 3210 mg daily of goldenseal for 14 days (Gurley et al. 2007). No effects on indinavir were observed after administration of 2280 mg daily of goldenseal for 14 days (Sandhu et al. 2003).

Case Reports of Suspected Drug or Supplement Interactions

No case reports of suspected drug or supplement interactions were identified.

Animal Trials of Drug or Supplement Interactions

The compound berberine has been shown to significantly prolong pentobarbital-induced sleeping time in rats (Janbaz and Gilani 2000).

II. Adverse Events

Adverse Events Reported in Clinical Trials

No adverse events were reported in the identified clinical trials of goldenseal (Gurley et al. 2005; Gurley et al. 2007).

Case Reports of Adverse Events

No case reports of adverse events were identified.

III. Pharmacology and Pharmacokinetics

Human Pharmacological Studies

Significant inhibition of the drug-metabolizing isoenzymes CYP2D6 and CYP3A4/5 was observed after oral administration of goldenseal for 14 (3210 or 3960 mg daily) or 28 days (2700 mg daily). No effect was observed on isoenzymes CYP1A2 and CYP2E1 (Gurley et al. 2005; Gurley et al. 2008a, 2008b). Goldenseal administered for 14 days (3210 mg daily) had no significant effect on P-glycoprotein function (Gurley et al. 2007).

Intravenous administration of the compound berberine (0.2 mg/kg/min) caused a significant decrease in blood pressure (Marin-Neto et al. 1988).

Animal Pharmacological Studies

Chronic administration of berberine to rats was shown to significantly decrease bilirubin protein binding at doses of 10 to 29 mg/kg (administered intraperitoneally), although at doses of 2 mg/kg the displacement was not significant (Chan 1993). Berberine is speculated to have possible interactions with drugs that displace bilirubin protein binding (Mills and Bone 2005).

Berberine was shown to promote blood coagulation in mice and rats (Ziablitskii et al. 1996).

In Vitro Pharmacological Studies

Two in vitro studies indicated that goldenseal tincture and the compound berberine significantly inhibit the drug-metabolizing isoenzyme CYP3A4 (Foster et al. 2003; Janbaz and Gilani 2000).

At non-relaxant doses, goldenseal extract was shown to potentiate the relaxant effects of isoprenaline in isolated guinea pig trachea (Abdel-Haq et al. 2000).

Berberine has been shown to increase the efflux of paclitaxel and rhodamine in human digestive tract and liver cancer cells by inducing P-glycoprotein (Chatterjee and Franklin 2003). Berberine may interfere with the action of tetracycline in the treatment of cholera (Khin et al. 1985).

IV. Pregnancy and Lactation

In rats administered daily doses of goldenseal at 65 times the human dose on either gestation days 1 through 8 or 8 through 15, no signs of maternal toxicity or effects on fertility, and no adverse effects in fetuses were observed

(Yao et al. 2005). In a related ex vivo study, dose-dependent adverse effects on development were observed in fetuses prematurely removed and subsequently treated with goldenseal extract. Fetuses treated with 2, 4, or 6 μg/ml had decreased length, number of somites, yolk sac size, brain size, and forelimb size (Yao et al. 2005). The authors of the study noted that the lack of adverse effects observed in vivo indicated that goldenseal may not be absorbed and available to fetuses (Yao et al. 2005).

No adverse effects were observed in pregnant mice or their fetuses administered doses of 2 g/kg goldenseal (5% berberine) daily on gestational days 6 to 17. A reduction in fetal weight was reported in offspring of mice administered 7.7 g/kg daily of goldenseal and an increase in maternal liver weight was reported in animals consuming more than 2 g/kg. The 7.7 g/kg dose rate is approximately 300 times the estimated average human dose (26 mg/kg) (NTP 2002).

No adverse effects were observed in pregnant rats administered goldenseal (5% berberine) at doses of 200 mg/kg on gestational days 6 to 20 or 0 to 20. At doses of 415 mg/kg or more, a dose-related increase was noted in maternal liver weight. At doses up to 1215 mg/kg daily, no effects were seen on prenatal mortality, average live litter size, average fetal body weight, or percent male fetuses per litter. No statistically significant dose-response patterns for malformations were seen, although a limited number of fetuses had multiple malformations; based on the absence of a dose-response relationship, however, these findings were not believed to be treatment-related (Price 2003).

In pregnant rats fed the compound berberine on gestational days 6 to 20, some reduction in maternal weight gain was observed, with a lowest-observed-adverse-effect level (LOAEL) of 530 mg/kg daily. Only a mild reduction in fetal weights was observed, and the LOAEL based on fetal weight reduction was 1000 mg/kg (Jahnke et al. 2006). Similarly, in mice administered berberine on gestational days 6 to 17, the maternal LOAEL was determined to be 531 mg/kg daily, and the developmental toxicity level was 1000 mg/kg daily (Jahnke et al. 2006).

A number of cases of kernicterus (brain damage caused by neonatal jaundice) were reported in south Asian countries in the 1970s and 1980s. Chinese goldthread (*Coptis chinensis*) was a common ingredient in traditional formulas used in pregnant women and newborns and has been shown to prevent bilirubin from binding to serum protein. Chinese goldthread contains berberine (also present in goldenseal), the compound believed to be responsible for the jaundice (Upton 2001), as berberine was shown to displace bilirubin in rats (*see* Animal pharmacological studies). A review of traditional Chinese medicine use in cases of neonatal jaundice indicated that no association could be made between maternal use of Chinese goldthread during pregnancy and neonatal jaundice (Fok 2001).

Berberine has been shown to stimulate uterine contractions in both pregnant and nonpregnant mice (Furuya 1957; Imaseki et al. 1961). A study of various berberine-containing herbal extracts on isolated uteruses, however, indicated that relaxation or stimulation of the uterus did not correlate with the concentration of berberine in the extract, suggesting that not all berberine-containing herbs will have the same effect on the uterus (Haginiwa and Harada 1962).

Berberine has been shown to be present in the breast milk of lactating women who have taken berberine-containing plants (Chan 1993).

V. Toxicity Studies

Acute Toxicity

In mice, the oral LD_{50} of goldenseal is 1.62 g/kg, whereas that of berberine is 329 mg/kg (Haginiwa and Harada 1962). The LD_{50} of orally administered berberine sulfate in rats could not be determined at doses up to 1000 mg/kg (Kowalewski et al. 1975).

Short-Term Toxicity

In a two week toxicity study, rats and mice were fed diets containing 0.156, 0.3121, 0.625, 1.25, 2.5, or 5% goldenseal. Liver weights of male rats administered 0.625% or higher and female rats administered 1.25% or higher were significantly greater than those of the controls. Minimal to moderate hepatocellular hypertrophy occurred in several males and all females administered 2.5% diets and in all males and females at the 5% dose (NTP 2010). Increases in liver weights occurred in male mice exposed at the 2.5 or 5% diets and in females on the 5% diet. Absolute and relative thymus weights of male mice in the two highest dose groups decreased. Minimal hypertrophy of centrilobular hepatocytes occurred in all males and females at the highest goldenseal concentration. The left epidydimal weight in male mice was significantly decreased relative to controls. The incidences of hepatocyte hypertrophy were significantly increased in male and female mice exposed to 12,500 ppm or greater (NTP 2010).

Subchronic Toxicity

In a 90-day toxicity study, rats and mice were fed diets containing 0.156, 0.3121, 0.625, 1.25, 2.5, or 5% goldenseal. Liver weights were significantly increased in male rats fed diets containing 0.625% or greater and in all exposed groups of females. The incidences of hepatocyte hypertrophy were increased in the liver of male and female rats fed concentrations of 1.25% or more. In mice, liver weights were increased in males fed diets containing 1.25% or more goldenseal and in females fed diets containing 2.5 or 5% goldenseal (NTP 2010).

Chronic Toxicity

In a two-year toxicity study, rats and mice were fed diets containing 0.3, 0.9, or 2.5% goldenseal, equivalent to average daily doses of approximately 135, 400, or 1175 mg/kg for male rats; 150, 470, or 1340 mg/kg for female rats; 375, 1120, or 3275 mg/kg for male mice; and 330, 1000, or 2875 mg/kg for female mice. These doses and the treatment duration are approximately equivalent to human exposure of 18, 54, and 150 g, given daily for 70 years (Hayes 2007). The standard oral dose for humans is 2 g daily (Upton 2001), and there are no uses recorded for goldenseal root that would recommend daily use over an entire lifetime. Survival was significantly greater than that of controls in female rats fed at the 0.9% level and significantly less in female mice at the same level, although the final report of the study states that survival "of all exposed groups of animals was similar to their controls." The rates of adenomas (benign tumors) of the liver was increased in male and female rats receiving 2.5% goldenseal, and male mice receiving goldenseal had increased rates of liver blastomas and multiple adenomas. The report concluded that, there was "no evidence of carcinogenic activity" of goldenseal in female mice at any of the feeding levels; "some evidence of carcinogenic activity" in male mice "based upon the increased incidence of hepatoblastoma in the 25,000 ppm [2.5%] group and the increased incidences of multiple hepatocellular adenoma in all exposed groups"; and "clear evidence of carcinogenic activity" in male and female rats as the study recorded, at the highest dosage level, that incidences of hepatocellular adenoma or hepatocellular carcinoma combined (note that the results "combined" a single carcinoma and several adenomas) in males and of hepatocellular adenoma in females "exceeded the current and historical control rates (NTP 2010)." A review of the report indicated, however, that these latter conclusions were "not appropriate" for a number of reasons, including that the one hepatocellular carcinoma at the high dose in a male rat "is within historic control incidence" (Beyer et al. 2009). Some potential protective roles of goldenseal root were recorded in this two-year feeding study. For example, incidences of cardiomyopathy were "significantly decreased in all exposed groups" of male rats and in the highest-dosed female rats, and the incidences of fibroadenoma of the mammary gland "occurred with a negative trend and were significantly decreased in all exposed groups" of female rats (NTP 2010).

H

Genotoxicity

No mutagenic activity of goldenseal was observed in *Salmonella typhimurium* strains TA100 or TA98 or *Escherichia coli* strain WP2 *uvrA* pKM101 with or without metabolic activation with S9. No increase in the frequency of micronucleated normochromatic (mature) erythrocytes was observed in peripheral blood samples from male or female B6C3F1 mice exposed to goldenseal root powder in feed (3121 to 50,000 ppm) for 3 months, and no significant exposure-related changes in the percentages of polychromatic (immature) erythrocytes were observed in peripheral blood of these mice, suggesting that no exposure-related bone marrow toxicity of goldenseal occurred (NTP 2010).

Literature Cited

Abdel-Haq, H., M.F. Cometa, M. Palmery, et al. 2000. Relaxant effects of *Hydrastis canadensis* L. and its major alkaloids on guinea pig isolated trachea. *Pharmacol. Toxicol.* 87(5):218-222.

Beyer, L., M. Seeley, and L. Rhomberg. 2009. Memorandum: Comments on NTP technical report on the toxicology and carcinogenesis of goldenseal root powder (*Hydrastis canadensis*) in F344/N rats and B6C3F1 mice. *Cited in* McGuffin, M. Report on the status of NTP's technical report on goldenseal. *AHPA Report* 25(1)12-14.

Bradley, P.R. 1992. *British herbal compendium: A handbook of scientific information on widely used plant drugs.* Bournemouth, UK: British Herbal Medicine Association.

Brinker, F. 2001. *Herb contraindications and drug interactions.* 3rd ed. Sandy, OR: Eclectic Medical Publications.

Chan, E. 1993. Displacement of bilirubin from albumin by berberine. *Biol. Neonate* 63(4):201-208.

Chatterjee, P., and M.R. Franklin. 2003. Human cytochrome P450 inhibition and metabolic-intermediate complex formation by goldenseal extract and its methylenedioxyphenyl components. *Drug Metab. Dispos.* 31(11):1391-1397.

Fok, T.F. 2001. Neonatal jaundice—Traditional Chinese medicine approach. *J. Perinatol.* 21(Suppl. 1):S98-S100; discussion S104-S107.

Foster, B.C., S. Vandenhoek, J. Hana, et al. 2003. In vitro inhibition of human cytochrome P450-mediated metabolism of marker substrates by natural products. *Phytomedicine* 10(4):334-342.

Furuya, T. 1957. Pharmacological action, including toxicity and excretion of berberine hydrochloride and its oxidation product. *Bull. Osaka Med. School* 3:62-67. *Cited in* De Smet, P.A.G.M. 1992. *Adverse effects of herbal drugs, Volume 1.* New York: Springer.

Gurley, B.J., S.F. Gardner, M.A. Hubbard, et al. 2005. *In vivo* effects of goldenseal, kava kava, black cohosh, and valerian on human cytochrome P450 1A2, 2D6, 2E1, and 3A4/5 phenotypes. *Clin. Pharmacol. Ther.* 77(5):415-426.

Gurley, B.J., A. Swain, G.W. Barone, et al. 2007. Effect of goldenseal (*Hydrastis canadensis*) and kava kava (*Piper methysticum*) supplementation on digoxin pharmacokinetics in humans. *Drug Metab. Dispos.* 35(2):240-245.

Gurley, B.J., A. Swain, M.A. Hubbard, et al. 2008a. Clinical assessment of CYP2D6-mediated herb-drug interactions in humans: Effects of milk thistle, black cohosh, goldenseal, kava kava, St. John's wort, and echinacea. *Mol. Nutr. Food Res.* 52(7):755.

Gurley, B.J., A. Swain, M.A. Hubbard, et al. 2008b. Supplementation with goldenseal (*Hydrastis canadensis*), but not kava kava (*Piper methysticum*), inhibits human CYP3A activity in vivo. *Clin. Pharmacol. Ther.* 83(1):61.

Haginiwa, J., and M. Harada. 1962. Pharmacological studies on crude drugs. V. Comparison of berberine type alkaloid-containing plants on their components and several pharmacological actions.. *Jpn. J. Pharmacol.* 82:726-731.

Imaseki, I., Y. Kitabatake, and T. Taguchi. 1961. Studies on the effect of berberine alkaloids on intestine and uterus in mice. *Yakugaku Zasshi* 81:1281-1284.

Jahnke, G.D., C.J. Price, M.C. Marr, C.B. Myers, and J.D. George. 2006. Developmental toxicity evaluation of berberine in rats and mice. *Birth Defects Res. B Dev. Reprod. Toxicol.* 77(3):195-206.

Janbaz, K.H., and A.H. Gilani. 2000. Studies on preventive and curative effects of berberine on chemical-induced hepatotoxicity in rodents. *Fitoterapia* 71(1):25-33.

Kettmann, V., D. Kosfalova, S. Jantova, M. Cernakova, and J. Drimal. 2004. In vitro cytotoxicity of berberine against HeLa and L1210 cancer cell lines. *Pharmazie* 59(7):548-551.

Khin, U., K. Myo, W. Nyunt, K. Aye, and U. Tin. 1985. Clinical trial of berberine in acute watery diarrhoea. *Br. Med. J.* 291:1601-1605.

Kim, H.R., H.Y. Min, Y.H. Jeong, et al. 2005. Cytotoxic constituents from the whole plant of *Corydalis pallida*. *Arch. Pharm. Res.* 28(11):1224-1247.

Kim, S.A., Y. Kwon, J.H. Kim, M.T. Muller, and I.K. Chung. 1998. Induction of topoisomerase II-mediated DNA cleavage by a protoberberine alkaloid, berberrubine. *Biochemistry* 37(46):16316-16324.

Kowalewski, Z., A. Mrozikiewicz, T. Bobkiewicz, K. Drost, and B. Hladon. 1975. Toxicity of berberine sulfate. *Acta Pol. Pharm.* 32(1):113-120.

Hayes, A.W. (Ed.) 2007. *Principles and methods of toxicology, 5th edition*. Boca Raton, FL: CRC Press.

List, P.H., and H. Hörhammer. 1973. *Hagers handbuch der pharmazeutischen praxis*. Berlin: Springer.

Mantena, S.K., S.D. Sharma, and S.K. Katiyar. 2006. Berberine inhibits growth, induces G_1 arrest and apoptosis in human epidermoid carcinoma A431 cells by regulating Cdki-Cdk-cyclin cascade, disruption of mitochondrial membrane potential and cleavage of caspase 3 and PARP. *Carcinogenesis* 27(10):2018-2027.

Marin-Neto, J.A., B.C. Maciel, A.L. Secches, and L. Gallo, Jr. 1988. Cardiovascular effects of berberine in patients with severe congestive heart failure. *Clin. Cardiol.* 11(4):253-260.

Mills, E., J. Dugoua, D. Perri, and G. Koren. 2006. *Herbal medicines in pregnancy and lactation—An evidence-based approach.* London: Taylor & Francis.

Mills, S., and K. Bone. 2005. *The essential guide to herbal safety*. St. Louis: Elsevier.

NTP. 2002. Final study report: Developmental toxicity evaluation for goldenseal (*Hydrastis canadensis*) root powder administered in the feed to Swiss (CD-1) mice on gestational days 6–17. Research Triangle Park, NC: National Toxicology Program.

NTP. 2010. NTP technical report on the toxicology and carcinogenesis studies of goldenseal root powder (*Hydrastis canadensis*) in F344/N rats and B6C3F1 mice. Research Triangle Park, NC: National Toxicology Program.

Orfila, L., M. Rodriguez, T. Colman, et al. 2000. Structural modification of berberine alkaloids in relation to cytotoxic activity in vitro. *J. Ethnopharmacol.* 71(3):449-456.

Price, C.J. 2003. Final study report on the developmental toxicity evaluation for goldenseal root powder (*'Hydrastis canadensis'*) administered in the feed to Sprague-Dawley rats on gestational days 6 to 20. *Govt. Rep. Announc. Index* 18:130.

Sandhu, R.S., R.P. Prescilla, T.M. Simonelli, and D.J. Edwards. 2003. Influence of goldenseal root on the pharmacokinetics of indinavir. *J. Clin. Pharmacol.* 43(11):1283-1288.

Stermitz, F.R., P. Lorenz, J.N. Tawara, L.A. Zenewicz, and K. Lewis. 2000. Synergy in a medicinal plant: Antimicrobial action of berberine potentiated by 5′-methoxyhydnocarpin, a multidrug pump inhibitor. *Proc. Natl. Acad. Sci. U.S.A.* 97(4):1433-1437.

Upton, R. 2001. *Goldenseal root:* Hydrastis canadensis; *Standards of analysis, quality control, and therapeutics. American Herbal Pharmacopoeia and therapeutic compendium.* Santa Cruz, CA: American Herbal Pharmacopoeia.

Yao, M., H.E. Ritchie, and P.D. Brown-Woodman. 2005. A reproductive screening test of goldenseal. *Birth Defects Res. B Dev. Reprod. Toxicol.* 74(5):399-404.

Ziablitskii, V.M., V.N. Romanovskaia, R.Z. Umurzakova, A.N. Starosel'skaia, and T. Mikhal'skaia. 1996. Modification to the functional status of the hemostatic system with the use of berberine sulfate. *Eksp. Klin. Farmakol.* 59(1):37-39.

Hypericum perforatum L. — Clusiaceae

SCN: St. John's wort

Part: flowering top, herb

Quick Reference Summary

Safety Class: 2d

Interaction Class: C*

Contraindications

St. John's wort should not be used during phototherapy (laser or ultraviolet) (Beattie et al. 2005).

Other Precautions

Avoidance of excessive exposure to sunlight during St. John's wort use is warranted in fair-skinned persons (Martindale and Reynolds 1996; Weiss and Meuss 2001; Wichtl 2004).

* Low-hyperforin and hyperforin-free extracts are commercially available and are believed to be less likely to interact with drugs than extracts with standard amounts of hyperforin (Madabushi et al. 2006).

Drug and Supplement Interactions

St. John's wort has been shown to induce certain drug-metabolizing enzymes in the CYP450 enzyme system and the drug transporter protein P-glycoprotein (P-gp). Such induction may decrease the blood levels of certain orally administered drugs through increased metabolism or efflux, resulting in decreased therapeutic activity of these drugs. In human studies, significant induction of the enzyme CYP3A4 (Bauer et al. 2003; Dresser et al. 2003; Gurley et al. 2002, 2005; Wang et al. 2001) and some evidence of induction of the enzyme CYP2C19 have been shown (Wang et al. 2004a, 2004b), leading to reduced plasma levels of drugs metabolized by these enzymes or transported by P-gp. *See* CYP450 in Appendix 3 for more information.

The compound hyperforin is believed to be primarily responsible for the effects of St. John's wort on CYP450 enzymes and P-gp (Gerhard 2005; Mueller et al. 2004).

Human studies and case series have demonstrated that St. John's wort may decrease plasma levels of the following drugs (Borrelli and Izzo 2009):

Immunosuppressants: Cyclosporine (Bauer et al. 2003; Mai et al. 2004), tacrolimus (Hebert et al. 2004; Mai et al. 2003)
Anticoagulants: Warfarin (Jiang et al. 2004, 2006), phenprocoumon (Maurer et al. 1999)
Antiarrhythmics: Digoxin (Durr et al. 2000; Johne et al. 1999; Mueller et al. 2004), verapamil (Tannergren et al. 2004)
Calcium channel blockers: Nifedipine (Smith et al. 2001), verapamil (Tannergren et al. 2004)
Anti-anginals: Ivabradine (Portoles et al. 2006)
Hormonal contraceptives: Ethinylestradiol, norethindrone (Hall et al. 2003; Murphy et al. 2005; Pfrunder et al. 2003)
Anxiolytics: Quazepam (Kawaguchi et al. 2004), midazolam (Dresser et al. 2003; Gurley et al. 2002, 2005; Mueller et al. 2006; Xie et al. 2005), alprazolam (Markowitz et al. 2000, 2003)
Antidepressants (tricyclic): Amitriptyline (Johne et al. 2002)
Antivirals: Indinavir (Piscitelli et al. 2000), nevirapine (de Maat et al. 2001; L'homme et al. 2006)
Statins: Simvastatin (Sugimoto et al. 2001), atorvastatin (Andrén et al. 2007)
Anticancer drugs (chemotherapy and other) Irinotecan (Mathijssen et al. 2002), imatinib (Frye et al. 2004; Smith 2004)
Beta-adrenergic blockers: Talinolol (Schwarz et al. 2007)
Hypoglycemics: Gliclazide (Xu et al. 2008)
Antiulcer agents: Omeprazole (Wang et al. 2004a)
Antifungals: Voriconazole (Rengelshausen et al. 2005) (increase of plasma levels with single St. John's wort dose, decrease after continued use)
Anticonvulsants: Mephenytoin (Wang et al. 2004b)
Skeletal muscle relaxants: Chlorzoxazone (Gurley et al. 2002, 2005)
Antifungals: Voriconazole (Rengelshausen et al. 2005)
Antihistamines: Fexofenadine (studies have indicated mixed results, with some reporting that plasma levels are increased while others indicating that plasma levels are decreased) (Dresser et al. 2003; Hamman et al. 2001; Wang et al. 2002)

St. John's wort may decrease plasma levels of other drugs metabolized by the isoenzyme CYP3A4. *See* CYP450 in Appendix 3 for more information.

Human studies have shown a lack of interaction between St. John's wort and some drugs including prednisone (Bell et al. 2007), mycophenolic acid (Mai et al. 2003), pravastatin (Sugimoto et al. 2001), tolbutamide (Wang et al. 2001), dextromethorphan (Markowitz et al. 2000, 2003; Wang et al. 2001; Wenk et al. 2004), carbamazepine (Burstein et al. 2000), and theophylline (Morimoto et al. 2004).

Due to the photosensitizing potential of St. John's wort, concomitant use with photosensitizing drugs is not advised (Fiume 2001).

Notice

Photosensitizing (Brockmoller et al. 1997; Cotterill 2001; Schempp et al. 2001, 2003); *see* Appendix 2.

Adverse Events and Side Effects

A systematic review of clinical trials indicated that St. John's wort has been well tolerated in trials, with no serious adverse effects reported. Patient dropout rates and adverse effects reports were similar to those of placebo (Knuppel and Linde 2004).

Evidence from clinical trials suggests that concerns regarding phototoxicity of St. John's wort are not a problem in the general population, but caution regarding excessive exposure to sunlight during St. John's wort use is warranted in fair-skinned persons. St. John's wort should not be used in persons undergoing phototherapy (laser or ultraviolet) or those taking photosensitizing drugs (Schempp et al. 2001; Schulz 2001; Woelk et al. 1994). Hypericin is the compound primarily responsible for phototoxic effects (Gulick et al. 1999; Jacobson et al. 2001). Cases of phototoxicity have been reported after oral and topical uses of St. John's wort products. Severe cases have been reported in persons undergoing laser or UV therapy and in persons taking the compound hypericin (Bove 1998; Cotterill 2001; Golsch et al. 1997; Gulick et al. 1999; Jacobson et al. 2001; Lane-Brown 2000).

Cases of adverse events that have been reported in two or more persons are hypomania and psychosis (Fahmi et al. 2002; Guzelcan et al. 2001; Khawaja et al. 1999; Laird and Webb 2001; Lal and Iskandar 2000; Moses and Mallinger 2000; Nierenberg et al. 1999; O'Breasail and Argouarch 1998; Shimizu et al. 2004; Shuster 1999) (note that in many of these reports, patients had a history of psychiatric illness or concomitant diagnoses of psychiatric disorders), hypertension (Patel et al. 2002; Zullino and Borgeat 2003), and sexual dysfunction (Assalian 2000; Bhopal 2001). The relationship between these events and St. John's wort use is not known.

Pharmacological Considerations

St. John's wort has been shown to induce the drug-metabolizing isoenzyme CYP3A4 and the drug transporter P-glycoprotein (P-gp), leading to lower plasma levels of drugs metabolized by CYP3A4 or transported by P-gp (Borrelli and Izzo 2009). The effect on CYP3A4 lasts for approximately 7 days after St. John's wort use (Imai et al. 2008).

Pregnancy and Lactation

Studies on the effect of prenatal consumption of St. John's wort on pregnancy in mice and rats were generally associated with normal gestation and offspring development (Borges et al. 2005; Cada et al. 2001; Ferguson et al. 1999; Rayburn et al. 2000, 2001a, 2001b). A limited number of human case reports indicated healthy pregnancies and infants when St. John's wort was used prenatally (Gregoretti et al. 2004).

In studies of nursing mothers, the compound hyperforin was detected in low concentrations in mother's milk while the compound hypericin was not detected in milk (Klier et al. 2002, 2006).

Review Details

H

I. Drug and Supplement Interactions

Clinical Trials of Drug or Supplement Interactions

Immunosuppressants

In kidney transplant patients orally administered 600 mg of St. John's wort daily for 2 weeks in addition to their standard dose of cyclosporine, an increase in the clearance of cyclosporine was observed. The maximum plasma concentration of cyclosporine was reduced by an average of 42% (Bauer et al. 2003). Similar effects were observed in healthy volunteers administered 2.5 mg/kg cyclosporine after daily administration of 900 mg St. John's wort for 14 days (Dresser et al. 2003).

In kidney transplant patients orally administered 900 mg daily of a low-hyperforin (0.6 mg hyperforin daily) or regular (42 mg hyperforin daily) St. John's wort preparation in addition to the standard dose of cyclosporine, no clinically relevant changes in cyclosporine levels were observed during the low-hyperforin treatment period. During treatment with the regular St. John's wort extract, a 45% reduction in plasma levels ($AUC_{0\text{-}12}$) of cyclosporine was observed (Mai et al. 2004). Cyclosporine is metabolized by CYP3A4 (Dresser et al. 2003).

In healthy volunteers orally administered 0.1 mg/kg of tacrolimus before or during 18 days of St. John's wort administration at a dose of 900 mg daily, a significant decrease in blood levels of tacrolimus was observed after St. John's wort dosing. Induction of CYP3A4 and P-gp were proposed as the mechanism for this interaction (Hebert et al. 2004).

In kidney transplant patients orally administered 600 mg St. John's wort daily for 2 weeks in addition to the regular regimen of tacrolimus and mycophenolic acid, a significant decrease in plasma levels of tacrolimus was observed. No changes in mycophenolic acid levels were observed (Mai et al. 2003).

No changes in plasma levels of prednisone were observed in healthy volunteers orally administered 900 mg St. John's wort daily for 4 weeks before or after single doses of prednisone (Bell et al. 2007).

Anticoagulants

A significant decrease in plasma levels of (*S*)-warfarin was observed in healthy volunteers orally administered 900 mg daily St. John's wort for 18 days. No change in the efficacy of (*S*)-warfarin was observed (Jiang et al. 2004, 2006). Both (*S*)-warfarin and (*R*)-warfarin were affected (Jiang et al. 2004).

In healthy volunteers orally administered 12 mg phenprocoumon before or after administration of 900 mg St. John's wort daily for 11 days, a significant reduction in plasma levels of phenprocoumon was observed after St. John's wort dosing (Maurer et al. 1999).

Antiarrhythmics

Clinical trials of standard St. John's wort and digoxin showed a significant decrease in plasma levels of digoxin both after a single dose (Durr et al. 2000) and after continued dosing of St. John's wort (Durr et al. 2000; Johne et al. 1999; Mueller et al. 2004). Digoxin is a P-gp substrate (Durr et al. 2000). In trials with low-hyperforin (3.5 mg hyperforin daily) St. John's wort products, the St. John's wort did not affect digoxin blood levels (Gerhard 2005; Mueller et al. 2004).

Calcium channel blockers

In healthy volunteers orally administered 10 mg nifedipine before or after 18 days of 900 mg St. John's wort daily, a significant decrease in the nifedipine plasma concentration was observed after St. John's wort dosing (Smith et al. 2001).

Concomitant use of 900 mg St. John's wort and 120 mg/l (*R*)- and (*S*)-verapamil (administered, via a tube, directly to the small intestine) caused a 77% decrease in the maximum plasma concentration of the (*R*)- and (*S*)-verapamil. Verapamil is primarily metabolized by CYP3A4 (Tannergren et al. 2004).

Anti-anginals

In healthy volunteers orally administered a single oral dose of ivabradine before or after 900 mg St. John's wort daily, a significant decrease in plasma levels of ivabradine was observed after St. John's wort dosing. Ivabradine is metabolized by CYP3A4 (Portoles et al. 2006).

Hormonal contraceptives

In healthy women using a hormonal contraceptive (ethinyl estradiol and norethindrone) and orally administered 900 mg St. John's wort daily for three menstrual cycles, a decrease in serum levels of ethinyl estradiol and norethindrone were observed. Testing with the CYP3A4 substrate midazolam on the last day of St. John's wort treatment confirmed activity of St. John's wort. Serum concentrations of follicle-stimulating hormone, luteinizing hormone, and progesterone were not significantly affected by St. John's wort dosing. Breakthrough bleeding occurred in 2 of 12 women in the control phase compared with 7 of 12 women in the St. John's wort phase. The oral clearance of midazolam after St. John's wort dosing was greater in women who had breakthrough bleeding than in those who did not (Hall et al. 2003). In another study with the same St. John's wort dosing regimen, a 15% decrease in exposure to ethinyl estradiol and norethindrone was observed along with an increase in intracyclic bleeding and evidence of follicle growth (Murphy et al. 2005). Participants in a third trial reported increased intracyclic bleeding as compared with contraceptive alone, but had no significant change in follicle maturation or serum estradiol and progesterone concentrations (Pfrunder et al. 2003).

In healthy volunteers who had taken a low-dose oral contraceptive (0.02 mg ethinylestradiol and 0.15 mg desogestrel) for at least 3 months, oral administration of 500 mg of a low-hyperforin (less than 1 mg hyperforin) St. John's wort extract daily for 14 days produced no significant effects on serum levels of the hormones or their metabolites (Will-Shahab et al. 2009).

Anxiolytics

In healthy volunteers orally administered a single dose of quazepam before or after oral administration of 900 mg St. John's wort daily for 14 days, a significant reduction in plasma levels of quazepam was observed after St. John's wort dosing. In subjective testing of volunteers, no significant effects of St. John's wort on quazepam activity were observed. Quazepam is metabolized by CYP3A4 (Kawaguchi et al. 2004).

A number of human studies have shown that St. John's wort decreases plasma levels of alprazolam (Markowitz et al. 2000, 2003) and midazolam (Dresser et al. 2003; Gurley et al. 2002, 2005; Wang et al. 2001; Xie et al. 2005) in healthy volunteers. These effects are not observed with low hyperforin-containing extracts (Arold et al. 2005) or after a short duration (3 days or less) of treatment (Markowitz et al. 2000).

Antidepressants (tricyclic)

Reduced serum levels of amitriptyline and its metabolite, nortriptyline, were observed in depressed patients taking amitriptyline along with 900 mg St. John's wort daily for 12 to 14 days (Johne et al. 2002).

Antivirals

A significant reduction in serum levels of indinavir was observed after oral administration of 800 mg indinavir to healthy volunteers who had been taking 900 mg St. John's wort daily for 2 weeks, as compared to indinavir administered prior to St. John's wort. Indinavir is metabolized by CYP3A4 (Piscitelli et al. 2000).

In HIV patients being treated with nevirapine, self-treatment with varying doses of St. John's wort led to a reduction in plasma levels of nevirapine. Nevirapine induces CYP3A4 (de Maat et al. 2002). In healthy volunteers orally administered 2 g of St. John's wort as an infusion daily for 14 days, before or after administration of single doses of nevirapine, a significant reduction in the half-life of nevirapine was observed after St. John's wort dosing (L'homme et al. 2006).

Statins

In healthy volunteers orally administered 900 mg St. John's wort daily for 14 days prior to oral administration of 10 mg simvastatin and 20 mg pravastatin, lowered plasma concentrations of simvastatin and its active metabolite were observed. Pravastatin concentrations were not significantly altered. Simvastatin is metabolized by CYP3A4 (Sugimoto et al. 2001).

In patients with hypercholesterolemia treated with a stable dose of atorvastatin (10–40 mg/day) for at least 3 months, administration of 600 mg of St. John's wort daily for 4 weeks resulted in an increased serum level of LDL cholesterol and total cholesterol. No significant changes were observed in HDL cholesterol or in triglycerides (Andrén et al. 2007). Atorvastatin is metabolized by CYP3A4 and is also a substrate for P-glycoprotein.

Anticancer drugs

In cancer patients undergoing treatment with irinotecan (350 mg/m^2, intravenously), oral administration of 900 mg daily of St. John's wort decreased plasma concentrations of the irinotecan active metabolite by 42%. Irinotecan is metabolized by CYP3A4 (Mathijssen et al. 2002).

In healthy volunteers orally administered 400 mg imatinib before and after oral administration of 900 mg St. John's wort daily for 14 days, a significant reduction in plasma levels of imatinib was observed after St. John's wort treatment. Imatinib is metabolized by CYP3A4 (Frye et al. 2004). Similar effects were observed in a second study with imatinib (Smith 2004).

Beta-adrenergic blockers

In healthy volunteers orally administered 900 mg St. John's wort daily for 12 days before or after single doses of talinolol, a significant reduction in plasma levels of talinolol was observed (Schwarz et al. 2007).

Hypoglycemics

In healthy volunteers orally administered 80 mg glicazide with or without 900 mg St. John's wort daily for 15 days, a reduction in plasma levels of glicazide was observed in the St. John's wort group (Xu et al. 2008).

Antiulcer drugs

In healthy volunteers orally administered 900 mg St. John's wort daily for 14 days, before or after single doses of omeprazole, a significant reduction in plasma levels of omeprazole was observed (Wang et al. 2004a).

Skeletal muscle relaxants

In healthy volunteers orally administered 900 mg St. John's wort daily for 28 days before or after single doses of 500 mg chlorzoxazone, a reduction in plasma levels of chlorzoxazone was observed (Gurley et al. 2002, 2005).

Anticonvulsants

In healthy volunteers orally administered 900 mg St. John's wort daily for 14 days, before or after single doses of mephenytoin, a significant increase in excretion of a mephenytoin metabolite was observed after St. John's wort dosing. The effect was observed in volunteers with the CYP2C19 wild genotype, but not in those who are CYP2C19 poor metabolizers (Wang et al. 2004b).

In healthy volunteers orally administered carbamazepine before or after 900 mg St. John's wort daily for 11 days, no changes in plasma levels of carbamazepine were observed. Plasma levels of hypericin and pseudohypericin from St. John's wort were also monitored, and a slight decrease in pseudohypericin levels was observed (Johne et al. 2004).

No significant changes in plasma levels of carbamazepine were observed in healthy volunteers orally administered 900 mg St. John's wort with 400 mg carbamazepine daily for 14 days (Burstein et al. 2000).

Antifungals

In healthy volunteers orally administered 900 mg St. John's wort daily for 15 days, voriconazole was administered before and on days 1 and 15 of St. John's wort consumption. Administration of voriconazole on day 1 resulted in an elevation of voriconazole plasma levels, although the elevation was considered clinically irrelevant. Plasma levels of voriconazole were significantly decreased when administered on day 15 of St. John's wort consumption (Rengelshausen et al. 2005).

Antihistamines

Findings on the use of St. John's wort and fexofenadine, a P-gp substrate, are mixed. In healthy volunteers orally administered single doses of 60 mg fexofenadine before and after a single dose of 900 mg St. John's wort or the same dose repeated for 14 days, the single dose of St. John's wort was found to significantly increase plasma concentrations of fexofenadine. After repeated dosing with St. John's wort, no significant changes in fexofenadine disposition were observed as compared to the non-St. John's wort portion of the experiment (Wang et al. 2002). An increase in oral clearance, with no changes in elimination half-life, was observed in healthy volunteers orally administered a single dose of 60 mg fexofenadine after oral administration of 900 mg St. John's wort daily for 10 days (Xie et al. 2005). A significant increase in oral clearance of fexofenadine was observed in healthy volunteers orally administered 180 mg fexofenadine after oral administration of 900 mg of St. John's wort daily for 14 days (Dresser et al. 2003).

Bronchodilators

No significant changes in plasma levels of theophylline were observed in healthy volunteers orally administered 900 mg daily St. John's wort for 15 days before and after dosing with 400 mg theophylline (Morimoto et al. 2004).

Other

Administration of St. John's wort was associated with significantly reduced methadone plasma levels in patients being treated in a methadone clinic (Eich-Hochli et al. 2003).

Also see Human pharmacological studies for this entry.

Case Reports of Suspected Drug or Supplement Interactions

Immunosuppressants

Cyclosporine and tacrolimus are substrates of the drug-metabolizing isoenzyme CYP3A4. Multiple case reports have demonstrated that St. John's wort decreases blood levels of cyclosporine (Ahmed et al. 2001; Alscher and Klotz 2003; Barone et al. 2000; Beer and Ostermann 2001; Breidenbach et al. 2000a, 2000b; Karliova et al. 2000; Mai et al. 2000; Mandelbaum et al. 2000; Moschella and Jaber 2001; Ruschitzka et al. 2000; Turton-Weeks et al. 2001). St. John's wort consumption for as little as 3 days was correlated with decreased cyclosporine blood levels (Mandelbaum et al. 2000).

One case report noted a decrease in the blood level of tacrolimus in a patient taking St. John's wort (Bolley et al. 2002).

Anticoagulants

One publication reported seven cases of decreased INR (a standardized scale used to report the results of blood coagulation tests; decreased INR indicates accelerated blood clotting) in patients taking warfarin and St. John's wort. Several of the cases reported a return to normal INR levels after cessation of St. John's wort; results of other cases were not reported (Yue et al. 2000).

Hormonal contraceptives

Several cases of intermenstrual bleeding and one case of unwanted pregnancy have been reported in women taking oral contraceptives and St. John's wort (Schwarz et al. 2003; Yue et al. 2000).

Antivirals

In patients taking St. John's wort and nevirapine, mild to moderate increases in oral clearance of nevirapine were recorded (de Maat et al. 2001, 2002).

Antidepressants (SSRI)

Several cases of suspected serotonin syndrome (Barbenel et al. 2000; Demott 1998; Gordon 1998; Lantz et al. 1999; Waksman et al. 2000) and one report of hypomania (Spinella and Eaton 2002) were reported in patients taking both St. John's wort and SSRIs (selective serotonin reuptake inhibitors).

Antipsychotics and anxiolytics

One case of possible serotonin syndrome was reported in a patient taking St. John's wort and buspirone (Dannawi 2002).

Bronchodilators

One case of lowered theophylline level was reported in a patient taking St. John's wort, theophylline, and numerous other drugs. Theophylline level returned to normal after St. John's wort was stopped (Nebel et al. 1999).

Animal Trials of Drug or Supplement Interactions

Animal trials of drug or supplement interactions were identified but omitted due to the availability of human data.

II. Adverse Events

Adverse Events Reported in Clinical Trials

A systematic review of adverse events reported in clinical trials of St. John's wort indicated that data from 35 double-blind randomized trials showed that dropout and adverse event rates in patients receiving St. John's wort extracts were similar to placebo, lower than with older antidepressants, and somewhat lower than with SSRI antidepressants. Dropout rates due to adverse events ranged from 0 to 5.7% in 17 observational studies that included 35,562 patients. No serious adverse events were reported in any of the studies (Knuppel and Linde 2004).

Case Reports of Adverse Events

A 35-year-old woman who had been taking 500 mg St. John's wort daily for 4 weeks developed subacute polyneuropathy after exposure to sunlight. Symptoms were restricted to areas of skin that were exposed to sunlight and abated after St. John's wort was discontinued (Bove 1998).

A 45-year-old woman developed a severe phototoxic reaction to laser treatment after taking St. John's wort (dose, duration, and product not specified). No adverse effects occurred during the same treatment prior to using or after discontinuation of St. John's wort (Cotterill 2001).

A 61-year-old woman developed recurring elevated itching erythematous lesions in light-exposed areas after taking six tablets daily of St. John's wort for 3 years. Routine patch testing did not reveal any relevant reactions and photo patch testing was negative. Using a systemic oral photoprovocation test with St. John's wort, a decrease of the MED-UVB (<0.039 J/cm^2) was observed and was reversible after discontinuation of St. John's wort (Golsch et al. 1997).

A 52-year-old woman with a history of cutaneous lupus developed a severe phototoxic reaction after using St. John's wort oil topically and orally three times daily for 2 weeks (Lane-Brown 2000).

A 63-year-old man with a history of psoriasis had a severe phototoxic reaction within 30 minutes of UVB phototherapy. The man had been taking six capsules of St. John's wort daily for an unspecified amount of time (Lane-Brown 2000).

A 45-year-old woman who had been using a topical St. John's wort cream for 3 weeks developed a phototoxic reaction after spending a day at the beach (Lane-Brown 2000).

Data from the 2002 U.S. National Health Interview Survey indicated a correlation between St. John's wort use and development of cataracts (Booth and McGwin 2009).

Cases of hypomania (Fahmi et al. 2002; Guzelcan et al. 2001; Nierenberg et al. 1999; O'Breasail and Argouarch 1998; Shuster 1999), mania (Moses and Mallinger 2000), psychosis (Laird and Webb 2001; Lal and Iskandar 2000; Shimizu et al. 2004), and delirium (Khawaja et al. 1999) have been associated with the use of St. John's wort. In many of these reports, patients were noted to have psychiatric histories or concomitant diagnoses of psychiatric disorders.

St. John's wort has been associated with hypertension in patients with no previous history of hypertension (Patel et al. 2002; Zullino and Borgeat 2003).

Sexual dysfunction was reported in a patient who had experienced similar dysfunction while on sertraline (Assalian 2000). Diminished libido was reported in a patient with a history of depression and anxiety (Bhopal 2001).

Other adverse events associated with the use of St. John's wort include the following: erythematous eruption (Holme and Roberts 2000), hair loss (Parker et al. 2001), cardiovascular collapse during anesthesia (Irefin and Sprung 2000), serotonin syndrome (Parker et al. 2001), bone marrow necrosis (Demiroglu et al. 2005), reduced TSH levels (Ferko and Levine 2001), and nausea and related physical or mental symptoms (Brown 2000; Dean et al. 2003).

III. Pharmacology and Pharmacokinetics

Human Pharmacological Studies

In trials of commercial St. John's wort extracts, humans ingesting up to 3600 mg of St. John's wort extract (11.25 mg of total hypericin) did not experience phototoxicity (Brockmoller et al. 1997). In patients taking the compound hypericin, doses as low as 0.05 mg/kg were associated with phototoxic reactions (Gulick et al. 1999; Jacobson et al. 2001).

St. John's wort ingestion has been associated with changes in P-gp expression. Studies have shown that single doses of St. John's wort inhibit P-gp expression (Hamman et al. 2001; Wang et al. 2002) but that doses taken over multiple days induce P-gp expression (Durr et al. 2000; Hennessy et al. 2002; Johne et al. 1999; Mueller et al. 2004; Schwarz et al. 2002; Wang et al. 2002; Xie et al. 2005). Conversely, one study showed that multiple doses had no effect on P-gp expression (Hamman et al. 2001). A hyperforin-free St. John's wort extract was shown to have a nonsignificant effect on P-gp expression as compared to an extract containing hyperforin (Mueller et al. 2004).

Multiple studies have demonstrated that St. John's wort significantly induces the drug-metabolizing enzyme CYP3A4 (Bauer et al. 2002; Dresser et al. 2003; Durr et al. 2000; Frye et al. 2004; Gurley et al. 2002; Kawaguchi et al. 2004; Markowitz et al. 2003; Roby et al. 2000; Wang et al. 2001; Xie et al. 2005). Clearance of probe drugs such as midazolam, cyclosporine, and imatinib increased by 40 to 90% (Dresser et al. 2003; Frye et al. 2004; Gurley et al. 2002; Wang et al. 2001). Most studies were conducted with a dose of 300 mg St. John's wort three times daily for 2 weeks.

In healthy men orally administered 5 mg of midazolam before or during 900 mg daily of St. John's wort, clearance of midazolam increased significantly during St. John's wort and returned to normal 7 days after cessation of St. John's wort (Imai et al. 2008).

The effects of St. John's wort ingestion on CYP2C19 were shown to vary from no significant effect (Burstein et al. 2000) to significant induction (Rengelshausen et al. 2005; Wang et al. 2004a, 2004b). Among the studies that showed induction, two indicated that all 2C19 genotypes (including wild and mutant) produced similar results (Wang et al. 2004a, 2004b), while a third study indicated that individuals with wild-type CYP2C19 were more affected than others (Rengelshausen et al. 2005).

In two studies, St. John's wort ingestion significantly induced the activity of CYP2E1 (Gurley et al. 2002, 2005).

Trials gave conflicting results on the effect of St. John's wort on CYP1A2 and CYP2D6 activity. Results indicated no effect (Gerhard 2005; Gurley et al. 2008; Markowitz et al. 2000; Wang et al. 2004b) and significant induction (Gurley et al. 2002).

In one study, St. John's wort demonstrated nonsignificant induction of CYP2C9 (Gerhard 2005).

In healthy volunteers orally administered St. John's wort capsules at the manufacturer's recommended dose (amount not specified) daily for two weeks, no effects on platelet function or other hematological parameters were observed, including prothrombin time, partial thromboplastin time, thrombin time, bleeding time, the collagen/epinephrine assay, or the collagen/adenosine diphosphate assay. Aspirin (325 mg daily) was used as a positive control and markedly inhibited platelet function (Beckert et al. 2007).

Animal Pharmacological Studies

Animal pharmacological studies were identified but omitted due to the availability of human data.

In Vitro Pharmacological Studies

In vitro pharmacological studies were identified but omitted due to the availability of human data.

IV. Pregnancy and Lactation

A number of studies on the effect of prenatal consumption of St. John's wort on pregnancy and offspring development in mice and rats were associated with normal gestation and offspring development (Borges et al. 2005; Cada et al. 2001; Ferguson et al. 1999; Rayburn et al. 2000, 2001a, 2001b). Morphological abnormalities were noted in rat embryos excised and exposed to high concentrations of hypericin (71.0 or 142.0 ng/ml) (Chan et al. 2001). One study noted hepatic and renal lesions in lactating rats that had ingested 100 or 1000 mg/kg/day St. John's wort (Gregoretti et al. 2004).

Two cases of healthy human pregnancy and baby were reported in mothers taking St. John's wort (Grush et al. 1998).

In nursing mothers taking 900 mg St. John's wort daily, levels of hypericin and hyperforin in mothers' milk and in mother and infant plasma were as follows: in mother's milk, hypericin was below quantification limits, hyperforin ranged from 0.58 to 18.20 ng/ml; in mother's plasma, hypericin was 10.71 ng/ml; hyperforin was 22.8–151 ng/ml; and in infant plasma, hypericin was below quantification limit and hyperforin was from below detection limits to 0.1 ng/ml (Klier et al. 2002, 2006). In a prospective study of breastfeeding mothers, no significant difference between the St. John's wort and control groups were reported (Lee et al. 2003).

V. Toxicity Studies

Acute Toxicity

The LD_{50} of orally administered St. John's wort extract (up to 5% in olive oil) could not be determined at doses up to 20 ml/kg (Fiume 2001).

The LD_{50} of intraperitoneally administered extract fractions of St. John's wort administered to mice is 780 mg/kg for the polyphenol fraction, 4300 mg/kg for the lipophile fraction, and 2800 mg/kg for the water-soluble fraction (Fiume 2001; Yevstifeyeva and Sibiryak 1996).

Short-Term Toxicity

Sheep fed St. John's wort flowers at doses of 4 to 16 g/kg for 14 days exhibited some signs of toxicity including restlessness, photophobia, tachycardia, and erythema of the exposed parts of the tail and legs (Fiume 2001; Kako et al. 1993).

Chronic Toxicity

In rats fed St. John's wort as 5% of their daily diet, significant weight loss was reported as compared to control. No other toxicities were reported (Garrett et al. 1982).

Genotoxicity

St. John's wort (*Hypericum perforatum*) contains hypericin and hypericin-like substances as well as flavonoids, of which particularly quercetin has generated a widespread controversial discussion with respect to mutagenic action.

No mutagenic activity of a hydroethanolic extract of St. John's wort was observed in in vivo and in vitro studies, including the fur spot test in mice, the chromosome aberration test with bone marrow cells of the Chinese hamster, the hypoxanthine guanidine phosphoribosyltransferase test, the cell transformation test using Syrian hamster embryo cells, and the unscheduled DNA synthesis test (Okpanyi et al. 1990).

In the Ames test for mutagenicity with *Salmonella typhimurium* strains TA98 and TA100 with or without metabolic activation from S9, a tincture of St. John's wort increased the number of revertants in TA98 with and without metabolic activation and in TA100 with metabolic activation (Goggelmann and Schimmer 1986).

Ethanol, chloroform, and ethyl acetate extracts of St. John's wort were assayed at concentrations of 10 or 40 µl in the Ames test for mutagenicity with *Salmonella typhimurium* strains TA98 and TA100 with or without metabolic activation by S9. The ethanol and ethyl acetate extracts had mutagenic activity with and without metabolic activation. Testing of fractions of the extracts indicated that mutagenicity of the full extract was found exclusively in quercetin and that hypericin was not mutagenic (Poginsky et al. 1988). The compound quercetin is recognized to have mutagenic activity in vitro but is regarded as safe in humans (Harwood et al. 2007).

Literature Cited

Ahmed, S.M., N.R. Banner, and S.W. Dubrey. 2001. Low cyclosporin-A level due to Saint John's wort in heart transplant patients. *J. Heart Lung Transplant.* 20(7):795.

Alscher, D.M., and U. Klotz. 2003. Drug interaction of herbal tea containing St. John's wort with cyclosporine. *Transplant. Int.* 16(7):543-544.

Andrén, L., Å. Andreasson, and R. Eggertsen. 2007. Interaction between a commercially available St. John's wort product (Movina) and atorvastatin in patients with hypercholesterolemia. *Eur. J. Clin. Pharmacol.* 63(10):913-916.

Arold, G., F. Donath, A. Maurer, et al. 2005. No relevant interaction with alprazolam, caffeine, tolbutamide, and digoxin by treatment with a low-hyperforin St. John's wort extract. *Planta Med.* 71(4):331-337.

Assalian, P. 2000. Sildenafil for St. John wort-induced sexual dysfunction. *J. Sex Marital Ther.* 26(4):357-358.

Barbenel, D.M., B. Yusufi, D. O'Shea, and C.J. Bench. 2000. Mania in a patient receiving testosterone replacement postorchidectomy taking St. John's wort and sertraline. *J. Psychopharmacol.* 14(1):84-86.

Barone, G.W., B.J. Gurley, B.L. Ketel, M.L. Lightfoot, and S.R. Abul-Ezz. 2000. Drug interaction between St. John's wort and cyclosporine. *Ann. Pharmacother.* 34(9):1013.

Bauer, S., E. Starmer, A. Johne, et al. 2003. Alterations in cyclosporin A pharmacokinetics and metabolism during treatment with St. John's wort in renal transplant patients. *Br. J. Clin. Pharmacol.* 55(2):203-211.

Bauer, S., E. Stormer, R. Kerb, et al. 2002. Differential effects of Saint John's wort (*Hypericum perforatum*) on the urinary excretion of D-glucaric acid and 6β-hydroxycortisol in healthy volunteers. *Eur. J. Clin. Pharmacol.* 58(9):581-585.

Beattie, P.E., R.S. Dawe, N.J. Traynor, et al. 2005. Can St. John's wort (hypericin) ingestion enhance the erythemal response during high-dose ultraviolet A1 therapy? *Br. J. Dermatol.* 153(6):1187-1191.

Beckert, B.W., M.J. Concannon, S.L. Henry, D.S. Smith, and C.L. Puckett. 2007. The effect of herbal medicines on platelet function: An in vivo experiment and review of the literature. *Plast. Reconstr. Surg.* 120 (7):2044-2050.

Beer, A.M., and T. Ostermann. 2001. St. John's wort: Interaction with cyclosporine increases risk of rejection for the kidney transplant and raises daily cost of medication. *Med. Klin.* 96(8):480-483.

Bell, E.C., W.R. Ravis, H.M. Chan, and Y.J. Lin. 2007. Lack of pharmacokinetic interaction between St. John's wort and prednisone. *Ann. Pharmacother.* 41(11):1819.

Bhopal, J.S. 2001. St. John's wort-induced sexual dysfunction. *Can. J. Psychiatr. Rev. Can. Psychiatr.* 46(5):456-457.

Bolley, R., C. Zulke, M. Kammerl, M. Fischereder, and B.K. Kramer. 2002. Tacrolimus-induced nephrotoxicity unmasked by induction of the CYP3A4 system with St. John's wort. *Transplantation* 73 (6):1009.

Booth, J.N., 3rd, and G. McGwin. 2009. The association between self-reported cataracts and St. John's wort. *Curr. Eye Res.* 34(10):863-866.

Borges, L.V., J.C. do Carmo Cancino, V.M. Peters, L. Las Casas, and M. de Oliveira Guerra. 2005. Development of pregnancy in rats treated with *Hypericum perforatum*. *Phytother. Res.* 19(10):885-887.

Borrelli, F., and A.A. Izzo. 2009. Herb-drug interactions with St. John's wort (*Hypericum perforatum*): An update on clinical observations. *AAPS J.* 11(4):710-727.

Bove, G.M. 1998. Acute neuropathy after exposure to sun in a patient treated with St. John's wort. *Lancet* 352(Oct.):1121-1122.

Breidenbach, T., M.W. Hoffmann, T. Becker, H. Schlitt, and J. Klempnauer. 2000a. Drug interaction of St. John's wort with cyclosporin [comment]. *Lancet* 355(9218):1912.

Breidenbach, T., V. Kliem, M. Burg, et al. 2000b. Profound drop of cyclosporin A whole blood trough levels caused by St. John's wort (*Hypericum perforatum*). *Transplantation* 69(10):2229-2230.

Brockmoller, J., T. Reum, S. Bauer, et al. 1997. Hypericin and pseudohypericin: Pharmacokinetics and effects on photosensitivity in humans. *Pharmacopsychiatry* 30(Suppl. 2):94-101.

Brown, T.M. 2000. Acute St. John's wort toxicity. *Am. J. Emerg. Med.* 18(2):231-232.

Burstein, A.H., R.L. Horton, T. Dunn, et al. 2000. Lack of effect of St. John's wort on carbamazepine pharmacokinetics in healthy volunteers. *Clin. Pharmacol. Ther.* 68(6):605-612.

Cada, A.M., D.K. Hansen, J.B. LaBorde, and S.A. Ferguson. 2001. Minimal effects from developmental exposure to St. John's wort (*Hypericum perforatum*) in Sprague-Dawley rats. *Nutr. Neurosci.* 4(2):135-141.

Chan, L., P. Chiu, and T. Lau. 2001. A study of hypericin-induced teratogenicity during organogenesis using a whole rat embryo culture model. *Fertil. Steril.* 76(5):1073-1074.

Cotterill, J.A. 2001. Severe phototoxic reaction to laser treatment in a patient taking St. John's wort. *J. Cosmet. Laser Ther.* 3(3):159-160.

Dannawi, M. 2002. Possible serotonin syndrome after combination of buspirone and St. John's wort. *J. Psychopharmacol.* 16(4):401.

de Maat, M.M., R.M. Hoetelmans, R.A. Math, et al. 2001. Drug interaction between St. John's wort and nevirapine. *AIDS* 15(3):420-421.

de Maat, M.M.R., A.D.R. Huitema, J.W. Mulder, et al. 2002. Population pharmacokinetics of nevirapine in an unselected cohort of HIV-1-infected individuals. *Br. J. Clin. Pharmacol.* 54(4):378-385.

Dean, A.J., G.M. Moses, and J.M. Vernon. 2003. Suspected withdrawal syndrome after cessation of St. John's wort. *Ann. Pharmacother.* 37(1):150.

Demiroglu, Y.Z., T.T. Yeter, C. Boga, et al. 2005. Bone marrow necrosis: A rare complication of herbal treatment with *Hypericum perforatum* (St. John's wort). *Acta Med. (Hradec Kralove)* 48(2):91-94.

Demott, K. 1998. St. John's wort tied to serotonin syndrome. *Clin. Psychiatr. News* 26:28.

Dresser, G.K., U.I. Schwarz, G.R. Wilkinson, and R.B. Kim. 2003. Coordinate induction of both cytochrome P4503A and MDR1 by St. John's wort in healthy subjects. *Clin. Pharmacol. Ther.* 73(1):41-50.

Durr, D., B. Stieger, G.A. Kullak-Ublick, et al. 2000. St. John's wort induces intestinal P-glycoprotein/MDR1 and intestinal and hepatic CYP3A4. *Clin. Pharmacol. Ther.* 68(6):598-604.

Eich-Hochli, D., R. Oppliger, K.P. Golay, P. Baumann, and C.B. Eap. 2003. Methadone maintenance treatment and St. John's wort—A case report. *Pharmacopsychiatry* 36(1):35-37.

Fahmi, M., C. Huang, and I. Schweitzer. 2002. A case of mania induced by *Hypericum*. *World J. Biol. Psychiatr.* 3(1):58-59.

Ferguson, S.A., A.M. Cada, E.P. Gray, and D.K. Hansen. 1999. Developmental treatment with St. John's wort (SJW) results in minimal neurobehavioral toxicity in rats. *Abstr. Soc. Neurosci.* 25(Pt. 2):1826.

Ferko, N., and M.A. Levine. 2001. Evaluation of the association between St. John's wort and elevated thyroid-stimulating hormone. *Pharmacotherapy* 21(12):1574-1578.

Fiume, M.Z. 2001. Final report on the safety assessment of *Hypericum perforatum* extract and *Hypericum perforatum* oil. *Int. J. Toxicol.* 20(Suppl. 2):31-39.

Frye, R.F., S.M. Fitzgerald, T.F. Lagattuta, M.W. Hruska, and M.J. Egorin. 2004. Effect of St. John's wort on imatinib mesylate pharmacokinetics. *Clin. Pharmacol. Ther.* 76(4):323-329.

Garrett, B.J., P.R. Cheeke, C.L. Miranda, D.E. Goeger, and D.R. Buhler. 1982. Consumption of poisonous plants (*Senecio jacobaea, Symphytum officinale, Pteridium aquilinum, Hypericum perforatum*) by rats: Chronic toxicity, mineral metabolism, and hepatic drug-metabolizing enzymes. *Toxicol. Lett.* 10:2-3.

Gerhard, A. 2005. No relevant interaction with alprazolam, caffeine, tolbutamide, and digoxin by treatment with a low-hyperforin St. John's wort extract. *Planta Med.* 71(4):331-337.

Goggelmann, W., and C. Schimmer. 1986. Mutagenic activity of phytotherapeutical drugs. *Prog. Clin. Biol. Res.* 206:63-72.

Golsch, S., E. Vocks, J. Rakoski, K. Brockow, and J. Ring. 1997. Reversible increase in photosensitivity to UV-B caused by St. John's wort extract. *Hautarzt* 48(4):249-252.

Gordon, J.B. 1998. SSRIs and St. John's wort: Possible toxicity? *Am. Fam. Physician* 57(5):950, 953.

Gregoretti, B., M. Stebel, L. Candussio, et al. 2004. Toxicity of *Hypericum perforatum* (St. John's wort) administered during pregnancy and lactation in rats. *Toxicol. Appl. Pharmacol.* 200(3):201-205.

Grush, L.R., A. Nierenberg, B. Keefe, and L.S. Cohen. 1998. St. John's wort during pregnancy. *J. Am. Med. Assoc.* 280(18):1566.

Gulick, R.M., V. McAuliffe, J. Holden-Wiltse, et al. 1999. Phase I studies of hypericin, the active compound in St. John's wort, as an antiretroviral agent in HIV-infected adults. AIDS Clinical Trials Group Protocols 150 and 258. *Ann. Intern. Med.* 130(6):510-514.

Gurley, B.J., S.F. Gardner, M.A. Hubbard, et al. 2002. Cytochrome P450 phenotypic ratios for predicting herb-drug interactions in humans. *Clin. Pharmacol. Ther.* 72(3):276-287.

Gurley, B.J., S.F. Gardner, M.A. Hubbard, et al. 2005. Clinical assessment of effects of botanical supplementation on cytochrome P450 phenotypes in the elderly: St. John's wort, garlic oil, *Panax ginseng* and *Ginkgo biloba*. *Drugs Aging* 22(6):525-539.

Gurley, B.J., A. Swain, M.A. Hubbard, et al. 2008. Clinical assessment of CYP2D6-mediated herb-drug interactions in humans: Effects of milk thistle, black cohosh, goldenseal, kava kava, St. John's wort, and echinacea. *Mol. Nutr. Food Res.* 52(7):755-763.

Guzelcan, Y., W.F. Scholte, J. Assies, and H.E. Becker. 2001. Mania during the use of a combination preparation with St. John's wort (*Hypericum perforatum*). *Ned. Tijdschr. Geneeskd.* 145(40):1943-1945.

Hall, S.D., Z. Wang, S.M. Huang, et al. 2003. The interaction between St. John's wort and an oral contraceptive. *Clin. Pharmacol. Ther.* 74(6):525-535.

Hamman, M., Z. Wang, P. Honig, et al. 2001. Effects of acute and chronic Saint John's wort (SJW) administration on fexofenadine (FEX) disposition PII-83 [abstract]. *Clin. Pharmacol. Ther.* 69(2).

Harwood, M., B. Danielewska-Nikiel, J.F. Borzelleca, et al. 2007. A critical review of the data related to the safety of quercetin and lack of evidence of in vivo toxicity, including lack of genotoxic/carcinogenic properties. *Food Chem. Toxicol.* 45(11):2179-2205.

Hebert, M.F., M.P. Jeong, C. Yu-Luan, A. Shahzad, and M.L. Anne. 2004. Effects of St. John's wort (*Hypericum perforatum*) on tacrolimus pharmacokinetics in healthy volunteers. *J. Clin. Pharmacol.* 44(1):89-94.

Hennessy, M., D. Kelleher, J.P. Spiers, et al. 2002. St. John's wort increases expression of P-glycoprotein: Implications for drug interactions. *Br. J. Clin. Pharmacol.* 53(1):75-82.

Holme, S.A., and D.L. Roberts. 2000. Erythroderma associated with St. John's wort. *Br. J. Dermatol.* 143(5):1127-1128.

Imai, H., T. Kotegawa, K. Tsutsumi, et al. 2008. The recovery time-course of CYP3A after induction by St. John's wort administration. *Br. J. Clin. Pharmacol.* 65(5):701-707.

Irefin, S., and J. Sprung. 2000. A possible cause of cardiovascular collapse during anesthesia: Long-term use of St. John's wort. *J. Clin. Anesth.* 12(6):498-499.

Jacobson, J.M., L. Feinman, L. Liebes, et al. 2001. Pharmacokinetics, safety, and antiviral effects of hypericin, a derivative of St. John's wort plant, in patients with chronic hepatitis C virus infection. *Antimicrob. Agents Chemother.* 45(2):517-524.

Jiang, X., E. Blair, and A.J. McLachlan. 2006. Investigation of the effects of herbal medicines on warfarin response in healthy subjects: A population pharmacokinetic-pharmacodynamic modeling approach. *J. Clin. Pharmacol.* 46(11):1370-1378.

Jiang, X., K.M. Williams, W.S. Liauw, et al. 2004. Effect of St. John's wort and ginseng on the pharmacokinetics and pharmacodynamics of warfarin in healthy subjects. *Br. J. Clin. Pharmacol.* 57(5):592-599.

Johne, A., J. Brockmoller, S. Bauer, et al. 1999. Pharmacokinetic interaction of digoxin with an herbal extract from St. John's wort (*Hypericum perforatum*). *Clin. Pharmacol. Ther.* 66(Oct.):338-345.

Johne, A., E. Perloff, S. Bauer, et al. 2004. Impact of cytochrome P-450 inhibition by cimetidine and induction by carbamazepine on the kinetics of hypericin and pseudohypericin in healthy volunteers. *Eur. J. Clin. Pharmacol.* 60(9):617-622.

Johne, A., J. Schmider, J. Brockmoller, et al. 2002. Decreased plasma levels of amitriptyline and its metabolites on comedication with an extract from St. John's wort (*Hypericum perforatum*). *J. Clin. Psychopharmacol.* 22(1):46-54.

Kako, M., M. Al-Sultan, and N. Saleem. 1993. Studies of sheep experimentally poisoned with *Hypericum perforatum. Vet. Human Toxicol.* 35(4 Suppl.):298-300.

Karliova, M., U. Treichel, M. Malago, et al. 2000. Interaction of *Hypericum perforatum* (St. John's wort) with cyclosporin A metabolism in a patient after liver transplantation. *J. Hepatol.* 33(5):853-855.

Kawaguchi, A., M. Ohmori, S. Tsuruoka, et al. 2004. Drug interaction between St. John's wort and quazepam. *Br. J. Clin. Pharmacol.* 58(4):403-410.

Khawaja, I.S., R.F. Marotta, and S. Lippmann. 1999. Herbal medicines as a factor in delirium. *Psychiatr. Serv.* 50(7):969-970.

Klier, C.M., M.R. Schafer, B. Schmid-Siegel, G. Lenz, and M. Mannel. 2002. St. John's wort (*Hypericum perforatum*)—Is it safe during breastfeeding? *Pharmacopsychiatry* 35(1):29-30.

Klier, C.M., B. Schmid-Siegel, M.R. Schafer, et al. 2006. St. John's wort (*Hypericum perforatum*) and breastfeeding: Plasma and breast milk concentrations of hyperforin for 5 mothers and 2 infants. *J. Clin. Psychiatr.* 67(2):305-309.

Knuppel, L., and K. Linde. 2004. Adverse effects of St. John's wort: A systematic review. *J. Clin. Psychiatr.* 65(11):1470-1479.

L'homme, R.F., T. Dijkema, A.J. van der Ven, and D.M. Burger. 2006. Brief report: Enzyme inducers reduce elimination half-life after a single dose of nevirapine in healthy women. *J. Acquir. Immune Defic. Syndr.* 43:193-196.

Laird, R.D., and M. Webb. 2001. Psychotic episode during use of St. John's wort. *J. Herbal Pharmacother.* 1(2):81-87.

Lal, S., and H. Iskandar. 2000. St. John's wort and schizophrenia. *Can. Med. Assoc. J.* 163(3):262-263.

Lane-Brown, M.M. 2000. Photosensitivity associated with herbal preparations of St. John's wort (*Hypericum perforatum*). *Med. J. Austr.* 172(6):302.

Lantz, M., E. Buchalter, and V. Giambanco. 1999. St. John's wort and antidepressant drug interactions in the elderly. *J. Geriatr. Psychiatr. Neurol.* 12(1):7-10.

Lee, A., R. Minhas, N. Matsuda, M. Lam, and S. Ito. 2003. The safety of St. John's wort (*Hypericum perforatum*) during breastfeeding. *J. Clin. Psychiatr.* 64(8):966-968.

Madabushi, R., B. Frank, B. Drewelow, H. Derendorf, and V. Butterweck. 2006. Hyperforin in St. John's wort drug interactions. *Eur. J. Clin. Pharmacol.* 62(3):225-233.

Mai, I., S. Bauer, E.S. Perloff, et al. 2004. Hyperforin content determines the magnitude of the St. John's wort-cyclosporine drug interaction. *Clin. Pharmacol. Ther.* 76(4):330-340.

Mai, I., K. Budde, E. Starmer, et al. 2003. Impact of St. John's wort treatment on the pharmacokinetics of tacrolimus and mycophenolic acid in renal transplant patients. *Nephrol. Dial. Transplant.* 18(4):819-822.

Mai, I., H. Kruger, K. Budde, et al. 2000. Hazardous pharmacokinetic interaction of Saint John's wort (*Hypericum perforatum*) with the immunosuppressant cyclosporin. *Int. J. Clin. Pharmacol. Ther.* 38(10):500-502.

Mandelbaum, A., F. Pertzborn, M. Martin-Facklam, and M. Wiesel. 2000. Unexplained decrease of cyclosporin trough levels in a compliant renal transplant patient. *Nephrol. Dial. Transplant.* 15(9):1473-1474.

Markowitz, J.S., C.L. DeVane, D.W. Boulton, et al. 2000. Effect of St. John's wort (*Hypericum perforatum*) on cytochrome P-450 2D6 and 3A4 activity in healthy volunteers. *Life Sci.* 66(9):PL133-139.

Markowitz, J.S., J.L. Donovan, C.L. DeVane, et al. 2003. Effect of St. John's wort on drug metabolism by induction of cytochrome P450 3A4 enzyme. *J. Am. Med. Assoc.* 290(11):1500-1504.

Martindale, W., and J.E.F. Reynolds. 1996. *The extra pharmacopoeia*. 31st ed. London: Pharmaceutical Press.

Mathijssen, R.H., J. Verweij, P. de Bruijn, W.J. Loos, and A. Sparreboom. 2002. Effects of St. John's wort on irinotecan metabolism. *J. Natl. Cancer Inst.* 94(16):1247-1249.

Maurer, A., A. Johne, and S. Bauer. 1999. Interaction of St. John's wort extract with phenprocoumon [abstract]. *Eur. J. Clin. Pharmacol.* 55:A22.

Morimoto, T., T. Kotegawa, K. Tsutsumi, et al. 2004. Effect of St. John's wort on the pharmacokinetics of theophylline in healthy volunteers. *J. Clin. Pharmacol.* 44(1):95-101.

Moschella, C., and B.L. Jaber. 2001. Interaction between cyclosporine and *Hypericum perforatum* (St. John's wort) after organ transplantation. *Am. J. Kidney Dis.* 38(5):1105-1107.

Moses, E.L., and A.G. Mallinger. 2000. St. John's wort: Three cases of possible mania induction. *J. Clin. Psychopharmacol.* 20(1):115-117.

Mueller, S.C., J. Majcher-Peszynska, B. Uehleke, et al. 2006. The extent of induction of CYP3A by St. John's wort varies among products and is linked to hyperforin dose. *Eur. J. Clin. Pharmacol.* 62(1):29-36.

Mueller, S.C., B. Uehleke, H. Woehling, et al. 2004. Effect of St. John's wort dose and preparations on the pharmacokinetics of digoxin. *Clin. Pharmacol. Ther.* 75(6):546-557.

Murphy, P.A., S.E. Kern, F.Z. Stanczyk, and C.L. Westhoff. 2005. Interaction of St. John's wort with oral contraceptives: Effects on the pharmacokinetics of norethindrone and ethinyl estradiol, ovarian activity and breakthrough bleeding. *Contraception* 71(6):402-408.

Nebel, A., B.J. Schneider, R.K. Baker, and D.J. Kroll. 1999. Potential metabolic interaction between St. John's wort and theophylline. *Ann. Pharmacother.* 33(4):502.

Nierenberg, A.A., T. Burt, J. Matthews, and A.P. Weiss. 1999. Mania associated with St. John's wort. *Biol. Psychiatr.* 46(12):1707-1708.

O'Breasail, A.M., and S. Argouarch. 1998. Hypomania and St. John's wort. *Can. J. Psychiatr.* 43(7):746-747.

Okpanyi, S.N., H. Lidzba, B.C. Scholl, and H.G. Miltenburger. 1990. [Genotoxicity of a standardized *Hypericum* extract.] *Arzneimittelforschung* 40(8):851-855.

Parker, V., A.H. Wong, H.S. Boon, and M.V. Seeman. 2001. Adverse reactions to St. John's wort. *Can. J. Psychiatr.* 46(1):77-79.

Patel, S., R. Robinson, and M. Burk. 2002. Hypertensive crisis associated with St. John's wort. *Am. J. Med.* 112(6):507-508.

Pfrunder, A., M. Schiesser, S. Gerber, et al. 2003. Interaction of St. John's wort with low-dose oral contraceptive therapy: A randomized controlled trial. *Br. J. Clin. Pharmacol.* 56(6):683-690.

Piscitelli, S.C., A.H. Burstein, D. Chaitt, R.M. Alfaro, and J. Falloon. 2000. Indinavir concentrations and St. John's wort [erratum in *Lancet* 2001 Apr. 14; 357(9263):1210]. *Lancet* 355(9203):547-548.

Poginsky, B., J. Westendorf, N. Prosenc, M. Kuppe, and H. Marquardt. 1988. St. John's wort (*Hypericum perforatum* L.) genotoxicity induced by quercetin content. *Dtsch. Apoth. Ztg.* 128:13464-13466.

Portoles, A., A. Terleira, A. Calvo, I. Martinez, and G. Resplandy. 2006. Effects of *Hypericum perforatum* on ivabradine pharmacokinetics in healthy volunteers: An open-label, pharmacokinetic interaction clinical trial. *J. Clin. Pharmacol.* 46(10):1188-1194.

Rayburn, W.F., H.D. Christensen, and C.L. Gonzalez. 2000. Effect of antenatal exposure to Saint John's wort (*Hypericum*) on neurobehavior of developing mice. *Am. J. Obstet. Gynecol.* 183(5):1225-1231.

Rayburn, W.F., C.L. Gonzalez, H.D. Christensen, T.L. Harkins, and T.C. Kupiec. 2001a. Impact of *Hypericum* (St.-John's-wort) given prenatally on cognition of mice offspring. *Neurotoxicol. Teratol.* 23(6):629-637.

Rayburn, W.F., C.L. Gonzalez, H.D. Christensen, and J.D. Stewart. 2001b. Effect of prenatally administered *Hypericum* (St. John's wort) on growth and physical maturation of mouse offspring. *Am. J. Obstet. Gynecol.* 184(2):191-195.

Rengelshausen, J., M. Banfield, K. Riedel, et al. 2005. Opposite effects of short-term and long-term St. John's wort intake on voriconazole pharmacokinetics. *Clin. Pharmacol. Ther.* 78(1):25-33.

Roby, C.A., G.D. Anderson, E. Kantor, D.A. Dryer, and A.H. Burstein. 2000. St. John's wort: Effect on CYP3A4 activity. *Clin. Pharmacol. Ther.* 67(5):451-457.

Ruschitzka, F., P.J. Meier, M. Turina, T.F. Luscher, and G. Noll. 2000. Acute heart transplant rejection due to Saint John's wort [see comment]. *Lancet* 355(9203):548-549.

Schempp, C.M., K. Muller, B. Winghofer, J. Schulte-Monting, and J.C. Simon. 2001. Single-dose and steady-state administration of *Hypericum perforatum* extract (St. John's wort) does not influence skin sensitivity to UV radiation, visible light, and solar-simulated radiation. *Arch. Dermatol.* 137(4):512-513.

Schempp, C.M., B. Winghofer, K. Muller, et al. 2003. Effect of oral administration of *Hypericum perforatum* extract (St. John's wort) on skin erythema and pigmentation induced by UVB, UVA, visible light and solar simulated radiation. *Phytother. Res.* 17(2):141-146.

Schulz, V. 2001. Incidence and clinical relevance of the interactions and side effects of *Hypericum* preparations. *Phytomedicine* 8(2):152-160.

Schwarz, U., H. Hanso, G. Dresser, et al. 2002. St. John's wort reduces oral bioavailability of talinolol in healthy volunteers. *Clin. Pharmacol. Ther.* 71(2):33.

Schwarz, U.I., B. Buschel, and W. Kirch. 2003. Unwanted pregnancy on self-medication with St. John's wort despite hormonal contraception. *Br. J. Clin. Pharmacol.* 55(1):112-113.

Schwarz, U.I., H. Hanso, R. Oertel, et al. 2007. Induction of intestinal P-glycoprotein by St. John's wort reduces the oral bioavailability of talinolol. *Clin. Pharmacol. Ther.* 81(5):669-678.

Shimizu, K., M. Nakamura, K. Isse, and P.J. Nathan. 2004. First-episode psychosis after taking an extract of *Hypericum perforatum* (St. John's wort). *Human Psychopharmacol.* 19(4):275-276.

Shuster, J. 1999. Hypomanic episode with St. John's wort. *Hosp. Pharm.* 34(Jun.):693-694.

Smith, M., K. Lin, and Y. Zheng. 2001. An open trial of nifedipine-herb interactions: Nifedipine with St. John's wort, ginseng or *Ginko biloba*. *Clin. Pharmacol. Ther.* 69(2):abstr. PIII-89.

Smith, P. 2004. The influence of St. John's wort on the pharmacokinetics and protein binding of imatinib mesylate. *Pharmacotherapy* 24(11):1508-1514.

Spinella, M., and L.A. Eaton. 2002. Hypomania induced by herbal and pharmaceutical psychotropic medicines following mild traumatic brain injury. *Brain Injury* 16(4):359-367.

Sugimoto, K., M. Ohmori, S. Tsuruoka, et al. 2001. Different effects of St. John's wort on the pharmacokinetics of simvastatin and pravastatin. *Clin. Pharmacol. Ther.* 70(6):518-524.

Tannergren, C., H. Engman, L. Knutson, et al. 2004. St. John's wort decreases the bioavailability of *R*- and *S*-verapamil through induction of the first-pass metabolism. *Clin. Pharmacol. Ther.* 75(4):298-309.

Turton-Weeks, S.M., G.W. Barone, B.J. Gurley, et al. 2001. St. John's wort: A hidden risk for transplant patients. *Prog. Transplant.* 11(2):116-120.

Waksman, J., K. Heard, and H. Joliff. 2000. Serotonin syndrome associated with the use of St. John's wort (*Hypericum perforatum*) and paroxetine [abstract]. *J. Toxicol. Clin. Toxicol.* 38:521.

Wang, L.S., G. Zhou, B. Zhu, et al. 2004a. St. John's wort induces both cytochrome P450 3A4-catalyzed sulfoxidation and 2C19-dependent hydroxylation of omeprazole. *Clin. Pharmacol. Ther.* 75(3):191-197.

Wang, L.S., B. Zhu, A.M. El-Aty, et al. 2004b. The influence of St. John's wort on CYP2C19 activity with respect to genotype. *J. Clin. Pharmacol.* 44(6):577-581.

Wang, Z., J.C. Gorski, M.A. Hamman, et al. 2001. The effects of St. John's wort (*Hypericum perforatum*) on human cytochrome P450 activity. *Clin. Pharmacol. Ther.* 70(4):317-326.

Wang, Z., M.A. Hamman, S.M. Huang, L.J. Lesko, and S.D. Hall. 2002. Effect of St. John's wort on the pharmacokinetics of fexofenadine. *Clin. Pharmacol. Ther.* 71(6):414-420.

Weiss, R.F., and A.R. Meuss. 2001. *Weiss's herbal medicine*. Classic ed. New York: Thieme.

Wenk, M., L. Todesco, and S. Krahenbuhl. 2004. Effect of St. John's wort on the activities of CYP1A2, CYP3A4, CYP2D6, N-acetyltransferase 2, and xanthine oxidase in healthy males and females. *Br. J. Clin. Pharmacol.* 57(4):495-499.

Wichtl, M. 2004. *Herbal drugs and phytopharmaceuticals: A handbook for practice on a scientific basis.* 3rd ed. Boca Raton, FL: CRC Press.

Will-Shahab, L., S. Bauer, U. Kunter, I. Roots, and A. Brattstrom. 2009. St. John's wort extract (Ze 117) does not alter the pharmacokinetics of a low-dose oral contraceptive. *Eur. J. Clin. Pharmacol.* 65(3):287-294.

Woelk, H., G. Burkard, and J. Gruenwald. 1994. Benefits and risks of the *Hypericum* extract LI 160: Drug monitoring study with 3250 patients. *J. Geriatr. Psychiatr. Neurol.* 7(Suppl. 1):S34-S38.

Xie, R., L.H. Tan, E.C. Polasek, et al. 2005. CYP3A and P-glycoprotein activity induction with St. John's wort in healthy volunteers from 6 ethnic populations. *J. Clin. Pharmacol.* 45(3):352-356.

Xu, H., K.M. Williams, W.S. Liauw, et al. 2008. Effects of St. John's wort and CYP2C9 genotype on the pharmacokinetics and pharmacodynamics of gliclazide. *Br. J. Pharmacol.* 153(7):1579-1586.

Yevstifeyeva, T.A., and S.V. Sibiryak. 1996. Immunotropic properties of biologically active products obtained from St. John's wort. *Eksp. Klin. Farmakol.* 59:54.

Yue, Q.Y., C. Bergquist, and B. Gerden. 2000. Safety of St. John's wort (*Hypericum perforatum*). *Lancet* 355(9203):576-577.

Zullino, D., and F. Borgeat. 2003. Hypertension induced by St. John's wort—A case report. *Pharmacopsychiatry* 36(1):32.

Hyssopus officinalis L. — Lamiaceae

SCN: hyssop

Part: herb

Quick Reference Summary

Safety Class: 2b
Interaction Class: A

Contraindications

Not for use in pregnancy except under the supervision of a qualified healthcare practitioner (Kuhn and Winston 2007; Madaus 1976; Riddle 1997).

Other Precautions

None known.

Drug and Supplement Interactions

None known.

Notice

Emmenagogue (Kuhn and Winston 2007); *see* Appendix 2.

Thujone (trace amounts) (Kerrola et al. 1994) ; *see* Appendix 1.

Adverse Events and Side Effects

None known.

Pharmacological Considerations

None known.

Pregnancy and Lactation

Hyssop has traditionally been used as both an emmenagogue and an abortifacient (Kuhn and Winston 2007; Riddle 1997). Based on this information, use during pregnancy is not recommended except under the supervision of a qualified healthcare practitioner.

No information on the safety of hyssop during lactation was identified. While this review did not identify any concerns for use while nursing, safety has not been conclusively established.

Review Details

I. Drug and Supplement Interactions

Clinical Trials of Drug or Supplement Interactions

No clinical trials of drug or supplement interactions were identified.

Case Reports of Suspected Drug or Supplement Interactions

No case reports of suspected drug or supplement interactions were identified.

Animal Trials of Drug or Supplement Interactions

No animal trials of drug or supplement interactions were identified.

II. Adverse Events

Case Reports of Adverse Events

No case reports of adverse events were identified.

III. Pharmacology and Pharmacokinetics

Human Pharmacological Studies

No relevant human pharmacological studies were identified.

Animal Pharmacological Studies

No relevant animal pharmacological studies were identified.

In Vitro Pharmacological Studies

No relevant in vitro pharmacological studies were identified.

IV. Pregnancy and Lactation

Hyssop has traditionally been used as both an emmenagogue and an abortifacient (Kuhn and Winston 2007; Riddle 1997).

No information on the safety of hyssop during lactation was identified.

V. Toxicity Studies

Acute Toxicity

The LD_{50} of intraperitoneally administered hyssop essential oil in rats could not be determined at doses up to 1.6 ml/kg, while that of pinocamphone and isopinocamphone is 0.05 ml/kg and that of thujone is 0.2 ml/kg (Steinmetz et al. 1980).

Genotoxicity

Some genotoxic activity of an alcohol extract of hyssop at concentrations of 0.05 and 0.1% was observed in human lymphocytes (Hadziselimovic and Sofradzija 1998).

Literature Cited

Hadziselimovic, R., and A. Sofradzija. 1998. Mutagenicity of the extract of *Hyssopus officinalis* L. Labiatae. *Cytogenet. Cell Genet.* 81(2):160.

Kerrola, K., B. Galambosi, and H. Kallio. 1994. Volatile components and odor intensity of four phenotypes of hyssop (*Hyssopus officinalis* L.). *J. Agric. Food. Chem.* 42 (3):776-781.

Kuhn, M.A., and D. Winston. 2007. *Herbal therapy and supplements: A scientific and traditional approach.* Philadelphia: Lippincott.

Madaus, G. 1976. *Lehrbuch der biologischen heilmittel.* Hildesheim: New York.

Riddle, J. 1997. *Eve's herbs: A history of contraception and abortion in the West.* Cambridge, MA: Harvard University Press.

Steinmetz, M.D., P. Tognetti, M. Mourgue, J. Jouglard, and Y. Millet. 1980. The toxicity of certain essential oils of commerce: Oil of hyssop and oil of sage. *Plant Med. Phytother.* 14(1):34-45.

Ilex paraguariensis A. St.-Hil. Aquifoliaceae

SCN: mate
OCN: Paraguay tea; yerba mate

Part: leaf

Quick Reference Summary

Safety Class: 1
Interaction Class: C*

Contraindications

None known.

Other Precautions

Mate (pronounced in two syllables) contains caffeine, a nervous system stimulant. If taken in large amounts, mate products containing caffeine can cause insomnia, nervousness, and the other well-known symptoms of excess caffeine intake (Donovan and DeVane 2001).

Due to the central nervous system (CNS) stimulant effects of caffeine, use of caffeine-containing products is cautioned in persons with heart disorders, as excessive caffeine consumption may increase heart rate or exacerbate arrhythmias; use is also cautioned in psychological disorders, as caffeine may aggravate depression or induce anxiety (Brinker 2001).

Drug and Supplement Interactions

Use of caffeine with other CNS stimulants, including bronchodilators or adrenergic drugs, may cause excessive central nervous system stimulation resulting in nervousness, irritability, insomnia, and possibly convulsions or cardiac arrhythmias (PDR 2006).

Notice

Caffeine (0.3–1.7%) (List and Hörhammer 1973; Wichtl 2004); *see* Appendix 1.

Diuretic (Brunton et al. 2006); *see* Appendix 2.

Tannins (4.0–16.0%) (List and Hörhammer 1973; Wichtl 2004); *see* Appendix 1.

Editors' Note

The American Herbal Products Association has established a trade requirement (AHPA 2011) that dietary supplement products that contain caffeine, whether as a direct ingredient or as a constituent of herbal ingredients, be labeled to disclose the presence of caffeine in the product and the quantity of added caffeine if greater than 25 mg; be formulated and labeled in a manner to recommend a maximum of 200 mg of caffeine per serving, not more often than every 3 to 4 hours; and bear the following or similar statement on the label of any dietary supplement that contains caffeine in sufficient quantity to warrant such labeling:

> Too much caffeine may cause nervousness, irritability, sleeplessness, and, occasionally, rapid heartbeat. Not recommended for use by children under 18 years of age.

See Appendix 1 for more specific details on this AHPA trade requirement.

Adverse Events and Side Effects

Epidemiological studies in regions where mate is frequently consumed have suggested an association between mate consumption and increased risk of developing certain cancers, such as esophageal, oral, lung, bladder, renal, and other cancers of the head and neck (Bates et al. 2007; De Stefani et al. 1996, 1998, 2007; Goldenberg et al. 2003; Pintos et al. 1994). The reason for the increased cancer rates is unknown, and several hypotheses have been suggested, including thermal injury from hot beverages, increased absorption of environmental carcinogens due to the temperature of mate at the time of consumption, or the presence of polycyclic aromatic hydrocarbons (compounds like those found in tobacco smoke and grilled meat) from smoke introduced during the traditional roasting process (new roasting processes used by some producers exclude smoke and thus the polycyclic aromatic hydrocarbons) (Heck and de Mejia 2007). Several of these studies also examined coffee, green tea, and other dietary factors such as red meat consumption, all of which had an increased risk of various cancers similar to or greater than that seen for mate (De Stefani et al. 1991, 1998; Heck and de Mejia 2007). In contrast to epidemiological data, a number of animal and in vitro studies have demonstrated anticancer effects of mate (Heck and de Mejia 2007).

Pharmacological Considerations

See Other Precautions and Drug and Supplement Interactions above.

* For caffeine-free preparations, no interactions are expected.

Pregnancy and Lactation

An epidemiological study on use of mate during pregnancy indicated no adverse effects of mate on pregnancy duration or birth weight (Santos et al. 2005).

Pregnant women are advised to limit use of mate to 300 mg caffeine daily (PDR 2006). Nursing women are advised to limit use of caffeinated mate products to 150 mg caffeine daily (AAP 2001).

Review Details

I. Drug and Supplement Interactions

Clinical Trials of Drug or Supplement Interactions

No clinical trials of drug or supplement interactions were identified.

Case Reports of Suspected Drug or Supplement Interactions

No case reports of suspected drug or supplement interactions were identified.

Animal Trials of Drug or Supplement Interactions

No animal trials of drug or supplement interactions were identified.

II. Adverse Events

Adverse Events Reported in Clinical Trials

Epidemiological studies in countries in which mate is frequently consumed have suggested an association between mate consumption and increased risk of developing certain cancers, such as esophageal, oral, lung, bladder, renal, and other cancers of the head and neck (Bates et al. 2007; De Stefani et al. 1996, 1998, 2007; Goldenberg et al. 2003; Pintos et al. 1994). The reason for the increased cancer rates is unknown, and several hypotheses have been suggested (Heck and de Mejia 2007). Studies have suggested that drinking more than one liter of mate daily can increase the risk of head and neck cancers by three to five times, and that some association between mate consumption and lung cancer exists (De Stefani et al. 1996; Sewram et al. 2003; Vassallo et al. 1985).

Some research has suggested that the increased risk of oral cancers may be due to the temperature of the mate at the time of consumption, as thermal injury may be responsible for the increased risk of cancers (Rolon et al. 1995). The custom of drinking very hot mate through a metal straw (bombilla) can cause repeated scalding to the back of the throat and esophagus. Consumption of hot coffee or hot green tea is reported to increase the risk of oral cancers by two to four times (Heck and de Mejia 2007).

Studies of mate drinking by smokers showed that, among mate drinkers, bladder cancer was more prevalent in smokers than in nonsmokers. The study also indicated that black tea and coffee consumption were risk factors for bladder cancer (De Stefani et al. 1991). Another similar study indicated an increased risk of bladder cancer in mate drinkers who smoked but no risk in nonsmokers (Bates et al. 2007).

Another hypothesis regarding higher oral cancer rates among mate drinkers that smoke tobacco or drink alcohol is that, due to the high temperature at time of consumption, mate may increase the absorption of the carcinogens found in cigarette smoke and other environmental contaminants that are carcinogens or cancer promoters (Goldenberg et al. 2004).

Mate contains polycyclic aromatic hydrocarbons (PAHs) including benzo[*a*]pyrene, compounds like those found in tobacco smoke and grilled meat that have carcinogenic properties and have been associated with increased risk of esophageal cancer (Heck and de Mejia 2007). One human study showed a dose-related correlation between the amount of mate consumed and urinary levels of PAHs (Fagundes et al. 2006). A study in Brazil indicated that PAH levels in beverages were 10.12 μg/kg for coffee and 0.70 μg/kg for mate (Rojo de Camargo and Toledo 2002). The PAH content of roasted mate comes from the traditional processing method of drying the leaves over a smoky wood fire (Heck and de Mejia 2007). Dried, unsmoked mate is also sold commercially.

Examining incidences of renal cell carcinoma in Uruguay, researchers indicated that, after adjusting for major covariates, heavy mate consumption was associated with a 3-fold increase in risk of renal cell carcinoma, while consumption of red meat was associated with a 3.4-fold increase in risk (De Stefani et al. 1998).

In contrast to epidemiological data, a number of animal and in vitro studies have demonstrated anticancer effects of mate (Heck and de Mejia 2007).

Case Reports of Adverse Events

No case reports of adverse events were identified.

III. Pharmacology and Pharmacokinetics

Human Pharmacological Studies

Caffeine is a substrate of the drug-metabolizing isoenzyme CYP1A2 (Nordmark et al. 1999).

Animal Pharmacological Studies

Mate has been observed to significantly inhibit absorption of nonheme iron in rats (Gutnisky et al. 1992).

No significant increases in cancers of the aerodigestive tract were seen in rats administered mate aqueous extract as drinking water for 5 months (Pereira Jotz et al. 2006).

In Vitro Pharmacological Studies

Catalytic topoisomerase inhibition by an extract of mate was seen for topoisomerase II but not topoisomerase I in vitro in an oral carcinoma cell line (Gonzalez de Mejia et al. 2005). Mate has been identified as a potent inhibitor of topoisomerase II, showing significant cancer cell growth inhibition, even at low concentrations (Heck and de Mejia 2007).

IV. Pregnancy and Lactation

In an epidemiological study of the effects of mate drinking in pregnancy on birth weight and the risk of preterm birth, no changes in duration of pregnancy and no harmful effects on intrauterine growth were detected. A total of 5189 single births were analyzed, and 68% of mothers reported drinking mate at least once a week during pregnancy (Santos et al. 2005).

A premature infant newborn of a mother who reported drinking one liter of mate tea (~930 mg caffeine) daily during pregnancy exhibited increased jitteriness and irritability, hypertonia in the limbs, and brisk tendon reflexes consistent with neonatal withdrawal syndrome. Symptoms progressively disappeared at 84 hours of age. High concentrations of caffeine and theobromine were detected in maternal and neonatal biological matrices, including the placenta, cord serum, neonatal urine, maternal and neonatal hair, meconium, and breast milk. The authors of the report indicated that while mate is frequently consumed by pregnant women in South America, neonatal withdrawal syndrome is rare, as infants are usually breast-fed within 4 hours after delivery, providing a dose of caffeine (Martin et al. 2007).

Caffeine is in the FDA pregnancy category C and has been shown to cross the placenta and achieve blood and tissue concentrations in the fetus. Excessive intake of caffeine by pregnant women has been associated with fetal arrhythmias. Pregnant women are advised to limit caffeine intake to less than 300 mg daily (PDR 2006).

Caffeine is listed as a "Maternal Medication Usually Compatible with Breastfeeding" by the American Academy of Pediatrics Committee on Drugs. The Committee noted that maternal consumption of caffeine may cause irritability and poor sleeping patterns in nursing infants, and that maternal consumption of caffeinated beverages should be limited to two to three cups daily (AAP 2001).

Epidemiological studies have indicated an association between high caffeine intake during pregnancy and an increased risk of spontaneous abortions. An analysis concluded that methodological flaws in many of the studies led to biased results and that a causal link between caffeine consumption and abortion cannot yet be confirmed (Signorello and McLaughlin 2004).

V. Toxicity Studies

Acute Toxicity

The LD_{50} of orally administered caffeine in rats is 335 mg/kg (Mills and Bone 2005).

Genotoxicity

No clastogenic or aneugenic activity of an aqueous extract of mate, without metabolic activation, was observed in a cytokinesis-blocked micronucleus assay (Alves et al. 2008).

No clastogenic activity of mate was observed in vivo in rat bone marrow cells, although an increased frequency of chromosomal aberrations was reported in vitro in human peripheral lymphocytes (da Fonseca et al. 1994). Genotoxic activity of an aqueous extract of mate was observed in *Escherichia coli* and *Salmonella typhimurium* without metabolic activation, but the addition of S9 microsomal fraction, catalase, thiourea, or dipyridyl counteracted the genotoxic activity of mate (da Fonseca et al. 1994; Leitão and Braga 1994).

Cytotoxicity

An extract of mate was cytotoxic against human liver cancer HepG2 cells in vitro with an IC_{50} value of the total polyphenol content of 12.01 μg/ml (Gonzalez de Mejia et al. 2005).

Literature Cited

AAP. 2001. The transfer of drugs and other chemicals into human milk. American Academy of Pediatrics Committee on Drugs. *Pediatrics* 108(3):776-789.

AHPA. July 2011. Code of Ethics & Business Conduct. Silver Spring, MD: American Herbal Products Association.

Alves, R.J.V., G.P. Jotz, V.S. do Amaral, et al. 2008. The evaluation of mate (*Ilex paraguariensis*) genetic toxicity in human lymphocytes by the cytokinesis-block in the micronucleus assay. *Toxicol. In Vitro* 22:695-698.

Bates, M.N., C. Hopenhayn, O.A. Rey, and L.E. Moore. 2007. Bladder cancer and mate consumption in Argentina: A case-control study. *Cancer Lett.* 246(1-2):268-273.

Brinker, F. 2001. *Herb contraindications and drug interactions.* 3rd ed. Sandy, OR: Eclectic Medical Publications.

Brunton, L.L., J.S. Lazo, and K.L. Parker. 2006. *Goodman & Gilman's the pharmacological basis of therapeutics.* 11th ed. New York: McGraw-Hill.

da Fonseca, C.A., J. Leal, S.S. Costa, and A.C. Leitao. 1994. Genotoxic and mutagenic effects of guarana (*Paullinia cupana*) in prokaryotic organisms. *Mutat. Res.* 321(3):165-173.

De Stefani, E., P. Boffetta, H. Deneo-Pellegrini, et al. 2007. Non-alcoholic beverages and risk of bladder cancer in Uruguay. *BMC Cancer* 7:57.

De Stefani, E., P. Correa, L. Fierro, et al. 1991. Black tobacco, mate, and bladder cancer. A case-control study from Uruguay. *Cancer* 67(2):536-540.

De Stefani, E., L. Fierro, P. Correa, et al. 1996. Mate drinking and risk of lung cancer in males: A case-control study from Uruguay. *Cancer Epidemiol. Biomarkers Prev.* 5(7):515.

De Stefani, E., L. Fierro, M. Mendilaharsu, et al. 1998. Meat intake, 'mate' drinking and renal cell cancer in Uruguay: A case-control study. *Br. J. Cancer* 78(9):1239-1243.

Donovan, J.L., and C.L. DeVane. 2001. A primer on caffeine pharmacology and its drug interactions in clinical psychopharmacology. *Psychopharmacol. Bull.* 35(3):30-48.

Fagundes, R.B., C.C. Abnet, P.T. Strickland, et al. 2006. Higher urine 1-hydroxy pyrene glucuronide (1-OHPG) is associated with tobacco smoke exposure and drinking mate in healthy subjects from Rio Grande do Sul, Brazil. *BMC Cancer* 6:139.

Goldenberg, D., A. Golz, and H.Z. Joachims. 2003. The beverage mate: A risk factor for cancer of the head and neck. *Head Neck* 25(7):595-601.

Goldenberg, D., J. Lee, W.M. Koch, et al. 2004. Habitual risk factors for head and neck cancer. *Otolaryngol. Head Neck Surg.* 131(6):986-993.

Gonzalez de Mejia, E., Y.S. Song, M.V. Ramirez-Mares, and H. Kobayashi. 2005. Effect of yerba mate (*Ilex paraguariensis*) tea on topoisomerase inhibition and oral carcinoma cell proliferation. *J. Agric. Food Chem.* 53(6):1966-1973.

Gutnisky, A., N. Rizzo, M.E. Castro, and G. Garbossa. 1992. The inhibitory action of chlorogenic acid on the intestinal iron absorption in rats. *Acta Physiol. Pharmacol. Ther. Latinoam.* 42(3):139-146.

Heck, C.I., E.G. de Mejia. 2007. Yerba Mate Tea (*Ilex paraguariensis*): A comprehensive review on chemistry, health implications, and technological considerations. *J. Food Sci.* 72(9):R138-51.

Leitão, A.C., and R.S. Braga. 1994. Mutagenic and genotoxic effects of mate (*Ilex paraguariensis*) in prokaryotic organisms. *Braz. J. Med. Biol. Res.* 27(7):1517-1525.

List, P.H., and H. Hörhammer. 1973. *Hagers handbuch der pharmazeutischen praxis*. Berlin: Springer.

Martin, I., M.A. Lopez-Vilchez, A. Mur, O. Garcia-Algar, et al. 2007. Neonatal withdrawal syndrome after chronic maternal drinking of mate. *Ther. Drug Monit.* 29(1):127-129.

Mills, S., and K. Bone. 2005. *The essential guide to herbal safety*. St. Louis: Elsevier.

Nordmark, A., S. Lundgren, S. Cnattingius, and A. Rane. 1999. Dietary caffeine as a probe agent for assessment of cytochrome P4501A2 activity in random urine samples. *Br. J. Clin. Pharmacol.* 47(4):397.

PDR. 2006. *Physicians' desk reference for nonprescription drugs and dietary supplements*. 27th ed. Montvale, NJ: Medical Economics Co.

Pereira Jotz, G., H. Sampaio Menezes, C. Galleano Zetter, et al. 2006. Mate (*Ilex paraguariensis*) as an etiological agent of neoplasia in the aerodigestive tract. *Int. Arch. Otorhinolaryngol. Sao Paulo* 10:306-311.

Pintos, J., E.L. Franco, B.V. Oliveira, et al. 1994. Mate, coffee, and tea consumption and risk of cancers of the upper aerodigestive tract in southern Brazil. *Epidemiology* 5(6):583-590.

Rojo de Camargo, M.C., and M.C.F. Toledo. 2002. Coffee and mate tea as a dietary source of polycyclic aromatic hydrocarbons (PAHs) in Campinas. *Cien. Tecnol. Aliment.* 22:49-53.

Rolon, P.A., X. Castellsague, M. Benz, and N. Munoz. 1995. Hot and cold mate drinking and esophageal cancer in Paraguay. *Cancer Epidemiol. Biomarkers Prev.* 4(6):595-605.

Santos, I.S., A. Matijasevich, and N.C. Valle. 2005. Mate drinking during pregnancy and risk of preterm and small for gestational age birth. *J. Nutr.* 135(5):1120-1123.

Sewram, V., E. De Stefani, P. Brennan, and P. Boffetta. 2003. Mate consumption and the risk of squamous cell esophageal cancer in Uruguay. *Cancer Epidemiol. Biomarkers Prev.* 12(6):508-513.

Signorello, L.B., and J.K. McLaughlin. 2004. Maternal caffeine consumption and spontaneous abortion: A review of the epidemiologic evidence. *Epidemiology* 15 (2):229-239.

Vassallo, A., P. Correa, E. Destefani, et al. 1985. Esophageal cancer in Uruguay—A case-control study. *J. Natl. Cancer Inst.* 75(6):1005-1009.

Wichtl, M. 2004. *Herbal drugs and phytopharmaceuticals: A handbook for practice on a scientific basis*. 3rd ed. Boca Raton, FL: CRC Press.

Illicium verum Hook. f. — Illiciaceae

SCN: star anise
AN: *takkola*
OCN: Chinese star anise
Part: fruit

Quick Reference Summary

Safety Class: 1
Interaction Class: A

Contraindications

None known.

Other Precautions

None known.

Drug and Supplement Interactions

None known.

Editors' Notes

Star anise (*Illicium verum*) is commonly confused with Japanese star anise, also known as shikimi, (*I. anisatum*), a species known to have neurological and gastrointestinal toxicity (Ize-Ludlow et al. 2004; Joshi et al. 2005; Small 1996). The fruit of the Japanese star anise is generally smaller than that of *I. verum* and consists of 6 to 8 follicles arranged in a star pattern and terminating in an upward curving tip. This curve is in contrast to the nearly straight beak of the standard species. Japanese star anise is reported to have a bitter flavor and smells of sassafras, whereas star anise is not bitter and smells of anise (Wichtl 2004; Youngken 1921).

Analytical methods to microscopically or chemically differentiate the species have been published (Howes et al. 2009; Joshi et al. 2005; Lederer et al. 2006; Techen et al. 2009; Upton 2006).

Seizures have been reported in infants administered teas labeled as star anise (Garzo Fernandez et al. 2002; Gil Campos et al. 2002; Ize-Ludlow et al. 2004; Minodier et al. 2003; Montoya-Cabrera 1990), but analysis of the tea products in several of the case reports indicated adulteration with Japanese star anise (*I. anisatum*). General malaise, nausea, and vomiting were reported in adults who consumed a tea product labeled as star anise that was later found to contain Japanese star anise (Johanns et al. 2002).

Adverse Events and Side Effects

None known.

Pharmacological Considerations

None known.

Pregnancy and Lactation

No information on the safety of star anise in pregnancy or lactation was identified in the scientific or traditional literature. Although this review did not identify any concerns for use while pregnant or nursing, safety has not been conclusively established.

Review Details

I. Drug and Supplement Interactions

Clinical Trials of Drug or Supplement Interactions

No clinical trials of drug or supplement interactions were identified.

Case Reports of Suspected Drug or Supplement Interactions

No case reports of suspected drug or supplement interactions were identified.

Animal Trials of Drug or Supplement Interactions

No animal trials of drug or supplement interactions were identified.

II. Adverse Events

Case Reports of Adverse Events

Seizures have been reported in infants administered teas labeled as star anise (Garzo Fernandez et al. 2002; Gil Campos et al. 2002; Ize-Ludlow et al. 2004; Minodier et al. 2003; Montoya-Cabrera 1990); however, analysis of the tea products in several of the case reports indicated adulteration with Japanese star anise (*I. anisatum*), a botanically similar species that has been documented to have neurological and gastrointestinal toxicity (Ize-Ludlow et al. 2004).

General malaise, nausea, and vomiting were reported in adults who consumed a tea product labeled as star anise that was later found to contain Japanese star anise (Johanns et al. 2002).

An anaphylactic reaction to the drug oseltamivir phosphate was reported in a patient with sensitization to star anise and celery-carrot-mugwort-spice syndrome (Hirschfeld et al. 2008).

III. Pharmacology and Pharmacokinetics

Human Pharmacological Studies

No relevant human pharmacological studies were identified.

Animal Pharmacological Studies

No relevant animal pharmacological studies were identified.

In Vitro Pharmacological Studies

No relevant in vitro pharmacological studies were identified.

IV. Pregnancy and Lactation

No information on the safety of star anise during pregnancy or lactation was identified.

V. Toxicity Studies

Acute Toxicity

The compounds veranisatins A, B, and C, isolated from star anise, caused convulsions and were lethal at doses of 3 mg/kg after oral administration to mice. At lower doses, the compounds caused hypothermia (Nakamura et al. 1996; Okuyama et al. 1993).

Literature Cited

Garzo Fernandez, C., P. Gomez Pintado, A. Barrasa Blanco, et al. 2002. Cases of neurological symptoms associated with star anise consumption used as a carminative. *An. Esp. Pediatr.* 57(4):290-294.

Gil Campos, M., J.L. Perez Navero, and I. Ibarra De La Rosa. 2002. Convulsive status secondary to star anise poisoning in a neonate. *An. Esp. Pediatr.* 57(4):366-368.

Hirschfeld, G., L. Weber, A. Renkl, K. Scharffetter-Kochanek, and J.M. Weiss. 2008. Anaphylaxis after oseltamivir (Tamiflu) therapy in a patient with sensitization to star anise and celery-carrot-mugwort-spice syndrome. *Allergy* 63(2):243-244.

Howes, M.J., G.C. Kite, and M.S. Simmonds. 2009. Distinguishing Chinese star anise from Japanese star anise using thermal desorption-gas chromatography-mass spectrometry. *J. Agric. Food Chem.* 57(13):5783-5789.

Ize-Ludlow, D., S. Ragone, I.S. Bruck, et al. 2004. Neurotoxicities in infants seen with the consumption of star anise tea. *Pediatrics* 114(5):e653-e656.
Johanns, E.S., L.E. van der Kolk, H.M. van Gemert, et al. 2002. An epidemic of epileptic seizures after consumption of herbal tea. *Ned. Tijdschr. Geneeskd.* 146(17):813-816.
Joshi, V.C., P.V. Srinivas, and I.A. Khan. 2005. Rapid and easy identification of *Illicium verum* Hook. f. and its adulterant *Illicium anisatum* Linn. by fluorescent microscopy and gas chromatography. *J. AOAC Int.* 88(3):703-706.
Lederer, I., G. Schulzki, J. Gross, and J.P. Steffen. 2006. Combination of TLC and HPLC-MS/MS methods. Approach to a rational quality control of Chinese star anise. *J. Agric. Food Chem.* 54(6):1970-1974.
Minodier, P., P. Pommier, E. Moulene, et al. 2003. Star anise poisoning in infants. *Arch. Pediatr.* 10(7):619-621.
Montoya-Cabrera, M.A. 1990.Poisoning by star anise (*Illicium verum*) tea.] *Gac. Med. Mex.* 126(4):341-342.
Nakamura, T., E. Okuyama, and M. Yamazaki. 1996. Neurotropic components from star anise (*Illicium verum* Hook. fil.). *Chem. Pharm. Bull. (Tokyo)* 44(10):1908-1914.
Okuyama, E., T. Nakamura, and M. Yamazaki. 1993. Convulsants from star anise (*Illicium verum* Hook.F.). *Chem. Pharm. Bull. (Tokyo)* 41(9):1670-1671.
Small, E. 1996. Confusion of common names for toxic and edible "star anise" (*Illicium*) species. *Econ. Bot.* 50(3):337-339.
Techen, N., Z. Pan, B.E. Scheffler, and I.A. Khan. 2009. Detection of *Illicium anisatum* as adulterant of *Illicium verum*. *Planta Med.* 75(4):392-395.
Upton, R. 2006. *Differentiation between star anise (Illicium verum) and the toxic adulterant shikimi (Illicium anisatum)*. Scotts Valley, CA: American Herbal Pharmacopoeia.
Wichtl, M. 2004. *Herbal drugs and phytopharmaceuticals: A handbook for practice on a scientific basis*. 3rd ed. Boca Raton, FL: CRC Press.
Youngken, H.W. 1921. *A textbook of pharmacognosy*. Philadelphia: Blakiston's Son & Co.

Inula helenium L. — Asteraceae

SCN: elecampane
PN: *tu mu xiang* (root)
Part: root

Quick Reference Summary

Safety Class: 1
Interaction Class: A

Contraindications

None known.

Other Precautions

Persons with allergies to other members of the Asteraceae family (such as feverfew, chamomile, or *Echinacea*) should exercise caution with elecampane, as allergic cross-reactivity to Asteraceae plants is common (Paulsen 2002).

Drug and Supplement Interactions

None known.

Notice

Emmenagogue (Felter and Lloyd 1898); *see* Appendix 2.

Adverse Events and Side Effects

Allergic reactions, primarily contact dermatitis, to elecampane have been reported (Mateo et al. 1995; Paulsen 2002; Pazzaglia et al. 1995).

Large doses (standard dose listed as tea of 1 g taken 3 to 4 times daily) may cause vomiting, diarrhea, cramping, and symptoms of paralysis (Roth et al. 1984; Wichtl 2004).

Pharmacological Considerations

Sensitization studies in animals, and anecdotal reports from human use, have indicated that compounds in elecampane are relatively common skin sensitizers (Alanso Blasi et al. 1992; Paulsen 2002; Stampf et al. 1982).

Pregnancy and Lactation

A British herbal text indicates that elecampane should not be used during pregnancy or lactation but does not indicate the reason for these contraindications (Bradley 1992). A text on traditional Chinese medicine indicates that elecampane is sometimes used in cases of threatened miscarriage (PPRC 2005). Based on the available evidence, the editors of this text believe that elecampane is generally safe for use in pregnancy and lactation.

Review Details

I. Drug and Supplement Interactions

Clinical Trials of Drug or Supplement Interactions

No clinical trials of drug or supplement interactions were identified.

Case Reports of Suspected Drug or Supplement Interactions

No case reports of suspected drug or supplement interactions were identified.

Animal Trials of Drug or Supplement Interactions

No animal trials of drug or supplement interactions were identified.

II. Adverse Events

Case Reports of Adverse Events

An erythema-multiforme-like eruption following allergic contact dermatitis was reported after contact with elecampane, with the effects attributed to sesquiterpene lactone compounds (Mateo et al. 1995).

Contact dermatitis due to a massage oil containing elecampane extract was reported (Pazzaglia et al. 1995).

III. Pharmacology and Pharmacokinetics

Human Pharmacological Studies

Among plants of the Asteraceae family that cause allergic reactions and contact sensitization, elecampane has been recognized as one of the species that, relatively frequently, causes sensitization (Paulsen 2002).

Animal Pharmacological Studies

In sensitization tests in guinea pigs, the compounds alantolactone and isoalantolactone showed sensitizing activity after intradermal injection, with no sensitization shown for helenin (a crystalline portion of steam-distilled elecampane extract) (Stampf et al. 1982).

The compound alantolactone demonstrated sensitizing activity in an epicutaneous test in mice. No sensitizing activity of isoalantolactone was observed (Alanso Blasi et al. 1992).

In Vitro Pharmacological Studies

No relevant in vitro pharmacological studies were identified.

IV. Pregnancy and Lactation

A British herbal text indicates that elecampane should not be used during pregnancy or lactation, but does not indicate the reason for these contraindications (Bradley 1992). A text on traditional Chinese medicine indicates that elecampane is sometimes used in cases of threatened miscarriage (PPRC 2005). Based on the available evidence, the editors of this text believe that elecampane is generally safe for use in pregnancy and lactation.

V. Toxicity Studies

Acute Toxicity

The LD_{50} of orally administered helenin (a crystalline portion of steam-distilled elecampane extract) in rabbits is 1.2 g/kg (Blaschek et al. 2002).

Literature Cited

Alanso Blasi, N., R. Fraginals, J.P. Lepoittevin, and C. Benezra. 1992. A murine in vitro model of allergic contact dermatitis to sesquiterpene alpha-methylene gamma-butyrolactones. *Arch. Dermatol. Res.* 284:297-302.

Blaschek, W., S. Ebel, E. Hackenthal, et al. 2002. *Hagers handbuch der drogen und arzneistoffe*. HagerROM 2002. Heidelberg: Springer.

Bradley, P.R. 1992. *British herbal compendium: A handbook of scientific information on widely used plant drugs*. Bournemouth, UK: British Herbal Medicine Association.

Felter, H.W., and J.U. Lloyd. 1898. *King's American dispensatory*. 18th ed., 3rd rev. 2 vols. Cincinnati: Ohio Valley Co.

Mateo, M.P.G., M. Velasco, F.J. Miquel, and J. De La Cuadra. 1995. Erythema-multiforme-like eruption following allergic contact dermatitis from sesquiterpene lactones in herbal medicine. *Contact Dermat.* 33(6):449-450.

Paulsen, E. 2002. Contact sensitization from Compositae-containing herbal remedies and cosmetics. *Contact Dermat.* 47(4):189-198.

Pazzaglia, M., N. Venturo, G. Borda, and A. Tosti. 1995. Contact dermatitis due to a massage liniment containing *Inula helenium* extract. *Contact Dermat.* 33(4):267.

PPRC. 2005. *Pharmacopoeia of the People's Republic of China*. Beijing: People's Medical Publishing House.

Roth, L., M. Daunderer, and K. Kormann. 1984. *Giftpflanzen-pflanzengifte: Vorkommen, wirkung, therapie*. Landsberg: Ecomed.

Stampf, J.L., C. Benezra, G. Klecak, et al. 1982. The sensitizing capacity of helenin and of 2 of its main constituents, the sesquiterpene lactones alantolactone and iso-alantolactone: A comparison of epicutaneous and intradermal sensitizing methods in different strains of guinea-pig. *Contact Dermat.* 8(1):16-24.

Wichtl, M. 2004. *Herbal drugs and phytopharmaceuticals: A handbook for practice on a scientific basis*. 3rd ed. Boca Raton, FL: CRC Press.

Inula spp. — Asteraceae

Inula britannica L.
SCN: British elecampane
PN: *xuan fu hua* (flower)
OCN: British inula

Inula japonica **Thunb.**
SCN: Japanese elecampane
Syn: *Inula britannica* L. var. *japonica* (Thunb.) Franch. & Sav.
PN: *xuan fu hua* (flower)
Part: flower

Quick Reference Summary

Safety Class: 1
Interaction Class: A

Contraindications

None known.

Other Precautions

Persons with allergies to other members of the Asteraceae family (such as feverfew, chamomile, or *Echinacea*) should exercise caution with British or Japanese elecampane, as allergic cross-reactivity to Asteraceae plants is common (Paulsen 2002).

Drug and Supplement Interactions

None known.

Editors' Note

Dried flowers of British or Japanese elecampane should be wrapped in cheesecloth or a similar filter prior to decocting to remove the small hairs from the flowers that may cause irritation of the throat or digestive tract (Chen and Chen 2004).

Adverse Events and Side Effects

Allergic reactions, primarily contact dermatitis, to Japanese and British elecampane have been reported (Bensky et al. 2004; Chen and Chen 2004).

Diarrhea has been associated with the use of Japanese elecampane (Chen and Chen 2004).

Pharmacological Considerations

A sensitization study in animals indicated that at least one compound in Japanese elecampane is a relatively common sensitizer (Stampf et al. 1982; Yong-Ming et al. 2006).

Pregnancy and Lactation

No information on the safety of Japanese and British elecampane during pregnancy or lactation was identified in the scientific or traditional literature. Although this review did not identify any concerns for pregnant or lactating women, safety has not been conclusively established.

Review Details

I. Drug and Supplement Interactions

Clinical Trials of Drug or Supplement Interactions

No clinical trials of drug or supplement interactions were identified.

Case Reports of Suspected Drug or Supplement Interactions

No case reports of suspected drug or supplement interactions were identified.

Animal Trials of Drug or Supplement Interactions

No animal trials of drug or supplement interactions were identified.

II. Adverse Events

Case Reports of Adverse Events

Repeated skin contact with British or Japanese elecampane flowers or stalks can cause allergic reactions such as contact dermatitis and pruritus (Bensky et al. 2004).

III. Pharmacology and Pharmacokinetics

Human Pharmacological Studies

No relevant human pharmacological studies were identified.

Animal Pharmacological Studies

In sensitization tests in guinea pigs, the compound isoalantolactone showed sensitizing activity after intradermal injection (Stampf et al. 1982).

In Vitro Pharmacological Studies

No relevant in vitro pharmacological studies were identified.

IV. Pregnancy and Lactation

No information on the safety of Japanese and British elecampane during pregnancy or lactation was identified.

V. Toxicity Studies

Acute Toxicity

The LD_{50} of a 15% decoction of Japanese elecampane intraperitoneally administered to mice is 22.5 g/kg (Chen and Chen 2004).

Literature Cited

Bensky, D., S. Clavey, and E. Stöger. 2004. *Chinese herbal medicine: Materia medica*. 3rd ed. Seattle: Eastland Press.

Chen, J.K., and T.T. Chen. 2004. *Chinese medical herbology and pharmacology*. City of Industry, CA: Art of Medicine Press.

Paulsen, E. 2002. Contact sensitization from Compositae-containing herbal remedies and cosmetics. *Contact Dermat.* 47(4):189-198.

Stampf, J.L., C. Benezra, G. Klecak, et al. 1982. The sensitizing capacity of helenin and of 2 of its main constituents, the sesquiterpene lactones alantolactone and iso-alantolactone: A comparison of epicutaneous and intradermal sensitizing methods in different strains of guinea-pig. *Contact Dermat.* 8(1):16-24.

Yong-Ming, Z., Z. Man-Li, S. Qing-Wen, and K. Hiromasa. 2006. Chemical constituents of plants from the genus *Inula*. *Chem. Biodivers.* 3(4):371-384.

Ipomoea purga (Wender.) Hayne — Convolvulaceae

SCN: jalap

Part: root

Quick Reference Summary

Safety Class: 3
Interaction Class: A

Contraindications

Not for use except under the supervision of an expert qualified in the appropriate use of this substance (Felter and Lloyd 1898; Wood and LaWall 1918).

Other Precautions

None known.

Drug and Supplement Interactions

See Pharmacological Considerations below.

Notice

Emetic (Felter and Lloyd 1898; Wood and LaWall 1918); *see* Appendix 2.

Stimulant laxative (BPC 1911; Felter and Lloyd 1898; Pereda-Miranda et al. 2006; Wood and LaWall 1918); *see* Appendix 2.

Adverse Events and Side Effects

Jalap is an irritant and cathartic that produces profuse and liquid stools, often accompanied by griping (BPC 1911; Felter and Lloyd 1898; Pereda-Miranda et al. 2006; Wood and LaWall 1918). Large doses may produce violent hypercatharsis (Felter and Lloyd 1898; Wood and LaWall 1918), which has led to fatalities in some cases (Felter and Lloyd 1898).

Pharmacological Considerations

The aqueous extract is reported to have a moderate purgative action, while the "portion not taken up by water" is reported to cause severe griping (Wood and LaWall 1918).

Concomitant use of stimulant laxatives such as jalap is cautioned with antiarrhythmic drugs and botanicals containing cardiac glycosides, as long-term use of jalap as a laxative can cause potassium loss, leading to increased toxicity of these drugs and botanicals (Brinker 2001; De Smet 1993; ESCOP 2003).

Concomitant internal use of stimulant laxatives such as jalap is cautioned with thiazide diuretics, corticosteroids, or licorice, and long-term use of jalap as a laxative may increase the potassium loss induced by these drugs and botanicals (Brinker 2001; De Smet 1993; ESCOP 2003).

Use of stimulant laxatives, such as jalap, may reduce the gastrointestinal transit time and thus reduce the absorption of orally administered drugs (Brinker 2001; De Smet 1993).

Pregnancy and Lactation

While no information on the use of jalap during pregnancy or lactation was identified in the scientific or traditional literature, the use of purgatives during pregnancy is contraindicated (Bensky et al. 2004; Chen and Chen 2004; Maciocia 1998).

While this review did not identify any concerns for use while nursing, safety has not been conclusively established, and this substance is not recommended for use except under the supervision of an expert qualified in its appropriate use.

Review Details

I. Drug and Supplement Interactions

Clinical Trials of Drug or Supplement Interactions

No clinical trials of drug or supplement interactions were identified.

Case Reports of Suspected Drug or Supplement Interactions

No case reports of suspected drug or supplement interactions were identified.

Animal Trials of Drug or Supplement Interactions

No animal trials of drug or supplement interactions were identified.

II. Adverse Events

Case Reports of Adverse Events

A reference text indicates that "large doses produce violent hypercatharsis, sometimes terminating fatally" (Felter and Lloyd 1898).

III. Pharmacology and Pharmacokinetics

Human Pharmacological Studies

No relevant human pharmacological studies were identified.

Animal Pharmacological Studies

No relevant animal pharmacological studies were identified.

In Vitro Pharmacological Studies

No relevant in vitro pharmacological studies were identified.

IV. Pregnancy and Lactation

No information on the safety of jalap during pregnancy or lactation was identified.

V. Toxicity Studies

No toxicity studies were identified.

Literature Cited

BPC. 1911. *British pharmaceutical codex*. London: Pharmaceutical Press

Bensky, D., S. Clavey, and E. Stöger. 2004. *Chinese herbal medicine: Materia medica*. 3rd ed. Seattle: Eastland Press.

Brinker, F. 2001. *Herb contraindications and drug interactions*. 3rd ed. Sandy, OR: Eclectic Medical Publications.

Chen, J.K., and T.T. Chen. 2004. *Chinese medical herbology and pharmacology*. City of Industry, CA: Art of Medicine Press.

De Smet, P.A.G.M. 1993. *Adverse effects of herbal drugs, Volume 2*. Berlin: Springer.

ESCOP. 2003. *ESCOP monographs: The scientific foundation for herbal medicinal products*. 2nd ed. Exeter, U.K.: European Scientific Cooperative on Phytotherapy.

Felter, H.W., and J.U. Lloyd. 1898. *King's American dispensatory*. 18th ed., 3rd rev. 2 vols. Cincinnati: Ohio Valley Co.

Maciocia, G. 1998. *Obstetrics & gynecology in Chinese medicine*. New York: Churchill Livingstone.

Pereda-Miranda, R., M. Fragoso-Serrano, E. Escalante-Sanchez, et al. 2006. Profiling of the resin glycoside content of Mexican Jalap roots with purgative activity. *J. Nat. Prod.* 69(10):1460-1466.

Wood, H., and C. LaWall. 1918. *The dispensatory of the United States of America*. 21st ed. Philadelphia: Lippincott.

Iris spp. — Iridaceae

Iris germanica L. var. ***florentina*** Dykes
SCN: orris
OCN: *fleur-de-lis*; Florentine iris

Iris pallida **Lam.**
SCN: orris
OCN: Dalmatian iris; sweet iris
Part: rhizome, root

Quick Reference Summary

Safety Class: 1
Interaction Class: A

Contraindications

None known.

Other Precautions

None known.

Drug and Supplement Interactions

None known.

Adverse Events and Side Effects

Fresh orris root may cause irritation of the gastrointestinal tract (Chadha 1988; Felter and Lloyd 1898). Large doses may cause nausea or vomiting (Wood and LaWall 1918).

Pharmacological Considerations

None known.

Pregnancy and Lactation

No information on the use of orris during pregnancy or lactation was identified in the scientific or traditional literature. While this review did not identify any concerns for pregnant or lactating women, safety has not been conclusively established.

Review Details

I. Drug and Supplement Interactions

Clinical Trials of Drug or Supplement Interactions

No clinical trials of drug or supplement interactions were identified.

Case Reports of Suspected Drug or Supplement Interactions

No case reports of suspected drug or supplement interactions were identified.

Animal Trials of Drug or Supplement Interactions

No animal trials of drug or supplement interactions were identified.

II. Adverse Events

Case Reports of Adverse Events

No case reports of adverse events were identified.

III. Pharmacology and Pharmacokinetics

Human Pharmacological Studies

No sensitization was reported in healthy volunteers using 2% orris absolute in petrolatum in a repeated insult patch test with eleven 24-hour exposures (Opdyke 1979). No irritation of orris absolute was observed in healthy volunteers in a 48-hour closed patch test using 3% orris absolute in petrolatum (Opdyke 1979).

Animal Pharmacological Studies

No relevant animal pharmacological studies were identified.

In Vitro Pharmacological Studies

No relevant in vitro pharmacological studies were identified.

I

IV. Pregnancy and Lactation

No information on the safety of orris during pregnancy or lactation was identified.

V. Toxicity Studies

Acute Toxicity

The LD_{50} of orris absolute orally administered to rats is 9.4 g/kg (Opdyke 1979).

Literature Cited

Chadha, Y. 1988. *The wealth of India: A dictionary of Indian raw materials and industrial products*. Delhi: Council of Scientific and Industrial Research.

Felter, H.W., and J.U. Lloyd. 1898. *King's American dispensatory*. 18th ed., 3rd rev. 2 vols. Cincinnati: Ohio Valley Co.

Opdyke, D.L.J. 1979. *Monographs on fragrance raw materials*. New York: Pergamon Press.

Wood, H., and C. LaWall. 1918. *The dispensatory of the United States of America*. 21st ed. Philadelphia: Lippincott.

Iris spp. — Iridaceae

Iris versicolor L.
SCN: blue flag
OCN: larger blue flag

Iris virginica L.
SCN: blue flag
Syn: *Iris caroliniana* S. Watson
OCN: southern blue flag
Part: rhizome, root

Quick Reference Summary

Safety Class: 2b, 2d
Interaction Class: A

Contraindications

Not for use in pregnancy except under the supervision of a qualified healthcare practitioner (McGuffin et al. 1997).

Do not exceed recommended dose (Felter and Lloyd 1898; List and Hörhammer 1973; Wood and LaWall 1918).

Other Precautions

May cause nausea or vomiting (Felter and Lloyd 1898; List and Hörhammer 1973; Wood and LaWall 1918).

Drug and Supplement Interactions

See Pharmacological Considerations below.

Standard Dose

The standard dose is 10–20 drops tincture, 2 to 3 times daily (Smith 2008; Winston 2010).

Notice

Emetic (List and Hörhammer 1973; Wood and LaWall 1918); *see* Appendix 2.

Stimulant laxative (Felter and Lloyd 1898; Wood and LaWall 1918); *see* Appendix 2.

Adverse Events and Side Effects

Fresh blue flag may cause irritation of the gastrointestinal tract that can lead to vomiting, watery diarrhea, intestinal burning, and colic (Felter and Lloyd 1898).

Blue flag is reported to have caused neuralgia of the face, head, and extremities (Felter and Lloyd 1898).

Pharmacological Considerations

Concomitant use of stimulant laxatives such as blue flag is cautioned with antiarrhythmic drugs and botanicals containing cardiac glycosides, as long-term use of blue flag as a laxative can cause potassium loss, leading to increased toxicity of these drugs and botanicals (Brinker 2001; De Smet 1993; ESCOP 2003).

Concomitant internal use of stimulant laxatives such as blue flag is cautioned with thiazide diuretics, corticosteroids, or licorice, and long-term use of blue flag as a laxative may increase the potassium loss induced by these drugs and botanicals (Brinker 2001; De Smet 1993; ESCOP 2003).

Use of stimulant laxatives, such as blue flag, may reduce the gastrointestinal transit time and thus reduce the absorption of orally administered drugs (Brinker 2001; De Smet 1993).

Pregnancy and Lactation

No information on the use of blue flag during pregnancy or lactation was identified in the scientific or traditional literature. While this review did not identify any concerns for pregnant or lactating women, safety has not been conclusively established.

Review Details

I. Drug and Supplement Interactions

Clinical Trials of Drug or Supplement Interactions

No clinical trials of drug or supplement interactions were identified.

Case Reports of Suspected Drug or Supplement Interactions

No case reports of suspected drug or supplement interactions were identified.

Animal Trials of Drug or Supplement Interactions

No animal trials of drug or supplement interactions were identified.

II. Adverse Events

Case Reports of Adverse Events

No case reports of adverse events were identified.

III. Pharmacology and Pharmacokinetics

Human Pharmacological Studies

No relevant human pharmacological studies were identified.

Animal Pharmacological Studies

No relevant animal pharmacological studies were identified.

In Vitro Pharmacological Studies

No relevant in vitro pharmacological studies were identified.

IV. Pregnancy and Lactation

No information on the safety of blue flag during pregnancy or lactation was identified.

V. Toxicity Studies

No toxicity studies were identified.

Literature Cited

Brinker, F. 2001. *Herb contraindications and drug interactions*. 3rd ed. Sandy, OR: Eclectic Medical Publications.

De Smet, P.A.G.M. 1993. *Adverse effects of herbal drugs, Volume 2*. Berlin: Springer.

ESCOP. 2003. *ESCOP monographs: The scientific foundation for herbal medicinal products*. 2nd ed. Exeter, U.K.: European Scientific Cooperative on Phytotherapy.

Felter, H.W., and J.U. Lloyd. 1898. *King's American dispensatory*. 18th ed., 3rd rev. 2 vols. Cincinnati: Ohio Valley Co.

List, P.H., and H. Hörhammer. 1973. *Hagers handbuch der pharmazeutischen praxis*. Berlin: Springer.

McGuffin, M., C. Hobbs, R. Upton, and A. Goldberg. 1997. *Botanical safety handbook*. Boca Raton, FL: CRC Press.

Smith, E. 2008. *Therapeutic herb manual*. Williams, OR: self-published.

Winston, D. 2010. *Winston's botanical materia medica*. Broadway, NJ: David Winston's Center for Herbal Studies

Wood, H., and C. LaWall. 1918. *The dispensatory of the United States of America*. 21st ed. Philadelphia: Lippincott.

I

Isatis spp.

Brassicaceae

Isatis indigotica Fortune
SCN: isatis
PN: *ban lan gen* (root)
OCN: indigo woad

Isatis tinctoria L.
SCN: dyer's wood
OCN: woad
Part: root

Quick Reference Summary

Safety Class: 1
Interaction Class: A

Contraindications

None known.

Other Precautions

Use with caution in persons with allergy to sulfonylurea (e.g., tolbutamide, glipizide, glyburide) or sulfonamide drugs (e.g., sulfadiazine, sulfisoxazole, sulfamethoxazole, trimethoprim-sulfamethoxazole) (Chen and Chen 2004).

Drug and Supplement Interactions

None known.

Editors' Note

Several plants are known by the common name *ban lan gen* in China, creating difficulty in understanding the identification and use of the different species. While one text on traditional Chinese medicine indicates that *Isatis indigotica* is the preferred species but *Baphicacanthus cusia* is a commonly traded acceptable alternate (Bensky et al. 2004), another text indicates that *Isatis tinctoria*, *Isatis indigotica*, *Baphicacanthus cusia*, *Polygonum tinctorium*, and *Clerodendron cyrtophyllum* are all acceptable species (Chen and Chen 2004).

Adverse Events and Side Effects

None known.

Pharmacological Considerations

None known.

Pregnancy and Lactation

No information on the safety of dyer's woad or isatis during pregnancy or lactation was identified in the scientific or traditional literature. While this review did not identify any concerns for pregnant or lactating women, safety has not been conclusively established.

Review Details

I. Drug and Supplement Interactions

Clinical Trials of Drug or Supplement Interactions

No clinical trials of drug or supplement interactions were identified.

Case Reports of Suspected Drug or Supplement Interactions

No case reports of suspected drug or supplement interactions were identified.

Animal Trials of Drug or Supplement Interactions

No animal trials of drug or supplement interactions were identified.

II. Adverse Events

Case Reports of Adverse Events

A case of bleeding in the upper gastrointestinal tract was reported in association with isatis use. Information on the product, dose, duration, concomitant medications, and relevant medical history was not provided (Bensky et al. 2004).

III. Pharmacology and Pharmacokinetics

Human Pharmacological Studies

No relevant human pharmacological studies were identified.

Animal Pharmacological Studies

Inhibition of platelet aggregation was observed in rabbits administered dyer's woad (dose, duration, and type of extract not specified in available English language translation) (Chen and Chen 2004).

In Vitro Pharmacological Studies

No relevant in vitro pharmacological studies were identified.

IV. Pregnancy and Lactation

No information on the safety of dyer's woad or isatis in pregnancy or lactation was identified.

V. Toxicity Studies

Acute Toxicity

No adverse reactions were observed in mice orally administered 5 g/kg of the compound indirubin (Chen and Chen 2004).

Literature Cited

Bensky, D., S. Clavey, and E. Stöger. 2004. *Chinese herbal medicine: Materia medica*. 3rd ed. Seattle: Eastland Press.

Chen, J.K., and T.T. Chen. 2004. *Chinese medical herbology and pharmacology*. City of Industry, CA: Art of Medicine Press.

Jasminum officinale L. — Oleaceae

SCN: jasmine
Syn: *Jasminum grandiflorum* L.
AN: *jati*

OCN: poet's jasmine
Part: flower

Quick Reference Summary

Safety Class: 1
Interaction Class: A

Contraindications

None known.

Other Precautions

None known.

Drug and Supplement Interactions

None known.

Adverse Events and Side Effects

None known.

Pharmacological Considerations

None known.

Pregnancy and Lactation

An animal study indicated that large doses of jasmine had some anti-implantation activity but no adverse effects on developing fetuses (Iqbal et al. 1993).

No information on the safety of jasmine during lactation was identified. While this review did not identify any concerns for use while nursing, safety has not been conclusively established.

J

Review Details

I. Drug and Supplement Interactions

Clinical Trials of Drug or Supplement Interactions

No clinical trials of drug or supplement interactions were identified.

Case Reports of Suspected Drug or Supplement Interactions

No case reports of suspected drug or supplement interactions were identified.

Animal Trials of Drug or Supplement Interactions

No animal trials of drug or supplement interactions were identified.

II. Adverse Events

Case Reports of Adverse Events

No case reports of adverse events were identified.

III. Pharmacology and Pharmacokinetics

Human Pharmacological Studies

In a maximization test in healthy volunteers, 3% jasmine absolute in petrolatum produced sensitization in 2 of 25 volunteers when tested with other sensitizing agents but produced no sensitization when tested alone in another 25 healthy volunteers (Opdyke 1979).

No irritation was observed after 3% jasmine absolute was tested in a 48-hour closed patch test in healthy volunteers (Opdyke 1979).

Animal Pharmacological Studies

No irritation was observed after undiluted jasmine absolute was applied to the backs of hairless mice and pigs or to intact or abraded rabbit skin (Opdyke 1979).

In Vitro Pharmacological Studies

No relevant in vitro pharmacological studies were identified.

IV. Pregnancy and Lactation

In rats orally administered 250 or 500 mg/kg of an aqueous extract of jasmine daily, a dose-dependent anti-implantation effect was observed, although neither dose produced complete infertility. A decrease in serum progesterone levels was observed on day 5 of pregnancy. No abortifacient activity was observed in rats administered the extract on days 8 to 20 of pregnancy, and no abnormalities were observed in offspring (Iqbal et al. 1993).

No information on the safety of jasmine during lactation was identified.

V. Toxicity Studies

Acute Toxicity

The LD_{50} of jasmine absolute orally administered to rats and topically administered to rabbits could not be determined at doses up to 5 g/kg (Opdyke 1979).

Literature Cited

Iqbal, M., A.K.M. Ghosh, and A.K. Saluja. 1993. Antifertility activity of the floral buds of *Jasminum officinale* var. *grandiflorum* in rats. *Phytother. Res.* 7(1):5-8.

Opdyke, D.L.J. 1979. *Monographs on fragrance raw materials*. New York: Pergamon.

Juglans cinerea L. — Juglandaceae

SCN: butternut

Part: inner bark of root

Quick Reference Summary

Safety Class: 1
Interaction Class: A

Contraindications

None known.

Other Precautions

None known.

Drug and Supplement Interactions

None known.

Adverse Events and Side Effects

Large doses can be mildly cathartic (Felter and Lloyd 1898; Wood and LaWall 1918).

Pharmacological Considerations

None known.

Pregnancy and Lactation

No information on the use of butternut during pregnancy or lactation was identified in the scientific or traditional literature. While this review did not identify any concerns for pregnant or lactating women, safety has not been conclusively established.

Review Details

I. Drug and Supplement Interactions

Clinical Trials of Drug or Supplement Interactions

No clinical trials of drug or supplement interactions were identified.

Case Reports of Suspected Drug or Supplement Interactions

No case reports of suspected drug or supplement interactions were identified.

Animal Trials of Drug or Supplement Interactions

No animal trials of drug or supplement interactions were identified.

II. Adverse Events

Case Reports of Adverse Events

No case reports of adverse events were identified.

III. Pharmacology and Pharmacokinetics

Human Pharmacological Studies

No relevant human pharmacological studies were identified.

Animal Pharmacological Studies

No relevant animal pharmacological studies were identified.

In Vitro Pharmacological Studies

No relevant in vitro pharmacological studies were identified.

IV. Pregnancy and Lactation

No information on the safety of butternut during pregnancy or lactation was identified.

V. Toxicity Studies

No toxicity studies were identified.

Literature Cited

Felter, H.W., and J.U. Lloyd. 1898. *King's American dispensatory*. 18th ed., 3rd rev. 2 vols. Cincinnati: Ohio Valley Co.

Wood, H., and C. LaWall. 1918. *The dispensatory of the United States of America*. 21st ed. Philadelphia: Lippincott.

Juglans nigra L. — Juglandaceae

SCN: black walnut

Part: hull, leaf

Quick Reference Summary

Safety Class: 2d
Interaction Class: A

Contraindications

Not for long-term use (Blumenthal et al. 1998).

Other Precautions

None known.

Drug and Supplement Interactions

None known.

Notice

Tannins (~10% in leaf) (Blumenthal et al. 2000); *see* Appendix 1.

Adverse Events and Side Effects

Although the German Commission E reported that daily use of juglone-containing preparations of walnut bark has been associated with an increased occurrence of cancer of the tongue and leukoplakia of the lips (Blumenthal et al. 1998), no epidemiological studies or case reports investigating the association of exposure to juglone and cancer risk in humans were identified (TRI 1999). Walnut bark, however, does contain tannins, and tannins have been associated with increased incidences of certain cancers (Chung et al. 1998).

Topical application of walnut hull preparations may result in a temporary yellow or brown discoloration of the skin at the site of application (Blumenthal et al. 1998).

Contact dermatitis in reaction to contact with English walnut (*Juglans regia*) hulls has been reported (Bonamonte et al. 2001; Neri et al. 2006). The compound juglone, present in the juice of fresh walnut hulls, is regarded as a strong skin irritant and may cause itching or burning sensations (Neri et al. 2006).

Pharmacological Considerations

None known.

Pregnancy and Lactation

No information on the use of black walnut during pregnancy or lactation was identified in the scientific or traditional literature. While this review did not identify any concerns for pregnant or lactating women, safety has not been conclusively established.

Review Details

I. Drug and Supplement Interactions

Clinical Trials of Drug or Supplement Interactions

No clinical trials of drug or supplement interactions were identified.

Case Reports of Suspected Drug or Supplement Interactions

No case reports of suspected drug or supplement interactions were identified.

Animal Trials of Drug or Supplement Interactions

No animal trials of drug or supplement interactions were identified.

II. Adverse Events

Case Reports of Adverse Events

Contact dermatitis of the finger webs was reported in a man who had been picking black walnuts. The reaction was believed to be due to exposure to the juice of the hull, and no reaction occurred on rechallenge (Siegel 1954). Contact dermatitis to English walnut (*Juglans regia*) hulls has been reported (Bonamonte et al. 2001; Neri et al. 2006).

III. Pharmacology and Pharmacokinetics

Human Pharmacological Studies

No relevant human pharmacological studies were identified.

Animal Pharmacological Studies

A depressant effect of ether and petroleum ether extracts of black walnut hulls was observed in frogs, fish, mice, and rabbits (Auyong et al. 1963; Westfall et al. 1961). In rabbits, an intravenously administered dose of 0.07 mg/kg juglone was reported to induce mild sedation; the toxicity is such that the dose reported to induce profound sleep was very close to the lethal dose (Westfall et al. 1961).

Horses exposed to black walnut sawdust or shavings develop laminitis, a hemodynamic dysfunction of the laminar vasculature of the hoof (Peroni et al. 2005).

In Vitro Pharmacological Studies

No relevant in vitro pharmacological studies were identified.

IV. Pregnancy and Lactation

No information on the safety of black walnut in pregnancy or lactation was identified.

V. Toxicity Studies

Acute Toxicity

The LD_{50} of orally administered juglone in mice is 2.5 mg/kg (Westfall et al. 1961).

In dogs intravenously administered the compound juglone at a dose of 5 mg/kg, histopathological changes in the lungs and liver were observed, indicating that juglone is toxic to the cell membrane and increases capillary permeability (Boelkins et al. 1968).

Genotoxicity

Mutagenic activity of the compounds juglone and plumbagin was observed in *Salmonella typhimurium* strain TA2637 with, but not without, metabolic activation (Matsushima et al. 1986; Tikkanen et al. 1983). No mutagenic activity of juglone was observed in *S. typhimurium* strains TA98 or TA100 with or without activation (Edenharder and Tang 1997; Matsushima et al. 1986; Tikkanen et al. 1983).

Carcinogenicity

Topical application of the compound juglone to mice promoted skin tumors induced with the carcinogen DMBA (7,12-dimethylbenz[*a*]anthracene) (Monks et al. 1990; Van Duuren et al. 1978), although no tumors were observed in mice treated with juglone alone (Van Duuren et al. 1978).

Cytotoxicity

HaCaT keratinocyte cells exposed to the compounds juglone or plumbagin at concentrations of 0 to 20 μ*M* exhibited dose-dependent decrease in cell viability. The cytotoxicity of these compounds is due to two different mechanisms, namely, redox cycling and reaction with glutathione (Inbaraj and Chignell 2004).

Literature Cited

Auyong, T.K., B.A. Westfall, and R.L. Russell. 1963. Pharmacological aspects of juglone. *Toxicon* 1: 235-239.

Blumenthal, M., W. Busse, A. Goldberg, et al. 1998. *The complete German Commission E monographs*. Austin, TX: American Botanical Council.

Blumenthal, M., A. Goldberg, and J. Brinckmann. 2000. *Herbal medicine: Expanded Commission E monographs*. Newton, MA: Integrative Medicine.

Boelkins, J.N., L.K. Everson, and T.K. Auyong. 1968. Effects of intravenous juglone in the dog. *Toxicon* 6(2):99-102.

Bonamonte, D., C. Foti, and G. Angelini. 2001. Hyperpigmentation and contact dermatitis due to *Juglans regia*. *Contact Dermat.* 44(2):101-102.

Chung, K.T., T.Y. Wong, C.I. Wei, Y.W. Huang, and Y. Lin. 1998. Tannins and human health: A review. *Crit. Rev. Food Sci. Nutr.* 38(6):421-464.

Edenharder, R., and X. Tang. 1997. Inhibition of the mutagenicity of 2-nitrofluorene, 3-nitrofluoranthene and 1-nitropyrene by flavonoids, coumarins, quinones and other phenolic compounds. *Food Chem. Toxicol.* 35(3-4):357-372.

Inbaraj, J.J., and C.F. Chignell. 2004. Cytotoxic action of juglone and plumbagin: A mechanistic study using HaCaT keratinocytes. *Chem. Res. Toxicol.* 17(1):55-62.

Matsushima, T., M. Muramatsu, O. Yagame, et al. 1986. Mutagenicity and chemical structure relations of naturally occurring mutagens from plants. In *Progress in clinical and biological research. Genetic toxicology of environmental chemicals, Part B: Genetic effects and applied mutagenesis*, edited by Ramel, C., B. Lambert, and J. Magnusson. New York: Liss.

Monks, T.J., S.E. Walker, L.M. Flynn, C.J. Conti, and J. DiGiovanni. 1990. Epidermal ornithine decarboxylase induction and mouse skin tumor promotion by quinones. *Carcinogenesis* 11(10):1795-1801.

Neri, I., F. Bianchi, F. Giacomini, and A. Patrizi. 2006. Acute irritant contact dermatitis due to *Juglans regia*. *Contact Dermat.* 55(1):62-63.

Peroni, J.F., W.E. Harrison, J.N. Moore, et al. 2005. Black walnut extract-induced laminitis in horses is associated with heterogeneous dysfunction of the laminar microvasculature. *Equine Vet. J.* 37(6):546-551.

Siegel, J.M. 1954. Dermatitis due to black walnut juice. *AMA Arch. Derm. Syphilol.* 70(4):511-513.

Tikkanen, L., T. Matsushima, S. Natori, and K. Yoshihira. 1983. Mutagenicity of natural naphthoquinones and benzoquinones in the *Salmonella*/microsome test. *Mutat. Res.* 124(1):25-34.

TRI. 1999. Summary of data for chemical selection: Juglone: Technical Resources International, prepared for the National Toxicology Program.

Van Duuren, B.L., A. Segal, S.S. Tseng, et al. 1978. Structure and tumor-promoting activity of analogs of anthralin (1,8-dihydroxy-9-anthrone). *J. Med. Chem.* 21(1):26-31.

Westfall, B.A., R.L. Russell, and T.K. Auyong. 1961. Depressant agent from walnut hulls. *Science* 134:1617.

Juniperus spp. — Cupressaceae

Juniperus communis **L.**
SCN: juniper
AN: *hapusha*
OCN: common juniper

Juniperus monosperma **(Engelm.) Sarg.**
SCN: one-seed juniper
OCN: cherrystone juniper

Juniperus osteosperma **(Torr.) Little**
SCN: Utah juniper

Juniperus oxycedrus **L.**
SCN: cade juniper
OCN: prickly juniper
Part: fruit (fleshy cone)

Quick Reference Summary

Safety Class: 2b, 2d
Interaction Class: A

Contraindications

Not for use in pregnancy except under the supervision of a qualified healthcare practitioner (Weiss and Meuss 2001; Wichtl 2004; Williamson 2003).

Not for use exceeding 6 weeks in succession (Weiss and Meuss 2001).

Other Precautions

Use with caution in persons with inflammatory kidney disease (Bone 1995; Yarnell 2002).

Drug and Supplement Interactions

None known.

Notice

Diuretic (Janků et al. 1957, 1960; Yarnell 2002); *see* Appendix 2.

Editors' Notes

Although some herbal references indicate that juniper has been associated with kidney toxicity (Blumenthal et al. 1998; Wichtl 2004), analyses of the literature and studies of juniper essential oil indicate that toxicity concerns are based on case reports likely involving misidentified or adulterated juniper essential oil (Bone 1995; ESCOP 2003; Schilcher et al. 1993; Schilcher and Heil 1994; Wichtl 2004; Yarnell 2002). Some of these analyses suggest that *Juniperus* species should be used with caution in cases of acute kidney inflammation (Bone 1995; Weiss and Fintelmann 2000; Yarnell 2002).

Adverse Events and Side Effects

No adverse events associated with oral use of *Juniperus* species were identified, although one infant suffered pulmonary edema and cardiovascular collapse after a rectal enema of cade juniper essential oil (Rahmani et al. 2004).

Pharmacological Considerations

An animal study demonstrated that juniper may modify glucose regulation (De Medina et al. 1994). People with diabetes are advised to monitor their blood sugar closely and discuss the use of this herb with a qualified healthcare practitioner prior to use.

Pregnancy and Lactation

An animal study indicated anti-implantation and abortifacient activity of juniper (Agrawal et al. 1980).

No information on the safety of *Juniperus* species in lactation was identified. While this review did not identify any concerns for use while nursing, safety has not been conclusively established.

Review Details

I. Drug and Supplement Interactions

Clinical Trials of Drug or Supplement Interactions

No clinical trials of drug or supplement interactions were identified.

Case Reports of Suspected Drug or Supplement Interactions

No case reports of suspected drug or supplement interactions were identified.

Animal Trials of Drug or Supplement Interactions

No animal trials of drug or supplement interactions were identified.

II. Adverse Events

Case Reports of Adverse Events

A 32-year-old man with a history of nephrolithiasis developed fever, severe hypotension, renal failure, hepatotoxicity, and severe cutaneous burns on the face after ingestion of homemade "*Juniper oxycedrus* tar." No information on dose or further details on the product consumed was reported (Koruk et al. 2005).

Pulmonary edema and cardiovascular collapse was reported in a 4-month-old after a rectal enema of cade juniper essential oil. No information on dose was provided (Rahmani et al. 2004).

III. Pharmacology and Pharmacokinetics

Human Pharmacological Studies

In an irritation test, irritation reactions were observed in 2 of 20 volunteers patch tested with pure juniper essential oil for 24 h. No irritation reactions were observed in volunteers treated with 8% juniper oil in petrolatum in a 48-hour closed patch test (Johnson 2001).

Animal Pharmacological Studies

A reduction in glycemic levels in healthy rats was observed after administration of 250 mg/kg of a decoction of juniper. In diabetic rats, a reduction in blood glucose levels was observed after oral administration of a decoction of juniper equivalent to 125 mg/kg of berries daily for 24 days (De Medina et al. 1994).

In Vitro Pharmacological Studies

No adverse effects on normal renal mammalian fibroblasts or tubular epithelial cells were observed after treatment with methanol, water-methanol, or ethyl acetate extracts of juniper at concentrations of 1250 µg/ml (Wojcikowski et al. 2009).

IV. Pregnancy and Lactation

A dose-dependent reduction in the number of implantation sites was observed on gestational day 10 in rats orally administered 300 or 500 mg/kg of an ethanol extract of juniper daily on days 1 to 7 of pregnancy. In animals administered the same doses on days 1 to 7 and 14 to 16 of pregnancy, a dose-dependent reduction in implantations was observed on gestational day 18, with animals at the 500 mg/kg dose having no implants on gestational day 18. In animals administered the same doses only on days 14 to 16 of pregnancy, fetuses were aborted and no pups were born. No implantations were reported when female rats from the first experiment that had no implantations on gestational day 10 were mated after a 2-month rest period. In pups that were born from these experiments, no teratogenic effects were observed (Agrawal et al. 1980).

A reduction of implantation sites was observed in rats orally administered 200 mg/kg of an acetone extract of juniper daily on days 1 to 7 of pregnancy (Prakash 1986).

No information on the safety of *Juniperus* species during lactation was identified.

V. Toxicity Studies

Acute Toxicity

The LD_{50} of an intraperitoneally administered lyophilized aqueous extract of juniper to mice is 3 g/kg (Lasheras et al. 1986). No adverse effects were observed in rats orally administered 2.5 g/kg of an 80% ethanolic extract of juniper (Mascolo et al. 1987).

The LD_{50} of juniper oil in rats is 6.28 g/kg after oral administration and could not be determined at doses up to 5 g/kg after topical administration (Opdyke 1979).

Toxicity testing of 10% juniper oil (species unspecified, but likely *J. communis*) in corn oil gave LD_{50} values of 750 mg/kg after intraperitoneal administration to mice, 1200 mg/kg after intra-abdominal administration to guinea pigs, 700 mg/kg after intramuscular administration to mice, and 1440 mg/kg after subcutaneous administration to guinea pigs (Mambetsadykov et al. 1990).

The LD_{50} of the compound terpinen-4-ol is 0.75 ml/kg after subcutaneous administration in mice, 0.78 ml/kg after intramuscular administration in mice, and 1.5 ml/kg after intramuscular administration in rats (Janků et al. 1960).

Short-Term Toxicity

In nephrotoxicity testing, rats were orally administered 100, 333, or 1000 mg/kg of an essential oil with an α-pinene + β-pinene to terpinen-4-ol ratio of 3:1, 100, 300, or 900 mg/kg of an essential oil with an α-pinene + β-pinene to terpinen-4-ol ratio of 5:1, or 400 mg/kg of the compound terpinen-4-ol daily for 28 days. Pathological and histological investigations indicated no toxic effects at any of the dose levels tested (Schilcher and Leuschnerb 1997).

Amelioration of tacrolimus-induced nephrotoxicity was observed in rats fed diets containing 5% juniper oil for 5 weeks, with tacrolimus administered the last 2 weeks. In rats administered juniper oil, a complete reversal of the decrease in inulin clearance seen with tacrolimus was observed, along with relatively high levels of urinary prostaglandin $F_{2\alpha}$ excretion (Butani et al. 2003).

Literature Cited

Agrawal, O.P., S. Bharadwai, and R. Mathur. 1980. Antifertility effects of fruits of *Juniperus communis*. *Planta Med.* 40(Suppl.):98-101.

Blumenthal, M., W. Busse, A. Goldberg, et al. 1998. *The Complete German Commission E monographs*. Austin, TX: American Botanical Council.

Bone, K. 1995. Juniper berry is not a kidney irritant *Br. J. Phytother.* 4:47-48.

Butani, L., A. Afshinnik, J. Johnson, et al. 2003. Amelioration of tacrolimus-induced nephrotoxicity in rats using juniper oil. *Transplantation* 76(2):306-311.

De Medina, F.S., M.J. Gamez, I. Jimenez, et al. 1994. Hypoglycemic activity of juniper 'berries'. *Planta Med.* 60(3):197-200.

ESCOP. 2003. *ESCOP monographs: The scientific foundation for herbal medicinal products*. 2nd ed. Exeter, U.K.: European Scientific Cooperative on Phytotherapy.

Janků, I., M. Háva, R. Kraus, and O. Motl. 1960. The diuretic principle of juniper. *Naunyn-Schmiedeberg's Arch. Pharmacol.* 238(1):112-113.

Janků, I., M. Hava, and O. Motl. 1957. Diuretic substance from juniper (*Juniperus communis* L.). *Experientia* 13(6):255.

Johnson, W. 2001. Final report on the safety assessment of *Juniperus communis* extract, *Juniperus oxycedrus* extract, *Juniperus oxycedrus* tar, *Juniperus phoenicea* extract, and *Juniperus virginiana* extract. *Int. J. Toxicol.* 20(Suppl. 2):41-56.
Koruk, S.T., E. Ozyilkan, P. Kaya, et al. 2005. Juniper tar poisoning. *Clin. Toxicol.* 43(1):47-49.
Lasheras, B., P. Turillas, and E. Cenarruzabeitia. 1986. Preliminary pharmacological study of *Prunus spinosa* L., *Amelanchier ovalis* Medikus, *Juniperus communis* L., and *Urtica dioica* L. *Plant. Med. Phytother.* 20:219-226.
Mambetsadykov, M.B., E.S. Matyev, M.A. Orozov, et al. 1990. Chemical composition and pharmacological properties of common juniper essential oil. *Khim. Farm. Zh.* 24(9):59-60.
Mascolo, N., G. Autore, F. Capasso, A. Menghini, and M.P. Fasulo. 1987. Biological screening of Italian medicinal plants for anti-inflammatory activity. *Phytother. Res.* 1(1):28-31.
Opdyke, D.L.J. 1979. *Monographs on fragrance raw materials*. New York: Pergamon.
Prakash, A.O. 1986. Potentialities of some indigenous plants for antifertility activity. *Int. J. Crude Drug. Res.* 24(1):19-24.
Rahmani, H., S. Leonhardt, D. Beladdale, et al. 2004. Severe acute lung oedema after rectal enema with cade oil. *J. Toxicol. Clin. Toxicol.* 42(4):487.
Schilcher, H., D. Emmrich, and C. Koehler. 1993. GLC comparison of commercially available juniper oils and their toxicological evaluation. *PZ Wissenschaft* 138(3-4):85-91.
Schilcher, H., and B.M. Heil. 1994. Nephrotoxicity of juniper berry preparations: A critical review of the literature from 1844 to 1993. *Z. Phytother.* 15 (4):205-208+211.
Schilcher, H., and F. Leuschnerb. 1997. Studies of potential nephrotoxic effects of essential juniper oil. *Arz. Forsch.* 47(7):855-858.
Weiss, R.F., and V. Fintelmann. 2000. *Herbal medicine*. 2nd ed. New York: Thieme.
Weiss, R.F., and A.R. Meuss. 2001. *Weiss's herbal medicine*. Classic ed. Stuttgart: New York.
Wichtl, M. 2004. *Herbal drugs and phytopharmaceuticals: A handbook for practice on a scientific basis*. 3rd ed. Boca Raton, FL: CRC Press.
Williamson, E.M. 2003. *Potter's herbal cyclopedia*. Saffron Walden, Essex: C.W. Daniel Co.
Wojcikowski, K., H. Wohlmuth, D.W. Johnson, M. Rolfe, and G. Gobe. 2009. An in vitro investigation of herbs traditionally used for kidney and urinary system disorders: Potential therapeutic and toxic effects. *Nephrology* 14(1):70-79.
Yarnell, E. 2002. Botanical medicines for the urinary tract. *World J. Urol.* 20(5):285-293.

Juniperus virginiana L. — Cupressaceae

SCN: eastern red cedar
OCN: pencil cedar; Virginia redcedar
Part: leaf, berry

Quick Reference Summary

Safety Class: 2b, 2d
Interaction Class: A

Contraindications

Not for use in pregnancy except under the supervision of a qualified healthcare practitioner (Moerman 1998; Wood and LaWall 1926).

Not for use exceeding six weeks in succession (Weiss and Meuss 2001).

Other Precautions

Use with caution in persons with inflammatory kidney disease (Bone 1995; Yarnell 2002).

Drug and Supplement Interactions

None known.

Notice

Abortifacient (Moerman 1998; Wood and LaWall 1926); *see* Appendix 2.

Editors' Note

Information on eastern red cedar is limited. Based on chemical similarity, the data and classifications that apply to other species of juniper (*Juniperus* spp.) are believed to be applicable to eastern red cedar.

Adverse Events and Side Effects

None known.

Pharmacological Considerations

None known.

Pregnancy and Lactation

Eastern red cedar has been reported to have abortifacient activity (Moerman 1998; Wood and LaWall 1926).

No information on the safety of eastern red cedar during lactation was identified. While this review did not identify any concerns for use while nursing, safety has not been conclusively established.

Review Details

I. Drug and Supplement Interactions

Clinical Trials of Drug or Supplement Interactions

No clinical trials of drug or supplement interactions were identified.

Case Reports of Suspected Drug or Supplement Interactions

No case reports of suspected drug or supplement interactions were identified.

Animal Trials of Drug or Supplement Interactions

No animal trials of drug or supplement interactions were identified.

II. Adverse Events

Case Reports of Adverse Events

Overdose after ingestion of eastern red cedar essential oil has been associated with burning in the stomach, vomiting, convulsions, coma, and a slow pulse. Fatal cases have been reported, although details on the amount consumed are lacking (Wood and LaWall 1926).

III. Pharmacology and Pharmacokinetics

Human Pharmacological Studies

Several sensitization studies with 5 to 10% eastern red cedar essential oil indicated sensitizing activity in zero to 2% of volunteers (Johnson 2001; Opdyke 1979).

Several skin irritation studies with 0.2 to 20% eastern red cedar essential oil indicated no irritant activity (Johnson 2001; Opdyke 1979).

Animal Pharmacological Studies

An increase in hexobarbital metabolism was observed in mice exposed to corncob bedding that had been sprayed with an ether extract of eastern red cedar (Wade et al. 1968).

In Vitro Pharmacological Studies

No relevant in vitro pharmacological studies were identified.

IV. Pregnancy and Lactation

Eastern red cedar has been reported to have abortifacient activity. No details were listed on the part used, dose, or whether the plant was used singly or in formula (Moerman 1998; Wood and LaWall 1926).

No information on the safety of eastern red cedar during lactation was identified.

V. Toxicity Studies

Acute Toxicity

The LD_{50} of eastern red cedar essential oil could not be determined at orally or topically administered doses up to 5 g/kg (Opdyke 1979).

Literature Cited

Bone, K. 1995. Juniper berry is not a kidney irritant. *Brit. J. Phytother.* 4:47-48.

Johnson, W. 2001. Final report on the safety assessment of *Juniperus communis* extract, *Juniperus oxycedrus* extract, *Juniperus oxycedrus* tar, *Juniperus phoenicea* extract, and *Juniperus virginiana* extract. Int. J. Toxicol. 20 Suppl 2:41-56.

Moerman, D.E. 1998. *Native American ethnobotany*. Portland, OR: Timber Press.

Opdyke, D.L.J. 1979. *Monographs on fragrance raw materials*. New York: Pergamon Press.

Wade, A.E., Holl, J.E., Hilliard, C.C., Molton, E., Greene, F.E. 1968. Alteration of drug metabolism in rats and mice by an environment of cedarwood. *Pharmacol.* 1(5):317-328.

Weiss, R.F., Meuss, A.R. 2001. *Weiss's herbal medicine*. Stuttgart: New York.

Wood, H., LaWall, C. 1926. *The dispensatory of the United States of America*. Philadelphia: J.B. Lippincott.

Yarnell, E. 2002. Botanical medicines for the urinary tract. *World J. Urol.* 20(5):285-293.

Kaempferia galanga L. — Zingiberaceae

SCN: *Kaempferia galanga*
PN: *shan nai* (rhizome)
OCN: resurrection lily
Part: rhizome

Quick Reference Summary

Safety Class: 1
Interaction Class: A

Contraindications

None known.

Other Precautions

None known.

Drug and Supplement Interactions

None known.

Adverse Events and Side Effects

None known.

Pharmacological Considerations

None known.

Pregnancy and Lactation

No information on the safety of *Kaempferia galanga* during pregnancy or lactation was identified in the scientific or traditional literature. While this review did not identify any concerns for pregnant or lactating women, safety has not been conclusively established.

K

Review Details

I. Drug and Supplement Interactions

Clinical Trials of Drug or Supplement Interactions

No clinical trials of drug or supplement interactions were identified.

Case Reports of Suspected Drug or Supplement Interactions

No case reports of suspected drug or supplement interactions were identified.

Animal Trials of Drug or Supplement Interactions

No animal trials of drug or supplement interactions were identified.

II. Adverse Events

Case Reports of Adverse Events

No case reports of adverse events were identified.

III. Pharmacology and Pharmacokinetics

Human Pharmacological Studies

No relevant human pharmacological studies were identified.

Animal Pharmacological Studies

In an irritation study, no signs of irritation were observed in rabbits topically treated with the hexane fraction of *Kaempferia galanga* (Kanjanapothi et al. 2004).

In Vitro Pharmacological Studies

No relevant in vitro pharmacological studies were identified.

IV. Pregnancy and Lactation

No information on the use of *Kaempferia galanga* during pregnancy or lactation was identified.

V. Toxicity Studies

Acute Toxicity

The LD_{50} of orally administered ethanol extract of *Kaempferia galanga* in rats could not be determined at doses up to 5 g/kg (Kanjanapothi et al. 2004).

Short-Term Toxicity

No changes in organ weights or serum chemistry or histopathological changes were observed in rats orally administered 25, 50, or 100 mg/kg of an ethanol extract of *Kaempferia galanga* daily for 28 days. In the 50 and 100 mg/kg groups, a slight decrease in lymphocyte counts was observed (Kanjanapothi et al. 2004).

Literature Cited

Kanjanapothi, D., A. Panthong, N. Lertprasertsuke, et al. 2004. Toxicity of crude rhizome extract of *Kaempferia galanga* L. (Proh Hom). *J. Ethnopharmacol.* 90(2-3):359-365.

Krameria spp. — Krameriaceae

Krameria argentea Mart. ex Spreng.
SCN: rhatany
OCN: Brazilian krameria; Brazilian rhatany; brown rhatany
Krameria lappacea (Dombey) Burdet & B.B. Simpson
SCN: rhatany
Syn: *Krameria triandra* Ruiz & Pav.
OCN: Peruvian krameria; Peruvian rhatany
Part: root

Quick Reference Summary

Safety Class: 1
Interaction Class: A

Contraindications

None known.

Other Precautions

None known.

Drug and Supplement Interactions

None known.

Notice

Tannins (10.0–15.0% in *Krameria lappacea*) (Wichtl 2004); *see* Appendix 1.

Adverse Events and Side Effects

Undiluted rhatany tincture may cause local irritation (De Smet 1993; Wichtl 2004).

Allergic reactions of the mucous membranes have been reported and confirmed in persons using topical products containing rhatany (Bujan et al. 1998; Grolnick 1938; Wichtl 2004).

Pharmacological Considerations

None known.

Pregnancy and Lactation

No information on the use of rhatany during pregnancy or lactation was identified in the scientific or traditional literature. While this review did not identify any concerns for pregnant or lactating women, safety has not been conclusively established.

Review Details

I. Drug and Supplement Interactions

Clinical Trials of Drug or Supplement Interactions

No clinical trials of drug or supplement interactions were identified.

Case Reports of Suspected Drug or Supplement Interactions

No case reports of suspected drug or supplement interactions were identified.

Animal Trials of Drug or Supplement Interactions

No animal trials of drug or supplement interactions were identified.

II. Adverse Events

Case Reports of Adverse Events

Allergic reactions of the mucous membranes, with patch tests confirming rhatany as the allergen, have been reported in persons using topical products containing rhatany (Bujan et al. 1998; Grolnick 1938; Wichtl 2004).

III. Pharmacology and Pharmacokinetics

Human Pharmacological Studies

No relevant human pharmacological studies were identified.

Animal Pharmacological Studies

No relevant animal pharmacological studies were identified.

In Vitro Pharmacological Studies

No relevant in vitro pharmacological studies were identified.

IV. Pregnancy and Lactation

No information on the use of rhatany during pregnancy or lactation was identified.

V. Toxicity Studies

No toxicity studies were identified.

Literature Cited

Bujan, J.J.G., J.M.O. Morante, I.Y. Bayona, M.G. Guemes, and R.S. Arechavala. 1998. Allergic contact dermatitis from *Krameria triandra* extract. *Contact Dermat.* 38(2):120.

De Smet, P.A.G.M. 1993. *Adverse effects of herbal drugs, Volume 2.* Berlin: Springer.

Grolnick, M. 1938. Dermatitis due to hemorrhoidal ointment containing krameria and oil of cade. *J. Am. Med. Assoc.* 110(13):951.

Wichtl, M. 2004. *Herbal drugs and phytopharmaceuticals: A handbook for practice on a scientific basis.* 3rd ed. Boca Raton, FL: CRC Press.

Lactuca spp. Asteraceae

Lactuca quercina L.
SCN: wild lettuce

Lactuca serriola L.
SCN: wild lettuce
OCN: prickly lettuce

Lactuca virosa L.
SCN: wild lettuce
Part: herb

Quick Reference Summary

Safety Class: 1
Interaction Class: A

Contraindications
None known.

Other Precautions
None known.

Drug and Supplement Interactions
None known.

Editors' Notes
Lactucarium, the dried latex of wild lettuce (typically, *L. virosa*), is a brown substance that physically resembles opium and is sometimes referred to as "lettuce opium." While several historical texts have noted "narcotic" properties of lactucarium (Felter and Lloyd 1898; Scudder 1898), the U.S. Dispensatory notes, "we believe that the general experience is in accord with our own in finding it to be almost devoid of narcotic properties" (Wood and LaWall 1918). Although early investigation indicated the presence of small amounts of hyoscyamine-like alkaloids (Farr and Wright 1904), more recent studies have not identified any alkaloids (Frohne and Pfänder 2000; Stojakowska et al. 1999). Wild lettuce is acrid when fresh and becomes less acrid on drying.

Adverse Events and Side Effects
Intravenous administration of wild lettuce extract led to fevers, chills, abdominal and back pain, neck stiffness, headache, elevated white blood cell counts, and mild liver function abnormalities (Mullins and Horowitz 1998). Such effects are not expected after oral use of wild lettuce.

Pharmacological Considerations
None known.

Pregnancy and Lactation
No information on the use of wild lettuce during pregnancy or lactation was identified in the scientific or traditional literature. While this review did not identify any concerns for pregnant or lactating women, safety has not been conclusively established.

Review Details

I. Drug and Supplement Interactions

Clinical Trials of Drug or Supplement Interactions
No clinical trials of drug or supplement interactions were identified.

Case Reports of Suspected Drug or Supplement Interactions
No case reports of suspected drug or supplement interactions were identified.

Animal Trials of Drug or Supplement Interactions
No animal trials of drug or supplement interactions were identified.

II. Adverse Events

Case Reports of Adverse Events
In a case series of nine patients, ages 12 to 38, reported to have eaten wild lettuce (*L. virosa*), adverse events were reported including mydriasis, dizziness, anxiety, urinary retention, decreased bowel sounds, and sympathetic overactivity. Amounts of the plant material ingested were not specified. A personal communication with the author indicated that the plant material was obtained and positively identified (Besharat et al. 2009).

Several young adult intravenous drug users had fevers, chills, abdominal and back pain, neck stiffness, headache, leukocytosis, and mild liver function abnormalities after intravenous administration of approximately 1 ml of an aqueous extract of wild lettuce powder. One also drank a

L

cup of the aqueous extract, and one also injected an evaporated ethanol extract of valerian. All three of the patients also took 2 to 6 pain relief tablets (325 mg acetaminophen, 30 mg caffeine, and 8 mg codeine), which one used chronically for headache relief. The authors of the report indicated that the plant material was labeled as *Lactuca virosa* but the identity was not confirmed (Mullins and Horowitz 1998).

III. Pharmacology and Pharmacokinetics

Human Pharmacological Studies

No relevant human pharmacological studies were identified.

Animal Pharmacological Studies

No relevant animal pharmacological studies were identified.

In Vitro Pharmacological Studies

No relevant in vitro pharmacological studies were identified.

IV. Pregnancy and Lactation

No information on the use of wild lettuce during pregnancy or lactation was identified.

V. Toxicity Studies

Acute Toxicity

The LD_{50} values of the compound lactucin administered to mice are 900 mg/kg after oral administration, 50 mg/kg after subcutaneous administration, and 15 mg/kg after intravenous administration. For the compound lactucopicrin, the LD_{50} values are 12 to 20 g/kg after oral administration, 3 g/kg after subcutaneous administration, and 1.5 to 2 g/kg after intravenous administration. Sublethal doses of these compounds had no significant effect on respiration or blood pressure, and toxic doses had no effect on the motor nerves or intestinal motility (Forst 1940).

Literature Cited

Besharat, S., M. Besharat, and A. Jabbari. 2009. Wild lettuce (*Lactuca virosa*) toxicity. *BMJ Case Rep.* 2009.

Farr, E.H., and R. Wright. 1904. The doubted presence of a mydriatic alkaloid in *Lactuca virosa*. *Pharm. J.* 18:186-187.

Felter, H.W., and J.U. Lloyd. 1898. *King's American dispensatory*. 18th ed., 3rd rev. 2 vols. Cincinnati: Ohio Valley Co.

Forst, A.W. 1940. [Pharmacological investigation of *Lactuca virosa*.] *Arch. Exp. Pathol. Pharmakol.* 195:1-25.

Frohne, D., and H.J. Pfänder. 2000. *A colour atlas of poisonous plants: A handbook for pharmacists, doctors, toxicologists, biologists and veterinarians*. 2nd ed. London: Manson.

Mullins, M.E., and B.Z. Horowitz. 1998. The case of the salad shooters: Intravenous injection of wild lettuce extract. *Vet. Hum. Toxicol.* 40(5):290-291.

Scudder, J.M. 1898. *American Eclectic materia medica and therapeutics*. Cincinnati: The Scudder Brothers Company.

Stojakowska, A., J. Malarz, and W. Kisiel. 1999. Culture and production of sesquiterpene lactones. In *Medicinal and aromatic plants*, edited by Bajaj, Y.P.S. Berlin: Springer.

Wood, H., and C. LaWall. 1918. *The dispensatory of the United States of America*. 21st ed. Philadelphia: Lippincott.

Laminaria spp., *Nereocystis* sp. — Laminariaceae, Lessoniaceae

Laminaria digitata (Huds.) J.V. Lamour.
SCN: kelp
OCN: horsetail kelp; seawand; tangle

Laminaria hyperborea **(Gunnerus) Foslie**
SCN: kelp
Syn: *Laminaria cloustonii* Edmonston

Laminaria japonica **Aresch.**
SCN: kombu
PN: *kun bu* (thallus)
OCN: Japanese kelp; Japanese sea tangle

Laminaria setchellii **P.C. Silva**
SCN: kelp
Syn: *Laminaria dentigera* Kjellm.

Laminaria sinclairii **(Harv.) Farl. et al.**
SCN: kelp

Nereocystis luetkeana **(Mert.) Postels & Rupr.**
SCN: bull kelp
Part: thallus

Quick Reference Summary

Safety Class: 2d
Interaction Class: A

Contraindications

Not for use in persons with hyperthyroidism (Ishizuki et al. 1989; Shilo and Hirsch 1986; Teas et al. 2004).

Other Precautions

Take with at least 250 ml (8 oz) of liquid (CFR 2011a).

Drug and Supplement Interactions

None known, although use with thyroid medications should be under the supervision of a qualified healthcare practitioner (Ishizuki et al. 1989; Shilo and Hirsch 1986; Shimizu et al. 2003; Teas et al. 2004).

Other drugs should be taken 1 hour prior to consumption of kelp, bull kelp, or kombu, or several hours after consumption, as mucilaginous plants such as kelp, bull kelp, or kombu may slow the absorption of orally administered drugs (Brinker 2001; De Smet 1993; Mills and Bone 2005).

Notice

Mucilages (Felter and Lloyd 1898); *see* Appendix 3.

Editors' Notes

Sections of kelp stems (known as laminaria tents) are used to dilate the cervix to induce labor or for surgical procedures including abortions (Boulvain et al. 2001). For use in this manner, the stem sections are inserted into the cervix. Safety of that use of kelp is not covered in this entry.

A number of seaweeds have been found to contain heavy metal residues (Almela et al. 2006; Rose et al. 2007). Total arsenic has been measured in kelp and kombu products at levels ranging from 40 to 105 mg/kg (Almela et al. 2006), though the inorganic form of arsenic is reported at much lower levels of 0.14–0.47 mg/kg (Almela et al. 2006) or as not detectable at a 0.3 mg/kg limit of detection (Rose et al. 2007). Evidence about the negative impact of inorganic arsenic on fetal health and infant development is emerging (EFSA 2009). Caution is therefore recommended in pregnancy and for young children.

U.S. regulations require that kelp directed for use as a dietary supplement must contain 225 µg or less of iodine per daily dose for supplements labeled without reference to age or physiological state. For supplements labeled for use by specific populations, iodine daily dose levels should not exceed 45 µg for infants, 105 µg for children under 4 years, 225 µg for adults and children over 4 years, and 300 µg for pregnant or nursing women (CFR 2011b).

Specific labeling is required in the United States for all over-the-counter drug products containing kelp and kombu (CFR 2011a); *see* Bulk-forming laxatives in Appendix 2.

Adverse Events and Side Effects

Cases of hyperthyroidism, hypothyroidism, and iodine-induced thyroid toxicity have been reported in association with excessive use of kelp and kombu (De Smet et al. 1990; Eliason 1998; Ishizuki et al. 1989; Mussig et al. 2006; Salas Coronas et al. 2002; Shilo and Hirsch 1986; Shimizu et al. 2003).

Kelp and kombu contain the element iodine (0.14–0.44% dry weight) (Dawczynski et al. 2007; Teas et al. 2004). Iodine idiosyncrasy, hyperthyroidism, and thyroid toxicity have been indicated as possible side effects after long-term, uncontrolled use of iodine-containing products, such as kelp, bull kelp, or kombu (Wichtl 2004). Excessive consumption of iodine can lead to goiter (swelling of the thyroid gland) (Baker 2004; Pennington 1990).

Pharmacological Considerations

None known.

Pregnancy and Lactation

Infants are reported to be particularly sensitive to the effects of iodine (Baker 2004), and cases of infant goiter after maternal use of iodine during pregnancy or lactation have been reported (Pennington 1990). *Also see* Editors' Notes for this entry.

Review Details

I. Drug and Supplement Interactions

Clinical Trials of Drug or Supplement Interactions

No clinical trials of drug or supplement interactions were identified.

Case Reports of Suspected Drug or Supplement Interactions

No case reports of suspected drug or supplement interactions were identified.

Animal Trials of Drug or Supplement Interactions

No animal trials of drug or supplement interactions were identified.

II. Adverse Events

Case Reports of Adverse Events

Cases of hyperthyroidism have been reported in association with use of kelp (De Smet et al. 1990; Eliason 1998; Salas Coronas et al. 2002; Shilo and Hirsch 1986). In one case, hyperthyroidism developed in a woman taking six 200 mg tablets of kelp daily (De Smet et al. 1990). In a second case, a 72-year-old woman ingested kelp tablets with an iodine dose of 2.8 to 4.2 mg daily for 6 months (Shilo and Hirsch 1986). The U.S. recommended dietary allowance for iodine is 150 µg/day (IOM 2001).

Hypothyroidism was reported in a 79-year-old woman who had been ingesting approximately 30 g of kombu weekly for approximately 1 year. Thyroid function tests indicated lowered levels of triiodothyronine (T_3) and thyroxine (T_4) and an increased level of thyroid-stimulating hormone (TSH), along with high blood levels of iodine. Symptoms resolved after cessation of kombu (Shimizu et al. 2003).

Iodine-induced thyroid toxicity was reported in a 39-year-old woman who was taking a combination of Chinese herbs that contained kombu and two other seaweeds in addition to other herbs. Based on the formula, the estimated iodine intake was 580 to 990 µg daily (Mussig et al. 2006). Two women, ages 42 and 59, developed iodine-induced thyroid toxicity 1 month and 1 year, respectively, after having eaten foods containing 28–140 mg/day of iodine, calculated from their daily diet. Both patients had high concentrations of serum T_3, low ratios of serum T_3/T_4, and high blood levels of iodine. Kombu was part of the diet of these women, although their average intake was not reported in the available English language case synopses. Their thyrotoxic signs and symptoms disappeared and thyroid hormone levels normalized 1 month after the prohibition of kombu intake (Ishizuki et al. 1989).

Arsenic poisoning was reported in a 54-year-old woman taking two to four 41 mg capsules of kelp supplement daily for several months. Analysis of the supplement being taken indicated that the supplement contained an arsenic concentration of 8.5 mg/kg, although the amounts of organic and inorganic arsenic were not reported (Amster et al. 2007). Commenting on this case report, other authors noted that the patient was taking two to four times the recommended dose, and that iodine toxicity may have been the cause of the symptoms (McGuffin and Dentali 2007), although the reporting physicians replied that iodine toxicity was not believed to be the cause of the reported symptoms (Schenker et al. 2007).

III. Pharmacology and Pharmacokinetics

Human Pharmacological Studies

In healthy volunteers with normal thyroid function orally administered two or four kelp capsules (containing 660 or 1320 µg of iodine) daily for 4 weeks, dose-related increases in TSH levels and urinary iodine levels were observed. In the high-dose group, a decrease in triiodothyronine and increase in poststimulation TSH response in the thyrotropin-releasing hormone stimulation test were observed. No changes in free thyroxine were reported. Two weeks after cessation of treatment, TSH levels in the high-dose kelp group were decreased, while all other thyroid values were normal for all groups (Clark et al. 2003).

In healthy volunteers ingesting 15 or 30 g of kombu (35 or 70 mg iodine) daily for 7 to 10 days, a significant increase in TSH concentrations was observed. In some cases, TSH levels exceeded normal limits. Serum levels of T_4 and T_3 were slightly decreased and within normal limits. During long-term administration (55–87 days), TSH levels were elevated and sustained while the T_4 and T_3 levels were essentially unchanged. Urinary excretion of iodine significantly increased during ingestion of kombu and returned to normal 7 to 40 days after cessation of kombu (Miyai et al. 2008).

Animal Pharmacological Studies

No relevant animal pharmacological studies were identified.

In Vitro Pharmacological Studies

An aqueous extract of kelp exhibited negative chronotropic effects on isolated right atria but no effect on atrial tension (Chiu and Fung 1997).

IV. Pregnancy and Lactation

Infants are reported to be particularly sensitive to the effects of iodine (Baker 2004), and cases of infant goiter after maternal use of iodine during pregnancy or lactation have been reported (Pennington 1990). *Also see* Editors' Notes for this entry.

V. Toxicity Studies

No toxicity studies were identified.

Literature Cited

Almela, C., M.J. Clemente, D. Velez, and R. Montoro. 2006. Total arsenic, inorganic arsenic, lead and cadmium contents in edible seaweed sold in Spain. *Food Chem. Toxicol.* 44(11):1901-1908.

Amster, E., A. Tiwary, and M.B. Schenker. 2007. Case report: Potential arsenic toxicosis secondary to herbal kelp supplement. *Environ. Health Perspect.* 115(4):606-608.

Baker, D.H. 2004. Iodine toxicity and its amelioration. *Exp. Biol. Med.* 229(6):473-478.

Boulvain, M., A. Kelly, C. Lohse, C. Stan, and O. Irion. 2001. Mechanical methods for induction of labour. *Cochrane Database Syst. Rev.* 4:CD001233.

Brinker, F. 2001. *Herb contraindications and drug interactions.* 3rd ed. Sandy, OR: Eclectic Medical Publications.

CFR. 2011a. *Code of federal regulations*, Title 21 Part 201.319, 2011 ed. Specific labeling requirements for specific drug products. Water-soluble gums, hydrophilic gums, and hydrophilic mucilloids (including, but not limited to agar, alginic acid, calcium polycarbophil, carboxymethylcellulose sodium, carrageenan, chondrus, glucomannan ((B-1,4 linked) polymannose acetate), guar gum, karaya gum, kelp, methylcellulose, plantago seed (psyllium), polycarbophil tragacanth, and xanthan gum) as active ingredients; required warnings and directions. Washington, DC: U.S. Government Printing Office.

CFR. 2011b. *Code of federal regulations*, Title 21 Part 172.365, 2011 ed. Food additives permitted for direct addition to food for human consumption. Special dietary and nutritional additives. Kelp. Washington, DC: U.S. Government Printing Office.

Chiu, K.W., and A.Y. Fung. 1997. The cardiovascular effects of green beans (*Phaseolus aureus*), common rue (*Ruta graveolens*), and kelp (*Laminaria japonica*) in rats. *Gen. Pharmacol.* 29(5):859-862.

Clark, C.D., B. Bassett, and M.R. Burge. 2003. Effects of kelp supplementation on thyroid function in euthyroid subjects. *Endocr. Pract.* 9(5):363-369.

Dawczynski, C., U. Schaefer, M. Leiterer, and G. Jahreis. 2007. Nutritional and toxicological importance of macro, trace, and ultra-trace elements in algae food products. *J. Agric. Food Chem.* 55(25):10470-10475.

De Smet, P.A., B.H. Stricker, F. Wilderink, and W.M. Wiersinga. 1990. Hyperthyroidism during treatment with kelp tablets. *Ned. Tijdschr. Geneeskd.* 134(21):1058-1059.

De Smet, P.A.G.M. 1993. *Adverse effects of herbal drugs, Volume 2.* Berlin: Springer.

EFSA. 2009. Scientific opinion on arsenic in food. *EFSA J.* 7(10):1351-1550.

Eliason, B.C. 1998. Transient hyperthyroidism in a patient taking dietary supplements containing kelp. *J. Am. Board Fam. Pract.* 11(6):478-480.

Felter, H.W., and J.U. Lloyd. 1898. *King's American dispensatory*. 18th ed., 3rd rev. 2 vols. Cincinnati: Ohio Valley Co.

IOM. 2001. *Iodine. Dietary reference intakes for vitamin A, vitamin K, boron, chromium, copper, iodine, iron, manganese, molybdenum, nickel, silicon, vanadium, and zinc.* Food and Nutrition Board, Institute of Medicine. Washington, DC: National Academy Press.

Ishizuki, Y., K. Yamauchi, and Y. Miura. 1989. Transient thyrotoxicosis induced by Japanese kombu. *Nippon Naibunpi Gakkai Zasshi* 65(2):91-98.

McGuffin, M., and S. Dentali. 2007. Safe use of herbal kelp supplements. *Environ. Health Perspect.* 115(12):A575-A576; author reply A576-A577.

Mills, S., and K. Bone. 2005. *The essential guide to herbal safety*. St. Louis: Elsevier.

Miyai, K., T. Tokushige, and M. Kondo. 2008. Suppression of thyroid function during ingestion of seaweed "kombu" (*Laminaria japonica*) in normal Japanese adults. *Endocrine J.* 55(6):1103.

Mussig, K., C. Thamer, R. Bares, et al. 2006. Iodine-induced thyrotoxicosis after ingestion of kelp-containing tea. *J. Gen. Intern. Med.* 21(6):C11-14.

Pennington, J.A. 1990. A review of iodine toxicity reports. *J. Am. Diet. Assoc.* 90(11):1571.

Rose, M., J. Lewis, N. Langford, et al. 2007. Arsenic in seaweed—Forms, concentration and dietary exposure. *Food Chem. Toxicol.* 45(7):1263-1267.

Salas Coronas, J., G. Cruz Caparros, F. Laynez Bretones, and F. Diez Garcia. 2002. [Hyperthyroidism secondary to kelp tablets ingestias.] *Med. Clin. (Barc.)* 118(20):797-798.

Schenker, M., E. Amster, and A. Tiwary. 2007. Arsenic in herbal kelp supplements: Schenker et al. respond. *Environ. Health Perspect.* 115(12):A576.

Shilo, S., and H.J. Hirsch. 1986. Iodine-induced hyperthyroidism in a patient with a normal thyroid gland. *Postgrad. Med. J.* 62(729):661-662.

Shimizu, C., K. Yamaji, K. Sugi, et al. 2003. A case of hypothyroidism by a large intake of kombu. *Jap. J. Clin. Exp. Med.* 80(3):460-462.

Teas, J., S. Pino, A. Critchley, and L.E. Braverman. 2004. Variability of iodine content in common commercially available edible seaweeds. *Thyroid* 14(10):836-841.

Wichtl, M. 2004. *Herbal drugs and phytopharmaceuticals: A handbook for practice on a scientific basis*. 3rd ed. Boca Raton, FL: CRC Press.

Lamium album L. — Lamiaceae

SCN: white nettle
OCN: dead nettle

Part: herb

Quick Reference Summary

Safety Class: 1
Interaction Class: A

Contraindications

None known.

Other Precautions

None known.

Drug and Supplement Interactions

None known.

Editors' Note

A reference text on British herbal medicine notes that white nettle is generally considered harmless and that the tender young leaves can be eaten like spinach (Bradley 2006).

Adverse Events and Side Effects

None known.

Pharmacological Considerations

None known.

Pregnancy and Lactation

No information on the safety of white nettle during pregnancy or lactation was identified in the scientific or traditional literature. While this review did not identify any concerns for pregnant or lactating women, safety has not been conclusively established.

Review Details

I. Drug and Supplement Interactions

Clinical Trials of Drug or Supplement Interactions

No clinical trials of drug or supplement interactions were identified.

Case Reports of Suspected Drug or Supplement Interactions

No case reports of suspected drug or supplement interactions were identified.

Animal Trials of Drug or Supplement Interactions

No animal trials of drug or supplement interactions were identified.

II. Adverse Events

Case Reports of Adverse Events

No case reports of adverse events were identified.

III. Pharmacology and Pharmacokinetics

Human Pharmacological Studies

No relevant human pharmacological studies were identified.

Animal Pharmacological Studies

No relevant animal pharmacological studies were identified.

In Vitro Pharmacological Studies

No relevant in vitro pharmacological studies were identified.

IV. Pregnancy and Lactation

No information on the safety of white nettle during pregnancy or lactation was identified.

V. Toxicity Studies

No toxicity studies were identified.

Literature Cited

Bradley, P.R. 2006. *British herbal compendium: A handbook of scientific information on widely used plant drugs*. Bournemouth, UK: British Herbal Medicine Association.

Larrea tridentata (Sessé & Moç. ex DC.) Coville — Zygophyllaceae

SCN: chaparral
Syn: *Larrea mexicana* Moric.
OCN: creosote bush
Part: leaf

Quick Reference Summary

Safety Class: 2b, 2d
Interaction Class: A

Contraindications

Not for use in pregnancy except under the supervision of a qualified healthcare practitioner (Heron and Yarnell 2001).

Not for use in persons with preexisting kidney disease or liver conditions, such as hepatitis or cirrhosis (De Smet 1993; McGuffin et al. 1997).

Other Precautions

None known.

Drug and Supplement Interactions

None known.

Editors' Notes

Reports of acute liver toxicity associated with consumption of chaparral surfaced from 1990 through 1997, leading to the issuance of a warning by the FDA to cease consumption of chaparral (FDA 1992). The American Herbal Products Association (AHPA) initiated a review of four cases (Watt et al. 1994) and found the reported toxicity to be due to idiosyncratic reactions in persons with preexisting liver conditions. The authors concluded, and AHPA recommended in 1995, that products containing chaparral should be labeled with the following cautionary statement: Seek advice from a health care practitioner before use if you have any history of liver disease. Discontinue use if nausea, fever, fatigue, or jaundice occur (e.g., dark urine or yellow discoloration of the eyes).

Adverse Events and Side Effects

Case reports have indicated a correlation between chaparral consumption and liver damage. Amounts of chaparral taken range from 0.3 to 6 g daily, and the duration of use has been from 20 days to "many years," with chaparral being taken as capsules or tablets in most of the cases. The product taken was analyzed in only one case. Although one patient with toxic liver damage required a liver transplant, other cases resolved on cessation of chaparral (IOM 2001). A review of those case reports and associated toxicity studies noted that the severity of liver damage was not related to the dose or duration of chaparral use (IOM 2001).

A case of autoimmune hemolytic anemia was reported in association with chaparral use (Tregellas and South 1980).

Chaparral-induced dermatitis has been reported after contact with living plants or chaparral compresses (Leonforte 1986).

Pharmacological Considerations

Traditional use and animal studies have indicated that chaparral and the compound nordihydroguaiaretic acid (NDGA) may modify glucose regulation (Lambert et al. 2004; Luo et al. 1998; Reed et al. 1999). People with diabetes are advised to monitor their blood sugar closely and discuss the use of this herb with a qualified healthcare practitioner prior to use.

Pregnancy and Lactation

An animal study indicated that large doses (700 mg/kg) of chaparral extracts exhibited anti-implantation activity (Konno et al. 1987). In this work, the contraindication for use in pregnancy is based on concerns regarding the cases of hepatotoxicity reported in association with chaparral use, as the implications of these case reports and possible mechanisms of hepatotoxicity have yet to be fully understood (Heron and Yarnell 2001).

No information on the safety of chaparral in lactation was identified. While this review did not identify any concerns for use while nursing, safety has not been conclusively established.

Review Details

I. Drug and Supplement Interactions

Clinical Trials of Drug or Supplement Interactions

No clinical trials of drug or supplement interactions were identified.

Case Reports of Suspected Drug or Supplement Interactions

No case reports of suspected drug or supplement interactions were identified.

Animal Trials of Drug or Supplement Interactions

No animal trials of drug or supplement interactions were identified.

II. Adverse Events

Case Reports of Adverse Events

Case reports have indicated a correlation between chaparral consumption and liver damage. Amounts of chaparral taken range from 0.3 to 6 g daily, and the duration of use has been from 20 days to "many years," with chaparral being taken as capsules or tablets in most of the cases. The product taken was analyzed in only one case. Although one patient with toxic liver damage required a liver transplant, other cases resolved on cessation of chaparral (IOM 2001). A voluntary rechallenge in one case led to a return of symptoms (Batchelor et al. 1995). A review of the case reports and associated toxicity studies noted that the severity of liver damage was not related to the dose or duration of chaparral use (IOM 2001).

Toxic liver damage with elevated aminotransferase levels and joint stiffness in the right hand were reported in a 38-year-old woman who had taken 400 mg chaparral daily for "many years." The patient had a history of drug abuse, alcohol use, and previous exposure to hepatitis C. The reporting authors indicated that chaparral might have potentiated or exacerbated underlying liver disease. The patient eventually required a liver transplant (Sheikh et al. 1997).

Toxic liver damage with jaundice and abdominal pain was reported in a 41-year-old woman who had taken tablets containing 259 mg/day of chaparral over an 11-week period. The patient retuned to normal after discontinuation of chaparral (Clark and Reed 1992; Sheikh et al. 1997).

Toxic liver damage with fatigue, jaundice, dark urine, nausea, abdominal pain, and diarrhea was reported in a 44-year-old woman who had taken capsules containing 2400 mg chaparral daily for 10 days and then 800 mg daily for another 10 days (Clark and Reed 1992; Sheikh et al. 1997).

Toxic liver damage with symptoms of cholecystitis and elevated aminotransferase levels were reported in a 60-year-old woman who had taken an unspecified amount of capsules containing chaparral (Sheikh et al. 1997).

Hepatic dysfunction with scleral icterus and diffuse jaundice was reported in a 42-year-old man who had consumed 1440 mg chaparral daily for 6 weeks. The patient returned to normal 1 month after discontinuing chaparral (Clark and Reed 1992; Sheikh et al. 1997).

Toxic liver damage with fatigue, jaundice, and dark urine was reported in a 25-year-old man who consumed 3830 mg chaparral capsules daily for 2 to 3 weeks, then 5760 mg daily for 10 weeks. After cessation of chaparral, the patient recovered within 2 weeks (Sheikh et al. 1997).

Toxic liver damage, confirmed by liver biopsy, with fatigue, jaundice, abdominal pain, light stools, and pruritus was reported in a 57-year-old woman who had consumed 480 mg of chaparral daily for 8 weeks. The woman had used conjugated estrogens in the past (a possible hepatotoxin with persistent effects, but noted as unlikely as the causative agent in the case) (IOM 2001; Sheikh et al. 1997).

Jaundice with possible toxic liver damage was reported in a 71-year-old man who had been taking an unspecified amount of chaparral capsules daily for an unspecified amount of time. Symptoms of flu-like illness, ascites, and jaundice abated 2 months after cessation of chaparral. The man had a history of alcohol use (14 oz wine daily). One month after restarting chaparral use, the man developed jaundice, ascites, scleral icterus, and nausea. Liver biopsy indicated diffuse necrosis with inflammation, portal tract expansion, mild cholestasis, and mild fibrous septation. A biopsy 3 months later indicated marked improvement (Batchelor et al. 1995).

Drug-induced hepatotoxicity was reported in a 33-year-old woman who had been taking 15 tablets (dose unspecified) of chaparral daily for 3 to 4 months. The patient had presented with jaundice, ascites, abdominal pain, fatigue, scleral icterus, anorexia, and pedal edema due to subacute hepatic necrosis. The symptoms resolved after cessation of chaparral (Katz and Saibil 1990).

Subacute liver damage was reported in a 43-year-old woman who had taken approximately three tablets (dose unspecified) of chaparral daily for 6 weeks. The woman was also taking aspirin regularly (Batchelor et al. 1995).

Liver damage was reported in a 54-year-old woman who had taken 1600 mg of chaparral daily for 3 weeks. The patient recovered within 6 weeks of stopping chaparral intake (Sheikh et al. 1997).

Liver damage was reported in a 39-year-old woman who drank tea made from 4 tea bags of chaparral daily for approximately 1.5 years. The patient recovered within one week after stopping chaparral intake (Sheikh et al. 1997).

An additional two cases of subacute liver damage and one case of hepatotoxicity associated with consumption of chaparral have been reported, but without information on doses taken (Sheikh et al. 1997).

A 40-year-old man had a positive direct antiglobulin test (DAT) (due to the presence of IgG antibodies) leading to a diagnosis of autoimmune hemolytic anemia after consumption of four tablets (dose unspecified) of chaparral daily. Chaparral was discontinued and the DAT became weaker over the course of 19 weeks, eventually becoming negative. After 8 weeks of negative DAT, the patient began taking chaparral again and the DAT test became positive after 5 weeks. During the follow-up study, haptoglobin and hematocrit tests showed no evidence of decreased red blood cell survival (Tregellas and South 1980).

Acute dermatitis, with patch tests confirming reaction to chaparral, was reported in six men. The lesions were primarily on sun-exposed sites but were also on the legs and scrotum. Contact was with the living plant in two men, from moist compresses or baths in two men, and as a result of burning the bushes in two men (Leonforte 1986).

III. Pharmacology and Pharmacokinetics

Human Pharmacological Studies

No relevant human pharmacological studies were identified.

Animal Pharmacological Studies

A reduction in plasma glucose concentrations with no changes in plasma insulin was observed in diabetic mice orally administered 300 mg/kg of the compound NDGA daily for 12 days (Luo et al. 1998).

A reduction in steady-state plasma glucose levels and glucose levels after an infusion of glucose was observed in diabetic rats orally administered 300 mg/kg of NDGA daily for 4 days (Reed et al. 1999).

Decreases in fasting plasma glucose levels and glucose response to an oral glucose load were observed in diabetic mice orally administered 300 mg/kg of the compound NDGA daily for 12 days (Luo et al. 1998).

In Vitro Pharmacological Studies

In vitro pharmacological studies were identified but omitted due to the availability of substantial human data.

IV. Pregnancy and Lactation

Anti-implantation activity was observed in rats orally administered 0.7 g/kg of a methanol extract, 0.58 g/kg of a chloroform extract, or 0.52 g/kg of a phenolic extract of chaparral daily on days 1 to 20 of pregnancy. The chloroform extract exhibited the strongest anti-implantation activity (Konno et al. 1987).

No information on the safety of chaparral during lactation was identified.

V. Toxicity Studies

Acute Toxicity

The oral LD_{50} of the compound nordihydroguaiaretic acid (NDGA) is 5.5 g/kg in rats, 4 g/kg in mice, and 0.8 g/kg in guinea pigs (Lehman et al. 1951). The LD_{50} of NDGA intraperitoneally administered to mice is 0.1 to 0.8 g/kg (Fujii et al. 1970; Kozubik et al. 1993; Madrigal-Bujaidar et al. 1998).

Short-Term Toxicity

Toxicity testing in mice examined the effects of chaparral phenolic resin as 0.5 to 5% of the diet, as well as oral gavage administered doses of 100 mg chaparral resin or 15 or 100 mg of the compound NDGA daily for 4 days. In animals administered chaparral as part of the diet, reductions in body mass and food intake were observed. Mice tolerated a maximum of 100 mg of phenolic resin in the diet, but oral gavage administered doses of 100 mg of resin or 15 mg of NDGA were lethal. No significant differences in detoxification, measured as glucuronic acid conjugates in urine, were detected among orally administered treatments (Rios et al. 2008).

In rats fed a diet containing 2% of the compound NDGA daily for 3 weeks and injected with endotoxin- or non-endotoxin-containing bacteria, a synergistic provocation of renal damage was observed in the animals injected with endotoxin-containing bacteria. Renal lesions were not characteristic of classic endotoxin treatment, and the results excluded bacterial colonization and intrarenal accumulation of NDGA as causes of nephropathy (Gardner et al. 1987).

Subchronic Toxicity

In hamsters fed a diet containing 4% of a hydroethanolic extract of chaparral for 70 days, marked growth retardation, pronounced irritability and aggressiveness, and a marked hypoplasia of testicles and accessory sex glands were observed (Granados and Cardenas 1994).

In rats fed a diet containing 2% of the compound NDGA daily for 1 to 24 weeks, the glomerular filtration rate was decreased as compared to control animals. Early during NDGA exposure, small polyps developed along the outer medullary segments of the collecting tubules in the kidney. At 2 months of NDGA exposure, kidneys were infiltrated by polymorphonuclear leukocytes and macrophages. Basement membrane thickening, fibrosis, tubular atrophy, and eventually proximal tubular cell necrosis were characterized adjacent to these infiltrated areas. By 6 months of NDGA exposure, cysts were found throughout the kidneys (Evan and Gardner 1979).

In rats fed diets containing 2% of the compound NDGA for 99 days, lesions in the kidneys were observed, including hydropic changes in tubular epithelial cells, tubular necrosis, proliferation of lysosomes in number and size, and invasion by macrophages (Goodman et al. 1970).

Chronic Toxicity

In rats fed diets containing 0.5 or 1.0% of the compound NDGA for 74 weeks, cysts of the mesenteric lymph nodes at the ileocecal junction were observed. In one rat, the cystic nodes were invaded by a malignant reticulum cell sarcoma. Mean body weight was lower in the NDGA group as compared to the control group (Grice et al. 1968).

In rats fed diets containing 0.5% of the compound NDGA for 2 years, growth inhibition was observed after 6 months. After 2 years, inflammatory cecal lesions, massive cecal hemorrhages with single and multiple cysts in the mesentery near the cecum, and slight cystic enlargement of the paracecal lymph nodes were observed (Lehman et al. 1951).

Genotoxicity

In human lymphocytes, NDGA produced a dose-dependent increase in sister-chromatid exchanges, and, at the highest dose tested (27 μ*M*), a decrease in the cell cycle delay (Madrigal-Bujaidar et al. 1998). In bone marrow cells of mice orally administered 8.8, 17.6, 35.3, and 70.7 mg/kg of NDGA, an increase in sister-chromatid exchanges was observed at the two highest dose levels (Madrigal-Bujaidar et al. 1998).

Conversely, NDGA exhibited protective activity against genotoxic damage induced by the compound norgestrel. Study parameters included sister-chromatid exchanges, chromosomal aberrations, mitotic index, and replication index (Siddique et al. 2006).

Literature Cited

Batchelor, W.B., J. Heathcole, and I.R. Wanless. 1995. Chaparral-induced hepatic injury. *Am. J. Gastroenterol.* 90(5):831-833.

Clark, F., and R. Reed. 1992. Chaparral-induced toxic hepatitis: California and Texas, 1992. *J. Am. Med. Assoc.* 268:3295-3298.

De Smet, P.A.G.M. 1993. *Adverse effects of herbal drugs, Volume 2.* Berlin: Springer.

Evan, A.P., and K.D. Gardner. 1979. Nephron obstruction in nordihydroguaiaretic acid-induced renal cystic disease. *Kidney Int.* 15(1):7-19.

FDA. 1992. U.S. Food and Drug Administration. FDA Advice to Chaparral Users. Online. December 11. News release.

Fujii, K., H. Jaffe, Y. Bishop, et al. 1970. Structure-activity relations for methylenedioxyphenyl and related compounds on hepatic microsomal enzyme function, as measured by prolongation of hexobarbital narcosis and zoxazolamine paralysis in mice. *Toxicol. Appl. Pharmacol.* 16(2):482-494.

Gardner, K.D., W.P. Reed, A.P. Evan, et al. 1987. Endotoxin provocation of experimental renal cystic disease. *Kidney Int.* 32(3):329-334.

Goodman, T., H.C. Grice, G.C. Becking, and F.A. Salem. 1970. A cystic nephropathy induced by nordihydroguaiaretic acid in the rat. Light and electron microscopic investigations. *Lab. Invest.* 23(1):93-107.

Granados, H., and R. Cardenas. 1994. Biliary calculi in the golden hamster. XXXVII. The prophylactic action of the creosote bush (*Larrea tridentata*) in pigmented cholelithiasis produced by vitamin A. *Rev. Gastroenterol. Mex.* 59(1):31-35.

Grice, H.C., G. Becking, and T. Goodman. 1968. Toxic properties of nordihydroguaiaretic acid. *Food Chem. Toxicol.* 6(2):155.

Heron, S., and E. Yarnell. 2001. The safety of low-dose *Larrea tridentata* (DC.) Coville (creosote bush or chaparral): A retrospective clinical study. *J. Altern. Complement. Med.* 7(2):175-185.

IOM. 2001. Committee on the Framework for Evaluating the Safety of Dietary Supplements. Safety review: Draft prototype monograph on chaparral. Washington DC: Institute of Medicine.

Katz, M., and F. Saibil. 1990. Herbal hepatitis: Subacute hepatic necrosis secondary to chaparral leaf. *J. Clin. Gastroenterol.* 12(2):203.

Konno, C., A. Martin, B.X. Ma, et al. 1987. Search for fertility regulating agents from *Larrea tridentata*. Paper read at Proceedings: The First Princess Chulabhorn Science Congress, International Congress on Natural Products.

Kozubik, A., J. Hofmanova, J. Hola, and J. Netikova. 1993. The effect of nordihydroguaiaretic acid, an inhibitor of prostaglandin and leukotriene biosynthesis, on hematopoiesis of gamma-irradiated mice. *Exp. Hematol.* 21(1):138-142.

Lambert, J., R. Dorr, and B. Timmermann. 2004. Nordihydroguaiaretic acid: A review of its numerous and varied biological activities. *Pharm. Biol.* 42(2):149-158.

Lehman, A.J., O.G. Fitzhugh, A.A. Nelson, and G. Woodard. 1951. The pharmacological evaluation of antioxidants. *Adv. Food Res.* 3:197-208.

Leonforte, J.F. 1986. Contact dermatitis from *Larrea* (creosote bush). *J. Am. Acad. Dermatol.* 14(2, Part 1):202-207.

Luo, J., T. Chuang, J. Cheung, et al. 1998. Masoprocol (nordihydroguaiaretic acid): A new antihyperglycemic agent isolated from the creosote bush (*Larrea tridentata*). *Eur. J. Pharmacol.* 346(1):77-79.

Madrigal-Bujaidar, E., S. D. Barriga, M. Cassani, D. Molina, and G. Ponce. 1998. In vivo and in vitro induction of sister-chromatid exchanges by nordihydroguaiaretic acid. *Mutat. Res.* 412(2):139-144.

McGuffin, M., C. Hobbs, R. Upton, and A. Goldberg. 1997. *Botanical safety handbook.* Boca Raton, FL: CRC Press.

Reed, M.J., K. Meszaros, L.J. Entes, et al. 1999. Effect of masoprocol on carbohydrate and lipid metabolism in a rat model of type II diabetes. *Diabetologia* 42(1):102-106.

Rios, J.M., A.M. Mangione, and J.C. Gianello. 2008. Effects of natural phenolic compounds from a desert dominant shrub *Larrea divaricata* Cav. on toxicity and survival in mice. *Rev. Chil. Hist. Nat.* 81(2):293–302.

Sheikh, N.M., R.M. Philen, and L.A. Love. 1997. Chaparral-associated hepatotoxicity. *Arch. Intern. Med.* 157(8):913-919.

Siddique, Y.H., T. Beg, and M. Afzal. 2006. Protective effect of nordihydroguaiaretic acid (NDGA) against norgestrel induced genotoxic damage. *Toxicol. In Vitro* 20(2):227-233.

Tregellas, W.M., and S.F. South. 1980. Autoimmune syndrome induced by chaparral ingestion. *Transfusion* 20:647-648.

Laurus nobilis L. — Lauraceae

SCN: bay
OCN: bay laurel; Grecian laurel; sweet bay; true bay

Part: leaf

Quick Reference Summary

Safety Class: 1
Interaction Class: A

Contraindications

None known.

Other Precautions

None known.

Drug and Supplement Interactions

None known.

Adverse Events and Side Effects

Allergic contact dermatitis, confirmed by patch testing, has been reported after contact with foods or topical products containing bay leaf or bay essential oil (Adisen and Onder 2007; Foussereau et al. 1975; Jirasek and Skach 1962; Opdyke 1979; Ozden et al. 2001).

Pharmacological Considerations

None known.

Pregnancy and Lactation

No information on the safety of bay during pregnancy or lactation was identified in the scientific or traditional literature. While this review did not identify any concerns for pregnant or lactating women, safety has not been conclusively established.

Review Details

I. Drug and Supplement Interactions

Clinical Trials of Drug or Supplement Interactions

No clinical trials of drug or supplement interactions were identified.

Case Reports of Suspected Drug or Supplement Interactions

No case reports of suspected drug or supplement interactions were identified.

Animal Trials of Drug or Supplement Interactions

No animal trials of drug or supplement interactions were identified.

II. Adverse Events

Case Reports of Adverse Events

Allergic contact dermatitis, confirmed by patch testing, has been reported after contact with foods or topical products containing bay leaf or bay essential oil (Adisen and Onder 2007; Foussereau et al. 1975; Jirasek and Skach 1962; Opdyke 1979; Ozden et al. 2001).

III. Pharmacology and Pharmacokinetics

Human Pharmacological Studies

No relevant human pharmacological studies were identified.

Animal Pharmacological Studies

No relevant animal pharmacological studies were identified.

In Vitro Pharmacological Studies

The compound cinnamtannin B-1 reduced thrombin-induced aggregation in platelets from type 2 diabetic subjects (Bouaziz et al. 2007). The same compound reduced thrombin-evoked microtubular remodeling and activation of the tyrosine kinases Btk and pp60(src), which leads to inhibition of platelet aggregation (Ben Amor et al. 2007).

IV. Pregnancy and Lactation

No information on the safety of bay during pregnancy or lactation was identified.

V. Toxicity Studies

Acute Toxicity

The LD_{50} of bay essential oil is 3.95 g/kg after oral administration to rats, and could not be determined at doses up to 5 g/kg after topical administration to rabbits (Opdyke 1979).

See *Eucalyptus globulus* essential oil Toxicity Studies for information on the toxicity of the compound 1,8-cineol (eucalyptol), the primary component of bay leaf essential oil (36–46%) (Kovacevic et al. 2007).

Literature Cited

Adisen, E., and M. Onder. 2007. Allergic contact dermatitis from *Laurus nobilis* oil induced by massage. *Contact Dermat.* 56(6):360-361.

Ben Amor, N., A. Bouaziz, C. Romera-Castillo, et al. 2007. Characterization of the intracellular mechanisms involved in the antiaggregant properties of cinnamtannin B-1 from bay wood in human platelets. *J. Med. Chem.* 50(16):3937-3944.

Bouaziz, A., S. Salido, P.J. Linares-Palomino, et al. 2007. Cinnamtannin B-1 from bay wood reduces abnormal intracellular Ca^{2+} homeostasis and platelet hyperaggregability in type 2 diabetes mellitus patients. *Arch. Biochem. Biophys.* 457(2):235-242.

Foussereau, J., J.C. Muller, and C. Benezra. 1975. Contact allergy to *Frullania* and *Laurus nobilis*: Cross-sensitization and chemical structure of the allergens. *Contact Dermat.* 1(4):223-230.

Jirasek, L., and M. Skach. 1962. Perioral contact eczema with eczematous stomatitis after the use of bay leaves (*Laurus nobilis* L.) in food. *Cesk. Dermatol.* 37:18-21.

Kovacevic, N.N., M.D. Simic, and M.S. Ristic. 2007. Essential oil of *Laurus nobilis* from Montenegro. *Chem. Nat. Compd.* 43(4):408-411.

Opdyke, D.L.J. 1979. *Monographs on fragrance raw materials.* New York: Pergamon.

Ozden, M.G., P. Oztas, M.O. Oztas, and M. Onder. 2001. Allergic contact dermatitis from *Laurus nobilis* (laurel) oil. *Contact Dermat.* 45(3):178.

Lavandula spp.

Lamiaceae

Lavandula angustifolia Mill.
SCN: English lavender
Syn: *Lavandula officinalis* Chaix.; *Lavandula spica* L.; *Lavandula vera* DC.
OCN: common lavender
Note: English lavender from France is often traded as "French lavender."

***Lavandula intermedia* Emeric ex Loisel.**
SCN: lavandin
OCN: Dutch lavender

***Lavandula latifolia* Medic.**
SCN: spike lavender

***Lavandula stoechas* L.**
SCN: Spanish lavender
OCN: French lavender
Part: flower

Quick Reference Summary

Safety Class: 1
Interaction Class: A

Contraindications

None known.

Other Precautions

None known.

Drug and Supplement Interactions

None known.

Editors' Notes

A series of case reports and an associated in vitro study suggested that lavender essential oil has estrogenic activity and that lavender as an ingredient in personal care products (i.e., shampoo, styling gel, and soap) was the cause of gynecomastia in several teenage boys (Henley et al. 2007). Those reports have been the subject of criticism (Dean 2007; Kalyan 2007; Kemper et al. 2007), with letters to the editor regarding the cases noting that estrogen levels reported were normal, and that traditional use of and research on lavender essential oil has not indicated estrogenic activity, although estrogenic activity has been shown for other essential oils (Kemper et al. 2007).

Adverse Events and Side Effects

Allergic contact dermatitis, confirmed by patch testing, has been reported in individuals exposed to lavender essential oil (Coulson and Ali Khan 1999; Rademaker 1994; Schaller and Korting 1995; Sugiura et al. 2000).

Pharmacological Considerations

None known.

Pregnancy and Lactation

A 1935 study indicated that lavender stimulated uterine contractions in isolated pregnant guinea pig uteruses (Superbi and Crispolti 1935). No other information on the safety of lavender in pregnancy was identified.

No information on the safety of lavender during lactation was identified in the scientific or traditional literature. While this review did not identify any concerns for use while nursing, safety has not been conclusively established.

Review Details

I. Drug and Supplement Interactions

Clinical Trials of Drug or Supplement Interactions

No clinical trials of drug or supplement interactions were identified.

Case Reports of Suspected Drug or Supplement Interactions

No case reports of suspected drug or supplement interactions were identified.

Animal Trials of Drug or Supplement Interactions

No animal trials of drug or supplement interactions were identified.

II. Adverse Events

Case Reports of Adverse Events

Allergic contact dermatitis, confirmed by patch testing, has been reported in individuals exposed to lavender essential oil (Coulson and Ali Khan 1999; Rademaker 1994; Schaller and Korting 1995; Sugiura et al. 2000).

Confusion and deep drowsiness were reported in an 8-month-old boy who ingested a "small amount" of a "handmade extract" (no details provided) of lavandin, identified as a hybrid of *L. angustifolia* and *L. latifolia* (Landelle et al. 2008).

A case series of gynecomastia associated with the use of products scented with "lavender" includes cases of a 4-year-old who had been treated with a "healing balm containing lavender oil," a 10-year-old boy who used a styling gel daily and a shampoo regularly, both of which contained lavender essential oil, and a 7-year-old who had used "lavender-scented" soap and "lavender-scented" commercial skin lotions. Gynecomastia resolved after discontinuation of these products. In all cases, other ingredients, product name, amount used, and frequency and duration of application were not specified (Henley et al. 2007). Letters to the editor regarding the case report noted that estrogen levels reported in the cases were normal, and that traditional use of and research on lavender essential oil have not indicated estrogenic activity, although estrogenic activity has been shown for other essential oils (Kemper et al. 2007). Letters also indicated that other ingredients or packaging (notably phthalates) could have been responsible for the effects (Kemper et al. 2007), as could other environmental or dietary factors (Dean 2007; Kalyan 2007).

III. Pharmacology and Pharmacokinetics

Human Pharmacological Studies

No relevant human pharmacological studies were identified.

Animal Pharmacological Studies

In a mouse model of pulmonary thromboembolism induced by intravenous injection of a collagen–epinephrine mixture, administration of 100 mg/kg of lavandin essential oil daily significantly reduced thrombotic events without inducing prohemorrhagic complications at variance with acetylsalicylic acid used as a reference drug (Ballabeni et al. 2004).

In Vitro Pharmacological Studies

Inhibition of platelet aggregation induced by arachidonic acid, U46619, collagen, and ADP (with IC_{50} values of 51, 84, 191, and 640 μg/ml, respectively) was observed in guinea pig platelet-rich plasma treated with lavandin essential oil. Lavandin oil also destabilized clot retraction (IC_{50} of 149 μg/ml) induced by thrombin on rat platelet-rich plasma (Ballabeni et al. 2004).

In estrogen receptor-positive (MCF-7) and androgen receptor-positive (MDA-kb2) human breast cancer cells, lavender essential oil stimulated ERE-dependent luciferase activity in a dose-dependent manner, with the maximum activity observed at 0.025% (v/v), corresponding to ~50% of the activity elicited by 1 n*M* 17β-estradiol. The pure estrogen receptor antagonist fulvestrant inhibited transactivation of the reporter plasmid (Henley et al. 2007). A critique of the study indicated that the effects of lavender were "very weak" (Kalyan 2007).

IV. Pregnancy and Lactation

An older study indicated that a hot aqueous extract of lavender flower (dose not specified) stimulated uterine contractions in isolated pregnant guinea pig uteri (Superbi and Crispolti 1935).

No information on the safety of lavender during lactation was identified.

V. Toxicity Studies

Acute Toxicity

The LD_{50} of lavandin essential oil could not be determined at doses up to 5 g/kg after oral administration in rats or topical application in rabbits (Opdyke 1979).

The LD_{50} of lavender absolute is 4.25 g/kg after oral administration in rats, whereas the dermal LD_{50} could not be determined at doses up to 5 g/kg in guinea pigs (Opdyke 1979).

The LD_{50} of spike lavender essential oil is 3.8 g/kg after oral administration in rats, whereas the dermal LD_{50} could not be determined at doses up to 2 g/kg in rabbits (Opdyke 1979).

Literature Cited

Ballabeni, V., M. Tognolini, M. Chiavarini, et al. 2004. Novel antiplatelet and antithrombotic activities of essential oil from *Lavandula hybrida* Reverchon "grosso." *Phytomedicine* 11(7-8):596-601.

Coulson, I.H., and A.S. Ali Khan. 1999. Facial 'pillow' dermatitis due to lavender oil allergy. *Contact Dermat.* 41(2):111.

Dean, C.J. 2007. Prepubertal gynecomastia linked to lavender and tea tree oils. *N. Engl. J. Med.* 356(24):2543; author reply 2543-2544.

Henley, D.V., N. Lipson, K.S. Korach, and C.A. Bloch. 2007. Prepubertal gynecomastia linked to lavender and tea tree oils. *N. Engl. J. Med.* 356(5):479-485.

Kalyan, S. 2007. Prepubertal gynecomastia linked to lavender and tea tree oils. *N. Engl. J. Med.* 356(24):2542; author reply 2543-2544.

Kemper, K.J., A.J. Romm, and P. Gardiner. 2007. Prepubertal gynecomastia linked to lavender and tea tree oils. *N. Engl. J. Med.* 356(24):2541-2542; author reply 2543-2544.

Landelle, C., G. Francony, N.F. Sam-Lai, et al. 2008. Poisoning by lavandin extract in a 18-month-old boy. *Clin. Toxicol.* 46(4):279-281.

Opdyke, D.L.J. 1979. *Monographs on fragrance raw materials*. New York: Pergamon.

Rademaker, M. 1994. Allergic contact dermatitis from lavender fragrance in Difflam gel. *Contact Dermat.* 31(1):58-59.

Schaller, M., and H.C. Korting. 1995. Allergic airborne contact dermatitis from essential oils used in aromatherapy. *Clin. Exp. Dermatol.* 20:143-145.

Sugiura, M., R. Hayakawa, Y. Kato, K. Sugiura, and R. Hashimoto. 2000. Results of patch testing with lavender oil in Japan. *Contact Dermat.* 43(3):157-160.

Superbi, C., and E. Crispolti. 1935. Effect on the uterine muscle of infusions and extracts of certain herbs used by the natives of Tripoli. *Ann. Ostetr. Ginecol.* 57:253-267.

Lawsonia inermis L. — Lythraceae

SCN: henna
Syn: *Lawsonia alba* Lam.
AN: *madayanti*
Part: leaf

Quick Reference Summary

Safety Class: 2a, 2d
Interaction Class: A

Contraindications

For external use only (McMillan et al. 2004; Zinkham and Oski 1996).

Not for use on infants or children with glucose-6-phosphate dehydrogenase (G6PD) deficiency (Devecioglu et al. 2001; Kandil et al. 1996; Kök et al. 2004; Raupp et al. 2001; Zinkham and Oski 1996).

Other Precautions

Use with caution on adults with G6PD deficiency (McMillan et al. 2004; Rund et al. 2007; Zinkham and Oski 1996).

Drug and Supplement Interactions

None known.

Editors' Notes

Some henna, particularly that marketed as "black henna," is mixed with the compound *para*-phenylenediamine (PPD), a black hair dye that frequently causes skin sensitization and has been responsible for numerous cases of allergic contact dermatitis, some of which have caused permanent scarring (Akhras and Ostlere 2005; Kang and Lee 2006). PPD is not a compound found in *Lawsonia inermis*.

Adverse Events and Side Effects

In children with glucose-6-phosphate dehydrogenase (G6PD) deficiency, a hereditary disease also known as favism, hemolytic anemia and hyperbilirubinemia have been reported after topical application of henna (Devecioglu et al. 2001; Kandil et al. 1996; Kök et al. 2004; Raupp et al. 2001; Zinkham and Oski 1996). In some cases, the patients also developed acute renal failure, sometimes fatal (Devecioglu et al. 2001; Kök et al. 2004). No case reports for such adverse effects in adults were identified after topical application. Relative thickness of skin and higher body mass to skin surface area ratio may account for a lack of case reports in adults.

Pharmacological Considerations

Animal and in vitro studies and human case reports indicated that the compound lawsone can produce hemolytic anemia in persons with G6PD deficiency (McMillan et al. 2004; Zinkham and Oski 1996).

In an in vitro study of skin absorption of henna, 0.3 to 1.3% of a henna hair coloring paste was absorbed into the skin after 1 hour of treatment. Extended absorption studies indicated that most of the product remained on the skin and a relatively small amount was absorbed (Kraeling et al. 2007).

Pregnancy and Lactation

An animal study indicated no developmental toxicity of henna but some maternal toxicity at the highest dose (20 mg/kg) tested (Nawaf et al. 2003). Henna is reported to have been used internally as an abortifacient (List and Hörhammer 1973).

No information on the safety of henna in lactation was identified. While this review did not identify any concerns for use while nursing, safety has not been conclusively established.

Review Details

I. Drug and Supplement Interactions

Clinical Trials of Drug or Supplement Interactions

No clinical trials of drug or supplement interactions were identified.

Case Reports of Suspected Drug or Supplement Interactions

No case reports of suspected drug or supplement interactions were identified.

Animal Trials of Drug or Supplement Interactions

No animal trials of drug or supplement interactions were identified.

L

II. Adverse Events

Adverse Events Reported in Clinical Trials

In a test of henna as a substance for creating temporary marks on patients undergoing external radiotherapy, no adverse reactions were reported in any of the 158 patients marked with henna (Wurstbauer et al. 2001).

Case Reports of Adverse Events

In children with glucose-6-phosphate dehydrogenase (G6PD) deficiency, a hereditary disease also known as favism, hemolytic anemia and hyperbilirubinemia were reported after topical application of henna (Devecioglu et al. 2001; Kandil et al. 1996; Kök et al. 2004; Raupp et al. 2001; Zinkham and Oski 1996). In some cases, the patients also developed acute renal failure, sometimes fatal (Devecioglu et al. 2001; Kök et al. 2004).

Some henna, particularly that marketed as "black henna," is mixed with the compound *para*-phenylenediamine (PPD), a black hair dye that frequently causes skin sensitization and has been responsible for numerous cases of allergic contact dermatitis (Akhras and Ostlere 2005; Kang and Lee 2006; Kazandjieva et al. 2007). Although the majority of allergic reactions to "henna" reported in the literature are reactions to PPD, allergic reactions to pure henna have been reported and confirmed by patch testing (Belhadjali et al. 2008; Garcia Ortiz et al. 1997; Gupta et al. 1986; Nigam and Saxena 1988; Oztas et al. 2001; Perez et al. 2003; Polat et al. 2009).

A young woman committed suicide by ingesting an unknown amount of henna. Autopsy findings were consistent with anaphylaxis, including laryngeal edema and pulmonary congestion (Kök et al. 2005).

Hemolysis, renal failure, and cardiac ischemia were reported in a 69-year-old man of northern Iraqi origin who mistakenly ingested two teaspoons of henna powder. The authors noted that, based on the man's origin, G6PD deficiency was likely but could not be tested for while the man was under care, due to inaccurate enzyme quantitation caused by the hemolysis (Rund et al. 2007).

Acute chemical colitis was reported in a 45-year-old woman with a history of ulcerative colitis who had reportedly ingested an unknown amount of henna mixed in 1 liter of water. The product was not verified and in the discussion section of the case report, the reporting physicians, citing a case of PPD-induced allergic contact dermatitis (Nawaf et al. 2003), indicated that henna was sometimes responsible for contact dermatitis (Uygur-Bayramicli et al. 2005), bringing into question authors' knowledge of the product ingested.

Newly applied henna is reported to block the passage of visible light to the skin. A case report indicated that a pulse oximeter was unable to provide readings on skin dyed with henna (Nirmalan and Baldwin 1997).

III. Pharmacology and Pharmacokinetics

Human Pharmacological Studies

No relevant human pharmacological studies were identified.

Animal Pharmacological Studies

In a study on the role of oxidant stress in lawsone-induced hemolytic anemia, radiolabeled erythrocytes exposed in vitro to lawsone or its hydroquinone form, 1,2,4-trihydroxynaphthalene (THN), were intravenously administered in rats. Neither lawsone nor THN were directly hemolytic or methemoglobinemic, even at high concentrations (>3 m*M*). Lawsone had no effect on erythrocytic glutathione (GSH levels), whereas THN (3 m*M*) induced a modest depletion. In contrast, ortho-substituted 1,4-naphthoquinones without a 2-hydroxy group, such as 2-methyl- and 2-methoxy-1,4-naphthoquinone, were redox active, were able to deplete GSH, and were direct-acting hemolytic agents. The authors indicated that lawsone is a weak direct-acting hemolytic agent that does not require extraerythrocytic metabolism to cause hemotoxicity and that the hemolytic response to henna may be restricted to individuals with compromised antioxidant defenses (McMillan et al. 2004).

In Vitro Pharmacological Studies

In blood from healthy and G6PD-deficient adults incubated with the compound lawsone (2.8×10^{-3} mol/l), methemoglobin (MHb) levels were 7.4% in the blood from healthy patients, and 40–44% in the blood from G6PD-deficient patients. The authors of the study noted that lawsone is capable of inducing oxidative injury in G6PD-normal red blood cells and produces more significant damage in G6PD-deficient red blood cells. The concentrations of lawsone producing oxidative damage are similar to those reported for the naphthalene metabolites α- and β-naphthol and α- and β-naphthoquinone, compounds documented to cause severe hemolysis in G6PD-deficient subjects (Zinkham and Oski 1996).

In a study of absorption of the compound lawsone, the principal color ingredient in henna, on human skin, two hair coloring products and two shampoo products, all containing henna, were tested on nonviable human skin. The shampoo was left on the skin for 5 min, while the coloring was left on for 1 h. For the henna hair paste products, 0.3 and 1.3% of the applied dose was absorbed into the receptor fluid in 24 hours while 2.2 and 4.0% remained in the skin. For both henna shampoo products, 0.3% of the applied dose was absorbed into the receptor fluid at 24 hours while 3.6 and 6.8% remained in the skin. For all products, most of the lawsone applied was washed from the surface of the skin (83–102%) at the end of the exposure period. Extended absorption studies were conducted for 72 hours to determine if skin levels of lawsone in the 24-hour studies might eventually be percutaneously absorbed. The studies determined that the majority of the lawsone remained in the skin with only a small increase in receptor fluid values (Kraeling et al. 2007).

IV. Pregnancy and Lactation

In rats orally administered 2, 7, or 20 mg/kg of the compound lawsone on days 6 to 15 of pregnancy, no adverse effects on pregnancy or embryo development were observed. In the 20 mg/kg group, one mouse aborted at day 15, but the authors indicated that this was not considered treatment related. Slight maternal toxicity, but no embryotoxicity, was observed in the 20 mg/kg dose group, and the no-observed-adverse-effect level (NOAEL) was established as 7 mg/kg daily for maternal toxicity (Nawaf et al. 2003).

No information on the safety of henna during lactation was identified.

V. Toxicity Studies

Acute Toxicity

The LD_{50} of the compound lawsone orally administered in rats is 570 mg/kg for females and 500 to 2000 mg/kg for males (SCCNFP 2002).

Short-Term Toxicity

Chemopreventive activity against chemically induced skin cancer was observed in mice topically administered 150 or 300 mg/kg daily of an 80% ethanolic extract of henna for 21 days (Dasgupta et al. 2003).

Subchronic Toxicity

In rats orally administered 8, 20, or 50 mg/kg of the compound lawsone daily for 13 weeks, no mortalities occurred, although evidence of hemolytic anemia was observed in the 50 mg/kg dose group. All female dose groups exhibited a significant but not dose-related decrease in clotting time. Dose-related decreases in blood urea, creatinine and albumin/globulin ratio, and increased bilirubin, were seen in females at the 20 and 50 mg/kg doses, and similar changes were seen in males at the high dose (SCCNFP 2002).

In a similar study, rats were orally administered 2, 7, or 20 mg/kg of the compound lawsone daily for 13 weeks. A dose-related decrease in erythrocyte count was observed in females, which was significantly lower than control at the 7 and 20 mg/kg dose levels. A significant decrease was also seen in males at the top dose only. Dose-related decreases in blood urea (mid and high dose), creatinine (high dose only), and albumin/globulin ratio (all doses), and increased bilirubin, were seen in females. Macroscopic abnormalities were observed in the forestomach and kidneys of animals at the 20 mg/kg dose. Dose-related increases in kidney and spleen weights were apparent for both sexes, and for livers of females. The female relative kidney weights were significantly higher than control at 7 and 20 mg/kg. The authors concluded that treatment-related effects occurred at 20 mg/kg, affecting mainly the kidneys, forestomach, and spleen, and that the no-observed-adverse-effect level (NOAEL) was 7 mg/kg (SCCNFP 2002).

Genotoxicity

Moderate mutagenic activity of an ethanol extract of henna was observed in the Ames test for mutagenicity using *Salmonella typhimurium* strains TA98 and TA102 (Mahmoud et al. 1992).

No evidence of chromosomal or other damage leading to micronucleus formation was observed in the mouse bone marrow micronucleus assay in mice orally administered 300 mg/kg henna and sacrificed 24, 48, or 72 hours after administration (SCCNFP 2002).

In the Ames test for mutagenicity in *S. typhimurium* strain TA98, the compound lawsone was mutagenic at concentrations of 500 to 1000 μg/plate without metabolic activation, whereas an extract of henna demonstrated no mutagenic activity at concentrations up to 1000 μg/plate (Stamberg et al. 1979).

In 2002, the EU Scientific Committee on Cosmetics and Non-Food Products (SCCNFP) evaluated the safety of the compound lawsone as a coloring agent in hair dye products of the European Union (EU). Based on in vitro and in vivo studies, the SCCNFP concluded that lawsone was mutagenic and not suitable for use as a hair coloring agent (SCCNFP 2002). More recent in vivo and in vitro studies have indicated a lack of mutagenic activity of lawsone (Kirkland and Marzin 2003; Marzin and Kirkland 2004).

Literature Cited

Akhras, V., and L. Ostlere. 2005. Is patch testing for PPD routinely necessary in patients with reactions to henna tattoos? *Contact Dermat.* 53(4):238.

Belhadjali, H., N. Ghannouchi, C. Amri, et al. 2008. Contact dermatitis to henna used as a hair dye. *Contact Dermat.* 58(3):182.

Dasgupta, T., A.R. Rao, and P.K. Yadava. 2003. Modulatory effect of henna leaf (*Lawsonia inermis*) on drug metabolising phase I and phase II enzymes, antioxidant enzymes, lipid peroxidation and chemically induced skin and forestomach papillomagenesis in mice. *Mol. Cell. Biochem.* 245(1-2):11-22.

Devecioglu, C., S. Katar, O. Dogru, and M.A. Tas. 2001. Henna-induced hemolytic anemia and acute renal failure. *Turk. J. Pediatr.* 43(1):65-66.

Garcia Ortiz, J.C., M. Terron, and J. Bellido. 1997. Contact allergy to henna. *Int. Arch. Allergy Immunol.* 114(3):298-299.

Gupta, B.N., A.K. Mathur, C. Agarwal, and A. Singh. 1986. Contact sensitivity to henna. *Contact Dermat.* 15(5):303-304.

Kandil, H.H., M.M. al-Ghanem, M.A. Sarwat, and F.S. al-Thallab. 1996. Henna (*Lawsonia inermis* Linn.) inducing haemolysis among G6PD-deficient newborns. A new clinical observation. *Ann. Trop. Paediatr.* 16(4):287-291.

Kang, I.J., and M.H. Lee. 2006. Quantification of *para*-phenylenediamine and heavy metals in henna dye. *Contact Dermat.* 55(1):26-29.

Kazandjieva, J., I. Grozdev, and N. Tsankov. 2007. Temporary henna tattoos. *Clin. Dermatol.* 25(4):383-387.

Kirkland, D., and D. Marzin. 2003. An assessment of the genotoxicity of 2-hydroxy-1,4-naphthoquinone, the natural dye ingredient of henna. *Mutat. Res.* 537(2):183-199.

Kök, A.N., M.V. Ertekin, V. Ertekin, and B. Avci. 2004. Henna (*Lawsonia inermis* Linn.) induced haemolytic anaemia in siblings. *Int. J. Clin. Pract.* 58(5):530-532.

Kök, A.N., V. Ertekin, Y. Bilge, and A.F. Isik. 2005. An unusual cause of suicide: Henna (*Lawsonia inermis* Linn.). *J. Emerg. Med.* 29(3):343-344.

Kraeling, M.E., R.L. Bronaugh, and C.T. Jung. 2007. Absorption of lawsone through human skin. *Cutan. Ocul. Toxicol.* 26(1):45-56.

List, P.H., and H. Hörhammer. 1973. *Hagers handbuch der pharmazeutischen praxis*. Berlin: Springer.

Mahmoud, I., A. Alkofahi, and A. Abdelaziz. 1992. Mutagenic and toxic activities of several spices and some Jordanian medicinal plants. *Int. J. Pharmacogn.* 30(2):81-85.

Marzin, D., and D. Kirkland. 2004. 2-Hydroxy-1,4-naphthoquinone, the natural dye of henna, is non-genotoxic in the mouse bone marrow micronucleus test and does not produce oxidative DNA damage in Chinese hamster ovary cells. *Mutat. Res.* 560(1):41-47.

McMillan, D.C., S.D. Sarvate, J.E. Oatis, Jr., and D.J. Jollow. 2004. Role of oxidant stress in lawsone-induced hemolytic anemia. *Toxicol. Sci.* 82(2):647-655.

Nawaf, A.M., A. Joshi, and O. Nour-Eldin. 2003. Acute allergic contact dermatitis due to *para*-phenylenediamine after temporary henna painting. *J. Dermatol.* 30(11):797-800.

Nigam, P.K., and A.K. Saxena. 1988. Allergic contact dermatitis from henna. *Contact Dermat.* 18(1):55-56.

Nirmalan, M., and J. Baldwin. 1997. Anaesthetic implications of henna. *Eur. J. Anaesthesiol.* 14(6):665-666.

Oztas, M.O., M. Onder, P. Oztas, and C. Atahan. 2001. Contact allergy to henna. *J. Eur. Acad. Dermatol. Venereol.* 15(1):91-92.

Perez, R.G., R. Gonzalez, M. Gonzalez, and R. Soloeta. 2003. Palpebral eczema due to contact allergy to henna used as a hair dye. *Contact Dermat.* 48(4):238.

Polat, M., M. Dikilita, P. Ozta, and N. Alli. 2009. Allergic contact dermatitis to pure henna. *Dermatol. Online J.* 15(1):15.

Raupp, P., J.A. Hassan, M. Varughese, and B. Kristiansson. 2001. Henna causes life threatening haemolysis in glucose-6-phosphate dehydrogenase deficiency. *Arch. Dis. Child.* 85(5):411-412.

Rund, D., T. Schaap, N. Da'as, D. Ben Yehuda, and J. Kalish. 2007. Plasma exchange as treatment for lawsone (henna) intoxication. *J. Clin. Apher.* 22(4):243-245.

SCCNFP. 2002. Opinion of the Scientific Committee on Cosmetic Products and Non-Food Products intended for consumers concerning lawsone. COLIPA No. C146. Brussels: Directorate-General Health and Consumer Protection of the European Union.

Stamberg, J., R. Werczberger, and Y. Koltin. 1979. Nonmutagenicity of the hair dye henna in the Ames test. *Mutat. Res.* 62(2):383-388.

Uygur-Bayramicli, O., R. Dabak, and M. Sargin. 2005. Acute chemical colitis resulting from oral intake of henna. *J. Clin. Gastroenterol.* 39(10):920-921.

Wurstbauer, K., F. Sedlmayer, and H.D. Kogelnik. 2001. Skin markings in external radiotherapy by temporary tattooing with henna: Improvement of accuracy and increased patient comfort. *Int. J. Radiat. Oncol. Biol. Phys.* 50(1):179-181.

Zinkham, W.H., and F.A. Oski. 1996. Henna: A potential cause of oxidative hemolysis and neonatal hyperbilirubinemia. *Pediatrics* 97(5):707-709.

Lentinula edodes (Berk.) Pegler — Tricholomataceae

SCN: shiitake
Syn: *Lentinus edodes* (Berk.) Singer
PN: *xiang xun* (fruiting body); *xiang gu* (fruiting body)
OCN: Japanese forest mushroom
Part: fruiting body, mycelium

Quick Reference Summary

Safety Class: 1
Interaction Class: A

Contraindications

None known.

Other Precautions

None known.

Drug and Supplement Interactions

None known.

Adverse Events and Side Effects

A condition known as shiitake dermatitis or toxicoderma occurs in some individuals after the ingestion of raw or partially cooked shiitakes. The clinical features are linear dermatitis on the trunk, and sometimes on the extremities, neck, face, and head, typically occurring up to 48 hours after shiitake ingestion (Garg and Cockayne 2008; Lippert et al. 2002; Nakamura 1992). An analysis of case reports suggested that some cases involved photodermatitis (development of a rash after exposure to sunlight) (Hanada and Hashimoto 1998).

Occupational contact dermatitis from shiitake has been reported (Aalto-Korte et al. 2005; Curnow and Tam 2003; Tarvainen et al. 1990, 1991; Ueda et al. 1992).

Pharmacological Considerations

Studies have indicated immunomodulating effects of shiitake, suggesting that shiitake should be used with caution in persons taking immune-suppressant medications (Lee et al. 2009a, 2009b; Zheng et al. 2005).

Pregnancy and Lactation

No information on the safety of shiitake during pregnancy or lactation was identified in the scientific or traditional literature. While this review did not identify any concerns for pregnant or lactating women, safety has not been conclusively established.

Review Details

I. Drug and Supplement Interactions

Clinical Trials of Drug or Supplement Interactions

No clinical trials of drug or supplement interactions were identified.

Case Reports of Suspected Drug or Supplement Interactions

No case reports of suspected drug or supplement interactions were identified.

Animal Trials of Drug or Supplement Interactions

No animal trials of drug or supplement interactions were identified.

II. Adverse Events

Case Reports of Adverse Events

A condition known as shiitake dermatitis or toxicoderma occurs in some individuals after the ingestion of raw or partially cooked shiitakes. The clinical features are linear groups of pruritic erythematous papules or flagellate dermatitis on the trunk, and sometimes on the extremities, neck, face, and head, typically occurring up to 48 hours after shiitake ingestion. The pathogenicity of the dermatitis is not known; however, patients generally do not test positive to skin-prick or patch testing with shiitake extract (Garg and Cockayne 2008; Lippert et al. 2002; Nakamura 1992). A similar case was reported in a patient who had taken the compound lentinan (Shimizu 1990). In some cases, photosensitive skin lesions have been associated with shiitake dermatitis (Hanada and Hashimoto 1998).

Occupational exposure to shiitake has resulted in contact dermatitis, with shiitake-specific immunoglobulin E reported in at least one case (Aalto-Korte et al. 2005; Curnow and Tam 2003; Tarvainen et al. 1990, 1991; Ueda et al. 1992).

Respiratory and immunological reactions, thought to be due to airborne spores, have been reported among shiitake workers (Sastre et al. 1990).

III. Pharmacology and Pharmacokinetics

Human Pharmacological Studies

In a phase I clinical safety and tolerability trial with 26 healthy volunteers, subjects were orally administered 9 g of an active hexose correlated compound extracted from shiitake daily for 14 days. Adverse effects of nausea, diarrhea, bloating, headache, fatigue, and foot cramps occurred in a total of 6 subjects but were mild and transient, and 2 subjects left the study due to nausea and intolerance of the shiitake extract. There were no laboratory abnormalities (Spierings et al. 2007).

Animal Pharmacological Studies

A study was conducted of the combined immune-potential effectiveness of intranasal application of the compound lentinan before and after administration in guinea pigs of BCG vaccination (Bacille Calmette-Guérin, a vaccine for tuberculosis), with doses of 1 mg/kg of lentinan three times at 2-day intervals. The results indicated that intranasal application of BCG alone or in combination with lentinan induced a high level of alveolar macrophage activation. Pretreatment with the lentinan enhanced the local immunohistological response to BCG in lung and reduced the generalized side effects of BCG (Drandarska et al. 2005).

A decrease in expression of the CYP1A drug-metabolizing isoenzymes was observed in mice orally or intraperitoneally administered 10 mg/kg of the compound lentinan every other day for four doses (Okamoto et al. 2004). In mice intraperitoneally administered 10 mg/kg of the compound lentinan every other day for four doses, suppression of constitutive and 3-methylcholanthrene-induced CYP1A expression and ethoxyresorufin-O-deethylation activity in the liver was observed (Hashimoto et al. 2002).

A reduction in plasma glucose levels was observed in diabetic rats orally administered 50 mg/kg exopolymers from shiitake daily for 7 days (Kim et al. 2001).

In Vitro Pharmacological Studies

In mouse splenocytes and human peripheral blood mononuclear cells treated with an extract of shiitake, the expression levels of IL-2 and TNF-α genes were augmented in both sets of cells. The production of IL-2 was augmented in the mouse splenocytes, and the production of TNF-α was augmented in murine peritoneal exudate macrophages. The production of IL-2 and TNF-α was augmented in the human mononuclear cells (Liu et al. 1998).

In the lymphocyte transformation test, aqueous extracts of shiitake were able to enhance the proliferation of rat thymocytes directly and act as costimulators in the presence of the T-mitogen phytohaemagglutinin (PHA) (Israilides et al. 2008).

The compound lenthionine inhibited platelet aggregation induced by arachidonic acid and U-46619 (Shimada et al. 2004).

IV. Pregnancy and Lactation

No information on the safety of shiitake during pregnancy or lactation was identified.

V. Toxicity Studies

Acute Toxicity

No adverse effects were observed in rats orally administered 0.5 or 2 g/kg of an extract of shiitake mycelium (Koike et al. 2002a).

In mice orally administered 3, 6, or 9 g/kg of dried shiitake daily for 5 days, an increase in plasma creatine kinase activity was observed at the 9 g/kg dose. No other changes in blood chemistry were observed (Nieminen et al. 2009).

Short-Term Toxicity

No adverse effects, including changes in body or organ weights, hematological or histopathological changes, or changes in blood chemistry, were observed in rats orally administered 0.5 or 1 g/kg of a shiitake mycelium extract daily for 28 days (Koike et al. 2002b).

No adverse effects, including changes in body or organ weights, hematological or histopathological changes, or changes in blood chemistry, were observed in rats given the compound lentinan in water (concentration of 150 g/kg) as the sole source of drinking water for 18 hours a day for 28 days (Odagiri et al. 2006).

Genotoxicity

In the Ames test for mutagenicity in *Salmonella typhimurium*, mutagenic activity of shiitake was observed without metabolic activation in *Salmonella* tester strains sensitive to base-pair substitution mutagens. Mutagenicity of other commonly consumed edible mushrooms, including *Agaricus bisporus* and *Boletus edulis*, was observed (Von Wright et al. 1982).

In a chromosomal aberration test in cultured mammalian cells, no aberrations were observed after continuous (up to 48 h) or short (6 h) treatment with an extract of shiitake mycelium at concentrations up to 4000 µg/ml (Miwa et al. 2002).

Literature Cited

Aalto-Korte, K., P. Susitaival, R. Kaminska, and S. Makinen-Kiljunen. 2005. Occupational protein contact dermatitis from shiitake mushroom and demonstration of shiitake-specific immunoglobulin E. *Contact Dermat.* 53(4):211-213.

Curnow, P., and M. Tam. 2003. Contact dermatitis to shiitake mushroom. *Australas. J. Dermatol.* 44(2):155-157.

Drandarska, I., V. Kussovski, S. Nikolaeva, and N. Markova. 2005. Combined immunomodulating effects of BCG and lentinan after intranasal application in guinea pigs. *Int. Immunopharmacol.* 5(4):795-803.

Garg, S., and S.E. Cockayne. 2008. Shiitake dermatitis diagnosed after 16 years! *Arch. Dermatol.* 144(9):1241-1242.

Hanada, K., and I. Hashimoto. 1998. Flagellate mushroom (shiitake) dermatitis and photosensitivity. *Dermatology* 197(3):255-257.

Hashimoto, T., Y. Nonaka, K. Minato, et al. 2002. Suppressive effect of polysaccharides from the edible and medicinal mushrooms, *Lentinus edodes* and *Agaricus blazei*, on the expression of cytochrome P450s in mice. *Biosci. Biotech. Biochem.* 66(7):1610-1614.

Israilides, C., D. Kletsas, D. Arapoglou, et al. 2008. In vitro cytostatic and immunomodulatory properties of the medicinal mushroom *Lentinula edodes*. *Phytomedicine* 15(6-7):512-519.

Kim, D.H., B.K. Yang, S.C. Jeong, et al. 2001. A preliminary study on the hypoglycemic effect of the exo-polymers produced by five different medicinal mushrooms. *J. Microbiol. Biotechnol.* 11(1):167-171.

Koike, T., H. Ihota, T. Fujimura, et al. 2002a. Single dose toxicity study of extract of cultured *Lentinus edodes* mycelia by oral administration in rats. *Oyo Yakuri* 62(1):1-3.

Koike, T., H. Ihota, T. Nagase, et al. 2002b. Repeated dose toxicity study of extract of cultured *Lentinus edodes* mycelia by 28-day oral administration in rats. *Oyo Yakuri* 62(1):5-12.

Lee, H.H., J.S. Lee, J.Y. Cho, Y.E. Kim, and E.K. Hong. 2009a. Structural characteristics of immunostimulating polysaccharides from *Lentinus edodes*. *J. Microbiol. Bioltechnol.* 19(5):455-461.

Lee, H.H., J.S. Lee, J.Y. Cho, Y.E. Kim, and E.K. Hong. 2009b. Study on immunostimulating activity of macrophage treated with purified polysaccharides from liquid culture and fruiting body of *Lentinus edodes*. *J. Microbiol. Biotechnol.* 19(6):566.

Lippert, U., V. Martin, C. Schwertfeger, and T. Fuchs. 2002. Shiitake (*Lentinula edodes*) dermatitis. *Allergologie* 25(9):484-488.

Liu, M., J. Li, F. Kong, J. Lin, and Y. Gao. 1998. Induction of immunomodulating cytokines by a new polysaccharide-peptide complex from culture mycelia of *Lentinus edodes*. *Immunopharmacology* 40(3):187-198.

Miwa, Y., R. Kobayashi, K. Hasegawa, and H. Nagaoka. 2002. Chromosomal aberration test of extract of cultured *Lentinus edodes* mycelia. *Oyo Yakuri* 62(1):13-18.

Nakamura, T. 1992. Shiitake *Lentinus edodes* dermatitis. *Contact Dermat.* 27(2):65-70.

Nieminen, P., V. Karja, and A.M. Mustonen. 2009. Myo- and hepatotoxic effects of cultivated mushrooms in mice. *Food Chem. Toxicol.* 47(1):70-74.

Odagiri, Y., N. Watari, T. Suga, and T. Masuyama. 2006. Four-week oral toxicity study of functional food containing superfine dispersed lentinan (β-1,3-glucan) in rats. *Biotherapy* 20(6):568-577.

Okamoto, T., R. Kodoi, Y. Nonaka, et al. 2004. Lentinan from shiitake mushroom (*Lentinus edodes*) suppresses expression of cytochrome P450 1A subfamily in the mouse liver. *Biofactors* 21(1-4):407-409.

Sastre, J., M.D. Ibanez, M. Lopez, and S.B. Lehrer. 1990. Respiratory and immunological reactions among shiitake (*Lentinus edodes*) mushroom workers. *Clin. Exp. Allergy* 20(1):13-20.

Shimada, S., K. Komamura, H. Kumagai, and H. Sakurai. 2004. Inhibitory activity of shiitake flavor against platelet aggregation. *Biofactors* 22(1-4):177-179.

Shimizu, R. 1990. A case of drug eruption caused by lentinan. *Rinsho Derma* 32:1065-1068.

Spierings, E.L., H. Fujii, B. Sun, and T. Walshe. 2007. A Phase I study of the safety of the nutritional supplement, active hexose correlated compound, AHCC, in healthy volunteers. *J. Nutr. Sci. Vitaminol.* 53(6):536-539.

Tarvainen, K., J.P. Salonen, L. Kanerva, et al. 1991. Allergy and toxicodermia from shiitake mushrooms. *J. Am. Acad. Dermatol.* 24(1):64-66.

Tarvainen, K., J.P. Salonen, L. Kanerva, T. Estlander, and T. Rantanen. 1990. Allergy to shiitake mushroom. *Contact Dermat.* 23(4):302-303.

Ueda, A., K. Obama, K. Aoyama, et al. 1992. Allergic contact dermatitis in shiitake (*Lentinus edodes* (Berk.) Sing.) growers. *Contact Dermat.* 26(4):228-233.

Von Wright, A., J. Knuutinen, and S. Lindroth. 1982. The mutagenicity of some edible mushrooms in the Ames test. *Food Chem. Toxicol.* 20(3):265-267.

Zheng, R., S. Jie, D. Hanchuan, and W. Moucheng. 2005. Characterization and immunomodulating activities of polysaccharide from *Lentinus edodes*. *Int. Immunopharmacol.* 5(5):811-820.

Leonurus cardiaca L. — Lamiaceae

SCN: motherwort
OCN: common motherwort

Part: herb

Quick Reference Summary

Safety Class: 2b
Interaction Class: A

Contraindications

Not for use in pregnancy except under the supervision of a qualified healthcare practitioner (Bradley 1992; Chadha 1988; List and Hörhammer 1973).

Other Precautions

None known.

Drug and Supplement Interactions

None known.

Notice

Emmenagogue (Chadha 1988; Felter and Lloyd 1898; List and Hörhammer 1973); *see* Appendix 2.

Adverse Events and Side Effects

None known.

Pharmacological Considerations

None known.

Pregnancy and Lactation

Information on the safety of motherwort in pregnancy is limited. Early studies of motherwort on uterine tissue provided conflicting results on uterine stimulant activity (Erspamer 1948; Pilcher 1916; Pilcher and Mauer 1918).

No information on the safety of motherwort in lactation was identified. While this review did not identify any concerns for use while nursing, safety has not been conclusively established.

Review Details

I. Drug and Supplement Interactions

Clinical Trials of Drug or Supplement Interactions

No clinical trials of drug or supplement interactions were identified.

Case Reports of Suspected Drug or Supplement Interactions

No case reports of suspected drug or supplement interactions were identified.

Animal Trials of Drug or Supplement Interactions

No animal trials of drug or supplement interactions were identified.

II. Adverse Events

Case Reports of Adverse Events

No case reports of adverse events were identified.

III. Pharmacology and Pharmacokinetics

Human Pharmacological Studies

No relevant human pharmacological studies were identified.

Animal Pharmacological Studies

No relevant animal pharmacological studies were identified.

In Vitro Pharmacological Studies

An ethanol extract of motherwort stimulated growth of estrogen receptor-positive MCF-7 breast cancer cells half as much as estradiol (Zava 1998).

IV. Pregnancy and Lactation

In experiments on the effects of motherwort extract on isolated uteruses or uterine tissue, one study indicated a mild stimulating effect (Erspamer 1948), while two others showed no effect (Pilcher 1916; Pilcher and Mauer 1918).

The compound leonurine was observed to increase uterine tone and contractions in isolated uteruses (Kubota and Nakashima 1930).

No information on the safety of motherwort during lactation was identified.

V. Toxicity Studies

Acute Toxicity

Irritation, convulsions, and respiratory paralysis were observed in mice intraperitoneally administered the compound leonurine (dose unspecified in available translated material). The same compound administered to cats in "small" doses caused respiratory stimulation, while "excessive" doses produced respiratory distress (Kubota and Nakashima 1930).

Genotoxicity

No mutagenic activity was observed for motherwort in the Ames mutagenicity test (Schimmer et al. 1994).

Literature Cited

Bradley, P.R. 1992. *British herbal compendium: A handbook of scientific information on widely used plant drugs*. Bournemouth, UK: British Herbal Medicine Association.

Buzogany, K., and V. Cucu. 1983. Studiu comparativ intre speciile *Leonurus cardiaca* L. si *Leonurus quinquelobatus* Gilib. II. Continutul in iridoide. *Clujul Med.* 56(4):S385-S388.

Chadha, Y. 1988. *The wealth of India: A dictionary of Indian raw materials and industrial products*. Delhi: Council of Scientific and Industrial Research.

Erspamer, V. 1948. Ricerche farmacologiche sul *Leonurus cardiaca* L. (e sul *Leonurus marrubiastrum* L). *Arch. Int. Pharmacodynam. Ther.* 76(1):132-152.

Felter, H.W., and J.U. Lloyd. 1898. *King's American dispensatory*. 18th ed., 3rd rev. 2 vols. Cincinnati: Ohio Valley Co.

Gulubov, A. 1970. Structure of alkaloids from *Leonurus cardiaca*. *Nauch. Tr. Vissh. Predagog. Inst. Plovdiv. Mat. Fiz. Khim. Biol.* 8:129-132.

Kubota, S., and S. Nakashima. 1930. The study of *Leonurus sibericus* L. II. Pharmacological study of the alkaloid leonurin isolated from *Leonurus sibericus* L. *Folia Pharmacol. Jap.* 11:159-167.

List, P.H., and H. Hörhammer. 1973. *Hagers handbuch der pharmazeutischen praxis*. Berlin: Springer.

Pilcher, J.D. 1916. The action of certain drugs on the excised uterus of the guineapig. *J. Pharm. Exp. Ther.* 8:110-111.

Pilcher, J.D., and R.T. Mauer. 1918. The action of "female remedies" on intact uteri of animals. *Surg. Gynecol. Obstet.* 27:97-99.

Schimmer, O., A. Kruger, H. Paulini, and F. Haefele. 1994. An evaluation of 55 commercial plant extracts in the Ames mutagenicity test. *Pharmazie* 49(6):448-451.

Zava, D.T. 1998. Estrogen and progestin bioactivity of foods, herbs, and spices. *Proc. Soc. Exp. Biol. Med.* 217(3):369-378.

Leonurus spp.

Lamiaceae

Leonurus japonicus Houtt.
SCN: Chinese motherwort
Syn: *Leonurus artemisia* (Lour.) S.Y. Hu; *Leonurus heterophyllus* Sweet
PN: *yi mu cao* (above-ground parts, herb)

Leonurus sibiricus L.
SCN: Siberian motherwort
PN: *yi mu cao* (above-ground parts, herb)
Part: above-ground parts, herb

Quick Reference Summary

Safety Class: 2b
Interaction Class: A

Contraindications

Not for use in pregnancy except under the supervision of a qualified healthcare practitioner (Bensky et al. 2004; Chen and Chen 2004).

Other Precautions

None known.

Drug and Supplement Interactions

See Pharmacological Considerations.

Notice

Emmenagogue and uterine stimulant (Bensky et al. 2004; Chen and Chen 2004; Tang and Eisenbrand 1992); *see* Appendix 2.

Adverse Events and Side Effects

Overdose (standard dose is a decoction of 9–15 g; 30 g is listed as a large dose) of Chinese or Siberian motherwort has been associated with symptoms such as generalized weakness, stiffness and paralysis, generalized body pain, a sense of oppression in the chest, excessive sweating, low blood pressure, and cold extremities (Bensky et al. 2004).

Pharmacological Considerations

Animal and in vitro studies have indicated that Chinese motherwort and compounds in Chinese motherwort have antiplatelet activity (Chen and Chen 2004; Lee et al. 1991). Based on these studies, one text on traditional Chinese medicine notes that Chinese or Siberian motherwort should be used with caution in patients taking anticoagulant or antiplatelet drugs, noting that this caution is theoretical and interactions have not been observed clinically (Chen and Chen 2004).

Pregnancy and Lactation

A reference text on traditional Chinese medicine indicates that neither Chinese nor Siberian motherwort should be used during pregnancy (Bensky et al. 2004), while a second text indicates that these herbs may be used during pregnancy with extreme caution (Chen and Chen 2004). Human and animal studies and traditional use have indicated that Chinese and Siberian motherwort have a stimulating effect on the uterus (Bensky et al. 2004; Chen and Chen 2004).

No information on the safety of Chinese or Siberian motherwort during lactation was identified. While this review did not identify any concerns for use while nursing, safety has not been conclusively established.

Review Details

I. Drug and Supplement Interactions

Clinical Trials of Drug or Supplement Interactions

No clinical trials of drug or supplement interactions were identified.

Case Reports of Suspected Drug or Supplement Interactions

No case reports of suspected drug or supplement interactions were identified.

Animal Trials of Drug or Supplement Interactions

No animal trials of drug or supplement interactions were identified.

II. Adverse Events

Case Reports of Adverse Events

Overdose (standard dose is a decoction of 9–15 g; 30 g is listed as a large dose) of Chinese or Siberian motherwort is reported as slightly toxic, with adverse effects appearing 4 to 6 hours after ingestion. Associated symptoms included generalized weakness, stiffness and paralysis, generalized body pain, a sense of oppression in the chest, excessive sweating, low blood pressure, and cold extremities. Severe cases may also include shock, cyanosis, and respiratory paralysis (Bensky et al. 2004). Details on doses, durations, products used, and relevant patient medical history were not listed in the available English language translation.

III. Pharmacology and Pharmacokinetics

Human Pharmacological Studies

No relevant human pharmacological studies were identified.

Animal Pharmacological Studies

An alcohol extract of Chinese motherwort significantly inhibited platelet aggregation after intravenous administration (animal and dose not specified in available English language translation) (Chen and Chen 2004).

In Vitro Pharmacological Studies

The compound prehispanolone was found to be a platelet-activating factor receptor antagonist (Lee et al. 1991).

IV. Pregnancy and Lactation

A reference text on traditional Chinese medicine indicates that Chinese or Siberian motherwort should not be used during pregnancy (Bensky et al. 2004), while a second text indicates that these herbs may be used during pregnancy but that extreme caution must be used (Chen and Chen 2004).

Overdose of Chinese or Siberian motherwort (standard dose is decoction of 9–15 g; 30 g is listed as a large dose) has been associated with miscarriage (Bensky et al. 2004).

A uterine-stimulating effect was observed in rabbits, cats, dogs, and guinea pigs orally administered water or alcohol extracts of Chinese motherwort (doses not specified in available English language translation). The effect was observed in both pregnant and nonpregnant animals, and early and late in pregnancy (Chen and Chen 2004).

Contraction of the uterus was recorded in nonpregnant women after ingestion of 15 to 20 g of a decoction or other extract of Chinese motherwort. The strength of the contraction was not dependent on the dose of herb taken (Chen and Chen 2004).

No information on the safety of Chinese or Siberian motherwort during lactation was identified.

V. Toxicity Studies

Acute Toxicity

The LD_{50} of orally administered Siberian motherwort in rats could not be determined at doses up to 3 or 5 g/kg (Oga et al. 1986; Pin et al. 2009).

The LD_{50} of intravenously administered Chinese motherwort in mice is between 30 and 60 g/kg (Chen and Chen 2004).

Subchronic Toxicity

In rats fed diets containing 0.5, 5, or 25 g/kg body weight of Siberian motherwort daily for 90 days, effects on various systems were observed in the medium- and high-dose groups. In those groups, plasma creatinine and liver enzymes were elevated, hematological analysis indicated mild anemia, and histopathological changes in the liver and kidneys were observed. No adverse effects were reported in the 0.5 g/kg group (Pin et al. 2009).

Literature Cited

Bensky, D., S. Clavey, and E. Stöger. 2004. *Chinese herbal medicine: Materia medica*. 3rd ed. Seattle: Eastland Press.

Chen, J.K., and T.T. Chen. 2004. *Chinese medical herbology and pharmacology*. City of Industry, CA: Art of Medicine Press.

Lee, C.M., L.M. Jiang, H.S. Shang, et al. 1991. Prehispanolone, novel platelet activating factor receptor antagonist from *Leonurus heterophyllus*. *Br. J. Pharmacol.* 103(3):1719-1724.

Oga, S., G. Akisue, E.A. Lopes, G. Kose, and S.Y. Yokoto. 1986. Effect of the extract of *Leonurus sibiricus* L. on CNS of rats. *Rev. Farm. Bioquim. Univ. Sao Paulo* 22(Jul-Dec):131-139.

Pin, C.H., A. Abdullah, and M. Murugaiyah. 2009. Toxicological evaluation of dried kacangma herb (*Leonurus sibiricus*) in rats. *Sains Malaysiana* 38(4):499-509.

Tang, W., and G. Eisenbrand. 1992. *Chinese drugs of plant origin: Chemistry, pharmacology, and use in traditional and modern medicine*. New York: Springer.

Lepidium meyenii Walp. — Brassicaceae

SCN: maca

Part: hypocotyl, root

Quick Reference Summary

Safety Class: 1
Interaction Class: A

Contraindications

None known.

Other Precautions

None known.

Drug and Supplement Interactions

None known.

Adverse Events and Side Effects

None known.

Pharmacological Considerations

Human and animal studies have indicated that maca does not affect levels of hormones such as testosterone, estrogen, follicle-stimulating hormone, or prolactin (Brooks et al. 2008; Gonzales et al. 2001, 2003a, 2003b, 2005).

A study of maca in patients taking selective serotonin reuptake inhibitors (SSRI) indicated no changes in SSRI efficacy (Dording et al. 2008).

Pregnancy and Lactation

No adverse effects on pregnancy or fetal development have been observed in animal studies with daily doses up to 1 g/kg (D'Arrigo et al. 2004; Kuo et al. 2003; Oshima et al. 2003; Ruiz-Luna et al. 2005).

No information on the safety of maca during lactation was identified. While this review did not identify any concerns for use while nursing, safety has not been conclusively established.

Review Details

I. Drug and Supplement Interactions

Clinical Trials of Drug or Supplement Interactions

No clinical trials of drug or supplement interactions were identified.

Case Reports of Suspected Drug or Supplement Interactions

No case reports of suspected drug or supplement interactions were identified.

Animal Trials of Drug or Supplement Interactions

No animal trials of drug or supplement interactions were identified.

II. Adverse Events

Case Reports of Adverse Events

No case reports of adverse events were identified.

III. Pharmacology and Pharmacokinetics

Human Pharmacological Studies

In a randomized, double-blind, placebo-controlled study, healthy postmenopausal women ingested 3.5 g of maca powder or placebo daily for 6 weeks. No differences were seen in serum concentrations of estradiol, follicle-stimulating hormone, luteinizing hormone, and sex hormone-binding globulin between baseline, maca treatment, and placebo (Brooks et al. 2008).

In a randomized, double-blind, placebo-controlled study, healthy men ingested 1.5 or 3 g of maca daily for 12 weeks. Testing at baseline and at weeks 2, 4, 8, and 12 of the study indicated that serum levels of luteinizing hormone, follicle-stimulating hormone, prolactin, 17α-hydroxyprogesterone, testosterone, and 17β-estradiol were unchanged (Gonzales et al. 2003a).

No changes in hormone levels, including luteinizing hormone, follicle-stimulating hormone, prolactin, testosterone, and estradiol, were observed in healthy men who ingested 1.5 or 3 g of maca daily for 4 months (Gonzales et al. 2001).

In a study on the effects of maca for sexual dysfunction induced by selective serotonin reuptake inhibitors (SSRI), patients who were stable on an SSRI (patients were taking escitalopram, citalopram, sertraline, venlafexine, fluoxetine, paroxetine, duloxetine, or fluvoxamine) were orally administered 1.5 or 3 g of maca daily for 12 weeks. No adverse effects of maca, including effects on SSRI efficacy, were reported, and patients in the high dose group had a modest improvement in depression (Dording et al. 2008).

Animal Pharmacological Studies

No changes in serum levels of testosterone or estradiol were observed in male rats orally administered 2 g/kg of red, yellow, or black ecotypes of maca daily for 7 days. Red maca, but neither yellow nor black maca, significantly reduced the ventral prostate size in treated animals (Gonzales et al. 2005).

In male rats orally administered 2 g/kg of red maca daily for 42 days with or without intramuscular injections of testosterone enanthate on days 1 and 7, maca prevented the prostate weight increase induced by testosterone treatment (Gonzales et al. 2005).

In male rats orally administered 48 or 96 mg/kg of maca daily for 7, 14, or 21 days, no changes in serum testosterone levels were observed. An increase in sperm count was observed after 7 days, while the sperm count was reduced at 14 and 21 days (Gonzales et al. 2003b).

In rats orally administered 0.01 to 5 g/kg of maca daily for 7 days, seminal vesicle weight was reduced at the 0.01 and 0.10 g/kg doses. Maca increased the length of stages VII–VIII of the seminiferous tubules in a dose-response fashion, with highest response at 1.0 g/kg. Cauda epididymal sperm count, sperm motility, and serum estradiol level were not affected at any of the doses studied (Chung et al. 2005).

Prevention of estrogen-deficient bone loss was observed in ovariectomized rats orally administered 240 mg/kg of an ethanol extract of maca daily for 28 weeks but not in animals administered 96 mg/kg (Zhang et al. 2006).

In hereditary hypertriglyceridemic rats, a strain used as a model for insulin resistance, improvement in glucose

L

tolerance was observed after administration of a diet containing 1% maca daily for 2 weeks (Vecera et al. 2007).

An increase in uterine weight was observed in ovariectomized mice orally administered 1 g/kg of a freeze-dried aqueous extract of maca daily for 42 days (Ruiz-Luna et al. 2005).

In Vitro Pharmacological Studies

No estrogenic or androgenic activity of aqueous or methanolic extracts of maca was observed at concentrations up to 4 mg/ml in yeast β-galactosidase reporter assays (Brooks et al. 2008). Conversely, estrogenic activity of maca was reported in an assay with estrogen receptor-positive human breast cancer cells (MCF-7) with an IC_{50} of 100 to 200 μg/ml (Valentova et al. 2006).

In a set of assays (luciferase and β-galactosidase) selected to assess whether maca was able to bind to the human androgen receptor, no regulation of the glucocorticoid response element activation was observed in tests with methanol, ethanol, hexane, and chloroform extracts of maca (Bogani et al. 2006).

IV. Pregnancy and Lactation

In female mice provided with a maca extract (5 g maca in 100 ml water) as the sole source of drinking water for 30 days (estimated intake not listed), increases in progesterone and testosterone levels were observed, but with no changes in levels of 17β-estradiol. After mating, no differences in the rate of embryo implantation were observed (Oshima et al. 2003).

No adverse effects on implantation or on fetal development were observed in mice orally administered 1 g/kg of a freeze-dried aqueous extract of maca on days 1 to 4 of pregnancy (D'Arrigo et al. 2004).

No adverse effects on fetal development or offspring growth were observed in mice fed diets containing 30% maca daily throughout pregnancy and lactation. The litter size was larger in treated animals than in controls (Kuo et al. 2003).

An increase in litter size was observed in mice orally administered 1 g/kg of a freeze-dried aqueous extract of maca daily 15 days prior to mating, throughout gestation, and for 21 days after birth. No adverse effects on fetal or pup development were observed (Ruiz-Luna et al. 2005).

No information on the safety of maca during lactation was identified.

V. Toxicity Studies

Acute Toxicity

No adverse effects were observed in rats orally administered 5 g/kg maca daily for 7 days (Chung et al. 2005).

In the brine shrimp lethality assay, the LC_{50} of micropulverized maca is 822.68 g/ml (Valerio and Gonzales 2005).

Subchronic Toxicity

In rats orally administered 1 g/kg of maca daily for 84 days, no histological effects on the liver or spleen or changes in DNA levels in the testis were observed (Gasco et al. 2007).

Literature Cited

Bogani, P., F. Simonini, M. Iriti, et al. 2006. *Lepidium meyenii* (maca) does not exert direct androgenic activities. *J. Ethnopharmacol.* 104(3):415-417.

Brooks, N.A., G. Wilcox, K.Z. Walker, et al. 2008. Beneficial effects of *Lepidium meyenii* (maca) on psychological symptoms and measures of sexual dysfunction in postmenopausal women are not related to estrogen or androgen content. *Menopause* 15(6):1157-1162.

Chung, F., J. Rubio, C. Gonzales, M. Gasco, and G.F. Gonzales. 2005. Dose-response effects of *Lepidium meyenii* (maca) aqueous extract on testicular function and weight of different organs in adult rats. *J. Ethnopharmacol.* 98(1-2):143-147.

D'Arrigo, G., V. Benavides, and J. Pino. 2004. Preliminary evaluation effect of *Lepidium meyenii* Walp. on the embryonic development of mouse. *Rev. Peru Biol.* 11(1):103-106.

Dording, C.M., L. Fisher, G. Papakostas, et al. 2008. A double-blind, randomized, pilot dose-finding study of maca root (*L. meyenii*) for the management of SSRI-induced sexual dysfunction. *CNS Neurosci. Ther.* 14(3):182-191.

Gasco, M., J. Aguilar, and G.F. Gonzales. 2007. Effect of chronic treatment with three varieties of *Lepidium meyenii* (maca) on reproductive parameters and DNA quantification in adult male rats. *Andrologia* 39(4):151-158.

Gonzales, G.F., A. Cordova, C. Gonzales, et al. 2001. *Lepidium meyenii* (maca) improved semen parameters in adult men. *Asian J. Androl.* 3(4):301-303.

Gonzales, G.F., A. Cordova, K. Vega, et al. 2003a. Effect of *Lepidium meyenii* (maca), a root with aphrodisiac and fertility-enhancing properties, on serum reproductive hormone levels in adult healthy men. *J. Endocrinol.* 176(1):163-168.

Gonzales, G.F., S. Miranda, J. Nieto, et al. 2005. Red maca (*Lepidium meyenii*) reduced prostate size in rats. *Reprod. Biol. Endocrinol.* 3:5.

Gonzales, G.F., J. Rubio, A. Chung, M. Gasco, and L. Villegas. 2003b. Effect of alcoholic extract of *Lepidium meyenii* (maca) on testicular function in male rats. *Asian J. Androl.* 5(4):349-352.

Kuo, T.-F., M.-H. Chang, and M.-Y. Liau. 2003. Effects of *Lepidium meyenii* Walp. (maca) on fecundity and puppy growth in mice. *Taiwan Vet. J.* 29(1):1-8.

Oshima, M., Y. Gu, and S. Tsukada. 2003. Effects of *Lepidium meyenii* Walp. and *Jatropha macrantha* on blood levels of estradiol-17 beta, progesterone, testosterone and the rate of embryo implantation in mice. *J. Vet. Med. Sci.* 65(10):1145-1146.

Ruiz-Luna, A.C., S. Salazar, N.J. Aspajo, et al. 2005. *Lepidium meyenii* (maca) increases litter size in normal adult female mice. *Reprod. Biol. Endocrinol.* 3:16.

Valentova, K., D. Buckiova, V. Kren, et al. 2006. The in vitro biological activity of *Lepidium meyenii* extracts. *Cell Biol. Toxicol.* 22(2):91-99.

Valerio, L.G., Jr., and G.F. Gonzales. 2005. Toxicological aspects of the South American herbs cat's claw (*Uncaria tomentosa*) and maca (*Lepidium meyenii*): A critical synopsis. *Toxicol. Rev.* 24(1):11-35.

Vecera, R., J. Orolin, N. Skottova, et al. 2007. The influence of maca (*Lepidium meyenii*) on antioxidant status, lipid and glucose metabolism in rat. *Plant Foods Hum. Nutr.* 62(2):59-63.

Zhang, Y., L. Yu, M. Ao, and W. Jin. 2006. Effect of ethanol extract of *Lepidium meyenii* Walp. on osteoporosis in ovariectomized rat. *J. Ethnopharmacol.* 105(1-2):274-279.

Ligusticum porteri J.M. Coult. & Rose — Apiaceae

SCN: osha
OCN: Porter's lovage
Part: rhizome

Quick Reference Summary

Safety Class: 2b
Interaction Class: A

Contraindications

Not for use in pregnancy except under the supervision of a qualified healthcare practitioner (Conway and Slocumb 1979; Moore 1979).

Other Precautions

None known.

Drug and Supplement Interactions

None known.

Adverse Events and Side Effects

None known.

Pharmacological Considerations

None known.

Pregnancy and Lactation

Osha has been used as an emmenagogue (Conway and Slocumb 1979). While a 1979 text on medicinal plants of the western United States indicated that large amounts (decoction of ~55 g) of osha have reportedly been used as an abortifacient (Moore 1979), a more recent edition of the same text notes that the dried root is thought to be safe "in any but the most difficult pregnancies" (Moore 2003).

No information on the safety of osha during lactation was identified. While this review did not identify any concerns for use while nursing, safety has not been conclusively established.

Review Details

I. Drug and Supplement Interactions

Clinical Trials of Drug or Supplement Interactions

No clinical trials of drug or supplement interactions were identified.

Case Reports of Suspected Drug or Supplement Interactions

No case reports of suspected drug or supplement interactions were identified.

Animal Trials of Drug or Supplement Interactions

No animal trials of drug or supplement interactions were identified.

II. Adverse Events

Case Reports of Adverse Events

No case reports of adverse events were identified.

III. Pharmacology and Pharmacokinetics

Human Pharmacological Studies

No relevant human pharmacological studies were identified.

Animal Pharmacological Studies

No relevant animal pharmacological studies were identified.

In Vitro Pharmacological Studies

No relevant in vitro pharmacological studies were identified.

IV. Pregnancy and Lactation

Osha has been used as an emmenagogue, and large amounts (decoction of ~55 g) have reportedly been used as an abortifacient (Conway and Slocumb 1979; Moore 1979).

No information on the safety of osha during lactation was identified.

V. Toxicity Studies

Acute Toxicity

The LD_{50} of an orally administered dichloromethane and methanol extract of osha in mice is 1085 mg/kg. In the brine shrimp lethality assay, the LC_{50} of the same extract is 778 μg/ml (Deciga-Campos et al. 2007).

Literature Cited

Conway, G.A., and J.C. Slocumb. 1979. Plants used as abortifacients and emmenagogues by Spanish New Mexicans. *J. Ethnopharmacol.* 1(3):241-261.

Deciga-Campos, M., I. Rivero-Cruz, M. Arriaga-Alba, et al. 2007. Acute toxicity and mutagenic activity of Mexican plants used in traditional medicine. *J. Ethnopharmacol.* 110(2):334-342.

Moore, M. 1979. *Medicinal plants of the Mountain West*. Santa Fe: Museum of New Mexico Press.

Moore, M. 2003. *Medicinal plants of the Mountain West*. Santa Fe: Museum of New Mexico Press.

Ligusticum spp. — Apiaceae

Ligusticum sinense Oliv. 'Chuanxiong'
SCN: Sichuan lovage
Syn: *Ligusticum chuanxiong* Hortorum ex Qiu, et al.
PN: *chuan xiong* (rhizome)

Ligusticum wallichii (Benth. & Hook. f.) Franch.
SCN: *Ligusticum wallichii*
PN: *chuan xiong* (rhizome)
Part: rhizome

Quick Reference Summary

Safety Class: 2b
Interaction Class: A

Contraindications

Not for use in pregnancy except under the supervision of a qualified healthcare practitioner (Bensky et al. 2004).

Other Precautions

Use with caution in excessive menstruation (Bensky et al. 2004; Chen 1987).

Drug and Supplement Interactions

See Pharmacological Considerations.

Adverse Events and Side Effects

Allergic skin reactions to Sichuan lovage or *Ligusticum wallichii* have been reported after oral ingestion (Bensky et al. 2004).

Pharmacological Considerations

Based on animal studies demonstrating anticoagulant and antiplatelet activity (Chen and Chen 2004; Zhu 1998), a text on traditional Chinese medicine suggests that Sichuan lovage or *Ligusticum wallichii* should be used with caution in patients taking anticoagulant or antiplatelet drugs, noting that this caution is theoretical, and interactions have not been observed clinically (Chen and Chen 2004).

Pregnancy and Lactation

A reference text on traditional Chinese medicine indicated that Sichuan lovage and *Ligusticum wallichii* should be used with caution during pregnancy (Bensky et al. 2004). Based on this information, use during pregnancy is not recommended except under the supervision of a qualified healthcare practitioner.

An animal study on the compounds tetramethylpyrazine and ferulic acid indicated that the compounds inhibited uterine contractions (Ozaki and Ma 1990).

No information on the safety of Sichuan lovage and *Ligusticum wallichii* during lactation was identified. While this review did not identify any concerns for use while nursing, safety has not been conclusively established.

Review Details

I. Drug and Supplement Interactions

Clinical Trials of Drug or Supplement Interactions

No clinical trials of drug or supplement interactions were identified.

Case Reports of Suspected Drug or Supplement Interactions

No case reports of suspected drug or supplement interactions were identified.

Animal Trials of Drug or Supplement Interactions

No animal trials of drug or supplement interactions were identified.

II. Adverse Events

Case Reports of Adverse Events

Ingestion of Sichuan lovage or *Ligusticum wallichii* has been associated with tingling pain in the abdomen, urinary frequency, pain on urinating, intense headache, and vomiting. Information on doses and products used was not listed in the available English language translation (Bensky et al. 2004).

Allergic skin reactions to Sichuan lovage or *Ligusticum wallichii* have been reported after oral ingestion (Bensky et al. 2004).

III. Pharmacology and Pharmacokinetics

Human Pharmacological Studies

The compound tetramethylpyrazine inhibited ADP- or collagen-induced platelet aggregation in healthy volunteers and in patients with coronary heart disease. Information on dose was not provided in the available English language translation (Zhu 1998).

Animal Pharmacological Studies

A dose-dependent reduction in plasma glucose levels was observed in diabetic mice intraperitoneally administered 10, 25, or 50 mg/kg of the compound tetramethylpyrazine (Lee et al. 2002).

In male rats orally administered 1 g (average animal weight 250 g) of an ethanol extract of Chinese lovage, a 2.3-fold increase in progestogenic activity was observed. After subcutaneous administration of 2 g of the same extract, a 4.5-fold increase was observed (Lim et al. 2006).

In rats orally administered water or alcohol extracts of Chinese lovage at doses equivalent to 3 g/kg of the dried herb daily, mild inhibition of the drug-metabolizing isoenzymes CYP1A2, CYP2C9, CYP2D6, CYP2E1, and CYP3A were observed (Tang et al. 2006).

A text on traditional Chinese medicine indicates that animal studies have demonstrated marked antiplatelet and anticoagulant effects. Details on doses and preparations used were not listed in the available English language translation (Chen and Chen 2004). The compound tetramethylpyrazine inhibited ADP- or collagen-induced platelet aggregation in rabbits. Information on dose was not provided in the available English language translation (Zhu 1998).

In Vitro Pharmacological Studies

The compound tetramethylpyrazine, at concentrations of 50 to 200 μ*M*, stimulated nitric oxide production in human platelets (Sheu et al. 2000).

Proestrogenic activity of Sichuan lovage was observed in HeLa cells with progesterone receptors (PR) and a PR-driven promoter (Lim et al. 2006).

IV. Pregnancy and Lactation

Inhibition of uterine contractions was observed in rats orally administered 100 or 300 mg/kg or intravenously administered 10 or 30 mg/kg of the compounds tetramethylpyrazine or ferulic acid. Synergistic activity was observed when the compounds were administered intravenously together, at doses individually insufficient to inhibit contractions (1 mg/kg tetramethylpyrazine, 3 mg/kg ferulic acid) (Ozaki and Ma 1990).

No information on the safety of Sichuan lovage or *Ligusticum wallichii* during lactation was identified.

V. Toxicity Studies

Acute Toxicity

The LD_{50} of Sichuan lovage in mice is 65.86 g/kg after intraperitoneal administration and 66.42 g/kg after intravenous administration (Chen and Chen 2004).

The LD_{50} of the compound tetramethylpyrazine intravenously administered in mice is 239 mg/kg (Zhu 1998).

Short-Term Toxicity

No adverse effects on liver or kidney function or changes in serum chemistry were observed in mice orally administered 5 or 10 mg/kg of the compound chuanxiongzine daily for 4 weeks (Chen and Chen 2004).

L

Literature Cited

Bensky, D., S. Clavey, and E. Stöger. 2004. *Chinese herbal medicine: Materia medica*. 3rd ed. Seattle: Eastland Press.

Chen, G.J. 1987. Pharmacological studies on the neutral oil of gao ben (*Ligusticum sinense*) (II): The inhibitory effect on the smooth muscle of the intestines and uterus. *Zhong Yao Tong Bao* 12(4):48-51.

Chen, J.K., and T.T. Chen. 2004. *Chinese medical herbology and pharmacology*. City of Industry, CA: Art of Medicine Press.

Lee, L.M., C.F. Liu, and P.P. Yang. 2002. Effect of tetramethylpyrazine on lipid peroxidation in streptozotocin-induced diabetic mice. *Am. J. Chin. Med.* 30(4):601-608.

Lim, L.S., P. Shen, Y.H. Gong, L.S. Lee, and E.L. Yong. 2006. Dynamics of progestogenic activity in serum following administration of *Ligusticum chuanxiong*. *Life Sci.* 79(13):1274-1280.

Ozaki, Y., and J.P. Ma. 1990. Inhibitory effects of tetramethylpyrazine and ferulic acid on spontaneous movement of rat uterus in situ. *Chem. Pharm. Bull.* 38(6):1620-1623.

Sheu, J.R., Y.C. Kan, W.C. Hung, C.H. Lin, and M.H. Yen. 2000. The antiplatelet activity of tetramethylpyrazine is mediated through activation of NO synthase. *Life Sci.* 67(8):937-947.

Tang, J.C., J.N. Zhang, Y.T. Wu, and Z.X. Li. 2006. Effect of the water extract and ethanol extract from traditional Chinese medicines *Angelica sinensis* (Oliv.) Diels, *Ligusticum chuanxiong* Hort. and *Rheum palmatum* L. on rat liver cytochrome P450 activity. *Phytother. Res.* 20(12):1046-1051.

Zhu, Y.-P. 1998. *Chinese materia medica: Chemistry, pharmacology and applications.* Amsterdam: Harwood Academic Publishers.

Ligusticum spp. — Apiaceae

Ligusticum jeholense (Nakai & Kitag.) Nakai & Kitag.
SCN: Chinese lovage
Syn: *Cnidium jeholense* Nakai & Kitag.
PN: *gao ben* (root and rhizome)

Ligusticum sinense Oliv.
SCN: Chinese lovage
PN: *gao ben* (root and rhizome)
Part: rhizome, root

Quick Reference Summary

Safety Class: 1
Interaction Class: A

Contraindications

None known.

Other Precautions

Long-term use of Chinese lovage is not recommended (Chen and Chen 2004).

Drug and Supplement Interactions

None known.

Adverse Events and Side Effects

An allergic skin reaction has been reported after ingestion of Chinese lovage (Bensky et al. 2004).

Pharmacological Considerations

None known.

Pregnancy and Lactation

No information on the safety of Chinese lovage in pregnancy or lactation was identified in the scientific or traditional literature. Although this review did not identify any concerns for use while pregnant or nursing, safety has not been conclusively established.

Review Details

I. Drug and Supplement Interactions

Clinical Trials of Drug or Supplement Interactions

No clinical trials of drug or supplement interactions were identified.

Case Reports of Suspected Drug or Supplement Interactions

No case reports of suspected drug or supplement interactions were identified.

Animal Trials of Drug or Supplement Interactions

No animal trials of drug or supplement interactions were identified.

II. Adverse Events

Case Reports of Adverse Events

An allergic skin reaction has been reported after ingestion of Chinese lovage (Bensky et al. 2004).

III. Pharmacology and Pharmacokinetics

Human Pharmacological Studies

No relevant human pharmacological studies were identified.

Animal Pharmacological Studies

No relevant animal pharmacological studies were identified.

In Vitro Pharmacological Studies

No relevant in vitro pharmacological studies were identified.

IV. Pregnancy and Lactation

No information on the safety of Chinese lovage during pregnancy or lactation was identified.

V. Toxicity Studies

Acute Toxicity

The LD_{50} of orally administered essential oil of Chinese lovage in mice is the equivalent of 70.17 g/kg of the dried herb (Chen and Chen 2004).

Literature Cited

Bensky, D., S. Clavey, and E. Stöger. 2004. *Chinese herbal medicine: Materia medica*. 3rd ed. Seattle: Eastland Press.

Chen, J.K., and T.T. Chen. 2004. *Chinese medical herbology and pharmacology*. City of Industry, CA: Art of Medicine Press.

Ligustrum lucidum W.T. Aiton — Oleaceae

SCN: ligustrum
PN: *nu zhen zi* (fruit)
OCN: glossy privet; white wax tree
Part: fruit

Quick Reference Summary

Safety Class: 1
Interaction Class: A

Contraindications

None known.

Other Precautions

None known.

Drug and Supplement Interactions

None known.

Adverse Events and Side Effects

None known.

Pharmacological Considerations

Traditional use and animal studies have indicated that ligustrum may modify glucose regulation (Chen and Chen 2004; Gao et al. 2007). People with diabetes are advised to monitor their blood sugar closely and discuss the use of this herb with a qualified healthcare practitioner prior to use.

Pregnancy and Lactation

No information on the safety of ligustrum in pregnancy or lactation was identified in the scientific or traditional literature. Although this review did not identify any concerns for use while pregnant or nursing, safety has not been conclusively established.

Review Details

I. Drug and Supplement Interactions

Clinical Trials of Drug or Supplement Interactions

No clinical trials of drug or supplement interactions were identified.

Case Reports of Suspected Drug or Supplement Interactions

No case reports of suspected drug or supplement interactions were identified.

Animal Trials of Drug or Supplement Interactions

No animal trials of drug or supplement interactions were identified.

II. Adverse Events

Case Reports of Adverse Events

No case reports of adverse events were identified.

III. Pharmacology and Pharmacokinetics

Human Pharmacological Studies

No relevant human pharmacological studies were identified.

Animal Pharmacological Studies

In diabetic rats orally administered 100 or 200 mg/kg of the compound oleanolic acid daily for 40 days, changes in blood glucose levels and in oral glucose tolerance tests showed that hypoglycemia was more pronounced in treated groups than in the control group. Rats treated with oleanolic acid also had increased serum insulin levels, while no changes in thyroid hormone or thyroid-stimulating hormone were observed (Gao et al. 2007).

In Vitro Pharmacological Studies

No relevant in vitro pharmacological studies were identified.

IV. Pregnancy and Lactation

No information on the safety of ligustrum during pregnancy or lactation was identified.

V. Toxicity Studies

Acute Toxicity

No adverse effects were reported in rabbits that ingested 75 g of ligustrum (animal weight not specified in available English language translation) (Chen and Chen 2004).

Subchronic Toxicity

A protective effect against diabetes-induced adverse effects on spermatogenesis was observed in diabetic rats orally administered an extract of ligustrum equivalent to 30 g/kg of the herb daily for 110 days (Feng et al. 2001).

Literature Cited

Chen, J.K., and T.T. Chen. 2004. *Chinese medical herbology and pharmacology*. City of Industry, CA: Art of Medicine Press.

Feng, S.L., S.H. Li, Y. Wang, C.C. Chen, and B. Gao. 2001. Effect of ligustrum fruit extract on reproduction in experimental diabetic rats. *Asian J. Androl.* 3(1):71-73.

Gao, D., Q. Li, Y. Li, et al. 2007. Antidiabetic potential of oleanolic acid from *Ligustrum lucidum* Ait. *Can. J. Physiol. Pharmacol.* 85(11):1076-1083.

Lilium spp. — Liliaceae

Lilium brownii F.E. Br. ex Miellez var. ***viridulum*** Baker
SCN: Brown's lily
PN: *bai he* (bulb)

Lilium lancifolium **Thunb.**
SCN: tiger lily
Syn: *Lilium tigrinum* Ker Gawl.
PN: *bai he* (bulb)

Lilium pumilum **DC.**
SCN: coral lily
PN: *bai he* (bulb)
Part: bulb

Quick Reference Summary

Safety Class: 1
Interaction Class: A

Contraindications

None known.

Other Precautions

None known.

Drug and Supplement Interactions

None known.

Adverse Events and Side Effects

Allergic reactions to the *Lilium* species included in this entry have been reported (Bensky et al. 2004).

Pharmacological Considerations

None known.

Pregnancy and Lactation

No information on the safety of Brown's lily, tiger lily, or coral lily in pregnancy or lactation was identified in the scientific or traditional literature. Although this review did not identify any concerns for use while pregnant or nursing, safety has not been conclusively established.

Review Details

I. Drug and Supplement Interactions

Clinical Trials of Drug or Supplement Interactions

No clinical trials of drug or supplement interactions were identified.

Case Reports of Suspected Drug or Supplement Interactions

No case reports of suspected drug or supplement interactions were identified.

Animal Trials of Drug or Supplement Interactions

Oral administration of lily bulb was reported to prolong sleeping time induced by barbiturates. Study details were not listed in the available English language translation (Chen and Chen 2004).

II. Adverse Events

Case Reports of Adverse Events

Allergic reactions to the *Lilium* species included in this entry have been reported in the literature on traditional Chinese medicine (Bensky et al. 2004).

Although the above-ground parts (bulbs are the part used medicinally) of tiger lily are toxic to cats, causing acute kidney damage and sometimes kidney failure (Berg et al. 2007; Gulledge et al. 1997; Langston 2002; Rumbeiha et al. 2004), no toxic effects have been identified in rats or rabbits, and dogs develop a mild, self-limiting gastroenteritis after ingesting large quantities of plant material (Hall 2001). No cases of ingestion of lily bulbs by cats or other animals were identified in the literature.

III. Pharmacology and Pharmacokinetics

Human Pharmacological Studies

No relevant human pharmacological studies were identified.

Animal Pharmacological Studies

No relevant animal pharmacological studies were identified.

In Vitro Pharmacological Studies

Inhibition of monoamine oxidase B (MAO-B) was observed in rat brain homogenates treated with an extract of Brown's lily (*L. brownii* var. *colchesteri*) with an IC_{50} value of 0.40 mg/ml (Lin et al. 2003).

IV. Pregnancy and Lactation

No information on the safety of Brown's lily, tiger lily, or coral lily during pregnancy or lactation was identified.

V. Toxicity Studies

No toxicity studies were identified.

Literature Cited

Bensky, D., S. Clavey, and E. Stöger. 2004. *Chinese herbal medicine: Materia medica*. 3rd ed. Seattle: Eastland Press.

Berg, R.I., T. Francey, and G. Segev. 2007. Resolution of acute kidney injury in a cat after lily (*Lilium lancifolium*) intoxication. *J. Vet. Intern. Med.* 21(4):857-859.

Chen, J.K., and T.T. Chen. 2004. *Chinese medical herbology and pharmacology*. City of Industry, CA: Art of Medicine Press.

Gulledge, L., D. Boos, and R. Wachsstock. 1997. Acute renal failure in a cat secondary to tiger lily (*Lilium tigrinum*) toxicity. *Feline Practice* 25(5-6):38-39.

Hall, J.O. 2001. Lily nephrotoxicity. In *Consultations in feline internal medicine 4*. Philadelphia: Saunders.

Langston, C.E. 2002. Acute renal failure caused by lily ingestion in six cats. *J. Am. Vet. Med. Assoc.* 220(1):49-52, 36.

Lin, R.D., W.C. Hou, K.Y. Yen, and M.H. Lee. 2003. Inhibition of monoamine oxidase B (MAO-B) by Chinese herbal medicines. *Phytomedicine* 10(8):650-656.

Rumbeiha, W.K., J.A. Francis, S.D. Fitzgerald, et al. 2004. A comprehensive study of Easter lily poisoning in cats. *J. Vet. Diagn. Invest.* 16(6):527-541.

Linum usitatissimum L. — Linaceae

SCN: flax
AN: *atasi*
PN: *ya ma zi* (seed)
OCN: linseed (seed)
Part: seed

Quick Reference Summary

Safety Class: 1
Interaction Class: A

Contraindications

None known.

Other Precautions

When taken as a bulk-forming laxative, flax should not be used in persons with bowel obstruction or with abnormal esophageal or intestinal narrowing (Wichtl 2004).

Flaxseed should be taken with at least 200 ml (6 oz) of water (Wichtl 2004).

Drug and Supplement Interactions

Other drugs should be taken 1 hour prior to consumption of flax or several hours after consumption, as flax may reduce the absorption of certain drugs due to the mucilage content and the increased speed of passage through the intestines. *See* Mucilages (Appendix 3).

Notice

Bulk-forming laxative (Wichtl 2004); *see* Appendix 2.

Mucilages (5–9%) (Bhatty 1993; Fedeniuk and Biliaderis 1994; Mazza and Biliaderis 1989); *see* Appendix 3.

Cyanogenic glycosides (0.1–1.5%) (Oomah et al. 1992; Wichtl 2004); *see* Appendix 1.

Editors' Note

Flaxseed contains the compound secoisolariciresinol diglycoside (SDG), which is converted by bacteria in the colon to the mammalian lignans enterodiol and enterolactone, compounds that have phytoestrogenic activity (Power and Thompson 2007).

Adverse Events and Side Effects

Allergic reactions, including anaphylactic reactions, to flaxseed have been reported (Alonso et al. 1996; Leon et al. 2003; Lezaun et al. 1998).

Pharmacological Considerations

In cases of bowel inflammation, flaxseed should be preswollen with fluids prior to consumption (Wichtl 2004). Overweight persons using flaxseed should consume the whole, unground seed to avoid caloric absorption (Wichtl 2004).

Studies in healthy women with normal menstrual cycles have generally shown no significant hormonal effects from ingestion of 5 to 10 g of flaxseed daily (Frische et al. 2003; Phipps et al. 1993). One study indicated changes in urinary ratios of some estrogen metabolites after ingestion of 5 or 10 g of flaxseed daily (Haggans et al. 1999). A longer luteal phase, with an increased progesterone/estradiol ratio, was observed in one study (Phipps et al. 1993). An animal study showed that a diet of 2.5 to 10% flaxseed increased in the length of estrus cycle, while animals fed flax with tamoxifen (an estrogen receptor agonist) showed an increase in irregular cycles (Orcheson et al. 1998).

Studies show a decrease in estrogen receptor-positive breast cancer tumor size in animals administered diets of 2.5 to 10% flaxseed (Bergman Jungestrom et al. 2007; Chen et al. 2007b; Dabrosin et al. 2002; Saarinen et al. 2006; Serraino and Thompson 1992; Thompson et al. 1996).

Pregnancy and Lactation

A human study of flaxseed oil in pregnancy indicated no difference in delivery time as compared to a control group (Knudsen et al. 2006).

Animal studies have indicated no adverse effects on fetal development in offspring of animals fed diets of 10% flaxseed during pregnancy (Collins et al. 2003; Flynn et al. 2003; Sankaran et al. 2006; Sprando et al. 2000). One study indicated conflicting effects of different dietary levels of flaxseed in pregnancy and lactation on hormone levels in offspring (Tou et al. 1998). Several animal studies indicated that exposure to flaxseed during gestation and lactation provided a protective effect against breast cancer later in life (Thompson et al. 2003; Tou and Thompson 1999), although one study showed the opposite effect (Khan et al. 2007).

A reduction in pregnancy losses was observed in cows administered a diet containing 10% flaxseed as compared to those fed a diet containing sunflower seed (Ambrose et al. 2006).

Animal studies have indicated no adverse effects on fetal development in offspring of animals fed diets of 10% flaxseed during pregnancy and lactation (Collins et al. 2003; Ward et al. 2001a, 2001b).

Review Details

I. Drug and Supplement Interactions

Clinical Trials of Drug or Supplement Interactions

No clinical trials of drug or supplement interactions were identified.

Case Reports of Suspected Drug or Supplement Interactions

Hematuria (blood in the urine) was reported in a man on chronic aspirin approximately 1 month after starting to take 1 tablespoon of flaxseed daily (Gruver 2003). Nosebleeds were reported after a man taking baby aspirin daily began to take flaxseed oil daily (dose unspecified) (Gruver 2003).

Animal Trials of Drug or Supplement Interactions

Coadministration of diets containing 5 or 10% flaxseed with tamoxifen (an estrogen receptor agonist) for 6 or 16 weeks to athymic rats with estrogen receptor-positive human breast cancer cells (MCF-7) caused a regression in tumor size. In rats fed flax but not tamoxifen, no effects on tumor growth were observed (Chen et al. 2004, 2007a, 2007b).

II. Adverse Events

Case Reports of Adverse Events

Anaphylactic reactions to flaxseed, confirmed by skin prick testing, have been reported (Alonso et al. 1996; Leon et al. 2003; Lezaun et al. 1998).

III. Pharmacology and Pharmacokinetics

Human Pharmacological Studies

In healthy women that ingested 10 g flaxseed in baked goods daily for 2 menstrual cycles, urinary lignan excretion significantly increased, but no changes in serum hormone concentrations or sex hormone-binding globulin were observed (Frische et al. 2003).

In healthy postmenopausal women, addition of 5 or 10 g daily of flaxseed dose-dependently increased urinary 2-hydroxyestrogen excretion and the ratio of urinary 2-hydroxyestrogen to 16α-hydroxyestrone. There were no significant differences in urinary 16α-hydroxyestrogen excretion (Haggans et al. 1999).

In a study of healthy, normally cycling women administered a diet containing 10 g flaxseed daily for 3 menstrual cycles, flax consumption was associated with longer luteal phase (LP) lengths. There were no significant differences between flax and control cycles for concentrations of either estradiol or estrone during the early follicular phase, midfollicular phase, or luteal phase. Although flaxseed ingestion had no significant effect on LP progesterone concentrations, the LP progesterone/estradiol ratios were significantly higher during the flax cycles. Midfollicular phase testosterone concentrations were slightly higher during flax cycles (Phipps et al. 1993).

Animal Pharmacological Studies

No stimulation of tumor growth was observed in athymic rats with estrogen receptor-positive MCF-7 human breast cancer cells. Administration of diets containing 5 or 10% flaxseed for 16 weeks did not stimulate tumor growth (Chen et al. 2007b). A reduction in tumor growth was observed in mice with established estrogen receptor-negative human breast cancer tumors (MDA-MB-435 cell line) administered a diet containing 10% flaxseed for 6 weeks (Dabrosin et al. 2002). In ovariectomized athymic mice with estrogen receptor-positive human breast cancer tumors, a diet of 10% flaxseed caused a regression in tumor size and reduced the tumor growth-stimulating effect that was induced by a diet containing 20% soy protein (Saarinen et al. 2006). In rats with induced mammary tumors, a diet of 5% flaxseed reduced tumor size by an average of 66.7%. Flaxseed feeding at the tumor initiation stage also tended to reduce the number of tumors per tumor-bearing animal (Serraino and Thompson 1992).

In rats with established mammary tumors, administration of diets containing 1.82% flaxseed oil (equivalent to 5% flaxseed) or 2.5% or 5% flaxseed for 7 weeks led to 50% smaller tumor volume as compared to control animals. There was no relationship between new or total tumor development and urinary lignan levels (Thompson et al. 1996).

In ovariectomized mice treated continuously with estradiol, a diet containing 10% flaxseed counteracted the estradiol-induced growth and angiogenesis in solid estrogen receptor-positive MCF-7 tumors. Extracellular vascular endothelial growth factor was significantly decreased compared with tumors in the basal diet group (Bergman Jungestrom et al. 2007).

An increase in expression of *Brca1* and *Brca2* genes was observed in ovariectomized rats administered a diet containing 10% flaxseed (~2% lignans) for 90 days (Vissac-Sabatier et al. 2003).

In immature rats fed a diet of 5% defatted flaxseed for 7 days, cadmium exposure through the diet, reportedly from flaxseed, was sixfold higher than allowed for humans by the World Health Organization, and cadmium significantly accumulated in the liver and kidneys of the rats (Khan et al. 2007).

A dose-related cessation or lengthening of estrous cycles was observed in approximately 66% of rats administered a diet containing 2.5, 5, or 10% flaxseed. When animals on the same diet were administered tamoxifen, 83% of the animals had irregular menstrual cycles or were in persistent diestrus (Orcheson et al. 1998).

No adverse effects on the hemopoietic system, including red or white blood cells, granulocytes, lymphocytes, monocytes, and platelets, were observed in normal or hypercholesterolemic rabbits orally administered 40 mg/kg daily of a flax-derived lignan complex for 2 months (Prasad 2005).

In obese rats administered diets containing 20% flax, no change in plasma glucose levels was observed, although plasma insulin levels were lower as compared to groups administered equal amounts of casein or soy protein concentrate (Velasquez et al. 2003).

An oil-in-water nanoemulsion made with flaxseed oil and saquinavir resulted in a significant increase in bioavailability of saquinavir in mice after oral or intravenous administration (Vyas et al. 2008).

In Vitro Pharmacological Studies

A dose-dependent decrease in trophoblast tumor cell proliferation was observed in estrogen receptor-positive Jeg3 cells treated with flaxseed extract or certain flaxseed extract fractions, with two fractions producing a 58 to 86% decrease in the production of progesterone. Some extract fractions showed a stimulating effect on hormone production and cell proliferation (Waldschlager et al. 2005).

IV. Pregnancy and Lactation

No differences in timing of spontaneous delivery were observed in pregnant women administered capsules containing 4 g flax oil (2.2 g α-linolenic acid) daily on weeks 17–27 of gestation until delivery, as compared to control. No adverse events were reported (Knudsen et al. 2006).

In rat pups whose mothers were fed a diet containing 10% flaxseed on postnatal days 0 to 21, exposure to flaxseed during suckling enhanced mammary gland morphogenesis through modulation of epidermal growth factor receptor and estrogen receptors (ER-α and ER-β), which resulted in more differentiated mammary glands at postnatal days 49 to 51. This suppressed chemically induced mammary tumorigenesis without causing adverse reproductive effects in dams or their offspring (Thompson et al. 2003).

A reduction in pregnancy losses was observed in pregnant cows administered a diet containing 9.8% flaxseed as compared with those fed a diet with 27% sunflower seed (Ambrose et al. 2006).

In a study examining the effects of exposure to flax during gestation only or during gestation and maturation in a lifetime study, no adverse effects on development were observed in 20-day-old fetuses of rats administered diets of 20 or 40% flaxseed or 13 or 26% defatted flaxseed meal during pregnancy. In the lifetime study group, the flaxseed but not defatted flax meal caused an increase in the anogenital index of females. Defatted flaxseed meal, but not flaxseed, caused delayed puberty in male offspring. Both diets caused a dose-dependent increase in the number of females with irregular estrous cycles, and a decrease in the thymus/body weight and thymus/brain ratios (Collins et al. 2002, 2003).

No significant embryotoxic effects were observed in rat embryos cultured for 45 hours in serum of pregnant rats that had been administered a diet of 20 or 40% flaxseed or 13 or 26% defatted flaxseed meal (Flynn et al. 1999, 2003).

L

In offspring of rats administered a diet of 5 or 10% flax during pregnancy or lactation, a shortened mammary tumor latency was observed in animals with chemically induced tumors, and 10% flaxseed exposure increased tumor multiplicity, as compared to the controls. When assessed in 8-week-old rats, offspring of mothers fed a 10% diet in pregnancy had increased lobular ER-α protein levels, and both in utero and postnatal flaxseed exposures dose-dependently reduced ER-β protein levels in the terminal end bud lobules and ducts (Khan et al. 2007).

In a study of the effects of flax on development of rat mammary gland structures, rats were administered diets of 5 or 10% flaxseed or secoisolariciresinol diglycoside (SDG) during the whole lifetime, during gestation and lactation, or after weaning. The results of the study indicated that lifetime or gestation and lactation exposure to 5 or 10% flaxseed induced structural changes in the mammary gland that may potentially reduce mammary cancer risk (Tou and Thompson 1999).

In rats exposed to diets of 5 or 10% flaxseed during pregnancy and lactation, or after weaning, perinatal exposure to 10% flaxseed increased serum estradiol and produced premature irregular cycles in female offspring, and elevated testosterone levels and increased the prostate weight in male offspring. Conversely, perinatal exposure to 5% flaxseed delayed puberty and produced persistent diestrus in female offspring (although sex hormone levels were unaffected), and elevated the estradiol/testosterone ratio and decreased the prostate weight in male offspring (Tou et al. 1998).

In rats administered 5 or 10% flaxseed diet starting at weaning on postnatal day (PND) 21 or continuously from gestation to PND 132 for lifetime exposure, compared to the control, exposure to flaxseed after weaning produced no marked reproductive effects, whereas lifetime flaxseed exposure produced some dose-dependent changes in the reproductive organs of offspring. In female rats, lifetime exposure to 5% flaxseed caused delayed puberty onset. In contrast, lifetime exposure to 10% flaxseed caused earlier puberty onset, higher relative ovarian weight, higher serum estradiol levels, and lengthened estrous cycles. In male rats, lifetime 10% flaxseed exposure raised serum testosterone and estradiol levels and produced higher relative sex organ weights and prostate cell proliferation. In contrast, lifetime exposure to 5% flaxseed reduced adult relative prostate weight and cell proliferation (Tou et al. 1999).

Flax seed contains the compound SDG, which is converted by bacteria in the colon to the mammalian lignans enterodiol and enterolactone. No significant effects on any reproductive indices were observed in male and female rats exposed, via maternal ingestion of a diet containing 10% flax or the equivalent isolated SDG, during suckling only or during suckling through the postsuckling period (Ward et al. 2001a).

No adverse effects on bone strength were observed after female mice were exposed to SDG purified from flaxseed during suckling via mother's milk or continuously to adolescence (Ward et al. 2001b).

No significant effects on reproduction were observed in male and female suckling rats whose mothers were fed a diet containing 10% flaxseed during lactation and then themselves were administered the same diet until postnatal day 132. The reproductive indices observed were anogenital distance, age and body weight at puberty onset, estrous cycle length, reproductive organ weights, and histological analysis of reproductive organs (Ward et al. 2001a).

In offspring of rats with inherited cystic kidney disease administered a diet of 7% flaxseed oil during pregnancy and lactation, with offspring then administered the same diet until 10 weeks of age, decreases in creatinine clearance and proteinuria were observed. The authors concluded that exposure to flax oil during pregnancy and lactation produced beneficial changes in adult renal disease (Sankaran et al. 2006).

No significant adverse effects on testis structure or spermatogenesis were observed in mice whose mothers were administered a diet containing up to 40% flaxseed during pregnancy and who were administered the same diet for 70 days after weaning (Sprando et al. 2000).

V. Toxicity Studies

Acute Toxicity

No toxic effects of flaxseed were observed in the brine shrimp lethality assay (Mahmoud et al. 1992).

Short-Term Toxicity

No adverse effects on growth, development, or behavior were observed in rats administered diets of 10% flaxseed from age 18 days to 86 days. There were no signs of toxicity, and plasma levels of alanine aminotransferase and γ-glutamyltranspeptidase (GGT) were the same as those of animals on a standard diet. The activity of GGT was increased in the livers of flax chow-fed rats after puberty, with the effect being more pronounced in males than females (Hemmings and Barker 2004).

Chronic Toxicity

A number of studies on the effects of lifetime exposure to flaxseed have indicated no signs of toxicity in rats administered diets of 10% flax daily for over 130 days (Tou et al. 1999; Tou and Thompson 1999; Ward et al. 2001a). *Also see* Pregnancy and Lactation for this entry.

Genotoxicity

Some genotoxic activity of flaxseed oil was observed in a *Drosophila* somatic mutation and recombination test after animals were administered media containing 6 or 12% flaxseed oil (Rojas-Molina et al. 2005).

No mutagenic effects of flaxseed were observed in the Ames test with *Salmonella typhimurium* strains TA98 and TA102 (Mahmoud et al. 1992).

Cytotoxicity

Flaxseed enhanced proliferation in Chinese hamster ovary cells at doses in the micromolar range (Sujjavanich and Gibson 2006).

Literature Cited

Alonso, L., M.L. Marcos, and J.M. Blanco. 1996. Anaphylaxis caused by linseed (flaxseed) intake. *J. Allergy Clin. Immunol.* 98:469-470.

Ambrose, D.J., J.P. Kastelic, R. Corbett, P.A. Pitney, H.V. Petit, J.A. Small, and P. Zalkovic. 2006. Lower pregnancy losses in lactating dairy cows fed a diet enriched in alpha-linolenic acid. *J. Dairy Sci.* 89(8):3066-3074.

Bergman Jungestrom, M., L.U. Thompson, and C. Dabrosin. 2007. Flaxseed and its lignans inhibit estradiol-induced growth, angiogenesis, and secretion of vascular endothelial growth factor in human breast cancer xenografts in vivo. *Clin. Cancer Res.* 13(3):1061-1067.

Bhatty, R.S. 1993. Further compositional analyses of flax: Mucilage, trypsin inhibitors and hydrocyanic acid. *J. Am. Oil Chem. Soc.* 70(9):899-904.

Chen, J., E. Hui, T. Ip, and L.U. Thompson. 2004. Dietary flaxseed enhances the inhibitory effect of tamoxifen on the growth of estrogen-dependent human breast cancer (MCF-7) in nude mice. *Clin. Cancer Res.* 10(22):7703-7711.

Chen, J., K.A. Power, J. Mann, A. Cheng, and L.U. Thompson. 2007a. Dietary flaxseed interaction with tamoxifen induced tumor regression in athymic mice with MCF-7 xenografts by downregulating the expression of estrogen related gene products and signal transduction pathways. *Nutr. Cancer* 58(2):162-170.

Chen, J., K.A. Power, J. Mann, A. Cheng, and L.U. Thompson. 2007b. Flaxseed alone or in combination with tamoxifen inhibits MCF-7 breast tumor growth in ovariectomized athymic mice with high circulating levels of estrogen. *Exp. Biol. Med. (Maywood)* 232(8):1071-1080.

Collins, T.F., R.L. Sprando, T.N. Black, N. Olejnik, P.W. Wiesenfeld, U.S. Babu, M. Bryant, T.J. Flynn, and D.I. Ruggles. 2002. Effects of flaxseed on reproduction and development of rats (abstract). *Teratology* 65(6):331.

Collins, T.F., R.L. Sprando, T.N. Black, N. Olejnik, P.W. Wiesenfeld, U.S. Babu, M. Bryant, T.J. Flynn, and D.I. Ruggles. 2003. Effects of flaxseed and defatted flaxseed meal on reproduction and development in rats. *Food Chem. Toxicol.* 41(6):819-834.

Dabrosin, C., J. Chen, L. Wang, and L.U. Thompson. 2002. Flaxseed inhibits metastasis and decreases extracellular vascular endothelial growth factor in human breast cancer xenografts. *Cancer Lett.* 185(1):31-37.

Fedeniuk, R.W., and C.G. Biliaderis. 1994. Composition and physicochemical properties of linseed (*Linum usitatissimum* L.) mucilage. *J. Agric. Food Chem.* 42:240-247.

Flynn, T.J., T.F. Collins, R.L. Sprando, T. Black, P. Wiesenfeld, U.S. Babu, and D. Ruggles. 1999. Effects of serum from flaxseed-fed rats on cultured rat embryos (abstract). *Teratology* 59(6):402.

Flynn, T.J., T.F. Collins, R.L. Sprando, T.N. Black, D.I. Ruggles, P.W. Wiesenfeld, and U.S. Babu. 2003. Developmental effects of serum from flaxseed-fed rats on cultured rat embryos. *Food Chem. Toxicol.* 41(6):835-840.

Frische, E.J., A.M. Hutchins, M.C. Martini, W. Thomas, and J.L. Slavin. 2003. Effect of flaxseed and wheat bran on serum hormones and lignan excretion in premenopausal women. *J. Am. Coll. Nutr.* 22(6):550-554.

Gruver, D.I. 2003. Does flaxseed interfere with the clotting system? *Plast. Reconstr. Surg.* 112(3):934.

Haggans, C.J., A.M. Hutchins, B.A. Olson, W. Thomas, M.C. Martini, and J.L. Slavin. 1999. Effect of flaxseed consumption on urinary estrogen metabolites in postmenopausal women. *Nutr. Cancer* 33(2):188-195.

Hemmings, S.J., and L. Barker. 2004. The effects of dietary flaxseed on the Fischer 344 rat: I. Development, behaviour, toxicity and the activity of liver gamma-glutamyltranspeptidase. *Cell Biochem. Funct.* 22(2):113-121.

Khan, G., P. Penttinen, A. Cabanes, A. Foxworth, A. Chezek, K. Mastropole, B. Yu, A. Smeds, T. Halttunen, C. Good, et al. 2007. Maternal flaxseed diet during pregnancy or lactation increases female rat offspring's susceptibility to carcinogen-induced mammary tumorigenesis. *Reprod. Toxicol.* 23(3):397-406.

Knudsen, V.K., H.S. Hansen, M.L. Osterdal, T.B. Mikkelsen, H. Mu, and S.F. Olsen. 2006. Fish oil in various doses or flax oil in pregnancy and timing of spontaneous delivery: A randomised controlled trial. *BJOG* 113(5):536-543.

Leon, F., M. Rodriguez, and M. Cuevas. 2003. Anaphylaxis to *Linum*. *Allergol. Immunopathol. (Madrid)* 31(1):47-49.

Lezaun, A., J. Fraj, C. Colas, F. Duce, M.A. Dominguez, M. Cuevas, and P. Eiras. 1998. Anaphylaxis from linseed. *Allergy* 53(1):105-106.

Mahmoud, I., A. Alkofahi, and A. Abdelaziz. 1992. Mutagenic and toxic activities of several spices and some Jordanian medicinal plants. *Int. J. Pharmacog.* 30(2):81-85.

Mazza, G., and C.G. Biliaderis. 1989. Functional properties of flax seed mucilage. *J. Food Sci.* 54:1302-1305.

Oomah, B.D., G. Mazza, and E.O. Kenaschuk. 1992. Cyanogenic compounds in flaxseed. *J. Agric. Food Chem.* 40 (8):1346-1348.

Orcheson, L.J., S.E. Rickard, M.M. Seidl, and L.U. Thompson. 1998. Flaxseed and its mammalian lignan precursor cause a lengthening or cessation of estrous cycling in rats. *Cancer Lett.* 125(1-2):69-76.

Phipps, W.R., M. Martini, J. Lampe, J. Slavin, and M. Kurzer. 1993. Effect of flax seed ingestion on the menstrual cycle. *J. Clin. Endocrinol. Metab.* 77(5):1215-1219.

Power, K.A., and L.U. Thompson. 2007. Can the combination of flaxseed and its lignans with soy and its isoflavones reduce the growth stimulatory effect of soy and its isoflavones on established breast cancer? *Mol. Nutr. Food Res.* 51(7):845-856.

Prasad, K. 2005. Effect of chronic administration of lignan complex isolated from flaxseed on the hemopoietic system. *Mol. Cell. Biochem.* 270(1-2):139-145.

Rojas-Molina, M., J. Campos-Sanchez, M. Analla, A. Munoz-Serrano, and A. Alonso-Moraga. 2005. Genotoxicity of vegetable cooking oils in the *Drosophila* wing spot test. *Environ. Mol. Mutagen.* 45(1):90-95.

Saarinen, N.M., K. Power, J. Chen, and L.U. Thompson. 2006. Flaxseed attenuates the tumor growth stimulating effect of soy protein in ovariectomized athymic mice with MCF-7 human breast cancer xenografts. *Int. J. Cancer* 119(4):925-931.

Sankaran, D., N. Bankovic-Calic, C.Y. Peng, M.R. Ogborn, and H.M. Aukema. 2006. Dietary flax oil during pregnancy and lactation retards disease progression in rat offspring with inherited kidney disease. *Pediatr. Res.* 60(6):729-733.

Serraino, M., and L.U. Thompson. 1992. The effect of flaxseed supplementation on the initiation and promotional stages of mammary tumorigenesis. *Nutr. Cancer* 17(2):153-159.

Sprando, R.L., T.F.X. Collins, P. Wiesenfeld, U.S. Babu, C. Rees, T. Black, N. Olejnik, and J. Rorie. 2000. Testing the potential of flaxseed to affect spermatogenesis: Morphometry. *Food Chem. Toxicol.* 38(10):887-892.

Sujjavanich, D.N., and J.E. Gibson. 2006. In vitro toxicity in CHO cells of herbs commonly used by postmenopausal women. *Toxicol. Sci.* 90(1):289.

Thompson, L.U., J. Chen, K.P. Tan, and W.E. Ward. 2003. Early exposure to flaxseed and its lignan reduced mammary cancer risk at adulthood. *J. Nutr.* 133(11, Suppl.):3860S-3861S.

Thompson, L.U., S.E. Rickard, L.J. Orcheson, and M.M. Seidl. 1996. Flaxseed and its lignan and oil components reduce mammary tumor growth at a late stage of carcinogenesis. *Carcinogenesis* 17:1373-1376.

Tou, J.C., J. Chen, and L.U. Thompson. 1998. The importance of timing of exposure to flaxseed on reproductive development of rats. *FASEB J.* 12(5 Pt 2):A768.

Tou, J.C., J. Chen, and L.U. Thompson. 1999. Dose, timing, and duration of flaxseed exposure affect reproductive indices and sex hormone levels in rats. *J. Toxicol. Environ. Health A* 56(8):555-570.

Tou, J.C., and L.U. Thompson. 1999. Exposure to flaxseed or its lignan component during different developmental stages influences rat mammary gland structures. *Carcinogenesis* 20(9):1831-1835.

Velasquez, M.T., S.J. Bhathena, T. Ranich, A.M. Schwartz, D.E. Kardon, A.A. Ali, C.C. Haudenschild, and C.T. Hansen. 2003. Dietary flaxseed meal reduces proteinuria and ameliorates nephropathy in an animal model of type II diabetes mellitus. *Kidney Int.* 64(6):2100-2107.

Vissac-Sabatier, C., V. Coxam, P. Dechelotte, C. Picherit, M.N. Horcajada, M.J. Davicco, P. Lebecque, Y.J. Bignon, and D. Bernard-Gallon. 2003. Phytoestrogen-rich diets modulate expression of *Brca1* and *Brca2* tumor suppressor genes in mammary glands of female Wistar rats. *Cancer Res.* 63(20):6607-6612.

Vyas, T.K., A. Shahiwala, and M.M. Amiji. 2008. Improved oral bioavailability and brain transport of saquinavir upon administration in novel nanoemulsion formulations. *Int. J. Pharm.* 347(1-2):93-101.

Waldschlager, J., C. Bergemann, W. Ruth, U. Effmert, U. Jeschke, R.U. Richter, U. Kragl, B. Piechulla, and V. Briese. 2005. Flaxseed extracts with phytoestrogenic effects on a hormone receptor-positive tumour cell line. *Anticancer Res.* 25(3A):1817-1822.

Ward, W.E., J. Chen, and L.U. Thompson. 2001a. Exposure to flaxseed or its purified lignan during suckling only or continuously does not alter reproductive indices in male and female offspring. *J. Toxicol. Environ. Health A* 64(7):567-577.

Ward, W.E., Y.V. Yuan, A.M. Cheung, and L.U. Thompson. 2001b. Exposure to purified lignan from flaxseed (*Linum usitatissimum*) alters bone development in female rats. *Br. J. Nutr.* 86(4):499-505.

Wichtl, M. 2004. *Herbal drugs and phytopharmaceuticals: A handbook for practice on a scientific basis.* 3rd ed. Boca Raton, FL: CRC Press.

Lobelia inflata L. — Campanulaceae

SCN: lobelia
OCN: Indian tobacco; puke weed

Part: herb

Quick Reference Summary

Safety Class: 2b
Interaction Class: A

Contraindications

Not for use in pregnancy except under the supervision of a qualified healthcare practitioner (Ellingwood 1919; Felter and Lloyd 1898).

Other Precautions

May cause nausea or vomiting (Bradley 1992; Felter and Lloyd 1898; Kaufmann and Bensimon 1960; Leung and Foster 1996; List and Hörhammer 1973; Rapp and Olen 1955; Scott et al. 1962; Williamson 2003).

Standard Dose

As an expectorant: 100 mg of the leaf; 0.6 to 2.0 ml of the tincture (Leung and Foster 1996; Martindale and Reynolds 1996; Osol and Farrar 1955).

Notice

Emetic (Bradley 1992; Felter and Lloyd 1898; Leung and Foster 1996; List and Hörhammer 1973; Williamson 2003); *see* Appendix 2.

Drug and Supplement Interactions

None known.

Editors' Notes

Controversy over the safety of lobelia dates to the early 19th century (Bergner 1998b). Ellingwood reported that "death has occurred in a very few cases from excessive doses of the remedy, but toxic effects are not apparent where the medicinal dose is prescribed" (Ellingwood 1919), and Felter and Lloyd state that "the emetic action is so prompt and decided, that the contained alkaloid could not, under ordinary circumstances, produce fatal results" (Felter and Lloyd 1898). Reviews note that information was insufficient to establish lobelia as being the likely cause of death in the alleged fatalities (Bergner 1998a, 1998b).

Adverse Events and Side Effects

In clinical trials of the compound lobeline, side effects of treatment included nausea and a burning sensation in the mouth and throat when taken at doses of 0.5 mg (London 1963). Higher doses were associated with nausea, vomiting, vertigo, and increased heart rate (tachycardia) (Kaufmann and Bensimon 1960; Rapp and Olen 1955; Scott et al. 1962).

Pharmacological Considerations

An increase in blood pressure and variable effects of the compound lobeline on heart rate were reported in one human study (Butler et al. 2001). Animal studies have shown transient and minor effects on the cardiovascular system (Dwoskin and Crooks 2002; Korczyn et al. 1969; Sloan et al. 1988).

Pregnancy and Lactation

Historical herbal texts note that lobelia promotes normal uterine contraction after the os is dilated, and relaxes the perineal tissues (Ellingwood 1919; Felter and Lloyd 1898). Based on this information, use during pregnancy is not recommended except under the supervision of a qualified healthcare practitioner.

No adverse effects were reported after pregnant women were injected with a single dose of 0.03 mg/kg of the compound lobeline (Klauer 1959).

No information on the safety of lobelia in lactation was identified. While this review did not identify any concerns for use while nursing, safety has not been conclusively established.

L

Review Details

I. Drug and Supplement Interactions

Clinical Trials of Drug or Supplement Interactions

No clinical trials of drug or supplement interactions were identified.

Case Reports of Suspected Drug or Supplement Interactions

No case reports of suspected drug or supplement interactions were identified.

Animal Trials of Drug or Supplement Interactions

No animal trials of drug or supplement interactions were identified.

II. Adverse Events

Adverse Events Reported in Clinical Trials

Nausea (14% lobeline group, 28% control group) and a burning sensation in the mouth or throat (50% lobeline group, 6% control group) were reported as side effects of treatment in a trial of the compound lobeline sulfate in chronic smokers. Lobeline sulfate was orally administered in pastilles containing 0.5 mg of the compound at a rate of one every 1 to 2 hours for the first week, one every 3 hours the second week, one every 4 hours the third week, and one every 4 to 6 hours the fourth week. The side effects occurred during the first week of treatment, decreased in frequency the second week, and were absent during the third and fourth weeks of treatment (London 1963).

A review of early clinical trials of the compound lobeline for smoking cessation indicated that doses of 2 mg or more were poorly tolerated, with a significant portion of trial participants refusing to complete the studies (London 1963). Side effects at the 2 mg dose were reported as nausea, vomiting, vertigo, and tachycardia (Kaufmann and Bensimon 1960; Rapp and Olen 1955; Scott et al. 1962). In a trial of lobeline sulfate with tablets containing 8 mg of the compound, few patients were willing to take more than 3 doses daily for more than 3 days (Von Wright et al. 1982). A more recent systematic review indicated that a large percentage of trials of lobeline for smoking cessation were not controlled and that in studies that were placebo controlled, descriptions of appropriate methods of randomization were frequently missing (Stead and Hughes 2000).

Case Reports of Adverse Events

Use of more than 0.5 ml of tincture three times daily may cause nausea in some people. Up to 1 ml of tincture three times daily may be needed in a larger adult and may not cause adverse effects. Mild nausea is not an indication for total discontinuation (Yarnell 1999). Severe overdose is reported to cause prostration, convulsions, coma, respiratory depression, and death from respiratory failure (Ellingwood 1919).

One controversial death sometimes cited as being a result of lobelia toxicity is that of Ezra Lovett, who allegedly died by lobelia poisoning at the hands of Samuel Thomson in Massachusetts in 1807. The death was the basis of murder charges 2 years later, but the judge and jury hearing the case found that no basis had been established for charges of murder or manslaughter. An analysis of the case and of the pharmacokinetics of lobeline indicated that death from lobelia was unlikely and may instead have been due to conventional medical treatment or another cause (Bergner 1998b). A second death attributed to lobelia, that of T.G. French in New York City in 1837, was reported, although no details of the case, including the dose of lobelia administered or the nature of the illness being treated, are available (Bergner 1998a).

III. Pharmacology and Pharmacokinetics

Human Pharmacological Studies

In human lung transplant patients intravenously administered 18 to 60 μg/kg of the compound lobeline, varying effects on heart rate were observed, with a rate increase in some patients and a decrease in others. A dose-dependent decrease in mean arterial pressure was also observed (Butler et al. 2001). At doses above 18 μg/kg, intravenously administered lobeline induced cough in healthy patients (Gandevia et al. 1998).

Lobeline-induced cough has been observed in humans intravenously administered the compound at doses of 35 μg/kg or more (Jaju et al. 1998; Raj et al. 2005). The cough is believed to be due to stimulation of the J receptors (Paintal 1995; Raj et al. 2005).

Animal Pharmacological Studies

The compound lobeline is reported to be a cardiovascular stimulant, producing effects through the carotid chemoreceptors (Cambar et al. 1969; Korczyn et al. 1969). Cardiovascular effects of the compound lobeline have been described as transient and minor, resulting from primary stimulant and secondary depressant effects on sympathetic and parasympathetic ganglia and the adrenal medulla (Dwoskin and Crooks 2002; Sloan et al. 1988). After administration of lobeline to dogs, an initial slowing of heart rate was observed, with a subsequent increase in blood pressure. Long-term administration, however, resulted in a decrease in blood pressure (Dwoskin and Crooks 2002; Korczyn et al. 1969).

In cats and dogs intramuscularly administered the compound lobeline sulfate at doses of 0.25 to 0.6 mg/kg, a dose of 0.5 mg/kg caused all cats and dogs to vomit (Laffan and Borison 1957).

In monkeys injected with 142 μg/kg of the compound lobeline, either apnea or excitation of breathing was reported, but no change in heart rate (Deep et al. 2001). Transitory hyperpnea with an increase in tidal volume and respiratory rate was induced in horses administered the compound lobeline. The increase in ventilation lasted for about 90 seconds and was accompanied by a sharp rise in the respiratory peak airflows, especially the expiratory flows (Art et al. 1991). Similar activity was reported in pigs and rabbits (Bredeck et al. 1961; Utashiro 1941).

In mice, administration of the nicotinic antagonist chlorisondamine stopped the physiological effects of nicotine but not lobeline (Decker et al. 1994).

In Vitro Pharmacological Studies

The compound lobeline has been shown to have a high affinity for nicotinic acetylcholine receptors and inhibits the function of vesicular monoamine and dopamine transporters (Felpin and Lebreton 2004; Flammia et al. 1999; Reavill et al. 1990; Teng et al. 1998). In guinea pig brain homogenates, lobeline functioned as a mu opioid receptor antagonist (Miller et al. 2007).

The compound lobeline reversed multidrug resistance in human colorectal adenocarcinoma cells in vitro by inhibiting P-glycoprotein (P-gp) activity. Lobeline, however, did not block breast cancer resistance protein-dependent mitoxantrone efflux (Ma and Wink 2008).

IV. Pregnancy and Lactation

Historical herbal texts note that lobelia promotes normal uterine contraction after the os is dilated, and relaxes the perineal tissues (Ellingwood 1919; Felter and Lloyd 1898).

No adverse events were reported after pregnant women were injected with a single dose of 0.03 mg/kg of the compound lobeline (Klauer 1959).

No information on the safety of lobelia during lactation was identified. While this review did not identify any concerns for use while nursing, safety has not been conclusively established.

V. Toxicity Studies

Acute Toxicity

The LD_{50} of intraperitoneally administered lobeline in mice is 107 mg/kg (Ganellin et al. 1997).

A lethal dose of 10 mg per animal was reported for the compound lobeline in rabbits (animal weight not specified in English language abstract) (Utashiro 1941).

Literature Cited

Art, T., D. Desmecht, H. Amory, and P. Lekeux. 1991. Lobeline-induced hyperpnea in equids. Comparison with rebreathing bag and exercise. *Zentralbl. Veterinarmed. A* 38(2):148-152.

Bergner, P. 1998a. Lobelia toxicity: A literature review. *Med. Herbalism* 10(1-2):15-26.

Bergner, P. 1998b. Lobelia: The controversial death of Ezra Lovett? *Med. Herbalism* 10(1-2):27-28.

Bradley, P.R. 1992. *British herbal compendium: A handbook of scientific information on widely used plant drugs*. Bournemouth, UK: British Herbal Medicine Association.

Bredeck, H.E., R.A. Herin, and N.H. Booth. 1961. Chemoceptor reflexes in swine. *Am. J. Physiol.* 201(1):89.

Butler, J.E., A. Anand, M.R. Crawford, et al. 2001. Changes in respiratory sensations induced by lobeline after human bilateral lung transplantation. *J. Physiol.* 534(Pt. 2):583-593.

Cambar, P.J., S.R. Shore, and D.M. Aviado. 1969. Bronchopulmonary and gastrointestinal effects of lobeline. *Arch. Int. Pharmacodyn. Ther.* 177(1):1-27.

Decker, M.W., M.J. Buckley, and J.D. Brioni. 1994. Differential effects of pretreatment with nicotine and lobeline on nicotine-induced changes in body temperature and locomotor activity in mice. *Drug Dev. Res.* 31(1):52-58.

Deep, V., M. Singh, and K. Ravi. 2001. Role of vagal afferents in the reflex effects of capsaicin and lobeline in monkeys. *Respir. Physiol.* 125(3):155-168.

Dwoskin, L.P., and P.A. Crooks. 2002. A novel mechanism of action and potential use for lobeline as a treatment for psychostimulant abuse. *Biochem. Pharmacol.* 63(2):89-98.

Ellingwood, F. 1919. *American materia medica, therapeutics, and pharmacognosy*. Sandy, OR: Eclectic Medical Publications (1998 reprint).

Felpin, F., and J. Lebreton. 2004. History, chemistry and biology of alkaloids from *Lobelia inflata*. *Tetrahedron* 60(45):10127-10153.

Felter, H.W., and J.U. Lloyd. 1898. *King's American dispensatory*. 18th ed., 3rd rev. 2 vols. Cincinnati: Ohio Valley Co.

Flammia, D., M. Dukat, M.I. Damaj, B. Martin, and R.A. Glennon. 1999. Lobeline: Structure-affinity investigation of nicotinic acetylcholinergic receptor binding. *J. Med. Chem.* 42(18):3726-3731.

Gandevia, S.C., J.E. Butler, J.L. Taylor, and M.R. Crawford. 1998. Absence of viscerosomatic inhibition with injections of lobeline designed to activate human pulmonary C fibres. *J. Physiol.* 511(Pt 1):289-300.

Ganellin, C., F. MacDonald, and D. Triggle. 1997. *Dictionary of pharmacological agents*. Boca Raton, FL: CRC Press.

Jaju, D.S., M.B. Dikshit, M.J. Agrawal, and N.A. Gupte. 1998. Comparison of respiratory sensations induced by J receptor stimulation with lobeline in left handers and right handers. *Indian J. Med. Res.* 108:291-295.

Kaufmann, H., and L. Bensimon. 1960. Le sevrage du tabac. *Vie Med.* 41:1139.

Klauer, D. 1959. Lobeline in the determination of circulatory rate in the healthy pregnant woman. *Obstet. Ginecol. Lat. Am.* 17(Sept.-Oct.):442-452.

Korczyn, A.D., I. Bruderman, and K. Braun. 1969. Cardiovascular effects of lobeline. *Arch. Int. Pharmacodyn. Ther.* 182(2):370-375.

Laffan, R.J., and H.L. Borison. 1957. Emetic action of nicotine and lobeline. *J. Pharmacol. Exp. Ther.* 121(4):468-476.

Leung, A.Y., and S. Foster. 1996. *Encyclopedia of common natural ingredients used in food, drugs, and cosmetics*. 2nd ed. New York: Wiley.

List, P.H., and H. Hörhammer. 1973. *Hagers handbuch der pharmazeutischen praxis*. Berlin: Springer.

London, S.J. 1963. Clinical evaluation of a new lobeline smoking deterrent. *Curr. Ther. Res. Clin. Exp.* 5(4):167.

Ma, Y., and M. Wink. 2008. Lobeline, a piperidine alkaloid from *Lobelia* can reverse P-gp dependent multidrug resistance in tumor cells. *Phytomedicine* 15(9):754-758.

Martindale, W., and J.E.F. Reynolds. 1996. *The extra pharmacopoeia*. 31st ed. London: Pharmaceutical Press.

Miller, D.K., J.R. Lever, K.R. Rodvelt, et al. 2007. Lobeline, a potential pharmacotherapy for drug addiction, binds to mu opioid receptors and diminishes the effects of opioid receptor agonists. *Drug Alcohol Depend.* 89(2-3):282-291.

Osol, A., and G. Farrar. 1955. *The dispensatory of the United States of America*. 25th ed. Philadelphia: Lippincott.

Paintal, A.S. 1995. Some recent advances in studies on J receptors. *Adv. Exp. Med. Biol.* 381:15-25.

Raj, H., G.S. Bakshi, R.R. Tiwari, A. Anand, and A.S. Paintal. 2005. How does lobeline injected intravenously produce a cough? *Respir. Physiol. Neurobiol.* 145(1):79-90.

Rapp, G.W., and A.A. Olen. 1955. A critical evaluation of a lobeline based smoking deterrent. *Am. J. Med. Sci.* 230(1):9.

Reavill, C., B. Walther, I.P. Stolerman, and B. Testa. 1990. Behavioural and pharmacokinetic studies on nicotine, cytisine and lobeline. *Neuropharmacology* 29(7):619-624.

Scott, G.W., A.G.C. Cox, K.S. Maclean, T.M.L. Price, and N. Southwell. 1962. Buffered lobeline as a smoking deterrent. *Lancet* 1(7219):54.

Sloan, J.W., W.R. Martin, M. Bostwick, R. Hook, and E. Wala. 1988. The comparative binding characteristics of nicotinic ligands and their pharmacology. *Pharmacol. Biochem. Behav.* 30(1):255-267.

Stead, L.F., and J.R. Hughes. 2000. Lobeline for smoking cessation. *Cochrane Database Syst. Rev.*

Teng, L., P.A. Crooks, and L.P. Dwoskin. 1998. Lobeline displaces [^{3}H]dihydrotetrabenazine binding and releases [^{3}H]dopamine from rat striatal synaptic vesicles: Comparison with *d*-amphetamine. *J. Neurochem.* 71(1):258-265.

Utashiro, S. 1941. Respiratory action of lobeline. *Nagoya Igakkai Zasshi* 54:603-609.

Von Wright, A., J. Knuutinen, and S. Lindroth. 1982. The mutagenicity of some edible mushrooms in the Ames test. *Food Chem. Toxicol.* 20(3):265-267.

Williamson, E.M. 2003. *Potter's herbal cyclopedia*. Saffron Walden, Essex: C.W. Daniel Co.

Yarnell, E. 1999. Misunderstood toxic herbs. *Altern. Complement. Ther.* 5(Feb.):6-11.

Lobelia siphilitica L. — Campanulaceae

SCN: blue lobelia
OCN: great blue lobelia

Part: herb

Quick Reference Summary

Safety Class: 2b
Interaction Class: A

Contraindications

Not for use in pregnancy except under the supervision of a qualified healthcare practitioner.

Other Precautions

May cause nausea or vomiting (Felter and Lloyd 1898).

Drug and Supplement Interactions

None known.

Notice

Emetic (Felter and Lloyd 1898); *see* Appendix 2.

Editors' Notes

Blue lobelia is reported to contain similar alkaloids to lobelia (*L. inflata*). The compound lobeline has been identified in blue lobelia, although the quantity is not known (Kesting et al. 2009). *See* Editors' Notes for *L. inflata*.

Adverse Events and Side Effects

Blue lobelia may cause nausea or vomiting (Felter and Lloyd 1898).

Pharmacological Considerations

None known.

Pregnancy and Lactation

Although no information on the safety of blue lobelia during pregnancy was identified, based on the similarity of this species to other species of *Lobelia,* use during pregnancy is not recommended unless under the supervision of a qualified healthcare practitioner.

L

Review Details

I. Drug and Supplement Interactions

Clinical Trials of Drug or Supplement Interactions

No clinical trials of drug or supplement interactions were identified.

Case Reports of Suspected Drug or Supplement Interactions

No case reports of suspected drug or supplement interactions were identified.

Animal Trials of Drug or Supplement Interactions

No animal trials of drug or supplement interactions were identified.

II. Adverse Events

Case Reports of Adverse Events

No case reports of adverse events were identified.

III. Pharmacology and Pharmacokinetics

Human Pharmacological Studies

No relevant human pharmacological studies were identified.

Animal Pharmacological Studies

No relevant animal pharmacological studies were identified.

In Vitro Pharmacological Studies

No relevant in vitro pharmacological studies were identified.

IV. Pregnancy and Lactation

No information on the safety of blue lobelia during pregnancy or lactation was identified.

V. Toxicity Studies

Acute Toxicity

The LD_{50} of intraperitoneally administered lobeline in mice is 107 mg/kg (Ganellin et al. 1997).

A lethal dose of 10 mg per animal was reported for the compound lobeline in rabbits (animal weight not specified in English language abstract) (Utashiro 1941).

Literature Cited

Felter, H.W., and J.U. Lloyd. 1898. *King's American dispensatory.* 18th ed., 3rd rev. 2 vols. Cincinnati: Ohio Valley Co.

Ganellin, C., F. MacDonald, and D. Triggle. 1997. *Dictionary of pharmacological agents.* Boca Raton, FL: CRC Press.

Kesting, J.R., I.L. Tolderlund, A.F. Pedersen, et al. 2009. Piperidine and tetrahydropyridine alkaloids from *Lobelia siphilitica* and *Hippobroma longiflora*. *J. Nat. Prod.* 72(2):312-315.

Utashiro, S. 1941. Respiratory action of lobeline. *Nagoya Igakkai Zasshi* 54:603-609.

Lomatium dissectum (Nutt.) Mathias & Constance — Apiaceae

SCN: lomatium
Syn: *Leptotaenia multifida* Nutt.; *Ferula multifida* (Nutt.) A. Gray
OCN: biscuit root; desert parsley; fern-leaf lomatium; Indian balsam
Part: root

Quick Reference Summary

Safety Class: 2b
Interaction Class: A

Contraindications

Not for use in pregnancy except under the supervision of a qualified healthcare practitioner (McGuffin et al. 1997).

Other Precautions

None known.

Drug and Supplement Interactions

None known.

Adverse Events and Side Effects

Skin rashes have been associated with internal use of lomatium. Rashes have been associated more with use of the fresh root than the dried root (Moore 1993). Mild fevers have been reported in some individuals after long-term consumption.

Pharmacological Considerations

None known.

Pregnancy and Lactation

The professional experience of the editors of the first edition of this text indicates that lomatium should not be used in pregnancy except under the supervision of a qualified healthcare practitioner (McGuffin et al. 1997).

No information on the safety of lomatium during lactation was identified. While this review did not identify any concerns for use while nursing, safety has not been conclusively established.

Review Details

I. Drug and Supplement Interactions

Clinical Trials of Drug or Supplement Interactions

No clinical trials of drug or supplement interactions were identified.

Case Reports of Suspected Drug or Supplement Interactions

No case reports of suspected drug or supplement interactions were identified.

Animal Trials of Drug or Supplement Interactions

No animal trials of drug or supplement interactions were identified.

II. Adverse Events

Case Reports of Adverse Events

Rashes have been reported as an occasional side effect of lomatium use. Rashes were associated more with use of the fresh root than the dried root. Some individuals that reacted to fresh root preparations did not react to dried root (Moore 1993).

III. Pharmacology and Pharmacokinetics

Human Pharmacological Studies

No relevant human pharmacological studies were identified.

Animal Pharmacological Studies

No relevant animal pharmacological studies were identified.

In Vitro Pharmacological Studies

No relevant in vitro pharmacological studies were identified.

IV. Pregnancy and Lactation

The professional experience of the editors of the first edition of this text indicates that lomatium should not be used in pregnancy except under the supervision of a qualified healthcare practitioner (McGuffin et al. 1997).

No information on the safety of lomatium during lactation was identified.

V. Toxicity Studies

No toxicity studies were identified.

Literature Cited

McGuffin, M., C. Hobbs, R. Upton, and A. Goldberg. 1997. *Botanical safety handbook*. Boca Raton, FL: CRC Press.

Moore, M. 1993. *Medicinal plants of the Pacific West*. Santa Fe: Red Crane Books.

Lycium spp. — Solanaceae

Lycium barbarum **L.**
SCN: lycium
PN: *gou qi zi* (fruit)
OCN: Barbary wolfberry; matrimony vine; goji

Lycium chinense **Mill.**
SCN: lycium
PN: *gou qi zi* (fruit)
OCN: Chinese boxthorn; Chinese wolfberry; goji
Part: fruit

Quick Reference Summary

Safety Class: 1
Interaction Class: A

Contraindications

None known.

Other Precautions

None known.

Drug and Supplement Interactions

See Pharmacological Interactions.

Editors' Notes

Although lycium fruit has been reported to contain the alkaloid atropine up to 0.95% (Harsh 1989), the presence of tropane alkaloids has been disputed (Frohne and Pfänder 1997; Merz and Stolte 1960), and a more recent analysis indicates that the atropine content of dried lycium fruit is very low, generally less than 10 parts per billion (ppb), with the highest level of analyzed fruit containing 19 ppb. Severe atropine intoxication with the examined lycium fruit samples would require the unrealistic consumption of several tons of berries (Adams et al. 2006).

Adverse Events and Side Effects

Allergic reaction to lycium fruit has been reported (Bensky et al. 2004).

Pharmacological Considerations

One case report indicated increased INR (a standardized scale used to report the results of blood coagulation tests; increased INR indicates slowed blood clotting) in a woman previously stable on warfarin who drank lycium fruit tea for several days. After cessation of the tea, INR levels returned to normal (Lam et al. 2001). The woman was also taking other drugs, at least one of which has been associated with increased INR levels in persons on warfarin (Trilli et al. 1996).

Pregnancy and Lactation

Information on the safety of lycium fruit in pregnancy is limited. While one Chinese medicine reference text advises caution for use in pregnancy, based on a study in rabbits showing that lycium fruit had a stimulating effect on the uterus (Chen and Chen 2004), another text lists no concerns for use in pregnancy (Bensky et al. 2004).

No information on the safety of lycium fruit during lactation was identified in the scientific or traditional literature. While this review did not identify any concerns for use while nursing, safety has not been conclusively established.

Review Details

I. Drug and Supplement Interactions

Clinical Trials of Drug or Supplement Interactions

No clinical trials of drug or supplement interactions were identified.

Case Reports of Suspected Drug or Supplement Interactions

An elevated (4.1) INR (a standardized scale used to report the results of blood coagulation tests; increased INR indicates slowed blood clotting) level was reported in a woman previously stable on warfarin. Her other medical conditions were hypertension, hypercholesterolemia, and tricuspid regurgitation, and other medications were benazepril, atenolol, digoxin, and fluvastatin. The woman had consumed 3 to 4 cups daily of a lycium fruit tea prior to her clinic visit. After cessation of the tea, INR levels returned to normal within 7 days. An in vitro study completed in conjunction with this case report indicated that only weak effects of lycium fruit extract were observed on activity of the drug-metabolizing isoenzyme CYP2C9, the isoenzyme that metabolizes warfarin (Lam et al. 2001). Fluvastatin has been associated with increased INR in patients taking warfarin (Trilli et al. 1996).

Animal Trials of Drug or Supplement Interactions

No animal trials of drug or supplement interactions were identified.

II. Adverse Events

Case Reports of Adverse Events

A Chinese herbal text indicates that lycium fruit is generally regarded as safe, with no adverse effects expected at the normal dose. Allergic reactions, including urticaria-like or papular rashes, have been reported (Bensky et al. 2004).

III. Pharmacology and Pharmacokinetics

Human Pharmacological Studies

No relevant human pharmacological studies were identified.

Animal Pharmacological Studies

A marked decrease in fasting plasma insulin levels was observed in streptozotocin-induced diabetic rats administered 10 mg/kg daily polysaccharides isolated from lycium fruit for 3 weeks. In an oral glucose tolerance test, administration of 2 g/kg significantly reduced postprandial glucose levels at 30 minutes and increased insulin-sensitive index values (Zhao et al. 2005).

In Vitro Pharmacological Studies

No relevant in vitro pharmacological studies were identified.

IV. Pregnancy and Lactation

A decoction of lycium fruit had a stimulating effect on the uterus in rabbits (dose and route of administration not specified in available translation) (Chen and Chen 2004). Based on that study, one reference text on Chinese herbal medicine indicates that lycium fruit should be used with caution in pregnancy (Chen and Chen 2004), while another text does not list any cautions in pregnancy (Bensky et al. 2004).

No information on the safety of lycium fruit during lactation was identified.

V. Toxicity Studies

Acute Toxicity

The LD_{50} of orally or intraperitoneally administered aqueous lycium fruit extract in mice is 83.2 g/kg (Chen and Chen 2004).

L

Literature Cited

Adams, M., M. Wiedenmann, G. Tittel, and R. Bauer. 2006. HPLC-MS trace analysis of atropine in *Lycium barbarum* berries. *Phytochem. Anal.* 17(5):279-283.

Bensky, D., S. Clavey, and E. Stöger. 2004. *Chinese herbal medicine: Materia medica.* 3rd ed. Seattle: Eastland Press.

Chen, J.K., and T.T. Chen. 2004. *Chinese medical herbology and pharmacology.* City of Industry, CA: Art of Medicine Press.

Frohne, D., and H.J. Pfänder. 1997. *Giftpflanzen. 4th ed.* Stuttgart: Wissenschaftliche Verlagsgesellschaft.

Harsh, M. 1989. Tropane alkaloids from *Lycium barbarum* L., in vivo and in vitro. *Curr. Sci.* 58:117-118.

Lam, A.Y., G.W. Elmer, and M.A. Mohutsky. 2001. Possible interaction between warfarin and *Lycium barbarum* L. *Ann. Pharmacother.* 35(10):1199-1201.

Merz, K., and H. Stolte. 1960. Constituents of *Lycium* species. *Planta Med.* 8:121-126.

Trilli, L.E., C.L. Kelley, S.L. Aspinall, and B.A. Kroner. 1996. Potential interaction between warfarin and fluvastatin. *Ann. Pharmacother.* 30(12):1399-1402.

Zhao, R., Q. Li, and B. Xiao. 2005. Effect of *Lycium barbarum* polysaccharide on the improvement of insulin resistance in NIDDM rats. *Yakugaku Zasshi* 125(12):981-988.

Lycium spp. — Solanaceae

Lycium barbarum L.
SCN: lycium
PN: *di gu pi* (root bark)
OCN: Barbary wolfberry; matrimony vine

Lycium chinense **Mill.**
SCN: lycium
PN: *di gu pi* (root bark)
OCN: Chinese boxthorn; Chinese wolfberry
Part: root bark

Quick Reference Summary

Safety Class: 1
Interaction Class: A

Contraindications

None known.

Other Precautions

None known.

Drug and Supplement Interactions

None known.

Adverse Events and Side Effects

None known.

Pharmacological Considerations

Several animal studies have shown that lycium root bark causes a decrease in fasting blood glucose levels and an increase in serum insulin levels in diabetic animals (Gao et al. 2007a, 2007b; Kim et al. 1994). Diabetic persons are advised to discuss the use of this herb with a qualified healthcare practitioner prior to use.

Pregnancy and Lactation

No information on the safety of lycium root bark in pregnancy or lactation was identified in the scientific or traditional literature. Although this review did not identify any concerns for use while pregnant or nursing, safety has not been conclusively established.

Review Details

I. Drug and Supplement Interactions

Clinical Trials of Drug or Supplement Interactions

No clinical trials of drug or supplement interactions were identified.

Case Reports of Suspected Drug or Supplement Interactions

No case reports of suspected drug or supplement interactions were identified.

Animal Trials of Drug or Supplement Interactions

No animal trials of drug or supplement interactions were identified.

L

II. Adverse Events

Case Reports of Adverse Events

A traditional Chinese medicine text indicates that no adverse effects are expected at standard therapeutic doses, although ingestion of a high dose (50 g) was reported to cause vertigo, palpitations, nausea, vomiting, and premature contractions (tissue or organ not specified, presumably premature uterine contractions) (Bensky et al. 2004). Case details were not available.

III. Pharmacology and Pharmacokinetics

Human Pharmacological Studies

No relevant human pharmacological studies were identified.

Animal Pharmacological Studies

Dose-dependent decreases in fasting blood glucose levels and increases in serum insulin levels were observed in streptozotocin-induced diabetic mice administered 100 or 200 mg/kg of a dried aqueous extract of lycium root bark extract daily for 28 days (Gao et al. 2007a, 2007b; Kim et al. 1994).

In Vitro Pharmacological Studies

No effects on the drug-metabolizing isoenzymes CYP1A2 or CYP3A4 were observed after treatment with an ethanolic extract of lycium root bark (Brandin et al. 2007).

IV. Pregnancy and Lactation

No information on the safety of lycium root bark in pregnancy or lactation was identified. Reference texts on Chinese medicine do not indicate any precautions regarding use in pregnancy (Bensky et al. 2004; Chen and Chen 2004).

V. Toxicity Studies

Acute Toxicity

The LD_{50} of an intraperitoneally administered extract of lycium root bark in mice is 12.8 g/kg (Chen and Chen 2004). The LD_{50} of an orally administered dried aqueous extract of lycium root bark in mice could not be determined at doses up to 2 g/kg. No toxic effects were seen at this dose (Gao et al. 2007a).

Literature Cited

Bensky, D., S. Clavey, and E. Stöger. 2004. *Chinese herbal medicine: Materia medica*. 3rd ed. Seattle: Eastland Press.

Brandin, H., E. Viitanen, O. Myrberg, and A.K. Arvidsson. 2007. Effects of herbal medicinal products and food supplements on induction of CYP1A2, CYP3A4 and MDR1 in the human colon carcinoma cell line LS180. *Phytother. Res.* 21(3):239-244.

Chen, J.K., and T.T. Chen. 2004. *Chinese medical herbology and pharmacology*. City of Industry, CA: Art of Medicine Press.

Gao, D., Q. Li, Z. Liu, et al. 2007a. Effects of *Lycium barbarum* L. root bark extract on alloxan-induced diabetic mice. *Therapy* 4(5):547-553.

Gao, D., Q. Li, Z. Liu, et al. 2007b. Hypoglycemic effects and mechanisms of action of *Cortex Lycii Radicis* on alloxan-induced diabetic mice. *Yakugaku Zasshi* 127(10):1715-1721.

Kim, N.J., W.G. Youn, and N.D. Hong. 1994. Pharmacological effects of *Lycium chinensis*. *Kor. J. Pharmacog.* 25(3):264-271.

Lycopus spp.

Lamiaceae

Lycopus americanus Muhl. ex W.P.C. Barton
SCN: American bugleweed
OCN: water horehound

Lycopus europaeus **L.**
SCN: European bugleweed
OCN: water horehound

Lycopus virginicus **L.**
SCN: bugleweed
OCN: Virginia bugleweed
Part: herb

Quick Reference Summary

Safety Class: 2b, 2c, 2d
Interaction Class: B

Contraindications

Not for use in pregnancy or lactation except under the supervision of a qualified healthcare practitioner (Beer et al. 2008; De Smet 1993; Mills and Bone 2005; Winterhoff et al. 1994).

Not for use in persons with hypothyroidism or thyroid enlargement (Beer et al. 2008; Vonhoff et al. 2006).

Other Precautions

None known.

Drug and Supplement Interactions

Bugleweeds should not be used with thyroid medications except under the supervision of a qualified healthcare practitioner (Beer et al. 2008; Winterhoff et al. 1994).

Adverse Events and Side Effects

Long-term use of "high" doses (standard dose listed as 1–2 g as a tea) of bugleweed or European bugleweed has been associated with enlargement of the thyroid, while sudden discontinuation of bugleweeds has been reported to cause increased symptoms of hyperthyroidism (Blumenthal et al. 1998).

Pharmacological Considerations

Bugleweeds have traditionally been used for hyperthyroidism. Several animal studies and one human study have indicated the effects of bugleweeds on thyroid hormones (Beer et al. 2008; Vonhoff et al. 2006; Winterhoff et al. 1994).

Pregnancy and Lactation

A recent reference on herbal safety suggests that bugleweeds should not be used during pregnancy or lactation due to effects on thyroid hormones in pregnancy and the possible transmission of antithyroid compounds during lactation (Mills and Bone 2005). An older study indicated that European bugleweed reduced the number of offspring in mice and rats (De Smet 1993). Based on this information, use during pregnancy is not recommended except under the supervision of a qualified healthcare practitioner.

L

Review Details

I. Drug and Supplement Interactions

Clinical Trials of Drug or Supplement Interactions

No clinical trials of drug or supplement interactions were identified.

Case Reports of Suspected Drug or Supplement Interactions

No case reports of suspected drug or supplement interactions were identified.

Animal Trials of Drug or Supplement Interactions

No animal trials of drug or supplement interactions were identified.

II. Adverse Events

Case Reports of Adverse Events

Long-term use of "high" doses (standard dose listed as 1–2 g as a tea) of bugleweed or European bugleweed has been associated with enlargement of the thyroid, while sudden discontinuation of bugleweeds has been reported to cause increased symptoms of hyperthyroidism (Blumenthal et al. 1998).

III. Pharmacology and Pharmacokinetics

Human Pharmacological Studies

In an open label study, patients with a basal *thyroid-stimulating hormone* (TSH) value less than 1.0 mU/l and hyperthyroidism-associated symptoms were administered one tablet containing 20 mg of European bugleweed daily for approximately 15 weeks. In treated patients, an increase in urinary thyroxine (T_4) excretion was observed, along with a reduction in symptoms related to hyperthyroidism, notably, a reduction in heart rate in the morning. In this study, European bugleweed was generally well tolerated, with only minor adverse events reported in the study. Of these, only one adverse event, subjective "disturbances of the cardiac rhythm," was reported after 7 weeks of treatment (Beer et al. 2008).

Animal Pharmacological Studies

In hyperthyroid rats treated with an ethanol extract equivalent to 400 mg/kg of European bugleweed alone or 10 or 400 mg/kg of European bugleweed with 0.7 mg/kg T_4 daily for 56 days, no significant changes of thyroid hormone concentrations or TSH levels were observed. Treatment reduced the increase in heart rate and blood pressure induced by T_4 administration, alleviated cardiac hypertrophy, and reduced the density of β-adrenoceptors in heart tissue (Vonhoff et al. 2006).

In rats orally administered 1 g/kg of an aqueous extract of European bugleweed, a decrease in triiodothyronine (T_3) levels (lasting over 24 h), not related to TSH levels, and due to peripheral T_4 deiodination, was observed. The decrease in T_3 levels was accompanied by a decrease in luteinizing hormone (Winterhoff et al. 1994).

In Vitro Pharmacological Studies

No relevant in vitro pharmacological studies were identified.

IV. Pregnancy and Lactation

Changes in the estrous cycle in mice and rats and a reduction in the number of offspring in the animals were reported after administration of European bugleweed. Extracts, doses, and duration of administration were not listed in the available English language translation (De Smet 1993).

Although one reference text indicates that bugleweeds should be avoided in pregnancy and lactation, due to "pronounced antiprolactin effects" (De Smet 1993), a more recent animal study indicated no effects of European bugleweed on prolactin levels (Beer et al. 2008). Another reference contraindicates bugleweeds during lactation due to the theoretical passage of antithyroid compounds through breast milk (Mills and Bone 2005).

V. Toxicity Studies

Acute Toxicity

In mice, intravenous administration of a dose of 3 ml of pressed juice of Virginia bugleweed was lethal, while a dose of 1 ml orally administered did not cause any toxic symptoms (De Smet 1993).

Literature Cited

L

Beer, A.M., K.R. Wiebelitz, and H. Schmidt-Gayk. 2008. *Lycopus europaeus* (gypsywort): Effects on the thyroidal parameters and symptoms associated with thyroid function. *Phytomedicine* 15(1-2):16-22.

Blumenthal, M., W. Busse, A. Goldberg, et al. 1998. *The complete German Commission E monographs*. Austin, TX: American Botanical Council.

De Smet, P.A.G.M. 1993. *Adverse effects of herbal drugs, Volume 2*. New York: Springer.

Mills, S., and K. Bone. 2005. *The essential guide to herbal safety*. St. Louis: Elsevier.

Vonhoff, C., A. Baumgartner, M. Hegger, et al. 2006. Extract of *Lycopus europaeus* L. reduces cardiac signs of hyperthyroidism in rats. *Life Sci.* 78(10):1063-1070.

Winterhoff, H., H.G. Gumbinger, U. Vahlensieck, et al. 1994. Endocrine effects of *Lycopus europaeus* L. following oral application. *Arzneim. Forsch.* 44(1):41-45.

Magnolia officinalis Rehder & E.H. Wilson — Magnoliaceae

SCN: magnolia
PN: *hou po* (bark, root bark)

Part: bark, root bark

Quick Reference Summary

Safety Class: 2b
Interaction Class: A

Contraindications

Not for use in pregnancy except under the supervision of a qualified healthcare practitioner (Bensky et al. 2004; Chen and Chen 2004).

Other Precautions

None known.

Drug and Supplement Interactions

None known.

Editors' Note

Magnolia bark contains trace amounts of the alkaloids magnocurarine and tubocurarine (Huang 1993). Tubocurarine has also been isolated from curare, the arrow poison used by indigenous hunters of South America that can produce death by asphyxiation when administered intravenously (Blubaugh and Linegar 1948). Intravenous use of magnolia bark is noted as exhibiting some curare-like effects, though the herb is considered generally safe for oral use (Bensky and Gamble 1993), and curare itself has long been recognized as innocuous when taken orally (Blubaugh and Linegar 1948).

Adverse Events and Side Effects

None known.

Pharmacological Considerations

Although compounds from magnolia bark have demonstrated anticoagulant activity in vitro (Teng et al. 1988), animal studies indicated no effects of magnolia bark extract on coagulation (Liu et al. 2007).

Pregnancy and Lactation

Reference texts on traditional Chinese medicine indicate that magnolia bark and root bark should be used with caution in pregnancy (Bensky et al. 2004; Chen and Chen 2004). Based on this information, use during pregnancy is not recommended except under the supervision of a qualified healthcare practitioner.

No information on the safety of magnolia bark or root bark in lactation was identified. While this review did not identify any concerns for use while nursing, safety has not been conclusively established.

M

Review Details

I. Drug and Supplement Interactions

Clinical Trials of Drug or Supplement Interactions

No clinical trials of drug or supplement interactions were identified.

Case Reports of Suspected Drug or Supplement Interactions

No case reports of suspected drug or supplement interactions were identified.

Animal Trials of Drug or Supplement Interactions

An increase in anxiolytic activity was observed in mice administered 0.2 mg/kg of the compound honokiol with diazepam as compared to diazepam alone (Maruyama and Kuribara 2000).

II. Adverse Events

Case Reports of Adverse Events

No case reports of adverse events were identified.

III. Pharmacology and Pharmacokinetics

Human Pharmacological Studies

No relevant human pharmacological studies were identified.

Animal Pharmacological Studies

No effects on coagulation were observed in rats fed diets containing up to 480 mg/kg of magnolia bark extract daily for 21 days or up to 240 mg/kg for 90 days (Liu et al. 2007).

In Vitro Pharmacological Studies

No estrogenic activity of an ethanol extract of magnolia bark was observed in a yeast assay system with human estrogen receptors (Shin et al. 2001).

The compounds magnolol and honokiol inhibited aggregation and ATP release of rabbit platelet-rich plasma induced by collagen and arachidonic acid without affecting that induced by ADP, platelet-activating factor (PAF), or thrombin. Aggregation of washed platelets was more markedly inhibited than that of platelet-rich plasma, while the aggregation of whole blood was least affected by both inhibitors (Teng et al. 1988).

IV. Pregnancy and Lactation

Reference texts on traditional Chinese medicine indicate that magnolia bark and root bark should be used with caution in pregnancy (Bensky et al. 2004; Chen and Chen 2004).

No information on the safety of magnolia bark and root bark in lactation was identified. While this review did not identify any concerns for use while nursing, safety has not been conclusively established.

V. Toxicity Studies

Acute Toxicity

The LD_{50} values of a magnolia bark decoction are 6.12 g/kg after intraperitoneal administration in mice, 4.25 g/kg after intravenous administration in cats, and could not be determined at doses up to 60 g/kg after oral administration in mice (Chen and Chen 2004).

Short-Term Toxicity

In rats fed diets containing 0, 60, 120, 240, or 480 mg/kg of magnolia bark extract daily for 21 days, no treatment-related effects in clinical observations, macroscopic or microscopic findings, or hematology, clinical chemistry, urinalysis, or organ weight measurements were observed, and there were no deaths or significant differences in body weight and weight gain (Liu et al. 2007).

Subchronic Toxicity

In rats fed diets containing 0, 60, 120, or 240 mg/kg of magnolia bark extract daily for 90 days, no mortality, ophthalmic abnormalities, or treatment-related changes in clinical observations, hematology, coagulation, organ weight measurements, or macroscopic or microscopic evaluations were found (Liu et al. 2007).

Genotoxicity

In a bacterial reverse mutation assay and an in vivo micronucleus test, no genotoxic activity of magnolia bark extract was observed (Li et al. 2007).

No genotoxic activity of magnolia bark extract was observed in chromosomal aberration assays with Chinese hamster ovary cells and Chinese hamster lung tissue (Zhang et al. 2008).

M

Literature Cited

Bensky, D., and A. Gample. 1993. *Chinese herbal medicine: Materia medica*. 2nd ed. Seattle: Eastland Press.

Bensky, D., S. Clavey, and E. Stöger. 2004. *Chinese herbal medicine: Materia medica*. 3rd ed. Seattle: Eastland Press.

Blubaugh, L.V., and C.R. Linegar. 1948. Curare and modern medicine. *Econ. Bot.* 2(1):73-82.

Chen, J.K., and T.T. Chen. 2004. *Chinese medical herbology and pharmacology*. City of Industry, CA: Art of Medicine Press.

Huang, K.C. 1993. *The pharmacology of Chinese herbs*. Boca Raton, FL: CRC Press.

Li, N., Y. Song, W. Zhang, et al. 2007. Evaluation of the in vitro and in vivo genotoxicity of magnolia bark extract. *Reg. Toxicol. Pharmacol.* 49(3):154-159.

Liu, Z., X. Zhang, W. Cui, et al. 2007. Evaluation of short-term and subchronic toxicity of magnolia bark extract in rats. *Reg. Toxicol. Pharmacol.* 49(3):160-171.

Maruyama, Y., and H. Kuribara. 2000. Overview of the pharmacological features of honokiol. *CNS Drug Rev.* 6(1):35-44.

Shin, T.Y., D.K. Kim, B.S. Chae, and E.J. Lee. 2001. Antiallergic action of *Magnolia officinalis* on immediate hypersensitivity reaction. *Arch. Pharmacol. Res.* 24(3):249-255.

Teng, C.M., C.C. Chen, F.N. Ko, et al. 1988. Two antiplatelet agents from *Magnolia officinalis*. *Thromb. Res.* 50(6):757-765.

Zhang, B., T. Maniatis, Y. Song, et al. 2008. Evaluation of magnolia bark extract in chromosomal aberration assays. *Mutat. Res.* 654(2):133-137.

Magnolia spp. Magnoliaceae

Magnolia biondii **Pamp.**
SCN: magnolia
PN: *xin yi hua* (flower bud)
OCN: yulan; yulan magnolia

Magnolia denudata **Desr. in Lam.**
SCN: magnolia
PN: *xin yi hua* (flower bud)

Magnolia sprengeri **Pamp.**
SCN: magnolia
PN: *xin yi hua* (flower bud)
Part: flower bud

Quick Reference Summary

Safety Class: 2b
Interaction Class: A

Contraindications

Not for use in pregnancy except under the supervision of a qualified healthcare practitioner (Bensky et al. 2004).

Other Precautions

None known.

Drug and Supplement Interactions

None known.

Adverse Events and Side Effects

Allergic reactions to magnolia flower bud have been reported (Bensky et al. 2004).

Pharmacological Considerations

Overdose of magnolia flower bud may cause dizziness, thirst, and dry nose (Bensky et al. 2004).

Pregnancy and Lactation

Magnolia flower bud should be used with caution in pregnancy (Bensky et al. 2004). Uterine stimulant activity of magnolia flower bud has been observed in an animal study (Chen and Chen 2004). Use during pregnancy is therefore not recommended except under the supervision of a qualified healthcare practitioner.

No information on the safety of magnolia flower bud during lactation was identified. While this review did not identify any concerns for use while nursing, safety has not been conclusively established.

Review Details

I. Drug and Supplement Interactions

Clinical Trials of Drug or Supplement Interactions

No clinical trials of drug or supplement interactions were identified.

Case Reports of Suspected Drug or Supplement Interactions

No case reports of suspected drug or supplement interactions were identified.

Animal Trials of Drug or Supplement Interactions

No animal trials of drug or supplement interactions were identified.

II. Adverse Events

Case Reports of Adverse Events

Allergic reactions to magnolia flower bud have been reported (Bensky et al. 2004).

Overdose of magnolia flower bud is reported to cause dizziness, thirst, and dry nose (Bensky et al. 2004).

III. Pharmacology and Pharmacokinetics

Human Pharmacological Studies

No relevant human pharmacological studies were identified.

Animal Pharmacological Studies

No relevant animal pharmacological studies were identified.

In Vitro Pharmacological Studies

Inhibition of the drug-metabolizing isoenzyme CYP2D6 was observed in rat liver microsomes treated with an extract of magnolia flower (Yu et al. 2007).

IV. Pregnancy and Lactation

Uterine stimulant activity of magnolia flower bud has been observed in dogs and rabbits. Details on product, dose, and route of administration were not listed in the available English language translation (Chen and Chen 2004).

Reference texts on traditional Chinese medicine suggest that magnolia flower bud be used with caution in pregnancy (Bensky et al. 2004) and cite a study in which uterine stimulant activity was observed in rabbits and dogs after ingestion of magnolia flower (Chen and Chen 2004).

No information on the safety of magnolia flower bud during lactation was identified.

V. Toxicity Studies

Acute Toxicity

The LD_{50} of intravenously administered magnolia flower bud could not be determined at doses up to 1 g/kg in dogs and 4.75 g/kg in rabbits (Chen and Chen 2004).

Short-Term Toxicity

No adverse effects were observed in rats orally administered 18 g/kg of an alcohol extract or 30 g/kg of a water extract of magnolia flower bud daily for 30 days (Chen and Chen 2004).

Literature Cited

Bensky, D., S. Clavey, and E. Stöger. 2004. *Chinese herbal medicine: Materia medica.* 3rd ed. Seattle: Eastland Press.

Chen, J.K., and T.T. Chen. 2004. *Chinese medical herbology and pharmacology.* City of Industry, CA: Art of Medicine Press.

Yu, W., L. Huang, and D. Zhu. 2007. Screening of Chinese materia medica possessed inhibitory effect on cytochrome P4502D6 in liver microsomes of rats. *Chin. Trad. Herbal Drugs* 38(3):397-401.

Magnolia virginiana L. — Magnoliaceae

SCN: sweetbay magnolia
Syn: *Magnolia glauca* (L.) L.
OCN: laurel magnolia; swamp laurel; sweetbay
Part: bark

Quick Reference Summary

Safety Class: 2b
Interaction Class: A

Contraindications

Sweetbay magnolia bark is not contraindicated in pregnancy in prominent 19th and early 20th century therapeutic texts in which it is listed (Ellingwood 1919; Felter and Lloyd 1898; Shoemaker 1893). Due to its close botanical relationship to other species of *Magnolia* which are recognized as not for use in pregnancy, however, it is recommended that sweetbay magnolia bark be avoided in pregnancy except under the supervision of a qualified healthcare practitioner.

Other Precautions

None known.

Drug and Supplement Interactions

None known.

Editors' Note

No cautions for use of sweetbay magnolia are reported in historical American medical texts (Felter and Lloyd 1898; Remington and Wood 1918).

Adverse Events and Side Effects

None known.

Pharmacological Considerations

None known.

Pregnancy and Lactation

Although no information on the safety of sweetbay magnolia during pregnancy was identified, based on the similarity of this species to other species of *Magnolia,* use during pregnancy is not recommended.

No information on the safety of sweetbay magnolia during lactation was identified. While this review did not identify any concerns for use while nursing, safety has not been conclusively established.

Review Details

I. Drug and Supplement Interactions

Clinical Trials of Drug or Supplement Interactions

No clinical trials of drug or supplement interactions were identified.

Case Reports of Suspected Drug or Supplement Interactions

No case reports of suspected drug or supplement interactions were identified.

Animal Trials of Drug or Supplement Interactions

No animal trials of drug or supplement interactions were identified.

II. Adverse Events

Case Reports of Adverse Events

No case reports of adverse events were identified.

III. Pharmacology and Pharmacokinetics

Human Pharmacological Studies

No relevant human pharmacological studies were identified.

Animal Pharmacological Studies

No relevant animal pharmacological studies were identified.

In Vitro Pharmacological Studies

No relevant in vitro pharmacological studies were identified.

IV. Pregnancy and Lactation

Although no information on the safety of sweetbay magnolia during pregnancy was identified, based on the similarity of this species to other species of *Magnolia*, use during pregnancy is not recommended.

No information on the safety of sweetbay magnolia during lactation was identified.

V. Toxicity Studies

No toxicity studies were identified.

Literature Cited

Ellingwood, F. 1919. *The American materia medica, therapeutics and pharmacognosy*. Evanston, IL: Ellingwood' Therapeutist.

Felter, H.W., and J.U. Lloyd. 1898. *King's American dispensatory*. 18th ed., 3rd rev. 2 vols. Cincinnati: Ohio Valley Co.

Remington, J.P., and H.C. Wood. 1918. *The dispensatory of the United States of America*. 20th ed. Philadelphia: Lippincott.

Shoemaker, J.V. 1893. *A practical treatise on materia medica and therapeutics*. 2nd ed. Philadelphia: The F.A. Davis Co.

Mahonia spp. — Berberidaceae

Mahonia aquifolium (Pursh) Nutt.
SCN: Oregon grape
Syn: *Berberis aquifolium* Pursh
OCN: holly-leaf barberry; mountain grape; Oregon grapeholly; Oregon barberry

***Mahonia nervosa* (Pursh) Nutt.**
SCN: Oregon grape
OCN: Oregon grapeholly; Oregon barberry

***Mahonia repens* (Lindl.) G. Don**
SCN: Oregon grape
OCN: creeping barberry; Oregon grapeholly; Oregon barberry
Part: root

M

Quick Reference Summary

Safety Class: 1
Interaction Class: A

Contraindications

None known.

Other Precautions

None known.

Drug and Supplement Interactions

None known.

Notice

Berberine (Leung and Foster 1996; List and Hörhammer 1973); *see* Appendix 1. The total isoquinoline alkaloid content is reported as 7 to 16% in the root bark, and 2.4 to 4.5% in the stem bark (Fleming 2000), although the typically traded portion is the whole root, which contains a significantly smaller percentage of berberine than the root bark.

Editors' Note

Most safety concerns reported for Oregon grape are based on studies of the compound berberine and other similar alkaloids. Data regarding isolated compounds may not apply directly to products or extracts made from Oregon grape.

Adverse Events and Side Effects

None known.

Pharmacological Considerations

None known.

Pregnancy and Lactation

The use of berberine-containing herbs is contraindicated or cautioned against in pregnancy in several contemporary texts on herbal safety (Brinker 2001; Mills and Bone 2005; Mills et al. 2006). These contraindications are based primarily on the uterine stimulant activity of the isolated compound berberine in excised mouse uteruses (Furuya 1957; Imaseki et al. 1961), and the potential ability of berberine to displace bilirubin and cause neonatal jaundice (Chan 1993). Although definitive data confirming the safety of Oregon grape during pregnancy is lacking, several reproductive toxicity studies on the isolated compound berberine in mice and rats have shown no adverse effects at doses of 530 mg/

kg in rats and mice (Jahnke et al. 2006; Marr et al. 2005; Price and George 2003).

Some concerns exist for use of berberine-containing plants during pregnancy, including uterine stimulation (Furuya 1957; Imaseki et al. 1961), although no uterine stimulation was noted in rats administered high doses of berberine during pregnancy (Jahnke et al. 2006), and a study of berberine-containing herbal extracts on isolated uteri showed no correlation between uterine stimulation and berberine concentration (Haginiwa and Harada 1962).

Although other berberine-containing plants in this text are classified as contraindicated in pregnancy except under supervision of a qualified healthcare practitioner, the lower percentage of berberine in Oregon grape root as compared to other berberine-containing botanicals led to the less restrictive classification (De Smet 1992; Fleming 2000; Leung and Foster 1996; List and Hörhammer 1973; Upton 2001).

Berberine has been shown to be present in the breast milk of lactating women who have taken berberine-containing plants (Chan 1993).

Review Details

I. Drug and Supplement Interactions

Clinical Trials of Drug or Supplement Interactions

No clinical trials of drug or supplement interactions were identified.

Case Reports of Suspected Drug or Supplement Interactions

No case reports of suspected drug or supplement interactions were identified.

Animal Trials of Drug or Supplement Interactions

No animal trials of drug or supplement interactions were identified.

II. Adverse Events

Adverse Events Reported in Clinical Trials

No drug-related adverse events were reported in a 12-week trial of a topically applied Oregon grape ointment containing 0.1% berberine (Gulliver 2004). Itching and burning were reported as adverse events in 4% of patients topically administered a preparation of Oregon grape. In that study, patients were simultaneously treated with placebo on one arm and Oregon grape on the other (Wiesenauer 1996). Other similar studies of topically applied Oregon grape preparations note adverse events of itching and burning in up to 1.2% of trial participants (Bernstein et al. 2006; Gieler et al. 1995).

Case Reports of Adverse Events

No case reports of adverse events were identified.

III. Pharmacology and Pharmacokinetics

Human Pharmacological Studies

No relevant human pharmacological studies were identified.

Animal Pharmacological Studies

Intraperitoneal administration of 0.02 mg/kg of the compound berberine daily for 1 week to adult rats resulted in a significant decrease in mean bilirubin serum protein binding due to a displacement effect (Chan 1993).

In Vitro Pharmacological Studies

The compound berberine was found to be 10 times more potent in vitro than phenylbutazone, a known displacer of bilirubin, and approximately 100 times more potent than papaverine (Chan 1993).

IV. Pregnancy and Lactation

In pregnant rats fed the compound berberine on gestational days (GD) 6 to 20, some reduction in maternal weight gain was observed, with a lowest-observed-adverse-effect level (LOAEL) of 530 mg/kg daily. Only a mild reduction in fetal weights was observed, and the LOAEL based on fetal weight reduction was 1000 mg/kg (Jahnke et al. 2006). Similarly, in mice administered berberine on GD 6 to 17 at doses up to 1155 mg/kg daily, the maternal LOAEL was determined to be 531 mg/kg daily, and the developmental toxicity level was 1000 mg/kg daily. In mice, 33% of the treated females died. Surviving animals had increased relative water intake, and average fetal body weight per litter decreased 5–6% with no change in live litter size (Jahnke et al. 2006).

Berberine has been shown to stimulate uterine contractions in both pregnant and nonpregnant mice (Furuya 1957; Imaseki et al. 1961). A study of various berberine-containing herbal extracts on isolated uteruses, however, indicated that relaxation or stimulation of the uterus did not correlate with the concentration of berberine in the extract, suggesting that not all berberine-containing herbs will have the same effect on the uterus (Haginiwa and Harada 1962).

Berberine has been shown to be present in the breast milk of lactating women who have taken berberine-containing plants (Chan 1993).

V. Toxicity Studies

Acute Toxicity

The LD_{50} of orally administered berberine in mice is 329 mg/kg (Haginiwa and Harada 1962). The LD_{50} of orally administered berberine sulfate in rats is greater than 1000 mg/kg (Kowalewski et al. 1975).

Genotoxicity

Fractions of Oregon grape extracts showed antimutagenic activity using *Euglena gracilis* as an eukaryotic test model. Both bisbenzylisoquinoline and protoberberine alkaloid fractions were active in this respect, but only the protoberberine derivatives jatrorrhizine and berberine showed significant concentration-dependent antimutagenic activity (Cernakova et al. 2002).

No mutagenic activity of berberine was observed in *Salmonella typhimurium* TA100 and TA98 with or without metabolic activation by S9 mix. Berberine hydrochloride was weakly mutagenic to strain TA98 without S9 mix, but showed no mutagenic activity in TA100 without S9 mix (Nozaka et al. 1990).

No genotoxic, mutagenic, or recombinogenic activity of berberine with or without metabolic activation was observed in the SOS chromotest. Berberine did not induce significant cytotoxic, mutagenic, or recombinogenic effects during treatments performed under nongrowth conditions; in dividing cells, however, the alkaloid induced cytotoxic and cytostatic effects in proficient and repair-deficient strains of *Saccharomyces cerevisiae*. In dividing cells, the induction of frameshift and mitochondrial mutations, as well as crossing over, indicated that berberine is not a potent mutagenic agent (Pasqual et al. 1993).

Literature Cited

Bernstein, S., H. Donsky, W. Gulliver, et al. 2006. Treatment of mild to moderate psoriasis with relieva, a *Mahonia aquifolium* extract—A double-blind, placebo-controlled study. *Am. J. Ther.* 13(2):121.

Brinker, F. 2001. *Herb contraindications and drug interactions*. 3rd ed. Sandy, OR: Eclectic Medical Publications.

Cernakova, M., D. Kost'alova, V. Kettmann, et al. 2002. Potential antimutagenic activity of berberine, a constituent of *Mahonia aquifolium*. *BMC Complement. Altern. Med.* 2:2.

Chan, E. 1993. Displacement of bilirubin from albumin by berberine. *Biol. Neonate* 63(4):201-208.

De Smet, P.A.G.M. 1992. *Adverse effects of herbal drugs, Volume 1*. Berlin: Springer.

Fleming, T.E. 2000. *PDR for herbal medicines*. Montvale, NJ: Medical Economics Company.

Furuya, T. 1957. Pharmacological action, including toxicity and excretion of berberine hydrochloride and its oxidation product. *Bull. Osaka Med. School* 3:62-67. *Cited in* De Smet, P.A.G.M. 1992. *Adverse effects of herbal drugs, Volume 1*. New York: Springer.

Gieler, U., A. Von der Weth, and M. Heger. 1995. *Mahonia aquifolium*—A new type of topical treatment for psoriasis. *J. Dermatol. Treat.* 6(1):31-34.

Gulliver, W. 2004. A report on three recent clinical trials utilizing *Mahonia aquifolium* 10% topical cream and worldwide clinical experience with *Mahonia aquifolium* for the treatment of plaque psoriasis. Part 1: Safety. Oldsmar, FL: Apollo Pharmaceuticals.

Haginiwa, J., and M. Harada. 1962. Pharmacological studies on crude drugs. V. Comparison of berberine type alkaloid-containing plants on their components and several pharmacological actions. *Jpn. J. Pharmacol.* 82:726-731.

Imaseki, I., Y. Kitabatake, and T. Taguchi. 1961. Studies on the effect of berberine alkaloids on intestine and uterus in mice. *Yakugaku Zasshi* 81:1281-1284.

Jahnke, G.D., C.J. Price, M.C. Marr, C.B. Myers, and J.D. George. 2006. Developmental toxicity evaluation of berberine in rats and mice. *Birth Defects Res. B Dev. Reprod. Toxicol.* 77(3):195-206.

Kowalewski, Z., A. Mrozikiewicz, T. Bobkiewicz, K. Drost, and B. Hladon. 1975. Toxicity of berberine sulfate. *Acta Pol. Pharm.* 32(1):113-120.

Leung, A.Y., and S. Foster. 1996. *Encyclopedia of common natural ingredients used in food, drugs, and cosmetics*. 2nd ed. New York: Wiley.

List, P.H., and H. Hörhammer. 1973. *Hagers handbuch der pharmazeutischen praxis*. Berlin: Springer.

Marr, M.C., C.J. Price, J.D. George, C.B. Myers, and G.D. Jahnke. 2005. Developmental toxicity evaluation of berberine chloride dihydrate (BCD) administered in the feed and by gavage to Swiss (CD-1) mice. *Birth Defects Res. A Clin. Mol. Teratol.* 73(5):357.

Mills, S., and K. Bone. 2005. *The essential guide to herbal safety*. St. Louis: Elsevier.

Mills, S., J. Duguoa, D. Perri, and G. Koren. 2006. *Herbal medicines in pregnancy and lactation: An evidence based approach*. New York: Taylor & Francis.

Nozaka, T., F. Watanabe, S.I. Tadaki, et al. 1990. Mutagenicity of isoquinoline alkaloids, especially of the aporphine type. *Mutat. Res.* 240(4):267-279.

Pasqual, M.S., C.P. Lauer, P. Moyna, and J.A. Henriques. 1993. Genotoxicity of the isoquinoline alkaloid berberine in prokaryotic and eukaryotic organisms. *Mutat. Res.* 286(2):243-252.

Price, C.J., and J.D. George. 2003. Final study report on the developmental toxicity evaluation for berberine chloride dihydrate (CAS No. 5956-60-5) administered in the feed to Swiss (CD-1) mice on gestational days 6 through 17. *Govt. Rep. Announcements Index* (20):112.

Upton, R. 2001. *Goldenseal root: Hydrastis canadensis; Standards of analysis, quality control, and therapeutics. American Herbal Pharmacopoeia and therapeutic compendium.* Santa Cruz, CA: American Herbal Pharmacopoeia.

Wiesenauer, L. 1996. *Mahonia aquifolium* in patients with psoriasis vulgaris—An intra-individual study. *Phytomedicine* 3:231-235.

Malva sylvestris L. — Malvaceae

SCN: high mallow
OCN: common mallow; malva

Part: flower, leaf

Quick Reference Summary

Safety Class: 1
Interaction Class: A

Contraindications

None known.

Other Precautions

Should be taken with at least 250 ml (8 oz) of liquid.

Drug and Supplement Interactions

Other drugs should be taken 1 hour prior to consumption of high mallow or several hours after consumption, as the mucilage content of high mallow may slow the absorption of orally administered drugs (Classen and Blaschek 1998; Franz 1966; Gonda et al. 1990; Tomoda et al. 1989).

Notice

Mucilages (5 to 12%) (Classen and Blaschek 1998; Franz 1966; Gonda et al. 1990; Tomoda et al. 1989); *see* Appendix 3.

Adverse Events and Side Effects

None known.

Pharmacological Considerations

None known.

Pregnancy and Lactation

No information on the safety of high mallow in pregnancy or lactation was identified in the scientific or traditional literature. Although this review did not identify any concerns for use while pregnant or nursing, safety has not been conclusively established.

Review Details

I. Drug and Supplement Interactions

Clinical Trials of Drug or Supplement Interactions

No clinical trials of drug or supplement interactions were identified.

Case Reports of Suspected Drug or Supplement Interactions

No case reports of suspected drug or supplement interactions were identified.

Animal Trials of Drug or Supplement Interactions

No animal trials of drug or supplement interactions were identified.

II. Adverse Events

Case Reports of Adverse Events

No case reports of adverse events were identified.

III. Pharmacology and Pharmacokinetics

Human Pharmacological Studies

No relevant human pharmacological studies were identified.

Animal Pharmacological Studies

No relevant animal pharmacological studies were identified.

In Vitro Pharmacological Studies

No relevant in vitro pharmacological studies were identified.

IV. Pregnancy and Lactation

No studies on the use of high mallow during pregnancy or lactation were identified.

V. Toxicity Studies

Cytotoxicity

No cytotoxic activity of a dichloromethane and methanol extract of the aerial parts of high mallow was observed in mouse fibrosarcoma cells (L929sA) (Kaileh et al. 2007).

Literature Cited

Classen, B., and W. Blaschek. 1998. High molecular weight acidic polysaccharides from *Malva sylvestris* and *Alcea rosea*. *Planta Med.* 64(7):640-644.

Franz, G. 1966. Die Schleimpolysaccharide von *Althea officinalis* und *Malva sylvestris*. *Planta Med.* 14:89-110.

Gonda, R., M. Tomoda, N. Shimizu, and H. Yamada. 1990. Structure and anticomplementary activity of an acidic polysaccharide from the leaves of *Malva sylvestris* var. *mauritiana*. *Carbohydr. Res.* 198(2):323-329.

Kaileh, M., W.V. Berghe, E. Boone, T. Essawi, and G. Haegeman. 2007. Screening of indigenous Palestinian medicinal plants for potential anti-inflammatory and cytotoxic activity. *J. Ethnopharmacol.* 113(3):510-516.

Tomoda, M., R. Gonda, N. Shimizu, and H. Yamada. 1989. Plant mucilages. XLII. An anticomplementary mucilage from the leaves of *Malva sylvestris* var. *mauritiana*. *Chem. Pharm. Bull. (Tokyo)* 37(11):3029-3032.

Maranta arundinacea L. — Marantaceae

SCN: arrowroot
OCN: St. Vincent arrowroot; West Indian arrowroot

Part: root

Quick Reference Summary

Safety Class: 1
Interaction Class: A

Contraindications
None known.

Other Precautions
None known.

Drug and Supplement Interactions
None known.

Editors' Note
A flour made from arrowroot is commonly used in cooking, notably in biscuits made for infants (Felter and Lloyd 1898; Kay and Gooding 1987).

Adverse Events and Side Effects
None known.

Pharmacological Considerations
None known.

Pregnancy and Lactation
No information on the safety of arrowroot in pregnancy or lactation was identified in the scientific or traditional literature. Although this review did not identify any concerns for use while pregnant or nursing, safety has not been conclusively established.

Review Details

I. Drug and Supplement Interactions

Clinical Trials of Drug or Supplement Interactions
No clinical trials of drug or supplement interactions were identified.

Case Reports of Suspected Drug or Supplement Interactions
No case reports of suspected drug or supplement interactions were identified.

Animal Trials of Drug or Supplement Interactions
No animal trials of drug or supplement interactions were identified.

II. Adverse Events

Case Reports of Adverse Events
No case reports of adverse events were identified.

III. Pharmacology and Pharmacokinetics

Human Pharmacological Studies
No relevant human pharmacological studies were identified.

Animal Pharmacological Studies
No relevant animal pharmacological studies were identified.

In Vitro Pharmacological Studies
No relevant in vitro pharmacological studies were identified.

IV. Pregnancy and Lactation

No information on the safety of arrowroot during pregnancy or lactation was identified.

V. Toxicity Studies

No toxicity studies were identified.

Literature Cited

Felter, H.W., and J.U. Lloyd. 1898. *King's American dispensatory*. 18th ed., 3rd rev. 2 vols. Cincinnati: Ohio Valley Co.

Kay, D.E., and G.B. Gooding. 1987. *Root crops*. London: Tropical Development and Research Institute.

Marrubium vulgare L. — Lamiaceae

SCN: horehound
OCN: white horehound

Part: herb

Quick Reference Summary

Safety Class: 2b
Interaction Class: A

Contraindications

Not for use in pregnancy except under the supervision of a qualified healthcare practitioner (Kchouck and Chadli 1963; List and Hörhammer 1973).

Other Precautions

None known.

Drug and Supplement Interactions

None known.

Notice

Emmenagogue (List and Hörhammer 1973); *see* Appendix 2.

Adverse Events and Side Effects

None known.

Pharmacological Considerations

Studies in rabbits and rats have indicated that horehound may help to regulate blood sugar levels (Novaes et al. 2001; Roman Ramos et al. 1992).

"Large doses" may have laxative effects (Chadha 1988; Osol and Farrar 1955).

Pregnancy and Lactation

Horehound has traditionally been used as an emmenagogue (List and Hörhammer 1973). Based on this information, use during pregnancy is not recommended except under the supervision of a qualified healthcare practitioner.

No information on the safety of horehound in lactation was identified in the scientific or traditional literature. While this review did not identify any concerns for use while nursing, safety has not been conclusively established.

M

Review Details

I. Drug and Supplement Interactions

Clinical Trials of Drug or Supplement Interactions

No clinical trials of drug or supplement interactions were identified.

Case Reports of Suspected Drug or Supplement Interactions

No case reports of suspected drug or supplement interactions were identified.

Animal Trials of Drug or Supplement Interactions

No animal trials of drug or supplement interactions were identified.

II. Adverse Events

Case Reports of Adverse Events

No case reports of adverse events were identified.

III. Pharmacology and Pharmacokinetics

Human Pharmacological Studies

No relevant human pharmacological studies were identified.

Animal Pharmacological Studies

Extracts of horehound helped decrease experimentally induced hyperglycemia in rabbits and in alloxan-induced diabetic rats (Novaes et al. 2001; Roman Ramos et al. 1992).

One reference indicates that a standard dose of horehound has a normalizing effect on extrasystolic arrythmias, whereas larger doses may disturb the heart rhythm in a manner that can be counteracted by atropine (List and Hörhammer 1973).

In Vitro Pharmacological Studies

No relevant in vitro pharmacological studies were identified.

IV. Pregnancy and Lactation

Horehound has traditionally been used as an emmenagogue (List and Hörhammer 1973).

In pregnant rats, mice, and guinea pigs subcutaneously, intraperitoneally, or orally administered approximately 2 ml/kg of a horehound decoction either once or twice during pregnancy, some abortifacient activity in rats was observed, with a lesser effect on mice and guinea pigs (Kchouck and Chadli 1963). This study did not include control groups, making results difficult to interpret.

No information on the safety of horehound in lactation was identified.

V. Toxicity Studies

Acute Toxicity

The LD_{50} of the compound marrubiin orally administered to rats is 370 mg/kg (Krejci and Zadina 1959).

Literature Cited

Chadha, Y. 1988. *The wealth of India: A dictionary of Indian raw materials and industrial products*. Delhi: Council of Scientific and Industrial Research.

Kchouck, M., and A. Chadli. 1963. On the abortive properties of white horehound (*Marrubium vulgare* L.). *Arch. Inst. Pasteur Tunis* 59(1):23-31.

Krejci, I., and R. Zadina. 1959. Die Gallentreibende Wirkung von Marrubiin und Marrabinsäure. *Planta Med.* 7(1).

List, P.H., and H. Hörhammer. 1973. *Hagers handbuch der pharmazeutischen praxis*. Berlin: Springer.

Novaes, A.P., C. Rossi, C. Poffo, et al. 2001. Preliminary evaluation of the hypoglycemic effect of some Brazilian medicinal plants. *Therapie* 56(4):427-430.

Osol, A., and G. Farrar. 1955. *The dispensatory of the United States of America*. 25th ed. Philadelphia: Lippincott.

Roman Ramos, R., F. Alarcon-Aguilar, A. Lara-Lemus, and J.L. Flores-Saenz. 1992. Hypoglycemic effect of plants used in Mexico as antidiabetics. *Arch. Med. Res.* 23(1):59-64.

Matricaria chamomilla L.

Asteraceae

SCN: chamomile
Syn: *Chamomilla recutita* (L.) Rauschert; *Matricaria recutita* L.
OCN: German chamomile; Hungarian chamomile; mayweed; sweet false chamomile; true chamomile
Part: flower

Quick Reference Summary

Safety Class: 1
Interaction Class: A

Contraindications

None known.

Other Precautions

Persons with allergies to other members of the Asteraceae family (such as feverfew or *Echinacea*) should exercise caution with chamomile, as allergic cross-reactivity is common to Asteraceae plants (Upton 2007).

Drug and Supplement Interactions

None known.

Editors' Note

The German Standard License, as presented by Wichtl (2004), requires the following label warning: The infusion should not be used near the eyes.

Adverse Events

Allergic reactions to chamomile, including anaphylactic reactions, have been reported (Benner and Lee 1973; Foti et al. 2000; Giordano-Labadie et al. 2000; Jensen-Jarolim et al. 1998; Lundh et al. 2006; McGeorge and Steele 1991; Pereira et al. 1997; Rodriguez-Serna et al. 1998; Rudzki et al. 2003; Subiza et al. 1989; Thien 2001). No other types of adverse events or interactions have been reported in association with chamomile.

Highly concentrated hot tea is noted as emetic (Chadha 1988).

Pharmacological Considerations

Chamomile has been indicated as an herb with the theoretical potential for interacting with warfarin due to the coumarin content of the herb (Heck et al. 2000). The coumarin content, however, consists of the coumarin derivatives herniarin and umbelliferone. Herniarin has been shown to have hemostatic activity (Ahmad and Misra 1997) and umbelliferone has shown no evidence of anticoagulant activity (Egan et al. 1990; Feuer 1974; Pelkonen et al. 1997). The flavonoid apigenin has been shown to inhibit platelet aggregation in vitro (Landolfi et al. 1984; Teng et al. 1985).

Coadministration of chamomile tea and iron-fortified bread reduced absorption of the iron (Hurrell et al. 1999).

Pregnancy and Lactation

Limited information on the safety of chamomile in pregnancy or lactation was identified in the scientific and traditional literature. Although this review did not identify any concerns for use while pregnant or nursing, safety has not been conclusively established.

Review Details

I. Drug and Supplement Interactions

Clinical Trials of Drug or Supplement Interactions

No clinical trials of drug or supplement interactions were identified.

Case Reports of Suspected Drug or Supplement Interactions

One woman taking warfarin had an increased INR, from 3.6 to 7.9 (a standardized scale used to report the results of blood coagulation tests; increased INR indicates slowed blood clotting), and pelvic and abdominal ecchymoses after consuming 4 or 5 cups of chamomile tea daily and applying a chamomile-containing skin lotion. The woman denied any dietary changes (Segal and Pilote 2006).

Animal Trials of Drug or Supplement Interactions

No relevant animal trials of drug or supplement interactions were identified.

II. Adverse Events

Case Reports of Adverse Events

Allergic reactions to chamomile including anaphylaxis (Benner and Lee 1973; Jensen-Jarolim et al. 1998; Subiza et al. 1989; Thien 2001) and contact dermatitis have been reported (Foti et al. 2000; Giordano-Labadie et al. 2000; Lundh et al. 2006; McGeorge and Steele 1991; Pereira et al. 1997; Rodriguez-Serna et al. 1998; Rudzki et al. 2003).

III. Pharmacology and Pharmacokinetics

Human Pharmacological Studies

Coadministration of iron-fortified bread and chamomile tea reduced absorption of the iron by 47%, an effect that was less pronounced than that of black tea, cocoa, or tea made with peppermint, European pennyroyal, European vervain, or linden flower (Hurrell et al. 1999).

Animal Pharmacological Studies

Administration of a 2% chamomile tea to rats for 4 weeks resulted in significant inhibition of the drug-metabolizing isoenzyme CYP1A2 (Maliakal and Wanwimolruk 2001).

In Vitro Pharmacological Studies

An extract of chamomile demonstrated inhibitory activity on the drug-metabolizing isoenzyme CYP3A4 in vitro (Budzinski et al. 2000).

IV. Pregnancy and Lactation

No adverse effects on prenatal development or signs of teratogenicity were observed after long-term oral administration of a commercial chamomile extract in rats (Homburg Pharma 1986).

No information on the safety of chamomile in lactation was identified.

V. Toxicity Studies

Acute Toxicity

The LD_{50} of orally administered chamomile essential oil in rats could not be determined at doses up to 5 g/kg (Opdyke 1974).

Chronic Toxicity

Long-term oral administration of chamomile extract in rats and dogs produced no signs of toxicity. Rat behavior was affected only at 500 mg/kg, the highest dose tested (Fundaro and Cassone 1980).

M

Literature Cited

Ahmad, A., and L. Misra. 1997. Isolation of a coumarin herniarin in German chamomile flowers. *Int. J. Pharmacog.* 35:121-125.

Benner, M.H., and H.J. Lee. 1973. Anaphylactic reaction to chamomile tea. *J. Allergy Clin. Immunol.* 52(5):307-308.

Budzinski, J.W., B.C. Foster, S. Vandenhoek, and J.T. Arnason. 2000. An in vitro evaluation of human cytochrome P450 3A4 inhibition by selected commercial herbal extracts and tinctures. *Phytomedicine* 7(4):273-282.

Chadha, Y. 1988. *The wealth of India: A dictionary of Indian raw materials and industrial products.* Delhi: Council of Scientific and Industrial Research.

Egan, D., R. O'Kennedy, E. Moran, et al. 1990. The pharmacology, metabolism, analysis, and applications of coumarin and coumarin-related compounds. *Drug Metab. Rev.* 22(5):503-529.

Feuer, G. 1974. The metabolism and biological action of coumarin. *Prog. Med. Chem.* 10:85-158.

Foti, C., E. Nettis, R. Panebianco, et al. 2000. Contact urticaria from *Matricaria chamomilla*. *Contact Dermat.* 42(6):360-361.

Fundaro, A., and M.C. Cassone. 1980. Action of essential oils of chamomile, cinnamon, absinthium, mace and origanum on operant conditioning behavior of the rat. *Boll. Soc. Ital. Biol. Sper.* 56(22):2375-2380.

Giordano-Labadie, F., H.P. Schwarze, and J. Bazex. 2000. Allergic contact dermatitis from camomile used in phytotherapy. *Contact Dermat.* 42(4):247.

Heck, A.M., B.A. DeWitt, and A.L. Lukes. 2000. Potential interactions between alternative therapies and warfarin. *Am. J. Health Syst. Pharm.* 57(13):1221-1227; quiz 1228-1230.

Homburg Pharma. Kamillosan Scientific Information. Homburg Pharma, Division of Degussa. *Cited in* Mann, C., and E. Staba. 1986. The chemistry, pharmacology and commercial formulations of chamomile. In Craker, L., and J. Simon, eds. *Herbs, spices, and medicinal plants: Recent advances, Volume 1.* Phoenix: Oryx Press.

Hurrell, R.F., M. Reddy, and J.D. Cook. 1999. Inhibition of non-haem iron absorption in man by polyphenolic-containing beverages. *Br. J. Nutr.* 81(4):289-295.

Jensen-Jarolim, E., N. Reider, R. Fritsch, and H. Breiteneder. 1998. Fatal outcome of anaphylaxis to camomile-containing enema during labor: A case study. *J. Allergy Clin. Immunol.* 102(6, Pt. 1):1041-1042.

Landolfi, R., R. Mower, and M. Steiner. 1984. Modification of platelet function and arachidonic acid metabolism by bioflavonoids. *Biochem. Pharmacol.* 33:1525-1530.

Lundh, K., M. Hindsen, B. Gruvberger, et al. 2006. Contact allergy to herbal teas derived from Asteraceae plants. *Contact Dermat.* 54(4):196-201.

Maliakal, P.P., and S. Wanwimolruk. 2001. Effect of herbal teas on hepatic drug metabolizing enzymes in rats. *J. Pharm. Pharmacol.* 53(10):1323-1329.

McGeorge, B.C., and M.C. Steele. 1991. Allergic contact dermatitis of the nipple from Roman chamomile ointment. *Contact Dermat.* 24(2):139-140.

Opdyke, D. 1974. Fragrance raw materials monographs. Chamomile oil German. *Food Cosmet. Toxicol.* 12(Suppl.):851-852.

Pelkonen, O., H. Raunio, A. Rautio, and M. Pasanen. 1997. The metabolism of coumarin. In O'Kennedy, K., and R.D. Thornes, eds. *Coumarins: Biology, applications and mode of action.* Hoboken, NJ: Wiley.

Pereira, F., R. Santos, and A. Pereira. 1997. Contact dermatitis from chamomile tea. *Contact Dermat.* 36(6):307.

Rodriguez-Serna, M., J.M. Sanchez-Motilla, R. Ramon, and A. Aliaga. 1998. Allergic and systemic contact dermatitis from *Matricaria chamomilla* tea. *Contact Dermat.* 39(4):192-193.

Rudzki, E., P. Rapiejko, and P. Rebandel. 2003. Occupational contact dermatitis, with asthma and rhinitis, from camomile in a cosmetician also with contact urticaria from both camomile and lime flowers. *Contact Dermat.* 49(3):162.

Segal, R., and L. Pilote. 2006. Warfarin interaction with *Matricaria chamomilla. Can. Med. Assoc. J.* 174(9):1281-1282.

Subiza, J., J.L. Subiza, M. Hinojosa, et al. 1989. Anaphylactic reaction after the ingestion of chamomile tea: A study of cross-reactivity with other composite pollens. *J. Allergy Clin. Immunol.* 84(3):353-358.

Teng, C., L. Lee, S. Ko, and T.F. Huang. 1985. Inhibition of platelet aggregation by apigenin from *Apium graveolens. Asia Pac. J. Pharmacol.* 3:85.

Thien, F.C. 2001. Chamomile tea enema anaphylaxis. *Med. J. Aust.* 175(1):54.

Upton, R. 2007. *Feverfew aerial parts: Tanacetum parthenium (L.) Schultz Bip: Standards of analysis, quality control, and therapeutics.* Santa Cruz, CA: American Herbal Pharmacopoeia.

Wichtl, M. 2004. *Herbal drugs and phytopharmaceuticals: A handbook for practice on a scientific basis.* 3rd ed. Boca Raton, FL: CRC Press.

Matricaria discoidea DC.

Asteraceae

SCN: pineapple weed
Syn: *Chamomilla suaveolens* (Pursh) Rydb.

OCN: disc mayweed
Part: flowering top

Quick Reference Summary

Safety Class: 1
Interaction Class: A

Contraindications

None known.

Other Precautions

Persons with allergies to other members of the Asteraceae family (such as feverfew or *Echinacea*) should exercise caution with chamomile, as allergic cross-reactivity is common to Asteraceae plants (Upton 2007).

Drug and Supplement Interactions

None known.

Adverse Events and Side Effects

None known.

Pharmacological Considerations

None known.

Pregnancy and Lactation

No information on the safety of pineapple weed in pregnancy or lactation was identified in the scientific or traditional literature. Although this review did not identify any concerns for use while pregnant or nursing, safety has not been conclusively established.

Review Details

I. Drug and Supplement Interactions

Clinical Trials of Drug or Supplement Interactions

No clinical trials of drug or supplement interactions were identified.

Case Reports of Suspected Drug or Supplement Interactions

No case reports of suspected drug or supplement interactions were identified.

Animal Trials of Drug or Supplement Interactions

No animal trials of drug or supplement interactions were identified.

II. Adverse Events

Case Reports of Adverse Events

No case reports of adverse events were identified.

III. Pharmacology and Pharmacokinetics

Human Pharmacological Studies

No relevant human pharmacological studies were identified.

Animal Pharmacological Studies

No relevant animal pharmacological studies were identified.

In Vitro Pharmacological Studies

No relevant in vitro pharmacological studies were identified.

IV. Pregnancy and Lactation

No information on the safety of pineapple weed during pregnancy or lactation was identified.

V. Toxicity Studies

No toxicity studies were identified.

Literature Cited

Upton, R. 2007. *Feverfew aerial parts: Tanacetum parthenium (L.) Schultz Bip: Standards of analysis, quality control, and therapeutics.* Scotts Valley, CA: American Herbal Pharmacopoeia.

Medicago sativa L. — Fabaceae

SCN: alfalfa
OCN: lucerne

Part: herb

Quick Reference Summary

Safety Class: 1
Interaction Class: A

Contraindications

None known.

Other Precautions

None known.

Drug and Supplement Interactions

None known.

Adverse Events and Side Effects

None known.

Pharmacological Considerations

Alfalfa is sometimes cited as a potential anticoagulant due to the presence of coumarin compounds and confusion between coumarins and the anticoagulant drug coumadin (sometimes referred to as coumarin). Different naturally occurring coumarins have been shown to have varying activities on coagulation, with some having anticoagulant activity, some having procoagulant activity, and others having no effects on coagulation (Booth et al. 2004). An in vitro study of alfalfa demonstrated some anticoagulant activity (Pierre et al. 2005), although the relevance of those in vitro data to human use is not known.

Animal studies of alfalfa seed and seed sprouts have indicated some association between those products and the autoimmune disease lupus. The amount of alfalfa used in the studies, however, was highly excessive, with animals being fed diets containing 40% alfalfa sprouts or 45% alfalfa seed (Malinow et al. 1982; Montanaro and Bardana 1991). Related studies implicated the compound L-canavanine as being responsible for the effects of alfalfa. Although the compound is present in alfalfa seed and sprouts, it is not present in the mature herb that is the subject of this entry (Brown 2000; Farnsworth 1995; Malinow et al. 1982; Whittam et al. 1995).

Pregnancy and Lactation

No information on the safety of alfalfa in pregnancy or lactation was identified in the scientific or traditional literature. Although this review did not identify any concerns for use while pregnant or nursing, safety has not been conclusively established.

Review Details

I. Drug and Supplement Interactions

Clinical Trials of Drug or Supplement Interactions

No clinical trials of drug or supplement interactions were identified.

Case Reports of Suspected Drug or Supplement Interactions

No case reports of suspected drug or supplement interactions were identified.

Animal Trials of Drug or Supplement Interactions

No animal trials of drug or supplement interactions were identified.

II. Adverse Events

Case Reports of Adverse Events

In two patients with well-controlled systemic lupus erythematosus (SLE), consumption of alfalfa tablets was associated with exacerbation of SLE symptoms. In one case, involving a 40-year-old woman with a 26-year history of SLE, exacerbation was observed after the woman had been taking 15 tablets of alfalfa daily for 9 months. In the second case, a 50-year-old woman with a 25-year history of SLE experienced an increase in symptoms over the course of 18 months. She had been taking 8 tablets of alfalfa daily for 2.5 years (Roberts and Hayashi 1983).

The compound L-canavanine has been suggested as the causative agent in SLE exacerbation. L-canavanine is found in minor amounts in the seed and sprouts of alfalfa but not in the mature plant. Alfalfa tablets from two manufacturers tested negative for L-canavanine (Brown 2000; Farnsworth 1995; Malinow et al. 1982; Whittam et al. 1995).

III. Pharmacology and Pharmacokinetics

Human Pharmacological Studies

No relevant human pharmacological studies were identified.

Animal Pharmacological Studies

In cynomolgus macaques fed diets containing 45% ground alfalfa seed, three of five monkeys developed autoimmune hemolytic anemia associated with systemic findings. All three of those animals developed significant antinuclear antibodies along with elevated antibodies to double-stranded DNA. During a 20-month observational period, one of the animals developed clinical signs consistent with the appearance of systemic lupus erythematosus temporally related to the ingestion of alfalfa seed. Withdrawal of alfalfa seed from the diet was associated with the normalization of immunologic parameters and serum complements. Antinuclear antibodies and antidouble-stranded DNA antibodies persisted for 2 years (Montanaro and Bardana 1991).

Hematological and serological abnormalities similar to those observed in human systemic lupus erythematosus (SLE) developed in cynomolgus macaques fed a diet containing 40% alfalfa sprouts. In animals that developed hematological and serological abnormalities, placement on a diet containing 1% of the compound L-canavanine sulfate for 4 weeks reactivated previously observed abnormalities associated with ingestion of alfalfa sprouts (Malinow et al. 1982).

In Vitro Pharmacological Studies

In an estrogen-dependent MCF-7 breast cancer cell proliferation assay, a methanol extract of alfalfa sprouts exhibited binding activity to estrogen receptor (ER) β with an IC_{50} value of 198 μg/ml. The pure estrogen antagonist, ICI 182,780, suppressed cell proliferation induced by the extracts. The ER-β binding activity was significantly less than that of kudzu root, red clover blossom, soy beans, and red clover sprouts. No activity of the alfalfa sprout extract was observed on ER-α (Boue et al. 2003).

The average content of the compound coumestrol in alfalfa tablets sold as supplements was 20 to 190 ppm (average of 99 ppm) (Elakovich and Hampton 1984). Other reports indicate that alfalfa meal contains 25 to 65 ppm or 80 to 100 ppm of the compound coumestrol (Livingston et al. 1961; Saloniemi et al. 1995). The amount of "estrogenic substances" in alfalfa meal was reported as 1.1 to 1.8 μg/g, which was greater than that of soybean meal (Kato et al. 2004).

An aqueous extract of alfalfa inhibited ADP- and collagen-induced human platelet aggregation but had no effect on arachadonic acid- or thrombin-induced platelet aggregation. Alfalfa inhibited thromboxane B_2 synthesis induced by ADP or collagen, and whole blood aggregation induced by collagen (Pierre et al. 2005).

The compound L-canavanine demonstrated dose-related effects on human immunoregulatory cells, including abrogation of concanavalin A–induced suppressor cell function and diminution of the mitogenic response to both phytohemagglutinin and concanavalin A but not to pokeweed mitogen (Alcocer-Varela et al. 1985).

IV. Pregnancy and Lactation

No information on the safety of alfalfa during pregnancy or lactation was identified.

V. Toxicity Studies

Subchronic Toxicity

No adverse effects in any of the observed parameters, including changes in liver, spleen, stomach, and kidney, were seen in rats fed diets containing 1 or 2% alfalfa saponins for 6 months (Malinow et al. 1981).

Genotoxicity

In the Ames test for mutagenicity with *Salmonella typhimurium* strains TA98 and TA100, no mutagenic activity of an acetone extract of alfalfa was observed with or without metabolic activation (White et al. 1983).

Literature Cited

Alcocer-Varela, J., A. Iglesias, L. Llorente, and D. Alarcon-Segovia. 1985. Effects of L-canavanine on T cells may explain the induction of systemic lupus erythematosus by alfalfa. *Arthritis Rheum.* 28(1):52-57.

Booth, N.L., D. Nikolic, R.B. van Breemen, et al. 2004. Confusion regarding anticoagulant coumarins in dietary supplements. *Clin. Pharmacol. Ther* .76(6):511-516.

Boue, S.M., T.E. Wiese, S. Nehls, et al. 2003. Evaluation of the estrogenic effects of legume extracts containing phytoestrogens. *J. Agric. Food Chem.* 51(8):2193-2199.

Brown, A.C. 2000. Lupus erythematosus and nutrition: A review of the literature. *J. Ren. Nutr.* 10(4):170-183.

Elakovich, S.D., and J.M. Hampton. 1984. Analysis of coumestrol, a phytoestrogen, in alfalfa tablets sold for human consumption. *J. Agric. Food Chem.* 32(1):173-175.

Farnsworth, N.R. 1995. Alfalfa pills and autoimmune diseases. *Am. J. Clin. Nutr.* 62(5):1026-1028.

Kato, H., T. Iwata, Y. Katsu, et al. 2004. Evaluation of estrogenic activity in diets for experimental animals using *in vitro* assay. *J. Agric. Food Chem.* 52(5):1410-1414.

Livingston, A.L., E.M. Bickoff, J. Guggolz, and C.R. Thompson. 1961. Alfalfa estrogens, quantitative determination of coumestrol in fresh and dried alfalfa. *J. Agric. Food Chem.* 9(2):135-137.

Malinow, M.R., E.J. Bardana, Jr., B. Pirofsky, S. Craig, and P. McLaughlin. 1982. Systemic lupus erythematosus-like syndrome in monkeys fed alfalfa sprouts: Role of a nonprotein amino acid. *Science* 216(4544):415-417.

Malinow, M.R., W.P. McNulty, and P. McLaughlin. 1981. The toxicity of alfalfa saponins in rats. *Food Cosmet. Toxicol.* 19(4):443-445.

Montanaro, A., and E.J. Bardana, Jr. 1991. Dietary amino acid-induced systemic lupus erythematosus. *Rheum. Dis. Clin. North Am.* 17(2):323-332.

Pierre, S., L. Crosbie, and A.K. Duttaroy. 2005. Inhibitory effect of aqueous extracts of some herbs on human platelet aggregation in vitro. *Platelets* 16(8):469-473.

Roberts, J.L., and J.A. Hayashi. 1983. Exacerbation of SLE associated with alfalfa ingestion. *N. Engl. J. Med.* 308(22):1361.

Saloniemi, H., K. Wahala, P. Nykanen-Kurki, K. Kallela, and I. Saastamoinen. 1995. Phytoestrogen content and estrogenic effect of legume fodder. *Proc. Soc. Exp. Biol. Med.* 208(1):13-17.

White, R.D., P.H. Krumperman, P.R. Cheeke, and D.R. Buhler. 1983. An evaluation of acetone extracts from six plants in the Ames mutagenicity test. *Toxicol. Lett.* 15(1):25-31.

Whittam, J., C. Jensen, and T. Hudson. 1995. Alfalfa, vitamin E, and autoimmune disorders. *Am. J. Clin. Nutr.* 62(5):1025-1026.

Melia azedarach L. — Meliaceae

SCN: melia
Syn: *Melia toosendan* Siebold & Zucc.
AN: *mahanimba*
PN: *chuan lian zi*
OCN: Chinaberry; Chinatree; pagoda tree
Part: fruit

Quick Reference Summary

Safety Class: 3
Interaction Class: A

Contraindications

Not for use except under the supervision of an expert qualified in the appropriate use of this substance (Bensky et al. 2004; Chen and Chen 2004).

Other Precautions

None known.

Drug and Supplement Interactions

None known.

Adverse Events and Side Effects

The entire melia plant is reported to be toxic, with the fruit listed as the most toxic part (Chen and Chen 2004). Eating six to eight fresh melia fruits may cause dizziness, vomiting, diarrhea, difficulty breathing, palpitations, and muscle spasms (Chen and Chen 2004). Onset of symptoms generally occurs 4 to 6 hours after ingestion, although they may appear as soon as 30 minutes (Chen and Chen 2004; Phau 2007). More severe toxicity may include symptoms of numbness, muscle cramps, tremors, arrhythmia, respiratory distress, convulsions, and unconsciousness or altered consciousness (Bensky et al. 2004; Chen and Chen 2004).

Pharmacological Considerations

None known.

Pregnancy and Lactation

No information on the safety of melia fruit in pregnancy or lactation was identified. While this review did not identify any concerns for use while pregnant or nursing, safety has not been conclusively established, and this substance is not recommended for use except under the supervision of an expert qualified in its appropriate use.

Editors' Notes

The cautions and concerns stated here are for raw, unprocessed melia fruit. Heat treated melia fruit is the primary article in trade, however, and processing by dry-frying is reported to minimize the toxicity of the fruit (Chen and Chen 2004).

Although *M. toosendan* is considered to be a taxonomic synonym for *M. azerdarach* (McGuffin et al. 2000), contemporary references list *M. azedarach* fruit as an adulterant of or potentially inappropriate substitute for *M. toosendan* fruit and regard the former as more toxic (Bensky et al. 2004; Chen and Chen 2004).

Review Details

I. Drug and Supplement Interactions

Clinical Trials of Drug or Supplement Interactions

No clinical trials of drug or supplement interactions were identified.

Case Reports of Suspected Drug or Supplement Interactions

No case reports of suspected drug or supplement interactions were identified.

Animal Trials of Drug or Supplement Interactions

No animal trials of drug or supplement interactions were identified.

II. Adverse Events

Case Reports of Adverse Events

Eating six to eight fresh melia fruits may cause dizziness, vomiting, diarrhea, difficulty breathing, palpitations, and muscle spasms (Chen and Chen 2004). Adverse reactions are reported to begin 4 to 6 hours after ingestion, although they may appear as soon as 30 minutes after ingestion (Chen and Chen 2004). More severe toxicity may include symptoms of numbness, muscle cramps, tremors, arrhythmia, respiratory distress, convulsions, and unconsciousness or altered consciousness (Bensky et al. 2004; Chen and Chen 2004).

In addition to the above-listed symptoms, overdose of melia fruit may result in jaundice, toxic hepatitis, arrhythmia, damage to the heart muscle, nosebleeds, bloody stools or urine, low blood pressure, and shock (Bensky et al. 2004).

Melia fruit poisoning has been reported in a number of animal species, including pigs, dogs, goats, cattle, poultry, rats, and guinea pigs. Symptoms of poisoning in animals are reported to appear 2 to 4 hours after ingestion and include nausea, vomiting, diarrhea, constipation, colic, convulsions, ataxia, depression, respiratory distress, muscle impairment, and coma (Hare 1998; Hare et al. 1997; Hothi et al. 1976). Muscle tremors, kicking, and respiratory distress were reported in ostriches that consumed melia fruit. Postmortem examination of a 7-month-old ostrich revealed a hemorrhage in the gut and enlargement of the liver (Cooper 2007).

III. Pharmacology and Pharmacokinetics

Human Pharmacological Studies

No relevant human pharmacological studies were identified.

Animal Pharmacological Studies

No relevant animal pharmacological studies were identified.

In Vitro Pharmacological Studies

No relevant in vitro pharmacological studies were identified.

IV. Pregnancy and Lactation

No information on the safety of melia fruit in pregnancy or lactation was identified.

V. Toxicity Studies

Acute Toxicity

The LD_{50} of melia fruit extract in rats is 1.03 g/kg after intraperitoneal administration but could not be determined at doses up to 16 g/kg after oral administration (Carpinella et al. 1999).

The LD_{50} of a hydroalcoholic extract of melia fruit is 700 mg/kg in mice and 925 mg/kg in rats after intravenous administration. After oral administration of the same extract, no toxic effects were reported at doses up to 1.5 g/kg (Zakir Ur et al. 1991).

In calves orally administered melia fruit at doses of 5 to 30 g/kg, half of the animals that received 15 g/kg and all animals that received 25 or 30 g/kg died. Clinical signs, which appeared 4 to 24 hours after administration, included depression, ruminal stasis, anorexia, diarrhea, incoordination, muscle tremors, difficulty standing, sternal recumbence, hypothermia, and dyspnea. Serum levels of aspartate aminotransferase (AST) and creatine phosphokinase (CPK) were elevated (Mendez et al. 2002). Feeding studies in pigs and sheep indicated that the toxic dose of melia fruit was approximately 5 g/kg (Hare 1998; Kingsbury 1964).

The LD_{50} of orally administered toosendanin is 244 mg/kg in mice, 120 mg/kg in rats, and 3.5 mg/kg in cats (Chen and Chen 2004).

M

Literature Cited

Bensky, D., S. Clavey, and E. Stöger. 2004. *Chinese herbal medicine: Materia medica*. 3rd ed. Seattle: Eastland Press.

Carpinella, M.C., S. Fulginiti, S. Britos, et al. 1999. Acute toxicity of fruit extracts from *Melia azedarach* L. in rats. *Rev. Toxicol.* 16(1):22-24.

Chen, J.K., and T.T. Chen. 2004. *Chinese medical herbology and pharmacology*. City of Industry, CA: Art of Medicine Press.

Cooper, R.G. 2007. Poisoning in ostriches following ingestion of toxic plants—Field observations. *Trop. Anim. Health Prod.* 39(6):439-442.

del Mendez, M.C., F. Elias, M. Aragao, E.J. Gimeno, and F. Riet-Correa. 2002. Intoxication of cattle by the fruits of *Melia azedarach*. *Vet. Hum. Toxicol.* 44(3):145-148.

Hare, W.R. 1998. Chinaberry (*Melia azedarach*) poisoning in animals. In Garland, T. and C. Barr, eds. *Toxic plants and other natural toxicants*. New York: CAB International.

Hare, W.R., H. Schutzman, B.R. Lee, and M.W. Knight. 1997. Chinaberry poisoning in two dogs. *J. Am. Vet. Med. Assoc.* 210(11):1638-1640.

Hothi, D.S., B. Singh, M.S. Kwatra, and R.S. Chawla. 1976. A note on the comparative toxicity of *Melia azedarach* (Dhrek) berries to piglets, buffalo calves, rabbits and fowls. *J. Res. Punjab Agric. Univ.* 13(2):232-234.

Kingsbury, J.M. 1964. *Poisonous plants of the United States and Canada*. Englewood Cliffs, NJ: Prentice-Hall.

McGuffin, M., J. Kartesz, A. Leung, and A.O. Tucker. 2000. *Herbs of commerce*. 2nd ed. Silver Spring, MD: American Herbal Products Association.

Phau, D.H., W-J. Tsai, J. Ger, J-F Deng, and C-C Yang. 2007. Human *Melia azedarach* poisoning. *Clin Tox* 46:1067-1070.

Zakir Ur, R., S. Ahmad, S. Qureshi, R. Atiq Ur, and Y. Badar. 1991. Toxicological studies of *Melia azedarach* L. flowers and berries. *Pak. J. Pharm. Sci.* 4(2):153-158.

Melia azedarach L. — Meliaceae

SCN: melia
Syn: *Melia toosendan* Siebold & Zucc.
AN: *mahanimba*
PN: *ku lian pi* (stem and root bark)
OCN: Chinaberry; Chinatree; pagoda tree
Part: bark, root bark

Quick Reference Summary

Safety Class: 3
Interaction Class: A

Contraindications

Not for use except under the supervision of an expert qualified in the appropriate use of this substance (Bensky et al. 2004; Chen and Chen 2004; Felter and Lloyd 1898).

Other Precautions

None known.

Drug and Supplement Interactions

None known.

Notice

Emetic (Felter and Lloyd 1898); *see* Appendix 2.

Adverse Events and Side Effects

Melia bark or root bark may cause side effects within the normal dosage range including dizziness, headache, nausea, vomiting, drowsiness, and abdominal pain. These symptoms disappear spontaneously when use of the herb is discontinued (Bensky et al. 2004; Chen and Chen 2004).

Allergic skin reactions to melia bark or root bark have been reported (Bensky et al. 2004).

Pharmacological Considerations

None known.

Pregnancy and Lactation

A reference text on traditional Chinese medicine indicates that melia bark or root bark should not be used during pregnancy (Bensky et al. 2004). Animal studies have indicated that melia root significantly reduces implantation and causes an increase in fetal resorptions (Keshri et al. 2003, 2004).

No information on the safety of melia bark or root bark in lactation was identified in the scientific or traditional literature. While this review did not identify any concerns for use while nursing, safety has not been conclusively established, and this substance is not recommended for use except under the supervision of an expert qualified in its appropriate use.

Review Details

I. Drug and Supplement Interactions

Clinical Trials of Drug or Supplement Interactions

No clinical trials of drug or supplement interactions were identified.

Case Reports of Suspected Drug or Supplement Interactions

No case reports of suspected drug or supplement interactions were identified.

Animal Trials of Drug or Supplement Interactions

No animal trials of drug or supplement interactions were identified.

II. Adverse Events

Case Reports of Adverse Events

Melia bark or root bark may cause side effects within the normal dosage range including dizziness, headache, nausea, vomiting, drowsiness, and abdominal pain. These symptoms will disappear spontaneously when use of the herb is discontinued (Bensky et al. 2004; Chen and Chen 2004; Felter and Lloyd 1898).

Overdose of melia bark or root bark may cause severe vomiting, diarrhea, abdominal distention, respiratory paralysis, palpitations, tachycardia, and blue coloration of the lips (Bensky et al. 2004; Chen and Chen 2004; Felter and Lloyd 1898).

In cases of severe overdose, melia bark or root bark may cause gastrointestinal bleeding, jaundice, hepatomegaly, elevated liver enzymes, toxic hepatitis, visual impairment, and respiratory or circulatory failure (Bensky et al. 2004; Chen and Chen 2004).

Melia bark poisoning was reported in a couple (77-year-old man and 66-year-old woman) who had been drinking an unspecified amount of melia root bark boiled in water daily for 3 weeks. Both the man and woman noted drooling after ingestion of larger amounts (~500 or 1000 ml of extract), later experiencing muscle weakness, ptosis, and elevated liver enzymes. The couple recovered after several days of supportive management (Phua et al. 2008).

An 18-year-old woman became comatose after ingesting an aqueous extract of an unspecified amount of melia bark. After remaining in a comatose state for 3 days, with doctors unsuccessfully attempting treatment, the woman died (Toh 1969).

Allergic skin reactions to melia bark or root bark have been reported (Bensky et al. 2004).

III. Pharmacology and Pharmacokinetics

Human Pharmacological Studies

No relevant human pharmacological studies were identified.

Animal Pharmacological Studies

No relevant animal pharmacological studies were identified.

In Vitro Pharmacological Studies

No estrogenic activity of an ethanol extract of melia bark was observed in a recombinant yeast system with a human estrogen receptor expression plasmid and a reporter plasmid (Kim et al. 2008).

IV. Pregnancy and Lactation

A 60 to 75% reduction in implantation was observed in rats orally administered 250 or 500 mg/kg of an ethanol extract of melia root bark daily on days 1 to 10 after mating. In animals that became pregnant, there was also a significant reduction in the number of implantations, and all implantations exhibited signs of resorption (Keshri et al. 2003).

An 83% reduction in implantation was observed in rats orally administered 250 mg/kg of a chloroform extract of melia root bark daily on days 1 to 7 after mating (Keshri et al. 2004).

A reference text on traditional Chinese medicine indicates that melia bark or root bark should not be used during pregnancy (Bensky et al. 2004).

No information on the safety of melia bark or root bark in lactation was identified.

V. Toxicity Studies

Acute Toxicity

The LD_{50} of orally administered toosendanin is 244 mg/kg in mice, 120 mg/kg in rats, and 3.5 mg/kg in cats (Chen and Chen 2004).

M

Literature Cited

Bensky, D., S. Clavey, and E. Stöger. 2004. *Chinese herbal medicine: Materia medica*. 3rd ed. Seattle: Eastland Press.

Chen, J.K., and T.T. Chen. 2004. *Chinese medical herbology and pharmacology*. City of Industry, CA: Art of Medicine Press.

Felter, H.W., and J.U. Lloyd. 1898. *King's American dispensatory*. 18th ed., 3rd rev. 2 vols. Cincinnati: Ohio Valley Co.

Keshri, G., M. Bajpai, V. Lakshmi, B.S. Setty, and G. Gupta. 2004. Role of energy metabolism in the pregnancy interceptive action of *Ferula assafoetida* and *Melia azedarach* extracts in rat. *Contraception* 70(5):429-432.

Keshri, G., V. Lakshmi, and M.M. Singh. 2003. Pregnancy interceptive activity of *Melia azedarach* Linn. in adult female Sprague-Dawley rats. *Contraception* 68(4):303-306.

Kim, I.G., S.C. Kang, K.C. Kim, E.S. Choung, and O.P. Zee. 2008. Screening of estrogenic and antiestrogenic activities from medicinal plants. *Environ. Toxicol. Pharmacol.* 25(1):75-82.

Phua, D.H., W.J. Tsai, J. Ger, J.F. Deng, and C.C. Yang. 2008. Human *Melia azedarach* poisoning. *Clin. Toxicol.* 46(10):1067-1070.

Toh, K.K. 1969. *Melia azedarach* poisoning. *Singapore Med. J.* 10(1):24-28.

Melissa officinalis L. — Lamiaceae

SCN: lemon balm
OCN: balm; bee balm; melissa; melissa balm

Part: leaf

Quick Reference Summary

Safety Class: 1
Interaction Class: A

Contraindications

None known.

Other Precautions

None known.

Drug and Supplement Interactions

None known.

Adverse Events and Side Effects

Clinical trials of orally and topically administered lemon balm preparations have demonstrated that they have been generally well tolerated with few adverse events (Akhondzadeh et al. 2003; Buchner et al. 1974; Kennedy et al. 2002, 2004; Koytchev et al. 1999; Wolbling and Leonhardt 1994).

Pharmacological Considerations

Initial animal and in vitro studies indicate that lemon balm may affect thyroid hormone levels and inhibit binding of thyroid-stimulating hormone (TSH) to TSH receptors (Auf'mkolk et al. 1985; Santini et al. 2003; Sourgens et al. 1982), although no human cases of thyroid effects have been reported.

Pregnancy and Lactation

No information on the safety of lemon balm in pregnancy or lactation was identified in the scientific or traditional literature. Although this review did not identify any concerns for use while pregnant or nursing, safety has not been conclusively established.

Review Details

M

I. Drug and Supplement Interactions

Clinical Trials of Drug or Supplement Interactions

No clinical trials of drug or supplement interactions were identified.

Case Reports of Suspected Drug or Supplement Interactions

No case reports of suspected drug or supplement interactions were identified.

Animal Trials of Drug or Supplement Interactions

Animal studies indicated that extracts of lemon balm may increase the sedative activity of pentobarbital and hexobarbital (Soulimani et al. 1991; Wagner and Sprinkmeyer 1973).

II. Adverse Events

Adverse Events Reported in Clinical Trials

Reviews (Brendler et al. 2006; ESCOP 2003) of clinical trials of lemon balm indicated that both internal (Akhondzadeh et al. 2003; Buchner et al. 1974; Kennedy et al. 2002, 2004) and topical administration (Koytchev et al. 1999; Wolbling and Leonhardt 1994) are generally very well tolerated. Headache and heart palpitations were reported in one early clinical trial of lemon balm (Buchner et al. 1974). Local reddening, burning sensation, paresthesia, residual pigmentation, and dermal irritation have been reported in clinical studies of topically applied lemon balm products (Wolbling and Leonhardt 1994; Wolbling and Milbradt 1984).

Case Reports of Adverse Events

Contact dermatitis was reported after topical use of a cosmetic containing lemon balm. Patch testing confirmed sensitivity to lemon balm and to other ingredients in the cosmetic (West and Maibach 1995).

III. Pharmacology and Pharmacokinetics

Human Pharmacological Studies

In healthy volunteers administered single doses of up to 900 mg of a standardized lemon balm extract, self-rated "calmness" was elevated while "alertness" was significantly reduced (Kennedy et al. 2002).

Animal Pharmacological Studies

A very weak sensitizing effect was observed in guinea pigs topically treated with lemon balm (Hausen and Schulze 1986).

Intravenous administration of 25 mg/kg of a freeze-dried extract of lemon balm to normal rats resulted in reduced serum and pituitary gland levels of TSH. In rats with goiter, no change in TSH levels was observed after administration of lemon balm. No changes in serum prolactin were observed in either the normal or goitrous rats (Sourgens et al. 1982).

In Vitro Pharmacological Studies

Incubation of thyroid-stimulating immunoglobulin G (IgG) found in the blood of patients with Graves' disease

(IgG resembles TSH in the ability of IgG to bind to the thyroid plasma membrane, probably at the TSH receptor) with a freeze-dried lemon balm extract decreased the TSH-binding activity of IgG in a dose-dependent manner (Auf'mkolk et al. 1985).

In Chinese hamster ovary cells transfected with the recombinant human TSH receptor, lemon balm produced a dose-dependent inhibition of TSH-stimulated adenylate cyclase activity, inhibited the cAMP production stimulated by TSH receptor antibody, and produced a significant inhibition of TSH binding to its receptor and of antibody binding to TSH (Santini et al. 2003).

IV. Pregnancy and Lactation

No information on the safety of lemon balm during pregnancy or lactation was identified.

V. Toxicity Studies

Acute Toxicity

No adverse effects were observed after intravenous injection of 25 mg/kg of lemon balm extract to rats (Sourgens et al. 1982).

Genotoxicity

No mutagenic effects of alcoholic extracts of lemon balm were observed in the Ames mutagenicity test with *Salmonella typhimurium* strains TA98 or TA100 with or without metabolic activation, or in *Aspergillus nidulans* using a plate incorporation assay (Ramos Ruiz et al. 1996; Schimmer et al. 1994). The aqueous extract also showed no genotoxic activity in *Aspergillus nidulans* using a plate incorporation assay (Schimmer et al. 1994).

Literature Cited

Akhondzadeh, S., M. Noroozian, M. Mohammadi, et al. 2003. *Melissa officinalis* extract in the treatment of patients with mild to moderate Alzheimer's disease: A double blind, randomised, placebo controlled trial. *J. Neurol. Neurosurg. Psychiatr.* 74(7):863-866.

Auf'mkolk, M., J.C. Ingbar, K. Kubota, S.M. Amir, and S.H. Ingbar. 1985. Extracts and auto-oxidized constituents of certain plants inhibit the receptor-binding and the biological activity of Graves' immunoglobulins. *Endocrinology* 116(5):1687-1693.

Brendler, T., J. Gruenwald, B. Kligler, et al. 2006. Lemon balm (*Melissa officinalis* L.): An evidence-based systematic review by the Natural Standard Research Collaboration. *J. Herbal Pharmacother.* 5(4):71-114.

Buchner, K.H., H. Hellings, M. Huber, et al. 1974. Double blind study as evidence of the therapeutic effect of Melissengeist on psycho-vegetative syndromes. *Med. Klin.* 69(23):1032-1036.

ESCOP. 2003. *ESCOP monographs: The scientific foundation for herbal medicinal products.* 2nd ed. Exeter, UK: European Scientific Cooperative on Phytotherapy.

Hausen, B.M., and R. Schulze. 1986. Comparative studies of the sensitizing capacity of drugs used in herpes simplex. *Derm. Beruf Umwelt* 34(6):163-170.

Kennedy, D.O., W. Little, and A.B. Scholey. 2004. Attenuation of laboratory-induced stress in humans after acute administration of *Melissa officinalis* (lemon balm). *Psychosom. Med.* 66(4):607-613.

Kennedy, D.O., A.B. Scholey, N.T.J. Tildesley, E.K. Perry, and K.A. Wesnes. 2002. Modulation of mood and cognitive performance following acute administration of *Melissa officinalis* (lemon balm). *Pharmacol. Biochem. Behav.* 72(4):953.

Koytchev, R., R.G. Alken, and S. Dundarov. 1999. Balm mint extract (Lo-701) for topical treatment of recurring herpes labialis. *Phytomedicine* 6(4):225-230.

Ramos Ruiz, A., R.A. De la Torre, N. Alonso, et al. 1996. Screening of medicinal plants for induction of somatic segregation activity in *Aspergillus nidulans. J. Ethnopharmacol.* 52(3):123-127.

Santini, F., P. Vitti, G. Ceccarini, et al. 2003. In vitro assay of thyroid disruptors affecting TSH-stimulated adenylate cyclase activity. *J. Endocrinol. Invest.* 26(10):950-955.

Schimmer, O., A. Kruger, H. Paulini, and F. Haefele. 1994. An evaluation of 55 commercial plant-extracts in the Ames mutagenicity test. *Pharmazie* 49(6):448-451.

Soulimani, R., J. Fleurentin, F. Mortier, et al. 1991. Neurotropic action of the hydroalcoholic extract of *Melissa officinalis* in the mouse. *Planta Med.* 57(2):105-109.

Sourgens, H., H. Winterhoff, H.G. Gumbinger, and F.H. Kemper. 1982. Anti-hormonal effects of plant-extracts: TSH-suppressing and prolactin-suppressing properties of *Lithospermum officinale* and other plants. *Planta Med.* 45(2):78-86.

Wagner, H., and L. Sprinkmeyer. 1973. Pharmacological effect of balm spirit. *Dtsch. Apoth. Ztg.* 113:1159-66.

West, I., and H.I. Maibach. 1995. Contact urticaria syndrome from multiple cosmetic components. *Contact Dermat.* 32(2):121.

Wolbling, R., and K. Leonhardt. 1994. Local therapy of herpes simplex with dried extract from *M. officinalis. Phytomedicine* 1:25-31.

Wolbling, R., and R. Milbradt. 1984. Clinical manifestations and treatment of herpes simplex infections. *Therapiewoche* 34:1193-1200.

Mentha piperita L. — Lamiaceae

SCN: peppermint

Part: leaf

Quick Reference Summary

Safety Class: 1
Interaction Class: A

Contraindications

None known.

Other Precautions

Use with caution in persons with gastroesophageal reflux or with a hiatal hernia, as peppermint may decrease esophageal sphincter pressure (Brinker 2001; Hiki et al. 2003; Hills and Aaronson 1991; Mills and Bone 2005).

Use with caution in persons with gastrointestinal ulcers or significant gastrointestinal inflammation (Mills and Bone 2005).

Drug and Supplement Interactions

None known.

Adverse Events and Side Effects

A review of botanicals used in the treatment of dyspepsia indicated that peppermint leaf had an "encouraging safety profile" (Thompson Coon and Ernst 2002).

Pharmacological Concerns

Peppermint essential oil has been shown to relax the smooth muscles of the gastrointestinal tract (McKay and Blumberg 2006), which may exacerbate gastroesophageal reflux or hiatal hernias.

Coadministration of peppermint tea and iron-fortified bread reduced absorption of the iron (Hurrell et al. 1999).

Pregnancy and Lactation

Limited information on the safety of peppermint leaf in pregnancy or lactation was identified in the scientific and traditional literature. Although this review did not identify any concerns for use while pregnant or nursing, safety has not been conclusively established.

M

Review Details

I. Drug and Supplement Interactions

Clinical Trials of Drug or Supplement Interactions

No human trials of drug or supplement interactions were identified.

Case Reports of Suspected Drug or Supplement Interactions

No case reports of suspected drug or supplement interactions were identified.

Animal Trials of Drug or Supplement Interactions

No animal trials of drug or supplement interactions were identified.

II. Adverse Events

Case Reports of Adverse Events

No case reports of adverse events with peppermint leaf were identified.

III. Pharmacology and Pharmacokinetics

Human Pharmacological Studies

Coadministration of iron-fortified bread and peppermint tea reduced absorption of the iron by 84%. This observed inhibition compared to a similar effect associated with black tea at a rate of 79–94% (Hurrell et al. 1999).

Animal Pharmacological Studies

Administration of peppermint leaf tea (ad libitum) to male rats for 30 days caused an increase in follicle-stimulating hormone and luteinizing hormone and a decrease in testosterone, and a decrease in iron absorption was observed in rats administered the tea (Akdogan et al. 2004).

A reduction in activity of the drug-metabolizing isoenzymes CYP1A2 and CYP2E was observed in rats administered a 2% tea of peppermint leaf for 4 weeks (Maliakal and Wanwimolruk 2001).

In Vitro Pharmacological Studies

Aqueous extracts of peppermint leaf showed a dose-dependent relaxation effect on isolated rabbit duodenum (Mahmood et al. 2003).

IV. Pregnancy and Lactation

See *Mentha piperita* essential oil entry for information on the safety of peppermint during pregnancy and lactation.

V. Toxicity Studies

See *Mentha piperita* essential oil entry for additional toxicity studies on peppermint.

Short-Term Toxicity

In rats administered peppermint leaf tea for 30 days as the sole source of drinking water, no significant histopathological changes were observed in the kidneys (Akdogan et al. 2003).

Literature Cited

Akdogan, M., I. Kilinc, M. Oncu, E. Karaoz, and N. Delibas. 2003. Investigation of biochemical and histopathological effects of *Mentha piperita* L. and *Mentha spicata* L. on kidney tissue in rats. *Hum. Exp. Toxicol.* 22(4):213-219.

Akdogan, M., M. Ozguner, A. Kocak, M. Oncu, and E. Cicek. 2004. Effects of peppermint teas on plasma testosterone, follicle-stimulating hormone, and luteinizing hormone levels and testicular tissue in rats. *Urology* 64(2):394-398.

Brinker, F. 2001. *Herb contraindications and drug interactions.* 3rd ed. Sandy, OR: Eclectic Medical Publications.

Hiki, N., H. Kurosaka, Y. Tatsutomi, et al. 2003. Peppermint oil reduces gastric spasm during upper endoscopy: A randomized, double-blind, double-dummy controlled trial. *Gastrointest. Endosc.* 57(4):475-482.

Hills, J.M., and P.I. Aaronson. 1991. The mechanism of action of peppermint oil on gastrointestinal smooth muscle. An analysis using patch clamp electrophysiology and isolated tissue pharmacology in rabbit and guinea pig. *Gastroenterology* 101(1):55-65.

Hurrell, R.F., M. Reddy, and J.D. Cook. 1999. Inhibition of non-haem iron absorption in man by polyphenolic-containing beverages. *Br. J. Nutr.* 81(4):289-295.

Mahmood, S., N. Abbas, and R. Rojas. 2003. Effects of aqueous extracts of peppermint, fennel, dill and cumin on isolated rabbit duodenum. *U. Aden J. Nat. Appl. Sci.* 7:377-383.

Maliakal, P.P., and S. Wanwimolruk. 2001. Effect of herbal teas on hepatic drug metabolizing enzymes in rats. *J. Pharm. Pharmacol.* 53(10):1323-1239.

McKay, D., and J. Blumberg. 2006. A review of the bioactivity and potential health benefits of peppermint tea (*Mentha piperita* L.). *Phytother. Res.* 20(8):619-633.

Mills, S., and K. Bone. 2005. *The essential guide to herbal safety.* St. Louis: Elsevier.

Thompson Coon, J., and E. Ernst. 2002. Systematic review: Herbal medicinal products for non-ulcer dyspepsia. *Aliment. Pharmacol. Ther.* 16(10):1689-1899.

Mentha piperita L.

Lamiaceae

SCN: peppermint

Part: leaf essential oil

Quick Reference Summary

Safety Class: 2b
Interaction Class: A

Contraindications

Not for use in pregnancy except under the supervision of a qualified healthcare practitioner (Tiran and Mack 2000).

Other Precautions

Peppermint leaf essential oil may cause heartburn in sensitive individuals (Nash et al. 1986; Somerville et al. 1984). Use is cautioned in persons with gastroesophageal reflux or with a hiatal hernia, as peppermint may decrease esophageal sphincter pressure (Brinker 2001; Hiki et al. 2003; Hills and Aaronson 1991; Mills and Bone 2005).

Use is cautioned in persons with gastrointestinal ulcers or significant gastrointestinal inflammation (Mills and Bone 2005).

Enteric-coated peppermint leaf essential oil capsules may cause anal burning, especially in patients with diarrhea (Grigoleit and Grigoleit 2005; Mills and Bone 2005).

Drug and Supplement Interactions

None known.

Standard Dose

The standard dose is 180–350 mg in enteric-coated capsules, three times daily (Grigoleit and Grigoleit 2005; Pittler and Ernst 1998).

Adverse Events and Side Effects

A review of botanicals used in the treatment of dyspepsia indicated that peppermint leaf oil had an "encouraging safety profile" (Thompson Coon and Ernst 2002).

Side effects reported in clinical trials of peppermint leaf essential oil include heartburn, and anal or perianal burning or discomfort (Grigoleit and Grigoleit 2005; Nash et al. 1986; Pittler and Ernst 1998; Somerville et al. 1984).

Allergic reactions, including contact dermatitis, to peppermint leaf essential oil have been reported (Foti et al. 2003; Morton et al. 1995; Sainio and Kanerva 1995; Wilkinson and Beck 1994).

Cases of mucosal ulcers have been associated with the oral ingestion of products containing peppermint, including mouthwashes and candies (Moghadam et al. 1999; Rogers and Pahor 1995).

Pharmacological Considerations

Peppermint leaf essential oil has been shown to slow intestinal transit, which may slow the absorption rate or increase the total absorption of coadministered drugs (Goerg and Spilker 2003).

Peppermint leaf essential oil has been shown to relax the smooth muscles of the gastrointestinal tract (McKay and Blumberg 2006), which may exacerbate gastroesophageal reflux or hiatal hernias.

Preliminary human and animal data suggest that large doses of peppermint leaf essential oil may inhibit the drug-metabolizing isoenzyme CYP3A4, leading to increased plasma levels of drugs metabolized by that isoenzyme (Dresser et al. 2002; Wacher et al. 2002).

Pregnancy and Lactation

Due to the highly concentrated nature of peppermint leaf essential oil, until further safety data is available, internal use during pregnancy should only be under the supervision of a qualified healthcare practitioner (Tiran and Mack 2000).

A study on the use of topical peppermint gel applied to nipples of nursing mothers indicated that no change occurred in infant breast-feeding behavior (Sayyah Melli et al. 2007).

Review Details

I. Drug and Supplement Interactions

Clinical Trials of Drug or Supplement Interactions

No clinical trials of drug or supplement interactions were identified.

Case Reports of Suspected Drug or Supplement Interactions

No case reports of suspected drug or supplement interactions were identified.

Animal Trials of Drug or Supplement Interactions

Coadministration of cyclosporine (25 mg/kg) and peppermint leaf oil (100 mg/kg) in rats significantly increased serum levels of cyclosporine (Wacher et al. 2002).

M

II. Adverse Events

Adverse Events Reported in Clinical Trials

In a systematic review of botanical products for non-ulcer dyspepsia, peppermint leaf essential oil was noted as having an "encouraging safety profile" (Thompson Coon and Ernst 2002).

Enteric-coated peppermint leaf essential oil capsules may produce anal burning in patients with diarrhea, due to excretion of peppermint oil. Use of enteric-coated capsules has been associated with adverse effects such as rashes, headache, bradycardia, muscle tremors, and ataxia (Mills and Bone 2005). Non-enteric-coated preparations of peppermint leaf essential oil have been reported to cause heartburn in sensitive individuals (Nash et al. 1986; Somerville et al. 1984).

A meta-analysis of trials on peppermint leaf essential oil for irritable bowel syndrome indicated that five of eight trials reported adverse effects including heartburn, perianal burning, blurred vision, nausea, and vomiting while taking the peppermint oil. The effects were observed in 11 to 36% (mean 20%) of the patients studied. The occurrence of adverse events in placebo groups was not reported (Pittler and Ernst 1998).

A review of 16 clinical trials on 180–200 mg enteric-coated peppermint leaf essential oil capsules in the treatment of irritable bowel syndrome indicated that adverse events reported in trials were generally mild and transient, but very specific, with peppermint leaf essential oil causing heartburn and anal or perianal burning or discomfort sensations (Grigoleit and Grigoleit 2005).

Case Reports of Adverse Events

Ingestion of a large dose (40 drops) of peppermint leaf essential oil resulted in a chemical burn of the oral cavity and pharynx with tachycardia, tachypnea, and edema of the lips, tongue, and uvula (Tamir et al. 2005). Several cases of mucosal ulcers have been associated with the oral ingestion of products containing peppermint, including mouthwashes and candies (Moghadam et al. 1999; Rogers and Pahor 1995).

Allergic reactions, including contact dermatitis, to peppermint leaf essential oil have been reported (Foti et al. 2003; Morton et al. 1995; Sainio and Kanerva 1995; Wilkinson and Beck 1994).

III. Pharmacology and Pharmacokinetics

Human Pharmacological Studies

Coadministration of peppermint leaf essential oil (600 mg) and felodipine moderately increased the plasma concentration of felodipine, possibly through inhibition of the drug-metabolizing isoenzyme CYP3A4 (Dresser et al. 2002).

Peppermint leaf essential oil has been shown to slow intestinal transit, which may slow the absorption rate or increase the total absorption of coadministered drugs (Goerg and Spilker 2003; Mills and Bone 2005).

Animal Pharmacological Studies

No relevant animal pharmacological studies were identified.

In Vitro Pharmacological Studies

No relevant in vitro pharmacological studies were identified.

IV. Pregnancy and Lactation

No teratogenic effects of the compound menthol were observed in mice fed 190 mg/kg, rats fed 220 mg/kg, hamsters fed 400 mg/kg, or rabbits fed 430 mg/kg (FAO/WHO 1999). The compound menthol has been reported to cause jaundice in newborn babies. This jaundice has been correlated with the hereditary disease glucose-6-phosphate dehydrogenase deficiency (Owa 1989).

No changes in breast-feeding frequency or duration were observed in infants of mothers using a topical application of peppermint gel daily for 14 days to prevent nipple cracking (Sayyah Melli et al. 2007).

V. Toxicity Studies

Acute Toxicity

The LD_{50} of orally administered peppermint leaf oil in mice is 2.4 g/kg, and in rats is 4.4 g/kg (Della Loggia et al. 1990). No toxic effects were observed after mice were administered 4 g/kg of a peppermint extract (Della Loggia et al. 1990).

The estimated lethal dose for the compound menthol has been reported to be 2 g, although individuals have survived doses up to 9 g (De Smet 1993).

Short-Term Toxicity

No adverse effects were observed in mice fed 20 to 500 mg/kg daily (containing 1 to 2% pulegone) of peppermint leaf oil for 5 weeks (Mengs and Stotzem 1989).

Subchronic Toxicity

Some histopathological changes were noted in the cerebellum of mice fed 100 mg/kg peppermint leaf essential oil for 90 days. In male rats, some nephrotoxicity was noted. No adverse effects were observed in mice fed 10 or 40 mg/kg doses, and a no-observed-adverse-effect level (NOAEL) of 40 mg/kg per day was suggested (Spindler and Madsen 1992). In a similar study, peppermint leaf essential oil was associated with lesions in the cerebellum at doses of 40 and 100 mg/kg. No lesions or other adverse effects were observed at a dose of 20 mg/kg (Thorup et al. 1983).

Genotoxicity

Genotoxicity assays indicated that peppermint leaf essential oil induced sister-chromatid exchanges in a dose-independent manner. In the SMART test, peppermint leaf essential oil induced mutations in a dose-independent manner (Lazutka et al. 2001).

Literature Cited

Brinker, F. 2001. *Herb contraindications and drug interactions*. 3rd ed. Sandy, OR: Eclectic Medical Publications.

De Smet, P.A.G.M. 1993. *Adverse effects of herbal drugs, Volume 2*. Berlin: Springer.

Della Loggia, R., A. Tubaro, and T. Lunder. 1990. Evaluation of some pharmacological activities of a peppermint extract. *Fitoterapia* 61:215-221.

Dresser, G.K., V. Wacher, S. Wong, H.T. Wong, and D.G. Bailey. 2002. Evaluation of peppermint oil and ascorbyl palmitate as inhibitors of cytochrome P4503A4 activity in vitro and in vivo. *Clin. Pharmacol. Ther.* 72(3):247-255.

FAO/WHO. 1999. *51st Meeting of the Joint FAO/WHO Expert Committee on Food Additives.* WHO Additives Series no 42. Geneva: WHO.

Foti, C., A. Conserva, A. Antelmi, L. Lospalluti, and G. Angelini. 2003. Contact dermatitis from peppermint and menthol in a local action transcutaneous patch. *Contact Dermat.* 49(6):312-313.

Goerg, K.J., and T. Spilker. 2003. Effect of peppermint oil and caraway oil on gastrointestinal motility in healthy volunteers: A pharmacodynamic study using simultaneous determination of gastric and gall-bladder emptying and orocaecal transit time. *Aliment. Pharmacol. Ther.* 17(3):445-451.

Grigoleit, H.G., and P. Grigoleit. 2005. Peppermint oil in irritable bowel syndrome. *Phytomedicine* 12(8):601-606.

Hiki, N., H. Kurosaka, Y. Tatsutomi, et al. 2003. Peppermint oil reduces gastric spasm during upper endoscopy: A randomized, double-blind, double-dummy controlled trial. *Gastrointest. Endosc.* 57(4):475-482.

Hills, J.M., and P.I. Aaronson. 1991. The mechanism of action of peppermint oil on gastrointestinal smooth muscle. An analysis using patch clamp electrophysiology and isolated tissue pharmacology in rabbit and guinea pig. *Gastroenterology* 101(1):55-65.

Lazutka, J.R., J. Mierauskiene, G. Slapsyte, and V. Dedonyte. 2001. Genotoxicity of dill (*Anethum graveolens* L.), peppermint (*Mentha piperita* L.) and pine (*Pinus sylvestris* L.) essential oils in human lymphocytes and *Drosophila melanogaster*. *Food Chem. Toxicol.* 39(5):485-492.

McKay, D., and J. Blumberg. 2006. A review of the bioactivity and potential health benefits of peppermint tea (*Mentha piperita* L.). *Phytother. Res.* 20(8):619-633.

Mengs, U., and C. Stotzem. 1989. Toxicological evaluation of peppermint oil in rodents and dogs. *Med. Sci. Res.* 17(11):499-500.

Mills, S., and K. Bone. 2005. *The essential guide to herbal safety*. St. Louis: Elsevier.

Moghadam, B.K., R. Gier, and T. Thurlow. 1999. Extensive oral mucosal ulcerations caused by misuse of a commercial mouthwash. *Cutis* 64(2):131-134.

Morton, C.A., J. Garioch, P. Todd, P.J. Lamey, and A. Forsyth. 1995. Contact sensitivity to menthol and peppermint in patients with intra-oral symptoms. *Contact Dermat.* 32(5):281-284.

Nash, P., S.R. Gould, and D.E. Bernardo. 1986. Peppermint oil does not relieve the pain of irritable bowel syndrome. *Br. J. Clin. Pract.* 40(7):292-293.

Owa, J.A. 1989. Relationship between exposure to icterogenic agents, glucose-6-phosphate dehydrogenase deficiency and neonatal jaundice in Nigeria. *Acta Paediatr. Scand.* 78(6):848-852.

Pittler, M.H., and E. Ernst. 1998. Peppermint oil for irritable bowel syndrome: A critical review and meta-analysis. *Am. J. Gastroenterol.* 93(7):1131-1135.

Rogers, S.N., and A.L. Pahor. 1995. A form of stomatitis induced by excessive peppermint consumption. *Dent. Update* 22(1):36-37.

Sainio, E.L., and L. Kanerva. 1995. Contact allergens in toothpastes and a review of their hypersensitivity. *Contact Dermat.* 33(2):100-105.

M

Sayyah Melli, M., M.R. Rashidi, A. Nokhoodchi, et al. 2007. A randomized trial of peppermint gel, lanolin ointment, and placebo gel to prevent nipple crack in primiparous breastfeeding women. *Med. Sci. Monit.* 13(9):CR406-411.

Somerville, K.W., C.R. Richmond, and G.D. Bell. 1984. Delayed release peppermint oil capsules (Colpermin) for the spastic colon syndrome: A pharmacokinetic study. *Br. J. Clin. Pharmacol.* 18(4):638-640.

Spindler, P., and C. Madsen. 1992. Subchronic toxicity study of peppermint oil in rats. *Toxicol. Lett.* 62(2-3):215-220.

Tamir, S., Z. Davidovich, P. Attal, and R. Eliashar. 2005. Peppermint oil chemical burn. *Otolaryngol. Head Neck Surg.* 133(5):801-802.

Thompson Coon, J., and E. Ernst. 2002. Systematic review: Herbal medicinal products for non-ulcer dyspepsia. *Aliment. Pharmacol. Ther.* 16(10):1689-1699.

Thorup, I., G. Würtzen, J. Carstensen, and P. Olsen. 1983. Short term toxicity study in rats dosed with peppermint oil. *Toxicol. Lett.* 19(3):211-215.

Tiran, D., and S. Mack. 2000. *Complementary therapies for pregnancy and childbirth.* New York: Elsevier Health Sciences.

Wacher, V.J., S. Wong, and H.T. Wong. 2002. Peppermint oil enhances cyclosporine oral bioavailability in rats: Comparison with D-alpha-tocopheryl poly(ethylene glycol 1000) succinate (TPGS) and ketoconazole. *J. Pharm. Sci.* 91(1):77-90.

Wilkinson, S.M., and M.H. Beck. 1994. Allergic contact dermatitis from menthol in peppermint. *Contact Dermat.* 30(1):42-43.

Mentha pulegium L. — Lamiaceae

SCN: European pennyroyal

Part: herb

Quick Reference Summary

Safety Class: 2b, 2c, 2d
Interaction Class: A

Contraindications

Not for use during pregnancy or lactation (Anderson et al. 1996; Chadha 1988; Ciganda and Laborde 2003; Gordon et al. 1987; List and Hörhammer 1973).

Not for use by women with a heavy menstrual flow (Kuhn and Winston 2007).

Other Precautions

Not recommended for use in persons with liver or kidney disease (Brinker 2001; Gordon et al. 1987; Mizutani et al. 1987; Speijers 2001; Sztajnkrycer et al. 2003).

Not recommended for use in children or infants (Bakerink et al. 1996; Brinker 2001).

Drug and Supplement Interactions

None known.

Notice

Emmenagogue (Chadha 1988); *see* Appendix 2.

Editors' Note

European pennyroyal leaf contains 1.0–2.0% essential oil consisting of 80.0–94.0% pulegone (List and Hörhammer 1973), a compound that causes severe liver toxicity when administered to rats in high doses (Gordon et al. 1987; Mizutani et al. 1987; Speijers 2001; Sztajnkrycer et al. 2003).

European pennyroyal and American pennyroyal (*Hedeoma pulegioides*) are historically interchangeable as the source for pennyroyal oil (De Smet 1992).

Adverse Events and Side Effects

Transient and generally mild adverse effects, including nausea, dizziness, and abdominal cramping, have been reported in association with the use of pennyroyal herb tea, even at doses as low as 1–3 cups. Much more serious side effects, including fatalities, are recorded with consumption of pennyroyal essential oil (*see* next entry) (Anderson et al. 1996).

Case reports also appear in the literature of significant adverse events with the use of incompletely described pennyroyal preparations to induce abortion. These include an epileptic event after use of 9–12 "pennyroyal tablets" daily for 4 days (Early 1961), and two deaths. One of the recorded deaths occurred after use of "a bottle of pennyroyal mixture" on two subsequent days (Vallance 1955). The other death was associated with ingestion of European pennyroyal herb extract in 48–56% alcohol by a 24-year-old diagnosed with a possibly ruptured ectopic pregnancy (Anderson et al. 1996). Two cases of coma associated with consumption of "essence of pennyroyal" are also reported (Anderson et al. 1996); the latter of these speculated that the material used was dilute pennyroyal oil (Braithwaite 1906).

Pharmacological Considerations

Coadministration of European pennyroyal tea and iron-fortified bread reduced absorption of the iron (Hurrell et al. 1999).

Pregnancy and Lactation

Pennyroyal essential oil (*see* next entry), has historically been used by women attempting to induce abortion (De Smet 1992; Williamson 2003). Although animal or other studies on the use of European pennyroyal herb during pregnancy and lactation are lacking, the content of the potentially toxic compound pulegone and traditional use of the oil suggest that European pennyroyal herb should not be used during pregnancy or when nursing.

Review Details

I. Drug and Supplement Interactions

Clinical Trials of Drug or Supplement Interactions

No clinical trials of drug or supplement interactions were identified.

Case Reports of Suspected Drug or Supplement Interactions

No case reports of suspected drug or supplement interactions were identified.

Animal Trials of Drug or Supplement Interactions

No animal trials of drug or supplement interactions were identified.

II. Adverse Events

Case Reports of Adverse Events

A review of pennyroyal toxicity identified 22 adverse events from 1883 to 1996, generally without differentiation between American pennyroyal (*Hedeoma pulegioides*) and European pennyroyal. Two of the identified events were associated with pennyroyal herb tea alone; two with use of pennyroyal herb tea and one or more additional preparations of pennyroyal herb; one with an alcohol extract of the herb; and the balance with pennyroyal essential oil (*see* next entry for further information on the case reports for essential oil). In women who drank pennyroyal tea, one or two cups of tea were associated with dizziness, weakness, and abdominal cramping. Consumption of the tea along with other preparations, including the essential oil, was associated with more severe symptoms including vomiting, stupor, and coma (Anderson et al. 1996). The case involving the extract, identified as "pennyroyal herb, 48% to 56% in an alcohol base" (probably an extract of the herb in 48–56% alcohol), was associated with the death of a 24-year-old woman with an undiagnosed ectopic pregnancy (Anderson et al. 1996; Young 1995).

Acute hepatic and neurological injury was observed in two Hispanic infants administered tea made from homegrown plants believed to be mint. An 8-week-old male developed fulminant liver failure with cerebral edema and necrosis, resulting in death, following a single dose of 120 ml of tea brewed from the plant's leaves. A 6-month-old male experienced hepatic dysfunction and severe epileptic encephalopathy after thrice weekly ingestion of 90 ml of leaf tea. Upon examination of samples of the plants involved, it was suggested that the subspecies used contained pennyroyal oil, and serum analyses identified pulegone in one infant and its metabolite menthofuran in both. (Bakerink et al. 1996).

A case of contact dermatitis was reported in a woman who had recently picked European pennyroyal leaves (Roe et al. 2005).

III. Pharmacology and Pharmacokinetics

Human Pharmacological Studies

Coadministration of iron-fortified bread and European pennyroyal tea reduced absorption of the iron by 73%, an effect that was less pronounced than that of black tea or peppermint tea, but more inhibitory than cocoa or tea made with European vervain, linden flower, or chamomile (Hurrell et al. 1999).

Animal Pharmacological Studies

No relevant animal pharmacological studies were identified.

In Vitro Pharmacological Studies

Menthofuran, a metabolite of the compound pulegone, has been shown to be a significant inhibitor of the drug-metabolizing isoenzyme CYP2A6 (Khojasteh-Bakht et al. 1998).

IV. Pregnancy and Lactation

Pennyroyal oil has been traditionally used as an abortifacient (De Smet 1992; Gordon et al. 1987; Williamson 2003). Although animal or other studies on the use of pennyroyal herb during pregnancy and lactation are lacking, the content of the potentially toxic compound pulegone (*see* Toxicity Studies) and traditional use of the oil suggest that pennyroyal herb should not be taken during pregnancy or while nursing.

V. Toxicity Studies

Toxicity testing of pulegone in rats indicated that the no-observed-adverse-effect level (NOAEL) for the compound ranges between 20 mg/kg (administered by gavage) and 250 mg/kg (administered in food) daily (Imaizumi et al. 1985; Thorup et al. 1983).

The compound pulegone is metabolized to form menthofuran, a recognized hepatotoxin (Gordon et al. 1987; Mizutani et al. 1987; Speijers 2001; Sztajnkrycer et al. 2003).

Acute Toxicity

The LD_{50} of orally administered pennyroyal essential oil in rats is 220 to 580 mg/kg (Opdyke 1972).

Short-Term Toxicity

Rats administered pulegone at doses of 80 or 160 mg/kg daily had induced atonia, decreased blood creatinine content, lowered terminal body weight, and histopathological changes in the liver and in the white matter of the cerebellum. Rats administered 20 mg/kg daily did not exhibit these signs of toxicity (Thorup et al. 1983).

Genotoxicity

The compound pulegone tested negative for genotoxicity in a *Salmonella typhimurium* test and was weakly positive in the *Drosophila melanogaster* wing spot test (Franzios et al. 1997).

Immunotoxicity

In a screening study for immunotoxicity, mice treated orally with the compound isopulegol at doses up to 500 mg/kg daily for 5 days exhibited no signs of toxicity (Vollmuth et al. 1989).

Literature Cited

Anderson, I.B., W.H. Mullen, J.E. Meeker, et al. 1996. Pennyroyal toxicity: Measurement of toxic metabolite levels in two cases and review of the literature. *Ann. Intern. Med.* 124(8):726-734.

Bakerink, J.A., S.M. Gospe, Jr., R.J. Dimand, and M.W. Eldridge. 1996. Multiple organ failure after ingestion of pennyroyal oil from herbal tea in two infants. *Pediatrics* 98(5):944-947.

Braithwaite, P.F. 1906. A case of poisoning by pennyroyal: Recovery. *Br. Med J.* 2388:865.

Brinker, F. 2001. *Herb contraindications and drug interactions.* 3rd ed. Sandy, OR: Eclectic Medical Publications.

Chadha, Y. 1988. *The wealth of India: A dictionary of Indian raw materials and industrial products.* Delhi: Council of Scientific and Industrial Research.

Ciganda, C., and A. Laborde. 2003. Herbal infusions used for induced abortion. *J. Toxicol. Clin. Toxicol.* 41(3):235-239.

De Smet, P.A.G.M. 1992. *Adverse effects of herbal drugs, Volume 1.* Berlin: Springer.

Early, D.F. 1961. Pennyroyal: A rare case of epilepsy. *Lancet* 278:580-581.

Franzios, G., M. Mirotsou, E. Hatziapostolou, et al. 1997. Insecticidal and genotoxic activities of mint essential oils. *J. Agric. Food Chem.* 45(7):2690-2694.

Gordon, W.P., A.C. Huitric, C.L. Seth, R.H. McClanahan, and S.D. Nelson. 1987. The metabolism of the abortifacient terpene, (*R*)-(+)-pulegone, to a proximate toxin, menthofuran. *Drug Metab. Dispos.* 15(5):589-594.

Hurrell, R.F., M. Reddy, and J.D. Cook. 1999. Inhibition of non-haem iron absorption in man by polyphenolic-containing beverages. *Br. J. Nutr.* 81(4):289-295.

Imaizumi, K., K. Hanada, K. Mawartari, and M. Sugano. 1985. Effect of essential oils on the concentration of serum lipids and apolipoproteins in rats. *J. Agric. Biol. Chem.* 49:2795-2796.

Khojasteh-Bakht, S.C., L.L. Koenigs, R.M. Peter, W.F. Trager, and S.D. Nelson. 1998. (*R*)-(+)-Menthofuran is a potent, mechanism-based inactivator of human liver cytochrome P450 2A6. *Drug Metab. Dispos.* 26(7):701-704.

Kuhn, M., and D. Winston. 2007. *Herbal therapy & supplements.* 2nd ed. St. Louis: Lippincott, Williams & Wilkins.

List, P.H., and H. Hörhammer. 1973. *Hagers handbuch der pharmazeutischen praxis.* Berlin: Springer.

Mizutani, T., H. Nomura, K. Nakanishi, and S. Fujita. 1987. Effects of drug metabolism modifiers on pulegone-induced hepatotoxicity in mice. *Res. Commun. Chem. Pathol. Pharmacol.* 58(1):75-83.

Opdyke, D. 1972. Monographs on fragrance raw materials. *Food Cosmet. Toxicol.* 12:949-950.

Roe, E., E. Serra-Baldrich, J. Dalmau, et al. 2005. *Mentha pulegium* contact dermatitis. *Contact Dermat.* 53(6):355.

Speijers, G. 2001. WHO Food Additives Series 46. Pulegone and related substances. Bilthoven, Netherlands: National Institute of Public Health and the Environment.

Sztajnkrycer, M.D., E.J. Otten, G.R. Bond, C.J. Lindsell, and R.J. Goetz. 2003. Mitigation of pennyroyal oil hepatotoxicity in the mouse. *Acad. Emerg. Med.* 10(10):1024-1028.

Thorup, I., G. Wurtzen, J. Carstensen, and P. Olsen. 1983. Short term toxicity study in rats dosed with pulegone and menthol. *Toxicol. Lett.* 19(3):207-210.

Vallance, W.B. 1955. Pennyroyal poisoning: A fatal case. *Lancet* 266:850-851.

Vollmuth, T.A., J.D. Heck, H.V. Ratajczak, and P.T. Thomas. 1989. Immunotoxicity assessment of flavoring ingredients using a rapid and economical screen. *Toxicologist* 9:206.

Williamson, E.M. 2003. *Potter's herbal cyclopedia.* Saffron Walden, Essex: C.W. Daniel Co.

Young, G. 1995. Lifestyle on trial. *Metro: Silicon Valley's Weekly Newspaper.* December 14-20, 1995 issue.

M

Mentha pulegium L. — Lamiaceae

SCN: European pennyroyal

Part: herb essential oil

Quick Reference Summary

Safety Class: 2a, 2b, 2c
Interaction Class: A

Contraindications

Not to be used during pregnancy (Anderson et al. 1996; Chadha 1988; Ciganda and Laborde 2003; Gordon et al. 1987; List and Hörhammer 1973).

For external use only (Anderson et al. 1996; Sullivan et al. 1979; Vallance 1955).

Other Precautions

If applying pennyroyal essential oil topically in humans or animals, the oil must be diluted in a carrier oil to minimize risk of adverse reactions. To avoid ingestion by companion animals, topical application of diluted essential oil should only be on a collar and not directly on fur or skin.

Not recommended for use in persons with liver or kidney disease (Gordon et al. 1987; Mizutani et al. 1987; Speijers 2001; Sztajnkrycer et al. 2003).

Not recommended for use in children or infants (Bakerink et al. 1996).

Not recommended for nursing women due to the theoretical potential for transfer of the compound pulegone through breast milk (Brinker 2001).

Drug and Supplement Interactions

None known.

Notice

Emmenagogue (Chadha 1988); *see* Appendix 2.

Abortifacient (De Smet 1992; Gordon et al. 1987; Williamson 2003); *see* Appendix 2.

Editors' Notes

Pennyroyal essential oil consists of 80.0 to 94.0% pulegone (List and Hörhammer 1973), a compound that causes severe liver toxicity when administered to rats in high doses (Gordon et al. 1987; Mizutani et al. 1987; Speijers 2001; Sztajnkrycer et al. 2003).

European pennyroyal and American pennyroyal (*Hedeoma pulegioides*) have been historically interchangeable as the source for pennyroyal oil (Furia and Bellanca 1971). The cautions presented here are relevant to the essential oil derived from either plant.

Adverse Events and Side Effects

Cases of toxicity have been reported in persons consuming pennyroyal essential oil (Anderson et al. 1996; Anonymous 1978; Macht 1913; Vallance 1955). Death was reported after consumption of as little as 15.0 ml (1/2 oz) of the oil and, in one instance, following consumption of an alcohol extract of pennyroyal over a 2-week period. Symptoms of toxicity ranged from nausea, dizziness, and dilation of pupils to respiratory depression, generalized seizures, and coma (Anderson et al. 1996).

Pharmacological Considerations

See Adverse Events and Side Effects.

Pregnancy and Lactation

Pennyroyal essential oil has been traditionally used as an abortifacient (Conway and Slocumb 1979; De Smet 1992; Williamson 2003). Although animal or other studies on the use of pennyroyal essential oil during pregnancy and lactation are lacking, the content of the potentially toxic compound pulegone and traditional use indicate that it should not be taken orally or used externally during pregnancy or while nursing.

Review Details

I. Drug and Supplement Interactions

Clinical Trials of Drug or Supplement Interactions

No clinical trials of drug or supplement interactions were identified.

Case Reports of Suspected Drug or Supplement Interactions

No case reports of suspected drug or supplement interactions were identified.

Animal Trials of Drug or Supplement Interactions

No animal trials of drug or supplement interactions were identified.

II. Adverse Events

Case Reports of Adverse Events

A review of pennyroyal toxicity identified 22 adverse events from 1883 to 1996, most of which involved oral consumption of the essential oil at doses greater than 10 ml. Four cases had fatal outcomes; the remaining cases had mild to severe outcomes ranging from nausea, dizziness, and dilation of pupils to respiratory depression, generalized seizures, and coma (Anderson et al. 1996). *See* above entry on European pennyroyal herb for a detailed discussion of this reference.

Toxicosis followed by death was reported in a dog treated topically with pennyroyal oil (Sudekum et al. 1992).

III. Pharmacology and Pharmacokinetics

Human Pharmacological Studies

No relevant human pharmacological studies were identified.

Animal Pharmacological Studies

No relevant animal pharmacological studies were identified.

In Vitro Pharmacological Studies

Menthofuran, a metabolite of the compound pulegone, has been shown to be a significant inhibitor of the drug-metabolizing isoenzyme CYP2A6 (Khojasteh-Bakht et al. 1998).

IV. Pregnancy and Lactation

Pennyroyal essential oil has been traditionally used as an abortifacient (Conway and Slocumb 1979; De Smet 1992; Williamson 2003). Although animal or other studies on the use of pennyroyal essential oil during pregnancy and lactation are lacking, the content of the potentially toxic compound pulegone and traditional use indicate that it should not be taken orally or used externally during pregnancy or while nursing.

V. Toxicity Studies

Toxicity testing of the compound pulegone in rats indicated that the no-observed-adverse-effect level (NOAEL) for the compound ranges between 20 mg/kg (administered by gavage) and 250 mg/kg (administered in food) daily (Imaizumi et al. 1985; Thorup et al. 1983).

The compound pulegone is metabolized to form menthofuran, a recognized hepatotoxin (Gordon et al. 1987; Mizutani et al. 1987; Speijers 2001; Sztajnkrycer et al. 2003).

Acute Toxicity

The LD_{50} of orally administered pennyroyal essential oil in rats is 220 to 580 mg/kg (Opdyke 1972).

Short-Term Toxicity

Rats administered pulegone at doses of 80 or 160 mg/kg daily had induced atonia, decreased blood creatinine content, lowered terminal body weight, and histopathological changes in the liver and in the white matter of the cerebellum. Rats administered 20 mg/kg daily did not exhibit those signs of toxicity (Thorup et al. 1983).

Genotoxicity

The compound pulegone tested negative for genotoxicity in a *Salmonella typhimurium* test and was weakly positive in the *Drosophila melanogaster* wing spot test (Franzios et al. 1997).

Immunotoxicity

In a screening study for immunotoxicity, mice treated orally with the compound isopulegol at doses up to 500 mg/kg daily for 5 days exhibited no signs of toxicity (Vollmuth et al. 1989).

Literature Cited

Anderson, I.B., W.H. Mullen, J.E. Meeker, et al. 1996. Pennyroyal toxicity: Measurement of toxic metabolite levels in two cases and review of the literature. *Ann. Intern. Med.* 124(8):726-734.

Anonymous. 1978. Fatality and illness associated with consumption of pennyroyal oil—Colorado. *Morb. Mortal. Wkly. Rep.* 27(51):511-513.

Bakerink, J.A., S.M. Gospe, Jr., R.J. Dimand, and M.W. Eldridge. 1996. Multiple organ failure after ingestion of pennyroyal oil from herbal tea in two infants. *Pediatrics* 98(5):944-947.

Brinker, F. 2001. *Herb contraindications and drug interactions.* 3rd ed. Sandy, OR: Eclectic Medical Publications.

Chadha, Y. 1988. *The wealth of India: A dictionary of Indian raw materials and industrial products.* Delhi: Council of Scientific and Industrial Research.

Ciganda, C., and A. Laborde. 2003. Herbal infusions used for induced abortion. *J. Toxicol. Clin. Toxicol.* 41(3):235-239.

Conway, G.A., and J.C. Slocumb. 1979. Plants used as abortifacients and emmenagogues by Spanish New Mexicans. *J. Ethnopharmacol.* 1(3):241-261.

Franzios, G., M. Mirotsou, E. Hatziapostolou, et al. 1997. Insecticidal and genotoxic activities of mint essential oils. *J. Agric. Food Chem.* 45(7):2690-2694.

Furia, T., and N. Bellanca. 1971. *Feranoli's handbook of flavor ingredients.* Cleveland: The Chemical Rubber Company.

Gordon, W.P., A.C. Huitric, C.L. Seth, R.H. McClanahan, and S.D. Nelson. 1987. The metabolism of the abortifacient terpene, (*R*)-(+)-pulegone, to a proximate toxin, menthofuran. *Drug Metab. Dispos.* 15(5):589-994.

Imaizumi, K., K. Hanada, K. Mawartari, and M. Sugano. 1985. Effect of essential oils on the concentration of serum lipids and apolipoproteins in rats. *J. Agric. Biol. Chem.* 49:2795-2796.

Khojasteh-Bakht, S.C., L.L. Koenigs, R.M. Peter, W.F. Trager, and S.D. Nelson. 1998. (*R*)-(+)-Menthofuran is a potent, mechanism-based inactivator of human liver cytochrome P450 2A6. *Drug Metab. Dispos.* 26(7):701-704.

List, P.H., and H. Hörhammer. 1973. *Hagers handbuch der pharmazeutischen praxis.* Berlin: Springer.

Macht, D.I. 1913. The action of so-called emmenagogue oils on the isolated uterus: With a report of a case of pennyroyal poisoning. *J. Am. Med. Assoc.* 61(2):105-107.

Mizutani, T., H. Nomura, K. Nakanishi, and S. Fujita. 1987. Effects of drug metabolism modifiers on pulegone-induced hepatotoxicity in mice. *Res. Commun. Chem. Pathol. Pharmacol.* 58(1):75-83.

Opdyke, D. 1972. Monographs on fragrance raw materials. *Food Cosmet. Toxicol.* 12:949-950.

Speijers, G. 2001. WHO Food Additives Series 46. Pulegone and related substances. Bilthoven, Netherlands: National Institute of Public Health and the Environment.

Sudekum, M., R.H. Poppenga, N. Raju, and W.E. Braselton, Jr. 1992. Pennyroyal oil toxicosis in a dog. *J. Am. Vet. Med. Assoc.* 200(6):817-818.

Sullivan, J.B., Jr., B.H. Rumack, H. Thomas, Jr., R.G. Peterson, and P. Bryson. 1979. Pennyroyal oil poisoning and hepatotoxicity. *J. Am. Med. Assoc.* 242(26):2873-4.

Sztajnkrycer, M.D., E.J. Otten, G.R. Bond, C.J. Lindsell, and R.J. Goetz. 2003. Mitigation of pennyroyal oil hepatotoxicity in the mouse. *Acad. Emerg. Med.* 10(10):1024-1028.

Thorup, I., G. Wurtzen, J. Carstensen, and P. Olsen. 1983. Short term toxicity study in rats dosed with pulegone and menthol. *Toxicol. Lett.* 19(3):207-210.

Vallance, W.B. 1955. Pennyroyal poisoning: A fatal case. *Lancet* 269(6895):850-851.

Vollmuth, T.A., J.D. Heck, H.V. Ratajczak, and P.T. Thomas. 1989. Immunotoxicity assessment of flavoring ingredients using a rapid and economical screen. *Toxicologist* 9:206.

M

Mentha spicata L.

Lamiaceae

SCN: spearmint
Syn: *Mentha viridis* L.

Part: leaf

Quick Reference Summary

Safety Class: 1
Interaction Class: A

Contraindications

None known.

Other Precautions

None known.

Drug and Supplement Interactions

None known.

Editors' Notes

Although spearmint leaf is the subject of this entry, spearmint essential oil is also available and is a highly concentrated extract. Safety considerations for spearmint essential oil are different than those for spearmint leaf.

Adverse Events and Side Effects

Cases of contact allergy to spearmint essential oil have been reported (Andersen 1978; Clayton and Orton 2004; Skrebova et al. 1998).

Spearmint essential oil may cause sensitivity reactions in certain individuals and especially in children (Mills and Bone 2005).

Pharmacological Considerations

Some changes in reproductive hormone levels have been reported in one human and two animal studies (Akdogan et al. 2004b, 2004c, 2007).

Pregnancy and Lactation

No information on the safety of spearmint in pregnancy or lactation was identified in the scientific or traditional literature. Although this review did not identify any concerns for use while pregnant or nursing, safety has not been conclusively established.

Review Details

I. Drug and Supplement Interactions

Clinical Trials of Drug or Supplement Interactions

No clinical trials of drug or supplement interactions were identified.

Case Reports of Suspected Drug or Supplement Interactions

No case reports of suspected drug or supplement interactions were identified.

Animal Trials of Drug or Supplement Interactions

No animal trials of drug or supplement interactions were identified.

II. Adverse Events

Adverse Events Reported in Clinical Trials

In a clinical trial of spearmint essential oil, some of the trial participants ingesting 500 mg of oil daily experienced chest pain, heartburn, or regurgitation (Bulat et al. 1999).

Case Reports of Adverse Events

Cases of contact allergy to spearmint essential oil have been reported (Andersen 1978; Clayton and Orton 2004; Skrebova et al. 1998). A case of allergic contact dermatitis to spearmint essential oil was reported (Bonamonte et al. 2001).

III. Pharmacology and Pharmacokinetics

Human Pharmacological Studies

No effects on lower esophageal sphincter tone were observed in healthy volunteers administered up to 500 mg spearmint essential oil, indicating that spearmint has no adverse effects on acid reflux (Bulat et al. 1999).

In women with hirsutism, drinking one cup of spearmint tea daily for 5 days was associated with a decrease in free testosterone and an increase in luteinizing hormone, follicle-stimulating hormone, and estradiol (Akdogan et al. 2007).

Animal Pharmacological Studies

A decrease in testosterone and an increase in luteinizing hormone and follicle-stimulating hormone were observed in male rats that consumed spearmint tea ad libitum for 30 days. Some adverse effects on testicular cells were observed (Akdogan et al. 2004c). A decrease in iron absorption was observed in rats administered spearmint tea (Akdogan et al. 2004a).

In Vitro Pharmacological Studies

No relevant in vitro pharmacological studies were identified.

IV. Pregnancy and Lactation

No studies on the safety of spearmint during pregnancy or lactation were identified.

V. Toxicity Studies

Acute Toxicity

The LD_{50} of orally administered spearmint essential oil in rats is 5000 mg/kg (Spectrum 2003).

Short-Term Toxicity

In rats administered spearmint tea for 30 days as the sole source of drinking water, some histopathological changes were observed in the kidneys (Akdogan et al. 2003). Similarly, in female rats administered spearmint tea for 30 days as the sole source of drinking water, some histopathological changes were observed in uterine tissues (Guney et al. 2006).

In rats given spearmint tea ad libitum for 30 days, liver enzyme changes were observed. Decreases in superoxide dismutase (SOD), glutathione peroxidase (GSH-Px), and catalase (CAT) and an increase in thiobarbituric acid-reactive substance (TBARS) were noted (Akdogan et al. 2004b).

Genotoxicity

Spearmint essential oil showed some mutagenic activity in *Drosophila melanogaster* (Franzios et al. 1997).

Literature Cited

Akdogan, M., F. Gultekin, and M. Yontem. 2004a. Effect of *Mentha piperita* (Labiatae) and *Mentha spicata* (Labiatae) on iron absorption in rats. *Toxicol. Ind. Health* 20(6-10):119-122.

Akdogan, M., I. Kilinc, M. Oncu, E. Karaoz, and N. Delibas. 2003. Investigation of biochemical and histopathological effects of *Mentha piperita* L. and *Mentha spicata* L. on kidney tissue in rats. *Hum. Exp. Toxicol.* 22(4):213-219.

Akdogan, M., M. Ozguner, G. Aydin, and O. Gokalp. 2004b. Investigation of biochemical and histopathological effects of *Mentha piperita* Labiatae and *Mentha spicata* Labiatae on liver tissue in rats. *Hum. Exp. Toxicol.* 23(1):21-28.

Akdogan, M., M. Ozguner, A. Kocak, M. Oncu, and E. Cicek. 2004c. Effects of peppermint teas on plasma testosterone, follicle-stimulating hormone, and luteinizing hormone levels and testicular tissue in rats. *Urology* 64(2):394-398.

Akdogan, M., M.N. Tamer, E. Cure, et al. 2007. Effect of spearmint (*Mentha spicata* Labiatae) teas on androgen levels in women with hirsutism. *Phytother. Res.* 21(5):444-447.

Andersen, K.E. 1978. Contact allergy to toothpaste flavors. *Contact Dermat.* 4(4):195-198.

Bonamonte, D., L. Mundo, M. Daddabbo, and C. Foti. 2001. Allergic contact dermatitis from *Mentha spicata* (spearmint). *Contact Dermat.* 45(5):298.

Bulat, R., E. Fachnie, U. Chauhan, Y. Chen, and G. Tougas. 1999. Lack of effect of spearmint on lower oesophageal sphincter function and acid reflux in healthy volunteers. *Aliment. Pharmacol. Ther.* 13(6):805-812.

Clayton, R., and D. Orton. 2004. Contact allergy to spearmint oil in a patient with oral lichen planus. *Contact Dermat.* 51(5-6):314-315.

Franzios, G., M. Mirotsou, E. Hatziapostolou, et al. 1997. Insecticidal and genotoxic activities of mint essential oils. *J. Agric. Food Chem.* 45(7):2690-2694.

Guney, M., B. Oral, N. Karahanli, T. Mungan, and M. Akdogan. 2006. The effect of *Mentha spicata* Labiatae on uterine tissue in rats. *Toxicol. Ind. Health* 22(8):343-348.

Mills, S., and K. Bone. 2005. *The essential guide to herbal safety*. St. Louis: Elsevier.

Skrebova, N., K. Brocks, and T. Karlsmark. 1998. Allergic contact cheilitis from spearmint oil. *Contact Dermat.* 39(1):35.

Spectrum. 2003. *Spearmint oil material safety data sheet.* Gardena, CA: Spectrum Chemical Manufacturing Corp.

Menyanthes trifoliata L. — Menyanthaceae

SCN: bog bean
OCN: buck bean; marsh trefoil

Part: leaf

Quick Reference Summary

Safety Class: 1
Interaction Class: A

Contraindications

None known.

Other Precautions

Use with caution in persons with diarrhea, dysentery, or colitis (Bradley 1993).

Drug and Supplement Interactions

None known.

Adverse Events and Side Effects

Bog bean may cause gastrointestinal irritation, especially in high doses (standard dose listed as a tea made from 0.5 or 1 g of herb in a cup of water) (Weiss and Meuss 2001; Wichtl 2004).

Pharmacological Considerations

None known.

M

Pregnancy and Lactation

No information on the safety of bog bean in pregnancy or lactation was identified in the scientific or traditional literature. Although this review did not identify any concerns for use while pregnant or nursing, safety has not been conclusively established.

Review Details

I. Drug and Supplement Interactions

Clinical Trials of Drug or Supplement Interactions

No clinical trials of drug or supplement interactions were identified.

Case Reports of Suspected Drug or Supplement Interactions

No case reports of suspected drug or supplement interactions were identified.

Animal Trials of Drug or Supplement Interactions

No animal trials of drug or supplement interactions were identified.

II. Adverse Events

Case Reports of Adverse Events

No case reports of adverse events were identified.

III. Pharmacology and Pharmacokinetics

Human Pharmacological Studies

No relevant human pharmacological studies were identified.

Animal Pharmacological Studies

No relevant animal pharmacological studies were identified.

In Vitro Pharmacological Studies

No relevant in vitro pharmacological studies were identified.

IV. Pregnancy and Lactation

No information on the safety of bog bean during pregnancy or lactation was identified.

V. Toxicity Studies

No toxicity studies were identified.

Literature Cited

Bradley, P.R. 1993. *British herbal compendium: A handbook of scientific information on widely used plant drugs*. Bournemouth, UK: British Herbal Medicine Association.

Weiss, R.F., and A.R. Meuss. 2001. *Weiss's herbal medicine*. Classic ed. New York: Thieme.

Wichtl, M. 2004. *Herbal drugs and phytopharmaceuticals: A handbook for practice on a scientific basis*. 3rd ed. Boca Raton, FL: CRC Press.

M

Mitchella repens L. — Rubiaceae

SCN: partridge berry
OCN: squawvine
Part: herb

Quick Reference Summary

Safety Class: 1
Interaction Class: A

Contraindications

None known.

Other Precautions

None known.

Drug and Supplement Interactions

None known.

Adverse Events and Side Effects

None known.

Pharmacological Considerations

None known.

Pregnancy and Lactation

Partridge berry has traditionally been used to prevent threatened abortions and taken in the last few weeks before delivery (Cook 1869; Ellingwood 1919; Felter 1922; Felter and Lloyd 1898; Noe et al. 2002).

No information on the safety of partridge berry in lactation was identified in the scientific or traditional literature. While this review did not identify any concerns for use while nursing, safety has not been conclusively established.

Review Details

I. Drug and Supplement Interactions

Clinical Trials of Drug or Supplement Interactions

No clinical trials of drug or supplement interactions were identified.

Case Reports of Suspected Drug or Supplement Interactions

No case reports of suspected drug or supplement interactions were identified.

Animal Trials of Drug or Supplement Interactions

No animal trials of drug or supplement interactions were identified.

II. Adverse Events

Case Reports of Adverse Events

No case reports of adverse events were identified.

III. Pharmacology and Pharmacokinetics

Human Pharmacological Studies

No relevant human pharmacological studies were identified.

Animal Pharmacological Studies

No relevant animal pharmacological studies were identified.

In Vitro Pharmacological Studies

No relevant in vitro pharmacological studies were identified.

IV. Pregnancy and Lactation

Partridge berry has traditionally been used to prevent threatened abortions and taken in the last few weeks before delivery (Cook 1869; Ellingwood 1919; Felter 1922; Felter and Lloyd 1898; Noe et al. 2002).

No information on the safety of partridge berry in lactation was identified.

V. Toxicity Studies

No toxicity studies were identified.

Literature Cited

Cook, W. 1869. *The physiomedical dispensatory*. Cincinnati, OH: W.H. Cook.

Ellingwood, F. 1919. *The American materia medica, therapeutics and pharmacognosy*. Evanston, IL: Ellingwood's Therapeutist.

Felter, H.W. 1922. *The Eclectic materia medica, pharmacology and therapeutics*. Cincinnati, OH: Scudder.

Felter, H.W., and J.U. Lloyd. 1898. *King's American dispensatory*. 18th ed., 3rd rev. 2 vols. Cincinnati: Ohio Valley Co.

Noe, J.E., M. Bove, and K. Janel. 2002. Herbal tonic formulas for naturopathic obstetrics. *Altern. Complement. Ther.* 8(6):327-335.

M

Monarda spp.

Lamiaceae

Monarda clinopodia L.
SCN: beebalm
OCN: white bergamot; wild bergamot

Monarda didyma **L.**
SCN: Oswego tea
OCN: Oswego beebalm; scarlet beebalm; scarlet monarda

Monarda fistulosa **L.**
SCN: wild bergamot beebalm
OCN: beebalm; Oswego tea

Monarda pectinata Nutt.
SCN: spotted beebalm
OCN: plains beebalm; pony beebalm

Monarda punctata **L.**
SCN: horsemint
OCN: American horsemint; monarda; origanum; spotted beebalm
Part: herb

Quick Reference Summary

Safety Class: 2b
Interaction Class: A

Contraindications

Not for use in pregnancy except under the supervision of a qualified healthcare practitioner (Felter and Lloyd 1898; List and Hörhammer 1973).

Other Precautions

None known.

Drug and Supplement Interactions

None known.

Notice

Emmenagogue (Felter and Lloyd 1898; List and Hörhammer 1973; Williamson 2003); *see* Appendix 2.

Adverse Events and Side Effects

None known.

Pharmacological Considerations

None known.

Pregnancy and Lactation

Species of *Monarda* have traditionally been used as an emmenagogue (Felter and Lloyd 1898; List and Hörhammer 1973; Williamson 2003). Based on this information, use during pregnancy is not recommended except under the supervision of a qualified healthcare practitioner.

No information on the safety of *Monarda* during lactation was identified in the scientific or traditional literature. While this review did not identify any concerns for use while nursing, safety has not been conclusively established.

Review Details

I. Drug and Supplement Interactions

Clinical Trials of Drug or Supplement Interactions

No clinical trials of drug or supplement interactions were identified.

Case Reports of Suspected Drug or Supplement Interactions

No case reports of suspected drug or supplement interactions were identified.

Animal Trials of Drug or Supplement Interactions

No animal trials of drug or supplement interactions were identified.

II. Adverse Events

Case Reports of Adverse Events

No case reports of adverse events were identified.

III. Pharmacology and Pharmacokinetics

Human Pharmacological Studies

No relevant human pharmacological studies were identified.

Animal Pharmacological Studies

No relevant animal pharmacological studies were identified.

In Vitro Pharmacological Studies

Some inhibition of arachidonic acid-induced platelet aggregation in guinea pig and rat plasma was observed after treatment with Oswego tea essential oil (Tognolini et al. 2006).

IV. Pregnancy and Lactation

Species of *Monarda* have traditionally been used as an emmenagogue (Felter and Lloyd 1898; List and Hörhammer 1973; Williamson 2003).

No information on the safety of *Monarda* in lactation was identified.

V. Toxicity Studies

No toxicity studies were identified.

M

Literature Cited

Felter, H.W., and J.U. Lloyd. 1898. *King's American dispensatory*. 18th ed., 3rd rev. 2 vols. Cincinnati: Ohio Valley Co.

List, P.H., and H. Hörhammer. 1973. *Hagers handbuch der pharmazeutischen praxis*. Berlin: Springer.

Tognolini, M., E. Barocelli, V. Ballabeni, et al. 2006. Comparative screening of plant essential oils: Phenylpropanoid moiety as basic core for antiplatelet activity. *Life Sci.* 78(13):1419-1432.

Williamson, E.M. 2003. *Potter's herbal cyclopedia*. Saffron Walden, Essex: C.W. Daniel Co.

Morella spp. — Myricaceae

Morella cerifera (L.) Sm.
SCN: bayberry
Syn: *Myrica cerifera* L.
OCN: candle berry; southern bayberry; wax myrtle

Morella pensylvanica **(Mirb.) Kartesz, comb. nov. ined.**
SCN: bayberry
Syn: *Myrica pensylvanica* Mirb.
OCN: northern bayberry
Part: bark, root

Quick Reference Summary

Safety Class: 1
Interaction Class: A

Contraindications

None known.

Other Precautions

None known.

Drug and Supplement Interactions

None known.

Adverse Events and Side Effects

Excessive doses of bayberry may cause vomiting (Felter and Lloyd 1898; Remington and Wood 1918).

Pharmacological Considerations

None known.

Pregnancy and Lactation

No information on the safety of bayberry in pregnancy or lactation was identified in the scientific or traditional literature. Although this review did not identify any concerns for use while pregnant or nursing, safety has not been conclusively established.

Review Details

M

I. Drug and Supplement Interactions

Clinical Trials of Drug or Supplement Interactions

No clinical trials of drug or supplement interactions were identified.

Case Reports of Suspected Drug or Supplement Interactions

No case reports of suspected drug or supplement interactions were identified.

Animal Trials of Drug or Supplement Interactions

No animal trials of drug or supplement interactions were identified.

II. Adverse Events

Case Reports of Adverse Events

No case reports of adverse events were identified.

III. Pharmacology and Pharmacokinetics

Human Pharmacological Studies

No relevant human pharmacological studies were identified.

Animal Pharmacological Studies

No relevant animal pharmacological studies were identified.

In Vitro Pharmacological Studies

No relevant in vitro pharmacological studies were identified.

IV. Pregnancy and Lactation

No information on the safety of bayberry during pregnancy or lactation was identified.

V. Toxicity Studies

No toxicity studies were identified.

Literature Cited

Felter, H.W., and J.U. Lloyd. 1898. *King's American dispensatory.* 18th ed., 3rd rev. 2 vols. Cincinnati: Ohio Valley Co.

Remington, J.P., and H.C. Wood. 1918. *The dispensatory of the United States of America.* 20th ed. Philadelphia: Lippincott.

Morinda citrifolia L. Rubiaceae

SCN: noni
OCN: Indian mulberry

Part: fruit

Quick Reference Summary

Safety Class: 2b
Interaction Class: A

Contraindications

Not for use in pregnancy except under the supervision of a qualified healthcare practitioner.

Other Precautions

None known.

Drug and Supplement Interactions

None known.

Adverse Events and Side Effects

Several cases of hepatotoxicity have been reported in persons taking noni juice (Lopez-Cepero Andrada et al. 2007; Millonig et al. 2005; Stadlbauer et al. 2005, 2008; Yuce et al. 2006). In some of the cases, the patients were taking other drugs or supplements. In one case, noni was being taken with a drug that commonly causes liver dysfunction (Francis et al. 2003; Yuce et al. 2006). Animal and human studies of noni juice have not shown any elevated levels of liver enzymes (Davies and Mugglestone 2003; Glerup 2000; Kalandakanond et al. 2004; Mancebo et al. 2002; Product Safety Labs 2000), and a 2006 review of case reports and toxicological data by the European Food Safety Authority indicated that no convincing evidence exists for a causal relationship between noni consumption and the reported cases of hepatotoxicity (EFSA 2006).

Hyperkalemia was reported in a man with chronic kidney insufficiency who was taking noni. Analysis of a noni juice indicated potassium levels similar to those of other common fruit juices, all of which are typically restricted in persons with kidney insufficiency (Mueller et al. 2000).

Pharmacological Considerations

A dose escalation study of capsules containing noni fruit indicated that it was well tolerated at doses up to 10 g daily for 28 days (Issell et al. 2005).

Pregnancy and Lactation

Animal studies on the use of noni provide conflicting information, with one study showing no adverse effects on development at very large (6 g/kg) doses, while another study indicated that use of noni juice in pregnancy caused a delay in fetal bone formation (Marques et al. 2010; West et al. 2008).

One ethnobotanical text indicated that ingestion of "a large amount" of noni fruit was reported to cause an abortion (Cambie and Brewis 1997), although this activity has not been confirmed experimentally (Pawlus and Kinghorn 2007). No other information on the safety of noni in pregnancy or lactation was identified.

In this text, the contraindication for use in pregnancy is based on concerns regarding the recent cases of hepatotoxicity reported in association with noni use, as the implications of these case reports and possible mechanisms of hepatotoxicity have yet to be fully understood.

No information on the safety of noni fruit lactation was identified in the scientific or traditional literature. While this review did not identify any concerns for use while nursing, safety has not been conclusively established.

M

Review Details

I. Drug and Supplement Interactions

Clinical Trials of Drug or Supplement Interactions

No clinical trials of drug or supplement interactions were identified.

Case Reports of Suspected Drug or Supplement Interactions

Acquired coumadin resistance and decreased INR (a standardized scale used to report the results of blood coagulation tests; decreased INR indicates faster blood clotting) were reported in a 41-year-old woman who had been consuming "two small glasses" daily of noni juice for an unspecified length of time. The reporting physicians indicated that the product used might have been fortified with vitamin K, which causes acquired coumadin resistance (Carr et al. 2004).

Animal Trials of Drug or Supplement Interactions

No animal trials of drug or supplement interactions were identified.

II. Adverse Events

Case Reports of Adverse Events

Hepatotoxicity was reported in a 45-year-old man who had been consuming "a glass" of noni juice daily for approximately 3 weeks. Highly elevated transaminase and gamma-glutamyltransferase (GGT), elevated lactate dehydrogenase, and slightly elevated direct bilirubin levels were reported, and liver damage was confirmed by biopsy. Following cessation of noni juice, transaminase levels returned to normal after 1 month (Millonig et al. 2005). Anthraquinones were suggested as the possible cause of the hepatotoxicity (Millonig et al. 2005), although subsequent independent analysis of noni juice purchased from the manufacturer of the product implicated in this case report indicated that no anthraquinones were present in the juice (Jensen et al. 2006) and that absence of anthraquinones was a prerequisite for approval as a novel food by the European Commission (EC 2003; Jensen et al. 2006).

A 29-year-old man with a history of toxic hepatitis associated with small doses of paracetamol developed subacute hepatic failure following consumption of 1.5 liters of noni juice over a 3-week period. He was also taking approximately 7 g daily of a Chinese herbal remedy containing *Bupleurum*, *Pinellia*, *Scutellaria*, *Codonopsis*, *Glycyrrhiza*, *Schizonepeta*, and *Paeonia*. Liver function tests indicated elevated levels of bilirubin, aspartate aminotransferase (AST), alanine aminotransferase (ALT), gamma-glutamyltransferase (GGT), and alkaline phosphatase (ALP) (Stadlbauer et al. 2005).

A 62-year-old woman with no history of liver disease developed an episode of self-limited acute hepatitis following consumption of 2 liters of noni juice over a 3-month period. A percutaneous liver biopsy revealed acute hepatitis consistent with an idiosyncratic drug reaction. Elevated levels of AST, ALT, and bilirubin were observed and returned to normal after cessation of noni. The woman had been treated the previous year with fludarabine for chronic B-cell leukemia (Stadlbauer et al. 2005).

M

Acute hepatitis was reported in a 24-year-old woman with multiple sclerosis who had been consuming noni juice (dose and frequency not specified) for 4 weeks. The woman had been treated with interferon b1a (IFN) for 6 weeks, a drug that commonly causes hepatic dysfunction of varying severity in multiple sclerosis patients (Francis et al. 2003). IFN was initially suspected as the cause of hepatotoxicity in this case. IFN administration was ceased, and approximately 1 week later noni was ceased. Liver enzyme levels (ALT and bilirubin) increased dramatically for 2 weeks after cessation of IFN and then returned to normal levels (Yuce et al. 2006).

Hepatotoxicity was reported in a 43-year-old man who had been recently diagnosed with a malignant brain tumor (glioblastoma). The man had consumed 20 ml daily of noni juice for 2 weeks and was also taking levetiracetam twice daily (Stadlbauer et al. 2008). Hepatotoxicity was reported in a 33-year-old woman who had consumed several doses of noni (amount unspecified) over a period of several weeks (Lopez-Cepero Andrada et al. 2007).

A case of hyperkalemia was reported in a man with chronic renal insufficiency. The man had consumed approximately 4.5 oz of noni juice daily for an unspecified length of time. The reporting physicians ordered and submitted for analysis a separate noni juice product that also contained grape juice, natural flavors, and benzoic acid. This juice was found to contain approximately 56 mEq/l of potassium, similar to that of orange, tomato, and grapefruit juices, all of which are usually restricted in the diets of patients with end-stage renal disease (Mueller et al. 2000).

III. Pharmacology and Pharmacokinetics

Human Pharmacological Studies

A dose escalation study of capsules containing 500 mg of an extract of ripe noni fruit in patients with advanced cancers showed that noni was well tolerated at doses up to 10 g daily for 28 days. That dose corresponds to approximately 200 ml of noni fruit juice. No adverse events were observed in the study (Issell et al. 2005).

In a clinical study, healthy volunteers were administered noni juice at three dose levels of up to 750 ml daily for 28 days. No significant changes in hematological, biochemical, or urological measurements were observed on examination at weeks 0, 2, 4, and 6. Biochemical analyses included alkaline phosphatase, ALT, AST, total bilirubin, lipids, creatine kinase, creatinine, gamma-glutamyltransferase, glucose, total protein, and uric acid (Davies and Mugglestone 2003).

Animal Pharmacological Studies

No allergic reactions were observed in guinea pigs subjected to sensitization and challenge tests with noni juice (Kaaber 2000; Product Safety Labs 2000).

In Vitro Pharmacological Studies

In human liver cells (HepG2), freeze-dried noni fruit puree did not decrease viability or induce neutral lipid accumulation or phospholipidosis (West et al. 2009).

IV. Pregnancy and Lactation

In rats orally administered 1.72, 3.43, or 6.86 g/kg freeze-dried noni puree daily on days 1 to 21 of pregnancy, no adverse effects on fetal development were observed, including decreased number of live fetuses, resorptions, decreased fetal weight and length, skeletal abnormalities, or gross malformations (West et al. 2008).

In rats orally administered 0.4, 2, or 20 ml/kg of noni aqueous extract or 0.4, 2, or 20 ml/kg of noni juice on days 7 to 15 of pregnancy, ossification in fetuses was delayed. No signs of maternal toxicity were observed (Marques et al. 2010).

In rats orally administered 7.5, 75, or 750 mg/kg of an aqueous extract of noni daily, no adverse effects on pregnancy duration, number of implants, litter size, or live birth index were observed. At the 7.5 mg/kg dose, a reduction in the parturition index was observed (Müller et al. 2009).

One ethnobotanical text indicates that ingestion of "a large amount" of noni fruit has been reported to cause an abortion (details on dose, duration, and preparation not available) (Cambie and Brewis 1997), although this activity has not been confirmed experimentally (Pawlus and Kinghorn 2007).

No information on the safety of noni during lactation was identified.

V. Toxicity Studies

Acute Toxicity

The LD_{50} of an intraperitoneally administered methanol extract of noni fruit in mice is 3.5 g/kg (Chearskul et al. 2004; Nakanishi et al. 1965), whereas that of an aqueous extract is 7.5 g/kg (Chearskul et al. 2004).

Short-Term Toxicity

No changes in clinical chemistry measurements, including serum glutamic pyruvic transaminase (SGPT) and serum glutamic oxaloacetic transaminase (SGOT), alkaline phosphatase, blood urea nitrogen, creatinine, Na^+, K^+, and Cl–, were observed in rats administered noni juice daily (dose not specified in English language translation) for 30 days (Kalandakanond et al. 2004).

No signs of toxicity or behavioral changes were observed in rats orally administered 15 g/kg of pureed noni fruit (Product Safety Labs 2000). No changes in biochemical parameters and no histopathological findings were observed in rats administered 1 g/kg of noni fruit extract daily for 28 days (Mancebo et al. 2002).

Subchronic Toxicity

No changes in histology, clinical chemistry (including liver enzymes), hematology, weight gain, or behavior were observed in rats administered 80 ml/kg noni juice daily for 13 weeks (Wang et al. 2002). Based on this study, a no-observable-adverse-effect level (NOAEL) was determined to be greater than 80 ml/kg body weight (Glerup 2000).

In rats orally administered 1.72, 3.43, or 6.86 g/kg of a freeze-dried noni fruit puree daily for 90 days, no histopathological changes or evidence of dose-related responses in hematological and clinical chemistry measurements, including liver function tests, were observed. Based on the results the NOAEL for freeze-dried noni fruit puree was determined as greater than 6.86 g/kg body weight, equivalent to approximately 90 ml/kg of noni fruit juice (West et al. 2009).

Genotoxicity

A slight mutagenic effect of noni juice extract was observed in the *Salmonella* microsome assay in strain TA1537 but not in strains TA98 or TA100. The activity was attributed to the presence of flavonoids (Westendorf et al. 2007). No mutagenicity of noni juice was observed in the mammalian mutagenicity test with V79 Chinese hamster fibroblasts (Westendorf et al. 2007). Rats treated with a noni juice concentrate did not show DNA repair synthesis in primary rat hepatocytes, nor could DNA adducts or DNA strand breaks be observed (Westendorf et al. 2007).

No clastogenic activity of noni was observed in mice orally administered a single dose of 10 g/kg of the dehydrated juice. Observation of bone marrow showed no increase in micronuclei related to ingestion of noni juice (Edwards 2002).

No activity of noni juice at concentrations of 625, 1250, 2500, and 5000 μg/ml (dry weight) was observed in the chromosomal aberration test with human lymphocytes with or without metabolic activation from S9 mix (Edwards 2003).

Literature Cited

Cambie, R.C., and A.A. Brewis. 1997. *Anti-fertility plants of the Pacific*. Collingwood, Victoria, Australia: CSIRO Publishing.

Carr, M.E., J. Klotz, and M. Bergeron. 2004. Coumadin resistance and the vitamin supplement "Noni." *Am. J. Hematol.* 77(1):103.

Chearskul, S., S. Koopiwut, S. Chatchawalvanit, et al. 2004. *Morinda citrifolia* has very weak estrogenic activity *in vivo*. *Thai J. Physiol. Sci.* 17:22-29.

Davies, C., and C. Mugglestone. 2003. A single centre, double-blind, three dose level, parallel group, placebo controlled safety study with Tahitian noni juice in healthy subjects Surrey, UK: BIBRA International Ltd.

EC. 2003. EU Commission Decision of 5 June 2003 authorising the placing on the market of "noni juice" (juice of the fruit of *Morinda citrifolia* L.) as a novel food ingredient under Regulation (EC) No 258/97 of the European Parliament and of the Council.

Edwards, C. 2002. Tahitian noni juice—Mouse micronucleus test. Test report. Lille Skensved, Denmark: Scantox Biologisk Laboratorium.

Edwards, C. 2003. In vitro mammalian chromosome aberration test performed with human lymphocytes. Test Report. Lille Skensved, Denmark: Scantox Biologisk Laboratorium.

EFSA. 2006. Opinion on request from the Commission related to safety of noni juice (juice of the fruits of *Morinda citrifolia*). *EFSA J.* 376:1-12.

Francis, G.S., Y. Grumser, E. Alteri, et al. 2003. Hepatic reactions during treatment of multiple sclerosis with interferon-β-1a: Incidence and clinical significance. *Drug Saf.* 26(11):815.

Glerup, P. 2000. Tahitian noni juice—A 13-wk oral (gavage) toxicity study in rats. Lille Skensved, Denmark: Scantox Biologisk Laboratorium.

Issell, B.F., C. Gotay, I. Pagano, and A. Franke. 2005. Quality of life measures in a phase I trial of noni. *J. Clin. Oncol.* 23(16):8217.

Jensen, C.J., J. Westendorf, M.Y. Wang, and D.P. Wadsworth. 2006. Noni juice protects the liver. *Eur. J. Gastroenterol. Hepatol.* 18(5):575-577.

Kaaber, K. 2000. Tahitian noni juice: Active systemic anaphylaxis test in the guinea pig. Lille Skensved, Denmark: Scantox Biologisk Laboratorium.

Kalandakanond, S., J. Pandaranandaga, S. Komolvanich, and S. Poonyachoti. 2004. A study on the anxiolytic effect of juice from the fruit of noni (*Morinda citrifolia* L. Rubiaceae) on Wistar rats. *Thai J. Pharm.* 34(1):99-105.

Lopez-Cepero Andrada, J.M., S. Lerma Castilla, M.D. Fernandez Olvera, and A. Amaya Vidal. 2007. Hepatotoxicity caused by a noni (*Morinda citrifolia*) preparation. *Rev. Esp. Enferm. Dig.* 99(3):179-181.

Mancebo, A., I. Scull, Y. Gonzalez, et al. 2002. Repeated dose oral toxicity assay (28 days) of the aqueous extract of *Morinda citrifolia* in Sprague-Dawley rats. *Rev. Toxicol.* 19(2):73-78.

Marques, N.F., A.P. Marques, A.L. Iwano, et al. 2010. Delayed ossification in Wistar rats induced by *Morinda citrifolia* L. exposure during pregnancy. *J. Ethnopharmacol.* 128 (1):85-91.

Millonig, G., S. Stadlmann, and W. Vogel. 2005. Herbal hepatotoxicity: Acute hepatitis caused by a noni preparation (*Morinda citrifolia*). *Eur. J. Gastroenterol. Hepatol.* 17(4):445-447.

Mueller, B.A., M.K. Scott, K.M. Sowinski, and K.A. Prag. 2000. Noni juice (*Morinda citrifolia*): Hidden potential for hyperkalemia? *Am. J. Kidney Dis.* 35(2):310-312.

Müller, J.C., G.G. Botelho, A.C. Bufalo, et al. 2009. *Morinda citrifolia* Linn. (noni): In vivo and in vitro reproductive toxicology. *J. Ethnopharmacol.* 121(2):229-233.

Nakanishi, K., S. Sasaki, A.K. Kiang, et al. 1965. Phytochemical survey of Malaysian plants: Preliminary chemical and pharmacological screening. *Chem. Pharm. Bull. (Tokyo)* 13(7):882-890.

Pawlus, A.D., and A.D. Kinghorn. 2007. Review of the ethnobotany, chemistry, biological activity and safety of the botanical dietary supplement *Morinda citrifolia* (noni). *J. Pharm. Pharmacol.* 59(12):1587-1609.

Product Safety Labs. 2000. *Cited in* West, B. J., C. J. Jensen, et al. (2006). A safety review of noni fruit juice. *J. Food Sci.* 71(8):R100-106.

Stadlbauer, V., P. Fickert, C. Lackner, et al. 2005. Hepatotoxicity of noni juice: Report of two cases. *World J. Gastroenterol.* 11(30):4758-4760.

Stadlbauer, V., S. Weiss, F. Payer, and R. Stauber. 2008. Herbal does not at all mean innocuous: The sixth case of hepatotoxicity associated with *Morinda citrifolia* (noni). *Am. J. Gastroenterol.* 103(9):2406-2407.

Wang, M.Y., B.J. West, C.J. Jensen, et al. 2002. *Morinda citrifolia* (noni): A literature review and recent advances in noni research. *Acta Pharmacol. Sin.* 23(12):1127-1141.

West, B.J., C.X. Su, and C.J. Jensen. 2008. Prenatal toxicity test of *Morinda citrifolia* (noni) fruit. *J. Toxicol. Sci.* 33(5):647-649.

West, B.J., C.X. Su, and C.J. Jensen. 2009. Hepatotoxicity and subchronic toxicity tests of *Morinda citrifolia* (noni) fruit. Review of noni update. *J. Toxicol. Sci.* 34(5):581-585.

Westendorf, J., K. Effenberger, H. Iznaguen, and S. Basar. 2007. Toxicological and analytical investigations of noni (*Morinda citrifolia*) fruit juice. *J. Agric. Food Chem.* 55(2):529-537.

Yuce, B., V. Gulberg, J. Diebold, and A.L. Gerbes. 2006. Hepatitis induced by noni juice from *Morinda citrifolia*: A rare cause of hepatotoxicity or the tip of the iceberg? *Digestion* 73(2-3):167-170.

Morus alba L. — Moraceae

SCN: white mulberry
PN: *sang shen zi* (fruit)

Part: fruit

Quick Reference Summary

Safety Class: 1
Interaction Class: A

Contraindications

None known.

Other Precautions

None known.

Drug and Supplement Interactions

None known.

Adverse Events and Side Effects

Allergic reactions to white mulberry fruit, including an anaphylactic reaction, have been reported (Bensky et al. 2004; Navarro et al. 1997).

Pharmacological Considerations

None known.

Pregnancy and Lactation

No information on the safety of white mulberry fruit in pregnancy or lactation was identified in the scientific or traditional literature. Although this review did not identify any concerns for use while pregnant or nursing, safety has not been conclusively established.

Review Details

I. Drug and Supplement Interactions

Clinical Trials of Drug or Supplement Interactions

No clinical trials of drug or supplement interactions were identified.

Case Reports of Suspected Drug or Supplement Interactions

No case reports of suspected drug or supplement interactions were identified.

Animal Trials of Drug or Supplement Interactions

No animal trials of drug or supplement interactions were identified.

II. Adverse Events

Case Reports of Adverse Events

A reference text on traditional Chinese medicine indicates that "high doses" of white mulberry fruit (standard dose is listed as a decoction of 9–15 g of dried fruit) can cause hemorrhagic colitis in children, and that two such cases have been reported in the literature. In one of the cases, a 3-year-old boy ate three handfuls of fruit (no indication of whether the fruit was fresh or dried). No details on this case or the second case were noted (Bensky et al. 2004). Conversely, another text on traditional Chinese medicine indicates no such concerns and notes that white mulberry fruit has a gentle and subtle effect and is appropriate as a long-term tonic (Chen and Chen 2004).

Allergic skin reactions to mulberry have been reported (Bensky et al. 2004). An anaphylactic reaction after eating a mulberry was reported in a woman with hypersensitivity to a number of tree pollens. The allergy to white mulberry was confirmed by skin prick testing (Navarro et al. 1997).

A small study in patients with multiple sensitizations to food allergens (mostly fruit) indicated that hypersensitivity to figs and mulberries might be associated as the result of allergen cross-reactivity (Caiaffa et al. 2003).

III. Pharmacology and Pharmacokinetics

Human Pharmacological Studies

No relevant human pharmacological studies were identified.

Animal Pharmacological Studies

No relevant animal pharmacological studies were identified.

In Vitro Pharmacological Studies

No estrogenic activity of an ethanol extract of white mulberry fruit was observed in a recombinant yeast system with a human estrogen receptor expression plasmid and a reporter plasmid (Kim et al. 2008).

IV. Pregnancy and Lactation

No information on the safety of white mulberry fruit during pregnancy or lactation was identified.

V. Toxicity Studies

No toxicity studies were identified.

M

Literature Cited

Bensky, D., S. Clavey, and E. Stöger. 2004. *Chinese herbal medicine: Materia medica*. 3rd ed. Seattle: Eastland Press.

Caiaffa, M.F., V.M. Cataldo, A. Tursi, and L. Macchia. 2003. Fig and mulberry cross-allergy. *Ann. Allergy Asthma Immunol.* 91(5):493-495.

Chen, J.K., and T.T. Chen. 2004. *Chinese medical herbology and pharmacology*. City of Industry, CA: Art of Medicine Press.

Kim, I.G., S.C. Kang, K.C. Kim, E.S. Choung, and O.P. Zee. 2008. Screening of estrogenic and antiestrogenic activities from medicinal plants. *Environ. Toxicol. Pharmacol.* 25(1):75-82.

Navarro, A.M., J.C. Orta, M.C. Sanchez, et al. 1997. Primary sensitization to *Morus alba*. *Allergy* 52(11):1144-1145.

Morus alba L. — Moraceae

SCN: white mulberry
PN: *sang ye* (leaf)
Part: leaf

Quick Reference Summary

Safety Class: 1
Interaction Class: A

Contraindications

None known.

Other Precautions

None known.

Drug and Supplement Interactions

None known.

Adverse Events and Side Effects

None known.

Pharmacological Considerations

Human and animal studies have demonstrated that white mulberry leaf may modify glucose regulation (Abou-Seif and Kamel 2008; Butt et al. 2008; Kimura et al. 2007; Mudra et al. 2007; Musabayane et al. 2006; Naowaboot et al. 2009). People with diabetes are advised to monitor their blood sugar closely and discuss the use of this herb with a qualified healthcare practitioner prior to use.

An animal study indicated that antidopaminergic activity of white mulberry leaf extract was observed in mice injected with relatively large doses (up to 200 mg/kg) (Yadav et al. 2008). The relevance of that finding to oral use at standard therapeutic doses is not known.

Pregnancy and Lactation

A study in pregnant sheep indicated no adverse effects of white mulberry leaf on the birth weight of offspring (Prasad et al. 1995).

No information on the safety of white mulberry leaf during lactation was identified. While this review did not identify any concerns for use while nursing, safety has not been conclusively established.

Review Details

I. Drug and Supplement Interactions

Clinical Trials of Drug or Supplement Interactions

No clinical trials of drug or supplement interactions were identified.

Case Reports of Suspected Drug or Supplement Interactions

No case reports of suspected drug or supplement interactions were identified.

M

Animal Trials of Drug or Supplement Interactions

No animal trials of drug or supplement interactions were identified.

II. Adverse Events

Case Reports of Adverse Events

No case reports of adverse events were identified.

III. Pharmacology and Pharmacokinetics

Human Pharmacological Studies

In healthy volunteers orally administered single doses of 0.4, 0.8, or 1.2 g of a white mulberry leaf powder enriched with the compound 1-deoxynojirimycin to a concentration of 1.5%, doses of 0.8 or 1.2 g of powder significantly suppressed the elevation of postprandial blood glucose and secretion of insulin (Kimura et al. 2007).

A reduction in blood glucose increases and a decrease in peak to trough fluctuations in blood glucose were observed in type 2 diabetes patients (without complications and with use of oral hypoglycemic agents) orally administered single doses of 1 g white mulberry leaf extract. Patients were allowed to continue taking any medication except acarbose (Mudra et al. 2007).

A survey indicated that white mulberry leaf is commonly used in the management of type 1 diabetes in Turkey (Arykan et al. 2009). White mulberry leaf has also been used in the management of diabetes in Chile, Thailand, and other countries (Lemus et al. 1999; Naowaboot et al. 2009).

Animal Pharmacological Studies

Amelioration of hyperglycemia was observed in diabetic rats orally administered 200 mg/kg of an aqueous extract of white mulberry leaf daily for 2 weeks (Abou-Seif and Kamel 2008). A reduction in blood glucose levels was observed in diabetic rats orally administered 200 mg/kg of a white mulberry leaf extract daily for 5 weeks. No effects on glucose levels in healthy rats were observed (Musabayane et al. 2006). A 22% reduction in blood glucose levels was observed in diabetic rats orally administered 1 g/kg white mulberry leaf aqueous extract daily for 6 weeks (Naowaboot et al. 2009). Several other animal studies have indicated similar findings (Chen et al. 1995; Devi and Urooj 2008; Hosseinzadeh and Sadeghi 1999; Miyahara et al. 2004).

In mice intraperitoneally administered doses of 50, 100, or 200 mg/kg of a methanol extract of white mulberry leaf, a reduction in the number of fights and an increased latency to fights in foot shock induced-aggression was observed, along with a dose-dependent decrease in amphetamine-induced stereotyped behavior and a prolongation of sleeping time induced by phenobarbitone. The authors of the study indicated that the results suggest antidopaminergic activity of white mulberry leaf methanol extract (Yadav et al. 2008).

In Vitro Pharmacological Studies

No estrogenic activity of an ethanol extract of white mulberry leaf was observed in a recombinant yeast system with a human estrogen receptor expression plasmid and a reporter plasmid (Kim et al. 2008).

IV. Pregnancy and Lactation

No effects on birth weight were observed in lambs from pregnant sheep (6 to 8 weeks pregnant) fed a hay containing 30 or 50% white mulberry leaf in addition to a base diet for 3 months (Prasad et al. 1995).

No information on the safety of white mulberry leaf during lactation was identified.

V. Toxicity Studies

Acute Toxicity

The LD_{50} of an intraperitoneally administered aqueous extract of white mulberry is 4 g/kg in mice and 5 g/kg in rats. The LD_{50} of the same extract orally administered to mice and rats could not be determined at doses up to 5 g/kg (Trabsung 2004).

Short-Term Toxicity

No adverse effects were observed in rats orally administered up to 3 g/kg daily of an aqueous extract of white mulberry leaf for 60 days. Observed parameters included blood chemistry, hematological values, and microscopic evaluation of major organs (Trabsung 2004).

No adverse effects were observed in rats fed diets containing 1 or 5% powdered white mulberry leaf daily for 4 weeks. Observed parameters included organ weights, hematologic and biochemical values, and pathological examination (Mitsuya et al. 2001).

No adverse effects, including signs of toxicity or changes in organ weights, were observed in rats fed diets containing 25% white mulberry leaf powder for 28 days (Srivastava et al. 2003).

No adverse effects in milking cows or changes in the yield or butterfat content of their milk were observed after animals were fed up to 6 kg of white mulberry leaf daily for 100 days (Srivastava et al. 2003).

Subchronic Toxicity

No adverse effects were observed in rats fed diets containing up to 1% white mulberry leaf extract (approximate daily mulberry dose of 884.5 mg/kg for males and 995.7 mg/kg for females) daily for 90 days. Observed parameters included body weight gain, hematology and blood chemistry, and microscopic examination of major organs (Miyazawa et al. 2003).

Literature Cited

Abou-Seif, M.A.M., and E.S.M. Kamel. 2008. Hypoglycemic and metabolic activity of aqueous extract of *Morus alba* in streptozotocin-diabetic rats. *Biosci. Biotechnol. Res. Asia* 5(1):139-144.

Arykan, D., S.K. Sivrikaya, and N. Olgun. 2009. Complementary alternative medicine use in children with type 1 diabetes mellitus in Erzurum, Turkey. *J. Clin. Nurs.* 18(15):2136-2144.

Butt, M.S., A. Nazir, M.T. Sultan, and K. Schroën. 2008. *Morus alba* L.: Nature's functional tonic. *Trends Food Sci. Technol.* 19(10):505-512.

Chen, F., N. Nakashima, I. Kimura, and M. Kimura. 1995. Hypoglycemic activity and mechanisms of extracts from mulberry leaves (*folium mori*) and cortex mori radicis in streptozotocin-induced diabetic mice. *Yakugaku Zasshi* 115(6):476-482.

Devi, V.D., and A. Urooj. 2008. Hypoglycemic potential of *Morus indica* L. and *Costus igneus* Nak.—A preliminary study. *Indian J. Exp. Biol.* 46(8):614-616.

Hosseinzadeh, H., and A. Sadeghi. 1999. Antihyperglycemic effects of *Morus nigra* and *Morus alba* in mice. *Pharm. Pharmacol. Lett.* 9(2):63-65.

Kim, I.G., S.C. Kang, K.C. Kim, E.S. Choung, and O.P. Zee. 2008. Screening of estrogenic and antiestrogenic activities from medicinal plants. *Environ. Toxicol. Pharmacol.* 25(1):75-82.

Kimura, T., K. Nakagawa, H. Kubota, et al. 2007. Food-grade mulberry powder enriched with 1-deoxynojirimycin suppresses the elevation of postprandial blood glucose in humans. *J. Agric. Food Chem.* 55(14):5869-5874.

Lemus, I., R. Garcia, E. Delvillar, and G. Knop. 1999. Hypoglycaemic activity of four plants used in Chilean popular medicine. *Phytother. Res.* 13(2):91-94.

Mitsuya, M., N. Suegara, Y. Kojima, et al. 2001. Four-week oral toxicity studies of the leaf powder of mulberry (*Morus alba* L.) in rats. *Pharmacometrics* 61(1):169-176.

Miyahara, C., M. Miyazawa, S. Satoh, A. Sakai, and S. Mizusaki. 2004. Inhibitory effects of mulberry leaf extract on postprandial hyperglycemia in normal rats. *J. Nutr. Sci. Vitaminol. (Tokyo)* 50 (3):161-164.

Miyazawa, M., C. Miyahara, S. Satoh, and A. Sakai. 2003. Ninety-day dietary toxicity study of mulberry leaf extract in rats. *Shokuhin Eiseigaku Zasshi* 44(4):191-197.

Mudra, M., N. Ercan-Fang, L. Zhong, J. Furne, and M. Levitt. 2007. Influence of mulberry leaf extract on the blood glucose and breath hydrogen response to ingestion of 75 g sucrose by type 2 diabetic and control subjects. *Diabetes Care* 30(5):1272-1274.

Musabayane, C.T., P.T. Bwititi, and J.A. Ojewole. 2006. Effects of oral administration of some herbal extracts on food consumption and blood glucose levels in normal and streptozotocin-treated diabetic rats. *Methods Find. Exp. Clin. Pharmacol.* 28(4):223-228.

Naowaboot, J., P. Pannangpetch, V. Kukongviriyapan, B. Kongyingyoes, and U. Kukongviriyapan. 2009. Antihyperglycemic, antioxidant and antiglycation activities of mulberry leaf extract in streptozotocin-induced chronic diabetic rats. *Plant Foods Hum. Nutr.* 64(2):116-121.

Prasad, P.E., D.N. Reddy, M.R. Reddy, and G.V.N. Reddy. 1995. Effect of feeding mulberry (*Morus alba*) hay in the rations to pregnant ewes. *Indian J. Anim. Nutr.* 12(2):109-111.

Srivastava, S., R. Kapoor, A. Thathola, and R.P. Srivastava. 2003. Mulberry (*Morus alba*) leaves as human food: A new dimension of sericulture. *Int. J. Food Sci. Nutr.* 54(6):411-416.

Trabsung, A. 2004. The toxicity study of *Morus alba* L. leaf extract. Faculty of Graduate Studies, Mahidol University, Nakhon Pathom, Thailand.

Yadav, A.V., L.A. Kawale, and V.S. Nade. 2008. Antidopaminergic effect of methanolic extract of *Morus alba* Linn. leaves. *PharmacologyOnLine* 1:218-232.

Morus alba L. — Moraceae

SCN: white mulberry
PN: *sang bai pi* (root bark)

Part: root bark

Quick Reference Summary

Safety Class: 1
Interaction Class: A

Contraindications

None known.

Other Precautions

None known.

Drug and Supplement Interactions

None known.

Adverse Events and Side Effects

None known.

Pharmacological Considerations

An animal study demonstrated that white mulberry root bark may modify glucose regulation (Singab et al. 2005). People with diabetes are advised to monitor their blood sugar closely and discuss the use of this herb with a qualified healthcare practitioner prior to use.

Pregnancy and Lactation

No information on the safety of white mulberry root bark in pregnancy or lactation was identified in the scientific or traditional literature. Although this review did not identify any concerns for use while pregnant or nursing, safety has not been conclusively established.

Review Details

I. Drug and Supplement Interactions

Clinical Trials of Drug or Supplement Interactions

No clinical trials of drug or supplement interactions were identified.

Case Reports of Suspected Drug or Supplement Interactions

No case reports of suspected drug or supplement interactions were identified.

Animal Trials of Drug or Supplement Interactions

No animal trials of drug or supplement interactions were identified.

II. Adverse Events

Case Reports of Adverse Events

No case reports of adverse events were identified.

III. Pharmacology and Pharmacokinetics

Human Pharmacological Studies

No relevant human pharmacological studies were identified.

Animal Pharmacological Studies

Reduction in glucose levels and increase in insulin levels were observed in diabetic rats orally administered 200 or 400 mg/kg of a hydroalcoholic extract of white mulberry root bark daily for 10 days (Singab et al. 2005).

In Vitro Pharmacological Studies

In assays to assess the immunomodulatory activity of a polysaccharide isolated from white mulberry root bark, the compound was found to enhance proliferation of splenic lymphocytes in a synergistic manner in the presence of mitogens. However, the compound suppressed primary IgM antibody production from B cells, which was activated with lipopolysaccharide, a polyclonal activator, or immunized with a T-cell-dependent antigen, sheep red blood cells (Kim et al. 2000).

No estrogenic activity of an ethanol extract of white mulberry root was observed in a recombinant yeast system with a human estrogen receptor expression plasmid and a reporter plasmid (Kim et al. 2008).

IV. Pregnancy and Lactation

No information on the use of white mulberry root bark during pregnancy or lactation was identified.

V. Toxicity Studies

Acute Toxicity

The LD_{50} values of water or alcohol extracts of white mulberry root bark in mice are 10 g/kg after oral administration and 5 g/kg after intravenous administration (Chen and Chen 2004).

Literature Cited

Chen, J.K., and T.T. Chen. 2004. *Chinese medical herbology and pharmacology.* City of Industry, CA: Art of Medicine Press.

Kim, H.M., S.B. Han, K.H. Lee, et al. 2000. Immunomodulating activity of a polysaccharide isolated from Mori Cortex Radicis. *Arch. Pharm. Res.* 23(3):240-242.

Kim, I.G., S.C. Kang, K.C. Kim, E.S. Choung, and O.P. Zee. 2008. Screening of estrogenic and antiestrogenic activities from medicinal plants. *Environ. Toxicol. Pharmacol.* 25(1):75-82.

Singab, A.N., H.A. El-Beshbishy, M. Yonekawa, T. Nomura, and T. Fukai. 2005. Hypoglycemic effect of Egyptian *Morus alba* root bark extract: Effect on diabetes and lipid peroxidation of streptozotocin-induced diabetic rats. *J. Ethnopharmacol.* 100(3):333-338.

Morus alba L.

Moraceae

SCN: white mulberry
PN: *sang zhi* (twig)

Part: twig

Quick Reference Summary

Safety Class: 1
Interaction Class: A

Contraindications

None known.

Other Precautions

None known.

Drug and Supplement Interactions

None known.

Adverse Events and Side Effects

None known.

Pharmacological Considerations

None known.

Pregnancy and Lactation

Texts on traditional Chinese medicine do not indicate any cautions for use of white mulberry twig during pregnancy or lactation (Bensky et al. 2004; Chen and Chen 2004). Although this review did not identify any concerns for use while pregnant or nursing, safety has not been conclusively established.

Review Details

I. Drug and Supplement Interactions

Clinical Trials of Drug or Supplement Interactions

No clinical trials of drug or supplement interactions were identified.

Case Reports of Suspected Drug or Supplement Interactions

No case reports of suspected drug or supplement interactions were identified.

Animal Trials of Drug or Supplement Interactions

No animal trials of drug or supplement interactions were identified.

II. Adverse Events

Case Reports of Adverse Events

No case reports of adverse events were identified.

III. Pharmacology and Pharmacokinetics

Human Pharmacological Studies

No relevant human pharmacological studies were identified.

Animal Pharmacological Studies

No relevant animal pharmacological studies were identified.

In Vitro Pharmacological Studies

No relevant in vitro pharmacological studies were identified.

IV. Pregnancy and Lactation

Texts on traditional Chinese medicine do not indicate any cautions for use of white mulberry twig during pregnancy or lactation (Bensky et al. 2004; Chen and Chen 2004).

V. Toxicity Studies

No toxicity studies were identified.

Literature Cited

Bensky, D., S. Clavey, and E. Stöger. 2004. *Chinese herbal medicine: Materia medica*. 3rd ed. Seattle: Eastland Press.

Chen, J.K., and T.T. Chen. 2004. *Chinese medical herbology and pharmacology*. City of Industry, CA: Art of Medicine Press.

Mucuna pruriens (L.) DC. — Fabaceae

SCN: velvet bean
AN: *atmagupta; kapikacchu*
OCN: buffalo bean; cowage; cowitch
Part: root, seed

Quick Reference Summary

Safety Class; 2b

Interaction Class: A

Contraindications

Not for use in pregnancy except under the supervision of a qualified healthcare practitioner (Nath et al. 1992).

Other Precautions

None known.

Drug and Supplement Interactions

None known.

M

Editors' Notes

Velvet bean contains trace amounts of certain hallucinogenic compounds. While the compound DMT (*N,N*-dimethyltryptamine) has been reported in velvet bean (Ghosal et al. 1971), a recent analysis did not find any DMT, but indicated the presence of the related compounds bufotenin (5-hydroxydimethyltryptamine) at a concentration of 1.2 to 1.5 ppm and 5-MeODMT (5-methoxydimethyltryptamine) at a concentration of 0.34 to 0.63 ppm (Szabo 2003).

Velvet bean contains 3.6 to 7% of the compound L-dopa (3,4-dihydroxyphenylalanine), an amino acid used in the treatment of Parkinson's disease (Pugalenthi et al. 2005).

Adverse Events and Side Effects

Contact with the hairs on the velvet bean seed pod can cause severe itching (pruritis) (Davidson et al. 2007; Kosteletzky et al. 2009; MMWR 1985).

Pharmacological Considerations

Animal studies have indicated that velvet bean may modify glucose regulation (Akhtar et al. 1990; Bhaskar et al. 2008; Rathi et al. 2002). People with diabetes are advised to monitor their blood sugar closely and discuss the use of this herb with a qualified healthcare practitioner prior to use.

Pregnancy and Lactation

An increase in fetal malformations was observed in an animal study with velvet bean (Nath et al. 1992). Based on this information, use during pregnancy is not recommended except under the supervision of a qualified healthcare practitioner.

No information on the safety of velvet bean during lactation was identified in the scientific or traditional literature. While this review did not identify any concerns for use while nursing, safety has not been conclusively established.

Review Details

I. Drug and Supplement Interactions

Clinical Trials of Drug or Supplement Interactions

No clinical trials of drug or supplement interactions were identified.

Case Reports of Suspected Drug or Supplement Interactions

No case reports of suspected drug or supplement interactions were identified.

Animal Trials of Drug or Supplement Interactions

No animal trials of drug or supplement interactions were identified.

II. Adverse Events

Case Reports of Adverse Events

An outbreak of cases of acute toxic psychosis was reported in a village in Mozambique during a period of famine and disruption of agriculture, when village residents were forced to subsist on wild plants, including velvet bean. Symptoms included severe headache and heart palpitations, confusion, agitation, hallucinations, and paranoid delusions. The average amount of velvet bean consumed by each person was not reported. Although the bean was traditionally boiled in repeated changes of water prior to consumption, a water shortage during the outbreak may have contributed to insufficient processing of the beans (Infante et al. 1990).

Contact with the hairs on the outside of velvet bean seed pods can cause severe itching (pruritis). An outbreak of pruritis was reported after someone maliciously placed velvet bean pods in a couple's bed. The couple, and people who came in physical contact with the couple for the few hours after contact, all developed moderate to severe pruritis (MMWR 1985).

III. Pharmacology and Pharmacokinetics

Human Pharmacological Studies

In a trial of velvet bean in patients with Parkinson's disease, patients were administered single doses of 200 mg L-dopa with 50 mg decarboxylase inhibitor (carbidopa) (LD/CD), 15 g velvet bean powder (containing 500 mg of L-dopa), or 30 g of velvet bean powder (containing 1000 mg of L-dopa). After treatment, one patient experienced vomiting after 30 g velvet bean; mild nausea was reported in two patients after LD/CD and in two after 15 g velvet bean. Dizziness occurred in one patient each after each LD/CD and 15 g velvet bean. No clinically relevant changes in hematology or biochemistry parameters were observed (Katzenschlager et al. 2004).

Animal Pharmacological Studies

In healthy and diabetic rabbits orally administered 0.5, 1, or 2 g/kg of velvet bean, a reduction in blood glucose levels was observed in healthy animals at all doses and in diabetic animals at doses of 1 g/kg or more (Akhtar et al. 1990). Reductions in blood glucose levels were reported in diabetic rats orally administered 100, 200, or 400 mg/kg of an alcohol extract of velvet bean daily for 21 days. No significant changes in blood sugar levels were observed in diabetic mice administered the same extract at the same doses for 60 days (Rathi et al. 2002). In diabetic rats orally administered 100 or 200 mg/kg of an aqueous extract of velvet bean daily for 21 days, a reduction in blood glucose levels was observed (Bhaskar et al. 2008).

In Vitro Pharmacological Studies

No relevant in vitro pharmacological studies were identified.

IV. Pregnancy and Lactation

An increase in fetal malformations was observed in the fetuses of rats orally administered 175 mg/kg of an aqueous extract of velvet bean on days 0 to 10 of pregnancy (Nath et al. 1992).

No information on the safety of velvet bean during lactation was identified in the scientific or traditional literature. While this review did not identify any concerns for use while nursing, safety has not been conclusively established.

V. Toxicity Studies

Short-Term Toxicity

No adverse effects were observed in rats orally administered 600 mg/kg of an alcohol extract of velvet bean daily for 30 days (Tripathi and Upadhyay 2002).

A reduction in growth rate was observed in chicks fed diets containing 10, 20, or 30% velvet bean for 28 days. At the 30% rate, feed intake declined significantly and animals weighed 39% of those in a control group (Del Carmen et al. 1999). After being processed by cracking, soaking, and cooking prior to being used as a feed, velvet beans could be used as up to 25% of chicken feed without adversely affecting growth or animal health (Emenalom et al. 2005).

In sheep fed diets containing 330, 670, or 1000 g/kg velvet bean (24 g/kg L-dopa) for 42 days, no changes in concentrations of blood ceruloplasmin or L-dopa, or concentrations of L-dopa metabolites, were observed (Chikagwa-Malunga et al. 2009).

Genotoxicity

No mutagenic activity of raw or lightly roasted velvet bean seed was observed in the Ames test for mutagenicity with *Salmonella typhimurium* strains TA97a, TA98, TA100, TA102, and TA1535 with or without metabolic activation (Burgess et al. 2003).

Literature Cited

Akhtar, M.S., A.Q. Qureshi, and J. Iqbal. 1990. Antidiabetic evaluation of *Mucuna pruriens* Linn. seeds. *J. Pak. Med. Assoc.* 40(7):147-150.

Bhaskar, A., V.G. Vidhya, and M. Ramya. 2008. Hypoglycemic effect of *Mucuna pruriens* seed extract on normal and streptozotocin-diabetic rats. *Fitoterapia* 79(7-8):539-543.

Burgess, S., A. Hemmer, and R. Myhrman. 2003. Examination of raw and roasted *Mucuna pruriens* for tumerogenic substances. *Trop. Subtrop. Agroecosys.* 1(2-3):287-294.

Chikagwa-Malunga, S.K., A.T. Adesogan, L.E. Sollenberger, et al. 2009. Nutritional characterization of *Mucuna pruriens* 4. Does replacing soybean meal with *Mucuna pruriens* in lamb diets affect ruminal, blood and tissue L-dopa concentrations? *Anim. Feed Sci. Technol.* 148(2-4):124-137.

Davidson, S., X. Zhang, C.H. Yoon, et al. 2007. The itch-producing agents histamine and cowhage activate separate populations of primate spinothalamic tract neurons. *J. Neurosci.* 27(37):10007-10014.

Del Carmen, J., A.G. Gernat, R. Myhrman, and L.B. Carew. 1999. Evaluation of raw and heated velvet beans (*Mucuna pruriens*) as feed ingredients for broilers. *Poult. Sci.* 78(6):866-872.

Emenalom, O.O., A.B.I. Udedibie, B.O. Esonu, and E.B. Etuk. 2005. Evaluation of processed velvet bean (*Mucuna pruriens*) as a feed ingredient in starter diets for broiler chickens. *J. Poult. Sci.* 42(4):301-307.

Ghosal, S., S. Singh, and S.K. Bhattacharya. 1971. Alkaloids of *Mucuna pruriens*: Chemistry and pharmacology. *Plant Med.* 19(3):279-284.

Infante, M.E., A.M. Perez, M.R. Simao, et al. 1990. Outbreak of acute toxic psychosis attributed to *Mucuna pruriens*. *Lancet* 336(8723):1129.

Katzenschlager, R., A. Evans, A. Manson, et al. 2004. *Mucuna pruriens* in Parkinson's disease: A double blind clinical and pharmacological study. *J. Neurol. Neurosurg. Psychiat.* 75(12):1672-1677.

Kosteletzky, F., B. Namer, C. Forster, and H.O. Handwerker. 2009. Impact of scratching on itch and sympathetic reflexes induced by cowhage (*Mucuna pruriens*) and histamine. *Acta Derm. Venereol.* 89(3):271-277.

MMWR. 1985. *Mucuna pruriens*-associated pruritus—New Jersey. *MMWR Morb. Mortal. Wkly. Rep.* 34(48):732-734.

Nath, D., N. Sethi, R.K. Singh, and A.K. Jain. 1992. Commonly used Indian abortifacient plants with special reference to their teratologic effects in rats. *J. Ethnopharmacol.* 36(2):147-154.

Pugalenthi, M., V. Vadivel, and P. Siddhuraju. 2005. Alternative food/feed perspectives of an underutilized legume *Mucuna pruriens* var. *utilis*—A review. *Plant. Foods Hum. Nutr.* 60(4):201-218.

Rathi, S.S., J.K. Grover, and V. Vats. 2002. The effect of *Momordica charantia* and *Mucuna pruriens* in experimental diabetes and their effect on key metabolic enzymes involved in carbohydrate metabolism. *Phytother. Res.* 16(3):236-243.

Szabo, N.J. 2003. Indolealkylamines in *Mucuna* species. *Trop. Subtrop. Agroecosys.* 1:295-307.

Tripathi, Y.B., and A.K. Upadhyay. 2002. Effect of the alcohol extract of the seeds of *Mucuna pruriens* on free radicals and oxidative stress in albino rats. *Phytother. Res.* 16(6):534-538.

Myrcia multiflora (Lam.) DC. — Myrtaceae

SCN: pedra hume
Syn: *Myrcia sphaerocarpa* DC.

Part: herb

Quick Reference Summary

Safety Class: 1
Interaction Class: A

Contraindications

None known.

Other Precautions

None known.

Drug and Supplement Interactions

None known.

Adverse Events and Side Effects

None known.

Pharmacological Considerations

Traditional human use and in vitro studies indicate that pedra hume may modify glucose regulation (Grune 1979; Matsuda 2002; Yoshikawa et al. 1998). People with diabetes are advised to monitor their blood sugar closely and discuss the use of this herb with a qualified healthcare practitioner prior to use.

Pregnancy and Lactation

No information on the safety of pedra hume in pregnancy or lactation was identified in the scientific or traditional literature. Although this review did not identify any concerns for use while pregnant or nursing, safety has not been conclusively established.

Review Details

I. Drug and Supplement Interactions

Clinical Trials of Drug or Supplement Interactions

No clinical trials of drug or supplement interactions were identified.

Case Reports of Suspected Drug or Supplement Interactions

No case reports of suspected drug or supplement interactions were identified.

Animal Trials of Drug or Supplement Interactions

No animal trials of drug or supplement interactions were identified.

II. Adverse Events

Case Reports of Adverse Events

No case reports of adverse events were identified.

III. Pharmacology and Pharmacokinetics

Human Pharmacological Studies

No relevant human pharmacological studies were identified.

Animal Pharmacological Studies

No relevant animal pharmacological studies were identified.

In Vitro Pharmacological Studies

Compounds from pedra hume have demonstrated aldose reductase inhibition (Matsuda 2002; Yoshikawa et al. 1998).

IV. Pregnancy and Lactation

No information on the safety of pedra hume during pregnancy or lactation was identified.

V. Toxicity Studies

No toxicity studies were identified.

Literature Cited

Grune, U. 1979. Sobre o principio antidiabetico da pedra-hume-caá, *Myrcia multiflora* (Lam). Federal University of Rio de Janeiro.

Matsuda, H. 2002. Antidiabetic principles of natural medicines. V. Aldose reductase inhibitors from *Myrcia multiflora* DC. (2): Structures of myrciacitrins III, IV, and V. *Chem. Pharm. Bull.* 50(3):429-431.

Yoshikawa, M., H. Shimada, N. Nishida, et al. 1998. Antidiabetic principles of natural medicines. II. Aldose reductase and alpha-glucosidase inhibitors from Brazilian natural medicine, the leaves of *Myrcia multiflora* DC.(Myrtaceae): Structures of myrciacitrins I and II and myrciaphenones A and B. *Chem. Pharm. Bull.* 46(1):113-119.

Myristica fragrans Houtt. — Myristicaceae

SCN: mace (aril); nutmeg (seed)
Syn: *Myristica moschata* Thunb.; *Myristica officinalis* L. f.
AN: *jatiphala*
PN: *rou dou kou* (seed)
Part: aril, seed

Quick Reference Summary

Safety Class: 2b, 2d
Interaction Class: A

Contraindications

Not for use in pregnancy except under the supervision of a qualified healthcare practitioner (Weil 1965; Zaki et al. 1987).

Do not exceed recommended dose (Barceloux 2008; Bensky et al. 2004; Forrest and Heacock 1972; Stein et al. 2001; Weil 1965).

Other Precautions

None known.

Drug and Supplement Interactions

None known.

Standard Dose

The standard dose is 0.5 to 1 g unroasted seed (Kapoor 1989); 3 to 10 g of properly roasted seed as a decoction (Bensky et al. 2004; Chen and Chen 2004); or 1.5 to 3 g of powdered, properly roasted seed (Chen and Chen 2004).

Amounts of 5 g or more of unroasted nutmeg have been associated with nutmeg poisoning, including symptoms listed below (Barceloux 2008).

Editors' Notes

The classifications and concerns for this herb are based on relatively higher doses used for therapeutic purposes in contrast to lower amounts generally used in cooking, and have not been associated with its use as a spice.

In traditional Chinese medicine, nutmeg is almost always roasted before using. The roasting process reduces the toxicity of nutmeg (Bensky et al. 2004; Chen and Chen 2004; PPRC 2005). An Eclectic medical text includes a nutmeg-based preparation that required roasting of the seed prior to use (Felter and Lloyd 1898). Nutmeg sold for use as a spice is not roasted.

In Finland, cases of nutmeg poisoning and abuse led authorities to discourage the sale of whole nutmeg and recommend that nutmeg be sold only as a powder with the packages labeled "Only for spice use. Harmful for health if consumed as such" (Evira 2009).

Adverse Events and Side Effects

The majority of cases of nutmeg intoxication have been reported in persons attempting to use nutmeg as a recreational hallucinogen (Abernethy and Becker 1992; Forrester 2005; Sangalli and Chiang 2000). A review of the literature on nutmeg indicates that use of nutmeg as a hallucinogen has been relatively limited due to the need for a large unpalatable dose, a high risk (unpleasantness)-to-benefit ratio, a lack of effectiveness as compared to other hallucinogens, and an unpredictable response (Sangalli and Chiang 2000).

Clinical features of nutmeg poisoning are similar to those of belladonna (anticholinergic), with facial flushing, tachycardia, hypertension, dry mouth, and blurred vision. Initial symptoms include nausea, vomiting, abdominal pain, chest pain, restlessness, agitation, tremor, ataxia, involuntary eye movement, vertigo, hallucinations, and a feeling of doom or giddiness. These symptoms occur with alternating periods of lethargy and delirium (Barceloux 2009). Although a lack of pupil dilation in nutmeg intoxication is sometimes noted as a way to differentiate between nutmeg and anticholinergic poisoning (Barceloux 2009), pupil dilation does occur in some cases of nutmeg poisoning (Ahmad and Thompson 1975; McKenna et al. 2004; Stein et al. 2001).

A review on nutmeg poisoning indicated that the last fatality reported in the medical literature related solely to nutmeg was in 1908 (Barceloux 2009; Cushny 1908).

Pharmacological Considerations

Although 6 g of nutmeg did not produce any significant effects in healthy young people (Beattie 1968), 5 g is typically listed as a toxic dose (Barceloux 2008). Ingestion of 25 to 28 g has been associated with more severe poisoning symptoms including tachycardia, palpitations, and anticholinergic symptoms (Barceloux 2008; Stein et al. 2001).

Pregnancy and Lactation

Nutmeg has been used by many women to bring on menstruation or to attempt to induce abortion, although the literature suggests that nutmeg is not efficacious as an abortifacient (Weil 1965). An increase in maternal and fetal heart rate was observed in a pregnant woman who consumed cookies containing an excessive amount of nutmeg. The baby was delivered healthy at term (Lavy 1987).

Animal studies have provided conflicting results, with no adverse effects of the essential oil at doses up to 400 mg/kg in rabbits (FDRL 1974) but some abnormalities in rats administered 300 mg/kg of nutmeg (Zaki et al. 1987). An animal study indicated that after administration of mace to lactating mice, physiological effects were observed on both mothers and nursing offspring (Chhabra and Rao 1994).

Based on this information, use during pregnancy is not recommended except under the supervision of a qualified healthcare practitioner.

Review Details

I. Drug and Supplement Interactions

Clinical Trials of Drug or Supplement Interactions

No clinical trials of drug or supplement interactions were identified.

Case Reports of Suspected Drug or Supplement Interactions

No case reports of suspected drug or supplement interactions were identified.

Animal Trials of Drug or Supplement Interactions

No animal trials of drug or supplement interactions were identified.

II. Adverse Events

Case Reports of Adverse Events

Clinical features of nutmeg poisoning are similar to those of belladonna (anticholinergic), with facial flushing, tachycardia, hypertension, dry mouth, and blurred vision. However, dilation of the pupils is uncommon in nutmeg intoxication. Symptoms usually begin about 3 to 6 hours after ingestion and resolve within 24 to 36 hours. Initial symptoms include nausea, vomiting, abdominal pain, chest pain, restlessness, agitation, tremor, ataxia, involuntary eye movement, vertigo, and a feeling of doom. These symptoms occur with alternating periods of lethargy and delirium. Other effects of nutmeg intoxication documented in case reports include the sensation of warmth and coldness of the extremities, distortion of space and colors, auditory and tactile hallucinations, headache, and generalized weakness (Barceloux 2009).

A review of nutmeg poisoning cases reported to a poison control center indicated that doses of 14 to 80 g elicited symptoms ranging from weakness or restlessness to nausea, vomiting, and tachycardia. None of the cases was fatal (Stein et al. 2001).

Chronic delusions, hallucinations, and disturbed behavior have been reported in persons who routinely ingested nutmeg. In one case, a 23-year-old man had been taking 5 g of nutmeg daily for several months and had increased intake several months prior to psychiatric referral (Kelly et al. 2003). In a similar case, a 25-year-old man had been taking 120 to 650 mg of nutmeg daily for an unspecified period of time (Brenner et al. 1993).

The only reported fatality attributed solely to nutmeg was of an 8-year-old boy who ate two whole nutmegs, became comatose, and died within 24 hours (Cushny 1908).

Numerous other cases of nutmeg poisoning have been reported in the literature (Abernethy and Becker 1992; Dinakar 1977; Faguet and Rowland 1978; Forrester 2005; Green 1959; INCHEM 1991; McCord and Jervey 1962; McKenna et al. 2004; Payne 1963; Perez Valdivieso 2007; Sjoholm et al. 1998; Wallace 1903).

An autopsy of a 55-year-old woman indicated toxic levels of flunitrazepam, a finding that alone could not account for the woman's death. Further examination indicated a nutmeg-like smell of the stomach contents, with a blood analysis indicating 4.0 μg/ml of the compound myristicin (Stein et al. 2001).

Occupational asthma has been reported in individuals working with mace in the food industry. Skin prick tests elicited positive reactions to mace (Sastre et al. 1996; van Toorenenbergen and Dieges 1985).

III. Pharmacology and Pharmacokinetics

Human Pharmacological Studies

In healthy volunteers orally administered a single dose of 400 mg of the compound myristicin, alertness, "a feeling of irresponsibility," a sense of euphoria in two subjects, and an unpleasant reaction (anxiety, fear, tremors, tachycardia, and nausea) in two subjects were reported. The dose of myristicin used was equivalent to approximately 40 g of nutmeg, although the author noted that myristicin does not reproduce the entire activities of whole nutmeg (Truitt et al. 1960).

Administration of 6 g of nutmeg to healthy students did not significantly alter performance on neuropsychiatric tests (Beattie 1968).

Animal Pharmacological Studies

No relevant animal pharmacological studies were identified.

In Vitro Pharmacological Studies

Inhibition of platelet aggregation induced by arachidonic acid was observed in rabbit platelets treated with nutmeg essential oil (Rasheed et al. 1984).

IV. Pregnancy and Lactation

An increase in maternal and fetal heart rate was observed in a pregnant woman who consumed cookies containing an excessive amount of nutmeg. Adverse effects subsided within 24 hours, and the baby was delivered healthy at term (Lavy 1987).

In rabbits orally administered 4 to 400 mg/kg of nutmeg essential oil daily on days 6 to 18 of pregnancy, no adverse effects on fetal survival or soft or skeletal tissue abnormalities were observed (FDRL 1974).

Administration of 300 mg/kg nutmeg to rats on day 8 of pregnancy resulted in deformed fetuses accompanied by a decrease in body weight (Zaki et al. 1987).

Nutmeg has been used by many women to bring on menstruation or attempt to induce abortion, although the literature suggests that nutmeg is not efficacious as an abortifacient (Weil 1965).

In mice orally administered an aqueous suspension of mace at doses of 0.025 or 0.1 g per animal daily on lactation days 1 to 14 or 21, mice receiving mace and their offspring had significantly elevated hepatic sulfhydryl content, glutathione S-transferase, and glutathione reductase activity and cytochrome b_5 content (Chhabra and Rao 1994).

V. Toxicity Studies

Acute Toxicity

The LD_{50} of orally administered nutmeg essential oil in rats is 2.63 g/kg (Jenner et al. 1964).

The LD_{50} of the compound myristicin could not be determined at doses up to 1 mg/kg after intraperitoneal administration to rats (Truitt et al. 1960).

Short-Term Toxicity

No adverse effects were observed in rats fed diets containing myristicin at amounts equivalent to 10 mg/kg daily for 26 days. Observed parameters included body weight and histological examination of the livers and kidneys (Hallstrom and Thuvander 1997).

In mice intraperitoneally administered 20, 40, or 80 mg/kg of an aqueous extract of nutmeg daily for 6 weeks, alterations in indicators of liver function were observed, including reduction of total protein and serum albumin and elevation of serum glutamic pyruvic transaminase (SGPT) and serum glutamic oxaloacetic transaminase (SGOT), and elevation of blood urea. Examination of the liver revealed hydropic and fatty degeneration. Myocardial indices including SGOT and lactate dehydrogenase (LDH) were elevated. No adverse effects on kidneys were observed (Al-Hazmi et al. 2004).

Genotoxicity

No genotoxic activity was observed in the mouse micronucleus assay in mice fed diets containing 0.5 or 2% mace daily (Kumari 1992).

Some mutagenic activity of nutmeg and other common spices was observed in assays with *Salmonella typhimurium* strains TA98 and TA102 (Mahmoud et al. 1992).

In pregnant mice orally administered the compound myristicin as a single dose on day 16 of pregnancy, or one dose each on days 16 and 17 of pregnancy, liver adducts were reported in the mothers, and significantly smaller amounts of liver adducts were observed in the fetuses. The number of adducts observed in pregnant mice was higher than that observed in nonpregnant treated mice (Randerath et al. 1993).

Literature Cited

Abernethy, M.K., and L.B. Becker. 1992. Acute nutmeg intoxication. *Am. J. Emerg. Med.* 10(5):429.

Ahmad, A., and H.S. Thompson. 1975. Nutmeg mydriasis. *J. Am. Med. Assoc.* 234:274.

Al-Hazmi, M.A., A.L. Assaggaf, G.N.E. Al-Sayed, and Y.S. Bin-Naser. 2004. Effect of acute and subchronic administration of nutmeg seed's extract on mice behaviour, histological structure and biochemical functions. *Saudi J. Biol. Sci.* 11(2):177-188.

Barceloux, D.G. 2008. Nutmeg (*Myristica fragrans* Houtt.). In *Medical toxicology of natural substances: Foods, fungi, medicinal herbs, plants, and venomous animals*. Hoboken, NJ: Wiley.

Barceloux, D.G. 2009. Nutmeg (*Myristica fragrans* Houtt.). *Dis. Mon.* 55(6):373-379.

Beattie, R.T. 1968. Nutmeg as a psychoactive agent. *Addiction* 63(1-2):105-109.

Bensky, D., S. Clavey, and E. Stöger. 2004. *Chinese herbal medicine: Materia medica*. 3rd ed. Seattle: Eastland Press.

Brenner, N., O.S. Frank, and E. Knight. 1993. Chronic nutmeg psychosis. *J. Roy. Soc. Med.* 86(3):179.

Chen, J.K., and T.T. Chen. 2004. *Chinese medical herbology and pharmacology*. City of Industry, CA: Art of Medicine Press.

Chhabra, S.K., and A.R. Rao. 1994. Transmammary modulation of xenobiotic metabolizing enzymes in liver of mouse pups by mace (*Myristica fragrans* Houtt.). *J. Ethnopharmacol.* 42(3):169-177.

Cushny, A.R. 1908. Nutmeg poisoning. *Proc. Roy. Soc Med.* (1):39-44.

Dinakar, H.S. 1977. Acute psychosis associated with nutmeg toxicity. *Med. Times* 105(12):63-64.

Evira. 2009. Nutmeg powder used as such is harmful for health. April 14 2009. Helsinki: Evira, Finnish Food Safety Authority.

Faguet, R.A., and K.F. Rowland. 1978. " Spice cabinet" intoxication. *Am. J. Psychiat.* 135(7):860.

FDRL. 1974. Food and Drug Research Labs. Teratologic evaluation of oil of nutmeg in rabbits. In *Nutmeg oil nutmeg powder*. University of California San Francisco Lorillard collection. Legacy Tobacco Documents Library.

Felter, H.W., and J.U. Lloyd. 1898. *King's American dispensatory*. 18th ed., 3rd rev. 2 vols. Cincinnati: Ohio Valley Co.

Forrest, J.E., and R.A. Heacock. 1972. Nutmeg and mace, the psychotropic spices from *Myristica fragrans*. *Lloydia* 35(4):440-449.

Forrester, M.B. 2005. Nutmeg intoxication in Texas, 1998–2004. *Hum. Exp. Toxicol.* 24(11):563-566.

Green, R.C. 1959. Nutmeg poisoning. *J. Am. Med. Assoc.* 171(10):166-168.

Hallstrom, H., and A. Thuvander. 1997. Toxicological evaluation of myristicin. *Nat. Toxins* 5(5):186-192.

INCHEM. 1991. *Myristica fragrans* Houtt. International Programme on Chemical Safety.

Jenner, P.M., E.C. Hagan, J.M. Taylor, E.L. Cook, and O.G. Fitzhugh. 1964. Food flavoring and compounds of related structure: I. Acute oral toxicity. *Food Cosmet. Toxicol.* 2:327-343.

Kapoor, L.D. 1989. *CRC handbook of Ayurvedic medicinal plants*. Boca Raton, FL: CRC Press.

Kelly, B.D., B.E. Gavin, M. Clarke, A. Lane, and C. Larkin. 2003. Nutmeg and psychosis. *Schizophr. Res.* 60(1):95-96.

Kumari, M.V. 1992. Modulatory influences of mace (*Myristica fragrans*, Houtt.) on hepatic detoxification systems and bone marrow genotoxicity in male Swiss albino mice. *Nutr. Res.* 12(3):385-394.

Lavy, G. 1987. Nutmeg intoxication in pregnancy. A case report. *J. Repro. Med.* 32(1):63.

Mahmoud, I., A. Alkofahi, and A. Abdelaziz. 1992. Mutagenic and toxic activities of several spices and some Jordanian medicinal plants. *Int. J. Pharmacogn.* 30(2):81-85.

McCord, J.A., and L.P. Jervey. 1962. Nutmeg (myristicin) poisoning. *J. South Carolina Med. Assoc.* 58:436-439.

McKenna, A., S.P. Nordt, and J. Ryan. 2004. Acute nutmeg poisoning. *Eur. J. Emerg. Med.* 11:240-241.

Payne, R.B. 1963. Nutmeg intoxication. *N. Engl. J. Med.* 269(1):36-88.

Perez Valdivieso, J.R. 2007. Acute nutmeg intoxication in Spain. *Rev. Esp. Anestesiol. Reanim.* 54(10):633-634.

PPRC. 2005 *Pharmacopoeia of the People's Republic of China*. Beijing: People's Medical Publishing House.

Randerath, K., K.L. Putman, and E. Randerath. 1993. Flavor constituents in cola drinks induce hepatic DNA adducts in adult and fetal mice. *Biochem. Biophys. Res. Commun.* 192(1):61-68.

Rasheed, A., G.M. Laekeman, A.J. Vlietinck, et al. 1984. Pharmacological influence of nutmeg and nutmeg constituents on rabbit platelet function. *Planta Med.* 50(3):222-226.

Sangalli, B., and W. Chiang. 2000. Toxicology of nutmeg abuse. *J. Toxicol. Clin. Toxicol.* 38(6):671-678.

Sastre, J., M. Olmo, A. Novalvos, D. Ibanez, and C. Lahoz. 1996. Occupational asthma due to different spices. *Allergy* 51(2):117-120.

Sjoholm, A., A. Lindberg, and M. Personne. 1998. Acute nutmeg intoxication. *J. Intern. Med.* 243(4):329.

Stein, U., H. Greyer, and H. Hentschel. 2001. Nutmeg (myristicin) poisoning—Report on a fatal case and a series of cases recorded by a poison information centre. *Forensic Sci. Int.* 118(1):87-90.

Truitt, E.B., Jr., E. Calloway, III, M.C. Braude, and J.C. Krantz, Jr. 1960. The pharmacology of myristicin. A contribution to the psychopharmacology of nutmeg. *J. Neuropsychiatr.* 2:205-210.

van Toorenenbergen, A.W., and P.H. Dieges. 1985. Immunoglobulin E antibodies against coriander and other spices. *J. Allerg. Clin. Immunol.* 76(3):477-481.

Wallace, G.B. 1903. On nutmeg poisoning. In *Contributions to medical research: University of Michigan Department of Surgery*. Ann Arbor, MI: George Wair.

Weil, A.T. 1965. Nutmeg as a narcotic. *Econ. Bot.* 19(3):194-217.

Zaki, N.G., N.M. El-Malkh, S.A. Abdelbaset, and A.A. Saeid. 1987. Teratogenicity of nutmeg in fetuses of rats. *Bull. Fac. Sci. Cairo Univ.* 55(1):105-124.

Myroxylon spp.

Fabaceae

Myroxylon balsamum (L.) Harms var. ***balsamum***
SCN: tolu balsam tree
Syn: *Myroxylon toluiferum* Kunth
OCN: opobalsam; Peruvian balsam; tolu

***Myroxylon balsamum* (L.) Harms var. *pereirae* (Royle) Harms**
SCN: balsam-of-Peru tree
OCN: Peruvian balsam
Part: oleoresin

Quick Reference Summary

Safety Class: 2d
Interaction Class: A

Contraindications

Not for use in persons with fevers (Felter and Lloyd 1898).

Other Precautions

May cause kidney irritation (Felter and Lloyd 1898).

Drug and Supplement Interactions

None known.

Adverse Events and Side Effects

Allergic reactions and sensitization to tolu balsam tree and balsam-of-Peru tree are relatively common and have been reported after topical and internal use (Pfutzner et al. 2003; Salam and Fowler 2001; Wichtl 2004; Zug et al. 2009).

Pharmacological Considerations

None known.

Pregnancy and Lactation

No information on the safety of tolu balsam tree and balsam-of-Peru in pregnancy or lactation was identified in the scientific or traditional literature. Although this review did not identify any concerns for use while pregnant or nursing, safety has not been conclusively established.

M

Review Details

I. Drug and Supplement Interactions

Clinical Trials of Drug or Supplement Interactions

No clinical trials of drug or supplement interactions were identified.

Case Reports of Suspected Drug or Supplement Interactions

No case reports of suspected drug or supplement interactions were identified.

Animal Trials of Drug or Supplement Interactions

No animal trials of drug or supplement interactions were identified.

II. Adverse Events

Case Reports of Adverse Events

Exacerbation of chronic eczema was reported in a 51-year-old brewery worker who showed delayed-type patch test reactions to balsam-of-Peru tree and fragrance-mix. Repeated oral challenges with balsam-of-Peru tree led to exacerbation of the eczema (Pfutzner et al. 2003).

III. Pharmacology and Pharmacokinetics

Human Pharmacological Studies

In a retrospective analysis of North American patients with suspected allergic contact dermatitis, balsam-of-Peru tree was the second most common allergen, with 11.6% of people tested producing positive reactions (Zug et al. 2009). Similarly, a retrospective study of patch test data from Denmark between 1985 and 2007 indicated that of the 16,173 patients tested, approximately 4% tested positively to balsam-of-Peru tree (Thyssen et al. 2008).

Animal Pharmacological Studies

No relevant animal pharmacological studies were identified.

In Vitro Pharmacological Studies

No relevant in vitro pharmacological studies were identified.

IV. Pregnancy and Lactation

No information on the safety of tolu balsam tree and balsam-of-Peru tree during pregnancy or lactation was identified.

V. Toxicity Studies

No toxicity studies were identified.

Literature Cited

Felter, H.W., and J.U. Lloyd. 1898. *King's American dispensatory.* 18th ed., 3rd rev. 2 vols. Cincinnati: Ohio Valley Co.

Pfutzner, W., A. Niedermeier, P. Thomas, and B. Przybilla. 2003. [Systemic contact eczema against balsam of Peru.] *J. Dtsch. Dermatol. Ges.* 1(9):719-721.

Salam, T.N., and J.F. Fowler. 2001. Balsam-related systemic contact dermatitis. *J. Am. Acad. Dermatol.* 45(3):377-381.

Thyssen, J.P., B.C. Carlsen, T. Menné, and J.D. Johansen. 2008. Trends of contact allergy to fragrance mix I and *Myroxylon pereirae* among Danish eczema patients tested between 1985 and 2007. *Contact Dermat.* 59(4):238-244.

Wichtl, M. 2004. *Herbal drugs and phytopharmaceuticals: A handbook for practice on a scientific basis*. 3rd ed. Boca Raton, FL: CRC Press.

Zug, K.A., E.M. Warshaw, J.F. Fowler Jr., et al. 2009. Patch-test results of the North American Contact Dermatitis Group 2005-2006. *Dermatitis* 20(3):149-160.

M

Nardostachys jatamansi (D. Don) DC. — Valerianaceae

SCN: jatamansi
Syn: *Nardostachys grandiflora* DC.
AN: *jatamansi*
PN: *gan song* (root and rhizome)
OCN: Indian spikenard; nard; spoon-leaf nardostachys
Part: rhizome, root

Quick Reference Summary

Safety Class: 2b
Interaction Class: A

Contraindications

Not for use in pregnancy except under the supervision of a qualified healthcare practitioner (Pole 2006).

Other Precautions

None known.

Drug and Supplement Interactions

None known.

Notice

Emmenagogue (Chadha 1988); *see* Appendix 2.
Diuretic (Chadha 1988); *see* Appendix 2.

Adverse Events and Side Effects

None known.

Pharmacological Considerations

An animal study on the antidepressant activity of jatamansi indicated that activity was due to MAO inhibition and/or interaction with GABA-B receptors (Dhingra and Goyal 2008).

Pregnancy and Lactation

Based on traditional use as an emmenagogue, use of jatamansi in pregnancy is not recommended (Chadha 1988; Pole 2006).

No information on the safety of jatamansi during lactation was identified in the scientific or traditional literature. While this review did not identify any concerns for use while nursing, safety has not been conclusively established.

N

Review Details

I. Drug and Supplement Interactions

Clinical Trials of Drug or Supplement Interactions

No clinical trials of drug or supplement interactions were identified.

Case Reports of Suspected Drug or Supplement Interactions

No case reports of suspected drug or supplement interactions were identified.

Animal Trials of Drug or Supplement Interactions

No animal trials of drug or supplement interactions were identified.

II. Adverse Events

Case Reports of Adverse Events

No case reports of adverse events were identified.

III. Pharmacology and Pharmacokinetics

Human Pharmacological Studies

No relevant human pharmacological studies were identified.

Animal Pharmacological Studies

Antidepressant-like activity of jatamansi was observed in rats orally administered 100, 200, or 400 mg/kg of an ethanol extract daily for 14 days. The antidepressant activity was attributed to a reduction in whole brain MAO-A and MAO-B activities and an increase in the level of monoamines. Pretreatment of animals with a GABA-B agonist resulted in a partial inhibition of the antidepressant effect, indicating a potential effect on GABA-B receptors (Dhingra and Goyal 2008).

In Vitro Pharmacological Studies

Methanol and water extracts of jatamansi exhibited acetylcholinesterase inhibitory activity at concentrations of 47.21 μg/ml, with the methanol extract demonstrating more activity than the water extract. Similar activity was observed at a lower concentration for ashwagandha and other botanicals (Vinutha et al. 2007).

IV. Pregnancy and Lactation

Based on traditional use as an emmenagogue, use of jatamansi in pregnancy is not recommended (Chadha 1988; Pole 2006).

No information on the safety of jatamansi during lactation was identified.

V. Toxicity Studies

Acute Toxicity

The LD_{50} of orally administered aqueous extract of jatamansi in mice could not be determined at doses up to 2 g/kg (Debelmas and Hache 1976).

The LD_{50} of an ethanol extract of jatamansi in rats is 569 mg/kg after intraperitoneal administration but could not be determined at doses up to 1000 mg/kg after oral administration (Rao et al. 2005).

Literature Cited

Chadha, Y. 1988. *The wealth of India: A dictionary of Indian raw materials and industrial products*. Delhi: Council of Scientific and Industrial Research.

Debelmas, A.M., and J. Hache. 1976. Pharmacologic study of some medicinal plants of Nepal. Acute toxicity, behavioural study and action on the central nervous system. *Plant. Med. Phytother.* 10(2):128-138.

Dhingra, D., and P.K. Goyal. 2008. Inhibition of MAO and GABA: Probable mechanisms for antidepressant-like activity of *Nardostachys jatamansi* DC. in mice. *Indian J. Exp. Biol.* 46(4):212-218.

Pole, S. 2006. *Ayurvedic medicine: The principles of traditional practice*. New York: Churchill Livingstone.

Rao, V.S., A. Rao, and K.S. Karanth. 2005. Anticonvulsant and neurotoxicity profile of *Nardostachys jatamansi* in rats. *J. Ethnopharmacol.* 102(3):351-356.

Vinutha, B., D. Prashanth, K. Salma, et al. 2007. Screening of selected Indian medicinal plants for acetylcholinesterase inhibitory activity. *J. Ethnopharmacol.* 109(2):359-363.

Nelumbo nucifera Gaertn. — Nymphaeaceae

N

SCN: sacred lotus
Syn: *Nelumbium speciosum* Willd.; *Nymphaea nelumbo* L.
AN: *kamala; padma*
PN: *lian zi* (seed); *lian zi xin* (green seed embryo)
OCN: East Indian lotus; Hindu lotus; oriental lotus
Part: seed

Quick Reference Summary

Safety Class: 2d
Interaction Class: A

Contraindications

Not for use in persons with constipation, dry stools, or abdominal distention (Bensky et al. 2004; Chen and Chen 2004).

Other Precautions

None known.

Drug and Supplement Interactions

None known.

Editors' Note

Sacred lotus seed as used in traditional Chinese medicine is prepared by removing the seed embryo (PPRC 2005).

Adverse Events and Side Effects

None known.

Pharmacological Considerations

None known.

Pregnancy and Lactation

Reference texts on traditional Chinese medicine do not indicate any cautions for the use of sacred lotus in pregnancy (Bensky et al. 2004; Chen and Chen 2004).

Oral and intraperitoneally administered extracts of sacred lotus seed exhibited effects on the estrus cycle in rats and mice (Mazumder et al. 1992; Mutreju et al. 2008).

An unreferenced statement in one published article asserts that sacred lotus seed is used as an antifertility agent by tribal women near Udaipur, India (Mutreju et al. 2008), but none of the classic texts on Indian materia medica indicate such usage (Chopra 1933; Dutt 1922; Kirtikar and Basu 1935; Nadkarni 1954). One contemporary ethnobotanical review reports that the seed is given in pregnancy (Singh et al. 2010).

No information on the safety of sacred lotus in lactation was identified. While this review did not identify any concerns for use while nursing, safety has not been conclusively established.

Review Details

I. Drug and Supplement Interactions

Clinical Trials of Drug or Supplement Interactions

No clinical trials of drug or supplement interactions were identified.

Case Reports of Suspected Drug or Supplement Interactions

No case reports of suspected drug or supplement interactions were identified.

Animal Trials of Drug or Supplement Interactions

No animal trials of drug or supplement interactions were identified.

II. Adverse Events

Case Reports of Adverse Events

No case reports of adverse events were identified.

III. Pharmacology and Pharmacokinetics

Human Pharmacological Studies

No relevant human pharmacological studies were identified.

Animal Pharmacological Studies

In rats orally administered 800 mg/kg of an ethanol extract of sacred lotus daily for 41 days, a decrease in the weights of the uterus, ovaries, and vagina was observed, along with a prolongation of the estrus cycle (Mutreju et al. 2008).

In mice intraperitoneally administered 3 mg/kg of a petroleum ether extract of sacred lotus every other day for 15 days, arrest of the estrus cycle was observed along with a decrease in uterine weight (Mazumder et al. 1992).

In diabetic rats orally administered 5 g/kg of the compound neferine daily for 3 weeks, a decrease in fasting blood glucose, insulin, triglycerides, and TNF-alpha and an enhancement of insulin sensitivity in insulin-resistant rats were observed (Pan et al. 2009).

In Vitro Pharmacological Studies

The compound neferine inhibited rabbit platelet aggregation induced by ADP, collagen, arachidonic acid, and platelet-activating factor (Yu and Hu 1997).

IV. Pregnancy and Lactation

References on traditional Chinese medicine (Bensky et al. 2004; Chen and Chen 2004) and classic texts on Indian materia medica (Chopra 1933; Dutt 1922; Dymock 1890; Kirtikar and Basu 1935; Nadkarni 1954) do not indicate any cautions for the use of sacred lotus in pregnancy.

Although one recent article reports that sacred lotus seed has traditionally been used as an antifertility agent by women in one tribal area in northern India (Mutreju et al. 2008), a contemporary ethnobotanical study records that the seed is given in pregnancy (Singh et al. 2010).

No information on the safety of sacred lotus in lactation was identified.

Also see Animal pharmacological studies for this entry.

V. Toxicity Studies

Acute Toxicity

The LD_{50} value of compounds from sacred lotus intravenously administered to mice is 34.9 mg/kg for liensinine, 26.0 mg/kg for neferine, and 20.0 mg/kg for the total alkaloids (Zhu 1998).

N

Literature Cited

Bensky, D., S. Clavey, and E. Stöger. 2004. *Chinese herbal medicine: Materia medica*. 3rd ed. Seattle: Eastland Press.

Chen, J.K., and T.T. Chen. 2004. *Chinese medical herbology and pharmacology*. City of Industry, CA: Art of Medicine Press.

Chopra, R.N. 1933. *Indigenous drugs of India*. Calcutta: The Art Press.

Dutt, U.C. 1922. *The materia medica of the Hindus*. rev. ed. Calcutta: Adi-Ayurveda Machine Press.

Kirtikar, K.R. and B.D. Basu. 1935 (2008 reprint). *Indian medicinal plants*. 2nd ed. Dehra Dun, India: Bishen Singh Mahendra Pal Singh.

Mazumder, U.K., M. Gupta, G. Pramanik, R.K. Mukhopadhyay, and S. Sarkar. 1992. Antifertility activity of seed of *Nelumbo nucifera* in mice. *Indian J. Exp. Biol.* 30(6):533-534.

Mutreju, A., M. Agarwal, S. Kushwaha, and A. Chauhan. 2008. Effect of *Nelumbo nucifera* seeds on the reproductive organs of female rats. *Iran. J. Reprod. Med.* 6(1):7-11.

Nadkarni, A.K. 1954. *Dr. K.M. Nadkarni's Indian materia medica*. 3rd ed. Bombay: Dhootapapeshwar Prakashan Ltd.

Pan, Y., B. Cai, K. Wang, et al. 2009. Neferine enhances insulin sensitivity in insulin resistant rats. *J. Ethnopharmacol.* 124(1):98-102.

PPRC. 2005 *Pharmacopoeia of the People's Republic of China*. Beijing: People's Medical Publishing House.

Singh, P.K., V. Kumar, R.K. Tiwari, A. Sharma, C.V. Rao, and R.H. Singh. 2010. Medico-ethnobotany of 'Chatara' block of District Sonebhadra, Uttar Pradesh, India. *Advances Biol. Res.* 4(1):65-80.

Yu, J., and W.S. Hu. 1997. Effects of neferine on platelet aggregation in rabbits. *Yao Xue Xue Bao* 32(1):1-4.

Zhu, Y.-P. 1998. *Chinese materia medica: Chemistry, pharmacology and applications*. Amsterdam: Harwood Academic Publishers.

Nepeta cataria L. — Lamiaceae

SCN: catnip
PN: *jia jing jie* (whole plant)

OCN: catmint
Part: herb

Quick Reference Summary

Safety Class: 1
Interaction Class: A

Contraindications
None known.

Other Precautions
None known.

Drug and Supplement Interactions
None known.

Notice
Emmenagogue (Felter and Lloyd 1898); *see* Appendix 2.

Adverse Events and Side Effects
None known.

Pharmacological Considerations
None known.

Pregnancy and Lactation
An animal study indicated that a diet of 10% catnip during pregnancy resulted in weight loss in mothers and fetuses but had no other effects on fetal health or survival (Bernardi et al. 1998). Catnip has been used as an emmenagogue (Felter and Lloyd 1898).

No information on the safety of catnip in lactation was identified. While this review did not identify any concerns for use while nursing, safety has not been conclusively established.

Review Details

I. Drug and Supplement Interactions

Clinical Trials of Drug or Supplement Interactions
No clinical trials of drug or supplement interactions were identified.

Case Reports of Suspected Drug or Supplement Interactions
No case reports of suspected drug or supplement interactions were identified.

Animal Trials of Drug or Supplement Interactions
No animal trials of drug or supplement interactions were identified.

II. Adverse Events

Case Reports of Adverse Events
A 19-month-old boy became lethargic after eating an unknown amount of raisins that had been soaked in a catnip tea that had been brewing in the refrigerator for 3 weeks (Osterhoudt et al. 1997).

III. Pharmacology and Pharmacokinetics

Human Pharmacological Studies
No relevant human pharmacological studies were identified.

Animal Pharmacological Studies
No relevant animal pharmacological studies were identified.

In Vitro Pharmacological Studies
No relevant in vitro pharmacological studies were identified.

IV. Pregnancy and Lactation
In mice fed a diet containing 10% catnip on days 6 to 18 of pregnancy, a decrease in maternal body weight at 18 days of pregnancy and a reduction of fetal and placental weights was observed. No differences were observed between the number of living and dead fetuses, in the number of resorptions, or in the ratio of maternal body weight to fetus weight plus placental weight of both groups. Catnip prenatal exposure reduced the body weight of male and female pups at birth while only females showed a decrease in this parameter at 7 days of lactation. The female eye, ear, and vaginal openings and male testis descent were delayed in the treatment group as compared to control (Bernardi et al. 1998).

Older herbal texts list catnip as an emmenagogue (Cook 1869; Felter and Lloyd 1898), although one text notes that the emmenagogue activity is weak (Cook 1869).

No information on the safety of catnip in lactation was identified.

V. Toxicity Studies

Acute Toxicity
The LD_{50} of catnip essential oil in mice is 1.3 g/kg (route of administration not listed) (De Vincenzi et al. 1996).

A decrease in CNS activity was observed in rats intraperitoneally administered up to 750 mg/kg catnip essential oil (Harney et al. 1978).

Literature Cited

Bernardi, M.M., S. Fernandes, A.L. Zodi, et al. 1998. Toxic effects of catnip (*Nepeta cataria*) exposure during embryogenetic period in mice. *Toxicon* 36(9):1261-1262.

Cook, W. 1869. *Physio-medical dispensatory*. Cincinnati, OH: W.H. Cook.

De Vincenzi, M., E. Mancini, and M.R. Dessi. 1996. Monographs on botanical flavouring substances used in foods. Part V. *Fitoterapia* 67(3):241-251.

Felter, H.W., and J.U. Lloyd. 1898. *King's American dispensatory*. 18th ed., 3rd rev. 2 vols. Cincinnati: Ohio Valley Co.

Harney, J.W., I.M. Barofsky, and J.D. Leary. 1978. Behavioral and toxicological studies of cyclopentanoid monoterpenes from *Nepeta cataria*. *Lloydia* 41(4):367-374.

Osterhoudt, K.C., S.K. Lee, J.M. Callahan, and F.M. Henretig. 1997. Catnip and the alteration of human consciousness. *Vet. Hum. Toxicol.* 39(6):373-375.

Notopterygium incisum K.C. Ting ex H.T. Chang — Apiaceae

SCN: notopterygium
PN: *qiang huo* (root and rhizome)

Part: root and rhizome

Quick Reference Summary

Safety Class: 1
Interaction Class: A

Contraindications

None known.

Other Precautions

None known.

Drug and Supplement Interactions

None known.

Adverse Events and Side Effects

Overdose (standard dose listed as a decoction of 3–10 g) of notopterygium may injure the stomach and cause nausea or vomiting (Chen and Chen 2004).

Pharmacological Considerations

None known.

Pregnancy and Lactation

No information on the safety of notopterygium in pregnancy or lactation was identified in the scientific or traditional literature. Although this review did not identify any concerns for use while pregnant or nursing, safety has not been conclusively established.

Review Details

I. Drug and Supplement Interactions

Clinical Trials of Drug or Supplement Interactions

No clinical trials of drug or supplement interactions were identified.

Case Reports of Suspected Drug or Supplement Interactions

No case reports of suspected drug or supplement interactions were identified.

Animal Trials of Drug or Supplement Interactions

No animal trials of drug or supplement interactions were identified.

II. Adverse Events

Case Reports of Adverse Events

Overdose (standard dose listed as a decoction of 3–10 g) of notopterygium may injure the stomach and cause nausea or vomiting (Chen and Chen 2004).

III. Pharmacology and Pharmacokinetics

Human Pharmacological Studies

No relevant human pharmacological studies were identified.

Animal Pharmacological Studies

No relevant animal pharmacological studies were identified.

In Vitro Pharmacological Studies

Inhibition of the drug-metabolizing isoenzyme CYP3A4 was observed in human liver microsomes treated with a decoction or an ethanolic extract of notopterygium (Guo et al. 2001).

IV. Pregnancy and Lactation

No information on the safety of notopterygium during pregnancy or lactation was identified.

V. Toxicity Studies

Acute Toxicity

The LD_{50} of orally administered notopterygium essential oil in mice is 2.83 g/kg (Chen and Chen 2004). The LD_{50} of an aqueous extract of notopterygium orally administered in mice could not be determined at doses up to 12 g/kg (Chen and Chen 2004).

Genotoxicity

No mutagenic activity of an aqueous extract of notopterygium was observed in the Ames mutagenicity assay with *Salmonella typhimurium* strains TA98 or TA100 with or without metabolic activation by S9 (Yin et al. 1991).

In mouse micronucleus and chromosomal aberration assays, mice were intraperitoneally administered an aqueous extract of notopterygium at doses of 0.1, 0.5, 1, or 2 g/kg. An increase in the incidence of chromosomal aberrations was observed at the 2 g/kg dose level. An increase of polychromatic erythrocytes was observed in the micronucleus assay at the 1 and 2 g/kg dose levels (Yin et al. 1991).

Literature Cited

Chen, J.K., and T.T. Chen. 2004. *Chinese medical herbology and pharmacology*. City of Industry, CA: Art of Medicine Press.

Guo, L.Q., M. Taniguchi, Q.Y. Chen, K. Baba, and Y. Yamazoe. 2001. Inhibitory potential of herbal medicines on human cytochrome P450-mediated oxidation: Properties of umbelliferous or citrus crude drugs and their relative prescriptions. *Jpn. J. Pharmacol.* 85(4):399-408.

Yin, X.J., D.X. Liu, H.C. Wang, and Y. Zhou. 1991. A study on the mutagenicity of 102 raw pharmaceuticals used in Chinese traditional medicine. *Mutat. Res.* 260(1):73-82.

N

Ocimum basilicum L. — Lamiaceae

SCN: basil
OCN: sweet basil

PN: *jiu ceng ta*
Part: leaf

Quick Reference Summary

Safety Class: 2b
Interaction Class: A

Contraindications

Not for use in pregnancy except under the supervision of a qualified healthcare practitioner (Chen and Chen 2004).

Other Precautions

None known.

Drug and Supplement Interactions

None known.

Notice

Alkenylbenzenes (estragole as 70–85% of essential oil; safrole as a minor constituent) (De Vincenzi et al. 2000; Leung and Foster 1996; Siano et al. 2003); *see* Appendix 1.

Editors' Note

The classifications and concerns for this herb are based on relatively higher doses used for therapeutic purposes in contrast to lower amounts generally used in cooking, and have not been associated with its use as a spice.

Adverse Events and Side Effects

An allergic reaction to basil has been reported (Vartholomaios et al. 2007).

Pharmacological Considerations

An animal study demonstrated that basil may modify glucose regulation (Zeggwagh et al. 2007). People with diabetes are advised to monitor their blood sugar closely and discuss the use of this herb with a qualified healthcare practitioner prior to use.

Pregnancy and Lactation

A text on traditional Chinese medicine indicates that basil should not be used during pregnancy (Chen and Chen 2004).

No information on the safety of basil during lactation was identified in the scientific or traditional literature. While this review did not identify any concerns for use while nursing, safety has not been conclusively established.

Review Details

I. Drug and Supplement Interactions

Clinical Trials of Drug or Supplement Interactions

No clinical trials of drug or supplement interactions were identified.

Case Reports of Suspected Drug or Supplement Interactions

No case reports of suspected drug or supplement interactions were identified.

Animal Trials of Drug or Supplement Interactions

No animal trials of drug or supplement interactions were identified.

II. Adverse Events

Case Reports of Adverse Events

An allergic reaction to basil has been reported (Vartholomaios et al. 2007).

III. Pharmacology and Pharmacokinetics

Human Pharmacological Studies

No relevant human pharmacological studies were identified.

Animal Pharmacological Studies

A reduction in blood glucose levels was observed in healthy and diabetic rats orally administered an aqueous extract of basil in a single dose of 20 mg/kg or a dose of 20 mg/kg daily for 15 days. Plasma insulin levels were unchanged. The hypoglycemic effect was greater in diabetic than in healthy animals (Zeggwagh et al. 2007).

In Vitro Pharmacological Studies

An extract of basil inhibited the drug-metabolizing isoenzyme CYP1A2 in a screening of several plant extracts (Jeurissen et al. 2007).

IV. Pregnancy and Lactation

A text on traditional Chinese medicine indicates that basil should not be used during pregnancy (Chen and Chen 2004).

V. Toxicity Studies

Acute Toxicity

The LD_{50} of an alcohol extract of holy basil orally administered in mice is 1.54 mg/kg (Lagarto Parra et al. 2001).

In the brine shrimp lethality assay, the LC_{50} of an alcohol extract of holy basil is 18.75 µg/ml (Lagarto Parra et al. 2001).

Literature Cited

Agrawal, P., V. Rai, and R.B. Singh. 1996. Randomized placebo-controlled, single blind trial of holy basil leaves in patients with noninsulin-dependent diabetes mellitus. *Int. J. Clin. Pharmacol. Ther.* 34(9):406-409.

Ahmed, M., R.N. Ahamed, R.H. Aladakatti, and M.G. Ghosesawar. 2002. Reversible anti-fertility effect of benzene extract of *Ocimum sanctum* leaves on sperm parameters and fructose content in rats. *J. Basic Clin. Physiol. Pharmacol.* 13(1):51-59.

Ahmed, M., R. Nazeer Ahamed, and R.H. Aladakatti. 2009. Effect of benzene extract of *Ocimum sanctum* leaves on cauda epididymis of albino rats. *J. Basic Clin. Physiol. Pharmacol.* 20(1):29-41.

Chattopadhyay, R.R. 1993. Hypoglycemic effect of *Ocimum sanctum* leaf extract in normal and streptozotocin diabetic rats. *Indian J. Exp. Biol.* 31(11):891-893.

Gholap, S., and A. Kar. 2004. Hypoglycaemic effects of some plant extracts are possibly mediated through inhibition in corticosteroid concentration. *Pharmazie* 59(11):876-878.

Kantak, N.M., and M.G. Gogate. 1992. Effect of short term administration of tulsi (*Ocimum sanctum* Linn.) on reproductive behaviour of adult male rats. *Indian J. Physiol. Pharmacol.* 36(2):109-111.

Kar, A., B.K. Choudhary, and N.G. Bandyopadhyay. 2003. Comparative evaluation of hypoglycaemic activity of some Indian medicinal plants in alloxan diabetic rats. *J. Ethnopharmacol.* 84(1):105-108.

Khanna, S., S.R. Gupta, and J.K. Grover. 1986. Effect of long term feeding of tulsi (*Ocimum sanctum* Linn) on reproductive performance of adult albino rats. *Indian J. Exp. Biol.* 24(5):302-304.

Lagarto Parra, A., R. Silva Yhebra, I. Guerra Sardinas, and L. Iglesias Buela. 2001. Comparative study of the assay of *Artemia salina* L. and the estimate of the median lethal dose (LD_{50} value) in mice, to determine oral acute toxicity of plant extracts. *Phytomedicine* 8(5):395-400.

Narendhirakannan, R.T., S. Subramanian, and M. Kandaswamy. 2006. Biochemical evaluation of antidiabetogenic properties of some commonly used Indian plants on streptozotocin-induced diabetes in experimental rats. *Clin. Exp. Pharmacol. Physiol.* 33(12):1150-1157.

Panda, S., and A. Kar. 1998. *Ocimum sanctum* leaf extract in the regulation of thyroid function in the male mouse. *Pharmacol. Res.* 38(2):107-110.

Rai, V., U. Iyer, and U.V. Mani. 1997. Effect of tulasi (*Ocimum sanctum*) leaf powder supplementation on blood sugar levels, serum lipids and tissue lipids in diabetic rats. *Plant Foods Hum. Nutr.* 50(1):9-16.

Sardessai, S.R., A.S. Borker, and M.E. Abraham. 1999. Effects of short term administration of tulsi leaves on sexual behaviour in female rats *Indian J. Physiol. Pharmacol.* 43(3):398-400.

Seth, S.D., N. Johri, and K.R. Sundaram. 1981. Antispermatogenic effect of *Ocimum sanctum*. *Indian J. Exp. Biol.* 19(10):975-976.

Vohora, S.B., S.K. Garg, and R.R. Chaudhury. 1969. Antifertility screening of plants. 3. Effect of six indigenous plants on early pregnancy in albino rats. *Indian J. Med. Res.* 57(5):893-899.

Oenothera biennis L. — Onagraceae

SCN: evening primrose

Part: seed, seed oil

Quick Reference Summary

Safety Class: 1

Interaction Class: A

Contraindications

None known.

Other Precautions

None known.

Drug and Supplement Interactions

None known.

Adverse Events and Side Effects

Evening primrose oil has been characterized as generally well tolerated in clinical trials and postmarketing surveillance studies (Gateley et al. 2001; Horrobin 1992). Headaches, abdominal pain, nausea, and loose stools have been reported as adverse events in clinical trials of evening primrose oil (Bamford et al. 1985; Barber 1988).

Pharmacological Considerations

None known.

Pregnancy and Lactation

No adverse effects on mothers or on pregnancy outcomes have been observed in three clinical trials of evening primrose oil use during pregnancy (Laivuori et al. 1993; Moodley and Norman 1989; O'Brien and Broughton Pipkin 1983). In one retrospective study, evening primrose oil use was associated with an increased need for obstetric intervention during labor and delivery (Dove and Johnson 1999).

Gamma-linolenic acid (GLA), one of the primary constituents of evening primrose oil, is a compound naturally formed in the human body and is a component of human breast milk (Carter 1988).

Review Details

I. Drug and Supplement Interactions

Clinical Trials of Drug or Supplement Interactions

No clinical trials of drug or supplement interactions were identified.

Case Reports of Suspected Drug or Supplement Interactions

No case reports of suspected drug or supplement interactions were identified.

Animal Trials of Drug or Supplement Interactions

No animal trials of drug or supplement interactions were identified.

II. Adverse Events

Adverse Events Reported in Clinical Trials

A 1992 review of evening primrose oil use indicated that no significant adverse effects were reported in approximately 500,000 users in the United Kingdom (Horrobin 1992). Adverse events reported in clinical trials of evening primrose oil include headache, abdominal pain, nausea, and loose stools (Bamford et al. 1985; Barber 1988). Evening primrose oil was reported as well tolerated at doses up to 3 g daily for 1 year (Gateley et al. 2001).

In a clinical trial of evening primrose oil (4 g daily for 4 months) and vitamins in chronic schizophrenics, 3 of 23 people enrolled in the study experienced grand mal seizures, 2 of which were in the evening primrose oil group. Both subjects taking evening primrose oil were also taking phenothiazine (Holman and Bell 1983), an antipsychotic medication that can lower seizure thresholds (Tobias et al. 2006). Other clinical trials of evening primrose oil in psychiatric patients, including schizophrenic patients, were not associated with any increase in adverse effects at daily doses up to 540 mg of the compound gamma-linolenic acid (GLA) (Vaddadi et al. 1989; Wolkin et al. 1986). A more recent review of the safety of evening primrose oil in persons with epilepsy indicated that there is no evidence that intake of the omega-6 fatty acids linoleic acid and gamma-linolenic acid may cause seizures and concluded, "It is therefore suggested that formularies should now remove seizures or epilepsy as a side-effect of evening primrose oil, and should remove a history of seizures or epilepsy as a contraindication to taking evening primrose oil supplementation or evening primrose oil-containing products" (Puri 2007).

Case Reports of Adverse Events

Three hospitalized schizophrenics who had failed to respond adequately to standard drug therapy were administered evening primrose oil (8 g daily in one patient, no dose stated for the other two patients). The condition of the patients worsened and electroencephalogram (EEG) features of temporal lobe epilepsy became apparent. The patients had taken or were taking phenothiazine (Vaddadi 1981), an antipsychotic medication that can lower seizure thresholds and may induce a discharge pattern in EEG (Tobias et al. 2006).

III. Pharmacology and Pharmacokinetics

Human Pharmacological Studies

Studies on the effects of evening primrose oil on blood pressure have yielded opposing results, with one study demonstrating a decrease in blood pressure in overweight subjects (Garcia et al. 1986) and a second study indicating no change in blood pressure of hyperlipidemic subjects (Viikari and Lehtonen 1986).

One study of evening primrose oil use in persons with Raynaud's phenomenon indicated some antiplatelet effect of evening primrose oil (Belch et al. 1985). Two other studies, however, one in multiple sclerosis patients and a second in patients with hypertriglyceridemia, indicated no effects on platelet function (Boberg et al. 1986; McGregor et al. 1989).

Animal Pharmacological Studies

Animal pharmacological studies were identified but omitted due to presence of human data.

In Vitro Pharmacological Studies

In vitro pharmacological studies were identified but omitted due to presence of human data.

IV. Pregnancy and Lactation

In women with preeclampsia treated with 4 g daily evening primrose oil, no adverse effects on perinatal outcome were observed (Moodley and Norman 1989). No adverse effects were reported in two small studies of evening primrose oil supplementation in pregnant women for 1 or 5 weeks (Laivuori et al. 1993; O'Brien and Broughton Pipkin 1983).

In a survey of American nurse midwives, 30% of respondents indicated that they used evening primrose oil to stimulate labor (McFarlin et al. 1999). A retrospective study on evening primrose use during pregnancy indicated that oral administration of evening primrose oil

Animal Pharmacological Studies

A reduction in blood glucose levels was observed in rabbits orally administered 0.2 g/kg of an aqueous or alcoholic extract of ophiopogon. At doses of 0.5 g/kg of ophiopogon daily for 4 days, a hypoglycemic effect was observed in diabetic rabbits (Zhu 1998). A single oral dose of 100 mg/kg of polysaccharides from ophiopogon produced a 54% reduction in blood sugar in 11 hours in diabetic mice (Zhu 1998).

A reduction in blood glucose levels was observed in normal and diabetic mice intraperitoneally administered 100 mg/kg of an *n*-butanol extract of ophiopogon (Kako et al. 1995).

In Vitro Pharmacological Studies

No relevant in vitro pharmacological studies were identified.

IV. Pregnancy and Lactation

No information on the safety of ophiopogon during pregnancy or lactation was identified.

V. Toxicity Studies

Acute Toxicity

The LD_{50} of intraperitoneally administered ophiopogon extract in mice was 20.6 g/kg. Intravenous administration of 1 ml of the same ophiopogon extract (equivalent to 2 g of the herb and over 100 times the human dose) produced no toxic reactions (Zhu 1998).

Literature Cited

Bensky, D., S. Clavey, and E. Stöger. 2004. *Chinese herbal medicine: Materia medica*. 3rd ed. Seattle: Eastland Press.

Kako, M., T. Miura, M. Usami, A. Kato, and S. Kadowaki. 1995. Hypoglycemic effect of the rhizomes of ophiopogonis tuber in normal and diabetic mice. *Biol. Pharm. Bull.* 18(5):785-787.

Zhu, Y.-P. 1998. *Chinese materia medica: Chemistry, pharmacology and applications*. Amsterdam: Harwood Academic Publishers.

Oplopanax horridus (Sm.) Miq. — Araliaceae

SCN: devil's club

Syn: *Echinopanax horridus* (Sm.) Decne. & Planch, nom. inval.

Part: root

O

Quick Reference Summary

Safety Class: 1

Interaction Class: A

Contraindications

None known.

Other Precautions

None known.

Drug and Supplement Interactions

None known.

Adverse Events and Side Effects

None known.

Pharmacological Considerations

Use by indigenous people of devil's club in diabetes has been recorded, though several human case studies and animal studies have provided equivocal results in evaluating effects on blood sugar levels, and no contemporary substantiation of efficacy in diabetes was found (Lantz et al. 2004; Smith 1983; Stuhr and Henry 1944; Thommassen et al. 1990).

Pregnancy and Lactation

No information on the safety of devil's club in pregnancy or lactation was identified in the scientific or traditional literature. Although this review did not identify any concerns for use while pregnant or nursing, safety has not been conclusively established.

Review Details

I. Drug and Supplement Interactions

Clinical Trials of Drug or Supplement Interactions

No clinical trials of drug or supplement interactions were identified.

Case Reports of Suspected Drug or Supplement Interactions

No case reports of suspected drug or supplement interactions were identified.

Animal Trials of Drug or Supplement Interactions

No animal trials of drug or supplement interactions were identified.

II. Adverse Events

Case Reports of Adverse Events

No case reports of adverse events were identified.

III. Pharmacology and Pharmacokinetics

Human Pharmacological Studies

In a hospital-based study, one insulin-dependent diabetes patient and one newly diagnosed non-insulin-dependent diabetes patient were orally administered 80 ml of a devil's claw decoction three to four times daily for several days. No significant effects on glucose levels were observed in the insulin-dependent patient, while a gradual reduction in glucose levels was observed in the non-insulin-dependent patient. No changes were observed in glucose levels in healthy volunteers (Thommassen et al. 1990).

Animal Pharmacological Studies

No relevant animal pharmacological studies were identified.

In Vitro Pharmacological Studies

No relevant in vitro pharmacological studies were identified.

IV. Pregnancy and Lactation

No information on the safety of devil's club during pregnancy or lactation was identified.

V. Toxicity Studies

No toxicity studies were identified.

Literature Cited

Lantz, T.C., K. Swerhun, and N.J. Turner. 2004. Devil's club (*Oplopanax horridus*): An ethnobotanical review. *HerbalGram* 62:33-48.

Smith, G.W. 1983. Arctic pharmacognosia. Part 2. Devil's club, *Oplopanax horridus*. *J. Ethnopharmacol.* 7(May):313-320.

Stuhr, E.T., and F.B. Henry. 1944. An investigation of the root bark of *Fatsia horrida*. *Pharm. Arch.* 15(9):6.

Thommassen, H.V., R.A. Wilson, and R.G. McIlwain. 1990. Effect of devil's club tea on blood glucose levels in diabetes mellitus. *Can. Fam. Phys.* 36:62–65.

Origanum majorana L. — Lamiaceae

SCN: sweet marjoram

Part: leaf

Quick Reference Summary

Safety Class: 1

Interaction Class: A

Contraindications

None known.

Other Precautions

None known.

Drug and Supplement Interactions

None known.

Adverse Events

None known.

Pharmacological Considerations

None known.

Pregnancy and Lactation

No information on the safety of sweet marjoram in pregnancy or lactation was identified in the scientific or traditional literature. Although this review did not identify any concerns for use while pregnant or nursing, safety has not been conclusively established.

Review Details

I. Drug and Supplement Interactions

Clinical Trials of Drug or Supplement Interactions

No clinical trials of drug or supplement interactions were identified.

Case Reports of Suspected Drug or Supplement Interactions

No case reports of suspected drug or supplement interactions were identified.

Literature Cited

Bruynzeel, D.P. 1989. Contact dermatitis due to *Paeonia* (peony). *Contact Dermat.* 20(2):152-153.

Madari, H., and R.S. Jacobs. 2004. An analysis of cytotoxic botanical formulations used in the traditional medicine of ancient Persia as abortifacients. *J. Nat. Prod.* 67(8):1204-1210.

Timmermans, M.W., S.E. Pentinga, T. Rustemeyer, and D.P. Bruynzeel. 2009. Contact dermatitis due to *Paeonia* (peony): A rare sensitizer? *Contact Dermat.* 60(4):232-233.

Paeonia suffruticosa Andrews — Paeoniaceae

SCN: tree peony
Syn: *Paeonia moutan* Sims
PN: *mu dan pi* (root bark)
OCN: mountain peony
Part: root bark

Quick Reference Summary

Safety Class: 2b, 2d
Interaction Class: A

Contraindications

Not for use in pregnancy except under the supervision of a qualified healthcare practitioner (Bensky et al. 2004; Chen and Chen 2004).

Not for use in excessive menstruation (Bensky et al. 2004; Chen and Chen 2004).

Other Precautions

None known.

Drug and Supplement Interactions

None known.

Adverse Events and Side Effects

Dizziness and nausea have been reported as occasional side effects of tree peony (Bensky et al. 2004).

Pharmacological Considerations

Human and in vitro studies have indicated that tree peony and compounds from tree peony may inhibit platelet aggregation (Hirai et al. 1983).

Traditional use and animal studies have demonstrated that tree peony and compounds in tree peony may modify glucose regulation (Jung et al. 2006; Lau et al. 2007; Suzuki et al. 1983). People with diabetes are advised to monitor their blood sugar closely and discuss the use of this herb with a qualified healthcare practitioner prior to use.

Pregnancy and Lactation

Reference texts on traditional Chinese medicine indicate that tree peony should not be used during pregnancy (Bensky et al. 2004; Chen and Chen 2004).

No information on the safety of tree peony during lactation was identified in the scientific or traditional literature. While this review did not identify any concerns for use while nursing, safety has not been conclusively established.

Review Details

I. Drug and Supplement Interactions

Clinical Trials of Drug or Supplement Interactions

No clinical trials of drug or supplement interactions were identified.

Case Reports of Suspected Drug or Supplement Interactions

No case reports of suspected drug or supplement interactions were identified.

Animal Trials of Drug or Supplement Interactions

No animal trials of drug or supplement interactions were identified.

II. Adverse Events

Case Reports of Adverse Events

Dizziness and nausea have been reported as occasional side effects of tree peony (Bensky et al. 2004).

III. Pharmacology and Pharmacokinetics

Human Pharmacological Studies

In platelets taken from healthy volunteers who orally ingested an aqueous extract of tree peony at a dose of 3 g of the herb daily for 1 week, a significant reduction in platelet aggregation and thromboxane B_2 formation induced by collagen, epinephrine, and ADP was observed (Hirai et al. 1983).

Animal Pharmacological Studies

Decreases in blood glucose levels were observed in rabbits, rats, and cats intravenously administered 50 to 100 mg/kg of an aqueous extract of tree peony (Suzuki et al. 1983).

A reduction in blood glucose levels was observed in healthy rats orally administered 100 mg/kg of an ethanol extract of tree peony, while a slight elevation of glucose levels was observed in diabetic rats administered the same extract (Jung et al. 2006).

The compound paeonol, orally administered at doses of 200 or 400 mg/kg, improved oral glucose tolerance in diabetic rats (Lau et al. 2007).

In Vitro Pharmacological Studies

Extracts of tree peony inhibited the drug-metabolizing isoenzymes CYP1A1/2, CYP2B1/2, and CYP2E1 (Jeong et al. 2002).

No estrogenic activity of tree peony was observed in a recombinant yeast system featuring a human estrogen receptor expression plasmid and a reporter plasmid (Kim et al. 2008).

A synergistic growth-inhibitory effect on an esophageal cancer cell line was observed with a combination of the compound paeonol and the drug cisplatin as compared to the activity of either substance alone (Wan et al. 2008).

The compound paeonol was found to cause dose-dependent inhibition of ADP- and collagen-induced platelet aggregation in human blood platelets (Hirai et al. 1983).

IV. Pregnancy and Lactation

Reference texts on traditional Chinese medicine indicate that tree peony should not be used during pregnancy (Bensky et al. 2004; Chen and Chen 2004).

No information on the safety of tree peony during lactation was identified.

V. Toxicity Studies

Acute Toxicity

The LD_{50} of an aqueous extract of tree peony in mice is 4.07 g/kg after intraperitoneal administration, 8 g/kg after subcutaneous administration, and could not be determined at doses up to 12.5 g/kg after oral administration (Suzuki et al. 1983).

The LD_{50} of the compound paeonol in mice is 3430 mg/kg after oral administration, 782 mg/kg after intraperitoneal administration, and 196 mg/kg after intravenous administration (Chen and Chen 2004).

Genotoxicity

In the mouse micronucleus assay, some chromosomal aberrations were observed in mice orally administered 3.0 g/kg of an aqueous extract of tree peony, but not in mice that were administered 0.15 to 1.5 g/kg of the same extract. An increase in the number of micronucleated polychromatic erythrocytes was observed in mice administered 0.75, 1.5, or 3.0, but not 0.15 g/kg of the same extract (Yin et al. 1991).

No mutagenic activity of the essential oil of tree peony was observed in *Salmonella typhimurium* strain TA100 with or without metabolic activation (Park 2002). Similarly, no mutagenic activity of an aqueous extract of tree peony was observed in *Salmonella typhimurium* strains TA98 or TA100 with or without metabolic activation (Yin et al. 1991).

An extract of tree peony and compounds from tree peony inhibited oxidative DNA damage and cleavage caused by phenylhydroquinone and *tert*-butylhydroquinone (Okubo et al. 2000).

Literature Cited

Bensky, D., S. Clavey, and E. Stöger. 2004. *Chinese herbal medicine: Materia medica*. 3rd ed. Seattle: Eastland Press.

Chen, J.K., and T.T. Chen. 2004. *Chinese medical herbology and pharmacology*. City of Industry, CA: Art of Medicine Press.

Hirai, A., T. Terano, T. Hamazaki, et al. 1983. Studies on the mechanism of antiaggregatory effect of *Moutan Cortex*. *Thromb. Res.* 31(1):29-40.

Jeong, H.G., H.J. You, Y.S. Chang, et al. 2002. Inhibitory effects of medicinal herbs on cytochrome P450 drug metabolizing enzymes. *Kor. J. Pharmacog.* 33(1):35-41.

Jung, C.H., S. Zhou, G.X. Ding, et al. 2006. Antihyperglycemic activity of herb extracts on streptozotocin-induced diabetic rats. *Biosci. Biotechnol. Biochem.* 70(10):2556-2559.

Kim, I.G., S.C. Kang, K.C. Kim, E.S. Choung, and O.P. Zee. 2008. Screening of estrogenic and antiestrogenic activities from medicinal plants. *Environ. Toxicol. Pharmacol.* 25(1):75-82.

Lau, C.H., C.M. Chan, Y.W. Chan, et al. 2007. Pharmacological investigations of the anti-diabetic effect of *Cortex Moutan* and its active component paeonol. *Phytomedicine* 14(11):778-784.

Okubo, T., F. Nagai, T. Seto, et al. 2000. The inhibition of phenylhydroquinone-induced oxidative DNA cleavage by constituents of *Moutan Cortex* and *Paeoniae Radix*. *Biol. Pharm. Bull.* 23(2):199-203.

Park, H.J. 2002. Mutagenicity of the essential oils in Ames test. *Kor. J. Pharmacog.* 33(4):372-375.

Suzuki, Y., K. Kajiyama, and K. Taguchi. 1983. Pharmacological studies on *Moutan Cortex*. (1) General pharmacological effect of water extract. *Pharmacometrics* 25(3):392-404.

Wan, X.A., G.P. Sun, H. Wang, et al. 2008. Synergistic effect of paeonol and cisplatin on oesophageal cancer cell lines. *Dig. Liver Dis.* 40(7):531-539.

Yin, X.J., D.X. Liu, H.C. Wang, and Y. Zhou. 1991. A study on the mutagenicity of 102 raw pharmaceuticals used in Chinese traditional medicine. *Mutat. Res.* 260(1):73-82.

Palmaria palmata (L.) Kuntze — Palmariaceae

SCN: dulse
Syn: *Rhodymenia palmata* (L.) Grev.
OCN: dillisk
Part: thallus

Quick Reference Summary

Safety Class: 2d
Interaction Class: A

Contraindications

Not for use in persons with hyperthyroidism (Lee et al. 2007; Teas et al. 2004).

Other Precautions

None known.

Drug and Supplement Interactions

None known.

Editors' Notes

A number of seaweeds have been found to contain heavy metal residues (Almela et al. 2002; Rose et al. 2007). Total arsenic has been measured in dulse products at levels ranging from 5 to 12 ppm (Almela et al. 2002; McSheehy and Szpunar 2000; Phaneuf et al. 1999).

Adverse Events and Side Effects

None known.

Pharmacological Considerations

None known.

Pregnancy and Lactation

No information on the safety of dulse in pregnancy or lactation was identified in the scientific or traditional literature. Although this review did not identify any concerns for use while pregnant or nursing, safety has not been conclusively established.

Review Details

I. Drug and Supplement Interactions

Clinical Trials of Drug or Supplement Interactions

No clinical trials of drug or supplement interactions were identified.

Case Reports of Suspected Drug or Supplement Interactions

No case reports of suspected drug or supplement interactions were identified.

Animal Trials of Drug or Supplement Interactions

No animal trials of drug or supplement interactions were identified.

II. Adverse Events

Case Reports of Adverse Events

No case reports of adverse events were identified.

III. Pharmacology and Pharmacokinetics

Human Pharmacological Studies

No relevant human pharmacological studies were identified.

Animal Pharmacological Studies

No effect on serum glucose levels or insulin responses were observed in pigs fed diets containing 5% dulse polysaccharides (Vaugelade et al. 2000).

In Vitro Pharmacological Studies

No relevant in vitro pharmacological studies were identified.

IV. Pregnancy and Lactation

No information on the safety of dulse during pregnancy or lactation was identified.

V. Toxicity Studies

No toxicity studies were identified.

Literature Cited

Almela, C., S. Algora, V. Benito, et al. 2002. Heavy metal, total arsenic, and inorganic arsenic contents of algae food products. *J. Agric. Food Chem.* 50(4):918-923.

Lee, S.M., J. Lewis, D.H. Buss, G.D. Holcombe, and P.R. Lawrance. 2007. Iodine in British foods and diets. *Br. J. Nutr.* 72(3):435-446.

McSheehy, S., and J. Szpunar. 2000. Speciation of arsenic in edible algae by bi-dimensional size-exclusion anion exchange HPLC with dual ICP-MS and electrospray MS/MS detection. *J. Anal. Atom. Spectrom.* 15(1):79-87.

Phaneuf, D., I. Côté, P. Dumas, L.A. Ferron, and A. LeBlanc. 1999. Evaluation of the contamination of marine algae (seaweed) from the St. Lawrence River and likely to be consumed by humans. *Environ. Res.* 80(2):S175-S182.

Rose, M., J. Lewis, N. Langford, et al. 2007. Arsenic in seaweed—Forms, concentration and dietary exposure. *Food Chem. Toxicol.* 45(7):1263-1267.

Teas, J., S. Pino, A. Critchley, and L.E. Braverman. 2004. Variability of iodine content in common commercially available edible seaweeds. *Thyroid* 14(10):836-841.

Vaugelade, P., C. Hoebler, F. Bernard, et al. 2000. Non-starch polysaccharides extracted from seaweed can modulate intestinal absorption of glucose and insulin response in the pig. *Reprod. Nutr. Devel.* 40(1):33-47.

Panax ginseng C.A. Mey. — Araliaceae

SCN: Asian ginseng
Syn: *Panax schinseng* T. Nees
PN: *ren shen* (root)
OCN: Chinese ginseng; Korean ginseng
Part: root

Quick Reference Summary

Safety Class: 1
Interaction Class: A

Contraindications

None known.

Other Precautions

Concurrent use with stimulants is not advised, as overstimulation may occur in susceptible persons (Mills and Bone 2005).

Drug and Supplement Interactions

Case reports of decreased blood clotting in patients taking warfarin and Asian ginseng have been reported (Janetzky and Morreale 1997; Rosado 2003), although trials of warfarin and Asian ginseng in humans and in rats have not shown any interaction (Jiang et al. 2004, 2006; Lee et al. 2008; Zhu et al. 1999).

Editors' Note

The use of red ginseng (processed by steaming) may potentiate the effects of caffeine and other stimulants (Bradley 1992).

Adverse Events and Side Effects

A systematic review of case reports of adverse events and adverse events in clinical trials indicated the relative safety of Asian ginseng. Adverse events in clinical trials were reported as similar for both Asian ginseng and placebo groups (Coon and Ernst 2002).

Cases of mastalgia, irregular menstrual bleeding, postmenopausal bleeding, hypertension, stroke, arteritis, and a coagulation disorder have been reported in people taking Asian ginseng (Coon and Ernst 2002). None of these cases has been causally attributed to Asian ginseng.

Allergic reactions to Asian ginseng have been reported (Dega et al. 1996; Wiwanitkit and Taungjaruwinai 2004).

Wichtl (2004) reports "relatively rare" side effects, "only with high doses and/or use over very long periods of time." These include sleeplessness, nervousness, diarrhea (particularly in the morning), menopausal bleeding, and hypertony. Martindale and Reynolds (1996) present a litany of side effects attributed to Siegel, whose work has been refuted due to methodological flaws (Blumenthal 1991; Buettner et al. 2006; De Smet 1992; Siegel 1979).

Pharmacological Considerations

A systematic review of human studies on the effects of Asian ginseng on blood pressure indicated mixed results, with most studies showing either no substantial change or slight decreases in blood pressure (Buettner et al. 2006).

Human studies have demonstrated that Asian ginseng may modify glucose regulation (Reay et al. 2006; Sotaniemi et al. 1995; Vuksan et al. 2008; Xie et al. 2005). People with diabetes are advised to monitor their blood sugar closely and discuss the use of this herb with a qualified healthcare practitioner prior to use.

No significant effects of Asian ginseng on the drug metabolizing isoenzymes CYP3A4 or CYP2D6 were observed in human studies (Anderson et al. 2003; Gurley et al. 2002, 2005).

No changes in platelet function or blood clotting parameters were observed in healthy volunteers taking Asian ginseng (Beckert et al. 2007).

Pregnancy and Lactation

Two retrospective studies of Asian ginseng use by pregnant women have not indicated any adverse effects of Asian ginseng on mother or fetus (Chin 1991; Chuang et al. 2006).

Although a British herbal text contraindicates Asian ginseng during pregnancy (Bradley 1992), no concerns for use in pregnancy are listed in texts on traditional Chinese medicine (Bensky et al. 2004; Chin 1991).

Animal Pharmacological Studies

No relevant animal pharmacological studies were identified.

In Vitro Pharmacological Studies

One study showed that an American ginseng extract significantly induced the growth of an estrogen receptor-positive breast cancer cell line (MCF-7) in vitro. The extract, however, did not show any estrogenic activity in the α- or β-estrogen receptors, and no increase in uterine weight was observed after 4 days of administration of the extract to mice (Amato et al. 2002).

IV. Pregnancy and Lactation

No information on the safety of American ginseng in pregnancy or lactation was identified.

V. Toxicity Studies

Subacute Toxicity

No significant toxic effects were observed after intraperitoneal administration of 450 mg/kg daily for 7 days in laboratory animals (Chen and Chen 2004).

Literature Cited

Amato, P., S. Christophe, and P. Mellon. 2002. Estrogenic activity of herbs commonly used as remedies for menopausal symptoms. *Menopause* 9(2):145-150.

Andrade, A.S., C. Hendrix, T.L. Parsons, et al. 2008. Pharmacokinetic and metabolic effects of American ginseng (*Panax quinquefolius*) in healthy volunteers receiving the HIV protease inhibitor indinavir. *BMC Complement. Altern. Med.* 8:50.

Bensky, D., S. Clavey, and E. Stöger. 2004. *Chinese herbal medicine: Materia medica*. 3rd ed. Seattle: Eastland Press.

Chen, J.K., and T.T. Chen. 2004. *Chinese medical herbology and pharmacology*. City of Industry, CA: Art of Medicine Press.

Hsu, C.C., M.C. Ho, L.C. Lin, B. Su, and M.C. Hsu. 2005. American ginseng supplementation attenuates creatine kinase level induced by submaximal exercise in human beings. *World J. Gastroenterol.* 11(34):5327-5231.

Lee, L.S., S.D. Wise, C. Chan, et al. 2008. Possible differential induction of phase 2 enzyme and antioxidant pathways by American ginseng, *Panax quinquefolius*. *J. Clin. Pharmacol.* 48 (5):599-609.

McElhaney, J.E., S. Gravenstein, S.K. Cole, et al. 2004. A placebo-controlled trial of a proprietary extract of North American ginseng (CVT-E002) to prevent acute respiratory illness in institutionalized older adults. *J. Am. Geriatr. Soc.* 52(1):13-19.

Predy, G.N., V. Goel, R. Lovlin, et al. 2005. Efficacy of an extract of North American ginseng containing poly-furanosyl-pyranosyl-saccharides for preventing upper respiratory tract infections: A randomized controlled trial. *Can. Med. Assoc. J.* 173(9):1043-1048.

Sievenpiper, J.L., J.T. Arnason, L.A. Leiter, and V. Vuksan. 2004. Decreasing, null and increasing effects of eight popular types of ginseng on acute postprandial glycemic indices in healthy humans: The role of ginsenosides. *J. Am. Coll. Nutr.* 23(3):248-258.

Sotaniemi, E.A., E. Haapakoski, and A. Rautio. 1995. Ginseng therapy in non-insulin-dependent diabetic patients. *Diabetes Care* 18(10):1373-1375.

Stavro, P.M., M. Woo, T.F. Heim, L.A. Leiter, and V. Vuksan. 2005. North American ginseng exerts a neutral effect on blood pressure in individuals with hypertension. *Hypertension* 46(2):406-411.

Stavro, P.M., M. Woo, L.A. Leiter, et al. 2006. Long-term intake of North American ginseng has no effect on 24-hour blood pressure and renal function. *Hypertension* 47(4):791-796.

Vuksan, V., J.L. Sievenpiper, V.Y. Koo, et al. 2000a. American ginseng (*Panax quinquefolius* L) reduces postprandial glycemia in nondiabetic subjects and subjects with type 2 diabetes mellitus. *Arch. Intern. Med.* 160(7):1009-1013.

Vuksan, V., M.P. Stavro, J.L. Sievenpiper, et al. 2000b. Similar postprandial glycemic reductions with escalation of dose and administration time of American ginseng in type 2 diabetes. *Diabetes Care* 23(9):1221-1226.

Yuan, C.S., G. Wei, L. Dey, et al. 2004. Brief communication: American ginseng reduces warfarin's effect in healthy patients: A randomized, controlled trial. *Ann. Intern. Med.* 141(1):23-27.

Parietaria spp. — **Urticaceae**

Parietaria judaica L.
SCN: pellitory-of-the-wall
Syn: *Parietaria diffusa* Mert. & W.D.J. Koch
OCN: spreading pellitory

Parietaria officinalis **L.**
SCN: pellitory-of-the-wall
OCN: upright pellitory
Part: herb

Quick Reference Summary

Safety Class: 1
Interaction Class: A

Contraindications

None known.

Drug and Supplement Interactions

None known.

Notice

Diuretic (Giachetti et al. 1986; Wood and LaWall 1918); *see* Appendix 2.

Adverse Events and Side Effects

The pollen of pellitory-of-the-wall is a major airborne pollen allergen in Europe and the Mediterranean (Ferrer et al. 2005; Pajno et al. 2004; Passalacqua et al. 1999).

Pharmacological Considerations

None known.

Pregnancy and Lactation

No information on the safety of pellitory-of-the-wall in pregnancy or lactation was identified in the scientific or traditional literature. Although this review did not identify any concerns for use while pregnant or nursing, safety has not been conclusively established.

Review Details

I. Drug and Supplement Interactions

Clinical Trials of Drug or Supplement Interactions

No clinical trials of drug or supplement interactions were identified.

Case Reports of Suspected Drug or Supplement Interactions

No case reports of suspected drug or supplement interactions were identified.

Animal Trials of Drug or Supplement Interactions

No animal trials of drug or supplement interactions were identified.

II. Adverse Events

Case Reports of Adverse Events

The pollen of pellitory-of-the-wall is one of the major airborne pollen allergens in Europe and the Mediterranean (Ferrer et al. 2005; Pajno et al. 2004; Passalacqua et al. 1999).

III. Pharmacology and Pharmacokinetics

Human Pharmacological Studies

No relevant human pharmacological studies were identified.

Animal Pharmacological Studies

No relevant animal pharmacological studies were identified.

In Vitro Pharmacological Studies

No relevant in vitro pharmacological studies were identified.

IV. Pregnancy and Lactation

No information on the safety of pellitory-of-the-wall during pregnancy or lactation was identified.

V. Toxicity Studies

No toxicity studies were identified.

P

Literature Cited

Ferrer, M., E. Burches, A. Pelaez, et al. 2005. Double-blind, placebo-controlled study of immunotherapy with *Parietaria judaica*: Clinical efficacy and tolerance. *J. Investig. Allergol. Clin. Immunol.* 15(4):283-292.

Giachetti, D., E. Taddei, and I. Taddei. 1986. Diuretic and uricosuric activity of *Parietaria judaica* L. *Boll. Soc. Ital. Biol. Sper.* 62(2):197-202.

Pajno, G.B., G. Passalacqua, D. Vita, et al. 2004. Sublingual immunotherapy abrogates seasonal bronchial hyperresponsiveness in children with *Parietaria*-induced respiratory allergy: A randomized controlled trial. *Allergy* 59(8):883-887.

Passalacqua, G., M. Albano, A. Riccio, et al. 1999. Clinical and immunologic effects of a rush sublingual immunotherapy to *Parietaria* species: A double-blind, placebo-controlled trial. *J. Allergy Clin. Immunol.* 104(5):964-968.

Wood, H., and C. LaWall. 1918. *The dispensatory of the United States of America*. 21st ed. Philadelphia: Lippincott.

Parthenium integrifolium L. — Asteraceae

SCN: prairie dock
OCN: American feverfew; Missouri snakeroot; parthenium

Part: root

Quick Reference Summary

Safety Class: 1
Interaction Class: A

Contraindications

None known.

Other Precautions

None known.

Drug and Supplement Interactions

None known.

Editors' Note

No cautions for use of prairie dock are reported in historical American medical or pharmaceutical literature (Cook 1869; Felter and Lloyd 1898; Meyer 1881).

Adverse Events and Side Effects

None known.

Pharmacological Considerations

None known.

Pregnancy and Lactation

No information on the safety of prairie dock in pregnancy or lactation was identified in the scientific or traditional literature. Although this review did not identify any concerns for use while pregnant or nursing, safety has not been conclusively established.

Review Details

I. Drug and Supplement Interactions

Clinical Trials of Drug or Supplement Interactions

No clinical trials of drug or supplement interactions were identified.

Case Reports of Suspected Drug or Supplement Interactions

No case reports of suspected drug or supplement interactions were identified.

Animal Trials of Drug or Supplement Interactions

No animal trials of drug or supplement interactions were identified.

II. Adverse Events

Case Reports of Adverse Events

No case reports of adverse events were identified.

III. Pharmacology and Pharmacokinetics

Human Pharmacological Studies

No relevant human pharmacological studies were identified.

Animal Pharmacological Studies

No relevant animal pharmacological studies were identified.

In Vitro Pharmacological Studies

No relevant in vitro pharmacological studies were identified.

IV. Pregnancy and Lactation

No information on the safety of prairie dock during pregnancy or lactation was identified.

V. Toxicity Studies

No toxicity studies were identified.

Literature Cited

Cook, W.H. 1869. *Physio-medical dispensatory*. Cincinnati: Wm. H. Cook.

Felter, H.W., and J.U. Lloyd. 1898. *King's American dispensatory*. 18th ed., 3rd rev. 2 vols. Cincinnati: Ohio Valley Co.

Meyer, F. 1881. *Parthenium Integrifolium*, Lin. *Am. J. Pharm.* 53:494-495.

Passiflora incarnata L. — Passifloraceae

SCN: passionflower
OCN: apricot vine; maypop; wild passionflower

Part: herb

Quick Reference Summary

Safety Class: 1
Interaction Class: A

Contraindications

None known.

Other Precautions

None known.

Drug and Supplement Interactions

None known.

Editors' Notes

Passionflower contains trace amounts of β-carboline alkaloids (e.g., harmane, harmine, and harmol). Analyses of commercial passionflower samples indicated that harmane content ranged from 0 to 0.1 μg/g and harmine ranged from 0 to 0.27 μg/g (Abourashed et al. 2003; Grice et al. 2001; Tsuchiya et al. 1999); another study indicated that these alkaloids were undetectable in all but 1 of 17 commercial samples (Rehwald et al. 1995).

Adverse Events and Side Effects

A meta-analysis of clinical trials of passionflower in the treatment of anxiety indicated that no serious adverse events were reported and that passionflower was generally well tolerated (Miyasaka et al. 2007).

Allergic reactions to passionflower have been reported (Echechipia et al. 1996; Giavina-Bianchi et al. 1997; Smith et al. 1993).

One case of gastrointestinal and cardiovascular disturbances was reported in a woman who had taken passionflower for several days (Fisher et al. 2000).

Pharmacological Considerations

None known.

Pregnancy and Lactation

One animal study showed no adverse effects of passionflower on fetal development (Hirakawa et al. 1981). No other information on the safety of passionflower in pregnancy or lactation was identified. While this review did not identify any concerns for use while nursing, safety has not been conclusively established.

Review Details

I. Drug and Supplement Interactions

Clinical Trials of Drug or Supplement Interactions

No clinical trials of drug or supplement interactions were identified.

Case Reports of Suspected Drug or Supplement Interactions

No case reports of suspected drug or supplement interactions were identified.

Animal Trials of Drug or Supplement Interactions

No animal studies of drug or supplement interactions were identified.

II. Adverse Events

Adverse Events Reported in Clinical Trials

A meta-analysis of clinical trials of passionflower for the treatment of anxiety indicated that no serious adverse events were reported in either of the trials; allergic reaction and dizziness were reported as adverse events, although the incidence of those was similar between those taking passionflower and those taking benzodiazepines (Miyasaka et al. 2007).

Case Reports of Adverse Events

An IgE-mediated case of anaphylactic reaction to an herbal product containing passionflower was reported in an 8-year-old boy with a history of latex allergy. Patch testing confirmed the allergy to passionflower (Echechipia et al. 1996). Vasculitis and urticaria caused by a hypersensitivity reaction were reported in association with passionflower use (Smith et al. 1993).

Occupational allergic asthma and rhinitis to passionflower, confirmed by patch testing, were reported in a pharmacy worker who prepared herbal products (Giavina-Bianchi et al. 1997).

Severe nausea, vomiting, drowsiness, prolonged QT_c, and episodes of nonsustained ventricular tachycardia were reported in a 34-year-old woman who had taken three or four 500 mg capsules of passionflower daily for several days. A sample of the product taken by the patient was analyzed and compared to other capsules produced by the same manufacturer and to other commercial samples of passionflower, and also tested for the presence of two

P

pharmaceutical drugs. Results indicated that the product taken was passionflower, and no adulteration with the tested drugs was found (Fisher et al. 2000).

III. Pharmacology and Pharmacokinetics

Human Pharmacological Studies

No relevant human pharmacological studies were identified.

Animal Pharmacological Studies

No relevant animal pharmacological studies were identified.

In Vitro Pharmacological Studies

Ethanol and aqueous extracts of passionflower inhibited GABA transaminase and glutamic acid decarboxylase in vitro. The half-maximal inhibitory concentration (IC_{50}) values were between 1.2 and 2.5 mg/ml (Awad et al. 2007).

IV. Pregnancy and Lactation

No adverse effects on development were observed in offspring of rats administered 400 mg/kg of a passionflower extract on days 7 to 17 of pregnancy (route of administration not stated in English language translations) (Hirakawa et al. 1981).

No information on the safety of passionflower during lactation was identified.

V. Toxicity Studies

Acute Toxicity

No toxic effects were observed in mice intraperitoneally administered passionflower extracts at doses of 500 or 900 mg/kg (Aoyagi et al. 1974; Speroni and Minghetti 1988).

Short-Term Toxicity

No adverse effects were observed in rats orally administered 5 g/kg daily of an aqueous-ethanolic extract of passionflower for 21 days (Sopranzi et al. 1990).

Genotoxicity

No mutagenic activity of a fluid extract of passionflower was observed in the somatic segregation assay (Ramos Ruiz et al. 1996).

Literature Cited

Abourashed, E.A., J. Vanderplank, and I.A. Khan. 2003. High-speed extraction and HPLC fingerprinting of medicinal plants—II. Application to harman alkaloids of genus *Passiflora*. *Pharmaceutical Biology* 41(2):100-106.

Aoyagi, N., R. Kimura, and T. Murata. 1974. Studies on *Passiflora incarnata* dry extract. I. Isolation of maltol and pharmacological action of maltol and ethyl maltol. *Chem. Pharm. Bull.* 22(5):1008-1013.

Awad, R., D. Levac, P. Cybulska, et al. 2007. Effects of traditionally used anxiolytic botanicals on enzymes of the gamma-aminobutyric acid (GABA) system. *Can. J. Physiol. Pharmacol.* 85(9):933-942.

Echechipia, S., B. Garcia, and M. Alvarez. 1996. *Passiflora* hypersensitivity in a latex allergic patient: Cross-reactivity study. *Allergy* 51(Suppl. 31):49.

Fisher, A.A., P. Purcell, and D.G. Le Couteur. 2000. Toxicity of *Passiflora incarnata* L. *J. Toxicol. Clin. Toxicol.* 38(1):63-66.

Giavina-Bianchi, P.F., Jr., F.F. Castro, M.L. Machado, and A.J. Duarte. 1997. Occupational respiratory allergic disease induced by *Passiflora alata* and *Rhamnus purshiana*. *Ann. Allergy Asthma Immunol.* 79(5):449-454.

Grice, I.D., L.A. Ferreira, and L.R. Griffiths. 2001. Identification and simultaneous analysis of harmane, harmine, harmol, isovitexin, and vitexin in *Passiflora incarnata* extracts with a novel HPLC method. *J. Liquid Chromatogr. Relat. Technol.* 24(16):2513-2523.

Hirakawa, T., T. Suzuki, Y. Sano, T. Kamata, and M. Nakamura. 1981. Reproductive studies of *P. incarnata* extract teratological study. *Kiso To Rinsho* 15:3431-3451.

Miyasaka, L.S., A.N. Atallah, and B.G. Soares. 2007. *Passiflora* for anxiety disorder. *Cochrane Database Syst. Rev.* 1:CD004518.

Ramos Ruiz, A., R.A. De la Torre, N. Alonso, et al. 1996. Screening of medicinal plants for induction of somatic segregation activity in *Aspergillus nidulans*. *J. Ethnopharmacol.* 52(3):123-127.

Rehwald, A., O. Sticher, and B. Meier. 1995. Trace analysis of harman alkaloids in *Passiflora incarnata* by reversed-phase high-performance liquid chromatography. *Phytochem. Anal.* 6(2):96-100.

Smith, G.W., T.M. Chalmers, and G. Nuki. 1993. Vasculitis associated with herbal preparation containing *Passiflora* extract. *Br. J. Rheumatol.* 32(1):87-88.

Sopranzi, N., G. De Feo, G. Mazzanti, and L. Tolu. 1990. Biological and electroencephalographic parameters in rats in relation to *Passiflora incarnata* L. *Clin. Ter.* 132(5):329-333.

Speroni, E., and A. Minghetti. 1988. Neuropharmacological activity of extracts from *Passiflora incarnata*. *Planta Med.* 54(6):488-491.

Tsuchiya, H., H. Hayashi, M. Sato, H. Shimizu, and M. Iinuma. 1999. Quantitative analysis of all types of beta-carboline alkaloids in medicinal plants and dried edible plants by high performance liquid chromatography with selective fluorometric detection. *Phytochem. Anal.* 10(5):247-253.

P

Paullinia cupana Kunth

Sapindaceae

SCN: guaraná

Part: seed

Quick Reference Summary

Safety Class: 1
Interaction Class: C*

Contraindications

None known.

Other Precautions

Guaraná extracts typically contain caffeine, a nervous system stimulant. If taken in large amounts, guaraná products containing caffeine can cause insomnia, nervousness, and the other well-known symptoms of excess caffeine intake (PDR 2006).

Due to the central nervous system (CNS) stimulant effects of caffeine, use of caffeine-containing products is cautioned in persons with heart disorders, as excessive caffeine consumption may increase heart rate or exacerbate arrhythmias; caution is also indicated in psychological disorders, as caffeine may aggravate depression or induce anxiety (Brinker 2001).

Drug and Supplement Interactions

Use of caffeine with other CNS stimulants, including bronchodilators or adrenergic drugs, may cause excessive central nervous system stimulation resulting in nervousness, irritability, insomnia, and possibly convulsions or cardiac arrhythmias (PDR 2006).

Caffeine is metabolized by the isoenzyme CYP1A2. Drugs that inhibit this isoenzyme (including fluvoxamine, ciprofloxacin, cimetidine, amiodarone, fluoroquinolones, furafylline, interferon, methoxsalen, and mibefradil) may slow the metabolism of caffeine, resulting in high blood levels of caffeine in persons drinking multiple cups of guaraná daily (Carrillo and Benitez 2000).

Notice

Caffeine (2.6–7.0%) (Leung and Foster 1996; List and Hörhammer 1973); *see* Appendix 1. A typical cup of guaraná contains approximately 200 to 400 mg caffeine, as compared to coffee, which contains 100 to 200 mg per cup.

Diuretic (Brunton et al. 2006); *see* Appendix 2.

Tannins (12%) (Leung and Foster 1996); *see* Appendix 1.

Editors' Note

The American Herbal Products Association has established a trade requirement (AHPA 2011) that dietary supplement products that contain caffeine, whether as a direct ingredient or as a constituent of herbal ingredients, be labeled to disclose the presence of caffeine in the product and the quantity of added caffeine if greater than 25 mg; be formulated and labeled in a manner to recommend a maximum of 200 mg of caffeine per serving, not more often than every 3 to 4 hours; and bear the following or similar statement on the label of any dietary supplement that contains caffeine in sufficient quantity to warrant such labeling:

> Too much caffeine may cause nervousness, irritability, sleeplessness, and, occasionally, rapid heartbeat. Not recommended for use by children under 18 years of age.

See Appendix 1 for more specific details on this AHPA trade requirement.

Adverse Events and Side Effects

Adverse events, primarily cardiovascular in nature, have been reported in persons consuming "energy drinks" and weight-loss supplements that contain guaraná and multiple other botanical extracts, added caffeine, and other stimulant compounds (Pittler et al. 2005).

Pharmacological Considerations

Caffeine is a central nervous system (CNS) stimulant, and excessive use may result in insomnia and nervousness (PDR 2006).

Guaraná has been reported to inhibit platelet aggregation in mice and in vitro (Bydlowski et al. 1988, 1991).

Pregnancy and Lactation

No studies on the safety of guaraná during pregnancy or lactation were identified. Caffeine is in FDA pregnancy category C and has been shown to cross the placenta and achieve blood and tissue concentrations in the fetus. Pregnant women are advised to limit caffeine intake to 300 mg or less daily (PDR 2006).

Caffeine is listed as a "Maternal Medication Usually Compatible with Breastfeeding" by the American Academy of Pediatrics Committee on Drugs. The Committee noted that maternal consumption of caffeine may cause irritability and poor sleeping patterns in nursing infants, and that maternal consumption of caffeinated beverages should be limited to 150 mg caffeine daily (AAP 2001).

* For caffeine-free preparations, no interactions are expected.

Review Details

I. Drug and Supplement Interactions

Clinical Trials of Drug or Supplement Interactions

No clinical trials of drug or supplement interactions were identified.

Case Reports of Suspected Drug or Supplement Interactions

No case reports of suspected drug or supplement interactions were identified.

Animal Trials of Drug or Supplement Interactions

No animal trials of drug or supplement interactions were identified.

II. Adverse Events

Adverse Events Reported in Clinical Trials

In a trial of 500 mg daily guaraná (2.1% caffeine) in healthy volunteers over age 60 for 5 months, one case of tachycardia and three cases of gastrointestinal upset were reported in the guaraná group, and one case each of gastrointestinal upset and insomnia were reported in the placebo group (Galduroz and Carlini 1996).

No adverse events were reported in a human study of single doses of 37.5, 75, 150, or 300 mg of standardized guaraná extract (11 –12% caffeine) administered to healthy adults (Haskell et al. 2007).

Case Reports of Adverse Events

Cardiovascular and other adverse events have been reported in persons consuming "energy drinks" and weight-loss supplements that contain guaraná and multiple other botanical extracts, added caffeine, and other stimulant compounds (Pittler et al. 2005). These adverse events are consistent with the known actions of stimulant compounds taken in excessive doses.

One case of acute tubular necrosis was reported in a woman who had taken guaraná and occasionally used injectable nonsteroidal painkillers (Vagasi et al. 2007). The use of nonsteroidal anti-inflammatory drugs is associated with kidney disease (Abuelo 1995).

III. Pharmacology and Pharmacokinetics

Human Pharmacological Studies

No relevant human pharmacological studies were identified.

Animal Pharmacological Studies

A decrease in platelet aggregation in response to ADP or arachidonate was observed in rabbits administered 1 ml of guaraná intravenously or 20 ml of guaraná through a nasogastric tube (Bydlowski et al. 1988).

Administration of 500 mg/kg of an aqueous extract of guaraná administered to glycogenolytic mice suppressed exercise-induced hypoglycemia, increased blood glucose and decreased liver glycogen after oral administration of maltose (Miura et al. 1998).

In Vitro Pharmacological Studies

A dose-dependent decrease in platelet aggregation was observed in vitro after the application of aqueous guaraná extracts to human or rabbit platelet-rich plasma (Bydlowski et al. 1988, 1991).

IV. Pregnancy and Lactation

No studies on the safety of guaraná during pregnancy or lactation were identified.

Caffeine is in the FDA pregnancy category C and has been shown to cross the placenta and achieve blood and tissue concentrations in the fetus. Excessive intake of caffeine by pregnant women has been associated with fetal arrhythmias. Pregnant women are advised to limit caffeine intake to less than 300 mg daily (PDR 2006).

Caffeine is listed as a "Maternal Medication Usually Compatible with Breastfeeding" by the American Academy of Pediatrics Committee on Drugs. The committee noted that maternal consumption of caffeine may cause irritability and poor sleeping patterns in nursing infants, and that maternal consumption of caffeinated beverages should be limited to two to three cups daily (AAP 2001).

Epidemiological studies have indicated an association between high caffeine intake during pregnancy and an increased risk of spontaneous abortions. An analysis concluded that methodological flaws in many of the studies led to biased results, and that a causal link between caffeine consumption and abortion cannot yet be confirmed (Signorello and McLaughlin 2004).

V. Toxicity Studies

Acute Toxicity

No toxic effects were observed in mice treated with guaraná at oral doses of 2 g/kg and at intraperitoneal doses of 1 or 2 g/kg when compared to control groups. Histopathological examination did not detect any differences between the guaraná and control groups in a number of organs, including liver, kidneys, and stomach (Mattei et al. 1998).

Chronic Toxicity

No adverse effects were observed in mice administered a 3 mg/ml aqueous-ethanolic extract of guaraná as the sole source of drinking water for 12 months (Mattei et al. 1998).

Cytotoxicity

No cytotoxicity was observed in three in vitro assays at concentrations of guaraná up to 40 mg/ml (Santa Maria et al. 1998).

Genotoxicity

In mice orally administered 133, 265, and 530 mg/kg daily of an aqueous extract of guaraná for 7 days, changes in the frequency of micronuclei in the femoral cells and inductions of testicular chromosomal aberrations were observed. An increase of malondialdehyde and depletion of nonprotein sulfhydryl, RNA, and DNA in both hepatic and testicular cells were observed (Al-Majed 2006).

No genotoxic effects of guaraná were observed in bacterial cells in the presence of S9 microsomal fraction, catalase, superoxide dismutase, or thiourea, although genotoxic effects were observed in the absence of these compounds (da Fonseca et al. 1994).

Literature Cited

AAP. 2001. The transfer of drugs and other chemicals into human milk. American Academy of Pediatrics Committee on Drugs. *Pediatrics* 108(3):776-789.

AHPA. July 2011. Code of Ethics & Business Conduct. Silver Spring, MD: American Herbal Products Association.

Abuelo, G.J. 1995. Diagnosing vascular causes of renal failure. *Ann. Intern. Med.* 123(8):601-614.

Al-Majed, A.A. 2006. Genetic and biochemical toxicity of guarana after sub-acute treatment in somatic and germ cells of Swiss albino mice. *Int. J. Pharmacol.* 2(2):226-232.

Brinker, F. 2001. *Herb contraindications and drug interactions*. 3rd ed. Sandy, OR: Eclectic Medical Publications.

Brunton, L.L., J.S. Lazo, and K.L. Parker. 2006. *Goodman & Gilman's the pharmacological basis of therapeutics*. 11th ed. New York: McGraw-Hill.

Bydlowski, S.P., E.A. D'Amico, and D.A. Chamone. 1991. An aqueous extract of guarana (*Paullinia cupana*) decreases platelet thromboxane synthesis. *Braz. J. Med. Biol. Res.* 24(4):421-424.

Bydlowski, S.P., R.L. Yunker, and M.T. Subbiah. 1988. A novel property of an aqueous guarana extract (*Paullinia cupana*): Inhibition of platelet aggregation *in vitro* and *in vivo*. *Braz. J. Med. Biol. Res.* 21(3):535-538.

Carrillo, J.A., and J. Benitez. 2000. Clinically significant pharmacokinetic interactions between dietary caffeine and medications. *Clin. Pharmacokin.* 39(2):127-153.

da Fonseca, C.A., J. Leal, S.S. Costa, and A.C. Leitao. 1994. Genotoxic and mutagenic effects of guarana (*Paullinia cupana*) in prokaryotic organisms. *Mutat. Res.* 321(3):165-173.

Galduroz, J.C., and E.A. Carlini. 1996. The effects of long-term administration of guarana on the cognition of normal, elderly volunteers. *Sao Paulo Med. J.* 114(1):1073-1078.

Haskell, C.F., D.O. Kennedy, K.A. Wesnes, A.L. Milne, and A.B. Scholey. 2007. A double-blind, placebo-controlled, multi-dose evaluation of the acute behavioural effects of guarana in humans. *J. Psychopharmacol.* 21(1):65-70.

Leung, A.Y., and S. Foster. 1996. *Encyclopedia of common natural ingredients used in food, drugs, and cosmetics*. 2nd ed. New York: Wiley.

List, P.H., and H. Hörhammer. 1973. *Hagers handbuch der pharmazeutischen praxis*. Berlin: Springer.

Mattei, R., R.F. Dias, E.B. Espinola, E.A. Carlini, and S.B.M. Barros. 1998. Guarana (*Paullinia cupana*): Toxic behavioral effects in laboratory animals and antioxidant activity *in vitro*. *J. Ethnopharmacol.* 60(2):111-116.

Miura, T., M. Tatara, K. Nakamura, and I. Suzuki. 1998. Effect of guarana on exercise in normal and epinephrine-induced glycogenolytic mice. *Biol. Pharm. Bull.* 21(6):646-648.

PDR. 2006. *Physicians' desk reference for nonprescription drugs and dietary supplements*. 27th ed. Montvale, NJ: Medical Economics Co.

Pittler, M.H., K. Schmidt, and E. Ernst. 2005. Adverse events of herbal food supplements for body weight reduction: Systematic review. *Obes. Rev.* 6(2):93-111.

Santa Maria, A., A. Lopez, M.M. Diaz, D. Munoz-Mingarro, and J.M. Pozuelo. 1998. Evaluation of the toxicity of guarana with *in vitro* bioassays. *Ecotoxicol. Environ. Saf.* 39(3):164-167.

Signorello, L.B., and J.K. McLaughlin. 2004. Maternal caffeine consumption and spontaneous abortion: A review of the epidemiologic evidence. *Epidemiology* 15(2):229-239.

Vagasi, K., P. Degrell, I. Kesoi, et al. 2007. Acute renal failure caused by plant extract. *Orv. Hetil.* 148(9):421-424.

Pausinystalia johimbe (K. Schum.) Pierre ex Beille — Rubiaceae

SCN: yohimbe
Syn: *Corynanthe yohimbe* K. Schum.
OCN: johimbe
Part: bark

Quick Reference Summary

Safety Class: 2b, 2c, 2d
Interaction Class: B

Contraindications

Not for use in pregnancy except under the supervision of a qualified healthcare practitioner (De Smet 1997). Use in lactation is also not recommended.

Price, L.H., D.S. Charney, and G.R. Heninger. 1984. Three cases of manic symptoms following yohimbine administration. *Am. J. Psychiatr.* 141(10):1267-1268.

Roth, L., M. Daunderer, and K. Kormann. 1984. *Giftpflanzen-pflanzengifte: Vorkommen, wirkung, therapie*. Landsberg: Ecomed.

RTECS. 1987. Registry of Toxic Effects of Chemical Substances. 1985-1986 Edition. U.S. Department of Health and Human Services, Public Health Service, Centers for Disease Control, National Institute for Occupational Safety and Health 5132-5133.

Ruck, B., R.D. Shih, and S.M. Marcus. 1999. Hypertensive crisis from herbal treatment of impotence. *Am. J. Emerg. Med.* 17(3):317-318.

Sandler, B., and P. Aronson. 1993. Yohimbine-induced cutaneous drug eruption, progressive renal failure, and lupus-like syndrome. *Urology* 41(4):343-345.

Tam, S.W., M. Worcel, and M. Wyllie. 2001. Yohimbine: A clinical review. *Pharmacol. Ther.* 91(3):215-243.

Ulbricht, C.E., and E.M. Basch. 2005. *Natural standard herb & supplement reference*. St. Louis: Elsevier Mosby.

Pelargonium sidoides DC. — Geraniaceae

SCN: *Pelargonium sidoides*

Part: root

Quick Reference Summary

Safety Class: 1
Interaction Class: A

Contraindications

None known.

Other Precautions

None known.

Drug and Supplement Interactions

None known.

Adverse Events and Side Effects

Allergic reactions, including anaphylactic reactions, to *Pelargonium sidoides* have been reported (de Boer et al. 2007; Timmer et al. 2008).

Pharmacological Considerations

An animal test indicated no interaction between *Pelargonium sidoides* and warfarin (Koch and Biber 2007). A human study indicated no interaction between *Pelargonium sidoides* and penicillin V (Roots et al. 2004).

A systematic review of clinical trials with *Pelargonium sidoides* in adults and children indicated that adverse events were rarely reported but were slightly more common with *Pelargonium sidoides* than control groups. Adverse events were generally gastrointestinal in nature, and none were severe (Timmer et al. 2008).

Pregnancy and Lactation

No information on the safety of *Pelargonium sidoides* in pregnancy or lactation was identified in the scientific or traditional literature. Although this review did not identify any concerns for use while pregnant or nursing, safety has not been conclusively established.

Review Details

I. Drug and Supplement Interactions

Clinical Trials of Drug or Supplement Interactions

No significant changes in serum levels of penicillin V were observed in healthy volunteers orally administered 30 drops of a *Pelargonium sidoides* extract three times daily and penicillin V at a dosage of 1,200,000 IU three times over 8 days (Roots et al. 2004).

Case Reports of Suspected Drug or Supplement Interactions

No case reports of suspected drug or supplement interactions were identified.

Animal Trials of Drug or Supplement Interactions

No changes in the anticoagulant activity (thromboplastin time or partial thromboplastin time) of warfarin was observed in rats orally coadministered single doses of 500 mg/kg of *Pelargonium sidoides* and 0.05 mg/kg of warfarin or in rats orally administered 500 mg/kg daily for 2 weeks prior to administration of a single dose of warfarin. The authors noted that coumarin-type anticoagulants inhibit the synthesis of vitamin K-dependent coagulation factors via identical mechanisms in rats and humans, and have a similar pattern of metabolism in both species (Koch and Biber 2007).

II. Adverse Events

Adverse Events Reported in Clinical Trials

A systematic review of eight clinical trials with *Pelargonium sidoides* in adults and children with a total of 1565 patients indicated that adverse events were rarely reported but were slightly more common with *Pelargonium sidoides* than control groups. Adverse events were generally gastrointestinal in nature, including nausea, vomiting, diarrhea and heartburn, and there were also allergic skin reactions with pruritus and urticaria. None of the adverse events were severe (Timmer et al. 2008).

Case Reports of Adverse Events

Between 2002 and 2006, 34 cases of hypersensitivity reactions to *Pelargonium sidoides* were reported to the WHO Uppsala Monitoring Centre. The case reports are anonymous and heterogeneous and vary with regard to their source, documentation, and likely relationship to the drug administered. All original reports were requested and reviewed individually. In 10 reports, concomitant use of other drugs was noted, but none of the other medications were reported as being cosuspect. In 15 of the 34 reports, the description and timing of the event, notably the combination of a skin rash with itching, urticaria, angioedema, and/or systemic involvement were suggestive of a Coombs and Gell Type I acute hypersensitivity reaction. Two of the patients needed treatment for circulatory failure or anaphylactic shock. A skin prick test confirmed allergy to *Pelargonium sidoides* in a patient who experienced life-threatening acute urticaria and circulatory failure (de Boer et al. 2007).

III. Pharmacology and Pharmacokinetics

Human Pharmacological Studies

No relevant human pharmacological studies were identified.

Animal Pharmacological Studies

No effects on thromboplastin time, partial thromboplastin time, or thrombin time were observed in rats orally administered 10, 75, or 500 mg/kg of *Pelargonium sidoides* daily for 2 weeks. The authors noted that the coumarins identified in *Pelargonium sidoides* do not posses the structural characteristics needed for anticoagulant activity, and that it is unlikely that *Pelargonium sidoides* would increase the tendency for hemorrhage in persons taking this herb (Koch and Biber 2007).

In Vitro Pharmacological Studies

No relevant in vitro pharmacological studies were identified.

IV. Pregnancy and Lactation

No information on the safety of *Pelargonium sidoides* during pregnancy or lactation was identified.

V. Toxicity Studies

A review of *Pelargonium sidoides* indicated that a full set of toxicity studies had been completed, including acute short-term toxicity studies in rats, 2-week dose finding and 13-week toxicity studies in dogs, the Ames test for mutagenicity, the chromosomal aberration test, the mouse micronucleus test, tumor promotion studies, immunotoxicity studies, and reproductive toxicology studies. The review reported that all studies yielded unremarkable results. Information on products and doses used were not reported in the review (Conrad et al. 2007).

Toxicity studies in rats and dogs indicated that the no-observed-adverse-effect level (NOAEL) is greater than 750 mg/kg (Loew et al. 2009).

P

Literature Cited

Conrad, A., H. Kolodziej, and V. Schulz. 2007. *Pelargonium sidoides*-extract (EPs 7630): Registration confirms efficacy and safety. *Wien Med. Wochenschr.* 157(13-14):331-336.

de Boer, H.J., U. Hagemann, J. Bate, and R.H. Meyboom. 2007. Allergic reactions to medicines derived from *Pelargonium* species. *Drug Saf.* 30(8):677-680.

Koch, E., and A. Biber. 2007. Treatment of rats with the *Pelargonium sidoides* extract EPs 7630 has no effect on blood coagulation parameters or on the pharmacokinetics of warfarin. *Phytomedicine* 14(Suppl. 6):40-45.

Loew, D., H. Hauer, and E. Koch. 2009. Differentiated risk consideration: Coumarins in phytopharmaceuticals. *Pharm. Ztg.* 154(7).

Roots, I., G. Arold, A. Dienel, et al. 2004. Placebokontrollierte doppelblinde Interaktionsstudie mit *Pelargonium-sidoides*-Extrakt und Penicillin V bei gesunden Probanden [abstract]. *Phytopharm. Phytother.* 22.

Timmer, A., J. Gunther, G. Rucker, et al. 2008. *Pelargonium sidoides* extract for acute respiratory tract infections. *Cochrane Database Syst. Rev.* 3:CD006323.

Petroselinum crispum (Mill.) Nyman ex A.W. Hill

Apiaceae

SCN: parsley
Syn: *Petroselinum sativum* Hoffm.

Part: leaf

Quick Reference Summary

Safety Class: 1
Interaction Class: A

Contraindications

None known.

Other Precautions

None known.

Drug and Supplement Interactions

None known.

Notice

Emmenagogue (Chadha 1988; Watt and Breyer-Brandwijk 1962); *see* Appendix 2.

Editors' Note

Concerns for this herb are based on relatively higher doses used for therapeutic purposes in contrast to lower amounts generally used in cooking, and have not been associated with its use as a spice.

Adverse Events and Side Effects

Parsley leaf contains furanocoumarins, compounds recognized to cause photodermatitis after topical exposure followed by exposure to sun or other sources of ultraviolet radiation (Beier and Ivie 1985; Zaynoun et al. 1985). Cases of dermatitis (phytophotodermatitis) have been reported in farm workers harvesting parsley (Lagey et al. 1995; Smith 1985). No cases of photosensitivity after ingestion of parsley were identified (Zaynoun et al. 1985).

Pharmacological Considerations

Animal studies and traditional use have indicated that parsley may modify glucose regulation (Ozsoy-Sacan et al. 2006; Tahraoui et al. 2007; Yanardağ et al. 2003). People with diabetes are advised to monitor their blood sugar closely and discuss the use of this herb with a qualified healthcare practitioner prior to use.

Pregnancy and Lactation

Several references indicate that parsley has historically been used as an emmenagogue and abortifacient. None of the references, however, indicates whether the leaf, seed, or root were the parts typically used (Ciganda and Laborde 2003; Laborde and Ciganda 1998; Mele 1968; Riddle 1997). Apiol, a compound found primarily in the seed but also in lesser concentrations in the leaf and root of parsley, is reported as abortifacient in high doses (Wichtl 2004). Parsley essential oil and parsley seed should not be used during pregnancy, although moderate use of parsley leaf is thought to be safe.

No information on the safety of parsley leaf during lactation was identified. While this review did not identify any concerns for use while nursing, safety has not been conclusively established.

P

Review Details

I. Drug and Supplement Interactions

Clinical Trials of Drug or Supplement Interactions

No clinical trials of drug or supplement interactions were identified.

Case Reports of Suspected Drug or Supplement Interactions

No case reports of suspected drug or supplement interactions were identified.

Animal Trials of Drug or Supplement Interactions

No animal trials of drug or supplement interactions were identified.

II. Adverse Events

Case Reports of Adverse Events

Phytophotodermatitis has been reported in persons commercially harvesting parsley (Lagey et al. 1995; Smith 1985).

Phytophotodermatitis was reported in pigs allowed to graze in a field of parsley. Vesicles with erythema and skin fissures were reported on the snouts, ears, and mammary gland teats (Griffiths and Douglas 2000). Acute photosensitivity was reported in a flock of ostriches exposed to parsley. Avian photosensitivity was confirmed by experimental reproduction of the typical lesions in ducks (Perelman and Kuttin 1988).

III. Pharmacology and Pharmacokinetics

Human Pharmacological Studies

No relevant human pharmacological studies were identified.

Animal Pharmacological Studies

In diabetic rats orally administered 2 g/kg aqueous extract of parsley leaf daily for 28 days, a reduction in blood glucose, serum alkaline phosphatase activity, sialic acid, uric acid, potassium, and sodium levels, and liver lipid peroxidation and non-enzymatic glycosylation levels was observed, along with an increase in GSH levels (Ozsoy-Sacan et al. 2006). A reduction in blood glucose levels was observed in diabetic rats orally administered 2 g/kg of an aqueous extract of parsley leaf daily for 28 days (Yanardağ et al. 2003).

In Vitro Pharmacological Studies

Dose-dependent inhibition of thrombin- and ADP-induced platelet aggregation was observed in rat platelets treated with an aqueous extract of parsley leaf. The IC_{50} was approximately 6.5 mg/ml (Mekhfi et al. 2004).

A methanol extract of aerial parts of parsley exhibited estrogenic activity in human estrogen receptor-positive breast cancer cells (MCF-7), with activity reported as similar to that of soy isoflavones (Shimoda et al. 2000; Yoshikawa et al. 2000).

IV. Pregnancy and Lactation

Several references indicate that parsley has historically been used as an emmenagogue and abortifacient. None of the references, however, indicate whether the leaf, seed, or root were the parts typically used (Ciganda and Laborde 2003; Laborde and Ciganda 1998; Mele 1968; Riddle 1997). Apiol, a compound found in the seed but also in lesser amounts in parsley leaf and root, has been reported as an abortifacient in high doses (Wichtl 2004).

No information on the safety of parsley leaf during lactation was identified.

V. Toxicity Studies

Genotoxicity

No mutagenic activity of an ethanolic extract of parsley leaf was observed in the Ames test for mutagenicity with *Salmonella typhimurium* strains TA98 and TA102 (Mahmoud et al. 1992).

Literature Cited

Beier, R.C., and G.W. Ivie. 1985. Linear furanocoumarins in the common herb parsley *Petroselinum sativum* biologically active compounds [abstract]. *Phytochemistry* 36:869-872.

Chadha, Y. 1988. *The wealth of India: A dictionary of Indian raw materials and industrial products*. Delhi: Council of Scientific and Industrial Research.

Ciganda, C., and A. Laborde. 2003. Herbal infusions used for induced abortion. *J. Toxicol. Clin. Toxicol.* 41(3):235-239.

Griffiths, I.B., and R.G. Douglas. 2000. Phytophotodermatitis in pigs exposed to parsley (*Petroselinum crispum*). *Vet. Rec.* 146(3):73-74.

Laborde, A., and C. Ciganda. 1998. Poisoning by herbal infusions ingested as abortifacient agents [abstract]. *J. Toxicol. Clin. Toxicol.* 36(5):454-455.

Lagey, K., L. Duinslaeger, and A. Vanderkelen. 1995. Burns induced by plants. *Burns* 21(7):542-543.

Mahmoud, I., A. Alkofahi, and A. Abdelaziz. 1992. Mutagenic and toxic activities of several spices and some Jordanian medicinal plants. *Pharm. Biol.* 30(2):81-85.

Mekhfi, H., M. El Haouari, A. Legssyer, et al. 2004. Platelet anti-aggregant property of some Moroccan medicinal plants. *J. Ethnopharmacol.* 94(2-3):317-322.

Mele, V. 1968. [On poisoning with parsley used as an abortifacient.] *Folia Med.* 51(8):601.

Ozsoy-Sacan, O., R. Yanardag, H. Orak, et al. 2006. Effects of parsley (*Petroselinum crispum*) extract versus glibornuride on the liver of streptozotocin-induced diabetic rats. *J. Ethnopharmacol.* 104(1-2):175-181.

Perelman, B., and E.S. Kuttin. 1988. Parsley-induced photosensitivity in ostriches and ducks. *Avian Pathol.* 17(1):183-192.

Peroutka, R., V. Schulzova, P. Botek, and J. Hajslova. 2007. Analysis of furanocoumarins in vegetables (Apiaceae) and citrus fruits (Rutaceae). *J. Sci. Food Agric.* 87(11):2152-2163.

Riddle, J. 1997. *Eve's herbs: A history of contraception and abortion in the West*. Cambridge, MA: Harvard University Press.

Shimoda, H., Y. Kawahara, and M. Yoshikawa. 2000. Bioactive constituents of medicinal herb: Phytoestrogens from parsley (*Petroselinum crispum*). *Aroma Res.* 1(2):67-74.

Smith, D.M. 1985. Occupational photodermatitis from parsley. *Practitioner* 229:673-675.

Tahraoui, A., J. El-Hilaly, Z.H. Israili, and B. Lyoussi. 2007. Ethnopharmacological survey of plants used in the traditional treatment of hypertension and diabetes in south-eastern Morocco (Errachidia province). *J. Ethnopharmacol.* 110(1):105-117.

Watt, J.M., and M.G. Breyer-Brandwijk. 1962. *The medicinal and poisonous plants of southern and eastern Africa*. 2nd ed. Edinburgh: E. & S. Livingstone.

Wichtl, M. 2004. *Herbal drugs and phytopharmaceuticals: A handbook for practice on a scientific basis*. 3rd ed. Boca Raton, FL: CRC Press.

Yanardağ, R., S. Bolkent, A. Tabakoglu-Oguz, and O. Ozsoy-Sacan. 2003. Effects of *Petroselinum crispum* extract on pancreatic B cells and blood glucose of streptozotocin-induced diabetic rats. *Biol. Pharm. Bull.* 26:1206-1210.

Yoshikawa, M., T. Uemura, H. Shimoda, et al. 2000. Medicinal foodstuffs. XVIII. Phytoestrogens from the aerial part of *Petroselinum crispum* MIll. (parsley) and structures of 6″-acetylapiin and a new monoterpene glycoside, petroside. *Chem. Pharm. Bull.* 48(7):1039-1044.

Zaynoun, S., L.A. Ali, K. Tenekjian, and A. Kurban. 1985. The bergapten content of garden parsley *Petroselinum sativum* and its significance in causing cutaneous photosensitization. *Clin. Exp. Dermatol.* 10(4):328-331.

Petroselinum crispum (Mill.) Nyman ex A. W. Hill — Apiaceae

SCN: parsley
Syn: *Petroselinum sativum* Hoffm.

Part: root

Quick Reference Summary

Safety Class: 1
Interaction Class: A

Contraindications

None known.

Other Precautions

None known.

Drug and Supplement Interactions

None known.

Notice

Diuretic (Felter and Lloyd 1898); *see* Appendix 2.

Adverse Events and Side Effects

Parsley root contains furanocoumarins, compounds recognized to cause photodermatitis after topical exposure followed by exposure to sun or other sources of ultraviolet radiation (Nitz et al. 1990).

Allergic skin reactions to parsley root have been reported (Wichtl 2004).

Pharmacological Considerations

None known.

Pregnancy and Lactation

Several references indicate that parsley has historically been used as an emmenagogue and abortifacient. None of the references, however, indicate whether the leaf, seed, or root were the parts typically used (Ciganda and Laborde 2003; Laborde and Ciganda 1998; Mele 1968; Riddle 1997). Apiol, a compound found primarily in the seed but also in lesser concentrations in the leaf and root of parsley, is reported as abortifacient in high doses (Wichtl 2004). Parsley essential oil and parsley seed should not be used during pregnancy, although moderate use of parsley root in pregnancy is thought to be safe.

No information on the safety of parsley root during lactation was identified. While this review did not identify any concerns for use while nursing, safety has not been conclusively established.

Review Details

I. Drug and Supplement Interactions

Clinical Trials of Drug or Supplement Interactions

No clinical trials of drug or supplement interactions were identified.

Case Reports of Suspected Drug or Supplement Interactions

No case reports of suspected drug or supplement interactions were identified.

Animal Trials of Drug or Supplement Interactions

No animal trials of drug or supplement interactions were identified.

II. Adverse Events

Case Reports of Adverse Events

Rare cases of allergic reactions of the skin and mucous membranes to parsley root have been reported (Wichtl 2004).

III. Pharmacology and Pharmacokinetics

Human Pharmacological Studies

No relevant human pharmacological studies were identified.

Animal Pharmacological Studies

No relevant animal pharmacological studies were identified.

In Vitro Pharmacological Studies

In mice orally administered 10 ml/kg of parsley root juice, prolongation of pentobarbital-induced (40 mg/kg intraperitoneally) sleeping time and of the analgesic action time of aminopyrine (60 mg/kg intraperitoneally) and paracetamol (80 mg/kg intraperitoneally) were observed. After a single dose of the juice, a decrease in CYP450 liver homogenates of treated mice was observed (Jakovljevic et al. 2002).

IV. Pregnancy and Lactation

Several references indicate that parsley has historically been used as an emmenagogue and abortifacient. None of the references, however, indicate whether the leaf, seed, or root were the parts typically used (Ciganda and Laborde

2003; Laborde and Ciganda 1998; Mele 1968; Riddle 1997). Apiol, a compound found primarily in the seed but also in lesser concentrations in the leaf and root of parsley, is reported to be abortifacient in high doses (Wichtl 2004).

No information on the safety of parsley root during lactation was identified.

V. Toxicity Studies

No toxicity studies were identified.

Literature Cited

Ciganda, C., and A. Laborde. 2003. Herbal infusions used for induced abortion. *J. Toxicol. Clin. Toxicol.* 41(3):235-239.

Felter, H.W., and J.U. Lloyd. 1898. *King's American dispensatory.* 18th ed., 3rd rev. 2 vols. Cincinnati: Ohio Valley Co.

Jakovljevic, V., A. Raskovic, M. Popovic, and J. Sabo. 2002. The effect of celery and parsley juices on pharmacodynamic activity of drugs involving cytochrome P450 in their metabolism. *Eur. J. Drug Metab. Pharmacokinet.* 27(3):153-156.

Laborde, A., and C. Ciganda. 1998. Poisoning by herbal infusions ingested as abortifacient agents [abstract]. *J. Toxicol. Clin. Toxicol.* 36(5):454-455.

Mele, V. 1968. On poisoning with parsley used as an abortifacient. *Folia Med.* 51(8):601.

Nitz, S., M.H. Spraul, and F. Drawert. 1990. C_{17} polyacetylenic alcohols as the major constituents in roots of *Petroselinum crispum* Mill. ssp. *tuberosum. J. Agric. Food Chem.* 38:1445-1447.

Riddle, J. 1997. *Eve's herbs: A history of contraception and abortion in the West.* Cambridge, MA: Harvard University Press.

Wichtl, M. 2004. *Herbal drugs and phytopharmaceuticals: A handbook for practice on a scientific basis.* 3rd ed. Boca Raton, FL: CRC Press.

Peumus boldus Molina — Monimiaceae

SCN: boldo

Part: leaf

Quick Reference Summary

Safety Class: 2d

Interaction Class: A

Contraindications

Not for use in persons with obstructions of the bile duct (ESCOP 2003; Wichtl 2004) or serious liver conditions (Wichtl 2004).

In persons with gallstones, boldo should only be used under the supervision of a qualified practitioner (Wichtl 2004).

Other Precautions

None known.

Drug and Supplement Interactions

None known.

Adverse Events and Side Effects

An anaphylactic reaction to boldo has been reported (Monzon et al. 2004).

Hepatotoxicity was reported in an elderly man who had been taking an herbal stimulant laxative for many years. The hepatotoxicity appeared shortly after the laxative was reformulated to contain boldo (Piscaglia et al. 2005). An animal study reported a protective effect of boldo against chemical-induced liver injury (Lanhers et al. 1991).

Pharmacological Considerations

None known.

Pregnancy and Lactation

An animal study indicated no adverse effects of boldo on fetal development at doses up to 500 mg/kg. At higher doses (800 mg/kg), equivalent to a human dose of approximately 58 g, some incidences of malformations of fetuses were observed (Almeida et al. 2000).

No information on the safety of boldo during lactation was identified in the scientific or traditional literature. While this review did not identify any concerns for use while nursing, safety has not been conclusively established.

Review Details

I. Drug and Supplement Interactions

Clinical Trials of Drug or Supplement Interactions

No clinical trials of drug or supplement interactions were identified.

Case Reports of Suspected Drug or Supplement Interactions

No case reports of suspected drug or supplement interactions were identified.

Animal Trials of Drug or Supplement Interactions

No animal trials of drug or supplement interactions were identified.

II. Adverse Events

Case Reports of Adverse Events

An anaphylactic reaction to boldo has been reported (Monzon et al. 2004).

An 82-year-old man with fatty liver who regularly used laxatives (herbal and nonherbal) over the course of the prior 20 years developed hepatotoxicity with elevated levels of liver enzymes after the commercial herbal laxative that he had been taking changed formulas to include an extract of boldo as one of the ingredients in a multi-ingredient product. Liver enzymes returned to normal after cessation of the herbal laxative (Piscaglia et al. 2005).

III. Pharmacology and Pharmacokinetics

Human Pharmacological Studies

No relevant human pharmacological studies were identified.

Animal Pharmacological Studies

No relevant animal pharmacological studies were identified.

In Vitro Pharmacological Studies

The compound boldine prevented ferric ATP-induced inactivation of the drug-metabolizing isoenzyme CYP2E1 but not inactivation induced by carbon tetrachloride (Kringstein and Cederbaum 1995).

P

IV. Pregnancy and Lactation

In rats orally administered 500 or 800 mg/kg daily of either an ethanolic extract of boldo or the same doses of the compound boldine on days 1 to 5 or days 7 to 12 of pregnancy, an increase in the number of fetal resorptions was observed at the 800 mg/kg level for both products in both treatment periods, with no adverse effects observed at the 500 mg/kg dose in either treatment period with either product. A timing-dependent increase in fetal malformations was observed in 3% of animals administered 800 mg/kg boldine on days 7 to 12, while malformations were observed in 1.5% of animals administered the same dose of boldine on days 1 to 5 of pregnancy (Almeida et al. 2000).

No information on the safety of boldo during lactation was identified.

V. Toxicity Studies

Acute Toxicity

No signs of toxicity were observed in rats orally administered 3 g/kg of a hydroalcoholic extract of boldo (Magistretti 1980). The oral LD_{50} of the compound boldine is 500 mg/kg in mice and 1000 mg/kg in guinea pigs (Kreitmair 1952).

Subchronic Toxicity

In rats orally administered 500 or 800 mg/kg of an ethanol extract of boldo or of the compound boldine daily for 90 days, increases in the levels of alanine aminotransferase (ALT), aspartate aminotransferase (AST), and cholesterol were observed in the 800 mg/kg groups for both products after 30 or 60 days, but levels returned to normal after 90 days. A decrease in bilirubin was observed in the 800 mg/kg groups for both products. No changes in liver enzymes or cholesterol levels were observed in the 500 mg/kg group (Almeida et al. 2000).

Genotoxicity

No significant effects on sister-chromatid exchanges were observed in the mouse bone marrow micronucleus assay in animals orally administered 225, 450, or 900 mg/kg of the compound boldine (Tavares and Takahashi 1994). Similarly, no significant effects on sister-chromatid exchanges were observed in vitro on human peripheral blood lymphocytes treated with the compound boldine at concentrations of 10, 20, and 40 μg/ml (Tavares and Takahashi 1994).

No genotoxic activity of the compound boldine was observed in the SOS chromotest and Ames test for mutagenicity with *Salmonella typhimurium* strains TA100, TA98, and TA102 with or without metabolic activation. Boldine weakly induced mitotic recombinational events such as crossing-over and gene conversion in diploid yeast cells. Weak induction of cytoplasmic *petite* mutation in haploid yeast cells was also observed after treatment with boldine (Moreno et al. 1991).

Literature Cited

Almeida, E.R., A.M. Melo, and H. Xavier. 2000. Toxicological evaluation of the hydro-alcohol extract of the dry leaves of *Peumus boldus* and boldine in rats. *Phytother. Res.* 14(2):99-102.

ESCOP. 2003. *ESCOP monographs: The scientific foundation for herbal medicinal products*. 2nd ed. Exeter, UK: European Scientific Cooperative on Phytotherapy.

Kreitmair, H. 1952. Pharmakologische wirkung des alkaloids aus *Peumus boldus* Molina. *Pharmazie* 7:507-511.

Kringstein, P., and A.I. Cederbaum. 1995. Boldine prevents human liver microsomal lipid peroxidation and inactivation of cytochrome P4502E1. *Free Radical Biol. Med.* 18(3):559-563.

Lanhers, M.C., M. Joyeux, R. Soulimani, et al. 1991. Hepatoprotective and anti-inflammatory effects of a traditional medicinal plant of Chile, *Peumus boldus*. *Planta Med.* 57(2):110-115.

Magistretti, M.J. 1980. Remarks on the pharmacological examination of plant extracts. *Fitoterapia* 51:67-79.

Monzon, S., A. Lezaun, D. Saenz, et al. 2004. Anaphylaxis to boldo infusion, a herbal remedy. *Allergy* 59(9):1019-1020.

Moreno, P.R., V.M. Vargas, H.H. Andrade, A.T. Henriques, and J.A. Henriques. 1991. Genotoxicity of the boldine aporphine alkaloid in prokaryotic and eukaryotic organisms. *Mutat. Res.* 260(2):145-152.

Piscaglia, F., S. Leoni, A. Venturi, et al. 2005. Caution in the use of boldo in herbal laxatives: A case of hepatotoxicity. *Scand. J. Gastroenterol.* 40(2):236-239.

Tavares, D.C., and C.S. Takahashi. 1994. Evaluation of the genotoxic potential of the alkaloid boldine in mammalian cell systems in vitro and in vivo. *Mutat. Res.* 321(3):139-145.

Wichtl, M. 2004. *Herbal drugs and phytopharmaceuticals: A handbook for practice on a scientific basis*. 3rd ed. Boca Raton, FL: CRC Press.

Pfaffia paniculata (Mart.) Kuntze — Amaranthaceae

SCN: suma
OCN: pfaffia

Part: root

Quick Reference Summary

Safety Class: 1
Interaction Class: A

Contraindications

None known.

Other Precautions

None known.

Drug and Supplement Interactions

None known.

Adverse Events and Side Effects

None known.

Pharmacological Considerations

None known.

Pregnancy and Lactation

No information on the use of suma in pregnancy or lactation was identified in the scientific or traditional literature. Although this review did not identify any concerns for use while pregnant or nursing, safety has not been conclusively established.

Review Details

I. Drug and Supplement Interactions

Clinical Trials of Drug or Supplement Interactions

No clinical trials of drug or supplement interactions were identified.

Case Reports of Suspected Drug or Supplement Interactions

No case reports of suspected drug or supplement interactions were identified.

Animal Trials of Drug or Supplement Interactions

No animal trials of drug or supplement interactions were identified.

II. Adverse Events

Case Reports of Adverse Events

Occupational asthma was reported in a worker routinely exposed to suma dust. Sensitivity to the dust was confirmed by immediate skin test reactivity, a positive bronchial challenge (immediate response), and the presence of specific IgE detected by ELISA (enzyme-linked immunosorbent assay) (Subiza et al. 1991).

III. Pharmacology and Pharmacokinetics

Human Pharmacological Studies

No relevant human pharmacological studies were identified.

Animal Pharmacological Studies

In rats provided with suma extract (5 g suma/100 ml water, estimated dose per animal not reported) as the sole source of drinking water for 30 days, increases in plasma concentrations of estradiol-17β and progesterone in female mice, and of testosterone in male mice, were observed (Oshima and Gu 2003).

In Vitro Pharmacological Studies

No relevant in vitro pharmacological studies were identified.

IV. Pregnancy and Lactation

No information on the use of suma during pregnancy or lactation was identified.

V. Toxicity Studies

Short-Term Toxicity

No adverse effects were observed in mice orally administered 400 mg/kg of suma daily for 10 days. Observed parameters included alanine aminotransferase levels and histology of the liver, spleen, and kidneys (Matsuzaki et al. 2003).

In mice orally administered 250, 500, or 1000 mg/kg of a methanol extract of suma daily for 10 days, no histological changes were observed in liver, lung, brain, cerebellum, eye, or kidney tissues. An increase in body weight was observed at the 250 mg/kg dose, and a decrease in body weight was observed at the 1000 mg/kg dose as compared to control animals (Carneiro et al. 2007).

Genotoxicity

In the Ames test for mutagenicity with *Salmonella typhimurium* strains TA97a, TA98, TA100, and TA104, with or without metabolic activation, an aqueous extract (1:5) of suma exhibited some mutagenic activity in TA100 with and without metabolic activation but showed no mutagenic activity in other strains (Rivera et al. 1994).

Literature Cited

Carneiro, C.S., F.A. Costa-Pinto, A.P. da Silva, et al. 2007. *Pfaffia paniculata* (Brazilian ginseng) methanolic extract reduces angiogenesis in mice. *Exp. Toxicol. Pathol.* 58(6):427-431.

Matsuzaki, P., G. Akisue, S.C.S. Oloris, S.L. Górniak, and M.L.Z. Dagli. 2003. Effect of *Pfaffia paniculata* (Brazilian ginseng) on the Ehrlich tumor in its ascitic form. *Life Sci.* 74(5):573-579.

Oshima, M., and Y. Gu. 2003. *Pfaffia paniculata*-induced changes in plasma estradiol-17beta, progesterone and testosterone levels in mice. *J. Reprod. Dev.* 49(2):175-180.

Rivera, I.G., M.T. Martins, P.S. Sanchez, et al. 1994. Genotoxicity assessment through the Ames test of medicinal plants commonly used in Brazil. *Environ. Toxicol. Water Qual.* 9(2):87-93.

Subiza, J., J.L. Subiza, P.M. Escribano, et al. 1991. Occupational asthma caused by Brazil ginseng dust. *J. Allergy Clin. Immunol.* 88(5):731-736.

Phellodendron spp. — Rutaceae

P

Phellodendron amurense Rupr.
SCN: phellodendron
PN: *huang bai* (bark); *huang bo* (bark)
OCN: Amur corktree

Phellodendron chinense **Schneid.**
SCN: phellodendron
PN: *huang bai* (bark); *huang bo* (bark)
OCN: Chinese corktree
Part: bark

Quick Reference Summary

Safety Class: 1
Interaction Class: A

Contraindications

None known.

Other Precautions

None known.

Drug and Supplement Interactions

None known.

Notice

Berberine (0.6–2.5% in *P. amurense*; 4.0–8.0% in *P. chinense*) (Bensky et al. 2004; De Smet 1992; Leung and Foster 1996); *see* Appendix 1.

Adverse Events and Side Effects

One case of rash associated with ingestion of phellodendron has been recorded (Bensky et al. 2004).

Pharmacological Considerations

None known.

Pregnancy and Lactation

No information on the safety of phellodendron in pregnancy or lactation was identified in the scientific or traditional literature. Although this review did not identify any concerns for use while pregnant or nursing, safety has not been conclusively established.

See Berberis vulgaris for information on the safety of the compound berberine in pregnancy and lactation.

Review Details

I. Drug and Supplement Interactions

Clinical Trials of Drug or Supplement Interactions

No clinical trials of drug or supplement interactions were identified.

Case Reports of Suspected Drug or Supplement Interactions

No case reports of suspected drug or supplement interactions were identified.

Animal Trials of Drug or Supplement Interactions

No animal trials of drug or supplement interactions were identified.

II. Adverse Events

Case Reports of Adverse Events

One case of rash associated with ingestion of phellodendron has been recorded (Bensky et al. 2004).

III. Pharmacology and Pharmacokinetics

Human Pharmacological Studies

No relevant human pharmacological studies were identified.

Animal Pharmacological Studies

No relevant animal pharmacological studies were identified.

In Vitro Pharmacological Studies

Purified polysaccharide fractions of phellodendron extract demonstrated immunopotentiating activity in mice administered doses up to 100 mg/kg (Park et al. 2004). Immunesuppressant effects of the compounds magnoflorine and phellodendrine were observed after mice were intraperitoneally administered the compounds at doses of 5 to 20 mg/kg (Mori et al. 1994).

An ethanol extract of phellodendron did not show estrogenic activity in a recombinant yeast system (Kim et al. 2008).

IV. Pregnancy and Lactation

No information on the safety of phellodendron in pregnancy or lactation was identified, although texts on traditional Chinese medicine do not contraindicate phellodendron during pregnancy or lactation.

See Berberis vulgaris for information on the safety of the compound berberine in pregnancy and lactation.

V. Toxicity Studies

See Berberis vulgaris for toxicity studies on the compound berberine.

Acute Toxicity

The LD_{50} of intraperitoneally administered phellodendron extract in mice is 2.7 g/kg (Chen and Chen 2004).

Literature Cited

Bensky, D., S. Clavey, and E. Stöger. 2004. *Chinese herbal medicine: Materia medica*. 3rd ed. Seattle: Eastland Press.

Chen, J.K., and T.T. Chen. 2004. *Chinese medical herbology and pharmacology*. City of Industry, CA: Art of Medicine Press.

De Smet, P.A.G.M. 1992. *Adverse effects of herbal drugs, Volume 1*. Berlin: Springer.

Kim, I.G., S.C. Kang, K.C. Kim, E.S. Choung, and O.P. Zee. 2008. Screening of estrogenic and antiestrogenic activities from medicinal plants. *Environ. Toxicol. Pharmacol.* 25(1):75-82.

Leung, A.Y., and S. Foster. 1996. *Encyclopedia of common natural ingredients used in food, drugs, and cosmetics*. 2nd ed. New York: Wiley.

Mori, H., M. Fuchigami, N. Inoue, et al. 1994. Principle of the bark of *Phellodendron amurense* to suppress the cellular immune response. *Planta Med.* 60(5):445-449.

Park, S.D., Y.S. Lai, and C.H. Kim. 2004. Immunopotentiating and antitumor activities of the purified polysaccharides from *Phellodendron chinense* Schneid. *Life Sci.* 75(22):2621-2632.

P

Phoradendron leucarpum (Raf.) Reveal & M.C. Johnst. — Viscaceae

SCN: American mistletoe
Syn: *Phoradendron flavescens* Nutt. ex Engelm.; *Phoradendron serotinum* (Raf.) M.C. Johnst.

OCN: oak mistletoe
Part: herb

Quick Reference Summary

Safety Class: 3
Interaction Class: A

Contraindications

Not for use except under the supervision of an expert qualified in the appropriate use of this substance (Hall et al. 1986; Krenzelok et al. 1997; Moore 1963; Spiller et al. 1996).

Other Precautions

None known.

Drug and Supplement Interactions

None known.

Adverse Events and Side Effects

Ingestion of several leaves or berries generally caused gastrointestinal upset or did not produce symptoms. Intentional cases of self-poisoning, including one attempted abortion, have been fatal. Details on doses, preparations, and part used are lacking (Hall et al. 1986; Moore 1963; Spiller et al. 1996).

American mistletoe has been used as an emetic for ritual purposes. No information on doses used is available (Moerman 1998).

Pharmacological Considerations

None known.

Pregnancy and Lactation

A reference ethnobotanical text indicates that American mistletoe has been used as an abortifacient and as a "medicine for pregnant women" (Moerman 1998). In the early 1900s, American mistletoe was reportedly used to stimulate uterine contractions (Wood and LaWall 1926).

No information on the safety of American mistletoe during lactation was identified in the scientific or traditional literature. While this review did not identify any concerns for use while nursing, safety has not been conclusively established, and this substance is not recommended for use except under the supervision of an expert qualified in its appropriate use.

Review Details

I. Drug and Supplement Interactions

Clinical Trials of Drug or Supplement Interactions

No clinical trials of drug or supplement interactions were identified.

Case Reports of Suspected Drug or Supplement Interactions

No case reports of suspected drug or supplement interactions were identified.

Animal Trials of Drug or Supplement Interactions

No animal trials of drug or supplement interactions were identified.

P

II. Adverse Events

Case Reports of Adverse Events

Information on case reports of intentional or accidental exposure to mistletoe is primarily from reviews of information reported to poison control centers and generally lacks details on amount and part ingested and specific symptoms. A review of 92 cases of American mistletoe exposure indicated that 14 of these were symptomatic with 6 cases of gastrointestinal upset, 2 cases of mild drowsiness, and 1 case each of ataxia and seizure (in an infant). None of the cases included any arrhythmias or cardiovascular changes. Amount ingested ranged from one berry or leaf to more than five leaves or 20 berries. In cases of leaf ingestion (1–5 leaves), three of five patients reported gastrointestinal upset, while a patient who ate five leaves was asymptomatic. The case of seizure was reported in an infant with an unwitnessed exposure, found with both berries and leaves in the crib. A case of ingestion of 20 berries in a 2-year-old resulted in one episode of vomiting prior to administration of activated charcoal, with no subsequent symptoms (Spiller et al. 1996).

Of the approximately 1860 exposures to American mistletoe listed in reviews of information reported to poison control centers, none were fatal (Hall et al. 1986; Krenzelok et al. 1997; Spiller et al. 1996). Other literature reports two fatal cases. One patient drank an unspecified amount of American mistletoe (part not specified), although symptoms and events leading to the fatality were not available. A woman who ingested an unspecified amount of berries in an attempted abortion developed abdominal pain, hypoventilation, and died from cardiovascular collapse (Hall et al. 1986; Moore 1963).

III. Pharmacology and Pharmacokinetics

Human Pharmacological Studies

No relevant human pharmacological studies were identified.

Animal Pharmacological Studies

Studies completed in the early 1900s indicated that injections of American mistletoe produced a hypertensive effect (Crawford 1911; Wood and LaWall 1926). An ethnobotanical text reports use of American mistletoe as a hypotensive (Moerman 1998).

The compounds viscotoxin A3 and phoratoxin produced reflex bradycardia, negative inotropic effect on the heart, and, in high doses, vasoconstriction of vessels in skin and skeletal muscle in cats (Rosell and Samuelsson 1966).

In Vitro Pharmacological Studies

No relevant in vitro pharmacological studies were identified.

IV. Pregnancy and Lactation

A reference ethnobotanical text indicates that American mistletoe has been used as an abortifacient and as a "medicine for pregnant women" (Moerman 1998). In the early

1900s, American mistletoe was reportedly used to stimulate uterine contractions (Wood and LaWall 1926).

No information on the safety of American mistletoe during lactation was identified.

V. Toxicity Studies

No toxicity studies were identified.

Literature Cited

Crawford, A.C. 1911. The pressor action of an American mistletoe. *J. Am. Med. Assoc.* 58(11):865-868.

Hall, A.H., D.G. Spoerke, and B.H. Rumack. 1986. Assessing mistletoe toxicity. *Ann. Emerg. Med.* 15(11):1320-1323.

Krenzelok, E.P., T.D. Jacobsen, and J. Aronis. 1997. American mistletoe exposures. *Am. J. Emerg. Med.* 15(5):516-520.

Moerman, D.E. 1998. *Native American ethnobotany*. Portland, OR: Timber Press.

Moore, H.W. 1963. Mistletoe poisoning. A review of the available literature, and the report of a case of probable fatal poisoning. *J. South Carolina Med. Assoc.* 59(8):269-271.

Rosell, S., and G. Samuelsson. 1966. Effect of mistletoe viscotoxin and phoratoxin on blood circulation. *Toxicon* 4(2):107-108.

Spiller, H.A., D.B. Willias, S.E. Gorman, and J. Sanftleban. 1996. Retrospective study of mistletoe ingestion. *J. Toxicol. Clin. Toxicol.* 34(4):405-408.

Wood, H., and C. LaWall. 1926. *The dispensatory of the United States of America*. Philadelphia: Lippincott.

Phyllanthus emblica L.

Euphorbiaceae

SCN: amla
Syn: *Emblica officinalis* Gaertn.
AN: *amalaki*
PN: *yu gan zi* (fruit)
OCN: emblic myrobalan; Indian gooseberry
Part: fruit

Quick Reference Summary

Safety Class: 1
Interaction Class: A

Contraindications

None known.

Other Precautions

None known.

Drug and Supplement Interactions

None known.

Adverse Events and Side Effects

None known.

Pharmacological Considerations

None known.

Pregnancy and Lactation

No information on the safety of amla in pregnancy or lactation was identified in the scientific or traditional literature. Although this review did not identify any concerns for use while pregnant or nursing, safety has not been conclusively established.

Review Details

I. Drug and Supplement Interactions

Clinical Trials of Drug or Supplement Interactions

No clinical trials of drug or supplement interactions were identified.

Case Reports of Suspected Drug or Supplement Interactions

No case reports of suspected drug or supplement interactions were identified.

Animal Trials of Drug or Supplement Interactions

No animal trials of drug or supplement interactions were identified.

II. Adverse Events

Case Reports of Adverse Events

No case reports of adverse events were identified.

III. Pharmacology and Pharmacokinetics

Human Pharmacological Studies

No relevant human pharmacological studies were identified.

Animal Pharmacological Studies

No relevant animal pharmacological studies were identified.

P

In Vitro Pharmacological Studies

No relevant in vitro pharmacological studies were identified.

IV. Pregnancy and Lactation

No information on the safety of amla in pregnancy or lactation was identified.

V. Toxicity Studies

Subchronic Toxicity

No adverse effects on hematology, biochemistry, or histology were observed in rats orally administered doses up to 2 g/kg daily of a standardized extract of amla for 90 days (Antony et al. 2007).

Literature Cited

Antony, B., B. Merina, and V. Sheeba. 2007. Toxicity studies of Amlamax—Purified standardized extract of *Emblica officinalis*. *Indian J. Nat. Prod.* 23(2):14-17.

Phyllanthus spp.

Euphorbiaceae

Phyllanthus amarus Schumach.
SCN: *Phyllanthus amarus*
OCN: carry-me-seed

Phyllanthus fraternus **G.L. Webster**
SCN: phyllanthus

Phyllanthus niruri **L.**
SCN: phyllanthus
AN: *bhumyamalaki*
OCN: *niruri*
Part: above-ground parts, whole plant

Quick Reference Summary

Safety Class: 2b
Interaction Class: A

Contraindications

Not for use in pregnancy except under the supervision of a qualified healthcare practitioner (Martins et al. 1994; Rao and Alice 2001; Taylor 2005).

Other Precautions

None known.

Drug and Supplement Interactions

None known.

Notice

Diuretic (Srividya and Periwal 1995; Taylor 2005); *see* Appendix 2.

Adverse Events and Side Effects

None known.

Pharmacological Considerations

Animal studies have demonstrated that phyllanthus may modify glucose regulation (Adeneye et al. 2006; Garg et al. 2008; Kumar et al. 1989; Okoli et al. 2010). People with diabetes are advised to monitor their blood sugar closely and discuss the use of this herb with a qualified healthcare practitioner prior to use.

An animal study indicated that *Phyllanthus amarus* inhibited the drug-metabolizing isoenzymes CYP1A1, CYP1A2, and CYP2B1/2 (Hari Kumar and Kuttan 2006).

Pregnancy and Lactation

An animal study demonstrated temporary antifertility effects of *Phyllanthus amarus* in female mice (Rao and Alice 2001). Some sources indicate that "large" doses of phyllanthus or *Phyllanthus amarus* have been used to produce abortion (Martins et al. 1994; Taylor 2005). Based on this information, use during pregnancy is not recommended except under the supervision of a qualified healthcare practitioner.

No information on the safety of *Phyllanthus amarus* during lactation was identified in the scientific or traditional literature. While this review did not identify any concerns for use while nursing, safety has not been conclusively established.

P

Review Details

I. Drug and Supplement Interactions

Clinical Trials of Drug or Supplement Interactions

No clinical trials of drug or supplement interactions were identified.

Case Reports of Suspected Drug or Supplement Interactions

No case reports of suspected drug or supplement interactions were identified.

Animal Trials of Drug or Supplement Interactions

No animal trials of drug or supplement interactions were identified.

II. Adverse Events

Adverse Events Reported in Clinical Trials

A review of human clinical trials with *Phyllanthus amarus* indicated a lack of reported adverse events or side effects (Calixto et al. 1998).

Case Reports of Adverse Events

No case reports of adverse events were identified.

III. Pharmacology and Pharmacokinetics

Human Pharmacological Studies

A study in 40- to 60-year-old subjects indicated that, after 10 days of administration of *Phyllanthus amarus,* an increase in urinary volume and urinary sodium levels was observed, along with reductions in blood pressure (Srividya and Periwal 1995). A review of the study noted that the control group was not well matched to the study group, making the results difficult to interpret (Wright et al. 2007).

Animal Pharmacological Studies

In diabetic rats orally administered 200 or 400 mg/kg of a methanol extract of phyllanthus, a dose-dependent reduction in fasting blood glucose levels and postprandial glucose levels was observed (Okoli et al. 2010). Hypoglycemic activity was observed in diabetic rabbits orally administered 250 mg/kg of an alcohol extract of phyllanthus (Kumar et al. 1989).

Dose-dependent decreases in fasting blood glucose levels were observed in healthy mice orally administered 150, 300, or 600 mg/kg of aqueous leaf and seed extracts of *Phyllanthus amarus* (Adeneye et al. 2006).

In diabetic rats orally administered 500 mg/kg of water, ethanol, or petroleum ether extracts of phyllanthus daily for 21 days, a reduction in blood sugar levels was observed. The ethanol extract was the most active, followed by the aqueous extract and then the petroleum ether extract (Garg et al. 2008).

Inhibition of the drug-metabolizing isoenzymes CYP1A1, CYP1A2, and CYP2B1/2 was observed in rats orally administered 250 mg/kg of a methanol extract of *Phyllanthus amarus* (Hari Kumar and Kuttan 2006).

In Vitro Pharmacological Studies

The compound methyl brevifolincarboxylate, isolated from phyllanthus, inhibited platelet aggregation induced by ADP and collagen (Iizuka et al. 2007).

An ethanol extract of *Phyllanthus amarus* was found to induce the drug-metabolizing isoenzymes CYP3A5 and CYP3A7 (Agbonon et al. 2010).

IV. Pregnancy and Lactation

After being orally administered 100 mg/kg of an alcohol extract of *Phyllanthus amarus* daily for 30 days, female mice were unable to become pregnant. On withdrawal of *Phyllanthus amarus* for 45 days, mice were successfully mated. Reductions in 3β- and 17β-hydroxysteroid dehydrogenase (HSD) levels were noted as probably affecting hormonal conversions in the ovaries (Rao and Alice 2001).

No information on the safety of *Phyllanthus amarus* during lactation was identified.

V. Toxicity Studies

Acute Toxicity

The LD_{50} of a methanol extract of phyllanthus orally administered to mice is 471 mg/kg (Okoli et al. 2010).

The LC_{50} of phyllanthus for brine shrimp is 581 μg/ml (Nascimenlo et al. 2008).

Short-Term Toxicity

In mice orally administered different fractions of *Phyllanthus amara* at doses of 400, 800, or 1600 mg/kg daily for 14 days, changes in serum biochemistry were observed. Changes included elevations in aspartate aminotransferase (AST) and alkaline phosphatase (ALP), with differing effects on protein, total bilirubin, creatinine, and other liver enzyme levels. The changes were generally not dose-dependent (Adedapo et al. 2005).

P

Literature Cited

Adedapo, A.A., M.O. Abatan, S.O. Idowu, and O.O. Olorunsogo. 2005. Toxic effects of chromatographic fractions of *Phyllanthus amarus* on the serum biochemistry of rats. *Phytother. Res.* 19(9):812-815.

Adeneye, A.A., O.O. Amole, and A.K. Adeneye. 2006. Hypoglycemic and hypocholesterolemic activities of the aqueous leaf and seed extract of *Phyllanthus amarus* in mice. *Fitoterapia* 77(7-8):511-514.

Agbonon, A., K. Eklu-Gadegbeku, K. Aklikokou, et al. 2010. In vitro inhibitory effect of West African medicinal and food plants on human cytochrome P450 3A subfamily. *J. Ethnopharmacol.* 128(2):390-394.

Calixto, J.B., A.R.S. Santos, V. Cechinel Filho, and R.A. Yunes. 1998. A review of the plants of the genus *Phyllanthus*: Their chemistry, pharmacology, and therapeutic potential. *Med. Res. Rev.* 18(4):225-258.

Garg, M., V.J. Dhar, and A.N. Kalia. 2008. Antidiabetic and antioxidant potential of *Phyllanthus fraternus* in alloxan induced diabetic animals. *Pharmacog. Mag.* 4(14):138-143.

Hari Kumar, K.B., and R. Kuttan. 2006. Inhibition of drug metabolizing enzymes (cytochrome P450) in vitro as well as in vivo by *Phyllanthus amarus* Schum. & Thonn. *Biol. Pharm. Bull.* 29(7):1310-1313.

Iizuka, T., M. Nagai, A. Taniguchi, H. Moriyama, and K. Hoshi. 2007. Inhibitory effects of methyl brevifolincarboxylate isolated from *Phyllanthus niruri* L. on platelet aggregation. *Biol. Pharm. Bull.* 30(2):382-384.

Kumar, N.G., A.M.C. Nair, V.R. Raghunandanan, and M.K. Rajagopalan. 1989. Hypoglycemic effect of *Phyllanthus niruri* leaves in rabbits. *Kerala J. Vet. Sci.* 20(1):77-80.

Martins, E.R., D.M. Castro, D.C. Castellani, and J.E. Dias. 1994. *Plantas medicinais*. Viçosa: UFV.

Nascimenlo, J.E., A.F.M. Melo, T.C. Lima E Silva, et al. 2008. Phytochemical screening and toxicological bioassay with brine shrimp larvae (*Artemia salina*) of three medicinal species of the genus *Phyllanthus* (Phyllanthaceae). *Rev. Cien. Farm. Basica Apl.* 29(2):143-148.

Okoli, C.O., A.F. Ibiam, A.C. Ezike, P.A. Akah, and T.C. Okoye. 2010. Evaluation of antidiabetic potentials of *Phyllanthus niruri* in alloxan diabetic rats. *Afr. J. Biotechnol.* 9(2):248-259.

Rao, M.V., and K.M. Alice. 2001. Contraceptive effects of *Phyllanthus amarus* in female mice. *Phytother. Res.* 15(3):265-267.

Srividya, N., and S. Periwal. 1995. Diuretic, hypotensive and hypoglycaemic effect of *Phyllanthus amarus*. *Indian J. Exp. Biol.* 33(11):861-864.

Taylor, L. 2005. *The healing power of rainforest herbs*. Garden City Park, NY: Square One Publishers.

Wright, C.I., L. Van-Buren, C.I. Kroner, and M.M. Koning. 2007. Herbal medicines as diuretics: A review of the scientific evidence. *J. Ethnopharmacol.* 114(1):1-31.

Phytolacca americana L.

Phytolaccaceae

P

SCN: poke
Syn: *Phytolacca decandra* L.
OCN: pokeweed
PN: *shang lu*
Part: root

Quick Reference Summary

Safety Class: 3
Interaction Class: A

Contraindications

Not for use except under the supervision of an expert qualified in the appropriate use of this substance (Brooke et al. 2001; De Smet 1993; French 1900; Frohne and Pfänder 2000; Lawrence 1990; Lewis and Smith 1979; Roberge et al. 1986).

Other Precautions

Topically applied poke root preparations are sometimes used to treat mastitis. To avoid exposure of nursing infants to poke, such preparations should not be applied on or near the nipple (Mills and Bone 2005).

Drug and Supplement Interactions

None known.

Editors' Notes

In the southern United States, the young shoots of poke are traditionally consumed as a green vegetable, after cooking in two to three changes of water. Poke berry is used medicinally. Even with proper preparation, both poke shoots and berries may cause diarrhea or nausea. Improperly prepared shoots and berries can cause gastroenteritis and other symptoms characteristic of poke poisoning.

Caution is advised for persons handling poke on a commercial scale. To avoid nausea, headaches, and diarrhea associated with dermal and respiratory exposure in commercial facilities, the use of gloves, a respirator, and other appropriate protective gear is recommended.

Adverse Events and Side Effects

In overdose, poisoning with characteristic symptoms of gastroenteritis, severe abdominal cramping, nausea, repeated vomiting, salivation, sweating, generalized weakness, and frothy diarrhea has been reported after ingestion of fresh or dried poke root (Brooke et al. 2001; French 1900; Lawrence 1990; Lewis and Smith 1979; Roberge et al. 1986).

Pharmacological Considerations

None known.

Pregnancy and Lactation

Poke is contraindicated in pregnancy (Bensky et al. 2004). *Also see* Other Precautions for this entry.

No information on the safety of poke root during lactation was identified. While this review did not identify any concerns for use while nursing, safety has not been conclusively established, and poke root is not recommended for use except under the supervision of an expert qualified in its appropriate use.

Review Details

I. Drug and Supplement Interactions

Clinical Trials of Drug or Supplement Interactions

No clinical trials of drug or supplement interactions were identified.

Case Reports of Suspected Drug or Supplement Interactions

No case reports of suspected drug or supplement interactions were identified.

Animal Trials of Drug or Supplement Interactions

Saponins isolated from poke root enhanced the absorption of heparin in mice by increasing the activated partial thromboplastin time (APTT) and thrombin time, modulating the transport of heparin via the intestinal route (Cho et al. 2003).

II. Adverse Events

Case Reports of Adverse Events

Poisonings with characteristic symptoms of gastroenteritis, severe abdominal cramping, nausea, repeated vomiting, salivation, sweating, generalized weakness, and frothy diarrhea have been reported after ingestion of fresh, dried, or cooked poke root, leaf, or stem (Callahan et al. 1981; Guthrie 1887; Hamilton et al. 1995; Jaeckle and Freemon 1981; Stein 1979). Poisoning from the berry was also reported (Edwards and Rodgers 1982; Mack 1982).

An 18-year-old man died after eating 4 to 5 inches of freshly dug poke root that he had mistaken for parsnip. He developed epigastric pain and vomiting 45 minutes after ingesting the root, collapsed, and was found to be in ventricular fibrillation 2 hours after ingestion. Treatment attempts were unsuccessful (Brooke et al. 2001). A 65-year-old woman experienced severe abdominal cramps, protracted vomiting, and profuse watery diarrhea after chewing on raw poke root (Roberge et al. 1986). After ingesting approximately 1 teaspoon of poke root that had been mistaken for horseradish root, a man experienced a warm sensation of the stomach followed by severe stomach cramps and vomiting, accompanied by a weak pulse and labored breathing, that continued for approximately 10 hours (French 1900). Grated fresh poke root, mistaken for horseradish, was served in a salad and as a side dish to 32 adults. All of the patients experienced symptoms ranging from dry throat to severe nausea, vomiting, and diarrhea, and most experienced extreme thirst. Symptoms began within 30 minutes to 6 hours after ingestion (Lawrence 1990).

A 43-year-old woman developed nausea, vomiting, cramping, generalized weakness, hematemesis, bloody diarrhea, and hypotension after ingesting one cup of poke root tea (~1 g of root) (Lewis and Smith 1979).

III. Pharmacology and Pharmacokinetics

Human Pharmacological Studies

Pokeweed contains compounds known as pokeweed antiviral proteins, ribosome-inactivating proteins that are currently under investigation in human and animal studies for use in the treatment of HIV and in several cancers, notably leukemia, lymphoma, and Hodgkin's disease (Ek et al. 1998; Hertler and Frankel 1989; Irvin and Uckun 1992; Messinger et al. 1999; Uckun et al. 1999; Waurzyniak et al. 1997).

Animal Pharmacological Studies

No relevant animal pharmacological studies were identified.

In Vitro Pharmacological Studies

Mitogenic activity of lectin compounds from poke root has been observed (Barker et al. 1965; Farnes et al. 1964; Kino et al. 1995).

IV. Pregnancy and Lactation

Poke root is contraindicated in pregnancy (Bensky et al. 2004).

No adverse effects on reproductive ability, neonatal survival, or pup development were observed in mice that had been treated prior to pregnancy with an intravaginally administered gel formulation of pokeweed antiviral protein (PAP) at doses of 0, 0.025, 0.05, or 0.1% PAP, 5 days per week for 13 weeks (D'Cruz et al. 2004). No other information on the safety of poke during pregnancy was identified.

One reference text indicates, with no details, that preparations of poke root have been used as abortifacients (Watt and Breyer-Brandwijk 1962).

No information on the safety of poke root during lactation was identified.

V. Toxicity Studies

Acute Toxicity

Cats intraperitoneally administered 1 g/kg of a hydroalcoholic extract of poke root experienced discomfort, retching, emesis, weakness of the legs, stupor, profound narcosis, and death from respiratory failure (Golstein et al. 1973).

No acute toxic effects were observed in rats intraperitoneally administered 3.6 ml/kg of ether, chloroform, water, or Prollius fluid extract of poke root. The extracts were prepared such that 1 ml was equivalent to 1 g of dried poke root. In rats administered an alcohol extract at the same dose, catharsis and uresis occurred, followed by drowsiness, labored breathing, paralysis of the lower limbs, failure of respiration, and death (Ahmed et al. 1949).

The LD_{50} of intraperitoneally administered saponins from poke root is 181 mg/kg in mice and 208 mg/kg in rats (Woo et al. 1976).

Intraperitoneal administration of 0.25 ml of a saline suspension of poke root was lethal to mice (animal weight not specified) (Macht 1937). The lethal dose of the same extract in rats and guinea pigs was 5 ml/kg (Macht 1937).

Subchronic Toxicity

No adverse effects on blood chemistry, histology, or reproductive performance were observed in a subchronic and reproductive toxicity screening of an intravaginally administered gel formulation of pokeweed antiviral protein (PAP) in mice. Animals were administered gel containing 0, 0.025, 0.05, or 0.1% PAP, 5 days a week for 13 weeks (D'Cruz et al. 2004).

Literature Cited

Ahmed, Z.F., C.J. Zufall, and G.L. Jenkins. 1949. A contribution to the chemistry and toxicology of the root of *Phytolacca americana*, L. *J. Am. Pharm. Assoc.* 38(8):443-448.

Barker, B.E., P. Farnes, and H. Fanger. 1965. Mitogenic activity in *Phytolacca americana* (pokeweed). *Lancet* 285(7377):170.

Bensky, D., S. Clavey, and E. Stöger. 2004. *Chinese herbal medicine: Materia medica*. 3rd ed. Seattle: Eastland Press.

Brooke, J., C. Obar, and L. Courtemanche. 2001. A fatality from *Phytolacca americana* (pokeweed) root ingestion. *J. Toxicol. Clin. Toxicol.* 39:549-550.

Callahan, R., F. Piccola, K. Gensheimer, et al. 1981. Plant poisonings—New Jersey. *Morbid. Mortal. Wkly. Rep.* 30(6):65-67.

Cho, S.Y., J.S. Sim, S.S. Kang, et al. 2003. Enhancement of heparin and heparin disaccharide absorption by the *Phytolacca americana* saponins. *Arch. Pharm. Res.* 26(12):1102-1108.

D'Cruz, O.J., B. Waurzyniak, and F.M. Uckun. 2004. A 13-week subchronic intravaginal toxicity study of pokeweed antiviral protein in mice. *Phytomedicine* 11(4):342-351.

De Smet, P.A.G.M. 1993. *Adverse effects of herbal drugs, Volume 2.* Berlin: Springer.

Edwards, N., and G.C. Rodgers. 1982. Pokeberry pancake breakfast—or—it's gonna be a great day. *Vet. Hum. Toxicol.* 24(Suppl.):135-137.

Ek, O., B. Waurzyniak, D.E. Myers, and F.M. Uckun. 1998. Antitumor activity of TP3(anti-p80)-pokeweed antiviral protein immunotoxin in hamster cheek pouch and severe combined immunodeficient mouse xenograft models of human osteosarcoma. *Clin. Cancer Res.* 4(7):1641-1647.

Farnes, P., B.E. Barker, L.E. Brownhill, and H. Fanger. 1964. Mitogenic activity in *Phytolacca americana* (pokeweed). *Lancet* 2(7369):1100-1101.

French, C. 1900. Pokeroot poisoning. *N. York Med. J.* 72: 653-654.

Frohne, D., and H.J. Pfänder. 2000. *A colour atlas of poisonous plants: A handbook for pharmacists, doctors, toxicologists, biologists and veterinarians.* 2nd ed. London: Manson.

Golstein, S., G. Jenkins, and M. Thompson. 1973. *Cited in* De Smet, P.A.G.M. 1993. *Adverse effects of herbal drugs, Volume 2.* Berlin: Springer Verlag.

Guthrie, A. 1887. Poisoning by poke root. *J. Am. Med. Assoc.* 9:125.

Hamilton, R.J., R.D. Shih, and R.S. Hoffman. 1995. Mobitz type I heart block after pokeweed ingestion. *Vet. Hum. Toxicol.* 37(1):66-67.

Hertler, A.A., and A.E. Frankel. 1989. Immunotoxins: A clinical review of their use in the treatment of malignancies. *J. Clin. Oncol.* 7(12):1932-1942.

Irvin, J.D., and F.M. Uckun. 1992. Pokeweed antiviral protein: Ribosome inactivation and therapeutic applications. *Pharmacol. Ther.* 55(3):279-302.

Jaeckle, K.A., and F.R. Freemon. 1981. Pokeweed (*Phytolacca americana*) poisoning. *South. Med. J.* 74(5):639-640.

Kino, M., K. Yamaguchi, H. Umekawa, and G. Funatsu. 1995. Purification and characterization of three mitogenic lectins from the roots of pokeweed (*Phytolacca americana*). *Biosci. Biotechnol. Biochem.* 59(4):683-688.

Lawrence, R.A. 1990. The clinical effect of pokeweed root ingestion upon 32 adults. American Association of Poison Control Centers, American Academy of Clinical Toxicology, American Board of Medical Toxicology, Canadian Association of Poison Control Centers Scientific Meeting, Tucson, Arizona, USA, September 14-18, 1990. *Vet. Hum. Toxicol.* 32(4):369.

Lewis, W.H., and P.R. Smith. 1979. Poke root herbal tea poisoning. *J. Am. Med. Assoc.* 242(Dec.):2759-2760.

Macht, D. 1937. A pharmacological study of *Phytolacca*. *J. Am. Pharm. Assoc. Sci. Ed.* 26:594-599.

Mack, R.B. 1982. Toxic encounters of the dangerous kind. Pokeweed. *N. C. Med. J.* 43(5):365.

Messinger, Y., G.H. Reaman, O. Ek, and F.M. Uckun. 1999. Evaluation of temozolomide in a SCID mouse model of human B-cell precursor leukemia. *Leuk. Lymphoma* 33(3-4):289-293.

Mills, S., and K. Bone. 2005. *The essential guide to herbal safety.* St. Louis: Elsevier.

Roberge, R., E. Brader, M.L. Martin, et al. 1986. The root of evil—Pokeweed intoxication. *Ann. Emerg. Med.* 15(4):470-473.

Stein, Z.L.G. 1979. Pokeweed induced gastroenteritis. *Am. J. Hosp. Pharm.* 36(Oct.):1303.

P

Uckun, F.M., K. Bellomy, K. O'Neill, et al. 1999. Toxicity, biological activity, and pharmacokinetics of TXU (anti-CD7)-pokeweed antiviral protein in chimpanzees and adult patients infected with human immunodeficiency virus. *J. Pharmacol. Exp. Ther.* 291(3):1301-1307.

Watt, J.M., and M.G. Breyer-Brandwijk. 1962. *The medicinal and poisonous plants of southern and eastern Africa*. 2nd ed. Edinburgh: E. & S. Livingstone.

Waurzyniak, B., E.A. Schneider, N. Tumer, et al. 1997. *In vivo* toxicity, pharmacokinetics, and antileukemic activity of TXU (anti-CD7)-pokeweed antiviral protein immunotoxin. *Clin. Cancer Res.* 3(6):881-890.

Woo, W.S., K. Shin, and S.S. Kang. 1976. Constituents of *Phytolacca* species I. Antiinflammatory saponins. *Soul Taehakkyo Saengyak Yonguso Opjukjip* 15:103-106.

Picrasma excelsa (Sw.) Planch. — Simaroubaceae

SCN: Jamaica quassia
Syn: *Quassia excelsa* Sw.
OCN: bitterwood
Part: bark, root, wood

Quick Reference Summary

Safety Class: 2b
Interaction Class: A

Contraindications

Not for use in pregnancy except under the supervision of a qualified healthcare practitioner (Bradley 1992; List and Hörhammer 1973).

Other Precautions

None known.

Drug and Supplement Interactions

None known.

Standard Dose

As infusion (tea) of 0.5 g (~0.25 tsp) of the bark 30 minutes prior to meals (Wichtl 2004), or decoction of 1–2 g daily (Merck 1930).

Editors' Note

Jamaica quassia should not be confused with quassia (*Quassia amara*) or senna (*Senna* spp., formerly classified as *Cassia* spp.).

Adverse Events and Side Effects

Consumption of large amounts of Jamaica quassia can irritate the mucous membrane of the stomach and lead to vomiting (Bradley 1992; List and Hörhammer 1973; Wichtl 2004; Wood and LaWall 1918).

Pharmacological Considerations

None known.

Pregnancy and Lactation

Several herbal reference texts indicate that Jamaica quassia should not be used during pregnancy (Bradley 1992; List and Hörhammer 1973).

No information on the safety of Jamaica quassia during lactation was identified. While this review did not identify any concerns for use while nursing, safety has not been conclusively established.

P

Review Details

I. Drug and Supplement Interactions

Clinical Trials of Drug or Supplement Interactions

No clinical trials of drug or supplement interactions were identified.

Case Reports of Suspected Drug or Supplement Interactions

No case reports of suspected drug or supplement interactions were identified.

Animal Trials of Drug or Supplement Interactions

No animal trials of drug or supplement interactions were identified.

II. Adverse Events

Case Reports of Adverse Events

An Eclectic medical text reports that a strong infusion, administered as an enema to a 4-year-old, caused collapse and related symptoms. No further details on the dose or the relevant medical history were provided (Felter and Lloyd 1898). The relevance of this information to oral use at standard therapeutic levels is not known.

Pimenta dioica (L.) Merr. — Myrtaceae

SCN: allspice
Syn: *Eugenia pimenta* DC.; *Pimenta officinalis* Lindl.
OCN: Jamaica pepper; myrtle pepper; pimenta
Part: unripe fruit

Quick Reference Summary

Safety Class: 1
Interaction Class: A

Contraindications

None known.

Other Precautions

None known.

Drug and Supplement Interactions

None known.

Adverse Events and Side Effects

Occupational allergic contact dermatitis from allspice, confirmed by patch testing, was reported in a food service worker (Kanerva et al. 1996).

Pharmacological Considerations

None known.

Pregnancy and Lactation

No information on the safety of allspice in pregnancy or lactation was identified in the scientific or traditional literature. Although this review did not identify any concerns for use while pregnant or nursing, safety has not been conclusively established.

Review Details

I. Drug and Supplement Interactions

Clinical Trials of Drug or Supplement Interactions

No clinical trials of drug or supplement interactions were identified.

Case Reports of Suspected Drug or Supplement Interactions

No case reports of suspected drug or supplement interactions were identified.

Animal Trials of Drug or Supplement Interactions

No animal trials of drug or supplement interactions were identified.

II. Adverse Events

Case Reports of Adverse Events

Among 1000 food service workers that visited a dermatological clinic between 1991 and 1995, five had occupational allergic contact dermatitis from spices. Allspice was confirmed as one of the causative spices (Kanerva et al. 1996).

III. Pharmacology and Pharmacokinetics

Human Pharmacological Studies

No relevant human pharmacological studies were identified.

Animal Pharmacological Studies

No relevant animal pharmacological studies were identified.

In Vitro Pharmacological Studies

In an estrogen receptor expression assay and in reporter and endogenous gene assays, an extract of allspice exhibited estrogenic activity which was inhibited by the addition of an estrogen antagonist (Doyle et al. 2009).

A hydromethanolic extract of allspice enhanced CYP3A4 promoter activity but had no effect on the pregnane-X receptor (Kluth et al. 2007).

IV. Pregnancy and Lactation

No information on the safety of allspice during pregnancy or lactation was identified.

V. Toxicity Studies

Acute Toxicity

No adverse effects were observed in mice orally administered up to 7.5 g/kg of an aqueous suspension of allspice (Al-Rehaily et al. 2002).

Genotoxicity

Allspice oleoresin gave a positive result in the DNA-repair test but not in the Ames *Salmonella* reversion assay or the *Bacillus subtilis rec* assay (Sekizawa and Shibamoto 1982). Antimutagenicity via antioxidant activity of allspice was observed in the DPPH radical reduction assay and lipid peroxidation inhibition testing (Ramos et al. 2003).

Literature Cited

Al-Rehaily, A.J., M.S. Al-Said, M.A. Al-Yahya, J.S. Mossa, and S. Rafatullah. 2002. Ethnopharmacological studies on allspice (*Pimenta dioica*) in laboratory animals. *Pharm. Biol.* 40(3):200-205.

Doyle, B.J., J. Frasor, L.E. Bellows, et al. 2009. Estrogenic effects of herbal medicines from Costa Rica used for the management of menopausal symptoms. *Menopause* 16(4):748-755.

Kanerva, L., T. Estlander, and R. Jolanki. 1996. Occupational allergic contact dermatitis from spices. *Contact Dermat.* 35(3):157-162.

Kluth, D., A. Banning, I. Paur, R. Blomhoff, and R. Brigelius-Flohe. 2007. Modulation of pregnane X receptor- and electrophile responsive element-mediated gene expression by dietary polyphenolic compounds. *Free Radicals Biol. Med.* 42(3):315-325.

Ramos, A., A. Visozo, J. Piloto, et al. 2003. Screening of antimutagenicity via antioxidant activity in Cuban medicinal plants. *J. Ethnopharmacol.* 87(2-3):241-246.

Sekizawa, J., and T. Shibamoto. 1982. Genotoxicity of safrole-related chemicals in microbial test systems. *Mutat. Res.* 101(2):127-140.

Pimpinella anisum L.

Apiaceae

SCN: anise
Syn: *Anisum vulgare* Gaertn.

Part: fruit (commonly known as "seed")

Quick Reference Summary

Safety Class: 1
Interaction Class: A

Contraindications

None known.

Other Precautions

None known.

Drug and Supplement Interactions

None known.

Notice

Alkenylbenzenes (estragole as 1–3% of essential oil) (De Vincenzi et al. 2000; Wichtl 2004); *see* Appendix 1.

Adverse Events and Side Effects

Allergic reactions to anise have been reported (Fraj et al. 1996; Garcia-Gonzalez et al. 2002; Gazquez Garcia et al. 2007; Wichtl 2004).

Pharmacological Considerations

None known.

Pregnancy and Lactation

The European Scientific Cooperative on Phytotherapy indicates that infusions of anise may be used in pregnancy but that alcohol extracts or the essential oil should not be used (ESCOP 2003).

An animal study suggested some anti-implantation activity of anise, but no adverse effects on fetal development (Dhar 1995).

No information on the safety of anise during lactation was identified. While this review did not identify any concerns for use while nursing, safety has not been conclusively established.

Review Details

I. Drug and Supplement Interactions

Clinical Trials of Drug or Supplement Interactions

No clinical trials of drug or supplement interactions were identified.

Case Reports of Suspected Drug or Supplement Interactions

No case reports of suspected drug or supplement interactions were identified.

Animal Trials of Drug or Supplement Interactions

No animal trials of drug or supplement interactions were identified.

II. Adverse Events

Case Reports of Adverse Events

Allergic reactions to anise have been reported (Wichtl 2004). A case of occupational allergy to anise associated with rhinoconjunctivitis and gastrointestinal symptoms was reported and confirmed by skin prick testing. Skin prick tests showed a positive immediate response to anise, asparagus, caraway, coriander, cumin, dill, and fennel extracts, and an intense late response to anise. Skin prick tests to celery, carrot, birch pollen, and mugwort pollen extracts were negative (Garcia-Gonzalez et al. 2002).

Occupational asthma induced by anise was observed in a butcher. Skin prick testing confirmed anise as the causative agent (Fraj et al. 1996). A case of anise-induced tongue angioedema was reported and confirmed by skin prick testing (Gazquez Garcia et al. 2007).

III. Pharmacology and Pharmacokinetics

Human Pharmacological Studies

No relevant human pharmacological studies were identified.

Animal Pharmacological Studies

In a study in mice intraperitoneally administered 0.125 to 0.5 ml/kg of anise essential oil, assays with GABA receptor agonists indicated that anise essential oil activated the GABAergic mechanism that depends on the GABA-A receptor subtype (Sahraei et al. 2002).

In Vitro Pharmacological Studies

In estrogen receptor-positive human breast cancer cells (MCF-7), antiestrogenic activity of an aqueous extract of anise was observed (Kassi et al. 2004). Some estrogenic activity of anise essential oil was observed in a yeast estrogen screening assay (Tabanca et al. 2004).

IV. Pregnancy and Lactation

P

Anise is used in herbal formulas for women in the first trimester of pregnancy (Noe et al. 2002).

Anise is one of 39 species listed as ingredients in 6 traditional abortifacient formulas. These species include commonly consumed nontoxic plants (e.g., kidney bean and pomegranate) as well as with other plants recognized to be toxic or inappropriate for use in pregnancy (e.g., pennyroyal). Parts and doses of botanicals used in these formulas were not listed (Madari and Jacobs 2004).

In rats orally administered 50, 70, or 80 mg/kg of the compound *trans*-anethole on days 1 to 10 of pregnancy, inhibition of implantation was 33, 66, and 100%, respectively (Dhar 1995).

In rats administered 80 mg/kg *trans*-anethole on days 1 to 2 of pregnancy, normal implantation and delivery occurred. After administration of *trans*-anethole on days 3 to 5, implantation was completely inhibited; and on days 6 to 10, three of five rats failed to deliver at term. No gross malformations were observed in any of the groups (Dhar 1995).

No information on the safety of anise during lactation was identified.

V. Toxicity Studies

Acute Toxicity

The LD_{50} of orally administered anise essential oil in rats is reported as 2.25 g/kg and 2.7 g/kg (Opdyke 1979; von Skramlik 1959).

The LD_{50} of orally administered *trans*-anethole is 3.41 g/kg in mice, 2.65 g/kg in rats, and 2.16 g/kg in guinea pigs (Lin 1991).

Genotoxicity

No mutagenic activity of an aqueous extract of anise was observed in the Ames test for mutagenicity with *Salmonella typhimurium* strains TA98, TA100, and TA102 (Al-Bataina et al. 2003).

A dried ethanolic extract of anise exhibited mutagenic activity at high concentrations (5 mg/plate) in a streptomycin-dependent strain of *Salmonella typhimurium* TA98 (Shashkanth and Hosono 1986).

No mutagenic activity of an anise ethanol extract was observed at concentrations up to 0.1 mg/ml in chromosomal aberration tests using a Chinese hamster fibroblast cell line (Ishidate et al. 1984).

Literature Cited

Al-Bataina, B.A., A.O. Maslat, and M.M. Al Kofahi. 2003. Element analysis and biological studies on ten oriental spices using XRF and Ames test. *J. Trace Elem. Med. Biol.* 17(2):85-90.

De Vincenzi, M., M. Silano, F. Maialetti, and B. Scazzocchio. 2000. Constituents of aromatic plants: II. Estragole. *Fitoterapia* 71(6):725-729.

Dhar, S.K. 1995. Anti-fertility activity and hormonal profile of *trans*-anethole in rats. *Indian J. Physiol. Pharmacol.* 39(1):63-67.

ESCOP. 2003. *ESCOP monographs: The scientific foundation for herbal medicinal products.* 2nd ed. Exeter, UK: European Scientific Cooperative on Phytotherapy.

Fraj, J., A. Lezaun, C. Colas, et al. 1996. Occupational asthma induced by aniseed. *Allergy* 51(5):337-339.

Garcia-Gonzalez, J.J., B. Bartolome-Zavala, S. Fernandez-Melendez, et al. 2002. Occupational rhinoconjunctivitis and food allergy because of aniseed sensitization. *Ann. Allergy Asthma Immunol.* 88(5):518-522.

Gazquez Garcia, V., P. Gaig Jane, and B. Bartolome Zavala. 2007. Aniseed-induced nocturnal tongue angioedema. *J. Investig. Allergol. Clin. Immunol.* 17(6):406-408.

Ishidate, M., T. Sofuni, K. Yoshikawa, et al. 1984. Primary mutagenicity screening of food additives currently used in Japan. *Food Chem. Toxicol.* 22(8):623-636.

Kassi, E., Z. Papoutsi, N. Fokialakis, et al. 2004. Greek plant extracts exhibit selective estrogen receptor modulator (SERM)-like properties. *J. Agric. Food Chem.* 52(23):6956-6961.

Lin, F.S.D. 1991. *trans*-Anethole. Joint FAO/WHO Expert Committee on Food Additives. WHO Food Additives Series 28. Geneva: World Health Organization.

Madari, H., and R.S. Jacobs. 2004. An analysis of cytotoxic botanical formulations used in the traditional medicine of ancient Persia as abortifacients. *J. Nat. Prod.* 67(8):1204-1210.

Noe, J.E., M. Bove, and K. Janel. 2002. Herbal tonic formulas for naturopathic obstetrics. *Altern. Complement. Ther.* 8(6):327-335.

Opdyke, D.L.J. 1979. *Monographs on fragrance raw materials*. New York: Pergamon.

Sahraei, H., H. Ghoshooni, S. Hossein Salimi, et al. 2002. The effects of fruit essential oil of the *Pimpinella anisum* on acquisition and expression of morphine induced conditioned place preference in mice. *J. Ethnopharmacol.* 80(1):43-47.

Shashkanth, K.N., and A. Hosono. 1986. *In vitro* mutagenicity of tropical spices to *Streptomycin*-dependent strains of *Salmonella typhimurium* TA98. *Agric. Biol. Chem.* 50(11):2947-2948.

Tabanca, N., S.I. Khan, E. Bedir, et al. 2004. Estrogenic activity of isolated compounds and essential oils of *Pimpinella* species from Turkey, evaluated using a recombinant yeast screen. *Planta Med.* 70(8):728-735.

von Skramlik, E. 1959. Ober die Giftigkeit und Vertraglichkeit von atherischen Olen. *Pharmazie* 14:435-445.

Wichtl, M. 2004. *Herbal drugs and phytopharmaceuticals: A handbook for practice on a scientific basis*. 3rd ed. Boca Raton, FL: CRC Press.

Pinellia ternata (Thunb.) Makino ex Breit. — Araceae

SCN: pinellia
PN: *ban xia* (rhizome); *jiang ban xia* (ginger-cured rhizome); *fa ban xia* (licorice-cured rhizome)

Part: prepared rhizome

Quick Reference Summary

Safety Class: 3
Interaction Class: A

Contraindications

Not for use except under the supervision of an expert qualified in the appropriate use of this substance (Bensky et al. 2004; Chen and Chen 2004).

Other Precautions

None known.

Drug and Supplement Interactions

None known.

Standard Dose

The standard dose is 3 to 10 g as a decoction (up to 60 g under professional supervision) (Bensky et al. 2004; Chen and Chen 2004).

Editors' Notes

Unprocessed pinellia is generally considered toxic and inappropriate for internal use. Proper processing reduces toxicity (Bensky et al. 2004; PPRC 2005; Wu et al. 1999, 2007). A review of research on pinellia processing and toxicity indicated that if unprocessed pinellia is cooked for an extended period of time, or if unprocessed pinellia is in a formula with fresh ginger, the toxicity is minimized (Chen and Chen 2004). The prepared rhizome, that has been processed to reduce toxicity, is the subject of this entry.

The unprocessed tuber of *Pinellia ternata* has been reported to contain ephedrine at a concentration of 0.002 to 0.003% (Oshio et al. 1978; Wu et al. 1996). This low concentration of ephedrine is further reduced by several traditional processing steps (Bensky et al. 2004; PPRC 2005; Wu et al. 1996, 1999, 2007). The U.S. Food and Drug Administration declared in 2004 that dietary supplements containing ephedrine alkaloids are adulterated, and identified *Pinellia ternata* as containing ephedrine alkaloids (FDA 2004). In establishing this ban for ephedrine in dietary supplements, however, the FDA stated that it does not apply to "traditional Asian medicine" that contains ephedrine alkaloids (FDA 2004).

Pinellia is sometimes adulterated with *Typhonium* spp. including *T. flagelliforme*, *T. divaricatum*, and *T. trilobatum* (Bensky et al. 2004).

Adverse Events and Side Effects

Ingestion of fresh pinellia or overdosage of processed pinellia can cause severe irritation of the mucosa of the mouth, pharynx, and gastrointestinal tract (Bensky et al. 2004; Chen and Chen 2004).

Pharmacological Considerations

A text on traditional Chinese herbs indicates that pinellia may act synergistically with barbiturates. No further details were reported (Bensky et al. 2004).

Pregnancy and Lactation

A text on traditional Chinese medicine reports that while a classic text on Chinese medicine indicates that pinellia may cause miscarriage, modern practitioners commonly use

P

pinellia to treat nausea and vomiting during pregnancy (Chen and Chen 2004).

Some effects on development were observed in the fetuses of rats administered high doses (2 g/kg) of pinellia during pregnancy (Shin et al. 2007). The compound pinellin has demonstrated anti-implantation activity in rabbits (dose unspecified in available translation) (Chen et al. 1984).

The compound pinellin has been shown to have abortifacient activity (Lu et al. 1986).

No information on the safety of pinellia during lactation was identified. While this review did not identify any concerns for use while nursing, safety has not been conclusively established, and this substance is not recommended for use except under the supervision of an expert qualified in its appropriate use.

Review Details

I. Drug and Supplement Interactions

Clinical Trials of Drug or Supplement Interactions

No clinical trials of drug or supplement interactions were identified.

Case Reports of Suspected Drug or Supplement Interactions

No case reports of suspected drug or supplement interactions were identified.

Animal Trials of Drug or Supplement Interactions

No animal trials of drug or supplement interactions were identified.

II. Adverse Events

Case Reports of Adverse Events

Occupational asthma and rhinitis have been reported and confirmed by skin prick testing. In one case, the patient also tested positively to *Dioscorea batatas*, and in a second case, the patient tested positive to unspecified species of *Dioscorea* and *Cnidium* (Lee et al. 2001; Park et al. 1994).

P

III. Pharmacology and Pharmacokinetics

Human Pharmacological Studies

No relevant human pharmacological studies were identified.

Animal Pharmacological Studies

No relevant animal pharmacological studies were identified.

In Vitro Pharmacological Studies

No estrogenic activity of an ethanol extract of pinellia was observed in a recombinant yeast system with a human estrogen receptor expression plasmid and a reporter plasmid (Kim et al. 2008).

IV. Pregnancy and Lactation

In pregnant rats orally administered 20, 200, or 2000 mg/kg of pinellia extract on days 6 to 15 of pregnancy, some adverse effects on development were observed in the 2000 mg/kg dose group. In that group, the rates of ureteric dilatation and renal malposition in fetuses were increased. Rates of skeletal malformations and variations of fetuses were also increased in the high-dose group, including asymmetric alignment of ribs, dumbbell ossification of thoracic centrum, and 14th supernumerary ribs. No adverse effects on the mothers were observed at any dose. Observed parameters in mothers included clinical signs, body weight gain, feed and water consumption, fertility parameters, and placental and organ weights (Shin et al. 2007).

The compound pinellin has demonstrated anti-implantation activity in rabbits. Doses and route of administration used were not listed in available English language translations (Chen et al. 1984).

A reference text on traditional Chinese medicine reports that while a classic text on Chinese medicine indicates that pinellia may cause miscarriage, modern practitioners commonly use pinellia to treat nausea and vomiting during pregnancy (Chen and Chen 2004).

The compound pinellin has been shown to have abortifacient activity (Lu et al. 1986).

No information on the safety of pinellia during lactation was identified.

V. Toxicity Studies

Acute Toxicity

The LD_{50} of unprocessed pinellia in mice is 0.325 g/kg after intraperitoneal administration and 42.7 g/kg after oral administration. No signs of toxicity were reported in mice orally administered 16 g/kg of processed pinellia every 3 hours for five doses (Chen and Chen 2004).

Literature Cited

Bensky, D., S. Clavey, and E. Stöger. 2004. *Chinese herbal medicine: Materia medica*. 3rd ed. Seattle: Eastland Press.

Chen, H., T. Song, and Z.J. Tao. 1984. Anti-implantation effect of pinellin in rabbits. *Acta Physiol. Sin.* 36(4):388-392.

Chen, J.K., and T.T. Chen. 2004. *Chinese medical herbology and pharmacology*. City of Industry, CA: Art of Medicine Press.

FDA. 2004. Final rule declaring dietary supplements containing ephedrine alkaloids adulterated because they present an unreasonable risk. *Federal Register* 69(28):6788-6854.

Kim, I.G., S.C. Kang, K.C. Kim, E.S. Choung, and O.P. Zee. 2008. Screening of estrogenic and antiestrogenic activities from medicinal plants. *Environ. Toxicol. Pharmacol.* 25(1):75-82.

Lee, S.K., H.K. Cho, S.H. Cho, et al. 2001. Occupational asthma and rhinitis caused by multiple herbal agents in a pharmacist. *Ann. Allergy Asthma Immunol.* 86(4):469-474.

Lu, Z.X., Q.L. Shi, and Q.Y. Xu. 1986. Conformational changes of pinellin in solution using intrinsic fluorescence and CD as probes. *Biopolymers* 25(3):393-405.

Oshio, H., M. Tsukui, and T. Matsuoka. 1978. Isolation of *l*-ephedrine from pinelliae tuber. *Chem. Pharm. Bull.* 26(7):2096-2097.

Park, H.S., M.J. Kim, and H.B. Moon. 1994. Occupational asthma caused by two herb materials, *Dioscorea batatas* and *Pinellia ternata*. *Clin. Exp. Allergy* 24(6):575-581.

PPRC. 2005 *Pharmacopoeia of the People's Republic of China*. Beijing: People's Medical Publishing House.

Shin, S., D. Park, J.H. Jeon, et al. 2007. Effect of *Pinellia ternata* extract on fetal development of rats. *Reprod. Toxicol.* 24(1):71.

Wu, H., W. Li, H. Han, R. Ji, and D.J. Ye. 1999. Studies on stimulating components of raw *Pinellia ternata* (Thunb.) (banxia). *Zhongguo Zhong Yao Za Zhi* 24(12):725-730, 763.

Wu, H., X. Tan, B. Cai, and D. Ye. 1996. Effect of ginger-processing on *l*-ephedrine contents in *Rhizoma Pinelliae*. *Zhongguo Zhong Yao Za Zhi* 21(3):157-158.

Wu, H., L.Y. Zhong, W. Li, and D.J. Ye. 2007. Study on processing mechanism of *Pinellia ternate*. *Zhongguo Zhong Yao Za Zhi* 32(14):1402-1406.

Pinus strobus L.

Pinaceae

SCN: white pine
OCN: eastern white pine

Part: bark

Quick Reference Summary

Safety Class: 1
Interaction Class: A

Contraindications

None known.

Other Precautions

None known.

Drug and Supplement Interactions

None known.

Adverse Events and Side Effects

None known.

Pharmacological Considerations

None known.

Pregnancy and Lactation

No information on the safety of white pine in pregnancy or lactation was identified in the scientific or traditional literature. Although this review did not identify any concerns for use while pregnant or nursing, safety has not been conclusively established.

Review Details

I. Drug and Supplement Interactions

Clinical Trials of Drug or Supplement Interactions

No clinical trials of drug or supplement interactions were identified.

Case Reports of Suspected Drug or Supplement Interactions

No case reports of suspected drug or supplement interactions were identified.

Animal Trials of Drug or Supplement Interactions

No animal trials of drug or supplement interactions were identified.

II. Adverse Events

Case Reports of Adverse Events

No case reports of adverse events were identified.

III. Pharmacology and Pharmacokinetics

Human Pharmacological Studies

No relevant human pharmacological studies were identified.

Animal Pharmacological Studies

No relevant animal pharmacological studies were identified.

In Vitro Pharmacological Studies

No relevant in vitro pharmacological studies were identified.

IV. Pregnancy and Lactation

No information on the safety of white pine during pregnancy or lactation was identified.

P

V. Toxicity Studies

Genotoxicity

In human peripheral blood lymphocytes, without metabolic activation, treated with condensate from white pine wood, a dose-response curve was observed using the chromosome aberration assay and sister-chromatid exchange analysis (Mark et al. 1996).

Literature Cited

Mark, H.F., R. Naram, J.T. Singer, et al. 1996. Wood-drying condensate from eastern white pine induced cytotoxicity and genotoxicity in vitro. *Ann. Clin. Lab. Sci.* 26(1):64-70.

Piper cubeba L. f. — Piperaceae

SCN: cubeb
AN: *kankola*

Part: unripe fruit

Quick Reference Summary

Safety Class: 1
Interaction Class: A

Contraindications

None known.

Other Precautions

Use with caution in persons with kidney inflammation (Felter and Lloyd 1898).

Drug and Supplement Interactions

See Pharmacological Considerations.

Notice

Piperine (Parmar et al. 1997); *see* Appendix 3.

Adverse Events and Side Effects

An Eclectic medical text indicated that, in "large" doses (standard dose listed as 0.6–4.0 g), cubeb occasionally causes nausea and vomiting, burning pain, griping, or even purging. Cubeb has been associated with skin rashes in some individuals (Felter and Lloyd 1898; Wood and LaWall 1926).

Pharmacological Considerations

The compound piperine has been shown to enhance the bioavailability of certain drugs (Hu et al. 2005; Srinivasan 2007; Wattanathorn et al. 2008).

Pregnancy and Lactation

No information on the safety of cubeb in pregnancy or lactation was identified in the scientific or traditional literature. Although this review did not identify any concerns for use while pregnant or nursing, safety has not been conclusively established.

Review Details

I. Drug and Supplement Interactions

Clinical Trials of Drug or Supplement Interactions

No clinical trials of drug or supplement interactions were identified.

Case Reports of Suspected Drug or Supplement Interactions

No case reports of suspected drug or supplement interactions were identified.

Animal Trials of Drug or Supplement Interactions

No animal trials of drug or supplement interactions were identified.

II. Adverse Events

Case Reports of Adverse Events

An Eclectic medical text indicated that, in "large" doses (standard dose listed as 0.6–4.0 g), cubeb occasionally causes nausea and vomiting, burning pain, griping, or even purging. Cubeb has been associated with skin rashes in some individuals (Felter and Lloyd 1898; Wood and LaWall 1926).

III. Pharmacology and Pharmacokinetics

Human Pharmacological Studies

No relevant human pharmacological studies were identified.

Animal Pharmacological Studies

No relevant animal pharmacological studies were identified.

In Vitro Pharmacological Studies

The ethyl acetate fraction of an aqueous extract of cubeb inhibited the drug-metabolizing isoenzyme CYP3A4 (Usia et al. 2005b, 2006). Subsequent testing indicated that methylenedioxyphenyl lignan compounds produced mechanism-based inhibition of CYP3A4 (Usia et al. 2005a).

No significant effects of a methanol extract of cubeb were observed on the drug-metabolizing isoenzymes CYP2D6 and CYP3A4 in human liver microsomes (Subehan et al. 2006).

An extract of cubeb demonstrated antagonistic activity in recombinant wild-type androgen receptors (Yam et al. 2008a).

An ethanolic extract of cubeb significantly inhibited growth induced by β-estradiol in estrogen receptor-positive human breast cancer cells (MCF-7) (Yam et al. 2008b).

IV. Pregnancy and Lactation

No information on the safety of cubeb during pregnancy or lactation was identified.

V. Toxicity Studies

No toxicity studies were identified.

Literature Cited

Felter, H.W., and J.U. Lloyd. 1898. *King's American dispensatory*. 18th ed., 3rd rev. 2 vols. Cincinnati: Ohio Valley Co.

Parmar, V.S., S.C. Jain, K.S. Bisht, et al. 1997. Phytochemistry of the genus *Piper*. *Phytochemistry* 46(4):597-673.

Subehan, T. Usia, H. Iwata, S. Kadota, and Y. Tezuka. 2006. Mechanism-based inhibition of CYP3A4 and CYP2D6 by Indonesian medicinal plants. *J. Ethnopharmacol.* 105(3):449-455.

Usia, T., H. Iwata, A. Hiratsuka, et al. 2006. CYP3A4 and CYP2D6 inhibitory activities of Indonesian medicinal plants. *Phytomedicine* 13(1-2):67-73.

Usia, T., T. Watabe, S. Kadota, and Y. Tezuka. 2005a. Metabolite-cytochrome P450 complex formation by methylenedioxyphenyl lignans of *Piper cubeba*: Mechanism-based inhibition. *Life Sci.* 76(20):2381-2391.

Usia, T., T. Watabe, S. Kadota, and Y. Tezuka. 2005b. Potent CYP3A4 inhibitory constituents of *Piper cubeba*. *J. Nat. Prod.* 68(1):64-68.

Wood, H., and C. LaWall. 1926. *The dispensatory of the United States of America*. Philadelphia: Lippincott.

Yam, J., M. Kreuter, and J. Drewe. 2008a. *Piper cubeba* targets multiple aspects of the androgen-signalling pathway. A potential phytotherapy against prostate cancer growth? *Planta Med.* 74(1):33-38.

Yam, J., A. Schaab, M. Kreuter, and J. Drewe. 2008b. *Piper cubeba* demonstrates anti-estrogenic and anti-inflammatory properties. *Planta Med.* 74(2):142-146.

Piper longum L. — Piperaceae

SCN: long pepper
AN: *pippali*
PN: *bi ba* (fruit)
OCN: jaborandi pepper
Part: fruit

Quick Reference Summary

Safety Class: 2b
Interaction Class: B

Contraindications

Not for use in pregnancy except under the supervision of a qualified healthcare practitioner (Adhikary et al. 1990; Daware et al. 2000; Kholkute et al. 1979; Lakshmi et al. 2006).

Other Precautions

None known.

Drug and Supplement Interactions

See Pharmacological Considerations.

Notice

Piperine (4–5%) (Tapadiya et al. 2009); *see* Appendix 3.

Adverse Events and Side Effects

None known.

Pharmacological Considerations

The compound piperine has been shown to enhance the bioavailability of certain drugs (Hu et al. 2005; Srinivasan 2007; Wattanathorn et al. 2008). Some studies have indicated

Other Precautions

Concomitant use of alcohol and large doses of kava (1 g/kg) may produce additive inhibition of cognitive and motor function induced by alcohol (Foo and Lemon 1997).

Kava has been associated with a potential risk of rare but severe liver injury (Schmidt et al. 2005). Patients who have or have had liver problems, frequently use alcohol, or are taking any medications, especially those metabolized by the liver, are advised to consult with a healthcare professional prior to use.

Drug and Supplement Interactions

Kava may inhibit the drug-metabolizing isoenzyme CYP2E1, resulting in a slowed clearance and increased plasma levels of drugs metabolized by this enzyme (Gurley et al. 2005). *See* Cytochrome P450 in Appendix 3.

Editors' Note

The American Herbal Products Association has established a trade requirement (AHPA 2011) that products containing kava bear the following or significantly similar statement:

> Caution: U.S. FDA advises that a potential risk of rare, but severe, liver injury may be associated with kava-containing dietary supplements. Ask a healthcare professional before use if you have or have had liver problems, frequently use alcoholic beverages, or are taking any medication. Stop use and see a doctor if you develop symptoms that may signal liver problems (e.g., unexplained fatigue, abdominal pain, loss of appetite, fever, vomiting, dark urine, pale stools, yellow eyes or skin). Not for use by persons under 18 years of age, or by pregnant or breastfeeding women. Not for use with alcoholic beverages. Excessive use, or use with products that cause drowsiness, may impair your ability to operate a vehicle or dangerous equipment.

Standard Dose

The standard dose is 2 to 4 g as a decoction, up to three times daily (BHMA 1983); 60–600 mg kavalactones (kavapyrones) per day (Blumenthal and Busse 1998; Dentali 1997).

Adverse Events and Side Effects

Kava is reported as being generally well tolerated in clinical trials (Pittler and Ernst 2003). Cases of liver toxicity have been reported in association with kava use (Brauer et al. 2001, 2003; Bujanda et al. 2002; Campo et al. 2002; Gow et al. 2003; Humberston et al. 2003; Musch et al. 2006; Russmann et al. 2003; Sass et al. 2001; Schmidt et al. 2005; Stoller 2000). These reports have been subject to critical analysis by a number of organizations and individuals (Schmidt 2007; Schmidt et al. 2005; TMEC 2002; Waller 2002). A review of liver toxicity cases indicated that the reported cases represented unpredictable reactions, and that cases were best characterized as idiosyncratic reactions of the metabolic type (Teschke et al. 2008). Of 82 cases of liver toxicity reported through 2002, one case was assessed as being "certain" or "probable" by multiple reviewers, with up to 12 others being noted as "probable" (BfArM 2002; Schmidt 2007)

Excessive or extended consumption of kava is reported to cause a scaly, yellowing skin condition that resolves when use is discontinued (Blumenthal and Busse 1998; Bone 1993/1994; Grace 2005; Jappe et al. 1998; Lewin 1931; Norton and Ruze 1994; Pfeiffer et al. 1967; Schmidt and Boehncke 2000).

Pharmacological Considerations

Human studies have provided conflicting information on the effects of kava on the drug-metabolizing isoenzyme CYP1A2, with one study showing inhibition of CYP1A2 in heavy kava users (Russmann et al. 2005) and another study showing no effect (Gurley et al. 2005). Kava showed inhibition of CYP2E1 but had no effect on CYP2D6 or CYP3A4/5 (Gurley et al. 2005, 2008a, 2008b).

A human study showed no interaction between digoxin and kava (Gurley et al. 2007).

Trials with concomitantly administered bromazepam indicated no additive adverse effects of kava on motor or cognitive functioning (Herberg 1996). Trials with concomitantly administered alcohol gave mixed results on the additive adverse effects of kava on cognitive functioning, with one study showing a reduction in visual-motor function after high doses of kava and another study showing no additive effects after low doses of kava (Foo and Lemon 1997; Herberg 1993).

Pregnancy and Lactation

Animal studies have indicated no adverse effects of several compounds from kava administered during pregnancy (Hansel and Woelk 1994), and no information on the safety of kava during lactation was identified in the scientific or traditional literature.

In this work, the contraindications for use in pregnancy and lactation are based on concerns regarding the cases of hepatotoxicity reported in association with kava use, as the implications of these case reports and possible mechanisms of hepatotoxicity have yet to be fully understood.

Review Details

I. Drug and Supplement Interactions

Clinical Trials of Drug or Supplement Interactions

A study of bromazepam (9 mg/day) and kava (400 mg/day standardized extract) indicated that taking the two products together did not produce significant differences in well-being or mental performance as compared with bromazepam alone (Herberg 1996). In a study of ethanol (0.05 blood alcohol concentration) and kava (210 mg/

day), no additive effects of the kava were observed on safety-related parameters as compared with alcohol alone (Herberg 1993). Another study of alcohol (0.05 blood alcohol concentration) and a relatively high dose of kava (1 g/kg) showed that kava and alcohol taken together produced more significant decreases in cognitive and visual-motor functions than alcohol alone (Foo and Lemon 1997).

A trial of single doses of digoxin (0.5 mg) before and after 14 days supplementation with kava (1227 mg/day) indicated no statistically significant effects of kava on digoxin pharmacokinetics (Gurley et al. 2007).

Case Reports of Suspected Drug or Supplement Interactions

A 44-year-old man taking kava, valerian, paroxetine, and pantoprazole was reported to have a fever, malaise, headache, and confusion (Rubin et al. 2006).

A 54-year-old man taking kava, alprazolam, cimetidine, and terazosin was reported to have lethargy and disorientation (Almeida and Grimsley 1996). Cimetidine has been shown to slow the metabolism of alprazolam, a benzodiazepine that may cause sleepiness, dizziness, and confusion if overdosed (Greenblatt and Wright 1993).

A 52-year-old woman on long-term treatment with benzodiazepines experienced confusion and auditory and visual hallucinations within 2 days of starting to take kava (Cartledge and Rutherford 2001). Several case reports of possible dopamine antagonism suggest that kava may interact with central dopamine agonists or antagonists (Mills and Bone 2005; Schelosky et al. 1995).

Animal Trials of Drug or Supplement Interactions

A study of ethanol (3.5 g/kg) and kava (200 mg/kg) in mice showed that the two products taken together produced a significant increase in sleeping time compared to ethanol alone (Jamieson and Duffield 1990). The applicability of those animal data to humans is unknown.

II. Adverse Events

Adverse Events Reported in Clinical Trials

In a meta-analysis of 12 double-blind, randomized controlled trials (n = 700), adverse events associated with kava use were reported as being mild, transient, and infrequent (Pittler and Ernst 2003). Doses of up to 6 g/day of kava caused yellowing of the skin in some individuals after several weeks of administration (Pfeiffer et al. 1967).

Case Reports of Adverse Events

A number of case reports of hepatotoxicity in association with kava use have been reported in the medical literature and to pharmacovigilance centers. Reports include cases of fulminant liver failure, severe liver damage, necrotizing hepatitis, cholestatic hepatitis, liver cell impairment, and an increase in liver enzymes (Brauer et al. 2001, 2003; Bujanda et al. 2002; Campo et al. 2002; Gow et al. 2003; Humberston et al. 2003; Musch et al. 2006; Russmann et al. 2003; Sass et al. 2001; Schmidt et al. 2005; Stoller 2000). These reports have been subject to critical analysis by a number of organizations and individuals (Schmidt 2007; Schmidt et al. 2005; TMEC 2002; Waller 2002). Details of case reports recorded by pharmacovigilance centers can be found in reviews by Schmidt et al. (2005) and Schmidt (2007). The analyses include assessments of causality for the case reports, with reviewers using assessment schemes that generally categorize causality of the adverse event as certain, probable, possible, unlikely, conditional, or unassessable in relation to kava ingestion (Schmidt et al. 2005; Teschke et al. 2008). Reviewers have included varying numbers of case reports in reviews, with some looking only at cases reported to German pharmacovigilance centers and others examining all known cases. German and Swiss cases of hepatotoxicity assessed as "probable" range from 0% (Waller 2002) to 42% (BfArM 2002) of cases analyzed. Information was noted as insufficient to assess causality in a number of the case reports and, in multiple cases, patients were taking other medications that have been associated with liver damage (BfArM 2002; Schmidt et al. 2005; TMEC 2002; Waller 2002). Of the 82 cases of hepatotoxicity reported through 2002, one case was assessed as being "certain" or "probable" by multiple reviewers with up to 12 others being noted as "probable" (BfArM 200; Schmidt 2007).

Based on sales data for 2000 to 2002, the total number of daily doses of kava in Germany has been estimated at 70 million annually. With 12 cases of hepatotoxicity reported in Germany and Switzerland assessed by some reviewers as probable (BfArM 2002), the incidence of reported adverse events is 0.24 per 1 million daily doses (Schmidt et al. 2005).

Traditional daily use of aqueous extracts of kava in the Pacific Islands has not been associated with any increased incidence of hepatotoxicity (Schmidt et al. 2005). Hypotheses for causes of hepatotoxicity reported outside of, but not in, Pacific Island populations include differences in solvent extraction (acetone and ethanol extracts versus traditional aqueous preparations), improper plant parts being included in manufactured products (i.e., the inclusion of stem peelings that may contain toxic compounds), increased levels of other hepatotoxic substances in populations taking manufactured kava extracts (Currie and Clough 2003), and genetic differences (Russmann et al. 2001). A review of hepatotoxicity cases indicated that cases were reported after the use of aqueous, ethanolic, and acetonic kava extracts, suggesting a lack of association between the solvent and hepatotoxicity (Teschke et al. 2009).

Use of kava in Fijian and Australian aboriginal communities has been recorded to range from 100 to 500 g, and occasionally to 900 g, per week. Adverse reactions associated with kava use in these communities include scaly skin, weight loss, headache, and decreases in blood lymphocytes (Schmidt et al. 2005). A condition known as kava dermopathy has been reported in heavy users of kava, with symptoms including yellow, scaly, leprosy-like eruptions of the skin and inflammation of the eyes (Norton and Ruze 1994). This condition is reported to resolve on discontinuation of kava (Blumenthal and Busse 1998; Bone 1993/1994; Lewin 1931).

Kava use has been associated with several other types of adverse reactions. Persistent Parkinsonism was reported in a 54-year-old woman with a family history of essential tremors after 10 days of kava use (65 mg/day); symptoms abated as kava was discontinued and medical treatment was started (Meseguer et al. 2002).

Several patients were reported to have episodes of dystonia and dyskinesia, and one patient with Parkinson's disease experienced an increase in motor oscillations. The reporting physicians indicated that dopamine antagonism was believed to be responsible for the adverse events (Schelosky et al. 1995).

Additional adverse reactions have been reported to include vertigo, nausea, and vomiting experienced by a 37-year-old man who drank several cups of "strong" kava tea (Perez and Holmes 2005); choreoathetosis in a 27-year-old man after occasions of consumption of "large quantities" of kava beverage (Spillane et al. 1997); acute urinary retention in a 61-year-old man who drank 1 liter of kava beverage (Leung 2004); and a case of "intoxication" in a 34-year-old man who drank up to 40 bowls of kava per day (Chanwai 2000).

III. Pharmacology and Pharmacokinetics

Human Pharmacological Studies

Conflicting information exists showing significant inhibition of CYP1A2 in chronic kava consumers (Russmann et al. 2005) or showing no effect after 28 days of 1.18 g daily kava in a non-kava drinking population (Gurley et al. 2005). In healthy volunteers orally administered 0.4 or 2 g kava for 28 days, a significant reduction in phenotypic ratios of CYP2E1 was observed, with no effects on CYP2D6 (Gurley et al. 2005, 2008b). No changes in CYP3A4 activity were observed in healthy volunteers orally administered 0.4 or 3.6 g kava daily for 14 days (Gurley et al. 2005, 2008b).

No effect was observed on pharmacokinetics of single doses of digoxin (0.5 mg) before or after 14 days of kava consumption, suggesting that kava does not effect P-gp activity (Gurley et al. 2007).

In a study of a kava-drinking population, recent users of kava were observed to have higher liver enzyme levels of γ-glutamyl transferase (GGT) and alkaline phosphatase (ALP) but not alanine aminotransferase (ALT) or bilirubin. Baselines were reported at normal levels after abstinence from kava (Clough et al. 2003).

Animal Pharmacological Studies

Animal studies were identified but omitted due to presence of human data.

In Vitro Pharmacological Studies

In vitro studies were identified but omitted due to presence of human data.

IV. Pregnancy and Lactation

Administration of synthetic kavain to rats (100 and 500 mg/kg) and rabbits (20 and 200 mg/kg) on days 6 to 17 of gestation did not result in any fetal abnormalities. Similarly, in rats administered dihydromethysticin (50 mg/kg) intraperitoneally for 3 months, no fetal abnormalities were noted in the first two generations of offspring (Hansel and Woelk 1994).

No information on the safety of kava during lactation was identified.

V. Toxicity Studies

Acute Toxicity

The oral LD_{50} of a standardized kava extract (70% kavalactones) is 180 mg/kg in mice and 1600 mg/kg in rats. The intraperitoneal LD_{50} is 380 mg/kg in mice and 370 mg/kg in rats (Hansel and Woelk 1994).

Short-Term Toxicity

In mice and rats fed a corn oil extract of kava at doses to 2 g/kg daily for 11 days, abnormal breathing, ataxia, and lethargy were observed. Hepatic cellular hypertrophy was noted in some animals (Sparrow et al. 2004).

In mice orally administered 0.125, 0.25, 0.5, 1.0, or 2.0 g/kg kava extract daily for 14 weeks, analyses of gene functions and pathways indicated that the levels of expression of a number of genes involving drug metabolism were changed, and that the pathways involving xenobiotics metabolism, Nrf2-mediated oxidative stress response, mitochondrial functions, and others were altered (Guo et al. 2010).

In rats orally administered kava containing 38 or 380 mg/kg kavalactones daily (approximately 10 and 100 times the suggested human dose) for 8 days, expression of CYP1A1 mRNA was markedly enhanced in the high-dose group along with ethoxyresorufin O-deethylase activities and CYP1A1 immunoreactivities. A moderate increase in CYP1A2 mRNA expression was also observed in the high-dose group (Yamazaki et al. 2008).

In rats administered kava aqueous extract (200 or 500 mg/kg kavalactones) for 2 or 4 weeks, no elevated levels of any liver enzymes were observed. After 2 weeks of treatment, levels of ALT and AST were decreased, and after 4 weeks of treatment, ALT levels were decreased (Singh and Devkota 2003).

In rats orally administered 10 mg/kg pipermethystine, 100 mg/kg acetone-water extracts of kava, or both pipermethystine and the kava extract daily for 2 weeks, no changes in liver enzyme levels were observed, although levels of hepatic glutathione, cytosolic superoxide dismutase,

TNF-α, mRNA expression, CYP2E1, and CYP1A2 were increased (Lim et al. 2007).

Subchronic Toxicity

In rats fed ethanol or acetone extracts of kava at 31.25, 62.5, or 133 mg/kg added to the diet for 3 months, no evidence of liver injury was observed, including changes in serum markers of liver damage and serum lipid peroxide readings. After intraperitoneal administration of the hepatotoxin galactosamine, kava showed no signs of enhancing liver injury induced by galactosamine (DiSilvestro et al. 2007).

Chronic Toxicity

Rats fed 7.3 or 73 mg/kg kava extract for 3 to 6 months showed no signs of hepatotoxicity or other adverse effects (Sorrentino et al. 2006).

Cytotoxicity

An in vitro assay of aqueous and methanolic extracts of kava (root, leaf, and stem peeling) on human liver carcinoma cells indicated that the methanolic extract displayed significantly greater cytotoxicity than the aqueous extract. The compound flavokavain B was believed to be responsible for the toxic effects (Jhoo et al. 2006).

Literature Cited

AHPA. July 2011. Code of Ethics & Business Conduct. Silver Spring, MD: American Herbal Products Association.

Almeida, J.C., and E.W. Grimsley. 1996. Coma from the health food store: Interaction between kava and alprazolam. *Ann. Intern. Med.* 125(11):940-941.

BfArM. 2002. Rejection of drug risks, Step II: Kava-kava (*Piper methysticum*)-containing, and kavain-containing drugs, including homeopathic preparations with a final concentration up to, and including D4. Bonn: Bundesinstitut für Arzneimittel und Medizinprodukte.

BHMA. 1983. *British herbal pharmacopoeia.* Consolidated ed. London: Scientific Committee of the British Herbal Medicine Association.

Blumenthal, M., and W. Busse. 1998. *The complete German Commission E monographs.* Austin, TX: American Botanical Council.

Bone, K. 1993/1994. Kava—A safe herbal treatment for anxiety. *Br. J. Phytother.* 3(4):147-153.

Brauer, R.B., R. Pfab, K. Becker, H. Berger, and M. Stangl. 2001. Fulminantes Leberversagen nach Einnahme des pflanzlichen Heilmittels Kava-Kava. *Z. Gastroenterol.* 39:491.

Brauer, R.B., M. Stangl, J.R. Stewart, R. Pfab, and K. Becker. 2003. Acute liver failure after administration of herbal tranquilizer kava-kava (*Piper methysticum*). *J. Clin. Psychiatr.* 64(2):216-218.

Bujanda, L., A. Palacios, R. Silvarino, A. Sanchez, and C. Munoz. 2002. Kava-induced acute icteric hepatitis. *Gastroenterol. Hepatol.* 25(6):434-435.

Campo, J.V., J. McNabb, J.M. Perel, et al. 2002. Kava-induced fulminant hepatic failure. *J. Am. Acad. Child Adolesc. Psychiatr.* 41(6):631-632.

Cartledge, A., and J. Rutherford. 2001. Kava and benzodiazepines—Worsening withdrawal [Rapid Response letter]. *BMJ.* http://www.bmj.com/rapid-response/2011/10/28/kava-and-benzodiazepines-worsening-withdrawal. Published August 31, 2001. Accessed September 7, 2012.

Chanwai, L.G. 2000. Kava toxicity. *Emerg. Med.* 12(2):142-145.

Clough, A.R., R.S. Bailie, and B. Currie. 2003. Liver function test abnormalities in users of aqueous kava extracts. *J. Toxicol. Clin. Toxicol.* 41(6):821-829.

Currie, B.J., and A.R. Clough. 2003. Kava hepatotoxicity with Western herbal products: Does it occur with traditional kava use? *Med. J. Aust.* 178(9):421-422.

Dentali, S.J. 1997. *Herb safety review: Kava, Piper methysticum Forster f. (Piperaceae).* Boulder, CO: Herb Research Foundation.

DiSilvestro, R.A., W. Zhang, and D.J. DiSilvestro. 2007. Kava feeding in rats does not cause liver injury nor enhance galactosamine-induced hepatitis. *Food Chem. Toxicol.* 45(7):1293-1300.

Foo, H., and J. Lemon. 1997. Acute effects of kava, alone or in combination with alcohol, on subjective measures of impairment and intoxication and on cognitive performance. *Drug Alcohol Rev.* 16(2):147-155.

Gow, P.J., N.J. Connelly, R.L. Hill, P. Crowley, and P.W. Angus. 2003. Fatal fulminant hepatic failure induced by a natural therapy containing kava. *Med. J. Aust.* 178(9):442-443.

Grace, R. 2005. Kava-induced urticaria. *J. Am. Acad. Dermatol.* 53(5):906.

Greenblatt, D.J., and C.E. Wright. 1993. Clinical pharmacokinetics of alprazolam. Therapeutic implications. *Clin. Pharmacokinet.* 24(6):453-471.

Guo, L., Q. Shi, S. Dial, et al. 2010. Gene expression profiling in male B6C3F1 mouse livers exposed to kava identifies changes in drug metabolizing genes and potential mechanisms linked to kava toxicity. *Food Chem. Toxicol.* 48(2):686-696.

Gurley, B.J., S.F. Gardner, M.A. Hubbard, et al. 2005. *In vivo* effects of goldenseal, kava kava, black cohosh, and valerian on human cytochrome P450 1A2, 2D6, 2E1, and 3A4/5 phenotypes. *Clin. Pharmacol. Ther.* 77(5):415-426.

Gurley, B.J., A. Swain, G.W. Barone, et al. 2007. Effect of goldenseal (*Hydrastis canadensis*) and kava kava (*Piper methysticum*) supplementation on digoxin pharmacokinetics in humans. *Drug Metab. Dispos.* 35(2):240-245.

Gurley, B.J., A. Swain, M.A. Hubbard, et al. 2008a. Supplementation with goldenseal (*Hydrastis canadensis*), but not kava kava (*Piper methysticum*), inhibits human CYP3A activity in vivo. *Clin. Pharmacol. Ther.* 83(1):61-69.

Gurley, B.J., A. Swain, M.A. Hubbard, et al. 2008b. Clinical assessment of CYP2D6-mediated herb-drug interactions in humans: Effects of milk thistle, black cohosh, goldenseal, kava kava, St. John's wort, and echinacea. *Mol. Nutr. Food Res.* 52(7):755.

Hansel, R., and H. Woelk. 1994. *Spektrum kava-kava.* Arzneimitteltherapie heute: Phytopharmaka Band 6, pp. 40-41. Basel: Aesopus Verlag.

Herberg, K. 1996. Safety-related performance after intake of kava-extract, bromazepam and their combination. *Ztschr. Allgemeinmed.* 72:973-977.

Herberg, K.W. 1993. Effect of kava-special extract WS 1490 combined with ethyl alcohol on safety-relevant performance parameters. *Blutalkohol* 30(2):96-105.

P

Humberston, C.L., J. Akhtar, and E.P. Krenzelok. 2003. Acute hepatitis induced by kava kava. *J. Toxicol. Clin. Toxicol.* 41(2):109-113.

Jamieson, D.D., and P.H. Duffield. 1990. Positive interaction of ethanol and kava resin in mice. *Clin. Exp. Pharmacol. Physiol.* 17(7):509-514.

Jappe, U., I. Franke, D. Reinhold, and H.P. Gollnick. 1998. Sebotropic drug reaction resulting from kava-kava extract therapy: A new entity? *J. Am. Acad. Dermatol.* 38(1):104-106.

Jhoo, J.W., J.P. Freeman, T.M. Heinze, et al. 2006. *In vitro* cytotoxicity of nonpolar constituents from different parts of kava plant (*Piper methysticum*). *J. Agric. Food Chem.* 54(8):3157-3162.

Leung, N. 2004. Acute urinary retention secondary to kava ingestion. *Emerg. Med. Australas.* 16(1):94.

Lewin, L. 1931. *Phantastica; narcotic and stimulating drugs, their use and abuse*. London: Paul, Trench, Trubner.

Lim, S.T., K. Dragull, C.S. Tang, et al. 2007. Effects of kava alkaloid, pipermethystine, and kavalactones on oxidative stress and cytochrome P450 in F-344 rats. *Toxicol. Sci.* 97(1):214-221.

Meseguer, E., R. Taboada, V. Sanchez, et al. 2002. Life-threatening parkinsonism induced by kava-kava. *Mov. Disord.* 17(1):195-196.

Mills, S., and K. Bone. 2005. *The essential guide to herbal safety*. St. Louis: Elsevier.

Musch, E., A. Chrissafidou, and M. Malek. 2006. Acute hepatitis due to kava-kava and St John's wort: An immune-mediated mechanism? *Dtsch. Med. Wochenschr.* 131(21):1214-1217.

Norton, S.A., and P. Ruze. 1994. Kava dermopathy. *J. Am. Acad. Dermatol.* 31(1):89-97.

Perez, J., and J.F. Holmes. 2005. Altered mental status and ataxia secondary to acute kava ingestion. *J. Emerg. Med.* 28(1):49-51.

Pfeiffer, C.C., H.B. Murphree, and L. Goldstein. 1967. Effect of kava in normal subjects and patients. *Psychopharmacol. Bull.* 4(3):12.

Pittler, M.H., and E. Ernst. 2003. Kava extract for treating anxiety. *Cochrane Database Syst. Rev.* 1:CD003383.

Rubin, D., B. McGovern, and R.I. Kopelman. 2006. Back to basics. *Am. J. Med.* 119(6):482-483.

Russmann, S., Y. Barguil, P. Cabalion, et al. 2003. Hepatic injury due to traditional aqueous extracts of kava root in New Caledonia. *Eur. J. Gastroenterol. Hepatol.* 15(9):1033-1036.

Russmann, S., B.H. Lauterburg, Y. Barguil, et al. 2005. Traditional aqueous kava extracts inhibit cytochrome P450 1A2 in humans: Protective effect against environmental carcinogens? *Clin. Pharmacol. Ther.* 77(5):453-454.

Russmann, S., B.H. Lauterburg, and A. Helbling. 2001. Kava hepatotoxicity. *Ann. Intern. Med.* 135(1):68-69.

Sass, M., S. Schnabel, J. Kröger, S. Liebe, and W. Schareck. 2001. Akutes Leberversagen durch Kava-Kava—Eine seltene Indikation zur Lebertransplantation. *Z. Gastroenterol.* 39:491.

Schelosky, L., C. Raffauf, K. Jendroska, and W. Poewe. 1995. Kava and dopamine antagonism. *J. Neurol. Neurosurg. Psychiatr.* 58(5):639-640.

Schmidt, M. 2007. Is kava really hepatotoxic? Accessed September 7, 2012: http://www.uni-muenster.de/imperia/md/content/pharmazeutische_biologie/_v/review.pdf.

Schmidt, M., M. Morgan, K. Bone, and J. McMillan. 2005. Kava: A risk benefit assessment. In *The essential guide to herbal safety*, edited by Mills, S. and K. Bone. St. Louis: Elsevier.

Schmidt, P., and W.H. Boehncke. 2000. Delayed-type hypersensitivity reaction to kava-kava extract. *Contact Dermat.* 42(6):363-364.

Singh, Y.N., and A.K. Devkota. 2003. Aqueous kava extracts do not affect liver function tests in rats. *Planta Med.* 69(6):496-499.

Sorrentino, L., A. Capasso, and M. Schmidt. 2006. Safety of ethanolic kava extract: Results of a study of chronic toxicity in rats. *Phytomedicine* 13(8):542-549.

Sparrow, B., M. Hejtmancik, M. Ryan, et al. 2004. Toxicity evaluation of kava kava extract in Fisher 344 rats and B6C3F1 mice following repeat dosing by oral gavage. *Toxicologist* 78(1 Suppl.):163.

Spillane, P.K., D.A. Fisher, and B.J. Currie. 1997. Neurological manifestations of kava intoxication. *Med. J. Aust.* 167(3):172-173.

Stoller, R. 2000. Leberschadigungen unter kava-extrakten. *Schweizerische Arztezeitung* 81(24):1335-1336.

Teschke, R., A. Genthner, and A. Wolff. 2009. Kava hepatotoxicity: Comparison of aqueous, ethanolic, acetonic kava extracts and kava-herbs mixtures. *J. Ethnopharmacol.* 123(3):378-384.

Teschke, R., A. Schwarzenboeck, and K.H. Hennermann. 2008. Kava hepatotoxicity: A clinical survey and critical analysis of 26 suspected cases. *Eur. J. Gastroenterol. Hepatol.* 20(12):1182.

TMEC. 2002. Response to concerns about *Piper methysticum* Forst. f., Kava: A submission prepared by the Traditional Medicines Evaluation Committee (TMEC), a submcommittee of the European Herbal Practitioners Association. Accessed September 7, 2012 at http://www.users.globalnet.co.uk/~ehpa/pdfs/kava11_01_02.pdf.

Waller, D. 2002. Report on kava and liver damage. Silver Spring, MD: American Herbal Products Association.

Yamazaki, Y., H. Hashida, A. Arita, K. Hamaguchi, and F. Shimura. 2008. High dose of commercial products of kava (*Piper methysticum*) markedly enhanced hepatic cytochrome P450 1A1 mRNA expression with liver enlargement in rats. *Food Chem. Toxicol.* 46(12):3732-3738.

P

Piper nigrum L.

Piperaceae

SCN: pepper
AN: *maricha*
PN: *hu jiao*
Part: fruit

Quick Reference Summary

Safety Class: 1
Interaction Class: B

Contraindications

None known.

Other Precautions

None known.

Drug and Supplement Interactions

See Pharmacological Considerations.

Notice

Contains piperine (5–9%) (Bhardwaj et al. 2002; Srinivasan 2007); *see* Appendix 3.

Alkenylbenzenes (safrole 0.01–0.09%) (Ames et al. 1990; Farag and Abo-Zeid 1997); *see* Appendix 1.

Editors' Note

The classifications and concerns for this herb are based on relatively higher doses used for therapeutic purposes in contrast to lower amounts generally used in cooking, and have not been associated with its use as a spice.

Adverse Events and Side Effects

Large doses of pepper may irritate the urinary tract. Overdoses of pepper may cause abdominal heat and burning, thirst, vomiting, fever, and sometimes convulsions (Felter and Lloyd 1898).

Pharmacological Considerations

The compound piperine has been shown to enhance the bioavailability of certain drugs (Hu et al. 2005; Srinivasan 2007; Wattanathorn et al. 2008). Some studies have indicated that coadministration of piperine with certain drugs allowed for a reduction in the required dose of the drugs (Hu et al. 2005; Srinivasan 2007; Wattanathorn et al. 2008).

An animal study suggested that an alcohol extract of pepper increased triiodothyronine (T_3) and thyroxine (T_4) levels, while an aqueous extract had no effects on T_3 and T_4 levels (Panda and Kar 2003).

Pregnancy and Lactation

Animal studies of pepper and the compound piperine have indicated some anti-implantation activity of these materials (Alkofahi et al. 1996; Daware et al. 2000). Another animal study indicated that piperine improved fertilization rates in artificially inseminated animals (Piyachaturawat and Pholpramool 1997).

No information on the safety of pepper during lactation was identified in the scientific or traditional literature. While this review did not identify any concerns for use while nursing, safety has not been conclusively established.

P

Review Details

I. Drug and Supplement Interactions

Clinical Trials of Drug or Supplement Interactions

No clinical trials of drug or supplement interactions were identified.

Case Reports of Suspected Drug or Supplement Interactions

No case reports of suspected drug or supplement interactions were identified.

Animal Trials of Drug or Supplement Interactions

No animal trials of drug or supplement interactions were identified.

II. Adverse Events

Case Reports of Adverse Events

A 4-year-old boy with a history of pica (a pattern of eating nonfood materials) experienced respiratory arrest, severe anoxia, and death after aspiration of an unknown amount of ground pepper (Sheahan et al. 1988).

Large doses of pepper may irritate the urinary tract. Overdoses of pepper may cause abdominal heat and burning, thirst, vomiting, fever, and sometimes convulsions (Felter and Lloyd 1898).

Urticaria has been associated with pepper use (Felter and Lloyd 1898).

III. Pharmacology and Pharmacokinetics

Human Pharmacological Studies

No relevant human pharmacological studies were identified.

Animal Pharmacological Studies

In mice orally administered 4 mg/kg of water or alcohol extracts of pepper daily for 15 days, no adverse effects on liver or thyroid functions were observed in the group

Plantago spp. — Plantaginaceae

Plantago lanceolata **L.**
SCN: English plantain
OCN: lance-leaf plantain; narrow-leaf plantain; ribgrass

Plantago major **L.**
SCN: plantain
PN: *che qian cao* (whole plant)
OCN: broad-leaf plantain; greater plantain

Plantago media **L.**
SCN: hoary plantain
OCN: plantain
Part: leaf

Quick Reference Summary

Safety Class: 1
Interaction Class: A

Contraindications

None known.

Other Precautions

None known.

Drug and Supplement Interactions

None known.

Editors' Note

Instances of adulteration of *P. lanceolata* leaves with those of *Digitalis lanata* have been rarely reported (Slifman et al. 1998; Whitmore 1997).

Adverse Events and Side Effects

None known.

Pharmacological Considerations

None known.

Pregnancy and Lactation

No information on the safety of plantain, English plantain, or hoary plantain leaf in pregnancy or lactation was identified in the scientific or traditional literature. Although this review did not identify any concerns for use while pregnant or nursing, safety has not been conclusively established.

Review Details

I. Drug and Supplement Interactions

Clinical Trials of Drug or Supplement Interactions

No clinical trials of drug or supplement interactions were identified.

Case Reports of Suspected Drug or Supplement Interactions

No case reports of suspected drug or supplement interactions were identified.

Animal Trials of Drug or Supplement Interactions

No animal trials of drug or supplement interactions were identified.

II. Adverse Events

Adverse Events Reported in Clinical Trials

No adverse effects on the gastrointestinal tract, liver, kidneys, or hemopoiesis were observed in bronchitis patients administered a plantain leaf preparation for 25 days (Matev et al. 1982).

Case Reports of Adverse Events

No case reports of adverse events were identified.

III. Pharmacology and Pharmacokinetics

Human Pharmacological Studies

No relevant human pharmacological studies were identified.

Animal Pharmacological Studies

No relevant animal pharmacological studies were identified.

In Vitro Pharmacological Studies

No relevant in vitro pharmacological studies were identified.

IV. Pregnancy and Lactation

No information on the safety of plantain leaf use during pregnancy or lactation was identified.

V. Toxicity Studies

Acute Toxicity

The LD_{50} of an orally administered plantain leaf extract in rats could not be determined at doses up to 4 g/kg, whereas the LD_{50} after intraperitoneal administration is 1 g/kg (Angelov et al. 1980).

An ethanol extract of plantain leaf was found to be toxic in the brine shrimp assay (Schmeda-Hirschmann et al. 1992).

Genotoxicity

Assays of the genotoxicity of plantain leaf extracts have provided mixed results. In the Ames test, a decoction of plantain had some mutagenic activity (Lim-Sylianco and Shier 1985), while a saline extract showed no mutagenic activity (Basaran et al. 1996). No toxicity of an alcohol extract of plantain leaf was shown in the plate incorporation assay with *Aspergillus nidulans* (Ruiz et al. 1996).

Literature Cited

Angelova, A., I. Lambev, M. Markov, et al. 1980. Study of acute and chronical toxicity of dispergue of *Plantago major*. *Med. Arch.* 18:47-52.

Basaran, A.A., T.W. Yu, M.J. Plewa, and D. Anderson. 1996. An investigation of some Turkish herbal medicines in *Salmonella typhimurium* and in the COMET assay in human lymphocytes. *Teratogen. Carcinogen. Mutagen.* 16:125-138.

Lim-Sylianco, C.Y., and W.T. Shier. 1985. Mutagenic and antimutagenic activities in Philippine medicinal and food plants. *J. Toxicol. Toxin Rev.* 4:71-105.

Matev, M., I. Angelova, A. Koichev, M. Leseva, and G. Stefanov. 1982. Clinical trial of a *Plantago major* preparation in the treatment of chronic bronchitis. *Vutr. Boles.* 21(2):133-137.

Ruiz, A.R., R.A. De la Torre, N. Alonso, et al. 1996. Screening of medicinal plants for induction of somatic segregation activity in *Aspergillus nidulans*. *J. Ethnopharmacol.* 52:123-127.

Schmeda-Hirschmann, G., J.I. Loyola, S.R. Retamal, and J. Rodriguez. 1992. Hypotensive effect and enzyme inhibition activity of Mapuche medicinal plant extracts. *Phytother. Res.* 6:184-188.

Slifman, N.R., W.R. Obermeyer, B.K. Aloi, et al. 1998. Contamination of botanical dietary supplements by *Digitalis lanata*. *N. Engl. J. Med.* 339(12):806-811.

Whitmore, A. 1997. FDA warns consumers against dietary supplement products that may contain digitalis mislabeled as "plantain." U.S. Food and Drug Administration.

Plantago spp.

Plantaginaceae

Plantago afra L.
SCN: *Plantago afra*
Syn: *Plantago psyllium* auct.
OCN: African psyllium (seed); black psyllium (seed)

Plantago arenaria **Waldst. & Kit.**
SCN: *Plantago arenaria*
Syn: *Plantago psyllium* L.
OCN: black psyllium (seed); French psyllium (seed); Spanish psyllium (seed)

Plantago asiatica **L.**
SCN: Asian plantain
PN: *che qian zi* (seed)
OCN: Asian psyllium (seed)

Plantago ovata Forssk.
SCN: Indian plantain
Syn: *Plantago ispaghula* Roxb. ex Fleming
OCN: blonde psyllium (seed); Indian psyllium; ispaghula (seed)
Note: The SCN of the seed of *P. afra*, *P. arenaria*, *P. asiatica*, and *P. ovata* is psyllium.
Part: seed, seed husk

P

Quick Reference Summary

Safety Class: 2d
Interaction Class: A

Contraindications

Contraindicated in bowel obstruction, esophageal stenosis, and abnormal intestinal narrowing (Bradley 1992; Brinker 2001; Wichtl 2004).

Other Precautions

Take with at least 250 ml (8 oz.) of liquid (CFR 2011; Wichtl 2004).

Drug and Supplement Interactions

Other drugs should be taken 1 hour prior to consumption of psyllium or several hours after consumption, as psyllium may reduce the absorption of certain drugs due to the mucilage content and the increased speed of passage through the intestines (Bradley 1992; Brinker 2001); *see* Mucilages in Appendix 3.

Notice

Bulk-forming laxative (Bradley 1992; Leung and Foster 1996; Martindale and Reynolds 1996; Wichtl 2004; Williamson 2003); *see* Appendix 2.

Mucilages (10–30%) (Leung and Foster 1996; Wichtl 2004); *see* Appendix 3.

Standard Dose

The standard adult dose is 2.5–7.5 g up to 3 times daily (Bradley 1992; Federal Register 1986).

Editors' Note

Specific labeling is required in the United States for all over-the-counter drug products containing psyllium (CFR 2011a, Federal Register 1986), and psyllium in granular form is not permitted in these products (CFR 2011b). *See* Bulk-forming laxatives in Appendix 2.

Adverse Events and Side Effects

Allergic sensitivity has been reported for psyllium products (Drake et al. 1991; Ford et al. 1992; Freeman 1994; Gauss et al. 1985; Hoffman 2006; Khalili et al. 2003; Lantner et al. 1990; Machado et al. 1979; Pozner et al. 1986; Schwartz et al. 1989; Scott 1987; Seggev et al. 1984; Suhonen et al. 1983; Sussman and Dorian 1990; Terho ja Martti Torkko 1980; Vaswani et al. 1996; Zaloga et al. 1984). Such sensitivity is most common among healthcare workers who have had repeated respiratory exposure to airborne psyllium powder (Freeman 1994).

Pharmacological Considerations

Human studies have demonstrated that psyllium may modify glucose regulation. People with diabetes are advised to monitor their blood sugar closely and discuss the use of this herb with a qualified healthcare practitioner prior to use (Anderson et al. 1999; Frati Munari et al. 1998; Pastors et al. 1991).

Pregnancy and Lactation

Over-the-counter psyllium products are categorized as an FDA category B, indicating that the products are generally considered safe during pregnancy although no adequate and well-controlled studies in pregnant humans have been completed (PDR 2006). Psyllium is generally considered to be safe for use by nursing mothers (Ulbricht and Basch 2005).

Review Details

I. Drug and Supplement Interactions

Clinical Trials of Drug or Supplement Interactions

No relevant clinical trials of drug or supplement interactions were identified.

Case Reports of Suspected Drug or Supplement Interactions

The addition of psyllium husk to the daily regimen of a woman taking lithium was associated with decreased blood levels of lithium (Perlman 1990).

Animal Trials of Drug or Supplement Interactions

No relevant animal trials of drug or supplement interactions were identified.

II. Adverse Events

Adverse Events Reported in Clinical Trials

Meta-analyses of psyllium preparations have not reported any adverse events associated with clinical trials of psyllium (Anderson et al. 2000; Brown et al. 1999; Olson et al. 1997).

Case Reports of Adverse Events

The bulking action of psyllium has resulted in gastrointestinal tract and, in rare occasions, respiratory tract obstructions (Angueira and Kadakia 1993; Berman and Schultz 1980; Frohna 1992; Herrle et al. 2004; Noble and Grannis 1984; Sauerbruch et al. 1980; Schapira et al. 1995; Schneider 1989). Some cases of obstructions have been reported in persons with some type of narrowing or restriction of the digestive tract (Angueira and Kadakia 1993; Herrle et al. 2004).

Reports of allergic reactions to psyllium included anaphylaxis, rashes, and asthma (Drake et al. 1991; Ford et al. 1992; Freeman 1994; Gauss et al. 1985; Hoffman 2006; Khalili et al. 2003; Lantner et al. 1990; Machado et al. 1979; Pozner et al. 1986; Schwartz et al. 1989; Scott 1987; Seggev et al. 1984; Suhonen et al. 1983; Sussman and Dorian 1990; Terho ja Martti Torkko 1980; Vaswani et al. 1996; Zaloga et al. 1984). Such reactions occur most commonly in healthcare workers, as sensitization to psyllium is thought to be caused by repeated respiratory exposure to powdered psyllium (Freeman 1994).

III. Pharmacology and Pharmacokinetics

Human Pharmacological Studies

In type 2 diabetes mellitus and non-insulin-dependent diabetes mellitus (NIDDM) subjects, administration of single doses of psyllium before a meal significantly reduced post-meal glucose levels (Anderson et al. 1999; Frati Munari et al. 1998; Pastors et al. 1991).

In NIDDM subjects, administration of 5.1 g psyllium twice daily for 8 weeks significantly reduced proximate and postmeal glucose levels (Pastors et al. 1991).

Animal Pharmacological Studies

No relevant animal pharmacological studies were identified.

In Vitro Pharmacological Studies

No relevant in vitro pharmacological studies were identified.

IV. Pregnancy and Lactation

Over-the-counter psyllium products are categorized as an FDA category B, indicating that the products are generally considered safe during pregnancy, although no adequate and well-controlled studies in pregnant humans have been completed (PDR 2006).

No evidence of adverse effects on reproduction was reported in rats or mice fed psyllium husk as 0, 1.25, or 5% of the diet for three generations (EMEA 2007). No evidence of adverse effects was observed in rabbits fed psyllium husk as

0, 2.5 5, or 10% of the diet on days 2 through 20 of gestation (EMEA 2007).

Psyllium-containing products are considered generally safe for use by nursing mothers (Ulbricht and Basch 2005).

V. Toxicity Studies

Acute Toxicity

No LD_{50} of orally administered psyllium husk could be determined at doses up to 2940 mg/kg in mice or in rats at doses up to 3360 mg/kg (EMEA 2007).

Short-Term and Subchronic Toxicity

In rats fed psyllium husk as 10% of the diet for 28 days or 13 weeks, with psyllium consumption ranging from 3876 to 11,809 mg/kg daily, lower serum total protein, albumin, globulin, total iron-binding capacity, calcium, potassium, and cholesterol, and higher aspartate transaminase (AST) and alanine transaminase (ALT) activities relative to control were observed (EMEA 2007). The reasons for the lower serum total protein, albumin, and globulin were not clear, but the absence of any increase in urinary protein, any evidence of gastrointestinal pathology which could account for protein loss, and any differences in growth or feed efficiency in psyllium fed rats suggests that psyllium does not adversely affect protein metabolism (EMEA 2007).

Literature Cited

Anderson, J.W., L.D. Allgood, A. Lawrence, et al. 2000. Cholesterol-lowering effects of psyllium intake adjunctive to diet therapy in men and women with hypercholesterolemia: Meta-analysis of 8 controlled trials. *Am. J. Clin. Nutr.* 71(2):472-479.

Anderson, J.W., L.D. Allgood, J. Turner, P.R. Oeltgen, and B.P. Daggy. 1999. Effects of psyllium on glucose and serum lipid responses in men with type 2 diabetes and hypercholesterolemia. *Am. J. Clin. Nutr.* 70(4):466-473.

Angueira, C., and S. Kadakia. 1993. Esophageal and duodenal bezoars from Perdiem. *Gastrointest. Endosc.* 39(1):110-111.

Berman, J.I., and M.J. Schultz. 1980. Bulk laxative ileus. *J. Am. Geriatr. Soc.* 28(5):224-226.

Bradley, P.R. 1992. *British herbal compendium: A handbook of scientific information on widely used plant drugs*. Bournemouth, UK: British Herbal Medicine Association.

Brinker, F. 2001. *Herb contraindications and drug interactions*. 3rd ed. Sandy, OR: Eclectic Medical Publications.

Brown, L., B. Rosner, W.W. Willett, and F.M. Sacks. 1999. Cholesterol-lowering effects of dietary fiber: A meta-analysis. *Am. J. Clin. Nutr.* 69(1):30-42.

CFR. 2011a. *Code of federal regulations*, Title 21 Part 201.319, 2011 ed. Specific labeling requirements for specific drug products. Water-soluble gums, hydrophilic gums, and hydrophilic mucilloids (including, but not limited to, agar, alginic acid, calcium polycarbophil, carboxymethylcellulose sodium, carrageenan, chondrus, glucomannan ((B-1,4 linked) polymannose acetate), guar gum, karaya gum, kelp, methylcellulose, plantago seed (psyllium), polycarbophil tragacanth, and xanthan gum) as active ingredients; required warnings and directions. Washington, DC: U.S. Government Printing Office.

CFR. 2011b. *Code of federal regulations*, Title 21 Part 310.545, 2011 ed. Requirements for specific new drugs or devices. Drug products containing certain active ingredients offered over-the-counter (OTC) for certain uses. Bulk laxatives-Approved as of March 29, 2007. Washington, DC: U.S. Government Printing Office.

Drake, C.L., E.S. Moses, and D. Tandberg. 1991. Systemic anaphylaxis after ingestion of a psyllium-containing breakfast cereal. *Am. J. Emerg. Med.* 9(5):449-451.

EMEA. 2007. European Medicines Agency. Overview of comments received on 'Community Herbal Monograph on *Plantago afra* L. et *Plantago indica* L., semen' (emea/hmpc/340865/2005). Doc. Ref. EMEA/HMPC/65063/2006.

Federal Register. 1986. 51 FR 35137. (October 1).

Ford, M.A., G. Cristea, Jr., W.D. Robbins, et al. 1992. Delayed psyllium allergy in three nurses. *Hosp. Pharm.* 27(12):1061-1062.

Frati Munari, A.C., W. Benitez Pinto, C. Raul Ariza Andraca, and M. Casarrubias. 1998. Lowering glycemic index of food by acarbose and *Plantago psyllium* mucilage. *Arch. Med. Res.* 29(2):137-141.

Freeman, G.L. 1994. Psyllium hypersensitivity. *Ann. Allergy* 73(6):490-492.

Frohna, W.J. 1992. Metamucil bezoar: An unusual cause of small bowel obstruction. *Am. J. Emerg. Med.* 10(4):393-395.

Gauss, W.F., J.P. Alarie, and M.H. Karol. 1985. Workplace allergenicity of a psyllium-containing bulk laxative. *Allergy* 40(1):73-76.

Herrle, F., T. Peters, C. Lang, et al. 2004. Bolus obstruction of pouch outlet by a granular bulk laxative after gastric banding. *Obes. Surg.* 14(7):1022-1024.

Hoffman, D. 2006. Psyllium: Keeping this boon for patients from becoming a bane for providers. *J. Fam. Pract.* 55(9):770-772.

Khalili, B., E.J. Bardana, Jr., and J.W. Yunginger. 2003. Psyllium-associated anaphylaxis and death: A case report and review of the literature. *Ann. Allergy Asthma Immunol.* 91(6):579-584.

Lantner, R.R., B.R. Espiritu, P. Zumerchik, and M.C. Tobin. 1990. Anaphylaxis following ingestion of a psyllium-containing cereal. *J. Am Med. Assoc.* 264(19):2534-2536.

Leung, A.Y., and S. Foster. 1996. *Encyclopedia of common natural ingredients used in food, drugs, and cosmetics*. 2nd ed. New York: Wiley.

Machado, L., O. Zetterstrom, and E. Fagerberg. 1979. Occupational allergy in nurses to a bulk laxative. *Allergy* 34(1):51-55.

Martindale, W., and J.E.F. Reynolds. 1996. *The extra pharmacopoeia*. 31st ed. London: Pharmaceutical Press.

Noble, J.A., and F.W. Grannis, Jr. 1984. Acute esophageal obstruction by a psyllium-based bulk laxative. *Chest* 86(5):800.

Olson, B.H., S.M. Anderson, M.P. Becker, et al. 1997. Psyllium-enriched cereals lower blood total cholesterol and LDL cholesterol, but not HDL cholesterol, in hypercholesterolemic adults: Results of a meta-analysis. *J. Nutr.* 127(10):1973-1980.

Pastors, J.G., P.W. Blaisdell, T.K. Balm, C.M. Asplin, and S.L. Pohl. 1991. Psyllium fiber reduces rise in postprandial glucose and insulin concentrations in patients with non-insulin-dependent diabetes. *Am. J. Clin. Nutr.* 53(6):1431-1435.

P

PDR. 2006. *PDR for nonprescription drugs, dietary supplements, and herbs: The definitive guide to OTC medications.* 26th ed. Montvale, NJ: Thomson PDR.

Perlman, B.B. 1990. Interaction between lithium salts and ispaghula husk. *Lancet* 335(8686):416.

Pozner, L.H., C. Mandarano, M.J. Zitt, M. Frieri, and N.S. Weiss. 1986. Recurrent bronchospasm in a nurse. *Ann. Allergy* 56(1):14-15, 44-47.

Sauerbruch, T., O. Kuntzen, and W. Unger. 1980. Agiolax bolus in the esophagus. Report of two cases. *Endoscopy* 12(2):83-85.

Schapira, M., J. Henrion, P. Jonard, et al. 1995. Esophageal bezoar: Report of five more cases. *Endoscopy* 27(4):342.

Schneider, R.P. 1989. Perdiem causes esophageal impaction and bezoars. *South. Med. J.* 82(11):1449-1450.

Schwartz, H.J., J.L. Arnold, and K.P. Strohl. 1989. Occupational allergic rhinitis reaction to psyllium. *J. Occup. Med.* 31(7):624-626.

Scott, D. 1987. Psyllium-induced asthma. Occupational exposure in a nurse. *Postgrad. Med.* 82(8):160-161.

Seggev, J.S., K. Ohta, and W.R. Tipton. 1984. IgE mediated anaphylaxis due to a psyllium-containing drug. *Ann. Allergy* 53(4):325-326.

Suhonen, R., I. Kantola, and F. Bjorksten. 1983. Anaphylactic shock due to ingestion of psyllium laxative. *Allergy* 38(5):363-365.

Sussman, G.L., and W. Dorian. 1990. Psyllium anaphylaxis. *Allergy Proc.* 11(5):241-242.

Terho ja Martti Torkko, E.O. 1980. Occupational asthma from psyllium laxatives. *Duodecim* 96(18):1213-1216.

Ulbricht, C.E., and E.M. Basch. 2005. *Natural standard herb & supplement reference.* St. Louis: Elsevier Mosby.

Vaswani, S.K., R.G. Hamilton, M.D. Valentine, and N.F. Adkinson, Jr. 1996. Psyllium laxative-induced anaphylaxis, asthma, and rhinitis. *Allergy* 51(4):266-268.

Wichtl, M. 2004. *Herbal drugs and phytopharmaceuticals: A handbook for practice on a scientific basis.* 3rd ed. Boca Raton, FL: CRC Press.

Williamson, E.M. 2003. *Potter's herbal cyclopedia.* Saffron Walden, Essex: C.W. Daniel Co.

Zaloga, G.P., U.R. Hierlwimmer, and R.J. Engler. 1984. Anaphylaxis following psyllium ingestion. *J. Allergy Clin. Immunol.* 74(1):79-80.

Platycladus orientalis (L.) Franco

Cupressaceae

SCN: oriental arborvitae
Syn: *Biota orientalis* (L.) Endl.; *Thuja orientalis* L.

PN: *ce bai ye* (leaf and branch tip)
Part: cacumen (leafy twig)

Quick Reference Summary

Safety Class: 2d
Interaction Class: A

Contraindications

Not for long-term use; do not exceed recommended dose (Bensky et al. 2004; Chen and Chen 2004).

Other Precautions

None known.

Drug and Supplement Interactions

None known.

Standard Dose

The standard dose is 6 to 15 g, raw or charred, daily as a tea (Bensky et al. 2004; Chen and Chen 2004).

Notice

Thujone (50–60% of the essential oil) (Bensky et al. 2004; Chizzola et al. 2004); *see* Appendix 1.

Adverse Events and Side Effects

Allergic reactions to oriental arborvitae twig have been reported (Bensky et al. 2004).

Pharmacological Considerations

None known.

Pregnancy and Lactation

No information on the safety of oriental arborvitae twig in pregnancy or lactation was identified in the scientific or traditional literature. Although this review did not identify any concerns for use while pregnant or nursing, safety has not been conclusively established.

Review Details

I. Drug and Supplement Interactions

Clinical Trials of Drug or Supplement Interactions

No clinical trials of drug or supplement interactions were identified.

Case Reports of Suspected Drug or Supplement Interactions

No case reports of suspected drug or supplement interactions were identified.

Animal Trials of Drug or Supplement Interactions

No animal trials of drug or supplement interactions were identified.

II. Adverse Events

Case Reports of Adverse Events

Overdose (standard dose is listed as a decoction of 6–15 g) or long-term use of oriental arborvitae twig has been associated with gastric discomfort, nausea, dizziness, and decreased appetite. In severe cases of overdose, vomiting of blood may occur (Bensky et al. 2004; Chen and Chen 2004).

Allergic reactions to oriental arborvitae twig have been reported (Bensky et al. 2004).

III. Pharmacology and Pharmacokinetics

Human Pharmacological Studies

No relevant human pharmacological studies were identified.

Animal Pharmacological Studies

No relevant animal pharmacological studies were identified.

In Vitro Pharmacological Studies

In assays with rabbit platelets, an aqueous extract of oriental arborvitae twig was found to be a platelet-activating factor (PAF) receptor binding antagonist (Yang et al. 1995). Related studies indicated that the compound pinusolide is responsible for this activity (Han et al. 1998; Yang and Han 1998). Pinusolide inhibited ^{3}H-serotonin release from rabbit platelets induced by PAF, but showed no effect on ^{3}H-serotonin release induced by ADP, collagen, or thrombin (Kim et al. 1999).

IV. Pregnancy and Lactation

No information on the safety of oriental arborvitae twig during pregnancy or lactation was identified.

V. Toxicity Studies

Acute Toxicity

No adverse effects were observed in mice orally administered up to 60 g/kg of a decoction of oriental arborvitae twig (Chen and Chen 2004). The LD_{50} of a decoction of oriental arborvitae twig intraperitoneally administered in mice is 15.2 g/kg (Chen and Chen 2004; Zhu 1998).

Short-Term Toxicity

Reduced appetite and activity were observed in rats orally administered up to 48 g/kg of oriental arborvitae twig daily for 6 weeks (Zhu 1998).

Literature Cited

Bensky, D., S. Clavey, and E. Stöger. 2004. *Chinese herbal medicine: Materia medica*. 3rd ed. Seattle: Eastland Press.

Chen, J.K., and T.T. Chen. 2004. *Chinese medical herbology and pharmacology*. City of Industry, CA: Art of Medicine Press.

Chizzola, R., W. Hochsteiner, and S. Hajek. 2004. GC analysis of essential oils in the rumen fluid after incubation of *Thuja orientalis* twigs in the Rusitec system. *Res. Vet. Sci.* 76(1):77-82.

Han, B.H., H.O. Yang, Y.H. Kang, et al. 1998. *In vitro* platelet-activating factor receptor binding inhibitory activity of pinusolide derivatives: A structure-activity study. *J. Med. Chem.* 41(14):2626-2630.

Kim, K.A., T.C. Moon, S.W. Lee, et al. 1999. Pinusolide from the leaves of *Biota orientalis* as potent platelet activating factor antagonist. *Planta Med.* 65(1):39-42.

Yang, H.O., and B.H. Han. 1998. Pinusolidic acid: A platelet-activating factor inhibitor from *Biota orientalis*. *Planta Med.* 64(1):72-74.

Yang, H.O., D.Y. Suh, and B.H. Han. 1995. Isolation and characterization of platelet-activating factor receptor binding antagonists from *Biota orientalis*. *Planta Med.* 61(1):37-40.

Zhu, Y.-P. 1998. *Chinese materia medica: Chemistry, pharmacology and applications.* Amsterdam: Harwood Academic Publishers.

P

Platycladus orientalis (L.) Franco

Cupressaceae

SCN: oriental arborvitae
Syn: *Biota orientalis* (L.) Endl.; *Thuja orientalis* L.

PN: *bai zi ren* (seed)
Part: seed

Quick Reference Summary

Safety Class: 1
Interaction Class: A

Contraindications

None known.

Other Precautions

Use with caution in persons with loose stools (Bensky et al. 2004; Chen and Chen 2004).

Drug and Supplement Interactions

None known.

Editors' Note

Contamination by aflatoxins, toxic compounds produced by fungi that grow in grains, nuts, and other products, has been reported in oriental arborvitae seed (Bensky et al. 2004). Proper processing and storage can prevent aflatoxin formation (Kabak et al. 2006).

Adverse Events and Side Effects

None known.

Pharmacological Considerations

None known.

Pregnancy and Lactation

No information on the safety of oriental arborvitae seed in pregnancy or lactation was identified in the scientific or traditional literature. Although this review did not identify any concerns for use while pregnant or nursing, safety has not been conclusively established.

Review Details

I. Drug and Supplement Interactions

Clinical Trials of Drug or Supplement Interactions

No clinical trials of drug or supplement interactions were identified.

Case Reports of Suspected Drug or Supplement Interactions

No case reports of suspected drug or supplement interactions were identified.

Animal Trials of Drug or Supplement Interactions

No animal trials of drug or supplement interactions were identified.

II. Adverse Events

Case Reports of Adverse Events

No case reports of adverse events were identified.

III. Pharmacology and Pharmacokinetics

Human Pharmacological Studies

No relevant human pharmacological studies were identified.

Animal Pharmacological Studies

No relevant animal pharmacological studies were identified.

In Vitro Pharmacological Studies

No relevant in vitro pharmacological studies were identified.

IV. Pregnancy and Lactation

No information on the safety of oriental arborvitae seed during pregnancy or lactation was identified.

V. Toxicity Studies

No toxicity studies were identified.

Literature Cited

Bensky, D., S. Clavey, and E. Stöger. 2004. *Chinese herbal medicine: Materia medica*. 3rd ed. Seattle: Eastland Press.

Chen, J.K., and T.T. Chen. 2004. *Chinese medical herbology and pharmacology*. City of Industry, CA: Art of Medicine Press.

Kabak, B., A.D.W. Dobson, and I. Var. 2006. Strategies to prevent mycotoxin contamination of food and animal feed: A review. *Crit. Rev. Food Sci. Nutr.* 46(8):593-619.

Platycodon grandiflorum (Jacq.) A. DC. — Campanulaceae

SCN: platycodon
PN: *jie geng* (root)
OCN: balloon flower; Chinese bellflower; Japanese bellflower
Part: root

Quick Reference Summary

Safety Class: 2d
Interaction Class: A

Contraindications

Not for use in persons with a tendency to cough up blood (Bensky et al. 2004; Chen and Chen 2004).

Other Precautions

None known.

Drug and Supplement Interactions

None known.

Adverse Events and Side Effects

None known.

Pharmacological Considerations

Animal studies have demonstrated that platycodon may modify glucose regulation (Kwon et al. 2009; Zheng et al. 2007). People with diabetes are advised to monitor their blood sugar closely and discuss the use of this herb with a qualified healthcare practitioner prior to use.

Pregnancy and Lactation

No information on the safety of platycodon in pregnancy or lactation was identified in the scientific or traditional literature. Although this review did not identify any concerns for use while pregnant or nursing, safety has not been conclusively established.

Review Details

I. Drug and Supplement Interactions

Clinical Trials of Drug or Supplement Interactions

No clinical trials of drug or supplement interactions were identified.

Case Reports of Suspected Drug or Supplement Interactions

No case reports of suspected drug or supplement interactions were identified.

Animal Trials of Drug or Supplement Interactions

No animal trials of drug or supplement interactions were identified.

II. Adverse Events

Case Reports of Adverse Events

No adverse effects have been reported within the normal dosage range (decoction of 3–9 g) of platycodon. Overdose may cause nausea and vomiting and, in severe cases, sweating on the limbs, exhaustion, and restlessness (Bensky et al. 2004).

III. Pharmacology and Pharmacokinetics

Human Pharmacological Studies

No relevant human pharmacological studies were identified.

Animal Pharmacological Studies

In diabetic mice orally administered 2 g/kg of platycodon or 0.2 g/kg of saponins from platycodon daily for 8 weeks, improvement of glucose homeostasis was observed. The saponin fraction was more active than the crude extract (Kwon et al. 2009).

In diabetic mice orally administered 150 or 300 mg/kg of an aqueous-ethanol extract of platycodon, a decrease in serum glucose levels was observed after a single dose and after daily dosing for 4 weeks. No changes in serum insulin levels were observed (Zheng et al. 2007).

Several animal studies have indicated that extracts of platycodon or saponins from platycodon prevented liver injury and changes in liver enzyme levels induced by ethanol or carbon tetrachloride. These studies indicated that inhibition of the drug-metabolizing isoenzyme CYP2E1 was responsible for this protection. Doses used in these studies were 10 to 100 mg/kg of an aqueous extract or 0.5 to 2 mg/kg of saponins (Khanal et al. 2009; Kim et al. 2007; Lee et al. 2008; Lee and Jeong 2002).

In Vitro Pharmacological Studies

No relevant in vitro pharmacological studies were identified.

IV. Pregnancy and Lactation

No information on the safety of platycodon during pregnancy or lactation was identified.

V. Toxicity Studies

Acute Toxicity

The LD_{50} of orally administered platycodon decoction in mice is 24 g/kg. The oral LD_{50} of the saponin fraction of platycodon is 420 mg/kg in mice and over 800 mg/kg in rats (Zhu 1998).

Genotoxicity

No mutagenic activity of an aqueous extract of platycodon was observed in the Ames mutagenicity assay with *Salmonella typhimurium* strains TA98 or TA100 with or without metabolic activation by S9 (Yin et al. 1991).

In the mouse micronucleus and chromosomal aberration assays, dose-dependent increases in polychromatic erythrocytes and in the incidences of chromosomal aberrations were observed in mice intraperitoneally administered an aqueous extract of platycodon at doses of 0.25, 0.5, 1.0, or 2.0 g/kg (Yin et al. 1991).

Literature Cited

Bensky, D., S. Clavey, and E. Stöger. 2004. *Chinese herbal medicine: Materia medica*. 3rd ed. Seattle: Eastland Press.

Chen, J.K., and T.T. Chen. 2004. *Chinese medical herbology and pharmacology*. City of Industry, CA: Art of Medicine Press.

Khanal, T., J.H. Choi, Y.P. Hwang, Y.C. Chung, and H.G. Jeong. 2009. Saponins isolated from the root of *Platycodon grandiflorum* protect against acute ethanol-induced hepatotoxicity in mice. *Food Chem. Toxicol.* 47(3):530-535.

Kim, H.K., D.S. Kim, and H.Y. Cho. 2007. Protective effects of Platycodi radix on alcohol-induced fatty liver. *Biosci. Biotechnol. Biochem.* 71(6):1550-1552.

Kwon, D.Y., Y.S. Kim, S.M. Hong, and S. Park. 2009. Long-term consumption of saponins derived from Platycodi radix (22 years old) enhances hepatic insulin sensitivity and glucose-stimulated insulin secretion in 90% pancreatectomized diabetic rats fed a high-fat diet. *Br. J. Nutr.* 101(3):358-366.

Lee, K.J., J.H. Choi, H.G. Kim, et al. 2008. Protective effect of saponins derived from the roots of *Platycodon grandiflorum* against carbon tetrachloride induced hepatotoxicity in mice. *Food Chem. Toxicol.* 46(5):1778-1785.

Lee, K.J., and H.G. Jeong. 2002. Protective effect of Platycodi radix on carbon tetrachloride-induced hepatotoxicity. *Food Chem. Toxicol.* 40(4):517-525.

Yin, X.J., D.X. Liu, H.C. Wang, and Y. Zhou. 1991. A study on the mutagenicity of 102 raw pharmaceuticals used in Chinese traditional medicine. *Mutat. Res.* 260(1):73-82.

Zheng, J., J. He, B. Ji, Y. Li, and X. Zhang. 2007. Antihyperglycemic effects of *Platycodon grandiflorum* (Jacq.) A. DC. extract on streptozotocin-induced diabetic mice. *Plant Foods Hum. Nutr.* 62(1):7-11.

Zhu, Y.-P. 1998. *Chinese materia medica: Chemistry, pharmacology and applications.* Amsterdam: Harwood Academic Publishers.

Podophyllum hexandrum Royle — Berberidaceae

SCN: Himalayan mayapple
Syn: *Podophyllum emodi* Wall. ex Hook. f. & Thomson
PN: *tao er qi* (root and rhizome)
Part: root and rhizome

Quick Reference Summary

Safety Class: 3
Interaction Class: A

Contraindications

Not for use except under the supervision of an expert qualified in the appropriate use of this substance (Didcock et al. 1952; Longstaff and von Krogh 2001).

Other Precautions

None known.

Drug and Supplement Interactions

None known.

Notice

Abortifacient (Didcock et al. 1952; Longstaff and von Krogh 2001); *see* Appendix 2.

Emetic (Felter and Lloyd 1898); *see* Appendix 2.

Stimulant laxative (Felter and Lloyd 1898); *see* Appendix 2.

Adverse Events and Side Effects

Toxicity has been reported after topical and internal administration of products (podophyllin and podophyllotoxin) derived from Himalayan mayapple and included nausea, vomiting, peripheral neuropathy, difficulty breathing, lethargy, and coma. Some cases of toxicity have been fatal (Cassidy et al. 1982; McFarland and McFarland 1981).

Pharmacological Considerations

None known.

Pregnancy and Lactation

Case reports and animal studies have indicated that compounds from Himalayan mayapple have abortifacient activity (Chamberlain et al. 1972; Didcock et al. 1952; Longstaff and von Krogh 2001). A safety assessment of the compounds podophyllotoxin and podophyllin indicated that the topical use of these compounds during pregnancy is strongly contraindicated (Longstaff and von Krogh 2001).

An animal study with aqueous extract of Himalayan mayapple indicated no adverse effects during pregnancy (Sajikumar and Goel 2003).

No information on the safety of Himalayan mayapple in lactation was identified. While this review did not identify any concerns for use while nursing, safety has not been conclusively established, and this substance is not recommended for use except under the supervision of an expert qualified in its appropriate use.

Review Details

I. Drug and Supplement Interactions

Clinical Trials of Drug or Supplement Interactions

No clinical trials of drug or supplement interactions were identified.

Case Reports of Suspected Drug or Supplement Interactions

No case reports of suspected drug or supplement interactions were identified.

Animal Trials of Drug or Supplement Interactions

No animal trials of drug or supplement interactions were identified.

II. Adverse Events

Case Reports of Adverse Events

Adverse effects from topical application of podophyllin have been reported and include erythema, edema, chemical burns, and allergic sensitivity. Systemic toxicity may cause nausea, vomiting, respiratory stimulation, peripheral neuropathy, fever, acute confusional states, tachycardia, oliguria, adynamic ileus, leukopenia, elevation of liver enzymes, coma, and death (Miller 1985; Rudrappa and Vijaydeva 2002).

A 2-year-old girl experienced bouts of vomiting, convulsions, and altered perception followed by a grade III coma after being accidentally administered 1 teaspoon of a 20% podophyllin preparation intended for topical treatment of warts. The girl's heart rate and blood pressure were normal, liver function tests on day 4 after poisoning indicated a mild increase in liver enzymes, and examination revealed no neurological effects. After supportive therapy for four weeks, the child made a full recovery (Rudrappa and Vijaydeva 2002).

A 22-year-old man developed a severe sensorimotor neuropathy following ingestion of podophyllin that had been prescribed for topical application for genital condylomata. The initial toxic symptoms were vomiting and diarrhea, followed by peripheral neuropathy. The neuropathy was still present 18 months later. Nerve conduction studies and sural nerve biopsy confirmed the presence of axonal degeneration (O'Mahony et al. 1990).

III. Pharmacology and Pharmacokinetics

Human Pharmacological Studies

No relevant human pharmacological studies were identified.

Animal Pharmacological Studies

No relevant animal pharmacological studies were identified.

In Vitro Pharmacological Studies

No relevant in vitro pharmacological studies were identified.

IV. Pregnancy and Lactation

Intravenous administration of 200 mg (per 200–250 g animal) of an aqueous extract of Himalayan mayapple to rats on day 16 of pregnancy demonstrated a protective effect against neuronal damage in offspring induced by exposure to radiation on day 17 of pregnancy (Sajikumar and Goel 2003).

A review of cases of adverse events associated with podophyllin use during pregnancy indicated that the adverse events were generally reported in young women with recent surgical procedures with or without general anesthesia, or in women who were taking other prescription or nonprescription drugs in addition to podophyllin use. The review concluded that podophyllin may be used topically during pregnancy without concern (Bargman 1988). A more recent review of the safety of the compounds podophyllin and podophyllotoxin indicates that these compounds are strongly contraindicated during pregnancy (Longstaff and von Krogh 2001).

A normal infant was born at term 3 months after the mother was accidentally administered 1 g of podophyllum resin orally (Balucani and Zellers 1964).

The child of a woman who had been topically treated with podophyllin on five occasions between weeks 23 and 29 of pregnancy was born with a simian crease and a preauricular skin tag. The mother had taken chloroquine and primaquine phosphate between weeks 19 and 22 of pregnancy (Karol et al. 1980). A comment on this case indicated that the podophyllin use was too late in pregnancy to cause the observed abnormalities (Bargman 1988; Fraser 1981).

In studies with rabbits, rats, and mice, administration of the compound podophyllotoxin to pregnant animals was associated with fetal resorption and abortifacient activity. Effects were observed in rats intravenously administered 2.5 mg/kg or subcutaneously administered 6 mg/kg, with no maternal mortality at these doses. In rabbits, subcutaneous doses of 18 mg/kg terminated pregnancy in two of five animals, while a dose of 35 mg/kg terminated pregnancy in

P

three of four animals with no maternal mortality reported in either group (Didcock et al. 1952).

No information on the safety of Himalayan mayapple in lactation was identified.

V. Toxicity Studies

Acute Toxicity

The LD_{50} of a fractionated extract (with podophyllin and podophyllotoxin removed) of Himalayan mayapple in mice is 250 mg/kg after oral administration and 90 mg/kg after intraperitoneal administration (Gupta et al. 2008a, 2008b; Lata et al. 2009).

The LD_{50} of the compound podophyllotoxin orally administered to rats was less than 100 mg/kg. After intravenous administration, the LD_{50} of podophyllotoxin is 10 mg/kg in rats and could not be determined at doses up to 20 mg/kg in mice. The dermal LD_{50} of podophyllotoxin is greater than 200 mg/kg in rabbits and greater than 500 mg/kg in rats (Longstaff and von Krogh 2001).

Genotoxicity

The compound podophyllin increased bacterial revertants and abnormal chromosomal structures in a concentration-dependent manner, both with and without metabolic activation, and increased the incidence of micronuclei in imprinted control region mouse reticulocytes (Lin et al. 2009).

Literature Cited

Balucani, M., and D.D. Zellers. 1964. *Podophyllum* resin poisoning with complete recovery. *J. Am. Med. Assoc.* 189(8):639-640.

Bargman, H. 1988. Is podophyllin a safe drug to use and can it be used during pregnancy? *Arch. Dermatol.* 124(11):1718-1720.

Cassidy, D.E., J. Drewry, and J.P. Fanning. 1982. *Podophyllum* toxicity: A report of a fatal case and a review of the literature. *J. Toxicol. Clin. Toxicol.* 19(1):35-44.

Chamberlain, M.J., A.L. Reynolds, and W.B. Yeoman. 1972. Toxic effect of *Podophyllum* application in pregnancy. *Br. Med. J.* 3:391-392.

Didcock, K.A., C.W. Picard, and J.M. Robson. 1952. The action of podophyllotoxin on pregnancy. *J. Physiol.* 117(4):65P-66P.

Felter, H.W., and J.U. Lloyd. 1898. *King's American dispensatory.* 18th ed., 3rd rev. 2 vols. Cincinnati: Ohio Valley Co.

Fraser, F.C. 1981. Letter to the editor. *Mod. Med. Can.* 36:1508.

Gupta, M.L., P.K. Agrawala, P. Kumar, et al. 2008a. Modulation of gamma radiation-inflicted damage in Swiss albino mice by an alcoholic fraction of *Podophyllum hexandrum* rhizome. *J. Med. Food* 11(3):486-492.

Gupta, M.L., S. Sankhwar, S. Verma, et al. 2008b. Whole body protection to lethally irradiated mice by oral administration of semipurified fraction of *Podophyllum hexandrum* and post irradiation treatment of *Picrorhiza kurroa. Tokai J. Exp. Clin. Med.* 33(1):6-12.

Karol, M.D., C.S. Conner, A.S. Watanabe, and K.J. Murphrey. 1980. *Podophyllum*: Suspected teratogenicity from topical application. *Clin. Toxicol.* 16(3):283-286.

Lata, M., J. Prasad, S. Singh, et al. 2009. Whole body protection against lethal ionizing radiation in mice by REC-2001: A semi-purified fraction of *Podophyllum hexandrum. Phytomedicine* 16(1):47-55.

Lin, M.C., H.W. Cheng, Y.C. Tsai, et al. 2009. Podophyllin, but not the constituents quercetin or kaempferol, induced genotoxicity *in vitro* and *in vivo* through ROS production. *Drug Chem. Toxicol.* 32(1):68-76.

Longstaff, E., and G. von Krogh. 2001. Condyloma eradication: Self-therapy with 0.15 –0.5% podophyllotoxin versus 20–25% podophyllin preparations—an integrated safety assessment. *Regul. Toxicol. Pharmacol.* 33(2):117-137.

McFarland, M.F., and J. McFarland. 1981. Accidental ingestion of *Podophyllum. Clin. Toxicol.* 18(8):973-977.

Miller, R.A. 1985. Podophyllin. *Int. J. Dermatol.* 24(8):491-498.

O'Mahony, S., C. Keohane, J. Jacobs, D. O'Riordain, and M. Whelton. 1990. Neuropathy due to podophyllin intoxication. *J. Neurol.* 237(2):110-112.

Rudrappa, S., and L. Vijaydeva. 2002. Podophyllin poisoning. *Indian Pediatr.* 39(6):598-599.

Sajikumar, S., and H.C. Goel. 2003. *Podophyllum hexandrum* prevents radiation-induced neuronal damage in postnatal rats exposed *in utero. Phytother. Res.* 17(7):761-766.

Podophyllum peltatum L. — Berberidaceae

SCN: mayapple
OCN: American mandrake

Part: root

Quick Reference Summary

Safety Class: 3
Interaction Class: A

Contraindications

Not for use except under the supervision of an expert qualified in the appropriate use of this substance (Cassidy et al. 1982; Longstaff and von Krogh 2001; McFarland and McFarland 1981).

Other Precautions

None known.

Drug and Supplement Interactions

None known.

Notice

Abortifacient (Chamberlain et al. 1972; Didcock et al. 1952; Longstaff and von Krogh 2001); *see* Appendix 2.

Emetic (Felter and Lloyd 1898); *see* Appendix 2.

Stimulant laxative (Felter and Lloyd 1898; Rosenstein et al. 1976; Wood and LaWall 1926); *see* Appendix 2.

Adverse Events and Side Effects

Mayapple toxicity has been reported after topical and internal use of mayapple and included nausea, vomiting, peripheral neuropathy, difficulty breathing, lethargy, and coma. Some cases of toxicity have been fatal (Cassidy et al. 1982; McFarland and McFarland 1981).

An Eclectic medical text indicated that mayapple may cause drooling and gastrointestinal pain. The fresh or recently dried root is reported to act as an "irritant poison" causing hypercatharsis, hyperemesis, griping, and other unpleasant symptoms. After inappropriate use, mayapple may cause prolonged gastrointestinal irritation and inflammation (Felter and Lloyd 1898).

Repeated topical application of the fresh or dried root may cause irritation followed by suppuration (Felter and Lloyd 1898).

Pharmacological Considerations

Although mayapple acts as a stimulant laxative, the plant is generally considered obsolete for this use (Rosenstein et al. 1976).

Pregnancy and Lactation

Case reports and animal studies have indicated that mayapple and compounds from mayapple have abortifacient activity (Chamberlain et al. 1972; Didcock et al. 1952; Longstaff and von Krogh 2001). A safety assessment of the compounds podophyllotoxin and podophyllin indicated that the topical use of these compounds during pregnancy is strongly contraindicated (Longstaff and von Krogh 2001).

No information on the safety of mayapple during lactation was identified in the scientific or traditional literature. While this review did not identify any concerns for use while nursing, safety has not been conclusively established, and this substance is not recommended for use except under the supervision of an expert qualified in its appropriate use.

Review Details

I. Drug and Supplement Interactions

Clinical Trials of Drug or Supplement Interactions

No clinical trials of drug or supplement interactions were identified.

Case Reports of Suspected Drug or Supplement Interactions

No case reports of suspected drug or supplement interactions were identified.

Animal Trials of Drug or Supplement Interactions

No animal trials of drug or supplement interactions were identified.

II. Adverse Events

Case Reports of Adverse Events

A review of cases of poisonings associated with mayapple use indicated that toxicity may occur after topical use or ingestion. Typical symptoms include nausea, vomiting, peripheral neuropathy, difficulty breathing, lethargy or coma, leukocytosis, and pancytopenia. Renal failure was reported in two fatal cases. Cases reviewed had topical doses of 150 to 1880 mg. Fatal poisoning was reported in a 60-year-old woman after application of 325 mg, while neuropathy was reported in an 18-year-old woman after application of 1800 mg. In persons who ingested 350 to 2800 mg of mayapple extracts, neuropathy with leukocytosis was observed after 350 mg, while coma with neuropathy was reported after a dose of 2800 mg (Cassidy et al. 1982; McFarland and McFarland 1981).

An Eclectic medical text indicated that mayapple may cause drooling, gastrointestinal pain, and griping. In overdose, violent vomiting and catharsis (diarrhea) may occur (Felter and Lloyd 1898). The fresh or recently dried root is reported to act as an "irritant poison" causing hypercatharsis, hyperemesis, griping, and other unpleasant symptoms. After inappropriate use, mayapple may cause prolonged gastrointestinal irritation and even inflammation (Felter and Lloyd 1898). Overdoses of mayapple have been lethal (Felter and Lloyd 1898). Repeated topical application of the fresh or dried root may cause irritation followed by suppuration (Felter and Lloyd 1898).

A 54-year-old alcoholic man with suicidal intent ingested a bottle of wine followed by a 10 to 11 g bottle of 25% mayapple in benzoin tincture. Treatment including ipecac, activated charcoal, and hemoperfusion was unsuccessful, and the poisoning was fatal (Cassidy et al. 1982).

A 20-year-old woman experienced nausea, vomiting, and an extended period of severe peripheral neuropathy after ingestion of approximately 4 g of mayapple in tincture of benzoin (McFarland and McFarland 1981).

A 22-year-old man developed a severe sensorimotor neuropathy following ingestion of podophyllin that had been prescribed for genital condylomata. The initial toxic symptoms were vomiting and diarrhea, followed by peripheral neuropathy. The neuropathy was still present 18 months later. Nerve conduction studies and sural nerve biopsy confirmed the presence of axonal degeneration (O'Mahony et al. 1990).

Irritation of the mucous membranes, including conjunctival inflammation, has been reported in workers processing dried mayapple or mayapple resin (Felter and Lloyd 1898).

III. Pharmacology and Pharmacokinetics

Human Pharmacological Studies

No relevant human pharmacological studies were identified.

Animal Pharmacological Studies

No relevant animal pharmacological studies were identified.

In Vitro Pharmacological Studies

No relevant in vitro pharmacological studies were identified.

IV. Pregnancy and Lactation

An 18-year-old woman in her 34th week of pregnancy experienced severe peripheral neuropathy followed by intrauterine death of the fetus after professional application of a 25% solution of podophyllum resin in tincture of benzoin on vulval warts. A total of 7.5 ml (equivalent to 1.88 g podophyllum) was applied. The woman was under general anesthesia at the time of application. After 3 months, the woman was walking without aid and was able to type, but very slowly and clumsily. Complete recovery from neuropathy and a subsequent healthy pregnancy was reported (Chamberlain et al. 1972).

P

A review of cases of adverse events associated with podophyllin use during pregnancy indicated that the adverse events were generally reported in young women with recent surgical procedures with or without general anesthesia, or in women who were taking other prescription or nonprescription drugs. The review concluded that podophyllin may be used topically during pregnancy without concern (Bargman 1988). A more recent review of the safety of the compounds podophyllin and podophyllotoxin indicates that these compounds are strongly contraindicated during pregnancy (Longstaff and von Krogh 2001).

A normal infant was born at term 3 months after the mother was accidentally administered 1 g of podophyllum resin orally (Balucani and Zellers 1964).

The child of a woman who had been topically treated with podophyllin on five occasions between weeks 23 and 29 of pregnancy was born with a simian crease and a preauricular skin tag. The mother had taken chloroquine and primaquine phosphate between weeks 19 and 22 of pregnancy (Karol et al. 1980). A comment on the case indicated that the podophyllin use was too late in pregnancy to cause the observed abnormalities (Bargman 1988; Fraser 1981).

In studies with rabbits, rats, and mice, administration of the compound podophyllotoxin to pregnant animals was associated with fetal resorption and abortifacient activity. Effects were observed in rats intravenously administered 2.5 mg/kg or subcutaneously administered 6 mg/kg, with no maternal mortality at these doses. In rabbits, subcutaneous dose of 18 mg/kg terminated pregnancy in two of five animals, while a dose of 35 mg/kg terminated pregnancy in three of four animals with no maternal mortality in either group (Didcock et al. 1952).

No information on the safety of mayapple during lactation was identified.

V. Toxicity Studies

Acute Toxicity

The LD_{50} of the compound podophyllotoxin orally administered in rats and rabbits could not be determined at doses up to 100 mg/kg. After intravenous administration, the LD_{50} of podophyllotoxin is 10 mg/kg in rats and could not be determined at doses up to 20 mg/kg in mice. The dermal LD_{50} of podophyllotoxin is greater than 200 mg/kg in rabbits and greater than 500 mg/kg in rats (Longstaff and von Krogh 2001).

Genotoxicity

The compound podophyllin increased bacterial revertants and abnormal chromosomal structures in a concentration-dependent manner, both with and without metabolic activation, and increased the incidence of micronuclei in imprinted control region mouse reticulocytes (Lin et al. 2009).

Literature Cited

Balucani, M., and D.D. Zellers. 1964. Podophyllum resin poisoning with complete recovery. *J. Am. Med. Assoc.* 189(8):639-640.

Bargman, H. 1988. Is podophyllin a safe drug to use and can it be used during pregnancy? *Arch. Dermatol.* 124(11):1718-1720.

Cassidy, D.E., J. Drewry, and J.P. Fanning. 1982. Podophyllum toxicity: A report of a fatal case and a review of the literature. *J. Toxicol. Clin. Toxicol.* 19(1):35-44.

Chamberlain, M.J., A.L. Reynolds, and W.B. Yeoman. 1972. Toxic effect of podophyllum application in pregnancy. *Br. Med. J.* 3:391-392.

Didcock, K.A., C.W. Picard, and J.M. Robson. 1952. The action of podophyllotoxin on pregnancy. *J. Physiol.* 117(4):65P-66P.
Felter, H.W., and J.U. Lloyd. 1898. *King's American dispensatory*. 18th ed., 3rd rev. 2 vols. Cincinnati: Ohio Valley Co.
Fraser, F.C. 1981. Letter to the editor. *Mod. Med. Can.* 36:1508.
Karol, M.D., C.S. Conner, A.S. Watanabe, and K.J. Murphrey. 1980. Podophyllum: Suspected teratogenicity from topical application. *Clin. Toxicol.* 16(3):283-286.
Lin, M.C., H.W. Cheng, Y.C. Tsai, et al. 2009. Podophyllin, but not the constituents quercetin or kaempferol, induced genotoxicity *in vitro* and *in vivo* through ROS production. *Drug Chem. Toxicol.* 32(1):68-76.
Longstaff, E., and G. von Krogh. 2001. Condyloma eradication: Self-therapy with 0.15–0.5% podophyllotoxin versus 20–25% podophyllin preparations—an integrated safety assessment. *Regul. Toxicol. Pharmacol.* 33(2):117-137.
McFarland, M.F., and J. McFarland. 1981. Accidental ingestion of podophyllum. *Clin. Toxicol.* 18(8):973-977.
O'Mahony, S., C. Keohane, J. Jacobs, D. O'Riordain, and M. Whelton. 1990. Neuropathy due to podophyllin intoxication. *J. Neurol.* 237(2):110-112.
Rosenstein, G., H. Rosenstein, M. Freeman, and N. Weston. 1976. Podophyllum a dangerous laxative. *Pediatrics* 57(3):419-421.
Wood, H., and C. LaWall. 1926. *The dispensatory of the United States of America*. Philadelphia: Lippincott.

Pogostemon cablin (Blanco) Benth.

Lamiaceae

SCN: patchouli
Syn: *Pogostemon patchouly* Pellet.
PN: *guang huo xiang* (herb)
Part: herb

Quick Reference Summary

Safety Class: 1
Interaction Class: A

Contraindications

None known.

Other Precautions

None known.

Drug and Supplement Interactions

None known.

Adverse Events and Side Effects

None known.

Pharmacological Considerations

None known.

Pregnancy and Lactation

A reference text on traditional Chinese medicine does not indicate any cautions for use of patchouli during pregnancy or lactation (Bensky et al. 2004). Although this review did not identify any concerns for use while pregnant or nursing, safety has not been conclusively established.

Review Details

I. Drug and Supplement Interactions

Clinical Trials of Drug or Supplement Interactions

No clinical trials of drug or supplement interactions were identified.

Case Reports of Suspected Drug or Supplement Interactions

No case reports of suspected drug or supplement interactions were identified.

Animal Trials of Drug or Supplement Interactions

No animal trials of drug or supplement interactions were identified.

II. Adverse Events

Case Reports of Adverse Events

No case reports of adverse events were identified.

III. Pharmacology and Pharmacokinetics

Human Pharmacological Studies

No relevant human pharmacological studies were identified.

Animal Pharmacological Studies

No relevant animal pharmacological studies were identified.

In Vitro Pharmacological Studies

No relevant in vitro pharmacological studies were identified.

IV. Pregnancy and Lactation

A reference text on traditional Chinese medicine does not indicate any cautions for use of patchouli during pregnancy or lactation (Bensky et al. 2004).

V. Toxicity Studies

No studies on the toxicity of patchouli were identified.

Literature Cited

Bensky, D., S. Clavey, and E. Stöger. 2004. *Chinese herbal medicine: Materia medica*. 3rd ed. Seattle: Eastland Press.

Polygala senega L. Polygalaceae

SCN: Seneca snakeroot
OCN: senega snakeroot

Part: root

Quick Reference Summary

Safety Class: 2b, 2d
Interaction Class: A

Contraindications

Not for use in pregnancy except under the supervision of a qualified healthcare practitioner (Felter and Lloyd 1898).

Not for long-term use; do not exceed recommended dose (Felter and Lloyd 1898; Wichtl 2004).

Not for use in persons with gastritis or gastric ulcers (Bradley 1992).

Other Precautions

None known.

Drug and Supplement Interactions

None known.

Notice

Diuretic (Yamahara et al. 1979); *see* Appendix 2.

Emmenagogue (Felter and Lloyd 1898); *see* Appendix 2.

Standard Dose

The standard dose is 0.5–1.0 g three times daily (Bradley 1992); 1–3 g daily (Merck 1930; Wichtl 2004).

Adverse Events and Side Effects

Prolonged use or overdose of Seneca snakeroot may cause gastrointestinal irritation, nausea, vomiting, retching, and diarrhea (Felter and Lloyd 1898; Wichtl 2004).

Pharmacological Considerations

Animal studies have demonstrated that Seneca snakeroot and compounds from Seneca snakeroot may modify glucose regulation (Kako et al. 1996, 1997; Yoshikawa et al. 1995, 1996). People with diabetes are advised to monitor their blood sugar closely and discuss the use of this herb with a qualified healthcare practitioner prior to use.

Pregnancy and Lactation

Seneca snake root has traditionally been used as an emmenagogue (Felter and Lloyd 1898). Based on this information, use during pregnancy is not recommended except under the supervision of a qualified healthcare practitioner.

No information on the safety of Seneca snake root in lactation was identified. While this review did not identify any concerns for use while nursing, safety has not been conclusively established.

Review Details

I. Drug and Supplement Interactions

Clinical Trials of Drug or Supplement Interactions

No clinical trials of drug or supplement interactions were identified.

Case Reports of Suspected Drug or Supplement Interactions

No case reports of suspected drug or supplement interactions were identified.

Animal Trials of Drug or Supplement Interactions

No animal trials of drug or supplement interactions were identified.

II. Adverse Events

Case Reports of Adverse Events

No case reports of adverse events were identified.

III. Pharmacology and Pharmacokinetics

Human Pharmacological Studies

No relevant human pharmacological studies were identified.

Animal Pharmacological Studies

In rats orally administered 2 g/kg of a hydromethanolic extract of Seneca snakeroot, a 62% reduction in congestive edema was observed along with a significant increase in 24-hour urine volume (Yamahara et al. 1979).

A reduction in blood glucose levels was observed in healthy and in non-insulin-dependent diabetic mice intraperitoneally administered 5 mg/kg of the *n*-butanol fraction of Seneca snakeroot. No changes in glucose levels were observed in insulin-dependent diabetic mice (Kako et al. 1996). A reduction in blood glucose levels was observed in healthy and in non-insulin-dependent diabetic mice intraperitoneally administered the compounds senegin II and senegin III (Kako et al. 1997).

In the oral glucose tolerance test in rats, after oral administration of 100 mg/kg of the compounds *E*,*Z*-senegasaponins a or b, a reduction in plasma glucose levels was observed (Yoshikawa et al. 1995). Similar inhibition of glucose elevation was observed after animals were administered *E*,*Z*-senegins II, III, and IV and *E*,*Z*-senegasaponin c (Yoshikawa et al. 1996).

In Vitro Pharmacological Studies

In high-throughput enzyme inhibition screening assays, an ethanol extract of Seneca snakeroot inhibited the drug-metabolizing isoenzyme CYP2C19 (Scott et al. 2006).

IV. Pregnancy and Lactation

Based on the traditional use of Seneca snake root as an emmenagogue (Felter and Lloyd 1898), use during pregnancy is not recommended except under the supervision of a qualified healthcare practitioner.

No information on the safety of Seneca snakeroot during lactation was identified.

V. Toxicity Studies

Genotoxicity

No mutagenic activity of aqueous or methanolic extracts of Seneca snakeroot was observed in the Ames mutagenicity test with *Salmonella typhimurium* strains TA98 and TA100 (Morimoto et al. 1982).

Literature Cited

Bradley, P.R. 1992. *British herbal compendium: A handbook of scientific information on widely used plant drugs*. Bournemouth, UK: British Herbal Medicine Association.

Felter, H.W., and J.U. Lloyd. 1898. *King's American dispensatory*. 18th ed., 3rd rev. 2 vols. Cincinnati: Ohio Valley Co.

Kako, M., T. Miura, Y. Nishiyama, et al. 1996. Hypoglycemic effects of the rhizomes of *Polygala senega* in normal and diabetic mice and its main component, the triterpenoid glycoside senegin-II. *Planta Med.* 62(5):440-443.

Kako, M., T. Miura, Y. Nishiyama, et al. 1997. Hypoglycemic activity of some triterpenoid glycosides. *J. Nat. Prod.* 60(6):604-605.

Merck, E. 1930. *Merck's index*. Darmstadt: E. Merck.

Morimoto, I., F. Watanabe, T. Osawa, T. Okitsu, and T. Kada. 1982. Mutagenicity screening of crude drugs with *Bacillus subtilis* rec-assay and *Salmonella*/microsome reversion assay. *Mutat. Res.* 97(2):81.

Scott, I.M., R.I. Leduc, A.J. Burt, et al. 2006. The inhibition of human cytochrome P450 by ethanol extracts of North American botanicals. *Pharm. Biol.* 44(5):315-327.

Wichtl, M. 2004. *Herbal drugs and phytopharmaceuticals: A handbook for practice on a scientific basis*. 3rd ed. Boca Raton, FL: CRC Press.

Yamahara, J., Y. Takagi, T. Sawada, et al. 1979. Effects of crude drugs on congestive edema. *Chem. Pharm. Bull.* 27(6):1464.

Yoshikawa, M., T. Murakami, H. Matsuda, et al. 1996. Bioactive saponins and glycosides. II. *Senegae Radix*.(2): Chemical structures, hypoglycemic activity, and ethanol absorption-inhibitory effect of *E*-senegasaponin c, *Z*-senegasaponin c, and *Z*-senegins II, III, and IV. *Chem. Pharm. Bull.* 44(7):1305-1313.

Yoshikawa, M., T. Murakami, T. Ueno, et al. 1995. *E*-Senegasaponins A and B, *Z*-senegasaponins A and B, *Z*-senegins II and III, new type inhibitors of ethanol absorption in rats from *Senegae Radix*, the roots of *Polygala senega* L. var. *latifolia* Torrey et Gray. *Chem. Pharm. Bull.* 43(2):350-352.

P

Polygala spp.

Polygalaceae

Polygala sibirica L.
SCN: polygala
PN: *yuan zhi* (root)
OCN: Siberian milkwort; Siberian polygala

Polygala tenuifolia **Willd.**
SCN: polygala
PN: *yuan zhi* (root)
OCN: thin-leaf polygala
Part: root

Quick Reference Summary

Safety Class: 2b, 2d
Interaction Class: A

Contraindications

Not for use in pregnancy except under the supervision of a qualified healthcare practitioner (Chen and Chen 2004).

Not for use in persons with gastritis or ulcers (Bensky et al. 2004; Chen et al. 2004).

Other Precautions

None known.

Drug and Supplement Interactions

None known.

Notice

Diuretic (Chen and Chen 2004; Yamahara et al. 1979); *see* Appendix 2.

Adverse Events and Side Effects

Polygala may be irritating to the gastrointestinal tract (Bensky et al. 2004; Chen et al. 2004).

Allergic reactions to polygala have been reported (Bensky et al. 2004; Park et al. 2005).

Pregnancy and Lactation

A reference text on traditional Chinese medicine indicates that polygala should be used with caution during pregnancy, as water and alcohol extracts of polygala have exhibited uterine-stimulating activity in pregnant and nonpregnant animals (Chen and Chen 2004). Based on this information, use during pregnancy is not recommended except under the supervision of a qualified healthcare practitioner.

No information on the safety of polygala during lactation was identified. While this review did not identify any concerns for use while nursing, safety has not been conclusively established.

Review Details

I. Drug and Supplement Interactions

Clinical Trials of Drug or Supplement Interactions

No clinical trials of drug or supplement interactions were identified.

Case Reports of Suspected Drug or Supplement Interactions

No case reports of suspected drug or supplement interactions were identified.

Animal Trials of Drug or Supplement Interactions

No animal trials of drug or supplement interactions were identified.

II. Adverse Events

Case Reports of Adverse Events

Overdose (standard dose is listed as a decoction of 6–15 g) of polygala may cause nausea and vomiting (Bensky et al. 2004).

Allergic reactions to polygala have been reported (Bensky et al. 2004).

Occupational asthma and rhinitis induced by polygala were reported in a patient who had worked in an herb processing facility for 8 years. A skin prick test confirmed polygala as the causative agent (Park et al. 2005).

III. Pharmacology and Pharmacokinetics

Human Pharmacological Studies

No relevant human pharmacological studies were identified.

Animal Pharmacological Studies

In rats orally administered 2 g/kg of a hydromethanolic extract of polygala, a 100% reduction in congestive edema was observed along with a significant increase in 24-hour urine volume (Yamahara et al. 1979).

Prolongation of hexobarbital-induced sleeping times was observed in mice intraperitoneally administered 6.25 to 50 mg/kg of the butanol-soluble fraction of a methanolic extract of polygala (Nikaido et al. 1982).

In Vitro Pharmacological Studies

No estrogenic activity of an ethanol extract of polygala was observed in a recombinant yeast assay system with a human estrogen receptor expression plasmid and a reporter plasmid (Kim et al. 2008).

The compound tetrahydrocolumbamine exhibited dopamine receptor binding activity in several in vitro assays (Shen et al. 1994).

IV. Pregnancy and Lactation

Water and alcohol extracts of polygala have exhibited uterine-stimulating activity in pregnant and nonpregnant rabbits, cats, and dogs. Information on dose and route of administration was not reported in the available English language translation (Chen and Chen 2004).

No information on the safety of polygala during lactation was identified.

V. Toxicity Studies

Acute Toxicity

The LD_{50} of polygala orally administered in mice is 16.95 g/kg for the whole root, 10.03 g/kg for the root bark, and over 75 g/kg for the root core (Chang and But 1986).

Genotoxicity

No mutagenic activity of aqueous or methanolic extracts of polygala was observed in the Ames mutagenicity test with *Salmonella typhimurium* strains TA98 and TA100 (Morimoto et al. 1982).

Literature Cited

Bensky, D., S. Clavey, and E. Stöger. 2004. *Chinese herbal medicine: Materia medica*. 3rd ed. Seattle: Eastland Press.

Chang, H.-M., and P.P.H. But. 1986. *Pharmacology and applications of Chinese materia medica*. English ed. Philadelphia: World Scientific.

Chen, J.K., and T.T. Chen. 2004. *Chinese medical herbology and pharmacology*. City of Industry, CA: Art of Medicine Press.

Chen, Y.L., C.L. Hsieh, P.H.B. Wu, and J.G. Lin. 2004. Effect of *Polygala tenuifolia* root on behavioral disorders by lesioning nucleus basalis magnocellularis in rat. *J. Ethnopharmacol.* 95(1):47-55.

Kim, I.G., S.C. Kang, K.C. Kim, E.S. Choung, and O.P. Zee. 2008. Screening of estrogenic and antiestrogenic activities from medicinal plants. *Environ. Toxicol. Pharmacol.* 25(1):75-82.

Morimoto, I., F. Watanabe, T. Osawa, T. Okitsu, and T. Kada. 1982. Mutagenicity screening of crude drugs with *Bacillus subtilis* rec-assay and *Salmonella*/microsome reversion assay. *Mutat. Res.* 97(2):81.

Nikaido, T., T. Ohmoto, H. Saitoh, et al. 1982. Inhibitors of cyclic adenosine monophosphate phosphodiesterase in *Polygala tenuifolia*. *Chem. Pharm. Bull.* 30(6):2020-2024.

Park, H.K., S.G. Jeon, T.B. Kim, et al. 2005. Occupational asthma and rhinitis induced by a herbal medicine, wonji (*Polygala tenuifolia*). *J. Kor. Med. Sci.* 20(1):46-49.

Shen, X.L., M.R. Witt, K. Dekermendjian, and M. Nielsen. 1994. Isolation and identification of tetrahydrocolumbamine as a dopamine receptor ligand from *Polygala tenuifolia* Willd. *Yao Xue Xue Bao* 29(12):887-890.

Yamahara, J., Y. Takagi, T. Sawada, et al. 1979. Effects of crude drugs on congestive edema. *Chem. Pharm. Bull.* 27(6):1464.

Polygonatum biflorum (Walter) Elliott — Liliaceae

SCN: Solomon's seal
OCN: small Solomon's seal
Part: root

Quick Reference Summary

Safety Class: 1
Interaction Class A

Contraindications
None known.

Other Precautions
None known.

Drug and Supplement Interactions
None known.

Editors' Note
In addition to medicinal use, the rhizomes of Solomon's seal were used by the Cherokee people as a food, being ground into a flour and made into bread or boiled and then eaten (Moerman 1998).

Adverse Events and Side Effects
None known.

Pharmacological Considerations
None known.

Pregnancy and Lactation
No information on the safety of Solomon's seal in pregnancy or lactation was identified in the scientific or traditional literature. Although this review did not identify any concerns for use while pregnant or nursing, safety has not been conclusively established.

Review Details

I. Drug and Supplement Interactions

Clinical Trials of Drug or Supplement Interactions
No clinical trials of drug or supplement interactions were identified.

Case Reports of Suspected Drug or Supplement Interactions
No case reports of suspected drug or supplement interactions were identified.

Animal Trials of Drug or Supplement Interactions
No animal trials of drug or supplement interactions were identified.

II. Adverse Events

Case Reports of Adverse Events
No case reports of adverse events were identified.

P

III. Pharmacology and Pharmacokinetics

Human Pharmacological Studies

No relevant human pharmacological studies were identified.

Animal Pharmacological Studies

No relevant animal pharmacological studies were identified.

In Vitro Pharmacological Studies

No relevant in vitro pharmacological studies were identified.

IV. Pregnancy and Lactation

No information on the safety of Solomon's seal during pregnancy or lactation was identified.

V. Toxicity Studies

No toxicity studies were identified.

Literature Cited

Moerman, D.E. 1998. *Native American ethnobotany*. Portland, OR: Timber Press.

Polygonatum odoratum (Mill.) Druce — Liliaceae

SCN: aromatic Solomon's seal
Syn: *Polygonatum officinale* All.
PN: *yu zhu* (rhizome)
OCN: fragrant Solomon's seal
Part: rhizome

Quick Reference Summary

Safety Class: 1
Interaction Class: A

Contraindications

None known.

Other Precautions

None known.

Drug and Supplement Interactions

None known.

Adverse Events and Side Effects

An allergic reaction to aromatic Solomon's seal has been reported (Bensky et al. 2004).

Pharmacological Considerations

Animal studies have demonstrated that aromatic Solomon's seal may modify glucose regulation (Choi and Park 2002; Kato and Miura 1994; Miura and Kato 1995). People with diabetes are advised to monitor their blood sugar closely and discuss the use of this herb with a qualified healthcare practitioner prior to use.

Pregnancy and Lactation

No information on the safety of aromatic Solomon's seal in pregnancy or lactation was identified in the scientific or traditional literature. Although this review did not identify any concerns for use while pregnant or nursing, safety has not been conclusively established.

Review Details

I. Drug and Supplement Interactions

Clinical Trials of Drug or Supplement Interactions

No clinical trials of drug or supplement interactions were identified.

Case Reports of Suspected Drug or Supplement Interactions

No case reports of suspected drug or supplement interactions were identified.

Animal Trials of Drug or Supplement Interactions

No animal trials of drug or supplement interactions were identified.

II. Adverse Events

Case Reports of Adverse Events

Overdose of aromatic Solomon's seal (standard dose listed as a decoction of 6–15 g) may cause nausea, vomiting, or diarrhea (Bensky et al. 2004).

An allergic reaction to aromatic Solomon's seal, including generalized pruritus and red papular rashes, has been reported (Bensky et al. 2004).

III. Pharmacology and Pharmacokinetics

Human Pharmacological Studies

No relevant human pharmacological studies were identified.

Animal Pharmacological Studies

In healthy and in diabetic mice intraperitoneally administered 800 mg/kg of a methanol extract of aromatic Solomon's seal, a reduction in blood glucose levels was observed (Kato and Miura 1994; Miura and Kato 1995).

In 90% pancreatectomized rats orally administered 30 mg/kg of steroidal glycoside compounds from aromatic Solomon's seal daily for 13 weeks, no change in insulin secretion from pancreatic beta-cells was observed; however, mice were found to have an increased insulin sensitivity (Choi and Park 2002).

In Vitro Pharmacological Studies

No estrogenic activity of an ethanol extract of aromatic Solomon's seal was observed in a recombinant yeast assay system with a human estrogen receptor expression plasmid and a reporter plasmid (Kim et al. 2008).

IV. Pregnancy and Lactation

No information on the safety of aromatic Solomon's seal during pregnancy or lactation was identified.

V. Toxicity Studies

Acute Toxicity

No adverse effects were observed in mice orally administered 10 g/kg of an aqueous extract of aromatic Solomon's seal (equivalent to 64.5 g/kg crude material) twice at 6 hour intervals (Chen et al. 2001).

The LD_{50} of an injectable preparation of aromatic Solomon's seal intraperitoneally administered in mice is 112.5 g/kg (Zhu 1998).

In a sperm aberration test, sexually mature male mice were orally administered 2.5, 5, or 10 g/kg of an aqueous extract of aromatic Solomon's seal daily for 5 days. Examination of sperm 30 days after the final aromatic Solomon's seal dose indicated no significant difference in the rates of sperm aberration between the control group and the extract-treated groups (Chen et al. 2001).

Subchronic Toxicity

In rats orally administered 4 or 16 g/kg of an aqueous extract of aromatic Solomon's seal (equivalent to 12.9 and 51.6 g/kg crude material) daily for 6 months, softened feces were noted in the 16 g/kg group. No signs of toxicity, including changes in behavior, appearance, diet, hematology, blood biochemistry, and histopathology, were observed (Chen et al. 2001).

Genotoxicity

In the mouse bone marrow micronucleus assay, mice were orally administered 2.5, 5, or 10 g/kg of an aqueous extract of aromatic Solomon's seal twice at a 24-hour interval. No significant changes in the rates of micronucleus cells in treated groups were observed (Chen et al. 2001).

In the Ames test for mutagenicity with *Salmonella typhimurium* strains TA97, TA98, TA100, or TA102 tested with and without metabolic activation by S9, no mutagenic activity of 0.2, 1, or 5 mg/plate of an aqueous extract of aromatic Solomon's seal was observed (Chen et al. 2001).

Literature Cited

Bensky, D., S. Clavey, and E. Stöger. 2004. *Chinese herbal medicine: Materia medica*. 3rd ed. Seattle: Eastland Press.

Chen, H., R. Feng, Y. Guo, et al. 2001. Toxicity studies of *Rhizoma Polygonati Odorati*. *J. Ethnopharmacol.* 74(3):221-224.

Choi, S.B., and S. Park. 2002. A steroidal glycoside from *Polygonatum odoratum* (Mill.) Druce. improves insulin resistance but does not alter insulin secretion in 90% pancreatectomized rats. *Biosci. Biotechnol. Biochem.* 66(10):2036-2043.

Kato, A., and T. Miura. 1994. Hypoglycemic action of the rhizomes of *Polygonatum officinale* in normal and diabetic mice. *Planta Med.* 60(3):201-203.

Kim, I.G., S.C. Kang, K.C. Kim, E.S. Choung, and O.P. Zee. 2008. Screening of estrogenic and antiestrogenic activities from medicinal plants. *Environ. Toxicol. Pharmacol.* 25 (1):75-82.

Miura, T., and A. Kato. 1995. The difference in hypoglycemic action between *polygonati rhizoma* and *polygonati officinalis rhizoma*. *Biol. Pharm. Bull.* 18(11):1605-1606.

Zhu, Y.-P. 1998. *Chinese materia medica: Chemistry, pharmacology and applications.* Amsterdam: Harwood Academic Publishers.

Polygonatum sibiricum F. Delaroche

Liliaceae

SCN: polygonatum
PN: *huang jing* (rhizome)
OCN: Siberian Solomon's seal
Part: rhizome

Quick Reference Summary

Safety Class: 1
Interaction Class: A

Contraindications

None known.

Other Precautions

None known.

Drug and Supplement Interactions

None known.

Adverse Events and Side Effects

None known.

Pharmacological Considerations

Animal studies have demonstrated that polygonatum may modify glucose regulation (Kato and Miura 1993; Miura and Kato 1995). People with diabetes are advised to monitor their blood sugar closely and discuss the use of this herb with a qualified healthcare practitioner prior to use.

Pregnancy and Lactation

No information on the safety of polygonatum in pregnancy or lactation was identified in the scientific or traditional literature. Although this review did not identify any concerns for use while pregnant or nursing, safety has not been conclusively established.

Review Details

I. Drug and Supplement Interactions

Clinical Trials of Drug or Supplement Interactions

No clinical trials of drug or supplement interactions were identified.

Case Reports of Suspected Drug or Supplement Interactions

No case reports of suspected drug or supplement interactions were identified.

Animal Trials of Drug or Supplement Interactions

No animal trials of drug or supplement interactions were identified.

II. Adverse Events

Case Reports of Adverse Events

No case reports of adverse events were identified.

III. Pharmacology and Pharmacokinetics

Human Pharmacological Studies

No relevant human pharmacological studies were identified.

Animal Pharmacological Studies

In healthy and in diabetic mice intraperitoneally administered 800 mg/kg of a methanol extract of polygonatum, a reduction in blood glucose levels was observed, with the effect being more pronounced in diabetic animals. Serum insulin levels did not change in either group (Kato and Miura 1993; Miura and Kato 1995).

In Vitro Pharmacological Studies

No relevant in vitro pharmacological studies were identified.

IV. Pregnancy and Lactation

No information on the safety of polygonatum during pregnancy or lactation was identified.

V. Toxicity Studies

Acute Toxicity

In mice orally administered an aqueous extract of polygonatum at a dose corresponding to 450 g/kg, all mice administered an extract of the unprocessed herb died, but all of the mice administered an extract of the steam-processed root survived (Zhu 1998).

Literature Cited

Kato, A., and T. Miura. 1993. Hypoglycemic activity of *Polygonati Rhizoma* in normal and diabetic mice. *Biol. Pharm. Bull.* 16(11):1118-1120.

Miura, T., and A. Kato. 1995. The difference in hypoglycemic action between *polygonati rhizoma* and *polygonati officinalis rhizoma*. *Biol. Pharm. Bull.* 18(11):1605-1606.

Zhu, Y.-P. 1998. *Chinese materia medica: Chemistry, pharmacology and applications.* Amsterdam: Harwood Academic Publishers.

Polygonum bistorta L. — Polygonaceae

SCN: bistort
PN: *quan shen* (rhizome)
OCN: English serpentary
Part: rhizome

Quick Reference Summary

Safety Class: 1
Interaction Class: A

Contraindications

None known.

Other Precautions

None known.

Drug and Supplement Interactions

None known.

Notice

Tannins (15 21%) (Remington and Wood 1918; Wattiez 1920); *see* Appendix 1.

Adverse Events and Side Effects

Mild side effects of bistort have been reported, including abdominal pain and diarrhea (Bensky et al. 2004).

Pharmacological Considerations

None known.

Pregnancy and Lactation

No information on the safety of bistort in pregnancy or lactation was identified in the scientific or traditional literature. Although this review did not identify any concerns for use while pregnant or nursing, safety has not been conclusively established.

Review Details

I. Drug and Supplement Interactions

Clinical Trials of Drug or Supplement Interactions

No clinical trials of drug or supplement interactions were identified.

Case Reports of Suspected Drug or Supplement Interactions

No case reports of suspected drug or supplement interactions were identified.

Animal Trials of Drug or Supplement Interactions

No animal trials of drug or supplement interactions were identified.

II. Adverse Events

Case Reports of Adverse Events

Mild side effects of bistort have been reported, including abdominal pain and diarrhea. These effects are thought to be due to irritation of the intestinal mucosa (Bensky et al. 2004).

III. Pharmacology and Pharmacokinetics

Human Pharmacological Studies

No relevant human pharmacological studies were identified.

Animal Pharmacological Studies

No relevant animal pharmacological studies were identified.

In Vitro Pharmacological Studies

An ethanol extract of bistort induced activity of a substance showing interferon-like activity in cultures of monkey kidney cells (Smolarz and Skwarek 1999).

IV. Pregnancy and Lactation

No information on the safety of bistort during pregnancy or lactation was identified.

V. Toxicity Studies

No toxicity studies were identified.

Literature Cited

Bensky, D., S. Clavey, and E. Stöger. 2004. *Chinese herbal medicine: Materia medica*. 3rd ed. Seattle: Eastland Press.

Remington, J.P., and H.C. Wood. 1918. *The dispensatory of the United States of America*. 20th ed. Philadelphia: Lippincott.

Smolarz, H.D., and T. Skwarek. 1999. The investigations into the interferon-like activity of *Polygonum* L. genus. *Acta Pol. Pharm.* 56(6):459-462.

Wattiez, N. 1920. Contribution to the study of *Polygonum bistorta*. Localization of tannin. Its employment as a substitute for *Krameria triandra*. *Ann. Bull. Soc. R. Med. Nat.* 4:121-128.

P

Populus balsamifera L. ssp. *balsamifera* Salicaeae

SCN: balsam poplar
Syn: *Populus candicans* Aiton; *Populus tacamahaca* Mill.
OCN: balm-of-gilead (leaf bud); tacamahac
Part: resinous leaf buds

Quick Reference Summary

Safety Class: 1
Interaction Class: A

Contraindications

None known.

Other Precautions

None known.

Drug and Supplement Interactions

None known.

Notice

Salicylates (List and Hörhammer 1973); *see* Appendix 1.

Adverse Events and Side Effects

Hypersensitivity and allergic skin reactions to poplar buds have been reported and noted as rarely occurring (Blumenthal et al. 1998; Bradley 1992).

Pharmacological Considerations

None known.

Pregnancy and Lactation

No information on the safety of balsam poplar in pregnancy or lactation was identified in the scientific or traditional literature. Although this review did not identify any concerns for use while pregnant or nursing, safety has not been conclusively established.

Review Details

I. Drug and Supplement Interactions

Clinical Trials of Drug or Supplement Interactions

No clinical trials of drug or supplement interactions were identified.

Case Reports of Suspected Drug or Supplement Interactions

No case reports of suspected drug or supplement interactions were identified.

Animal Trials of Drug or Supplement Interactions

No animal trials of drug or supplement interactions were identified.

II. Adverse Events

Case Reports of Adverse Events

No case reports of adverse events were identified.

III. Pharmacology and Pharmacokinetics

Human Pharmacological Studies

No relevant human pharmacological studies were identified.

Animal Pharmacological Studies

No relevant animal pharmacological studies were identified.

In Vitro Pharmacological Studies

No relevant in vitro pharmacological studies were identified.

IV. Pregnancy and Lactation

No information on the safety of balsam poplar during pregnancy or lactation was identified.

V. Toxicity Studies

No toxicity studies were identified.

Literature Cited

Blumenthal, M., W. Busse, A. Goldberg, et al. 1998. *The complete German Commission E monographs*. Austin, TX: American Botanical Council.

Bradley, P.R. 1992. *British herbal compendium: A handbook of scientific information on widely used plant drugs*. Bournemouth, UK: British Herbal Medicine Association.

List, P.H., and H. Hörhammer. 1973. *Hagers handbuch der pharmazeutischen praxis*. Berlin: Springer.

P

Portulaca oleracea L. — Portulacaceae

SCN: purslane
PN: *ma chi xian* (above-ground parts)
Part: above-ground parts

Quick Reference Summary

Safety Class: 2b
Interaction Class: A

Contraindications

Not for use in pregnancy except under the supervision of a qualified healthcare practitioner (Bensky et al. 2004).

Other Precautions

Use with caution in persons with a history of kidney stones (McGuffin et al. 1997).

Drug and Supplement Interactions

None known.

Notice

Diuretic (Pagar et al. 2007; Yasuye and Honda 1944); *see* Appendix 2.

Editors' Note

Purslane is commonly consumed as a vegetable in many regions of the world (Ezekwe et al. 1999; Omara-Alwala et al. 1991).

Adverse Events and Side Effects

None known.

Pharmacological Considerations

Animal studies have demonstrated that purslane may modify glucose regulation (Cui et al. 2005; Li et al. 2009). People with diabetes are advised to monitor their blood sugar closely and discuss the use of this herb with a qualified healthcare practitioner prior to use.

Pregnancy and Lactation

Fresh purslane is commonly consumed as a vegetable in many regions of the world with no cautions listed for use as a food during pregnancy (Ezekwe et al. 1999; Omara-Alwala et al. 1991). Animal studies have provided conflicting data on the activity of purslane on the uterus (Chen and Chen 2004). Of two reference texts on traditional Chinese medicine, one indicates that purslane should not be used during pregnancy (standard dose listed is 9–15 grams of dried purslane) (Bensky et al. 2004) while the other text does not (Chen and Chen 2004). Texts on traditional medicine from India do not indicate any concern for use of purslane in pregnancy (Khory and Katrak 1887; Nadkarni 1954).

No information on the safety of purslane during lactation was identified in the scientific or traditional literature. While this review did not identify any concerns for use while nursing, safety has not been conclusively established.

P

Review Details

I. Drug and Supplement Interactions

Clinical Trials of Drug or Supplement Interactions

No clinical trials of drug or supplement interactions were identified.

Case Reports of Suspected Drug or Supplement Interactions

No case reports of suspected drug or supplement interactions were identified.

Animal Trials of Drug or Supplement Interactions

No animal trials of drug or supplement interactions were identified.

II. Adverse Events

Case Reports of Adverse Events

No case reports of adverse events were identified.

III. Pharmacology and Pharmacokinetics

Human Pharmacological Studies

No relevant human pharmacological studies were identified.

Animal Pharmacological Studies

In diabetic and healthy rats orally administered 0.5, 2.5, or 5 g/kg of purslane powder daily for 21 days, a dose-dependent reduction in blood glucose levels was observed in diabetic but not in healthy rats (Cui et al. 2005).

In diabetic mice orally administered 200 or 400 mg/kg of polysaccharides from purslane daily for 28 days, a decrease in fasting blood glucose and increase in serum insulin levels were observed (Li et al. 2009).

In rats administered 200 or 400 mg/kg of a hydroethanolic extract of purslane, an increase in the volume of urine and urinary concentration of sodium, potassium, and chloride ions was observed at the 400 mg/kg dose. At the 200 mg/kg dose, no change in urine volume or urinary concentration of chloride ions was observed, although urinary concentrations of sodium and potassium levels were elevated (Pagar et al. 2007).

In Vitro Pharmacological Studies

No relevant in vitro pharmacological studies were identified.

IV. Pregnancy and Lactation

Of two authoritative English-language texts on traditional Chinese medicine, one indicates that purslane should not be used during pregnancy (standard dose listed is 9–15 grams of dried purslane, approx. 195–326 g fresh) (Bensky et al. 2004) while the second text does not (Chen and Chen 2004). Texts on traditional medicine from India do not indicate any concern for use of purslane in pregnancy (Khory and Katrak 1887; Nadkarni 1954).

An extract of purslane was reported to have a stimulating effect on the smooth muscles of the intestines and uteruses of rats, rabbits, and guinea pigs. In another study, purslane was reported to have mixed effects on the uterus, with an extract of the stem stimulating the uterus and an extract of the leaves relaxing the uterus. Details on the extracts, doses, and routes of administration were not reported in the available English language translation (Chen and Chen 2004).

No information on the safety of purslane during lactation was identified.

V. Toxicity Studies

Acute Toxicity

The LD_{50} of a methanol extract of purslane intraperitoneally administered in mice is 1.87 g/kg (Musa et al. 2007).

Genotoxicity

No mutagenic activity of an aqueous extract of purslane was observed in the Ames test for mutagenicity with *Salmonella typhimurium* strains TA98 and TA100 with or without metabolic activation by S9 (Yen et al. 2001).

Literature Cited

Bensky, D., S. Clavey, and E. Stöger. 2004. *Chinese herbal medicine: Materia medica*. 3rd ed. Seattle: Eastland Press.

Chen, J.K., and T.T. Chen. 2004. *Chinese medical herbology and pharmacology*. City of Industry, CA: Art of Medicine Press.

Cui, M.Z., H. Liu, and C.Y. Li. 2005. Changes of blood glucose in diabetic rats and the interventional effect of purslane. *Chin. J. Clin. Rehab.* 9(27):92-93.

Ezekwe, M.O., T.R. Omara-Alwala, and T. Membrahtu. 1999. Nutritive characterization of purslane accessions as influenced by planting date. *Plant Food Hum. Nutr.* 54(3):183-191.

Khory, R. and N.N. Katrak 1887. Bombay materia medica. Delhi: Neeraj Publishing House, Delhi (1981 reprint).

Li, F., Q. Li, D. Gao, Y. Peng, and C. Feng. 2009. Preparation and antidiabetic activity of polysaccharide from *Portulaca oleracea* L. *Afr. J. Biotechnol.* 8(4):569-573.

McGuffin, M., C. Hobbs, R. Upton, and A. Goldberg. 1997. *Botanical safety handbook*. Boca Raton, FL: CRC Press.

Musa, K.Y., A. Ahmed, G. Ibrahim, et al. 2007. Toxicity studies on the methanolic extract of *Portulaca oleracea* L. (fam. Portulacaceae). *J. Biol. Sci.* 7(7):1293-1295.

Nadkarni, A.K., and K.M. Nadkarni, 1954. Dr. K.M. Nadkarni's Indian materia medica. Bombay: Popular Prakashan.

Omara-Alwala, T.R., T. Mebrahtu, D.E. Prior, and M.O. Ezekwe. 1991. Omega-three fatty acids in purslane (*Portulaca oleracea*) tissues. *J. Am. Oil Chem. Soc.* 68(3):198-199.

Pagar, H.J., T.M. Jyothi, S.V. Rajendra, et al. 2007. A study on preliminary phytochemical and diuretic activity of leaves of *Portulaca oleracea*. *Pharmacog. Mag.* 3(12):264-266.

Yasuye, M., and Y. Honda. 1944. Components of Portulaceae plants. *Yakugaku Zasshi* 64:177-178.

Yen, G.C., H.Y. Chen, and H.H. Peng. 2001. Evaluation of the cytotoxicity, mutagenicity and antimutagenicity of emerging edible plants. *Food Chem. Toxicol.* 39(11):1045-1053.

P

Potentilla erecta (L.) Räusch. — Rosaceae

SCN: cinquefoil
Syn: *Potentilla tormentilla* Stokes
OCN: erect cinquefoil; tormentil
Part: rhizome

Quick Reference Summary

Safety Class: 1
Interaction Class: A

Contraindications

None known.

Other Precautions

None known.

Drug and Supplement Interactions

None known.

Notice

Tannins (15.0–20.0%) (Tomczyk and Latte 2009; Wichtl 2004); *see* Appendix 1.

Adverse Events and Side Effects

Cinquefoil may cause gastric discomfort in sensitive individuals (Huber et al. 2007; Wichtl 2004).

Pharmacological Considerations

None known.

Pregnancy and Lactation

No information on the safety of cinquefoil in pregnancy or lactation was identified in the scientific or traditional literature. Although this review did not identify any concerns for use while pregnant or nursing, safety has not been conclusively established.

Review Details

I. Drug and Supplement Interactions

Clinical Trials of Drug or Supplement Interactions

No clinical trials of drug or supplement interactions were identified.

Case Reports of Suspected Drug or Supplement Interactions

No case reports of suspected drug or supplement interactions were identified.

Animal Trials of Drug or Supplement Interactions

No animal trials of drug or supplement interactions were identified.

II. Adverse Events

Adverse Events Reported in Clinical Trials

No adverse effects were reported in a randomized, double-blind, placebo-controlled study of cinquefoil in children ages 3 months to 7 years with rotavirus diarrhea. Patients were orally administered 3 drops cinquefoil per year of age (~0.01 ml/kg), three times daily, for up to 5 days (Subbotina et al. 2003).

Case Reports of Adverse Events

No case reports of adverse events were identified.

III. Pharmacology and Pharmacokinetics

Human Pharmacological Studies

In an open-label dose escalation study in patients with ulcerative colitis, patients were orally administered 1200 mg, 1800 mg, 2400 mg, or 3000 mg of capsules containing an ethanolic dried extract of cinquefoil (containing 15 to 22% tannins) daily, divided into three doses. Adverse effects reported as likely related to the cinquefoil were stomach discomfort, heartburn, and nausea or fullness. Other adverse events reported as possible or "unclear" in relation to cinquefoil causality were abdominal cramps, rash on the arms and legs, and exacerbation of the ulcerative colitis (Huber et al. 2007).

Animal Pharmacological Studies

No relevant animal pharmacological studies were identified.

In Vitro Pharmacological Studies

No relevant in vitro pharmacological studies were identified.

IV. Pregnancy and Lactation

No information on the safety of cinquefoil during pregnancy or lactation was identified.

V. Toxicity Studies

Acute Toxicity

After oral administration, no adverse effects were observed in rats administered 2.5 g/kg or mice administered 6.8 g/kg of cinquefoil extract. After intraperitoneal administration, no adverse effects were observed in rats administered 3.8 g/kg or mice administered 14.5 g/kg of cinquefoil extract (Subbotina et al. 2003).

Literature Cited

Huber, R., A.V. Ditfurth, F. Amann, et al. 2007. Tormentil for active ulcerative colitis: An open-label, dose-escalating study. *J. Clin. Gastroenterol.* 41(9):834-838.

Subbotina, M.D., V.N. Timchenko, M.M. Vorobyov, et al. 2003. Effect of oral administration of tormentil root extract (*Potentilla tormentilla*) on rotavirus diarrhea in children: A randomized, double blind, controlled trial. *Pediatr. Infect. Dis. J.* 22(8):706-711.

Tomczyk, M., and K.P. Latte. 2009. Potentilla—A review of its phytochemical and pharmacological profile. *J. Ethnopharmacol.* 122(2):184-204.

Wichtl, M. 2004. *Herbal drugs and phytopharmaceuticals: A handbook for practice on a scientific basis*. 3rd ed. Boca Raton, FL: CRC Press.

Primula veris L. — Primulaceae

SCN: cowslip
Syn: *Primula officinalis* (L.) Hill

Part: flower, root

Quick Reference Summary

Safety Class: 2d
Interaction Class: A

Contraindications

Not for use in persons with gastritis or gastric ulcers (ESCOP 2003).

Other Precautions

Use with caution in persons with hypersensitivity to *Primula* species (EMEA 2008).

Drug and Supplement Interactions

None known.

Adverse Events and Side Effects

Occasional stomach discomfort and nausea has been reported with the use of cowslip root or flower (ESCOP 2003; Wichtl 2004).

Cowslip flower may cause skin reactions in individuals sensitive to species of *Primula* (Wichtl 2004).

Pharmacological Considerations

None known.

Pregnancy and Lactation

No information on the safety of cowslip in pregnancy or lactation was identified in the scientific or traditional literature. Although this review did not identify any concerns for use while pregnant or nursing, safety has not been conclusively established.

Review Details

I. Drug and Supplement Interactions

Clinical Trials of Drug or Supplement Interactions

No clinical trials of drug or supplement interactions were identified.

Case Reports of Suspected Drug or Supplement Interactions

No case reports of suspected drug or supplement interactions were identified.

Animal Trials of Drug or Supplement Interactions

No animal trials of drug or supplement interactions were identified.

II. Adverse Events

Case Reports of Adverse Events

Overdose of cowslip root (standard dose: decoction of 0.5–1.5 g, 2–4 g of tincture) may cause upset stomach, nausea, vomiting, or diarrhea (Hänsel et al. 1993; Wichtl 2004).

The leaf of various species of *Primula* contains the compound primin, a potent sensitizer that causes contact dermatitis in sensitive persons (Aplin and Lovell 2001; Ingber and Menne 1990). Primula dermatitis typically manifests as linear streaks of erythema with vesicles and bullae on the forearms, vesicles on the fingers, and sometimes on the face (Zachariae et al. 2007). The underground parts of cowslip do not contain primin, and thus no cutaneous effects are expected after oral use of cowslip root (EMEA 2008; Hausen 1978).

III. Pharmacology and Pharmacokinetics

Human Pharmacological Studies

No relevant human pharmacological studies were identified.

Animal Pharmacological Studies

No relevant animal pharmacological studies were identified.

In Vitro Pharmacological Studies

No relevant in vitro pharmacological studies were identified.

IV. Pregnancy and Lactation

No information on the safety of cowslip during pregnancy or lactation was identified.

V. Toxicity Studies

Acute Toxicity

The LD_{50} of the saponin fraction of cowslip root intraperitoneally administered in mice is 24.5 mg/kg, whereas that of primula acid intravenously administered to rats is 1.2 mg/kg (Hänsel et al. 1993). The toxic effects of cowslip are generally attributed to the saponin content (EMEA 2008).

Literature Cited

Aplin, C.G., and C.R. Lovell. 2001. Contact dermatitis due to hardy *Primula* species and their cultivars. *Contact Dermat.* 44(1):23-29.

EMEA. 2008. Assessment report on *Primula veris* L., *Primula elatior* (L.) Hill, *radix*. London: European Medicines Agency. Original edition, EMEA/HMPC/144474/2006.

ESCOP. 2003. *ESCOP monographs: The scientific foundation for herbal medicinal products*. 2nd ed. Exeter, UK: European Scientific Cooperative on Phytotherapy.

Hänsel, R., K. Keller, H. Rimpler, and G. Schneider, eds. 1993. *Hagers handbuch der pharmazeutischen praxis.* 5th ed. Berlin: Springer

Hausen, B.M. 1978. On the occurrence of the contact allergen primin and other quinoid compounds in species of the family of Primulaceae. *Arch. Dermatol. Res.* 261:311-321.

Ingber, A., and T. Menne. 1990. Primin standard patch testing: 5 years experience. *Contact Dermat.* 23(1):15-19.

Wichtl, M. 2004. *Herbal drugs and phytopharmaceuticals: A handbook for practice on a scientific basis.* 3rd ed. Boca Raton, FL: CRC Press.

Zachariae, C., K. Engkilde, J.D. Johansen, and T. Menne. 2007. Primin in the European standard patch test series for 20 years. *Contact Dermat.* 56(6):344-346.

Prunella vulgaris L. — Lamiaceae

SCN: heal all
PN: *xia ku cao* (fruit spike)
OCN: self heal
Part: fruit spike

Quick Reference Summary

Safety Class: 1
Interaction Class: A

Contraindications

None known.

Other Precautions

None known.

Drug and Supplement Interactions

None known.

Adverse Events and Side Effects

Allergic reactions to heal all have been reported (Bensky et al. 2004).

Pharmacological Considerations

An animal study demonstrated that heal all may modify glucose regulation (Zheng et al. 2007). People with diabetes are advised to monitor their blood sugar closely and discuss the use of this herb with a qualified healthcare practitioner prior to use.

In animal studies, antiestrogenic activity, with no adverse effects on fertility, was observed after treatment with heal all (Collins et al. 2009).

Pregnancy and Lactation

Limited information on the safety of heal all in pregnancy or lactation was identified in the scientific or traditional literature. Although this review did not identify any concerns for use while pregnant or nursing, safety has not been conclusively established.

Review Details

I. Drug and Supplement Interactions

Clinical Trials of Drug or Supplement Interactions

No clinical trials of drug or supplement interactions were identified.

Case Reports of Suspected Drug or Supplement Interactions

No case reports of suspected drug or supplement interactions were identified.

Animal Trials of Drug or Supplement Interactions

No animal trials of drug or supplement interactions were identified.

II. Adverse Events

Case Reports of Adverse Events

Rare cases of allergic reactions to heal all have been reported, with rashes, swelling of the throat, lips, and tongue, and gastrointestinal symptoms including nausea, vomiting, and diarrhea (Bensky et al. 2004). In an animal study, heal all demonstrated inhibition of immediate-type allergic reactions (Shin et al. 2001).

III. Pharmacology and Pharmacokinetics

Human Pharmacological Studies

No relevant human pharmacological studies were identified.

Animal Pharmacological Studies

A dose-dependent reduction in blood glucose levels was observed in an acute glucose tolerance test with diabetic mice orally administered 50, 75, 100, or 125 mg/kg of an aqueous-ethanolic extract of heal all. No increase in plasma levels of insulin was observed. A synergistic reduction of blood glucose levels was observed in mice administered a 100 mg/kg dose of the same extract along with 5 mg/kg of the drug glibenclamide (Zheng et al. 2007).

Antiestrogenic activity of heal all was observed in ovariectomized mice administered an aqueous extract of heal all as the sole source of drinking water (estimated daily intake not specified). The mice were implanted with human endometrial xenografts that remain only when treated with estrogen. Mice treated with estrogen and heal all for 1 month had fewer and smaller xenograft implants compared with their estrogen-treated counterparts that drank only water (Collins et al. 2009).

No adverse effects on fertility were observed in mice administered an aqueous extract of heal all as the sole source of drinking water (estimated daily intake not specified) for 2 weeks prior to mating (Collins et al. 2009).

In Vitro Pharmacological Studies

In a hormone-responsive endometrial cell line (ECC-1), a methanolic extract of heal all exhibited antiestrogenic activity. Specifically, a reduction in alkaline phosphatase activity and cell proliferation in response to estrogen in a dose-dependent manner was observed. The expression of an estrogen-induced protein was blocked in the cell line both by a standard antiestrogen compound and by the heal all extract (Collins et al. 2009).

In an in vitro screening system, an extract of heal all exhibited some inhibition of the drug-metabolizing isoenzyme CYP3A4 (Lee et al. 2007).

IV. Pregnancy and Lactation

In excised rabbit uteruses, a decoction of heal all produced tonic contractions in uteruses from nonpregnant animals but had only a weak effect on uteruses from pregnant animals (Zhu 1998).

No information on the safety of heal all during lactation was identified.

V. Toxicity Studies

Acute Toxicity

No adverse effects were observed in mice orally administered 10 g/kg of an extract of heal all (Chen and Chen 2004).

Literature Cited

Bensky, D., S. Clavey, and E. Stöger. 2004. *Chinese herbal medicine: Materia medica*. 3rd ed. Seattle: Eastland Press.

Chen, J.K., and T.T. Chen. 2004. *Chinese medical herbology and pharmacology*. City of Industry, CA: Art of Medicine Press.

Collins, N.H., E.C. Lessey, C.D. DuSell, et al. 2009. Characterization of antiestrogenic activity of the Chinese herb, *Prunella vulgaris*, using in vitro and in vivo (mouse xenograft) models. *Biol. Reprod.* 80(2):375-383.

Lee, S.S., B. Zhang, M.L. He, V.S.C. Chang, and H.F. Kung. 2007. Screening of active ingredients of herbal medicine for interaction with CYP450 3A4. *Phytother. Res.* 21(11):1096-1099.

Shin, T.Y., Y.K. Kim, and H.M. Kim. 2001. Inhibition of immediate-type allergic reactions by *Prunella vulgaris* in a murine model. *Immunopharmacol. Immunotoxicol.* 23(3):423-435.

Zheng, J., J. He, B. Ji, Y. Li, and X. Zhang. 2007. Antihyperglycemic activity of *Prunella vulgaris* L. in streptozotocin-induced diabetic mice. *Asia Pac. J. Clin. Nutr.* 16(Suppl. 1):427-431.

Zhu, Y.-P. 1998. *Chinese materia medica: Chemistry, pharmacology and applications.* Amsterdam: Harwood Academic Publishers.

P

Prunus armeniaca L. — Rosaceae

SCN: apricot
Syn: *Armeniaca vulgaris* Lam.
PN: *ku xing ren* (seed)
OCN: Chinese bitter almond
Part: seed

Quick Reference Summary

Safety Class: 3
Interaction Class: A

Contraindications

Not for use except under the supervision of an expert qualified in the appropriate use of this substance (Bensky et al. 2004; Femenia et al. 1995).

Other Precautions

None known.

Drug and Supplement Interactions

None known.

Notice

Cyanogenic glycosides (up to 8% amygdalin) (Chen and Chen 2004; Encarna et al. 1998; FAO/WHO 2009; Femenia et al. 1995; Gunders et al. 1969); *see* Appendix 1.

Editors' Notes

"Sweet" varieties of apricot kernel, with very low levels of cyanogenic glycosides (<0.000005% cyanide) are available commercially (Bensky et al. 2004). Concerns stated here refer to the bitter apricot kernel (0.2–0.4% cyanide) and not the sweet apricot kernel.

Bitter varieties of almonds (*Prunus dulcis*), formerly marketed as a supplement for persons with cancer, also contain the cyanogenic glycoside amygdalin (0.2% cyanide), while commonly consumed sweet varieties of almonds have only trace amounts of the compound (Dicenta et al. 2002). Concerns listed for bitter apricot also apply to bitter almond (Shragg et al. 1982).

Oil extracted from apricot kernel does not contain cyanogenic glycosides (Alpaslan and Hayta 2006; Beyer and Melton 1990).

The U.K. Food Standards Agency suggested that a daily dose of 0.005 mg/kg of cyanide would be unlikely to cause acute effects (FSA 2006). A discussion paper from the FAO and WHO indicated that this estimate may be overly conservative and does not account for the different toxicokinetics for amygdalin (the cyanogenic glycoside present in apricot kernels) that involves bacterial enzymatic conversion to hydrocyanic acid once ingested (FAO/WHO 2009). Other organizations have indicated provisional tolerable daily intake levels of cyanide at 0.012 to 0.108 mg/kg (FAO/WHO 2009).

Bitter apricot kernel should not be sold in retail in bulk; labels on all products should state: Not for use by children (McGuffin et al. 1997).

Adverse Events and Side Effects

Cases of poisoning, both fatal and nonfatal, have been reported in children and adults after ingestion of apricot kernels (Bensky et al. 2004; Gunders et al. 1969; Lasch and Shawa 1981; Sayre and Kaymakcalan 1964; Suchard et al. 1998; Townsend and Boni 1975). The toxic dose for children has been reported as 10 to 20 seeds, while for adults 40 to 60 seeds may cause toxicity (Bensky et al. 2004; Chen and Chen 2004). Poisoning from the cyanogenic glycosides results in symptoms such as salivation, gastric discomfort, nausea, vomiting, abdominal pain, diarrhea, headache, dizziness, weakness, difficulty breathing, restlessness, terror, and palpitations. In severe cases, symptoms may include coma, convulsions, cyanosis, dilated pupils, and death by respiratory failure (Bensky et al. 2004; Chen and Chen 2004).

Pharmacological Considerations

None known.

Pregnancy and Lactation

An animal study indicted that in rats fed diets containing high-amygdalin apricot kernel, offspring survival, lactation indices, and weaning weights were reduced (Miller et al. 1981).

No information on the safety of apricot kernel during lactation was identified in the scientific or traditional literature. While this review did not identify any concerns for use while nursing, safety has not been conclusively established, and this substance is not recommended for use except under the supervision of an expert qualified in its appropriate use.

Review Details

I. Drug and Supplement Interactions

Clinical Trials of Drug or Supplement Interactions

No clinical trials of drug or supplement interactions were identified.

Case Reports of Suspected Drug or Supplement Interactions

No case reports of suspected drug or supplement interactions were identified.

Animal Trials of Drug or Supplement Interactions

No animal trials of drug or supplement interactions were identified.

II. Adverse Events

Case Reports of Adverse Events

The toxic dose for children has been reported as 10 to 20 seeds; for adults, 40 to 60 seeds may cause toxicity. Ingestion of 50 to 120 seeds has resulted in fatal poisoning. Poisoning from the cyanogenic glycoside content results in symptoms such as salivation, gastric discomfort, nausea, vomiting, abdominal pain, diarrhea, headache, dizziness, weakness, difficulty breathing, restlessness, terror, and palpitations. In severe cases, symptoms may include coma, cyanosis, dilated pupils, and death by respiratory failure (Bensky et al. 2004; Chen and Chen 2004).

Within the normal dosage range (3–10 g as a decoction) and taken as a decoction, no toxic side effects are expected.

A suspension of the powdered kernel is considered four to five times as toxic as a decoction of the seed. Boiling reduces the toxicity of the kernel (Bensky et al. 2004).

Fatal cyanide poisoning occurred in a 3-year-old girl who ingested an unspecified number of apricot kernels. Analysis of apricot kernels from the same lot that caused the poisoning showed an average of 2.15 mg cyanide per kernel (Gunders et al. 1969).

A 41-year-old woman became weak and had difficulty breathing after ingesting 30 apricot kernels (estimated total, 15 g) purchased at a health food store. She became comatose and hypothermic but responded promptly to antidotal therapy for cyanide poisoning (Suchard et al. 1998).

Nonfatal cyanide poisoning was reported in a 49-year-old woman with lymphoma who had eaten 20 to 40 bitter apricot seeds (Suchard et al. 1998).

Other cases of cyanide poisoning in children and adults have been reported (Lasch and Shawa 1981; Sayre and Kaymakcalan 1964; Townsend and Boni 1975).

III. Pharmacology and Pharmacokinetics

Human Pharmacological Studies

No relevant human pharmacological studies were identified.

Animal Pharmacological Studies

No relevant animal pharmacological studies were identified.

In Vitro Pharmacological Studies

No relevant in vitro pharmacological studies were identified.

P

IV. Pregnancy and Lactation

In pregnant rats fed diets containing 10% ground apricot kernel from cultivars of apricots containing low amygdalin (<500 mg/kg cyanide), moderate amygdalin (1000–2000 mg/kg cyanide), or high amygdalin (2000 mg/kg cyanide) for 18 weeks, 3-day survival indices of offspring were lower in the high-amygdalin group than the low-amygdalin group. Lactation indices and weaning weights were also lower in the high-dose group than in the low-dose group (Miller et al. 1981).

No information on the safety of apricot kernel during lactation was identified.

V. Toxicity Studies

Acute Toxicity

The oral LD_{50} of an aqueous extract of apricot kernel (amygdalin content not specified in English language abstract) in mice is 2.25 g/kg, whereas that of the compound amygdalin is 443 mg/kg (Yamashita et al. 1987).

The LD_{50} of the compound amygdalin in rats is 25 g/kg after intravenous injection, 8 g/kg after intraperitoneal administration, and 0.6 g/kg after oral administration. Hydrocyanic acid is produced from amygdalin in the presence of gastric acid; thus the oral route is the most toxic (Chen and Chen 2004).

Short-Term Toxicity

In rats fed diets containing 30% powdered pomace of apricot kernels (containing "low" levels of amygdalin) for 4 weeks, poor weight gain and food efficiency ratio were observed. If water was added to the meal prior to feeding, these effects were ameliorated. No histological manifestations were observed in the organs examined (Gandhi et al. 1997).

Subchronic Toxicity

In a toxicity test, rats were fed diets containing 10% ground apricot kernel from cultivars of apricots containing low amygdalin (<500 mg/kg cyanide), moderate amygdalin (1000–2000 mg/kg cyanide), or high amygdalin (2000 mg/kg cyanide) for 18 weeks. In female but not male rats, liver rhodanese activity and thiocyanate (SCN) blood levels increased with the high-amygdalin diet. Both males and females efficiently excreted thiocyanate, indicating efficient detoxification and clearance of cyanide hydrolyzed from dietary amygdalin. No changes in blood chemistry were observed (Miller et al. 1981).

Literature Cited

Alpaslan, M., and M. Hayta. 2006. Apricot kernel: Physical and chemical properties. *J. Am. Oil Chem. Soc.* 83(5):469-471.

Bensky, D., S. Clavey, and E. Stöger. 2004. *Chinese herbal medicine: Materia medica*. 3rd ed. Seattle: Eastland Press.

Beyer, R., and L.D. Melton. 1990. Composition of New Zealand apricot kernels. *N.Z. J. Crop Hort. Sci.* 18(1):39-42.

Chen, J.K., and T.T. Chen. 2004. *Chinese medical herbology and pharmacology*. City of Industry, CA: Art of Medicine Press.

Dicenta, F., P. Martinez-Gomez, N. Grane, et al. 2002. Relationship between cyanogenic compounds in kernels, leaves, and roots of sweet and bitter kernelled almonds. *J. Agric. Food Chem.* 50(7):2149-2152.

Encarna, G., B. Lorenzo, S. Constanza, and M. Josefa. 1998. Amygdalin content in the seeds of several apricot cultivars. *J. Sci. Food Agric.* 77(2):184-186.

FAO/WHO. 2009. Discussion paper on cyanogenic glycosides. Third Session, March 2009. Joint FAO/WHO Food Standards Programme Codex Committee on Contaminants in Foods.

Femenia, A., C. Rossello, A. Mulet, and J. Canellas. 1995. Chemical composition of bitter and sweet apricot kernels. *J. Agric. Food Chem.* 43(2):356-361.

FSA. 2006. Statement on cyanogenic glycosides in bitter apricot kernels. Committee on Toxicity of Chemicals in Food, Consumer Products and the Environment. U.K. Food Standards Agency. TOX-2006-13.

Gandhi, V.M., B. Mukerji, V.J. Iyer, and K.M. Cherian. 1997. Nutritional and toxicological evaluation of wild apricot pomace. *J. Food Sci. Technol.* 34(2):132-135.

Gunders, A.E., A. Abrahamov, E. Weisenberg, S. Gertner, and S. Shafran. 1969. Cyanide poisoning following ingestion of apricot (*Prunus armeniaca*) kernels. *Harefuah* 76(12):536-538.

Lasch, E.E., and R.E. Shawa. 1981. Multiple cases of cyanide poisoning by apricot kernels in children from Gaza. *Pediatrics* 68(1):5.

McGuffin, M., C. Hobbs, R. Upton, and A. Goldberg. 1997. *Botanical safety handbook*. Boca Raton, FL: CRC Press.

Miller, K.W., J.L. Anderson, and G.S. Stoewsand. 1981. Amygdalin metabolism and effect on reproduction of rats fed apricot (*Prunus armeniaca*) kernels. *J. Toxicol. Environ. Health* 7(3-4):457-468.

Sayre, J.W., and S. Kaymakcalan. 1964. Cyanide poisoning from apricot seeds among children in Central Turkey. *N. Engl. J. Med.* 270:1113-1115.

Shragg, T.A., T.E. Albertson, and C.J. Fisher, Jr. 1982. Cyanide poisoning after bitter almond ingestion. *West. J. Med.* 136(1):65-67.

Suchard, J.R., K.L. Wallace, and R.D. Gerkin. 1998. Acute cyanide toxicity caused by apricot kernel ingestion. *Ann. Emerg. Med.* 32(6):742-744.

Townsend, W.A., and B. Boni. 1975. Cyanide poisoning from ingestion of apricot kernels. *Morbid. Mortal. Weekly Rep.* 24:427.

Yamashita, M., S. Minematsu, Y. Kobayashi, N. Kiuchi, and M. Aburada. 1987. Acute toxicity of apricot kernel (Armeniacae Semen) in mice. *Pharm. Mon.* 29(Jun):1291-1294.

Prunus mume Siebold & Zucc. — Rosaceae

SCN: Japanese apricot
Syn: *Armeniaca mume* Siebold
PN: *wu mei* (smoked fruit)
OCN: mume; ume
Part: unripe fruit

Quick Reference Summary

Safety Class: 1
Interaction Class: A

Contraindications

None known.

Other Precautions

None known.

Drug and Supplement Interactions

None known.

Adverse Events and Side Effects

None known.

Pharmacological Considerations

None known.

Pregnancy and Lactation

No information on the safety of Japanese apricot in pregnancy or lactation was identified in the scientific or traditional literature. Although this review did not identify any concerns for use while pregnant or nursing, safety has not been conclusively established.

Review Details

I. Drug and Supplement Interactions

Clinical Trials of Drug or Supplement Interactions

No clinical trials of drug or supplement interactions were identified.

Case Reports of Suspected Drug or Supplement Interactions

No case reports of suspected drug or supplement interactions were identified.

Animal Trials of Drug or Supplement Interactions

No animal trials of drug or supplement interactions were identified.

II. Adverse Events

Case Reports of Adverse Events

No case reports of adverse events were identified.

III. Pharmacology and Pharmacokinetics

Human Pharmacological Studies

No relevant human pharmacological studies were identified.

Animal Pharmacological Studies

No relevant animal pharmacological studies were identified.

P

In Vitro Pharmacological Studies

No significant effects of Japanese apricot juice were observed on the drug-metabolizing isoenzyme CYP3A4 (Kim et al. 2006).

IV. Pregnancy and Lactation

No information on the safety of Japanese apricot during pregnancy or lactation was identified.

V. Toxicity Studies

No toxicity studies were identified.

Literature Cited

Kim, H., Y.-J. Yoon, J.-H. Shon, et al. 2006. Inhibitory effects of fruit juices on CYP3A activity. *Drug Metab. Dispos.* 34(4):521-523.

Prunus persica (L.) Batsch

Rosaceae

SCN: peach
Syn: *Amygdalus persica* L.
PN: *tao ren* (seed)
Part: leaf, seed, twig

Quick Reference Summary

Safety Class: 3
Interaction Class: A

Contraindications

Not for use except under the supervision of an expert qualified in the appropriate use of this substance (Bensky et al. 2004; Holzbecher et al. 1984).

Other Precautions

None known.

Drug and Supplement Interactions

None known.

Notice

Cyanogenic glycosides (2–6% amygdalin in seed, ~0.5 to 1.5% in leaf) (Holzbecher et al. 1984; List and Hörhammer 1973; Machel and Dorsett 1970); *see* Appendix 1.

Editors' Notes

Peach kernel is generally taken as a decoction. If taken directly, in pills or powders, the kernels should be processed by peeling after being boiled in water for a short time. This processing reduces the toxicity (Bensky et al. 2004). The leaf and twig are less toxic than the kernel (Machel and Dorsett 1970).

Peach leaf and twig should be completely dried before use. The partially dried material contains higher levels of hydrocyanic acid than the fresh or fully dried material. Complete drying reduces the hydrocyanic acid content (Kingsbury 1964; Radostits et al. 2000).

Adverse Events and Side Effects

Overdose (standard dose is a decoction of 4.5–10 g of the kernel) of peach kernel, twig, or leaf may result in poisoning. In children under five, 5 to 10 kernels have caused adverse effects, and 20 kernels have been lethal (Bensky et al. 2004). Symptoms of overdose are similar to cyanide poisoning, including nausea, vomiting, irritation of the gastrointestinal mucosa, headache, dizziness, weakness, blurred vision, increased heart rate, and difficulty breathing (Bensky et al. 2004).

Pharmacological Considerations

None known.

Pregnancy and Lactation

Texts on traditional Chinese medicine indicate that peach kernel should not be used during pregnancy (Bensky et al. 2004; Chen and Chen 2004).

No information on the safety of peach kernel during lactation was identified in the scientific or traditional literature. While this review did not identify any concerns for use while nursing, safety has not been conclusively established, and this substance is not recommended for use except under the supervision of an expert qualified in its appropriate use.

Review Details

I. Drug and Supplement Interactions

Clinical Trials of Drug or Supplement Interactions

No clinical trials of drug or supplement interactions were identified.

Case Reports of Suspected Drug or Supplement Interactions

No case reports of suspected drug or supplement interactions were identified.

Animal Trials of Drug or Supplement Interactions

No animal trials of drug or supplement interactions were identified.

II. Adverse Events

Case Reports of Adverse Events

Overdose of peach kernel, twig, or leaf (standard dose is a decoction of 4.5–10 g of the kernel) may result in poisoning. In children under five, 5 to 10 kernels have caused adverse effects, and 20 kernels have been lethal (Bensky et al. 2004). Symptoms of overdose are similar to cyanide poisoning, including nausea, vomiting, irritation of the gastrointestinal tract, headache, dizziness, weakness, blurred vision, increased heart rate, and difficulty breathing. In more severe cases, symptoms may include incontinence of the urine or stool, loss of consciousness, dilation of pupils, absence of pupillary reflexes, severe spasms, cyanosis, and shock (Bensky et al. 2004).

III. Pharmacology and Pharmacokinetics

Human Pharmacological Studies

No relevant human pharmacological studies were identified.

Animal Pharmacological Studies

In rats orally administered an aqueous extract (boiled for 3 hours) of peach kernel, inhibition of cholinesterase activity in the brain and plasma was observed. The inhibitory dose (ID50) for brain cholinesterase activity was 2.7 g/kg; for plasma cholinesterase activity inhibition, the ID50 was 18.6 g/kg (Suh et al. 2006).

In Vitro Pharmacological Studies

No estrogenic activity of ethanol extracts of peach kernel was observed in a recombinant yeast system with a human estrogen receptor expression plasmid and a reporter plasmid (Kang et al. 2006; Kim et al. 2008).

IV. Pregnancy and Lactation

Texts on traditional Chinese medicine indicate that peach kernel should not be used during pregnancy (Bensky et al. 2004; Chen and Chen 2004).

No information on the safety of peach kernel during lactation was identified.

V. Toxicity Studies

The LD_{50} of an intramuscularly injected aqueous extract of peach kernel in mice is 222 g/kg (Chen and Chen 2004).

The oral LD_{50} of an aqueous extract containing the compound amygdalin in mice is 443 mg/kg (Yamashita et al. 1987).

Literature Cited

Bensky, D., S. Clavey, and E. Stöger. 2004. *Chinese herbal medicine: Materia medica*. 3rd ed. Seattle: Eastland Press.

Chen, J.K., and T.T. Chen. 2004. *Chinese medical herbology and pharmacology*. City of Industry, CA: Art of Medicine Press.

Holzbecher, M.D., M.A. Moss, and H.A. Ellenberger. 1984. The cyanide content of laetrile preparations, apricot, peach and apple seeds. *Clin. Toxicol.* 22(4):341-347.

Kang, S.C., C.M. Lee, H. Choi, et al. 2006. Evaluation of oriental medicinal herbs for estrogenic and antiproliferative activities. *Phytother. Res.* 20(11):1017-1019.

Kim, I.G., S.C. Kang, K.C. Kim, E.S. Choung, and O.P. Zee. 2008. Screening of estrogenic and antiestrogenic activities from medicinal plants. *Environ. Toxicol. Pharmacol.* 25(1):75-82.

Kingsbury, J.M. 1964. *Poisonous plants of the United States and Canada*. Englewood Cliffs, NJ: Prentice-Hall.

List, P.H., and H. Hörhammer. 1973. *Hagers handbuch der pharmazeutischen praxis*. Berlin: Springer.

Machel, A.R., and C.I. Dorsett. 1970. Cyanide analyses of peaches. *Econ. Bot.* 24:5-12.

Radostits, O.M., C.C. Gay, D.C. Blood, and K.W. Hinchcliff. 2000. *Veterinary medicine*. 9th ed. Edinburgh: Saunders.

Suh, S.J., B.S. Koo, U.H. Jin, et al. 2006. Pharmacological characterization of orally active cholinesterase inhibitory activity of *Prunus persica* L. Batsch in rats. *J. Mol. Neurosci.* 29(2):101-107.

Yamashita, M., S. Minematsu, Y. Kobayashi, N. Kiuchi, and M. Aburada. 1987. [Acute toxicity of apricot kernel (Armeniacae Semen) in mice.] *Pharm. Mon.* 29(Jun.):1291-1294.

Prunus serotina Ehrh. — Rosaceae

SCN: black cherry
OCN: wild black cherry

Part: dried bark

Quick Reference Summary

Safety Class: 1
Interaction Class: A

Contraindications

None known.

P

Other Precautions

None known.

Drug and Supplement Interactions

None known.

Standard Dose

The standard dose is 2.0–4.0 g (Osol and Farrar 1955).

Notice

Cyanogenic glycosides (prunasin, yielding up to 0.15% hydrocyanic acid) (List and Hörhammer 1973); *see* Appendix 1.

Editors' Notes

Black cherry bark should be completely dried before use. The partially dried bark contains higher levels of hydrocyanic acid than the fresh or fully dried bark, while complete drying reduces the hydrocyanic acid content (Kingsbury 1964; Radostits et al. 2000).

Adverse Events and Side Effects

None known.

Pharmacological Considerations

None known.

Pregnancy and Lactation

No information on the safety of black cherry in pregnancy or lactation was identified in the scientific or traditional literature. Although this review did not identify any concerns for use while pregnant or nursing, safety has not been conclusively established.

Review Details

I. Drug and Supplement Interactions

Clinical Trials of Drug or Supplement Interactions

No clinical trials of drug or supplement interactions were identified.

Case Reports of Suspected Drug or Supplement Interactions

No case reports of suspected drug or supplement interactions were identified.

Animal Trials of Drug or Supplement Interactions

No animal trials of drug or supplement interactions were identified.

II. Adverse Events

Case Reports of Adverse Events

No case reports of adverse events were identified.

III. Pharmacology and Pharmacokinetics

Human Pharmacological Studies

No relevant human pharmacological studies were identified.

Animal Pharmacological Studies

No relevant animal pharmacological studies were identified.

In Vitro Pharmacological Studies

No relevant in vitro pharmacological studies were identified.

IV. Pregnancy and Lactation

No information on the safety of black cherry during pregnancy or lactation was identified.

V. Toxicity Studies

No toxicity studies were identified.

Literature Cited

Kingsbury, J.M. 1964. *Poisonous plants of the United States and Canada*. Englewood Cliffs, N.J.: Prentice-Hall.

List, P.H., and H. Hörhammer. 1973. *Hagers handbuch der pharmazeutischen praxis*. Berlin: Springer.

Osol, A., and G. Farrar. 1955. *The dispensatory of the United States of America*. 25th ed. Philadelphia: Lippincott Company.

Radostits, O.M., C.C. Gay, D.C. Blood, and K.W. Hinchcliff. 2000. *Veterinary medicine*, 9th ed. Edinburgh: Saunders.

Prunus spinosa L. — Rosaceae

SCN: sloe

OCN: blackthorn

Part: flower, fruit

Quick Reference Summary

Safety Class: 1

Interaction Class: A

Contraindications

None known.

Other Precautions

None known.

Drug and Supplement Interactions

None known.

Notice

Cyanogenic glycosides (seed and fresh flower) (List and Hörhammer 1973); *see* Appendix 1.

Diuretic (Wichtl 2004); *see* Appendix 2.

Adverse Events and Side Effects

None known.

Pharmacological Considerations

None known.

Pregnancy and Lactation

No information on the safety of sloe in pregnancy or lactation was identified in the scientific or traditional literature. Although this review did not identify any concerns for use while pregnant or nursing, safety has not been conclusively established.

Review Details

I. Drug and Supplement Interactions

Clinical Trials of Drug or Supplement Interactions

No clinical trials of drug or supplement interactions were identified.

Case Reports of Suspected Drug or Supplement Interactions

No case reports of suspected drug or supplement interactions were identified.

Animal Trials of Drug or Supplement Interactions

No animal trials of drug or supplement interactions were identified.

II. Adverse Events

Case Reports of Adverse Events

No case reports of adverse events were identified.

III. Pharmacology and Pharmacokinetics

Human Pharmacological Studies

No relevant human pharmacological studies were identified.

Animal Pharmacological Studies

No relevant animal pharmacological studies were identified.

In Vitro Pharmacological Studies

No relevant in vitro pharmacological studies were identified.

IV. Pregnancy and Lactation

No information on the safety of sloe during pregnancy or lactation was identified.

V. Toxicity Studies

Acute Toxicity

In the brine shrimp lethality assay, the LC_{50} of a methanol extract of sloe fruit is 230 mg/ml (Kumarasamy et al. 2004).

P

Literature Cited

Kumarasamy, Y., P.J. Cox, M. Jaspars, L. Nahar, and S.D. Sarker. 2004. Comparative studies on biological activities of *Prunus padus* and *P. spinosa*. *Fitoterapia* 75(1):77-80.

List, P.H., and H. Hörhammer. 1973. *Hagers handbuch der pharmazeutischen praxis*. Berlin: Springer.

Wichtl, M. 2004. *Herbal drugs and phytopharmaceuticals: A handbook for practice on a scientific basis*. 3rd ed. Boca Raton, FL: CRC Press.

Pterocarpus santalinus L. f. — Fabaceae

SCN: red saunders

AN: *rakta chandana*

OCN: red sandalwood

Part: heartwood

Quick Reference Summary

Safety Class: 1

Interaction Class: A

Contraindications

None known.

Other Precautions

None known.

Drug and Supplement Interactions

None known.

Editors' Note

Use of this species as a food additive in the U.S. is limited to its function as a flavoring substance in alcoholic beverages (CFR 2011). Dietary ingredients for use in dietary supplements, however, are specifically excluded from the federal food additive definition (U.S.C. 2010).

Adverse Events and Side Effects

Allergic contact dermatitis from red saunders has been reported (Sandra et al. 1996).

Pharmacological Considerations

An animal study demonstrated that red saunders may modify glucose regulation (Kameswara Rao et al. 2001). People with diabetes are advised to monitor their blood sugar closely and discuss the use of this herb with a qualified healthcare practitioner prior to use.

Pregnancy and Lactation

No information on the safety of red saunders in pregnancy or lactation was identified in the scientific or traditional literature. Although this review did not identify any concerns for use while pregnant or nursing, safety has not been conclusively established.

Review Details

I. Drug and Supplement Interactions

Clinical Trials of Drug or Supplement Interactions

No clinical trials of drug or supplement interactions were identified.

Case Reports of Suspected Drug or Supplement Interactions

No case reports of suspected drug or supplement interactions were identified.

Animal Trials of Drug or Supplement Interactions

No animal trials of drug or supplement interactions were identified.

P

II. Adverse Events

Case Reports of Adverse Events

Allergic contact dermatitis from red saunders has been reported (Sandra et al. 1996).

III. Pharmacology and Pharmacokinetics

Human Pharmacological Studies

No relevant human pharmacological studies were identified.

Animal Pharmacological Studies

In healthy or diabetic rats orally administered 0.25, 0.5, 0.75, or 2.0 g/kg of the water, ethanol, or hexane fractions of red saunders bark, a reduction in blood glucose levels was observed in diabetic but not in healthy animals. The ethanol extract produced the maximum antihyperglycemic activity (Kameswara Rao et al. 2001).

In Vitro Pharmacological Studies

No relevant in vitro pharmacological studies were identified.

IV. Pregnancy and Lactation

No information on the safety of red saunders during pregnancy or lactation was identified.

V. Toxicity Studies

No toxicity studies were identified.

Literature Cited

Kameswara Rao, B., R. Giri, M.M. Kesavulu, and C. Apparao. 2001. Effect of oral administration of bark extracts of *Pterocarpus santalinus* L. on blood glucose level in experimental animals. *J. Ethnopharmacol.* 74(1):69-74.

Sandra, A., S.D. Shenoi, and C.R. Srinivas. 1996. Allergic contact dermatitis from red sandalwood (*Pterocarpus santalinus*). *Contact Dermat.* 34(1):69.

U.S.C. 2010. United States Code, Title 21, Part 321 (s)(6). Current as of January 7, 2011. Washington, DC: U.S. Government Printing Office.

Ptychopetalum spp.

Olacaceae

Ptychopetalum olacoides Benth.
SCN: muira puama
OCN: marapuama; potency wood

Ptychopetalum uncinatum **Anselmino**
SCN: *Ptychopetalum uncinatum*
OCN: marapuama; muira puama
Part: root, wood

Quick Reference Summary

Safety Class: 1
Interaction Class: A

Contraindications

None known.

Other Precautions

None known.

Drug and Supplement Interactions

None known.

Adverse Events and Side Effects

None known.

Pharmacological Considerations

None known.

Pregnancy and Lactation

No information on the safety of muira puama or *Ptychopetalum uncinatum* in pregnancy or lactation was identified in the scientific or traditional literature. Although this review did not identify any concerns for use while pregnant or nursing, safety has not been conclusively established.

Review Details

I. Drug and Supplement Interactions

Clinical Trials of Drug or Supplement Interactions

No clinical trials of drug or supplement interactions were identified.

Case Reports of Suspected Drug or Supplement Interactions

No case reports of suspected drug or supplement interactions were identified.

Animal Trials of Drug or Supplement Interactions

No animal trials of drug or supplement interactions were identified.

II. Adverse Events

Case Reports of Adverse Events

No case reports of adverse events were identified.

III. Pharmacology and Pharmacokinetics

Human Pharmacological Studies

No relevant human pharmacological studies were identified.

Animal Pharmacological Studies

No relevant animal pharmacological studies were identified.

In Vitro Pharmacological Studies

In rat brains treated with different concentrations of an ethanol extract of muira puama, dose- and time-dependent inhibition of acetylcholinesterase was observed in the rat frontal cortex, hippocampus, and striatum. Acetylcholinesterase inhibition in the same brain regions was observed in ex vivo brains of aged mice that had been intraperitoneally administered 100 mg/kg of an ethanol extract of muira puama (Siqueira et al. 2003).

IV. Pregnancy and Lactation

No information on the safety of muira puama or *Ptychopetalum uncinatum* during pregnancy or lactation was identified.

V. Toxicity Studies

No toxicity studies were identified.

Literature Cited

Siqueira, I.R., C. Fochesatto, A.L. da Silva, et al. 2003. *Ptychopetalum olacoides*, a traditional Amazonian "nerve tonic," possesses anticholinesterase activity. *Pharmacol. Biochem. Behav.* 75(3):645-650.

P

Pueraria montana (Lour.) Merr. — Fabaceae

Pueraria montana **(Lour.) Merr. var. *chinense* Maesen & S.M. Almeida**
SCN: kudzu
Syn: *Pueraria thomsonii* Benth.
PN: *ge gen* (root)

Pueraria montana **(Lour.) Merr. var. *lobata* (Willd.) Maesen & S.M. Almeida**
SCN: kudzu
Syn: *Pueraria lobata* (Willd.) Ohwi
PN: *ge gen* (root)
Part: root

Quick Reference Summary

Safety Class: 1
Interaction Class: A

Contraindications

None known.

Other Precautions

None known.

Drug and Supplement Interactions

See Pharmacological Considerations.

Adverse Events and Side Effects

None known.

Pharmacological Considerations

In vitro studies with kudzu and the compound puerarin, and animal and in vitro studies of the compound puerarin, have shown that these substances may inhibit platelet aggregation (Choo et al. 2002; Pan et al. 2003; Zhu 1998). Although no cases of interactions have been reported, kudzu should be used with caution in persons taking anticoagulant or antiplatelet drugs (Chen and Chen 2004).

Several in vitro studies have indicated estrogenic activity of kudzu (Boue et al. 2003; Kang et al. 2006; Kim et al. 2008; Zhang et al. 2005), while another in vitro study and a human clinical trial indicated no estrogenic activity (Cherdshewasart et al. 2004; Malaivijitnond et al. 2006).

An animal study indicated that high doses of kudzu (2 or 4 g/kg) slowed the elimination of and increased exposure to the drug methotrexate (Chiang et al. 2005).

Pregnancy and Lactation

No information on the safety of kudzu in pregnancy or lactation was identified in the scientific or traditional literature. Although this review did not identify any concerns for use while pregnant or nursing, safety has not been conclusively established.

Review Details

I. Drug and Supplement Interactions

Clinical Trials of Drug or Supplement Interactions

No clinical trials of drug or supplement interactions were identified.

Case Reports of Suspected Drug or Supplement Interactions

No case reports of suspected drug or supplement interactions were identified.

Animal Trials of Drug or Supplement Interactions

A decrease in the elimination rate and increase in exposure to the drug methotrexate were observed in rats administered 1 mg/kg intravenously or 5 mg/kg orally of methotrexate with 2 or 4 g/kg of orally administered kudzu (Chiang et al. 2005).

II. Adverse Events

Adverse Events Reported in Clinical Trials

In menopausal women orally administered an aqueous extract of kudzu containing 100 to 200 mg of isoflavones daily for 3 months, slight reductions in estradiol and follicle-stimulating hormone were observed, with no changes in luteinizing hormone (Woo et al. 2003).

Case Reports of Adverse Events

Large overdose of kudzu (standard dose is a decoction of 9–21 g) has been associated with arrhythmias (Bensky et al. 2004).

III. Pharmacology and Pharmacokinetics

Human Pharmacological Studies

No relevant human pharmacological studies were identified.

Animal Pharmacological Studies

In ovariectomized rats orally administered 10, 100, or 1000 mg/kg of kudzu daily for 14 days, an increase in vaginal cornification but no changes in vaginal epithelium were observed in animals in the 100 and 1000 mg/kg dose groups. An increase in uterine weight was observed in animals at the 1000 mg/kg dose (Malaivijitnond et al. 2006).

In rats intravenously administered the compound puerarin (dose not specified in available English language translation), a reduction in the viscosity of whole blood and plasma, blood yield stress, and the maximum rate of platelet aggregation was observed in animals with acute blood stasis (Pan et al. 2003).

In diabetic and healthy rats intravenously administered 15 mg/kg of the compound puerarin, a reduction in plasma glucose levels was observed. The effect was greater in diabetic than in healthy animals (Hsu et al. 2003).

In rats orally administered 100 or 200 mg/kg of the compound puerarin or 700 or 1400 mg/kg of kudzu, a "complex pattern of modulation" of drug-metabolizing isoenzymes was observed, with induction of the drug-metabolizing isoenzymes CYP1A2, CYP3A1, and CYP2B1 in rats treated with kudzu and induction of CYP2A1, 1CYPA1/2, CYP3A1, and CYP2C11 in animals administered puerarin. Both treatments were also reported to inactivate CYP3A, CYP2E1, and CYP2B1 (Guerra et al. 2000).

In Vitro Pharmacological Studies

Inhibition of ADP-induced platelet aggregation was observed in rat platelets treated with 0.25, 0.5, or 1 mg/ml of the compound puerarin. At concentrations of 0.5 to 3.0 mg/ml, puerarin inhibited rabbit, sheep, and human platelet aggregation induced by ADP or 5-HT (Zhu 1998).

Inhibition of ADP- and collagen-induced platelet aggregation was observed in the blood of rats that had been orally administered 1 g/kg of kudzu or 0.25 or 0.5 mg/kg of the compound puerarin (Choo et al. 2002).

An increase in cell proliferation was observed in estrogen receptor-positive human breast cancer cells (MCF-7) treated with a methanol extract of kudzu. The addition of an estrogen antagonist (ICI 182,780) suppressed cell proliferation induced by kudzu (Boue et al. 2003).

In estrogen receptor-positive human breast cancer cells (MCF-7) treated with kudzu, no proliferation and a mild antiproliferative effect was observed (Cherdshewasart et al. 2004).

Competitive binding of a methanol extract of kudzu to estrogen receptor β was observed in a competitive binding assay (Boue et al. 2003).

In assays using recombinant yeast systems with a human estrogen receptor expression plasmid and a reporter plasmid, estrogenic activity of kudzu was observed (Kang et al. 2006; Kim et al. 2008; Zhang et al. 2005).

The compound puerarin blocked L-type calcium channels and potassium channels in isolated guinea pig ventricular myocytes (Sun et al. 2007).

IV. Pregnancy and Lactation

No information on the safety of kudzu during pregnancy or lactation was identified.

V. Toxicity Studies

Acute Toxicity

No evidence of toxicity was observed in mice orally administered up to 20 g/kg of an ethanol extract of kudzu daily for 3 days (Zhu 1998).

The LD_{50} of intravenously administered puerarin in mice is 738 mg/kg (Zhu 1998).

Short-Term Toxicity

In mice orally administered 2 g/kg of an ethanol extract of kudzu daily for 60 days, no pathological changes in organs were observed (Zhu 1998).

No toxic effects were observed in hypertensive dogs orally administered 2 g/kg of an ethanol extract of kudzu daily for 14 days (Zhu 1998).

P

Literature Cited

Bensky, D., S. Clavey, and E. Stöger. 2004. *Chinese herbal medicine: Materia medica*. 3rd ed. Seattle: Eastland Press.

Boue, S.M., T.E. Wiese, S. Nehls, et al. 2003. Evaluation of the estrogenic effects of legume extracts containing phytoestrogens. *J. Agric. Food Chem.* 51(8):2193-2199.

Chen, J.K., and T.T. Chen. 2004. *Chinese medical herbology and pharmacology*. City of Industry, CA: Art of Medicine Press.

Cherdshewasart, W., W. Cheewasopit, and P. Picha. 2004. The differential anti-proliferation effect of white (*Pueraria mirifica*), red (*Butea superba*), and black (*Mucuna collettii*) kwao krua plants on the growth of MCF-7 cells. *J. Ethnopharmacol.* 93(2-3):255-260.

Chiang, H.M., S.H. Fang, K.C. Wen, et al. 2005. Life-threatening interaction between the root extract of *Pueraria lobata* and methotrexate in rats. *Toxicol. Appl. Pharmacol.* 209(3):263-268.

Choo, M.K., E.K. Park, H.K. Yoon, and D.H. Kim. 2002. Antithrombotic and antiallergic activities of daidzein, a metabolite of puerarin and daidzin produced by human intestinal microflora. *Biol. Pharm. Bull.* 25(10):1328-1332.

Guerra, M.C., E. Speroni, M. Broccoli, et al. 2000. Comparison between chinese medical herb *Pueraria lobata* crude extract and its main isoflavone puerarin antioxidant properties and effects on rat liver CYP-catalysed drug metabolism. *Life Sci.* 67(24):2997-3006.

Hsu, F.L., I.M. Liu, D.H. Kuo, et al. 2003. Antihyperglycemic effect of puerarin in streptozotocin-induced diabetic rats. *J. Nat. Prod.* 66(6):788-792.

Kang, S.C., C.M. Lee, H. Choi, et al. 2006. Evaluation of oriental medicinal herbs for estrogenic and antiproliferative activities. *Phytother. Res.* 20(11):1017-1019.

Kim, I.G., S.C. Kang, K.C. Kim, E.S. Choung, and O.P. Zee. 2008. Screening of estrogenic and antiestrogenic activities from medicinal plants. *Environ. Toxicol. Pharmacol.* 25(1):75-82.

Malaivijitnond, S., K. Chansri, P. Kijkuokul, N. Urasopon, and W. Cherdshewasart. 2006. Using vaginal cytology to assess the estrogenic activity of phytoestrogen-rich herb. *J. Ethnopharmacol.* 107(3):354-360.

Pan, H.P., J.Z. Yang, L.L. Li, et al. 2003. Experimental study of puerarin injection on the hemorheology in acute blood-stasis model rats. *Zhongguo Zhong Yao Za Zhi* 28(12):1178-1180.

Sun, X.-H., J.-P. Ding, H. Li, et al. 2007. Activation of large-conductance calcium-activated potassium channels by puerarin: The underlying mechanism of puerarin-mediated vasodilation. *J. Pharmacol. Exp. Ther.* 323(1):391-397.

Woo, J., E. Lau, S.C. Ho, et al. 2003. Comparison of *Pueraria lobata* with hormone replacement therapy in treating the adverse health consequences of menopause. *Menopause* 10 (4):352-361.

Zhang, C.Z., S.X. Wang, Y. Zhang, J.P. Chen, and X.M. Liang. 2005. *In vitro* estrogenic activities of Chinese medicinal plants traditionally used for the management of menopausal symptoms. *J. Ethnopharmacol.* 98(3):295-300.

Zhu, Y.-P. 1998. *Chinese materia medica: Chemistry, pharmacology and applications.* Amsterdam: Harwood Academic Publishers.

Pulmonaria officinalis L. — Boraginaceae

SCN: lungwort

Part: herb

Quick Reference Summary

Safety Class: 1
Interaction Class: A

Contraindications

None known.

Other Precautions

None known.

Drug and Supplement Interactions

None known.

Editors' Notes

Although the presence of pyrrolizidine alkaloids is common in the Boraginaceae family, and the presence of those compounds in lungwort has been suspected, analyses of plants containing pyrrolizidine alkaloids failed to detect any pyrrolizidine alkaloids in lungwort (Dobler et al. 2000; Williamson 2003).

A set of anticholinergic poisonings, improperly attributed to lungwort, was reported (Baca-Garcia et al. 2007). Lungwort does not contain compounds that would cause anticholinergic poisoning.

Adverse Events and Side Effects

None known.

Pharmacological Considerations

None known.

Pregnancy and Lactation

No information on the safety of lungwort in pregnancy or lactation was identified in the scientific or traditional literature. Although this review did not identify any concerns for use while pregnant or nursing, safety has not been conclusively established.

Review Details

I. Drug and Supplement Interactions

Clinical Trials of Drug or Supplement Interactions

No clinical trials of drug or supplement interactions were identified.

Case Reports of Suspected Drug or Supplement Interactions

No case reports of suspected drug or supplement interactions were identified.

Animal Trials of Drug or Supplement Interactions

No animal trials of drug or supplement interactions were identified.

II. Adverse Events

Case Reports of Adverse Events

A set of anticholinergic poisonings, improperly attributed to lungwort, was reported in three members of a family who consumed tea from what were believed at the time to be lungwort leaves. The identity of the leaves was not confirmed or analyzed for the presence of adulterants (Baca-Garcia et al. 2007). Lungwort does not contain compounds that would cause anticholinergic poisoning.

III. Pharmacology and Pharmacokinetics

Human Pharmacological Studies

No relevant human pharmacological studies were identified.

Animal Pharmacological Studies

No relevant animal pharmacological studies were identified.

In Vitro Pharmacological Studies

No relevant in vitro pharmacological studies were identified.

IV. Pregnancy and Lactation

No information on the safety of lungwort during pregnancy or lactation was identified.

V. Toxicity Studies

No toxicity studies were identified.

Literature Cited

Baca-Garcia, E., H. Blasco-Fontecilia, C. Blanco, et al. 2007. Acute atropine intoxication with psychiatric symptoms by herbal infusion of *Pulmonaria officinalis* (lungwort). *Eur. J. Psychiatr.* 21 (2):93-97.

Dobler, S., W. Haberer, L. Witte, and T. Hartmann. 2000. Selective sequestration of pyrrolizidine alkaloids from diverse host plants by *Longitarsus* flea beetles. *J. Chem. Ecol.* 26(5):1281-1298.

Williamson, E.M. 2003. *Potter's herbal cyclopedia*. Saffron Walden, Essex: C.W. Daniel Co.

Punica granatum L.

Punicaceae

SCN: pomegranate
AN: *dadima*

PN: *shi liu pi* (fruit rind)
Part: fruit husk

Quick Reference Summary

Safety Class: 2b, 2d
Interaction Class: A

Contraindications

Not for use in pregnancy except under the supervision of a qualified healthcare practitioner (Gujral et al. 1960; Prakash et al. 1985).

Not for use in the early stages of diarrhea or dysentery, although one of the common therapeutic uses is in chronic diarrhea and dysentery (Bensky et al. 2004; Chen and Chen 2004).

Not for use in excess of 14 days.

Other Precautions

None known.

Drug and Supplement Interactions

None known.

Notice

Tannins (up to 28.0%) (List and Hörhammer 1973); *see* Appendix 1.

Editors' Note

Pomegranate fruit husk should not be mixed with oils or fats when taken to kill parasites (Bensky et al. 2004; Chen and Chen 2004).

Adverse Events and Side Effects

Allergic reactions to pomegranate fruit have been reported (Gaig et al. 1992, 1999; Igea et al. 1991).

In the normal dose range (decoction made from 3–10 g of husk), nausea, vomiting, diarrhea, abdominal pain, dizziness, tinnitus, and tremors have been reported as side effects (no further details available) (Bensky et al. 2004; Chen and Chen 2004).

Pharmacological Considerations

Extracts of whole pomegranate fruit have shown varying activity on drug-metabolizing enzymes in vitro (Subehan et al. 2006; Usia et al. 2006). The relevance of those data to human use of preparations made from pomegranate husk is not known.

Pregnancy and Lactation

Animal studies have indicated anti-implantation and antifertility activity of pomegranate fruit husk (Gujral et al. 1960; Prakash et al. 1985). Based on this information, use during pregnancy is not recommended except under the supervision of a qualified healthcare practitioner.

No information on the safety of pomegranate fruit husk during lactation was identified. While this review did not identify any concerns for use while nursing, safety has not been conclusively established.

Review Details

I. Drug and Supplement Interactions

Clinical Trials of Drug or Supplement Interactions

No clinical trials of drug or supplement interactions were identified.

Case Reports of Suspected Drug or Supplement Interactions

No case reports of suspected drug or supplement interactions were identified.

Animal Trials of Drug or Supplement Interactions

No animal trials of drug or supplement interactions were identified.

II. Adverse Events

Case Reports of Adverse Events

Allergic reactions after ingestion of pomegranate fruit, confirmed by skin prick testing, have been reported (Gaig et al. 1992, 1999; Igea et al. 1991). Anaphylactic reactions to pomegranate fruit have been reported. Skin prick testing has indicated that the compound mannitol, naturally present in pomegranate, is responsible for this reaction (Hegde et al. 2002; Hegde and Venkatesh 2004).

In the normal dose range (decoction made from 3–9 g of husk), nausea, vomiting, diarrhea, abdominal pain, dizziness, tinnitus, and tremors have been reported as side effects (case details not available) (Bensky et al. 2004; Chen and Chen 2004). In overdose, pomegranate husk has been associated with vertigo, headache, blurred vision, tinnitus, weakness, and tremors. In severe cases, death due to respiratory paralysis has been reported (case details not available) (Bensky et al. 2004).

III. Pharmacology and Pharmacokinetics

Human Pharmacological Studies

No relevant human pharmacological studies were identified.

Animal Pharmacological Studies

Hypoglycemic activity of pomegranate husk extract was observed in diabetic rats administered doses of 2.5 g/kg (Zafar 1990).

In animal studies, gross overdose of pomegranate peel led to respiratory suppression (dose, product, and route of administration not reported in English language translation) (Chen and Chen 2004).

In Vitro Pharmacological Studies

No significant effects of 0.5 mg/ml of a methanolic extract of pomegranate fruit on CYP3A4 or CYP2D6 were observed in human liver microsomes (Subehan et al. 2006). In another study, however, a methanolic extract of the fruit significantly inhibited CYP3A4, with an inhibitory concentration (IC_{50}) of 35 μg/ml, and CYP2D6, with an IC_{50} of 32 μg/ml (Usia et al. 2006). The studies used an extract of the whole fruit, including juice, and findings may or may not apply to products made from pomegranate husk.

IV. Pregnancy and Lactation

Some antifertility activity was observed in rats and guinea pigs fed a diet supplemented with pomegranate husk (Gujral et al. 1960). Anti-implantation activity of acetone, aqueous, and methanol extracts of pomegranate seed, root, or whole plant was observed in rats (Prakash et al. 1985).

Early studies indicated that the fruit pulp of pomegranate had a stimulatory effect on isolated rat uteri (Dhawan and Saxena 1958).

No information on the safety of pomegranate fruit husk during lactation was identified.

V. Toxicity Studies

Acute Toxicity

See Punica granatum juice entry (next entry).

Short-Term Toxicity

See Punica granatum juice entry.

Subchronic Toxicity

See Punica granatum juice entry.

Genotoxicity

In two human lymphoma cell lines, administration of a decoction of pomegranate husk did not induce chromosomal aberrations but did induce apoptotic DNA fragmentation (Settheetham and Ishida 1995).

Literature Cited

Bensky, D., S. Clavey, and E. Stöger. 2004. *Chinese herbal medicine: Materia medica*. 3rd ed. Seattle: Eastland Press.

Chen, J.K., and T.T. Chen. 2004. *Chinese medical herbology and pharmacology*. City of Industry, CA: Art of Medicine Press.

Dhawan, B.N., and P.N. Saxena. 1958. Evaluation of some indigenous drugs for stimulant effect on the rat uterus: A preliminary report. *Indian J. Med. Res.* 46:808.

Gaig, P., B. Bartolome, R. Lleonart, et al. 1999. Allergy to pomegranate (*Punica granatum*). *Allergy* 54(3):287-288.

Gaig, P., J. Botey, V. Gutierrez, et al. 1992. Allergy to pomegranate (*Punica granatum*). *J. Investig. Allergol. Clin. Immunol.* 2(4):216-218.

Gujral, M.L., D.R. Varma, and K.N. Sareen. 1960. Preliminary observations on the antifertility effect of some indigenous drugs. *Indian J. Med. Res.* 48:46-51.

P

Hegde, V.L., P.A. Mahesh, and Y.P. Venkatesh. 2002. Anaphylaxis caused by mannitol in pomegranate (*Punica granatum*). *Allergy Clin. Immunol. Int.* 14(1):37-39.

Hegde, V.L., and Y.P. Venkatesh. 2004. Anaphylaxis to excipient mannitol: Evidence for an immunoglobulin E-mediated mechanism. *Clin. Exp. Allergy* 34(10):1602-1609.

Igea, J.M., J. Cuesta, M. Cuevas, et al. 1991. Adverse reaction to pomegranate ingestion. *Allergy* 46 (6):472-474.

List, P.H., and H. Hörhammer. 1973. *Hagers handbuch der pharmazeutischen praxis*. Berlin: Springer.

Prakash, A.O., V. Saxena, S. Shukla, et al. 1985. Anti-implantation activity of some indigenous plants in rats. *Acta Eur. Fertil.* 16(6):441-448.

Settheetham, W., and T. Ishida. 1995. Study of genotoxic effects of antidiarrheal medicinal herbs on human cells *in vitro*. *Southeast Asian J. Trop. Med. Public Health* 26(Suppl. 1):306-310.

Subehan, T. Usia, H. Iwata, S. Kadota, and Y. Tezuka. 2006. Mechanism-based inhibition of CYP3A4 and CYP2D6 by Indonesian medicinal plants. *J. Ethnopharmacol.* 105(3):449-455.

Usia, T., H. Iwata, A. Hiratsuka, et al. 2006. CYP3A4 and CYP2D6 inhibitory activities of Indonesian medicinal plants. *Phytomedicine* 13(1-2):67-73.

Zafar, R. and J. Singh. 1990. Antidiabetic activity of *Punica granatum* L. *Sci. Culture* 56(7):3.

Punica granatum L. — Punicaceae

SCN: pomegranate
AN: *dadima*
PN: *shi liu*
Part: fruit juice

Quick Reference Summary

Safety Class: 1
Interaction Class: A

Contraindications

None known.

Other Precautions

None known.

Drug and Supplement Interactions

See Pharmacological Considerations.

Adverse Events and Side Effects

A review of six clinical studies of a pomegranate juice product indicated that no serious adverse events or changes in blood chemistry were observed (McCutcheon et al. 2008).

Pharmacological Considerations

Animal (Faria et al. 2007; Hidaka et al. 2005; Nagata et al. 2007) and in vitro (Farkas et al. 2007; Hidaka et al. 2005; Kim et al. 2006; Nagata et al. 2007; Subehan et al. 2006; Usia et al. 2006) studies have provided conflicting data on the effects of pomegranate juice on drug-metabolizing isoenzymes CYP3A, CYP2C9, and CYP2D6, with some studies showing inhibition of the enzymes and others showing no effects. One human study comparing pomegranate juice and grapefruit juice (a known CYP3A4 inhibitor), however, showed no effect of pomegranate juice on CYP3A4 (Farkas et al. 2007).

In vitro studies have indicated that pomegranate juice competitively binds to estrogen receptors (Kajiya et al. 2005; Maru et al. 2001). The relevance of the in vitro data to human use is not known.

Pregnancy and Lactation

An animal study of pomegranate juice during pregnancy and lactation indicated a strong protective effect of pomegranate juice in experimentally induced brain damage due to lack of oxygen and blood flow (hypoxia-ischemia) in the fetuses and nursing offspring (Loren et al. 2005).

No information on the safety of pomegranate fruit juice during lactation was identified. While this review did not identify any concerns for use while nursing, safety has not been conclusively established.

Review Details

I. Drug and Supplement Interactions

Clinical Trials of Drug or Supplement Interactions

See Human pharmacological studies.

Case Reports of Suspected Drug or Supplement Interactions

Rhabdomyolysis was reported in a 48-year-old man with possible underlying myopathy. The man had been successfully treated with ezetimibe 10 mg/day and rosuvastatin 5 mg every other day for 17 months. Three weeks before presentation, he began drinking 200 ml pomegranate juice twice weekly. Muscle pain and weakness is reported to occur in 1 to 5% of patients taking statin drugs such as rosuvastatin, and rhabdomyolysis also occurs (Wooltorton 2004). Ezetimibe has also been associated with rhabdomyolysis (Health Canada 2005). The authors of the

P

report indicated that pomegranate juice may increase the risk of rhabdomyolysis from rosuvastatin through inhibition of CYP3A4, although rosuvastatin is not metabolized by CYP3A4 (AstraZeneca 2003; Sorokin et al. 2006).

Animal Trials of Drug or Supplement Interactions

See Animal pharmacological studies.

II. Adverse Events

Adverse Events Reported in Clinical Trials

No serious adverse events were reported in a phase II study of pomegranate juice in men being treated for prostate cancer. The men were administered 8 oz of pomegranate juice daily for 33 months (Pantuck et al. 2006). No changes in blood chemistry were observed in patients with atherosclerosis administered 50 ml of pomegranate juice daily for 1 to 3 years (Aviram et al. 2004).

Case Reports of Adverse Events

Allergic reactions after ingestion of pomegranate fruit, confirmed by skin prick testing, have been reported (Gaig et al. 1992, 1999; Igea et al. 1991). Anaphylactic reactions to pomegranate fruit have been reported. Skin prick testing has indicated that the compound mannitol, naturally present in pomegranate, is responsible for this reaction (Hegde et al. 2002; Hegde and Venkatesh 2004).

III. Pharmacology and Pharmacokinetics

Human Pharmacological Studies

In healthy volunteers pretreated with pomegranate juice or grapefruit juice, no effects of pomegranate juice were seen on the activity of the drug-metabolizing isoenzyme CYP3A4. Grapefruit juice, however, significantly inhibited CYP3A4 (Farkas et al. 2007).

No adverse events were reported in a safety study of a pomegranate ellagitannin-enriched polyphenol extract in overweight patients administered capsules containing 710 mg (435 mg of gallic acid equivalents, GAEs) or 1420 mg (870 mg of GAEs) daily for 28 days (Heber et al. 2007). No serious adverse events were reported in a safety study of pomegranate juice in men with erectile dysfunction administered 8 oz of pomegranate juice daily for 28 days. Nonserious adverse events were similar in the pomegranate and control groups (Forest et al. 2007).

Animal Pharmacological Studies

In rats administered 3 ml of pomegranate juice, a significant increase in the area under the time-concentration curve of orally administered tolbutamide was observed, although the elimination half-life of tolbutamide was not affected, indicating that pomegranate juice impairs the function of intestinal but not hepatic CYP2C9 (Nagata et al. 2007).

In rats administered 2 ml of pomegranate juice, a significant increase in the area under the concentration-time curve of orally administered carbamazepine was observed, although the elimination half-life of carbamazepine was not affected, indicating that pomegranate juice impairs the function of intestinal but not hepatic CYP3A (Hidaka et al. 2005).

In mice administered pomegranate juice (made from whole fruit including seed) as the sole source of drinking water for 4 weeks, a decrease in total hepatic cytochrome P450 content and expression of CYP1A2 and CYP3A were observed (Faria et al. 2007).

An increase in uterine weight was reported in ovariectomized rats administered pomegranate juice (dose and duration not specified in English-language abstract) (Maru et al. 2001).

In Vitro Pharmacological Studies

In human liver microsomes preincubated with pomegranate juice, no inhibition of the drug-metabolizing isoenzyme CYP3A4 was observed. Without preincubation, some inhibition was observed (Farkas et al. 2007). In another test in human liver microsomes, pomegranate juice was shown to inhibit CYP3A. The effect was less than that of grapefruit, black mulberry, wild grape, and black raspberry fruit juices (Kim et al. 2006).

Addition of pomegranate juice to human liver microsomes resulted in a dose-dependent inhibition of CYP3A and CYP2C9 activity, with a 5% (by volume) concentration providing almost complete inhibition of both enzymes. The effects on CYP3A were similar to that of grapefruit juice (Hidaka et al. 2005; Nagata et al. 2007).

No significant effects of 0.5 mg/ml of a methanolic extract of pomegranate fruit on CYP3A4 or CYP2D6 were observed in human liver microsomes (Subehan et al. 2006). In another study, however, a methanolic extract of the fruit significantly inhibited CYP3A4, with an inhibitory concentration (IC_{50}) of 35 μg/ml, and CYP2D6, with an IC_{50} of 32 μg/ml (Usia et al. 2006).

Pomegranate juice was reported to have competitive binding activity with 17β-estradiol for the estrogen receptor and to stimulate proliferation of the estrogen receptor-positive human breast cancer cell line MCF-7 (active concentrations not reported in English language abstract) (Maru et al. 2001). Another study of competitive binding indicated that pomegranate juice was more active in estrogen receptor (ER)-α than ER-β, and that the methanolic fraction of pomegranate juice showed more competitive binding against 17β-estradiol than the water fraction (Kajiya et al. 2005). A phytochemical analysis of pomegranate seed and juice failed to find any steroid-like compounds in either product (Choi et al. 2006).

IV. Pregnancy and Lactation

A neuroprotective effect of pomegranate juice was observed in a mouse model of neonatal hypoxic-ischemic brain injury in offspring of mothers administered pomegranate juice beginning at the third trimester of pregnancy and continuing through lactation. The juice was provided alongside drinking water for consumption ad libitum in low, medium, or high concentrations (8, 16, or 32 μmol polyphenols daily estimated intake), with the high dose equivalent to a human dose of two 8 oz glasses of pomegranate juice daily. Ellagic acid, a polyphenolic component in pomegranate juice, was detected in plasma from treated but not control pups (Loren et al. 2005).

Early studies indicated that the fruit pulp of pomegranate had a stimulatory effect on isolated rat uteri (Dhawan and Saxena 1958).

No information on the safety of pomegranate fruit juice during lactation was identified.

V. Toxicity Studies

Acute Toxicity

The LD_{50} of intraperitoneally administered whole plant (excluding roots) extract in rats is 250 mg/kg (Dhawan et al. 1977). The LD_{50} of an intraperitoneally administered standardized whole pomegranate fruit extract (30% punicalagins) in rats is 217 mg/kg and in mice is 187 mg/kg. The LD_{50} of an orally administered standardized whole pomegranate fruit extract (30% punicalagins) in rats and mice could not be determined at doses up to 5 g/kg (Patel et al. 2008). The LD_{50} of an intraperitoneally administered pomegranate whole fruit hydroalcoholic extract is 731 mg/kg in mice (Vidal et al. 2003).

In chick embryos, a dose of 0.1 mg pomegranate whole fruit hydroalcoholic extract per embryo was nontoxic (Vidal et al. 2003).

Short-Term Toxicity

No adverse effects were observed in rats intranasally administered 0.4 and 1.2 mg/kg pomegranate whole fruit hydroalcoholic extract daily for up to 35 days (Vidal et al. 2003).

In rats fed a diet containing 6% of the compound punicalagin (present in pomegranate juice at concentrations of 2 g/l) daily for 37 days, a decrease in food intake and growth rate was observed, attributed to the palatability of the supplemented food. Blood chemistry was normal with the exception of urea and triglycerides, which remained at low values throughout the experiment. Histopathological analysis of liver and kidney indicated no adverse effects on the organs (Cerda et al. 2003).

Subchronic Toxicity

No adverse effects were observed in rats orally administered a standardized whole pomegranate fruit extract (30% punicalagins) at doses of 60, 240 and 600 mg/kg daily for 90 days. Based on the results of the study, the authors indicated that the no-observed-adverse-effect level (NOAEL) for the standardized pomegranate fruit extract was 600 mg/kg daily (Patel et al. 2008).

Genotoxicity

A hydroalcoholic extract of whole pomegranate fruit showed genotoxic activity in *Salmonella typhimurium* strain TA100 and *Saccharomyces cerevisiae* with or without activation by S9 (Sanchez-Lamar et al. 2008). Administration of a hydroalcoholic extract of whole pomegranate fruit to mice at doses of 7 to 700 mg/kg (intraperitoneal) daily for 2 days resulted in some mutagenic activity in the bone marrow micronucleus assay and the sperm shape abnormality assay at doses over 70 mg/kg (Sanchez-Lamar et al. 2008). Antimutagenic effects of pomegranate extracts have also been shown in *Salmonella typhimurium* and in micronucleus assays in mice and rats (Alekperov 2002; Bala and Grover 1992).

Literature Cited

Alekperov, U.K. 2002. Plant antimutagens and their mixtures in inhibition of genotoxic effects of xenobiotics and aging processes. *Eur. J. Cancer Prev.* 11:S8-S11.

AstraZeneca. 2003. Crestor Product Information. Wilmington, DE: AstraZeneca Pharmaceuticals.

Aviram, M., M. Rosenblat, D. Gaitini, et al. 2004. Pomegranate juice consumption for 3 years by patients with carotid artery stenosis reduces common carotid intima-media thickness, blood pressure and LDL oxidation. *Clin. Nutr.* 23(3):423-433.

Bala, S., and I. Grover. 1992. Antimutagenic effect of pomegranate (*Punica granatum* var. *anardana*) fruit extracts on direct acting and S9 dependent mutagens in *Salmonella typhimurium*. *Plant Sci. Res.* 8:14-16.

Cerda, B., J.J. Ceron, F.A. Tomas-Barberan, and J.C. Espin. 2003. Repeated oral administration of high doses of the pomegranate ellagitannin punicalagin to rats for 37 days is not toxic. *J. Agric. Food Chem.* 51(11):3493-3501.

Choi, D.W., J.Y. Kim, S.H. Choi, et al. 2006. Identification of steroid hormones in pomegranate (*Punica granatum*) using HPLC and GC-mass spectrometry. *Food Chem.* 96(4):562-571.

Dhawan, B.N., G.K. Patnaik, R.P. Rastogi, K.K. Singh, and J.S. Tandon. 1977. Screening of Indian plants for biological activity: Part VI. *Indian J. Exp. Biol.* 15(3):208-219.

Dhawan, B.N., and P.N. Saxena. 1958. Evaluation of some indigenous drugs for stimulant effect on the rat uterus: A preliminary report. *Indian J. Med. Res.* 46:808.

Faria, A., R. Monteiro, I. Azevedo, and C. Calhau. 2007. Pomegranate juice effects on cytochrome P450s expression: *In vivo* studies. *J. Med. Food* 10(4):643-649.

Farkas, D., L.E. Oleson, Y. Zhao, et al. 2007. Pomegranate juice does not impair clearance of oral or intravenous midazolam, a probe for cytochrome P450-3A activity: Comparison with grapefruit juice. *J. Clin. Pharmacol.* 47(3):286-294.

Forest, C.P., H. Padma-Nathan, and H.R. Liker. 2007. Efficacy and safety of pomegranate juice on improvement of erectile dysfunction in male patients with mild to moderate erectile dysfunction: A randomized, placebo-controlled, double-blind, crossover study. *Int. J. Impot. Res.* 19(6):564-567.

Gaig, P., B. Bartolome, R. Lleonart, et al. 1999. Allergy to pomegranate (*Punica granatum*). *Allergy* 54(3):287-288.

Gaig, P., J. Botey, V. Gutierrez, et al. 1992. Allergy to pomegranate (*Punica granatum*). *J. Investig. Allergol. Clin. Immunol.* 2(4):216-218.

Health Canada. 2005. Association of Ezetrol (ezetimibe) with myalgia, rhabdomyolysis, hepatitis, pancreatitis, and thrombocytopenia. February. Ottawa.

Heber, D., N.P. Seeram, H. Wyatt, et al. 2007. Safety and antioxidant activity of a pomegranate ellagitannin-enriched polyphenol dietary supplement in overweight individuals with increased waist size. *J. Agric. Food Chem.* 55(24):10050-10054.

Hegde, V.L., P.A. Mahesh, and Y.P. Venkatesh. 2002. Anaphylaxis caused by mannitol in pomegranate (*Punica granatum*). *Allergy Clin. Immunol. Int.* 14(1):37-39.

Hegde, V.L., and Y.P. Venkatesh. 2004. Anaphylaxis to excipient mannitol: Evidence for an immunoglobulin E-mediated mechanism. *Clin. Exp. Allergy* 34(10):1602-1609.

Hidaka, M., M. Okumura, K.I. Fujita, et al. 2005. Effects of pomegranate juice on human cytochrome P450 3A (CYP3A) and carbamazepine pharmacokinetics in rats. *Drug Metab. Dispos.* 33(5):644-648.

Igea, J.M., J. Cuesta, M. Cuevas, et al. 1991. Adverse reaction to pomegranate ingestion. *Allergy* 46(6):472-474.

Kajiya, H., N. Ikeda, Y. Nakanishi, et al. 2005. An *in vitro* and *in vivo* study of the estrogenic action of pomegranate juice. *Jpn. J. Clin. Physiol.* 35(2):89-99.

Kim, H., Y.J. Yoon, J.H. Shon, et al. 2006. Inhibitory effects of fruit juices on CYP3A activity. *Drug Metab. Dispos.* 34(4):521-523.

Loren, D.J., N.P. Seeram, R.N. Schulman, and D.M. Holtzman. 2005. Maternal dietary supplementation with pomegranate juice is neuroprotective in an animal model of neonatal hypoxic-ischemic brain injury. *Pediatr. Res.* 57(6):858-864.

Maru, I., J. Ohnishi, S. Yamaguchi, et al. 2001. An estrogen-like activity in pomegranate juice. *Nippon Shokuhin Kagaku Kogaku Kaishi* 48(2):146-149.

McCutcheon, A., J. Udani, and D. Brown. 2008. Scientific and Clinical Monograph: POM Wonderful pomegranate juice. Austin, TX: American Botanical Council.

Nagata, M., M. Hidaka, H. Sekiya, et al. 2007. Effects of pomegranate juice on human cytochrome P450 2C9 and tolbutamide pharmacokinetics in rats. *Drug Metab. Dispos.* 35(2):302-305.

Pantuck, A.J., J.T. Leppert, N. Zomorodian, et al. 2006. Phase II study of pomegranate juice for men with rising prostate-specific antigen following surgery or radiation for prostate cancer. *Clin. Cancer Res.* 12(13):4018-4026.

Patel, C., P. Dadhaniya, L. Hingorani, and M.G. Soni. 2008. Safety assessment of pomegranate fruit extract: Acute and subchronic toxicity studies. *Food Chem. Toxicol.* 46(8):2718-2735.

Sanchez-Lamar, A., G. Fonseca, J.L. Fuentes, et al. 2008. Assessment of the genotoxic risk of *Punica granatum* L. (Punicaceae) whole fruit extracts. *J. Ethnopharmacol.* 115(3):416-422.

Sorokin, A.V., B. Duncan, R. Panetta, and P.D. Thompson. 2006. Rhabdomyolysis associated with pomegranate juice consumption. *Am. J. Cardiol.* 98(5):705-706.

Subehan, T. Usia, H. Iwata, S. Kadota, and Y. Tezuka. 2006. Mechanism-based inhibition of CYP3A4 and CYP2D6 by Indonesian medicinal plants. *J. Ethnopharmacol.* 105(3):449-455.

Usia, T., H. Iwata, A. Hiratsuka, et al. 2006. CYP3A4 and CYP2D6 inhibitory activities of Indonesian medicinal plants. *Phytomedicine* 13(1-2):67-73.

Vidal, A., A. Fallarero, B.R. Pena, et al. 2003. Studies on the toxicity of *Punica granatum* L. (Punicaceae) whole fruit extracts. *J. Ethnopharmacol.* 89(2-3):295-300.

Wooltorton, E. 2004. Rosuvastatin (Crestor) and rhabdomyolysis. *Can. Med. Assoc. J.* 171(2):129.

Punica granatum L. Punicaceae

SCN: pomegranate
AN: *dadima*
ON: *shi liu zi* (seed)
Part: seed

Quick Reference Summary

Safety Class: 1
Interaction Class: A

Contraindications

None known.

Other Precautions

None known.

Drug and Supplement Interactions

None known.

Adverse Events and Side Effects

Allergic reactions to pomegranate fruit have been reported (Gaig et al. 1992, 1999; Igea et al. 1991). Ingestion of dried crushed seed has been associated with an increased risk of esophageal cancer in Iranian woman. The risk is believed to be due to repeated mechanical irritation of the esophagus, which is a recognized cause of esophageal cancer (Ghadirian et al. 1992).

Pharmacological Considerations

None known.

Pregnancy and Lactation

Information on the safety of pomegranate seed use during pregnancy and lactation is limited. One study indicated that pomegranate seed is traditionally consumed by pregnant women in Iran (Ghadirian et al. 1992). Although this review did not identify any concerns for use while pregnant or nursing, safety has not been conclusively established.

Review Details

I. Drug and Supplement Interactions

Clinical Trials of Drug or Supplement Interactions

No clinical trials of drug or supplement interactions were identified.

Case Reports of Suspected Drug or Supplement Interactions

No case reports of suspected drug or supplement interactions were identified.

Animal Trials of Drug or Supplement Interactions

No animal trials of drug or supplement interactions were identified.

II. Adverse Events

Case Reports of Adverse Events

An increased risk of esophageal cancer in Iranian women was associated with the consumption of crushed pomegranate seed. As traditionally prepared for consumption, the seed pieces have sharp edges that mechanically irritate or damage the esophagus, which is a recognized cause of esophageal cancer (Ghadirian et al. 1992).

Allergic reactions after ingestion of pomegranate fruit, confirmed by skin prick testing, have been reported (Gaig et al. 1992, 1999; Igea et al. 1991). Anaphylactic reactions to pomegranate fruit have been reported. Skin prick testing indicated that the compound mannitol, naturally present in pomegranate, is responsible for the reaction (Hegde et al. 2002; Hegde and Venkatesh 2004).

III. Pharmacology and Pharmacokinetics

Human Pharmacological Studies

No relevant human pharmacological studies were identified.

Animal Pharmacological Studies

A significant reduction of blood glucose levels was observed in diabetic rats orally administered an extract of pomegranate seed at doses of 150, 300, or 600 mg/kg (Das et al. 2001).

Pomegranate seed oil administered orally or intramuscularly increased the uterine weight in ovariectomized mice and increased vaginal cornification in immature female rabbits (Sharaf 1969; Sharaf and Nigm 1964).

In Vitro Pharmacological Studies

A polyphenol-rich fraction of pomegranate seed oil inhibited 17β-hydroxysteroid dehydrogenase type 1 at concentrations ranging from 100 to 1000 μg/ml. The oil also inhibited proliferation of estrogen receptor (ER)-positive human breast cancer cells (MCF-7), inhibited invasion of MCF-7 across a membrane, and induced apoptosis in ER-negative human breast cancer cells (MDA-MB-435) (Kim et al. 2002). An older study indicated that dried pomegranate seed contains the compound estrone at a concentration of 17 mg/kg (Heftmann et al. 1966); a more recent analysis indicated that no estrone was present (Choi et al. 2006).

IV. Pregnancy and Lactation

Pomegranate seed is traditionally consumed by pregnant women in Iran (Ghadirian et al. 1992). No other information on the safety of pomegranate seed use during pregnancy or lactation was identified.

V. Toxicity Studies

Acute Toxicity

The LD_{50} of an intraperitoneally administered whole plant extract (excluding roots) in rats is 250 mg/kg (Dhawan et al. 1977).

Pomegranate seed oil was not toxic to brine shrimp larvae (Fatope et al. 2002).

Chronic Toxicity

A lack of adverse effects was noted in rats administered a diet containing 0.01 to 1% pomegranate seed oil for up to 32 weeks (Kohno et al. 2004).

P

Literature Cited

Choi, D.W., J.Y. Kim, S.H. Choi, et al. 2006. Identification of steroid hormones in pomegranate (*Punica granatum*) using HPLC and GC-mass spectrometry. *Food Chem.* 96(4):562-571.

Das, A.K., S.C. Mandal, S.K. Banerjee, et al. 2001. Studies on the hypoglycaemic activity of *Punica granatum* seed in streptozotocin induced diabetic rats. *Phytother. Res.* 15(7):628-629.

Dhawan, B.N., G.K. Patnaik, R.P. Rastogi, K.K. Singh, and J.S. Tandon. 1977. Screening of Indian plants for biological activity: Part VI. *Indian J. Exp. Biol.* 15(3):208-219.

Fatope, M.O., S.K. Al Burtomani, and Y. Takeda. 2002. Monoacylglycerol from *Punica granatum* seed oil. *J. Agric. Food Chem.* 50:357-360.

Gaig, P., B. Bartolome, R. Lleonart, et al. 1999. Allergy to pomegranate (*Punica granatum*). *Allergy* 54(3):287-288.

Gaig, P., J. Botey, V. Gutierrez, et al. 1992. Allergy to pomegranate (*Punica granatum*). *J. Investig. Allergol. Clin. Immunol.* 2(4):216-218.

Ghadirian, P., J.M. Ekoe, and J.P. Thouez. 1992. Food habits and esophageal cancer: An overview. *Cancer Detect. Prev.* 16(3):163-168.

Heftmann, E., S. Ko, and R. Bennet. 1966. Identification of estrone in pomegranate seeds. *Phytochemistry* 5:1337-1340.

Hegde, V.L., P.A. Mahesh, and Y.P. Venkatesh. 2002. Anaphylaxis caused by mannitol in pomegranate (*Punica granatum*). *Allergy Clin. Immunol. Int.*14(1):37-39.

Hegde, V.L., and Y.P. Venkatesh. 2004. Anaphylaxis to excipient mannitol: Evidence for an immunoglobulin E-mediated mechanism. *Clin. Exp. Allergy* 34(10):1602-1609.

Igea, J.M., J. Cuesta, M. Cuevas, et al. 1991. Adverse reaction to pomegranate ingestion. *Allergy* 46(6):472-474.

Kim, N.D., R. Mehta, W. Yu, et al. 2002. Chemopreventive and adjuvant therapeutic potential of pomegranate (*Punica granatum*) for human breast cancer. *Breast Cancer Res. Treat.* 71(3):203-217.

Kohno, H., R. Suzuki, Y. Yasui, et al. 2004. Pomegranate seed oil rich in conjugated linolenic acid suppresses chemically induced colon carcinogenesis in rats. *Cancer Sci.* 95(6):481-486.

Sharaf, A. 1969. Food plants as a possible factor in fertility control. *Qual. Plant. Mat. Veg.* 153-160.

Sharaf, A., and S.A.R. Nigm. 1964. The oestrogenic activity of pomegranate seed oil. *J. Endocrinol.* 29:91-92.

Quassia amara L.

Simaroubaceae

SCN: quassia
OCN: Surinam quassia

Part: bark, root, wood

Quick Reference Summary

Safety Class: 2b, 2d
Interaction Class: A

Contraindications

Not for use in pregnancy except under the supervision of a qualified healthcare practitioner (Felter and Lloyd 1898; Remington and Wood 1918).

Not for use in children under 6 years of age (Felter and Lloyd 1898).

Not for internal use by men attempting to conceive (Faisal et al. 2006; Parveen et al. 2003; Raji and Bolarinwa 1997).

Other Precautions

None known.

Drug and Supplement Interactions

None known.

Standard Dose

The standard dose is 1–2 g per day of the wood as a decoction; average dose, 0.5 g, 2 to 3 times daily (Merck 1930; Wichtl 2004).

Editors' Note

Quassia should not be confused with Jamaica quassia (*Picrasma excelsa*) or senna (*Senna* spp., formerly classified as *Cassia* spp.).

Adverse Events and Side Effects

Consumption of large amounts of quassia can irritate the mucous membrane of the stomach and lead to vomiting (List and Hörhammer 1973; Remington and Wood 1918; Wichtl 2004).

Pharmacological Considerations

Animal studies with quassia have demonstrated changes in sperm count, shape, and viability (Faisal et al. 2006; Parveen et al. 2003; Raji and Bolarinwa 1997). Changes were shown to return to normal several weeks after cessation of treatment (Raji and Bolarinwa 1997). No human studies on the safety or efficacy of quassia as a male contraceptive have been completed.

Pregnancy and Lactation

Although no information on the safety of quassia during pregnancy was identified, traditional use in the treatment of parasites suggests that quassia should not be used during pregnancy (Felter and Lloyd 1898; Remington and Wood 1918).

No information on the safety of quassia during lactation was identified. While this review did not identify any concerns for use while nursing, safety has not been conclusively established.

Q

Review Details

I. Drug and Supplement Interactions

Clinical Trials of Drug or Supplement Interactions

No clinical trials of drug or supplement interactions were identified.

Case Reports of Suspected Drug or Supplement Interactions

No case reports of suspected drug or supplement interactions were identified.

Animal Trials of Drug or Supplement Interactions

No animal trials of drug or supplement interactions were identified.

II. Adverse Events

Case Reports of Adverse Events

No case reports of adverse events were identified.

III. Pharmacology and Pharmacokinetics

Human Pharmacological Studies

No relevant human pharmacological studies were identified.

Animal Pharmacological Studies

A reduction in sperm counts and motility and an increase in sperm abnormalities were observed in mice intraperitoneally administered 25, 50, or 100 mg/kg of dried methanol extract of quassia (Faisal et al. 2006).

Rehmannia glutinosa (Gaertn.) Steud. — Orobanchaceae

SCN: rehmannia
PN: *sheng di huang* (unprocessed tuber)

Part: dried tuber

Quick Reference Summary

Safety Class: 2d
Interaction Class: A

Contraindications

Not for use in persons with diarrhea (Bensky et al. 2004; Chen and Chen 2004).

Other Precautions

None known.

Drug and Supplement Interactions

None known.

Editors' Notes

Two forms of rehmannia tuber are used in traditional Chinese medicine, the dried tuber (*sheng di huang*), which is the subject of this entry, and the processed tuber (*shu di huang*), which is the subject of a separate entry. Processing reduces the content of the compound catalpol (an iridoid glycoside) such that the catalpol content of prepared rehmannia root is only 1/20–1/30 that of fresh rehmannia (Wang et al. 1997). Methods of proper processing and resulting chemical profiles are described in other texts and articles (Bensky et al. 2004; Wang et al. 1997).

Adverse Events and Side Effects

Occasional side effects of rehmannia tuber include diarrhea, abdominal pain, dizziness, fatigue, and heart palpitations (Bensky et al. 2004).

Allergic reactions to rehmannia tuber, including an anaphylactic reaction after a very high dose, have been reported (Bensky et al. 2004).

Pharmacological Considerations

None known.

Pregnancy and Lactation

Limited information on the use of rehmannia tuber in pregnancy or lactation was identified in the scientific and traditional literature. Although this review did not identify any concerns for use while pregnant or nursing, safety has not been conclusively established.

Review Details

I. Drug and Supplement Interactions

Clinical Trials of Drug or Supplement Interactions

No clinical trials of drug or supplement interactions were identified.

Case Reports of Suspected Drug or Supplement Interactions

No case reports of suspected drug or supplement interactions were identified.

Animal Trials of Drug or Supplement Interactions

No animal trials of drug or supplement interactions were identified.

II. Adverse Events

Case Reports of Adverse Events

Allergic reaction to rehmannia tuber, including an anaphylactic reaction after a 100 g dose, has been reported (Bensky et al. 2004).

Occasional side effects of rehmannia tuber include diarrhea, abdominal pain, dizziness, fatigue, and heart palpitations. Ingestion of very high doses (standard dose is a decoction of 9–15 g) has been associated with headache, dizziness, constricted pupils and loss of pupil reflexes, blue discoloration of the lips, low blood pressure, irregular heartbeat, and coma (Bensky et al. 2004).

III. Pharmacology and Pharmacokinetics

Human Pharmacological Studies

No relevant human pharmacological studies were identified.

Animal Pharmacological Studies

Several different extracts of rehmannia tuber, including an ethanol fraction of an aqueous extract, rehmannia oligosaccharide, and rehmannia glycoside D, showed hypoglycemic activity in healthy and diabetic rats and mice (Kiho et al. 1992; Kitagawa et al. 1971; Oshio and Inouye 1982).

In Vitro Pharmacological Studies

No relevant in vitro pharmacological studies were identified.

IV. Pregnancy and Lactation

A decrease in litter numbers was observed in mice subcutaneously administered 0.1 to 0.4 ml daily of an aqueous extract of rehmannia (condition of herbal material not stated) (Matsui et al. 1967).

No information on the safety of dried (unprepared) rehmannia tuber during lactation was identified.

V. Toxicity Studies

Acute Toxicity

No deaths were reported in mice orally administered 60 g/kg daily of an aqueous or alcoholic extract of dried rehmannia tuber for 3 days (Chen and Chen 2004).

Short-Term Toxicity

No deaths were reported in rats orally administered 18 g/kg daily of an aqueous or alcoholic extract of dried rehmannia for 15 days. No adverse effects on internal organs were seen (Chen and Chen 2004).

Genotoxicity

Dried rehmannia tuber did not exhibit mutagenic activity in the Ames mutagenicity test or the micronucleus assays in mice (Yin et al. 1991).

Literature Cited

Bensky, D., S. Clavey, and E. Stöger. 2004. *Chinese herbal medicine: Materia medica*. 3rd ed. Seattle: Eastland Press.

Chen, J.K., and T.T. Chen. 2004. *Chinese medical herbology and pharmacology*. City of Industry, CA: Art of Medicine Press.

Kiho, T., T. Watanabe, K. Nagai, and S. Ukai. 1992. Hypoglycemic activity of polysaccharide fraction from rhizome of *Rehmannia glutinosa* Libosch F-Hueichingensis Hsiao and the effect on carbohydrate-metabolism in normal mouse-liver. *Yakugaku Zasshi* 112(6):393-400.

Kitagawa, I., T. Nishimura, A. Furubaya, and I. Yosioka. 1971. Constituents of rhizome of *Rehmannia glutinosa* Libosch forma-Hueichingensis Hsiao. *Yakugaku Zasshi* 91(5):593-597.

Matsui, A., J. Rogers, and Y. Woo. 1967. Effects of some natural products on fertility in mice. *Med. Pharmacol. Exp.* 16(5):414-424.

Oshio, H., and H. Inouye. 1982. Iridoid glycosides of *Rehmannia glutinosa*. *Phytochemistry* 21(1):133-138.

Wang, H., B.L. Bian, and J. Yang. 1997. A study on catalpol content changes in *Rehmannia glutinosa* (Gaertn.) Libosch. under certain conditions. *China J Chin. Mat. Med.* 22:408-409.

Yin, X.J., D.X. Liu, H.C. Wang, and Y. Zhou. 1991. A study on the mutagenicity of 102 raw pharmaceuticals used in Chinese traditional medicine. *Mutat. Res.* 260(1):73-82.

Rehmannia glutinosa (Gaertn.) Steud. — Orobanchaceae

R

SCN: rehmannia
PN: *shu di huang* (prepared root tuber)
Part: prepared root tuber

Quick Reference Summary

Safety Class: 1
Interaction Class: A

Contraindications

None known.

Other Precautions

None known.

Drug and Supplement Interactions

None known.

Editors' Notes

Two forms of rehmannia tuber are used in traditional Chinese medicine, the processed or prepared tuber (*shu di huang*), which is the subject of this entry, and the dried tuber (*sheng di huang*), which is the subject of a separate entry. Processing reduces the content of the compound catalpol such that the catalpol content of prepared rehmannia root is only 1/20 to 1/30 that of fresh rehmannia (Wang et al. 1997). Methods of proper processing and resulting chemical profiles are described in other texts and articles (Bensky et al. 2004; Wang et al. 1997).

Adverse Events and Side Effects

An allergic skin reaction after use of rehmannia tuber has been reported (Bensky et al. 2004).

Pharmacological Considerations

None known.

Pregnancy and Lactation

Limited information on the use of prepared rehmannia tuber in pregnancy or lactation was identified in the scientific and traditional literature. Although this review did not identify any concerns for use while pregnant or nursing, safety has not been conclusively established.

Review Details

I. Drug and Supplement Interactions

Clinical Trials of Drug or Supplement Interactions

No clinical trials of drug or supplement interactions were identified.

Case Reports of Suspected Drug or Supplement Interactions

No case reports of suspected drug or supplement interactions were identified.

Animal Trials of Drug or Supplement Interactions

No animal trials of drug or supplement interactions were identified.

II. Adverse Events

Case Reports of Adverse Events

An allergic skin reaction after use of rehmannia tuber has been reported (Bensky et al. 2004).

III. Pharmacology and Pharmacokinetics

Human Pharmacological Studies

No relevant human pharmacological studies were identified.

Animal Pharmacological Studies

Several different extracts of rehmannia tuber, including an ethanol fraction of an aqueous extract, rehmannia oligosaccharide, and rehmannia glycoside D, showed hypoglycemic activity in healthy and diabetic rats and mice (Kiho et al. 1992; Kitagawa et al. 1971; Oshio and Inouye 1982).

In Vitro Pharmacological Studies

No relevant in vitro pharmacological studies were identified.

IV. Pregnancy and Lactation

A decrease in litter numbers was observed in mice subcutaneously administered 0.1 to 0.4 ml daily of an aqueous extract of rehmannia (herbal processing not specified) (Matsui et al. 1967).

No information on the safety of rehmannia prepared root tuber during lactation was identified.

V. Toxicity Studies

Genotoxicity

Processed rehmannia demonstrated mutagenic potential in the chromosomal aberration and micronucleus assays in mice intraperitoneally administered doses of 2 to 4 g/kg, although no mutagenic activity was observed in the Ames test (Yin et al. 1991).

An aqueous extract of rehmannia exhibited a protective effect against the mutagenic activity of benzo[*a*]pyrene in *Salmonella typhimurium* (Sakai et al. 1988).

Literature Cited

Bensky, D., S. Clavey, and E. Stöger. 2004. *Chinese herbal medicine: Materia medica*. 3rd ed. Seattle: Eastland Press.

Kiho, T., T. Watanabe, K. Nagai, and S. Ukai. 1992. Hypoglycemic activity of polysaccharide fraction from rhizome of *Rehmannia glutinosa* Libosch F-Hueichingensis Hsiao and the effect on carbohydrate-metabolism in normal mouse-liver. *Yakugaku Zasshi* 112(6):393-400.

Kitagawa, I., T. Nishimura, A. Furubaya, and I. Yosioka. 1971. Constituents of rhizome of *Rehmannia glutinosa* Libosch forma-Hueichingensis Hsiao. *Yakugaku Zasshi* 91(5):593.

Matsui, A., J. Rogers, and Y. Woo. 1967. Effects of some natural products on fertility in mice. *Med. Pharmacol. Exp.* 16(5):414-424.

Oshio, H., and H. Inouye. 1982. Iridoid glycosides of *Rehmannia glutinosa*. *Phytochemistry* 21(1):133-138.

Sakai, Y., H. Nagase, Y. Ose, et al. 1988. Effects of medicinal plant extracts from Chinese herbal medicines on the mutagenic activity of benzo-*a*-pyrene. *Mutat. Res.* 206(3):327-334.

Wang, H., B.L. Bian, and J. Yang. 1997. A study on catalpol content changes in *Rehmannia glutinosa* (Gaertn.) Libosch. under certain conditions. *China J. Chin. Mat. Med.* 22:408-409.

Yin, X.J., D.X. Liu, H.C. Wang, and Y. Zhou. 1991. A study on the mutagenicity of 102 raw pharmaceuticals used in Chinese traditional medicine. *Mutat. Res.* 260(1):73-82.

Reynoutria multiflora (Thunb.) Moldenke — Polygonaceae

SCN: fo-ti
Syn: *Fallopia multiflora* (Thunb.) Haraldson; *Polygonum multiflorum* Thunb.
PN: *he shou wu* (cured root tuber)
OCN: fleeceflower
Part: prepared root tuber

Quick Reference Summary

Safety Class: 1
Interaction Class: A

Contraindications

None known.

Other Precautions

Use with caution in persons with loose stools or diarrhea (Chen and Chen 2004).

Drug and Supplement Interactions

None known.

Editors' Notes

Two forms of fo-ti tuber are used in traditional Chinese medicine, the processed or prepared tuber (*zhi he shou wu*), which is the subject of this entry, and the dried tuber (*sheng he shou wu*), which is the subject of a separate entry (*see* next entry). The prepared tuber is used almost exclusively, while the dried tuber is not commonly used and is difficult to find commercially. The dried tuber contains stilbene glycosides and anthraquinone glycosides and is traditionally used for short periods of time as a stimulant laxative; the processed tuber is used as a tonic and has significantly lower anthraquinone glycoside content (Bensky et al. 2004; Leung and Foster 1996). The processing is responsible for this change in constituents, reducing the concentrations of free and conjugated anthraquinones by 42 to 96% (Avula et al. 2007; Leung and Foster 1996; Zhang et al. 2006). Methods of proper processing and resulting chemical profiles are described in other texts and articles (Avula et al. 2007; Bensky et al. 2004; Liu et al. 1991; PPRC 2005; Sionneau and Flaws 1995; Yi et al. 2007; Zhang et al. 2006).

Adverse Events and Side Effects

Several cases of acute hepatitis have been reported in persons taking fo-ti (Battinelli et al. 2004; But et al. 1996; Cardenas et al. 2006; Laird et al. 2008; Panis et al. 2005; Park et al. 2001). Although products in the case reports were marketed as processed fo-ti, analyses of some products indicated elevated levels of anthraquinone glycosides and stilbene glycosides (Battinelli et al. 2004; But et al. 1996; Laird et al. 2008; Panis et al. 2005). The hepatitis is believed to be due to long-term use of unprocessed or incompletely processed tuber, which is indicated only for short-term use (*see* Editors' Notes). The stilbene glycosides (Panis et al. 2005) and anthraquinone glycosides (Cardenas et al. 2006; Stickel et al. 2005; Vanderperren et al. 2005) in the unprocessed or incompletely processed tuber have been indicated as the compounds responsible for the hepatotoxicity, although the mechanism of action is not yet understood.

Pharmacological Considerations

In vitro, a mild estrogenic effect (Kang et al. 2006; Zhang et al. 2005) and inhibition of some CYP enzymes have been observed (Unger and Frank 2004).

Pregnancy and Lactation

No information on the safety of fo-ti in pregnancy or lactation was identified in the scientific or traditional literature. Although this review did not identify any concerns for use while pregnant or nursing, safety has not been conclusively established.

Review Details

I. Drug and Supplement Interactions

Clinical Trials of Drug or Supplement Interactions

No clinical trials of drug or supplement interactions were identified.

Case Reports of Suspected Drug or Supplement Interactions

No case reports of suspected drug or supplement interactions were identified.

Animal Trials of Drug or Supplement Interactions

No animal trials of drug or supplement interactions were identified.

II. Adverse Events

Case Reports of Adverse Events

Cases of acute hepatitis have been reported as follows: a woman who had taken 2 tablets daily of shen-min tablets (20% 12:1 extract of fo-ti and 40% fo-ti powder) for 8 weeks (Cardenas et al. 2006); a man who had taken fo-ti tablets at the recommended dose daily for 1 month (Battinelli et al. 2004); and a woman who had taken fo-ti tablets for 2 weeks at the recommended dose (Park et al. 2001). In one case, the herbal product was analyzed and found to contain 0.14% by weight anthraquinones (Battinelli et al. 2004). The author of one of the reports indicated that the hepatotoxicity was an idiosyncratic reaction because the patient developed a fine maculopapular rash and mild eosinophilia, and did not overdose (Cardenas et al. 2006).

Recurrent hepatitis was reported in a 5-year-old girl who had taken 3 tablets daily of shou wu pian, a product made from fo-ti tuber, for 4 months. Elevated serum levels of bilirubin and liver enzymes were observed, and viral markers were used to rule out a number of diseases. Symptoms disappeared and liver function tests were normal 1 month after cessation of the product. The girl experienced hepatitis again after being administered 2 tablets daily of the same product for 1 month (Panis et al. 2005). The product identity was confirmed by nuclear magnetic resonance analysis and showed 2,3,5,4′-tetrahydroxystilbene-2-*O*-β-D-glucopyranoside, a stilbene glycoside, to be the main constituent. The anthraquinone emodin was present in trace amounts (Panis et al. 2005). Such a chemical composition indicates that raw or incompletely processed fo-ti was used.

Hepatitis was diagnosed in a 31-year-old pregnant woman who had been taking shou wu pian (dose unspecified), a proprietary product made from fo-ti tuber, for several weeks prior to the hepatitis. Elevated plasma levels of liver enzymes were observed, and viral hepatitis was ruled out. Liver enzyme levels returned to normal 3 weeks after cessation of the product. The identity of the product was confirmed by thin-layer chromatography, and two anthraquinones, emodin and physcion, were identified. The patient described a case of hepatitis that she had 2 years prior after taking a liquid extract prepared from fo-ti (But et al. 1996).

Acute hepatitis mimicking iron overload syndrome was reported in a 35-year-old man who had been taking fo-ti (dose and duration unspecified). Laboratory studies included alanine transferase 2714 U/l (normal <50 U/l), aspartate aminotransferase 1170 U/l (normal <50 U/l), AP 137 U/l (normal <130 U/l), total bilirubin 4.6 mg/dl (normal <1.4 mg/dl), direct bilirubin 3.0 mg/dl (normal <0.4 mg/dl), and ferritin 13,862 ng/ml (normal 8 to 282 ng/ml) and a fasting transferrin saturation of 86% (normal 20% to 60%). Analysis of the herbal supplement identified extracts from fo-ti including the anthraquinones emodin and physcion. The patient recovered after cessation of the herbal products, and liver function tests 4 months after hospitalization were normal (Laird et al. 2008).

A traditional Chinese medicine text lists the following adverse events reported in association with processed fo-ti tuber products (no details on cases are provided): mild abdominal pain, nausea, vomiting, mild jaundice, abnormal liver function studies, bleeding from the upper digestive tract, diarrhea, and melena. In extremely high doses, stimulation, agitation, tachycardia, clonic or tetanic spasms, and even death due to respiratory paralysis have been reported. A repetitive, malarial-like sequence of drug fevers, chills, sweating, and weakness has been observed. Allergic reactions have been reported (Bensky et al. 2004).

III. Pharmacology and Pharmacokinetics

Human Pharmacological Studies

No relevant human pharmacological studies were identified.

Animal Pharmacological Studies

No relevant animal pharmacological studies were identified.

In Vitro Pharmacological Studies

At concentrations greater than 0.1 μg/ml, a mild estrogenic effect of fo-ti was observed in vitro (Kang et al. 2006). The EC_{50} for fo-ti estrogenic activity in the recombinant yeast bioassay was 80 μg/ml (Zhang et al. 2005).

Inhibition of several CYP450 enzymes was observed in vitro at concentrations of 500 μg/ml, but not at concentrations of 20 or 100 μg/ml. The authors of the study indicated that the relevance of the in vitro results to in vivo situations is unknown (Unger and Frank 2004).

Fo-ti inhibited DNA synthesis and growth in human vascular epithelial cells under normal culture conditions at a concentration of 100 μg/ml (Ling et al. 2008).

IV. Pregnancy and Lactation

No studies were identified on the use of fo-ti during pregnancy or lactation. No contraindications for use during pregnancy or lactation are listed in two traditional Chinese medicine texts (Bensky et al. 2004; Chen and Chen 2004).

R

V. Toxicity Studies

Acute Toxicity

The LD_{50} of intraperitoneally administered alcohol extract of processed fo-ti tuber in mice is 164.9 g/kg (Leung and Foster 1996; Shen et al. 1982), whereas that of the intragastrically administered extract is over 1000 g/kg, with no deaths reported at that dose (Chen and Chen 2004; Shen et al. 1982).

Genotoxicity

No mutagenic activity of fo-ti aqueous extracts was reported in the Ames test for mutagenicity (Kam 1981). Antimutagenic activity of fo-ti was observed in a *Tradescantia* micronucleus assay (Zhang et al. 1999).

Literature Cited

Avula, B., V.C. Joshi, Y.H. Wang, and I.A. Khan. 2007. Simultaneous identification and quantification of anthraquinones, polydatin, and resveratrol in *Polygonum multiflorum*, various *Polygonum* species, and dietary supplements by liquid chromatography and microscopic study of *Polygonum* species. *J. AOAC Int.* 90(6):1532-1538.

Battinelli, L., C. Daniele, G. Mazzanti, et al. 2004. New case of acute hepatitis following the consumption of shou wu pian, a chinese herbal product derived from *Polygonum multiflorum*. *Ann. Intern. Med.* 140(7):589-590.

Bensky, D., S. Clavey, and E. Stöger. 2004. *Chinese herbal medicine: Materia medica.* 3rd ed. Seattle: Eastland Press.

But, P.P., B. Tomlinson, and K.L. Lee. 1996. Hepatitis related to the Chinese medicine *shou-wu-pian* manufactured from *Polygonum multiflorum*. *Vet. Hum. Toxicol.* 38(4):280-282.

Cardenas, A., J.C. Restrepo, F. Sierra, and G. Correa. 2006. Acute hepatitis due to shen-min: A herbal product derived from *Polygonum multiflorum*. *J. Clin. Gastroenterol.* 40(7):629-632.

Chen, J.K., and T.T. Chen. 2004. *Chinese medical herbology and pharmacology*. City of Industry, CA: Art of Medicine Press.

Kam, J.K. 1981. Mutagenic activity of *ho shao wu* (*Polygonum multiflorum* Thunb). *Am. J. Chin. Med.* 9(3):213-215.

Kang, S.C., C.M. Lee, H. Choi, et al. 2006. Evaluation of oriental medicinal herbs for estrogenic and antiproliferative activities. *Phytother. Res.* 20(11):1017-1019.

Laird, A.R., N. Ramchandani, E.M. Degoma, et al. 2008. Acute hepatitis associated with the use of an herbal supplement (*Polygonum multiflorum*) mimicking iron-overload syndrome. *J. Clin. Gastroenterol.* 42(7):861-862.

Leung, A.Y., and S. Foster. 1996. *Encyclopedia of common natural ingredients used in food, drugs, and cosmetics.* 2nd ed. New York: Wiley.

Ling, S., L. Nheu, A. Dai, Z. Guo, and P. Komesaroff. 2008. Effects of four medicinal herbs on human vascular endothelial cells in culture. *Int. J. Cardiol.* 128(3):350-358.

Liu, C., Q. Zhang, and Q. Zhou. 1991. Assay of stilbene glucoside in *Polygonum multiflorum* Thunb. and its processed products. *Zhongguo Zhong Yao Za Zhi* 16(8):469-472, 511.

Panis, B., D.R. Wong, P.M. Hooymans, P.A.G.M. De Smet, and P.P.R. Rosias. 2005. Recurrent toxic hepatitis in a Caucasian girl related to the use of shou-wu-pian, a Chinese herbal preparation. *J. Pediatr. Gastroenterol. Nutr.* 41(2):256-258.

Park, G.J., S.P. Mann, and M.C. Ngu. 2001. Acute hepatitis induced by shou-wu-pian, a herbal product derived from *Polygonum multiflorum*. *J. Gastroenterol. Hepatol.* 16(1):115-117.

PPRC. 2005. *Pharmacopoeia of the People's Republic of China.* Beijing: Chemical Industry Press.

Shen, D., Y. Gu, and X. Ren. 1982. The pharmacological study on the processed products of *Radix Polygoni Multiflori. [Chinese Traditional Patent Medicine]* 1(1):21.

Sionneau, P., and B. Flaws. 1995. *Pao zhi: An introduction to the use of processed Chinese medicinals.* Boulder, CO: Blue Poppy Enterprises, Inc.

Stickel, F., E. Patsenker, and D. Schuppan. 2005. Herbal hepatotoxicity. *J. Hepatol.* 43(5):901-910.

Unger, M., and A. Frank. 2004. Simultaneous determination of the inhibitory potency of herbal extracts on the activity of six major cytochrome P450 enzymes using liquid chromatography/mass spectrometry and automated online extraction. *Rapid Commun. Mass Spectrom.* 18(19):2273-2281.

Vanderperren, B., M. Rizzo, L. Angenot, et al. 2005. Acute liver failure with renal impairment related to the abuse of senna anthraquinone glycosides. *Ann. Pharmacother.* 39(7-8):1353-1357.

Yi, T., K.S. Leung, G.H. Lu, H. Zhang, and K. Chan. 2007. Identification and determination of the major constituents in traditional Chinese medicinal plant *Polygonum multiflorum* Thunb. by HPLC coupled with PAD and ESI/MS. *Phytochem. Anal.* 18(3):181-187.

Zhang, C.Z., S.X. Wang, Y. Zhang, J.P. Chen, and X.M. Liang. 2005. *In vitro* estrogenic activities of Chinese medicinal plants traditionally used for the management of menopausal symptoms. *J. Ethnopharmacol.* 98(3):295-300.

Zhang, H., B.S. Jeong, and T.H. Ma. 1999. Antimutagenic property of an herbal medicine, *Polygonum multiflorum* Thunb., detected by the *Tradescantia* micronucleus assay. *J. Environ. Pathol. Toxicol. Oncol.* 18(2):127-130.

Zhang, Z.G., T.S. Lu, and Q.Q. Yao. 2006. Effect of preparation on the major chemical constituents of *Polygonum multiflorum*. *Zhong Yao Cai* 29(10):1017-1019.

Reynoutria multiflora (Thunb.) Moldenke — Polygonaceae

SCN: fo-ti
Syn: *Fallopia multiflora* (Thunb.) Haraldson; *Polygonum multiflorum* Thunb.
PN: *sheng he shou wu* (dried root tuber)
OCN: fleeceflower
Part: unprocessed root tuber

Quick Reference Summary

Safety Class: 3
Interaction Class: A

Contraindications

Not for use except under the supervision of an expert qualified in the appropriate use of this substance (Battinelli et al. 2004; But et al. 1996; Laird et al. 2008; Panis et al. 2005).

Other Precautions

None known.

Drug and Supplement Interactions

None known.

Notice

Stimulant laxative (Leung and Foster 1996); *see* Appendix 2.

Editors' Notes

Two forms of fo-ti tuber are used in traditional Chinese medicine, the dried tuber (*sheng he shou wu*), which is the subject of this entry, and the processed tuber (*zhi he shou wu*), which is the subject of a separate entry (*see* previous entry). The processed tuber is used almost exclusively, while the dried tuber is not commonly used and is difficult to find commercially. The dried tuber contains stilbene glycosides and anthraquinone glycosides and is traditionally used for short periods of time as a stimulant laxative (Avula et al. 2007; Leung and Foster 1996; Zhang et al. 2006).

Adverse Events and Side Effects

Although no cases of adverse events associated explicitly with unprocessed fo-ti have been reported in the English language literature, analysis of fo-ti products marketed as processed fo-ti and associated with cases of hepatitis showed levels of anthraquinone glycosides or stilbene glycosides that suggested incomplete processing (Battinelli et al. 2004; But et al. 1996; Laird et al. 2008; Panis et al. 2005).

Pharmacological Considerations

None known.

Pregnancy and Lactation

No studies on the use of unprocessed fo-ti during pregnancy or lactation were identified. No contraindications for use during pregnancy or lactation are listed in two traditional Chinese medicine texts (Bensky et al. 2004; Chen and Chen 2004). Based on stimulant laxative activity and an incomplete understanding of reports of hepatotoxicity associated with the incompletely processed tuber, use during pregnancy or lactation is not recommended except under the supervision of a qualified healthcare practitioner.

Review Details

I. Drug and Supplement Interactions

Clinical Trials of Drug or Supplement Interactions

No clinical trials of drug or supplement interactions were identified.

Case Reports of Suspected Drug or Supplement Interactions

No case reports of suspected drug or supplement interactions were identified.

Animal Trials of Drug or Supplement Interactions

No animal trials of drug or supplement interactions were identified.

II. Adverse Events

Case Reports of Adverse Events

Although no cases of adverse events associated explicitly with unprocessed fo-ti have been reported in the English language literature, analysis of fo-ti products marketed as processed fo-ti and associated with cases of hepatitis showed levels of anthraquinone glycosides or stilbene glycosides that suggested incomplete processing (Battinelli et al. 2004; But et al. 1996; Laird et al. 2008; Panis et al. 2005).

III. Pharmacology and Pharmacokinetics

Human Pharmacological Studies

No relevant human pharmacological studies were identified.

Animal Pharmacological Studies

In rats fed peroxidized corn oil, administration of stilbene components (100 mg/kg piceid or 50 or 100 mg/kg 2,3,5,4′-tetrahydroxystilbene-2-*O*-D-glucoside) isolated from fo-ti tubers was found to partly inhibit the deposition of lipid peroxides in the liver, providing a hepatoprotective effect, and to reduce the elevation of glutamic oxaloacetic transaminase (GOT) and glutamic pyruvate transaminase (GPT) levels in the serum (Kimura et al. 1983).

In Vitro Pharmacological Studies

No relevant in vitro pharmacological studies were identified.

IV. Pregnancy and Lactation

No studies on the use of unprocessed fo-ti during pregnancy or lactation were identified. No contraindications for use during pregnancy or lactation are listed in two traditional Chinese medicine texts (Bensky et al. 2004; Chen and Chen 2004). Based on stimulant laxative activity and an incomplete understanding of reports of hepatotoxicity associated with the incompletely processed tuber, use during pregnancy or lactation is not recommended except under the supervision of a qualified healthcare practitioner.

V. Toxicity Studies

Acute Toxicity

The LD_{50} of an intraperitoneally administered alcohol extract of unprocessed fo-ti tuber in mice is 2.7 g/kg, whereas that of an intragastrically administered extract is 50 g/kg (Leung and Foster 1996; Shen et al. 1982).

Literature Cited

Avula, B., V.C. Joshi, Y.H. Wang, and I.A. Khan. 2007. Simultaneous identification and quantification of anthraquinones, polydatin, and resveratrol in *Polygonum multiflorum*, various *Polygonum* species, and dietary supplements by liquid chromatography and microscopic study of *Polygonum* species. *J. AOAC Int.* 90(6):1532-1538.

Battinelli, L., C. Daniele, G. Mazzanti, et al. 2004. New case of acute hepatitis following the consumption of shou wu pian, a chinese herbal product derived from *Polygonum multiflorum*. *Ann. Intern. Med.* 140(7):589-590.

Bensky, D., S. Clavey, and E. Stöger. 2004. *Chinese herbal medicine: Materia medica*. 3rd ed. Seattle: Eastland Press.

But, P.P., B. Tomlinson, and K.L. Lee. 1996. Hepatitis related to the Chinese medicine *shou-wu-pian* manufactured from *Polygonum multiflorum*. *Vet. Hum. Toxicol.* 38(4):280-282.

Chen, J.K., and T.T. Chen. 2004. *Chinese medical herbology and pharmacology*. City of Industry, CA: Art of Medicine Press.

Kimura, Y., H. Ohminami, and H. Okuda. 1983. Effects of stilbene components of roots of *Polygonum* spp. on liver injury in peroxidized oil-fed rats. *Planta Med.* 49(1):51-54.

Laird, A.R., N. Ramchandani, E.M. Degoma, et al. 2008. Acute hepatitis associated with the use of an herbal supplement (*Polygonum multiflorum*) mimicking iron-overload syndrome. *J. Clin. Gastroenterol.* 42(7):861-862.

Leung, A.Y., and S. Foster. 1996. *Encyclopedia of common natural ingredients used in food, drugs, and cosmetics*. 2nd ed. New York: Wiley.

Panis, B., D.R. Wong, P.M. Hooymans, P.A.G.M. De Smet, and P.P.R. Rosias. 2005. Recurrent toxic hepatitis in a Caucasian girl related to the use of shou-wu-pian, a Chinese herbal preparation. *J. Pediatr. Gastroenterol. Nutr.* 41(2):256-258.

Shen, D., Y. Gu, and X. Ren. 1982. The pharmacological study on the processed products of *Radix Polygoni Multiflori. [Chinese Traditional Patent Medicine]* 1(1):21.

Zhang, Z.G., T.S. Lu, and Q.Q. Yao. 2006. Effect of preparation on the major chemical constituents of *Polygonum multiflorum*. *Zhong Yao Cai* 29(10):1017-1019.

Rhamnus cathartica L. — Rhamnaceae

SCN: buckthorn

Part: fruit

Quick Reference Summary

Safety Class: 2b, 2c, 2d
Interaction Class: A

Contraindications

Not for use in pregnancy or lactation except under the supervision of a qualified healthcare practitioner (Bradley 1992; Chadha 1988; Roth et al. 1984; Wichtl 2004).

Not for use in persons with intestinal obstruction, abdominal pain of unknown origin, or any inflammatory condition of the intestines (i.e., appendicitis, colitis, Crohn's disease, irritable bowel syndrome, and melanosis coli) (Bradley 1992; De Smet 1993; Wichtl 2004).

Not for use in children less than 12 years of age (Bradley 1992; De Smet 1993).

Not for use in excess of 8 consecutive days (Bradley 1992; De Smet 1993; Leung and Foster 1996; Weiss and Meuss 2001; Wichtl 2004).

Other Precautions

None known.

Drug and Supplement Interactions

See Pharmacological Considerations.

Notice

Stimulant laxative (Bradley 1992; Chadha 1988; De Smet 1993; Felter and Lloyd 1898; Leung and Foster 1996; Martindale and Reynolds 1996; Roth et al. 1984; Weiss and Meuss 2001; Wichtl 2004); *see* Appendix 2.

Standard Dose

The individually correct dose is the smallest dose necessary to produce a comfortable, soft stool (ESCOP 2003).

Editors' Note

The American Herbal Products Association has established a trade requirement (AHPA 2011) that products containing this herb in sufficient quantity to warrant such labeling bear the following label statement:

> NOTICE: Do not use this product if you have abdominal pain or diarrhea. Consult a health care provider prior to use if you are pregnant or nursing. Discontinue use in the event of diarrhea or watery stools. Do not exceed recommended dose. Not for long-term use.

Adverse Events and Side Effects

Discoloration of the urine by buckthorn metabolites may occur but is not clinically significant. Abdominal spasms and pain have been reported (ESCOP 2003).

Pharmacological Considerations

Concomitant use of buckthorn is cautioned with antiarrhythmic drugs and botanicals containing cardiac glycosides, as long-term use of buckthorn as a laxative can cause potassium loss, leading to increased toxicity of those drugs and botanicals (Brinker 2001; De Smet 1993; ESCOP 2003).

Concomitant internal use of buckthorn is cautioned with thiazide diuretics, corticosteroids, or licorice and long-term use of buckthorn as a laxative may increase the potassium loss induced by those drugs and botanicals (Brinker 2001; De Smet 1993; ESCOP 2003).

Use of stimulant laxatives, such as buckthorn, may reduce the gastrointestinal transit time and thus reduce the absorption of orally administered drugs (Brinker 2001; De Smet 1993).

Pregnancy and Lactation

While most stimulant laxatives have traditionally been contraindicated in pregnancy due to concerns regarding stimulation of the uterus, a number of stimulant laxatives, including buckthorn, have shown a lack of adverse effects on pregnancy or the fetus when used according to the recommended dosage schedule (De Smet 1993; ESCOP 2003). Thus, these laxatives are now considered appropriate for use during pregnancy (De Smet 1993; ESCOP 2003; Prather 2004). Due to the potential genotoxicity of certain anthraquinones, however, it is recommended that use of certain anthranoid laxatives, including buckthorn, should be avoided in the first trimester of pregnancy or used under professional supervision (ESCOP 2003).

No information on the safety of buckthorn during lactation was identified in the scientific or traditional literature. While this review did not identify any concerns for use while nursing, safety has not been conclusively established.

Review Details

I. Drug and Supplement Interactions

Clinical Trials of Drug or Supplement Interactions

No clinical trials of drug or supplement interactions were identified.

Case Reports of Suspected Drug or Supplement Interactions

No case reports of suspected drug or supplement interactions were identified.

Animal Trials of Drug or Supplement Interactions

No animal trials of drug or supplement interactions were identified.

II. Adverse Events

Case Reports of Adverse Events

No case reports of adverse events were identified.

III. Pharmacology and Pharmacokinetics

Human Pharmacological Studies

No relevant human pharmacological studies were identified.

Animal Pharmacological Studies

No relevant animal pharmacological studies were identified.

In Vitro Pharmacological Studies

No relevant in vitro pharmacological studies were identified.

IV. Pregnancy and Lactation

While most stimulant laxatives have traditionally been contraindicated in pregnancy due to concerns regarding stimulation of the uterus, a number of stimulant laxatives, including buckthorn, have shown a lack of adverse effects on pregnancy or the fetus when used according to the recommended dosage schedule (De Smet 1993; ESCOP 2003). Thus, these laxatives are now considered appropriate for use during pregnancy (De Smet 1993; ESCOP 2003; Prather 2004). Due to the potential genotoxicity of certain anthraquinones, however, it is recommended that use of certain anthranoid laxatives, including buckthorn, should be avoided in the first trimester of pregnancy or used under professional supervision (ESCOP 2003).

No information on the safety of buckthorn during lactation was identified.

V. Toxicity Studies

No toxicity studies on buckthorn were identified.

R

Literature Cited

AHPA. July 2011. Code of Ethics & Business Conduct. Silver Spring, MD: American Herbal Products Association.

Bradley, P.R. 1992. *British herbal compendium: A handbook of scientific information on widely used plant drugs*. Bournemouth, UK: British Herbal Medicine Association.

Brinker, F. 2001. *Herb contraindications and drug interactions*. 3rd ed. Sandy, OR: Eclectic Medical Publications.

Chadha, Y. 1988. *The wealth of India: A dictionary of Indian raw materials and industrial products*. Delhi: Council of Scientific and Industrial Research.

De Smet, P.A.G.M. 1993. *Adverse effects of herbal drugs, Volume 2*. Berlin: Springer.

ESCOP. 2003. *ESCOP monographs: The scientific foundation for herbal medicinal products*. 2nd ed. Exeter, UK: European Scientific Cooperative on Phytotherapy.

Felter, H.W., and J.U. Lloyd. 1898. *King's American dispensatory*. 18th ed., 3rd rev. 2 vols. Cincinnati: Ohio Valley Co.

Leung, A.Y., and S. Foster. 1996. *Encyclopedia of common natural ingredients used in food, drugs, and cosmetics*. 2nd ed. New York: Wiley.

Martindale, W., and J.E.F. Reynolds. 1996. *The extra pharmacopoeia*. 31st ed. London: Pharmaceutical Press.

Roth, L., M. Daunderer, and K. Kormann. 1984. *Giftpflanzen-pflanzengifte: Vorkommen, wirkung, therapie*. Landsberg: Ecomed.

Weiss, R.F., and A.R. Meuss. 1998. *Weiss's herbal medicine*. Classic ed. New York: Thieme.

Wichtl, M. 2004. *Herbal drugs and phytopharmaceuticals: A handbook for practice on a scientific basis*. 3rd ed. Boca Raton, FL: CRC Press.

Rheum spp. — Polygonaceae

Rheum officinale **Baill.**
SCN: Chinese rhubarb
PN: *da huang* (root and rhizome)
OCN: Turkey rhubarb

Rheum palmatum **L.**
SCN: Chinese rhubarb
PN: *da huang* (root and rhizome)
OCN: Turkey rhubarb

Rheum palmatum **L. var.** ***tanguticum*** **Maxim. ex Regel**
SCN: Chinese rhubarb
Syn: *Rheum tanguticum* Maxim. ex Balf.
PN: *da huang* (root and rhizome)
OCN: Turkey rhubarb
Part: rhizome, root

Quick Reference Summary

Safety Class: 2b, 2c, 2d
Interaction Class: A

Contraindications

Not for use in pregnancy or lactation except under the supervision of a qualified healthcare practitioner (Bensky et al. 2004; Chen and Chen 2004).

Not for use in intestinal obstruction, abdominal pain of unknown origin, or any inflammatory condition of the intestines (i.e., appendicitis, colitis, Crohn's disease, irritable bowel syndrome, and melanosis coli) (Bradley 1992; De Smet 1993; Martindale and Reynolds 1996; Roth et al. 1984; Wichtl 2004), or during menstruation (Bensky et al. 2004; Chen and Chen 2004).

Not for use in excess of 8 consecutive days (Bradley 1992; De Smet 1993; Leung and Foster 1996; Weiss and Meuss 2001; Wichtl 2004).

Not for use in children less than 12 years of age (Bradley 1992; De Smet 1993).

Other Precautions

None known.

Drug and Supplement Interactions

See Pharmacological Considerations.

Standard Dose

The standard dose is 1 to 5 g daily, directly consumed or as a tea (Bradley 1992). Traditional Chinese medicine texts indicate that the standard dose is 3 to 10 g daily as a tea, with up to 20 g in severe cases, and that the dose should be reduced by half when the powdered product is taken internally (Bensky et al. 2004; Chen and Chen 2004).

Notice

Stimulant laxative (Bensky et al. 2004; Bradley 1992; De Smet 1993; ESCOP 2003; Felter and Lloyd 1898; Leung and Foster 1996; Weiss and Meuss 2001); *see* Appendix 2.

Tannins (4.0–11.0%) (Bradley 1992; Tang and Eisenbrand 1992); *see* Appendix 1.

Editors' Notes

Although not used in medicine, the leaf blade of Chinese rhubarb contains oxalic acid, which can cause precipitation of calcium oxalate in the renal tubules, thus causing kidney failure (Bensky et al. 2004).

Some compounds in Chinese rhubarb produce a purgative action. These compounds are highly sensitive to heat and may be removed by different forms of processing (Chen and Chen 2004).

The American Herbal Products Association has established a trade requirement (AHPA 2011) that products containing this herb in sufficient quantity to warrant such labeling bear the following label statement:

> NOTICE: Do not use this product if you have abdominal pain or diarrhea. Consult a health care provider prior to use if you are pregnant or nursing. Discontinue use in the event of diarrhea or watery stools. Do not exceed recommended dose. Not for long-term use.

Adverse Events and Side Effects

Gastrointestinal cramping may occur during Chinese rhubarb use (Wichtl 2004). Metabolites of Chinese rhubarb may cause the urine to have a red-brown or bright yellow coloration (Wichtl 2004). Side effects of nausea, vomiting, and poor appetite have been reported (Bensky et al. 2004).

R

Pharmacological Considerations

Long-term use of Chinese rhubarb and other anthranoid laxatives may cause pseudomelanosis coli, a brownish pigmentation of the colon. Although pseudomelanosis coli has been regarded as harmless (Leng-Peschlow 1992; Wichtl 2004), some researchers have suggested a possible relationship between the condition and an increased risk of development of colon cancer (Siegers et al. 1993). The risk, however, may be due to other confounding factors such as chronic constipation or diet (Sonnenberg and Müller 1993; van Gorkom et al. 1999).

Concomitant use of Chinese rhubarb is cautioned with antiarrhythmic drugs and botanicals containing cardiac glycosides, as long-term use of Chinese rhubarb as a laxative can cause potassium loss, leading to increased toxicity of those drugs and botanicals (Brinker 2001; De Smet 1993; ESCOP 2003).

Concomitant internal use of Chinese rhubarb is cautioned with thiazide diuretics, corticosteroids, or licorice, and long-term use of Chinese rhubarb as a laxative may increase the potassium loss induced by those drugs and botanicals (Brinker 2001; De Smet 1993; ESCOP 2003; Wichtl 2004).

Use of stimulant laxatives, such as Chinese rhubarb, may reduce the gastrointestinal transit time and thus reduce the absorption of orally administered drugs (Brinker 2001; De Smet 1993).

Pregnancy and Lactation

A reference text on traditional Chinese medicine indicated that Chinese rhubarb should be used with great caution and only when absolutely necessary during pregnancy and lactation (Bensky et al. 2004). Based on this information, use during pregnancy or lactation is not recommended except under the supervision of a qualified healthcare practitioner.

Review Details

I. Drug and Supplement Interactions

Clinical Trials of Drug or Supplement Interactions

No clinical trials of drug or supplement interactions were identified.

Case Reports of Suspected Drug or Supplement Interactions

Acute renal failure was reported in a 23-year-old woman after 6 weeks of use of an herbal slimming pill that contained Chinese rhubarb (amount unspecified). The woman had begun taking diclofenac, a nonsteroidal anti-inflammatory drug (NSAID) that has been associated with kidney failure (Juhlin et al. 2004; Revai and Harmos 1999; Rossi et al. 1985; Rubio Garcia and Tellez Molina 1992). The authors of the report indicated that although NSAIDs are well recognized as a major cause of kidney dysfunction, such dysfunction is usually associated with dehydration, with consequent hemodynamically mediated renal impairment (Kwan et al. 2006).

Animal Trials of Drug or Supplement Interactions

No animal trials of drug or supplement interactions were identified.

II. Adverse Events

Case Reports of Adverse Events

Overdoses of Chinese rhubarb can cause a toxic reaction, especially if the fresh herb is used. Symptoms include diarrhea, nausea, vomiting, dizziness, abdominal colic, and jaundice. Long-term use may cause severe damage to the large intestine, cirrhosis of the liver, and potassium loss (Bensky et al. 2004).

Allergic reactions to Chinese rhubarb have also been reported, usually after large doses (30 g), and include flushing, pruritus, papular rashes, asthma, and tachypnea (Bensky et al. 2004).

III. Pharmacology and Pharmacokinetics

Human Pharmacological Studies

No relevant human pharmacological studies were identified.

Animal Pharmacological Studies

No relevant animal pharmacological studies were identified.

In Vitro Pharmacological Studies

In guinea pig stomachs, an aqueous extract of Chinese rhubarb dose-dependently increased the resting tension and contractile frequency of gastric body circular muscle. The authors indicated that the exciting action of Chinese rhubarb was partly mediated via the cholinergic M receptor, the cholinergic N receptor, and the L-type calcium channel (Yu et al. 2005).

The compound rhapontin exhibited growth inhibition and induced apoptosis in human stomach cancer cells (Hibasami et al. 2007). The compound rhein induced apoptosis and blocked cell cycle progression in the G_1 phase in human liver cancer cells (Kuo et al. 2004).

An extract of Chinese rhubarb induced estrogenic activity at a concentration of 0.1 mg/ml in a yeast-based estrogen receptor activity assay (Kang et al. 2006).

IV. Pregnancy and Lactation

A reference text on traditional Chinese medicine indicates that Chinese rhubarb should be used with great caution, and only when absolutely necessary, during pregnancy and lactation (Bensky et al. 2004).

V. Toxicity Studies

Acute Toxicity

The LD_{50} of an orally administered Chinese rhubarb decoction in mice is 153 g/kg (Chen and Chen 2004).

Subchronic Toxicity

In rats orally administered 140, 794, or 4500 mg/kg daily of anthraquinones isolated from rhubarb for 13 weeks, nephrotoxicity developed in the group administered 4500 mg/kg, and tubule epithelial cells swelled and denatured (Yan et al. 2006).

Literature Cited

AHPA. July 2011. Code of Ethics & Business Conduct. Silver Spring, MD: American Herbal Products Association.

Bensky, D., S. Clavey, and E. Stöger. 2004. *Chinese herbal medicine: Materia medica*. 3rd ed. Seattle: Eastland Press.

Bradley, P.R. 1992. *British herbal compendium: A handbook of scientific information on widely used plant drugs*. Bournemouth, UK: British Herbal Medicine Association.

Brinker, F. 2001. *Herb contraindications and drug interactions. 3rd ed.* Sandy, OR: Eclectic Medical Publications.

Chen, J.K., and T.T. Chen. 2004. *Chinese medical herbology and pharmacology*. City of Industry, CA: Art of Medicine Press.

De Smet, P.A.G.M. 1993. *Adverse effects of herbal drugs, Volume 2.* Berlin: Springer.

ESCOP. 2003. *ESCOP monographs: The scientific foundation for herbal medicinal products*. 2nd ed. Exeter, UK: European Scientific Cooperative on Phytotherapy.

Felter, H.W., and J.U. Lloyd. 1898. *King's American dispensatory*. 18th ed., 3rd rev. 2 vols. Cincinnati: Ohio Valley Co.

Hibasami, H., K. Takagi, T. Ishii, et al. 2007. Induction of apoptosis by rhapontin having stilbene moiety, a component of rhubarb (*Rheum officinale* Baillon) in human stomach cancer KATO III cells. *Oncol. Rep.* 18(2):347-351.

Juhlin, T., S. Bjorkman, B. Gunnarsson, et al. 2004. Acute administration of diclofenac, but possibly not long term low dose aspirin, causes detrimental renal effects in heart failure patients treated with ACE-inhibitors. *Eur. J. Heart Fail.* 6(7):909-916.

Kang, S.C., C.M. Lee, H. Choi, et al. 2006. Evaluation of oriental medicinal herbs for estrogenic and antiproliferative activities. *Phytother. Res.* 20(11):1017-1019.

Kuo, P.L., Y.L. Hsu, L.T. Ng, and C.C. Lin. 2004. Rhein inhibits the growth and induces the apoptosis of Hep G2 cells. *Planta Med.* 70(1):12-16.

Kwan, T.H., M.K. Tong, K.T. Leung, et al. 2006. Acute renal failure associated with prolonged intake of slimming pills containing anthraquinones. *Hong Kong Med. J.* 12(5):394-397.

Leng-Peschlow, E. 1992. Senna and pseudomelanosis coli. *Pharmacology* 44 (Suppl. 1):33-35.

Leung, A.Y., and S. Foster. 1996. *Encyclopedia of common natural ingredients used in food, drugs, and cosmetics*. 2nd ed. New York: Wiley.

Martindale, W., and J.E.F. Reynolds. 1996. *The extra pharmacopoeia*. 31st ed. London: Pharmaceutical Press.

Revai, T., and G. Harmos. 1999. Nephrotic syndrome and acute interstitial nephritis associated with the use of diclofenac. *Wien Klin. Wochenschr.* 111(13):523-524.

Rossi, E., G.F. Ferraccioli, F. Cavalieri, et al. 1985. Diclofenac-associated acute renal failure. Report of 2 cases. *Nephron* 40(4):491-493.

Roth, L., M. Daunderer, and K. Kormann. 1984. *Giftpflanzen-pflanzengifte: Vorkommen, wirkung, therapie*. Landsberg: Ecomed.

Rubio Garcia, J.A., and M.J. Tellez Molina. 1992. Acute renal failure and nephrotic syndrome associated with treatment with diclofenac. *Rev. Clin. Esp.* 191(5):289-290.

Siegers, C.P., E. von Hertzberg-Lottin, M. Otte, and B. Schneider. 1993. Anthranoid laxative abuse—A risk for colorectal cancer? *Gut* 34(8):1099-1101.

Sonnenberg, A., and A.D. Müller. 1993. Constipation and cathartics as risk factors of colorectal cancer: A meta-analysis. *Pharmacology* 47(1):224-233.

Tang, W., and G. Eisenbrand. 1992. *Chinese drugs of plant origin: Chemistry, pharmacology, and use in traditional and modern medicine*. New York: Springer.

van Gorkom, B.A.P., E.G.E. de Vries, A. Karrenbeld, and J.H. Kleibeuker. 1999. Review article: Anthranoid laxatives and their potential carcinogenic effects. *Aliment. Pharmacol. Ther.* 13(4):443-452.

Weiss, R.F., and A.R. Meuss. 1998. *Weiss's herbal medicine*. Classic ed. New York: Thieme.

Wichtl, M. 2004. *Herbal drugs and phytopharmaceuticals: A handbook for practice on a scientific basis*. 3rd ed. Boca Raton, FL: CRC Press.

Yan, M., L.Y. Zhang, L.X. Sun, Z.Z. Jiang, and X.H. Xiao. 2006. Nephrotoxicity study of total rhubarb anthraquinones on Sprague Dawley rats using DNA microarrays. *J. Ethnopharmacol.* 107(2):308-311.

Yu, M., Y.L. Luo, J.W. Zheng, et al. 2005. Effects of rhubarb on isolated gastric muscle strips of guinea pigs. *World J. Gastroenterol.* 11(17):2670-2673.

Rhus spp.

Anacardiaceae

Rhus copallinum **L.**
SCN: winged sumac
OCN: shining sumac

Rhus coriaria **L.**
SCN: Sicilian sumac
OCN: tanner's sumac

Rhus glabra **L.**
SCN: smooth sumac
OCN: scarlet sumac
Part: fruit

Quick Reference Summary

Safety Class: 1
Interaction Class: A

Contraindications

None known.

Other Precautions

None known.

Drug and Supplement Interactions

None known.

Adverse Events and Side Effects

None known.

Pharmacological Considerations

None known.

Pregnancy and Lactation

No information on the use of Sicilian sumac or smooth sumac in pregnancy or lactation was identified in the scientific or traditional literature. Although this review did not identify any concerns for use while pregnant or nursing, safety has not been conclusively established.

Review Details

I. Drug and Supplement Interactions

Clinical Trials of Drug or Supplement Interactions

No clinical trials of drug or supplement interactions were identified.

Case Reports of Suspected Drug or Supplement Interactions

No case reports of suspected drug or supplement interactions were identified.

Animal Trials of Drug or Supplement Interactions

No animal trials of drug or supplement interactions were identified.

II. Adverse Events

Case Reports of Adverse Events

No case reports of adverse events were identified.

III. Pharmacology and Pharmacokinetics

Human Pharmacological Studies

No relevant human pharmacological studies were identified.

Animal Pharmacological Studies

No relevant animal pharmacological studies were identified.

In Vitro Pharmacological Studies

An ethyl acetate extract of Sicilian sumac inhibited alpha-amylase, a glycoside hydrolase, indicating hypoglycemic activity of this extract. The IC_{50} was 28.7 μg/ml (Giancarlo et al. 2006).

IV. Pregnancy and Lactation

No information on the use of Sicilian sumac or smooth sumac during pregnancy or lactation was identified.

V. Toxicity Studies

No toxicity studies were identified.

Literature Cited

Giancarlo, S., L.M. Rosa, F. Nadjafi, and M. Francesco. 2006. Hypoglycaemic activity of two spices extracts: *Rhus coriaria* L. and *Bunium persicum* Boiss. *Nat. Prod. Res.* 20(9):882-886.

Ribes nigrum L. — Grossulariaceae

SCN: black currant
OCN: cassis

Part: fruit, leaf

Quick Reference Summary

Safety Class: 1
Interaction Class: A

Contraindications
None known.

Other Precautions
None known.

Drug and Supplement Interactions
None known.

Notice
Diuretic (leaf) (Rácz-Kotilla and Rácz 1977; Wichtl 2004); *see* Appendix 2.

Adverse Events and Side Effects
None known.

Pharmacological Considerations
None known.

Pregnancy and Lactation
No information on the safety of black currant in pregnancy or lactation was identified in the scientific or traditional literature. Although this review did not identify any concerns for use while pregnant or nursing, safety has not been conclusively established.

Review Details

I. Drug and Supplement Interactions

Clinical Trials of Drug or Supplement Interactions
No clinical trials of drug or supplement interactions were identified.

Case Reports of Suspected Drug or Supplement Interactions
No case reports of suspected drug or supplement interactions were identified.

Animal Trials of Drug or Supplement Interactions
No animal trials of drug or supplement interactions were identified.

II. Adverse Events

Case Reports of Adverse Events
No case reports of adverse events were identified.

III. Pharmacology and Pharmacokinetics

Human Pharmacological Studies
No relevant human pharmacological studies were identified.

Animal Pharmacological Studies
In rats orally administered a fluid extract of black currant leaf at a dose equivalent to 1.5 g/kg of the dried leaf, a diuretic action (diuretic quotient of 1.56) was observed and was similar to a 50 mg/kg dose of furosemide (Rácz-Kotilla and Rácz 1977).

In Vitro Pharmacological Studies
No relevant in vitro pharmacological studies were identified.

IV. Pregnancy and Lactation
No information on the safety of black currant during pregnancy or lactation was identified.

V. Toxicity Studies

Acute Toxicity
No adverse effects were observed in mice orally administered 2 g/kg of an extract of the outer layer of black currant fruit (Kyerematen and Sandberg 1986).

The LD_{50} of a fluid extract (1:1) of black currant leaf in mice is 49 g/kg (Rácz-Kotilla and Rácz 1977).

The LD_{50} of a freeze-dried hydroethanolic extract of black currant leaf in mice is 1.09 g/kg after intraperitoneal administration and could not be determined at doses up to 3 g/kg after oral administration (Mongold et al. 1993).

Short-Term Toxicity
No adverse effects, including gastric ulceration or signs of toxicity, were observed in rats orally administered a freeze-dried hydroethanolic extract of black currant leaf (1 g of extract equivalent to 1.8 g of leaf) at doses of 2 g/kg daily for 21 days or 1.34 g/kg daily for 28 days.

Literature Cited

Kyerematen, G., and F. Sandberg. 1986. Preliminary pharmacological studies of Pecarin, new preparation from *Ribes nigrum* fruits. *Acta Pharm. Suec.* 23(2):101-106.

Mongold, J.J., P. Susplugas, C. Taillade, and J.J. Serrano. 1993. Anti-inflammatory activity of *Ribes nigrum* leaf extract in rats. *Plant. Med Phytother.* 26(2):109-116.

Rácz-Kotilla, E., and G. Rácz. 1977. Salidiuretische und hypotensive Wirkung der Auszüge von *Ribes* Blättern. *Planta Med.* 32:110-114.

Wichtl, M. 2004. *Herbal drugs and phytopharmaceuticals: A handbook for practice on a scientific basis*. 3rd ed. Boca Raton, FL: CRC Press.

Ricinus communis L.

Euphorbiaceae

SCN: castor
AN: *eranda*
PN: *bi ma you* (seed oil)
OCN: palma christi
Part: seed oil

Quick Reference Summary

Safety Class: 2b, 2d
Interaction Class: A

Contraindications

Not for use in pregnancy except under the supervision of a qualified healthcare practitioner (Lippert and Mueck 2002; Martindale and Reynolds 1996; McFarlin et al. 1999; Nabors 1958; Osol and Farrar 1955).

Not for use in intestinal obstruction, abdominal pain of unknown origin, or any inflammatory condition of the intestines (i.e., appendicitis, colitis, Crohn's disease, irritable bowel syndrome, and melanosis coli) (Bradley 1992; De Smet 1993; Martindale and Reynolds 1996; Roth et al. 1984; Wichtl 2004).

Not for use in excess of 8 consecutive days.

Other Precautions

None known.

Drug and Supplement Interactions

See Pharmacological Considerations.

Standard Dose

Adults: 5.0–20.0 ml per dose, not to exceed 60.0 ml per day. Children: 4.0–12.0 ml (Leung and Foster 1996; Osol and Farrar 1955; Williamson 2003).

Notice

Stimulant laxative (Chadha 1988; Felter and Lloyd 1898; Leung and Foster 1996; List and Hörhammer 1973; Martindale and Reynolds 1996); *see* Appendix 2.

Editors' Notes

Castor seed contains the highly toxic compound ricin. Ricin is removed in the process of oil extraction and thus is not present in castor oil (Audi et al. 2005; Doan 2004; Lord et al. 2003; Wilbur 2007). Several hundred cases of poisoning after ingestion of castor seed have been reported in the literature. Symptoms of poisoning include acute gastroenteritis, fluid and electrolyte depletion, gastrointestinal bleeding, hemolysis, and hypoglycemia (Challoner and McCarron 1990).

Adverse Events and Side Effects

Adverse events and side effects reported in association with castor oil include thrombosed hemorrhoids, precipitous labor, nausea, vomiting, diarrhea, intestinal colic, flatulence, disturbances of electrolyte balance, dehydration, hemorrhagic gastritis, hyperemia of the pelvic organs, hemolysis, and liver cell necrosis (Lippert and Mueck 2002; McFarlin et al. 1999).

Allergic reactions to castor oil, including cheilitis and dermatitis, confirmed by patch testing, have been reported after topical application of personal care products containing castor oil (Aplin and Eliseo 1997; Brandle et al. 1983; Di Berardino and Della Torre 2003; Fisher 1991; Kalavala et al. 2007; le Coz and Ball 2000; Lodi et al. 1992; Sai 1983; Taghipour et al. 2008; Tan et al. 1997; Wakelin et al. 1996; Wilbur 2007).

Pharmacological Considerations

Concomitant use of castor oil is cautioned with antiarrhythmic drugs and botanicals containing cardiac glycosides, as long-term use of castor oil as a laxative can cause potassium loss, leading to increased toxicity of those drugs and botanicals (Brinker 2001; De Smet 1993; ESCOP 2003).

Concomitant internal use of castor oil is cautioned with thiazide diuretics, corticosteroids, or licorice, and long-term use of castor oil as a laxative may increase the potassium loss induced by those drugs and botanicals (Brinker 2001; De Smet 1993; ESCOP 2003; Wichtl 2004).

Use of stimulant laxatives, such as castor oil, may reduce the gastrointestinal transit time and thus reduce the absorption of orally administered drugs (Brinker 2001; De Smet 1993).

Pregnancy and Lactation

Castor oil has traditionally been used for the induction of labor (McFarlin et al. 1999). Human studies have indicated that castor oil can be used safely for labor induction, although the studies provide mixed results for the need for cesarean sections in castor oil-treated groups versus control groups (Davis 1984; Garry et al. 2000). Possible side effects and case reports of adverse events have led some reviewers to caution against the use of castor oil as a labor inducer (Lippert and Mueck 2002; Nabors 1958), although castor oil is the most commonly used natural product for labor induction among certified nurse midwives that use natural products for labor induction (McFarlin et al. 1999). Based on this information, use during pregnancy is not recommended except under the supervision of a qualified healthcare practitioner.

An Eclectic medical text notes that castor oil is "the most suitable laxative for constipation of pregnant women" (Felter and Lloyd 1898), and a more recent survey of gastroenterologists and obstetricians indicated that 26% of gastroenterologists and 38% of obstetricians would prescribe castor oil as a laxative to pregnant patients (Vinod et al. 2007). Other texts note that castor oil must be used with caution in pregnancy (Martindale and Reynolds 1996; Osol and Farrar 1955).

A case of an amniotic fluid embolism causing a maternal cardiorespiratory arrest was temporally associated with ingestion of castor oil by a woman at week 40 of gestation, although causality could not be determined (Steingrub et al. 1988).

Review Details

I. Drug and Supplement Interactions

Clinical Trials of Drug or Supplement Interactions

No clinical trials of drug or supplement interactions were identified.

Case Reports of Suspected Drug or Supplement Interactions

No case reports of suspected drug or supplement interactions were identified.

Animal Trials of Drug or Supplement Interactions

In dogs intravenously administered toxic doses of methyprylon with or without a large oral dose of castor oil, no significant difference was observed in the sleep times of the dogs treated with castor oil, or in the methyprylon pharmacokinetics, as compared to controls (Gwilt et al. 1982).

R

II. Adverse Events

Case Reports of Adverse Events

Cutaneous reactions to castor oil, including cheilitis and dermatitis, confirmed by patch testing, have been reported after topical application of personal care products containing castor oil (Aplin and Eliseo 1997; Brandle et al. 1983; Di Berardino and Della Torre 2003; Fisher 1991; Kalavala et al. 2007; le Coz and Ball 2000; Lodi et al. 1992; Sai 1983; Taghipour et al. 2008; Tan et al. 1997; Wakelin et al. 1996; Wilbur 2007).

An anaphylactic reaction to polyethoxylated castor oil, confirmed by intradermal testing, was reported in a woman who had been intravenously administered cyclosporine with a vehicle of polyethoxylated castor oil (Ebo et al. 2001).

A case of epoxydicarboxylic aciduria was reported in a woman excreting large amounts of 3,6-epoxyoctanedioic, 3,6-epoxydecanedioic, and 3,6-epoxydodecanedioic acids, the presence of which was traced to factitious ingestion of castor oil (Hagenfeldt et al. 1986).

Also see Pregnancy and Lactation for this entry.

III. Pharmacology and Pharmacokinetics

Human Pharmacological Studies

No relevant human pharmacological studies were identified.

Animal Pharmacological Studies

Skin irritation tests on rats, guinea pigs, and pigs have shown that topical application of castor oil for 10 to 56 days is generally well tolerated (Butcher 1951; Meyer et al. 1976; Motoyoshi et al. 1979; Wilbur 2007).

In rabbits, one irritation test showed severe irritation after several doses of castor oil (Motoyoshi et al. 1979), while other studies indicated that daily application for up to 54 days was well tolerated (Guillot et al. 1979; Rantuccio et al. 1981).

Castor oil did not promote chemically induced skin tumors in mice treated topically with castor oil twice a week for 20 weeks (Shubik 1950).

Slight epidermal hyperplasia was observed in mice treated topically with castor oil 4 times in a 2-week period (Shubik 1950).

In Vitro Pharmacological Studies

No relevant in vitro pharmacological studies were identified.

IV. Pregnancy and Lactation

In a study of pregnant women with premature rupture of the membranes, who were between weeks 37 and 42 of gestation, administration of castor oil increased the rates of spontaneous labor, decreased the need for cesarean section, and decreased the need for labor induction with oxytocin as compared to women not administered castor oil. No significant adverse events were reported in the mothers. The authors of the study concluded that castor oil is safe for the induction of labor (Davis 1984).

No adverse outcomes for mother or fetus were identified in a study of pregnant women administered a single 60 ml dose of castor oil to induce labor. The rate of cesarean sections was 19% in the group receiving castor oil and 8% in the untreated control group (Garry et al. 2000).

In pregnant women between gestational weeks 38 and 41 orally administered single doses of 71 ml of castor oil, an increase in the contractile work of the uterus was observed. Effects on fetal health were not reported (Mathie and Dawson 1959).

In a study of amniotic fluid at the time of rupture of the membranes, meconium passage was significantly more common among women who had recently taken castor oil as compared with women that did not use castor oil (Mitri et al. 1987).

A survey of American certified nurse midwives that used natural products to induce labor (90 midwives responded) indicated that castor oil was the most commonly used product for the induction of labor, with 93% of midwives using castor oil. Adverse events reported in association with castor oil use were thrombosed hemorrhoids, precipitous labor, nausea, vomiting, diarrhea, and flatulence, and two midwives reported an increase in meconium-stained fluid (McFarlin et al. 1999).

A survey of attending gastroenterologists and obstetricians at a New York hospital (75 physicians responded) indicated that 26% of gastroenterologists and 38% of obstetricians would prescribe castor oil as a laxative to pregnant patients (Vinod et al. 2007).

A review of the use of castor oil for labor induction lists possible side effects of castor oil as follows: nausea, vomiting, intestinal colic, diarrhea, disturbances of electrolyte balance, dehydration, hemorrhagic gastritis, hyperemia of the pelvic organs, hemolysis, and liver cell necrosis (Lippert and Mueck 2002).

A woman at 39 weeks of gestation, with a history of a previous cesarean delivery, had severe abdominal pains and rupture of membranes shortly after ingesting 5 ml of castor oil. Forty-five minutes later, repetitive variable decelerations prompted a cesarean delivery. At surgery, a portion of the umbilical cord was protruding from a 2 cm rupture of the lower transverse scar. The reporting authors indicated that the risk of uterine rupture in women with a low transverse incision from a previous cesarean section is 0.2 to 1.5% (Sicuranza and Figueroa 2003).

A case of an amniotic fluid embolism causing a cardiorespiratory arrest was temporally associated with ingestion of 30 ml castor oil by a woman at week 40 of gestation. Approximately 1 hour after ingestion, the woman experienced spontaneous rupture of the membranes and suffered cardiorespiratory arrest. The reporting authors noted that the association between castor oil ingestion and the adverse events was temporal but that causality could not be determined (Steingrub et al. 1988).

V. Toxicity Studies

Acute Toxicity

Damage to the duodenal and jejunal mucosa was observed from 1 to 7 hours after oral administration in rats of single doses of 2 ml of castor oil per 160 to 180 g animal. Injury was not observed at 0.5 or 9 hours after dosing (Capasso et al. 1994).

In ponies administered 2.5 ml/kg castor oil via nasogastric tubes, hyperemia of the intestinal mucosa was observed 24, 48, and 72 hours postdosing. Dosing was associated with transient diarrhea, neutrophilic inflammation, and mucosal erosion in the cecum and ventral colon (Johnson et al. 1993).

No toxic effects on the renal tubular epithelium were observed in rats in which castor oil was injected into proximal tubules (Langer et al. 1968).

Short-Term Toxicity

No adverse effects or catharsis were observed in rats fed a diet containing 10% castor oil daily for 5 weeks (Masri et al. 1962).

Subchronic Toxicity

No toxicity to any specific tissue, organ system, or organ was observed in mice and rats fed a diet of up to 10% castor oil daily for 13 weeks. A dose-related increase in alkaline phosphatase was observed on days 5, 12, and at the end of the study (NTP 1992). The no-observed-adverse-effect levels (NOAEL) in the study were estimated as 5 g/kg daily for rats and 7.5 g/kg daily for mice (Burdock et al. 2006).

Genotoxicity

No evidence of induction of micronuclei in peripheral erythrocytes was observed in mice fed diets of up to 10% castor oil for 13 weeks (NTP 1992).

No induction of sister-chromatid exchanges was observed in Chinese hamster ovary cells incubated with castor oil at concentrations up to 5000 µg/ml with or without metabolic activation (NTP 1992).

No mutagenic activity was observed in *Salmonella* strains TA97, TA98, TA100, TA102, TA1535, and TA1537 with or without metabolic activation with castor oil at concentrations of up to 10,000 µg/plate (Hachiya 1987; Zeiger et al. 1988).

No genotoxic activity of polyoxyethylene hydrogenated castor oil was observed in the micronucleus test in mice, or in the reverse mutation test in bacteria and the chromosome aberration test in vitro, with or without metabolic activation (Hirai et al. 1994).

Cytotoxicity

At concentrations of 100 and 5000 µg/plate, with and without metabolic activation, castor oil exhibited cytotoxicity against an *E. coli* strain (Hachiya 1987).

Literature Cited

Aplin, P.J., and T. Eliseo. 1997. Ingestion of castor oil plant seeds. *Med. J. Aust.* 167(5):260-261.

Audi, J., M. Belson, M. Patel, J. Schier, and J. Osterloh. 2005. Ricin poisoning: A comprehensive review. *J. Am. Med. Assoc.* 294(18):2342-2351.

Bradley, P.R. 1992. *British herbal compendium: A handbook of scientific information on widely used plant drugs.* Bournemouth, UK: British Herbal Medicine Association.

Brandle, I., A. Boujnah-Khouadja, and J. Foussereau. 1983. Allergy to castor oil. *Contact Dermat.* 9(5):424-425.

Brinker, F. 2001. *Herb contraindications and drug interactions.* 3rd ed. Sandy, OR: Eclectic Medical Publications.

Burdock, G.A., I.G. Carabin, and J.C. Griffiths. 2006. Toxicology and pharmacology of sodium ricinoleate. *Food Chem. Toxicol.* 44(10):1689-1698.

Butcher, E.O. 1951. The effects of applications of various substances on the epidermis of the rat. *J. Invest. Dermatol.* 16(2):85-90.

Capasso, F., N. Mascolo, A.A. Izzo, and T.S. Gaginella. 1994. Dissociation of castor oil-induced diarrhoea and intestinal mucosal injury in rat: Effect of N$^{\$1}$-nitro-L-arginine methyl ester. *Br. J. Pharmacol.* 113(4):1127.

Chadha, Y. 1988. *The wealth of India: A dictionary of Indian raw materials and industrial products.* Delhi: Council of Scientific and Industrial Research.

Challoner, K.R., and M.M. McCarron. 1990. Castor bean intoxication. *Ann. Emerg. Med.* 19(10):1177-1183.

Davis, L. 1984. The use of castor oil to stimulate labor in patients with premature rupture of membranes. *J. Nurse Midwifery* 29(6):366-370.

De Smet, P.A.G.M. 1993. *Adverse effects of herbal drugs, Volume 2.* Berlin: Springer.

Di Berardino, L., and F. Della Torre. 2003. Side effects to castor oil. *Allergy* 58(8):826.

Doan, L.G. 2004. Ricin: Mechanism of toxicity, clinical manifestations, and vaccine development. A review. *J. Toxicol. Clin. Toxicol.* 42(2):201-208.

Ebo, D.G., G.C. Piel, V. Conraads, and W.J. Stevens. 2001. IgE-mediated anaphylaxis after first intravenous infusion of cyclosporine. *Ann. Allergy Asthma Immunol.* 87(3):243-245.

ESCOP. 2003. *ESCOP monographs: The scientific foundation for herbal medicinal products.* 2nd ed. Exeter, UK: European Scientific Cooperative on Phytotherapy.

Felter, H.W., and J.U. Lloyd. 1898. *King's American dispensatory.* 18th ed., 3rd rev. 2 vols. Cincinnati: Ohio Valley Co.

Fisher, A.A. 1991. Allergic cheilitis due to castor oil in lipsticks. *Cutis* 47(6):389-390.

Garry, D., R. Figueroa, J. Guillaume, and V. Cucco. 2000. Use of castor oil in pregnancies at term. *Altern. Ther. Health Med.* 6(1):77-101.

Guillot, J.P., J.Y. Giauffret, and M.C. Martini. 1979. Etude de tolerances oculaire et cutanee chez la lapin, de differentes matieres premieres utilisees en cosmetologie, et provenant de fabrications diverses II. *Int. J. Cosmet. Sci.* 1:27-57.

Gwilt, P.R., M.C. Pankaskie, and J.J. Mitala. 1982. The effect of oral castor oil on the disposition of methyprylon in intoxicated dogs. *Can. Anaesth. Soc. J.* 29(4):381-383.

Hachiya, N. 1987. Evaluation of genotoxicity by a series of short-term tests. *Akita J. Med.* 14:269-292.

Hagenfeldt, L., L. Blomquist, and T. Midtvedt. 1986. Epoxydicarboxylic aciduria resulting from the ingestion of castor oil. *Clin. Chim. Acta* 161(2):157-163.

Hirai, O., Y. Miyamae, K. Zaizen, et al. 1994. Mutagenicity tests of polyoxyethylene hydrogenated castor oil 60 (HCO-60). *J. Toxicol. Sci.* 19(2):89-96.

Johnson, C.M., J.M. Cullen, and M.C. Roberts. 1993. Morphologic characterization of castor oil-induced colitis in ponies. *Vet. Pathol. Online* 30(3):248-255.

Kalavala, M., T.M. Hughes, and N.M. Stone. 2007. Allergic contact dermatitis to polyethylene glycol-7 hydrogenated castor oil. *Contact Dermat.* 56(5):287-288.

Langer, K.H., W. Thoenes, and M. Wiederho. 1968. Light and electron microscopic investigations of proximal convoluted tubules of rat kidney after intraluminal injection of oil. *Pflugers Arch. Eur. J. Physiol.* 302(2):149.

le Coz, C.J., and C. Ball. 2000. Recurrent allergic contact dermatitis and cheilitis due to castor oil. *Contact Dermat.* 42(2):114-115.

Leung, A.Y., and S. Foster. 1996. *Encyclopedia of common natural ingredients used in food, drugs, and cosmetics.* 2nd ed. New York: Wiley.

Lippert, T.H., and A.O. Mueck. 2002. Labour induction with alternative drugs? *J. Obstet. Gynaecol.* 22(4):343.

List, P.H., and H. Hörhammer. 1973. *Hagers handbuch der pharmazeutischen praxis.* Berlin: Springer.

Lodi, A., S. Leuchi, L. Mancini, G. Chiarelli, and C. Crosti. 1992. Allergy to castor oil and colophony in a wart remover. *Contact Dermat.* 26(4):266-267.

Lord, M.J., N.A. Jolliffe, C.J. Marsden, et al. 2003. Ricin. Mechanisms of cytotoxicity. *Toxicol. Rev.* 22(1):53-64.

Martindale, W., and J.E.F. Reynolds. 1996. *The extra pharmacopoeia.* 31st ed. London: Pharmaceutical Press.

Masri, M.S., L.A. Goldblatt, F. De Eds, and G.O. Kohler. 1962. Relation of cathartic activity to structural modifications of ricinoleic acid of castor oil. *J. Pharm. Sci.* 51:999-1002.

Mathie, J.G., and B.H. Dawson. 1959. Effect of castor oil, soap enema, and hot bath on the pregnant human uterus near term. *Br. Med. J.* 1(5130):1162.

McFarlin, B.L., M.H. Gibson, J. O'Rear, and P. Harman. 1999. A national survey of herbal preparation use by nurse-midwives for labor stimulation. Review of the literature and recommendations for practice. *J. Nurse Midwifery* 44(3):205-216.

Meyer, F.U., C. Wollmann, N. Exner, and G. Exner. 1976. Comparative studies on the skin of guinea pigs, swine and man following the effect of various external agents. *Dermatol. Monatsschr.* 162:986-991.

Mitri, F., G.J. Hofmeyr, and C.J. van Gelderen. 1987. Meconium during labour—Self-medication and other associations. *S. Afr. Med. J.* 71(7):431-433.

Motoyoshi, K., Y. Toyoshima, M. Sato, and M. Yoshimura. 1979. Comparative studies on the irritancy of oils and synthetic perfumes to the skin of rabbit, guinea pig, rat, miniature swine and man. *Cosmet. Toiletries* 94:41-48.

Nabors, G. 1958. Castor oil as an adjunct to induction of labor: Critical re-evaluation. *Am. J. Obstet. Gynecol.* 75:36-39.

NTP. 1992. Toxicity studies of castor oil in F344/N rats and B6C3F1 mice. Report NTIS PB93-151439. Research Triangle Park, NC: National Toxicology Program.

Osol, A., and G. Farrar. 1955. *The dispensatory of the United States of America.* 25th ed. Philadelphia: Lippincott.

Rantuccio, F., D. Sinist, A. Scardigno, and C. Coviello. 1981. Histological changes in rabbits after application of medicaments and cosmetic bases. II. *Contact Dermat.* 7(2):94-97.

Roth, L., M. Daunderer, and K. Kormann. 1984. *Giftpflanzen-pflanzengifte: Vorkommen, wirkung, therapie.* Landsberg: Ecomed.

Sai, S. 1983. Lipstick dermatitis caused by castor oil. *Contact Dermat.* 9(1):75.

Shubik, P. 1950. Studies on the promoting phase in the stages of carcinogenesis in mice, rats, rabbits, and guinea pigs. *Cancer Res.* 10(1):13-17.

Sicuranza, G.B., and R. Figueroa. 2003. Uterine rupture associated with castor oil ingestion. *J. Matern. Fetal Neonat. Med.* 13(2):133-134.

Steingrub, J.S., T. Lopez, D. Teres, and R. Steingart. 1988. Amniotic fluid embolism associated with castor oil ingestion. *Crit. Care Med.* 16(6):642-643.

Taghipour, K., F. Tatnall, and D. Orton. 2008. Allergic axillary dermatitis due to hydrogenated castor oil in a deodorant. *Contact Dermat.* 58(3):168-169.

Tan, B.B., A.L. Noble, M.E. Roberts, J.T. Lear, and J.S.C. English. 1997. Allergic contact dermatitis from oleyl alcohol in lipstick cross-reacting with ricinoleic acid in castor oil and lanolin. *Contact Dermat.* 37(1):41-42.

Vinod, J., J. Bonheur, B.I. Korelitz, and G. Panagopoulos. 2007. Choice of laxatives and colonoscopic preparation in pregnant patients from the viewpoint of obstetricians and gastroenterologists. *World J. Gastroenterol.* 13(48):6549-6552.

Wakelin, S.H., A.J. Harris, and S. Shaw. 1996. Contact dermatitis from castor oil in zinc and castor oil cream. *Contact Dermat.* 35(4):259.

Wichtl, M. 2004. *Herbal drugs and phytopharmaceuticals: A handbook for practice on a scientific basis.* 3rd ed. Boca Raton, FL: CRC Press.

Wilbur, J. 2007. Final report on the safety assessment of *Ricinus communis* (castor) seed oil, hydrogenated castor oil, glyceryl ricinoleate, glyceryl ricinoleate se, ricinoleic acid, potassium ricinoleate, sodium ricinoleate, zinc ricinoleate, cetyl ricinoleate, ethyl ricinoleate, glycol ricinoleate, isopropyl ricinoleate, methyl ricinoleate, and octyldodecyl ricinoleate. *Int. J. Toxicol.* 26(Suppl. 3):31-77.

Williamson, E.M. 2003. *Potter's herbal cyclopedia.* Saffron Walden, Essex: C.W. Daniel Co.

Zeiger, E., B. Anderson, S. Haworth, T. Lawlor, and K. Mortelmans. 1988. *Salmonella* mutagenicity tests. 4. Results from the testing of 300 chemicals. *Environ. Mol. Mutagen.* 11:1-158.

Rorippa nasturtium-aquaticum (L.) Hayek — Brassicaceae

SCN: watercress
Syn: *Nasturtium officinale* W.T. Aiton; *Sisymbrium nasturtium-aquaticum* L.
PN: *xi yang cai gan* (herb)
Part: herb

Quick Reference Summary

Safety Class: 1
Interaction Class: A

Contraindications

None known.

Other Precautions

Use with caution in cases of gastric and duodenal ulcers, inflammatory kidney disorders, and in children under 4 years of age (Wichtl 2004).

Drug and Supplement Interactions

A decrease in blood levels of acetaminophen was observed after pretreatment with a large dose (50 g) of watercress (Chen et al. 1996).

No significant changes in coumarin metabolism were observed after several large doses (57 g each) of watercress (Murphy et al. 2001).

Adverse Events and Side Effects

Irritation of the gastric mucosa may develop if watercress is taken in large amounts or for an extended time (Wichtl 2004).

Pharmacological Considerations

In human studies, watercress inhibited the drug-metabolizing isoenzyme CYP2E1 in one study but had no effect on this isoenzyme in a second study (Desager et al. 2002; Leclercq et al. 1998).

Pregnancy and Lactation

Limited information on the safety of watercress in pregnancy or lactation was identified in the scientific and traditional literature. Although this review did not identify any concerns for use while pregnant or nursing, safety has not been conclusively established.

Review Details

I. Drug and Supplement Interactions

Clinical Trials of Drug or Supplement Interactions

In healthy volunteers orally administered coumarin before or after ingestion of four doses of 56.8 g each of fresh watercress over the course of 24 hours, a marginal inhibitory effect on coumarin metabolism was observed. In approximately 30% of volunteers, a decrease in urinary excretion of coumarin was observed in the first 2 hours after coumarin administration, although the mean excretion of coumarin during that time for all test subjects was not significantly different from the control. Total excretion of the coumarin metabolite was not affected by watercress (Murphy et al. 2001).

In healthy volunteers orally administered 50 g of watercress homogenates 10 hours prior to a single dose of acetaminophen, a reduction in serum levels of acetaminophen was observed. Total urinary excretion of Cys-acetaminophen was also reduced, although pharmacokinetic processes and the urinary excretion of acetaminophen, acetaminophen glucuronide, and acetaminophen sulfate were not altered (Chen et al. 1996).

Case Reports of Suspected Drug or Supplement Interactions

No case reports of suspected drug or supplement interactions were identified.

Animal Trials of Drug or Supplement Interactions

No animal trials of drug or supplement interactions were identified.

II. Adverse Events

Case Reports of Adverse Events

No case reports of adverse events were identified.

III. Pharmacology and Pharmacokinetics

Human Pharmacological Studies

In healthy volunteers orally administered chlorzoxazone (a substrate of the drug-metabolizing isoenzyme CYP2E1) before or after 50 g fresh watercress homogenate, an increase in chlorzoxazone plasma concentration was observed. Similarly, chlorzoxazone elimination half-life was prolonged after watercress ingestion. The effects of watercress were one-third to one-half the effects of the CYP2E1 inhibitor isoniazid (Leclercq et al. 1998).

In healthy volunteers orally administered 50 g fresh watercress homogenate 10 hours or 1 hour prior to ingestion of ethanol (a substrate of the drug-metabolizing isoenzyme CYP2E1), no significant effects on ethanol metabolism were observed, although ethanol absorption was delayed by ingestion of watercress just prior to ethanol consumption. Ingestion of watercress was associated with a weak inhibition of acetaldehyde (a metabolite of ethanol) metabolism (Desager et al. 2002).

Although the compound phenethyl isothiocyanate, found in watercress and vegetables such as cabbage, was found to inhibit the drug-metabolizing isoenzyme CYP2D6 in vitro, a study in healthy human volunteers indicated no effects on CYP2D6. In the human study, phenotyping was performed on healthy volunteers before and after ingestion of 50 g of fresh watercress. No significant changes in CYP2D6 metabolism were observed (Caporaso et al. 1994).

Animal Pharmacological Studies

No relevant animal pharmacological studies were identified.

In Vitro Pharmacological Studies

No relevant in vitro pharmacological studies were identified.

IV. Pregnancy and Lactation

In rats orally administered 200, 400, or 800 mg/kg of an aqueous extract of watercress daily on days 1 to 6 of pregnancy, a reduction in the number of viable fetuses and increase in the number of resorptions were observed (Elbetieha et al. 1996).

No information on the safety of watercress during lactation was identified.

V. Toxicity Studies

No toxicity studies were identified.

Literature Cited

Caporaso, N., J. Whitehouse, S. Monkman, et al. 1994. *In vitro* but not *in vivo* inhibition of CYP2D6 by phenethyl isothiocyanate (PEITC), a constituent of watercress. *Pharmacogenetics* 4(5):275-280.

Chen, L., S.N. Mohr, and C.S. Yang. 1996. Decrease of plasma and urinary oxidative metabolites of acetaminophen after consumption of watercress by human volunteers. *Clin. Pharmacol. Ther.* 60(6):651-660.

Desager, J.P., J.L. Golnez, C. De Buck, and Y. Horsmans. 2002. Watercress has no importance for the elimination of ethanol by CYP2E1 inhibition. *Phamacol. Toxicol.* 91(3):103-105.

Elbetieha, A., M.H. Al-Hamood, and A. Alkofahi. 1996. Anti-implantation potential of some medicinal plants in female rats. *Arch. STD/HIV Res.* 10(3):181-188.

Leclercq, I., J.P. Desager, and Y. Horsmans. 1998. Inhibition of chlorzoxazone metabolism, a clinical probe for CYP2E1, by a single ingestion of watercress. *Clin. Pharmacol. Ther.* 64(2):144-149.

Murphy, S.E., L.M. Johnson, L.M. Losey, S.G. Carmella, and S.S. Hecht. 2001. Consumption of watercress fails to alter coumarin metabolism in humans. *Drug Metab. Dispos.* 29(6):786-788.

Wichtl, M. 2004. *Herbal drugs and phytopharmaceuticals: A handbook for practice on a scientific basis*. 3rd ed. Boca Raton, FL: CRC Press.

Rosa spp. — Rosaceae

Rosa alba L.
SCN: white rose

Rosa canina **L.**
SCN: dog rose
OCN: dog brier; brier rose

Rosa centifolia **L.**
SCN: cabbage rose
AN: *shatapatri; shatapatrika*
OCN: Provence rose

Rosa damascena **Mill.**
SCN: damask rose

Rosa gallica **L.**
SCN: French rose
OCN: apothecary rose; rosa mundi

Rosa rugosa **Thunb.**
SCN: rugose rose
OCN: ramanas rose; Turkestan rose
Part: fruit ("hips")

Quick Reference Summary

Safety Class: 1
Interaction Class: A

Contraindications

None known.

Other Precautions

None known.

Drug and Supplement Interactions

None known.

Editors' Notes

Small hairs on fresh or dried whole rose hips may cause mechanical irritation in the mouth, throat, and other areas. Commercial processing removes the hairs.

Adverse Events and Side Effects

A systematic review of clinical trials on rose fruit products indicated that rose fruit was generally well tolerated with no significant adverse events reported in the trials (Chrubasik et al. 2006). No case reports of adverse events were identified for any of the rose species.

Allergic reactions to rose fruit have been reported (Chrubasik et al. 2008; Lleonart et al. 2007).

Pharmacological Considerations

None known.

Pregnancy and Lactation

No information on the safety of *Rosa* species in pregnancy or lactation was identified in the scientific or traditional literature. Although this review did not identify any concerns for use while pregnant or nursing, safety has not been conclusively established.

Review Details

I. Drug and Supplement Interactions

Clinical Trials of Drug or Supplement Interactions

No clinical trials of drug or supplement interactions were identified.

Case Reports of Suspected Drug or Supplement Interactions

No case reports of suspected drug or supplement interactions were identified.

Animal Trials of Drug or Supplement Interactions

No animal trials of drug or supplement interactions were identified.

II. Adverse Events

Adverse Events Reported in Clinical Trials

A systematic review of clinical trials on dog rose fruit (5 g daily for over 3 months) did not indicate any adverse events in clinical trials (Chrubasik et al. 2006).

Case Reports of Adverse Events

No case reports of adverse events were identified.

III. Pharmacology and Pharmacokinetics

Human Pharmacological Studies

No relevant human pharmacological studies were identified.

Animal Pharmacological Studies

No relevant animal pharmacological studies were identified.

In Vitro Pharmacological Studies

No relevant in vitro pharmacological studies were identified.

IV. Pregnancy and Lactation

No studies on the use of rose fruit during pregnancy or lactation were identified.

V. Toxicity Studies

Acute Toxicity

No toxicity of ethyl acetate, *n*-butanol, or ethanol extracts of dog rose fruit was observed after oral administration of 919 mg/kg to mice (Orhan et al. 2007).

Literature Cited

Chrubasik, C., R.K. Duke, and S. Chrubasik. 2006. The evidence for clinical efficacy of rose hip and seed: A systematic review. *Phytother. Res.* 20(1):1-3.

Chrubasik, C., B.D. Roufogalis, U. Müller-Ladner, and S. Chrubasik. 2008. A systematic review on the *Rosa canina* effect and efficacy profiles. *Phytother. Res.* 22(6):725-733.

Lleonart, R., M. Corominas, and M. Lombardero. 2007. Tea infusion, another source of Rosaceae allergy. *Allergy* 62:89-90.

Orhan, D.D., A. Hartevioglu, E. Kupeli, and E. Yesilada. 2007. *In vivo* anti-inflammatory and antinociceptive activity of the crude extract and fractions from *Rosa canina* L. fruits. *J. Ethnopharmacol.* 112(2):394-400.

Rosmarinus officinalis L. — Lamiaceae

SCN: rosemary

Part: leaf

Quick Reference Summary

Safety Class: 1

Interaction Class: A

Contraindications

None known.

Other Precautions

None known.

Drug and Supplement Interactions

None known.

Notice

Emmenagogue (Chadha 1988); *see* Appendix 2.

Editors' Note

Concerns for this herb are based on relatively higher doses used for therapeutic purposes in contrast to lower amounts generally used in cooking, and have not been associated with its use as a spice.

Adverse Events and Side Effects

Topical allergic reactions to rosemary, confirmed by patch testing, have been reported (Armisen et al. 2003; Fernandez et al. 1997; Guin 2001; Inui and Katayama 2005; Klarman 1958; Serra et al. 2005).

Pharmacological Considerations

Absorption of nonheme iron was reduced in healthy women who ingested a phenolic-rich extract of rosemary with a test meal (Samman et al. 2001).

Animal studies have indicated that rosemary reduced serum glucose levels in both healthy and diabetic animals (Bakirel et al. 2008; Erenmemisoglu et al. 1997). An ethnobotanical survey indicated that rosemary is traditionally used in Morocco to treat diabetes and hypertension (Tahraoui et al. 2007). Thus, diabetic persons using insulin or oral hypoglycemic drugs should continue to monitor glucose levels during rosemary use.

Induction of the drug-metabolizing isoenzymes CYP1A1, CYP1A2, and CYP2B1,2 was observed in rats administered an aqueous extract of rosemary or rosemary essential oil (Debersac et al. 2001a, 2001b).

Inhibition of platelet aggregation was observed in one animal and two in vitro studies (Lee et al. 2007; Yamamoto et al. 2005).

Pregnancy and Lactation

Information on the safety of rosemary during pregnancy is conflicting. While some older references indicate that rosemary (perhaps referring to rosemary essential oil) was used to induce abortions (Casey 1960; Chadha 1988; Greshoff 1913), animal studies of rosemary or compounds isolated from rosemary have not indicated such activity. One animal study showed no significant adverse effects of a rosemary aqueous extract administered to pregnant rats, although a small, statistically insignificant increase in preimplantation loss was observed (Lemonica et al. 1996). No teratogenic effects of the compounds D-camphor or 1,8-cineol were observed in other studies (Jori and Briatico 1973; Leuschner 1997). In rats administered the compound 1,8-cineol during pregnancy, 1,8-cineol was shown to cross the placenta, suggesting that caution is warranted for use during pregnancy

(Jori and Briatico 1973). Although dried rosemary or rosemary tea may be safe in pregnancy, the essential oil and other concentrated extracts should not be used.

In rats administered the compound 1,8-cineol during or after pregnancy, 1,8-cineol was shown to cross the placenta but was reported not to cross into breast milk (Jori and Briatico 1973).

Review Details

I. Drug and Supplement Interactions

Clinical Trials of Drug or Supplement Interactions

No clinical trials of drug or supplement interactions were identified.

Case Reports of Suspected Drug or Supplement Interactions

No case reports of suspected drug or supplement interactions were identified.

Animal Trials of Drug or Supplement Interactions

No animal trials of drug or supplement interactions were identified.

II. Adverse Events

Case Reports of Adverse Events

Several cases of topical allergic reactions, including cheilitis and eczema, after ingestion of rosemary or topical application of products containing rosemary, confirmed by patch testing, have been reported (Armisen et al. 2003; Fernandez et al. 1997; Guin 2001; Inui and Katayama 2005; Klarman 1958; Serra et al. 2005). In two cases, the patients also tested positive to thyme (*Thymus vulgaris*) in skin patch tests (Armisen et al. 2003; Martinez-Gonzalez et al. 2007). Photopatch tests produced stronger reactions than standard tests (Armisen et al. 2003). A case of occupational contact dermatitis was reported, with the compound carnosol identified as the major allergen (Hjorther et al. 1997).

III. Pharmacology and Pharmacokinetics

Human Pharmacological Studies

Absorption of nonheme iron was significantly reduced in healthy women who ingested a phenolic-rich extract of rosemary with a test meal. The effect of rosemary was less than that of green tea (Samman et al. 2001).

An ethnobotanical survey of diabetic and hypertensive people in Morocco indicated that rosemary is commonly used in the treatment of both conditions (Tahraoui et al. 2007).

Animal Pharmacological Studies

In healthy and diabetic rabbits orally administered an ethanolic extract of rosemary at doses of 50, 100, or 200 mg/kg, the 200 mg/kg dose caused a lowering of blood glucose in both healthy and diabetic animals, and the 100 and 200 mg/kg doses caused increases in serum insulin levels in diabetic animals (Bakirel et al. 2008). A decrease in serum glucose levels was observed in healthy or diabetic mice provided aqueous extracts of rosemary ad libitum for 3 months. No changes in liver enzymes were observed in either the treatment or control groups (Erenmemisoglu et al. 1997).

Antithrombotic activity in the laser-induced thrombosis test was observed in mice orally administered a juice extract of fresh rosemary at doses of 3.85 ml/kg. No effect on vasodilation was observed (Yamamoto et al. 2005).

In male rats orally administered 250 or 500 mg/kg of a concentrated ethanol extract of rosemary daily for 63 days, a significant decline in spermatogenesis in testes was observed due to a decrease in the number of spermatocytes and spermatids in the high-dose group, which was attributed to a decrease in testosterone. Sperm motility and density also decreased in this group. No changes in the weights of testes were observed, although a weight decrease in some other reproductive tissues was noted. The treated male rats were mated with untreated females, and an increase in the number of fetal resorptions was observed in the females that had been mated with males from the high dose group (Nusier et al. 2007).

An increased weight of seminal vesicles was reported in male rats orally administered 582 mg/kg of rosemary aqueous extract daily for 5 days. No other changes in organ weights or sperm production were observed at that dose, and no adverse effects were observed in rats administered 291 mg/kg (Waseem et al. 2006).

The compound 1,8-cineol was detected in the blood of rats after inhalation of 0.5 ml of rosemary essential oil (Kovar et al. 1987).

In Vitro Pharmacological Studies

Rosemary extract inhibited the drug-metabolizing isoenzyme CYP1A1 by 70 to 90% (Offord et al. 1995).

The addition of a methanol-extracted fraction of rosemary at a concentration of 16.5–82 mg/ml significantly increased the intracellular accumulation of doxorubicin and vinblastine in drug-resistant P-glycoprotein-expressing MCF-7 cells (Plouzek et al. 1999).

An aqueous extract of rosemary inhibited angiotensin-converting enzyme (ACE) activity in vitro (Kwon et al. 2006).

An aqueous extract of rosemary enhanced the permeability of Caco-2 cells to furosemide but not verapamil, metoprolol, ketoprofen, or paracetamol (Laitinen et al. 2004).

The juice extracted from fresh rosemary inhibited platelet aggregation as assessed by a shear-induced in vitro platelet function test (Yamamoto et al. 2005). The compound carnosic acid produced a dose-dependent inhibition of collagen-, arachidonic acid-, U46619- and thrombin-induced washed rabbit platelet aggregation in a concentration-dependent manner, with IC_{50} values of 29 to 48 μM (Lee et al. 2007).

IV. Pregnancy and Lactation

In rats orally administered an aqueous extract of rosemary leaf, flower, and stem at doses of 26 mg (13 mg/ml of solids) on days 1 to 6 (preimplantation) or 6 to 15 (organogenic period) of pregnancy, no differences between the rosemary and control groups were observed for postimplantation loss or number of malformations, although a nonsignificant increase in preimplantation loss was observed in the rosemary group (Lemonica et al. 1996).

No teratogenic effects of the compound D-camphor were observed in rats orally administered the compound at doses up to 1000 mg/kg daily during the organogenesis period of pregnancy, or in pregnant rabbits administered doses up to 681 mg/kg daily (Leuschner 1997).

Some references indicate that rosemary has been used as an abortifacient. Details on that use are lacking, including whether crude extracts or the essential oil were used (Casey 1960; Chadha 1988; Greshoff 1913).

In rats subcutaneously administered the compound 1,8-cineol (eucalyptol) on gestational days 10 to 14, the last 4 days of pregnancy, or on postnatal days 2 through 6, treatment enhanced microsomal enzyme activity of the mothers and induced the enzyme activity in fetal livers, but it did not induce microsomal activity of the suckling newborn rats, suggesting that 1,8-cineol is able to cross the placenta but does not cross into breast milk (Jori and Briatico 1973).

V. Toxicity Studies

Acute Toxicity

The LD_{50} of orally administered rosemary essential oil in rats is 5 ml/kg (Opdyke 1974).

No toxic effects of an alcoholic rosemary extract were observed in rats and mice intraperitoneally administered 2 g/kg (Mongold et al. 1991). In rats orally administered 2 g/kg of a supercritical fluid extract of rosemary, no adverse effects, including changes in hematological and serum chemistry values, organ weights, or gross histological characteristics, were noted during a 2-week observation period (Anadon et al. 2008).

Toxicity studies on the compound eucalyptol (1,8-cineol) are reviewed in the *Eucalyptus globulus* leaf essential oil entry.

Subchronic Toxicity

A rosemary extract inhibited tumor promotion in mice with initiated skin tumors topically treated with 3.6 mg of rosemary acetone extract twice a week for 19 weeks (Ho et al. 1994). A similar inhibition of tumor growth was observed in rats with induced mammary carcinoma fed a diet of 1% rosemary extract daily for 21 weeks (Singletary and Nelshoppen 1991).

Hepatotoxicity

Significant hepatoprotective activity of orally or intraperitoneally administered aqueous and ethanolic extracts of rosemary have been observed in mice treated with the hepatotoxic compounds carbon tetrachloride or azathioprine (Fahim et al. 1999; Hoefler et al.; Rusu et al. 2005; Sotelo-Felix et al. 2002; Waseem et al. 2006).

No change in liver enzymes was observed in male rats administered 250 or 500 mg/kg of rosemary daily for 63 days (Nusier et al. 2007).

Genotoxicity

Antimutagenic activity of rosemary extracts has been observed in the Ames mutagenicity assay and in rats (Fahim et al. 1999; Minnunni et al. 1992; Wolleb et al. 1992).

Literature Cited

Anadon, A., M.R. Martinez-Larranaga, M.A. Martinez, et al. 2008. Acute oral safety study of rosemary extracts in rats. *J. Food Prot.* 71(4):790-795.

Armisen, M., V. Rodriguez, and C. Vidal. 2003. Photoaggravated allergic contact dermatitis due to *Rosmarinus officinalis* cross-reactive with *Thymus vulgaris*. *Contact Dermat.* 48(1):52-53.

Bakirel, T., U. Bakirel, O.U. Keles, S.G. Ulgen, and H. Yardibi. 2008. *In vivo* assessment of antidiabetic and antioxidant activities of rosemary (*Rosmarinus officinalis*) in alloxan-diabetic rabbits. *J. Ethnopharmacol.* 116(1):64-73.

Casey, R. 1960. Alleged antifertility plants of India. *Indian J. Med. Sci.* 14:590-600.

Chadha, Y. 1988. *The wealth of India: A dictionary of Indian raw materials and industrial products.* Delhi: Council of Scientific and Industrial Research.

Debersac, P., J.M. Heydel, M.J. Amiot, et al. 2001a. Induction of cytochrome P450 and/or detoxication enzymes by various extracts of rosemary: Description of specific patterns. *Food Chem. Toxicol.* 39(9):907-918.

Debersac, P., M.F. Vernevaut, M.J. Amiot, M. Suschetet, and M.H. Siess. 2001b. Effects of a water-soluble extract of rosemary and its purified component rosmarinic acid on xenobiotic-metabolizing enzymes in rat liver. *Food Chem. Toxicol.* 39(2):109-117.

Erenmemisoglu, A., R. Saraymen, and H. Ustun. 1997. Effect of a *Rosmarinus officinalis* leaf extract on plasma glucose levels in normoglycemic and diabetic mice. *Pharmazie* 52(Aug.):645-646.

Fahim, F.A., A.Y. Esmat, H.M. Fadel, and K.F. Hassan. 1999. Allied studies on the effect of *Rosmarinus officinalis* L. on experimental hepatotoxicity and mutagenesis. *Int. J. Food Sci. Nutr.* 50(6):413-427.

Fernandez, L., S. Duque, I. Sanchez, et al. 1997. Allergic contact dermatitis from rosemary (*Rosmarinus officinalis* L.). *Contact Dermat.* 37(5):248-249.

Greshoff, M. 1913. Buitenzorg Med. Dep. Landb. *Cited in* Watt, J.M., and M.G. Breyer-Brandwijk. 1962. *The medicinal and poisonous plants of southern and eastern Africa.* Edinburgh: E. & S. Livingstone.

Guin, J.D. 2001. Rosemary cheilitis: One to remember. *Contact Dermat.* 45(1):63.

Hjorther, A.B., C. Christophersen, B.M. Hausen, and T. Menne. 1997. Occupational allergic contact dermatitis from carnosol, a naturally-occurring compound present in rosemary. *Contact Dermat.* 37(3):99-100.

Ho, C.-T., T. Ferraro, Q. Chen, R. Rosen, and M.-T. Huang. 1994. Phytochemicals in teas and rosemary and their cancer-preventative properties. In *Food phytochemicals for cancer prevention II: Teas, spices, and herbs. ACS Symposium Series 547*, edited by Ho, C.-T., T. Osawa, M.-T. Huang, and R. Rosen. Washington, DC: American Chemical Society.

Hoefler, C., J. Fleurentin, F. Mortier, J.M. Pelt, and J. Guillemain. Comparative choleretic and hepatoprotective properties of young sprouts and total plant extracts of *Rosmarinus officinalis* in rats. *J. Ethnopharmacol.* 19(2):133.

Inui, S., and I. Katayama. 2005. Allergic contact dermatitis induced by rosemary leaf extract in a cleansing gel. *J. Dermatol.* 32(8):667-669.

Jori, A., and G. Briatico. 1973. Effect of eucalyptol on microsomal enzyme activity of foetal and newborn rats. *Biochem. Pharmacol.* 22(4):543-544.

Klarman, E. 1958. Perfume dermatitis. *Ann. Allergy* 16:425-434.

Kovar, K.A., B. Gropper, D. Friess, and H.P.T. Ammon. 1987. Blood levels of 1,8-cineole and locomotor activity of mice after inhalation and oral administration of rosemary oil. *Planta Med.* 53(4):315-318.

Kwon, Y.I., D.A. Vattem, and K. Shetty. 2006. Evaluation of clonal herbs of Lamiaceae species for management of diabetes and hypertension. *Asia Pac. J. Clin. Nutr.* 15(1):107-118.

Laitinen, L.A., P.S.M. Tammela, A. Galkin, et al. 2004. Effects of extracts of commonly consumed food supplements and food fractions on the permeability of drugs across Caco-2 cell monolayers. *Pharm. Res.* 21(10):1904-1916.

Lee, J.J., Y.R. Jin, J.H. Lee, et al. 2007. Antiplatelet activity of carnosic acid, a phenolic diterpene from *Rosmarinus officinalis. Planta Med.* 73(2):121-127.

Lemonica, I.P., D.C. Damasceno, and L.C. Di-Stasi. 1996. Study of the embryotoxic effects of an extract of rosemary (*Rosmarinus officinalis* L.). *Braz. J. Med. Biol. Res.* 29(2):223-227.

Leuschner, J. 1997. Reproductive toxicity studies of D-camphor in rats and rabbits. *Arzneimittelforschung* 47(2):124-128.

Martinez-Gonzalez, M.C., J.J. Goday Bujan, W. Martinez Gomez, and E. Fonseca Capdevila. 2007. Concomitant allergic contact dermatitis due to *Rosmarinus officinalis* (rosemary) and *Thymus vulgaris* (thyme). *Contact Dermat.* 56(1):49-50.

Minnunni, M., U. Wolleb, O. Mueller, A. Pfeifer, and H.U. Aeschbacher. 1992. Natural antioxidants as inhibitors of oxygen species induced mutagenicity. *Mutat. Res.* 269(2):193-200.

Mongold, J., S. Camillieri, P. Susplugas, et al. 1991. The cholagogue/choleretic properties of a lyophilised extract of *Rosmarinus officinalis* L. *Planta Med. Phytother.* 25:6-11.

Nusier, M.K., H.N. Bataineh, and H.M. Daradkah. 2007. Adverse effects of rosemary (*Rosmarinus officinalis* L.) on reproductive function in adult male rats. *Exp. Biol. Med. (Maywood)* 232(6):809-813.

Offord, E.A., K. Macé, C. Ruffieux, A. Malnoë, and A. Pfeifer. 1995. Rosemary components inhibit benzo[*a*]pyrene-induced genotoxicity in human bronchial cells. *Carcinogenesis* 16(9):2057.

Opdyke, D. 1974. Fragrance raw material monographs: Rosemary oil. *Food Cosmet. Toxicol.* 12:977-978.

Plouzek, C.A., H.P. Ciolino, R. Clarke, and G.C. Yeh. 1999. Inhibition of P-glycoprotein activity and reversal of multidrug resistance *in vitro* by rosemary extract. *Eur. J. Cancer* 35(10):1541-1545.

Rusu, M.A., M. Tamas, C. Puica, I. Roman, and M. Sabadas. 2005. The hepatoprotective action of ten herbal extracts in CCl_4 intoxicated liver. *Phytother. Res.* 19(9):744-749.

Samman, S., B. Sandstrom, M.B. Toft, et al. 2001. Green tea or rosemary extract added to foods reduces nonheme-iron absorption. *Am. J. Clin. Nutr.* 73(3):607-612.

Serra, E., A. Vila, L. Peramiquel, et al. 2005. Allergic contact dermatitis due to rosemary. *Contact Dermat.* 53(3):179-180.

Singletary, K.W., and J.M. Nelshoppen. 1991. Inhibition of 7,12-dimethylbenz[*a*]anthracene (DMBA)-induced mammary tumorigenesis and of *in vivo* formation of mammary DMBA-DNA adducts by rosemary extract. *Cancer Lett.* 60(2):169-175.

Sotelo-Felix, J.I., D. Martinez-Fong, P. Muriel, et al. 2002. Evaluation of the effectiveness of *Rosmarinus officinalis* (Lamiaceae) in the alleviation of carbon tetrachloride-induced acute hepatotoxicity in the rat. *J. Ethnopharmacol.* 81(2):145-154.

Tahraoui, A., J. El-Hilaly, Z.H. Israili, and B. Lyoussi. 2007. Ethnopharmacological survey of plants used in the traditional treatment of hypertension and diabetes in south-eastern Morocco (Errachidia province). *J. Ethnopharmacol.* 110(1):105-117.

Waseem, M., M.A.U. Shah, R.A. Qureshi, et al. 2006. Ethnopharmacological survey of plants used for the treatment of stomach, diabetes, and ophthalmic diseases in Sudhan Gali, Kashmir, Pakistan. *Acta Bot. Yunnan.* 28(5):535.

Wolleb, U., O. Mueller, A. Pfeifer, and H.U. Aeschbacher. 1992. Natural antioxidants as inhibitors of oxygen species induced mutagenicity. *Mutat. Res.* 269(2):193-200.

Yamamoto, J., K. Yamada, A. Naemura, T. Yamashita, and R. Arai. 2005. Testing various herbs for antithrombotic effect. *Nutrition* 21(5):580-587.

Rubia cordifolia L. — Rubiaceae

SCN: Indian madder
AN: *manjishtha*
PN: *qian cao gen* (root)
OCN: Bengal madder
Part: root

Quick Reference Summary

Safety Class: 1
Interaction Class: A

Contraindications

None known.

Other Precautions

None known.

Drug and Supplement Interactions

None known.

Adverse Events and Side Effects

None known.

Pharmacological Considerations

Animal studies have demonstrated that Indian madder may modify glucose regulation (Patil et al. 2006; Somani et al. 2007). People with diabetes are advised to monitor their blood sugar closely and discuss the use of this herb with a qualified healthcare practitioner prior to use.

Pregnancy and Lactation

Although one animal study showed some anti-implantation activity of relatively high doses (250 mg/kg) of Indian madder (Sharma et al. 1983), reference texts on traditional Chinese medicine do not caution against the use of Indian madder during pregnancy (Bensky et al. 2004; Chen and Chen 2004).

No information on the safety of Indian madder during lactation was identified in the scientific or traditional literature. While this review did not identify any concerns for use while nursing, safety has not been conclusively established.

Review Details

I. Drug and Supplement Interactions

Clinical Trials of Drug or Supplement Interactions

No clinical trials of drug or supplement interactions were identified.

Case Reports of Suspected Drug or Supplement Interactions

No case reports of suspected drug or supplement interactions were identified.

Animal Trials of Drug or Supplement Interactions

No animal trials of drug or supplement interactions were identified.

II. Adverse Events

Case Reports of Adverse Events

No case reports of adverse events were identified.

III. Pharmacology and Pharmacokinetics

Human Pharmacological Studies

No relevant human pharmacological studies were identified.

Animal Pharmacological Studies

A reduction in blood sugar was observed in healthy and diabetic mice orally administered single or repeated (for 2 weeks) doses of an alcohol extract of Indian madder (Somani et al. 2007). A reduction in blood sugar was observed in diabetic mice intraperitoneally administered 100, 200, or 400 mg/kg of an alcohol extract of Indian madder (Patil et al. 2006).

In Vitro Pharmacological Studies

A partially purified fraction of Indian madder inhibited PAF-induced, but not thrombin-induced, platelet aggregation in rabbit platelets (Tripathi et al. 1993).

IV. Pregnancy and Lactation

In rats orally administered relatively large doses (250 mg/kg) of an ethanolic extract of Indian madder on days 1 to 7 of pregnancy, a reduction in the number of implanted fetuses was observed (Sharma et al. 1983).

No information on the safety of Indian madder during lactation was identified.

V. Toxicity Studies

Acute Toxicity

No deaths were reported in mice orally administered 150 g/kg of an Indian madder decoction. At a dose of 175 g/kg, one of five mice died (Zhu 1998).

The LD_{50} of an alcohol extract of Indian madder intraperitoneally administered to mice could not be determined at doses up to 1 g/kg (Patil et al. 2006).

The LD_{50} of the compound rubidate in mice is 3 g/kg after intraperitoneal administration (Zhu 1998).

Chronic Toxicity

The compound 1-hydroxyanthraquinone is a metabolite of the compound alizarin primeveroside that is found in Indian madder. In rats fed diets containing 1% 1-hydroxy-anthraquinone daily for 480 days, an increase in incidences of tumors in the colon and intestines was observed (Mori et al. 1990).

Literature Cited

Bensky, D., S. Clavey, and E. Stöger. 2004. *Chinese herbal medicine: Materia medica*. 3rd ed. Seattle: Eastland Press.

Chen, J.K., and T.T. Chen. 2004. *Chinese medical herbology and pharmacology*. City of Industry, CA: Art of Medicine Press.

Mori, H., N. Yoshimi, H. Iwata, et al. 1990. Carcinogenicity of naturally occurring 1-hydroxyanthraquinone in rats: Induction of large bowel, liver and stomach neoplasms. *Carcinogenesis* 11(5):799-802.

Patil, R.A., S.C. Jagdale, and S.B. Kasture. 2006. Antihyperglycemic, antistress and nootropic activity of roots of *Rubia cordifolia* Linn. *Indian J. Exp. Biol.* 44(12):987-992.

Sharma, B.B., M.D. Varshney, D.N. Gupta, and A.O. Prakash. 1983. Antifertility screening of plants. Part I. Effect of ten indigenous plants on early pregnancy in albino rats. *Int. J. Crude Drug Res.* 21(4):183-187.

Somani, R.S., K.S. Jain, and A.K. Singhai. 2007. Hypoglycaemic activity of roots of *Rubia cordifolia* in normal and diabetic rats. *PharmacologyOnline* 1:162-169.

Tripathi, Y.B., S. Pandey, and S.D. Shukla. 1993. Anti-platelet activating factor property of *Rubia cordifolia* Linn. *Indian J. Exp. Biol.* 31(6):533-535.

Zhu, Y.-P. 1998. *Chinese materia medica: Chemistry, pharmacology and applications.* Amsterdam: Harwood Academic Publishers.

Rubus spp.

Rosaceae

Rubus chingii **Hu**
SCN: palm-leaf raspberry
Syn: *Rubus officinalis* Koidz.
PN: *fu pen zi* (fruit)

Rubus suavissimus **S.K. Lee**
SCN: Chinese blackberry
PN: *tian cha* (leaf)
OCN: sweet tea
Part: fruit

Quick Reference Summary

Safety Class: 1
Interaction Class: A

Contraindications

None known.

Other Precautions

None known.

Drug and Supplement Interactions

None known.

Editors' Note

The fruits of palm-leaf raspberry are eaten fresh and are also used for making jam, jelly, and various beverages (Wu et al. 2003).

Adverse Events and Side Effects

None known.

Pharmacological Considerations

None known.

Pregnancy and Lactation

No information on the safety of palm-leaf raspberry or Chinese blackberry in pregnancy or lactation was identified in the scientific or traditional literature. Although this review did not identify any concerns for use while pregnant or nursing, safety has not been conclusively established.

Review Details

I. Drug and Supplement Interactions

Clinical Trials of Drug or Supplement Interactions

No clinical trials of drug or supplement interactions were identified.

Case Reports of Suspected Drug or Supplement Interactions

No case reports of suspected drug or supplement interactions were identified.

Animal Trials of Drug or Supplement Interactions

No animal trials of drug or supplement interactions were identified.

II. Adverse Events

Case Reports of Adverse Events

No case reports of adverse events were identified.

III. Pharmacology and Pharmacokinetics

Human Pharmacological Studies

No relevant human pharmacological studies were identified.

Animal Pharmacological Studies

No relevant animal pharmacological studies were identified.

In Vitro Pharmacological Studies

No relevant in vitro pharmacological studies were identified.

IV. Pregnancy and Lactation

No information on the safety of palm-leaf raspberry or Chinese blackberry during pregnancy or lactation was identified.

V. Toxicity Studies

No toxicity studies were identified.

Literature Cited

Wu, Z.Y., P. H. Raven, and D. Y. Hong, eds. 2003. *Flora of China. Vol. 9 (Pittosporaceae through Connaraceae).* Beijing: Science Press, and St. Louis: Missouri Botanical Garden Press.

Rubus spp. — Rosaceae

Rubus fruticosus L.
SCN: blackberry
OCN: bramble; shrubby blackberry

Rubus idaeus L. ssp. *idaeus*
SCN: raspberry
OCN: red raspberry

Rubus idaeus L. ssp. *strigosus* (Michx.) Focke
SCN: raspberry
Syn: *Rubus strigosus* Michx.
OCN: American raspberry; red raspberry
Part: leaf

Quick Reference Summary

Safety Class: 1
Interaction Class: A

Contraindications

None known.

Other Precautions

None known.

Drug and Supplement Interactions

None known.

Notice

Tannins (8–14% in *R. fruticosus*) (Mullen et al. 2002; Wichtl 2004); *see* Appendix 1.

Adverse Events and Side Effects

None known.

Pharmacological Considerations

Animal studies have demonstrated that blackberry leaf may modify glucose regulation (Alonso et al. 1980; Jouad et al. 2002). People with diabetes are advised to monitor their blood sugar closely and discuss the use of this herb with a qualified healthcare practitioner prior to use.

Pregnancy and Lactation

Human studies have indicated no adverse effects of raspberry leaf on fetal or maternal health with doses of 2.4 g daily from week 32 until birth in a double-blind, randomized, placebo-controlled trial, and in a retrospective study with women taking one to eight raspberry leaf tablets or one to six cups of raspberry leaf tea daily (Parsons et al. 1999; Simpson et al. 2001).

An animal study indicated a lengthening of gestation in animals treated with raspberry leaf (Johnson et al. 2009).

No information on the safety of blackberry or raspberry leaf during lactation was identified in the scientific or traditional literature. While this review did not identify any concerns for use while nursing, safety has not been conclusively established.

Review Details

I. Drug and Supplement Interactions

Clinical Trials of Drug or Supplement Interactions

No clinical trials of drug or supplement interactions were identified.

Case Reports of Suspected Drug or Supplement Interactions

No case reports of suspected drug or supplement interactions were identified.

Animal Trials of Drug or Supplement Interactions

No animal trials of drug or supplement interactions were identified.

II. Adverse Events

Case Reports of Adverse Events

No case reports of adverse events were identified.

III. Pharmacology and Pharmacokinetics

Human Pharmacological Studies

No relevant human pharmacological studies were identified.

Animal Pharmacological Studies

A reduction in blood glucose levels was observed in healthy and diabetic rabbits orally administered 5 g/kg of an aqueous extract of blackberry leaf (Alonso et al. 1980). Similar effects were observed in healthy and diabetic rats orally administered 100 mg/kg of an aqueous extract of blackberry leaf as a single dose or the same dose daily for 9 days (Jouad et al. 2002).

In Vitro Pharmacological Studies

Extracts of raspberry leaf prepared with different solvents (*n*-hexane, ethyl acetate, chloroform, and methanol) exhibited a relaxant effect on isolated guinea pig ileum. The methanol extract was the most active (Rojas-Vera et al. 2002).

IV. Pregnancy and Lactation

In a double-blind, randomized, placebo-controlled trial, healthy pregnant women were orally administered tablets containing 2.4 g raspberry leaf daily from 32 weeks of gestation until birth. No adverse effects on fetal or maternal health were observed, including length of gestation, medical augmentation of labor, need for pain relief during labor, time of the stages of labor (shorter second stage in the raspberry leaf group), blood loss during labor, and diastolic blood pressure in the mother. Other observed parameters included meconium stained fluid, Apgar score at 5 minutes of age, birth weight, transfer of baby to special care, and rate of emergency caesarean sections. Slightly more women in the placebo group had forceps or vacuum-assisted birth (Simpson et al. 2001).

In a retrospective study of mothers who used raspberry leaf during pregnancy, no adverse effects on pregnancy or on fetal or maternal health were observed. Doses ingested ranged from one to six cups of raspberry leaf tea daily or one to eight tablets of raspberry leaf (tablet size not specified) daily. Findings from this study suggest that raspberry leaf might decrease the likelihood of preterm and post-term gestation. Women who ingested raspberry leaf were less likely to receive an artificial rupture of their membranes or require a caesarean section, forceps birth, or vacuum birth than the women in the control group (Parsons et al. 1999).

In rats orally administered 10 mg/kg raspberry leaf daily for the full gestational period, gestation was an average of 1.6 days longer in the raspberry group as compared to control. In offspring of treated animals, the time to vaginal opening was 2 days earlier than control group animals (Johnson et al. 2009).

Studies examining the effects of intravenous administration of raspberry leaf extract to cats and rabbits, and the effects of the same extract on isolated uterine strips from dogs, cats, rabbits, and guinea pigs, indicated that raspberry leaf toned smooth muscles that were relaxed and relaxed muscles that were contracted (Burn and Withell 1941).

In isolated uterine strips from pregnant and nonpregnant rats and pregnant and nonpregnant humans (pathological uterine tissue, removed for therapeutic purposes), treatment with an aqueous extract of raspberry leaf had little or no effect on nonpregnant uteruses. In strips from pregnant uteruses, inhibition of contractions was observed along with a more regular rhythm of contractions (Bamford et al. 1970).

Raspberry leaf extracts applied to isolated uterine tissue produced a spasmolytic effect (Beckett et al. 1954).

No information on the safety of blackberry or raspberry leaf during lactation was identified.

V. Toxicity Studies

Acute Toxicity

The LD_{50} of an orally administered blackberry leaf aqueous extract in mice is 8.1 g/kg (Jouad et al. 2002).

R

Literature Cited

Alonso, R., I. Cadavid, and J.M. Calleja. 1980. A preliminary study of hypoglycemic activity of *Rubus fruticosus*. *Planta Med.* (Suppl.):102-106.

Bamford, D.S., R.C. Percival, and A.U. Tothill. 1970. Raspberry leaf tea: A new aspect to an old problem [abstract]. *Br. J. Pharmacol.* 40(1):161.

Beckett, A.H., F.W. Belthle, and K.R. Fell. 1954. The active constituents of raspberry leaves; a preliminary investigation. *J. Pharm. Pharmacol.* 6(11):785-796.

Burn, J.H., and E.R. Withell. 1941. A principle in raspberry leaves which relaxes uterine muscle. *Lancet* 238(6149):1-3.

Johnson, J.R., E. Makaji, and S. Ho. 2009. Effect of maternal raspberry leaf consumption in rats on pregnancy outcome and the fertility of the female offspring. *Reprod. Sci.* 16(6):605.

Jouad, H., M. Maghrani, and M. Eddouks. 2002. Hypoglycaemic effect of *Rubus fructicosis* L. and *Globularia alypum* L. in normal and streptozotocin-induced diabetic rats. *J. Ethnopharmacol.* 81(3):351-356.

Mullen, W., J. McGinn, M.E. Lean, et al. 2002. Ellagitannins, flavonoids, and other phenolics in red raspberries and their contribution to antioxidant capacity and vasorelaxation properties. *J Agric. Food Chem.* 50(18):5191-5196.

Parsons, M., M. Simpson, and T. Ponton. 1999. Raspberry leaf and its effect on labour: Safety and efficacy. *Aust. Coll. Midwives Inc. J.* 12(3):20-25.

Rojas-Vera, J., A.V. Patel, and C.G. Dacke. 2002. Relaxant activity of raspberry (*Rubus idaeus*) leaf extract in guinea-pig ileum *in vitro. Phytother. Res.* 16(7):665-668.

Simpson, M., M. Parsons, J. Greenwood, and K. Wade. 2001. Raspberry leaf in pregnancy: Its safety and efficacy in labor. *J. Midwifery Womens Health* 46(2):51-59.

Wichtl, M. 2004. *Herbal drugs and phytopharmaceuticals: A handbook for practice on a scientific basis*. 3rd ed. Boca Raton, FL: CRC Press.

Rumex hymenosepalus Torr. — Polygonaceae

SCN: canaigre

Part: root

Quick Reference Summary

Safety Class: 1

Interaction Class: A

Contraindications

None known.

Other Precautions

None known.

Drug and Supplement Interactions

None known.

Notice

Tannins (11–35%) (Krochmal and Paur 1951; Moore 1989); *see* Appendix 1.

Adverse Events and Side Effects

None known.

Pharmacological Considerations

None known.

Pregnancy and Lactation

No information on the use of canaigre in pregnancy or lactation was identified in the scientific or traditional literature. Although this review did not identify any concerns for use while pregnant or nursing, safety has not been conclusively established.

Review Details

I. Drug and Supplement Interactions

Clinical Trials of Drug or Supplement Interactions

No clinical trials of drug or supplement interactions were identified.

Case Reports of Suspected Drug or Supplement Interactions

No case reports of suspected drug or supplement interactions were identified.

Animal Trials of Drug or Supplement Interactions

No animal trials of drug or supplement interactions were identified.

II. Adverse Events

Case Reports of Adverse Events

No case reports of adverse events were identified.

III. Pharmacology and Pharmacokinetics

Human Pharmacological Studies

No relevant human pharmacological studies were identified.

Animal Pharmacological Studies

No relevant animal pharmacological studies were identified.

In Vitro Pharmacological Studies

No relevant in vitro pharmacological studies were identified.

IV. Pregnancy and Lactation

No information on the use of canaigre during pregnancy or lactation was identified.

V. Toxicity Studies

No toxicity studies were identified.

Literature Cited

Krochmal, A., and S. Paur. 1951. Canaigre—A desert source of tannin. *Econ. Bot.* 5(4):367-377.

Moore, M. 1989. *Medicinal plants of the desert and canyon west*. Santa Fe: Museum of New Mexico Press.

Rumex spp.

Polygonaceae

Rumex acetosa **L.**
SCN: sorrel
PN: *suan mo ye* (leaf)
OCN: garden sorrel

Rumex acetosella **L.**
SCN: sheep sorrel
OCN: sour grass
Part: leaf

Quick Reference Summary

Safety Class: 1
Interaction Class: A

Contraindications

None known.

Other Precautions

Use with caution in persons with a history of kidney stones (McGuffin et al. 1997).

Drug and Supplement Interactions

None known.

Notice

Tannins (7–15%) (List and Hörhammer 1973); *see* Appendix 1.

Adverse Events and Side Effects

None known.

Pharmacological Considerations

None known.

Pregnancy and Lactation

No information on the safety of sorrel or sheep sorrel in pregnancy or lactation was identified in the scientific or traditional literature. Although this review did not identify any concerns for use while pregnant or nursing, safety has not been conclusively established.

Review Details

I. Drug and Supplement Interactions

Clinical Trials of Drug or Supplement Interactions

No clinical trials of drug or supplement interactions were identified.

Case Reports of Suspected Drug or Supplement Interactions

No case reports of suspected drug or supplement interactions were identified.

Animal Trials of Drug or Supplement Interactions

No animal trials of drug or supplement interactions were identified.

II. Adverse Events

Case Reports of Adverse Events

No case reports of adverse events were identified.

III. Pharmacology and Pharmacokinetics

Human Pharmacological Studies

No relevant human pharmacological studies were identified.

Animal Pharmacological Studies

No relevant animal pharmacological studies were identified.

In Vitro Pharmacological Studies

No relevant in vitro pharmacological studies were identified.

IV. Pregnancy and Lactation

No information on the safety of sorrel or sheep sorrel during pregnancy or lactation was identified.

V. Toxicity Studies

No toxicity studies were identified.

Literature Cited

List, P.H., and H. Hörhammer. 1973. *Hagers handbuch der pharmazeutischen praxis*. Berlin: Springer.

McGuffin, M., C. Hobbs, R. Upton, and A. Goldberg. 1997. *Botanical safety handbook*. Boca Raton, FL: CRC Press.

Rumex spp. — Polygonaceae

Rumex crispus L.
SCN: yellow dock
OCN: curled dock; curly dock; dock

Rumex obtusifolius L.
SCN: broad-leaf dock
OCN: bitter dock
Part: root

Quick Reference Summary

Safety Class: 1
Interaction Class: A

Contraindications

None known.

Other Precautions

Use with caution in persons with a history of kidney stones (McGuffin et al. 1997).

Drug and Supplement Interactions

None known.

Notice

Tannins (12–20%) (List and Hörhammer 1973); *see* Appendix 1.

Editors' Note

Although yellow dock and broad-leaf dock contain small amounts of anthraquinone glycosides (0.35–4.0%), these species have, at most, a mild laxative effect (Demirezer 1994; Demirezer and Kuruuzum 1995; List and Hörhammer 1973; Mills and Bone 2005).

Adverse Events and Side Effects

None known.

Pharmacological Considerations

None known.

Pregnancy and Lactation

No information on the safety of yellow dock or broad-leaf dock in pregnancy or lactation was identified in the scientific or traditional literature. Although this review did not identify any concerns for use while pregnant or nursing, safety has not been conclusively established.

R

Review Details

I. Drug and Supplement Interactions

Clinical Trials of Drug or Supplement Interactions

No clinical trials of drug or supplement interactions were identified.

Case Reports of Suspected Drug or Supplement Interactions

No case reports of suspected drug or supplement interactions were identified.

Animal Trials of Drug or Supplement Interactions

No animal trials of drug or supplement interactions were identified.

II. Adverse Events

Case Reports of Adverse Events

Fatal oxalate poisoning was reported in a 53-year-old insulin-dependent diabetic man who had ingested 1 kg of yellow dock (part not specified) along with his family members. Other family members who ingested smaller amounts of yellow dock had only mild symptoms of poisoning and recovered within several days (Reig et al. 1990).

Oxalate poisoning was reported in a flock of sheep within 40 hours of being set to graze in a lot that contained significant amounts of yellow dock leaf. Clinical signs of toxicosis included excess salivation, tremors, ataxia, and recumbency. Affected ewes were markedly hypocalcemic and azotemic. Samples of yellow dock contained 6.6 to 11.1% oxalic acid on a dry-weight basis, a concentration comparable with that in other oxalate-containing plants that have caused acute oxalate toxicosis (Panciera et al. 1990).

III. Pharmacology and Pharmacokinetics

Human Pharmacological Studies

No relevant human pharmacological studies were identified.

Animal Pharmacological Studies

No relevant animal pharmacological studies were identified.

In Vitro Pharmacological Studies

No relevant in vitro pharmacological studies were identified.

IV. Pregnancy and Lactation

No information on the safety of yellow dock or broad-leaf dock during pregnancy or lactation was identified.

V. Toxicity Studies

No toxicity studies were identified.

Literature Cited

Demirezer, L.O. 1994. Anthraquinone derivatives in *Rumex gracilescens* and *R. crispus*. *Pharmazie* 49:378-379.

Demirezer, L.O., and A. Kuruuzum. 1995. Determination of the cytotoxicity of *Rumex crispus* during the vegetation period using a brine shrimp bioassay. *Z. Naturforsch.* 50(5-6):461-462.

List, P.H., and H. Hörhammer. 1973. *Hagers handbuch der pharmazeutischen praxis*. Berlin: Springer.

McGuffin, M., C. Hobbs, R. Upton, and A. Goldberg. 1997. *Botanical safety handbook*. Boca Raton, FL: CRC Press.

Mills, S., and K. Bone. 2005. *The essential guide to herbal safety*. St. Louis: Elsevier.

Panciera, R.J., T. Martin, G.E. Burrows, D.S. Taylor, and L.E. Rice. 1990. Acute oxalate poisoning attributable to ingestion of curly dock (*Rumex crispus*) in sheep. *J. Am. Vet. Med. Assoc.* 196(12):1981-1984.

Reig, R., P. Sanz, C. Blanche, et al. 1990. Fatal poisoning by rumex-crispus (curled dock): Pathological findings and application of scanning electron microscopy. *Vet. Hum. Toxicol.* 32(5):468-470.

Ruscus aculeatus L. — Lilaceae

SCN: butcher's broom
OCN: box holly
Part: root and rhizome

Quick Reference Summary

Safety Class: 1
Interaction Class: A

Contraindications

None known.

Other Precautions

None known.

Drug and Supplement Interactions

None known.

Adverse Events and Side Effects

A review of the literature on butcher's broom indicated that no side effects or other undesirable effects are expected (ESCOP 2003).

Contact allergy to butcher's broom and extracts of butcher's broom has been reported and confirmed by patch testing (Breuil et al. 1989; Elbadir et al. 1998; Landa et al. 1990; Ramirez-Hernandez et al. 2006).

Pharmacological Considerations

None known.

Pregnancy and Lactation

Information on the safety of butcher's broom during pregnancy is limited. A small human study indicated that suppositories containing ruscogenin were well tolerated, with no adverse effects on pregnancy or on fetal or maternal health (Anger and Neietsch 1981).

No information on the safety of butcher's broom during lactation was identified in the scientific or traditional literature. While this review did not identify any concerns for use while nursing, safety has not been conclusively established.

Review Details

I. Drug and Supplement Interactions

Clinical Trials of Drug or Supplement Interactions

No clinical trials of drug or supplement interactions were identified.

Case Reports of Suspected Drug or Supplement Interactions

No case reports of suspected drug or supplement interactions were identified.

R

Animal Trials of Drug or Supplement Interactions

No animal trials of drug or supplement interactions were identified.

II. Adverse Events

Case Reports of Adverse Events

Contact allergy to butcher's broom and extracts of butcher's broom has been reported and confirmed by patch testing (Breuil et al. 1989; Elbadir et al. 1998; Landa et al. 1990; Ramirez-Hernandez et al. 2006).

III. Pharmacology and Pharmacokinetics

Human Pharmacological Studies

No relevant human pharmacological studies were identified.

Animal Pharmacological Studies

No relevant animal pharmacological studies were identified.

In Vitro Pharmacological Studies

In a human COX inhibitor screening assay, a methanol extract of butcher's broom exhibited COX-1 inhibition at 2.09 mg/ml (IC_{50}) and COX-2 inhibition at 6.83 mg/ml (IC_{50}) (Seaver and Smith 2004).

IV. Pregnancy and Lactation

Ruscogenin-containing suppositories were well tolerated in a study with 30 pregnant women. No adverse effects on newborns were reported (Anger and Neietsch 1981).

No information on the safety of butcher's broom during lactation was identified.

V. Toxicity Studies

Acute Toxicity

The LD_{50} of orally administered butcher's broom fluid extract in mice is 4.6 g/kg, and in rats could not be determined at doses up to 4.6 g/kg (Seidenberger et al. 1974). The LD_{50} of ruscogenins (combined ruscogenin and neoruscogenin) orally administered to mice and rats could not be determined at doses up to 3 g/kg (Capra 1972).

The oral LD_{50} of a fluid extract of butcher's broom rhizome is 2.2 ml/kg in rats and 29.21 ml/kg in mice. After intraperitoneal administration, the toxicity is 10- to 20-fold higher. An extract of the root was reportedly more toxic than that of the rhizome (Boucard et al. 1967).

An ethanol extract of butcher's broom intravenously administered to dogs was fatal at doses from 830 to 1800 mg/kg. In guinea pigs intraperitoneally administered the same extract, no toxic signs were observed at doses up to 1.5 g/kg, whereas doses over 2 g/kg were fatal (Caujolle et al. 1953; Moscarella 1953).

Subchronic Toxicity

In rabbits orally administered 2 or 5 g/kg of a butcher's broom extract daily for 26 weeks, no changes in body weight or blood count were observed (Roux 1969).

In rats orally administered 300 mg/kg of ruscogenin or saponins from butcher's broom daily for 8 weeks, no signs of toxicity were observed in any of the parameters measured, including body and organ weights, blood glucose and liver function levels, and histological examination of organs (Capra 1972).

Literature Cited

R

Anger, H., and P. Neietsch. 1981. Ruscorectal bei Analerkrankungen. Ergebnisse aus Klinik und Praxis. *Med. Welt.* 33(41):1450-1452.

Boucard, M., I.S. Beaulaton, and C. Reboul. 1967. Study of the acute toxicity of various fluid extracts of the thorny holly (*Ruscus aculeatus* L.). *Trav. Soc. Pharm. Montpellier* 27(3):187-191.

Breuil, K., F. Patte, J.C. Meurice, and B. Vandel. 1989. Allergie de contact à une pommade aux extraits de petit houx. *Rev. Fr. Allergol. Immunol. Clin.* 29(4):215.

Capra, C. 1972. Pharmacology and toxicology of some components of *Ruscus aculeatus* L. *Fitoterapia* 4:99-113.

Caujolle, F., P. Mériel, and E. Stanislas. 1953. Sur les propriétés pharmacologiques de l'extrait de *Ruscus aculeatus*. *Ann. Pharm. Fr.* 11:109-120.

Elbadir, S., F. El Sayed, and F. Renaud. 1998. L'allergie de contact aux ruscogenines. *Rev. Fr. Allergol. Immunol. Clin.* 38:37-40.

ESCOP. 2003. *ESCOP monographs: The scientific foundation for herbal medicinal products*. 2nd ed. Exeter, UK: European Scientific Cooperative on Phytotherapy.

Landa, N., A. Aguirre, J. Goday, J.A. Ratón, and J.L. Díaz-Pérez. 1990. Allergic contact dermatitis from a vasoconstrictor cream. *Contact Dermat.* 22(5):290.

Moscarella, C. 1953. Contribution a l'étude pharmacodynamique du *Ruscus aculeatus* L. Thesis, Université de Toulouse.

Ramirez-Hernandez, M., J. Garcia-Selles, C. Merida-Fernandez, and J.A. Martinez-Escribano. 2006. Allergic contact dermatitis to ruscogenins. *Contact Dermat.* 54(1):60.

Roux, G. 1969. *Cited in* ESCOP. 2003. *ESCOP monographs: The scientific foundation for herbal medicinal products*. 2nd ed. Exeter, UK: European Scientific Cooperative on Phytotherapy.

Seaver, B., and J.R. Smith. 2004. Inhibition of COX isoforms by nutraceuticals. *J. Herb. Pharmacother.* 4(2):11-18.

Seidenberger, A.V., I. Müller, and H.J.H. Heindl. 1974. *Cited in* Mills, S., and K. Bone. 2005. *The essential guide to herbal safety.* St. Louis: Elsevier.

Ruta graveolens L. — Rutaceae

SCN: rue
PN: *chou cao* (herb)
OCN: common rue; herb-of-grace
Part: herb

Quick Reference Summary

Safety Class: 2b
Interaction Class: A

Contraindications

Not for use in pregnancy except under the supervision of a qualified healthcare practitioner (Al-Mahmoud et al. 2003; Ciganda and Laborde 2003; de Freitas et al. 2005; Wood and LaWall 1918).

Other Precautions

Avoid prolonged exposure to sunlight after ingestion (McGuffin et al. 1997; Zobel and Brown 1990).

Drug and Supplement Interactions

None known.

Notice

Abortifacient (Ciganda and Laborde 2003; Conway and Slocumb 1979; Felter and Lloyd 1898; Wood and LaWall 1918); *see* Appendix 2.

Emmenagogue (Chadha 1988; Conway and Slocumb 1979; Felter and Lloyd 1898); *see* Appendix 2.

Photosensitizing (Milesi et al. 2001; Zobel and Brown 1990); *see* Appendix 2.

Editors' Notes

Rue is identified as generally recognized as safe by the U.S. FDA when present in all categories of food at concentrations that do not exceed 2 ppm (CFR 2011a). Although an associated regulation states that other uses in foods would require a food additive regulation (CFR 2011b), dietary ingredients for use in dietary supplements are specifically excluded from the federal food additive definition (U.S.C. 2010).

Adverse Events and Side Effects

Rue contains furanocoumarin compounds known to cause phytophotodermatitis after topical application followed by exposure to sunlight or UVA light (Furniss and Adams 2007; Milesi et al. 2001; Schempp et al. 1999).

At standard therapeutic doses, ingestion of rue has been associated with melancholic mood, sleep disorders, drowsiness, dizziness, and cramps (Weiss and Fintelmann 2000). Rue may cause gastrointestinal irritation (Felter and Lloyd 1898; Remington and Wood 1918).

A review of cases of women who attempted to use rue as an abortifacient (doses not listed, but likely represent overdose) indicated that adverse effects of rue included abdominal pain, vomiting, genital hemorrhage, anemia, jaundice, liver enlargement, neurological depression, respiratory distress, and elevated liver enzyme levels (Ciganda and Laborde 2003).

Pharmacological Considerations

None known.

Pregnancy and Lactation

Rue has traditionally been used as an abortifacient (Ciganda and Laborde 2003; Conway and Slocumb 1979). Animal studies have indicated that rue decreased birth rates and increased fetal resorption rates (Al-Mahmoud et al. 2003; de Freitas et al. 2005). Based on this information, use during pregnancy is not recommended except under the supervision of a qualified healthcare practitioner.

No information on the safety of rue during lactation was identified in the scientific or traditional literature. While this review did not identify any concerns for use while nursing, safety has not been conclusively established.

R

Review Details

I. Drug and Supplement Interactions

Clinical Trials of Drug or Supplement Interactions

No clinical trials of drug or supplement interactions were identified.

Case Reports of Suspected Drug or Supplement Interactions

No case reports of suspected drug or supplement interactions were identified.

Animal Trials of Drug or Supplement Interactions

No animal trials of drug or supplement interactions were identified.

II. Adverse Events

Case Reports of Adverse Events

A 78-year-old woman with a history of heart problems developed bradycardia, acute renal failure with hyperkalemia, and coagulopathy after 3 days of consuming a decoction made from rue. Each dose was made from 50 g of fresh rue in 1000 ml of water, made into 250 ml of decoction taken twice daily. The woman had a 5-year history of non-obstructive type hypertrophic cardiomyopathy, and was taking bisoprolol, diltiazem, and amiloride. Her heart rate was around 76 beats per minute in the previous 6 months (Seak and Lin 2007).

Phytophotodermatitis from rue has been reported after contact with the fresh plant (Asefi et al. 1999; Gawkrodger and Savin 1983; Heskel et al. 1983) or after topical application of infusions or ointments containing rue (Arias-Santiago et al. 2009; Morais et al. 2008; Wessner et al. 1999). Some cases have produced severe bullae (Eickhorst et al. 2007; Schempp et al. 1999).

A 2-year-old developed acute phytophotodermatitis with systemic upset after contact with rue (Furniss and Adams 2007).

III. Pharmacology and Pharmacokinetics

Human Pharmacological Studies

No relevant human pharmacological studies were identified.

Animal Pharmacological Studies

No relevant animal pharmacological studies were identified.

In Vitro Pharmacological Studies

In sperm immobilization tests with human sperm, dose-dependent immobilization of sperm was observed after treatment with a freeze-dried aqueous extract of rue, with 100% immobilization at a concentration of 100 mg/ml. No effects were observed on viability of cells, DNA status, or mitochondrial activity. After washing, motility was returned to approximately 30% of sperm cells. The part of the extract responsible for immobilization of the sperm cells was stable on boiling (Harat et al. 2008).

IV. Pregnancy and Lactation

A review of cases of herbal infusion ingestion with abortive intent received by a poison control center in Uruguay indicated that rue was one of the most commonly used herbs. In women who attempted to use rue as an abortifacient, reported symptoms of toxicity included abdominal pain, vomiting, genital hemorrhage, anemia, jaundice, liver enlargement, a decrease in urine production, neurological depression, respiratory distress, potassium levels less than 3.5 mEq/l, thrombocytes lower than 140,000, leukocytes higher than 10,000, elevated levels of aspartate aminotransferase (AST) and alanine aminotransferase (ALT), and increased creatinine. No information on doses used or method of preparation was given (Ciganda and Laborde 2003).

In mice orally administered 1 g/kg of a dried hydroalcoholic extract of rue daily on days 1 to 3, 4 to 6, or 7 to 9 of pregnancy, no significant differences in preimplantation or in fetal resorptions were observed in treated animals as compared to control. The fetal death rate was 8% in animals treated on days 7 to 9, and similar to control for other treatment periods. Birth rates were 91% for animals treated on pregnancy days 1 to 3, 82% in animals treated on days 4 to 6, 64% for days 7 to 9, and 76% for the control group (de Freitas et al. 2005).

In rats orally administered 800 mg/kg of water, methanol, ethanol, hexane, ether, or dichloromethane extracts of rue daily on days 1 to 6 of pregnancy, no anti-implantation activity was observed for the water, methanol, or ethanol extracts; significant anti-implantation activity was observed for the hexane, ether, and dichloromethane extracts. At the 800 mg/kg dose, the ether extract produced severe toxicity in the mothers. At 400 mg/kg, an increase in the number of fetal resorptions was observed in animals treated with the water, ethanol, or hexane extracts. A reduction in fetal weight gain was observed after treatment with water, methanol, or dichloromethane extracts. After administration of the same extracts and doses on days 6 to 15 of pregnancy, an increase in fetal mortality was observed (Al-Mahmoud et al. 2003).

To determine the effects of rue on embryo implantation and development, superovulated mated female mice were provided with aqueous extracts of rue at concentrations of 5, 10, or 20% as the sole source of drinking water for 4 days after mating. Examination of embryos 98 hours after superovulation treatment indicated that in the 10% group, 37% of embryos were abnormal while in the 20% group, 64% were abnormal. Cell number was diminished and embryo transport was slightly delayed in the high-dose group (Gutierrez-Pajares et al. 2003).

Rue is listed as an ingredient in four of six abortifacient formulas from ancient Persia (Madari and Jacobs 2004).

No information on the safety of rue during lactation was identified.

V. Toxicity Studies

Acute Toxicity

In goats orally administered 5 g/kg of rue leaf daily, tremor, dyspnea, frequent urination, incoordination of movement, ataxia, and recumbency, with death after 1 to 7 days, were observed (el Agraa et al. 2002).

Short-Term Toxicity

In male rats orally administered 500 mg/kg of an aqueous extract of rue daily for 60 days, decreases in the weight of reproductive organs, sperm motility, density in cauda epididymides and testicular ducts, and spermatogenesis in the seminiferous tubules were observed. Ingestion of the extract was associated with a suppression of sexual behavior in adult males (Khouri and El-Akawi 2005).

In goats orally administered 1 g/kg of rue leaf daily for 40 days, animals had pallor of the visible mucous membranes and loss in condition, with one animal dying on day 17. Changes in serum aspartate transaminase, creatine kinase, total protein, cholesterol, urea and other serum constituents were observed (el Agraa et al. 2002).

R

In rats orally administered 500 mg/kg of rue daily for 3 weeks, significant increases in body weight with remarkable increase in total food intake and protein efficiency ratio were observed. No significant changes were observed in other biochemical and nutritional parameters except an elevation of alkaline phosphatase (Al-Okbi et al. 2002).

Genotoxicity

In mice orally administered 400 or 1000 mg/kg of a hydromethanolic extract of rue daily for 30 days, damage to cellular DNA was observed in the comet assay at the 1000 mg/kg dose. Damage was observed at day 30 but not at day 20 or earlier. In the chromosomal aberration test with animals on the same treatment regimen, a dose-dependent increase in aberrations was observed (Preethi et al. 2008).

In the Ames test for mutagenicity in *Salmonella typhimurium* strains TA98 and TA100, a strong mutagenic effect was observed in TA98 without metabolic activation. With metabolic activation, only a weak mutagenic response was observed. Moderate mutagenic effects were detected with and without metabolic activation by S9 mix (Paulini et al. 1987).

Compounds isolated from rue have demonstrated mutagenic activity and photomutagenic activity (Paulini and Schimmer 1989; Schimmer et al. 1991; Schimmer and Kuhne 1990).

Literature Cited

Al-Mahmoud, M.S., A. Elbetieha, and R.A. Al-Muhur. 2003. Anticonceptive and antifertility activities of various *Ruta graveolens* extracts in female rats. *Acta Pharm. Turc.* 45(3):203-212.

Al-Okbi, S.Y., E.M. El-Sayed, N.M. Ammar, N.K. El-Sayed, and L.T. Abou-El Kassem. 2002. Effect of *Ruta graveolens* L. and *Euphorbia peplus* L. anti-inflammatory extracts on nutritional status of rats and the safety of their use. *Indian J. Exp. Biol.* 40(1):45-48.

Arias-Santiago, S.A., M.A. Fernandez-Pugnaire, F.M. Almazan-Fernandez, C. Serrano-Falcon, and S. Serrano-Ortega. 2009. Phytophotodermatitis due to *Ruta graveolens* prescribed for fibromyalgia. *Rheumatology* 48(11):1401.

Asefi, M., E. Schopf, and M. Augustin. 1999. Bullous phototoxic dermatitis caused by *Ruta graveolens* (garden rue). *Aktuelle Dermatol.* 25(7):230-232.

CFR. 2011a. *Code of federal regulations*, Title 21 Part 184.1698, 2011 ed. Direct food substances affirmed as generally recognized as safe. Listing of specific substances affirmed as GRAS. Rue. Washington, DC: U.S. Government Printing Office.

CFR. 2011b. *Code of federal regulations*, Title 21 Part 184.1(b)(2), 2011 ed. Direct food substances affirmed as generally recognized as safe. Substances added directly to human food affirmed as generally recognized as safe (GRAS). Washington, DC: U.S. Government Printing Office.

Chadha, Y. 1988. *The wealth of India: A dictionary of Indian raw materials and industrial products.* Delhi: Council of Scientific and Industrial Research.

Ciganda, C., and A. Laborde. 2003. Herbal infusions used for induced abortion. *J. Toxicol. Clin. Toxicol.* 41(3):235-239.

Conway, G.A., and J.C. Slocumb. 1979. Plants used as abortifacients and emmenagogues by Spanish New Mexicans. *J. Ethnopharmacol.* 1(3):241-261.

de Freitas, T.G., P.M. Augusto, and T. Montanari. 2005. Effect of *Ruta graveolens* L. on pregnant mice. *Contraception* 71(1):74-77.

Eickhorst, K., V. DeLeo, and J. Csaposs. 2007. Rue the herb: *Ruta graveolens*-associated phytophototoxicity. *Dermatitis* 18(1):52-55.

el Agraa, S.E., S.M. el Badwi, and S.E. Adam. 2002. Preliminary observations on experimental *Ruta graveolens* toxicosis in Nubian goats. *Trop. Anim. Health Prod.* 34(4):271-281.

Felter, H.W., and J.U. Lloyd. 1898. *King's American dispensatory.* 18th ed., 3rd rev. 2 vols. Cincinnati: Ohio Valley Co.

Furniss, D., and T. Adams. 2007. Herb of grace: An unusual cause of phytophotodermatitis mimicking burn injury. *J. Burn Care Res.* 28(5):767-769.

Gawkrodger, D.J., and J.A. Savin. 1983. Phytophotodermatitis due to common rue (*Ruta graveolens*). *Contact Dermat.* 9(3):224.

Gutierrez-Pajares, J.L., L. Zuniga, and J. Pino. 2003. *Ruta graveolens* aqueous extract retards mouse preimplantation embryo development. *Reprod. Toxicol.* 17(6):667-672.

Harat, Z.N., M.R. Sadeghi, H.R. Sadeghipour, M. Kamalinejad, and M.R. Eshraghian. 2008. Immobilization effect of *Ruta graveolens* L. on human sperm: A new hope for male contraception. *J. Ethnopharmacol.* 115(1):36-41.

Heskel, N.S., R.B. Amon, F.J. Storrs, and C.R. White, Jr. 1983. Phytophotodermatitis due to *Ruta graveolens*. *Contact Dermat.* 9(4):278-280.

Khouri, N.A., and Z. El-Akawi. 2005. Antiandrogenic activity of *Ruta graveolens* L. in male albino rats with emphasis on sexual and aggressive behavior. *Neuroendocrinol. Lett.* 26(6):823-829.

Madari, H., and R.S. Jacobs. 2004. An analysis of cytotoxic botanical formulations used in the traditional medicine of ancient Persia as abortifacients. *J. Nat. Prod.* 67(8):1204-1210.

McGuffin, M., C. Hobbs, R. Upton, and A. Goldberg. 1997. *Botanical safety handbook.* Boca Raton, FL: CRC Press.

Milesi, S., B. Massot, E. Gontier, F. Bourgaud, and A. Guckert. 2001. *Ruta graveolens* L.: A promising species for the production of furanocoumarins. *Plant Sci.* 161(1):189-199.

Morais, P., A. Mota, A.P. Cunha, L. Peralta, and F. Azevedo. 2008. Phytophotodermatitis due to homemade ointment for *Pediculosis capitis*. *Contact Dermat.* 59(6):373-374.

Paulini, H., U. Eilert, and O. Schimmer. 1987. Mutagenic compounds in an extract from rutae herba (*Ruta graveolens* L.). I. Mutagenicity is partially caused by furoquinoline alkaloids. *Mutagenesis* 2(4):271-273.

Paulini, H., and O. Schimmer. 1989. Mutagenicity testing of rutacridone epoxide and rutacridone, alkaloids in *Ruta graveolens* L., using the *Salmonella*/microsome assay. *Mutagenesis* 4(1):45-50.

Preethi, K.C., C.K. Nair, and R. Kuttan. 2008. Clastogenic potential of *Ruta graveolens* extract and a homeopathic preparation in mouse bone marrow cells. *Asian Pac. J. Cancer Prev.* 9(4):763-769.

Remington, J.P., and H.C. Wood. 1918. *The dispensatory of the United States of America.* 20th ed. Philadelphia: Lippincott.

Schempp, C.M., E. Schopf, and J.C. Simon. 1999. Bullous phototoxic contact dermatitis caused by *Ruta graveolens* L. (garden rue), Rutaceae. Case report and review of the literature. *Hautarzt* 50(6):432-434.

Schimmer, O., J. Kiefer, and H. Paulini. 1991. Inhibitory effects of furocoumarins in *Salmonella typhimurium* TA98 on the mutagenicity of dictamnine and rutacridone, promutagens from *Ruta graveolens* L. *Mutagenesis* 6(6):501-506.

Schimmer, O., and I. Kuhne. 1990. Mutagenic compounds in an extract from rutae herba (*Ruta graveolens* L.). II. UV-A mediated mutagenicity in the green alga *Chlamydomonas reinhardtii* by furoquinoline alkaloids and furocoumarins present in a commercial tincture from rutae herba. *Mutat. Res.* 243(1):57-62.

Seak, C.J., and C.C. Lin. 2007. *Ruta graveolens* intoxication. *Clin. Toxicol.* 45(2):173-175.

U.S.C. 2010. United States Code, Title 21, Part 321 (s)(6). Current as of January 7, 2011. Washington, DC: U.S. Government Printing Office.

Weiss, R.F., and V. Fintelmann. 2000. *Herbal medicine.* 2nd ed. New York: Thieme.

Wessner, D., H. Hofmann, and J. Ring. 1999. Phytophotodermatitis due to *Ruta graveolens* applied as protection against evil spells. *Contact Dermat.* 41(4):232.

Wood, H., and C. LaWall. 1918. *The dispensatory of the United States of America.* 21st ed. Philadelphia: Lippincott.

Zobel, A.M., and S.A. Brown. 1990. Dermatitis-inducing furanocoumarins on leaf surfaces of eight species of rutaceous and umbelliferous plants. *J. Chem. Ecol.* 16(3):693-700.

R

Salix spp.

Salicaceae

Salix alba **L.**
SCN: white willow
PN: *bai liu pi* (bark)

Salix daphnoides **Vill.**
SCN: violet willow
OCN: Daphne willow

Salix fragilis **L.**
SCN: brittle willow
OCN: crack willow

Salix pentandra **L.**
SCN: laurel willow
OCN: bay willow

Salix purpurea **L.**
SCN: purple willow
PN: *shui yang pi* (bark)
OCN: basket willow; purple osier
Part: bark

Quick Reference Summary

Safety Class: 1
Interaction Class: A

Contraindications

None known.

Other Precautions

Use in persons with sensitivity to aspirin or other salicylate-containing drugs is cautioned (ESCOP 2003; Mills and Bone 2005; Wichtl 2004).

Use in persons with G6PD deficiency (a hereditary disease also known as favism) is cautioned (Baker and Thomas 1987; Brinker 2001; Mills and Bone 2005).

Drug and Supplement Interactions

White willow has been hypothesized to increase the risk of bleeding in persons on anticoagulant medication. Human studies, however, have indicated that the antiplatelet effect of white willow is mild compared to that of aspirin (Higgs et al. 1987; Krivoy et al. 2001; Meier and Liebi 1990).

Notice

Salicylates (1.5–11.0%) (Bradley 1992; List and Hörhammer 1973; Wichtl 2004; Williamson 2003); *see* Appendix 1.

Tannins (8.0–20.0%) (List and Hörhammer 1973; Wichtl 2004); *see* Appendix 1.

Editors' Note

Other species of *Salix* are used in trade, with salicylate content ranging from 1.5 to 11% (Bradley 1992; Wichtl 2004).

Adverse Events

Preparations of white willow have been generally well tolerated in clinical trials, with only mild and transient adverse events reported (ESCOP 2003; Marz and Kemper 2002).

Pharmacological Considerations

Theoretical concerns have been reported for the use of white willow bark in children or adolescents with viral infections due to the possibility of Reye's syndrome, an acute inflammatory disease in children that has been linked with use of aspirin and other salicylate-containing medications during acute viral infections (Clauson et al. 2005; Mills and Bone 2005; Upton 1999). Data to support or refute such a concern are lacking, and the U.S. Food and Drug Administration has indicated that no association has been identified between nonaspirin salicylates and Reye's syndrome (FDA 2003).

Only mild inhibition of platelet aggregation has been observed, indicating that white willow cannot be used as an antiplatelet medication, and that white willow is not likely to present any additive effects on antiplatelet medications (Krivoy et al. 2001; Meier and Liebi 1990).

Pregnancy and Lactation

No information on the safety of willow in pregnancy or lactation was identified in the scientific or traditional literature. Although this review did not identify any concerns for use while pregnant or nursing, safety has not been conclusively established.

Review Details

I. Drug and Supplement Interactions

Clinical Trials of Drug or Supplement Interactions

No clinical trials of drug or supplement interactions were identified.

Case Reports of Suspected Drug or Supplement Interactions

No case reports of suspected drug or supplement interactions were identified.

Animal Trials of Drug or Supplement Interactions

No animal trials of drug or supplement interactions were identified.

II. Adverse Events

Adverse Events Reported in Clinical Trials

A review of clinical trials on three different white willow extracts indicated that mild adverse events were reported in 3.7% of persons in the white willow groups (ESCOP 2003). Another review of clinical trials of 120 to 240 mg daily of white willow indicated only mild adverse events associated with white willow use. Adverse events were reported by 3.8 to 35.8% in treatment groups and 2.8 to 35.2% of persons in the placebo groups (Marz and Kemper 2002). Adverse events in these two reviews included nausea, stomach ache, dizziness, fatigue, sweating, skin rash, and allergic reactions (ESCOP 2003; Marz and Kemper 2002).

Case Reports of Adverse Events

A massive intravascular hemolysis was reported in a woman with G6PD deficiency who had taken an herbal product containing *Salix caprea*, a willow species that contains salicin (Baker and Thomas 1987). In patients with G6PD deficiency, aspirin can cause hemolytic anemia (Colonna 1981).

An anaphylactic reaction to an herbal product containing white willow bark was reported in a woman with a history of allergy to acetylsalicylic acid (Boullata et al. 2003).

S

III. Pharmacology and Pharmacokinetics

Human Pharmacological Studies

A clinical trial of white willow bark extract (240 mg daily salicin) for 4 weeks resulted in a mild inhibition of platelet aggregation, as compared to control and 100 mg daily acetylsalicylate (Krivoy et al. 2001). A review of salicin-containing plants indicated that salicin does not elicit the same irreversible inhibition of platelet aggregation that acetylsalicylic acid in aspirin does (Meier and Liebi 1990).

No gastric injury was observed after administration of a single dose of 5 mmol/kg salicin to rats. In the same study, saligenin and sodium salicylate induced severe gastric lesions (Akao et al. 2002).

Animal Pharmacological Studies

Administration of salicylate (200 mg/kg) or aspirin (200 mg/kg) to rats demonstrated that salicylate had a much weaker effect than aspirin on thromboxane B_2 production in blood clotting, indicating that salicylate is not an effective inhibitor of platelet aggregation (Higgs et al. 1987).

In Vitro Pharmacological Studies

No relevant in vitro pharmacological studies were identified.

IV. Pregnancy and Lactation

No information on the safety of willow in pregnancy or lactation was identified.

Low concentrations of salicylates are excreted into breast milk (Bailey et al. 1982; Bennett 1988). Although the World Health Organization Working Group on Human Lactation has advised against the use of acetylsalicylic acid (a salicylate) during lactation (Bennet 1988), the American Academy of Pediatrics has recommended that "aspirin (salicylates)" should be given to nursing mothers with caution (Ressel 2002).

V. Toxicity Studies

Acute Toxicity

The LD_{50} of an orally administered white willow ethanolic extract in mice is 28 ml/kg (Leslie 1978).

Literature Cited

Akao, T., T. Yoshino, K. Kobashi, and M. Hattori. 2002. Evaluation of salicin as an antipyretic prodrug that does not cause gastric injury. *Planta Med.* 68(8):714-718.

Bailey D.N., R.T. Weibert, A.J. Naylor, and R.F Shaw 1982. A study of salicylate and caffeine excretion in the breast milk of two nursing mothers. *J. Anal. Toxicol.* 6(2):64-68.

Baker, S., and P. Thomas. 1987. Herbal medicine precipitating massive haemolysis. *Lancet* 1:1039-1040.

Bennett, P.N. 1988. *Drugs and human lactation*. New York: Elsevier.

Boullata, J.I., P.J. McDonnell, and C.D. Oliva. 2003. Anaphylactic reaction to a dietary supplement containing willow bark. *Ann. Pharmacother.* 37(6):832-835.

Bradley, P.R. 1992. *British herbal compendium: A handbook of scientific information on widely used plant drugs*. Bournemouth, UK: British Herbal Medicine Association.

Brinker, F. 2001. *Herb contraindications and drug interactions*. 3rd ed. Sandy, OR: Eclectic Medical Publications.

Clauson, K.A., M.L. Santamarina, C.M. Buettner, and J.S. Cauffield. 2005. Evaluation of presence of aspirin-related warnings with willow bark. *Ann. Pharmacother.* 39(7-8):1234-1237.

Colonna, P. 1981. Aspirin and glucose-6-phosphate dehydrogenase deficiency. *Br. Med. J.* 283(6300):1189.
ESCOP. 2003. *ESCOP monographs: The scientific foundation for herbal medicinal products.* 2nd ed. Exeter, UK: European Scientific Cooperative on Phytotherapy.
FDA. 2003. U.S. Food and Drug Administration. Labeling for oral and rectal over the counter drug products containing aspirin and nonaspirin salicylates; Reye's syndrome warning. Final rule. *Fed. Regist.* 68:18861-18869.
Higgs, G.A., J.A. Salmon, B. Henderson, and J.R. Vane. 1987. Pharmacokinetics of aspirin and salicylate in relation to inhibition of arachidonate cyclooxygenase and antiinflammatory activity. *Proc. Natl. Acad. Sci. U.S.A.* 84(5):1417-1420.
Krivoy, N., E. Pavlotzky, S. Chrubasik, E. Eisenberg, and G. Brook. 2001. Effect of salicis cortex extract on human platelet aggregation. *Planta Med.* 67(3):209-212.
Leslie, G. 1978. A pharmacometric evaluation of nine Bio-Strath herbal remedies. *Medica* 10:31-37.
List, P.H., and H. Hörhammer. 1973. *Hagers handbuch der pharmazeutischen praxis.* Berlin: Springer.
Marz, R.W., and F. Kemper. 2002. Willow bark extract—Effects and effectiveness. Status of current knowledge regarding pharmacology, toxicology and clinical aspects. *Wien Med. Wochenschr.* 152(15-16):354-359.
Meier, B., and M. Liebi. 1990. Salicinhaltige Pflanzliche Arzneimittel. *Z. Phytother.* 11:50-58.
Mills, S., and K. Bone. 2005. *The essential guide to herbal safety.* St. Louis: Elsevier.
Ressel, G. 2002. AAP updates statement for transfer of drugs and other chemicals into breast milk. *Am. Fam. Physician* 65(5):979-980.
Upton, R. 1999. *Willow bark, Salix spp.: Analytical, quality control, and therapeutic monograph.* Santa Cruz, CA: American Herbal Pharmacopoeia.
Wichtl, M. 2004. *Herbal drugs and phytopharmaceuticals: A handbook for practice on a scientific basis.* 3rd ed. Boca Raton, FL: CRC Press.
Williamson, E.M. 2003. *Potter's herbal cyclopedia.* Saffron Walden, Essex: C.W. Daniel Co.

Salvia miltiorrhiza Bunge — Lamiaceae

SCN: Chinese salvia
PN: *dan shen* (root)
OCN: Chinese sage; red-root sage
Part: root

Quick Reference Summary

Safety Class: 2b
Interaction Class: C

Contraindications

Not for use in pregnancy except under the supervision of a qualified healthcare practitioner (Bensky et al. 2004).

Other Precautions

Chinese salvia should be used with caution in any situation associated with bleeding, including menstruation, nosebleeds, or blood in the urine, or if blood is expectorated during coughing (Chen and Chen 2004).

Drug and Supplement Interactions

Chinese salvia has been shown to slow the metabolism of warfarin, increasing plasma levels of the drug (Chan et al. 1995; Izzat et al. 1998; Lo et al. 1992).

A review of the literature on the use of Chinese salvia suggested that this herb may act synergistically with digitalis, cardiac glycosides, and hypotensive drugs, and that the dosage of these drugs may need to be modified if used concurrently with Chinese salvia (Wu et al. 2008).

Adverse Events and Side Effects

A review of Chinese salvia studies and case reports indicated that adverse events are rare relative to the extent of use of this herb. Adverse events reported in association with Chinese salvia and fufang danshen (a combination of Chinese salvia, *Panax notoginseng*, and *Cinnamomum camphora*) were anaphylactic reactions, abdominal discomfort, loss of appetite, pruritus, low blood pressure, dizziness, and headache (Wu et al. 2008).

A small percentage of persons taking Chinese salvia are reported to experience dry mouth, dizziness, nausea, weakness, shortness of breath, numbing or coldness of the hands, anxiety, or tachycardia (Bensky et al. 2004).

Pharmacological Considerations

Case reports and animal studies have indicated that concomitant use of Chinese salvia and warfarin slows the metabolism and increases blood levels of warfarin (Chan 1995; Chan et al. 1995; Izzat et al. 1998; Lo et al. 1992; Yu et al. 1997).

A human study indicated no interaction between Chinese salvia and theophylline (Qiu et al. 2008). One animal study indicated that Chinese salvia can decrease plasma levels of diazepam (Qiao et al. 2003).

Inhibition of platelet aggregation and a decrease in blood viscosity have been reported in animal studies and in vitro (Hou et al. 2007; Liu et al. 2002; Qiu et al. 2008; Wang et al. 1982).

S

Pregnancy and Lactation

A reference text on traditional Chinese medicine indicates that Chinese salvia is contraindicated in pregnancy (Bensky et al. 2004).

No information on the safety of Chinese salvia during lactation was identified in the scientific or traditional literature. While this review did not identify any concerns for use while nursing, safety has not been conclusively established.

Review Details

I. Drug and Supplement Interactions

Clinical Trials of Drug or Supplement Interactions

No effects on theophylline metabolism were observed in healthy volunteers before and after administration of 12 tablets (1 g each) of Chinese salvia for 14 days (Qiu et al. 2008).

Case Reports of Suspected Drug or Supplement Interactions

Several published case reports have suggested an interaction between Chinese salvia and warfarin. In all cases, the patients were on long-term warfarin therapy and were taking other drugs and/or herbs. They all experienced increased INR (a standardized scale used to report the results of blood coagulation tests; increased INR indicates slowed blood clotting and increased risk of bleeding) temporally related to the consumption of Chinese salvia. None of the cases report the dose of Chinese salvia taken by the patients, and some suggest that patients were using other herbs, but no details were provided (Chan 2001).

Case one was a 66-year-old man also taking digoxin and propanolol (Tam et al. 1995). Five days prior to admission due to melena (dark-colored feces) and chest pain, and 2 days prior to treatment, the man had self-treated with a topically applied medicated oil that contained 15% methyl salicylate. Methyl salicylate has been shown to augment the anticoagulant effect of warfarin (Chan 1995, 1998; Chow et al. 1989; Joss and LeBlond 2000; Yip et al. 1990).

Case two was a 48-year-old woman also taking furosemide, digoxin, mefenamic acid, and theophylline (Yu et al. 1997). She had consumed Chinese salvia and "other herbs" every other day for approximately 1 month prior to elevated INR.

Case three was a 62-year-old man also taking digoxin, captopril, and furosemide (Izzat et al. 1998). The man had consumed a decoction of Chinese salvia daily for 2 weeks prior to hospital admission for a massive pleural effusion and large pericardial effusion.

Animal Trials of Drug or Supplement Interactions

A significant increase in the plasma levels of (*R*)- and (*S*)-warfarin was observed in rats stabilized on warfarin (2 mg/kg daily) and then intraperitoneally administered 5 g/kg of a Chinese salvia extract twice daily for 3 days. A corresponding significant increase in prothrombin time was observed (Chan et al. 1995).

In rats administered warfarin (2 mg/kg) with or without pretreatment with intraperitoneal administration of 5 g/kg of an aqueous extract of Chinese salvia twice daily for 3 days, the absorption rate, volume of distribution, and elimination half-life of warfarin were significantly decreased, while the maximum concentration was significantly increased in the group treated with Chinese salvia. No significant changes in prothrombin time were observed (Lo et al. 1992).

An increase in clearance and decrease in plasma levels of orally administered diazepam (15 mg/kg single dose) were observed in rats orally pretreated with a Chinese salvia extract at a dose of 100 mg/kg daily for 15 days (Qiao et al. 2003).

Cyclosporine-induced nephropathy was reduced by administration of Chinese salvia extract to rats (Peng et al. 2006; Qiao et al. 2001). Extracts of Chinese salvia attenuated adriamycin-induced cardiac and hepatic toxicity in healthy rats (You et al. 2007) and vanadium-induced gastrointestinal stress and metal accumulation in diabetic rats (Zhang et al. 2008).

II. Adverse Events

Adverse Events Reported in Clinical Trials

A meta-analysis of clinical trials on Chinese salvia and fufang danshen (a standard formula containing Chinese salvia, *Panax notoginseng*, and *Cinnamomum camphora*) indicated that "literally thousands" of studies on Chinese salvia for the treatment of heart attack had been published, and adverse events were rarely reported. The route of administration in the studies reviewed included intravenous and oral. Reviewing the literature, adverse events reported in association with Chinese salvia and fufang danshen were anaphylactic reactions, abdominal discomfort, loss of appetite, pruritus, low blood pressure, dizziness, headache, and, with excessive use, an increased risk of bleeding. Bleeding events were bleeding in the skin or mucous membranes or excessive menstrual bleeding. An increase in serum aminotransferase has also been reported. Details on doses and preparations associated with the specific adverse events were not included with the review (Wu et al. 2008).

No differences in adverse events between treatment and control groups were reported in a meta-analysis of Chinese salvia for acute ischemic stroke. Compound dan shen injection, a formula derived from an aqueous extract of Chinese salvia, was the primary intervention used, usually administered intravenously at doses of 20 or 30 ml, and sometimes administered orally (Wu et al. 2007).

S

Case Reports of Adverse Events

A small percentage of persons taking Chinese salvia may experience dry mouth, dizziness, nausea, weakness, shortness of breath, numbing or coldness of the hands, anxiety, or tachycardia. Allergic reactions to Chinese salvia have been reported. One case of liver damage and two cases of shock have been reported after injection of preparations made from Chinese salvia (details on dose and product were not reported) (Bensky et al. 2004).

III. Pharmacology and Pharmacokinetics

Human Pharmacological Studies

Extracts of Chinese salvia have been administered intravenously for the treatment of various neurological and cardiovascular conditions. No mention of adverse events was made in a report on the intravenous therapeutic use of Chinese salvia in 100 people (ages 5 to 60 years) with nerve deafness. The standard dose used was 20 to 30 ml in solution four times daily for two to four treatment courses of 2 weeks each, separated by 3-day intervals (Hu et al. 1992).

In a dosing study of Chinese salvia extract injections in patients with hepatitis B, doses of 8, 16, or 24 ml were administered for 60 days. Chinese salvia treatment was associated with a decrease in liver enzyme levels, including alanine aminotransferase and total bilirubin. All doses were apparently well tolerated, and no adverse events were reported in the English language abstract of this study (Ye et al. 2005).

Animal Pharmacological Studies

In aging guinea pigs fed a diet containing 75, 100, or 150 mg/kg daily of a water-soluble extract of Chinese salvia for 28 days, blood biochemical parameters of the aging guinea pigs remained unaffected, except that the fibrinogen levels of the group fed the high dose decreased. The medium dose significantly reduced erythrocyte membrane malondialdehyde (MDA) levels, although no improvement in erythrocyte aggregation, blood viscosity, and blood viscoelasticity could be observed. At the high dose, a significant decrease in whole blood viscosity was observed at high, medium, and low shear rates (Hou et al. 2007). A review of studies published in the Chinese literature indicated that intravenous administration of Chinese salvia extracts has been associated with antiplatelet activity in healthy rabbits, rats, and mice (doses not specified in English translations) (Chen and Chen 2004).

Inhibition of platelet aggregation was observed in rabbits intragastrically administered an extract of Chinese salvia (dose and product not specified in English language abstract) (Liu et al. 2002). A combination of Chinese salvia and *Panax notoginseng* increased inhibition of platelet aggregation as compared to Chinese salvia alone (dose and product not specified in English language abstract) (Liu et al. 2002). Bioavailability of orally administered Chinese salvia (5 g/kg) was increased by coadministration of an extract of fragrant rosewood (*Dalbergia odorifera*) (2.5 g/kg) (Zheng et al. 2007).

An increase in cytochrome P450 activity was observed in the livers of rats that had been orally administered 20 or 100 mg/kg of an aqueous extract of Chinese salvia daily for 15 days. The cytochrome activity was primarily induction of the drug-metabolizing isoenzyme CYP3A (Jinping et al. 2003). Induction of the drug-metabolizing isoenzymes CYP1A, CYP2C, and CYP3A was observed in mice orally administered an ethyl acetate extract of Chinese salvia, although no effects were observed in mice administered a water extract of Chinese salvia (Kuo et al. 2006).

In rats with bile duct ligation, administration of 0.4 g daily of an aqueous extract of Chinese salvia for 4 weeks significantly reduced histological grades of fibrosis and ameliorated the portal hypertensive state as compared with control. Treatment had no effect on plasma biochemical profiles of either bile duct ligated or normal rats (Huang et al. 2001).

The compound tanshinone IIA crossed the blood-brain barrier at a greater rate than that for sucrose, and the brain penetration was increased in the presence of a P-glycoprotein or the multidrug-resistance-associated protein (Mrp1/2) inhibitor (Chen et al. 2007).

Inhibition of calcium channels has been suggested as the mechanism of action for the vasorelaxant activity of Chinese salvia (Lam et al. 2005, 2006, 2007). Inhibition of cardiac aldosterone activity was observed in rats treated with Chinese salvia (dose and preparation not specified in English language abstract) (Han et al. 2002). Studies in rats and dogs have indicated a therapeutic effect of Chinese salvia in animal models of myocardial infarction (Liu and Lu 1999; Sun et al. 2005).

In rats administered Chinese salvia, differing effects on digoxin assays were observed, with one assay (microparticle enzyme immunoassay) showing falsely lower digoxin concentrations, a second showing falsely elevated levels (fluorescence polarization immunoassay), and a third showing no change (chemiluminescence assay) (Dasgupta et al. 2002).

In Vitro Pharmacological Studies

Dose-dependent inhibition of platelet aggregation by ADP or epinephrine was observed in platelets treated with an extract of Chinese salvia prepared for injectable use (Wang et al. 1982).

An ethyl acetate extract of Chinese salvia exhibited modest digoxin-like immunoreactivity with the fluorescence polarization immunoassay, but no apparent digoxin activity was exhibited in Roche and Beckman brand digoxin immunoassays (Chow et al. 2003).

Inhibition of estrogen receptor-positive breast cancer cells (MCF-7) was observed in cells treated with an ethanolic extract of Chinese salvia (Hsieh and Wu 2006).

IV. Pregnancy and Lactation

A reference text on traditional Chinese medicine indicates that Chinese salvia is contraindicated for use during pregnancy (Bensky et al. 2004).

No information on the safety of Chinese salvia in lactation was identified.

V. Toxicity Studies

Acute Toxicity

The LD_{50} of orally administered water-soluble extract of Chinese salvia in mice is 25.8 g/kg (TTPG 1998). No fatalities were reported in mice intraperitoneally administered 43 g/kg of a decoction of Chinese salvia (Chen and Chen 2004).

Short-Term Toxicity

No adverse effects were observed in mice intraperitoneally administered 2.4 g/kg of a decoction of Chinese salvia daily for 14 days (Chen and Chen 2004).

Subchronic Toxicity

No toxic effects were observed in rats orally administered 2.5 g/kg of a water-soluble extract of Chinese salvia daily for 90 days (TTPG 1998).

Literature Cited

Bensky, D., S. Clavey, and E. Stöger. 2004. *Chinese herbal medicine: Materia medica*. 3rd ed. Seattle: Eastland Press.

Chan, K., A.C. Lo, J.H. Yeung, and K.S. Woo. 1995. The effects of danshen (*Salvia miltiorrhiza*) on warfarin pharmacodynamics and pharmacokinetics of warfarin enantiomers in rats. *J. Pharm. Pharmacol.* 47(5):402-406.

Chan, T.Y. 1995. Adverse interactions between warfarin and nonsteroidal antiinflammatory drugs: Mechanisms, clinical significance, and avoidance. *Ann. Pharmacother.* 29(12):1274-1283.

Chan, T.Y. 1998. Drug interactions as a cause of overanticoagulation and bleedings in Chinese patients receiving warfarin. *Int. J. Clin. Pharmacol. Ther.* 36(7):403-405.

Chan, T.Y. 2001. Interaction between warfarin and danshen (*Salvia miltiorrhiza*). *Ann. Pharmacother.* 35(4):501-504.

Chen, J.K., and T.T. Chen. 2004. *Chinese medical herbology and pharmacology*. City of Industry, CA: Art of Medicine Press.

Chen, X., Z.W. Zhou, C.C. Xue, X.X. Li, and S.F. Zhou. 2007. Role of P-glycoprotein in restricting the brain penetration of tanshinone IIA, a major active constituent from the root of *Salvia miltiorrhiza* Bunge, across the blood-brain barrier. *Xenobiotica* 37(6):635-678.

Chow, L., M. Johnson, A. Wells, and A. Dasgupta. 2003. Effect of the traditional Chinese medicines chan su, lu-shen-wan, dan shen, and Asian ginseng on serum digoxin measurement by Tina-quant (Roche) and Synchron LX system (Beckman) digoxin immunoassays. *J. Clin. Lab. Anal.* 17(1):22-27.

Chow, W.H., K.L. Cheung, H.M. Ling, and T. See. 1989. Potentiation of warfarin anticoagulation by topical methyl-salicylate ointment. *J. Roy. Soc. Med.* 82(8):501-502.

Dasgupta, A., J.K. Actor, M. Olsen, A. Wells, and P. Datta. 2002. *In vivo* digoxin-like immunoreactivity in mice and interference of Chinese medicine danshen in serum digoxin measurement: Elimination of interference by using a chemiluminescent assay. *Clin. Chim. Acta* 317(1-2):231-234.

Han, S., Z. Zheng, and D. Ren. 2002. [Effect of *Salvia miltiorrhiza* on left ventricular hypertrophy and cardiac aldosterone in spontaneously hypertensive rats.] *J. Huazhong Univ. Sci. Technol. Med. Sci.* 22(4):302-304.

Hou, W.C., H.S. Tsay, H.J. Liang, et al. 2007. Improving abnormal hemorheological parameters in aging guinea pigs by water-soluble extracts of *Salvia miltiorrhiza* Bunge. *J. Ethnopharmacol.* 111(3):483-489.

Hsieh, T.C., and J.M. Wu. 2006. Differential control of growth, cell cycle progression, and gene expression in human estrogen receptor positive MCF-7 breast cancer cells by extracts derived from polysaccharopeptide I'm-Yunity and danshen and their combination. *Int. J. Oncol.* 29(5):1215-1222.

Hu, Y., Y. Ge, Y. Zhang, et al. 1992. Treatment of 100 cases of nerve deafness with injectio *radix salviae miltiorrhizae*. *J. Trad. Chin. Med.* 12(4):256-258.

Huang, Y.T., T.Y. Lee, H.C. Lin, et al. 2001. Hemodynamic effects of *Salvia miltiorrhiza* on cirrhotic rats. *Can. J. Physiol. Pharmacol.* 79(7):566-572.

Izzat, M.B., A.P.C. Yim, and M.H. El-Zufari. 1998. A taste of Chinese medicine! *Ann. Thorac. Surg.* 66(3):941-942.

Jinping, Q., H. Peiling, L. Yawei, and Z. Abliz. 2003. Effects of the aqueous extract from *Salvia miltiorrhiza* Bge on the pharmacokinetics of diazepam and on liver microsomal cytochrome P450 enzyme activity in rats. *J. Pharm. Pharmacol.* 55(8):1163-1167.

Joss, J.D., and R.F. LeBlond. 2000. Potentiation of warfarin anticoagulation associated with topical methyl salicylate. *Ann. Pharmacother.* 34(6):729-733.

Kuo, Y.H., Y.L. Lin, M.J. Don, R.M. Chen, and Y.F. Ueng. 2006. Induction of cytochrome P450-dependent monooxygenase by extracts of the medicinal herb *Salvia miltiorrhiza*. *J. Pharm. Pharmacol.* 58(4):521-527.

Lam, F.F., J.H. Yeung, K.M. Chan, and P.M. Or. 2007. Relaxant effects of danshen aqueous extract and its constituent danshensu on rat coronary artery are mediated by inhibition of calcium channels. *Vasc. Pharmacol.* 46(4):271-277.

Lam, F.F., J.H. Yeung, and J.H. Cheung. 2005. Mechanisms of the dilator action of danshen (*Salvia miltiorrhiza*) on rat isolated femoral artery. *J. Cardiovasc. Pharmacol.* 46(3):361-368.

Lam, F.F., J.H. Yeung, J.H. Cheung, and P.M. Or. 2006. Pharmacological evidence for calcium channel inhibition by danshen (*Salvia miltiorrhiza*) on rat isolated femoral artery. *J. Cardiovasc. Pharmacol.* 47(1):139-145.

Liu, Q., and Z. Lu. 1999. Effect of *Salvia miltiorrhiza* on coronary collateral circulation in dogs with experimental acute myocardial infarction. *J. Tongji Med. Univ.* 19(1):40-41, 69.

Liu, T., C.L. Qin, Y. Zhang, et al. 2002. [Effect of dan-shen, san-qi of different proportion on platelet aggregation and adhesion in normal rabbits.] *Zhongguo Zhong Yao Za Zhi* 27(8):609-611.

S

Lo, A.C., K. Chan, J.H. Yeung, and K.S. Woo. 1992. The effects of danshen (*Salvia miltiorrhiza*) on pharmacokinetics and pharmacodynamics of warfarin in rats. *Eur. J. Drug Metab. Pharmacokinet.* 17(4):257-262.

Peng, B., M. Li, T. Niu, and S. He. 2006. Impact of Lotensin and Salviae on the changes of TGF-beta1 and its receptors in a rat model of chronic cyclosporine-induced nephropathy. *Transplant. Proc.* 38(7):2183-2186.

Qiao, B.P., X.D. Tang, and Q. Ruan. 2001. [Experimental study of compound salvia injection in preventing and treating chronic nephrotoxicity induced by cyclosporine A in rats.] *Zhongguo Zhong Xi Yi Jie He Za Zhi* 21(8):611-614.

Qiao, J., P. Hou, Y. Li, and A. Zeper. 2003. Effects of the aqueous extract from *Salvia miltiorrhiza* Bge on the pharmacokinetics of diazepam and on liver microsomal cytochrome P450 enzyme activity in rats. *J. Pharm. Pharmacol.* 55(8):1163-1167.

Qiu, F., G. Wang, Y. Zhao, et al. 2008. Effect of danshen extract on pharmacokinetics of theophylline in healthy volunteers. *Br. J. Clin. Pharmacol.* 65(2):270-274.

Sun, J., S.H. Huang, B.K. Tan, et al. 2005. Effects of purified herbal extract of *Salvia miltiorrhiza* on ischemic rat myocardium after acute myocardial infarction. *Life Sci.* 76(24):2849-2860.

Tam, L.S., T.Y.K. Chan, W.K. Leung, and J. Critchley. 1995. Warfarin interactions with Chinese traditional medicines: Danshen and methyl salicylate medicated oil. *Intern. Med. J.* 25(3):258.

TTPG. 1998. Tianjin Talisco Pharmaceutical Group Co. Approval of compound danshen dripping pill by FDA through pre-IND for clinical trials. *Cited in* Zhou, L., Z. Zuo, and M.S. Chow. 2005. Danshen: An overview of its chemistry, pharmacology, pharmacokinetics, and clinical use. *J. Clin. Pharmacol.* 45(12):1345-1359.

Wang, W.C. 1993. [Effect of *Salvia miltiorrhizae* in the treatment of 36 infantile acute toxic myocarditis.] *Zhongguo Zhong Xi Yi Jie He Za Zhi* 13(11):665-666, 645.

Wang, Z., J.M. Roberts, P.G. Grant, R.W. Colman, and A.D. Schreiber. 1982. The effect of a medicinal Chinese herb on platelet function. *Thromb. Haemost.* 48(3):301-306.

Wu, B., M. Liu, and S. Zhang. 2007. Dan shen agents for acute ischaemic stroke. *Cochrane Database Syst. Rev.* 2:CD004295.

Wu, T., J. Ni, and J. Wu. 2008. Danshen (Chinese medicinal herb) preparations for acute myocardial infarction. *Cochrane Database Syst. Rev.* 2:CD004465.

Ye, F., Y. Liu, G. Qiu, Y. Zhao, and M. Liu. 2005. [Clinical study on treatment of cirrhosis by different dosages of salvia injection.] *Zhong Yao Cai* 28(9):850-854.

Yip, A.S., W.H. Chow, Y.T. Tai, and K.L. Cheung. 1990. Adverse effect of topical methylsalicylate ointment on warfarin anticoagulation: An unrecognized potential hazard. *Postgrad. Med. J.* 66(775):367-369.

You, J.S., T.L. Pan, and Y.S. Lee. 2007. Protective effects of danshen (*Salvia miltiorrhiza*) on adriamycin-induced cardiac and hepatic toxicity in rats. *Phytother. Res.* 21(12):1146-1152.

Yu, C.M., J.C.N. Chan, and J.E. Sanderson. 1997. Chinese herbs and warfarin potentiation by 'danshen.' *J. Intern. Med.* 241:337-339.

Zhang, L., Y. Zhang, Q. Xia, et al. 2008. Effective control of blood glucose status and toxicity in streptozotocin-induced diabetic rats by orally administration of vanadate in an herbal decoction. *Food Chem. Toxicol.* 46:2996-3002.

Zheng, X., X. Zhao, S. Wang, et al. 2007. Co-administration of *Dalbergia odorifera* increased bioavailability of *Salvia miltiorrhizae* in rabbits. *Am. J. Chin. Med.* 35(5):831-840.

Salvia officinalis L. — Lamiaceae

SCN: sage
OCN: Dalmatian sage; garden sage; common sage

Part: leaf

Quick Reference Summary

Safety Class: 2b, 2d
Interaction Class: A

Contraindications

Not for use in pregnancy except under the supervision of a qualified healthcare practitioner (Wichtl 2004).

Do not exceed recommended dose (Mills and Bone 2005).

Other Precautions

None known.

Drug and Supplement Interactions

None known.

Standard Dose

The standard dose is 3 to 12 g of dried leaf as a tea (Mills and Bone 2005; Wichtl 2004).

Notice

Thujone (0.5–1.5%) (Farrell 1990; Länger et al. 1996; Leung and Foster 1996); *see* Appendix 1. The essential oil contains up to 60% α-thujone and up to 25% β-thujone (Länger et al. 1996).

Editors' Note

The classifications and concerns for this herb are based on relatively higher doses used for therapeutic purposes in contrast to lower amounts generally used in cooking, and have not been associated with its use as a spice.

Adverse Events and Side Effects

Overdose of sage essential oil may cause vomiting, excessive salivation, or seizures (Burkhard et al. 1999; Millet et al. 1981).

Pharmacological Considerations

None known.

Pregnancy and Lactation

Information on the use of sage in pregnancy and lactation is limited. The German Commission E contraindicates the use of the alcohol extract or essential oil of sage in pregnancy (Wichtl 2004). Sage is traditionally used to stop lactation and should not be used by women who are lactating and wish to continue to do so (Leung and Foster 1996; Mills and Bone 2005).

Review Details

I. Drug and Supplement Interactions

Clinical Trials of Drug or Supplement Interactions

No clinical trials of drug or supplement interactions were identified.

Case Reports of Suspected Drug or Supplement Interactions

No case reports of suspected drug or supplement interactions were identified.

Animal Trials of Drug or Supplement Interactions

No animal trials of drug or supplement interactions were identified.

II. Adverse Events

Case Reports of Adverse Events

A 54-year-old woman with no history of epilepsy experienced dystonic movements of the tongue, followed by a typical generalized tonic-clonic seizure and loss of consciousness, after ingesting a "large" (unmeasured accidental overdose) amount of sage essential oil. The woman had previously taken smaller amounts (reported as "a mouthful") of sage oil weekly for the past several years. With each intake she had experienced gastric burning, faintness, profuse sweating, dizziness, and rapid breathing, all of which receded after 10 minutes (Burkhard et al. 1999).

A 53-year-old man developed a generalized tonic-clonic seizure and loss of consciousness, followed by muscle aches, after ingesting 12 drops of sage essential oil. A follow-up 2 years later indicated no recurrence of seizures (Burkhard et al. 1999).

Overdose of sage essential oil may result in episodes of vomiting, excessive salivation, tonic and/or clonic convulsions, and cyanosis (blue discoloration of the skin). The episodes may be separated by periods of atonicity. The dose at which such effects may occur was not specified (Millet et al. 1981).

III. Pharmacology and Pharmacokinetics

Human Pharmacological Studies

No relevant human pharmacological studies were identified.

Animal Pharmacological Studies

A methanolic extract of sage decreased serum glucose levels in diabetic but not in healthy rats after intraperitoneal administration of doses of 100 to 500 mg/kg. No effects on serum glucose levels were observed in rats administered 0.042 to 0.4 ml/kg of sage essential oil (Eidi et al. 2005).

In Vitro Pharmacological Studies

A sage-based product induced the drug-metabolizing isoenzyme CYP3A4 but had no effect on CYP1A2. Induction of the drug transporter protein MDR1 was also observed (Brandin et al. 2007). An ethanolic extract of sage induced the drug-metabolizing isoenzymes CYP1A2, CYP2D6, and CYP3A4 (Hellum et al. 2007).

The essential oil of sage showed mild inhibition of platelet aggregation induced by arachidonic acid but not by thrombin, ADP, or the thromboxane A_2 agonist U46619 (Tognolini et al. 2006).

An extract of sage inhibited α-glucosidase activity (Kwon et al. 2006).

IV. Pregnancy and Lactation

The German Commission E contraindicates the use of the alcohol extract or essential oil of sage in pregnancy (Wichtl 2004).

In an anti-implantation study, no adverse effects were observed in rats orally administered 250 mg/kg aqueous or ethanolic extracts of sage on postcoital days 1 to 10 (Kamboj and Dhawan 1982).

Sage is traditionally used to stop lactation and should not be used by women who are lactating and wish to continue to do so (Leung and Foster 1996; Mills and Bone 2005).

V. Toxicity Studies

Acute Toxicity

The LD_{50} of an orally administered sage essential oil in rats is 2600 mg/kg (Opdyke 1974). Convulsions in rats intraperitoneally administered sage essential oil were observed at doses between 300 and 500 mg/kg. Doses over 3200 mg/kg were lethal to the rats (Millet et al. 1981).

The LD_{50} of an intraperitoneally administered methanolic extract of sage in rats was 4000 mg/kg (Eidi et al. 2005).

Hepatotoxicity

An aqueous extract of sage, provided as the sole source of drinking water, increased liver damage from the liver toxin carbon tetrachloride (CCl_4) in rats. Sage ingestion increased GST activity, glutathione peroxidase, and the CYP2E1 protein (CCl_4 toxicity results from bioactivation mainly by CYP2E1) (Lima et al. 2007a).

Hepatoprotective activity of aqueous and methanolic extracts of sage was observed in human liver cells treated

S

with sage extract and/or *tert*-butyl hydroperoxide (*t*-BHP). The methanolic extract had a higher content of phenolic compounds than the water extract and conferred better protection in this in vitro model of oxidative stress. Both extracts significantly prevented *t*-BHP-induced lipid peroxidation and GSH depletion but not DNA damage as assessed by the comet assay (Lima et al. 2007b).

A protective effect against oxidative liver damage produced by azathioprine was observed in rats treated with an aqueous extract of sage (Amin and Hamza 2005).

Genotoxicity

A terpenoid fraction of sage exhibited a dose-dependent protective effect against mutagenicity induced by mitomycin C in rats (Vujosevic and Blagojevic 2004).

Cytotoxicity

Sage used at a concentration of 100 µl/kg was cytotoxic to mammalian cells (Vujosevic and Blagojevic 2004). No significant cytotoxicity of sage essential oil was observed in human melanoma, renal, and adenocarcinoma cell lines, although some inhibition of tumor growth was observed (Loizzo et al. 2007).

Literature Cited

Amin, A., and A.A. Hamza. 2005. Hepatoprotective effects of *Hibiscus, Rosmarinus* and *Salvia* on azathioprine-induced toxicity in rats. *Life Sci.* 77(3):266-278.

Brandin, H., E. Viitanen, O. Myrberg, and A.K. Arvidsson. 2007. Effects of herbal medicinal products and food supplements on induction of CYP1A2, CYP3A4 and MDR1 in the human colon carcinoma cell line LS180. *Phytother. Res.* 21(3):239-244.

Burkhard, P.R., K. Burkhardt, C.A. Haenggeli, and T. Landis. 1999. Plant-induced seizures: Reappearance of an old problem. *J. Neurol.* 246(8):667-670.

Eidi, M., A. Eidi, and H. Zamanizadeh. 2005. Effect of *Salvia officinalis* L. leaves on serum glucose and insulin in healthy and streptozotocin-induced diabetic rats. *J. Ethnopharmacol.* 100(3):310-313.

Farrell, K. 1990. *Spices, condiments and seasonings*. 2nd ed. New York: Van Nostrand Reinhold.

Hellum, B.H., Z. Hu, and O.G. Nilsen. 2007. The induction of CYP1A2, CYP2D6 and CYP3A4 by six trade herbal products in cultured primary human hepatocytes. *Basic Clin. Pharmacol. Toxicol.* 100(1):23-30.

Kamboj, V.P., and B.N. Dhawan. 1982. Research on plants for fertility regulation in India. *J. Ethnopharmacol.* 6(2):191-226.

Kwon, Y.I., D.A. Vattem, and K. Shetty. 2006. Evaluation of clonal herbs of Lamiaceae species for management of diabetes and hypertension. *Asia Pac. J. Clin. Nutr.* 15(1):107-118.

Länger, R., C. Mechtler, and J. Jurenitsch. 1996. Composition of the essential oils of commercial samples of *Salvia officinalis* L. and *S. fruticosa* Miller: A comparison of oils obtained by extraction and steam distillation. *Phytochem. Anal.* 7(6):289-293.

Leung, A.Y., and S. Foster. 1996. *Encyclopedia of common natural ingredients used in food, drugs, and cosmetics*. 2nd ed. New York: Wiley.

Lima, C.F., M. Fernandes-Ferreira, and C. Pereira-Wilson. 2007a. Drinking of *Salvia officinalis* tea increases CCl_4-induced hepatotoxicity in mice. *Food Chem. Toxicol.* 45(3):456-464.

Lima, C.F., P.C. Valentao, P.B. Andrade, et al. 2007b. Water and methanolic extracts of *Salvia officinalis* protect HepG2 cells from *t*-BHP induced oxidative damage. *Chem. Biol. Interact.* 167(2):107-115.

Loizzo, M.R., R. Tundis, F. Menichini, et al. 2007. Cytotoxic activity of essential oils from Labiatae and Lauraceae families against *in vitro* human tumor models. *Anticancer Res.* 27(5A):3293-3299.

Millet, Y., J. Jouglard, M.D. Steinmetz, et al. 1981. Toxicity of some essential plant oils—Clinical and experimental study. *Clin. Toxicol.* 18(12):1485-1498.

Mills, S., and K. Bone. 2005. *The essential guide to herbal safety*. St. Louis: Elsevier.

Opdyke, D. 1974. Sage oil Dalmatian. *Food Cosmet. Toxicol.* 12:987-988.

Tognolini, M., E. Barocelli, V. Ballabeni, et al. 2006. Comparative screening of plant essential oils: Phenylpropanoid moiety as basic core for antiplatelet activity. *Life Sci.* 78(13):1419-1432.

Vujosevic, M., and J. Blagojevic. 2004. Antimutagenic effects of extracts from sage (*Salvia officinalis*) in mammalian system *in vivo*. *Acta Vet. Hung.* 52(4):439-443.

Wichtl, M. 2004. *Herbal drugs and phytopharmaceuticals: A handbook for practice on a scientific basis*. 3rd ed. Boca Raton, FL: CRC Press.

S

Salvia sclarea L. — Lamiaceae

SCN: clary sage
OCN: clary; muscatel sage

Part: herb

Quick Reference Summary

Safety Class: 1
Interaction Class: A

Contraindications

None known.

Other Precautions

None known.

Drug and Supplement Interactions

None known.

Adverse Events and Side Effects

None known.

Pharmacological Considerations

None known.

Pregnancy and Lactation

No information on the safety of clary sage in pregnancy or lactation was identified in the scientific or traditional literature. Although this review did not identify any concerns for use while pregnant or nursing, safety has not been conclusively established.

Review Details

I. Drug and Supplement Interactions

Clinical Trials of Drug or Supplement Interactions

No clinical trials of drug or supplement interactions were identified.

Case Reports of Suspected Drug or Supplement Interactions

No case reports of suspected drug or supplement interactions were identified.

Animal Trials of Drug or Supplement Interactions

No animal trials of drug or supplement interactions were identified.

II. Adverse Events

Case Reports of Adverse Events

No case reports of adverse events were identified.

III. Pharmacology and Pharmacokinetics

Human Pharmacological Studies

In 24- and 48-hour closed patch test studies, no positive reactions were observed after use of full strength clary sage essential oil alone or in 8% petrolatum. Further, there were no sensitization reactions observed in volunteers tested with 8% clary oil in petrolatum (Opdyke 1979).

Animal Pharmacological Studies

No relevant animal pharmacological studies were identified.

In Vitro Pharmacological Studies

The essential oil of clary sage showed mild inhibition of platelet aggregation induced by arachidonic acid but not by thrombin, ADP, or the thromboxane A_2 agonist U46619 (Tognolini et al. 2006).

IV. Pregnancy and Lactation

No information on the safety of clary sage in pregnancy or lactation was identified.

V. Toxicity Studies

Acute Toxicity

The LD_{50} of orally administered clary sage essential oil in rats has been reported as 5.0 and 6.3 g/kg. The acute dermal toxicity in rabbits could not be determined at doses up to 2 g/kg (Opdyke 1979).

Genotoxicity

No significant genotoxic activity of clary sage was observed in the *Bacillus subtilis rec* assay or the *Salmonella* microsome reversion assay (Zani et al. 1991).

S

Literature Cited

Opdyke, D.L.J. 1979. *Monographs on fragrance raw materials.* New York: Pergamon Press.

Tognolini, M., E. Barocelli, V. Ballabeni, et al. 2006. Comparative screening of plant essential oils: Phenylpropanoid moiety as basic core for antiplatelet activity. *Life Sci.* 78(13):1419-1432.

Zani, F., G. Massimo, S. Benvenuti, et al. 1991. Studies on the genotoxic properties of essential oils with *Bacillus subtilis* rec-assay and *Salmonella*/microsome reversion assay. *Planta Med.* 57(3):237-241.

Salvia spp.

Lamiaceae

Salvia columbariae Benth.
SCN: chia
OCN: California chia; California sage

Salvia hispanica L.
SCN: chia
OCN: Spanish sage
Part: seed

Quick Reference Summary

Safety Class: 1
Interaction Class: A

Contraindications

None known.

Other Precautions

None known.

Drug and Supplement Interactions

None known.

Adverse Events and Side Effects

None known.

Pharmacological Considerations

A study with diabetic patients indicated no adverse effects of chia on safety parameters including liver, kidney, and hemostatic function, or on fasting blood glucose levels (Vuksan et al. 2007).

Pregnancy and Lactation

No information on the safety of chia in pregnancy or lactation was identified in the scientific or traditional literature. Although this review did not identify any concerns for use while pregnant or nursing, safety has not been conclusively established.

Review Details

I. Drug and Supplement Interactions

Clinical Trials of Drug or Supplement Interactions

No clinical trials of drug or supplement interactions were identified.

Case Reports of Suspected Drug or Supplement Interactions

No case reports of suspected drug or supplement interactions were identified.

Animal Trials of Drug or Supplement Interactions

No animal trials of drug or supplement interactions were identified.

II. Adverse Events

Case Reports of Adverse Events

No case reports of adverse events were identified.

III. Pharmacology and Pharmacokinetics

Human Pharmacological Studies

In type 2 diabetes patients given supplementation of diets with 33 to 41 g of chia daily for 12 weeks, no changes in safety parameters including liver, kidney, or hemostatic function, or in fasting blood glucose levels were observed (Vuksan et al. 2007).

Animal Pharmacological Studies

In dyslipemic rats fed diets containing 33% chia seed for 3 weeks, the onset of dyslipidemia and insulin resistance was prevented, with no change in glycemia (Chicco et al. 2008).

In Vitro Pharmacological Studies

No relevant in vitro pharmacological studies were identified.

IV. Pregnancy and Lactation

No information on the safety of chia during pregnancy or lactation was identified.

V. Toxicity Studies

No toxicity studies were identified.

Literature Cited

Chicco, A.G., M.E. D'Alessandro, G.J. Hein, M.E. Oliva, and Y.B. Lombardo. 2008. Dietary chia seed (*Salvia hispanica* L.) rich in γ-linolenic acid improves adiposity and normalises hypertriacylglycerolaemia and insulin resistance in dyslipaemic rats. *Br. J. Nutr.* 101(1):41-50.

Vuksan, V., D. Whitham, J.L. Sievenpiper, et al. 2007. Supplementation of conventional therapy with the novel grain salba (*Salvia hispanica* L.) improves major and emerging cardiovascular risk factors in type 2 diabetes: Results of a randomized controlled trial. *Diabetes Care* 30(11):2804-2810.

S

Sambucus spp. — Caprifoliaceae

Sambucus nigra L.
SCN: European elder
OCN: black elder

Sambucus nigra **L. ssp.** ***canadensis*** **(L.) Bolli**
SCN: American elder
Syn: *Sambucus canadensis* L.
OCN: sweet elder
Part: flower

Quick Reference Summary

Safety Class: 1
Interaction Class: A

Contraindications

None known.

Other Precautions

None known.

Drug and Supplement Interactions

None known.

Editors' Notes

The leaf, bark, seed, and raw unripe fruit of European elder, American elder, and related *Sambucus* species contain the cyanogenic glycoside sambunigrin, ingestion of which may cause vomiting or severe diarrhea (Buhrmester et al. 2000; Frohne and Pfander 2000; Nelson et al. 2006; Weiss and Meuss 2001; Wichtl 2004). Cooking or drying reduces the content of cyanogenic glycosides (FSANZ 2004).

Extracts of cooked elder flowers are traditionally used as a beverage flavoring (Leung and Foster 1996).

Adverse Events and Side Effects

None known.

Pharmacological Considerations

None known.

Pregnancy and Lactation

No information on the safety of elder flower in pregnancy or lactation was identified in the scientific or traditional literature. Although this review did not identify any concerns for use while pregnant or nursing, safety has not been conclusively established.

Review Details

I. Drug and Supplement Interactions

Clinical Trials of Drug or Supplement Interactions

No clinical trials of drug or supplement interactions were identified.

Case Reports of Suspected Drug or Supplement Interactions

No case reports of suspected drug or supplement interactions were identified.

Animal Trials of Drug or Supplement Interactions

Oral administration of 2 ml/kg of an aqueous extract of European elder flower reduced sleep onset time and increased duration of sleeping time induced by orally administered pentobarbitone (30 mg/kg, administered subcutaneously) but had no significant effects on the analgesic activity of morphine (5 mg/kg, administered subcutaneously) (Jakovljevic et al. 2001).

II. Adverse Events

Case Reports of Adverse Events

No case reports of adverse events were identified.

III. Pharmacology and Pharmacokinetics

Human Pharmacological Studies

No relevant human pharmacological studies were identified.

Animal Pharmacological Studies

No relevant animal pharmacological studies were identified.

In Vitro Pharmacological Studies

An aqueous extract of European elder flower exhibited insulin-like and insulin-releasing actions in isolated mouse abdominal muscles (Gray et al. 2000).

IV. Pregnancy and Lactation

No studies on the use of elder flower during pregnancy or lactation were identified.

V. Toxicity Studies

Short-Term Toxicity

No significant toxic effects were observed in rabbits intragastrically administered 39 mg/kg daily of an ethanolic extract of European elder flower for 3 days (Chibanguza et al. 1984).

Literature Cited

Buhrmester, R.A., J.E. Ebingerla, and D.S. Seigler. 2000. Sambunigrin and cyanogenic variability in populations of *Sambucus canadensis* L. (Caprifoliaceae). *Biochem. Syst. Ecol.* 28(7):689-695.

Chibanguza, V.G., R. Marz, and W. Sterner. 1984. Zur Wirksamkeit und Toxizität Eines Pflanzlichen Sekretolytikum und Seiner Einzeldrogen. *Arzneimittelforschung* 34:32.

Frohne, D., and H.J. Pfander. 2000. *A colour atlas of poisonous plants: A handbook for pharmacists, doctors, toxicologists, biologists and veterinarians.* 2nd ed. London: Manson.

FSANZ. 2004. Cyanogenic glycosides in cassava and bamboo shoots. Food Standards Australia New Zealand. A Human Health Risk Assessment, Technical Report Series No. 28.

Gray, A.M., Y.H. Abdel-Wahab, and P.R. Flatt. 2000. The traditional plant treatment, *Sambucus nigra* (elder), exhibits insulin-like and insulin-releasing actions *in vitro*. *J. Nutr.* 130(1):15-20.

Jakovljevic, V., M. Popovic, N. Mimica-Dukic, and J. Sabo. 2001. Interaction of *Sambucus nigra* flower and berry decoctions with the actions of centrally acting drugs in rats. *Pharm. Biol.* 39(2):142-145.

Leung, A.Y., and S. Foster. 1996. *Encyclopedia of common natural ingredients used in food, drugs, and cosmetics.* 2nd ed. New York: Wiley.

Nelson, L., R.D. Shih, M.J. Balick, and K.F. Lampe. 2006. *Handbook of poisonous and injurious plants.* 2nd ed. New York: Springer.

Weiss, R.F., and A.R. Meuss. 2001. *Weiss's herbal medicine.* Classic ed. Stuttgart: Thieme.

Wichtl, M. 2004. *Herbal drugs and phytopharmaceuticals: A handbook for practice on a scientific basis.* 3rd ed. Boca Raton, FL: CRC Press.

Sambucus spp. — Caprifoliaceae

Sambucus nigra L.
SCN: European elder
OCN: black elder

Sambucus nigra **L. ssp.** ***canadensis*** **(L.) Bolli**
SCN: American elder
Syn: *Sambucus canadensis* L.
OCN: sweet elder
Part: ripe fruit

Quick Reference Summary

Safety Class: 1
Interaction Class: A

Contraindications

None known.

Other Precautions

None known.

Drug and Supplement Interactions

None known.

Notice

Cyanogenic glycosides (Buhrmester et al. 2000; Frohne and Pfander 2000; Nelson et al. 2006; Weiss and Meuss 2001; Wichtl 2004); *see* Appendix 1.

Editors' Notes

The leaf, bark, seed, and raw unripe fruit of European elder, American elder, and related *Sambucus* species contain the cyanogenic glycoside sambunigrin, ingestion of which may cause vomiting or severe diarrhea (Buhrmester et al. 2000; Frohne and Pfander 2000; Nelson et al. 2006; Weiss and Meuss 2001; Wichtl 2004). Cooking or drying reduces the content of cyanogenic glycosides (FSANZ 2004).

Ripe elder fruit is used as a food and beverage ingredient (Jagendorf 1963; Nichols 1972; Osol and Farrar 1955; Rombauer and Becker 1975).

Adverse Events and Side Effects

No adverse events were reported in clinical trials of European elder fruit extracts (Murkovic et al. 2004; Zakay-Rones et al. 1995, 2004).

Nausea, vomiting, abdominal cramps, and weakness were reported after consumption of a juice made from elder fruit, leaf, and branches (MMWR 1984).

Pharmacological Considerations

None known.

Pregnancy and Lactation

No information on the safety of elder fruit in pregnancy or lactation was identified in the scientific or traditional literature. Although this review did not identify any concerns for use while pregnant or nursing, safety has not been conclusively established.

S

Review Details

I. Drug and Supplement Interactions

Clinical Trials of Drug or Supplement Interactions

No clinical trials of drug or supplement interactions were identified.

Case Reports of Suspected Drug or Supplement Interactions

No case reports of suspected drug or supplement interactions were identified.

Animal Trials of Drug or Supplement Interactions

Oral administration of 2 ml/kg of an aqueous extract of European elder fruit reduced sleep onset time and increased duration of sleeping time induced by orally administered pentobarbitone (30 mg/kg, administered subcutaneously) but had no significant effects on the analgesic activity of morphine (5 mg/kg, administered subcutaneously) (Jakovljevic et al. 2001).

II. Adverse Events

Adverse Events Reported in Clinical Trials

No adverse events were reported in clinical trials of European elder fruit extracts at doses of 60 ml or 1200 mg daily (Murkovic et al. 2004; Zakay-Rones et al. 1995, 2004).

Case Reports of Adverse Events

Consumption of an improperly prepared juice, made with the fruit, leaf, and branches of wild elder plants in California (believed to be *S. mexicana*), resulted in acute gastrointestinal and neurological symptoms in 11 individuals. Symptoms of poisoning included nausea, vomiting, abdominal cramps, and weakness (MMWR 1984).

III. Pharmacology and Pharmacokinetics

Human Pharmacological Studies

No relevant human pharmacological studies were identified.

Animal Pharmacological Studies

No relevant animal pharmacological studies were identified.

In Vitro Pharmacological Studies

No relevant in vitro pharmacological studies were identified.

IV. Pregnancy and Lactation

No studies on the use of elder fruit during pregnancy or lactation were identified.

V. Toxicity Studies

No toxicity studies were identified.

Literature Cited

Buhrmester, R.A., J.E. Ebingerla, and D.S. Seigler. 2000. Sambunigrin and cyanogenic variability in populations of *Sambucus canadensis* L. (Caprifoliaceae). *Biochem. Syst. Ecol.* 28(7):689-695.

Frohne, D., and H.J. Pfander. 2000. *A colour atlas of poisonous plants: A handbook for pharmacists, doctors, toxicologists, biologists and veterinarians*. 2nd ed. London: Manson.

FSANZ. 2004. Cyanogenic glycosides in cassava and bamboo shoots. Food Standards Australia New Zealand. A Human Health Risk Assessment, Technical Report Series No. 28.

Jagendorf, M.A. 1963. *Folk wines, cordials, & brandies*. New York: The Vanguard Press, Inc.

Jakovljevic, V., M. Popovic, N. Mimica-Dukic, and J. Sabo. 2001. Interaction of *Sambucus nigra* flower and berry decoctions with the actions of centrally acting drugs in rats. *Pharm. Biol.* 39(2):142-145.

MMWR. 1984. Poisoning from elderberry juice—California. *MMWR Morbid. Mortal. Wkly. Rep.* 33(13):173-174.

Murkovic, M., P.M. Abuja, A.R. Bergmann, et al. 2004. Effects of elderberry juice on fasting and postprandial serum lipids and low-density lipoprotein oxidation in healthy volunteers: A randomized, double-blind, placebo-controlled study. *Eur. J. Clin. Nutr.* 58(2):244-249.

Nelson, L., R.D. Shih, M.J. Balick, and K.F. Lampe. 2006. *Handbook of poisonous and injurious plants*. 2nd ed: New York: Springer.

Nichols, N.B., ed. 1972. *Farm Journal's country cookbook*, revised, enlarged edition. Garden City, NY: Doubleday & Co., Inc.

Osol, A., and G. Farrar. 1955. *The dispensatory of the United States of America*. 25th ed. Philadelphia: Lippincott.

Rombauer, I.S., and M.R. Becker. 1975. *Joy of cooking*. Indianapolis: Bobbs-Merrill Company, Inc.

Weiss, R.F., and A.R. Meuss. 2001. *Weiss's herbal medicine*. Classic ed. Stuttgart: New York.

Wichtl, M. 2004. *Herbal drugs and phytopharmaceuticals: A handbook for practice on a scientific basis*. 3rd ed. Boca Raton, FL: CRC Press.

Zakay-Rones, Z., E. Thom, T. Wollan, and J. Wadstein. 2004. Randomized study of the efficacy and safety of oral elderberry extract in the treatment of influenza A and B virus infections. *J. Int. Med. Res.* 32(2):132-140.

Zakay-Rones, Z., N. Varsano, M. Zlotnik, et al. 1995. Inhibition of several strains of influenza virus *in vitro* and reduction of symptoms by an elderberry extract (*Sambucus nigra* L.) during an outbreak of influenza B Panama. *J. Altern. Complement. Med.* 1(4):361-369.

S

Sanguinaria canadensis L. — Papaveraceae

SCN: bloodroot
OCN: red puccoon; red root

Part: rhizome, root

Quick Reference Summary

Safety Class: 2b
Interaction Class: A

Contraindications

Not for use in pregnancy except under the supervision of a qualified healthcare practitioner (McGuffin et al. 1997).

Other Precautions

May cause nausea and vomiting (Felter and Lloyd 1898; Osol and Farrar 1955; Remington and Wood 1918; Scudder 1898).

Drug and Supplement Interactions

None known.

Notice

Berberine (Salmore and Hunter 2001); *see* Appendix 1.

Emetic (Felter and Lloyd 1898; Osol and Farrar 1955; Remington and Wood 1918; Scudder 1898); *see* Appendix 2.

Editors' Note

Bloodroot contains the compound berberine, which is present in significantly smaller quantities than other alkaloids in the plant (Salmore and Hunter 2001).

Adverse Events and Side Effects

Vomiting has been caused by as little as 1 g of bloodroot (Felter and Lloyd 1898; Osol and Farrar 1955; Remington and Wood 1918; Scudder 1898). The presence of the toxic alkaloid sanguinarine suggests that the plant should not be used in large amounts (Leung and Foster 1996; List and Hörhammer 1973; Martindale and Reynolds 1996). Large doses of bloodroot may cause irritation of the gastrointestinal tract (Felter and Lloyd 1898; Scudder 1898).

An association between increased risk of oral leukoplakia (white patches that develop on a mucous membrane, sometimes precancerous) and use of a bloodroot-containing commercial toothpaste has been recognized and been the subject of significant study and debate (Allen 1999; Damm et al. 1999; Eversole et al. 2000; Mascarenhas et al. 2001, 2002; Munro et al. 1999). A review of human studies on oral care products that contain bloodroot extracts indicates that the products were generally well tolerated, with cases of soft tissue irritation being reported in a small number of trial participants (Munro et al. 1999). A more recent analysis of leukoplakia tissue samples from patients that used bloodroot-containing toothpaste indicated that the leukoplakia was precancerous (Anderson et al. 2005).

Bloodroot has traditionally been used as an ingredient in topical escharotic (an agent that destroys tissue and causes sloughing) preparations for the treatment of skin cancers, although bloodroot itself does not have escharotic activity. Cases of late recurrences and metastasis of cancers, due to incomplete removal of cancerous cells, in patients who are self-treating skin cancers have been reported (Affleck and Varma 2007; Laub 2008; McDaniel and Goldman 2002). Scarring, inflammation, escharotic activity, ulceration, and acute pain have been reported after repeated topical applications of escharotic products that contain bloodroot (Affleck and Varma 2007; Jellinek and Maloney 2005; Laub 2008; McDaniel and Goldman 2002; Moran and Helm 2008).

Pharmacological Considerations

In vitro studies on the compound sanguinarine have indicated antiplatelet, antiangiogenic, and apoptotic activity, and have shown either inhibition or no effect on the drug-metabolizing isoenzyme CYP1A1 (Adhami et al. 2004; Basini et al. 2007; Eun and Koh 2004; Hussain et al. 2007; Jeng et al. 2007; Karp et al. 2005; Zdarilová et al. 2006). The relevance of those in vitro findings to human use is not known.

Pregnancy and Lactation

Animal studies indicated no adverse effects on fetal development or on fertility in animals administered up to 100 mg/kg bloodroot extract daily. At a dose of 60 mg/kg in rats and 25 mg/kg in rabbits, some toxicity was observed in the pregnant animals. Reduced body weights in the offspring during lactation were observed at 100 mg/kg treatment level concomitant with maternal toxicity (Keller and Meyer 1989).

Review Details

I. Drug and Supplement Interactions

Clinical Trials of Drug or Supplement Interactions

No clinical trials of drug or supplement interactions were identified.

Case Reports of Suspected Drug or Supplement Interactions

No case reports of suspected drug or supplement interactions were identified.

S

Animal Trials of Drug or Supplement Interactions

No animal trials of drug or supplement interactions were identified.

II. Adverse Events

Adverse Events Reported in Clinical Trials

A review of human studies on oral care products that contain bloodroot extracts indicates that multiple 14-day, 28-day, 90-day, and 6-month studies on these products have been completed and that the products were generally well tolerated, with cases of soft tissue irritation being reported in a small number of trial participants (Munro et al. 1999).

Case Reports of Adverse Events

Bloodroot has traditionally been used in topical preparations as an escharotic (an agent that destroys tissue and causes sloughing) for the treatment of skin cancers. This use is based on a technique developed in the 1930s by physician Frederic Mohs. Dr. Mohs used a salve containing zinc chloride, antimony trisulfide, and bloodroot (reportedly included as an "organic stabilizer") as an escharotic and fixative prior to surgical removal of tumors (McDaniel and Goldman 2002).

While anecdotal evidence indicates successful elimination of cancerous cells in some patients, case reports indicate late recurrences and metastasis of cancers, due to incomplete removal of cancerous cells, in patients who were self-treating skin cancers (Affleck and Varma 2007; Laub 2008; McDaniel and Goldman 2002). Scarring, inflammation, escharotic activity, ulceration, and acute pain have been reported after repeated topical applications of escharotic products that contain bloodroot (Affleck and Varma 2007; Jellinek and Maloney 2005; Laub 2008; McDaniel and Goldman 2002; Moran and Helm 2008).

A 27-year-old man with a history of abnormal moles presented with necrotic ulcers after use of "black salve," a topical preparation containing bloodroot that is used as an escharotic in the treatment of skin cancers. Histological examination and excisional biopsy showed ulceration and extensive tissue necrosis (involving the epidermis, dermis, and subcutaneous tissue). A diffuse mixed neutrophilic and lymphocytic infiltrate was present in all layers of the skin and in subcutaneous tissues. Information on dose and duration of use was not reported (Moran and Helm 2008).

Internally, overdose of bloodroot extracts is reported to cause burning in the epigastrium with vomiting, tormenting thirst, faintness, vertigo, dimness of vision, and prostration (Remington and Wood 1918).

The manufacturers of Viadent, a toothpaste that, at the time, contained 0.075% bloodroot extract, commissioned a safety and toxicology review of bloodroot and the compound sanguinarine by an expert panel. The panel concluded that the available toxicology data confirmed the safety of bloodroot at the exposure level of Viadent (Frankos et al. 1990). Bloodroot is no longer an ingredient in Viadent, and all references here to adverse events are in relation to the formulation that included bloodroot.

More recent studies and case reports, however, have suggested an association between increased risk of oral leukoplakia (white patches that develop on a mucous membrane, sometimes precancerous) and use of Viadent (Allen et al. 2001), although this association has been the focus of much study and debate (Allen 1999; Damm et al. 1999; Eversole et al. 2000; Mascarenhas et al. 2001, 2002; Munro et al. 1999).

In a retrospective study of immunohistochemical assessment of Viadent-associated leukoplakia, normal, dysplastic, and Viadent-associated tissue samples from an oral pathology archive were evaluated. Biomarker profiles evaluated in this study consistently showed intermediate staining intensities for Viadent-related specimens relative to dysplastic and normal. Based on biomarker profiles, unique histology, and clinical behavior, the authors of this study concluded that Viadent-associated lesions are preneoplastic in nature (Anderson et al. 2005).

III. Pharmacology and Pharmacokinetics

Human Pharmacological Studies

No relevant human pharmacological studies were identified.

Animal Pharmacological Studies

Dermal, ocular, and oral cheek pouch sensitization studies of the compound sanguinarine chloride in rabbits and dogs at concentrations of 0.1 to 1.0% indicated that the compound caused slight dermal irritation in rabbits (0.1% concentration), mild ocular irritation in rabbits (0.1% concentration), and was nonirritating in the oral cheek pouch test in dogs (1.0% concentration for 28 days) (Frankos et al. 1990).

No effect on ventricular pressure, diastolic pressure, or cardiac output and no changes in lead II EKG were observed after intra-arterial administration of 0.075 mg/kg of the compound sanguinarine to beagle dogs (Frankos et al. 1990; Schwartz 1986). In monkeys orally administered 0, 10, 30, and 60 mg/kg daily for 90 days, no electrocardiographic evidence of a drug effect was noted at any of the dose levels (Frankos et al. 1990).

In Vitro Pharmacological Studies

Sanguinarine inhibited platelet aggregation induced by arachidonic acid, collagen, U46619, and a subthreshold concentration of thrombin with IC_{50} (the half-maximal inhibitory concentration) concentrations of 4.4 to 8.3 μ*M*. Sanguinarine also inhibited platelet thromboxane B_2 production, but not platelet aggregation induced by higher concentration of thrombin. SQ22536, an adenylate cyclase inhibitor, attenuated the inhibitory effect of sanguinarine toward arachidonic acid-induced platelet calcium mobilization and aggregation (Jeng et al. 2007).

The compound sanguinarine inhibited both vascular endothelial growth factor (VEGF) production and VEGF-induced Akt activation in swine granulosa cells, and blocked vessel growth induced by VEGF. The authors of the study suggested that supplementation of livestock feed with sanguinarine should be carefully considered, since sanguinarine could be detrimental to follicular angiogenesis (Basini et al. 2007).

The compound sanguinarine was shown to activate polycyclic aromatic hydrocarbon-associated signaling and metabolic pathways, inducing AhR-associated gene expression and inhibiting CYP1A1 microsomal oxidative activity (Karp et al. 2005). Other studies have shown that sanguinarine did not activate aryl hydrocarbon receptor signaling pathways in rat hepatoma cells (Dvorák et al. 2006) and had no direct effect on CYP1A expression (Zdarilová et al. 2006).

The compound sanguinarine inhibited cell proliferation and induced apoptosis in a dose-dependent manner in several lymphoma cell lines (Hussain et al. 2007). Sanguinarine has been shown to inhibit angiogenesis in the membrane of chick embryos (Eun and Koh 2004). In human prostate cancer cell lines, the compound sanguinarine caused cell cycle blockade and apoptosis by modulation of cyclin kinase inhibitor-cyclin-cyclin-dependent kinase machinery (Adhami et al. 2004).

In rat liver microsomes, the compound sanguinarine was metabolized to a species that caused DNA adduct formation in a dose-dependent manner (Stiborová et al. 2002). In human colon cancer cells treated with the compound sanguinarine, a relative abundance of rapidly initiated DNA double strand breaks was observed and were believed to be the result of, rather than the cause of, apoptotic cell death. The results of the study indicated that apoptosis induced by sanguinarine is independent of p53 and most likely independent of DNA damage (Matkar et al. 2008). Several studies have shown that the compound sanguinarine intercalates with DNA (Faddeeva and Beliaeva 1997; Piehler et al. 1997; Saran et al. 1995; Schmeller et al. 1997).

The compound sanguinarine had a positive inotropic effect followed by contracture in rat ventricular and atrial strips. Sanguinarine also dose-dependently reduced the rate of spontaneous contractions of isolated right atria (Hu et al. 2005). Positive inotropic effects were also observed in isolated guinea pig atria treated with the compound sanguinarine (Seifen et al. 1979). A dose-dependent response in ventricular refractoriness was observed in anesthetized pigs after vascular infusion of up to 4 mg/kg of the compound sanguinarine (Whittle et al. 1980).

IV. Pregnancy and Lactation

In a set of reproductive and developmental toxicity studies, female rats were orally administered bloodroot extract (containing ~68% alkaloids) at doses of up to 100 mg/kg daily 14 days prior to mating, through weaning, or doses of up to 60 mg/kg on gestational days (GD) 6 to 15, or the same dose on GD 15 through lactation day 20. Rabbits were orally administered up to 75 mg/kg daily on GD 6 to 18. No developmental toxicity, including teratogenicity, was observed in the fetuses of rats following maternal administration of the 5 to 60 mg/kg doses. An increase in post-implantation loss was observed at maternally toxic dosage levels of 50 and 75 mg/kg in rabbits. Oral administration of bloodroot in perinatal and postnatal studies in rats caused no adverse effects on litter size, parturition, or lactation of female rats nor on survival or growth of their offspring at dosage levels of 5 to 60 mg/kg. Maternal oral toxicity thresholds were 60 mg/kg daily in rats and 25 mg/kg daily in rabbits (Keller and Meyer 1989).

No adverse effects on estrous cycling or female copulatory and fertility indices or gestation/lactation parameters were observed in rats administered 10 to 100 mg/kg of bloodroot daily. Reduced body weights in the offspring during lactation were observed at the 100 mg/kg treatment level concomitant with maternal toxicity (Keller and Meyer 1989).

V. Toxicity Studies

Acute Toxicity

The LD_{50} of orally administered bloodroot extract in rats is 1440 mg/kg (Frankos et al. 1990). No mortalities were reported in rabbits dermally administered undiluted bloodroot extract at a dose of 200 mg/kg (Becci et al. 1987; Schwartz 1986).

The LD_{50} of the compound sanguinarine in rats has been reported as 1525 mg/kg or 1658 mg/kg after oral administration (Becci et al. 1987; Frankos et al. 1990; Schwartz 1986) and 29 mg/kg after intravenous administration (Becci et al. 1987; Schwartz 1986).

In rats intraperitoneally administered a single dose of 10 mg/kg of the compound sanguinarine, increased levels of liver enzymes (SGPT and SGOT)) were observed (Dalvi 1985). In mice intraperitoneally administered the same dose, significant decreases in liver glutathione and P450 enzyme activities were observed along with increased serum levels of alanine aminotransferase (Williams et al. 2000).

Short-Term Toxicity

No toxic effects were observed in rats fed up to 150 ppm of the compound sanguinarine in the diet for 14 days, or in rats orally administered up to 0.6 mg/kg daily for 30 days (Becci et al. 1987; Schwartz 1986).

In rats administered a bloodroot extract as part of the diet, with estimated daily doses of 0 to 405 mg/kg for 28 days, no mortality was observed. At the highest dose level, a reduction in body weight was observed (Frankos et al. 1990).

In cynomolgus monkeys orally administered bloodroot extract at doses of 0 to 200 mg/kg for 28 days, emesis, diarrhea, inappetence, reduction of body weight, and mortality was observed. The publicly available data from the study do not specify at which doses these adverse effects were observed (Frankos et al. 1990). Dose-related emesis and diarrhea (due to gastrointestinal irritation) were reported in cynomolgus monkeys orally administered the compound sanguinarine chloride at doses of 0 to 60 mg/kg daily for 90 days. The emesis occurred only during the first 5 weeks of the study and was not observed in the last 8 weeks (Frankos et al. 1990).

Subchronic Toxicity

No changes in liver enzymes were observed in pigs administered the compounds sanguinarine and chelerythrine (3:1 ratio) as 0.0002 to 0.01% of the diet for 90 days (Kosina et al. 2004).

Genotoxicity

In mice intraperitoneally administered single doses of 1.35 to 21.6 mg/kg of the compound sanguinarine, a dose-dependent increase in DNA damage in blood and bone marrow cells was observed 24 hours after treatment (Ansari et al. 2005).

In the Ames mutagenicity assay, an extract of bloodroot tested weakly positive in *Salmonella typhimurium* (Ames assay) with metabolic activation, while no mutagenic activity of the compound sanguinarine was observed with or without metabolic activation. No mutagenic activity was observed in *E. coli* in the unscheduled DNA synthesis assay in rat hepatocytes (Frankos et al. 1990).

Literature Cited

Adhami, V.M., M.H. Aziz, S.R. Reagan-Shaw, et al. 2004. Sanguinarine causes cell cycle blockade and apoptosis of human prostate carcinoma cells via modulation of cyclin kinase inhibitor-cyclin-cyclin-dependent kinase machinery. *Mol. Cancer Ther.* 3(8):933-940.

Affleck, A.G., and S. Varma. 2007. A case of do-it-yourself Mohs' surgery using bloodroot obtained from the Internet. *Br. J. Dermatol.* 157(5):1078-1079.

Allen, C.L., J. Loudon, and A.K. Mascarenhas. 2001. *Sanguinaria*-related leukoplakia: Epidemiologic and clinicopathologic features of a recently described entity. *Gen. Dent.* 49(6):608-614.

Allen, C.M. 1999. Viadent-related leukoplakia—The tip of the iceberg? *Oral Surg. Oral Med. Oral Pathol. Oral Radiol. Endodon.* 87(4):393.

Anderson, K.M., G.D. Stoner, H.W. Fields, et al. 2005. Immunohistochemical assessment of Viadent®-associated leukoplakia. *Oral Oncol.* 41(2):200-207.

Ansari, K.M., A. Dhawan, S.K. Khanna, and M. Das. 2005. *In vivo* DNA damaging potential of sanguinarine alkaloid, isolated from argemone oil, using alkaline Comet assay in mice. *Food Chem. Toxicol.* 43(1):147-153.

Basini, G., S.E. Santini, S. Bussolati, and F. Grasselli. 2007. The plant alkaloid sanguinarine is a potential inhibitor of follicular angiogenesis. *J. Reprod. Dev.* 53(3):573-579.

Becci, P.J., H. Schwartz, H.H. Barnes, and G.L. Southard. 1987. Short-term toxicity studies of sanguinarine and of two alkaloid extracts of *Sanguinaria canadensis* L. *J. Toxicol. Environ. Health* 20(1-2):199-208.

Dalvi, R.R. 1985. Sanguinarine: Its potential as a liver toxic alkaloid present in the seeds of *Argemone mexicana Cell. Mol. Life Sci.* 41(1):77-78.

Damm, D.D., A. Curran, D.K. White, and J.F. Drummond. 1999. Leukoplakia of the maxillary vestibule—An association with Viadent? *Oral Surg. Oral Med. Oral Pathol. Oral Radiol. Endodon.* 87(1):61.

Dvorák, Z., I. Sovadinová, L. Bláha, J.P. Giesy, and J. Ulrichová. 2006. Quaternary benzo[*c*]phenanthridine alkaloids sanguinarine and chelerythrine do not affect transcriptional activity of aryl hydrocarbon receptor: Analyses in rat hepatoma cell line H4IIE.luc. *Food Chem. Toxicol.* 44(9):1466-1473.

Eun, J.P., and G.Y. Koh. 2004. Suppression of angiogenesis by the plant alkaloid, sanguinarine. *Biochem. Biophys. Res. Commun.* 317(2):618-624.

Eversole, L.R., G.M. Eversole, and J. Kopcik. 2000. *Sanguinaria*-associated oral leukoplakia: Comparison with other benign and dysplastic leukoplakic lesions. *Oral Surg. Oral Med. Oral Pathol. Oral Radiol. Endodon.* 89(4):455-464.

Faddeeva, M.D., and T.N. Beliaeva. 1997. Sanguinarine and ellipticine cytotoxic alkaloids isolated from well-known antitumor plants. Intracellular targets of their action. *Tsitologiia* 39(2-3):181-208.

Felter, H.W., and J.U. Lloyd. 1898. *King's American dispensatory*. 18th ed., 3rd rev. 2 vols. Cincinnati: Ohio Valley Co.

Frankos, V.H., D.J. Brusick, E.M. Johnson, et al. 1990. Safety of *Sanguinaria* extract as used in commercial toothpaste and oral rinse products. *J. Can. Dent. Assoc.* 56(7, Suppl.):41-47.

Hu, C.M., Y.W. Cheng, J.W. Liao, H.W. Cheng, and J.J. Kang. 2005. Induction of contracture and extracellular Ca^{2+} influx in cardiac muscle by sanguinarine: A study on cardiotoxicity of sanguinarine. *J. Biomed. Sci.* 12(2):399-407.

Hussain, A.R., N.A. Al-Jomah, A.K. Siraj, et al. 2007. Sanguinarine-dependent induction of apoptosis in primary effusion lymphoma cells. *Cancer Res.* 67(8):3888-3897.

Jellinek, N., and M.E. Maloney. 2005. Escharotic and other botanical agents for the treatment of skin cancer: A review. *J. Am. Acad. Dermatol.* 53(3):487-495.

Jeng, J.H., H.L. Wu, B.R. Lin, et al. 2007. Antiplatelet effect of sanguinarine is correlated to calcium mobilization, thromboxane and cAMP production. *Atherosclerosis* 191(2):250-258.

Karp, J.M., K.A. Rodrigo, P. Pei, et al. 2005. Sanguinarine activates polycyclic aromatic hydrocarbon associated metabolic pathways in human oral keratinocytes and tissues. *Toxicol. Lett.* 158(1):50-60.

Keller, K.A., and D.L. Meyer. 1989. Reproductive and developmental toxicological evaluation of *Sanguinaria* extract. *J. Clin. Dent.* 1(3):59-66.

Kosina, P., D. Walterova, J. Ulrichova, et al. 2004. Sanguinarine and chelerythrine: Assessment of safety on pigs in ninety days feeding experiment. *Food Chem. Toxicol.* 42(1):85-91.

Laub, D.R., Jr. 2008. Death from metastatic basal cell carcinoma: Herbal remedy or just unlucky? *J. Plast. Reconstr. Aesthet. Surg.* 61(7):846-848.

Leung, A.Y., and S. Foster. 1996. *Encyclopedia of common natural ingredients used in food, drugs, and cosmetics*. 2nd ed. New York: Wiley.

List, P.H., and H. Hörhammer. 1973. *Hagers handbuch der pharmazeutischen praxis*. Berlin: Springer.

Martindale, W., and J.E.F. Reynolds. 1996. *The extra pharmacopoeia*. 31st ed. London: Pharmaceutical Press.

Mascarenhas, A.K., C.M. Allen, and J. Loudon. 2001. The association between Viadent use and oral leukoplakia. *Epidemiology* 12(6):741.

Mascarenhas, A.K., C.M. Allen, and M.L. Moeschberger. 2002. The association between Viadent use and oral leukoplakia—Results of a matched case-control study. *J. Pub. Health Dent.* 62(3):158-162.

Matkar, S.S., L.A. Wrischnik, and U. Hellmann-Blumberg. 2008. Sanguinarine causes DNA damage and p53-independent cell death in human colon cancer cell lines. *Chem. Biol. Interact.* 172(1):63-71.

McDaniel, S., and G.D. Goldman. 2002. Consequences of using escharotic agents as primary treatment for nonmelanoma skin cancer. *Arch. Dermatol.* 138(12):1593-1596.

McGuffin, M., C. Hobbs, R. Upton, and A. Goldberg. 1997. *Botanical safety handbook*. Boca Raton, FL: CRC Press.

Moran, A.M., and K.F. Helm. 2008. Histopathologic findings and diagnostic difficulties posed with use of escharotic agents for treatment of skin lesions: A case report and review of the literature. *J. Cutan. Pathol.* 35(4):404-406.

Munro, I.C., E.S. Delzell, E.R. Nestmann, and B.S. Lynch. 1999. Viadent usage and oral leukoplakia: A spurious association. *Regul. Toxicol. Pharmacol.* 30(3):182-196.

Osol, A., and G. Farrar. 1955. *The dispensatory of the United States of America*. 25th ed. Philadelphia: Lippincott.

Piehler, J., A. Brecht, G. Gauglitz, et al. 1997. Specific binding of low molecular weight ligands with direct optical detection. *Biosens. Bioelectron.* 12(6):531-538.

Remington, J.P., and H.C. Wood. 1918. *The dispensatory of the United States of America*. 20th ed. Philadelphia: Lippincott.

Salmore, A.K., and M.D. Hunter. 2001. Environmental and genotypic influences on isoquinoline alkaloid content in *Sanguinaria canadensis*. *J. Chem. Ecol.* 27(9):1729-1747.

Saran, A., S. Srivastava, E. Coutinho, and M. Maiti. 1995. ^{1}H NMR investigation of the interaction of berberine and sanguinarine with DNA. *Indian J. Biochem. Biophys.* 32(2):74-77.

Schmeller, T., B. Latz-Bruning, and M. Wink. 1997. Biochemical activities of berberine, palmatine and sanguinarine mediating chemical defence against microorganisms and herbivores. *Phytochemistry* 44(2):257-266.

Schwartz, H.G. 1986. Safety profile of sanguinarine and *Sanguinaria* extract. *Compend. Contin. Educ. Dent. Suppl.* 7:S212-S217.

Scudder, J. 1898. *The American Eclectic materia medica and therapeutics*. Cincinnati, OH: The Scudder Brothers Company.

Seifen, E., R.J. Adams, and R.K. Riemer. 1979. Sanguinarine: A positive inotropic alkaloid which inhibits cardiac Na^+,K^+-ATPase. *Eur. J. Pharmacol.* 60(4):373-377.

Stiborová, M., V. Šimánek, E. Frei, P. Hobza, and J. Ulrichová. 2002. DNA adduct formation from quaternary benzo[*c*] phenanthridine alkaloids sanguinarine and chelerythrine as revealed by the ^{32}P-postlabeling technique. *Chem. Biol. Interact.* 140(3):231-242.

Whittle, J.A., J.K. Bissett, K.D. Straub, J.E. Doherty, and J.R. McConnell. 1980. Effect of sanguinarine on ventricular refractoriness. *Res. Commun. Chem. Pathol. Pharmacol.* 29(2):377-380.

Williams, M.K., S. Dalvi, and R.R. Dalvi. 2000. Influence of 3-methylcholanthrene pretreatment on sanguinarine toxicity in mice. *Vet. Human Toxicol.* 42(4):196-198.

Zdarilová, A., R. Vrzal, M. Rypka, J. Ulrichová, and Z. Dvorák. 2006. Investigation of sanguinarine and chelerythrine effects on CYP1A1 expression and activity in human hepatoma cells. *Food Chem. Toxicol.* 44(2):242-249.

Santalum album L. — Santalaceae

SCN: sandalwood
AN: *chandana; shveta chandana*
PN: *tan xiang* (heartwood)
OCN: East Indian sandalwood; white saunders; yellow sandalwood; yellow saunders
Part: wood

S

Quick Reference Summary

Safety Class: 1
Interaction Class: A

Contraindications

None known.

Other Precautions

None known.

Drug and Supplement Interactions

None known.

Adverse Events and Side Effects

Among patients with allergic reactions to fragrances, positive patch test results to sandalwood were observed in less than 3% of patients (An et al. 2005; Larsen et al. 1996; Romaguera et al. 1983; Trattner and David 2003).

Pharmacological Considerations

None known.

Pregnancy and Lactation

No information on the safety of sandalwood in pregnancy was identified in the scientific or traditional literature.

While this review did not identify any concerns for pregnant women, safety has not been conclusively established.

Sandalwood essential oil is a commonly sold, highly concentrated extract of sandalwood. An animal study demonstrated that sandalwood oil crosses into breast milk (Chhabra and Rao 1993).

Review Details

I. Drug and Supplement Interactions

Clinical Trials of Drug or Supplement Interactions

No clinical trials of drug or supplement interactions were identified.

Case Reports of Suspected Drug or Supplement Interactions

No case reports of suspected drug or supplement interactions were identified.

Animal Trials of Drug or Supplement Interactions

No animal trials of drug or supplement interactions were identified.

II. Adverse Events

Case Reports of Adverse Events

Among 422 patients with suspected contact allergy in Korea, 9.6% tested positive to a fragrance mix and 2.4% tested positive to sandalwood essential oil (An et al. 2005).

Among 167 patients with suspected fragrance allergy in a worldwide investigation of fragrance contact dermatitis, 6.6% were allergic to sandalwood, while 1.8% had irritation reactions to sandalwood (Larsen et al. 1996). Among 641 Israeli patients with eczema, none tested positive to sandalwood (Trattner and David 2003).

In an analysis of patch test reactions in 58,128 patients, 5539 patients were patch tested and 80 tested positive to the fragrance mix. Among these 80, only one tested positive to sandalwood essential oil (Romaguera et al. 1983).

III. Pharmacology and Pharmacokinetics

Human Pharmacological Studies

No relevant human pharmacological studies were identified.

Animal Pharmacological Studies

No relevant animal pharmacological studies were identified.

In Vitro Pharmacological Studies

No significant activity of a methanol extract or ethyl acetate fraction of sandalwood was observed on the drug-metabolizing isoenzymes CYP3A4 or CYP2D6 in human liver microsomes (Subehan et al. 2006; Usia et al. 2006).

IV. Pregnancy and Lactation

No information on the safety of sandalwood in pregnancy was identified.

Increases in hepatic glutathione S-transferase, glutathione reductase, and glutathione peroxidase activity were observed in the nursing pups of rats that were orally administered 5 or 10 μl of sandalwood essential oil daily for 14 or 21 days. A decrease in hepatic CYP450 content was observed in pups and dams treated with 10 μl of sandalwood oil for 21 days (Chhabra and Rao 1993).

V. Toxicity Studies

Acute Toxicity

The oral LD_{50} of sandalwood essential oil in rats is 5.58 g/kg (Opdyke 1979). The dermal LD_{50} of sandalwood essential oil in rabbits could not be determined at doses up to 5 g/kg (Opdyke 1979).

Genotoxicity

No genotoxicity of sandalwood essential oil was observed in *Bacillus subtilis* spore *rec* assays with H17 Rec^+ and M45 Rec^- with or without metabolic activation by S9 (Burdock and Carabin 2008; Ishizaki et al. 1985; Watanabe 1994).

S

Literature Cited

An, S., A.Y. Lee, C.H. Lee, et al. 2005. Fragrance contact dermatitis in Korea: A joint study. *Contact Dermat.* 53(6):320-333.

Burdock, G.A., and I.G. Carabin. 2008. Safety assessment of sandalwood oil (*Santalum album* L.). *Food Chem. Toxicol.* 46(2):421-432.

Chhabra, S.K., and A.R. Rao. 1993. Postnatal modulation of xenobiotic metabolizing enzymes in liver of mouse pups following translactational exposure to sandalwood oil. *Nutr. Res.* 13(10):1191-1202.

Ishizaki, M., S. Ueno, N. Oyamada, K. Kubota, and M. Noda. 1985. The DNA-damaging activity of natural food additives (III). *J. Food Hyg. Soc. Jap.* 26:523-527.

Larsen, W., H. Nakayama, M. Lindberg, et al. 1996. Fragrance contact dermatitis: A worldwide multicenter investigation (Part I). *Am. J. Contact Dermat.* 7:77-83.

Opdyke, D.L.J. 1979. *Monographs on fragrance raw materials.* New York: Pergamon.

Romaguera, C., J.M. Camarasa, A. Alomar, and F. Grimalt. 1983. Patch tests with allergens related to cosmetics. *Contact Dermat.* 9(2):167-168.

Subehan, T. Usia, H. Iwata, S. Kadota, and Y. Tezuka. 2006. Mechanism-based inhibition of CYP3A4 and CYP2D6 by Indonesian medicinal plants. *J. Ethnopharmacol.* 105(3):449-455.

Trattner, A., and M. David. 2003. Patch testing with fine fragrances: Comparison with fragrance mix, balsam of Peru and a fragrance series. *Contact Dermat.* 49(6):287-289.

Usia, T., H. Iwata, A. Hiratsuka, et al. 2006. CYP3A4 and CYP2D6 inhibitory activities of Indonesian medicinal plants. *Phytomedicine* 13(1-2):67-73.

Watanabe, S. 1994. A simple screening test for chemical compounds to induce delayed allergic contact dermatitis: Use of *Bacillus subtilis* spore rec-assay in place of animal methods. *Oyo Yakuri* 47(3):177-198.

Saposhnikovia divaricata (Turcz.) Schischk. — Apiaceae

SCN: siler
Syn: *Ledebouriella divaricata* (Turcz.) Hiroe; *Ledebouriella seseloides* (Hoffm.) Wolff; *Siler divaricatum* (Turcz.) Benth. & Hook. f.
PN: *fang feng* (root)
Part: root

Quick Reference Summary

Safety Class: 1
Interaction Class: A

Contraindications

None known.

Other Precautions

None known.

Drug and Supplement Interactions

None known.

Adverse Events and Side Effects

Allergic reactions to siler, affecting the skin and digestive system, have been reported (Bensky et al. 2004).

Pharmacological Considerations

None known.

Pregnancy and Lactation

No information on the safety of siler in pregnancy or lactation was identified in the scientific or traditional literature. Although this review did not identify any concerns for use while pregnant or nursing, safety has not been conclusively established.

Review Details

I. Drug and Supplement Interactions

Clinical Trials of Drug or Supplement Interactions

No clinical trials of drug or supplement interactions were identified.

Case Reports of Suspected Drug or Supplement Interactions

No case reports of suspected drug or supplement interactions were identified.

Animal Trials of Drug or Supplement Interactions

No animal trials of drug or supplement interactions were identified.

II. Adverse Events

Case Reports of Adverse Events

Allergic reactions to siler, affecting the skin and digestive system, have been reported (Bensky et al. 2004).

III. Pharmacology and Pharmacokinetics

Human Pharmacological Studies

No relevant human pharmacological studies were identified.

Animal Pharmacological Studies

No relevant animal pharmacological studies were identified.

In Vitro Pharmacological Studies

Inhibition of the drug-metabolizing isoenzyme CYP3A was observed in human liver microsomes treated with a decoction or infusion of siler (Guo et al. 2001).

No estrogenic activity of an ethanolic extract of siler was observed in a recombinant yeast system featuring a human estrogen receptor expression plasmid and a reporter plasmid (Kim et al. 2008).

IV. Pregnancy and Lactation

No information on the safety of siler during pregnancy or lactation was identified.

V. Toxicity Studies

Acute Toxicity

The LD_{50} of orally administered siler in mice is 213.8 g/kg (type of extract not specified in available English language translation) (Chen and Chen 2004).

Literature Cited

Bensky, D., S. Clavey, and E. Stöger. 2004. *Chinese herbal medicine: Materia medica*. 3rd ed. Seattle: Eastland Press.

Chen, J.K., and T.T. Chen. 2004. *Chinese medical herbology and pharmacology*. City of Industry, CA: Art of Medicine Press.

Guo, L.Q., M. Taniguchi, Q.Y. Chen, K. Baba, and Y. Yamazoe. 2001. Inhibitory potential of herbal medicines on human cytochrome P450-mediated oxidation: Properties of umbelliferous or citrus crude drugs and their relative prescriptions. *Jpn. J. Pharmacol.* 85(4):399-408.

Kim, I.G., S.C. Kang, K.C. Kim, E.S. Choung, and O.P. Zee. 2008. Screening of estrogenic and antiestrogenic activities from medicinal plants. *Environ. Toxicol. Pharmacol.* 25(1):75-82.

Sassafras albidum (Nutt.) Nees — Lauraceae

SCN: sassafras
Syn: *Sassafras officinale* T. Nees & C.H. Eberm.

Part: root

Quick Reference Summary

*Safety Class:** 2b, 2d
Interaction Class: A

Contraindications

Not for use in pregnancy except under the supervision of a qualified healthcare practitioner (Vesselinovitch et al. 1979).

Not for long-term use; do not exceed recommended dose (Wichtl 2004).

Other Precautions

None known.

Drug and Supplement Interactions

None known.

Standard Dose

The standard dose is 4–7 g of the root bark as a tea (Remington and Wood 1918).

Notice

Alkenylbenzenes (safrole 5.0–8.0% in the root bark, <1.0% in the root, 80% of the essential oil) (Cook and Martin 1948; Craig 1953; JECFA 1981; List and Hörhammer 1973); *see* Appendix 1.

Editors' Notes

U.S. regulations allow the use of sassafras in foods only if safrole-free (CFR 2011).

Animal studies in the 1960s raised concern about the cancer-causing potential of the purified compound safrole when fed in relatively high amounts (0.01 to 0.1% of the diet) for extended periods of time (2 years, equivalent to approximately 68 years of human exposure) (Abbott et al. 1961; Hagan et al. 1965, 1967; Long et al. 1963), leading to the ban on safrole in foods (CFR 2011c). Although safrole is listed as insoluble in water, one study indicated that approximately 11% of the safrole is extracted in sassafras tea, and a much higher percentage is extracted by alcohol (Carlson and Thompson 1997; Merck 1999). The content of safrole in sassafras tea made from 3 g sassafras root powder or root bark powder is 90 to 4120 μg per cup (Carlson and Thompson 1997). Safrole is also present in black pepper. With an estimated intake of 400 mg of black pepper per person daily in the United States, daily individual intake of safrole is approximately 36 μg (Ames et al. 1990; Bhardwaj et al. 2002; Farag and Abo-Zeid 1997; Gold et al. 2001). While occasional consumption of sassafras tea is thought to be safe, use of alcohol extracts is discouraged.

Adverse Events and Side Effects

None known.

Pharmacological Considerations

None known.

Pregnancy and Lactation

No information on the safety of sassafras during pregnancy or lactation was identified. Studies with the compound safrole in pregnant and lactating rats indicated higher incidences of tumors in offspring of treated pregnant or lactating mice as compared to control groups (Vesselinovitch et al. 1979). Based on this information, use during pregnancy is not recommended except under the supervision of a qualified healthcare practitioner.

* Safrole-free extracts of sassafras are commercially available. Concerns regarding the use of sassafras do not apply to safrole-free products.

S

Review Details

I. Drug and Supplement Interactions

Clinical Trials of Drug or Supplement Interactions

No clinical trials of drug or supplement interactions were identified.

Case Reports of Suspected Drug or Supplement Interactions

No case reports of suspected drug or supplement interactions were identified.

Animal Trials of Drug or Supplement Interactions

No animal trials of drug or supplement interactions were identified.

II. Adverse Events

Case Reports of Adverse Events

Hot flashes and sweating were reported in a 72-year-old woman with a history of hypertension. The woman was taking furosemide, potassium chloride, and aspirin in addition to drinking up to 10 cups of sassafras tea daily for an unspecified amount of time. The hot flashes and sweating resolved after cessation of the tea (Haines 1991).

Vomiting, tachycardia, and tremulousness were reported in a 47-year-old woman who inadvertently ingested one teaspoon of sassafras essential oil (Grande and Dannewitz 1987). A fatal case of poisoning was reported in young man after ingestion of one teaspoon of sassafras essential oil. Symptoms of poisoning included vomiting, collapse, dilated pupils, and stupor (Cincinnati 1888). Several other cases of sassafras oil poisoning have been reported, although details are lacking. A review of essential oil poisonings reported that vomiting and signs of shock are common in sassafras poisoning (Craig 1953).

III. Pharmacology and Pharmacokinetics

Human Pharmacological Studies

No relevant human pharmacological studies were identified.

Animal Pharmacological Studies

In rats subcutaneously administered 15 mg of an extract of sassafras (produced with a series of solvents that omitted safrole) once a week for up to 78 weeks, tumors developed in 11 of 15 males and 9 of 15 females. The types of tumors were not specified (Kapadia et al. 2002).

In partially hepatectomized rats, an increase in liver regeneration was observed after subcutaneous administration of 15 or 375 mg of sassafras essential oil or after administration of a diet containing 1.5 or 7% sassafras tea for 7 days postoperatively (Gershbein 1977).

In Vitro Pharmacological Studies

No relevant in vitro pharmacological studies were identified.

IV. Pregnancy and Lactation

Offspring of pregnant mice orally administered 120 μg/g of the compound safrole on gestational days 12, 14, 16, and 18 were sacrificed at 94 weeks of age. Among animals exposed in utero to safrole, renal epithelial tumors were observed in 7% of female offspring, with none in male offspring (Vesselinovitch et al. 1979).

After oral administration of 120 μg/g of the compound safrole to lactating mice every other day for 12 weeks, 34% of male offspring had hepatocellular tumors, but no tumors were observed in female offspring. This incidence was significantly lower than observed in 4-week-old offspring orally administered the same dose of safrole twice a week for 90 weeks. In these animals, hepatocellular tumors were observed in 48% of females and 8% of males (Vesselinovitch et al. 1979).

V. Toxicity Studies

Acute Toxicity

The oral LD_{50} of the compound safrole is 2350 mg/kg in mice and 1950 mg/kg in rats (Jenner et al. 1964).

Chronic Toxicity

In rats fed sassafras essential oil as 0.117% of the diet or safrole as 0.039% of the diet for 2 years, equivalent to approximately 68 years of human exposure, hepatocarcinomas were found in all animals along with changes in the kidneys, adrenals, thyroid, pituitary, and testes or ovaries. In animals sacrificed at 22 months, kidneys showed evidence of congestion, but no liver tumors were observed (Abbott et al. 1961).

In rats fed the compound safrole as 0, 0.1, 0.25, 0.5, or 1% of the diet for 2 years, growth was depressed at levels of 0.25% and above. None of the animals in the 1% group survived beyond 62 weeks. These animals showed testicular atrophy, changes in the stomach, and changes in the liver, including tumor formation. Similar types of liver changes were seen with the lower doses. Liver damage was slight at 0.1% and lacked tumors and cirrhosis, moderate at 0.25% but lacked cirrhosis, and severe at 0.5%, with cysts and tumors (Hagan et al. 1965, 1967).

In rats fed diets containing 0.01, 0.05, 0.1, or 0.5% of the compound safrole for 2 years, changes in the liver, including benign and malignant tumors, were observed. The liver injury was rated as very slight at the 0.01% safrole level, slight at 0.05% safrole, slight to moderate at 0.1% safrole, and moderate to severe at 0.5% safrole. Tumor incidence was significantly increased at 0.5%. Weight gain and survival were significantly decreased at the 0.5% level (Long et al. 1963).

Genotoxicity

A review of mutagenicity studies on the compound safrole indicated that safrole was generally inactive in mutagenicity studies in various strains of *Salmonella typhimurium* with or without metabolic activation, but has occasionally been reported to be weakly positive (JECFA 1981).

Safrole was positive in the following short-term tests for mutagenesis: mammalian cell transformation in culture, Rabin's test of degranulation of rough endoplasmic reticulum from rat liver, and mouse sebaceous gland suppression test. The compound was negative in the tetrazolium reduction and tissue reaction to subcutaneous implants in mice. Safrole was positive in mutagenicity assays with *Escherichia coli* and *Saccharomyces cerevisiae* and in the intraperitoneal host-mediated assay with *Salmonella typhimurium* strain TA1535 or *Saccharomyces cerevisiae*. Safrole was positive in the host-mediated assay with *Salmonella typhimurium* strains TA1950 and TA1952 (JECFA 1981).

Literature Cited

Abbott, D.D., E.W. Packman, J.W.E. Harrisson, and B.M. Wagner. 1961. Chronic oral toxicity of oil of sassafras and safrole. *Pharmacologist* 3:62.

Ames, B., M. Profet, and L.S. Gold. 1990. Dietary pesticides (99.9% natural). *Proc. Natl. Acad. Sci. U.S.A.* 87:7777-7781.

Bhardwaj, R.K., H. Glaeser, L. Becquemont, et al. 2002. Piperine, a major constituent of black pepper, inhibits human P-glycoprotein and CYP3A4. *J. Pharmacol. Exp. Ther.* 302(2):645-650.

Carlson, M., and R.D. Thompson. 1997. Liquid chromatographic determination of safrole in sassafras-derived herbal products. *J. AOAC Int.* 80(5):1023-1028.

CFR. 2011. *Code of federal regulations*, Title 21 Part 189.180, 2011 ed. Substances generally prohibited from direct addition or use as human food. Safrole. Washington, DC: U.S. Government Printing Office.

Cincinnati. 1888. Cincinnati Lancet-Clinic, Dec. 1888. *Cited in* Craig, J.O. 1953. Poisoning by the volatile oils in childhood. *Arch. Dis. Child.* 28(142):475-483.

Cook, E.F., and E.W. Martin. 1948. *Remington's practice of pharmacy.* 9th ed. Easton, PA: Mack Publishing Company.

Craig, J.O. 1953. Poisoning by the volatile oils in childhood. *Arch. Dis. Child.* 28(142):475-483.

Farag, S.E.A., and M. Abo-Zeid. 1997. Degradation of the natural mutagenic compound safrole in spices by cooking and irradiation. *Nahrung* 41:359-361.

Gershbein, L.L. 1977. Regeneration of rat liver in the presence of essential oils and their components. *Food Cosmet. Toxicol.* 15(3):173-181.

Gold, J.L., D.A. Laxer, J.M. Dergal, K.L. Lanctot, and P.A. Rochon. 2001. Herbal-drug therapy interactions: A focus on dementia. *Curr. Opin. Clin. Nutr. Metab. Care* 4(1):29-34.

Grande, G.A., and S.R. Dannewitz. 1987. Symptomatic sassafras oil ingestion. *Vet. Human Toxicol.* 29(6):447.

Hagan, E.C., W.H. Hansen, O.G. Fitzhugh, et al. 1967. Food flavourings and compounds of related structure. II. Subacute and chronic toxicity. *Food Cosmet. Toxicol.* 5(2):141-157.

Hagan, E.C., P.M. Jenner, W.I. Jones, et al. 1965. Toxic properties of compounds related to safrole. *Toxicol. Appl. Pharmacol.* 7(1):18-24.

Haines, J.D. 1991. Sassafras tea and diaphoresis. *Postgrad. Med.* 90(4):75-76.

JECFA. 1981. Safrole. WHO Food Additives Series 16. Joint FAO/WHO Expert Committee on Food Additives. Geneva WHO/FAO.

Jenner, P.M., E.C. Hagan, J.M. Taylor, E.L. Cook, and O.G. Fitzhugh. 1964. Food flavourings and compounds of related structure I. Acute oral toxicity. *Food Cosmet. Toxicol.* 2:327-343.

Kapadia, G.J., M.A. Azuine, H. Tokuda, et al. 2002. Inhibitory effect of herbal remedies on 12-*O*-tetradecanoylphorbol-13-acetate-promoted Epstein-Barr virus early antigen activation. *Pharmacol. Res.* 45(3):213-220.

List, P.H., and H. Hörhammer. 1973. *Hagers handbuch der pharmazeutischen praxis*. Berlin: Springer.

Long, E.L., A.A. Nelson, O.G. Fitzhugh, and W.H. Hansen. 1963. Liver tumors produced in rats by feeding safrole. *Arch. Pathol.* 75(6):595-604.

Merck. 1999. *The Merck index*. Whitehouse Station, NJ: Merck.

Remington, J.P., and H.C. Wood. 1918. *The dispensatory of the United States of America*. 20th ed. Philadelphia: Lippincott.

Vesselinovitch, S.D., K.V.N. Rao, and N. Mihailovich. 1979. Transplacental and lactational carcinogenesis by safrole. *Cancer Res.* 39:4378-4380.

Wichtl, M. 2004. *Herbal drugs and phytopharmaceuticals: A handbook for practice on a scientific basis*. 3rd ed. Boca Raton, FL: CRC Press.

Satureja spp. — Lamiaceae

Satureja hortensis L.
SCN: summer savory
OCN: annual savory

Satureja montana L.
SCN: winter savory
Part: leaf

Quick Reference Summary

Safety Class: 1
Interaction Class: A

Contraindications

None known.

Other Precautions

None known.

Drug and Supplement Interactions

None known.

Notice

Diuretic (*S. hortensis*) (Hajhashemi et al. 2002; Stanic and Samaržija 1993); *see* Appendix 2.

Emmenagogue (*S. hortensis*) (Felter and Lloyd 1898); *see* Appendix 2.

Editors' Note

The classifications and concerns for these herbs are based on relatively higher doses used for therapeutic purposes in contrast to lower amounts generally used in cooking, and have not been associated with their use as spices.

Adverse Events and Side Effects

None known.

Pharmacological Considerations

None known.

Pregnancy and Lactation

No information on the safety of summer savory or winter savory in pregnancy or lactation was identified in the scientific or traditional literature. Although this review did not identify any concerns for use while pregnant or nursing, safety has not been conclusively established.

Review Details

I. Drug and Supplement Interactions

Clinical Trials of Drug or Supplement Interactions

No clinical trials of drug or supplement interactions were identified.

Case Reports of Suspected Drug or Supplement Interactions

No case reports of suspected drug or supplement interactions were identified.

Animal Trials of Drug or Supplement Interactions

No animal trials of drug or supplement interactions were identified.

II. Adverse Events

Case Reports of Adverse Events

No case reports of adverse events were identified.

III. Pharmacology and Pharmacokinetics

Human Pharmacological Studies

No relevant human pharmacological studies were identified.

Animal Pharmacological Studies

In rats orally administered 50 ml/kg of winter savory, an initial decrease in diuresis with a subsequent mild stimulation of diuresis was observed. Administration of a 10% aqueous extract or 0.1% essential oil solution resulted in more pronounced diuretic activity. Tests with the essential oil indicated that a 0.1% solution of essential oil was considered to be a diuretic dose, whereas 0.5% and 1% solutions caused toxic symptoms (Stanic and Samaržija 1993).

In Vitro Pharmacological Studies

No relevant in vitro pharmacological studies were identified.

IV. Pregnancy and Lactation

No information on the safety of summer savory or winter savory during pregnancy or lactation was identified.

V. Toxicity Studies

Acute Toxicity

The LD_{50} of summer savory essential oil in rats is 1.37 g/kg after oral administration (Opdyke 1979).

Genotoxicity

No mutagenic activity of the essential oils of winter savory and summer savory was observed in the *Bacillus subtilis rec* assay and *Salmonella* microsome reversion assay (Zani et al. 1991).

S

Literature Cited

Felter, H.W., and J.U. Lloyd. 1898. *King's American dispensatory*. 18th ed., 3rd rev. 2 vols. Cincinnati: Ohio Valley Co.

Hajhashemi, V., A. Ghannadi, and S.K. Pezeshkian. 2002. Antinociceptive and anti-inflammatory effects of *Satureja hortensis* L. extracts and essential oil. *J. Ethnopharmacol.* 82(2-3):83-87.

Opdyke, D.L.J. 1979. *Monographs on fragrance raw materials*. New York: Pergamon.

Stanic, G., and I. Samaržija. 1993. Diuretic activity of *Satureja montana* subsp. *montana* extracts and oil in rats. *Phytother. Res.* 7(5):363-366.

Zani, F., G. Massimo, S. Benvenuti, et al. 1991. Studies on the genotoxic properties of essential oils with *Bacillus subtilis* rec-assay and *Salmonella*/microsome reversion assay. *Planta Med.* 57(3):237-241.

Saussurea costus (Falc.) Lipsch. — Asteraceae

SCN: costus
Syn: *Aucklandia costus* Falc.; *Aucklandia lappa* Decne.; *Saussurea lappa* (Decne.) C.B. Clarke
AN: *kushtha*
PN: *mu xiang* (root)
OCN: aucklandia
Part: root

Quick Reference Summary

Safety Class: 1
Interaction Class: A

Contraindications
None known.

Other Precautions
None known.

Drug and Supplement Interactions
None known.

Editors' Notes
Costus, which is sold as *mu xiang*, is sometimes confused with the aristolochic acid-containing *Aristolochia* species, which are sometimes traded as *qing mu xiang* (Shum et al. 2007). Methods for differentiation of *Saussurea* and *Aristolochia* species and for detection of aristolochic acid have been developed (Shum et al. 2007; Upton 2006; Yamasaki et al. 2009).

Adverse Events and Side Effects
Allergic reactions to costus have been reported (Bensky et al. 2004; Cheminat et al. 1981).

Pharmacological Considerations
None known.

Pregnancy and Lactation
No information on the safety of costus in pregnancy or lactation was identified in the scientific or traditional literature. Although this review did not identify any concerns for use while pregnant or nursing, safety has not been conclusively established.

Review Details

I. Drug and Supplement Interactions

Clinical Trials of Drug or Supplement Interactions
No clinical trials of drug or supplement interactions were identified.

Case Reports of Suspected Drug or Supplement Interactions
No case reports of suspected drug or supplement interactions were identified.

Animal Trials of Drug or Supplement Interactions
No animal trials of drug or supplement interactions were identified.

II. Adverse Events

Case Reports of Adverse Events
Costus essential oil is used in perfumery and has been responsible for a number of cases of contact dermatitis (Cheminat et al. 1981). Sesquiterpene lactone compounds are believed to be responsible for the reaction (Cheminat et al. 1981; Pandey et al. 2007; Robinson et al. 2008; Sun et al. 2003).

An allergic reaction was reported in a person who ingested a decoction containing 10 g of costus. The reaction was confirmed with an oral rechallenge (Bensky et al. 2004).

III. Pharmacology and Pharmacokinetics

Human Pharmacological Studies
No relevant human pharmacological studies were identified.

Animal Pharmacological Studies
No relevant animal pharmacological studies were identified.

In Vitro Pharmacological Studies
No estrogenic activity of costus was observed in a recombinant yeast system featuring a human estrogen receptor expression plasmid and a reporter plasmid (Kim et al. 2008).

IV. Pregnancy and Lactation
No information on the safety of costus during pregnancy or lactation was identified.

V. Toxicity Studies

Acute Toxicity
The LD_{50} of the compound costuslactone is 300 mg/kg in rats after intraperitoneal administration (Chen and Chen 2004).

Subchronic Toxicity

No adverse effects were observed in rats orally administered 1.77 mg/kg (males) or 2.17 mg/kg (females) of costus daily for 90 days (Chen and Chen 2004).

Genotoxicity

In a mutagenicity assay with *Salmonella typhimurium* strains TA98, TA100, TA1535, and TA1537, an extract of costus exhibited some mutagenic activity in TA98 (Riazuddin et al. 1987).

Literature Cited

Bensky, D., S. Clavey, and E. Stöger. 2004. *Chinese herbal medicine: Materia medica*. 3rd ed. Seattle: Eastland Press.

Cheminat, A., J.L. Stampf, C. Benezra, M.J. Farrall, and J.M. Frechet. 1981. Allergic contact dermatitis to costus: Removal of haptens with polymers. *Acta Derm. Venereol.* 61(6):525-529.

Chen, J.K., and T.T. Chen. 2004. *Chinese medical herbology and pharmacology*. City of Industry, CA: Art of Medicine Press.

Kim, I.G., S.C. Kang, K.C. Kim, E.S. Choung, and O.P. Zee. 2008. Screening of estrogenic and antiestrogenic activities from medicinal plants. *Environ. Toxicol. Pharmacol.* 25(1):75-82.

Pandey, M.M., S. Rastogi, and A.K. Rawat. 2007. *Saussurea costus*: Botanical, chemical and pharmacological review of an ayurvedic medicinal plant. *J. Ethnopharmacol.* 110(3):379-390.

Riazuddin, S., M.M. Malik, and A. Nasim. 1987. Mutagenicity testing of some medicinal herbs. *Environ. Mol. Mutagen.* 10(2):141-148.

Robinson, A., T.V. Kumar, E. Sreedhar, et al. 2008. A new sesquiterpene lactone from the roots of *Saussurea lappa*: Structure-anticancer activity study. *Bioorg. Med. Chem. Lett.* 18(14):4015-4017.

Shum, K.C., F. Chen, S.L. Li, et al. 2007. Authentication of *Radix Aucklandiae* and its substitutes by GC-MS and hierarchical clustering analysis. *J. Sep. Sci.* 30(18):3233-3239.

Sun, C.M., W.J. Syu, M.J. Don, J.J. Lu, and G.H. Lee. 2003. Cytotoxic sesquiterpene lactones from the root of *Saussurea lappa*. *J. Nat. Prod.* 66(9):1175-1180.

Upton, R. 2006. Characterization of selected plants that may contain or be adulterated with aristolochic acid. Scotts Valley, CA: American Herbal Pharmacopoeia.

Yamasaki, K., T. Tagami, M. Kawaguchi, et al. 2009. Simple and rapid analysis of aristolochic acid contained in crude drugs and Kampo formulations with solid-phase extraction and HPLC photodiode-array detection. *J. Nat. Med.* 63(4):451-458.

Schinus spp. — Anacardiaceae

Schinus molle L.
SCN: Peruvian peppertree
OCN: California pepper tree; molle

Schinus terebinthifolius **Raddi**
SCN: Brazilian peppertree
OCN: Christmasberry; pink pepper
Part: bark

Quick Reference Summary

Safety Class: 1
Interaction Class: A

Contraindications

None known.

Other Precautions

None known.

Drug and Supplement Interactions

None known.

Notice

Tannins (10–14%) (Morton 1978); *see* Appendix 1.

Adverse Events and Side Effects

Rashes have been reported in persons coming in contact with the resin, wood, and, less frequently, the fruit of Brazilian peppertree (Morton 1978).

Pharmacological Considerations

None known.

Pregnancy and Lactation

Emmenagogue properties are attributed to the bark and leaf of the Peruvian peppertree in Argentina and also to the fruit and leaf in Iraq (Hieronymus 1882; Dellacassa 2010).

No information on the safety of Brazilian or Peruvian peppertree bark in lactation was identified in the scientific or traditional literature. While this review did not identify any concerns for use while nursing, safety has not been conclusively established.

S

Review Details

I. Drug and Supplement Interactions

Clinical Trials of Drug or Supplement Interactions

No clinical trials of drug or supplement interactions were identified.

Case Reports of Suspected Drug or Supplement Interactions

No case reports of suspected drug or supplement interactions were identified.

Animal Trials of Drug or Supplement Interactions

No animal trials of drug or supplement interactions were identified.

II. Adverse Events

Case Reports of Adverse Events

Rashes have been reported in persons coming in contact with the resin, wood, and, less frequently, the fruit of Brazilian peppertree (Morton 1978).

III. Pharmacology and Pharmacokinetics

Human Pharmacological Studies

No relevant human pharmacological studies were identified.

Animal Pharmacological Studies

No relevant human pharmacological studies were identified.

In Vitro Pharmacological Studies

No relevant in vitro pharmacological studies were identified.

IV. Pregnancy and Lactation

A review of the economic botany of Argentina records that emmenagogue properties are attributed to the bark and leaf of the Peruvian peppertree (Hieronymus 1882).

No information on the safety of Brazilian or Peruvian peppertree bark in lactation was identified.

V. Toxicity Studies

Genotoxicity

In a series of cell-free and bacterial assays, an extract of Brazilian peppertree bark was negative in a cell-free plasmid DNA test, indicating that it did not directly break DNA. Positive results were obtained in the SOS chromotest, in a forward mutagenesis assay employing strains of *Escherichia coli*, and in the *Salmonella* reversion assay with strains TA97, TA98, TA100, and TA102. All the bacterial tests were performed without metabolic activation. The authors indicated that results suggest that Brazilian peppertree bark produces DNA damage and mutation in bacteria, and that oxidative damage may be responsible for the genotoxicity (de Carvalho et al. 2003).

Literature Cited

de Carvalho, M.C., F.N. Barca, L.F. Agnez-Lima, and S.R. de Medeiros. 2003. Evaluation of mutagenic activity in an extract of pepper tree stem bark (*Schinus terebinthifolius* Raddi). *Environ. Mol. Mutagen.* 42(3):185-191.

Dellacassa, E. 2010. *Normalización de productos naturales obtenidos de especies de la flora aromática latinoamericana.* Porto Alegre, Brasil: Editora Universitária da PUCRS.

Hieronymus, J. 1882. *Plantae Diaphoricae – Florae Argentinae.* Buenos Aires: Guillermo Kraft.

Morton, J.F. 1978. Brazilian pepper—Its impact on people, animals and the environment. *Econ. Bot.* 32(4):353-359.

Stahl, E., K. Keller, and C. Blinn. 1983. Cardanol, a skin irritant in pink pepper. *Planta Med.* 48(5):5-9.

S

Schinus spp. — Anacardiaceae

Schinus molle L.
SCN: Peruvian peppertree
OCN: California pepper tree; molle

Schinus terebinthifolius **Raddi**
SCN: Brazilian peppertree
OCN: Christmasberry; pink pepper
Part: fruit

Quick Reference Summary

Safety Class: 2d
Interaction Class: A

Contraindications

Consumption of the fruit should not exceed small amounts for use as a condiment (Morton 1978; Watt and Breyer-Brandwijk 1962).

Other Precautions

None known.

Drug and Supplement Interactions

None known.

Adverse Events and Side Effects

Gastrointestinal irritation after ingestion of Brazilian and Peruvian peppertree fruit has been reported (Morton 1978; Watt and Breyer-Brandwijk 1962).

Rashes have been reported in persons coming in contact with the resin, wood, and, less frequently, the fruit of Brazilian peppertree (Morton 1978). An irritant compound, cardanol, was identified in Brazilian peppertree fruit (Stahl et al. 1983).

Pharmacological Considerations

None known.

Pregnancy and Lactation

An animal study with an ethanolic extract of fruit of Peruvian peppertree indicated stimulant activity in isolated rabbit and rat uterus (Zaidi et al. 1970). Emmenagogue properties are attributed to the fruit and leaf of the Peruvian peppertree in Iraq (Dellacassa 2010).

No information on the safety of Brazilian or Peruvian peppertree fruit in lactation was identified in the scientific or traditional literature. While this review did not identify any concerns for use while nursing, safety has not been conclusively established.

Review Details

I. Drug and Supplement Interactions

Clinical Trials of Drug or Supplement Interactions

No clinical trials of drug or supplement interactions were identified.

Case Reports of Suspected Drug or Supplement Interactions

No case reports of suspected drug or supplement interactions were identified.

Animal Trials of Drug or Supplement Interactions

No animal trials of drug or supplement interactions were identified.

II. Adverse Events

Case Reports of Adverse Events

Eaten "in quantity" (no more specific information was given), the fruit of Peruvian peppertree has been reported to cause gastrointestinal irritation in children, accompanied by vomiting, diarrhea, headache and fatigue, and the same effects but with "more severe exhaustion" have been reported in association with eating Brazilian peppertree fruit (Watt and Breyer-Brandwijk 1962). Cases of poisoning from eating the fruit of the Peruvian peppertree were reported to include irritation in the throat, vomiting and purging (Watt and Breyer-Brandwijk 1962). Eating more than a few of the ripe or unripe fruits of Brazilian peppertree was reported to cause digestive upset and vomiting (Morton 1978).

Rashes have been reported in persons coming in contact with the resin, wood, and, less frequently, the fruit of Brazilian peppertree (Morton 1978). A researcher preparing extracts of fresh ripe Brazilian peppertree fruit experienced headaches, swollen eyelids, shortness of breath, and chest pains (Morton 1978).

III. Pharmacology and Pharmacokinetics

Human Pharmacological Studies

No relevant human pharmacological studies were identified.

Animal Pharmacological Studies

An irritant compound, cardanol, was identified in Brazilian peppertree fruit. A skin test showed that the substance had a strong skin irritating effect after a relatively long latency period (Stahl et al. 1983).

In Vitro Pharmacological Studies

No relevant in vitro pharmacological studies were identified.

IV. Pregnancy and Lactation

A study with a specially prepared extract of the fruit of Peruvian peppertree indicated stimulant activity on isolated rabbit and rat gravid uterus, but not in virgin uterus. The extract used was concentrated at 1 g/ml. Although described as an aqueous extract, it was prepared with ethanol (95%), then dried, mixed with chloroform and distilled water in several steps, neutralized to a pH of 6.8 with dilute ammonia solution, and subjected to charcoal filtration (Zaidi et al. 1970). Emmenagogue properties are attributed to the fruit and leaf of the Peruvian peppertree in Iraq (Dellacassa 2010).

No information on the safety of Brazilian or Peruvian peppertree fruit in lactation was identified.

V. Toxicity Studies

Acute Toxicity

The LD_{50} of Brazilian peppertree fruit in mice is 3.5 g/kg after intraperitoneal administration and could not be determined at doses up to 5 g/kg after oral administration (Pires et al. 2004).

No toxic effects were observed in rats fed 2 g/kg of an ethanol extract of Peruvian peppertree fruit in the diet for 1 day (Ferrero et al. 2007).

Short-Term Toxicity

No toxic effects or histopathological changes in the brain, liver, kidney, lung, heart, stomach, or intestines were observed in rats fed diets containing a daily dose of 1 g/kg of an ethanol extract of Peruvian peppertree fruit for 14 days (Ferrero et al. 2007).

Genotoxicity

Acute Toxicity

The LD_{50} of Brazilian peppertree in mice is 3.5 g/kg after intraperitoneal administration and could not be determined at doses up to 5 g/kg after oral administration (Pires et al. 2004).

No toxic effects were observed in rats fed 2 g/kg of an ethanol extract of Peruvian peppertree in the diet for 1 day (Ferrero et al. 2007).

Short-Term Toxicity

No toxic effects or histopathological changes in the brain, liver, kidney, lung, heart, stomach, or intestines were observed in rats fed diets containing a daily dose of 1 g/kg of an ethanol extract of Peruvian peppertree for 14 days (Ferrero et al. 2007).

Literature Cited

Dellacassa, E. 2010. *Normalización de productos naturales obtenidos de especies de la flora aromática latinoamericana.* Porto Alegre, Brasil: Editora Universitária da PUCRS.

Ferrero, A., A. Minetti, C. Bras, and N. Zanetti. 2007. Acute and subacute toxicity evaluation of ethanolic extract from fruits of *Schinus molle* in rats. *J. Ethnopharmacol.* 113(3):441-447.

Morton, J.F. 1978. Brazilian pepper—Its impact on people, animals and the environment. *Econ. Bot.* 32(4):353-359.

Pires, O.C., A.V. Corsi Taquemasa, G. Akisue, F. De Oliveira, and C.E. Pulz Araujo. 2004. Preliminary comparative analysis of the acute toxicity and median lethal dose (LD_{50}) of the fruit of the Brazilian black pepper (*Schinus terebinthifolius* Raddi) and black pepper (*Piper nigrum* L.). *Acta Farm. Bon.* 23(2):176-182.

Stahl, E., K. Keller, and C. Blinn. 1983. Cardanol, a skin irritant in pink pepper. *Planta Med.* 48(5):5-9.

Watt, J.M., and M.G. Breyer-Brandwijk. 1962. *The medicinal and poisonous plants of southern and eastern Africa.* Edinburgh: E. & S. Livingstone.

Zaidi, S., A. Hanan, and S. Babar. 1970. Some preliminary studies of the pharmacological activities of *Schinus molle. Pak. J. Sci. Ind. Res.* 13:53.

Schisandra spp. — Schisandraceae

Schisandra chinensis (Turcz.) Baill.
SCN: schisandra
PN: *bei wu wei zi* (fruit)
OCN: northern schisandra; schisandra

Schisandra sphenanthera **Rehder & E.H. Wilson**
SCN: southern schisandra
PN: *nan wu wei zi* (fruit)
OCN: schizandra
Part: fruit

S

Quick Reference Summary

Safety Class: 1
Interaction Class: C

Contraindications

None known.

Other Precautions

None known.

Drug and Supplement Interactions

In human studies, schisandra has been shown to increase the plasma levels of tacrolimus and talinolol, and is thought to inhibit the drug metabolizing isoenzyme CYP3A and the drug transporter protein P-gp, leading to increased plasma levels of drugs metabolized by CYP3A or transported by P-gp (Fan et al. 2009; Jiang et al. 2010; Xin et al. 2007; Xin et al. 2009). *See* Cytochrome P450 in Appendix 3.

Adverse Events and Side Effects

Heartburn has been reported as a side effect in some individuals. Allergic reactions to schisandra have been reported (Bensky et al. 2004).

Pharmacological Considerations

Large doses (500 mg/kg) of schisandra increased the metabolism rate of intravenously administered warfarin in rats (Mu et al. 2006). Animal studies have indicated that schisandra berry induced the drug-metabolizing isoenzyme CYP1A2 and inhibited CYP3A4 (Makino et al. 2006; Zhang et al. 2002).

Pregnancy and Lactation

A reduction in postpartum hemorrhaging and no adverse effects on maternal or fetal health were reported in pregnant women that took a schisandra extract during pregnancy (Gaistruk and Taranovskij 1968). In one study, schisandra was used to induce labor in women with prolonged labor (Trifonova 1954).

No information on the safety of schisandra during lactation was identified in the scientific or traditional literature. While this review did not identify any concerns for use while nursing, safety has not been conclusively established.

Review Details

I. Drug and Supplement Interactions

Clinical Trials of Drug or Supplement Interactions

In healthy patients orally administered 100 mg talinolol before and after oral administration of 600 mg schisandra extract daily for 14 days, an increase in plasma levels of the talinolol was observed after the schisandra administration (51% increase in the maximum plasma concentration). Talinolol is a substrate of the drug transporter protein P-glycoprotein (P-gp), and inhibition of P-gp was proposed as the mechanism of interaction (Fan et al. 2009).

In liver transplant patients taking tacrolimus, a dose dependent increase in blood levels of tacrolimus was observed after treatment with two capsules of schisandra (each containing 11.25 mg deoxyschizandrin) extract along with one of two doses of tacrolimus, 0.1 to 0.15 mg/kg daily or 0.5 to 3 mg/person daily. Increases in the maximum plasma concentration of tacrolimus were 339% for the 0.1 to 0.15 mg/kg dose and 262% for the 0.5 to 3 mg/person dose. A decrease in diarrhea and agitation, two side effects typically observed with tacrolimus treatment, was noted along with improvement in liver function (Jiang et al. 2010).

Increased bioavailability of midazolam was observed in healthy volunteers orally administered three capsules schisandra extract (each containing 11.25 mg deoxyschizandrin) twice daily for seven days. As compared to pretreatment levels, an 85% increase in the maximum plasma concentration of midazolam and a 52% decrease in the oral clearance of this drug was observed after administration of a single dose of 15 mg midazolam. The authors indicated that inhibition of CYP3A was the likely mechanism of interaction (Xin et al. 2009).

An increase in the bioavailability of tacrolimus was observed in healthy volunteers orally administered three capsules schisandra extract (each containing 11.25 mg deoxyschizandrin) twice daily for 13 days. The maximum plasma concentration of tacrolimus increased by 227% as compared to tacrolimus administered prior to treatment with schisandra. Tacrolimus is metabolized by CYP3A4 and CYP3A5, and P-gp is important in the absorption metabolism of this drug. The authors suggested that inhibition of CYP3A and P-gp by schisandra were likely responsible for the observed interaction (Xin et al. 2007).

Case Reports of Suspected Drug or Supplement Interactions

No case reports of suspected drug or supplement interactions were identified.

Animal Trials of Drug or Supplement Interactions

In rats orally administered an aqueous extract of schisandra at a dose equivalent to 500 mg/kg of dried fruit daily for 5 days, the clearance rate of intravenously administered warfarin (2 mg/kg) was significantly larger than that seen in the control group. Related assays indicated that the interaction was due to induction of the pregnane X receptor (Mu et al. 2006).

The compounds schisandrol A and B and schisandrin B and C increased pentobarbital- and barbital-induced sleeping time in mice (Bao et al. 1980; Liu 1991).

II. Adverse Events

Case Reports of Adverse Events

Schisandra may occasionally cause heartburn (Bensky et al. 2004). Allergic rashes affecting the eyelids, back of the hands, chest, and lower back have been reported (Bensky et al. 2004; Sandberg 1993). Overdoses of schisandra (standard dose listed as a decoction of 3–6 g) have been associated with abdominal discomfort and burning, cold and sore sensations in the epigastrium, stomach pain, and reduced appetite (Bensky et al. 2004).

III. Pharmacology and Pharmacokinetics

Human Pharmacological Studies

No relevant human pharmacological studies were identified.

Animal Pharmacological Studies

Administration of an aqueous extract of schisandra (dose not specified in English language abstract) in rats for 3 or 6 days increased the metabolism of imipramine, a CYP1A2 substrate (Zhang et al. 2002). An aqueous extract of schisandra berry inhibited the drug-metabolizing isoenzyme CYP3A4 to a greater extent than grapefruit juice (Makino et al. 2006). In rats orally administered a single dose of schisandra equivalent to 3 g/kg of dried herb, examination of excised livers indicated an increase in total CYP450 expression (Mu et al. 2006).

In a study of the potential of schisandra to restore hepatic drug metabolism in rats with liver damage, rats were orally administered the hepatotoxin carbon tetrachloride with or without pretreatment with 160 mg/kg of a lignan fraction of schisandra. In rats administered carbon tetrachloride alone, metabolism of antipyrine was significantly reduced, but rats pretreated with schisandra metabolized antipyrine at rates similar to those of control animals, indicating a protective effect of schisandra on phase I metabolism (Zhu et al. 1999).

Schisandra exhibited a significant protective effect against adriamycin-induced cardiotoxicity in rats (You et al. 2006).

In Vitro Pharmacological Studies

An ethanol extract of schisandra induced the drug-metabolizing isoenzymes CYP3A4 and CYP1A2, and MDR1 (Brandin et al. 2007). Aqueous extracts of schisandra induced the drug-metabolizing isoenzymes CYP3A4 and CYP2C9 in human liver cells (Mu et al. 2006). Dose-dependent induction of CYP3A4 via the pregnane X receptor was observed in human liver cells (Hua et al. 2007; Mu et al. 2006).

A study on different compounds isolated from schisandra berry indicated that the compound gomisin C is a mechanism-based inhibitor that competitively inhibits and irreversibly inactivates CYP3A4 with inhibitory effects greater than those of the drug ketoconazole (Iwata et al. 2004). The compounds schisandrin A and B and schisantherin A inhibited P-glycoprotein (Pan et al. 2006; Qiangrong et al. 2005).

An extract of schisandra activated the estrogen-responsive luciferase gene in cells transiently transfected with estrogen receptors and reporter plasmids (Lee et al. 2004).

IV. Pregnancy and Lactation

Administration of 20 to 25 drops of schisandra tincture three times daily, during three consecutive hours, for three consecutive days was effective in inducing labor in women with prolonged labor. No adverse effects were observed on blood pressure, elimination of the placenta, or postnatal health of mother and infant (Trifonova 1954).

S

In hypotensive or normotensive pregnant women administered 30 to 40 drops of schisandra tincture three times daily for an unspecified number of days, no adverse effects on maternal or fetal health were observed. Blood pressure was increased in hypotensive women and unchanged in normal women. A reduction in postpartum hemorrhaging was observed in women treated with schisandra, with 2.3% of women in the schisandra group and 12.4% of women in the control group experiencing hemorrhaging (Gaistruk and Taranovskij 1968).

Increases in uterine tension and the amplitude of uterine contractions were observed in rabbits subcutaneously administered 0.1 ml/kg of a schisandra tincture (Trifonova 1954).

No information on the safety of schisandra during lactation was identified.

V. Toxicity Studies

Acute Toxicity

No adverse effects were observed in mice orally administered 15 or 21 g/kg of schisandra extracts (Chen and Chen 2004; Hancke et al. 1999). The LD_{50} of a petroleum-ether extract containing 10% of the compound schisandrin in mice is 10.5 g/kg after oral administration and 4.4 g/kg after intraperitoneal administration (Volicer et al. 1966). The LD_{50} of a petroleum-ether extract orally administered in mice is 2.8 g/kg for an extract containing 40% schisandrin and 1.4 g/kg for an extract containing 80% schisandrin (Volicer et al. 1966).

The LD_{50} of orally administered schisandra essential oil in mice is 8.75 mg/kg (Chen and Chen 2004). Another reference indicates that oral administration of 280 mg/kg of the essential oil to mice caused depression, dyspnea, ataxia, and death (Chang and But 1986).

The LD_{50} of the compound schizandrol B in mice is 878 mg/kg after oral administration and 855 mg/kg after subcutaneous administration (Hänsel et al. 1994). In mice orally administered compounds isolated from schisandra, no deaths occurred after administration of 2 g/kg of schisandrins A, B, and C, 2 g/kg schisandrer A, 0.5 g/kg schizandrol A, or 0.25 g/kg of schizandrol B or schisandrer B (Bao et al. 1980). Another study indicated that mice died 7 days after oral administration of 2 g/kg of schisandrins A and C (Hänsel et al. 1994).

Short-Term Toxicity

No adverse effects were observed after oral administration of the compound schisandrin B for 30 days in mice at a dose of 200 mg/kg daily, or in dogs at a dose of 10 mg/kg daily (Chang and But 1986).

Oral administration of 1.3 g/kg of schisandra berries daily to mice for 10 days resulted in only mild toxic effects, including piloerection, apathy, and increased body weight (Sandberg 1993).

Subchronic Toxicity

No adverse effects, including changes in blood chemistry, were observed in piglets orally administered a standardized extract of schisandra at doses of 70, 360, or 720 mg/kg daily for 90 days. No changes in the liver, heart, kidneys, intestine, lungs, spleen, and gonads were seen (Burgos and Hancke 1992).

Hepatotoxicity

Significant hepatoprotective activity of extracts of schisandra and compounds isolated from schisandra have been observed in mice, rats, and rabbits treated with hepatotoxic compounds (Hancke et al. 1999; Ko et al. 2002; Nakagiri et al. 2003; Zhu et al. 1999, 2000).

Genotoxicity

No mutagenic effects of the compound schisandrin B were observed in the Ames mutagenicity test. Schisandrin B decreased the mutagenic effects of benzo[*a*]pyrene and 2-acetylaminofluorene in this assay (Liu 1991). However, in mice administered a diet containing 5% of an ethanol extract of schisandra, increased mutagenicity of benzo[*a*]pyrene and aflatoxin were observed in liver microsomes from the animals (Hendrich and Bjeldanes 1986).

Literature Cited

Bao, T., G. Liu, Z. Song, G. Xu, and R. Sun. 1980. A comparison of the pharmacologic actions of 7 constituents isolated from *Fructus Schizandrae*. *Chin. Med. J.* 93(1):41-47.

Bensky, D., S. Clavey, and E. Stöger. 2004. *Chinese herbal medicine: Materia medica.* 3rd ed. Seattle: Eastland Press.

Brandin, H., E. Viitanen, O. Myrberg, and A.K. Arvidsson. 2007. Effects of herbal medicinal products and food supplements on induction of CYP1A2, CYP3A4 and MDR1 in the human colon carcinoma cell line LS180. *Phytother. Res.* 21(3):239-244.

Burgos, R., and J. Hancke. 1992. Toxicological studies on *S. chinensis*. *Cited in* Hancke, J.L., R.A. Burgos, and F. Ahumada. 1999. *Schisandra chinensis* (Turcz.) Baill. *Fitoterapia* 70(5):451-471.

Chang, H.-M., and P.P.H. But. 1986. *Pharmacology and applications of Chinese materia medica*. English ed. Philadelphia: World Scientific.

Chen, J.K., and T.T. Chen. 2004. *Chinese medical herbology and pharmacology*. City of Industry, CA: Art of Medicine Press.

Fan, L., X.Q. Mao, G.Y. Tao, et al. 2009. Effect of *Schisandra chinensis* extract and *Ginkgo biloba* extract on the pharmacokinetics of talinolol in healthy volunteers. *Xenobiotica* 39 (3):249-254.

Gaistruk, A., and K. Taranovskij. 1968. The treatment of arterial hypotension in pregnant women using *Schizandra chinensis*. *Urg. Prob. Obstet. Gynecol. L'vov.* 1:183-186.

Hancke, J.L., R.A. Burgos, and F. Ahumada. 1999. *Schisandra chinensis* (Turcz.) Baill. *Fitoterapia* 70 (5):451-471.

Hänsel, R., K. Keller, H. Rimpler, and G. Schneider. 1994. *Hagers handbuch der pharmazeutischen praxis*. Berlin: Springer.

Hendrich, S., and L.F. Bjeldanes. 1986. Effects of dietary *Schizandra chinensis*, Brussels sprouts and *Illicium verum* extracts on carcinogen metabolism systems in mouse liver. *Food Chem. Toxicol.* 24(9):903-912.

Hua, Z.T., X.N. Liu, Y.H. Guo, and C. Meng. 2007. Transcriptional regulation of some antihyperglycemic agents on cytochrome P450 3A4. *Pharm. Biotechnol.* 14(3):182-185, 195.

Iwata, H., Y. Tezuka, S. Kadota, A. Hiratsuka, and T. Watabe. 2004. Identification and characterization of potent CYP3A4 inhibitors in *Schisandra* fruit extract. *Drug Metab. Dispos.* 32(12):1351-1358.

Jiang, W., X. Wang, X. Xu, and L. Kong. 2010. Effect of *Schisandra sphenanthera* extract on the concentration of tacrolimus in the blood of liver transplant patients. *Int. J. Clin. Pharmacol. Ther.* 48 (3):224-229.

Ko, K.M., M.K.T. Poon, S.P. Ip, K. Wu, and R. Ko. 2002. Protection against carbon tetrachloride liver toxicity by enantiomers of schisandrin B associated with differential changes in hepatic glutathione antioxidant system in mice. *Pharm. Biol.* 40(4):298-301.

Lee, Y.J., J.Y. Cho, J.H. Kim, et al. 2004. Extracts from *Schizandra chinensis* fruit activate estrogen receptors: A possible clue to its effects on nitric oxide-mediated vasorelaxation. *Biol. Pharm. Bull.* 27(7):1066-1069.

Liu, G. 1991. Pharmacological actions and clinical uses of *Fructus schizandrae*. In *Recent advances in Chinese herbal drugs—Actions and uses*, edited by Zhou, J., G. Liu, and J. Chen. Beijing: Science Press.

Makino, T., F. Mizuno, and H. Mizukami. 2006. Does a kampo medicine containing *Schisandra* fruit affect pharmacokinetics of nifedipine like grapefruit juice? *Biol. Pharm. Bull.* 29(10):2065-2069.

Mu, Y., J. Zhang, S. Zhang, et al. 2006. Traditional Chinese medicines *wu wei zi* (*Schisandra chinensis* Baill.) and *gan cao* (*Glycyrrhiza uralensis* Fisch) activate pregnane X receptor and increase warfarin clearance in rats. *J. Pharmacol. Exp. Ther.* 316(3):1369-1377.

Nakagiri, R., H. Oda, and T. Kamiya. 2003. Small scale rat hepatocyte primary culture with applications for screening hepatoprotective substances. *Biosci. Biotechnol. Biochem.* 67(8):1629-1635.

Pan, Q., Q. Lu, K. Zhang, and X. Hu. 2006. Dibenzocyclooctadiene lignans: A class of novel inhibitors of P-glycoprotein. *Cancer Chemother. Pharmacol.* 58(1):99-106.

Qiangrong, P., T. Wang, Q. Lu, and X. Hu. 2005. Schisandrin B—A novel inhibitor of P-glycoprotein. *Biochem. Biophys. Res. Commun.* 335(2):406-411.

Sandberg, F. 1993. *Schisandrae fructus—Wu wei zi.* Gothenburg. *Cited in* Upton, R. 1999. Schisandra berry: Analytical, quality control and therapeutic monograph. Santa Cruz, CA: American Herbal Pharmacopoeia.

Trifonova, A. 1954. Stimulation of labor activity using *Schizandra chinensis*. *Obstet. Gynecol.* 4:19-22.

Volicer, L., M. Sramka, C. Janku, R. Smetana, and V. Ditteova. 1966. Some pharmacological effects of *Schizandra chinensis*. *Arch. Int. Pharmacodyn. Ther.* 163(2):249-262.

Xin, H.W., X.C. Wu, Q. Li, et al. 2007. Effects of *Schisandra sphenanthera* extract on the pharmacokinetics of tacrolimus in healthy volunteers. *Br. J. Clin. Pharmacol.* 64(4):469-475.

Xin, H.W., X.C. Wu, Q. Li, A.R. Yu, and L. Xiong. 2009. Effects of *Schisandra sphenanthera* extract on the pharmacokinetics of midazolam in healthy volunteers. *Br. J. Clin. Pharmacol.* 67(5):541-546.

You, J.S., T.L. Pan, and Y.C. Hou. 2006. *Schisandra chinensis* protects against adriamycin-induced cardiotoxicity in rats. *Chang Gung Med. J.* 29(1):63-70.

Zhang, J.N., Y.W. Li, Y.X. Xu, and S.L. Yan. 2002. Induction effects of *Glycyrrhiza uralensis* Fisch. and *Schisandra chinensis* Baill. on the hepatic microsomal cytochrome P450 in rats. *Chin. Pharm. J.* 37(6):424-426.

Zhu, M., K.F. Lin, R.Y. Yeung, and R.C. Li. 1999. Evaluation of the protective effects of *Schisandra chinensis* on phase I drug metabolism using a CCl_4 intoxication model. *J. Ethnopharmacol.* 67(1):61-68.

Zhu, M., R.Y. Yeung, K.F. Lin, and R.C. Li. 2000. Improvement of phase I drug metabolism with *Schisandra chinensis* against CCl_4 hepatotoxicity in a rat model. *Planta Med.* 66(6):521-525.

Scrophularia spp.

Scrophulariaceae

Scrophularia marilandica L.
SCN: figwort
OCN: carpenter's square; eastern figwort

Scrophularia nodosa **L.**
SCN: figwort
Part: herb, root

Quick Reference Summary

Safety Class: 1
Interaction Class: A

Contraindications

None known.

Other Precautions

None known.

Drug and Supplement Interactions

None known.

Adverse Events and Side Effects

None known.

Pharmacological Considerations

While some earlier literature cautions against the use of figwort in persons with ventricular tachycardia (Mitchell 1983), this concern is believed to be theoretical in nature. No studies have been completed to support or refute this concern.

Pregnancy and Lactation

No information on the safety of figwort in pregnancy or lactation was identified in the scientific or traditional literature. Although this review did not identify any concerns for use while pregnant or nursing, safety has not been conclusively established.

Review Details

I. Drug and Supplement Interactions

Clinical Trials of Drug or Supplement Interactions

No clinical trials of drug or supplement interactions were identified.

Case Reports of Suspected Drug or Supplement Interactions

No case reports of suspected drug or supplement interactions were identified.

Animal Trials of Drug or Supplement Interactions

No animal trials of drug or supplement interactions were identified.

II. Adverse Events

Case Reports of Adverse Events

No case reports of adverse events were identified.

III. Pharmacology and Pharmacokinetics

Human Pharmacological Studies

No relevant human pharmacological studies were identified.

Animal Pharmacological Studies

No relevant animal pharmacological studies were identified.

In Vitro Pharmacological Studies

The compound harpagoside demonstrated negative chronotropic and positive inotropic effects in isolated rabbit hearts (Circosta et al. 1984; Sesterhenn et al. 2007).

IV. Pregnancy and Lactation

No information on the safety of figwort during pregnancy or lactation was identified.

V. Toxicity Studies

No toxicity studies were identified.

Literature Cited

Circosta, C., F. Occhiuto, S. Ragusa, et al. 1984. A drug used in traditional medicine: *Harpagophytum procumbens* DC. II. Cardiovascular activity. *J. Ethnopharmacol.* 11(3):259-274.

Mitchell, H. 1983. *British herbal pharmacopoeia*. Bournemouth, U.K.: British Herbal Medicine Association.

Sesterhenn, K., M. Distl, and M. Wink. 2007. Occurrence of iridoid glycosides in in vitro cultures and intact plants of *Scrophularia nodosa* L. *Plant. Cell Rep.* 26(3):365-371.

Scutellaria baicalensis Georgi — Lamiaceae

SCN: Chinese skullcap
PN: *huang qin* (root)
OCN: Baikal skullcap; scute
Part: root

Quick Reference Summary

Safety Class: 1
Interaction Class: B

Contraindications

None known.

Other Precautions

None known.

Drug and Supplement Interactions

Chinese skullcap may decrease plasma levels of the immune-suppressant drug cyclosporine (Lai et al. 2004).

Adverse Events and Side Effects

Allergic reactions to Chinese skullcap have been reported (Bensky et al. 2004). Diarrhea and stomach discomfort have been reported after ingestion of Chinese skullcap (Bensky et al. 2004).

Pharmacological Considerations

Coadministration of Chinese skullcap and cyclophosphamide potentiated the antimetastatic effect of cyclophosphamide in mice (Kaplya et al. 2004; Razina et al. 1987).

In vitro studies indicate that Chinese skullcap can inhibit the drug-metabolizing isoenzyme CYP1A2 (Kim et al. 2001; Kim et al. 2002).

The compound baicalin was found to decrease plasma levels of rosuvastatin and to induce the drug metabolizing isoenzyme CYP2B6 (Fan et al. 2008, 2009).

Pregnancy and Lactation

In traditional Chinese medicine, Chinese skullcap is commonly used during pregnancy (Chen and Chen 2004).

Very large doses (25 g/kg) of Chinese skullcap produced some developmental abnormalities in offspring of pregnant mice administered Chinese skullcap during pregnancy. Lower doses (0.25 and 12.5 g/kg) did not produce any adverse effects (Kim et al. 1993).

No information on the safety of Chinese skullcap during lactation was identified in the scientific or traditional literature. While this review did not identify any concerns for use while nursing, safety has not been conclusively established.

Review Details

I. Drug and Supplement Interactions

Clinical Trials of Drug or Supplement Interactions

In healthy volunteers, the compound baicalin was found to decrease plasma levels of rosuvastatin. Volunteers were orally administered 20 mg rosuvastatin before or after oral administration of 150 mg baicalin daily for 14 days. After baicalin treatment, the area under the plasma concentration-time curve (AUC_{0-72}) of rosuvastatin decreased by 1.7 to 47%, depending on the haplotype of the volunteer. Rosuvastatin is a substrate of the organic anion-transporting polypeptide 1B1 (OATP1B1), which was proposed as the mechanism of interaction (Fan et al. 2008).

Case Reports of Suspected Drug or Supplement Interactions

No case reports of suspected drug or supplement interactions were identified.

Animal Trials of Drug or Supplement Interactions

A reduction in the plasma levels of orally administered cyclosporine was observed in rats orally administered a decoction of Chinese skullcap at doses of 1 or 2 g/kg. No changes in cyclosporine levels were observed after intravenous administration of the drug (Lai et al. 2004).

In rats administered the compounds baicalin or baicalein at doses of 112 mmol/kg, an increase in serum cyclosporine levels was observed (Lai et al. 2004).

Coadministration of 1 ml/kg of Chinese skullcap extract for 12 days and a single dose of 125 mg/kg cyclophosphamide potentiated the antimetastatic effect of cyclophosphamide in mice with grafted tumors. The effect was attributed to modulated cytotoxic activity of natural killer cells and peritoneal macrophages during tumor growth (Kaplya et al. 2004; Razina et al. 1987).

II. Adverse Events

Case Reports of Adverse Events

Diarrhea and stomach discomfort have been reported after ingestion or injection of Chinese skullcap. Details on dose and preparation were not reported (Bensky et al. 2004).

Allergic reactions to Chinese skullcap have been reported, including flushing of the skin and macular erythema, mainly in the face and on the exposed areas of the limbs (Bensky et al. 2004).

III. Pharmacology and Pharmacokinetics

Human Pharmacological Studies

In healthy volunteers, the compound baicalin was found to induce the drug metabolizing isoenzyme CYP2B6. Volunteers were orally administered 150 mg of the CYP2B6 substrate bupropion before and after receiving 1.5 g baicalin daily for seven days. After treatment with baicalin, the maximum plasma concentration of hydroxybupropion (the metabolized form of bupropion) was increased by 73%, and the area under the time-concentration time curve ($AUC_{0-\infty}$) of hydroxybupropion increased by 87%, with no change in the elimination half-life of hydroxybupropion (Fan et al. 2009).

Animal Pharmacological Studies

No relevant animal pharmacological studies were identified.

In Vitro Pharmacological Studies

In human liver microsomes, individual flavonoid compounds isolated from Chinese skullcap inhibited the drug-metabolizing isoenzyme CYP1A2 at concentrations ranging from 0.7 to 51.3 μM. No effects on the drug-metabolizing isoenzymes CYP2B1, CYP2C19, CYP2D6, or CYP2E1 were observed. The compound baicalein inhibited CYP3A4 and the compound oroxylin A inhibited CYP2C9 (Kim et al. 2002).

In human and rat liver microsomes, an aqueous extract of Chinese skullcap inhibited the drug-metabolizing isoenzyme CYP1A2 (Kim et al. 2001). An extract of Chinese skullcap inhibited the drug-metabolizing isoenzyme CYP3A4 (Lee et al. 2007).

Platelet aggregation induced by platelet-activating factor was attenuated by the compound baicalein (Michibayashi 2002). Baicalein has been shown to inhibit platelet lipoxygenase (Sekiya and Okuda 1982).

IV. Pregnancy and Lactation

Offspring of rats orally administered 0.25, 12.49, or 24.98 g/kg of a water extract concentrate of Chinese skullcap on gestational days 7 to 11 demonstrated a dose-dependent increase in the incidence of skeletal variations, while urinary system abnormalities were observed in the two higher dose groups, with an equivalent number of abnormalities in each group. No changes in hematological or other developmental parameters were observed at any of the dose levels (Kim et al. 1993).

In traditional Chinese medicine, Chinese skullcap is commonly used during pregnancy (Chen and Chen 2004).

No information on the safety of Chinese skullcap in lactation was identified.

V. Toxicity Studies

Acute Toxicity

No nausea or vomiting was observed in dogs orally administered a Chinese skullcap extract at a dose of 15 g/kg (Chen and Chen 2004).

The LD_{50} of intraperitoneally administered baicalein in mice is 3 g/kg (Chen and Chen 2004). The LD_{50} of intravenously administered wogonin in rats is 286 mg/kg (Qi et al. 2009).

Short-Term Toxicity

Loose stools were observed in dogs orally administered 5 g/kg of Chinese skullcap extract daily for 8 weeks (Chen and Chen 2004).

Subchronic Toxicity

In rats intravenously administered 30, 60, or 120 mg/kg of the compound wogonin daily for 90 days, no macroscopic changes in heart, liver, spleen, lung, kidney, adrenal gland, thymus, thyroid gland, brain, uterus, testis, ovary, or prostate were observed. In the high-dose group, microscopic investigation revealed myofibrosis cordis and interstitial fibroblast, histiocyte, and inflammatory cell infiltrate in the hearts of rats (Qi et al. 2009).

Hepatotoxicity

The compound baicalein has demonstrated hepatoprotective effects against damage induced by *tert*-butyl hydroperoxide, carbon tetrachloride, and acetaminophen (Hwang et al. 2005; Jang et al. 2003; Park et al. 2008).

Genotoxicity

Chinese skullcap provided a protective effect against liver mutagenesis induced by aflatoxin B1 in rats (de Boer et al. 2005).

S

Literature Cited

Bensky, D., S. Clavey, and E. Stöger. 2004. *Chinese herbal medicine: Materia medica*. 3rd ed. Seattle: Eastland Press.

Chen, J.K., and T.T. Chen. 2004. *Chinese medical herbology and pharmacology*. City of Industry, CA: Art of Medicine Press.

de Boer, J.G., B. Quiney, P.B. Walter, et al. 2005. Protection against aflatoxin-B1-induced liver mutagenesis by *Scutellaria baicalensis*. *Mutat. Res.* 578(1-2):15-22.

Fan, L., J.-C. Wang, F. Jiang, et al. 2009. Induction of cytochrome P450 2B6 activity by the herbal medicine baicalin as measured by bupropion hydroxylation. *Eur. J. Clin. Pharmacol.* 65:403-409.

Fan, L., W. Zhang, D. Guo, et al. 2008. The effect of herbal medicine baicalin on pharmacokinetics of rosuvastatin, substrate of organic anion-transporting polypeptide 1B1. *Clin. Pharmacol. Ther.* 83 (3):471-476.

Hwang, J.M., C.J. Wang, F.P. Chou, et al. 2005. Protective effect of baicalin on *tert*-butyl hydroperoxide-induced rat hepatotoxicity. *Arch. Toxicol.* 79(2):102-109.

Jang, S.I., H.J. Kim, K.M. Hwang, et al. 2003. Hepatoprotective effect of baicalin, a major flavone from *Scutellaria radix*, on acetaminophen-induced liver injury in mice. *Immunopharmacol. Immunotoxicol.* 25(4):585-594.

Kaplya, O.A., E.Y. Sherstoboev, E.P. Zueva, et al. 2004. Effect of baikal skullcap extract administered alone or in combination with cyclophosphamide on natural cytotoxicity system in mice with Lewis lung carcinoma. *Bull. Exp. Biol. Med.* 137(5):471-474.

Kim, B.R., D.H. Kim, R. Park, et al. 2001. Effect of an extract of the root of *Scutellaria baicalensis* and its flavonoids on aflatoxin B1 oxidizing cytochrome P450 enzymes. *Planta Med.* 67(5):396-399.

Kim, J.-Y., S. Lee, D.-H. Kim, et al. 2002. Effects of flavonoids isolated from *Scutellariae radix* on cytochrome P-450 activities in human liver microsomes. *J. Toxicol. Environ. Health A* 65(5-6):373-381.

Kim, S.H., Y. Kim, S.S. Han, and J. Roh. 1993. Teratogenicity study of *Scutellariae radix* in rats. *Reprod. Toxicol.* 7(1):73-79.

Lai, M.Y., S.L. Hsiu, Y.C. Hou, S.Y. Tsai, and P.D. Chao. 2004. Significant decrease of cyclosporine bioavailability in rats caused by a decoction of the roots of *Scutellaria baicalensis*. *Planta Med.* 70(2):132-137.

Lee, S.S., B. Zhang, M.L. He, V.S.C. Chang, and H.F. Kung. 2007. Screening of active ingredients of herbal medicine for interaction with CYP450 3A4. *Phytother. Res.* 21(11):1096-1099.

Michibayashi, T. 2002. Platelet aggregating response to platelet activating factor participates in activation of the 12-lipoxygenase pathway in platelets from rabbits. *Int. Angiol.* 21(3):260-267.

Park, S.W., C.H. Lee, S.K. Yeong, et al. 2008. Protective effect of baicalin against carbon tetrachloride-induced acute hepatic injury in mice. *J. Pharmacol. Sci.* 106(1):136-143.

Qi, Q., J. Peng, W. Liu, et al. 2009. Toxicological studies of wogonin in experimental animals. *Phytother Res.* 23:417-422.

Razina, T.G., S.N. Udintsev, T.P. Prishchep, and K.V. Iaremenko. 1987. Enhancement of the selectivity of the action of the cytostatics cyclophosphane and 5-fluorouracil by using an extract of the Baikal skullcap in an experiment. *Voprosy Onkol.* 33(2):80-84.

Sekiya, K., and H. Okuda. 1982. Selective inhibition of platelet lipoxygenase by baicalein. *Biochem. Biophys. Res. Commun.* 105(3):1090-1095.

Scutellaria lateriflora L. — Lamiaceae

SCN: skullcap
OCN: blue skullcap; scullcap
Part: herb

Quick Reference Summary

Safety Class: 1
Interaction Class: A

Contraindications

None known.

Other Precautions

None known.

Drug and Supplement Interactions

None known.

Editors' Notes

Cases of toxicity reported in association with use of products labeled as skullcap are likely due to adulteration with germander (*Teucrium* spp.), which has been reported to cause liver toxicity. A text on herbal safety cites an "increasing number of case reports to suggest that the ingestion of skullcap-containing preparations can induce hepatoxic reactions." Yet an appended clarification acknowledges the adulteration of skullcap with species of *Teucrium*. This adulteration, along with more recent reports of hepatitis associated with *Teucrium* consumption, led to the conclusion that it is "unclear at the moment, whether the hepatotoxic effects that have been associated with preparations containing skullcap should be attributed to *Scutellaria*, *Teucrium*, or both" (De Smet 1993).

A more recent analysis of the literature (Mills and Bone 2005) on liver damage caused by diterpenes isolated from germander species and hepatic effects of related diterpenes isolated from skullcap indicated that, although both sets of diterpenes demonstrated hepatotoxic activity in mice, adverse effects of the skullcap diterpenes were limited to a few liver cells in a small percentage of tested animals, whereas germander diterpenes caused toxicity that was characterized as extensive and severe. Skullcap neoclerodane diterpenes lack the furan ring that is a structural requirement for the hepatotoxicity caused by the compound teucrin A from germander (Fau et al. 1997; Furbee et al. 2006; Haouzi et al. 2000; Lekehal et al. 1996).

Adverse Events and Side Effects

None known.

Pharmacological Considerations

None known.

Pregnancy and Lactation

No information on the safety of skullcap in pregnancy or lactation was identified in the scientific or traditional literature. Although this review did not identify any concerns for use while pregnant or nursing, safety has not been conclusively established.

Review Details

I. Drug and Supplement Interactions

Clinical Trials of Drug or Supplement Interactions

No clinical trials of drug or supplement interactions were identified.

Case Reports of Suspected Drug or Supplement Interactions

No case reports of suspected drug or supplement interactions were identified.

Animal Trials of Drug or Supplement Interactions

No animal trials of drug or supplement interactions were identified.

II. Adverse Events

Adverse Events Reported in Clinical Trials

In a trial of freeze-dried skullcap preparations at doses of up to 400 mg, no adverse events were reported in the skullcap group, and one adverse event was reported in the placebo group (Wolfson and Hoffmann 2003).

Case Reports of Adverse Events

Several cases of hepatotoxicity have been reported in persons taking products labeled to contain skullcap, usually in combination with other herbs. In most cases, skullcap was suggested as the herb responsible for the hepatotoxicity (Enlow 1996; Hullar et al. 1999; MacGregor et al. 1989); however, skullcap is botanically similar to two species of germander (*Teucrium canadense* and *T. chamaedrys*), plants that contain a hepatotoxic compound. A number of sources acknowledge that germander species have been common adulterants of skullcap (Applequist 2006; De Smet 1993; Gafner et al. 2003; Mills and Bone 2005; Sundaresan et al. 2006). The hepatotoxicity in the case reports above is most likely not due to skullcap but instead to germander, other herbs in the consumed formulas, such as mistletoe (*Viscum album*), or other causes (Wolfson and Hoffmann 2003).

S

III. Pharmacology and Pharmacokinetics

Human Pharmacological Studies

No relevant human pharmacological studies were identified.

Animal Pharmacological Studies

No relevant animal pharmacological studies were identified.

In Vitro Pharmacological Studies

In male rat hepatocytes incubated with 100 μg/ml of diterpenoids from skullcap, the compounds caused apoptosis. Reactive metabolites formed by CYP3A depleted cellular thiols, increasing cellular calcium and opening the mitochondrial permeability transition pore (MPTP). Cyclosporine A, an inhibitor of MPTP opening, prevented cytochrome *c* release, caspase activation, and apoptosis, and caspase inhibitors also prevented apoptosis (Haouzi et al. 2000).

Ethanol, glycerin, and water extracts of skullcap showed significant inhibition of the drug-metabolizing isoenzyme CYP3A4 (Awad et al. 2003).

IV. Pregnancy and Lactation

No information on the safety of skullcap in pregnancy or lactation was identified.

V. Toxicity Studies

Acute Toxicity

In mice administered a single 20 mg dose (per 28–30 g animal) of diterpenoids isolated from skullcap, an increase in hepatic caspases was observed as compared to controls. In one skullcap-treated mouse, apoptosis affected a few hepatocytes, while in two other skullcap-treated mice, necrosis affected a few hepatocytes. No adverse effects were observed in the 13 other skullcap-treated mice. No apoptosis or necrosis was observed in the control group, nor in a group administered skullcap and pretreated with a caspase inhibitor (Haouzi et al. 2000).

Literature Cited

Applequist, W. 2006. *The identification of medicinal plants: A handbook of the morphology of botanicals in commerce.* Austin, TX: American Botanical Council.

Awad, R., J.T. Arnason, V. Trudeau, et al. 2003. Phytochemical and biological analysis of skullcap (*Scutellaria lateriflora* L.): A medicinal plant with anxiolytic properties. *Phytomedicine* 10(8):640-649.

De Smet, P.A.G.M. 1993. *Adverse effects of herbal drugs, Volume 2.* Berlin: Springer.

Enlow, M. 1996. Herbal hepatotoxicity. *Pharmacol. Toxicol.* 21:162.

Fau, D., M. Lekehal, G. Farrell, et al. 1997. Diterpenoids from germander, an herbal medicine, induce apoptosis in isolated rat hepatocytes. *Gastroenterology* 113(4):1334-1346.

Furbee, R.B., K.S. Barlotta, M.K. Allen, and C.P. Holstege. 2006. Hepatotoxicity associated with herbal products. *Clin. Lab. Med.* 26(1):227-241.

Gafner, S., C. Bergeron, L.L. Batcha, et al. 2003. Analysis of *Scutellaria lateriflora* and its adulterants *Teucrium canadense* and *Teucrium chamaedrys* by LC-UV/MS, TLC, and digital photomicroscopy. *J. AOAC Int.* 86(3):453-460.

Haouzi, D., M. Lekehal, A. Moreau, et al. 2000. Cytochrome P 450-generated reactive metabolites cause mitochondrial permeability transition, caspase activation, and apoptosis in rat hepatocytes. *Hepatology* 32(2):303-311.

Hullar, T.E., B.L. Sapers, P.M. Ridker, et al. 1999. Herbal toxicity and fatal hepatic failure. *Am. J. Med.* 106(2):267-268.

Lekehal, M., D. Pessayre, J.M. Lereau, et al. 1996. Hepatotoxicity of the herbal medicine germander: Metabolic activation of its furano diterpenoids by cytochrome P 450 3A depletes cytoskeleton-associated protein thiols and forms plasma membrane blebs in rat hepatocytes. *Hepatology* 24(1):212-218.

MacGregor, F.B., V.E. Abernethy, S. Dahabra, I. Cobden, and P.C. Hayes. 1989. Hepatotoxicity of herbal remedies. *Br. Med. J.* 299(6708):1156.

Mills, S., and K. Bone. 2005. *The essential guide to herbal safety*. St. Louis: Elsevier.

Sundaresan, P.R., S.A. Slavoff, E. Grundel, et al. 2006. Isolation and characterisation of selected germander diterpenoids from authenticated *Teucrium chamaedrys* and *T. canadense* by HPLC, HPLC-MS and NMR. *Phytochem. Anal.* 17:243-250.

Wolfson, P., and D.L. Hoffmann. 2003. An investigation into the efficacy of *Scutellaria lateriflora* in healthy volunteers. *Altern. Ther. Health Med.* 9(2):74-78.

Selenicereus grandiflorus (L.) Britton & Rose — Cactaceae

SCN: night-blooming cereus
Syn: *Cactus grandiflorus* L.; *Cereus grandiflorus* (L.) Mill.
OCN: queen-of-the-night
Part: flower, stem

Quick Reference Summary

Safety Class: 2d
Interaction Class: B

Contraindications

Do not exceed recommended dose (Felter and Lloyd 1898; Mitchell 2003; Yarnell and Abascal 2003).

Other Precautions

None known.

Drug and Supplement Interactions

Historical records of the cardiac activity of night-blooming cereus (Felter and Lloyd 1898) suggests that it be used with caution in persons on heart medications.

Notice

Diuretic (Felter and Lloyd 1898; Parke Davis 1894); *see* Appendix 2.

Standard Dose

The standard dose is 10 to 20 drops of tincture twice daily (Mitchell 2003).

Editors' Notes

Although one reference (Williamson 2003) reports a mild positive inotropic effect for the alkaloid hordenine, another reference indicates that hordenine is noncumulative (List and Hörhammer 1973). The long history of use and the lack of reports of human toxicity suggest that such an effect is likely to be minimal.

Adverse Events and Side Effects

None known.

Pharmacological Considerations

None known.

Pregnancy and Lactation

No information on the safety of night-blooming cereus in pregnancy or lactation was identified in the scientific or traditional literature. Although this review did not identify any concerns for use while pregnant or nursing, safety has not been conclusively established.

Review Details

I. Drug and Supplement Interactions

Clinical Trials of Drug or Supplement Interactions

No clinical trials of drug or supplement interactions were identified.

Case Reports of Suspected Drug or Supplement Interactions

No case reports of suspected drug or supplement interactions were identified.

Animal Trials of Drug or Supplement Interactions

No animal trials of drug or supplement interactions were identified.

II. Adverse Events

Case Reports of Adverse Events

No case reports of adverse events were identified.

III. Pharmacology and Pharmacokinetics

Human Pharmacological Studies

No relevant human pharmacological studies were identified.

S

Literature Cited

Aaronov, D., D. Tasher, A. Levine, et al. 2008. Natural history of food allergy in infants and children in Israel. *Ann. Allergy Asthma Immunol.* 101(6):637-640.

Agne, P.S., E. Bidat, P.S. Agne, F. Rance, and E. Paty. 2004. Sesame seed allergy in children. *Eur. Ann. Allergy Clin. Immunol.* 36(8):300-305.

Asero, R., G. Mistrello, D. Roncarolo, P.L. Antoniotti, and P. Falagiani. 1999. A case of sesame seed-induced anaphylaxis. *Allergy* 54(5):526-527.

Caminiti, L., D. Vita, G. Passalacqua, et al. 2006. Tahini, a little known sesame-containing food, as an unexpected cause of severe allergic reaction. *J. Invest. Allergol. Clin. Immunol.* 16(5):308-310.

Dalal, I., I. Binson, A. Levine, et al. 2003. The pattern of sesame sensitivity among infants and children. *Pediatr. Allergy Immunol.* 14(4):312-316.

Derby, C.J., M.H. Gowland, and J.O. Hourihane. 2005. Sesame allergy in Britain: A questionnaire survey of members of the Anaphylaxis Campaign. *Pediatr. Allergy Immunol.* 16(2):171-175.

Gangur, V., C. Kelly, and L. Navuluri. 2005. Sesame allergy: A growing food allergy of global proportions? *Ann. Allergy Asthma Immunol.* 95(1):4-11.

James, C., A. Williams-Akita, Y.A. Rao, L.T. Chiarmonte, and A.T. Scheider. 1991. Sesame seed anaphylaxis. *N.Y. State Med. J.* 91:457-458.

Morisset, M., D.A. Moneret-Vautrin, G. Kanny, et al. 2003. Thresholds of clinical reactivity to milk, egg, peanut and sesame in immunoglobulin E-dependent allergies: Evaluation by double-blind or single-blind placebo-controlled oral challenges. *Clin. Exp. Allergy* 33(8):1046-1051.

Panizzolo, C., M. Tura, and A. Barbato. 2005. Anaphylaxis to sesame paste. *Eur. Ann. Allergy Clin. Immunol.* 37(1):34-35.

Sacco, S.M., K.A. Power, J. Chen, W.E. Ward, and L.U. Thompson. 2007. Interaction of sesame seed and tamoxifen on tumor growth and bone health in athymic mice. *Exp. Biol. Med.* 232(6):754-761.

Takeuchi, H., L.Y. Mooi, Y. Inagaki, and P. He. 2001. Hypoglycemic effect of a hot-water extract from defatted sesame (*Sesamum indicum* L.) seed on the blood glucose level in genetically diabetic KK-Ay mice. *Biosci. Biotechnol. Biochem.* 65(10):2318-2321.

Sida cordifolia L. — Malvaceae

SCN: heart-leaf sida
AN: *bala*
OCN: country mallow; fanpetals; flannelweed; llima
Part: leaf, root, seed

Quick Reference Summary

Safety Class: 1
Interaction Class: A

Contraindications

None known.

Other Precautions

None known.

Drug and Supplement Interactions

None known.

Editors' Notes

Heart-leaf sida contains 0.29 to 0.97% of the compound ephedrine (Gunatilaka et al. 1980; Marchei et al. 2006). The U.S. Food and Drug Administration declared in 2004 that dietary supplements containing ephedrine alkaloids are adulterated, and identified heart-leaf sida as containing ephedrine alkaloids (FDA 2004). In establishing this ban for ephedrine in dietary supplements, however, FDA stated that it does not apply to "traditional Asian medicine" that contains ephedrine alkaloids (FDA 2004).

Adverse Events and Side Effects

None known.

Pharmacological Considerations

None known.

Pregnancy and Lactation

No information on the safety of heart-leaf sida in pregnancy or lactation was identified in the scientific or traditional literature. Although this review did not identify any concerns for use while pregnant or nursing, safety has not been conclusively established.

Review Details

I. Drug and Supplement Interactions

Clinical Trials of Drug or Supplement Interactions

No clinical trials of drug or supplement interactions were identified.

Case Reports of Suspected Drug or Supplement Interactions

No case reports of suspected drug or supplement interactions were identified.

Lekehal, M., D. Pessayre, J.M. Lereau, et al. 1996. Hepatotoxicity of the herbal medicine germander: Metabolic activation of its furano diterpenoids by cytochrome P 450 3A depletes cytoskeleton-associated protein thiols and forms plasma membrane blebs in rat hepatocytes. *Hepatology* 24(1):212-218.

MacGregor, F.B., V.E. Abernethy, S. Dahabra, I. Cobden, and P.C. Hayes. 1989. Hepatotoxicity of herbal remedies. *Br. Med. J.* 299(6708):1156.

Mills, S., and K. Bone. 2005. *The essential guide to herbal safety*. St. Louis: Elsevier.

Sundaresan, P.R., S.A. Slavoff, E. Grundel, et al. 2006. Isolation and characterisation of selected germander diterpenoids from authenticated *Teucrium chamaedrys* and *T. canadense* by HPLC, HPLC-MS and NMR. *Phytochem. Anal.* 17:243-250.

Wolfson, P., and D.L. Hoffmann. 2003. An investigation into the efficacy of *Scutellaria lateriflora* in healthy volunteers. *Altern. Ther. Health Med.* 9(2):74-78.

Selenicereus grandiflorus (L.) Britton & Rose — Cactaceae

SCN: night-blooming cereus
Syn: *Cactus grandiflorus* L.; *Cereus grandiflorus* (L.) Mill.
OCN: queen-of-the-night
Part: flower, stem

Quick Reference Summary

Safety Class: 2d
Interaction Class: B

Contraindications

Do not exceed recommended dose (Felter and Lloyd 1898; Mitchell 2003; Yarnell and Abascal 2003).

Other Precautions

None known.

Drug and Supplement Interactions

Historical records of the cardiac activity of night-blooming cereus (Felter and Lloyd 1898) suggests that it be used with caution in persons on heart medications.

Notice

Diuretic (Felter and Lloyd 1898; Parke Davis 1894); *see* Appendix 2.

Standard Dose

The standard dose is 10 to 20 drops of tincture twice daily (Mitchell 2003).

Editors' Notes

Although one reference (Williamson 2003) reports a mild positive inotropic effect for the alkaloid hordenine, another reference indicates that hordenine is noncumulative (List and Hörhammer 1973). The long history of use and the lack of reports of human toxicity suggest that such an effect is likely to be minimal.

Adverse Events and Side Effects

None known.

Pharmacological Considerations

None known.

Pregnancy and Lactation

No information on the safety of night-blooming cereus in pregnancy or lactation was identified in the scientific or traditional literature. Although this review did not identify any concerns for use while pregnant or nursing, safety has not been conclusively established.

S

Review Details

I. Drug and Supplement Interactions

Clinical Trials of Drug or Supplement Interactions

No clinical trials of drug or supplement interactions were identified.

Case Reports of Suspected Drug or Supplement Interactions

No case reports of suspected drug or supplement interactions were identified.

Animal Trials of Drug or Supplement Interactions

No animal trials of drug or supplement interactions were identified.

II. Adverse Events

Case Reports of Adverse Events

No case reports of adverse events were identified.

III. Pharmacology and Pharmacokinetics

Human Pharmacological Studies

No relevant human pharmacological studies were identified.

Animal Pharmacological Studies

No relevant animal pharmacological studies were identified.

In Vitro Pharmacological Studies

No relevant in vitro pharmacological studies were identified.

IV. Pregnancy and Lactation

No information on the safety of night-blooming cereus during pregnancy or lactation was identified.

V. Toxicity Studies

No toxicity studies were identified.

Literature Cited

Felter, H.W., and J.U. Lloyd. 1898. *King's American dispensatory*. 18th ed., 3rd rev. 2 vols. Cincinnati: Ohio Valley Co.

List, P.H., and H. Hörhammer. 1973. *Hagers handbuch der pharmazeutischen praxis*. Berlin: Springer.

Mitchell, W.A. 2003. *Plant medicine in practice: Using the teachings of John Bastyr*. New York: Churchill Livingstone.

Parke Davis. 1894. *Descriptive catalogue of the laboratory products of Parke, Davis & Company*. Detroit: Press of Parke, Davis & Company.

Williamson, E.M. 2003. *Potter's herbal cyclopedia*. Saffron Walden, Essex: C.W. Daniel Co.

Yarnell, E., and K. Abascal. 2003. Botanicals for regulating heart rhythms. *Altern. Complement. Ther.* 9(3):125-129.

Senna spp. — Fabaceae

Senna alexandrina **Mill.**
SCN: senna
Syn: *Cassia acutifolia* Delile; *Cassia angustifolia* Vahl; *Cassia lanceolata* Forssk.; *Cassia senna* L.; *Senna acutifolia* (Delile) Batka; *Senna angustifolia* (Vahl) Batka
AN: *svarnapatri*
OCN: Alexandrian senna; Indian senna; Tinnevelly senna; true senna

Senna obtusifolia **(L.) H.S. Irwin & Barneby**
SCN: sickle-pod senna
Syn: *Cassia obtusifolia* L.
PN: *jue ming zi* (seed)

Senna tora **(L.) Roxb.**
SCN: sickle-pod senna
Syn: *Cassia tora* L.
OCN: foetid cassia; wild senna
Part: fruit (pod), leaf

Quick Reference Summary

Safety Class: 2d
Interaction Class: A

Contraindications

Not for use in persons with intestinal obstruction, abdominal pain of unknown origin, or any inflammatory condition of the intestines (i.e., appendicitis, colitis, Crohn's disease, irritable bowel syndrome, and melanosis coli) (Bradley 1992; De Smet 1993; Martindale and Reynolds 1996; Wichtl 2004).

Not for use in excess of 8 consecutive days (Bradley 1992; De Smet 1993; Leung and Foster 1996; List and Hörhammer 1973; Weiss and Meuss 2001; Wichtl 2004).

Other Precautions

None known.

Drug and Supplement Interactions

See Pharmacological Considerations.

Use of stimulant laxatives, such as senna or sickle-pod senna, may reduce the gastrointestinal transit time and thus reduce the absorption of orally administered drugs (Brinker 2001; De Smet 1993).

Standard Dose

Senna leaf: 0.5–3.0 g infused in hot water for 10–15 minutes (Bensky et al. 2004; Weiss and Meuss 2001; Wichtl 2004).

Senna fruit or leaf containing the following dose of the compounds sennosides A and B: Adults and children 12 years of age and over, oral dosage is 12–50 mg of sennosides once or twice daily; children 6 to under 12 years of age, oral dosage is 6–25 mg of sennosides once or twice daily; children 2 to under 6 years of age, oral dosage is 3–12.5 mg of sennosides once or twice daily (FDA 1985).

Senna leaf typically contains 1.5–3% dianthrone glucosides, consisting primarily of sennosides A and B (Khan and Abourashed 2011).

Pharmacopeias of Europe, India, and China specify that senna leaf must contain a minimum of 2.5% hydroxyanthracene glycosides, calculated as sennoside B. Senna fruit typically contains 3–5% sennosides. Pharmacopeias of Europe and India recognize *Senna alexandrina* as two species, *S. angustifolia*, which must contain a minimum of 3.4%, and *S. acutifolia*, which must contain a minimum of 2.2% hydroxyanthracene glycosides, calculated as sennoside B.

The proportion of anthraquinone glycosides is higher in the leaves than in the pods, with the composition of the laxative compounds differing in the leaf and pod (Wichtl 2004).

Notice

Stimulant laxative (Bradley 1992; De Smet 1993; FDA 1985; Leng-Peschlow 1992a; Leung and Foster 1996; List and Hörhammer 1973; Martindale and Reynolds 1996; Weiss and Meuss 2001; Wichtl 2004; Williamson 2003); *see* Appendix 2.

Editors' Notes

Records of use, case reports, research papers, and recommendations for appropriate use of senna products often do not differentiate between the fruit (pod) or the leaf, and numerous references are addressed to the sennosides contained in both of these plant parts. Unless otherwise specified, the information in this entry is relevant to both senna fruit and senna leaf.

Several authorities note that preparations of senna fruit act more gently than senna leaf (Weiss and Meuss 2001; Wichtl 2004). However, concerns regarding the long-term use of stimulant laxatives are relevant to both the fruit and the leaf.

The American Herbal Products Association has established a trade requirement (AHPA 2011) that products containing this herb in sufficient quantity to warrant such labeling bear the following label statement:

> NOTICE: Do not use this product if you have abdominal pain or diarrhea. Consult a health care provider prior to use if you are pregnant or nursing. Discontinue use in the event of diarrhea or watery stools. Do not exceed recommended dose. Not for long-term use.

Adverse Events and Side Effects

In overdose, senna may cause griping and severe diarrhea with consequent loss of fluid and electrolytes (ESCOP 2003).

Cases of hepatitis have been associated with chronic use of senna fruit and leaf (Beuers et al. 1991; Seybold et al. 2004; Soyuncu et al. 2008).

Accidental exposure to senna-containing laxatives in children under 6 years old has resulted in severe diaper rash, blistering, and skin sloughing after ingestion of doses containing 15 mg of sennosides (the standard adult dose) or more (Spiller et al. 2003).

Allergic reactions to senna have been reported in workers routinely exposed to senna pod dust (Marks et al. 1991) and to senna used in hair dyes (Helin and Makinen-Kiljunen 1996), such as *Senna italica* (syn. *S. obovata*).

Pharmacological Considerations

Long-term use of senna and other anthraquinone laxatives may cause pseudomelanosis coli, a brownish pigmentation of the colon. Although pseudomelanosis coli has been regarded as a harmless condition (Leng-Peschlow 1992c), some researchers have suggested a possible relationship between the condition and an increased risk of the development of colon cancer (Siegers et al. 1993b). This risk, however, may be due to other confounding factors such as chronic constipation or diet (Sonnenberg and Müller 1993; van Gorkom et al. 1999).

Concomitant use of senna or sickle-pod senna is cautioned with antiarrhythmic drugs, thiazide diuretics, corticosteroids, licorice, and botanicals containing cardiac glycosides, as long-term use of senna or sickle-pod senna as a laxative can cause or exacerbate potassium loss (Brinker 2001; De Smet 1993; ESCOP 2003).

Use of stimulant laxatives, such as senna or sickle-pod senna, may reduce the gastrointestinal transit time and thus reduce the absorption of orally administered drugs (Brinker 2001; De Smet 1993).

Pregnancy and Lactation

Some references advise cautious or supervised use of senna fruit or the contained seed and senna leaf during pregnancy (Bradley 1992; Chen and Chen 2004; ESCOP 2003).

A review of human studies of senna in pregnancy indicated that the use of senna for 2 weeks to 9 months showed good laxative efficacy with few side effects. Senna was not associated with stimulating uterine contractions, even in high-risk pregnancies, including those with a tendency to premature labor or to bleeding late in pregnancy (Leng-Peschlow 1992b). Epidemiological studies have shown a lack of birth defects in infants whose mothers used senna-based laxatives during pregnancy (Ács et al. 2009; Ács et al. 2010).

An evaluation of the safety of a variety of laxatives concluded that senna, without differentiation between the fruit and the leaf, is the laxative of choice in pregnancy and lactation (Gattuso and Kamm 1994).

Animal studies have indicated a lack of abortifacient, teratogenic, or fetotoxic effects, and one study in sheep indicated that senna did not stimulate uterine contractions (Garcia-Villar 1988; Mengs 1986).

The American Academy of Pediatrics lists senna as a product that is usually compatible with breast-feeding (AAP 2001). Although laxative compounds from senna have been shown to cross into breast milk, no laxative effects have been observed in nursing infants whose mothers took sennosides (Baldwin 1963; Faber and Strenge-Hesse 1988; Werthmann and Krees 1973).

Review Details

I. Drug and Supplement Interactions

Clinical Trials of Drug or Supplement Interactions

No clinical trials of drug or supplement interactions were identified.

Case Reports of Suspected Drug or Supplement Interactions

Near fatal bleeding was reported in a woman who had been stable on warfarin for 7 years and had initiated use of a "senna-based laxative" (plant part not stated) several months prior to the bleeding event. An increase in INR (a standardized scale used to report the results of blood coagulation tests; increased INR indicates slowed blood clotting) was reported in conjunction with the bleeding and was associated with senna-induced diarrhea subsequent to a significant increase in senna use for three weeks. The reporting physicians indicated that diarrhea can reduce absorption of vitamin K and increase the risk of bleeding (Kittisupamongkol et al. 2008).

Animal Trials of Drug or Supplement Interactions

No animal trials of drug or supplement interactions were identified.

II. Adverse Events

Case Reports of Adverse Events

A 52-year-old woman developed severe hepatotoxicity after ingestion of 1 liter per day of tea made from 70 g of dried senna fruit for over 3 years (Vanderperren et al. 2005). High levels of cadmium and mercury were found in the woman's urine, though analysis of the senna product indicated no cadmium or mercury contamination (Vanderperren et al. 2005).

Hepatitis was reported in a 28-year-old woman with a variant of the drug-metabolizing isoenzyme CYP2D6. The woman had recently been consuming 3 to 4 liters of beer per week, and had also been consuming an herbal tea containing senna leaf (dose, duration, and presence of other ingredients not specified). Re-exposure to the tea led to elevated liver enzyme levels (Seybold et al. 2004).

Hepatitis was reported in a 26-year-old woman who had been taking 10 g senna leaf tea twice a week for an unspecified length of time, and began also taking senna fruit containing 100 mg sennoside B daily 1 month prior to the hepatitis. This combined dose of senna products was noted as being 10 times the recommended dose. Liver enzyme levels returned to normal after cessation of senna products (Beuers et al. 1991).

Portal vein thrombosis was reported in a 42-year-old woman who had taken 200 ml of senna leaf tea (amount of senna used to make tea was not specified) daily for 2 years (Soyuncu et al. 2008).

Paralytic ileum was reported in an 85-year-old man who had been taking a tea with senna leaf and fruit (with licorice, mallow, fennel, "balum," and caraway) every other day for 2 months. At the time the condition was diagnosed, the man was being hospitalized for transitory ischemic stroke. He had a history of high blood pressure, chronic obstructive bronchopneumopathy, and cholelithiasis, and a 5-year history of constipation, for which lactulose, bisacodyl, and sodium picosulfate were taken, with enemas used every 15 to 30 days (Sossai et al. 2007).

A study of cases of accidental exposure to senna-containing laxatives in children under 6 years of age reported that 33% experienced severe diaper rash, and 10% developed blisters and skin sloughing. The mean dose ingested was 105 mg (presumably measured as sennosides), with a range of 15 to 375 mg. Approximately 12% of children ingested only a single dose, and several of those children experienced severe diaper rash or developed blisters and skin sloughing. Symptoms were significantly worse in children who wore diapers as compared to those who were fully toilet trained (Spiller et al. 2003).

Clubbing of fingers and toes has been reported in association with long-term abuse of senna, primarily in patients with anorexia nervosa (Armstrong et al. 1981; Levine et al. 1981; Lim et al. 2008; Malmquist et al. 1980; Prior and White 1978; Silk et al. 1975).

IgE-mediated occupational asthma, allergy, and rhinoconjunctivitis have been reported in workers routinely exposed to senna pod dust (Marks et al. 1991) and to species of senna used in hair colors (Helin and Makinen-Kiljunen 1996).

III. Pharmacology and Pharmacokinetics

Human Pharmacological Studies

A prospective cohort study of the risk of anthraquinone laxatives use for the development of colorectal neoplasms indicated no statistically significant risk of anthranoid use for the development of colorectal adenomas or carcinomas. Macroscopic and high-grade microscopic melanosis coli were not significant risk factors for the development of adenomas or carcinomas (Nusko et al. 2000).

In a retrospective study of 2277 patients who underwent colonoscopy, the correlation between patients' laxative use or melanosis coli and colorectal neoplasms was examined. Increases in colorectal adenomas were observed in patients with melanosis coli and in patients with a history of laxative use (Nusko et al. 1993).

Animal Pharmacological Studies

No effects of senna on intestinal platelet-activating factor were observed in rats, mice, and guinea pigs orally administered senna in single or repeated doses of 60 to 240 mg/kg (Capasso et al. 1993; Mascolo et al. 1992).

S

No chronic changes in colonic motility were observed in rats orally administered 10 or 40 mg/kg sennosides daily for 23 weeks (Fioramonti et al. 1993).

No significant damage of myenteric neurons in the colon of rats or mice was observed after animals were given purgative doses of sennosides in drinking water for 4 or 5 months (Kiernan and Heinicke 1989).

In guinea pigs intragastrically administered sennosides daily for 14 days, the mucosa of the cecum and upper colon became brown, and microscopic examination revealed cytoplasmic degeneration and increased apoptosis in the colonic surface epithelium (Mengs and Rudolph 1993).

No induction of aberrant crypt foci (considered putative preneoplastic lesions) or increases in incidence of chemically induced aberrant crypt foci were observed in rats fed diets containing up to 0.2% anthraquinone glycosides from senna for 56 days (Mereto et al. 1996).

In Vitro Pharmacological Studies

No relevant in vitro pharmacological studies were identified.

IV. Pregnancy and Lactation

Some references advise cautious or supervised use of senna fruit or the contained seed and senna leaf during pregnancy (Bradley 1992; Chen and Chen 2004; ESCOP 2003).

A review of the safety of senna in pregnancy indicated that nine human studies on senna or a combination of various senna preparations and psyllium seed for 2 weeks to 9 months showed good laxative efficacy with few side effects. Senna was not associated with stimulating uterine contractions, even in high-risk pregnancies, including those with a tendency to premature labor or bleeding late in pregnancy (Leng-Peschlow 1992b).

In an epidemiological study of congenital malformations, 22,843 infants with congenital malformations were included. Of these, 506 had mothers who had taken senna during pregnancy. An additional 38,151 normal infants were identified whose mothers also used senna during pregnancy. Daily dosage ranged between 10 and 30 mg of sennoside A and B, and most pregnant women used 20 mg per day. The authors concluded that senna treatment was not associated with a higher risk of congenital abnormalities in the offspring of pregnant women with constipation (Ács et al. 2009).

No abortifacient, teratogenic, or fetotoxic effects were observed in rats or rabbits administered sennosides, even when large doses were administered (Mengs 1986). Electromyographic studies in pregnant sheep did not show any stimulation of uterine contractions by sennosides, and a slight inhibition of contraction frequency was observed in some animals (Garcia-Villar 1988).

The American Academy of Pediatrics lists senna as a product that is usually compatible with breast-feeding (AAP 2001).

In lactating women administered doses of senna containing 15 mg sennosides daily for 3 days, approximately 0.007% of the sennoside dose (calculated as rhein) was detected in breast milk. No effects of the treatment on infant bowel habits were observed (Faber and Strenge-Hesse 1988).

In lactating women administered a single dose of senna containing 8.6 mg of sennosides A and B, some infants had diarrhea. Doubling the dose administered to the women, however, did not cause diarrhea, suggesting that the diarrhea observed was due to a cause other than maternal laxative ingestion (Werthmann and Krees 1973).

No effects on infant bowel habits were observed after nursing mothers were administered 1 teaspoon of a granular product made from senna, described as equivalent to the total active constituents of 450 mg of senna fruit. The authors indicated that this study is not indicative of whether laxative compounds are excreted in breast milk, but only indicates the lack of effect on infant bowel habits (Baldwin 1963).

V. Toxicity Studies

Acute Toxicity

The LD_{50} of orally administered sennosides in mice and rats is 5 g/kg, with death attributed to loss of water and electrolytes after severe diarrhea (Mengs 1988).

Short-Term Toxicity

No significant adverse effects were observed in rats orally administered sennosides at doses of 5, 10, or 20 mg/kg daily for 4 weeks. In the 20 mg/kg group, a low-grade laxative effect was observed, along with an increase in kidney weight (Mengs 1988).

No adverse effects were observed in dogs orally administered up to 500 mg/kg sennosides daily for 4 weeks (Mengs 1988).

In goats orally administered fresh fruit and leaf of senna at doses of 1, 5, and 10 g/kg daily, several goats died within 30 days. Serum levels of aspartate aminotransferase, ammonia, urea, and total cholesterol were increased, while total protein was decreased (El Sayed et al. 1983).

Subchronic Toxicity

In rats administered senna pods via oral gavage at doses of 100, 300, 750, or 1500 mg/kg daily for up to 13 weeks, dose- and treatment-related clinical signs included abnormal feces, which were seen to varying degrees in animals administered 300 mg/kg or more. In animals receiving 750 or 1500 mg/kg, notable changes in electrolytes in both serum and urine were observed, along with an increase in absolute and relative kidney weights, and a darkening of the kidneys. However, there were no indications in laboratory parameters of any renal dysfunction (Mengs et al. 2004).

No promotion of chemically induced colorectal tumors was observed in mice administered a diet containing 0.03% sennosides daily for 20 weeks, and no significant changes in serum electrolytes or parameters of hepatotoxicity and nephrotoxicity were observed (Siegers et al. 1993a).

Chronic Toxicity

No induction of aberrant crypt foci was observed in rats administered 30 or 60 mg/kg of senna fruit extract daily for 2 years. Administration of the senna fruit extract inhibited tumor development in animals treated with a tumor initiating agent (Borrelli et al. 2005).

An increase in the appearance of chemically induced tumors was observed in rats administered a dose of 100 mg/kg senna fruit extract (a dose that induced chronic diarrhea) daily for 13 to 28 weeks. No induction of aberrant crypt foci was observed in rats administered 10 mg/kg of senna fruit extract daily (a dose that produced laxation), and no promotion of chemically induced tumors was seen at that dose (Mascolo et al. 1999).

In rats administered 5, 15, or 25 mg/kg of a purified senna extract via the drinking water daily for 2 years, no increases in tumor incidences in the gastrointestinal tract, liver, kidney, or adrenal gland were observed (Lyden-Sokolowski et al. 1993).

In male rats orally administered 25 or 100 mg/kg sennosides daily for 6 months, the low dose caused slightly softer stools than normal, while the high dose resulted in severe diarrhea. No significant hematological or urinary differences were observed (Mengs 1988).

Genotoxicity

No elevated levels of micronuclei in bone marrow cells were observed in mice orally administered 2 g/kg of a senna extract (equivalent to 119 mg/kg rhein, 5.74 mg/kg aloe-emodin, and 0.28 mg/kg emodin) (Mengs et al. 1999).

No mutagenic effects of sennosides were observed in the Ames mutagenicity test (concentration up to 5000 μg/plate), the mouse lymphoma forward mutation assay (up to 5000 μg/ml without metabolic activation and 4000 μg/ml with activation), or the mice micronucleus assay (doses up to 2.5 mg/kg) (Mengs 1988).

In the *Salmonella typhimurium* reversion assay, senna glycosides were inactive in all strains, except for a slight but significant increase in mutant frequency in TA102 in the absence and presence of liver microsomes. Extracts of senna fruit and senna leaf demonstrated weak activity in TA97a, TA100, and TA102 in the presence of liver microsomes, and in TA97a and TA102 in the absence of liver microsomes. A strong increase in mutant frequency (three- to five-fold above background frequency) was observed with all extracts in TA98 in the presence of liver microsomes. This activity increased further following enzymatic hydrolysis with hesperidinase of extracts of senna fruit from one source, and could be correlated to the release of kaempferol and quercetin, suggesting that mutagenicity cannot be attributed solely to the anthraquinone content of these plant materials. A strong increase in mutant frequency (three- to five-fold above background frequency) was observed with all extracts in TA98 in the presence of liver microsomes (Sandnes et al. 1992).

Genotoxicity tests were performed with senna fruit, senna leaf extract, sennosides, rhein, and aloe-emodin. Senna fruit, sennosides, and rhein did not increase mutation frequencies in the following test systems: bacterial systems, mammalian cell culture tests, mouse lymphoma test, chromosome aberration test with Chinese hamster ovary (CHO) cells, bone marrow micronucleus test, chromosome aberration tests, and melanoblast cell test. With aloe-emodin, mutagenic effects were observed only in vitro in the chromosome aberration test with CHO cells and in the *Salmonella* reverse mutation test. In the in vitro gene mutation test with V79 cells, no mutagenic potential of aloe-emodin was observed. In vivo studies indicated no mutagenic activity of aloe-emodin, and aloe-emodin did not induce unscheduled DNA synthesis in an ex vivo assay performed with hepatocytes of male rats (Heidemann et al. 1993).

An aqueous extract of senna leaf was tested in four different assays for genotoxic and mutagenic effects: *E. coli* cultures, bacterial growth inhibition, reverse mutation test, and DNA strand break analysis in plasmid DNA. The extract produced single and double strand breaks in plasmid DNA in a cell-free system but was not cytotoxic or mutagenic in the *E. coli* strains tested (Silva et al. 2008).

In mice orally administered anthraquinones from senna or their equivalent amount in senna extracts, no increase in the number of chromosomal aberrations or aberrant cells in bone marrow was observed (Mukhopadhyay et al. 1998).

No clastogenic activity of the compound chrysophanol was observed in the Chinese hamster ovary cell assay at concentrations up to the limit of solubility, with or without metabolic activation (Mengs et al. 2001).

S

Literature Cited

AAP. 2001. Transfer of drugs and other chemicals into human milk. American Academy of Pediatrics, Committee on Drugs. *Pediatrics* 108(3):776-789.

AHPA. July 2011. Code of Ethics & Business Conduct. Silver Spring, MD: American Herbal Products Association.

Ács, N., F. Bánhidy, E.H. Puhó, and A.E. Czeizel. 2009. Senna treatment in pregnant women and congenital abnormalities in their offspring: A population-based case-control study. *Repro. Toxicol.* 28 (1):100-104.

Ács, N., F. Bánhidy, E.H. Puhó, and A.E. Czeizel. 2010. No association between severe constipation with related drug treatment in pregnant women and congenital abnormalities in their offspring: A population based case control study. *Congen. Anom.* 50 (1):15-20.

Armstrong, R.D., A.J. Crisp, R. Grahame, and D.L. Woolf. 1981. Hypertrophic osteoarthropathy and purgative abuse. *Br. Med. J.* 282(6279):1836.

Baldwin, W.F. 1963. Clinical study of senna administration to nursing mothers: Assessment of effects on infant bowel habits. *Can. Med. Assoc. J.* 89:566-568.

Bensky, D., S. Clavey, and E. Stöger. 2004. *Chinese herbal medicine: Materia medica*. 3rd ed. Seattle: Eastland Press.

Beuers, U., U. Spengler, and G.R. Pape. 1991. Hepatitis after chronic abuse of senna. *Lancet* 337(8737):372-373.

Borrelli, F., R. Capasso, G. Aviello, et al. 2005. Senna and the formation of aberrant crypt foci and tumors in rats treated with azoxymethane. *Phytomedicine* 12(6-7):501-505; discussion 505.

Bradley, P.R. 1992. *British herbal compendium: A handbook of scientific information on widely used plant drugs*. Bournemouth, UK: British Herbal Medicine Association.

Brinker, F. 2001. *Herb contraindications and drug interactions*. 3rd ed. Sandy, OR: Eclectic Medical Publications.

Capasso, F., A.A. Izzo, N. Mascolo, G. Autore, and G. Di Carlo. 1993. Effect of senna is not mediated by platelet-activating factor. *Pharmacology* 47(Suppl. 1):58-63.

Chen, J.K., and T.T. Chen. 2004. *Chinese medical herbology and pharmacology*. City of Industry, CA: Art of Medicine Press.

De Smet, P.A.G.M. 1993. *Adverse effects of herbal drugs, Volume 2*. Berlin: Springer.

El Sayed, N.Y., E.M. Abdelbari, O.M. Mahmoud, and S.E. Adam. 1983. The toxicity of cassia senna to Nubian goats. *Vet. Q.* 5(2):80-85.

ESCOP. 2003. *ESCOP monographs: The scientific foundation for herbal medicinal products*. 2nd ed. Exeter, UK: European Scientific Cooperative on Phytotherapy.

Faber, P., and A. Strenge-Hesse. 1988. Relevance of rhein excretion into breast milk. *Pharmacology* 36(Suppl. 1):212-220.

FDA. 1985. 21 CFR Part 334: Laxative products for over-the-counter human use; tentative final monograph. U.S. Food and Drug Administration. *Federal Register.* 15 January 1985; 50(10):2124-2158.

Fioramonti, J., C. Dupuy, and L. Bueno. 1993. In vivo motility of rat colon chronically pretreated with sennosides. *Pharmacology* 47(Suppl. 1):155-161.

Garcia-Villar, R. 1988. Evaluation of the effects of sennosides on uterine motility in the pregnant ewe. *Pharmacology* 36 (Suppl. 1):203-211.

Gattuso, J.M., and M.A. Kamm. 1994. Adverse effects of drugs used in the management of constipation and diarrhoea. *Drug Saf.* 10(1):47-65.

Heidemann, A., H.G. Miltenburger, and U. Mengs. 1993. The genotoxicity status of senna. *Pharmacology* 47(Suppl. 1):178-186.

Helin, T., and S. Makinen-Kiljunen. 1996. Occupational asthma and rhinoconjunctivitis caused by senna. *Allergy* 51(3):181-184.

Khan, I.A., and E.A. Abourashed. 2011. *Leung's encyclopedia of common natural ingredients: Used in food, drugs and cosmetics.* Hoboken, NJ: John Wiley & Sons.

Kiernan, J.A., and E.A. Heinicke. 1989. Sennosides do not kill myenteric neurons in the colon of the rat or mouse. *Neuroscience* 30(3):837-842.

Kittisupamongkol, W., V. Nilaratanakul, and W. Kulwichit. 2008. Near-fatal bleeding, senna, and the opposite of lettuce. *Lancet* 371(9614):784.

Leng-Peschlow, E. 1992a. Classification of senna as a laxative. *Pharmacology* 44(Suppl. 1):6-9.

Leng-Peschlow, E. 1992b. Risk assessment for senna during pregnancy. *Pharmacology* 44(Suppl. 1):20-22.

Leng-Peschlow, E. 1992c. Senna and pseudomelanosis coli. *Pharmacology* 44(Suppl. 1):33-35.

Leung, A.Y., and S. Foster. 1996. *Encyclopedia of common natural ingredients used in food, drugs, and cosmetics*. 2nd ed. New York: Wiley.

Levine, D., A.W. Goode, and D.L. Wingate. 1981. Purgative abuse associated with reversible cachexia, hypogammaglobulinaemia, and finger clubbing. *Lancet* 1(8226):919-920.

Lim, A.K., D.H. Hooke, and P.G. Kerr. 2008. Anorexia nervosa and senna misuse: Nephrocalcinosis, digital clubbing and hypertrophic osteoarthropathy. *Med. J. Aust.* 188(2):121-122.

List, P.H., and H. Hörhammer. 1973. *Hagers handbuch der pharmazeutischen praxis*. Berlin: Springer.

Lyden-Sokolowski, A., A. Nilsson, and P. Sjoberg. 1993. Two-year carcinogenicity study with sennosides in the rat: Emphasis on gastro-intestinal alterations. *Pharmacology* 47(Suppl. 1):209-215.

Malmquist, J., B. Ericsson, M.B. Hulten-Nosslin, J.O. Jeppsson, and O. Ljungberg. 1980. Finger clubbing and aspartylglucosamine excretion in a laxative-abusing patient. *Postgrad. Med. J.* 56(662):862-864.

Marks, G.B., C.M. Salome, and A.J. Woolcock. 1991. Asthma and allergy associated with occupational exposure to ispaghula and senna products in a pharmaceutical work force. *Am. Rev. Respir. Dis.* 144(5):1065-1069.

Martindale, W., and J.E.F. Reynolds. 1996. *The extra pharmacopoeia*. 31st ed. London: Pharmaceutical Press.

Mascolo, N., E. Mereto, F. Borrelli, et al. 1999. Does senna extract promote growth of aberrant crypt foci and malignant tumors in rat colon? *Dig. Dis. Sci.* 44(11):2226-2230.

Mengs, U. 1986. Reproductive toxicological investigations with sennosides. *Arzneimittelforschung* 36:1355-1358.

Mengs, U. 1988. Toxic effects of sennosides in laboratory animals and in vitro. *Pharmacology* 36:180-187.

Mengs, U., W. Grimminger, G. Krumbiegel, et al. 1999. No clastogenic activity of a senna extract in the mouse micronucleus assay. *Mutat. Res.* 444(2):421-426.

Mengs, U., J. Mitchell, S. McPherson, R. Gregson, and J. Tigner. 2004. A 13-week oral toxicity study of senna in the rat with an 8-week recovery period. *Arch. Toxicol.* 78(5):269-275.

Mengs, U., and R.L. Rudolph. 1993. Light and electron-microscopic changes in the colon of the guinea pig after treatment with anthranoid and non-anthranoid laxatives. *Pharmacology* 47(Suppl. 1):172-177.

Mengs, U., D. Schuler, and R.R. Marshall. 2001. No induction of chromosomal aberrations in Chinese hamster ovary cells by chrysophanol. *Mutat. Res.* 492(1-2):69-72.

Mereto, E., M. Ghia, and G. Brambilla. 1996. Evaluation of the potential carcinogenic activity of senna and cascara glycosides for the rat colon. *Cancer Lett.* 101(1):79-83.

Mukhopadhyay, M.J., A. Saha, A. Dutta, B. De, and A. Mukherjee. 1998. Genotoxicity of sennosides on the bone marrow cells of mice. *Food Chem. Toxicol.* 36(11):937-940.

Nusko, G., B. Schneider, G. Muller, J. Kusche, and E.G. Hahn. 1993. Retrospective study on laxative use and melanosis-coli as risk-factors for colorectal neoplasm. *Pharmacology* 47:234-241.

Nusko, G., B. Schneider, I. Schneider, C. Wittekind, and E.G. Hahn. 2000. Anthranoid laxative use is not a risk factor for colorectal neoplasia: Results of a prospective case control study. *Gut* 46(5):651-655.

Prior, J., and I. White. 1978. Tetany and clubbing in patient who ingested large quantities of senna. *Lancet* 2(8096):947.

Rikans, L., and T. Yamano. 2001. Mechanisms of cadmium-mediated acute hepatotoxicity. *J. Biochem. Mol. Toxicol.* 14(2):110-117.

Sandnes, D., T. Johansen, G. Teien, and G. Ulsaker. 1992. Mutagenicity of crude senna and senna glycosides in *Salmonella typhimurium*. *Pharmacol. Toxicol.* 71(3, Pt. 1):165-172.

Seybold, U., N. Landauer, S. Hillebrand, and F.D. Goebel. 2004. Senna-induced hepatitis in a poor metabolizer. *Ann. Intern. Med.* 141(8):650-651.

Siegers, C.P., J. Siemers, and G. Baretton. 1993a. Sennosides and aloin do not promote dimethylhydrazine-induced colorectal tumors in mice. *Pharmacology* 47(Suppl. 1):205-208.

Siegers, C.P., E. Vonhertzberglottin, M. Otte, and B. Schneider. 1993b. Anthranoid laxative abuse—A risk for colorectal-cancer. *Gut* 34(8):1099-1101.

Silk, D.B., J.A. Gibson, and C.R. Murray. 1975. Reversible finger clubbing in a case of purgative abuse. *Gastroenterology* 68(4, Pt. 1):790-794.

Silva, C.R., M.R. Monteiro, H.M. Rocha, et al. 2008. Assessment of antimutagenic and genotoxic potential of senna (*Cassia angustifolia* Vahl.) aqueous extract using in vitro assays. *Toxicol. In Vitro* 22(1):212-218.

Sonnenberg, A., and A.D. Müller. 1993. Constipation and cathartics as risk factors of colorectal cancer: A meta-analysis. *Pharmacology* 47(Suppl. 1):224-233.

Sossai, P., C. Nasone, and F. Cantalamessa. 2007. Are herbs always good for you? A case of paralytic ileum using a herbal tisane. *Phytother. Res.* 21(6):587-588.

Soyuncu, S., Y. Cete, and A. Nokay. 2008. Portal vein thrombosis related to *Cassia angustifolia*. *Clin. Toxicol. (Phila.)* 46(8):774-777.

Spiller, H.A., M.L. Winter, J.A. Weber, et al. 2003. Skin breakdown and blisters from senna-containing laxatives in young children. *Ann. Pharmacother.* 37(5):636-639.

van Gorkom, B., E. de Vries, and K. Kleibeuker. 1999. Review article: Anthranoid laxatives and their potential carcinogenic effects. *Aliment Pharmacol. Ther.* 13(4):443-452.

Vanderperren, B., M. Rizzo, L. Angenot, et al. 2005. Acute liver failure with renal impairment related to the abuse of senna anthraquinone glycosides. *Ann. Pharmacother.* 39(7-8):1353-1357.

Weiss, R.F., and A.R. Meuss. 2001. *Weiss's herbal medicine.* Classic ed. Stuttgart: Thieme.

Werthmann, M.W., Jr., and S.V. Krees. 1973. Quantitative excretion of Senokot in human breast milk. *Med. Ann. Dist. Columbia* 42(1):4-5.

Wichtl, M. 2004. *Herbal drugs and phytopharmaceuticals: A handbook for practice on a scientific basis.* 3rd ed. Boca Raton, FL: CRC Press.

Williamson, E.M. 2003. *Potter's herbal cyclopedia.* Saffron Walden, Essex: C.W. Daniel Co.

Serenoa repens (W. Bartram) Small

Arecaceae

SCN: saw palmetto

Syn: *Sabal serrulata* (Michx.) Nutt. ex Schult. & Schult. f.; *Serenoa serrulata* (Michx.) G. Nichols.

OCN: sabal palm

Part: fruit

S

Quick Reference Summary

Safety Class: 1

Interaction Class: A

Contraindications

None known.

Other Precautions

None known.

Drug and Supplement Interactions

None known.

Editors' Note

Most contemporary research on saw palmetto fruit is specific to one or another brand of extract, usually standardized to 85% or more of the fruit's naturally occurring fatty acids. Unless otherwise stated, the references cited below identify standardized saw palmetto fruit extracts.

Adverse Events and Side Effects

Systematic reviews and meta-analyses of saw palmetto clinical trials and case reports indicate that saw palmetto was generally well tolerated, with adverse events reported similar to those of placebo (Agbabiaka et al. 2009; Tacklind et al. 2009; Wilt et al. 2002). A human study on the safety of saw palmetto indicated no significant differences between saw palmetto and treatment groups for adverse events, bilirubin levels, urine tests, or prostate-specific antigen levels (Avins et al. 2008).

An allergic reaction to topical application of "an unlabelled extemporaneous preparation that [the patient] was told contained saw palmetto" has been reported (Sinclair et al. 2002). Single cases of pancreatitis (Jibrin et al. 2006) and hemorrhage during surgery (Cheema et al. 2001) have been reported in persons taking saw palmetto (no additional description provided). A case of cholestatic hepatitis was associated with a multi-ingredient product including saw palmetto, which the treating physician described as "presumably the most active ingredient" due to its "estrogenic and antiandrogenic effect" (Hamid et al. 1997).

Pharmacological Considerations

No effects of saw palmetto on the drug-metabolizing isoenzymes CYP3A4, CYP2D6, CYP1A2, or CYP2E1 were observed in human studies (Gurley et al. 2004; Markowitz et al. 2003).

Pregnancy and Lactation

No information on the safety of saw palmetto in pregnancy or lactation was identified in the scientific or traditional literature. Although this review did not identify any concerns for use while pregnant or nursing, safety has not been conclusively established.

The typical use of saw palmetto is for prostate health concerns; thus use by pregnant or lactating women is very uncommon.

Review Details

I. Drug and Supplement Interactions

Clinical Trials of Drug or Supplement Interactions

No clinical trials of drug or supplement interactions were identified.

Case Reports of Suspected Drug or Supplement Interactions

No case reports of suspected drug or supplement interactions were identified.

Animal Trials of Drug or Supplement Interactions

No animal trials of drug or supplement interactions were identified.

II. Adverse Events

Adverse Events Reported in Clinical Trials

In a meta-analysis of saw palmetto clinical trials, data from 3139 men in 21 randomized trials lasting 4 to 48 weeks were assessed. Adverse effects in the saw palmetto groups were reported as generally minor and similar to those of placebo. Withdrawal rates in the studies were 8.9% for saw palmetto, 7.1% for placebo, and 9.0% for the drug finasteride. Specific adverse events were reported in 11 of the clinical trials. The most common event was impotence, with incidences reported at 1.1% for saw palmetto, 0.7% for placebo, and 4.9% for finasteride. Gastrointestinal side effects were reported in 1.3% of men on saw palmetto, 0.9% on placebo, and 1.5% on finasteride (Wilt et al. 2002).

In a study on the safety of 320 mg saw palmetto daily for 1 year in men with benign prostatic hyperplasia, no significant differences in serious or nonserious adverse events were reported between the saw palmetto and placebo groups. No clinically significant differences in sexual functioning, bilirubin levels, urine tests, or prostate-specific antigen levels were observed between the saw palmetto and placebo groups (Avins et al. 2008).

Case Reports of Adverse Events

A systematic review of adverse events reported in association with saw palmetto indicated that saw palmetto is well tolerated by most users and is not associated with serious adverse events. Reported adverse events were characterized as mild, infrequent, and reversible and included abdominal pain, diarrhea, nausea, fatigue, headache, decreased libido, and rhinitis (Agbabiaka et al. 2009).

One case of hemorrhage during surgery was reported in a man taking saw palmetto (Cheema et al. 2001). Single cases of pancreatitis (Jibrin et al. 2006) and protracted cholestatic hepatitis (Hamid et al. 1997) have been reported. A single case of allergic reaction, confirmed by patch testing, was reported (Sinclair et al. 2002).

III. Pharmacology and Pharmacokinetics

Human Pharmacological Studies

In human clinical trials, saw palmetto extracts did not affect the drug-metabolizing cytochrome P450 isoenzymes CYP3A4, CYP2D6, CYP1A2, or CYP2E1 (Gurley et al. 2004; Markowitz et al. 2003). In a trial on the effects of saw palmetto on testosterone, follicle-stimulating hormone, and luteinizing hormone in men, saw palmetto had no effect on hormone levels (Casarosa et al. 1988).

In healthy volunteers orally administered saw palmetto capsules at the manufacturer's recommended dose (amount not specified) daily for two weeks, no effects on platelet function or other hematological parameters were observed, including prothrombin time, partial thromboplastin time, thrombin time, bleeding time, the collagen/epinephrine assay, or the collagen/adenosine diphosphate assay. Aspirin (325 mg daily) was used as a positive control and markedly inhibited platelet function (Beckert et al. 2007).

Animal Pharmacological Studies

No relevant animal pharmacological studies were identified.

In Vitro Pharmacological Studies

In high-throughput *in vitro* screening, saw palmetto extract caused significant inhibition of CYP3A4, CYP2D6, and CYP2C9 (Yale and Glurich 2005).

IV. Pregnancy and Lactation

The typical use of saw palmetto is for prostate health concerns; thus use by pregnant or lactating women is not expected.

V. Toxicity Studies

Acute Toxicity

In toxicity tests in mice, rats, and guinea pigs, the LD_{50} could not be determined at doses up to 10 g/kg of an ethanolic extract of saw palmetto (Mills and Bone 2005).

Short-Term Toxicity

In a hepatotoxicity test in rats, rats fed 9 or 23 mg/kg/day of a hexane extract of saw palmetto for 2 weeks did not develop any signs of hepatotoxicity (Singh et al. 2007).

Chronic Toxicity

Chronic toxicity tests of an ethanolic extract of saw palmetto in rats for 6 weeks at 360 times the standard human dose and 6 months at 80 times the standard human dose did not cause any adverse effects. In the 6-month toxicity trial, fertility was reported as unaffected (Mills and Bone 2005). In the Ames test, no evidence of mutagenicity was found (Degenring et al. 2001).

Literature Cited

Agbabiaka, T.B., M.H. Pittler, B. Wider, and E. Ernst. 2009. *Serenoa repens* (saw palmetto): A systematic review of adverse events. *Drug Saf.* 32(8):637-647.

Avins, A.L., S. Bent, S. Staccone, et al. 2008. A detailed safety assessment of a saw palmetto extract. *Complement. Ther. Med.* 16(3):147-154.

Beckert, B.W., M.J. Concannon, S.L. Henry, D.S. Smith, and C.L. Puckett. 2007. The effect of herbal medicines on platelet function: An in vivo experiment and review of the literature. *Plast. Reconstr. Surg.* 120 (7):2044-2050.

Casarosa, C., M. Cosci di Coscio, and M. Fratta. 1988. Lack of effects of a lyposterolic extract of *Serenoa repens* on plasma levels of testosterone, follicle-stimulating hormone, and luteinizing hormone. *Clin. Ther.* 10(5):585-588.

Cheema, P., O. El-Mefty, and A.R. Jazieh. 2001. Intraoperative haemorrhage associated with the use of extract of saw palmetto herb: A case report and review of literature. *J. Intern. Med.* 250(2):167-169.

Degenring, F., A. Sokolowski, A. Suter, and M. Weber. 2001. *Salmonella typhimurium* reverse mutation assay with the *Serenoa repens* extract Prostasan®. *Eur. Phytoj.* 2.

Gurley, B.J., S.F. Gardner, M.A. Hubbard, et al. 2004. *In vivo* assessment of botanical supplementation on human cytochrome P450 phenotypes: *Citrus aurantium, Echinacea purpurea*, milk thistle, and saw palmetto. *Clin. Pharmacol. Ther.* 76(5):428-440.

Hamid, S., S. Rojter, and J. Vierling. 1997. Protracted cholestatic hepatitis after the use of Prostata. *Ann. Intern. Med.* 127(July):169-170.

Jibrin, I., A. Erinle, A. Saidi, and Z.Y. Aliyu. 2006. Saw palmetto-induced pancreatitis. *South. Med. J.* 99(6):611-612.

Markowitz, J.S., J.L. Donovan, C.L. Devane, et al. 2003. Multiple doses of saw palmetto (*Serenoa repens*) did not alter cytochrome P450 2D6 and 3A4 activity in normal volunteers. *Clin. Pharmacol. Ther.* 74(6):536-542.

Mills, S., and K. Bone. 2005. *The essential guide to herbal safety*. St. Louis: Elsevier.

Sinclair, R.D., R.S. Mallari, and B. Tate. 2002. Sensitization to saw palmetto and minoxidil in separate topical extemporaneous treatments for androgenetic alopecia. *Australas. J. Dermatol.* 43(4):311-312.

Singh, Y.N., A.K. Devkota, D.C. Sneeden, K.K. Singh, and F. Halaweish. 2007. Hepatotoxicity potential of saw palmetto (*Serenoa repens*) in rats. *Phytomedicine* 14(2):204-208.

Tacklind, J., R. MacDonald, I. Rutks, and T.J. Wilt. 2009. *Serenoa repens* for benign prostatic hyperplasia. *Cochrane Database Syst. Rev.* 2:CD001423.

Wilt, T., A. Ishani, and R. Mac Donald. 2002. *Serenoa repens* for benign prostatic hyperplasia. *Cochrane Database of Syst. Rev.* 3:CD001423.

Yale, S.H., and I. Glurich. 2005. Analysis of the inhibitory potential of *Ginkgo biloba, Echinacea purpurea*, and *Serenoa repens* on the metabolic activity of cytochrome P450 3A4, 2D6, and 2C9. *J. Altern. Complement. Med.* 11(3):433-439.

S

Sesamum indicum L. — Pedaliaceae

SCN: sesame
Syn: *Sesamum orientale* L.
AN: *tila*
Part: seed

Quick Reference Summary

Safety Class: 1
Interaction Class: A

Contraindications

None known.

Other Precautions

None known.

Drug and Supplement Interactions

None known.

Adverse Events and Side Effects

Sesame is a relatively common cause of food allergy (Aaronov et al. 2008; Agne et al. 2004; Gangur et al. 2005). Severe allergic reactions, including anaphylactic reactions, to sesame have been reported and confirmed by skin prick testing (Agne et al. 2004; Asero et al. 1999; Caminiti et al. 2006; Dalal et al. 2003; Derby et al. 2005; Gangur et al. 2005; James et al. 1991; Morisset et al. 2003; Panizzolo et al. 2005).

Pharmacological Considerations

An animal study indicated that a diet very high in sesame (10% of diet) counteracted the effects of the estrogen receptor antagonist tamoxifen in mice with estrogen receptor-positive human breast cancer tumors (Sacco et al. 2007).

Pregnancy and Lactation

No information on the safety of sesame in pregnancy or lactation was identified in the scientific or traditional literature. Although this review did not identify any concerns for use while pregnant or nursing, safety has not been conclusively established.

Review Details

I. Drug and Supplement Interactions

Clinical Trials of Drug or Supplement Interactions

No clinical trials of drug or supplement interactions were identified.

Case Reports of Suspected Drug or Supplement Interactions

No case reports of suspected drug or supplement interactions were identified.

Animal Trials of Drug or Supplement Interactions

Ovariectomized athymic mice under premenopausal-simulated conditions and estrogen receptor-positive human breast cancer cells (MCF-7) were fed a diet containing 10% sesame seed and implanted with tamoxifen pellets (5 mg; 60-day release). Results indicated that sesame did not inhibit tumor growth and tended to negate the tumor inhibitory effect of tamoxifen. Sesame alone and combined with tamoxifen enhanced femur biomechanical strength but caused no differences in bone mineral content or bone mineral density in either the femur or lumbar vertebrae (Sacco et al. 2007).

II. Adverse Events

Case Reports of Adverse Events

Sesame has been characterized as a relatively common food allergen. A review on sesame allergy indicated that most sesame allergy was presented in one of two forms. The first form was an immediate hypersensitivity, often expressed as systemic anaphylaxis, associated with positive skin prick test and/or immunoglobulin E (IgE) antibody test results to sesame proteins. The second form was delayed hypersensitivity to lignin-like compounds in sesame oil, which appeared clinically as allergic contact dermatitis. Several cases of immediate hypersensitivity to sesame were negative in skin prick and/or IgE antibody tests but were confirmed by oral challenge tests (Gangur et al. 2005).

Anaphylactic reactions to sesame have been reported (Agne et al. 2004; Asero et al. 1999; Dalal et al. 2003; Derby et al. 2005; James et al. 1991; Morisset et al. 2003; Panizzolo et al. 2005).

Severe allergic reactions, including vomiting, tightness in the chest, cough, sneezing, generalized itching, facial erythema, breathlessness, and laryngeal edema, were reported in a woman with a known allergy to sesame who accidentally consumed sesame paste (tahini) on two occasions (Caminiti et al. 2006).

In Israeli children with suspected food allergies, sesame was one of the four most common allergens among those tested, with 30 of 234 children testing positive for sesame allergy. Of these 30, 70% had an atopic background. Reactions to sesame were generally IgE-mediated (Aaronov et al. 2008; Dalal et al. 2003).

III. Pharmacology and Pharmacokinetics

Human Pharmacological Studies

No relevant human pharmacological studies were identified.

Animal Pharmacological Studies

A reduction in plasma glucose and serum insulin levels was observed in diabetic mice fed diets containing 4% of a hot water extract of defatted sesame daily for 4 weeks (Takeuchi et al. 2001).

In Vitro Pharmacological Studies

No relevant in vitro pharmacological studies were identified.

IV. Pregnancy and Lactation

No information on the safety of sesame during pregnancy or lactation was identified.

V. Toxicity Studies

No toxicity studies were identified.

Literature Cited

Aaronov, D., D. Tasher, A. Levine, et al. 2008. Natural history of food allergy in infants and children in Israel. *Ann. Allergy Asthma Immunol.* 101(6):637-640.

Agne, P.S., E. Bidat, P.S. Agne, F. Rance, and E. Paty. 2004. Sesame seed allergy in children. *Eur. Ann. Allergy Clin. Immunol.* 36(8):300-305.

Asero, R., G. Mistrello, D. Roncarolo, P.L. Antoniotti, and P. Falagiani. 1999. A case of sesame seed-induced anaphylaxis. *Allergy* 54(5):526-527.

Caminiti, L., D. Vita, G. Passalacqua, et al. 2006. Tahini, a little known sesame-containing food, as an unexpected cause of severe allergic reaction. *J. Invest. Allergol. Clin. Immunol.* 16(5):308-310.

Dalal, I., I. Binson, A. Levine, et al. 2003. The pattern of sesame sensitivity among infants and children. *Pediatr. Allergy Immunol.* 14(4):312-316.

Derby, C.J., M.H. Gowland, and J.O. Hourihane. 2005. Sesame allergy in Britain: A questionnaire survey of members of the Anaphylaxis Campaign. *Pediatr. Allergy Immunol.* 16(2):171-175.

Gangur, V., C. Kelly, and L. Navuluri. 2005. Sesame allergy: A growing food allergy of global proportions? *Ann. Allergy Asthma Immunol.* 95(1):4-11.

James, C., A. Williams-Akita, Y.A. Rao, L.T. Chiarmonte, and A.T. Scheider. 1991. Sesame seed anaphylaxis. *N.Y. State Med. J.* 91:457-458.

Morisset, M., D.A. Moneret-Vautrin, G. Kanny, et al. 2003. Thresholds of clinical reactivity to milk, egg, peanut and sesame in immunoglobulin E-dependent allergies: Evaluation by double-blind or single-blind placebo-controlled oral challenges. *Clin. Exp. Allergy* 33(8):1046-1051.

Panizzolo, C., M. Tura, and A. Barbato. 2005. Anaphylaxis to sesame paste. *Eur. Ann. Allergy Clin. Immunol.* 37(1):34-35.

Sacco, S.M., K.A. Power, J. Chen, W.E. Ward, and L.U. Thompson. 2007. Interaction of sesame seed and tamoxifen on tumor growth and bone health in athymic mice. *Exp. Biol. Med.* 232(6):754-761.

Takeuchi, H., L.Y. Mooi, Y. Inagaki, and P. He. 2001. Hypoglycemic effect of a hot-water extract from defatted sesame (*Sesamum indicum* L.) seed on the blood glucose level in genetically diabetic KK-Ay mice. *Biosci. Biotechnol. Biochem.* 65(10):2318-2321.

Sida cordifolia L. — Malvaceae

SCN: heart-leaf sida
AN: *bala*
OCN: country mallow; fanpetals; flannelweed; llima
Part: leaf, root, seed

Quick Reference Summary

Safety Class: 1
Interaction Class: A

Contraindications

None known.

Other Precautions

None known.

Drug and Supplement Interactions

None known.

Editors' Notes

Heart-leaf sida contains 0.29 to 0.97% of the compound ephedrine (Gunatilaka et al. 1980; Marchei et al. 2006). The U.S. Food and Drug Administration declared in 2004 that dietary supplements containing ephedrine alkaloids are adulterated, and identified heart-leaf sida as containing ephedrine alkaloids (FDA 2004). In establishing this ban for ephedrine in dietary supplements, however, FDA stated that it does not apply to "traditional Asian medicine" that contains ephedrine alkaloids (FDA 2004).

Adverse Events and Side Effects

None known.

Pharmacological Considerations

None known.

Pregnancy and Lactation

No information on the safety of heart-leaf sida in pregnancy or lactation was identified in the scientific or traditional literature. Although this review did not identify any concerns for use while pregnant or nursing, safety has not been conclusively established.

Review Details

I. Drug and Supplement Interactions

Clinical Trials of Drug or Supplement Interactions

No clinical trials of drug or supplement interactions were identified.

Case Reports of Suspected Drug or Supplement Interactions

No case reports of suspected drug or supplement interactions were identified.

S

Animal Trials of Drug or Supplement Interactions

No animal trials of drug or supplement interactions were identified.

II. Adverse Events

Case Reports of Adverse Events

No case reports of adverse events were identified.

III. Pharmacology and Pharmacokinetics

Human Pharmacological Studies

No relevant human pharmacological studies were identified.

Animal Pharmacological Studies

In rats intravenously administered 5 to 40 mg/kg of the aqueous fraction of a hydroalcoholic extract of heart-leaf sida, a reduction in heart rate and blood pressure was observed (Medeiros et al. 2006). The activity has been traced to the compound vasicine, which caused hypotension and a marked reduction in heart rate after intravenous administration to rats at doses of 2.5 to 10 mg/kg (Silveira et al. 2003).

In Vitro Pharmacological Studies

No relevant in vitro pharmacological studies were identified.

IV. Pregnancy and Lactation

No information on the use of heart-leaf sida during pregnancy or lactation was identified.

V. Toxicity Studies

Acute Toxicity

The LD_{50} of an aqueous extract of heart-leaf sida orally administered to mice could not be determined at doses up to 3 g/kg (Franzotti et al. 2000).

The LD_{50} of a hydroalcoholic extract of heart-leaf sida in mice is 2.639 g/kg after intraperitoneal administration and could not be determined at doses up to 5 g/kg after oral administration (Franco et al. 2005). Similarly, the oral LD_{50} of a methanol extract could not be determined at doses up to 5 g/kg in mice (Philip and Venkataraman 2001).

Chronic Toxicity

No adverse effects were reported in mice chronically administered an oral dose of 500 mg/kg of a methanol extract of heart-leaf sida (Philip and Venkataraman 2001).

Literature Cited

FDA. 2004. Final rule declaring dietary supplements containing ephedrine alkaloids adulterated because they present an unreasonable risk. *Fed. Reg.* 69(28):6788-6854.

Franco, C.I., L.C. Morais, L.J. Quintans-Junior, R.N. Almeida, and A.R. Antoniolli. 2005. CNS pharmacological effects of the hydroalcoholic extract of *Sida cordifolia* L. leaves. *J. Ethnopharmacol.* 98(3):275-279.

Franzotti, E.M., C.V. Santos, H.M. Rodrigues, et al. 2000. Anti-inflammatory, analgesic activity and acute toxicity of *Sida cordifolia* L. (Malva-branca). *J. Ethnopharmacol.* 72(1-2):273-277.

Gunatilaka, A.A.L., S. Sotheeswaran, S. Balasubramaniam, A.I. Chandrasekara, and H.T. Badrasriyani. 1980. Studies on medicinal plants of Sri Lanka 3. Pharmacologically important alkaloids of some *Sida* species. *Planta Med.* 39(1):66-72.

Marchei, E., M. Pellegrini, R. Pacifici, P. Zuccaro, and S. Pichini. 2006. A rapid and simple procedure for the determination of ephedrine alkaloids in dietary supplements by gas chromatography-mass spectrometry. *J. Pharm. Biomed. Anal.* 41(5):1633-1641.

Medeiros, I.A., M.R. Santos, N.M. Nascimento, and J.C. Duarte. 2006. Cardiovascular effects of *Sida cordifolia* leaves extract in rats. *Fitoterapia* 77(1):19-27.

Philip, B.K., and S. Venkataraman. 2001. Evaluation of acute and chronic toxicity profile of *Sida cordifolia* Linn. in mice with respect to biochemical and hematological parameters. *Biomedicine* 21(2-3):65-70.

Silveira, A.L., M.A.S. Gomes, M.R.V. Santos, et al. 2003. Evaluation of the cardiovascular effects of vasicine, an alkaloid isolated from the leaves of *Sida cordifolia* L. (Malvaceae). *Rev. Bras. Farmacogn.* 13:37-39.

Silybum marianum (L.) Gaertn. — Asteraceae

SCN: milk thistle
OCN: Mary's thistle

Part: seed

Quick Reference Summary

Safety Class: 1
Interaction Class: A

Contraindications

None known.

Other Precautions

None known.

Drug and Supplement Interactions

None known.

Adverse Events and Side Effects

Allergic reactions, including anaphylactic reactions, to milk thistle products have been reported (Geier et al. 1990; Mironets and Krasovskaia 1990).

Pharmacological Considerations

Human studies have demonstrated that milk thistle may modify glucose regulation (Huseini et al. 2006; Velussi et al. 1997). People with diabetes are advised to monitor their blood sugar closely and discuss the use of this herb with a qualified healthcare practitioner prior to use.

Human studies have indicated no interactions between milk thistle and indinavir (DiCenzo et al. 2003; Mills et al. 2005; Piscitelli et al. 2002), digoxin (Gurley et al. 2006b), or irinotecan (van Erp et al. 2005).

Human studies have shown a lack of clinically significant effects of milk thistle on the CYP450 drug-metabolizing isoenzymes (Fuhr et al. 2007; Gurley et al. 2004, 2006a, 2008; Rajnarayana et al. 2004; Rao et al. 2007). Human studies on the effects of the compound silymarin on the drug-metabolizing isoenzyme CYP3A4 have given conflicting results, with one study showing induction of the enzyme and another study showing no effect (Fuhr et al. 2007; Rao et al. 2007).

Pregnancy and Lactation

Limited human and animal studies on use of the compound silymarin during pregnancy have indicated no adverse effects (Gonzalez et al. 1988; Hahn et al. 1968). No adverse effects were reported in a study of nursing women taking the compound silymarin (Di Pierro et al. 2008).

Review Details

I. Drug and Supplement Interactions

Clinical Trials of Drug or Supplement Interactions

No interaction between digoxin and milk thistle was observed in healthy volunteers orally administered a single dose of 0.4 mg of digoxin before or after 900 mg of a standardized extract of milk thistle daily for 14 days (Gurley et al. 2006b).

No interaction between indinavir and milk thistle was observed in healthy volunteers orally administered four 800 mg doses of indinavir before or after oral administration of 525 mg milk thistle (containing 459 mg silymarin) daily for 21 days (Piscitelli et al. 2002) or before or after oral administration of 1350 mg milk thistle daily for 30 days (Mills et al. 2005).

No interaction between indinavir and the compound silymarin was observed in healthy volunteers orally administered four 800 mg doses of indinavir before or after oral administration of 480 mg silymarin daily for 14 days (DiCenzo et al. 2003).

No interaction between irinotecan and milk thistle was observed in patients undergoing chemotherapy with irinotecan (125 mg/m^2) once a week along with 200 mg of orally administered milk thistle daily for 14 days (van Erp et al. 2005).

Case Reports of Suspected Drug or Supplement Interactions

No case reports of suspected drug or supplement interactions were identified.

Animal Trials of Drug or Supplement Interactions

No animal trials of drug or supplement interactions were identified.

II. Adverse Events

Adverse Events Reported in Clinical Trials

In systematic reviews of clinical trials, adverse events reported in the milk thistle groups were noted as being low in frequency, minor in nature, and indistinguishable from adverse events reported in placebo groups (Jacobs et al. 2002; Mulrow et al. 2000).

Case Reports of Adverse Events

One woman experienced repeated attacks of sweating, nausea, colicky abdominal pain, diarrhea, vomiting, weakness, and collapse after taking a milk thistle product (Adverse Drug Reactions Advisory Committee 1999). Other cases of adverse events, reported without details, include a case of abdominal pain, nausea, listlessness, and insomnia and one case of thrombocytopenia (Adverse Drug Reactions Advisory Committee 1999). The symptoms in these reports are similar to those associated with liver disease, for which the patients may have been taking milk thistle.

A case of an anaphylactic reaction after milk thistle tea consumption and a case of urticaria after consumption of a silymarin product have been reported (Geier et al. 1990; Mironets and Krasovskaia 1990).

Exacerbation of hemochromatosis (an inherited disease in which excessive iron accumulates in the body) was reported in a 68-year-old woman with a history of type 2 diabetes, asthma, hypothyroidism, borderline hypertension, borderline diastolic dysfunction, and a fatty liver. The women had been taking 200 mg of milk thistle daily for over a year, along with two extra-strength acetaminophen pills every 2 to 3 days, and a can of cola (presumably a diet cola) every day (Whittington 2007). A letter regarding this case noted that elevated liver enzymes, reported as a sign of hemochromatosis exacerbation, may occur with regular acetaminophen use (Kidd 2008).

III. Pharmacology and Pharmacokinetics

Human Pharmacological Studies

In two human studies, the compound silymarin (600 mg daily) significantly decreased fasting blood glucose levels in alcoholic cirrhosis patients with non-insulin-dependent diabetes mellitus and in type 2 diabetic patients (Huseini et al. 2006; Velussi et al. 1997).

Human studies indicated no significant effects of milk thistle on the drug-metabolizing isoenzymes CYP1A2, CYP2D6, CYP2E1, and CYP3A4 at doses of 900 mg milk thistle daily for 14 days or 350 mg milk thistle daily for 28 days (Gurley et al. 2004, 2006a, 2008).

Human studies of the compound silymarin on the drug-metabolizing isoenzyme CYP3A4 have yielded conflicting results. One study found that administration of 140 mg silymarin daily for 9 days resulted in significant induction of CYP3A4 (using metronidazole as the substrate) (Rajnarayana et al. 2004), while no effect of silymarin on CYP3A4 (using nifedipine as the substrate) was observed after administration of two doses of 280 mg each or after 7 days of administration of 420 mg daily (Fuhr et al. 2007; Rao et al. 2007).

In a dose escalation study of the compound silymarin in patients with noncirrhotic hepatitis C, no adverse effects were reported after oral administration of up to 2.1 g silymarin daily (Hawke et al. 2010).

Animal Pharmacological Studies

In healthy and diabetic rats orally administered 100 mg/kg daily of silymarin or 1.2 g/kg daily of an ethanol extract of milk thistle, non-significant decreases in blood glucose levels were observed for both silymarin and the milk thistle extracts in diabetic rats (Vessal et al 2010).

In Vitro Pharmacological Studies

A number of in vitro studies on the effects of milk thistle and compounds isolated from milk thistle on CYP450 drug-metabolizing isoenzymes have been completed and provide conflicting information on activity. In human hepatocytes, milk thistle has been shown to inhibit CYP2C8 and CYP3A4 (Etheridge et al. 2007), but it has also been shown to inhibit CYP3A4 at low concentrations while inducing the enzyme at high concentrations (Budzinski et al. 2007). The compound silybin has shown significant inhibition of CYP2C9 (Sridar et al. 2004) and no effect to significant inhibition of CYP3A4 (Kosina et al. 2005; Sridar et al. 2004; Zuber et al. 2002). Silybin moderately inhibited CYP1A1 (Awad et al. 2003), CYP2C8 (Dvorak et al. 2006; Jancova et al. 2007), CYP2D6 (Zuber et al. 2002), and CYP2E (Zuber et al. 2002). No effect to moderate inhibition from silybin was observed for CYP1A2 (Jancova et al. 2007; Kosina et al. 2005). Inhibition of CYP3A4 was observed after treatment with the compound silibinin (Budzinski et al. 2007).

A study with the compounds silybin A and silybin B indicated a possible interaction between these compounds and warfarin, with CYP2C9 noted as the mechanism of interaction (Brantley et al. 2010).

The compound silybin has been shown to have significant iron-chelating activity in vitro (Borsari et al. 2001). The implications of this to human clinical use are unclear. The binding may enhance or impair iron absorption (Mills and Bone 2005).

IV. Pregnancy and Lactation

In a study of women with cholestasis of pregnancy, no adverse effects were noted when the compound silymarin (210 mg daily) was taken for 15 days (Gonzalez et al. 1988). No teratogenic effects were observed after pregnant rabbits were administered the compound silymarin on days 8 to 17 of gestation, and when it was administered to pregnant rats on days 8 to 12 of gestation (Hahn et al. 1968).

No adverse effects were reported in nursing women taking 420 mg daily of the compound silymarin for 63 days during lactation (Di Pierro et al. 2008).

V. Toxicity Studies

Acute Toxicity

No adverse effects were observed in dogs administered a single dose of 1 g/kg of the compound silymarin (Hahn et al. 1968).

Short-Term Toxicity

No adverse effects were observed in mice administered the compound silymarin, 20 g/kg daily for 7 days. No adverse effects were observed in rats administered silymarin, 1 g/kg daily for 2 weeks (Hahn et al. 1968).

Chronic Toxicity

No adverse effects were observed in rats administered the compound silymarin at 100 mg/kg daily for 22 weeks (Hahn et al. 1968).

S

Literature Cited

Adverse Drug Reactions Advisory Committee. 1999. An adverse reaction to the herbal medication milk thistle (*Silybum marianum*). *Med. J. Aust.* 170(5):218-219.

Borsari, M., C. Gabbi, F. Ghelfi, et al. 2001. Silybin, a new iron-chelating agent. *J. Inorg. Biochem.* 85(2-3):123-129.

Brantley, S.J., N.H. Oberlies, D.J. Kroll, and M.F. Paine. 2010. Two flavonolignans from milk thistle (*Silybum marianum*) inhibit CYP2C9-mediated warfarin metabolism at clinically achievable concentrations. *J. Pharmacol. Exp. Ther.* 332(3):1081-1087.

Budzinski, J.W., V.L. Trudeau, C.E. Drouin, et al. 2007. Modulation of human cytochrome P450 3A4 (CYP3A4) and P-glycoprotein (P-gp) in Caco-2 cell monolayers by selected commercial-source milk thistle and goldenseal products. *Can. J. Physiol. Pharmacol.* 85(9):966-978.

Di Pierro, F., A. Callegari, D. Carotenuto, and M.M. Tapia. 2008. Clinical efficacy, safety and tolerability of BIO-C (micronized silymarin) as a galactagogue. *Acta Biomed.* 79(3):205-210.

DiCenzo, R., M. Shelton, K. Jordan, et al. 2003. Coadministration of milk thistle and indinavir in healthy subjects. *Pharmacotherapy* 23(7):866-870.

Dvorak, Z., R. Vrzal, and J. Ulrichova. 2006. Silybin and dehydrosilybin inhibit cytochrome P450 1A1 catalytic activity: A study in human keratinocytes and human hepatoma cells. *Cell Biol. Toxicol.* 22(2):81-90.

Etheridge, A.S., S.R. Black, P.R. Patel, J. So, and J.M. Mathews. 2007. An *in vitro* evaluation of cytochrome P450 inhibition and P-glycoprotein interaction with goldenseal, *Ginkgo biloba*, grape seed, milk thistle, and ginseng extracts and their constituents. *Planta Med.* 73(8):731-741.

Fuhr, U., S. Beckmann-Knopp, A. Jetter, H. Luck, and U. Mengs. 2007. The effect of silymarin on oral nifedipine pharmacokinetics. *Planta Med.* 73(14):1429-1435.

Geier, J., T. Fuchs, and R. Wahl. 1990. Anaphylactic shock due to an extract of *Silybum marianum* in a patient with immediate-type allergy to kiwi fruit. *Allergologie* 13:387-388.

Gonzalez, M., H. Reyes, J. Ribalta, et al. 1988. Effects of sylimarin [*sic*] on pruritis of cholestasis. *Hepatology* 8(5):1356.

Gurley, B., M.A. Hubbard, D.K. Williams, et al. 2006a. Assessing the clinical significance of botanical supplementation on human cytochrome P450 3A activity: Comparison of a milk thistle and black cohosh product to rifampin and clarithromycin. *J. Clin. Pharmacol.* 46(2):201-213.

Gurley, B.J., G.W. Barone, D.K. Williams, et al. 2006b. Effect of milk thistle (*Silybum marianum*) and black cohosh (*Cimicifuga racemosa*) supplementation on digoxin pharmacokinetics in humans. *Drug Metab. Dispos.* 34(1):69-74.

Gurley, B.J., S.F. Gardner, M.A. Hubbard, et al. 2004. *In vivo* assessment of botanical supplementation on human cytochrome P450 phenotypes: *Citrus aurantium, Echinacea purpurea*, milk thistle, and saw palmetto. *Clin. Pharmacol. Ther.* 76(5):428-440.

Gurley, B.J., A. Swain, M.A. Hubbard, et al. 2008. Clinical assessment of CYP2D6-mediated herb-drug interactions in humans: Effects of milk thistle, black cohosh, goldenseal, kava kava, St. John's wort, and echinacea. *Mol. Nutr. Food Res.* 52(7):755-763.

Hahn, G., H.D. Lehmann, M. Kurten, H. Uebel, and G. Vogel. 1968. On the pharmacology and toxicology of silymarin, an antihepatotoxic active principle from *Silybum marianum* (L.) Gaertn. *Arzneimittelforschung* 18(6):698-704. In Mills, S., and K. Bone. 2005. *The essential guide to herbal safety*. St. Louis: Elsevier.

Hawke, R.L., S.J. Schrieber, T.A. Soule, et al. 2010. Silymarin ascending multiple oral dosing phase I study in noncirrhotic patients with chronic hepatitis C. *J. Clin. Pharmacol.* 50(4):434-449.

Huseini, H.F., B. Larijani, R. Heshmat, et al. 2006. The efficacy of *Silybum marianum* (L.) Gaertn. (silymarin) in the treatment of type II diabetes: A randomized, double-blind, placebo-controlled, clinical trial. *Phytother. Res.* 20(12):1036-1039.

Jacobs, B.P., C. Dennehy, G. Ramirez, J. Sapp, and V.A. Lawrence. 2002. Milk thistle for the treatment of liver disease: A systematic review and meta-analysis. *Am. J. Med.* 113(6):506-515.

Jancova, P., E. Anzenbacherova, B. Papouskova, et al. 2007. Silybin is metabolized by cytochrome P450 2C8 *in vitro*. *Drug Metab. Dispos.* 35(11):2035-2039.

Kidd, R. 2008. Exacerbation of hemochromatosis by ingestion of milk thistle. *Can. Fam. Physician* 54(2):182; author reply 182-183.

Kosina, P., P. Maurel, J. Ulrichova, and Z. Dvorak. 2005. Effect of silybin and its glycosides on the expression of cytochromes P450 1A2 and 3A4 in primary cultures of human hepatocytes. *J. Biochem. Mol. Toxicol.* 19(3):149-153.

Mills, E., K. Wilson, M. Clarke, et al. 2005. Milk thistle and indinavir: A randomized controlled pharmacokinetics study and meta-analysis. *Eur. J. Clin. Pharmacol.* 61(1):1-7.

Mills, S., and K. Bone. 2005. *The essential guide to herbal safety*. St. Louis: Elsevier.

Mironets, V., and E. Krasovskaia. 1990. A case of urticaria during carsil treatment. *Vrach Delo* 7:86-87.

Mulrow, C., V. Lawrence, B. Jacobs, et al. 2000. Milk thistle: Effects on liver disease and cirrhosis and clinical adverse effects. *Evid Rep. Technol. Assess. (Summ.)* 21:1-3.

Piscitelli, S.C., E. Formentini, A.H. Burstein, et al. 2002. Effect of milk thistle on the pharmacokinetics of indinavir in healthy volunteers. *Pharmacotherapy* 22(5):551-556.

Rajnarayana, K., M. Reddy, J. Vidyasagar, and D. Krishna. 2004. Study on the influence of silymarin pretreatment on metabolism and disposition of metronidazole. *Arzneimittelforschung* 54(2):109-113.

Rao, B.N., M. Srinivas, Y.S. Kumar, and Y.M. Rao. 2007. Effect of silymarin on the oral bioavailability of ranitidine in healthy human volunteers. *Drug Metabol. Drug Interact.* 22(2-3):175-185.

Sridar, C., T.C. Goosen, U.M. Kent, J.A. Williams, and P.F. Hollenberg. 2004. Silybin inactivates cytochromes P450 3A4 and 2C9 and inhibits major hepatic glucuronosyltransferases. *Drug Metab. Dispos.* 32(6):587-594.

van Erp, N.P., S.D. Baker, M. Zhao, et al. 2005. Effect of milk thistle (*Silybum marianum*) on the pharmacokinetics of irinotecan. *Clin. Cancer Res.* 11(21):7800-7806.

Velussi, M., A.M. Cernigoi, A. De Monte, et al. 1997. Long-term (12 months) treatment with an anti-oxidant drug (silymarin) is effective on hyperinsulinemia, exogenous insulin need and malondialdehyde levels in cirrhotic diabetic patients. *J. Hepatol.* 26(4):871-879.

Vessal G., M. Akmali, P. Najafi, M.R. Moein, and M.M. Sagheb. 2010. Silymarin and milk thistle extract may prevent the progression of diabetic nephropathy in streptozotocin-induced diabetic rats. *Renal Failure* 32(6):733-739.

Whittington, C. 2007. Exacerbation of hemochromatosis by ingestion of milk thistle. *Can. Fam. Physician* 53(10):1671-1673.

Zuber, R., M. Modriansky, Z. Dvorak, et al. 2002. Effect of silybin and its congeners on human liver microsomal cytochrome P450 activities. *Phytother. Res.* 16(7):632-638.

Smilax spp. — Smilacaceae

Smilax aristolochiifolia Mill.
SCN: sarsaparilla
Syn: *Smilax medica* Schltdl. & Cham.; *Smilax ornata* Lem.
OCN: gray sarsaparilla; Mexican sarsaparilla; Vera Cruz sarsaparilla

Smilax febrifuga **Kunth**
SCN: sarsaparilla
OCN: Ecuadorian sarsaparilla

Smilax regelii **Killip & C.V. Morton**
SCN: sarsaparilla
Syn: *Smilax officinalis* Kunth; *Smilax ornata* Hook., nom. illeg.; *Smilax utilis* Hemsl.
OCN: brown sarsaparilla; Honduran sarsaparilla; Jamaican sarsaparilla
Part: root

Quick Reference Summary

Safety Class: 1
Interaction Class: A

Contraindications

None known.

Other Precautions

None known.

Drug and Supplement Interactions

None known.

Adverse Events and Side Effects

None known.

Pharmacological Considerations

Although the German Commission E reported that sarsaparilla could increase the absorption of digitalis glycosides, bismuth, or hypnotic drugs (Blumenthal et al. 1998), no information was found to support or refute this assertion.

Pregnancy and Lactation

No information on the safety of sarsaparilla in pregnancy or lactation was identified in the scientific or traditional literature. Although this review did not identify any concerns for use while pregnant or nursing, safety has not been conclusively established.

Review Details

I. Drug and Supplement Interactions

Clinical Trials of Drug or Supplement Interactions

No clinical trials of drug or supplement interactions were identified.

Case Reports of Suspected Drug or Supplement Interactions

No case reports of suspected drug or supplement interactions were identified.

Animal Trials of Drug or Supplement Interactions

No animal trials of drug or supplement interactions were identified.

II. Adverse Events

Case Reports of Adverse Events

No case reports of adverse events were identified.

III. Pharmacology and Pharmacokinetics

Human Pharmacological Studies

No relevant human pharmacological studies were identified.

Animal Pharmacological Studies

No relevant animal pharmacological studies were identified.

In Vitro Pharmacological Studies

No relevant in vitro pharmacological studies were identified.

IV. Pregnancy and Lactation

No information on the use of sarsaparilla during pregnancy or lactation was identified.

V. Toxicity Studies

Acute Toxicity

No adverse effects were observed in rats orally administered up to 3 g/kg of a sarsaparilla extract (Rafatullah et al. 1991).

Subchronic Toxicity

No adverse effects were observed in rats orally administered up to 100 mg/kg of a sarsaparilla extract daily for 90 days. Observed parameters included body weight and hematological values (Rafatullah et al. 1991).

Literature Cited

Blumenthal, M., W. Busse, A. Goldberg, et al. 1998. *The complete German Commission E monographs*. Austin, TX: American Botanical Council.

Rafatullah, S., J.S. Mossa, A.M. Ageel, M.A. Al-Yahya, and M. Tariq. 1991. Hepatoprotective and safety evaluation studies on sarsaparilla. *Int. J. Pharmacog.* 29(4):296-301.

Solidago spp. — Asteraceae

Solidago canadensis L. var. ***lepida*** (DC.) Cronquist
SCN: Canadian goldenrod
Syn: *Solidago lepida* DC.

Solidago gigantea **Aiton**
SCN: early goldenrod
Syn: *Solidago serotina* Aiton
OCN: giant goldenrod

Solidago virgaurea **L.**
SCN: European goldenrod
OCN: virgaurea
Part: herb

Quick Reference Summary

Safety Class: 1
Interaction Class: A

Contraindications

None known.

Other Precautions

Use with caution in persons with hypersensitivity to the active substance or to other plants of the Asteraceae family (EMEA 2008).

Drug and Supplement Interactions

None known.

Notice

Diuretic (Chodera et al. 1991; Felter and Lloyd 1898; Remington and Wood 1918); *see* Appendix 2.

Adverse Events and Side Effects

None known.

Pharmacological Considerations

None known.

Pregnancy and Lactation

No information on the safety of Canadian goldenrod, early goldenrod, or European goldenrod in pregnancy or lactation was identified in the scientific or traditional literature. Although this review did not identify any concerns for use while pregnant or nursing, safety has not been conclusively established.

Review Details

I. Drug and Supplement Interactions

Clinical Trials of Drug or Supplement Interactions

No clinical trials of drug or supplement interactions were identified.

Case Reports of Suspected Drug or Supplement Interactions

No case reports of suspected drug or supplement interactions were identified.

Animal Trials of Drug or Supplement Interactions

No animal trials of drug or supplement interactions were identified.

II. Adverse Events

Case Reports of Adverse Events

No case reports of adverse events were identified.

III. Pharmacology and Pharmacokinetics

Human Pharmacological Studies

No relevant human pharmacological studies were identified.

Animal Pharmacological Studies

No relevant animal pharmacological studies were identified.

In Vitro Pharmacological Studies

In a screening of different herbal products for effects on the cytochrome P450 drug-metabolizing isoenzymes CYP1A2 and CYP3A4 and the transporter protein MDR1, no effects of a hydroethanolic extract of European goldenrod were observed on CYP1A2 or MDR1. An inducing effect on CYP3A4 was observed (Brandin et al. 2007).

IV. Pregnancy and Lactation

No information on the safety of Canadian goldenrod, early goldenrod, or European goldenrod during pregnancy or lactation was identified.

V. Toxicity Studies

Acute Toxicity

The LD_{50} of intraperitoneally administered European goldenrod leaf extract (solvent not specified) in mice is 11.2 (g/kg presumably, dose units not specified for toxicity test), while that of the flower extract is 20.8 (g/kg presumably) (Racz-Kotilla and Racz 1978).

Literature Cited

Brandin, H., E. Viitanen, O. Myrberg, and A.K. Arvidsson. 2007. Effects of herbal medicinal products and food supplements on induction of CYP1A2, CYP3A4 and MDR1 in the human colon carcinoma cell line LS180. *Phytother. Res.* 21(3):239-244.

Chodera, A., K. Dabrowska, A. Sloderbach, L. Skrzypczak, and J. Budzianowski. 1991. Effect of flavonoid fractions of *Solidago virgaurea* L. on diuresis and levels of electrolytes. *Acta Pol. Pharm.* 48(5-6):35.

EMEA. 2008. Community herbal monograph on *Solidago virgaurea* L., herba. London: European Medicines Agency, Committee on Herbal Medicinal Products.

Felter, H.W., and J.U. Lloyd. 1898. *King's American dispensatory.* 18th ed., 3rd rev. 2 vols. Cincinnati: Ohio Valley Co.

Racz-Kotilla, E., and G. Racz. 1978. Hypotensive and sedative effect of extracts obtained from *Solidago virgaurea* L. [abstract]. *Planta Med.* 33(3):300.

Remington, J.P., and H.C. Wood. 1918. *The dispensatory of the United States of America.* 20th ed. Philadelphia: Lippincott.

Sophora flavescens Aiton — Fabaceae

SCN: shrubby sophora
PN: *ku shen* (root)
OCN: light-yellow sophora
Part: root

Quick Reference Summary

Safety Class: 1
Interaction Class: A

Contraindications

None known.

Other Precautions

None known.

Drug and Supplement Interactions

None known.

Adverse Events and Side Effects

Occasional side effects such as mild dizziness, nausea, vomiting, and constipation may occur (Bensky et al. 2004; Chen and Chen 2004). The nausea and vomiting are usually related to the very bitter taste of shrubby sophora (Chen and Chen 2004).

Allergic reactions to shrubby sophora have been reported (Bensky et al. 2004).

Pharmacological Considerations

None known.

Pregnancy and Lactation

No information on the safety of shrubby sophora in pregnancy or lactation was identified in the scientific or traditional literature. Although this review did not identify any concerns for use while pregnant or nursing, safety has not been conclusively established.

Review Details

I. Drug and Supplement Interactions

Clinical Trials of Drug or Supplement Interactions

No clinical trials of drug or supplement interactions were identified.

Case Reports of Suspected Drug or Supplement Interactions

No case reports of suspected drug or supplement interactions were identified.

Animal Trials of Drug or Supplement Interactions

No animal trials of drug or supplement interactions were identified.

II. Adverse Events

Case Reports of Adverse Events

Occasional side effects such as mild dizziness, nausea, vomiting, and constipation may occur. In cases of overdose (standard dose is a decoction of 3–10 g), symptoms may include irritability, spasms or cramps, disturbance of speech, and irregular breathing. In cases of gross overdose, respiratory failure has been reported (Bensky et al. 2004; Chen and Chen 2004).

Allergic reactions to shrubby sophora have been reported (Bensky et al. 2004).

III. Pharmacology and Pharmacokinetics

Human Pharmacological Studies

No relevant human pharmacological studies were identified.

Animal Pharmacological Studies

In rats with arrhythmias induced by myocardial infarction (MI), the compound matrine exhibited antiarrhythmic activity along with a shortening of the MI-induced prolonged action potential duration (Li et al. 2009).

In Vitro Pharmacological Studies

Some estrogenic activity of an ethanolic extract of shrubby sophora was observed in a recombinant yeast assay system with a human estrogen receptor expression plasmid and a reporter plasmid (Kang et al. 2006; Kim et al. 2008).

A methanol extract of shrubby sophora inhibited mouse brain monoamine oxidase (MAO) (Hwang et al. 2005).

Compounds isolated from shrubby sophora had no activity on the drug transporter protein P-gp (Choi et al. 1999).

An extract of shrubby sophora inhibited the drug-metabolizing isoenzyme CYP3A4 (Lee et al. 2007).

In aortic smooth muscle tissue, the compound matrine was found to inhibit phenylephrine-induced contractions by inhibiting activation of α-adrenoceptors and interfering with the release of intracellular calcium and the influx of extracellular calcium (Zheng et al. 2009).

IV. Pregnancy and Lactation

No information on the safety of shrubby sophora during pregnancy or lactation was identified.

V. Toxicity Studies

Acute Toxicity

The LD_{50} of shrubby sophora in mice is 14.5 g/kg after oral administration and 14.4 g/kg after intramuscular injection (Chen and Chen 2004).

The LD_{50} of the total alkaloids from shrubby sophora in mice is 1.18 g/kg after oral administration (Zhu 1998). The LD_{50} of the compound oxymatrine in mice is 521 mg/kg after intraperitoneal administration (Zhu 1998).

Short-Term Toxicity

No significant damage to the heart, spleen, or kidneys was observed in mice intraperitoneally administered 100 to 500 mg/kg of the compound oxymatrine daily for 2 to 4 weeks (Zhu 1998).

Genotoxicity

In the mouse micronucleus and chromosomal aberration assays, mutagenic activity was observed in mice intraperitoneally administered an aqueous extract of shrubby sophora at doses of 0.24 to 3.6 or 0.1 to 2 g/kg. A dose-dependent increase in the incidence of chromosomal aberrations was observed at doses of 0.24 to 3.6 g/kg. In the micronucleus assay, an increase of polychromatic erythrocytes was observed at the 0.5 to 2 g/kg dose levels (Yin et al. 1991).

No mutagenic activity of an aqueous extract of shrubby sophora was observed in the Ames mutagenicity assay with *Salmonella typhimurium* strains TA98 or TA100 with or without metabolic activation by S9 (Yin et al. 1991).

S

Literature Cited

Bensky, D., S. Clavey, and E. Stöger. 2004. *Chinese herbal medicine: Materia medica.* 3rd ed. Seattle: Eastland Press.

Chen, J.K., and T.T. Chen. 2004. *Chinese medical herbology and pharmacology.* City of Industry, CA: Art of Medicine Press.

Choi, S.U.N., K.H. Kim, E.J. Choi, et al. 1999. P-glycoprotein (Pgp) does not affect the cytotoxicity of flavonoids from *Sophora flavescens,* which also have no effects on Pgp action. *Anticancer Res.* 19(3A):2035-2040.

Hwang, J.S., S.A. Lee, S.S. Hong, et al. 2005. Monoamine oxidase inhibitory components from the roots of *Sophora flavescens. Arch. Pharm. Res.* 28(2):190-194.

Kang, S.C., C.M. Lee, H. Choi, et al. 2006. Evaluation of oriental medicinal herbs for estrogenic and antiproliferative activities. *Phytother. Res.* 20(11):1017-1019.

Kim, I.G., S.C. Kang, K.C. Kim, E.S. Choung, and O.P. Zee. 2008. Screening of estrogenic and antiestrogenic activities from medicinal plants. *Environ. Toxicol. Pharmacol.* 25(1):75-82.

Lee, S.S., B. Zhang, M.L. He, V.S. Chang, and H.F. Kung. 2007. Screening of active ingredients of herbal medicine for interaction with CYP450 3A4. *Phytother. Res.* 21(11):1096-1099.

Li, X., W. Chu, J. Liu, et al. 2009. Antiarrhythmic properties of long-term treatment with matrine in arrhythmic rat induced by coronary ligation. *Biol. Pharm. Bull.* 32(9):1521-1526.

Yin, X.J., D.X. Liu, H.C. Wang, and Y. Zhou. 1991. A study on the mutagenicity of 102 raw pharmaceuticals used in Chinese traditional medicine. *Mutat. Res.* 260(1):73-82.

Zheng, J.I.E., P. Zheng, X.U. Zhou, et al. 2009. Relaxant effects of matrine on aortic smooth muscles of guinea pigs. *Biomed. Environ. Sci.* 22(4):327-332.

Zhu, Y.-P. 1998. *Chinese materia medica: Chemistry, pharmacology and applications.* Amsterdam: Harwood Academic Publishers.

Spigelia marilandica (L.) L.

Loganiaceae

SCN: spigelia
OCN: Indian pink; pinkroot; woodland pinkroot; wormgrass

Part: root

Quick Reference Summary

Safety Class: 2b, 2d
Interaction Class: A

Contraindications

Not for use in pregnancy except under the supervision of a qualified healthcare practitioner (Felter and Lloyd 1898; Wood and LaWall 1918).

Not for long-term use (McGuffin et al. 1997).

Do not exceed recommended dose (McGuffin et al. 1997).

Other Precautions

None known.

Drug and Supplement Interactions

None known.

Standard Dose

Adults: 2 to 5 g, morning and evening, given with a strong purgative, such as senna (Wren et al. 1988).

Children over 4 years: 0.5 to 4 g, morning and evening, given with a strong purgative, such as senna (Wren et al. 1988).

Editors' Note

Historically, adulteration of spigelia with species of *Ruellia* or *Phlox* has been reported (Grieve 1971). No recent cases of adulteration have been reported.

Adverse Events and Side Effects

Texts from the early 1900s report that in standard doses, and when catharsis is produced either by spigelia or a coadministered stimulant laxative, no adverse effects are expected (Felter and Lloyd 1898; Osol and Farrar 1955; Wood and LaWall 1918). In large doses and without catharsis, spigelia may have a purgative effect, also producing various unpleasant symptoms such as increased action of the heart and arteries, dizziness, dilation of the pupils, and muscular spasms, often terminating in convulsions. Spasmodic twitching of the eyelids has been reported as one of the more frequent effects (Felter and Lloyd 1898; Wood and LaWall 1918). A later text reported that spigelia was commonly used throughout the United States as a vermifuge, and side effects were "almost unheard of" (Osol and Farrar 1955).

Pharmacological Considerations

None known.

Pregnancy and Lactation

Although no information on the safety of spigelia during pregnancy was identified, traditional use as a vermifuge (used to expel worms) suggests that spigelia should not be used during pregnancy (Felter and Lloyd 1898; Wood and LaWall 1918).

No information on the safety of spigelia during lactation was identified. While this review did not identify any concerns for use while nursing, safety has not been conclusively established.

S

Review Details

I. Drug and Supplement Interactions

Clinical Trials of Drug or Supplement Interactions

No clinical trials of drug or supplement interactions were identified.

Case Reports of Suspected Drug or Supplement Interactions

No case reports of suspected drug or supplement interactions were identified.

Animal Trials of Drug or Supplement Interactions

No animal trials of drug or supplement interactions were identified.

II. Adverse Events

Case Reports of Adverse Events

Texts from the early 1900s report that in standard doses, and when catharsis is produced either by spigelia or a coadministered stimulant laxative, no adverse effects are expected (Felter and Lloyd 1898; Osol and Farrar 1955; Wood and LaWall 1918). In large doses and without catharsis, spigelia may have a purgative effect, also producing various unpleasant symptoms such as increased action of the heart and arteries, dizziness, dilation of the pupils, and muscular spasms, often terminating in convulsions. Spasmodic twitching of the eyelids has been reported as one of the more frequent effects (Felter and Lloyd 1898; Wood and LaWall 1918). A later text reported that spigelia was commonly used throughout the United States as a vermifuge, and side effects were "almost unheard of" (Osol and Farrar 1955).

III. Pharmacology and Pharmacokinetics

Human Pharmacological Studies

No relevant human pharmacological studies were identified.

Animal Pharmacological Studies

No relevant animal pharmacological studies were identified.

In Vitro Pharmacological Studies

No relevant in vitro pharmacological studies were identified.

IV. Pregnancy and Lactation

Although no information on the safety of spigelia during pregnancy was identified, traditional use as a vermifuge (used to expel worms) suggests that spigelia should not be used during pregnancy (Felter and Lloyd 1898; Wood and LaWall 1918).

No information on the safety of spigelia during lactation was identified.

V. Toxicity Studies

No toxicity studies were identified.

Literature Cited

Felter, H.W., and J.U. Lloyd. 1898. *King's American dispensatory.* 18th ed., 3rd rev. 2 vols. Cincinnati: Ohio Valley Co.

Grieve, M. 1971. *A modern herbal.* New York: Dover.

McGuffin, M., C. Hobbs, R. Upton, and A. Goldberg. 1997. *Botanical safety handbook.* Boca Raton, FL: CRC Press.

Osol, A., and G. Farrar. 1955. *The dispensatory of the United States of America.* 25th ed. Philadelphia: Lippincott.

Wood, H., and C. LaWall. 1918. *The dispensatory of the United States of America.* 21st ed. Philadelphia: Lippincott.

Wren, R.C., E.M. Williamson, and F.J. Evans. 1988. *Potter's new cyclopaedia of botanical drugs and preparations.* Essex: C.W. Daniel Co. Ltd.

Spilanthes spp. — Asteraceae

Spilanthes acmella (L.) L.
SCN: spilanthes
OCN: para cress; toothache plant

Spilanthes oleracea **L.**
SCN: spilanthes
OCN: para cress; spotflower; toothache plant
Part: herb

Quick Reference Summary

Safety Class: 1
Interaction Class: A

Contraindications

None known.

Other Precautions

None known.

Drug and Supplement Interactions

None known.

Adverse Events and Side Effects

None known.

Pharmacological Considerations

Rats injected with doses of 100 mg/kg or more of spilanthes had tonic-clonic seizures (Moreira et al. 1989, 1991), although mice orally administered doses up to 1500 mg/kg experienced no adverse effects (Ratnasooriya et al. 2004).

Pregnancy and Lactation

Although no studies on the safety of spilanthes in pregnancy or lactation were identified, traditional use of spilanthes in south Asia as a cooked vegetable or salad ingredient provides some indication of the relative safety of this herb (Burdock and Fenaroli 2005 ; Facciola 1990).

Review Details

I. Drug and Supplement Interactions

Clinical Trials of Drug or Supplement Interactions

No clinical trials of drug or supplement interactions were identified.

Case Reports of Suspected Drug or Supplement Interactions

No case reports of suspected drug or supplement interactions were identified.

Animal Trials of Drug or Supplement Interactions

No animal trials of drug or supplement interactions were identified.

II. Adverse Events

Case Reports of Adverse Events

No case reports of adverse events were identified.

III. Pharmacology and Pharmacokinetics

Human Pharmacological Studies

No relevant human pharmacological studies were identified.

Animal Pharmacological Studies

In rats intraperitoneally administered doses of 50 to 150 mg/kg of a hexane extract of spilanthes, minor behavioral changes (grooming and "wet dog shakes") were observed in animals administered 50 or 75 mg/kg. At doses of 100 to 150 mg/kg, full tonic-clonic seizures were induced in a dose-dependent manner (Moreira et al. 1989). Testing of the convulsant activity of different parts of spilanthes plants indicated that the flower was most active, followed by root and then a leaf and stem combination (Moreira et al. 1991).

A diuretic effect of a cold water extract of spilanthes flower was observed in rats orally administered doses of 1500 mg/kg. Doses of 500 or 1000 mg/kg did not significantly alter urine output. The extract caused a marked increase in urinary sodium and potassium levels and a reduction in the osmolarity of urine, suggesting that the activity is as a loop diuretic (Ratnasooriya et al. 2004).

In Vitro Pharmacological Studies

An ethanol extract of spilanthes inhibited CYP2E1, with alkylamides isolated from spilanthes showing greater activity (Raner et al. 2007).

IV. Pregnancy and Lactation

Although no studies on the safety of spilanthes in pregnancy or lactation were identified, traditional use of spilanthes in south Asia as a cooked vegetable or salad ingredient provides some indication of the relative safety of this herb (Burdock and Fenaroli 2005 ; Facciola 1990).

V. Toxicity Studies

Acute Toxicity

The LD_{50} of an intraperitoneally administered hexane extract of spilanthes in rats is 150 mg/kg (Moreira et al. 1989).

Literature Cited

Burdock, G.A., and G. Fenaroli. 2005 *Fenaroli's handbook of flavor ingredients.* Boca Raton, FL: CRC Press.

Facciola, S. 1990. *Cornucopia: A source book of edible plants.* Vista, CA: Kampong Publications.

Moreira, V.M., J.G. Maia, J.M. de Souza, Z.A. Bortolotto, and E.A. Cavalheiro. 1989. Characterization of convulsions induced by a hexanic extract of *Spilanthes acmella* var. *oleracea* in rats. *Braz. J. Med. Biol. Res.* 22(1):65-67.

Moreira, V.M.T., L. Calderazzo, and E.A. Cavalheiro. 1991. Comparison of convulsant effect of extracts of different parts of *Spilanthes oleracea* jack jambu. 19th International Epilepsy Congress, Rio De Janeiro, Brazil, October 14-19, 1991. *Epilepsia* 32(Suppl. 1):47.

Raner, G.M., S. Cornelious, K. Moulick, et al. 2007. Effects of herbal products and their constituents on human cytochrome P450(2E1) activity. *Food Chem. Toxicol.* 45(12):2359-2365.

Ratnasooriya, W.D., K.P. Pieris, U. Samaratunga, and J.R. Jayakody. 2004. Diuretic activity of *Spilanthes acmella* flowers in rats. *J. Ethnopharmacol.* 91(2-3):317-320.

S

Stachys officinalis (L.) Trevis.

Lamiaceae

SCN: wood betony
Syn: *Betonica officinalis* L.; *Stachys betonica* Benth. nom. illeg.

Part: herb

Quick Reference Summary

Safety Class: 1
Interaction Class: A

Contraindications

None known.

Other Precautions

None known.

Drug and Supplement Interactions

None known.

Editors' Note

No cautions for use of wood betony are reported in historical medical texts (Madaus 1938; Sayre 1917).

Adverse Events and Side Effects

None known.

Pharmacological Considerations

None known.

Pregnancy and Lactation

No information on the safety of wood betony in pregnancy or lactation was identified in the scientific or traditional literature. Although this review did not identify any concerns for use while pregnant or nursing, safety has not been conclusively established.

Review Details

I. Drug and Supplement Interactions

Clinical Trials of Drug or Supplement Interactions

No clinical trials of drug or supplement interactions were identified.

Case Reports of Suspected Drug or Supplement Interactions

No case reports of suspected drug or supplement interactions were identified.

Animal Trials of Drug or Supplement Interactions

No animal trials of drug or supplement interactions were identified.

II. Adverse Events

Case Reports of Adverse Events

No case reports of adverse events were identified.

III. Pharmacology and Pharmacokinetics

Human Pharmacological Studies

No relevant human pharmacological studies were identified.

Animal Pharmacological Studies

No relevant animal pharmacological studies were identified.

In Vitro Pharmacological Studies

No relevant in vitro pharmacological studies were identified.

IV. Pregnancy and Lactation

No information on the safety of wood betony during pregnancy or lactation was identified.

V. Toxicity Studies

No toxicity studies were identified.

Literature Cited

Madaus, G. 1938. *Lehrbuch der Biologischen Heilmittel*. Leipzig: G. Thieme.

Sayre, L.E. 1917. *A manual of organic materia medica*. Philadelphia: P. Blakiston's Son & Co.

Stellaria media (L.) Vill. — Caryophyllaceae

SCN: chickweed

Part: herb

S

Quick Reference Summary

Safety Class: 1

Interaction Class: A

Contraindications

None known.

Other Precautions

None known.

Drug and Supplement Interactions

None known.

Adverse Events and Side Effects

None known.

Pharmacological Considerations

None known.

Pregnancy and Lactation

No information on the safety of chickweed in pregnancy or lactation was identified in the scientific or traditional literature. Although this review did not identify any concerns for use while pregnant or nursing, safety has not been conclusively established.

Review Details

I. Drug and Supplement Interactions

Clinical Trials of Drug or Supplement Interactions

No clinical trials of drug or supplement interactions were identified.

Case Reports of Suspected Drug or Supplement Interactions

No case reports of suspected drug or supplement interactions were identified.

Animal Trials of Drug or Supplement Interactions

No animal trials of drug or supplement interactions were identified.

II. Adverse Events

Case Reports of Adverse Events

A case report indicated a positive patch test result to chickweed in a patient allergic to a number of common weeds. Analysis of the plant material used for the testing indicated the presence of borneol, menthol, linalool, and 1,8-cineole, all compounds that are not known to be present in chickweed, bringing into doubt the identity of the material tested (Jovanovi et al. 2003).

There is one report of an alleged case of nitrate toxicity associated with chickweed that resulted in a mild form of paralysis (Chadha 1988). The standard nitrate content of chickweed is 0.1% (Guil et al. 1997), suggesting that the toxicity case was due to environmental factors, such as harvesting from fields where synthetic fertilizers had been used.

III. Pharmacology and Pharmacokinetics

Human Pharmacological Studies

No relevant human pharmacological studies were identified.

Animal Pharmacological Studies

No relevant animal pharmacological studies were identified.

In Vitro Pharmacological Studies

No relevant in vitro pharmacological studies were identified.

IV. Pregnancy and Lactation

No information on the safety of chickweed during pregnancy or lactation was identified.

V. Toxicity Studies

Acute Toxicity

The LD_{50} of an ethanol extract of chickweed intraperitoneally administered to mice is 1 g/kg (Sharma et al. 1978).

Literature Cited

Chadha, Y. 1988. *The wealth of India: A dictionary of Indian raw materials and industrial products*. Delhi: Council of Scientific and Industrial Research.

Guil, J.L., I. Rodriguez-Garci, and E. Torija. 1997. Nutritional and toxic factors in selected wild edible plants. *Plant Food Human Nutr.* 51(2):99-107.

Jovanovi, M., N. Mimica-Dukic, M. Poljacki, and P. Boza. 2003. Erythema multiforme due to contact with weeds: A recurrence after patch testing. *Contact Dermat.* 48(1):17-25.

Sharma, M.L., N. Chandokhe, B.J. Ray Ghatak, et al. 1978. Pharmacological screening of Indian medicinal plants. *Indian J. Exp. Biol.* 16:228-240.

Stephania tetrandra S. Moore — Menispermaceae

S

SCN: stephania
PN: *han fang ji* (root)

Part: root

Quick Reference Summary

Safety Class: 1
Interaction Class: A

Contraindications

None known.

Other Precautions

None known.

Drug and Supplement Interactions

None known.

Notice

Diuretic (Chen and Chen 2004); *see* Appendix 2.

Editors' Notes

Cases of nephrotoxicity were reported from ingestion of weight-loss products sold in Europe that listed *Stephania* as an ingredient (Vanherweghem et al. 1993). These were analyzed and found to contain the nephrotoxic compound aristolochic acid (Schmeiser et al. 1996), which is not a constituent of stephania (Chen et al. 1990; Wu et al. 2007; Zhu and Phillipson 1996). Stephania, which is sold as *han fang ji*, is sometimes confused with the aristolochic acid-containing herb *Aristolochia fangchi*, which is traded as *guang fang ji* (Bensky et al. 2004; Chen and Chen 2004; Wu et al. 2007). Methods for differentiation of *Stephania* and *Aristolochia* species and for detection of aristolochic acid have been developed (Joshi et al. 2008; Koh et al. 2006; Sorenson and Sullivan 2007).

Adverse Events and Side Effects

None known.

Pharmacological Considerations

Animal studies have demonstrated that stephania may modify glucose regulation. People with diabetes are advised to monitor their blood sugar closely and discuss the use of this herb with a qualified healthcare practitioner prior to use (Tsutsumi et al. 2003, 2008).

Pregnancy and Lactation

A compound from stephania has been demonstrated in an animal study to have a therapeutic effect on fetal lung malformations (Xu et al. 2009). No other information on the safety of this herb in pregnancy or lactation was identified in the scientific or traditional literature. Although this review did not identify any concerns for use while pregnant or nursing, safety has not been conclusively established.

Review Details

I. Drug and Supplement Interactions

Clinical Trials of Drug or Supplement Interactions

No clinical trials of drug or supplement interactions were identified.

Case Reports of Suspected Drug or Supplement Interactions

No case reports of suspected drug or supplement interactions were identified.

Animal Trials of Drug or Supplement Interactions

No animal trials of drug or supplement interactions were identified.

II. Adverse Events

Case Reports of Adverse Events

No adverse effects are expected within the normal dosage range (4.5–9 g). Overdoses of stephania (15–30 g) have been associated with symptoms including headache, dizziness, numbness and tremor of the upper extremities, fainting, an oppressive feeling in the chest, and palpitations (Bensky et al. 2004).

S

III. Pharmacology and Pharmacokinetics

Human Pharmacological Studies

No relevant human pharmacological studies were identified.

Animal Pharmacological Studies

A reduction in blood glucose levels was observed in diabetic rats orally administered 60 mg/kg of an aqueous extract of stephania (Tsutsumi et al. 2008).

A dose-dependent reduction in blood glucose levels was observed in diabetic mice orally administered 0.3 to 3 mg/kg of the compound fangchinoline. The activity of fangchinoline was greater than that of an aqueous extract of stephania (Tsutsumi et al. 2003).

In Vitro Pharmacological Studies

The compounds tetrandrine and fangchinoline inhibited platelet-activating factor (PAF)-induced human platelet aggregation. The IC_{50} of tetrandrine was 28.6 μ*M*, whereas that of fangchinoline was 21.7 μ*M*. Tetrandrine and fangchinoline also inhibited PAF-, thrombin- and arachidonic acid-induced thromboxane B_2 formation in human washed platelets. In the PAF-receptor binding assay, neither compound showed any inhibitory effects on the specific binding of PAF to its receptor (Kim et al. 1999).

The compound tetrandrine had a synergistic effect on the cytotoxicity of the chemotherapeutic agents 5-fluorouracil, oxaliplatin, and docetaxel in two gastric cancer cell lines (Wei et al. 2007).

IV. Pregnancy and Lactation

In rat fetuses with a model of congenital diaphragmatic hernia, oral administration of 30 mg/kg (based on maternal weight) of the compound tetrandrine to the pregnant mothers led to an improvement of morphological parameters (percentage of alveoli area, counting bronchus) in treated animals as compared to controls (Xu et al. 2009).

No information on the safety of stephania during lactation was identified.

V. Toxicity Studies

Acute Toxicity

The LD_{50} of the total alkaloids of stephania intraperitoneally administered to mice is 113 mg/kg (Zhu 1998).

The LD_{50} of the compound tetrandrine in mice is 82.5 mg/kg after intravenous administration, 1450 mg/kg after intramuscular administration, and 3700 mg/kg after oral administration (Zhu 1998).

Literature Cited

Bensky, D., S. Clavey, and E. Stöger. 2004. *Chinese herbal medicine: Materia medica*. 3rd ed. Seattle: Eastland Press.

Chen, H.A., G.J. Xu, R. Jin, and L.S. Xu. 1990. Hong Kong samples of the traditional Chinese medicine "Fang Ji" contain aristolochic acid toxins. *China J. Chin. Mat. Med.* 15:707-708.

Chen, J.K., and T.T. Chen. 2004. *Chinese medical herbology and pharmacology*. City of Industry, CA: Art of Medicine Press.

Joshi, V.C., B. Avula, and I.A. Khan. 2008. Authentication of *Stephania tetrandra* S. Moore (fang ji) and differentiation of its common adulterants using microscopy and HPLC analysis. *J. Nat. Med.* 62(1):117-121.

Kim, H.S., Y.H. Zhang, L.H. Fang, Y.P. Yun, and H.K. Lee. 1999. Effects of tetrandrine and fangchinoline on human platelet aggregation and thromboxane B_2 formation. *J. Ethnopharmacol.* 66(2):241-246.

Koh, H.L., H. Wang, S. Zhou, E. Chan, and S.O. Woo. 2006. Detection of aristolochic acid I, tetrandrine and fangchinoline in medicinal plants by high performance liquid chromatography and liquid chromatography/mass spectrometry. *J. Pharm. Biomed. Anal.* 40(3):653-661.

Schmeiser, H.H., C.A. Bieler, M. Wiessler, C. van Ypersele de Strihou, and J.P. Cosyns. 1996. Detection of DNA adducts formed by aristolochic acid in renal tissue from patients with Chinese herbs nephropathy. *Cancer Res.* 56(9):2025.

Sorenson, W.R., and D. Sullivan. 2007. Determination of aristolochic acid I in botanicals and dietary supplements potentially contaminated with aristolochic acid I using LC-UV with confirmation by LC/MS: Collaborative study. *J. AOAC Int.* 90(4):925-933.

Tsutsumi, T., N. Hagino, X.C. Liang, S.S. Guo, and S. Kobayashi. 2008. Effects of oral administration of *Stephania tetrandra* S. Moore on neovascularization of retinal and choroidal capillaries of diabetes in rats. *Phytother. Res.* 22(5):591-596.

Tsutsumi, T., S. Kobayashi, Y.Y. Liu, and H. Kontani. 2003. Antihyperglycemic effect of fangchinoline isolated from *Stephania tetrandra Radix* in streptozotocin-diabetic mice. *Biol. Pharm. Bull.* 26(3):313-317.

Vanherweghem, J.L., M. Depierreux, C. Tielemans, et al. 1993. Rapidly progressive interstitial renal fibrosis in young women: Association with slimming regimen including Chinese herbs. *Lancet* 341(8842):387-391.

Wei, J., B. Liu, L. Wang, et al. 2007. Synergistic interaction between tetrandrine and chemotherapeutic agents and influence of tetrandrine on chemotherapeutic agent-associated genes in human gastric cancer cell lines. *Cancer Chemother. Pharmacol.* 60(5):703-711.

Wu, K.M., J.G. Farrelly, R. Upton, and J. Chen. 2007. Complexities of the herbal nomenclature system in traditional Chinese medicine (TCM): Lessons learned from the misuse of *Aristolochia*-related species and the importance of the pharmaceutical name during botanical drug product development. *Phytomedicine* 14(4):273-279.

Xu, C., W. Liu, Z. Chen, et al. 2009. Effect of prenatal tetrandrine administration on transforming growth factor-beta1 level in the lung of nitrofen-induced congenital diaphragmatic hernia rat model. *J. Pediatr. Surg.* 44(8):1611-1620.

Zhu, M., and J.D. Phillipson. 1996. Hong Kong samples of the traditional Chinese medicine "fang ji" contain aristolochic acid toxins. *Int. J. Pharmacog.* 34:283-289.

Zhu, Y.-P. 1998. *Chinese materia medica: Chemistry, pharmacology and applications.* Amsterdam: Harwood Academic Publishers.

Stevia rebaudiana (Bertoni) Bertoni — Asteraceae

SCN: stevia
Syn: *Eupatorium rebaudianum* Bertoni
OCN: candyleaf; Paraguayan sweet herb; sweetleaf
Part: leaf

S

Quick Reference Summary

Safety Class: 1
Interaction Class: A

Contraindications

None known.

Other Precautions

None known.

Drug and Supplement Interactions

None known.

Editors' Note

Stevia has been extensively used as a sweetener for beverages, originally in Paraguay and, in the twentieth century, in other South American countries and Asia (Leung and Foster 1996).

Adverse Events and Side Effects

Reviews of stevia and compounds from stevia have indicated a lack of adverse effects (Carakostas et al. 2008; Chatsudthipong and Muanprasat 2009).

Pharmacological Considerations

Although two animal studies reported reductions in male or female fertility after large doses of stevia (Mazzei

Planas and Kuc 1968; Melis 1999), no adverse effects on fertility were observed in four other studies (Akashi and Yokoyama 1975; Mori et al. 1981; Sinchomi and Marcorities 1989; Yodyingyuad and Bunyawong 1991).

Pregnancy and Lactation

Animal studies have shown no adverse effects of large doses (up to 3 g/kg) of the compound stevioside on pregnancy or fetal and maternal health (Mori et al. 1981; Takanaka et al. 1991; Usami et al. 1995; Yodyingyuad and Bunyawong 1991).

No information on the safety of stevia during lactation was identified in the scientific or traditional literature. While this review did not identify any concerns for use while nursing, safety has not been conclusively established.

Review Details

I. Drug and Supplement Interactions

Clinical Trials of Drug or Supplement Interactions

No clinical trials of drug or supplement interactions were identified.

Case Reports of Suspected Drug or Supplement Interactions

No case reports of suspected drug or supplement interactions were identified.

Animal Trials of Drug or Supplement Interactions

No animal trials of drug or supplement interactions were identified.

II. Adverse Events

Adverse Events Reported in Clinical Trials

Adverse events reported in three clinical trials of stevia leaf or leaf extract were similar in the stevia and placebo groups (Chan et al. 2000; Ferri et al. 2006; Hsieh et al. 2003).

Case Reports of Adverse Events

No case reports of adverse events were identified.

III. Pharmacology and Pharmacokinetics

Human Pharmacological Studies

No relevant human pharmacological studies were identified.

S

Animal Pharmacological Studies

In female rats orally administered an aqueous extract of 5 g stevia daily (~40 ml/kg) for 18 days, a reduction in fertility was observed. Fertility remained reduced during the 60-day post-treatment period (Mazzei Planas and Kuc 1968). A number of subsequent studies have not shown any effects of stevia, or the compound stevioside, on fertility or reproduction (Akashi and Yokoyama 1975; Mori et al. 1981; Sinchomi and Marcorities 1989; Yodyingyuad and Bunyawong 1991).

In male rats orally administered an aqueous extract of 2.66 g stevia daily for 60 days, a decrease in the concentration of spermatozoa stored in the cauda epididymides was observed along with decreases in the weights of the cauda epididymides and the seminal vesicles (Melis 1999).

In Vitro Pharmacological Studies

No relevant in vitro pharmacological studies were identified.

IV. Pregnancy and Lactation

In three successive generations of hamsters orally administered 0.5, 1, or 2.5 g/kg of the compound stevioside daily, no adverse effects on growth or reproduction were observed. Histological examinations of reproductive tissues from all three generations revealed no evidence of abnormality (Yodyingyuad and Bunyawong 1991).

In rats fed diets containing 0.15, 0.75, or 3% (equivalent to 150, 750, or 3000 mg/kg body weight daily) of the compound stevioside for 14 days before mating and for 7 days during gestation, no effects on fertility or fetal development were observed. At the highest dose, a slight increase in the number of fetal resorptions was observed (Mori et al. 1981).

In pregnant rats orally administered 250, 500, or 1000 mg/kg of the compound stevioside daily on gestational days 6 to 15, no adverse effects on fetal development or maternal health were observed (Takanaka et al. 1991; Usami et al. 1995).

No information on the safety of stevia during lactation was identified.

V. Toxicity Studies

Reviews of safety and toxicity information on stevia and the compound stevioside indicate that both are generally safe, with no adverse effects noted (Geuns 2002, 2003).

Acute Toxicity

The LD_{50} of orally administered stevia extract in rats is 17 g/kg (Mitusuhashi 1976). The LD_{50} of the compound stevioside orally administered to mice, rats, and hamsters could not be determined at doses up to 15 g/kg (Toskulkao et al. 1997).

Subchronic Toxicity

An increase in renal plasma flow was observed in rats administered stevia (4 ml of extract daily) for 60 days; no effects on mean arterial pressure were observed. The authors of the study suggested that stevia may induce vasodilation, low blood pressure, and increased urination (Melis 1995).

A decrease in the weight of seminal vesicles, but no effects on thyroid function or other parameters, were observed in male rats administered stevia (53 g/kg) for 60 days (Oliveira-Filho et al. 1989). A second study with similar stevia dose and duration indicated similar results but with some decrease in male fertility noted (Melis 1999).

In rats fed diets containing 25,000, 50,000, 75,000, or 100,000 ppm (~3, 6, 9, or 12 g/kg body weight) of the compound rebaudioside A for 13 weeks, a reduction in body weight gain, attributed to taste aversion and lower caloric density of the diet, was observed in the high-dose groups. Inconsistent reductions in serum bile acids and cholesterol were attributed to physiological changes in bile acid metabolism due to excretion of high levels of rebaudioside A. All other liver function test results and liver histopathology were within normal limits. No significant changes were observed in other clinical pathology results, including organ weights, and macroscopic and microscopic examinations of all organs, including testes and kidneys. The no-observed-adverse-effect level (NOAEL) was determined to be 4.4 g/kg (Curry and Roberts 2008).

In rats orally administered 500, 1000, or 2000 mg/kg of the compound rebaudioside A daily for 90 days, no treatment-related adverse effects were observed in clinical observations, hematology, serum chemistry, or urinalysis. Macroscopic and microscopic findings revealed no treatment-related effects on any organ evaluated. A reduction in body weight gain was noted in males at the 2000 mg/kg dose level (Nikiforov and Eapen 2008).

Chronic Toxicity

In rats fed diets containing the compound stevioside at a concentration of 0.2, 0.6, or 1.2% (equivalent to 100, 300, or 600 mg/kg body weight daily) for 2 years, no treatment-related changes were observed in hematological, urinary, or clinical biochemical values. The incidence and severity of non-neoplastic and neoplastic changes were unrelated to the concentration of stevioside in the diet. The no-observed-effect level was equivalent to 600 mg/kg per day (Xili et al. 1992).

Genotoxicity

A review of the genetic toxicity of the compounds steviol and stevioside indicated that the majority of the findings show no evidence of genotoxic activity for either of the compounds, and that neither stevioside nor steviol have been shown to react directly with DNA or demonstrate genotoxic damage in assays relevant to human risk (Brusick 2008).

Literature Cited

Akashi, H., and Y. Yokoyama. 1975. Security of dried-leaf extracts of stevia. Toxicological tests. *Food Ind.* 18:34-43.

Brusick, D.J. 2008. A critical review of the genetic toxicity of steviol and steviol glycosides. *Food Chem. Toxicol.* 46(Suppl. 7):S83-S91.

Carakostas, M.C., L.L. Curry, A.C. Boileau, and D.J. Brusick. 2008. Overview: The history, technical function and safety of rebaudioside A, a naturally occurring steviol glycoside, for use in food and beverages. *Food Chem. Toxicol.* 46(Suppl. 7):S1-S10.

Chan, P., B. Tomlinson, Y.J. Chen, J.C. Liu, M.H. Hsieh, and J.T. Cheng. 2000. A double-blind placebo-controlled study of the effectiveness and tolerability of oral stevioside in human hypertension. *Br. J. Clin. Pharmacol.* 50(3):215-220.

Chatsudthipong, V., and C. Muanprasat. 2009. Stevioside and related compounds: Therapeutic benefits beyond sweetness. *Pharmacol. Ther.* 121(1):41-54.

Curry, L.L., and A. Roberts. 2008. Subchronic toxicity of rebaudioside A. *Food Chem. Toxicol.* 46(Suppl. 7):S11-S20.

Ferri, L.A., W. Alves-Do-Prado, S.S. Yamada, S. Gazola, M.R. Batista, and R.B. Bazotte. 2006. Investigation of the antihypertensive effect of oral crude stevioside in patients with mild essential hypertension. *Phytother. Res.* 20(9):732-736.

Geuns, J. 2002. Safety evaluation of *Stevia* and stevioside. *Stud. Nat. Prod. Chem.* 27:299-319.

Geuns, J.M. 2003. Stevioside. *Phytochemistry* 64(5):913-921.

Hsieh, M.H., P. Chan, Y.M. Sue, J.C. Liu, T.H. Liang, T.Y. Huang, B. Tomlinson, M.S. Chow, P.F. Kao, and Y.J. Chen. 2003. Efficacy and tolerability of oral stevioside in patients with mild essential hypertension: A two-year, randomized, placebo-controlled study. *Clin. Ther.* 25(11):2797-2808.

Leung, A.Y., and S. Foster. 1996. *Encyclopedia of common natural ingredients used in food, drugs, and cosmetics.* New York: Wiley.

Mazzei Planas, G., and J. Kuc. 1968. Contraceptive properties of *Stevia rebaudiana*. *Science* 162(857):1007.

Melis, M.S. 1995. Chronic administration of aqueous extract of *Stevia rebaudiana* in rats: Renal effects. *J. Ethnopharmacol.* 47(3):129-134.

Melis, M.S. 1999. Effects of chronic administration of *Stevia rebaudiana* on fertility in rats. *J. Ethnopharmacol.* 67(2):157-161.

Mitusuhashi, H. 1976. Safety of stevioside. In Tama Biochemical Co. Ltd. Report on safety of stevia. *Cited in* Geuns, J. 2002. Safety evaluation of *Stevia* and stevioside. *Stud. Nat. Prod. Chem.* 27:299-319.

Mori, N., M. Sakanoue, M. Takeuchi, K. Shimpo, and T. Tanabe. 1981. Effect of stevioside on fertility in rats. *Shokuhin Eiseiga Ku Zasshi* 22:409-414.

Nikiforov, A.I., and A.K. Eapen. 2008. A 90-day oral (dietary) toxicity study of rebaudioside A in Sprague-Dawley rats. *Int. J. Toxicol.* 27(1):65-80.

Oliveira-Filho, R.M., O.A. Uehara, C.A.S.A. Minetti, and L.B.S. Valle. 1989. Chronic administration of aqueous extract of *Stevia rebaudiana* Bertoni in rats: Endocrine effects. *Gen. Pharmacol.* 20(2):187-192.

Sinchomi, D., and P. Marcorities. 1989. Etude de l'activité anti-androgénique d'un extrait de *Stevia rebaudiana* Bertoni. *Plant. Med. Phytother.* 23(4):282-287.

Takanaka, T., K. Kawashima, M. Usami, and K. Sakami. 1991. A teratological study of stevioside administered orally to rats. *Cited in* JECFA. 1999. Stevioside. WHO Food Additives Series 42. Geneva: Joint FAO/WHO Expert Committee on Food Additives.

S

Toskulkao, C., L. Chaturat, P. Temcharoen, and T. Glinsukon. 1997. Acute toxicity of stevioside, a natural sweetener, and its metabolite, steviol, in several animal species. *Drug Chem. Toxicol.* 20:31-44.

Usami, M., K. Sakemi, K. Kawashima, M. Tsuda, and Y. Ohno. 1995. Teratogenicity study of stevioside in rats. *Bull. Nat. Inst. Hyg. (Jap.)* 113:31-35.

Xili, L., B. Chengjiany, X. Eryi, S. Reiming, W. Yuengming, S. Haodong, and H. Zhiyian. 1992. Chronic oral toxicity and carcinogenicity study of stevioside in rats. *Food Chem. Toxicol.* 30(11):957-965.

Yodyingyuad, V., and S. Bunyawong. 1991. Effect of stevioside on growth and reproduction. *Human Reprod.* 6(1):158-165.

Stillingia sylvatica Garden ex L. — Euphorbiaceae

SCN: stillingia
OCN: queen's delight; queen's root; yaw root

Part: root

Quick Reference Summary

Safety Class: 3
Interaction Class: A

Contraindications

Not for use except under the supervision of an expert qualified in the appropriate use of this substance (Felter and Lloyd 1898; Winston 2010; Wood and LaWall 1918).

Other Precautions

None known.

Drug and Supplement Interactions

None known.

Standard Dose

The standard dose is 2 to 5 drops of tincture, 2 to 3 times daily (Winston 2010).

Adverse Events and Side Effects

Overdose of stillingia may cause symptoms such as vomiting and purging, sometimes accompanied by a burning sensation in the stomach or other part of the gastrointestinal tract (Felter and Lloyd 1898; Wood and LaWall 1918). Stillingia use has been associated with tachycardia (rapid heart rate) (Winston 2010).

Pharmacological Considerations

None known.

Pregnancy and Lactation

Although no information on the safety of stillingia during pregnancy or lactation was identified, the side effects listed above suggest that stillingia should not be used during pregnancy or lactation (Felter and Lloyd 1898; McGuffin et al. 1997; Wood and LaWall 1918).

Review Details

I. Drug and Supplement Interactions

Clinical Trials of Drug or Supplement Interactions

No clinical trials of drug or supplement interactions were identified.

Case Reports of Suspected Drug or Supplement Interactions

No case reports of suspected drug or supplement interactions were identified.

Animal Trials of Drug or Supplement Interactions

No animal trials of drug or supplement interactions were identified.

II. Adverse Events

Case Reports of Adverse Events

Overdose of stillingia may cause symptoms such as vomiting and purging, sometimes accompanied by a burning sensation in the stomach or other part of the gastrointestinal tract (Felter and Lloyd 1898; Wood and LaWall 1918).

III. Pharmacology and Pharmacokinetics

Human Pharmacological Studies

No relevant human pharmacological studies were identified.

Animal Pharmacological Studies

No relevant animal pharmacological studies were identified.

In Vitro Pharmacological Studies

No relevant in vitro pharmacological studies were identified.

IV. Pregnancy and Lactation

Although no information on the safety of stillingia during pregnancy or lactation was identified, the side effects listed above suggest that stillingia should not be used during pregnancy or lactation (Felter and Lloyd 1898; Wood and LaWall 1918).

V. Toxicity Studies

No toxicity studies were identified.

Literature Cited

Felter, H.W., and J.U. Lloyd. 1898. *King's American dispensatory*. 18th ed., 3rd rev. 2 vols. Cincinnati: Ohio Valley Co.

McGuffin, M., C. Hobbs, R. Upton, and A. Goldberg. 1997. *Botanical safety handbook*. Boca Raton, FL: CRC Press.

Winston, D. 2010. *Winston's botanical materia medica*. Broadway, NJ: David Winston's Center for Herbal Studies.

Wood, H., and C. LaWall. 1918. *The dispensatory of the United States of America*. 21st ed. Philadelphia: Lippincott.

Styrax spp.

Styracaeae

Styrax benzoin Dryand.
SCN: benzoin tree
OCN: Benjamin tree; Sumatra benzoin
PN: *an xi xiang*

Styrax paralleloneurum **Perkins**
SCN: benzoin tree
OCN: haminjon toba; Sumatra benzoin

Styrax tonkinensis **(Pierre) Craib ex Hartwich**
SCN: benzoin tree
OCN: Siam styrax
PN: *an xi xiang*
Part: gum resin

Quick Reference Summary

Safety Class: 1
Interaction Class: A

Contraindications

None known.

Other Precautions

None known.

Drug and Supplement Interactions

None known.

Editors' Note

Among other products, benzoin is used in tincture of benzoin, which contains benzoin resin and alcohol, and compound tincture of benzoin, which contains benzoin resin, aloe resin, storax resin, and balsam of tolu resin in alcohol (Scardamaglia et al. 2003).

Adverse Events and Side Effects

Allergic contact dermatitis to benzoin has been reported and confirmed by patch testing (Coskey 1978; Hoffman and Adams 1978; Klein et al. 2009; Spott and Shelley 1970).

Pharmacological Considerations

None known.

Pregnancy and Lactation

No information on the safety of benzoin in pregnancy or lactation was identified in the scientific or traditional literature. Although this review did not identify any concerns for use while pregnant or nursing, safety has not been conclusively established.

S

Review Details

I. Drug and Supplement Interactions

Clinical Trials of Drug or Supplement Interactions

No clinical trials of drug or supplement interactions were identified.

Case Reports of Suspected Drug or Supplement Interactions

No case reports of suspected drug or supplement interactions were identified.

Animal Trials of Drug or Supplement Interactions

No animal trials of drug or supplement interactions were identified.

II. Adverse Events

Case Reports of Adverse Events

Allergic contact dermatitis to benzoin has been reported and confirmed by patch testing (Coskey 1978; Hoffman and Adams 1978; Klein et al. 2009; Spott and Shelley 1970).

III. Pharmacology and Pharmacokinetics

Human Pharmacological Studies

No relevant human pharmacological studies were identified.

Animal Pharmacological Studies

No relevant animal pharmacological studies were identified.

In Vitro Pharmacological Studies

No relevant in vitro pharmacological studies were identified.

IV. Pregnancy and Lactation

No information on the safety of benzoin during pregnancy or lactation was identified.

V. Toxicity Studies

Acute Toxicity

The LD_{50} of orally administered benzoin in rats is 10 g/kg (Opdyke 1979).

Genotoxicity

A review of genotoxicity testing of benzoin indicated that the product showed mutagenic activity in 2 of 12 assays with rodents (Ashby and Shelby 1989).

Literature Cited

Ashby, J., and M.D. Shelby. 1989. Overview of the genetic toxicity of caprolactam and benzoin. *Mutat. Res.* 224(3):321-324.

Coskey, R.J. 1978. Contact dermatitis owing to tincture of benzoin. *Arch. Dermatol.* 114(1):128.

Hoffman, T.E., and R.M. Adams. 1978. Contact dermatitis to benzoin in greasepaint makeup. *Contact Dermat.* 4(6):379-380.

Klein, T.G., H.J. Woehlck, and P.S. Pagel. 2009. Severe allergic contact dermatitis resulting from occupational exposure to tincture of benzoin aerosol spray in an anesthesiologist. *J. Anesthes.* 23(2):292-294.

Opdyke, D.L.J. 1979. *Monographs on fragrance raw materials.* New York: Pergamon.

Scardamaglia, L., R. Nixon, and J. Fewings. 2003. Compound tincture of benzoin: A common contact allergen? *Australas. J. Dermatol.* 44(3):180-184.

Spott, D.A., and W.B. Shelley. 1970. Exanthem due to contact allergen (benzoin) absorbed through skin. *J. Am. Med. Assoc.* 214(10):1881-1882.

Symphytum officinale L. — Boraginaceae

SCN: comfrey
OCN: common comfrey; healing-herb; knitbone

Part: leaf, root

Quick Reference Summary

Safety Class:* 2a, 2b, 2c
Interaction Class: A

Contraindications

For external use only (Bradley 1992; De Smet 1992; Wichtl 2004).

Not for use during pregnancy or lactation (Chan et al. 1994; De Smet 1992; Panter and James 1990).

Other Precautions

None known.

Drug and Supplement Interactions

None known.

* Extracts of comfrey that have had the pyrrolizidine alkaloids removed (PA-free comfrey) are commercially available. Concerns regarding the internal use of comfrey products do not apply to PA-free products.

Notice

Pyrrolizidine alkaloids (0.012–0.16% in leaf; 0.3–0.6% in dried root) (Altamirano et al. 2005; Betz et al. 1994; Cao et al. 2008; Couet et al. 1996; Dąbska et al. 1980; Mattocks 1980; Oberlies et al. 2004; Roitman 1981; Stickel and Seitz 2000; Tittel et al. 1979; Vollmer et al. 1988); *see* Appendix 1.

Editors' Notes

Pyrrolizidine alkaloid compounds in comfrey have been associated with cases of liver toxicity (see Adverse Events and Side Effects below). The American Herbal Products Association has established a trade requirement (AHPA 2011) that all products with botanical ingredients that contain toxic pyrrolizidine alkaloids, including comfrey, are not offered for sale for internal use and display the following cautionary label: "For external use only. Do not apply to broken or abraded skin. Do not use when nursing."

Considering the lower concentration of alkaloids in comfrey leaf and the absence of echimidine in most samples of the species (Couet et al. 1996; Roitman 1981), the limitation to external use recommended above may be overly cautious for the leaf. One reference states, without differentiation of the species, that chronic internal use of *Symphytum* preparations may cause severe hepatic damage (De Smet 1992).

Adulteration of comfrey may occur with *Symphytum* species that contain echimidine (one of the most toxic of the pyrrolizidine alkaloids), such as *S. asperum* and *S. uplandicum* (Wichtl 2004). A chemical analysis of *Symphytum* species indicated a lack of echimidine in *S. officinale* (Huizing et al. 1982), although a more recent analysis reported echimidine to be present in one-quarter of tested samples of *S. officinale,* in a range reported as from a "generally very small quantity" up to "a rather high echimidine content comparable to concentrations found in *Symphytum asperum*" (Awang et al. 1993).

German authorities have recommended that external application of comfrey be limited to 4 to 6 weeks per year with a daily exposure at or below 100 μg unsaturated pyrrolizidine alkaloids (Wichtl 2004).

Adverse Events and Side Effects

Cases of veno-occlusive liver disease (blockage of some very small veins in the liver) have been reported in persons taking comfrey leaf or root (Bach et al. 1989; Huxtable et al. 1986; Ridker et al. 1985; Weston et al. 1987; Yeong et al. 1990). This type of liver disease is caused by consumption of unsaturated pyrrolizidine alkaloids (Cao et al. 2008).

The leaves of comfrey and foxglove (*Digitalis purpurea*) are similar in appearance. Several cases of poisoning have been reported in persons who accidentally ingested foxglove instead of comfrey (Awang et al. 1993; Bain 1985; Routledge and Spriggs 1989; Turley and Muir 2008).

Pharmacological Considerations

Rats fed diets containing 2 or 8% comfrey root for 12 weeks did not develop liver toxicity but did have a higher mutation rate in liver and lung cell genes, and expressed genes that were involved in liver injury and abnormalities, including liver fibrosis and cancer development (Mei et al. 2005, 2006; Mei and Chen 2007).

In contrast to case reports of liver disease, no evidence of liver damage was found in regular consumers (1 to 10 years of regular use) of comfrey (species not identified) leaf (Anderson and McLean 1989).

In one animal study, topical administration of comfrey resulted in excretion of PA metabolites (*N*-oxides) 20 to 50 times lower than that seen after oral administration (Brauchli et al. 1982).

Pregnancy and Lactation

Data from animal studies indicates that pyrrolizidine alkaloid compounds present in comfrey can cross the placenta and are present in the breast milk of animals that have consumed comfrey (Chan et al. 1994; Panter and James 1990). Based on this information, use during pregnancy is not recommended except under the supervision of a qualified healthcare practitioner.

Review Details

I. Drug and Supplement Interactions

Clinical Trials of Drug or Supplement Interactions

No clinical trials of drug or supplement interactions were identified.

Case Reports of Suspected Drug or Supplement Interactions

Support for the cytochrome P450-mediated activation of pyrrolizidines into their corresponding toxins comes from the observation that potent microsomal enzyme inducers, such as the anticonvulsant phenobarbital, can enhance the toxicity of pyrrolizidines (McLean 1974).

Animal Trials of Drug or Supplement Interactions

No animal trials of drug or supplement interactions were identified.

II. Adverse Events

Case Reports of Adverse Events

Consumption of comfrey tea has been associated with liver damage, primarily veno-occlusive disease, a nonthrombotic obliteration of small hepatic veins leading to cirrhosis and eventually liver failure. Patients have presented with either acute or chronic clinical signs, with portal hypertension, hepatomegaly, and abdominal pain as the main features (Stickel and Seitz 2000). In the case reports below, the species of comfrey consumed was generally not reported.

A 49-year-old woman was diagnosed with veno-occlusive disease, a form of Budd-Chiari syndrome. The patient had portal hypertension associated with obliteration of the smaller hepatic venules. A liver biopsy specimen showed centrilobular necrosis and congestion. The woman had been regularly taking two products that contained comfrey, one for 6 months and the other for 4 months. Analyses of products taken indicated an estimated intake of pyrrolizidine alkaloids of 0.49 to 1.45 μg/kg daily. Other supplements being taken by the woman included a number of vitamins, minerals, and sterotrophic adrenal bovine extract (Ridker et al. 1985).

Veno-occlusive disease was reported in a 13-year-old boy with Crohn's disease who was "regularly" administered comfrey leaf tea (dose, duration, and frequency not specified) (Weston et al. 1987). Veno-occlusive disease was reported in a 47-year-old woman who had been consuming

10 cups of comfrey tea daily for approximately 4 years (part and dose used not specified) (Bach et al. 1989).

Hepatic veno-occlusive disease leading to death was reported in a 23-year-old man who ingested 4 to 5 cooked comfrey leaves daily for 1 to 2 weeks (Yeong et al. 1990). Two cases of veno-occlusive disease were reported in women who took comfrey-pepsin tablets for 6 months. The daily doses included 280 mg/kg total pyrrolizidine alkaloids in one patient and 988 mg/kg total pyrrolizidine alkaloids in the other patient (Huxtable 1987).

The leaves of comfrey and foxglove (*Digitalis purpurea*) are similar in appearance. Several cases of poisoning have been reported in persons who accidentally ingested foxglove instead of comfrey (Awang et al. 1993; Bain 1985; Routledge and Spriggs 1989; Turley and Muir 2008).

III. Pharmacology and Pharmacokinetics

Human Pharmacological Studies

No evidence of liver damage was found in regular consumers of comfrey leaf (species not identified). Liver function tests showed slightly elevated levels of bilirubin in two patients and aspartate aminotransferase (AST) in one patient. Most volunteers (72%) had used comfrey for 1 to 10 years (mean intake 3.0 g dry leaf/day); 17% used it for 11 to 20 years (mean intake 2.6 g dry leaf/day); and 10% used it for 21 to 30 years (mean intake 11 g dry leaf/day). The estimated intake of pyrrolizidine alkaloids was 0.015 to 0.15 mg/kg daily (Anderson and McLean 1989).

Animal Pharmacological Studies

In rats topically administered a dose of an ethanolic comfrey extract corresponding to 194 mg/kg of alkaloid *N*-oxides (with 7-acetylintermedine and 7-acetyllycopsamine as the main constituents along with lycopsamine, intermedine, symphytine, and traces of two unidentified alkaloids), the excretion of *N*-oxides in the urine was in the range of 0.1 to 0.4% of the dose, indicating that the dermally absorbed *N*-oxides are either not, or only to a small extent, converted to free alkaloids. Oral administration led to a 20 to 50 times higher excretion of *N*-oxides and free alkaloids in the urine (Brauchli et al. 1982).

In rats orally administered 1 mg/kg of the compound riddelliine (a pyrrolizidine alkaloid) 5 days per week or fed a daily diet of 8% comfrey root for 12 weeks, gene expression and biological processes observed were significantly different between the comfrey and riddelliine groups, although a number of genes related to carcinogenesis were expressed in both groups (Guo et al. 2007).

Unsaturated pyrrolizidine alkaloids, such as those found in comfrey, are metabolized by the CYP3A drug-metabolizing isoenzymes (Fu et al. 2004; Wang et al. 2005; Yarnell and Abascal 2007).

In Vitro Pharmacological Studies

No relevant in vitro pharmacological studies were identified.

S

IV. Pregnancy and Lactation

Offspring of rats administered the pyrrolizidine alkaloid riddelliine during pregnancy and lactation had lower weight than control animals. The authors of the study indicated that riddelliine may cross the placenta and may be found in milk (Chan et al. 1994).

Pyrrolizidine alkaloids, including those present in comfrey, have been shown to be present in milk of animals that have consumed plants that contain these compounds (Panter and James 1990). Cases of veno-occlusive disease have been reported in infants of lactating women who consumed species of plants containing pyrrolizidine alkaloids (Sperl et al. 1995).

V. Toxicity Studies

Acute Toxicity

The LD_{50} of intraperitoneally administered symphytine is 300 mg/kg in mice (Culvenor et al. 1980) and 130 mg/kg in rats (Hirono et al. 1979).

Short-Term Toxicity

In rats administered a diet consisting of 5, 10, or 30% comfrey (part unspecified) for 21 days, those on the 30% comfrey diet gained less weight than control animals. Rats fed 10 or 30% comfrey or 30% alfalfa (*Medicago sativa*) in the diet exhibited an increase in hepatic aminopyrine N-demethylase activity but had no change in hepatic glutathione S-transferase or epoxide hydrolase activities, as compared to rats fed a control diet (Garrett et al. 1982).

Chronic Toxicity

A dose-dependent induction of liver tumors was observed in rats fed a diet of 8, 16, or 33% comfrey leaf for 600 days or a diet of 33% comfrey for 480 days. Liver tumors were also observed in rats fed a diet of 1 to 8% comfrey root for varying periods of time. The 8% comfrey root diet was administered until the rats died, while the 1, 2, and 4% diets were administered for 180 to 275 days with a reduced content of comfrey root (0 to 2%) administered for a total of 245 days or until the rats died (Hirono et al. 1978).

Hepatotoxicity

In rats fed a diet of 2% comfrey root daily for 12 weeks, a higher mutation rate was observed in liver and lung *cII* genes than in rats fed a control diet. No signs of liver toxicity were observed (Mei et al. 2005; Mei and Chen 2007). Riddelliine, one of the pyrrolizidine alkaloids of comfrey, was previously shown to dose-dependently induce mutations in the *cII* gene after oral administration to rats at doses of 0.1, 0.3, and 1.0 mg/kg (Mei et al. 2004). A gene expression profile of rats administered a diet of 8% comfrey root indicated that differentially expressed genes in the livers of treated animals were involved in metabolism, injury of endothelial cells, and liver injury and abnormalities, including liver fibrosis and cancer development (Mei et al. 2006).

In rats orally administered riddelliine at doses up to 10 mg/kg and in mice administered riddelliine at doses up to 25 mg/kg 5 days a week for 13 weeks, dose-related hepatopathy and intravascular macrophage accumulation in rats and hepatocytomegaly in mice were observed. Some animals were given a 14-week recovery period after treatment, and during that time, hepatic foci of cellular alteration in male rats and bile duct proliferation in female rats and male and female mice increased in severity (Chan et al. 1994).

Genotoxicity

Formation of DNA adducts was observed in female rats orally administered extracts of comfrey root but not comfrey leaf. The compound riddelliine caused formation of a significantly larger number of adducts than comfrey root (Chou and Fu 2006).

In the Ames mutagenicity test, comfrey produced "toxic responses" in *Salmonella typhimurium* strains TA98 and TA100 in the absence of microsomal activation, but showed no adverse effects in the presence of microsomal activation (White et al. 1983).

An alkaloid extract of comfrey induced sister-chromatid exchanges (SCE) and chromosomal aberrations at concentrations of 140 and 1400 μg/ml but not at concentrations of 1.4 or 14 μg/ml. In the presence of S9 mix, the clastogenic activity and SCE-inducing effect were enhanced (Behninger et al. 1989).

No mutagenic effects of comfrey leaf were observed in *Drosophila* (Clark 1982).

Literature Cited

AHPA. July 2011. Code of Ethics & Business Conduct. Silver Spring, MD: American Herbal Products Association.

Altamirano, J., S.R. Gratz, and K.A. Wolnik. 2005. Investigation of pyrrolizidine alkaloids and their *N*-oxides in commercial comfrey-containing products and botanical materials by liquid chromatography electrospray ionization mass spectrometry. *J. AOAC Int.* 88(2):406-412.

Anderson, P.C., and A.E.M. McLean. 1989. Comfrey and liver damage. *Human Toxicol.* 8:68-69.

Awang, D.V.C., B.A. Dawson, J. Fillion, M. Girad, and D. Klindack. 1993. Echimidine content of commercial comfrey. *J. Herbs Spices Med. Plants* 2(1):21-34.

Bach, N., S.N. Thung, and F. Schaffner. 1989. Comfrey herb tea-induced hepatic veno-occlusive disease. *Am. J. Med.* 87(1):97-99.

Bain, R.J.I. 1985. Accidental digitalis poisoning due to drinking herbal tea. *Br. Med. J.* 290:1624.

Behninger, C., G. Abel, E. Roder, V. Neuberger, and W. Goggelmann. 1989. Studies on the effect of an alkaloid extract of *Symphytum officinale* on human lymphocyte cultures. *Planta Med.* 55(6):518-522.

Betz, J.M., R.M. Eppley, W.C. Taylor, and D. Andrzejewski. 1994. Determination of pyrrolizidine alkaloids in commercial comfrey products (*Symphytum* sp.). *J. Pharm. Sci.* 83(5):649-653.

Bradley, P.R. 1992. *British herbal compendium: A handbook of scientific information on widely used plant drugs*. Bournemouth, UK: British Herbal Medicine Association.

Brauchli, J., J. Luthy, U. Zweifel, and C. Schlatter. 1982. Pyrrolizidine alkaloids from *Symphytum officinale* L. and their percutaneous absorption in rats. *Experientia* 38(9):1085-1087.

Cao, Y., S.M. Colegate, and J.A. Edgar. 2008. Safety assessment of food and herbal products containing hepatotoxic pyrrolizidine alkaloids: Interlaboratory consistency and the importance of *N*-oxide determination. *Phytochem Anal.* 19(6):526-533.

Chan, P.C., J. Mahler, J.R. Bucher, G.S. Travlos, and J.B. Reid. 1994. Toxicity and carcinogenicity of riddelliine following 13 weeks of treatment to rats and mice. *Toxicon* 32(8):891-908.

Chou, M.W., and P.P. Fu. 2006. Formation of DHP-derived DNA adducts in vivo from dietary supplements and Chinese herbal plant extracts containing carcinogenic pyrrolizidine alkaloids. *Toxicol. Ind. Health* 22 (8):321-327.

Clark, A.M. 1982. The use of larval stages of *Drosophila* in screening for some naturally occurring mutagens. *Mutat. Res.* 92(1):89-97.

Couet, C.E., C. Crews, and A.B. Hanley. 1996. Analysis, separation, and bioassay of pyrrolizidine alkaloids from comfrey (*Symphytum officinale*). *Nat. Toxins* 4(4):163-167.

Culvenor, C.C., M. Clarke, J.A. Edgar, et al. 1980. Structure and toxicity of the alkaloids of Russian comfrey (*Symphytum* × *uplandicum* Nyman), a medicinal herb and item of human diet. *Experientia* 36(4):377-379.

De Smet, P.A.G.M. 1992. *Adverse effects of herbal drugs, Volume 1.* Berlin: Springer.

Dąbska, W., A. Owczarska, and R. Madalihska. 1980. *Radix symphyti* for lasiocarpine content. *Herb. Pol.* 26:47-52.

Fu, P.P., Q. Xia, G. Lin, and M.W. Chou. 2004. Pyrrolizidine alkaloids—Genotoxicity, metabolism enzymes, metabolic activation, and mechanisms. *Drug Metab. Rev.* 36(1):1-55.

Garrett, B.J., P.R. Cheeke, and C.L. Miranda. 1982. Consumption of poisonous plants (*Senecio jacobaea, Symphytum officinale, Pteridium aquilinum, Hypericum perforatum*) by rats: Chronic toxicity, mineral metabolism, and hepatic drug-metabolizing enzymes. *Toxicol. Lett.* 10(2-3):183-188.

Guo, L., N. Mei, S. Dial, J. Fuscoe, and T. Chen. 2007. Comparison of gene expression profiles altered by comfrey and riddelliine in rat liver. *BMC Bioinformatics* 8(Suppl. 7):S22.

Hirono, I., M. Haga, M. Fujii, et al. 1979. Induction of hepatic tumors in rats by senkirkine and symphytine. *J. Natl. Cancer Inst.* 63(2):469-472.

Hirono, I., H. Mori, and M. Haga. 1978. Carcinogenic activity of *Symphytum officinale*. *J. Natl. Cancer Inst.* 61(3):865-869.

Huizing, H.J., T.W.J. Gadella, and E. Kliphuis. 1982. Chemotaxonomical investigations of the *Symphytum officinale* polyploid complex and *S. asperum* (Boraginaceae): The pyrrolizidine alkaloids. *Plant Sys. Evol.* 140(4):279-292.

Huxtable, R.J. 1987. Pyrrolizidine alkaloid poisoning in the United States [abstract]. *Pharmacologist* 29(3):155.

Huxtable, R.J., J. Luthy, and U. Zweifel. 1986. Toxicity of comfrey-pepsin preparations. *N. Engl. J. Med.* 315(17):1095.

Mattocks, A.R. 1980. Toxic pyrrolizidine alkaloids in comfrey. *Lancet* 2(8204):1136-1137.

McLean, E. 1974. *Senecio* and other plants as liver poisons. *Isr. J. Med. Sci.* 436:40.
Mei, N., L. Guo, P.P. Fu, R.H. Heflich, and T. Chen. 2005. Mutagenicity of comfrey (*Symphytum officinale*) in rat liver. *Br. J. Cancer* 92(5):873-875.
Mei, N., L. Guo, L. Zhang, et al. 2006. Analysis of gene expression changes in relation to toxicity and tumorigenesis in the livers of Big Blue transgenic rats fed comfrey (*Symphytum officinale*). *BMC Bioinformatics* 7(Suppl. 2):S16.
Mei, N., R.H. Heflich, M.W. Chou, and T. Chen. 2004. Mutations induced by the carcinogenic pyrrolizidine alkaloid riddelliine in the liver *cII* gene of transgenic Big Blue rats. *Chem. Res. Toxicol.* 17(6):814-818.
Mei, X.B., and T. Chen. 2007. [The mutant frequencies and types of mutations induced by comfrey in the lungs of transgenic Big Blue rats.] *J. Food Drug Anal.* 15(4):458-465.
Oberlies, N.H., N.C. Kim, D.R. Brine, et al. 2004. Analysis of herbal teas made from the leaves of comfrey (*Symphytum officinale*): Reduction of *N*-oxides results in order of magnitude increases in the measurable concentration of pyrrolizidine alkaloids. *Public Health Nutr.* 7(7):919-924.
Panter, K.E., and L.F. James. 1990. Natural plant toxicants in milk: A review. *J. Anim. Sci.* 68(3):892-904.
Ridker, P.M., S. Ohkuma, and W.V. McDermott. 1985. Hepatic veno-occlusive disease associated with the consumption of pyrrolizidine-containing dietary supplements. *Gastroenterol.* 88 (4):1050-1054.
Roitman, J.N. 1981. Comfrey and liver damage. *Lancet* 1(8226):944.
Routledge, P.A., and T.L. Spriggs. 1989. Atropine as possible contaminant of comfrey tea. *Lancet* 1(8644):963-964.
Sperl, W., H. Stuppner, I. Gassner, et al. 1995. Reversible hepatic veno-occlusive disease in an infant after consumption of pyrrolizidine-containing herbal tea. *Eur. J. Pediatr.* 154(2):112-116.
Stickel, F., and H.K. Seitz. 2000. The efficacy and safety of comfrey. *Public Health Nutr.* 3(4A):501-508.
Tittel, G., H. Hinz, and H. Wagner. 1979. Quantitative bestimmung der pyrrolizidinalkaloide in *Symphyti radix* durch HPLC. *Planta Med.* 37(1).
Turley, A.J., and D.F. Muir. 2008. ECG for physicians: A potentially fatal case of mistaken identity. *Resuscitation* 76(3):323-324.
Vollmer, J.J., N.C. Steiner, G.Y. Larsen, K.M. Muirhead, and R.J. Molyneux. 1988. Pyrrolizidine alkaloids: Testing for toxic constituents of comfrey. *J. Chem. Educ.* 64(12):1027-1030.
Weston, C.F., B.T. Cooper, J.D. Davies, and D.F. Levine. 1987. Veno-occlusive disease of the liver secondary to ingestion of comfrey. *Br. Med. J. (Clin. Res. Ed.)* 295(6591):183.
White, R.D., P.H. Krumperman, P.R. Cheeke, and D.R. Buhler. 1983. An evaluation of acetone extracts from six plants in the Ames mutagenicity test. *Toxicol. Lett.* 15(1):25-31.
Wichtl, M. 2004. *Herbal drugs and phytopharmaceuticals: A handbook for practice on a scientific basis*. 3rd ed. Boca Raton, FL: CRC Press.
Yarnell, E., and K. Abascal. 2007. Interaction of herbal constituents with cytochrome P450 enzymes. *Altern. Complement. Ther.* 13(5):239-247.
Yeong, M.L., B. Swinburn, M. Kennedy, and G. Nicholson. 1990. Hepatic veno-occlusive disease associated with comfrey ingestion. *J. Gastroenterol. Hepatol.* 5(2):211-214.

Symphytum spp. — Boraginaceae

Symphytum asperum Lepechin
SCN: prickly comfrey
OCN: rough comfrey

Symphytum uplandicum **Nyman**
SCN: Russian comfrey
OCN: prickly comfrey; Quaker comfrey
Part: leaf, root

Quick Reference Summary

Safety Class:* 2a, 2b, 2c
Interaction Class: A

Contraindications

For external use only (Bradley 1992; De Smet 1992; Wichtl 2004).

Not for use during pregnancy or lactation (Chan et al. 1994; De Smet 1992; Panter and James 1990).

Other Precautions

None known.

Drug and Supplement Interactions

None known.

Notice

Pyrrolizidine alkaloids (0.01–0.15% in leaf; up to 0.37% in root) (Culvenor et al. 1980; De Smet 1992; Mattocks 1980; Roitman 1981; Stickel and Seitz 2000; Wuilloud et al. 2004); *see* Appendix 1.

Editors' Notes

Pyrrolizidine alkaloid compounds in comfrey have been associated with cases of liver toxicity (see Adverse Events and Side Effects below). The American Herbal Products Association has established a trade requirement (AHPA 2011) that all products with botanical ingredients that contain toxic pyrrolizidine alkaloids, including prickly comfrey

* Extracts of prickly comfrey and Russian comfrey that have had the pyrrolizidine alkaloids removed are commercially available. Concerns regarding the internal use of comfrey products do not apply to PA-free products.

and Russian comfrey, are not offered for sale for internal use and display the following cautionary label: "For external use only. Do not apply to broken or abraded skin. Do not use when nursing."

Along the alkaloids found in different species of comfrey, the pyrrolizidine alkaloid echimidine has exhibited a more toxic effect than other alkaloids (Culvenor et al. 1980). Russian comfrey and prickly comfrey contain echimidine in greater amounts than comfrey (*S. officinale*) (Awang et al. 1993; Jaarsma et al. 1989). Pyrrolizidine alkaloids and their metabolites are of particular interest, since these compounds are associated with toxicity (Stickel and Seitz 2000). Reports indicate that prickly comfrey contains from 0.14 to 0.37% pyrrolizidine alkaloids (Mattocks 1980), while the concentration in a commercial extract of Russian comfrey was reported as 0.0001% (Cao et al. 2008).

Symphytum uplandicum is a hybrid of *S. asperum* and *S. officinale* (Culvenor et al. 1980).

Adverse Events and Side Effects

Cases of liver disease have been reported in persons taking the leaf or root of *Symphytum* species (Bach et al. 1989; Huxtable et al. 1986; Ridker et al. 1985; Weston et al. 1987; Yeong et al. 1990). These case reports generally do not identify the species of comfrey consumed. Liver disease is caused by consumption of unsaturated pyrrolizidine alkaloids (Cao et al. 2008).

The leaves of comfrey and foxglove (*Digitalis purpurea*) are similar in appearance. Several cases of poisoning have been reported in persons who accidentally ingested foxglove instead of comfrey (Awang and Kindack 1989; Bain 1985; Routledge and Spriggs 1989; Turley and Muir 2008).

Pharmacological Considerations

In contrast to case reports of liver disease, no evidence of liver damage was found in regular consumers (1 to 10 years of regular use) of comfrey (species not identified) leaf (Anderson and McLean 1989).

Dose-dependent impairment of liver function has been observed in rats orally administered mixtures of alkaloids extracted from Russian comfrey (Culvenor et al. 1980; Yeong et al. 1991).

Pregnancy and Lactation

Data from animal studies indicates that pyrrolizidine compounds present in comfrey can cross the placenta and are present in the breast milk of animals that have consumed comfrey (Chan et al. 1994; Panter and James 1990). Based on this information, use during pregnancy is not recommended except under the supervision of a qualified healthcare practitioner.

Review Details

I. Drug and Supplement Interactions

Clinical Trials of Drug or Supplement Interactions

No clinical trials of drug or supplement interactions were identified.

Case Reports of Suspected Drug or Supplement Interactions

No case reports of suspected drug or supplement interactions were identified.

Animal Trials of Drug or Supplement Interactions

No animal trials of drug or supplement interactions were identified.

II. Adverse Events

Case Reports of Adverse Events

The leaves of comfrey and foxglove (*Digitalis purpurea*) are similar in appearance. Several cases of poisoning have been reported in persons who accidentally ingested foxglove instead of *Symphytum* species (Awang and Kindack 1989; Bain 1985; Routledge and Spriggs 1989; Turley and Muir 2008).

Also see Case reports of adverse events in entry for *Symphytum officinale* leaf and root.

III. Pharmacology and Pharmacokinetics

Human Pharmacological Studies

No evidence of liver damage was found in regular consumers of comfrey leaf (species not identified). Liver function tests showed slightly elevated levels of bilirubin in two patients and AST in one patient. Most volunteers (72%) had used comfrey for 1 to 10 years (mean intake 3.0 g dry leaf/day); 17% used it for 11 to 20 years (mean intake 2.6 g dry leaf/day); and 10% used it for 21 to 30 years (mean intake 11 g dry leaf/day). The estimated intake of pyrrolizidine alkaloids was 0.015 to 0.15 mg/kg daily (Anderson and McLean 1989).

Animal Pharmacological Studies

In rats orally administered 1 mg/kg of the compound riddelliine 5 days per week or fed a daily diet of 8% comfrey root for 12 weeks, gene expression and biological processes observed were significantly different between the comfrey and riddelliine groups, although a number of genes related to carcinogenesis were expressed in both groups. The study authors concluded that the pyrrolizidine alkaloids contained in comfrey are the main active components responsible for carcinogenicity of the plant (Guo et al. 2007).

Unsaturated pyrrolizidine alkaloids, such as those found in comfrey, are metabolized by the CYP3A drug-metabolizing isoenzymes (Fu et al. 2004; Wang et al. 2005; Yarnell and Abascal 2007).

In Vitro Pharmacological Studies

No relevant in vitro pharmacological studies were identified.

IV. Pregnancy and Lactation

Offspring of rats administered the pyrrolizidine alkaloid riddelliine during pregnancy and lactation had lower weight than control animals. The authors of the study indicated that riddelliine may cross the placenta and may be found in milk (Chan et al. 1994). Based on this information, use during pregnancy is not recommended except under the supervision of a qualified healthcare practitioner.

Pyrrolizidine alkaloids, including those present in comfrey, have been shown to be present in the milk of animals that have consumed plants that contain these compounds (Panter and James 1990). Cases of veno-occlusive disease have been reported in infants of lactating women who consumed species of plants containing pyrrolizidine alkaloids (Sperl et al. 1995).

V. Toxicity Studies

Acute Toxicity

The LD_{50} of an intraperitoneally administered extract of the alkaloids of Russian comfrey in young rats is 284 mg/kg (Culvenor et al. 1980).

The LD_{50} of the compound echimidine administered intraperitoneally to rats is 200 mg/kg (Bull et al. 1968). The LD_{50} of intraperitoneally administered symphytine is 300 mg/kg in mice (Furuya and Araki 1968) and 130 mg/kg in rats (Hirono et al. 1979).

Short-Term Toxicity

In rats orally administered 50 mg/kg daily of Russian comfrey-derived alkaloids for 6 weeks, sinusoidal fibrosis, sloughing of endothelial cells, and hepatocyte membrane damage were observed (Yeong et al. 1993).

In rats orally administered pyrrolizidine alkaloids isolated from Russian comfrey at a single dose of 200 mg/kg, or 100 mg/kg 3 days per week for 3 weeks, or 50 mg/kg 3 days per week for 3 weeks, all rats showed evidence of liver damage, the severity of which was dose dependent (Yeong et al. 1991).

In rats administered alkaloids derived from Russian comfrey, a dose of 71 mg/kg administered intraperitoneally three times per week resulted in severely impaired liver function and death within 3 to 4 weeks. Plasma protein and the albumin/globulin ratio were reduced by 40% or more, with only small increases in liver enzymes (Culvenor et al. 1980).

Literature Cited

AHPA. July 2011. Code of Ethics & Business Conduct. Silver Spring, MD: American Herbal Products Association.

Anderson, P.C., and A.E.M. McLean. 1989. Comfrey and liver damage. *Human Toxicol.* 8:68-69.

Awang, D.V., and D.G. Kindack. 1989. Atropine as possible contaminant of comfrey tea. *Lancet* 2(8653):44.

Awang, D.V.C., B.A. Dawson, J. Fillion, M. Girad, and D. Klindack. 1993. Echimidine content of commercial comfrey. *J. Herbs Spices Med. Plants* 2:1.

Bach, N., S.N. Thung, and F. Schaffner. 1989. Comfrey herb tea-induced hepatic veno-occlusive disease. *Am. J Med.* 87(1):97-99.

Bain, R.J.I. 1985. Accidental digitalis poisoning due to drinking herbal tea. *Br. Med. J.* 290:1624.

Bradley, P.R. 1992. *British herbal compendium: A handbook of scientific information on widely used plant drugs*. Bournemouth, UK: British Herbal Medicine Association.

Bull, L., C.C. Culvenor, and A. Dick. 1968. *The pyrrolizidine alkaloids*. Amsterdam: North Holland.

Cao, Y., S.M. Colegate, and J.A. Edgar. 2008. Safety assessment of food and herbal products containing hepatotoxic pyrrolizidine alkaloids: Interlaboratory consistency and the importance of *N*-oxide determination. *Phytochem. Anal.* 19(6):526-533.

Chan, P.C., J. Mahler, J.R. Bucher, G.S. Travlos, and J.B. Reid. 1994. Toxicity and carcinogenicity of riddelliine following 13 weeks of treatment to rats and mice. *Toxicon* 32(8):891-908.

Culvenor, C.C., M. Clarke, J.A. Edgar, et al. 1980. Structure and toxicity of the alkaloids of Russian comfrey (*Symphytum* × *uplandicum* Nyman), a medicinal herb and item of human diet. *Experientia* 36(4):377-379.

De Smet, P.A.G.M. 1992. *Adverse effects of herbal drugs, Volume 1*. Berlin: Springer.

Fu, P.P., Q. Xia, G. Lin, and M.W. Chou. 2004. Pyrrolizidine alkaloids—Genotoxicity, metabolism enzymes, metabolic activation, and mechanisms. *Drug Metab. Rev.* 36(1):1-55.

Furuya, T., and K. Araki. 1968. Studies on constituents of crude drugs. I. Alkaloids of *Symphytum officinale* L. *Chem. Pharm. Bull.* 16(12):2512-2516.

Guo, L., N. Mei, S. Dial, J. Fuscoe, and T. Chen. 2007. Comparison of gene expression profiles altered by comfrey and riddelliine in rat liver. *BMC Bioinformatics* 8(Suppl. 7):S22.

Hirono, I., M. Haga, M. Fujii, et al. 1979. Induction of hepatic tumors in rats by senkirkine and symphytine. *J. Natl. Cancer Inst.* 63(2):469-472.

Huxtable, R.J., J. Luthy, and U. Zweifel. 1986. Toxicity of comfrey-pepsin preparations. *N. Engl. J. Med.* 315(17):1095.

Jaarsma, T.A., E. Lohmanns, T.W.J. Gadella, and T.M. Malingre. 1989. Chemotaxonomy of the *Symphytum officinale* agg. (Boraginaceae). *Plant Sys. Evol.* 167(3-4).

Mattocks, A.R. 1980. Toxic pyrrolizidine alkaloids in comfrey. *Lancet* 2 (8204):1136-1137.

Panter, K.E., and L.F. James. 1990. Natural plant toxicants in milk: A review. *J. Anim. Sci.* 68 (3):892-904.

S

Ridker, P.M., S. Ohkuma, and W.V. McDermott. 1985. Hepatic veno-occlusive disease associated with the consumption of pyrrolizidine-containing dietary supplements. *Gastroenterology* 88(4):1050-1054.
Roitman, J.N. 1981. Comfrey and liver damage. *Lancet* 1(8226):944.
Routledge, P.A., and T.L. Spriggs. 1989. Atropine as possible contaminant of comfrey tea. *Lancet* 1(8644):963-964.
Sperl, W., H. Stuppner, I. Gassner, et al. 1995. Reversible hepatic veno-occlusive disease in an infant after consumption of pyrrolizidine-containing herbal tea. *Eur. J. Pediatr.* 154(2):112-116.
Stickel, F., and H.K. Seitz. 2000. The efficacy and safety of comfrey. *Public Health Nutr.* 3(4A):501-508.
Turley, A.J., and D.F. Muir. 2008. ECG for physicians: A potentially fatal case of mistaken identity. *Resuscitation* 76(3):323-324.
Wang, Y.P., J. Yan, P.P. Fu, and M.W. Chou. 2005. Human liver microsomal reduction of pyrrolizidine alkaloid *N*-oxides to form the corresponding carcinogenic parent alkaloid. *Toxicol. Lett.* 155(3):411-420.
Weston, C.F., B.T. Cooper, J.D. Davies, and D.F. Levine. 1987. Veno-occlusive disease of the liver secondary to ingestion of comfrey. *Br. Med. J. (Clin. Res. Ed.)* 295(6591):183.
Wichtl, M. 2004. *Herbal drugs and phytopharmaceuticals: A handbook for practice on a scientific basis*. 3rd ed. Boca Raton, FL: CRC Press.
Wuilloud, J.C.A., S.R. Gratz, B.M. Gamble, and K.A. Wolnik. 2004. Simultaneous analysis of hepatotoxic pyrrolizidine alkaloids and *N*-oxides in comfrey root by LC-ion trap mass spectrometry. *Analyst* 129(2):150-156.
Yarnell, E., and K. Abascal. 2007. Interaction of herbal constituents with cytochrome P450 enzymes. *Altern. Complement. Ther.* 13(5):239-247.
Yeong, M.L., S.P. Clark, J.M. Waring, R.D. Wilson, and S.J. Wakefield. 1991. The effects of comfrey derived pyrrolizidine alkaloids on rat liver. *Pathology* 23(1):35-38.
Yeong, M.L., B. Swinburn, M. Kennedy, and G. Nicholson. 1990. Hepatic veno-occlusive disease associated with comfrey ingestion. *J. Gastroenterol. Hepatol.* 5(2):211-214.
Yeong, M.L., S.J. Wakefield, and H.C. Ford. 1993. Hepatocyte membrane injury and bleb formation following low dose comfrey toxicity in rats. *Int. J. Exp. Pathol.* 74(2):211-217.

Symplocarpus foetidus (L.) Salisb. ex Nutt. — Araceae

SCN: skunk cabbage
Syn: *Dracontium foetidum* L.
Part: herb, root

Quick Reference Summary

Safety Class: 2d
Interaction Class: A

Contraindications

Do not exceed recommended dose (Felter and Lloyd 1898; Wood and LaWall 1918).

Other Precautions

Individuals with a history of kidney stones should use skunk cabbage with caution (McGuffin et al. 1997).

Drug and Supplement Interactions

None known.

Standard Dose

Tincture of the fresh root: 3 to 8 drops two or three times daily, diluted in a full cup of juice or water (Winston 2010).

Tincture of the dried root: 15 to 30 drops two or three times daily, diluted in a full cup of juice or water (Winston 2010).

Adverse Events and Side Effects

Large doses of skunk cabbage have been associated with nausea and vomiting, headache, dizziness, and dimness of vision (Felter and Lloyd 1898; Wood and LaWall 1918).

Skunk cabbage contains calcium oxalate crystals, microscopic needlelike structures that mechanically irritate the skin and mucous membranes, causing a painful burning sensation of the lips and mouth after ingestion (Genua and Hillson 1985; Higley 1880; Keating 2004; Nelson et al. 2006; Rowlee and Nichols 1896). Ingestion may cause an inflammatory reaction, often with edema and blistering, sometimes resulting in hoarseness and difficulty swallowing (Nelson et al. 2006).

Pharmacological Considerations

None known.

Pregnancy and Lactation

No information on the safety of skunk cabbage in pregnancy or lactation was identified in the scientific or traditional literature. Although this review did not identify any concerns for use while pregnant or nursing, safety has not been conclusively established.

Review Details

I. Drug and Supplement Interactions

Clinical Trials of Drug or Supplement Interactions

No clinical trials of drug or supplement interactions were identified.

Case Reports of Suspected Drug or Supplement Interactions

No case reports of suspected drug or supplement interactions were identified.

Animal Trials of Drug or Supplement Interactions

No animal trials of drug or supplement interactions were identified.

II. Adverse Events

Case Reports of Adverse Events

No case reports of adverse events were identified.

III. Pharmacology and Pharmacokinetics

Human Pharmacological Studies

No relevant human pharmacological studies were identified.

Animal Pharmacological Studies

No relevant animal pharmacological studies were identified.

In Vitro Pharmacological Studies

No relevant in vitro pharmacological studies were identified.

IV. Pregnancy and Lactation

No information on the safety of skunk cabbage during pregnancy or lactation was identified.

V. Toxicity Studies

No toxicity studies were identified.

Literature Cited

Felter, H.W., and J.U. Lloyd. 1898. *King's American dispensatory*. 18th ed., 3rd rev. 2 vols. Cincinnati: Ohio Valley Co.

Genua, J.M., and C.J. Hillson. 1985. The occurrence, type and location of calcium oxalate crystals in the leaves of fourteen species of Araceae. *Ann. Bot.* 56(3):351.

Higley, W.K. 1880. On the microscopic crystals contained in plants. *Am. Naturalist* 14(10):720-725.

Keating, R.C. 2004. Systematic occurrence of raphide crystals in Araceae. *Ann. Missouri Bot. Gar.* 91(3):495-504.

McGuffin, M., C. Hobbs, R. Upton, and A. Goldberg. 1997. *Botanical safety handbook*. Boca Raton, FL: CRC Press.

Nelson, L., R.D. Shih, M.J. Balick, and K.F. Lampe. 2006. *Handbook of poisonous and injurious plants*. 2nd ed. New York: Springer.

Rowlee, W.W., and M.A. Nichols. 1896. Contributions to the life-history of *Symplocarpus fœtidus*. *Trans. Am. Microsc. Soc.* 17:157-164.

Winston, D. 2010. *Winston's botanical materia medica*. Broadway, NJ: David Winston's Center for Herbal Studies.

Wood, H., and C. LaWall. 1918. *The dispensatory of the United States of America*. 21st ed. Philadelphia: Lippincott.

Syzygium aromaticum (L.) Merr. & L.M. Perry — Myrtaceae

S

SCN: clove

Syn: *Caryophyllus aromaticus* L.; *Eugenia aromatica* (L.) Baill., nom. illeg.; *Eugenia caryophyllata* Thunb.

AN: *lavanga*

PN: *ding xiang* (flower bud)

Part: flower bud

Quick Reference Summary

Safety Class: 2d

Interaction Class: A

Contraindications

Do not exceed recommended dose (Bensky et al. 2004).

Other Precautions

None known.

Drug and Supplement Interactions

None known.

Standard Dose

The standard dose is 1 to 5 g as a decoction (Bensky et al. 2004; Chen and Chen 2004), 120 to 300 mg of powder (Martindale and Reynolds 1967).

Editors' Notes

The classifications and concerns for this herb are based on relatively higher doses used for therapeutic purposes in contrast to lower amounts generally used in cooking, and have not been associated with its use as a spice.

Clove should not be confused with clove essential oil, which is also available commercially. Although the oil has long been used as a topical pain reliever for toothaches and is regarded as safe for this use (Alqareer et al. 2006), ingestion of large amounts may cause toxicity (Hartnoll et al. 1993).

Adverse Events and Side Effects

None known.

Pharmacological Considerations

The compounds eugenol and acetyleugenol inhibited platelet aggregation in vitro (Laekeman et al. 1990; Saeed et al. 1995; Srivastava and Malhotra 1991).

In one animal study, some changes in sperm formation were reported after administration of clove extract (Mishra and Singh 2008).

Pregnancy and Lactation

Limited information on the safety of clove in pregnancy or lactation was identified in the scientific and traditional literature. Although this review did not identify any concerns for use while pregnant or nursing, safety has not been conclusively established.

Review Details

I. Drug and Supplement Interactions

Clinical Trials of Drug or Supplement Interactions

No clinical trials of drug or supplement interactions were identified.

Case Reports of Suspected Drug or Supplement Interactions

No case reports of suspected drug or supplement interactions were identified.

Animal Trials of Drug or Supplement Interactions

No animal trials of drug or supplement interactions were identified.

II. Adverse Events

Case Reports of Adverse Events

No adverse reactions to clove have been reported in the normal dosage range (a decoction of 1–3 g). Overdose of clove has resulted in symptoms of nausea, diarrhea, vomiting, and upper gastrointestinal hemorrhage. Cases of severe overdoses have been reported to cause changes in liver function, difficulty breathing, loss of consciousness, or death (Bensky et al. 2004).

A 2-year-old boy that ingested 5 to 10 ml of clove essential oil experienced coma, fits, a coagulopathy, and acute liver damage (Hartnoll et al. 1993).

III. Pharmacology and Pharmacokinetics

Human Pharmacological Studies

No relevant human pharmacological studies were identified.

Animal Pharmacological Studies

An increase in muscle propulsion in the gastrointestinal tract was observed in fasted mice and rats orally administered 300 or 700 mg/kg of an aqueous extract of clove (Agbaje 2008).

In mice orally administered 15, 30, or 60 mg/kg of a hexane extract of cloves daily for 35 days, the 15 mg/kg dose increased the activities of hydroxysteroid dehydrogenases and serum level of testosterone. At the 30 and 60 mg/kg dose, inhibition of these parameters was observed, along with induction of nonuniform degenerative changes in the seminiferous tubules associated with a decrease in daily sperm production and depletion of round and elongated spermatid populations (Mishra and Singh 2008).

In Vitro Pharmacological Studies

A methanol extract of clove inhibited the human drug-metabolizing isoenzyme CYP3A4 in a radiometric assay (Usia et al. 2006).

In human plasma, the compound eugenol strongly inhibited platelet aggregation induced by platelet-activating factor. A weaker inhibition was observed on platelet aggregation induced by arachidonic acid or collagen (Saeed et al. 1995). Inhibition of arachidonic acid-induced platelet inhibition was also observed in rabbit plasma (Laekeman et al. 1990).

The compounds eugenol and acetyleugenol inhibited platelet aggregation induced by arachidonic acid, adrenaline, and collagen in platelet-rich human plasma (IC_{50} values of 0.8 μM for eugenol and 2 μM for acetyleugenol), with effects on arachidonic acid-induced aggregation being greater than that of aspirin (IC_{50} of 28 μM). A combination of the two compounds had an additive effect (Srivastava and Malhotra 1991).

Two polysaccharides isolated from clove were found to have antithrombotic activity (Lee et al. 2001).

IV. Pregnancy and Lactation

In mice fed diets containing 0.25% clove essential oil (estimated daily intake of 375 mg/kg) for 2 weeks and then mated, examination of 4-day-old embryos indicated an increase in cell death rate. The number of embryos and the rate of pregnancy were higher in the clove group than in the control group (Domaracky et al. 2007).

No information on the safety of clove during lactation was identified.

V. Toxicity Studies

Acute Toxicity

The LD_{50} of clove essential oil orally administered in rats is reported as 2.65 and 3.72 g/kg (Opdyke 1979). The dermal LD_{50} for clove essential oil is 5 g/kg (Opdyke 1979).

The oral LD_{50} of the compound eugenol is 3 g/kg in mice, 1.9 or 2.7 g/kg in rats, and 2.1 g/kg in guinea pigs (JECFA 1982). Clove essential oil is approximately 90% eugenol (Chaieb et al. 2007).

The Joint FAO/WHO Expert Committee on Food Additives determined that the acceptable daily intake for the compound eugenol is 2.5 mg/kg. This group also indicated that 250 mg/kg was the level causing no effect in the diet of rats (JECFA 1982).

Short-Term Toxicity

In rats orally administered 35 or 70 mg of clove essential oil daily for 8 weeks, the doses were reported as well tolerated. At a dose of 105 mg, severe liver and kidney damage was observed after 2 to 3 weeks. A single dose of 140 mg was fatal (Opdyke 1979).

Genotoxicity

In the Ames test for mutagenicity with *Salmonella typhimurium* strains TA98 and TA100, an ethanol extract of clove exhibited strong mutagenic activity in strain TA100 and weak activity in TA98 (Mahmoud et al. 1992).

An ethanol extract of clove showed no mutagenic activity in the Ames test but gave a strong positive result in the *Bacillus subtilis rec* assay (Morimoto et al. 1982).

In *Salmonella typhimurium* strain TA100, clove essential oil was mutagenic without metabolic activation, but not mutagenic after metabolic activation by S9 (Park 2002). Another study with TA100 indicated that no mutagenic activity of clove essential oil was observed without metabolic activation and that mutagenic effect was stronger after the addition of S9 (Shoeibi et al. 2009).

An aqueous extract of clove showed no mutagenic activity in the SOS chromotest and in the *Salmonella reversion* assay using strains TA97a, TA98, TA100, and TA102. In a forward mutagenesis assay, an increase in mutagenesis was observed with the CC104 *mutMmutY* strain, suggesting that oxidative DNA damage occurred (dos Santos et al. 2008).

In the mouse bone marrow micronucleus assay, mice fed diets containing 0.5 or 2% clove for 10 days did not have any induced micronuclei (Kumari 1991).

Phenylpropanoid compounds from clove exhibited antimutagenic activity after induction of the SOS response in *Salmonella typhimurium* TA1535/pSK1002 treated with various mutagens (Miyazawa and Hisama 2003). A methanol extract of clove showed a suppressive effect on the SOS-inducing activity of furylfuramide in the *Salmonella typhimurium* TA1535/pSK1002 umu test (Miyazawa and Hisama 2001).

Literature Cited

Agbaje, E.O. 2008. Gastrointestinal effects of *Syzigium aromaticum* (L.) Merr. & Perry (Myrtaceae) in animal models. *Nig. Q. J. Hosp. Med.* 18(3):137-141.

Alqareer, A., A. Alyahya, and L. Andersson. 2006. The effect of clove and benzocaine versus placebo as topical anesthetics. *J. Dentist.* 34(10):747-750.

Bensky, D., S. Clavey, and E. Stöger. 2004. *Chinese herbal medicine: Materia medica.* 3rd ed. Seattle: Eastland Press.

Chaieb, K., H. Hajlaoui, T. Zmantar, et al. 2007. The chemical composition and biological activity of clove essential oil, *Eugenia caryophyllata* (*Syzigium aromaticum* L. Myrtaceae): A short review. *Phytother. Res.* 21(6):501-506.

Chen, J.K., and T.T. Chen. 2004. *Chinese medical herbology and pharmacology.* City of Industry, CA: Art of Medicine Press.

Domaracky, M., P. Rehak, S. Juhas, and J. Koppel. 2007. Effects of selected plant essential oils on the growth and development of mouse preimplantation embryos in vivo. *Physiol. Res.* 56(1):97-104.

dos Santos, P.E., L.C. Monte Egito, S.R. Batistuzzo de Medeiros, and L.F. Agnez-Lima. 2008. Genotoxicity induced by *Eugenia caryophyllata* infusion. *J. Toxicol. Environ. Health A* 71(7):439-444.

Hartnoll, G., D. Moore, and D. Douek. 1993. Near fatal ingestion of oil of cloves. *Arch. Dis. Child.* 69(3):392-393.

JECFA. 1982. Eugenol. WHO Food Additives Series 17: FAO/WHO Joint Expert Committee on Food Additives.

Kumari, M.V. 1991. Modulatory influences of clove (*Caryophyllus aromaticus*, L.) on hepatic detoxification systems and bone marrow genotoxicity in male Swiss albino mice. *Cancer Lett.* 60(1):67-73.

Laekeman, G.M., L.V. Hoof, A. Haemers, et al. 1990. Eugenol, a valuable compound for in vitro experimental research and worthwhile for further *in vivo* investigation. *Phytother. Res.* 4 (3):90-96.

Lee, J.I., H.S. Lee, W.J. Jun, et al. 2001. Purification and characterization of antithrombotics from *Syzygium aromaticum* (L.) Merr. and Perry. *Biol. Pharm. Bull.* 24(2):181-187.

Mahmoud, I., A. Alkofahi, and A. Abdelaziz. 1992. Mutagenic and toxic activities of several spices and some Jordanian medicinal plants. *Int. J. Pharmacog.* 30(2):81-85.

Martindale, W., and J.E.F. Reynolds. 1967. *The extra pharmacopoeia.* 25th ed. London: Pharmaceutical Press.

Mishra, R.K., and S.K. Singh. 2008. Safety assessment of *Syzygium aromaticum* flower bud (clove) extract with respect to testicular function in mice. *Food Chem. Toxicol.* 46(10):3333-3338.

Miyazawa, M., and M. Hisama. 2001. Suppression of chemical mutagen-induced SOS response by alkylphenols from clove (*Syzygium aromaticum*) in the *Salmonella typhimurium* TA1535/pSK1002 umu test. *J. Agric. Food Chem.* 49(8):4019-4025.

Miyazawa, M., and M. Hisama. 2003. Antimutagenic activity of phenylpropanoids from clove (*Syzygium aromaticum*). *J. Agric. Food Chem.* 51(22):6413-6422.

S

Morimoto, I., F. Watanabe, T. Osawa, T. Okitsu, and T. Kada. 1982. Mutagenicity screening of crude drugs with *Bacillus subtilis* rec-assay and *Salmonella*/microsome reversion assay. *Mutat. Res.* 97:81-102.

Opdyke, D.L.J. 1979. *Monographs on fragrance raw materials*. New York: Pergamon.

Park, H.J. 2002. Mutagenicity of the essential oils in Ames test. *Kor. J. Pharmacog.* 33(4):372-375.

Saeed, S.A., R.U. Simjee, G. Shamim, and A.H. Gilani. 1995. Eugenol: A dual inhibitor of platelet activating factor and arachidonic acid metabolism. *Phytomedicine* 2:23-28.

Shoeibi, S., N. Rahimifard, B. Pirouz, et al. 2009. Mutagenicity of four natural flavors: Clove, cinnamon, thyme and *Zataria multiflora* Boiss. *J. Med. Plant* 8 (Suppl. 5):89-96.

Srivastava, K.C., and N. Malhotra. 1991. Acetyl eugenol, a component of oil of cloves (*Syzygium aromaticum* L.) inhibits aggregation and alters arachidonic acid metabolism in human blood platelets. *Prostaglandins Leukot. Essent. Fatty Acids* 42(1):73-81.

Usia, T., H. Iwata, A. Hiratsuka, et al. 2006. CYP3A4 and CYP2D6 inhibitory activities of Indonesian medicinal plants. *Phytomedicine* 13(1-2):67-73.

Syzygium cumini (L.) Skeels — Myrtaceae

SCN: jambolan
Syn: *Eugenia jambolana* Lam.; *Syzygium jambolana* DC.
AN: *jambu*
OCN: Java plum; jumbul
Part: bark, seed

Quick Reference Summary

Safety Class: 1
Interaction Class: A

Contraindications

None known.

Other Precautions

None known.

Drug and Supplement Interactions

None known.

Adverse Events and Side Effects

None known.

Pharmacological Considerations

Animal studies have demonstrated that jambolan seed and bark may modify glucose regulation (Mallick et al. 2006; Pandey and Khan 2002; Ravi et al. 2004; Saravanan and Pari 2008; Sharma et al. 2003, 2008; Sridhar et al. 2005; Villaseñor and Lamadrid 2006). People with diabetes are advised to monitor their blood sugar closely and discuss the use of this herb with a qualified healthcare practitioner prior to use.

Pregnancy and Lactation

No information on the safety of jambolan in pregnancy or lactation was identified in the scientific or traditional literature. Although this review did not identify any concerns for use while pregnant or nursing, safety has not been conclusively established.

Review Details

I. Drug and Supplement Interactions

Clinical Trials of Drug or Supplement Interactions

No clinical trials of drug or supplement interactions were identified.

Case Reports of Suspected Drug or Supplement Interactions

No case reports of suspected drug or supplement interactions were identified.

Animal Trials of Drug or Supplement Interactions

No animal trials of drug or supplement interactions were identified.

II. Adverse Events

Case Reports of Adverse Events

No case reports of adverse events were identified.

III. Pharmacology and Pharmacokinetics

Human Pharmacological Studies

No relevant human pharmacological studies were identified.

Animal Pharmacological Studies

In diabetic rats orally administered 800 mg/kg of a methanolic extract of jambolan seed daily for 14 days, corrections in fasting blood glucose levels and plasma insulin levels were observed (Mallick et al. 2006).

In diabetic rats orally administered 100 mg/kg of an ethanol extract of jambolan seed, a significant decrease in

S

blood glucose levels and glucose tolerance was observed (Ravi et al. 2004).

In rats with mild or severe diabetes, oral administration of 500 mg/kg of a flavonoid-rich extract of jambolan seed resulted in a reduction in fasting blood glucose and peak blood glucose levels (Sharma et al. 2008).

In rabbits with moderate or severe diabetes, oral administration of 100 mg/kg of an ethanol extract of jambolan seed daily for 15 days resulted in a significant reduction in fasting blood glucose and peak blood glucose and an increase in serum insulin levels (Sharma et al. 2003).

In diabetic rats fed diets containing 15% jambolan seed powder daily for 21 days, significantly lowered blood glucose levels were observed along with improved oral glucose tolerance (Pandey and Khan 2002).

In diabetic rats orally administered 250, 500, or 1000 mg/kg of jambolan seed powder daily for 15 days, decreases in fasting blood glucose and in post-treatment fasting and peak blood glucose were observed at the 500 and 1000 mg/kg doses (Sridhar et al. 2005).

In nondiabetic mice orally administered 250 mg/kg of a methanol extract of jambolan bark, a significant decrease in the blood glucose level was observed in the oral glucose tolerance test (Villaseñor and Lamadrid 2006).

In diabetic rats orally administered 75, 150, or 300 mg/kg of an aqueous extract of jambolan bark daily for 45 days, significant decreases in levels of blood glucose and urine sugar were observed. The activity of the 300 mg/kg dose was greater than that of the 75 and 150 mg/kg doses (Saravanan and Pari 2008).

In Vitro Pharmacological Studies

No relevant in vitro pharmacological studies were identified.

IV. Pregnancy and Lactation

No information on the safety of jambolan during pregnancy or lactation was identified.

V. Toxicity Studies

Acute Toxicity

No mortality or abnormalities were observed in rats orally administered single doses of 2.5 or 5 g/kg of jambolan seed powder (Sridhar et al. 2005).

Short-Term Toxicity

In diabetic rats orally administered 800 mg/kg of a methanolic extract of jambolan daily for 14 days, decreases in hepatic and renal GOT and GPT activity were observed (Mallick et al. 2006).

Literature Cited

Mallick, C., R. Maiti, and D. Ghosh. 2006. Antidiabetogenic effects of separate and composite extract of seed of jamun (*Eugenia jambolana*) and root of kadali (*Musa paradisiaca*) in streptozotocin-induced diabetic male albino rat: A comparative study. *Int. J. Pharmacol.* 2(5):492-503.

Pandey, M., and A. Khan. 2002. Hypoglycaemic effect of defatted seeds and water soluble fibre from the seeds of *Syzygium cumini* (Linn.) Skeels in alloxan diabetic rats. *Indian J. Exp. Biol.* 40(10):1178-1182.

Ravi, K., K. Sivagnanam, and S. Subramanian. 2004. Anti-diabetic activity of *Eugenia jambolana* seed kernels on streptozotocin-induced diabetic rats. *J. Med. Food* 7(2):187-191.

Saravanan, G., and L. Pari. 2008. Hypoglycaemic and antihyperglycaemic effect of *Syzygium cumini* bark in streptozotocin-induced diabetic rats. *J. Pharmacol. Toxicol.* 3(1):1-10.

Sharma, B., C. Balomajumder, and P. Roy. 2008. Hypoglycemic and hypolipidemic effects of flavonoid rich extract from *Eugenia jambolana* seeds on streptozotocin induced diabetic rats. *Food Chem. Toxicol.* 46(7):2376-2383.

Sharma, S.B., A. Nasir, K.M. Prabhu, P.S. Murthy, and G. Dev. 2003. Hypoglycaemic and hypolipidemic effect of ethanolic extract of seeds of *Eugenia jambolana* in alloxan-induced diabetic rabbits. *J. Ethnopharmacol.* 85(2-3):201-206.

Sridhar, S.B., U.D. Sheetal, M.R. Pai, and M.S. Shastri. 2005. Preclinical evaluation of the antidiabetic effect of *Eugenia jambolana* seed powder in streptozotocin-diabetic rats. *Braz. J. Med. Biol. Res.* 38(3):463-468.

Villaseñor, I.M., and M.R. Lamadrid. 2006. Comparative anti-hyperglycemic potentials of medicinal plants. *J. Ethnopharmacol.* 104(1-2):129-131.

S

Tabebuia impetiginosa (Mart. ex DC.) Standl. — Bignoniaceae

SCN: pau d'arco
Syn: *Tabebuia avellanedae* Lorentz ex Griseb.; *Tabebuia heptaphylla* (Vell.) Toledo; *Tecoma impetiginosa* Mart. ex DC.

OCN: ipe roxo; lapacho; taheebo
Part: bark

Quick Reference Summary

Safety Class: 1
Interaction Class: A

Contraindications

None known.

Other Precautions

None known.

Drug and Supplement Interactions

None known.

Editors' Notes

Much of the research related to pau d'arco has been done on the compound lapachol, which is reported by one reference to be present in the heartwood at a concentration of 2 to 7%, with less in the bark (Taylor 2005). An analysis of commercial pau d'arco wood and bark products, however, indicated that the lapachol content in the wood was 0.001%, with no lapachol detected in bark products (Awang et al. 1994). Similarly, no lapachol was identified in an aqueous extract of pau d'arco inner bark (Steinert et al. 1996). Publications on lapachol, including those cited here, may therefore have little or no relevance to products that contain pau d'arco.

Adverse Events and Side Effects

Occupational asthma and allergic contact dermatitis have been reported in woodworkers exposed to pau d'arco dust (Algranti et al. 2005; Estlander et al. 2001).

At doses of the compound lapachol over 1.5 g/day, nausea and vomiting were reported as adverse events in one clinical trial (Block et al. 1974).

Pharmacological Considerations

At doses of 2 g daily of the compound lapachol, a prolonged thrombin time was observed, while other tests of clotting gave normal results (Block et al. 1974).

Pregnancy and Lactation

While animal studies indicated mixed effects of the compound lapachol (*see* Editors' Note) in pregnant animals, no information on the safety of pau d'arco in pregnancy or lactation was identified in the scientific or traditional literature. Although this review did not identify any concerns for use while pregnant or nursing, safety has not been conclusively established.

Review Details

I. Drug and Supplement Interactions

Clinical Trials of Drug or Supplement Interactions

No clinical trials of drug or supplement interactions were identified.

Case Reports of Suspected Drug or Supplement Interactions

No case reports of suspected drug or supplement interactions were identified.

Animal Trials of Drug or Supplement Interactions

No animal trials of drug or supplement interactions were identified.

II. Adverse Events

Case Reports of Adverse Events

Occupational asthma and allergic contact dermatitis have been reported in woodworkers exposed to pau d'arco wood dust (Algranti et al. 2005; Estlander et al. 2001).

Adverse Events Reported in Clinical Trials

At doses of 2 g daily of the compound lapachol, thrombin time was prolonged and required correction with vitamin K. Other tests of clotting gave normal results (Block et al. 1974). No toxicity was observed at doses up to 1.5 g daily. At doses over 1.5 g/day, nausea and vomiting were reported as adverse events (Block et al. 1974).

III. Pharmacology and Pharmacokinetics

Human Pharmacological Studies

No relevant human pharmacological studies were identified.

Animal Pharmacological Studies

The compound lapachol had contact-sensitizing effects in guinea pigs (Schulz et al. 1977).

An increase in thrombin time but no change in clotting time were observed in dogs administered the compound lapachol (Morrison et al. 1970).

A decrease in weight of seminal vesicles but no changes in body weight or other organs were observed in male rats administered 100 mg/kg daily of hydroalcoholic solution of lapachol for 5 days (de Cassia da Silveira and Guerra 2007).

Also see Pregnancy and Lactation for this entry.

In Vitro Pharmacological Studies

Lapachol and lapachol derivatives exhibited cytotoxicity against A549 human breast cancer cells in vitro. The IC_{50} of lapachol was 0.78 m*M* (Oliveira et al. 2002).

The compounds lapachol and β-lapachone have been shown to be cytotoxic to cancer cells in animal studies and at micromolar concentrations in vitro in certain human cancer cell lines (Li et al. 1999; Menacho-Marquez and Murguia 2006; Ough et al. 2005; Pardee et al. 2002; Perez-Sacau et al. 2007; Queiroz et al. 2008). β-lapachone has been shown to have inhibitory activity on DNA topoisomerase (Lee et al. 2005), and cell death may also be caused by activation of a futile cycling of the drug by the cytoplasmic two-electron reductase NAD(P)H:quinone oxidoreductase (Ough et al. 2005; Pardee et al. 2002).

IV. Pregnancy and Lactation

No studies on the safety of pau d'arco in pregnancy or lactation were identified though a number of studies have examined the effects of the compound lapachol on reproduction in mice and rats.

T

In rats orally administered doses of 55 mg/kg or 110 mg/kg of lapachol on days 8 to 12 of pregnancy, 99 and 100% fetal mortality rates, respectively, were observed at the two dose levels, although no maternal toxicity was seen (Guerra et al. 1999, 2001). Fetal growth retardation but no effects on implantation or occurrences of resorption were observed in rats orally administered 100 mg/kg lapachol on days 17 to 20 of pregnancy (Felicio et al. 2002). Inhibition of pregnancy was observed in mice intramuscularly administered 20 mg/kg lapachol daily for 2 to 7 days. A 100% inhibition of pregnancy was observed in mice treated with lapachol on days 1 to 7 of pregnancy, while a 71% inhibition was observed in mice treated on days 1 to 3 (Sareen et al. 1995). No fetal resorption was observed in rats orally administered 100 mg/kg of lapachol daily for days 14 to 19 of pregnancy, although the same product and dose administered on days 7 to 12 of pregnancy resulted in a 79% fetal resorption rate (Rodrigues de Almeida et al. 1988). No effects on implantation and no resorptions were observed in rats orally administered lapachol at doses of 100 to 200 mg/kg daily on days 3 to 5 of pregnancy (Almeida et al. 1999).

V. Toxicity Studies

Acute Toxicity

No LD_{50} of pau d'arco concentrated aqueous extract could be determined at doses up to 5 g/kg orally administered to mice (de Miranda et al. 2001).

In tumor-bearing mice treated with β-lapachone, doses of 2 and 5 mg/kg had no effects on survival. Toxicity was evident at the 5 mg/kg dose with the death of 33% of treated tumor-bearers in the first 24 hours after β-lapachone administration, as compared to the 17-day delay in the onset of death of nontreated tumor-bearing mice (Queiroz et al. 2008).

The LD_{50} of orally administered lapachol in mice is 621 mg/kg, and in rats is 2.4 g/kg (Morrison et al. 1970). The LD_{50} for intraperitoneally administered lapachol in mice is 1.6 g/kg (de Santana et al. 1968).

In monkeys orally administered lapachol at doses of 0.0625, 0.125, 0.25, 0.5, or 1 g/kg daily, death occurred after six doses of 0.5 g/kg and after five doses of 1 g/kg. Severe anemia and elevated alkaline phosphatase levels were observed at the higher doses (Morrison et al. 1970).

Short-Term Toxicity

In dogs orally administered lapachol at doses of 0.25, 0.5, 1, or 2 g/kg daily six times per week for 4 weeks, no lethal effects were observed, but severe anemia and elevated alkaline phosphatase levels occurred (Morrison et al. 1970).

Genotoxicity

In the Ames test, lapachol had no mutagenic activity (Hakura et al. 1994).

Literature Cited

Algranti, E., E.M. Mendonca, S.A. Ali, C.M. Kokron, and V. Raile. 2005. Occupational asthma caused by ipe (*Tabebuia* spp.) dust. *J. Investig. Allergol. Clin. Immunol.* 15(1):81-83.

Almeida, M., M. Brandão, M. Guerra, and V. Peters. 1999. Avaliação preliminar do efeito interceptivo do lapachol em ratas Wistar. *Bol. Centr. Biol. Reprod.* 18:37-48.

Awang, D., B. Dawson, J. Ethier, et al. 1994. Naphthoquinone constituents of commercial lapacho/pau d'arco/taheebo products. *J. Herbs Spices Med. Plants* 2(4):27-43.

Block, J.B., A.A. Serpick, W. Miller, and P.H. Wiernik. 1974. Early clinical studies with lapachol (Nsc-11905). *Cancer Chemother. Rep. Part 2 Suppl.* 4(4):27-28.

de Cassia da Silveira, E.S.R., and M.O. Guerra. 2007. Reproductive toxicity of lapachol in adult male Wistar rats submitted to short-term treatment. *Phytother. Res.* 21(7):658-662.

de Miranda, F., J. Vilar, I. Alves, S. Cavalcanti, and A. Antoniolli. 2001. Antinociceptive and antiedematogenic properties and acute toxicity of *Tabebuia avellanedae* Lor. ex Griseb. inner bark aqueous extract. *BMC Pharmacol.* 1(6).

de Santana, C.F., O. de Lima, I.L. d' Albuquerque, A.L. Lacerda, and D.G. Martins. 1968. Antitumoral and toxicological properties of extracts of bark and various wood components of pau d'arco (*Tabebuia avellanedae*). *Rev. Inst. Antibiot. (Recife)* 8(1):89-94.

Estlander, T., R. Jolanki, K. Alanko, and L. Kanerva. 2001. Occupational allergic contact dermatitis caused by wood dusts. *Contact Dermat.* 44(4):213-217.

Felicio, A.C., C.V. Chang, M.A. Brandao, V.M. Peters, and M.D. Guerra. 2002. Fetal growth in rats treated with lapachol. *Contraception* 66(4):289-293.

Guerra, M.O., A.S. Mazoni, M.A. Brandao, and V.M. Peters. 1999. Interceptive effect of lapachol in rats. *Contraception* 60(5):305-307.

Guerra, M.O., A.S. Mazoni, M.A. Brandao, and V.M. Peters. 2001. Toxicology of lapachol in rats: Embryolethality. *Braz. J. Biol.* 61(1):171-174.

Hakura, A., H. Mochida, Y. Tsutsui, and K. Yamatsu. 1994. Mutagenicity and cytotoxicity of naphthoquinones for Ames *Salmonella* tester strains. *Chem. Res. Toxicol.* 7(4):559-567.

Lee, J.H., J. Cheong, Y.M. Park, and Y.H. Choi. 2005. Down-regulation of cyclooxygenase-2 and telomerase activity by beta-lapachone in human prostate carcinoma cells. *Pharmacol. Res.* 51(6):553-560.

Li, C.J., Y.Z. Li, A.V. Pinto, and A.B. Pardee. 1999. Potent inhibition of tumor survival *in vivo* by beta-lapachone plus taxol: Combining drugs imposes different artificial checkpoints. *Proc. Natl. Acad. Sci. U.S.A.* 96(23):13369-13374.

Menacho-Marquez, M., and J.R. Murguia. 2006. Beta-lapachone activates a Mre11p-Tel1p G1/S checkpoint in budding yeast. *Cell Cycle* 5(21):2509-2516.

Morrison, R.K., D.E. Brown, J.J. Oleson, and D.A. Cooney. 1970. Oral toxicology studies with lapachol. *Toxicol. Appl. Pharmacol.* 17(1):1-11.

Oliveira, M.F., T.G. Lemos, M.C. de Mattos, et al. 2002. New enamine derivatives of lapachol and biological activity. *An. Acad. Bras. Cienc.* 74(2):211-221.

Ough, M., A. Lewis, E.A. Bey, et al. 2005. Efficacy of beta-lapachone in pancreatic cancer treatment: Exploiting the novel, therapeutic target NQO1. *Cancer Biol. Ther.* 4(1):95-102.

Pardee, A.B., Y.Z. Li, and C.J. Li. 2002. Cancer therapy with beta-lapachone. *Curr. Cancer Drug Targets* 2(3):227-242.

Perez-Sacau, E., R.G. Diaz-Penate, A. Estevez-Braun, et al. 2007. Synthesis and pharmacophore modeling of naphthoquinone derivatives with cytotoxic activity in human promyelocytic leukemia HL-60 cell line. *J. Med. Chem.* 50(4):696-706.

Queiroz, M.L.S., M.C. Valadares, C.O. Torello, et al. 2008. Comparative studies of the effects of *Tabebuia avellanedae* bark extract and β-lapachone on the hematopoietic response of tumour-bearing mice. *J. Ethnopharmacol.* 117(2):228-235.

Rodrigues de Almeida, E., E. Santos, A. Filho, and C. Lopes. 1988. The action of 2-hydroxy-3-(3-methyl-2-butenyl)-1,4-naphthoquinone (lapachol) in pregnant rats. *Rev. Port. Farm.* 38:21-23.

Sareen, V., S. Jain, and A. Narula. 1995. Evaluation of estrogenicity and pregnancy interceptory efficacy of lapachol (2-hydroxy-3-(3-methyl-2-butenyl)-1,4-naphthoquinone) in the mouse. *Phytother. Res.* 9(2):139-141.

Schulz, K.H., I. Garbe, B.M. Hausen, and M.H. Simatupang. 1977. The sensitizing capacity of naturally occurring quinones. Experimental studies in guinea pigs. I. Naphthoquinones and related compounds. *Arch. Dermatol. Res.* 258(1):41-52.

Steinert, J., H. Khalaf, and M. Rimpler. 1996. High-performance liquid chromatographic separation of some naturally occurring naphthoquinones and anthraquinones. *J. Chromatogr. A* 723(1):206-209.

Taylor, L. 2005. *The healing power of rainforest herbs*. Garden City Park, NY: Square One Publishers.

Tanacetum parthenium (L.) Sch. Bip. — Asteraceae

SCN: feverfew
Syn: *Chrysanthemum parthenium* (L.) Bernh.

Part: herb

T

Quick Reference Summary

Safety Class: 2b
Interaction Class: A

Contraindications

Not for use in pregnancy except under the supervision of a qualified healthcare practitioner (De Smet 1992; Mitchell 1983; Yao et al. 2006).

Other Precautions

Persons with allergies to other members of the Asteraceae family (such as chamomile) should exercise caution with feverfew, as allergic cross-reactivity is common to Asteraceae plants (Hausen and Osmundsen 1983; Upton 2007).

Persons discontinuing feverfew should reduce the dosage gradually over a 1-month period to avoid "post-feverfew syndrome" (Johnson et al. 1985; Mills and Bone 2005; Upton 2007).

Drug and Supplement Interactions

None known.

Adverse Events and Side Effects

Systematic reviews of clinical trials on feverfew indicate that the herb is generally well tolerated with few adverse effects (Ernst and Pittler 2000; Pittler and Ernst 2004). No adverse effects were observed in a dose escalation study of a standardized feverfew extract at doses up to 1 g daily (Curry et al. 2004).

Daily consumption of fresh feverfew leaves, but not the CO_2 extract of the leaf (Diener et al. 2005; Pfaffenrath et al. 2002; Upton 2007), may cause canker sores that have been shown to resolve after discontinuation of the herb (Johnson 1984).

"Post-feverfew syndrome" has been reported after cessation of feverfew in approximately 10% of long-term feverfew users, with symptoms of aches and pains, joint and muscle stiffness, anxiety, and poor sleep (Johnson et al. 1985).

Pharmacological Considerations

While some in vitro studies have indicated that feverfew can inhibit platelet aggregation (Loesche et al. 1988; Makheja and Bailey 1982), no effects on aggregation were observed in the one human study of feverfew and aggregation (Biggs et al. 1982), and no cases of bleeding in association with feverfew have been reported (Stargrove et al. 2008).

Pregnancy and Lactation

Limited information on the safety of feverfew during pregnancy and lactation is available. Animal studies with high doses of feverfew in pregnant rats or on isolated rat fetuses have shown some adverse effects on the fetus (Yao et al. 2006). Based on this information, use during pregnancy is not recommended except under the supervision of a qualified healthcare practitioner.

Review Details

I. Drug and Supplement Interactions

Clinical Trials of Drug or Supplement Interactions

No clinical trials of drug or supplement interactions were identified.

Case Reports of Suspected Drug or Supplement Interactions

No case reports of suspected drug or supplement interactions were identified.

Animal Trials of Drug or Supplement Interactions

No animal trials of drug or supplement interactions were identified.

II. Adverse Events

Adverse Events Reported in Clinical Trials

A systematic review of clinical trials of feverfew monopreparations, including six clinical trials, characterized feverfew as generally well tolerated with only mild and transient adverse events. In two of the six trials, adverse events were higher in the placebo group, and in a third trial, adverse events were equivalent in the placebo and feverfew groups (Ernst and Pittler 2000). One study indicated a higher incidence of mouth ulcers in the placebo group than in the feverfew group (Murphy et al. 1988).

In a survey of 270 migraine patients who had been self-treating their headaches by chewing several feverfew leaves daily for 2 to 4 years, 18% of respondents reported adverse effects. The most frequent complaint was canker sores reported by 11.3% of respondents, although only 7% discontinued therapy. Other complaints included sore tongue, abdominal pain, indigestion, tingling sensation, urinary problems, and headaches (Johnson 1983). Canker sores are believed to be a systemic effect that resolves within a week of discontinuing feverfew (Johnson 1984). Canker sores and ulceration have not been reported in any clinical trials using a CO_2 extract of feverfew (Diener et al. 2005; Pfaffenrath et al. 2002; Upton 2007).

"Post-feverfew syndrome" has been reported after cessation of feverfew in approximately 10% of long-term feverfew users. Symptoms included aches and pains, joint and muscle stiffness, anxiety, and poor sleep (Johnson et al. 1985).

Case Reports of Adverse Events

Chewing fresh leaves may induce a generalized inflammation of oral mucosa (Johnson et al. 1985), which is likely due to the sesquiterpene lactone compounds in the plant. These compounds were suggested as the cause for contact allergy to live feverfew plants (Hausen and Osmundsen 1983).

Occasional side effects, such as mouth ulceration or gastric disturbance, have been observed in 6 to 15% of users, usually in the first week of use (Bradley 1992; De Smet 1992; Martindale and Reynolds 1996; Williamson 2003). No adverse effects have been reported from long-term consumption (Bradley 1992).

III. Pharmacology and Pharmacokinetics

Human Pharmacological Studies

A small study of long-term feverfew users (3.5 to 8 years of use) indicated that aggregation responses to ADP and thrombin were identical to a control group that had not taken feverfew for at least 6 months (Biggs et al. 1982). A review on feverfew indicated that the herb does not appear to affect blood pressure, heart rate, or other hematological parameters (Johnson et al. 1985).

In a dose escalation trial of a standardized feverfew product (500 μg parthenolide per 125 mg capsule), doses up to 4 mg of parthenolide daily, taken orally, were well tolerated with no dose-limiting toxicity observed during 4 weeks of treatment (Curry et al. 2004).

Animal Pharmacological Studies

No relevant animal pharmacological trials were identified.

In Vitro Pharmacological Studies

A methanolic extract of feverfew indicated some inhibition of CYP1A2, CYP2C9, CYP2C8, CYP2C19, CYP2D6, and CYP3A4 (Unger and Frank 2004). An ethanolic extract exhibited very low inhibitory action on CYP3A4, while the compound parthenolide significantly inhibited CYP3A4 (Budzinski et al. 2000).

Extracts of feverfew have inhibited human platelet aggregation in vitro (Groenewegen and Heptinstall 1990; Heptinstall et al. 1987, 1988; Loesche et al. 1988; Makheja and Bailey 1982).

IV. Pregnancy and Lactation

When pregnant rats were administered 839 mg/kg of feverfew daily on gestational days 1–8 or 8–15, fetuses from the group that received feverfew later in pregnancy were smaller than fetuses in the control group (Yao et al. 2006). The dose used in this study was 60 times the recommended human dose (Upton 2007). Feverfew demonstrated toxicity when 10-day-old fetuses were cultured for 26 hours in rat serum with feverfew extract (Yao et al. 2006). Low doses of feverfew used prophylactically against migraines might not elicit any adverse effects, but prudence suggests that feverfew should not be used in pregnancy until more evidence is available (Upton 2007).

No information on the safety of feverfew during lactation was identified.

V. Toxicity Studies

Acute Toxicity

No LD_{50} of feverfew has been reported. The highest nontoxic acute dose in rats was reported at 860 mg/kg (Yao and Brown-Woodman 2001).

Short-Term Toxicity

No toxic effects were reported in guinea pigs or rats fed 100 to 150 times the daily human dose for 5 to 7 weeks (Johnson 1983).

Genotoxicity

A study of the genotoxicity of feverfew in humans examined the number of sister-chromatid exchanges in feverfew users compared with nonusers and indicated that no adverse effects of feverfew were found. The mutagenicity of urine samples from feverfew users was tested using the Ames mutagenicity test. The frequency of chromosomal aberrations was lower in the feverfew group than in the control group (Johnson et al. 1987).

Literature Cited

Biggs, M.J., E.S. Johnson, N.P. Persaud, and D.M. Ratcliffe. 1982. Platelet aggregation in patients using feverfew for migraine. *Lancet* 2(8301):776.

Bradley, P.R. 1992. *British herbal compendium: A handbook of scientific information on widely used plant drugs*. Bournemouth, UK: British Herbal Medicine Association.

Budzinski, J.W., B.C. Foster, S. Vandenhoek, and J.T. Arnason. 2000. An *in vitro* evaluation of human cytochrome P450 3A4 inhibition by selected commercial herbal extracts and tinctures. *Phytomedicine* 7(4):273-282.

Curry, E.A., 3rd, D.J. Murry, C. Yoder, et al. 2004. Phase I dose escalation trial of feverfew with standardized doses of parthenolide in patients with cancer. *Invest. New Drugs* 22(3):299-305.

De Smet, P.A.G.M. 1992. *Adverse effects of herbal drugs, Volume 1*. Berlin: Springer.

Diener, H.C., V. Pfaffenrath, J. Schnitker, M. Friede, and H.H. Henneicke-von Zepelin. 2005. Efficacy and safety of 6.25 mg t.i.d. feverfew CO_2-extract (MIG-99) in migraine prevention—A randomized, double-blind, multicentre, placebo-controlled study. *Cephalalgia* 25(11):1031-1041.

Ernst, E., and M.H. Pittler. 2000. The efficacy and safety of feverfew (*Tanacetum parthenium* L.): An update of a systematic review. *Public Health Nutr.* 3(4A):509-514.

Groenewegen, W.A., and S. Heptinstall. 1990. A comparison of the effects of an extract of feverfew and parthenolide, a component of feverfew, on human platelet activity *in vitro*. *J. Pharm. Pharmacol.* 42(8):553-557.

Hausen, B.M., and P.E. Osmundsen. 1983. Contact allergy to parthenolide in *Tanacetum parthenium* (L.) Schulz-Bip. (feverfew, Asteraceae) and cross-reactions to related sesquiterpene lactone containing Compositae species. *Acta Derm. Venereol.* 63(4):308-314.

Heptinstall, S., W.A. Groenewegen, P. Spangenberg, and W. Loesche. 1987. Extracts of feverfew may inhibit platelet behaviour via neutralization of sulphydryl groups. *J. Pharm. Pharmacol.* 39(6):459-465.

Heptinstall, S., W.A. Groenewegen, P. Spangenberg, and W. Losche. 1988. Inhibition of platelet behaviour by feverfew: A mechanism of action involving sulphydryl groups. *Folia Haematol. Int. Mag. Klin. Morphol. Blutforsch.* 115(4):447-449.

Johnson, E.S. 1983. Patients who chew chrysanthemum leaves. *MIMS Mag.* (May 15):32-35.

Johnson, E.S. 1984. *Feverfew: A traditional herbal remedy for migraine and arthritis*. London: Sheldon Press.

Johnson, E.S., N.P. Kadam, D. Anderson, et al. 1987. Investigation of possible genetoxic effects of feverfew in migraine patients. *Human Toxicol.* 6(6):533-534.

Johnson, E.S., N.P. Kadam, D.M. Hylands, and P.J. Hylands. 1985. Efficacy of feverfew as prophylactic treatment of migraine. *Br. Med. J. (Clin. Res. Ed.)* 291(6495):569-573.

Loesche, W., W.A. Groenewegen, S. Krause, P. Spangenberg, and S. Heptinstall. 1988. Effects of an extract of feverfew (*Tanacetum parthenium*) on arachidonic acid metabolism in human blood platelets. *Biomed. Biochim. Acta* 47(10-11):S241-S243.

Makheja, A.N., and J.M. Bailey. 1982. A platelet phospholipase inhibitor from the medicinal herb feverfew (*Tanacetum parthenium*). *Prostaglandins Leukot. Med.* 8(6):653-660.
Martindale, W., and J.E.F. Reynolds. 1996. *The extra pharmacopoeia*. 31st ed. London: Pharmaceutical Press.
Mills, S., and K. Bone. 2005. *The essential guide to herbal safety*. St. Louis: Elsevier.
Mitchell, H. 1983. *British herbal pharmacopoeia*. Bournemouth, U.K.: British Herbal Medicine Association.
Murphy, J.J., S. Heptinstall, and J.R. Mitchell. 1988. Randomised double-blind placebo-controlled trial of feverfew in migraine prevention. *Lancet* 2(8604):189-192.
Pfaffenrath, V., H.C. Diener, M. Fischer, M. Friede, and H.H. Henneicke-von Zepelin. 2002. The efficacy and safety of *Tanacetum parthenium* (feverfew) in migraine prophylaxis—A double-blind, multicentre, randomized placebo-controlled dose-response study. *Cephalalgia* 22(7):523-532.
Pittler, M.H., and E. Ernst. 2004. Feverfew for preventing migraine. *Cochrane Database Syst. Rev.* 1:CD002286.
Stargrove, M., J. Treasure, and D. McKee. 2008. *Herb, nutrient, and drug interactions: Clinical implications and therapeutic solutions*. St. Louis: Elsevier.
Unger, M., and A. Frank. 2004. Simultaneous determination of the inhibitory potency of herbal extracts on the activity of six major cytochrome P450 enzymes using liquid chromatography/mass spectrometry and automated online extraction. *Rapid Commun. Mass Spectrom.* 18(19):2273-2281.
Upton, R. 2007. *Feverfew aerial parts: Tanacetum parthenium (L.) Schultz Bip*. Santa Cruz, CA: American Herbal Pharmacopoeia.
Williamson, E.M. 2003. *Potter's herbal cyclopedia*. Saffron Walden, Essex: C.W. Daniel Co.
Yao, M., and P.D. Brown-Woodman. 2001. Do herbal remedies have an adverse effect on pregnancy outcome in the rat? *Teratology* 64:323-324.
Yao, M., H.E. Ritchie, and P.D. Brown-Woodman. 2006. A reproductive screening test of feverfew: Is a full reproductive study warranted? *Reprod. Toxicol.* 22(4):688-693.

Tanacetum vulgare L. — Asteraceae

SCN: tansy
Syn: *Chrysanthemum vulgare* (L.) Bernh.
Part: herb

Quick Reference Summary

Safety Class: 2b, 2c, 2d
Interaction Class: A

Contraindications

Not for use in pregnancy or lactation except under the supervision of a qualified healthcare practitioner (Felter and Lloyd 1898; Wood and LaWall 1918).

Do not exceed recommended dose (Felter and Lloyd 1898; Opdyke 1979; Wood and LaWall 1918).

Other Precautions

None known.

Drug and Supplement Interactions

None known.

Standard Dose

1 to 2 g as an infusion up to 3 times daily (Mitchell 1983).

1 to 2 ml of the fluid extract (1:1) up to 3 times daily (Mitchell 1983).

0.6 to 1.5 ml of the tincture (1:5) up to 3 times daily (Mills and Bone 2005).

Notice

Abortifacient (Felter and Lloyd 1898; Wood and LaWall 1918); *see* Appendix 2.

Emmenagogue (Felter and Lloyd 1898; Wood and LaWall 1918); *see* Appendix 2.

Diuretic (Lahlou et al. 2007); *see* Appendix 2.

Thujone in some chemotypes (0–73.5% α-thujone and 0–97.7% β-thujone in the essential oil) (Dragland et al. 2005; Holopainen et al. 1987; Rohloff et al. 2004); *see* Appendix 1.

Editors' Notes

A number of different chemotypes (plants that are botanically identical but chemically different) of tansy have been identified. The compounds in these chemotypes vary widely, with α-thujone ranging from 0 to 73.5% of the essential oil and β-thujone ranging from 0 to 97.7% of the essential oil. The essential oil content of the plant ranges from 0.04 to 0.19 ml/kg (Dragland et al. 2005; Holopainen et al. 1987; Rohloff et al. 2004). This chemical variation leads to challenges for safe dosing of tansy (Blumenthal et al. 1998; List and Hörhammer 1973).

Adverse Events and Side Effects

Contact dermatitis from tansy has been reported and is attributed to sesquiterpene lactone compounds in the plant. Cross-reactivity has been reported between sesquiterpene lactone-containing plants in the Asteraceae family including tansy, dandelion, feverfew, and yarrow (Guin and Skidmore 1987; Hausen 1996; Hausen and Osmundsen 1983; Killoran et al. 2007; Mark et al. 1999; Opdyke 1979; Paulsen et al. 1993, 2001).

Pharmacological Considerations

An animal study indicated that tansy has diuretic activity (Lahlou et al. 2007).

Pregnancy and Lactation

Tansy and tansy essential oil have been used in attempted abortions. Several texts report that such use is "highly dangerous and generally ineffectual," with some fatal cases of tansy essential oil use reported (Felter and Lloyd 1898; Whitehill 1906; Wood and LaWall 1918). Based on this information, use during pregnancy is not recommended except under the supervision of a qualified healthcare practitioner.

No information on the safety of tansy during lactation was identified in the scientific or traditional literature. While this review did not identify any concerns for use while nursing, safety has not been conclusively established.

Review Details

I. Drug and Supplement Interactions

Clinical Trials of Drug or Supplement Interactions

No clinical trials of drug or supplement interactions were identified.

Case Reports of Suspected Drug or Supplement Interactions

No case reports of suspected drug or supplement interactions were identified.

Animal Trials of Drug or Supplement Interactions

No animal trials of drug or supplement interactions were identified.

II. Adverse Events

Case Reports of Adverse Events

Allergic contact dermatitis from tansy has been reported. Patch testing has indicated cross-reactions to a number of sesquiterpene lactone-containing plants in the Asteraceae family including tansy, dandelion, feverfew, and yarrow. The dermatitis is generally attributed to the sesquiterpene lactone compounds in tansy (Guin and Skidmore 1987; Hausen 1996; Hausen and Osmundsen 1983; Killoran et al. 2007; Mark et al. 1999; Paulsen et al. 1993, 2001).

In overdose, tansy and tansy essential oil have been reported to cause abdominal pain, vomiting, convulsions, seizures, coma, dilated pupils, irregular respiration, and fast and weak pulse. Some cases of overdose have been fatal (Felter and Lloyd 1898; Opdyke 1979; Wood and LaWall 1918).

Also see Pregnancy and Lactation for this entry.

III. Pharmacology and Pharmacokinetics

Human Pharmacological Studies

No relevant human pharmacological studies were identified.

Animal Pharmacological Studies

In rats orally administered single doses of 100 mg/kg of an aqueous extract of tansy or 10 mg/kg of the drug furosemide, an increase in urine output was observed. Increases in urinary levels of sodium and potassium were also observed in the tansy group, while furosemide increased urinary levels of sodium and decreased urinary levels of potassium. Despite changes in urinary excretion of the electrolytes, plasma levels of sodium and potassium were unaffected. In an 8-day study, both substances induced significant diuresis and natriuresis, but only tansy increased urinary potassium excretion (Lahlou et al. 2007).

In Vitro Pharmacological Studies

No relevant in vitro pharmacological studies were identified.

IV. Pregnancy and Lactation

Tansy and tansy essential oil have been used in attempted abortions. Several texts report that such use is "highly dangerous and generally ineffectual," with some fatal cases of tansy essential oil use reported (Felter and Lloyd 1898; Wood and LaWall 1918).

Convulsions and seizures were reported in an 18-year-old woman who had taken two or three doses of tansy essential oil over the course of one evening in an attempted abortion. The pregnancy was not interrupted (Whitehill 1906).

No information on the safety of tansy during lactation was identified.

V. Toxicity Studies

Acute Toxicity

The oral LD_{50} of tansy essential oil is 1.15 g/kg in rats and 3 g/kg in dogs (Opdyke 1979).

The LD_{50} of a freeze-dried aqueous extract of tansy leaf in mice is 9.9 g/kg after oral administration and 2.8 g/kg after intraperitoneal administration (Lahlou et al. 2008).

Subchronic Toxicity

In mice orally administered 100, 300, or 600 mg/kg of a freeze-dried aqueous extract of tansy leaf daily for 90 days, no significant adverse effects or changes in biological or hematological parameters were observed (Lahlou et al. 2008).

Literature Cited

Blumenthal, M., W. Busse, A. Goldberg, et al. 1998. *The complete German Commission E monographs*. Austin, TX: American Botanical Council.

Dragland, S., J. Rohloff, R. Mordal, and T.-H. Iversen. 2005. Harvest regimen optimization and essential oil production in five tansy (*Tanacetum vulgare* L.) genotypes under a northern climate. *J. Agric. Food Chem.* 53(12):4946-4953.

Felter, H.W., and J.U. Lloyd. 1898. *King's American dispensatory*. 18th ed., 3rd rev. 2 vols. Cincinnati: Ohio Valley Co.

Guin, J.D., and G. Skidmore. 1987. Compositae dermatitis in childhood. *Arch. Dermatol.* 123(4):500-502.

Hausen, B.M. 1996. A 6-year experience with compositae mix. *Am. J. Contact Dermat.* 7(2):94-99.

Hausen, B.M., and P.E. Osmundsen. 1983. Contact allergy to parthenolide in *Tanacetum parthenium* (L.) Schulz-Bip. (feverfew, Asteraceae) and cross-reactions to related sesquiterpene lactone containing Compositae species. *Acta Derm. Venereol.* 63(4):308-314.

Holopainen, M., R. Hiltunen, and M. Von Schantz. 1987. A study on tansy chemotypes. *Planta Med.* 53(3):284.

Killoran, C.E., G.H. Crawford, and A. Pedvis-Leftick. 2007. Two cases of Compositae dermatitis exacerbated by moisturizer containing feverfew. *Dermatitis* 18(4):225-229.

Lahlou, S., Z.H. Israili, and B. Lyoussi. 2008. Acute and chronic toxicity of a lyophilised aqueous extract of *Tanacetum vulgare* leaves in rodents. *J. Ethnopharmacol.* 117(2):221-227.

Lahlou, S., A. Tahraoui, Z. Israili, and B. Lyoussi. 2007. Diuretic activity of the aqueous extracts of *Carum carvi* and *Tanacetum vulgare* in normal rats. *J. Ethnopharmacol.* 110(3):458-463.

List, P.H., and H. Hörhammer. 1973. *Hagers handbuch der pharmazeutischen praxis*. Berlin: Springer.

Mark, K.A., R.R. Brancaccio, N.A. Soter, and D.E. Cohen. 1999. Allergic contact and photoallergic contact dermatitis to plant and pesticide allergens. *Arch. Dermatol.* 135(1):67-70.

Mills, S., and K. Bone. 2005. *The essential guide to herbal safety*. St. Louis: Elsevier.

Mitchell, H. 1983. *British herbal pharmacopoeia*. Bournemouth, UK: British Herbal Medicine Association.

Opdyke, D.L.J. 1979. *Monographs on fragrance raw materials*. New York: Pergamon.

Paulsen, E., K.E. Andersen, and B.M. Hausen. 1993. Compositae dermatitis in a Danish dermatology department in one year: (I). Results of routine patch testing with the sesquiterpene lactone mix supplemented with aimed patch testing with extracts and sesquiterpene lactones of Compositae plants. *Contact Dermat.* 29(1):6-10.

Paulsen, E., K.E. Andersen, and B.M. Hausen. 2001. Sensitization and cross-reaction patterns in Danish Compositae-allergic patients. *Contact Dermat.* 45(4):197-204.

Rohloff, J., R. Mordal, and S. Dragland. 2004. Chemotypical variation of tansy (*Tanacetum vulgare* L.) from 40 different locations in Norway. *J. Agric. Food Chem.* 52(6):1742-1748.

Whitehill, N. 1906. Poisoning from oil of tansy. *J. Am. Med. Assoc.* 47(7):509.

Wood, H., and C. LaWall. 1918. *The dispensatory of the United States of America*. 21st ed. Philadelphia: Lippincott.

Taraxacum officinale Weber ex F.H. Wigg. — Asteraceae

SCN: dandelion
Syn: *Taraxacum dens-leonis* Desf.; *Taraxacum vulgare* (Lam.) Schrank
OCN: lion's tooth
Part: leaf, root

Quick Reference Summary

T

Safety Class: 1
Interaction Class: A

Contraindications

None known.

Other Precautions

None known.

Drug and Supplement Interactions

None known.

Notice

Diuretic (Clare et al. 2009; Mitchell 1983; Racz-Kotilla et al. 1974; Schutz et al. 2006); *see* Appendix 2.

Adverse Events and Side Effects

None known.

Pharmacological Considerations

Human and animal studies have indicated that dandelion leaf is a diuretic (Clare et al. 2009; Racz-Kotilla et al. 1974).

Due to the choleretic activity of dandelion, caution has been suggested for persons with gallstones (Mills and Bone 2005; Schutz et al. 2006; Wichtl 2004).

Pregnancy and Lactation

No information on the safety of dandelion in pregnancy or lactation was identified in the scientific or traditional literature. Although this review did not identify any concerns for use while pregnant or nursing, safety has not been conclusively established.

Review Details

I. Drug and Supplement Interactions

Clinical Trials of Drug or Supplement Interactions

No clinical trials of drug or supplement interactions were identified.

Case Reports of Suspected Drug or Supplement Interactions

No case reports of suspected drug or supplement interactions were identified.

Animal Trials of Drug or Supplement Interactions

No animal trials of drug or supplement interactions were identified.

II. Adverse Events

Case Reports of Adverse Events

Contact dermatitis from fresh dandelion leaf has been reported and confirmed by patch testing. These reactions are generally attributed to the sesquiterpene lactone compounds in dandelion (Guin and Skidmore 1987; Hausen 1982; Ingber 2000; Jovanovic et al. 2003; Mark et al. 1999; Paulsen et al. 2008; Wakelin et al. 1997).

III. Pharmacology and Pharmacokinetics

Human Pharmacological Studies

In healthy volunteers orally administered 8 ml of a hydroethanolic extract of dandelion leaf every 5 hours for three doses, an increase in the frequency of urination was observed after the first dose, and an increase in the excretion ratio (urination volume:fluid intake) was observed after the second dose. No changes were observed after the third dose (Clare et al. 2009).

Animal Pharmacological Studies

No significant changes in urine volume were reported in saline-loaded mice orally administered 50 ml/kg of different extract fractions of dandelion root. Excretion of sodium and potassium were higher in animals given certain extract fractions (petroleum ether and chloroform) than in control animals. A crude methanol extract increased potassium concentration in the urine similar to levels observed with furosemide treatment (37.5 mg/kg), although sodium excretion was significantly lower after dandelion than furosemide (Hook et al. 1993).

In studies with mice and rats, increases in urine output and salt excretion were observed after administration of dandelion leaf fluid extract. A dose corresponding to 8 g/kg of the dried herb demonstrated activity equivalent to 80 mg/kg of the diuretic furosemide (Racz-Kotilla et al. 1974).

Inhibition of the drug-metabolizing isoenzyme CYP2E1 was observed in rats provided with drinking water containing 2% dandelion root tea (Maliakal and Wanwimolruk 2001).

In Vitro Pharmacological Studies

No relevant in vitro pharmacological studies were identified.

IV. Pregnancy and Lactation

No information on the use of dandelion during pregnancy or lactation was identified.

V. Toxicity Studies

Acute Toxicity

The LD_{50} of intraperitoneally administered dandelion leaf fluid extract (1:1) in mice is 28.8 g/kg, while that of the root is 36.6 g/kg (Racz-Kotilla et al. 1974).

No adverse effects were observed in rabbits orally administered doses equivalent to 6 g/kg of dried whole dandelion (Akhtar et al. 1985). An ethanol extract of dandelion root demonstrated very low toxicity in mice and rats orally administered a dose equivalent to 10 g/kg of the dried herb or intraperitoneally administered a dose equivalent to 4 g/kg of the dried herb (Tita et al. 1993).

Literature Cited

Akhtar, M.S., Q.M. Khan, and T. Khaliq. 1985. Effects of *Portulaca oleracea* (kulfa) and *Taraxacum officinale* (dhudhal) in normoglycaemic and alloxan-treated hyperglycaemic rabbits. *J. Pak. Med. Assoc.* 35:207-210.

Clare, B.A., R.S. Conroy, and K. Spelman. 2009. The diuretic effect in human subjects of an extract of *Taraxacum officinale* folium over a single day. *J. Altern. Complement. Med.* 15(8):929-934.

Guin, J.D., and G. Skidmore. 1987. Compositae dermatitis in childhood. *Arch. Dermatol.* 123(4):500-502.

Hausen, B.M. 1982. Taraxinsäure-1′-*O*-β-D-glucopyranosid, das Kontaktallergen des Löwenzahns (*Taraxacum officinale* Wiggers). *Dermatosensitivity* 30:51-53.

Hook, I., A. McGee, and M. Henman. 1993. Evaluation of dandelion for diuretic activity and variation in potassium content. *Int. J. Pharmacog.* 31:29-34.

Ingber, A. 2000. Seasonal allergic contact dermatitis from *Taraxacum officinale* (dandelion) in an Israeli florist. *Contact Dermat.* 43(1):49.

Jovanovic, M., N. Mimica-Dukic, M. Poljacki, and P. Boza. 2003. Erythema multiforme due to contact with weeds: A recurrence after patch testing. *Contact Dermat.* 48(1):17-25.

Maliakal, P.P., and S. Wanwimolruk. 2001. Effect of herbal teas on hepatic drug metabolizing enzymes in rats. *J. Pharm. Pharmacol.* 53(10):1323-1329.

Mark, K.A., R.R. Brancaccio, N.A. Soter, and D.E. Cohen. 1999. Allergic contact and photoallergic contact dermatitis to plant and pesticide allergens. *Arch. Dermatol.* 135(1):67-70.

Mills, S., and K. Bone. 2005. *The essential guide to herbal safety*. St. Louis: Elsevier.

Mitchell, H. 1983. *British herbal pharmacopoeia*. Bournemouth, U.K.: British Herbal Medicine Association.

Paulsen, E., A. Otkjaer, and K.E. Andersen. 2008. Sesquiterpene lactone dermatitis in the young: Is atopy a risk factor? *Contact Dermat.* 59(1):1-6.

Racz-Kotilla, E., G. Racz, and A. Solomon. 1974. The action of *Taraxacum officinale* extracts on the body weight and diuresis of laboratory animals. *Planta Med.* 26(3):212-217.

Schutz, K., R. Carle, and A. Schieber. 2006. *Taraxacum*—A review on its phytochemical and pharmacological profile. *J. Ethnopharmacol.* 107(3):313-323.

Tita, B., U. Bello, P. Faccendini, R. Bartolini, and P. Bolle. 1993. *Taraxacum officinale* W.: Pharmacological effect of ethanol extract. *Pharmacol. Res.* 27:23-24.

Wakelin, S.H., P. Marren, E. Young, and S. Shaw. 1997. Compositae sensitivity and chronic hand dermatitis in a seven-year-old boy. *Br. J. Dermatol.* 137(2):289-291.

Wichtl, M. 2004. *Herbal drugs and phytopharmaceuticals: A handbook for practice on a scientific basis*. 3rd ed. Boca Raton, FL: CRC Press.

Taxus brevifolia Nutt. — Taxaceae

SCN: Pacific yew

Part: needles

Quick Reference Summary

Safety Class: 2b
Interaction Class: A

Contraindications

Not for use in pregnancy except under the supervision of a qualified healthcare practitioner (Krag 1976).

Other Precautions

None known.

Drug and Supplement Interactions

None known.

Notice

Emmenagogue (Krag 1976); *see* Appendix 2.

Editors' Notes

Pacific yew should not be confused with botanically similar species, *T. baccata* (English yew) and *T. cuspidata* (Japanese yew), that are recognized as toxic due to effects on heart rate and rhythm (Rowinsky et al. 1990; Vance et al. 2001). Effects on the heart are caused by taxine compounds, present at 0.5 to 1% in *T. baccata* as compared to 0.0007% in Pacific yew (Jenniskens et al. 1996; Tyler 1960).

Pacific yew is a source of the compound taxol, used in chemotherapy of several types of cancers. The concentration of taxol in Pacific yew needles is approximately 0.006% and 0.01% in the bark (Witherup et al. 1990).

Adverse Events and Side Effects

None known.

Pharmacological Considerations

None known.

Pregnancy and Lactation

Pacific yew has traditionally been used as an emmenagogue (Krag 1976).

No information on the safety of Pacific yew during lactation was identified in the scientific or traditional literature. While this review did not identify any concerns for use while nursing, safety has not been conclusively established.

Review Details

I. Drug and Supplement Interactions

Clinical Trials of Drug or Supplement Interactions

No clinical trials of drug or supplement interactions were identified.

Case Reports of Suspected Drug or Supplement Interactions

No case reports of suspected drug or supplement interactions were identified.

Animal Trials of Drug or Supplement Interactions

No animal trials of drug or supplement interactions were identified.

II. Adverse Events

Case Reports of Adverse Events

No case reports of adverse events were identified.

III. Pharmacology and Pharmacokinetics

Human Pharmacological Studies

No relevant human pharmacological studies were identified.

Animal Pharmacological Studies

No relevant animal pharmacological studies were identified.

In Vitro Pharmacological Studies

No relevant in vitro pharmacological studies were identified.

IV. Pregnancy and Lactation

Pacific yew has traditionally been used as an emmenagogue (Krag 1976).

No information on the safety of Pacific yew during lactation was identified.

V. Toxicity Studies

Acute Toxicity

The LD_{50} of Pacific yew powder orally administered to rats could not be determined at doses up to 5 g/kg (PSL 1999).

Literature Cited

Jenniskens, L.H.D., E.L.M. van Rozendaal, T.A. van Beek, P.H.G. Wiegerinck, and H.W. Scheeren. 1996. Identification of six taxine alkaloids from *Taxus baccata* needles. *J. Nat. Prod.* 59(2):117-123.

Krag, K.J. 1976. Plants used as contraceptives by the North American Indians—An ethnobotanical study. Cambridge, MA: Botanical Museum, Harvard University.

PSL. 1999. Montana yew tip powder acute oral toxicity test in rats. Unpublished report. E90601-5D. Dayton, NJ: Product Safety Labs.

Rowinsky, E.K., L.A. Cazenave, and R.C. Donehower. 1990. Taxol: A novel investigational antimicrotubule agent. *J. Natl. Cancer Inst.* 82(15):1247-1259.

Tyler, V.E. 1960. Note on the occurrence of taxine in *Taxus brevifolia*. *J. Am. Pharm. Assoc.* 49(10):683-684.

Vance, N., M. Borsting, D. Pilz, and J. Freed. 2001. Special forest products: Species information guide for the Pacific Northwest. General Technical Report PNW-GTR-513. Portland, OR: U.S. Forest Service.

Witherup, K.M., S.A. Look, M.W. Stasko, et al. 1990. *Taxus* spp. needles contain amounts of taxol comparable to the bark of *Taxus brevifolia*: Analysis and isolation. *J. Nat. Prod.* 53(5):1249-1255.

Terminalia arjuna (Roxb. ex DC.) Wight & Arn. — Combretaceae

SCN: arjuna
AN: *arjuna*

Part: bark

Quick Reference Summary

Safety Class: 2b
Interaction Class: A

Contraindications

Not for use in pregnancy except under the supervision of a qualified healthcare practitioner (Gupta et al. 1989; Pole 2006).

Other Precautions

None known.

Drug and Supplement Interactions

None known.

Notice

Tannins (12–24%) (Chadha 1988; Dwivedi 2007); *see* Appendix 1.

Adverse Events and Side Effects

None known.

Pharmacological Considerations

None known.

Pregnancy and Lactation

An animal study indicated that some extracts of arjuna reduced the number of viable fetuses (Gupta et al. 1989). A text on Ayurvedic medicine indicates that arjuna should not be used during pregnancy (Pole 2006). Based on this information, use during pregnancy is not recommended except under the supervision of a qualified healthcare practitioner.

No information on the safety of arjuna during lactation was identified in the scientific or traditional literature. While this review did not identify any concerns for use while nursing, safety has not been conclusively established.

Review Details

I. Drug and Supplement Interactions

Clinical Trials of Drug or Supplement Interactions

In a phase II open label study of arjuna in patients with severe refractory heart failure, patients stabilized on digitalis and/or diuretic and vasodilator drugs were orally administered 500 mg of a hydroalcoholic extract of arjuna bark every 8 hours for 4 months. Biochemical, electrocardiographic, radiological, and echocardiographic parameters were monitored. No adverse events were reported, and in all cases patients were able to reduce the diuretic dosage to the minimum necessary to keep symptom free (Bharani et al. 1995).

Case Reports of Suspected Drug or Supplement Interactions

No case reports of suspected drug or supplement interactions were identified.

Animal Trials of Drug or Supplement Interactions

No relevant animal studies of drug or supplement interactions were identified.

II. Adverse Events

Adverse Events Reported in Clinical Trials

A review of clinical trials of arjuna bark and bark extracts in patients with cardiovascular disorders indicated that arjuna bark is generally well tolerated at doses of 1 to 2 g daily. At those doses, only minor adverse events were reported, including mild gastritis, headache, and constipation (Dwivedi 2007). *Also see* Clinical trials of drug or supplement interactions above.

Case Reports of Adverse Events

No case reports of adverse events were identified.

III. Pharmacology and Pharmacokinetics

Human Pharmacological Studies

No relevant pharmacological studies were identified.

Animal Pharmacological Studies

A dose-dependent reduction in blood glucose levels was observed in diabetic rats orally administered 250 or 500 mg/kg of an ethanol extract of arjuna daily for 30 days (Ragavan and Krishnakumari 2006).

In Vitro Pharmacological Studies

In platelets from healthy persons and patients with coronary artery disease (CAD), the effect of an ethanol extract of arjuna on platelet function indices was determined by incubating the platelets with arjuna extract in a time- and dose-dependent (concentrations of 25 to 100 µg/ml) manner in the presence and absence of adenosine diphosphate (ADP). Arjuna was able to significantly inhibit platelet aggregation in platelets from CAD patients and from healthy controls. Significant attenuation of intracellular free calcium release and expression of CD62P was also observed after treatment with arjuna (Malik et al. 2009).

IV. Pregnancy and Lactation

In rats orally administered 200 or 400 mg/kg of a freeze-dried ethanol extract of arjuna on gestational days 1 to 7, no anti-implantation activity was observed. In animals administered the same dose on days 12 to 14, a dose-dependent increase in the number of fetal resorptions was observed. Hexane (10 mg/kg) and butanol (100 mg/kg) fractions of the extract and the compounds arjunolone (5 mg/kg) and baicalein (25 mg/kg), but not chloroform (50 mg/kg) or benzene (50 mg/kg) fractions, also produced an increase in the number of fetal resorptions (Gupta et al. 1989).

A text on Ayurvedic medicine indicates that arjuna should not be used during pregnancy (Pole 2006).

No information on the safety of arjuna during lactation was identified.

V. Toxicity Studies

Short-Term Toxicity

No signs of toxicity were observed in diabetic rats orally administered up to 500 mg/kg of an ethanol extract of arjuna daily for 30 days (Ragavan and Krishnakumari 2006).

T

Literature Cited

Bharani, A., A. Ganguly, and K.D. Bhargava. 1995. Salutary effect of *Terminalia arjuna* in patients with severe refractory heart failure. *Int. J. Cardiol.* 49(3):191-199.

Chadha, Y. 1988. *The wealth of India: A dictionary of Indian raw materials and industrial products.* Delhi: Council of Scientific and Industrial Research.

Dwivedi, S. 2007. *Terminalia arjuna* Wight & Arn.—A useful drug for cardiovascular disorders. *J. Ethnopharmacol.* 114(2):114-129.

Gupta, D.N., G. Keshri, V. Lakshmi, and R.S. Kapil. 1989. Post coital contraceptive efficacy of *Terminalia arjuna* in albino rats. *Fitoterapia* 60(3):275-276.

Malik, N., V. Dhawan, A. Bahl, and D. Kaul. 2009. Inhibitory effects of *Terminalia arjuna* on platelet activation in vitro in healthy subjects and patients with coronary artery disease. *Platelets* 20(3):183-190.

Pole, S. 2006. *Ayurvedic medicine: The principles of traditional practice.* New York: Elsevier.

Ragavan, B., and S. Krishnakumari. 2006. Hypoglycemic and hypolipidemic activities of *Terminalia arjuna* stem bark in alloxan induced diabetic rats. *J. Nat. Remedies* 6(2):124-130.

Terminalia bellerica (Gaertn.) Roxb. — Combretaceae

SCN: belleric myrobalan
AN: *bibhitaki*

Part: fruit

Quick Reference Summary

Safety Class: 1
Interaction Class: A

Contraindications

None known.

Other Precautions

None known.

Drug and Supplement Interactions

None known.

Notice

Tannins (up to 21%) (Chadha 1988; List and Hörhammer 1973); *see* Appendix 1.

Adverse Events and Side Effects

None known.

Pharmacological Considerations

None known.

Pregnancy and Lactation

No information on the safety of belleric myrobalan in pregnancy or lactation was identified in the scientific or traditional literature. Although this review did not identify any concerns for use while pregnant or nursing, safety has not been conclusively established.

Review Details

I. Drug and Supplement Interactions

Clinical Trials of Drug or Supplement Interactions

No clinical trials of drug or supplement interactions were identified.

Case Reports of Suspected Drug or Supplement Interactions

No case reports of suspected drug or supplement interactions were identified.

Animal Trials of Drug or Supplement Interactions

No animal trials of drug or supplement interactions were identified.

II. Adverse Events

Case Reports of Adverse Events

No case reports of adverse events were identified.

III. Pharmacology and Pharmacokinetics

Human Pharmacological Studies

No relevant human pharmacological studies were identified.

Animal Pharmacological Studies

No relevant animal pharmacological studies were identified.

In Vitro Pharmacological Studies

No relevant in vitro pharmacological studies were identified.

IV. Pregnancy and Lactation

No information on the safety of belleric myrobalan during pregnancy or lactation was identified.

V. Toxicity Studies

No toxicity studies were identified.

T

Literature Cited

Chadha, Y. 1988. *The wealth of India: A dictionary of Indian raw materials and industrial products*. Delhi: Council of Scientific and Industrial Research.

List, P.H., and H. Hörhammer. 1973. *Hagers handbuch der pharmazeutischen praxis*. Berlin: Springer.

Terminalia chebula Retz. — Combretaceae

SCN: chebulic myrobalan
AN: *haritaki*
PN: *he zi* (fruit)
OCN: black myrobalan; true myrobalan
Part: fruit

Quick Reference Summary

Safety Class: 1
Interaction Class: A

Contraindications

None known.

Other Precautions

None known.

Drug and Supplement Interactions

None known.

Notice

Tannins (25–32%) (Chadha 1988; List and Hörhammer 1973); *see* Appendix 1.

Adverse Events and Side Effects

None known.

Pharmacological Considerations

Animal studies and traditional human use have demonstrated that chebulic myrobalan may modify glucose regulation (Kumar et al. 2006; Murali et al. 2007; Rao and Nammi 2006). People with diabetes are advised to monitor their blood sugar closely and discuss the use of this herb with a qualified healthcare practitioner prior to use.

Pregnancy and Lactation

No information on the safety of chebulic myrobalan in pregnancy or lactation was identified in the scientific or traditional literature. Although this review did not identify any concerns for use while pregnant or nursing, safety has not been conclusively established.

Review Details

I. Drug and Supplement Interactions

Clinical Trials of Drug or Supplement Interactions

No clinical trials of drug or supplement interactions were identified.

Case Reports of Suspected Drug or Supplement Interactions

No case reports of suspected drug or supplement interactions were identified.

Animal Trials of Drug or Supplement Interactions

No animal trials of drug or supplement interactions were identified.

II. Adverse Events

Case Reports of Adverse Events

No case reports of adverse events were identified.

III. Pharmacology and Pharmacokinetics

Human Pharmacological Studies

No relevant human pharmacological studies were identified.

Animal Pharmacological Studies

In diabetic rats orally administered 200 mg/kg of an ethanol extract of chebulic myrobalan daily for 30 days, a significant reduction in blood glucose and glycosylated hemoglobin was observed. Determination of plasma insulin levels revealed the insulin-stimulating action of the fruit extract (Kumar et al. 2006). In diabetic rats orally administered 100, 200, or 300 mg/kg of a chloroform extract of chebulic myrobalan seed daily for up to 8 weeks, a dose-dependent reduction in blood glucose levels was observed (Rao and Nammi 2006). A reduction in blood glucose levels was observed in diabetic rats orally administered 200 mg/kg of arjuna aqueous extract daily for 2 months. Related in vitro studies with pancreatic islets showed that the insulin release was nearly two times more in treated than in untreated diabetic animals (Murali et al. 2007).

In Vitro Pharmacological Studies

No relevant in vitro pharmacological studies were identified.

IV. Pregnancy and Lactation

No information on the safety of chebulic myrobalan during pregnancy or lactation was identified.

V. Toxicity Studies

Acute Toxicity

The LD_{50} of an orally administered aqueous extract of arjuna in rats could not be determined at doses up to 3 g/kg. At doses up to 3 g/kg, no changes in blood parameters for liver and kidney function were observed including total protein, alkaline phosphatase, SGPT, SGOT, total bilirubin, and creatinine (Murali et al. 2007).

Short-Term Toxicity

In rats fed a diet containing 25% chebulic myrobalan seed powder daily for 10 to 14 days, early centrilobular vein disruption and centrilobular sinusoidal congestion were observed (Arseculeratne et al. 1985).

Genotoxicity

In the Vitotox assay, no mutagenic activity of chebulic myrobalan was observed, whereas an increase in DNA damage was observed in the comet assay with an extract of chebulic myrobalan at a concentration of 500 ppm. The authors noted that these differing results are not considered contradictory because DNA damage in the alkaline comet assay may not be permanent and hence may not necessarily result in mutations; chebulic myrobalan was previously found to have significant antimutagenic activity in the Ames assay (Arora et al. 2005).

Literature Cited

Arora, S., E. Brits, S. Kaur, et al. 2005. Evaluation of genotoxicity of medicinal plant extracts by the comet and VITOTOX tests. *J. Environ. Pathol. Toxicol. Oncol.* 24(3):193-200.

Arseculeratne, S.N., A.A. Gunatilaka, and R.G. Panabokke. 1985. Studies of medicinal plants of Sri Lanka. Part 14: Toxicity of some traditional medicinal herbs. *J. Ethnopharmacol.* 13(3):323-335.

Chadha, Y. 1988. *The wealth of India: A dictionary of Indian raw materials and industrial products.* Delhi: Council of Scientific and Industrial Research.

Kumar, G.P.S., P. Arulselvan, D.S. Kumar, and S.P. Subramanian. 2006. Anti-diabetic activity of fruits of *Terminalia chebula* on streptozotocin induced diabetic rats. *J. Health Sci.* 52(3):283-291.

List, P.H., and H. Hörhammer. 1973. *Hagers handbuch der pharmazeutischen praxis.* Berlin: Springer.

Murali, Y.K., P. Anand, V. Tandon, et al. 2007. Long-term effects of *Terminalia chebula* Retz. on hyperglycemia and associated hyperlipidemia, tissue glycogen content and in vitro release of insulin in streptozotocin induced diabetic rats. *Exp. Clin. Endocrinol. Diabetes* 115(10):641-646.

Rao, N.K., and S. Nammi. 2006. Antidiabetic and renoprotective effects of the chloroform extract of *Terminalia chebula* Retz. seeds in streptozotocin-induced diabetic rats. *BMC Complement. Altern. Med.* 6:17.

Ternstroemia pringlei (Rose) Standl. — Theaceae

SCN: tilia estrella
Syn: *Taonabo pringlei* Rose

OCN: star tilia
Part: flower

Quick Reference Summary

Safety Class: 1
Interaction Class: A

Contraindications

None known.

Other Precautions

None known.

Drug and Supplement Interactions

None known.

Adverse Events and Side Effects

None known.

Pharmacological Considerations

None known.

Pregnancy and Lactation

No information on the safety of tilia estrella in pregnancy or lactation was identified in the scientific or traditional literature. Although this review did not identify any concerns for use while pregnant or nursing, safety has not been conclusively established.

Review Details

I. Drug and Supplement Interactions

Clinical Trials of Drug or Supplement Interactions

No clinical trials of drug or supplement interactions were identified.

Case Reports of Suspected Drug or Supplement Interactions

No case reports of suspected drug or supplement interactions were identified.

T

Animal Trials of Drug or Supplement Interactions

No animal trials of drug or supplement interactions were identified.

II. Adverse Events

Case Reports of Adverse Events

No case reports of adverse events were identified.

III. Pharmacology and Pharmacokinetics

Human Pharmacological Studies

No relevant human pharmacological studies were identified.

Animal Pharmacological Studies

No relevant animal pharmacological studies were identified.

In Vitro Pharmacological Studies

No relevant in vitro pharmacological studies were identified.

IV. Pregnancy and Lactation

No information on the safety of tilia estrella during pregnancy or lactation was identified.

V. Toxicity Studies

Acute Toxicity

The LD_{50} of a tilia estrella methanol extract intraperitoneally administered to mice is 1.05 g/kg (Aguilar-Santamaria and Tortoriello 1996).

Literature Cited

Aguilar-Santamaria, L., and J. Tortoriello. 1996. Anticonvulsant and sedative effects of crude extracts of *Ternstroemia pringlei* and *Ruta chalepensis*. *Phytother. Res.* 10(6):531-533.

Tetradium ruticarpum (A. Juss.) T.G. Hartley — Rutaceae

SCN: evodia
Syn: *Euodia ruticarpa* (A. Juss.) Benth.
PN: *wu zhu yu* (unripe fruit)
Part: unripe fruit

Quick Reference Summary

Safety Class: 2d
Interaction Class: B

Contraindications

Not for long-term use (Bensky et al. 2004; Chen and Chen 2004).

Do not exceed recommended dose (Bensky et al. 2004; Chen and Chen 2004).

Other Precautions

None known.

Drug and Supplement Interactions

See Pharmacological Considerations.

Editors' Notes

Unprocessed evodia is sometimes used topically, whereas material for internal use should be processed. Processing of evodia, which reduces the toxicity, is typically done by cooking the unripe fruit with licorice (Bensky et al. 2004; Chen and Chen 2004; Hong et al. 2008). Other methods of processing are sometimes used (Bensky et al. 2004).

Adverse Events and Side Effects

Large doses (standard dose listed as a decoction of 1.5–5 g) have a stimulating effect on the central nervous system and can lead to visual disturbances and hallucinations (Bensky et al. 2004; Chen and Chen 2004).

Pharmacological Considerations

Animal studies have indicated that evodia and compounds from evodia may affect the metabolism of theophylline, acetaminophen, chlorzoxazone, and caffeine (Bista et al. 2008; Jan et al. 2005; Lee et al. 2007; Tsai et al. 2005). In all of these studies, evodia or a compound from evodia was administered orally, while the drugs were administered intravenously. The relevance of those studies to human oral use is not known.

Animal studies have indicated that evodia and compounds from evodia may induce the drug-metabolizing isoenzymes CYP1A1, CYP1A2, and CYP2E1 (Bista et al. 2008; Jan et al. 2005, 2006; Lee et al. 2004, 2007; Tsai et al. 2005; Ueng et al. 2001, 2002a, 2002c).

Pregnancy and Lactation

Limited information on the safety of evodia in pregnancy or lactation was identified in the scientific or traditional literature. Although this review did not identify any concerns for use while pregnant or nursing, safety has not been conclusively established.

Review Details

I. Drug and Supplement Interactions

Clinical Trials of Drug or Supplement Interactions

No clinical trials of drug or supplement interactions were identified.

Case Reports of Suspected Drug or Supplement Interactions

No case reports of suspected drug or supplement interactions were identified.

Animal Trials of Drug or Supplement Interactions

A decrease in serum levels and a dose-dependent increase in the rate of clearance of the drug theophylline was observed in rats orally administered 1 or 2 g/kg of an ethanol extract of evodia or 50 mg/kg of the compound rutaecarpine daily for 3 days prior to intravenous administration of 2 mg/kg theophylline, a CYP1A2 substrate (Jan et al. 2005; Ueng et al. 2005).

A decrease in serum levels and an increase in the clearance rate of caffeine was observed in rats orally administered 1 g/kg daily of a hydroalcoholic extract of evodia, or 25 mg/kg daily of the compound rutaecarpine daily, for 3 days prior to the intravenous administration of 5 mg/kg caffeine, a CYP1A2 substrate (Tsai et al. 2005).

A decrease in serum levels of chlorzoxazone was observed in rats orally administered 80 mg/kg of the compound rutaecarpine daily for 3 days prior to the intravenous administration of 20 mg/kg chlorzoxazone, a CYP2E1 substrate (Bista et al. 2008).

A decrease in serum levels of acetaminophen was observed in rats orally administered 40 or 80 mg/kg of the compound rutaecarpine daily for 3 days prior to intravenous administration of acetaminophen (Lee et al. 2007).

II. Adverse Events

Case Reports of Adverse Events

Overdoses (standard dose is a decoction of 1.5–5 g) of evodia or internal use of the unprocessed herb have been associated with abdominal pain, diarrhea, visual disturbances, dizziness, hair loss, headache, a sense of oppression in the chest, and rashes (Bensky et al. 2004; Chen and Chen 2004).

Allergic reactions, primarily in the form of scarlatiniform rashes, have been reported (Bensky et al. 2004).

III. Pharmacology and Pharmacokinetics

Human Pharmacological Studies

No relevant human pharmacological studies were identified.

Animal Pharmacological Studies

In mice orally administered 2 g/kg of methanolic or aqueous extracts of evodia daily, immunoblot analyses showed that the methanol extract increased the levels of the drug-metabolizing isoenzymes CYP1A1, CYP1A2, and CYP2B and GSTYb-immunoreactive proteins while the aqueous extract increased only CYP1A2 protein levels (Ueng et al. 2002c).

A significant increase in activity of the drug-metabolizing isoenzymes CYP1A1 and CYP1A2 was observed in mice orally administered 50 mg/kg of the compound rutaecarpine daily for 3 days. No effect on CYP3A was observed (Ueng et al. 2001).

Induction of the drug-metabolizing isoenzyme CYP1A2 was observed in mice orally administered 3.5 mg/kg daily of the compound rutaecarpine or 0.77 g/kg per day of a concentrate of evodia; the compounds evodiamine and dehydroevodiamine had no significant effects on CYP1A2 activity (Ueng et al. 2002a).

Induction of the drug-metabolizing isoenzymes CYP1A and CYP2B was observed in mice orally administered 20, 40, or 80 mg/kg of the compound rutaecarpine daily for 3 days (Lee et al. 2004).

The drug-metabolizing isoenzymes CYP1A and CYP2B, but not CYP3A, were found to play major roles in the hydroxylation of rutaecarpine in mice (Jan et al. 2006).

Also see Animal trials of drug or supplement interactions for this entry.

In Vitro Pharmacological Studies

Significant inhibition of the drug-metabolizing isoenzyme CYP1A2 was observed in mouse and human liver microsomes treated with the compound rutaecarpine. The compounds evodiamine and dehydroevodiamine were also tested, but no significant activity of the compounds on CYP1A2 was observed (Ueng et al. 2002b).

Mechanism-based inhibition of the drug-metabolizing isoenzyme CYP3A4 was observed in human liver microsomes treated with the compounds rutaecarpine or limonin, with activity reported as preincubation time dependent (Iwata et al. 2005).

Moderate modulation of the drug transporter P-glycoprotein (P-gp) was observed in the calcein assay in porcine brain capillary endothelial cells treated with the compounds rutaecarpine or evodiamine. The compound evocarpine had no effect on P-gp (Adams et al. 2007).

The essential oil of evodia enhanced percutaneous absorption of ibuprofen (Luo et al. 2007).

In human platelet-rich plasma, inhibition of aggregation induced by collagen, ADP, adrenaline, and arachidonic acid was observed after treatment with the compound rutaecarpine (40–200 μM) (Sheu et al. 1996).

IV. Pregnancy and Lactation

In an in vitro uterotonic assay, the compounds rutaecarpine and dehydroevodiamine demonstrated uterotonic activity. No uterotonic activity of evodiamine hydrochloride was observed (King et al. 1980).

No information on the safety of evodia during lactation was identified in the scientific or traditional literature. While this review did not identify any concerns for use while nursing, safety has not been conclusively established.

V. Toxicity Studies

Acute Toxicity

The LD_{50} of orally administered aqueous and ethanolic extracts of evodia could not be determined at doses up to 10 g/kg (Yang 2008). The LD_{50} of aqueous extracts of processed (cooked with a licorice decoction) or unprocessed evodia could not be determined at doses up to 40 g/kg (route of administration not specified in English language abstract) (Hong et al. 2008).

The LD_{50} values for compounds isolated from evodia and intravenously administered to mice are 65.0 mg/kg for rutaecarpine, 77.8 mg/kg for evodiamine, 64.9 mg/kg for 1-methyl-2-undecyl-4(1*H*)-quinoline, 36.0 mg/kg for 2-undecyl-4(1*H*)-quinolone, and 47.6 mg/kg for 1-methyl-2-undecanone-10′-4(1*H*)-quinolone. The LD_{50} of the total quinolones of evodia is 14 mg/kg after intravenous administration (Yang et al. 2006).

In the mouse sperm aberration test, no abnormalities were observed in the sperm of mice orally administered up to 5 g/kg of aqueous or ethanolic extracts of evodia daily for 5 days (Yang 2008).

Genotoxicity

No mutagenic activity of aqueous or ethanolic extracts of evodia was observed in the Ames test for mutagenicity or in the mouse bone marrow micronucleus test. The Ames test examined effects in *Salmonella typhimurium* strains TA97, TA98, TA100, or TA102. In the mouse bone marrow micronucleus test, animals were administered up to 5 g/kg of the evodia extracts daily for 2 days (Yang 2008).

Literature Cited

Adams, M., A. Mahringer, O. Kunert, et al. 2007. Cytotoxicity and P-glycoprotein modulating effects of quinolones and indoloquinazolines from the Chinese herb *Evodia rutaecarpa*. *Planta Med.* 73(15):1554-1557.

Bensky, D., S. Clavey, and E. Stöger. 2004. *Chinese herbal medicine: Materia medica*. 3rd ed. Seattle: Eastland Press.

Bista, S.R., S.K. Lee, D. Thapa, et al. 2008. Effects of oral rutaecarpine on the pharmacokinetics of intravenous chlorzoxazone in rats. *Toxicol. Res.* 24(3):195-199.

Chen, J.K., and T.T. Chen. 2004. *Chinese medical herbology and pharmacology*. City of Industry, CA: Art of Medicine Press.

Hong, Y.M., W.H. Wang, Z.M. Wang, and J.H. Wang. 2008. Study of liquorice processing fructus. *Zhongguo Zhong Yao Za Zhi* 33(8):884-888.

Iwata, H., Y. Tezuka, S. Kadota, A. Hiratsuka, and T. Watabe. 2005. Mechanism-based inactivation of human liver microsomal CYP3A4 by rutaecarpine and limonin from evodia fruit extract. *Drug Metab. Pharmacokinet.* 20(1):34-45.

Jan, W.C., M.J. Don, L.K. Ho, C.F. Chen, and Y.F. Ueng. 2006. Characterization of mouse cytochrome P450-catalyzed oxidative metabolism of rutaecarpine, an alkaloid in the herbal medicine *Evodia rutaecarpa*. *J. Food Drug. Anal.* 14(2):159-165.

Jan, W.C., L.C. Lin, C. Chieh Fu, and T.H. Tsai. 2005. Herb-drug interaction of *Evodia rutaecarpa* extract on the pharmacokinetics of theophylline in rats. *J. Ethnopharmacol.* 102(3):440-445.

King, C.L., Y.C. Kong, N.S. Wong, et al. 1980. Uterotonic effect of *Evodia rutaecarpa* alkaloids. *J. Nat. Prod.* 43 (5):577-582.

Lee, S.K., S.R. Bista, H. Jeong, et al. 2007. The effects of rutaecarpine on the pharmacokinetics of acetaminophen in rats. *Arch. Pharm. Res.* 30(12):1629-1634.

Lee, S.K., N.H. Kim, J. Lee, et al. 2004. Induction of cytochrome P450s by rutaecarpine and metabolism of rutaecarpine by cytochrome P450s. *Planta Med.* 70(8):753-757.

Luo, X.Q., Y.H. Gu, and Z.Y. Wu. 2007. Comparison of the effect of eight kinds of volatile oil of Chinese materia medica on percutaneous absorption of ibuprofen in vitro. *Zhong Yao Cai* 30(5):571-573.

Sheu, J.R., W.C. Hung, Y.M. Lee, and M.H. Yen. 1996. Mechanism of inhibition of platelet aggregation by rutaecarpine, an alkaloid isolated from *Evodia rutaecarpa*. *Eur. J. Pharmacol.* 318(2-3):469-475.

Tsai, T.H., C.H. Chang, and L.C. Lin. 2005. Effects of *Evodia rutaecarpa* and rutaecarpine on the pharmacokinetics of caffeine in rats. *Planta Med.* 71(7):640-645.

Ueng, Y.F., M.J. Don, H.C. Peng, et al. 2002a. Effects of *wu chu yu tang* and its component herbs on drug-metabolizing enzymes. *Jpn. J. Pharmacol.* 89(3):267-273.

Ueng, Y.F., W.C. Jan, L.C. Lin, et al. 2002b. The alkaloid rutaecarpine is a selective inhibitor of cytochrome P450 1A in mouse and human liver microsomes. *Drug Metab. Dispos.* 30(3):349-353.

Ueng, Y.F., H.C. Ko, C.F. Chen, J.J. Wang, and K.T. Chen. 2002c. Modulation of drug-metabolizing enzymes by extracts of a herbal medicine *Evodia rutaecarpa* in C57BL/6J mice. *Life Sci.* 71(11):1267-1277.

Ueng, Y.F., T.H. Tsai, M.J. Don, R.M. Chen, and T.L. Chen. 2005. Alteration of the pharmacokinetics of theophylline by rutaecarpine, an alkaloid of the medicinal herb *Evodia rutaecarpa*, in rats. *J. Pharm. Pharmacol.* 57(2):227-232.

Ueng, Y.F., J.J. Wang, L.C. Lin, S.S. Park, and C.F. Chen. 2001. Induction of cytochrome P450-dependent monooxygenase in mouse liver and kidney by rutaecarpine, an alkaloid of the herbal drug *Evodia rutaecarpa*. *Life Sci.* 70(2):207-217.

Yang, X.W. 2008. [Toxicological assessment on safety of water and 70% ethanolic extracts of nearly ripe fruit of *Evodia rutaecarpa*.] *Zhongguo Zhong Yao Za Zhi* 33(11):1317-1321.

Yang, X.W., H. Zhang, M. Li, et al. 2006. Studies on the alkaloid constituents of *Evodia rutaecarpa* (Juss.) Benth. var. *bodinaieri* (Dode) Huang and their acute toxicity in mice. *J. Asian Nat. Prod. Res.* 8(8):697-703.

T

Thuja occidentalis L. — Cupressaceae

SCN: thuja
OCN: eastern white cedar; eastern arborvitae; northern white cedar; swamp cedar

Part: leaf

Quick Reference Summary

Safety Class: 2b, 2d
Interaction Class: A

Contraindications

Not for use in pregnancy except under the supervision of a qualified healthcare practitioner (Felter and Lloyd 1898; Osol and Farrar 1955).

Not for long-term use; do not exceed recommended dose (Wood and LaWall 1926).

Other Precautions

None known.

Drug and Supplement Interactions

None known.

Standard Dose

The standard dose is 1–3 g as a tea (Remington and Wood 1918).

Notice

Abortifacient (Felter and Lloyd 1898; Osol and Farrar 1955); *see* Appendix 2.

Thujone (0.76–2.4% of herb; essential oil contains approximately 55% α-thujone and 16% β-thujone) (Naser et al. 2005; Svajdlenka et al. 1999); *see* Appendix 1.

Editors' Notes

Use of thuja as a food additive in the United States is subject to a limitation that the finished food or beverage is thujone-free (CFR 2011). Dietary ingredients for use in dietary supplements, however, are specifically excluded from the federal food additive definition (U.S.C. 2010).

Adverse Events and Side Effects

Seizures have been reported after overdose of thuja essential oil (Friesen and Phillips 2006; Millet et al. 1981).

Pharmacological Considerations

None known.

Pregnancy and Lactation

Thuja has historically been used as an abortifacient, although more recent reviews have indicated that such use was ineffective and often highly toxic (Felter and Lloyd 1898; Harnischfeger and Stolze 1983; Naser et al. 2005; Osol and Farrar 1955). Thuja and thuja essential oil should not be used internally or externally during pregnancy.

No information on the safety of thuja during lactation was identified in the scientific or traditional literature. While this review did not identify any concerns for use while nursing, safety has not been conclusively established.

Review Details

I. Drug and Supplement Interactions

Clinical Trials of Drug or Supplement Interactions

No clinical trials of drug or supplement interactions were identified.

Case Reports of Suspected Drug or Supplement Interactions

No case reports of suspected drug or supplement interactions were identified.

Animal Trials of Drug or Supplement Interactions

No animal trials of drug or supplement interactions were identified.

II. Adverse Events

Case Reports of Adverse Events

Seizures were reported in a 2-year-old girl who ingested up to 15 ml of 0.1% thuja essential oil. The two seizures, each lasting 5 min, occurred within 20 minutes of ingestion (Friesen and Phillips 2006). A tonic seizure was reported in a 50-year-old woman who had taken 20 drops of thuja essential oil twice daily for 5 days (Millet et al. 1981).

In children who have eaten the fresh shoots or cones of thuja, gastrointestinal pain was reported. In some instances, the pain was accompanied by bloody vomiting. Quantities ingested were not reported in the available English language translation (Schulte 2002).

Allergic contact dermatitis to thuja has been reported (Grimm 1991).

T

III. Pharmacology and Pharmacokinetics

Human Pharmacological Studies

No relevant human pharmacological studies were identified.

Animal Pharmacological Studies

No relevant animal pharmacological studies were identified.

In Vitro Pharmacological Studies

No relevant in vitro pharmacological studies were identified.

IV. Pregnancy and Lactation

Thuja has historically been used as an abortifacient, although more recent reviews have indicated that such use was ineffective and often highly toxic (Felter and Lloyd 1898; Harnischfeger and Stolze 1983; Naser et al. 2005; Osol and Farrar 1955).

No information on the safety of thuja during lactation was identified.

V. Toxicity Studies

Acute Toxicity

The LD_{50} of the compound thujone (mixture of α- and β-thujone) is 500 mg/kg after oral administration to rats, 87.5 mg/kg after subcutaneous administration to mice, 240 mg/kg after intraperitoneal administration to rats, and 0.031 mg/kg after intravenous administration to rabbits (EMEA 1999; Naser et al. 2005).

In the brine shrimp lethality assay, the LC_{50} of an alcohol extract of thuja is 11.94 µg/ml (Lagarto Parra et al. 2001).

Genotoxicity

No mutagenic activity of thuja tinctures was observed in the *Salmonella*/mammalian microsome assay and the SOS-chromotest (Valsa and Felzenszwalb 2001).

Literature Cited

CFR. 2011. *Code of federal regulations*, Title 21 Part 172.510, 2011 ed. Food additives permitted for direct addition to food for human consumption. Flavoring agents and related substances. Natural flavoring substances and natural substances used in conjunction with flavors. Washington, DC: U.S. Government Printing Office.

EMEA. 1999. *Thuja occidentalis* Summary Report. EMEA/MRL/602/99, edited by European Agency for Evaluation of Medicinal Products, V.M.E.U. London.

Felter, H.W., and J.U. Lloyd. 1898. *King's American dispensatory*. 18th ed., 3rd rev. 2 vols. Cincinnati: Ohio Valley Co.

Friesen, M.S., and B. Phillips. 2006. Status epilepticus following pediatric ingestion of thuja essential oil. *Clin. Toxicol.* 44(4):557.

Grimm, I. 1991. Allergic contact dermatitis from a *Thuja occidentalis* extract. *Allergologie* 14(7):272-274.

Harnischfeger, G., and H. Stolze. 1983. *Bewährte Pflanzendrogen in Wissenschaft und Medizin*. Bad Homburg/Melsungen: Notamed Verlag.

Lagarto Parra, A., R. Silva Yhebra, I. Guerra Sardinas, and L. Iglesias Buela. 2001. Comparative study of the assay of *Artemia salina* L. and the estimate of the medium lethal dose (LD_{50} value) in mice, to determine oral acute toxicity of plant extracts. *Phytomedicine* 8(5):395-400.

Millet, Y., J. Jouglard, M.D. Steinmetz, et al. 1981. Toxicity of some essential plant oils. Clinical and experimental study. *Clin. Toxicol.* 18(12):1485-1498.

Naser, B., C. Bodinet, M. Tegtmeier, and U. Lindequist. 2005. *Thuja occidentalis* (arbor vitae): A review of its pharmaceutical, pharmacological and clinical properties. *Evid. Based Complement. Alternat. Med.* 2(1):69-78.

Osol, A., and G. Farrar. 1955. *The dispensatory of the United States of America*. 25th ed. Philadelphia: Lippincott.

Remington, J.P., and H.C. Wood. 1918. *The dispensatory of the United States of America*. 20th ed. Philadelphia: Lippincott.

Schulte, U. 2002. Der Wacholder, Baum des Jahres. *Dtsch. Apoth. Ztg.* 142(4):64-66.

Svajdlenka, E., P. Mártonfi, I. Tomasko, D. Grancai, and M. Nagy. 1999. Essential oil composition of *Thuja occidentalis* L. samples from Slovakia. *J. Essen. Oil Res.* 11:532-536.

U.S.C. 2010. United States Code, Title 21, Part 321 (s)(6). Current as of January 7, 2011. Washington, DC: U.S. Government Printing Office.

Valsa, J.O., and I. Felzenszwalb. 2001. Genotoxic evaluation of the effect of *Thuja occidentalis* tinctures. *Braz. J. Biol.* 61(2):329-332.

Wood, H., and C. LaWall. 1926. *The dispensatory of the United States of America*. 21st ed. Philadelphia: Lippincott.

T

Thymus vulgaris L. — Lamiaceae

SCN: thyme
OCN: common thyme; garden thyme

Part: herb

Quick Reference Summary

Safety Class: 2b
Interaction Class: A

Contraindications

Not for use in pregnancy except under the supervision of a qualified healthcare practitioner (Chadha 1988; Felter and Lloyd 1898; Watt and Breyer-Brandwijk 1962; Wichtl 2004).

Other Precautions

None known.

Drug and Supplement Interactions

None known.

Notice

Emmenagogue (Chadha 1988; Felter and Lloyd 1898; Watt and Breyer-Brandwijk 1962); *see* Appendix 2.

Editors' Note

The classifications and concerns for this herb are based on relatively higher doses used for therapeutic purposes in contrast to lower amounts generally used in cooking, and have not been associated with its use as a spice.

Adverse Events and Side Effects

Allergic reactions to thyme have been reported, sometimes with cross-sensitivity to rosemary or oregano (Armisen et al. 2003; Benito et al. 1996; Le Roy et al. 1981; Lorenzi et al. 1995; Martinez-Gonzalez et al. 2007).

Pharmacological Considerations

None known.

Pregnancy and Lactation

Several authors list thyme as an emmenagogue or abortifacient and note that thyme essential oil should not be used in pregnancy (Chadha 1988; Felter and Lloyd 1898; Watt and Breyer-Brandwijk 1962; Wichtl 2004). Large doses of the compound thymol (which comprises approximately 51% of thyme essential oil) have been associated with abortifacient activity (Gillespie 1973; Sollman 1948). Based on this information, use during pregnancy is not recommended except under the supervision of a qualified healthcare practitioner.

No information on the safety of thyme during lactation was identified in the scientific or traditional literature. While this review did not identify any concerns for use while nursing, safety has not been conclusively established.

Review Details

I. Drug and Supplement Interactions

Clinical Trials of Drug or Supplement Interactions

No clinical trials of drug or supplement interactions were identified.

Case Reports of Suspected Drug or Supplement Interactions

No case reports of suspected drug or supplement interactions were identified.

Animal Trials of Drug or Supplement Interactions

No animal trials of drug or supplement interactions were identified.

II. Adverse Events

Case Reports of Adverse Events

Allergic reactions to thyme, confirmed by patch testing, have been reported (Benito et al. 1996; Le Roy et al. 1981; Lorenzi et al. 1995). In some cases, patients exhibited cross-sensitivity to oregano or rosemary (Armisen et al. 2003; Benito et al. 1996; Martinez-Gonzalez et al. 2007).

Relative to urban-dwelling control groups, respiratory ailments and sensitivity to microbial allergens were increased in thyme farmers regularly exposed to airborne thyme dust (Golec et al. 2003, 2005). Occupational airborne contact dermatitis caused by thyme dust was reported in some farmers during threshing of thyme (Spiewak et al. 2001).

III. Pharmacology and Pharmacokinetics

Human Pharmacological Studies

No relevant human pharmacological studies were identified.

Animal Pharmacological Studies

Induction of phase I and phase II enzymes was observed in mice fed a diet containing up to 2% thyme or treated orally with thymol (200 mg/kg) or carvacrol (up to 200 mg/kg) daily for 7 days. Bioactivity was attributed to the thymol and carvacrol content of thyme (Sasaki et al. 2005).

In Vitro Pharmacological Studies

Thyme essential oil inhibited arachidonic acid-induced platelet aggregation at a concentration of 7 μM, and ADP-induced aggregation at a concentration of 100 μM, but had no effect on U46619-induced aggregation (Tognolini et al. 2006). The compound thymol inhibited platelet aggregation induced by collagen, ADP, arachidonic acid, and thrombin (Okazaki et al. 2002).

An aqueous-ethanolic extract of thyme bound competitively to estradiol and progesterone receptors (Zava et al. 1998).

IV. Pregnancy and Lactation

No adverse effects on development were observed in 4-day-old embryos from pregnant mice that were administered a diet of 0.25% thyme essential oil (approximate dose of 375 mg/kg daily) for 2 weeks prior to pregnancy until day 4 of pregnancy (Domaracky et al. 2007).

Large doses (over 1 g) of the compound thymol may cause abortion (Sollman 1948). In the 1970s, a combination of thymol, potassium iodide, soap, and astringents was applied as a topical irritant to produce abortion. This practice was considered high risk, and a number of maternal deaths were associated with the use of this combination (Gillespie 1973).

No information on the safety of thyme during lactation was identified.

V. Toxicity Studies

Acute Toxicity

The LD_{50} of orally administered thyme essential oil in rats is reported as 2.48 and 4.7 g/kg (Opdyke 1974; Skramlik 1959).

In mice orally administered doses of 0.5 to 3 g/kg of a concentrated thyme extract (equivalent to 4.3 to 25.8 g/kg thyme), a decrease in locomotor activity and modest slowing of the respiratory system were observed (Qureshi et al. 1991).

Short-Term Toxicity

No significant adverse effects, including differences in weight gain, were observed in piglets administered a diet of 1% dried thyme (duration of feeding not specified in English language abstract) (Jugl-Chizzola et al. 2005).

Subchronic Toxicity

In mice administered 100 mg/kg of a concentrated thyme extract (equivalent to 900 mg/kg thyme) daily for 3 months, an increase in liver and testes weight was noted, but no spermatotoxic effects were observed (Qureshi et al. 1991).

Chronic Toxicity

No significant adverse effects were observed in rats fed a diet supplemented with thyme essential oil (providing 42.5 mg/kg daily) for 28 months (Youdim and Deans 1999).

Genotoxicity

No mutagenic effects of thyme essential oil or the compound thymol were observed in the Ames mutagenicity assay (Azizan and Blevins 1995; Zani et al. 1991). Similarly, no mutagenic effects of thyme essential oil were observed in the *Bacillus subtilis rec* assay (Zani et al. 1991).

Literature Cited

Armisen, M., V. Rodriguez, and C. Vidal. 2003. Photoaggravated allergic contact dermatitis due to *Rosmarinus officinalis* cross-reactive with *Thymus vulgaris*. *Contact Dermat.* 48(1):52-53.

Azizan, A., and R.D. Blevins. 1995. Mutagenicity and antimutagenicity testing of six chemicals associated with the pungent properties of specific spices as revealed by the Ames *Salmonella*/microsomal assay. *Arch. Environ. Contam. Toxicol.* 28(2):248-258.

Benito, M., G. Jorro, C. Morales, A. Pelaez, and A. Fernandez. 1996. Labiatae allergy: Systemic reactions due to ingestion of oregano and thyme. *Ann. Allergy Asthma Immunol.* 76(5):416-418.

Chadha, Y. 1988. *The wealth of India: A dictionary of Indian raw materials and industrial products*. Delhi: Council of Scientific and Industrial Research.

Domaracky, M., P. Rehak, S. Juhas, and J. Koppel. 2007. Effects of selected plant essential oils on the growth and development of mouse preimplantation embryos *in vivo*. *Physiol. Res.* 56(1):97-104.

Felter, H.W., and J.U. Lloyd. 1898. *King's American dispensatory*. 18th ed., 3rd rev. 2 vols. Cincinnati: Ohio Valley Co.

Gillespie, A. 1973. Techniques of abortion. *Br. J. Hosp. Med.* Mar.:309-316.

Golec, M., C. Skorska, B. Mackiewicz, and J. Dutkiewicz. 2003. Health effects of exposure to thyme dust in a group of thyme growing farmers. *Ann. Univ. Mariae Curie Sklodowska [Med.]* 58(1):195-203.

Golec, M., C. Skorska, B. Mackiewicz, A. Gora, and J. Dutkiewicz. 2005. Respiratory effects of exposure to dust from herbs. *Ann. Agric. Environ. Med.* 12(1):5-10.

Jugl-Chizzola, M., J. Spergser, F. Schilcher, et al. 2005. Effects of *Thymus vulgaris* L. as feed additive in piglets and against haemolytic *E. coli in vitro*. *Berl. Munch. Tierarztl. Wochenschr.* 118(11-12):495-501.

Le Roy, R., E. Grosshans, and J. Foussereau. 1981. Investigation of contact allergies in 100 cases of ulcus cruris. *Derm. Beruf. Umwelt.* 29(6):168-170.

Lorenzi, S., F. Placucci, C. Vincenzi, F. Bardazzi, and A. Tosti. 1995. Allergic contact dermatitis due to thymol. *Contact Dermat.* 33(6):439-440.

Martinez-Gonzalez, M.C., J.J. Goday Bujan, W. Martinez Gomez, and E. Fonseca Capdevila. 2007. Concomitant allergic contact dermatitis due to *Rosmarinus officinalis* (rosemary) and *Thymus vulgaris* (thyme). *Contact Dermat.* 56(1):49-50.

Okazaki, K., K. Kawazoe, and Y. Takaishi. 2002. Human platelet aggregation inhibitors from thyme (*Thymus vulgaris* L.). *Phytother. Res.* 16(4):398-399.

Opdyke, D.L.J. 1974. Thyme oil, red. *Food Cosmet. Toxicol.* 12:1003-1004.

Qureshi, S., A. Shah, and M. Al-Yahya. 1991. Toxicity of *Achillea fragrantissima* and *Thymus vulgaris* in mice. *Fitoterapia* 62(4):319-323.

Sasaki, K., K. Wada, Y. Tanaka, et al. 2005. Thyme (*Thymus vulgaris* L.) leaves and its constituents increase the activities of xenobiotic-metabolizing enzymes in mouse liver. *J. Med. Food* 8(2):184-189.

Skramlik, E.V. 1959. On the toxicity and tolerance of ethereal oils. *Pharmazie* 14:435-445.

Sollman, T. 1948. *A manual of pharmacology*. 7th ed. Philadelphia: Saunders.

Spiewak, R., C. Skorska, and J. Dutkiewicz. 2001. Occupational airborne contact dermatitis caused by thyme dust. *Contact Dermat.* 44(4):235-239.

Tognolini, M., E. Barocelli, V. Ballabeni, et al. 2006. Comparative screening of plant essential oils: Phenylpropanoid moiety as basic core for antiplatelet activity. *Life Sci.* 78(13):1419-1432.

Watt, J.M., and M.G. Breyer-Brandwijk. 1962. *The medicinal and poisonous plants of southern and eastern Africa*. 2nd ed. Edinburgh: E. & S. Livingstone.

Wichtl, M. 2004. *Herbal drugs and phytopharmaceuticals: A handbook for practice on a scientific basis*. 3rd ed. Boca Raton, FL: CRC Press.

Youdim, K.A., and S.G. Deans. 1999. Dietary supplementation of thyme (*Thymus vulgaris* L.) essential oil during the lifetime of the rat: Its effects on the antioxidant status in liver, kidney and heart tissues. *Mech. Ageing Dev.* 109(3):163-175.

Zani, F., G. Massimo, S. Benvenuti, et al. 1991. Studies on the genotoxic properties of essential oils with *Bacillus subtilis* rec-assay and *Salmonella*/microsome reversion assay. *Planta Med.* 57(3):237-241.

Zava, D.T., C.M. Dollbaum, and M. Blen. 1998. Estrogen and progestin bioactivity of foods, herbs, and spices. *Proc. Soc. Exp. Biol. Med.* 217(3):369-378.

Tilia spp.

Tiliaceae

Tilia cordata **Mill.**
SCN: linden
OCN: small-leaf lime tree

Tilia platyphyllos **Scop.**
SCN: linden
OCN: large-leaf linden; tilia

Tilia vulgaris **Hayne**
SCN: linden
Syn: *Tilia europaea* L.
OCN: European linden; European lime tree; tilia
Part: flower, leaf

Quick Reference Summary

Safety Class: 1
Interaction Class: A

Contraindications

None known.

Other Precautions

None known.

Drug and Supplement Interactions

None known.

Adverse Events and Side Effects

An allergic reaction to linden pollen in a linden flower tea was reported and confirmed by patch testing (De Smet 1993; Faillers 1989). A case of contact urticaria after use of a linden-containing shampoo was reported and confirmed by patch testing (Picardo et al. 1988).

Pharmacological Considerations

Coadministration of linden flower tea and iron-fortified bread reduced absorption of the iron (Hurrell et al. 1999).

Pregnancy and Lactation

No information on the safety of linden in pregnancy or lactation was identified in the scientific or traditional literature. Although this review did not identify any concerns for use while pregnant or nursing, safety has not been conclusively established.

Review Details

I. Drug and Supplement Interactions

Clinical Trials of Drug or Supplement Interactions

No clinical trials of drug or supplement interactions were identified.

Case Reports of Suspected Drug or Supplement Interactions

No case reports of suspected drug or supplement interactions were identified.

Animal Trials of Drug or Supplement Interactions

No animal trials of drug or supplement interactions were identified.

II. Adverse Events

Case Reports of Adverse Events

Contact urticaria was reported after use of a shampoo containing a linden extract. Patch testing with linden elicited a positive reaction (Picardo et al. 1988).

An allergic reaction to linden pollen in a linden flower tea was reported in a woman allergic to multiple types of tree pollen. The reaction was confirmed by patch testing (De Smet 1993; Faillers 1989).

III. Pharmacology and Pharmacokinetics

Human Pharmacological Studies

Coadministration of iron-fortified bread and linden flower tea reduced absorption of the iron by 52%, an effect that was less pronounced than that of black tea, cocoa, or teas made with peppermint, European pennyroyal, or European vervain, but greater than the inhibitory effect of chamomile tea (Hurrell et al. 1999).

Animal Pharmacological Studies

No relevant animal pharmacological studies were identified.

In Vitro Pharmacological Studies

No relevant in vitro pharmacological studies were identified.

IV. Pregnancy and Lactation

No information on the use of linden during pregnancy or lactation was identified.

V. Toxicity Studies

Genotoxicity

No mutagenic activity of an aqueous extract of linden was observed in the somatic mutation and recombination test (SMART) in *Drosophila melanogaster* (Romero-Jimenez et al. 2005).

Literature Cited

De Smet, P.A.G.M. 1993. *Adverse effects of herbal drugs. Volume 2.* Berlin: Springer.

Faillers, C.J. 1989. Pine nut allergy in perspective. *Ann. Allerg.* 62:186-189.

Hurrell, R.F., M. Reddy, and J.D. Cook. 1999. Inhibition of non-haem iron absorption in man by polyphenolic-containing beverages. *Br. J. Nutr.* 81(04):289-295.

Picardo, M., R. Rovina, A. Cristaudo, C. Cannistraci, and B. Santucci. 1988. Contact urticaria from *Tilia* (lime). *Contact Dermat.* 19(1):72-73.

Romero-Jimenez, M., J. Campos-Sanchez, M. Analla, A. Munoz-Serrano, and A. Alonso-Moraga. 2005. Genotoxicity and anti-genotoxicity of some traditional medicinal herbs. *Mutat. Res.* 585(1-2):147-155.

Tinospora cordifolia (Willd.) Miers

Menispermaceae

SCN: Indian tinospora
AN: *guduchi*

Part: leaf, root, starch, stem

Quick Reference Summary

Safety Class: 1
Interaction Class: A

Contraindications

None known.

Other Precautions

None known.

Drug and Supplement Interactions

None known.

Notice

Diuretic (Kapoor 2001; Nayampalli et al. 1988; Remington and Wood 1918); *see* Appendix 2.

Adverse Events and Side Effects

None known.

Pharmacological Considerations

Animal studies have demonstrated that Indian tinospora may modify glucose regulation (Grover et al. 2000; Kar et al. 2003; Panchabhai et al. 2008; Stanely et al. 2000, 2003). People with diabetes are advised to monitor their blood sugar closely and discuss the use of this herb with a qualified healthcare practitioner prior to use.

Pregnancy and Lactation

No information on the safety of Indian tinospora in pregnancy or lactation was identified in the scientific or traditional literature. Although this review did not identify any concerns for use while pregnant or nursing, safety has not been conclusively established.

T

Review Details

I. Drug and Supplement Interactions

Clinical Trials of Drug or Supplement Interactions

No clinical trials of drug or supplement interactions were identified.

Case Reports of Suspected Drug or Supplement Interactions

No case reports of suspected drug or supplement interactions were identified.

Animal Trials of Drug or Supplement Interactions

No animal trials of drug or supplement interactions were identified.

II. Adverse Events

Case Reports of Adverse Events

No case reports of adverse events were identified.

III. Pharmacology and Pharmacokinetics

Human Pharmacological Studies

In healthy male volunteers orally administered 50 to 1520 mg of an aqueous extract of Indian tinospora daily for 4 days, an increase in urine volume was observed. A reduction in serum potassium and chloride levels was observed in one of the ten study subjects. No significant changes in liver enzyme levels or other serum biochemical parameters were observed. All study subjects noted an increase in appetite (Nayampalli et al. 1988).

Animal Pharmacological Studies

A dose-dependent reduction in blood glucose levels was observed in diabetic rats orally administered 25, 50, or 100 mg/kg of an alcohol extract of Indian tinospora daily for 42 days (Stanely et al. 2003). Similarly, a reduction in blood glucose levels was observed in diabetic rats orally administered 2.5 or 5 g/kg of an aqueous extract of Indian tinospora daily for 42 days (Stanely et al. 2000). In diabetic mice and rats, a reduction in blood sugar levels was observed after administration of 400 mg/kg of an aqueous extract of Indian tinospora daily for 15 weeks (Grover et al. 2000). A reduction in blood glucose levels was observed in diabetic rats orally administered a single dose of 250 mg/kg of an ethanol extract of Indian tinospora (Kar et al. 2003).

In male rats orally administered 100 mg/animal of a methanolic extract of Indian tinospora daily for 60 days, decreases in the weight of testis, epididymis, seminal vesicle, and ventral prostate were observed. Sperm motility and density were significantly reduced, resulting in complete infertility (Gupta and Sharma 2003).

In mice with cyclophosphamide-induced immunosuppression, oral administration of 100 mg/kg of an aqueous extract of Indian tinospora resulted in an increase in the number of leukocytes. Indian tinospora was found to inhibit cyclophosphamide-induced immunosuppression (Manjrekar et al. 2000).

In rats orally administered 100 mg/kg of an aqueous extract of Indian tinospora, an increase in urine output was observed. As compared to 2.5 mg/kg of hydrochlorothiazide, Indian tinospora produced a greater loss of sodium and potassium (Nayampalli et al. 1988).

In Vitro Pharmacological Studies

No relevant in vitro pharmacological studies were identified.

IV. Pregnancy and Lactation

No information on the safety of Indian tinospora during pregnancy or lactation was identified.

V. Toxicity Studies

Short-Term Toxicity

In mice orally administered 400 mg/kg of the freeze-dried powder of an aqueous extract of Indian tinospora daily for 60 days, no significant changes were observed in white and red blood cell counts, hemoglobin, mean corpuscular volume, or hematocrit (Grover et al. 2000).

Genotoxicity

No genotoxicity of an aqueous extract of Indian tinospora was observed in any of four different genotoxicity tests. Tests included the Ames test with *Salmonella typhimurium* strains TA97a, TA98, TA100, TA102, and TA1535 at concentrations of extract up to 5000 µg/plate, the chromosome aberration assay, the rodent bone marrow micronucleus assay, and the comet assay. For the bone marrow micronucleus and comet assays, mice were orally administered the extract for 7 days at doses of 150, 200, and 250 mg/kg (Chandrasekaran et al. 2009).

T

Literature Cited

Chandrasekaran, C.V., L.N. Mathuram, P. Daivasigamani, and U. Bhatnagar. 2009. *Tinospora cordifolia*, a safety evaluation. *Toxicol. In Vitro* 23(7):1220-1226.

Grover, J.K., V. Vats, and S.S. Rathi. 2000. Anti-hyperglycemic effect of *Eugenia jambolana* and *Tinospora cordifolia* in experimental diabetes and their effects on key metabolic enzymes involved in carbohydrate metabolism. *J. Ethnopharmacol.* 73(3):461-470.

Gupta, R.S., and A. Sharma. 2003. Antifertility effect of *Tinospora cordifolia* (Willd.) stem extract in male rats. *Indian J. Exp. Biol.* 41(8):885-889.

Kapoor, L.D. 2001. *Handbook of Ayurvedic medicinal plants*. Boca Raton, FL: CRC Press.

Kar, A., B.K. Choudhary, and N.G. Bandyopadhyay. 2003. Comparative evaluation of hypoglycaemic activity of some Indian medicinal plants in alloxan diabetic rats. *J. Ethnopharmacol.* 84(1):105-108.

Manjrekar, P.N., C.I. Jolly, and S. Narayanan. 2000. Comparative studies of the immunomodulatory activity of *Tinospora cordifolia* and *Tinospora sinensis*. *Fitoterapia* 71(3):254-257.

Nayampalli, S.S., S.S. Ainapure, B.D. Samant, et al. 1988. A comparative study of diuretic effects of *Tinospora cordifolia* and hydrochlorothiazide in rats and a preliminary phase I study in human volunteers. *J. Postgrad. Med.* 34(4):233-236.

Panchabhai, T.S., U.P. Kulkarni, and N.N. Rege. 2008. Validation of therapeutic claims of *Tinospora cordifolia*: A review. *Phytother. Res.* 22(4):425-441.

Remington, J.P., and H.C. Wood. 1918. *The dispensatory of the United States of America.* 20th ed. Philadelphia: Lippincott.

Stanely, P., M. Prince, and V.P. Menon. 2000. Hypoglycaemic and other related actions of *Tinospora cordifolia* roots in alloxan-induced diabetic rats. *J. Ethnopharmacol.* 70(1):9-15.

Stanely, P., M. Prince, and V.P. Menon. 2003. Hypoglycaemic and hypolipidaemic action of alcohol extract of *Tinospora cordifolia* roots in chemical induced diabetes in rats. *Phytother. Res.* 17(4):410-413.

Tribulus terrestris L. — Zygophyllaceae

SCN: tribulus
AN: *gokshura*
PN: *bai ji li* (fruit)
OCN: puncturevine caltrop; small caltrops
Part: fruit

Quick Reference Summary

Safety Class: 2b
Interaction Class: A

Contraindications

Not for use in pregnancy except under the supervision of a qualified healthcare practitioner (Bensky et al. 2004; Chen and Chen 2004).

Other Precautions

None known.

Drug and Supplement Interactions

None known.

Notice

Diuretic (Chen and Chen 2004; Wright et al. 2007); *see* Appendix 2.

Editors' Note

Tribulus fruit contains trace amounts of β-carboline alkaloids (e.g., harmane, harmine, and harmol). Analyses indicate that the content of these alkaloids in the fruit is approximately 0.000045% (Tsuchiya et al. 1999).

Adverse Events and Side Effects

Topical application of tribulus fruit has been associated with rash and itching (Chen and Chen 2004).

Allergic reactions, including anaphylactic reactions, to tribulus fruit have been reported (Bensky et al. 2004; Chen and Chen 2004).

Pharmacological Considerations

A 20-day human study on a tribulus product reported to contain both tribulus fruit extract and tribulus aerial parts showed a temporary increase in testosterone levels after the first 10 days (Milašius et al. 2010).

Pregnancy and Lactation

A text on traditional Chinese medicine indicates that tribulus fruit should be used with caution during pregnancy (Bensky et al. 2004), while another text indicates that it may cause miscarriage and should not be used during pregnancy (Chen and Chen 2004). Due to the above cautions listed in the traditional Chinese literature, use should be discontinued at the time of attempted conception.

No information on the safety of tribulus fruit during lactation was identified in the scientific or traditional literature. While this review did not identify any concerns for use while nursing, safety has not been conclusively established.

Review Details

I. Drug and Supplement Interactions

Clinical Trials of Drug or Supplement Interactions

No clinical trials of drug or supplement interactions were identified.

Case Reports of Suspected Drug or Supplement Interactions

No case reports of suspected drug or supplement interactions were identified.

Animal Trials of Drug or Supplement Interactions

No animal trials of drug or supplement interactions were identified.

II. Adverse Events

Case Reports of Adverse Events

Overdose of tribulus fruit (standard dose is a decoction of 6–15 g) has been associated with dizziness, weakness, drowsiness, nausea, vomiting, heart palpitations, increased breathing and heart rates, and blue coloration of the skin. In severe cases of overdose, pulmonary edema and respiratory failure have been reported (Bensky et al. 2004).

III. Pharmacology and Pharmacokinetics

Human Pharmacological Studies

In healthy volunteers orally administered 625 mg of a tribulus product three times daily for 20 days, an increase in blood levels of testosterone was observed after the first 10 days of treatment, but the levels returned to near baseline at the end of the study (Milašius et al. 2010). The tested product is reported to contain a proprietary blend of tribulus fruit extract and tribulus aerial parts (Optimum Nutrition 2012).

Animal Pharmacological Studies

Hypotensive activity of tribulus was observed in rats with high blood pressure orally administered 10 mg/kg of a freeze-dried aqueous extract of tribulus fruit daily for 4 weeks (Sharifi et al. 2003).

Diuretic activity of tribulus was observed in rats orally administered 5 g/kg of an aqueous extract of a combination of tribulus fruit and leaf (Al-Ali et al. 2003).

In Vitro Pharmacological Studies

A methanol extract of tribulus fruit was shown to inhibit prostaglandin E_2 production (Hong et al. 2002).

IV. Pregnancy and Lactation

A text on traditional Chinese medicine indicates that tribulus fruit should be used with caution during pregnancy (Bensky et al. 2004), while another text indicates that it may cause miscarriage and should not be used during pregnancy (Chen and Chen 2004).

No information on the safety of tribulus fruit during lactation was identified.

V. Toxicity Studies

No toxicity studies were identified.

Literature Cited

Al-Ali, M., S. Wahbi, H. Twaij, and A. Al-Badr. 2003. *Tribulus terrestris*: Preliminary study of its diuretic and contractile effects and comparison with *Zea mays*. *J. Ethnopharmacol.* 85(2-3):257-260.

Bensky, D., S. Clavey, and E. Stöger. 2004. *Chinese herbal medicine: Materia medica*. 3rd ed. Seattle: Eastland Press.

Chen, J.K., and T.T. Chen. 2004. *Chinese medical herbology and pharmacology*. City of Industry, CA: Art of Medicine Press.

Hong, C.H., S.K. Hur, O.J. Oh, et al. 2002. Evaluation of natural products on inhibition of inducible cyclooxygenase (COX-2) and nitric oxide synthase (iNOS) in cultured mouse macrophage cells. *J. Ethnopharmacol.* 83(1-2):153-159.

Milašius, K., M. Pečiukonienė, R. Dadelienė, and J. Skernevičius. 2010. Efficacy of the tribulus food supplement used by athletes. *Acta Med. Lituanica.* 17(1-2):65-70.

Optimum Nutrition. 2012. Product label accessed April 19, 2012: http://www.optimumnutrition.com/products/images/tribulus-facts.jpg.

Sharifi, A.M., R. Darabi, and N. Akbarloo. 2003. Study of antihypertensive mechanism of *Tribulus terrestris* in 2K1C hypertensive rats: Role of tissue ACE activity. *Life Sci.* 73(23):2963-2971.

Tsuchiya, H., H. Shimizu, and M. Iinuma. 1999. Beta-carboline alkaloids in crude drugs. *Chem. Pharm. Bull.* 47(3):440-443.

Wright, C.I., L. Van-Buren, C.I. Kroner, and M.M.G. Koning. 2007. Herbal medicines as diuretics: A review of the scientific evidence. *J. Ethnopharmacol.* 114(1):1-31.

T

Tribulus terrestris L. — Zygophyllaceae

SCN: tribulus
AN: *gokshura*
OCN: puncturevine caltrop; small caltrops
Part: above-ground parts

Quick Reference Summary

Safety Class: 2b
Interaction Class: A

Contraindications

None known.

Other Precautions

None known.

Drug and Supplement Interactions

None known.

Adverse Events and Side Effects

Animals grazing on live tribulus stems and leaves sometimes develop a condition known as geeldikkop, which includes photosensitivity, jaundice, and facial edema (Aslani et al. 2004; Smith and Henderson 1991). No such activity has been associated with human consumption of tribulus.

Pharmacological Considerations

Human studies on the effect of tribulus on reproductive hormone levels (testosterone and related hormones) have provided conflicting results. One study reported to contain both tribulus fruit extract and tribulus aerial parts showed a temporary increase in testosterone levels after the first 10 days (Milašius et al. 2010). Other human studies on products derived just from the aerial parts have shown no effects on hormone levels (Neychev and Mitev 2005; Saudan et al. 2008). Animal studies of tribulus aerial parts have also provided conflicting results (Gauthaman and Ganesan 2008; Martino-Andrade et al. 2010).

A 20-day human study on a tribulus product reported to contain both tribulus fruit extract and tribulus aerial parts showed a temporary increase in testosterone levels after the first 10 days (Milašius et al. 2010; Optimum Nutrition 2012).

Pregnancy and Lactation

Tribulus is reported to sometimes be used to promote fertility (Romm 2010).

No information on the safety of tribulus aboveground parts during lactation was identified in the scientific or traditional literature. While this review did not identify any concerns for use while nursing, safety has not been conclusively established.

Review Details

I. Drug and Supplement Interactions

Clinical Trials of Drug or Supplement Interactions

No clinical trials of drug or supplement interactions were identified.

Case Reports of Suspected Drug or Supplement Interactions

No case reports of suspected drug or supplement interactions were identified.

Animal Trials of Drug or Supplement Interactions

No animal trials of drug or supplement interactions were identified.

II. Adverse Events

Case Reports of Adverse Events

A case of gynecomastia was reported in a 21-year-old male, identified as a "keen weight-trainer," who was reportedly taking tribulus. The specific product used, dose, and duration of use were not specified, though the product was described as derived from the aboveground parts of the plant. Testing revealed decreased levels of follicle-stimulating hormone, luteinizing hormone, and testosterone that returned to normal after cessation of the product. The man had a history of atypical ductal hyperplasia, although it was not clear from the case report whether the hyperplasia began during or prior to tribulus use (Jameel et al. 2004).

T

III. Pharmacology and Pharmacokinetics

Human Pharmacological Studies

In healthy young men orally administered 10 or 20 mg/kg daily for 4 weeks of a tribulus extract standardized to 60% steroid saponins, no changes in serum hormone levels were observed, including testosterone, androstenedione, or luteinizing hormone (Neychev and Mitev 2005).

Conversely, oral administration in thirty-two 20- to 22-year-old athletes of 625 mg of a tribulus product three times daily for 20 days measured an increase in blood levels of testosterone after the first 10 days of treatment; the levels returned to near baseline at the end of the study (Milašius et al. 2010). The tested product is reported to contain a proprietary blend of tribulus fruit extract and tribulus aerial parts (Optimum Nutrition 2012).

No changes in urinary levels of hormones were observed in two healthy female volunteers who ingested 500 mg of a tribulus extract three times daily for 2 days (Saudan et al. 2008).

Animal Pharmacological Studies

In baboons and rhesus monkeys intravenously administered single doses of 7.5, 15, or 30 mg/kg of an extract of the aerial parts of tribulus, a brief rise in testosterone level 15 minutes after administration was observed, followed by a return to near but above control levels. A similar pattern was observed for dihydrotestosterone, whereas serum levels of dehydroepiandrosterone sulfate were generally higher than control. For all three hormones, the effects of the tribulus extract were not dose dependent. No significant changes in blood pressure occurred. Electrocardiograms indicated no significant changes in heart rate or beat (Gauthaman and Ganesan 2008).

In rats and rabbits orally administered 2.5, 5, or 10 mg/kg of the same tribulus extract daily for 8 weeks, an increase in dihydrotestosterone was observed in rabbits at the 5 and 10 mg/kg doses. In castrated rats orally administered 5.5 mg/kg of the tribulus extract daily for 8 weeks, an increase in serum testosterone levels was observed (Gauthaman and Ganesan 2008).

No changes in levels of androgen hormones were observed in male rats or ovariectomized female rats orally administered 11, 42, or 110 mg/kg of a dry ethanolic extract of tribulus aerial parts daily for 28 days, or in castrated male rats for 7 days (Martino-Andrade et al. 2010).

Sheep and goats grazing on live tribulus plants sometimes develop a condition known as geeldikkop, which includes photosensitivity, jaundice, and facial edema. Biliary obstruction occurs due to accumulation of microliths high in sulfur and lower in sodium and potassium (Aslani et al. 2004; Smith and Henderson 1991).

Diuretic activity of tribulus was observed in rats orally administered 5 g/kg of an aqueous extract of a combination of tribulus fruit and leaf (Al-Ali et al. 2003).

In Vitro Pharmacological Studies

No relevant in vitro pharmacological studies were identified.

IV. Pregnancy and Lactation

In sheep fed hay containing tribulus herb on days 103–112 of pregnancy, no adverse effects on maternal health or pregnancy outcome were observed. The incidences of fetal breathing movements were lower in the tribulus group as compared to the control group, and the breathing movements did not show the typical day-night variation (Walker et al. 1992).

No information on the safety of tribulus aboveground parts during lactation was identified.

V. Toxicity Studies

No toxicity studies were identified.

Literature Cited

Al-Ali, M., S. Wahbi, H. Twaij, and A. Al-Badr. 2003. *Tribulus terrestris*: Preliminary study of its diuretic and contractile effects and comparison with *Zea mays*. *J. Ethnopharmacol.* 85(2-3):257-260.

Aslani, M.R., A.R. Movassaghi, M. Mohri, V. Ebrahim-Pour, and A.N. Mohebi. 2004. Experimental *Tribulus terrestris* poisoning in goats. *Small Rumin. Res.* 51(3):261-267.

Gauthaman, K., and A.P. Ganesan. 2008. The hormonal effects of *Tribulus terrestris* and its role in the management of male erectile dysfunction—An evaluation using primates, rabbit and rat. *Phytomedicine* 15(1-2):44-54.

Jameel, J.K., P.J. Kneeshaw, V.S. Rao, and P.J. Drew. 2004. Gynaecomastia and the plant product *"Tribulis terrestris."* *Breast* 13(5):428-430.

Martino-Andrade, A.J., R.N. Morais, K.M. Spercoski, et al. 2010. Effects of *Tribulus terrestris* on endocrine sensitive organs in male and female Wistar rats. *J. Ethnopharmacol.* 127(1):165-170.

Milašius, K., M. Pečiukonienė, R. Dadelienė, and J. Skernevičius. 2010. Efficacy of the tribulus food supplement used by athletes. *Acta Med. Lituanica.* 17(1-2):65-70.

Neychev, V.K., and V.I. Mitev. 2005. The aphrodisiac herb *Tribulus terrestris* does not influence the androgen production in young men. *J. Ethnopharmacol.* 101(1-3):319-323.

Optimum Nutrition. 2012. Product label accessed April 19, 2012: http://www.optimumnutrition.com/products/images/tribulus-facts.jpg.

Romm, A.J. 2010. *Botanical medicine for women's health.* New York: Churchill Livingstone.

Saudan, C., N. Baume, C. Emery, E. Strahm, and M. Saugy. 2008. Short term impact of *Tribulus terrestris* intake on doping control analysis of endogenous steroids. *Forensic Sci. Int.* 178(1):e7-e10.

Smith, J.E., and R.S. Henderson. 1991. *Mycotoxins and animal foods.* Boca Raton, FL: CRC Press.

Walker, D., A. Bird, T. Flora, and B. O'Sullivan. 1992. Some effects of feeding *Tribulus terrestris*, *Ipomoea lonchophylla* and the seed of *Abelmoschus ficulneus* on fetal development and the outcome of pregnancy in sheep. *Reprod. Fertil. Dev.* 4(2):135-144.

T

Trichosanthes kirilowii Maxim. — Cucurbitaceae

SCN: trichosanthes
PN: *gua lou* (fruit); *gua lou pi* (fruit rind); *gua lou ren* (seed)
OCN: Chinese cucumber; Mongolian snakegourd
Part: fruit, fruit rind, seed

Quick Reference Summary

Safety Class: 1
Interaction Class: A

Contraindications

None known.

Other Precautions

None known.

Drug and Supplement Interactions

None known.

Adverse Events and Side Effects

Occasional cases of mild diarrhea and gastric discomfort have been reported in persons taking tablets of trichosanthes fruit (Chang and But 1986).

Unprocessed trichosanthes seed may cause nausea. The seed is typically processed by dry-frying to reduce the frequency of nausea (Chen and Chen 2004).

Pharmacological Considerations

None known.

Pregnancy and Lactation

No information on the safety of trichosanthes in pregnancy or lactation was identified in the scientific or traditional literature. Although this review did not identify any concerns for use while pregnant or nursing, safety has not been conclusively established.

Review Details

I. Drug and Supplement Interactions

Clinical Trials of Drug or Supplement Interactions

No clinical trials of drug or supplement interactions were identified.

Case Reports of Suspected Drug or Supplement Interactions

No case reports of suspected drug or supplement interactions were identified.

Animal Trials of Drug or Supplement Interactions

No animal trials of drug or supplement interactions were identified.

II. Adverse Events

Case Reports of Adverse Events

Occasional cases of mild diarrhea and gastric discomfort have been reported in persons taking tablets of trichosanthes fruit (Chang and But 1986).

Overdose of the seed (standard dose is a decoction of 10–15 g) may result in gastric discomfort, nausea, vomiting, abdominal pain, and diarrhea (Chang and But 1986).

Unprocessed trichosanthes seed may cause nausea. The seed is typically processed by dry-frying to reduce the frequency of nausea (Chen and Chen 2004).

III. Pharmacology and Pharmacokinetics

Human Pharmacological Studies

No relevant human pharmacological studies were identified.

Animal Pharmacological Studies

No relevant animal pharmacological studies were identified.

In Vitro Pharmacological Studies

No relevant in vitro pharmacological studies were identified.

IV. Pregnancy and Lactation

No information on the safety of trichosanthes during pregnancy or lactation was identified.

V. Toxicity Studies

Acute Toxicity

The LD_{50} of trichosanthes fruit in mice is 363 g/kg after intraperitoneal administration and 306 g/kg after intravenous administration (Chen and Chen 2004; Zhu 1998).

Literature Cited

Bensky, D., S. Clavey, and E. Stöger. 2004. *Chinese herbal medicine: Materia medica*. 3rd ed. Seattle: Eastland Press.

Chang, H.-M., and P.P.H. But. 1986. *Pharmacology and applications of Chinese materia medica*. English ed. Philadelphia: World Scientific.

Chen, J.K., and T.T. Chen. 2004. *Chinese medical herbology and pharmacology*. City of Industry, CA: Art of Medicine Press.

Zhu, Y.-P. 1998. *Chinese materia medica: Chemistry, pharmacology and applications.* Amsterdam: Harwood Academic Publishers.

Trifolium pratense L.

Fabaceae

SCN: red clover

Part: flower, herb

Quick Reference Summary

Safety Class: 1

Interaction Class: A

Contraindications

None known.

Other Precautions

None known.

Drug and Supplement Interactions

None known.

Editors' Notes

Red clover herb and flower contain approximately 0.17% isoflavones (DerMarderosian and Beutler 2005), with the herb generally containing a higher percent isoflavone content than the flower (Booth et al. 2006). Commercially available red clover products include products with this normal percentage of isoflavones and those modified to contain up to 100% isoflavones. Products with modified concentrations of selected compounds may be expected to have different physiological effects than traditional preparations of the herb. Information on both red clover and red clover derived isoflavones is provided in this entry.

Adverse Events and Side Effects

Adverse events reported in clinical trials of red clover products were minor and similar in the placebo and red clover groups (Coon et al. 2007; Lethaby et al. 2007; Low Dog 2005; Nelson et al. 2006).

Pharmacological Considerations

Red clover is commonly cited as a potential anticoagulant due to the presence of coumarins and confusion between coumarins and the anticoagulant drug coumadin (sometimes referred to as coumarin) (Abebe 2002; Booth et al. 2004; Fugh-Berman and Kronenberg 2001; Heck et al. 2000). However, a screening of red clover for 17 naturally occurring coumarins identified 5 coumarin compounds in red clover, 1 with anticoagulant activity, 1 with procoagulant activity, and 3 with no activity reported (Booth et al. 2004). No changes in bleeding or blood clotting were observed in women after 30 days of red clover supplementation (Booth et al. 2004).

Red clover contains isoflavones, compounds that are structurally similar to the human hormone estradiol and capable of binding to estrogen receptors (Umland et al. 2000). Some studies have indicated that red clover isoflavones have a greater affinity for the estrogen receptor β (found primarily in bone, brain, heart, and vascular system) than estrogen receptor α (found primarily in uterus, breast, ovaries, and adrenal glands) (Beck et al. 2005; Dornstauder et al. 2001).

In human studies, conflicting data on the estrogenic activity of red clover have been reported. Some studies have shown no effect on endometrial thickness or vaginal cytology, indicating a lack of estrogenic activity (Baber et al. 1999; Clifton-Bligh et al. 2001; Powles et al. 2008), whereas others showed a decrease in (Imhof et al. 2006) or no effect on (Hale et al. 2001) endometrial thickness or a significant improvement in vaginal cytology (Hidalgo et al. 2005).

Pregnancy and Lactation

Limited information on the safety of red clover in pregnancy or lactation was identified in the scientific and traditional literature. Although this review did not identify any concerns for use while pregnant or nursing, safety has not been conclusively established.

Review Details

I. Drug and Supplement Interactions

Clinical Trials of Drug or Supplement Interactions

No clinical trials of drug or supplement interactions were identified.

Case Reports of Suspected Drug or Supplement Interactions

No case reports of suspected drug or supplement interactions were identified.

Animal Trials of Drug or Supplement Interactions

No animal trials of drug or supplement interactions were identified.

II. Adverse Events

Adverse Events Reported in Clinical Trials

In a meta-analysis of clinical trials of monopreparations containing red clover isoflavones, the reviewers noted that there was no apparent evidence of adverse events during short-term use of red clover, but long-term studies were lacking. Doses of products in the meta-analysis ranged from 40 to 82 mg daily (Coon et al. 2007).

A second meta-analysis, including six placebo-controlled red clover isoflavone trials, with doses ranging from 40 to 80 mg daily, and study duration from 12 to 16 weeks, concluded that adverse events did not differ between isoflavone and placebo groups, although the events were not well characterized in several trials. Gastrointestinal symptoms were generally the most common adverse events in both isoflavone and placebo groups (Nelson et al. 2006).

A systematic review of phytoestrogens for vasomotor menopausal symptoms, including seven studies on red clover-derived isoflavone extracts, concluded that phytoestrogen products do not appear to have an estrogen agonistic effect on the endometrium when given for up to 1 year. The authors noted that the long-term endometrial safety of high doses of phytoestrogen supplements has not been fully established (Lethaby et al. 2007). Similarly, a review of red clover clinical trials indicated that no adverse effects have been reported, but that the question of safety in hormone-sensitive tissue is important, especially when considering high doses over long periods of time (Low Dog 2005).

Case Reports of Adverse Events

No case reports of adverse reactions were identified.

III. Pharmacology and Pharmacokinetics

Human Pharmacological Studies

No increase in mammographic breast density and no significant effects were observed on estradiol, follicle-stimulating hormone, or luteinizing hormone levels in a trial of women, ages 49 to 65, taking 44 mg red clover-derived isoflavones daily for 1 year (Atkinson et al. 2004).

No significant differences in vaginal cytology, endometrial thickness, serum estradiol, follicle-stimulating hormone, or sex hormone-binding globulin were observed after administration of 40 mg daily of a red clover-based isoflavone extract (Baber et al. 1999). No increase in endometrial thickness was observed after administration of doses up to 85.5 mg daily of a red clover isoflavone preparation for 6 months (Clifton-Bligh et al. 2001). A decrease in endometrial thickness and increase in plasma testosterone levels were observed in postmenopausal women after administration of 80 mg red clover-derived isoflavones daily for 90 days (Imhof et al. 2006). After administration of 50 g daily of a red clover-derived isoflavone extract to perimenopausal women for 3 months, no evidence of an antiproliferative effect in the Ki-67 proliferative marker of endometrial biopsies was found (Hale et al. 2001). In postmenopausal women administered 80 mg daily of a red clover isoflavone supplement for 90 days, significant improvement in all vaginal cytology indices (karyopyknotic index, cornification index, maturation index), as compared to placebo, was observed (Hidalgo et al. 2005).

In women aged 35 to 70 with at least one first-degree relative with breast cancer, administration of 40 mg of red clover isoflavones daily for 3 years did not result in any significant differences in breast density, endometrial thickness, serum cholesterol, follicle-stimulating hormone levels, or bone mineral density, as compared to placebo. In postmenopausal women, some differences in bone marker levels were seen between active and placebo groups at 6 and 12 months (Powles et al. 2008).

In a noncontrolled clinical study of postmenopausal women administered 80 mg daily of red clover-derived isoflavone extract for 6 months, some changes in endometrial activity but no changes in endometrial thickness were observed. Of the 32 study participants, 6 presented with vaginal bleeding and 3 presented with endometrial alteration as compared to the initial exams, 2 developed endometrial cell proliferation, and 1 developed endometrial hyperplasia (Wolff et al. 2006).

A review of studies on red clover and soy-derived isoflavones concluded that 2 mg/kg was a safe daily dose of isoflavones for most populations (Barnes 2003). A second review of studies on red clover and soy-derived isoflavones indicated that 40–50 mg of isoflavones is recommended as the daily dose. This recommendation was based on the daily intake of isoflavones in a traditional Japanese diet (Beck et al. 2005).

No change in prothrombin time or INR time was found after administration of 400 mg daily of red clover extract to postmenopausal women for 30 days (Booth et al. 2004).

Animal Pharmacological Studies

In ovariectomized rats administered 250, 500, or 750 mg daily of a red clover extract (15% isoflavones) for 21 days, a dose-dependent increase in uterine weight and differentiated vaginal cells were observed at the two higher doses, but no stimulation of cell proliferation was observed in the mammary glands. Neither antiestrogenic nor additive estrogenic properties were observed in any of the tissues studied (Burdette et al. 2002).

In ovariectomized rats administered 20 or 40 mg daily of red clover-derived total isoflavones for 14 weeks, treatment with isoflavones significantly increased bone mineral content, mechanical strength of the tibia, femoral weight, and femoral density and prevented the rise of serum alkaline phosphatase levels. In addition, the treatment with isoflavones significantly reduced the number of osteoclasts compared with the ovariectomized control rats (Occhiuto et al. 2007).

In Vitro Pharmacological Studies

In vitro assays of red clover extracts in endometrial cells and MCF-7 (estrogen receptor-positive) breast cancer cells, differential estrogenic activity was observed in the endometrial cells, whereas nondifferential activity was observed in the MCF-7 cells, indicating the significance of the type of bioassay used to determine the estrogenic activity of red clover (Booth et al. 2006).

An extract of a red clover isoflavone preparation increased MCF-7 breast cancer cell proliferation rates (Bodinet and Freudenstein 2004). A standardized red clover isoflavone extract (9% isoflavones by dry weight) showed an affinity for both estrogen receptor (ER) α and β with a significantly stronger affinity for the ERβ receptor in a yeast two-plasmid system (Dornstauder et al. 2001).

A red clover-derived extract of 30% isoflavone aglycones was shown to bind to mu- and delta-opiate receptors in Chinese hamster ovaries (Nissan et al. 2007).

An ethanolic extract of a commercial red clover preparation inhibited the drug-metabolizing isoenzyme CYP3A4 in a fluorometric microtiter plate assay (Budzinski et al. 2000). Genistein and daidzein, metabolites of the predominant red clover isoflavones biochanin A and formononetin, were shown to inhibit the enzyme CYP1B1 (Roberts et al. 2004).

IV. Pregnancy and Lactation

In a study of amniotic fluid samples in women between weeks 15 and 23 of pregnancy, isoflavonoids were detected in 92% of samples. The isoflavonoids daidzein, genistein, formononetin, biochanin A, and coumestrol were detected (Foster et al. 2002).

No information was identified on the safety of red clover use during lactation.

V. Toxicity Studies

Short-Term Toxicity

In ovariectomized sheep, feeding with 3.5 kg daily of red clover silage for 14 days as the sole source of food (81–95 mg/kg daily total phytoestrogens) resulted in increased plasma concentrations of T_3 and T_4 and increased thyroid follicle size as compared to sheep fed hay (Madej et al. 2002).

Literature Cited

Abebe, W. 2002. Herbal medication: Potential for adverse interactions with analgesic drugs. *J. Clin. Pharm. Ther.* 27(6):391-401.

Atkinson, C., R.M. Warren, E. Sala, et al. 2004. Red-clover-derived isoflavones and mammographic breast density: A double-blind, randomized, placebo-controlled trial. *Breast Cancer Res.* 6(3):R170-R179.

Baber, R.J., C. Templeman, T. Morton, G.E. Kelly, and L. West. 1999. Randomized placebo-controlled trial of an isoflavone supplement and menopausal symptoms in women. *Climacteric* 2(2):85-92.

Barnes, S. 2003. Phyto-oestrogens and osteoporosis: What is a safe dose? *Br. J. Nutr.* 89(Suppl. 1):S101-S108.

Beck, V., U. Rohr, and A. Jungbauer. 2005. Phytoestrogens derived from red clover: An alternative to estrogen replacement therapy? *J. Steroid Biochem. Mol. Biol.* 94(5):499-518.

Bodinet, C., and J. Freudenstein. 2004. Influence of marketed herbal menopause preparations on MCF-7 cell proliferation. *Menopause* 11(3):281-289.

Booth, N.L., D. Nikolic, R.B. van Breemen, et al. 2004. Confusion regarding anticoagulant coumarins in dietary supplements. *Clin. Pharmacol. Ther.* 76(6):511-516.

Booth, N.L., C.R. Overk, P. Yao, et al. 2006. Seasonal variation of red clover (*Trifolium pratense* L., Fabaceae) isoflavones and estrogenic activity. *J. Agric. Food Chem.* 54(4):1277-1282.

Budzinski, J.W., B.C. Foster, S. Vandenhoek, and J.T. Arnason. 2000. An *in vitro* evaluation of human cytochrome P450 3A4 inhibition by selected commercial herbal extracts and tinctures. *Phytomedicine* 7(4):273-282.

Burdette, J.E., J. Liu, D. Lantvit, et al. 2002. *Trifolium pratense* (red clover) exhibits estrogenic effects *in vivo* in ovariectomized Sprague-Dawley rats. *J. Nutr.* 132(1):27-30.

Clifton-Bligh, P.B., R.J. Baber, G.R. Fulcher, M.L. Nery, and T. Moreton. 2001. The effect of isoflavones extracted from red clover (Rimostil) on lipid and bone metabolism. *Menopause* 8(4):259-265.

Coon, J.T., M.H. Pittler, and E. Ernst. 2007. *Trifolium pratense* isoflavones in the treatment of menopausal hot flushes: A systematic review and meta-analysis. *Phytomedicine* 14(2-3):153-159.

DerMarderosian, A., and J. Beutler. 2005. *The review of natural products.* 4th ed. St Louis: Facts and Comparisons.

Dornstauder, E., E. Jisa, I. Unterrieder, et al. 2001. Estrogenic activity of two standardized red clover extracts (Menoflavon) intended for large scale use in hormone replacement therapy. *J. Steroid Biochem. Mol. Biol.* 78(1):67-75.

Foster, W.G., S. Chan, L. Platt, and C.L. Hughes, Jr. 2002. Detection of phytoestrogens in samples of second trimester human amniotic fluid. *Toxicol. Lett.* 129(3):199-205.

Fugh-Berman, A., and F. Kronenberg. 2001. Red clover (*Trifolium pratense*) for menopausal women: Current state of knowledge. *Menopause* 8(5):333-337.

Hale, G.E., C.L. Hughes, S.J. Robboy, S.K. Agarwal, and M. Bievre. 2001. A double-blind randomized study on the effects of red clover isoflavones on the endometrium. *Menopause* 8(5):338-346.

Heck, A.M., B.A. DeWitt, and A.L. Lukes. 2000. Potential interactions between alternative therapies and warfarin. *Am. J. Health Syst. Pharm.* 57(13):1221-1227; quiz 1228-1230.

Hidalgo, L.A., P.A. Chedraui, N. Morocho, S. Ross, and G. San Miguel. 2005. The effect of red clover isoflavones on menopausal symptoms, lipids and vaginal cytology in menopausal women: A randomized, double-blind, placebo-controlled study. *Gynecol. Endocrinol.* 21(5):257-264.

Imhof, M., A. Gocan, F. Reithmayr, et al. 2006. Effects of a red clover extract (MF11RCE) on endometrium and sex hormones in postmenopausal women. *Maturitas* 55(1):76-81.

Lethaby, A., J. Brown, J. Marjoribanks, et al. 2007. Phytoestrogens for vasomotor menopausal symptoms. *Cochrane Database Syst. Rev.* 4:CD001395.

Low Dog, T. 2005. Menopause: A review of botanical dietary supplements. *Am. J. Med.* 118(Suppl. 12B):98-108.

Madej, A., E. Persson, T. Lundh, and Y. Ridderstrale. 2002. Thyroid gland function in ovariectomized ewes exposed to phytoestrogens. *J. Chromatogr. B Anal. Technol. Biomed. Life Sci.* 777(1-2):281-287.

Nelson, H.D., K.K. Vesco, E. Haney, et al. 2006. Nonhormonal therapies for menopausal hot flashes: Systematic review and meta-analysis. *J. Am. Med. Assoc.* 295(17):2057-2071.

Nissan, H.P., J. Lu, N.L. Booth, et al. 2007. A red clover (*Trifolium pratense*) phase II clinical extract possesses opiate activity. *J. Ethnopharmacol.* 112(1):207-210.

Occhiuto, F., R. De Pasquale, G. Guglielmo, et al. 2007. Effects of phytoestrogenic isoflavones from red clover (*Trifolium pratense* L.) on experimental osteoporosis. *Phytother. Res.* 21(2):130-134.

Powles, T.J., A. Howell, D.G. Evans, et al. 2008. Red clover isoflavones are safe and well tolerated in women with a family history of breast cancer. *Menopause Int.* 14(1):6-12.

Roberts, D.W., D.R. Doerge, M.I. Churchwell, et al. 2004. Inhibition of extrahepatic human cytochromes P450 1A1 and 1B1 by metabolism of isoflavones found in *Trifolium pratense* (red clover). *J. Agric. Food Chem.* 52(21):6623-6632.

Umland, E.M., J.S. Cauffield, J.K. Kirk, and T.E. Thomason. 2000. Phytoestrogens as therapeutic alternatives to traditional hormone replacement in postmenopausal women. *Pharmacotherapy* 20(8):981-990.

Wolff, L.P., M.R. Martins, A.J. Bedone, and I.M. Monteiro. 2006. [Endometrial evaluation in menopausal women after six months of isoflavones]. *Rev. Assoc. Med. Bras.* 52(6):419-423.

Trigonella foenum-graecum L. — Fabaceae

SCN: fenugreek
AN: *methi; methika*
PN: *hu lu ba* (seed)
Part: seed

Quick Reference Summary

Safety Class: 2b
Interaction Class: A

Contraindications

Not for use in pregnancy except under the supervision of a qualified healthcare practitioner (Friedman et al. 2007; Kassem et al. 2006; Petropoulos 2002; Sethi et al. 1990).

Other Precautions

None known.

Drug and Supplement Interactions

Other drugs should be taken 1 hour prior to consumption of fenugreek or several hours after consumption, as the mucilage present in fenugreek may slow the absorption of orally administered drugs (Leung and Foster 1996; Petropoulos 2002).

Notice

Mucilages (17–22%) (Leung and Foster 1996; Petropoulos 2002); *see* Appendix 3.

Editors' Notes

The classifications and concerns for this herb are based on relatively higher doses used for therapeutic purposes in contrast to lower amounts generally used in cooking, and have not been associated with its use as a spice.

Fenugreek contains steroidal saponins (0.28–2.2%), notably diosgenin and yamogenin (Taylor et al. 2002; Petropoulos 2002).

Adverse Events and Side Effects

Infants administered large quantities of fenugreek, either directly or through maternal ingestion, have been observed to smell of maple syrup, which has led in some cases to false suspicion of maple syrup urine disease, a metabolic disease characterized by maple-scented urine (Bartley et al. 1981; Korman et al. 2001; Sewell et al. 1999). The compound sotolon is responsible for the maple scent (Korman et al. 2001).

Allergic reactions to fenugreek have been reported (Patil et al. 1997).

Pharmacological Considerations

Human studies have demonstrated that fenugreek may modify glucose regulation (Bordia et al. 1997; Gupta et al. 2001; Madar et al. 1988; Ragurham et al. 1994; Sharma and Raghuram 1990; Sharma et al. 1990). People with diabetes are advised to monitor their blood sugar closely and discuss the use of this herb with a qualified healthcare practitioner prior to use.

Purported anticoagulant activity of fenugreek has been widely cited due to the presence of coumarins (Abebe 2002; Czigle et al. 2006; Heck et al. 2000; Holbrook et al. 2005; Izzo 2005; Izzo et al. 2005; Juurlink 2007; Myers 2002). This effect has not been observed in human or animal studies at standard therapeutic doses (Bordia et al. 1997; Sharma et al. 1996), although some anticoagulant effects have been observed in rats at doses many times the standard therapeutic dose (Bordia et al. 1997; Hannan et al. 2003; Xue et al. 2007).

Pregnancy and Lactation

Information on the use of fenugreek during pregnancy is conflicting. No adverse effects were observed in rats fed fenugreek as 5 or 20% of the diet (Mital and Gopaldas 1986), while some adverse effects on the fetus were observed in rabbits fed fenugreek as 30% of the diet (Kassem et al. 2006). Some antifertility or abortifacient properties have been attributed to fenugreek in surveys of herbs used in India as antifertility and abortifacient agents (Casey 1960; Nath et al. 1997), although no such activities are recognized in the traditional Chinese medicine literature (Bensky et al. 2004; Chen and Chen 2004). Fenugreek contains steroidal saponins (0.28–2.2%), notably diosgenin and yamogenin (Friedman et al. 2007; Petropoulos 2002).

Fenugreek has traditionally been used in a number of cultures to stimulate milk production in lactating women (Tiran 2003).

Also see Adverse Events and Side Effects for this entry.

T

Review Details

I. Drug and Supplement Interactions

Clinical Trials of Drug or Supplement Interactions

No clinical trials of drug or supplement interactions were identified.

Case Reports of Suspected Drug or Supplement Interactions

No case reports of interactions in patients taking fenugreek monopreparations were identified. One case of increased INR was reported in a patient taking warfarin, boldo (*Peumus boldus*), and fenugreek (Lambert and Cormier 2001). This case report and the coumarin content of fenugreek seed have led to frequent warnings regarding concomitant use of warfarin and fenugreek (Abebe 2002; Czigle et al. 2006; Heck et al. 2000; Holbrook et al. 2005; Izzo 2005; Izzo et al. 2005; Juurlink 2007; Myers 2002).

Animal Trials of Drug or Supplement Interactions

No animal trials of drug or supplement interactions were identified.

II. Adverse Events

Adverse Events Reported in Clinical Trials

In a systematic review of human studies on the effects of fenugreek seed powder and defatted seed powder in persons with diabetes mellitus, adverse events were listed as "not reported" in three studies, while no side effects were observed in a fourth study (Yeh et al. 2003).

Case Reports of Adverse Events

The scent of maple syrup has been observed from the skin and urine of several infants, in some cases leading to a false suspicion of maple syrup urine disease, a metabolic disease characterized by maple-scented urine. Two infants were administered teas containing fenugreek seed, and the mother of a third infant had ingested a fenugreek seed paste at the onset of labor (Bartley et al. 1981; Korman et al. 2001; Sewell et al. 1999). Persons consuming fenugreek sometimes have the characteristic odor of maple syrup due to the presence of the compound sotolon, an aromatic compound present in fenugreek that is used in the production of artificial maple syrup (Korman et al. 2001).

Two cases of allergic reaction to fenugreek have been reported. In the first case, inhalation of fenugreek seed powder resulted in rhinorrhea, wheezing, and fainting. In the second case, a patient with chronic asthma developed numbness of the head, facial angioedema, and wheezing after application of fenugreek paste to her scalp. Skin scratch tests indicated that both patients had a strong sensitivity to fenugreek and chickpeas (Patil et al. 1997).

III. Pharmacology and Pharmacokinetics

Human Pharmacological Studies

Studies on the effects of fenugreek seed and defatted fenugreek seed have shown beneficial effects on blood sugar levels in persons with type 1 or 2 diabetes mellitus. Increases in glucose tolerance test (GTT) scores and serum clearance rates of glucose were observed in persons with type 2 diabetes controlled on glibenclamide and administered 25 g/day of fenugreek seed powder (added to the diet in the form of fiber) for 15 days (Ragurham et al. 1994). Improvements in fasting blood glucose (FBG), GTT, and 24-hour urinary glucose excretion test were observed in type 2 diabetes patients in a study in which 100 g/day of defatted fenugreek seed powder was administered for 10 days (Sharma and Raghuram 1990). In persons with type 2 diabetes, administration of 1 g daily of a hydroalcoholic extract of fenugreek seed or standard dietary discretion for 2 months resulted in a decrease in FBG in both the fenugreek and dietary controlled groups (Gupta et al. 2001). A review of several case series on the efficacy of fenugreek for glucose control in type 2 diabetes patients indicated that fenugreek may improve glucose control in this population (Basch et al. 2003). In healthy patients administered 5 g/day of fenugreek seed for 3 months, no effect on blood sugar was observed. In the same study, patients with mild cases of type 2 diabetes, however, experienced a significant reduction in postprandial glucose (PPG) and FBG, while in patients with severe diabetes only a slight reduction in PPG and FBG was observed (Bordia et al. 1997). A significant reduction in PPG was observed in type 2 diabetes patients administered a single 15 g dose of fenugreek seed soaked in water. A nonsignificant reduction in plasma insulin was also observed (Madar et al. 1988).

In patients with type 1 diabetes, decreases in 24-hour urine glucose levels, GTT scores, and fasting serum glucose were observed after administration of 100 g/day of fenugreek seed powder in meals (Sharma et al. 1990).

No changes in platelet aggregation, fibrinolytic activity, or fibrinogen were observed in patients with coronary artery disease administered 5 g/day of fenugreek for 3 months (Bordia et al. 1997). In type 2 diabetic persons administered diets containing 25 g fenugreek seed daily for 24 weeks, no changes in hematological parameters were observed (Sharma et al. 1996).

Fenugreek is reported as being rich in iron (6.5 mg/100 g) and has been promoted to prevent anemia (Gopaldas 1995), although a study of anemic preschool children correlated the consumption of fenugreek (part and doses unspecified) with an increased risk of anemia. The authors of the study indicated that the mucilage content of fenugreek inhibits iron absorption (Adish et al. 1999).

Animal Pharmacological Studies

Modification of glucose regulation has been observed in a number of studies of fenugreek preparations in rats and mice (Ajabnoor and Tilmisany 1988; Hannan et al. 2003, 2007; Khosla et al. 1995; Mohammad et al. 2006; Mondal et al. 2004; Puri et al. 2002; Raju et al. 2001; Vijayakumar et al. 2005; Xue et al. 2007; Zia et al. 2001).

No significant effect on platelet aggregation was found in type 2 model diabetic rats administered 1 g/kg daily of fenugreek soluble dietary fiber fraction for 28 days, although a tendency to lower aggregation was observed (Hannan et al. 2003). In diabetic rats orally administered fenugreek extract at doses of 0.44, 0.87, or 1.74 g/kg daily for 6 weeks, the plasma viscosity, whole blood viscosity of high and low shear rates, erythrocyte sedimentation rate, whole blood reduction viscosity, and platelet conglutination were significantly reduced in diabetic rats treated with high and middle doses of fenugreek extract but not in those treated with the low dose of the extract (Xue et al. 2007).

Administration of a dried hydroalcoholic extract of fenugreek seed in male mice and rats at doses of 110 mg/kg for 15 days resulted in decreases in the serum triiodothyronine (T_3) level and T_3:T_4 ratio but caused an increase in serum thyroxine (T_4) level and body weight. The authors indicated that inhibition of T_4 to T_3 conversion was not peroxidation mediated and the inhibition in SOD activity could be the result of a decrease in the protein anabolic hormone, T_3 (Panda et al. 1999).

In Vitro Pharmacological Studies

In vitro studies were identified but omitted due to the availability of similar human and animal studies.

IV. Pregnancy and Lactation

Fenugreek was listed as one of nearly 300 reported antifertility plants identified in a review of the Ayurvedic literature (Casey 1960). Abortifacient activity was attributed to fenugreek, although reviewers of the research have noted that the definition of abortifacient was broad and included emmenagogues, ecbolics (uterine contractors), and "antimetabolites" (Mills and Bone 2005). A survey of plants used to promote abortion in northeastern India indicated that fenugreek seed was one of 14 plants commonly used; no information on efficacy was reported (Nath et al. 1997). Antifertility activity was observed in rats administered a petroleum extract of fenugreek at doses of 0.5 to 1.25 g/kg daily for days 1 to 10 of pregnancy (Adhikary et al. 1990).

In pregnant rats, intake of fenugreek seed as 5 or 20% of the diet produced no adverse effects on fertility or growth and development of the fetus (Mital and Gopaldas 1986). In pregnant rats orally administered 175 mg/kg fenugreek powder on days 1 to 10 of pregnancy, some increase in fetal abnormalities was observed (Sethi et al. 1990). Aqueous and alcoholic extracts of fenugreek produced a stimulating effect on isolated guinea pig uteri (Abdo and al-Kafawi 1969).

In male rabbits fed a diet of 30% fenugreek seed for 3 months, testis weight was reduced, and damage to the seminiferous tubules and interstitial tissues was observed. Plasma concentration of androgen hormones and sperm concentrations were halved in the treated animals. In female rabbits fed the same diet for 3 months prior to conception, a significant reduction in the number of developing fetuses was observed at 20 days of gestation. The circulating plasma progesterone concentrations at 10 and 20 days of gestation were significantly increased, with no significant effect on prebreeding estrogen levels. Histopathological changes were observed in endometrial glands (Kassem et al. 2006).

Fenugreek is considered a galactagogue and has been traditionally used by lactating women in Ayurvedic, traditional Chinese medicine, and Western medicine traditions (Tiran 2003). An increase in prolactin levels was observed in rats administered fenugreek during different phases of reproductive life, including during pregnancy and lactation. No adverse effects were reported in offspring of animals administered fenugreek (Basha et al. 1987).

Also see Case reports of adverse events for this entry.

V. Toxicity Studies

Acute Toxicity

The LD_{50} of an intraperitoneally administered ethanol extract of fenugreek in rats is 500 mg/kg (Sharma et al. 1978). The LD_{50} of the compound trigonelline orally administered to rats is 5 g/kg (Mishkinsky et al. 1974).

No signs of toxicity or mortality were observed in mice orally administered acute doses of 2 g/kg debitterized fenugreek powder or rats orally administered 5 g/kg defatted fenugreek powder (Muralidhara et al. 1999).

Short-Term Toxicity

In chicks administered crude seed saponins of fenugreek at doses of 10 mg/kg intramuscularly, 50 mg/kg intraperitoneally, 50 mg/kg subcutaneously, or 500 mg/kg orally in drinking water daily for 21 days, reductions in body weight, pathological changes in the liver and kidneys, and varying degrees of hemorrhage in the thigh and breast were observed (Nakhla et al. 1991).

In rats orally administered fenugreek extract at doses of 1.0, 1.5, or 2.0 g/kg twice a week for 4 weeks, a necrotic effect on the liver and the kidney tissues was observed at all dose levels (Effraim et al. 1999).

Subchronic Toxicity

In rats fed a diet of 0, 1, 5, or 10% debitterized fenugreek powder daily for 90 days, no effects on food intake, growth, vital organ weights, liver biochemistry, or hematological parameters were observed (Muralidhara et al. 1999). Similarly, in rats fed diets of 0, 5, 10, or 20% fenugreek seed for 90 days, no significant differences in body weight, food intake, feed efficiency, hematological parameters, or liver function tests were observed between the fenugreek and control groups (Udayasekhara Rao et al. 1996).

Chronic Toxicity

In type 2 diabetic persons administered diets containing 25 g fenugreek seed daily for 24 weeks, no renal or hepatic toxicity or changes in hematological parameters were observed, although blood urea levels were seen to decrease after week 12 (Sharma et al. 1996).

Genotoxicity

No genotoxic effects of a fenugreek seed extract were observed in standard genotoxicity assays including the reverse mutation assay, mouse lymphoma forward mutation assay, and the mouse micronucleus assay (Flammang et al. 2004).

Literature Cited

Abdo, M., and A. al-Kafawi. 1969. Experimental studies on the effect of *Trigonella foenum-graecum*. *Planta Med.* 17:14-18.

Abebe, W. 2002. Herbal medication: Potential for adverse interactions with analgesic drugs. *J. Clin. Pharm. Ther.* 27(6):391-401.

Adhikary, P., J. Banerji, D. Choudhury, et al. 1990. Anti-implantation activity of some indigenous plants in adult female rats. *Indian J. Pharmacol.* 22(1):24-25.

Adish, A.A., S.A. Esrey, T.W. Gyorkos, and T. Johns. 1999. Risk factors for iron deficiency anaemia in preschool children in northern Ethiopia. *Public Health Nutr.* 2(3):243-252.

Ajabnoor, M.A., and A.K. Tilmisany. 1988. Effect of *Trigonella foenum-graceum* on blood glucose levels in normal and alloxan-diabetic mice. *J. Ethnopharmacol.* 22(1):45-49.

Bartley, G.B., M.D. Hilty, B.D. Andreson, A.C. Clairmont, and S.P. Maschke. 1981. "Maple-syrup" urine odor due to fenugreek ingestion. *N. Engl. J. Med.* 305(8):467.

Basch, E., C. Ulbricht, G. Kuo, P. Szapary, and M. Smith. 2003. Therapeutic applications of fenugreek. *Altern. Med. Rev.* 8(1):20-27.

Basha, L.A., R.M. Hussein, M.M. Badawi, and A.M. Abdalla. 1987. Influence of *Trigonella foenum-graecum* on prolactin release in female albino rats during different phases of reproductive life. *J. Drug Res.* 17(1-2):9-16.

Bensky, D., S. Clavey, and E. Stöger. 2004. *Chinese herbal medicine: Materia medica*. 3rd ed. Seattle: Eastland Press.

Bordia, A., S.K. Verma, and K.C. Srivastava. 1997. Effect of ginger (*Zingiber officinale* Rosc.) and fenugreek (*Trigonella foenum-graecum* L.) on blood lipids, blood sugar and platelet aggregation in patients with coronary artery disease. *Prostaglandins Leukot. Essent. Fatty Acids* 56(5):379-384.

Casey, R.C. 1960. 298 Alleged anti-fertility plants of India. *Indian J. Med. Sci.* 14:590-600.

Chen, J.K., and T.T. Chen. 2004. *Chinese medical herbology and pharmacology*. City of Industry, CA: Art of Medicine Press.

Czigle, S., J. Toth, and D. Kostalova. 2006. Herbal medicine and cardiovascular therapy—The risk of drug interactions. *Farm. Obzor.* 75(2-3):35-43.

Effraim, K.D., T.W. Jaxks, and P.A. Nwafor. 1999. Investigation of the toxicity potential of *Trigonella foenum-graecum* (Linn). *Pak. Vet. J.* 19(1):13-16.

Flammang, A.M., M.A. Cifone, G.L. Erexson, and L.F. Stankowski, Jr. 2004. Genotoxicity testing of a fenugreek extract. *Food Chem. Toxicol.* 42(11):1769-1775.

Gopaldas, T. 1995. India's control programs for iron deficiency anemia in preschool children: Past, present, and future. In Nestel, P., Ed. *Proceedings of conference on iron interventions for child survival, opportunities for micronutrient interventions*. London, United Kingdom: John Snow Inc.

Gupta, A., R. Gupta, and B. Lal. 2001. Effect of *Trigonella foenum-graecum* (fenugreek) seeds on glycaemic control and insulin resistance in type 2 diabetes mellitus: A double blind placebo controlled study. *J. Assoc. Physicians India* 49:1057-1061.

Hannan, J.M., L. Ali, B. Rokeya, et al. 2007. Soluble dietary fibre fraction of *Trigonella foenum-graecum* (fenugreek) seed improves glucose homeostasis in animal models of type 1 and type 2 diabetes by delaying carbohydrate digestion and absorption, and enhancing insulin action. *Br. J. Nutr.* 97(3):514-521.

Hannan, J.M., B. Rokeya, O. Faruque, et al. 2003. Effect of soluble dietary fibre fraction of *Trigonella foenum-graecum* on glycemic, insulinemic, lipidemic and platelet aggregation status of type 2 diabetic model rats. *J. Ethnopharmacol.* 88(1):73-77.

Heck, A.M., B.A. DeWitt, and A.L. Lukes. 2000. Potential interactions between alternative therapies and warfarin. *Am. J. Health Syst. Pharm.* 57(13):1221-1227; quiz 1228-1230.

Holbrook, A.M., J.A. Pereira, R. Labiris, et al. 2005. Systematic overview of warfarin and its drug and food interactions. *Arch. Intern. Med.* 165(10):1095-1106.

Izzo, A.A. 2005. Herb-drug interactions: An overview of the clinical evidence. *Fund. Clin. Pharmacol.* 19(1):1-16.

Izzo, A.A., G. Di Carlo, F. Borrelli, and E. Ernst. 2005. Cardiovascular pharmacotherapy and herbal medicines: The risk of drug interaction. *Int. J. Cardiol.* 98(1):1-14.

Juurlink, D.N. 2007. Drug interactions with warfarin: What clinicians need to know. *Can. Med. Assoc. J.* 177(4):369-371.

Kassem, A., A. Al-Aghbari, M. Al-Habori, and M. Al-Mamary. 2006. Evaluation of the potential antifertility effect of fenugreek seeds in male and female rabbits. *Contraception* 73(3):301-306.

Khosla, P., D.D. Gupta, and R.K. Nagpal. 1995. Effect of *Trigonella foenum-graecum* (fenugreek) on blood glucose in normal and diabetic rats. *Indian J. Physiol. Pharmacol.* 39(2):173-174.

Korman, S.H., E. Cohen, and A. Preminger. 2001. Pseudo-maple syrup urine disease due to maternal prenatal ingestion of fenugreek. *J. Paediatr. Child Health* 37(4):403-404.

Lambert, J.P., and J. Cormier. 2001. Potential interaction between warfarin and boldo-fenugreek. *Pharmacotherapy* 21(4):509-512.

Leung, A.Y., and S. Foster. 1996. *Encyclopedia of common natural ingredients used in food, drugs, and cosmetics*. 2nd ed. New York: Wiley.

Madar, Z., R. Abel, S. Samish, and J. Arad. 1988. Glucose-lowering effect of fenugreek in non-insulin dependent diabetics. *Eur. J. Clin. Nutr.* 42(1):51-54.

Mills, S., and K. Bone. 2005. *The essential guide to herbal safety*. St. Louis: Elsevier.

Mishkinsky, J.S., A. Goldschmied, B. Joseph, Z. Ahronson, and F.G. Sulman. 1974. Hypoglycaemic effect of *Trigonella foenum-graecum* and *Lupinus termis* (Leguminosae) seeds and their major alkaloids in alloxan-diabetic and normal rats. *Arch. Int. Pharmacodyn. Ther.* 210(1):27-37.

Mital, N., and T. Gopaldas. 1986. Effect of fenugreek (*Trigonella foenum-graecum*) seed based diets on the lactational performance in albino-rats. *Nutr. Rep. Int.* 33(3):477-484.

Mohammad, S., A. Taha, R.N. Bamezai, and N.Z. Baquer. 2006. Modulation of glucose transporter (GLUT4) by vanadate and *Trigonella* in alloxan-diabetic rats. *Life Sci.* 78(8):820-824.

Mondal, D.K., B.M. Yousuf, L.A. Banu, et al. 2004. Effect of fenugreek seeds on the fasting blood glucose level in the streptozotocin induced diabetic rats. *Mymensingh Med. J.* 13(2):161-164.

Muralidhara, K. Narasimhamurthy, S. Viswanatha, and B.S. Ramesh. 1999. Acute and subchronic toxicity assessment of debitterized fenugreek powder in the mouse and rat. *Food Chem. Toxicol.* 37(8):831-838.

Myers, S.P. 2002. Interactions between complementary medicines and warfarin. *Austr. Prescriber* 25(3):54-56.

Nakhla, H.B., O.S. Mohamed, I.M. Abu, A.L. Fatuh, and S.E. Adam. 1991. The effect of *Trigonella foenum-graecum* (fenugreek) crude saponins on Hisex-type chicks. *Vet. Human Toxicol.* 33(6):561-564.

Nath, D., N. Sethi, and S. Srivastav. 1997. Survey on indigenous medicinal plants used for abortion in some districts of Uttar Pradesh. *Fitoterapia* 68(3):223-225.

Panda, S., P. Tahiliani, and A. Kar. 1999. Inhibition of triiodothyronine production by fenugreek seed extract in mice and rats. *Pharmacol. Res.* 40(5):405-409.

Patil, S.P., P.V. Niphadkar, and M.M. Bapat. 1997. Allergy to fenugreek (*Trigonella foenum-graecum*). *Ann. Allergy Asthma Immunol.* 78(3):297-300.

Petropoulos, G. 2002. Cultivation. In Petropoulos, G.A. *Fenugreek: The genus Trigonella*. Boca Raton, FL: CRC Press.

Puri, D., K.M. Prabhu, and P.S. Murthy. 2002. Mechanism of action of a hypoglycemic principle isolated from fenugreek seeds. *Indian J. Physiol. Pharmacol.* 46(4):457-462.

Ragurham, T., R. Sharma, B. Sivakumar, and B. Sahay. 1994. Effect of fenugreek seeds on intravenous glucose disposition in non-insulin dependent diabetic patients. *Phytother. Res.* 8(2):83-86.

Raju, J., D. Gupta, A.R. Rao, P.K. Yadava, and N.Z. Baquer. 2001. *Trigonella foenum-graecum* (fenugreek) seed powder improves glucose homeostasis in alloxan diabetic rat tissues by reversing the altered glycolytic, gluconeogenic and lipogenic enzymes. *Mol. Cell. Biochem.* 224(1-2):45-51.

Sethi, N., D. Nath, R.K. Singh, and R.K. Srivastava. 1990. Antifertility and teratogenic activity of some indigenous medicinal plants in rats. *Fitoterapia* 61(1):64-67.

Sewell, A.C., A. Mosandl, and H. Bohles. 1999. False diagnosis of maple syrup urine disease owing to ingestion of herbal tea. *N. Engl. J. Med.* 341(10):769.

Sharma, M.L., N. Chandokhe, B.J. Ghatak, et al. 1978. Pharmacological screening of Indian medicinal plants. *Indian J. Exp. Biol.* 16(2):228-240.

Sharma, R., and T. Raghuram. 1990. Hypoglycemic effect of fenugreek seeds in non-insulin dependent diabetic subjects. *Nutr. Res.* 10:731-739.

Sharma, R.D., T.C. Raghuram, and N.S. Rao. 1990. Effect of fenugreek seeds on blood glucose and serum lipids in type I diabetes. *Eur. J. Clin. Nutr.* 44(4):301-306.

Sharma, R.D., A. Sarkar, D.K. Hazra, et al. 1996. Toxicological evaluation of fenugreek seeds: A long term feeding experiment in diabetic patients. *Phytother. Res.* 10(6):519-520.

Taylor, W.G., H.J. Zulyniak, K.W. Richards, et al. 2002. Variation in diosgenin levels among 10 accessions of fenugreek seeds produced in western Canada. *J. Agric. Food. Chem.* 50 (21):5994-5997.

Tiran, D. 2003. The use of fenugreek for breast feeding women. *Complement. Ther. Nurs. Midwifery* 9(3):155-156.

Udayasekhara Rao, P., B. Sesikeran, P. Srinivasa Rao, et al. 1996. Short term nutritional and safety evaluation of fenugreek. *Nutr. Res.* 16(9):1495-1505.

Vijayakumar, M.V., S. Singh, R.R. Chhipa, and M.K. Bhat. 2005. The hypoglycaemic activity of fenugreek seed extract is mediated through the stimulation of an insulin signalling pathway. *Br. J. Pharmacol.* 146(1):41-48.

Xue, W.L., X.S. Li, J. Zhang, et al. 2007. Effect of *Trigonella foenum-graecum* (fenugreek) extract on blood glucose, blood lipid and hemorheological properties in streptozotocin-induced diabetic rats. *Asia Pac. J. Clin. Nutr.* 16(Suppl. 1):422-426.

Yeh, G.Y., D.M. Eisenberg, T.J. Kaptchuk, and R.S. Phillips. 2003. Systematic review of herbs and dietary supplements for glycemic control in diabetes. *Diabetes Care* 26(4):1277-1294.

Zia, T., S.N. Hasnain, and S.K. Hasan. 2001. Evaluation of the oral hypoglycaemic effect of *Trigonella foenum-graecum* L. (methi) in normal mice. *J. Ethnopharmacol.* 75(2-3):191-195.

Trillium erectum L.

Liliaceae

SCN: bethroot

OCN: birth root; purple trillium; red trillium; wakerobin

Part: root

T

Quick Reference Summary

Safety Class: 2b

Interaction Class: A

Contraindications

Not for use in pregnancy except under the supervision of a qualified healthcare practitioner (Cook 1869; Foster and Johnson 2008; Wood and LaWall 1918).

Other Precautions

None known.

Drug and Supplement Interactions

None known.

Adverse Events and Side Effects

Trillium may cause gastrointestinal irritation (McGuffin et al. 1997).

Pharmacological Considerations

None known.

Pregnancy and Lactation

Bethroot has been used to facilitate childbirth (Cook 1869; Foster and Johnson 2008; Wood and LaWall 1918). Based on this information, use during pregnancy is not recommended except under the supervision of a qualified healthcare practitioner.

No information on the safety of bethroot during lactation was identified in the scientific or traditional literature. While this review did not identify any concerns for use while nursing, safety has not been conclusively established.

Review Details

I. Drug and Supplement Interactions

Clinical Trials of Drug or Supplement Interactions

No clinical trials of drug or supplement interactions were identified.

Case Reports of Suspected Drug or Supplement Interactions

No case reports of suspected drug or supplement interactions were identified.

Animal Trials of Drug or Supplement Interactions

No animal trials of drug or supplement interactions were identified.

II. Adverse Events

Case Reports of Adverse Events

No case reports of adverse events were identified.

III. Pharmacology and Pharmacokinetics

Human Pharmacological Studies

No relevant human pharmacological studies were identified.

Animal Pharmacological Studies

No relevant animal pharmacological studies were identified.

In Vitro Pharmacological Studies

No relevant in vitro pharmacological studies were identified.

IV. Pregnancy and Lactation

Bethroot has been used to facilitate childbirth (Cook 1869; Foster and Johnson 2008; Wood and LaWall 1918).

No information on the safety of bethroot during lactation was identified.

V. Toxicity Studies

No toxicity studies were identified.

Literature Cited

Cook, W. 1869. *Physio-medical dispensatory*. Cincinnati, OH: W.H. Cook.

Foster, S., and R. Johnson. 2008. *Desk reference to nature's medicine*. Washington, DC: National Geographic Society.

McGuffin, M., C. Hobbs, R. Upton, and A. Goldberg. 1997. *Botanical safety handbook*. Boca Raton, FL: CRC Press.

Wood, H., and C. LaWall. 1918. *The dispensatory of the United States of America*. 21st ed. Philadelphia: Lippincott.

Turnera diffusa Willd. ex Schult. var. *diffusa* — Turneraceae

SCN: damiana

Part: leaf

Quick Reference Summary

Safety Class: 1

Interaction Class: A

Contraindications

None known.

Other Precautions

None known.

Drug and Supplement Interactions

None known.

Notice

Cyanogenic glycosides (0.26% tetraphyllin B) (Spencer and Seigler 1981) *see* Appendix 1.

Adverse Events and Side Effects

None known.

Pharmacological Considerations

Studies in animals have yielded conflicting results on the effects of damiana on glucose levels in healthy and diabetic animals, with one study indicating a reduction in blood sugar and another study indicating no effect (Alarcon-Aguilar et al. 2002; Alarcon-Aguilara et al. 1998).

Pregnancy and Lactation

No information on the safety of damiana in pregnancy or lactation was identified in the scientific or traditional literature. Although this review did not identify any concerns for use while pregnant or nursing, safety has not been conclusively established.

Review Details

I. Drug and Supplement Interactions

Clinical Trials of Drug or Supplement Interactions

No clinical trials of drug or supplement interactions were identified.

Case Reports of Suspected Drug or Supplement Interactions

No case reports of suspected drug or supplement interactions were identified.

Animal Trials of Drug or Supplement Interactions

No animal trials of drug or supplement interactions were identified.

II. Adverse Events

Case Reports of Adverse Events

After consumption of 8 oz of damiana extract, a person with a history of alcohol abuse experienced tetanus-like convulsions and sudden outbursts. Further details on this case are lacking (Kumar et al. 2005).

III. Pharmacology and Pharmacokinetics

Human Pharmacological Studies

No relevant human pharmacological studies were identified.

Animal Pharmacological Studies

In healthy rabbits orally administered 4 ml/kg of a decoction of damiana, a reduction in peak blood glucose levels and a mild reduction in the area under the glucose tolerance curve was observed in the glucose tolerance test (Alarcon-Aguilara et al. 1998).

No significant hypoglycemic activity was observed in healthy or diabetic mice orally administered 500 mg/kg of a hydroethanolic extract of damiana (Alarcon-Aguilar et al. 2002).

In Vitro Pharmacological Studies

In a yeast estrogen screen assay, damiana exhibited estrogenic activity. In a tritiated-water release assay, a methanol extract of damiana dose-dependently inhibited aromatase (Zhao et al. 2008).

IV. Pregnancy and Lactation

No information on the safety of damiana during pregnancy or lactation was identified.

V. Toxicity Studies

Genotoxicity

No mutagenic activity of a hydroethanolic extract of damiana was observed in the Ames test for mutagenicity with *Salmonella typhimurium* strain TA98 with or without metabolic activation by S9 (Schimmer et al. 1994).

Literature Cited

Alarcon-Aguilar, F.J., R. Roman-Ramos, J.L. Flores-Saenz, and F. Aguirre-Garcia. 2002. Investigation on the hypoglycaemic effects of extracts of four Mexican medicinal plants in normal and alloxan-diabetic mice. *Phytother. Res.* 16(4):383-386.

Alarcon-Aguilara, F.J., R. Roman-Ramos, S. Perez-Gutierrez, et al. 1998. Study of the anti-hyperglycemic effect of plants used as antidiabetics. *J. Ethnopharmacol.* 61(2):101-110.

Kumar, S., R. Taneja, and A. Sharma. 2005. The genus *Turnera*: A review update. *Pharm. Biol.* 43(5):383-391.

Schimmer, O., A. Kruger, H. Paulini, and F. Haefele. 1994. An evaluation of 55 commercial plant extracts in the Ames mutagenicity test. *Pharmazie* 49(6):448-451.

Spencer, K.C., and D.S. Seigler. 1981. Tetraphyllin B from *Turnera diffusa*. *Plant Med.* 43(10):175-178.

Zhao, J., A.K. Dasmahapatra, S.I. Khan, and I.A. Khan. 2008. Anti-aromatase activity of the constituents from damiana (*Turnera diffusa*). *J. Ethnopharmacol.* 120(3):387-393.

Tussilago farfara L.

Asteraceae

SCN: coltsfoot
PN: *kuan dong hua* (flower bud)

Part: flower bud

Quick Reference Summary

Safety Class: * 2b, 2c, 2d

Interaction Class: A

* Extracts of coltsfoot that have had the pyrrolizidine alkaloids removed (PA-free coltsfoot) are commercially available. Concerns regarding the internal use of coltsfoot products do not apply to PA-free products.

Contraindications

Not for use in pregnancy or lactation except under the supervision of a qualified healthcare practitioner (Blumenthal et al. 1998; Wichtl 2004).

Not for use in excess of 6 weeks per year (Wichtl 2004).

Other Precautions

Use is cautioned in those with coughing of blood or pus and blood (Bensky et al. 2004).

Drug and Supplement Interactions

None known.

Notice

Pyrrolizidine alkaloids (0.004–0.03%, calculated as senkirkine) (Culvenor et al. 1976; De Smet 1992; Lebada et al. 2000; Wichtl 2004; Williamson 2003; Yu et al. 2005); *see* Appendix 1.

Editors' Notes

The American Herbal Products Association has established a trade requirement (AHPA 2011) that all products with botanical ingredients that contain toxic pyrrolizidine alkaloids, including coltsfoot flower bud, are not offered for sale for internal use and display the following cautionary label: "For external use only. Do not apply to broken or abraded skin. Do not use when nursing."

Coltsfoot flower buds of European origin are reported to contain a smaller amount of pyrrolizidine alkaloids than those of Chinese origin (De Smet 1992).

Tussilagone and isotussilagone are pyrrolizidine alkaloids that have a saturated necine ring and are thus not toxic. Senkirkine and senecionine, however, have shown hepatotoxic and/or carcinogenic effects (Wichtl 2004).

Adverse Events and Side Effects

A Chinese herbal reference text indicates that although no adverse reactions to coltsfoot are reported in the literature, long-term administration of the herb in high doses may present a risk of liver cancer (Bensky et al. 2004), presumably due to the presence of pyrrolizidine alkaloids.

Overdose of coltsfoot flower bud may cause restlessness, excitation, irritability, and increased respiration (Chen and Chen 2004).

Pharmacological Considerations

None known.

Pregnancy and Lactation

No information on the safety of coltsfoot flower bud in pregnancy or lactation was identified (Hirono et al. 1979). Data from animal studies indicate that pyrrolizidine alkaloids can cross the placenta and may be transferred through breast milk to nursing infants (Cheeke 1988; Panter and James 1990; Schoental 1968). Based on this information, use during pregnancy or lactation is not recommended except under the supervision of a qualified healthcare practitioner.

Review Details

I. Drug and Supplement Interactions

Clinical Trials of Drug or Supplement Interactions

No clinical trials of drug or supplement interactions were identified.

Case Reports of Suspected Drug or Supplement Interactions

No case reports of suspected drug or supplement interactions were identified.

Animal Trials of Drug or Supplement Interactions

No animal trials of drug or supplement interactions were identified.

II. Adverse Events

Case Reports of Adverse Events

Overdose of coltsfoot flower bud may cause restlessness, excitation, irritability, and increased respiration (Chen and Chen 2004).

III. Pharmacology and Pharmacokinetics

Human Pharmacological Studies

No relevant human pharmacological studies were identified.

Animal Pharmacological Studies

The compound tussilagone was found to be a potent cardiovascular and respiratory stimulant. In anesthetized dogs, cats, and rats administered the compound intravenously, tussilagone produced an instant and dose-dependent pressor effect similar to that of dopamine; however, no tachyphylaxis was observed (Li and Wang 1987, 1988).

In Vitro Pharmacological Studies

No relevant in vitro pharmacological studies were identified.

IV. Pregnancy and Lactation

No information on the safety of coltsfoot flower bud in pregnancy or lactation was identified. Data from animal studies indicate that pyrrolizidine alkaloids can cross the placenta and may be transferred through breast milk to nursing infants (Cheeke 1988; Panter and James 1990; Schoental 1968).

V. Toxicity Studies

Acute Toxicity

The LD_{50} values of different preparations of coltsfoot flower bud in mice are 124 g/kg after oral administration of a decoction, 112 g/kg after intraperitoneal administration of an alcohol extract, and 43 g/kg after intraperitoneal administration of an ether extract (Chen and Chen 2004). Overdose of the flower bud led to restlessness, excitation, mania, increased respiration, muscle stiffness, tremor, and muscle spasms (Chen and Chen 2004).

The LD_{50} of intravenously administered tussilagone in mice is 28.9 mg/kg (Li and Wang 1988). The LD_{50} of intraperitoneally administered senkirkine in rats is 220 mg/kg (Hirono et al. 1979).

Chronic Toxicity

In rats administered coltsfoot flower buds as 32% of the diet for 4 days then as 16% of the diet for 380 days, 8 of 12 rats developed hemangioendothelial sarcoma in the liver. Adverse effects were attributed to the compound senkirkine (Hirono et al. 1976).

In rats administered coltsfoot flower buds as 4 or 8% of the diet for 600 days, 1 of 10 rats in the 8% group developed hemangioendothelial sarcoma in the liver, and no tumors were observed in the 4% group (Hirono et al. 1976).

Genotoxicity

No increase in the number of structural chromosomal aberrations was observed in human lymphocytes treated with the compounds senkirkine and tussilagone at concentrations up to 1000 μ*M*. In contrast, heliotrine, a pyrrolizidine alkaloid used for comparison, induced chromosomal aberrations when tested at a concentration of 100 μ*M* (Kraus et al. 1985).

Literature Cited

AHPA. July 2011. Code of Ethics & Business Conduct. Silver Spring, MD: American Herbal Products Association.

Bensky, D., S. Clavey, and E. Stöger. 2004. *Chinese herbal medicine: Materia medica*. 3rd ed. Seattle: Eastland Press.

Blumenthal, M., W. Busse, A. Goldberg, et al. 1998. *The complete German Commission E monographs*. Austin, TX: American Botanical Council.

Cheeke, P.R. 1988. Toxicity and metabolism of pyrrolizidine alkaloids. *J. Anim. Sci.* 66(9):2343.

Chen, J.K., and T.T. Chen. 2004. *Chinese medical herbology and pharmacology*. City of Industry, CA: Art of Medicine Press.

Culvenor, C.C.J., J.A. Edgar, L.W. Smith, and I. Hirono. 1976. The occurrence of senkirkine in *Tussilago farfara*. *Aust. J. Chem.* 29.

De Smet, P.A.G.M. 1992. *Adverse effects of herbal drugs, Volume 1*. Berlin: Springer.

Hirono, I., M. Haga, M. Fujii, et al. 1979. Induction of hepatic tumors in rats by senkirkine and symphytine. *J. Natl. Cancer Inst.* 63(2):469-472.

Hirono, I., H. Mori, and C.C.J. Culvenor. 1976. Carcinogenic activity of coltsfoot, *Tussilago farfara* L. *Gann* 67(1):125-129.

Kraus, C., G. Abel, and O. Schimmer. 1985. Studies on the chromosome damaging effect of some pyrrolizidine alkaloids in human lymphocytes *in vitro*. *Planta Med.* 51(2):89-91.

Lebada, R., A. Schreier, S. Scherz, et al. 2000. Quantitative analysis of the pyrrolizidine alkaloids senkirkine and senecionine in *Tussilago farfara* L. by capillary electrophoresis. *Phytochem. Anal.* 11(6):366-369.

Li, Y.P., and Y.M. Wang. 1987. The effects of tussilagone on the hemodynamics of conscious dogs and dogs during hemorrhagic shock. *Yaoxue Xuebao* 22(7):486-490.

Li, Y.P., and Y.M. Wang. 1988. Evaluation of tussilagone: A cardiovascular-respiratory stimulant isolated from Chinese herbal medicine. *Gen. Pharmacol.* 19(2):261-263.

Panter, K.E., and L.F. James. 1990. Natural plant toxicants in milk: A review. *J. Anim. Sci.* 68(3):892-904.

Schoental, R. 1968. Toxicology and carcinogenic action of pyrrolizidine alkaloids. *Cancer Res.* 28(11):2237.

Wichtl, M. 2004. *Herbal drugs and phytopharmaceuticals: A handbook for practice on a scientific basis*. 3rd ed. Boca Raton, FL: CRC Press.

Williamson, E.M. 2003. *Potter's herbal cyclopedia*. Saffron Walden, Essex: C.W. Daniel Co.

Yu, L., Y. Xu, H. Feng, and S.F. Li. 2005. Separation and determination of toxic pyrrolizidine alkaloids in traditional Chinese herbal medicines by micellar electrokinetic chromatography with organic modifier. *Electrophoresis* 26(17):3397-3404.

Tussilago farfara L. Asteraceae

SCN: coltsfoot

Part: leaf

Quick Reference Summary

*Safety Class:** 2b, 2c, 2d
Interaction Class: A

Contraindications

Not for use in pregnancy or lactation except under the supervision of a qualified healthcare practitioner (Blumenthal et al. 1998; Wichtl 2004).

Not for use in excess of 6 weeks per year (Wichtl 2004).

Other Precautions

None known.

Drug and Supplement Interactions

None known.

Standard Dose

The standard dose is 1.5 to 6 g of fresh or dried leaf or its equivalent in finished preparations; daily consumption of unsaturated pyrrolizidine alkaloids must not exceed 10 μg for teas, or 1 μg for extracts or juice pressed from fresh leaves (Blumenthal et al. 1998; Wichtl 2004).

Notice

Pyrrolizidine alkaloids (0–0.007%, primarily senkirkine, with smaller amounts of senecionene and tussilagone and isotussilagone) (Bartkowski et al. 1997; Culvenor et al. 1976; De Smet 1992; Lebada et al. 2000; Miething and Steinbach 1990; Wichtl 2004); *see* Appendix 1.

* Extracts of coltsfoot that have had the pyrrolizidine alkaloids removed (PA-free coltsfoot) are commercially available. Concerns regarding the internal use of coltsfoot products do not apply to PA-free products.

Editors' Notes

The American Herbal Products Association maintains as a trade requirement that all products with botanical ingredients that contain toxic pyrrolizidine alkaloids, including coltsfoot leaf, are not offered for sale for internal use and display the following cautionary label: "For external use only. Do not apply to broken or abraded skin. Do not use when nursing."

Alone among plants containing unsaturated pyrrolizidine alkaloids, coltsfoot leaf is allowed for internal use by the German Commission E, with duration of use limited to 4 to 6 weeks per year (Wichtl 2004).

Tussilagone and isotussilagone are saturated pyrrolizidine alkaloids and are thus not toxic. Senkirkine and senecionine, however, are unsaturated and have shown hepatotoxic and carcinogenic effects (Wichtl 2004).

The leaves of coltsfoot, *Petasites,* and *Adenostyles* are similar in appearance. Cases of poisoning have been reported in persons who accidentally ingested plants of those species instead of coltsfoot (Roulet et al. 1988; Sperl et al. 1995).

Adverse Events and Side Effects

None known.

Pharmacological Considerations

None known.

Pregnancy and Lactation

No information on the safety of coltsfoot leaf in pregnancy or lactation was identified. Data from animal studies indicate that pyrrolizidine alkaloids can cross the placenta and may be transferred through breast milk to nursing infants (Cheeke 1988; Panter and James 1990; Schoental 1968). Based on this information, use during pregnancy or lactation is not recommended except under the supervision of a qualified healthcare practitioner.

Review Details

I. Drug and Supplement Interactions

Clinical Trials of Drug or Supplement Interactions

No clinical trials of drug or supplement interactions were identified.

Case Reports of Suspected Drug or Supplement Interactions

No case reports of suspected drug or supplement interactions were identified.

Animal Trials of Drug or Supplement Interactions

No animal trials of drug or supplement interactions were identified.

II. Adverse Events

Case Reports of Adverse Events

Two cases of adverse effects have been reported from the use of plants mistaken for coltsfoot. In both cases, the plants used (*Petasites hybridus* and *Adenostyles alliariae*) contained pyrrolizidine alkaloids (Roulet et al. 1988; Sperl et al. 1995).

III. Pharmacology and Pharmacokinetics

Human Pharmacological Studies

No relevant human pharmacological studies were identified.

Animal Pharmacological Studies

Coltsfoot demonstrated a weak sensitizing capacity in a guinea pig sensitization test (Zeller et al. 1985).

The compound tussilagone was found to be a potent cardiovascular and respiratory stimulant. In anesthetized dogs, cats, and rats administered the compound intravenously, tussilagone produced an instant and dose-dependent pressor effect similar to that of dopamine; however, no tachyphylaxis has been observed (Li and Wang 1987, 1988).

In Vitro Pharmacological Studies

No relevant in vitro pharmacological studies were identified.

IV. Pregnancy and Lactation

No information on the safety of coltsfoot leaf in pregnancy or lactation was identified. Data from animal studies indicate that pyrrolizidine alkaloids can cross the placenta and may be transferred through breast milk to nursing infants (Cheeke 1988; Panter and James 1990; Schoental 1968).

V. Toxicity Studies

Acute Toxicity

The LD_{50} of intravenously administered tussilagone in mice is 28.9 mg/kg (Li and Wang 1988). The LD_{50} of intraperitoneally administered senkirkine in rats is 220 mg/kg.

Genotoxicity

No increase in the number of structural chromosomal aberrations was observed in human lymphocytes treated with the compounds senkirkine and tussilagone at concentrations up to 1000 μ*M*. In contrast, heliotrine, a pyrrolizidine alkaloid used for comparison, induced chromosomal aberrations when tested at a concentration of 100 μ*M* (Kraus et al. 1985).

Literature Cited

Bartkowski, J.-P.B., H. Wiedenfeld, and E. Roeder. 1997. Quantitative photometric determination of senkirkine in farfarae folium. *Phytochem. Anal.* 8(1):1-4.

Blumenthal, M., W. Busse, A. Goldberg, et al. 1998. *The complete German Commission E monographs*. Austin, TX: American Botanical Council.

Cheeke, P.R. 1988. Toxicity and metabolism of pyrrolizidine alkaloids. *J. Anim. Sci.* 66(9):2343.

Culvenor, C.C.J., J.A. Edgar, L.W. Smith, and I. Hirono. 1976. The occurrence of senkirkine in *Tussilago farfara*. *Aust. J. Chem.* 29.

De Smet, P.A.G.M. 1992. *Adverse effects of herbal drugs, Volume 1*. Berlin: Springer.

Kraus, C., G. Abel, and O. Schimmer. 1985. Studies on the chromosome damaging effect of some pyrrolizidine alkaloids in human lymphocytes *in vitro*. *Planta Med.* 51(2):89-91.

Lebada, R., A. Schreier, S. Scherz, et al. 2000. Quantitative analysis of the pyrrolizidine alkaloids senkirkine and senecionine in *Tussilago farfara* L. by capillary electrophoresis. *Phytochem. Anal.* 11(6):366-369.

Li, Y.P., and Y.M. Wang. 1987. The effects of tussilagone on the hemodynamics of conscious dogs and dogs during hemorrhagic shock. *Yaoxue Xuebao* 22(7):486-490.

Li, Y.P., and Y.M. Wang. 1988. Evaluation of tussilagone: A cardiovascular-respiratory stimulant isolated from Chinese herbal medicine. *Gen. Pharmacol.* 19(2):261-263.

Miething, H., and R.A. Steinbach. 1990. Evaluation of the senkirkine in aqueous preparations of coltsfoot leaves. *Pharm. Ztg. Wissensch.* 135(4):153-155.

Panter, K.E., and L.F. James. 1990. Natural plant toxicants in milk: A review. *J. Anim. Sci.* 68(3):892-904.

Roulet, M., R. Laurini, L. Rivier, and A. Calame. 1988. Hepatic veno-occlusive disease in newborn infant of a woman drinking herbal tea. *J. Pediatr.* 112(3):433-436.

Schoental, R. 1968. Toxicology and carcinogenic action of pyrrolizidine alkaloids. *Cancer Res.* 28(11):2237.

Sperl, W., H. Stuppner, I. Gassner, et al. 1995. Reversible hepatic veno-occlusive disease in an infant after consumption of pyrrolizidine-containing herbal tea. *Eur. J. Pediatr.* 154(2):112-116.

Wichtl, M. 2004. *Herbal drugs and phytopharmaceuticals: A handbook for practice on a scientific basis*. 3rd ed. Boca Raton, FL: CRC Press.

Zeller, W., M. de Gols, and B.M. Hausen. 1985. The sensitizing capacity of Compositae plants. VI. Guinea pig sensitization experiments with ornamental plants and weeds using different methods. *Arch. Dermatol. Res.* 277(1):28-35.

T

Ulmus rubra Muhl.

Ulmaceae

SCN: slippery elm
Syn: *Ulmus fulva* Michx.

Part: bark

Quick Reference Summary

Safety Class: 1
Interaction Class: A

Contraindications

None known.

Other Precautions

Slippery elm should be taken with at least 250 ml (8 oz) of liquid.

Drug and Supplement Interactions

Other drugs should be taken 1 hour prior to or several hours after consumption of slippery elm, as the mucilage may slow the absorption of orally administered drugs (Brinker 1997; Evans 2002).

Notice

Mucilages (Brinker 1997; Evans 2002); *see* Appendix 3.

Adverse Events and Side Effects

Bladder stones have been reported in women after vaginal insertion of slippery elm bark in attempted abortions (Williams 1954). Such events would not occur during normal oral administration of slippery elm.

Pharmacological Considerations

None known.

Pregnancy and Lactation

No information on the safety of slippery elm in pregnancy or lactation was identified in the scientific or traditional literature. Although this review did not identify any concerns for use while pregnant or nursing, safety has not been conclusively established.

Review Details

I. Drug and Supplement Interactions

Clinical Trials of Drug or Supplement Interactions

No clinical trials of drug or supplement interactions were identified.

Case Reports of Suspected Drug or Supplement Interactions

No case reports of suspected drug or supplement interactions were identified.

Animal Trials of Drug or Supplement Interactions

No animal trials of drug or supplement interactions were identified.

II. Adverse Events

Case Reports of Adverse Events

Two cases of bladder calculus have been reported in pregnant women who had vaginally administered pieces of slippery elm bark in an attempt to induce abortion. In both cases, the bark pieces had passed into the bladder and remained to form a stone while the pregnancy continued (Williams 1954). Such events would not occur during normal oral administration of slippery elm.

III. Pharmacology and Pharmacokinetics

Human Pharmacological Studies

No relevant human pharmacological studies were identified.

Animal Pharmacological Studies

No relevant animal pharmacological studies were identified.

In Vitro Pharmacological Studies

No relevant in vitro pharmacological studies were identified.

IV. Pregnancy and Lactation

See Adverse Events. No other information was identified on the safety of slippery elm in pregnancy or lactation.

V. Toxicity Studies

No information on the toxicity of slippery elm was identified.

Literature Cited

Brinker, F. 1997. Interactions of pharmaceutical and botanical medicines. *J. Naturopathic Med.* 7(2):14-20.

Evans, W. 2002. *Trease and Evans' pharmacognosy.* 15th ed. New York: Saunders.

Williams, B. 1954. Two cases of slippery elm bladder calculus in pregnancy. *J. Obstet. Gynaecol. Br. Emp.* 61(4):499-500.

Uncaria gambir (Hunter) Roxb. — Rubiaceae

SCN: gambir (dried extract of leaf and twig)
PN: *zong er cha*
OCN: brown cutch; pale catechu; white cutch
Part: leaf, twig

Quick Reference Summary

Safety Class: 1
Interaction Class: A

Contraindications

None known.

Other Precautions

None known.

Drug and Supplement Interactions

None known.

Notice

Tannins (calculated as catechin, 30–35%) (Bradley 1992; Leung and Foster 1996; List and Hörhammer 1973); *see* Appendix 1.

Adverse Events and Side Effects

None known.

Pharmacological Considerations

None known.

Pregnancy and Lactation

No information on the safety of gambir in pregnancy or lactation was identified in the scientific or traditional literature. Although this review did not identify any concerns for use while pregnant or nursing, safety has not been conclusively established.

Review Details

I. Drug and Supplement Interactions

Clinical Trials of Drug or Supplement Interactions

No clinical trials of drug or supplement interactions were identified.

Case Reports of Suspected Drug or Supplement Interactions

No case reports of suspected drug or supplement interactions were identified.

Animal Trials of Drug or Supplement Interactions

No animal trials of drug or supplement interactions were identified.

II. Adverse Events

Case Reports of Adverse Events

No case reports of adverse events were identified.

III. Pharmacology and Pharmacokinetics

Human Pharmacological Studies

No relevant human pharmacological studies were identified.

Animal Pharmacological Studies

Intravenous administration of the compound gambirine at doses of 0.2 to 10.0 mg/kg caused a dose-related fall in both systolic and diastolic blood pressures as well as heart rate in rats. At all doses, gambirine showed a prompt onset of action; at doses over 5 mg/kg, marked persistence of hypotension accompanied by severe bradycardia was observed (Merlini et al. 1967; Mok et al. 1992).

In Vitro Pharmacological Studies

No relevant in vitro pharmacological studies were identified.

IV. Pregnancy and Lactation

No information on the use of gambir during pregnancy or lactation was identified.

V. Toxicity Studies

No toxicity studies were identified.

U

Literature Cited

Bradley, P.R. 1992. *British herbal compendium: A handbook of scientific information on widely used plant drugs.* Bournemouth, UK: British Herbal Medicine Association.

Leung, A.Y., and S. Foster. 1996. *Encyclopedia of common natural ingredients used in food, drugs, and cosmetics.* 2nd ed. New York: Wiley.

List, P.H., and H. Hörhammer. 1973. *Hagers handbuch der pharmazeutischen praxis.* Berlin: Springer.

Merlini, L., R. Mendelli, G. Nasini, and M. Hesse. 1967. Gambirine, a new indole alkaloid from *Uncaria gambier* Roxb. *Tetrahedron Lett.* 16:1571-1574.

Mok, J.S., P. Chang, K.H. Lee, T.S. Kam, and S.H. Goh. 1992. Cardiovascular responses in the normotensive rat produced by intravenous injection of gambirine isolated from *Uncaria callophylla* Bl. ex Korth. *J. Ethnopharmacol.* 36(3):219-223.

Uncaria tomentosa (Willd.) DC. — Rubiaceae

SCN: cat's claw
OCN: *uña de gato*

Part: root bark, stem bark

Quick Reference Summary

Safety Class: 2b
Interaction Class: A

Contraindications

Not for use during pregnancy or in women attempting to become pregnant except under the supervision of a qualified healthcare practitioner (De Feo 1992; Jones 1995; Reinhard 1999; Valerio and Gonzales 2005).

Other Precautions

None known.

Drug and Supplement Interactions

None known.

Editors' Note

Although some researchers have reported that two chemotypes (chemically distinct types of a plant species) of cat's claw are recognized, one containing pentacyclic oxindole alkaloids (POA) and the other containing tetracyclic oxindole alkaloids (TOA) (Laus et al. 1997), others have indicated that such chemotypes are not typically found and that cat's claw contains both POAs and TOAs (Taylor 2002a, 2002b).

Adverse Events and Side Effects

In a trial of HIV patients taking a cat's claw extract high in POAs, a small number of cases of a mild erythrocytosis, constipation, loose stool, and aggravation of preexisting acne were reported (Immodal Pharmaka 1996).

One case of kidney failure was reported in a woman taking cat's claw (Hilepo et al. 1997).

Pharmacological Considerations

Several references have indicated a theoretical caution against the use of cat's claw in persons with autoimmune diseases (Jones 1995; Immodal Pharmaka 1996; Mills and Bone 2005). Jones lists a wide range of contraindications, e.g., patients undergoing skin grafts and organ transplants; hemophiliacs prescribed fresh blood plasma; simultaneous administration of certain vaccines, hormone therapies, thymus extracts, and insulin; and children under 3 years of age (Jones 1995). Concerns regarding cat's claw use in these conditions are theoretical, and definitive data supporting or refuting these concerns are lacking. One clinical trial of a high-POA content cat's claw extract in persons with rheumatoid arthritis (an inflammatory autoimmune disorder) demonstrated a reduction in joint pain and swelling (Mur et al. 2002).

The anticancer effect of cat's claw can be partly attributed to the pentacyclic oxindole alkaloids that are likely accompanied by tetracyclic oxindole alkaloids, some of which are calcium channel blockers and hence hypotensive (Bacher et al. 2006; Garcia Prado et al. 2007; Laus et al. 1997; Reinhard 1999; Shi et al. 2003).

Pregnancy and Lactation

Traditional usage of cat's claw as a contraceptive suggests that the herb should be avoided in pregnancy and in women attempting to conceive (De Feo 1992; Jones 1995).

No information on the safety of cat's claw during lactation was identified in the scientific or traditional literature. While this review did not identify any concerns for use while nursing, safety has not been conclusively established.

Review Details

I. Drug and Supplement Interactions

Clinical Trials of Drug or Supplement Interactions

In a year-long clinical trial examining the effects of a high-POA content cat's claw extract in patients with active rheumatoid arthritis undergoing sulfasalazine or hydroxychloroquine treatment, a reduction in the number of painful and swollen joints was observed (Mur et al. 2002).

Case Reports of Suspected Drug or Supplement Interactions

No case reports of suspected drug or supplement interactions were identified.

Animal Trials of Drug or Supplement Interactions

No animal trials of drug or supplement interactions were identified.

II. Adverse Events

Adverse Events Reported in Clinical Trials

In a trial with HIV patients taking a cat's claw extract high in pentacyclic oxindole alkaloids, a small number of cases of a mild erythrocytosis, constipation, loose stool, and aggravation of preexisting acne were reported. In rare cases, a rise in uric acid values due to increased cellular immune system activity was also reported (Immodal Pharmaka 1996).

Case Reports of Adverse Events

Acute renal failure was reported in a 35-year-old woman with systemic lupus erythematosus taking four capsules daily of a cat's claw product (species unspecified) for an unspecified length of time. The woman was also taking other medications including prednisone, atenolol, metolazone, furosemide, and nifedipine. Biochemical parameters for renal function were reported to return to normal after cessation of the cat's claw product (Hilepo et al. 1997).

III. Pharmacology and Pharmacokinetics

U

Human Pharmacological Studies

No relevant human pharmacological studies were identified.

Animal Pharmacological Studies

No relevant animal pharmacological studies were identified.

In Vitro Pharmacological Studies

One in vitro study indicated that cat's claw inhibited the drug-metabolizing isoenzyme CYP3A4, although the applicability of this finding to human use is unknown (Budzinski et al. 2000).

IV. Pregnancy and Lactation

In pregnant mice administered 0.125 to 0.5 mg/ml of cat's claw extract in drinking water for 72 hours after copulation, a significant number of abnormal embryos were observed (Iziga et al. 1998).

Oral administration of 6.25 mg/kg of a tannin-free extract of cat's claw to female mice prevented pregnancy (Keplinger 1982).

High doses of cat's claw preparations have traditionally been used as contraceptives in South America (De Feo 1992; Jones 1995).

No information on the safety of cat's claw during lactation was identified.

V. Toxicity Studies

Acute Toxicity

The LD_{50} of orally administered aqueous cat's claw extract in rats is greater than 8 g/kg. The LD_{50} of a cat's claw powder is over 2 g/kg and that of an aqueous/ethanolic extract (4% alkaloids) is over 5 g/kg (Sheng et al. 1999). The LD_{50} of an orally administered freeze-dried aqueous cat's claw root extract in mice is greater than 16 g/kg (Kynoch and Lloyd 1975). The LD_{50} of an intraperitoneally administered aqueous cat's claw extract in mice is greater than 2 g/kg (Kreutzkamp 1984).

Short-Term Toxicity

In rats orally administered 1000 mg/kg daily of an aqueous cat's claw extract (0.75% oxindole alkaloids) for 28 days, increases in lymphocytes and kidney weight and a decrease in neutrophil granulocytes were observed (Svendsen and Skydsgaard 1986).

In rats orally administered 5 to 80 mg/kg of an aqueous extract of cat's claw for 56 days, an increase in white blood cells was observed at the highest dose, and no significant signs of toxicity were noted (Sheng et al. 2000).

Cytotoxicity

No cytotoxicity of cat's claw extract at doses up to 100 mg/ml was observed in Chinese hamster ovary cells using the tetrazolium assay and a Microtox bacterial test in vitro (Santa Maria et al. 1997).

Genotoxicity

No genotoxicity of an aqueous extract of cat's claw was observed in a somatic mutation and recombination test in *Drosophila melanogaster* (Romero-Jimenez et al. 2005).

Mutagenicity

No mutagenicity of cat's claw was observed in tests on *Salmonella typhimurium* strains with or without metabolic activation at doses up to 100 μg/plate (Rizzi et al. 1993).

Literature Cited

Bacher, N., M. Tiefenthaler, S. Sturm, et al. 2006. Oxindole alkaloids from *Uncaria tomentosa* induce apoptosis in proliferating, G_0/G_1-arrested and bcl-2-expressing acute lymphoblastic leukaemia cells. *Br. J. Haematol.* 132(5):615-622.

Budzinski, J.W., B.C. Foster, S. Vandenhoek, and J.T. Arnason. 2000. An *in vitro* evaluation of human cytochrome P450 3A4 inhibition by selected commercial herbal extracts and tinctures. *Phytomedicine* 7(4):273-282.

De Feo, V. 1992. Medicinal and magical plants in the northern Peruvian Andes. *Fitoterapia* 63:417-440.

Garcia Prado, E., M.D. Garcia Gimenez, R. De la Puerta Vazquez, J.L. Espartero Sanchez, and M.T. Saenz Rodriguez. 2007. Antiproliferative effects of mitraphylline, a pentacyclic oxindole alkaloid of *Uncaria tomentosa* on human glioma and neuroblastoma cell lines. *Phytomedicine* 14(4):280-284.

Hilepo, J.N., A.G. Bellucci, and R.T. Mossey. 1997. Acute renal failure caused by cat's claw herbal remedy in a patient with systemic lupus erythematosus. *Nephron* 77(3):361.

Immodal Pharmaka. 1996. Krallendorn. *Uncaria tomentosa* (Willd.) DC. mod. pent. root extract. Report on experiences with probands. Austria: Volders.

Iziga, R., J. Gutierrez-Pajares, and J. Pino. 1998. Efecto *in vivo* de *Uncaria tomentosa* (Willd.) DC. (Rubiaceae) "uña de gato" en el desarrollo de embriones preimplantacionales de ratón de 72 h.p.c. *Bol. Soc. Biol. Concep.* 69:141-145.

Jones, K. 1995. *Cat's claw: Healing vine of Peru.* Seattle: Sylvan Press.

Keplinger, K. 1982. Cytostat, contraceptive, and antiinflammatory agent from *Uncaria tomentosa* roots. PCT Int. Appl. WO 821130 A1.

Kreutzkamp, B. 1984. Niedermolekulare Inhaltsstoffe mit immunstimulierenden Eigenschaften aus *Uncaria tomentosa* und *Okoubaka aubrevillei* und anderen Drogen [dissertation]. Munich: University of Munich. *Cited in* Valerio, L.G., Jr., and G.F. Gonzales. 2005. Toxicological aspects of the South American herbs cat's claw (*Uncaria tomentosa*) and maca (*Lepidium meyenii*): A critical synopsis. *Toxicol. Rev.* 24(1):11-35.

Kynoch, S., and G. Lloyd. 1975. Acute oral toxicity of substance E-2919. Huntington, UK: Huntington Research Centre. *Cited in* Valerio, L.G., Jr., and G.F. Gonzales. 2005. Toxicological aspects of the South American herbs cat's claw (*Uncaria tomentosa*) and maca (*Lepidium meyenii*): A critical synopsis. *Toxicol. Rev.* 24(1):11-35.

Laus, G., D. Brossner, and K. Keplinger. 1997. Alkaloids of Peruvian *Uncaria tomentosa. Phytochemistry* 45:855-860.

Mills, S., and K. Bone. 2005. *The essential guide to herbal safety.* St. Louis: Elsevier.

Mur, E., F. Hartig, G. Eibl, and M. Schirmer. 2002. Randomized double blind trial of an extract from the pentacyclic alkaloid-chemotype of *Uncaria tomentosa* for the treatment of rheumatoid arthritis. *J. Rheumatol.* 29(4):678-681.

Reinhard, K.H. 1999. *Uncaria tomentosa* (Willd.) DC.: Cat's claw, una de gato, or saventaro. *J. Altern. Complement. Med.* 5(2):143-151.

Rizzi, R., F. Re, A. Bianchi, et al. 1993. Mutagenic and antimutagenic activities of *Uncaria tomentosa* and its extracts. *J. Ethnopharmacol.* 38(1):63-77.

Romero-Jimenez, M., J. Campos-Sanchez, M. Analla, A. Munoz-Serrano, and A. Alonso-Moraga. 2005. Genotoxicity and antigenotoxicity of some traditional medicinal herbs. *Mutat. Res.* 585(1-2):147-155.

Santa Maria, A., A. Lopez, M.M. Diaz, et al. 1997. Evaluation of the toxicity of *Uncaria tomentosa* by bioassays *in vitro*. *J. Ethnopharmacol.* 57(3):183-187.

Sheng, Y., C. Bryngelsson, and R. Pero. 1999. Enhanced DNA repair, immune function, and reduced toxicity of C-MED, a novel aqueous extract from *Uncaria tomentosa*. *J. Ethnopharmacol.* 69:115-126.

Sheng, Y., R. Pero, and H. Wagner. 2000. Treatment of chemotherapy-induced leucopenia in a rat model with aqueous extract from *Uncaria tomentosa. Phytomedicine* 7:137-143.

Shi, J.S., J.X. Yu, X.P. Chen, and R.X. Xu. 2003. Pharmacological actions of *Uncaria* alkaloids, rhynchophylline and isorhynchophylline. *Acta Pharmacol. Sin.* 24(2):97-101.

Svendsen, O., and K. Skydsgaard. 1986. Test report (extraction *radicis Uncariae tomentosae*). Denmark: Scantox Biological Laboratory Ltd.

Taylor, L. 2002a. Cat's claw. In *Herbal secrets of the rainforest.* Garden City Park, NY: Sage Press.

Taylor, L. 2002b. *The cat's claw TOA/POA controversy.* Carson City, NV: Rain Tree Nutrition.

Valerio, L.G., Jr., and G.F. Gonzales. 2005. Toxicological aspects of the South American herbs cat's claw (*Uncaria tomentosa*) and maca (*Lepidium meyenii*): A critical synopsis. *Toxicol. Rev.* 24(1):11-35.

Urtica dioica L. ssp. *dioica* — Urticaceae

SCN: stinging nettle
OCN: nettle

Part: leaf

Quick Reference Summary

Safety Class: 1
Interaction Class: A

Contraindications

None known.

Other Precautions

Fresh stinging nettle leaf causes skin irritation and should be used with care. Uncooked fresh nettle should not be consumed.

Drug and Supplement Interactions

None known.

Notice

Diuretic (Kirchhoff 1983; Tahri et al. 2000); *see* Appendix 2.

Adverse Events and Side Effects

The stems and leaves of stinging nettle have tiny tubular hairs (trichomes) that contain biologically active substances including histamine, acetylcholine, serotonin, leukotrienes, oxalic acid, and tartaric acid (Collier and Chesher 1956; Czarnetzki et al. 1990; Emmelin and Feldberg 1947; Fu et al. 2006; Ganora 2009; Oliver et al. 1991). On contact with skin, hollow hairs on live plants will release their contents into skin, causing a range of cutaneous reactions, from mild to severe burning and stinging sensation and irritant and allergic contact dermatitis (Lovell 1993; Oliver et al. 1991). Although properly dried whole or chopped stinging nettle may cause skin irritation due to preservation of some of the contents of the stinging trichomes, no such reaction occurs after contact with or ingestion of powdered, extracted, or cooked stinging nettle leaf.

A review of clinical trials of stinging nettle leaf preparations indicated that no serious adverse events were reported in any of the trials. Mild gastrointestinal discomfort or allergic reaction was reported in 1.2 to 2.7% of the 10,368 participants (ESCOP 2003).

Pharmacological Considerations

Animal studies have demonstrated that stinging nettle leaf may modify glucose regulation (Bnouham et al. 2003; Farzami et al. 2003; Swanston-Flatt et al. 1989). People with diabetes are advised to monitor their blood sugar closely and discuss the use of this herb with a qualified healthcare practitioner prior to use.

Several in vitro studies have shown some inhibition of platelet aggregation (El Haouari et al. 2006; Goun et al. 2002; Mekhfi et al. 2004; Sajid et al. 1991), although one study showed enhancement of platelet aggregation by phospholipids isolated from stinging nettle (Antonopoulou et al. 1996). The relevance of these in vitro studies to human use is unknown.

Pregnancy and Lactation

Limited information on the safety of stinging nettle leaf in pregnancy or lactation was identified in the scientific and traditional literature. Although this review did not identify any concerns for use while pregnant or nursing, safety has not been conclusively established.

Review Details

I. Drug and Supplement Interactions

Clinical Trials of Drug or Supplement Interactions

No clinical trials of drug or supplement interactions were identified.

Case Reports of Suspected Drug or Supplement Interactions

No case reports of suspected drug or supplement interactions were identified.

Animal Trials of Drug or Supplement Interactions

No animal trials of drug or supplement interactions were identified.

II. Adverse Events

Adverse Events Reported in Clinical Trials

A review of clinical studies of stinging nettle leaf indicated that no serious adverse events were reported in five clinical studies with a total of 10,368 participants taking 670 mg twice daily of a dried hydroethanolic extract (~9.7 g dried leaf) for 3 to 52 weeks. Minor gastrointestinal upset or allergic reaction occurred in 1.2 to 2.7% of participants (ESCOP 2003). Abdominal gas was reported in 3 of 19 patients administered 50 g daily of a stinging nettle leaf puree for 14 days (Chrubasik et al. 1997).

Case Reports of Adverse Events

Immediate and delayed hypersensitivity to stinging nettle leaf was reported in a boy who fell into a stinging nettle patch. Urticaria was present for 12 hours, and a fine vesicular rash erupted at 48 hours (Edwards 1992). A case of severe tongue edema was reported in a woman who placed a fresh leaf of dwarf stinging nettle (*U. urens*) on her tongue (Caliskaner et al. 2004).

Gynecomastia was reported in a man who had consumed two cups of stinging nettle tea (part unspecified) daily for 1 month prior to the onset of the gynecomastia. Testing ruled out malnutrition, hepatic and renal diseases, gonadal insufficiency, testicular tumors, paraneoplastic syndromes, and hyperthyroidism. No medications were noted, and the gynecomastia had resolved by a follow-up visit, 2 months after cessation of stinging nettle consumption (Sahin et al. 2007).

In a woman with an 18-month history of galactorrhea (inappropriate secretion of breast milk), tests revealed a complicated breast cyst, very high levels of estrogen (543 pg/ml), and low levels of follicle-stimulating hormone (1.2 mIU/ml) and luteinizing hormone (1.7 mIU/ml). The woman had begun drinking stinging nettle tea (dose and part used unspecified) daily for 1 month prior to seeing the reporting physician (17 months after diagnosis of galactorrhea). After cessation of stinging nettle, estrogen levels decreased (45 pg/ml) and follicle-stimulating hormone (5.9 mIU/ml) and luteinizing hormone (2.9 mIU) levels increased (Sahin et al. 2007).

Atropine poisoning was reported in a woman who had consumed a tea labeled as stinging nettle. On analysis, the tea was shown to be contaminated with belladonna (*Atropa belladonna*), a plant known to contain atropine (Scholz et al. 1980).

III. Pharmacology and Pharmacokinetics

Human Pharmacological Studies

An open study in patients with myocardial or chronic venous insufficiency indicated that administration of 45 ml stinging nettle leaf juice daily caused an increase in the volume of urine of patients. The effect was more pronounced in patients with chronic venous insufficiency, with a 24% increase in urine volume in this group as compared to a 9% increase in the myocardial insufficiency group (Kirchhoff 1983).

Animal Pharmacological Studies

In healthy rats, 250 mg/kg of an aqueous extract of stinging nettle leaf showed a strong glucose-lowering effect when administered 30 minutes before an oral glucose tolerance test. In contrast, no hypoglycemic effects were observed in alloxan-induced diabetic rats (Bnouham et al. 2003). An increase in serum insulin and a decrease in blood sugar levels were reported in rats intraperitoneally administered an extract of stinging nettle leaf. The results showed that the blood sugar-lowering effect of the extract was due to the enhancement of insulin secretion by pancreatic islets (Farzami et al. 2003). No effect on blood sugar in healthy rats was seen after administration of nettle aqueous extract as the sole source of drinking water or nettle leaf as 6.25% of the diet for 28 days. In diabetic rats, ingestion of the same diet was reported to adversely affect glucose homeostasis (Swanston-Flatt et al. 1989).

Hypotensive, diuretic, and natriuretic activity of an aqueous extract of stinging nettle leaf was observed in anesthetized rats intravenously administered a continuous perfusion at a rate of 4 mg/kg per hour or 24 mg/kg per hour for 1.25 h. The decrease in arterial blood pressure was reduced by 15% at the lower dose and 38% at the higher dose (Tahri et al. 2000). A rapid but transient blood pressure decrease of 37% in rats was observed after intravenous administration of an aqueous extract of 25 mg/kg stinging nettle leaf (Lasheras 1986). In cats, administration of 26.6 mg/kg produced a hypotensive effect and bradycardia that was not compensated by administration of adrenaline (Broncano 1983).

A fraction of a water-methanol extract of stinging nettle leaf produced a marked decrease of inotropic activity in spontaneously beating atria of guinea pigs as well as marked but transient hypotensive activity on the blood pressure of anesthetized rats (Testai et al. 2002).

Case reports indicate that horses exposed to stinging nettle leaf sometimes become distressed and develop ataxia and muscle weakness. Such incidents of stinging nettle exposure are often accompanied by the characteristic nettle rash (Bathe 1994; Conwell and Findlay 2008).

In Vitro Pharmacological Studies

Weak inhibition of thrombin- and ADP-induced platelet aggregation was observed, with IC_{50} values of an aqueous extract of the aerial parts of stinging nettle of 15.5 mg/ml for thrombin- and 12.8 mg/ml for ADP-induced aggregation (Mekhfi et al. 2004). Lipophilic extracts of stinging nettle leaf had more potent antithrombotic effects than aqueous extracts (El Haouari et al. 2006), although a methanolic extract had weak antithrombotic activity (Goun et al. 2002). Inhibition of adrenaline-induced platelet aggregation was observed with an IC_{50} of 2.17 mg/ml of a stinging nettle extract (Sajid et al. 1991). Conversely, a phospholipid from stinging nettle produced a dose-dependent induction of platelet aggregation (Antonopoulou et al. 1996).

IV. Pregnancy and Lactation

No adverse effects on implantation were observed in rats orally administered 250 mg/kg of an ethanol extract of stinging nettle leaf (Sharma et al. 1983). No other information on the safety of stinging nettle leaf during pregnancy or lactation was identified, although stinging nettle leaf is traditionally used as a nutritional supplement for pregnant and lactating women (Yarnell 1998).

V. Toxicity Studies

Acute Toxicity

Minimal toxicity of an ethanol extract of stinging nettle was observed in mice and rats orally and intraperitoneally administered doses up to 2 g/kg (dried herb equivalent) (Tita et al. 1993), although other studies have shown LD_{50} values lower than that dose. The LD_{50} of orally administered stinging nettle leaf infusion in rats is reported as 1.31 g/kg (Baraibar et al. 1983). In mice intraperitoneally administered extracts of stinging nettle, the LD_{50} of the leaf infusion is reported as 1.92 g/kg (Baraibar et al. 1983), 3.5 g/kg (Bnouham et al. 2003), and 3.625 g/kg (Lasheras 1986), whereas the decoction has an LD_{50} value of 1.72 g/kg (Baraibar et al. 1983).

Genotoxicity

No mutagenic activity of saline, aqueous, or chloroform extracts of the aerial parts of stinging nettle was observed in *Salmonella typhimurium* strains TA98 or TA100 with or without metabolic activation (Basaran et al. 1996). The same extracts showed some activity in the comet assay, although the relevance of results of that test to humans is difficult to determine (Basaran et al. 1996). Weak genotoxic activity, similar to that of the flavonoids rutin and quercetin, of a stinging nettle aqueous extract was observed in the *Drosophila* wing somatic mutation and recombination test (Graf et al. 1994).

Literature Cited

Antonopoulou, S., C.A. Demopoulos, and N.K. Andrikopoulos. 1996. Lipid separation from *Urtica dioica*: Existence of platelet-activating factor. *J. Agric. Food Chem.* 44:3052-3056.

Baraibar, C., F.J. Broncano, M.J. Lazaro-Carrasco, M. Rebuelta, and L. Villanua. 1983. Toxicity study of *Urtica dioica* L. nettles. *An. Bromatol.* 35(1):99-104.

Basaran, A.A., T.W. Yu, M.J. Plewa, and D. Anderson. 1996. An investigation of some Turkish herbal medicines in *Salmonella typhimurium* and in the COMET assay in human lymphocytes. *Teratog. Carcinog. Mutagen.* 16(2):125-138.

Bathe, A.P. 1994. An unusual manifestation of nettle rash in three horses. *Vet. Rec.* 134(1):11-12.

Bnouham, M., F.Z. Merhfour, A. Ziyyat, et al. 2003. Antihyperglycemic activity of the aqueous extract of *Urtica dioica*. *Fitoterapia* 74(7-8):677-681.

Broncano, F.J. 1983. Etude de l'effet sur le centre cardiovasculaire de quelques préparations de l'*Urtica dioica* L. *Planta Med.* 17:222-229.

Caliskaner, Z., M. Karaayvaz, and S. Ozturk. 2004. Misuse of a herb: Stinging nettle (*Urtica urens*) induced severe tongue oedema. *Complement. Ther. Med.* 12(1):57-58.

Chrubasik, S., W. Enderlein, R. Bauer, and W. Grabner. 1997. Evidence for antirheumatic effectiveness of *Herba Urticae dioicae* in acute arthritis: A pilot study. *Phytomedicine* 4(2):105-108.

Collier, H.O.J., and G.B. Chesher. 1956. Identification of 5-hydroxytryptamine in the sting of the nettle (*Urtica dioica*). *Br. J. Pharmacol. Chemother.* 11(2):186.

Conwell, R., and C. Findlay. 2008. Nettle reaction in a horse. *Vet. Rec.* 162(8):256.

Czarnetzki, B.M., T. Thiele, and T. Rosenbach. 1990. Immunoreactive leukotrienes in nettle plants (*Urtica urens*). *Int. Arch. Allergy Appl. Immunol.* 91(1):43-46.

Edwards, E. 1992. Immediate and delayed hypersensitivity to the nettle plant. *Contact Dermat.* 27(4):264-265.

El Haouari, M., M. Bnouham, M. Bendahou, et al. 2006. Inhibition of rat platelet aggregation by *Urtica dioica* leaves extracts. *Phytother. Res.* 20(7):568-572.

Emmelin, N., and W. Feldberg. 1947. The mechanism of the sting of the common nettle (*Urtica urens*). *J. Physiol.* 106(4):440.

ESCOP. 2003. *ESCOP monographs: The scientific foundation for herbal medicinal products*. 2nd ed. Exeter, UK: European Scientific Cooperative on Phytotherapy.

Farzami, B., D. Ahmadvand, S. Vardasbi, F.J. Majin, and S. Khaghani. 2003. Induction of insulin secretion by a component of *Urtica dioica* leave extract in perifused islets of Langerhans and its *in vivo* effects in normal and streptozotocin diabetic rats. *J. Ethnopharmacol.* 89(1):47-53.

Fu, H.Y., S.J. Chen, R.F. Chen, et al. 2006. Identification of oxalic acid and tartaric acid as major persistent pain-inducing toxins in the stinging hairs of the nettle, *Urtica thunbergiana*. *Ann. Bot.* 98(1):57-65.

Ganora, L. 2009. *Herbal constituents: Foundations of phytochemistry*. Louisville, CO: Herbalchem Press.

Goun, E.A., V.M. Petrichenko, S.U. Solodnikov, et al. 2002. Anticancer and antithrombin activity of Russian plants. *J. Ethnopharmacol.* 81(3):337-342.

Graf, U., A. Alonso Moraga, R. Castro, and E. Diaz Carrillo. 1994. Genotoxicity testing of different types of beverages in the *Drosophila* wing somatic mutation and recombination test. *Food Chem. Toxicol.* 32(5):423-430.

Kirchhoff, H.W. 1983. *Urtica* juice as a diuretic. *Z. Phytother.* 4:621-626.

Lasheras, B. 1986. Etude pharmacologique préliminaire de *Prunus spinosa* L., *Amelanchier ovalis* medikus, *Juniperus communis* L. et *Urtica dioica* L. *Plant. Med. Phytother.* 20:219-226.

Lovell, C.R. 1993. *Plants and the skin*. Boston: Blackwell Scientific Publications.

Mekhfi, H., M. El Haouari, A. Legssyer, et al. 2004. Platelet anti-aggregant property of some Moroccan medicinal plants. *J. Ethnopharmacol.* 94(2-3):317-322.

Oliver, F., E.U. Amon, A. Breathnach, et al. 1991. Contact urticaria due to the common stinging nettle (*Urtica dioica*)—Histological, ultrastructural and pharmacological studies. *Clin. Exp. Dermatol.* 16(1):1-7.

Sahin, M., H. Yilmaz, A. Gursoy, et al. 2007. Gynaecomastia in a man and hyperoestrogenism in a woman due to ingestion of nettle (*Urtica dioica*). *N. Z. Med. J.* 120(1265):U2803.

Sajid, T.M., S. Rashid, and S.A. Saeed. 1991. Inhibition of adrenaline-induced aggregation of human platelets by Pakistani medicinal plants. *Pak. J. Pharm. Sci.* 4(2):145-152.

Scholz, H., S. Kascha, and H. Zingerle. 1980. Atropine poisoning from "health tea." *Fortschr. Med.* 98(39):1525-1526.

Sharma, B.B., M.D. Varshney, D.N. Gupta, and A.O. Prakash. 1983. Antifertility screening of plants. Part I. Effect of ten indigenous plants on early pregnancy in Albino rats. *Pharm. Biol.* 21(4):183-187.

Swanston-Flatt, S.K., C. Day, P.R. Flatt, B.J. Gould, and C.J. Bailey. 1989. Glycaemic effects of traditional European plant treatments for diabetes. Studies in normal and streptozotocin diabetic mice. *Diabetes Res.* 10(2):69-73.

Tahri, A., S. Yamani, A. Legssyer, et al. 2000. Acute diuretic, natriuretic and hypotensive effects of a continuous perfusion of aqueous extract of *Urtica dioica* in the rat. *J. Ethnopharmacol.* 73(1-2):95-100.

Testai, L., S. Chericoni, V. Calderone, et al. 2002. Cardiovascular effects of *Urtica dioica* L. (Urticaceae) roots extracts: *In vitro* and *in vivo* pharmacological studies. *J. Ethnopharmacol.* 81(1):105-109.

Tita, B., P. Faccendini, U. Bello, L. Martinoli, and P. Bolle. 1993. *Urtica dioica* L.: Pharmacological effect of ethanol extract. *Pharmacol. Res.* 27 (Suppl. 1):21-22.

Yarnell, E. 1998. Stinging nettle: A modern view of an ancient healing plant. *Altern. Complement. Ther.* 4:180-186.

U

Urtica dioica L. ssp. *dioica* — Urticaceae

SCN: stinging nettle
OCN: nettle

Part: root

Quick Reference Summary

Safety Class: 1
Interaction Class: A

Contraindications

None known.

Other Precautions

None known.

Drug and Supplement Interactions

None known.

Adverse Events and Side Effects

A review of 15 clinical studies, with a total of over 16,000 participants, indicated that stinging nettle root was generally well tolerated with no serious adverse events reported (ESCOP 2003).

Pharmacological Considerations

None known.

Pregnancy and Lactation

No information on the safety of stinging nettle root in pregnancy or lactation was identified in the scientific or traditional literature. Although this review did not identify any concerns for use while pregnant or nursing, safety has not been conclusively established.

Review Details

I. Drug and Supplement Interactions

Clinical Trials of Drug or Supplement Interactions

No clinical trials of drug or supplement interactions were identified.

Case Reports of Suspected Drug or Supplement Interactions

No case reports of suspected drug or supplement interactions were identified.

Animal Trials of Drug or Supplement Interactions

No animal trials of drug or supplement interactions were identified.

II. Adverse Events

Adverse Events Reported in Clinical Trials

A review of 15 clinical studies of stinging nettle root, with a total of over 16,000 participants, indicated that no serious adverse events were reported in any of the studies. Daily doses taken by participants were up to 756 mg of hydroalcoholic dry native extract for up to 6 months, although several studies included doses of 300 mg daily for 24 months. Adverse events reported in the studies were primarily mild gastrointestinal upset, with fewer than 5% of participants experiencing such events (ESCOP 2003). In a study of stinging nettle root extract, adverse events deemed probably or possibly related to treatment occurred in approximately 1% of the 1319 participants (Kaldewey 1995).

Case Reports of Adverse Events

No case reports of adverse events related to stinging nettle root were identified.

III. Pharmacology and Pharmacokinetics

Human Pharmacological Studies

No relevant human pharmacological studies were identified.

Animal Pharmacological Studies

No relevant animal pharmacological studies were identified.

In Vitro Pharmacological Studies

No relevant in vitro pharmacological studies were identified.

IV. Pregnancy and Lactation

No information on the safety of stinging nettle root in pregnancy or lactation was identified.

V. Toxicity Studies

Acute Toxicity

The LD_{50} of orally administered stinging nettle root extract in rats could not be determined at doses up to 30 g/kg (Chrubasik et al. 2007).

Genotoxicity

No mutagenic activity of a stinging nettle root extract was observed in the Ames mutagenicity test with

Salmonella typhimurium strains TA98 and TA100 at concentrations of extract up to 5000 μg/plate (de Meester and Leonard 1988).

Cytotoxicity

No effects on cell viability were observed in mouse white blood cells (splenocytes and peritoneal macrophages) treated with an aqueous extract of stinging nettle root (Harput et al. 2005).

Literature Cited

Chrubasik, J.E., B.D. Roufogalis, H. Wagner, and S. Chrubasik. 2007. A comprehensive review on the stinging nettle effect and efficacy profiles. Part II: *Urticae radix*. *Phytomedicine* 14(7-8):568-579.

de Meester, C., and A. Leonard. 1988. Ames reversion mutation test with *Salmonella typhimurium*. Brussels: Laboratory of Teratogenesis and Mutagenesis, Université Catholique de Louvani.

ESCOP. 2003. *ESCOP monographs: The scientific foundation for herbal medicinal products*. 2nd ed. Exeter, UK: European Scientific Cooperative on Phytotherapy.

Harput, U.S., I. Saracoglu, and Y. Ogihara. 2005. Stimulation of lymphocyte proliferation and inhibition of nitric oxide production by aqueous *Urtica dioica* extract. *Phytother. Res.* 19(4):346-348.

Kaldewey, W. 1995. Behandlung der benignen Prostatahyperplasie und der Prostatitis mit einem standardisierten *Urticae-radix*-Extrakt. *Urologe B* 35:430-433.

Usnea barbata (L.) F.H. Wigg.

Usneaceae

SCN: usnea

OCN: usnea lichen; old man's beard

Part: lichen

Quick Reference Summary

Safety Class: 1

Interaction Class: A

Contraindications

None known.

Other Precautions

None known.

Drug and Supplement Interactions

None known.

Editors' Note

Most safety concerns reported for usnea are based on studies of the compound usnic acid, which is present at a concentration of 1 to 3% (Cansaran et al. 2006, 2008; Cocchietto et al. 2002). Data regarding this isolated compound may not apply directly to products or extracts made from usnea.

Adverse Events and Side Effects

A case of liver failure was reported in a woman who had taken 500 mg of the compound usnic acid daily for 2 weeks (Durazo et al. 2004). Other cases of liver toxicity have been reported in persons taking different multiherbal products containing the compound usnic acid (Durazo et al. 2004; Favreau et al. 2002; Neff et al. 2004; Sanchez et al. 2006). Daily doses of usnic acid in the products used ranged from 300 to 1350 mg (Guo et al. 2008).

Pharmacological Considerations

None known.

Pregnancy and Lactation

No information on the safety of usnea in pregnancy or lactation was identified in the scientific or traditional literature. Although this review did not identify any concerns for use while pregnant or nursing, safety has not been conclusively established.

Review Details

I. Drug and Supplement Interactions

Clinical Trials of Drug or Supplement Interactions

No clinical trials of drug or supplement interactions were identified.

Case Reports of Suspected Drug or Supplement Interactions

No case reports of suspected drug or supplement interactions were identified.

U

Animal Trials of Drug or Supplement Interactions

No animal trials of drug or supplement interactions were identified.

II. Adverse Events

Case Reports of Adverse Events

Fulminant liver failure requiring a transplant was reported in a 28-year-old woman who had taken 500 mg of the compound usnic acid daily for 2 weeks, followed by a 2-week break, and then again for 4 days. The woman had not taken any other drugs or supplements, aside from two doses of a weight-loss product that contained albumin, oligopeptides, gelatin, cellulose, and magnesium 3 weeks prior to the onset of symptoms (Durazo et al. 2004).

III. Pharmacology and Pharmacokinetics

Human Pharmacological Studies

No relevant human pharmacological studies were identified.

Animal Pharmacological Studies

No relevant animal pharmacological studies were identified.

In Vitro Pharmacological Studies

Usnic acid was tested for activity with the drug-metabolizing isoenzymes CYP1A2, CYP2B6, CYP2C8, CYP2C9, CYP2C19, CYP2D6, and CYP3A4/5. Results suggested that the oxidative metabolism of usnic acid was mainly mediated by CYP1A2. Usnic acid was a weak inhibitor of CYP2D6, a potent inhibitor of CYP2C19 and CYP2C9, and a less potent inhibitor of CYP2C8 and CYP2C18 (Foti et al. 2008).

In a study of usnic acid in isolated rat hepatocytes, treatment with 100 or 1000 μM usnic acid for 1 hour induced the release of hepatic transaminases (AST and ALT), decreased the content of reduced glutathione, and caused loss of cell membrane integrity. The hepatotoxin carbon tetrachloride and usnic acid exhibited similar cellular responses, suggesting that the compounds elicited activity via the same mechanisms (Pramyothin et al. 2004).

In mouse primary hepatocytes, treatment with 5 μM usnic acid for 16 hours resulted in 98% cell death that appeared to be associated with cell necrosis rather than apoptosis. Results indicated that usnic acid was associated with inhibition and uncoupling of the electron transport chain in mitochondria. Usnic acid triggered oxidative stress by increasing free radical generation, and the oxidative stress appears to be critical in usnic acid-induced hepatotoxicity (Han et al. 2004).

IV. Pregnancy and Lactation

No information on the safety of usnea in pregnancy or lactation was identified.

V. Toxicity Studies

Acute Toxicity

The LD_{50} of a hydroalcoholic extract of usnea in rats is 22.53 g/kg after intraperitoneal administration, 7.43 g/kg after intravenous administration, and could not be determined at doses up to 32 g/kg after oral administration (Dobrescu et al. 1993).

In male rats intraperitoneally administered 50 or 200 mg/kg of the compound usnic acid daily for 5 days, swelling of the liver mitochondria and endoplasmic reticulum were observed by electron microscopy. Serum transaminase activity was unchanged (Pramyothin et al. 2004).

In rats orally administered single doses of 500, 1000, or 2000 mg/kg of the compound usnic acid, no signs of toxicity were observed at doses up to 1000 mg/kg, while unspecified "toxic effects" were observed at the 2000 mg/kg dose (Odabasoglu et al. 2006).

Short-Term Toxicity

In mice and rats orally administered 5, 30, 100, or 200 mg/kg of the compound usnic acid daily for 14 days, liver toxicity was observed at doses of 100 mg/kg or higher (Guo et al. 2008).

In mice intraperitoneally administered 15 mg/kg of usnic acid daily for 15 days, elevated levels of serum transaminase activity and extensive liver necrosis were observed. No signs of toxicity in other organs, such as the kidney or spleen, were detected. A similar pattern of toxicity was revealed in tumor-bearing mice (da Silva Santos et al. 2006; Ribeiro-Costa et al. 2004).

In sheep fed diets containing usnic acid at doses of 323 to 776 mg/kg daily for up to 9 days, lethargy and anorexia, and in some cases death, were observed. The estimated median toxic dose was between 485 and 647 mg/kg. Serum lactate dehydrogenase, aspartate aminotransferase, and creatine kinase were elevated at those doses. Complete postmortem examination revealed that pathological changes in the sheep occurred exclusively in skeletal muscles (Dailey et al. 2008). A review of usnic acid toxicology noted that the skeletal muscle toxicity observed in sheep is in sharp contrast to findings in mice, rats, and humans, in which the liver is considered to be the most vulnerable organ to usnic acid (Guo et al. 2008).

A reduction in weight gain was observed in female guinea pigs with tuberculosis subcutaneously administered 20 mg/animal of usnic acid daily for 6 days. No apparent organ-specific toxicities in the liver, spleen, or lung were observed (Marshak and Kuschner 1950).

Genotoxicity

In the Ames test for mutagenicity with *Salmonella typhimurium* strains TA98 and TA100, no mutagenic activity of the compound usnic acid was observed in either strain without or with activation by S9 (Shibamoto and Wei 1984).

No genotoxic activity of the compounds (+)-usnic acid and (–)-usnic acid was observed in the cytokinesis-blocked micronucleus assay with lymphocytes from two healthy male donors (Koparal et al. 2006).

Literature Cited

Cansaran, D., S. Aras, and O. Atakol. 2008. Determination of usnic acid content in some lichen species found in Anatolia. *J. Appl. Biol. Sci.* 2(3):41-44.

Cansaran, D., D. Kahya, E. Yurdakulol, and O. Atakol. 2006. Identification and quantitation of usnic acid from the lichen *Usnea* species of Anatolia and antimicrobial activity. *Z. Naturfor.* 61(11/12):773.

Cocchietto, M., N. Skert, P.L. Nimis, and G. Sava. 2002. A review on usnic acid, an interesting natural compound. *Naturwissenschaften* 89:137-146.

da Silva Santos, N.P., S.C. Nascimento, M.S.O. Wanderley, et al. 2006. Nanoencapsulation of usnic acid: An attempt to improve antitumour activity and reduce hepatotoxicity. *Eur. J. Pharm. Biopharm.* 64(2):154-160.

Dailey, R.N., D.L. Montgomery, J.T. Ingram, et al. 2008. Toxicity of the lichen secondary metabolite (+)-usnic acid in domestic sheep. *Vet. Pathol. Online* 45(1):19.

Dobrescu, D., M. Tanasescu, A. Mezdrea, et al. 1993. Contributions to the complex study of some lichens—*Usnea* genus. Pharmacological studies on *Usnea barbata* and *Usnea hirta* species. *Rom. J. Physiol.* 30(1-2):101-107.

Durazo, F.A., C. Lassman, S.H.B. Han, et al. 2004. Fulminant liver failure due to usnic acid for weight loss. *Am. J. Gastroenterol.* 99(5):950-952.

Favreau, J.T., M.L. Ryu, G. Braunstein, et al. 2002. Severe hepatotoxicity associated with the dietary supplement LipoKinetix. *Ann. Int. Med.* 136(8):590-595.

Foti, R.S., L.J. Dickmann, J.A. Davis, et al. 2008. Metabolism and related human risk factors for hepatic damage by usnic acid containing nutritional supplements. *Xenobiotica* 38(3):264-280.

Guo, L., Q. Shi, J.L. Fang, et al. 2008. Review of usnic acid and *Usnea barbata* toxicity. *J. Environ. Sci. Health C Environ. Carcinog. Ecotoxicol. Rev.* 26(4):317-338.

Han, D., K. Matsumaru, D. Rettori, and N. Kaplowitz. 2004. Usnic acid-induced necrosis of cultured mouse hepatocytes: Inhibition of mitochondrial function and oxidative stress. *Biochem. Pharmacol.* 67 (3):439-451.

Koparal, A.T., B.A. Tüylü, and H. Türk. 2006. In vitro cytotoxic activities of (+)-usnic acid and (–)-usnic acid on V79, A549, and human lymphocyte cells and their non-genotoxicity on human lymphocytes. *Nat. Prod. Res.* 20(14):1300.

Marshak, A., and M. Kuschner. 1950. The action of streptomycin and usnic acid on the development of tuberculosis in guinea pigs. *Public Health Rep.* 65(5):131-162.

Neff, G.W., K. Rajender Reddy, F.A. Durazo, et al. 2004. Severe hepatotoxicity associated with the use of weight loss diet supplements containing ma huang or usnic acid. *J. Hepatol.* 41(6):1062-1064.

Odabasoglu, F., A. Cakir, H. Suleyman, et al. 2006. Gastroprotective and antioxidant effects of usnic acid on indomethacin-induced gastric ulcer in rats. *J. Ethnopharmacol.* 103(1):59-65.

Pramyothin, P., W. Janthasoot, N. Pongnimitprasert, S. Phrukudom, and N. Ruangrungsi. 2004. Hepatotoxic effect of (+) usnic acid from *Usnea siamensis* Wainio in rats, isolated rat hepatocytes and isolated rat liver mitochondria. *J. Ethnopharmacol.* 90(2-3):381-387.

Ribeiro-Costa, R.M., A.J. Alves, N.P. Santos, et al. 2004. In vitro and in vivo properties of usnic acid encapsulated into PLGA-microspheres. *J. Microencaps.* 21(4):371-384.

Sanchez, W., J.T. Maple, L.J. Burgart, and P.S. Kamath. 2006. Severe hepatotoxicity associated with use of a dietary supplement containing usnic acid. *Mayo Clin. Proc.* 81:541-544.

Shibamoto, T., and C.I. Wei. 1984. Mutagenicity of lichen constituents. *Environ. Mutagen.* 6:757-762.

U

Vaccinium myrtillus L. — Ericaceae

SCN: bilberry
OCN: European blueberry; huckleberry; whortleberry

Part: fruit

Quick Reference Summary

Safety Class: 1
Interaction Class: A

Contraindications

None known.

Other Precautions

None known.

Drug and Supplement Interactions

None known.

Editors' Note

Bilberry fruit is traditionally consumed as a food (Upton 2001).

Adverse Events and Side Effects

A systematic review of clinical trials of bilberry or anthocyanosides of bilberry indicated that no significant adverse events were reported in any of the trials (Canter and Ernst 2004).

A postmarketing survey of persons taking bilberry fruit extract indicated that respondents reported adverse events involving the gastrointestinal tract, skin, or nervous system, although no causal association between bilberry and the symptoms could be established (Eandi 1987).

Pharmacological Considerations

A study on the use of bilberry fruit extract prior to surgery showed reduced bleeding during surgery (Gentile 1987). One ex vivo human study, one animal study, and two in vitro studies indicated that bilberry fruit may reduce blood coagulation (Gomez-Serranillos et al. 1983; Morazzoni and Magistretti 1990; Pulliero et al. 1989; Zaragoza et al. 1985).

Pregnancy and Lactation

A limited number of human studies of bilberry fruit extract used during pregnancy have not shown any adverse effects on mother or fetus (Eandi 1987; Grismondi 1980; Pourrat et al. 1967; Zaragoza et al. 1985). No information was identified on the safety of bilberry fruit or leaf during lactation, although traditional consumption as a food suggests that no adverse effects are expected from bilberry fruit (Upton 2001).

Review Details

I. Drug and Supplement Interactions

Clinical Trials of Drug or Supplement Interactions

No clinical trials of drug or supplement interactions were identified.

Case Reports of Suspected Drug or Supplement Interactions

No case reports of suspected drug or supplement interactions were identified.

Animal Trials of Drug or Supplement Interactions

No animal trials of drug or supplement interactions were identified.

II. Adverse Events

Adverse Events Reported in Clinical Trials

A systematic review of placebo-controlled studies on anthocyanosides of bilberry fruit indicated that no adverse effects were reported in any of the clinical trials on bilberry extracts (Canter and Ernst 2004). In a postmarketing survey of persons taking bilberry fruit extract, 94 of the 2295 survey respondents noted side effects, mostly related to the gastrointestinal, skin, or nervous system. Due to the nature of the survey, a causal association between bilberry and the symptoms could not be established (Eandi 1987).

Case Reports of Adverse Events

No case reports of adverse events were identified.

III. Pharmacology and Pharmacokinetics

Human Pharmacological Studies

After oral administration of bilberry fruit extract (173 mg daily) to healthy volunteers, platelet aggregation was inhibited in blood samples tested ex vivo after 30 days of bilberry administration and was further inhibited after 60 days of consumption (Pulliero et al. 1989). In contrast, preoperative treatment with bilberry fruit standardized extract (160–320 mg daily) in patients undergoing ear, nose, or throat surgery significantly reduced the incidence and severity of intraoperative and postoperative bleeding and hemorrhagic complications (Gentile 1987). One older study reported that bilberry anthocyanins had a tendency to reduce hemorrhagic retinopathy due to anticoagulant therapy (Scharrer and Ober 1981).

Animal Pharmacological Studies

A bilberry fruit extract high in anthocyanosides administered to rats in single oral doses from 2.5 to 400 mg/kg significantly increased bleeding time at doses of 5 mg/kg or more (Morazzoni and Magistretti 1990).

In Vitro Pharmacological Studies

Two in vitro studies of bilberry fruit anthocyanosides indicated that inhibition of platelet aggregation by bilberry was greater than that of aspirin (Gomez-Serranillos et al. 1983; Zaragoza et al. 1985).

IV. Pregnancy and Lactation

No treatment-related adverse events were reported in a study of pregnant women taking a standardized bilberry fruit extract (320 mg anthocyanins daily) for 60 to 90 days, starting in the sixth month of pregnancy (Grismondi 1980), nor in pregnant women taking a standardized bilberry fruit extract (160–320 mg daily) for 90 days (Teglio et al. 1987).

A standardized bilberry fruit extract demonstrated no adverse effects on fertility in rats (Eandi 1987). Administration of anthocyanins or standardized bilberry fruit extracts did not produce any teratogenic activity in a single generation of rats or in three generations of rats and rabbits (Eandi 1987; Pourrat et al. 1967).

No information on the safety of bilberry fruit during lactation was identified. Based on the traditional safe consumption of bilberry fruit as a food, no adverse effects are expected (Upton 2001).

V. Toxicity Studies

Acute Toxicity

The LD_{50} of intraperitoneally administered standardized bilberry fruit extract in rats is 2.4 g/kg, and 0.24 g/kg for intravenously administered extract. No lethal dose could be determined for orally administered standardized extract at doses up to 20 g/kg in rats (Pourrat et al. 1967). In mice, the LD_{50} of intraperitoneally administered standardized bilberry fruit extract is 4.1 g/kg, and 0.84 g/kg for intravenously administered extract. No lethal dose could be determined for an orally administered standardized extract at doses up to 25 g/kg in mice (Pourrat et al. 1967). Other acute toxicity studies indicated that no signs of toxicity were observed for orally administered bilberry fruit extract at doses over 2 g/kg in mice and rats and over 3 g/kg in dogs. A darkening of urine and feces was noted in the animals (Eandi 1987).

Short-Term Toxicity

No toxic effects were observed in guinea pigs administered bilberry fruit extract at doses up to 43 mg/kg daily for 2 weeks, nor in rats fed the same dose for 6 weeks (Pourrat et al. 1967).

Similarly, no toxic effects were observed in rats administered up to 36 mg/kg bilberry fruit extract intravenously every day for 4 weeks or in dogs administered 12 mg/kg bilberry extract intravenously every day for 13 weeks, although dark blue pigmentation of urine, skin, eyes, and sometimes liver, kidneys, and ovaries was observed in the rats and dogs (Eandi 1987).

Chronic Toxicity

No changes in urinary, hematological, gross, or microscopic parameters were observed after oral administration of standardized bilberry fruit extract in rats at doses of 125 to 500 mg/kg daily nor in dogs administered 80 to 320 mg/kg daily for 6 months (Eandi 1987).

Literature Cited

Canter, P.H., and E. Ernst. 2004. Anthocyanosides of *Vaccinium myrtillus* (bilberry) for night vision—A systematic review of placebo-controlled trials. *Surv. Ophthalmol.* 49(1):38-50.

Eandi, M. 1987. Relazione del'esperto sulla Documentazione Farmacologica e Tossicologica Relative alla Specialità Tegens. Inverni della Beffa SpA. *Cited in* Morazzoni, P., and E. Bombardelli. 1996. *Vaccinium myrtillus* L. *Fitoterapia* 67(1):3-29.

Gentile, A. 1987. Relazione clinica nel' impiego preventivo a scopo antiemorragico degli antociandioli del mirtillo (Tegens Inverni della Beffa) in chirurgia otorinolaringoiatrica. Milan, Italy: Indena. Unpublished study. *Cited in* Morazzoni, P., and E. Bombardelli. 1996. *Vaccinium myrtillus* L. *Fitoterapia* 67(1):3-29.

Gomez-Serranillos, F., F. Zaragoza, and P. Alvarez. 1983. Efectos sobre la agregacion plaquetaria 'in vitro' de los antocianosidos del *Vaccinium myrtillus* L. *An. R. Acad. Farm.* 49:79.

Grismondi, G. 1980. Contributo al trattamento delle flebopatie da stasi in gravidanza. *Min. Gin.* 32(1):1-10.

Morazzoni, P., and M.J. Magistretti. 1990. Activity of bilberry, an anthocyanoside complex from *Vaccinium myrtillus* (VMA), on platelet aggregation and adhesiveness. *Fitoterapia* 61(1):13-21.

Pourrat, H., P. Bastide, P. Dorier, A. Pourrat, and A. Tronche. 1967. Préparation et activité thérapeutique de quelques glycosides d'anthocyanes. *Chim. Ther.* 2:33-38.

Pulliero, G., S. Montin, V. Bettini, et al. 1989. Ex vivo study of the inhibitory effects of *Vaccinium myrtillus* anthocyanosides on human platelet aggregation. *Fitoterapia* 60:69-74.

Scharrer, A., and M. Ober. 1981. Anthocyanosides in the treatment of retinopathies. *Klin. Monatsbl. Augenheilkd.* 178:386-389.

Teglio, L., C. Mazzanti, R. Tronconi, and E. Guerresi. 1987. *Vaccinium myrtillus* anthocyanosides (Tegens) in the treatment of venous insufficiency of lower limbs and acute piles in pregnancy [in Italian]. *Q. Clin. Ostet. Ginecol.* 42:221. *Cited in* Morazzoni, P., and E. Bombardelli. 1996. *Vaccinium myrtillus* L. *Fitoterapia* 67(1):3-29.

Upton, R. 2001. *Bilberry fruit: Vaccinium myrtillus L.: Standards of analysis, quality control, and therapeutics. American Herbal Pharmacopoeia and therapeutic compendium.* Santa Cruz, CA: American Herbal Pharmacopoeia.

Zaragoza, F., I. Iglesias, and J. Benedi. 1985. Comparative study of the anti-aggregation effects of anthocyanosides and other agents. *Arch. Farmacol. Toxicol.* 11(3):183-188.

Vaccinium spp. — Ericaceae

Vaccinium angustifolium Aiton
SCN: blueberry
OCN: low-bush blueberry

Vaccinium corymbosum **L.**
SCN: blueberry
OCN: giant whortleberry; high-bush blueberry

Vaccinium pallidum **Aiton**
SCN: blueberry
Part: fruit

Quick Reference Summary

Safety Class: 1
Interaction Class: A

Contraindications

None known.

Other Precautions

None known.

Drug and Supplement Interactions

None known.

Editors' Note

Blueberry fruit is traditionally consumed as a food (McGee 2004; Rombauer et al. 1997).

Adverse Events

No adverse events were reported in recent clinical trials of blueberry fruit (Hiraishi et al. 1995; Kay and Holub 2002; Pedersen et al. 2000).

Pharmacological Considerations

None known.

Pregnancy and Lactation

No information was identified on the safety of blueberry fruit during pregnancy or lactation, although traditional consumption as a food suggests that no adverse effects are expected.

Review Details

I. Drug and Supplement Interactions

Clinical Trials of Drug or Supplement Interactions

No clinical trials of drug or supplement interactions were identified.

Case Reports of Suspected Drug or Supplement Interactions

No case reports of suspected drug or supplement interactions were identified.

Animal Trials of Drug or Supplement Interactions

No animal trials of drug or supplement interactions were identified.

II. Adverse Events

Adverse Events Reported in Clinical Trials

No adverse events were reported in recent clinical trials of blueberry fruit (Hiraishi et al. 1995; Kay and Holub 2002; Pedersen et al. 2000).

Case Reports of Adverse Events

No case reports of adverse events were identified.

III. Pharmacology and Pharmacokinetics

Human Pharmacological Studies

No relevant human pharmacological studies were identified.

Animal Pharmacological Studies

No relevant animal pharmacological studies were identified.

In Vitro Pharmacological Studies

No relevant in vitro pharmacological studies were identified.

IV. Pregnancy and Lactation

No information on the use of blueberry fruit during pregnancy or lactation was identified, although traditional consumption of blueberry as a food suggests that no adverse reactions are expected.

V. Toxicity Studies

No toxicological studies of blueberry fruit or leaf were identified.

Literature Cited

Hiraishi, K., I. Narabayashi, O. Fujita, et al. 1995. Blueberry juice: Preliminary evaluation as an oral contrast agent in gastrointestinal MR imaging. *Radiology* 194(1):119-123.

Kay, C.D., and B.J. Holub. 2002. The effect of wild blueberry (*Vaccinium angustifolium*) consumption on postprandial serum antioxidant status in human subjects. *Br. J. Nutr.* 88(4):389-398.

McGee, H. 2004. *On food and cooking: The science and lore of the kitchen*. New York: Simon and Schuster.

Pedersen, C.B., J. Kyle, A.M. Jenkinson, et al. 2000. Effects of blueberry and cranberry juice consumption on the plasma antioxidant capacity of healthy female volunteers. *Eur. J. Clin. Nutr.* 54(5):405-408.

Rombauer, I.S., M.R. Becker, E. Becker, and M. Guarnaschelli. 1997. *Joy of cooking*. New York: Simon and Schuster.

Vaccinium spp. — Ericaceae

Vaccinium angustifolium Aiton
SCN: blueberry
OCN: low-bush blueberry

Vaccinium myrtillus L.
SCN: bilberry
OCN: European blueberry; huckleberry; whortleberry

Vaccinium pallidum **Aiton**
SCN: blueberry
Part: leaf

Quick Reference Summary

Safety Class: 2d
Interaction Class: A

Contraindications

Not for long-term use (Wichtl 2004).

Other Precautions

None known.

Drug and Supplement Interactions

None known.

Adverse Events and Side Effects

None known.

Pharmacological Considerations

None known.

Pregnancy and Lactation

No information on the safety of bilberry or blueberry leaf in pregnancy or lactation was identified in the scientific or traditional literature. Although this review did not identify any concerns for use while pregnant or nursing, safety has not been conclusively established.

V

Review Details

I. Drug and Supplement Interactions

Clinical Trials of Drug or Supplement Interactions

No clinical trials of drug or supplement interactions were identified.

Case Reports of Suspected Drug or Supplement Interactions

No case reports of suspected drug or supplement interactions were identified.

Animal Trials of Drug or Supplement Interactions

No animal trials of drug or supplement interactions were identified.

II. Adverse Events

Case Reports of Adverse Events

No case reports of adverse events were identified.

III. Pharmacology and Pharmacokinetics

Human Pharmacological Studies

No relevant human pharmacological studies were identified.

Animal Pharmacological Studies

No relevant animal pharmacological studies were identified.

In Vitro Pharmacological Studies

No relevant in vitro pharmacological studies were identified.

IV. Pregnancy and Lactation

No information on the safety of bilberry or blueberry leaf during pregnancy or lactation was identified.

V. Toxicity Studies

Chronic Toxicity

In animal studies, chronic high doses (up to 1.5 g/kg) of bilberry leaf have been associated with symptoms of chronic intoxication such as muscular atrophy, weight loss, anemia, yellow discoloration of the skin, and acute states of excitation (Wichtl 2004).

Literature Cited

Wichtl, M. 2004. *Herbal drugs and phytopharmaceuticals: A handbook for practice on a scientific basis*. 3rd ed. Boca Raton, FL: CRC Press.

Vaccinium spp.

Ericaceae

Vaccinium macrocarpon Aiton
SCN: cranberry
OCN: American cranberry; large cranberry

Vaccinium oxycoccos L.
SCN: cranberry
OCN: small cranberry
Part: fruit

Quick Reference Summary

Safety Class: 1
Interaction Class: A

Contraindications

None known.

Other Precautions

None known.

Drug and Supplement Interactions

None known.

Editors' Note

The relatively small number of adverse events reported to date in comparison to the widespread consumption of cranberry juice as a beverage (USDA 2007) provides an indication of the relative low risk of adverse effects.

Adverse Events and Side Effects

Single cases of high levels of potassium, metabolic acidosis, and kidney stones have been reported in persons consuming cranberry juice (Garcia-Calatayud et al. 2002; Terris et al. 2001; Thomson and Perry 2001). The relationship between cranberry juice and those events has not been determined.

Pharmacological Considerations

Although several case reports have indicated a concern for a potential interaction between cranberry juice and warfarin (CSM 2003; Grant 2004; Rindone and Murphy 2006; Suvarna et al. 2003), human clinical trials have indicated no interaction at amounts up to 480 ml (~2 cups) daily (Ansell et al. 2009; Li et al. 2006; Lilja et al. 2007; Mellen et al. 2010). A review of clinical trials and published and unpublished case reports indicated that there is limited evidence to support an interaction between cranberry juice and warfarin (Zikria et al. 2010). A study with cranberry juice concentrate indicated a possible interaction with warfarin (Abdul et al. 2008).

Human studies have provided conflicting data on the association of cranberry extract consumption with calcium oxalate stone formation (Brinkley et al. 1981; Gettman et al. 2005; Kahn et al. 1967; Leahy et al. 2001; Light et al. 1973; Massey et al. 1993; Terris et al. 2001).

Pregnancy and Lactation

No information was identified on the safety of cranberry during pregnancy or lactation, although traditional consumption as a food and beverage suggests that no adverse effects are expected.

Review Details

I. Drug and Supplement Interactions

Clinical Trials of Drug or Supplement Interactions

Two studies in patients stable on warfarin indicated no significant changes in prothrombin time or INR (INR, or international normalized ratio, is a system used to report the results of blood coagulation tests, with elevated INR indicating a long blood clotting time) after administration of 250 or 480 ml of cranberry juice daily for 7 or 14 days (Li et al. 2006; Mellen et al. 2010). In another study of patients stable on warfarin, ingestion of 240 ml cranberry juice daily for 2 weeks resulted in a mild increase in INR in 8 of 30 patients. The mean INR level was increased only on the 12th day of treatment. Cranberry juice had no effect on plasma levels of warfarin (Ansell et al. 2009).

In a study of cranberry juice (200 ml daily) and warfarin in healthy volunteers, no clinically significant effects of cranberry juice on warfarin metabolism were observed after 10 days of treatment (Lilja et al. 2007).

Increased INR was observed in healthy volunteers orally administered 1000 mg cranberry juice concentrate (equivalent to 57 g of cranberries) daily, before or after single doses of 25 mg warfarin. Plasma levels of warfarin were unchanged (Abdul et al. 2008).

A study of cranberry juice (240 ml) and cyclosporine indicated that a single dose of cranberry juice had no effect on cyclosporine metabolism, whereas a single dose of a citrus fruit juice (pomelo) had a significant effect (Grenier et al. 2006).

Consumption of cranberry juice (100–150 ml daily) for 7 weeks was reported to interfere with urinary dipstick tests for glucose and hemoglobin (Kilbourn 1987).

Case Reports of Suspected Drug or Supplement Interactions

Several detailed cases of possible interactions between cranberry juice and warfarin have been reported in the literature. In all cases, the patients had elevated INR levels. In two cases, the patients were in poor health and consumed almost nothing except cranberry juice for approximately 2 weeks (Griffiths et al. 2008; Suvarna et al. 2003). In other cases, the patients were drinking approximately 700 or 2000 ml daily of cranberry juice, and INR levels returned to normal after cessation of cranberry (Grant 2004; Rindone and Murphy 2006). One case of elevated INR was in a patient who had been consuming turkey sandwiches with 113 g of cranberry sauce daily during the week following the Thanksgiving holiday (Mergenhagen and Sherman 2008). Between 1999 and 2003, five cases of increased INR due to possible interactions between cranberry and warfarin, including one reported death, were reported to the British Committee on the Safety of Medicines. Details of the cases were not published (CSM 2003).

In a systematic review of interactions between warfarin and selected foods, the interaction potential for cranberry was ranked as "possible" on a scale that included the rankings of "highly probable, probable, possible, and highly improbable" (Holbrook et al. 2005). A review of case reports and clinical trial data on the potential interaction between cranberry and warfarin concluded that the available information does not support a clinically relevant interaction, but that patients taking warfarin should be cautioned about the potential interaction (Pham and Pham 2007). A third review on cranberry and warfarin suggested that large amounts of cranberry juice may adversely affect patients on warfarin, while small amounts of juice are not expected to cause an interaction (Aston et al. 2006).

Animal Trials of Drug or Supplement Interactions

No relevant animal trials of drug or supplement interactions were identified.

II. Adverse Events

Adverse Events Reported in Clinical Trials

A review of clinical trials on cranberry juice indicated a high dropout rate for many of the trials, although no significant adverse events were reported in the trials reviewed (Jepson et al. 2007).

Case Reports of Adverse Events

One case of hyperkalemia was reported in a man who consumed 2 liters of cranberry juice daily for several days. The man was on other medications and had a history of traumatic injury and urinary problems (Thomson and Perry 2001).

One case of intoxication was reported in a 4-month-old infant fed 180 ml of cranberry juice. Symptoms of intoxication included diarrhea, hyperglycemia, and metabolic acidosis (Garcia-Calatayud et al. 2002).

One case of kidney stones was reported in a man who had been taking cranberry concentrate tablets for 6 months. The man had passed two kidney stones 6 years prior (Terris et al. 2001).

III. Pharmacology and Pharmacokinetics

Human Pharmacological Studies

In a clinical trial in healthy volunteers, no effects of cranberry juice (200 ml daily for 10 days or 2 doses of 240 ml) were observed on the drug-metabolizing isoenzymes cytochrome P450 CYP2C9, CYP1A2, or CYP3A4 (Greenblatt et al. 2006; Lilja et al. 2007).

Cranberry concentrate tablets taken for 7 days were shown in one study to significantly increase urinary oxalate levels and urinary calcium, phosphate, and sodium along with magnesium and potassium, both inhibitors of urinary stone formation (Terris et al. 2001). The study has been criticized for methodological flaws such as failing to measure oxalate content of the studied tablets and not assessing dietary intake of calcium and vitamin C, contributors to urinary oxalates (Leahy et al. 2001).

A single dose of 500 ml cranberry juice had no effect on urinary oxalate secretion but significantly increased mean urinary calcium levels (Brinkley et al. 1981). A significant increase in urinary calcium excretion was observed in a small study after consumption of 2 pints of cranberry juice (Kahn et al. 1967). Consumption of 2 pints daily of cranberry juice for 1 month reduced urinary ionized calcium by 50% (Light et al. 1973). A study on dietary intake and urinary excretion of oxalates indicated that cranberry juice had no effect on oxalate excretion (Massey et al. 1993).

In a study of cranberry juice (1 liter daily for 7 days) on urinary tract stone formation risk in normal subjects and in subjects with a history of calcium oxalate stone formation, significant increases in urinary calcium and oxalate levels and a decrease in urinary pH were observed. These results suggest that cranberry juice may increase the risk of calcium oxalate and uric acid stone formation but reduce the risk of brushite stone formation (Gettman et al. 2005).

In patients with a history of struvite kidney stones, consumption of 2 pints of cranberry juice daily for 9 years resulted in no recurrence in 60% of participants, no increase in stone size in 32%, and new stone formation or an increase in stone size in 6%. The study did not include a control group (Zinsser et al. 1968).

Animal Pharmacological Studies

No relevant animal pharmacological studies were identified.

In Vitro Pharmacological Studies

An in vitro trial of cranberry juice on human liver and rat small intestine microsomes demonstrated that cranberry juice significantly inhibited the activity of human and rat drug-metabolizing cytochrome P450 isoenzyme CYP3A (Uesawa and Mohri 2006).

IV. Pregnancy and Lactation

No information on the safety of cranberry fruit during pregnancy or lactation was identified. Based on the traditional safe consumption of cranberry as a food, no adverse effects are expected (Mills and Bone 2005; Upton 2002).

V. Toxicity Studies

No toxicity studies on cranberry were identified.

Literature Cited

Abdul, M.I.M., X. Jiang, K.M. Williams, et al. 2008. Pharmacodynamic interaction of warfarin with cranberry but not with garlic in healthy subjects. *Br. J. Pharmacol.* 154(8):1691.

Ansell, J., M. McDonough, Y. Zhao, J.S. Harmatz, and D.J. Greenblatt. 2009. The absence of an interaction between warfarin and cranberry juice: A randomized, double-blind trial. *J. Clin. Pharmacol.* 49(7):824.

Aston, J.L., A.E. Lodolce, and N.L. Shapiro. 2006. Interaction between warfarin and cranberry juice. *Pharmacotherapy* 26(9):1314-1319.

Brinkley, L., J. McGuire, J. Gregory, and C.Y. Pak. 1981. Bioavailability of oxalate in foods. *Urology* 17(6):534-538.

CSM. 2003. Possible interaction between warfarin and cranberry juice. Committee on the Safety of Medicines. *Curr. Prob. Pharmacovigilance* 29:8.

Garcia-Calatayud, S., J.J. Larreina Cordoba, and M.J. Lozano De La Torre. 2002. Severe cranberry juice poisoning. *An. Esp. Pediatr.* 56(1):72-73.

Gettman, M.T., K. Ogan, L.J. Brinkley, et al. 2005. Effect of cranberry juice consumption on urinary stone risk factors. *J. Urol.* 174(2):590-594; quiz 801.

Grant, P. 2004. Warfarin and cranberry juice: An interaction? *J. Heart Valve Dis.* 13(1):25-26.

Greenblatt, D.J., L.L. von Moltke, E.S. Perloff, et al. 2006. Interaction of flurbiprofen with cranberry juice, grape juice, tea, and fluconazole: In vitro and clinical studies. *Clin. Pharmacol. Ther.* 79(1):125-133.

Grenier, J., C. Fradette, G. Morelli, et al. 2006. Pomelo juice, but not cranberry juice, affects the pharmacokinetics of cyclosporine in humans. *Clin. Pharmacol. Ther.* 79(3):255-262.

Griffiths, A.P., A. Beddall, and S. Pegler. 2008. Fatal haemopericardium and gastrointestinal haemorrhage due to possible interaction of cranberry juice with warfarin. *J. Roy. Soc. Prom. Health* 128(6):324-326.

Holbrook, A.M., J.A. Pereira, R. Labiris, et al. 2005. Systematic overview of warfarin and its drug and food interactions. *Arch. Intern. Med.* 165(10):1095-1106.

Jepson, R.G., L. Mihaljevic, and J. Craig. 2007. Cranberries for preventing urinary tract infections. *Cochrane Database Syst. Rev.* 2:CD001321.

Kahn, H., V. Panariello, J. Saeli, J. Sampson, and E. Schwartz. 1967. Implications for therapy of urinary tract infection and calculi: Effect of cranberry juice on urine. *J. Am. Diet. Assoc.* 51:251-254.

Kilbourn, J.P. 1987. Interference with dipstick tests for glucose and hemoglobin in urine by ascorbic acid in cranberry juice. *Clin. Chem.* 33(7):1297.

Leahy, M., R. Roderick, and K. Brilliant. 2001. The cranberry—Promising health benefits, old and new. *Nutr. Today* 36(5):254-265.

Li, Z., N.P. Seeram, C.L. Carpenter, et al. 2006. Cranberry does not affect prothrombin time in male subjects on warfarin. *J. Am. Diet. Assoc.* 106(12):2057-2061.

Light, I., E. Gursel, and H. Zinnser. 1973. Urinary ionized calcium in urolithiasis. *Urology* 1(1):67-70.

Lilja, J.J., J.T. Backman, and P.J. Neuvonen. 2007. Effects of daily ingestion of cranberry juice on the pharmacokinetics of warfarin, tizanidine, and midazolam—Probes of CYP2C9, CYP1A2, and CYP3A4. *Clin. Pharmacol. Ther.* 81(6):833-839.

Massey, L.K., H. Roman-Smith, and R.A. Sutton. 1993. Effect of dietary oxalate and calcium on urinary oxalate and risk of formation of calcium oxalate kidney stones. *J. Am. Diet. Assoc.* 93(8):901-906.

Mellen, C.K., M. Ford, and J.P. Rindone. 2010. Effect of high dose cranberry juice on the pharmacodynamics of warfarin in patients. *Br. J. Clin. Pharmacol.* 70(1):139-142.

Mergenhagen, K.A., and O. Sherman. 2008. Elevated international normalized ratio after concurrent ingestion of cranberry sauce and warfarin. *Am. J. Health Syst. Pharm.* 65(22):2113-2116.

Mills, S., and K. Bone. 2005. *The essential guide to herbal safety.* St. Louis: Elsevier.

Pham, D.Q., and A.Q. Pham. 2007. Interaction potential between cranberry juice and warfarin. *Am. J. Health Syst. Pharm.* 64(5):490-494.

Rindone, J.P., and T.W. Murphy. 2006. Warfarin-cranberry juice interaction resulting in profound hypoprothrombinemia and bleeding. *Am. J. Ther.* 13(3):283-284.

Suvarna, R., M. Pirmohamed, and L. Henderson. 2003. Possible interaction between warfarin and cranberry juice. *Br. Med. J.* 327(7429):1454.

Terris, M.K., M.M. Issa, and J.R. Tacker. 2001. Dietary supplementation with cranberry concentrate tablets may increase the risk of nephrolithiasis. *Urology* 57(1):26-29.

Thomson, F., and L. Perry. 2001. Hyperkalaemia associated with cranberry juice. *Pharm. Prac.* 11(7):215-216.

Uesawa, Y., and K. Mohri. 2006. Effects of cranberry juice on nifedipine pharmacokinetics in rats. *J. Pharm. Pharmacol.* 58(8):1067-1072.

Upton, R. 2002. *Cranberry fruit: Vaccinium macrocarpon Aiton: Standards of analysis, quality control, and therapeutics. American Herbal Pharmacopoeia and therapeutic compendium.* Santa Cruz, CA: American Herbal Pharmacopoeia.

USDA. 2007. Fruit juices: Per capita consumption. In *Food availability (per capita) data system.* Washington, DC: U.S. Dept. of Agriculture, Economic Research Service.

Zikria, J., R. Goldman, and J. Ansell. 2010. Cranberry juice and warfarin: When bad publicity trumps science. *Am. J. Med.* 123(5):384-392.

Zinsser, H.H., H. Seneca, I. Light, et al. 1968. Management of infected stones with acidifying agents. *N.Y. State J. Med.* 68(23):3001-3010.

Valeriana spp. — Valerianaceae

Valeriana edulis Nutt. ex Torr. & A. Gray ssp. ***procera*** (Kunth) F.G. Mey.
SCN: Mexican valerian
Syn: *Valeriana procera* Kunth

Valeriana jatamansi Jones
SCN: Indian valerian
Syn: *Nardostachys jatamansi* (Jones) DC.; *Valeriana wallichii* DC.
AN: *sugandhabala; tagara*

Valeriana officinalis L.
SCN: valerian
Syn: *Valeriana exaltata* J.C. Mikan
AN: *sugandhbala; tagara*
OCN: garden heliotrope; garden valerian

Valeriana sitchensis Bong.
SCN: Pacific valerian
Syn: *Valeriana scouleri* Rydb.
OCN: mountain heliotrope; Sitka valerian
Part: rhizome, root

Quick Reference Summary

Safety Class: 1
Interaction Class: B

Contraindications

None known.

V

Other Precautions

None known.

Drug and Supplement Interactions

Caution is advised during use of barbiturates, benzodiazepines, and other sedative drugs, as valerian has been speculated to potentiate the effects of some sedatives (Brinker 2001).

Editors' Note

Although reports of toxicity of compounds called valepotriates have been published, such reports are not considered to be clinically relevant, as valepotriates are absorbed poorly and degrade quickly into less toxic metabolites (Bradley 1992; Leung and Foster 1996).

Adverse Events and Side Effects

A systematic review of valerian clinical trials indicated that adverse events in valerian groups were reported as diarrhea in one study and headache and gastrointestinal disturbance in another study. In the remaining studies reviewed, adverse events were either similar to placebo or were not observed (Bent et al. 2006).

Cases of liver-associated adverse effects have been reported for 2 women taking products that apparently contained standardized valerian extracts in the U.S. (Cohen and del Toro 2008) and Greece (Vassiliadis et al. 2009) and one other in France who used valerian as a tea (Mennecier et al. 1999). The identity of the valerian ingredient was not verified in any of these three reports. Each case resolved after the ingested product was discontinued. The authors of the U.S. event expressed their belief "that this represents a case of valerian-associated hepatotoxicity," while acknowledging that, "there is no specific diagnostic test to confirm that valerian was the cause of the liver abnormalities in this case" (Cohen and del Toro 2008). The report from Greece concludes, "We believe that the present case represents an idiosyncratic reaction to valerian herb manifested as acute hepatitis" (Vassiliadis et al. 2009).

Pharmacological Considerations

Although some sources have indicated caution for the use of valerian while driving or operating heavy machinery (Bradley 1992), clinical trials have shown no impairment of cognitive or psychomotor performance at doses up to 1800 mg (Albrecht et al. 1995; Gutierrez et al. 2004; Hallam et al. 2003; Mills and Bone 2005).

Although valerian is generally recognized to have sedative and anxiolytic effects, clinical herbalists have reported that valerian may have stimulant effects, including agitation and increased heart rate, in a small percentage (~2–3%) of the population (Kuhn and Winston 2007).

Human studies have indicated a lack of effect of valerian on the CYP450 drug-metabolizing isoenzymes (Donovan et al. 2004; Gurley et al. 2005).

Pregnancy and Lactation

Animal studies and human case reports have indicated no adverse effects of relatively high doses (2.8 g/kg) of valerian in pregnancy (Czeizel et al. 1997; Tufik et al. 1994; Yao et al. 2003, 2007).

No information on the safety of valerian during lactation was identified in the scientific or traditional literature. While this review did not identify any concerns for use while nursing, safety has not been conclusively established.

Review Details

I. Drug and Supplement Interactions

Clinical Trials of Drug or Supplement Interactions

No clinical trials of drug or supplement interactions were identified.

Case Reports of Suspected Drug or Supplement Interactions

No case reports of suspected drug or supplement interactions were identified.

Animal Trials of Drug or Supplement Interactions

Coadministration of thiopental and 2 mg/kg of an aqueous alkaline extract of valerian to mice increased thiopental-induced sleeping time by a factor of 1.6. Administration of 200 mg/kg increased the thiopental-induced sleeping time by a factor of 7.6 (Leuschner et al. 1993).

In rats provided with drinking water containing 1% valerian tincture (estimated dose 200–250 mg/kg daily) for 12 weeks with intramuscular injection of haloperidol (equivalent to 1 mg/kg daily), an increase in lipid peroxidation levels and dichlorofluorescein-reactive species production was observed in the hepatic tissue. In the liver and kidneys, δ-aminolevulinate dehydratase activity was inhibited. Serum alanine aminotransferase (ALT) was elevated, as compared to controls, while aspartate aminotransferase (AST) levels were unchanged (Dalla Corte et al. 2008).

In rats administered isoflurane alone (5%), isoflurane plus valerian (30 mg/kg), isoflurane plus midazolam (2 mg/kg), or isoflurane plus valerian (30 mg/kg) and midazolam (2 mg/kg), no significant differences in emergence time from anesthesia were observed in the isoflurane plus valerian group. Delayed emergence was observed in the isoflurane plus midazolam group, with a greater delay observed in the group receiving isoflurane, valerian, and midazolam combination (Chaplin et al. 2007).

II. Adverse Events

Adverse Events Reported in Clinical Trials

In a systematic review of 16 clinical trials of valerian, one study was noted to have a significant increase in diarrhea in the valerian group as compared to the control group; in other studies, adverse events were either similar to placebo or were not observed (Bent et al. 2006). Adverse events reported in one clinical trial of valerian included headache and gastrointestinal disturbance (Seifert 1988).

Case Reports of Adverse Events

A case of hepatotoxicity was reported in a 27-year-old woman who had been taking 300 mg daily of valerian for the prior 3 months. Liver enzyme levels returned to normal within 4 weeks after valerian was discontinued (Cohen and del Toro 2008). Elevated liver enzyme levels were observed in a 50-year-old woman who had been taking 5 ml of valerian extract three times per week for 3 weeks along with 10 tablets (unclear as to whether this was a single or total dose) containing 125 mg valerian extract each. Liver enzyme levels returned to normal within 2 months after discontinuation of valerian use (Vassiliadis et al. 2009). A case of acute hepatitis was reported in a woman who had been drinking valerian tea for several weeks (Mennecier et al. 1999). The authors of the 2008 report express their "belie[f] that this represents a case of valerian-associated hepatotoxicity," while acknowledging that, "there is no specific diagnostic test to confirm that valerian was the cause of the liver abnormalities in this case" (Cohen and del Toro 2008). The report from 2009 concludes, "We believe that the present case represents an idiosyncratic reaction to valerian herb manifested as acute hepatitis" (Vassiliadis et al. 2009).

In an attempted suicide, a woman took 20 g of valerian powder. Thirty minutes after ingestion, she experienced fatigue, abdominal pain, chest tightness, tremors, and lightheadedness. Her blood pressure was low, and her liver enzymes were normal. The woman was treated with activated charcoal, and all symptoms resolved within 24 hours (Willey et al. 1995). One case of chest and abdominal pain was reported in a woman who had intravenously administered herself a homemade aqueous extract of valerian (Wells 1995).

A case of heart failure with delirium was reported in a man with a significant history of cardiovascular disease who had taken valerian (500–2000 mg daily) for 5 years and was also taking numerous other medications and supplements (isosorbide dinitrate, digoxin, furosemide, benazepril, aspirin, lovastatin, ibuprofen, potassium, zinc, and vitamins). The reporting physicians stated that valerian is considered to exert a benzodiazepine-like action through the enhancement of γ-aminobutyric acid neurotransmission; they suggested that symptoms reported in this case were similar to benzodiazepine withdrawal symptoms and hypothesized that the cardiac failure was due to valerian withdrawal (Garges et al. 1998).

Delirium was reported in a woman who had taken valerian regularly for 6 months, along with St. John's wort and loperamide (Khawaja et al. 1999).

A syncopal episode and paranoia were reported in a woman who had been taking valerian for 2 years and prior to the episode consumed 3.5 liters of wine, over 1.25 liters of vodka, and an unspecified amount of ginkgo (Chen et al. 2002).

III. Pharmacology and Pharmacokinetics

Human Pharmacological Studies

In healthy volunteers orally administered 375 mg of valerian daily for 28 days, no significant changes in activity of the drug-metabolizing isoenzymes CYP1A2, CYP2D6, CYP2E1, or CYP3A4/5 were observed. Probe drugs used in the study were midazolam (CYP3A4/5), caffeine (CYP1A2), chlorzoxazone (CYP2E1), and debrisoquin (CYP2D6) (Gurley et al. 2005).

In healthy volunteers orally administered 1000 mg valerian nightly for 14 nights, mild, clinically insignificant inhibition of the drug-metabolizing isoenzyme CYP3A4 was observed, with no changes in CYP2D6 activity (Donovan et al. 2004).

In studies of valerian (500–1800 mg) on cognitive and psychomotor performance, no significant effects were observed on any of the parameters measured, including reaction time and concentration (Albrecht et al. 1995; Gutierrez et al. 2004; Hallam et al. 2003).

Animal Pharmacological Studies

In rats and mice orally administered alcohol extracts of valerian at doses up to 1000 mg/kg, anxiolytic and antidepressant effects were observed without sedative or muscle relaxant activity (Hattesohl et al. 2008).

In Vitro Pharmacological Studies

In cultured human hepatocytes, an extract of valerian was found to weakly induce the drug-metabolizing isoenzyme CYP2C19, inhibit CYP3A4 and CYP2D6, and had no effect on CYP2E1 (Hellum et al. 2007, 2009).

In tests of hepatotoxicity, human hepatoma cells incubated in vitro with 2.0 and 20 mg/ml valerian (no further description provided) showed some cell death only at the higher concentration (Vo et al. 2003).

IV. Pregnancy and Lactation

In pregnant rats orally administered 2790 mg/kg (~65 times the human dose) of an ethanolic valerian extract on gestational days 1 to 8 or 8 to 15, no adverse effects on fetal development and no signs of maternal toxicity were observed (Yao et al. 2007).

Observations of pregnant rats fed varying concentrations of valerian on days 1 through 8 of pregnancy indicated that the highest nontoxic dose of valerian was 2.8 g/kg. Doses above 2.8 g/kg resulted in reduced placental weight but not reduced birth weight as compared to control (Yao et al. 2003).

In female rats fed a mixture of valepotriates, 6 to 24 mg/kg daily, for 30 days before pregnancy or on days 1 through 19 of pregnancy, no effects on fertility or signs of fetotoxicity were observed. An increased number of fetuses with slowed bone formation were observed with the higher doses (12 and 24 mg/kg) of valerian (Tufik et al. 1994).

No adverse effects of valerian were noted in three cases termed by the report authors as "intentional valerian overdose" (2 to 5 g) during weeks 3 to 10 of pregnancy (Czeizel et al. 1997). Authoritative sources list the standard dose of valerian as 1-5 g daily (ESCOP 2003; Wichtl 2004).

No information on the safety of valerian during lactation was identified.

V. Toxicity Studies

Acute Toxicity

The LD_{50} of an intraperitoneally administered valerian extract (9.5:1 ethanol extract) in mice is 3.3 g/kg (Rosecrans et al. 1961). The LD_{50} of orally administered valerian essential oil in rats is 15 g/kg (Skramlik 1959). The LD_{50} of orally administered Mexican valerian extract is 3.8 g/kg (Deciga-Campos et al. 2007).

Short-Term Toxicity

No remarkable adverse effects were observed in rats fed 600 mg/kg valerian extract daily for 30 days (Fehri et al. 1991).

Subchronic Toxicity

In rats intraperitoneally administered 400 to 600 mg/kg of valerian daily for 45 days, no significant changes in weight, blood, or urine were observed as compared to controls (Rosecrans et al. 1961).

In rats fed valerian at daily doses of 3.1 g/kg for 28 days or a single oral dose of 18.6 g/kg, no changes in bile flow, liver enzyme activity, or liver histology were observed (Vo et al. 2003).

Genotoxicity

In the Ames test for mutagenicity, a tincture of Mexican valerian did not show signs of mutagenicity, although a methanol-dichloromethane extract induced mutations in *S. typhimurium* TA100, in the presence of human liver S9 fractions, at the highest concentration tested (Deciga-Campos et al. 2007).

In a test for genotoxicity, valerian showed no mutagenic activity in fruit flies (*Drosophila melanogaster*) exposed to concentrations of valerian aqueous extract up to 55.2 mg/ml (Romero-Jimenez et al. 2005).

Literature Cited

Albrecht, M., W. Berger, and P. Laux. 1995. Psychopharmaceuticals and safety in traffic. *Z. Allgemeinmed.* 71:1215-1225.

Bent, S., A. Padula, D. Moore, M. Patterson, and W. Mehling. 2006. Valerian for sleep: A systematic review and meta-analysis. *Am. J. Med.* 119(12):1005-1012.

Bradley, P.R. 1992. *British herbal compendium: A handbook of scientific information on widely used plant drugs.* Bournemouth, UK: British Herbal Medicine Association.

Brinker, F. 2001. *Herb contraindications and drug interactions.* 3rd ed. Sandy, OR: Eclectic Medical Publications.

Chaplin, R.L., Jr., J. Jedynak, D. Johnson, et al. 2007. The effects of valerian on the time course of emergence from general anesthesia in Sprague-Dawley rats (*Rattus norvegicus*). *Am. Assoc. Nurse Anesthetists J.* 75(6):431-435.

Chen, D., J. Klesmer, A. Giovanniello, and J. Katz. 2002. Mental status changes in an alcohol abuser taking valerian and *Ginkgo biloba*. *Am. J. Addict.* 11(1):75-77.

Cohen, D.L., and Y. del Toro. 2008. A case of valerian-associated hepatotoxicity. *J. Clin. Gastroenterol.* 42(8):961-962.

Czeizel, A.E., M. Tomcsik, and L. Timar. 1997. Teratologic evaluation of 178 infants born to mothers who attempted suicide by drugs during pregnancy. *Obstet. Gynecol.* 90(2):195-201.

Dalla Corte, C.L., R. Fachinetto, D. Colle, et al. 2008. Potentially adverse interactions between haloperidol and valerian. *Food Chem. Toxicol.* 46(7):2369-2375.

Deciga-Campos, M., I. Rivero-Cruz, M. Arriaga-Alba, et al. 2007. Acute toxicity and mutagenic activity of Mexican plants used in traditional medicine. *J. Ethnopharmacol.* 110(2):334-342.

Donovan, J.L., C.L. DeVane, K.D. Chavin, et al. 2004. Multiple night-time doses of valerian (*Valeriana officinalis*) had minimal effects on CYP3A4 activity and no effect on CYP2D6 activity in healthy volunteers. *Drug Metab. Dispos.* 32(12):1333-1336.

ESCOP. 2003. *ESCOP monographs: The scientific foundation for herbal medicinal products.* 2nd ed. Exeter, U.K.: European Scientific Cooperative on Phytotherapy.

Fehri, B., J.M. Aiache, K. Boukef, A. Memmi, and B. Hizaoui. 1991. [*Valeriana officinalis* and *Crataegus oxyacantha*: Toxicity from repeated administration and pharmacologic investigations.] *J. Pharm. Belg.* 46(3):165-176.

Garges, H.P., I. Varia, and P.M. Doraiswamy. 1998. Cardiac complications and delirium associated with valerian root withdrawal. *J. Am. Med. Assoc.* 280(18):1566-1567.

Gurley, B.J., S.F. Gardner, M.A. Hubbard, et al. 2005. In vivo effects of goldenseal, kava kava, black cohosh, and valerian on human cytochrome P450 1A2, 2D6, 2E1, and 3A4/5 phenotypes. *Clin. Pharmacol. Ther.* 77(5):415-426.

Gutierrez, S., M.K. Ang-Lee, D.J. Walker, and J.P. Zacny. 2004. Assessing subjective and psychomotor effects of the herbal medication valerian in healthy volunteers. *Pharmacol. Biochem. Behav.* 78(1):57-64.

Hallam, K.T., J.S. Olver, C. McGrath, and T.R. Norman. 2003. Comparative cognitive and psychomotor effects of single doses of *Valeriana officianalis* and triazolam in healthy volunteers. *Human Psychopharmacol.* 18(8):619-625.

Hattesohl, M., B. Feistel, H. Sievers, et al. 2008. Extracts of *Valeriana officinalis* L. s.l. show anxiolytic and antidepressant effects but neither sedative nor myorelaxant properties. *Phytomedicine* 15(1-2):2-15.

Hellum, B.H., Z. Hu, and O.G. Nilsen. 2007. The induction of CYP1A2, CYP2D6 and CYP3A4 by six trade herbal products in cultured primary human hepatocytes. *Basic Clin. Pharmacol. Toxicol.* 100(1):23-30.

Hellum, B.H., Z. Hu, and O.G. Nilsen. 2009. Trade herbal products and induction of CYP2C19 and CYP2E1 in cultured human hepatocytes. *Basic Clin. Pharmacol. Toxicol.* 105(1):58-63.

Khawaja, I.S., R.F. Marotta, and S. Lippmann. 1999. Herbal medicines as a factor in delirium. *Psychiatr. Serv.* 50(7):969-970.

Kuhn, M.A., and D. Winston. 2007. *Herbal therapy and supplements: A scientific and traditional approach.* Philadelphia: Lippincott, Williams & Wilkins.

Leung, A.Y., and S. Foster. 1996. *Encyclopedia of common natural ingredients used in food, drugs, and cosmetics.* 2nd ed. New York: Wiley.

Leuschner, J., J. Müller, and M. Rudmann. 1993. Characterization of the central nervous depressant activity of a commercially available valerian root extract. *Arzneim. Forsch. Drug Res.* 43[Suppl. 1 (6)]:638-644.

Mennecier, D., T. Saloum, P.M. Dourthe, et al. 1999. Acute hepatitis after phytotherapy. *Presse Med.* 28(18):966.

Mills, S., and K. Bone. 2005. *The essential guide to herbal safety.* St. Louis: Elsevier.

Romero-Jimenez, M., J. Campos-Sanchez, M. Analla, A. Munoz-Serrano, and A. Alonso-Moraga. 2005. Genotoxicity and antigenotoxicity of some traditional medicinal herbs. *Mutat. Res.* 585(1-2):147-155.

Rosecrans, J.A., J.J. Defeo, and H.W. Youngken, Jr. 1961. Pharmacological investigation of certain *Valeriana officinalis* L. extracts. *J. Pharm. Sci.* 50:240-244.

Seifert, T. 1988. Therapeutische Effekte von Baldrian bei nervösen Störungen. *Klin. Praxis Ther.* 2:94-98.

Skramlik, E.V. 1959. On the toxicity and tolerance of ethereal oils. *Pharmazie* 14:435-445.

Tufik, S., K. Fujita, L. Seabra Mde, and L.L. Lobo. 1994. Effects of a prolonged administration of valepotriates in rats on the mothers and their offspring. *J. Ethnopharmacol.* 41(1-2):39-44.

Vassiliadis, T., P. Anagnostis, K. Patsiaoura, et al. 2009. Valeriana hepatotoxicity. *Sleep Med.* 10(8):935.

Vo, L.T., D. Chan, and R.G. King. 2003. Investigation of the effects of peppermint oil and valerian on rat liver and cultured human liver cells. *Clin. Exp. Pharmacol. Physiol.* 30(10):799-804.

Wells, S.R. 1995. Intentional intravenous administration of a crude valerian root extract. North American Congress of Clinical Toxicology Annual Meeting, Rochester, New York, September 16-19, 1995. *J. Toxicol. Clin. Toxicol.* 33(5):542.

Wichtl, M. 2004. *Herbal drugs and phytopharmaceuticals: A handbook for practice on a scientific basis*. 3rd ed. Boca Raton, FL: CRC Press.

Willey, L.B., S.P. Mady, D.J. Cobaugh, and P.M. Wax. 1995. Valerian overdose: A case report. *Vet. Human Toxicol.* 37(4):364-365.

Yao, M., P.D. Brown-Woodman, and H. Ritchie. 2003. Do the herbal remedies feverfew and valerian have an adverse effect on pregnancy outcome in the rat? *Birth Defects Res. A Clin. Mol. Teratol.* 67(2):145-146.

Yao, M., H.E. Ritchie, and P.D. Brown-Woodman. 2007. A developmental toxicity-screening test of valerian. *J. Ethnopharmacol.* 113(2):204-209.

Vanilla spp. — Orchidaceae

Vanilla planifolia Jacks.
SCN: vanilla
Syn: *Vanilla fragrans* (Salisb.) Ames
OCN: Bourbon vanilla; Mexican vanilla; Madagascar vanilla

Vanilla tahitensis **J.W. Moore**
SCN: vanilla
OCN: Tahitian vanilla
Part: fruit

Quick Reference Summary

Safety Class: 1
Interaction Class: A

Contraindications

None known.

Other Precautions

None known.

Drug and Supplement Interactions

None known.

Adverse Events and Side Effects

None known.

Pharmacological Considerations

None known.

Pregnancy and Lactation

No information on the safety of vanilla in pregnancy or lactation was identified in the scientific or traditional literature, although traditional consumption as a food suggests that no adverse effects are expected.

Review Details

I. Drug and Supplement Interactions

Clinical Trials of Drug or Supplement Interactions

No clinical trials of drug or supplement interactions were identified.

Case Reports of Suspected Drug or Supplement Interactions

No case reports of suspected drug or supplement interactions were identified.

Animal Trials of Drug or Supplement Interactions

No animal trials of drug or supplement interactions were identified.

II. Adverse Events

Case Reports of Adverse Events

No case reports of adverse events were identified.

III. Pharmacology and Pharmacokinetics

Human Pharmacological Studies

No relevant human pharmacological studies were identified.

Animal Pharmacological Studies

No relevant animal pharmacological studies were identified.

In Vitro Pharmacological Studies

No relevant in vitro pharmacological studies were identified.

IV. Pregnancy and Lactation

No information on the safety of vanilla during pregnancy or lactation was identified.

V. Toxicity Studies

Acute Toxicity

The oral LD_{50} of the compound vanillin is 3 g/kg in rabbits, 1.58 to 2.8 g/kg in rats, and 1.4 g/kg in guinea pigs

(Deichmann and Kitzmiller 1940; Hake and Rowe 1963; Hodge and Downs 1961; Jenner et al. 1964; Taylor et al. 1964).

The intravenous LD_{50} of the compound vanillin in dogs is 1.32 g/kg (Caujolle and Meynier 1954).

The intraperitoneal LD_{50} of the compound vanillin is 0.47 to 0.78 g/kg in mice, 1.19 g/kg in guinea pigs, and 1.16 g/kg in rats (Caujolle and Meynier 1954; Caujolle et al. 1956; NIOSH 1975).

No deaths and no macroscopic changes in livers were observed in rats orally administered 530 mg/kg of vanillin daily for 4 days (Taylor et al. 1964).

Subchronic Toxicity

No adverse effects were observed in rats orally administered 300 mg/kg of the compound vanillin twice weekly for 14 weeks (Deichmann and Kitzmiller 1940).

Rats fed diets containing doses of 20 mg/kg of vanillin for 18 weeks had no adverse effects, while doses of 64 mg/kg daily for 10 weeks resulted in growth depression and damage to the myocardium, liver, kidney, lung, spleen, and stomach (Deichmann and Kitzmiller 1940).

In rats fed diets containing vanillin for 13 weeks, growth depression and enlargement of the liver, kidney, and spleen were observed at the 5% dietary level (2500 mg/kg daily), while mild changes were observed at the 1% level (500 mg/kg daily), and no changes were observed at the 0.3% level (150 mg/kg daily) (Deichmann and Kitzmiller 1940).

In rats fed diets containing 3000, 10,000, or 50,000 ppm of the compound vanillin (dose equivalent of 150, 500, or 2500 mg/kg daily), no adverse effects were observed at the 3000 ppm level, while mild adverse effects were observed at 10,000 ppm and growth depression and enlargement of the liver, kidneys, and spleen were observed at 50,000 ppm (Hake and Rowe 1963).

Chronic Toxicity

No adverse effects on growth or hematology or macroscopic or microscopic changes in tissues were observed in rats fed diets containing the compound vanillin at 10,000 ppm (500 mg/kg daily) for 16 weeks, 1000 ppm (50 mg/kg daily) for 27 weeks, 20,000 or 50,000 ppm (1000 or 2500 mg/kg daily) for 1 year, or 5000, 10,000, or 20,000 ppm (250, 500, or 1000 mg/kg daily) for 2 years (Hagan et al. 1967).

Literature Cited

Caujolle, F., and D. Meynier. 1954. Comparative toxicity of orthovaniline and of ethylorthovaniline. *C. R. Hebd. Seances Acad. Sci.* 238(26):2576-2578.

Caujolle, F., D. Meynier, and J.M. et Farthouat. 1956. Sur la toxicité des acides et diethylamides de la série vanillique. *C. R. Hebd. Seances Acad. Sci.* 243:609.

Deichmann, W., and K.V. Kitzmiller. 1940. On the toxicity of vanillin and ethyl vanillin for rabbits and rats. *J. Am. Pharm. Assoc.* 29(10):425-428.

Hagan, E.C., W.H. Hansen, O.G. Fitzhugh, et al. 1967. Food flavourings and compounds of related structure. II. Subacute and chronic toxicity. *Food Cosmet. Toxicol.* 5(2):141.

Hake, C.L., and V.K. Rowe. 1963. Ethers. In Patty, F.A., ed., *Industrial hygiene and toxicology.* New York: Wiley.

Hodge, H.C., and W.L. Downs. 1961. The approximate oral toxicity in rats of selected household products. *Toxicol. Appl. Pharmacol.* 3(6):689-695.

Jenner, P.M., E.C. Hagan, J.M. Taylor, E.L. Cook, and O.G. Fitzhugh. 1964. Food flavourings and compounds of related structure. I. Acute oral toxicity. *Food Cosmet. Toxicol.* 2:327.

NIOSH. 1975. Entry No. YW57750. In Christensen, A.E., and T.T. Luginbyhl, eds., *Registry of toxic effects of chemical substances.* Washington, DC: National Institute for Occupational Safety and Health.

Taylor, J.M., P.M. Jenner, and W.I. Jones. 1964. A comparison of the toxicity of some allyl, propenyl, and propyl compounds in the rat. *Toxicol. Appl. Pharmacol.* 6(4):378-387.

Veratrum viride Aiton **Liliaceae**

SCN: American hellebore
OCN: American white hellebore; false hellebore; Indian poke

Part: root

V

Quick Reference Summary

Safety Class: 3
Interaction Class: A

Contraindications

Not for use except under the supervision of an expert qualified in the appropriate use of this substance (Felter and Lloyd 1898; Freis et al. 1949; Prince and Stork 2000; Remington and Wood 1918).

Other Precautions

See Adverse Events and Side Effects below.

Drug and Supplement Interactions

Caution is required for use with blood pressure or cardiac medications (Coe et al. 1950; Elek et al. 1953; Freis et al. 1949).

Adverse Events and Side Effects

Side effects are common with the therapeutic use of American hellebore and include nausea, vomiting, flushing of the face, sweating, and burning of the mouth and throat (Alban et al. 1952; Assali 1950; Assali et al. 1950). Larger doses may produce a slowed heart rate (Crummett et al. 1985; Freis et al. 1949; Jaffe et al. 1990; Prince and Stork 2000).

Symptoms of American hellebore poisoning include vomiting accompanied by nausea, a slowed heart rate (bradycardia), muscular weakness, reduced body temperature, cold sweat, dizziness, faintness, failure of sight, dilation of the pupils, slow and shallow breathing, syncope, and atrioventricular block (interruption of the electrical signal in the heart) (Crummett et al. 1985; Felter and Lloyd 1898; Freis et al. 1949; Jaffe et al. 1990; Prince and Stork 2000; Remington and Wood 1918; Zumoff 1954).

Pharmacological Considerations

Human studies and clinical use have indicated that American hellebore can rapidly reduce blood pressure (Coe et al. 1950; Elek et al. 1953; Freis et al. 1949). American hellebore was used in the 1930s through the 1960s in the treatment of hypertension. Effects were observed after both oral and intravenous administration. Use was discontinued due to the relatively severe side effects (*see* Adverse Events and Side Effects above).

Pregnancy and Lactation

American hellebore was formerly used to treat preeclampsia and eclampsia (pregnancy-induced high blood pressure). Due in part to side effects noted above, use was discontinued in the 1960s (Alban et al. 1952; Assali 1950; Assali et al. 1950, 1954; Bryant and Fleming 1962; Remington and Wood 1918).

Animal studies have reported teratogenic effects of some compounds isolated from American hellebore (Omnell et al. 1990; Young et al. 1976).

No information on the safety of American hellebore during lactation was identified in the scientific or traditional literature. While this review did not identify any concerns for use while nursing, safety has not been conclusively established, and this substance is not recommended for use except under the supervision of an expert qualified in its appropriate use.

Review Details

I. Drug and Supplement Interactions

Clinical Trials of Drug or Supplement Interactions

No clinical trials of drug or supplement interactions were identified.

Case Reports of Suspected Drug or Supplement Interactions

No case reports of suspected drug or supplement interactions were identified.

Animal Trials of Drug or Supplement Interactions

No animal trials of drug or supplement interactions were identified.

II. Adverse Events

Case Reports of Adverse Events

Medical texts from the late 1800s and early 1900s indicate that American hellebore poisoning includes vomiting accompanied by severe nausea and much retching, a slowed pulse, muscular weakness, reduced body temperature, cold sweat, dizziness, faintness, failure of sight, dilation of the pupils, slow and shallow breathing, sleepiness, coma, and unconsciousness, sometimes with stertorous breathing. The rapid onset of emesis from American hellebore reportedly prevents lethal effects (Felter and Lloyd 1898; Remington and Wood 1918).

Several cases of American hellebore poisoning have been reported in persons who mistook the plant for ramps (*Allium tricoccum*), a type of wild edible leek. Cases typically involved vomiting, slowed heart rate, and reduced blood pressure. Some cases include first- and second-degree atrioventricular block (Crummett et al. 1985; Jaffe et al. 1990; Prince and Stork 2000).

A number of cases of poisoning were reported in persons accidentally poisoned by sneezing powders that contained alkaloids from American hellebore (Carlier et al. 1983; Charles et al. 1984). Of nine cases reported together, all in boys aged 7 to 18, one ingested the sneezing powder, five both ingested and inhaled the powder, and three only inhaled the powder. Aside from sneezing, vomiting was reported in all nine patients, epigastric pain was reported in two patients, syncope in three patients, a slowed heart rate in seven patients, and decreased blood pressure in five patients (Carlier et al. 1983).

Applied to the skin, American hellebore is rubefacient, and applied to the nose, causes sneezing (Felter and Lloyd 1898).

III. Pharmacology and Pharmacokinetics

Human Pharmacological Studies

No relevant human pharmacological studies were identified.

Animal Pharmacological Studies

No relevant animal pharmacological studies were identified.

In Vitro Pharmacological Studies

No relevant in vitro pharmacological studies were identified.

IV. Pregnancy and Lactation

In the early 1900s through the 1960s, alkaloids isolated from American hellebore were used in the treatment of pre-eclampsia and eclampsia (pregnancy-induced high blood pressure). Preparations were originally used orally, but later use, as preparations and technologies were developed, was intravenous. Side effects of treatment were common and included nausea, vomiting, flushing of the face, sweating, and burning of the mouth and throat (Alban et al. 1952; Assali 1950; Assali et al. 1950, 1954; Bryant and Fleming 1962; Jaffe et al. 1990; Remington and Wood 1918).

In mice orally administered a single dose of 70, 150, or 300 mg/kg of the compound jervine on day 8, 9, or 10 of pregnancy, teratogenic effects were observed in some strains of mice (C57BL/6J and A/J) but not in others (Swiss N:GP[S] and Swiss Webster). Teratogenic effects included cleft lip, cleft palate, isolated cleft palate, and mandibular and limb malformations. Fetal teratogenicity and maternal and fetal toxicity were highly correlated (Omnell et al. 1990).

Teratogenic effects were observed in hamsters orally administered steroidal alkaloids (cyclopamine and jervine) from American hellebore. The doses used were not reported in the published abstract (Young et al. 1976).

No information on the safety of American hellebore during lactation was identified.

V. Toxicity Studies

Acute Toxicity

The LD_{50} of American hellebore orally administered to rabbits is 300 mg/kg. In rats, the LD_{50} of American hellebore is 118 mg/kg after intraperitoneal administration but could not be determined at doses up to 8 g/kg after oral administration (Perdue et al. 1950).

Subchronic Toxicity

In rats fed diets containing 0.66, 1.8, 2.0, 2.9, or 3.7% American hellebore for 60 days, severe toxicity was observed at the 2.9% level and above. Toxic effects reported were attributed to lack of nutrition (presumably due to vomiting), and no significant adverse effects on organ weights or blood chemistry were observed (Perdue et al. 1950).

Chronic Toxicity

Weight loss and emesis were reported in adult dogs fed diets containing a purified extract of American hellebore. The daily dose was dependent on each dog's recent history of emesis and diet refusal. The mean daily intake of the American hellebore preparation was 0.19 mg/kg, a limit determined by emesis and failure to eat. No organ-specific toxicities were observed after 6 or 13 months of dosing (Gourzis and Bauer 1951).

Growth retardation was observed in rats fed diets containing 7.3, 11, or 19% of a purified extract of American hellebore daily for 12 months. No pathological changes in organs were observed (Gourzis and Bauer 1951).

Literature Cited

Alban, E.J., M.S. Dennis, and C.N. Swanson. 1952. Intravenous *Veratrum viride* in the treatment of toxemia of pregnancy. *Am. J. Obstet. Gynecol.* 64(5):1083-1092.

Assali, N.S. 1950. Studies on *Veratrum viride*: Standardization of intravenous technique and its clinical application in the treatment of toxemia of pregnancy. *Am. J. Obstet. Gynecol.* 60(2):387-394.

Assali, N.S., A.A. Brust, S.T. Garber, and E.B. Ferris. 1950. Comparative study of the effects of tetraethyl-ammonium chloride and *Veratrum viride* on blood pressure in normal and toxemic pregnancy. *J. Clin. Invest.* 29(3):290-296.

Assali, N.S., B. Neme, and J.G. Rosenkrantz. 1954. *Veratrum viride* alkaloids; the hypotensive effects of the mixture (deravine) in human subjects. *Obstet. Gynecol.* 3(3):270-273.

Bryant, R.D., and J.G. Fleming. 1962. *Veratrum viride* in the treatment of eclampsia. III. *Obstet. Gynecol.* 19:372-383.

Carlier, P., M.L. Efthymiou, and R. Garnier. 1983. Poisoning with *Veratrum*-containing sneezing powders. *Human Toxicol.* 2(2):321-325.

Charles, M.H., R. Grimee, and F. Crucke. 1984. Toxicity of sneezing powders. *J. Pharm. Belg.* 39(6):371-382.

Coe, W.S., M.M. Best, and J.M. Kinsman. 1950. *Veratrum viride* in the treatment of hypertensive vascular disease. *J. Am. Med. Assoc.* 143(1):5-7.

Crummett, D., D. Bronstein, and Z. Weaver, 3rd. 1985. Accidental *Veratrum viride* poisoning in three "ramp" foragers. *N.C. Med. J.* 46 (9):469-471.

Elek, S.R., J. Douglas McNair, and G.C. Griffith. 1953. *Veratrum viride*; hypotensive and cardiac effects of intravenous use. *Calif. Med.* 79(4):300-305.

Felter, H.W., and J.U. Lloyd. 1898. *King's American dispensatory*. 18th ed., 3rd rev. 2 vols. Cincinnati: Ohio Valley Co.

Freis, E.D., J.R. Stanton, J.W. Culbertson, et al. 1949. The hemodynamic effects of hypotensive drugs in man. I. *Veratrum viride*. *J. Clin. Invest.* 28(2):353-368.

Gourzis, J., and R.O. Bauer. 1951. Chronic toxicity of a purified extract of *Veratrum viride*. *Proc. Soc. Exp. Biol. Med.* 76(4):767-770.

Jaffe, A.M., D. Gephardt, and L. Courtemanche. 1990. Poisoning due to ingestion of *Veratrum viride* (false hellebore). *J. Emerg. Med.* 8(2):161-167.

Omnell, M.L., F.R.P. Sim, R.F. Keeler, L.C. Harne, and K.S. Brown. 1990. Expression of *Veratrum* alkaloid teratogenicity in the mouse. *Teratology* 42(2):105-119.

Perdue, A.S., F. Bell, et al. 1950. Acute and chronic toxicity of *Veratrum viride*. *J. Am. Pharm. Assoc. A (Sci. Ed.)* 39(2):91-94.
Prince, L.A., and C.M. Stork. 2000. Prolonged cardiotoxicity from poison lily (*Veratrum viride*). *Vet. Human Toxicol.* 42(5):282-285.
Remington, J.P., and H.C. Wood. 1918. *The dispensatory of the United States of America.* 20th ed. Philadelphia: Lippincott.
Young, S., R.F. Keeler, and D. Brown. 1976. Teratologic effects in hamsters of *Veratrum* and *Solanum* steroidal alkaloids [abstract]. *Teratology* 13(2):41A.
Zumoff, B. 1954. Temporary atrioventricular conduction disturbance associated with ingestion of *Veratrum viride*. *Am. Heart J.* 47(4):630-633.

Verbascum spp. — Scrophulariaceae

Verbascum densiflorum Bertol.
SCN: mullein

Verbascum phlomoides L.
SCN: mullein
OCN: orange mullein

Verbascum thapsus L.
SCN: mullein
OCN: Aaron's rod
Part: flower, leaf

Quick Reference Summary

Safety Class: 1
Interaction Class: A

Contraindications

None known.

Other Precautions

None known.

Drug and Supplement Interactions

None known.

Editors' Note

Small hairs on mullein leaf may cause mechanical irritation in the mouth and throat if not filtered out of extracts prior to consumption.

Adverse Events and Side Effects

None known.

Pharmacological Considerations

None known.

Pregnancy and Lactation

No information on the safety of mullein in pregnancy or lactation was identified in the scientific or traditional literature. Although this review did not identify any concerns for use while pregnant or nursing, safety has not been conclusively established.

Review Details

I. Drug and Supplement Interactions

Clinical Trials of Drug or Supplement Interactions

No clinical trials of drug or supplement interactions were identified.

Case Reports of Suspected Drug or Supplement Interactions

No case reports of suspected drug or supplement interactions were identified.

Animal Trials of Drug or Supplement Interactions

No animal trials of drug or supplement interactions were identified.

II. Adverse Events

Case Reports of Adverse Events

No case reports of adverse events were identified.

III. Pharmacology and Pharmacokinetics

Human Pharmacological Studies

No relevant human pharmacological studies were identified.

Animal Pharmacological Studies

No relevant animal pharmacological studies were identified.

In Vitro Pharmacological Studies

No relevant in vitro pharmacological studies were identified.

IV. Pregnancy and Lactation

No information on the safety of mullein in pregnancy or lactation was identified.

V. Toxicity Studies

Acute Toxicity

The LD_{50} of an intraperitoneally administered ethanol extract of mullein in mice is 1 g/kg (Bhakuni et al. 1969).

In the brine shrimp bioassay, extracts of mullein are toxic at concentrations around 1 g/l, with decoctions showing a greater toxicity than infusions (Turker and Camper 2002).

Literature Cited

Bhakuni, O.S., M.L. Dhar, M.M. Dhar, B.N. Dhawan, and B.N. Mehrotra. 1969. Screening of Indian plants for biological activity: Part II. *Indian J. Exp. Biol.* 7:250-262.

Turker, A.U., and N.D. Camper. 2002. Biological activity of common mullein, a medicinal plant. *J. Ethnopharmacol.* 82(2-3):117-125.

Verbena spp.

Verbenaceae

Verbena hastata L.
SCN: blue vervain
OCN: American blue vervain; wild hyssop

Verbena officinalis L. ssp. ***officinalis***
SCN: European vervain
PN: *ma bian cao*
Part: herb

Quick Reference Summary

Safety Class: 2b
Interaction Class: A

Contraindications

Not for use in pregnancy except under the supervision of a qualified healthcare practitioner (Bensky et al. 2004; Chen and Chen 2004).

Other Precautions

None known.

Drug and Supplement Interactions

None known.

Adverse Events and Side Effects

Allergic contact dermatitis from European vervain has been reported (Del Pozo et al. 1994).

Pharmacological Considerations

Coadministration of European vervain tea and iron-fortified bread reduced absorption of the iron (Hurrell et al. 1999).

Pregnancy and Lactation

Texts on traditional Chinese medicine indicate that European vervain should be used with caution during pregnancy (Bensky et al. 2004; Chen and Chen 2004). Based on this information, use during pregnancy is not recommended except under the supervision of a qualified healthcare practitioner.

No information on the safety of European vervain during lactation was identified in the scientific or traditional literature. While this review did not identify any concerns for use while nursing, safety has not been conclusively established.

Review Details

I. Drug and Supplement Interactions

Clinical Trials of Drug or Supplement Interactions

No clinical trials of drug or supplement interactions were identified.

Case Reports of Suspected Drug or Supplement Interactions

No case reports of suspected drug or supplement interactions were identified.

Animal Trials of Drug or Supplement Interactions

No animal trials of drug or supplement interactions were identified.

II. Adverse Events

Case Reports of Adverse Events

Allergic contact dermatitis from European vervain has been reported (Del Pozo et al. 1994).

Overdose of European vervain (standard dose listed as 4.5–30 g as an infusion) may cause abdominal pain, diarrhea, nausea, vomiting, vertigo, and headache. Symptoms reportedly disappear after cessation of the herb (Bensky et al. 2004).

III. Pharmacology and Pharmacokinetics

Human Pharmacological Studies

Coadministration of iron-fortified bread and European vervain tea reduced absorption of the iron by 59%, an effect that was less pronounced than that of black tea, cocoa, or teas made with peppermint or European pennyroyal, but greater than the inhibitory effects of linden flower tea or chamomile tea (Hurrell et al. 1999).

Animal Pharmacological Studies

No relevant animal pharmacological studies were identified.

In Vitro Pharmacological Studies

In an in vitro model of digestion and dialysis used to determine the effects of an infusion of European vervain on iron absorption by infants, two gastric pHs were used: pH 4, as in the first week of life, and pH 2.5, as in older infants. At pH 4 and at pH 2.5, iron availability was decreased by vervain alone and increased by ascorbic acid alone. Iron absorption was increased by the administration of European vervain and ascorbic acid together, but the combined activity was not as high as ascorbic acid alone (Zaida et al. 2006).

In a competitive binding assay in human breast cancer cells with estrogen and progesterone receptors, European vervain bound to both receptor types and was one of the six highest binding herbs of 150 species of herbs tested (Zava et al. 1998).

IV. Pregnancy and Lactation

Texts on traditional Chinese medicine indicate that European vervain should be used with caution during pregnancy (Bensky et al. 2004; Chen and Chen 2004).

Texts from India from the 1950s indicate that European vervain has been listed as an abortifacient. Details on use, such as part and doses used, and whether the plant was used singly or as part of a formula, are lacking (Casey 1960).

No information on the safety of European vervain during lactation was identified.

V. Toxicity Studies

No toxicity studies were identified.

Literature Cited

Bensky, D., S. Clavey, and E. Stöger. 2004. *Chinese herbal medicine: Materia medica.* 3rd ed. Seattle: Eastland Press.

Casey, R.C.D. 1960. Alleged anti-fertility plants of India. *Indian J. Med. Sci.* 14:590-600.

Chen, J.K., and T.T. Chen. 2004. *Chinese medical herbology and pharmacology.* City of Industry, CA: Art of Medicine Press.

Del Pozo, M.D., G. Gastaminza, J.A. Navarro, et al. 1994. Allergic contact dermatitis from *Verbena officinalis* L. *Contact Dermat.* 31(3):200-201.

Hurrell, R.F., M. Reddy, and J.D. Cook. 1999. Inhibition of nonhaem iron absorption in man by polyphenolic-containing beverages. *Br. J. Nutr.* 81(04):289-295.

Zaida, F., F. Bureau, S. Guyot, et al. 2006. Iron availability and consumption of tea, vervain and mint during weaning in Morocco. *Ann. Nutr. Metab.* 50(3):237-241.

Zava, D.T., C.M. Dollbaum, and M. Blen. 1998. Estrogen and progestin bioactivity of foods, herbs, and spices. *Proc. Soc. Exp. Biol. Med.* 217(3):369-378.

Veronicastrum virginicum (L.) Farw. — Scrophulariaceae

SCN: Culver's root
Syn: *Leptandra virginica* (L.) Nutt.
OCN: blackroot
Part: dried root

Quick Reference Summary

V

Safety Class: 1
Interaction Class: A

Contraindications

None known.

Other Precautions

None known.

Drug and Supplement Interactions

None known.

Editors' Notes

Fresh Culver's root is strongly cathartic and has produced violent emesis and bloody purging, accompanied by vertigo. The dried root, which is the subject of this entry, acts as a mild laxative (Felter and Lloyd 1898; Remington and Wood 1918).

Adverse Events and Side Effects

None known.

Pharmacological Considerations

None known.

Pregnancy and Lactation

No information on the safety of Culver's root in pregnancy or lactation was identified in the scientific or traditional literature. Although this review did not identify any concerns for use while pregnant or nursing, safety has not been conclusively established.

Review Details

I. Drug and Supplement Interactions

Clinical Trials of Drug or Supplement Interactions

No clinical trials of drug or supplement interactions were identified.

Case Reports of Suspected Drug or Supplement Interactions

No case reports of suspected drug or supplement interactions were identified.

Animal Trials of Drug or Supplement Interactions

No animal trials of drug or supplement interactions were identified.

II. Adverse Events

Case Reports of Adverse Events

No case reports of adverse events were identified.

III. Pharmacology and Pharmacokinetics

Human Pharmacological Studies

No relevant human pharmacological studies were identified.

Animal Pharmacological Studies

No relevant animal pharmacological studies were identified.

In Vitro Pharmacological Studies

No relevant in vitro pharmacological studies were identified.

IV. Pregnancy and Lactation

No information on the safety of Culver's root during pregnancy or lactation was identified.

V. Toxicity Studies

No toxicity studies were identified.

Literature Cited

Felter, H.W., and J.U. Lloyd. 1898. *King's American dispensatory.* 18th ed., 3rd rev. 2 vols. Cincinnati: Ohio Valley Co.

Remington, J.P., and H.C. Wood. 1918. *The dispensatory of the United States of America.* 20th ed. Philadelphia: Lippincott.

Viburnum opulus L. — Caprifoliaceae

SCN: cramp bark
OCN: guelder rose; high-bush cranberry
Part: bark, root bark

Quick Reference Summary

Safety Class: 1
Interaction Class: A

Contraindications

None known.

Other Precautions

None known.

Drug and Supplement Interactions

None known.

Adverse Events and Side Effects

None known.

Pharmacological Considerations

None known.

Pregnancy and Lactation

Cramp bark is traditionally used during pregnancy to prevent miscarriage and premature labor, and is taken during the third trimester as a partus preparator (Felter and Lloyd 1898; Upton 2000).

No information on the safety of cramp bark during lactation was identified in the scientific or traditional literature. While this review did not identify any concerns for use while nursing, safety has not been conclusively established.

Review Details

I. Drug and Supplement Interactions

Clinical Trials of Drug or Supplement Interactions

No clinical trials of drug or supplement interactions were identified.

Case Reports of Suspected Drug or Supplement Interactions

No case reports of suspected drug or supplement interactions were identified.

Animal Trials of Drug or Supplement Interactions

No animal trials of drug or supplement interactions were identified.

II. Adverse Events

Case Reports of Adverse Events

No case reports of adverse events were identified.

III. Pharmacology and Pharmacokinetics

Human Pharmacological Studies

No relevant human pharmacological studies were identified.

Animal Pharmacological Studies

No relevant animal pharmacological studies were identified.

In Vitro Pharmacological Studies

A hydroalcoholic extract of cramp bark showed no binding to estrogen or progesterone receptors in human breast cancer cells (Zava et al. 1998).

IV. Pregnancy and Lactation

Cramp bark is traditionally used during pregnancy to prevent miscarriage and premature labor, and is taken during the third trimester as a partus preparator (Felter and Lloyd 1898; Upton 2000).

No information on the safety of cramp bark during lactation was identified.

V. Toxicity Studies

No toxicity studies on cramp bark were identified.

Literature Cited

Felter, H.W., and J.U. Lloyd. 1898. *King's American dispensatory.* 18th ed., 3rd rev. 2 vols. Cincinnati: Ohio Valley Co.

Upton, R. 2000. *Cramp bark: Viburnum opulus: Analytical, quality control, and therapeutic monograph. American Herbal Pharmacopoeia and therapeutic compendium.* Santa Cruz, CA: American Herbal Pharmacopoeia.

Zava, D.T., C.M. Dollbaum, and M. Blen. 1998. Estrogen and progestin bioactivity of foods, herbs, and spices. *Proc. Soc. Exp. Biol. Med.* 217(3):369-378.

Viburnum prunifolium L.

Caprifoliaceae

SCN: black haw
OCN: nannybush

Part: bark

Quick Reference Summary

V

Safety Class: 1
Interaction Class: A

Contraindications

None known.

Other Precautions

Black haw should be used with caution in persons with a history of kidney stones (McGuffin et al. 1997).

Drug and Supplement Interactions

None known.

Adverse Events and Side Effects

None known.

Pharmacological Considerations

None known.

Pregnancy and Lactation

Black haw is traditionally used during pregnancy to prevent miscarriage and premature labor, and is taken during the third trimester as a partus preparator (Felter and Lloyd 1898; Upton 2000).

No information on the safety of black haw during lactation was identified in the scientific or traditional literature. While this review did not identify any concerns for use while nursing, safety has not been conclusively established.

Review Details

I. Drug and Supplement Interactions

Clinical Trials of Drug or Supplement Interactions

No clinical trials of drug or supplement interactions were identified.

Case Reports of Suspected Drug or Supplement Interactions

No case reports of suspected drug or supplement interactions were identified.

Animal Trials of Drug or Supplement Interactions

No animal trials of drug or supplement interactions were identified.

II. Adverse Events

Case Reports of Adverse Events

No case reports of adverse events were identified.

III. Pharmacology and Pharmacokinetics

Human Pharmacological Studies

No relevant human pharmacological studies were identified.

Animal Pharmacological Studies

No relevant animal pharmacological studies were identified.

In Vitro Pharmacological Studies

No relevant in vitro pharmacological studies were identified.

IV. Pregnancy and Lactation

Black haw is traditionally used during pregnancy to prevent miscarriage and premature labor, and is taken during the third trimester as a partus preparator (Felter and Lloyd 1898; Upton 2000).

No information on the safety of black haw during lactation was identified.

V. Toxicity Studies

No toxicity studies on black haw were identified.

Literature Cited

Felter, H.W., and J.U. Lloyd. 1898. *King's American dispensatory*. 18th ed., 3rd rev. 2 vols. Cincinnati: Ohio Valley Co.

McGuffin, M., C. Hobbs, R. Upton, and A. Goldberg. 1997. *Botanical safety handbook*. Boca Raton, FL: CRC Press.

Upton, R. 2000. *Black haw bark: Viburnum prunifolium: Analytical, quality control, and therapeutic monograph. American Herbal Pharmacopoeia and therapeutic compendium.* Santa Cruz, CA: American Herbal Pharmacopoeia.

Vinca minor L. — Apocynaceae

SCN: lesser periwinkle

Part: herb

Quick Reference Summary

Safety Class: 1

Interaction Class: A

Contraindications

None known.

Other Precautions

None known.

Drug and Supplement Interactions

None known.

Editors' Notes

Lesser periwinkle contains small amounts (0.02–0.13%) of the compound vincamine (Karabaev et al. 1972; Proksa and Grossmann 1991; Vachnadze et al. 2001). Vincamine and the closely related compound vinpocetine have been widely studied, and vinpocetine is sold as a dietary supplement (EMEA 1999; NCI 2003; Szatmari and Whitehouse 2003).

Adverse Events and Side Effects

None known.

Pharmacological Considerations

None known.

Pregnancy and Lactation

No information on the safety of lesser periwinkle in pregnancy or lactation was identified in the scientific or traditional literature. Although this review did not identify any concerns for use while pregnant or nursing, safety has not been conclusively established.

Review Details

I. Drug and Supplement Interactions

Clinical Trials of Drug or Supplement Interactions

No clinical trials of drug or supplement interactions were identified.

Case Reports of Suspected Drug or Supplement Interactions

No case reports of suspected drug or supplement interactions were identified.

Animal Trials of Drug or Supplement Interactions

No animal trials of drug or supplement interactions were identified.

II. Adverse Events

Case Reports of Adverse Events

No case reports of adverse events were identified.

III. Pharmacology and Pharmacokinetics

Human Pharmacological Studies

No relevant human pharmacological studies were identified.

Animal Pharmacological Studies

No relevant animal pharmacological studies were identified.

In Vitro Pharmacological Studies

No relevant in vitro pharmacological studies were identified.

IV. Pregnancy and Lactation

No information on the safety of lesser periwinkle during pregnancy or lactation was identified.

V. Toxicity Studies

Acute Toxicity

The oral LD_{50} for the compound vincamine is reported as 810 or 1200 mg/kg in rats and 460, 825, or 1000 mg/kg in mice (EMEA 1999; NCI 2003).

In animal studies, administration of lesser periwinkle has been reported to cause leukocytopenia, lymphocytopenia, and reductions in α-, α_2-, and γ-globulin levels. Information on doses used, duration of study, and the type of animal subject were not reported in available English language translations (Blumenthal et al. 1998).

Ventricular extrasystole was observed in rabbits after oral administration of 20 mg/kg vincamine (EMEA 1999).

Short-Term Toxicity

No signs of toxicity were observed in rats orally administered 30 or 100 mg/kg of vincamine daily for 6 weeks (EMEA 1999).

In rats orally administered 120 mg/kg of vincamine daily for 6 days, no changes in liver enzyme levels were observed (Porquet et al. 1992).

Subchronic Toxicity

No signs of toxicity were observed in rats orally administered 6.6, 20, or 62 mg/kg of vincamine or in guinea pigs given 2.5 mg/kg daily for 3 months (EMEA 1999).

In dogs orally administered 1, 7, or 20 mg/kg of vincamine daily for 3 months, behavioral changes were observed at the 20 mg/kg dose, with no effects reported at lower doses (EMEA 1999).

V

Literature Cited

Blumenthal, M., W. Busse, A. Goldberg, et al. 1998. *The complete German Commission E monographs*. Austin, TX: American Botanical Council.

EMEA. 1999. Committee for Veterinary Medicinal Products; Vincamine Summary Report. EMEA/MRL/587/99. London: European Agency for the Evaluation of Medicinal Products.

Karabaev, S.S., K.N. Aripov, and T.T. Shakirov. 1972. Isolation of vincamine from *Vinca minor*. *Chem. Nat. Compd.* 8(5):674.

NCI. 2003. Summary of data for chemical selection: Vincamine. Vincamine dietary supplements. 1617-90-9. Technical Resources International for the National Cancer Institute.

Porquet, D., M. Appel, T. Fournier, et al. 1992. Evaluation of the hepatotoxicological effects of a drug in an in vivo/in vitro model. *Experientia* 48:257-261.

Proksa, B., and E. Grossmann. 1991. High performance liquid chromatographic determination of alkaloids from *Vinca minor* L. *Phytochem. Anal.* 2(2):74-76.

Szatmari, S.Z., and P.J. Whitehouse. 2003. Vinpocetine for cognitive impairment and dementia. *Cochrane Database Syst. Rev.* CD003119.

Vachnadze, V.Y., E.Z. Dzhakeli, Z.V. Robakidze, et al. 2001. Chemical composition and pharmacological activity of alkaloids from the common periwinkle cultured in Georgia. *Pharm. Chem. J.* 35(5):268-270.

Viola odorata L. — Violaceae

SCN: sweet violet
OCN: English violet; garden violet; sweet blueviolet

Part: leaf

Quick Reference Summary

Safety Class: 1
Interaction Class: A

Contraindications

None known.

Other Precautions

None known.

Drug and Supplement Interactions

None known.

Adverse Events and Side Effects

None known.

Pharmacological Considerations

An animal study indicated that sweet violet exhibited diuretic activity (Rebuelta et al. 1983).

Pregnancy and Lactation

No information on the safety of sweet violet in pregnancy or lactation was identified in the scientific or traditional literature. Although this review did not identify any concerns for use while pregnant or nursing, safety has not been conclusively established.

Review Details

I. Drug and Supplement Interactions

Clinical Trials of Drug or Supplement Interactions

No clinical trials of drug or supplement interactions were identified.

Case Reports of Suspected Drug or Supplement Interactions

No case reports of suspected drug or supplement interactions were identified.

Animal Trials of Drug or Supplement Interactions

No animal trials of drug or supplement interactions were identified.

II. Adverse Events

Case Reports of Adverse Events

No case reports of adverse events were identified.

III. Pharmacology and Pharmacokinetics

Human Pharmacological Studies

No relevant human pharmacological studies were identified.

Animal Pharmacological Studies

Diuretic activity of violet leaf extracts was observed after oral administration in rats of 77.4 mg/animal of an aqueous infusion or dried methanolic extract of violet leaf or 34.3 mg/animal of violet leaf ash (animal weight not specified in available English language translation). The diuretic effects were reportedly greater than those observed for 5 mg/kg of theophylline. Minor natriuretic and kaliuretic activity was also observed (Rebuelta et al. 1983).

In Vitro Pharmacological Studies

No relevant in vitro pharmacological studies were identified.

IV. Pregnancy and Lactation

No information on the safety of sweet violet during pregnancy or lactation was identified.

V. Toxicity Studies

Acute Toxicity

Hexane, chloroform, and aqueous extracts of sweet violet are well tolerated in rabbits after oral administration of doses of 1.2 g/kg (Khattak et al. 1985).

Cytotoxicity

Cyclotide compounds from sweet violet have demonstrated cytotoxic activity in human tumor lines in vitro (Lindholm et al. 2002). Cycloviolacin O2, isolated from sweet violet, exhibited significant cytotoxicity in human tumor cell lines, with an IC_{50} of 0.1 to 0.3 μ*M* (Lindholm et al. 2002).

Literature Cited

Khattak, S.G., S.N. Gilani, and M. Ikram. 1985. Antipyretic studies on some indigenous Pakistani medicinal plants. *J. Ethnopharmacol.* 14(1):45-51.

Lindholm, P., U. Goransson, S. Johansson, et al. 2002. Cyclotides: A novel type of cytotoxic agents. *Mol. Cancer Ther.* 1(6):365-369.

Rebuelta, M., J.M. Vivas, L. San Roman, and G. Serranillo-Fdez. 1983. Study of the diuretic effect of various preparations of the leaves of *Viola odorata. Plant. Med. Phytother.* 17:215-221.

Viola sororia Willd. — Violaceae

SCN: violet
Syn: *Viola papilionacea* Pursh
AN: *banafsha*
OCN: common blue violet; downy blueviolet; hooded blueviolet
Part: leaf

Quick Reference Summary

Safety Class: 1
Interaction Class: A

Contraindications

None known.

Other Precautions

None known.

Drug and Supplement Interactions

None known.

Adverse Events and Side Effects

None known.

Pharmacological Considerations

None known.

Pregnancy and Lactation

No information on the safety of violet in pregnancy or lactation was identified in the scientific or traditional literature. Although this review did not identify any concerns for use while pregnant or nursing, safety has not been conclusively established.

Review Details

I. Drug and Supplement Interactions

Clinical Trials of Drug or Supplement Interactions

No clinical trials of drug or supplement interactions were identified.

Case Reports of Suspected Drug or Supplement Interactions

No case reports of suspected drug or supplement interactions were identified.

Animal Trials of Drug or Supplement Interactions

No animal trials of drug or supplement interactions were identified.

II. Adverse Events

Case Reports of Adverse Events

No case reports of adverse events were identified.

III. Pharmacology and Pharmacokinetics

Human Pharmacological Studies

No relevant human pharmacological studies were identified.

Animal Pharmacological Studies

No relevant animal pharmacological studies were identified.

In Vitro Pharmacological Studies

No relevant in vitro pharmacological studies were identified.

IV. Pregnancy and Lactation

No information on the safety of violet during pregnancy or lactation was identified.

V. Toxicity Studies

Cytotoxicity

Cyclotide compounds from violet have demonstrated cytotoxic activity in human tumor lines in vitro (Goransson et al. 2003; Lindholm et al. 2002).

Literature Cited

Goransson, U., A.M. Broussalis, and P. Claeson. 2003. Expression of *Viola* cyclotides by liquid chromatography-mass spectrometry and tandem mass spectrometry sequencing of intercysteine loops after introduction of charges and cleavage sites by aminoethylation. *Anal. Biochem.* 318(1):107-117.

Lindholm, P., U. Goransson, S. Johansson, et al. 2002. Cyclotides: A novel type of cytotoxic agents. *Mol. Cancer Ther.* 1(6):365-369.

Viola tricolor L.

Violaceae

SCN: heartsease
OCN: European wild pansy; Johnny-jump-up; wild violet

Part: herb

Quick Reference Summary

Safety Class: 1
Interaction Class: A

Contraindications

None known.

Other Precautions

None known.

Drug and Supplement Interactions

None known.

Editors' Notes

Use of this species as a food additive in the U.S. is limited to its function as a flavoring substance in alcoholic beverages (CFR 2011). Dietary ingredients for use in dietary supplements, however, are specifically excluded from the federal food additive definition (U.S.C. 2010).

Adverse Events and Side Effects

An infant with glucose-6-phosphate dehydrogenase (G6PD) deficiency (a hereditary disease also known as favism) developed moderate hemolysis after ingesting water extract of heartsease (Behmanesh and Abdollahi 2002).

Pharmacological Considerations

None known.

Pregnancy and Lactation

No information on the safety of heartsease in pregnancy or lactation was identified in the scientific or traditional literature. Although this review did not identify any concerns for use while pregnant or nursing, safety has not been conclusively established.

Review Details

I. Drug and Supplement Interactions

Clinical Trials of Drug or Supplement Interactions

No clinical trials of drug or supplement interactions were identified.

Case Reports of Suspected Drug or Supplement Interactions

No case reports of suspected drug or supplement interactions were identified.

Animal Trials of Drug or Supplement Interactions

No animal trials of drug or supplement interactions were identified.

II. Adverse Events

Case Reports of Adverse Events

A 9-month-old infant with a history of G6PD deficiency (a hereditary disease also known as favism) developed moderate hemolysis after ingesting half a cup of boiled heartsease (Behmanesh and Abdollahi 2002).

III. Pharmacology and Pharmacokinetics

Human Pharmacological Studies

No relevant human pharmacological studies were identified.

Animal Pharmacological Studies

No relevant animal pharmacological studies were identified.

In Vitro Pharmacological Studies

No relevant in vitro pharmacological studies were identified.

IV. Pregnancy and Lactation

No information on the safety of heartsease during pregnancy or lactation was identified.

V. Toxicity Studies

Cytotoxicity

Cyclotide compounds, such as those from heartsease, have demonstrated cytotoxic activity in human tumor lines in vitro (Goransson et al. 2003; Lindholm et al. 2002). In human cancer cell lines, the most active cyclotides isolated from heartsease had IC_{50} values of 0.6 to 6 μM (Svangård et al. 2004).

Literature Cited

Behmanesh, Y., and M. Abdollahi. 2002. Haemolysis after consumption of *Viola tricolor*. *WHO Drug Info.* 16(1):15-16.

CFR. 2011. *Code of federal regulations*, Title 21 Part 172.510, 2011 ed. Food additives permitted for direct addition to food for human consumption. Flavoring agents and related substances. Natural flavoring substances and natural substances used in conjunction with flavors. Washington, DC: U.S. Government Printing Office.

Goransson, U., A.M. Broussalis, and P. Claeson. 2003. Expression of *Viola* cyclotides by liquid chromatography-mass spectrometry and tandem mass spectrometry sequencing of intercysteine loops after introduction of charges and cleavage sites by aminoethylation. *Anal. Biochem.* 318(1):107-117.

Lindholm, P., U. Goransson, S. Johansson, et al. 2002. Cyclotides: A novel type of cytotoxic agents. *Mol. Cancer Ther.* 1(6):365-369.

Svangård, E., U. Goransson, Z. Hocaoglu, et al. 2004. Cytotoxic cyclotides from *Viola tricolor*. *J. Nat. Prod.* 67(2):144-147.

U.S.C. 2010. United States Code, Title 21, Part 321 (s)(6). Current as of January 7, 2011. Washington, DC: U.S. Government Printing Office.

Viscum album L. — Viscaceae

SCN: European mistletoe

Part: herb

Quick Reference Summary

Safety Class: 2b, 2d
Interaction Class: A

Contraindications

Not for use in pregnancy except under the supervision of a qualified healthcare practitioner (Büssing 1996).

Do not exceed recommended dose (McGuffin et al. 1997).

Other Precautions

None known.

Drug and Supplement Interactions

None known.

Standard Dose

The standard dose is 2.5 g infused in cold water for 10 to 12 hours, up to two times daily (Wichtl 2004).

Adverse Events and Side Effects

Systematic reviews of several dozen clinical trials (all subcutaneous injections of 1 to 200 mg) indicated that European mistletoe is generally very well tolerated, with no serious adverse events reported. Minor localized reactions at the injection site are sometimes associated with injections of European mistletoe, but such effects are not expected after oral use (Horneber et al. 2009; Kienle et al. 2009; Stein and Berg 2000).

Allergic reactions, including anaphylactic reactions, to European mistletoe have been reported (Bauer et al. 2005; Horneber et al. 2009; Hutt et al. 2001; Kienle et al. 2009).

Pharmacological Considerations

Animal studies have demonstrated that European mistletoe may modify glucose regulation (Eno et al. 2008; Nwaegerue et al. 2007; Ohiri et al. 2003). People with diabetes are advised to monitor their blood sugar closely and discuss the use of this herb with a qualified healthcare practitioner prior to use.

Some authorities suggest that blood pressure should be checked regularly in persons consuming European mistletoe tea (Wichtl 2004).

Pregnancy and Lactation

No information on the safety of European mistletoe during pregnancy or lactation was identified in the scientific or traditional literature. Due to the relative toxicity of some compounds in European mistletoe, use during pregnancy is not recommended unless under the care of a qualified healthcare practitioner (Büssing 1996).

Review Details

I. Drug and Supplement Interactions

Clinical Trials of Drug or Supplement Interactions

No clinical trials of drug or supplement interactions were identified.

Case Reports of Suspected Drug or Supplement Interactions

No case reports of suspected drug or supplement interactions were identified.

Animal Trials of Drug or Supplement Interactions

No animal trials of drug or supplement interactions were identified.

II. Adverse Events

Adverse Events Reported in Clinical Trials

Systematic reviews of clinical trials indicate that European mistletoe is generally well tolerated, with very few adverse events reported. The reviews examined 46 clinical trials that met inclusion criteria, with a total of 9949 patients subcutaneously administered standardized extracts of European mistletoe at doses ranging from 1 to 200 mg daily for a time span of several weeks to over a year (treatment longer than several months usually consisted of cycles of treatment and breaks) (Horneber et al. 2009; Kienle et al. 2009). Localized reactions at the injection site, including redness, swelling, itching, prurigo, and induration (a hardening at the site), occurred in approximately one-third of patients. Reactions were reported to be dose-dependent and may intensify during concomitant chemotherapy (Horneber et al. 2009; Kienle et al. 2009; Stein and Berg 2000). Flulike symptoms and eosinophilia have also been reported after injection of European mistletoe (Horneber et al. 2009; Kienle et al. 2009; Stein and Berg 2000). These reactions are generally attributed to the route of administration and are not expected after oral use.

An increase in eosinophil (a type of white blood cell) counts has been reported after administration of lectin-rich European mistletoe preparations (Huber et al. 2005; Stein and Berg 2000).

Case Reports of Adverse Events

Allergic reactions, including anaphylactic reactions, to European mistletoe have been reported (Bauer et al. 2005; Horneber et al. 2009; Hutt et al. 2001; Kienle et al. 2009).

III. Pharmacology and Pharmacokinetics

Human Pharmacological Studies

In a study of mistletoe on T lymphocytes in patients with prostate, breast, or colorectal cancer, a decrease in T lymphocyte function was observed in colorectal and prostate cancer patients on the "swift" dose escalation scheme. Patients on a "slow" dose escalation scheme had less of a decline in T lymphocytes. An escalation of dose over the course of treatment is a standard practice for European mistletoe. In this study, the subcutaneous doses began at 0.01 mg and went to 20 mg (Bussing et al. 2007).

Animal Pharmacological Studies

In diabetic and healthy rats intravenously administered 100 mg/kg of European mistletoe, a reduction in blood glucose levels was observed in diabetic animals but not in healthy animals. An increase in insulin secretion was observed in both healthy and diabetic animals (Eno et al. 2008).

In diabetic mice and rabbits intraperitoneally administered 200 or 400 mg/kg of an aqueous extract of European mistletoe, a reduction in blood glucose levels was observed (Ohiri et al. 2003).

In diabetic and healthy rats administered 250, 500, 750, or 1000 mg/kg of a hydroethanolic extract of European mistletoe, a dose-dependent reduction in glucose levels was observed (Nwaegerue et al. 2007).

In Vitro Pharmacological Studies

Some inhibition of the drug-metabolizing isoenzyme CYP3A4 was observed after treatment with European mistletoe, although the authors indicated that clinically relevant systemic or intestinal interactions with CYP3A4 were considered unlikely (Engdal and Nilsen 2009).

An aqueous extract of European mistletoe was found to stimulate insulin secretion from clonal pancreatic B cells (Gray and Flatt 1999).

IV. Pregnancy and Lactation

No information on the safety of European mistletoe during pregnancy or lactation was identified.

V. Toxicity Studies

Acute Toxicity

The LD_{50} of an intraperitoneally administered aqueous extract of European mistletoe in mice is 4.18 g/kg (Ohiri et al. 2003).

The LD_{50} of intraperitoneally administered standardized extracts of European mistletoe is 700 mg/kg in mice and 378 mg/kg in rats (Stein 2000). The LD_{50} after intravenous administration to mice is 500 mg/kg and after subcutaneous administration to mice is 1200 mg/kg (Luther et al. 1986).

The LD_{50} of viscotoxins is 0.5 mg/kg after intraperitoneal administration to mice and 0.1 mg/kg after intravenous administration to cats. Sublethal doses were reported to cause hypotension, bradycardia, and negative inotropic effects (Samuelsson 1974).

Short-Term Toxicity

In rats intravenously administered up to 5 mg/kg of a standardized extract of European mistletoe daily for 28 days, no relevant organ toxicity was observed, and no changes in clinical signs, organ weights, hematology, histology, or other parameters were seen (Mengs 1998).

Genotoxicity

No genotoxic activity of European mistletoe was observed in a variety of assays, including a bacterial mutation assay, mammalian cell gene mutation assay, in vitro cytogenetic assay, cell transformation assay, and the Ames test (Mengs 1998; Mengs et al. 1997).

Literature Cited

Bauer, C., T. Oppel, F. Rueff, and B. Przybilla. 2005. Anaphylaxis to viscotoxins of mistletoe (*Viscum album*) extracts. *Ann. Allergy Asthma Immunol.* 94(1):86-89.

Büssing, A. 1996. Induction of apoptosis by the mistletoe lectins: A review on the mechanisms of cytotoxicity mediated by *Viscum album* L. *Apoptosis* 1(1):25-32.

Bussing, A., C. Stumpf, W. Troger, and M. Schietzel. 2007. Course of mitogen-stimulated T lymphocytes in cancer patients treated with *Viscum album* extracts. *Anticancer Res.* 27(4C):2903-2910.

Engdal, S., and O.G. Nilsen. 2009. *In vitro* inhibition of CYP3A4 by herbal remedies frequently used by cancer patients. *Phytother. Res.* 23(7):906-912.

Eno, A.E., O.E. Ofem, C.O. Nku, E.J. Ani, and E.H. Itam. 2008. Stimulation of insulin secretion by *Viscum album* (mistletoe) leaf extract in streptozotocin-induced diabetic rats. *Afr. J. Med. Med. Sci.* 37(2):141.

Gray, A.M., and P.R. Flatt. 1999. Insulin-secreting activity of the traditional antidiabetic plant *Viscum album* (mistletoe). *J. Endocrinol.* 160(3):409.

Horneber, M., G. Bueschel, R. Huber, K. Linde, and M. Rostock. 2009. Mistletoe therapy in oncology (review). *Cochrane Lib.* 2:1-93.

Huber, R., M. Rostock, R. Goedl, et al. 2005. Mistletoe treatment induces GM-CSF- and IL-5 production by PBMC and increases blood granulocyte- and eosinophil counts: A placebo controlled randomized study in healthy subjects. *Eur. J. Med. Res.* 10(10):411-418.

Hutt, N., M. Kopferschmitt-Kubler, J. Cabalion, et al. 2001. Anaphylactic reactions after therapeutic injection of mistletoe (*Viscum album* L.). *Allergol. Immunopathol.* 29(5):201-203.

Kienle, G.S., A. Glockmann, M. Schink, and H. Kiene. 2009. *Viscum album* L. extracts in breast and gynaecological cancers: A systematic review of clinical and preclinical research. *J. Exp. Clin. Cancer Res.* 28:79.

Luther, P., G. Uhlenbruck, H. Reuigen, et al. 1986. Are lectins from *Viscum album* interesting tools in lung disease? *Z. Erkrank. Atm. Org.* 166:247-256.

McGuffin, M., C. Hobbs, R. Upton, and A. Goldberg. 1997. *Botanical safety handbook*. Boca Raton, FL: CRC Press.

Mengs, U. 1998. Toxicity of an aqueous mistletoe extract: Acute and subchronic toxicity in rats, genotoxicity *in vitro*. In Bardocz, S., U. Pfueller, and A. Pusztai, eds. *Effects of antinutrients on the nutritional value of legume diets.* Luxembourg: European Commission.

Mengs, U., C.B. Clare, and J.A. Poiley. 1997. Genotoxicity testing of an aqueous mistletoe extract *in vitro*. *Arzneimittelforschung* 47:316-319.

Nwaegerue, E., I.N. Nweke, C.C. Ezeala, and P.C. Unekwe. 2007. Glucose lowering effect of leaf extracts of *Viscum album* in normal and diabetic rats. *J. Res. Med. Sci.* 12(5):235-240.

Ohiri, F.C., C.O. Esimone, S.V. Nwafor, C.O. Okoli, and O.O. Ndu. 2003. Hypoglycemic properties of *Viscum album* (mistletoe) in alloxan-induced diabetic animals. *Pharm. Biol.* 41(3):184-187.

Samuelsson, G. 1974. Mistletoe toxins. *Syst. Zool.* 22:566-569.

Stein, G.M. 2000. Toxicology of mistletoe extracts and their components. In Bussing, A., ed. *Mistletoe: The genus Viscum*. Boca, Raton, FL: CRC Press.

Stein, G.M., and P.A. Berg. 2000. Adverse effects during therapy with mistletoe extracts. In Bussing, A., ed. *Mistletoe: The genus Viscum*. Boca Raton, FL: CRC Press.

Wichtl, M. 2004. *Herbal drugs and phytopharmaceuticals: A handbook for practice on a scientific basis*. 3rd ed. Boca Raton, FL: CRC Press.

Vitex agnus-castus L.

Verbenaceae

V

SCN: chastetree

AN: *renuka*

OCN: agnus-castus; chasteberry; monk's pepper

Part: fruit (berry)

Quick Reference Summary

Safety Class: 1

Interaction Class: A

Contraindications

None known.

Other Precautions

Chaste tree is not recommended for use with hormonal contraceptives (Hobbs 1990; Krochmal et al. 2004; Mills and Bone 2000; Stargrove et al. 2008).

Drug and Supplement Interactions

None known.

Adverse Events and Side Effects

Reviews of clinical trials and case reports of chaste tree products have concluded that chaste tree is generally well tolerated and that any adverse events reported were mild and transient (Daniele et al. 2005).

Pharmacological Considerations

Human studies have indicated that chaste tree may inhibit prolactin secretion (Upton 2001).

Pregnancy and Lactation

Herbal practitioners have traditionally used chaste tree to prevent miscarriages in women with a history of prior miscarriages. In studies where pregnancy has been achieved while taking chaste tree, no follow-up studies have been completed on pregnancy outcomes or newborn health (Romm 2010).

Chaste tree has traditionally been used as a galactagogue (a substance that promotes lactation) (Brown 1994; Upton 2001). No recent human or animal studies on the safety of chaste tree during lactation were identified, although a review of human studies indicated that chaste tree may decrease prolactin secretion (Upton 2001).

Review Details

I. Drug and Supplement Interactions

Clinical Trials of Drug or Supplement Interactions

No clinical trials of drug or supplement interactions were identified.

Case Reports of Suspected Drug or Supplement Interactions

No case reports of suspected drug or supplement interactions were identified.

Animal Trials of Drug or Supplement Interactions

No animal trials of drug or supplement interactions were identified.

II. Adverse Events

Adverse Events Reported in Clinical Trials

A systematic review of adverse events reported in clinical trials, case reports, pharmacovigilance data, and postmarketing surveillance studies concluded that although further studies are needed to assess the safety of chaste tree, the available data indicate that chaste tree is not associated with serious risk to health and that the vast majority of adverse events reported in association with chaste tree use have been mild and transient (Daniele et al. 2005). A review of human studies on chaste tree products indicated that adverse events were generally similar in the chaste tree and placebo groups and that adverse events reported were often the same as the symptoms being treated. The most common adverse events were listed as gastrointestinal disturbance (notably nausea) and skin conditions, with side effects occasionally reported including headache, fatigue, and hormone-related symptoms (Upton 2001).

Case Reports of Adverse Events

Occasional minor skin irritations have been reported in persons taking chaste tree preparations (Leung and Foster 1996).

III. Pharmacology and Pharmacokinetics

Human Pharmacological Studies

No effects on blood pressure, heart rate, testosterone, follicle-stimulating hormone, or luteinizing hormone were observed in healthy men administered 120 to 480 mg chaste tree extract daily for 14 days (Loew et al. 1996; Merz et al. 1996).

Compounds in chaste tree are reported to bind to the estrogen receptor β (ERβ) in the heart, vasculature, bone, and bladder (Wuttke et al. 2003).

Animal Pharmacological Studies

A reduction in prolactin secretion in rats administered chaste tree extract was reported (Winterhoff et al. 1991).

In Vitro Pharmacological Studies

Chaste tree has been shown to bind to estrogen receptor β (ERβ) but not ERα in one in vitro study (Jarry et al. 2003), whereas a second study showed that a methanol extract of chaste tree was found to competitively bind to ERα and ERβ (Liu et al. 2001). Chaste tree has also been shown to bind to the μ-opiate receptor in Chinese hamster ovaries (Webster et al. 2006).

In high concentrations, chaste tree has agonistic effects on the pituitary dopamine (D2) receptors (Jarry et al. 1994; Wuttke 1996).

In cultured rat pituitary cells, a chaste tree extract was observed to inhibit basal as well as thyrotropin-releasing hormone-stimulated prolactin secretion (Sliutz et al. 1993).

IV. Pregnancy and Lactation

Herbal practitioners have traditionally used chaste tree to prevent miscarriages in women with a history of prior miscarriages. Increased progesterone levels as a result of improved luteal function associated with use of chaste tree have been hypothesized as a mechanism of action for miscarriage prevention. In studies where pregnancy has been achieved while taking chaste tree, no follow-up studies have been completed on pregnancy outcomes or newborn health (Romm 2010).

No reduction in number of fetuses was observed in pregnant rats orally administered chaste tree fruit at doses of 1 to 2 g/kg daily for days 1 through 10 of pregnancy (Lal et al. 1985).

Chaste tree has been traditionally used as a galactagogue (Brown 1994; Upton 2001). Although older human studies have indicated that chaste tree helps to increase milk production (Mohr 1954; Noack 1943), a review of nine human studies (including placebo-controlled and open studies) on the effects of chaste tree on prolactin secretion indicated that four studies showed a significant decrease in prolactin, one showed a significant increase, one showed a nonsignificant increase, and three showed no effect (Upton 2001).

A review of the safety of chaste tree during lactation concluded that "the low toxicity profile and tolerability of chaste tree makes it unlikely to be toxic for the newborn, especially after filtration and dilution through the mother" (Dugoua et al. 2008).

V. Toxicity Studies

Acute Toxicity

In acute toxicity tests, no deaths were observed in rats and mice orally administered up to 2 g/kg of a standardized chaste tree extract (Bionorica 2001).

Short-Term Toxicity

In rats and mice orally administered a standardized chaste tree extract for 28 days, the no-observed-adverse-effect level (NOAEL) was 50 mg/kg. At higher doses, a slight increase in hemopoietic activity in the spleen and an increase in liver weight were observed. These changes were noted as being transient and interpreted as adaptive rather than toxic (Bionorica 2001).

Subchronic Toxicity

In a chronic oral toxicity study of a standardized chaste tree extract orally administered to mice and rats, the NOAEL was 40 mg/kg after 26 weeks of administration (Bionorica 2001).

Genotoxicity

No genotoxic effects of a standardized chaste tree extract were observed in a mammalian cell line assay, a DNA-breakage assay, or a chromosomal breakage assay (Bionorica 2001).

Literature Cited

Bionorica. 2001. Unpublished data. *Cited in* Upton, R. 2001. *Chaste tree fruit: Vitex agnus-castus*. Santa Cruz, CA: American Herbal Pharmacopoeia.

Brown, D. 1994. *Vitex agnus-castus* clinical monograph. *Q. Rev. Nat. Med.* Summer:111-121.

Daniele, C., J. Thompson Coon, M.H. Pittler, and E. Ernst. 2005. *Vitex agnus-castus*: A systematic review of adverse events. *Drug Saf.* 28(4):319-332.

Dugoua, J.J., D. Seely, D. Perri, G. Koren, and E. Mills. 2008. Safety and efficacy of chastetree (*Vitex agnus-castus*) during pregnancy and lactation. *Can. J. Clin. Pharmacol.* 15(1):e74-e79.

Hobbs, C. 1990. *Vitex: The women's herb*. Santa Cruz: Botanica Press.

Jarry, H., S. Leonhardt, C. Gorkow, and W. Wuttke. 1994. *In vitro* prolactin but not LH and FSH release is inhibited by compounds in extracts of *agnus-castus*: Direct evidence for a dopaminergic principle by the dopamine receptor assay. *Exp. Clin. Endocrinol.* 102(6):448-454.

Jarry, H., B. Spengler, A. Porzel, et al. 2003. Evidence for estrogen receptor beta-selective activity of *Vitex agnus-castus* and isolated flavones. *Planta Med.* 69(10):945-947.

Krochmal, R., M. Hardy, S. Bowerman, et al. 2004. Phytochemical assays of commercial botanical dietary supplements. *Evid. Based Complement. Alternat. Med.* 1(3):305-313.

Lal, R., A. Sankaranarayanan, V.S. Mathur, and P.L. Sharma. 1985. Antifertility and oxytocic activity of *Vitex agnus-castus* (seeds) in female albino rats. *Bull. Postgrad. Inst. Med. Educ. Res. Chandigarh* 19(2):44-47.

Leung, A.Y., and S. Foster. 1996. *Encyclopedia of common natural ingredients used in food, drugs, and cosmetics*. 2nd ed. New York: Wiley.

Liu, J., J.E. Burdette, H. Xu, et al. 2001. Evaluation of estrogenic activity of plant extracts for the potential treatment of menopausal symptoms. *J. Agric. Food Chem.* 49 (5):2472-2479.

Loew, D., C. Gorkow, A. Schrödter, et al. 1996. Zur dosisabhaengigen Vertraeglichkeit eines *Agnus-castus*-Spezialextraktes. *Z. Phytother.* 17(4):237-243.

Merz, P.G., C. Gorkow, A. Schroedter, et al. 1996. The effects of a special *agnus-castus* extract (BP1095E1) on prolactin secretion in healthy male subjects. *Exp. Clin. Endrocrin. Diabetes* 104(6):447-453.

Mills, S., and K. Bone. 2000. *Principles and practice of phytotherapy: Modern herbal medicine*. New York: Churchill Livingstone.

Mohr, H. 1954. Clinical investigations of means to increase lactation. *Dtsch. Med. Wochenschr.* 79(41):1513-1516.

Noack, M. 1943. Unsere Erfalwungen mit *Agnus castus* Oligoplex bei der Lalctationssteigerung. *Dtsch. Med. Wochenschr.* 204-206.

Romm, A. 2010. *Botanical medicine for women's health*. London: Churchill Livingstone.

Sliutz, G., P. Speiser, A.M. Schultz, J. Spona, and R. Zeillinger. 1993. *Agnus-castus* extracts inhibit prolactin secretion of rat pituitary cells. *Horm. Metab. Res.* 25(5):253-255.

Stargrove, M., J. Treasure, and D. McKee. 2008. *Herb, nutrient, and drug interactions: Clinical implications and therapeutic solutions*. St. Louis, MO: Elsevier.

V

Upton, R. 2001. *Chaste tree fruit: Vitex agnus-castus: Standards of analysis, quality control, and therapeutics. American Herbal Pharmacopoeia and therapeutic compendium.* Santa Cruz, CA: American Herbal Pharmacopoeia.

Webster, D.E., J. Lu, S.N. Chen, N.R. Farnsworth, and Z.J. Wang. 2006. Activation of the mu-opiate receptor by *Vitex agnus-castus* methanol extracts: Implication for its use in PMS. *J. Ethnopharmacol.* 106(2):216-221.

Winterhoff, H., C. Gorkow, and B. Behr. 1991. Reduced lactation in rats following application of *agnus-castus*-extract. An indirect evidence for the prolactin inhibiting activity. *Ztschr. Phytother.* 12(6):175-179.

Wuttke, W. 1996. Dopaminergic action of extracts of agnus castus. *Forsch. Komplementarmed.* 3:329-330.

Wuttke, W., H. Jarry, V. Christoffel, B. Spengler, and D. Seidlova-Wuttke. 2003. Chaste tree (*Vitex agnus-castus*)—Pharmacology and clinical indications. *Phytomedicine* 10(4):348-357.

Withania somnifera (L.) Dunal

Solanaceae

SCN: ashwagandha
AN: *ashvagandha*
OCN: winter cherry
Part: root

Quick Reference Summary

Safety Class: 2b
Interaction Class: A

Contraindications

Not for use in pregnancy except under the supervision of a qualified healthcare practitioner (Chadha 1988; Upton 2000).

Other Precautions

None known.

Drug and Supplement Interactions

None known.

Adverse Events and Side Effects

"Large" but undefined doses of ashwagandha have been reported to cause irritation of mucous and serous membranes, resulting in gastrointestinal upset, diarrhea, and vomiting (Chadha 1976).

Pharmacological Considerations

Human and animal studies have indicated that ashwagandha may modify glucose levels, suggesting that diabetic persons taking ashwagandha should continue to monitor blood glucose levels (Andallu and Radhika 2000).

Several older studies suggest that ashwagandha may potentiate the effects of barbiturates (Malhotra et al. 1960; Rao and Karanth 1990; Singh et al. 1979), although one study showed that ashwagandha reduced the effects of barbiturates (Singh et al. 1978).

In vitro, falsely elevated digoxin levels were observed in some types of digoxin assays when ashwagandha was added to tested serum (Dasgupta and Reyes 2005; Dasgupta et al. 2008).

Pregnancy and Lactation

In the ethnobotanical literature, there are references of ashwagandha use as both an "abortifacient" (Badhwar and Chopra 1946; Casey 1960; Chadha 1976) and as a "pregnancy tonic" (Kapoor 1990; Tirtha 1998; Upton 2000). The strength or accuracy of this information is difficult to assess given the lack of detail regarding dose, duration, and plant part used. In a review of the scientific literature, an animal study failed to detect any decreased number of pregnancies, change in litter size, or fetal loss in rats when administered 100 mg/kg/day of an aqueous root extract for 8 months (Sharma et al. 1986). Based upon this data, it may be that low doses of ashwagandha pose little danger; however, clinicians should discuss the strengths and limitations of the evidence when counseling about its use in pregnancy.

In Ayurvedic medicine, ashwagandha is traditionally used to promote lactation (Kapoor 1990).

Review Details

I. Drug and Supplement Interactions

Clinical Trials of Drug or Supplement Interactions

No clinical trials of drug or supplement interactions were identified.

Case Reports of Suspected Drug or Supplement Interactions

No case reports of suspected drug or supplement interactions were identified.

Animal Trials of Drug or Supplement Interactions

Ashwagandha was reported to potentiate the activity of pentobarbital sodium in mice after intraperitoneal administration of doses of 700 mg/kg or more (Malhotra et al. 1960). Studies of administration of 300 mg/kg ashwagandha report a barbiturate potentiation (Rao and Karanth 1990) or a reduction in barbiturate-induced sleeping time (Singh et al. 1978). A fourth study indicated that a dose of 1 g/kg of ashwagandha produced a potentiation of barbiturates (Singh et al. 1979).

II. Adverse Events

Case Reports of Adverse Events

No case reports of adverse events were identified.

W

III. Pharmacology and Pharmacokinetics

Human Pharmacological Studies

In contrast to animal studies that reported a decrease in libido and sexual performance, an increase in sexual performance was reported in healthy men (age 50 to 59) administered 3 g daily of ashwagandha for one year (Kuppurajan et al. 1980).

In patients with mild non-insulin-dependent diabetes, administration of 3 g powdered ashwagandha root daily for 30 days resulted in a decrease in blood glucose that was comparable to that of an oral hypoglycemic drug (Andallu and Radhika 2000).

Animal Pharmacological Studies

In rats with non-insulin-dependent diabetes, oral administration of 200 or 400 mg/kg ashwagandha daily for 5 weeks reduced elevated levels of blood glucose, HbA(1)c, and insulin. Improved glucose tolerance was observed in an oral glucose tolerance test, and treatment significantly improved the insulin sensitivity index (Anwer et al. 2008; Jain et al. 2006).

A decrease in fasting blood glucose levels was observed in diabetic rats intraperitoneally administered 20 mg/kg of an alcohol extract of ashwagandha (Jain et al. 2006).

An increase in serum T_3 and T_4 levels was observed in mice intragastrically administered 1.4 g/kg of ashwagandha root daily for 20 days (Panda and Kar 1998).

Ashwagandha has been shown to bind to the GABA receptor in mice, with studies showing a reduction in effective dose of diazepam (Kulkarni et al. 1998) and an increased seizure threshold for pentylenetetrazol-induced seizures (Kulkarni et al. 2008; Kulkarni and Dhir 2008).

In male rats administered 470 mg/kg of a lyophilized aqueous extract of ashwagandha, an increase in testicular weight was observed, and histological examination revealed an apparent increase in the diameter of seminiferous tubules and the number of seminiferous tubular cell layers in the testes of treated rats, as compared with control animals. A decrease in serum testosterone and FSH levels, and an increase in LH levels was also observed (Abdel-Magied et al. 2001).

In male rats orally administered 3 g/kg of a methanol extract of ashwagandha daily for 7 days, impairment in libido, sexual performance, and sexual vigor was observed, along with penile erectile dysfunction. The effects were partly reversible on cessation of treatment, and were not due to changes in testosterone levels or toxicity, but instead were attributed to hyperprolactinemic, GABAergic, serotonergic or sedative activities of the extract (Ilayperuma et al. 2002).

A decrease in litter size and some infertility were observed in male and female mice orally administered 25 mg/kg daily of ashwagandha for 10 days (Garg and Parasar 1965).

In Vitro Pharmacological Studies

Falsely elevated levels were observed in the Digoxin III digoxin assay in serum treated with 25 or 50 µl/ml of an aqueous/ethanolic extract of ashwagandha, but not after treatment with 10 µl/ml. In the Tina-quant and FPIA digoxin assays, falsely elevated digoxin levels were observed in serum treated with 50 µl/ml but not in serum treated with 10 or 25 µl/ml (Dasgupta and Reyes 2005; Dasgupta et al. 2008).

The compound withaferin A has been shown to induce apoptosis in leukemia cells (Malik et al. 2007; Mandal et al. 2008; Oh et al. 2008; Senthil et al. 2007).

IV. Pregnancy and Lactation

Offspring from pregnant mice administered 100 mg/kg ashwagandha decoction daily for 8 months had a higher birthweight than offspring from control animals (Sharma et al. 1986).

Information on the history of ashwagandha use in pregnancy is conflicting. While some literature indicates that ashwagandha has been used for abortion (Casey 1960; Chadha 1976; Svoboda 1992), other studies indicate that ashwagandha is used to prevent miscarriage and stabilize the fetus (Kapoor 1990; Tirtha 1998). The strength or accuracy of this information is difficult to assess given the lack of detail regarding plant part used, dose, and duration. One of the articles citing abortifacient activity uses a broad definition of abortifacients, including botanicals used for amenorrhea, those that have ecbolic (stimulating uterine contractions) activity, anti-metabolites, and "others."

In Ayurvedic medicine, ashwagandha is traditionally used to promote lactation (Kapoor 1990).

V. Toxicity Studies

Acute Toxicity

The LD_{50} of orally administered ashwagandha in mice was approximately 7.7 g/kg (Singh et al. 1982). In mice intraperitoneally administered single doses of 1.1, 1.2, 1.3, 1.4, or 1.5 g/kg of an ethanol extract of ashwagandha, the 1.1 g/kg dose did not kill any animals, while the 1.5 g/kg dose was lethal to all animals. Based on these results, the LD_{50} of the extract was estimated to be 1.26 g/kg (Sharada et al. 1993). No toxic effects were observed in mice intraperitoneally administered 1 g/kg of an ethanol extract of ashwagandha (Sharada et al. 1993).

The LD_{50} of intraperitoneally administered sitoindosides VII, VIII, IX, and X in mice were 1076, 1564, 518, and 808 mg/kg, respectively (Bhattacharya et al. 1987; Ghosal et al. 1989). No deaths were reported in mice orally administered up to 1 g/kg of sitoindosides IX and X (Ghosal et al. 1989).

The LD_{50} of intraperitoneally administered alcohol extract of ashwagandha in mice was 1.26 g/kg (Sharada et al. 1993).

Short-Term Toxicity

In rats administered ashwagandha as 25% of the diet (approximate dose of 5 g per mouse daily) for 10 to 14 days, no macroscopic abnormalities were observed, but microscopic examination revealed centrilobular hydropic degeneration in the liver, peribronchial and perivenous edema in the lungs, intertubular vascular congestion, tubular casts, and tubular degeneration in the kidneys (Arseculeratne et al. 1985).

In mice intraperitoneally administered 100 mg/kg of an ethanol extract of ashwagandha daily for 30 days, a significant reduction in the weights of spleen, thymus, and adrenal glands was observed in male rats, and a significant increase in blood acid phosphatase activity was observed in animals of both sexes (Sharada et al. 1993).

Other Toxicity

Treatment of neonatal and adult rat cardiomyocytes with 100 µl of a 10% extract of ashwagandha resulted in cessation of beating in neonatal cardiomyocytes due to calcium overload, while sequential dilutions revealed that treatment with a low dose (0.01% v/v, 0.1 µl/ml of the medium) resulted in constant, regular beats (transients), and a slight elevation of diastolic calcium without overload. Addition of ashwagandha to sparking, calcium-tolerant adult cardiomyocytes resulted in initiation of calcium transients, and adult cells were able to tolerate exposure to high concentrations of extract (Poindexter et al. 2006).

Literature Cited

Abdel-Magied, E.M., H.A. Abdel-Rahman, and F.M. Harraz. 2001. The effect of aqueous extracts of *Cynomorium coccineum* and *Withania somnifera* on testicular development in immature Wistar rats. *J. Ethnopharmacol.* 75 (1):1-4.

Andallu, B., and B. Radhika. 2000. Hypoglycemic, diuretic and hypocholesterolemic effect of winter cherry (*Withania somnifera*, Dunal) root. *Indian J. Exp. Biol.* 38 (6):607-9.

Anwer, T., M. Sharma, K.K. Pillai, and M. Iqbal. 2008. Effect of *Withania somnifera* on insulin sensitivity in non-insulin-dependent diabetes mellitus rats. *Basic Clin. Pharmacol. Toxicol.* 102 (6):498-503.

Arseculeratne, S.N., A.A. Gunatilaka, and R.G. Panabokke. 1985. Studies of medicinal plants of Sri Lanka. Part 14: Toxicity of some traditional medicinal herbs. *J. Ethnopharmacol.* 13 (3):323-35.

Badhwar, R.L., and I.C. Chopra. 1946. Reputed abortifacient plants of India. *Indian J. Agric. Sci.* 16:342-355.

Bhattacharya, S.K., R.K. Goel, R. Kaur, and S. Ghosal. 1987. Anti-stress activity of Sitoindosides VII and VIII, new acylsteryl-glucosides from *Withania somnifera*. *Phytother. Res.* 1 (1):32-37.

Casey, R. 1960. Alleged antifertility plants of India. *Indian J. Med. Res.* 14:590-600.

Chadha, Y. 1976. *The Wealth of India: A dictionary of Indian raw materials and industrial products*. Vol. 10. New Delhi: Council of Scientific and Industrial Research.

Chadha, Y. 1988. *The Wealth of India: A dictionary of Indian raw materials and industrial products*. Delhi: Council of Scientific and Industrial Research.

Dasgupta, A., and M.A. Reyes. 2005. Effect of Brazilian, Indian, Siberian, Asian, and North American ginseng on serum digoxin measurement by immunoassays and binding of digoxin-like immunoreactive components of ginseng with Fab fragment of antidigoxin antibody (Digibind). *Am. J. Clin. Pathol.* 124 (2):229-236.

Dasgupta, A., G. Tso, and A. Wells. 2008. Effect of Asian ginseng, Siberian ginseng, and Indian ayurvedic medicine Ashwagandha on serum digoxin measurement by Digoxin III, a new digoxin immunoassay. *J. Clin. Lab. Anal.* 22 (4):295-301.

Garg, L.C., and G.C. Parasar. 1965. Effect of *Withania somnifera* on reproduction in mice. *Planta Medica* 13 (1):46-47.

Ghosal, S., J. Lal, R. Srivastava, et al. 1989. Immunomodulatory and CNS effects of sitoindosides IX and X, two new glycowithanolides from *Withania somnifera*. *Phytother. Res.* 3 (5):201-206.

Ilayperuma, I., W.D. Ratnasooriya, and T.R. Weerasooriya. 2002. Effect of *Withania somnifera* root extract on the sexual behaviour of male rats. *Asian J. Androl.* 4 (4):295-298.

Jain, S., P. Pandhi, A.P. Singh, and S. Malhotra. 2006. Efficacy of standardised herbal extracts in type 1 diabetes—An experimental study. *Afric. J. Trad., Complemen. Altern. Med.* 3 (4):23-33.

Kapoor, L. 1990. *CRC handbook of Ayurvedic medicinal plants*. Boca Raton CRC Press.

Kulkarni, S.K., K.K. Akula, and A. Dhir. 2008. Effect of *Withania somnifera* Dunal root extract against pentylenetetrazol seizure threshold in mice: Possible involvement of GABAergic system. *Indian J. Exp. Biol.* 46 (6):465-469.

Kulkarni, S.K., and A. Dhir. 2008. *Withania somnifera*: An Indian ginseng. *Prog. Neuropsychopharmacol. Biol. Psychiatry* 32 (5):1093-1105.

Kulkarni, S.K., B. George, and R. Mathur. 1998. Protective effect of *Withania somnifera* root extract on electrographic activity in lithium palocarpine model of status epiteplicus. *Phytother. Res.* 12:451-453.

Kuppurajan, K., S. Rajagopalan, R. Sitaraman, et al. 1980. Effect of ashwagandha (*Withania somnifera* Dunal) on the process of aging in human volunteers. *J. Res. Ayurveda Siddha* 1:247-258.

Malhotra, C.L., P.K. Das, and N.S. Dhalla. 1960. Studies on *Withania ashwagandha*. (Part II): Effect of total extract on cardiovascular system, respiration and skeletal muscle. *Indian J. Physiol. Pharmacol.* 4:49-64.

Malik, F., A. Kumar, S. Bhushan, et al. 2007. Reactive oxygen species generation and mitochondrial dysfunction in the apoptotic cell death of human myeloid leukemia HL-60 cells by a dietary compound withaferin A with concomitant protection by N-acetyl cysteine. *Apoptosis* 12 (11):2115-2133.

Mandal, C., A. Dutta, A. Mallick, et al. 2008. Withaferin A induces apoptosis by activating p38 mitogen-activated protein kinase signaling cascade in leukemic cells of lymphoid and myeloid origin through mitochondrial death cascade. *Apoptosis* 13 (12):1450-1464.

Oh, J.H., T.J. Lee, S.H. Kim, et al. 2008. Induction of apoptosis by withaferin A in human leukemia U937 cells through down-regulation of Akt phosphorylation. *Apoptosis* 13 (12):1494-1504.

Panda, S., and A. Kar. 1998. Changes in thyroid hormone concentrations after administration of ashwagandha root extract to adult male mice. *J. Pharm. Pharmacol.* 50 (9):1065-1068.

Poindexter, B.J., A.W. Allison, R.J. Bick, and A. Dasgupta. 2006. Ginseng: Cardiotonic in adult rat cardiomyocytes, cardiotoxic in neonatal rat cardiomyocytes. *Life Sci.* 79 (25):2337-2344.

Rao, A., and K. Karanth. 1990. Neuropharmacological activity of *Withania somnifera*. *Fitoterapia* 61 (3):237-40.

Senthil, V., S. Ramadevi, V. Venkatakrishnan, et al. 2007. Withanolide induces apoptosis in HL-60 leukemia cells via mitochondria mediated cytochrome c release and caspase activation. *Chem. Biol. Interact.* 167 (1):19-30.

Sharada, A.C., F.E. Solomon, and P.U. Devi. 1993. Toxicity of *Withania somnifera* root extract in rats and mice. *Int. J. Pharmacog.* 31 (3):205-212.

Sharma, S., S. Dahanukar, and S.M. Karandikar. 1986. Effects of long-term administration of the roots of ashwagandha (*Withania somnifera*) and shatavari (*Asparagus racemosus*) in rats. *Indian Drugs* 23:133-139.

Singh, N., R. Nath, A. Lata, et al. 1982. *Withania somnifera* (Ashwagandha), a rejuvenating herbal drug which enhances survival during stress (an adaptogen). *Pharmaceutical Biol.* 20 (1):29-35.

Singh, N., R. Nath, D. Singh, M. Gupta, and R. Kohli. 1978. An experimental evaluation of protective effects of some indigenous drugs on carbon tetrachloride-induced hepatotoxicity in mice and rats. *Int. J. Crude Drug Res.* 16 (1):8-16.

Singh, R., P. Malviya, F. Sarkar, and K. Udupa. 1979. Studies on the psychotropic effect of Indian indigenous drug, asvagandha [*Withania somnifera* Dunal.]. Part II: Experimental studies. *J. Res. Indian Med. Yoga Homeop.* 14 (1):49-54.

Svoboda, R. 1992. *Ayurveda: Life, health and longevity*. London: Arkana Penguin.

Tirtha, S. 1998. *The Ayurveda encyclopedia*. Bayville, NY: Ayurvedic Holistic Center.

Upton, R. 2000. *Ashwagandha root: Withania somnifera: Analytical, quality control, and therapeutic monograph, American Herbal Pharmacopoeia and therapeutic compendium;* Santa Cruz, CA: American Herbal Pharmacopoeia.

Wolfiporia cocos (F.A. Wolf) Ryvarden & Gilb. — Polyporaceae

SCN: poria
Syn: *Poria cocos* F.A. Wolf
PN: *fu ling* (sclerotium)
OCN: hoelen; Indian bread (sclerotium); polyporus
Part: sclerotium

Quick Reference Summary

Safety Class: 1
Interaction Class: A

Contraindications

None known.

Other Precautions

None known.

Drug and Supplement Interactions

None known.

Adverse Events and Side Effects

None known.

Pharmacological Considerations

None known.

Pregnancy and Lactation

No information on the safety of poria in pregnancy or lactation was identified in the scientific or traditional literature. Although this review did not identify any concerns for use while pregnant or nursing, safety has not been conclusively established.

Review Details

I. Drug and Supplement Interactions

Clinical Trials of Drug or Supplement Interactions

No clinical trials of drug or supplement interactions were identified.

Case Reports of Suspected Drug or Supplement Interactions

No case reports of suspected drug or supplement interactions were identified.

Animal Trials of Drug or Supplement Interactions

No animal trials of drug or supplement interactions were identified.

II. Adverse Events

Case Reports of Adverse Events

Allergic reactions to poria have been reported. In sensitive individuals, contact with powdered poria may elicit allergic asthma, allergic rhinitis, rapid breathing, and cold sweats (Bensky et al. 2004).

III. Pharmacology and Pharmacokinetics

Human Pharmacological Studies

No relevant human pharmacological studies were identified.

Animal Pharmacological Studies

No relevant animal pharmacological studies were identified.

In Vitro Pharmacological Studies

In a human hepatocellular liver cancer cell line, an extract of poria was found to induce expression of gene transcription of the drug metabolizing isoenzyme CYP3A4 through activation of the pregnane X receptor (Dong et al. 2008a).

In rat liver microsomes, no effect of poria was observed on the drug metabolizing isoenzymes CYP3A, CYP3A1, or CYP3A2 (Dong et al. 2008b).

In a screening on the effects of poria on chemotherapy drug uptake in human cervical carcinoma cells, an aqueous extract of poria enhanced the target cell sensitivity to the p-glycoprotein (P-gp) substrate paclitaxel, with no changes in sensitivity to 5-fluorouracil, which is not a substrate for P-gp. Poria increased cellular uptake of rhodamine 123 (Takara et al. 2005).

In a monocyte culture medium containing 10% poria extract, significant inhibition of secretion of TNF-alpha, IL-beta, IL-6, and GM-CSF from the monocytes was observed (Tseng and Chang 1992).

No significant anti-cholinesterase activity of a methanol extract of poria was observed in a screening of plants for anti-cholinesterase activity (Oh et al. 2004).

IV. Pregnancy and Lactation

No information on the safety of poria during pregnancy or lactation was identified.

V. Toxicity Studies

Acute Toxicity

The LD_{50} of aqueous extracts of poria in mice could not be determined at doses up to 10 g/kg after oral administration and 2 g/kg after intraperitoneal administration (Zhu 1998).

Literature Cited

Bensky, D., S. Clavey, and E. Stöger. 2004. *Chinese herbal medicine: Materia medica*. 3rd ed. Seattle: Eastland Press.

Dong, H.Y., J.W. Shao, J.F. Chen, et al. 2008a. Transcriptional regulation of cytochrome P450 3A4 by four kinds of traditional Chinese medicines. *Zhongguo Zhongyao Zazhi* 33 (9):1014-1017+1089.

Dong, H.Y., J.W. Shao, T. Wang, Y.H. Guo, and L.Y. Yan. 2008b. Effects on the activities and mRNA expression of CYP3A in rat's liver by four kinds of extracts from anti-cancer traditional Chinese medicines. *Zhong Yao Cai* 31 (1):68-71.

Oh, M.H., P.J. Houghton, W.K. Whang, and J.H. Cho. 2004. Screening of Korean herbal medicines used to improve cognitive function for anti-cholinesterase activity. *Phytomedicine* 11 (6):544-548.

Takara, K., S. Horibe, Y. Obata, et al. 2005. Effects of 19 herbal extracts on the sensitivity to paclitaxel or 5-fluorouracil in HeLa cells. *Bio. Pharm. Bull.* 28 (1):138-142.

Tseng, J., and J.G. Chang. 1992. Suppression of tumor necrosis factor-alpha, interleukin-1 beta, interleukin-6 and granulocyte-monocyte colony stimulating factor secretion from human monocytes by an extract of *Poria cocos*. *Zhonghua Min Guo Wei Sheng Wu Ji Mian Yi Xue Za Zhi* 25 (1):1-11.

Zhu, Y.-P. 1998. *Chinese materia medica: Chemistry, pharmacology and applications*. Amsterdam: Harwood Academic Publishers.

Yucca spp.

Agavaceae

Yucca aloifolia L.
SCN: yucca
OCN: aloe yucca; amole; dagger plant; Spanish bayonet

Yucca brevifolia Engelm.
SCN: Joshua tree
OCN: yucca

Yucca glauca Nutt.
SCN: yucca
OCN: amole; soapweed; soapwell

Yucca schidigera Roezl ex Ortgies
SCN: Mojave yucca
OCN: amole

Yucca whipplei Torr.
SCN: our Lord's candle
OCN: amole
Part: root

Quick Reference Summary

Safety Class: 2d
Interaction Class: A

Contraindications

Do not exceed the recommended dose.

Other Precautions

None known.

Drug and Supplement Interactions

None known.

Standard Dose

The standard dose is 2 to 4 g of the herb daily (Moore 2003).

Adverse Events and Side Effects

Allergic reactions to yucca have been reported and confirmed by skin prick testing (Kanerva et al. 2001).

Pharmacological Considerations

None known.

Pregnancy and Lactation

In pigs, a reduction in stillbirths and improvement in piglet health were observed after feeding of Mojave yucca powder to pregnant sows for 3 weeks prior to giving birth (Herpin et al. 2004).

No information on the safety of any of the listed *Yucca* species during lactation was identified in the scientific or traditional literature. While this review did not identify any concerns for use while nursing, safety has not been conclusively established.

Review Details

I. Drug and Supplement Interactions

Clinical Trials of Drug or Supplement Interactions

No clinical trials of drug or supplement interactions were identified.

Case Reports of Suspected Drug or Supplement Interactions

No case reports of suspected drug or supplement interactions were identified.

Animal Trials of Drug or Supplement Interactions

No animal trials of drug or supplement interactions were identified.

II. Adverse Events

Adverse Events Reported in Clinical Trials

No significant adverse effects were observed in a clinical trial of yucca saponins in approximately 700 patients with arthritis (Bingham et al. 1975).

Case Reports of Adverse Events

Occupational contact dermatitis was reported in an atopic worker routinely exposed to yucca (*Yucca aloifolia*) leaves, weeping fig (*Ficus benjamina*), and spathe flower (*Spathiphyllum wallisii*). Skin prick tests were positive to all three plants, and IgE antibodies were found to weeping fig and spathe flower (Kanerva et al. 2001).

Positive reactions to skin prick testing with yucca (*Yucca aloifolia*) were observed in 5.8% of approximately 600 individuals (Kanerva et al. 2001).

III. Pharmacology and Pharmacokinetics

Human Pharmacological Studies

No relevant human pharmacological studies were identified.

Animal Pharmacological Studies

No relevant animal pharmacological studies were identified.

In Vitro Pharmacological Studies

The compound sarsasapogenin has been shown to have hemolytic activity in human blood samples (Zhang et al. 1999). Saponin compounds such as sarsasapogenin are recognized to have hemolytic activity in vitro, although the activity is reduced when given intravenously or orally. Saponins are thought to be poorly absorbed in the gastrointestinal tract and are typically regarded as a safe food additive (Bingham et al. 1975; Oakenfull 1981).

In pig blood platelets, application of phenolic compounds, including *trans*-3,3′,5,5′-tetrahydroxy-4′-methoxystilbene and yuccaols A and C, as a pretreatment slightly reduced platelet aggregation stimulated by ADP. With thrombin-induced platelet aggregation, the selected compounds exhibited significant inhibition of aggregation (Olas et al. 2002).

Phenolic compounds isolated from Mojave yucca, including resveratrol, yuccaol C, and other compounds, have been shown to inhibit NF-κB (Cheeke et al. 2006).

IV. Pregnancy and Lactation

In pregnant pigs fed 120 mg/kg Mojave yucca powder daily from day 107 of pregnancy to day 21 of lactation, a reduction in the number of stillbirths was observed. Piglets from sows fed Mojave yucca powder exhibited improved thermoregulatory abilities after birth and tended to have a lower incidence of stillbirth and preweaning mortality (Herpin et al. 2004).

No information on the safety of any of the listed *Yucca* species during lactation was identified.

V. Toxicity Studies

Short-Term Toxicity

In cats orally administered 150 mg/kg Mojave yucca extract daily for 40 days, no adverse effects or changes in hematological parameters, including red and white blood cell count, hematocrit, hemoglobin, or blood urea nitrogen, were observed (Lowe and Kershaw 1997).

No adverse effects were observed in cows (average animal weight 730 kg) orally administered 60 g of Mojave yucca extract (10% saponins) daily for 28 days (Benchaar et al. 2008).

In lambs intraruminally administered 55 mg/kg Mojave yucca saponins for 11 days, no signs of toxicity were observed (Flaoyen et al. 2002).

In a toxicity study, three groups of lambs were administered varying amounts of Mojave yucca extract or Mojave yucca juice daily for 21 days. In group 1, animals were administered 1.5 g/kg Mojave yucca extract (63 mg sapogenin); in group 2, animals were administered 1.5 g/kg Mojave yucca juice. In group 3, the lambs were dosed as follows: 1.5 g/kg juice on days 1 and 2, 3.0 g/kg on days 3 and 4, 4.5 g/kg on day 5, 3.0 g/kg on days 6 and 7, 1.5 g/kg days 8, 9 and 10, 3.0 g/kg on days 11 to 21. During the first 11 days of the experiment, six lambs from group 2 and six lambs from group 3 were found dead or had to be euthanized for ethical reasons. In these animals, diarrhea and dehydration were the main signs, and these signs were typically accompanied by elevated levels of creatinine and urea, acute tubular necrosis in the kidneys, and watery content in the gastrointestinal tract. Saponins were found in the liver and kidney samples from all lambs that were dosed with the juice of Mojave yucca (Wisloff et al. 2008).

Genotoxicity

No mutagenic activity of compounds isolated from Mojave yucca, including yuccaols A, B, and C, *trans*-resveratrol, and *trans*-3,3′,5,5′-tetrahydroxy-4′-methoxystilbene, was observed in the Ames test for mutagenicity in *Salmonella typhimurium* strains TA97, TA98, TA100, or TA102 with or without metabolic activation by S9 (Czeczot et al. 2003).

Literature Cited

Benchaar, C., T.A. McAllister, and P.Y. Chouinard. 2008. Digestion, ruminal fermentation, ciliate protozoal populations, and milk production from dairy cows fed cinnamaldehyde, quebracho condensed tannin, or *Yucca schidigera* saponin extracts. *J. Dairy Sci.* 91(12):4765-4777.

Bingham, R., B.A. Bellow, and J.G. Bellow. 1975. Yucca plant saponin in the management of arthritis. *J. Appl. Nutr.* 27:45-50.

Cheeke, P.R., S. Piacente, and W. Oleszek. 2006. Anti-inflammatory and anti-arthritic effects of *Yucca schidigera*: A review. *J. Inflamm.* 3:6.

Czeczot, H., M. Podsiad, M. Skrzycki, A. Stochmal, and W. Oleszek. 2003. Evaluation of the mutagenic activity of phenolics from the bark of *Yucca schidigera* Roezl. *Acta Pol. Pharm.* 60(5):357-362.

Flaoyen, A., A.L. Wilkins, and M. Sandvik. 2002. Ruminal metabolism in sheep of saponins from *Yucca schidigera*. *Vet. Res. Commun.* 26(2):159-169.

Herpin, P., A. Vincent, and P.R. Cheeke. 2004. Effect of feeding *Yucca schidigera* (DK powder) to the sow on piglet blood oxygenation and survival. *Proc. West. Sect. Am. Soc. An. Sci.* 55:145-150.

Y

Kanerva, L., T. Estlander, L. Petman, and S. Makinen-Kiljunen. 2001. Occupational allergic contact urticaria to yucca (*Yucca aloifolia*), weeping fig (*Ficus benjamina*), and spathe flower (*Spathiphyllum wallisii*). *Allergy* 56(10):1008-1011.

Lowe, J.A., and S.J. Kershaw. 1997. The ameliorating effect of *Yucca schidigera* extract on canine and feline faecal aroma. *Res. Vet. Sci.* 63(1):61-66.

Moore, M. 2003. *Medicinal plants of the Mountain West*. Revised and expanded edition. Santa Fe: Museum of New Mexico Press.

Oakenfull, D. 1981. Saponins in food—A review. *Food Chem.* 7(1):19-40.

Olas, B., B. Wachowicz, A. Stochmal, and W. Oleszek. 2002. Antiplatelet effects of different phenolic compounds from *Yucca schidigera* Roezl. bark. *Platelets* 13(3):167-173.

Wisloff, H., S. Uhlig, E. Scheie, et al. 2008. Toxicity testing of saponin-containing *Yucca schidigera* Roetzl. juice in relation to hepato- and nephrotoxicity of *Narthecium ossifragum* (L.) Huds. *Toxicon* 51(1):140-150.

Zhang, J., Z. Meng, M. Zhang, et al. 1999. Effect of six steroidal saponins isolated from *Anemarrhenae rhizoma* on platelet aggregation and hemolysis in human blood. *Clin. Chim. Acta* 289(1-2):79-88.

Zanthoxylum spp.

Rutaceae

Zanthoxylum americanum Mill.
SCN: prickly ash
OCN: northern prickly ash; toothache tree

Zanthoxylum clava-herculis L.
SCN: southern prickly ash
OCN: Hercules' club
Part: bark

Quick Reference Summary

Safety Class: 2b
Interaction Class: A

Contraindications

Not for use in pregnancy except under the supervision of a qualified healthcare practitioner (Bradley 1992).

Other Precautions

None known.

Drug and Supplement Interactions

None known.

Notice

Emmenagogue (Felter and Lloyd 1898); *see* Appendix 2.

Adverse Events and Side Effects

Prickly ash and southern prickly ash are toxic to cattle and fish (Bowen et al. 1996; Yin et al. 2007). No cases of toxicity associated with human use were identified.

Pharmacological Considerations

None known.

Pregnancy and Lactation

A British reference text contraindicates the use of prickly ash and southern prickly ash during pregnancy (Bradley 1992).

No information on the safety of prickly ash during lactation was identified in the scientific or traditional literature. While this review did not identify any concerns for use while nursing, safety has not been conclusively established.

Review Details

I. Drug and Supplement Interactions

Clinical Trials of Drug or Supplement Interactions

No clinical trials of drug or supplement interactions were identified.

Case Reports of Suspected Drug or Supplement Interactions

No case reports of suspected drug or supplement interactions were identified.

Animal Trials of Drug or Supplement Interactions

No animal trials of drug or supplement interactions were identified.

II. Adverse Events

Case Reports of Adverse Events

No case reports of adverse events were identified.

III. Pharmacology and Pharmacokinetics

Human Pharmacological Studies

No relevant human pharmacological studies were identified.

Animal Pharmacological Studies

No relevant animal pharmacological studies were identified.

In Vitro Pharmacological Studies

Prickly ash and southern prickly ash have caused toxic and sometimes fatal reactions in cattle that grazed on the bark (Bowen et al. 1996). The bark is also toxic to fish (Yin et al. 2007). Testing in nerves isolated from rats indicated that southern prickly ash extract appeared to exert action on neuromuscular transmission, probably through blockade of postjunctional, end-plate receptors or enhanced release of neurotransmitters (Bowen et al. 1996).

IV. Pregnancy and Lactation

A British reference text contraindicates the use of prickly ash and southern prickly ash during pregnancy (Bradley 1992).

No information on the safety of prickly ash during lactation was identified.

V. Toxicity Studies

No toxicity studies were identified.

Literature Cited

Bowen, J.M., R.J. Cole, D. Bedell, and D. Schabdach. 1996. Neuromuscular effects of toxins isolated from southern prickly ash (*Zanthoxylum clava-herculis*) bark. *Am. J. Vet. Res.* 57(8):1239-1244.

Bradley, P.R. 1992. *British herbal compendium: A handbook of scientific information on widely used plant drugs, Volume 1*. Bournemouth, UK: British Herbal Medicine Association.

Felter, H.W., and J.U. Lloyd. 1898. *King's American dispensatory*. 18th ed., 3rd rev. 2 vols. Cincinnati: Ohio Valley Co.

Yin, Y., W. Fu, M. Fu, G. He, and L. Traore. 2007. The immune effects of edible fungus polysaccharides compounds in mice. *Asia Pac. J. Clin. Nutr.* 16(Suppl. 1):258-260.

Zanthoxylum spp. — Rutaceae

Zanthoxylum bungeanum Maxim.
SCN: Sichuan pepper
PN: *hua jiao* (pericarp)
OCN: Bunge's prickly ash; Sichuan peppercorn

Zanthoxylum schinifolium Siebold & Zucc.
SCN: Sichuan pepper
PN: *hua jiao* (pericarp)
OCN: Sichuan peppercorn

Zanthoxylum simulans Hance
SCN: Sichuan pepper
PN: *hua jiao* (pericarp)
OCN: wild Sichuan pepper
Part: fruit pericarp

Quick Reference Summary

Safety Class: 2b
Interaction Class: A

Contraindications

Not for use in pregnancy except under the supervision of a qualified healthcare practitioner (Bensky et al. 2004; Chen and Chen 2004).

Other Precautions

None known.

Drug and Supplement Interactions

None known.

Adverse Events and Side Effects

None known.

Pharmacological Considerations

None known.

Pregnancy and Lactation

Texts on traditional Chinese medicine indicate that Sichuan pepper should be used with caution during pregnancy (Bensky et al. 2004; Chen and Chen 2004). Based on this information, use during pregnancy is not recommended except under the supervision of a qualified healthcare practitioner.

No information on the safety of Sichuan pepper during lactation was identified in the scientific or traditional literature. While this review did not identify any concerns for use while nursing, safety has not been conclusively established.

Review Details

I. Drug and Supplement Interactions

Clinical Trials of Drug or Supplement Interactions

No clinical trials of drug or supplement interactions were identified.

Case Reports of Suspected Drug or Supplement Interactions

No case reports of suspected drug or supplement interactions were identified.

Animal Trials of Drug or Supplement Interactions

No animal trials of drug or supplement interactions were identified.

II. Adverse Events

Case Reports of Adverse Events

Overdose of Sichuan pepper (standard dose is an aqueous extract of 3–6 g) may cause nausea, vomiting, dry mouth, and vertigo. Severe cases of overdose have been associated with delirium, spasms, difficulty breathing, loss of consciousness, and respiratory failure (Bensky et al. 2004).

Z

III. Pharmacology and Pharmacokinetics

Human Pharmacological Studies

No relevant human pharmacological studies were identified.

Animal Pharmacological Studies

No relevant animal pharmacological studies were identified.

In Vitro Pharmacological Studies

The compound chelerythrine chloride inhibited platelet aggregation induced by ADP, arachidonic acid, PAF, collagen, ionophore A23187, and thrombin in washed rabbit platelets. Less inhibition was observed in platelet-rich plasma. Treatment of washed platelets with chelerythrine chloride decreased thromboxane B_2 formation induced by arachidonic acid, collagen, ionophore A23187, and thrombin (Ko et al. 1990).

IV. Pregnancy and Lactation

Texts on traditional Chinese medicine indicate that Sichuan pepper should be used with caution during pregnancy (Bensky et al. 2004; Chen and Chen 2004).

No information on the safety of Sichuan pepper during lactation was identified.

V. Toxicity Studies

No toxicity studies were identified.

Literature Cited

Bensky, D., S. Clavey, and E. Stöger. 2004. *Chinese herbal medicine: Materia medica*. 3rd ed. Seattle: Eastland Press.

Chen, J.K., and T.T. Chen. 2004. *Chinese medical herbology and pharmacology*. City of Industry, CA: Art of Medicine Press.

Ko, F.N., I.S. Chen, S.J. Wu, et al. 1990. Antiplatelet effects of chelerythrine chloride isolated from *Zanthoxylum simulans*. *Biochim. Biophys. Acta* 1052(3):360-365.

Zea mays L.

Poaceae

SCN: corn
PN: *yu mi xu* (stigma)
OCN: maize
Part: stigma

Quick Reference Summary

Safety Class: 1
Interaction Class: A

Contraindications

None known.

Other Precautions

None known.

Drug and Supplement Interactions

None known.

Notice

Diuretic (Chen and Chen 2004; Felter and Lloyd 1898; Remington and Wood 1918; Velazquez et al. 2005; Wichtl 2004); *see* Appendix 2.

Adverse Events and Side Effects

None known.

Pharmacological Considerations

None known.

Pregnancy and Lactation

No information on the safety of corn stigma during pregnancy or lactation was identified in the scientific or traditional literature. While this review did not identify any concerns for pregnant or lactating women, safety has not been conclusively established.

Review Details

I. Drug and Supplement Interactions

Clinical Trials of Drug or Supplement Interactions

No clinical trials of drug or supplement interactions were identified.

Case Reports of Suspected Drug or Supplement Interactions

No case reports of suspected drug or supplement interactions were identified.

Animal Trials of Drug or Supplement Interactions

No animal trials of drug or supplement interactions were identified.

II. Adverse Events

Case Reports of Adverse Events

No case reports of adverse events were identified.

III. Pharmacology and Pharmacokinetics

Human Pharmacological Studies

No relevant human pharmacological studies were identified.

Animal Pharmacological Studies

In water-loaded rats orally administered single doses of 25, 50, 200, 350, and 500 mg/kg corn stigma extract, an increase in potassium excretion was observed at doses of 350 and 500 mg/kg. At the 500 mg/kg dose, diuretic activity was observed. In rats administered a single dose of 500 mg/kg of the same extract, glomerular filtration and filtered load decreased without affecting proximal tubular function, sodium, or uric acid excretion (Velazquez et al. 2005).

In Vitro Pharmacological Studies

No relevant in vitro pharmacological studies were identified.

IV. Pregnancy and Lactation

No information on the safety of corn stigma in pregnancy or lactation was identified.

V. Toxicity Studies

Acute Toxicity

The LD_{50} of intravenously administered corn stigma in rabbits is 250 mg/kg (Chen and Chen 2004). The LD_{50} of an intraperitoneally administered corn stigma aqueous extract in rats is 14.5 g/kg (Al-Ali et al. 2003).

Short-Term Toxicity

In rats orally administered 50, 100, or 150 mg/kg of an aqueous extract of corn stigma daily for 21 days, no significant toxicity was observed at the 50 and 100 mg/kg doses. Reductions in levels of hemoglobin, red blood corpuscles, clotting time, mean corpuscular volume, hematocrit, blood urea nitrogen, aspartate transaminase, alanine transaminase, calcium, total protein, total albumin, and total acid phosphatase were reported. Increases were reported in white blood corpuscles, mean corpuscular hemoglobin, alkaline phosphatase, and creatinine (Garg and Goyal 1992).

Subchronic Toxicity

In rats administered corn silk as 0.5%, 2.0% and 8.0% (w/w) of the diet for 90 days, no adverse changes in body weight, food consumption, hematology, blood chemistry, organ weights, or gross and microscopic appearance of tissues were observed. The authors of this study indicated that the no-observed-adverse-effect level (NOAEL) of corn silk is at least 8.0%, which corresponds to a mean daily corn silk intake of approximately 9.354 g/kg for males and 10.308 g/kg for females (Wang et al 2011).

Literature Cited

Al-Ali, M., S. Wahbi, H. Twaij, and A. Al-Badr. 2003. *Tribulus terrestris*: Preliminary study of its diuretic and contractile effects and comparison with *Zea mays*. *J. Ethnopharmacol.* 85(2-3):257-260.

Chen, J.K., and T.T. Chen. 2004. *Chinese medical herbology and pharmacology*. City of Industry, CA: Art of Medicine Press.

Felter, H.W., and J.U. Lloyd. 1898. *King's American dispensatory*. 18th ed., 3rd rev. 2 vols. Cincinnati: Ohio Valley Co.

Garg, D.K., and R.N. Goyal. 1992. Haematological and hepatotoxic effects of silken styles of corn in albino rats. *J. Appl. Toxicol.* 12(5):359-363.

Remington, J.P., and H.C. Wood. 1918. *The dispensatory of the United States of America*. 20th ed. Philadelphia: Lippincott.

Velazquez, D.V., H.S. Xavier, J.E. Batista, and C. de Castro-Chaves. 2005. *Zea mays* L. extracts modify glomerular function and potassium urinary excretion in conscious rats. *Phytomedicine* 12(5):363-369.

Wang, C., T. Zhang, J. Liu, S. Lu, C. Zhang, E. Wang, Z. Wang, Y. Zhang, J. Liu. 2011. Subchronic toxicity study of corn silk with rats. *J. Ethnopharmacol.* 137(1):36-43.

Wichtl, M. 2004. *Herbal drugs and phytopharmaceuticals: A handbook for practice on a scientific basis*. 3rd ed. Boca Raton, FL: CRC Press.

Zingiber officinale Roscoe

Zingiberaceae

SCN: ginger
AN: *ardraka* (fresh rhizome); *shunthi* (dried rhizome)

PN: *sheng jiang* (fresh rhizome); *gan jiang* (dried rhizome); *pao jiang* (prepared rhizome); *sheng jiang pi* (peel)
Part: rhizome

Z

Quick Reference Summary

Safety Class: 1

Interaction Class: B

Contraindications

None known.

Other Precautions

In persons with active gallstone disease, use of ginger should be under the supervision of a qualified healthcare practitioner (Mills and Bone 2005; Srinivasan and Sambaiah 1991; Yamahara et al. 1985).

Drug and Supplement Interactions

A human trial indicated that nifedipine with ginger had a greater antiplatelet action than nifedipine alone (Young et al. 2006).

Adverse Events and Side Effects

Gastrointestinal symptoms, such as heartburn, are occasionally associated with ginger use, especially with large doses of ginger (Chittumma et al. 2007; Sripramote and Lekhyananda 2003; Vutyavanich et al. 2001; Wigler et al. 2003; Willetts et al. 2003). Allergic reactions to ginger have been reported, primarily in spice factory workers (Futrell and Rietschel 1993; Kanerva et al. 1996; Stager et al. 1991; Zuskin et al. 1988).

Excessive doses of preparations made from fresh ginger have been reported to cause dry mouth, sore throat, nosebleeds, and kidney inflammation (Bensky et al. 2004).

Pharmacological Considerations

Although ginger has been reported to have antiplatelet activity (Srivastava 1984), thus slowing blood clotting, human studies have not demonstrated any antiplatelet activity (Bordia et al. 1997; Janssen et al. 1996; Lumb 1994; Srivastava 1989). Bile-stimulating effects have been reported in animal studies (Srinivasan and Sambaiah 1991; Yamahara et al. 1985).

No effects on coagulation were observed in a human study of ginger and warfarin (Jiang et al. 2005, 2006).

Pregnancy and Lactation

Although the use of dried ginger in pregnancy is cautioned in texts on traditional Chinese medicine (Bensky et al. 2004; Chen and Chen 2004), data from several clinical trials, including over 900 pregnant women, have not reported any adverse effects of 1 to 2 g ginger daily on pregnancy outcomes (Chittumma et al. 2007; Fischer-Rasmussen et al. 1991; Keating and Chez 2002; Smith et al. 2004; Sripramote and Lekhyananda 2003; Vutyavanich et al. 2001; Willetts et al. 2003).

No information on the safety of ginger during lactation was identified in the scientific or traditional literature. While this review did not identify any concerns for use while nursing, safety has not been conclusively established.

Review Details

I. Drug and Supplement Interactions

Clinical Trials of Drug or Supplement Interactions

No effect on warfarin (25 mg) was observed after administration of ginger (3.6 g daily) for 7 days (Jiang et al. 2005, 2006).

Administration of 1 g ginger daily for 7 days along with 10 mg of nifedipine daily exhibited a more significant inhibition of platelet aggregation than ginger or nifedipine separately, suggesting a synergistic effect of the two agents. The effect was more pronounced in subjects with normal blood pressure than in those with hypertension. While nifedipine is a calcium channel blocker typically used to treat high blood pressure, this drug has also been shown to decrease platelet aggregation (Young et al. 2006).

Case Reports of Suspected Drug or Supplement Interactions

Two case reports of suspected drug interactions with phenprocoumon and celecoxib were identified, but the reports did not identify the doses or type of products used (Handler 2003; Kruth et al. 2004).

Animal Trials of Drug or Supplement Interactions

No relevant animal trials of drug or supplement interactions were identified.

II. Adverse Events

Adverse Events Reported in Clinical Trials

A systematic review of randomized controlled studies on the use of ginger during pregnancy, including a total of 675 participants, indicated a lack of significant side effects or adverse effects on pregnancy outcomes. The review concluded that, "observational studies, with a larger sample size, are needed to confirm the encouraging preliminary data on ginger safety" (Borrelli et al. 2005). Heartburn has been reported as an adverse effect of ginger in several clinical trials (Chittumma et al. 2007; Sripramote and Lekhyananda 2003; Vutyavanich et al. 2001; Wigler et al. 2003; Willetts et al. 2003).

Case Reports of Adverse Events

Cases of allergic reactions to ginger have been reported; these cases have primarily been observed in allergic patch testing of spice factory or food service workers (Futrell and Rietschel 1993; Kanerva et al. 1996; Stager et al. 1991; Zuskin et al. 1988). A correlation has been suggested between ginger allergy and birch-mugwort-celery allergy syndrome, although the correlation has been challenged (Moneret-Vautrin et al. 2002; Stager et al. 1991).

Gastrointestinal symptoms, such as heartburn, are occasionally associated with ginger use, especially with large doses of ginger (De Smet 1997; Desai et al. 1990).

III. Pharmacology and Pharmacokinetics

Human Pharmacological Studies

Although ginger has been reported to have antiplatelet activity (Srivastava 1984), data from human studies has not substantiated such concerns. No effects on platelet aggregation or fibrinolytic activity were observed in volunteers with coronary artery disease or non-insulin-dependent diabetes mellitus who had taken 4 g of powdered ginger daily for 3 months (Bordia et al. 1997), although a significant reduction in platelet aggregation was observed in men after a single dose of 10 g of powdered ginger (Bordia et al. 1997). No effects were observed after a single dose of 2 g of powdered ginger (Lumb 1994). Likewise, no significant change in thromboxane B_2 formation was observed in healthy women after consumption of 5 g raw ginger daily for 7 days (Srivastava 1989). No ex vivo effects on thromboxane production were observed in blood from healthy volunteers who had taken 15 g of raw ginger or 40 g of cooked ginger daily for 2 weeks (Janssen et al. 1996).

Animal Pharmacological Studies

An acetone extract of ginger administered intraduodenally resulted in an increase in bile secretion of rats. In the same study, an aqueous ginger extract had no effects on bile secretion (Yamahara et al. 1985). In rats, ginger has been shown to stimulate hepatic cholesterol 7α-hydroxylase, an enzyme responsible for the conversion of cholesterol to bile acids (Srinivasan and Sambaiah 1991).

Other animal pharmacological studies were identified but omitted due to the availability of human data.

In Vitro Pharmacological Studies

Ginger was shown to lower blood pressure in several in vitro models (Ghayur and Gilani 2005). Estrogenic effects of ginger extracts were observed at high concentrations in one in vitro yeast assay study (Kang et al. 2006).

IV. Pregnancy and Lactation

A number of clinical trials and epidemiological studies evaluating the use of ginger in morning sickness have been completed. A systematic review of randomized controlled studies on the use of ginger during pregnancy, including a total of 675 participants, indicated that no adverse events were reported during the studies. The review concluded that, "observational studies, with a larger sample size, are needed to confirm the encouraging preliminary data on ginger safety" (Borrelli et al. 2005).

Clinical trials of ginger in pregnancy have generally been limited to 4 days of treatment, although some studies lasted 2 or 3 weeks (Keating and Chez 2002; Smith et al. 2004). Doses in the studies ranged from 1 to 2 g daily, with 1 g daily divided into three doses being the most common regimen. Most products studied were capsules of dried ginger, although one study used a ginger syrup. No significant adverse effects were observed on mothers, on pregnancy, or on fetuses (Chittumma et al. 2007; Fischer-Rasmussen et al. 1991; Keating and Chez 2002; Smith et al. 2004; Sripramote and Lekhyananda 2003; Vutyavanich et al. 2001; Willetts et al. 2003). In one study with women taking 1.95 g daily of ginger, heartburn was reported in 12% of the participants (Chittumma et al. 2007). In a prospective study, no significant differences were found between pregnant women self-administering ginger-containing products and a control group of pregnant women (Portnoi et al. 2003).

Surveys of pregnant women in Australia and Canada indicated a relatively high incidence of ginger use, with approximately 12 to 30% of women responding to the survey having used ginger for relief of nausea during pregnancy (Forster et al. 2006; Hollyer et al. 2002; Maats and Crowther 2002).

In rats administered tea of fresh ginger on days 6 to 15 of gestation, a higher embryonic loss was observed in the treatment group as compared to the untreated control group, although fetuses that survived were heavier and had more advanced skeletal development as compared to controls. No fetal malformations or signs of maternal toxicity were observed (Wilkinson 2000).

No information on the safety of ginger during lactation was identified in the scientific or traditional literature. While this review did not identify any concerns for use while nursing, safety has not been conclusively established.

V. Toxicity Studies

Acute Toxicity

The LD_{50} of orally administered ginger in rats is over 250 g/kg (Wu et al. 1990).

Literature Cited

Bensky, D., S. Clavey, and E. Stöger. 2004. *Chinese herbal medicine: Materia medica*. 3rd ed. Seattle: Eastland Press.

Bordia, A., S.K. Verma, and K.C. Srivastava. 1997. Effect of ginger (*Zingiber officinale* Rosc.) and fenugreek (*Trigonella foenumgraecum* L.) on blood lipids, blood sugar and platelet aggregation in patients with coronary artery disease. *Prostaglandins Leukot. Essent. Fatty Acids* 56(5):379-384.

Borrelli, F., R. Capasso, G. Aviello, M.H. Pittler, and A.A. Izzo. 2005. Effectiveness and safety of ginger in the treatment of pregnancy-induced nausea and vomiting. *Obstet. Gynecol.* 105(4):849-856.

Chen, J.K., and T.T. Chen. 2004. *Chinese medical herbology and pharmacology*. City of Industry, CA: Art of Medicine Press.

Z

Chittumma, P., K. Kaewkiattikun, and B. Wiriyasiriwach. 2007. Comparison of the effectiveness of ginger and vitamin B_6 for treatment of nausea and vomiting in early pregnancy: A randomized double-blind controlled trial. *J. Med. Assoc. Thai.* 90(1):15-20.

De Smet, P.A.G.M. 1997. *Adverse effects of herbal drugs, Volume 3.* Berlin: Springer.

Desai, H.G., R.H. Kalro, and A.P. Choksi. 1990. Effect of ginger and garlic on DNA content of gastric aspirate. *Indian J. Med. Res.* 92:139-141.

Fischer-Rasmussen, W., S.K. Kjaer, C. Dahl, and U. Asping. 1991. Ginger treatment of hyperemesis gravidarum. *Eur. J. Obstet. Gynecol. Reprod. Biol.* 38(1):19-24.

Forster, D.A., A. Denning, G. Wills, M. Bolger, and E. McCarthy. 2006. Herbal medicine use during pregnancy in a group of Australian women. *BMC Pregnancy Childbirth* 6:21.

Futrell, J.M., and R.L. Rietschel. 1993. Spice allergy evaluated by results of patch tests. *Cutis* 52(5):288-290.

Ghayur, M.N., and A.H. Gilani. 2005. Ginger lowers blood pressure through blockade of voltage-dependent calcium channels. *J. Cardiovasc. Pharmacol.* 45(1):74-80.

Handler, J. 2003. Drug-induced hypertension. *J. Clin. Hypertens. (Greenwich)* 5(1):83-85.

Hollyer, T., H. Boon, A. Georgousis, M. Smith, and A. Einarson. 2002. The use of CAM by women suffering from nausea and vomiting during pregnancy. *BMC Complement. Altern. Med.* 2:5.

Janssen, P.L., S. Meyboom, W.A. van Staveren, F. de Vegt, and M.B. Katan. 1996. Consumption of ginger (*Zingiber officinale* Roscoe) does not affect *ex vivo* platelet thromboxane production in humans. *Eur. J. Clin. Nutr.* 50(11):772-774.

Jiang, X., E.Y. Blair, and A.J. McLachlan. 2006. Investigation of the effects of herbal medicines on warfarin response in healthy subjects: A population pharmacokinetic-pharmacodynamic modeling approach. *J. Clin. Pharmacol.* 46(11):1370-1378.

Jiang, X., K.M. Williams, W.S. Liauw, et al. 2005. Effect of ginkgo and ginger on the pharmacokinetics and pharmacodynamics of warfarin in healthy subjects. *Br. J. Clin. Pharmacol.* 59(4):425-432.

Kanerva, L., T. Estlander, and R. Jolanki. 1996. Occupational allergic contact dermatitis from spices. *Contact Dermat.* 35(3):157-162.

Kang, S.C., C.M. Lee, H. Choi, et al. 2006. Evaluation of oriental medicinal herbs for estrogenic and antiproliferative activities. *Phytother. Res.* 20(11):1017-1019.

Keating, A., and R.A. Chez. 2002. Ginger syrup as an antiemetic in early pregnancy. *Altern. Ther. Health Med.* 8(5):89-91.

Kruth, P., E. Brosi, R. Fux, K. Morike, and C.H. Gleiter. 2004. Ginger-associated overanticoagulation by phenprocoumon. *Ann. Pharmacother.* 38(2):257-260.

Lumb, A.B. 1994. Effect of dried ginger on human platelet function. *Thromb. Haemost.* 71(1):110-111.

Maats, F.H., and C.A. Crowther. 2002. Patterns of vitamin, mineral and herbal supplement use prior to and during pregnancy. *Aust. N.Z. J. Obstet. Gynaecol.* 42(5):494-496.

Mills, S., and K. Bone. 2005. *The essential guide to herbal safety.* St. Louis: Elsevier.

Moneret-Vautrin, D.A., M. Morisset, P. Lemerdy, A. Croizier, and G. Kanny. 2002. Food allergy and IgE sensitization caused by spices: CICBAA data (based on 589 cases of food allergy). *Allerg. Immunol. (Paris)* 34(4):135-140.

Portnoi, G., L-A. Chng, L. Karimi-Tabesh, et al. 2003. Prospective comparative study of the safety and effectiveness of ginger for the treatment of nausea and vomiting in pregnancy. *Am. J. Obstet. Gynecol.* 189(5):1374-1377.

Smith, C., C. Crowther, K. Willson, N. Hotham, and V. McMillian. 2004. A randomized controlled trial of ginger to treat nausea and vomiting in pregnancy. *Obstet. Gynecol.* 103(4):639-645.

Srinivasan, K., and K. Sambaiah. 1991. The effect of spices on cholesterol 7 alpha-hydroxylase activity and on serum and hepatic cholesterol levels in the rat. *Int. J. Vitam. Nutr. Res.* 61(4):364-369.

Sripramote, M., and N. Lekhyananda. 2003. A randomized comparison of ginger and vitamin B_6 in the treatment of nausea and vomiting of pregnancy. *J. Med. Assoc. Thai.* 86:846-853.

Srivastava, K.C. 1984. Aqueous extracts of onion, garlic and ginger inhibit platelet aggregation and alter arachidonic acid metabolism. *Biomed. Biochim. Acta* 43(8-9):S335-S346.

Srivastava, K.C. 1989. Effect of onion and ginger consumption on platelet thromboxane production in humans. *Prostaglandins Leukot. Essent. Fatty Acids* 35(3):183-185.

Stager, J., B. Wuthrich, and S.G. Johansson. 1991. Spice allergy in celery-sensitive patients. *Allergy* 46(6):475-478.

Vutyavanich, T., T. Kraisarin, and R. Ruangsri. 2001. Ginger for nausea and vomiting in pregnancy: Randomized, double-masked, placebo-controlled trial. *Obstet. Gynecol.* 97(4):577-582.

Wigler, I., I. Grotto, D. Caspi, and M. Yaron. 2003. The effects of Zintona EC (a ginger extract) on symptomatic gonarthritis. *Osteoarthritis Cartilage* 11(11):783-789.

Wilkinson, J.M. 2000. Effect of ginger tea on the fetal development of Sprague-Dawley rats. *Reprod. Toxicol.* 14(6):507-512.

Willetts, K.E., A. Ekangaki, and J.A. Eden. 2003. Effect of a ginger extract on pregnancy-induced nausea: A randomised controlled trial. *Aust. N.Z. J. Obstet. Gynaecol.* 43(2):139-144.

Wu, H., D. Ye, Y. Bai, and Y. Zhao. 1990. Effect of dry ginger and roasted ginger on experimental gastric ulcers in rats. *Zhongguo Zhong Yao Za Zhi* 15(5):278-280, 317-318.

Yamahara, J., K. Miki, T. Chisaka, et al. 1985. Cholagogic effect of ginger and its active constituents. *J. Ethnopharmacol.* 13(2):217-225.

Young, H.Y., J.C. Liao, Y.S. Chang, et al. 2006. Synergistic effect of ginger and nifedipine on human platelet aggregation: A study in hypertensive patients and normal volunteers. *Am. J. Chin. Med.* 34(4):545-551.

Zuskin, E., B. Kanceljak, Z. Skuric, et al. 1988. Immunological and respiratory findings in spice-factory workers. *Environ. Res.* 47(1):95-108.

Ziziphus jujuba Mill. — Rhamnaceae

SCN: jujube
Syn: *Ziziphus vulgaris* Lam.
AN: *badara*
PN: *da zao* (fruit)

OCN: Chinese date; Chinese jujube; jujube date

Part: fruit

Quick Reference Summary

Safety Class: 1
Interaction Class: A

Contraindications
None known.

Other Precautions
None known.

Drug and Supplement Interactions
None known.

Editors' Note
Jujube fruit, often referred to as jujube date, is a sweet dried fruit commonly used in Asian cooking (Chen and Chen 2004).

Adverse Events and Side Effects
None known.

Pharmacological Considerations
Allergic reactions to jujube fruit have been reported (Bensky et al. 2004).

Pregnancy and Lactation
No information on the safety of jujube fruit in pregnancy or lactation was identified in the scientific or traditional literature. Although this review did not identify any concerns for use while pregnant or nursing, safety has not been conclusively established.

Review Details

I. Drug and Supplement Interactions

Clinical Trials of Drug or Supplement Interactions
No clinical trials of drug or supplement interactions were identified.

Case Reports of Suspected Drug or Supplement Interactions
No case reports of suspected drug or supplement interactions were identified.

Animal Trials of Drug or Supplement Interactions
No animal trials of drug or supplement interactions were identified.

II. Adverse Events

Case Reports of Adverse Events
Allergic reactions to jujube fruit have been reported (Bensky et al. 2004).

III. Pharmacology and Pharmacokinetics

Human Pharmacological Studies
No relevant human pharmacological studies were identified.

Animal Pharmacological Studies
No relevant animal pharmacological studies were identified.

In Vitro Pharmacological Studies
No relevant in vitro pharmacological studies were identified.

IV. Pregnancy and Lactation
No information on the safety of jujube fruit during pregnancy or lactation was identified.

V. Toxicity Studies
No toxicity studies were identified.

Literature Cited

Bensky, D., S. Clavey, and E. Stöger. 2004. *Chinese herbal medicine: Materia medica*. 3rd ed. Seattle: Eastland Press.

Chen, J.K., and T.T. Chen. 2004. *Chinese medical herbology and pharmacology*. City of Industry, CA: Art of Medicine Press.

Ziziphus jujuba Mill. var. *spinosa* (Bunge) Hu ex H.F. Chow — Rhamnaceae

SCN: ziziphus
Syn: *Ziziphus spinosa* (Bunge) Hu ex Chen
PN: *suan zao ren* (seed)
OCN: sour date; sour jujube

Part: seed

Quick Reference Summary

Safety Class: 2b
Interaction Class: A

Contraindications

Not for use in pregnancy except under the supervision of a qualified healthcare practitioner (Chen and Chen 2004).

Other Precautions

Use with caution in persons with diarrhea (Bensky et al. 2004; Chen and Chen 2004).

Drug and Supplement Interactions

None known.

Notice

Uterine stimulant (Chen and Chen 2004); *see* Appendix 2.

Adverse Events and Side Effects

An acute serotonin reaction was reported in a woman taking ziziphus who subsequently began to take venlafaxine (Stewart 2004).

Allergic reactions to ziziphus seed have been reported (Bensky et al. 2004).

Pharmacological Considerations

Ziziphus seed may cause drowsiness and sedation (Chen and Chen 2004; Jiang et al. 2007; Ma et al. 2008; Peng et al. 2000).

An in vitro study indicated that ziziphus seed had an antagonistic effect on serotonin receptors and dopamine receptors (Koetter et al. 2009).

Pregnancy and Lactation

A text on traditional Chinese medicine indicates that ziziphus seed should be used with caution during pregnancy due to uterine-stimulating activity (Chen and Chen 2004). Based on this information, use during pregnancy is not recommended except under the supervision of a qualified healthcare practitioner.

No information on the safety of zizyphus seed during lactation was identified. While this review did not identify any concerns for use while nursing, safety has not been conclusively established.

Review Details

I. Drug and Supplement Interactions

Clinical Trials of Drug or Supplement Interactions

No clinical trials of drug or supplement interactions were identified.

Case Reports of Suspected Drug or Supplement Interactions

A severe acute serotonin reaction with some anaphylactic features was reported in a 40-year-old woman. The woman had been taking 0.5 g of ziziphus seed daily for several weeks and was subsequently prescribed 37.5 mg daily of venlafaxine. Approximately 1 hour after ingesting the first dose of venlafaxine and ziziphus at bedtime, the patient became agitated, restless, nauseated, dizzy, and ataxic, and subsequently collapsed. On examination, she was pale, drooling, unable to sit, profusely diaphoretic, tachypneic, trembling, and shivering. The patient was instructed to stop taking ziziphus and instead take 150 mg venlafaxine daily, which she did for 1 month without adverse effects (Stewart 2004).

Animal Trials of Drug or Supplement Interactions

No animal trials of drug or supplement interactions were identified.

II. Adverse Events

Case Reports of Adverse Events

Allergic reactions to ziziphus seed have been reported. Symptoms included pruritus, urticaria, numbness of the mouth and lips, shortness of breath, dizziness, nausea, vomiting, facial pallor, and cold sweat (Bensky et al. 2004).

While one text on traditional Chinese medicine indicates that overdose (standard dose listed as an extract of 9–15 g) of ziziphus seed should be avoided (Bensky et al. 2004), another text indicates that up to 30 g may be used and that 70 g has been used under professional supervision without toxic reactions (Chen and Chen 2004).

III. Pharmacology and Pharmacokinetics

Human Pharmacological Studies

No relevant human pharmacological studies were identified.

Animal Pharmacological Studies

In mice orally administered 1 g/kg of an ethanol extract of ziziphus seed, a prolongation of hexobarbital-induced sleeping time was observed (Peng et al. 2000). A cyclopeptide alkaloid fraction of ziziphus seed shortened sleeping onset and prolonged sleeping time induced by hypnotic doses of pentobarbital (42 mg/kg) and increased the falling asleep rate and duration of sleeping time at a subhypnotic dosage of pentobarbital (28 mg/kg) (Ma et al. 2008). Saponin compounds from ziziphus seed were also found to prolong pentobarbital-induced sleeping time in mice (Jiang et al. 2007).

In Vitro Pharmacological Studies

A hydroethanolic extract of ziziphus seed had an antagonistic effect with the serotonin 5-HT_{1B} receptor, with 23% inhibition of control (10 μ*M* imipramine) binding at 100 μg/ml. Antagonism of the dopamine D1 receptor, with inhibition of control (10 μ*M* dopamine) binding at 100 μg/ml was reported (Koetter et al. 2009).

IV. Pregnancy and Lactation

A text on traditional Chinese medicine indicates that ziziphus seed should be used with caution during pregnancy, due to uterine-stimulating activity (Chen and Chen 2004).

No information on the safety of zizyphus seed during lactation was identified.

V. Toxicity Studies

Acute Toxicity

The LD_{50} of intraperitoneally administered ziziphus seed in mice is 14.3 g/kg (Chen and Chen 2004). Intraperitoneal administration of 10 to 15 g/kg of an aqueous extract of ziziphus seed was lethal to guinea pigs (Zhu 1998).

Short-Term Toxicity

No signs of toxicity were observed in rats orally administered 20 g/kg of an aqueous extract of ziziphus seed daily for 30 days (Zhu 1998).

Literature Cited

Bensky, D., S. Clavey, and E. Stöger. 2004. *Chinese herbal medicine: Materia medica.* 3rd ed. Seattle: Eastland Press.

Chen, J.K., and T.T. Chen. 2004. *Chinese medical herbology and pharmacology.* City of Industry, CA: Art of Medicine Press.

Jiang, J.G., X.J. Huang, and J. Chen. 2007. Separation and purification of saponins from *Semen Ziziphus jujuba* and their sedative and hypnotic effects. *J. Pharm. Pharmacol.* 59(8):1175-1180.

Koetter, U., M. Barrett, S. Lacher, A. Abdelrahman, and D. Dolnick. 2009. Interactions of *Magnolia* and *Ziziphus* extracts with selected central nervous system receptors. *J. Ethnopharmacol.* 124(3):421-425.

Ma, Y., H. Han, S.Y. Nam, et al. 2008. Cyclopeptide alkaloid fraction from Zizyphi Spinosi Semen enhances pentobarbital-induced sleeping behaviors. *J. Ethnopharmacol.* 117(2):318-224.

Peng, W.H., M.T. Hsieh, Y.S. Lee, Y.C. Lin, and J. Liao. 2000. Anxiolytic effect of seed of *Ziziphus jujuba* in mouse models of anxiety. *J. Ethnopharmacol.* 72(3):435-441.

Stewart, D.E. 2004. Venlafaxine and sour date nut. *Am. J. Psychiatr.* 161(6):1129-1130.

Zhu, Y.P. 1998. *Chinese materia medica: Chemistry, pharmacology and applications.* Boca Raton, FL: CRC Press.

APPENDIX 1: HERBAL CONSTITUENT PROFILES

Alkenylbenzenes

Written by Soaring Bear, Ph.D.

Alkenylbenzenes, also known as allylbenzenes or allylphenols, contribute to the flavor of some plants. Alkenylbenzenes found in medicinal plants include, among others, asarone, estragole, safrole, and methyleugenol. β-asarone is found in species of *Acorus* and *Asarum*. Traces of estragole can be found in herbs such as tarragon (*Artemisia dracunculus*), basil (*Ocimum basilicum*), and fennel (*Foeniculum vulgare*) (EMEA 2005). Safrole is a minor component of aromatic oils of nutmeg (*Myristica fragrans*), cinnamon leaf (*Cinnamomum verum*), and camphor (*Cinnamomum camphora*), and a major constituent of sassafras (*Sassafras albidum*) essential oil (Keeler and Tu 1983). Safrole is also found in black pepper (*Piper nigrum*) and in trace amounts in basil (*Ocimum basilicum*) (Farag and Abo-Zeid 1997; Leung and Foster 1996).

Adverse Effects

There is some controversy regarding the safety of the use of plants containing alkenylbenzenes, with some of them limited in usage by regulatory authorities. Extract of sassafras was used for many years as a flavoring agent in the soft-drink industry, providing one of the familiar natural root beer flavors. In 1960, however, researchers began to question the safety of safrole (Barceloux 2008). Animal studies showed an increase in liver tumors in animals fed the purified compound safrole in relatively high amounts (0.01 to 0.1 % of the diet) for extended periods of time (2 years, equivalent to approximately 68 years of human exposure) (Abbott et al. 1961; Hagan et al. 1967; Hagan et al. 1965; Long et al. 1963).

A 1961 report, "Toxic and possible carcinogenic effects of 4-allyl-1,2-methylenedioxybenzene (safrole) in rats," led to further in vitro studies that culminated in a ban by the FDA (CFR 2011; Homburger et al. 1961). Safrole's potential damage of DNA has not been confirmed in humans, yet it is of such substantial consequence that public health agencies are inclined to err on the side of safety. Safrole is also used as a precursor in the synthesis of the insecticide synergist piperonyl butoxide and for the clandestine manufacture of MDMA (ecstasy) (Barceloux 2008), which raises suspicions about politicization of the regulation.

Animal studies with isolated estragole raised similar concerns with regulatory agencies regarding the association between estragole and liver cancer, although estragole has been regarded as a "weak inducer" in regards to liver cancer. Metabolic studies indicate that in high doses (150–600 mg/kåg), the production of 1′-hydroxyestragole, expressed as percentage of the dose, is about 5–10 times higher than that at lower doses (0.05–50 mg/kg). European authorities have recommended on precautionary grounds that the content of estragole and methyleugenol in foods be reduced as far as possible (BGVV 2002). The Committee on Herbal Medicinal Products of the EMEA issued a public statement on the use of herbal medicinal products containing estragole and concluded that, "The present exposure to estragole resulting from consumption of herbal medicinal products (short time use in adults at recommended posology) does not pose a significant cancer risk" (EMEA 2005).

Generally in toxicology, there is some threshold dose below which toxicity is inconsequential and above which our normal detoxification systems are overwhelmed. There is evidence that small quantities of alkenylbenzenes are quickly broken down by the cytosolic and microsomal epoxide hydrolases of the liver and that the potential hazard to humans of low doses of allylbenzenes (e.g., β-asarone, estragole, and safrole) is minimal. The question of exactly how much is too much has not yet been answered.

Mechanism of Action

Alkenylbenzenes are not directly hepatotoxic or hepatocarcinogenic. Cytochrome P450 enzymes in the liver oxidize the double bond of alkenylbenzenes to an epoxide, which is mostly conjugated by glutathione for excretion, but at levels exceeding the detoxification capacity, the overflow can be reactive electrophilic and mutagenic sulfuric acid esters that give rise to DNA adducts. The propenyl analogues isosafrole, anethole, and methylisoeugenol, which cannot undergo 1-hydroxylation, are not genotoxic (Hasheminejad and Caldwell 1994).

Relative to other carcinogens, the hazard of alkenylbenzenes is small yet present. One study compared the number of liver tumors (hepatomas) induced in mice by a set of compounds with well known carcinogenic effects. Diethylnitrosamine and aflatoxin B1 respectively induced 1100 and 350 hepatomas per micromole per gram of body weight, whereas the estragole and safrole hydroxyl-metabolites respectively induced 32 and 20 hepatomas per micromole per gram of body weight (Wiseman et al. 1987).

Listed Herbs

Herbs listed in the *Botanical Safety Handbook* that contain alkenylbenzenes:

- *Acorus calamus* rhizome of the asarone-containing triploid or tetraploid varieties
- *Acorus gramineus* rhizome
- *Artemisia dracunculus* herb
- *Cinnamomum camphora* wood distillate
- *Foeniculum vulgare* fruit
- *Pimpinella anisum* fruit
- *Piper nigrum* fruit
- *Ocimum basilicum* leaf
- *Ocimum gratissimum* aboveground parts
- *Ocimum tenuiflorum* leaf
- *Sassafras albidum* root

Literature Cited

Abbott, D.D., E.W. Packman, J.W.E. Harrisson, and B.M. Wagner. 1961. Chronic oral toxicity of oil of sassafras and safrole. *Pharmacologist* 3:62.

Barceloux, D.G. 2008. *Medical toxicology of natural substances: Foods, fungi, medicinal herbs.* New York: John Wiley and Sons.

BGVV. 2002. Reduce estragole and methyleugenol contents in foods. German Federal Institute for Health Protection of Consumers and Veterinary Medicine (BgVV). Berlin.

CFR. 2011. *Code of federal regulations*, Title 21 Part 189.180, 2011 ed. Substances generally prohibited from direct addition or use as human food. Safrole. Washington, DC: U.S. Government Printing Office.

EMEA. 2005. Final position paper on the use of herbal medicinal products containing estragole. European Agency for the Evaluation of Medicinal Products, Committee on Herbal Medicinal Products. EMEA/HMPC/137212/2005.

Farag, S.E.A., and M. Abo-Zeid. 1997. Degradation of the natural mutagenic compound safrole in spices by cooking and irradiation. *Nahrung* 41:359–361.

Hagan, E.C., W.H. Hansen, O.G. Fitzhugh, et al. 1967. Food flavourings and compounds of related structure. II. Subacute and chronic toxicity. *Food Cosmet. Toxicol.* 5 (2):141-157.

Hagan, E.C., P.M. Jenner, W.I. Jones, et al. 1965. Toxic properties of compounds related to safrole. *Toxicol. Appl. Pharmacol.* 7 (1):18-24.

Hasheminejad, G., and J. Caldwell. 1994. Genotoxicity of the alkenylbenzenes α- and β-asarone, myristicin and elemicin as determined by the UDS assay in cultured rat hepatocytes. *Food Chem. Toxicol.* 32 (3):223-231.

Homburger, F., T. Kelley, G. Friedler, and A.B. Russfield. 1961. Toxic and possible carcinogenic effects of 4-allyl-1,2-methylenedioxybenzene (safrole) in rats on deficient diets. *Med. Exp. Int. J. Exp. Med.* 4:1-11.

Keeler, R.F., and A.T. Tu. 1983. *Plant and fungal toxins.* New York: Marcel Dekker.

Leung, A.Y., and S. Foster. 1996. *Encyclopedia of common natural ingredients used in food, drugs, and cosmetics.* 2nd ed. New York: Wiley.

Long, E.L., A.A. Nelson, O.G. Fitzhugh, and W.H. Hansen. 1963. Liver tumors produced in rats by feeding safrole. *Arch. Pathol.* 75 (6):595–604.

Wiseman, R.W., E.C. Miller, J.A. Miller, and A. Liem. 1987. Structure-activity studies of the hepatocarcinogenicities of alkenylbenzene derivatives related to estragole and safrole on administration to preweanling male C57BL/6J. x C3H/HeJ F_1 mice. *Canc. Res.* 47 (9):2275-2283.

Berberine

Written by Lisa Ganora

Berberine is a bitter, yellow compound belonging to the subclass of the isoquinoline alkaloids known as the protoberberines. Berberine has a positively-charged quaternary amine group and thus is quite soluble in water. Berberine is found in a number of medicinal plants including goldthread (*Coptis* spp.), goldenseal (*Hydrastis canadensis*), barberry (*Berberis* spp.), Oregon grape (*Mahonia* spp.), phellodendron (*Phellodendron* spp.), California poppy (*Eschscholzia californica*), celandine (*Chelidonium majus*), and bloodroot (*Sanguinaria canadensis*). Traditionally, berberine-rich herbs have been used as bitter choleretic and cholagogue, astringent, anti-inflammatory, antimicrobial, anticarcinogenic, and antidiabetic agents. Berberine-rich herbs have been used for conditions involving the mucous membranes in the digestive, reproductive, ocular, and respiratory systems. Berberine also has activity on the cardiovascular system, with antihypertensive, anti-atherosclerotic, antiarrhythmic, and anti-aggregatory effects.

Adverse Effects

Berberine is not usually considered to be harmful at clinical doses (Imanshahidi and Hosseinzadeh 2008). However, some authors suggest that berberine-rich herbs should be contraindicated during pregnancy and lactation, based on the fact that higher doses of berberine can strongly displace bilirubin both from human serum albumin *in vitro* and at a dosage over 2 mg/kg intraperitoneally administered to rats (Chan 1993). High concentrations of unconjugated bilirubin can accumulate in and cause damage to human brain tissue. If an excessive dose of berberine were to be ingested during pregnancy, this could be of concern especially for neonates with pre-existing jaundice or hereditary diseases (such as Gilbert's syndrome and Crigler–Najjar syndrome) which also involve hyperbilirubinemia.

There is some controversy regarding the clinical significance of berberine's bilirubin-displacing effect. Chinese literature from the 1970s and 1980s reported an association between the maternal and neonatal use of formulas containing *Coptis chinensis* (approx. 7–9% berberine content) and an increased incidence of kernicterus in infants with neonatal jaundice (Chan 1993; Upton 2001). There is also a widespread belief in China that formulas containing *Coptis* can be hazardous to infants born with an erythrocyte glucose-6-phosphate dehydrogenase (G6PD) deficiency, in whom such formulas could cause hemolytic anemia. However, it is unclear if *Coptis* or some other substance in the formulas commonly given to neonates could be responsible for such an effect (Yeung et al. 1990). A review of traditional Chinese medicine use in cases of neonatal jaundice identified one case that associated exposure to coptis with fatal hemolysis and kernicterus in a baby (Fok 2001). There are no reports in the contemporary literature of these conditions being associated with berberine-containing Western herbs such as goldenseal, barberry, and Oregon grape. Mills and Bone, however, recommend that berberine-containing herbs not be used during pregnancy except with professional supervision (Mills and Bone 2005).

Another rationale for contraindicating berberine during pregnancy is based on a few older reports claiming that it could induce uterine contractions in mice (Furuya 1957; Imaseki et al. 1961). A recent literature search found no contemporary reports of such activity ascribed to the use of berberine-containing herbs. Some studies (using several types of isolated tissue) have found berberine to have a contractile effect on smooth muscle, while others have found it to be antispasmodic and relaxant (Tice 1997). Hydrastine, a related alkaloid found along with berberine in goldenseal, was historically employed as a uterine astringent and hemostatic by the Eclectic physicians (Felter and Lloyd 1898). At least one historical source notes that goldenseal was not observed to cause or enhance contractions when used for this purpose (Shoemaker 1906).

In high doses, isolated berberine salts are moderately toxic; the LD_{50} for intraperitoneally administered berberine chloride dihydrate (BCD) was reported to be 30 mg/kg in the mouse and 205 mg/kg in the rat (Jahnke et al. 2006). The LD_{50} of orally administered berberine sulfate was reported to be greater than 1000 mg/kg in the rat (Kowalewski et al. 1975). These dosages are far beyond what one would obtain from the clinical use of berberine-containing herbs, which typically contain concentrations ranging from 0.5 to 6% in goldenseal root and 4 to 7% in coptis (Chang and But 1986; Upton 2001).

In an evaluation of reproductive toxicity, isolated berberine salts were given by oral gavage to pregnant mice over the course of eleven days. The maternal lowest-observed-adverse-effect-level (LOAEL) was determined to be 841 mg/kg daily of BCD (equivalent to approximately 698 mg of pure berberine). No signs of developmental toxicity were observed until the dosage reached approximately 938 mg BCD/kg/day, and these were limited to a 5 to 6% decrease in average fetal body weight; there was no evidence of teratogenicity (Price and George 2003). Another evaluation found no adverse effects at dosages up to 1000 mg/kg/day of BCD in rats. The authors noted that the no-observed-adverse-effect-level (NOAEL) for both rats and mice was approximately 500 times greater than the amount of berberine that one would obtain from herbs used as dietary supplements (Jahnke et al. 2006).

A 2005 reproductive screening in rats, which found no adverse effects from a hydroethanolic extract of goldenseal at a dosage of 1.86 g/kg daily (reported as 65 times the recommended human dose), concluded that toxic levels of berberine were unlikely to be reached in the plasma due to poor intestinal absorption (Yao et al. 2005). In humans, symptoms of berberine overdose are reported to include hypotension, bradycardia, dyspnea, and gastrointestinal disturbances (Lau et al. 2001).

Mechanism of Action

Berberine has demonstrated a number of anti-inflammatory and anti-cancer properties in numerous different cell lines and tissue types. A recent investigation identified NF-κβ modulation as a major mechanism underlying these effects. Berberine was found to suppress NF-κβ activation when induced by several different pro-inflammatory and carcinogenic agents. This activity led to the down-regulation of gene products responsible for blocking apoptosis in cancer cells, for promoting inflammation via COX-2 induction, and for enabling tumor metastasis (Pandey et al. 2008).

In a mechanism distinct from that of the statin drugs, berberine can significantly reduce plasma levels of LDL cholesterol. The mechanism involves upregulation of the low-density lipoprotein receptors (LDLR) in the liver. The LDL receptor system coordinates cholesterol metabolism, allowing excessive LDL cholesterol to be cleared from the bloodstream (Goldstein and Brown 2009). Berberine was found to extend the half-life of LDLR mRNA (without having an effect on gene transcription), resulting in a strong increase in LDLR protein expression (Abidi et al. 2006). A clinical study of hypercholesterolemic patients in China found that an oral dose of one gram of berberine/day for three months lowered total cholesterol by 29%, LDL by 25%, and triglycerides by 35% (Kong et al. 2004).

Listed Herbs

Herbs listed in the *Botanical Safety Handbook* that contain berberine:

- *Berberis vulgaris* root, root bark
- *Chelidonium majus* herb
- *Coptis chinensis* rhizome
- *Coptis groenlandica* rhizome
- *Corydalis yanhusuo* tuber
- *Hydrastis canadensis* rhizome, root
- *Mahonia aquifolium* root
- *Mahonia nervosa* root
- *Mahonia repens* root
- *Phellodendron amurense* bark
- *Phellodendron chinense* bark
- *Sanguinaria canadensis* rhizome, root

Literature Cited

Abidi, P., W. Chen, F.B. Kraemer, H. Li, and J.W. Liu. 2006. The medicinal plant goldenseal is a natural LDL-lowering agent with multiple bioactive components and new action mechanisms. *J. Lipid Res.* 47 (10):2134-2147.

Chan, E. 1993. Displacement of bilirubin from albumin by berberine. *Biol. Neonate* 63 (4):201-208.

Chang, H.-M., and P.P.H. But. 1986. *Pharmacology and applications of Chinese materia medica*. English ed. Singapore: Philadelphia, PA, USA.

Felter, H.W., and J.U. Lloyd. 1898. *King's American dispensatory*. Cincinnati: Ohio Valley Co.

Fok, T.F. 2001. Neonatal jaundice—traditional Chinese medicine approach. *J. Perinatol.* 21 Suppl 1:S98-S100, 104-107.

Furuya, T. 1957. Pharmacological action, including toxicity and excretion of berberine hydrochloride and its oxidation product. *Bull. Osaka Med. School* 3:62-67.

Goldstein J.L. and M.S. Brown. 2009. The LDL receptor. *Arterioscler. Thromb. Vasc. Biol.* 29(4):431-438.

Imanshahidi, M., and H. Hosseinzadeh. 2008. Pharmacological and therapeutic effects of *Berberis vulgaris* and its active constituent, berberine. *Phytother. Res.* 22 (8):999-1012.

Imaseki, I., Y. Kitabatakea, and T. Taguchi. 1961. Studies on the effect of berberine alkaloids on intestine and uterus in mice. *Yakugaku Zasshi* 81:1281-1284.

Jahnke, G.D., C.J. Price, M.C. Marr, C.B. Myers, and J.D. George. 2006. Developmental toxicity evaluation of berberine in rats and mice. *Birth Defects Res. B* 77 (3):195-206.

Kong, W., J. Wei, P. Abidi, et al. 2004. Berberine is a novel cholesterol-lowering drug working through a unique mechanism distinct from statins. *Nature Medicine* 10:1344-1351.

Kowalewski, Z., A. Mrozikiewicz, T. Bobkiewicz, K. Drost, and B. Hladon. 1975. Studies of toxicity of berberine sulfate. *Acta Polon. Pharmaceut.* 32 (1):113-120.

Lau, C.W., X.Q. Yao, Z.Y. Chen, W.H. Ko, and Y. Huang. 2001. Cardiovascular actions of berberine. *Cardiovasc. Drug Rev.* 19 (3):234-244.

Mills, S., and K. Bone. 2005. *The essential guide to herbal safety*. St. Louis: Elsevier

Pandey, M.K., B. Sung, A.B. Kunnumakkara, et al. 2008. Berberine modifies cysteine 179 of I kappa B alpha kinase, suppresses nuclear factor-kappa B-regulated antiapoptotic gene products, and potentiates apoptosis. *Cancer Res.* 68 (13):5370-5379.

Price, C.J., and J.D. George. 2003. Final study report on the developmental toxicity evaluation for berberine chloride dihydrate (CAS no. 5956-60-5) administered in the feed to Swiss (cd-1) mice on gestational days 6 through 17. *Gov. Rep. Announce. Index* (20):112.

Shoemaker, J. 1906. *A practical treatise on materia medica and therapeutics: With especial reference to the clinical application of drugs*. 6th ed. Philadelphia: F.A. Davis.

Tice, R. 1997. Goldenseal (*Hydrastis canadensis* L.) and two of its constituent alkaloids berberine and hydrastine; review of toxicological literature. Research Triangle Park, NC: Integrated Laboratory Systems.

Upton, R. 2001. *Goldenseal root: Hydrastis canadensis; standards of analysis, quality control, and therapeutics*. Santa Cruz, CA: American Herbal Pharmacopoeia.

Yao, M., H.E. Ritchie, and P.D. Brown-Woodman. 2005. A reproductive screening test of goldenseal. *Birth Defects Res. B* 74 (5):399-404.

Yeung, C.Y., F.T. Lee, and H.N. Wong. 1990. Effect of a popular Chinese herb on neonatal bilirubin protein-binding. *Biol. Neonate* 58 (2):98-103.

Caffeine

Written by Zoë Gardner, Ph.D.(c)

Caffeine is an alkaloid classified as a methylxanthine, a group of closely related compounds including caffeine, theophylline, and theobromine, that have similar physiological effects. Caffeine is the most widely consumed and researched psychoactive substance in the world. The worldwide average daily intake of caffeine is 159 mg per person, with Americans consuming approximately 168 mg daily, while the Dutch are the heaviest consumers at an average of 414 mg per day (Fredholm et al. 1999).

Adverse Effects

Ingestion of caffeine results in a number of physiological effects including central nervous system stimulation, acute elevation of blood pressure, increased metabolic rate, increased gastric and colonic activity, and diuretic activity (Higdon and Frei 2006; James 2000). Long-term use of caffeine usually results in tolerance to some of the physiological and behavioral effects (Griffiths and Mumford 1996). *See* Appendix 2 for more information on the diuretic activity of caffeine.

Overdose of caffeine may result in caffeine intoxication with symptoms including nervousness, anxiety, restlessness, insomnia, gastrointestinal upset, tremors, and a rapid heart rate (APA 1994). Symptoms of caffeine intoxication may be similar to those of anxiety or other mood disorders (Greden 1974). In rare cases, caffeine overdose can be fatal, although such cases are generally from intentional self-poisoning with caffeine pills or tablets rather than from drinking caffeine-containing beverages (Holmgren et al. 2004; Mrvos et al. 1989).

Regular use of caffeine produces physical dependence on caffeine, and withdrawal symptoms are common with reduction or cessation of caffeine (Hughes et al. 1998; Strain et al. 1994). Withdrawal symptoms begin to occur 12 to 24 hours after abstaining from caffeine consumption. The most common symptom of withdrawal is headache, with fatigue, decreased energy, decreased alertness, a depressed mood, irritability, and other related symptoms also being commonly reported (Juliano and Griffiths 2004).

Studies on the effects of caffeine on blood pressure indicate that caffeine causes an acute rise in blood pressure, usually occurring 30 minutes to 4 hours after ingestion (Nurminen et al. 1999). The blood-pressure raising effects may be more pronounced in persons with high blood pressure (Nurminen et al. 1999). With routine consumption of caffeine, most individuals develop tolerance to the blood-pressure raising effects, while some do not (James 1994; Lovallo et al. 2004). Studies on the effects of caffeine or coffee on blood pressure have mixed results. Some studies show a mild elevation of blood pressure after caffeine consumption (James 2004), while others show no effect or a habituation to the effect. In addition, the effects of coffee on blood pressure may be different than those of caffeine, since coffee contains other compounds, such as polyphenols, soluble fiber, and potassium, which typically have a beneficial effect on the cardiovascular system (Geleijnse 2008).

The effects of caffeine on human reproduction and pregnancy have been widely studied. While current reviews suggest a lack of adverse effects of caffeine on fetal development and pregnancy outcomes (Christian and Brent 2001; Peck et al. 2010), women are generally advised to limit caffeine intake to approximately 300 mg daily during pregnancy and 200 to 300 mg daily while nursing (AAP 2001; ADA 2008; PDR 2006).

The American Herbal Products Association has established a trade requirement (AHPA 2011) that dietary supplement products that contain caffeine,* whether as a direct ingredient or as a constituent of herbal ingredients, conform to all of the following:

1. The label of caffeine-containing dietary supplements discloses the presence of caffeine in the product.
2. The label or labeling of caffeine containing dietary supplements, except for such supplements as are described in paragraph 3 below, discloses the quantity of caffeine per recommended serving of the dietary supplement, stated in both milligrams per serving and in equivalent approximate cups of coffee, where 100 mg of caffeine represents one cup of coffee.
3. The label of caffeine-containing dietary supplements discloses the presence of caffeine, but not necessarily the quantity of caffeine per recommended serving, if at least one of the following conditions is met:
   The caffeine-containing dietary ingredient is an herb or herbal source ingredient that is less concentrated than a 1:1 weight/weight or weight/volume concentration ratio of raw herb to dietary ingredient; or

* Consisting of caffeine and all so-called caffeine analogues that include, but are not limited to, the following terms: caffeine, guaranine, mateina, mateine, methyltheobromine, thein, theine, 1,3,7-trimethylxanthine, 1,3,7-trimethyl-2,6-dioxopurine, and 7-methyltheophylline.

The amount of caffeine per recommended serving of the caffeine-containing dietary supplement is less than 25 mg.

4. Caffeine-containing dietary supplements are formulated and labeled in a manner to recommend a maximum of 200 mg of caffeine per serving, not more often than every 3 to 4 hours.
5. The following or similar statement is included on the label of any dietary supplement that contains caffeine in sufficient quantity to warrant such labeling:

Too much caffeine may cause nervousness, irritability, sleeplessness, and, occasionally, rapid heartbeat. Not recommended for use by children under 18 years of age.

Drug Interactions

Caffeine is metabolized by the isoenzyme CYP1A2. Drugs that inhibit this isoenzyme (including fluvoxamine, ciprofloxacin, cimetidine, amiodarone, fluoroquinolones, furafylline, interferon, methoxsalen, and mibefradil) may slow the metabolism of caffeine. In persons drinking multiple cups of coffee daily, high levels of caffeine could accumulate (Carrillo and Benitez 2000).

Mechanism of Action

Methylxanthines, including caffeine, stimulate the central nervous system and the heart, elicit a diuretic effect in the kidneys, and relax smooth muscles.

Caffeine works in part by competing with adenosine, a neurotransmitter that accumulates in the brain during periods of wakefulness and helps to induce sleep. Caffeine binds to the adenosine receptors, effectively blocking the adenosine, thus promoting alertness and reducing the ability to fall asleep. Changes in motor activity are due to the effect of caffeine on neurotransmitters in the basal ganglia, an area of the brain responsible for motor control and other activities (Fisone et al. 2004).

Average amount of caffeine per 8 oz. cup

Green tea	25–40 mg
Black tea	25–55 mg
Espresso (single shot)	60–75 mg
Brewed coffee	70–125 mg
Cola beverage	23–31 mg

(China et al. 2008; McCusker et al. 2003)

Listed Herbs

Herbs listed in the *Botanical Safety Handbook* that contain caffeine:

- *Camellia sinensis* leaf, stem
- *Coffea arabica* seed kernel
- *Cola acuminata* seed
- *Cola nitida* seed
- *Ilex paraguariensis* leaf
- *Paullinia cupana* seed

Literature Cited

AAP. 2001. The transfer of drugs and other chemicals into human milk. American Academy of Pediatrics Committee on Drugs. *Pediatrics* 108 (3):776-789.

ADA. 2008. Position of the American Dietetic Association: Nutrition and lifestyle for a healthy pregnancy outcome. *J. Am. Diet. Assoc.* 108:553-561.

APA. 1994. *Diagnostic and statistical manual of mental disorders: DSM-IV*. Washington DC: American Psychiatric Association.

Carrillo, J.A., and J. Benitez. 2000. Clinically significant pharmacokinetic interactions between dietary caffeine and medications. *Clin. Pharmacokin.* 39 (2):127-153.

China, J.M., M.L. Merves, B.A. Goldberger, A. Sampson-Cone, and E.J. Cone. 2008. Caffeine content of brewed teas. *J. Analyt. Toxicol.* 32 (8):702-704.

Christian, M.S., and R.L. Brent. 2001. Teratogen update: Evaluation of the reproductive and developmental risks of caffeine. *Teratol.* 64 (1):51-78.

Fisone, G., A. Borgkvist, and A. Usiello. 2004. Caffeine as a psychomotor stimulant: Mechanism of action. *Cell Molec. Life Sci.* 61 (7):857-872.

Fredholm, B.B., K. Bättig, J. Holmén, A. Nehlig, and E.E. Zvartau. 1999. Actions of caffeine in the brain with special reference to factors that contribute to its widespread use. *Pharmacol. Rev.* 51 (1):83-133.

Geleijnse, J.M. 2008. Habitual coffee consumption and blood pressure: An epidemiological perspective. *Vasc. Health Risk Manag.* 4 (5):963-970.

Greden, J.F. 1974. Anxiety or caffeinism: A diagnostic dilemma. *Am J. Psych.* 131 (10):1089.

Griffiths, R.R., and G.K. Mumford. 1996. Caffeine reinforcement, discrimination, tolerance, and physical dependence in laboratory animals and humans. In *Pharmacological aspects of drug dependence: Toward an integrated neurobehavioral approach*, edited by Schuster, C.R. and M.J. Kuhar. New York: Springer.

Higdon, J.V., and B. Frei. 2006. Coffee and health: A review of recent human research. *Crit. Rev. Food Sci. Nutr.* 46 (2):101-123.

Holmgren, P., L. Nordén-Pettersson, and J. Ahlner. 2004. Caffeine fatalities—four case reports. *Forensic Sci. Int.* 139 (1):71-73.

Hughes, J.R., A.H. Oliveto, A. Liguori, J. Carpenter, and T. Howard. 1998. Endorsement of DSM-IV dependence criteria among caffeine users. *Drug Alc. Depend.* 52 (2):99-107.

James, J.E. 1994. Chronic effects of habitual caffeine consumption on laboratory and ambulatory blood pressure levels. *J. Cardiovasc. Risk* 1:159-164.

James, J.E. 2000. Acute and chronic effects of caffeine on performance, mood, headache, and sleep. *Neuropsychobiology* 38 (1):32-41.

James, J.E. 2004. Critical review of dietary caffeine and blood pressure: A relationship that should be taken more seriously. *Psychosom. Med.* 66 (1):63-71.

Juliano, L.M., and R.R. Griffiths. 2004. A critical review of caffeine withdrawal: Empirical validation of symptoms and signs, incidence, severity, and associated features. *Psychopharmacology* 176 (1):1-29.

Lovallo, W.R., M.F. Wilson, and A.S. Vincent. 2004. Blood pressure response to caffeine shows incomplete tolerance after short-term regular consumption. *Hypertension* 43:760-765.

McCusker, R.R., B.A. Goldberger, and E.J. Cone. 2003. Technical note: Caffeine content of specialty coffees. *J. Analyt. Toxicol.* 27 (7):520-522.

Mrvos, R.M., P.E. Reilly, B.S. Dean, and E.P. Krenzelok. 1989. Massive caffeine ingestion resulting in death. *Vet. Hum. Toxicol.* 31 (6):571-572.

Nurminen, M.L., L. Niittynen, R. Korpela, and H. Vapaatalo. 1999. Coffee, caffeine and blood pressure: A critical review. *Eur. J. Clin. Nutr.* 53:831–839.

PDR. 2006. *Physicians' desk reference for nonprescription drugs and dietary supplements.* 27th ed. Montvale, NJ: Medical Economics Co.

Peck, J.D., A. Leviton, and L.D. Cowan. 2010. A review of the epidemiologic evidence concerning the reproductive health effects of caffeine consumption: A 2000–2009 Update. *Food Chem. Toxicol.* 48 (10):2549-2576.

Strain, E.C., G.K. Mumford, K. Silverman, and R.R. Griffiths. 1994. Caffeine dependence syndrome: Evidence from case histories and experimental evaluations. *JAMA.* 272 (13):1043.

Cyanogenic Glycosides

Written by Michael McGuffin; revised by Zoë Gardner, Ph.D.(c)

Cyanogenic glycosides are sugar-containing compounds with a nitrile group (hydrogen triple-bonded to nitrogen). After being metabolized, these compounds can release cyanide (in the form of hydrocyanic acid), a substance that can, in significant amounts, be toxic to humans and other animals. The best-known cyanogenic glycoside, amygdalin, is found in the seeds of many common fruits of the Rosaceae family, such as cherries, apples, peaches, apricots, almonds and pears (Vetter 2000).

Several common food plants, including bamboo shoots, cassava, and lima beans, contain cyanogenic glycosides (FSANZ 2004; Ologhobo et al. 1984). Medicinally, a number of plants containing cyanogenic glycosides, including black cherry bark (*Prunus serotina*) and loquat leaf (*Eriobotrya japonica*), have traditionally been used as cough remedies (Mills and Bone 2000). Foods and botanicals containing low levels of cyanogenic glycosides are generally not dangerous to consume.

Adverse Effects

Cyanide is released during the metabolism of cyanogenic glycosides. The best established and probably most important toxic action of cyanide is incapacitation of the cell's mechanism for using oxygen, resulting in chemical asphyxiation (oxygen deprivation) (Nelson 2006).

Of the plants included in this text that contain this class of glycosides, the seeds of several species of *Prunus* present the most, and possibly the only, concern. Peach kernels contain 2 to 6% amygdalin, while apricot kernels contain up to 8% amygdalin (Encarna et al. 1998; Femenia et al. 1995; Gunders et al. 1969; Holzbecher et al. 1984; Machel and Dorsett 1970). The toxic dose for apricot seeds has been reported as 10 to 20 seeds in children and 40 to 60 seeds in adults, though removal of the seed skin and heating of the seeds reduce the amygdalin content (Bensky et al. 2004; Chen and Chen 2004). Levels of cyanogenic glycosides in other species listed below are generally not of toxicological concern.

The LD_{50} of hydrocyanic acid is 3.7 mg/kg in mice and 4 mg/kg in dogs, while the LD_{50} of amygdalin, the compound found in apricot seeds, is 522 mg/kg in rats (Milne 1995; Newton et al. 1981). Based on the concentration of amygdalin in apricot seeds, a toxic dose would be equivalent to 34–39 g of amygdalin in an adult of normal weight. An adult would need to consume 425–480 grams of apricot seeds, for example, in order to reach this toxic intake level.

The symptoms of cyanide poisoning are well known from industrial exposure or exposure to cyanide in smoke from residential or industrial fires. Early signs and symptoms of acute cyanide poisoning include attempts of the respiratory, neurologic, and cardiovascular systems to overcome tissue hypoxia (whole body oxygen deprivation). These include transient increases in blood pressure and heart rate, hyperventilation, shortness of breath, heart palpitations, and headache. Late symptoms or symptoms of severe poisoning include neurologic, respiratory, and cardiovascular depression, as tissues fail to compensate for their inability to use oxygen (Borron 2006; Nelson 2006). A number of sources review the available treatment protocols for cyanide poisoning (Cummings 2004; Goldfrank and Flomenbaum 2006; Hall et al. 2009).

Mechanism of Action

Plants containing cyanogenic glycosides do not contain detectable free hydrocyanic acid. Instead, the glycosides and enzymes that break down the glycosides are stored separately until the plant tissue is crushed, chewed, wilted, or otherwise disturbed, at which time the glycosides and enzymes come together, and cyanide (in the form of hydrocyanic acid) is released (Ganora 2009; Thayer and Conn 1981).

Cyanide is a normal waste product of protein degradation, and humans are able to detoxify about 1 mg/kg of cyanide per hour (Aminlari et al. 2007; Nelson 2006). Additionally, the acidic environment of the human stomach is not optimal for β-glucosidase, the main enzyme that liberates hydrocyanic acid from cyanogenic glycosides. Ruminant animals, such as cows, are more susceptible to poisoning from plants containing cyanogenic glycosides due to the relatively neutral pH of the ruminant digestive tract (Ganora 2009; Majak 1992).

Different methods of processing have been developed for reducing the cyanide content in food products. Cassava, a root crop widely used as a staple food in tropical countries, contains two cyanogenic glycosides and must be processed prior to consumption. Processing is done through a combination of several steps that may include crushing, soaking, drying, fermenting, or roasting (Cardoso et al. 2005; Lancaster et al. 1982).

Listed Herbs

Herbs listed in the *Botanical Safety Handbook* that contain cyanogenic glycosides:

- *Eriobotrya japonica* leaf (0.06% amygdalin)
- *Hydrangea arborescens* root (1–3% hydrangin)
- *Linum usitatissimum* seeds (0.1 to 1.5% linustatin and neolinustatin)
- *Prunus armeniaca* seed (up to 8% amygdalin)
- *Prunus persica* seed (2–6% amygdalin; leaf, 0.5 to 1.5% amygdalin)
- *Prunus serotina* dried bark (prunasin yielding up to 0.15% hydrocyanic acid)
- *Prunus spinosa* seeds and fresh flowers (minor amounts)
- *Sambucus canadensis* leaves, bark, seeds, and raw unripe fruits (minor amounts)
- *Sambucus nigra* leaves, bark, seeds, and raw unripe fruits (minor amounts)
- *Turnera diffusa* leaf (0.26% tetraphyllin B)

Literature Cited

Aminlari, M., A. Malekhusseini, F. Akrami, and H. Ebrahimnejad. 2007. Cyanide-metabolizing enzyme rhodanese in human tissues: Comparison with domestic animals. *Compar. Clin. Pathol.*16 (1):47-51.

Bensky, D., S. Clavey, and E. Stöger. 2004. *Chinese herbal medicine: Materia medica.* 3rd ed. Seattle: Eastland Press.

Borron, S.W. 2006. Recognition and treatment of acute cyanide poisoning. *J. Emerg. Nurs.* 32 (4 Suppl):S12-S18.

Cardoso, A.P., E. Mirione, M. Ernesto, et al. 2005. Processing of cassava roots to remove cyanogens. *J. Food Comp. Anal.* 18 (5):451-460.

Chen, J.K., and T.T. Chen. 2004. *Chinese medical herbology and pharmacology.* City of Industry, CA: Art of Medicine Press.

Cummings, T.F. 2004. The treatment of cyanide poisoning. *Occ. Med.* 54 (2):82.

Encarna, G., B. Lorenzo, S. Constanza, and M. Josefa. 1998. Amygdalin content in the seeds of several apricot cultivars. *J. Sci. Food Agric.* 77 (2):184-186.

Femenia, A., C. Rossello, A. Mulet, and J. Canellas. 1995. Chemical composition of bitter and sweet apricot kernels. *J. Agric. Food Chem.* 43 (2):356-361.

FSANZ. 2004. Cyanogenic glycosides in cassava and bamboo shoots. Technical Report Series No. 28, Food Standards Australia New Zealand. Canberra.

Ganora, L. 2009. *Herbal constituents: Foundations of phytochemistry.* Louisville, CO: HerbalChem Press.

Goldfrank, L.R., and N. Flomenbaum. 2006. *Goldfrank's toxicologic emergencies.* New York: McGraw-Hill Professional.

Gunders, A.E., A. Abrahamov, E. Weisenberg, S. Gertner, and S. Shafran. 1969. Cyanide poisoning following ingestion of apricot (*Prunus armeniaca*) kernels. *Harefuah* 76 (12):536-538.

Hall, A.H., J. Saiers, and F. Baud. 2009. Which cyanide antidote? *Crit. Rev. Toxicol.* 39 (7):541-552.

Holzbecher, M.D., M.A. Moss, and H.A. Ellenberger. 1984. The cyanide content of laetrile preparations, apricot, peach and apple seeds. *Clin. Toxicol.* 22 (4):341-347.

Lancaster, P.A., J.S. Ingram, M.Y. Lim, and D.G. Coursey. 1982. Traditional cassava-based foods: Survey of processing techniques. *Econ. Bot.* 36 (1):12-45.
Machel, A.R., and C.I. Dorsett. 1970. Cyanide analyses of peaches. *Econ. Bot.* 24:51-52.
Majak, W. 1992. Metabolism and absorption of toxic glycosides by ruminants. *J. Range Manag.* 45 (1):67-71.
Mills, S., and K. Bone. 2000. *Principles and practice of phytotherapy: Modern herbal medicine*. New York: Churchill Livingstone.
Milne, G.W.A. 1995. *CRC handbook of pesticides*. Boca Raton, FL: CRC Press.
Nelson, L. 2006. Acute cyanide toxicity: Mechanisms and manifestations. *J. Emerg. Nurs.* 32:S8-11.
Newton, G.W., E.S. Schmidt, J.P. Lewis, R. Lawrence, and E. Conn. 1981. Amygdalin toxicity studies in rats predict chronic cyanide poisoning in humans. *West. J. Med.* 134 (2):97.
Ologhobo, A.D., B.L. Fetuga, and O.O. Tewe. 1984. The cyanogenic glycoside contents of raw and processed limabean varieties. *Food Chem.* 13 (2):117-128.
Thayer, S.S., and E.E. Conn. 1981. Subcellular localization of dhurrin β-glucosidase and hydroxynitrile lyase in the mesophyll cells of sorghum leaf blades. *Plant Physiol.* 67 (4):617.
Vetter, J. 2000. Plant cyanogenic glycosides. *Toxicon* 38 (1):11-36.

Pyrrolizidine Alkaloids

Written by Michael McGuffin; revised by Zoë Gardner, Ph.D.(c)

Pyrrolizidine alkaloids (PAs) are compounds found in a number of plant species that have been associated with liver toxicity. Based on their chemistry, different PAs may be saturated or unsaturated (the difference between the two is determined by whether a chemical bond between two particular carbons in the central ring structure is double or single). Saturated PAs, such as those found in *Euphrasia* spp. and *Echinacea* spp., are nontoxic. Unsaturated PAs, such as those in *Senecio* species, are recognized as causing liver toxicity when ingested in sufficient amounts. Certain unsaturated PAs are more toxic than others.

Adverse Effects

Initial concern regarding PAs was probably based on cases of livestock poisoning due to consumption of *Senecio* and *Amsinckia* (Cheeke 1988; Johnson et al. 1985). Supplies of grain have been contaminated by PA-containing weeds in grain fields, leading to outbreaks of PA toxicity, causing acute cases of liver damage in persons eating the contaminated grain (Prakash et al. 1999). Serious liver damage has also occurred after chronic consumption of PA-containing medicinal plants that have traditionally been used for therapeutic purposes.

The herbs most widely used in the United States that contain PAs are comfrey root and leaf (*Symphytum* spp.), coltsfoot leaf and flower (*Tussilago farfara*), and borage leaf (*Borago officinale*). The amounts and relative safety of PAs in these plants and in the various plant parts vary widely. For example, in comfrey, the concentration of alkaloids is measured at about 10 times higher in the root than in the leaf (Tyler 1994). Moreover, echimidine, the most toxic of the alkaloids found in comfrey, is present in *Symphytum asperum* and *S.* x *uplandicum* but is absent in most samples of *S. officinale* (Awang et al. 1993; Huizing et al. 1982; Jaarsma et al. 1989). Although a number of species of *Eutrochium* (recently reclassified from *Eupatorium*) are known to contain PAs (Zhang et al. 2008), the PA content of several species used in the U.S. (*E. fistulosum, E. purpureum*, and *E. maculatum*) has not been adequately investigated. For other plants, such as borage, the amounts of PAs are generally cited to be "low," although reliable information on the concentration of PAs in these plants is lacking.

While some of these alkaloids have shown carcinogenic and mutagenic properties, and kidney toxicity has been reported (Fu et al. 2004), the primary concern for use of these herbs is the potential for serious liver damage, specifically hepatic veno-occlusive disease (a condition in which veins in the liver become blocked). This potentially fatal condition manifests symptoms such as abdominal pain, swelling of the liver and spleen, accumulation of fluid in the abdominal cavity, elevated levels of bilirubin, jaundice, cirrhosis of the liver, and liver failure (Chen and Huo 2010; McDermott and Ridker 1990).

Cautious restrictions on the use of all of the herbs containing unsaturated (toxic) PAs have been recommended by the American Herbal Products Association, with suggestions to limit use to external application on unbroken skin only, and to refrain from use while nursing (AHPA 2011). All use is contraindicated in pregnancy and in persons with a history of liver disease.

Mechanism of Action

PAs are metabolized in the liver by the drug metabolizing isoenzyme CYP3A4 to form N-oxides and conjugated dienic pyrroles (alkylating compounds that are highly reactive with proteins and nucleic acids). The complex of pyrroles with proteins and nucleic acids may remain in tissues and cause chronic injury, while the N-oxides may be transformed into epoxides and toxic necines. Substances that induce CYP3A4 may enhance the toxicity of PAs, while inhibitors of this isoenzyme may reduce the toxicity. The development of veno-occlusive disease remains poorly understood, although studies suggest that endothelial cell injury, cytokines, and hemostatic derangement are all involved. A strict dose-dependent association between PA consumption and veno-occlusive disease development may not be present, and not all persons taking PAs develop the disease (Chen and Huo 2010).

An animal study demonstrated that the systemic bioavailability of PAs after external use is about 20 to 50 times lower than that after oral ingestion, although absorption may be increased after application to inflamed, cut, or abraded skin (Brauchli et al. 1982).

Listed Herbs

Herbs listed in the *Botanical Safety Handbook* that contain unsaturated pyrrolizidine alkaloids*:

- *Alkanna tinctoria* root
- *Borago officinalis*† herb
- *Eutrochium fistulosum* herb, root, and rhizome
- *Eutrochium maculatum* herb, root, and rhizome
- *Eutrochium purpureum*‡ herb, root, and rhizome
- *Symphytum asperum*§ leaf, root
- *Symphytum officinale*§ leaf, root
- *Symphytum* x *uplandicum*§ leaf, root
- *Tussilago farfara*§ flower, leaf.

* Note that although medicinal species of *Petasites*, including purple butterbur (*Petasites hybridus*) and Arctic butterbur (*Petasites frigidus*), are not listed in this text, these species also contain pyrrolizidine alkaloids. PA-free extracts of *Petasites* species are available, and PA-free products are considered appropriate for internal use.

† Processing of borage seed oil eliminates PAs.

‡ Presence and type of PAs has not been confirmed.

§ PA-free extracts of *Symphytum* spp., *Tussilago farfara*, and other botanicals are available commercially.

Literature Cited

AHPA. July 2011. Code of Ethics & Business Conduct. Silver Spring, MD: American Herbal Products Association.

Awang, D.V.C., B.A. Dawson, J. Fillion, M. Girad, and D. Klindack. 1993. Echimidine content of commercial comfrey. *J. Herbs Spices Med. Plants* 2 (1):21-34.

Brauchli, J., J. Luthy, U. Zweifel, and C. Schlatter. 1982. Pyrrolizidine alkaloids from *Symphytum officinale* L. and their percutaneous absorption in rats. *Experientia* 38 (9):1085-1087.

Cheeke, P.R. 1988. Toxicity and metabolism of pyrrolizidine alkaloids. *J. Animal Sci.* 66 (9):2343-2350.

Chen, Z., and R.-H. Huo. 2010. Hepatic veno-occlusive disease associated with toxicity of pyrrolizidine alkaloids in herbal preparations. *Neth. J. Med.* 68 (6):252-260.

Fu, P.P., Q. Xia, G. Lin, and M.W. Chou. 2004. Pyrrolizidine alkaloids—Genotoxicity, metabolism enzymes, metabolic activation, and mechanisms. *Drug Metab. Rev.* 36 (1):1-55.

Huizing, H.J., T.W.J. Gadella, and E. Kliphuis. 1982. Chemotaxonomical investigations of the *Symphytum officinale* polyploid complex and *S. asperum* (Boraginaceae): The pyrrolizidine alkaloids. *Plant Systemat. Evol.* 140 (4):279-292.

Jaarsma, T.A., E. Lohmanns, T.W.J. Gadella, and T.M. Malingre. 1989. Chemotaxonomy of the *Symphytum officinale* agg. (Boraginaceae). *Plant Sys. Evol.* 167 (3-4).

Johnson, A.E., R.J. Molyneux, and G.B. Merrill. 1985. Chemistry of toxic range plants. Variation in pyrrolizidine alkaloid content of *Senecio*, *Amsinckia*, and *Crotalaria* species. *J. Agric. Food Chem.* 33 (1):50-55.

McDermott, W.V., and P.M. Ridker. 1990. The Budd-Chiari syndrome and hepatic veno-occlusive disease: Recognition and treatment. *Arch. Surg.* 125 (4):525-527.

Prakash, A.S., T.N. Pereira, P.E.B. Reilly, and A.A. Seawright. 1999. Pyrrolizidine alkaloids in human diet. *Mutat. Res.* 443 (1-2):53-67.

Tyler, V. 1994. *Herbs of choice*. Binghamton, NY: Pharmaceutical Products Press.

Zhang, M.L., M. Wu, J.J. Zhang, et al. 2008. Chemical constituents of plants from the genus *Eupatorium*. *Chem. Biodivers.* 5 (1):40-55.

Salicylates

Written by Michael McGuffin; revised by Zoë Gardner, Ph.D.(c)

Salicylates are phenolic acids derived from salicylic acid, and include salicin (in *Salix* species), populin (in *Populus* species), methyl salicylate (in *Gaultheria* and *Betula* species), and acetylsalicylic acid (aspirin). Salicylic acid was first synthesized in 1860, and the salicylate-containing plants were soon supplanted by the synthetic analog, acetylsalicylic acid (Weissmann 1991). Salicylates are commonly consumed for pain relief, especially for low intensity pain, with an estimated 40,000 metric tons of aspirin being consumed worldwide every year (Warner and Mitchell 2002).

Adverse Effects

Concern regarding the consumption of the salicin-containing plants is addressed here primarily to assure that the known adverse effects of aspirin have been examined in relationship to these naturally occurring related compounds. While persons with known sensitivity to aspirin and other salicylates should exercise caution with these plants, there is no evidence that the types of reactions known to be associated with the pharmaceutical salicylates is observed with *Salix* or any other salicin-rich plant. A study using serum from human volunteers taking willow bark extract (providing 240 mg of salicin daily for 28 days) did note a modest effect on platelet aggregation, but it was less than the effect seen with 100 mg per day aspirin (acetylsalicylic acid) (Krivoy, et al. 2001). At this dose, there is reassurance that salicin will not adversely affect bleeding, though there have not been studies conducted in patients with disorders of thrombotic function. It also suggests that salicin should not be used as a substitute for aspirin in the prevention of heart attacks and strokes.

The concentration of salicylates in most botanicals listed here is quite low, and salicylate overdose is unlikely except in the case of wintergreen essential oil (which contains 98% methyl salicylate), for which multiple cases of overdose have been reported after oral and topical use (Chan 1996; Chyka et al. 2007; Stevenson 1937). Symptoms of mild salicylate poisoning (serum concentrations of 30–50 mg/dl) include deep breathing (hyperpnea), nausea, vomiting, tinnitus, and dizziness. Moderate poisoning (serum concentrations of 50–70 mg/dl) can produce symptoms of rapid breathing (tachypnea), fever, sweating, dehydration, incoordination, and listlessness. With severe intoxication (> 75mg/dl), symptoms may include coma, seizures, hallucinations, stupor, cerebral edema, dysrhythmias, heart failure, low blood pressure, decreased urine production (oliguria), or kidney failure (Pearlman and Gambhir 2009).

Mechanism of Action

Most research has focused on the ability of salicylates to suppress the synthesis of prostaglandins, hormones thought to play an integral role in pain, inflammation, and fever. Two specific enzymes, cyclooxygenase 1 and 2 (COX1 and COX2), are considered to be predominant in this process. COX1 occurs in platelets, blood vessels, and other organs; COX2 acts primarily in inflamed tissue.

Aspirin is the most commonly used salicylate. It blocks the synthesis of prostaglandins through the acetylation of cyclooxygenase, especially COX1, by an irreversible transfer of the acetyl group into the enzyme (Hardman and Limbird 1996). Salicylic acid and salicylates (such as salicin) that lack an acetyl group are not as effective as aspirin in inhibiting platelet aggregation. Therefore, there is little concern for salicin-containing plants causing hematological disturbances. Conversely, these plants are not appropriate as a preventative treatment against stroke, a benefit associated with aspirin consumption.

Listed Herbs

Herbs listed in the *Botanical Safety Handbook* that contain salicylates:

- *Betula lenta* leaf and bark
- *Filipendula ulmaria* herb
- *Gaultheria procumbens* leaf
- *Populus balsamifera* ssp. *balsamifera* leaf buds
- *Salix alba* bark
- *Salix daphnoides* bark
- *Salix fragilis* bark
- *Salix pentandra* bark
- *Salix purpurea* bark

Literature Cited

Chan, T.Y. 1996. Potential dangers from topical preparations containing methyl salicylate. *Hum. Exp. Toxicol.* 15 (9):747-50.

Chyka, P.A., A.R. Erdman, G. Christianson, et al. 2007. Salicylate poisoning: An evidence-based consensus guideline for out-of-hospital management. *Clin. Toxicol.* 45 (2):95-131.

Hardman, J.G., and L.E. Limbird, eds. 1996. *Goodman and Gilman's the pharmacological basis of therapeutics.* New York: McGraw Hill.

Krivoy, N., E. Pavlotzky, S. Chrubasik, E. Eisenberg, and G. Brook. 2001. Effect of salicis cortex extract on human platelet aggregation. *Planta Med.* 2001; 67(3): 209-212.

Pearlman, B.L., and R. Gambhir. 2009. Salicylate intoxication: A clinical review. *Postgrad. Med.* 121 (4):162-168.

Stevenson, C.S. 1937. Oil of wintergreen (methyl salicylate) poisoning: Report of three cases, one with autopsy, and a review of the literature. *Am. J. Med. Sci.* 193 (6):772-788.

Warner, T.D., and J.A. Mitchell. 2002. Cyclooxygenase-3 (COX-3): Filling in the gaps toward a COX continuum? *P.N.A.S. U.S.* 99 (21):13371-13373.

Weissmann, G. 1991. Aspirin. *Sci. Amer.* 264 (1):84-90.

Tannins

Written by Michael McGuffin

Tannins are a broad class of complex phenolic compounds that are comprised of two chemical groups: the hydrolyzable tannins (gallotannins) and the condensed tannins (proanthocyanidins). Tannins bind to and precipitate proteins, producing the astringent activity of tannin-containing herbs. Tannins are natural components of many herbs and common foods, and some tannins are used in the processing of foods, alcoholic beverages, and medicines. Condensed tannins are found in grapes (*Vitis vinifera*), green tea (*Camellia sinensis*), hawthorn (*Crataegus* spp.), and many other plants, while hydrolyzable tannins are found in pomegranate (*Punica granatum*), green and black tea (*Camellia sinensis*), white oak (*Quercus alba*), witch hazel (*Hamamelis virginiana*), and cranesbill (*Geranium maculatum*). Both types of tannins have astringent properties, providing the basis for many of the historical medicinal uses of the plants containing them.

Adverse Effects

Tannins are broadly distributed throughout the plant kingdom, occurring in the barks, roots, leaves, fruits, seeds, and other parts of many different species. Only those plants which are reported to contain at least 10% tannins have been identified as relevant to this discussion of the potential adverse effects of tannin consumption.

Tannins have been shown to reduce the availability of certain nutrients. In the digestive tract, tannins form complexes with proteins, starch, and digestive enzymes, thereby reducing the nutritional values of ingested foods. Condensed tannins, in particular, inhibit digestive enzymes, although the major effects of condensed tannins within the digestive tract are thought to be due to the formation of less digestible complexes with dietary proteins, rather than by inhibition of digestive enzymes (Chung et al. 1998a). Tannins are also known to reduce the absorption of certain vitamins and minerals, notably iron (Chung et al. 1998a; Disler et al. 1975; Salunkhe et al. 1990). To optimize nutrient absorption, supplements or beverages that contain tannins should be taken separately from meals.

Most of the known adverse effects related to tannins are specifically recorded for consumption of tannic acid, an ethereal or hydroalcoholic extract of nutgalls (from *Quercus* spp.), and include gastrointestinal disturbances and kidney damage, as well as severe necrotic conditions in the liver (Gilman et al. 1985; Osol and Farrar 1955). While these concerns may be theoretically relevant to the use of high tannin content herbs, only the digestive irritating properties of tannins are traditionally associated with the consumption of these other plants.

Both carcinogenic and anti-cancer properties of tannins have been reported in experimental settings that measured the effect of tannins on laboratory animals (Chung et al. 1998a; Chung et al. 1998b). Condensed tannins, also called proanthocyanidins, are recognized to have significant anti-oxidant activity and potential anti-cancer activity (Nandakumar et al. 2008).

Mechanism of Action

The therapeutic activities of tannins are associated with their ability to bind with and precipitate proteins and to force dehydration of mucosal tissues. In external use, these actions allow the formation of a protective layer of harder, constricted cells; internally, both normal and pathologic secretions of all types are reduced. During internal use, tannins alter the fluidity of the bowel contents, hence their use as anti-diarrheal remedies.

Listed Herbs

Herbs listed in the *Botanical Safety Handbook* that contain over 10% tannins:

- *Agrimonia eupatoria* herb
- *Alchemilla xanthochlora* herb
- *Arctostaphylos uva-ursi* leaf
- *Camellia sinensis* leaf and stem
- *Castanea dentata* leaf
- *Corylus avellana* leaf and bark
- *Corylus cornuta* leaf and bark

- *Epilobium parviflorum* herb
- *Eucalyptus globulus* leaf
- *Euphrasia rostkoviana* herb
- *Euphrasia stricta* herb
- *Filipendula ulmaria* herb
- *Geranium maculatum* root
- *Hamamelis virginiana* bark and leaf
- *Heuchera micrantha* root
- *Ilex paraguariensis* leaf
- *Juglans nigra* leaf
- *Krameria argentea* root
- *Krameria lappacea* root
- *Paullinia cupana* seed
- *Polygonum bistorta* root
- *Potentilla erecta* rhizome
- *Punica granatum* fruit husk
- *Quercus alba* bark
- *Quercus petraea* bark
- *Quercus robur* bark
- *Rheum officinale* rhizome and root
- *Rheum palmatum* rhizome and root
- *Rheum tanguticum* rhizome and root
- *Rubus fruticosus* leaf
- *Rumex acetosa* leaf
- *Rumex acetosella* leaf
- *Rumex crispus* root
- *Rumex hymenosepalus* root
- *Rumex obtusifolius* root
- *Salix alba* bark
- *Salix daphnoides* bark
- *Salix fragilis* bark
- *Salix pentandra* bark
- *Salix purpurea* bark
- *Schinus molle* bark
- *Schinus terebinthifolius* bark
- *Terminalia arjuna* bark
- *Terminalia bellerica* fruit
- *Terminalia chebula* fruit
- *Uncaria gambir* leaf and twig

Literature Cited

Chung, K.T., C.I. Wei, and M.G. Johnson. 1998a. Are tannins a double-edged sword in biology and health? *Trends Food Sci. Tech.* 9 (4):168-175.

Chung, K.T., T.Y. Wong, C.I. Wei, Y.W. Huang, and Y. Lin. 1998b. Tannins and human health: A review. *Crit. Rev. Food Sci. Nutr.* 38 (6):421-464.

Disler, P.B., S.R. Lynch, R.W. Charlton, et al. 1975. The effect of tea on iron absorption. *Gut* 16 (3):193-200.

Gilman, A.G., L.S. Goodman, T.W. Rall, and F. Murad, eds. 1985. *Goodman and Gilman's the pharmacological basis of therapeutics.* New York: Macmillan Publishing Company.

Nandakumar, V., T. Singh, and S.K. Katiyar. 2008. Multi-targeted prevention and therapy of cancer by proanthocyanidins. *Cancer Lett.* 269 (2):378-387.

Osol, A., and G. Farrar. 1955. *The dispensatory of the United States of America, 25th ed.* Philadelphia: J.B. Lippincott Company.

Salunkhe, D.K., J.K. Chavan, and S.S. Kadam. 1990. *Dietary tannins: Consequences and remedies.* Boca Raton, FL: CRC Press.

Thujone

Written by Lisa Ganora

Thujone (which occurs as both α-thujone and its isomer, β-thujone) is a bicyclic monoterpene ketone found as a constituent of certain volatile oils. α-Thujone is a modulator of $GABA_A$ and $5\text{-}HT_3$ receptors; high doses are neurotoxic and cause epileptiform convulsions in mammals (Dettling et al. 2004). β-Thujone is less active in this respect. Because of toxicological concerns, isolated thujone is banned in many countries as a food additive. Contemporary EU regulations limit its content in sage-containing foods to 25 mg/kg, while bitters may contain 35 mg/L of thujone (ECSCF 2003). In a draft document issued in January 2011, the European Medicines Agency recommended that thujone intake from herbal medicines be limited to 6 mg per day, and indicated that higher amounts may be acceptable if deemed appropriate on a case by case basis (EMA 2011).

One analysis found that the total thujone content in essential oil of common garden sage (*Salvia officinalis*) ranged from 9 to 44% (Perry et al. 1999), while wormwood (*Artemisia absinthium*) essential oil has been reported to contain anywhere from 0 to 90% total thujone (Lachenmeier et al. 2006). Thuja (*Thuja occidentalis*) oil may have up to 73% total thujone (Naser et al. 2005), tansy oil (*Tanacetum vulgare*) up to 81% (Rohloff et al. 2004), and yarrow (*Achillea millefolium*) oil from 0 to 27% α-thujone and 0 to 11% β-thujone (Orav et al. 2006). Depending on any given plant's developmental stage, chemotype, and geographical origin, there can be wide variations in total thujone content as well as in the proportion of α-thujone to β-thujone.

It was formerly assumed that thujone (from *A. absinthium*) was responsible for the alleged psychotropic activity and toxicity of absinthe; this notion has recently been refuted by multiple analyses demonstrating that insignificant concentrations of the compound are present in both historical and contemporary examples of the beverage. It is now generally believed that absinthe's high ethanol content, and perhaps the presence of chemical adulterants (e.g., copper salts added as green dyes to inferior grades of absinthe) or other potential toxins, were responsible for any actual neurological effects (Lachenmeier et al. 2008).

Adverse Effects

Thujone has very low solubility in water; therefore little can be found in aqueous preparations (e.g., teas); however, it can be present in hydroethanolic extracts having a high percentage of ethanol, and especially in distilled products (Tegtmeier and Harnischfeger 1994).

In a toxicological assessment of thujone in mice, no adverse effects were found at concentrations below 5 mg/kg body weight, given orally for fourteen weeks (Council of Europe 1999). The LD_{50} for orally administered thujone in the rat has been reported as 192 to 500 mg/kg (ECSCF 2003).

Numerous investigations have established that essential oil of wormwood can cause convulsions in animals (Padosch et al. 2006). A case report relates that the ingestion of approximately 10 ml of the oil produced seizures, mental confusion, and agitation in a 31-year-old man; this was followed by apparent rhabdomyolysis and subsequent acute renal failure which resolved after treatment (Weisbord et al. 1997). In another case, a 2-year-old ingested up to 15 ml of dilute *Thuja* oil; the resulting seizures responded to treatment with lorazepam and phenytoin, and she was released after fifteen hours in the hospital with no apparent adverse sequelae (Friesen and Phillips 2006).

In a recent investigation, an ethanol drink high in thujone (100 mg/l) was demonstrated to have a negative effect on attention performance in human volunteers and to counteract the anxiolytic effects of ethanol alone; a low-thujone (10 mg/l) preparation did not have these properties (Dettling et al. 2004).

Mechanism of Action

α-Thujone, which binds at a non-competitive blocker site, has been established as a reversible modulator of $GABA_A$ receptors. This monoterpene has an analeptic effect similar to picrotoxinin and the pesticide dieldrin, both $GABA_A$ receptor antagonists. In the case of all three compounds, binding and toxicity is blocked by diazepam, phenobarbital, and ethanol. β-Thujone was found to have a 2.3-fold lower binding affinity and has demonstrated lesser toxicity in mice (Hold et al. 2000). It seems likely that thujone's activity on $GABA_A$ receptors is largely responsible for its seizure-promoting effects.

One study has reported that α-thujone reduced the activity of cloned human $5\text{-}HT_3$ receptors, resulting in an inhibition of serotonergic responses. It is not yet known if this mechanism contributes to the observed neurological effects of the compound (Deiml et al. 2004).

It was formerly suggested that thujone might interact with cannabinoid receptors in the CNS to bring about a psychotropic effect. Despite internet marketing claims, this idea has been discredited by a study which found that thujone has low affinity for cannabinoid receptors and does not demonstrate cannabimimetic properties (Meschler and Howlett 1999).

In chick embryo liver cells, thujone has demonstrated porphyrogenic activity; on this basis, it has been suggested that it could be hazardous to patients with acquired or genetic defects of heme synthesis in the liver (Bonkovsky et al. 1992).

In human cells, the thujone isomers are metabolized by several cytochrome P450 enzymes, including CYP2D6 and CYP3A4. 7-Hydroxy-α-thujone and 7-hydroxy-β-thujone are the major metabolites, followed by their 4-hydroxylated congeners (Jiang et al. 2006). These metabolites have significantly reduced $GABA_A$ binding affinity compared to their parent compounds and are therefore considered to be less toxic (Hold et al. 2000). Little is known about the pharmacokinetics of thujone in humans.

Listed Herbs

Herbs listed in the *Botanical Safety Handbook* that contain thujone:

- *Achillea millefolium* herb
- *Artemisia absinthium* herb
- *Artemisia capillaris* herb
- *Artemisia douglasiana* herb
- *Artemisia lactiflora* herb
- *Artemisia scoparia* herb
- *Artemisia vulgaris* herb
- *Evernia furfuracea* thallus*
- *Evernia prunastri* thallus*
- *Hyssopus officinalis* herb
- *Platycladus orientalis* cacumen
- *Salvia officinalis* leaf

* May contain thujone.

- *Tanacetum vulgare* herb*
- *Thuja occidentalis* leaves

* Some chemotypes contain thujone.

Literature Cited

Bonkovsky, H.L., E.E. Cable, J.W. Cable, et al. 1992. Porphyrogenic properties of the terpenes camphor, pinene, and thujone (with a note on historic implications for absinthe and the illness of Vincent van Gogh). *Biochem. Pharmacol.* 43 (11):2359-2368.

Council of Europe, 1999. Revised detailed datasheet on thujone. Document RD 4.2/14-44.

Deiml, T., R. Haseneder, W. Zieglgansberger, et al. 2004. Alpha-thujone reduces 5-HT_3 receptor activity by an effect on the agonist-induced desensitization. *Neuropharmacol.* 46 (2):192-201.

Dettling, A., H. Grass, A. Schuff, et al. 2004. Absinthe: Attention performance and mood under the influence of thujone. *J. Stud. Alcohol* 65 (5):573-581.

ECSCF. 2003. Opinion of the Scientific Committee on Food on thujone. European Commission Scientific Committee on Food. SCF/CS/FLAV/FLAVOUR/23 ADD2 Final.

EMA. 2011. Public statement on the use of herbal medicinal products containing thujone: Draft. European Medicines Agency, Committee on Herbal Medicinal Products. EMA/HMPC/732886/2010.

Friesen, M., and B. Phillips. 2006. Status epilepticus following pediatric ingestion of Thuja essential oil. LCLT abstracts of the European Association of Poisons Centres and Clinical Toxicologists XXVI International Congress. 219.

Hold, K.M., N.S. Sirisoma, T. Ikeda, T. Narahashi, and J.E. Casida. 2000. Alpha-thujone (the active component of absinthe): Gamma-aminobutyric acid type A receptor modulation and metabolic detoxification. *P.N.A.S. U.S.* 97 (8):3826-3831.

Jiang, Y.Y., X. He, and P.R.O. de Montellano. 2006. Radical intermediates in the catalytic oxidation of hydrocarbons by bacterial and human cytochrome P450 enzymes. *FASEB J.* 20 (4):A42-A42.

Lachenmeier, D.W., D. Nathan-Maister, T.A. Breaux, et al. 2008. Chemical composition of vintage preban absinthe with special reference to thujone, fenchone, pinocamphone, methanol, copper, and antimony concentrations. *J. Agric. Food Chem.* 56 (9):3073-3081.

Lachenmeier, D.W., S.G. Walch, S.A. Padosch, and L.U. Kroner. 2006. Absinthe—A review. *Crit. Rev. Food Sci. Nutr.* 46 (5):365-377.

Meschler, J.P., and A.C. Howlett. 1999. Thujone exhibits low affinity for cannabinoid receptors but fails to evoke cannabimimetic responses. *Pharmacol. Biochem. Behav.* 62 (3):473-480.

Naser, B., C. Bodinet, M. Tegtmeier, and U. Lindequist. 2005. Thuja occidentalis (Arbor vitae): A review of its pharmaceutical, pharmacological and clinical properties. *Evid.-Based Compl. Altern. Med.* 2 (1):69-78.

Orav, A., E. Arak, and A. Raal. 2006. Phytochemical analysis of the essential oil of *Achillea millefolium* L. from various European countries. *Nat. Prod. Res.* 20 (12):1082-1088.

Padosch, S.A., D.W. Lachenmeier, and L.U. Kroner. 2006. Absinthism: A fictitious 19th century syndrome with present impact. *Subst Abuse Treat Prev Policy* 1 (1):14.

Perry, N.B., R.E. Anderson, N.J. Brennan, et al. 1999. Essential oils from Dalmatian sage (*Salvia officinalis* L.): Variations among individuals, plant parts, seasons, and sites. *J. Agric. Food Chem.* 47 (5):2048-2054.

Rohloff, J., R. Mordal, and S. Dragland. 2004. Chemotypical variation of tansy (*Tanacetum vulgare* L.) from 40 different locations in Norway. *J. Agric. Food Chem.* 52 (6):1742-1748.

Tegtmeier, M., and G. Harnischfeger. 1994. Methods for the reduction of thujone content in pharmaceutical preparations of artemisia, salvia and thuja. *Eur. J. Pharm. Biopharm.* 40 (5):337-340.

Weisbord, S.D., J.B. Soule, and P.L. Kimmel. 1997. Poison on line—Acute renal failure caused by oil of wormwood purchased through the internet. *NEJM.* 337 (12):825-827.

APPENDIX 2. HERBAL ACTION PROFILES

Abortifacients

Written by Michael McGuffin, revised by Aviva Romm, M.D. and Tieraona Low Dog, M.D.

Abortifacients are agents that are used to induce abortion and terminate pregnancy. Herbal abortion is not a recommended method of intentional pregnancy termination.

Adverse Effects

There is a long history of use of select botanicals as abortifacients. Research on the use of botanicals to induce abortion is extremely limited, and most information on the topic comes from historical or empirical reports. Little reliable data exists on the effectiveness, toxic levels, or possible effects of these plants on the developing fetus.

Terminating a fetus in a healthy human being is a difficult and potentially dangerous process. When using botanicals for this purpose, the dose of the herb required to produce an abortion is generally extremely high and may pose toxicity risks to the mother or may have negative effects on the developing fetus should the attempt fail. Because of this, a follow-up medical abortion should be considered if an attempt to abort with an herbal agent is not effective. To date, no reports of failed abortions resulting in fetal damage have been found, although cases of maternal toxicity have been reported. Such cases include a single fatal event of a mother recorded in 1978 following consumption of an extremely high dose (one ounce) of pennyroyal essential oil (identified as *Mentha pulegium* or *Hedeoma pulegioides*) in an attempted abortion. The victim suffered two heart attacks, liver and kidney failure, and disseminated vascular coagulation before her death (Sullivan et al. 1979). Additionally, a case of nicotinic poisoning was reported after an abortion attempt using blue cohosh (*Caulophyllum thalictroides*) in excessive doses (Rao and Hoffman 2002; Rao et al. 1998).

Abortifacient or potential harmful effects to a fetus typically require very high amounts of a botanical or botanical formula over a continuous period of time. Thus, ingesting small amounts of the herbs listed below during pregnancy, except in rare cases, is not a cause for alarm. In addition, some of the herbs listed below as abortifacients, such as saffron (*Crocus sativus*), safflower (*Carthamus tinctorius*), and Roman chamomile (*Chamaemelum nobile*), may be safely used during pregnancy in amounts typically consumed in foods or beverages.

Mechanism of Action

Abortifacients have many different mechanisms of action. Some abortifacients act indirectly, meaning that they induce abortion through peripheral systems such as the endocrine, cardiovascular, gastrointestinal, or nervous systems. Others are direct-acting abortifacients that target the uterus, endometrium, and/or fetus, causing abortion to commence. It is not possible to generalize the action, efficacy, or safety of plants listed in this text as abortifacients, since the mechanisms of action of these plants have not been well studied (Bingel and Farnsworth 1980).

Certain abortifacients have drastic purgative effects or are gastrointestinal irritants that can produce reflex uterine contraction. Many volatile oils, such as oil of tansy (*Tanacetum vulgare*), and saponin glycosides, such as those found in bethroot (*Trillium erectum*), act in this manner.

Listed Herbs

Herbs listed in the *Botanical Safety Handbook* with potential abortifacient action:

- *Andrographis paniculata* herb
- *Carthamus tinctorius* flower
- *Catharanthus roseus* herb
- *Caulophyllum thalictroides* root
- *Chamaemelum nobile* flower
- *Chrysopogon zizanoides* root

- *Crocus sativus* stigma
- *Cytisus scoparius* flowering tops
- *Gossypium herbaceum* root bark
- *Gossypium hirsutum* root bark
- *Juniperus virginiana* leaf, berry
- *Mentha pulegium* leaf, essential oil
- *Podophyllum peltatum* root
- *Podophyllum hexandrum* root
- *Ruta graveolens* herb
- *Tanacetum vulgare* herb
- *Thuja occidentalis* leaves

Literature Cited

Bingel, A., and N. Farnsworth. 1980. Botanical sources of fertility regulating agents: Chemistry and pharmacology. In *Progress in hormone biochemistry and pharmacology* edited by Briggs, M. and A. Corbin. St. Albans, VT: Eden Medical Research.

Rao, R.B., and R.S. Hoffman. 2002. Nicotinic toxicity from tincture of blue cohosh (*Caulophyllum thalictroides*) used as an abortifacient. *Vet. Hum. Toxicol.* 44 (4):221-222.

Rao, R.B., R.S. Hoffman, R. Desiderio, et al. 1998. Nicotinic toxicity from tincture of blue cohosh (*Caulophyllum thalictroides*) used as an abortifacient. *J. Toxicol. Clin. Toxicol.* 36 (5):455.

Sullivan, J.B., B.H. Rumack, H. Thomas, Jr., R.G. Peterson, and P. Bryson. 1979. Pennyroyal oil poisoning and hepatotoxicity. *JAMA* 242(26):2873-2874.

Bulk-forming Laxatives

Written by Michael McGuffin; revised by Eric Yarnell, N.D.

Bulk-forming laxatives are substances that promote bowel evacuation by increasing the bulk volume and water content of the stool. These are generally considered to be the safest laxative agents.

Adverse Effects

Bulk-forming laxatives are contraindicated in bowel obstruction and must be taken with adequate liquid to avoid paradoxical constipation and esophageal or bowel obstruction (Frohna 1992; Herrle et al. 2004; Noble and Grannis 1984; Schapira et al. 1995). Cases of esophageal or bowel obstruction typically occur in persons with abnormal esophageal or intestinal narrowing or in persons who took the laxative product with insufficient liquid (Angueira and Kadakia 1993; Frohna 1992; Herrle et al. 2004).

The U.S. Food and Drug Administration (FDA) requires specific labeling of products classified as over-the-counter drug products that contain certain of the herbs listed here, including agar (*Gelidiella acerosa* and *Gelidium* spp.), guar gum (*Cyamopsis tetragonolobus*), and psyllium (*Plantago* spp.) if these are marketed in a "dry or incompletely hydrated form," such as capsules and powders. The designated labeling is as follows:

> **Choking** [highlighted in bold type]: Taking this product without adequate fluid may cause it to swell and block your throat or esophagus and may cause choking. Do not take this product if you have difficulty in swallowing. If you experience chest pain, vomiting, or difficulty in swallowing or breathing after taking this product, seek immediate medical attention (CFR 2011).

Additional language is required under the directions for use, as follows:

> **Directions** [highlighted in bold type]: (Select one of the following, as appropriate: "Take" or "Mix") this product (child or adult dose) with at least 8 ounces (a full glass) of water or other fluid. Taking this product without enough liquid may cause choking. See choking warning.

Although the above labeling is not specifically required by FDA for dietary supplements, it is suggested that the above or significantly similar language be included on the label of dietary supplements that contain bulk-forming laxative ingredients and posted at the point of sale in any retail setting where any of these are sold in bulk.

Bulk-forming laxatives may inhibit the absorption of other drugs. The drugs usually associated with this consideration are aspirin, digitalis and other cardiac glycosides, antibiotics, thyroid hormones, and anticoagulants. To ensure complete absorption of drugs, bulk-forming laxatives and other drugs should be taken several hours apart

(Brunton et al. 2006). The absorption of dietary nutrients, including calcium, iron, zinc, sodium, and potassium, can also be inhibited (ESCOP 2003). Appropriate supplementation must therefore be considered when using bulk-forming laxatives for extended periods. For individuals accustomed to a low-fiber diet, gradual use of less refined fiber in the diet is recommended before using bulk fiber agents.

Mechanism of Action

Bulk-forming laxative herbs include gel-forming fibers such as psyllium husk (*Plantago* spp.) and flax seed (*Linum usitatissimum*). Gel-forming fibers contain a form of starch called mucilage, composed of mucopolysaccharides, and roughage or indigestible plant fiber called cellulose. These plant starches are hydrophilic, absorbing water or other liquid to form a mucilaginous or gel-like substance. Because these herbs also expand on contact with liquid, they add moisture and bulk to stools in the colon (Brunton et al. 2006; Williams et al. 2006).

Bulk-forming laxatives that contain mucilage have additional minor benefits complementing their primary effect of relieving constipation. Mucilaginous herbs are demulcent, meaning that they are soothing to inflamed mucosal surfaces (Brunton et al. 2006). Demulcents form a temporary gelatinous barrier which protects the intestinal wall from irritation caused by caustic material in the intestines, thus allowing repair of the adjoining tissues.

Besides providing the demulcent properties associated with these plants' mucilage content, the indigestible cellulose fiber of bulk-forming laxatives plays additional roles related to diet and digestion. Fiber absorbs dietary fats, decreasing the absorption of cholesterol into the bloodstream. In addition, since dietary fiber from bulk laxatives cannot be digested, these herbs add a feeling of fullness without calories. Fiber also slows the release of dietary sugar from the digestive tract into the blood stream, assisting in the stabilization of blood sugar (Brennan 2005; Singh 2007; Sirtori et al. 2009).

Listed Herbs

Herbs listed in the *Botanical Safety Handbook* as bulk forming laxatives:

- *Gelidiella acerosa* (thallus)
- *Gelidium amansii* (thallus)
- *Gelidium cartilagineum* (thallus)
- *Gelidium crinale* (thallus)
- *Gelidium divaricatum* (thallus)
- *Gelidium pacificum* (thallus)
- *Gelidium vagum* (thallus)
- *Linum usitatissimum* (seed)
- *Plantago arenaria* (seed, seed husk)
- *Plantago asiatica* (seed, seed husk)
- *Plantago ovata* (seed, seed husk)

Literature Cited

Angueira, C., and S. Kadakia. 1993. Esophageal and duodenal bezoars from Perdiem. *Gastrointest. Endosc.* 39 (1):110-111.

Brennan, C.S. 2005. Dietary fibre, glycaemic response, and diabetes. *Molec. Nutr. Food Res.* 49 (6):560-570.

Brunton, L.L., J.S. Lazo, and K.L. Parker. 2006. *Goodman & Gilman's the pharmacological basis of therapeutics*, 11th ed. New York: McGraw-Hill.

CFR. 2011. *Code of federal regulations*, Title 21 Part 201.319, 2011 ed. Specific labeling requirements for specific drug products. Water-soluble gums, hydrophilic gums, and hydrophilic mucilloids (including, but not limited to agar, alginic acid, calcium polycarbophil, carboxymethylcellulose sodium, carrageenan, chondrus, glucomannan ((B-1,4 linked) polymannose acetate), guar gum, karaya gum, kelp, methylcellulose, plantago seed (psyllium), polycarbophil tragacanth, and xanthan gum) as active ingredients; required warnings and directions. Washington, DC: U.S. Government Printing Office.

ESCOP. 2003. *ESCOP monographs: The scientific foundation for herbal medicinal products*. 2nd ed. New York: Thieme.

Frohna, W.J. 1992. Metamucil bezoar: An unusual cause of small bowel obstruction. *Am. J. Emerg. Med.* 10 (4):393-395.

Herrle, F., T. Peters, C. Lang, et al. 2004. Bolus obstruction of pouch outlet by a granular bulk laxative after gastric banding. *Obes. Surg.* 14 (7):1022-1024.

Noble, J.A., and F.W. Grannis, Jr. 1984. Acute esophageal obstruction by a psyllium-based bulk laxative. *Chest* 86 (5):800.

Schapira, M., J. Henrion, P. Jonard, et al. 1995. Esophageal bezoar: Report of five more cases. *Endoscopy* 27 (4):342.

Singh, B. 2007. Psyllium as therapeutic and drug delivery agent. *Int. J. Pharmaceut.* 334 (1-2):1-14.

Sirtori, C.R., C. Galli, J.W. Anderson, E. Sirtori, and A. Arnoldi. 2009. Functional foods for dyslipidaemia and cardiovascular risk prevention. *Nutr. Res. Rev.* 22 (02):244-261.

Williams, P.A., G.O. Phillips, A.M. Stephen, and S.C. Churms. 2006. Gums and mucilages. In *Food polysaccharides and their applications*, edited by Stephen, A.M., G.O. Phillips, and P.A. Williams. Boca Raton, FL: CRC Press.

Diuretics

Written by Zoë Gardner, Ph.D.(c), reviewed by Roy Upton, RH (AHG), DAyu and Lana Dvorkin-Camiel, Pharm.D., R.Ph.

A diuretic is a substance that increases the volume and/or rate of urinary output.

Botanicals with diuretic activity can be divided into two general categories: those that induce urinary output and thus fluid loss through excretion of sodium ions (natriuretics), and those that induce urinary output without impacting electrolyte balance (aquaretics).

In conventional medicine, diuretics are employed to assist the kidneys in eliminating excess fluid from the body, fluid that accumulates in conditions such as congestive heart failure, pulmonary edema, and liver failure (Brunton et al. 2006). To be clinically useful for such indications, a diuretic must also cause excretion of sodium or sodium chloride ions in order to cause substantial fluid output (Wright 2007). Diuretics may also be employed for relief of conditions such as mild primary hypertension or to increase the flow of urine in cases of urinary tract infections and other conditions of the urinary tract (Brunton et al. 2006; Yarnell 2001).

The listing of herbs as diuretics in this text is generally based on traditional use and clinical observation. While there are a number of botanicals for which diuretic activity has been confirmed, formal research is lacking on the specific activity (i.e., natriuretic or aquaretic) and degree of effect of most of these (Wright et al. 2007). This makes it difficult to provide clear guidance on which species and what dosage may cause the level of diuresis that invokes the concerns typically ascribed to diuretics. At the same time, there is evidence to suggest that select botanical diuretics (e.g., dandelion leaf; *Taraxacum officinale*) do not result in the potassium loss common to many conventional diuretics (Racz-Kotilla et al. 1974). The magnitude of diuretic activity of the species listed here is considered by many to be generally more mild than those of pharmaceutical diuretics (Racz-Kotilla et al. 1974; Wright et al. 2007). Therefore, formal investigation would be helpful in determining the clinical efficacy, potential adverse effects profile, and potential advantage of using botanical as compared to conventional diuretics.

Adverse Effects

The potential adverse effects associated with diuretics are generally related to shifts in electrolyte balances. Although botanical diuretics are generally not as strong as pharmaceutical diuretics, and have not been associated with many of the adverse effects (i.e., hypotension, dehydration, and significant electrolyte loss) of pharmaceutical diuretics, botanical diuretics may theoretically cause electrolyte imbalances (Brunton et al. 2006; Wright et al. 2007). Therefore, individuals with conditions that may cause electrolyte imbalances (i.e., congestive heart failure, liver failure, kidney failure, etc.) should be cautious when taking diuretic herbs or do so under medical supervision, as shifts in electrolyte balance may exacerbate the disease state.

Use of diuretics is generally cautioned in persons taking drugs with narrow therapeutic ranges (small differences between the effective and toxic doses), such as warfarin, steroids (i.e., prednisone), digoxin, tacrolimus, cyclosporine, valproic acid, phenytoin, and carbemazepine, as shifts in serum levels of sodium and potassium may affect serum levels of these drugs. Serum electrolyte shifts, notably sodium, can also cause an increase in serum lithium levels, and may result in lithium toxicity (Finley et al. 1995). Use of lithium with diuretics is not recommended, but if taken concomitantly, serum drug and electrolyte levels need to be monitored closely. Corticosteroids or licorice may amplify the potassium-depleting effects of diuretics (Brunton et al. 2006; Isbrucker and Burdock 2006).

Concomitant use of diuretic herbs with prescription loop diuretics, thiazide diuretics, osmotic diuretics, and potassium-sparing diuretics may cause excessive fluid loss. Reductions in potassium levels caused by diuretics may increase the toxicity of cardiac glycosides, such as digoxin, and their combination should be avoided (Anon 2010).

Diuretics may irritate or exacerbate symptoms of kidney stones, and professional guidance is recommended when employing these therapies for this condition (Chitme et al. 2010). Some naturopathic treatment protocols suggest that kidney stones smaller than 5 mm in size may be passed with the assistance of herbal diuretics, but such protocols should only be followed under the supervision of a qualified health professional (Yarnell 2001).

Mechanism of Action

Due to lack of research, it is not yet known whether botanical diuretics have the same or different mechanisms of action as non-botanical diuretic drugs.

Listed Herbs

Herbs listed in the *Botanical Safety Handbook* as diuretics:

- *Agathosma betulina* leaf
- *Agathosma crenulata* leaf
- *Agathosma serratifolia* leaf
- *Alisma plantago-aquatica* rhizome
- *Anethum graveolens* herb and fruit
- *Apocynum androsaemifolium* root
- *Apocynum cannabinum* root
- *Asparagus officinalis* rhizome
- *Asparagus racemosus* rhizome
- *Betula pendula* leaf
- *Betula pubescens* leaf
- *Boerhavia diffusa* root
- *Camellia sinensis* leaf and stem
- *Chamaesyce hirta* herb
- *Coffea arabica* seed kernel
- *Cola acuminata* seed
- *Cola nitida* seed
- *Daucus carota* fruit
- *Equisetum arvense* herb
- *Equisetum hyemale* herb
- *Equisetum telmateia* herb
- *Gossypium herbaceum* root bark
- *Gossypium hirsutum* root bark
- *Ilex paraguariensis* leaf
- *Juniperus communis* fruit
- *Juniperus monosperma* fruit
- *Juniperus osteosperma* fruit
- *Juniperus oxycedrus* fruit
- *Nardostachys jatamansi* rhizome and root
- *Parietaria judaica* herb
- *Parietaria officinalis* herb
- *Paullinia cupana* seed
- *Petroselinum crispum* root
- *Phyllanthus amarus* whole plant
- *Phyllanthus fraternus* whole plant
- *Phyllanthus niruri* whole plant
- *Polygala senega* root
- *Polygala sibirica* root
- *Polygala tenuifolia* root
- *Portulaca oleracea* herb
- *Prunus spinosa* seeds and fresh flowers
- *Ribes nigrum* leaf
- *Satureja hortensis* leaf
- *Satureja montana* leaf
- *Selenicereus grandiflorus* flowers and stem
- *Solidago canadensis* var. *lepida* herb
- *Solidago gigantea* herb
- *Solidago virgaurea* herb
- *Stephania tetrandra* root
- *Tanacetum vulgare* herb
- *Taraxacum officinale* leaf, root
- *Tinospora cordifolia* root, stem, and leaf
- *Tribulus terrestris* fruit
- *Urtica dioica* leaf
- *Zea mays* stigma

Literature Cited

Anon. 2010. Digoxin: Serious drug interactions. *Prescrire Int.* 19 (106):68-70.

Brunton, L.L., J.S. Lazo, and K.L. Parker. 2006. *Goodman & Gilman's the pharmacological basis of therapeutics*, 11th ed. New York: McGraw-Hill.

Chitme, H.R., S. Alok, S.K. Jain, and M. Sabharwal. 2010. Herbal treatment for urinary stones. *Int. J. Pharm. Sci. Res.* 1 (24-31).

Combest, W., M. Newton, A. Combest, and J.H. Kosier. 2005. Effects of herbal supplements on the kidney. *Urol. Nurs.* 25 (5):381-386.

Finley, P.R., M.D. Warner, and C.A. Peabody. 1995. Clinical relevance of drug interactions with lithium. *Clin. Pharmacokin.* 29 (3):172-191.

Isbrucker, R.A., and G.A. Burdock. 2006. Safety and risk assessment on the consumption of licorice root. *Regul. Toxicol. Pharmacol.* 46:168-192.

Racz-Kotilla, E., G. Racz,, and A. Solomon. 1974. The action of *Taraxacum officinale* extracts on the body weight and diuresis of laboratory animals. *Planta Med* 26(3):212-217.

Supuran, C.T., A. Scozzafava, and J. Conway, eds. 2004. *Carbonic anhydrase—Its inhibitors and activators.* Boca Raton, FL: CRC Press.

Wright, C.I., L. Van-Buren, C.I. Kroner, and M.M.G. Koning. 2007. Herbal medicines as diuretics: A review of the scientific evidence. *J. Ethnopharmacol.* 114 (1):1-31.

Yarnell, E. 2001. *Naturopathic urology and men's health.* Wenatchee, WA: Healing Mountain Publishing.

Emetics

Written by Michael McGuffin; revised by Zoë Gardner, Ph.D.(c)

An emetic is a substance that induces vomiting in a sufficient dose. Emetics have two primary functions. The first is to empty the stomach, especially in cases of ingestion of non-caustic poisons; the second is as an expectorant for respiratory diseases, primarily to expel mucus and phlegm from the bronchioles. Historically, use in respiratory conditions was mainly in children with conditions such as asthma, bronchitis, and diphtheria before they were able to expectorate effectively. Doses for this latter use are significantly smaller than doses used to empty the stomach.

Adverse Effects

The concerns related to herbs with emetic potential are of importance only when they are consumed in dosage levels sufficient to produce vomiting. Such use is typically limited in duration to one or several doses, and use of emetics for more than three to four days can produce dehydration and severe electrolyte imbalances. Continual retching action from chronic vomiting will strain the abdominal and stomach muscles and the diaphragm, causing severe cramping and potential development of hernias (Rakel 1996; Hardman & Limbird 1996; Katcher et al. 1983). Chronic use by persons with eating disorders may lead to generalized muscle weakness, diarrhea, mild tremors, fluid retention, dehydration, and metabolic disturbances (hypokalemia, hypochloremic acidosis, elevation of creatinine phosphokinase) (Manno and Manno 1977; Quang and Woolf 2000).

Emetics are contraindicated in those with aneurysms, hernia, arteriosclerosis, or in cases of hemorrhage. The use of emetics is often associated with depression of central motor functions, and so is best used under appropriate medical supervision.

The following guidelines have been developed for the use of ipecac (*Cephaelis ipecacuanha*), one of the most common botanical emetics used, but are applicable to any botanical being used for emetic purposes in the case of poisonings. Ipecac (and other emetics) should not be used unless directed by a qualified healthcare professional (physician, poison control center, or other professional), and should not be used if:

- A patient is comatose or has altered mental status and the risk of breathing in the stomach contents is high.
- The patient is having convulsions.
- The substance ingested is capable of causing altered mental status or convulsions.
- The substance ingested is a caustic or corrosive agent.
- The substance ingested is a low viscosity petroleum distillate with the potential for pulmonary aspiration.
- The patient has a medical condition that may be exacerbated by vomiting (e.g., severe high blood pressure, slow heart rate, or tendency for bleeding) (Manoguerra and Cobaugh 2005).

Other guidelines list additional "relative contraindications" beyond the "absolute contraindications" listed above. These guidelines indicate that ipecac should not be used if:

- The patient is already vomiting.
- More than one hour has passed since ingestion of the product of concern.
- The patient is susceptible to bleeding or hemorrhaging (bleeding diathesis).
- An oral antidote to the consumed poison is available.
- The patient is less than 6 months of age.
- The patient is elderly or has a history of heart disease.
- The patient ingested cardiotoxic drugs (i.e., calcium channel blockers, beta blockers) (Quang and Woolf 2000).

While emetics such as ipecac were, in the past, remedies of choice for emptying the stomach in the case of poisonings, more recent recommendations indicate that emetics should not be used in certain circumstances, and that other treatment methods may be preferable (Manoguerra and Cobaugh 2005; Quang and Woolf 2000). Activated charcoal and emetics should not be administered together since the charcoal can absorb the emetic substance and reduce the emetic effect and because the charcoal will be expelled through vomiting (Hardman and Limbird 1996).

Ipecac contains the compound emetine, which may adversely affect the heart. In the event that vomiting does not occur after administration of ipecac, gastric lavage (stomach pumping) should be performed to avoid a toxic reaction to emetine (Manno and Manno 1977).

Mechanism of Action

There are two primary classes of emetics: those that work on the vomiting centers in the medulla (central emetics), and those that act directly on the stomach itself (gastric emetics). There are botanical emetics that fall into each class, although the mechanism of action of most botanical emetics is not fully understood. Central emetics act by affecting a section of the brain stem known as the chemoreceptor trigger zone. This zone is affected by certain chemical abnormalities in the body and sends a signal to the vomiting centers, which stimulate and coordinate the process of vomiting. Gastric emetics, such as ipecac, act as irritants in the gastrointestinal tract and signal the vomiting center via the vagus nerve (Hardman and Limbird 1996).

Listed Herbs

Herbs listed in the *Botanical Safety Handbook* as emetics:

Apocynum androsaemifolium root
Apocynum cannabinum root
Asclepias tuberosa root
Cephaelis ipecacuanha rhizome
Genista tinctoria herb and flower
Ipomoea purga root
Iris versicolor rhizome and root
Iris virginica rhizome and root
Lobelia inflata herb
Lobelia siphilitica herb
Melia azedarach fruit, root bark
Podophyllum peltatum root
Podophyllum hexandrum root
Sanguinaria canadensis root

Literature Cited

Hardman, J.G., and L.E. Limbird, eds. 1996. *Goodman & Gilman's the pharmacological basis of therapeutics*. New York: McGraw-Hill.

Katcher, B.S., L.Y. Young, and M.A. Koda-Kimble. 1983. *Applied therapeutics — The clinical use of drugs*, 3rd ed. Spokane, WA: Applied Therapeutics.

Manno, B.R., and J.E. Manno. 1977. Toxicology of ipecac: A review. *Clin. Toxicol.* 10 (2):221-242.

Manoguerra, A.S., and D.J. Cobaugh. 2005. Guideline on the use of ipecac syrup in the out-of-hospital management of ingested poisons. *Clin. Toxicol.* 43 (1):1-10.

Quang, L.S., and A.D. Woolf. 2000. Past, present, and future role of ipecac syrup. *Curr. Opin. Pediatr.* 12 (2):153-162.

Rakel R, editor. 1996. *Conn's Current Therapy*. Philadelphia: W.B. Saunders Co.

Emmenagogues and Uterine Stimulants

Written by Tieraona Low Dog, M.D.

Emmenagogues are agents used to promote menstrual flow. Uterine stimulants are used to cause a woman's uterus to contract, or to increase the frequency and intensity of the contractions.

As a category, emmenagogues are not well defined, and their mechanisms of action have not been well studied or understood. Emmenagogues are principally used for amenorrhea, a condition that can be due to a wide variety of causes. Logically thinking through the etiology of amenorrhea, the category could include herbs with inherent nutritive value, herbs that have a calming effect on the nervous system, herbs that enhance or promote circulation, agents that induce or increase uterine contractions (uterine stimulants), and herbs that interact with the hypothalamic-adrenal-gonadal axis (part of the neuroendocrine system). Examples might include the following:

Amenorrhea secondary to anemia or malnourishment might be treated with nutritive herbs (i.e., nettles, red clover, dong quai).

Amenorrhea secondary to trauma or great emotional stress might be treated with nervines (i.e., chamomile, motherwort, bupleurum).

Amenorrhea due to pregnancy might be addressed using herbs with abortifacient or uterine stimulant activity (e.g., pennyroyal, artemisia, cotton root bark). Abortifacient activity may involve reduction in maternal progestogen and testosterone levels, as well as an increase in immunoreactive cells (Al-Dissi et al. 2001; Boareto et al. 2008; Mukherjee et al. 1996; Talwar et al. 1997).

In traditional Chinese medicine, blood-moving herbs, or "herbs which invigorate the blood," are thought to increase menstrual blood flow by regulating the blood vessels in the uterus or stimulating the general blood circulation (e.g., safflower (*Carthamus tinctorius*) flower and myrhh (*Commiphora wightii*) gum resin (Bensky et al. 2004; Chen and Chen 2004).

The broad variation in herbs used to "bring about menstruation" makes it impossible to generalize mechanistic activity.

Adverse Effects

Depending upon the plant and the intended use, one could predict to some degree the adverse effects that might occur. Herbs that are widely known to be used as abortifacients and/or uterine stimulants should be avoided during pregnancy and in women with menorrhagia (heavy menstrual bleeding) (Chalker and Downer 1992).

Uterine stimulants for labor induction should always be used under the direct supervision of a qualified and experienced individual. If used improperly, these herbs could potentially lead to such complications as uterine hypercontractility, uterine rupture, and maternal hypotension (Kelsey and Prevost 1994).

Adverse effects in pregnancy from emmenagogues that do not have abortifacient activity, such as chamomile or catmint, are unlikely when they are used at normal and customary doses. The same is true for the culinary herbs, such as thyme, parsley, and rosemary, when they are used to flavor food.

The editors of this text acknowledge that there is a lack of consensus regarding which herbs should be listed as emmenagogues, and that safety is a grey area when it comes to their use in pregnancy. The editors actively discourage the use of any emmenagogue as an abortifacient due to the potential for serious and significant risk to both mother and fetus; they encourage the herbal, medical and research communities to explore the category more fully and conduct studies that would help to clarify mechanisms of action and safety.

Listed Herbs

Herbs listed in the *Botanical Safety Handbook* as emmenagogues or uterine stimulants:

- Emmenagogues
  - *Angelica archangelica* root and fruit
  - *Angelica atropurpurea* root and fruit
  - *Anthriscus cerefolium* herb
  - *Artemisia abrotanum* herb
  - *Artemisia douglasiana* herb
  - *Artemisia lactiflora* herb
  - *Artemisia vulgaris* herb
  - *Caulophyllum thalictroides* root
  - *Chamaemelum nobile* flower
  - *Ferula assa-foetida* oleo-gum-resin
  - *Ferula foetida* oleo-gum-resin
  - *Forsythia suspensa* fruit
  - *Gentiana lutea* root
  - *Hyssopus officinalis* herb
  - *Inula helenium* root
  - *Leonurus cardiaca* herb
  - *Leonurus heterophyllus* herb
  - *Leonurus sibiricus* herb
  - *Marrubium vulgare* herb
  - *Mentha pulegium* leaf and essential oil
  - *Monarda clinopodia* herb
  - *Monarda didyma* herb
  - *Monarda fistulosa* herb
  - *Monarda pectinata* herb
  - *Monarda punctata* herb
  - *Nardostachys jatamansi* rhizome, root
  - *Nepeta cataria* herb
  - *Petroselinum crispum* leaf
  - *Polygala senega* root
  - *Rosmarinus officinalis* leaf
  - *Ruta graveolens* herb
  - *Satureja hortensis* leaf
  - *Satureja montana* leaf
  - *Tanacetum vulgare* herb
  - *Taxus brevifolia* needles
  - *Thymus vulgaris* herb
  - *Zanthoxylum americanum* bark
  - *Zanthoxylum clava-herculis* bark
- Uterine stimulants
  - *Achyranthes bidentata* root
  - *Capsella bursa-pastoris* herb
  - *Carthamus tinctorius* flower
  - *Commiphora mukul* gum resin
  - *Commiphora wightii* gum resin
  - *Corydalis yanhusuo* tuber
  - *Cytisus scoparius* flowering tops
  - *Gossypium herbaceum* root bark
  - *Gossypium hirsutum* root bark
  - *Leonurus heterophyllus* herb
  - *Leonurus sibiricus* herb
  - *Ziziphus jujuba* var. *spinosa* seed

Literature Cited

Al-Dissi, N.M., A.S. Salhab, and H.A. Al-Hajj. 2001. Effects of *Inula viscosa* leaf extracts on abortion and implantation in rats. *J. Ethnopharmacol.* 77 (1):117-121.

Bensky, D., S. Clavey, and E. Stöger. 2004. *Chinese herbal medicine: Materia medica*, 3rd ed. Seattle: Eastland Press.

Boareto, A.C., J.C. Muller, A.C. Bufalo, et al. 2008. Toxicity of artemisinin (*Artemisia annua* L.) in two different periods of pregnancy in Wistar rats. *Repro. Toxicol.* 25 (2):239-246.

Chalker, R., and C. Downer. 1992. *A woman's book of choices: Abortion, menstrual extraction, RU-486*. New York: Four Walls Eight Windows.

Chen, J.K., and T.T. Chen. 2004. *Chinese medical herbology and pharmacology*. City of Industry, CA: Art of Medicine Press.

Kelsey, J.J., and R.R. Prevost. 1994. Drug-therapy during labor and delivery. *Am. J. Hosp. Pharm.* 51 (19):2394-2402.

Mukherjee, S., N.K. Lohiya, R. Pal, M.G. Sharma, and G.P. Talwar. 1996. Purified neem (*Azadirachta indica*) seed extracts (Praneem) abrogate pregnancy in primates. *Contraception* 53 (6):375-378.

Talwar, G.P., S. Shah, S. Mukherjee, and R. Chabra. 1997. Induced termination of pregnancy by purified extracts of Azadirachta Indica (Neem): Mechanisms involved. *Am. J. Reprod. Immunol.* 37 (6):485.

Photosensitizing Agents

Written by Zoë Gardner, Ph.D.(c)

Photosensitizing substances cause the development of abnormally heightened reactivity of the skin or eyes to sunlight. Exposure to either the substance or light alone is not sufficient to induce the reaction, which may include rashes, blistering, skin irritation, swelling, or hyperpigmentation. Such reactions may occur after topical application or ingestion of the photosensitizing substance (Moore 2002).

A number of plants that may cause reactions after topical exposure do not pose any risk after internal use. For example, furanocoumarin compounds, such as those present in fragrant angelica (*Angelica dahurica*), have photosensitizing effects after contact with skin, although no cases of photosensitivity after ingestion of fragrant angelica are known (Bensky et al. 2004). While certain plants elicit photosensitivity in most people, the majority of photosensitizing plants listed in this text affect sensitive or otherwise predisposed people (i.e., those who are fair skinned) or those undergoing phototherapy (laser or UV treatment).

Adverse Effects

There are two distinct types of photosensitizing reactions associated with the use of photosensitizing agents: phototoxic reactions and photoallergic reactions. Phototoxic reactions are dose dependent and may occur several minutes or hours after exposure to sunlight. These reactions are limited to sun-exposed skin and generally resemble severe sunburns, sometimes accompanied by erythema (reddening of the skin), edema (swelling and fluid retention), and blistering. Skin pigmentation may also occur (Stein and Scheinfeld 2007). Phototoxic reactions are generally associated with products containing psoralens (linear furocoumarins), although other compounds, such as hypericin in St. John's wort (*Hypericum perforatum*), may contribute to photosensitivity (Brockmöller et al. 1997; Stein and Scheinfeld 2007). Psoralens are found in more than two dozen plant sources, including species of the Rutaceae (i.e., *Ruta graveolens, Citrus aurantifolia,* and *C. bergamia*), Apiaceae (i.e., *Ammi majus* and *Apium graveolens*), Fabaceae (*Psoralea* spp.), and Moraceae (*Ficus carica*) families (Bollero et al. 2001; Egan and Sterling 1993; Eickhorst et al. 2007; Maso et al. 1991; Thomson et al. 2007; Wagner et al. 2002; Wang et al. 2002).

Photoallergic reactions occur after light causes a substance to change from one that is non-allergenic to one that is allergenic. These reactions are less common than phototoxic reactions and are not dependent on the dose (which may be very small) of the substance taken or the amount of light exposure, but do require prior sensitization. Photoallergic reactions typically occur 24 hours or more after initial exposure and present as an eczema-like rash that may spread beyond the sun-exposed skin (Nigel et al. 2003; Stein and Scheinfeld 2007). Photoallergic reactions are exhibited as skin conditions characteristic of allergic contact dermatitis, with the reaction limited to sun-exposed areas of the body. However, when the reactions are severe or prolonged, they may extend into covered areas of skin (Stein and Scheinfeld 2007).

Photosensitivity associated with the use of common herbs such as St. John's wort generally occurs at doses many times higher than standard recommended dosages, though there are exceptions. While photosensitivity reactions may occur after oral use of some botanicals, reactions are more common after topical use or accidental topical exposure. Fair-skinned individuals are more susceptible to developing photosensitivity than others. Severe reactions may occur during exposure to high levels of ultraviolet (UV) light, and especially during therapeutic UV treatment (Beattie et al. 2005).

Mechanism of Action

Phototoxic reactions occur by light activation of plant compounds such as psoralens. The psoralen or other compound in the skin absorbs energy which increases the energy state of the compound's electrons, creating an excited state. As the electrons return to ground state, energy is released that incites an inflammatory response and damages cellular molecules and organelles. Damage may occur from formation of radicals or by production of singlet oxygen, which then oxidizes cell structures (Stein and Scheinfeld 2007).

Photoallergic reactions occur after light causes a conversion from a non-allergenic substance to an allergenic substance (Nigel et al. 2003; Stein and Scheinfeld 2007). Radiant energy converts a compound (from a botanical or drug) in the skin into a photoactive compound, initiating an immunologic, cell-mediated hypersensitivity reaction. This can occur through the production of stable photodrugs, one of which acts as a hapten that conjugates with a carrier molecule to form an antigen (a substance that elicits production of antibodies). Alternatively, the compound may be converted by radiant energy to a higher energy state and, upon return to resting state, the released energy promotes conjugation of a compound to a carrier protein, forming a completely new antigen (Stein and Scheinfeld 2007).

Listed Herbs

Herbs listed in the *Botanical Safety Handbook* with potential photosensitizing action:

- *Angelica pubescens* root
- *Apium graveolens* fruit
- *Citrus* × *aurantifolia* peel
- *Citrus bergamia* peel
- *Cullen corylifolia* seed
- *Hypericum perforatum* herb
- *Ruta graveolens* herb

Literature Cited

Beattie, P.E., R.S. Dawe, N.J. Traynor, et al. 2005. Can St John's wort (hypericin) ingestion enhance the erythemal response during high-dose ultraviolet A1 therapy? *Br. J. Dermatol.* 153 (6):1187-1191.

Bensky, D., S. Clavey, and E. Stöger. 2004. *Chinese herbal medicine: Materia medica,* 3rd ed. Seattle: Eastland Press.

Bollero, D., M. Stella, A. Rivolin, et al. 2001. Fig leaf tanning lotion and sun-related burns: Case reports. *Burns* 27 (7):777-779.

Brockmöller, J., T. Reum, S. Bauer, et al. 1997. Hypericin and pseudohypericin: Pharmacokinetics and effects on photosensitivity in humans. *Pharmacopsych.* 30:94-101.

Egan, C.L., and G. Sterling. 1993. Phytophotodermatitis: A visit to Margaritaville. *Cutis* 51 (1):41.

Eickhorst, K., V. DeLeo, and J. Csaposs. 2007. Rue the herb: *Ruta graveolens*-associated phytophototoxicity. *Dermatitis* 18 (1):52-55.

Maso, M.J., A.M. Ruszkowski, J. Bauerle, V.A. DeLeo, and F.P. Gasparro. 1991. Celery phytophotodermatitis in a chef. *Arch. Dermatol.* 127 (6):912.

Moore, D.E. 2002. Drug-induced cutaneous photosensitivity: Incidence, mechanism, prevention and management. *Drug Safety* 25 (5):345-372.

Nigel, S., S.R. Knowles, and N.H. Shear. 2003. Drug eruptions: Approaching the diagnosis of drug-induced skin diseases. *J. Drugs Dermatol.* 2 (3):278-299.

Stein, K.R., and N.S. Scheinfeld. 2007. Drug-induced photoallergic and phototoxic reactions. *Exp. Opin. Drug Safe.* 6 (4):431-443.

Thomson, M.A., P.W. Preston, L. Prais, and I.S. Foulds. 2007. Lime dermatitis from gin and tonic with a twist of lime. *Contact Derm.* 56 (2):114-115.

Wagner, A.M., J.J. Wu, R.C. Hansen, H.N. Nigg, and R.C. Beiere. 2002. Bullous phytophotodermatitis associated with high natural concentrations of furanocoumarins in limes. *Am. J. Cont. Derm.* 13 (1):10-14.

Wang, L., B. Sterling, and P. Don. 2002. Berloque dermatitis induced by "Florida water." *Cutis* 70 (1):29-30.

Stimulant Laxatives

Written by Michael McGuffin,
updated by Eric Yarnell, N.D.

Stimulant laxatives are agents used to relieve constipation by local stimulation and contraction of the smooth muscle of the lower bowel. This results in increased peristalsis that empties stools more quickly.

Adverse Effects

Short-term side effects of stimulant laxative consumption may include intestinal cramps, uterine contractions, and watery diarrhea. Continuous use for more than 10 days can cause dependency, resulting in colonic atonicity requiring the aid of stimulant laxatives to have bowel movements. When used in excess or for long periods, the resultant loss of fluids and electrolytes, especially potassium, can cause pathological alterations to the colon, kidney malfunction, or heart palpitations. Patients taking cardiac glycosides are particularly susceptible to cardiotoxicity (De Smet 1993).

The American Herbal Products Association (AHPA) recommends the following labeling for products that contain *Aloe* spp. latex, *Frangula alnus* bark, *Frangula purshiana* bark, *Rhamnus catahrtica* fruit, *Rheum* spp. root/rhizome, and *Senna* spp. fruit (pod) and leaf:

> NOTICE: Do not use this product if you have abdominal pain or diarrhea. Consult a health care provider prior to use if you are pregnant or nursing a baby. Discontinue use in the event of diarrhea or watery stools. Do not exceed recommended dose. Not for long-term use.

The State of California has established labeling requirements that supersede the AHPA recommendation for products sold in California. All dietary supplements that contain any of the above listed ingredients are required to bear the following label (California 2010):

> NOTICE: This product contains [name of substance(s) and common name(s) if different]. Read and follow directions carefully. Do not use if you have or develop diarrhea, loose stools, or abdominal pain because [insert common name] may worsen these conditions and be harmful to your health. Consult your physician if you have frequent diarrhea or if you are pregnant, nursing, taking medication, or have a medical condition.

Beyond these regulations, stimulant laxatives must not be given to patients with eating disorders, and every attempt should be made to determine if such patients are abusing them. Chronic abuse by such patients has resulted in severe muscle damage, kidney failure, and death. Urine tests exist to detect the presence of anthraquinones in such patients (Roerig et al. 2010).

Cathartic laxatives can turn the urine and/or stool a red or dark color (Roerig et al. 2010). While this does not represent any health problem, people taking laxatives should be warned of this, so they do not mistake the color for blood and seek unnecessary health care.

The majority of evidence supports that, although anthraquinone-containing glycosides cause dark patches in the colon that can last for months (known as pseudomelanosis coli), this is not a precursor to nor does it increase the risk of colorectal cancer (Sonnenberg and Müller 1993).

While most stimulant laxatives have traditionally been contraindicated in pregnancy due to concerns regarding stimulation of the uterus, senna (*Senna alexandrina*) and sickle-pod senna (*Senna obtusifolia, S. tora*) have shown a lack of adverse effects on pregnancy or the fetus when used according to the recommended dosage schedule (Ács et al. 2010; ESCOP 2003). Thus, senna laxatives are now considered appropriate for use during the second and third trimesters of pregnancy (ESCOP 2003; Prather 2004). Due to the potential genotoxicity of certain anthraquinones, however, it is recommended that use of senna be avoided in the first trimester of pregnancy or used under professional supervision (ESCOP 2003).

Mechanism of Action

The action of stimulant laxative herbs is, in most cases, due primarily to their content of anthraquinones. The one exception among the herbs listed in this category is castor oil (from *Ricinus communis*), the action of which is due to ricinoleic acid (Brunton et al. 2006).

Stimulant laxatives increase the motility of the colon, induce changes in the surface cells of the colon, and cause the loss of water and electrolytes. Although intensive research has been performed, the exact mechanism of action is still unclear. However, herbs that contain anthraquinones do affect the colonic mucosa and produce a laxative effect. Anthraquinones disturb the equilibrium between the absorption of water from the intestinal lumen (via an active sodium transport) and the secretion of water into the lumen by the hydrostatic blood pressure or a prostaglandin-dependent chloride secretion. Anthraquinone glycosides utilize the intestinal flora to produce a laxative effect and are stronger in action than anthraquinone aglycones, which are absorbed in the stomach and duodenum (De Smet 1993). Both aglycones and glycosides occur in *Aloe*, *Rhamnus*, and *Rheum*.

Listed Herbs

Herbs listed in the *Botanical Safety Handbook* as stimulant laxatives:

- *Aloe ferox* latex
- *Aloe perryi* latex
- *Aloe vera* latex
- *Frangula alnus* bark
- *Frangula purshiana* bark
- *Ipomoea purga* root
- *Iris versicolor* rhizome, root
- *Iris virginica* rhizome, root
- *Podophyllum peltatum* root
- *Podophyllum hexandrum* root
- *Reynoutria multiflora* unprocessed root tuber
- *Rhamnus cathartica* fruit
- *Rheum officinale* rhizome, root
- *Rheum palmatum* rhizome, root
- *Rheum tanguticum* rhizome, root
- *Ricinus communis* seed oil
- *Senna alexandrina* fruit (pod), leaf
- *Senna obtusifolia* fruit (pod), leaf
- *Senna tora* fruit (pod), leaf

Literature Cited

Ács, N., F. Bánhidy, E.H. Puhó, and A.E. Czeizel. 2010. No association between severe constipation with related drug treatment in pregnant women and congenital abnormalities in their offspring: A population based case control study. *Congen. Anom.* 50 (1):15-20.

Brunton, L.L., J.S. Lazo, and K.L. Parker. 2006. *Goodman & Gilman's the pharmacological basis of therapeutics*, 11th ed. New York: McGraw-Hill.

California. 2010. State of California, Title 17, California Code of Regulations, Section 10750.

De Smet, P.A.G.M. 1993. *Adverse effects of herbal drugs, volume* 2. Berlin: Springer.

ESCOP. 2003. *ESCOP monographs: The scientific foundation for herbal medicinal products*. 2nd ed., completely rev. and expand ed. New York: Thieme.

Prather, C.M. 2004. Pregnancy-related constipation. *Curr. Gastroenterol. Reports* 6 (5):402-404.

Roerig, J.L., K.J. Steffen, J.E. Mitchell, and C. Zunker. 2010. Laxative abuse: Epidemiology, diagnosis and management. *Drugs* 70 (12):1487-1503.

Sonnenberg, A., and A. Müller. 1993. Constipation and cathartics as risk factors of colorectal cancer: A meta-analysis. *Pharmacology* 47 (Supp 1):224-233.

APPENDIX 3. HERBAL INTERACTION PROFILES

Pharmacokinetic Drug Interactions: CYP450 and P-Glycoprotein

Written by Bill Gurley, Ph.D. and Zoë Gardner, Ph.D.(c)

Drug interactions are generally split into two classes. The first is pharmacodynamic interactions, which are interactions that happen due to additive or opposing effects of two different substances. Pharmacodynamic interactions can be predicted based on the biological activities of the different drugs or herbs (i.e., stimulants and sedatives have opposing effects). The second type of interactions, pharmacokinetic interactions, are unrelated to the therapeutic activity of the herbs and drugs being taken. Pharmacokinetic interactions involve enzymes or transporter proteins that metabolize drugs or compounds from herbs and foods or transport these substances into and out of cells. Several herbs and foods, and a number of drugs, can change the activity of these enzymes and proteins, which can be inhibited or induced, in turn influencing the blood levels of specific drugs or other compounds. While some changes in activity may be minor, with little change in the efficacy of herbs or drugs involved, other changes may produce clinically relevant interactions. In pharmacokinetic interactions, the severity of an interaction is based on the potential toxicity of the drug being used (in the case of increased blood levels of a drug) or the consequences of the therapeutic dose not being achieved (in the case of decreased blood levels of a drug). Until identified through testing or well-documented case reports, these interactions generally cannot be predicted. Once identified, however, these interactions can be easily avoided.

Once understood, herb-drug interactions may also be used therapeutically. Inhibition of drug metabolizing enzymes or transporter proteins, for example, can help increase or maintain levels of certain drugs in the blood or in cells, allowing for a reduced dose (and sometimes reduced side effects) of drugs or increased therapeutic activity (Dresser et al. 2000; Padowski and Pollack 2010). Therapeutic interactions are an emerging area of research, although few human studies have been completed (Bano et al. 1987; Kasibhatta and Naidu 2007).

While some foods and herbs, such as grapefruit (*Citrus × paradisi*), St. John's wort (*Hypericum perforatum*), and schisandra (*Schisandra sphenanthera*), have been shown to affect CYP450 enzymes or P-glycoprotein in a clinically significant manner, some other botanicals, such as black cohosh and milk thistle, have shown a lack of metabolic interactions in human studies (Fuhr et al. 2007; Gurley et al. 2006a; Gurley et al. 2006b; Gurley et al. 2008; Rajnarayana et al. 2004; Rao et al. 2007).

Cytochrome P450

The cytochrome P450 enzymes are known as a "superfamily" of enzymes found in nearly all living organisms. These enzymes are important in metabolizing a variety of compounds from foods, medicines, environmental contaminants, and compounds produced in the body. In humans, cytochrome P450 enzymes are primarily responsible for the initial metabolism (phase I) of many drugs to prepare them for conjugation (phase II) and then elimination. These enzymes are important for understanding potential drug interactions and an individual's response to certain drugs (Danielson 2002).

CYP-Based Drug Interactions

Certain CYP enzymes may be induced or inhibited by conventional drugs or by compounds in plants, leading to an increase or decrease in the activity of the enzyme. If a CYP enzyme is induced, plasma levels of drugs metabolized by that enzyme may be decreased, which, in turn, may result in plasma levels too low to be effective. For example, the compound hyperforin in St. John's wort induces CYP3A4 and reduces plasma levels of cyclosporine, a drug used for immune system suppression in patients who have had organ transplants (Mai et al. 2004). This interaction has resulted in sub-therapeutic levels of cyclosporine, causing organ rejection in transplant patients (Barone et al. 2000; Breidenbach et al. 2000). Conversely, if a CYP enzyme is inhibited, metabolism of a drug metabolized by that enzyme will be slowed and plasma levels will increase, leading to potential overexposure and drug toxicity. Grapefruit juice is a well-known inhibitor of CYP3A4 (David et al. 1998) and thus should not be consumed by patients taking drugs metabolized by CYP3A4.

Induction or inhibition of CYP enzymes occurs at different levels, with some inhibitors or inducers producing weak effects and others producing strong effects. For drugs with narrow therapeutic indices (a small difference between the effective dose and the toxic dose), a potentially "weak" induction or inhibition could result in significant adverse side effects (Huang et al. 2007). By understanding the effects of herbal products and drugs on different CYP enzymes and cross-referencing with known substrates, inducers, and inhibitors, potentially dangerous interactions can be avoided.

Effects on CYP enzymes have been shown to last several days, with patients drinking a single glass of grapefruit juice showing normal enzyme activity after three days, while those taking St. John's wort for two weeks had normal enzyme activity one week after stopping St. John's wort (Greenblatt et al. 2003; Imai et al. 2008). Such a multiday time to return to normal enzyme activity suggests that simply separating ingestion of a CYP-metabolized drug and a CYP inducer or inhibitor by several hours is not sufficient to prevent interactions, and that products affecting CYP enzymes should not be used during treatment with CYP-metabolized drugs.

While most concern has been placed on herbs that can modify the effects of drugs, drugs may also modify the activity of herbs or herbal compounds. For example, drugs that inhibit CYP1A2 may slow the metabolism of caffeine (Carrillo and Benitez 2000; Christensen et al. 2002). While this is not likely to pose any risk with average caffeine consumption, persons taking a drug that inhibits CYP1A2 and drinking large amounts of coffee (10 cups or more daily) may experience prolonged or more pronounced effects of caffeine.

P-Glycoprotein

P-glycoprotein (P-gp), also known as ABCB1, or multi-drug resistance protein 1 (MDR1), is a multi-drug efflux pump (a protein that removes drugs and related compounds from cells) that can move a wide variety of compounds across cellular membranes. P-gp is concentrated in the excretory tissues (liver and kidney) and in barrier tissues (intestines, blood-brain barrier, placental barrier, blood-testes barrier, and blood-ovary barrier), helping to detoxify or protect certain organs and the fetus (Cordon-Cardo et al. 1989; Fojo et al. 1987; Thiebaut et al. 1987).

P-gp is also concentrated in tumor cells and is primarily responsible for multi-drug resistance found in some patients undergoing chemotherapy (Bellamy 1996).

P-Glycoprotein-Based Drug Interactions

Like CYP450 enzymes, P-gp can be both inhibited and induced. Induction of P-gp can lead to low plasma levels of P-gp-transported drugs, potentially reducing the efficacy of the drug. For example, St. John's wort has been shown to induce P-gp, leading to reduced plasma levels of digoxin, a drug transported by P-gp and used to treat heart failure and abnormal heart rhythms (Durr et al. 2000; Johne et al. 1999). Inhibition of P-gp can lead to higher intracellular (inside the cell) levels of drugs transported by P-gp, resulting in potential drug toxicity.

Differences in Human, Animal, and In Vitro Study Findings

To determine potential CYP450 interactions, in vitro, animal, and human studies are used, each with varying levels of accuracy for predicting clinically relevant interactions. While most research to date has been in vitro, the results of *in vitro* studies are often quite different from animal and human studies. Such differences may be caused by chemicals added to in vitro studies to promote uptake of some compounds that can exaggerate experimental findings by changing solubility that would otherwise limit uptake. *In vitro* studies sometimes use phytochemical concentrations that exceed those achieved in humans or other animals, or focus on isolated compounds that do not reflect the phytochemical complexity typical of extracts that may contribute to the bioactivity. Many compounds and extracts also undergo extensive metabolism in the intestines or liver, which may also change the bioactivity (Brinker 2009; Hines 1999; Markowitz et al. 2008; Venkataramanan et al. 2006). While animal studies may be more accurate in predicting clinically relevant drug interactions, these studies cannot be taken as conclusive, since large, non-physiological doses are often administered and, due to species variation in metabolism and transport, results are rarely generalizable to humans (Brinker 2009; Venkataramanan et al. 2006). In considering interaction ratings for herbs in this text, the types of evidence (human, animal, or in vitro) available were important for determining the level of concern for a potential interaction.

Literature Cited

Bano, G., V. Amla, R.K. Raina, U. Zutshi, and C.L. Chopra. 1987. The effect of piperine on pharmacokinetics of phenytoin in healthy volunteers. *Planta Med.* 53 (6):568-569.

Barone, G.W., B.J. Gurley, B.L. Ketel, M.L. Lightfoot, and S.R. Abul-Ezz. 2000. Drug interaction between St. John's wort and cyclosporine. *Ann. Pharmacother.* 34 (9):1013-1016.

Bellamy, W.T. 1996. P-Glycoproteins and multidrug resistance. *Ann. Rev. Pharmacol. Toxicol.* 36 (1):161-183.

Breidenbach, T.H., V. Kliem, M. Burg, et al. 2000. Profound drop of cyclosporin A whole blood trough levels caused by St. John's wort (*Hypericum perforatum*). *Transplantation* 69 (10):2229-2230.

Brinker, F. 2009. Managing and interpreting the complexities of botanical research. *HerbalGram* 82:42-49.

Carrillo, J.A., and J. Benitez. 2000. Clinically significant pharmacokinetic interactions between dietary caffeine and medications. *Clin. Pharmacokinet.* 39:127-153.

Christensen, M., G. Tybring, K. Mihara, et al. 2002. Low daily 10-mg and 20-mg doses of fluvoxamine inhibit the metabolism of both caffeine (cytochrome P4501A2) and omeprazole (cytochrome P4502C19). *Clin. Pharmacol. Ther.* 71 (3):141-152.

Cordon-Cardo, C., J.P. O'Brien, D. Casals, et al. 1989. Multidrug-resistance gene (P-glycoprotein) is expressed by endothelial cells at blood-brain barrier sites. *P.N.A.S. U.S.* 86 (2):695.

Danielson, P.B. 2002. The cytochrome P450 superfamily: Biochemistry, evolution and drug metabolism in humans. *Curr. Drug Metab.* 3 (6):561-597.

David, G.B., J. Malcolm, O. Arnold, and J.D. Spence. 1998. Grapefruit juice-drug interactions. *Br. J. Clin. Pharmacol.* 46 (2):101-110.

Dresser, G.K., J.D. Spence, and D.G. Bailey. 2000. Pharmacokinetic-pharmacodynamic consequences and clinical relevance of cytochrome P450 3A4 inhibition. *Clin. Pharmacokinet.* 38 (1):41-57.

Durr, D., B. Stieger, G.A. Kullak-Ublick, et al. 2000. St John's Wort induces intestinal P-glycoprotein/MDR1 and intestinal and hepatic CYP3A4. *Clin. Pharmacol. Ther.* 68 (6):598-604.

Fojo, A.T., K. Ueda, D.J. Slamon, et al. 1987. Expression of a multidrug-resistance gene in human tumors and tissues. *P.N.A.S. U.S.* 84 (1):265.

Fuhr, U., S. Beckmann-Knopp, A. Jetter, H. Luck, and U. Mengs. 2007. The effect of silymarin on oral nifedipine pharmacokinetics. *Planta Med.* 73 (14):1429-1435.

Greenblatt, D.J., L.L. von Moltke, J.S. Harmatz, et al. 2003. Time course of recovery of cytochrome p450 3A function after single doses of grapefruit juice. *Clin. Pharmacol. Ther.* 74 (2):121-129.

Gurley, B., M.A. Hubbard, D.K. Williams, et al. 2006a. Assessing the clinical significance of botanical supplementation on human cytochrome P450 3A activity: Comparison of a milk thistle and black cohosh product to rifampin and clarithromycin. *J. Clin. Pharmacol* 46 (2):201-213.

Gurley, B.J., G.W. Barone, D.K. Williams, et al. 2006b. Effect of milk thistle (*Silybum marianum*) and black cohosh (*Cimicifuga racemosa*) supplementation on digoxin pharmacokinetics in humans. *Drug Metab. Dispos.* 34 (1):69-74.

Gurley, B.J., A. Swain, M.A. Hubbard, et al. 2008. Clinical assessment of CYP2D6-mediated herb-drug interactions in humans: Effects of milk thistle, black cohosh, goldenseal, kava kava, St. John's wort, and Echinacea. *Mol. Nutr. Food Res.* 52 (7):755-763.

Hines, E. 1999. Standardizing botanical extracts: Can the part exceed the whole? *Pharmaceut. Formulat. Qual.* 3:28-33.

Huang, S.M., R. Temple, D.C. Throckmorton, and L.J. Lesko. 2007. Drug interaction studies: Study design, data analysis, and implications for dosing and labeling. *Clin. Pharmacol. Ther.* 81 (2):298-304.

Imai, H., T. Kotegawa, K. Tsutsumi, et al. 2008. The recovery time-course of CYP3A after induction by St John's wort administration. *Br. J. Clin. Pharmacol* 65 (5):701-707.

Johne, A., J. Brockmoller, S. Bauer, et al. 1999. Pharmacokinetic interaction of digoxin with an herbal extract from St. John's wort (*Hypericum perforatum*). *Clin. Pharmacol. Ther.* 66 (Oct):338-345.

Kasibhatta, R., and M.U.R. Naidu. 2007. Influence of piperine on the pharmacokinetics of nevirapine under fasting conditions: A randomised, crossover, placebo-controlled study. *Drugs R&D* 8 (6):383-391.

Mai, I., S. Bauer, E.S. Perloff, et al. 2004. Hyperforin content determines the magnitude of the St John's Wort–cyclosporine drug interaction. *Clin. Pharmacol. Ther.* 76 (4):330-340.

Markowitz, J.S., L.L. von Moltke, and J.L. Donovan. 2008. Predicting interactions between conventional medications and botanical products on the basis of in vitro investigations. *Molec. Nutr. Food Res.* 52 (7):747-754.

Padowski, J.M., and G.M. Pollack. 2010. Pharmacokinetic and pharmacodynamic implications of P-glycoprotein modulation. *Meth. Mol. Biol.* 596:359-384.

Rajnarayana, K., M. Reddy, J. Vidyasagar, and D. Krishna. 2004. Study on the influence of silymarin pretreatment on metabolism and disposition of metronidazole. *Arzneim-Forsch* 54 (2):109-113.

Rao, B.N., M. Srinivas, Y.S. Kumar, and Y.M. Rao. 2007. Effect of silymarin on the oral bioavailability of ranitidine in healthy human volunteers. *Drug Metabol. Drug Interact.* 22 (2-3):175-185.

Thiebaut, F., T. Tsuruo, H. Hamada, et al. 1987. Cellular localization of the multidrug-resistance gene product P-glycoprotein in normal human tissues. *P.N.A.S. U.S.* 84 (21):7735.

Venkataramanan, R., B. Komoroski, and S. Strom. 2006. In vitro and in vivo assessment of herb drug interactions. *Life Sci.* 78 (18):2105-2115.

Mucilages

Written by Zoë Gardner, Ph.D.(c)

Definition

Mucilages are polysaccharides that form gels or viscous solutions when mixed with water. These compounds are highly branched and form large hydrophilic (water-loving) cage-like structures that are capable of trapping water (Williams et al. 2006). After being mixed with water, mucilaginous herbs swell to many times their original size as the water is absorbed (Mills and Bone 2000).

Mucilage-Based Drug Interactions

Mucilages are a source of dietary fiber, and, like certain other sources of soluble fiber, are capable of inhibiting the absorption of some drugs (Brunton et al. 2006). To ensure complete absorption of drugs, mucilaginous plants should be taken at least one hour after other drugs or supplements (Wichtl 2004).

Mucilages may also inhibit the absorption of certain nutrients, so diabetics taking mucilaginous herbs should continue to monitor blood sugar levels (Ziai et al. 2005).

Drinking a glass of water with mucilaginous plants (especially when taken in powdered or granular form) is important to prevent the plant material from swelling in clumps and causing an obstruction in the esophagus or intestines (Angueira and Kadakia 1993; Frohna 1992; Herrle et al. 2004). Many mucilaginous herbs are also used as bulk-forming laxatives. See Appendix 2 for further information.

Herbs listed in the *Botanical Safety Handbook* containing mucilages:*

- *Alcea rosea* root
- *Aloe ferox* leaf gel
- *Aloe perryi* leaf gel
- *Aloe vera* leaf gel
- *Althaea officinalis* root, leaf, and flower
- *Laminaria digitata* thallus
- *Laminaria hyperborea* thallus
- *Laminaria japonica* thallus
- *Laminaria setchellii* thallus
- *Laminaria sinclairii* thallus
- *Linum usitatissimum* seed
- *Malva sylvestris* leaf and flower
- *Nereocystis luetkeana* thallus
- *Plantago arenaria* seed, seed husk
- *Plantago asiatica* seed, seed husk
- *Plantago ovata* seed, seed husk
- *Trigonella foenum-graecum* seed
- *Ulmus rubra* bark

* Many other herbs, such as mullein (*Verbascum thapsus*) and cinnamon (*Cinnamomum verum*), contain small amounts of mucilages that are unlikely to interfere with drug or nutrient absorption.

Literature Cited

Angueira, C., and S. Kadakia. 1993. Esophageal and duodenal bezoars from Perdiem. *Gastrointest. Endosc.* 39 (1):110-111.

Brunton, L.L., J.S. Lazo, and K.L. Parker. 2006. *Goodman & Gilman's the pharmacological basis of therapeutics,* 11th ed. New York: McGraw-Hill.

Frohna, W.J. 1992. Metamucil bezoar: An unusual cause of small bowel obstruction. *Am J. Emerg. Med.* 10 (4):393-395.

Herrle, F., T. Peters, C. Lang, et al. 2004. Bolus obstruction of pouch outlet by a granular bulk laxative after gastric banding. *Obes. Surg.* 14 (7):1022-1024.

Mills, S., and K. Bone. 2000. *Principles and practice of phytotherapy: Modern herbal medicine.* New York: Churchill Livingstone.

Wichtl, M. 2004. *Herbal drugs and phytopharmaceuticals: A handbook for practice on a scientific basis.* 3rd ed. Boca Raton, FL: CRC Press.

Williams, P.A., G.O. Phillips, A.M. Stephen, and S.C. Churms. 2006. Gums and mucilages. In *Food polysaccharides and their applications,* edited by Stephen, A.M., G.O. Phillips and P.A. Williams. Boca Raton, FL: CRC Press.

Ziai, S.A., B. Larijani, S. Akhoondzadeh, et al. 2005. Psyllium decreased serum glucose and glycosylated hemoglobin significantly in diabetic outpatients. *J. Ethnopharmacol.* 102 (2):202-207.

Piperine

Written by Zoë Gardner, Ph.D.(c)

Piperine is an alkaloid found in black pepper (*Piper nigrum*), long pepper (*Piper longum*), cubeb (*Piper cubeba*) and grains-of-paradise (*Aframomum melegueta*), and is responsible in part for the pungent taste of these spices. In addition to being found naturally in these botanicals, isolated piperine is sometimes added to other herbal products to help increase absorption and bioavailability.

Piperine-Based Drug Interactions

The compound piperine has been shown to increase the bioavailability of a number of drugs and other substances. Human studies have indicated that piperine generally increases the absorption and plasma concentrations and reduces the elimination rate of drugs including phenytoin (Bano et al. 1987; Pattanaik et al. 2006; Velpandian et al. 2001), carbamazepine (Pattanaik et al. 2009), propranolol (Bano et al. 1991), theophylline (Bano et al. 1991), rifampicin (Zutshi et al. 1985), nevirapine (Kasibhatta and Naidu 2007), and the supplements coenzyme Q_{10} (Badmaev et al. 2000) and curcumin (Shoba et al. 1998). Increases in the plasma levels of these substances have ranged from 30 to 120%. The standard dose of piperine used in human studies is 20 mg daily, though one study measured a delay in elimination of phenytoin in human subjects fed soup with black pepper (Velpandian et al. 2001). In many cases, increases in bioavailability can result in enhanced therapeutic effects, while in the case of drugs with a narrow therapeutic window (a small difference between the effective dose and the toxic dose), piperine and piperine-containing herbs may need to be used with caution.

Animal studies have shown that piperine can increase the bioavailability of the drug fexofenadine (Jin and Han 2010), the compounds curcumin, from turmeric (Shoba et al. 1998), and EGCG, from green tea (Lambert et al. 2004).

Mechanism of Action

Piperine's mechanism of enhancing bioavailability is not yet fully understood, although it is generally attributed to increased absorption, which may be due to alteration in membrane lipid dynamics and changes in the conformation of enzymes in the intestine (Khajuria et al. 2002).

An animal study demonstrated that piperine slowed the gastric emptying time and gastrointestinal transit time of solid foods (Bajad et al. 2001). An in vitro study suggested that piperine inhibited both the drug metabolizing isoenzyme CYP3A4 and the drug transporter protein P-gp (Bhardwaj et al. 2002; Han et al. 2008). Several studies have suggested that piperine acts as an MAO inhibitor (Rahman and Rahmatullah 2010).

Herbs listed in the *Botanical Safety Handbook* that contain piperine:

- *Aframomum melegueta* fruit, seed
- *Piper cubeba* unripe fruit
- *Piper longum* fruit
- *Piper nigrum* fruit

Literature Cited

Badmaev, V., M. Majeed, and L. Prakash. 2000. Piperine derived from black pepper increases the plasma levels of coenzyme q10 following oral supplementation. *J. Nutr. Biochem.* 11 (2):109-113.

Bajad, S., K.L. Bedi, A.K. Singla, and R.K. Johri. 2001. Piperine inhibits gastric emptying and gastrointestinal transit in rats and mice. *Planta Med.* 67 (2):176-179.

Bano, G., V. Amla, R.K. Raina, U. Zutshi, and C.L. Chopra. 1987. The effect of piperine on pharmacokinetics of phenytoin in healthy volunteers. *Planta Med.* 53 (6):568-569.

Bano, G., R.K. Raina, U. Zutshi, et al. 1991. Effect of piperine on bioavailability and pharmacokinetics of propranolol and theophylline in healthy volunteers. *Eur. J. Clin. Pharmacol.* 41 (6):615-617.

Bhardwaj, R.K., H. Glaeser, L. Becquemont, et al. 2002. Piperine, a major constituent of black pepper, inhibits human P-glycoprotein and CYP3A4. *J. Pharmacog. Exp. Ther.* 302 (2):645-650.

Han, Y., T.M. Chin Tan, and L.Y. Lim. 2008. In vitro and in vivo evaluation of the effects of piperine on P-gp function and expression. *Toxicol. Appl. Pharmacol.* 230 (3):283-289.

Jin, M.J., and H.K. Han. 2010. Effect of piperine, a major component of black pepper, on the intestinal absorption of fexofenadine and its implication on food-drug interaction. *J. Food Sci.* 75 (3):H93-6.

Kasibhatta, R., and M.U.R. Naidu. 2007. Influence of piperine on the pharmacokinetics of nevirapine under fasting conditions: A randomised, crossover, placebo-controlled study. *Drugs R&D* 8 (6):383-391.

Khajuria, A., N. Thusu, and U. Zutshi. 2002. Piperine modulates permeability characteristics of intestine by inducing alterations in membrane dynamics: Influence on brush border membrane fluidity, ultrastructure and enzyme kinetics. *Phytomed.* 9 (3):224-231.

Lambert, J.D., J. Hong, D.H. Kim, V.M. Mishin, and C.S. Yang. 2004. Piperine enhances the bioavailability of the tea polyphenol (-)-epigallocatechin-3-gallate in mice. *J. Nutr.* 134 (8):1948.

Pattanaik, S., D. Hota, S. Prabhakar, P. Kharbanda, and P. Pandhi. 2006. Effect of piperine on the steady-state pharmacokinetics of phenytoin in patients with epilepsy. *Phytother. Res.* 20 (8):683-686.

Pattanaik, S., D. Hota, S. Prabhakar, P. Kharbanda, and P. Pandhi. 2009. Pharmacokinetic interaction of single dose of piperine with steady-state carbamazepine in epilepsy patients. *Phytother. Res.* 23 (9):1281-1286.

Rahman, T., and M. Rahmatullah. 2010. Proposed structural basis of interaction of piperine and related compounds with monoamine oxidases. *Bioorg. Med. Chem. Lett.* 20 (2):537-540.

Shoba, G., D. Joy, T. Joseph, et al. 1998. Influence of piperine on the pharmacokinetics of curcumin in animals and human volunteers. *Planta Med.* 64:353-356.

Velpandian, T., R. Jasuja, R.K. Bhardwaj, J. Jaiswal, and S.K. Gupta. 2001. Piperine in food: Interference in the pharmacokinetics of phenytoin. *Eur. J. Drug Metab. Pharmacokinet.* 26 (4):241-247.

Zutshi, R.K., R. Singh, U. Zutshi, R.K. Johri, and C.K. Atal. 1985. Influence of piperine on rifampicin blood levels in patients of pulmonary tuberculosis. *J. Assoc. Physicians India* 33 (3):223-224.

APPENDIX 4. SAFETY OF BOTANICALS IN PREGNANCY AND LACTATION

Written by Aviva Romm

Introduction

Herbs have been used by women during pregnancy, to help prepare for birth, and to promote lactation since time immemorial, with texts and treatises dating at least back to ancient Egypt (O'Dowd 2001). Herbal medicines are still commonly used by childbearing women for a variety of reasons, for example, ginger to treat "morning sickness," raspberry leaf (*Rubus* spp.) as a uterine tonic, or echinacea (*Echinacea* spp.) for colds (Ernst 2002; Gibson et al. 2001; Hepner et al. 2002; Pinn and Pallett 2002). Studies suggest the safety of these commonly used herbs during pregnancy. Two human clinical trials evaluated raspberry leaf for its effects on labor outcome and did not demonstrate any adverse effects (Parsons et al. 1999; Simpson et al. 2001). An observational study showed no harmful effects of echinacea when used during various stages of pregnancy (Gallo et al. 2000), and numerous studies have evaluated ginger root (*Zingiber officinale*) for the reduction of nausea and vomiting of pregnancy, finding it both safe and effective at the doses used (dried rhizome 1000–1500 mg per day) (Bryer 2005). There is considerable evidence that modern obstetric professionals such as midwives and obstetricians are recommending herbs during pregnancy, particularly for labor stimulation and for other concerns as well (Allaire et al. 2000; Chez and Jonas 1999; Hardy 2000; Hepner et al. 2002; Low Dog 2009; Pinn and Pallett 2002; Romm 2009b). Because pregnancy is the most sensitive time in human development, and many substances cross the placenta, questions have been raised about the safety of herbal medicines taken during pregnancy.

Similar safety concerns are raised about the safety of herbal remedies used by breastfeeding mothers. About one percent of any substance a lactating mother ingests will be passed to her baby through the breast milk; thus, while many herbs would likely be safe, we must also be mindful about the maternal ingestion of some herbal products during lactation. Consideration should also be given to the age of the infant, as there are considerable differences between the liver and kidney function of a 4-week-old and a 12-month-old infant (Humphrey and McKenna 1997).

Most herbs have not been evaluated for safety in the childbearing cycle; thus, there is little scientific evidence to support or refute their safety during this time. Ethical considerations surrounding experimentation on pregnant women and the need for large sample sizes severely limit human clinical investigation during pregnancy (Hepner et al. 2002; Low Dog 2009; Romm 2009b). This chapter elucidates some of the key issues surrounding the use of herbs during pregnancy and lactation, and explains the careful process involved in creating the pregnancy and lactation safety ratings found in this book.

Contraindicated Herb Lists and Botanical Safety Classifications

The herbal literature is rife with lists of herbs contraindicated in pregnancy and lactation. There are limitations inherent in most lists, particularly in their lack of specifics as to how, when, and why each herb is contraindicated. Herbs may sometimes be broadly contraindicated in pregnancy yet in actuality be only contextually contraindicated. For example, some herbs are absolutely contraindicated during the first and second trimesters but may be reasonably used during labor, or they may be safe in small doses for a very limited duration. Culinary herbs, appearing on many contraindicated lists, represent no harm to the fetus or mother when ingested in the small amounts typically used as food seasonings. Herbs such as goldenseal root (*Hydrastis canadensis*) may be used topically with no risk but should be avoided for internal use, yet are contraindicated on such lists with no differentiation, leading to confusion about safety. Certain contraindications have become pervasive myths; for example, the frequent contraindication of chamomile (*Matricaria recutita*) in pregnancy due to its alleged action as an abortifacient (McKenna et al. 2002). In fact, chamomile provides an excellent example of how misapplication of a scientific finding can lead to unjustified contraindication of a safe herb. A study conducted in 1979 found that a concentrated extract of the compound a-bisabolol caused birth defects at high doses. No birth defects were seen at lower doses, and the dose of a-bisabolol required to cause defects is far greater than would be possible to ingest by someone drinking the tea. However, based on this single study, chamomile continues to be improperly contraindicated for consumption during pregnancy (Low Dog 2004).

Finally, herbs may be contraindicated based on limited or conflicting information. For example, information on traditional use indicates that ashwagandha (*Withania somnifera*) is used as both an "abortifacient" (Badhwar and Chopra 1946; Casey 1960; Chadha 1976) and as a "pregnancy tonic" (Kapoor 1990; Tirtha 1998; Upton 2000). However, no details are available regarding plant part, doses used, and duration of use. A more recent study in animals indicated a lack of adverse effects of ashwagandha in pregnancy (Sharma et al. 1986). To complicate matters, certain herbs that are contraindicated by western herbalists for use during pregnancy may be routinely used in traditional medicines of non-western cultures. For example, dong quai is prescribed in blood tonic formulas for pregnant women in China, and listed in official Chinese and Japanese texts for the prevention of miscarriage, yet is considered contraindicated for use in pregnancy by some western herbalists (Brinker 2001; Mills and Bone 2005).

Safety of Herbs in Pregnancy and Lactation

There is a lack of human clinical trials on the safety or efficacy for many botanical therapies during pregnancy. Risks associated with the use of herbs during pregnancy include:

- Toxicity to the mother which might indirectly affect the embryo/fetus
- Direct teratogenicity, mutagenicity, or fetal toxicity
- Abortifacient activity
- Poor neonatal outcomes
- Delayed administration of necessary medical therapy in favor of herbs, regardless of their safety (Mills and Bone 2005; Romm 2009b)

In general, most herbs on the market in the U.S. have a relatively high track record of safety, with few case reports of adverse effects (Hansten 2000). Negative outcomes have been reported for only a very limited number of herbal products used by women in or about to be in labor (Beal 1998; Ernst 2002; Mabina et al. 1997; Romm 2009a). When apparent adverse events have occurred, cause and effect have been difficult to establish due to a wide range of confounding factors (Ernst and Schmidt 2002). Adverse events have typically involved the consumption of known toxic herbs, contaminated or adulterated herbal products, or inappropriate use or dosage of specific botanical therapies. However, lack of proof of harm is not synonymous with proof of safety. Some of the harmful effects of herbs may not be readily apparent until after they have been discontinued, or may only occur with cumulative use. Some researchers therefore believe that, in the absence of scientific proof of safety, herbs that are used for therapeutic reasons should be avoided during pregnancy (Ernst 2002). However, many practitioners, including herbalists, naturopaths, and midwives, continue to use herbs based on historical data and observational evidence, tempered by the knowledge that many pharmaceutical preparations recommended during pregnancy also carry unknown risks (Low Dog 2004; Romm 2009b).

After much careful consideration, the editors of this text deemed that the most reasonable standard on which to base pregnancy safety ratings is a combination of the best available evidence including:

- Clinical evidence in the medical/scientific literature
- Animal and in vitro studies
- Case reports
- Known and suspected mechanisms of action of the herb
- Safety and bioavailability of the herb in the form most likely to be used (i.e., powder, tea, tincture, concentrated extract, etc.)
- Dose of the herb typically ingested and typical duration of use
- Clinical (empiric, observational) consensus of safety/risk
- Chemical constituent profile
  - teratogenicity
  - mutagenicity
  - toxicity
  - extractability
  - bioavailability
- Historical information and traditional use as derived from a circumscribed set of approved texts and sources

There are virtually no formal studies that can definitively make a determination that a particular herb, or drug for that matter, is absolutely safe during pregnancy. There are also a number of botanicals for which there are little data or clinical experience regarding their use in pregnancy and lactation. It is the opinion of the editors that the absence of formal data and clinical experience regarding the use of a botanical in pregnancy or lactation, in and of itself, is not justification to contraindicate the botanical in pregnancy and lactation. In cases where there is neither substantial historical nor scientific data, the editors have used their best judgment in looking at the totality of the data available to make the most appropriate determination. However, consumers and health professionals should be cautioned about taking any botanical in pregnancy and lactation regardless of its classification. This is especially true for those botanicals for which there is insufficient traditional or scientific evidence of safety. For such botanicals, the following Editors' Note was included to highlight the relative lack of data specifically regarding use in pregnancy and lactation, but also a lack of data to suggest that any safety concern exists:

> No information on the use of wild lettuce during pregnancy or lactation was identified in the scientific or traditional literature. While this review did not identify any concerns for pregnant or lactating women, safety has not been conclusively established.

Herbal medicines may be used during lactation for a variety of concerns, ranging from increasing the quantity of breast milk (lactagogues) to the treatment of any number of postpartum or common general conditions. The risks of using herbs during lactation appear to be less significant than in pregnancy, as only very small quantities of plant constituents ingested by the mother actually pass into the breast milk, and risks of birth defects and abortifacient activity are no longer present (Humphrey 2009). However, quantification of chemical constituents from herbal medicines appearing in breast milk has not been conducted for most herbs, so some element of caution is still required. As with pregnancy, it is prudent to avoid known toxic herbs or those with possible effects on still developing biological systems, such as the nervous and endocrine systems. The editors of this text have taken these points into consideration when assigning safety classifications to herbs in pregnancy and lactation.

Emmenagogues, Uterine Stimulants, Abortifacients, and Partus Preparator

Much of the literature on these categories comes from late 19th century references, such as Felter & Lloyd's *King's American Dispensatory* (Felter and Lloyd 1898), as well as the experience of contemporary practicing herbalists.

Historically, as in modern times, the term emmenagogue has referred to a plant or substance that was used with the intention of bringing about menstruation (Santow 2001). For this reason, many of these herbs have been contraindicated during pregnancy. While some of these herbs were undoubtedly used as abortifacients, it is also quite plausible that a number were used to nourish and strengthen a woman who might have lost her menstrual cyclicity due to anemia, malnourishment, or severe stress. This has led to considerable disagreement about what should and shouldn't be called an emmenagogue. Those that are considered to be unsafe during pregnancy include rue (*Ruta graveolens*), Scotch broom (*Cytisus scoparius*), tansy (*Tanacetum vulgare*), thuja (*Thuja occidentalis*), wormwood (*Artemisia absinthium*), and pennyroyal (*Mentha pulegium*).

Uterine tonics have been used historically with the belief that they improve the strength and tone of the uterine muscle, while uterine stimulants were thought to bring on menses or induce labor. Although both of these are important categories in modern herbal therapeutics, there is little understanding or research regarding their mechanism of action, effective doses, or safety. Some herbs, such as dong quai (*Angelica sinensis*), are considered by some sources to be emmenagogues but are also believed to have both uterine relaxant and uterine stimulatory activity, and are used in Chinese and Japanese herbal formulas for the prevention of miscarriage (Chen and Chen 2004; Liang 2004; Upton 2003; Zhu 1998). Clearly, this is an area where more research is needed. Herbs that are intentionally being used to induce labor should be administered under the direct supervision of a qualified and experienced individual, usually a midwife, knowledgeable in such therapies.

Abortifacients are herbs used to facilitate a miscarriage or induce abortion. How effective these herbs are for inducing abortion is unknown; however, the amount required to terminate pregnancy is likely enough to pose significant risk to the mother's health, including kidney and liver damage, and may not result in a successful abortion attempt. However, because the risks of these herbs to the fetus are unknown, women who have attempted to abort unsuccessfully may be well advised to obtain a clinical abortion. Abortifacients should be entirely avoided during pregnancy, and herbal abortion is not a recommended method of intentional pregnancy termination (also see Abortifacients in Appendix 2).

Partus preparators are herbs generally taken during the last few weeks of pregnancy and were historically used to facilitate a timely delivery. Blue cohosh (*Caulophyllum thalictroides*) is the herb most commonly used for this purpose. Case reports of adverse events, along with the activity of different compounds present in blue cohosh (see entry for *Caulophyllum thalictroides*) suggest that it should be avoided as a partus preparator, with use limited to short-term application to induce or augment labor, and then only with the proper guidance and monitoring of an expert qualified in the appropriate use of the selected herb or formula (Jones and Lawson 1998; Wright 1999).

Teratogens and Mutagens

Teratogens are substances that cause structural abnormalities and other birth defects in the fetus. There is extremely limited knowledge about which herbs may act as teratogens. Most of what is known is derived primarily from animal studies, observation of teratogenesis in grazing animals, and from human ingestion of suspected harmful herbal products. The primary means for identifying the propensity for this type of reaction is through toxicological screenings or through pharmacovigilance programs (monitoring of drugs and supplements for adverse effects).

Mutagens are substances that cause genetic mutations in cells. The potential for mutagenicity is typically discerned through in vitro assays followed by animal studies. In both types of tests, isolated compounds rather than whole herbs are most often investigated. While in vitro tests can be useful, they are generally quite limited for extrapolation to human use. The presence of a mutagen in an herb does not automatically contraindicate its use in pregnancy. There are many commonly consumed plants such as basil (*Ocimum basilicum*), black pepper (*Piper nigrum*), coffee (*Coffea arabica*), tomatoes (*Solanum lycopersicum*), and potatoes (*Solanum tuberosum*) that contain compounds with mutagenic potential (ACHS 1996). However, consumption of these in pregnancy is not contraindicated when consumed as a normal part of the diet, and the presence of such compounds in a food or herb does not mean that this substance crosses the placenta or reaches the fetus in appreciable or clinically significant quantities.

Phytoestrogens

Phytoestrogens are weak forms of plant estrogens found in numerous plants, including common foods such as beans and other legumes (most notably soy foods) and dark leafy greens (Franke et al. 1994; Kuiper et al. 1998). Some traditional Asian diets, which include large amounts of tofu and other soy products, are particularly high in phytoestrogens, but abnormal levels of fetal and neonatal problems have not been observed when compared to other populations. Nonetheless, there is concern that consumption of concentrated doses of phytoestrogens (i.e., isolated soy isoflavones) during pregnancy may exert abnormal hormonal effects on the developing embryo or fetus, particularly in female embryogenesis, as occurred with the drug diethylstilbestrol (DES), a synthetic nonsteroidal estrogen. Although the amounts found in plant sources are incomparably minute compared to DES (Kuiper et al. 1998), it is generally recommended that women not supplement with concentrated phytoestrogen products such as soy isoflavone concentrates during pregnancy. Additionally, herbs with known or suspected estrogenic activity or activity on the endocrine system, including hops, alfalfa, or red clover, are best avoided for long-term use or in large doses during pregnancy.

Nervous System Stimulants or Depressants

Herbs which strongly affect the nervous system, including stimulants such as ephedra (*Ephedra sinica*), guarana (*Paullinia cupana*), coffee (*Coffea arabica*), or green and black teas (*Camellia sinensis*), may have adverse effects on pregnancy and the developing fetal nervous system. Caffeine should be avoided or limited to no more than 300 mg (approx. 3 cups coffee) per day during pregnancy (ADA 2008). Kava (*Piper methysticum*), a strong anxiolytic and sedative herb, has been implicated in a number of cases of liver toxicity (Teschke et al. 2008). While causality is uncertain, kava should be avoided during pregnancy and lactation until the relationship between kava and liver toxicity is fully understood.

Summary

Women may turn to herbs for the relief of common complaints and concerns that arise during pregnancy, childbirth, and breastfeeding. The power of herbs should be respected and, therefore, used with caution during the childbearing cycle. However, a number of herbs have been contraindicated in modern herbal literature on the basis of very limited findings and erroneous or incomplete reports. Consumers, manufacturers, and practitioners need to be educated when it comes to the safe use of herbal medicine during pregnancy and lactation. Further research needs to be conducted not only into the safety and efficacy of herbs for both the mother and her baby, but also a comparative safety analysis should be done on herbs and pharmaceuticals commonly used during pregnancy. The editors of this text have done their best to shine light on a complicated subject by providing a safety ranking to help healthcare professionals, manufacturers, and consumers differentiate those herbs which are generally considered safe from those which should be avoided during pregnancy and lactation.

Literature Cited

ACHS. 1996. Does nature know best? Natural carcinogens and anticarcinogens in America's food. New York: American Council on Science and Health.

Ács, N., F. Bánhidy, E.H. Puhó, and A.E. Czeizel. 2009. Senna treatment in pregnant women and congenital abnormalities in their offspring: A population-based case-control study. *Repro. Toxicol.* 28 (1):100-104.

ADA. 2008. Position of the American Dietetic Association: Nutrition and lifestyle for a healthy pregnancy outcome. *J. Am. Diet. Assoc.* 108:553-561.

Allaire, A.D., M.K. Moos, and S.R. Wells. 2000. Complementary and alternative medicine in pregnancy: A survey of North Carolina certified nurse-midwives. *Obstet. Gynecol.* 95 (1):19-23.

Badhwar, R.L., and I.C. Chopra. 1946. Reputed abortifacient plants of India. *Ind. J. Agric. Sci.* 16:342-355.

Beal, M.W. 1998. Women's use of complementary and alternative therapies in reproductive health care. *J. Nurse-Midwifery* 43 (3):224-234.

Brinker, F. 2001. *Herb contraindications and drug interactions.* 3rd ed. Sandy, OR: Eclectic Medical Publications.

Bryer, E. 2005. A literature review of the effectiveness of ginger in alleviating mild-to-moderate nausea and vomiting of pregnancy. *J. Midwifery Womens Health* 50 (1):e1-3.

Casey, R. 1960. Alleged antifertility plants of India. *Indian J. Med. Res.* 14:590-600.

Chadha, Y. 1976. *The wealth of India; a dictionary of Indian raw materials and industrial products.* New Delhi: Council of Scientific and Industrial Research.

Chen, J.K., and T.T. Chen. 2004. *Chinese medical herbology and pharmacology.* City of Industry, CA: Art of Medicine Press.

Chen, Z., and J.R. Huo. 2010. Hepatic veno-occlusive disease associated with toxicity of pyrrolizidine alkaloids in herbal preparations. *Neth. J. Med.* 68 (6):252-260.

Chez, R.A., and W.B. Jonas. 1999. Complementary and alternative medicine. Part I: Clinical studies in obstetrics. *Obstet. Gynecol. Survey* 54 (11):118-122.

Ernst, E. 2002 Herbal medicinal products during pregnancy: Are they safe? *B.J.O.G.* 109 (3):227-235.

Ernst, E., and K. Schmidt. 2002. Health risks over the internet: Advice offered by "medical herbalists" to a pregnant woman. *Wien. Med. Wschr.* 152 (78):190-192.

Felter, H.W., and J.U. Lloyd. 1898. *King's American dispensatory.* Cincinnati: Ohio Valley Co.

Franke, A.A., L.J. Custer, C.M. Cerna, and K.K. Narala. 1994. Quantitation of phytoestrogens in legumes by HPLC. *J. Agric. Food Chem.* 42 (9):1905-1913.

Gallo, M., M. Sarkar, W. Au, et al. 2000. Pregnancy outcome following gestational exposure to echinacea: A prospective controlled study. *Arch. Int. Med.* 160 (20):3141-3143.

Gibson, P.S., R. Powrie, and J. Star. 2001. Herbal and alternative medicine use during pregnancy: A cross-sectional survey. *Obstet. Gynecol.* 97 (4):S44-S45.

Hansten, P. 2000. Managing drug interactions with herbal products. A.S.H.P. Midyear Clinical Meeting. 35:102.

Hardy, M.L. 2000. Herbs of special interest to women. *J. Am. Pharm. Assoc.* 40 (2):234-239.

Hepner, D.L., M. Harnett, S. Segal, et al. 2002. Herbal medicine use in parturients. *Anesth. Analg.* 94 (3):690.

Humphrey, S. 2009. A comprehensive review of safety considerations and breastfeeding concerns for the mother-infant dyad. In Romm, A., *Botanical medicines for women's health.* St. Louis: Churchill Livingstone Elsevier.

Humphrey, S.I., and D.J. McKenna. 1997. Herbs and breastfeeding. *Breastfeed. Abstr.* 17 (2):11-12.

Jones, T.K., and B.M. Lawson. 1998. Profound neonatal congestive heart failure caused by maternal consumption of blue cohosh herbal medication. *J. Pediatr.* 132 (3):550-552.

Kapoor, L. 1990. *CRC handbook of Ayurvedic medicinal plants.* Boca Raton, FL: CRC Press.

Kuiper, G.G.J.M., J.G. Lemmen, B.O. Carlsson, et al. 1998. Interaction of estrogenic chemicals and phytoestrogens with estrogen receptor β. *Endocrinol.* 139 (10):4252-4263.

Liang, L. 2004. *Acupuncture & IVF.* Boulder, CO: Blue Poppy Enterprises.

Low Dog, T. 2004. *Women's health in complementary and integrative medicine: A clinical guide.* St Louis: Elsevier.

Low Dog, T. 2009. The use of botanicals during pregnancy and lactation. *Altern. Ther. Health. Med.* 15 (1):54-58.

Mabina, M.H., S.B. Pitsoe, and J. Moodley. 1997. The effect of traditional herbal medicines on pregnancy outcome. *S.A.J.M.* 87 (8):1008-1010.

McKenna, D., K. Jones, K. Hughes, and S. Humphrey. 2002. *Botanical medicines: The desk reference for major herbal supplements*. New York: Haworth Press.

Mills, S., and K. Bone. 2005. *The essential guide to herbal safety*. St Louis: Elsevier.

O'Dowd, M. 2001. *The history of medications for women: Materia medica woman*. New York: Parthenon Publishing Group.

Parsons, M., M. Simpson, and T. Ponton. 1999. Raspberry leaf and its effect on labour: Safety and efficacy. *Austral. Coll. Midwives J.* 12 (3):20-25.

Pinn, G., and L. Pallett. 2002. Herbal medicine in pregnancy. *Comp. Ther. Nurs. Midwif.* 8 (2):77-80.

Romm, A. 2009a. Blue cohosh: History, science, safety, and midwife prescribing of a potentially fetotoxic herb. New Haven: Yale University School of Medicine.

Romm, A. 2009b. *Botanical medicines for women's health*. St Louis: Elsevier.

Santow, G. 2001. Emmenagogues and abortifacients in the twentieth century: An issue of ambiguity. In *Regulating menstruation: Beliefs, practices, interpretations*. In Van de Walle, É. and E.P. Renne. Chicago: University of Chicago Press.

Sharma, S.S., S. Dahanukar, and S. Karandikar. 1986. Effects of long-term administration of the roots of ashwagandha (*Withania somnifera*) and shatavari (*Asparagus racemosus*) in rats. *Indian Drugs* 23:133-139.

Simpson, M., M. Parsons, J. Greenwood, and K. Wade. 2001. Raspberry leaf in pregnancy: Its safety and efficacy in labor. *J. Midwif. Womens Health* 46 (2):51-59.

Teschke, R., A. Schwarzenboeck, and K.H. Hennermann. 2008. Kava hepatotoxicity: A clinical survey and critical analysis of 26 suspected cases. *Eur. J. Gastroenterol. Hepatol.* 20 (12):1182-1193.

Tirtha, S. 1998. *The Ayurveda encyclopedia*. Bayville, NY: Ayurvedic Holistic Center.

Upton, R. 2000. *Ashwagandha root:* Withania somnifera*: Analytical, quality control, and therapeutic monograph*. Santa Cruz, CA: American Herbal Pharmacopoeia.

Upton, R. 2003. *Dang gui root:* Angelica sinensis *(Oliv.) Diels: Standards of analysis, quality control, and therapeutics*. Scotts Valley, CA: American Herbal Pharmacopoeia.

Wright, I.M. 1999. Neonatal effects of maternal consumption of blue cohosh. *J. Pediatr.* 134 (3):384-385.

Zhu, Y. P. 1998. *Chinese materia medica: Chemistry, pharmacology and applications*. Amsterdam: Harwood Academic Publishers.

APPENDIX 5. HERB LISTINGS BY CLASSIFICATION

Lists are provided below of all of the plants classified in the main section of the text in either Safety Class 2a, 2b, 2c, or 3, or in Interaction Class B or C. Note that the restrictions described for plants in each of these Safety Classes (as well as in Class 2d, not listed here) apply unless otherwise directed by an expert qualified in the use of the described substance.

The lists presented here are provided as a compilation of the classifications found at individual entries in the text. Additional information, including in some cases exceptions to one of these classifications, may be found at specific entries.

Safety Class 2a: For External Use Only

- *Alkanna tinctoria* root
- *Borago officinalis* herb
- *Eutrochium fistulosum* herb, rhizome, root
- *Eutrochium maculatum* herb, rhizome, root
- *Eutrochium purpureum* herb, rhizome, root
- *Lawsonia inermis* leaf
- *Mentha pulegium* herb essential oil
- *Symphytum asperum* leaf, root
- *Symphytum officinale* leaf, root
- *Symphytum × uplandicum* leaf, root

Safety Class 2b: Not To Be Used during Pregnancy

- *Achyranthes bidentata* root
- *Actaea racemosa* rhizome
- *Adiantum capillus-veneris* herb
- *Adiantum pedatum* herb
- *Agathosma betulina* leaf
- *Agathosma crenulata* leaf
- *Agathosma serratifolia* leaf
- *Albizia julibrissin* bark
- *Alkanna tinctoria* root
- *Aloe ferox* latex
- *Aloe perryi* latex
- *Aloe vera* latex
- *Andrographis paniculata* herb
- *Angelica archangelica* fruit, root
- *Angelica atropurpurea* fruit, root
- *Anthriscus cerefolium* herb
- *Apium graveolens* fruit
- *Aralia racemosa* rhizome
- *Artemisia abrotanum* herb
- *Artemisia absinthium* herb
- *Artemisia annua* aboveground parts
- *Artemisia douglasiana* herb
- *Artemisia lactiflora* herb
- *Artemisia vulgaris* herb
- *Asclepias asperula* root
- *Asclepias tuberosa* root
- *Baptisia tinctoria* root
- *Berberis vulgaris* root, root bark
- *Boswellia sacra* gum resin
- *Boswellia serrata* gum resin
- *Capsella bursa-pastoris* herb
- *Carica papaya* leaf
- *Carthamus tinctorius* flower
- *Catharanthus roseus* herb
- *Caulophyllum thalictroides* root
- *Chamaemelum nobile* flower
- *Changium smyrnioides* root
- *Chelidonium majus* herb
- *Chrysopogon zizanioides* root
- *Cinchona calisaya* bark
- *Cinchona officinalis* bark
- *Cinchona pubescens* bark
- *Cinnamomum aromaticum* bark
- *Cinnamomum camphora* distillate of the wood
- *Cinnamomum verum* bark
- *Coix lacryma-jobi* seed
- *Commiphora madagascariensis* gum resin
- *Commiphora molmol* gum resin
- *Commiphora myrrha* gum resin
- *Commiphora wightii* gum resin
- *Coptis chinensis* rhizome
- *Coptis trifolia* rhizome
- *Corydalis yanhusuo* tuber
- *Crocus sativus* stigma
- *Cullen corylifolium* seed
- *Curculigo orchioides* rhizome
- *Curcuma zedoaria* rhizome
- *Cyathula officinalis* root
- *Daemonorops draco* resin
- *Daucus carota* ssp. *carota* seed
- *Equisetum hyemale* aboveground parts
- *Eschscholzia californica* whole plant in flower

- *Eucalyptus globulus* essential oil
- *Euonymus atropurpureus* root bark
- *Eutrochium fistulosum* herb, rhizome, root
- *Eutrochium maculatum* herb, rhizome, root
- *Eutrochium purpureum* herb, rhizome, root
- *Ferula assa-foetida* oleo gum resin
- *Ferula foetida* oleo gum resin
- *Fouquieria splendens* stem
- *Frangula alnus* bark
- *Frangula purshiana* bark
- *Fraxinus americana* bark
- *Genista tinctoria* flower, herb
- *Glycyrrhiza echinata* rhizome, root
- *Glycyrrhiza glabra* rhizome, root
- *Glycyrrhiza uralensis* rhizome, root
- *Gossypium herbaceum* root bark
- *Gossypium hirsutum* root bark
- *Hedeoma pulegioides* herb
- *Hepatica nobilis* var. *obtusa* herb
- *Hydrastis canadensis* rhizome, root
- *Hyssopus officinalis* herb
- *Iris versicolor* rhizome, root
- *Iris virginica* rhizome, root
- *Juniperus communis* fruit
- *Juniperus monosperma* fruit
- *Juniperus osteosperma* fruit
- *Juniperus oxycedrus* fruit
- *Juniperus virginiana* berry, leaf
- *Larrea tridentata* leaf
- *Leonurus cardiaca* herb
- *Leonurus japonicus* aboveground parts, herb
- *Leonurus sibiricus* aboveground parts, herb
- *Ligusticum porteri* rhizome
- *Ligusticum sinense* 'Chuanxiong' rhizome
- *Ligusticum wallichii* rhizome
- *Lobelia inflata* herb
- *Lobelia siphilitica* herb
- *Lomatium dissectum* root
- *Lycopus americanus* herb
- *Lycopus europaeus* herb
- *Lycopus virginicus* herb
- *Magnolia biondii* flower bud
- *Magnolia denudata* flower bud
- *Magnolia officinalis* bark, root bark
- *Magnolia sprengeri* flower bud
- *Magnolia virginiana* bark
- *Marrubium vulgare* herb
- *Mentha* × *piperita* essential oil
- *Mentha pulegium* essential oil
- *Mentha pulegium* herb
- *Monarda clinopodia* herb
- *Monarda didyma* herb
- *Monarda fistulosa* herb
- *Monarda pectinata* herb
- *Monarda punctata* herb
- *Morinda citrifolia* fruit
- *Mucuna pruriens* root, seed
- *Myristica fragrans* aril, seed
- *Nardostachys jatamansi* rhizome, root
- *Ocimum basilicum* leaf
- *Ocimum gratissimum* aboveground parts
- *Paeonia suffruticosa* root bark
- *Pausinystalia johimbe* bark
- *Phyllanthus amarus* aboveground parts, whole plant
- *Phyllanthus fraternus* aboveground parts, whole plant
- *Phyllanthus niruri* aboveground parts, whole plant
- *Picrasma excelsa* bark, root, wood
- *Piper longum* fruit
- *Piper methysticum* rhizome, root
- *Polygala senega* root
- *Polygala sibirica* root
- *Polygala tenuifolia* root
- *Portulaca oleracea* aboveground parts
- *Punica granatum* fruit husk
- *Quassia amara* bark, root, wood
- *Rhamnus cathartica* fruit
- *Rheum officinale* rhizome, root
- *Rheum palmatum* rhizome, root
- *Rheum palmatum* var. *tanguticum* rhizome, root
- *Ricinus communis* seed oil
- *Ruta graveolens* herb
- *Salvia miltiorrhiza* root
- *Salvia officinalis* leaf
- *Sanguinaria canadensis* rhizome, root
- *Sassafras albidum* root
- *Spigelia marilandica* root
- *Symphytum asperum* leaf, root
- *Symphytum officinale* leaf, root
- *Symphytum* × *uplandicum* leaf, root
- *Tanacetum parthenium* herb
- *Tanacetum vulgare* herb
- *Taxus brevifolia* needles
- *Terminalia arjuna* bark
- *Thuja occidentalis* leaf
- *Thymus vulgaris* herb
- *Tribulus terrestris* aboveground parts
- *Tribulus terrestris* fruit
- *Trigonella foenum-graecum* seed
- *Trillium erectum* root
- *Tussilago farfara* flower bud
- *Tussilago farfara* leaf
- *Uncaria tomentosa* root bark, stem bark
- *Verbena hastata* herb
- *Verbena officinalis* ssp. *officinalis* herb
- *Viscum album* herb
- *Withania somnifera* root

- *Zanthoxylum americanum* bark
- *Zanthoxylum bungeanum* fruit pericarp
- *Zanthoxylum clava-herculis* bark
- *Zanthoxylum schinifolium* fruit pericarp
- *Zanthoxylum simulans* fruit pericarp
- *Ziziphus jujuba* var. *spinosa* seed

Safety Class 2c: Not To Be Used while Nursing

- *Alkanna tinctoria* root
- *Aloe ferox* latex
- *Aloe perryi* latex
- *Aloe vera* latex
- *Artemisia absinthium* herb
- *Chelidonium majus* herb
- *Euonymus atropurpureus* root bark
- *Eutrochium fistulosum* herb, rhizome, root
- *Eutrochium maculatum* herb, rhizome, root
- *Eutrochium purpureum* herb, rhizome, root
- *Frangula alnus* bark
- *Frangula purshiana* bark
- *Hedeoma pulegioides* herb
- *Lycopus americanus* herb
- *Lycopus europaeus* herb
- *Lycopus virginicus* herb
- *Mentha pulegium* essential oil
- *Mentha pulegium* herb
- *Pausinystalia johimbe* bark
- *Piper methysticum* rhizome, root
- *Rhamnus cathartica* fruit
- *Rheum officinale* rhizome, root
- *Rheum palmatum* rhizome, root
- *Rheum palmatum* var. *tanguticum* rhizome, root
- *Symphytum asperum* leaf, root
- *Symphytum officinale* leaf, root
- *Symphytum* × *uplandicum* leaf, root
- *Tanacetum vulgare* herb
- *Tussilago farfara* flower bud
- *Tussilago farfara* leaf

Safety Class 3: For Use Only under the Supervision of Qualified Expert

- *Aconitum carmichaelii* prepared root
- *Acorus calamus* rhizome
- *Acorus gramineus* rhizome
- *Apocynum androsaemifolium* root
- *Apocynum cannabinum* root
- *Arisaema amurense* prepared rhizome
- *Arisaema erubescens* prepared rhizome
- *Arisaema heterophyllum* prepared rhizome
- *Arnica latifolia* root, rhizome, whole plant
- *Arnica montana* root, rhizome, whole plant
- *Atropa belladonna* leaf
- *Buxus sempervirens* leaf
- *Cephaelis ipecacuanha* rhizome
- *Convallaria majalis* entire plant
- *Cytisus scoparius* flowering top
- *Digitalis purpurea* leaf
- *Digitalis lanata* leaf
- *Dryopteris filix-mas* rhizome
- *Ipomoea purga* root
- *Melia azedarach* bark, fruit, root bark
- *Phoradendron leucarpum* herb
- *Phytolacca americana* root
- *Pilocarpus jaborandi* leaf
- *Pilocarpus microphyllus* leaf
- *Pilocarpus pennatifolius* leaf
- *Pinellia ternata* prepared rhizome
- *Podophyllum hexandrum* rhizome, root
- *Podophyllum peltatum* root
- *Prunus armeniaca* seed
- *Prunus persica* leaf, seed, twig
- *Reynoutria multiflora* unprocessed root tuber
- *Stillingia sylvatica* root
- *Veratrum viride* root

Interaction Class B: Herbs for Which Clinically Relevant Interactions Are Biologically Plausible

- *Apocynum androsaemifolium* root
- *Apocynum cannabinum* root
- *Astragalus mongholicus* root
- *Carthamus tinctorius* flower
- *Commiphora wightii* gum resin
- *Convallaria majalis* entire plant
- *Cytisus scoparius* flowering top
- *Ginkgo biloba* leaf
- *Glycyrrhiza echinata* rhizome, root
- *Glycyrrhiza glabra* rhizome, root
- *Glycyrrhiza uralensis* rhizome, root
- *Lycopus americanus* herb
- *Lycopus europaeus* herb
- *Lycopus virginicus* herb
- *Panax quinquefolius* root
- *Pausinystalia johimbe* bark

- *Pilocarpus jaborandi* leaf
- *Pilocarpus microphyllus* leaf
- *Pilocarpus pennatifolius* leaf
- *Piper longum* fruit
- *Piper methysticum* rhizome, root
- *Piper nigrum* fruit
- *Scutellaria baicalensis* root
- *Selenicereus grandiflorus* flower, stem
- *Tetradium ruticarpum* unripe fruit
- *Valeriana edulis* ssp. *procera* rhizome, root
- *Valeriana jatamansi* rhizome, root
- *Valeriana officinalis* rhizome, root
- *Valeriana sitchensis* rhizome, root
- *Zingiber officinale* rhizome

Interaction Class C: Herbs for Which Clinically Relevant Interactions Are Known to Occur

- *Allium sativum* bulb
- *Angelica sinensis* root
- *Atropa belladonna* leaf
- *Camellia sinensis* leaf, stem
- *Citrus × aurantium* fruit
- *Coffea arabica* roasted seed kernel
- *Cola acuminata* seed
- *Cola nitida* seed
- *Digitalis purpurea* leaf
- *Digitalis lanata* leaf
- *Hydrastis canadensis* root, rhizome
- *Hypericum perforatum* flowering top, herb
- *Ilex paraguariensis* leaf
- *Paullinia cupana* seed
- *Salvia miltiorrhiza* root
- *Schisandra chinensis* fruit
- *Schisandra sphenanthera* fruit

INDEX

B

C

E

F

G

H

I

L

M

N

O

P

U

V

Y

Z